T0263707

Meyler's Side Effects of Cardiovascular Drugs

Meyler's Side Effects of Cardiovascular Drugs

Editor

J K Aronson, MA, DPhil, MBChB, FRCP, FBPharmacolS, FFPM (Hon)

Oxford, United Kingdom

ELSEVIER

AMSTERDAM • BOSTON • HEIDELBERG • LONDON • NEW YORK • OXFORD
PARIS • SAN DIEGO • SAN FRANCISCO • SINGAPORE • SYDNEY • TOKYO

Elsevier
Radarweg 29, PO Box 211, 1000 AE Amsterdam, The Netherlands
The Boulevard, Langford Lane, Kidlington, Oxford OX5 1GB, UK
525 B Street, Suite 1900, San Diego, CA 92101-4495, USA

British Library Cataloguing in Publication Data
A catalogue record for this book is available from the British Library

Library of Congress Catalog Number: 2008933971

ISBN: 978-044-453268-8

For information on all Elsevier publications
visit our web site at http://www.elsevierdirect.com

Typeset by Integra Software Services Pvt. Ltd, Pondicherry, India www.integra-india.com

Contents

Preface

This volume covers the adverse effects of drugs used in managing cardiovascular disorders. The material has been collected from *Meyler's Side Effects of Drugs: The International Encyclopedia of Adverse Drug Reactions and Interactions* (15th edition, 2006, in six volumes), which was itself based on previous editions of *Meyler's Side Effects of Drugs*, and from the *Side Effects of Drugs Annuals* (SEDA) 28, 29, and 30. The main contributors of this material were M Andréjak, JK Aronson, JJ Coleman, A del Favero, MG Franzosi, V Gras, J Harenberg, GD Johnston, P Joubert, DM Keeling, R Latini, PO Lim, TM MacDonald, AP Maggioni, U Martin, GT McInnes, K Peerlinck, DA Sica, R Verhaeghe, P Verhamme, J Vermylen, and F Zannad. For contributors to earlier editions of *Meyler's Side Effects of Drugs* and the *Side Effects of Drugs Annuals*, see http://www.elsevier.com/wps/find/bookseriesdescription.cws_home/BS_SED/description.

A brief history of the Meyler series

Leopold Meyler was a physician who was treated for tuberculosis after the end of the Nazi occupation of The Netherlands. According to Professor Wim Lammers, writing a tribute in Volume VIII (1975), Meyler got a fever from para-aminosalicylic acid, but elsewhere Graham Dukes has written, based on information from Meyler's widow, that it was deafness from dihydrostreptomycin; perhaps it was both. Meyler discovered that there was no single text to which medical practitioners could look for information about unwanted effects of drug therapy; Louis Lewin's text "Die Nebenwirkungen der Arzneimittel" ("The Untoward Effects of Drugs") of 1881 had long been out of print (SEDA-27, xxv-xxix). Meyler therefore determined to make such information available and persuaded the Netherlands publishing firm of Van Gorcum to publish a book, in Dutch, entirely devoted to descriptions of the adverse effects that drugs could cause. He went on to agree with the Elsevier Publishing Company, as it was then called, to prepare and issue an English translation. The first edition of 192 pages (*Schadelijke Nevenwerkingen van Geneesmiddelen*) appeared in 1951 and the English version (*Side Effects of Drugs*) a year later.

The book was a great success, and a few years later Meyler started to publish what he called surveys of unwanted effects of drugs. Each survey covered a period of two to four years. They were labelled as volumes rather than editions, and after Volume IV had been published Meyler could no longer handle the task alone. For subsequent volumes he recruited collaborators, such as Andrew Herxheimer. In September 1973 Meyler died unexpectedly, and Elsevier invited Graham Dukes to take over the editing of Volume VIII.

Dukes persuaded Elsevier that the published literature was too large to be comfortably encompassed in a four-yearly cycle, and he suggested that the volumes should be produced annually instead. The four-yearly volume could then concentrate on providing a complementary critical encyclopaedic survey of the entire field. The first *Side Effects of Drugs Annual* was published in 1977. The first encyclopaedic edition of *Meyler's Side Effects of Drugs*, which appeared in 1980, was labelled the ninth edition, and since then a new encyclopaedic edition has appeared every four years. The 15th edition was published in 2006, in both hard and electronic versions.

Monograph structure

This volume is in six sections:

- drugs used to treat hypertension, heart failure, and angina pectoris;
- diuretics—a general introduction to their adverse effects, followed by monographs on individual drugs;
- antidysrhythmic drugs— a general introduction to their adverse effects, followed by monographs on individual drugs;
- drugs that act on the cerebral and peripheral circulations;
- anticoagulants, thrombolytic agents, and anti-platelet drugs;
- cardiovascular adverse effects of non-cardiovascular drugs.

In each monograph in the Meyler series the information is organized into sections as shown below (although not all the sections are covered in each monograph).

DoTS classification of adverse drug reactions

A few adverse effects have been classified using the system known as DoTS. In this system adverse reactions are classified according to the ***Dose*** at which they usually occur, the ***Time-course*** over which they occur, and the ***Susceptibility factors*** that make them more likely, as follows:

- ***Relation to Dose***
  - Toxic reactions—reactions that occur at supratherapeutic doses
  - Collateral reactions—reactions that occur at standard therapeutic doses
  - Hypersusceptibility reactions—reactions that occur at subtherapeutic doses in susceptible individuals
- ***Time course***
  - Time-independent reactions—reactions that occur at any time during a course of therapy
  - Time-dependent reactions
    - Immediate or rapid reactions—reactions that occur only when a drug is administered too rapidly
    - First-dose reactions—reactions that occur after the first dose of a course of treatment and not necessarily thereafter

- Early reactions—reactions that occur early in treatment then either abate with continuing treatment (owing to tolerance) or persist
- Intermediate reactions—reactions that occur after some delay but with less risk during longer term therapy, owing to the "healthy survivor" effect
- Late reactions—reactions the risk of which increases with continued or repeated exposure
- Withdrawal reactions—reactions that occur when, after prolonged treatment, a drug is withdrawn or its effective dose is reduced
- Delayed reactions—reactions that occur some time after exposure, even if the drug is withdrawn before the reaction appears

- ***Susceptibility factors***
  - Genetic
  - Age
  - Sex
  - Physiological variation
  - Exogenous factors (for example drug–drug or drug–food interactions, smoking)
  - Diseases

Drug names

Drugs have usually been designated by their recommended or proposed International Non-proprietary Names (rINN or pINN); when these are not available, chemical names have been used. In some cases brand names have been used.

Spelling

For indexing purposes, American spelling has been used, e.g. anemia, estrogen rather than anaemia, oestrogen.

Cross-references

The various editions of *Meyler's Side Effects of Drugs* are cited in the text as SED-13, SED-14, etc; the *Side Effects of Drugs Annuals* are cited as SEDA-1, SEDA-2, etc.

J K Aronson
Oxford, August 2008

Organization of material in monographs in the Meyler series (not all sections are included in each monograph)

General information

Drug studies
- Observational studies
- Comparative studies
- Drug-combination studies
- Placebo-controlled studies
- Systematic reviews

Organs and systems
- Cardiovascular
- Respiratory
- Ear, nose, throat
- Nervous system
- Neuromuscular function
- Sensory systems
- Psychological
- Psychiatric
- Endocrine
- Metabolism
- Nutrition
- Electrolyte balance
- Mineral balance
- Metal metabolism
- Acid-base balance
- Fluid balance
- Hematologic
- Mouth
- Teeth
- Salivary glands
- Gastrointestinal
- Liver
- Biliary tract
- Pancreas
- Urinary tract
- Skin
- Hair
- Nails
- Sweat glands
- Serosae
- Musculoskeletal
- Sexual function
- Reproductive system
- Breasts
- Immunologic
- Autacoids
- Infection risk
- Body temperature
- Multiorgan failure
- Trauma
- Death

Long-term effects
- Drug abuse
- Drug misuse
- Drug tolerance
- Drug resistance
- Drug dependence
- Drug withdrawal
- Genotoxicity
- Cytotoxicity
- Mutagenicity
- Tumorigenicity

Second-generation effects
- Fertility
- Pregnancy
- Teratogenicity
- Fetotoxicity
- Lactation
- Breast feeding

Susceptibility factors
- Genetic factors
- Age
- Sex
- Physiological factors
- Disease
- Other features of the patient

Drug administration
- Drug formulations
- Drug additives
- Drug contamination and adulteration
- Drug dosage regimens
- Drug administration route
- Drug overdose

Interactions
- Drug-drug interactions
- Food-drug interactions
- Drug-device interactions
- Smoking
- Other environmental interactions

Interference with diagnostic tests

Diagnosis of adverse drug reactions

Management of adverse drug reactions

Monitoring therapy

References

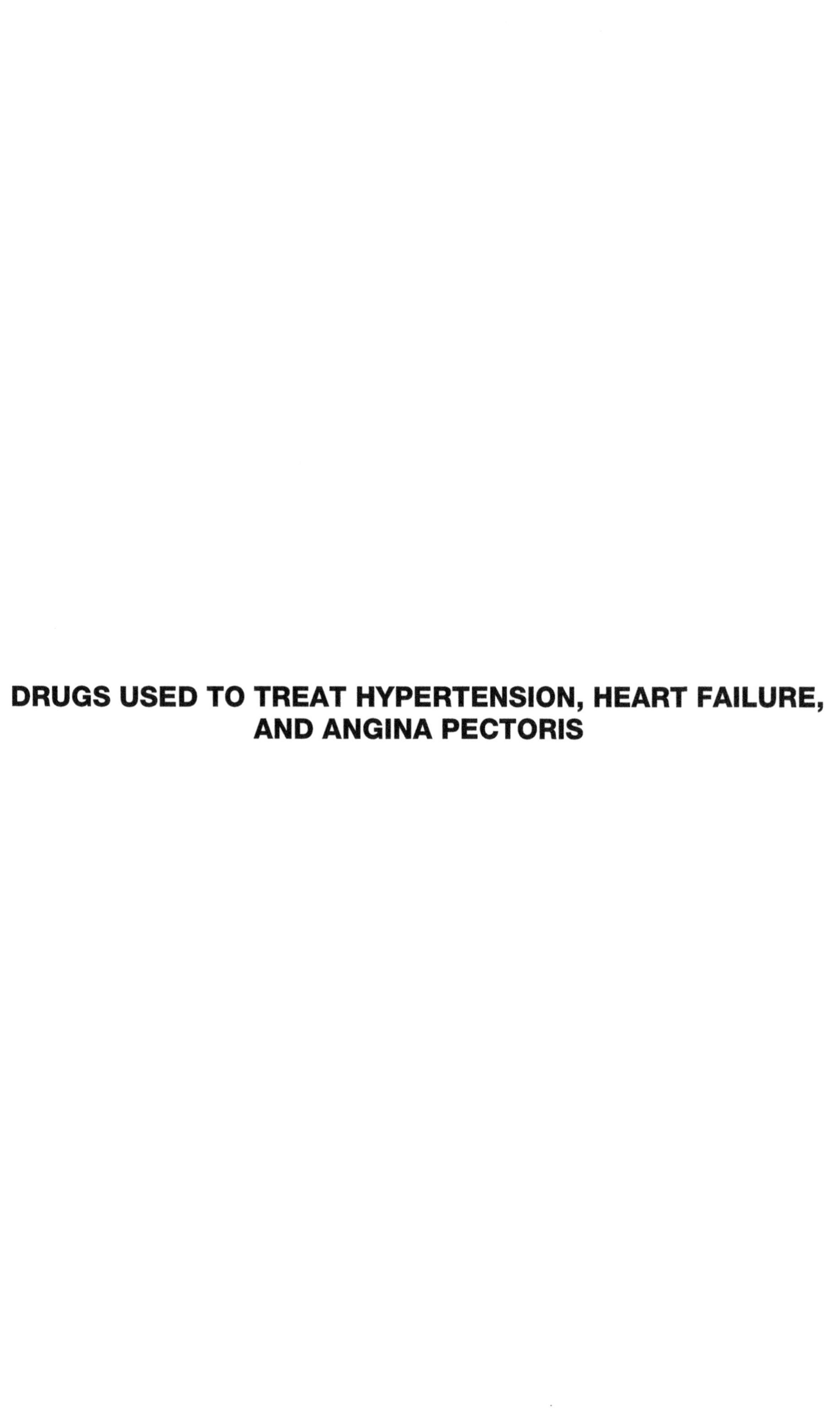

DRUGS USED TO TREAT HYPERTENSION, HEART FAILURE, AND ANGINA PECTORIS

Antihypertensive Drugs

General information

Moving targets and patterns of prescribing antihypertensive drugs

The landscape of hypertension management has changed considerably, and changes in treatment are reviewed every few years by national and international groups with interests in cardiovascular disease. In 2003 the Joint National Committee on prevention, detection, evaluation, and treatment of high blood pressure produced its seventh report (1). On the basis of data on the lifetime risk of hypertension and the risks of cardiovascular disease in patients with hypertension, their report emphasized the targets of disease treatment and pointed to new patterns of prescribing. Guidelines from the European Society of Hypertension and the European Society of Cardiology, also published in 2003 (2), gave similar perspectives.

In 2004 the British Society of Hypertension produced a comprehensive set of guidelines, endorsing the A(B)/CD algorithm (3). This strategy targets the renin–angiotensin–aldosterone system in younger Caucasian patients with angiotensin-converting enzyme (ACE) inhibitors or angiotensin receptor antagonists, while first-line treatment in older Caucasian or black patients of any age is with calcium channel blockers or thiazide diuretics; beta-blockers take a less important initial role in the absence of compelling indications. There are also concerns regarding the possible adverse metabolic consequences of long-term therapy with thiazide diuretics and beta-blockers.

Since the hypertension guidelines were published new evidence that strengthens this argument has appeared. Conventional blood pressure-lowering therapy (atenolol + bendroflumethiazide) has been compared with a more contemporary regimen of drugs (amlodipine + perindopril) in a large randomized controlled trial (4). The Anglo-Scandinavian Cardiac Outcomes Trial—Blood Pressure Lowering Arm (ASCOT-BPLA) has shown that treating hypertension with amlodipine and additional perindopril as required was associated with a reduction in the incidence of all types of cardiovascular events compared with atenolol + a thiazide. The overall incidence of adverse effects was similar in the two groups, but not surprisingly the specific adverse effects profiles were different. Cough, joint swelling, and peripheral edema were more common with amlodipine + perindopril, and bradycardia, dizziness, diarrhea, dyspnea, erectile dysfunction, fatigue, and cold extremities were more common with atenolol + a thiazide. Moreover, the amlodipine-based regimen caused new-onset diabetes in significantly fewer patients than the atenolol-based regimen did.

What implications does this newer evidence have on the current pattern of prescribing in hypertension? The combination of a calcium channel blocker with an ACE inhibitor (or an angiotensin receptor antagonist) has not previously been used as often as other combinations. Fixed-dose combinations are therefore not generally available, although they are likely to become more widely available. The ASCOT-BPLA study reaffirmed that most hypertensive patients require two or more agents to reach blood pressure targets. This endorses the latest guidelines, which propose that combination treatment should be considered for patients who present with a systolic blood pressure of 160 mmHg or more or a diastolic blood pressure of 100 mmHg or more.

Monitoring therapy

The publication of clear and explicit guidance on monitoring therapy in order to maximize efficacy and minimize adverse drug reactions is rare. The publication of practical recommendations for the use of ACE inhibitors, beta-blockers, aldosterone antagonists, and angiotensin receptor antagonists in heart failure may also be helpful in the safer administration of these drugs in hypertension (5). These guidelines provide advice about how these drugs should be used safely, including what advice should be given to the patient and what monitoring needs to be undertaken. Of equal value are the recommendations about the actions to be taken if problems occur, for example what to do in the event of electrolyte imbalance or renal dysfunction in patients taking ACE inhibitors.

Choice of antihypertensive drugs in patients with diabetes and hypertension

The choice of drugs in patients with diabetes and hypertension is important because antihypertensive drugs affect the development of complications such as albuminuria and the development of nephropathy, and because the metabolic effects of antihypertensive drugs can complicate treatment or enhance the development of diabetes.

The authors of a review of the treatment of combined diabetes and hypertension pointed out the importance of tight blood pressure control (aiming for a blood pressure below 130/80 for all diabetics and below 125/75 in the presence of significant proteinuria) for the prevention of cardiovascular mortality and morbidity, and the development and progression of diabetic nephropathy (6). Adequate control of blood pressure is more important than the choice of drug, and multiple drugs are often required.

The general consensus is that ACE inhibitors should be the first-line choice, angiotensin II receptor blockers being a reasonable alternative.

Thiazide diuretics impair glucose tolerance. On the other hand the increase in renin that they cause enhances the effects of ACE inhibitors and angiotensin II receptor blockers. It also appears that the adverse effect on blood glucose can be eliminated by avoiding hypokalemia.

Beta-blockers reduce proteinuria and cardiovascular mortality. They can worsen glycemic control, reduce awareness of hypoglycemia, and adversely affect lipid profiles. However, in patients with diabetes and hypertension and a history of myocardial infarction, the benefits may outweigh the risks.

Calcium channel blockers combined with ACE inhibitors appear to provide additional renoprotection.

The LIFE study

Further commentaries on the LIFE study in over 9000 patients (7) have appeared in 2003.

The key findings alluded to in a commentary (8), in terms of hypertension and diabetes, were that atenolol or losartan as monotherapy reduced blood pressure in patients with diabetes and hypertension, but not to the target blood pressure, suggesting that more intensive therapy is required than was used in the LIFE study. The data suggest that the onset of diabetes can be prevented or delayed by losartan, and losartan is also more effective than atenolol in reducing cardiovascular mortality and morbidity in patients with diabetes taking suboptimal treatment. In another commentary (9) it was suggested that losartan is clearly better and that elderly patients with hypertension should not be exposed to beta-blockers.

The ALPINE study

In a 1-year study, 392 newly diagnosed patients with hypertension were randomized to either candesartan 16 mg/day or hydrochlorothiazide 25 mg/day; if the blood pressure did not fall below 135/85 in patients aged under 65 years or 140/90 in patients aged 65 years or older, extended-release felodipine 2.5–5.0 mg was added to candesartan or atenolol 50–100 mg to hydrochlorothiazide (10). The fall in blood pressure was similar in the two groups and most patients required two drugs. Fasting insulin and glucose concentrations increased in the hydrochlorothiazide + atenolol group, but were unaffected in the candesartan + felodipine group. Eight patients in the thiazide group developed diabetes mellitus compared with one in the candesartan group.

Other studies in diabetes

In 463 patients with type II diabetes and hypertension, a combination of atenolol + chlortalidone produced worse metabolic control (HbA_{1c}), whereas metabolic control was minimally affected with verapamil + trandolapril (11). Both regimens produced similar suboptimal falls in mean blood pressure.

In 457 patients with type II diabetes, hypertension, and albuminuria, the effect of daily perindopril 2 mg + indapamide 0.625 mg was compared with the effect of daily enalapril 10 mg (12). Based on blood pressure, doses could be increased to a maximum of 8.0 mg of perindopril + 2.5 mg of indapamide or 40 mg of enalapril. The combination produced a statistically significant greater fall in blood pressure, but it is difficult to see this as clinically relevant (3.0 and 1.5 mm more for systolic and diastolic pressures respectively). There was a significantly greater reduction in albuminuria with the combination (−40%), than with monotherapy (−27%).

Combination therapy

Several smaller studies have suggesting that monotherapy is usually not optimal for patients with diabetes and hypertension, and that combination therapy would be required in most cases. In 24 patients with diabetes and hypertension, dual renin-angiotensin blockade with lower doses of an ACE inhibitor and an angiotensin II receptor blocker was superior to maximal doses of either alone (13). In 38 patients with diabetes and hypertension benazepril + amlodipine produced better reduction in blood pressure and a more favorable effect on fibrinolytic balance than either drug alone (14).

References

1. Chobanian AV, Bakris GL, Black HR, Cushman WC, Green LA, Izzo JL Jr, Jones DW, Materson BJ, Oparil S, Wright JT Jr, Roccella EJ; Joint National Committee on Prevention, Detection, Evaluation, and Treatment of High Blood Pressure. National Heart, Lung, and Blood Institute; National High Blood Pressure Education Program Coordinating Committee. Seventh report of the Joint National Committee on Prevention, Detection, Evaluation, and Treatment of High Blood Pressure. Hypertension 2003;42(6):1206–52.
2. European Society of Hypertension–European Society of Cardiology guidelines for the management of arterial hypertension. J Hypertens 2003;21(6):1011–53.
3. Williams B, Poulter NR, Brown MJ, Davis M, McInnes GT, Potter JF, Sever PS, Thom SMcG; British Hypertension Society. Guidelines for management of hypertension: report of the Fourth Working Party of the British Hypertension Society, 2004—BHS IV. J Hum Hypertens 2004;18(3):139–85.
4. Dahlof B, Sever PS, Poulter NR, Wedel H, Beevers DG, Caulfield M, Collins R, Kjeldsen SE, Kristinsson A, McInnes GT, Mehlsen J, Nieminen M, O'Brien E, Ostergren J; ASCOT Investigators. Prevention of cardiovascular events with an antihypertensive regimen of amlodipine adding perindopril as required versus atenolol adding bendroflumethiazide as required, in the Anglo-Scandinavian Cardiac Outcomes Trial-Blood Pressure Lowering Arm (ASCOT-BPLA): a multicentre randomised controlled trial. Lancet 2005;366(9489):895–906.
5. McMurray J, Cohen-Solal A, Dietz R, Eichhorn E, Erhardt L, Hobbs FD, Krum H, Maggioni A, McKelvie RS, Pina IL, Soler-Soler J, Swedberg K. Practical recommendations for the use of ACE inhibitors, beta-blockers, aldosterone antagonists and angiotensin receptor blockers in heart failure: putting guidelines into practice. Eur J Heart Fail 2005;7(5):710–21.
6. Padilla R, Estacio RO. New insights into the combined burden of type 2 diabetes and hypertension. Heart Drug 2003;3:25–33.
7. Dahlop B, Devereux RB, Kjeldsen SE. Cardiovascular mortality and morbidity in the Losartan Intervention For Endpoint reduction in hypertension study (LIFE): a randomized trial against atenolol. Lancet 2002;359:995–1003.
8. Nadar I, Lim HS, Lip GYH. Implications of the LIFE trial. Exp Opin Investig Drugs 2003;12:871–7.
9. Messerli FH. The LIFE study: the straw that should break the camel's back. Eur Heart J 2003;24:487–9.
10. Lindholm LH, Persson M, Alaupovic P, Carlberg B, Svensson A, Samuelsson O. Metabolic outcome during 1 year in newly detected hypertensives: results of the Antihypertensive treatment and Lipid Profile In a North of Sweden Efficacy evaluation (ALPINE study). J Hypertens 2003;21:1563–74.
11. Holzgreve H, Nakov R, Beck K, Janka HU. Antihypertensive therapy with verapamil SR plus trandolapril versus atenolol plus chlorthalidone on glycaemic control. Am J Hypertens 2003;16:381–6.

12. Morgensen CE, Viberti G, Halimi Í, Ritz E, Ruilope L, Jermendy G, Widimsky J, Sarelli, P, Taton J, Rull J, Erdogan G, De Leeuw PW, Ribeiro A, Sanchez R, Mechmeche R, Nolan J, Sirotiokova J, Hamani A, Scheen A, Hess B, Luger A, Thomas SM. Effect of low-dose perindopril/indapamide on albuminuria in diabetes. Hypertension 2003;41:1063–71.
13. Jacobsen P, Andersen S, Rossing K, Jensen BR, Parving H. Dual blockade of the renin angiotensin system versus maximal recommended dose of ACE inhibition in diabetic nephropathy. Kidney Int 2003;63:1874–80.
14. Fogari R, Preti P, Lazzari P, Corradi L, Zoppi A, Fogari E, Mugellini A. Effect of benazepril amlodipine combination on fibrinolysis in hypertensive diabetic patients. Eur J Clin Pharmacol 2003;59:271–3.

Acebutolol

See also Beta-adrenoceptor antagonists

General Information

Acebutolol is a beta-adrenoceptor antagonist with membrane-stabilizing activity that is sometimes cited as being cardioselective but has considerable effects on bronchioles and peripheral blood vessels.

Organs and Systems

Respiratory

Bronchiolitis obliterans has been attributed to acebutolol (1).

Liver

Six cases of reversible hepatitis have been attributed to acebutolol (2).

Skin

Various drugs can cause a lupus-like syndrome. Beta-adrenoceptor antagonists have been implicated only infrequently and there have been no cases of subacute cutaneous lupus erythematosus associated with the use of beta-adrenoceptor antagonists. Subacute cutaneous lupus erythematosus has been attributed to acebutolol (3).

- A 57-year-old woman with hypertension developed a cutaneous eruption taking acebutolol for 1 month. She had no history of photosensitivity, photodermatosis, or immunological diseases. A complete blood cell count, liver and kidney tests, rheumatoid factor, and complement fractions were all within the refrence ranges. There was a positive titer of antinuclear antibodies. A biopsy specimen showed atrophy of the epidermis. A positive lupus band test was found at direct immunofluorescence. Acebutolol was withdrawn, and she was given chloroquine sulfate associated with photoprotection. The cutaneous eruption resolved progressively. After 4 months the skin lesions had completely cleared. A Seroly test was negative for antihistone antibodies.

While several cases of subacute cutaneous lupus erythematosus have been described with other antihypertensive agents, such as captopril, calcium channel blockers, and hydrochlorothiazide, this seems to have been the first case described in a patient taking a beta-adrenoceptor antagonist. This case and its evolution suggest a link between acebutolol therapy and the onset of a lupus-like syndrome, whose pathogenesis is unclear.

Immunologic

Patients taking acebutolol relatively commonly develop antinuclear antibodies (4,5).

Drug Administration

Drug overdose

The membrane-stabilizing activity of beta-blockers can play a major role in toxicity. Of 208 deaths in subjects who had taken beta-blockers, 206 occurred with drugs that have membrane-stabilizing activity. This quinidine-like effect can be reversed by sodium bicarbonate, which is also used to counteract the cardiotoxic effects of cyclic antidepressants, which also have membrane-stabilizing activity.

- An overdose of acebutolol (6.4 mg) in a 48-year-old man caused cardiac arrest with ventricular tachycardia (6). An intravenous bolus of sodium bicarbonate 50 mmol produced sinus rhythm.

References

1. Camus P, Lombard JN, Perrichon M, Piard F, Guerin JC, Thivolet FB, Jeannin L. Bronchiolitis obliterans organising pneumonia in patients taking acebutolol or amiodarone. Thorax 1989;44(9):711–5.
2. Tanner LA, Bosco LA, Zimmerman HJ. Hepatic toxicity after acebutolol therapy. Ann Intern Med 1989;111(6):533–4.
3. Fenniche S, Dhaoui A, Ben Ammar F, Benmously R, Marrak H, Mokhtar I. Acebutolol-induced subacute cutaneous lupus erythematosus. Skin Pharmacol Physiol 2005;18:230–3.
4. Booth RJ, Bullock JY, Wilson JD. Antinuclear antibodies in patients on acebutolol. Br J Clin Pharmacol 1980;9(5):515–7.
5. Cody RJ Jr, Calabrese LH, Clough JD, Tarazi RC, Bravo EL. Development of antinuclear antibodies during acebutolol therapy. Clin Pharmacol Ther 1979;25(6):800–5.
6. Donovan KD, Gerace RV, Dreyer JF. Acebutolol-induced ventricular tachycardia reversed with sodium bicarbonate. J Toxicol Clin Toxicol 1999;37(4):481–4.

Alfuzosin

See also Alpha-adrenoceptor antagonists

General Information

Alfuzosin is a uroselective $alpha_1$-adrenoceptor antagonist used to relieve the symptoms of prostatic hyperplasia (1). Its safety has been investigated in a large prospective 3-year open trial in 3228 patients with benign prostatic hyperplasia. There were no unexpected adverse effects. Only 4.2% of the patients dropped out owing to adverse effects.

In a large database of 7093 patients with lower urinary tract symptoms related to benign prostatic hyperplasia treated for up to 3 years with alfuzosin in general practice, adverse events were reported in a very complex and uninformative way (2). In another paper, the same authors reported on a subcohort of 2829 patients, with special focus on effects on quality of life. Adverse events occurred in 15% of the patients, 1.7% died during the study, and 5.2% had serious effects, which the authors did not detail, but which they stated were not related to treatment. Most adverse effects occurred during the first 3 months of treatment (3). In another database of 3095 Spanish patients taking alfuzosin 5 mg bd for 60 days, adverse events were reported in 3.3% of the patients, and led to drug withdrawal in 1.6%; postural hypotension occurred in 1.8% (4).

In a systematic review 11 trials of alfuzosin in 3901 men were analysed (5). Alfuzosin was safe and well tolerated. Most of the reported adverse events, such as dizziness and syncope, were related to its vasodilatory action.

Organs and Systems

Nervous system

Dizziness, headache, postural hypertension, and other symptoms familiar from the older alpha-blockers occur primarily during the first 2 weeks of treatment with alfuzosin (1).

Liver

Hepatitis potentially related to alfuzosin has been reported.

- A 63-year-old man, who had taken amiloride and alfuzosin for 9 months for hypertension and benign prostatic hyperplasia, became jaundiced (6). His aspartate transaminase was 3013 IU/l, alanine transaminase 2711 IU/l, alkaline phosphatase 500 IU/l, and total bilirubin 415 μmol/l. Viral causes, autoimmune hepatitis, and biliary obstruction were excluded. After withdrawal of alfuzosin, his liver function tests gradually returned to normal within 6 months.
- An 80-year-old man with chronic liver disease due to hepatitis B virus took alfuzosin for 3 weeks for benign prostatic hyperplasia and developed raised liver enzymes, which settled rapidly on withdrawal of alfuzosin (7).

Immunologic

Dermatomyositis has been attributed to alfuzosin.

- A 75-year-old man, who had taken alfuzosin for 1 year, developed muscle pain and weakness over 4 days, accompanied by tenderness and swelling of the deltoid muscles (8). There was erythema, with rash, periungual purpura, and erythematous plaques over the finger joints. Serum CK, LDH, and transaminase activities were raised and ANA was positive. An MRI scan showed findings consistent with inflammation of muscle and a biopsy confirmed the diagnosis of dermatomyositis. Three days after drug withdrawal there was no improvement, so prednisone was started and he recovered within a few days. The temporal relation in this case was weak.
- Dermatomyositis, with typical clinical effects, biochemical tests, electromyography, and muscle biopsy, occurred in a 75-year-old man who had taken alfuzosin for 1 year (9). There was no malignancy and he recovered fully after alfuzosin withdrawal (timing not given).

Drug Administration

Drug formulations

The pharmacology, including the tolerability and drug-interaction potential, of a modified-release formulation of alfuzosin, relating mainly to studies in patients symptomatic benign prostatic hyperplasia, has been reviewed (10).

References

1. McKeage K, Plosker GL. Alfuzosin: a review of the therapeutic use of the prolonged-release formulation given once daily in the management of benign prostatic hyperplasia. Drugs 2002;62(4):633–53.
2. Lukacs B, Grange JC, Comet D, McCarthy C. History of 7,093 patients with lower urinary tract symptoms related to benign prostatic hyperplasia treated with alfuzosin in general practice up to 3 years. Eur Urol 2000;37(2):183–90.
3. Lukacs B, Grange JC, Comet D. One-year follow-up of 2829 patients with moderate to severe lower urinary tract symptoms treated with alfuzosin in general practice according to IPSS and a health-related quality-of-life questionnaire. BPM Group in General Practice. Urology 2000;55(4):540–6.
4. Sanchez-Chapado M, Guil M, Alfaro V, Badiella L, Fernandez-Hernando N. Safety and efficacy of sustained-release alfuzosin on lower urinary tract symptoms suggestive of benign prostatic hyperplasia in 3,095 Spanish patients evaluated during general practice. Eur Urol 2000;37(4):421–7.
5. MacDonald R, Wilt TJ. Alfuzosin for treatment of lower urinary tract symptoms compatible with benign prostatic hyperplasia: a systematic review of efficacy and adverse effects. Urology 2005;66(4):780–8.
6. Zabala S, Thomson C, Valdearcos S, Gascon A, Pina MA. Alfuzosin-induced hepatotoxicity. J Clin Pharm Ther 2000;25(1):73–4.
7. Yolcu OF, Koklu S, Koksal AS, Yuksel O, Beyazit Y, Basar O. Alfuzosin-induced acute hepatitis in a patient with chronic liver disease. Ann Pharmacother 2004;38(9):1443–5.
8. Vela-Casasempere P, Borras-Blasco J, Navarro-Ruiz A. Alfuzosin-associated dermatomyositis. Br J Rheumatol 1998;37(10):1135–6.
9. Schmutz J-L, Barbaud A, Trechot PH. Alfuzosine, inducteur de dermatomyosite. [Alfuzosine-induced dermatomyositis.] Ann Dermatol Venereol 2000;127(4):449.
10. Guay DR. Extended-release alfuzosin hydrochloride: a new alpha-adrenergic receptor antagonist for symptomatic benign prostatic hyperplasia. Am J Geriatr Pharmacother 2004;2(1):14–23.

Alpha-adrenoceptor antagonists

See also Alfuzosin, Doxazosin, Indoramin, Prazosin, Terazosin

General Information

The postsynaptic alpha-adrenoceptor antagonists, indoramin, prazosin, and related quinazoline derivatives, block alpha$_1$-adrenoceptor-mediated vasoconstriction of peripheral blood vessels (both arterial and venous) and are effectively peripheral vasodilators (1,2). Qualitatively and quantitatively common adverse effects are generally similar, although indoramin has additional effects on other neurotransmitter systems and therefore tends to be considered separately. Their use in benign prostatic hyperplasia has been reviewed (3,4).

Several recent articles have reviewed the pharmacology, pharmacokinetics, mode of action, use, efficacy, and adverse effects of the selective alpha$_1$-adrenoceptor blockers doxazosin, prazosin, and terazosin in benign prostatic hyperplasia (5).

The frequencies and the profile of adverse effects of five major classes of antihypertensive agents have been assessed in an unselected group of 2586 chronically drug-treated hypertensive patients (6). This was accompanied by a questionnaire-based survey among patients attending a general practitioner. The percentage of patients who reported adverse effects spontaneously, on general inquiry, and on specific questioning were 16, 24, and 62% respectively. With alpha-blockers the figures were 15, 25, and 50%. The percentage of patients in whom discontinuation was due to adverse effects was 6.8% with alpha-blockers. Alpha-blockers were associated with less fatigue, cold extremities, sexual urge, and insomnia, and more bouts of palpitation than other antihypertensive drugs (RR = 2.5; CI = 1.2, 5.4). The authors did not find a significant effect of age on the pattern of adverse effects. Women reported more effects and effects that were less related to the pharmacological treatment.

The first-dose effect (profound postural hypotension and reflex tachycardia) is a well-recognized complication of the first dose of prazosin and related agents. This phenomenon is dose-related and can usually be avoided by using a low initial dosage taken at bedtime. During long-term treatment, orthostatic hypotension and dizziness is reported by about 10% of patients.

Current guidelines on the use of postsynaptic alpha-adrenoceptor antagonists have been reviewed (7).

Drug-drug interactions

Inhibitors of phosphodiesterase type V

Postsynaptic alpha-adrenoceptor antagonists are used both in hypertension and for urological conditions, and can cause orthostatic hypotension due to vasodilatation. This adverse effect can be potentiated considerably if they are co-administered with inhibitors of phosphodiesterase type V for the treatment of erectile dysfunction (8).

References

1. Grimm RH Jr. Alpha 1-antagonists in the treatment of hypertension. Hypertension 1989;13(5 Suppl):I131–6.
2. Luther RR. New perspectives on selective alpha 1 blockade. Am J Hypertens 1989;2(9):729–35.
3. Beduschi MC, Beduschi R, Oesterling JE. Alpha-blockade therapy for benign prostatic hyperplasia: from a nonselective to a more selective alpha$_{1A}$-adrenergic antagonist. Urology 1998;51(6):861–72.
4. Narayan P, Man In't Veld AJ. Clinical pharmacology of modern antihypertensive agents and their interaction with alpha-adrenoceptor antagonists. Br J Urol 1998; 81(Suppl 1):6–16.
5. Akduman B, Crawford ED. Terazosin, doxazosin, and prazosin: current clinical experience. Urology 2001;58(6 Suppl 1): 49–54.
6. Olsen H, Klemetsrud T, Stokke HP, Tretli S, Westheim A. Adverse drug reactions in current antihypertensive therapy: a general practice survey of 2586 patients in Norway. Blood Press 1999;8(2):94–101.
7. Sica DA. Alpha1-adrenergic blockers: current usage considerations. J Clin Hypertens (Greenwich) 2005;7(12):757–62.
8. Kloner RA. Pharmacology and drug interaction effects of the phosphodiesterase 5 inhibitors: focus on alpha-blocker interactions. Am J Cardiol 2005;96(12B):42M–46M.

Ambrisentan

General Information

Ambrisentan is an endothelin ET_A receptor antagonist (1). It has been used in pulmonary arterial hypertension and there have been one dose-ranging study, two randomized, double-blind, placebo-controlled studies, and one drug-conversion study. In the dose-ranging study, ambrisentan 1–10 mg produced significant improvements from baseline in walking distance at 12 weeks (2). In the placebo-controlled studies, ambrisentan 2.5–10 mg/day was associated with significant improvement in walking distance at 12 weeks and sustained for up to 1 year. The most common adverse effects associated with ambrisentan in clinical trials were peripheral edema (17%), nasal congestion (6%), palpitation (5%), constipation (4%), flushing (4%), abdominal pain (3%), nasopharyngitis (3%), and sinusitis (3%). In the placebo-controlled studies, the incidence of liver aminotransferase and bilirubin abnormalities at 12 weeks was lower with ambrisentan than with placebo (0.8% versus 2.3% respectively). Patients who had had raised serum transaminase activities during previous therapy with bosentan or sitaxsentan were switched to ambrisentan without further abnormalities in liver function. In a double-blind, dose-ranging study in 64 patients with pulmonary hypertension adverse events reflected those common to the endothelin receptor antagonist class, but two patients developed raised serum transaminase activities, one of whom required treatment withdrawal (3). Raised liver enzymes have been seen with bosentan and other drugs in this class.

References

1. Hussar DA. New drugs: ambrisentan, temsirolimus, and eculizumab. J Am Pharm Assoc (2003) 2007;47(5):664, 666–7, 669–71.
2. Cheng JW. Ambrisentan for the management of pulmonary arterial hypertension. Clin Ther 2008;30(5):825–33.
3. Galié N, Badesch D, Oudiz R, Simonneau G, McGoon MD, Keogh AM, Frost AE, Zwicke D, Naeije R, Shapiro S, Olschewski H, Rubin LJ. Ambrisentan therapy for pulmonary arterial hypertension. J Am Coll Cardiol 2005;46(3): 529–35.

Amlodipine

See also Calcium channel blockers

General Information

Amlodipine is a long-acting dihydropyridine calcium channel blocker. It has an adverse effects profile similar to those of other dihydropyridines, but at a lower frequency (1). Along with felodipine (2), but unlike other calcium channel blockers, it may also be safer in severe chronic heart failure when there is concurrent angina or hypertension (3).

The effects of amlodipine and isosorbide-5-mononitrate for 3 weeks on exercise-induced myocardial stunning have been compared in a randomized, double-blind, crossover study in 24 patients with chronic stable angina and normal left ventricular function (4). Amlodipine attenuated stunning, evaluated by echocardiography, significantly more than isosorbide, without difference in anti-ischemic action or hemodynamics. Amlodipine was better tolerated than isosorbide, mainly because of a lower incidence of headache (4).

Vasodilatory calcium channel blockers have been reported to improve exercise tolerance in some preliminary studies. A multicenter, randomized, placebo-controlled trial was therefore performed in 437 patients with mild to moderate heart failure to assess the effects of amlodipine 10 mg/day in addition to standard therapy (5). Over 12 weeks amlodipine did not improve exercise time and did not increase the incidence of adverse events.

Mental stress is a risk factor for cardiovascular disease. In 24 patients with mild to moderate hypertension, amlodipine reduced the blood pressure rise during mental stress compared with placebo, but increased plasma noradrenaline concentrations (6).

Hypertension leading to cardiac dysfunction is very frequent in patients with the inherited syndrome called Ribbing's disease, which is characterized by multiple epiphyseal dystrophy. In a randomized, double-blind comparison of amlodipine (10 mg/day) and enalapril (20 mg/day) in 50 patients for 6 months, both drugs significantly reduced blood pressure, but amlodipine increased heart rate and plasma concentrations of noradrenaline and angiotensin II (7). These undesired effects make ACE inhibitors a better choice for prevention of cardiac dysfunction.

Placebo-controlled studies

The efficacy and safety of amlodipine have been assessed in a multicenter, double-blind, placebo-controlled trial in 268 children with hypertension aged 6–16 years (8). Amlodipine produced significantly greater reductions in systolic blood pressure than placebo. Twelve patients withdrew from the study because of adverse events, six of which were attributed to the study drug: three cases of worsening hypertension, one of facial edema, one of finger edema and rash, and one of ventricular extra beats. The maximal dose, 5 mg/day, was not high, and the target to reduce blood pressure below the 95th centile was reached in 35% of children with systolic hypertension and in 55% of those with diastolic hypertension.

Organs and Systems

Nervous system

- A 35-year-old woman with benign intracranial hypertension and high blood pressure was given amlodipine, with good control of her blood pressure (9). However, her headache worsened and she developed papilledema. The CSF pressure was 30 cm. Her symptoms disappeared shortly after amlodipine withdrawal.

Fluid balance

Calcium channel blockers often cause peripheral edema, usually limited to the lower legs; periocular and perioral edema are less common. Occasionally edema can be more severe, and a case of anasarca has been reported in a 77-year-old woman with essential hypertension taking amlodipine 10 mg/day (10).

Hematologic

Thrombocytopenia has been attributed to amlodipine (11).

- A 79-year-old man developed epistaxis and gum bleeding; his platelet count was 1×109/l. Amlodipine was withdrawn and immunoglobulins and glucocorticoids were given. The platelet count returned to 204×109/l in 7 days. Amlodipine was restarted, and 2 days later bleeding recurred and resolved after amlodipine was withdrawn for the second time. ELISA (enzyme-linked immunosorbent assay) showed an IgG antibody reactive with patient's platelets only in the presence of amlodipine.

The authors suggested that drug-related thrombocytopenia can occur after long-term treatment with a drug, such as in this patient who had been taking amlodipine for 10 years before the event.

Liver

Hepatitis has been attributed to amlodipine.

- A 77-year-old man took amlodipine for 1 month and developed jaundice and raised aspartate transaminase, alanine transaminase, and bilirubin (12). A liver biopsy suggested a drug-induced hepatitis and the amlodipine was withdrawn. His symptoms and laboratory values normalized. Other drugs (metformin, fluindione, and omeprazole) were not withdrawn.
- A 69-year-old hypertensive man who had taken amlodipine for 10 months abruptly developed jaundice, choluria, raised serum bilirubin, and increased transaminases (13). After amlodipine withdrawal he progressively recovered in a few weeks without sequelae or relapses. However, after several months he presented again with jaundice and an enlarged liver, having started to take diltiazem 5 months before. He recovered completely in a few weeks after drug withdrawal.

In the second case the authors hypothesized an idiosyncratic mechanism.

- An 87-year-old woman who had taken amlodipine for several years for hypertension developed pruritus and 2 weeks later painless jaundice (14). She had a raised bilirubin concentration and raised aspartate and alanine transaminase activities. Infectious causes were not found and a liver biopsy suggested drug-induced liver damage. After withdrawal of amlodipine the transaminases and measures of cholestasis improved markedly within 2 weeks.

Skin

Recognized skin eruptions associated with amlodipine include erythematous and maculopapular rashes, skin discoloration, urticaria, dryness, alopecia, dermatitis, erythema multiforme, and lichen planus. A granuloma annulare-like eruption has been reported (15).

- A 64-year-old Caucasian woman, with a history of ankylosing spondylitis, hypertension, and osteoporosis, took amlodipine for 13 days and developed a rash on her lower legs. Amlodipine was withdrawn, but the rash progressed to involve both of her hands. The eruption consisted of multiple erythematous pruritic papules. Histology showed focal collagen degeneration and a significant interstitial histiocytic dermal infiltrate, suggestive of granuloma annulare. Within 3 months of withdrawal of amlodipine the reaction cleared and did not recur during follow-up for 3 years.

Amlodipine can cause generalized pruritus, which usually happens within 24 hours and resolves within 24 hours of withdrawal (16).

Photosensitivity presenting with telangiectasia can be caused by calcium channel blockers.

- A 57-year-old hypertensive man developed telangiectasia, initially on the forehead and rapidly extending to the upper back, shoulders, and chest, particularly during the summer (17). The eruption began 1 month after starting amlodipine and diminished considerably 3 months after withdrawal.
- A 3-year-old girl developed telangiectases on the cheeks and gingival hyperplasia while taking furosemide, captopril, and amlodipine for hypertension due to hemolytic–uremic syndrome (18). Both lesions disappeared on withdrawal of amlodipine.

Calcium channel blockers can cause lichen planus.

- A 56-year-old Nigerian woman, with a previous history of sickle cell trait, osteoarthritis, and non-insulin-dependent diabetes mellitus, took amlodipine 5 mg/day for hypertension for 2 weeks and developed a lichenoid eruption (19). Histological examination confirmed the diagnosis of lichen planus. Amlodipine was withdrawn and there was rapid symptomatic and clinical improvement after treatment with glucocorticoids and antihistamines.

Generalized hyperpigmentation has been reported (20).

- A 45-year-old Turkish man with a history of hypertension who had taken amlodipine 10 mg/day for 3 years developed Fitzpatrick's skin type III after a 2-year history of gradually increasing, asymptomatic, generalized hyperpigmentation. Although cutaneous hyperpigmentation was more prominent on the photoexposed areas, there was no history of previous photosensitivity, pruritus, or flushing. Photo protection and withdrawal of amlodipine was advised. The skin discoloration faded slightly 8 months after changing amlodipine to metoprolol and strict avoidance of sun exposure.

Nails

Longitudinal melanonychia is tan, brown, or black longitudinal streaking in the nail plate due to increased melanin deposition and Hutchinson's sign is periungual pigmentation. In a 75-year-old Indian man longitudinal melanonychia and periungual pigmentation affecting several fingernails and toenails were attributed to amlodipine, which he had taken for 2 years for hypertension (21).

Musculoskeletal

A patient presented with severe, generalized muscle stiffness, joint pain, and fatigue while taking amlodipine for hypertension and zafirlukast for asthma. Stopping zafirlukast did not change her symptoms; the dose of amlodipine was increased at different times up to 15 mg to control blood pressure better. The neurological symptoms worsened, in the absence of any evidence of immunological or neurological disorders, and so amlodipine was withdrawn: the symptoms disappeared within 4 days (22).

Reproductive system

Gynecomastia is not uncommon in men undergoing hemodialysis for end-stage renal disease. Two cases of gynecomastia have been reported in patients taking amlodipine 10 mg/day (23). In both cases the gynecomastia abated within a month or so of substituting amlodipine with an angiotensin receptor blocker. In one case, amlodipine was re-administered because of worsening of hypertension, and gynecomastia reappeared.

Second-Generation Effects

Pregnancy

Subcutaneous fat necrosis in a neonate has been attributed to maternal use of amlodipine during pregnancy (24).

- A boy weighing 4 kg was born by spontaneous normal delivery at 39 weeks to a 38-year-old Afro-Caribbean woman, whose pregnancy was complicated by essential hypertension treated with amlodipine. On day 1 the child developed firm, red, pea-sized nodular lesions on the face, buttocks, back, shoulders, and arms.

Subcutaneous fat necrosis of the newborn is relatively uncommon. It is said to be benign and painless and to resolve within a few weeks. However, in this case it was extremely painful and was relieved only by opiates. The skin changes persisted beyond the age of 6 months and remained extremely symptomatic until the age of 9 months, when the skin had become normal. Calcium abnormalities have often been reported in association with subcutaneous fat necrosis, and exposure to amlodipine during pregnancy may have resulted in impairment of enzyme systems dependent on calcium fluxes for their action; it may also have affected calcium homeostasis in the neonate. Since previous reports of teratogenicity in animals have been published, few women take calcium channel blockers during pregnancy and there are no reports to date of an association between these drugs and subcutaneous fat necrosis (24).

Drug Administration

Drug overdose

Amlodipine overdose has been reported (25).

- A 23-year-old woman took 60 tablets of amlodipine intentionally and developed tachycardia and severe hypotension. She did not improve with intensive therapy and developed left ventricular failure and oliguria and underwent hemodiafiltration. Her condition slowly improved over 4 days.

Drug–Drug Interactions

Chloroquine

A possible interaction of amlodipine with chloroquine has been reported (26).

- A 48-year-old hypertensive physician, who had optimal blood pressure control after taking oral amlodipine 5 mg/day for 3 months, developed a slight frontal headache and fever, thought that he had malaria, and took four tablets of chloroquine sulfate (total 600 mg base). Two hours later he became nauseated and dizzy and collapsed; his systolic blood pressure was 80 mmHg and his diastolic pressure was unrecordable, suggesting vasovagal syncope, which was corrected by dextrose–saline infusion.

There was no malaria parasitemia in this case, and hence the syncope may have resulted from the acute synergistic hypotensive, venodilator, and cardiac effects of chloroquine plus amlodipine, possibly acting via augmented nitric oxide production and calcium channel blockade. Since malaria fever is itself associated with orthostatic hypotension, this possible interaction may be unrecognized and unreported in these patients.

Ciclosporin

Ciclosporin increases the survival of allografts in man. However, it causes renal vasoconstriction and increases proximal tubular reabsorption, leading in some cases to hypertension (27). The concomitant use of calcium channel blockers can prevent most of these adverse effects of ciclosporin. However, some calcium channel blockers (verapamil, diltiazem, nicardipine) can increase plasma concentrations of ciclosporin up to three-fold through inhibition of cytochrome P450. Eight different studies have been performed on the combination of amlodipine and ciclosporin given for 1–6 months to kidney transplant recipients, and the results have been reviewed (28). In three studies, in a total of 41 patients, amlodipine increased ciclosporin concentrations, while in the others, a total of 85 patients, there was no evidence of an interaction.

In normotensive renal transplant recipients treated for 2 months with amlodipine there was a small but significant nephroprotective effect (29). Thus, amlodipine, in contrast to other calcium channel blockers, does not affect ciclosporin blood concentrations and can be safely added in transplant recipients.

Sildenafil

The effect of sildenafil on arterial pressure has been tested in 16 hypertensive men taking amlodipine 5–10 mg/day (30). Sildenafil did not affect amlodipine pharmacokinetics, but caused a further additive fall in blood pressure. Adverse events with the combination of sildenafil and amlodipine, headache, dyspepsia, and nausea, did not require drug withdrawal.

References

1. Osterloh I. The safety of amlodipine. Am Heart J 1989;118(5 Pt 2):1114–9.
2. Cohn JN, Ziesche S, Smith R, Anand I, Dunkman WB, Loeb H, Cintron G, Boden W, Baruch L, Rochin P, Loss LVasodilator-Heart Failure Trial (V-HeFT) Study Group. Effect of the calcium antagonist felodipine as supplementary vasodilator therapy in patients with chronic heart failure treated with enalapril: V-HeFT III. Circulation 1997;96(3):856–63.
3. Packer M, O'Connor CM, Ghali JK, Pressler ML, Carson PE, Belkin RN, Miller AB, Neuberg GW, Frid D, Wertheimer JH, Cropp AB, DeMets DLProspective Randomized Amlodipine Survival Evaluation Study Group. Effect of amlodipine on morbidity and mortality in severe chronic heart failure. N Engl J Med 1996;335(15):1107–14.

4. Rinaldi CA, Linka AZ, Masani ND, Avery PG, Jones E, Saunders H, Hall RJ. Randomized, double-blind crossover study to investigate the effects of amlodipine and isosorbide mononitrate on the time course and severity of exercise-induced myocardial stunning. Circulation 1998;98(8):749–56.
5. Udelson JE, DeAbate CA, Berk M, Neuberg G, Packer M, Vijay NK, Gorwitt J, Smith WB, Kukin ML, LeJemtel T, Levine TB, Konstam MA. Effects of amlodipine on exercise tolerance, quality of life, and left ventricular function in patients with heart failure from left ventricular systolic dysfunction. Am Heart J 2000;139(3):503–10.
6. Spence JD, Munoz C, Huff MW, Tokmakjian S. Effect of amlodipine on hemodynamic and endocrine responses to mental stress. Am J Hypertens 2000;13(5 Pt 1):518–22.
7. Cocco G, Ettlin T, Baumeler HR. The effect of amlodipine and enalapril on blood pressure and neurohumoral activation in hypertensive patients with Ribbing's disease (multiple epiphysal dystrophy). Clin Cardiol 2000;23(2):109–14.
8. Flynn JT, Newburger JW, Daniels SR, Sanders SP, Portman RJ, Hogg RJ, Saul JP, for the PATH-I Investigators. A randomized, placebo-controlled trial of amlodipine in children with hypertension. J Pediatr 2004;145:353–9.
9. Gurm HS, Farooq M. Calcium channel blockers and benign hypertension. Arch Intern Med 1999;159(9):1011.
10. Sener D, Halil M, Yavuz BB, Cankurtaran M, Ariogul S. Anasarca edema with amlodipine treatment. Ann Pharmacother 2005;39(4):761–3.
11. Garbe E, Meyer O, Andersohn F, Aslan T, Kiesewetter H, Salama A. Amlodipine-induced immune thrombocytopenia. Vox Sanguinis 2004;86:75–6.
12. Khemissa-Akouz F, Ouguergouz F, Sulem P, Tkoub el M, Vaucher E. Hepatite aiguë a l'amlodipine. [Amlodipine-induced acute hepatitis.] Gastroenterol Clin Biol 2002;26(6–7):637–8.
13. Lafuente NG, Egea AM. Calcium channel blockers and hepatotoxicity. Am J Gastroenterol 2000;95(8):2145.
14. Zinsser P, Meyer-Wyss B, Rich P. Hepatotoxicity induced by celecoxib and amlodipine. Swiss Med Wkly 2004;134(14):201.
15. Lim AC, Hart K, Murrell D. A granuloma annulare-like eruption associated with the use of amlodipine. Australas J Dermatol 2002;43(1):24–7.
16. Orme S, da Costa D. Generalised pruritus associated with amlodipine. BMJ 1997;315(7106):463.
17. Grabczynska SA, Cowley N. Amlodipine induced-photosensitivity presenting as telangiectasia. Br J Dermatol 2000;142(6):1255–6.
18. van der Vleuten CJ, Trijbels-Smeulders MA, van de Kerkhof PC. Telangiectasia and gingival hyperplasia as side-effects of amlodipine (Norvasc) in a 3-year-old girl. Acta Dermatol Venereol 1999;79(4):323–4.
19. Swale VJ, McGregor JM. Amlodipine-associated lichen planus. Br J Dermatol 2001;144(4):920–1.
20. Erbagci Z. Amlodipine associated hyperpigmentation. Saudi Med J 2004;25:103–5.
21. Sladden MJ, Mortimer NJ, Osborne JE. Longitudinal melanonychia and pseudo-Hutchinson sign associated with amlodipine. Br J Dermatol 2005;153(1):219–20.
22. Phillips BB, Muller BA. Severe neuromuscular complications possibly associated with amlodipine. Ann Pharmacother 1998;32(11):1165–7.
23. Komine N, Takeda Y, Nakamata T. Amlodipine-induced gynecomastia in two patients on long-term hemodialysis therapy. Clin Exp Nephrol 2003;7:85–6.
24. Rosbotham JL, Johnson A, Haque KN, Holden CA. Painful subcutaneous fat necrosis of the newborn associated with intra-partum use of a calcium channel blocker. Clin Exp Dermatol 1998;23(1):19–21.
25. Feldman R, Glinska-Serwin M. Gleboka hipotensja z przemijajaca oliguria oraz ciezka niewydolnosc serca W przebiegu ostrego zamierzonego zatrucia amlodypina. [Deep hypotension with transient oliguria and severe heart failure in course of acute intentional poisoning with amlodipine.] Pol Arch Med Wewn 2001;105(6):495–9.
26. Ajayi AA, Adigun AQ. Syncope following oral chloroquine administration in a hypertensive patient controlled on amlodipine. Br J Clin Pharmacol 2002;53(4):404–5.
27. Curtis JJ. Hypertension following kidney transplantation. Am J Kidney Dis 1994;23(3):471–5.
28. Schrama YC, Koomans HA. Interactions of cyclosporin A and amlodipine: blood cyclosporin A levels, hypertension and kidney function. J Hypertens Suppl 1998;16(4):S33–8.
29. Venkat Raman G, Feehally J, Coates RA, Elliott HL, Griffin PJ, Olubodun JO, Wilkinson R. Renal effects of amlodipine in normotensive renal transplant recipients. Nephrol Dial Transplant 1999;14(2):384–8.
30. Knowles S, Gupta AK, Shear NH. The spectrum of cutaneous reactions associated with diltiazem: three cases and a review of the literature. J Am Acad Dermatol 1998;38(2 Pt 1):201–6.

Angiotensin-converting enzyme inhibitors

See also Benazepril, Captopril, Cilazapril, Enalapril, Fosinopril, Imidapril, Lisinopril, Perindopril, Quinapril, Ramipril, Temocapril, Trandolapril, Zofenopril

General Information

Angiotensin-converting enzyme (ACE) inhibitors inhibit the conversion of angiotensin I to angiotensin II. The ACE is also a kininase, and so ACE inhibitors inhibit the breakdown of kinins. Some of the adverse effects of these drugs are related to these pharmacological effects. For example, cough is thought to be due to the action of kinins on axon fibers in the lungs and hypotension is due to vasodilatation secondary to reduced concentrations of the vasoconstrictor angiotensin II.

Our knowledge of the use of ACE inhibitors has expanded dramatically during recent years, thanks to the publication of the results of a number of large clinical trials (1).

The Heart Outcomes Prevention Evaluation (HOPE) study showed that virtually all patients with a history of cardiovascular disease, not only those who have had an acute myocardial infarction or who have heart failure, benefit from ACE inhibitor therapy (2). The authors selected 9297 patients at increased risk of cardiovascular disease, defined as a history of a cardiovascular event or evidence of disease, such as angina. People with diabetes but no indication of heart disease were included, but they had to have an additional risk factor. They were allocated to receive the ACE inhibitor ramipril 10 mg/day or placebo. The trial was stopped early, according to the predefined rules, because of an overwhelming effect of ramipril on the primary end-point, a 22% reduction in a composite measure of myocardial infarction, stroke, and death from cardiovascular causes. Significance was also achieved on outcomes as diverse as myocardial infarction, revascularization, heart failure, cardiac arrest, and worsening angina. Patients with diabetes had a similar 25% reduction for the composite cardiovascular end-point. Moreover, patients taking ramipril had 16% less overt nephropathy (defined as urine albumin over 300 mg/24 hours, or urine total protein excretion over 500 mg/24 hours, or a urine albumin/creatinine ratio over 36 mg/mmol). They also needed 22% less laser therapy for retinopathy. Since all the patients in the HOPE study were not hypertensive, and since the cardiovascular benefit was greater than that attributable to the fall in blood pressure, the authors suggested that ACE inhibitors are cardioprotective, vasculoprotective, and renal protective, independently of their blood pressure lowering effect.

Relative to the dosage issue, the dosage–plasma concentration relation for enalaprilat (the active metabolite of enalapril) in patients with heart failure and its relation to drug-related adverse effects has been investigated (3). In patients taking enalapril for more than 3 months, in dosages of 5–20 mg bd, there were highly variable trough concentrations of enalaprilat. They were affected by serum creatinine, the severity of heart failure, and body weight. Adverse effects, such as cough and rises in serum creatinine and potassium, were more common at high enalaprilat trough concentrations. The authors concluded that these results provide a rationale for individually adjusting ACE inhibitor doses in case of adverse effects.

Use in hypertension

In hypertension, the Captopril Prevention Project (CAPPP) trial evaluated an ACE inhibitor as an alternative first-line agent in mild to moderate hypertension. It was a prospective randomized open study with blinded end point evaluation (PROBE design), comparing an antihypertensive strategy based on either captopril or conventional therapy with a beta-blocker or a diuretic in patients with mild to moderate hypertension. At the end of follow-up the incidence of cardiovascular events was equal with the two strategies. However, imbalances in the assignment of treatment resulted in a 2 mmHg higher average diastolic blood pressure at entry in the group assigned to captopril. This difference in blood pressure alone would be sufficient to confer an excess of cardiovascular risk within this group, could mask real differences between the regimens in their effects on coronary events, and could explain the greater risk of stroke among patients who took captopril. The authors claimed that the overall results support the position that from now on one should consider ACE inhibitors as first-line agents, equal to diuretics and beta-blockers (4). The CAPPP study also reported a reduced risk of diabetes with captopril, which may be explained by the fact that thiazides and beta-blockers cause changes in glucose metabolism and by favorable effects of ACE inhibition on insulin responsiveness.

The second Swedish Trial in Old Patients with hypertension, STOP-2, was designed to compare the effects of conventional antihypertensive drugs on cardiovascular mortality and morbidity with those of newer antihypertensive drugs, including ACE inhibitors, in elderly patients (5). The study was prospective, randomized, and open, but with a blinded end-point evaluation. It included 6614 patients aged 70–84 years with hypertension (blood pressure over 180 mmHg systolic, or over 105 mmHg diastolic, or both). The patients were randomly assigned to conventional drugs (atenolol 50 mg/day, metoprolol 100 mg/day, pindolol 5 mg/day, or hydrochlorothiazide 25 mg/day plus amiloride 2.5 mg/day) or to newer drugs (enalapril 10 mg/day or lisinopril 10 mg/day, or felodipine 2.5 mg/day or isradipine 2.5 mg/day). Blood pressure fell similarly in all treatment groups. There were equal incidences of the primary end-points (fatal stroke, fatal myocardial infarction, and other fatal cardiovascular disease combined) in all groups (20 events per 1000 patient years). Subgroup analyses showed that conventional therapy, ACE inhibitors, and calcium antagonists had similar efficacy in preventing cardiovascular mortality and major morbidity. This finding argues against the hypothesis that some classes of antihypertensive drugs have efficacy advantages over others, at least in this population of elderly hypertensive patients. Therefore, the choice of antihypertensive treatment will be related to other factors, such as cost, co-existing disorders, and adverse effects. With respect to the reported adverse effects, since the study was open, causality cannot be established. Nevertheless, the size of the study and its naturalistic design allowed accurate assessment of the incidence of adverse effects in this population of elderly hypertensive patients. With ACE inhibitors the most frequently reported adverse effects were cough 30%, dizziness 28%, ankle edema 8.7%, headache 7.7%, shortness of breath 7.3%, and palpitation 5.5%. Actually, little detail was given in the section devoted to safety in the main publication of the results of the trial.

Use in heart failure

In heart failure much debate has been generated by the observation of general "under-use" of ACE inhibitors and the use of smaller doses than have been beneficial in clinical trials. This was partly related to concern about safety with the highest doses, especially in high-risk groups, such as the elderly and patients with renal insufficiency (6). Actually, outcome trials effectively excluded elderly patients (75–80 years and over) and usually patients with renal insufficiency. As elderly patients have poorer renal function, they are more likely to have vascular disease in their renal and carotid arteries, and may be more prone to symptomatic hypotension, it cannot be assumed that the benefit to harm balance observed in younger patients will be the same, at the same doses, in elderly people. The NETWORK trial, a comparison of small and large doses of enalapril in heart failure, was poorly designed and is not conclusive. However, it suggested that apart from a trend to more fatigue with higher doses (10 mg bd), the incidence of adverse effects, including symptomatic hypotension, was similar across the three dosages (2.5, 5, and 10 mg bd) (7).

In heart failure the issue of whether it is justified to use doses of ACE inhibitors substantially smaller than the target doses used in the large-scale studies that established the usefulness of these drugs has been examined in the ATLAS (Assessment of Treatment with Lisinopril and Survival) trial (8). This trial randomized 3164 patients with New York Heart Association (NYHA) class II–IV heart failure and ejection fractions less than 30% to double-blind treatment with either low doses (2.5–5.0 mg/day) or high doses (32.5–35 mg/day) of the ACE inhibitor lisinopril for 39–58 months, while background therapy for heart failure was continued. When compared with the low-dose group, patients in the high-dose group had a non-significant 8% lower risk of death but a significant 12% lower risk of death plus hospitalization for any reason and 24% fewer hospitalizations for heart failure. Dizziness and renal insufficiency were more frequent in the high-dose group, but the two groups were similar in the number of patients who required withdrawal of the study medication. These findings suggest that patients with heart failure should not generally be maintained on very low doses of an ACE inhibitor, unless higher doses cannot be tolerated. However, the ATLAS trial did not address this issue properly. The doses in the small-dose arm were actually very small, and much smaller than those used in routine practice, as reported in several other studies (9). The doses in the large-dose arm may have been unnecessarily high. The recommendation of using target doses proven to be effective in large-scale trials remains unchallenged.

In the studies of left ventricular dysfunction (SOLVD), adverse effects related to the long-term use of enalapril have been thoroughly investigated (10).

Use in myocardial infarction

In the acute infarction ramipril efficacy (AIRE) study, oral ramipril in 2006 patients with heart failure after acute myocardial infarction resulted in a substantial reduction in deaths within 30 days (11).

More trials during and after myocardial infarction have been published and subjected to meta-analysis (12). This very large database provides valuable information on the rate of the most common adverse effects. Of all trials of the effects of ACE inhibitors on mortality in acute myocardial infarction, only the CONSENSUS II trial did not show a positive effect. In this trial, enalaprilat was infused within 24 hours after the onset of symptoms, followed by oral enalapril. The reasons for the negative result of CONSENSUS II remain unresolved, but hypotension and a proischemic effect linked to a poorer prognosis have been suggested.

Use in nephropathy

The results of two trials in patients with chronic nephropathy have reinforced the benefit of ACE inhibitors in slowing the progression of chronic renal insufficiency due to renal diseases other than diabetic nephropathy (13–15) and have provided sufficient information on the safety profile of these agents in chronic renal insufficiency. This was found to be essentially the same as in patients with normal renal function. The current practice of avoiding ACE inhibitors in severe renal insufficiency, to prevent further renal impairment and hyperkalemia, is no longer justified, although careful monitoring should still be observed.

Ramipril has a renal protective effect in non-diabetic nephropathies with nephrotic and non-nephrotic proteinuria (14). It also improves cardiovascular morbidity and all-cause mortality in patients with some cardiovascular risk (2).

The Ramipril Efficacy in Nephropathy (REIN) trial was designed to test whether glomerular protein traffic, and its modification by an ACE inhibitor, influenced disease progression in non-diabetic chronic nephropathies (13). Patients were stratified before randomization by 24-hour proteinuria. Treatment with ramipril or placebo plus conventional antihypertensive therapy was targeted at the same bloodpressure control. At the second interim analysis, ramipril had slowed the fall in glomerular filtration rate (GFR) more than expected from the degree of blood pressure reduction. In the follow-up study GFR almost stabilized in patients who had been originally randomized to ramipril and had continued to take it for more than 36 months. The combined risk of doubling of the serum creatinine or end-stage renal insufficiency was half that found in those taking placebo plus conventional therapy. In patients with proteinuria of 1–3 g/day the fall in GFR per month was not significantly affected, but progression to end-stage renal insufficiency was significantly less common with ramipril (9/99 versus 18/87) for a relative risk of 2.72 (CI = 1.22, 6.08) (14); and so was progression to overt proteinuria (15/99 versus 27/87; RR = 2.40; CI = 1.27, 4.52).

The results of this trial show that ramipril was well tolerated and even protective in cases of advanced renal insufficiency. One major reason for the current practice of underprescription and of prescription of suboptimal doses of ACE inhibitors, especially in patients with heart failure, is the presence of renal insufficiency (16). In such patients, not only should ACE inhibitors no longer be avoided, they are indeed indicated for preservation of renal function.

General adverse effects

The commonest unwanted effects of ACE inhibitors are related to their pharmacological actions (that is inhibition of angiotensin-converting enzyme and kininase II): renal insufficiency, potassium retention, pronounced first-dose hypotension, cough, and the serious but less common angioedema. Skin rashes and taste disturbances are uncommon, but may be more likely with sulfhydryl-containing drugs, particularly captopril. Rare hypersensitivity reactions include rashes, bone-marrow suppression, hepatitis, and alveolitis. If administered in the second or third term of pregnancy, ACE inhibitors can cause a number of fetal anomalies, including growth retardation, renal impairment, oligohydramnios, hypocalvaria, fetal pulmonary hypoplasia, and fetal death. Neonatal anuria and neonatal death can also occur (17,18). Tumor-inducing effects have not been reported.

The frequencies and the profile of adverse effects of five major classes of antihypertensive agents have been assessed in an unselected group of 2586 chronically drug-treated hypertensive patients (19). This was accompanied by a questionnaire-based survey among patients visiting a general practitioner. The percentages of patients who reported adverse effects spontaneously, on general inquiry, and on specific questioning were 16, 24, and 62% respectively. With ACE inhibitors the figures were 15, 22, and 55%. The percentage of patients in whom discontinuation was due to adverse effects was 8.1% with ACE inhibitors (significantly higher than diuretics). Compared with beta-blockers, ACE inhibitors were associated with less fatigue (RR = 0.57; 95% CI = 0.38, 0.85), cold extremities (RR = 0.11; CI = 0.07, 0.18), sexual urge (RR = 0.52; CI = 0.33, 0.82), insomnia (RR = 0.10; CI = 0.04, 0.26), dyspnea (RR = 0.38; CI = 0.17, 0.85), and more coughing (RR = 13; CI = 5.6, 30). The authors did not find a significant effect of age on the pattern of adverse effects. Women reported more effects and effects that were less related to the pharmacological treatment.

Organs and Systems

Cardiovascular

Marked reductions in blood pressure, without any significant change in heart rate, can occur at the start of ACE inhibitor therapy. Such reductions, which are not orthostatic, are sometimes symptomatic but rarely fatal. The volume of evidence is greatest with the longer established agents, but continues to suggest that the problems of first-dose hypotension are most likely to occur in patients whose renin–angiotensin system is stimulated (renin-dependent states), such as in renovascular hypertension or other causes of renal hypoperfusion, dehydration, or previous treatment with other vasodilators (20). These conditions can co-exist, particularly in severe heart failure (21–23). Similar problems have occurred in the treatment of hypertensive neonates and infants (24), but again were particularly likely in the setting of high plasma renin activity associated with either renovascular disease or concurrent diuretic treatment.

The use of very low doses to avoid first-dose hypotension is common, although the rationale remains unclear (25). It is even less clear whether or not there are differences between different ACE inhibitors, that is whether first-dose hypotension is agent-specific or a class effect (26,27).

Respiratory

Cough

DoTS classification
Dose-relation: collateral effect
Time-course: time-independent
Susceptibility factors: genetic (polymorphisms of the bradykinin B_2 receptor gene and the ACE gene); sex (men); exogenous factors (non-smokers).

A non-productive irritant cough was reported as an adverse effect of ACE inhibitors in the mid 1980s. It can be distressing and inconvenient, leading to withdrawal of therapy. Certain susceptibility factors are clearly recognized (for example non-smoking and female sex), but racial group can also affect the incidence.

Frequency

In different studies there has been large variability in the absolute incidence of cough (0.7–48%), the discontinuation rate (1–10%), and the relative incidences with different ACE inhibitors (28). However, the placebo-controlled, randomized, HOPE study has provided a remarkable database, with the largest cohort and the longest follow-up ever reported with such therapy (over 9000 patients followed for 5 years on average). Compared with placebo, ACE inhibitor therapy with ramipril caused cough leading to drug withdrawal in 7.3% of patients (compared with 1.8% for placebo) (13).

HOPE TIPS was a prospective study of patients with high cardiovascular risk, in which the practicability and tolerability of ramipril titration was tested in 1881 patients (29). Cough occurred in 14% over a period of up to 3 months, and 4% discontinued ramipril as a result. The author of an accompanying editorial (30) pointed out that the true incidence of ramipril-induced cough had conceivably been overestimated in the study, owing to the large proportions of patients with type 2 diabetes (52%) and non-smokers (80%) and the high doses used. The authors suggested that cough was not necessarily more common in Asian patients (79% of the patients in this study), although within this broad category the differential susceptibility to cough may quite large, and the editorial examined this; on the balance of evidence, Chinese patients (and perhaps some other racial groups in Asian countries) probably develop cough more commonly with ACE inhibitors than Caucasian patients do.

Mechanism

The mechanism of this effect has been explored (31–33). It may be more complicated than just an increase in concentrations of bradykinin and substance P, increased microvasculature leakage, and stimulation of vagal C fibers (34). Sulindac and indometacin may abolish or reduce the intensity and frequency of cough, supposedly because of inhibition of prostaglandin synthesis (35,36). Common variant genetics of ACE, chymase, and the bradykinin B_2 receptor do not explain the occurrence of ACE inhibitor-related cough (37). In general, bronchial hyper-reactivity has been causally implicated and may also be associated with exaggerated dermal responses to histamine (31,33). However, in one report, airways hyper-responsiveness was not a consistent finding (38).

Susceptibility factors

Cough is more common in non-smokers (39) and in women (39,40). It has been speculated that the risk of cough is genetically predetermined. The possibility that polymorphisms of the human bradykinin B_2 receptor gene may be involved in ACE inhibitor-related cough has been investigated in a case-control study (41). The DNA of 60 subjects with and without cough who were treated with ACE inhibitors was compared with that of 100 patients with untreated essential hypertensive and 100 normotensive subjects. The frequencies of the TT genotype and T allele were significantly higher in the subjects with cough than in subjects without. These tendencies were more pronounced in women. Subjects with the CC genotype were less susceptible to cough. According to the authors, high transcriptional activity of the bradykinin B_2 receptor promoter may be related to the risk of ACE inhibitor-related cough. This is the first demonstration that a genetic variant is involved in ACE inhibitor-related cough. It may therefore be possible to predict the occurrence of cough related to ACE inhibitor use.

The genetic basis of ACE inhibitor-induced cough and its relation to bradykinin have been further explored in a study of the effect of cilazapril in two groups of healthy volunteers genotyped for ACE insertion/deletion (I/D) polymorphism (42). The cough threshold to inhaled capsaicin was significantly lower in the genotype II group than in the DD group. Skin responses to intradermal bradykinin were significantly enhanced in the genotype II group. There was no difference in responsiveness to intradermal substance P. The authors suggested that these findings provide further evidence of the link between ACE inhibitor-induced cough and I/D polymorphism of the ACE gene, and that this supports the hypothesis that ACE inhibitors cause cough by modulating tissue concentrations of bradykinin.

Chinese patients experience more cough from ACE inhibitors than Caucasians. A review of the pharmacokinetics and blood pressure-lowering efficacy of ACE inhibitors as well as of ACE and angiotensinogen gene polymorphism did not find significant differences between Chinese and Caucasians to account for the difference in cough incidence (43).

Management

ACE inhibitor-associated cough seems to be a class effect: switching to another ACE inhibitor rarely solves the problem, although there are occasional anecdotal reports (40,44). However, most patients who develop a cough related to an ACE inhibitor are able and willing to continue therapy. In a small randomized study inhaled sodium cromoglicate relieved the symptom (45). In those in whom the symptom is intolerable, a switch to an angiotensin receptor antagonist is justified.

Obstructive airways disease

It has been suggested that ACE inhibitors are also associated with an increased incidence of symptomatic obstructive airways disease, leading to bronchospasm and asthma (46). However, a prescription event monitoring study of more than 29 000 patients taking ACE inhibitors, compared with 278 000 patients taking other drugs, failed to confirm this association (47).

Endocrine

Gynecomastia has been reported in a patient taking captopril 75 mg/day; it resolved when captopril was withdrawn but recurred when the patient was given enalapril (48). This suggests that gynecomastia may not be simply attributable to the sulfhydryl group of captopril.

Metabolism

ACE inhibition has been associated with increased insulin sensitivity in diabetic patients, and it has therefore been hypothesized that ACE inhibitors can precipitate hypoglycemia in such patients. A Dutch case-control study suggested that among users of insulin or oral hypoglycemic drugs, the use of ACE inhibitors was significantly associated with an increased risk of hospital admission for hypoglycemia (49). However, a French case/non-case study from the pharmacovigilance database did not confirm this finding (50).

In a matched case-control study of 404 cases of hospitalization for hypoglycemia in diabetic patients and 1375 controls, the risk of hypoglycemia was greater in those who used insulin versus a sulfonylurea and was not influenced by the use of ACE inhibitors (51). However, the use of enalapril was associated with an increased risk of hypoglycemia (OR = 2.4; CI = 1.1, 5.3) in sulfonylurea users. Although the authors emphasized the fact that previous reports of ACE inhibitor-related hypoglycemia were more frequent with enalapril, it is unclear why only enalapril, and not ACE inhibitors as a class, was associated with a significantly increased risk of hypoglycemia, and why this occurred only in sulfonylurea users.

Conversely, it has been suggested that the protective effect of ACE inhibitors against severe hypoglycemia should be tested in high-risk patients with high ACE activity. About 10–20% of patients with type 1 diabetes mellitus have a risk of severe hypoglycemia. In 307 unselected consecutive diabetic outpatients, those with the ACE DD genotype had a relative risk of severe hypoglycemia of 3.2 (95% CI = 1.4, 7.4) compared with those with the genotype II (52). There was a significant relation between serum ACE activity and the risk of severe hypoglycemia.

Electrolyte balance

ACE inhibitors can cause hyperkalemia because they inhibit the release of aldosterone. The effect is usually not significant in patients with normal renal function. However, in patients with impaired kidney function and/or in patients taking potassium supplements (including salt substitutes) or potassium-sparing diuretics, and especially aldosterone antagonists, hyperkalemia can occur. In two cases, hypoaldosteronism with diabetes was implicated (53,54).

Hyponatremia, defined as a plasma sodium concentration of 133 mmol/l or under, has been investigated in a prospective study of elderly patients with hip fractures. ACE inhibitors were the most frequently used drugs (five of 14 cases) (55). Of course, this does not prove a cause and effect relation, since in elderly people ACE inhibitors are likely to be among the most frequently prescribed drugs. However, hypoaldosteronism would be a likely mechanism.

Hematologic

ACE inhibitors are used to treat erythrocytosis, for example after transplantation (56). Efficacy in treating erythrocytosis in chronic obstructive pulmonary disease has also been described with the angiotensin II receptor antagonist losartan (57). ACE inhibitors can also lower normal erythrocyte counts and cause anemia (58). This effect has been assessed in a retrospective study of 92 patients after transplantation with and without erythrocytosis, comparing patients taking the same anti-rejection therapy (steroids plus ciclosporin or steroids, ciclosporin, and azathioprine) taking ACE inhibitors with those not taking ACE inhibitors (59). There were significantly lower hemoglobin and erythropoietin concentrations in patients taking ACE inhibitors. When enalapril was given to those who had not previously taken an ACE inhibitor, the hemoglobin concentration fell by around 10% and erythropoietin by around 40%. These effects were not affected by the presence or absence of azathioprine. Although the hemoglobin-lowering effect of ACE inhibition is not a new finding, the lack of an influence of azathioprine adds some further understanding to the effect.

Liver

Hepatic injury is a rare adverse effect of the ACE inhibitors (60,61). Both acute and chronic hepatitis and cholestatic jaundice can occur (62,63), as can cross-reactivity, as identified in a report involving enalapril and captopril (64).

Pancreas

Acute pancreatitis has been reported with both enalapril and lisinopril (65,66).

Urinary tract

The ACE inhibitors can cause reversible impairment of renal function in the setting of reduced renal perfusion, whether due to bilateral renal artery stenosis, severe congestive heart failure, volume depletion, hyponatremia, high dosages of diuretics, combined treatment with NSAIDs, or diabetes mellitus (67). Beyond treatment of the cause, preventive measures include withholding diuretics for a few days, beginning therapy with very small doses of ACE inhibitors, and cautious dosage titration. Therapy involves increasing dietary sodium intake and reducing dosages of diuretics or temporarily withdrawing them. The ACE inhibitor may have to be given in reduced dosages or withdrawn for a time. Because they prolong survival in heart failure and after myocardial infarction, if withdrawal is deemed necessary ACE inhibitors should be reintroduced after a brief respite.

The agreed mechanism of renal function impairment with ACE inhibitors is as follows: when perfusion pressure or afferent arteriolar pressure is reduced in the glomerulus, glomerular filtration is maintained by efferent arteriolar vasoconstriction, an effect of angiotensin II. Blocking the formation of angiotensin II, and perhaps increasing the formation of bradykinin, causes selective efferent arteriolar vasodilatation and results in a reduction in glomerular filtration (68).

In a retrospective study of 64 patients, mean age 71 years, with acute renal insufficiency associated with an ACE inhibitor, over 85% presented with overt dehydration due to diuretics or gastrointestinal fluid loss (69). Bilateral renal artery stenosis or stenosis in a solitary kidney was documented in 20% of cases. In seven patients dialysis was required, but none became dialysis dependent. After resolution of acute renal insufficiency, the plasma creatinine concentration returned to baseline and renal function was not significantly worsened. Two-year mortality was the highest in a subgroup of patients with pre-existing chronic renal insufficiency.

Skin

There have been numerous reports of different rashes in association with ACE inhibitors. The most common skin reaction is a pruritic maculopapular eruption, which is reportedly more common with captopril (2–7%) than with enalapril (about 1.5%). This rash occurs in the usual dosage range and is more common in patients with renal insufficiency (70). Lichenoid reactions, bullous pemphigoid, exfoliative dermatitis, flushing and erythroderma, vasculitis/purpura, subcutaneous lupus erythematosus, and reversible alopecia have all been reported (70–72).

The ACE inhibitors can worsen psoriasis by a mechanism mediated by inhibition of the activity of leukotrienes, which are implicated in the pathogenesis of psoriasis (73).

The ACE inhibitor-related pemphigus has been reviewed in the light of two cases of pemphigus attributed to fosinopril and quinapril (74). Drug-related pemphigus can be classified into two major types, based on the clinical course: induced pemphigus and triggered pemphigus, in which endogenous factors are more important and the drug plays a secondary role. The first type is usually related to thiol drugs. It is impossible to distinguish drug-related pemphigus reliably from idiopathic pemphigus on the basis of clinical findings, histopathology, or immunofluorescence. Captopril tends to be associated with pemphigus foliaceus, whereas the non-thiol ACE inhibitors are more often associated with pemphigus vulgaris, although there are exceptions. A transition from pemphigus vulgaris to pemphigus foliaceus is more common than the reverse. Several mechanisms have been proposed to be involved in the induction of pemphigus: interaction of the thiol group with sulfur-containing groups on the keratinocyte membrane, leading to acantholysis by biochemical interference with adhesion mechanisms; antigen modification resulting in antibody formation; inhibition of suppressor T cells, resulting in pathogenic autoantibody formation by B cell clones; or enzyme activation or inhibition. The maximum latency to the development of pemphigus reported for ACE inhibitors is 2 years. It can take up to 17 months for lesions to resolve after drug withdrawal. A significant proportion of cases will not improve or resolve spontaneously on drug withdrawal alone. It is important to withdraw the offending drug, treat the bullous reaction appropriately, and advise avoidance of ACE inhibitors, although substitution of enalapril for captopril or vice versa has been successful in some cases.

Immunologic

There have been reports of a lupus-like syndrome with captopril (75) and lisinopril (76).

Autacoids

Angioedema

DoTS classification (BMJ 2003;327:1222-5):
Dose-relation: Collateral
Time-course: Intermediate
Susceptibility factors: Genetic (black Americans); sex (female); drugs (NSAIDs, vaccines, immunosuppressants); other exogenous factors (polyacrylonitrile membranes in hemodialysis); diseases (a history of angioedema)

Angioedema was first described many centuries ago, although many attribute the first description to Heinrich Quincke in the late 19th century. Cases of hereditary angioedema were reported by Sir William Osler in 1888 (77). Angioedema can be caused by autoimmunity, infection, or drugs; the hereditary form is due to deficiency or underfunctioning of the blood protein C1 esterase inhibitor. Angioedema is the most common term used (in 1325 titles of papers listed in Pubmed at the time of searching); other terms are angioneurotic edema (486 instances), Quincke's edema (111 instances), or giant urticaria (8 instances).

Drugs that have been reported as causing angioedema include non-steroidal anti-inflammatory drugs (NSAIDs), both selective (COX-2 inhibitors) and non-selective, and vaccines, although the most commonly implicated drugs are ACE inhibitors. Angiotensin receptor blockers have also been reported as causative agents, especially in patients who have previously had reactions to ACE inhibitors (78) or in those with co-existing immunological predisposition.

Angioedema is a potentially fatal complication that has been associated with several different ACE inhibitors, with a reported incidence of 0.1–0.5%.

Presentation

Angioedema due to ACE inhibitors can manifest as recurrent episodes of facial swelling, which resolves on withdrawal, or as acute oropharyngeal edema and airways obstruction, which requires emergency treatment with an antihistamine and corticosteroids. It may be life-threatening (79) and may need tracheostomy (80). It is occasionally fatal (81). An unusual presentation with subglottic stenosis has also been reported (82). A variant form is angioedema of the intestine, which tends to occur within the first 24–48 hours of treatment (83,84).

- Two patients presented with isolated visceral angioedema with episodes of recurrent abdominal symptoms (85). Each had undergone surgical procedures for symptoms that persisted after surgery and were ultimately relieved by withdrawal of their ACE inhibitors.
- Another similar case was diagnosed as angioedema of the small bowel after an abdominal CT scan (86). Angioedema occurred in a 58-year-old woman 3 hours after biopsy of a hypopharyngeal mass under general anesthesia and was accompanied by transient electrocardiographic features of anterior myocardial infarction with severe hypokinesis of the anterior wall regions on echocardiography but no significant change in creatinine kinase activity (87). Only T wave inversion persisted on follow-up. Repeat echocardiography showed significant spontaneous improvement and coronary angiography showed normal coronary arteries. Hypotension and hypoxemia did not seem to occur, and the authors could not therefore speculate on the mechanism of the concomitant cardiac changes.
- Recurrent episodes of tongue swelling have been reported with cilazapril (88) and perindopril (89).
- A 74-year-old man with a permanent latex condom catheter developed penile swelling that was non-pitting and involved the subcutaneous tissue of a normal scrotum, after taking lisinopril 5 mg/day for 6 days (90). Removal of the catheter had no effect. After other possible causes were ruled out, ACE inhibitor-induced angioedema was suspected and lisinopril was withdrawn. Within a few days, the swelling, which had not spread, resolved.

Urticaria and itching are not usual features, although drug-induced urticaria and angioedema can co-exist as adverse effects of the same drug. Intestinal edema is also described (91,92), and tends to occur within the first 24–48 hours of treatment (93,94).

ognition requires a high degree of suspicion, as the symptoms can manifest as non-specific gastrointestinal disturbance and/or abdominal pain; there have been several reports of repeated laparotomies before the correct diagnosis has been made (95).

- A 73-year-old woman developed unilateral tongue angioedema during treatment with enalapril for hypertension (96).

Five cases of angioedema in association with various ACE inhibitors have been reported (97). The patients were taking lisinopril, trandalopril, or ramipril, and all presented with increasing breathlessness and/or dysphagia associated with significant tongue edema. There was significant airway obstruction in all cases, requiring either tracheotomy or intubation. Withdrawal of the ACE inhibitor and standard supportive care, including glucocorticoids, antihistamines, and adrenaline in two cases, led to resolution of the angioedema.

Orolingual angioedema has been attributed to benazepril after recombinant tissue plasminogen activator (rtPA) treatment for acute stroke (98).

- A 58-year-old man taking amlodipine and benazepril received intravenous rtPA for an acute left middle cerebral artery territory stroke. He developed orolingual angioedema 5 minutes later. There was no airway compromise or hemodynamic instability to suggest an anaphylactic reaction. He was treated with dexamethasone and an antihistamine. The angioedema resolved completely over the next 48 hours, as did his neurological deficits.

Lisinopril-associated angioedema has been reported in a patient undergoing maxillofacial surgery (99).

Small bowel angioedema has been attributed to perindopril (100).

- Angioedema occurred in a 57-year-old man who had taken trandolapril for 2 days, not having occurred while he took ramipril for 3 years (101).

The authors suggested that this implied that angioedema is not a class effect of ACE inhibitors. However, the association could have been coincidental.

Incidence

The overall incidence of drug-induced angioedema is not known, but it is estimated that it could occur in 0.1–0.5% of patients taking ACE inhibitors (102). In a prospective placebo-controlled study in 12 557 patients with hypertension treated with enalapril maleate 5–40 mg/day, angioedema occurred in 86 (0.68%) (103).

Black Americans are at increased risk, with an adjusted relative risk of 4.5 (95% CI = 2.9, 6.8) compared with white users (104). This increase in risk was unrelated to the dosage of ACE inhibitor or the concurrent use of cardiovascular drugs. Since millions of patients take these agents worldwide, this represents one of the most common adverse drug reaction in terms of absolute numbers affected. It has been pointed out that the incidence of ACE inhibitor-induced angioedema seems to be on the increase (105).

In a large, double-blind, randomized comparison of enalapril maleate and omapatrilat, angioedema occurred in 86 (0.86%) of 12 557 patients who were randomized to enalapril (103).

The Australian Adverse Reactions Advisory Committee has issued a warning in its bulletin about the continued reporting of angioedema due to ACE inhibitors (106). Of over 7000 reports of angioedema since 1970, 13% had been attributable to ACE inhibitors. In some cases angioedema occurred episodically with long symptom-free intervals.

- An elderly woman, who had been taking ramipril for 1 year without adverse effects, had several episodes of unilateral swelling of the face, lips, jaw line, and cheek, each lasting 2–3 days over 4 months.

Dose relation

ACE inhibitor-induced angioedema can occur at any dose in the therapeutic range (a collateral reaction).

Time-course

The time course of angioedema is very variable in relation to the start of drug treatment. It can occur within a day after the start of treatment (107,109) and the risk is highest within the first month. In a study of enalapril (103), the incidence of angioedema was higher immediately after the start of therapy (3.6/1000 patients per month) and fell to 0.4/1000 patients per month. However, it can occur at any time and has occasionally been reported many months after the start of therapy (108,109) or even years after (104,110,111,112).

Susceptibility factors

One of the strongest independent risk factors is ethnicity—black patients are three or four times more susceptible to the adverse effect than non-black patients (104,113,114), with an overall rate of 1.6 per 1000 person years of ACE inhibitor use. Moreover, angioedema is more severe in black American users (104). In the study of enalapril mentioned above (103) stepwise logistic regression identified black race (OR = 2.88; 95%CI = 1.72, 4.82), a history of drug rash (OR = 3.78; 95%CI = 1.80, 7.92), age greater than 65 years (OR = 1.60; 95%CI = 1.02, 2.53), and seasonal allergies (OR = 1.79; 95%CI = 1.06, 3.00) as independent risk factors for angioedema.

Other susceptibility factors that have been identified include a history of angioedema, female sex, and co-existent use of NSAIDs (110). Recent initiation of ACE inhibitor therapy and the use of enalapril or lisinopril are also associated with a higher rate of angioedema (104).

Anaphylactoid reactions, with hypotension, and flushing, occasionally associated with abdominal cramping, diarrhea, nausea, and sweating, have been reported in patients taking ACE inhibitors undergoing hemodialysis. Angioedema can also occur in these patients; it is occasionally life-threatening and is usually associated with the concurrent use of ACE inhibitors and dialysers in which the membrane is made of polyacrylonitrile (also known as AN69) (115,116). The mechanism may be related to release of bradykinin. This combination should be avoided, since well-established alternatives are available.

For unclear reasons, ACE inhibitor-induced angioedema was more prevalent among immunosuppressed patients after cardiac or renal transplantation than among other patients (117). In 156 cardiac patients and 341 patients with renal transplants, this adverse effect was observed in 4.8 and 1% respectively, that is 24 times and 5 times higher than in the general population (0.1–0.2%).

In a series of 15 kidney transplant patients given a combination of treatments, including the ACE inhibitor ramipril and the immunosuppressant sirolimus, there were five cases of tongue edema (118). All of the patients had previously taken ramipril before renal transplantation without adverse effects. The tongue edema was observed only in those who took high doses of both ramipril (5 mg/day) and sirolimus. There was resolution of the edema after withdrawal of the ramipril, and rechallenge with lower doses of both ramipril (2.5 mg/day) and sirolimus did not result in the same adverse effects. The authors hypothesized that the two drugs act synergistically only when full doses of both are used.

The characteristics of treated hypertensive patients in the Antihypertensive and Lipid-Lowering treatment to prevent Heart Attack Trial (ALLHAT) who developed angioedema have been published (119). There were 42 418 participants, of whom 53 developed angioedema; 70% were assigned to the ACE inhibitor lisinopril. Susceptibility factors included, as expected, black ethnicity (55%), but unusually there was a male preponderance (60%). The timing of the adverse effect fitted with the known variable time-course: three cases (6%) within 1 day of randomization, 22% within the first week, 34% within the first month, and 68% within the first year.

Mechanism

The exact mechanism of angioedema associated with ACE inhibitors has not been determined. Although the reaction may be immune mediated, IgE antibodies or other specific antibodies have not been detected. Some authors have speculated that it may be related to a deficiency of carboxypeptidase N and complement components, because of its parallel role with that of ACE in enzymatic inactivation of bradykinin.

The putative pathophysiological mechanisms of ACE inhibitor-induced angioedema have been discussed in the context of 19 cases (120) and five other cases have been reported (121). New insights into the mechanisms have been reviewed with accompanying guidance regarding clinical management (122). The cutaneous manifestations result from subcutaneous and mucosal inflammation, arteriolar dilatation, vascular leakage, and localized swelling.

The pathophysiological mechanism has been related to bradykinin accumulation, release of IgE, and mast cell-mediated release of vasoactive mediators. However, immunoglobulin E (IgE) antibodies or other specific antibodies have not been detected. ACE inhibitors inactivate bradykinin. Some authors have speculated that it may be related to a deficiency of carboxypeptidase N and complement components, because of its parallel role with that of ACE in the enzymatic inactivation of bradykinin. Increased bradykinin concentrations have been found during acute attacks (123) and in a well-documented study, a reliable assay for specific measurement of plasma bradykinin, excluding other immunoreactive kinins, detected a very high concentration of bradykinin (47 pmol/l) during an acute attack of angioedema in a patient taking captopril (124). The concentrations fell to 3.2 pmol/l in remission after drug withdrawal. The concentration of bradykinin during chronic ACE inhibition with no angioedema was not reported. One major contribution of this paper was to demonstrate that plasma bradykinin concentrations were substantially increased in 22 patients with hereditary angioedema and 22 others with acquired angioedema, both conditions being associated with inadequate inhibition of the first component of human complement. The infusion of C1 esterase inhibitor immediately lowered bradykinin concentrations in patients with hereditary or acquired C1 esterase inhibitor deficiency. Infusion of C1 esterase inhibitor in ACE inhibitor-induced angioedema was not investigated.

In a case-control study nested within an 8-week open study of the use of quinapril for hypertension in 12 275 patients there were 22 cases of angioedema (125). They were matched with 48 controls taking quinapril. Patients with angioedema had significantly lower mean activities of serum carboxypeptidase N and C1 esterase inhibitor compared with controls, but all mean values were within the laboratory's reference range. Although this may support the involvement of low activities of carboxypeptidase in the pathogenesis of ACE inhibitor-induced angioedema, prior testing of patients for low enzyme activities is not likely to be helpful in screening for angioedema risk in patients in whom ACE inhibitor therapy is being considered. In this study it was also reported that a history of prior episodes of angioedema was associated with a six-fold increase in the subsequent risk of angioedema after ACE inhibitor therapy. Another anecdotal report has pointed out the risk of recurrence of angioedema, in relation to a case of coincident occurrence of angioedema on several occasions in one patient after the consecutive administration of captopril, fosinopril, and quinapril (126).

Not all patients develop the adverse effect, which suggests that other factors are important; it has been speculated that it may be related to different rates of degradation of endogenous bradykinin (127).

Individual drugs

Angioedema is generally regarded as a class effect and has been attributed to benazepril (109), captopril (128), cilazapril (129), enalapril (103), lisinopril (130), perindopril (131), quinapril (132), and ramipril (133).

The angiotensin receptor antagonists losartan, irbesartan, telmisartan, eprosartan, and valsartan have all also been implicated, and they are therefore not necessarily absolute safe substitutes for patients with ACE inhibitor-induced angioedema (78).

Management

All cases of angioedema should be evaluated to look for evidence of an offending drug. Withdrawal of the presumed causative agent and supportive therapy are then key to management. Most cases are self-limiting and will resolve over the first 24–48 hours. In view of the potential for airway compromise, which is the usual cause of death in fatal cases, patients should be carefully examined for any respiratory manifestations. Airway management with intubation or emergency tracheostomy may become necessary in severe cases. Other measures that may help to alleviate oropharyngeal or airway swelling include the administration of intramuscular adrenaline (epinephrine), in accordance with anaphylaxis guidelines, and antihistamine and glucocorticoid therapy. There is no evidence to support the routine use of C1-esterase inhibitor concentrate. Patients who have developed angioedema on treatment should be advised not to use any drugs in the class that precipitated the attack.

- A 43-year-old, white woman took ramipril, and after 3 weeks developed angioedema, which resolved with antihistamines, glucocorticoids, and one dose of adrenaline (134). A low dose of ramipril was restarted 4 days later, and increased over the next 4 days. A few months later she developed severe upper lip and tongue edema. Her C1 esterase inhibitor concentration was normal. After 4 days of treatment with antihistamines, glucocorticoids, adrenaline, leukotriene receptor antagonists, ciclosporin, and intravenous immunoglobulin, without effect, she responded to two units of intravenous fresh frozen plasma, and had no further recurrence.

Second-Generation Effects

Fetotoxicity

Enalapril, captopril, and lisinopril (and presumably other ACE inhibitors) cross the placenta in pharmacologically significant amounts (17). There is clear evidence of fetotoxicity when ACE inhibitors are used beyond the first trimester of pregnancy. Since continuation of treatment beyond the first trimester carries an excess risk of low fetal birth weight and other more severe complications, it is important to withdraw the ACE inhibitor at this time. Intrauterine growth retardation, oligohydramnios, and neonatal renal impairment, often with a serious outcome, are characteristic (135); failure of ossification of the skull or hypocalvaria also appear to be part of the pattern (17). There is also evidence that persistence of a patent ductus arteriosus is also more likely to occur.

Susceptibility Factors

Renal disease

Enalapril has been specifically studied in patients who are resistant to other drugs and intolerant of captopril (136) and in patients with collagen vascular disease and renal disease known to be at high risk of adverse effects (137). In the first study (136), the major reasons for discontinuing captopril were a low white blood cell count, proteinuria, taste disturbance, and rash. In the vast majority of the 281 patients, these adverse effects did not recur during enalapril treatment. The main adverse events that warranted withdrawal of enalapril were impairment of renal function (5%), hypotension (2%), and rashes (2%). The authors noted that patients with angioedema should not be given alternative ACE inhibitors.

In the second study (137) of 738 high-risk patients the main reasons for the withdrawal of enalapril were increases in serum creatinine (4%), hypotension (1%), and nausea (1%).

The long-term safety of enalapril in patients with severe renal insufficiency and hypertension has been evaluated in a pooled analysis of three similar, randomized, placebo-controlled clinical trials in 317 patients with renal insufficiency (138). Only patients without diabetes were included. Follow-up was for 2 and 3 years. One protocol used a fixed dose (5 mg/day) and the other two allowed titration up to 40 mg/day. Cough occurred in 18% of the patients taking enalapril and in 6.1% of those taking placebo. Hypotension (5.9 versus 1.2%) and paresthesia (7.8 versus 2.4%) were more frequent with enalapril. Angioedema (1.3 versus 0.6%) and first-dose hypotension (1.3 versus 0%) tended to occur more often with enalapril. Hyperkalemia, defined as any increase from baseline and left to the judgement of the investigators, was excessive in the enalapril-treated patients (28 versus 8.8%). Finally, the hematocrit fell more often with enalapril (7.1 versus 2.0%).

It is important to stress that although it is a risk factor, renal insufficiency is a good indication for ACE inhibitors, which slow the progression of chronic renal insufficiency (139). In trials in patients with chronic renal insufficiency (13–15,17), the safety profile was essentially the same as in patients with normal renal function. The current practice of avoiding ACE inhibitors in severe renal insufficiency, to prevent further renal impairment and hyperkalemia, seems no longer justified. However, patients with bilateral renal artery stenosis carry an excess risk of renal insufficiency when treated with ACE inhibitors. These agents are therefore contraindicated in such patients.

Drug–Drug Interactions

Aprotinin

Aprotinin, a proteolytic enzyme inhibitor acting on plasmin and kallidinogenase (kallikrein), is hypothesized to contribute significantly to a reduction in glomerular perfusion pressure when it is used in combination with ACE inhibitors. In a retrospective investigation of this combination in adults undergoing coronary artery bypass surgery, the combination of preoperative ACE inhibition and intraoperative aprotinin was associated with a significant increase in the incidence of acute renal insufficiency (OR = 2.9; 95% CI = 1.4, 5.8) (140). The authors concluded that this combination should be avoided in cardiac surgery.

Aspirin

Antagonistic effects of cyclo-oxygenase inhibitors (indometacin or aspirin) have been repeatedly reported both in hypertension and in heart failure, strongly suggesting that there may be prostaglandin participation in the clinical response to ACE inhibitors (141,142). In animals, although not in all experimental models, aspirin can attenuate the beneficial effects of ACE inhibitors on ventricular remodelling after myocardial infarction. However, there are conflicting reports on the clinical significance of this interaction (143).

Positive studies

From a post-hoc analysis of the SOLVD trial, it appears that in patients with left ventricular systolic dysfunction, the use of aspirin was associated with improved survival and reduced morbidity. In aspirin users, benefit from enalapril was retained but reduced (144).

The WASH pilot study (Warfarin/Aspirin Study in Heart Failure) compared the effects on cardiovascular events of warfarin and aspirin, and on antithrombotic therapy in patients with heart failure, most of whom were also taking an ACE inhibitor. Patients taking aspirin had more events and hospitalizations related to worsening heart failure than patients in the two other groups (unpublished data, reported at the 1999 annual meeting of the European Society of Cardiology, John Cleland, personal communication). The authors speculated that this may have been related to a negative interaction between ACE inhibitor therapy and aspirin, which would counteract the beneficial effects of ACE inhibitors. The Warfarin-Antiplatelet Trial in Chronic Heart Failure (WATCH) is indirectly addressing the issue. It is based on the hypothesis that warfarin or clopidogrel (an antiplatelet agent that acts by a pathway independent of cyclo-oxygenase) may be preferred to aspirin as antithrombotic therapy in patients with heart failure. It will randomize 4500 patients, most of whom will be taking ACE inhibitors. Meanwhile, it may be advisable to avoid aspirin in patients with heart failure and no clear indication for aspirin (no evidence of atherosclerosis), and to consider substituting warfarin or clopidogrel for aspirin in patients with refractory or rapidly progressive heart failure (145). In all other cases, because each drug is clearly associated with a substantial clinical benefit, it would be excessive to deny patients aspirin or ACE inhibitors.

In a series of studies of ACE inhibitor-induced improvement in pulmonary function, treatment with aspirin 325 mg/day for 8 weeks in patients with mild to moderate heart failure due to primitive dilated cardiomyopathy did not affect ventilation and peak oxygen consumption during exercise when the patients were not taking an ACE inhibitor but worsened pulmonary diffusion capacity and made the ventilatory response to exercise (tidal volume, ventilation to carbon dioxide production) less effective in those who were, regardless of the duration of ACE inhibition (146).

A systematic overview of major ACE inhibitor trials (CONSENSUS II, AIRE, TRACE, SMILE) found a trend toward less benefit from ACE inhibitors among aspirin users (147). Although the interaction was not statistically significant, the authors concluded that the data did not "refute the hypothesis of a major aspirin interaction with ACE inhibitors," especially because patients taking aspirin had only 60% of the benefit seen in patients not taking it.

GUSTO-1 and EPILOG, two different antithrombotic trials, GUSTO-1 and EPILOG, the first in acute myocardial infarction and the second during coronary stenting, compared the event rates in patients taking aspirin, an ACE inhibitor, or both (148). In each of these trials, events were more frequent in patients taking the combination than in those taking aspirin alone. The authors interpreted these findings as suggesting that ACE inhibitors may reduce the benefit of aspirin in these patients, whereas the results of the ACE inhibitor trials suggested that aspirin may interfere with the effect of ACE inhibitors.

Negative studies

A post-hoc analysis of the CATS trial database in patients with acute myocardial infarction suggested that aspirin does not attenuate the acute and long-term effects of captopril (149). Because of the demonstrated benefit on morbidity and mortality with each agent, textbooks and official guidelines do not recommend withholding either aspirin or ACE inhibitors in patients with heart failure or myocardial infarction. With no sufficient proof of lack of interaction, the use of small doses of aspirin (100 mg/day or less) is recommended.

In a study of the effects of aspirin 325 mg/day, both acute (4 hours after the dose) and chronic (6 weeks), in 62 patients with mild to moderate heart failure taking enalapril (more than 10 mg/day for at least 3 months), there were no significant changes in mean arterial pressure or in forearm blood flow and vascular resistance measured by venous plethysmography (150). In another arm of the study the same results were observed with ifetroban 250 mg/day, a thromboxane A_2 receptor antagonist.

Conclusions

Two post-hoc analyses of clinical trials (148) have added to the confusion engendered by these conflicting results. GUSTO-1 and EPILOG, two different antithrombotic trials, the first in acute myocardial infarction and the second during coronary stenting, compared the event rates in patients taking aspirin, an ACE inhibitor, or both (148). In each of these trials, events were more frequent in patients taking the combination than in those taking aspirin alone. The authors interpreted these findings as suggesting that ACE inhibitors may reduce the benefit of aspirin in these patients, whereas the results of the ACE inhibitor trials suggested that aspirin may interfere with the effect of ACE inhibitors.

These conflicting results may be partly explained by the various mechanisms of the vasodilatory action of ACE inhibitors, which may differ according to the regional peripheral circulation. In none of the studies was central hemodynamics or cardiac output assessed.

However, none of the trials post-hoc analysis, which suggested that there is an interaction between aspirin and ACE inhibitors, was specifically designed to examine this question. Post-hoc and subgroup analyses may be heavily biased, and multivariate adjustment may not have been able to account fully for confounding factors. Aspirin in itself may be harmful in certain patients, such as those with heart failure, because of its antiprostaglandin activity, rather than because it interferes with the actions of ACE inhibitors, a phenomenon that would also manifest as an aspirin–ACE inhibitor interaction.

The interaction between aspirin and ACE inhibitors in patients with heart failure is probably clinically important (143). Both drugs are often prescribed for a large number of patients with a variety of cardiovascular diseases. These agents have mechanisms of action that interact at the physiological level, and there are consequently many theoretical reasons to expect important clinical consequences. In animals, although not in all experimental models, aspirin can attenuate the beneficial effects of ACE inhibitors on ventricular remodelling after myocardial infarction. Some clinical studies have suggested that there is minimal, if any, adverse peripheral hemodynamic effect. In the most recent such study, the acute (4 hours after the dose) and chronic (6 weeks) effects of aspirin 325 mg/day were investigated in 62 patients with mild to moderate heart failure treated with enalapril (more than 10 mg/day for at least 3 months) (150). This did not produce significant changes in mean arterial pressure or in forearm blood flow and vascular resistance measured by venous plethysmography. In another arm of the study the same results were observed with ifetroban 250 mg/day, a thromboxane A_2 receptor antagonist. These conflicting results may be explained by the various mechanisms of the vasodilatory action of ACE inhibitors, which may differ according to the regional peripheral circulation. In none of the studies was central hemodynamics or cardiac output assessed. There is evidence that co-administration of aspirin and ACE inhibitors can be detrimental to renal function in patients with heart failure.

Beta-lactams

Intestinal absorption of beta-lactams occurs at least in part by an active mechanism involving a dipeptide carrier, and this pathway can be inhibited by dipeptides and tripeptides (151,152), which reduce the rate of absorption of the beta-lactams. ACE inhibitors, which have an oligopeptide structure, are absorbed by the same carrier (153) and interact with beta-lactams in isolated rat intestine (154).

A second potential site of interaction between ACE inhibitors and beta-lactams is the renal anionic transport system, and concomitant administration sometimes results in pronounced inhibition of the elimination of beta-lactams (155).

Diuretics

Because of dehydration, patients taking diuretics can be particularly sensitive to the hypotensive effect of ACE inhibitors (20).

Interferon alfa

An increased risk of severe and early but reversible neutropenia has been found in patients taking angiotensin-converting enzyme inhibitors (enalapril and captopril) with interferon alfa (156).

Potassium supplements, potassium-sparing diuretics, or salt substitutes

Concurrent administration of potassium supplements, potassium-sparing diuretics, or salt substitutes can precipitate hyperkalemia in ACE inhibitor-treated patients, in whom aldosterone is suppressed (SED-14, 674). Regular monitoring of serum potassium is essential in these patients, because of the risk of hyperkalemia in patients given potassium (or potassium-sparing diuretics) and ACE inhibitors or angiotensin receptor antagonists.

In a retrospective study, five patients developed extreme hyperkalemia (9.4–11 mmol/l) within 8–18 days of starting combination therapy with co-amilozide and an ACE inhibitor (157).

In eight healthy subjects, treatment with spironolactone and losartan increased mean plasma potassium concentration by 0.8 mmol/l (up to 5.0 mmol/l) and reduced mean urinary potassium excretion from 108 to 87 mmol/l (158).

Until more data are available, it is prudent to consider angiotensin II receptor antagonists similar to ACE inhibitors as risk factors for hyperkalemia in patients taking potassium-sparing diuretics.

Selective cyclo-oxygenase-2 (COX-2) inhibitors

Non-selective non-steroidal inflammatory drugs can attenuate the antihypertensive effects of ACE inhibitors and increase the risk of renal insufficiency. In 2278 patients taking NSAIDs, 328 taking ACE inhibitors, and 162 taking both, no nephrotoxicity was found in patients taking monotherapy, but there were three cases of reversible renal insufficiency in patients taking the combination (159).

This effect is more prominent in patients with low renin concentrations. The interaction of COX-2 inhibitors with ACE inhibitors has been much less well investigated. In a review of Phase II/III studies of COX-2 inhibitors, it was reported that the co-administration of rofecoxib 25 mg/day and benazepril 10–40 mg/day for 4 weeks was associated with an average increase in mean arterial pressure of about 3 mmHg compared with ACE inhibitor monotherapy (160).

One report has described a case of increased blood pressure in a patient taking rofecoxib and lisinopril (161).

- The blood pressure of a 59-year-old man with hypertension and normal renal function rose when rofecoxib 25 mg/day was added to lisinopril 10 mg/day (from an average of 135/80–85 to 168/98 mmHg within 5 weeks). Four days after rofecoxib was withdrawn the blood pressure was 127/78 mmHg. Rechallenge with the same dose of rofecoxib produced the same effect and the blood pressure fell when the dosage of lisinopril was increased to 20 mg/day on continuous rofecoxib. The authors did not report on the course of renal function.

The increase in blood pressure with COX-2 inhibitors from interaction with ACE inhibitors may be greater in some patients than has previously been reported.

References

1. Brown NJ, Vaughan DE. Angiotensin-converting enzyme inhibitors. Circulation 1998;97(14):1411–20.
2. Yusuf S, Sleight P, Pogue J, Bosch J, Davies R, Dagenais G. Effects of an angiotensin-converting-enzyme inhibitor, ramipril, on cardiovascular events in high-risk patients. The Heart Outcomes Prevention Evaluation Study Investigators. N Engl J Med 2000;342(3):145–53.
3. Brunner-La Rocca HP, Weilenmann D, Kiowski W, Maly FE, Follath F. Plasma levels of enalaprilat in chronic therapy of heart failure: relationship to adverse events. J Pharmacol Exp Ther 1999;289(1):565–71.
4. Hansson L, Lindholm LH, Niskanen L, Lanke J, Hedner T, Niklason A, Luomanmaki K, Dahlof B, de Faire U, Morlin C, Karlberg BE, Wester PO, Bjorck JE. Effect of angiotensin-converting-enzyme inhibition compared with conventional therapy on cardiovascular morbidity and mortality in hypertension: the Captopril Prevention Project (CAPPP) randomised trial. Lancet 1999;353(9153):611–6.
5. Hansson L, Lindholm LH, Ekbom T, Dahlof B, Lanke J, Schersten B, Wester PO, Hedner T, de Faire U. Randomised trial of old and new antihypertensive drugs in elderly patients: cardiovascular mortality and morbidity the Swedish Trial in Old Patients with Hypertension-2 study. Lancet 1999;354(9192):1751–6.
6. Cleland JG. ACE inhibitors for the prevention and treatment of heart failure: why are they "under-used"? J Hum Hypertens 1995;9(6):435–42.
7. Poole-Wilson PA. The NETWORK study. The effect of dose of an ACE inhibitor on outcome in patients with heart failure. J Am Coll Cardiol 1996;27(Suppl A):141A on behalf of the NETWORK investigators.

8. Packer M, Poole-Wilson PA, Armstrong PW, Cleland JG, Horowitz JD, Massie BM, Ryden L, Thygesen K, Uretsky BFATLAS Study Group. Comparative effects of low and high doses of the angiotensin-converting enzyme inhibitor, lisinopril, on morbidity and mortality in chronic heart failure. Circulation 1999;100(23):2312–8.
9. Gerstein HC, Yusuf S, Mann JFE. Effects of ramipril on cardiovascular and microvascular outcomes in people with diabetes mellitus: results of the HOPE study and MICRO-HOPE substudy. Heart Outcomes Prevention Evaluation Study Investigators. Lancet 2000;355(9200):253–9.
10. Kostis JB, Shelton B, Gosselin G, Goulet C, Hood WB Jr, Kohn RM, Kubo SH, Schron E, Weiss MB, Willis PW 3rd, Young JB, Probstfield J. Adverse effects of enalapril in the Studies of Left Ventricular Dysfunction (SOLVD). SOLVD Investigators. Am Heart J 1996;131(2):350–5.
11. The Acute Infarction Ramipril Efficacy (AIRE) Study Investigators. Effect of ramipril on mortality and morbidity of survivors of acute myocardial infarction with clinical evidence of heart failure. Lancet 1993;342(8875):821–8.
12. Latini R, Maggioni AP, Flather M, Sleight P, Tognoni G. ACE inhibitor use in patients with myocardial infarction. Summary of evidence from clinical trials. Circulation 1995;92(10):3132–7.
13. Ruggenenti P, Perna A, Gherardi G, Gaspari F, Benini R, Remuzzi G. Renal function and requirement for dialysis in chronic nephropathy patients on long-term ramipril: REIN follow-up trial. Gruppo Italiano di Studi Epidemiologici in Nefrologia (GISEN). Ramipril Efficacy in Nephropathy. Lancet 1998;352(9136):1252–6.
14. Ruggenenti P, Perna A, Gherardi G, Garini G, Zoccali C, Salvadori M, Scolari F, Schena FP, Remuzzi G. Renoprotective properties of ACE-inhibition in non-diabetic nephropathies with non-nephrotic proteinuria. Lancet 1999;354(9176):359–64.
15. Maschio G, Alberti D, Janin G, Locatelli F, Mann JF, Motolese M, Ponticelli C, Ritz E, Zucchelli P, Marai P, Marcelli D, Tentori F, Oldrizzi L, Rugiu C, Salvadeo A, Villa G, Picardi L, Borghi M, Moriggi M, et alThe Angiotensin-Converting-Enzyme Inhibition in Progressive Renal Insufficiency Study Group. Effect of the angiotensin-converting-enzyme inhibitor benazepril on the progression of chronic renal insufficiency. N Engl J Med 1996;334(15):939–45.
16. Echemann M, Zannad F, Briancon S, Juilliere Y, Mertes PM, Virion JM, Villemot JP. Determinants of angiotensin-converting enzyme inhibitor prescription in severe heart failure with left ventricular systolic dysfunction: the EPICAL study. Am Heart J 2000;139(4):624–31.
17. Barr M Jr. Teratogen update: angiotensin-converting enzyme inhibitors. Teratology 1994;50(6):399–409.
18. Sedman AB, Kershaw DB, Bunchman TE. Recognition and management of angiotensin converting enzyme inhibitor fetopathy. Pediatr Nephrol 1995;9(3):382–5.
19. Olsen H, Klemetsrud T, Stokke HP, Tretli S, Westheim A. Adverse drug reactions in current antihypertensive therapy: a general practice survey of 2586 patients in Norway. Blood Press 1999;8(2):94–101.
20. Scott RA, Barnett DB. Lower than conventional doses of captopril in the initiation of converting enzyme inhibition in patients with severe congestive heart failure. Clin Cardiol 1989;12(4):225–6.
21. Kjekshus J, Swedberg K. Enalapril for congestive heart failure. Am J Cardiol 1989;63(8):D26–32.
22. Francis GS, Rucinska EJ. Long-term effects of a once-a-day versus twice-a-day regimen of enalapril for congestive heart failure. Am J Cardiol 1989;63(8):D17–21.
23. Lewis GR. Comparison of lisinopril versus placebo for congestive heart failure. Am J Cardiol 1989;63(8):D12–6.
24. Perlman JM, Volpe JJ. Neurologic complications of captopril treatment of neonatal hypertension. Pediatrics 1989;83(1):47–52.
25. Reznik V, Griswold W, Mendoza S. Dangers of captopril therapy in newborns. Pediatrics 1989;83(6):1076.
26. MacFadyen RJ, Lees KR, Reid JL. Differences in first dose response to angiotensin converting enzyme inhibition in congestive heart failure: a placebo controlled study. Br Heart J 1991;66(3):206–11.
27. Mullen PJ. Unexpected first dose hypotensive reaction to enalapril. Postgrad Med J 1990;66(782):1087–8.
28. Kaplan NM. The CARE Study: a postmarketing evaluation of ramipril in 11,100 patients. The Clinical Altace Real-World Efficacy (CARE) Investigators. Clin Ther 1996;18(4):658–70.
29. Sharpe N; International HOPE TIPS Investigators. The HOPE TIPS: the HOPE study translated into practices. Cardiovasc Drugs Ther 2005;19(3):197–201.
30. Nicholls MG, Gilchrist NL. Cough with ACE inhibitors: a bigger problem in some racial groups? Cardiovasc Drugs Ther 2005;19(3):173–5.
31. Lindgren BR, Andersson RG. Angiotensin-converting enzyme inhibitors and their influence on inflammation, bronchial reactivity and cough. A research review. Med Toxicol Adverse Drug Exp 1989;4(5):369–80.
32. Kaufman J, Casanova JE, Riendl P, Schlueter DP. Bronchial hyperreactivity and cough due to angiotensin-converting enzyme inhibitors. Chest 1989;95(3):544–8.
33. Lindgren BR, Rosenqvist U, Ekstrom T, Gronneberg R, Karlberg BE, Andersson RG. Increased bronchial reactivity and potentiated skin responses in hypertensive subjects suffering from coughs during ACE-inhibitor therapy. Chest 1989;95(6):1225–30.
34. Emanueli C, Grady EF, Madeddu P, Figini M, Bunnett NW, Parisi D, Regoli D, Geppetti P. Acute ACE inhibition causes plasma extravasation in mice that is mediated by bradykinin and substance P. Hypertension 1998;31(6):1299–304.
35. Ohya Y, Kumamoto K, Fujishima M. Effects of crossover application of sulindac and azelastine on enalapril-induced cough. J Hum Hypertens 1992;6(1):81–2.
36. Fogari R, Zoppi A, Tettamanti F, Malamani GD, Tinelli C, Salvetti A. Effects of nifedipine and indomethacin on cough induced by angiotensin-converting enzyme inhibitors: a double-blind, randomized, cross-over study. J Cardiovasc Pharmacol 1992;19(5):670–3.
37. Zee RY, Rao VS, Paster RZ, Sweet CS, Lindpaintner K. Three candidate genes and angiotensin-converting enzyme inhibitor-related cough: a pharmacogenetic analysis. Hypertension 1998;31(4):925–8.
38. Boulet LP, Milot J, Lampron N, Lacourciere Y. Pulmonary function and airway responsiveness during long-term therapy with captopril. JAMA 1989;261(3):413–6.
39. Os I, Bratland B, Dahlof B, Gisholt K, Syvertsen JO, Tretli S. Female sex as an important determinant of lisinopril-induced cough. Lancet 1992;339(8789):372.
40. Israili ZH, Hall WD. Cough and angioneurotic edema associated with angiotensin-converting enzyme inhibitor therapy. A review of the literature and pathophysiology. Ann Intern Med 1992;117(3):234–42.
41. Mukae S, Aoki S, Itoh S, Iwata T, Ueda H, Katagiri T. Bradykinin B(2) receptor gene polymorphism is associated with angiotensin-converting enzyme inhibitor-related cough. Hypertension 2000;36(1):127–31.

42. Takahashi T, Yamaguchi E, Furuya K, Kawakami Y. The ACE gene polymorphism and cough threshold for capsaicin after cilazapril usage. Respir Med 2001;95(2):130–5.
43. Ding PY, Hu OY, Pool PE, Liao W. Does Chinese ethnicity affect the pharmacokinetics and pharmacodynamics of angiotensin-converting enzyme inhibitors? J. Hum Hypertens 2000;14(3):163–70.
44. Sharif MN, Evans BL, Pylypchuk GB. Cough induced by quinapril with resolution after changing to fosinopril. Ann Pharmacother 1994;28(6):720–2.
45. Hargreaves MR, Benson MK. Inhaled sodium cromoglycate in angiotensin-converting enzyme inhibitor cough. Lancet 1995;345(8941):13–6.
46. Lunde H, Hedner T, Samuelsson O, Lotvall J, Andren L, Lindholm L, Wiholm BE. Dyspnoea, asthma, and bronchospasm in relation to treatment with angiotensin converting enzyme inhibitors. BMJ 1994;308(6920):18–21.
47. Inman WH, Pearce G, Wilton L, Mann RD. Angiotensin Converting Enzyme Inhibitors and AsthmaSouthampton: Drug Safety Research Unit;. 1994.
48. Nakamura Y, Yoshimoto K, Saima S. Gynaecomastia induced by angiotensin converting enzyme inhibitor. BMJ 1990;300(6723):541.
49. Herings RM, de Boer A, Stricker BH, Leufkens HG, Porsius A. Hypoglycaemia associated with use of inhibitors of angiotensin converting enzyme. Lancet 1995;345(8959): 1195–8.
50. Moore N, Kreft-Jais C, Haramburu F, Noblet C, Andrejak M, Ollagnier M, Begaud B. Reports of hypoglycaemia associated with the use of ACE inhibitors and other drugs: a case/non-case study in the French pharmacovigilance system database. Br J Clin Pharmacol 1997;44(5):513–8.
51. Thamer M, Ray NF, Taylor T. Association between antihypertensive drug use and hypoglycemia: a case-control study of diabetic users of insulin or sulfonylureas. Clin Ther 1999;21(8):1387–400.
52. Pedersen-Bjergaard U, Agerholm-Larsen B, Pramming S, Hougaard P, Thorsteinsson B. Activity of angiotensin-converting enzyme and risk of severe hypoglycaemia in type 1 diabetes mellitus. Lancet 2001;357(9264):1248–53.
53. Uchida K, Azukizawa S, Nakano S, Kaneko M, Kigoshi T, Morimoto S, Matsui A. Reversible hyperkalemia during antihypertensive therapy in a hypertensive diabetic patient with latent hypoaldosteronism and mild renal failure. South Med J 1994;87(11):1153–5.
54. Bonnet F, Thivolet CH. Reversible hyperkalemia at the initiation of ACE inhibitors in a young diabetic patient with latent hyporeninemic hypoaldosteronism. Diabetes Care 1996;19(7):781.
55. Schwab M, Roder F, Morike K, Thon KP, Klotz U. Drug-induced hyponatraemia in elderly patients. Br J Clin Pharmacol 1999;48(1):105–6.
56. Mazzali M, Filho GA. Use of aminophylline and enalapril in posttransplant polycythemia. Transplantation 1998;65(11):1461–4.
57. Olger AF, Ozlem OK, Ozgur K, Peria A, Doganay A. Effects of losartan on the renin–angiotensin–aldosterone system and erythrocytosis in patients with chronic obstructive pulmonary diseases and systemic hypertension. Clin Drug Invest 2001;21:337–43.
58. Gossmann J, Kachel HG, Schoeppe W, Scheuermann EH. Anemia in renal transplant recipients caused by concomitant therapy with azathioprine and angiotensin-converting enzyme inhibitors. Transplantation 1993;56(3):585–9.
59. Montanaro D, Gropuzzo M, Tulissi P, Boscutti G, Risaliti A, Baccarani U, Mioni G. Angiotensin-converting enzyme inhibitors reduce hemoglobin concentrations, hematocrit, and serum erythropoietin levels in renal transplant recipients without posttransplant erythrocytosis. Transplant Proc 2001;33(1–2):2038–40.
60. Deira JL, Corbacho L, Bondia A, Lerma JL, Gascon A, Martin B, Garcia P, Tabernero JM. Captopril hepatotoxicity in a case of renal crisis due to systemic sclerosis. Nephrol Dial Transplant 1997;12(8):1717–8.
61. Nissan A, Spira RM, Seror D, Ackerman Z. Captopril-associated "pseudocholangitis." A case report and review of the literature Arch Surg 1996;131(6):670–1.
62. Valle R, Carrascosa M, Cillero L, Perez-Castrillon JL. Enalapril-induced hepatotoxicity. Ann Pharmacother 1993;27(11):1405.
63. Droste HT, de Vries RA. Chronic hepatitis caused by lisinopril. Neth J Med 1995;46(2):95–8.
64. Hagley MT, Benak RL, Hulisz DT. Suspected cross-reactivity of enalapril- and captopril-induced hepatotoxicity. Ann Pharmacother 1992;26(6):780–1.
65. Maringhini A, Termini A, Patti R, Ciambra M, Biffarella P, Pagliaro L. Enalapril-associated acute pancreatitis: recurrence after rechallenge. Am J Gastroenterol 1997;92(1):166–7.
66. Standridge JB. Fulminant pancreatitis associated with lisinopril therapy. South Med J 1994;87(2):179–81.
67. Packer M. Identification of risk factors predisposing to the development of functional renal insufficiency during treatment with converting-enzyme inhibitors in chronic heart failure. Cardiology 1989;76(Suppl 2):50–5.
68. Kon V, Fogo A, Ichikawa I. Bradykinin causes selective efferent arteriolar dilation during angiotensin I converting enzyme inhibition. Kidney Int 1993;44(3):545–50.
69. Wynckel A, Ebikili B, Melin JP, Randoux C, Lavaud S, Chanard J. Long-term follow-up of acute renal failure caused by angiotensin converting enzyme inhibitors. Am J Hypertens 1998;11(9):1080–6.
70. Kuechle MK, Hutton KP, Muller SA. Angiotensin-converting enzyme inhibitor-induced pemphigus: three case reports and literature review. Mayo Clin Proc 1994;69(12):1166–71.
71. Gilleaudeau P, Vallat VP, Carter DM, Gottlieb AB. Angiotensin-converting enzyme inhibitors as possible exacerbating drugs in psoriasis. J Am Acad Dermatol 1993;28(3):490–2.
72. Butt A, Burge SM. Pemphigus vulgaris induced by captopril. Dermatology 1993;186:315.
73. Ikai K. Exacerbation and induction of psoriasis by angiotensin-converting enzyme inhibitors. J Am Acad Dermatol 1995;32(5 Pt 1):819.
74. Ong CS, Cook N, Lee S. Drug-related pemphigus and angiotensin converting enzyme inhibitors. Australas J Dermatol 2000;41(4):242–6.
75. Sieber C, Grimm E, Follath F. Captopril and systemic lupus erythematosus syndrome. BMJ 1990; 301(6753):669.
76. Leak D. Absence of cross-reaction between lisinopril and enalapril in drug-induced lupus. Ann Pharmacother 1997;31(11):1406–7.
77. Chagas KdeN, Arruk VG, Andrade ME, Vasconcelos DdeM, Kirschfink M, Duarte AJ, Grumach AS. Angioedema hereditario: consideracoes sobre terapia. [Therapeutic approach of hereditary angioedema.] Rev Assoc Med Bras 2004;50(3):314–9.
78. Howes LG, Tran D. Can angiotensin receptor antagonists be used safely in patients with previous ACE inhibitor-induced angioedema? Drug Saf 2002;25(2):73–6.

79. Sadeghi N, Panje WR. Life-threatening perioperative angioedema related to angiotensin-converting enzyme inhibitor therapy. J Otolaryngol 1999;28(6):354–6.
80. Maestre ML, Litvan H, Galan F, Puzo C, Villar Landeira JM. Imposibilidad de intubacion por angioedema secundario a IECA. [Impossibility of intubation due to angioedema secondary to an angiotensin-converting enzyme inhibitor.] Rev Esp Anestesiol Reanim 1999;46(2):88–91.
81. Hedner T, Samuelsson O, Lunde H, Lindholm L, Andren L, Wiholm BE. Angio-oedema in relation to treatment with angiotensin converting enzyme inhibitors. BMJ 1992;304(6832):941–6.
82. Martin DJ, Grigg RG, Tomkinson A, Coman WB. Subglottic stenosis: an unusual presentation of ACE inhibitor-induced angioedema. Aust NZ J Surg 1999;69(4):320–1.
83. Jacobs RL, Hoberman LJ, Goldstein HM. Angioedema of the small bowel caused by an angiotensin-converting enzyme inhibitor. Am J Gastroenterol 1994;89(1):127–8.
84. Dupasquier E. Une forme clinique rare d'oedème angioneurotique sous énalapril: l'abdomen aigu. [A rare clinical form of angioneurotic edema caused by enalapril: acute abdomen.] Arch Mal Coeur Vaiss 1994;87(10):1371–4.
85. Byrne TJ, Douglas DD, Landis ME, Heppell JP. Isolated visceral angioedema: an underdiagnosed complication of ACE inhibitors? Mayo Clin Proc 2000;75(11):1201–4.
86. Chase MP, Fiarman GS, Scholz FJ, MacDermott RP. Angioedema of the small bowel due to an angiotensin-converting enzyme inhibitor. J Clin Gastroenterol 2000;31(3):254–7.
87. Blomberg PJ, Surks HK, Long A, Rebeiz E, Mochizuki Y, Pandian N. Transient myocardial dysfunction associated with angiotensin-converting enzyme inhibitor-induced angioedema: recognition by serial echocardiographic studies. J Am Soc Echocardiogr 1999;12(12):1107–9.
88. Kyrmizakis DE, Papadakis CE, Fountoulakis EJ, Liolios AD, Skoulas JG. Tongue angioedema after long-term use of ACE inhibitors. Am J Otolaryngol 1998;19(6):394–6.
89. Lapostolle F, Borron SW, Bekka R, Baud FJ. Lingual angioedema after perindopril use. Am J Cardiol 1998;81(4):523.
90. Henson EB, Bess DT, Abraham L, Bracikowski JP. Penile angioedema possibly related to lisinopril. Am J Health Syst Pharm 1999;56(17):1773–4.
91. Schmidt TD, McGrath KM. Angiotensin-converting enzyme inhibitor angioedema of the intestine: a case report and review of the literature. Am J Med Sci 2002;324(2):106–8.
92. Orr KK, Myers JR. Intermittent visceral edema induced by long-term enalapril administration. Ann Pharmacother 2004;38(5):825–7.
93. Jacobs RL, Hoberman LJ, Goldstein HM. Angioedema of the small bowel caused by an angiotensin-converting enzyme inhibitor. Am J Gastroenterol 1994;89(1):127–8.
94. Dupasquier E. Une forme clinique rare d'oedème angioneurotique sous énalapril: l'abdomen aigu. [RA rare clinical form of angioneurotic edema caused by enalapril: acute abdomen.] Arch Mal Coeur Vaiss 1994;87(10):1371–4.
95. Byrne TJ, Douglas DD, Landis ME, Heppell JP. Isolated visceral angioedema: an underdiagnosed complication of ACE inhibitors? Mayo Clin Proc 2000;75(11):1201–4.
96. Mlynarek A, Hagr A, Kost K. Angiotensin-converting enzyme inhibitor-induced unilateral tongue angioedema. Otolaryngol Head Neck Surg 2003;129(5):593–5.
97. Rai MR, Amen F, Idrees F. Angiotensin-converting enzyme inhibitor related angioedema and the anaesthetist. Anaesthesia 2004;59(3):283–9.
98. Rafii MS, Koenig M, Ziai WC. Orolingual angioedema associated with ACE inhibitor use after rtPA treatment of acute stroke. Neurology 2005;65(12):1906.
99. O'Ryan F, Poor DB, Hattori M. Intraoperative angioedema induced by angiotensin-converting enzyme inhibitors: overview and case report. J Oral Maxillofac Surg 2005;63(4):551–6.
100. Salloum H, Locher C, Chenard A, Bigorie B, Beroud P, Gatineau-Sailliant G, Glikmanas M. Angidème intestinal après prise de perindopril. Gastroenterol Clin Biol 2005;29(11):1180–1.
101. Karagiannis A, Pyrpasopoulou A, Tziomal;os K, Florentin M, Athyros V. Q J Med 2005;99:197–8.
102. Agostoni A, Cicardi M. Drug-induced angioedema without urticaria: incidence, prevention and management. Drug Saf 2001;24(8):599–606.
103. Kostis JB, Kim HJ, Rusnak J, Casale T, Kaplan A, Corren J, Levy E. Incidence and characteristics of angioedema associated with enalapril. Arch Intern Med 2005; 165(14):1637–42.
104. Brown NJ, Ray WA, Snowden M, Griffin MR. Black Americans have an increased rate of angiotensin converting enzyme inhibitor-associated angioedema. Clin Pharmacol Ther 1996;60(1):8–13.
105. Sondhi D, Lippmann M, Murali G. Airway compromise due to angiotensin-converting enzyme inhibitor-induced angioedema. Chest 2004;126:400–4.
106. Adverse Drug Reactions Advisory Committee (ADRAC). Angioedema—still a problem with ACE inhibitors. Aust Adv Drug React Bull 2005;24(2):7.
107. Jae Joo Cho, Woo Seok Koh, Bang Soon Kim. A case of angioedema probably induced by captopril. Korean J Dermatol 1999;37:404–6.
108. Mchaourab A, Sarantopoulos C, Stowe DF. Airway obstruction due to late-onset angioneurotic edema from angiotensin-converting enzyme inhibition. Can J Anaesth 1999;46(10):975–8.
109. Maliekal J, Del Rio G. Acute angioedema associated with long-term benazepril therapy. J Pharm Technol 1999;15:208–11.
110. Vleeming W, van Amsterdam JGC, Stricker BHCh, de Wildt DJ. ACE inhibitor-induced angioedema: incidence, prevention and management. Drug Saf 1998;18(3):171–88.
111. O'Mara NB, O'Mara EM. Delayed onset of angioedema with angiotensin-converting enzyme inhibitors: case report and review of the literature. Pharmacotherapy 1996; 16:675–9.
112. McHaourab A, Sarantopoulos C, Stowe DF. Airway obstruction due to late-onset angioneurotic edema from angiotensin-converting enzyme inhibition. Can J Anaesth 1999;46:975–8.
113. Burkhart DG, Brown NJ, Griffin MR, Ray WA, Hammerstrom T, Weiss S. Angiotensin converting enzyme inhibitor-associated angioedema: higher risk in blacks than whites. Pharmacoepidemiol Drug Saf 1996;5(3):149–54.
114. Coleman JJ, McDowell SE. Ethnicity and adverse drug reactions. Adv Drug React Bull 2005;234:899–902.
115. Committee on Safety of Medicines. Anaphylactoid reactions to high-flux polyacrylonitrile membranes in combination with ACE inhibitors. Curr Probl 1992;33.
116. Kammerl MC, Schaefer RM, Schweda F, Schreiber M, Riegger GA, Kramer BK. Extracorporal therapy with AN69 membranes in combination with ACE inhibition

causing severe anaphylactoid reactions: still a current problem? Clin Nephrol 2000;53(6):486–8.
117. Abbosh J, Anderson JA, Levine AB, Kupin WL. Angiotensin converting enzyme inhibitor-induced angioedema more prevalent in transplant patients. Ann Allergy Asthma Immunol 1999;82(5):473–6.
118. Stallone G, Infante B, Di Paolo S, Schena A, Grandaliano G, Gesualdo L, Schena FP. Sirolimus and angiotensin-converting enzyme inhibitors together induce tongue oedema in renal transplant recipients. Nephrol Dial Transplant 2004;19(11):2906–8.
119. Piller L, Ford C, Davis B, Nwachuku C, Black H, Oparil S, Gappy S, Retta T, Probstfield J. Angioedema in the antihypertensive and lipid-lowering treatment to prevent heart attack trial (ALLHAT). Am J Hypertens 2005;18(5 Suppl):A92.
120. Ducroix JP, Outurquin S, Benabes-Jezraoui B, Gras V, Chaby G, Strunski V, Salle V, Smail A, Lok C, Andrejak M. Angio-dèmes et inhibiteurs de l'enzyme de conversion de l'angiotensine: à propos de 19 cas. [Angioedema and angiotensin converting enzyme inhibitors: a report of 19 cases.] Rev Med Interne 2004;25(7):501–6.
121. Rai MR, Amen F, Idrees F. Angiotensin-converting enzyme inhibitor related angioedema and the anaesthetist. Anaesthesia 2004;59(3):283–9.
122. Dykewicz MS. Cough and angioedema from angiotensin-converting enzyme inhibitors: new insights into mechanisms and management. Curr Opin Allergy Clin Immunol 2004;4(4):267–70.
123. Cugno M, Nussberger J, Cicardi M, Agostoni A. Bradykinin and the pathophysiology of angioedema. Int Immunopharmacol 2003;3:311–7.
124. Nussberger J, Cugno M, Amstutz C, Cicardi M, Pellacani A, Agostoni A. Plasma bradykinin in angiooedema. Lancet 1998;351(9117):1693–7.
125. Van De Carr S, Sigler C, Annis K, Cooper K, Haber H. Examination of baseline levels of carboxypeptidase N and complement components as potential predictors of angioedema associated with the use of an angiotensin-converting enzyme inhibitor. Arch Dermatol 1997;133:972–5.
126. Ebo DG, Stevens WJ, Bosmans JL. An adverse reaction to angiotensin-converting enzyme inhibitors in a patient with neglected C1 esterase inhibitor deficiency. J Allergy Clin Immunol 1997;99:425–6.
127. Molinaro G, Cugno M, Perez M, Lepage Y, Gervais N, Agostoni A, Adam A. Angiotensin-converting enzyme inhibitor-associated angioedema is characterized by a slower degradation of des-arginine(9)-bradykinin. J Pharmacol Exp Ther 2002;303(1):232–7.
128. Jae Joo Cho, Woo Seok Koh, Bang Soon Kim. A case of angioedema probably induced by captopril. Korean J Dermatol 1999;37:404–6.
129. Kyrmizakis DE, Papadakis CE, Fountoulakis EJ, Liolios AD, Skoulas JG. Tongue angioedema after long-term use of ACE inhibitors. Am J Otoryngol 1998;19:394–6.
130. Abdelmalek MF, Douglas DD. Lisinopril-induced isolated visceral angioedema. Review of ACE-inhibitor-induced small bowel angioedema. Dig Dis Sci 1997;42:847–50.
131. Lapostolle F, Barron SW, Bekka R, Baud FJ. Lingual angioedema after perindopril use. Am J Cardiol 1998;81:523.
132. Cosano L, Gonzalez Ramallo VJ, Huertas AJ, Garcia Castano J. Edema angioneurotico y broncospasmo en el tratamiento con quinapril. [RAngioneurotic edema and bronchospasm in quinapril treatment] Med Clin (Barc) 1994;102(7):275.
133. Epeldo Gonzalo F, Boada Montagut L, Vecina ST. Angioedema caused by ramipril. Ann Pharmacother 1995;29(4):431–2.
134. Warrier MR, Copilevitz CA, Dykewicz MS, Slavin RG. Fresh frozen plasma in the treatment of resistant angiotensin-converting enzyme inhibitor angioedema. Ann Allergy Asthma Immunol 2004;92(5):573–5.
135. Pryde PG, Sedman AB, Nugent CE, Barr M Jr. Angiotensin-converting enzyme inhibitor fetopathy. J Am Soc Nephrol 1993;3(9):1575–82.
136. Rucinska EJ, Small R, Mulcahy WS, Snyder DL, Rodel PV, Rush JE, Smith RD, Walker JF, Irvin JD. Tolerability of long term therapy with enalapril maleate in patients resistant to other therapies and intolerant to captopril. Med Toxicol Adverse Drug Exp 1989;4(2): 144–52.
137. Rucinska EJ, Small R, Irvin J. High-risk patients treated with enalapril maleate: safety considerations. Int J Cardiol 1989;22(2):249–59.
138. Keane WF, Polis A, Wolf D, Faison E, Shahinfar S. The long-term tolerability of enalapril in hypertensive patients with renal impairment. Nephrol Dial Transplant 1997;12(Suppl 2):75–81.
139. Zanchetti A. Contribution of fixed low-dose combinations to initial therapy in hypertension. Eur Heart J 1999; 1(Suppl L):L5–9.
140. Kincaid EH, Ashburn DA, Hoyle JR, Reichert MG, Hammon JW, Kon ND. Does the combination of aprotinin and angiotensin-converting enzyme inhibitor cause renal failure after cardiac surgery? Ann Thorac Surg 2005;80(4):1388–93.
141. Guazzi MD, Campodonico J, Celeste F, Guazzi M, Santambrogio G, Rossi M, Trabattoni D, Alimento M. Antihypertensive efficacy of angiotensin converting enzyme inhibition and aspirin counteraction. Clin Pharmacol Ther 1998;63(1):79–86.
142. Spaulding C, Charbonnier B, Cohen-Solal A, Juilliere Y, Kromer EP, Benhamda K, Cador R, Weber S. Acute hemodynamic interaction of aspirin and ticlopidine with enalapril: results of a double-blind, randomized comparative trial. Circulation 1998;98(8):757–65.
143. Teerlink JR, Massie BM. The interaction of ACE inhibitors and aspirin in heart failure: torn between two lovers. Am Heart J 1999;138(2 Pt 1):193–7.
144. Al-Khadra AS, Salem DN, Rand WM, Udelson JE, Smith JJ, Konstam MA. Antiplatelet agents and survival: a cohort analysis from the Studies of Left Ventricular Dysfunction (SOLVD) trial. J Am Coll Cardiol 1998;31(2):419–25.
145. Massie BM, Teerlink JR. Interaction between aspirin and angiotensin-converting enzyme inhibitors: real or imagined. Am J Med 2000;109(5):431–3.
146. Guazzi M, Pontone G, Agostoni P. Aspirin worsens exercise performance and pulmonary gas exchange in patients with heart failure who are taking angiotensin-converting enzyme inhibitors. Am Heart J 1999;138(2 Pt 1):254–60.
147. Flather MD, Yusuf S, Kober L, Pfeffer M, Hall A, Murray G, Torp-Pedersen C, Ball S, Pogue J, Moye L, Braunwald E. Long-term ACE-inhibitor therapy in patients with heart failure or left-ventricular dysfunction: a systematic overview of data from individual patients. ACE-Inhibitor Myocardial Infarction Collaborative Group. Lancet 2000;355(9215):1575–81.
148. Peterson JG, Topol EJ, Sapp SK, Young JB, Lincoff AM, Lauer MS. Evaluation of the effects of aspirin combined with angiotensin-converting enzyme inhibitors in patients with coronary artery disease. Am J Med 2000;109(5):371–7.

149. Oosterga M, Anthonio RL, de Kam PJ, Kingma JH, Crijns HJ, van Gilst WH. Effects of aspirin on angiotensin-converting enzyme inhibition and left ventricular dilation one year after acute myocardial infarction. Am J Cardiol 1998;81(10):1178–81.
150. Katz SD, Radin M, Graves T, Hauck C, Block A, LeJemtel THIfetroban Study Group. Effect of aspirin and ifetroban on skeletal muscle blood flow in patients with congestive heart failure treated with Enalapril. J Am Coll Cardiol 1999;34(1):170–6.
151. Sugawara M, Toda T, Iseki K, Miyazaki K, Shiroto H, Kondo Y, Uchino J. Transport characteristics of cephalosporin antibiotics across intestinal brush-border membrane in man, rat and rabbit. J Pharm Pharmacol 1992;44(12):968–72.
152. Dantzig AH, Bergin L. Uptake of the cephalosporin, cephalexin, by a dipeptide transport carrier in the human intestinal cell line, Caco-2. Biochim Biophys Acta 1990;1027(3):211–7.
153. Friedman DI, Amidon GL. Intestinal absorption mechanism of dipeptide angiotensin converting enzyme inhibitors of the lysyl-proline type: lisinopril and SQ 29,852. J Pharm Sci 1989;78(12):995–8.
154. Hu M, Amidon GL. Passive and carrier-mediated intestinal absorption components of captopril. J Pharm Sci 1988;77(12):1007–11.
155. Padoin C, Tod M, Perret G, Petitjean O. Analysis of the pharmacokinetic interaction between cephalexin and quinapril by a nonlinear mixed-effect model. Antimicrob Agents Chemother 1998;42(6):1463–9.
156. Casato M, Pucillo LP, Leoni M, di Lullo L, Gabrielli A, Sansonno D, Dammacco F, Danieli G, Bonomo L. Granulocytopenia after combined therapy with interferon and angiotensin-converting enzyme inhibitors: evidence for a synergistic hematologic toxicity. Am J Med 1995;99(4):386–91.
157. Chiu TF, Bullard MJ, Chen JC, Liaw SJ, Ng CJ. Rapid life-threatening hyperkalemia after addition of amiloride HCl/hydrochlorothiazide to angiotensin-converting enzyme inhibitor therapy. Ann Emerg Med 1997;30(5):612–5.
158. Henger A, Tutt P, Hulter HM, Krapf R. Acid-base effects of inhibition of aldosterone and angiotensin II action in chronic metabolic acidosis in humans. J Am Soc Nephrol 1999;10:121A.
159. Seelig CB, Maloley PA, Campbell JR. Nephrotoxicity associated with concomitant ACE inhibitor and NSAID therapy. South Med J 1990;83(10):1144–8.
160. Kaplan-Machlis B, Klostermeyer BS. The cyclooxygenase-2 inhibitors: safety and effectiveness. Ann Pharmacother 1999;33(9):979–88.
161. Brown CH. Effect of rofecoxib on the antihypertensive activity of lisinopril. Ann Pharmacother 2000;34(12):1486.

Antianginal drugs

See also Beta-blockers, Calcium channel blockers, Molsidomine, Nicorandil, Nitrates, organic

General Information

Drugs that are used in the treatment of angina pectoris include agents from the following groups:

- Nitric oxide donors
- Beta-blockers
- Calcium channel blockers
- Potassium channel activators.

New agents, such as L-carnitine and trimetazidine, are also being studied (1).

Safety factors that govern the choice of antianginal drug

Insights into the epidemiology, physiology, cellular biology, molecular biology, and treatment of ischemic heart disease have shown that there are three major modifiable risk factors (hypertension, hypercholesterolemia, smoking) that are the main targets of our preventive strategies. However, evidence that intervention is beneficial has been considerably strengthened.

Hypertension

In high-risk elderly people, treatment of hypertension prevents coronary events (2,3). In the Hypertension Optimal Treatment (HOT) study, which included middle-aged and elderly subjects, good blood pressure control resulted in improved outcome, the optimal blood pressure in non-diabetic subjects being about 138/82 mmHg (4). The target blood pressure was achievable but most patients required more than one drug. A further finding was that low-dose aspirin was beneficial in high-risk subjects. The HOT study found particular benefits of rigorous blood pressure lowering in diabetic subjects, findings that were confirmed by the UK Prospective Diabetes Study (UKPDS) (5), which did not find an advantage in macrovascular complications with angiotensin converting enzyme inhibitors compared with beta-blockers (6). The CAPPP study similarly showed no benefit of captopril over other drugs (7). This originally led to the advice that drugs other than diuretics and beta-blockers should be used infrequently (8); however, subsequent recommendations for the choice of initial therapy have been increasingly based on indications for other conditions that often co-exist with hypertension (9). The Syst-Eur study of systolic hypertension gave further prominence to isolated systolic hypertension as a treatable risk factor and provided some reassuring data on the efficacy of calcium antagonists (10). Despite rather depressing news about inadequate blood pressure control in the UK (11) and USA (12), the increased use of antihypertensive drugs appears to have resulted in less left ventricular hypertrophy (13). This may account in part for the considerable fall in mortality from cardiovascular disease observed since the late 1960s.

Hypercholesterolemia

The landmark 4S study of cholesterol-lowering therapy has convincingly shown the effectiveness of treating patients with high serum cholesterol concentrations and ischemic heart disease (14). This has largely been confirmed by subsequent studies (15–17), and the beneficial effects of lowering serum cholesterol have been extended to primary prevention (18,19). Currently, statins are recommended in all those over the age of 55 years and/or in those with pre-existing cardiovascular disease (20). Statins reduce cardiac events if begun early after acute coronary syndromes (21) and may also be indicated in any patient with aortic stenosis, regardless of severity (22).

Smoking

The risks of smoking have been highlighted (23). Evidence has emerged that changing from high-tar to low-tar cigarettes is ineffective in reducing myocardial infarction (24), and the antismoking lobby has become more vocal (25,26). The role of cigarette smoking as a major factor in myocardial infarction has been further emphasized by a study of the survivors of the ISIS studies (27). Smoking increased the risk of a non-fatal myocardial infarction five-, three-, and two-fold in the age ranges 30–49, 50–59, and 70–79 respectively. In addition, the results of this study suggested that enforcement of a European Union upper limit of cigarette tar of 12 mg will result in only a modest reduction in myocardial infarction. Preventing adolescents from smoking seems to be the only strategy, as few adults start smoking after the age of 18 (28). Passive smoking (29) and cigar smoking (30) may be associated with coronary disease, making this initiative even more important. The US ruling in July 1995 that nicotine is an addictive substance (31) was a monumental advance in fighting smoking, since cigarettes are considered to be nicotine delivery systems. If the FDA gains legal jurisdiction in regulating tobacco, active steps will be taken to reduce smoking among children.

The roles of these modifiable risks have thus become more important, and other risks interventions, such as dietary antioxidants (32) and physical activity (33), have also been highlighted. In addition, observational studies have implicated hyperhomocysteinemia as a powerful risk for premature atherosclerotic coronary artery disease (34). Whether or not treating this risk factor is beneficial in reducing cardiac events is currently being tested (35). In the USA, grain is already enriched with folic acid (140 μg/100 g) (36), and this has resulted in reduced homocysteine concentrations in the middle-aged and older population from the Framingham Offspring Study cohort (37).

In recent years, potassium channel activators have been added to the antianginal armamentarium. In addition, the line between antianginal drugs and drugs used to prevent angina has become somewhat blurred. Aspirin has now become a mainstay drug in the treatment and prevention of ischemic heart disease (38), and another antiplatelet drug, clopidogrel, has also been licensed for this indication. Lipid-lowering drugs have also been advocated for both primary and secondary prevention, and the role of beta-blockers in the secondary prevention of myocardial infarction has been extended into the treatment of systolic heart failure. The angiotensin-converting-enzyme (ACE) inhibitor enalapril has gained a licence for the prevention of coronary ischemic events in patients with left ventricular dysfunction, following the Studies of Left Ventricular Dysfunction Trials (39). More recently, evidence on lisinopril and trandolapril has suggested that these drugs prolong the lives of subjects who have a myocardial infarction, either uncomplicated or complicated by impaired ventricular function (40). The benefits of ACE inhibitors in secondary prevention are now beyond doubt, and are especially marked in diabetic patients (41).

Our view of drug therapy has subtly changed with these findings. The physician's role was to use drugs to relieve the symptoms of angina. Nowadays, drugs must not only relieve symptoms but also, when possible, improve life expectancy (20). We are also more concerned with the quality of symptomatic relief. The notion of "well-being" has become important and has been assessed in "quality-of-life" comparisons of different agents (42). Subtle effects that individually might not be detected have been highlighted in such studies and can modify our view of a drug's adverse effects profile. As the subtle adverse effects of drugs become more important, serious drug toxicity becomes even more unacceptable.

Other susceptibility factors

The roles of other modifiable susceptibility factors have become more important, and other interventions, such as dietary antioxidants (32) and physical activity (33), have also been highlighted in the past. However, a meta-analysis of the effect of vitamin E supplements has suggested that doses over 400 IU/day can increase the risk of death from any cause (43).

Pharmacoeconomics

A further consideration, the economic evaluation of drugs, has become an issue, although not without criticism (44). In its most rigorous form the cost of drug therapy is measured against the aggregate number of years of improved health that such therapy might be expected to bring (45). These so-called "quality-adjusted life years" (QUALYs) are reduced if a drug has an adverse effect, so that the net benefit of a drug that prolongs life but makes that life miserable may well be a negative number of QUALYs. Such studies have put a price-tag on mild adverse events as well as on beneficial effects. In time we may pay increasing attention to the price of adverse effects when choosing a drug.

The drug treatment of angina is time-consuming, empirical, and often relatively unrewarding. Drug treatment simply palliates the underlying disease, and no symptomatic treatment improves survival or prevents myocardial infarction. This is in contrast to surgical intervention, which, in good hands, prolongs survival (46) and improves its quality. The three main comparisons of coronary surgery with medical therapy (47–49) are now out of date, since both medical therapy and surgical techniques have improved considerably since they were completed. However, long-term medical therapy is now more often being reserved for patients who for one reason or another are unsuitable for surgery or percutaneous coronary angioplasty, which, at least for single-vessel disease, is superior to antianginal drug treatment (50) but may be equivalent to lipid-lowering therapy (51). Preventive therapy must now be the dominant strategy.

References

1. Chierchia SL, Fragasso G. Metabolic management of ischaemic heart disease. Eur Heart J 1993;14(Suppl G):2–5.
2. Dahlof B, Lindholm LH, Hansson L, Schersten B, Ekbom T, Wester PO. Morbidity and mortality in the Swedish Trial in Old Patients with Hypertension (STOP-Hypertension). Lancet 1991;338(8778):1281–5.
3. SHEP Cooperative Research Group. Prevention of stroke by antihypertensive drug treatment in older persons with isolated systolic hypertension. Final results of the Systolic Hypertension in the Elderly Program (SHEP). JAMA 1991;265(24):3255–64.
4. Hansson L, Zanchetti A, Carruthers SG, Dahlof B, Elmfeldt D, Julius S, Menard J, Rahn KH, Wedel H, Westerling SHOT Study Group. Effects of intensive blood-pressure lowering and low-dose aspirin in patients with hypertension: principal results of the Hypertension Optimal Treatment (HOT) randomised trial. Lancet 1998;351(9118):1755–62.
5. UK Prospective Diabetes Study Group. Tight blood pressure control and risk of macrovascular and microvascular complications in type 2 diabetes: UKPDS 38. BMJ 1998;317(7160):703–13.
6. UKPDS 39. UK Prospective Diabetes Study Group. Efficacy of atenolol and captopril in reducing risk of macrovascular and microvascular complications in type 2 diabetes. BMJ 1998;317(7160):713–20.
7. Hansson L, Lindholm LH, Niskanen L, Lanke J, Hedner T, Niklason A, Luomanmaki K, Dahlof B, de Faire U, Morlin C, Karlberg BE, Wester PO, Bjorck JE. Effect of angiotensin-converting-enzyme inhibition compared with conventional therapy on cardiovascular morbidity and mortality in hypertension: the Captopril Prevention Project (CAPPP) randomised trial. Lancet 1999;353(9153):611–6.
8. Cutler J. Which drug for treatment of hypertension? Lancet 1999;353(9153):604–5.
9. European Society of Hypertension-European Society of Cardiology Guidelines Committee. 2003 European Society of Hypertension–European Society of Cardiology guidelines for the management of arterial hypertension. J Hypertens 2003;21(6):1011–53.
10. Staessen JA, Fagard R, Thijs L, Celis H, Arabidze GG, Birkenhager WH, Bulpitt CJ, de Leeuw PW, Dollery CT, Fletcher AE, Forette F, Leonetti G, Nachev C, O'Brien ET, Rosenfeld J, Rodicio JL, Tuomilehto J, Zanchetti A. Randomised double-blind comparison of placebo and active treatment for older patients with isolated systolic hypertension. The Systolic Hypertension in Europe (Syst-Eur) Trial Investigators. Lancet 1997;350(9080):757–64.
11. Colhoun HM, Dong W, Poulter NR. Blood pressure screening, management and control in England: results from the health survey for England 1994. J Hypertens 1998;16(6):747–52.
12. Berlowitz DR, Ash AS, Hickey EC, Friedman RH, Glickman M, Kader B, Moskowitz MA. Inadequate management of blood pressure in a hypertensive population. N Engl J Med 1998;339(27):1957–63.
13. Mosterd A, D'Agostino RB, Silbershatz H, Sytkowski PA, Kannel WB, Grobbee DE, Levy D. Trends in the prevalence of hypertension, antihypertensive therapy, and left ventricular hypertrophy from 1950 to 1989. N Engl J Med 1999;340(16):1221–7.
14. Scandinavian Simvastatin Survival Study Group. Randomized trial of cholesterol lowering in 4444 patients with coronary heart disease: the Scandinavian Simvastatin Survival Study (4S). Lancet 1994;344:1383.
15. Byington RP, Jukema JW, Salonen JT, Pitt B, Bruschke AV, Hoen H, Furberg CD, Mancini GB. Reduction in cardiovascular events during pravastatin therapy. Pooled analysis of clinical events of the Pravastatin Atherosclerosis Intervention Program. Circulation 1995; 92(9):2419–25.
16. Sacks FM, Pfeffer MA, Moye LA, Rouleau JL, Rutherford JD, Cole TG, Brown L, Warnica JW, Arnold JM, Wun CC, Davis BR, Braunwald E. The effect of pravastatin on coronary events after myocardial infarction in patients with average cholesterol levels. Cholesterol and Recurrent Events Trial investigators. N Engl J Med 1996;335(14):1001–9.
17. The Long-Term Intervention with Pravastatin in Ischaemic Disease (LIPID) Study Group. Prevention of cardiovascular events and death with pravastatin in patients with coronary heart disease and a broad range of initial cholesterol levels. N Engl J Med 1998;339(19):1349–57.
18. Shepherd J, Cobbe SM, Ford I, Isles CG, Lorimer AR, MacFarlane PW, McKillop JH, Packard CJWest of Scotland Coronary Prevention Study Group. Prevention of coronary heart disease with pravastatin in men with hypercholesterolemia. N Engl J Med 1995;333(20):1301–7.
19. Downs JR, Clearfield M, Weis S, Whitney E, Shapiro DR, Beere PA, Langendorfer A, Stein EA, Kruyer W, Gotto AM Jr. Primary prevention of acute coronary events with lovastatin in men and women with average cholesterol levels: results of AFCAPS/TexCAPS. Air Force/Texas Coronary Atherosclerosis Prevention Study. JAMA 1998;279(20):1615–22.
20. Wald NJ, Law MR. A strategy to reduce cardiovascular disease by more than 80%. BMJ 2003;326(7404):1419.
21. Cannon CP, Braunwald E, McCabe CH, Rader DJ, Rouleau JL, Belder R, Joyal SV, Hill KA, Pfeffer MA, Skene AM. for the Pravastatin or Atorvastatin Evaluation and Infection Therapy–Thrombolysis in Myocardial Infarction 22 Investigators. Intensive versus moderate lipid lowering with statins after acute coronary syndromes. N Engl J Med 2004;350(15):1495–504.
22. Rosenhek R, Rader F, Loho N, Gabriel H, Heger M, Klaar U, Schemper M, Binder T, Maurer G, Baumgartner H. Statins but not angiotensin-converting enzyme inhibitors delay progression of aortic stenosis. Circulation 2004;110(10):1291–5.
23. Bartecchi CE, MacKenzie TD, Schrier RW. The human costs of tobacco use (1). N Engl J Med 1994;330(13):907–12.
24. Negri E, Franzosi MG, La Vecchia C, Santoro L, Nobili A, Tognoni G. Tar yield of cigarettes and risk of acute myocardial infarction. GISSI-EFRIM Investigators. BMJ 1993;306(6892):1567–70.
25. Vickers A. Why cigarette advertising should be banned. BMJ 1992;304(6836):1195–6.
26. Anonymous. Enlightenment on the road to death. Lancet 1994;343(8906):1109–10.
27. Parish S, Collins R, Peto R, Youngman L, Barton J, Jayne K, Clarke R, Appleby P, Lyon V, Cederholm-Williams S, et al. Cigarette smoking, tar yields, and non-fatal myocardial infarction: 14,000 cases and 32,000 controls in the United Kingdom. The International Studies of Infarct Survival (ISIS) Collaborators. BMJ 1995;311(7003):471–7.
28. McNeill AD, Jarvis MJ, Stapleton JA, Russell MA, Eiser JR, Gammage P, Gray EM. Prospective study of factors predicting uptake of smoking in adolescents. J Epidemiol Community Health 1989;43(1):72–8.
29. He J, Vupputuri S, Allen K, Prerost MR, Hughes J, Whelton PK. Passive smoking and the risk of coronary

heart disease—a meta-analysis of epidemiologic studies. N Engl J Med 1999;340(12):920–6.
30. Iribarren C, Tekawa IS, Sidney S, Friedman GD. Effect of cigar smoking on the risk of cardiovascular disease, chronic obstructive pulmonary disease, and cancer in men. N Engl J Med 1999;340(23):1773–80.
31. Roberts J. Nicotine is addictive, rules FDA. BMJ 1995;311(6999):211.
32. Steinberg D. Antioxidant vitamins and coronary heart disease. N Engl J Med 1993;328(20):1487–9.
33. Lakka TA, Venalainen JM, Rauramaa R, Salonen R, Tuomilehto J, Salonen JT. Relation of leisure-time physical activity and cardiorespiratory fitness to the risk of acute myocardial infarction. N Engl J Med 1994;330(22):1549–54.
34. Boushey CJ, Beresford SA, Omenn GS, Motulsky AG. A quantitative assessment of plasma homocysteine as a risk factor for vascular disease. Probable benefits of increasing folic acid intakes. JAMA 1995;274(13):1049–57.
35. Homocysteine Lowering Trialists' Collaboration. Lowering blood homocysteine with folic acid based supplements: meta-analysis of randomised trials. BMJ 1998;316(7135):894–8.
36. Oakley GP Jr, Johnston RB Jr. Balancing benefits and harms in public health prevention programmes mandated by governments. BMJ 2004;329(7456):41–3.
37. Jacques PF, Selhub J, Bostom AG, Wilson PW, Rosenberg IH. The effect of folic acid fortification on plasma folate and total homocysteine concentrations. N Engl J Med 1999;340(19):1449–54.
38. Willard JE, Lange RA, Hillis LD. The use of aspirin in ischemic heart disease. N Engl J Med 1992;327(3):175–81.
39. Yusuf S, Pepine CJ, Garces C, Pouleur H, Salem D, Kostis J, Benedict C, Rousseau M, Bourassa M, Pitt B. Effect of enalapril on myocardial infarction and unstable angina in patients with low ejection fractions. Lancet 1992;340(8829):1173–8.
40. Torp-Pedersen C, Kober L. Effect of ACE inhibitor trandolapril on life expectancy of patients with reduced left-ventricular function after acute myocardial infarction. TRACE Study Group. Trandolapril Cardiac Evaluation. Lancet 1999;354(9172):9–12.
41. MacDonald TM, Butler R, Newton RW, Morris AD. Which drugs benefit diabetic patients for secondary prevention of myocardial infarction? DARTS/MEMO Collaboration. Diabet Med 1998;15(4):282–9.
42. Fitzpatrick R, Fletcher A, Gore S, Jones D, Spiegelhalter D, Cox D. Quality of life measures in health care. I: Applications and issues in assessment. BMJ 1992;305(6861):1074–7.
43. Miller ER 3rd, Pastor-Barriuso R, Dalal D, Riemersma RA, Appel LJ, Guallar E. Meta-analysis: high-dosage vitamin E supplementation may increase all-cause mortality. Ann Intern Med 2005;142(1):37–46.
44. Menard J. Oil and water? Economic advantage and biomedical progress do not mix well in a government guidelines committee. Am J Hypertens 1994;7(10 Pt 1):877–85.
45. Fletcher A. Pressure to treat and pressure to cost: a review of cost-effectiveness analysis. J Hypertens 1991;9(3):193–8.
46. Myers WO, Davis K, Foster ED, Maynard C, Kaiser GC. Surgical survival in the Coronary Artery Surgery Study. (CASS) registry. Ann Thorac Surg 1985;40(3):245–60.
47. Detre K, Murphy ML, Hultgren H. Effect of coronary bypass surgery on longevity in high and low risk patients. Report from the V.A. Cooperative Coronary Surgery Study Lancet 1977;2(8051):1243–5.
48. European Coronary Surgery Study Group. Long-term results of prospective randomised study of coronary artery bypass surgery in stable angina pectoris. Lancet 1982;2(8309):1173–80.
49. CASS Principal Investigators and their Associates. Coronary artery surgery study (CASS): a randomised trial of coronary artery bypass surgery. Survival data. Circulation 1983;68:939.
50. Parisi AF, Folland ED, Hartigan P. A comparison of angioplasty with medical therapy in the treatment of single-vessel coronary artery disease. Veterans Affairs ACME Investigators. N Engl J Med 1992;326(1):10–6.
51. Pitt B, Waters D, Brown WV, van Boven AJ, Schwartz L, Title LM, Eisenberg D, Shurzinske L, McCormick LS. Aggressive lipid-lowering therapy compared with angioplasty in stable coronary artery disease. Atorvastatin versus Revascularization Treatment Investigators. N Engl J Med 1999;341(2):70–6.

Angiotensin II receptor antagonists

See also Candesartan cilexetil, Eprosartan, Irbesartan, Losartan, Telmisartan, Valsartan

General Information

Inhibition of the renin–angiotensin system by ACE inhibitors has proved efficacious in the treatment of hypertension, cardiac failure, myocardial infarction, in secondary prevention after myocardial infarction, and for kidney protection in diabetic and non-diabetic nephropathy. The development of specific antagonists to subtype 1 of the angiotensin II receptor (AT_1) has provided a new tool for inhibiting the renin–angiotensin system.

Experimental data and preliminary clinical experience have suggested that the efficacy of AT_1 receptor antagonists in the treatment of hypertension is similar to that of ACE inhibitors. However, the two drug categories also have potential differences, because of their different mechanisms of action. Since they have an action that is exclusively targeted at angiotensin II, the AT_1 receptor antagonists should lack the effects of ACE inhibitors that are mediated through the accumulation of bradykinin and other peptides. Furthermore, since a significant amount of angiotensin II can be generated by enzymes other than ACE, particularly the chymases, AT_1 receptor antagonists achieve more complete blockade of the effects of angiotensin than ACE inhibitors do. Nevertheless, whether these differences are clinically important is still being investigated.

The first attempt at blocking the AT_1 receptor with the peptide analogues of angiotensin II resulted in the discovery of saralasin, an antagonist that lacked oral activity and had partial agonist properties. Losartan was the first agent in the new class of orally active AT_1 receptor antagonists. Other agents with different receptor affinities and binding kinetics, such as candesartan, eprosartan, irbesartan, losartan, tasosartan, telmisartan, and valsartan, have since become available (1). Variations in chemical structure may lead to marginal but potentially clinically significant differences in the time-effect profile. Non-competitive antagonists may have longer durations of action than competitive antagonists, because they bind irreversibly to angiotensin II receptors. There have been several reviews of this class of agents (2,3). All have emphasized their remarkable tolerance profile. In double-blind, placebo-controlled trials, the type and frequency of reported adverse effects were consistently no different from placebo (4).

The angiotensin II receptor antagonists are being considered for the treatment of diseases other than hypertension (heart failure with or without left ventricular systolic dysfunction, during and after acute myocardial infarction, diabetic nephropathy, other forms of glomerulopathy, restenosis after coronary angioplasty, and atherosclerosis).

Comparative studies

The RESOLVD (Randomized Evaluation of Strategies for Left Ventricular Dysfunction) trial was a pilot study investigating the effects of candesartan, enalapril, and their combination on exercise tolerance, ventricular function, quality of life, neurohormone concentrations, and tolerability in congestive heart failure (5). Candesartan alone was as effective, safe, and well tolerated as enalapril. The combination of candesartan with enalapril was more beneficial in preventing left ventricular remodelling than either alone. Although the trial was not powered to assess effects on cardiovascular events, it was terminated prematurely because of a trend toward a greater number of events in the candesartan alone and combination groups compared with enalapril alone.

In the ELITE study losartan produced greater survival benefit in elderly patients with heart failure than captopril (6). ELITE II was performed in order to confirm this. It included 3152 patients aged 60 years and over with NYHA class II–IV heart failure and was powered to detect a clinically significant effect on all-cause mortality. Median follow up was 555 days. Losartan was not superior to captopril in improving survival; however, it was better tolerated. Significantly fewer patients taking losartan withdrew because of adverse effects, including effects attributed to the study drug, or because of cough. Fewer patients discontinued treatment because of adverse effects (10 versus 15%), including effects attributed to the study drug (3 versus 8%) or because of cough (0.4 versus 2.8%) (6).

The results of SPICE (The Study of Patients Intolerant of Converting Enzyme Inhibitors) and of the previously published RESOLVD led to the design of the current CHARM trial, which is investigating the effect of candesartan in 6600 patients with heart failure in three different ways: versus an ACE inhibitor in patients with preserved left ventricular function; versus placebo in patients intolerant of ACE inhibitors; and in addition to ACE inhibitors in all other patients. While waiting for the results of this trial it is advisable to continue to use ACE inhibitors as the initial therapy for heart failure. In patients with documented intolerance of ACE inhibitors (which may represent 10–20% of patients with heart failure) angiotensin receptor antagonists may be useful as a substitute to block the renin–angiotensin–aldosterone system.

Placebo-controlled studies

The SPICE was a smaller trial (270 patients, 12 weeks follow-up) which evaluated the use of candesartan versus placebo in patients with heart failure and a history of intolerance of ACE inhibitors (most commonly because of cough, symptomatic hypotension, or renal insufficiency). Titration to the highest dose of candesartan 16 mg was possible in 69% of the patients (84% in the placebo group). Death and cardiovascular events tended to be lower with candesartan (7).

The results of another trial in heart failure with another angiotensin receptor antagonist (valsartan) are now available but are still to be published. In the VAL-HeFT (Valsartan in Heart Failure Trial) valsartan 160 mg bd was compared with placebo in 5010 patients with heart failure and left ventricular systolic dysfunction receiving optimal conventional therapy, including ACE inhibitors. The results showed a non-significant effect on mortality (19% on placebo, 20% on valsartan), but a highly significant effect on the primary endpoint of all-cause mortality and morbidity (32% on placebo, 29% on valsartan). A subgroup analysis suggested that the combination of valsartan, an ACE inhibitor, and a beta-blocker was not beneficial and might even be harmful, whereas the combination of valsartan and an ACE inhibitor caused a reduction of 45% (8).

General adverse effects

The safety profile of angiotensin II receptor antagonists is so far remarkably good. Except for hypotension, virtually no dose-related adverse effects have been reported. Headache, dizziness, weakness, and fatigue are the most common adverse effects. There have been reports of raised liver enzymes (9), cholestatic hepatitis (10), and pancreatitis (11) with losartan. Several cases of angioedema have been reported but no other obvious hypersensitivity reactions.

Organs and Systems

Cardiovascular

The action of angiotensin II receptor antagonists in interfering with the activity of the renin–angiotensin system is so similar to the action of the ACE inhibitor drugs that episodes of first-dose hypotension, renal impairment, and hyperkalemia would be anticipated. However, there have been only a few such reports to date, although this may reflect the careful screening of patients for clinical trials, from which, by inference, obviously high-risk patients may have been excluded.

Angiotensin II receptor blockers have provoked concerns about the risk of myocardial infarction. The use of this class as an alternative to ACE inhibitors for all indications has been suggested, but an incomplete analysis of some of the evidence led to a provocative suggestion that angiotensin II receptor antagonists confer a risk of harm (12). However, several meta-analyses have been performed to examine the association between angiotensin II receptor antagonists and the risk of myocardial infarction in a variety of clinical trial settings; there was no significant increase in the risk (13,14).

Respiratory

Cough has been specifically studied, because the mechanism of action of losartan differs from the ACE inhibitors, in that there is no accumulation of kinins, which have been implicated in the non-productive cough associated with ACE inhibitors (15). In 135 patients known to have ACE inhibitor-induced cough, lisinopril, losartan, or hydrochlorothiazide were given. Of the patients rechallenged with lisinopril, 72% developed a cough compared with only 29 and 34% challenged with losartan and hydrochlorothiazide respectively (16). However, in a Prescription Event Monitoring (PEM) study in four cohorts of 9000 patients exposed to losartan, enalapril, lisinopril, or perindopril, the rate of cough was high even with losartan. The authors attributed this to a carry-over effect, since presumably patients taking losartan have previously had cough with an ACE inhibitor (17).

A compilation of three controlled trials in 1200 patients showed incidence rates of cough of 3.6% with valsartan versus 9.5% with ACE inhibitors and 0.4% with placebo (1).

In a PEM study in 9000 patients cough occurred in 3.1% of patients taking losartan, compared with 3.9, 14, and 16% in patients taking enalapril, lisinopril, and perindopril respectively (17). The unusual low rate of cough with enalapril was surprising and emphasizes the limitations of PEM studies. Worthy of remark is a possible confounding effect, discussed by the authors, related to the fact that a large number of patients were given losartan because they had had cough with an ACE inhibitor. Because of a carry-over effect, cough may still be reported, especially in the first week when patients change to losartan. This carry-over effect may have been the cause of cough reported with losartan in other cases (18).

In a multicenter controlled study in patients with hypertension and a history of ACE inhibitor-induced cough, eprosartan significantly reduced the risk of cough by 88% compared with enalapril (19).

Hematologic

A Japanese group has studied the effects of various concentrations of angiotensin II receptor antagonists and ACE inhibitors on in vitro burst-forming erythroid units in seven healthy volunteers (40–47 years) and in 10 men (40–49 years) with chronic renal insufficiency undergoing hemodialysis, seven of whom required erythropoietin 42 185 IU/week to maintain a hematocrit of 30% (20). None was taking an angiotensin II receptor antagonist or an ACE inhibitor. The blood from healthy volunteers yielded about four times the number of burst-forming erythroid units than blood from patients. Angiotensin II significantly increased the number of burst-forming units. Losartan inhibited this effect dose-dependently in both healthy volunteers and patients, but enalaprilat and trandolaprilate had no effects. The authors conclude that angiotensin II receptor blockade causes direct inhibition of erythropoiesis and they suggested that hematocrit and hemoglobin should be monitored when angiotensin II receptor antagonists are given to patients with chronic renal insufficiency.

Immunologic

Because ACE inhibitor-induced anaphylaxis is thought to be related to accumulation of bradykinin, it was assumed that angiotensin II receptor antagonists would not cause this reaction. However, angioedema has been described within 30 minutes of a first dose of losartan 50 mg in a 52-year-old man (21). The author also referred to a single case of losartan-induced angioedema mentioned in the manufacturers' package insert from among 4058 patients treated with losartan. In an international safety update report based on 200 000 patients there were 13 cases of angioedema (22). Two had also taken an ACE inhibitor and three others had previously developed angioedema when taking ACE inhibitors.

Autacoids

DoTS classification (BMJ 2003;327:1222-5):
Reaction: Angioedema due to angiotensin II receptor antagonists
Dose-relation: ?Collateral
Time-course: Time-independent
Susceptibility factors: Previous angioedema with an ACE inhibitor

The introduction of the angiotensin II receptor antagonists has provided a new tool for inhibition of the renin–angiotensin system, and these drugs are used extensively in the treatment of hypertension and heart failure, in diabetic nephropathy, and after myocardial infarction. They are purported to be as efficacious as the ACE inhibitors, owing to downstream receptor antagonism, and were also originally thought to cause fewer adverse effects, such as cough (23). More recently it has become clear that some of the common adverse effects of ACE inhibitors also occur with angiotensin II receptor antagonists. In particular, angioedema has been reported both with initial therapy and when substituting treatment in patients who have had ACE inhibitor-induced angioedema. In a commentary on the use of angiotensin receptor antagonists in patients who previously had angioedema while taking ACE inhibitors has advised taking extreme caution when prescribing angiotensin receptor antagonists for patients with a history of ACE inhibitor-associated angioedema (24,25).

Angioedema is a self-limiting, but potentially fatal condition caused by non-pitting edema affecting the skin and mucous membranes, including the upper respiratory and intestinal epithelial linings.

Mechanisms

Angioedema may be bradykinin-mediated or dependent on mast cell degranulation or other mechanisms (26). It is thought that ACE inhibitors induce angioedema by increasing the availability of bradykinin, due to reduced enzymatic degradation by ACE (27). Angiotensin II receptor antagonists theoretically do not affect bradykinin, so they should be appropriate substitutes in patients with ACE inhibitor-associated angioedema. However, there have been several reports of angioedema associated with angiotensin II receptor antagonists, and it has been suggested that they cause it by potentiating the effects of bradykinin (28).

The antithypertensive effects of angiotensin II receptor antagonists are thought to be secondary to inhibition of the binding of angiotensin II to AT_1 receptors, which are the receptors responsible for most of the deleterious effects of angiotensin II, including increased blood pressure due to vasoconstriction and salt retention due to aldosterone synthesis. However, there is increasing evidence that unopposed activation of AT_2 receptors by angiotensin II during therapy with angiotensin II receptor antagonists can produce vasodilatation and increased vascular permeability through mechanisms that involve nitric oxide release and potentiation of the effects of bradykinin (28).

Under normal circumstances the angiotensin II receptor antagonists have a low affinity for AT_2 receptors compared with AT_1 receptors. Antagonism of AT_1 receptors causes a transient rise in angiotensin II concentrations, which may increase AT_2 receptor expression and activity (29). In animals angiotensin II receptor antagonists increase aortic and renal bradykinin concentrations, perhaps secondary to increased stimulation of AT_2 receptors (30,31). They may therefore precipitate angioedema by causing increased AT_2 receptor activity, leading to increased tissue bradykinin concentrations (32). Although these increased concentrations may not be detectable in the serum, in the tissues they may cause angioedema (29). However, increased serum bradykinin concentrations can sometimes be detected; in one study losartan increased the serum bradykinin concentration two-fold in hypertensive humans (33).

Other mechanisms have been proposed by which angiotensin II receptor antagonists could increase bradykinin, including inhibition of neutral endopeptidase, an enzyme involved in the breakdown of bradykinin (33); there was reduced neutral endopeptidase activity in homogenates of isolated heart components in rats that had been treated with angiotensin II receptor antagonists (34). It has also been proposed that increased plasma angiotensin II concentrations caused by angiotensin II receptor antagonists can cause negative feedback inhibition of ACE activity (35). Finally, an abnormality of degradation of an active metabolite of bradykinin, des-arginine(9)-bradykinin may be involved (36). All of these factors would favor the development of angioedema.

Two patients with hereditary angioedema not associated with C1 esterase inhibitor deficiency (so-called Type III hereditary angioedema) have been described as having severe exacerbations of attacks of angioedema in relation to treatment with angiotensin receptor antagonists (37).

Incidence

Angioedema has been reported with losartan (38,39,40,41,42,43,44,45,46), valsartan (29,35,47), candesartan (48), irbesartan (49), olmesartan (32), and telmisartan (50). The incidence of angioedema associated with the use of ACE inhibitors is 0.1–1% (28,51,52), whereas the incidence of angioedema due to angiotensin II receptor antagonists is reported to be 0.1–0.4% (53).

Dose-relation

There is little information about the relation to dose of this adverse effect. In one case valsartan-induced angioedema occurred when the dose of valsartan was increased from 160 to 320 mg/day (29). The patient taking the lower dose for 2 years and the symptoms of lip and tongue swelling occurred within 2 hours of taking the higher dose; once the dose was reduced to 160 mg/day no further episodes of angioedema occurred.

Time course

The time of onset varies with angiotensin II receptor antagonists, ranging from hours to years. For example, losartan has been associated with angioedema from within 30 minutes after administration to 3 years after starting therapy.

Susceptibility factors

There is evidence that angioedema due to angiotensin II receptor antagonists is more likely in patients who have had angioedema while taking ACE inhibitors (52,54). In one series, 32% of patients reported to have angioedema with angiotensin II receptor antagonists had had a prior episode of angioedema attributable to an ACE inhibitor (38). It has also been reported that almost half of those who have angioedema due to an angiotensin II receptor antagonist have also had it with an ACE inhibitor (55). Despite this, angioedema due to angiotensin II receptor antagonists has also been reported to occur in patients not previously exposed to an ACE inhibitor and in patients who have previously received an ACE inhibitor without developing angioedema (35). There is no information about factors that increase susceptibility.

Conclusions

Although angioedema due to angiotensin II receptor antagonists is uncommon, caution is advised when they are prescribed, particularly if there is a history of ACE inhibitor-associated angioedema.

Second-Generation Effects

Fetotoxicity

Angiotensin II receptor antagonists can cause a wide variety of fetotoxic effects, including oligohydramnios, fetal growth retardation, and pulmonary hypoplasia, abnormalities that are similar to those seen with ACE inhibitors. The available evidence has been reviewed, with the conclusion that pharmacological suppression of the fetal renin–angiotensin system through AT_1 receptor blockade disrupts fetal vascular perfusion and renal function, and that angiotensin II receptor antagonists should be withdrawn as soon as pregnancy is recognized (56). Furthermore, there has been a report of a child born with renal impairment after anhydramnios due to maternal exposure to valsartan and hydrochlorothiazide during the first 28 weeks of pregnancy (57).

- A hypertensive woman taking valsartan 80 mg/day, hydrochlorothiazide 12.5 mg/day, prazosin 10 mg/day, lysine acetylsalicylate 100 mg/day, and levothyroxine 250 micrograms/day became pregnant. At 28 weeks anhydramnios associated with high β_2-microglobulin concentrations in fetal cord blood was observed. On withdrawal of valsartan, fetal renal prognosis improved. At the age of 2.5 years the child had mild chronic renal insufficiency. Growth parameters were within the expected range, and there was no evidence of developmental delay.

In this case the Naranjo probability scale suggested that the child's renal impairment was probably related to valsartan.

Drug-drug interactions

Drug interactions with angiotensin II receptor blockers have been reviewed (58).

References

1. Birkenhager WH, de Leeuw PW. Non-peptide angiotensin type 1 receptor antagonists in the treatment of hypertension. J Hypertens 1999;17(7):873–81.
2. Bakris GL, Giles TD, Weber MA. Clinical efficacy and safety profiles of AT1 receptor antagonists. Cardiovasc Rev Rep 1999;20:77–100.
3. Burnier M. Angiotensin II type 1 receptor blockers. Circulation 2001;103(6):904–12.
4. Hedner T, Oparil S, Rasmussen K, Rapelli A, Gatlin M, Kobi P, Sullivan J, Oddou-Stock P. A comparison of the angiotensin II antagonists valsartan and losartan in the treatment of essential hypertension. Am J Hypertens 1999;12(4 Pt 1):414–7.
5. McKelvie RS, Yusuf S, Pericak D, Avezum A, Burns RJ, Probstfield J, Tsuyuki RT, White M, Rouleau J, Latini R, Maggioni A, Young J, Pogue J. Comparison of candesartan, enalapril, and their combination in congestive heart failure: randomized evaluation of strategies for left ventricular dysfunction (RESOLVD) pilot study. The RESOLVD Pilot Study Investigators. Circulation 1999;100(10):1056–64.
6. Pitt B, Poole-Wilson PA, Segal R, Martinez FA, Dickstein K, Camm AJ, Konstam MA, Riegger G, Klinger GH, Neaton J, Sharma D, Thiyagarajan B. Effect of losartan compared with captopril on mortality in patients with symptomatic heart failure: randomised trial—the Losartan Heart Failure Survival Study ELITE II. Lancet 2000;355(9215):1582–7.
7. Granger CB, Ertl G, Kuch J, Maggioni AP, McMurray J, Rouleau JL, Stevenson LW, Swedberg K, Young J, Yusuf S, Califf RM, Bart BA, Held P, Michelson EL, Sellers MA, Ohlin G, Sparapani R, Pfeffer MA. Randomized trial of candesartan cilexetil in the treatment of patients with congestive heart failure and a history of intolerance to angiotensin-converting enzyme inhibitors. Am Heart J 2000; 139(4):609–17.
8. Cohn JN, Tognoni G, Glazer RD, Spormann D, Hester A. Rationale and design of the Valsartan Heart Failure Trial: a large multinational trial to assess the effects of valsartan, an angiotensin-receptor blocker, on morbidity and mortality in chronic congestive heart failure. J Card Fail 1999;5(2):155–60.
9. Losartan potassium prescribing information. Merck & Co Inc., West Point, PA 19486, USA, April 1995.
10. Bosch X, Goldberg AL, Smith IS, Stephenson WP. Losartan-induced hepatotoxicity. JAMA 1997;278(19):1572.
11. Bosch X. Losartan-induced acute pancreatitis. Ann Intern Med 1997;127(11):1043–4.
12. Verma S, Strauss M. Angiotensin receptor blockers and myocardial infarction. BMJ 2004;329(7477):1248–9.
13. McDonald MA, Simpson SH, Ezekowitz JA, Gyenes G, Tsuyuki RT. Angiotensin receptor blockers and risk of myocardial infarction: systematic review. BMJ 2005;331(7521):873.
14. Volpe M, Mancia G, Trimarco B. Angiotensin II receptor blockers and myocardial infarction:deeds and misdeeds. J Hypertens 2005;23(12):2113–8.
15. Lacourciere Y, Brunner H, Irwin R, Karlberg BE, Ramsay LE, Snavely DB, Dobbins TW, Faison EP, Nelson EBLosartan Cough Study Group. Effects of modulators of the renin–angiotensin–aldosterone system on cough. J Hypertens 1994;12(12):1387–93.
16. Lacourciere Y, Lefebvre J. Modulation of the renin–angiotensin–aldosterone system and cough. Can J Cardiol 1995;11(Suppl F):F33–9.
17. Mackay FJ, Pearce GL, Mann RD. Cough and angiotensin II receptor antagonists: cause or confounding? Br J Clin Pharmacol 1999;47(1):111–4.
18. Conigliaro RL, Gleason PP. Losartan-induced cough after lisinopril therapy. Am J Health Syst Pharm 1999;56(9):914–5.
19. Oparil S. Eprosartan versus enalapril in hypertensive patients with angiotensin-converting enzyme inhibitor-induced cough. Curr Ther Res Clin Exp 1999;60:1–4.
20. Naito M, Kawashima A, Akiba T, Takanashi M. Effects of an angiotensin II receptor antagonist and angiotensin-converting enzyme inhibitors on burst-forming units-erythroid in chronic hemodialysis patients. Am J Nephrol 2003;23:287–93.
21. Acker CG, Greenberg A. Angioedema induced by the angiotensin II blocker losartan. N Engl J Med 1995; 333(23):1572.
22. Hansson L. Medical and cost-economy aspects of modern antihypertensive therapy—with special reference to 2 years of clinical experience with losartan. Blood Press Suppl 1997;1:52–5.
23. Pylypchuk GB. ACE inhibitor- versus angiotensin II blocker-induced cough and angioedema. Ann Pharmacother 1998;32(10):1060–6.
24. Fuchs SA, Meyboom RH, van Puijenbroek EP, Guchelaar HJ. Use of angiotensin receptor antagonists in patients with ACE inhibitor induced angioedema. Pharm World Sci 2004; 26(4):191–2.
25. Agostoni A, Cicardi M. Drug-induced angioedema without urticaria. Drug Saf 2001;24(8):599–606.
26. Kaplan AP, Greaves MW. Angioedema. J Am Acad Dermatol 2005;53(3):373–88.

27. Vleeming W, van Amsterdam JG, Stricker BH, de Wildt DJ. ACE inhibitor-induced angioedema. Incidence, prevention and management. Drug Saf 1998;18(3):171–88.
28. Howes LG, Tran D. Can angiotensin receptor antagonists be used safely in patients with previous ACE inhibitor-induced angioedema? Drug Saf 2002;25(2):73–6.
29. Irons BK, Kumar A. Valsartan-induced angioedema. Ann Pharmacother 2003;37(7–8):1024–7.
30. Siragy HM, de GM, El-Kersh M, Carey RM. Angiotensin-converting enzyme inhibition potentiates angiotensin II type 1 receptor effects on renal bradykinin and cGMP. Hypertension 2001;38(2):183–6.
31. Gohlke P, Pees C, Unger T. AT2 receptor stimulation increases aortic cyclic GMP in SHRSP by a kinin-dependent mechanism. Hypertension 1998;31(1 Pt 2):349–55.
32. Nykamp D, Winter EE. Olmesartan medoxomil-induced angioedema. Ann Pharmacother 2007;41(3):518–20.
33. Campbell DJ, Krum H, Esler MD. Losartan increases bradykinin levels in hypertensive humans. Circulation 2005;111(3):315–20.
34. Walther T, Siems WE, Hauke D, Spillmann F, Dendorfer A, Krause W, Schultheiss HP, Tschöpe C. AT1 receptor blockade increases cardiac bradykinin via neutral endopeptidase after induction of myocardial infarction in rats. FASEB J 2002;16(10):1237–41.
35. Arakawa M, Murata Y, Rikimaru Y, Sasaki Y. Drug-induced isolated visceral angioneurotic edema. Intern Med 2005;44(9):975–8.
36. Molinaro G, Cugno M, Perez M, Lepage Y, Gervais N, Agostoni A, Adam A. Angiotensin-converting enzyme inhibitor-associated angioedema is characterized by a slower degradation of des-arginine(9)-bradykinin. J Pharmacol Exp Ther 2002;303(1):232–7.
37. Bork K, Dewald G. Hereditary angioedema type III, angioedema associated with angiotensin II receptor antagonists, and female sex. Am J Med 2004 1;116(9):644–5.
38. Warner KK, Visconti JA, Tschampel MM. Angiotensin II receptor blockers in patients with ACE inhibitor-induced angioedema. Ann Pharmacother 2000;34(4):526–8.
39. van Rijnsoever EW, Kwee-Zuiderwijk WJ, Feenstra J. Angioneurotic edema attributed to the use of losartan. Arch Intern Med 1998;158(18):2063–5.
40. Acker CG, Greenberg A. Angioedema induced by the angiotensin II blocker losartan. N Engl J Med 1995;333(23):1572.
41. Boxer M. Accupril- and Cozaar-induced angioedema in the same patient. J Allergy Clin Immunol 1996;98(2):471.
42. Sharma PK, Yium JJ. Angioedema associated with angiotensin II receptor antagonist losartan. South Med J 1997;90(5):552–3.
43. Frye CB, Pettigrew TJ. Angioedema and photosensitive rash induced by valsartan. Pharmacotherapy 1998;18(4):866–8.
44. Rivera JO. Losartan-induced angioedema. Ann Pharmacother 1999;33(9):933–5.
45. Cha YJ, Pearson VE. Angioedema due to losartan. Ann Pharmacother 1999;33(9):936–8.
46. Chiu AG, Krowiak EJ, Deeb ZE. Angioedema associated with angiotensin II receptor antagonists: challenging our knowledge of angioedema and its etiology. Laryngoscope 2001;111(10):1729–31.
47. Tojo A, Onozato ML, Fujita T. Repeated subileus due to angioedema during renin–angiotensin system blockade. Am J Med Sci 2006;332(1):36–8.
48. Lo KS. Angioedema associated with candesartan. Pharmacotherapy 2002;22(9):1176–9.
49. Nielsen EW. Hypotensive shock and angio-oedema from angiotensin II receptor blocker: a class effect in spite of tripled tryptase values. J Intern Med 2005;258(4):385–7.
50. Borazan A, Ustun H, Yilmaz A. Angioedema induced by angiotensin II blocker telmisartan. Allergy 2003;58(5):454.
51. Israili ZH, Hall WD. Cough and angioneurotic edema associated with angiotensin-converting enzyme inhibitor therapy. A review of the literature and pathophysiology. Ann Intern Med 1992;117(3):234–42.
52. Cicardi M, Zingale LC, Bergamaschini L, Agostoni A. Angioedema associated with angiotensin-converting enzyme inhibitor use: outcome after switching to a different treatment. Arch Intern Med 2004;164(8):910–3.
53. Dickstein K, Kjekshus J. Effects of losartan and captopril on mortality and morbidity in high-risk patients after acute myocardial infarction: the OPTIMAAL randomised trial. Optimal Trial in Myocardial Infarction with Angiotensin II Antagonist Losartan. Lancet 2002;360(9335):752–60.
54. Fuchs SA, Meyboom RH, van Puijenbroek EP, Guchelaar HJ. Use of angiotensin receptor antagonists in patients with ACE inhibitor induced angioedema. Pharm World Sci 2004;26(4):191–2.
55. Abdi R, Dong VM, Lee CJ, Ntoso KA. Angiotensin II receptor blocker-associated angioedema: on the heels of ACE inhibitor angioedema. Pharmacotherapy 2002;22(9): 1173–5.
56. Alwan S, Polifka JE, Friedman JM. Angiotensin II receptor antagonist treatment during pregnancy. Birth Defects Res A Clin Mol Teratol 2005;73(2):123–30.
57. Bos-Thompson MA, Hillaire-Buys D, Muller F, Dechaud H, Mazurier E, Boulot P, Morin D. Fetal toxic effects of angiotensin II receptor antagonists: case report and follow-up after birth. Ann Pharmacother 2005;39(1):157–61.
58. Unger T, Kaschina E. Drug interactions with angiotensin blockers: a comparison with other antihypertensives. Drug Saf 2003;26:707–20.

Atenolol

See also Beta-adrenoceptor antagonists

General Information

Although atenolol, a hydrophilic cardioselective beta-adrenoceptor antagonist with no partial agonist activity, is generally regarded as one of the safest beta-blockers, severe adverse effects are occasionally reported. These include profound hypotension after a single oral dose (1), organic brain syndrome (2), cholestasis (3), and cutaneous vasculitis (4).

Organs and Systems

Nervous system

Atenolol is hydrophilic, making it less likely to cross the blood–brain barrier. However, nervous system effects are occasionally reported.

- A 54-year-old man developed progressive memory loss after taking atenolol 100 mg/day for 3 years. Four weeks after withdrawal, he completely recovered his memory (5).

Second-Generation Effects

Fetotoxicity

Atenolol-induced developmental toxicity has been reviewed, combining data from eight randomized clinical trials, three surveys, and one case series (6). The main indications for atenolol were pregnancy-induced hypertension and pre-existing chronic hypertension. The most frequent prenatal adverse events were reduced placental weight, reduced birth weight, and intrauterine growth retardation. The most likely reasons for these effects were placental and fetal hemodynamic disturbances, characterized by a reduction in umbilical and fetal aortic blood flow, with or without reduced fetal heart rate. However, there were no increases in embryo-fetal deaths or congenital abnormalities. The effects of atenolol prenatal toxicity were similar to those reported in animal studies (rats and rabbits) suggesting that animal experiments can predict developmental toxicity caused by beta adrenoceptor antagonists.

References

1. Kholeif M, Isles C. Profound hypotension after atenolol in severe hypertension. BMJ 1989;298(6667):161–2.
2. Arber N. Delirium induced by atenolol. BMJ 1988; 297(6655):1048.
3. Schwartz MS, Frank MS, Yanoff A, Morecki R. Atenolol-associated cholestasis. Am J Gastroenterol 1989;84(9):1084–6.
4. Wolf R, Ophir J, Elman M, Krakowski A. Atenolol-induced cutaneous vasculitis. Cutis 1989;43(3):231–3.
5. Ramanathan M. Atenolol induced memory impairment: a case report. Singapore Med J 1996;37(2):218–9.
6. Tabacova S, Kimmel CA, Wall K, Hansen D. Atenolol developmental toxicity: animal-to-human comparisons. Birth Defects Research (Part A) 2003; 67: 181–92.

Barnidipine

See also Calcium channel blockers

General Information

Barnidipine is a dihydropyridine with antihypertensive activity and tolerability similar to that of other calcium antagonists of the same class. The most frequent adverse events are edema, headache, and flushing, but barnidipine does not cause reflex tachycardia (1).

Reference

1. Malhotra HS, Plosker GL. Barnidipine. Drugs 2001; 61(7):989–96.

Benazepril

See also Angiotensin converting enzyme inhibitors

General Information

Safety data derived from the controlled trials submitted in the benazepril New Drug Application for hypertension have been reviewed (1). In more than 100 trials involving 6000 patients with hypertension or congestive heart failure, the types and incidences of adverse effects were comparable with those of other ACE inhibitors.

Organs and Systems

Respiratory

In a large open study of benazepril in Chinese patients with hypertension there was *cough* in up to one-fifth of the treated population, leading to frequent withdrawal of therapy and associated with poor adherence to therapy (2). The symptom was more common in women; 168 of 741 women (23%) compared with 196 of 1090 men (18%). The prevalence of cough in this population may relate to ethnicity; in another study East-Asian ethnicity (Chinese, Korean, or Japanese) was an independent risk factor for ACE inhibitor-induced cough (3).

Endocrine

Hypoaldosteronism with metabolic acidosis has been reported in a child taking an ACE inhibitor (4).

- A 4-year-old boy with minimal-change nephrotic syndrome since the age of 11 months had been treated with cyclophosphamide and glucocorticoids. After several relapses and the development of mild hypertension and proteinuria, he was given benazepril 0.3 mg/kg/day. He was admitted 4 months later with a metabolic acidosis (pH 7.28, base excess –15) and mild hyperchloremia (chloride 110 mmol/l). He had mild proteinuria and a normal creatinine clearance. The diagnosis was metabolic acidosis due to gastroenteritis and he was treated with intravenous saline and bicarbonate and discharged, but was re-admitted with anorexia and nausea and the same findings as before. A 24-hour urine sample showed high sodium and a low potassium and chloride excretion. The serum aldosterone concentration was below the limit of detection. The dose of benazepril was reduced to 0.2 mg/kg/day for 1 week and then withdrawn. Ten days later the aldosterone concentration was normal (29 pg/ml). After 9 months of follow-up, he still had mild proteinuria, but there had been no further episodes of metabolic acidosis.

The authors pointed out that there is evidence of ACE inhibitor-induced hypoaldosteronism in adults. This condition should be considered in children and adults taking ACE inhibitors who present with metabolic acidosis.

Pancreas

Benazepril has been associated with pancreatitis (5).

- A 70-year old man with type II diabetes had severe epigastric pain 30 minutes after taking his first dose of 5 mg benazepril and lasting 6–8 hours. The next day he had the same pain, complicated by vomiting. Benazepril was withdrawn and he was unable to eat for 4 days. He later developed severe epigastric pain, nausea, and vomiting 30 minutes after taking a third dose. Laboratory findings confirmed pancreatitis and imaging showed a mildly edematous inflamed pancreas. He improved progressively with bowel rest and pethidine. He was symptom-free 2 months after discharge.

References

1. MacNab M, Mallows S. Safety profile of benazepril in essential hypertension. Clin Cardiol 1991;14(8 Suppl 4):IV33–7.
2. Lu J, Lee L, Cao W, Zhan S, Zhu G, Dai L, Hu Y. Postmarketing surveillance study of benazepril in Chinese patients with hypertension: an open-label, experimental, epidemiologic study. Curr Ther Res Clin Exp 2004;65(3):300–19.
3. Morimoto T, Gandhi TK, Fiskio JM, Seger AC, So JW, Cook EF, Fukui T, Bates DW. An evaluation of risk factors for adverse drug events associated with angiotensin-converting enzyme inhibitors. J Eval Clin Pract 2004;10(4):499–509.
4. Bruno I, Pennesi M, Marchetti F. ACE-inhibitors-induced metabolic acidosis in a child with nephrotic syndrome. Pediatr Nephrol 2003;18:1293–4.
5. Muchnick JS, Mehta JL. Angiotensin-converting enzyme inhibitor-induced pancreatitis. Clin Cardiol 1999;22(1):50–1.

Beta-adrenoceptor antagonists

See also Acebutolol, Atenolol, Bevantolol, Bisoprolol, Carvedilol, Celiprolol, Dilevalol, Esmolol, Labetalol, Metoprolol, Pindolol, Practolol, Propranolol, Sotalol, Timolol, Xamoterol

General Information

Many beta-adrenoceptor antagonists (beta-blockers) have been developed, and their adverse effects have been comprehensively reviewed (1). The spectrum of adverse effects is broadly similar for all beta-blockers, despite differences in their pharmacological properties, notably cardioselectivity, partial agonist activity, membrane-stabilizing activity, and lipid solubility (see Table 1). The influence of these properties is mentioned in the general discussion when appropriate and is summarized at the end of the section. Individual differences in toxicity are largely unimportant but will be mentioned briefly.

Although beta-blockers have been available for many years, new members of this class with novel pharmacological profiles continue to be developed. These new drugs are claimed to have either greater cardioselectivity or vasodilatory and beta$_2$-agonist properties. The claimed advantages of these new drugs serve to highlight the supposed disadvantages of the older members of the class (their adverse constrictor effects on the airways and peripheral blood vessels). Strong commercial emphasis is being placed on these new properties, and papers extolling these effects often appear in non-peer-reviewed supplements or even in reputable journals (2).

Although the toxicity of the beta-adrenoceptor antagonists has been fairly well documented, there has been a subtle change in perceptions of their potential benefits and drawbacks. The cardioprotective effect of beta-blockers after myocardial infarction and their efficacy in reducing "silent" myocardial ischemia have persuaded some clinicians to use them preferentially. On the other hand, they can significantly impair the quality of life (3) and are contraindicated in some patients. A few patients cannot tolerate beta-blockade at all. These include patients with bronchial asthma, patients with second- or third-degree heart block, and those with seriously compromised limb perfusion causing claudication, ischemic rest pain, and pregangrene.

General adverse effects

The adverse effects of beta-blockers are usually mild, with occurrence rates of 10–20% for the most common in most studies. Most are predictable from the pharmacological and physicochemical properties of these drugs. Examples include fatigue, cold peripheries, bradycardia, heart failure, sleep disturbances, bronchospasm, and altered glucose tolerance. Gastrointestinal upsets are also relatively common. Serious adverse cardiac effects and even sudden death can follow abrupt withdrawal of therapy in patients with ischemic heart disease. Most severe adverse reactions can be avoided by careful selection of patients and consideration of individual beta-blockers. Hypersensitivity reactions have been relatively rare since the withdrawal of practolol. Tumor-inducing effects have not been established in man.

Fatigue

Fatigue is one of the most commonly reported adverse effects of beta-adrenoceptor antagonists, with reported occurrence rates of up to 20% or more, particularly in those who exert themselves. It has to be viewed alongside the ability to produce fatigue and lethargy by a possible effect on the nervous system. The precise cause of physical fatigue is not known, but hypotheses include impaired muscle blood supply, effects on intermediary metabolism, and a direct effect on muscle contractility (4).

Theoretically, beta$_1$-selective drugs are less likely to alter these variables, and might therefore have an advantage over non-selective drugs. However, this has not always been shown in single-dose studies in volunteers (5,6), although in two such studies atenolol produced less exercise intolerance than propranolol at comparable dosages (7,8). For an unexplained reason, cardioselectivity impaired performance relatively less in subjects with a high proportion of slow-twitch muscle fibers than it did in those whose muscle biopsy specimens showed a high percentage of fast-twitch fibers (9). The muscle fibers of long-distance runners are predominantly of the slow-twitch type, and this probably explains the superiority of

Table 1 Properties of beta-adrenoceptor antagonists (where known)

Drug	*Lipid solubility*[a]	*Cardioselectivity*	*Partial agonist activity*	*Membrane-stabilizing activity*
Acebutolol	0.7	±	+	+
Alprenolol	31	−	+	+
Amosulalol		−		
Arotinolol				
Atenolol	0.02	+	−	
Befunolol				
Betaxolol		+	−	±
Bevantolol	Low	+	−	+
Bisoprolol		++		−
Bopindolol				
Bucindolol				
Bufetolol				
Bufuralol	+	−	+	
Bunitrolol	+	−	++	±
Bupranolol				
Butofilolol				
Carazolol				
Carteolol	−	+		
Carvedilol	++	−	−	
Celiprolol		±	+[b]	−
Cetamolol				
Cicloprolol				
Cloranolol				
Dexpropranolol				
Diacetolol				
Dilevalol				
Draquinolol				
Epanolol	Minimal	+	+	
Esmolol	−	+	−	
Flestolol	−	−	−	
Indenolol				
Labetalol	+	−	−	
Levobetaxolol				
Levobunolol		−		
Levomoprolol				
Medroxalol				
Mepindolol				
Metipranolol				
Metoprolol	0.2	+	−	±
Moprolol				
Nadolol	0.03	−	−	−
Nebivolol		+	−	−
Nifenalol				
Nipradilol				
Oxprenolol	0.7	−	+	+
Penbutolol	−	−	+	+
Pindolol	0.2	−	++	±
Practolol	0.02	+	+	−
Pronethalol				
Propranolol	4.3	−	−	++
Sotalol	0.02	−	−	−
Talinolol		+		
Tertatolol				
Tilisolol				
Timolol	0.03	−	±	±
Xamoterol		+	++	

[a]Octanol:water partition coefficient
[b]Partial $beta_2$-adrenoceptor agonist

atenolol over propranolol when exercise performance was assessed in such subjects (7). The release of lactic acid from skeletal muscle cells is impaired to a greater extent by non-selective beta-blockers than by cardioselective drugs, and cardioselectivity was associated with a less marked fall in blood glucose during and after maximal and submaximal exercise (10,11). Partial agonist activity might have been the reason for the superiority of oxprenolol over propranolol in terms of exercise duration (12).

Differences among beta-blockers

Although there are now many different beta-adrenoceptor antagonists, and the number is still increasing, there are only a few important characteristics that distinguish them in terms of their physicochemical and pharmacological properties: lipid solubility, cardioselectivity, partial agonist activity, and membrane-stabilizing activity. The characteristics of the currently available compounds are shown in Table 1.

Lipid solubility

Lipid solubility (13) determines the extent to which a drug partitions between an organic solvent and water. Propranolol, oxprenolol, metoprolol, and timolol are the most lipid-soluble beta-adrenoceptor antagonists, and atenolol, nadolol, and sotalol are the most water-soluble; acebutolol and pindolol are intermediate (14).

The more lipophilic drugs are extensively metabolized in the gut wall and liver (first-pass metabolism). This first-pass clearance is variable and can result in 20-fold differences in plasma drug concentrations between patients who have taken the same dose. It also produces susceptibility to drug interactions with agents that alter hepatic drug metabolism, for example cimetidine, and can result in altered kinetics and hence drug response in patients with hepatic disease, particularly cirrhosis. Lipid-soluble drugs pass the blood–brain barrier more readily (15) and should be more likely to cause adverse nervous system effects, such as disturbance of sleep, but the evidence for this is not very convincing.

In contrast, water-soluble drugs are cleared more slowly from the body by the kidneys. These drugs therefore tend to accumulate in patients with renal disease, do not interact with drugs that affect hepatic metabolism, and gain access to the brain less readily.

Cardioselectivity

Cardioselectivity (16), or more properly $beta_1$-adrenoceptor selectivity, is the term used to indicate that there are at least two types of beta-adrenoceptors, and that while some drugs are non-selective (that is they are competitive antagonists at both $beta_1$- and $beta_2$-adrenoceptors), others appear to be more selective antagonists at $beta_1$-adrenoceptors, which are predominantly found in the heart. Bronchial tissue, peripheral blood vessels, the uterus, and pancreatic beta-cells contain principally $beta_2$-adrenoceptors. Thus, cardioselective beta-adrenoceptor antagonists, such as atenolol and metoprolol, might offer theoretical benefits to patients with bronchial asthma, peripheral vascular disease, and diabetes mellitus.

Cardioselective drugs may have relatively less effect on the airways, but they are in no way cardiospecific and they should be used with great care in patients with evidence of reversible obstructive airways disease.

The benefits of cardioselective drugs in patients with Raynaud's phenomenon or intermittent claudication have been difficult to prove. Because of vascular sparing, cardioselective agents may also be preferable in stress, when adrenaline is released.

Cardioselective drugs are less likely to produce adverse effects in patients with type I diabetes than non-selective drugs. At present, hypoglycemia in patients with type I diabetes mellitus is the only clinical problem in which cardioselectivity is considered important. Even there, any potential advantages of cardioselective drugs in minimizing adverse effects apply only at low dosages, since cardioselectivity is dose-dependent.

Partial agonist activity

Partial agonist activity (17,18) is the property whereby a molecule occupying the beta-adrenoceptor exercises agonist effects of its own at the same time as it competitively inhibits the effects of other extrinsic agonists. The effects of these drugs depend on the degree on endogenous tone of the sympathetic nervous system. When there is high endogenous sympathetic tone they tend to act as beta-blockers; when endogenous sympathetic tone is low they tend to act as beta-agonists. Thus, xamoterol had a beneficial effect in patients with mild heart failure (NYHA classes I and II), through a positive inotropic effect on the heart; however, in severe heart failure (NYHA classes III and IV), in which sympathetic tone is high, it acted as a beta-blocker and worsened the heart failure, through a negative inotropic effect (19).

Partial agonists, such as acebutolol, oxprenolol, pindolol, practolol, and xamoterol, produce less resting bradycardia. It has also been claimed that such agents cause a smaller increase in airways resistance in asthmatics, less reduction in cardiac output (and consequently a lower risk of congestive heart failure), and fewer adverse effects in patients with cold hands, Raynaud's phenomenon, or intermittent claudication. However, none of these advantages has been convincingly demonstrated in practice, and patients with bronchial asthma or incipient heart failure must be considered at risk with this type of compound.

Drugs with partial agonist activity can produce tremor (20).

Drugs that combine $beta_1$-antagonism or partial agonism with $beta_2$-agonism (celiprolol, dilevalol, labetalol, pindolol) or with alpha-antagonism (carvedilol, labetalol) have been developed (21). Both classes have significant peripheral vasodilating effects. Drugs with significant agonist activity at $beta_1$-adrenoceptors have poor antihypertensive properties (22).

Membrane-stabilizing activity

Drugs with membrane-stabilizing activity reduce the rate of rise of the cardiac action potential and have other electrophysiological effects. Membrane-stabilizing activity has only been shown in human cardiac muscle in vitro in concentrations 100 times greater than those produced by therapeutic doses (23). It is therefore likely to be of clinical relevance only if large overdoses are taken.

Use in heart failure

Traditionally, beta-blockers have been contraindicated in patients with heart failure. However, there are some patients with systolic heart failure who benefit from a beta-blocker (24). Early evidence was strongest for idiopathic dilated cardiomyopathy rather than ischemic heart disease (25). However, several randomized controlled trials of beta-blockers in patients with mild to moderate heart failure have been published. These include the CIBIS II trial with bisoprolol (26), the MERIT-HF trial with metoprolol (27), and the PRECISE trial with carvedilol (28). These have shown that cardiac mortality in these patients can be reduced by one-third, despite concurrent treatment with conventional therapies of proven benefit (that is ACE inhibitors).

Diastolic dysfunction can lead to congestive heart failure, even when systolic function is normal (29,30). Since ventricular filling occurs during diastole, failure of intraventricular pressure to fall appropriately during diastole leads to increased atrial pressure, which eventually leads to increased pulmonary and systemic venous pressures, causing a syndrome of congestive heart failure indistinguishable clinically from that caused by systolic pump failure (31). Diastolic dysfunction occurs in systemic arterial hypertension, hypertrophic obstructive cardiomyopathy, and infiltrative heart diseases, which reduce ventricular compliance or increase ventricular stiffness (32). As energy is required for active diastolic myocardial relaxation, a relative shortage of adenosine triphosphate in ischemic heart disease also often leads to co-existing diastolic and systolic dysfunction (33). Beta-blockers improve diastolic function in general, and this may be beneficial in patients with congestive heart failure associated with poor diastolic but normal systolic function.

Beta-blockade reduces mortality in patients with heart failure by at least a third when initiated carefully, with gradual dose titration, in those with stable heart failure (34,35). Similarly, beta-blocker prescribing should be encouraged in people with diabetes, since they have a worse outcome after cardiac events and beta-blockade has an independent secondary protective effect (36,37). The small risk of masking metabolic and autonomic responses to hypoglycemia, which was only a problem with non-selective agents in type I diabetes, is a very small price worth paying in diabetics with coronary heart disease.

Use in glaucoma

It has been more than a quarter of a century since the discovery that oral propranolol reduces intraocular pressure in patients with glaucoma. However, the use of propranolol for glaucoma was limited by its local anesthetic action (membrane-stabilizing activity).

Topical timolol was released for general use in 1978. That timolol is systemically absorbed was suggested by early reports of reduced intraocular pressure in the untreated eyes of patients using monocular treatment. About 80–90% of a topically administered drop drains through the nasolacrimal duct and enters the systemic circulation through the highly vascular nasal mucosa, without the benefit of first-pass metabolism in the liver; only a small fraction is swallowed. Thus, topical ophthalmic dosing is probably more akin to intravenous delivery than to oral dosing, and systemic adverse reactions are potentially serious. However, although patients may give their physicians a detailed list of current medications, they often fail to mention the use of eye-drops, about which physicians are often either unaware or do not have time to ask specific questions.

Betaxolol is a $beta_1$-selective adrenoceptor antagonist without significant membrane-stabilizing activity or intrinsic sympathomimetic activity. It may be no more effective than other drugs in reducing intraocular pressure, but it may be safer for some patients, particularly those with bronchospastic disease (but see the section on Respiratory under Drug administration route) (38).

Partial agonist activity of beta-blockers may help to prevent ocular nerve damage and subsequent visual field loss associated with glaucoma. Such damage may be related to a reduction in ocular perfusion, as might occur if an ocular beta-blocker caused local vasoconstriction. An agent with intrinsic sympathomimetic activity might preserve ocular perfusion through local vasodilatation or by minimizing local vasoconstriction. The data are sparse and inconclusive, but carteolol appears to have no effect on retinal blood flow or may even increase it, making it potentially suitable as a neuroprotective drug (38,39).

Organs and Systems

Cardiovascular

A randomized comparison of oral atenolol and bisoprolol in 334 patients with acute myocardial infarction was associated with drug withdrawal in 70 patients (21%) because of significant bradydysrhythmias, hypotension, heart failure, and abnormal atrioventricular conduction (40). Logistic regression analysis suggested that critical events were more likely to occur in patients who were pretreated with dihydropyridine calcium antagonists.

Heart failure

Beta-adrenoceptor antagonists reduce cardiac output through their negative inotropic and negative chronotropic effects. They can therefore cause worsening systolic heart failure or new heart failure in patients who depend on high sympathetic drive to maintain cardiac output. Plasma noradrenaline is increased in patients with heart failure, and the extent of this increase is directly related to the degree of ventricular impairment (41). Since the greatest effect on sympathetic activity occurs with the first (and usually the lowest) doses, heart failure associated with beta-blockade seems to be independent of dosage. Heart failure is one of the most serious adverse effects of the beta-adrenoceptor antagonists (42), but it is usually predictable and can be attenuated by pretreatment with diuretics and angiotensin-converting enzyme inhibitors in patients who are considered to be at risk.

It has been suggested that drugs with partial agonist activity (see Table 1), which have a minimal depressant effect on normal resting sympathetic tone, might cause less reduction in cardiac output (43) and thus protect against the development of cardiac decompensation (44). However, this has not been satisfactorily shown for drugs with high partial agonist activity, for example acebutolol, oxprenolol, and pindolol (45), and these drugs should therefore be given with the same caution as others in compromised patients. Xamoterol, a beta-antagonist with substantial $beta_1$ partial agonist activity, was hoped to be of benefit in mild congestive heart failure (46), but its widespread use by non-specialists in more severe degrees of heart failure resulted in many reports of worsening heart failure (47). Heart failure has also been produced by labetalol (48) and after the use of timolol eye-drops in the treatment of glaucoma (49).

Identification of the possible predictors of intolerance to beta-blockade in heart failure was the object of an analysis of a series of 236 patients (50). A B-type natriuretic peptide (BNP) concentration of over 1000 pg/ml in the first 8 days from the start of beta-blockade was a significant predictor of worsening heart failure. This neurohormone not only has powerful prognostic value, but it can also provide useful information in the selection of patients in whom beta-blockade should be started.

Hypotension

Beta-adrenoceptor antagonists lower blood pressure, probably by a variety of mechanisms, including reduced cardiac output. More severe reductions in blood pressure can occur and can be associated with syncope (42). It has been suggested that this is more likely to occur in old people, but comprehensive studies have stressed the safety of beta-blockers in this age group (51).

- Profound hypotension, resulting in renal insufficiency, has been reported in a single patient after the administration of atenolol 100 mg orally (52); however, large doses of furosemide and diazoxide were also given in this case, and this appears more likely to have been a consequence of a drug interaction.

Cardiac dysrhythmias and heart block

Beta-blockade can result in sinus bradycardia, because blockade of sympathetic tone allows unopposed parasympathetic activity. Drugs with partial agonist activity may prevent bradycardia (53). However, heart rates under 60/minute often worry the physician more than the patient: in a retrospective study of nearly 7000 patients taking beta-adrenoceptor antagonists, apart from dizziness in patients with heart rates under 40/minute (0.4% of the total group), slow heart rates were well tolerated (54).

All beta-blockers cause an increase in atrioventricular conduction time; this is most pronounced with drugs that have potent membrane-depressant properties and no partial agonist activity. Sotalol differs from other beta-blockers in that it increases the duration of the action potential in the cardiac Purkinje fibers and ventricular muscle at therapeutic doses. This is a class III antidysrhythmic effect, and because of this, sotalol has been used to treat ventricular (55–57) and supraventricular dysrhythmias (58). The main serious adverse effect of sotalol is that it is prodysrhythmic in certain circumstances, and can cause torsade de pointes (59,60).

Acute chest pain

Worsening of angina pectoris has been attributed to beta-blocker therapy. The reports include 35 cases in a series of 296 elderly patients admitted to hospital with suspected myocardial infarction; in these 35 the pain disappeared within 7 hours of withdrawing beta-blocker therapy (61).

Worsening of angina has been reported at very low heart rates (62). Propranolol resulted in vasotonic angina in six patients during a double-blind trial, with prolongation of the duration of pain and electrocardiographically assessed ischemia. It has been suggested that this reflects a reduction in coronary perfusion as a result of reduced cardiac output, and also coronary arterial spasm provoked by non-selective agents by inhibition of $beta_2$-mediated vasodilatation (63). The latter explanation is controversial, and the use of beta-blockers in patients with arteriographic evidence of coronary artery spasm has not consistently caused worsening of the disorder (64).

Unstable angina has also followed treatment for hypertension with cardioselective drugs such as betaxolol (65).

Peripheral vascular effects

Cold extremities or exacerbation of Raynaud's phenomenon are amongst the commonest adverse effects reported with beta-blockers (5.8% of nearly 800 patients taking propranolol) (42); Raynaud's phenomenon occurs in 0.5–6% of patients (66). The mechanism may be potentiation of the effects of a cold environment on an already abnormal circulation, but whether symptoms can be produced de novo is more difficult to determine. However, a retrospective questionnaire study in 758 patients taking antihypertensive drugs showed that 40% of patients taking beta-blockers noted cold extremities, compared with 18% of those taking diuretics; there were no significant differences among patients taking alprenolol, atenolol, metoprolol, pindolol, and propranolol (67). Similarly, a large randomized study showed that the incidence of Raynaud's phenomenon was the same for atenolol and pindolol (68). In another study, vasospastic symptoms improved when labetalol was substituted for a variety of beta-blockers (69). On the other hand, a small, double-blind, placebo-controlled study in patients with established Raynaud's phenomenon showed that the prevalence of symptoms with both propranolol and labetalol was no greater than that with placebo (70).

Intermittent claudication has also been reported to be worsened by beta-adrenoceptor antagonists, but has been difficult to document because of the difficulty of study design in patients with advanced atherosclerosis. As early as 1975 it was reported from one small placebo-controlled study that propranolol did not exacerbate symptoms in patients with intermittent claudication (71). This has subsequently been supported by the results of several large placebo-controlled trials of beta-blockers in mild hypertension and reports of trials of the secondary prevention of myocardial infarction, in which intermittent claudication was not mentioned as an adverse effect, even though it was not a specific contraindication to inclusion (72). In addition, a comprehensive study of the effects of beta-adrenoceptor antagonists in patients with intermittent claudication did not show beta-blockade to be an independent risk factor for the disease (73). In men with chronic stable intermittent claudication, atenolol (50 mg bd) had no effect on walking distance or foot temperature (74). These findings have been confirmed in a recent meta-analysis of 11 randomized, controlled trials to determine whether beta-blockers exacerbate intermittent claudication (SEDA-17, 234).

Patchy skin necrosis has been described in hypertensive patients with small-vessel disease in the legs who were taking beta-blockers. Characteristically, pedal pulses remained palpable and the lesions occurred during cold weather and healed on withdrawal of the drugs (75–77). Three cases have been reported in which long-lasting incipient gangrene of the leg was immediately overcome when a beta-blocker was withdrawn (78,79), showing how easily these drugs are overlooked in such circumstances. In several cases of beta-blocker-induced gangrene, recovery did not follow withdrawal of therapy, and amputation was necessary (80,81). Thus, when possible, other forms of therapy should be used in patients with critical ischemia or rest pain.

It has also been suggested that beta-blockade may compromise the splanchnic vasculature. Intravenous propranolol reduces splanchnic blood flow experimentally by 29% while reducing cardiac output by only 6% (82).

Five patients developed mesenteric ischemia, four with ischemic colitis, and one with abdominal angina, while taking beta-adrenoceptor antagonists (83). Although causation was not proven, it was possible.

Respiratory

The respiratory and cardiovascular adverse effects of topical therapy with timolol or betaxolol have been studied in a randomized, controlled trial in 40 elderly patients with glaucoma (84). Five of the 20 allocated to timolol discontinued treatment for respiratory reasons, compared with three of the 20 patients allocated to betaxolol. There were no significant differences in mean values of spirometry, pulse, or blood pressure between the groups. This study confirms that beta-blockers administered as eye-drops can reach the systemic circulation and that serious adverse respiratory events can occur in elderly people, even if they are screened before treatment for cardiac and respiratory disease. These events can occur using either the selective betaxolol agent or the non-selective timolol.

Airways obstruction

Since the introduction of propranolol, it has been recognized that patients with bronchial asthma treated with beta-adrenoceptor antagonists can develop severe airways obstruction (85), which can be fatal (86) or near fatal (87,88); this has even followed the use of eye-drops containing timolol (89). Beta-blockers upset the balance of bronchial smooth muscle tone by blocking the bronchial β_2-adrenoceptors responsible for bronchodilatation. They also promote degranulation of mast cells and depress central responsiveness to carbon dioxide (90,91).

Although beta$_1$-selective drugs are theoretically safer, there are reports of serious reductions in ventilatory function (92,93), even when used as eye-drops (94). However, it has been concluded that if beta-blockade is necessary in the treatment of glaucoma, cardioselective beta-blocking drugs should be preferred (95). While cardioselectivity is dose-dependent (96), and higher dosages might therefore be expected to produce adverse effects, metoprolol and bevantolol, even in dosages that are lower than those usually required for a therapeutic effect, may be poorly tolerated by patients with asthma (97).

Whether drugs with partial agonist activity confer any advantage is uncertain. Some of the evidence that patients with asthma tolerate beta-blockers is probably misleading, relating to patients with chronic obstructive airways disease who have irreversible changes and who do not respond to either bronchoconstricting or bronchodilating drugs (98). In contrast, a few patients who have never had asthma or chronic bronchitis develop severe bronchospasm when given a beta-blocker. Some, but not all, of these cases (99) may have been allergic reactions to the dyestuffs (for example tartrazine) that are used to color some formulations. Other patients, who need not have a history of chest disease, only develop increased airways resistance with beta-blockers during respiratory infections.

It is against this background that claims that some asthmatic subjects will tolerate certain beta-blockers (100) must be viewed. Some asthmatic patients may indeed tolerate either cardioselective beta-blockers (such as atenolol and metoprolol) or labetalol (101,102), and in patients taking atenolol beta$_2$-adrenoceptor agonists may continue to produce bronchodilatation (103), but in most instances other therapeutic options are preferable (104). Celiprolol is a beta$_1$-adrenoceptor antagonist that has partial beta$_2$-agonist activity. Small studies have suggested that it may be useful in patients with asthma (105), but worsening airways obstruction has been reported (106); it has been concluded that celiprolol has no advantage over existing beta-blockers in the treatment of hypertension (107).

Bronchospasm, which can be life-threatening, can be precipitated by beta-blocker eye-drops. Even beta$_1$-selective antagonists, such as betaxolol, can cause a substantial reduction in forced expiratory volume. Wheezing and dyspnea have been reported among patients using betaxolol: the symptoms resolved after withdrawal. A cross-sectional study has shown that ophthalmologists were more aware than chest physicians about the use of beta-blocker eye-drops by patients with obstructive airways disease; patient awareness was also poor (38,108).

Attention has also been drawn to the increased risks of the adverse effects of beta-blockers on respiratory function in old people (109).

Central ventilatory suppression

Reduced sensitivity of the respiratory center to carbon dioxide has been reported (90,110). The clinical significance of this is unknown, but lethal synergism between morphine and propranolol in suppressing ventilation in animals has been described (111).

Pneumonitis, pulmonary fibrosis, and pleurisy

Pulmonary fibrosis (112) and pleural fibrosis (113) have both been described as infrequent complications associated with practolol. Pulmonary fibrosis has also occurred during treatment with pindolol (114) and acebutolol (115,116). Pleuritic and pneumonitic reactions to acebutolol have been reported (117).

Ear, nose, throat

Nasal polyps, rhinitis, and sinusitis resistant to long courses of antibiotics and surgical intervention have been described in five patients taking non-selective beta-adrenoceptor antagonists (propranolol and timolol) (118). The symptoms resolved when the drugs were withdrawn and did not recur when $beta_1$-selective adrenoceptor blockers (metoprolol or atenolol) were given instead.

Nervous system

Some minor neuropsychiatric adverse effects, such as light-headedness, visual and auditory hallucinations, illusions, sleep disturbances, vivid dreams, and changes in mood and affect, have been causally related to long-term treatment with beta-adrenoceptor antagonists (119,120). Other occasional nervous system effects of beta-blockers include hearing impairment (121), episodic diplopia (122), and myotonia (123).

Although some migraine sufferers use beta-adrenoceptor antagonists prophylactically, there are also reports of the development of migraine on exposure to propranolol or rebound aggravation when the drug is withdrawn (124). Stroke, a rare complication of migraine, has been reported in three patients using propranolol for prophylaxis (125–127). Seizures have been reported with the short-acting beta-blocker esmolol, usually with excessive doses (118). Myasthenia gravis has been associated with labetalol (128), oxprenolol, and propranolol (129), and carpal tunnel syndrome has been reported with long-term beta-blockade, the symptoms gradually disappearing on withdrawal of therapy (130).

Propranolol and gabapentin are both effective in essential tremor. However, pindolol, which has substantial partial agonist activity, can cause tremor (131), and gabapentin can occasionally cause reversible movement disorders. A patient who developed dystonic movements after the combined use of gabapentin and propranolol has been described (132).

- A 68-year-old man with a 10-year history of essential tremor was initially treated with propranolol (120 mg/day), which was only slightly effective. Propranolol was replaced by gabapentin (900 mg/day). The tremor did not improve and propranolol (80 mg/day) was added. Two days later he developed paroxysmal dystonic movements in both hands. Between episodes, neurological examination was normal. When propranolol was reduced to 40 mg/day the abnormal movements progressively disappeared.

This case suggests that there is a synergistic effect between propranolol and gabapentin.

In addition, tiredness, fatigue, and lethargy, probably the commonest troublesome adverse effects of beta-blockers and often the reason for withdrawal (133), may have a contributory nervous system component, although they are probably primarily due to reduced cardiac output and altered muscle metabolism (66) (see also the section on Fatigue in this monograph). In general, a definite neurological association has been difficult to prove, and studies of patients taking beta-adrenoceptor antagonists for hypertension, which incorporated control groups of patients taking either other antihypertensive drugs or a placebo, appear to have shown that the incidence of symptoms that can be specifically attributed to beta-adrenoceptor antagonists is lower than anticipated (134).

The more lipophilic drugs, such as propranolol and oxprenolol, would be expected to pass the blood–brain barrier more readily than hydrophilic drugs, such as atenolol and nadolol, and there is some evidence that they do so (15). In theory, therefore, hydrophilic drugs might be expected to produce fewer neuropsychiatric adverse effects. A double-blind, placebo-controlled evaluation of the effects of four beta-blockers (atenolol, metoprolol, pindolol, and propranolol) on central nervous function (135) showed that disruption of sleep was similar with the three lipid-soluble drugs, averaging six to seven wakenings per night, compared with an average of three wakenings per night for atenolol and placebo. Only pindolol, which has a higher CSF/plasma concentration ratio than metoprolol and propranolol, significantly altered rapid eye movement sleep and latency (136). Patients who took pindolol and propranolol also had high depression scores.

In a placebo-controlled sleep laboratory study of atenolol, metoprolol, pindolol, and propranolol, the three lipophilic drugs reduced dreaming (equated with rapid eye movement sleep) but increased the recollection of dreaming and the amount of wakening; in contrast, although atenolol also reduced sleep, it had no effect on subjective measures of sleep (137).

The published data on the effects of beta-blockers on the nervous system have been extensively reviewed (138). The overall incidence of effects was low, and lowest with the hydrophilic drugs. However, a meta-analysis of 55 studies of the cognitive effects of beta-blockade did not show any firm evidence that lipophilic drugs caused more adverse effects than hydrophilic ones (139). Recent data confirming a correlation between lipophilicity and serum concentrations on the one hand and nervous system effects on the other (140) have fuelled this controversy.

Car-driving and other specialized skills

In view of the large numbers of people who take beta-adrenoceptor antagonists regularly for hypertension or ischemic heart disease, the question arises whether these drugs impair performance in tasks that require psychomotor coordination. The occupations under scrutiny include car-driving, the operation of industrial machinery, and the piloting of aeroplanes. The current evidence is conflicting and controversial. One report suggested that propranolol and pindolol given for 5 days impaired slalom driving in a manner comparable with the coordination defects caused by alcohol (141). In contrast, other studies have shown that driving skills were not impaired during long-term beta-blocker therapy and might even be improved (142,143). There is also a suggestion that tolerance to the central effects of these drugs can develop within 3 weeks of starting therapy, provided the dosage does not change (144). Until more information is available from well-controlled studies, it is advisable to inform patients who are starting treatment with beta-blockers that they should exercise special care in the performance of skills requiring psychomotor coordination for the first 1 or 2 weeks.

Sensory systems

Eyes

Keratopathy in association with the practolol syndrome is the major serious ocular effect ascribed to beta-adrenoceptor antagonists. Conjunctivitis and visual disturbances have also been reported, and a case of ocular pemphigoid has been described in a patient taking timolol eye-drops for glaucoma (145). Anterior uveitis has been reported in patients taking betaxolol (146) and metipranolol (147,148). Corneal anesthesia and epithelial sloughing with continuing use of topical beta-blockers have also been reported (149), as have ocular myasthenia and worsened sicca syndrome. Patients who lack CYP2D6 are more likely to have higher systemic concentrations of beta-blockers after topical application, making them susceptible to adverse effects.

- Recurrent retinal arteriolar spasm with associated visual loss has been described in a 68-year-old man with hypertension treated with atenolol (SEDA-17, 236).
- A 60-year-old man with open-angle glaucoma developed an allergic contact conjunctivitis and dermatitis from carteolol, a topical non-cardioselective beta-blocker (150). He had extensive cross-reactivity to other topical beta-blockers, such as timolol and levobunolol. Cross-reactivity among different beta-blockers is possibly due to a common lateral aliphatic chain.

Psychological, psychiatric

Disturbances of psychomotor function

Beta-adrenoceptor antagonists impair performance in psychomotor tests after single doses. These include effects of atenolol, oxprenolol, and propranolol on pursuit rotor and reaction times (151,152). However, other studies with the same drugs have failed to show significant effects (153–157), and the issue has remained controversial. A report that sotalol improved psychomotor performance in 12 healthy individuals in a dose of 320 mg/day but impaired performance at 960 mg/day (158) has been interpreted to indicate that the water-soluble beta-adrenoceptor antagonists would be less likely than the fat-soluble drugs to produce nervous system effects. Both atenolol and propranolol alter the electroencephalogram; atenolol affects body sway and alertness and propranolol impairs short-term memory and the ability to concentrate (159,160). These results suggest that both lipophilic and hydrophilic beta-adrenoceptor antagonists can affect the central nervous system, although the effects may be subtle and difficult to demonstrate.

In 27 hypertensive patients aged 65 years or more, randomized to continue atenolol treatment for 20 weeks or to discontinue atenolol and start cilazapril, there was a significant improvement in the choice reaction time in the patients randomized to cilazapril (161). This study has confirmed previous reports that chronic beta-blockade can determine adverse effects on cognition in elderly patients. Withdrawal of beta-blockers should be considered in any elderly patient who has signs of mental impairment.

In a placebo-controlled trial of propranolol in 312 patients with diastolic hypertension, 13 tests of cognitive function were assessed at baseline, 3 months, and 12 months (162). Propranolol had no significant effects on 11 of the 13 tests. Compared with placebo, patients taking propranolol had fewer correct responses at 3 months and made more errors of commission.

Psychoses

Bipolar affective disorder

Bipolar depression affects 1% of the general population, and treatment resistance is a significant problem. The addition of pindolol can lead to significant improvement in depressed patients who are resistant to antidepressant drugs, such as selective serotonin reuptake inhibitors or phenelzine. Of 17 patients with refractory bipolar depression, in whom pindolol was added to augment the effect of antidepressant drugs, eight responded favorably (163). However, two developed transient hypomania, and one of these became psychotic after the resolution of hypomanic symptoms. In both cases transient hypomanic symptoms resolved without any other intervention, while psychosis required pindolol withdrawal.

Anxiety and depression have been reported after the use of nadolol, which is hydrophilic (164). In a study of the co-prescribing of antidepressants in 3218 new users of beta-blockers (165), 6.4% had prescriptions for antidepressant drugs within 34 days, compared with 2.8% in a control population. Propranolol had the highest rate of co-prescribing (9.5%), followed by other lipophilic beta-blockers (3.9%) and hydrophilic beta-blockers (2.5%). In propranolol users, the risk of antidepressant use was 4.8 times greater than the control group, and was highest in those aged 20–39 (RR = 17; 95% CI = 14, 22).

Organic brain syndrome

The development of a severe organic brain syndrome has been reported in several patients taking beta-adrenoceptor antagonists regularly without a previous history of psychiatric illness (166–168). A similar phenomenon was seen in a young healthy woman who took propranolol 160 mg/day (169). The psychosis can follow initial therapy or dosage increases during long-term therapy (170). The symptoms, which include agitation, confusion, disorientation, anxiety, and hallucinations, may not respond to treatment with neuroleptic drugs, but subsides rapidly when the beta-blockers are withdrawn. Symptoms are also ameliorated by changing from propranolol to atenolol (171).

Schizophrenia

A schizophrenia-like illness has also been seen in close relation to the initiation of propranolol therapy (172).

Endocrine

Prolactin

- Reversible hyperprolactinemia with galactorrhea occurred in a 38-year-old woman taking atenolol for hypertension (173).

Thyroid

Propranolol inhibits the conversion of thyroxine (T4) to tri-iodothyronine (T3) by peripheral tissues (174), resulting in increased formation of inactive reverse T3. There have been several reports of hyperthyroxinemia in clinically euthyroid patients taking propranolol for non-thyroid reasons in high dosages (320–480 mg/day) (175,176). The incidence was considered to be higher than could be accounted for by the development of spontaneous hyperthyroidism, but the mechanism is unknown.

The effect of beta-adrenoceptor antagonists on thyroid hormone metabolism is unlikely to play a significant role in their use in hyperthyroidism. Since D-propranolol has similar effects on thyroxine metabolism to those seen with the racemic mixture, membrane-stabilizing activity may be involved (177).

In one case, beta-adrenoceptor blockade masked an unexpected thyroid crisis, resulting in severe cerebral dysfunction before the diagnosis was made (178).

Metabolism

Hypoglycemia and blood glucose control

Hypoglycemia, producing loss of consciousness in some cases, can occur in non-diabetic individuals who are taking beta-adrenoceptor antagonists, particularly those who undergo prolonged fasting (179) or severe exercise (180,181). Patients on maintenance dialysis are also at risk (182). It has been suggested that non-selective drugs are most likely to produce hypoglycemia and that cardioselective drugs are to be preferred in at-risk patients (183), but the same effect has been reported with atenolol under similar circumstances (181).

Two children in whom propranolol was used to treat attention deficit disorders and anxiety became unarousable, with low heart rates and respiratory rates, due to hypoglycemia (184). Hypoglycemia can be caused by reduced glucose intake (fasting), increased utilization (hyperinsulinemia), or reduced production (enzymatic defects). One or more of these mechanisms can be responsible for hypoglycemia secondary to drugs. Children treated with propranolol may be at increased risk of hypoglycemia, particularly if they are fasting. Concomitant treatment with methylphenidate can increase the risk of this metabolic disorder.

However, contrary to popular belief, beta-adrenoceptor antagonists do not by themselves increase the risk of hypoglycemic episodes in insulin-treated diabetics, in whom their use was concluded to be generally safe (185). Indeed, in 20 such patients treated with diet or diet plus oral hypoglycemic agents, both propranolol and metoprolol produced small but significant increases in blood glucose concentrations after 4 weeks (186). The rise was considered clinically important in only a few patients.

However, in insulin-treated diabetics who become hypoglycemic, non-selective beta-adrenoceptor antagonists can mask the adrenaline-mediated symptoms, such as palpitation, tachycardia, and tremor; they can cause a rise in mean and diastolic blood pressures, due to unopposed alpha-adrenoceptor stimulation from catecholamines, because the β_2-adrenoceptor-mediated vasodilator response is blocked (187); they can also impair the rate of rise of blood glucose toward normal (188). In contrast, cardioselective drugs mask hypoglycemic symptoms less (189); because of vascular sparing, they are less likely to be associated with a diastolic pressor response in the presence of catecholamines, although this has been reported with metoprolol (190); and delay in recovery from hypoglycemia is either less marked or undetectable with cardioselective drugs, such as atenolol or metoprolol. Thus, if insulin-requiring diabetics need to be treated with a beta-adrenoceptor antagonist, a cardioselective agent should always be chosen for reasons of safety, while allowing that this type of beta-blocker is associated with insulin resistance and can impair insulin sensitivity by 15–30% (SEDA-17, 235), and hence increase insulin requirements.

People with diabetes have a much worse outcome after acute myocardial infarction, with a mortality rate at least twice that in non-diabetics. However, tight control of blood glucose, with immediate intensive insulin treatment during the peri-infarct period followed by intensive subcutaneous insulin treatment, was associated with a 30% reduction in mortality at 1 year, as reported in the DIGAMI study. In addition, the use of beta-blockers in this group of patients had an independent secondary preventive effect (191). The use of beta-blockers in diabetics with ischemic heart disease should be encouraged (37).

In 686 hypertensive men treated for 15 years, beta-blockers were associated with a higher incidence of diabetes than thiazide diuretics (192). This was an uncontrolled study, but the observation deserves further study.

In a randomized controlled comparison of the effects of beta-adrenoceptor antagonists with different pharmacological profiles, namely metoprolol and carvedilol, on glycemic and metabolic control in 1235 hypertensive patients with type 2 diabetes already taking a blocker of the renin–angiotensin–aldosterone system, blood pressure reduction was similar in the two groups but the mean glycosylated hemoglobin increased significantly from baseline to the end of the study with metoprolol (0.15%;1 95%CI = 0.08, 0.22) but not carvedilol (0.02%)

(193). Insulin sensitivity improved with carvedilol (–9.1%) but not metoprolol (–2.0%). The between-group difefrence was –7.2%. Progression to microalbuminuria was more common with metoprol than with carvedilol. Even if both agents were effective in reducing blood pressure and well tolerated, the use of carvedilol in addition to blockers of the renin–angiotensin–aldosterone system seems to be associated with a better metabolic profile in diabetic patients.

Blood lipids

There is increasing evidence that beta-adrenoceptor antagonists increase total triglyceride concentrations in blood and reduce high-density lipoprotein (HDL) cholesterol. Comparisons of non-selective and cardioselective drugs have shown that lipid changes are less marked but still present with $beta_1$-selective agents (194). Current information suggests that $beta_1$-selective drugs may be preferable in patients with hypertriglyceridemia (195). Topical beta-blockers can cause rises in serum triglyceride concentrations and falls in serum high-density lipoprotein concentrations; this makes them less suitable in patients with coronary heart disease (38,196).

The importance of these effects for the long-term management of patients with hypertension or ischemic heart disease is unknown, but it is recognized that a high serum total cholesterol and a low HDL cholesterol are associated with an increased risk of ischemic heart disease. However, a significant reduction in HDL cholesterol after treatment for 1 year with timolol was of no prognostic significance and did not attenuate the protective effect of the drug (197). In a 4-year randomized, placebo-controlled study of six antihypertensive monotherapies, acebutolol produced only a small and probably clinically irrelevant (0.17 mmol/l) reduction in total cholesterol (198), which was not statistically different from four of the other antihypertensive drugs.

Obesity

It has been suggested that beta-blockers may predispose to obesity by reducing basal metabolic rate via beta-adrenoceptor blockade (199). Thermogenesis in response to heat and cold, meals, stress, and anxiety is also reduced by beta-adrenoceptor blockade, promoting weight gain (SEDA-16, 193). $Beta_3$-adrenoceptors have been implicated in this mechanism (200,201). Since propranolol blocks $beta_3$-receptors in vivo (202), it would be wise on theoretical grounds to avoid propranolol in obese patients; nadolol is another non-selective beta-blocker that does not act on $beta_3$-adrenoceptors.

A systematic review of eight prospective, randomized trials in 7048 patients with hypertension (3205 of whom were taking beta-blockers) confirmed that body weight was higher in those taking beta-blockers than in controls at the end of the studies (203). The median difference in body weight was 1.2 kg (range –0.4–3.5 kg). There was no relation between demographic characteristics and changes in body weight. The weight gain was observed in the first few months of treatment and thereafter there was no further weight gain compared with controls. This observation suggests that first-line use of beta-blockers in obese patients with hypertension should be considered with caution.

Electrolyte balance

Hypokalemia

Adrenaline by infusion produces a transient increase in plasma potassium, followed by a prolonged fall; pretreatment with beta-adrenoceptor antagonists results in a rise in plasma potassium (204). These effects may be mediated via $beta_2$-adrenoceptors (205), and cardioselective drugs should have smaller effects (204). In the Treatment of Mild Hypertension Study (TOMHS), acebutolol did not change serum potassium after 4 years (198).

It has been argued that drug combinations that contain a beta-adrenoceptor antagonist in combination with a thiazide diuretic minimize the hypokalemic effect of the latter; however, marked hypokalemia in the absence of primary hyperaldosteronism has been reported in a patient taking Sotazide (a combination of hydrochlorothiazide and the non-selective drug sotalol) (206). The use of a combination formulation of chlortalidone and atenolol has also produced hypokalemia (207), in one case complicated by ventricular fibrillation after myocardial infarction (208).

In addition to a rise in serum potassium, timolol increases plasma uric acid concentrations (209). In the TOMHS study, acebutolol increased serum urate by 7 μmol/l (198).

Mineral balance

A fall in serum calcium has been reported with atenolol (210), but whether this was causal has been disputed (211).

Hematologic

Thrombocytopenia has been reported in patients taking oxprenolol (212,213) and alprenolol (214,215); it can recur on rechallenge. This effect is presumed to have an immunological basis.

In the International Agranulocytosis and Aplastic Anemia Study the relation between cardiovascular drugs and agranulocytosis was examined: there was a relative risk of 2.5 (95% CI = 1.1, 6.1) for propranolol (216). Other beta-blockers did not increase risk and propranolol had no association with aplastic anemia. There are also anecdotal reports of this association (217).

Gastrointestinal

Mild gastrointestinal adverse effects, such as nausea, dyspepsia, constipation, or diarrhea, have been reported in 5–10% of patients taking beta-adrenoceptor antagonists (42). A reduction in dosage or a change to another member of the group will usually produce amelioration. Severe reactions of this type are very infrequent, but severe diarrhea, dehydration, hypokalemia, and weight loss, recurring after rechallenge, occurred with propranolol in a single case (218).

Nausea and vomiting have been attributed to timolol eye-drops (219).

- Severe nausea and vomiting occurred in a 77-year-old woman treated with timolol eye-drops for glaucoma. Her weight had fallen by 8 kg (13%). All physical, laboratory, and instrumental examinations were negative. Gastroduodenoscopy and duodenal biopsy were unremarkable and *Helicobacter pylori* was absent. When timolol was replaced by betaxolol, her complaints disappeared and she gained 2 kg. On rechallenge 3 months later she developed severe nausea, vomiting, and anorexia after some days of treatment. She immediately stopped taking the treatment and 4 days later the symptoms disappeared.

Since timolol has been satisfactorily used by millions of patients, the incidence of serious gastrointestinal events appears to be very low. Absence of symptoms after betaxolol therapy in this patient is in agreement with its lower risk of non-cardiac adverse reactions compared with the non-selective agent timolol.

Beta-adrenoceptor antagonists can cause non-anginal chest pain because of esophagitis (220), due to adherence of the tablet mass, resulting in esophageal spasm, inflammatory change, and even perforation.

Sclerosing peritonitis and retroperitoneal fibrosis

Sclerosing peritonitis was described as part of the practolol syndrome (221–223), and it can also occur with other beta-adrenoceptor antagonists (224,225).

Retroperitoneal fibrosis has been reported in patients taking oxprenolol (226), atenolol (227), propranolol (228), metoprolol (229), sotalol (230), and timolol (including eye-drops) (231,232). However, this disorder often occurs spontaneously and has been reported very infrequently in patients taking beta-blockers (233). Thus, in the absence of any causal relation it is most likely that it reflects the spontaneous incidence in patients taking a common therapy. This conclusion has been supported by an analysis of 100 cases of retroperitoneal fibrosis (234).

Liver

Many beta-adrenoceptor antagonists undergo substantial first-pass hepatic metabolism; these include alprenolol, metoprolol, oxprenolol, and propranolol. Hepatic cirrhosis, with consequent portosystemic shunting, can therefore result in increased systemic availability and higher plasma concentrations, perhaps resulting in adverse effects. Beta-blockers may also reduce liver blood flow and cause interactions with drugs with flow-dependent hepatic clearance.

The oxidative clearance of the lipophilic drugs, metoprolol, timolol, and bufuralol, is influenced by the debrisoquine hydroxylation gene locus, resulting in polymorphic metabolism (235). This might result in an increase in the adverse effects of these beta-blockers in poor metabolizers, but to date there is no objective evidence of such an association (236).

Beta-adrenoceptor antagonists, used in the prevention of bleeding from esophageal varices in patients with hepatic cirrhosis, have reportedly caused hepatic encephalopathy in several patients (237–241). Thus, extreme caution is required, particularly because resuscitation can be difficult when beta-blockers are given to patients with gastrointestinal bleeding or encephalopathy (242).

Biliary tract

Biliary cirrhosis was reported as part of the practolol syndrome (243), but there have been no comparable reports with other beta-adrenoceptor antagonists.

Urinary tract

Propranolol reduces renal blood flow and glomerular filtration rate after acute administration, associated with, and probably partly due to, falls in cardiac output and blood pressure (244,245). There has been some argument about whether these effects persist during long-term therapy (246). Despite early suggestions that renal function might be worsened by such therapy, particularly in patients with chronic renal insufficiency (247), the clinical significance of these changes is debatable (248). Claims that nadolol increases renal blood flow and that cardioselective drugs such as atenolol reduce renal blood flow less than non-selective agents in old people (249) are thus probably relatively unimportant. The vasodilating beta-blocker carvedilol maintains renal blood flow whilst reducing glomerular filtration rate, suggesting that renal vasodilatation occurs (250), although a single case of reversible renal insufficiency has been described in a clinical trial in patients with severe heart failure (251).

Skin

Rashes were part of the practolol (oculomucocutaneous) syndrome, but are infrequent with other beta-adrenoceptor antagonists. The eruptions can be urticarial, morbilliform, eczematous, vesicular, bullous, psoriasiform, or lichenoid (252–257).

Beta-blocker eye-drops can cause skin rashes (258).

- A 70-year-old woman treated with topical timolol for glaucoma developed a papular eruption on the arms and back, consistent with prurigo. All tests were within the reference ranges. There was no improvement after 1 month of topical corticosteroids. The eruption cleared completely within 1 month of timolol withdrawal. Betaxolol eye-drops were introduced and the eruption recurred within 1 week. When beta-blocker therapy was replaced by synthetic cholinergic eye-drops (drug unspecified) the eruption cleared completely without any recurrence a year later.
- Allergic contact dermatitis due to carteolol eye drops occurred in a 61-year-old woman (259). Withdrawal of carteolol and the use of timolol instead led to improvement within 10 days, suggesting that in some cases there is no cross-reactivity between different beta-blockers.

Although cutaneous adverse effects have been previously described after oral beta-blockers, including timolol, this observation further suggests a class effect of topical beta-blockers. This case also suggests a cross-reaction between timolol and betaxolol.

In a review of 588 patients with established psoriasis it was concluded that about two-thirds of such patients are likely to have a flare-up with a beta-adrenoceptor antagonist, regardless of the agent used (260). In patients with vitiligo, beta-blockers rarely exacerbate depigmentation (7/548 patients) (261). In a separate report, topical betaxolol used for glaucoma was associated with periocular cutaneous pigmentary changes (262).

Some patients have positive patch tests and/or a positive response to oral rechallenge. There can also be cross-sensitivity to other beta-blockers in compromised patients. The mechanisms appear to include both immunological and pharmacological effects; in the latter case the drug may modify growth regulation in the epidermis (263).

Contact allergy to topical beta-blockers can occur.

- A 68-year-old woman developed contact allergy after many years of using befunolol (264). Patch-testing showed cross-sensitivity to carteolol. Evidence of such cross-sensitivity has not previously been reported.

Sweat glands

Hyperhydrosis or sweating with beta-blockers has been reported with both oral formulations (sotalol and acebutolol) (265) and topical formulations (carteolol) (266), although the patients described in these reports were not rechallenged to ascertain the link. However, it was suggested that beta-blockade increased exercise-related sweating in healthy volunteers, more so with a non-selective beta-blocker (propranolol), than a selective one (atenolol) (267). The mechanism for this was uncertain, but was thought to be due to an imbalance between beta- and alpha-adrenergic activity. In some instances, clonidine, an alpha-adrenoceptor antagonist, was effective in treating hyperhydrosis (268), whereas in other cases, propranolol was paradoxically effective (269).

Hair

There have been single case reports of alopecia in association with propranolol (270) and metoprolol (271).

Musculoskeletal

It has been suggested that arthralgia is a not an uncommon adverse effect of beta-adrenoceptor antagonists, particularly metoprolol (272), although the association was not confirmed by rechallenge in any patient. A later case-control study in 127 patients attending a hypertension clinic who had arthropathy showed no significant relation between the arthropathy and the use of beta-blockers (273). On the other hand, five cases of metoprolol-associated arthralgia, most with negative serological tests for collagenases, have been reported to the FDA (274).

Muscle cramps have been reported in patients taking beta-blockers with partial agonist activity (275); it has been suggested (276) that this might be a beta$_2$-partial agonist effect, although this has not subsequently been supported (277). However, a crossover study in 78 hypertensive patients suggested that beta-blockers with partial agonist activity (pindolol and carteolol) caused muscle cramps in up to 40% of these patients, with an associated rise in serum CK and CK-MB, although the severity of the cramps did not correlate with the enzyme activities (278).

In a large case-control study of beta-blockers alone or in combination with thiazides in 30 601 patients with a fracture and 120 819 matched controls, patients who took beta-blockers alone had a 23% (95% CI = 17, 28) lower risk of fractures (279). Patients who took thiazides alone had a 20% (95% CI = 14, 26) risk reduction, and patients who took both had a risk reduction of 29% (95% CI = 21, 36). The data were adjusted for the main possible confounding variables. These findings seem to confirm the experimental evidence that beta-blockers cause increased bone formation. From a practical point of view, in elderly patients with hypertension at high risk of osteoporosis, a beta-blocker alone or in combination with a thiazide diuretic may be of potential benefit.

Sexual function

Erectile drys function

Uncontrolled studies of the effect of beta-adrenoceptor antagonists on sexual function have often shown a high incidence of absence of erections, reduced potency, and reduced libido (280). Several large controlled trials in hypertension and ischemic heart disease have provided more exact information. In a large-scale, prospective, placebo-controlled study, reduced sexual activity was an adverse effect of sufficient severity to lead to the withdrawal of some patients in the propranolol-treated group (281). In the TOMHS study, the incidence of difficulty in obtaining and maintaining an erection over 48 months was 17% with both acebutolol and placebo (198). In the TAIM study (282), a randomized, placebo-controlled study lasting 6 months, atenolol did not cause a significant increase in erectile problems in men (11%; 95% CI 2, 20%) compared with placebo (3%; 0, 9%). Loss of libido and difficulty in sustaining an erection can be induced in young healthy volunteers (135); although these effects may be more common with lipophilic drugs, such as propranolol (283) and pindolol (284), they have also been reported with atenolol (285).

Patients with different newly diagnosed cardiovascular diseases and without erectile dysfunction were randomized to take atenolol 50 mg/day blindly (n = 32, group A), or atenolol 50 mg/day with information about the treatment but not its adverse effects (n = 32, group B), or atenolol 50 mg/day with information about both the kind of drug and the possible adverse effects (n = 32, group C) (286). Erectile dysfunction occurred in 3.1%, 16%, and 31% of the patients in groups A, B, and C respectively. All patients who reported erectile dysfunction were then randomized to sildenafil 50 mg or placebo, which were equally effective in reversing erectile dysfunction in all but one patient. This study confirms how knowledge of the adverse effects of beta blockers can cause erectile dysfunction. This suggests that the problem is predominantly psychological in origin.

New-generation beta-blockers, such as nebivolol, seem to cause less sexual dysfunction, probably because they increase the release of nitric oxide. In a randomized, single-blind, multicenter trial, 131 patients with a new diagnosis of hypertension, without prior erectile dysfunction, were randomized to either atenolol or atenolol plus chlorthalidone or nebivolol (287). Over 12 weeks the patients who were allocated to nebivolol did not have any change in the number of satisfactory episodes of sexual intercourse, while those who took atenolol or atenolol plus chlorthalidone had a significant reduction from baseline. Given a similar effect in reducting blood pressure, nebivolol seems to maintain a more active sexual life in hypertensive men than atenolol. Increased release of nitric oxide due to nebivolol may counteract the detrimental effect of beta-adrenoceptor antagonists on sexual activity.

Peyronie's disease

Peyronie's disease is a fibrotic condition of the penis that has been associated with beta-blockers, such as propranolol (288), metoprolol (289), and labetalol (289). However, 100 consecutive cases of Peyronie's disease included only five men who had taken a beta-blocker before the onset of the condition (290); the authors concluded that the syndrome was likely to be associated with chronic degenerative arterial disease and not with beta-adrenoceptor antagonists.

Immunologic

Leukocytoclastic vasculitis has been reported with sotalol (291).

- A progressive cutaneous vasculitis occurred in a 66-year-old man taking sotalol for prevention of a symptomatic atrial fibrillation. After 7 days he noted a petechial eruption on his wrists and ankles. This progressed during the next days to palpable purpura on the hands, wrists, ankles, and feet. A biopsy specimen showed changes consistent with leukocytoclastic vasculitis. After withdrawal of sotalol the skin rash cleared completely without any other intervention.

Other beta-blockers associated with leukocytoclastic vasculitis include acebutolol, alprenolol, practolol, and propranolol.

Antinuclear antibodies in high titers were detected in a number of patients with the practolol oculomucocutaneous syndrome. Tests in patients taking acebutolol (292,293) and celiprolol (294) have also shown a high frequency of antinuclear antibodies. Positive lupus erythematosus cell preparations have been observed in patients taking acebutolol (293).

The lupus-like syndrome was part of the practolol syndrome and has also been attributed to acebutolol (295,296), atenolol (SEDA-16, 194), labetalol (297), pindolol (298), and propranolol (299). However, apart from practolol, it seems to be very rare during treatment with beta-adrenoceptor antagonists.

Anaphylactic reactions have been attributed to beta-adrenoceptor antagonists only very infrequently (300). However, it appears that anaphylactic reactions precipitated by other agents can be particularly severe in patients taking beta-blockers, especially non-selective drugs, and may require higher-than-usual doses of adrenaline for treatment (301–304). The authors of a brief review of the risk of anaphylaxis with beta-blockers concluded that the risk is not increased (305). The view that allergy skin testing or immunotherapy is inadvisable in patients taking beta-blockers (306) has been disputed, bearing in mind the low incidence of this adverse effect (307).

Long-Term Effects

Drug abuse

In an unusual case of Munchausen's syndrome, a female general practitioner repeatedly took high doses of beta-blockers in order to simulate symptomatic sick sinus syndrome (308).

Drug withdrawal

Interest in the possible effects of the sudden withdrawal of beta-adrenoceptor antagonists followed a 1975 report of two deaths and four life-threatening complications of coronary artery disease within 2 weeks of withdrawal of propranolol (309). Subsequent analyses did not always confirm these findings (310,311), and it has not been easy to distinguish between natural progression and deterioration caused by drug withdrawal under such circumstances. However, a case-control study in hypertensive patients showed a relative risk of 4.5 (95% CI = 1.1, 19) associated with recent withdrawal of beta-blockers and the development of myocardial infarction or angina (312).

The symptoms attributed to the sudden withdrawal of beta-adrenoceptor antagonists (severe exacerbation of angina pectoris, acute myocardial infarction, sudden death, malignant tachycardia, sweating, palpitation, and tremor) are consistent with transient adrenergic hypersensitivity. Unequivocal signs of rebound hypersensitivity have been observed after drug withdrawal in patients with ischemic heart disease (313), but not in hypertensive patients (314–316). The density of beta-adrenoceptors on human lymphocyte membranes increased by 40% during treatment with propranolol for 8 days (317), and hypersensitivity to isoprenaline can be shown in hypertensive patients after the withdrawal of different beta-adrenoceptor antagonists, including propranolol (318), metoprolol (319), and atenolol (320). This hypersensitivity occurs within 2 days of drug withdrawal, can persist for up to 14 days, and is presumed to reflect the up-regulation of beta-adrenoceptors that occurs with prolonged treatment. This phenomenon is said to be diminished by gradual withdrawal of therapy and by the use of drugs with partial agonist activity, such as pindolol (321). Whether this is directly relevant to the effects of the sudden withdrawal of beta-adrenoceptor antagonists in patients with ischemic heart disease is speculative.

Although there is evidence that abrupt withdrawal of long-acting beta-blockers is not associated with the development of the beta-blocker withdrawal syndrome (322), current information suggests that withdrawal of beta-adrenoceptor antagonists, particularly in patients with ischemic heart disease, should be accomplished by gradual dosage reduction over 10–14 days. However, even gradual withdrawal may not always prevent rebound effects (323).

Second-Generation Effects

Pregnancy

Great concern at one time accompanied the use of beta-adrenoceptor antagonists in pregnancy, particularly in the management of hypertension. On theoretical grounds beta-adrenoceptor antagonists might be expected to increase uterine contractions, impair placental blood flow, cause intrauterine growth retardation, accentuate fetal and neonatal distress, and increase the risk of neonatal hypoglycemia and perinatal mortality. There are many anecdotal reports of such complications attributed to beta-adrenoceptor antagonists, often propranolol. However, many of the adverse effects listed above are also potential complications of hypertension in pregnancy, and in the absence of a properly controlled trial of therapy, definite conclusions of cause and effect have been impossible on the basis of these anecdotes alone.

Many of the fears expressed were set aside by a double-blind, randomized, placebo-controlled trial of atenolol in pregnancy-associated hypertension in 120 women (323), which showed that babies in the placebo group had a higher morbidity, that atenolol reduced the occurrence of respiratory distress syndrome and intrauterine growth retardation, and that neonatal hypoglycemia and hyperbilirubinemia were equally common in the treated and placebo groups. Although bradycardia was more common with atenolol, it had no deleterious consequences. However, the offspring of women taking atenolol had lower body weights on follow up, the significance of which is unclear (323).

It is reasonable to consider that beta-adrenoceptor antagonists can be used in pregnancy without serious risks, provided patients are kept under careful clinical observation. Since all beta-blockers cross the placenta freely, major differences in effects or toxicity among the various drugs are unlikely, and in a review of beta-blockers in pregnancy it was concluded that no single beta-blocker is superior (324). However, in a small study in 51 women with pregnancy-induced hypertension the combination of hydralazine and propranolol was associated with lower blood glucose and weight at birth compared with the combination of hydralazine and pindolol (with partial agonist activity), despite similar blood pressure control. It was suggested that beta-adrenoceptor antagonists without partial agonist activity might reduce uteroplacental blood flow (325).

Fetotoxicity

The adverse effects of beta-adrenoceptor antagonists on the fetus have been reviewed (326). Beta blockers cross the placenta, and can have adverse maternal and fetal effects. Studies of the use of beta-blockers during pregnancy have generally been small, and the gestational age at the start of the study was generally 29–33 weeks, leaving substantially unanswered the possibility that treatment of more patients and/or longer treatment durations may reveal unrecognized adverse events. These observations underline the fact that the safety of beta-blockers remains uncertain and that they are therefore better not given before the third trimester.

Non-cardioselective beta-adrenoceptor antagonists

Observations derived from uncontrolled studies have shown an association between maternal use of propranolol and intrauterine growth retardation, neonatal respiratory depression, bradycardia, hypoglycemia, and increased perinatal mortality. However, in randomized, placebo-controlled studies of metoprolol and oxprenolol, there was no evidence of effects on birth weight.

- Two infants with features of severe beta-blockade (bradycardia, persistent hypotension), persistent hypoglycemia, pericardial effusion, and myocardial hypertrophy were born before term to mothers taking long-term oral labetalol for hypertension in pregnancy.

Although labetalol is considered to be generally safe in neonates, impaired urinary excretion and lower albumin binding in preterm infants can prolong the half-life of labetalol and increase its systemic availability and toxicity (327).

Cardioselective beta-adrenoceptor antagonists

There is reluctance to use atenolol in pregnancy, especially if treatment starts early. In placebo-controlled studies, birth weight was significantly lower with atenolol groups. The same was true when atenolol was compared with non-cardioselective agents: the weight of infants born to women taking atenolol was significantly lower. When atenolol was started later there was no difference in birth weight between infants born to women treated with atenolol or other beta-blockers, suggesting the relevance of the time of initiation of atenolol. Atenolol should therefore be avoided in the early stages of pregnancy and given with caution in the later stages.

- Fetal bradycardia and pauses after each two normal beats occurred at 21 weeks gestation in a 37-year-old woman using timolol eye-drops for glaucoma; when timolol was withdrawn, the fetal heart rate recovered (328).

The authors concluded that when a woman taking glaucoma therapy becomes pregnant, it is usually possible to interrupt therapy during pregnancy. Treatment may be deferred until delivery of the infant.

Lactation

The list of beta-adrenoceptor antagonists that have been detected in breast milk includes atenolol (329), acebutolol and its active *N*-acetyl metabolite (330), metoprolol (331), nadolol (332), oxprenolol and timolol (333), propranolol (334), and sotalol (335). Most authors have concluded that the estimated daily infant dose derived from breastfeeding is likely to be too low to produce untoward effects in the suckling infant, and indeed such effects were not noted in the above cases. However, in the case of acebutolol it was considered that clinically important amounts of drug could be transferred after increasing plasma concentrations were noted in two breastfed infants.

Susceptibility Factors

Genetic factors

Most beta-blockers undergo extensive oxidation (336). There have been anecdotal reports of high plasma concentrations of some beta-blockers in poor metabolizers of debrisoquine, and controlled studies have shown that debrisoquine oxidation phenotype is a major determinant of the metabolism, pharmacokinetics, and some of the pharmacological effects of metoprolol, bufuralol, timolol, and bopindolol. The poor metabolizer phenotype is associated with increased plasma drug concentrations, a prolonged half-life, and more intense and sustained beta-blockade. There are also phenotypic differences in the pharmacokinetics of the enantiomers of metoprolol and bufuralol.

Renal disease

The hydrophilic drugs atenolol and sotalol are eliminated largely unchanged in the urine; with deteriorating renal function their half-lives can be prolonged as much as 10-fold (337,338). Other beta-adrenoceptor antagonists, for example acebutolol and metoprolol, have active metabolites that can accumulate (339). Massive retention of the metabolite propranolol gluconate has also been reported in patients with renal insufficiency taking long-term oral propranolol (340); this metabolite is then deconjugated, and concentrations of propranolol can be significantly increased in these patients. Thus, in a patient with a low creatinine clearance, either dosage adjustment or a change of beta-blocker may be necessary.

Other features of the patient

Heart failure

Untreated congestive heart failure secondary to systolic pump failure is a contraindication to the use of beta-adrenoceptor antagonists. Patients in frank or incipient heart failure have reduced sympathetic drive to the heart, and acute life-threatening adverse effects can therefore follow beta-blockade. This is one of the recognized potentially serious complications of beta-blockers in the management of thyrotoxic crisis (341). However, patients with heart failure treated with ACE inhibitors and/or diuretics and digoxin may well gain long-term benefit from beta-adrenoceptor antagonists (34,35).

Heart block

Second-degree or third-degree heart block is a contraindication to beta-adrenoceptor blockade. If it is considered necessary for the control of dysrhythmias, a beta-blocker can be given after the institution of pacing.

Acute myocardial infarction

After many trials including thousands of patients, it is increasingly accepted that treatment of acute myocardial infarction with beta-adrenoceptor antagonists is beneficial. Given intravenously within 4–6 hours of the onset of the infarction these drugs can prevent ventricular dysrhythmias and cardiac rupture (342,343). When given orally during the first year after infarction, beta-adrenoceptors reduce mortality by about 25% (344) and probably more in diabetic subjects (37). Since heart failure, hypotension, and bradycardia are complications of both myocardial infarction and beta-adrenoceptor blockade, it might be assumed that these effects would be more common when the two are combined. However, reviews of the relevant studies (342,345–353) do not suggest that beta-adrenoceptor antagonists, given after acute myocardial infarction, either acutely intravenously or for secondary prophylaxis, increase the incidence of adverse effects or the risk of any particular adverse effect. Nevertheless, patients were rigorously selected for inclusion in these trials; less careful decisions to treat may carry increased risks.

Bronchial asthma

Beta-adrenoceptor antagonists should not be given to patients with bronchial asthma or obstructive airways disease, unless there are no other treatment options, because of the risk of precipitating bronchospasm resistant to bronchodilators. Celiprolol, a $beta_1$-antagonist with $beta_2$-agonist activity, has a theoretical but unproven advantage. Alternatively, cardioselective drugs should be chosen in the lowest possible dosages and in conjunction with a $beta_2$-adrenoceptor agonist, such as salbutamol or terbutaline, to minimize bronchoconstriction.

Smoking

Some common activities, such as mental effort (354), cigarette smoking, and coffee drinking (355,356), can produce stress associated with increased catecholamine secretion. In the presence of a non-selective beta-adrenoceptor antagonist, there can be a marked diastolic pressor response, due to mechanisms identical to those described above in hypoglycemia in diabetes. This effect may be smaller with cardioselective drugs.

Theoretically, frequent rises in diastolic blood pressure associated with smoking whilst taking a non-selective beta-adrenoceptor antagonist could be harmful; in a patient with ischemic heart disease or hypertension a cardioselective drug might offer advantages. There is no evidence of differences in morbidity or mortality in patients taking non-selective and cardioselective agents, but both the MRC and IPPPSH trials of mild hypertension showed increases in the incidence of coronary events in patients taking non-selective beta-adrenoceptor antagonists who were also cigarette smokers (357). The explanation that cigarette smoking increases the hepatic metabolism of beta-adrenoceptor blockers, reducing their effectiveness (358), does not extend to the use of cardioselective beta-blockers in smokers, as reported in the HAPPHY and MAPHY trials.

Insulin-treated diabetes
Beta-adrenoceptor antagonists may mask the symptoms of hypoglycemia, result in a catecholamine-mediated rise in diastolic blood pressure, and delay the return of blood glucose concentrations to normal. These effects are minimized or abolished by using a beta$_1$-selective drug, and this type of drug should always be used in preference to a non-selective drug in insulin-treated diabetes.

Anaphylaxis
Beta-blockers can make anaphylactic reactions more difficult to diagnose and treat (355). Even patients with spontaneous attacks of angioedema or urticaria can be at risk when given beta-blockers (359).

Drug Administration

Drug administration route

Beta-adrenoceptor antagonists are used as ocular tension-lowering drugs without notable effects on pupillary size or refraction. Their systemic effects are greater than one would expect, since there is no first-pass metabolism after ocular administration and the plasma concentration can therefore attain therapeutic concentrations (360).

The systemic adverse effects of ophthalmic beta-blockers have been reviewed (361). Symptomatic bradycardia from systemic or ophthalmic use of beta-blockers alone suggests underlying cardiac conduction disturbances. Beta$_2$-adrenoceptor blockade can exacerbate or trigger bronchospasm in patients with asthma or pulmonary disease associated with hyper-reactive airways. Occasionally, adverse systemic reactions can be severe enough to require drug withdrawal. Obtaining a careful medical history and checking pulse rate and rhythm and peak expiratory flow rate should identify the vast majority of patients with potential cardiac and respiratory contraindications.

Beta-blockers that are available as eye-drops include timolol, metipranolol, and levobunolol, which are non-selective beta$_1$- and beta$_2$-adrenoceptor antagonists, and betaxolol, a relatively cardioselective beta$_1$-adrenoceptor antagonist. Although selective beta$_1$-blockers are less likely to precipitate bronchospasm, this and other systemic effects can nevertheless occur (SED-12, 1200).

In 165 patients who used timolol 0.5% eye-drops, adverse effects were reported in 23%, including psychiatric effects (40%), cardiovascular effects (19%), respiratory effects (7%), and local effects (26%) (362).

Cardiovascular
Hemodynamic changes after the topical ocular use of beta-blockers sometimes include only small reductions in heart rate and resting pulse rate and an insignificant reduction in blood pressure. However, patients with cardiovascular disorders, especially those with an irregular heart rate and dysrhythmias, are certainly at risk (SEDA-4, 339). Bradycardia, cardiac arrest, heart block, hypotension, palpitations, syncope, and cerebral ischemia and stroke can occur (363). Rebound tachycardia has been reported after withdrawal of ophthalmic timolol (89,364). Continuous 24-hour monitoring of blood pressure has shown that beta-blocker eye-drops for glaucoma can increase the risk of nocturnal arterial hypotension (365).

Respiratory
Beta-blocker eye-drops can aggravate or precipitate bronchospasm (SEDA-4, 339) (SEDA-5, 426) and potentially life-threatening respiratory failure can occur.

- A 58-year-old patient using topical timolol maleate for open-angle glaucoma developed cough and dyspnea due to interstitial pneumonitis. Three months after withdrawal of the eye-drops, he was asymptomatic with normal lung function, chest X-ray, and thoracic CT scan (366).

Nervous system
Light-headedness, mental depression, weakness, fatigue, acute anxiety, dissociative behaviour, disorientation, and memory loss can develop a few days to some months after the start of timolol eye-drop therapy (89). Central nervous system complaints are most common in patients who have the greatest reduction in intraocular pressure (SEDA-5, 426). Patients may be unaware of the symptoms until the medication is stopped.

Sensory systems
Dry eyes have been reported after the systemic or ocular use of timolol (367). A sensation of dryness in the eyes can develop and is usually transitory. There can be a reduction in the Schirmer test and tear film break-up time. Symptomatic superficial punctate keratitis in association with complete corneal anesthesia has been observed (368).

Amaurosis fugax has been reported in association with topical timolol (369).

Endocrine
Ophthalmic beta-blockers can cause hypoglycemia in insulin-dependent diabetes (370). Conversely, in diabetic patients taking oral hypoglycemic drugs, hyperglycemia can develop because of impaired insulin secretion (SEDA-21, 487).

Electrolyte balance
Severe hyperkalemia has been reported in association with topical timolol, confirmed by rechallenge (371).

Skin
Cutaneous changes secondary to instillation of betaxolol have been described (262).

Musculoskeletal
Aggravation of myasthenia gravis has been observed during ophthalmic timolol therapy (372). Bilateral pigmentation of the fingernails and toenails, marked hyperkalemia, and arthralgia after ocular timolol have all been reported (SED-12, 1200).

Sexual function
Erectile impotence can occur after ophthalmic use of beta-blockers (89).

Susceptibility factors
All of the susceptibility factors that apply to systemically administered beta-blockers also apply to eye-drops. This particularly applies to asthma (262).

Eyes with potential angle closure require a miotic drug and should not be treated with beta-blockers alone. To exclude the risk of precipitating glaucoma in a susceptible individual, gonioscopy is recommended before starting topical beta-adrenoceptor antagonist therapy.

Drug tolerance
Tachyphylaxis can develop after treatment with ophthalmic beta-blockers (373). There are two forms, short-term "escape," which occurs over a few days, and long-term "drift," which occurs over months and years.

Drug overdose

The increasing use of beta-adrenoceptor antagonists appears to have resulted in more frequent reports of severe high-dose intoxication (374,375), in which beta-adrenoceptor antagonists are often taken in combination with sedatives or alcohol. There can be a very short latency from intake of the drug until fulminant symptoms occur (376). The clinical features are well established. Cardiovascular suppression results in bradycardia, heart block, and congestive heart failure, and intraventricular conduction abnormalities are common (377). Ventricular tachycardias with sotalol intoxication may reflect its class III antidysrhythmic properties, leading to prolongation of the QT interval (378) and torsade de pointes, which may respond to lidocaine (379). Bronchospasm and occasionally hypoglycemia can also occur. Coma and epileptiform seizures are often seen (377,380) and may not be secondary to circulatory changes. The outcome is seldom fatal, but 16 fatal cases of intoxication with talinolol (which is $beta_1$-selective) have been described. Deaths have also occurred with metoprolol and acebutolol. Acebutolol has membrane-stabilizing activity, and it has been suggested that drugs with this property carry greater risk when taken in overdose (381). Lipid solubility influences the rate of nervous system penetration of a drug, and overdosage with highly lipophilic drugs, such as oxprenolol and propranolol, has been associated with rapid loss of consciousness and coma (382–384).

Management
The management of self-poisoning with beta-blockers has been reviewed (385).

Treatment should include isoprenaline (although massive doses may be required), glucagon, and atropine. If a $beta_1$-selective antagonist has been taken, isoprenaline may reduce diastolic blood pressure by its unopposed vasodilator effect on $beta_2$-adrenoceptors (386). The $beta_1$-selective agonist dobutamine may be preferable in such patients (387). A temporary transvenous pacemaker should be inserted if significant heart block or bradycardia occur. Seizures in overdosage with a beta-adrenoceptor antagonist respond poorly to diazepam and barbiturates; muscle relaxants and artificial ventilation may be required. In general, the lipid-soluble drugs are highly protein-bound with a large apparent volume of distribution; forced diuresis or hemodialysis are therefore unlikely to be of use.

Propranolol intoxication can cause central nervous system depression in the absence of clinical signs of cardiac toxicity.

- A 16-year-old boy developed central nervous system depression and an acute dilated cardiomyopathy after taking 3200 mg of propranolol in a suicide attempt (388). He was treated with gastric lavage, activated charcoal, and mechanical ventilation. Echocardiography showed a poorly contracting severely dilated left ventricle. After intravenous isoprenaline hydrochloride and glucagon, echocardiography showed normal left ventricular size and function. He became fully alert 20 hours later and made a good recovery without sequelae.

Early echocardiographic evaluation is important in beta-blocker overdose and can prevent delay in the diagnosis and treatment of cardiac toxicity.

Two regional poison centers in the USA have reviewed 280 cases of beta-blocker overdose (389). All patients with symptoms developed them within 6 hours of ingestion. Four patients died as a result of overdosage. There was cardiovascular morbidity in 41 patients (15%), requiring treatment with cardioactive drugs. Propranolol, atenolol, and metoprolol were responsible for 87% of the cases and 84% of cardiovascular morbidity. Beta-blockers with membrane-stabilizing activity (acebutolol, labetalol, metoprolol, pindolol, and propranolol) accounted for 62% of beta-blocker exposures and 73% of cardiovascular morbidity. Symptomatic bradycardia (heart rate less than 60/minute) or hypotension (systolic blood pressure less than 90 mmHg) were observed in all cases classified as having cardiovascular morbidity. Beta-blocker exposure was complicated by a history of at least one co-ingestant in 73% of the cases, benzodiazepines and ethanol being the most frequent. Cardioactive co-ingestants were reported in 26% of cases: calcium channel blockers, cyclic antidepressants, neuroleptic drugs, and ACE inhibitors were the most common. Multivariate analysis showed that the only independent variable significantly associated with cardiovascular morbidity was the presence of another cardioactive drug.

When patients who took another cardioactive drug were excluded, the only variable associated with cardiovascular morbidity was the ingestion of a beta-blocker with membrane-stabilizing activity.

Two fatal cases of acebutolol intoxication (6 and 4 g) have been reported (390). In both cases, the onset of symptoms was sudden (within 2 hours of ingestion), with diminished consciousness, PR, QRS, and QT prolongation, and hypotension unresponsive to inotropic drugs. In both cases there were episodes of repetitive polymorphous ventricular tachycardia.

These cases have confirmed the potential toxicity of beta-blockers with membrane stabilizing activity; they predispose the patient to changes in ventricular repolarization, which can cause QT prolongation and serious ventricular dysrhythmias. This is generally not seen in cases of propranolol intoxication.

Drug–Drug Interactions

General

Drug interactions with beta-adrenoceptor antagonists can be pharmacokinetic or pharmacodynamic (391–393).

Pharmacokinetic interactions

Absorption interactions

The absorption of some beta-adrenoceptor antagonists is altered by aluminium hydroxide, ampicillin, and food; these are interactions of doubtful clinical relevance.

Metabolism interactions

Beta-adrenoceptor antagonists that are cleared predominantly by the liver (for example metoprolol, oxprenolol, propranolol, and timolol) are more likely to participate in drug interactions involving changes in liver blood flow, hepatic drug metabolism, or both. Thus, enzyme-inducing drugs, such as phenobarbital and rifampicin, increase the clearance of drugs such as propranolol and metoprolol and reduce their systemic availability (394,395). Similarly, the histamine H_2 receptor antagonist cimetidine increases the systemic availability of labetalol, metoprolol, and propranolol by inhibiting hepatic oxidation (396–399). The disposition of drugs with high extraction ratios, such as propranolol and metoprolol, is also affected by changes in liver blood flow, and this may be the mechanism by which hydralazine reduces the first-pass clearance of oral propranolol and metoprolol (400).

Lipophilic beta-adrenoceptor antagonists are metabolized to varying degrees by oxidation by liver microsomal cytochrome P450 (for example propranolol by CYP1A2 and CYP2D6 and metoprolol by CYP2D6). These agents can therefore reduce the clearance and increase the steady-state plasma concentrations of other drugs that undergo similar metabolism, potentiating their effects. Drugs that are affected in this way include theophylline (401), thioridazine (402), chlorpromazine (403), warfarin (404), diazepam (405), isoniazid (406), and flecainide (407). These interactions are most likely to be of clinical significance when the affected drug has a low therapeutic ratio, for example theophylline or warfarin.

Beta-blockers can also affect the clearance of high clearance drugs by altering hepatic blood flow. This occurs when propranolol is co-administered with lidocaine (408), but it appears that this interaction is due more to inhibition of enzyme activity than to a reduction in hepatic blood flow (409). Atenolol inhibits the clearance of disopyramide, but the mechanism is unknown (410). Conversely, quinidine doubles propranolol plasma concentrations in extensive but not poor metabolizers (411) and oral contraceptives increase metoprolol plasma concentrations (412).

Pharmacodynamic interactions

The pharmacodynamic interactions of the beta-adrenoceptor antagonists can mostly be predicted from their pharmacology.

Blood pressure

The antihypertensive effect of beta-blockers can be impaired by the concurrent administration of some non-steroidal anti-inflammatory drugs (NSAIDs), possibly because of inhibition of the synthesis of renal vasodilator prostaglandins. This interaction is probably common to all beta-blockers, but may not occur with all NSAIDs; for example, sulindac appears to affect blood pressure less than indometacin (413–415).

The hypertensive crisis that can follow the withdrawal of clonidine can be accentuated by beta-blockers. It has also been reported that when beta-blockers are used in conjunction with drugs that cause arterial vasoconstriction they can have an additional effect on peripheral perfusion, which can be hazardous. Thus, combining beta-blockers with ergot alkaloids, as has been recommended for migraine, can cause severe peripheral ischemia and even tissue necrosis (416).

The hypotensive effects of halothane and barbiturates can be exaggerated by beta-adrenoceptor antagonists. However, they are not contraindicated in anesthesia, provided the anesthetist is aware of what the patient is taking.

The combination of caffeine with beta-blockers causes a raised blood pressure (417).

Cardiac dysrhythmias

The bradycardia produced by digoxin can be enhanced by beta-adrenoceptor antagonists. Neostigmine enhances vagal activity and can aggravate bradycardia (418). An apparent interaction between sotalol and thiazide-induced hypokalemia, resulting in torsade de pointes (419), has prompted the withdrawal of the combination formulation Sotazide.

- The co-prescription of sotalol 80 mg bd with terfenadine 60 mg bd (both drugs that can prolong the QT interval) in a 71-year-old lady with hypertension, atrial fibrillation, and nasal congestion was complicated by recurrent torsade de pointes, causing dizzy spells and confusion after 8 days (420). She was treated with temporary pacing, but her symptoms resolved 72 hours after drug withdrawal.

Cardiac contractility

The negative inotropic effects of class I antidysrhythmic agents, such as disopyramide, procainamide, quinidine, and tocainide can be accentuated by beta-blockers; this is most pronounced in patients with pre-existing myocardial disease and can result in left ventricular failure or even asystole (421). Digoxin can obviate the negative inotropic effect of beta-blockers in patients with poor left ventricular function.

Adrenaline

Small quantities of adrenaline, such as are present as an additive in local anesthetic formulations, can be dangerously potentiated by beta-adrenoceptor blockers; propranolol should be discontinued at least 3 days in advance of administering such products for local anesthesia. A combined infusion of adrenaline and propranolol has been used for diagnosing insulin resistance, but it can evoke cardiac dysrhythmias, even in patients without signs of coronary disease (422).

Bepridil

The effect of a beta-blocker (metoprolol 30–40 mg/day or bisoprolol 2.5–5.0 mg/day for 1 month) on the change in QT interval, QT dispersion, and transmural dispersion of repolarization caused by bepridil has been studied in 10 patients with paroxysmal atrial fibrillation resistant to various antidysrhythmic drugs (423). Bepridil significantly prolonged the QTc interval from 0.42 to 0.50 seconds, QT dispersion from 0.07 to 0.14 seconds, and transmural dispersion of repolarization from 0.10 to 0.16 seconds. The addition of a beta-blocker shortened the QTc interval from 0.50 to 0.47 seconds, QTc dispersion from 0.14 to 0.06 seconds, and transmural dispersion of repolarization from 0.16 to 0.11 seconds. The authors therefore suggested that combined therapy with bepridil and a beta-blocker might be useful for intractable atrial fibrillation.

Bepridil does not interact with propranolol (424).

Calcium channel blockers

The greatest potential for serious mishap arises from interactions between calcium channel blockers (especially verapamil and related compounds) and beta-adrenoceptor antagonists (425,426). This combination can cause severe hypotension and cardiac failure, particularly in patients with poor myocardial function (427–429). The major risk appears to be associated with the intravenous administration of verapamil to patients who are already taking a beta-blocker (430), but a drug-like tiapamil, which closely resembles verapamil in its pharmacological profile, might be expected to carry a similar risk (431). Conversely, intravenous diltiazem does not produce deleterious hemodynamic effects in patients taking long-term propranolol (432). However, there have been instances when the combination of diltiazem with metoprolol caused sinus arrest and atrioventricular block (433).

The concurrent use of oral calcium channel blockers and beta-adrenoceptor antagonists in the management of angina pectoris or hypertension is less likely to result in heart block or other serious adverse effects (434), and these two drug groups are commonly used together. However, caution is still advised, and nifedipine or other dihydropyridine derivatives would be preferred in this type of combination (435,436,437). Nevertheless, the combination of nifedipine with atenolol in patients with stable intermittent claudication resulted in a reduction in walking distance and skin temperature, whereas either drug alone produced benefits (438).

Chlorpromazine

Lipophilic beta-adrenoceptor antagonists are metabolized to varying degrees by oxidation by liver microsomal cytochrome P450 (for example propranolol by CYP1A2 and CYP2D6 and metoprolol by CYP2D6). They can therefore reduce the clearance and increase the steady-state plasma concentrations of other drugs that undergo similar metabolism, potentiating their effects. Drugs that interact in this way include chlorpromazine (439).

- A schizophrenic patient experienced delirium, tonic-clonic seizures, and photosensitivity after the addition of propranolol to chlorpromazine, suggesting that chlorpromazine concentrations are increased by propranolol (440).

Although high dosages of propranolol (up to 2 g) have been used in combination with chlorpromazine to treat schizophrenia, the combination of propranolol or pindolol with chlorpromazine should be avoided if possible (441).

Diazepam

Lipophilic beta-adrenoceptor antagonists are metabolized to varying degrees by oxidation by liver microsomal cytochrome P450 (for example propranolol by CYP1A2 and CYP2D6 and metoprolol by CYP2D6). They can therefore reduce the clearance and increase the steady-state plasma concentrations of other drugs that undergo similar metabolism, potentiating their effects. Drugs that interact in this way include diazepam (442).

Flecainide

The combination of flecainide with propranolol results in additive hypotensive and negative inotropic effects (443).

Lipophilic beta-adrenoceptor antagonists are metabolized to varying degrees by oxidation by liver microsomal cytochrome P450 (for example propranolol by CYP1A2 and CYP2D6 and metoprolol by CYP2D6). They can therefore reduce the clearance and increase the steady-state plasma concentrations of other drugs that undergo similar metabolism, potentiating their effects. Drugs that interact in this way include flecainide (444).

Isoniazid

Lipophilic beta-adrenoceptor antagonists are metabolized to varying degrees by oxidation by liver microsomal cytochrome P450 (for example propranolol by CYP1A2 and CYP2D6 and metoprolol by CYP2D6). They can therefore reduce the clearance and increase the steady-state plasma concentrations of other drugs that undergo similar metabolism, potentiating their effects. Drugs that interact in this way include isoniazid (445).

Lidocaine

The combination of lidocaine with beta-adrenoceptor antagonists is associated with a slightly increased risk of some minor non-cardiac adverse events (dizziness, numbness, somnolence, confusion, slurred speech, and nausea and vomiting) (446). The combination is not associated with an increased risk of dysrhythmias.

Some beta-blockers reduce hepatic blood flow and inhibit microsomal enzymes, reducing the clearance of lidocaine; there is a clinically significant increase in the plasma concentration of lidocaine during concomitant propranolol therapy (447).

Neuroleptic drugs

The cardiac effects of neuroleptic drugs can be potentiated by propranolol (448).

In general, concurrent use of neuroleptic and antihypertensive drugs merits close patient monitoring (449).

Thioridazine

Lipophilic beta-adrenoceptor antagonists are metabolized to varying degrees by oxidation by liver microsomal cytochrome P450 (for example propranolol by CYP1A2 and CYP2D6 and metoprolol by CYP2D6). They can therefore reduce the clearance and increase the steady-state plasma concentrations of other drugs that undergo similar metabolism, potentiating their effects. Drugs that interact in this way include thioridazine (450).

The combination of propranolol or pindolol with thioridazine should be avoided if possible (451).

References

1. Cruickshank JM, Prichard BNC. Beta-blockers in Clinical PracticeEdinburgh: Churchill Livingstone;. 1988.
2. Brennan TA. Buying editorials. N Engl J Med 1994;331(10):673–5.
3. Croog SH, Levine S, Testa MA, Brown B, Bulpitt CJ, Jenkins CD, Klerman GL, Williams GH. The effects of antihypertensive therapy on the quality of life. N Engl J Med 1986;314(26):1657–64.
4. Anonymous. Fatigue as an unwanted effect of drugs. Lancet 1980;1(8181):1285–6.
5. Pearson SB, Banks DC, Patrick JM. The effect of beta-adrenoceptor blockade on factors affecting exercise tolerance in normal man. Br J Clin Pharmacol 1979;8(2):143–8.
6. Anderson SD, Bye PT, Perry CP, Hamor GP, Theobald G, Nyberg G. Limitation of work performance in normal adult males in the presence of beta-adrenergic blockade. Aust NZ J Med 1979;9(5):515–20.
7. Kaiser P. Running performance as a function of the dose–response relationship to beta-adrenoceptor blockade. Int J Sports Med 1982;3(1):29–32.
8. Kaijser L, Kaiser P, Karlsson J, Rossner S. Beta-blockers and running. Am Heart J 1980;100(6 Pt 1):943–4.
9. Bowman WC. Effect of adrenergic activators and inhibitors on the skeletal muscles. In: Szekeres L, editor. Adrenergic Activators and Inhibitors. Berlin: Springer Verlag, 1980:473.
10. Frisk-Holmberg M, Jorfeldt L, Juhlin-Dannfelt A. Metabolic effects in muscle during antihypertensive therapy with beta 1- and beta 1/beta 2-adrenoceptor blockers. Clin Pharmacol Ther 1981;30(5):611–8.
11. Koch G, Franz IW, Lohmann FW. Effects of short-term and long-term treatment with cardio-selective and non-selective beta-receptor blockade on carbohydrate and lipid metabolism and on plasma catecholamines at rest and during exercise. Clin Sci (Lond) 1981;61(Suppl 7):S433–5.
12. Franciosa JA, Johnson SM, Tobian LJ. Exercise performance in mildly hypertensive patients. Impairment by propranolol but not oxprenolol. Chest 1980;78(2):291–9.
13. McDevitt DG. Differential features of beta-adrenoceptor blocking drugs for therapy. In: Laragh J, Buhler F, editors. Frontiers in Hypertension Research. New York: Springer Verlag, 1981:473.
14. Woods PB, Robinson ML. An investigation of the comparative liposolubilities of beta-adrenoceptor blocking agents. J Pharm Pharmacol 1981;33(3):172–3.
15. Neil-Dwyer G, Bartlett J, McAinsh J, Cruickshank JM. Beta-adrenoceptor blockers and the blood–brain barrier. Br J Clin Pharmacol 1981;11(6):549–53.
16. McDevitt DG. Clinical significance of cardioselectivity: state of the art. Drugs 1983;25(Suppl 2):219.
17. McDevitt DG. Beta-adrenoceptor blocking drugs and partial agonist activity. Is it clinically relevant? Drugs 1983;25(4):331–8.
18. Cruickshank JM. Measurement and cardiovascular relevance of partial agonist activity (PAA) involving beta 1- and beta 2-adrenoceptors. Pharmacol Ther 1990;46(2):199–242.
19. Cruickshank JM. The xamoterol experience in the treatment of heart failure. Am J Cardiol 1993;71(9):C61–4.
20. McCaffrey PM, Riddell JG, Shanks RG. An assessment of the partial agonist activity of Ro 31–1118, flusoxolol and pindolol in man. Br J Clin Pharmacol 1987;24(5):571–80.
21. Prichard BN. Beta-blocking agents with vasodilating action. J Cardiovasc Pharmacol 1992;19(Suppl 1):S1–4.
22. Prichard BN, Owens CW. Mode of action of beta-adrenergic blocking drugs in hypertension. Clin Physiol Biochem 1990;8(Suppl 2):1–10.
23. Coltart DJ, Meldrum SJ, Hamer J. The effect of propranolol on the human and canine transmembrane action potential. Br J Pharmacol 1970;40(1):148P.
24. Barnett DB. Beta-blockers in heart failure: a therapeutic paradox. Lancet 1994;343(8897):557–8.
25. Waagstein F, Bristow MR, Swedberg K, Camerini F, Fowler MB, Silver MA, Gilbert EM, Johnson MR, Goss FG, Hjalmarson AMetoprolol in Dilated Cardiomyopathy (MDC) Trial Study Group. Beneficial effects of metoprolol in idiopathic dilated cardiomyopathy. Lancet 1993;342(8885):1441–6.
26. The Cardiac Insufficiency Bisoprolol Study II (CIBIS-II): a randomised trial. Lancet 1999;353(9146):9–13.
27. Merit-HF Study Group. Effect of metoprolol CR/XL in chronic heart failure: Metoprolol CR/XL Randomised Intervention Trial in Congestive Heart Failure (MERIT-HF). Lancet 1999;353(9169):2001–7.
28. Packer M, Colucci WS, Sackner-Bernstein JD, Liang CS, Goldscher DA, Freeman I, Kukin ML, Kinhal V, Udelson JE, Klapholz M, Gottlieb SS, Pearle D, Cody RJ, Gregory JJ, Kantrowitz NE, LeJemtel TH,

Young ST, Lukas MA, Shusterman NH. Double-blind, placebo-controlled study of the effects of carvedilol in patients with moderate to severe heart failure. The PRECISE Trial. Prospective Randomized Evaluation of Carvedilol on Symptoms and Exercise. Circulation 1996;94(11):2793–9.

29. Dougherty AH, Naccarelli GV, Gray EL, Hicks CH, Goldstein RA. Congestive heart failure with normal systolic function. Am J Cardiol 1984;54(7):778–82.
30. Wheeldon NM, MacDonald TM, Flucker CJ, McKendrick AD, McDevitt DG, Struthers AD. Echocardiography in chronic heart failure in the community. Q J Med 1993;86(1):17–23.
31. Clarkson P, Wheeldon NM, Macdonald TM. Left ventricular diastolic dysfunction. Q J Med 1994;87(3):143–8.
32. Wheeldon NM, Clarkson P, MacDonald TM. Diastolic heart failure. Eur Heart J 1994;15(12):1689–97.
33. Pouleur H. Diastolic dysfunction and myocardial energetics. Eur Heart J 1990;11(Suppl C):30–4.
34. Krumholz HM. Beta-blockers for mild to moderate heart failure. Lancet 1999;353(9146):2–3.
35. Sharpe N. Benefit of beta-blockers for heart failure: proven in 1999. Lancet 1999;353(9169):1988–9.
36. Gottlieb SS, McCarter RJ, Vogel RA. Effect of beta-blockade on mortality among high-risk and low-risk patients after myocardial infarction. N Engl J Med 1998;339(8):489–97.
37. MacDonald TM, Butler R, Newton RW, Morris AD. Which drugs benefit diabetic patients for secondary prevention of myocardial infarction? DARTS/MEMO Collaboration. Diabet Med 1998;15(4):282–9.
38. Frishman WH, Kowalski M, Nagnur S, Warshafsky S, Sica D. Cardiovascular considerations in using topical, oral, and intravenous drugs for the treatment of glaucoma and ocular hypertension: focus on beta-adrenergic blockade. Heart Dis 2001;3(6):386–97.
39. Girkin CA. Neuroprotection: does it work for any neurological diseases? Ophthalmic Pract 2001;19:298–302.
40. Van De Ven LLM, Spanjaard JN, De Jongste MJL, Hillege H, Verkenne P, Van Gilst WH, Lie KI. Safety of beta-blocker therapy with and without thrombolysis: A comparison or bisoprolol and atenolol in acute myocardial infarction. Curr Ther Res Clin Exp 1996;57:313.
41. Thomas JA, Marks BH. Plasma norepinephrine in congestive heart failure. Am J Cardiol 1978;41(2):233–43.
42. Greenblatt DJ, Koch-Weser J. Clinical toxicity of propranolol and practolol: a report from the Boston Collaborative Drug Surveillance Program. In: Avery GS, editor. Cardiovascular Drugs Vol 2. Sydney: Adis Press, 1977:179 Beta-Adrenoceptor Blocking Drugs, Chapter VIII.
43. Aellig WH. Pindolol—a beta-adrenoceptor blocking drug with partial agonist activity: clinical pharmacological considerations. Br J Clin Pharmacol 1982;13(Suppl 2):S187–92.
44. Imhof P. The significance of beta1-beta2-selectivity and intrinsic sympathomimetic activity in beta-blockers, with particular reference to antihypertensive treatment. Adv Clin Pharmacol 1976;11:26–32.
45. Davies B, Bannister R, Mathias C, Sever P. Pindolol in postural hypotension: the case for caution. Lancet 1981;2(8253):982–3.
46. Anonymous. Xamoterol: stabilising the cardiac beta receptor? Lancet 1988;2(8625):1401–2.
47. Anonymous. New evidence on xamoterol. Lancet 1990;336(8706):24.
48. Frais MA, Bayley TJ. Left ventricular failure with labetalol. Postgrad Med J 1979;55(646):567–8.
49. Britman NA. Cardiac effects of topical timolol. N Engl J Med 1979;300(10):566.
50. Hery E, Jourdain P, Funck F, Bellorini M, Loiret J, Thebault B, Guillard N, El Hallak A, Desnos M. Prediction of intolerance to beta blocker therapy in chronic heart failure patients using BNP. Ann Cardiol Angeiol 2004;53:298–304.
51. Wikstrand J, Berglund G. Antihypertensive treatment with beta-blockers in patients aged over 65. BMJ (Clin Res Ed) 1982;285(6345):850.
52. Montoliu J, Botey A, Darnell A, Revert L. Hipotension prolongada tras la primera dosis de atenolol. [Prolonged hypotension after the first dose of atenolol.] Med Clin (Barc) 1981;76(8):365–6.
53. McNeil JJ, Louis WJ. A double-blind crossover comparison of pindolol, metoprolol, atenolol and labetalol in mild to moderate hypertension. Br J Clin Pharmacol 1979;8(Suppl 2):S163–6.
54. Cruickshank JM. Beta-blockers, bradycardia and adverse effects. Acta Ther 1981;7:309.
55. Anastasiou-Nana MI, Anderson JL, Askins JC, Gilbert EM, Nanas JN, Menlove RL. Long-term experience with sotalol in the treatment of complex ventricular arrhythmias. Am Heart J 1987;114(2):288–96.
56. Obel IW, Jardine R, Haitus B, Millar RN. Efficacy of oral sotalol in reentrant ventricular tachycardia. Cardiovasc Drugs Ther 1990;4(Suppl 3):613–8.
57. Griffith MJ, Linker NJ, Garratt CJ, Ward DE, Camm AJ. Relative efficacy and safety of intravenous drugs for termination of sustained ventricular tachycardia. Lancet 1990;336(8716):670–3.
58. Juul-Moller S, Edvardsson N, Rehnqvist-Ahlberg N. Sotalol versus quinidine for the maintenance of sinus rhythm after direct current conversion of atrial fibrillation. Circulation 1990;82(6):1932–9.
59. Desoutter P, Medioni J, Lerasle S, Haiat R. Bloc auriculo-ventriculaire et torsade de pointes après surdosage par le sotalol. [Atrioventricular block and torsade de pointes following sotalol overdose.] Nouv Presse Méd 1982;11(52):3855.
60. Belton P, Sheridan J, Mulcahy R. A case of sotalol poisoning. Ir J Med Sci 1982;151(4):126–7.
61. Pathy MS. Acute central chest pain in the elderly. A review of 296 consecutive hospital admissions during 1976 with particular reference to the possible role of beta-adrenergic blocking agents in inducing substernal pain. Am Heart J 1979;98(2):168–70.
62. Warren V, Goldberg E. Intractable angina pectoris. Combined therapy with propranolol and permanent pervenous pacemaker. JAMA 1976;235(8):841–2.
63. Robertson RM, Wood AJ, Vaughn WK, Robertson D. Exacerbation of vasotonic angina pectoris by propranolol. Circulation 1982;65(2):281–5.
64. McMahon MT, McPherson MA, Talbert RL, Greenberg B, Sheaffer SL. Diagnosis and treatment of Prinzmetal's variant angina. Clin Pharm 1982;1(1):34–42.
65. Aubran M, Trigano JA, Allard-Laour G, Ebagosti A, Torresani J. Angor accéléré sous béta-bloquants. [Angina accelerated under betablockers.] Ann Cardiol Angeiol (Paris) 1986;35(2):99–101.
66. Hall PE, Kendall MJ, Smith SR. Beta blockers and fatigue. J Clin Hosp Pharm 1984;9(4):283–91.
67. Feleke E, Lyngstam O, Rastam L, Ryden L. Complaints of cold extremities among patients on antihypertensive treatment. Acta Med Scand 1983;213(5):381–5.
68. Greminger P, Vetter H, Boerlin JH, Havelka J, Baumgart P, Walger P, Lüscher T, Siegenthaler W,

Vetter W. A comparative study between 100 mg atenolol and 20 mg pindolol slow-release in essential hypertension Drugs 1983;25(Suppl 2):37–41.

69. Eliasson K, Danielson M, Hylander B, Lindblad LE. Raynaud's phenomenon caused by beta-receptor blocking drugs. Improvement after treatment with a combined alpha- and beta-blocker. Acta Med Scand 1984;215(4):333–9.
70. Steiner JA, Cooper R, Gear JS, Ledingham JG. Vascular symptoms in patients with primary Raynaud's phenomenon are not exacerbated by propranolol or labetalol. Br J Clin Pharmacol 1979;7(4):401–3.
71. Reichert N, Shibolet S, Adar R, Gafni J. Controlled trial of propranolol in intermittent claudication. Clin Pharmacol Ther 1975;17(5):612–5.
72. Breckenridge A. Which beta blocker? BMJ (Clin Res Ed) 1983;286(6371):1085–8.
73. Lepantalo M. Chronic effects of labetalol, pindolol, and propranolol on calf blood flow in intermittent claudication. Clin Pharmacol Ther 1985;37(1):7–12.
74. Solomon SA, Ramsay LE, Yeo WW, Parnell L, Morris-Jones W. Beta-blockade and intermittent claudication: placebo controlled trial of atenolol and nifedipine and their combination. BMJ 1991;303(6810):1100–4.
75. Gokal R, Dornan TL, Ledingham JGG. Peripheral skin necrosis complicating beta-blockage. BMJ 1979;1(6165):721–2.
76. Hoffbrand BI. Peripheral skin necrosis complicating beta-blockade. BMJ 1979;1(6170):1082.
77. Rees PJ. Peripheral skin necrosis complicating beta-blockade. BMJ 1979;1(6168):955.
78. O'Rourke DA, Donohue MF, Hayes JA. Beta-blockers and peripheral gangrene. Med J Aust 1979;2(2):88.
79. Fogoros RN. Exacerbation of intermittent claudication by propranolol. N Engl J Med 1980;302(19):1089.
80. Stringer MD, Bentley PG. Peripheral gangrene associated with beta-blockade. Br J Surg 1986;73(12):1008.
81. Dompmartin A, Le Maitre M, Letessier D, Leroy D. Nécrose digitales sous béta-bloquants. [Digital necroses induced by beta-blockers.] Ann Dermatol Venereol 1988;115(5):593–6.
82. Price HL, Cooperman LH, Warden JC. Control of the splanchnic circulation in man. Role of beta-adrenergic receptors. Circ Res 1967;21(3):333–40.
83. Schneider R. Do beta-blockers cause mesenteric ischemia? J Clin Gastroenterol 1986;8(2):109–10.
84. Diggory P, Cassels-Brown A, Vail A, Hillman JS. Randomised, controlled trial of spirometric changes in elderly people receiving timolol or betaxolol as initial treatment for glaucoma. Br J Ophthalmol 1998;82(2): 146–9.
85. McNeill RS. Effect of a beta-adrenergic-blocking agent, propranolol, on asthmatics. Lancet 1964;13:1101–2.
86. Harries AD. Beta-blockade in asthma. BMJ (Clin Res Ed) 1981;282(6272):1321.
87. Australian Adverse Drug Reactions Advisory Committee. Beta-blockers. Med J Aust 1980;2:130.
88. Raine JM, Palazzo MG, Kerr JH, Sleight P. Near-fatal bronchospasm after oral nadolol in a young asthmatic and response to ventilation with halothane. BMJ (Clin Res Ed) 1981;282(6263):548–9.
89. McMahon CD, Shaffer RN, Hoskins HD Jr, Hetherington J Jr. Adverse effects experienced by patients taking timolol. Am J Ophthalmol 1979;88(4):736–8.
90. Mustchin CP, Gribbin HR, Tattersfield AE, George CF. Reduced respiratory responses to carbon dioxide after propranolol: a central action. BMJ 1976;2(6046):1229–31.
91. Trembath PW, Taylor EA, Varley J, Turner P. Effect of propranolol on the ventilatory response to hypercapnia in man. Clin Sci (Lond) 1979;57(5):465–8.
92. Chang LC. Use of practolol in asthmatics: a plea for caution. Lancet 1971;2(7719):321.
93. Waal-Manning HJ, Simpson FO. Practolol treatment in asthmatics. Lancet 1971;2(7736):1264–5.
94. Harris LS, Greenstein SH, Bloom AF. Respiratory difficulties with betaxolol. Am J Ophthalmol 1986;102(2): 274–5.
95. Diggory P, Cassels-Brown A, Vail A, Abbey LM, Hillman JS. Avoiding unsuspected respiratory side-effects of topical timolol with cardioselective or sympathomimetic agents. Lancet 1995;345(8965):1604–6.
96. Formgrein H. The effect of metoprolol and practolol on lung function and blood pressure in hypertensive asthmatics. Br J Clin Pharmacol 1976;3:1007.
97. Wilcox PG, Ahmad D, Darke AC, Parsons J, Carruthers SG. Respiratory and cardiac effects of metoprolol and bevantolol in patients with asthma. Clin Pharmacol Ther 1986;39(1):29–34.
98. Nordstrom LA, MacDonald F, Gobel FL. Effect of propranolol on respiratory function and exercise tolerance in patients with chronic obstructive lung disease. Chest 1975;67(3):287–92.
99. Fraley DS, Bruns FJ, Segel DP, Adler S. Propranolol-related bronchospasm in patients without history of asthma. South Med J 1980;73(2):238–40.
100. Mue S, Sasaki T, Shibahara S, Takahashi M, Ohmi T, Yamauchi K, Suzuki S, Hida W, Takishima T. Influence of metoprolol on hemodynamics and respiratory function in asthmatic patients. Int J Clin Pharmacol Biopharm 1979;17(8):346–50.
101. Assaykeen TA, Michell G. Metoprolol in hypertension: an open evaluation. Med J Aust 1982;1(2):73–7.
102. Jackson SH, Beevers DG. Comparison of the effects of single doses of atenolol and labetalol on airways obstruction in patients with hypertension and asthma. Br J Clin Pharmacol 1983;15(5):553–6.
103. Ellis ME, Sahay JN, Chatterjee SS, Cruickshank JM, Ellis SH. Cardioselectivity of atenolol in asthmatic patients. Eur J Clin Pharmacol 1981;21(3):173–6.
104. Committee on Safety of Medicines. Fatal bronchospasm associated with beta-blockers. Curr Probl 1987;20:2.
105. van Zyl AI, Jennings AA, Bateman ED, Opie LH. Comparison of respiratory effects of two cardioselective beta-blockers, celiprolol and atenolol, in asthmatics with mild to moderate hypertension. Chest 1989;95(1):209–13.
106. Waal-Manning HJ, Simpson FO. Safety of celiprolol in hypertensives with chronic obstructive respiratory disease. NZ Med J 1990;103:222.
107. Anonymous. Celiprolol—a better beta blocker? Drug Ther Bull 1992;30(9):35–6.
108. Malik A, Memon AM. Beta blocker eye drops related airway obstruction. J Pak Med Assoc 2001;51(5):202–4.
109. Tattersfield AE. Respiratory function in the elderly and the effects of beta blockade. Cardiovasc Drugs Ther 1991;4(Suppl 6):1229–32.
110. Campbell SC, Lauver GL, Cobb RB Jr. Central ventilatory depression by oral propranolol. Clin Pharmacol Ther 1981;30(6):758–64.
111. Davis WM, Hatoum NS. Lethal synergism between morphine or other narcotic analgesics and propranolol. Toxicology 1979;14(2):141–51.
112. Erwteman TM, Braat MC, van Aken WG. Interstitial pulmonary fibrosis: a new side effect of practolol. BMJ 1977;2(6082):297–8.

113. Marshall AJ, Eltringham WK, Barritt DW, Davies JD, Griffiths DA, Jackson LK, Laszlo G, Read AE. Respiratory disease associated with practolol therapy. Lancet 1977;2(8051):1254–7.
114. Musk AW, Pollard JA. Pindolol and pulmonary fibrosis. BMJ 1979;2(6190):581–2.
115. Wood GM, Bolton RP, Muers MF, Losowsky MS. Pleurisy and pulmonary granulomas after treatment with acebutolol. BMJ (Clin Res Ed) 1982;285(6346):936.
116. Akoun GM, Herman DP, Mayaud CM, Perrot JY. Acebutolol-induced hypersensitivity pneumonitis. BMJ (Clin Res Ed) 1983;286(6361):266–7.
117. Akoun GM, Touboul JL, Mayaud CM, et al. Pneumopathie d'hypersensibilité à l'acébutolol: données en faveur d'un mécanisme immunologique et médiation cellulaire. Rev Fr Allergol 1985;75:85.
118. Das G, Ferris JC. Generalized convulsions in a patient receiving ultrashort-acting beta-blocker infusion. Drug Intell Clin Pharm 1988;22(6):484–5.
119. Fleminger R. Visual hallucinations and illusions with propranolol. BMJ 1978;1(6121):1182.
120. Greenblatt DJ, Shader RI. On the psychopharmacology of beta adrenergic blockade. Curr Ther Res Clin Exp 1972;14(9):615–25.
121. Faldt R, Liedholm H, Aursnes J. Beta blockers and loss of hearing. BMJ (Clin Res Ed) 1984;289(6457):1490–2.
122. Weber JC. Beta-adrenoreceptor antagonists and diplopia. Lancet 1982;2(8302):826–7.
123. Turkewitz LJ, Sahgal V, Spiro A. Propranolol-induced myotonia. Mt Sinai J Med 1984;51(2):207.
124. Robson RH. Recurrent migraine after propranolol. Br Heart J 1977;39(10):1157–8.
125. Prendes JL. Considerations on the use of propranolol in complicated migraine. Headache 1980;20(2):93–5.
126. Gilbert GJ. An occurrence of complicated migraine during propranolol therapy. Headache 1982;22(2):81–3.
127. Bardwell A, Trott JA. Stroke in migraine as a consequence of propranolol. Headache 1987;27(7):381–3.
128. Leys D, Pasquier F, Vermersch P, Gosset D, Michiels H, Kassiotis P, Petit H. Possible revelation of latent myasthenia gravis by labetalol chlorhydrate. Acta Clin Belg 1987;42(6):475–6.
129. Komar J, Szalay M, Szel I. Myasthenische Episode nach Einnahme grosser Mengen Beta-blocker. [A myasthenic episode following intake of large amounts of a beta blocker.] Fortschr Neurol Psychiatr 1987;55(6):201–2.
130. Emara MK, Saadah AM. The carpal tunnel syndrome in hypertensive patients treated with beta-blockers. Postgrad Med J 1988;64(749):191–2.
131. Hod H, Har-Zahav J, Kaplinsky N, Frankl O. Pindolol-induced tremor. Postgrad Med J 1980;56(655):346–7.
132. Palomeras E, Sanz P, Cano A, Fossas P. Dystonia in a patient treated with propranolol and gabapentin. Arch Neurol 2000;57(4):570–1.
133. Medical Research Council Working Party on Mild to Moderate Hypertension. Adverse reactions to bendrofluazide and propranolol for the treatment of mild hypertension. Lancet 1981;2(8246):539–43.
134. Bengtsson C, Lennartsson J, Lindquist O, Noppa H, Sigurdsson J. Sleep disturbances, nightmares and other possible central nervous disturbances in a population sample of women, with special reference to those on antihypertensive drugs. Eur J Clin Pharmacol 1980;17(3):173–7.
135. Kostis JB, Rosen RC. Central nervous system effects of beta-adrenergic-blocking drugs: the role of ancillary properties. Circulation 1987;75(1):204–12.
136. Patel L, Turner P. Central actions of beta-adrenoceptor blocking drugs in man. Med Res Rev 1981;1(4):387–410.
137. Betts TA, Alford C. Beta-blockers and sleep: a controlled trial. Eur J Clin Pharmacol 1985;28(Suppl):65–8.
138. McAinsh J, Cruickshank JM. Beta-blockers and central nervous system side effects. Pharmacol Ther 1990;46(2):163–97.
139. Dimsdale JE, Newton RP, Joist T. Neuropsychological side effects of beta-blockers. Arch Intern Med 1989;149(3):514–25.
140. Dahlof C, Dimenas E. Side effects of beta-blocker treatments as related to the central nervous system. Am J Med Sci 1990;299(4):236–44.
141. Braun P, Reker K, Friedel B, et al. Driving tests with beta-receptor blockers. Blutalkohol 1979;16:495.
142. Betts T. Effects of beta blockade on driving. Aviat Space Environ Med 1981;52(11 Pt 2):S40–5.
143. Panizza D, Lecasble M. Effect of atenolol on car drivers in a prolonged stress situation. Eur J Clin Pharmacol 1985;28(Suppl):97–9.
144. Broadhurst AD. The effect of propranolol on human psychomotor performance. Aviat Space Environ Med 1980;51(2):176–9.
145. Fiore PM, Jacobs IH, Goldberg DB. Drug-induced pemphigoid. A spectrum of diseases. Arch Ophthalmol 1987;105(12):1660–3.
146. Jain S. Betaxolol-associated anterior uveitis. Eye 1994;8(Pt 6):708–9.
147. Schultz JS, Hoenig JA, Charles H. Possible bilateral anterior uveitis secondary to metipranolol (optipranolol) therapy. Arch Ophthalmol 1993;111(12):1606–7.
148. O'Connor GR. Granulomatous uveitis and metipranolol. Br J Ophthalmol 1993;77(8):536–8.
149. Fraunfelder FT. Drug-induced ocular side effects. Folia Ophthalmol Jpn 1996;47:770.
150. Kellner U, Kraus H, Foerster MH. Multifocal ERG in chloroquine retinopathy: regional variance of retinal dysfunction. Graefes Arch Clin Exp Ophthalmol 2000;238(1):94–7.
151. Bryan PC, Efiong DO, Stewart-Jones J, Turner P. Propranolol on tests of visual function and central nervous activity. Br J Clin Pharmacol 1974;1:82.
152. Glaister DH, Harrison MH, Allnutt MF. Environmental influences on cardiac activity. In: Burley DM, Frier JH, Rondel RK, Taylor SH, editors. New Perspectives in Beta-blockade. Horsham, UK: Ciba Laboratories, 1973:241.
153. Landauer AA, Pocock DA, Prott FW. Effects of atenolol and propranolol on human performance and subjective feelings. Psychopharmacology (Berl) 1979;60(2):211–5.
154. Salem SA, McDevitt DG. Central effects of beta-adrenoceptor antagonists. Clin Pharmacol Ther 1983;33(1):52–7.
155. Ogle CW, Turner P, Markomihelakis H. The effects of high doses of oxprenolol and of propranolol on pursuit rotor performance, reaction time and critical flicker frequency. Psychopharmacologia 1976;46(3):295–9.
156. Turner P, Hedges A. An investigation of the central effects of oxprenolol. In: Burley DM, Frier JH, Rondel RK, Taylor SH, editors. New Perspectives in Beta-blockade. Horsham, UK: Ciba Laboratories, 1973:269.
157. Tyrer PJ, Lader MH. Response to propranolol and diazepam in somatic and psychic anxiety. BMJ 1974;2(909):14–6.
158. Greil W. Central nervous system effects. Curr Ther Res 1980;28:106.
159. Currie D, Lewis RV, McDevitt DG, Nicholson AN, Wright NA. Central effects of beta-adrenoceptor antagonists. I—Performance and subjective assessments of mood. Br J Clin Pharmacol 1988;26(2):121–8.

160. Nicholson AN, Wright NA, Zetlein MB, Currie D, McDevitt DG. Central effects of beta-adrenoceptor antagonists. II—Electroencephalogram and body sway. Br J Clin Pharmacol 1988;26(2):129–41.
161. Hearing SD, Wesnes KA, Bowman CE. Beta blockers and cognitive function in elderly hypertensive patients: withdrawal and consequences of ACE inhibitor substitution. Int J Geriatr Psychopharmacol 1999;2:13–7.
162. Perez-Stable EJ, Halliday R, Gardiner PS, Baron RB, Hauck WW, Acree M, Coates TJ. The effects of propranolol on cognitive function and quality of life: a randomized trial among patients with diastolic hypertension. Am J Med 2000;108(5):359–65.
163. Yatham LN, Lint D, Lam RW, Zis AP. Adverse effects of pindolol augmentation in patients with bipolar depression. J Clin Psychopharmacol 1999;19(4):383–4.
164. Russell JW, Schuckit NA. Anxiety and depression in patient on nadolol. Lancet 1982;2(8310):1286–7.
165. Thiessen BQ, Wallace SM, Blackburn JL, Wilson TW, Bergman U. Increased prescribing of antidepressants subsequent to beta-blocker therapy. Arch Intern Med 1990;150(11):2286–90.
166. Topliss D, Bond R. Acute brain syndrome after propranolol treatment. Lancet 1977;2(8048):1133–4.
167. Helson L, Duque L. Acute brain syndrome after propranolol. Lancet 1978;1(8055):98.
168. Kurland ML. Organic brain syndrome with propranolol. N Engl J Med 1979;300(7):366.
169. Gershon ES, Goldstein RE, Moss AJ, van Kammen DP. Psychosis with ordinary doses of propranolol. Ann Intern Med 1979;90(6):938–9.
170. Kuhr BM. Prolonged delirium with propanolol. J Clin Psychiatry 1979;40(4):198–9.
171. McGahan DJ, Wojslaw A, Prasad V, Blankenship S. Propranolol-induced psychosis. Drug Intell Clin Pharm 1984;18(7–8):601–3.
172. Steinhert J, Pugh CR. Two patients with schizophrenic-like psychosis after treatment with beta-adrenergic blockers. BMJ 1979;1(6166):790.
173. Lee ST. Hyperprolactinemia, galactorrhea, and atenolol. Ann Intern Med 1992;116(6):522.
174. Harrower AD, Fyffe JA, Horn DB, Strong JA. Thyroxine and triiodothyronine levels in hyperthyroid patients during treatment with propranolol. Clin Endocrinol (Oxf) 1977;7(1):41–4.
175. Cooper DS, Daniels GH, Ladenson PW, Ridgway EC. Hyperthyroxinemia in patients treated with high-dose propranolol. Am J Med 1982;73(6):867–71.
176. Mooradian A, Morley JE, Simon G, Shafer RB. Propranolol-induced hyperthyroxinemia. Arch Intern Med 1983;143(11):2193–5.
177. Heyma P, Larkins RG, Higginbotham L, Ng KW. D-propranolol and DL-propranolol both decrease conversion of L-thyroxine to L-triiodothyronine. BMJ 1980;281(6232):24–5.
178. Jones DK, Solomon S. Thyrotoxic crisis masked by treatment with beta-blockers. BMJ (Clin Res Ed) 1981;283(6292):659.
179. Gold LA, Merimee TJ, Misbin RI. Propranolol and hypoglycemia: the effects of beta-adrenergic blockade on glucose and alanine levels during fasting. J Clin Pharmacol 1980;20(1):50–8.
180. Uusitupa M, Aro A, Pietikainen M. Severe hypoglycaemia caused by physical strain and pindolol therapy. A case report. Ann Clin Res 1980;12(1):25–7.
181. Holm G, Herlitz J, Smith U. Severe hypoglycaemia during physical exercise and treatment with beta-blockers. BMJ (Clin Res Ed) 1981;282(6273):1360.
182. Zarate A, Gelfand M, Novello A, Knepshield J, Preuss HG. Propranolol-associated hypoglycemia in patients on maintenance hemodialysis. Int J Artif Organs 1981;4(3):130–4.
183. Belton P, O'Dwyer WF, Carmody M, Donohoe J. Propranolol associated hypoglycaemia in non-diabetics. Ir Med J 1980;73(4):173.
184. Chavez H, Ozolins D, Losek JD. Hypoglycemia and propranolol in pediatric behavioral disorders. Pediatrics 1999;103(6 Pt 1):1290–2.
185. Barnett AH, Leslie D, Watkins PJ. Can insulin-treated diabetics be given beta-adrenergic blocking drugs? BMJ 1980;280(6219):976–8.
186. Wright AD, Barber SG, Kendall MJ, Poole PH. Beta-adrenoceptor-blocking drugs and blood sugar control in diabetes mellitus. BMJ 1979;1(6157):159–61.
187. Davidson NM, Corrall RJ, Shaw TR, French EB. Observations in man of hypoglycaemia during selective and non-selective beta-blockade. Scott Med J 1977;22(1):69–72.
188. Deacon SP, Barnett D. Comparison of atenolol and propranolol during insulin-induced hypoglycaemia. BMJ 1976;2(6030):272–3.
189. Blohme G, Lager I, Lonnroth P, Smith U. Hypoglycemic symptoms in insulin-dependent diabetics. A prospective study of the influence of beta-blockade. Diabete Metab 1981;7(4):235–8.
190. Shepherd AM, Lin MS, Keeton TK. Hypoglycemia-induced hypertension in a diabetic patient on metoprolol. Ann Intern Med 1981;94(3):357–8.
191. Malmberg K, Ryden L, Hamsten A, Herlitz J, Waldenstrom A, Wedel H. Mortality prediction in diabetic patients with myocardial infarction: experiences from the DIGAMI study. Cardiovasc Res 1997;34(1):248–53.
192. Samuelsson O, Hedner T, Berglund G, Persson B, Andersson OK, Wilhelmsen L. Diabetes mellitus in treated hypertension: incidence, predictive factors and the impact of non-selective beta-blockers and thiazide diuretics during 15 years treatment of middle-aged hypertensive men in the Primary Prevention Trial Goteborg, Sweden. J Hum Hypertens 1994;8(4):257–63.
193. Bakris GL, Fonseca V, Katholi RE, McGill JB, Messerli FH, Phillips RA, Raskin P, Wright JT, Oakes R, Lukas MA, Anderson KM, Bell DSH, for the GEMINI Investigators. Metabolic effects of carvedilol vs metoprolol in patients with type 2 diabetes mellitus and hypertension. A randomized controlled trial. JAMA 2004;292:2227–36.
194. Van Brammelen P. Lipid changes induced by beta-blockers. Curr Opin Cardiol 1988;3:513.
195. Bielmann P, Leduc G, Jequier JC, et al. Changes in the lipoprotein composition after chronic administration of metoprolol and propranolol in hypertriglyceridemic-hypertensive subjects. Curr Ther Res 1981;30:956.
196. Gavalas C, Costantino O, Zuppardi E, Scaramucci S, Doronzo E, Aharrh-Gnama A, Nubile M, Di Nuzzo S, De Nicola GC. Variazioni della colesterolemia in pazienti sottoposti a terapia topica con il timololo. Ann Ottalmol Clin Ocul 2001;127:9–14.
197. Northcote RJ. Beta blockers, lipids, and coronary atherosclerosis: fact or fiction? BMJ (Clin Res Ed) 1988;296(6624):731–2.
198. Neaton JD, Grimm RH Jr, Prineas RJ, Stamler J, Grandits GA, Elmer PJ, Cutler JA, Flack JM, Schoenberger JA, McDonald R, et al. Treatment of Mild Hypertension Study. Final results. Treatment of Mild Hypertension Study Research Group. JAMA 1993; 270(6):713–24.

199. Astrup AV. Fedme og diabetes som bivirkninger til beta-blockkere. [Obesity and diabetes as side-effects of beta-blockers.] Ugeskr Laeger 1990;152(40):2905–8.
200. Connacher AA, Jung RT, Mitchell PE. Weight loss in obese subjects on a restricted diet given BRL 26830A, a new atypical beta adrenoceptor agonist. BMJ (Clin Res Ed) 1988;296(6631):1217–20.
201. Wheeldon NM, McDevitt DG, McFarlane LC, Lipworth BJ. Do beta 3-adrenoceptors mediate metabolic responses to isoprenaline. Q J Med 1993;86(9):595–600.
202. Emorine LJ, Marullo S, Briend-Sutren MM, Patey G, Tate K, Delavier-Klutchko C, Strosberg AD. Molecular characterization of the human beta 3-adrenergic receptor. Science 1989;245(4922):1118–21.
203. Sharma AM, Pischon T, Hardt S, Kunz I, Luft FC. Hypothesis: Beta-adrenergic receptor blockers and weight gain: a systematic analysis. Hypertension 2001; 37(2):250–4.
204. Saunders J, Prestwich SA, Avery AJ, Kilborn JR, Morselli PL, Sonksen PH. The effect of non-selective and selective beta-1-blockade on the plasma potassium response to hypoglycaemia. Diabete Metab 1981; 7(4):239–42.
205. Arnold JMO, Shanks RG, McDevitt DG. Beta-adrenoceptor antagonism of isoprenaline induced metabolic changes in man. Br J Clin Pharmacol 1983;16:621P.
206. Skehan JD, Barnes JN, Drew PJ, Wright P. Hypokalaemia induced by a combination of a beta-blocker and a thiazide. BMJ (Clin Res Ed) 1982;284(6309):83.
207. Walters EG, Horswill CE, Shelton JR, Ali Akbar F. Hazards of beta-blocker/diuretic tablets. Lancet 1985; 2(8448):220–1.
208. Odugbesan O, Chesner IM, Bailey G, Barnett AH. Hazards of combined beta-blocker/diuretic tablets. Lancet 1985;1(8439):1221–2.
209. Pedersen OL, Mikkelsen E. Serum potassium and uric acid changes during treatment with timolol alone and in combination with a diuretic. Clin Pharmacol Ther 1979;26(3):339–43.
210. Bushe CJ. Does atenolol have an effect on calcium metabolism? BMJ (Clin Res Ed) 1987;294(6583):1324–5.
211. Freestone S, MacDonald TM. Does atenolol have an effect on calcium metabolism? BMJ (Clin Res Ed) 1987;295(6589):53.
212. Dodds WN, Davidson RJ. Thrombocytopenia due to slow-release oxprenolol. Lancet 1978;2(8091):683.
213. Hare DL, Hicks BH. Thrombocytopenia due to oxprenolol. Med J Aust 1979;2(5):259.
214. Caviet NL, Klaassen CH. Trombocuytopenie veroorzaakt door alprenolol. [Thrombocytopenia caused by alprenolol.] Ned Tijdschr Geneeskd 1979;123(1):18–20.
215. Magnusson B, Rodjer S. Alprenolol-induced thrombocytopenia. Acta Med Scand 1980;207(3):231–3.
216. Kelly JP, Kaufman DW, Shapiro S. Risks of agranulocytosis and aplastic anemia in relation to the use of cardiovascular drugs: The International Agranulocytosis and Aplastic Anemia Study. Clin Pharmacol Ther 1991; 49(3):330–41.
217. Nawabi IU, Ritz ND. Agranulocytosis due to propranolol. JAMA 1973;223(12):1376–7.
218. Robinson JD, Burtner DE. Severe diarrhea secondary to propranolol. Drug Intell Clin Pharm 1981;15(1):49–50.
219. Wolfhagen FH, van Neerven JA, Groen FC, Ouwendijk RJ. Severe nausea and vomiting with timolol eye drops. Lancet 1998;352(9125):373.
220. Carlborg B, Kumlien A, Olsson H. Medikamentella esofagusstrikturen. [Drug-induced esophageal strictures.] Lakartidningen 1978;75(49):4609–11.
221. Windsor WO, Durrein F, Dyer NH. Fibrinous peritonitis: a complication of practolol therapy. BMJ 1975;2(5962):68.
222. Eltringham WK, Espiner HJ, Windsor CW, Griffiths DA, Davies JD, Baddeley H, Read AE, Blunt RJ. Sclerosing peritonitis due to practolol: a report on 9 cases and their surgical management. Br J Surg 1977;64(4):229–35.
223. Marshall AJ, Baddeley H, Barritt DW, Davies JD, Lee RE, Low-Beer TS, Read AE. Practolol peritonitis. A study of 16 cases and a survey of small bowel function in patients taking beta adrenergic blockers. Q J Med 1977;46(181):135–49.
224. Ahmad S. Sclerosing peritonitis and propranolol. Chest 1981;79(3):361–2.
225. Nillson BV, Pederson KG. Sclerosing peritonitis associated with atenolol. BMJ (Clin Res Ed) 1985;290:518.
226. McClusky DR, Donaldson RA, McGeown MG. Oxprenolol and retroperitoneal fibrosis. BMJ 1980;281(6253):1459–60.
227. Johnson JN, McFarland J. Retroperitoneal fibrosis associated with atenolol. BMJ 1980;280(6217):864.
228. Pierce JR Jr, Trostle DC, Warner JJ. Propranolol and retroperitoneal fibrosis. Ann Intern Med 1981;95(2):244.
229. Thompson J, Julian DG. Retroperitoneal fibrosis associated with metoprolol. BMJ (Clin Res Ed) 1982;284(6309):83–4.
230. Laakso M, Arvala I, Tervonen S, Sotarauta M. Retroperitoneal fibrosis associated with sotalol. BMJ (Clin Res Ed) 1982;285(6348):1085–6.
231. Rimmer E, Richens A, Forster ME, Rees RW. Retroperitoneal fibrosis associated with timolol. Lancet 1983;1(8319):300.
232. Benitah E, Chatelain C, Cohen F, Herman D. Fibrose retroperitonéale: effet systémique d'un collyre béta-bloquant?. [Retroperitoneal fibrosis: a systemic effect of beta-blocker eyedrops?.] Presse Méd 1987;16(8):400–1.
233. Bullimore DW. Retroperitoneal fibrosis associated with atenolol. BMJ 1980;281(6239):564.
234. Pryor JP, Castle WM, Dukes DC, Smith JC, Watson ME, Williams JL. Do beta-adrenoceptor blocking drugs cause retroperitoneal fibrosis? BMJ (Clin Res Ed) 1983;287(6393):639–41.
235. Mahgoub A, Idle JR, Dring LG, Lancaster R, Smith RL. Polymorphic hydroxylation of debrisoquine in man. Lancet 1977;2(8038):584–6.
236. Smith RL. Polymorphic metabolism of the beta-adrenoreceptor blocking drugs and its clinical relevance. Eur J Clin Pharmacol 1985;28(Suppl):77–84.
237. Sherlock S. In: Diseases of the Liver and Biliary System. 6th edn.. Oxford: Blackwell Scientific Publications, 1981:163.
238. Conn HO. Propranolol in the treatment of portal hypertension: a caution. Hepatology 1982;2(5):641–4.
239. Hayes PC, Shepherd AN, Bouchier IA. Medical treatment of portal hypertension and oesophageal varices. BMJ (Clin Res Ed) 1983;287(6394):733–6.
240. Tarver D, Walt RP, Dunk AA, Jenkins WJ, Sherlock S. Precipitation of hepatic encephalopathy by propranolol in cirrhosis. BMJ (Clin Res Ed) 1983;287(6392):585.
241. Watson P, Hayes JR. Cirrhosis, hepatic encephalopathy, and propranolol. BMJ (Clin Res Ed) 1983;287(6398):1067.
242. Anonymous. Beta-adrenergic blockers in cirrhosis. Lancet 1985;1(8442):1372–3.

243. Brown PJ, Lesna M, Hamlyn AN, Record CO. Primary biliary cirrhosis after long-term practolol administration. BMJ 1978;1(6127):1591.
244. Falch DK, Odegaard AE, Norman N. Decreased renal plasma flow during propranolol treatment in essential hypertension. Acta Med Scand 1979;205(1–2):91–5.
245. Bauer JH, Brooks CS. The long-term effect of propranolol therapy on renal function. Am J Med 1979;66(3):405–10.
246. Kincaid-Smith P, Fang P, Laver MC. A new look at the treatment of severe hypertension. Clin Sci Mol Med 1973;45(Suppl 1):s75–87.
247. Warren DJ, Swainson CP, Wright N. Deterioration in renal function after beta-blockade in patients with chronic renal failure and hypertension. BMJ 1974;2(912):193–4.
248. Wilkinson R. Beta-blockers and renal function. Drugs 1982;23(3):195–206.
249. Britton KE, Gruenwald SM, Nimmon CC. Nadolol and renal haemodynamics. In: International Experience With Nadolol, No. 37International Congress and Symposium Series . London: Royal Society of Medicine, 1981:77.
250. Dupont AG. Effects of carvedilol on renal function. Eur J Clin Pharmacol 1990;38(Suppl 2):S96–S100.
251. Krum H, Sackner-Bernstein JD, Goldsmith RL, Kukin ML, Schwartz B, Penn J, Medina N, Yushak M, Horn E, Katz SD, et al. Double-blind, placebo-controlled study of the long-term efficacy of carvedilol in patients with severe chronic heart failure. Circulation 1995;92(6):1499–506.
252. Hawk JL. Lichenoid drug eruption induced by propanolol. Clin Exp Dermatol 1980;5(1):93–6.
253. Guillet G, Chouvet V, Perrot H. Un accident des béta-bloquants: lichen induit par le pindolol avec anticorps pemphigus-like. Bordeaux Med 1981;14:95.
254. Faure M, Hermier C, Perrot H. Accidents cutanés provoqués par le propranolol. [Cutaneous reactions to propranolol.] Ann Dermatol Venereol 1979;106(2): 161–5.
255. Newman BR, Schultz LK. Epinephrine-resistant anaphylaxis in a patient taking propranolol hydrochloride. Ann Allergy 1981;47(1):35–7.
256. Kauppinen K, Idanpaan-Heikkila J. Cutaneous reactions to beta-blocking agents. In: Proceedings. XV International Congress of DermatologyMexico, 1977 1979:702.
257. Halevy S, Feuerman EJ. Psoriasiform eruption induced by propranolol. Cutis 1979;24(1):95–8.
258. Girardin P, Derancourt C, Laurent R. A new cutaneous side-effect of ocular beta-blockers. Clin Exp Dermatol 1998;23(2):95.
259. Sanchez-Perez J, Cordoba S, Bartolome B, Garcia-Diez A. Allergic contact dermatitis due to the beta-blocker carteolol in eyedrops. Contact Dermatitis 1999;41(5):298.
260. Gold MH, Holy AK, Roenigk HH Jr. Beta-blocking drugs and psoriasis. A review of cutaneous side effects and retrospective analysis of their effects on psoriasis. J Am Acad Dermatol 1988;19(5 Pt 1):837–41.
261. Schallreuter KU. Beta-adrenergic blocking drugs may exacerbate vitiligo. Br J Dermatol 1995;132(1):168–9.
262. Arnoult L, Bowman ZL, Kimbrough RL, Stewart RH. Periocular cutaneous pigmentary changes associated with topical betaxolol. J Glaucoma 1995;4:263–7.
263. Neumann HAM, Van Joost TH. Dermatitis as a side-effect of long-term treatment with beta-adrenoceptor blocking agents. Br J Dermatol 1980;103:566.
264. Nino M, Suppa F, Ayala F, Balato N. Allergic contact dermatitis due to the beta-blocker befunolol in eyedrops, with cross-sensitivity to carteolol. Contact Dermatitis 2001;44(6):369.
265. Schmutz JL, Houet C, Trechot P, Barbaud A, Gillet-Terver MN. Sweating and beta-adrenoceptor antagonists. Dermatology 1995;190(1):86.
266. Schmutz JL, Barbaud A, Reichert S, Vasse JP, Trechot P. First report of sweating associated with topical beta-blocker therapy. Dermatology 1997;194(2):197–8.
267. Gordon NF. Effect of selective and nonselective beta-adrenoceptor blockade on thermoregulation during prolonged exercise in heat. Am J Cardiol 1985;55(10):D74–8.
268. Feder R. Clonidine treatment of excessive sweating. J Clin Psychiatry 1995;56(1):35.
269. Tanner CM, Goetz CG, Klawans HL. Paroxysmal drenching sweats in idiopathic parkinsonism: response to propanolol. Neurology 1982;32(Suppl A):162.
270. Hilder RJ. Propranolol and alopecia. Cutis 1979;24(1): 63–64.
271. Graeber CW, Lapkin RA. Metoprolol and alopecia. Cutis 1981;28(6):633–4.
272. Savola J. Arthropathy induced by beta blockade. BMJ (Clin Res Ed) 1983;287(6401):1256–7.
273. Waller PC, Ramsay LE. Do beta blockers cause arthropathy? A case control study. BMJ (Clin Res Ed) 1985;291(6510):1684.
274. Sills JM, Bosco L. Arthralgia associated with beta-adrenergic blockade. JAMA 1986;255(2):198–9.
275. Zimlichman R, Krauss S, Paran E. Muscle cramps induced by beta-blockers with intrinsic sympathomimetic activity properties: a hint of a possible mechanism. Arch Intern Med 1991;151(5):1021.
276. Tomlinson B, Cruickshank JM, Hayes Y, Renondin JC, Lui JB, Graham BR, Jones A, Lewis AD, Prichard BN. Selective beta-adrenoceptor partial agonist effects of pindolol and xamoterol on skeletal muscle assessed by plasma creatine kinase changes in healthy subjects. Br J Clin Pharmacol 1990;30(5):665–72.
277. Wheeldon NM, Newnham DM, Fraser GC, McDevitt DG, Lipworth BJ. The effect of pindolol on creatine kinase is not due to beta 2-adrenoceptor partial agonist activity. Br J Clin Pharmacol 1991;31(6):723–4.
278. Imai Y, Watanabe N, Hashimoto J, Nishiyama A, Sakuma H, Sekino H, Omata K, Abe K. Muscle cramps and elevated serum creatine phosphokinase levels induced by beta-adrenoceptor blockers. Eur J Clin Pharmacol 1995;48(1):29–34.
279. Schlienger RG, Kraenzlin ME, Jick SS, Meier CR. Use of beta-blockers and risk of fractures. JAMA 2004;292: 1326–32.
280. Burnett WC, Chahine RA. Sexual dysfunction as a complication of propranolol therapy in man. Cardiovasc Med 1979;4:811.
281. Beta-blocker Heart Attack Trial Research Group. A randomized trial of propranolol in patients with acute myocardial infarction. I. Mortality results. JAMA 1982;247(12):1707–14.
282. Wassertheil-Smoller S, Blaufox MD, Oberman A, Davis BR, Swencionis C, Knerr MO, Hawkins CM, Langford HG. Effect of antihypertensives on sexual function and quality of life: the TAIM Study. Ann Intern Med 1991;114(8):613–20.
283. Croog SH, Levine S, Sudilovsky A, Baume RM, Clive J. Sexual symptoms in hypertensive patients. A clinical trial of antihypertensive medications. Arch Intern Med 1988;148(4):788–94.

284. Kostis JB, Rosen RC, Holzer BC, Randolph C, Taska LS, Miller MH. CNS side effects of centrally-active antihypertensive agents: a prospective, placebo-controlled study of sleep, mood state, and cognitive and sexual function in hypertensive males. Psychopharmacology (Berl) 1990;102(2):163–70.
285. Suzuki H, Tominaga T, Kumagai H, Saruta T. Effects of first-line antihypertensive agents on sexual function and sex hormones. J Hypertens Suppl 1988;6(4):S649–51.
286. Silvestri A, Galetta P, Cerquetani E, Marazzi G, Patrizi R, Fini M, Rosano GMC. Report of erectile dysfunction after therapy with beta-blockers is related to patient knowledge of side effects and is reversed by placebo. Eur Heart J 2003;24:1928–32.
287. Boydak B, Nalbantgil S, Fici F, Nalbantgil I, Zoghi M, Ozerkan F, Tengiz I, Ercan E, Yilmaz H, Yoket U, Onder R. A randomised comparison of the effects of nebivolol and atenolol with and without chlorthalidone on the sexual function of hypertensive men. Clin Drug Invest 2005;25(6):409–16.
288. Osborne DR. Propranolol and Peyronie's disease. Lancet 1977;1(8021):1111.
289. Kristensen BO. Labetalol-induced Peyronie's disease? A case report. Acta Med Scand 1979;206(6):511–2.
290. Pryor JP, Castle WM. Peyronie's disease associated with chronic degenerative arterial disease and not with beta-adrenoceptor blocking agents. Lancet 1982;1(8277):917.
291. Rustmann WC, Carpenter MT, Harmon C, Botti CF. Leukocytoclastic vasculitis associated with sotalol therapy. J Am Acad Dermatol 1998;38(1):111–2.
292. Booth RJ, Bullock JY, Wilson JD. Antinuclear antibodies in patients on acebutolol. Br J Clin Pharmacol 1980;9(5):515–7.
293. Cody RJ Jr, Calabrese LH, Clough JD, Tarazi RC, Bravo EL. Development of antinuclear antibodies during acebutolol therapy. Clin Pharmacol Ther 1979;25(6):800–5.
294. Huggins MM, Menzies CW, Quail D, Rumfitt IW. An open multicenter study of the effect of celiprolol on serum lipids and antinuclear antibodies in patient with mild to moderate hypertension. J Drug Dev 1991;4:125–33.
295. Bigot MC, Trenque T, Moulin M, Beguin J, Loyau G. Acebutolol-induced lupus syndrome. Therapie 1984;39:571–5.
296. Hourdebaigt-Larrusse P, Grivaux M. Une nouvelle obscuration de lupus induit par un bêta-bloquant. Sem Hop 1984;60:1515.
297. Griffiths ID, Richardson J. Lupus-type illness associated with labetalol. BMJ 1979;2(6188):496–7.
298. Clerens A, Guilmot-Bruneau MM, Defresne C, Bourlond A. Beta-blocking agents: side effects. Biomedicine 1979;31(8):219.
299. Harrison T, Sisca TS, Wood WH. Case report. Propranolol-induced lupus syndrome? Postgrad Med 1976;59(1):241–4.
300. Holzbach E. Ein Beta-blocker als Zusatztherapie beim Delirium tremens. [Beta-Blockers as adjuvant therapy in delirium tremens.] MMW Munch Med Wochenschr 1980;122(22):837–40.
301. Jacobs RL, Rake GW Jr, Fournier DC, Chilton RJ, Culver WG, Beckmann CH. Potentiated anaphylaxis in patients with drug-induced beta-adrenergic blockade. J Allergy Clin Immunol 1981;68(2):125–7.
302. Hannaway PJ, Hopper GD. Severe anaphylaxis and drug-induced beta-blockade. N Engl J Med 1983;308(25):1536.
303. Cornaille G, Leynadier F, Modiano, Dry J. Gravité du choc anaphylactic chez les malades traités par béta-bloqueurs. [Severity of anaphylactic shock in patients treated with beta-blockers.] Presse Méd 1985;14(14):790–1.
304. Raebel MA. Potentiated anaphylaxis during chronic beta-blocker therapy. DICP Ann Pharmacother 1988;22:720.
305. Miller MM, Miller MM. Beta-blockers and anaphylaxis: are the risks overstated? J Allergy Clin Immunol 2005;116(4):931–3.
306. Toogood JH. Beta-blocker therapy and the risk of anaphylaxis. CMAJ 1987;136(9):929–33.
307. Arkinstall WW, Toogood JH. Beta-blocker therapy and the risk of anaphylaxis. CMAJ 1987;137(5):370–1.
308. Steinwender C, Hofmann R, Kypta A, Leisch F. Recurrent symptomatic bradycardia due to secret ingestion of beta-blockers—a rare manifestation of cardiac Munchhausen syndrome. Wien Klin Wochenschr 2005;117(18):647–50.
309. Miller RR, Olson HG, Amsterdam EA, Mason DT. Propranolol-withdrawal rebound phenomenon. Exacerbation of coronary events after abrupt cessation of antianginal therapy. N Engl J Med 1975;293(9):416–8.
310. Myers MG, Wisenberg G. Sudden withdrawal of propranolol in patients with angina pectoris. Chest 1977;71(1):24–6.
311. Shiroff RA, Mathis J, Zelis R, Schneck DW, Babb JD, Leaman DM, Hayes AH Jr. Propranolol rebound—a retrospective study. Am J Cardiol 1978;41(4):778–80.
312. Psaty BM, Koepsell TD, Wagner EH, LoGerfo JP, Inui TS. The relative risk of incident coronary heart disease associated with recently stopping the use of beta-blockers. JAMA 1990;263(12):1653–7.
313. Olsson G, Hjemdahl P, Rehnqvist N. Rebound phenomena following gradual withdrawal of chronic metoprolol treatment in patients with ischemic heart disease. Am Heart J 1984;108(3 Pt 1):454–62.
314. Maling TJ, Dollery CT. Changes in blood pressure, heart rate, and plasma noradrenaline concentration after sudden withdrawal of propranolol. BMJ 1979;2(6186):366–7.
315. Lederballe Pedersen O, Mikkelsen E, Lanng Nielsen J, Christensen NJ. Abrupt withdrawal of beta-blocking agents in patients with arterial hypertension. Effect on blood pressure, heart rate and plasma catecholamines and prolactin. Eur J Clin Pharmacol 1979;15(3):215–7.
316. Webster J, Hawksworth GM, Barber HE, Jeffers TA, Petrie JC. Withdrawal of long-term therapy with atenolol in hypertensive patients. Br J Clin Pharmacol 1981;12(2):211–4.
317. Aarons RD, Nies AS, Gal J, Hegstrand LR, Molinoff PB. Elevation of beta-adrenergic receptor density in human lymphocytes after propranolol administration. J Clin Invest 1980;65(5):949–57.
318. Nattel S, Rangno RE, Van Loon G. Mechanism of propranolol withdrawal phenomena. Circulation 1979;59(6):1158–64.
319. Rangno RE, Langlois S, Lutterodt A. Metoprolol withdrawal phenomena: mechanism and prevention. Clin Pharmacol Ther 1982;31(1):8–15.
320. Walden RJ, Bhattacharjee P, Tomlinson B, Cashin J, Graham BR, Prichard BN. The effect of intrinsic sympathomimetic activity on beta-receptor responsiveness after beta-adrenoceptor blockade withdrawal. Br J Clin Pharmacol 1982;13(Suppl 2):S359–64.
321. Rangno RE, Langlois S. Comparison of withdrawal phenomena after propranolol, metoprolol and pindolol. Br J Clin Pharmacol 1982;13(Suppl 2):S345–51.
322. Krukemyer JJ, Boudoulas H, Binkley PF, Lima JJ. Comparison of hypersensitivity to adrenergic stimulation

after abrupt withdrawal of propranolol and nadolol: influence of half-life differences. Am Heart J 1990;120(3):572–9.
323. Rubin PC, Butters L, Clark DM, Reynolds B, Sumner DJ, Steedman D, Low RA, Reid JL. Placebo-controlled trial of atenolol in treatment of pregnancy-associated hypertension. Lancet 1983;1(8322):431–4.
324. Lowe SA, Rubin PC. The pharmacological management of hypertension in pregnancy. J Hypertens 1992;10(3):201–7.
325. Paran E, Holzberg G, Mazor M, Zmora E, Insler V. Beta-adrenergic blocking agents in the treatment of pregnancy-induced hypertension. Int J Clin Pharmacol Ther 1995;33(2):119–23.
326. Khedun SM, Maharaj B, Moodley J. Effects of antihypertensive drugs on the unborn child: what is known, and how should this influence prescribing? Paediatr Drugs 2000;2(6):419–36.
327. Crooks BN, Deshpande SA, Hall C, Platt MP, Milligan DW. Adverse neonatal effects of maternal labetalol treatment. Arch Dis Child Fetal Neonatal Ed 1998;79(2):F150–1.
328. Wagenvoort AM, van Vugt JM, Sobotka M, van Geijn HP. Topical timolol therapy in pregnancy: is it safe for the fetus? Teratology 1998;58(6):258–62.
329. White WB, Andreoli JW, Wong SH, Cohn RD. Atenolol in human plasma and breast milk. Obstet Gynecol 1984;63(Suppl 3):S42–4.
330. Boutroy MJ, Bianchetti G, Dubruc C, Vert P, Morselli PL. To nurse when receiving acebutolol: is it dangerous for the neonate? Eur J Clin Pharmacol 1986;30(6):737–9.
331. Sandstrom B, Regardh CG. Metoprolol excretion into breast milk. Br J Clin Pharmacol 1980;9(5):518–9.
332. Devlin RG, Duchin KL, Fleiss PM. Nadolol in human serum and breast milk. Br J Clin Pharmacol 1981;12(3):393–6.
333. Fidler J, Smith V, De Swiet M. Excretion of oxprenolol and timolol in breast milk. Br J Obstet Gynaecol 1983;90(10):961–5.
334. Smith MT, Livingstone I, Hooper WD, Eadie MJ, Triggs EJ. Propranolol, propranolol glucuronide, and naphthoxylactic acid in breast milk and plasma. Ther Drug Monit 1983;5(1):87–93.
335. O'Hare MF, Murnaghan GA, Russell CJ, Leahey WJ, Varma MP, McDevitt DG. Sotalol as a hypotensive agent in pregnancy. Br J Obstet Gynaecol 1980;87(9):814–20.
336. Lennard MS, Tucker GT, Woods HF. The polymorphic oxidation of beta-adrenoceptor antagonists. Clinical pharmacokinetic considerations. Clin Pharmacokinet 1986;11(1):1–17.
337. McAinsh J, Holmes BF, Smith S, Hood D, Warren D. Atenolol kinetics in renal failure. Clin Pharmacol Ther 1980;28(3):302–9.
338. Berglund G, Descamps R, Thomis JA. Pharmacokinetics of sotalol after chronic administration to patients with renal insufficiency. Eur J Clin Pharmacol 1980;18(4):321–6.
339. Verbeeck RK, Branch RA, Wilkinson GR. Drug metabolites in renal failure: pharmacokinetic and clinical implications. Clin Pharmacokinet 1981;6(5):329–45.
340. Stone WJ, Walle T. Massive retention of propranolol metabolites in maintenance hemodialysis patients. Clin Pharmacol Ther 1980;27:288.
341. McDevitt DG. Beta-adrenoceptor blockade in hyperthyroidism. In: Shanks RG, editor. Advanced Medicine: Topics in Therapeutics 3. London: Pitman Medical, 1977:100.
342. Rossi PR, Yusuf S, Ramsdale D, Furze L, Sleight P. Reduction of ventricular arrhythmias by early intravenous atenolol in suspected acute myocardial infarction. BMJ (Clin Res Ed) 1983;286(6364):506–10.
343. Ryden L, Ariniego R, Arnman K, Herlitz J, Hjalmarson A, Holmberg S, Reyes C, Smedgard P, Svedberg K, Vedin A, Waagstein F, Waldenstrom A, Wilhelmsson C, Wedel H, Yamamoto M. A double-blind trial of metoprolol in acute myocardial infarction. Effects on ventricular tachyarrhythmias. N Engl J Med 1983;308(11):614–8.
344. Anonymous. Long-term and short-term beta-blockade after myocardial infarction. Lancet 1982;1(8282):1159–61.
345. Baber NS, Evans DW, Howitt G, Thomas M, Wilson T, Lewis JA, Dawes PM, Handler K, Tuson R. Multicentre post-infarction trial of propranolol in 49 hospitals in the United Kingdom, Italy, and Yugoslavia. Br Heart J 1980;44(1):96–100.
346. Beta-Blocker Heart Attack Study Group. The Beta-blocker Heart Attack Trial. JAMA 1981;246(18):2073–4.
347. Wilhelmsson C, Vedin JA, Wilhelmsen L, Tibblin G, Werko L. Reduction of sudden deaths after myocardial infarction by treatment with alprenolol. Preliminary results. Lancet 1974;2(7890):1157–60.
348. Andersen MP, Bechsgaard P, Frederiksen J, Hansen DA, Jurgensen HJ, Nielsen B, Pedersen F, Pedersen-Bjergaard O, Rasmussen SL. Effect of alprenolol on mortality among patients with definite or suspected acute myocardial infarction. Preliminary results. Lancet 1979;2(8148):865–8.
349. Ahlmark G, Saetre H, Korsgren M. Reduction of sudden deaths after myocardial infarction. Lancet 1974;2(7896):1563.
350. Hjalmarson A, Elmfeldt D, Herlitz J, Holmberg S, Malek I, Nyberg G, Ryden L, Swedberg K, Vedin A, Waagstein F, Waldenstrom A, Waldenstrom J, Wedel H, Wilhelmsen L, Wilhelmsson C. Effect on mortality of metoprolol in acute myocardial infarction. A double-blind randomised trial. Lancet 1981;2(8251):823–7.
351. Norwegian Multicentre Study Group. Timolol-induced reduction in mortality and reinfarction in patients surviving acute myocardial infarction. N Engl J Med 1981;304(14):801–7.
352. Julian DG, Prescott RJ, Jackson FS, Szekely P. Controlled trial of sotalol for one year after myocardial infarction. Lancet 1982;1(8282):1142–7.
353. Hansteen V, Moinichen E, Lorentsen E, Andersen A, Strom O, Soiland K, Dyrbekk D, Refsum AM, Tromsdal A, Knudsen K, Eika C, Bakken J Jr, Smith P, Hoff PI. One year's treatment with propranolol after myocardial infarction: preliminary report of Norwegian multicentre trial. BMJ (Clin Res Ed) 1982;284(6310):155–60.
354. Heidbreder E, Pagel G, Rockel A, Heidland A. Beta-adrenergic blockade in stress protection. Limited effect of metoprolol in psychological stress reaction. Eur J Clin Pharmacol 1978;14(6):391–8.
355. Trap-Jensen J, Carlsen JE, Svendsen TL, Christensen NJ. Cardiovascular and adrenergic effects of cigarette smoking during immediate non-selective and selective beta adrenoceptor blockade in humans. Eur J Clin Invest 1979;9(3):181–3.
356. Freestone S, Ramsay LE. Effect of coffee and cigarette smoking in untreated and diuretic-treated hypertensive patients. Br J Clin Pharmacol 1981;11:428.
357. Ramsay LE. Antihypertensive drugs. Curr Opin Cardiol 1987;1:524.
358. Deanfield J, Wright C, Krikler S, Ribeiro P, Fox K. Cigarette smoking and the treatment of angina with propranolol, atenolol, and nifedipine. N Engl J Med 1984; 310(15):951–4.

359. Howard PJ, Lee MR. Beware beta-adrenergic blockers in patients with severe urticaria!. Scott Med J 1988; 33(5):344–5.
360. Korte JM, Kaila T, Saari KM. Systemic bioavailability and cardiopulmonary effects of 0.5% timolol eyedrops Graefes Arch Clin Exp Ophthalmol 2002;240(6):430–5.
361. Caballero F, Lopez-Navidad A, Cotorruelo J, Txoperena G. Ecstasy-induced brain death and acute hepatocellular failure: multiorgan donor and liver transplantation. Transplantation 2002;74(4):532–7.
362. Fraunfelder FT. Ocular beta-blockers and systemic effects. Arch Intern Med 1986;146(6):1073–4.
363. Stewart WC, Castelli WP. Systemic side effects of topical beta-adrenergic blockers. Clin Cardiol 1996;19(9):691–7.
364. Nelson WL, Fraunfelder FT, Sills JM, Arrowsmith JB, Kuritsky JN. Adverse respiratory and cardiovascular events attributed to timolol ophthalmic solution, 1978–1985. Am J Ophthalmol 1986;102(5):606–11.
365. Hayreh SS, Podhajsky P, Zimmerman MB. Beta-blocker eyedrops and nocturnal arterial hypotension. Am J Ophthalmol 1999;128(3):301–9.
366. Vandezande LM, Gallouj K, Lamblin C, Fourquet B, Maillot E, Wallaert B. Pneumopathie interstitielle induite par un collyre de timolol. [Interstitial lung disease induced by timolol eye solution.] Rev Mal Respir 1999;16(1):91–3.
367. Heel RC, Brogden RN, Speight TM, Avery GS. Timolol: a review of its therapeutic efficacy in the topical treatment of glaucoma. Drugs 1979;17(1):38–55.
368. Van Buskirk EM. Corneal anesthesia after timolol maleate therapy. Am J Ophthalmol 1979;88(4):739–43.
369. Coppeto JR. Transient ischemic attacks and amaurosis fugax from timolol. Ann Ophthalmol 1985;17(1):64–5.
370. Silverstone BZ, Marcus T. [Hypoglycemia due to ophthalmic timolol in a diabetic.]Harefuah 1990;118(12):693–4.
371. Swenson ER. Severe hyperkalemia as a complication of timolol, a topically applied beta-adrenergic antagonist. Arch Intern Med 1986;146(6):1220–1.
372. Shaivitz SA. Timolol and myasthenia gravis. JAMA 1979;242(15):1611–2.
373. Boger WP 3rd. Shortterm "escape" and longterm "drift." The dissipation effects of the beta adrenergic blocking agents Surv Ophthalmol 1983;28(Suppl):235–42.
374. Anonymous. Self-poisoning with beta-blockers. BMJ 1978;1(6119):1010–1.
375. Anonymous. Beta-blocker poisoning. Lancet 1980; 1(8172):803–4.
376. Tynan RF, Fisher MM, Ibels LS. Self-poisoning with propranolol. Med J Aust 1981;1(2):82–3.
377. Buiumsohn A, Eisenberg ES, Jacob H, Rosen N, Bock J, Frishman WH. Seizures and intraventricular conduction defect in propranolol poisoning. A report of two cases. Ann Intern Med 1979;91(6):860–2.
378. Neuvonen PJ, Elonen E, Vuorenmaa T, Laakso M. Prolonged Q-T interval and severe tachyarrhythmias, common features of sotalol intoxication. Eur J Clin Pharmacol 1981;20(2):85–9.
379. Assimes TL, Malcolm I. Torsade de pointes with sotalol overdose treated successfully with lidocaine. Can J Cardiol 1998;14(5):753–6.
380. Lagerfelt J, Matell G. Attempted suicide with 5.1 g of propranolol. A case report Acta Med Scand 1976; 199(6):517–8.
381. Henry JA, Cassidy SL. Membrane stabilising activity: a major cause of fatal poisoning. Lancet 1986;1(8495):1414–7.
382. Aura ED, Wexler LF, Wirtzburg RA. Massive propranolol overdose: successful treatment with high dose isoproterenol and glucagon. Am J Med 1986;80:755.
383. Weinstein RS. Recognition and management of poisoning with beta-adrenergic blocking agents. Ann Emerg Med 1984;13(12):1123–31.
384. Nicolas F, Villers D, Rozo L, Haloun A, Bigot A. Severe self-poisoning with acebutolol in association with alcohol. Crit Care Med 1987;15(2):173–4.
385. Ojetti V, Migneco A, Bononi F, De Lorenzo A, Gentiloni Silveri N. Calcium channel blockers, beta-blockers and digitalis poisoning: management in the emergency room. Eur Rev Med Pharmacol Sci 2005;9(4):241–6.
386. Richards DA, Prichard BN. Self-poisoning with beta-blockers. BMJ 1978;1(6127):1623–4.
387. Freestone S, Thomas HM, Bhamra RK, Dyson EH. Severe atenolol poisoning: treatment with prenalterol. Hum Toxicol 1986;5(5):343–5.
388. Lifshitz M, Zucker N, Zalzstein E. Acute dilated cardiomyopathy and central nervous system toxicity following propranolol intoxication. Pediatr Emerg Care 1999;15(4):262–3.
389. Love JN, Howell JM, Litovitz TL, Klein-Schwartz W. Acute beta blocker overdose: factors associated with the development of cardiovascular morbidity. J Toxicol Clin Toxicol 2000;38(3):275–81.
390. Love JN. Acebutolol overdose resulting in fatalities. J Emerg Med 2000;18(3):341–4.
391. McDevitt DG. Clinically important adverse drug interactions. In: Petrie JC, editor. Cardiovascular and Respiratory Disease Therapy 1. Amsterdam: Elsevier/North Holland, Biomedical Press, 1980:21.
392. Lewis RV, McDevitt DG. Adverse reactions and interactions with beta-adrenoceptor blocking drugs. Med Toxicol 1986;1(5):343–61.
393. Kendall MJ, Beeley L. Beta-adrenoceptor blocking drugs: adverse reactions and drug interactions. Pharmacol Ther 1983;21(3):351–69.
394. Alvan G, Piafsky K, Lind M, von Bahr C. Effect of pentobarbital on the disposition of alprenolol. Clin Pharmacol Ther 1977;22(3):316–21.
395. Bennett PN, John VA, Whitmarsh VB. Effect of rifampicin on metoprolol and antipyrine kinetics. Br J Clin Pharmacol 1982;13(3):387–91.
396. Feely J, Wilkinson GR, Wood AJ. Reduction of liver blood flow and propranolol metabolism by cimetidine. N Engl J Med 1981;304(12):692–5.
397. Daneshmend TK, Roberts CJ. Cimetidine and bioavailability of labetalol. Lancet 1981;1(8219):565.
398. Kirch W, Kohler H, Spahn H, Mutschler E. Interaction of cimetidine with metoprolol, propranolol, or atenolol. Lancet 1981;2(8245):531–2.
399. Sax MJ. Analysis of possible drug interactions between cimetidine (and ranitidine) and beta-blockers. Adv Ther 1988;5:210.
400. McLean AJ, Skews H, Bobik A, Dudley FJ. Interaction between oral propranolol and hydralazine. Clin Pharmacol Ther 1980;27(6):726–32.
401. Conrad KA, Nyman DW. Effects of metoprolol and propranolol on theophylline elimination. Clin Pharmacol Ther 1980;28(4):463–7.
402. Greendyke RM, Kanter DR. Plasma propranolol levels and their effect on plasma thioridazine and haloperidol concentrations. J Clin Psychopharmacol 1987;7(3):178–82.
403. Peet M, Middlemiss DN, Yates RA. Pharmacokinetic interaction between propranolol and chlorpromazine in schizophrenic patients. Lancet 1980;2(8201):978.
404. Bax ND, Lennard MS, Tucker GT, Woods HF, Porter NR, Malia RG, Preston FE. The effect of beta-adrenoceptor

antagonists on the pharmacokinetics and pharmacodynamics of warfarin after a single dose. Br J Clin Pharmacol 1984;17(5):553–7.
405. Ochs HR, Greenblatt DJ, Verburg-Ochs B. Propranolol interactions with diazepam, lorazepam, and alprazolam. Clin Pharmacol Ther 1984;36(4):451–5.
406. Santoso B. Impairment of isoniazid clearance by propranolol. Int J Clin Pharmacol Ther Toxicol 1985;23(3):134–6.
407. Lewis GP, Holtzman JL. Interaction of flecainide with digoxin and propranolol. Am J Cardiol 1984;53(5):B52–7.
408. Ochs HR, Carstens G, Greenblatt DJ. Reduction in lidocaine clearance during continuous infusion and by coadministration of propranolol. N Engl J Med 1980;303(7):373–7.
409. Bax ND, Tucker GT, Lennard MS, Woods HF. The impairment of lignocaine clearance by propranolol—major contribution from enzyme inhibition. Br J Clin Pharmacol 1985;19(5):597–603.
410. Bonde J, Bodtker S, Angelo HR, Svendsen TL, Kampmann JP. Atenolol inhibits the elimination of disopyramide. Eur J Clin Pharmacol 1985;28(1):41–3.
411. Leemann T, Dayer P, Meyer UA. Single-dose quinidine treatment inhibits metoprolol oxidation in extensive metabolizers. Eur J Clin Pharmacol 1986;29(6):739–41.
412. Kendall MJ, Jack DB, Quarterman CP, Smith SR, Zaman R. Beta-adrenoceptor blocker pharmacokinetics and the oral contraceptive pill. Br J Clin Pharmacol 1984;17(Suppl 1):S87–9.
413. Watkins J, Abbott EC, Hensby CN, Webster J, Dollery CT. Attenuation of hypotensive effect of propranolol and thiazide diuretics by indomethacin. BMJ 1980;281(6242):702–5.
414. Wong DG, Spence JD, Lamki L, Freeman D, McDonald JW. Effect of non-steroidal anti-inflammatory drugs on control of hypertension by beta-blockers and diuretics. Lancet 1986;1(8488):997–1001.
415. Lewis RV, Toner JM, Jackson PR, Ramsay LE. Effects of indomethacin and sulindac on blood pressure of hypertensive patients. BMJ (Clin Res Ed) 1986;292(6525):934–5.
416. Venter CP, Joubert PH, Buys AC. Severe peripheral ischaemia during concomitant use of beta blockers and ergot alkaloids. BMJ (Clin Res Ed) 1984;289(6440):288–9.
417. Smits P, Hoffmann H, Thien T, Houben H, van't Laar A. Hemodynamic and humoral effects of coffee after beta 1-selective and nonselective beta-blockade. Clin Pharmacol Ther 1983;34(2):153–8.
418. Eldor J, Hoffman B, Davidson JT. Prolonged bradycardia and hypotension after neostigmine administration in a patient receiving atenolol. Anaesthesia 1987;42(12):1294–7.
419. McKibbin JK, Pocock WA, Barlow JB, Millar RN, Obel IW. Sotalol, hypokalaemia, syncope, and torsade de pointes. Br Heart J 1984;51(2):157–62.
420. Feroze H, Suri R, Silverman DI. Torsades de pointes from terfenadine and sotalol given in combination. Pacing Clin Electrophysiol 1996;19(10):1519–21.
421. Ikram H. Hemodynamic and electrophysiologic interactions between antiarrhythmic drugs and beta blockers, with special reference to tocainide. Am Heart J 1980;100(6 Pt 2):1076–80.
422. Lampman RM, Santinga JT, Bassett DR, Savage PJ. Cardiac arrhythmias during epinephrine–propranolol infusions for measurement of in vivo insulin resistance. Diabetes 1981;30(7):618–20.
423. Yoshiga Y, Shimizu A, Yamagata T, Hayano T, Ueyama T, Ohmura M, Itagaki K, Kimura M, Matsuzaki M. Beta-blocker decreases the increase in QT dispersion and transmural dispersion of repolarization induced by bepridil. Circ J 2002;66(11):1024–8.
424. Frishman WH, Charlap S, Farnham DJ, Sawin HS, Michelson EL, Crawford MH, DiBianco R, Kostis JB, Zellner SR, Michie DD, et al. Combination propranolol and bepridil therapy in stable angina pectoris. Am J Cardiol 1985;55(7):C43–9.
425. Klieman RL, Stephenson SH. Calcium antagonists–drug interactions. Rev Drug Metab Drug Interact 1985;5(2–3):193–217.
426. Pringle SD, MacEwen CJ. Severe bradycardia due to interaction of timolol eye drops and verapamil. BMJ (Clin Res Ed) 1987;294(6565):155–6.
427. Opie LH, White DA. Adverse interaction between nifedipine and beta-blockade. BMJ 1980;281(6253):1462.
428. Staffurth JS, Emery P. Adverse interaction between nifedipine and beta-blockade. BMJ (Clin Res Ed) 1981;282(6259):225.
429. Anastassiades CJ. Nifedipine and beta-blocker drugs. BMJ 1980;281(6250):1251–2.
430. Young GP. Calcium channel blockers in emergency medicine. Ann Emerg Med 1984;13(9 Pt 1):712–22.
431. Saini RK, Fulmor IE, Antonaccio MJ. Effect of tiapamil and nifedepine during critical coronary stenosis and in the presence of adrenergic beta-receptor blockade in anesthetized dogs. J Cardiovasc Pharmacol 1982;4(5):770–6.
432. Rocha P, Baron B, Delestrain A, Pathe M, Cazor JL, Kahn JC. Hemodynamic effects of intravenous diltiazem in patients treated chronically with propranolol. Am Heart J 1986;111(1):62–8.
433. Kjeldsen SE, Syvertsen JO, Hedner T. Cardiac conduction with diltiazem and beta-blockade combined. A review and report on cases. Blood Press 1996;5(5):260–3.
434. Leon MB, Rosing DR, Bonow RO, Lipson LC, Epstein SE. Clinical efficacy of verapamil alone and combined with propranolol in treating patients with chronic stable angina pectoris. Am J Cardiol 1981;48(1):131–9.
435. DeWood MA, Wolbach RA. Randomized double-blind comparison of side effects of nicardipine and nifedipine in angina pectoris. The Nicardipine Investigators Group. Am Heart J 1990;119(2 Pt 2):468–78.
436. Sorkin EM, Clissold SP, Brogden RN. Nifedipine. A review of its pharmacodynamic and pharmacokinetic properties, and therapeutic efficacy, in ischaemic heart disease, hypertension and related cardiovascular disorders. Drugs 1985;30(3):182–274.
437. Goa KL, Sorkin EM. Nitrendipine. A review of its pharmacodynamic and pharmacokinetic properties, and therapeutic efficacy in the treatment of hypertension. Drugs 1987;33(2):123–55.
438. Solomon SA, Ramsay LE, Yeo WW, Parnell L, Morris-Jones W. Beta blockade and intermittent claudication: placebo controlled trial of atenolol and nifedipine and their combination. BMJ 1991;303(6810):1100–4.
439. Peet M, Middlemiss DN, Yates RA. Pharmacokinetic interaction between propranolol and chlorpromazine in schizophrenic patients. Lancet 1980;2(8201):978.
440. Miller FA, Rampling D. Adverse effects of combined propranolol and chlorpromazine therapy. Am J Psychiatry 1982;139(9):1198–9.
441. Markowitz JS, Wells BG, Carson WH. Interactions between antipsychotic and antihypertensive drugs. Ann Pharmacother 1995;29(6):603–9.
442. Ochs HR, Greenblatt DJ, Verburg-Ochs B. Propranolol interactions with diazepam, lorazepam, and alprazolam. Clin Pharmacol Ther 1984;36(4):451–5.

443. Almeyda J, Levantine A. Cutaneous reactions to cardiovascular drugs. Br J Dermatol 1973;88(3):313–9.
444. Lewis GP, Holtzman JL. Interaction of flecainide with digoxin and propranolol. Am J Cardiol 1984;53(5):B52–7.
445. Santoso B. Impairment of isoniazid clearance by propranolol. Int J Clin Pharmacol Ther Toxicol 1985;23(3):134–6.
446. Wyse DG, Kellen J, Tam Y, Rademaker AW. Increased efficacy and toxicity of lidocaine in patients on beta-blockers. Int J Cardiol 1988;21(1):59–70.
447. Naguib M, Magboul MM, Samarkandi AH, Attia M. Adverse effects and drug interactions associated with local and regional anaesthesia. Drug Saf 1998;18(4):221–50.
448. Ayd FJ Jr. Loxapine update: 1966–1976. Dis Nerv Syst 1977;38(11):883–7.
449. Markowitz JS, Wells BG, Carson WH. Interactions between antipsychotic and antihypertensive drugs. Ann Pharmacother 1995;29(6):603–9.
450. Greendyke RM, Kanter DR. Plasma propranolol levels and their effect on plasma thioridazine and haloperidol concentrations. J Clin Psychopharmacol 1987;7(3):178–82.
451. Markowitz JS, Wells BG, Carson WH. Interactions between antipsychotic and antihypertensive drugs. Ann Pharmacother 1995;29(6):603–9.

Bevantolol

See also Beta-adrenoceptor antagonists

General Information

Bevantolol, a hydrophilic cardioselective beta-blocker with membrane-stabilizing activity, may have a higher incidence of fatigue, headache, and dizziness than atenolol or propranolol (1,2).

References

1. Maclean D. Bevantolol vs propranolol: a double-blind controlled trial in essential hypertension. Angiology 1988;39(6):487–96.
2. Rodrigues EA, Lawrence JD, Dasgupta P, Hains AD, Lahiri A, Wilkinson PR, Raftery EB. Comparison of bevantolol and atenolol in chronic stable angina. Am J Cardiol 1988;61(15):1204–9.

Bisoprolol

See also Beta-adrenoceptor antagonists

General Information

Bisoprolol is a highly selective $beta_1$-adrenoceptor antagonist. Its adverse effects profile is similar to that of atenolol (1), and despite theoretical benefits there is no convincing clinical evidence that bisoprolol has an advantage (2).

Organs and Systems

Metabolism

Bisoprolol can increase serum triglycerides and reduce HDL cholesterol (3).

Urinary tract

The lower urinary tract symptoms of benign prostatic hyperplasia in a 60-year-old man worsened antihypertensive treatment with bisoprolol (4). Withdrawal resulted in relief whereas and rechallenge caused repeated deterioration.

References

1. Neutel JM, Smith DH, Ram CV, Kaplan NM, Papademetriou V, Fagan TC, Lefkowitz MP, Kazempour MK, Weber MA. Application of ambulatory blood pressure monitoring in differentiating between antihypertensive agents. Am J Med 1993;94(2):181–7.
2. Wheeldon NM, MacDonald TM, Prasad N, Maclean D, Peebles L, McDevitt DG. A double-blind comparison of bisoprolol and atenolol in patients with essential hypertension. QJM 1995;88(8):565–70.
3. Lancaster SG, Sorkin EM. Bisoprolol. A preliminary review of its pharmacodynamic and pharmacokinetic properties, and therapeutic efficacy in hypertension and angina pectoris. Drugs 1988;36(3):256–85.
4. Sein Anand J, Chodorowski Z, Hajduk A. Repeated intensification of lower urinary tract symptoms in the patient with benign prostatic hyperplasia during bisoprolol treatment. Przegl Lek 2005;62(6):522–3.

Bosentan

See also Endothelin receptor antagonists

General Information

Bosentan is an endothelin-A and endothelin-B receptor antagonist. It is effective in pulmonary arterial hypertension (1,2) and has been marketed for this indication. The studies showed an improvement in exercise capacity and dyspnea and an increased time to clinical worsening. Efficacy in pulmonary hypertension has also been reported in an open study with a selective endothelin-A receptor antagonist (3).

The effects of bosentan (62.5 mg bd for 4 weeks followed by either 125 or 250 mg bd for a minimum of 12 weeks) have been studied in a double-blind, placebo-controlled trial in 213 patients with pulmonary arterial hypertension (2). The patients who took bosentan had improved exercise capacity, less dyspnea, and delayed worsening. There were similar results in 32 patients who took bosentan for a minimum of 12 weeks (62.5 mg bd for 4 weeks then 125 mg bd) (4). Bosentan significantly reduced pulmonary vascular resistance. The number and nature of adverse events were similar with bosentan and placebo.

The pharmacokinetics and drug interactions of bosentan have been reviewed (5). The risk management strategies and post-marketing surveillance plans in place in the USA and Europe have been described in a paper that also briefly describes the potential hepatotoxicity and teratogenicity of bosentan (6).

Placebo-controlled studies

The BREATHE-2 trial examined the effects of a combination of intravenous epoprostenol with either bosentan or placebo in a randomized trial in 33 patients with pulmonary arterial hypertension (7). There were three deaths in the combination treated group during or soon after the end of the study, although the authors concluded that the severity of the patients' illnesses it precluded definite attribution of the deaths to the treatment. Other adverse effects were mainly those attributed to epoprostenol, although an excess number of patients developed leg edema in the combination group (27% versus 9%).

Organs and Systems

Cardiovascular

In a dose-finding study with bosentan (100, 500, 1000, and 2000 mg/day) in 293 hypertensive patients (8) there were statistically significant falls in diastolic blood pressure with the 500 and 2000 mg/day doses. The effects were similar to that of enalapril 20 mg/day. The lowering of blood pressure was not associated with any changes in heart rate, plasma noradrenaline concentrations, plasma renin activity, or angiotensin II concentrations.

In the ENABLE (Endothelin Antagonist Bosentan for Lowering Cardiac Events in Heart Failure) placebo-controlled study the effects of low-dose bosentan (125 mg bd) were evaluated in 1613 patients with severe heart failure (left ventricular ejection fraction < 35%, New York Heart Association classes IIIb-IV) (9). The primary endpoint of all-cause mortality or hospitalization for heart failure was reached in 321 of 808 patients who took placebo and 312 of 805 who took bosentan. Treatment with bosentan conferred an early risk of worsening heart failure necessitating hospitalization, because of fluid retention. These results throw doubt on the potential benefits of non-specific endothelin receptor blockade in heart failure. Preliminary data suggest that in heart failure selective endothelin-A receptor antagonists (BQ-123, sitaxsentan) may be more beneficial than non-selective antagonists, especially when there is associated pulmonary hypertension (10).

Liver

The major safety issue that has emerged with bosentan, the endothelin receptor antagonist that has been most extensively studied in man, has been dose-dependent reversible impairment of hepatic function (3% with 125 mg, 7% with 250 mg), manifesting as raised transaminases (2). The effect of bosentan on hepatocanalicular bile-salt transport has been studied in rats in conjunction with a re-examination of the safety database from two clinical trials (in hypertension and congestive cardiac failure) and measurement of bile-salt concentrations in stored blood samples from these trials (11). Hepatic injury was defined as a three-fold increase in alanine transaminase activity. In the hypertension trial there were no cases of hepatic injury with placebo or enalapril. With bosentan, the frequencies were 2, 4, 11, and 8% at dosages of 100, 500, 1000, and 2000 mg/day respectively. There was a dose-dependent increase in bile-salt concentrations. In the study in patients with heart failure (New York Heart Association classes III/IV), liver injury occurred in 4% of 126 patients taking placebo and 18% of 244 patients taking bosentan 500 mg bd. A subgroup analysis showed a higher incidence of hepatic injury in patients taking concomitant bosentan and glibenclamide. Patients with hepatic injury had raised bile-salt concentrations.

In rats, intravenous bosentan produced a dose-dependent increase in plasma bile salts. The effect was potentiated when glibenclamide was co-administered. In vitro studies in rat canalicular liver plasma membranes confirmed inhibition of bile-salt transport. Three bosentan metabolites were also investigated. The M2 metabolite was more potent than bosentan, whereas the M1 and M3 metabolites produced less inhibition of bile acid transport than bosentan. The effects of bosentan and its main metabolites, both of which are eliminated in the bile, on biliary secretion have been studied in rats with biliary fistulae and with or without a genetic defect in mrp2 (12). Intravenous bosentan 0.1–10 mg/kg caused a dose-dependent increase in biliary bilirubin excretion, and doses of 10 mg/kg or over caused a sustained increase in canalicular salt-independent bile flow, combined with significant increases in the concentrations and output of glutathione and of bicarbonate in the bile. Phospholipid and cholesterol secretion were profoundly inhibited and uncoupled from bile-salt secretion. In mrp2-deficient rats, the choleretic effect of bosentan was markedly reduced. Thus, bosentan alters canalicular bile formation mostly via mrp2-mediated mechanisms. Intermittent uncoupling of lipid from bile-salt secretion may contribute to its hepatic adverse effects.

These data suggest that bosentan causes cholestatic liver injury due to inhibition of bile-salt efflux and damage due to intracellular accumulation of bile salts.

Immunologic

Severe necrotizing leukocytoclastic vasculitis had been attributed to bosentan (13).

- A 47-year-old woman with idiopathic pulmonary arterial hypertension was given bosentan 62.5 mg bd plus spironolactone. Four weeks later, after some clinical improvement, the dosage of bosentan was increased to 125 mg bd, and 4 days later she developed itching of both legs. Within another 3 days typical vasculitic skin lesions appeared on both legs. A skin biopsy showed leukocytoclastic vasculitis and negative immune complex staining. Hematology, blood chemistry, and blood immunology were unremarkable. Bosentan was withdrawn immediately. Diuretics and acenocoumarol were continued. The vasculitic skin lesions improved slowly over the following weeks.

Long-Term Effects

Drug tolerance

Despite short-term benefits of bosentan in systemic hypertension and congestive heart failure, it increases plasma concentrations of endothelin-1, probably by inhibiting its clearance via endothelin-B receptors. This may mean that its effectiveness is reduced during long-term therapy (10).

Susceptibility Factors

Renal disease

The effects of severe renal insufficiency on the pharmacokinetics and metabolism of bosentan 125 mg have been studied in an open, parallel-group study in eight patients with creatinine clearances 17–27 ml/minute and in eight healthy subjects (creatinine clearances 99–135 ml/minute) (14). The pharmacokinetics of bosentan did not differ significantly, although the concentrations of the three CYP2C9- and CYP3A4-derived metabolites increased about two-fold in the patients with renal insufficiency. It is not necessary to adjust the dosage of bosentan in patients with any grade of renal insufficiency.

References

1. Channick RN, Simonneau G, Sitbon O, Robbins IM, Frost A, Tapson VF, Badesch DB, Roux S, Rainisio M, Bodin F, Rubin LJ. Effects of the dual endothelin-receptor antagonist bosentan in patients with pulmonary hypertension: a randomised placebo-controlled study. Lancet 2001;358(9288):1119–23.
2. Rubin LJ, Badesch DB, Barst RJ, Galie N, Black CM, Keogh A, Pulido T, Frost A, Roux S, Leconte I, Landzberg M, Simonneau G. Bosentan therapy for pulmonary arterial hypertension. N Engl J Med 2002;346(12):896–903.
3. Barst RJ, Rich S, Widlitz A, Horn EM, McLaughlin V, McFarlin J. Clinical efficacy of sitaxsentan, an endothelin-A receptor antagonist, in patients with pulmonary arterial hypertension: open-label pilot study. Chest 2002;121(6):1860–8.
4. Badesch DB, Bodin F, Channick RN, Frost A, Rainisio M, Robbins IM, Roux S, Rubin LJ, Simonneau G, Sitbon O, Tapson VF. Complete results of the first randomized, placebo-controlled study of bosentan, a dual endothelin receptor antagonist, in pulmonary arterial hypertension. Curr Ther Res Clin Exp 2002;63:227–46.
5. Dingemanse J, van Giersbergen PL. Clinical pharmacology of bosentan, a dual endothelin receptor antagonist. Clin Pharmacokinet 2004;43(15):1089–115.
6. Segal ES, Valette C, Oster L, Bouley L, Edfjall C, Herrmann P, Raineri M, Kempff M, Beacham S, van Lierop C. Risk management strategies in the postmarketing period:safety experience with the US and European bosentan surveillance programmes. Drug Saf 2005;28(11):971–80.
7. Humbert M, Barst RJ, Robbins IM, Channick RN, Galie N, Boonstra A, Rubin LJ, Horn EM, Manes A, Simonneau G. Combination of bosentan with epoprostenol in pulmonary arterial hypertension: BREATHE-2. Eur Respir J 2004;24(3):353–9.
8. Krum H, Viskoper RJ, Lacourciere Y, Budde M, Charlon V. The effect of an endothelin-receptor antagonist, bosentan, on blood pressure in patients with essential hypertension. Bosentan Hypertension Investigators. N Engl J Med 1998;338(12):784–90.
9. Kalra PR, Moon JC, Coats AJ. Do results of the ENABLE (Endothelin Antagonist Bosentan for Lowering Cardiac Events in Heart Failure) study spell the end for non-selective endothelin antagonism in heart failure? Int J Cardiol 2002;85(2–3):195–7.
10. Doggrell SA. The therapeutic potential of endothelin-1 receptor antagonists and endothelin-converting enzyme inhibitors on the cardiovascular system. Expert Opin Investig Drugs 2002;11(11):1537–52.
11. Fattinger K, Funk C, Pantze M, Weber C, Reichen J, Stieger B, Meier PJ. The endothelin antagonist bosentan inhibits the canalicular bile salt export pump: a potential mechanism for hepatic adverse reactions. Clin Pharmacol Ther 2001;69(4):223–31.
12. Fouassier L, Kinnman N, Lefevre G, Lasnier E, Rey C, Poupon R, Elferink RP, Housset C. Contribution of mrp2 in alterations of canalicular bile formation by the endothelin antagonist bosentan. J Hepatol 2002;37(2):184–91.
13. Gasser S, Kuhn M, Speich R. Severe necrotising leucocytoclastic vasculitis in a patient taking bosentan. BMJ 2004;329(7463):430.
14. Dingemanse J, van Giersbergen PL. Influence of severe renal dysfunction on the pharmacokinetics and metabolism of bosentan, a dual endothelin receptor antagonist. Int J Clin Pharmacol Ther 2002;40(7):310–6.

Calcium channel blockers

See also Amlodipine, Barnidipine, Diltiazem, Isradipine, Lercanidipine, Lidoflazine, Manidipine, Mibefradil, Nicardipine, Nifedipine, Nimodipine, Nitrendipine, Prenylamine, Verapamil

General Information

The calcium channel blockers block the movement of calcium across L-type calcium channels. The main drugs that share this action are verapamil (a phenylalkylamine), diltiazem (a benzthiazepine), and the dihydropyridines, which include amlodipine, darodipine, felodipine, isradipine, lacidipine, lercanidipine, manidipine, nicardipine, nifedipine, nimodipine, nisoldipine, and nitrendipine. Other agents, for example prenylamine and lidoflazine, are now rarely used, and perhexiline, having failed to reach the market at all in some countries, was withdrawn in the UK after continuing concerns about its safety (1). Mibefradil blocks T type calcium channels; it was withdrawn within 1 year of marketing because of multiple drug interactions, emphasizing the need for rigorous drug assessment before release and the importance of postmarketing surveillance of new drugs (2).

Although they are chemically heterogeneous, many adverse effects are common to all calcium channel blockers, predictable from their pharmacological actions. Calcium plays a role in the functions of contraction and conduction in the heart and in the smooth muscle of arteries; drugs that interfere with its availability (of which there are many, the calcium channel blockers being the most specific) will therefore act in all these tissues. A few idiosyncratic and hypersensitivity reactions have also been reported with individual calcium channel blockers.

The properties of these drugs vary widely (3). Nifedipine is said to have little negative inotropic effect and no effect on the atrioventricular node; verapamil is a potent cardiac depressant, with a marked effect on the atrioventricular node; and diltiazem has less cardiac depressant effect but inhibits atrioventricular nodal activity.

Controversy has surrounded the use of calcium channel blockers in the treatment of hypertension (4) and ischemic heart disease (5), with evidence for an association with unfavorable coronary outcomes compared with other therapies. Most of the evidence comes from the use of short-acting formulations, especially short-acting nifedipine. The hypothesis put forward to explain these findings was that short-acting formulations cause reflex activation of the sympathetic nervous system (6). Further observational studies showed that these drugs were also associated with gastrointestinal hemorrhage (7) and cancer (8). Claims of conflicts of interests amongst authors were published (9), often amid heated debate. Further evidence against the short-acting calcium channel blockers in hypertension has been forthcoming (10,11) and a worse than expected outcome with respect to coronary outcomes in diabetic patients has fuelled the debate (12). Further evidence of gastrointestinal bleeding has been published (13), but there is evidence against a link with cancer (14). All these scares have undoubtedly reduced the standing of this class of drugs in the eyes of physicians.

That calcium channel blockers are effective in relieving the symptoms of angina pectoris is beyond doubt. However, the Angina and Silent Ischemia Study (15), in which nifedipine, diltiazem, and propranolol were compared with placebo in a crossover study, produced conflicting results. Only diltiazem improved treadmill exercise time and only propranolol convincingly reduced the number of silent ischemic episodes during ambulatory monitoring. These findings are hard to explain (16). Beta-blockers may be cardioprotective and therefore preferable to calcium channel blockers (17). The clinically significant deterioration seen in patients with impaired left ventricular function taking calcium channel blockers is important, as many patients with angina have previously had a myocardial infarction or have poor left ventricular function. Calcium channel blockers cannot be assumed to be safe second-line drugs for angina in patients with poor cardiac reserve, although newer agents may prove to be safer (18).

Calcium channel blockers are very effective in controlling variant angina, and are often used during coronary angioplasty and after coronary artery surgery. They are also useful in patients who are intolerant of beta-blockers (19), or who have a poor response to nitrates, or who have concurrent hypertension.

General adverse effects

Throbbing headache, facial warmth and flushing, and dizziness are minor complaints associated with the use of calcium channel blockers; these effects are believed to be caused by inhibitory actions on smooth muscle (20). Palpitation, muscle cramps, and pedal edema also occur (21–27). Dizziness, facial flushing, leg edema, postural hypotension, and constipation have been reported in up to one-third of patients. They are rarely severe and often abate on continued therapy. More serious adverse effects, mainly those affecting cardiac conduction, are much less common, and only rarely is withdrawal necessary.

Organs and Systems

Cardiovascular

Cardiac failure

Although acute hemodynamic studies have suggested that calcium channel blockers can be beneficial in cardiac failure (28), long-term treatment has been associated with clinical deterioration. Calcium channel blockers should therefore be prescribed with caution for patients with impaired cardiac function, who should be regularly reassessed; treatment should be withdrawn if the signs or symptoms of cardiac failure appear. In some cases heart failure is predictable, as in the case of a patient with aortic stenosis who developed left ventricular failure after treatment with nifedipine (29). Increased sympathetic activity can also compensate for the myocardial suppressant effects of calcium channel blockers, and the combination of these drugs (particularly verapamil) with beta-adrenoceptor antagonists has therefore given cause for concern in the past, although this combination is now considered relatively safe for the majority of patients with normal cardiac function (30–32).

Myocardial ischemia

There have been many studies of the efficacy of calcium channel blockers in early and late intervention in myocardial infarction (33). These studies have failed to show convincing benefits. Indeed, in the nifedipine intervention studies there was a consistent trend towards higher mortality in the treated patients than in those taking placebo. A study in which patients were randomized to placebo or nifedipine within 48 hours of admission was terminated after 1358 patients had been recruited, because mortality at 6 months was 15.4% on nifedipine and 13.3% on placebo (34).

It has been argued that dihydropyridine calcium channel blockers, which increase heart rate, can all increase the risk of death and reinfarction (35,36). Early beneficial results with diltiazem in patients with non-Q-wave infarction (37) were not confirmed in the Multicenter Diltiazem Postinfarction Trial (38). In patients with pulmonary congestion, diltiazem was associated with an increase in cardiac events, and there was a similar result in patients with low ejection fractions. However, verapamil does appear to reduce reinfarction (39), a benefit that is more marked in those without heart failure (40). Nifedipine may also have a detrimental effect in unstable angina; it certainly appears to offer no benefit (41).

A retrospective case-control study (4) has sparked controversy concerning the use of short-acting calcium channel blockers in treating hypertension. The study involved 623 cases of fatal and non-fatal myocardial infarction over a period of 8 years, and 2032 age- and sex-matched controls. The risk of myocardial infarction in patients taking calcium channel blockers was 16 per 1000, compared with 10 per 1000 in patients taking beta-blockers or thiazides. However, this result may have been an example of confounding by indication, since patients exposed to calcium channel blockers will have been more likely to have had peripheral vascular disease, lung disease (a low forced expiratory volume being a risk factor for cardiovascular disease), higher serum cholesterol concentrations, and diabetes mellitus. Careful statistical analysis was carried out in an attempt to control for some of the confounding factors, but such confounding can only be properly controlled for in a randomized study. A meta-analysis of 16 randomized secondary prevention studies in patients with coronary heart disease showed that the use of short-acting nifedipine is associated with an increased mortality in a dose-related manner (dose, risk: 30–50 mg, 1.06; 60 mg, 1.18; 80 mg, 2.83) (5). However, the event rates in this study were relatively small. A prospective cohort study in 906 elderly hypertensive patients showed that short-acting nifedipine is associated with a relative mortality risk of 1.7 compared with beta-blockers (42). After the publication of these studies, the FDA recommended that short-acting nifedipine should no longer be used in hypertension or unstable angina (43).

Disturbances of cardiac rhythm

Calcium channel blockers differ in their effects on the myocardial conduction system. Both verapamil and diltiazem have significant inhibitory effects on both sinoatrial and atrioventricular nodal function, whereas nifedipine has little or no effect. Nevertheless, nifedipine can on occasion cause troublesome bradydysrhythmias (44,45).

Severe conduction disturbances can also occur if calcium channel blockers are used in hypertrophic cardiomyopathy (46), but these drugs are used in this condition (47).

Hypotension

There are many case reports of symptomatic hypotension, usually in hypertensive patients treated with large dosages of calcium channel blockers (48,49) or in patients with myocardial infarction (50). These may represent injudicious prescribing rather than true adverse drug effects. In the DAVIT II study, 1.9% of the verapamil-treated group versus 1.6% of the placebo-treated group developed hypotension or dizziness (40); the frequency of hypotension in a randomized study of diltiazem after infarction was 0.6% in the drug-treated group and 0.2% in the placebo-treated group (37).

Respiratory

Adverse respiratory effects are uncommon with calcium channel blockers. However, three cases of acute bronchospasm accompanied by urticaria and pruritus have been reported in patients taking verapamil (51), and a patient with Duchenne-type muscular dystrophy developed respiratory failure during intravenous verapamil therapy for supraventricular tachycardia (52). Recurrent exacerbations of asthma occurred in a 66-year-old lady with hypertension and bronchial asthma given modified-release verapamil (53).

In pulmonary hypertension, both verapamil and nifedipine increase mean right atrial pressure in association with hypotension, chest pain, dyspnea, and hypoxemia; the severe hemodynamic upset resulted in cardiac arrest in two patients after verapamil and death in another after nifedipine (54). A patient with pulmonary hypertension also developed pulmonary edema whilst taking nifedipine (55) and another seems to have developed this as an allergic reaction (56).

Nervous system

Calcium channel blockers can cause parkinsonism. Of 32 patients with this complication, only three had made a full recovery 18 months after withdrawal; patients under 73 years of age tended to have a better prognosis (57). It is not known if these patients would have developed parkinsonism in any case, and whether the drugs merely act as precipitants.

Calcium channel blockers can worsen myasthenic syndromes. Myasthenia gravis can deteriorate with oral verapamil (58). A patient with Lambert–Eaton syndrome and a small-cell carcinoma of the lung developed respiratory failure within hours of starting treatment with verapamil for atrial flutter, and required assisted ventilation (59). Only after verapamil had been withdrawn did breathing improve. Verapamil affects calcium channels in nerve membranes in animals, but the experimental concentrations used exceeded those found in clinical practice (59). Thus, the evidence for a drug-related effect is circumstantial. In another case, diltiazem triggered Lambert–Eaton syndrome, which improved with drug withdrawal (60).

Sensory systems

Eyes

Painful eyes occurred in 14% of patients taking nifedipine compared with 9% in captopril-treated patients in a post-marketing surveillance study (61). The mechanism is unknown but is not via ocular vasodilatation (62).

Taste

Transient disturbances of taste and smell, without other signs of neurological deficit, have been reported after nifedipine and diltiazem. The time to the onset of symptoms after nifedipine varied from days to months, and symptoms regressed within 24 hours of withdrawal (63). With diltiazem the effect gradually abated over 10 weeks, despite continuation of therapy (64).

Psychological, psychiatric

A patient taking diltiazem developed the signs and symptoms of mania (65) and another developed mania with psychotic features (66). There have also been reports that nifedipine can cause agitation, tremor, belligerence, and depression (67), and that verapamil can cause toxic delirium (68). Nightmares and visual hallucinations have been associated with nifedipine (69). Depression has been reported as a possible adverse effect of nifedipine (70).

Some reports have suggested that calcium channel blockers may be associated with an increased incidence of depression or suicide. However, there is a paucity of evidence from large-scale studies. A study of the rates of depression with calcium channel blockers, using data from prescription event monitoring, involved gathering information on symptoms or events in large cohorts of patients after the prescription of lisinopril, enalapril, nicardipine, and diltiazem by general practitioners (71). The crude overall rates of depression during treatment were 1.89, 1.92, and 1.62 per 1000 patient-months for the ACE inhibitors, diltiazem, and nicardipine respectively. Using the ACE inhibitors as the reference group, the rate ratios for depression were 1.07 (95% CI = 0.82, 1.40) and 0.86 (0.69, 1.08) for diltiazem and nicardipine respectively. This study does not support the hypothesis that calcium channel blockers are associated with depression.

Endocrine

In six hypertensive patients given nitrendipine 20 mg/day for 30 days, there was inhibition of aldosterone response but no significant change in ACTH secretion in response to corticotrophin-releasing hormone (72).

The calcium-dependent pathway of aldosterone synthesis in the zona glomerulosa is blocked by calcium channel blockers, producing a negative feedback increase in the pituitary secretion of ACTH, which in turn causes hyperplasia of the zona glomerulosa. This leads to increased production of androgenic steroid intermediate products and subsequently testosterone, which acts on gingival cells and matrix, giving rise to gingival hyperplasia (see the section on Mouth and teeth).

Metabolism

Calcium transport is essential for insulin secretion, which is therefore inhibited by calcium channel blockers (73). Despite this, calcium channel blockers generally have minimal effects on glucose tolerance in both healthy and diabetic subjects. Oral glucose tolerance is not affected by verapamil, and basal blood glucose concentrations were not altered during long-term verapamil administration (74). Similarly, neither nifedipine nor nicardipine produced significant hyperglycemic effects in either diabetic or non-diabetic patients (75–77). In 117 hypertensive patients nifedipine caused a significant rise in mean random blood glucose of only 0.3 mmol/l (78), an effect that was clearly of no clinical relevance. In the Treatment of Mild Hypertension Study, 4 years of monotherapy with amlodipine maleate caused no change compared with placebo in the serum glucose of 114 hypertensive patients (79). In a review (80) it was concluded that in usual dosages calcium channel blockers do not alter glucose handling. However, in a few patients diabetes appeared de novo or worsened considerably on starting nifedipine (78,81), so there may be a small risk in some individuals.

Fluid balance

Edema of the legs is a well-recognized reaction to nifedipine and also occurs with verapamil, diltiazem, and the long-acting dihydropyridines (82,83), suggesting that this is a class effect of calcium channel blockers.

Hematologic

Calcium channel blockers rarely cause hematological effects. A hemorrhagic diathesis, including impaired platelet function, develops in chronic renal insufficiency, in which calcium channel blockers are used widely as antihypertensive agents. In 156 patients with moderate to severe chronic renal insufficiency not on hemodialysis calcium channel blockers prolonged the bleeding time (OR = 3.52; 95% CI = 1.01, 12.3) (84). However, despite this effect, there were no clinically serious hemorrhagic events during the study. Among those taking calcium channel blockers, 21 patients with prolonged bleeding times were randomly assigned to two groups; in one group treatment was withdrawn and bleeding time shortened; in those who continued to take the treatment the bleeding time was unchanged.

Nifedipine has been reported to cause agranulocytosis (85) and leukopenia was attributed to diltiazem; the latter patient had scleroderma, active rheumatoid disease, and pulmonary fibrosis, but the white cell count fell after 3 weeks of diltiazem, recovered on withdrawal, and fell on rechallenge (86). Diltiazem has also been reported to cause immune thrombocytopenia in a 68-year-old man with angina (87).

Mouth and teeth

Gingival hyperplasia, similar to that seen with phenytoin and ciclosporin, is a rare but well-recognized adverse effect of nifedipine (88). It has also been reported in patients taking felodipine (89,90), nitrendipine (SEDA-16, 200), and verapamil (91), suggesting that this adverse effect is a class effect. Only one case of gingival hyperplasia related to calcium channel blockers was reported to the Norwegian Adverse Drug Reaction Committee up to 1991, despite their widespread use (92). However, subclinical gingival hyperplasia on tissue histology was found in 83 and 74% of patients taking nifedipine and diltiazem respectively (93). The reaction generally occurs within a few months of starting treatment, and in some cases drug withdrawal produces marked regression of clinical hyperplasia. The mechanism of this adverse effect is unclear, but has been proposed to involve a hormonal imbalance in the hypothalamic–pituitary–adrenal axis (94).

Periodontal disease has been assessed in 911 patients taking calcium channel blockers, of whom 442 were taking nifedipine, 181 amlodipine, and 186 diltiazem, and in 102 control subjects (95). There was significant gingival overgrowth in 6.3% of the subjects taking nifedipine, while the prevalence induced by amlodipine or diltiazem was not significantly different than in the controls. The severity of overgrowth in the nifedipine group was related to the amount of gingival inflammation and also to sex, men being three times as likely to develop overgrowth than women.

Gastrointestinal

Because of effects on smooth muscle, the calcium channel blockers (particularly verapamil (96) but also diltiazem) can cause constipation. This may be due to colonic motor activity inhibition (97). Gastroesophageal reflux can also occur, and the calcium channel blockers should be avoided in patients with symptoms suggestive of reflux esophagitis (98). Calcium channel blockers (verapamil, diltiazem, and nifedipine) can also be associated with an increased incidence of gastrointestinal bleeding, as reported in a prospective cohort study in 1636 older hypertensives, with a relative risk of 1.86 (95% CI = 1.22, 2.82) compared with beta-blockers (7). However, this finding was not confirmed in other retrospective studies (13,99,100).

Liver

Mild hepatic reactions have been observed in association with verapamil, nifedipine (101–104), and diltiazem (105,106). In some cases fever, chills, and sweating have been associated with right upper quadrant pain, hepatomegaly, and mild increases in serum bilirubin and transaminase activity; in others, patients have remained asymptomatic. One patient had granulomatous hepatitis with diltiazem (107); another had a periportal infiltrate rich in eosinophils while taking verapamil (108). The increase in liver enzyme activities is generally transient, although mild persistent abnormalities have been seen. Occasionally, extreme increases in hepatic enzyme activities have been reported (85,105). Their frequency appears to be low, and since the symptoms and signs are mild they could easily be overlooked.

Skin

Apart from minor flushing and leg erythema associated with edema, skin reactions with calcium channel blockers are infrequent; the frequency has been estimated at 1.3% for diltiazem (109).

An erythematous rash with painful edema has been described with nifedipine (110) and also with diltiazem, but without the edematous element (109).

Mild erythema multiforme and Stevens–Johnson syndrome have been reported as probable reactions to diltiazem (85) and long-acting nifedipine (111).

Nifedipine, verapamil, and diltiazem have all been implicated as possible causes of erythema multiforme and its variants, Stevens–Johnson syndrome and toxic epidermal necrolysis, and/or exfoliative dermatitis from FDA data (112).

Psoriasiform eruptions have been reported in patients taking verapamil and nicardipine (SEDA-16, 199).

Photo-induced annular or papulosquamous eruptions due to subacute cutaneous lupus erythematosus with positive antinuclear, anti-Ro, and anti-La antibodies have been reported with verapamil, nifedipine, and diltiazem (113). The association of calcium channel blockers with photo-damage has been assessed in 82 patients with renal transplants (114). Most of the patients (90%) had photo-damaged skin (50% mild, 24% moderate, and 13% severe) and 53 (65%) had used a calcium channel blocker (49 nifedipine and four amlodipine). There were strong associations between calcium channel blockers and the grade of photo-damage and the presence of telangiectasia, with a less marked association with solar elastosis. There was no convincing association between the grade of photo-damage and the duration of treatment.

Reproductive system

Calcium channel blockers can occasionally cause menorrhagia (115) and gynecomastia (116).

Immunologic

Verapamil, nifedipine, and diltiazem have all been associated with allergic reactions, including skin eruptions and effects on liver and kidney function. Nifedipine has also been reported to cause a febrile reaction (117), and diltiazem was associated with fever, lymphadenopathy, hepatosplenomegaly, an erythematous maculopapular rash, and eosinophilia in a 50-year-old man (118).

Long-Term Effects

Drug withdrawal

The possibility of a calcium antagonist withdrawal syndrome has been raised (119–128), as it has been reported that withdrawal of verapamil, nifedipine, and diltiazem can worsen angina or even cause myocardial infarction. However, in a randomized, double-blind study of withdrawal of nifedipine in 81 patients before coronary artery bypass surgery, angina at rest occurred only in patients who had experienced similar symptoms previously, and there were no early untoward effects of drug withdrawal (121). If a withdrawal syndrome does exist, it could be due to rebound coronary vasospasm, but the present weight of evidence suggests that withdrawal results in no more than the loss of a useful therapeutic effect or the unmasking of progressive disease (128).

Tumorigenicity

A retrospective cohort study in 5052 elderly subjects, of whom 451 were taking verapamil, diltiazem, or nifedipine, showed that these drugs were associated with a cancer risk of 1.72 (95% CI = 1.27, 2.34), and there was a significant dose–response relation (8). A small risk of cancer (RR = 1.27; 95% CI = 0.98, 1.63) with calcium channel blockers was reported in a nested case-control retrospective study involving 446 cases of cancers in hypertensive patients (129). However, the authors concluded that this finding may have been spurious, as there was no relation between the cancer risk and the duration of drug use. Another study did not show any excess cancer risk with short-acting nifedipine after myocardial infarction in patients followed up for 10 years, although there were only 22 cancer deaths in 2607 patients (130). Neither did the much larger Bezafibrate Infarction Prevention (BIP) Study, which reported cancer incidence data in 11 575 patients followed for a mean period of 5.2 years, with 246 incident cancer cases, 129 among users (2.3%) and 117 (2.1%) among non-users of calcium channel blockers (131). Others also failed to find a positive link between calcium channel blockers and cancer (14,132). However, elderly women taking estrogens and short-acting calcium channel blockers had a significantly increased risk of breast carcinoma (hazard ratio = 8.48; 95% CI = 2.99, 24) (133).

In another study, the responses of 975 women with invasive breast carcinoma were compared with the responses of 1007 women in a control group (134). Women who had ever used calcium channel blockers, beta-blockers, or ACE-inhibitors did not have an altered risk of breast carcinoma compared with women who had never used antihypertensive drugs. There was a modestly increased risk of breast carcinoma among users of immediate-release calcium channel blockers (OR = 1.5; 95% CI = 1.0, 2.1), thiazide diuretics (OR = 1.4; 95% CI = 1.1, 1.8), and potassium-sparing diuretics (OR = 1.6; 95% CI = 1.2, 2.1). No clear trends emerged from the analysis of the correlations between risk and duration of use. This controversy can perhaps only be resolved by prospective studies with longer follow-up periods (135), although ideal studies are unlikely ever to be conducted.

Second-Generation Effects

Pregnancy

The calcium channel blockers have had very limited use in pregnancy. The absence of reports of fetal deaths, malformations, or other maternal or neonatal adverse effects cannot therefore be construed as indicating safety. However, a comparison of nifedipine and hydralazine in 54 patients with severe pre-eclampsia showed that nifedipine is more effective, allowing delivery of more mature infants (136).

- Modified-release nifedipine 40 mg tds caused marked hypotension when used to delay preterm labor in a previously healthy 29-year-old woman who started contracting at 29 weeks; the hypotension may have precipitated an uncomplicated non-Q-wave myocardial infarction (137).

When nifedipine is combined with intravenous magnesium to delay preterm labor, colonic pseudo-obstruction can occur (138).

Lactation

Both verapamil (139) and diltiazem (140) are excreted in breast milk, but the risk to the suckling infant is unclear.

Susceptibility Factors

Patients with impaired function of the sinus node or impaired atrioventricular conduction can develop sinus bradycardia, sinus arrest, heart block, hypotension and shock, and even asystole, with verapamil (141) or diltiazem. These drugs should not be given to patients with aberrant conduction pathways associated with broad-complex tachydysrhythmias, and they can cause severe conduction disturbances in hypertrophic cardiomyopathy.

Similarly, verapamil should be used with caution in patients with heart failure, and both diltiazem and nifedipine can cause problems in patients with poor cardiac reserve. However, the PRAISE study (18) suggested that amlodipine may be used safely, even in the presence of severe heart failure optimally treated with diuretics, digoxin, and ACE inhibitors. In this study, amlodipine significantly reduced cardiac mortality by more than a third in non-ischemic dilated cardiomyopathy, without significantly affecting mortality in ischemic cardiomyopathy (142).

Calcium antagonists should be avoided when possible in the peri-infarction period.

The use of calcium channel blockers in patients with pulmonary hypertension has been associated with cardiac arrest and sudden death.

Caution should be exercised in using verapamil in patients with hepatic cirrhosis, as its metabolism is reduced, leading to high plasma concentrations and potential toxicity (SEDA-16, 198). Similarly, lower starting and maintenance doses of other calcium channel blockers should be used in the presence of liver impairment. This also applies to patients with chronic renal insufficiency, especially those taking the modified-release formulation of verapamil (SEDA-17, 238).

Drug Administration

Drug overdose

The treatment of overdosage with calcium channel blockers has been reviewed (143,144); other reports have reviewed poisoning with verapamil (145–147), and other calcium channel blockers (148,149).

The features appear to be arterial hypotension, bradycardia due to sinus node depression and atrioventricular block, and congestive cardiac failure and angina (150–152). Although the therapeutic effects are different according to the drug, in overdosage the effects are similar (148). Severe metabolic acidosis (usually lactic acidosis) and generalized convulsions can also occur (153) and hypoglycemia has been reported (154). Non-cardiogenic pulmonary edema has been reported with diltiazem (155) and verapamil (156). Several deaths have occurred with verapamil.

- An overdose of nifedipine 280 mg produced marked vasodilatation in a young patient with advanced renal insufficiency; it was successfully treated with intravenous calcium (157).

An overdose of a mixture of calcium channel blockers mimicked acute myocardial infarction (158).

- A 42-year-old man developed shortness of breath, weakness, sweating, and left bundle branch block. Coronary angiography showed only non-obstructive lesions, ruling out acute closure of a coronary artery, and his left ventriculogram showed no wall motion abnormalities, but rather a markedly hyperdynamic left ventricle with an ejection fraction of 80%. Despite this, he subsequently developed profound bradycardia and hypotension, which were refractory to standard treatments, including pressor agents, calcium, and transvenous pacing. He gradually improved over several days and made a full recovery. After extubation he admitted to having taken "several" tablets each of long-acting verapamil, diltiazem, and nifedipine, of unclear dosages, and over an unclear period time, trying to self-medicate for symptoms he related to life-long paroxysmal supraventricular tachycardia.

This case highlights the fact that calcium channel blocker overdose must be considered in the differential diagnosis of patients who present with apparent acute myocardial infarction.

Treatment consists of gastric lavage, activated charcoal, and cathartics. Contrary to popular belief, significant overdosage of immediate-release verapamil can be associated with delayed absorption, as suggested by a case report, the authors of which suggested the use of repeated doses of activated charcoal (145). In severe cases total gut lavage should be considered. Intravenous calcium gluconate (159), glucagon (160), pressor amines (isoprenaline, adrenaline, or dobutamine), artificial ventilation, and cardiac pacing may all be required. Hemoperfusion does not appear to influence the clinical course, but 4-aminopyridine reversed the features of a modest accidental overdose of verapamil in a patient on maintenance hemodialysis (161). The rationale for the use of aminopyridine, an antagonist of non-depolarizing neuromuscular blocking agents, supported by prior animal experiments, was the enhancement of transmembrane calcium flux and the facilitation of synaptic transmission. This is of potential value, in view of the apparent unresponsiveness of some patients to supportive measures.

Five cases of overdose of calcium channel blockers have been reported (162):

- a 34-year-old woman who took amlodipine 0.86 mg/kg;
- a 48-year-old man who took an unknown amount of modified-release diltiazem;
- a 5-month-old girl inadvertently given nifedipine 20 mg;
- a 14-year-old girl who took modified-release verapamil 30 mg/kg;
- a 31-year-old man who took modified-release verapamil 71 mg/kg.

All were successfully treated with hyperinsulinemia/euglycemia therapy. The authors described the mechanism of action of this form of therapy, which is mainly related to improvement in cardiac contractility and peripheral vascular resistance and reversal of acidosis. They proposed indications and dosing for this therapy consisting in most cases of intravenous glucose with an intravenous bolus dose of insulin 1 U/kg followed by an infusion of 0.5–1 U/kg/hour until the systolic blood pressure is over 100 mm/Hg and the heart rate over 50/minute. Hyperinsulinemia/euglycemia therapy is currently reserved as an adjunct to conventional therapy and is recommended only after an inadequate response to fluid resuscitation, high-dose calcium salts, and pressor agents.

- A 43-year old man took amlodipine 560 mg and failed to respond to fluid resuscitation, calcium salts, glucagon, and noradrenaline/adrenaline inotropic support (163). However, intravenous metaraminol 2 mg followed by 83 micrograms/minute produced an improvement in his blood pressure, cardiac output, and urine output.
- A 65-year-old man with aortic stenosis died after mistakenly taking six tablets of modified-release diltiazem SR 360 mg (164). He developed symptoms of toxicity within 7 hours and died after 17 hours. The diltiazem concentration in an antemortem blood sample 11.5 hours after ingestion was 2.9 μg/ml and in a postmortem sample of central blood 6 μg/ml.

Drug–Drug Interactions

Benzodiazepines

Diltiazem and verapamil compete for hepatic oxidative pathways that metabolize most benzodiazepines, as well as zolpidem, zopiclone, and buspirone (SEDA-22, 39) (SEDA-22, 41).

Beta-adrenoceptor antagonists

The greatest potential for serious mishap arises from interactions between calcium channel blockers (especially verapamil and related compounds) and beta-adrenoceptor antagonists (165,166). This combination can cause severe hypotension and cardiac failure, particularly in patients with poor myocardial function (167–169). The major risk appears to be associated with the intravenous administration of verapamil to patients who are already taking a beta-blocker (170), but a drug-like tiapamil, which closely resembles verapamil in its pharmacological profile, might be expected to carry a similar risk (171). Conversely, intravenous diltiazem does not produce deleterious hemodynamic effects in patients taking long-term propranolol (172). However, there have been instances when the combination of diltiazem with metoprolol caused sinus arrest and atrioventricular block (173).

The concurrent use of oral calcium channel blockers and beta-adrenoceptor antagonists in the management of angina pectoris or hypertension is less likely to result in heart block or other serious adverse effects (174), and these two drug groups are commonly used together. However, caution is still advised, and nifedipine or other dihydropyridine derivatives would be preferred in this type of combination (26,175,176). Nevertheless, the combination of nifedipine with atenolol in patients with stable intermittent claudication resulted in a reduction in walking distance and skin temperature, whereas either drug alone produced benefits (177).

Bupivacaine

Calcium channel blockers in combination with bupivacaine produce significant negative inotropic effects on the heart in animals, possibly due to reduced protein binding of the local anesthetic, as well as a generalized myocardial depressant effect (178,179).

However, bupivacaine cardiotoxicity was reduced in rats by pretreatment with low doses of calcium channel blockers (180). In vivo, the LD_{50} for bupivacaine was increased from 3.08 to 3.58 mg/kg after pretreatment with verapamil 150 micrograms/kg, and to 3.50 mg/kg after nimodipine 200 micrograms/kg. Of the rats that died, only one developed cardiac arrest first, whilst the majority developed respiratory arrest. In vitro, bupivacaine alone dose-dependently reduced heart rate, contractile force, and coronary perfusion pressure. Dysrhythmias were also noted: bradycardias, ventricular extra beats, and ventricular tachycardia were the most common. Verapamil made no difference to these adverse effects, but nimodipine significantly reduced the negative chronotropic and dysrhythmogenic effects of bupivacaine. These results, although interesting, cannot be used to reach any clinical conclusions, particularly as the mechanism of interaction between bupivacaine and calcium channel blockers has yet to be elucidated.

Buspirone

In a randomized placebo-controlled trial, the possible interactions of buspirone with verapamil and diltiazem were investigated. Both verapamil and diltiazem considerably increased plasma buspirone concentrations, probably by inhibiting CYP3A4. Thus, enhanced effects and adverse effects of buspirone are possible when it is used with verapamil, diltiazem, or other inhibitors of CYP3A4 (181).

Carbamazepine

A pharmacokinetic interaction has been described between carbamazepine and the calcium channel blockers verapamil (182) and diltiazem (183). With both drugs, inhibition of the hepatic metabolism of carbamazepine resulted in increased serum carbamazepine concentrations and neurotoxicity, with dizziness, nausea, ataxia, and diplopia. Adding nifedipine to carbamazepine was not associated with alterations in steady-state carbamazepine concentrations (183).

Cardiac glycosides

Calcium channel blockers interact with cardiac glycosides. The main mechanism is inhibition of digoxin renal tubular secretion by inhibition of P glycoprotein. In a review of the interactions of calcium channel blockers with digoxin, in which their clinical relevance was assessed, it was concluded that serious consequences can be prevented by careful monitoring, especially in patients whose serum digoxin concentration is already near the upper end of the therapeutic range (184).

Verapamil suppresses renal digoxin elimination acutely, but this suppression disappears over a few weeks (185). However, inhibition of the extrarenal clearance of digoxin persists, and the result of this complex interaction is an increase in steady-state plasma digoxin concentrations of less than 100%. Patients taking both drugs should be carefully monitored. However, the pharmacodynamic effects of digoxin are apparently reduced by verapamil (186), so that dosage adjustment may be unnecessary. Cardiovascular collapse and/or asystole has followed the use of intravenous verapamil in patients taking oral digoxin alone (187) or in combination with quinidine, propranolol, or disopyramide (171).

The interaction of digoxin with nifedipine increases plasma digoxin concentrations by only about 15% (188,189) and is less important.

Diltiazem increases digoxin concentrations by 20–50% (190,191).

Interactions of digoxin with nitrendipine (192) and bepridil (193) have also been described.

Verapamil and diltiazem, but not nifedipine, increase steady-state plasma digitoxin concentrations (194).

Calcium channel blockers have varying effects on the disposition of digoxin. The calcium channel blockers for which varying amounts of information are available include cinnarizine, diltiazem, felodipine, fendiline, gallopamil, isradipine, lidoflazine, mibefradil, nicardipine, nifedipine, nitrendipine, tiapamil, and verapamil (195).

Diltiazem

Studies of the effects of diltiazem on the pharmacokinetics of digoxin have yielded variable results. In one study, diltiazem reduced the steady-state total body clearance of beta-acetyldigoxin in 12 healthy men, perhaps because of reduced renal and non-renal clearances (SEDA-10, 145). In some studies diltiazem 120–240 mg/day increased steady-state plasma digoxin concentrations by about 20–40% (SEDA-14, 146) (196), although not in other studies (SEDA-14, 146), and reduced the total body clearance of digoxin, with changes in both renal and non-renal clearances (SEDA-11, 154), although others did not find this (197). In at least one case digoxin toxicity was attributed to this interaction (SEDA-17, 216). In eight patients with chronic heart failure taking digoxin 0.25 mg/day, diltiazem 180 mg/day increased the AUC and mean steady-state serum concentrations of digoxin by 50% and reduced its total clearance (198).

Mibefradil

In 40 healthy subjects mibefradil 50 or 100 mg/day for 6 days had no significant effects on the steady-state pharmacokinetics of digoxin, apart from a very small increase in the C_{max} (199).

Tiapamil

Tiapamil reversed digoxin-induced splanchnic vasoconstriction in healthy men (200), but this has no direct effect on systemic hemodynamics.

Verapamil

Verapamil increases plasma digoxin concentrations at steady state by inhibiting the active tubular secretion and non-renal clearance of digoxin (201,202). There is only anecdotal evidence that this can result in digitalis toxicity (202,203).

Verapamil reversed digoxin-induced splanchnic vasoconstriction in healthy men (200), but this has no direct effect on systemic hemodynamics.

Ciclosporin

Calcium channel blockers are given to transplant patients for their protective effect against ciclosporin-induced nephrotoxicity and to optimize ciclosporin immunosuppression in order to reduce early rejection of renal grafts. Nifedipine has been used to treat ciclosporin-induced hypertension, although amlodipine may be just as effective (204).

However, some calcium channel blockers have pharmacokinetic interactions: diltiazem, verapamil, nicardipine, and amlodipine increase ciclosporin concentrations, whereas nifedipine, felodipine, and isradipine do not (SED-14, 604) (SEDA-21, 210) (SEDA-21, 212) (SEDA-22, 216). Two confirmations of these observations have been published. In a retrospective study of 103 transplant patients verapamil and diltiazem, but not nifedipine or isradipine, caused a significant increase in plasma ciclosporin concentrations (205). The effect of verapamil and diltiazem on ciclosporin concentrations was independent of dosage. In a crossover comparison between verapamil, felodipine, and isradipine in 22 renal transplant recipients, verapamil interacted pharmacokinetically with ciclosporin but felodipine and isradipine did not (206).

Nine kidney transplant recipients had an increase in trough whole blood ciclosporin concentrations of 24–341% after introduction of nicardipine (207). A similar interaction has been reported with diltiazem (208) and verapamil (209).

A large amount of data has accumulated on the effects of various calcium channel blockers on ciclosporin metabolism or a possible renal protective effect. Diltiazem, nicardipine, or verapamil inhibit ciclosporin metabolism, and this has been investigated as a potential beneficial combination for ciclosporin-sparing effects, particularly for diltiazem or verapamil (210,211). Any change in the formulation of calcium channel blockers in patients previously stabilized should be undertaken cautiously because unpredictable changes in ciclosporin concentrations can occur (212). In contrast, nifedipine, isradipine, or felodipine do not significantly affect ciclosporin pharmacokinetics (SED-13, 1129). Results obtained with amlodipine are conflicting; some studies have shown no effect, while others indicate an increase of up to 40% in ciclosporin blood concentrations (SEDA-19, 351) (SEDA-20, 345). Co-administration of calcium channel blockers is also regarded as a valuable option in the treatment of ciclosporin-induced hypertension, or to prevent ciclosporin nephrotoxicity.

There are conflicting results from studies on the protective role of calcium channel blockers in patients taking ciclosporin in regard to blood pressure and preservation of renal graft function. In a multicenter, randomized, placebo-controlled study in 131 de novo recipients of cadaveric renal allografts, lacidipine improved graft function from 1 year onwards, but had no effect on acute rejection rate, trough blood ciclosporin concentrations, blood pressure, number of antihypertensive drugs, hospitalization rate, or rate of adverse events (213).

The combination of ciclosporin with nifedipine produces an additive on gingival hyperplasia, with an increased prevalence and/or severity (SED-13, 1127) in both children (214,215) and adults (216,217,218). In contrast, verapamil had no significant additional effects on the prevalence or severity of ciclosporin-induced gingival overgrowth (SEDA-21, 385).

Cimetidine

The histamine H_2 receptor antagonist cimetidine increases plasma concentrations of nifedipine and delays its elimination by inhibition of hepatic mono-oxygenases. Maximum plasma nifedipine concentrations and AUC can be increased by as much as 80%, and this results in a significant increase in the antihypertensive and antianginal effects of nifedipine and also toxicity (219,220).

Cimetidine also increases plasma concentrations of nitrendipine and nisoldipine (192,221).

Ranitidine, which inhibits the microsomal mono-oxygenase system only slightly, does not alter plasma dihydropyridine concentrations to the same extent (222).

Combinations of calcium channel blockers

Paralytic ileus has been attributed to the combined use of diltiazem and nifedipine (223).

- A 62-year-old man with chest pain underwent cardiac catheterization. The diagnosis was vasospastic angina and he was given nifedipine 20 mg bd; when his angina attacks persisted he was also given oral diltiazem 100 mg bd. After 2 days, although his angina was well controlled, abdominal distension and vomiting occurred, and an X-ray suggested intestinal ileus. The drugs were withdrawn and the ileus resolved. It recurred when the treatment was resumed and gradually resolved again after withdrawal.

The disorder was suspected to be due to enhanced pharmacodynamic effects caused by the combination of the two calcium channel blockers. However, plasma concentrations of nifedipine have also been reported to increase about three-fold when it is combined with diltiazem (224).

Corticosteroids—glucocorticoids

Methylprednisolone concentrations increased with the co-administration of diltiazem (2.6-fold) and mibefradil (3.8-fold) (225).

Dantrolene

Dantrolene interacts with verapamil and with diltiazem, causing myocardial depression and cardiogenic shock (SEDA-16, 199).

The combination of dantrolene with calcium channel blockers, such as verapamil, can result in severe cardiovascular depression and hyperkalemia (SEDA-12, 113) (226,227), so that extreme care is required.

Fluconazole

- Fluconazole enhanced the blood pressure-lowering effects of nifedipine by increasing its plasma concentrations in a 16-year-old patient with malignant pheochromocytoma taking chronic nifedipine for arterial hypertension who was given fluconazole for *Candida* septicemia (228).

Grapefruit juice

The ability of grapefruit to increase the plasma concentrations of some drugs was accidentally discovered when grapefruit juice was used as a blinding agent in a drug interaction study of felodipine and alcohol (229). It was noticed that plasma concentrations of felodipine were much higher when the drug was taken with grapefruit juice than those previously reported for the dose of drug administered. In other studies concurrent administration of grapefruit juice and felodipine increased the AUC, causing increased heart rate, and reduced diastolic blood pressure (230), or caused increased blood pressure and heart rate, headaches, flushing, and light-headedness (231). Grapefruit increases plasma concentrations of nifedipine (232) and nisoldipine (233) by increasing their systemic availability (234); with nisoldipine or nitrendipine there was an increase in heart rate.

Ketoconazole

The effects of ketoconazole 200 mg on the pharmacokinetics of nisoldipine 5 mg have been investigated in a randomized, cross-over trial (235). Pretreatment with and concomitant administration of ketoconazole resulted in 24-fold and 11-fold increases in the AUC and C_{max} of nisoldipine, respectively. The ketoconazole-induced increase in plasma concentrations of the metabolite M9 was of similar magnitude. Thus, ketoconazole and other potent inhibitors of CYP3A should not be used concomitantly with nisoldipine.

In an intestinal perfusion study of the effect of ketoconazole 40 μg/ml on the jejunal permeability and first-pass metabolism of (*R*)- and (*S*)-verapamil 120 μg/ml in six healthy volunteers, ketoconazole did not alter the jejunal permeability of the isomers, suggesting that it had no effect on the P-glycoprotein mediated efflux. However, the rate of absorption increased, suggesting inhibition by ketoconazole of the gut wall metabolism of (*R/S*)-verapamil by CYP3A4 (236).

Lithium

Lithium clearance is reduced by about 30% by nifedipine (237).

- A 30-year-old man required a reduction in lithium dosage from 1500 to 900 mg/day to maintain his serum lithium concentration in the target range shortly after he started to take nifedipine 60 mg/day (238).

There have been reports of neurotoxicity, bradycardia, and reduced lithium concentrations associated with verapamil (239–241).

Mibefradil

In a Prescription-Event Monitoring study in 3085 patients, mean age 65 years, one patient developed collapse and severe bradycardia after starting to take a dihydropyridine calcium channel blocker within 24 hours of stopping mibefradil (242).

Prazosin

An interaction of prazosin with nifedipine or verapamil resulted in acute hypotension (243,244). The mechanism appears to be partly kinetic (the systemic availability of prazosin increasing by 60%) and partly dynamic.

Sildenafil

Retrospective analysis of clinical trials has suggested that the concomitant use of antihypertensive drugs did not lead to an increase in adverse events in patients also taking sildenafil (245).

Hypertensive patients taking amlodipine, in contrast to glyceryl trinitrate, had only a minor supplementary fall in blood pressure when challenged with a single dose of sildenafil, and a few had a mild to moderate headache (246).

Diltiazem is metabolized by CYP3A4 and was held responsible for unanticipated prolonged hypotension after sublingual glyceryl trinitrate in a patient who underwent coronary angiography 2 days after last using sildenafil (247).

Simvastatin

In a meta-analysis of megatrials of simvastatin, the overall incidence of myopathy was 0.025%; the same proportion of those with myositis had used calcium channel blockers as the proportion overall, suggesting that there is no important interaction between these two groups of drugs (248).

However, diltiazem interacts with lovastatin although not with pravastatin (SEDA-24, 511), and an interaction has also been observed with simvastatin in a 75-year-old man who developed impaired renal function (249). He had extreme weakness and muscle pain.

Tacrolimus

A 3-day course of diltiazem 90 mg/day produced a four-fold increase in tacrolimus trough concentrations in a 68-year-old patient with a liver transplant (250).

In a non-randomized, pharmacokinetic study, four patients taking tacrolimus after kidney and liver transplantation were given diltiazem in seven incremental dosages of 0–180 mg at 2-week intervals (251). The mean tacrolimus-sparing effect was similar to the ciclosporin-sparing effect previously reported. This effect occurred at a lower dose of diltiazem in renal transplant patients than in liver transplant patients. Tacrolimus is metabolized by CYP3A4 and is also a substrate for P glycoprotein, and this interaction could have occurred by inhibition of these mechanisms.

A retrospective study has shown a significant improvement in kidney function and a 38% reduction in tacrolimus dosage requirements in patients taking both nifedipine and tacrolimus compared to patients not taking nifedipine (252).

Theophylline

Theophylline toxicity has been reported in several patients, apparently stabilized on theophylline, after the introduction of verapamil (253) or nifedipine (254).

Theophylline toxicity has been reported in several patients, apparently stabilized on theophylline, after the introduction of verapamil (255) or nifedipine (256).

Tubocurarine

Calcium channel blockers, such as verapamil and nifedipine, can potentiate neuromuscular blocking agents (257,258) and it has been suggested that in long-term use they can accumulate in muscle and make block-reversal difficult (259).

References

1. Committee on Safety of Medicines. Perhexiline maleate (Pexid): adverse reactions. Curr Probl 1983;11.
2. Po AL, Zhang WY. What lessons can be learnt from withdrawal of mibefradil from the market? Lancet 1998;351(9119):1829–30.
3. Wood AJ. Calcium antagonists. Pharmacologic differences and similarities. Circulation 1989;80(Suppl 6):IV184–8.
4. Psaty BM, Heckbert SR, Koepsell TD, Siscovick DS, Raghunathan TE, Weiss NS, Rosendaal FR, Lemaitre RN, Smith NL, Wahl PW, et al. The risk of myocardial infarction associated with antihypertensive drug therapies. JAMA 1995;274(8):620–5.
5. Furberg CD, Psaty BM, Meyer JV. Nifedipine. Dose-related increase in mortality in patients with coronary heart disease. Circulation 1995;92(5):1326–31.
6. Grossman E, Messerli FH. Calcium antagonists in cardiovascular disease: a necessary controversy but an unnecessary panic. Am J Med 1997;102(2):147–9.
7. Pahor M, Guralnik JM, Furberg CD, Carbonin P, Havlik R. Risk of gastrointestinal haemorrhage with calcium antagonists in hypertensive persons over 67 years old. Lancet 1996;347(9008):1061–5.
8. Pahor M, Guralnik JM, Ferrucci L, Corti MC, Salive ME, Cerhan JR, Wallace RB, Havlik RJ. Calcium-channel blockade and incidence of cancer in aged populations. Lancet 1996;348(9026):493–7.
9. Stelfox HT, Chua G, O'Rourke K, Detsky AS. Conflict of interest in the debate over calcium-channel antagonists. N Engl J Med 1998;338(2):101–6.
10. Alderman MH, Cohen H, Roque R, Madhavan S. Effect of long-acting and short-acting calcium antagonists on cardiovascular outcomes in hypertensive patients. Lancet 1997;349(9052):594–8.
11. McMurray J, Murdoch D. Calcium-antagonist controversy: the long and short of it? Lancet 1997;349(9052):585–6.
12. Estacio RO, Jeffers BW, Hiatt WR, Biggerstaff SL, Gifford N, Schrier RW. The effect of nisoldipine as compared with enalapril on cardiovascular outcomes in patients with non-insulin-dependent diabetes and hypertension. N Engl J Med 1998;338(10):645–52.
13. Garcia Rodriguez LA, Cattaruzzi C, Troncon MG, Agostinis L. Risk of hospitalization for upper gastrointestinal tract bleeding associated with ketorolac, other nonsteroidal anti-inflammatory drugs, calcium antagonists, and other antihypertensive drugs. Arch Intern Med 1998;158(1):33–9.
14. Rosenberg L, Rao RS, Palmer JR, Strom BL, Stolley PD, Zauber AG, Warshauer ME, Shapiro S. Calcium channel blockers and the risk of cancer. JAMA 1998;279(13):1000–4.
15. Stone PH, Gibson RS, Glasser SP, DeWood MA, Parker JD, Kawanishi DT, Crawford MH, Messineo FC, Shook TL, Raby K, et alThe ASIS Study Group. Comparison of propranolol, diltiazem, and nifedipine in the treatment of ambulatory ischemia in patients with stable angina. Differential effects on ambulatory ischemia, exercise performance, and anginal symptoms. Circulation 1990;82(6):1962–72.
16. Maseri A. Medical therapy of chronic stable angina pectoris. Circulation 1990;82(6):2258–62.
17. Psaty BM, Koepsell TD, LoGerfo JP, Wagner EH, Inui TS. Beta-blockers and primary prevention of coronary heart disease in patients with high blood pressure. JAMA 1989;261(14):2087–94.
18. Packer M, O'Connor CM, Ghali JK, Pressler ML, Carson PE, Belkin RN, Miller AB, Neuberg GW, Frid D, Wertheimer JH, Cropp AB, DeMets DLProspective Randomized Amlodipine Survival Evaluation Study Group. Effect of amlodipine on morbidity and mortality in severe chronic heart failure. N Engl J Med 1996;335(15):1107–14.
19. Vetrovec GW, Parker VE. Alternative medical treatment for patients with angina pectoris and adverse reactions to beta blockers. Usefulness of nifedipine. Am J Med 1986;81(4A):20–7.
20. Andersson KE. Effects of calcium and calcium antagonists on the excitation-contraction coupling in striated and smooth muscle. Acta Pharmacol Toxicol (Copenh) 1978;43(Suppl 1):5–14.
21. Jones RI, Hornung RS, Sonecha T, Raftery EB. The effect of a new calcium channel blocker nicardipine on 24-hour ambulatory blood pressure and the pressor response to isometric and dynamic exercise. J Hypertens 1983;1(1):85–9.
22. Stoepel K, Deck K, Corsing C, Ingram C, Vanov SK. Safety aspects of long-term nitrendipine therapy. J Cardiovasc Pharmacol 1984;6(Suppl 7):S1063–6.
23. Dubois C, Blanchard D, Loria Y, Moreau M. Clinical trial of new antihypertensive drug nicardipine: efficacy and tolerance in 29,104 patients. Curr Ther Res 1987;42:727.
24. Sorkin EM, Clissold SP. Nicardipine. A review of its pharmacodynamic and pharmacokinetic properties, and therapeutic efficacy, in the treatment of angina pectoris, hypertension and related cardiovascular disorders. Drugs 1987;33(4):296–345.

25. Sundstedt CD, Ruegg PC, Keller A, Waite R. A multicenter evaluation of the safety, tolerability, and efficacy of isradipine in the treatment of essential hypertension. Am J Med 1989;86(4A):98–102.
26. DeWood MA, Wolbach RA. Randomized double-blind comparison of side effects of nicardipine and nifedipine in angina pectoris. The Nicardipine Investigators Group. Am Heart J 1990;119(2 Pt 2):468–78.
27. Cheer SM, Mc Clellan K. Manidipine: a review of its use in hypertension. Drugs 2001;61(12):1777–99.
28. Matsumoto S, Ito T, Sada T, Takahashi M, Su KM, Ueda A, Okabe F, Sato M, Sekine I, Ito Y. Hemodynamic effects of nifedipine in congestive heart failure. Am J Cardiol 1980;46(3):476–80.
29. Gillmer DJ, Kark P. Pulmonary oedema precipitated by nifedipine. BMJ 1980;280(6229):1420–1.
30. Subramanian B, Bowles MJ, Davies AB, Raftery EB. Combined therapy with verapamil and propranolol in chronic stable angina. Am J Cardiol 1982;49(1):125–32.
31. Bassan M, Weiler-Ravell D, Shalev O. Additive antianginal effect of verapamil in patients receiving propranolol. BMJ (Clin Res Ed) 1982;284(6322):1067–70.
32. Terry RW. Nifedipine therapy in angina pectoris: evaluation of safety and side effects. Am Heart J 1982;104(3):681–9.
33. Yusuf S, Wittes J, Friedman L. Overview of results of randomized clinical trials in heart disease. I. Treatments following myocardial infarction. JAMA 1988; 260(14):2088–93.
34. Goldbourt U, Behar S, Reicher-Reiss H, Zion M, Mandelzweig L, Kaplinsky E. Early administration of nifedipine in suspected acute myocardial infarction. The Secondary Prevention Reinfarction Israel Nifedipine Trial 2 Study. Arch Intern Med 1993;153(3):345–53.
35. Held PH, Yusuf S, Furberg CD. Calcium channel blockers in acute myocardial infarction and unstable angina: an overview. BMJ 1989;299(6709):1187–92.
36. Held PH, Yusuf S. Effects of beta-blockers and calcium channel blockers in acute myocardial infarction. Eur Heart J 1993;14(Suppl F):18–25.
37. Gibson RS, Boden WE, Theroux P, Strauss HD, Pratt CM, Gheorghiade M, Capone RJ, Crawford MH, Schlant RC, Kleiger RE, et al. Diltiazem and reinfarction in patients with non-Q-wave myocardial infarction. Results of a double-blind, randomized, multicenter trial. N Engl J Med 1986;315(7):423–9.
38. The Multicenter Diltiazem Postinfarction Trial Research Group. The effect of diltiazem on mortality and reinfarction after myocardial infarction. N Engl J Med 1988;319(7):385–92.
39. The Danish Study Group on Verapamil in Myocardial Infarction. Verapamil in acute myocardial infarction. Eur Heart J 1984;5(7):516–28.
40. The Danish Verapamil Infarction Trial II—DAVIT II. Effect of verapamil on mortality and major events after acute myocardial infarction. Am J Cardiol 1990; 66(10):779–85.
41. Lubsen J, Tijssen JGP, Kerkkamp HJJ. Early treatment of unstable angina in the coronary care unit: a randomised, double blind, placebo controlled comparison of recurrent ischaemia in patients treated with nifedipine or metoprolol or both. Report of The Holland Interuniversity Nifedipine/Metoprolol Trial (HINT) Research Group. Br Heart J 1986;56(5):400–13.
42. Pahor M, Guralnik JM, Corti MC, Foley DJ, Carbonin P, Havlik RJ. Long-term survival and use of antihypertensive medications in older persons. J Am Geriatr Soc 1995; 43(11):1191–7.
43. Barnett AA. News. Lancet 1996;347:313.
44. Zangerle KF, Wolford R. Syncope and conduction disturbances following sublingual nifedipine for hypertension. Ann Emerg Med 1985;14(10):1005–6.
45. Villani GQ, del Giudice S, Arruzzoli S, Dieci G. Blocco seno-atriale dopo somministrazione orale di nifedipina. Descrizione di un caso. [Sinoatrial block after oral administration of nifedipine. Description of a case.] Minerva Cardioangiol 1985;33(9):557–9.
46. Epstein SE, Rosing DR. Verapamil: its potential for causing serious complications in patients with hypertrophic cardiomyopathy. Circulation 1981;64(3):437–41.
47. Hopf R, Rodrian S, Kaltenbach M. Behandlung der hypertrophen Kardiomyopathie mit Kalziumantagonisten. Therapiewoche 1986;36:1433.
48. Wachter RM. Symptomatic hypotension induced by nifedipine in the acute treatment of severe hypertension. Arch Intern Med 1987;147(3):556–8.
49. Schwartz M, Naschitz JE, Yeshurun D, Sharf B. Oral nifedipine in the treatment of hypertensive urgency: cerebrovascular accident following a single dose. Arch Intern Med 1990;150(3):686–7.
50. Shettigar UR, Loungani R. Adverse effects of sublingual nifedipine in acute myocardial infarction. Crit Care Med 1989;17(2):196–7.
51. Graham CF. Intravenous verapamil-isotopin (Calan): acute bronchospasm. ADR Highlights 1982;868:82.
52. Zalman F, Perloff JK, Durant NN, Campion DS. Acute respiratory failure following intravenous verapamil in Duchenne's muscular dystrophy. Am Heart J 1983;105(3):510–1.
53. Ben-Noun L. Acute asthma associated with sustained-release verapamil. Ann Pharmacother 1997;31(5):593–5.
54. Packer M, Medina N, Yushak M. Adverse hemodynamic and clinical effects of calcium channel blockade in pulmonary hypertension secondary to obliterative pulmonary vascular disease. J Am Coll Cardiol 1984;4(5):890–901.
55. Batra AK, Segall PH, Ahmed T. Pulmonary edema with nifedipine in primary pulmonary hypertension. Respiration 1985;47(3):161–3.
56. Hasebe N, Fijikana T, Wantanabe M, et al. A case of respiratory failure precipitated by injecting nifedipine. Kokya To Junkan 1988;36:1255.
57. Garcia-Ruiz PJ, Garcia de Yebenes J, Jimenez-Jimenez FJ, Vazquez A, Garcia Urra D, Morales B. Parkinsonism associated with calcium channel blockers: a prospective follow-up study. Clin Neuropharmacol 1992;15(1):19–26.
58. Swash M, Ingram DA. Adverse effect of verapamil in myasthenia gravis. Muscle Nerve 1992;15(3):396–8.
59. Krendel DA, Hopkins LC. Adverse effect of verapamil in a patient with the Lambert–Eaton syndrome. Muscle Nerve 1986;9(6):519–22.
60. Ueno S, Hara Y. Lambert–Eaton myasthenic syndrome without anti-calcium channel antibody: adverse effect of calcium antagonist diltiazem. J Neurol Neurosurg Psychiatry 1992;55(5):409–10.
61. Coulter DM. Eye pain with nifedipine and disturbance of taste with captopril: a mutually controlled study showing a method of postmarketing surveillance. BMJ (Clin Res Ed) 1988;296(6629):1086–8.
62. Kelly SP, Walley TJ. Eye pain with nifedipine. BMJ (Clin Res Ed) 1988;296(6633):1401.
63. Levenson JL, Kennedy K. Dysosmia, dysgeusia, and nifedipine. Ann Intern Med 1985;102(1):135–6.

64. Berman JL. Dysosmia, dysgeusia and diltiazem. Ann Intern Med 1985;103:154.
65. Brink DD. Diltiazem and hyperactivity. Ann Intern Med 1984;100(3):459–60.
66. Ahmad S. Nifedipine-induced acute psychosis. J Am Geriatr Soc 1984;32(5):408.
67. Palat GK, Hooker EA, Movahed A. Secondary mania associated with diltiazem. Clin Cardiol 1984;7(11):611–2.
68. Jacobsen FM, Sack DA, James SP. Delirium induced by verapamil. Am J Psychiatry 1987;144(2):248.
69. Pitlik S, Manor RS, Lipshitz I, Perry G, Rosenfeld J. Transient retinal ischaemia induced by nifedipine. BMJ (Clin Res Ed) 1983;287(6408):1845–6.
70. Eccleston D, Cole AJ. Calcium-channel blockade and depressive illness. Br J Psychiatry 1990;156:889–91.
71. Dunn NR, Freemantle SN, Mann RD. Cohort study on calcium channel blockers, other cardiovascular agents, and the prevalence of depression. Br J Clin Pharmacol 1999;48(2):230–3.
72. Rocco S, Mantero F, Boscaro M. Effects of a calcium antagonist on the pituitary–adrenal axis. Horm Metab Res 1993;25(2):114–6.
73. Malaisse WJ, Sener A. Calcium-antagonists and islet function-XII. Comparison between nifedipine and chemically related drugs. Biochem Pharmacol 1981;30(10):1039–41.
74. Giugliano D, Gentile S, Verza M, Passariello N, Giannetti G, Varricchio M. Modulation by verapamil of insulin and glucagon secretion in man. Acta Diabetol Lat 1981;18(2):163–71.
75. Donnelly T, Harrower AD. Effect of nifedipine on glucose tolerance and insulin secretion in diabetic and non-diabetic patients. Curr Med Res Opin 1980;6(10):690–3.
76. Abadie E, Passa PH. Diabetogenic effect of nifedipine. BMJ (Clin Res Ed) 1984;289(6442):438.
77. Collings WCJ, Cullen MJ, Feely J. The effect of therapy with dihydropyridine calcium channel blockers on glucose tolerance in non-insulin dependent diabetes. Br J Clin Pharmacol 1986;21:568.
78. Zezulka AV, Gill JS, Beevers DG. Diabetogenic effects of nifedipine. BMJ (Clin Res Ed) 1984;289(6442):437–8.
79. Neaton JD, Grimm RH Jr, Prineas RJ, Stamler J, Grandits GA, Elmer PJ, Cutler JA, Flack JM, Schoenberger JA, McDonald R, et al. Treatment of Mild Hypertension Study. Final results. Treatment of Mild Hypertension Study Research Group. JAMA 1993; 270(6):713–24.
80. Trost BN. Glucose metabolism and calcium antagonists. Horm Metab Res Suppl 1990;22:48–56.
81. Bhatnagar SK, Amin MMA, Al-Yusuf AR. Diabetogenic effects of nifedipine. BMJ (Clin Res Ed) 1984;289:19.
82. Lindenberg BS, Weiner DA, McCabe CH, Cutler SS, Ryan TJ, Klein MD. Efficacy and safety of incremental doses of diltiazem for the treatment of stable angina pectoris. J Am Coll Cardiol 1983;2(6):1129–33.
83. Petru MA, Crawford MH, Sorensen SG, Chaudhuri TK, Levine S, O'Rourke RA. Short- and long-term efficacy of high-dose oral diltiazem for angina due to coronary artery disease: a placebo-controlled, randomized, double-blind crossover study. Circulation 1983;68(1):139–47.
84. Hayashi K, Matsuda H, Honda M, Ozawa Y, Tokuyama H, Okubo K, Takamatsu I, Kanda T, Tatematsu S, Homma K, Saruta T. Impact of calcium antagonists on bleeding time in patients with chronic renal failure. J Hum Hypertens 2002;16(3):199–203.
85. Voth AJ, Turner RH. Nifedipine and agranulocytosis. Ann Intern Med 1983;99(6):882.
86. Quigley MA, White KL, McGraw BF. Interpretation and application of world-wide safety data on diltiazem. Acta Pharmacol Toxicol (Copenh) 1985;57(Suppl 2):61–73.
87. Baggott LA. Diltiazem-associated immune thrombocytopenia. Mt Sinai J Med 1987;54(6):500–4.
88. Ramon Y, Behar S, Kishon Y, Engelberg IS. Gingival hyperplasia caused by nifedipine—a preliminary report. Int J Cardiol 1984;5(2):195–206.
89. Lombardi T, Fiore-Donno G, Belser U, Di Felice R. Felodipine-induced gingival hyperplasia: a clinical and histologic study. J Oral Pathol Med 1991;20(2):89–92.
90. Young PC, Turiansky GW, Sau P, Liebman MD, Benson PM. Felodipine-induced gingival hyperplasia. Cutis 1998;62(1):41–3.
91. Cucchi G, Giustiniani S, Robustelli F. Gengivite ipertrofica da verapamil. [Hypertrophic gingivitis caused by verapamil.] G Ital Cardiol 1985;15(5):556–7.
92. Lokken P, Skomedal T. Kalsiumkanalblokkerindusert gingival hyperplasi. Sjelden, eller tusener av tilfeller i Norge?. [Gingival hyperplasia induced by calcium channel blockers. Rare or frequent in Norway?.] Tidsskr Nor Laegeforen 1992;112(15):1978–80.
93. Fattore L, Stablein M, Bredfeldt G, Semla T, Moran M, Doherty-Greenberg JM. Gingival hyperplasia: a side effect of nifedipine and diltiazem. Spec Care Dentist 1991;11(3):107–9.
94. Nyska A, Shemesh M, Tal H, Dayan D. Gingival hyperplasia induced by calcium channel blockers: mode of action. Med Hypotheses 1994;43(2):115–8.
95. Ellis JS, Seymour RA, Steele JG, Robertson P, Butler TJ, Thomason JM. Prevalence of gingival overgrowth induced by calcium channel blockers: a community-based study. J Periodontol 1999;70(1):63–7.
96. Hedback B, Hermann LS. Antihypertensive effect of verapamil in patients with newly discovered mild to moderate essential hypertension. Acta Med Scand Suppl 1984;681:129–35.
97. Bassotti G, Calcara C, Annese V, Fiorella S, Roselli P, Morelli A. Nifedipine and verapamil inhibit the sigmoid colon myoelectric response to eating in healthy volunteers. Dis Colon Rectum 1998;41(3):377–80.
98. Gaginella TS, Maxfield DL. Calcium-channel blocking agents and chest pain. Drug Intell Clin Pharm 1988;22(7–8):623–5.
99. Smalley WE, Ray WA, Daugherty JR, Griffin MR. No association between calcium channel blocker use and confirmed bleeding peptic ulcer disease. Am J Epidemiol 1998;148(4):350–4.
100. Suissa S, Bourgault C, Barkun A, Sheehy O, Ernst P. Antihypertensive drugs and the risk of gastrointestinal bleeding. Am J Med 1998;105(3):230–5.
101. Rotmensch HH, Roth A, Liron M, Rubinstein A, Gefel A, Livni E. Lymphocyte sensitisation in nifedipine-induced hepatitis. BMJ 1980;281(6246):976–7.
102. Davidson AR. Lymphocyte sensitisation in nifedipine-induced hepatitis. BMJ 1980;281(6251):1354.
103. Centrum Voor Geneesmiddelenbewaking. Nifedipine en hepatitis. Folia Pharmacother 1981;8:7.
104. Stern EH, Pitchon R, King BD, Wiener I. Possible hepatitis from verapamil. N Engl J Med 1982;306(10):612–3.
105. Tartaglione TA, Pepine CJ, Pieper JA. Diltiazem: a review of its clinical efficacy and use. Drug Intell Clin Pharm 1982;16(5):371–9.
106. McGraw BF, Walker SD, Hemberger JA, Gitomer SL, Nakama M. Clinical experience with diltiazem in Japan. Pharmacotherapy 1982;2(3):156–61.

107. Sarachek NS, London RL, Matulewicz TJ. Diltiazem and granulomatous hepatitis. Gastroenterology 1985;88(5 Pt 1):1260–2.
108. Guarascio P, D'Amato C, Sette P, Conte A, Visco G. Liver damage from verapamil. BMJ (Clin Res Ed) 1984;288(6414):362–3.
109. Wirebaugh SR, Geraets DR. Reports of erythematous macular skin eruptions associated with diltiazem therapy. DICP 1990;24(11):1046–9.
110. Grunwald Z. Painful edema, erythematous rash, and burning sensation due to nifedipine. Drug Intell Clin Pharm 1982;16(6):492.
111. Barker SJ, Bayliff CD, McCormack DG, Dilworth GR. Nifedipine-induced erythema multiforme. Can J Hosp Pharm 1996;49:160.
112. Stern R, Khalsa JH. Cutaneous adverse reactions associated with calcium channel blockers. Arch Intern Med 1989;149(4):829–32.
113. Crowson AN, Magro CM. Subacute cutaneous lupus erythematosus arising in the setting of calcium channel blocker therapy. Hum Pathol 1997;28(1):67–73.
114. Cooper SM, Wojnarowska F. Photo-damage in Northern European renal transplant recipients is associated with use of calcium channel blockers. Clin Exp Dermatol 2003;28:588–91.
115. Rodger JC, Torrance TC. Can nifedipine provoke menorrhagia? Lancet 1983;2(8347):460.
116. Clyne CAC. Unilateral gynaecomastia and nifedipine. BMJ (Clin Res Ed) 1986;292:380.
117. Carraway RD. Febrile reaction following nifedipine therapy. Am Heart J 1984;108(3 Pt 1):611.
118. Scolnick B, Brinberg D. Diltiazem and generalized lymphadenopathy. Ann Intern Med 1985;102(4):558.
119. Offerhaus L, Dunning AJ. Angina pectoris: variaties op het thema nifedipine. [Angina pectoris; variations on the nifedipine theme.] Ned Tijdschr Geneeskd 1980;124(45):1928–32.
120. Pedersen OL, Mikkelsen E, Andersson KE. Paradoks angina pectoris efter nifedipin. [Paradoxal angina pectoris following nifedipine.] Ugeskr Laeger 1980;142(29):1883–4.
121. Gottlieb SO, Gerstenblith G. Safety of acute calcium antagonist withdrawal: studies in patients with unstable angina withdrawn from nifedipine. Am J Cardiol 1985;55(12):E27–30.
122. Gottlieb SO, Ouyang P, Achuff SC, Baughman KL, Traill TA, Mellits ED, Weisfeldt ML, Gerstenblith G. Acute nifedipine withdrawal: consequences of preoperative and late cessation of therapy in patients with prior unstable angina. J Am Coll Cardiol 1984;4(2):382–8.
123. Kay R, Blake J, Rubin D. Possible coronary spasm rebound to abrupt nifedipine withdrawal. Am Heart J 1982;103(2):308.
124. Engelman RM, Hadji-Rousou I, Breyer RH, Whittredge P, Harbison W, Chircop RV. Rebound vasospasm after coronary revascularization in association with calcium antagonist withdrawal. Ann Thorac Surg 1984;37(6):469–72.
125. Lette J, Gagnon RM, Lemire JG, Morissette M. Rebound of vasospastic angina after cessation of long-term treatment with nifedipine. Can Med Assoc J 1984; 130(9):1169–74.
126. Mysliwiec M, Rydzewski A, Bulhak W. Calcium antagonist withdrawal syndrome. BMJ (Clin Res Ed) 1983; 286(6381):1898.
127. Schick EC Jr, Liang CS, Heupler FA Jr, Kahl FR, Kent KM, Kerin NZ, Noble RJ, Rubenfire M, Tabatznik B, Terry RW. Randomized withdrawal from nifedipine: placebo-controlled study in patients with coronary artery spasm. Am Heart J 1982;104(3):690–7.
128. Subramanian VB, Bowles MJ, Khurmi NS, Davies AB, O'Hara MJ, Raftery EB. Calcium antagonist withdrawal syndrome: objective demonstration with frequency-modulated ambulatory ST-segment monitoring. BMJ (Clin Res Ed) 1983;286(6364):520–1.
129. Jick H, Jick S, Derby LE, Vasilakis C, Myers MW, Meier CR. Calcium-channel blockers and risk of cancer. Lancet 1997;349(9051):525–8.
130. Jonas M, Goldbourt U, Boyko V, Mandelzweig L, Behar S, Reicher-Reiss H. Nifedipine and cancer mortality: ten-year follow-up of 2607 patients after acute myocardial infarction. Cardiovasc Drugs Ther 1998;12(2):177–81.
131. Braun S, Boyko V, Behar S, Reicher-Reiss H, Laniado S, Kaplinsky E, Goldbourt U. Calcium channel blocking agents and risk of cancer in patients with coronary heart disease. Benzafibrate Infarction Prevention (BIP) Study Research Group. J Am Coll Cardiol 1998;31(4):804–8.
132. Hole DJ, Gillis CR, McCallum IR, McInnes GT, MacKinnon PL, Meredith PA, Murray LS, Robertson JW, Lever AF. Cancer risk of hypertensive patients taking calcium antagonists. J Hypertens 1998;16(1):119–24.
133. Fitzpatrick AL, Daling JR, Furberg CD, Kronmal RA, Weissfeld JL. Use of calcium channel blockers and breast carcinoma risk in postmenopausal women. Cancer 1997;80(8):1438–47.
134. Li CI, Malone KE, Weiss NS, Boudreau DM, Cushing-Haugen KL, Daling JR. Relation between use of antihypertensive medications and risk of breast carcinoma among women ages 65-79 years. Cancer 2003;98:1504–13.
135. Howes LG, Edwards CT. Calcium antagonists and cancer. Is there really a link? Drug Saf 1998;18(1):1–7.
136. Fenakel K, Fenakel G, Appelman Z, Lurie S, Katz Z, Shoham Z. Nifedipine in the treatment of severe preeclampsia. Obstet Gynecol 1991;77(3):331–7.
137. Oei SG, Oei SK, Brolmann HA. Myocardial infarction during nifedipine therapy for preterm labor. N Engl J Med 1999;340(2):154.
138. Pecha RE, Danilewitz MD. Acute pseudo-obstruction of the colon (Ogilvie's syndrome) resulting from combination tocolytic therapy. Am J Gastroenterol 1996;91(6):1265–6.
139. Inove H. Excretion of verapamil in human milk. BMJ (Clin Res Ed) 1984;288(6417):645.
140. Okada M, Inoue H, Nakamura Y, Kishimoto M, Suzuki T. Excretion of diltiazem in human milk. N Engl J Med 1985;312(15):992–3.
141. Hagemeijer F. Verapamil in the management of supraventricular tachyarrhythmias occurring after a recent myocardial infarction. Circulation 1978;57(4):751–5.
142. O'Connor CM, Carson PE, Miller AB, Pressler ML, Belkin RN, Neuberg GW, Frid DJ, Cropp AB, Anderson S, Wertheimer JH, DeMets DL. Effect of amlodipine on mode of death among patients with advanced heart failure in the PRAISE trial. Prospective Randomized Amlodipine Survival Evaluation. Am J Cardiol 1998;82(7):881–7.
143. Kenny J. Treating overdose with calcium channel blockers. BMJ 1994;308(6935):992–3.
144. Ojetti V, Migneco A, Bononi F, De Lorenzo A, Gentiloni Silveri N. Calcium channel blockers, beta-blockers and digitalis poisoning: management in the emergency room. Eur Rev Med Pharmacol Sci 2005;9(4):241–6.
145. Buckley CD, Aronson JK. Prolonged half-life of verapamil in a case of overdose: implications for therapy. Br J Clin Pharmacol 1995;39(6):680–3.

146. Sauder P, Kopferschmitt J, Dahlet M, Tritsch L, Flesch F, Siard P, Mantz JM, Jaeger A. Les intoxications aiguës par le verapamil. A propos de 6 cas. Revue de la litterature. [Acute verapamil poisoning. 6 cases. Review of the literature.] J Toxicol Clin Exp 1990;10(4):261–70.
147. McMillan R. Management of acute severe verapamil intoxication. J Emerg Med 1988;6(3):193–6.
148. Ramoska EA, Spiller HA, Myers A. Calcium channel blocker toxicity. Ann Emerg Med 1990;19(6):649–53.
149. Howarth DM, Dawson AH, Smith AJ, Buckley N, Whyte IM. Calcium channel blocking drug overdose: an Australian series. Hum Exp Toxicol 1994;13(3):161–6.
150. Perkins CM. Serious verapamil poisoning: treatment with intravenous calcium gluconate. BMJ 1978;2(6145):1127.
151. Candell J, Valle V, Soler M, Rius J. Acute intoxication with verapamil. Chest 1979;75(2):200–1.
152. Kenney J. Calcium channel blocking agents and the heart. BMJ (Clin Res Ed) 1985;291:1150.
153. Borkje B, Omvik P, Storstein L. Fatal verapamilforgiftning. [Fatal verapamil poisoning.] Tidsskr Nor Laegeforen 1986;106(5):401–2.
154. Zogubi W, Schwartz JB. Verapamil overdose: report of a case and review of the literature. Cardiovasc Rev Rep 1984;5:356.
155. Humbert VH Jr, Munn NJ, Hawkins RF. Noncardiogenic pulmonary edema complicating massive diltiazem overdose. Chest 1991;99(1):258–9.
156. Brass BJ, Winchester-Penny S, Lipper BL. Massive verapamil overdose complicated by noncardiogenic pulmonary edema. Am J Emerg Med 1996;14(5):459–61.
157. Schiffl H, Ziupa J, Schollmeyer P. Clinical features and management of nifedipine overdosage in a patient with renal insufficiency. J Toxicol Clin Toxicol 1984;22(4):387–95.
158. Henrikson CA, Chandra-Strobos N. Calcium channel blocker overdose mimicking an acute myocardial infarction. Resuscitation 2003;59:361–4.
159. Pearigen PD, Benowitz NL. Poisoning due to calcium antagonists. Experience with verapamil, diltiazem and nifedipine. Drug Saf 1991;6(6):408–30.
160. Walter FG, Frye G, Mullen JT, Ekins BR, Khasigian PA. Amelioration of nifedipine poisoning associated with glucagon therapy. Ann Emerg Med 1993;22(7):1234–7.
161. ter Wee PM, Kremer Hovinga TK, Uges DR, van der Geest S. 4-Aminopyridine and haemodialysis in the treatment of verapamil intoxication. Hum Toxicol 1985;4(3):327–9.
162. Boyer EW, Duic PA, Evans A. Hyperinsulinemia/euglycemia therapy for calcium channel blocker poisoning. Pediatr Emerg Care 2002;18(1):36–7.
163. Wood DM, Wright KD, Jones AL, Dargan PI. Metaraminol (Aramine) in the management of a significant amlodipine overdose. Hum Exp Toxicol 2005;24(7):377–81.
164. Cantrell FL, Williams SR. Fatal unintentional overdose of diltiazem with antemortem and postmortem values. Clin Toxicol (Phila) 2005;43(6):587–8.
165. Klieman RL, Stephenson SH. Calcium antagonists–drug interactions. Rev Drug Metab Drug Interact 1985;5(2–3):193–217.
166. Pringle SD, MacEwen CJ. Severe bradycardia due to interaction of timolol eye drops and verapamil. BMJ (Clin Res Ed) 1987;294(6565):155–6.
167. Opie LH, White DA. Adverse interaction between nifedipine and beta-blockade. BMJ 1980;281(6253):1462.
168. Staffurth JS, Emery P. Adverse interaction between nifedipine and beta-blockade. BMJ (Clin Res Ed) 1981;282(6259):225.
169. Anastassiades CJ. Nifedipine and beta-blocker drugs. BMJ 1980;281(6250):1251–2.
170. Young GP. Calcium channel blockers in emergency medicine. Ann Emerg Med 1984;13(9 Pt 1):712–22.
171. Saini RK, Fulmor IE, Antonaccio MJ. Effect of tiapamil and nifedepine during critical coronary stenosis and in the presence of adrenergic beta-receptor blockade in anesthetized dogs. J Cardiovasc Pharmacol 1982;4(5):770–6.
172. Rocha P, Baron B, Delestrain A, Pathe M, Cazor JL, Kahn JC. Hemodynamic effects of intravenous diltiazem in patients treated chronically with propranolol. Am Heart J 1986;111(1):62–8.
173. Kjeldsen SE, Syvertsen JO, Hedner T. Cardiac conduction with diltiazem and beta-blockade combined. A review and report on cases. Blood Press 1996;5(5):260–3.
174. Leon MB, Rosing DR, Bonow RO, Lipson LC, Epstein SE. Clinical efficacy of verapamil alone and combined with propranolol in treating patients with chronic stable angina pectoris. Am J Cardiol 1981;48(1):131–9.
175. Sorkin EM, Clissold SP, Brogden RN. Nifedipine. A review of its pharmacodynamic and pharmacokinetic properties, and therapeutic efficacy, in ischaemic heart disease, hypertension and related cardiovascular disorders. Drugs 1985;30(3):182–274.
176. Goa KL, Sorkin EM. Nitrendipine. A review of its pharmacodynamic and pharmacokinetic properties, and therapeutic efficacy in the treatment of hypertension. Drugs 1987;33(2):123–55.
177. Solomon SA, Ramsay LE, Yeo WW, Parnell L, Morris-Jones W. Beta blockade and intermittent claudication: placebo controlled trial of atenolol and nifedipine and their combination. BMJ 1991;303(6810):1100–4.
178. Wulf H, Godicke J, Herzig S. Functional interaction between local anaesthetics and calcium antagonists in guineapig myocardium: 2. Electrophysiological studies with bupivacaine and nifedipine. Br J Anaesth 1994;73(3):364–70.
179. Herzig S, Ruhnke L, Wulf H. Functional interaction between local anaesthetics and calcium antagonists in guineapig myocardium: 1. Cardiodepressant effects in isolated organs. Br J Anaesth 1994;73(3):357–63.
180. Adsan H, Tulunay M, Onaran O. The effects of verapamil and nimodipine on bupivacaine-induced cardiotoxicity in rats: an in vivo and in vitro study. Anesth Analg 1998; 86(4):818–24.
181. Lamberg TS, Kivisto KT, Neuvonen PJ. Effects of verapamil and diltiazem on the pharmacokinetics and pharmacodynamics of buspirone. Clin Pharmacol Ther 1998;63(6):640–5.
182. Macphee GJ, McInnes GT, Thompson GG, Brodie MJ. Verapamil potentiates carbamazepine neurotoxicity: a clinically important inhibitory interaction. Lancet 1986;1(8483):700–3.
183. Brodie MJ, MacPhee GJ. Carbamazepine neurotoxicity precipitated by diltiazem. BMJ (Clin Res Ed) 1986;292(6529):1170–1.
184. De Vito JM, Friedman B. Evaluation of the pharmacodynamic and pharmacokinetic interaction between calcium antagonists and digoxin. Pharmacotherapy 1986;6(2):73–82.
185. Pedersen KE, Dorph-Pedersen A, Hvidt S, Klitgaard NA, Pedersen KK. The long-term effect of verapamil on plasma digoxin concentration and renal digoxin clearance in healthy subjects. Eur J Clin Pharmacol 1982;22(2): 123–127.
186. Schwartz JB, Keefe D, Kates RE, Kirsten E, Harrison DC. Acute and chronic pharmacodynamic interaction of verapamil and digoxin in atrial fibrillation. Circulation 1982; 65(6):1163–70.

187. Kounis NG. Asystole after verapamil and digoxin. Br J Clin Pract 1980;34(2):57–8.
188. Kleinbloesem CH, van Brummelen P, Hillers J, Moolenaar AJ, Breimer DD. Interaction between digoxin and nifedipine at steady state in patients with atrial fibrillation. Ther Drug Monit 1985;7(4):372–6.
189. Kirch W, Hutt HJ, Dylewicz P, Graf KJ, Ohnhaus EE. Dose-dependence of the nifedipine–digoxin interaction? Clin Pharmacol Ther 1986;39(1):35–9.
190. Oyama Y, Fujii S, Kanda K, Akino E, Kawasaki H, Nagata M, Goto K. Digoxin–diltiazem interaction. Am J Cardiol 1984;53(10):1480–1.
191. D'Arcy PF. Diltiazem–digoxin interactions. Pharm Int 1985;6:148.
192. Kirch W, Hutt HJ, Heidemann H, Ramsch K, Janisch HD, Ohnhaus EE. Drug interactions with nitrendipine. J Cardiovasc Pharmacol 1984;6(Suppl 7):S982–5.
193. Belz GG, Wistuba S, Matthews JH. Digoxin and bepridil: pharmacokinetic and pharmacodynamic interactions. Clin Pharmacol Ther 1986;39(1):65–71.
194. Kuhlmann J. Effects of verapamil, diltiazem, and nifedipine on plasma levels and renal excretion of digitoxin. Clin Pharmacol Ther 1985;38(6):667–73.
195. Pliakos ChC, Papadopoulos K, Parcharidis G, Styliadis J, Tourkantonis A. Effects of calcium channel blockers on serum concentrations of digoxin. Epitheorese Klin Farmakol Farmakokinetikes 1991;9:118–25.
196. North DS, Mattern AL, Hiser WW. The influence of diltiazem hydrochloride on trough serum digoxin concentrations. Drug Intell Clin Pharm 1986;20(6):500–3.
197. Halawa B, Mazurek W. Interakcje digoksyny z nifedypina i diltiazemem. [Interactions of digoxin with nifedipine and diltiazem.] Pol Tyg Lek 1990;45(23–24):467–9.
198. Mahgoub AA, El-Medany AH, Abdulatif AS. A comparison between the effects of diltiazem and isosorbide dinitrate on digoxin pharmacodynamics and kinetics in the treatment of patients with chronic ischemic heart failure. Saudi Med J 2002;23(6):725–31.
199. Peters J, Welker HA, Bullingham R. Pharmacokinetic and pharmacodynamic aspects of concomitant mibefradil–digoxin therapy at therapeutic doses. Eur J Drug Metab Pharmacokinet 1999;24(2):133–40.
200. Gasic S, Eichler HG, Korn A. Effect of calcium antagonists on basal and digitalis-dependent changes in splanchnic and systemic hemodynamics. Clin Pharmacol Ther 1987;41(4):460–6.
201. Pedersen KE, Dorph-Pedersen A, Hvidt S, Klitgaard NA, Nielsen-Kudsk F. Digoxin–verapamil interaction. Clin Pharmacol Ther 1981;30(3):311–6.
202. Klein HO, Lang R, Weiss E, Di Segni E, Libhaber C, Guerrero J, Kaplinsky E. The influence of verapamil on serum digoxin concentration. Circulation 1982;65(5):998–1003.
203. Zatuchni J. Verapamil–digoxin interaction. Am Heart J 1984;108(2):412–3.
204. Venkat-Raman G, Feehally J, Elliott HL, Griffin P, Moore RJ, Olubodun JO, Wilkinson R. Renal and haemodynamic effects of amlodipine and nifedipine in hypertensive renal transplant recipients. Nephrol Dial Transplant 1998;13(10):2612–6.
205. Jacob LP, Malhotra D, Chan L, Shapiro JI. Absence of a dose-response of cyclosporine levels to clinically used doses of diltiazem and verapamil. Am J Kidney Dis 1999;33(2):301–3.
206. Yildiz A, Sever MS, Turkmen A, Ecder T, Turk S, Akkaya V, Ark E. Interaction between cyclosporine A and verapamil, felodipine, and isradipine. Nephron 1999;81(1):117–8.
207. Bourbigot B, Guiserix J, Airiau J, Bressollette L, Morin JF, Cledes J. Nicardipine increases cyclosporin blood levels. Lancet 1986;1(8495):1447.
208. Pochet JM, Pirson Y. Cyclosporin–diltiazem interaction. Lancet 1986;1(8487):979.
209. Citterio F, Serino F, Pozzetto U, Fioravanti P, Caizzi P, Castagneto M. Verapamil improves Sandimmune immunosuppression, reducing acute rejection episodes. Transplant Proc 1996;28(4):2174–6.
210. Sketris IS, Methot ME, Nicol D, Belitsky P, Knox MG. Effect of calcium-channel blockers on cyclosporine clearance and use in renal transplant patients. Ann Pharmacother 1994;28(11):1227–31.
211. Smith CL, Hampton EM, Pederson JA, Pennington LR, Bourne DW. Clinical and medicoeconomic impact of the cyclosporine–diltiazem interaction in renal transplant recipients. Pharmacotherapy 1994;14(4):471–81.
212. Jones TE, Morris RG, Mathew TH. Formulation of diltiazem affects cyclosporin-sparing activity. Eur J Clin Pharmacol 1997;52(1):55–8.
213. Kuypers DR, Neumayer HH, Fritsche L, Budde K, Rodicio JL, Vanrenterghem YLacidipine Study Group. Calcium channel blockade and preservation of renal graft function in cyclosporine-treated recipients: a prospective randomized placebo-controlled 2-year study. Transplantation 2004;78(8):1204–11.
214. Wondimu B, Dahllof G, Berg U, Modeer T. Cyclosporin-A-induced gingival overgrowth in renal transplant children. Scand J Dent Res 1993;101(5):282–6.
215. Bokenkamp A, Bohnhorst B, Beier C, Albers N, Offner G, Brodehl J. Nifedipine aggravates cyclosporine A-induced gingival hyperplasia. Pediatr Nephrol 1994;8(2):181–5.
216. Thomason JM, Seymour RA, Ellis JS, Kelly PJ, Parry G, Dark J, Idle JR. Iatrogenic gingival overgrowth in cardiac transplantation. J Periodontol 1995;66(8):742–6.
217. Sooriyamoorthy M, Gower DB, Eley BM. Androgen metabolism in gingival hyperplasia induced by nifedipine and cyclosporin. J Periodontal Res 1990;25(1):25–30.
218. Thomason JM, Seymour RA, Rice N. The prevalence and severity of cyclosporin and nifedipine-induced gingival overgrowth. J Clin Periodontol 1993;20(1):37–40.
219. Kirch W, Janisch HD, Heidemann H, Ramsch K, Ohnhaus EE. Einfluss von Cimetidin und Ranitidin auf Pharmakokinetik und antihypertensiven Effect von Nifedipin. [Effect of cimetidine and ranitidine on the pharmacokinetics and anti-hypertensive effect of nifedipine.] Dtsch Med Wochenschr 1983;108(46):1757–61.
220. Dylewicz P, Kirch W, Benesch L, Ohnhaus EE. Influence of nifedipine with and without cimetidine on exercise tolerance in patients after myocardial infarction. Proceedings 6th International Adalat Symposium, Geneva, 1985. ICS 71Amsterdam: Excerpta Medica;. 1986.
221. van Harten J, van Brummelen P, Lodewijks MT, Danhof M, Breimer DD. Pharmacokinetics and hemodynamic effects of nisoldipine and its interaction with cimetidine. Clin Pharmacol Ther 1988;43(3):332–41.
222. Kirch W, Kleinbloesem CH, Belz GG. Drug interactions with calcium antagonists. Pharmacol Ther 1990;45(1):109–36.
223. Harada T, Ohtaki E, Sumiyoshi T, Hosoda S. Paralytic ileus induced by the combined use of nifedipine and diltiazem in the treatment of vasospastic angina. Cardiology 2002;97(2):113–4.
224. Toyosaki N, Toyo-oka T, Natsume T, Katsuki T, Tateishi T, Yaginuma T, Hosoda S. Combination therapy

with diltiazem and nifedipine in patients with effort angina pectoris. Circulation 1988;77(6):1370–5.
225. Varis T, Backman JT, Kivisto KT, Neuvonen PJ. Diltiazem and mibefradil increase the plasma concentrations and greatly enhance the adrenal-suppressant effect of oral methylprednisolone. Clin Pharmacol Ther 2000;67(3): 215–21.
226. Saltzman LS, Kates RA, Corke BC, Norfleet EA, Heath KR. Hyperkalemia and cardiovascular collapse after verapamil and dantrolene administration in swine. Anesth Analg 1984;63(5):473–8.
227. Rubin AS, Zablocki AD. Hyperkalemia, verapamil, and dantrolene. Anesthesiology 1987;66(2):246–9.
228. Kremens B, Brendel E, Bald M, Czyborra P, Michel MC. Loss of blood pressure control on withdrawal of fluconazole during nifedipine therapy. Br J Clin Pharmacol 1999;47(6):707–8.
229. Bailey DG, Spence JD, Edgar B, Bayliff CD, Arnold JM. Ethanol enhances the hemodynamic effects of felodipine. Clin Invest Med 1989;12(6):357–62.
230. Rodvold KA, Meyer J. Drug–food interactions with grapefruit juice. Infect Med 1996;13:868–912.
231. Feldman EB. How grapefruit juice potentiates drug bioavailability. Nutr Rev 1997;55(11 Pt 1):398–400.
232. Hashimoto Y, Kuroda T, Shimizu A, Hayakava M, Fukuzaki H, Morimoto S. Influence of grapefruit juice on plasma concentration of nifedipine. Jpn J Clin Pharmacol Ther 1996;27:599–606.
233. Azuma J, Yamamoto I, Wafase T, Orii Y, Tinigawa T, Terashima S, Yoshikawa K, Tanaka T, Kawano K. Effects of grapefruit juice on the pharmacokinetics of the calcium channel blockers nifedipine and nisoldipine. Curr Ther Res Clin Exp 1998;59:619–34.
234. Bailey DG, Spence JD, Munoz C, Arnold JM. Interaction of citrus juices with felodipine and nifedipine. Lancet 1991;337(8736):268–9.
235. Heinig R, Adelmann HG, Ahr G. The effect of ketoconazole on the pharmacokinetics, pharmacodynamics and safety of nisoldipine. Eur J Clin Pharmacol 1999;55(1): 57–60.
236. Sandstrom R, Knutson TW, Knutson L, Jansson B, Lennernas H. The effect of ketoconazole on the jejunal permeability and CYP3A metabolism of (R/S)-verapamil in humans. Br J Clin Pharmacol 1999;48(2):180–9.
237. Bruun NE, Ibsen H, Skott P, Toftdahl D, Giese J, Holstein-Rathlou NH. Lithium clearance and renal tubular sodium handling during acute and long-term nifedipine treatment in essential hypertension. Clin Sci (Lond) 1988;75(6):609–13.
238. Pinkofsky HB, Sabu R, Reeves RR. A nifedipine-induced inhibition of lithium clearance. Psychosomatics 1997; 38(4):400–1.
239. Price WA, Giannini AJ. Neurotoxicity caused by lithium-verapamil synergism. J Clin Pharmacol 1986;26(8):717–9.
240. Price WA, Shalley JE. Lithium–verapamil toxicity in the elderly. J Am Geriatr Soc 1987;35(2):177–8.
241. Dubovsky SL, Franks RD, Allen S. Verapamil: a new antimanic drug with potential interactions with lithium. J Clin Psychiatry 1987;48(9):371–2.
242. Riley J, Wilton LV, Shakir SA. A post-marketing observational study to assess the safety of mibefradil in the community in England. Int J Clin Pharmacol Ther 2002;40(6):241–8.
243. Jee LD, Opie LH. Acute hypotensive response to nifedipine added to prazosin in treatment of hypertension. BMJ (Clin Res Ed) 1983;287(6404):1514.
244. Pasanisi F, Meredith PA, Elliott HL, Reld JL. Verapamil and prazosin: pharmacodynamic and pharmacokinetic interactions in normal man. Br J Clin Pharmacol 1984;18:290.
245. Stauffer JC, Ruiz V, Morard JD. Subaortic obstruction after sildenafil in a patient with hypertrophic cardiomyopathy. N Engl J Med 1999;341(9):700–1.
246. Spencer CM, Gunasekara NS, Hills C. Zolmitriptan: a review of its use in migraine. Drugs 1999;58(2):347–74.
247. Khoury V, Kritharides L. Diltiazem-mediated inhibition of sildenafil metabolism may promote nitrate-induced hypotension. Aust NZ J Med 2000;30(5):641–2.
248. Gruer PJ, Vega JM, Mercuri MF, Dobrinska MR, Tobert JA. Concomitant use of cytochrome P450 3A4 inhibitors and simvastatin. Am J Cardiol 1999;84(7):811–5.
249. Peces R, Pobes A. Rhabdomyolysis associated with concurrent use of simvastatin and diltiazem. Nephron 2001;89(1):117–8.
250. Hebert MF, Lam AY. Diltiazem increases tacrolimus concentrations. Ann Pharmacother 1999;33(6):680–2.
251. Jones TE, Morris RG. Pharmacokinetic interaction between tacrolimus and diltiazem: dose-response relationship in kidney and liver transplant recipients. Clin Pharmacokinet 2002;41(5):381–8.
252. Seifeldin RA, Marcos-Alvarez A, Gordon FD, Lewis WD, Jenkins RL. Nifedipine interaction with tacrolimus in liver transplant recipients. Ann Pharmacother 1997;31(5):571–5.
253. Burnakis TG, Seldon M, Czaplicki AD. Increased serum theophylline concentrations secondary to oral verapamil. Clin Pharm 1983;2(5):458–61.
254. Parrillo SJ, Venditto M. Elevated theophylline blood levels from institution of nifedipine therapy. Ann Emerg Med 1984;13(3):216–7.
255. Burnakis TG, Seldon M, Czaplicki AD. Increased serum theophylline concentrations secondary to oral verapamil. Clin Pharm 1983;2(5):458–61.
256. Parrillo SJ, Venditto M. Elevated theophylline blood levels from institution of nifedipine therapy. Ann Emerg Med 1984;13(3):216–7.
257. Durant NN, Nguyen N, Katz RL. Potentiation of neuromuscular blockade by verapamil. Anesthesiology 1984;60(4):298–303.
258. Jones RM, Cashman JN, Casson WR, Broadbent MP. Verapamil potentiation of neuromuscular blockade: failure of reversal with neostigmine but prompt reversal with edrophonium. Anesth Analg 1985;64(10):1021–5.
259. Bikhazi GB, Leung I, Flores C, Mikati HM, Foldes FF. Potentiation of neuromuscular blocking agents by calcium channel blockers in rats. Anesth Analg 1988;67(1):1–8.

Candesartan cilexetil

See also Angiotensin II receptor antagonists

General Information

Candesartan cilexetil is the prodrug of candesartan, an angiotensin II type 1 (AT_1) receptor antagonist. Absorbed candesartan cilexetil is completely metabolized to candesartan. Candesartan has a half-life of about 9 hours (slightly longer in elderly people).

In an open study of 4531 hypertensive patients, the total incidence of adverse effects of candesartan was 6.1% (1). Individual adverse effects did not occur in more than 0.8% and were mainly dizziness and headache.

Sensory systems

Taste disorders have been reported with ACE inhibitors and now also with an angiotensin receptor antagonist (2).

- A renal transplant patient developed stomatitis and dysgeusia while taking candesartan cilexetil for hypertension. Although the patient was taking many concomitant medications, the chronology of the adverse effect suggested a causal relation with candesartan.

The potential for taste disturbances with candesartan has been investigated in a randomized, double-blind, placebo-controlled, crossover trial in 8 healthy subjects, measuring taste sensation by a paper disc method and also by electrogustometry (3). In view of the suggestion that taste disturbance is associated with low serum zinc concentrations, in that drugs may chelate or complex with zinc, serial serum and salivary zinc measurements were also made. The data suggested that candesartan subclinically reduces taste sensitivity after repeated dosage in healthy subjects but serum and salivary zinc concentrations were not affected, suggesting a zinc-independent effect.

Electrolyte balance

As angiotensin II receptor antagonists are often used for similar indications or substituted for ACE inhibitor therapy, the combination of an angiotensin II receptor antagonist with spironolactone will become more commonly used in treating cardiovascular and renal disease. Severe hyperkalemia, due to a combination of candesartan and spironolactone, has been described in a patient with hypertensive nephrosclerosis and only mild renal impairment (4). The authors reasonably advised that close monitoring of serum potassium should be mandatory during combination therapy, in order to prevent hyperkalemia. They also suggested that the transtubular potassium gradient and fractional excretion of potassium should be assessed before starting therapy to identify high-risk patients, and that a thiazide or loop diuretic be co-administered in order to reduce the risk of hyperkalemia; however, neither of these strategies has clear evidence to support it.

Liver

Jaundice and ductopenic hepatitis, similar to that described above under enalapril, has been attributed to candesartan (5). Hepatotoxicity has been reported with losartan, but this appears to be the first case associated with candesartan.

Biliary tract

Prolonged severe cholestasis induced by candesartan cilexetil improved after extracorporeal albumin dialysis (6).

Pancreas

A 75-year-old hypertensive man taking long-term candesartan cilexetil died after developing fulminant pancreatitis, complicated by acute renal insufficiency, respiratory insufficiency, and septic shock (7).

Skin

Erythema multiforme has been attributed to candesartan.

- A 50 year old man with hypertension taking candesartan cilexetil developed an ulcerative plaque on the upper lip after five weeks of treatment; biopsy showed erythema multiforme (8).

Immunologic

Vasculitis has been attributed to candesartan.

- A 73 year old man taking candesartan developed a vasculitic rash and nephritic syndrome (9). A skin biopsy showed a lymphocytic vasculitis. The rash and microscopic hematuria resolved within a week of stopping the drug.

The authors concluded that this reaction was consistent with a drug reaction.

Susceptibility Factors

Genetic

An 89-year-old hypertensive Japanese man with the CYP2C9*1/*3 slow metabolizer genotype had reduced clearance of candesartan, resulting in excessive blood pressure lowering (10). The authors pointed out that although losartan metabolism is also reduced in CYP2C9*1/*3 slow metabolizers, the inhibition results in reduced formation of the active metabolite E-3174, with no apparent effect on blood pressure lowering. In contrast, as candesartan has an inactive metabolite, its blood pressure lowering effect is enhanced.

Renal disease

In patients undergoing chronic hemodialysis, the safety profile did not differ from that reported in other populations, except for some rare cases of hypotension during hemodialysis. Hemodialysis does not affect the kinetics of candesartan. Because of the variability of oral clearance and the pronounced influence of hemodialysis-induced volume contraction on the hemodynamic effects of candesartan, careful monitoring is recommended (11).

References

1. Schulte KL, Fischer M, Lenz T, Meyer-Sabellek W. Efficacy and tolerability of candesartan cilexetil monotherapy or in combination with other antihypertensive drugs. Results of the AURA study. Clin Drug Invest 1999;18:453–60.
2. Chen C, Chevrot D, Contamin C, Romanet T, Allenet B, Mallaret M. Stomatite et agueusie induites par candésartan. [Stomatitis and ageusia induced by candesartan.] Nephrologie 2004;25(3):97–9.

3. Tsuruoka S, Wakaumi M, Nishiki K, Araki N, Harada K, Sugimoto K, Fujimura A. Subclinical alteration of taste sensitivity induced by candesartan in healthy subjects. Br J Clin Pharmacol 2004;57(6):807–12.
4. Fujii H, Nakahama H, Yoshihara F, Nakamura S, Inenaga T, Kawano Y. Life-threatening hyperkalemia during a combined therapy with the angiotensin receptor blocker candesartan and spironolactone. Kobe J Med Sci 2005; 51(1–2):1–6.
5. Basile G, Villari D, Gangemi S, Ferrara T, Accetta MG, Nicita-Mauro V. Candesartan cilexetil-induced severe hepatotoxicity. J Clin Gastroenterol 2003;36:273–5.
6. Sturm N, Hilleret MN, Dreyfus T, Barnoud D, Leroy V, Zarski JP. Hépatite sévère et prolongée secondaire à la prise de candesartan cilexitil (Atacand®) améliorée par le système MARS. Gastroenterol Clin Biol 2005;29(12): 1299–301.
7. Gill CJ, Jennings AE, Newton JB, Schwartz DE. Fatal acute pancreatitis in a patient chronically treated with candesartan. J Pharm Tech 2005;21(2):79–82.
8. Ejaz AA, Walsh JS, Wasiluk A. Erythema multiforme associated with candesartan cilexetil. South Med J 2004; 97(6):614–5.
9. Morton A, Muir J, Lim D. Rash and acute nephritic syndrome due to candesartan. BMJ 2004;328(7430):25.
10. Uchida I, Watanabe H, Nishio I, Hashimoto H, Yamasaki K, Hayashi H, Ohasi K. Altered pharmacokinetics and excessive hypotensive effect of candesartan in a patient with the CYP2C9*1/*3 genotype. Clin Pharmacol Ther 2003;74:505–8.
11. Pfister M, Schaedeli F, Frey FJ, Uehlinger DE. Pharmacokinetics and haemodynamics of candesartan cilexetil in hypertensive patients on regular haemodialysis. Br J Clin Pharmacol 1999;47(6):645–51.

Captopril

See also Angiotensin converting enzyme inhibitors

General Information

Captopril is a sulfhydryl-containing ACE inhibitor used in the management of hypertension and heart failure, after myocardial infarction, and in diabetic nephropathy.

Organs and Systems

Respiratory

Two cases of alveolitis have been reported with captopril, one associated with eosinophilia (1) and the other with a lymphocytic pulmonary infiltrate (2).

Sensory systems

A report from the Australian Drug Evaluation Committee has confirmed that dysgeusia (taste disturbance and taste loss) is more likely to complicate treatment with captopril than with other ACE inhibitors; captopril accounted for more than half the cases of taste loss (3). Taste loss was dose-related, in that more than 90% of reports detailed a daily dose of 50 mg or more. Most cases recovered on withdrawal, although the sense of taste had not returned 7 months after withdrawal in one case, and in two of the reports there was associated anosmia. Although there was no significant difference between captopril and lisinopril (4), it appears that captopril, presumably via its sulfhydryl group, is more likely to provoke dysgeusia.

Hematologic

Anecdotal reports of a bone-marrow suppressant effect of captopril have been published. Cases of neutropenia and agranulocytosis (5–8) have also been reported, although in some of these there were complicating issues, making a clear association difficult to establish. Usually the incidence is higher in patients with renal insufficiency or collagen vascular disease. In few cases the white cell count returned to normal when captopril was withdrawn. Although agranulocytosis has been attributed to the presence of the sulfhydryl group in captopril, it has also been reported with enalapril (9).

Agranulocytosis has also been reported in association with toxic epidermal necrolysis (10).

- A 59-year-old woman with long-standing hypertension took captopril 200 mg bd plus hydrochlorothiazide 50 mg od and 3 days later developed nausea, vomiting, malaise, and a severe, pruritic, erythematous rash, with severe dehydration and orthostatic hypotension. The diuretic was withdrawn and intravenous fluids and diphenhydramine were given. She recovered within 1 week and remained asymptomatic for 4 weeks. She then developed fever, malaise, a new rash, orthostatic hypotension, and diffuse erythroderma. Her posterior pharynx was erythematous. She had neutropenia (900×10^6/l) and captopril was withdrawn. She was treated empirically with antibiotics and intravenous fluids and subcutaneous granulocyte colony-stimulating factor (G-CSF) for 5 days. The erythroderma progressed to large coalescing blisters, which then ruptured, producing large areas of weeping skin over all her limbs. The bone-marrow and skin recovered fully within 2 weeks.

Autoimmune thrombocytopenia, gradually reversible on withdrawal of captopril, was reported in three patients taking captopril (11).

Hematologic

Pancytopenia has been reported in a premature newborn treated with captopril for renovascular hypertension (12). Trilineage bone marrow suppression leading to pancytopenia is a very rare complication, which may be dose-related in the therapeutic range (a collateral or toxic reaction). In this case the authors surmised that accumulation due to renal tubular dysfunction of prematurity, combined with renal artery stenosis, may have explained the adverse effect in the absence of overt renal dysfunction. They recommended that captopril should be used with caution in premature babies and neonates with underlying renal or renovascular disease, even if they do not have overt renal dysfunction.

Salivary glands

Sialadenitis has been reported in two patients who had taken captopril for hypertension and developed non-painful bilateral parotid and submandibular gland swelling (13). In one case there was associated upper chest and face erythema, and in the other conjunctival erythema. The authors concluded that the symptoms had probably been caused by the drug, and they suggested that inflammatory mediators or vasodilators (for example bradykinin) could have induced salivary gland obstruction by spasm or edema, or that the drug may have interfered with salivary secretion at a cellular level.

Pancreas

Two new cases of pancreatitis with captopril have been reported (14,15). It has been suggested that early detection of raised serum amylase and lipase activities can prevent the development of full-blown pancreatitis (15).

Urinary tract

Acute renal insufficiency with tubular necrosis has been described (16). It occurred within 24 hours of the first dose of captopril and required hemodialysis for 8 weeks. The diagnosis of ischemic tubular necrosis was confirmed by renal biopsy. Few previous cases of ACE inhibitor-induced nephropathy have been documented by renal biopsy.

Immunologic

Lupus-like syndrome has been reported with captopril (17). The authors believed their patient to be the fifth such published case.

- A 54-year-old Caucasian man presented with a 4-week history of chills, fever, malaise, and generalized arthralgia. Following an aortic valve replacement, he had taken aspirin, coumadin, and captopril 25 mg tds for 1 year. He was febrile (temperature 39.4°C), normotensive, with diffuse livedo reticularis, and the physical signs of aortic valve disease. Infective endocarditis was ruled out by appropriate investigations. He had a raised erythrocyte sedimentation rate (142 mm/hour) and a positive antinuclear antibody test (FANA 1:2560) with a negative antinative DNA test. Captopril was withdrawn and he was given prednisone for 5 days. His symptoms resolved rapidly and the livedo reticularis cleared within 2 days. The FANA and ESR returned to normal and remained so at follow-up 6 months later.

During early drug development, the occurrence of antinuclear antibodies in 10 of 37 patients taking high doses of captopril was described (18).

Drug–Drug Interactions

Acetylsalicylic acid

Aspirin is thought to reduce the antihypertensive effect of captopril (19).

Cardiac glycosides

There have been contradictory results in studies of the effects of captopril on digoxin pharmacokinetics. In some cases, captopril increased steady-state plasma digoxin concentrations (20), while in others there was no evidence of an interaction (21–24). In eight patients with NHYA Class IV congestive heart failure, captopril 12.5 mg tds for 1 week increased the peak digoxin concentration, reduced the time to peak, and increased the AUC during a single dosage interval at steady state; trough digoxin concentrations did not change (25). There are two possible explanations for these findings: that captopril reduced the clearance of digoxin or that it increased both the rate and extent of absorption of digoxin; unfortunately, the authors did not measure either the mean steady-state concentration or the half-life, which would have clarified this. On the whole, however, it is unlikely that this interaction is of any clinical significance, since in no study has there been evidence of digoxin toxicity during concomitant captopril therapy, and in one formal study (24) there was evidence of no pharmacodynamic interaction.

General anesthesia

The hazards of general anesthesia in the presence of ACE inhibitors have been recognized and reviewed (26). It has generally been accepted that ACE inhibitors should be continued up to the time of surgery. Patients should be monitored aggressively and hemodynamic instability treated appropriately.

- An 86-year-old man, who was taking captopril 25 mg bd and bendroflumethiazide 25 mg/day for hypertension, had a transurethral resection of the prostate under spinal anesthesia, and developed profound bradycardia and hypotension with disturbances of consciousness during transfer to the recovery room (27). Initial treatment with atropine produced rapid improvement in cardiovascular and cerebral function. A further hypotensive episode, without bradycardia, occurred about 1 hour later, but responded rapidly to methoxamine. He made a full recovery overnight.

The authors suggested that the concomitant administration of captopril may have contributed to the adverse effect, since the renin–angiotensin system is important in maintaining blood pressure during anesthesia. Because of a protective effect of angiotensin II in the presence of sympathetic blockade by presynaptic stimulation of adrenergic neurons, the authors hypothesized that hypotension and bradycardia in this patient may have been magnified by suppression of the renin–angiotensin system, resulting in an enhanced negative Bainbridge effect.

Iron and other transition metals

Captopril has been reported to react with iron and other transition metals. This interaction has been investigated in vivo in seven healthy adults (28). Co-administration of ferrous sulfate and captopril resulted in a 37% decrease in the AUC of unconjugated captopril, with no significant changes in C_{max} or t_{max}. The plasma AUC of total captopril was not altered. The authors suggested that the interaction may be specific to captopril among ACE inhibitors, because it contains a sulfhydryl group. Therefore, if iron salts were to be taken by a patient requiring an ACE inhibitor, an agent other than captopril might be considered.

Naloxone

Following the use of naloxone the hypotensive effect of captopril was abolished (29).

Interference with Diagnostic Tests

Legal reaction

Like other sulfhydryl-containing compounds, captopril can cause false-positive ketonuria when assessed with the Legal reaction (sodium nitroprusside reacting with acetoacetic acid and possibly with acetone). It has therefore been suggested that in patients with diabetes taking such drugs ketonuria should be assessed with the Acetest (30). Alternatively, a blood ketone test with Acetest and/ or enzymatic detection of beta-hydroxybutyric acid can be performed for confirmation.

References

1. Schatz PL, Mesologites D, Hyun J, Smith GJ, Lahiri B. Captopril-induced hypersensitivity lung disease. An immune-complex-mediated phenomenon. Chest 1989;95(3):685–7.
2. Kidney JC, O'Halloran DJ, FitzGerald MX. Captopril and lymphocytic alveolitis. BMJ 1989;299(6705):981.
3. Boyd I. Captopril-induced taste disturbance. Lancet 1993;342(8866):304.
4. Neil-Dwyer G, Marus A. ACE inhibitors in hypertension: assessment of taste and smell function in clinical trials. J Hum Hypertens 1989;3(Suppl 1):169–76.
5. Lacueva J, Enriquez R, Bonilla F, Cabezuelo JB. Agranulocitosis por captopril en insufficienca renal. Nefrologia 1992;12:76–7.
6. Ortega G, Molina Boix M, Vidal JB, de Paco M, del Bano MD, Ruiz F. Tratamiento con captopril de la hipertension arterial en la nefropatia lupica. [Treatment of arterial hypertension with captopril in lupus nephropathy.] An Med Interna 1992;9(2):72–5.
7. Fernandez Seara J, Dominguez Alvarez LM, Rodriguez Perez R, Pereira Jorge JA, Rodriguez Canal A. Agranulocitosis inducida por captopril tras cuarto mese de tratamiento. [Agranulocytosis induced by captopril after 4 months of treatment.] An Med Interna 1991;8(8):398–400.
8. Ortega G, Molina M, Rivera MD, Serrano C. Neutropenia inducida porbajas dosis de captopris en pacientes hipertensos sin enfermedad autoimmune asociada. [Neutropenia induced by low doses of captopril in hypertensive patients without associated autoimmune disease.] An Med Interna 1999;16(8):436.
9. Elis A, Lishner M, Lang R, Ravid M. Agranulocytosis associated with enalapril. DICP 1991;25(5):461–2.
10. Winfred RI, Nanda S, Horvath G, Elnicki M. Captopril-induced toxic epidermal necrolysis and agranulocytosis successfully treated with granulocyte colony-stimulating factor. South Med J 1999;92(9):918–20.
11. Pujol M, Duran-Suarez JR, Martin Vega C, Sanchez C, Tovar JL, Valles M. Autoimmune thrombocytopenia in three patients treated with captopril. Vox Sang 1989;57(3):218.
12. Tarcan A, Gurakan B, Ozbek N. Captopril-induced pancytopenia in a premature newborn. J Paediatr Child Health 2004;40(7):404–5.
13. Gislon Da Silva RM. Captopril-induced bilateral parotid and submandibular sialadenitis. Eur J Clin Pharmacol 2004;60(6):449–53.
14. Iliopoulou A, Giannakopoulos G, Pagoy H, Christos T, Theodore S. Acute pancreatitis due to captopril treatment. Dig Dis Sci 2001;46(9):1882–3.
15. Borgia MC, Celestini A, Caravella P, Catalano C. Angiotensin-converting-enzyme inhibitor administration must be monitored for serum amylase and lipase in order to prevent an acute pancreatitis: a case report. Angiology 2001;52(9):645–7.
16. Al Shohaib S, Raweily E. Acute tubular necrosis due to captopril. Am J Nephrol 2000;20(2):149–52.
17. Ratliff NB 3rd, Pieranna F, Manganelli P. Captopril induced lupus. J Rheumatol 2002;29(8):1807–8.
18. Reidenberg MM, Case DB, Drayer DE, Reis S, Lorenzo B. Development of antinuclear antibody in patients treated with high doses of captopril. Arthritis Rheum 1984;27(5):579–81.
19. Moore TJ, Crantz FR, Hollenberg NK, Koletsky RJ, Leboff MS, Swartz SL, Levine L, Podolsky S, Dluhy RG, Williams GH. Contribution of prostaglandins to the antihypertensive action of captopril in essential hypertension. Hypertension 1981;3(2):168–73.
20. Cleland JG, Dargie HJ, Pettigrew A, Gillen G, Robertson JI. The effects of captopril on serum digoxin and urinary urea and digoxin clearances in patients with congestive heart failure. Am Heart J 1986;112(1):130–5.
21. Douste-Blazy P, Blanc M, Montastruc JL, Conte D, Cotonat J, Galinier F. Is there any interaction between digoxin and enalapril? Br J Clin Pharmacol 1986;22(6):752–3.
22. Magelli C, Bassein L, Ribani MA, Liberatore S, Ambrosioni E, Magnani B. Lack of effect of captopril on serum digoxin in congestive heart failure. Eur J Clin Pharmacol 1989;36(1):99–100.
23. Miyakawa T, Shionoiri H, Takasaki I, Kobayashi K, Ishii M. The effect of captopril on pharmacokinetics of digoxin in patients with mild congestive heart failure. J Cardiovasc Pharmacol 1991;17(4):576–80.
24. de Mey C, Elich D, Schroeter V, Butzer R, Belz GG. Captopril does not interact with the pharmacodynamics and pharmacokinetics of digitoxin in healthy man. Eur J Clin Pharmacol 1992;43(4):445–7.
25. Kirimli O, Kalkan S, Guneri S, Tuncok Y, Akdeniz B, Ozdamar M, Guven H. The effects of captopril on serum digoxin levels in patients with severe congestive heart failure. Int J Clin Pharmacol Ther 2001;39(7):311–4.

26. Mets B. The renin angiotensin system and ACE inhibitors in the perioperative period. In: Skarvan K, editor. Arterial Hypertension. London: Bailliere Tindall, 1997:581–604.
27. Williams NE. Profound bradycardia and hypotension following spinal anaesthesia in a patient receiving an ACE inhibitor: an important "drug" interaction? Eur J Anaesthesiol 1999;16(11):796–8.
28. Schaefer JP, Tam Y, Hasinoff BB, Tawfik S, Peng Y, Reimche L, Campbell NR. Ferrous sulphate interacts with captopril. Br J Clin Pharmacol 1998;46(4):377–81.
29. Ajayi AA, Campbell BC, Rubin PC, Reid JL. Effect of naloxone on the actions of captopril. Clin Pharmacol Ther 1985;38(5):560–5.
30. Csako G, Elin RJ. Spurious ketonuria due to captopril and other free sulfhydryl drugs. Diabetes Care 1996;19(6):673–4.

Carvedilol

See also Beta-adrenoceptor antagonists

General Information

Carvedilol is a highly lipophilic non-selective beta-adrenoceptor antagonist with alpha$_1$-blocking action, promoting peripheral vasodilatation. It also has free radical scavenging and antimitogenic effects.

The safety and tolerability profile of carvedilol in heart failure appears to be reassuring (1). In trials in patients with congestive heart failure, carvedilol was withdrawn in only about 5% (2). The most common adverse reactions were edema, dizziness, bradycardia, hypotension, nausea, diarrhea, and blurred vision. The rate of drug withdrawal was not different among patients under 65 years and among older ones.

The largest trial of carvedilol in patients with heart failure documented only a 5% withdrawal rate in 1197 patients during the open phase that preceded randomization. The major reasons for withdrawal were worsening heart failure (2%), dizziness (1.1%), bradycardia (0.2%), and death (0.5%). Overall, 78% of the patients who entered the trial and were randomized to carvedilol achieved the target dose of 50 mg/day, and withdrawal rates were comparable with carvedilol and placebo (11–14% over 6.5 months of treatment).

The issue of tolerability is particularly important in patients with severe heart failure (NYHA class IV). Unfortunately, so far all trials with carvedilol have failed to recruit a large number of such patients. A retrospective analysis of the tolerability profile of carvedilol in 63 patients with NYHA class IV heart failure showed that non-fatal adverse events while taking carvedilol were more frequent than in patients with class II-III heart failure (43 versus 24%) and more often resulted in permanent withdrawal of the drug (25 versus 13%) (3). However, 59% of the patients with class IV heart failure improved by one or more functional class after 3 months of treatment. The conclusion was that carvedilol is a useful adjunctive therapy for patients with NYHA class IV heart failure, but these patients require close observation during the start of treatment and titration of the dose.

In patients with heart failure, amiodarone is often required for the treatment of serious ventricular dysrhythmias. The beneficial effects of carvedilol on left ventricular remodeling, systolic function, and symptomatic status were not altered by amiodarone in 80 patients with heart failure. Adverse effects that necessitated withdrawal of carvedilol were no more frequent in patients taking amiodarone than in those taking carvedilol alone (26 versus 25%) (4).

In 10 patients with gastroesophageal varices, none having bled, treated with oral carvedilol 12.5 mg/day for 4 weeks, hemodynamic measurements were performed before the first administration and at 1 hour and 4 weeks after (5). After acute administration, the hepatic venous pressure gradient was significantly reduced by 23% (from 16.4 to 12.6 mmHg), with a significant reduction in heart rate. After 4 weeks, the hepatic venous pressure gradient was further significantly reduced to 9.3 mmHg. Carvedilol was well tolerated, and only one patient had asymptomatic hypotension. The results of this study suggest that low-dose carvedilol significantly reduces portal pressure without significantly systemic hemodynamic effects. The reduction in portal pressure with carvedilol was larger than that obtained with propranolol, encouraging specific trials of carvedilol for the primary prevention of variceal hemorrhage.

Organs and Systems

Cardiovascular

Since the adrenergic system is activated to support reduced contractility of the failing heart, the administration of a beta-blocker in a patient with heart failure can cause myocardial depression. This effect is more marked with non-selective first-generation beta-blockers, such as propranolol. Carvedilol has an acceptable tolerability profile, reducing after-load and thus counteracting the negative inotropic properties of adrenoceptor blockade. While these vasodilatory properties can play a favorable role at the start of treatment, it is less likely that vasodilatation can make a substantial contribution to the long-term effects of third-generation beta-blockers.

There are concerns about the risks of starting beta-blockers in patients with heart failure, specifically those with severe impairment. The database of the COPERNICUS study, which tested the effects of carvedilol in patients with advanced heart failure, has been evaluated in order to assess the rates of death, hospitalizations, and permanent withdrawal of carvedilol in the first 8 weeks from the start of treatment (6). The patients allocated to carvedilol did not have any significant increase in cardiovascular risk compared with those randomized to placebo and carvedilol was associated with fewer deaths, hospitalizations, and drug withdrawals. These data suggest that in patients with advanced heart failure the benefit to harm balance in the first 2 months of treatment is similar to that observed during long-term therapy. These findings should encourage clinicians to use carvedilol in such patients.

Carvedilol has been evaluated in patients with heart failure (7). In more than 50% of the patients given carvedilol the dosage could be titrated up to the maximum (50 mg/day). The major reason for missing the target dose was hypotension followed by bradycardia. Carvedilol had to be withdrawn in 21/316 patients (6.6%), in 14 (4.4%) for cardiovascular reasons: six with worsening heart failure, five with hypotension, two with syncope, and one with bradycardia. The most frequent non-cardiovascular reason for withdrawal was chronic obstructive pulmonary disease. Age did not influence tolerability: the rate of discontinuation was similar in patients aged 70 years or older and in younger patients. These data have confirmed in a real clinical context what controlled studies have shown, that carvedilol is generally well tolerated and that the maximum target dose can be reached in a relevant number of patients irrespective of age and severity of heart failure.

As with other alpha-blockers, postural hypotension is quite common with carvedilol, especially in elderly people (8). In a comparison of carvedilol with pindolol in elderly patients, postural hypotension occurred even with small doses of carvedilol (12.5 mg) (9). The authors concluded that lower starting doses should be used in elderly people, in patients taking diuretics, and in patients with heart failure.

Respiratory

There have been many reports of drug-related pneumonitis caused by beta-adrenoceptor antagonists. Pneumonitis associated with carvedilol has been described (10).

- A 69-year-old woman, a non-smoker with a history of asthma, had had no respiratory exacerbation in the previous 12 months. She had taken carvedilol for hypertension and left ventricular hypertrophy for several months and fluoxetine for depression for the last 5 years. Over 20 days she developed progressive dyspnea on exercise, fever, cough, and then dyspnea at rest. Chest X-rays showed bilateral opacities. A high-resolution CT scan showed bilateral infiltrates diffusely distributed over all the lung fields. Pulmonary function tests showed deterioration in spirometry. Both carvedilol and fluoxetine were withdrawn and 3 days later she became afebrile, with clear improvement in her respiratory symptoms. Another high-resolution CT scan showed great improvement with only minimal residual findings. Fluoxetine was restarted and 40 days later she was completely asymptomatic, still taking fluoxetine.

Even in the absence of rechallenge in this case, the rapid improvement after carvedilol withdrawal suggested a causative relation between pneumonitis and carvedilol.

Metabolism

Severe diabetes mellitus has been described in a patient with heart failure treated with carvedilol and furosemide (11).

- A 37-year old man with a dilated cardiomyopathy was given furosemide, spironolactone, and candesartan. After 1 year carvedilol was introduced in a maintenance dose of 10 mg/day. HbA_{1c} was 5.1% at the beginning of carvedilol treatment. After 9 months of treatment, he started to feel extremely thirsty and lost 10 kg in 3 months. No viral infections or pancreatitis were detected. The HbA_{1c} concentration increased to 17%, the blood glucose concentration was 31 mmol/l (557 mg/dl). Furosemide was withdrawn and the blood glucose concentration fell within a week to 8.9 mmol/l (160 mg/dl). Carvedilol was then replaced by metoprolol and after 2 further weeks the fasting blood glucose concentration fell to 5.9 mmol/l (106 mg/dl). The patient was stable thereafter.

The mechanisms of carvedilol-induced hyperglycemia are not known, although alpha-blockade in the pancreas could impair insulin secretion. In this case furosemide could have increased insulin resistance.

Liver

Liver function abnormalities in trials of carvedilol occurred in 1.1% of patients taking carvedilol compared with 0.9% in patients taking placebo (2). However, in all trials the patients with pre-existing liver disease were excluded, and so information on the effects of carvedilol in these patients is not available.

Urinary tract

Renal insufficiency can occur in patients with heart failure treated with carvedilol, usually when pre-existing renal insufficiency, low blood pressure, or diffuse vascular disease are present (2). Patients at high risk of renal dysfunction should be carefully monitored, particularly at the beginning of treatment, and the drug should be withdrawn in case renal function worsens.

Skin

- Stevens–Johnson syndrome occurred in a 71-year-old man with stable ischemic cardiomyopathy taking carvedilol (12).

Sexual function

In a comparison of carvedilol with valsartan in 160 patients with hypertension (mean age 46 years) each treatment was continued for 16 weeks, with crossover after 4 weeks of placebo (13). Blood pressure was significantly lowered by both drugs (48% normalization with valsartan and 45% with carvedilol). In the first month of treatment, sexual activity (assessed as the number of episodes of sexual intercourse per month) fell with both treatments compared with baseline, although the change was statistically significant only with carvedilol. After the first month of treatment, sexual activity further worsened with carvedilol, but it improved or recovered fully with valsartan. The results were confirmed by the crossover. This confirms that beta-blockers can cause chronic worsening of sexual function.

Drug Administration

Drug overdose

Overdosage of carvedilol results mainly in hypotension and bradycardia (2). For excessive bradycardia, atropine has been used successfully, while to support ventricular function intravenous glucagon, dobutamine, or isoprenaline have been recommended. For severe hypotension, adrenaline or noradrenaline can be given.

References

1. Tang WHW, Fowler MB. Clinical trials of carvedilol in heart failure. Heart Fail Rev 1999;4:79–88.
2. Frishman WH. Carvedilol. N Engl J Med 1998;339(24):1759–65.
3. Macdonald PS, Keogh AM, Aboyoun CL, Lund M, Amor R, McCaffrey DJ. Tolerability and efficacy of carvedilol in patients with New York Heart Association class IV heart failure. J Am Coll Cardiol 1999;33(4):924–31.
4. Macdonald PS, Keogh AM, Aboyoun C, Lund M, Amor R, McCaffrey D. Impact of concurrent amiodarone treatment on the tolerability and efficacy of carvedilol in patients with chronic heart failure. Heart 1999;82(5):589–93.
5. Tripathi D, Therapondos G, Lui HF, Stanley AJ, Hayes PC. Haemodynamic effects of acute and chronic administration of low-dose carvedilol, a vasodilating beta-blocker, in patients with cirrhosis and portal hypertension. Aliment Pharmacol Ther 2002;16(3):373–80.
6. Krum H, Roecker EB, Mohacsi P, Rouleau JL, Tendera M, Coats AJS, Katus HA, Fowler MB, Packer M, for the Carvedilol Prospective Randomized Cumulative Survival (COPERNICUS) Study Group. Effects of initiating carvedilol in patients with severe chronic heart failure. Results from the COPERNICUS Study. J Am Med Assoc 2003;289:712–8.
7. Nul D, Zambrano C, Diaz A, Ferrante D, Varini S, Soifer S, Grancelli H, Doval H; Grupo de Estudio de la Sobrevida en la Insuficiencia Cardiaca en Argentina. Impact of a standardized titration protocol with carvedilol in heart failure: safety, tolerability, and efficacy—a report from the GESICA Registry. Cardiovasc Drugs Ther 2005;19:125–34.
8. Louis WJ, Krum H, Conway EL. A risk-benefit assessment of carvedilol in the treatment of cardiovascular disorders. Drug Saf 1994;11(2):86–93.
9. Krum H, Conway EL, Broadbear JH, Howes LG, Louis WJ. Postural hypotension in elderly patients given carvedilol. BMJ 1994;309(6957):775–6.
10. Markou N, Antzoulatos N, Haniotou A, Kanakaki M, Parissis J, Damianos A. A case of drug-induced pneumonitis caused by carvedilol. Respiration 2004;71:650–2.
11. Kobayacawa N, Sawaki D, Otani Y, Sekita G, Fukushima K, Takeuchi H, Aoyagi T. A case of severe diabetes mellitus occurred during management of heart failure with carvedilol and furosemide. Cardiovasc Drugs Ther 2003;17:295.
12. Kowalski BJ, Cody RJ. Stevens–Johnson syndrome associated with carvedilol therapy. Am J Cardiol 1997;80(5):669–70.
13. Fogari R, Zoppi A, Poletti L, Marasi G, Mugellini A, Corradi L. Sexual activity in hypertensive men treated with valsartan or carvedilol: a crossover study. Am J Hypertens 2001;14(1):27–31.

Celiprolol

See also Beta-adrenoceptor antagonists

General Information

Celiprolol is a $beta_1$-selective antagonist with partial $beta_2$-agonist activity. In healthy volunteers it caused "particularly unpleasant" subjective adverse effects, including headache, sleepiness, and feeling cold and generally unwell (1). However, it has beneficial effects on lipids, reducing total cholesterol by 6% and low-density lipoproteins by 10% (SEDA-17, 235). Whether this translates into clinical benefit is not known, and celiprolol has no convincing advantages over other beta-blockers (2).

Organs and Systems

Respiratory

There has been a report of hypersensitivity pneumonitis secondary to celiprolol (3).

References

1. Busst CM, Bush A. Comparison of the cardiovascular and pulmonary effects of oral celiprolol, propranolol and placebo in normal volunteers. Br J Clin Pharmacol 1989;27(4):405–10.
2. Anonymous. Celiprolol—a better beta blocker? Drug Ther Bull 1992;30(9):35–6.
3. Lombard JN, Bonnotte B, Maynadie M, Foucher P, Reybet Degat O, Jeannin L, Camus P. Celiprolol pneumonitis. Eur Respir J 1993;6(4):588–91.

Cilazapril

See also Angiotensin converting enzyme inhibitors

General Information

Cilazapril is a non-sulfhydryl ACE inhibitor (1). Like enalapril and ramipril it is a prodrug and is hydrolysed after absorption to cilazaprilat, which has a long half-life allowing once-daily administration (2).

Organs and Systems

Cardiovascular

Refractory hypotension occurred when the long-acting ACE inhibitor cilazapril was used in a patient undergoing spinal surgery (3).

- A 76-year-old woman, who had taken cilazapril 2.5 mg/day for 5 years for hypertension, underwent back surgery. On the morning of the procedure she took her usual dose of cilazapril, and 10 minutes after induction of anesthesia her blood pressure fell to 60/40 mmHg. She was successfully treated with ephedrine and fluid boluses. However, about 1 hour later her blood pressure fell to 60/30 mmHg when she was placed in the prone position. She was given ephedrine and fluids, with no effect on blood pressure. After a dopamine infusion her systolic blood pressure stabilized at 80–100 mmHg. Cilazapril was withdrawn. On the third postoperative day her blood pressure progressively increased, and she was weaned from the vasopressor. By the end of that day her blood pressure was 185/65 mmHg and cilazapril was restarted.

Urinary tract

Uremia has been reported in association with cilazapril (4).

- A 72-year-old man had taken captopril, cilazapril, and enalapril for heart disorders and developed uremia in the context of progressive symptoms of renal impairment and dehydration, presenting with renal insufficiency and hyperkalemia. The plasma potassium concentration normalized after hemodialysis but the creatinine concentration remained high and the eventual outcome was not clearly stated.

The association with cilazapril was relatively weak in this case.

Skin

Many cases of pemphigus have been reported with ACE inhibitors, related to the amide group they contain. This is the case for cilazapril.

- A 69-year-old white woman developed skin lesions of pemphigus foliaceus after she had taken cilazapril for 3 months; they resolved on withdrawal and the addition of prednisone and azathioprine (5).
- Pemphigus vulgaris possibly triggered by cilazapril has been reported (6).

References

1. Szucs T. Cilazapril. A review. Drugs 1991;41(Suppl 1):18–24.
2. Deget F, Brogden RN. Cilazapril. A review of its pharmacodynamic and pharmacokinetic properties, and therapeutic potential in cardiovascular disease. Drugs 1991;41(5):799–820.
3. Akinci SB, Ayhan B, Kanbak M, Aypar U. Refractory hypotension in a patient chronically treated with a long acting angiotensin-converting enzyme inhibitor. Anaesth Intensive Care 2004;32(5):722–3.
4. Fomin VV, Taronishvili OI, Shvetsov MIu, Shilov EM, Moiseev SV, Kushnir VV, Sorokin IuD. [Progressive azotemia provoked by ACE inhibitor in renal ischemia.]. Ter Arkh 2004;76(9):66–70.
5. Buzon E, Perez-Bernal AM, de la Pena F, Rios JJ, Camacho F. Pemphigus foliaceus associated with cilazapril. Acta Derm Venereol 1998;78(3):227.
6. Orion E, Gazit E, Brenner S. Pemphigus vulgaris possibly triggered by cilazapril. Acta Dermatol Venereol 2000; 80(3):220.

Clonidine and apraclonidine

General Information

The presynaptic alpha-adrenoceptor agonists, particularly methyldopa and clonidine, are agonists at presynaptic $alpha_2$-adrenoceptors. Guanabenz, guanfacine, and tiamenidine appear to be qualitatively similar to clonidine; clear evidence of quantitative differences remains to be confirmed.

The mechanism of action of these drugs depends on reducing sympathetic nervous outflow from the nervous system by interference with regulatory neurotransmitter systems in the brainstem. Because none of the drugs was selective or specific for circulatory control systems, the hypotensive effects were invariably accompanied by other nervous system effects. However, the identification of imidazoline receptors has suggested that it may be possible to develop drugs, such as moxonidine and relminidine, with greater selectivity for circulatory control mechanisms and a reduced likelihood of unwanted nervous system depressant effects (1).

Clonidine has been used to treat hypertension and migraine. It ameliorates the opioid withdrawal syndrome by reducing central noradrenergic activity. Its role in the treatment of psychiatric disorders has been the subject of an extensive review, but without new information on its safety (2). Clonidine is also used epidurally, in combination with opioids, neostigmine, and anesthetic and analgesic agents, to produce segmental analgesia, particularly for postoperative relief of pain after obstetrical and surgical procedures.

The use of clonidine in children has been comprehensively reviewed (3,4).

Apraclonidine (*para*-aminoclonidine) is a relatively nonspecific $alpha_1$- and $alpha_2$-adrenoceptor agonist, which is less likely to cross the blood–brain barrier than clonidine. Apraclonidine suppresses aqueous humor flow by 39–44% and lowers intraocular pressure by 20–23% (5).

Observational studies

on of clonidine is attributed to its central action as an $alpha_2$ adrenoceptor agonist, although this assumption may not explain all of its antihypertensive actions. In an experimental study of the hemodynamic response to clonidine in 28 young healthy men, the hemodynamic response suggested that the hypotensive action of clonidine is caused by both an immediate reduction in peripheral vascular resistance and a prolonged reduction in cardiac output (6). The study also showed that central blood pressure was reduced more than peripheral blood pressure, according to pulse wave velocity indices.

Organs and Systems

Cardiovascular

Clonidine causes sinus bradycardia and atrioventricular block, as illustrated by two cases, one a 10-year-old boy (7) and the other a 71-year-old woman (8), who developed Wenckebach's phenomenon. Clonidine was also studied in seven patients subjected to electrophysiological studies after 5 weeks of therapy (9). It slowed the sinus rate and increased the atrial pacing rate, producing Wenckebach's phenomenon, indicating depressed function of the sinus and AV nodes.

In three patients with chronic schizophrenia and primary polydipsia given clonidine in doses of up to 800 micrograms/day for 2–5 months, blood pressure and pulse fell significantly in a dose-dependent manner, but fluid intake, as assessed by measurements of weight and 24-hour urine volume, was not affected (10). Hypotension and bradycardia limited the extent to which the dose of clonidine could be increased. The lack of effect of clonidine on polydipsia in this small sample and the inconsistent results of two other recent studies have provided little overall support for using clonidine to treat primary polydipsia associated with schizophrenia.

Clonidine-induced hypertension has been reported in a patient with autonomic dysfunction (11).

- A 39-year-old quadriplegic man with poorly controlled pain had many features consistent with autonomic dysfunction (for example a C4 spinal lesion, orthostatic hypotension, hypertension). He routinely used transdermal clonidine and transdermal glyceryl trinitrate as needed for control of acute hypertensive episodes. The clonidine was discontinued, after which his blood pressure fell (maximum systolic and diastolic pressures by about 50 and 25 mmHg respectively).

Because clonidine relies on central alpha$_2$-adrenoceptor agonist activity for its hypotensive effects, it can cause hypertension in patients with autonomic dysfunction. It should therefore be used with great caution when autonomic dysfunction is suspected.

The effect of intrathecal clonidine has been evaluated in a prospective randomized study in 45 children aged 6–15 years, who were randomized to receive either 0.5% hyperbaric bupivacaine or 0.5% hyperbaric bupivacaine plus clonidine 2 micrograms/kg (12). Clonidine was associated with non-significant prolongation of motor block, from an average of 150–190 minutes. Postoperative analgesia was significantly longer with clonidine (490 versus 200 minutes). Clonidine was associated with higher incidences of hypotension (54 versus 36%) and bradycardia (30 versus 0%).

Respiratory

The effects of clonidine and other alpha$_2$-adrenoceptor agonists on respiratory function in asthmatics have been reviewed (13). Inhaled alpha$_2$-adrenoceptor agonists reduce the bronchial response to allergens, and if ingested they can aggravate the bronchial response to histamine.

Nervous system

Treatment with centrally acting agents is characterized by a relatively high incidence (up to 60% in some studies) of nervous system depressant effects (dizziness, drowsiness, tiredness, dry mouth, headache, depression), particularly during the initial period of treatment or after dosage increments. Sedation, lethargy, and tiredness are common with clonidine, particularly at the start of treatment (14).

In a placebo-controlled study of a single oral dose of clonidine 0.25–3 mg in 18 healthy men aged 18–21 years, clonidine caused more sleepiness than placebo; it significantly reduced stage 1 and rapid-eye-movement (REM) sleep and increased stage 2 sleep (15).

Clonidine 300 micrograms was given orally to 30 patients 60 minutes before intrathecal anesthesia with lidocaine 40 mg or 80 mg, and 60 other patients received either plain lidocaine 100 mg or lidocaine 40 mg or 80 mg with clonidine 100 micrograms intrathecally (16). Clonidine, both intrathecally and orally, prolonged the duration of spinal block and allowed a reduction in the dose of lidocaine needed for a given duration of block, but prolongation of motor block, exceeding the duration of sensory block, was a drawback with both routes and doses of clonidine. The smallest hemodynamic changes were seen with lidocaine 40 mg + clonidine 100 micrograms intrathecally, which provided adequate anesthesia for operations lasting up to 140 minutes. All doses of clonidine caused sedation.

The electroencephalographic effects of clonidine have been reported (17). There was electrophysiological sedation after parenteral clonidine, an effect that the authors concluded was likely to relate to the drug's analgesic properties rather than direct modulation of the nociceptive system.

Anecdotal reports have shown that clonidine can induce epileptic activity. In 22 patients with chronic, medically intractable, localization-related epilepsy, magnetoencephalography was used to investigate whether oral clonidine can aid localization of spike or sharp-wave activity as well as to gain knowledge about whether clonidine is epileptogenic (18). Oral clonidine-induced magnetoencephalographic activity was superior to sleep deprivation, and therefore may be an aid to localization as well as showing epileptogenic activity.

Endocrine

Clonidine stimulates the release of growth hormone and has been used as a provocation test of growth hormone reserve (19).

Clonidine reduces plasma renin activity and urinary aldosterone and catecholamine concentrations (20).

Mouth

Dry mouth is common with clonidine, and investigation of the mechanism has come from an experimental study in which the direct effect of clonidine on salivary amylase excretion from rat parotid dispersed cells was measured in vitro (21). Clonidine stimulated calcium influx into the cells and inhibited postsynaptic alpha$_1$-adrenoceptor responses.

Gastrointestinal

Constipation is frequent with clonidine, and one case of pseudo-obstruction of the bowel has been reported (22).

Urinary tract

Epidural clonidine is less likely to produce urinary retention than epidural opioids (23,24).

Skin

Rashes occur in about 4% of patients taking clonidine (25).

Transdermal clonidine formulated in adhesive skin patches has been used for long-term treatment with once-weekly application. When transdermal clonidine is used to lower blood pressure systemic adverse effects seem to be fewer than with oral clonidine (26). However, localized skin reactions are common, and the incidence increases with the dose and duration of use. Common signs include erythema, scaling, vesiculation, excoriation, and induration. Allergic contact dermatitis is less frequent but still common. Hyperpigmentation and depigmentation also occur. Pretreatment with 0.5% hydrocortisone is associated with less skin irritation and higher blood concentrations.

Immunologic

During long-term treatment of glaucoma with apraclonidine allergic reactions can occur. In a retrospective analysis of 64 patients who used apraclonidine 1% for more than 2 weeks, 31 (48%) developed an allergic reaction that led to withdrawal of treatment, with a mean latency of 4.7 months (27). Those who had allergic reactions tended to be older and female.

- A 46-year-old woman, who took clonidine 25 mg bd for menopausal flushing, developed depigmentation and swelling of her forearms (28). A skin biopsy showed a pattern consistent with immune complex disease, with IgG, IgM, Clq, C2c, and C4 complement between muscle fibers and at the dermo-epidermal junctions. All of these abnormalities disappeared after withdrawal.

Long-Term Effects

Drug tolerance

During long-term treatment of glaucoma, apraclonidine 0.5% is effective in reducing intraocular pressure, both in the short term and for up to 24 months. However, efficacy is lost in some patients. Of 174 patients followed for up to 24 months apraclonidine was ineffective in 38 patients (21%), at some point during the study (29).

Drug dependence

A case of clonidine dependence has been reported (30). The authors pointed out that abuse of clonidine in combination with other drugs, particularly opioids, is relatively common. The case they reported was unusual, as it involved clonidine as the sole drug of abuse.

- A 37-year-old man requested admission for voluntary detoxification from clonidine. He had started using alcohol and marijuana in his teens, became dependent on intravenous heroin at the age of 20, and switched to methadone at 27. He had several unsuccessful attempts at methadone withdrawal using clonidine, benzodiazepines, and other therapies. In a final attempt at the age of 35, he managed to self-detoxify using clonidine 0.8 mg 3–5 times per day. At the time of admission he was using 1.8–3.2 mg/day, compared with the recommended antihypertensive dose of 0.2–0.6 mg/day. Clonidine relieved his anxiety, controlled his anger, relieved headaches, and allowed him to sleep. When he ran out of clonidine he had severe headaches, chest pain, blurred vision, and irritability, with slurred speech and some paranoia. His blood pressure was 218/110 and his blood chemistry was normal. He was detoxified over a period of 3 days and his blood pressure was controlled. He was discharged taking benazepril 40 mg/day, gabapentin 900 mg tds, hydrochlorothiazide 25 mg/day, and a 0.3 mg clonidine patch to be removed the next day. On the same evening he took an unknown amount of clonidine and had a seizure. He was unconscious, with a blood pressure of 78/46 mmHg. He was resuscitated and had raised cardiac enzymes with no electrocardiographic evidence of myocardial infarction. He was discharged 4 days later taking diltiazem, benazepril, gabapentin, low-dose aspirin, paroxetine, and a clonidine patch 0.3 mg per day.

The authors commented that hyperactive central $alpha_2$ adrenergic neurons mediate symptoms such as anxiety, irritability, and insomnia, and that alleviation of these symptoms can result in clonidine dependence. They also quoted rodent evidence that $\alpha lpha_2$ adrenoceptors mediate GABA release, which could explain a benzodiazepine-like effect of clonidine.

Drug withdrawal

A withdrawal syndrome with marked rebound hypertension (attributed to increased sympathetic nervous activity as a result of drug-modified receptor responses) is an occasional but well-recognized complication of the abrupt withdrawal of $alpha_2$-adrenoceptor agonists, particularly with clonidine. In its most florid form, the clonidine withdrawal syndrome is characterized by a pronounced increase in blood pressure, tachycardia, tremulousness, and sweating (31). There is an associated marked increase in catecholamine output and the features are reminiscent of pheochromocytoma or accelerated hypertension. Milder cases can pass unnoticed, unless the patient is being carefully monitored. The syndrome begins within 48 hours of withdrawal, and although the exact incidence is unknown it is more likely if the patient has been taking a high dosage. There has even been a case report of rebound hypertension in association with transdermal clonidine (32). Treatment is by combined alpha- and beta-blockade, and it is important to avoid monotherapy with a beta-blocker, since this may exacerbate the syndrome by promoting unopposed alpha-adrenergic effects.

There have been 8 case reports of other serious complications arising from the acute withdrawal of clonidine. Neuropsychiatric disturbance with self-injury occurred in a child (33) and myocardial infarction developed in a patient with no history of ischemic heart disease (34). Acute myocardial infarction has also been reported as a complication of clonidine withdrawal in a patient with hypertension (35).

Drug Administration

Drug formulations

Transdermal clonidine (clonidine TTS) has been used with some success for the treatment of mild hypertension. Systemic adverse effects are similar to those seen after oral administration, but are less frequent and milder. They include dry mouth, drowsiness, headache, sexual disturbance, cold extremities, obstipation, and fatigue (36–38). These adverse effects rarely necessitate withdrawal of clonidine TTS.

A high percentage of patients who take clonidine (up to 38%) (39) develop contact allergic reactions, usually due to the active ingredient, at the patch application site (40). This has been reported with a frequency of 15% in 357 African–American hypertensive patients. It can lead to drug discontinuation in 4.2% of patients.

Recurrent maculopapular rash due to both topical and systemic administration of clonidine has been reported (41).

- A 47-year-old woman was given transdermal, oral, and intravenous clonidine at different times, separated by several months. The patches were withdrawn after 2 months because of pruritic erythematous vesiculation at the site of application. Oral clonidine (0.3 mg/day) had to be withdrawn when a generalized maculopapular rash appeared on the third day, and was promptly exacerbated by each dose. Intravenous clonidine 0.150 mg was followed within 30 minutes by a severe, generalized, maculopapular reaction requiring systemic steroids and antihistamines.

The authors interpreted these reactions as being allergic. A patch test was very positive to the commercial formulation. Clonidine is a weak sensitizer, but occlusion and prolonged skin contact during transdermal application can cause delayed hypersensitivity.

Drug administration route

The most common ocular adverse effects of topical apraclonidine are conjunctival hyperemia, itching, foreign body sensation, and lacrimation. The most frequent systemic adverse effects are dry mouth and unusual taste perception (42–44). With *para*-aminoclonidine, headaches, attributable to ocular vasoconstriction, as well as anxiety, vomiting, dry mouth, tremor, and pallor, have been reported (45).

Drug overdose

The effects of exposure to clonidine hydrochloride in children, as reported to US poison centers from 1993 to 1999, have been retrospectively reviewed (46). There were 10 060 reported exposures, of which 57% were in children under 6 years, 34% in children aged 6–12 years, and 9% in adolescents aged 13–18 years. In 1999 there were 2.5 times as many exposures as in 1993. Clonidine was the child's medication in 10% of those under 6 years, 35% of those aged 6–12 years, and 26% of the adolescents. Unintentional overdose was most common in those under 6 years, while therapeutic errors and suicide attempts predominated in those aged 6–12-years and the adolescents. In 6042 symptomatic children (60%), the most common symptoms were lethargy (80%), bradycardia (17%), hypotension (15%), and respiratory depression (5%). Most of the exposures resulted in no effect (40%) or minor effects (39%). There were moderate effects in 1907 children (19%), major effects in 230 (2%), and one death in a 23-month-old child.

In a prospective evaluation of clonidine overdose in children, all of the clinical effects occurred rapidly after ingestion, but all of the patients recovered fully with observation and supportive therapy (47). The authors suggested that direct medical evaluation should be undertaken for all children aged 4 years or younger who have taken an unintentional overdose of at least 0.1 mg, in children 5–8 years of age who have taken at least 0.2 mg, and in children older than 8 years of age who have taken at least 0.4 mg; and that observation for 4 hours may be sufficient to detect patients who will develop severe adverse effects.

Clonidine overdose with marked hypothermia has been reported (48).

- A 39-year-old woman taking multiple psychiatric medications became apneic and a core body temperature of 28.9°C and was thought to have died. Twenty doses of clonidine 0.2 mg and 20 of amitriptyline 25 mg were unaccounted for. After resuscitation she made a full recovery within 3 days.

A literature search revealed three cases of clonidine overdose with hypothermia and an extended recovery period.

- A hypertensive crisis and myocardial infarction occurred in a 62-year-old woman after a combined injection of hydromorphone 48 mg and clonidine 12 mg subcutaneously in an attempt to refill an implanted epidural infusion pump (49). She was immediately treated with naloxone, but she subsequently had accelerated hypertension, a brief tonic-clonic seizure, and an anteroseptal myocardial infarction. Cardiac catheterization showed no coronary narrowing or blockage, but an anterior infarct was confirmed.

It is believed that the reaction was secondary to the vasoconstricting effects of high-dose clonidine through stimulation of peripheral alpha-adrenoceptors.

Drug–Drug Interactions

Mianserin

Mianserin has alpha-adrenoceptor activity and so might interact with clonidine (50). In healthy volunteers, pretreatment with mianserin 60 mg/day for 3 days did not modify the hypotensive effects of a single 300 mg dose of clonidine. In 11 patients with essential hypertension, the addition of mianserin 60 mg/day (in divided doses) for 2 weeks did not reduce the hypotensive effect of clonidine. The results of this study appear to have justified the authors' conclusion that adding mianserin to treatment with clonidine will not result in loss of blood pressure control.

Mirtazapine

Hypertension has been reported with mirtazapine plus clonidine (51).

- A 20-year-old man had had Goodpasture's syndrome for 2.5 years, end-stage renal disease on chronic hemodialysis for 15 months, and hypertension controlled with metoprolol, losartan, and clonidine. He developed dyspnea and hypertension (blood pressure 178/115 mmHg) 2 weeks after his psychiatrist first gave him mirtazapine 15 mg at bedtime to treat depression. His blood pressure did not fall significantly, despite the addition of losartan and minoxidil and the use of intravenous glyceryl trinitrate and labetalol. Only after emergency dialysis and intravenous nitroprusside did his blood pressure fall to 150–180/80–100 mmHg. When mirtazapine was withdrawn, his blood pressure was controlled with minoxidil 5 mg, clonidine 0.1 mg, and metoprolol 10 mg, all bd.

The authors recognized that mirtazapine alone could have caused the hypertensive event. In postmarketing surveillance of mirtazapine, hypertension occurred in at least 1% of patients. However, it is likely that the patient lost antihypertensive control because mirtazapine antagonized the antihypertensive effect of clonidine. Mirtazapine, a tetracyclic antidepressant, stimulates the noradrenergic system through antagonism at central $alpha_2$ inhibitory receptors, which is precisely opposite to the effect of clonidine.

Neostigmine

Clonidine increases the analgesic effect of intrathecal neostigmine without enhancing its adverse effects (52).

References

1. Webster J, Koch HF. Aspects of tolerability of centrally acting antihypertensive drugs. J Cardiovasc Pharmacol 1996;27(Suppl 3):S49–54.
2. Ahmed I, Takeshita J. Clonidine: a critical review of its role in the treatment of psychiatric disorders. Drug Ther 1996;6:53–70.
3. Dollery CT. Advantages and disadvantages of alpha-adrenoceptor agonists for systemic hypertension. Am J Cardiol 1988;61(7):D1–5.
4. Nishina K, Mikawa K, Shiga M, Obara H. Clonidine in paediatric anaesthesia. Paediatr Anaesth 1999;9(3):187–202.
5. Schadlu R, Maus TL, Nau CB, Brubaker RF. Comparison of the efficacy of apraclonidine and brimonidine as aqueous suppressants in humans. Arch Ophthalmol 1998; 116(11):1441–4.
6. Mitchell A, Buhrmann S, Opazo Saez A, Rushentsova U, Schafers RF, Philipp T, Nurnberger J. Clonidine lowers blood pressure by reducing vascular resistance and cardiac output in young, healthy males. Cardiovasc Drugs Ther 2005;19(1):49–55.
7. Dawson PM, Vander Zanden JA, Werkman SL, Washington RL, Tyma TA. Cardiac dysrhythmia with the use of clonidine in explosive disorder. DICP 1989;23(6):465–6.
8. Marini M, Cavani E, Abbatangelo R, Mascelloni R. Periodismo di Luciani Wenckeback e clonidina. Presentazione di un caso. [Luciani-Wenckeback period and clonidine. Presentation of a case.] Clin Ter 1988;126(4):273–6.
9. Roden DM, Nadeau JH, Primm RK. Electrophysiologic and hemodynamic effects of chronic oral therapy with the alpha-agonists clonidine and tiamenidine in hypertensive volunteers. Clin Pharmacol Ther 1988;43(6):648–54.
10. Delva NJ, Chang A, Hawken ER, Lawson JS, Owen JA. Effects of clonidine in schizophrenic patients with primary polydipsia: three single case studies. Prog Neuropsychopharmacol Biol Psychiatry 2002;26(2):387–92.
11. Backo AL, Clause SL, Triller DM, Gibbs KA. Clonidine-induced hypertension in a patient with a spinal lesion. Ann Pharmacother 2002;36(9):1396–8.
12. Kaabachi O, Ben Rajeb A, Mebazaa M, Safi H, Jelel C, Ben Ghachem M, Ben Ammar M. La rachianesthésie chez l'enfant: étude comparative de la bupivacaïne hyperbare avec et sans clonidine. [Spinal anesthesia in children: comparative study of hyperbaric bupivacaine with or without clonidine.] Ann Fr Anesth Reanim 2002;21(8):617–21.
13. Rosen B, Ovsyshcher IA, Zimlichman R. Complete atrioventricular block induced by methyldopa. Pacing Clin Electrophysiol 1988;11(11 Pt 1):1555–8.
14. Schmitt H, Schwartz J, Blanchot P, Fritel D, Froment A, Traeger J. Expérience française de la clonidine. Nouv Presse Méd 1972;1:877.
15. Carskadon MA, Cavallo A, Rosekind MR. Sleepiness and nap sleep following a morning dose of clonidine. Sleep 1989;(4):338–44.
16. Dobrydnjov I, Samarutel J. Enhancement of intrathecal lidocaine by addition of local and systemic clonidine. Acta Anaesthesiol Scand 1999;43(5):556–62.
17. Bischoff P, Schmidt GN, Scharein E, Bromm B, Schulte am EJ. Clonidine induced sedation and analgesia—an EEG study. J Neurol 2004;251(2):219–21.
18. Kettenmann B, Feichtinger M, Tilz C, Kaltenhauser M, Hummel C, Stefan H. Comparison of clonidine to sleep deprivation in the potential to induce spike or sharp-wave activity. Clin Neurophysiol 2005;116(4):905–12.
19. Morris AH, Harrington MH, Churchill DL, Olshan JS. Growth hormone stimulation testing with oral clonidine: 90 minutes is the preferred duration for the assessment of growth hormone reserve. J Pediatr Endocrinol Metab 2001;14(9):1657–60.
20. Houston MC. Clonidine hydrochloride. South Med J 1982;75(6):713–9.
21. Yamada K, Tanaka K, Aoki M, Banno A, Nakagawa S, Yamaguchi M, Togari A, Matsumoto S. Effect of clonidine on amylase secretion from rat parotid gland. Asia Pacific J Pharmacol 1996;11:19–24.
22. Kellaway GS. Adverse drug reactions during treatment of hypertension. Drugs 1976;11(Suppl 1):91–9.
23. Gentili M, Bonnet F. Spinal clonidine produces less urinary retention than spinal morphine. Br J Anaesth 1996;76(6):872–3.
24. Batra YK, Gill PK, Vaidyanathan S, Aggarwal A. Effect of epidural buprenorphine and clonidine on vesical functions in women. Int J Clin Pharmacol Ther 1996;34(7):309–11.

25. Onesti G, Bock KD, Heimsoth V, Kim KE, Merguet P. Clonidine: a new antihypertensive agent. Am J Cardiol 1971;28(1):74–83.
26. Prisant LM. Transdermal clonidine skin reactions. J Clin Hypertens (Greenwich) 2002;4(2):136–8.
27. Butler P, Mannschreck M, Lin S, Hwang I, Alvarado J. Clinical experience with the long-term use of 1% apraclonidine. Incidence of allergic reactions. Arch Ophthalmol 1995;113(3):293–6.
28. Petersen HH, Hansen M, Albrectsen JM. Clonidine-induced immune complex disease. Acta Dermatol Venereol 1989;69(6):519–20.
29. Gross RL, Pinyero A, Orengo-Nania S. Clinical experience with apraclonidine 0.5% J Glaucoma 1997;6(5):298–302.
30. Lanford W, Myrick M, O'Bryan E. A severe case of clonidine dependence and withdrawal. J Psychiatr Pract 2003;9:167–9.
31. Hansson L, Hunyor SN, Julius S, Hoobler SW. Blood pressure crisis following withdrawal of clonidine (Catapres, Catapresan), with special reference to arterial and urinary catecholamine levels, and suggestions for acute management. Am Heart J 1973;85(5):605–10.
32. Schmidt GR, Schuna AA. Rebound hypertension after discontinuation of transdermal clonidine. Clin Pharm 1988;7(10):772–4.
33. Dillon JE. Self-injurious behavior associated with clonidine withdrawal in a child with Tourette's disorder. J Child Neurol 1990;5(4):308–10.
34. Berge KH, Lanier WL. Myocardial infarction accompanying acute clonidine withdrawal in a patient without a history of ischemic coronary artery disease. Anesth Analg 1991;72(2):259–61.
35. Simic J, Kishineff S, Goldberg R, Gifford W. Acute myocardial infarction as a complication of clonidine withdrawal. J Emerg Med 2003;25(4):399–402.
36. Weber MA, Drayer JI, Brewer DD, Lipson JL. Transdermal continuous antihypertensive therapy. Lancet 1984;1(8367):9–11.
37. Groth H, Vetter H, Knusel J, Boerlin HJ, Walger P, Baumgart P, Wehling M, Siegenthaler W, Vetter W. Clonidin-TTS bei essentieller Hypertonie: Wirkung und Vertr äglichkeit. [Clonidine transdermal therapeutic system in essential hypertension: effect and tolerance.] Schweiz Med Wochenschr 1983;113(49):1841–5.
38. Olivari MT, Cohn JN. Cutaneous administration of nitroglycerin: a review. Pharmacotherapy 1983;3(3):149–57.
39. Carmichael AJ. Skin sensitivity and transdermal drug delivery. A review of the problem. Drug Saf 1994;10(2):151–9.
40. Dias VC, Tendler B, Oparil S, Reilly PA, Snarr P, White WB. Clinical experience with transdermal clonidine in African–American and Hispanic–American patients with hypertension: evaluation from a 12-week prospective, open-label clinical trial in community-based clinics. Am J Ther 1999;6(1):19–24.
41. Crivellaro MA, Bonadonna P, Dama A, Senna G, Passalacqua G. Skin reactions to clonidine: not just a local problem. Case report. Allergol Immunopathol (Madr) 1999;27(6):318–9.
42. Araujo SV, Bond JB, Wilson RP, Moster MR, Schmidt CM Jr, Spaeth GL. Long term effect of apraclonidine. Br J Ophthalmol 1995;79(12):1098–101.
43. Robin AL, Ritch R, Shin D, Smythe B, Mundorf T, Lehmann RPThe Apraclonidine Maximum Tolerated Medical Therapy Study Group. Topical apraclonidine hydrochloride in eyes with poorly controlled glaucoma. Trans Am Ophthalmol Soc 1995;93:421–41.
44. Robin AL, Ritch R, Shin DH, Smythe B, Mundorf T, Lehmann RPApraclonidine Maximum-Tolerated Medical Therapy Study Group. Short-term efficacy of apraclonidine hydrochloride added to maximum-tolerated medical therapy for glaucoma. Am J Ophthalmol 1995;120(4):423–32.
45. Abrams DA, Robin AL, Pollack IP, deFaller JM, DeSantis L. The safety and efficacy of topical 1% ALO 2145 (p-aminoclonidine hydrochloride) in normal volunteers. Arch Ophthalmol 1987;105(9):1205–7.
46. Klein-Schwartz W. Trends and toxic effects from pediatric clonidine exposures. Arch Pediatr Adolesc Med 2002;156(4):392–6.
47. Spiller HA, Klein-Schwartz W, Colvin JM, Villalobos D, Johnson PB, Anderson DL. Toxic clonidine ingestion in children. J Pediatr 2005;146(2):263–6.
48. Quail MT, Shannon M, Severe hypothermia caused by clonidine. Am J Emerg Med 2003;21:86.
49. Frye CB, Vance MA. Hypertensive crisis and myocardial infarction following massive clonidine overdose. Ann Pharmacother 2000;34(5):611–5.
50. Elliott HL, Whiting B, Reid JL. Assessment of the interaction between mianserin and centrally-acting antihypertensive drugs. Br J Clin Pharmacol 1983;15(Suppl 2):S323–8.
51. Abo-Zena RA, Bobek MB, Dweik RA. Hypertensive urgency induced by an interaction of mirtazapine and clonidine. Pharmacotherapy 2000;20(4):476–8.
52. Hood DD, Mallak KA, Eisenach JC, Tong C. Interaction between intrathecal neostigmine and epidural clonidine in human volunteers. Anesthesiology 1996;85(2):315–25.

Diazoxide

General Information

Diazoxide is a direct vasodilator that acts on vascular smooth muscle to produce systemic vasodilatation. As a result there is baroreceptor-mediated activation of the sympathetic nervous system and the renin–angiotensin system.

Vasodilators (as monotherapy) are associated acutely with flushing, headache, dizziness, reflex tachycardia, and palpitation. Chronic treatment can be complicated by fluid retention.

Organs and Systems

Cardiovascular

Rapid administration of diazoxide intravenously can cause severe hypotension. In some cases this has resulted in abnormal neurological signs associated with ischemic damage (1).

Metabolism

Diazoxide can cause hyperglycemia and diabetes mellitus and has been used to treat hyperinsulinism in infancy, although its use may be hazardous (2,3).

Drug Administration

Drug overdose

Severe diazoxide toxicity has been reported when it was used to treat persistent hypoglycemia in a neonate (4).

References

1. Ledingham JG, Rajagopalan B. Cerebral complications in the treatment of accelerated hypertension. Q J Med 1979;48(189):25–41.
2. Low LC, Yu EC, Chow OK, Yeung CY, Young RT. Hyperinsulinism in infancy. Aust Paediatr J 1989;25(3):174–7.
3. Abu-Osba YK, Manasra KB, Mathew PM. Complications of diazoxide treatment in persistent neonatal hyperinsulinism. Arch Dis Child 1989;64(10):1496–500.
4. Silvani P, Camporesi A, Mandelli A, Wolfler A, Salvo I. A case of severe diazoxide toxicity. Paediatr Anaesth 2004; 14(7): 607–9.

Dilevalol

See also Beta-adrenoceptor antagonists

General Information

Dilevalol is one of the four enantiomers of labetalol. It was withdrawn from the market by its manufacturers in early 1991 because of an unacceptably high incidence of liver damage (1).

Reference

1. Harvengt C. Labetalol hepatoxicity. Ann Intern Med 1990;114:341.

Diltiazem

See also Calcium channel blockers

General Information

Diltiazem is a benzthiazepine calcium channel blocker.

Organs and Systems

Cardiovascular

Diltiazem can be associated with heart block (1,2). In a randomized trial in patients with non-Q-wave myocardial infarction, 38 of 287 patients who took diltiazem developed some degree of atrioventricular block at some time. Of these episodes, 32 were first-degree, eight were second-degree, and only two were third-degree. In the placebo-treated group there were 10 events in 289 patients; eight were first-degree, two second-degree, and none third-degree.

Atrioventricular block has been reported in three patients taking therapeutic doses of diltiazem; one died (3).

- A 47-year-old man taking furosemide for hypertension was given diltiazem 300 mg/day to achieve better blood pressure control; 1 month later he developed atrioventricular block, resolved by atropine.
- A 62-year-old hypertensive man with renal artery stenosis, an adrenal adenoma, peripheral artery disease, and an abdominal aortic aneurysm developed a hypertensive crisis with chest pain. He was treated with nitrates, heparin, aspirin, and nicardipine, which were afterwards replaced by diltiazem 200 mg/day, because of persistent chest pain. He developed atrioventricular block 2 hours after the second dose of diltiazem, and was successfully treated with a pacemaker.
- A 59-year-old woman with a previous myocardial infarction, hypertension, diabetes, and uterine cancer developed angina. She was treated with nitrates, aspirin, and heparin; diltiazem 200 mg/day was then added because of persistent chest pain. She developed atrioventricular block 72 hours later, and despite resuscitative efforts died in electromechanical dissociation.

Following the publication of a report of heart block (4), further cases (junctional bradycardia, sinoatrial block) have been reported (5).

Nervous system

Paresthesia of the hands and feet has been reported with diltiazem (SEDA-17, 237). Akathisia has been attributed to diltiazem (6). Parkinsonism associated with diltiazem has been reported (7).

- A 53-year-old man with hypertension took diltiazem 60–120 mg/day for 5 years and then developed parkinsonism. His neurological symptoms were treated without success and only after the substitution of diltiazem with an ACE inhibitor did his parkinsonian symptoms begin to regress, with eventual complete recovery.

Mouth and teeth

Gum hyperplasia has been attributed to diltiazem (8).

- A 49-year-old Afro-Caribbean man with resistant hypertension developed pronounced gum hyperplasia with bleeding, which gradually resolved on withdrawal of amlodipine 20 mg/day. Since the blood pressure was poorly controlled, he was given a non-dihydropyridine calcium channel blocker diltiazem 240 mg/day. The gum hyperplasia recurred 3 months later and slowly resolved after withdrawal of diltiazem.

This case suggests that the well-known adverse effect of gum hyperplasia is a class effect of calcium channel blockers, not just limited to the dihydropyridines.

Gingival enlargement has been studied in 46 patients taking diltiazem (n = 32) or verapamil (n = 14) compared with 49 cardiovascular controls who had never taken them

(9). There was more gingival enlargement in the patients taking diltiazem compared with controls: 31% as assessed by the vertical gingival overgrowth index and in 50% as assessed by the horizontal Miranda & Brunet index. The corresponding figures for verapamil were 21% and 36% (njot different from controls). The risk of gingival enlargement (odds ratio) associated with diltiazem therapy, adjusted for gingival index values, was 3.5 (1.0-12) the vertical gingival overgrowth index and 6.2 (1.9, 20) for the horizontal Miranda & Brunet index. Verapamil had no significant effect.

Gastrointestinal

Intestinal pseudo-obstruction has been attributed to diltiazem (10).

- A 74-year-old man with acute myelogenous leukemia and neutropenia was given five doses of intravenous diltiazem 5 mg every 5–10 minutes plus amiodarone for atrial fibrillation, followed by diltiazem 30 mg orally qds and a continuous infusion of amiodarone. Amiodarone was withdrawn after 1 day because of an adverse effect and the dose of diltiazem was increased to 120 mg qds orally by day 14. On day 15, he developed increasing abdominal distention with hyperactive bowel sounds. An abdominal X-ray showed multiple dilated loops in the small and large bowel and diltiazem was withdrawn, after which he recovered.

Medications rarely cause intestinal pseudo-obstruction, but calcium channel blockers cause smooth muscle relaxation, which was probably the causative mechanism in this case.

Urinary tract

Diltiazem was associated with the development of acute renal insufficiency in a patient being treated for severe retrosternal chest pain who had neither primary kidney disease nor urinary tract obstruction (11,12).

Acute interstitial nephritis has been attributed to diltiazem (12).

- A 53-year-old man was given diltiazem for precordial pain and about 2 hours later developed an erythematous maculopapular rash mainly on the trunk and lower limbs. Four days later he developed abdominal pain radiating to both renal angles, accompanied by dysuria and tenesmus and followed 6 days later by acute renal insufficiency associated with raised liver function test results.

In this case, the self-limiting resolution in 4–5 days without relapse, the presence of the skin rash, and the liver sequelae suggested a common immunoallergic mechanism. The clinical symptoms, the time relation between drug administration and the occurrence of the syndrome, the inability to explain the syndrome otherwise, and its disappearance on withdrawal of diltiazem support an association with the drug.

A retrospective analysis of postoperative renal function in patients undergoing cardiac operations has been conducted to evaluate whether the use of prophylactic intravenous diltiazem, in order to reduce the incidence of ischemia and dysrhythmias, was associated with increased renal dysfunction (13). The incidence of acute renal insufficiency requiring dialysis was 4.4% with diltiazem versus 0.7% in the controls. Logistic regression analysis suggested that the risk of acute renal insufficiency was strongly associated with intravenous diltiazem, age, baseline serum creatinine, the presence of left main coronary disease, and the presence of cerebrovascular disease.

Skin

Skin reactions ranging from exanthems to severe adverse events have been reported in association with diltiazem (SED-13, 513; SEDA-18, 215; SEDA-22, 216). Three cases of skin reactions (hypersensitivity syndrome reaction, pruritic exanthematous eruption, and acute generalized exanthematous pustulosis) possibly induced by diltiazem have been described and the literature on skin reactions associated with calcium antagonists has been reviewed. The number of diltiazem-induced cutaneous events was significantly greater than those induced by either nifedipine or verapamil. However, there was no difference in the proportion of serious cutaneous adverse events due to any of these three drugs (14).

Cutaneous vasculitis (15) and angioedema (16,17) have been reported with diltiazem. In 1988, the Federal German Health Authorities imposed a warning of dermal hypersensitivity reactions (including erythema multiforme) with diltiazem.

- Acute generalized exanthematous pustulosis in an 82-year-old woman was confirmed as being due to diltiazem by a positive patch test (18).
- Exfoliative dermatitis with fever occurred in a 69-year-old man with ischemic heart disease treated with mexiletine and diltiazem for three weeks; the rash resolved after withdrawal of both drugs and systemic corticosteroid therapy. Patch tests with mexiletine and diltiazem were positive. In addition to this case, 39 cases of drug eruption due to diltiazem have been reported in Japan (19).
- Lichenoid purpura of 6 months' duration occurred in a 65-year-old man with hypertension who had taken diltiazem for 1 month. Topical therapy with a very potent glucocorticoid was not effective and the eruption began to regress only after diltiazem withdrawal, after which it disappeared 3 weeks later (20).

Four cases of photodistributed hyperpigmentation associated with long-term administration of a modified-release formulation of diltiazem hydrochloride have been reported (21). All the patients were African-American women, mean age 62 (range 49–72) years. The duration of diltiazem administration before the development of hyperpigmentation was 6–11 months. The hyperpigmentation was slate-gray and reticulated. Phototesting during diltiazem therapy showed a reduced minimal erythema dose to UVA in one patient. Histological examination showed lichenoid dermatitis with prominent pigmentary incontinence. Electron microscopy showed multiple melanosome complexes. Withdrawal of diltiazem resulted in gradual resolution of the hyperpigmentation.

Acute generalized exanthematous pustular dermatitis has been reported after diltiazem (22).

Skin reactions to calcium channel blockers have been reviewed in the light of a report of three reactions to diltiazem (23).

- A 54-year-old man developed a generalized erythema-multiforme-like reaction followed by erythrodermia and exfoliative dermatitis after taking diltiazem for 6–7 days.
- An 80-year-old woman had a pruritic exanthematous eruption on her trunk 10 days after taking diltiazem, which evolved to generalized erythrodermia and superficial desquamation; it gradually improved in 10–12 days after withdrawal.
- A 79-year-old man developed erythema and pruritus initially on the back, and then affecting the thorax, limbs, and face after taking diltiazem for 3 days. Diltiazem was withdrawn and the skin improved, but verapamil was started 3 days later and the skin worsened again.

The authors carried out skin tests with different calcium channel blockers. Diltiazem was positive in all three patients; nifedipine was positive in patient 2 and verapamil in patient 3.

Four patients (one man and three women, mean age 60 years) who developed a maculopapular rash at 8–12 days after starting to take diltiazem underwent drug skin tests in order to determine the value of patch tests and cross-reactions among calcium channel blockers (24). In all cases patch tests were positive to diltiazem and there were no cross-reactions with nimodipine, nifedipine, nitrendipine, or nicardipine. There was a cross-reaction with diltiazem and verapamil in only one patient.

Drug Administration

Drug overdose

Several cases of accidental or deliberate self-poisoning with calcium antagonists have been described (SED-12, 452) (SEDA-16, 198) (SEDA-20, 186) and different approaches to the management of these patients have been proposed.

- A 52-year-old woman with essential hypertension erroneously took a 7-day supply of modified-release diltiazem, enalapril, and trichlormethiazide, and 30 minutes later complained of nausea (25). She was treated with gastric lavage, calcium gluconate, activated charcoal, and polyethylene glycol. Her blood pressure and heart rate fell progressively to 120/62 mmHg and 40/minute respectively, 14 hours after ingestion. Afterwards, her hemodynamic status gradually normalized without further treatment.
- A 54-year-old man with severe triple vessel coronary artery disease took six modified-release diltiazem tablets 180 mg following an episode of severe angina, and 10 hours later developed bradycardia, hypotension, and severe pulmonary edema, but was free of chest pain (26). After intensive hemodynamic monitoring and noradrenaline treatment, his renal, respiratory, and cardiac problems recovered to baseline over the next 48 hours. Diltiazem overdose was confirmed by a diltiazem serum concentration of 1230 ng/ml (usual target range 40–160 ng/ml).
- A 15-year-old woman intentionally took 10 modified-release tablets of diltiazem 200 mg. She developed hypertension, oliguria, pulmonary edema, and respiratory distress syndrome, and required mechanical ventilation for 3 days, besides intravenous calcium, dopamine, and noradrenaline. After 5 days in an intensive care unit, she was transferred to a psychiatric hospital in good physical condition (27).
- A 50-year-old man took 28 modified-release tablets of diltiazem 240 mg and 28 tablets of hydrochlorothiazide and 12–14 hours later became lethargic but oriented and complaining of nausea and dizziness; he had bradypnea, hypotension, and second-degree heart block with bradycardia (28). He was given activated charcoal, oxygen, atropine, glucagon, and calcium gluconate by prolonged infusion. His heart rate, blood pressure, and electrocardiogram recovered to baseline over the next 24 hours, with no further episodes of dysrhythmias or hypotension.
- A 38-year-old white man with a history of coronary artery disease, myocardial infarction, coronary artery by-pass, alcoholism, and depression took a combined massive overdose of diltiazem and atenolol (29). He underwent cardiopulmonary resuscitation because of cardiac arrest; bradycardia, hypotension, and oliguria followed and were resistant to intravenous pacing and multiple pharmacological interventions, including intravenous fluids, calcium, dopamine, dobutamine, adrenaline, prenalterol, and glucagon. Adequate mean arterial pressure and urine output were restored only after the addition of phenylephrine and transvenous pacing. He survived despite myocardial infarction and pneumonia.

Drug–Drug Interactions

Alfentanil

Diltiazem reduces the elimination of alfentanil and prolongs the time to tracheal extubation (30).

Amiodarone

Sinus arrest with hypotension has been reported in a patient with a congestive cardiomyopathy when diltiazem was added to amiodarone (31).

Cardiac glycosides

Studies of the effects of diltiazem on the pharmacokinetics of digoxin have yielded variable results. In one study, diltiazem reduced the steady-state total body clearance of beta-acetyldigoxin in 12 healthy men, perhaps because of reduced renal and non-renal clearances (SEDA-10, 145). In some studies diltiazem 120–240 mg/

day increased steady-state plasma digoxin concentrations by about 20–40% (SEDA-14, 146; 32), although not in other studies (SEDA-14, 146), and reduced the total body clearance of digoxin, with changes in both renal and non-renal clearances (SEDA-11, 154), although others did not find this (33). In at least one case digoxin toxicity was attributed to this interaction (SEDA-17, 216). In eight patients with chronic heart failure taking digoxin 0.25 mg/day, diltiazem 180 mg/day increased the AUC and mean steady-state serum concentrations of digoxin by 50% and reduced its total clearance (34).

Ciclosporin

Many studies have shown an interaction of ciclosporin with diltiazem: concomitant administration allows reduction of the daily dose of ciclosporin. However, according to a study in eight renal transplant recipients, the low systemic availability and high degree of variation in diltiazem metabolism within and between patients can give unpredictable results (35).

Diltiazem abolishes the acute renal hypoperfusion and vasoconstriction induced by ciclosporin in renal transplant patients. Plasma endothelin-1 may be a mediator of ciclosporin-induced renal hypoperfusion, but is not affected by diltiazem (36). This interaction has been confirmed with a new microemulsion formulation of ciclosporin in nine patients with renal transplants who took diltiazem 90–120 mg bd for 4 weeks (37). Diltiazem caused a 51% increase in the AUC of ciclosporin and a 34% increase in peak concentration, without altering the time to peak concentration. However, the ciclosporin microemulsion did not significantly affect the pharmacokinetics of diltiazem.

- Ciclosporin-induced encephalopathy was precipitated by diltiazem in a 76-year-old white woman with corticosteroid-resistant aplastic anemia and thrombocytopenia, type 2 diabetes, and coronary artery disease, who was taking diltiazem for hypertension (38). She became comatose after 13 days of therapy with ciclosporin, and clinical examination and electroencephalography showed diffuse encephalopathy of moderate severity. Ciclosporin was withdrawn and she regained consciousness after 36 hours.

Concomitant diltiazem without proper dosage adjustment of ciclosporin can cause adverse neurological events.

Cisapride

A possible interaction of cisapride 20 mg/day with diltiazem has been reported in a 45-year-old woman who developed near syncope and had QT interval prolongation (39). The QT interval returned to normal after withdrawal of cisapride. Rechallenge was not attempted. Diltiazem inhibits CYP3A4, and should therefore probably be avoided in combination with cisapride.

HMG coenzyme-A reductase inhibitors

Atorvastatin

Rhabdomyolysis and acute hepatitis have been reported in association with the co-administration of diltiazem and atorvastatin (40).

- A 60-year-old African-American man developed abdominal pain, a racing heart, and shortness of breath over 24 hours. He had also noticed increasing fatigue and reduced urine output over the previous 2–3 days. He had been taking several medications, including atorvastatin, for more than 1 year, but diltiazem had been added 3 weeks before for atrial fibrillation. On the basis of laboratory findings and physical examination, a diagnosis of acute hepatitis and rhabdomyolysis with accompanying acute renal insufficiency was made. His renal function gradually normalized and his CK activity reached a maximum of 2092 units/l on day 1 and fell to 623 units/l on discharge. His liver function tests returned to normal by 3 months.

While rhabdomyolysis from statins is rare, the risk is increased when they are used in combination with agents that share similar metabolic pathways. Atorvastatin is metabolized by CYP3A4, which is inhibited by diltiazem.

Lovastatin

The effects of co-administration of oral diltiazem, a potent inhibitor of CYP3A, on the pharmacokinetics of lovastatin have been evaluated in a randomized study in 10 healthy volunteers (41). Lovastatin is oxidized by CYP3A to active metabolites. Diltiazem significantly increased the oral AUC and maximum serum concentration of lovastatin, but did not alter its half-life. The magnitude of the increase of plasma concentration of lovastatin suggested that caution is necessary when co-administering diltiazem and lovastatin.

In another study by the same investigators, 10 healthy volunteers were randomized in a two-way, crossover study either to oral lovastatin or to intravenous diltiazem followed by oral lovastatin. Intravenous diltiazem did not significantly affect the pharmacokinetics of lovastatin (oral AUC, C_{max}, t_{max}, or half-life), suggesting that the interaction does not occur systemically and is primarily a first-pass effect (42). Drug interactions with diltiazem may therefore become evident when a patient is changed from intravenous to oral dosing.

Pravastatin

The effects of co-administration of diltiazem, a potent inhibitor of CYP3A, on the pharmacokinetics of pravastatin have been evaluated in a randomized study in 10 healthy volunteers (41). Pravastatin is active alone and is not metabolized by CYP3A. Diltiazem did not alter the oral AUC, maximum serum concentration, or half-life of pravastatin.

Simvastatin

The interaction of diltiazem with simvastatin has been investigated in 135 patients attending a hypertension clinic (43). Cholesterol reduction in the 19 patients taking diltiazem was 33% compared with 25% in the other 116 patients (median difference 8.6%; 95% CI = 1.1, 12). Multivariate analysis showed that concurrent diltiazem therapy, age, and the starting dose of simvastatin were independent predictors of percentage cholesterol response. The authors concluded that patients who take both diltiazem and simvastatin may need lower doses of simvastatin to achieve the recommended reduction in cholesterol.

Results from two clinical studies of the interaction of diltiazem with simvastatin showed that diltiazem increased the C_{max} of simvastatin (44) and enhanced its cholesterol-reducing effect (43).

In 10 healthy volunteers taking oral simvastatin 20 mg/day, diltiazem 120 mg bd for 2 weeks significantly increased the simvastatin C_{max} 3.6-fold, the AUC 5-fold, and the half-life 2.3-fold (44). There were no changes in the t_{max} of simvastatin or simvastatin acid.

In a prospective pharmacokinetic study, seven men and four women with hypercholesterolemia and hypertension were given oral simvastatin 5 mg/day, then oral simvastatin 5 mg/day combined with diltiazem 90mg/day, and then oral diltiazem 90 mg/day, each for 4 weeks (45). Diltiazem plus simvastatin resulted in a two-fold increase in the C_{max} and AUC of simvastatin, accompanied by an enhanced cholesterol-lowering effect. In contrast, co-administration reduced the C_{max} and AUC of diltiazem without altering its blood pressure lowering effects. Statins are metabolized by CYP3A4, which is inhibited by diltiazem; the mechanism whereby simvastatin reduced the AUC of diltiazem is unknown.

Of 135 patients attending a hypertension clinic who were taking simvastatin for primary or secondary prevention of coronary heart disease, 19 were also taking diltiazem (43). The cholesterol reduction in the 19 patients taking diltiazem was significantly higher than in the other 116 (33 versus 25%), with less interindividual variability. Concurrent diltiazem therapy, age, and the starting dose of simvastatin were significant independent predictors of the percentage cholesterol response.

Rhabdomyolysis due to an interaction of simvastatin with diltiazem has been reported (46).

- A 75-year-old-man taking simvastatin 80 mg/day and diltiazem 240 mg/day developed extreme weakness and diffuse muscle pain. All drugs were withdrawn and he underwent hemodialysis. Within 3 weeks his muscle pain disappeared and he regained function in his legs. The activities of creatine kinase and transaminases gradually returned to normal, but he continued to need hemodialysis.

Methylprednisolone

Co-administration of diltiazem with methylprednisolone increased plasma concentrations of methylprednisolone and its adrenal suppressant effects in nine healthy volunteers (47). Care should be taken when these two drugs are co-administered for a long period, even if the clinical relevance of this pharmacokinetic interaction still needs to be evaluated.

Midazolam

Diltiazem caused a 43% mean increase in the half-life of midazolam in 30 patients who underwent coronary artery bypass grafting (30). Similar effects were observed with alfentanil. The proposed mechanism was diltiazem-induced inhibition of benzodiazepine metabolism by CYP3A. Patients taking diltiazem had delayed early post-operative recovery as a result.

Moracizine

Moracizine is an enzyme inducer; it increases the rate of clearance of diltiazem (48). Conversely, diltiazem caused a doubling of the AUC of moracizine in healthy volunteers.

Sildenafil

Sildenafil is metabolized predominantly by CYP3A4, which diltiazem inhibits. An interaction of diltiazem with sildenafil has been reported (49).

- A 72-year-old man, who regularly took aspirin, metoprolol, diltiazem, and sublingual glyceryl trinitrate for stable angina, reported chest pain during elective prognostic coronary angiography, which resolved with half of a sublingual tablet of glyceryl trinitrate. Within 2 minutes he developed severe hypotension, with an unchanged electrocardiogram and no evidence of anaphylaxis. He had taken sildenafil 50 mg 48 hours before angiography.

The interval after which even short-acting nitrates can be safely given after the use of sildenafil is likely to be substantially longer than 24 hours when elderly patients are concurrently taking a CYP3A4 inhibitor, such as diltiazem.

Sirolimus

The pharmacokinetic interaction of a single oral dose of diltiazem 120 mg with a single oral dose of sirolimus 10 mg has been studied in 18 healthy subjects, 12 men and 6 women, 20–43 years old, in an open, three-period, randomized, crossover study (50). The whole-blood sirolimus AUC increased by 60% and the C_{max} by 43% with diltiazem co-administration; the apparent oral clearance and volume of distribution of sirolimus fell by 38 and 45% respectively, consistent with the change in half-life from 79 to 67 hours. Sirolimus had no effect on the pharmacokinetics of diltiazem or on the effects of diltiazem on either diastolic or systolic blood pressures or the electrocardiogram. Single-dose diltiazem co-administration leads to higher sirolimus exposure, presumably by the inhibition of first-pass metabolism. Because of pronounced intersubject variability in this interaction, whole-blood sirolimus concentrations should be monitored closely in patients taking the two drugs.

Sulfonylureas

The interaction of tolbutamide with diltiazem has been studied in eight healthy men (51). Tolbutamide had no effect on diltiazem, but diltiazem increased the AUC of tolbutamide by 10% without an effect on blood glucose concentrations.

Tacrolimus

Diltiazem can increase the blood concentration of the macrolide immunosuppressant tacrolimus (52).

- A 68-year-old man developed diarrhea, dehydration, and atrial fibrillation 4 months after liver transplantation. He was taking tacrolimus (blood concentration 13 ng/ml) and was given a continuous infusion of diltiazem for 1 day followed by oral therapy. Three days later he became delirious, confused, and agitated, and the blood concentration of tacrolimus was 55 ng/ml. His mental status gradually improved after withdrawal of both drugs.

In a retrospective study in 96 renal transplant recipients who took tacrolimus for immunosuppression, 64 took a mean dose of 214 mg/day of diltiazem for hypertension; tacrolimus concentrations increased (53). Besides its potential use in reducing tacrolimus dosage requirements, diltiazem has the potential to optimize allograft function by reducing ischemia–reperfusion injury. There was no difference in renal function between the groups over 2 years and no differences in graft or patient survival, which at 2 years were 97 and 98% with diltiazem and 100 and 100% without diltiazem. Acute rejection episodes were also similar between the two groups (15%). Diltiazem was discontinued in four patients because of adverse effects. There was no difference in tacrolimus-related adverse effects between the two groups. The authors concluded that diltiazem is efficacious and acceptably safe in renal transplant recipients taking with tacrolimus-based immunosuppressive therapy.

A dramatic increase in tacrolimus blood concentration has been attributed to simultaneous treatment with diltiazem and protease inhibitors (54).

- A 40-year-old man with diabetes, who took diltiazem 300 mg/day for hypertension and antiretroviral drugs, including lamivudine, stavudine, saquinavir, and ritonavir, underwent renal transplantation. He was also positive for hepatitis B and C. Four days after the graft, tacrolimus blood concentrations were 130 μg/l before the morning dose, 186 μg/l by the evening, and 202 μg/l the following day. Tacrolimus was withdrawn and the blood concentration slowly fell to 12 μg/l on day 26. Tacrolimus 0.5 mg bd was then reintroduced. There was no evidence of toxicity. Saquinavir and ritonavir were continued unchanged and the doses of lamivudine and stavudine were increased. Fifteen months after the graft, he was taking tacrolimus 0.5 mg once a week, which is 140 times lower than the initial dose, with a tacrolimus blood concentration of 10 μg/l. His renal function was normal.

As tacrolimus metabolism is mostly due to CYP3A4, coadministration of ritonavir and saquinavir, which are CYP3A4 substrates and inhibitors, may have altered the pharmacokinetics of tacrolimus. Diltiazem is also a CYP3A4 substrate and inhibitor, and the authors suggested that it may have contributed; however, it was withdrawn on the day of the graft and is unlikely to have played a part.

Thrombolytic agents

Clinical and experimental data suggested a possible increase of hemorrhage risk during thrombolytic therapy with alteplase in combination with diltiazem (55).

References

1. Hossack KF. Conduction abnormalities due to diltiazem. N Engl J Med 1982;307(15):953–4.
2. Schroeder JS, Feldman RL, Giles TD, Friedman MJ, DeMaria AN, Kinney EL, Mallon SM, Pit B, Meyer R, Basta LL, Curry RC Jr, Groves BM, MacAlpin RN. Multiclinic controlled trial of diltiazem for Prinzmetal's angina. Am J Med 1982;72(2):227–32.
3. Boujnah MR, Jaafari A, Boukhris B, Boussabah I, Thameur M. Bloc sino-auriculaire induit par le diltiazem aux doses thérapeutiques. A propos de trois observations. [Sinoatrial block induced by therapeutic doses of diltiazem. Report of 3 cases.] Tunis Med 2000;78(12):735–7.
4. Waller PC, Inman WH. Diltiazem and heart block. Lancet 1989;1(8638):617.
5. Nagle RE, Low-Beer T, Horton R. Diltiazem and heart block. Lancet 1989;1(8643):907.
6. Jacobs MB. Diltiazem and akathisia. Ann Intern Med 1983;99(6):794–5.
7. Remblier C, Kassir A, Richard D, Perault MC, Guibert S. Syndrome parkinsonien sous diltiazem. [Parkinson syndrome from diltiazem.] Therapie 2001;56(1):57–9.
8. Samarasinghe YP, Cox A, Feher MD. Calcium channel blocker induced gum hypertrophy: no class distinction. Heart 2004;90:16.
9. Miranda J, Brunet L, Roset P, Berini L, Farre M, Mendieta C. Prevalence and risk of gingival overgrowth in patients treated with diltiazem or verapamil. J Clin Periodontol. 2005;32(3):294–8.
10. Young RP, Wu H. Intestinal pseudo-obstruction caused by diltiazem in a neutropenic patient. Ann Pharmacother 2005;39(10):1749–51.
11. ter Wee PM, Rosman JB, van der Geest S. Acute renal failure due to diltiazem. Lancet 1984;2(8415):1337–8.
12. Abadin JA, Duran JA, Perez de Leon JA. Probable diltiazem-induced acute interstitial nephritis. Ann Pharmacother 1998;32(6):656–8.
13. Young EW, Diab A, Kirsh MM. Intravenous diltiazem and acute renal failure after cardiac operations. Ann Thorac Surg 1998;65(5):1316–9.
14. Knowles S, Gupta AK, Shear NH. The spectrum of cutaneous reactions associated with diltiazem: three cases and a review of the literature. J Am Acad Dermatol 1998;38(2 Pt 1):201–6.
15. Sheehan-Dare RA, Goodfield MJ. Widespread cutaneous vasculitis associated with diltiazem. Postgrad Med J 1988;64(752):467–8.
16. Romano A, Pietrantonio F, Garcovich A, Rumi C, Bellocci F, Caradonna P, Barone C. Delayed hypersensitivity to diltiazem in two patients. Am Allergy 1992;69(1):31–2.
17. Sadick NS, Katz AS, Schreiber TL. Angioedema from calcium channel blockers. J Am Acad Dermatol 1989;21(1):132–3.
18. Jan V, Machet L, Gironet N, Martin L, Machet MC, Lorette G, Vaillant L. Acute generalized exanthematous pustulosis induced by diltiazem: value of patch testing. Dermatology 1998;197(3):274–5.
19. Umebayashi Y. Drug eruption due to mexiletine and diltiazem. Nishinihon J Dermatol 2000;62:80–2.

20. Inui S, Itami S, Yoshikawa K. A case of lichenoid purpura possibly caused by diltiazem hydrochloride. J Dermatol 2001;28(2):100–2.
21. Scherschun L, Lee MW, Lim HW. Diltiazem-associated photodistributed hyperpigmentation: a review of 4 cases. Arch Dermatol 2001;137(2):179–82.
22. Lambert DG, Dalac S, Beer F, Chavannet P, Portier H. Acute generalized exanthematous pustular dermatitis induced by diltiazem. Br J Dermatol 1988;118(2):308–9.
23. Gonzalo Garijo MA, Pérez Calderón R, de Argila Fernández-Durán D, Rangel Mayoral JF. Cutaneous reactions due to diltiazem and cross reactivity with other calcium channel blockers. Allergol Immunopathol (Madr) 2005;33(4):238–40.
24. Cholez C, Trechot P, Schmutz JL, Faure G, Bene MC, Barbaud A. Maculopapular rash induced by diltiazem: allergological investigations in four patients and cross reactions between calcium channel blockers. Allergy 2003;58:1207–9.
25. Morimoto S, Sasaki S, Kiyama M, Hatta T, Moriguchi J, Miki S, Kawa T, Nakamura K, Itoh H, Nakata T, Takeda K, Nakagawa M. Sustained-release diltiazem overdose. J Hum Hypertens 1999;13(9):643–4.
26. Satchithananda DK, Stone DL, Chauhan A, Ritchie AJ. Unrecognised accidental overdose with diltiazem. BMJ 2000;321(7254):160–1.
27. Quispel R, Baur HJ. Tentamen suicidii door diltiazem met gereguleerde afgifte. [Attempted suicide with sustained release diltiazem.] Ned Tijdschr Geneeskd 2001;145(19):918–22.
28. Shah SJ, Quartin AA, Schein RMH. Diltiazem overdose—a case report. JK Pract 2001;8:40–2.
29. Snook CP, Sigvaldason K, Kristinsson J. Severe atenolol and diltiazem overdose. J Toxicol Clin Toxicol 2000;38(6):661–5.
30. Ahonen J, Olkkola KT, Salmenpera M, Hynynen M, Neuvonen PJ. Effect of diltiazem on midazolam and alfentanil disposition in patients undergoing coronary artery bypass grafting. Anesthesiology 1996;85(6):1246–52.
31. Lee TH, Friedman PL, Goldman L, Stone PH, Antman EM. Sinus arrest and hypotension with combined amiodarone–diltiazem therapy. Am Heart J 1985;109(1):163–4.
32. North DS, Mattern AL, Hiser WW. The influence of diltiazem hydrochloride on trough serum digoxin concentrations. Drug Intell Clin Pharm 1986;20(6):500–3.
33. Halawa B, Mazurek W. Interakcje digoksyny z nifedypina i diltiazemem. [Interactions of digoxin with nifedipine and diltiazem.] Pol Tyg Lek 1990;45(23–24): 467–9.
34. Mahgoub AA, El-Medany AH, Abdulatif AS. A comparison between the effects of diltiazem and isosorbide dinitrate on digoxin pharmacodynamics and kinetics in the treatment of patients with chronic ischemic heart failure. Saudi Med J 2002;23(6):725–31.
35. Morris RG, Jones TE. Diltiazem disposition and metabolism in recipients of renal transplants. Ther Drug Monit 1998;20(4):365–70.
36. Asberg A, Christensen H, Hartmann A, Berg KJ. Diltiazem modulates cyclosporin A induced renal hemodynamic effects but not its effect on plasma endothelin-1. Clin Transplant 1998;12(5):363–70.
37. Asberg A, Christensen H, Hartmann A, Carlson E, Molden E, Berg KJ. Pharmacokinetic interactions between microemulsion formulated cyclosporine A and diltiazem in renal transplant recipients. Eur J Clin Pharmacol 1999;55(5):383–7.
38. Jiang TT, Huang W, Patel D. Cyclosporine-induced encephalopathy predisposed by diltiazem in a patient with aplastic anemia. Ann Pharmacother 1999;33(6):750–1.
39. Thomas AR, Chan LN, Bauman JL, Olopade CO. Prolongation of the QT interval related to cisapride-diltiazem interaction. Pharmacotherapy 1998;18(2):381–5.
40. Lewin JJ 3rd, Nappi JM, Taylor MH, Lugo SI, Larouche M. Rhabdomyolysis with concurrent atorvastatin and diltiazem. Ann Pharmacother 2002;36(10):1546–9.
41. Azie NE, Brater DC, Becker PA, Jones DR, Hall SD. The interaction of diltiazem with lovastatin and pravastatin. Clin Pharmacol Ther 1998;64(4):369–77.
42. Masica AL, Azie NE, Brater DC, Hall SD, Jones DR. Intravenous diltiazem and CYP3A-mediated metabolism. Br J Clin Pharmacol 2000;50(3):273–6.
43. Yeo KR, Yeo WW, Wallis EJ, Ramsay LE. Enhanced cholesterol reduction by simvastatin in diltiazem-treated patients. Br J Clin Pharmacol 1999;48(4):610–5.
44. Mousa O, Brater DC, Sunblad KJ, Hall SD. The interaction of diltiazem with simvastatin. Clin Pharmacol Ther 2000;67(3):267–74.
45. Watanabe H, Kosuge K, Nishio S, Yamada H, Uchida S, Satoh H, Hayashi H, Ishizaki T, Ohashi K. Pharmacokinetic and pharmacodynamic interactions between simvastatin and diltiazem in patients with hypercholesterolemia and hypertension. Life Sci 2004;76:281–92.
46. Peces R, Pobes A. Rhabdomyolysis associated with concurrent use of simvastatin and diltiazem. Nephron 2001;89(1):117–8.
47. Varis T, Backman JT, Kivisto KT, Neuvonen PJ. Diltiazem and mibefradil increase the plasma concentrations and greatly enhance the adrenal-suppressant effect of oral methylprednisolone. Clin Pharmacol Ther 2000;67(3): 215–21.
48. Shum L, Pieniaszek HJ Jr, Robinson CA, Davidson AF, Widner PJ, Benedek IH, Flamenbaum W. Pharmacokinetic interactions of moricizine and diltiazem in healthy volunteers. J Clin Pharmacol 1996;36(12):1161–8.
49. Khoury V, Kritharides L. Diltiazem-mediated inhibition of sildenafil metabolism may promote nitrate-induced hypotension. Aust NZ J Med 2000;30(5):641–2.
50. Bottiger Y, Sawe J, Brattstrom C, Tollemar J, Burke JT, Hass G, Zimmerman JJ. Pharmacokinetic interaction between single oral doses of diltiazem and sirolimus in healthy volunteers. Clin Pharmacol Ther 2001;69(1): 32–40.
51. Dixit AA, Rao YM. Pharmacokinetic interaction between diltiazem and tolbutamide. Drug Metabol Drug Interact 1999;15(4):269–77.
52. Hebert MF, Lam AY. Diltiazem increases tacrolimus concentrations. Ann Pharmacother 1999;33(6):680–2.
53. Kothari J, Nash M, Zaltzman J, Ramesh Prasad GV, Diltiazem use in tacrolimus-treated renal transplant recipients. J Clin Pharm Ther 2004;29(5):425–30.
54. Hardy G, Stanke-Labesque F, Contamin C, Serre-Debeauvais F. Bayle F, Zaoui P, Bessard G. Protease inhibitors and diltiazem increase tacrolimus blood concentration in a patient with renal transplantation: a case report. Eur J Clin Pharmacol 2004;60:603–5.
55. Becker RC, Caputo R, Ball S, Corrao JM, Baker S, Gore JM. Hemorrhagic potential of combined diltiazem and recombinant tissue-type plasminogen activator administration. Am Heart J 1993;126(1):11–4.

Direct Renin Inhibitors

General Information

The development of novel antihypertensive agents is important, as the number of patients with hypertension continues to burgeon throughout the world. Identifying novel targets is an important prelude to developing appropriate leads, but the identification of the target for the latest class of antihypertensive agents, direct renin inhibitors, is not new. The idea of blocking renin at the proximal step of the renin–angiotensin–aldosterone pathway (Figure 1) was proposed several decades ago. However, early attempts at producing renin inhibitors were thwarted by difficulties in achieving either adequate systemic availability or acceptable blood pressure lowering activity. Previous peptide-like molecules were successful when given parenterally, but had little activity when given orally (1).

It was the advent of computational molecular modelling that made the non-peptide renin inhibitors a reality. Aliskiren is a novel compound that was developed by the use of X-ray crystallography of the active site of renin and subsequent computational modelling. Its non-peptide structure overcame some of the difficulties of previous attempts at producing renin inhibitors, although it still has some interesting pharmacokinetic properties. It is the first in class of the orally active, non-peptide, direct renin inhibitors for the treatment of hypertension to go into phase III trials. Here we outline current knowledge about the clinical effects and safety of this new drug.

Mechanism of action

The renin–angiotensin system plays a central role in hypertension, mediating its effects through the peptide hormone angiotensin II, which increases arterial tone, stimulates aldosterone release, activates sympathetic neurotransmission, and promotes renal sodium reabsorption. Overactivation of the renin–angiotensin system therefore contributes to hypertension and its associated end-organ damage. The system can be inhibited at various points: ACE inhibitors reduce the conversion of angiotensin I to angiotensin II and angiotensin II receptor blockers antagonize the interaction of angiotensin II with the type-1 angiotensin II (AT_1) receptor. However, both of these classes of agent interfere with the normal feedback mechanism to the juxtaglomerular apparatus in the kidneys and thus lead to a reactive rise in plasma renin activity, which can partially counteract their effects. Optimal blockade of the renin–angiotensin–aldosterone system should therefore be achievable by blocking the proximal step in the conversion of angiotensinogen to angiotensin I by directly inhibiting the action of renin, thus attenuating the reactive rise in plasma renin activity.

In healthy volunteers aliskiren produces dose-dependent reductions in plasma renin activity and angiotensin I and angiotensin II concentrations (2), which translates into effective blood pressure lowering.

Pharmacokinetics

The systemic availability of aliskiren is limited, less than 3% of the parent compound being absorbed (3). Its half-life is in the region of 40 hours, which allows effective once-daily dosing; steady-state plasma concentrations are achieved after 5–8 days (3). Aliskiren has little clinically significant interaction with cytochrome P450 liver enzymes and no significant hepatic metabolism. Mild to severe hepatic impairment had no significant effect on the single-dose pharmacokinetics of aliskiren (4). The main route of elimination is biliary excretion and consequent fecal elimination. Renal clearance plays a minor role, about 1% being excreted in the urine. Aliskiren is a substrate for P glycoprotein and co-administration of other drugs that interfere with P glycoprotein could theoretically alter exposure to aliskiren; however, in elderly

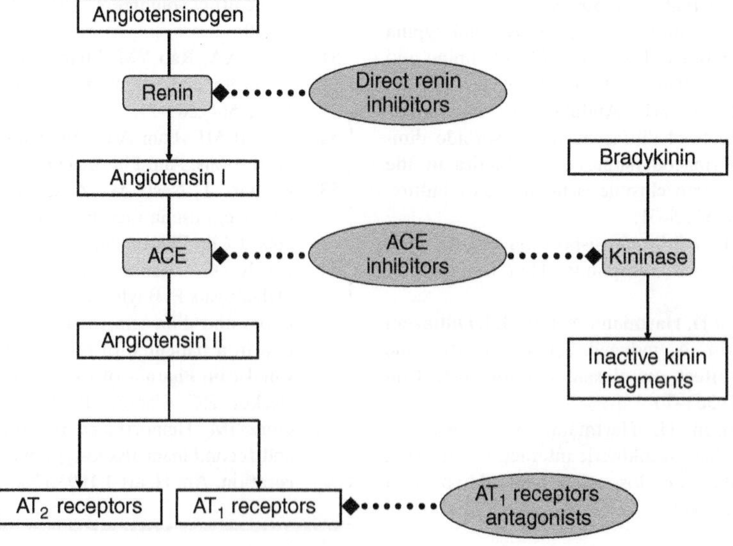

Figure 1

patients (65 years and over) there were only modest increases in drug exposure (5).

Clinical studies

The largest clinical studies to date with aliskiren have investigated the use of doses of up to 600 mg/day and have shown effective blood pressure lowering in patients with hypertension. In an 8-week study in 652 patients, aliskiren (150, 300, and 600 mg) was compared with placebo and irbesartan (6). Aliskiren 150 mg was as effective as irbesartan 150 mg, while higher doses lowered mean sitting diastolic blood pressure significantly more. The incidences of adverse events were comparable with aliskiren, irbesartan, and placebo.

In an 8-week placebo-controlled study in 672 hypertensive patients, aliskiren 150–600 mg/day was significantly superior to placebo in lowering mean sitting systolic and diastolic blood pressures at all doses; maximal or near-maximal reductions were achieved by week 4 (7). In a subgroup of patients evaluated with ambulatory blood pressure monitoring, the blood pressure lowering effect was sustained throughout the 24-hour dosing interval. There are many more clinical studies in progress or soon to be published.

Combination therapy

The favorable effect of aliskiren on plasma renin activity implies that combination treatment with antihypertensive agents that otherwise increase renin activity should be beneficial. Aliskiren has been investigated in open studies in combination with hydrochlorothiazide, ramipril, and irbesartan (8). Plasma renin activity did not increase compared with baseline, which suggests that combination therapy can achieve increased renin–angiotensin system suppression and improved blood pressure control.

In a comparison of aliskiren and ramipril, alone or in combination, the incidence of cough with aliskiren was lower than with ramipril, but the incidence with the combination was, surprisingly, lower still (5). Combination treatment with calcium channel blockers may also reduce the incidence of edema seen with agents such as amlodipine and provide an advantage to this combination. In a comparison of the combined use of aliskiren and valsartan against both agents alone or placebo in 1979 patients with hypertension the combination produced significantly greater reductions in blood pressure than either agent alone (9).

Adverse effects

The adverse effects of renin inhibitors compared with other renin–angiotensin blockers have been reviewed (10). The renin inhibitors should lack the effects of ACE inhibitors that are mediated through the accumulation of substance P, bradykinin, and other peptides, such as dry cough and angioedema. Aliskiren has been purported in trials to have "placebo-like" tolerability and a safety profile equivalent to that of angiotensin receptor antagonists (11). The most common adverse effects reported in the trials to date have been headache, diarrhea, and dizziness, all at low incidence rates. It has been suggested that gastrointestinal tolerability of aliskiren is dose-limiting above doses of 300 mg/day (12). Long-term adverse effects have not been studied, as trials reported to date have been principally short-term, with trials lasting up to 12 months in the more recent larger phase III studies.

Organs and systems

Electrolyte balance and renal function

One would expect aliskiren to have the same effects on electrolytes and renal function as those seen with angiotensin receptor antagonists or ACE inhibitors. Hyperkalemia and rises in creatinine have not been described in many of the early-phase trials published to date, most of which have reported that clinical laboratory values remained normal. However, combination therapy in the aliskiren/valsartan trial was associated with laboratory abnormalities—there were higher serum potassium concentrations and clinically relevant increases in serum creatinine in patients who took the combination (11). In an accompanying commentary it was pointed out that in the absence of trials with hard end-point data, the routine use of combinations of agents with dual inhibition of the renin–angiotensin system may expose patients to hyperkalemia and renal insufficiency, and this strategy is unlikely to become routine in conditions such as hypertension (13).

Second-Generation Effects

Pregnancy

The reproductive effects of renin inhibitors have not been reported, although it is likely that aliskiren will be contraindicated in pregnant women and those who are planning a pregnancy, as are ACE inhibitors and angiotensin receptor antagonists, because of concerns about potential teratogenic effects.

Susceptibility factors

Age

Aliskiren has not been evaluated in children, and this is therefore a contraindication pending clinical trials or further experience in younger subjects.

Renal disease

As with other inhibitors of the renin–angiotensin system, aliskiren is contraindicated in patients with bilateral renal artery stenosis or renal artery stenosis in a single functioning kidney.

Biliary disease

As the excretion of aliskiren depends on hepatobiliary routes, care should be taken in patients with biliary abnormalities.

Drug-drug interactions

Some of the most important potential drug interactions with aliskiren have been investigated in trials. There were no significant changes in drug exposure when it was combined with amlodipine, atenolol, celecoxib, cimetidine, hydrochlorothiazide, lovastatin, ramipril, or valsartan (14,15), or in single-dose studies with warfarin (16).

References

1. Staessen JA, Li Y, Richart T. Oral renin inhibitors. Lancet 2006;368(9545):1449–56.
2. Nussberger J, Wuerzner G, Jensen C, Brunner HR. Angiotensin II suppression in humans by the orally active renin inhibitor aliskiren (SPP100): comparison with enalapril. Hypertension 2002;39(1):E1–8.
3. Schmieder RE. Aliskiren: a clinical profile. J Renin Angiotensin Aldosterone Syst 2006;7(Suppl 2):S16–20.
4. Vaidyanathan S, Warren V, Yeh C, Bizot MN, Dieterich HA, Dole WP. Pharmacokinetics, safety, and tolerability of the oral renin inhibitor aliskiren in patients with hepatic impairment. J Clin Pharmacol 2007;47(2):192–200.
5. Vaidyanathan S, Reynolds C, Yeh CM, Bizot MN, Dieterich HA, Howard D, Dole WP. Pharmacokinetics, safety, and tolerability of the novel oral direct renin inhibitor aliskiren in elderly healthy subjects. J Clin Pharmacol 2007;47(4):453–60.
6. Gradman AH, Schmieder RE, Lins RL, Nussberger J, Chiang Y, Bedigian MP. Aliskiren, a novel orally effective renin inhibitor, provides dose-dependent antihypertensive efficacy and placebo-like tolerability in hypertensive patients. Circulation 2005;111(8):1012–8.
7. Oh BH, Mitchell J, Herron JR, Chung J, Khan M, Keefe DL. Aliskiren, an oral renin inhibitor, provides dose-dependent efficacy and sustained 24-hour blood pressure control in patients with hypertension. J Am Coll Cardiol 2007;49(11):1157–63.
8. O'Brien E, Barton J, Nussberger J, Mulcahy D, Jensen C, Dicker P, Stanton A. Aliskiren reduces blood pressure and suppresses plasma renin activity in combination with a thiazide diuretic, an angiotensin-converting enzyme inhibitor, or an angiotensin receptor blocker. Hypertension 2007; 49(2):276–84.
9. Oparil S, Yarows SA, Patel S, Zhang J, Satlin A. Efficacy and safety of combined use of aliskiren in patients with hypertension: a randomised double-blind trial. Lancet 2007;370:221–9.
10. Cheng H, Harris RC. Potential side effects of renin inhibitors—mechanisms based on comparison with other renin–angiotensin blockers. Expert Opin Drug Saf 2006; 5(5):631–41.
11. Van Tassell BW, Munger MA. Aliskiren for renin inhibition: a new class of antihypertensives. Ann Pharmacother 2007;41(3):456–64.
12. Krum H, Gilbert RE. Novel therapies blocking the renin–angiotensin–aldosterone system in the management of hypertension and related disorders. J Hypertens 2007; 25(1):25–35.
13. Birkenhäger WH, Staessen JA. Dual inhibition of the renin system by aliskiren and valsartan. Lancet 2007;370:195–6.
14. Dieterle W, Corynen S, Vaidyanathan S, Mann J. Pharmacokinetic interactions of the oral renin inhibitor aliskiren with lovastatin, atenolol, celecoxib and cimetidine. Int J Clin Pharmacol Ther 2005;43(11):527–35.
15. Vaidyanathan S, Valencia J, Kemp C, Zhao C, Yeh CM, Bizot MN, Denouel J, Dieterich HA, Dole WP. Lack of pharmacokinetic interactions of aliskiren, a novel direct renin inhibitor for the treatment of hypertension, with the antihypertensives amlodipine, valsartan, hydrochlorothiazide (HCTZ) and ramipril in healthy volunteers. Int J Clin Pract 2006;60(11):1343–56.
16. Dieterle W, Corynen S, Mann J. Effect of the oral renin inhibitor aliskiren on the pharmacokinetics and pharmacodynamics of a single dose of warfarin in healthy subjects. Br J Clin Pharmacol 2004;58(4):433–6.

Doxazosin

See also Alpha-adrenoceptor antagonists

General Information

The alpha$_1$-adrenoceptor antagonist doxazosin has similar adverse effects to those of prazosin. In a review of its clinical pharmacology and therapeutic uses, reports of tolerability included an update from previously unpublished pooled data on file (1). In 339 patients treated with doxazosin for hypertension compared with 336 treated with placebo, adverse effects were essentially the same as those reported in 665 patients being treated for benign prostatic hyperplasia compared with 300 treated with placebo. Dizziness (19 versus 9% and 16 versus 9%) and fatigue (12 versus 6% and 8 versus 2%) were reported significantly more often than with placebo (figures for patients with hypertension and benign prostate hyperplasia respectively, versus placebo). However, in hypertensive patients somnolence was reported more often than with placebo (5 versus 1%). In patients with benign prostatic hyperplasia the following adverse effects were reported significantly more often than with placebo: hypotension (17 versus 0%), edema (2.7 versus 0.7%), and dyspnea (2.6 versus 0.3%).

The Antihypertensive and Lipid-Lowering Treatment to Prevent Heart Attack Trial (ALLHAT) (2) was a multicenter comparison of drugs from each of three classes of antihypertensive agents (amlodipine, lisinopril, and doxazosin) with chlortalidone as an active control, in 24 335 adults with hypertension and at least one other coronary heart disease risk factor. In an interim analysis (median follow-up 3.3 years) there was no significant difference in the rates of fatal coronary heart disease or non-fatal myocardial infarction, or in total mortality between doxazosin and chlortalidone, but doxazosin was associated with higher rates of stroke (RR = 1.19; CI = 1.01, 1.40) and combined cardiovascular disease (RR = 1.24; CI = 1.17, 1.33). Considered separately, the risk of congestive heart failure was doubled (RR = 2.04; CI = 1.79, 2.32). Mean systolic blood pressure with doxazosin was about 2–3 mmHg higher than with chlortalidone and mean diastolic pressure was similar in the two groups. The authors concluded that the difference in blood pressure control did not account for the differences in cardiovascular end-points.

Doxazosin, and perhaps the whole class of alpha-blockers, should no longer be considered as first-line antihypertensive therapy. Doxazosin can still be used for symptom relief in patients with nocturia secondary to prostatic hyperplasia, although it should probably be avoided in patients with manifest or latent congestive heart failure (3). This issue has been intensely debated (4).

Phenoxybenzamine and doxazosin have been compared in 35 patients with pheochromocytoma (5). Hemodynamic, pharmacological, and biochemical indicators of alpha- and beta-adrenoceptor blockade were measured before, during, and after anesthesia and surgery in eight patients pretreated with phenoxybenzamine and in 27 patients pretreated with doxazosin. Doxazosin (2–16 mg/day) was as effective as phenoxybenzamine in controlling arterial blood pressure and heart rate before and during surgery and it caused fewer adverse effects.

Organs and Systems

Cardiovascular

Hypotension is a risk with doxazosin and has been reported to cause stroke (6).

- A 64-year-old man developed a right hemiparesis after taking one dose of doxazosin 4 mg for prostatic symptoms. A CT scan of the brain and carotid ultrasound studies were normal. He recovered most of his neurological function within a few days. Ambulatory blood pressure monitoring after treatment with doxazosin 2 mg showed striking blood pressure reduction during sleep.

Psychological, psychiatric

Doxazosin, 16 mg/day, has been reported to have caused an acute psychosis.

- A 71-year-old woman with type II diabetes and hypertension began to hear voices and to have auditory hallucinations. Doxazosin was progressively withdrawn over the next 14 days and by the time the dosage had been reduced to 8 mg a day the psychosis was much less severe; it disappeared completely after withdrawal (7).

Hematologic

Possible doxazosin-induced leukopenia with a positive response to dechallenge has been reported (8).

Drug Administration

Drug formulations

Doxazosin can cause a sharp fall in blood pressure at the start of therapy, and a modified-release formulation has been developed in an attempt to obviate this. In an open, non-comparative, sequential study in primary care, the ordinary formulation (1–16 mg/day for 3–6 months) was replaced by the modified-release formulation (4–8 mg/day for 3 months) in 3537 patients (9). The most common reasons for withdrawal from the study were loss to follow-up (37%) and adverse events (28%). Blood pressure fell from 160/95 to 139/82 mmHg with the ordinary formulation and to 135/79 mmHg when the modified-release formulation was used instead. The most common adverse events were weakness, headache, dizziness, and hypotension, all of which were more common with the ordinary formulation. However, these differences could have been due to a healthy survivor effect, in which those who had an adverse event early on (that is while taking the ordinary formulation) dropped out before taking the modified-release formulation, which would therefore appear to be safer. Thus, the lack of crossover in this study makes the results hard to interpret, although it appears that the two formulations were at worst no different from each other.

Drug overdose

Doxazosin overdose has been reported.

- Hypotension, bradycardia, and ST segment elevation on the electrocardiogram occurred in a patient who took doxazosin 40 mg (10). His blood pressure was 90/60 mmHg and his heart rate fell to 50/minute. Eight hours after aggressive saline infusion and gastric lavage the patient was awake, but his hypotension and bradycardia were corrected only after 96 hours and the administration of intravenous atropine 0.5 mg.
- A 19-year-old girl took doxazosin 60 mg and developed drowsiness, hypotension, and tachycardia (11). She was given supportive therapy, including activated charcoal to reduce gastrointestinal absorption, and her hypotension resolved with crystalloid infusion. She survived and was discharged after 48 hours.

References

1. Fulton B, Wagstaff AJ, Sorkin EM. Doxazosin. An update of its clinical pharmacology and therapeutic applications in hypertension and benign prostatic hyperplasia. Drugs 1995;49(2):295–320.
2. Davies BR, Furberg CD, Wright JT. Major cardiovascular events in hypertensive patients randomized to doxazosin vs chlorthalidone: the antihypertensive and lipid-lowering treatment to prevent heart attack trial (ALLHAT). ALLHAT Collaborative Research Group. JAMA 2000;283(15):1967–75.
3. Messerli FH. Implications of discontinuation of doxazosin arm of ALLHAT. Antihypertensive and Lipid-Lowering Treatment to Prevent Heart Attack Trial. Lancet 2000;355(9207):863–4.
4. Beevers DG, Lip GY. Do alpha blockers cause heart failure and stroke? Observations from ALLHAT. J Hum Hypertens 2000;14(5):287–9.
5. Prys-Roberts C, Farndon JR. Efficacy and safety of doxazosin for perioperative management of patients with pheochromocytoma. World J Surg 2002;26(8):1037–42.
6. Mansoor GA, Tendler BE. Stroke associated with alpha blocker therapy for benign prostatic hypertrophy. J Natl Med Assoc 2002;94(1):1–4.
7. Evans M, Perera PW, Donoghue J. Drug induced psychosis with doxazosin. BMJ 1997;314(7098):1869.
8. Ledger S, Mainra RR. Possible doxazosin-induced leukopenia. Can J Hosp Pharm 2004;57(5):297–8.

9. Anegon M, Esteban J, Jimenez-Garcia R, Sanz de Burgoa V, Martinez J, Gil de Miguel A. A postmarketing, open-label study to evaluate the tolerability and effectiveness of replacing standard-formulation doxazosin with doxazosin in the gastrointestinal therapeutic system formulation in adult patients with hypertension. Clin Ther 2002;24(5):786–97.
10. Gokel Y, Dokur M, Paydas S. Doxazosin overdosage. Am J Emerg Med 2000;18(5):638–9.
11. Satar S, Sebe A, Avci A, Yesilagac H, Gokel Y. Acute intoxication with doxazosin. Hum Exp Toxicol 2005;24(6):337–9.

Enalapril

See also Angiotensin converting enzyme inhibitors

General Information

Enalapril is an ACE inhibitor used in the treatment of hypertension and heart failure and prophylactically in patients with asymptomatic left ventricular dysfunction.

Organs and Systems

Cardiovascular

Of all studies of the effects of ACE inhibitors on mortality in acute myocardial infarction, only the CONSENSUS II trial did not show a positive effect. In this trial enalaprilat was infused within 24 hours after the onset of symptoms, followed by oral enalapril. The reasons for the negative result of CONSENSUS II remain unresolved, but hypotension linked to a poorer prognosis has been suggested as an explanation. In a small substudy of this large trial, 60 patients were investigated for residual ischemia before discharge, with exercise testing and Holter monitoring (1). Episodes of hypotension and predischarge ischemia were more common with enalapril than with placebo. The authors suggested that enalapril induced a proischemic effect in hypotension-prone patients, mediated through exacerbation of the hemodynamic response, since the initial blood pressure fall after myocardial infarction is related to residual ischemia and recurrent acute ischemic syndromes. This conclusion seems very speculative. The data were derived from a small substudy with multiple subanalyses and cannot support a cause-and-effect relation between the acute hemodynamic effect and the predischarge ischemic conditions. ACE inhibitors should still be used in acute myocardial infarction with cautious dose titration.

Neuromuscular function

Muscular weakness has been reported in a patient with mild renal impairment taking enalapril (2).

- A 78-year-old man, taking enalapril (10 mg/day), furosemide, and digoxin for cardiac failure due to ischemic heart disease, suddenly developed generalized muscle weakness. He had grade 3/5 weakness of all four limbs, his cranial nerves were intact, and there was no sensory impairment. His tendon reflexes were reduced and he had flexor plantar reflexes. The initial diagnosis was Guillain–Barré syndrome. Further investigation showed peaked T waves on his electrocardiogram, with a serum potassium of 9.4 mmol/l and a creatinine of 266 μmol/l. He was treated with glucose plus insulin, calcium gluconate, and sodium bicarbonate. Enalapril was withdrawn. His potassium concentration normalized.

Hyperkalemic muscle paralysis has been reported in renal insufficiency and trauma and in patients taking spironolactone and amiloride plus hydrochlorothiazide (co-amilozide). ACE inhibitors inhibit the release of aldosterone, reducing renal potassium loss, which can be enhanced by potassium-sparing diuretics or pre-existing renal insufficiency.

Enalapril

Sensory systems

ACE inhibitors can cause taste disturbances.

- A 66-year-old man developed severe dysgeusia and impaired quality of life while taking enalapril 5 mg/day, having previously been intolerant of quinapril because of cough (3).

This case illustrates the occurrence of different adverse effects of two antihypertensive drugs in the same class.

Endocrine

Cases of the syndrome of inappropriate secretion of antidiuretic hormone has been reported with enalapril (4,5).

Electrolyte imbalance

The management of hyperkalemia caused by inhibitors of the renin–angiotensin–aldosterone system has previously been reviewed (6). Volume depletion is a risk factor for the development of both renal impairment and subsequent hyperkalemia, because of the effects of ACE inhibitors on the renal vasculature. A further case of life-threatening hyperkalemia has been reported in a woman who developed a severe diarrheal illness on a background of widespread vascular disease, for which she was taking enalapril, bisoprolol, bumetanide, and isosorbide mononitrate (7).

Hematologic

Thrombocytopenia has been reported with enalapril in two elderly sisters (8). Both had the same HLA phenotype (B8DR3), which the authors postulated may reflect a genetic predisposition to this reaction. However, no further cases have been reported.

Enalapril-induced anemia has been reported in hemodialysis patients and in kidney transplant recipients. The anemia is usually normocytic, normochromic, and non-regenerative. Deficient bone-marrow erythropoiesis has occasionally been described. The anemia is reversible after withdrawal. Although the hypothesized mechanism of a reduction in erythropoietin blood concentrations by enalapril is controversial, in a prospective controlled study enalapril increased recombinant human erythropoietin requirements to maintain hemoglobin concentrations (9).

Enalapril-induced anemia has been reported in a child (10).

- A 7-year-old girl with segmental glomerulosclerosis and nephrotic syndrome failed to respond to prednisolone and cyclophosphamide, which were withdrawn. While her renal function was deteriorating she was given enalapril 2.5 mg bd. After 3 months, the enalapril was withdrawn because her hemoglobin fell from 12.7 to 6.2 g/dl. Ferritin, folate, and vitamin B_{12} were normal. Recovery was incomplete. Only 10 weeks after enalapril withdrawal her hemoglobin rose to 8.5 g/dl and remained unchanged thereafter.

The incomplete resolution of anemia in this case suggests that factors other than the speculated enalapril-related inhibition of erythropoietin may have contributed to the initial anemia (worsening renal function, frequent blood sampling, withdrawal of prednisolone).

An equivocal case of neutropenia has been ascribed to enalapril (11).

- A 52-year-old male renal transplant recipient on stable therapy (4.5 months) with ciclosporin, mycophenolate mofetil, and co-trimoxazole developed erythrocytosis and hypertension. His leukocyte count fell 19 days after he started to take enalapril. Enalapril was withdrawn, the dose of co-trimoxazole was halved, and the dose of mycophenolate was first reduced and then withdrawn on day 25. On day 28 the leukocyte and neutrophil counts were so low that granulocyte-stimulating factor had to be used. The leukocyte count normalized during continued treatment with ciclosporin and co-trimoxazole. CMV tests were negative.

The authors suggested a synergistic effect between enalapril and mycophenolate. As leukopenia is the commonest clinically significant adverse effect of mycophenolate, the evidence in this case is inadequate to support this hypothesis, particularly in the absence of information on mycophenolate blood concentrations.

Mouth

Stomatodynia (burning mouth) has previously been reported as an adverse effect of ACE inhibitors.

- A 50-year-old man developed burning of the lingual, palatal, and labial mucosa in the absence of xerostomia, attributed to treatment with enalapril (12);

Liver

Hepatitis has been attributed to enalapril.

- A 45-year-old man who had taken enalapril (dose not reported) for more than 2 years presented with hepatitis with negative viral serology (13). Biopsy showed necrosis and cholestasis. He recovered fully after drug withdrawal (timing not reported).
- A 46-year-old man had taken enalapril for hypertension for 3 years before he presented with jaundice and progressive liver failure (14). He was taking no other drugs and had a moderate daily consumption of alcohol. All known causes of acute liver failure were excluded by careful and extensive investigation. Analysis of liver biopsies showed a pathological pattern comparable to that observed in severe halothane hepatitis. Serological studies, including T cell stimulation with enalapril and a broad spectrum of tests for autoimmunity, were negative. The hepatitis persisted despite enalapril withdrawal and finally led to orthotopic liver transplantation and subsequently to death.

The mechanism of enalapril-induced liver injury in this case was obscure. The causal relation was unconvincing.

Enalapril-induced ductopenia with cholestatic hepatitis has been reported (15)

- A 58-year-old man with no previous history of liver disease, intravenous drug abuse, blood transfusion, or alcohol abuse, developed progressively worsening jaundice. He had been taking enalapril 20 mg/day for 2 years. He was afebrile, and had raised bilirubin, alkaline phosphatase, and transaminases. Serological tests for viral hepatitis, HIV, Epstein–Barr virus, cytomegalovirus, *Varicella*, Rickettsiae, and *Salmonella* were negative. Tests for a variety of antibodies and congenital liver disease were negative. A liver biopsy showed ductopenia and cholestasis with centrilobular hepatocyte necrosis. Enalapril was withdrawn, and the biochemical parameters improved over a period of 20 days. Ten days later he was readmitted with severe cholestatic jaundice. A second liver biopsy confirmed the presence of ductopenia (bile duct-to-portal tract ratio 0.5, normal 0.9–1.8). The jaundice progressively improved and resolved within 2 months.

Pancreas

There have been many reports of pancreatitis associated with ACE inhibitors, including enalapril, as in a case with recurrence after inadvertent rechallenge (16) and in a 74-year-old woman with chronic renal insufficiency on hemodialysis (17).

Urinary tract

Acute renal insufficiency has been reported in a neonate after treatment with enalapril for congestive heart failure (18). Although impaired renal function has been reported in young infants, cases in neonates are sparse.

- A 2.75 kg male neonate developed signs of congestive heart failure 3 days after birth and a diagnosis of a ventricular septal defect was made. Because of persistence of heart failure after digoxin and diuretic therapy, oral enalapril 0.1 mg/kg was started on day 6. The signs of congestive heart failure improved, but 3 days later oliguria developed. The child was normotensive, urinalysis was normal, and there was no evidence of infection or renal artery stenosis. Enalapril was withdrawn. He became anuric and had a dangerously raised serum potassium concentration (8.1 mmol/l). Peritoneal dialysis was started and digoxin and diuretics withdrawn. By day 20, the renal function tests were normal and dialysis was stopped. By week 3, congestive heart failure recurred and digoxin and diuretics were restarted. At follow-up at 1 year of age, the ventricular septal defect was closing spontaneously and renal function has remained normal.

The authors conclude that ACE inhibitors should be given with extreme caution to neonates, particularly in the presence of left-to-right shunts and congestive heart failure. Renal function should be carefully monitored.

Because of concerns over the adverse effects of prostaglandin synthesis inhibitors (for example non-steroidal anti-inflammatory drugs) in patients taking ACE inhibitors for heart failure, a human experimental study has been undertaken in 14 elderly healthy subjects, who were randomized to oral diclofenac or placebo with either no pre-treatment or after the administration of bendroflumethiazide and enalapril to activate the renin–angiotensin system (19). Diclofenac caused significant reductions in urine flow, the excretion rates of electrolytes, osmolality clearance, and free water clearance, although all of these were further reduced by pre-treatment with a diuretic and an ACE inhibitor. The authors deduced that in elderly patients with congestive cardiac failure, NSAIDs can counteract the beneficial effects of ACE inhibitors.

Skin

Adult Henoch-Schönlein purpura has been attributed to enalapril (20).

- A 41-year-old man with a history of hypertension and a recent stroke was admitted 10 days after starting to take enalapril (dose not reported). He complained of severe abdominal pain, myalgia, arthralgia, paresthesia, and Raynaud's phenomenon, and had a high erythrocyte sedimentation rate, but no leukocytosis or thrombocytopenia. He developed purpura-like cutaneous lesions with acute renal insufficiency, severe hematuria, and proteinuria. Liver tests became altered. There was a polyclonal rise in IgA concentrations. Skin biopsy suggested leukocytoclastic vasculitis, with deposition of IgA and complement. Other diagnostic procedures were negative. Enalapril was withdrawn and he was given glucocorticoids, after which he fully recovered (timing not reported).

The acantholytic potential of enalapril has been shown in vitro in skin cultures and in vivo acantholysis without pemphigus has been reported.

- A 66-year-old man, who had taken enalapril 10 mg/day for 1 year, had a basal cell carcinoma removed surgically (21). Histology showed suprabasal acantholytic clefts in the perilesional epidermis. Direct immunofluorescence did not show intracellular deposits of IgG, IgA, IgM, or C3. Indirect immunofluorescence showed serum anticellular antibodies (titer 1:80). He stopped taking enalapril, and 3 months later his anticellular antibody titer was 1:320. A further biopsy of apparently healthy skin on the back showed new foci of acantholysis. Another biopsy in the same location 2 months later showed no acantholytic changes. Direct immunofluorescence was negative and the anticellular antibody titer was 1:80. HLA typing showed pemphigus-predisposing antigens (DR14, DQ1), which has previously been reported as a predisposing factor for acantholytic changes in the epidermis exposed to enalapril.

As pemphigus vulgaris has been attributed to ACE inhibitors, it is not surprising that giving an ACE inhibitor to a patient with pemphigus could worsen the disease. Aggravation of pre-existing childhood pemphigus by enalapril has been reported (22).

- A 12-year-old boy with pemphigus vulgaris was treated with intravenous dexamethasone and prednisolone. He developed severe hypertension, which was unresponsive to atenolol, and he was given enalapril 2.5 mg bd. Although the blood pressure responded, the pemphigus deteriorated markedly over the next 2 weeks. Additional dexamethasone did not produce improvement. Enalapril was replaced by amlodipine, and the disease resolved over 3 weeks.

The evidence to date suggests that ACE inhibitors should be avoided in patients with pre-existing pemphigus.

Pemphigus has also been reported in a patient with pre-existing psoriasis.

- A 70-year-old man with plaque psoriasis for 35 years, treated topically, developed new erythematous scaly plaques on his trunk and scalp 2 months after starting to take enalapril and lacidipine (23). He also had a history of hypertension and congestive heart failure treated with digoxin, acenocoumarol, theophylline, amiloride, furosemide, isosorbide mononitrate, and pravastatin. The diagnosis was pemphigus foliaceus, based on the clinical picture and a skin biopsy. He was treated intermittently for the next 2 years with hostacycline and prednisone, with periods of improvement and relapse. When seen again with extensive skin lesions, a firm diagnosis of pemphigus foliaceus co-existing with psoriasis was made, based on a skin biopsy and immunofluoresence testing. Enalapril was withdrawn and he was given topical betamethasone. Four months later the skin condition had cleared completely.

The author speculated that in the presence of psoriasis, a T cell mediated disorder, enalapril acted as an exogenous trigger.

There has been a report of a lichenoid eruption in a patient taking enalapril.

- A 45-year-old woman developed a lichenoid eruption shortly after starting to take enalapril (24). The condition cleared when enalapril was withdrawn and subcutaneous enoxaparin given. The patient was not rechallenged.

Immunologic

Eosinophilic gastroenteritis after enalapril has been described (25). The authors briefly reviewed this rare condition, which is diagnosed on the basis of the presence of gastrointestinal symptoms, eosinophilic infiltration of the gastrointestinal tract, and the absence of parasitic or extra-intestinal disease. It has also been reported after clofazimine and naproxen.

- A 63-year-old hypertensive woman, who had a carcinoma of the distal esophagus resected 19 months earlier, developed chronic diarrhea. *Clostridium difficile* toxin was identified in her stools and the diarrhea resolved after treatment with metronidazole. Enalapril was added to her antihypertensive treatment, and 3 months later the diarrhea recurred. Stool examination was negative and there was no *Clostridium difficile* toxin. Her condition worsened and she lost 5 kg in weight. She had marked eosinophilia (2.4×10^9/l), and a small bowel biopsy showed mild chronic inflammation and edema, partial villous atrophy, and large clusters of eosinophils in the lamina propria with some focal infiltration of the epithelium. She stopped taking enalapril and her diarrhea promptly abated and the eosinophil count fell to 0.5×10^9/l at 3 weeks and 0.1×10^9/l at 2 months.

Body temperature

Fever has been attributed to enalapril (26).

- A 50-year-old man with heart failure and a valve prosthesis, taking digoxin, furosemide, and spironolactone, was given enalapril 5 mg/day. Two days later, after increasing the dose to 10 mg, he developed a fever with cough and clear sputum, with a normal chest X-ray. Enalapril was withdrawn and 24 hours later the fever resolved. It recurred immediately after rechallenge.

Second-Generation Effects

Fetotoxicity

The use of ACE inhibitors in pregnancy can have serious effects on the fetus. A further case of fatal neonatal renal insufficiency associated with maternal enalapril use during the third trimester has been reported (27).

- A woman with pregnancy-induced hypertension took enalapril 5 mg/day for 21 days before delivery, in addition to other antihypertensive drugs. The use of enalapril was temporally associated with oligohydramnios. The neonate had intrauterine growth retardation, hydrops, and oliguric renal insufficiency, which did not respond to furosemide, peritoneal dialysis, and exchange transfusion. Autopsy showed macroscopically and microscopically normal kidneys.

Susceptibility factors

Severe aortic stenosis is traditionally viewed as a contraindication to ACE inhibitor therapy. However, in a controlled trial, patients with symptomatic aortic stenosis taking ACE inhibitors had significantly better exercise tolerance and less dyspnea (28). The ACE inhibitors were generally well tolerated, although patients with congestive cardiac failure, poor left ventricular function, and resting hypotension were particularly likely to develop hypotension.

Drug Administration

Drug overdose

A single case of overdose of enalapril has been reported (29).

Drug-drug interactions

Sirolimus

An interaction of sirolimus with ramipril, causing tongue edema, has been described after renal transplantation (30) and in two renal graft patients, both of whom developed acute severe allergic reactions (31).

- A 54-year-old woman was given sirolimus to allow tacrolimus withdrawal 26 months after a renal transplant when she had also been taking enalapril for 2 months for difficult blood pressure control. Nine days later she developed an urticarial erythematous skin lesion with localized non-pitting edema on the face, neck, and upper thorax.
- A 61-year-old man developed non-pitting facial edema after ramipril was added to an immunosuppressive regimen.

In neither case was rechallenge performed.

References

1. Sogaard P, Thygesen KThe CONSENSUS II Holter Substudy Group. Potential proischemic effect of early enalapril in hypotension-prone patients with acute myocardial infarction. Cardiology 1997;88(3):285–91.
2. Dutta D, Fischler M, McClung A. Angiotensin converting enzyme inhibitor induced hyperkalaemic paralysis. Postgrad Med J 2001;77(904):114–5.
3. Unnikrishnan D, Murakonda P, Dharmarajan TS. If it is not cough, it must be dysgeusia: differing adverse effects of angiotensin-converting enzyme inhibitors in the same individual. J Am Med Dir Assoc 2004;5(2):107–10.
4. Castrillon JL, Mediavilla A, Mendez MA, Cavada E, Carrascosa M, Valle R. Syndrome of inappropriate antidiuretic hormone secretion (SIADH) and enalapril. J Intern Med 1993;233(1):89–91.
5. Fernandez Fernandez FJ, De la Fuente AJ, Vazquez TL, Perez FS. Síndrome de secreción inadecuada de hormona antidiurética causado por enalapril. Med Clin (Barc) 2004;123(4):159.
6. Palmer BF. Managing hyperkalemia caused by inhibitors of the renin–angiotensin–aldosterone system. N Engl J Med 2004;351(6):585–92.
7. McGuigan J, Robertson S, Isles C. Life threatening hyperkalaemia with diarrhoea during ACE inhibition. Emerg Med J 2005;22(2):154–5.
8. Grosbois B, Milton D, Beneton C, Jacomy D. Thrombocytopenia induced by angiotensin converting enzyme inhibitors. BMJ 1989;298(6667):189–90.

9. Albitar S, Genin R, Fen-Chong M, Serveaux MO, Bourgeon B. High dose enalapril impairs the response to erythropoietin treatment in haemodialysis patients. Nephrol Dial Transplant 1998;13(5):1206–10.
10. Sackey AH. Anaemia after enalapril in a child with nephrotic syndrome. Lancet 1998;352(9124):285–6.
11. Donadio C, Lucchesi A. Neutropenia after treatment of posttransplantation erythrocytosis with enalapril. Transplantation 2001;72(3):553–4.
12. Triantos D, Kanakis P. Stomatodynia (burning mouth) as a complication of enalapril therapy. Oral Dis 2004;10(4):244–5.
13. Quilez C, Palazon JM, Chulia T, Cordoba YC. Hepatatoxicided por enalapril. [Hepatoxicity by enalapril.] Gastroenterol Hepatol 1999;22(2):113–4.
14. Jeserich M, Ihling C, Allgaier HP, Berg PA, Heilmann C. Acute liver failure due to enalapril Herz 2000;25(7):689–93.
15. Macías FMJ, Campos FRR, Salguerro TP, Soria TP, Carrasco FG, Martin JMS. Ductopenic hepatitis related to enalapril. J Hepatol 2003;39:1091–2.
16. Carnovale A, Esposito P, Bassano P, Russo L, Uomo G. Enalapril-induced acute recurrent pancreatitis. Dig Liver Dis 2003;35:55–7.
17. Kishino T, Nakamura K, Mori H, Yamaguchi Y, Takahashi S, Ishida H, Saito S, Watanabe T. Acute pancreatitis during haemodialysis. Nephrol Dial Transplant 2005;20(9):2012–3.
18. Dutta I, Narang A. Enalapril-induced acute renal failure in a newborn infant. Pediatr Nephrol 2003;18:570–2.
19. Juhlin T, Bjorkman S, Hoglund P. Cyclooxygenase inhibition causes marked impairment of renal function in elderly subjects treated with diuretics and ACE-inhibitors. Eur J Heart Fail 2005;7(6):1049–56.
20. Goncalves R, Cortez Pinto H, Serejo F, Ramalho F. Adult Schönlein–Henoch purpura after enalapril. J Intern Med 1998;244(4):356–7.
21. Lo Schiavo A, Guerrera V, Cozzani E, Aurilia A, Ruocco E, Pinto F. In vivo enalapril-induced acantholysis. Dermatology 1999;198(4):391–3.
22. Thami GP, Kaur S, Kanwar AJ. Severe childhood pemphigus vulgaris aggravated by enalapril. Dermatology 2001;202(4):341.
23. Stavropoulos PG. Coexistence of psoriasis and pemphigus after enalapril intake. Dermatology 2003;207:337–8.
24. Villaverdi RR, Melguizo JB, Solano JL, Ortega SS. Lichen planus-like eruption due to enalapril. J Eur Acad Dermatol Venereol 2003;17:612–14.
25. Barak N, Hart J, Sitrin MD. Enalapril-induced eosinophilic gastroenteritis. J Clin Gastroenterol 2001;33(2):157–8.
26. Lorente L, Falguera M, Jover A, Rubio M. Fiebra secundaria a la administracion de enalapril. [Fever secondary to enalapril administration.] Med Clin (Barc) 1999;112(16):638–9.
27. Murki S, Kumar P, Dutta S, Narang A. Fatal neonatal renal failure due to maternal enalapril ingestion. J Matern Fetal Neonatal Med 2005;17(3):235–7.
28. Chockalingam A, Venkatesan S, Subramaniam T, Jagannathan V, Elangovan S, Alagesan R, Gnanavelu G, Dorairajan S, Krishna BP, Chockalingam V. Safety and efficacy of angiotensin-converting enzyme inhibitors in symptomatic severe aortic stenosis: Symptomatic Cardiac Obstruction-Pilot Study of Enalapril in Aortic Stenosis (SCOPE-AS). Am Heart J 2004;147(4):E19.
29. Newby DE, Lee MR, Gray AJ, Boon NA. Enalapril overdose and the corrective effect of intravenous angiotensin II. Br J Clin Pharmacol 1995;40(1):103–4.
30. Stallone G, Infante B, Di Paolo S, Schena A, Grandaliano G, Gesualdo L, Schena FP. Sirolimus and angiotensin-converting enzyme inhibitors together induce tongue oedema in renal transplant recipients. Nephrol Dial Transplant 2004;19(11):2906–8.
31. Burdese M, Rossetti M, Guarena C, Consiglio V, Mezza E, Soragna G, Gai M, Segoloni GP, Piccoli GB. Sirolimus and ACE-inhibitors: a note of caution. Transplantation 2005;79(2):251–2.

Endothelin receptor antagonists

See also Ambrisentan, Bosentan, Sitaxsentan, Tezosentan

General Information

It is disappointing that mechanistically novel antihypertensive drugs have not emerged during the last decade or two. The only real novelty was the addition of angiotensin II receptor antagonists as a refinement of the approach introduced with ACE inhibitors. The expectations for direct renin antagonists have not been realized. Some of the cornerstone classes of antihypertensive drugs, such as the diuretics, beta-adrenoceptor antagonists, calcium channel blockers, and direct vasodilators, have been around for several decades. Advances have predominantly been made in the pharmacokinetic properties and pharmacodynamic specificities of compounds in existing antihypertensive drug classes.

With the emergence of knowledge about endothelial factors such as nitric oxide and endothelin, there was much expectation that endothelin antagonists would become useful in the management of hypertension. Endothelin was discovered in 1988 (1) and is the most potent vasoconstrictor known. In terms of pathophysiology, endothelin receptor antagonists could play a role in a variety of diseases associated with vasoconstriction, such as hypertension, renal disease, occlusive vascular disease, pulmonary hypertension, and congestive heart failure (2). Apart from vasoconstriction, endothelin is also involved in the structural changes associated with these diseases, and it is now recognized that there is a wider target in hypertension treatment than lowering blood pressure. Vascular and myocardial remodelling may be key issues in determining long-term outcome (3) and effects on myocardial fibrosis and vascular compliance may be as important as lowering blood pressure. ACE inhibitor treatment can produce regression of hypertensive myocardial fibrosis in animal models (4) and in hypertensive patients (5).

Several endothelin receptor antagonists are in clinical use or currently in trials in the treatment of idiopathic pulmonary hypertension or pulmonary artery hypertension secondary to connective tissue disorders. Some agents block both endothelin A (ET_A) receptors and endothelin B (ET_B) receptors (bosentan); others block ET_A only (sitaxsentan and ambrisentan). They antagonize the effect of the powerful vasoconstrictor and smooth muscle mitogen hormone endothelin-1. Theoretically, the selective ET_A receptor antagonists should inhibit the vasoconstrictor effects of ET_A stimulation while preserving the natural vasodilator responses that are mediated through ET_B receptors. All of the endothelin receptor antagonists have adverse effects typical of vasodilatory agents, including headache, nasal congestion, dizziness, and peripheral edema.

Trials of the effects of endothelin receptor antagonists in patients with heart failure, coronary artery disease, arterial hypertension, and pulmonary hypertension have been reviewed (6), as have their uses in treating cancers (7) and cerebral vasospasm (8), and their potential uses in atherosclerosis, re-stenosis, myocarditis, shock, and portal hypertension (9)

As far as hypertension is concerned, there is a substantial body of preclinical evidence of the potential efficacy of endothelin antagonists in hypertension, and this has been extensively reviewed (10). Circulating endothelin concentrations are not increased in hypertension, but it is postulated that there is an imbalance between the vasodilatory effects of nitric oxide and the vasoconstrictor effects of endothelin at a local vascular level, resulting in increased endothelin vasoconstrictor tone and endothelin-mediated end-organ damage (11,12).

References

1. Yanagisawa M, Kurihara H, Kimura S, Tomobe Y, Kobayashi M, Mitsui Y, Yazaki Y, Goto K, Masaki T. A novel potent vasoconstrictor peptide produced by vascular endothelial cells. Nature 1988;332(6163):411–5.
2. Luscher TF, Barton M. Endothelins and endothelin receptor antagonists: therapeutic considerations for a novel class of cardiovascular drugs. Circulation 2000;102(19):2434–40.
3. Weber KT. Cardioreparation in hypertensive heart disease. Hypertension 2001;38(3 Pt 2):588–91.
4. Brilla CG, Matsubara L, Weber KT. Advanced hypertensive heart disease in spontaneously hypertensive rats. Lisinopril-mediated regression of myocardial fibrosis. Hypertension 1996;28(2):269–75.
5. Brilla CG, Funck RC, Rupp H. Lisinopril-mediated regression of myocardial fibrosis in patients with hypertensive heart disease. Circulation 2000;102(12):1388–93.
6. Neunteufl T, Berger R, Pacher R. Endothelin receptor antagonists in cardiology clinical trials. Expert Opin Investig Drugs 2002;11(3):431–43.
7. Wu-Wong JR. Endothelin receptor antagonists as therapeutic agents for cancer. Curr Opin Investig Drugs 2002;3(8):1234–9.
8. Chow M, Dumont AS, Kassell NF, Seifert V, Zimmermann M, Elliott JP, Awad IA, Dempsey RJ, Hodge CJ Jr. Endothelin receptor antagonists and cerebral vasospasm: an update. Neurosurgery 2002;51(6):1333–42.
9. Doggrell SA. The therapeutic potential of endothelin-1 receptor antagonists and endothelin-converting enzyme inhibitors on the cardiovascular system. Expert Opin Investig Drugs 2002;11(11):1537–52.
10. Moreau P. Endothelin in hypertension: a role for receptor antagonists? Cardiovasc Res 1998;39(3):534–42.
11. Taddei S, Virdis A, Ghiadoni L, Sudano I, Magagna A, Salvetti A. Role of endothelin in the control of peripheral vascular tone in human hypertension. Heart Fail Rev 2001;6(4):277–85.
12. Donckier JE. Therapeutic role of bosentan in hypertension: lessons from the model of perinephritic hypertension. Heart Fail Rev 2001;6(4):253–64.

Eprosartan

See also Angiotensin II receptor antagonists

General Information

Eprosartan is a non-biphenyl, non-tetrazole competitive antagonist at angiotensin II type 1 (AT_1) receptors that is chemically distinct from other angiotensin II receptor antagonists (1). It causes dual blockade of AT_1 receptors both presynaptically and postsynaptically, reducing sympathetic nerve activity significantly more than other receptor antagonists.

Its use in hypertension has been reviewed (2).

Organs and Systems

Respiratory

In a multicenter controlled study in patients with hypertension and a history of ACE inhibitor-induced cough, eprosartan significantly reduced the risk of cough by 88% compared with enalapril (3).

References

1. Puig JG, Lopez MA, Bueso TS, Bernardino JI, Jimenez RT. Grupo MAPA-MADRID Investigators. Clinical profile of eprosartan. Cardiovasc Drugs Ther 2002;16(6):543–9.
2. Robins GW, Scott LJ. Eprosartan: a review of its use in the management of hypertension. Drugs 2005;65(16):2355–77.
3. Oparil S. Eprosartan versus enalapril in hypertensive patients with angiotensin-converting enzyme inhibitor-induced cough. Curr Ther Res Clin Exp 1999;60:1–4.

Esmolol

See also Beta-adrenoceptor antagonists

General Information

Esmolol is a $beta_1$-selective adrenoceptor antagonist with an extremely short half-life (about 10 minutes), because of extensive metabolism by esterases in blood, liver, and other tissues.

Organs and Systems

Cardiovascular

Esmolol causes hypotension, sometimes symptomatic, in up to 44% of patients (1–3).

Drug Administration

Drug administration route

Irritation at the infusion site occurs in up to 9% of patients and depends on the duration of the infusion (4).

Drug–Drug Interactions

Digoxin

Esmolol increased digoxin concentrations by 10–20% in healthy volunteers (5).

Morphine

Co-administration of esmolol with morphine resulted in increased steady-state concentrations of esmolol (5).

References

1. Benfield P, Sorkin EM. Esmolol. A preliminary review of its pharmacodynamic and pharmacokinetic properties, and therapeutic efficacy. Drugs 1987;33(4):392–412.
2. Gray RJ, Bateman TM, Czer LS, Conklin CM, Matloff JM. Esmolol: a new ultrashort-acting beta-adrenergic blocking agent for rapid control of heart rate in postoperative supraventricular tachyarrhythmias. J Am Coll Cardiol 1985;5(6):1451–6.
3. Morganroth J, Horowitz LN, Anderson J, Turlapaty P. Comparative efficacy and tolerance of esmolol to propranolol for control of supraventricular tachyarrhythmia. Am J Cardiol 1985;56(11):F33–9.
4. Angaran DM, Schultz NJ, Tschida VH. Esmolol hydrochloride: an ultrashort-acting, beta-adrenergic blocking agent. Clin Pharm 1986;5(4):288–303.
5. Lowenthal DT, Porter RS, Saris SD, Bies CM, Slegowski MB, Staudacher A. Clinical pharmacology, pharmacodynamics and interactions with esmolol. Am J Cardiol 1985;56(11):F14–8.

Felodipine

See also Calcium channel blockers

General Information

Felodipine is a dihydropyridine derivative with diuretic properties (1). Its diuretic properties are not unique but are shared by other dihydropyridines. Its vasodilator-related adverse effects include flushing, headache, and tachycardia (2,3). Reduced arterial oxygen saturation has been seen in patients given intravenous felodipine for pulmonary hypertension (4,5). Along with amlodipine, but unlike other calcium channel blockers, felodipine may be safer in severe chronic heart failure accompanied by angina or hypertension.

Organs and Systems

Mouth

Felodipine-associated gingival enlargement has been reported in a patient with type 2 diabetes (6). The changes resolved after withdrawal of felodipine and control of the diabetes, and there were further improvements after scaling and root planing and instruction about oral hygiene. The histological features, elongated rete pegs, fibrous hyperplasia, a low-grade chronic inflammatory infiltrate, predominantly consisting of lymphocytes, and collagen bundle groups randomly distributed, were similar to those present in other cases of drug-associated gingival enlargement.

Skin

Different skin reactions have been reported with calcium antagonists, and in particular telangiectases in light-exposed areas of the skin with nifedipine and amlodipine (7).

- A 70-year-old woman took felodipine 10 mg/day and enalapril for about 1 year for hypertension. She developed telangiectatic lesions of both sides of the trunk. After excluding other causes, felodipine was withdrawn. After 2 months the lesions slightly abated, but never completely disappeared. The diagnosis was confirmed by histological evidence of enlarged capillaries parallel to the skin surface, in the absence of mast cells.
- Photodistributed telangiectases, made worse by solar radiation have been reported in a 67-year-old man taking felodipine 5 mg/day (8). Rosacea was also present. Felodipine was withdrawn and the telangiectases improved. Rechallenge was not performed.
- A 67-year-old man who had taken felodipine 5 mg/day for 4 years developed facial telangiectatic lesions that worsened with solar exposure for 9 months before felodipine was withdrawn; 2 months later the lesions had markedly diminished (8).

Drug–Drug Interactions

Erythromycin

The metabolism of felodipine is inhibited by erythromycin (SEDA-17, 239).

Nelfinavir

Nelfinavir inhibits CYP3A4, which metabolizes felodipine.

- A nurse who was taking felodipine 5 mg/day was accidentally exposed to blood infected with HIV; she was given nelfinavir and 3 days after treatment was started she developed bilateral leg edema, dizziness, fatigue, and orthostatic hypotension (9). The edema resolved after felodipine was withdrawn and her hypertension was controlled with a diuretic.

Oxcarbazepine

Oxcarbazepine reduces serum felodipine concentrations, although less than carbamazepine does (10).

Grapefruit juice

In 12 elderly people (aged 70 years and over), grapefruit juice increased the AUC of felodipine three-fold and its peak concentration four-fold (11). Blood pressure was lower with grapefruit juice after a single dose of felodipine, but not at steady state. Heart rate was higher with grapefruit juice after both single and multiple doses. Elderly patients should avoid taking grapefruit juice during treatment with felodipine because of this marked and unpredictable interaction.

References

1. Edgar B, Bengtsson B, Elmfeldt D, Lundborg P, Nyberg G, Raner S, Ronn O. Acute diuretic/natriuretic properties of felodipine in man. Drugs 1985;29(Suppl 2):176–84.
2. Sluiter HE, Huysmans FT, Thien TA, Koene RA. Haemodynamic effects of intravenous felodipine in normotensive and hypertensive subjects. Drugs 1985;29(Suppl 2):144–53.
3. Agner E, Rehling M, Trap-Jensen J. Haemodynamic effects of single-dose felodipine in normal man. Drugs 1985;29(Suppl 2):36–40.
4. Bratel T, Hedenstierna G, Nyquist O, Ripe E. The effect of a new calcium antagonist, felodipine, on pulmonary hypertension and gas exchange in chronic obstructive lung disease. Eur J Respir Dis 1985;67(4):244–53.
5. Cohn JN, Ziesche S, Smith R, Anand I, Dunkman WB, Loeb H, Cintron G, Boden W, Baruch L, Rochin P, Loss LVasodilator-Heart Failure Trial (V-HeFT) Study Group. Effect of the calcium antagonist felodipine as supplementary vasodilator therapy in patients with chronic heart failure treated with enalapril: V-HeFT III. Circulation 1997;96(3):856–63.
6. Fay AA, Satheesh K, Gapski R. Felodipine-influenced gingival enlargement in an uncontrolled type 2 diabetic patient. J Periodontol 2005;76(7):1217.
7. Karonen T, Stubb S, Keski-Oja J. Truncal telangiectases coinciding with felodipine. Dermatology 1998;196(2):272–3.
8. Silvestre JF, Albares MP, Carnero L, Botella R. Photodistributed felodipine-induced facial telangiectasia. J Am Acad Dermatol 2001;45(2):323–4.
9. Izzedine H, Launay-Vacher V, Deray G, Hulot JS. Nelfinavir and felodipine: a cytochrome P450 3A4-mediated drug interaction. Clin Pharmacol Ther 2004;75:362–3.
10. Zaccara G, Gangemi PF, Bendoni L, Menge GP, Schwabe S, Monza GC. Influence of single and repeated doses of oxcarbazepine on the pharmacokinetic profile of felodipine. Ther Drug Monit 1993;15(1):39–42.
11. Love JN. Acebutolol overdose resulting in fatalities. J Emerg Med 2000;18(3):341–4.

Fosinopril

See also Angiotensin converting enzyme inhibitors

General Information

Fosinopril is a phosphorus-containing ester prodrug of the ACE inhibitor fosinoprilat. Its clinical pharmacology, clinical use, and safety profile have been reviewed (1).

Organs and Systems

Respiratory

Claims that cough is less often observed with fosinopril than with other ACE inhibitors are primarily based on a single open unconvincing study (2).

Liver

Hepatic injury has been described from time to time with ACE inhibitors. Severe, prolonged cholestatic jaundice has been reported with fosinopril (3). The evidence of a link to fosinopril was convincing.

- A 61-year-old man developed weakness, severe jaundice, pruritus, and weight loss over 2 weeks. He had started to take metoprolol, fosinopril, and diazepam for hypertension 5 weeks before. He had raised hepatic transaminases and bilirubin. A liver biopsy showed cholestasis in a normal cellular architecture. A lymphocyte transformation assay showed reactivity to fosinopril but not diazepam or metoprolol. Bilirubin concentrations took 4 months to normalize and pruritus persisted for 6 months.

Skin

There is a recognized association between the use of ACE inhibitors and the development of pemphigus vulgaris, for example with enalapril (SEDA-26, 235). A case of worsening of pemphigus vulgaris with fosinopril, and a subsequent in vitro mechanistic study, has been published (4).

- A 64-year-old woman with insulin-dependent diabetes mellitus and pemphigus vulgaris controlled by deflazacort 12 mg/day was given fosinopril 10 mg/day for hypertension. Within 1 month her skin lesions worsened and an indirect immunofluorescence test became positive. Fosinopril was withdrawn and her skin lesions improved without modification of her steroid regimen; 10 months later the immunofluorescence test was negative.

Following this, normal human skin slices (obtained after informed consent from mammoplasty patients) were incubated with increasing concentrations of either fosinopril or captopril for 2–24 hours. Sera from patients with pemphigus vulgaris, containing antidesmoglein-3 antibodies (anti-Dsg3) were tested on the skin samples incubated with fosinopril and captopril, as well as control skin samples incubated with 0.9% saline. Indirect immunofluorescence testing showed that captopril at a concentration of 1.7×10^{-9} mmol/l blocked the binding of anti-Dsg3 to the keratinocyte surface, probably because captopril blocked adhesion molecules. In contrast, fosinopril only had this effect at a concentration of 1.7×10^{-2} mmol/l, a concentration much higher than would occur in vivo. The authors proposed that captopril produces acantholysis by blocking adhesion molecules, but that fosinopril does not have this effect, and that another mechanism must have been involved in the case that they reported.

References

1. Shionoiri H, Naruse M, Minamisawa K, Ueda S, Himeno H, Hiroto S, Takasaki I. Fosinopril. Clinical pharmacokinetics and clinical potential. Clin Pharmacokinet 1997;32(6): 460–480.
2. Punzi HA. Safety update: focus on cough. Am J Cardiol 1993;72(20):H45–8.
3. Nunes ACR, Amaro P, Maçoas F, Cipriano A, Martins I, Rosa A, Pimenta I, Donato A, Freitas D. Fosinopril-induced prolonged cholestatic jaundice and pruritus: first case report. Eur J Gastroenterol Hepatol 2001;13(3):279–82.
4. Parodi A, Cozzani E, Milesi G, Drosera M, Rebora A. Fosinopril as a possible pemphigus-inducing drug. Dermatology 2002;204(2):139–41.

Hydralazine

General Information

Hydralazine is a direct vasodilator that acts on vascular smooth muscle to produce systemic vasodilatation. As a result there is baroreceptor-mediated activation of the sympathetic nervous system and the renin–angiotensin system.

Vasodilators (as monotherapy) are associated acutely with flushing, headache, dizziness, reflex tachycardia, and palpitation. Chronic treatment can be complicated by fluid retention.

Organs and Systems

Cardiovascular

Aggravation of angina, presumably as a result of reflex tachycardia, has been reported with hydralazine.

- Pericarditis has been attributed to hydralazine in a 75-year-old man who developed a full-blown hydralazine-induced autoimmune syndrome (1).

The authors referred to a similar previously published case of late pericarditis after hydralazine treatment.

Nervous system

Peripheral neuropathy has been described in association with deficiency of pyridoxine (vitamin B6) in slow acetylators (2).

Metabolism

Salt and water retention mediated by secondary hyperaldosteronism can complicate hydralazine treatment, leading to weight gain and peripheral edema, loss of blood pressure control, and rarely cardiac failure (SED-8, 474).

Gastrointestinal

Anorexia, nausea, and vomiting can complicate hydralazine treatment, particularly initially (SED-8, 474).

Liver

Acute hepatitis (usually with negative antinuclear antibodies) has been attributed to hydralazine (3–7).

Urinary tract

Pauci-immune renal vasculitis in association with renal cell carcinoma has been reported in a patient taking hydralazine (8).

- A 61-year-old man taking hydralazine 50 mg bd for hypertension developed weight loss, proteinuria and hematuria, biochemical features of renal insufficiency, and a mass in the upper pole of the left kidney. Antinuclear antibodies were strongly positive. Biopsy of the right kidney showed focal proliferative glomerulonephritis. A left radical nephrectomy was performed and a renal clear cell carcinoma with no evidence of local or metastatic spread was found. He required five postoperative hemodialysis sessions and was entered into a permanent hemodialyis program 12 months later, at which time hydralazine was withdrawn. He remained well 1 year later, with no evidence of tumor recurrence, but antinuclear antibodies remain strongly positive and he did not recover renal function.

The authors cited several reports of an association between antinuclear antibody-positive renal vasculitis and either renal cell carcinoma or hydralazine therapy. The fact that the patient remained positive for antinuclear antibodies could have been due to the failure to withdraw hydralazine early enough or the presence of undetected residual tumor.

Skin

- A 44-year-old man taking hydralazine (dose not stated) developed a pruritic rash, attributed to a drug-induced lichenoid eruption, based on biopsy findings (9). The hydralazine was withdrawn and the rash resolved in a few weeks.

Immunologic

The lupus-like syndrome (SED-9, 318) with hydralazine occurs particularly in slow acetylators (and only rarely in fast acetylators) and in patients with the HLA-DR4 antigen. Blood dyscrasias and necrotizing vasculitis are additional features.

Current knowledge about the possible mechanisms of drug-induced lupus-like syndrome has been reviewed (10). Three mechanisms seem most plausible. One involves a change, possibly caused by a reactive metabolite, in the way that antigens are processed and presented to T cells, leading to the presentation of cryptic antigens. Another possibility is that a reactive metabolite binds to the class II major histocompatibility antigen and induces an autoimmune reaction analogous to a graft-versus-host reaction. A third possibility is that hydralazine inhibits DNA methylation, leading to an increase in DNA transcription and a generalized activation of the immune system.

A variety of vasculitic diseases, including Wegener's granulomatosis, microscopic polyangiitis, Churg–Strauss syndrome, and crescentic glomerulonephritis, are associated with antineutrophil cytoplasmic antibodies (ANCA) or leukocytoclastic vasculitis. In drug-induced ANCA-positive vasculitis, antimyeloperoxidase antibodies are most often found; they produce a perinuclear pattern of staining by indirect immunofluorescence (pANCA), but antiproteinase 3 (anti-PR3) antibodies can also occur (cANCA).

The possible drug causes of ANCA-positive vasculitis with high titers of antimyeloperoxidase antibodies in 30 new patients have been reviewed (11). The findings illustrate that this type of vasculitis is a predominantly drug-induced disorder. Only 12 of the 30 cases were not related to a drug. The most frequently implicated drug was hydralazine (10 cases); the remainder involved propylthiouracil (3 cases), penicillamine (2 cases), allopurinol (2 cases), and sulfasalazine.

Second-Generation Effects

Fetotoxicity

Thrombocytopenia has been reported in infants whose mothers were taking hydralazine with no evidence of the lupus-like syndrome (12).

The administration of hydralazine 25 mg bd at 34 weeks to a hypertensive pregnant woman was associated, within 1 week, with atrial extra beats in the fetus. These subsided when hydralazine was withdrawn, and the rest of the pregnancy and delivery was uneventful (13).

- A 29-year-old woman, 26 weeks pregnant, was treated with hydralazine for toxemia of pregnancy. She developed arthralgia and dyspnea and was subsequently found to be antinuclear antibody-positive. Following an induced labor, a low-birth-weight infant was born but died aged 36 hours. At autopsy the neonate was found to have a pericardial effusion and tamponade (14).

Although hydralazine was implicated as the cause of a maternal and neonatal lupus-like syndrome, the toxemia and low birth weight of the child make interpretation of the case difficult.

References

1. Franssen CF, el Gamal MI, Gans RO, Hoorntje SJ. Hydralazine-induced constrictive pericarditis. Neth J Med 1996;48(5):193–7.
2. Raskin NH, Fishman RA. Pyridoxine-deficiency neuropathy due to hydralazine. N Engl J Med 1965;273(22):1182–5.
3. Bartoli E, Massarelli G, Solinas A, Faedda R, Chiandussi L. Acute hepatitis with bridging necrosis due to hydralazine intake. Report of a case. Arch Intern Med 1979;139(6):698–9.
4. Forster HS. Hepatitis from hydralazine. N Engl J Med 1980;302(24):1362.
5. Itoh S, Ichinoe A, Tsukada Y, Itoh Y. Hydralazine-induced hepatitis. Hepatogastroenterology 1981;28(1):13–6.
6. Itoh S, Yamaba Y, Ichinoe A, Tsukada Y. Hydralazine-induced liver injury. Dig Dis Sci 1980;25(11):884–7.
7. Barnett DB, Hudson SA, Golightly PW. Hydrallazine-induced hepatitis? BMJ 1980;280(6224):1165–6.
8. Norris JH, Leeds J, Jeffrey RF. P-ANCA positive renal vasculitis in association with renal cell carcinoma and prolonged hydralazine therapy. Renal Fail 2003;25:311–4.
9. Bargout R, Malhotra A. A 44-year-old man with a pruritic skin rash. Cleve Clin J Med 2001;68(11):952–3.
10. Uetrecht JP. Drug induced lupus: possible mechanisms and their implications for prediction of which new drugs may induce lupus. Exp Opin Invest Drugs 1996;5:851–60.
11. Choi HK, Merkel PA, Walker AM, Niles JL. Drug-associated antineutrophil cytoplasmic antibody-positive vasculitis: prevalence among patients with high titers of antimyeloperoxidase antibodies. Arthritis Rheum 2000;43(2):405–13.
12. Widerlov E, Karlman I, Storsater J. Hydralazine-induced neonatal thrombocytopenia. N Engl J Med 1980; 303(21):1235.
13. Lodeiro JG, Feinstein SJ, Lodeiro SB. Fetal premature atrial contractions associated with hydralazine. Am J Obstet Gynecol 1989;160(1):105–7.
14. Yemini M, Shoham Z, Dgani R, Lancet M, Mogilner BM, Nissim F, Bar-Khayim Y. Lupus-like syndrome in a mother and newborn following administration of hydralazine; a case report. Eur J Obstet Gynecol Reprod Biol 1989;30(2):193–7.

Imidapril

See also Angiotensin converting enzyme inhibitors

General Information

Imidapril is a long-acting, non-sulfhydryl ACE inhibitor that has been used in patients with hypertension, congestive heart failure, acute myocardial infarction, and diabetic nephropathy. It is a prodrug, rapidly converted in the liver to its active metabolite imidaprilat, which has a half-life of about 15 hours. Imidapril and its metabolites are mainly excreted in the urine.

Organs and Systems

Respiratory

The incidences of cough with imidapril and enalapril have been compared in a poorly designed open study in 489 hypertensive patients (1). The authors claimed a significantly smaller incidence of cough with imidapril. However, because of important methodological flaws, this result cannot be trusted.

Mineral balance

Hyperkalemia is a well-known adverse effect of ACE inhibitors and can sometimes be severe (2).

- An 85-year-old woman with diabetes mellitus and a prior myocardial infarction, who was taking ioxoprofen and imidapril, lost consciousness owing to marked bradycardia caused by hyperkalemia (7.4 mmol/l). An electrocardiogram showed T wave changes compatible with hyperkalemia. After right ventricular pacing she promptly recovered consciousness. She was given glucose and insulin and her plasma potassium fell to 4.1 mmol/l within 3 hours. Simultaneously her heart rate became normal.

Skin

There has been a report of atenolol-induced lupus erythematosus followed 4 years later by imidapril-induced pemphigus foliaceus (3). The patient had HLA type DR4, which the authors surmised to be a predisposing factor to these two rare drug-induced immune-mediated dermatoses.

- A 67-year-old woman developed photo-distributed telangiectatic erythema together with periungual erythema and ragged posterior cuticles; this was attributed to atenolol, which she had taken for 1 year. She had a positive antinuclear antibody titer (1:160, homogeneous pattern), anti-dsDNA antibodies 125 IU (0–30), and antihistone antibodies 44 IU (0–5), which confirmed drug-induced lupus erythematosus. The photosensitive erythema improved and her autoantibodies returned to normal 3 months after atenolol was withdrawn. Four years later she developed crusted plaques and erosions over her upper back, arms, and scalp. Her medication included imidapril, which she had taken for 2.5 years, and thyroxine. The clinical presentation and skin histology and immunofluorescence were consistent with pemphigus foliaceus, which was thought to be drug-induced. Imidapril was withdrawn and she was given prednisolone 30 mg/day and hydroxychloroquine 200 mg bd. Within 4 months the pemphigus improved significantly.

Serosae

An unusual case of eosinophilic pleurisy has been described in association with imidapril (4). Eosinophilic lung diseases have been rarely reported and only with some ACE inhibitors (for example captopril and perindopril); this is the first reported case with imidapril. The authors made no formal assessment of the probability that this was drug-induced. Rechallenge was not attempted, but the result of dechallenge was dramatic, and improvements in both diagnostic imaging and laboratory findings made the association probable.

References

1. Saruta T, Arakawa K, Iimura O, Abe K, Matsuoka H, Nakano T, Nakagawa M, Ogihara T, Kajiyama G, Hiwada K, Fujishima M, Nakajima M. Difference in the incidence of cough induced by angiotensin converting enzyme inhibitors: a comparative study using imidapril hydrochloride and enalapril maleate. Hypertens Res 1999; 22(3):197–202.
2. Kurata C, Uehara A, Sugi T, Yamazaki K. Syncope caused by nonsteroidal anti-inflammatory drugs and angiotensin-converting enzyme inhibitors. Jpn Circ J 1999;63(12):1002–3.
3. Ah-Weng A, Natarajan S. A case report of drug-induced pemphigus foliaceus preceded by a history of drug-induced lupus erythematosus. J Am Acad Dermatol 2005;52(3 Suppl 1):42.
4. Yoshida H, Hasegawa R, Hayashi H, Irie Y. Imidapril-induced eosinophilic pleurisy. Case report and review of the literature. Respiration 2005;72(4):423–6.

Indoramin

See also Alpha-adrenoceptor antagonists

General Information

Indoramin is a postsynaptic selective $alpha_1$-adrenoceptor antagonist that is chemically distinct from the quinazolines. Unlike some other alpha-blockers, indoramin lowers blood pressure without a resulting reflex tachycardia or postural hypotension (1). However, it has largely been supplanted by more modern drugs, such as doxazosin, prazosin, and terazosin.

Organs and Systems

Nervous system

Indoramin penetrates the nervous system significantly and is reported to have a relatively high incidence of adverse effects, including sedation, dizziness, depression, headache, palpitation, dry mouth, and constipation (2). In a comparison with prazosin, prazosin produced a lower incidence of sedation, which is the most common adverse effect of indoramin, usually transient, in about 19% of cases (1). Other adverse effects that have sometimes led to withdrawal of indoramin have been dry mouth, dizziness, and failure of ejaculation. These adverse effects can be reduced by starting therapy with small doses and titrating gradually.

Drug–Drug Interactions

Alcohol

The sedative effect of indoramin is enhanced by alcohol (3).

References

1. Holmes B, Sorkin EM. Indoramin. A review of its pharmacodynamic and pharmacokinetic properties, and therapeutic efficacy in hypertension and related vascular, cardiovascular and airway diseases. Drugs 1986;31(6):67–99 Erratum in: Drugs 1986;32(4):preceding 291.

2. Marshall AJ, Kettle MA, Barritt DW. Evaluation of indoramin added to oxprenolol and bendrofluazide as a third agent in severe hypertension. Br J Clin Pharmacol 1980;10(3):217–21.
3. Abrams SM, Pierce DM, Johnston A, Hedges A, Franklin RA, Turner P. Pharmacokinetic interaction between indoramin and ethanol. Hum Toxicol 1989;8(3):237–41.

Irbesartan

See also Angiotensin II receptor antagonists

General Information

Irbesartan is a potent selective angiotensin II type 1 (AT_1) receptor antagonist. Its pharmacology is the same as that of other angiotensin II receptor antagonists (1). In the registration studies and other controlled trials adverse effects were not dose-related and not different from placebo.

An updated review of the pharmacology and therapeutic use of irbesartan in cardiovascular disorders has been published, including a brief section on drug tolerability, which referred to an unpublished postmarketing surveillance study in which 14% of the patients (1232 of 9009) had mostly mild adverse events (2). No further details were given on the nature of the adverse events.

There has been a large post-marketing assessment of irbesartan in 4500 patients followed for 6 months, of whom 2.2% had adverse effects, most commonly headache, epigastric pain, nausea, and vomiting (3). An allergic reaction and syncope due to hypotension were the only serious adverse events.

Organs and Systems

Sensory systems

Acute bilateral angle closure glaucoma is often due to an adverse drug reaction in patients who are predisposed by having narrowed angles in the anterior chambers of the eye. This has been reported with a wide variety of agents, including adrenergic agents, antidepressants, sulpha drugs, and ACE inhibitors.

- A 42-year-old woman gradually developed a severe, pulsatile, temporal headache, with nausea, blurred vision, and red eyes 2 weeks after starting to take a combination of irbesartan and hydrochlorothiazide for hypertension (4). She was also taking gabapentin and pantoprazole. She had severe bilateral intraocular hypertension and was given topical antiglaucoma treatment. The next day she had laser iridotomy, and her symptoms resolved.

The authors suggested that the timing made a drug-induced cause probable, although they did not specifically discuss the possible contribution of irbesartan. There have been several reports of angle closure glaucoma with hydrochlorothiazide, but only one with angiotensin II receptor antagonists. It is likely that the combination antihypertensive treatment was responsible, but unclear which of the agents (or both) was causative.

Liver

Cholestasis with irbesartan has been reported (5).

- A 62-year-old woman developed deep icterus and hepatomegaly 1 month after starting to take irbesartan 300 mg/day. She had been hypertensive for 15 years and had no history of liver disease or risk factors for liver disease. Her bilirubin was 403 μmol/l, alkaline phosphatase 3193 IU/l, and aspartate transaminase 177 IU/l. Serology and autoimmune screens were negative, as were liver ultrasonography and computerized tomography. Cholangiopancreatography was normal. Irbesartan was replaced by amlodipine and metoprolol, and 2 months later she remained jaundiced (bilirubin 324 μmol/l). Liver biopsy showed portal tract expansion, minimal inflammation, ectatic ductules, and cholestatic rosettes. Within 16 weeks she fully recovered and continued to be anicteric more than 1 year later.

The temporal profile in this case and the lack of an alternative cause for liver dysfunction suggested a drug reaction.

There have been a few reports of hepatitis in association with irbesartan.

- A 69-year-old man developed autoimmune cholestatic hepatitis while taking irbesartan for essential hypertension (6).
- A 51-year-old woman developed acute hepatitis while taking irbesartan (7).

Pancreas

An elderly patient developed pancreatitis soon after starting to take irbesartan 300 mg/day (8). The Naranjo criteria suggested a probable drug-induced event.

Urinary tract

Renal insufficiency has been attributed to irbesartan.

- A 77-year-old woman took captopril 25 mg tds and furosemide 40 mg/day for decompensated heart failure (9). Irbesartan 75 mg/day then replaced captopril, and 3 days later she developed renal insufficiency. Irbesartan was withdrawn and her renal function normalized within 5 days. It was not stated whether her blood pressure also changed.
- A 78-year-old Caucasian man with insulin-treated type II diabetes, hypertension, and stable mild renal insufficiency took captopril 25 mg/day and torasemide 150 mg/day (10). His general physician substituted irbesartan 150 mg/day for the captopril. The basal average serum creatinine rose from 220 to 294 μmol/l 10 days after beginning irbesartan, and to 752 μmol/l 3 weeks later, at which stage he was admitted with acute renal insufficiency, oliguria, and edema with a 6 kg weight

gain. Two days after withdrawal of irbesartan his creatinine reached a maximum of 907 μmol/l and then fell progressively to 570 μmol/l and never returned to basal values. The patient was started on chronic hemodialysis. Renal Doppler and MRI scans showed no renal artery stenosis, but there were signs of chronic renal ischemia, which may have contributed to the adverse drug reaction.

It is possible in the second case that irbesartan 150 mg/day produced abrupt and more pronounced inhibition of the renin–angiotensin system than the basal stable small dose of captopril 25 mg/day.

A further convincing case of ACE inhibitor-related renal insufficiency has been described with irbesartan and losartan (11).

- A 67-year-old woman with congestive heart failure developed oliguric renal insufficiency 2 days after the introduction of irbesartan. She rapidly recovered after withdrawal, and was then given losartan. The condition recurred shortly afterwards and subsided when losartan was stopped.

Skin

Maculopapular allergic skin lesions developed 5 days after introduction of irbesartan for hypertension in a 76-year-old man (12). Skin biopsy showed upper dermal lymphocytic infiltration resembling an early phase T cell lymphoma. This cleared rapidly and completely after withdrawal of irbesartan. Similar findings have been described with enalapril (13).

Drug–Drug Interactions

Lithium

Lithium toxicity developed in a 74-year-old woman with bipolar disorder, type 2 diabetes mellitus, hypertension, and ischemic heart disease (14). She was taking several potentially causative drugs, including two diuretics, an ACE inhibitor, and irbesartan, which had been started several weeks before. The authors suggested that the addition of the angiotensin II receptor antagonist may have contributed to lithium intoxication. Since ACE inhibitors enhance tubular reabsorption of lithium they proposed that the mechanism of this interaction is probably similar.

Simvastatin acid

A small well-designed study in 14 healthy subjects showed no significant effect of irbesartan on the single-dose pharmacokinetics of total simvastatin acid (15).

References

1. Johnston CI. Pharmacology of irbesartan. Expert Opin Investig Drugs 1999;8(5):655–70.
2. Markham A, Spencer CM, Jarvis B. Irbesartan: an updated review of its use in cardiovascular disorders. Drugs 2000;59(5):1187–206.
3. Morales-Olivas FJ, Aristegui I, Estan L, Rodicio JL, Moreno A, Gil V, Ferron G, Velasco O; KARTAN Study Investigators. The KARTAN study: a postmarketing assessment of irbesartan in patients with hypertension. Clin Ther 2004;26(2):232–44.
4. Rahim SA, Sahlas DJ, Shadowitz S. Blinded by pressure and pain. Lancet 2005;365(9478):2244.
5. Hariraj R, Stoner E, Jader S, Preston DM. Drug points: prolonged cholestasis associated with irbesartan. BMJ 2000;321(7260):547.
6. Annicchiarico BE, Siciliano M. Could irbesartan trigger autoimmune cholestatic hepatitis? Eur J Gastroenterol Hepatol 2005;17(2):247–8.
7. Peron JM, Robic MA, Bureau C, Vinel JP. Hépatite aiguë cytolytique du a la prise d'irbesartan (Aprovel): à propos d'un cas. Gastroenterol Clin Biol 2005;29(6–7):747–8.
8. Famularo G, Minisola G, Nicotra GC, De SC. Acute pancreatitis associated with irbesartan therapy. Pancreas 2005;31(3):294–5.
9. Anglada Pintado JC, Gallego Puerto P, Zapata Lopez A, Cayon Blanco M. Fracaso renal agudo asociado con irbesartan. [Acute renal failure associated with irbesartan.] Med Clin (Barc) 1999;113(9):358–9.
10. Descombes E, Fellay G. End-stage renal failure after irbesartan prescription in a diabetic patient with previously stable chronic renal insufficiency. Ren Fail 2000;22(6):815–21.
11. Lee HY, Kim CH. Acute oliguric renal failure associated with angiotensin II receptor antagonists. Am J Med 2001;111(2):162–3.
12. Gambini D, Sala F, Gianotti R, Cusini M. Exanthematous reaction to irbesartan. Eur Acad Dermatol Venereol 2003;17:469–90.
13. Furness PM, Goodfield MJ, MacLennan KA, Stevens A, Millard LG. Severe cutaneous reactions to captopril and analapril; histological study and comparison with early mycosis fungoides. J Clin Pathol 1986;39:902–7.
14. Spinewine A, Schoevaerdts D, Mwenge GB, Swine C, Dive A. Drug-induced lithium intoxication: a case report. J Am Geriatr Soc 2005;53(2):360–1.
15. Marino MR, Vachharajani NN, Hadjilambris OW. Irbesartan does not affect the pharmacokinetics of simvastatin in healthy subjects. J Clin Pharmacol 2000;40(8):875–9.

Isradipine

See also Calcium channel blockers

General Information

Isradipine is a dihydropyridine calcium channel blocker.

The MIDAS study (1) was a randomized trial of isradipine versus hydrochlorothiazide over 3 years in 883 patients, designed primarily to assess the effect on the rate of progression of medial intimal thickness in carotid arteries. The control of diastolic blood pressure was equivalent in both groups, but mean systolic blood pressure was 3.5 mmHg higher in isradipine group at 6 months, a significant difference that persisted throughout the study. This might explain the higher incidence of vascular events with isradipine, although there was no difference in the rate of progression of medial intimal thickness between the groups.

Drug–Drug Interactions

Ethanol

Studies in rodents and primates have suggested that calcium channel blockers, including darodipine, nifedipine, and verapamil, can attenuate the behavioral effects of ethanol. Isradipine seems to be the most effective. However, the results of published reports in man on the effects of calcium antagonists on the acute subject-rated, performance-impairing, and cardiovascular effects of ethanol have suggested that the behavioral effects of ethanol are not attenuated by verapamil, nifedipine, or nimodipine. The same conclusions have been reached from a study in nine healthy volunteers (2). Combined ethanol and isradipine produced increases in heart rate and reductions in blood pressure that were not observed with either drug alone. Isradipine significantly reduced peak breath alcohol concentrations, but it did not significantly alter the subject-rated performance-impairing effects of ethanol.

Phenytoin and fosphenytoin

Isradipine, an inhibitor of CYP450, provoked phenytoin intoxication in one patient (3).

Phenytoin

An interaction of isradipine with phenytoin has been reported (4).

- A 21-year-old white man, who was taking phenytoin and carbamazepine for seizures, developed severe lethargy, ataxia, and weakness after isradipine was prescribed for blood pressure control. The interaction was verified 2 months later, when he was rechallenged with the same dose of isradipine, 2.5 mg bd, and developed the same symptoms. Phenytoin plasma concentrations were at the upper limit of the usual target range (maximum assayed 100 μmol/l) and did not fall when the dose of phenytoin was reduced while he was taking isradipine.

Both pharmacokinetic and pharmacodynamic mechanisms can explain these observations. Isradipine inhibits CYP450 isoforms that are responsible for the metabolism of phenytoin, and both isradipine and phenytoin can bind to calcium channels in the brain and thereby affect neurological function.

References

1. Borhani NO, Mercuri M, Borhani PA, Buckalew VM, Canossa-Terris M, Carr AA, Kappagoda T, Rocco MV, Schnaper HW, Sowers JR, Bond MG. Final outcome results of the Multicenter Isradipine Diuretic Atherosclerosis Study (MIDAS). A randomized controlled trial. JAMA 1996;276(10):785–91.
2. Rush CR, Pazzaglia PJ. Pretreatment with isradipine, a calcium-channel blocker, does not attenuate the acute behavioral effects of ethanol in humans. Alcohol Clin Exp Res 1998;22(2):539–47.
3. Cachat F, Tufro A. Phenytoin/isradipine interaction causing severe neurologic toxicity. Ann Pharmacother 2002;36(9):1399–402.
4. Cachat F, Tufro A, Dalmady-Israel C, Laplante S. Phenytoin/isradipine interaction causing severe neurologic toxicity. Ann Pharmacother 2002;36(9):1399–402.

Labetalol

See also Beta-adrenoceptor antagonists

General Information

Labetalol is a non-selective alpha- and beta-adrenoceptor antagonist; these actions reside in four different enantiomers.

Labetalol is less likely to increase airways resistance in patients with bronchial asthma or to reduce peripheral blood flow. However, it can produce postural hypotension (SEDA-3, 166–7), and paresthesia of the scalp (1,2) and perioral numbness (3) have been described.

Organs and Systems

Cardiovascular

Labetalol is generally used effectively as an antihypertensive drug without reports of high vascular resistance, even in patients with pheochromocytoma. However, a case of pheochromocytoma with increased systemic vascular resistance and reduced cardiac index has been described (4).

- A 36-year-old man with a pheochromocytoma underwent adrenalectomy. After induction of anesthesia he was given intravenous labetalol 30 mg, and after intubation his blood pressure rose from 147/85 to 247/150 mmHg, his systemic vascular resistance index rose from 1958 to 3458 dynsec^{-1} m^2 cm^5, and his cardiac index fell to 3.6 l min^{-1} m^2. During tumor resection, he was given sodium nitroprusside to reduce his blood pressure. After tumor resection, his blood pressure fell to 77/52 mmHg and his systemic vascular resistance index to 1635 dynsec^{-1} m^2 cm^5. His blood pressure was effectively controlled with dobutamine.

This case affords a reminder that intravenous labetalol can increase the systemic vascular resistance and cause a hypertensive crisis. In these cases, labetalol should be replaced with a pure alpha-blocker or another vasodilator, to prevent the possibility of cardiovascular complications.

Neonatal cardiac arrest was precipitated in one case by labetalol (5).

Electrolyte balance

Three cases of severe hyperkalemia have been reported in renal transplant recipients taking labetalol for acute hypertension (6) and life-threatening hyperkalemia has been reported after intravenous labetalol (7).

- A 28-year-old man with severe hypertension and end-stage renal disease was given two intravenous doses of labetalol 20 mg 1 hour apart for malignant hypertension. The serum potassium concentration before treatment was 6.2 mmol/l, but 8 hours after labetalol it rose to 9.9 mmol/l and he developed left bundle branch block, ventricular tachycardia, and hypotension. He was given intravenous calcium gluconate, sodium bicarbonate, and lidocaine and reverted to sinus rhythm. The potassium concentration after 2 hours was 8.0 mmol/l. After hemodialysis the potassium concentration fell to 6.1 mmol/l.

In a retrospective chart review in 103 transplanted patients from January to November 1994, hypertension requiring perioperative treatment was observed in 51 cases (8). Treatment for hyperkalemia was necessary in 13 of the 38 patients who were treated with labetalol, compared with 11 of the 65 who were treated with another antihypertensive treatment.

Patients with end-stage renal disease on dialysis can have an enhanced hyperkalemic response to labetalol, which is partly attributable to electrochemical disturbances in the cells, characterized by an increase in intracellular sodium and chloride and a fall in intracellular potassium.

Liver

Eleven cases of hepatotoxicity have been reported with labetalol (9). Acute hepatitis has been described in an East-Asian man (10).

- A 50-year-old man with chronic hepatitis B infection, a hemiparesis due to a hemorrhage in the left basal ganglia, and a high systolic blood pressure was given intermittent intravenous labetalol followed by oral therapy (200 mg bd). After 1 week, his aspartate transaminase, alanine transaminase, and bilirubin started to rise. Ultrasonography of the liver and gallbladder was normal. Labetalol was withdrawn, and all the liver function tests normalized within a few days.

Prior hepatitis B infection may have predisposed this patient to labetalol-induced liver toxicity.

Musculoskeletal

Two children developed a proximal myopathy and rhabdomyolysis, which resolved on withdrawal of labetalol (11).

Immunologic

Antinuclear and antimitochondrial antibodies develop not uncommonly during long-term administration of labetalol (12).

Second-Generation Effects

Fetotoxicity

Labetalol has a high transplacental transfer rate.

- A single intravenous dose of labetalol 30 mg given to a woman with severe pregnancy-related hypertension 20 minutes before cesarean section was associated with significant neonatal beta-adrenoceptor blockade (hypoglycemia, bradycardia, and hypotension), and there were high labetalol concentrations in the umbilical cord blood (150–180 ng/ml) (13).
- An infant was born dead to a woman given intravenous labetalol 50 mg, which lowered her blood pressure from 170/110 to 115/85 mmHg before surgery (14).

Oral labetalol is well tolerated in pre-eclampsia, but if intravenous treatment is necessary, a small dose (5–10 mg) should be used initially and titrated up as necessary.

References

1. Scowen E. Scalp tingling on labetalol. Lancet 1978;1:98.
2. Hua AS, Thomas GW, Kincaid-Smith P. Scalp tingling in patients on labetalol. Lancet 1977;2(8032):295.
3. Gabriel R. Circumoral paraesthesiae and labetalol. BMJ 1978;1(6112):580.
4. Chung PC, Li AH, Lin CC, Yang MW. Elevated vascular resistance after labetalol during resection of a pheochromocytoma (brief report). Can J Anaesth 2002;49(2):148–50.
5. Sala X, Monsalve C, Comas C, Botet F, Nalda MA. Paro cardiaco en neonato de madre tratada con labetalol. [Cardiac arrest in newborn of mother treated with labetalol.] Rev Esp Anestesiol Reanim 1993;40(3):146–7.
6. Arthur S, Greenberg A. Hyperkalemia associated with intravenous labetalol therapy for acute hypertension in renal transplant recipients. Clin Nephrol 1990;33(6): 269–271.
7. Hamad A, Salameh M, Zihlif M, Feinfeld DA, Carvounis CP. Life-threatening hyperkalemia after intravenous labetolol injection for hypertensive emergency in a hemodialysis patient. Am J Nephrol 2001;21(3):241–4.
8. McCauley J, Murray J, Jordan M, Scantlebury V, Vivas C, Shapiro R. Labetalol-induced hyperkalemia in renal transplant recipients. Am J Nephrol 2002;22(4):347–51.
9. Clark JA, Zimmerman HJ, Tanner LA. Labetalol hepatotoxicity. Ann Intern Med 1990;113(3):210–3.
10. Marinella MA. Labetalol-induced hepatitis in a patient with chronic hepatitis B infection. J Clin Hypertens (Greenwich) 2002;4(2):120–1.
11. Willis JK, Tilton AH, Harkin JC, Boineau FG. Reversible myopathy due to labetalol. Pediatr Neurol 1990;6(4):275–6.
12. Kanto JH. Current status of labetalol, the first alpha- and beta-blocking agent. Int J Clin Pharmacol Ther Toxicol 1985;23(11):617–28.
13. Klarr JM, Bhatt-Mehta V, Donn SM. Neonatal adrenergic blockade following single dose maternal labetalol administration. Am J Perinatol 1994;11(2):91–3.
14. Olsen KS, Beier-Holgersen R. Fetal death following labetalol administration in pre-eclampsia. Acta Obstet Gynecol Scand 1992;71(2):145–7.

Lercanidipine

See also Calcium channel blockers

General Information

Lercanidipine is a third-generation dihydropyridine for once-daily dosing in hypertension. It has similar antihypertensive efficacy and tolerability to other calcium channel blockers (1).

Comparative studies

ACE inhibitors slow the progression of diabetic nephropathy. In a double-blind, randomized comparison of ramipril 10–20 mg/day and lercanidipine 10–20 mg/day, the two drugs were equally effective in reducing albumin excretion rate and blood pressure in 180 patients with microalbuminuria and type 2 diabetes and hypertension (2). The proportions of patients who had adverse events were similar: 29% with lercanidipine and 23% with rampril. Six patients taking lercanidipine and five taking ramipril withdrew because of one or more adverse events, as follows: lercanidipine—hypotension (n = 2), ankle edema (n = 1), tachycardia (n = 2), headache (n = 2), and epigastralgia (n = 1); ramipril—cough (n = 3), hypotension (n = 1), worsened peptic ulcer (n = 1).

Fluid balance

Lercanidipine has been tested in the COHORT double-blind trial in 828 hypertensives aged over 59 years (3). Lacidipine and lercanidipine were associated with a lower incidence of *edema* than amlodipine. In another double-blind study in 92 postmenopausal hypertensive women, lercanidipine was associated with a smaller increase in leg volume and a lower incidence of leg edema than amlodipine (4).

Susceptibility factors

Renal disease

Dihydropyridine calcium channel blockers often cause adverse effects that necessitate withdrawal, particularly in more fragile patients, such as those with renal dysfunction. Lercanidipine has been assessed in a multicenter trial in 203 hypertensive patients with chronic renal insufficiency (5). None had acchieved a target blood pressure (systolic pressure >130 mmHg or diastolic blood pressure >90 mmHg) despite treatment with a blocker of the renin-angiotensin system, either an ACE inhibitor or an angiotensin receptor blocker. Lercanidipine was added in a dosage of 10 mg/day for 6 months. In all, 43/203 patients (21%) withdrew from the study for several reasons, poor adherence and uncontrolled blood pressure being the most frequent. In three of these 43 the reason was possibly related to lercanidipine: erectile dysfunction in one, urinary incontinence in one, and dry mouth and eosinophilia in one. Biochemical values did not change significantly during treatment, with the exception of (a) creatinine clearance, which improved significantly after 6 months; (b) proteinuria, which improved significantly, and (c) cholesterol and trygliceride concentrations, which were significantly lower at the end of the treatment period.

References

1. McClellan KJ, Jarvis B. Lercanidipine: a review of its use in hypertension. Drugs 2000;60(5):1123–40.
2. Dalla Vestra M, Pozza G, Mosca A, Grazioli V, Lapolla A, Fioretto P, Crepaldi G. Effect of lercanidipine compared with ramipril on albumin excretion rate in hypertensive type 2 diabetic patients with microalbuminuria: DIAL Study (Diabete, Ipertensione, Albuminuria, Lercanidipina). Diab Nutr Metab 2004;17:259–66.
3. Zanchetti A. Emerging data on calcium-channel blockers: the COHORT study. Clin Cardiol 2003;26:II-17–20.
4. Lund-Johansen P, Stranden E, Helberg S, Wessel-Aas T, Risberg K, Ronnevik PK, Istad H, Madsbu S. Quantification of leg oedema in postmenopausal hypertensive patients treated with lercanidipine or amlodipine. J Hypertens 2003;21:1003–10.
5. Robles NR, Ocon J, Gomez CF, Manjon M, Pastor L, Herrera J, Villatoro J, Calls J, Torrijos J, Rodríguez VI, Rodriguez MM, Mendez ML, Morey A, Martinez FI, Marco J, Liebana A, Rincon B, Tornero F. Lercanidipine in patients with chronic renal failure: the ZAFRA Study. Renal Failure 2005;1:73–80.

Lidoflazine

See also Calcium channel blockers

General Information

Lidoflazine is a calcium channel blocker (1). The use of lidoflazine in patients with microvascular angina has been associated with malignant ventricular dysrhythmias (SEDA-16, 199).

Reference

1. Janssen PA. Pharmacologie et effets cliniques de la lidoflazine. [Pharmacology and clinical effects of lidoflazine.] Actual Pharmacol (Paris) 1970;23:135–61.

Lisinopril

See also Angiotensin converting enzyme inhibitors

General Information

Lisinopril is a non-sulfhydryl ACE inhibitor. It has been used in patients with hypertension, heart failure, myocardial infarction, and diabetic nephropathy.

Organs and Systems

Cardiovascular

An updated and comprehensive review of the use of lisinopril in congestive heart failure has been published (1), including a section on tolerability and details of the ATLAS (Assessment with Treatment with Lisinopril and Survival) trial. The tolerance of high doses of lisinopril (32.5–35 mg od) in heart failure, one of the issues addressed by this trial, was not significantly different from that of low doses (2.5–5 mg od). Since high doses were more effective than low doses, the authors recommended that more aggressive use of ACE inhibitors is warranted (2). However, this conclusion is valid only in the conditions of the trial, with careful and slow dose escalation. With such a strategy, most patients with heart failure can be titrated successfully to high maintenance doses.

Shock after myocardial infarction has been attributed to an ACE inhibitor (3).

- A 42-year-old woman suffered an acute anterior myocardial infarction, initially associated with pulmonary edema. After hemodynamic stabilization she was given lisinopril 10 mg orally. Two hours later she developed circulatory failure in conjunction with acute renal insufficiency. Right heart catheterization showed markedly reduced systemic vascular resistance but a normal cardiac index. After the usual causes of cardiogenic shock had been ruled out, repeated fluid challenges and intravenous noradrenaline failed to improve her hemodynamic status. She was therefore given angiotensin II intravenously (5–7.5 μg/minute), which immediately and markedly raised the systematic vascular resistance and resulted in subsequent regression of shock. She was discharged after an otherwise uneventful course.

Endocrine

The syndrome of inappropriate antidiuretic hormone secretion (SIADH) has been attributed to lisinopril (4).

- A 76-year-old woman taking lisinopril 20 mg/day and metoprolol for hypertension developed headaches, nausea, and a tingling sensation in her arms. Her serum sodium was 109 mmol/l, with a serum osmolality of 225 mosm/kg, urine osmolality of 414 mosm/kg, and urine sodium of 122 mmol/l. She had taken diclofenac 75 mg/day for arthritic pain for 6 years and naproxen for about 1 month. Propoxyphene napsylate and paracetamol had then been substituted and zolpidem had been started. A diagnosis of SIADH was postulated and thyroid and adrenal causes were excluded. Lisinopril was withdrawn and fluid was restricted to 100 ml/day. The serum sodium gradually corrected to 143 mmol/l.

The authors referred to three other similar cases, in two of which the diagnosis may have been confused by the concomitant use of diuretics in patients with heart failure. However, the present and one other case had occurred without co-existing risk factors for hyponatremia. They discussed a synergistic effect of zolpidem and/or diclofenac, and suggested a potential mechanism involving non-inhibition of brain ACE, which leaves brain angiotensin II receptors exposed to high circulating concentrations of angiotensin which would strongly stimulate thirst and the release of antidiuretic hormone.

Pancreas

Several cases of pancreatitis have been reported with ACE inhibitors, including lisinopril.

- A 67-year-old man without other risk factors developed acute pancreatitis only 3 hours after taking lisinopril. The originality of this case resides in the fact that the patient had experienced a similar but less severe reaction to the medication 3 months before. Thus, this case probably represents the first time a patient was rechallenged with lisinopril and had a more severe adverse reaction (5).
- A 42-year-old woman, who was taking lisinopril, nicardipine, metoprolol, simvastatin, omeprazole, an estrogen, and sublingual glyceryl trinitrate, presented with epigastric pain and vomiting (6). Pancreatitis was diagnosed on the grounds of raised activities of lipase (200 IU/l) and amylase (117 IU/l) and diffuse homogeneous enlargement of the head and body of the pancreas on CT scan. The pancreatic enzyme activities normalized 5 days after lisinopril and simvastatin withdrawal. Two months later she was taking estrogen and simvastatin, but not lisinopril. Her amylase and lipase were normal.

The temporal relation in the second case suggests that lisinopril was causative. The authors raised the possibility that the concomitant use of an estrogen and simvastatin, as well as a history of familial hypertriglyceridemia, may have predisposed her to pancreatitis.

A further case has been reported in a patient taking lisinopril and hydrochlorothiazide (7). There was associated radiological evidence of biliary tract pathology, and the authors concluded that in this case the drug may have aggravated the acute illness.

Pancreatitis has been reported in a patient taking lisinopril and atorvastatin (8). The relative contributions of each drug were not explored.

Skin

Dermatological adverse effects are relatively uncommon with ACE inhibitors but erythematous rashes, urticaria, alopecia, and lichenoid eruptions have all been reported. A pityriasis rosea-like eruption has been reported (9).

- A 72-year-old Caucasian man developed flat, round, or oval scaly patches, bright red to violet in color, on the trunk, abdomen, and proximal limbs. The scaly surface showed a typical marginal collarette. He reported severe itching, unresponsive to antihistamine treatment. The eruption had begun suddenly 1 month before, had been exanthematous from the start, and evolved in successive crops with no tendency to spontaneous remission. There were no general or prodromic symptoms. He had started to take lisinopril about 2 weeks before the onset of the eruption, was taking no other medications, and had never suffered adverse drug reactions. Lisinopril was withdrawn and the lesions ameliorated rapidly and healed in about a week, without specific treatment.

Pemphigus foliaceus has previously been reported with captopril, ramipril, enalapril, and fosinopril, as well as a wide variety of other drugs, and lisinopril has now also been implicated.

- A 66-year-old man developed pemphigus foliaceus while taking lisinopril for hypertension and ischemic heart disease (10). Response to dechallenge was rapid, with resolution of blister formation within 48 hours. There was no recurrence during 3 weeks without ACE inhibitor treatment.

Although there was co-existent rheumatoid arthritis, which can be associated with pemphigus foliaceus, the authors concluded that the time relation in the case suggested that lisinopril was the more important factor.

Worsening of pre-existing cold urticaria has been associated with lisinopril (11).

- A 43-year-old Caucasian man had a history of mild cold urticaria during the preceding 10 years. Shortly after starting to take lisinopril 20 mg/day and hydrochlorothiazide 25 mg/day for hypertension his condition worsened considerably and he had several bouts of severe cold urticaria. On one occasion, after swimming in water at 24°C, he developed severe urticaria and angioedema, and required emergency hospital admission. Standard ice-cube and cold-water immersion tests were strongly positive within 1 minute. Tests for a variety of causative factors were negative, and a diagnosis of acquired cold urticaria was made. Lisinopril was withdrawn, he was given valsartan 80 mg/day, and the dose of hydrochlorothiazide was reduced to 12.5 mg/day. During the next 24 weeks he had only mild symptoms of cold urticaria. Repeat ice-cube and cold-water immersion tests were negative at 1 and 3 minutes, but positive at 5 minutes.

The authors cited evidence that the kallikrein/kinin system is involved in cold urticaria, that ACE inhibitors increase wheal-and-flare reactions to cutaneously applied bradykinin, and that ACE inhibitor therapy is associated with raised plasma kinin concentrations. They therefore recommended avoidance of ACE inhibitors in patients with cold urticaria.

Erythema multiforme has been attributed to lisinopril (12).

- A 68-year-old man who had taken lisinopril 10 mg/day for several years stopped taking it temporarily because of a cough. It was then restarted in a dosage of 5 mg/day and a few days later he developed several bean-sized non-pruritic erythematous plaques with scales over the whole body, particularly on the chest. The appearance was of seborrheic dermatitis, but histology showed the features of erythema multiforme. The eruptions cleared completely a few days after lisinopril withdrawal. Two months later, a rechallenge test was positive.

Sexual function

Lisinopril 20 mg/day has been compared with atenolol 100 mg/day in a 16-week, double-blind, randomized, controlled trial in 90 hypertensive men aged 40–49 and without a history of sexual dysfunction. The number of occasions on which they had sexual intercourse fell during the first month in both groups (13). Subsequently, sexual activity tended to recover with lisinopril but not atenolol. The authors suggested that lisinopril may cause only a temporary reduction in sexual function.

Second-Generation Effects

Teratogenicity

An intact renin-angiotensin system is required for the normal intrauterine development of the kidneys. The use of ACE inhibitors during pregnancy can result in multiple organ failure, resulting in oligohydramnios, pulmonary hypoplasia, neonatal renal insufficiency, and bony abnormalities. In one such case lisinopril was successfully cleared from the neonate's blood by early peritoneal dialysis (14).

Drug–Drug Interactions

Clozapine

Raised clozapine blood concentrations have been reported after the introduction of lisinopril (15).

- A 39-year-old man with schizophrenia and diabetes, who had taken clozapine 300 mg/day and glipizide 10 mg/day for a year, took lisinopril 5 mg/day for newly diagnosed hypertension. On several occasions afterwards he had roughly a doubling of his blood concentrations of clozapine and norclozapine. He had typical effects of clozapine toxicity. After replacement of lisinopril by diltiazem, the blood concentrations of clozapine and norclozapine returned to the values that were present before lisinopril was introduced.

The information given here was sketchy and there was no information on the timing of blood samples relative to the dose of clozapine. Clozapine is metabolized by CYP1A2 and CYP3A4, but there is no evidence that lisinopril affects these pathways.

Lithium

Lithium toxicity occurred in a patient who took concomitant lisinopril (16).

- A 46-year-old man with a schizoaffective disorder took increasing doses of lithium for breakthrough hypomanic symptoms, in addition to neuropsychotherapies and antihypertensive drugs. Initially he had been taking atenolol and fosinopril; however, the latter was replaced by lisinopril because of supply shortage. A month after substitution, and from a stable lithium concentration and stable renal function, he developed symptoms of lithium toxicity, confirmed by serum lithium concentration measurement. Withdrawal of therapy and supportive measures resolved the problem.

Tizanidine

Hypotension followed the addition of tizanidine to lisinopril (17).

- A 10-year-old boy developed hypotension and reduced alertness. His blood pressure was 56/24 mmHg and his heart rate 88/minute. He had a history of a hypoxic ischemic insult to the central nervous system, subsequent hypertension, and spastic quadriplegia. His blood pressure had been controlled for the last 10 months with lisinopril (dose not stated). Tizanidine had been added 1 week before admission for spasticity. Lisinopril and tizanidine were withdrawn and his blood pressure rose to 149/89 mmHg over the next day. He was discharged and lisinopril was restarted but not tizanidine. He had no further problems with hypotension.

The authors interpreted the finding as a consequence of limited ability of the patient to respond to hypotension because of simultaneous blockade of the sympathetic system with the centrally acting $alpha_2$-adrenoceptor agonist tizanidine.

- A 48-year-old hypertensive woman, who was also taking lisinopril, was given tizanidine as a centrally acting antispastic agent for decerebrate rigidity after an intracerebral hemorrhage (18). Within 2 hours her blood pressure fell from 130/85 to 66/42 mmHg.

The authors discussed the possible mechanisms of this interaction, which was probably pharmacodynamic, since tizanidine has actions similar to those of clonidine, and stimulates central $alpha_2$ adrenoceptors and imidazoline-I receptors.

References

1. Simpson K, Jarvis B. Lisinopril: a review of its use in congestive heart failure. Drugs 2000;59(5):1149–67.
2. Massie BM, Armstrong PW, Cleland JG, Horowitz JD, Packer M, Poole-Wilson PA, Ryden L. Toleration of high doses of angiotensin-converting enzyme inhibitors in patients with chronic heart failure: results from the ATLAS trial. The Assessment of Treatment with Lisinopril and Survival. Arch Intern Med 2001;161(2):165–71.
3. Desachy A, Normand S, Francois B, Cassat C, Gastinne H, Vignon P. Choc refractaire après administration d'un inhibiteur de l'enzyme de conversion. Interêt d'un traitement par angiotensine. [Refractory shock after converting enzyme inhibitor administration. Usefulness of angiotensin II.] Presse Méd 2000;29(13):696–8.
4. Shaikh ZH, Taylor HC, Maroo PV, Llerena LA. Syndrome of inappropriate antidiuretic hormone secretion associated with lisinopril. Ann Pharmacother 2000;34(2):176–9.
5. Gershon T, Olshaker JS. Acute pancreatitis following lisinopril rechallenge. Am J Emerg Med 1998;16(5):523–4.
6. Miller LG, Tan G. Drug-induced pancreatitis (lisinopril). J Am Board Fam Pract 1999;12(2):150–3.
7. Tosun E, Oksuzoglu B, Topaloglu O. Relationship between acute pancreatitis and ACE inhibitors. Acta Cardiol 2004; 59(5):571–2.
8. Kanbay M, Sekuk H, Yilmaz U, Gur G, Boyacioglu S. Acute pancreatitis associated with combined lisinopril and atorvastatin therapy. Dig Dis 2005;23(1):92–4.
9. Atzori L, Ferreli C, Pinna AL, Aste N. "Pityriasis rosea-like" adverse reaction to lisinopril. J Eur Acad Dermatol Venereol 2004;18(6):743–5.
10. Patterson CR, Davies MG. Pemphigus foliaceus: an adverse reaction to lisinopril. J Dermatol Treat 2004;15(1):60–2.
11. Kranke B, Mayr-Kanhauser S. Cold urticaria and angiotensin converting enzyme inhibitor. Acta Derm Venereol 2002;82(2):149–50.
12. Horiuchi Y, Matsuda M. Eruptions induced by the ACE inhibitor, lisinopril. J Dermatol 1999;26(2):128–30.
13. Fogari R, Zoppi A, Corradi L, Mugellini A, Poletti L, Lusardi P. Sexual function in hypertensive males treated with lisinopril or atenolol: a cross-over study. Am J Hypertens 1998;11(10):1244–7.
14. Filler G, Wong H, Condella AS, Charbonneau C, Sinclair B, Kovesi T, Hutchison J. Early dialysis in a neonate with intrauterine lisinopril exposure. Arch Dis Child Fetal Neonatal Ed 2003;88:F154–6.
15. Abraham G, Grunberg B, Gratz S. Possible interaction of clozapine and lisinopril. Am J Psychiatry 2001;158(6):969.
16. Meyer JM, Dollarhide A, Tuan IL. Lithium toxicity after switch from fosinopril to lisinopril. Int Clin Psychopharmacol 2005;20(2):115–8.
17. Johnson TR, Tobias JD. Hypotension following the initiation of tizanidine in a patient treated with an angiotensin converting enzyme inhibitor for chronic hypertension. J Child Neurol 2000;15(12):818–9.
18. Kao CD, Chang JB, Chen JT, Wu ZA, Shan DE, Liao KK. Hypotension due to interaction between lisinopril and tizanidine. Ann Pharmacother 2004;38(11):1840–3.

Losartan

See also Angiotensin II receptor antagonists

General Information

Losartan is a non-peptide, selective angiotensin type 1 (AT_1) receptor antagonist. Because it has been suggested that there is a local renin–angiotensin system in the eye, losartan may be useful in treating glaucoma. In a small, well-designed, placebo-controlled, crossover study in four groups of subjects—controls, hypertensive patients with normal intraocular pressure, and patients with primary open-angle glaucoma with and without hypertension—a single oral dose of losartan 50 mg produced a drop in intraocular pressure in all subjects within 2–6 hours after drug intake, proportional to baseline eye pressure (1). Blood pressure fell only in the hypertensive subjects. Thus, the fall in intraocular pressure was independent of the systemic effect on blood pressure. Beyond the potential for using losartan as a therapy for glaucoma, ophthalmologists should take into account concomitant therapy with losartan (and possibly other angiotensin receptor antagonists) when measuring intraocular pressure.

Using the method of prescription event monitoring (PEM), the incidence densities of adverse effects per 1000 patient-months of exposure have been measured in 14 522 patients (2). Most were hypertensive (63%). During treatment months 2–6, the commonest adverse effects were cough (17%), malaise and lassitude (15%), dizziness (15%), headache or migraine (11%), nausea and vomiting (6.0%), rash (5.0%), dyspnea (4.9%), edema (4.8%), dyspepsia (4.2%), and diarrhea (3.7%). These were also the most common reasons for drug withdrawal. The seemingly high incidence of cough may have been due to carry over from previous use of ACE inhibitors. In 1378 patients with mild to moderate hypertension valsartan 80 mg/day has been compared with losartan 50 mg/day and placebo for 8 weeks (3). This study confirmed the excellent tolerance profile, not different from placebo, of both angiotensin II receptor antagonists.

An extensive review of the use of losartan, with special focus on elderly patients, has included an update on the tolerability profile, mainly in clinical trials, but with no significant new information (4). The clinical pharmacokinetics of losartan have also been reviewed (5).

Observational studies

Losartan 2.5–100 mg has been studied in 175 hypertensive children aged 6–16 years (6). There were non-severe adverse events in 14; *headache* was the most common. In a comparison of losartan and nebivolol, headache was a common symptom in patients taking losartan; there were no serious adverse events (7).

Organs and Systems

Nervous system

An isolated case of migraine occurring after a single dose of losartan was reported (8). The temporal association during challenge and rechallenge, although in unblinded conditions, argues for a possible causal relation.

Sensory systems

There have been reports of taste disturbance in patients taking ACE inhibitors, tentatively attributed to chelation of metal ions. However, no mechanism for taste loss due to angiotensin-II receptor blockers has been proposed. Reversible loss of taste discrimination has been reported with losartan (9). In two cases dysgeusia occurred some time after switching from an ACE inhibitor to losartan (1 week in one case and 3 months in the other). In both cases the dysgeusia disappeared after withdrawal of losartan (timing not reported in one case, 1 week after in the other) (10). The authors referred to previously reported cases, one with losartan (11) and one with valsartan (12), and to a personal communication from the manufacturer of 12 other cases within a large safety monitoring program.

Electrolyte balance

Severe hyperkalemia (8.4 mmol/l), which required hemodialysis, occurred in an 84-year-old woman taking losartan 50 mg/day and spironolactone 25 mg/day for hypertension and severe mitral insufficiency (13).

Metal metabolism

Zinc, magnesium, and nitric oxide have been studied in a prospective observational study in patients with hypertension taking losartan alone and in combination with hydrochlorothiazide (14). Losartan caused zinc deficiency with increased urinary zinc excretion, an effect that was exacerbated by co-administration of hydrochlorothiazide. There were no effects on magnesium or nitric oxide. Zinc deficiency can aggravate hypertension and is also reported to be responsible for some adverse drug effects, such as dysgeusia.

Hematologic

Losartan has been used after renal transplantation to control both blood pressure and post-transplant erythrocytosis. The interaction between the renin–angiotensin system and erythropoietic mechanisms is complex, and after transplantation ACE inhibition can cause anemia without erythrocytosis. Anemia in hemodialysis (15) and mainly in renal transplant patients has been reported (16–18). In a prospective controlled study renal transplant patients taking losartan had a significant fall in hemoglobin concentration compared with controls (19). The suggested mechanism is by an action on erythropoietin, similar to that of ACE inhibitors, which produce the same adverse effect. However, in a small uncontrolled but prospective study, losartan given for 3 months to 15 patients on chronic hemodialysis with anemia, neither altered plasma erythropoietin concentrations nor aggravated the anemia (20). In those taking losartan there was no need for higher doses of co-administered r-Hu Epo in order to correct anemia, in contrast to controls.

Liver

Hepatic injury due to losartan is a recognized rare adverse effect (21).

- A 52-year-old woman developed jaundice, right upper abdominal discomfort, and weakness. She had had a similar problem previously, after taking losartan 50 mg/day for 5 months for hypertension. At that time she was jaundiced with increased serum transaminase activities. The losartan was withdrawn and during the following 3 weeks she improved. Losartan was then reintroduced, and 2 weeks later her symptoms recurred. She was jaundiced, with raised transaminase, alkaline phosphatase, and lactic dehydrogenase activities; tests for viral hepatitis were negative. A liver biopsy showed chronic hepatitis with moderate to severe inflammation and mild fibrosis. Losartan was withdrawn and the laboratory tests returned to normal over a period of 4 months. She was healthy with normal laboratory tests 2 years later.

The authors made a strong case for a causal relation, based on the absence of other causes, the return to normality after withdrawal of losartan, and re-occurrence on rechallenge. A literature search revealed four previous reports of losartan-induced liver damage.

Pancreas

Pancreatitis can occur with losartan (22).

- A 42-year-old woman with insulin-dependent diabetes and chronic renal insufficiency developed pancreatitis 5 days after taking enalapril 2.5 mg and recovered after drug withdrawal. The pancreatitis relapsed one week after her general practitioner had prescribed losartan 50 mg/day, with full recovery after withdrawal. A rechallenge test with losartan was fully positive.

The case was well documented and causality seems to have been very convincingly established. No mechanistic speculation was given by the authors.

Urinary tract

Renal insufficiency has been reported with losartan (23).

- A 69-year-old man with hypertension and heart failure took losartan 25 mg/day, increasing to 50 mg/day after 2 weeks. He also took spironolactone 50 mg/day, furosemide 40 mg/day, digoxin 0.25 mg/day, acenocoumarol, and allopurinol. Two weeks later he developed acute renal insufficiency with a plasma creatinine concentration of 725 μmol/l (previously 115 μmol/l). Within 24 hours after losartan withdrawal (it was not stated whether spironolactone was also stopped) and hemodialysis, he recovered renal function (plasma creatinine 124 mg/l). He was later found to have bilateral renal artery stenosis, which is a contraindication to angiotensin II receptor antagonists.

The authors did not discuss the possible aggravating role of spironolactone.

- Anuric renal insufficiency occurred with losartan in a 70-year-old man with a solitary kidney (24). An interesting feature was that the patient's blood pressure could be controlled with captopril without the development of anuria.

This is in line with a study in dogs, in which there was a greater deterioration in glomerular filtration with angiotensin II receptor inhibition than with ACE inhibition, for the same effect on blood pressure (25).

Acute renal insufficiency has been attributed to losartan in a patient with bilateral renal artery stenosis, reinforcing the fact that the same caution is needed with angiotensin II receptor antagonists as with ACE inhibitors in such patients (26).

Skin

Two cases of atypical cutaneous lymphoid hyperplasia have been reported (27). Both occurred after 2 weeks of treatment with losartan 50 mg/day.

Immunologic

Angioedema has been reported with angiotensin II receptor antagonists (28,29).

- An African-American developed swelling of the lips and shortness of breath within 24 hours of starting losartan. However, similar reactions had been noted before, while the patient was taking captopril. It was therefore unclear whether the reaction was related to losartan or was a carry-over reaction to captopril.
- A 45-year-old white man, who had taken losartan, hydrochlorothiazide, allopurinol, and colchicine for 9 months, developed facial urticaria, eyelid swelling, shortness of breath, and upper chest tightness, which resolved quickly with famotidine, methylprednisolone, and adrenaline. He had a recurrence 7 hours later, not having taken another dose of losartan. The patient had no history of allergy and was well after losartan withdrawal.

The late onset and recurrence of symptoms after initial resolution are the unique original features of this case report. Patients with such reactions should be kept under observation after the resolution of initial symptoms.

Second-Generation Effects

Teratogenicity

In the study of 14 522 patients mentioned in the introduction to this monograph (2) three mothers were exposed to losartan during the first trimester of pregnancy. One pregnancy terminated in a spontaneous abortion. The two others were both complicated by hypertension. One baby died with extreme prematurity and growth retardation, and the second survived after having been born prematurely. A woman who took losartan from 17 to 31 weeks of gestation developed oligohydramnios and delivered a stillborn fetus with deformities of the extremities and face (30).

Susceptibility Factors

Renal disease

Losartan was evaluated in 406 patients with end-stage renal insufficiency undergoing hemodialysis (31). Only 15 patients discontinued losartan because of adverse effects. In seven the adverse reaction was hypotension. Two patients reported a possible anaphylactoid reaction on treatment with AN69 dialysis membranes. However, nine patients with a history of previous anaphylactoid reactions on treatment with AN69 have not shown this complication with losartan and AN69.

Drug Administration

Drug overdose

In a case of accidental overdose in an 80-year-old patient, who was taking 200 mg/day for 2 months, no ill effects were observed (2).

Drug-Drug Interactions

Erythromycin

Because losartan is metabolized by CYP2C9 and CYP3A4 to an active metabolite, E3174, which has greater antihypertensive activity than the parent compound, the effects of co-administration of losartan 50 mg/day and erythromycin, a moderate inhibitor of CYP3A4, have been investigated in a well-designed study in healthy volunteers. There was no significant effect of erythromycin.

Fluconazole

The pharmacokinetic interaction of fluconazole 200 mg/day with losartan 100 mg/day has been investigated in 32 healthy subjects (32). Fluconazole significantly increased the steady-state concentration of losartan by 66% and inhibited the formation of the active metabolite EXP-3174 by 34%.

Another very similar study investigated interactions with itraconazole, an inhibitor of CYP3A4, and fluconazole, which is more specific for CYP2C9 (33). Fluconazole reduced the mean peak plasma concentrations of losartan and E3174 to 30% and 47% of their control concentrations respectively. The half-life of losartan was prolonged by 67%. Itraconazole had no significant effect. The possibility of a reduced effect of losartan when co-administered with fluconazole should be expected.

Grapefruit juice

Administration of grapefruit juice (which inhibits both cytochrome P450 and P glycoprotein) to healthy volunteers resulted in an increased serum concentration ratio of losartan to its active metabolite E3174 (34). As both losartan and its metabolite contribute to the therapeutic effects, the absence of pharmacodynamic measurements in this study obviated conclusions about the clinical implications of this interaction.

Indometacin

Indometacin 50 mg bd for one week did not interfere with the antihypertensive efficacy of losartan 50–100 mg/day in patients with essential hypertension, despite the fact that indometacin caused significant increases in weight and extracellular fluid volume (35). Glomerular filtration rate remained unchanged. Indometacin did not adversely influence peripheral hemodynamics. This is in contrast to the reported effects of indometacin during ACE inhibition, leading to an increase in blood pressure. It suggests that prostaglandins in part mediate vasodilatation during ACE inhibition, a mechanism that is not shared by angiotensin II antagonists. However, the result of this study is in contrast with those of several similar interaction studies with other antihypertensive agents, in which prostaglandin inhibition for 1–4 weeks caused an increase in weight combined with an increase in blood pressure. If indometacin had been given with losartan for more than 1 week, it may have produced a similar increase in blood pressure. Thus, the results of this study should not be taken as an argument against the recommendation that non-steroidal anti-inflammatory drugs should not be combined with angiotensin II antagonists, a recommendation that should be re-emphasized.

Lithium

An interaction of losartan with lithium has been reported (36).

- A 77-year-old woman who had been taken lithium carbonate 625 mg/day with a stable lithium concentration of around 0.6 mmol/l started to take losartan for hypertension. Within 4 weeks she developed ataxia, dysarthria, and confusion, and her serum lithium concentration was 2.0 mmol/l. Her symptoms resolved on withdrawal of losartan.

The authors proposed that this effect had occurred by reduced aldosterone secretion, an effect which is greater with ACE inhibitors than with angiotensin II receptor antagonists. In healthy volunteers losartan did not alter the fractional secretion of lithium (37), so presumably this patient had some susceptibility that caused the interaction.

Rifampicin

A study in 10 healthy volunteers has suggested that rifampicin interacts with losartan (38). The AUCs of both losartan and its active metabolite were markedly reduced (about 30 and 40% respectively) when rifampicin 300 mg bd was given for 1 week.

Because losartan is metabolized by CYP2C9 and CYP3A4 to an active metabolite, E3174, which has greater antihypertensive activity than the parent compound, the effects of co-administration of losartan 50 mg/day and the potent enzyme inducer rifampicin have been investigated in a well-designed study in healthy volunteers. Rifampicin produced 35 and 40% reductions in the AUCs of losartan and E3174 respectively. Losartan oral clearance increased by 44% and the half-lives of both compounds were shortened by 50% (39). Given the magnitude of the effect, this interaction is likely to be of clinical relevance.

References

1. Costagliola C, Verolino M, De Rosa ML, Iaccarino G, Ciancaglini M, Mastropasqua L. Effect of oral losartan potassium administration on intraocular pressure in normotensive and glaucomatous human subjects. Exp Eye Res 2000;71(2):167–71.
2. Mann R, Mackay F, Pearce G, Freemantle S, Wilton L. Losartan: a study of pharmacovigilance data on 14 522 patients J Hum Hypertens 1999;13(8):551–7.
3. Hedner T, Oparil S, Rasmussen K, Rapelli A, Gatlin M, Kobi P, Sullivan J, Oddou-Stock P. A comparison of the angiotensin II antagonists valsartan and losartan in the treatment of essential hypertension. Am J Hypertens 1999;12(4 Pt 1):414–7.
4. Simpson KL, McClellan KJ. Losartan: a review of its use, with special focus on elderly patients. Drugs Aging 2000; 16(3):227–50.

5. Sica DA, Gehr TW, Ghosh S. Clinical pharmacokinetics of losartan. Clin Pharmacokinet 2005;44(8):797–814.
6. Shahinfar S, Cano F, Soffer BA, Ahmed T, Santoro EP, Zhang Z, Gleim G, Miller K, Vogt B, Blumer J, Briazgounov I. A double-blind, dose-response study of losartan in hypertensive children. Am J Hypertens 2005;18(2 Pt 1):183–90 [erratum 2006;19(6):658].
7. Van Bortel LM, Bulpitt CJ, Fici F. Quality of life and antihypertensive effect with nebivolol and losartan. Am J Hypertens 2005;18(8):1060–6.
8. Ahmad S. Losartan and severe migraine. JAMA 1995;274(16):1266–7.
9. Ohkoshi N, Shoji S. Reversible ageusia induced by losartan: a case report. Eur J Neurol 2002;9(3):315.
10. Heeringa M, van Puijenbroek EP. Reversible dysgeusia attributed to losartan. Ann Intern Med 1998;129(1):72.
11. Schlienger RG, Saxer M, Haefeli WE. Reversible ageusia associated with losartan. Lancet 1996;347(8999):471–2.
12. Stroeder D, Zeissing I, Heath R, Federlin K. Angiotensin-II-antagonist CGP 48933 (valsartan). Results of a double-blind, placebo-controlled multicenter study. Nieren Hochdruckkr 1994;23:217–20.
13. Kauffmann R, Orozco R, Venegas JC. Hiperkalemia grave asociada a drogas que actúan sobre el sistema renina, angiotensina, aldosterona: un problema que requiere atención. Caso clínico. Rev Med Chil 2005;133(8):947–52.
14. Koren-Michowitz M, Dishy V, Zaidenstein R, Yona O, Berman S, Weissgarten J, Golik A. The effect of losartan and losartan/hydrochlorothiazide fixed-combination on magnesium, zinc, and nitric oxide metabolism in hypertensive patients: a prospective open-label study. Am J Hypertens 2005;18(3):358–63.
15. Schwarzbeck A, Wittenmeier KW, Hallfritzsch U. Anaemia in dialysis patients as a side-effect of sartanes. Lancet 1998;352(9124):286.
16. Ducloux D, Saint-Hillier Y, Chalopin JM. Effect of losartan on haemoglobin concentration in renal transplant recipients—a retrospective analysis. Nephrol Dial Transplant 1997;12(12):2683–6.
17. Brantley RP, Mrug M, Barker CV. Blockade of AT_1 receptors lowers hematocrit in post-transplant erythrocytosis. J Am Soc Nephrol 1996;7:1939.
18. Horn S, Holzer H, Horina J. Losartan and renal transplantation. Lancet 1998;351(9096):111.
19. Ersoy A, Kahvecioglu S, Ersoy C, Cift A, Dilek K. Anemia due to losartan in hypertensive renal transplant recipients without posttransplant erythrocytosis. Transplant Proc 2005;37(5):2148–50.
20. Lang SM, Schiffl H. Losartan and anaemia of end-stage renal disease. Lancet 1998;352(9141):1708.
21. Tabak F, Mert A, Ozaras R, Biyikli M, Ozturk R, Ozbay G, Senturk H, Aktuglu Y. Losartan-induced hepatic injury. J Clin Gastroenterol 2002;34(5):585–6.
22. Birck R, Keim V, Fiedler F, van der Woude FJ, Rohmeiss P. Pancreatitis after losartan. Lancet 1998;351(9110):1178.
23. Cuxart M, Matas M, Sans R, Garci M. Acute renal failure due to losartan. Nefrologia 1999;19:374–5.
24. Maillard JO, Descombes E, Fellay G, Regamey C. Repeated transient anuria following losartan administration in a patient with a solitary kidney. Ren Fail 2001;23(1):143–7.
25. Brooks DP, DePalma PD, Ruffolo RR Jr. Effect of captopril and the nonpeptide angiotensin II antagonists, SK&F 108566 and EXP3174, on renal function in dogs with a renal artery stenosis. J Pharmacol Exp Ther 1992;263(2):422–7.
26. Wargo KA, Chong K, Chan CY. Acute renal failure secondary to angiotensin II receptor blockade in a patient with bilateral renal artery stenosis. Pharmacotherapy 2003; 23:1199–204.
27. Viraben R, Lamant L, Brousset P. Losartan-associated atypical cutaneous lymphoid hyperplasia. Lancet 1997; 350(9088):1366.
28. Cha YJ, Pearson VE. Angioedema due to losartan. Ann Pharmacother 1999;33(9):936–8.
29. Rivera JO. Losartan-induced angioedema. Ann Pharmacother 1999;33(9):933–5.
30. Saji H, Yamanaka M, Hagiwara A, Ijiri R. Losartan and fetal toxic effects. Lancet 2001;357(9253):363.
31. Saracho R, Martin-Malo A, Martinez I, Aljama P, Montenegro J. Evaluation of the Losartan in Hemodialysis (ELHE) Study. Kidney Int Suppl 1998;68:S125–9.
32. Kazierad DJ, Martin DE, Blum RA, Tenero DM, Ilson B, Boike SC, Etheredge R, Jorkasky DK. Effect of fluconazole on the pharmacokinetics of eprosartan and losartan in healthy male volunteers. Clin Pharmacol Ther 1997; 62(4):417–25.
33. Kaukonen KM, Olkkola KT, Neuvonen PJ. Fluconazole but not itraconazole decreases the metabolism of losartan to E-3174. Eur J Clin Pharmacol 1998;53(6):445–9.
34. Zaidenstein R, Soback S, Gips M, Avni B, Dishi V, Weissgarten Y, Golik A, Scapa E. Effect of grapefruit juice on the pharmacokinetics of losartan and its active metabolite E3174 in healthy volunteers. Ther Drug Monit 2001;23(4):369–73.
35. Olsen ME, Thomsen T, Hassager C, Ibsen H, Dige-Petersen H. Hemodynamic and renal effects of indomethacin in losartan-treated hypertensive individuals. Am J Hypertens 1999;12(2 Pt 1):209–16.
36. Blanche P, Raynaud E, Kerob D, Galezowski N. Lithium intoxication in an elderly patient after combined treatment with losartan. Eur J Clin Pharmacol 1997;52(6):501.
37. Burnier M, Rutschmann B, Nussberger J, Versaggi J, Shahinfar S, Waeber B, Brunner HR. Salt-dependent renal effects of an angiotensin II antagonist in healthy subjects. Hypertension 1993;22(3):339–47.
38. Strayhorn VA, Baciewicz AM, Self TH. Update on rifampin drug interactions, III. Arch Intern Med 1997;157(21):2453–8.
39. Williamson KM, Patterson JH, McQueen RH, Adams KF Jr, Pieper JA. Effects of erythromycin or rifampin on losartan pharmacokinetics in healthy volunteers. Clin Pharmacol Ther 1998;63(3):316–23.

Manidipine

See also Calcium channel blockers

General Information

Manidipine is a dihydropyridine calcium channel blocker that can be given once a day for hypertension. In a comparison of manidipine 10 mg/day and amlodipine 5 mg/day in a multicenter, randomized, double-blind study in 530 patients with mild-to-moderate hypertension, the two drugs had comparable antihypertensive effects, but manidipine was associated with a significantly lower incidence of ankle edema (1). Nevertheless, adverse events caused withdrawal from treatment in a similar number of patients, 23 with manidipine and 26 with amlodipine.

Reference

1. Zanchetti A, Omboni S, La Commare P, De Cesaris R, Palatini P. Efficacy, tolerability, and impact on quality of life of long-term treatment with manidipine or amlodipine in patients with essential hypertension. J Cardiovasc Pharmacol 2001;38(4):642–50.

Methyldopa

General Information

The presynaptic alpha-adrenoceptor agonists, particularly methyldopa and clonidine, are agonists at presynaptic alpha$_2$-adrenoceptors. Guanabenz, guanfacine, and tiamenidine appear to be qualitatively similar to clonidine; clear evidence of quantitative differences remains to be confirmed.

The mechanism of action of these drugs depends on reducing sympathetic nervous outflow from the nervous system by interference with regulatory neurotransmitter systems in the brainstem. Because none of the drugs was selective or specific for circulatory control systems, the hypotensive effects were invariably accompanied by other nervous system effects. However, the identification of imidazoline receptors has suggested that it may be possible to develop drugs, such as moxonidine and relminidine, with greater selectivity for circulatory control mechanisms and a reduced likelihood of unwanted nervous system depressant effects (1).

Organs and Systems

Cardiovascular

Although postural hypotension is often quoted as a complication of treatment with methyldopa, there is no convincing evidence that this unwanted effect is more likely with methyldopa than with other commonly used antihypertensive drugs.

Complete heart block, which resolved after the withdrawal of methyldopa but recurred on rechallenge, has been reported in two elderly patients and attributed to the central sympatholytic effect of methyldopa (2).

Nervous system

Treatment with centrally acting agents is characterized by a relatively high incidence (up to 60% in some studies) of nervous system depressant effects (dizziness, drowsiness, tiredness, dry mouth, headache, depression), particularly during the initial period of treatment or after dosage increments. Some tolerance develops during long-term treatment, but the insidious presence of these depressant effects may not be fully apparent until the patient is given alternative treatment.

Methyldopa should probably be avoided in patients with parkinsonism, because in theory the production of alpha-methyldopamine can worsen the disease, and there have indeed been reports of exacerbation (3).

Endocrine

Acutely methyldopa promotes the release of growth hormone and prolactin, but the long-term significance of this is unclear. However, there have been reports of hyperprolactinemia, leading to amenorrhea and galactorrhea (4).

Hematologic

Coombs' positivity (in about 20% of cases) and Coombs' positive hemolytic anemia (in about 2%) are well-recognized complications of methyldopa.

- Anemia, reticulocytosis, and a positive Coombs' test were seen in a 64-year-old woman taking methyldopa 1000 mg bd; these problems resolved after withdrawal (5).

Gastrointestinal

Dry mouth occurred twice as often with methyldopa (41%) as with atenolol (21%) in a crossover comparison in 69 hypertensive patients (6).

There is an association between methyldopa and diarrhea.

- An 84-year-old black woman over a period of 7 years was free of diarrhea during periods when methyldopa was stopped but suffered diarrhea as soon as methyldopa was re-introduced (7).
- Similar recurrences on repeated exposure occurred in a 71-year-old man who took methyldopa 250 mg bd (8).

Liver

Hepatotoxicity due to methyldopa is most often an acute hepatitis, which resolves when the drug is withdrawn. It can start at any time after the start of therapy, but it usually occurs within the first few weeks and ranges in severity from a minor disturbance of liver function tests to acute hepatic necrosis.

- Fatal toxic hepatitis occurred in a 22-year-old woman treated with methyldopa in late pregnancy (9).

Sometimes, a more chronic picture develops after treatment for 2 or 3 years, with features of chronic active hepatitis (10).

Immunologic

Lupus-like syndrome has been attributed to methyldopa, causing hemolytic anemia, arthritis, photosensitivity, and high titers of antinuclear antibody (1:256) and of IgG antibodies to class I histones in a 55-year-old man who took methyldopa 250 mg bd for 13 months; the syndrome resolved spontaneously when methyldopa was withdrawn (11).

Long-Term Effects

Drug withdrawal

A withdrawal syndrome with rebound hypertension has been reported with methyldopa (12). Although it is similar to that associated with clonidine, it is less well defined, less severe, and less frequent.

- Withdrawal of methyldopa has been associated with the development of an acute manic syndrome in a 62-year-old man who had had methyldopa withdrawn 4 weeks before; there had been no previous history of psychiatric disorder (13).

Drug–Drug Interactions

Haloperidol

Dementia occurred when methyldopa was combined with haloperidol (14).

Lithium

There have been occasional reports of neurotoxic symptoms when methyldopa was combined with lithium, both with and without an increase in serum lithium concentration (15).

Mianserin

Mianserin has alpha-adrenoceptor activity and so might interact with methyldopa (16). In 11 patients with essential hypertension, the addition of mianserin 60 mg/day (in divided doses) for 2 weeks did not reduce the hypotensive effect of methyldopa. In patients treated with methyldopa, there were additive hypotensive effects after the first dose of mianserin, but these were not significant after 1 or 2 weeks of combined treatment. The results of this study appear to have justified the authors' conclusion that adding mianserin to treatment with methyldopa will not result in loss of blood pressure control.

Mianserin lacks potential for peripheral adrenergic interactions, but since it has α-adrenoceptor activity, it might interact with the centrally acting α-adrenoceptor agonists clonidine and methyldopa. In healthy volunteers, pretreatment with mianserin 60 mg/day for 3 days did not modify the hypotensive effects of a single 300 mg dose of clonidine, and in 11 patients with essential hypertension, the addition of mianserin 60 mg/day (in divided doses) for 2 weeks did not reduce the hypotensive effects of clonidine or methyldopa (17). In patients taking methyldopa, there were additive hypotensive effects after the first dose of mianserin, but these were not significant after 1 or 2 weeks of combined treatment. The results of this study appear to have justified the authors' conclusion that combining mianserin with centrally-acting hypotensive agents will not result in loss of blood pressure control.

Naproxen

Naproxen reduces the actions of methyldopa (SEDA- 8, 107).

References

1. Webster J, Koch HF. Aspects of tolerability of centrally acting antihypertensive drugs. J Cardiovasc Pharmacol 1996;27(Suppl 3):S49–54.
2. Rosen B, Ovsyshcher IA, Zimlichman R. Complete atrioventricular block induced by methyldopa. Pacing Clin Electrophysiol 1988;11(11 Pt 1):1555–8.
3. Rosenblum AM, Montgomery EB. Exacerbation of parkinsonism by methyldopa. JAMA 1980;244(24):2727–8.
4. Arze RS, Ramos JM, Rashid HU, Kerr DN. Amenorrhoea, galactorrhoea, and hyperprolactinaemia induced by methyldopa. BMJ (Clin Res Ed) 1981;283(6285):194.
5. Egbert D, Hendricksen DK. Congestive heart failure and respiratory arrest secondary to methyldopa-induced hemolytic anemia. Ann Emerg Med 1988;17(5):526–8.
6. Glazer N, Goldstein RJ, Lief PD. A double blind, randomized, crossover study of adverse experiences among hypertensive patients treated with atenolol and methyldopa. Curr Ther Res 1989;45:782.
7. Gloth FM 3rd, Busby MJ. Methyldopa-induced diarrhea: a case of iatrogenic diarrhea leading to request for nursing home placement. Am J Med 1989;87(4):480–1.
8. Troster M, Sullivan SN. Acute colitis due to methyldopa. Can J Gastroenterol 1989;3:182.
9. Picaud A, Walter P, de Preville G, Nicolas P. Hépatite toxique mortelle au cours de la grossesse. [Fatal toxic hepatitis in pregnancy. A discussion of the role of methyldopa.] J Gynecol Obstet Biol Reprod (Paris) 1990;19(2):192–6.
10. Arranto AJ, Sotaniemi EA. Morphologic alterations in patients with alpha-methyldopa-induced liver damage after short- and long-term exposure. Scand J Gastroenterol 1981;16(7):853–63.
11. Nordstrom DM, West SG, Rubin RL. Methyldopa-induced systemic lupus erythematosus. Arthritis Rheum 1989; 32(2):205–8.
12. Burden AC, Alexander CP. Rebound hypertension after acute methyldopa withdrawal. BMJ 1976;1(6017):1056–7.
13. Labbate LA, Holzgang AJ. Manic syndrome after discontinuation of methyldopa. Am J Psychiatry 1989;146(8):1075–6.
14. Thornton WE. Dementia induced by methyldopa with haloperidol. N Engl J Med 1976;294(22):1222.
15. O'Regan JB. Letter: Adverse interaction of lithium carbonate and methyldopa. Can Med Assoc J 1976;115(5):385–6.
16. Elliott HL, Whiting B, Reid JL. Assessment of the interaction between mianserin and centrally-acting antihypertensive drugs. Br J Clin Pharmacol 1983;15(Suppl 2):S323–8.
17. Elliott HL, Whiting B, Reid JL. Assessment of the interaction between mianserin and centrally-acting antihypertensive drugs. Br J Clin Pharmacol 1983;15(Suppl 2):323S–8S.

Metoprolol

See also Beta-adrenoceptor antagonists

General Information

Metoprolol is a cardioselective beta-adrenoceptor antagonist with some membrane-stabilizing activity. It is metabolized by polymorphic hydroxylation (1).

Organs and Systems

Psychiatric

Delirium has been attributed to metoprolol (2).

- A 53-year-old man with a history of alcohol abuse suffered personality changes and multiple hallucinogenic episodes for 2 years, attributed to dementia probably caused by alcohol abuse. He also taking acenocoumerol, atorvastatin, quinapril, and metoprolol 50 mg bd for atrial flutter. Metoprolol, which had been started 2 years before, was withdrawn. Within 24 hours the delirium had disappeared completely and the liver enzymes fell.

Liver

Metoprolol was associated with hepatitis in a patient who was a poor metabolizer (3).

Musculoskeletal

A metoprolol-induced polymyalgia-like syndrome (4) has been described.

Susceptibility Factors

Genetic factors

The metabolism of metoprolol varies six-fold higher between poor metabolizers of debrisoquine and extensive metabolizers (1).

Drug Administration

Drug formulations

A modified-release formulation of metoprolol has been associated with more skin reactions, probably due to the succinate component, instead of the tartrate component used in the old fast-acting formulation (SEDA-16, 195).

References

1. Silas JH, McGourty JC, Lennard MS, Tucker GT, Woods HF. Polymorphic metabolism of metoprolol: clinical studies. Eur J Clin Pharmacol 1985;28(Suppl):85–8.
2. van der Vleuten PA, van den Brink E, Schoonderwoerd BA, van den Berg F, Tio RA, Zijlstra F. Delirant beeld, toegeschreven aan het gebruik van metoprolol. Ned Tijdschr Geneeskd 2005;149(39):2183–6.
3. Larrey D, Henrion J, Heller F, Babany G, Degott C, Pessayre D, Benhamou JP. Metoprolol-induced hepatitis: rechallenge and drug oxidation phenotyping. Ann Intern Med 1988;108(1):67–8.
4. Snyder S. Metoprolol-induced polymyalgia-like syndrome. Ann Intern Med 1991;114(1):96–7.

Mibefradil

See also Calcium channel blockers

General Information

Mibefradil is a member of a class of calcium antagonists that specifically block the T-type calcium channel. It was marketed as an antihypertensive and antianginal drug with an adverse effects profile similar to that of placebo (1) and a favorable pharmacokinetic profile, allowing once-a-day dosage (2).

Owing to its attractive pharmacological profile, namely an anti-ischemic action with little or no negative inotropism, mibefradil was tested in a large-scale mortality trial in heart failure (MACH-I, Mortality Assessment in Congestive Heart Failure Trial). However, early in June 1998, only a few months after its launch, Roche Laboratories withdrew mibefradil from the market, after several dangerous interactions with at least 25 drugs had been reported (3,4) associated with rhabdomyolysis and life-threatening cardiac dysrhythmias; these problems were detected by postmarketing surveillance (4). The FDA stated that: "since [mibefradil] has not been shown to offer special benefits (such as treating patients who do not respond to other anti-hypertensive and antianginal drugs), the drug's problems are viewed as an unreasonable risk to consumers."

In the MACH-1 (Mortality Assessment in Congestive Heart Failure) trial, 2590 patients with moderate to severe congestive heart failure were randomized to mibefradil or placebo (5). Mibefradil given for a maximum of 3 years did not affect mortality or morbidity. However, a subgroup analysis by concomitant drugs showed that digoxin, class I antidysrhythmic drugs, amiodarone, and other drugs associated with torsade de pointes increased the risk of death with mibefradil. These worrisome drug–drug interactions were consistent with the results of postmarketing surveillance, which prompted withdrawal of mibefradil from the market by its manufacturers in June 1998, even before complete results of the trial became available.

The safety and efficacy of mibefradil in association with beta-blockers was assessed in 205 patients with chronic stable angina, randomized to placebo or mibefradil 25 or 50 mg/day for 2 weeks (6). Besides an improvement in angina with mibefradil, it dose-dependently reduced heart rate and increased the PR interval. One patient taking mibefradil had an escape junctional rhythm 26 hours after the last dose of 50 mg. The nodal rhythm disappeared on withdrawal of mibefradil, but based on the overall results it was concluded that mibefradil was safe and effective when given for a short time with beta-blockers.

In a Prescription-Event Monitoring study in 3085 patients, mean age 65 years, the commonest reported adverse events and reasons for stopping were malaise/lassitude, dizziness, edema, and headache (7).

Organs and Systems

Cardiovascular

Mibefradil causes slight bradycardia associated with its hypotensive effect. In a Prescription-Event Monitoring study in 3085 patients, mean age 65 years, there were seven reports of serious bradycardia/collapse thought to be possible adverse drug reactions to mibefradil (7). All were in patients over 65 years and six were thought to have resulted from a drug interaction. In all, there were 11 possible drug interactions. Nine (eight reports of bradycardia and one of syncope) involved beta-blockers. Collapse and severe bradycardia occurred in a patient who had started to take a dihydropyridine calcium channel blocker within 24 hours of stopping mibefradil. Palpitation and dyspnea occurred in a patient taking digoxin and sotalol. None of the 53 deaths during the study was attributed to mibefradil.

Abrupt switch of therapy from mibefradil to other dihydropyridine calcium antagonist was reported to cause shock, fatal in one case, in four patients also taking beta-blockers (8).

Drug–Drug Interactions

General

Mibefradil inhibits CYP3A4 (2). Other drugs that are metabolized by this pathway accumulate as a result. Drugs that were commonly affected included amiodarone, astemizole, ciclosporin, cisapride, erythromycin, imipramine, lovastatin, propafenone, quinidine, simvastatin (9), tacrolimus (10), tamoxifen, terfenadine, thioridazine, and drugs that impair sinoatrial node function (for example beta-blockers) (6).

- Severe rhabdomyolysis occurred in an 83-year-old woman who developed progressive immobilizing myopathy, low back pain, and oliguria; she was taking simvastatin and mibefradil (9). The symptoms disappeared completely after 4 weeks of withdrawal.

Mibefradil was probably responsible for raising plasma concentrations of simvastatin to toxic concentrations.

Beta-blockers

In a Prescription-Event Monitoring study in 3085 patients, mean age 65 years, there were 11 possible drug interactions. Nine involved beta-blockers, with eight reports of bradycardia and one of syncope. One patient developed palpitation and dyspnea while also taking digoxin and sotalol.

Calcium channel blockers

In a Prescription-Event Monitoring study in 3085 patients, mean age 65 years, one patient developed collapse and severe bradycardia after starting to take a dihydropyridine calcium channel blocker within 24 hours of stopping mibefradil (7).

Cardiac glycosides

In 40 healthy subjects mibefradil 50 or 100 mg/day for 6 days had no significant effects on the steady-state pharmacokinetics of digoxin, apart from a very small increase in the C_{max} (11).

Tacrolimus

Mibefradil, a potent inhibitor of CYP3A, increased tacrolimus blood concentrations dramatically (12).

References

1. Braun S, van der Wall EE, Emanuelsson H, Kobrin IThe Mibefradil International Study Group. Effects of a new calcium antagonist, mibefradil (Ro 40-5967), on silent ischemia in patients with stable chronic angina pectoris: a multicenter placebo-controlled study. J Am Coll Cardiol 1996;27(2):317–22.
2. Billups SJ, Carter BL. Mibefradil: a new class of calcium-channel antagonists. Ann Pharmacother 1998;32(6):659–71.
3. SoRelle R. Withdrawal of Posicor from market. Circulation 1998;98(9):831–2.
4. Po AL, Zhang WY. What lessons can be learnt from withdrawal of mibefradil from the market? Lancet 1998;351(9119):1829–30.
5. Levine TB, Bernink PJ, Caspi A, Elkayam U, Geltman EM, Greenberg B, McKenna WJ, Ghali JK, Giles TD, Marmor A, Reisin LH, Ammon S, Lindberg E. Effect of mibefradil, a T-type calcium channel blocker, on morbidity and mortality in moderate to severe congestive heart failure: the MACH-1 study. Mortality Assessment in Congestive Heart Failure Trial. Circulation 2000;101(7):758–64.
6. Alpert JS, Kobrin I, DeQuattro V, Friedman R, Shepherd A, Fenster PE, Thadani U. Additional antianginal and anti-ischemic efficacy of mibefradil in patients pretreated with a beta blocker for chronic stable angina pectoris. Am J Cardiol 1997;79(8):1025–30.
7. Riley J, Wilton LV, Shakir SA. A post-marketing observational study to assess the safety of mibefradil in the community in England. Int J Clin Pharmacol Ther 2002;40(6):241–8.
8. Mullins ME, Horowitz BZ, Linden DH, Smith GW, Norton RL, Stump J. Life-threatening interaction of mibefradil and beta-blockers with dihydropyridine calcium channel blockers. JAMA 1998;280(2):157–8.
9. Schmassmann-Suhijar D, Bullingham R, Gasser R, Schmutz J, Haefeli WE. Rhabdomyolysis due to interaction of simvastatin with mibefradil. Lancet 1998;351(9120):1929–30.
10. Krahenbuhl S, Menafoglio A, Giostra E, Gallino A. Serious interaction between mibefradil and tacrolimus. Transplantation 1998;66(8):1113–5.
11. Peters J, Welker HA, Bullingham R. Pharmacokinetic and pharmacodynamic aspects of concomitant mibefradil–digoxin therapy at therapeutic doses. Eur J Drug Metab Pharmacokinet 1999;24(2):133–40.
12. Krahenbuhl S, Menafoglio A, Giostra E, Gallino A. Serious interaction between mibefradil and tacrolimus. Transplantation 1998;66(8):1113–5.

Minoxidil

General Information

Minoxidil (2,4-diamino-6-piperidinopyrimidine-3-oxide) is a potent vasodilator effective in severe hypertension irrespective of the cause. Isolated case reports have been published of hair growth in areas of male pattern baldness in patients treated with oral minoxidil, therefore topical minoxidil has been used for the treatment of alopecia areata and alopecia androgenica, with some success.

The pattern of adverse vasodilator effects with minoxidil is similar to, but more severe than, that of hydralazine and fluid retention can be troublesome.

Topical minoxidil stimulates new hair growth and arrests loss of hair in androgenic alopecia. Minoxidil is poorly absorbed through the skin (less than 4%) (1), and plasma concentrations of minoxidil are far less than 10% of the mean minoxidil concentration present 2 hours after oral ingestion of 5 mg, the lowest dose for the treatment of hypertension (2). Its adverse effects after topical application are therefore usually limited to the application site on the scalp. They include irritant contact dermatitis, allergic contact dermatitis, and exacerbation of seborrheic dermatitis. These effects are seen in 5.7% of patients who use a 5% solution and in 1.9% of those who use a 2% formulation (3).

Placebo-controlled studies

The results with a 2% solution of minoxidil are variable, and experience has suggested that higher concentrations may be more effective. Topical twice-daily minoxidil 2% and 5% have therefore been compared for 48 weeks in a double-blind, randomized, placebo-controlled, multicenter trial in 393 men (aged 18–49 years) with androgenic alopecia (4). The higher concentration of minoxidil was significantly more effective than both the lower concentration and placebo in terms of change from baseline in non-vellus hair count, patient rating of scalp coverage and treatment benefit, and investigator rating of scalp coverage; the response to treatment occurred earlier with 5% minoxidil. However, there was more pruritus and local irritation with 5% minoxidil.

Organs and Systems

Cardiovascular

In a 1-year observational study of more than 3 million subject-days of exposure to topical minoxidil solution, patients who use topical minoxidil were no more likely to have major medical events resulting in hospitalization or death than control subjects who had never been exposed to topical minoxidil. In addition, there was no difference in the rates of cardiovascular events between the two groups (5).

Minoxidil can cause pericardial effusion (6).

- A 52-year-old man with a history of chronic hypertension presented with worsening dyspnea and leg edema. He had been taking minoxidil for 10 years. The cardiac silhouette was markedly enlarged. Echocardiography and computed tomography showed a large pericardial effusion. Minoxidil was withdrawn and the effusion resolved within 1 month.

A hypertensive crisis occurred in one patient after withdrawal of topical minoxidil (SEDA-16, 158).

Respiratory

An isolated unilateral pleural effusion secondary to minoxidil has been reported in a patient undergoing peritoneal dialysis (7). It resolved on withdrawal and reappeared after the patient reintroduced the drug himself.

Endocrine

Pseudoacromegaly has been reported in a patient who had taken large oral doses of minoxidil for about 10 years (8). There have been no reports of pseudoacromegaly associated with topical minoxidil.

Skin

Local adverse effects of topical minoxidil (usually in an alcohol–propylene glycol base) include dryness, irritation, pruritus, contact allergy (SEDA-11, 139), and photocontact allergy (SEDA-11, 139). Allergic contact dermatitis has been reported in about 1–3% of patients who use topical minoxidil solution.

- A 72-year-old woman developed an allergic contact dermatitis during treatment with topical minoxidil for alopecia (9).
- A 24-year-old woman developed allergic contact dermatitis while taking minoxidil and again while taking *Serenoa repens* (saw palmetto) solution for androgenic alopecia (10).

In many cases, propylene glycol is the causative allergen, but not always.

- In a 40-year-old woman with a history of scalp desquamative dermatitis 1 week after starting 5% minoxidil solution, patch tests showed that minoxidil itself was the inciting agent (11).

Genital ulceration, probably due to inadvertent local contamination from topical minoxidil used for androgenic baldness, has been described (12).

Leukoderma has been reported in two men from India who used 2% minoxidil lotion for 2–3 months to treat baldness (13). The depigmentation was localized to the scalp. Other possibilities of leukoderma were ruled out, as was vitiligo. There was repigmentation of the leukodermic area within 3 months of minoxidil withdrawal in both cases.

Hair

A particular feature of minoxidil is excessive hair growth (14), which occurs in about 70% of patients who take oral minoxidil, usually within 2 months of the start of therapy. Severe hypertrichosis, unacceptable even to men, has complicated the otherwise successful antihypertensive treatment of six patients after renal transplantation, for which ciclosporin was also used. Since hypertrichosis has also been described with ciclosporin, there may be an additive pharmacodynamic interaction (15).

Topical minoxidil has been used to treat women with androgenic alopecia. Severe hypertrichosis of the face and the limbs occurred in three women after 2–3 months of treatment with 5% topical minoxidil (16). The hypertrichosis disappeared from the face and arms within 1–3 months of withdrawal and from the legs after 4–5 months.

A review of minoxidil topical solution clinical trials revealed that approximately 4% of women noted hypertrichosis/facial hair. Excessive hair growth was reported primarily on the face (including cheeks, chin, forehead, upper lip, sideburns and around the eyes), but also on the neck, chest, back and extremities. However, post-marketing data showed a lower occurrence (0.5%) of hypertrichosis/facial hair than in the clinical trials. A dose-related pattern of response was noted. The hypertrichotic effect of minoxidil is reversible, and does not necessarily require discontinuation of therapy (17).

Immunologic

A "polymyalgia syndrome," characterized by fatigue, anorexia, weight loss, and severe pain in the shoulders and the pelvic girdle, was attributed to minoxidil in four men (18).

In 11 patients allergic to topical minoxidil lotion, patch tests showed that four were positive to minoxidil itself (19). Propylene and butylene glycol are used as solvents for minoxidil in topical formulations. Nine of the 11 patients appeared to have positive patch tests to propylene glycol and one of the 11 reacted to its alternative butylene glycol.

- A 57-year-old man developed a pigmented contact dermatitis after using topical minoxidil 5% for 2 years (20). Patch tests were negative with the European standard series and with a textile and finishes series, but positive with minoxidil 5% on days 3 and 7. However, withdrawal of the minoxidil did not lead to improvement after 10 months.
- A 24-year-old woman with androgenic alopecia became sensitized to topical minoxidil after using minoxidil 4% with retinoic acid in a propylene glycol base (21). She subsequently also became sensitized to saw palmetto (*Serenoa repens*), a topical herbal extract commonly promoted for the treatment of hair loss.
- Allergic contact dermatitis occurred in a 54-year-old man who had used 1% minoxidil on the scalp for 8 months (22). He had positive patch tests to minoxidil in alcohol, but not to minoxidil in petrolatum, piperidine, pyrimidine, or diaminopyrimidine.

The authors of the last report suggested that the whole structure of minoxidil is required for sensitization and that propylene glycol in the formulation of minoxidil that is used therapeutically increases the penetration of minoxidil into the skin, enhancing the risk of a reaction.

Long-Term Effects

Drug withdrawal

A hypertensive crisis occurred in one patient after withdrawal of topical minoxidil (SEDA-16, 158).

Second-Generation Effects

Pregnancy

There was no indication of increased risk of adverse pregnancy outcomes in women who used topical minoxidil during their pregnancy. However, it should be noted that this study was not designed to determine whether subjects continued to use topical minoxidil solution after they became pregnant (5). In contrast to the apparent safety of minoxidil in pregnancy, a case report describes a 28-year-old pregnant woman who applied daily minoxidil 2% to her scalp because of hair loss. A routine ultrasound test at the 22nd gestational week showed significant brain, heart and vascular malformations of the fetus; pregnancy was interrupted. The most evident pathological event was formation of abnormal vessels (23).

Teratogenicity

Caudal regression syndrome is a rare anomaly, a continuum of congenital malformations ranging from isolated sacral agenesis to absence of the lumbosacral spine and major visceral anomalies. While the exact cause of this syndrome is unclear, maternal diabetes, genetic factors, teratogens, and vascular anomalies that alter blood flow have been hypothesized to play a role. A case has been attributed to maternal minoxidil exposure (24).

A fetus had an extremely hypotrophic caudal body pole, aplasia of the lower spine, and complete renal agenesis diagnosed in the second trimester by ultrasound. The mother had used minoxidil solution to prevent hair loss for 4 years before and during gestation. She had also taken co-trimoxazole during the first trimester. She was not diabetic and had no history of familial genetic disorders.

Drug Administration

Drug administration route

Despite the poor transdermal absorption of minoxidil, in 7 out of 30 non-hypertensive patients treated with 3% minoxidil solution twice daily, there was a "significant" fall in blood pressure (25).

Some patients using minoxidil solution had increased hair growth outside the area of drug application (SEDA-18, 175), which suggests a systemic effect (26).

The sudden death of a patient using topical minoxidil (27) was probably not drug-related but due to cardiovascular disease and hypertension. Nevertheless, in patients who are known to be hypertensive and who are also receiving other antihypertensive medication extra caution is warranted when topical minoxidil is used (28).

Smoking intolerance possibly related to topical minoxidil has been observed rarely (29).

Drug overdose

In some countries, an extra-strength solution of minoxidil is available over the counter for hair regrowth treatment in men. Ingestion of a large dose in a suicide attempt has been reported (30).

- A 26-year-old woman took 60 ml of minoxidil solution 5% and 1 hour later developed hypotension (75/40 mmHg) and tachycardia (130/minute). She was alert and oriented and physical examination was unremarkable. She was given a saline infusion, oral activated charcoal, and sorbitol, and had worsening hypotension (40 mmHg systolic) and bradycardia (20/minute), developed chest pain and T wave inversion on the electrocardiogram, and lost consciousness. Atropine and dopamine restored a heart rate of 120/minute, but she remained hypotensive (54/36 mmHg). Her blood ethanol concentration was 5.64 mmol/l and other toxicological screening tests were negative. The addition of phenylephrine up to a dose of 200 μg/minute restored her systolic blood pressure to 90–100 mmHg.

References

1. Franz TJ. Percutaneous absorption of minoxidil in man. Arch Dermatol 1985;121(2):203–6.
2. Vanderveen EE, Ellis CN, Kang S, Case P, Headington JT, Voorhees JJ, Swanson NA. Topical minoxidil for hair regrowth. J Am Acad Dermatol 1984;11(3):416–21.
3. Rogaine Extra Strength for Men slide lecture kit. Pharmacia & Upjohn Company, 1998:M-7909P.
4. Olsen EA, Dunlap FE, Funicella T, Koperski JA, Swinehart JM, Tschen EH, Trancik RJ. A randomized clinical trial of 5% topical minoxidil versus 2% topical minoxidil and placebo in the treatment of androgenetic alopecia in men. J Am Acad Dermatol 2002;47(3):377–85.
5. Shapiro J. Safety of topical minoxidil solution: a one-year, prospective, observational study. J Cutan Med Surg 2003;7:322–9.
6. Shirwany A, D'Cruz IA, Munir A. Very large pericardial effusion attributable to minoxidil: resolution without drainage of fluid. Echocardiography 2002;19(6):513–6.
7. Palomar R, Morales P, Sanz de Castro S, Tasis A, Rodrigo E, Pinera C, Ruiz JC, Fernandez-Fresnedo G, Martin de Francisco AL, Arias M. Pleural effusion secondary to minoxidil in a peritoneal dialysis patient. Nephrol Dial Transplant 2004;19(10):2688.
8. Nguyen K, Marks J. Pseudoacromegaly induced by the long-term use of minoxidil. J Am Acad Dermatol 2003; 48:962–5.
9. Hagemann T, Schlutter-Bohmer B, Allam JP, Bieber T, Novak N. Positive lymphocyte transformation test in a patient with allergic contact dermatitis of the scalp after short-term use of topical minoxidil solution. Contact Dermatitis 2005;53(1):53–5.
10. Sinclair R, Mallair R, Tate B. Sensitization to saw palmetto and minoxidil in separate topical extemporaneous treatments for androgenetic alopecia. Aust J Dermatol 2002; 43:311–12.
11. Carreno P, Martin E, Trabado A, Peral C, Timon S, Arbeiza FJ, Lopez RC. Allergic contact dermatitis due to minoxidil itself. Allergol Immunol Clin 2003;18:225–8.
12. Spenatto N, Alibert M, Cambon L, Viraben R. Ulc ration g nitale originale d'origine toxique. Nouv Dermatol 2004;23(3):111.
13. Malakar S, Dhar S. Leucoderma associated with the use of topical minoxidil: a report of two cases. Dermatology 2000;201(2):183–4.
14. Toriumi DM, Konior RJ, Berktold RE. Severe hypertrichosis of the external ear canal during minoxidil therapy. Arch Otolaryngol Head Neck Surg 1988;114(8):918–9.
15. Sever MS, Sonmez YE, Kocak N. Limited use of minoxidil in renal transplant recipients because of the additive side-effects of cyclosporine on hypertrichosis. Transplantation 1990;50(3):536.
16. Peluso AM, Misciali C, Vincenzi C, Tosti A. Diffuse hypertrichosis during treatment with 5% topical minoxidil. Br J Dermatol 1997;136(1):118–20.
17. Dawber R, Rundegren J. Hypertrichosis in females applying minoxidil topical solution and in normal controls. J Eur Acad Dermatol Venereol 2003;17:271–5.
18. Colamarino R, Dubost JJ, Brun P, Flori B, Tournilhac M, Eschalier A, Sauvezie B. Etats polyalgiques induits par le minoxidil topique. [Polymyalgia induced by topical minoxidil.] Ann Med Interne (Paris) 1990;141(5):425–8.
19. Friedman ES, Friedman PM, Cohen DE, Washenik K. Allergic contact dermatitis to topical minoxidil solution: etiology and treatment. J Am Acad Dermatol 2002; 46(2):309–12.
20. Trattner A, David M. Pigmented contact dermatitis from topical minoxidil 5%. Contact Dermatitis 2002; 46(4):246.
21. Sinclair RD, Mallari RS, Tate B. Sensitization to saw palmetto and minoxidil in separate topical extemporaneous treatments for androgenetic alopecia. Australas J Dermatol 2002;43(4):311–2.
22. Suzuki K, Suzuki M, Akamatsu H, Matsungaga K. Allergic contact dermatitis from minoxidil: study of the cross-reaction to minoxidil. Am J Contact Dermat 2002;13(1):45–6.
23. Smorlesi C, Caldarella A, Caramelli L, Di Lollo S, Moroni F. Topically applied minoxidil may cause fetal malformation: a case report. Birth Defects Res 2003;67:997–1001.
24. Rojansky N, Fasouliotis SJ, Ariel I, Nadjari M. Extreme caudal agenesis. Possible drug-related etiology? J Reprod Med 2002;47(3):241–5.
25. Ranchoff RE, Bergfeld WF. Topical minoxidil reduces blood pressure. J Am Acad Dermatol 1985;12(3):586–7.
26. Gonzalez M, Landa N, Gardeazabal J, Calderon MJ, Bilbao I, Diaz Perez JL. Generalized hypertrichosis after treatment with topical minoxidil. Clin Exp Dermatol 1994;19(2):157–8.
27. Baral J. Minoxidil and sudden death. J Am Acad Dermatol 1985;13(2 Pt 1):297–9.
28. Vanderveen EE. Minoxidil and sudden death: reply. J Am Acad Dermatol 1985;13(2 Pt 1):298.
29. Trattner A, Ingber A. Topical treatment with minoxidil 2% and smoking intolerance. Ann Pharmacother 1992; 26(2):198–9.
30. Farrell SE, Epstein SK. Overdose of Rogaine Extra Strength for Men topical minoxidil preparation. J Toxicol Clin Toxicol 1999;37(6):781–3.

Molsidomine

See also Antianginal drugs

General Information

Molsidomine is a vasodilator that acts as a nitric oxide donor and has been used in angina pectoris, heart failure, and after myocardial infarction.

In a multicenter, randomized, double-blind, crossover, placebo-controlled study of the efficacy of two regimens of molsidomine in 90 patients with stable angina, beneficial effects on exercise load did not seem to be reduced by 6 weeks of continuous therapy, suggesting lack of tolerance (1). The most frequently reported adverse effect of molsidomine was headache, as for all other nitric oxide donors.

Drug Administration

Drug formulations

A new form of molsidomine, a prolonged-release tablet containing 16 mg for once-a-day administration has been evaluated in 533 patients (2). Blood pressure and heart rate, electrocardiographic findings, and blood parameters were not affected by molsidomine. Drug-related adverse events occurred in respectively 4.7%, 5.4%, and 6.9% of patients allocated to placebo, molsidomine 8 mg, and molsidomine 16 mg. The difference was not statistically significant. The most commonly reported events were headache and hypotension; the latter was significantly associated with molsidomine 16 mg, although the hypotension episodes were reported as being non-serious.

References

1. Messin R, Karpov Y, Baikova N, Bruhwyler J, Monseu MJ, Guns C, Geczy J. Short- and long-term effects of molsidomine retard and molsidomine nonretard on exercise capacity and clinical status in patients with stable angina: a multicenter randomized double-blind crossover placebo-controlled trial. J Cardiovasc Pharmacol 1998;31(2):271–6.
2. Messin R, Opolski G, Fenyvesi T, Carreer-Bruhwyler F, Dubois C, Famaey JP, Géczy J. Efficacy and safety of molsidomine once-a-day in patients with stable angina pectoris. Int J Cardiol 2005;98:79–89.

Moxonidine

General Information

Moxonidine is a selective agonist at imidazoline I_1 receptors in the rostroventrolateral medulla and has very little alpha$_2$-adrenoceptor agonist activity (1). It has similar efficacy to other antihypertensive drugs. Its safety has been evaluated in a number of individual studies and in a systematic review of randomized trials (2).

Unlike clonidine, moxonidine does not appear to cause sedation or to impair psychomotor performance or cognitive function. However, possible potentiation of the effect of benzodiazepines can occur. There is no evidence of a withdrawal syndrome or rebound hypertension associated with sudden withdrawal.

There has been some controversy over the mechanisms of action of moxonidine in cardiovascular and other diseases, and excessive deaths in a trial of heart failure patients has fuelled this controversy. Its use as an antihypertensive drug is still important, and it is relatively well tolerated, sedation being the most frequent adverse effect. Its apparently favorable metabolic effects in clinical studies will be tested in the MARRIAGE study (Moxonidine And Ramipril Regarding Insulin And Glucose Evaluation), in which moxonidine and the ACE inhibitor ramipril, separately and in combination, will be studied with respect to metabolic and hemodynamic effects in patients with hypertension and impaired fasting glycemia (3).

Moxonidine has been used as add-on therapy in elderly patients with resistant hypertension (4). The results suggested that moxonidine is efficacious and well tolerated in patients who are already taking two or more antihypertensive agents.

Organs and Systems

Mouth

Dryness of the mouth occurs in about 10% of patients who take moxonidine, but it is usually mild and occurs early in therapy. The effect is dose-dependent but shows no tendency to accumulate over time. This effect is consistent with an action of mildly inhibiting salivary flow.

Liver

- Cholestatic hepatitis occurred in an 83-year-old man after 9 months of continuous therapy with moxonidine. No autoantibodies were detected and liver biopsy showed features compatible with drug-induced inflammatory intrahepatic cholestasis. The patient recovered fully and his biochemical data normalized 8 weeks after withdrawal.

In a postmarketing surveillance program none of more than 20 000 patients developed cholestatic hepatitis (5).

Drug Administration

Drug formulations

Whereas moxonidine appears to be generally well tolerated in hypertension, in congestive heart failure it was prematurely withdrawn because of adverse outcomes on survival in a major trial, the MOXCON trial, in patients with NYHA class II-IV heart failure who were treated with modified-release moxonidine 0.5–3.0 mg/day or placebo (6). An interim analysis of the first 2000 patients enrolled showed that 53 patients taking moxonidine, compared with 29 taking placebo, died in the first 6 months and the trial was stopped prematurely. Excess mortality in the moxonidine arm appeared to be related to sudden death. The reason for this adverse outcome has to be fully elucidated. Although it has not been established that moxonidine caused a higher mortality rate, it is no longer being used in patients with heart failure. From the large database available there is no concern with regard to any adverse effect on survival with the immediate-release formulation used in hypertension.

Drug–Drug Interactions

Lorazepam

When co-administered with lorazepam 0.4 mg, moxonidine increased impairment of attentional tasks (choice, simple reaction time and digit vigilance performance, memory tasks, immediate word recall, delayed word recall accuracy, and visual tracking). These effects should be considered when moxonidine is co-administered with lorazepam, although they were smaller than would have been produced by a single dose of lorazepam 2 mg alone (7).

When co-administered with lorazepam 0.4 mg, moxonidine increased impairment of attentional tasks (choice, simple reaction time and digit vigilance performance, memory tasks, immediate word recall, delayed word recall accuracy, and visual tracking). These effects should be considered when moxonidine is coadministered with lorazepam, although they were smaller than would have been produced by a single dose of lorazepam 2 mg alone(8).

References

1. Fenton C, Keating GM, Lyseng-Williamson KA. Moxonidine: a review of its use in essential hypertension. Drugs 2006;66(4):477–96.
2. Webster J, Koch HF. Aspects of tolerability of centrally acting antihypertensive drugs. J Cardiovasc Pharmacol 1996;27(Suppl 3):S49–54.
3. Rayner B. Selective imidazoline agonist moxonidine plus the ACE inhibitor ramipril in hypertensive patients with impaired insulin sensitivity: partners in a successful MARRIAGE? Curr Med Res Opin 2004;20(3):359–67.
4. Martin U, Hill C, O'Mahoney D. Use of moxonidine in elderly patients with resistant hypertension. J Clin Pharm Ther 2005;30(5):433–7.
5. Tamm M, Sieber C, Schnyder F, Haefeli WE. Moxonidine-induced cholestatic hepatitis. Lancet 1997;350(9094):1822.
6. Curel P. Paradoxical findings in halted MOXCON trial. Reactions 1999;770:3.
7. Wesnes K, Simpson PM, Jansson B, Grahnen A, Weimann HJ, Kuppers H. Moxonidine and cognitive function: interactions with moclobemide and lorazepam. Eur J Clin Pharmacol 1997;52(5):351–8.
8. Wesnes K, Simpson PM, Jansson B, Grahnen A, Weimann HJ, Kuppers H. Moxonidine and cognitive function: interactions with moclobemide and lorazepam. Eur. J Clin Pharmacol 1997;52(5):351–8.

Nicardipine

See also Calcium channel blockers

General Information

Nicardipine is a dihydropyridine calcium channel blocker that has been used intravenously for hypertensive crises in adults, although increase in intracranial pressure was observed in a case series (1).

Organs and Systems

Cardiovascular

Severe bradycardia has been attributed to nicardipine (2).

- A 75-year-old man with atrial fibrillation undergoing surgery for esophageal carcinoma developed hypothermia and bradycardia during anesthesia. In an attempt to reduce his blood pressure of over 190/100 mmHg, he was given intravenous nicardipine 1 mg twice, 3 minutes apart. After the second dose, his blood pressure fell and his heart rate fell to 10/minute. He recovered after cardiopulmonary resuscitation.

The authors suggested that direct suppression of the sinoatrial node by nicardipine in the presence of low sympathetic tone might have been the mechanism of this serious adverse reaction.

Invasive coronary procedures, such as rotational atherectomy and coronary artery bypass graft stenting, are associated with vasoconstriction, which can increase the risk of a bad outcome. Nicardipine is relatively cardiac and vascular selective with a rapid onset of action after intravenous administration, and it is used as an alternative to sodium nitroprusside and adenosine to prevent vasoconstriction during interventional coronary procedures. A review of its efficacy and safety (3) has shown that nicardipine is as effective as diltiazem, but because it has minimal negative inotropic and chronotropic effects it can readily be given with beta-blockers in postoperative therapy.

A 38-year-old parturient developed acute pulmonary edema more than 48 hours after tocolytic treatment with nicardipine and salbutamol at 30 weeks gestation (4). The authors discussed the question of whether a second tocolytic should be used after a first has failed.

Liver

Spider nevi, typical of chronic liver disease, have been seen in patients taking long-term nicardipine (5). The lesions disappeared within 2 weeks of drug withdrawal.

Second-Generation Effects

Pregnancy

Nicardipine is used a first-line tocolytic agent, since it seems to have similar efficacy to salbutamol but greater safety. Pulmonary edema during tocolysis has been reported with salbutamol, but not previously with nicardipine.

- A 27-year-old woman developed severe dyspnea and orthopnea after receiving an of infusion nicardipine 2 mg/hour for 3 days for preterm labor at 27 weeks of gestation (6). She had also received betametasone 12 mg/day intravenously for 2 days for fetal pulmonary maturation. There was no evidence of infection or pre-eclampsia, but radiographic evidence of pulmonary edema with a tachycardia of 156/minute and a blood pressure of 140/56 mmHg. The fetal heart rhythm and ultrasound were normal. The pulmonary edema quickly resolved with intravenous furosemide 40 mg and nasal oxygen. Echocardiography showed normal left ventricular function without evidence of valvular disease, and 2 weeks later she was asymptomatic. The pregnancy ended with a full-term vaginal delivery at 38 weeks. Peripartum cardiomyopathy was excluded based on the fast clinical recovery.

The mechanism in this case was unknown, but the patient may have had acute diastolic dysfunction due to tachycardia and volume overload aggravated by glucocorticoid therapy.

Five cases of pulmonary edema have also been reported in women given nicardipine 3–6 mg/hour by intravenous infusion) and intramuscular betametasone to prevent preterm labor at 26–38 weeks of gestation in the presence of contraindications to sympathomimetics (7). There was no evidence of myocardial infarction or other cardiac dysfunction. Tocolysis was immediately stopped and the pulmonary edema resolved within 48 hours with specific therapy. Clinicians should be aware of the risk of pulmonary edema during tocolysis with nicardipine, especially when associated with glucocorticoids and in high-risk maternal conditions, particularly carediovascular.

Nicardipine has been compared with magnesium sulfate in acute therapy of preterm labor in a randomized trial in 122 women (8). The women treated with nicardipine were less likely to have recurrent preterm labor and adverse effects. Neonatal outcomes were similar in the two groups. The most frequent adverse effect was nausea or vomiting with magnesium sulfate (12/65 women).

Susceptibility Factors

Age

Nicardipine has scarcely been used in children. Intravenous nicardipine was effective in controlling blood pressure in three cases of hypertensive emergency secondary to renal disease (9). Children aged 12–14 years received intravenous infusions of 1–3 μg/kg/minute for 3–27 days, which normalized blood pressure without significant adverse reactions.

Intravenous nicardipine reduced systolic (16%) and diastolic (23%) blood pressures in a retrospective uncontrolled series of 29 children aged 2 days to 18 years (10). Tachycardia was recorded in four patients, palpitation in one, and flushing in one.

References

1. Nishikawa T, Omote K, Namiki A, Takahashi T. The effects of nicardipine on cerebrospinal fluid pressure in humans. Anesth Analg 1986;65(5):507–10.
2. Arima H, Sobue K, Tanaka S, Morishima T, Ando H, Katsuya H. Profound sinus bradycardia after intravenous nicardipine. Anesth Analg 2002;95(1):53–5.
3. Fischell TA, Maheshwari A. Current applications for nicardipine in invasive and interventional cardiology. J Invasive Cardiol 2004;16:428–32.
4. Chapuis C, Menthonnex E, Debaty G, Koch FX, Rancurel E, Menthonnex P, Pons JC. Oedème aigu du poumon au decours d'une tocolyse par nicardipine et salbutamol lors d'une menace d'accouchement prématuré sur grossesse gémellaire. J Gynecol Obstet Biol Reprod (Paris) 2005;34(5):493–6.
5. Labadie H, Faucher F, Sangla S. Poussée d'angiomes stellaires sous traitement par inhibiteurs calciques. [Growth of stellar angiomas during treatment with calcium inhibitors.] Gastroenterol Clin Biol 1999;23(6–7):788–9.
6. Bal L, Thierry S, Brocas E, Adam M, Van de Louw A, Tenaillon A. Pulmonary edema induced by calcium-channel blockade for tocolysis. Anesth Analg 2004;99:910–1.
7. Vaast P, Dubreucq-Fossaert S, Houfflin-Debarge V, Provost-Helou N, Ducloy-Bouthors AS, Puech F, Subtil D. Acute pulmonary oedema during nicardipine therapy for premature labour. Report of five cases. Eur J Obstet Gynecol Reprod Biol 2004;113:98–9.
8. Larmon JE, Ross BS, May WL, Dickerson GA, Fischer RG, Morrison JC. Oral nicardipine versus intravenous magnesium sulfate for the treatment of preterm labor. Am J Obstet Gynecol 1999;181(6):1432–7.
9. Michael J, Groshong T, Tobias JD. Nicardipine for hypertensive emergencies in children with renal disease. Pediatr Nephrol 1998;12(1):40–2.
10. Flynn JT, Mottes TA, Brophy PD, Kershaw DB, Smoyer WE, Bunchman TE. Intravenous nicardipine for treatment of severe hypertension in children. J Pediatr 2001;139(1):38–43.

Nicorandil

See also Antianginal drugs

General Information

Nicorandil is a potassium channel activator. The potassium channel activators include cromakalim and its levorotatory isomer lemakalim, bimakalim, nicorandil, and pinacidil (1). Older drugs with this property include minoxidil and diazoxide. Nicorandil, a nicotinamide derivative, is the only drug in the group to be specifically designed as one of this fourth class of agents in the management of angina pectoris after organic nitrates, beta-blockers, and calcium antagonists. Apart from being an arterial/coronary vasodilator, through modulation of ATP-sensitive potassium channels, nicorandil also contains a nitrate moiety, giving it properties similar to organic nitrates but lacking the disadvantages of drug tolerance. Thus, nicorandil has two major hemodynamic actions. It causes venodilatation, reducing preload, by its nitrate-like effect, and systemic arterial vasodilatation, reducing systemic vascular resistance and afterload, by its potassium channel opening effect. In the heart it has a selective effect on coronary vasculature compared with myocardial muscle (2), and it does not appear to affect the sinus node or atrioventricular conduction in animals (3). It also has a vasospasmolytic action (4), and in animals is cardioprotective in induced myocardial ischemia (5).

Nicorandil is effective in controlling 69–80% of cases of chronic stable angina when used as monotherapy (6,7). However, clinical experience with nicorandil is still limited, although it has been in general use in Japan for over 8 years. A review of toxicity data has been published (8).

General adverse effects

The arterial and venous vasodilatory properties of nicorandil precipitate postural hypotension, leading to dizziness, syncope, palpitation, and headache, through a mechanism similar to that of organic nitrates. Other minor gastrointestinal symptoms, such as nausea, vomiting, abdominal pain, and diarrhea have been reported, as has flushing due to cutaneous vasodilatation. There have been no reports of hypersensitivity reactions. Tumor-inducing effects have not been reported in animals or man.

Organs and Systems

Cardiovascular

The systemic hemodynamic actions of nicorandil are occasionally associated with a transient increase in heart rate of up to 18% (9,10). Larger doses are associated with cardiac depression, a dose-dependent fall in sinus rate and atrioventricular conduction velocity (11), and shortening of the cardiac action potential duration (12). However, no prodysrhythmic effects have been observed in man (13,14). Single oral doses over 40 mg have been associated with severe postural hypotension, dizziness, and syncope (10,15).

Respiratory

Experimental asthma caused by IgE antibodies in animals is inhibited by nicorandil (16). The relevance of this to man is not known.

Nervous system

Headache occurs in up to one-third of cases in dosages of 10–20 mg bd (17,18), but it is usually mild to moderate, necessitating withdrawal in about 5% of cases. Headache is more common at the start of therapy and is dose-dependent; it gets better with continued therapy or progressive dose titration (8,19).

Endocrine

Nicorandil suppresses the release of insulin from isolated animal pancreatic cells (20). However, this effect is four times weaker than that of diazoxide, and its effect in man is not known.

Hematologic

Nicorandil has an inhibitory effect on platelet aggregation by increasing intracellular platelet cyclic GMP concentrations (21). Whether this effect will potentiate bleeding or protect against myocardial infarction in patients with coronary heart disease is not known

Gastrointestinal

Nausea, vomiting, gastralgia, diarrhea, and abdominal pains have been reported in some studies (6,10).

Skin

Nicorandil, like minoxidil, appears to enhance the incorporation of cysteine into hair shafts (22). However, hypertrichosis has not been reported.

Anal eczema, which resolved after the withdrawal of nicorandil, has been reported (19).

Sexual function

Nicorandil's potassium channel opening action and its ability to release nitric oxide promotes tumescence in isolated human penile corpus cavernosum (23). This effect has been exploited to provide treatment for erectile dysfunction by local injection. Whether oral nicorandil is associated with enhanced erectile function has not been assessed.

Death

Nicorandil was not associated with an excess of sudden deaths complicating acute myocardial infarction in a Japanese series of 1000 sudden deaths between 1983 and 1987 (24).

Long-Term Effects

Drug withdrawal

Adverse effects have not been observed with the withdrawal of nicorandil after long-term therapy. Unlike organic nitrates, there is no drug tolerance with prolonged use of nicorandil (25), and there does not appear to be any cross-tolerance (26). The effects of organic nitrates depend solely on the activation of cyclic GMP in smooth muscle, but this is not the case with nicorandil. In a comparison of intravenous nicorandil and glyceryl trinitrate it took twice as long for blood concentrations of cyclic GMP to return to baseline after the latter, but the hemodynamic effects of the former continued for longer (27). This may be because of the additional non-nitrate potassium channel opening effect of nicorandil when further formation of cyclic GMP ceases.

Susceptibility Factors

Renal disease

The use of nicorandil in elderly patients with chronic renal insufficiency and in patients with stable hepatic cirrhosis is not associated with any significant alteration in its pharmacokinetics. In addition, its metabolism is not affected by drugs that interfere with drug-metabolizing liver enzymes (28).

Hepatic disease

Nicorandil is denitrated in the liver. Its half-life was slightly prolonged in eight patients with cirrhotic liver impairment given a single intravenous bolus of nicorandil (0.1 mg/kg for 5 minutes, half-life 1.7 versus 1.1 hours in controls) (29). This is unlikely to have any significance in adjusting dosages in patients with stable liver disease.

Other features of the patient

There is a theoretical contraindication to nicorandil in patients with cardiogenic shock, acute left ventricular failure with low filling pressure, and hypotension. A sublingual dose of 20 mg in patients with coronary artery disease and normal left ventricular function was associated with a 12% fall in left ventricular end-systolic pressure, a 3% fall in left ventricular end-diastolic pressure, accentuated diastolic filling, a 13% reduction in mean aortic pressure, and a reduced cardiac output at rest (9,30). However, cardiac output may be augmented by up to 60% in patients with congestive cardiac failure or with a previous history of myocardial infarction, principally by reduced preload, reduced afterload, and improved myocardial oxygen supply (31–33).

Drug–Drug Interactions

Beta-blockers

The combination of nicorandil and beta-blockers (atenolol/propranolol) potentiates hypotension, with attenuation of reflex tachycardia (6).

Dipyridamole

The coronary vasodilator response of nicorandil is potentiated by dipyridamole (34).

Food–Drug Interactions

Food reduces the rate but not the extent of absorption of nicorandil after an oral dose (35). Its systemic availability is good and is generally greater than 75% in healthy volunteers, suggesting limited first-pass hepatic metabolism, unlike the organic nitrates (36).

Smoking

Cigarette exposure in animals suppresses and delays the absorption of nicorandil from the gastrointestinal tract (37), but this has yet to be investigated in man.

References

1. Hamilton TC, Weston AH. Cromakalim, nicorandil and pinacidil: novel drugs which open potassium channels in smooth muscle. Gen Pharmacol 1989;20(1):1–9.
2. Sakai K. Nicorandil: animal pharmacology. Am J Cardiol 1989;63(21):J2–J10.
3. Taira N, Satoh K, Yanagisawa T, Imai Y, Hiwatari M. Pharmacological profile of a new coronary vasodilator drug, 2-nicotinamidoethyl nitrate (SG-75). Clin Exp Pharmacol Physiol 1979;6(3):301–16.
4. Feldman RL. A review of medical therapy for coronary artery spasm. Circulation 1987;75(6 Pt 2):V96–V102.
5. Pieper GM, Gross GJ. Salutary action of nicorandil, a new antianginal drug, on myocardial metabolism during ischemia and on postischemic function in a canine preparation of brief, repetitive coronary artery occlusions: comparison with isosorbide dinitrate. Circulation 1987;76(4):916–28.
6. Krumenacker M, Roland E. Clinical profile of nicorandil: an overview of its hemodynamic properties and therapeutic efficacy. J Cardiovasc Pharmacol 1992;20(Suppl 3):S93–S102.
7. IONA Study group. Effect of nicorandil on coronary events in patients with stable angina: the Impact Of Nicorandil in Angina (IONA) randomised trial. Lancet 2002; 359(9314): 1269–75.
8. Roland E. Safety profile of an anti-anginal agent with potassium channel opening activity: an overview. Eur Heart J 1993;14(Suppl B):48–52.
9. Coltart DJ, Signy M. Acute hemodynamic effects of single-dose nicorandil in coronary artery disease. Am J Cardiol 1989;63(21):J34–9.
10. Camm AJ, Maltz MB. A controlled single-dose study of the efficacy, dose response and duration of action of nicorandil in angina pectoris. Am J Cardiol 1989;63(21):J61–5.
11. Taira N. Nicorandil as a hybrid between nitrates and potassium channel activators. Am J Cardiol 1989;63(21):J18–24.
12. Fenici RR, Melillo G. Effects of nicorandil on human cardiac electrophysiological parameters. Cardiovasc Drugs Ther 1991;5(Suppl 3):367.
13. Gross GJ, Auchampach JA. Role of ATP dependent potassium channels in myocardial ischaemia. Cardiovasc Res 1992;26(11):1011–6.
14. Mitrovic V, Neuss H, Kindler M. Elektrophysiologische Auswirkungen einer Vasodilatation durch Nicorandil. Herz/Kreislauf 1986;18:403.
15. Galie N, Varani E, Maiello L, Boriani G, Boschi S, Binetti G, Magnani B. Usefulness of nicorandil in congestive heart failure. Am J Cardiol 1990;65(5):343–8.
16. Nagai H, Kitagaki K, Goto S, Suda H, Koda A. Effect of three novel K+ channel openers, cromakalim, pinacidil and nicorandil on allergic reaction and experimental asthma. Jpn J Pharmacol 1991;56(1):13–21.
17. Doring G. Antianginal and anti-ischemic efficacy of nicorandil in comparison with isosorbide-5-mononitrate and isosorbide dinitrate: results from two multicenter, double-blind, randomized studies with stable coronary heart disease patients. J Cardiovasc Pharmacol 1992;20(Suppl 3):S74–81.
18. Ulvenstam G, Diderholm E, Frithz G, Gudbrandsson T, Hedback B, Hoglund C, Moelstad P, Perk J, Sverrisson JT. Antianginal and anti-ischemic efficacy of nicorandil compared with nifedipine in patients with angina pectoris and coronary heart disease: a double-blind, randomized, multicenter study. J Cardiovasc Pharmacol 1992;20(Suppl 3):S67–73.
19. Wagner G. Selected issues from an overview on nicorandil: tolerance, duration of action, and long-term efficacy. J Cardiovasc Pharmacol 1992;20(Suppl 3):S86–92.
20. Garrino MG, Plant TD, Henquin JC. Effects of putative activators of K+ channels in mouse pancreatic beta-cells. Br J Pharmacol 1989;98(3):957–65.
21. Jaraki O, Strauss WE, Francis S, Loscalzo J, Stamler JS. Antiplatelet effects of a novel antianginal agent, nicorandil. J Cardiovasc Pharmacol 1994;23(1):24–30.
22. Buhl AE, Waldon DJ, Conrad SJ, Mulholland MJ, Shull KL, Kubicek MF, Johnson GA, Brunden MN, Stefanski KJ, Stehle RG, et al. Potassium channel conductance: a mechanism affecting hair growth both in vitro and in vivo. J Invest Dermatol 1992;98(3):315–9.
23. Hedlund P, Holmquist F, Hedlund H, Andersson KE. Effects of nicorandil on human isolated corpus cavernosum and cavernous artery. J Urol 1994;151(4):1107–13.
24. Kambara H, Kinoshita M, Nakagawa M, Sakurai T, Kawai C. [Sudden death among 1,000 patients with

myocardial infarction: incidence and contributory factors. KYSMI Study Group.]J Cardiol 1995;25(2):55–61.
25. Guermonprez JL, Blin P, Peterlongo F. A double-blind comparison of the long-term efficacy of a potassium channel opener and a calcium antagonist in stable angina pectoris. Eur Heart J 1993;14(Suppl B):30–4.
26. Frampton J, Buckley MM, Fitton A. Nicorandil. A review of its pharmacology and therapeutic efficacy in angina pectoris. Drugs 1992;44(4):625–55.
27. Tsutamoto T, Miyauchi N, Kanamori T, Kinoshita M. Mechanism of nicorandil intolerance in patients with heart failure. Circulation 1991;84(Suppl II):II–58.
28. Frydman A. Pharmacokinetic profile of nicorandil in humans: an overview. J Cardiovasc Pharmacol 1992; 20(Suppl 3):S34–44.
29. Jungbluth GL, Della-Coletta AA, Blum RA, et al. Comparative pharmacokinetics and bioavailability of nicorandil in subjects with stabilized cirrhosis and matched healthy volunteers. Clin Pharmacol Ther 1991;49:181.
30. Suryapranata H, MacLeod D. Nicorandil and cardiovascular performance in patients with coronary artery disease. J Cardiovasc Pharmacol 1992;20(Suppl 3):S45–51.
31. Solal AC, Jaeger P, Bouthier J, Juliard JM, Dahan M, Gourgon R. Hemodynamic action of nicorandil in chronic congestive heart failure. Am J Cardiol 1989;63(21):J44–8.
32. Tice FD, Binkley PF, Cody RJ, Moeschberger ML, Mohrland JS, Wolf DL, Leier CV. Hemodynamic effects of oral nicorandil in congestive heart failure. Am J Cardiol 1990;65(20):1361–7.
33. Murakami M, Takeyama Y, Matsubara H, Hasegawa S, Nakamura N, Sekita S, Katagiri T. Effects of intravenous injection of nicorandil on systemic and coronary hemodynamics in patients with old myocardial infarction. A comparison with nifedipine and ISDN. Eur Heart J 1989;10(Suppl):426.
34. Sakai K, Shiraki Y, Nabata H. Cardiovascular effects of a new coronary vasodilator N-(2-hydroxyethyl)nicotinamide nitrate (SG-75): comparison with nitroglycerin and diltiazem. J Cardiovasc Pharmacol 1981;3(1):139–50.
35. Horii D, Ishibashi A, Iwamoto A. Bioavailability study of nicorandil before and after meals. Rinsho Yakuri 1984;15:489.
36. Frydman AM, Chapelle P, Diekmann H, Bruno R, Thebault JJ, Bouthier J, Caplain H, Ungethuem W, Gaillard C, Le Liboux A, et al. Pharmacokinetics of nicorandil. Am J Cardiol 1989;63(21):J25–33.
37. Gomita Y, Eto K, Furuno K, Mimaki Y, Araki Y. Influences of exposure to cigarette smoke on concentration of nicorandil in plasma of rats. J Pharm Sci 1992;81(3):228–31.

Nifedipine

See also Calcium channel blockers

General Information

Nifedipine is a dihydropyridine calcium channel blocker.

A significant fall in pulmonary vascular resistance with high doses of oral calcium channel blockers seems to be associated with an improved prognosis, and potential clinical efficacy in primary pulmonary hypertension. However, significant adverse effects have been reported during acute testing with calcium channel blockers. Therefore, to identify patients who may benefit from long-term calcium channel blockers, there is a need for a safe, potent, short-acting vasodilator. Two studies have been conducted to assess the use of inhaled nitric oxide, a selective pulmonary vasodilator, for predicting the safety of high-dose oral calcium channel blockers and acute hemodynamic responses to them in primary pulmonary hypertension. In one study, 17 patients with primary pulmonary hypertension undergoing a trial of nifedipine (20 mg hourly for 8 hours) were assessed for the hemodynamic response to inhaled nitric oxide, 80 parts per million for 5 minutes (1). All nitric oxide responders also responded to nifedipine, and nine of the ten nitric oxide non-responders were nifedipine non-responders. All nitric oxide responders tolerated a full trial of nifedipine without hypotension. There was a highly significant correlation between the effects of nitric oxide and nifedipine on pulmonary vascular resistance. In conclusion, the pulmonary vascular response to inhaled nitric oxide accurately predicted the acute hemodynamic response to nifedipine in primary pulmonary hypertension, and a positive response to nitric oxide was associated with a safe nifedipine trial. In patients comparable to those evaluated, a trial of nifedipine in nitric oxide non-responders appears unwarranted and potentially dangerous.

In another study the acute response to inhaled nitric oxide and high doses of oral nifedipine or verapamil was assessed in 33 consecutive patients with primary pulmonary hypertension (2). Ten patients responded acutely to nitric oxide, nine of whom responded acutely to calcium channel blockers, without any complications. The other 23 patients failed to respond to nitric oxide and calcium channel blockers. In these non-responders there were nine serious adverse effects with calcium channel blockers. There was no clinical or baseline hemodynamic feature that predicted the acute vasodilator response. Long-term oral treatment with calcium channel blockers was restricted to the nine acute responders, and there was a sustained clinical and hemodynamic improvement in only six patients. It was concluded that nitric oxide may be used as a screening agent for safely identifying patients with primary pulmonary hypertension who may benefit from long-term treatment with calcium channel blockers.

Organs and Systems

Cardiovascular

The long-term safety of dihydropyridine calcium channel blockers has been extensively debated since 1995, with reports of conflicting results from observational and randomized clinical studies about possible increases in cardiovascular mortality, myocardial infarction, and neoplastic diseases (SEDA-22, 214).

In the INSIGHT (Intervention as a Goal in Hypertension Treatment) study, a prospective, multicenter, double-blind study in 6321 hypertensive patients aged 55–80 years, long-acting nifedipine 30 mg was compared with co-amilozide (hydrochlorothiazide 25 mg plus amiloride 2.5 mg) (3). There were no differences between the two treatments in the primary end-points of cardiovascular death, myocardial infarction, heart failure, or stroke during follow-up for 4 years.

In the Canadian Study of Health and Aging, a population-based prospective study of people aged 65 years or more, 5-year follow-up of 837 subjects who reportedly used at least one antihypertensive or diuretic agent showed that the risk of all-cause and cardiac mortality was significantly higher among nifedipine users than beta-blocker users (4). Compared with beta-blockers, the hazard ratios (95% CI) were:

- loop diuretics 1.84 (1.21, 2.82);
- nifedipine 1.82 (1.09, 3.04);
- ACE inhibitors 0.98 (0.54, 1.78);
- diltiazem/verapamil 0.96 (0.58, 1.60).

Among nifedipine users, the risk of death increased with the average daily dose and with recent initiation of therapy and remained significant for long-acting formulations.

The ACTION trial started in 1996 following concerns about the long-term safety of nifedipine, particularly short-acting formulations. The study included 7665 patients with stable angina randomized to receive nifedipine or placebo on top of their conventional therapy. Over a mean follow-up period of 4.9 years there were no differences between the two groups in the primary endpoint, which was the combination of death, acute myocardial infarction, refractory angina, new heart failure, stroke, and peripheral revascularization. Although the results were disappointing with regard to efficacy, they offered reassurance about the safety of long-acting nifedipine (5).

Myocardial ischemia

Ischemic chest pain is an infrequent but well-documented adverse effect of initiating therapy with nifedipine (6). Its mechanism is unclear, but it may arise from reduced overall coronary blood flow or coronary steal (7).

Ischemic electrocardiographic changes without pain have also been reported (8).

There has been a meta-analysis of 60 published, randomized clinical trials that enrolled at least 10 patients with stable angina and that compared any nifedipine formulation, either as monotherapy or combination therapy, with a non-dihydropyridine active drug or a placebo control for at least 1 week (9). There were 3096 subjects and 5571 treatment exposures. There was an increased risk of cardiovascular events in patients allocated to nifedipine. The unadjusted odds ratios for nifedipine versus controls were 1.40 (CI = 0.56, 3.49) for major events (death, non-fatal myocardial infarction, stroke, revascularization procedures) and 1.75 (CI = 0.83, 3.67) for increased angina. Episodes of increased angina were more frequent with immediate-release nifedipine (OR = 4.19; CI = 1.41, 12.49) and with nifedipine monotherapy (OR = 2.61; CI = 1.30, 5.26). The odds ratio for immediate-release nifedipine was significantly higher than that for modified-release nifedipine, and the odds ratio for nifedipine monotherapy was significantly higher than that for nifedipine combination therapy. This meta-analysis suggests that the adverse effects of nifedipine on cardiovascular events in patients with stable angina are primarily due to more frequent episodes of angina when monotherapy with immediate-release formulations is used. Modified-release formulations and concurrent beta-blockade do not appear to be associated with an increased risk.

Hypotension

Sublingual nifedipine, given for hypertensive crises in elderly patients, can cause adverse effects associated with a precipitous fall in blood pressure, even at low doses. In 93 consecutive hypertensive patients without coronary heart disease, aged 65 years or over, nifedipine reduced blood pressure significantly, increased heart rate, and relieved symptoms associated with raised blood pressure (10). However, there were electrocardiographic changes consistent with myocardial ischemia in six of 55 patients with left ventricular hypertrophy and in one patient without left ventricular hypertrophy.

Increased sympathetic activity has been identified as a risk factor for cardiovascular events, but long-acting dihydropyridines should have little if any sympathoexcitatory effects when given over the long term. Treatment for 4 weeks with low-dose GITS (gastrointestinal therapeutic system) nifedipine 20 mg/day reduced the blood pressure in older patients with mild hypertension (mean age 67 years) but not in younger ones (mean age 45 years) (11). There was a reflex increase in sympathetic activity in the young patients, which could have attenuated the blood pressure response.

Rapid blood pressure reduction is not recommended in severe hypertension in older people because of adverse effects related to impaired cerebral autoregulation, although sublingual nifedipine is still used in children, in whom the cerebral circulation is more robust.

Cardiac dysrhythmias

A reported causal relation between torsade de pointes and nifedipine (12) has been questioned (13).

Ventricular dysrhythmias have been described in a teenager who was given nifedipine 10 mg sublingually for symptomatic, severe hypertension (blood pressure 180/120) (14). Within minutes she complained of palpitation and had a tachycardia of 100/minute with ventricular bigeminy and later ventricular extra beats. The authors suggested that reflex sympathetic activation following an abrupt drop in blood pressure may have caused the dysrhythmias, because of raised catecholamine concentrations, and that it may be more appropriate to treat severe hypertension in young people with intravenous antihypertensive agents that can be titrated to produce controlled reductions in blood pressure.

Nervous system

Rapid lowering of the blood pressure with nifedipine, particularly sublingually, can precipitate cerebral ischemia, with confusion, loss of consciousness, and stroke. Cases of cortical blindness with macular sparing secondary to occipital lobe infarction have been reported (SEDA-17, 238).

Transient cerebral ischemia, with aphasia and hemiparesis in one patient and cerebellar dysfunction and loss of consciousness in another, has been observed with nifedipine (15). In addition, rapidly progressive hemiparesis, aphasia, and confusion, accompanied by a substantial fall in blood pressure, occurred in a patient whose hypertension (230/136 mmHg) was treated with sublingual nifedipine (16). Transient retinal ischemia, which may be recurrent, has also been attributed to nifedipine (17,18). Hypotension or cerebral steal are the likely mechanisms.

Cerebral vasodilatation from nifedipine might have contributed to a subarachnoid hemorrhage in a 32-year-old woman with a berry aneurysm (19).

Paresthesia of the hands and feet has been reported with nifedipine (SEDA-17, 237).

Electrolyte balance

Nifedipine can increase urinary potassium loss in patients treated with thiazide diuretics (20), but it has no effect on adrenaline-induced hypokalemia (21). In the Treatment of Mild Hypertension Study, 4 years of monotherapy with amlodipine maleate caused no change compared with placebo in the serum potassium, uric acid, aspartate transaminase, or creatinine of 114 hypertensive patients (22).

- A 56-year-old man developed hypokalemia associated with muscle weakness after taking nifedipine for 2 weeks for angina; the potassium concentration rose on withdrawal, but fell rapidly again on rechallenge (23).

Hematologic

In a case-control surveillance of agranulocytosis and aplastic anemia conducted in the metropolitan area of Barcelona, where 178 cases of aplastic anemia were identified, nifedipine was associated with a significant relative risk of aplastic anemia, which translates into an absolute risk of 1.2 per patient-year. Among the 178 patients, 147 were interviewed and compared with 1295 controls. Six cases (4.1%) and 11 controls (0.8%) had been exposed to nifedipine during the window period, as all of them had taken nifedipine for at least 7 months. The multivariate odds ratio was 4.6 (95% CI = 1.5, 15). All six died within 5 months of diagnosis. The authors concluded that the risk of aplastic anemia associated with nifedipine is of a similar magnitude to that associated with chloramphenicol (1.7 per 100 000 patients) and that associated with phenylbutazone (2.2 per 100 000 patients) (24).

Mouth and teeth

Nifedipine and ciclosporin can cause gingival hyperplasia, which is associated with gingival inflammation and can be mitigated by careful oral hygiene, although this has not been clearly shown. In gingival biopsies from nine nifedipine-treated cardiac outpatients, 13 immunosuppressant-treated renal transplant recipients, nine of whom were also taking nifedipine, and 30 healthy individuals, there were significant differences in macrophage and lymphocyte subpopulations in gingival connective tissue from patients with nifedipine-associated gingival lesions compared with healthy individuals (25). For example, the proportion of CD8-labelled cells was significantly higher and the CD4/CD8 ratio was significantly lower in connective tissue beneath the sulcus epithelium in those taking nifedipine. The authors suggested that immune responsiveness may be altered in drug-induced gingival overgrowth.

In another study, gingival samples were collected from 19 healthy individuals, 12 nifedipine-treated cardiac patients, and 22 immunosuppressant-treated organ transplant recipients, 11 of whom were also taking nifedipine (26). Mitotic activity was measured in the basal cell layer, and the results suggested that the increased epithelial thickness observed in nifedipine- and ciclosporin-induced gingival overgrowth is associated with increased mitotic activity.

The true prevalence of gingival overgrowth induced by chronic treatment with calcium channel blockers is still uncertain, since most studies have been small. In a cross-sectional study in 65 patients taking nifedipine and 147 controls in a primary-care center in Barcelona there was a higher prevalence of gingival overgrowth in patients taking nifedipine (34 versus 4.1%) (27).

Gastrointestinal

Acute abdominal pain due to mesenteric ischemia has been described with nifedipine (28).

Two patients developed severe gastric mucosal damage probably due to nifedipine (29).

- A 62-year-old man with gastric hemorrhage had a deep fundal ulcer in which a tablet of nifedipine was firmly embedded. The tablet was removed endoscopically and he was given cimetidine and recovered.
- A 67-year-old man with diabetes and hypertension, taking glibenclamide and nifedipine, developed dysphagia. Gastroscopy showed severe damage to the mid-esophagus. He did not respond to omeprazole and sucralfate, but gradually improved after nifedipine withdrawal, and the lesion disappeared.

Modified-release formulations of nifedipine have been associated in case reports with the formation of bezoars, concretions of undigested material within the gastrointestinal tract, mostly in the stomach. An unusual case of tablet impaction in the duodenum, with gastric outlet obstruction, was discovered 1 year after the patient stopped taking a modified-release formulation of nifedipine (30).

- A 77-year-old woman developed constipation and weight loss. Esophagogastroduodenoscopy showed a deformed pylorus and an elongated duodenal bulb, with numerous impacted tablets and ulceration in the outlet. Since only a few tablets could be recovered by endoscopy, she underwent partial duodenal resection to remove the contents: 25 intact tablets that were confirmed to be modified-release nifedipine.

Retention of modified-release capsules can cause bezoar formation. In a patient with a enteric stricture, modified-release capsules of nifedipine caused recurrent small bowel obstruction (31).

Liver

Hepatic steatosis with Mallory inclusion bodies has been reported with nifedipine (32). In patients with liver cirrhosis, nifedipine increases portal pressure due to splanchnic vasodilatation (33); whether this increases the risk of variceal bleeding is unknown.

Urinary tract

Several reports have linked renal dysfunction with nifedipine. In a study of hypertensive diabetics with renal insufficiency, nifedipine increased proteinuria and worsened renal function (SEDA-16, 196). Others have reported mild reversible renal impairment in patients with chronic renal insufficiency taking nifedipine for angina or hypertension; a biopsy in one of the patients, who had heavy proteinuria, showed focal and segmental glomerulosclerosis (34). Immune-complex nephritis was reported in a patient taking nifedipine, but the proteinuria persisted (and indeed worsened) on changing to verapamil (35).

Nifedipine was associated with an apparent diuresis within a short time of the first dose in 14 of 24 patients (36) and with nocturia, which ceased or improved markedly on withdrawal of therapy in nine patients (37). The nocturia was considered to reflect an effect of calcium channel blockade on detrusor function.

Skin

Bullous eruptions secondary to nifedipine include bullous fixed drug eruptions, phototoxic bullous eruptions, erythema multiforme, and pemphigus foliaceus.

Two cases of pemphigus vulgaris, one aggravated and one induced by nifedipine, have been described (38).

- A 54-year-old woman with an 8-year history of pemphigus vulgaris had a flare-up of her disease, refractory to oral glucocorticoids, after taking nifedipine for 2 years for hypertension. The nifedipine was withdrawn and she was given high-dose immunosuppressant treatment with prednisone and azathioprine. The lesions subsided promptly and cleared totally within several weeks.
- A 68-year-old woman had a bullous eruption of pemphigus vulgaris after antihypertensive treatment with nifedipine. She was given high-dose systemic glucocorticoids. However, she was unresponsive and nifedipine was finally withdrawn. Although new lesions did not appear, she died a few days later with uncontrollable sepsis.

Probable nifedipine-induced pemphigoid has also been reported (39).

- A 70-year-old man with hypertension took nifedipine 10 mg bd and 3 months later developed a pruritic eruption on his back, which spread to cover his trunk, limbs, face, and scalp, with involvement of the mouth. His skin deteriorated after the dosage of nifedipine was increased to 20 mg bd 1 year later. Antibodies to the 230 kDa pemphigoid antigen were detected in his serum. His skin improved considerably after withdrawal of nifedipine and the use of topical steroids and oral minocycline.

The findings supported a diagnosis of pemphigoid.

Photosensitivity (40) and a generalized bullous eruption (41) have been attributed to nifedipine, as have telangiectasia and spider nevi (42).

Musculoskeletal

Severe muscle cramps in the legs and hands can occur during nifedipine treatment (43).

Immunologic

Anaphylaxis has been attributed to sublingual nifedipine (44).

- A 71-year-old man with prostatic adenocarcinoma and a pathological vertebral fracture received sublingual nifedipine for hypertension and 15 minutes later became stuporose and complained of pruritus, generalized erythema, dizziness, and nausea. His blood pressure had fallen to 60/40 mmHg, his pulse rate was 120/minute, and his respiratory rate was 30/minute. He had cyanosis and severe bronchospasm with no focal neurological abnormalities. After treatment with subcutaneous adrenaline, intravenous fluids, hydrocortisone, and aminophylline, his blood pressure increased to 125/70 mmHg. During the following days his neurological and pulmonary status rapidly improved.

Cutaneous lupus erythematosus has been associated with nifedipine (45).

- Nifedipine-induced subacute cutaneous lupus erythematosus has been reported in a 48-year-old white woman who had taken nifedipine for essential hypertension for 4 years. She developed a papulosquamous annular eruption in sun-exposed areas during the summer. She had taken no other drugs. Serological tests (antinuclear and antihistone antibodies), histopathology, and immunological tests (granular IgM deposits at the dermoepidermal junction) confirmed the diagnosis. Nifedipine withdrawal led to rapid improvement, with almost complete resolution of the skin lesions in 1 month. Antinuclear and antihistone antibodies titers fell within 6 months.

Second-Generation Effects

Pregnancy

Clinical guidelines have advocated nifedipine as a tocolytic of choice (46). Tocolytic therapy with nifedipine has been reported in several studies to be at least effective as ritodrine, terbutaline, or magnesium sulfate, with fewer maternal adverse effects (SEDA-18, 216; SEDA-20, 187; SEDA-22, 217; SEDA-23, 211). These observations have been confirmed in two trials in 102 and 54 pregnant women of under 34 weeks gestation, randomized to nifedipine or ritodrine. There were no differences in the time of delivery, but significantly fewer maternal adverse effects in those given nifedipine (47,48).

The possibility that calcium channel blockers can cause cardiovascular adverse effects in pregnancy has been widely debated (SED-14, 598; SEDA-20, 185; SEDA-21, 208; SEDA-22, 214). An uncomplicated non-Q wave myocardial infarction has been reported during nifedipine therapy for preterm labor (49).

- A 29-year-old woman was admitted because of preterm rupture of membranes and uterine contractions at 29 weeks. She was given ritodrine and 30 hours later reported chest pain. The electrocardiogram was normal. Ritodrine was withdrawn. Oral nifedipine was started and 4 hours later she reported chest pain. Her blood pressure was 70/50 mmHg and her heart rate 124/minute. The electrocardiogram showed T wave inversion in lead III and the serum creatine kinase activity was raised.

The authors concluded that even if beta$_2$-adrenoceptor agonists can cause myocardial ischemia in pregnant women by further increasing oxygen consumption, acute myocardial necrosis is rare; however, in this case nifedipine could have posed an extra risk by causing hypotension and reflex tachycardia.

In a review of almost 800 patients randomized to beta-adrenoceptor agonists or nifedipine, the latter was associated with more frequent prolongation of pregnancy, a lower incidence of respiratory distress syndrome, and lower incidences of maternal and fetal adverse effects (50).

Nifedipine is not without risks in pregnancy, particularly in women with underlying cardiac anomalies.

- A 31 year old woman presented at 32 weeks of gestation with threatened preterm labor (51). An asymptomatic ventricular septal defect had been incompletely repaired in childhood. Despite this, she had good pre-pregnancy exercise tolerance and no dyspnea. An echocardiogram undertaken in early pregnancy showed normal left ventricular size and function. She received betamethasone, nifedipine 20 mg orally three times 20 minutes apart for preterm labour tocolysis, and amoxicillin 1 g intravenously for prophylaxis of bacterial endocarditis. Within 10 minutes of the second dose of nifedipine, she became short of breath. Her symptoms worsened overnight and by 20 hours after nifedipine, her respiratory distress necessitated transfer to the intensive care unit. She was afebrile and had mild cyanosis, bilateral peripheral edema, a raised jugular venous pressure, a pulse rate of 120/minute, and a blood pressure of 130/65 mmHg. Her respiratory rate was 32/minute. There was a loud systolic murmur at the left sternal edge, consistent with her ventricular septal defect, a right ventricular heave, and reduced breath sounds at the lungs bases. With conservative management the unexplained dyspnea and hypoxia gradually settled over 7 days. Her delivery and postnatal recovery was uneventful and the caused of her hypoxia remained undiagnosed.

Hypocalcemia has been reported after treatment with magnesium sulfate for tocolysis.

- A 25-year-old Hispanic woman was given magnesium sulfate 132 mg intravenously over 55 hours during betamethasone administration, nifedipine 10 mg orally 30 minutes after magnesium discontinuation, which she tolerated well, and nifedipine 20 mg orally 4 hours later (52). Her blood pressure fell to 93/49 mmHg and she was given no further doses. She had bilateral hand contractures about 6 hours later and her serum calcium was 1.35 mmol/l. She was given 40 ml of 10% calcium gluconate intravenously and 4 hours later was moving her hands normally and the serum calcium was 2.45 mmol/l.

The association between magnesium sulfate and maternal hypocalcemia is well recognized, given its effect on calcium metabolism, and concern has been raised that the risk may be increased if a calcium channel blocker, such as nifedipine, is also used. Even if evidence of synergistic toxicity is scanty, caution is advised when these drugs are used simultaneously.

Severe hypotension followed by fetal death has been reported after the administration of nifedipine as a tocolytic (53).

- A 23-year-old woman of African origin went into preterm labor and was given tocolytic therapy with atosiban. The uterine contractions continued and indometacin was added until the contractions ceased after 24 hours. After 48 hours the atosiban was withdrawn and 5 hours later the uterine contractions started again. She was given oral nifedipine 10 mg but after the second dose the blood pressure fell to 73/30 mmHg. Ultrasound examination showed severe fetal bradycardia soon followed by death. Tocolytic therapy was withdrawn and a colloid infusion started. After 6 hours the blood pressure was again normal.

The authors concluded that care should be taken when administering nifedipine as a tocolytic in pregnancy. They did not deny that nifedipine is generally safe and effective in these circumstances, but others disagreed that nifedipine had caused severe hypotension in this case: "In our 15 years of personal experience with nifedipine for the management of preterm labour, we have never seen such severe hypotension in a previously normotensive 'euvolemic' patient." (54). They pointed out that the half-life of nifedipine is only 1.3 hours and that its adverse effects are generally mild and transient side. They cited several meta-analyses that have shown that nifedipine is associated with better neonatal outcomes and fewer adverse maternal outcomes than other tocolytic agents, especially beta-adren oceptor agonists.

However, others wrote to describe another case of severe hypotension, in a 24-year-old woman in her third pregnancy who had suspected preterm labour at 26 weeks of gestation and was given nifedipine tocolysis, 30 mg orally followed by two doses of 20 mg, after which she complained of severe headache, visual disturbances, flushing, dizziness, palpitation, nausea, and vomiting; her pulse rate was 120/minute and her blood pressure 80/40 mmHg (55). She required circulatory resuscitation using intravenous infusions of crystalloids and colloids, and her blood pressure returned to normal over the next 30 minutes. There was no evidence of fetal compromise during these events.

Furthermore, there is no doubt that severe hypotension can happen in these cases. Of 73 consultant obstetricians in the United Kingdom who replied to a survey, 53% used nifedipine as their first-line tocolytic and 40% used atosiban (56). Of these, 17 reported maternal hypotension of sufficient severity to cause fetal compromise necessitating cesarean section with nifedipine, compared with four with ritodrine and two with atosiban.

Others have pointed out that conclusions from meta-analyses that favor the use of nifedipine as a tocolytic agent are not supported by close examination of the data, since the tocolytic actions have been studied primarily in normal pregnancies (57). They have suggested that care should be taken in giving nifedipine when the maternal cardiovascular condition is compromised, such as in intrauterine infection, twin pregnancy, maternal hypertension, and cardiac disease, because of the risks of life-threatening pulmonary edema and/or cardiac failure, in which baroreceptor-mediated increases in sympathetic tone may not balance the cardiac depressant activity of nifedipine.

Fetotoxicity

The use of nifedipine during pregnancy and labor has been widely debated, although its effects on child development have not been well evaluated. In one study nifedipine did not affect the development and health of 190 children, aged 18 months, born to women with mild to moderate hypertension who had been randomized to nifedipine, given for 12–34 gestational weeks before delivery, or expectant management (58).

Susceptibility Factors

Treatment of hypertensive emergencies has been the object of a randomized comparison of isosorbide dinitrate aerosol and nifedipine in 60 adults (59). Two patients taking nifedipine had myocardial ischemia early after administration and four patients had rebound hypertension during the follow-up period. Authors who reviewed the role of nifedipine in hypertensive emergencies reached the conclusion that the use of short-acting sublingual or oral nifedipine is no longer recommended for the treatment of these urgencies because it can precipitate serious adverse reactions (60).

There has been a report of cardiovascular collapse after uncomplicated coronary artery bypass surgery in patients taking nifedipine preoperatively (61); the condition responded to intravenous calcium chloride in three patients and adrenaline in four. Nifedipine may facilitate cardiac arrest during induction of anesthesia by sensitizing the carotid sinus (62).

Drug Administration

Drug formulations

Crushing an extended-release nifedipine tablet (nifedipine XL) can alter its release characteristics.

- A 38-year-woman was treated for acute pulmonary edema and pneumonia and her medications were changed to oral hydralazine, labetalol, and nifedipine XL, crushed and administered through a nasogastric tube (63). She developed bradycardia, hypotension, and cardiac arrest. She was resuscitated, but died the next morning after worsening bradycardia with hypotension subsequent to a further dose of crushed labetalol and nifedipine XL.

The administration of crushed nifedipine XL resulted in severe hypotension and the concurrent administration of labetalol prevented a compensatory increase in heart rate. The release characteristics of oral modified-release formulations are destroyed when the formulation is crushed, resulting in rapid availability of the total dose, which can cause significant harm.

Drug overdose

Unlike verapamil and diltiazem, overdose with nifedipine is not usually fatal. Multiple case series of pediatric nifedipine ingestion have been published, and none has reported any deaths. However, two fatal cases of ingestion of long-acting nifedipine in children have been reported: a 24-month-old girl who took 20 tablets of nifedipine 10 mg and a 14-month-old girl who took a single tablet of nifedipine 10 mg; neither responded to aggressive supportive care (64).

Drug–Drug Interactions

Alcohol

There is a substantial increase in the systemic availability of nifedipine when alcohol is taken at the same time (65).

Atenolol

The combination of nifedipine with atenolol in patients with intermittent claudication resulted in a reduction in walking distance and skin temperature, whereas either drug alone produced benefits (66).

Beta-lactam antibiotics

An active dipeptide transport system that depends on hydrogen ions takes up non-ester amino-beta-lactams (penicillin, amoxicillin, and oral first-generation cephalosporins) (67–69) and specific cephalosporins that lack the alpha-amino group (cefixime, ceftibuten, cefdinir, cefprozil) (70,71). Nifedipine increases amoxicillin and cefixime absorption, probably by stimulating the dipeptide transport system, since the serum concentrations of passively absorbed drugs and intestinal blood flow did not change (72–74).

Ciclosporin

Ciclosporin significantly inhibits the metabolism of nifedipine, leading to increased effects (75).

Clarithromycin

Severe hypotension with reduced systemic vascular resistence occurred in a patient taking nifedipine and clarithromycin (76).

- A 77-year-old ex-smoker with a history of type 2 diabetes and hypertension took modified-release nifedipine 60 mg tds, doxazosin 8 mg tds, captopril 25 mg tds, metformin 850 mg tds, and aspirin 300 mg/day. Clarithromicin 500 mg tds was added for a presumed chest infection. Two days later he developed shock, heart block, and multiorgan failure, with reduced systemic vascular resistence, compatible with septic shock secondary to a respiratory infection. However, no causative micro-organisms were identified and the leukocyte count was normal. The antihypertensive agents were withdrawn and noradrenaline was given. He recovered.

The authors suggested that extreme hypotension had occurred secondary to a fall in systemic vascular resistence, because of excessive hypotension, due to an interaction of nifedipine with clarithromicin. The time course supported this hypothesis.

Fluconazole

A pharmacokinetic interaction between fluconazole and nifedipine has been reported (77).

- A 16-year-old man with neurofibromatosis type 1, a malignant pheochromocytoma with lung and bone metastases, and candidiasis of the gastrointestinal tract with fungemia was taking nifedipine for arterial hypertension. The fungal infection responded to fluconazole, but three attempts to withdraw the fluconazole resulted in recurrence of headache, palpitation, and increased blood pressure.

After careful pharmacokinetic assessment of the effects of fluconazole withdrawal on 24-hour ambulatory blood pressure and nifedipine plasma concentration, the authors concluded that fluconazole can enhance the blood-pressure lowering effects of nifedipine by increasing its plasma concentrations, most likely by inhibiting CYP3A4.

Lithium

Lithium clearance is reduced by about 30% by nifedipine (78).

Meglitinides

Concomitant treatment with the CYP3A4 substrate nifedipine altered the mean AUC and mean C_{max} of repaglinide by 11 and 3% respectively (79).

Melatonin

Melatonin has a hypotensive effect in both normotensive and hypertensive subjects. In a double-blind, randomized, crossover study designed to evaluate whether evening ingestion of melatonin potentiates the antihypertensive effect of nifedipine monotherapy in 50 patients with well-controlled mild to moderate hypertension aged 38–65 years (28 men, 22 women), there was a surprising significant increase in blood pressure and heart rate throughout 24 hours (80). The authors suggested that there was competition between melatonin and nifedipine, with impairment of the antihypertensive efficacy of the calcium channel blocker.

Nafcillin

Nine healthy men, aged 21–23 years, participated in a randomized, placebo-controlled, crossover study of the effects of 5 days pretreatment with nafcillin or placebo four times daily on the pharmacokinetics of an oral dose of nifedipine of 10 mg. Nafcillin markedly increased the clearance of nifedipine, suggesting that it is a potent enzyme inducer (81).

Nitrates, organic

- A case of laryngeal edema has been reported with the use of isosorbide dinitrate spray followed shortly by sublingual nifedipine in an attempt to reduced high blood pressure (230/130 mmHg) in a 65-year-old woman who presented with coma secondary to a large intracerebral hemorrhage; the same complication recurred when she was rechallenged with the same sequence of drugs (82).

Phenobarbital

A possible interaction of nifedipine with phenobarbital has been reported (83).

- A 67-year-old man with hypertension and seizures taking nifedipine, digoxin, ticlopidine, paroxetine, and clorazepate dipotassium also took phenobarbital 100 mg/day for 2 months, after which the serum phenobarbital concentration 24 hours after the last dose of phenobarbital was almost twice as high as expected.

Since nifedipine and ticlopidine inhibit CYP2C9 and CYP2C19 respectively, both involved in phenobarbital metabolism, the authors suggested that phenobarbital concentrations should be monitored when these drugs are co-administered.

Phenytoin

Increased serum phenytoin concentrations have been reported after the introduction of nifedipine (84,85).

Quinidine

Nifedipine may enhance the elimination of quinidine (86) and cause hypotension and a loss of antidysrhythmic effect (87).

Rifamycins

The elimination of nifedipine is enhanced by rifampicin (SEDA-14, 259).

Tacrolimus

Nifedipine can increase the blood concentration of the macrolide immunosuppressant tacrolimus (SEDA-21, 392).

Food–Drug Interactions

The effect of food on the systemic availability of two modified-release dosage forms of nifedipine for once-daily administration (Adalat OROS, a tablet with an osmotic push–pull system, and Slofedipine XL, a tablet with an acid-resistant coating) has been investigated in 24 healthy men in an open, randomized, crossover study (88). After fasted administration the systemic availability was slightly lower for Slofedipine XL than Adalat OROS, with a point estimate of 82%, mainly resulting from differences in nifedipine concentrations during the first 15 hours after administration. Maximum plasma concentrations were lower after Slofedipine XL compared with Adalat OROS (point estimate 84%). After a high-fat breakfast the differences in availability between the two products were greater than while fasting, with point estimates of 70% for AUC and 81% for C_{max}. Most striking was the lag time after food for Slofedipine XL which was more than 15 hours in 15 of 24 subjects. The availability of nifedipine from Slofedipine XL compared with Adalat OROS was only 28% over the intended dosing interval of 24 hours. The delay in nifedipine absorption when Slofedipine XL is administered may be explained by properties of the formulation, since the acid-resistant coating probably confers delayed absorption, due to prolongation of the gastric residence time, while the osmotic push–pull system is not sensitive to food. The same authors conducted a similar study in 24 healthy subjects, in which they compared Adalat OROS with Nifedicron, a product that consists of a capsule containing several mini-tablets (89). There was a higher rate of availability for Nifedicron in the fasted state, and after a high-fat breakfast the differences between the products became even more pronounced. The most important effect of concomitant food was reflected by a pronounced increase in Cmax after Nifedicron, which resulted in a more than three-fold higher mean concentration than Adalat OROS. This phenomenon may result in safety and tolerability problems.

Management of adverse drug reactions

Ankle oedema is often associated with nifedipine, as a consequence of local vasodilatation. Prevention studies of this adverse reaction with simultaneous administration of diuretics have yielded controversial results: furosemide did not prevent the acute increase in foot volume due by nifedipine in healthy volunteers (90), while other studies have shown that edema formation can be attenuated by diuretics (91). To investigate further whether diuretic pretreatment is effective in controlling foot swelling, the effect of nifedipine after premedication with amiloride or chlorthalidone was compared with the effect of nifedipine alone in 10 healthy volunteers (8 men, 2 women) aged 19–27 years (92). In four separate experiments they received: (a) nifedipine 20 mg/day with placebo pretreatment; (b) nifedipine 20 mg/day after treatment for 5 days with amiloride 5 mg bd; (c) nifedipine 20 mg/day after treatment for 5 days with chlortalidone 50 mg/day; (d) pre-treatment with placebo and placebo in place of nifedipine. Amiloride and chlortalidone pretreatment produced marked foot volume depletion, with a 2–3% reduction in body weight, a 5–10 increase in hematocrit, and a 14–23% increase in plasma colloid osmotic pressure. The mean foot volume after both chlortalidone (1282 ml) and amiloride (1289 ml) was significantly lower than without pretreatment (1315 ml). However, neither amiloride nor chlortalidone significantly altered the acute increase in foot volume with nifedipine, although foot volume remained lower after pretreatment. The authors therefore concluded that diuretics mitigate edema due to nifedipine but do not directly interfere with its formation.

References

1. Ricciardi MJ, Knight BP, Martinez FJ, Rubenfire M. Inhaled nitric oxide in primary pulmonary hypertension: a safe and effective agent for predicting response to nifedipine. J Am Coll Cardiol 1998;32(4):1068–73.
2. Sitbon O, Humbert M, Jagot JL, Taravella O, Fartoukh M, Parent F, Herve P, Simonneau G. Inhaled nitric oxide as a screening agent for safely identifying responders to oral calcium-channel blockers in primary pulmonary hypertension. Eur Respir J 1998;12(2):265–70.
3. Brown MJ, Palmer CR, Castaigne A, de Leeuw PW, Mancia G, Rosenthal T, Ruilope LM. Morbidity and mortality in patients randomised to double-blind treatment with a long-acting calcium-channel blocker or diuretic in the International Nifedipine GITS study: Intervention as a Goal in Hypertension Treatment (INSIGHT). Lancet 2000;356(9227):366–72.
4. Maxwell CJ, Hogan DB, Campbell NRC, Ebly EM. Nifedipine and mortality risk in the elderly: relevance of drug formulation, dose and duration. Pharmacoepidemiol Drug Saf 2000;9:11–23.
5. Poole-Wilson PA, Lubsen J, Kirwan BA, van Dalen FJ, Wagener G, Danchin N, Just H, Fox KA, Pocock SJ, Clayton TC, Motro M, Parker JD, Bourassa MG, Dart AM, Hildebrandt P, Hjalmarson Å, Kragten JA, Molhoek GP, Otterstad JE, Seabra-Gomes R, Soler-Soler J, Weber S. Effect of long-acting nifedipine on mortality and cardiovascular morbidity in patients with stable angina requiring treatment (ACTION trial): randomised controlled trial. Lancet 2004;364:849–57.
6. Committee on Safety of Medicines. Nifedipine (Adalat) and myocardial ischaemia. Curr Probl 1979;4:1.
7. Deanfield J, Wright C, Fox K. Treatment of angina pectoris with nifedipine: importance of dose titration. BMJ (Clin Res Ed) 1983;286(6376):1467–70.
8. Yagil Y, Kobrin I, Leibel B, Ben-Ishay D. Ischemic ECG changes with initial nifedipine therapy of severe hypertension. Am Heart J 1982;103(2):310–1.

9. Stason WB, Schmid CH, Niedzwiecki D, Whiting GW, Caubet JF, Cory D, Luo D, Ross SD, Chalmers TC. Safety of nifedipine in angina pectoris: a meta-analysis. Hypertension 1999;33(1):24–31.
10. Ishibashi Y, Shimada T, Yoshitomi H, Sano K, Oyake N, Umeno T, Sakane T, Murakami Y, Morioka S. Sublingual nifedipine in elderly patients: even a low dose induces myocardial ischaemia. Clin Exp Pharmacol Physiol 1999; 26(5–6):404–10.
11. Grassi G. Neuroadrenergic effects of calcium channel blockers: a developing concept. J Hypertens 2004;22:887–8.
12. Grayson HA, Kennedy JD. Torsades de pointes and nifedipine. Ann Intern Med 1982;97(1):144.
13. Krikler DM, Rowland E. Torsade de pointes and nifedipine. Ann Intern Med 1982;97(4):618–9.
14. Castaneda MP, Walsh CA, Woroniecki RP, Del Rio M, Flynn JT. Ventricular arrhythmia following short-acting nifedipine administration. Pediatr Nephrol 2005; 20(7):1000–2.
15. Nobile-Orazio E, Sterzi R. Cerebral ischaemia after nifedipine treatment. BMJ (Clin Res Ed) 1981;283(6297):948.
16. Ellrodt AG, Ault MJ, Riedinger MS, Murata GH. Efficacy and safety of sublingual nifedipine in hypertensive emergencies. Am J Med 1985;79(4A):19–25.
17. Pitlik S, Manor RS, Lipshitz I, Perry G, Rosenfeld J. Transient retinal ischaemia induced by nifedipine. BMJ (Clin Res Ed) 1983;287(6408):1845–6.
18. Bertel O, Conen LD. Treatment of hypertensive emergencies with the calcium channel blocker nifedipine. Am J Med 1985;79(4A):31–5.
19. Gill JS, Zezulka AV, Horrocks PM. Rupture of a cerebral aneurysm associated with nifedipine treatment. Postgrad Med J 1986;62(733):1029–30.
20. Murphy MB, Scriven AJ, Brown MJ, Causon R, Dollery CT. The effects of nifedipine and hydralazine induced hypotension on sympathetic activity. Eur J Clin Pharmacol 1982;23(6):479–82.
21. Struthers AD, Reid JL. Nifedipine does not influence adrenaline induced hypokalaemia in man. Br J Clin Pharmacol 1983;16(3):342–3.
22. Neaton JD, Grimm RH Jr, Prineas RJ, Stamler J, Grandits GA, Elmer PJ, Cutler JA, Flack JM, Schoenberger JA, McDonald R, et al. Treatment of Mild Hypertension Study. Final results. Treatment of Mild Hypertension Study Research Group. JAMA 1993; 270(6):713–24.
23. Tishler M, Armon S. Nifedipine-induced hypokalemia. Drug Intell Clin Pharm 1986;20(5):370–1.
24. Laporte JR, Ibanez L, Ballarin E, Perez E, Vidal X. Fatal aplastic anaemia associated with nifedipine. Lancet 1998;352(9128):619–20.
25. Pernu HE, Knuuttila ML. Macrophages and lymphocyte subpopulations in nifedipine- and cyclosporin A-associated human gingival overgrowth. J Periodontol 2001;72(2):160–6.
26. Nurmenniemi PK, Pernu HE, Knuuttila ML. Mitotic activity of keratinocytes in nifedipine- and immunosuppressive medication-induced gingival overgrowth. J Periodontol 2001;72(2):167–73.
27. Miranda J, Brunet L, Roset P, Berini L, Farre M, Mendieta C. Prevalence and risk of gingival enlargement in patients treated with nifedipine. J Periodontol 2001;72(5):605–11.
28. Goglin WK, Elliott BM, Deppe SA. Nifedipine-induced hypotension and mesenteric ischemia. South Med J 1989;82(2):274–5.
29. Lavy A. Corrosive effect of nifedipine in the upper gastrointestinal tract. Diagn Ther Endosc 2000;6:39–41.
30. Niezabitowski LM, Nguyen BN, Gums JG. Extended-release nifedipine bezoar identified one year after discontinuation. Ann Pharmacother 2000;34(7–8):862–4.
31. Yeen WC, Willis IH. Retention of extended release nifedipine capsules in a patient with enteric stricture causing recurrent small bowel obstruction. South Med J 2005;98(8):839–42.
32. Babany G, Uzzan F, Larrey D, Degott C, Bourgeois P, Rene E, Vissuzaine C, Erlinger S, Benhamou JP. Alcoholic-like liver lesions induced by nifedipine. J Hepatol 1989;9(2):252–5.
33. Ota K, Shijo H, Kokawa H, Kubara K, Kim T, Akiyoshi N, Yokoyama M, Okumura M. Effects of nifedipine on hepatic venous pressure gradient and portal vein blood flow in patients with cirrhosis. J Gastroenterol Hepatol 1995;10(2):198–204.
34. Diamond JR, Cheung JY, Fang LS. Nifedipine-induced renal dysfunction. Alterations in renal hemodynamics. Am J Med 1984;77(5):905–9.
35. Hall-Craggs M, Light PD, Peters RW. Development of immune complex nephritis during treatment with the calcium channel-blocking agent nifedipine. Hum Pathol 1984;15(7):691–4.
36. Groth H, Foerster EC, Neyses L, Kulhmann U, Vetter H, Vetter W. Nifedipine beim hypertensiven Notfall und bei schwerer Hypertonie. [Nifedipine in hypertensive emergencies and severe hypertension.] Schweiz Rundsch Med Prax 1984;73(2):45–9.
37. Williams G, Donaldson RM. Nifedipine and nocturia. Lancet 1986;1(8483):738.
38. Brenner S, Golan H, Bialy-Golan A, Ruocco V. Lesion topography in two cases of nifedipine-related pemphigus. J Eur Acad Dermatol Venereol 1999;13(2):123–6.
39. Ameen M, Harman KE, Black MM. Pemphigoid nodularis associated with nifedipine. Br J Dermatol 2000;142(3):575–7.
40. Wood TML. Photosensitivity reaction associated with nifedipine. BMJ (Clin Res Ed) 1986;292:992.
41. Alcalay J, David M. Generalised fixed drug eruption associated with nifedipine. BMJ (Clin Res Ed) 1986;292:450.
42. Tsele E, Chu AC. Nifedipine and telangiectasias. Lancet 1992;339(8789):365–6.
43. Keidar S, Binenboim C, Palant A. Muscle cramps during treatment with nifedipine. BMJ (Clin Res Ed) 1982;285(6350):1241–2.
44. Pedro-Botet J, Minguez S, Supervia A. Sublingual nifedipine-induced anaphylaxis. Arch Intern Med 1998;158(12):1379.
45. Gubinelli E, Cocuroccia B, Girolomoni G. Subacute cutaneous lupus erythematosus induced by nifedipine. J Cutan Med Surg 2003;7:243–6.
46. Royal College of Obstetricians and Gynaecologists. Clinical Green Top Guidelines. Tocolytic drugs for women in preterm labour 1(B). Oct 2002. http://www.rcog.org.uk/index.asp?PageID = 536.
47. Koks CA, Brolmann HA, de Kleine MJ, Manger PA. A randomized comparison of nifedipine and ritodrine for suppression of preterm labor. Eur J Obstet Gynecol Reprod Biol 1998;77(2):171–6.
48. Garcia-Velasco JA, Gonzalez Gonzalez A. A prospective, randomized trial of nifedipine vs. ritodrine in threatened preterm labor. Int J Gynaecol Obstet 1998;61(3):239–44.
49. Oei SG, Oei SK, Brolmann HA. Myocardial infarction during nifedipine therapy for preterm labor. N Engl J Med 1999;340(2):154.

50. Papatsonis DN, Lok CA, Bos JM, Geijn HP, Dekker GA. Calcium channel blockers in the management of preterm labor and hypertension in pregnancy. Eur J Obstet Gynecol Reprod Biol 2001;97(2):122–40.
51. Hodges R, Barkehall-Thomas A, Tippett C. Maternal hypoxia associated with nifedipine for threatened preterm labour. Br J Obstet Gynaecol 2004;111:380–1.
52. Koontz SL, Friedman SA, Schwartz ML. Symptomatic hypocalcemia after tocolytic therapy with magnesium sulfate and nifedipine. Am J Obstet Gynecol 2004;190:1773–6.
53. van Veen AJ, Pelinck MJ, van Pampus MG, Erwich JJ. Severe hypotension and fetal death due to tocolysis with nifedipine. BJOG 2005;112(4):509–10.
54. Papatsonis DN, Carbonne B, Dekker GA, Flenady V, King JF. Severe hypotension and fetal death due to tocolysis with nifedipine. BJOG 2005;112(11):1582–3.
55. Kandysamy V, Thomson AJ. Severe hypotension and fetal death due to tocolysis with nifedipine. BJOG 2005;112(11): 1583–4.
56. Johnson KA, Mason GC. Severe hypotension and fetal death due to tocolysis with nifedipine. BJOG 2005; 112(11):1583.
57. van Geijn HP, Lenglet JE, Bolte AC. Nifedipine trials: effectiveness and safety aspects. BJOG 2005;112 Suppl 1:79–83.
58. Bortolus R, Ricci E, Chatenoud L, Parazzini F. Nifedipine administered in pregnancy: effect on the development of children at 18 months. BJOG 2000;107(6):792–4.
59. Rubio-Guerra AF, Vargas-Ayala G, Lozano-Nuevo JJ, Narvaez-Rivera JL, Rodriguez-Lopez L. Comparison between isosorbide dinitrate aerosol and nifedipine in the treatment of hypertensive emergencies. J Hum Hypertens 1999;13(7):473–6.
60. Sunderji R, Shalansky KF, Beauchesne MF, Fung A. Is there a role for nifedipine in hypertensive urgencies? Can J Hosp Pharm 1999;52:167–70.
61. Goiti JJ. Calcium channel blocking agents and the heart. BMJ (Clin Res Ed) 1985;291(6507):1505.
62. Plotkin CN, Eckenbrecnt PD, Waldo DA. Consecutive cardiac arrests on induction of anesthesia associated with nifedipine-induced carotid sinus hypersensitivity. Anesth Analog 1989;68(3):402–5.
63. Schier JG, Howland MA, Hoffman RS, Nelson LS. Fatality from administration of labetalol and crushed extended-release nifedipine. Ann Pharmacother 2003;37:1420–3.
64. Lee DC, Greene T, Dougherty T, Pearigen P. Fatal nifedipine ingestions in children. J Emerg Med 2000;19(4):359–61.
65. Qureshi S, Laganiere S, McGilveray IJ, Lacasse Y, Calle G. Nifedipine–alcohol interaction. JAMA 1990;264(13):1660–1.
66. Solomon SA, Ramsay LE, Yeo WW, Parnell L, Morris-Jones W. Beta blockade and intermittent claudication: placebo controlled trial of atenolol and nifedipine and their combination. BMJ 1991;303(6810):1100–4.
67. Ganapathy ME, Prasad PD, Mackenzie B, Ganapathy V, Leibach FH. Interaction of anionic cephalosporins with the intestinal and renal peptide transporters PEPT 1 and PEPT 2. Biochim Biophys Acta 1997;1324(2):296–308.
68. Matsumoto S, Saito H, Inui K. Transcellular transport of oral cephalosporins in human intestinal epithelial cells, Caco-2: interaction with dipeptide transport systems in apical and basolateral membranes. J Pharmacol Exp Ther 1994;270(2):498–504.
69. Sugawara M, Iseki K, Miyazaki K, Shiroto H, Kondo Y, Uchino J. Transport characteristics of ceftibuten, cefixime and cephalexin across human jejunal brush-border membrane. J Pharm Pharmacol 1991;43(12):882–4.
70. Dantzig AH, Duckworth DC, Tabas LB. Transport mechanisms responsible for the absorption of loracarbef, cefixime, and cefuroxime axetil into human intestinal Caco-2 cells. Biochim Biophys Acta 1994;1191(1):7–13.
71. Winstanley PA, Orme ML. The effects of food on drug bioavailability. Br J Clin Pharmacol 1989;28(6):621–8.
72. Westphal JF, Trouvin JH, Deslandes A, Carbon C. Nifedipine enhances amoxicillin absorption kinetics and bioavailability in humans. J Pharmacol Exp Ther 1990; 255(1):312–7.
73. Duverne C, Bouten A, Deslandes A, Westphal JF, Trouvin JH, Farinotti R, Carbon C. Modification of cefixime bioavailability by nifedipine in humans: involvement of the dipeptide carrier system. Antimicrob Agents Chemother 1992;36(11):2462–7.
74. Deslandes A, Camus F, Lacroix C, Carbon C, Farinotti R. Effects of nifedipine and diltiazem on pharmacokinetics of cefpodoxime following its oral administration. Antimicrob Agents Chemother 1996;40(12):2879–81.
75. McFadden JP, Pantin JE, Parkes AV, et al. Cyclosporin decreases nifedipine metabolism. BMJ 1989;299:1224.
76. Gerònimo-Pardo M, Cuartero-del-Pozo AB, Jimènez-Vizuete JM, Cortinas-Sàez M, Peyrò-Garcia R. Clarithromycin-nifedipine interaction as possible cause of vasodilatory shock. Ann Pharmacother 2005;39:538–42.
77. Kremens B, Brendel E, Bald M, Czyborra P, Michel MC. Loss of blood pressure control on withdrawal of fluconazole during nifedipine therapy. Br J Clin Pharmacol 1999;47(6):707–8.
78. Bruun NE, Ibsen H, Skott P, Toftdahl D, Giese J, Holstein-Rathlou NH. Lithium clearance and renal tubular sodium handling during acute and long-term nifedipine treatment in essential hypertension. Clin Sci (Lond) 1988;75(6):609–13.
79. Hatorp V, Hansen KT, Thomsen MS. Influence of drugs interacting with CYP3A4 on the pharmacokinetics, pharmacodynamics, and safety of the prandial glucose regulator repaglinide. J Clin Pharmacol 2003;43(6):649–60.
80. Lusardi P, Piazza E, Fogari R. Cardiovascular effects of melatonin in hypertensive patients well controlled by nifedipine: a 24-hour study. Br J Clin Pharmacol 2000;49(5):423–7.
81. Lang CC, Jamal SK, Mohamed Z, Mustafa MR, Mustafa AM, Lee TC. Evidence of an interaction between nifedipine and nafcillin in humans. J Clin Pharmacol 2003;55:588–90.
82. Silfvast T, Kinnunen A, Varpula T. Laryngeal oedema after isosorbide dinitrate spray and sublingual nifedipine. BMJ 1995;311(6999):232.
83. Sánchez-Romero A, García-Delgado R, Durán-Quintana JA. ¿Puede el tratamiento asociado con ticlopidina y nifedipina aumentar los niveles séricos de fonobardital? Rev Neurol 2003;36:433–4.
84. Ahmad S. Nifedipine–phenytoin interaction. J Am Coll Cardiol 1984;3(6):1582.
85. Bahls FH, Ozuna J, Ritchie DE. Interactions between calcium channel blockers and the anticonvulsants carbamazepine and phenytoin. Neurology 1991;41(5):740–2.
86. Green JA, Clementi WA, Porter C, Stigelman W. Nifedipine–quinidine interaction. Clin Pharm 1983;2(5):461–5.
87. Farringer JA, Green JA, O'Rourke RA, Linn WA, Clementi WA. Nifedipine-induced alterations in serum quinidine concentrations. Am Heart J 1984;108(6):1570–2.
88. Schug BS, Brendel E, Chantraine E, Wolf D, Martin W, Schall R, Blume HH. The effect of food on the pharmacokinetics of nifedipine in two slow release formulations: pronounced lag-time after a high fat breakfast. Br J Clin Pharmacol 2002;53(6):582–8.
89. Schug BS, Brendel E, Wonnemann M, Wolf D, Wargenau M, Dingler A, Blume HH. Dosage form-related food interaction observed in a marketed once-daily nifedipine formulation after a high-fat American breakfast. Eur J Clin Pharmacol 2002;58(2):119–25.

90. Tvan Hamersvelt HW, van Kestern M, Kloke HJ, Wetzels JF, Valk I, Koene RA, Huysmans FT. Ankle edema with nifedipine is not primarily due to sodium retention. Jam Soc Nephrol 1993;4:541.
91. Luscher TF, Waeber B. Efficacy and safety of various combination therapies based on a calcium antagonists in essential hypertension: results of a placebo-controlled randomized trial. J Cardiovasc Pharmacol 1993;21:305–9.
92. van der Heijden AG, Huysmans FTM, van Hamersvelt HW. Foot volume increase on nifedipine is not prevented by pretreatment with diuretics. J Hypertens 2004;22:425–30.

Nimodipine

See also Calcium channel blockers

General Information

Nimodipine is a dihydropyridine calcium channel blocker.

In the absence of effective neuroprotective treatment for ischemic stroke, a double-blind, randomized, placebo-controlled trial has been performed in 454 patients in primary care (1). Nimodipine 30 mg/day or placebo was started within 6 hours after the onset of the stroke and continued for 10 days. Nimodipine had no effect on all-cause mortality or dependency in daily life. In patients with ischemic stroke documented by CT scan, nimodipine had a borderline significant adverse effect on outcome. Nimodipine was tolerated as well as placebo (7 versus 8 treatment withdrawals respectively), but the lack of benefit does not support the use of any voltage-sensitive calcium channel blocker in ischemic stroke.

It has been suggested that calcium channel blockers can be used to treat cocaine dependence, and some studies have shown reductions in cocaine-induced subjective and cardiovascular responses with nifedipine and diltiazem. The cardiovascular and subjective responses to cocaine have been evaluated in a double-blind, placebo-controlled, crossover study in five subjects pretreated with two dosages of nimodipine (2). Nimodipine 60 mg attenuated the systolic, but not the diastolic, blood pressure rise after cocaine. In three subjects, nimodipine 90 mg produced greater attenuation than 60 mg. The subjective effects of cocaine were not altered by either dosage of nimodipine.

Placebo-controlled studies

In a post-hoc analysis of the relation between blood pressure, nimodipine treatment, and outcome in 350 patients in a placebo-controlled trial in ischemic stroke, there was a higher fatality rate with nimodipine, together with a greater reduction in blood pressure (3). Stroke severity was the strongest predictor of a bad outcome at both 3 weeks and 3 months. A high initial blood pressure within the first 24 hours was associated with better survival in moderately severe strokes, but the reverse in severe stroke.

Organs and Systems

Cardiovascular

When nimodipine was given intravenously to 87 patients with proven subarachnoid hemorrhage in a maintenance dosage of 2 mg/hour, it caused significant hypotension in one-third of cases (4). With maintenance oral administration, the incidence of hypotension was much lower. It is therefore recommended that if nimodipine is prescribed, the oral route is preferable, even in intubated patients who should receive it via a nasogastric tube.

Nimodipine reportedly aggravated cardiac dysfunction in a patient with subarachnoid hemorrage (5).

- A 52-year-old woman with a subarachnoid hemorrage had severe myocardial depression with anterolateral ischemic electrocardiographic signs after an intravenous infusion of nimodipine 0.5–1.2 mg/h had been started. The cardiac index improved only after nimodipine had been withdrawn in the presence of inotropes. The cardiac index fell shortly after nimodipine had been reintroduced; nimodipine could only be restarted 48 hour later, whien it was well tolerated.

The authors suggested that nimodipine may aggravate cardiac dysfunction associated with the acute phase of subarachnoid hemorrage.

Nervous system

Nimodipine has been routinely used to reduce cerebral vasospasm in patients with subarachnoid hemorrhage. A randomized, double-blind, placebo-controlled trial was stopped early because of excess mortality (8/75 versus 1/74 deaths with nimodipine and placebo respectively), which was attributed to an increased incidence of surgical bleeding (6). In addition, its use for cerebral protection in a cardiac valve replacement trial involving cardiopulmonary bypass was also terminated prematurely because of increased cerebrovascular complications and excessive blood loss in a small subgroup of patients (7). In traumatic subarachnoid hemorrhage, a trial in 123 patients showed a significant reduction in unfavorable outcomes (death, vegetative survival, or severe disability) at 6 months with nimodipine (25 versus 46% for placebo) (8).

References

1. Horn J, de Haan RJ, Vermeulen M, Limburg M. Very Early Nimodipine Use in Stroke (VENUS): a randomized, double-blind, placebo-controlled trial. Stroke 2001;32(2):461–5.
2. Kosten TR, Woods SW, Rosen MI, Pearsall HR. Interactions of cocaine with nimodipine: a brief report. Am J Addict 1999;8(1):77–81.
3. Fogelholm R, Palomäki H, Erilä T, Rissanen A, Kaste M. Blood pressure, nimodipine, and outcome of ischemic stroke. Acta Neurol Scand 2004;109:200–4.
4. Porchet F, Chiolero R, de Tribolet N. Hypotensive effect of nimodipine during treatment for aneurysmal subarachnoid haemorrhage. Acta Neurochir (Wien) 1995;137(1–2):62–9.
5. Subramani K, Ghrew M. Severe myocardial depression following intravenous nimodipine for aneurysmal subarachnoid haemorrhage. Intensive Care Med 2004;30:1498–9.

6. Wagenknecht LE, Furberg CD, Hammon JW, Legault C, Troost BT. Surgical bleeding: unexpected effect of a calcium antagonist. BMJ 1995;310(6982):776–7.
7. Heininger K, Kuebler J. Use of nimodipine is safe. Stroke 1996;27(10):1911–3.
8. Harders A, Kakarieka A, Braakman RGerman tSAH Study Group. Traumatic subarachnoid hemorrhage and its treatment with nimodipine. J Neurosurg 1996;85(1):82–9.

Nisoldipine

See also Calcium channel blockers

General Information

Nisoldipine is a dihydropyridine calcium channel blocker (1).

Nisoldipine is almost completely absorbed, but its systemic availability is only 5% because of first-pass metabolism in the gut and liver (2). Its pharmacokinetics therefore vary with liver blood flow. It is over 99% protein bound and is eliminated by hepatic metabolism. Although it is administered as a racemic mixture, its plasma concentrations are almost entirely attributable to the active isomer, because of stereoselective clearance (3).

In a double-blind study in African Americans with hypertension, extended-release nisoldipine and amlodipine were equally effective (4). The safety profiles were comparable: most of the adverse effects were mild and transient and common to all vasodilators (i.e. headache, dizziness, and edema); three of 92 patients taking nisoldipine and five of 101 taking amlodipine withdrew because of adverse events.

Organs and Systems

Cardiovascular

In the Appropriate Blood Pressure Control in Diabetes (ABCD) Trial, a prospective, randomized, blind comparison of moderate control of blood pressure (target diastolic pressure, 80–89 mmHg) with intensive control (diastolic pressure 75 mmHg) in 470 patients with hypertension, nisoldipine was associated with a higher incidence of fatal and non-fatal myocardial infarctions (n = 24) than enalapril (n = 4) (RR = 9.5; 95%CI = 2.7, 34) (5).

Drug Administration

Drug formulations

Nisoldipine has been marketed as a modified-release formulation for once-daily therapy using a technology called coat-core (6), from which the drug is absorbed across the entire gastrointestinal tract, including the colon. This formulation is reportedly not associated with the proischemic effects that have been reported with the rapid-release formulation (7).

Susceptibility factors

Age

Plasma concentrations of nisoldipine increase with age (8).

Liver disease

Nisoldipine should not be used in patients with hepatic cirrhosis.

Drug-Drug Interactions

Cimetidine

Cimetidine inhibits the metabolism of nisoldipine, but the effect may not be clinically significant (9).

Grapefruit juice

Concomitant ingestion of nisoldipine with grapefruit juice should be avoided, since grapefruit inhibits the metabolism of nisoldipine (10).

Enzyme inducers

Phenytoin (11) and rifampicin (12,13), inducers of CYP3A4, reduce the systemic availability of nisoldipine and can reduce its efficacy.

Food-Drug Interactions

Concomitant intake of the nisoldipine with high fat, high calorie foods increased the maximum plasma concentrations of nisoldipine. This effect can be avoided by taking the drug up to 30 minutes before food.

References

1. Mitchell J, Frishman W, Heiman M. Nisoldipine: a new dihydropyridine calcium-channel blocker. J Clin Pharmacol 1993;33(1):46–52.
2. Heinig R. Clinical pharmacokinetics of nisoldipine coat-core. Clin Pharmacokinet 1998;35(3):191–208.
3. Tokuma Y, Noguchi H. Stereoselective pharmacokinetics of dihydropyridine calcium antagonists. J Chromatogr A 1995; 694(1):181–93.
4. White WB, Saunders E, Noveck RJ, Ferdinand K. Comparative efficacy and safety of nisoldipine extended-release (ER) and amlodipine (CESNA-III study) in African American patients with hypertension. Am J Hypertens 2003;16:739–45.
5. Estacio RO, Jeffers BW, Hiatt WR, Biggerstaff SL, Gifford N, Schrier RW. The effect of nisoldipine as compared with enalapril on cardiovascular outcomes in patients with non-insulin-dependent diabetes and hypertension. N Engl J Med 1998;338(10):645–52.
6. White WB. Pharmacologic agents in the management of hypertension—nisoldipine coat-core. J Clin Hypertens (Greenwich) 2007;9(4):259–66.
7. Hamilton SF, Houle LM, Thadani U. Rapid-release and coat-core formulations of nisoldipine in treatment of hypertension, angina, and heart failure. Heart Dis 1999;1(5):279–88.

8. Davidsson GK, Edwards JS, Davidson C. The effect of age and liver disease on the pharmacokinetics of the calcium antagonist, nisoldipine. Curr Med Res Opin 995;13(5):285–97.
9. van Harten J, van Brummelen P, Lodewijks MT, Danhof M, Breimer DD. Pharmacokinetics and hemodynamic effects of nisoldipine and its interaction with cimetidine. Clin Pharmacol Ther 1988;43(3):332–41.
10. Takanaga H, Ohnishi A, Murakami H, Matsuo H, Higuchi S, Urae A, Irie S, Furuie H, Matsukuma K, Kimura M, Kawano K, Orii Y, Tanaka T, Sawada Y. Relationship between time after intake of grapefruit juice and the effect on pharmacokinetics and pharmacodynamics of nisoldipine in healthy subjects. Clin Pharmacol Ther 2000;67(3):201–14.
11. Michelucci R, Cipolla G, Passarelli D, Gatti G, Ochan M, Heinig R, Tassinari CA, Perucca E. Reduced plasma nisoldipine concentrations in phenytoin-treated patients with epilepsy. Epilepsia 1996;37(11):1107–10.
12. Tada Y, Tsuda Y, Otsuka T, Nagasawa K, Kimura H, Kusaba T, Sakata T. Case report: nifedipine-rifampicin interaction attenuates the effect on blood pressure in a patient with essential hypertension. Am J Med Sci 1992;303(1):25–7.
13. Yoshimoto H, Takahashi M, Saima S. Influence of rifampicin on antihypertensive effects of dihydropiridine calcium-channel blockers in four elderly patients. Nippon Ronen Igakkai Zasshi 1996;33(9):692–6.

Nitrates, organic

See also Antianginal drugs

General Information

Organic nitrates have been in use for over 135 years, and their adverse effects are therefore well documented. The first compound of this class to be used was amyl nitrite, synthesized in 1844 by Ballard and used to treat angina pectoris by Brunton in 1869. Glyceryl trinitrate was first synthesized by Ascagne Sobrero in 1847 and was used by William Murrell to treat angina pectoris in 1879.

Nitrates have been produced in many chemical forms, but the current choice lies between the rapid-onset, short-acting glyceryl trinitrate (nitroglycerin) and the longer-acting isosorbide mononitrate and isosorbide dinitrate.

Mechanism of action

The nitrates act by releasing nitric oxide, which relaxes vascular smooth muscle. The discovery that endothelium-derived relaxing factor (EDRF) is nitric oxide (1) stimulated new interest in these drugs, as nitric oxide not only controls local vessel wall tension in response to shear stress, but also plays a role in regulating the interaction of platelets with blood vessel walls. The release of nitric oxide from the walls of atheromatous arteries is reduced, because of malfunctioning or absent endothelium. Atheromatous arteries behave differently from healthy arteries, in that these vessels vasoconstrict rather than vasodilate when stimulated by acetylcholine. This impairment of the acetylcholine vasomotor response appears to be related to serum cholesterol concentration (2).

Uses

Traditionally, nitrates have been first-line therapy in angina. They can be given to treat acute attacks, taken prophylactically before exertion or other stimuli known to provoke an attack, or used as continuous prophylaxis over 24 hours. Since angina is an episodic symptom, the first two methods are attractive, as continuous prophylaxis requires treatment over long periods when there are no symptoms. Tempering this argument is the realization that there may be silent myocardial ischemia, especially overnight, and morning angina on waking can be troublesome. In practice, a combination of continuous prophylaxis and as-required therapy is often used.

Nitrates are the drugs of choice in patients with left ventricular impairment, in whom they are of benefit when used in combination with hydralazine (3), and they should be used in preference to the calcium antagonists, which cause deterioration in myocardial function by an as yet unknown mechanism (4). In black patients with congestive heart failure taking ACE inhibitors and beta-blockers, a combination of isosorbide dinitrate plus hydralazine significantly reduced total mortality (5).

The potentially life-threatening hypotension caused by an interaction with sildenafil citrate (Viagra), a treatment for male erectile dysfunction, should be highlighted, especially in patients with coronary heart disease needing regular nitrates (6).

Observational studies

Glyceryl trinitrate can be used to treat anal fissure, as an alternative to surgery. Although glyceryl trinitrate causes resolution of symptoms in some cases, its local use as ointment is associated with frequent adverse reactions, mostly headache and anal burning (7,8). The difficulty in adjusting the dose and time of application, the high frequency of adverse reactions (up to 75% of patients), and the high rate of recurrence make the use of glyceryl trinitrate in non-surgical treatment of anal fissures questionable (9). In one case, a patient reported itching, dyspnea, and hypotension after applying glyceryl trinitrate. This unexplained systemic reaction to topical glyceryl trinitrate was reported as "anaphylactic," without a clear demonstration that it was an allergic reaction (10).

Besides its usual indications, topical glyceryl trinitrate has been used to prevent intravenous infusion failure because of phlebitis and extravasation, which occur in 30–60% of patients and can cause discomfort and harm, such as pulmonary embolism, septicemia, and increased mortality. Local application of glyceryl trinitrate patches was efficacious in several studies (11); the mechanism is probably vasodilatation and increased capillary flow. Glyceryl trinitrate is safe, since it causes only skin rashes and transient headache.

Transdermal glyceryl trinitrate is effective as an analgesic co-adjuvant for cancer pain. In combination with opiates it is well tolerated and reduces the daily consumption of morphine (12). Transdermal glyceryl trinitrate was safe in potentiating the analgesic effect of spinal sufentanil analgesia after knee surgery (13). The effects of glyceryl trinitrate include headache, a feeling of pressure in the head, nausea, vomiting, dizziness, tiredness, blackouts, and possibly penile erection and ageusia (14,15).

Placebo-controlled studies

In a multicenter, randomized, placebo-controlled, double-blind study of 0.2% glyceryl trinitrate ointment in 132 patients with anal fissures over at least 4 weeks, healing rates were similar with glyceryl trinitrate and placebo, but adverse events were more frequent in those who used glyceryl trinitrate: 34% complained of headache and 5.9% had orthostatic hypotension (16).

High-dose transdermal glyceryl trinitrate 50–100 mg given over a period of 12 hours each day for 3 months improved exercise tolerance and left ventricular systolic function in 29 patients with congestive heart failure taking a long-term ACE inhibitor (17). The most frequent adverse effects were skin irritation at the site of patch application and headache, which caused premature glyceryl trinitrate withdrawal in three patients.

Intravenous glyceryl trinitrate (starting rate 50 micrograms/minute uptitrated with blood pressure monitoring) for an average of 63 hours was very effective in preventing adverse ischemic events in 200 patients with unstable angina secondary to restenosis after coronary artery angioplasty, while heparin had no effect (18). In this acute setting, complications and adverse effects were not frequent, even if there was an excess of cases with headache or hypotension with glyceryl trinitrate, never leading to premature discontinuation. This study has shown that glyceryl trinitrate is safe and effective in reducing the need for urgent invasive procedures in these patients.

Comparative studies

Glyceryl trinitrate has been used as a tocolytic to prolong gestation, reducing the incidence of preterm labor. When compared with a widely used tocolytic, the beta$_2$-adrenoceptor agonist ritodrine, in a randomized, multicenter trial, transdermal patches of glyceryl trinitrate 10–20 mg had similar efficacy in 245 women with preterm labor. The main adverse effect of glyceryl trinitrate was headache, while with ritodrine palpitation and tachycardia were significantly more frequent. Headache with glyceryl trinitrate could usually be managed with paracetamol. Therefore, glyceryl trinitrate appears to be an alternative to ritodrine and is devoid of dangerous cardiac adverse effects (19).

In 90 patients randomly assigned to surgery (internal sphincterectomy) or topical glyceryl trinitrate, surgery led to a higher rate of healing than glyceryl trinitrate, with fewer adverse events: 29% of surgical patients versus 84% of the patients given glyceryl trinitrate (20). In conclusion, glyceryl trinitrate should not be used to treat anal fissure.

Variceal bleeding is a frequent and serious event in cirrhosis, and carries an increased risk of death (SEDA-22, 218). Therapy to prevent bleeding is therefore essential in these patients. Propranolol alone has been compared with propranolol plus isosorbide-5-mononitrate in a randomized, double-blind study in 95 patients (21). The combined treatment reduced the incidence of variceal bleeding compared with propranolol alone, but without any improvement in survival. Isosorbide-5-mononitrate added to propranolol appeared to be less well tolerated than propranolol alone, since seven patients had to be withdrawn from treatment because of adverse effects (four with feelings of faintness, two with headache, one with angina-like chest pain), compared with one with atrioventricular block taking propranolol alone.

Isosorbide-5-mononitrate has been tested with and without propranolol in a placebo-controlled study in 30 patients with liver cirrhosis and esophageal varices (22). The aim of the study was to assess the severity of previously reported adverse effects (that is renal dysfunction and hepatic encephalopathy) when vasoactive drugs are used to prevent variceal bleeding. Neither isosorbide-5-mononitrate nor propranolol alone or together had any adverse effect on subclinical hepatic encephalopathy or renal function in patients with well-compensated cirrhosis. Severe headache in those taking isosorbide caused three patients to withdraw.

Routes of administration

Nitrates can be administered by various routes. For example, glyceryl trinitrate is used not only as traditional sublingual tablets but also in the form of modified-release tablets, buccal tablets, aerosolized oral spray, intravenous injection, and topical ointment or skin patches for percutaneous absorption (23). These different formulations have been developed largely as a means of controlling the onset and duration of action of glyceryl trinitrate, since in conventional oral form its action is limited by marked hepatic first-pass metabolism.

General adverse effects

The adverse effects of the nitrates are unaffected by the chemical form or by the route of administration. The spectrum and frequency of adverse effects for the individual drugs, such as glyceryl trinitrate, isosorbide dinitrate, and isosorbide-5-mononitrate, appear to be similar.

Generalized vasodilatation predictably produces some common mild adverse effects. These include headache, flushing, and palpitation, which are most often experienced during the early treatment period and to which a degree of tolerance invariably occurs. More severe adverse effects, including syncope, transient cerebral ischemic attacks, and peripheral edema, are less common. Sublingual glyceryl trinitrate tablets produce a characteristic local burning or tingling sensation. Adverse cardiac effects can follow abrupt withdrawal of treatment after long-term exposure or high-dose therapy. Contact irritation and allergic contact dermatitis can occur with cutaneous glyceryl trinitrate in ointment or skin patches, and may be attributed to the drug itself or an inactive excipient, such as an adhesive agent. There was no excess risk of cancer in a historical cohort study in 2023 male fertilizer workers exposed to industrial nitrates at work in Norway during 1945–79 (24).

Amyl nitrite

Amyl nitrite, despite its name, is also used as a nitrate. It is a yellowish liquid taken by inhalation of its vapor for angina pectoris. It has also been used to treat cyanide poisoning, but is nowadays used exclusively as a substance of abuse. Although it is a prescription-only medicine in the UK, it is sold over the counter in glass ampoules to be used as a "room deodorizer" or "incense." It is known under numerous different nicknames and brand names, including snappers, poppers, Rush, Kix, Liquid gold, Locker room, Hardware, Ram, Thrust, and Rock Hard.

Amyl nitrate can cause headache, flushing, hypotension, tachycardia, dizziness, and relaxation of the anal sphincter. Less commonly it causes fainting, stupor, vertigo, incontinence, dyspnea, glaucoma, and methemoglobinemia.

Organs and Systems

Cardiovascular

Hypotension is a frequent adverse effect of nitrates (24), resulting from dilatation of arteries and veins. It can be associated with reflex tachycardia and palpitation, and even syncope on assuming an upright posture. Hypotension occurs particularly in old people, especially those with recurrent falls (SEDA-16, 196), and more commonly with glyceryl trinitrate oral spray than with sublingual tablets, because of its faster onset of action (24). It tends to occur more commonly in those who take nitrates infrequently.

Glyceryl trinitrate ointment has been used topically to improve blood perfusion of surgical wounds that are under tension, but it can cause unwanted bradycardia and hypotension.

- An 85-year-old man became hypotensive (BP 60/40 mmHg) following topical application of 2% glyceryl trinitrate paste to improve wound healing after knee surgery (25). The patient recovered within 4 hours after institution of a dopamine infusion and removal of the glyceryl trinitrate paste.

The authors recommended monitoring patients given topical glyceryl trinitrate for potentially dangerous hypotension.

Alternatively, bradycardia can accompany hypotension or syncope (26–28), presumably due to a vagal effect (it is responsive to atropine); complete heart block has also been reported (29). Occasionally, similar events occur in patients without ischemic heart disease (30), suggesting a mechanism similar to vasovagal syncope or coronary artery spasm (31,32) rather than coronary steal.

Nitrates can precipitate myocardial infarction in susceptible patients (33), and occasionally worsen myocardial infarction, producing asystole (34) or ventricular fibrillation (35,36), possibly due to coronary steal or reperfusion of ischemic myocardium.

Throbbing headache as a result of dilatation of cerebral vessels (37,38) is common, although it often abates as tolerance develops. The once-a-day modified-release formulations of oral isosorbide mononitrate are particularly likely to cause headache on first exposure and have accounted for premature withdrawal in about 1% of cases in some trials (39,40).

Of 118 patients (66 men and 52 women) with chest pain (mean age 63 years) 30 patients had normal coronary arteries or minimal or non-obstructive coronary artery disease and 88 had obstructive lesions defined as luminal narrowing greater than 50% in any one or more of the left or right coronary arteries or their major branches (41). Glyceryl trinitrate produced variable relief of chest pain within 10 minutes in all cases. Significant headache occurred in 36% and the patients with normal or minimally diseased coronary arteries had a greater risk of significant headache than those with obstructive lesions (73% versus 23%).

Respiratory

Transient hypoxemia can complicate the use of glyceryl trinitrate and has been described after sublingual (42) and intravenous administration (43). In patients with ischemic heart disease this can result in a paradoxical response, with further angina pectoris and ST segment depression on the electrocardiogram (42). A possible explanation is vasodilatation of pulmonary arteries, leading to relative hypoventilation in some areas of the lungs, and it may be especially common and troublesome when glyceryl trinitrate is being tried for pulmonary hypertensive disease (44,45).

Episodes of severe oxygen desaturation occurred in a 43-year-old man with obesity, hypertension, and diabetes mellitus after intravenous administration of glyceryl trinitrate on two occasions, which the authors attributed to inhibition of hypoxic pulmonary vasoconstriction (46).

Nervous system

Glyceryl trinitrate can reduce cerebral blood flow, and this may explain occasional reports of impaired cognitive function and neurological disturbances in patients with coronary heart disease. Continuous intravenous infusion of glyceryl trinitrate in 12 healthy volunteers reduced blood flow velocity in the middle cerebral artery and increased slow-wave power during electroencephalography (47). Patients with cognitive disturbances receiving glyceryl trinitrate should be carefully monitored.

Transient cerebral ischemic attacks, unrelated to blood pressure changes, occurred with isosorbide dinitrate and glyceryl trinitrate ointment in a patient with previous cerebrovascular disease, and resolved on withdrawal (48). Such changes may be more frequent than is realized and, added to the risk of hypotension, suggest that nitrates should be used with great caution in patients with cerebrovascular disease.

Glyceryl trinitrate has been associated with encephalopathy (49).

- A 27-year-old woman developed pre-eclampsia at 35 weeks and after urgent cesarean section was given glyceryl trinitrate sublingually in an attempt to reduce her blood pressure, which was associated with intractable headache and which had not responded to urapidil and metoprolol. She lost her sight and 6 days later had a tonic-clonic seizure. Posterior reversible encephalopathy syndrome was diagnosed. The seizures did not recur in the absence of anticonvulsants, and her blood pressure was controlled with verapamil.

The vasodilatory action of glyceryl trinitrate may have aggravated the encephalopathy.

In a patient with migraine triggered on two separate occasions by glyceryl trinitrate positron emission tomography showed activation in the primary visual area of the occipital cortex during the aura (50).

Isosorbide mononitrate precipitated pituitary apoplexy in a patient with an unsuspected pituitary macroadenoma (51). Intracranial hypertension associated with glyceryl trinitrate has been reported and attributed to cerebral vasodilatation (52). However, it may occasionally be a consequence of hyperosmolality, hemolysis, and lactic acidosis caused by the excipient ethylene glycol used with intravenous isosorbide dinitrate. The risk of this may be greatest in patients with renal impairment (53).

Psychological, psychiatric

In an elderly woman, isosorbide dinitrate caused visual hallucinations and subsequent suicidal ideation, thought to be due to hypotension and cerebral ischemia (54).

Sensory systems

Eyes

On the basis of animal experiments it has been suggested that nitrates could increase intraocular pressure and should therefore not be used in the presence of glaucoma (55). However, both intravenous glyceryl trinitrate and oral isosorbide dinitrate actually lower the intraocular pressure, both in healthy individuals and in patients with open-angle and closed-angle glaucoma (56).

Taste

Glyceryl trinitrate in sublingual tablets causes a burning sensation under the tongue, an indication of the activity of the formulation. Altered and later absent taste sensation developed in a patient taking long-term transdermal glyceryl trinitrate patches (15).

Fluid balance

Peripheral edema occurred in patients treated for left ventricular failure with isosorbide dinitrate (57). This disappeared on withdrawal. The mechanism is unclear, in common with fluid retention reported with other vasodilators.

Hematologic

Methemoglobinemia has been reported after overdosage of sublingual glyceryl trinitrate (58) but does not occur with usual doses; prolonged intravenous infusion (up to 128 micrograms/minute) was not associated with pathological concentrations of methemoglobin (59). Higher mean infusion rates (290 micrograms/minute) (60), or occasionally lower rates in a patient with reduced drug clearance (61), can cause the formation of methemoglobin and could conceivably cause a clinical problem; however, recovery is rapid after injection of methylene blue (62).

In contrast, amyl nitrite can cause methemoglobinemia when abused, since it has a 2000 times higher affinity for hemoglobin (63). In vitro both amyl nitrite and glyceryl trinitrate reacted immediately with oxyhemoglobin to effect oxidation to methemoglobin, while for sodium nitrite there was a lag phase before the reaction occurred. The affinity rate constants were:

- 11 400 M^{-1} minute^{-1} for sodium nitrite;
- 74 500 M^{-1} minute^{-1} for amyl nitrite;
- 35 M^{-1} minute^{-1} for glyceryl trinitrate.

Organic nitrates inhibit platelet aggregation in response to a wide variety of stimuli and prolong bleeding time.

Mouth and teeth

An increased frequency of dental caries with the use of buccal glyceryl trinitrate formulations has been reported in Sweden (SEDA-17, 237).

Skin

In a multicenter comparison of two transdermal nitrate delivery systems, there were skin problems at the application site in 20–39% of patients (64).

Allergic contact dermatitis has been reported with the use of an ointment containing glyceryl trinitrate (65), and its development should be suspected if persistent erythema, vesicles, papules, hyperpigmentation, or edema follow the use of such formulations. Allergy may be due to the drug itself or to an excipient in the ointment base. However, erythema and pruritus produced by glyceryl trinitrate ointment can also result from the pharmacological action of the drug. The use of transdermal nitrate patches has also resulted in allergic contact dermatitis which was attributed both to the active drug (66) and to an inactive component of the delivery system or adhesive (67).

A causal relation between rosacea and the facial flushing seen with vasodilators such as glyceryl trinitrate seems likely (68).

Long-Term Effects

Drug tolerance

There is progressive attenuation of the effects of nitrates within the first 24 hours of continuous exposure. In a review of a transdermal nitrate system, the short duration of action of the formulation was ascribed to tolerance (69).

At first it was thought that tolerance was due to depletion of intracellular endothelial sulfhydryl groups (70), which are necessary for the biotransformation of organic nitrates to nitric oxide. It seems likely that tolerance occurs through several mechanisms, and the precise mechanisms are still uncertain (71). However, one possible mechanism of tolerance has been elucidated (72). Organic nitrates are converted to *S*-nitrosothiols when combined with sulfhydryl moieties in smooth muscles. These compounds activate guanylate cyclase, leading to an increase in the intracellular concentration of cyclic GMP, with resultant vascular relaxation and vasodilatation. As the supply of sulfhydryl within the smooth muscles is limited, prolonged exposure to organic nitrates results in its depletion and hence tolerance (73). Other mechanisms of nitrate tolerance that have been proposed include neurohumoral adaptation and a plasma volume-dependent mechanism (74). A unifying hypothesis, involving the excess production of superoxide anions, has been proposed (75).

In an assessment of how nitrate tolerance might be avoided, the efficacy of thrice-daily buccal glyceryl trinitrate and that of four times a day oral isosorbide dinitrate were compared (76); over 2 weeks, prolonged antianginal efficacy was maintained only with thrice-daily buccal administration. Nevertheless, in another study, transdermal glyceryl trinitrate seemed to offer a prolonged effect, sufficient to wean patients with unstable angina from intravenous nitrate therapy (76).

It has been convincingly shown that a nitrate-free period reduces nitrate tolerance (77). The modern strategy is to have a nitrate-free period when the patient is asleep, although this results in reduced protection from ischemia in the early morning, the time when many episodes of silent ischemia and myocardial infarction occur. Reports that ACE inhibitors (78) or carvedilol (79) reduce nitrate tolerance are as yet unconfirmed (80).

Three published controlled trials of the use of vitamin C in treating nitrate tolerance in 77 subjects overall have been reviewed (81). Vitamin C (3–6 g/day orally or by intravenous infusion) reduced nitrate tolerance for up to 3 days, without causing adverse reactions. However, before making definitive recommendations, studies of longer duration should be done in a larger number of patients.

Drug withdrawal

Adverse reactions to industrial exposure to glyceryl trinitrate were first reported in 1914 (82), and there have been several subsequent reports of symptoms during the weekend resulting from withdrawal of the sustained vasodilator response of nitrates. Exposure to organic nitrates, for example in workers in the explosives industry, can cause profound vascular changes after departure from the environment (83). Vasoconstriction ensues and can cause non-exertional angina pectoris, or myocardial infarction (84), and peripheral ischemia, with Raynaud's phenomenon and gangrene of the extremities (85). The development of tolerance to the vasodilatory effects of glyceryl trinitrate, with arterial spasm when exposure ceases, has been postulated as the mechanism (86).

- A 41-year-old man employed in a munitions factory was admitted with crushing chest pain on a Sunday morning. There was evidence of an acute inferior myocardial infarction, and cardiac catheterization showed 80% narrowing of the proximal right coronary artery, which was reduced to 10% by intracoronary administration of glyceryl trinitrate. The electrocardiogram became normal.

Despite continuing reports of adverse reactions to withdrawal from chronic occupational exposure to nitrates, there is no clear evidence of similar events in patients taking chronic nitrate therapy. However, since this risk cannot be excluded, patients taking chronic nitrates should be carefully advised and monitored.

Tolerance can also occur in patients who are given long-term continuous treatment with nitrates in high dosages for angina pectoris. Harmful withdrawal symptoms can be avoided by gradual dosage reduction, but high doses should be avoided where possible (87). Likewise, abrupt withdrawal of intravenous glyceryl trinitrate in patients with unstable angina precipitates further angina (SEDA-16, 196).

Tumorigenicity

In view of the hypothesis that nitric oxide can have mutagenic or carcinogenic effects, chromosomal alterations have been sought in blood lymphocytes from patients taking chronic nitrates and after in vitro exposure to a nitric oxide donor (88). No structural alterations were found in vivo or in vitro, only an increased frequency of micronucleus formation, a nuclear morphological change that has been suggested to be associated with a risk of cancer. However, this first report of a genotoxic effect of nitrates probably has no clinical relevance.

Second-Generation Effects

Pregnancy

Glyceryl trinitrate has been used for its relaxing action on smooth muscle during cesarean delivery, with the aim of facilitating fetal extraction. After the publication of reports, a randomized, placebo-controlled trial was performed in 97 patients undergoing elective cesarean delivery after the 34th week of gestation (89). The patients were randomized to receive an intravenous bolus of glyceryl trinitrate (0.25 or 0.5 mg) or placebo. Glyceryl trinitrate did not improve fetal extraction. There were no adverse reactions attributable to glyceryl trinitrate in the neonate or mother.

The safety of glyceryl trinitrate in emergency or elective cesarean delivery has also been documented in a prospective series of 23 women, who received 400–800 micrograms of glyceryl trinitrate spray (90). However, no conclusions about the efficacy of glyceryl trinitrate were possible, owing to the lack of a control group.

Transdermal glyceryl trinitrate 5 mg for 16 hours each day has been compared with placebo in 25 mothers bearing intrauterine growth-retarded fetuses (91). It was applied from week 27 to week 35 of gestation until delivery. Three mothers were excluded from the analysis because they did not use the glyceryl trinitrate correctly. Placental flow, measured by Doppler, was improved by glyceryl trinitrate, and there was a better short-term neonatal outcome. Moreover, a biochemical marker of brain distress, the protein S100B, was normalized after glyceryl trinitrate, without an excess of adverse events in either the neonates or the mothers.

An authoritative overview of the published evidence on the use of glyceryl trinitrate to induce uterine relaxation in obstetric emergencies has shown that it is safe for fetus and mother (92).

Susceptibility Factors

The problems that can complicate the use of nitrates in patients with acute myocardial infarction, severe coronary atherosclerosis, and prior cerebrovascular disease have already been stressed. Hypotension, reflex tachycardia, or bradycardia can all be hazardous in individual patients.

Myocardial infarction

Although meta-analysis of studies in acute myocardial infarction suggested that nitrates protect against death and reinfarction (93), the ISIS-4 study found no such protection for 1 month after infarction (94). However, this study did show that nitrates are relatively safe in suspected myocardial infarction, even though 14% of patients who received nitrates had heart failure. Compared with placebo-treated patients there was an excess of hypotension considered severe enough to withdraw the drug in 15 per 1000 patients; in about half the cases it occurred soon after the start of treatment. The excess risk for headache was 18.7 per 1000 and of dizziness 1.6 per 1000, 0.8 per 1000 being associated with profound hypotension.

Congestive heart failure

The combination of isosorbide dinitrate with hydralazine is safe in systolic heart failure. Only headache and palpitation necessitated drug withdrawal and were more common than with enalapril (95). Headache and dizziness were more common with isosorbide dinitrate plus hydralazine than with placebo in patients taking ACE inhibitors and beta-blockers (5).

Pulmonary hypertension

A paradoxical increase in pulmonary vascular resistance has been reported after intravenous infusion of glyceryl trinitrate, and care is needed when giving potent vasodilators to patients with idiopathic pulmonary hypertension (96).

Hypertrophic obstructive cardiomyopathy

Organic nitrates should be used with caution in patients with hypertrophic obstructive cardiomyopathy, as they reduce afterload and venous return, which can exacerbate intraventricular obstruction and reduce left ventricular filling (SEDA-16, 196).

Exercise

A dramatic rise in plasma glyceryl trinitrate concentrations occurred after exercise in healthy subjects following transdermal administration (97). This altered systemic availability could cause adverse effects.

Drug Administration

Drug administration route

Transdermal

In one case a small explosion occurred when a patient was defibrillated with the paddle placed over a glyceryl trinitrate patch (98). The authors felt that this was probably due to arcing of the electrical current from the aluminium backing of the patch, but they advised that patches and gel should be removed before defibrillation.

An unusual skin burn resulted from the application of a glyceryl trinitrate transdermal patch in a patient who was subsequently exposed to radiation that leaked from a damaged microwave oven. The patient sustained a second-degree burn of the size and shape of his skin patch, probably due to a reaction with a metallic element in the adhesive strip (99).

Anal

Topical nitrates are used as an alternative to surgical treatment of chronic anal fissures. Their often limiting adverse effect is headache, due to vasodilatation, which is why glyceryl trinitrate ointment has been compared with diltiazem cream in a double-blind study in 60 patients with chronic anal fissures (100). After 8 weeks of treatment, 25 of 29 patients who received glyceryl trinitrate and 24 of 31 who received diltiazem had healed or improved; adverse effects were more common with glyceryl trinitrate (21/29) than diltiazem (13/31), the most frequent being headache. The authors suggested that diltiazem might be used as an alternative to glyceryl trinitrate in the treatment of chronic anal fissures, since it has similar efficacy with fewer adverse effects.

Drug–Drug Interactions

Acetylsalicylic acid

Aspirin in low dosages (under 300 mg/day) is widely used in cardiovascular prophylaxis, but its use is accompanied by an increased risk of gastrointestinal bleeding (SEDA-21, 100). Of particular interest therefore are data from a retrospective case-control study showing that nitrate therapy may reduce the risk of aspirin-induced gastrointestinal bleeding (101). As nitrates are often used in the same population of patients, such data merit further confirmation from larger prospective studies.

Angiotensin-converting enzyme (ACE) inhibitors

Headache associated with nitrates may be less frequent in patients taking ACE inhibitors (12/762, 1.6%) than in those not taking ACE inhibitors (24/775, 3.2%) according to an analysis of a database of elderly patients (102).

Cardiac glycosides

The effects of nitrates on the pharmacokinetics of digoxin are probably small.

In eight patients with chronic heart failure taking digoxin 0.25 mg/day, isosorbide dinitrate 10 mg tds caused only a small increase in C_{max} (15%) and had no effect on the mean steady-state serum concentration or AUC (103).

Nitroprusside increased the renal clearance of digoxin, perhaps by increasing renal blood flow and therefore renal tubular secretion (104). This caused a small fall in plasma digoxin concentration, the clinical significance of which is unclear.

Clonidine

The combination of isosorbide dinitrate with clonidine predisposes to postural hypotension (105).

Disopyramide

Disopyramide, by producing xerostomia, can prevent the dissolution of sublingual isosorbide tablets (106).

Ethanol

Alcohol potentiates the hypotensive effects of sublingual glyceryl trinitrate (107).

The presence of high concentrations of ethanol and glycerol in intravenous glyceryl trinitrate formulations has led to alcohol intoxication after high-dose therapy (108–110).

- A 74-year-old man developed Wernicke's encephalopathy, which required treatment with intravenous thiamine after discontinuation of the glyceryl trinitrate infusion.

Thus, high-dose intravenous therapy should be given with caution to patients with prior vitamin deficiency or a history of alcohol abuse (111).

Ethanol has also been implicated as the cause of acute gout in four patients with unstable angina who received intravenous infusions of glyceryl trinitrate (112).

Heparin

There have been reports of heparin resistance associated with glyceryl trinitrate (113) (SEDA-16, 196), but subsequent review of this interaction showed no clinical relevance, provided that heparin anticoagulation is monitored closely (114,115).

Hydralazine

Severe postural hypotension has been observed in patients given both isosorbide dinitrate and hydralazine for chronic heart failure (27), although this has not been a problem in large studies (100).

Lidocaine

Complete atrioventricular block has been reported after the use of sublingual nitrates in patients receiving lidocaine by infusion (116,117) and can result in asystole.

Metoprolol

Dizziness or faintness has been reported to follow the use of sublingual nitrate in patients taking the beta-adrenoceptor antagonist metoprolol (118).

Nifedipine

- A case of laryngeal edema has been reported with the use of isosorbide dinitrate spray followed shortly by sublingual nifedipine in an attempt to reduced high blood pressure (230/130 mmHg) in a 65-year-old woman who presented with coma secondary to a large intracerebral hemorrhage; the same complication recurred when she was rechallenged with the same sequence of drugs (119).

Phosphodiesterase type 5 inhibitors

Inhibitors of phosphodiesterase type 5 (PDE5), such as sildenafil and tadalafil, act synergistically with nitrates to cause falls in blood pressure, and nitrate use is contraindicated in patients taking such inhibitors (120). More than 3.6 million prescriptions for sildenafil were issued as of August 1998, and so far there have been 69 deaths, 12 of which were attributable to the interaction with nitrates, as reported to the FDA. This hemodynamic interaction lasts 24 hours, and in the case of tadalafil had completely disappeared 8 hours after treatment (121).

The pressure-lowering effect of a step-wise intravenous infusion of glyceryl trinitrate was significantly increased in 12 healthy volunteers treated for 4 days with sildenafil (Viagra) 25 mg tds; episodes of symptomatic hypotension were also more frequent with sildenafil (122). When sublingual glyceryl trinitrate was used, the reduction in systolic blood pressure was four-fold greater in association with sildenafil.

An elderly man with anginal pain died after treatment with glyceryl trinitrate (123). The family physician discovered that the man had also taken sildenafil, and he suggested suicidal intent, since he claimed that the patient was aware of the adverse cardiac events of the association of these two drugs.

The concomitant use of nitrates and sildenafil can precipitate a hypotensive reaction, and this combination should therefore be avoided (124,125). More than 3.6 million prescriptions for sildenafil were issued as of August 1998; of 69 deaths, 12 were attributable to the interaction with nitrates, as reported to the FDA.

Potentiation by sildenafil of the hypotensive effects of glyceryl trinitrate and amlodipine has been investigated. The fall in systolic blood pressure with glyceryl trinitrate was amplified four-fold by sildenafil in healthy subjects.

The pressure-lowering effect of a step-wise intravenous infusion of glyceryl trinitrate was significantly increased in 12 healthy volunteers treated for 4 days with sildenafil 25 mg tds; episodes of symptomatic hypotension were also more frequent with sildenafil (126). When sublingual glyceryl trinitrate was used, the reduction in systolic blood pressure was fourfold greater in association with sildenafil.

Management of adverse reactions

Abrupt cessation of intravenous glyceryl trinitrate after infusion for more than 24 hours often causes rebound vasoconstriction; this can be prevented by the use of oral isosorbide dinitrate (127).

References

1. Palmer RM, Ferrige AG, Moncada S. Nitric oxide release accounts for the biological activity of endothelium-derived relaxing factor. Nature 1987;327(6122):524–6.
2. Leung WH, Lau CP, Wong CK. Beneficial effect of cholesterol-lowering therapy on coronary endothelium-dependent relaxation in hypercholesterolaemic patients. Lancet 1993;341(8859):1496–500.

3. Cohn JN, Archibald DG, Ziesche S, Franciosa JA, Harston WE, Tristani FE, Dunkman WB, Jacobs W, Francis GS, Flohr KH, et al. Effect of vasodilator therapy on mortality in chronic congestive heart failure. Results of a Veterans Administration Cooperative Study. N Engl J Med 1986;314(24):1547–52.
4. Anonymous. Calcium antagonist caution. Lancet 1991; 337(8746):885–6.
5. Taylor AL, Ziesche S, Yancy C, Carson P, D'Agostino R Jr, Ferdinand K, Taylor M, Adams K, Sabolinski M, Worcel M, Cohn JN. African-American Heart Failure Trial Investigators. Combination of isosorbide dinitrate and hydralazine in blacks with heart failure. N Engl J Med 2004;351(20):2049–57.
6. Kloner RA, Jarow JP. Erectile dysfunction and sildenafil citrate and cardiologists. Am J Cardiol 1999;83(4):576–82.
7. Brisinda G, Maria G, Bentivoglio AR, Cassetta E, Gui D, Albanese A. A comparison of injections of botulinum toxin and topical nitroglycerin ointment for the treatment of chronic anal fissure. N Engl J Med 1999;341(2):65–9.
8. Dorfman G, Levitt M, Platell C. Treatment of chronic anal fissure with topical glyceryl trinitrate. Dis Colon Rectum 1999;42(8):1007–10.
9. Hyman NH, Cataldo PA. Nitroglycerin ointment for anal fissures: effective treatment or just a headache? Dis Colon Rectum 1999;42(3):383–5.
10. Pietroletti R, Navarra L, Simi M. Anaphylactic reaction caused by perianal application of glyceryl trinitrate ointment. Am J Gastroenterol 1999;94(1):292–3.
11. Tjon JA, Ansani NT. Transdermal nitroglycerin for the prevention of intravenous infusion failure due to phlebitis and extravasation. Ann Pharmacother 2000;34(10):1189–92.
12. Lauretti GR, Lima IC, Reis MP, Prado WA, Pereira NL. Oral ketamine and transdermal nitroglycerin as analgesic adjuvants to oral morphine therapy for cancer pain management. Anesthesiology 1999;90(6):1528–33.
13. Lauretti GR, de Oliveira R, Reis MP, Mattos AL, Pereira NL. Transdermal nitroglycerine enhances spinal sufentanil postoperative analgesia following orthopedic surgery. Anesthesiology 1999;90(3):734–9.
14. Olivari MT, Cohn JN. Cutaneous administration of nitroglycerin: a review. Pharmacotherapy 1983;3(3):149–57.
15. Ewing RC, Janda SM, Henann NE. Ageusia associated with transdermal nitroglycerin. Clin Pharm 1989;8(2):146–7.
16. Richard CS, Gregoire R, Plewes EA, Silverman R, Burul C, Buie D, Reznick R, Ross T, Burnstein M, O'Connor BI, Mukraj D, McLeod RS. Internal sphincterotomy is superior to topical nitroglycerin in the treatment of chronic anal fissure: results of a randomized, controlled trial by the Canadian Colorectal Surgical Trials Group. Dis Colon Rectum 2000;43(8):1048–58.
17. Elkayam U, Johnson JV, Shotan A, Bokhari S, Solodky A, Canetti M, Wani OR, Karaalp IS. Double-blind, placebo-controlled study to evaluate the effect of organic nitrates in patients with chronic heart failure treated with angiotensin-converting enzyme inhibition. Circulation 1999;99(20):2652–7.
18. Doucet S, Malekianpour M, Theroux P, Bilodeau L, Cote G, de Guise P, Dupuis J, Joyal M, Gosselin G, Tanguay JF, Juneau M, Harel F, Nattel S, Tardif JC, Lesperance J. Randomized trial comparing intravenous nitroglycerin and heparin for treatment of unstable angina secondary to restenosis after coronary artery angioplasty. Circulation 2000;101(9):955–61.
19. Lees CC, Lojacono A, Thompson C, Danti L, Black RS, Tanzi P, White IR, Campbell S. Glyceryl trinitrate and ritodrine in tocolysis: an international multicenter randomized study. GTN Preterm Labour Investigation Group. Obstet Gynecol 1999;94(3):403–8.
20. Altomare DF, Rinaldi M, Milito G, Arcana F, Spinelli F, Nardelli N, Scardigno D, Pulvirenti-D'Urso A, Bottini C, Pescatori M, Lovreglio R. Glyceryl trinitrate for chronic anal fissure—healing or headache? Results of a multicenter, randomized, placebo-controled, double-blind trial. Dis Colon Rectum 2000;43(2):174–81.
21. Gournay J, Masliah C, Martin T, Perrin D, Galmiche JP. Isosorbide mononitrate and propranolol compared with propranolol alone for the prevention of variceal rebleeding. Hepatology 2000;31(6):1239–45.
22. Silva G, Segovia R, Ponce R, Backhouse C, Palma M, Roblero JP, Abadal J, Quijada C, Troncoso M, Iturriaga H. Effects of 5-isosorbide mononitrate and propranolol on subclinical hepatic encephalopathy and renal function in patients with liver cirrhosis. Hepatogastroenterology 2002;49(47):1357–62.
23. Abrams J. Nitrate delivery systems in perspective. A decade of progress. Am J Med 1984;76(6A):38–46.
24. Sporl-Radun S, Betzien G, Kaufmann B, Liede V, Abshagen U. Effects and pharmacokinetics of isosorbide dinitrate in normal man. Eur J Clin Pharmacol 1980;18(3): 237–44.
25. Siddiqi M, Marco AP, Gorp CV. Postoperative hypotension from topical use of 2% nitroglycerin ointment after a total knee replacement procedure. J Clin Anesth 2004;16:77–8.
26. Nemerovski M, Shah PK. Syndrome of severe bradycardia and hypotension following sublingual nitroglycerin administration. Cardiology 1981;67(3):180–9.
27. Massie B, Kramer B, Haughom F. Postural hypotension and tachycardia during hydralazine–isosorbide dinitrate therapy for chronic heart failure. Circulation 1981;63(3):658–64.
28. Lamaud M, Duranton B, Verneyre H. Une syncope nitrée. Lyon Med 1985;253:39.
29. Viskin S, Heller K, Porat R, Belhassen B. Complete atrioventricular block due to sublingual isosorbide dinitrate. South Med J 1991;84(3):369–70.
30. Brandes W, Santiago T, Limacher M. Nitroglycerin-induced hypotension, bradycardia, and asystole: report of a case and review of the literature. Clin Cardiol 1990; 13(10):741–4.
31. Feldman RL, Pepine CJ, Conti CR. Unusual vasomotor coronary arterial responses after nitroglycerin. Am J Cardiol 1978;42(3):517–9.
32. Curry RC Jr. Coronary vasoconstriction following nitroglycerin. Cathet Cardiovasc Diagn 1980;6(2):211–2.
33. Scardi S, Zingone B, Pandullo C. Myocardial infarction following sublingual administration of isosorbide dinitrate. Int J Cardiol 1990;26(3):378–9.
34. Ong EA, Canlas C, Smith W. Nitroglycerin-induced asystole. Arch Intern Med 1985;145(5):954.
35. Quigley PJ, Maurer BJ. Ventricular fibrillation during coronary angiography: association with potassium-containing glyceryl trinitrate. Am J Cardiol 1985;56(1):191.
36. Berisso MZ, Cavallini A, Iannetti M. Sudden death during continuous Holter monitoring out of hospital after nitroglycerin consumption. Am J Cardiol 1984;54(6):677–9.
37. Iversen HK, Olesen J, Tfelt-Hansen P. Intravenous nitroglycerin as an experimental model of vascular headache. Basic characteristics. Pain 1989;38(1):17–24.
38. Askmark H, Lundberg PO, Olsson S. Drug-related headache. Headache 1989;29(7):441–4.
39. Herrmann H, Kuhl A, Maier-Lenz H. Der Einfluss des Dosierungszeitpunktes von Isosorbidmononitrat auf objektive und subjektive Parameter der Angina pectoris.

[The effect of time and dosage of isosorbide mononitrate on objective and subjective parameters in angina pectoris.] Arzneimittelforschung 1988;38(5):694–8.
40. Ankier SI, Fay L, Warrington SJ, Woodings DF. A multicentre open comparison of isosorbide-5-mononitrate and nifedipine given prophylactically to general practice patients with chronic stable angina pectoris. J Int Med Res 1989;17(2):172–8.
41. Hsi DH, Roshandel A, Singh N, Szombathy T, Meszaros ZS. Headache response to glyceryl trinitrate in patients with and without obstructive coronary artery disease. Heart 2005;91(9):1164–6.
42. Kopiman ES. Hypoxemia after sublingual nitroglycerin. Prim Cardiol 1981;7:105.
43. Jaffe AS, Roberts R. The use of intravenous nitroglycerin in cardiovascular disease. Pharmacotherapy 1982;2(5):273–80.
44. Pepke-Zaba J, Higenbottam TW, Dinh-Xuan AT, Stone D, Wallwork J. Inhaled nitric oxide as a cause of selective pulmonary vasodilatation in pulmonary hypertension. Lancet 1991;338(8776):1173–4.
45. Packer M, Halperin JL, Brooks KM, Rothlauf EB, Lee WH. Nitroglycerin therapy in the management of pulmonary hypertensive disorders. Am J Med 1984; 76(6A):67–75.
46. Karaaslan P, Dönmez A, Arslan G. Severe hypoxaemia following intravenous nitroglycerine administration in an obese patient: case report. Eur J Anaesthesiol 2005; 22(12):957–8.
47. Siepmann M, Kirch W. Effects of nitroglycerine on cerebral blood flow velocity, quantitative electroencephalogram and cognitive performance. Eur J Clin Invest 2000; 30(9):832–7.
48. Purvin VA, Dunn DW. Nitrate-induced transient ischemic attacks. South Med J 1981;74(9):1130–1.
49. Finsterer J, Schlager T, Kopsa W, Wild E. Nitroglycerin-aggravated pre-eclamptic posterior reversible encephalopathy syndrome (PRES). Neurology 2003;61:715–6.
50. Afridi S, Kaube H, Goadsby PJ. Occipital activation in glyceryl trinitrate induced migraine with visual aura. J Neurol Neurosurg Psychiatry 2005;76(8):1158–60.
51. Bevan JS, Oza AM, Burke CW, Adams CB. Pituitary apoplexy following isosorbide administration. J Neurol Neurosurg Psychiatry 1987;50(5):636–7.
52. Ahmad S. Nitroglycerin and intracranial hypertension. Am Heart J 1991;121(6 Pt 1):1850–1.
53. Demey HE, Daelemans RA, Verpooten GA, De Broe ME, Van Campenhout CM, Lakiere FV, Schepens PJ, Bossaert LL. Propylene glycol-induced side effects during intravenous nitroglycerin therapy. Intensive Care Med 1988;14(3):221–6.
54. Rosenthal R. Visual hallucinations and suicidal ideation attributed to isosorbide dinitrate. Psychosomatics 1987; 28(10):555–6.
55. Needleman P, Johnson EM Jr. Vasodilators in the treatment of angina. In: Gilman AG, Goodman LS, Gilman A, editors. The Pharmacological Basis of Therapeutics. 6th ed. New York: MacMillan, 1980:819.
56. Wizemann AJ, Wizemann V. Organic nitrate therapy in glaucoma. Am J Ophthalmol 1980;90(1):106–9.
57. Rodger JC. Peripheral oedema in patients treated with isosorbide dinitrate. BMJ (Clin Res Ed) 1981; 283(6303):1365–6.
58. Marshall JB, Ecklund RE. Methemoglobinemia from overdose of nitroglycerin. JAMA 1980;244(4):330.
59. Kaplan JA, Finlayson DC, Woodward S. Vasodilator therapy after cardiac surgery: a review of the efficacy and toxicity of nitroglycerin and nitroprusside. Can Anaesth Soc J 1980;27(3):254–9.
60. Kaplan KJ, Taber M, Teagarden JR, Parker M, Davison R. Association of methemoglobinemia and intravenous nitroglycerin administration. Am J Cardiol 1985;55(1):181–3.
61. Zurick AM, Wagner RH, Starr NJ, Lytle B, Estafanous FG. Intravenous nitroglycerin, methemoglobinemia, and respiratory distress in a postoperative cardiac surgical patient. Anesthesiology 1984;61(4):464–6.
62. Buenger JW, Mauro VF. Organic nitrate-induced methemoglobinemia. DICP 1989;23(4):283–8.
63. Tarburton JP, Metcalf WK. Kinetics of amyl nitrite-induced hemoglobin oxidation in cord and adult blood. Toxicology 1985;36(1):15–21.
64. Chinoy DA, Camp J, Elchahal S, Godoy C, Grossman W, Hai H, Hamilton W, Kushner M, McGreevy M, Mulvihill RJ. A multicenter comparison of adhesion, preference, tolerability, and safety characteristics of two transdermal nitroglycerin delivery systems: Transderm-Nitro and Deponit. Clin Ther 1989;11(5):678–84.
65. Hendricks AA, Dec GW Jr. Contact dermatitis due to nitroglycerin ointment. Arch Dermatol 1979;115(7):853–5.
66. Rosenfeld AS, White WB. Allergic contact dermatitis secondary to transdermal nitroglycerin. Am Heart J 1984;108(4 Pt 1):1061–2.
67. Letendre PW, Barr C, Wilkens K. Adverse dermatologic reaction to transdermal nitroglycerin. Drug Intell Clin Pharm 1984;18(1):69–70.
68. Wilkin JK. Vasodilator rosacea. Arch Dermatol 1980; 116(5):598.
69. Anonymous. Transdermal glyceryl trinitrate patches (Transiderm-Nitro). Drug Ther Bull 1986;24(2):5–6.
70. Cowan JC. Nitrate tolerance. Int J Cardiol 1986;12(1): 1–19.
71. Cowan JC. Antianginal drug therapy. Curr Opin Cardiol 1990;5:453.
72. Needleman P, Johnson EM Jr. Mechanism of tolerance development to organic nitrates. J Pharmacol Exp Ther 1973;184(3):709–15.
73. Elkayam U. Tolerance to organic nitrates: evidence, mechanisms, clinical relevance, and strategies for prevention. Ann Intern Med 1991;114(8):667–77.
74. Cowan JC. Avoiding nitrate tolerance. Br J Clin Pharmacol 1992;34(2):96–101.
75. Gori T, Parker JD. Nitrate tolerance: a unifying hypothesis. Circulation 2002;106(19):2510–3.
76. Parker JO, Vankoughnett KA, Farrell B. Comparison of buccal nitroglycerin and oral isosorbide dinitrate for nitrate tolerance in stable angina pectoris. Am J Cardiol 1985;56(12):724–8.
77. Abrams J. Management of myocardial ischemia: role of intermittent nitrate therapy. Am Heart J 1990;120(3):762–5.
78. Muiesan ML, Boni E, Castellano M, Beschi M, Cefis G, Cerri B, Verdecchia P, Porcellati C, Pollavini G, Agabiti-Rosei E. Effects of transdermal nitroglycerin in combination with an ACE inhibitor in patients with chronic stable angina pectoris. Eur Heart J 1993;14(12):1701–8.
79. Watanabe H, Kakihana M, Ohtsuka S, Sugishita Y. Randomized, double-blind, placebo-controlled study of carvedilol on the prevention of nitrate tolerance in patients with chronic heart failure. J Am Coll Cardiol 1998;32(5):1194–200.
80. Glasser SP. Prospects for therapy of nitrate tolerance. Lancet 1999;353(9164):1545–6.
81. Daniel TA, Nawarskas JJ. Vitamin C in the prevention of nitrate tolerance. Ann Pharmacother 2000;34(10): 1193–7.
82. Ebright GE. The effects of nitroglycerin on those engaged in its manufacture. JAMA 1914;62:201.

83. Hogstedt C, Axelson O. Nitroglycerine–nitroglycol exposure and the mortality in cardio-cerebrovascular diseases among dynamite workers. J Occup Med 1977;19(10): 675–8.
84. RuDusky BM. Acute myocardial infarction secondary to coronary vasospasm during withdrawal from industrial nitroglycerin exposure—a case report. Angiology 2001;52(2):143–4.
85. Lange RL, Reid MS, Tresch DD, Keelan MH, Bernhard VM, Coolidge G. Nonatheromatous ischemic heart disease following withdrawal from chronic industrial nitroglycerin exposure. Circulation 1972;46(4):666–78.
86. Danahy DT, Aronow WS. Hemodynamics and antianginal effects of high dose oral isosorbide dinitrate after chronic use. Circulation 1977;56(2):205–12.
87. Aronow WS. Nitrates as antianginal agents. Prim Cardiol 1980;6:46.
88. Andreassi MG, Picano E, Del Ry S, Botto N, Colombo MG, Giannessi D, Lubrano V, Vassalle C, Biagini A. Chronic long-term nitrate therapy: possible cytogenetic effect in humans? Mutagenesis 2001;16(6):517–21.
89. David M, Halle H, Lichtenegger W, Sinha P, Zimmermann T. Nitroglycerin to facilitate fetal extraction during cesarean delivery. Obstet Gynecol 1998;91(1):119–24.
90. Craig S, Dalton R, Tuck M, Brew F. Sublingual glyceryl trinitrate for uterine relaxation at Caeserean section—a prospective trial. Aust NZ J Obstet Gynaecol 1998; 38(1):34–9.
91. Gazzolo D, Bruschettini M, Di Iorio R, Marinoni E, Lituania M, Marras M, Sarli R, Bruschettini PL, Michetti F. Maternal nitric oxide supplementation decreases cord blood S100B in intrauterine growth-retarded fetuses. Clin Chem 2002;48(4):647–50.
92. Caponas G. Glyceryl trinitrate and acute uterine relaxation: a literature review. Anaesth Intensive Care 2001;29(2):163–77.
93. De Caterina R. Nitrate als Thrombocytenfunktionskemmer. [Nitrates as thrombocyte function inhibitors.] Z Kardiol 1994;83(7):463–73.
94. ISIS-4 (Fourth International Study of Infarct Survival) Collaborative Group. ISIS-4: a randomised factorial trial assessing early oral captopril, oral mononitrate, and intravenous magnesium sulphate in 58,050 patients with suspected acute myocardial infarction. Lancet 1995;345(8951):669–85.
95. Cohn JN, Johnson G, Ziesche S, Cobb F, Francis G, Tristani F, Smith R, Dunkman WB, Loeb H, Wong M, et al. A comparison of enalapril with hydralazine–isosorbide dinitrate in the treatment of chronic congestive heart failure. N Engl J Med 1991;325(5):303–10.
96. Hoit B, Gregoratos G, Shabetai R. Paradoxical pulmonary vasoconstriction induced by nitroglycerin in idiopathic pulmonary hypertension. J Am Coll Cardiol 1985; 6(2):490–2.
97. Lefebvre RA, Bogaert MG, Teirlynck O, Sioufi A, Dubois JP. Influence of exercise on nitroglycerin plasma concentrations after transdermal application. Br J Clin Pharmacol 1990;30(2):292–6.
98. Wrenn K. The hazards of defibrillation through nitroglycerin patches. Ann Emerg Med 1990;19(11):1327–8.
99. Murray KB. Hazard of microwave ovens to transdermal delivery system. N Engl J Med 1984;310(11):721.
100. Kocher HM, Steward M, Leather AJ, Cullen PT. Randomized clinical trial assessing the side-effects of glyceryl trinitrate and diltiazem hydrochloride in the treatment of chronic anal fissure. Br J Surg 2002;89(4):413–7.
101. Lanas A, Bajador E, Serrano P, Arroyo M, Fuentes J, Santolaria S. Effects of nitrate and prophylactic aspirin on upper gastrointestinal bleeding: a retrospective case-control study. J Int Med Res 1998;26(3):120–8.
102. Onder G, Pahor M, Gambassi G, Federici A, Savo A, Carbonin P, Bernabei R, on behalf of the GIFA Study. Association between ACE inhibitors use and headache caused by nitrates among hypertensive patients: results from the Italian Group of Pharmacoepidemiology in the Elderly (GIFA). Cephalalgia 2003;23:901–6.
103. Mahgoub AA, El-Medany AH, Abdulatif AS. A comparison between the effects of diltiazem and isosorbide dinitrate on digoxin pharmacodynamics and kinetics in the treatment of patients with chronic ischemic heart failure. Saudi Med J 2002;23(6):725–31.
104. Cogan JJ, Humphreys MH, Carlson CJ, Benowitz NL, Rapaport E. Acute vasodilator therapy increases renal clearance of digoxin in patients with congestive heart failure. Circulation 1981;64(5):973–6.
105. Tongia SK. Exaggerated tendency to postural hypotension with isosorbiddinitrate on clonidine's background activity. J Assoc Physicians India 1994;42(7):580.
106. Barletta MA, Eisen H. Isosorbide dinitrate–disopyramide phosphate interaction. Drug Intell Clin Pharm 1985;19(10):764.
107. Abrams J, Schroeder K, Raizada V. Potentially adverse effects of sublingual nitroglycerin during consumption of alcohol. J Am Coll Cardiol 1990;15:226A.
108. Korn SH, Comer JB. Intravenous nitroglycerin and ethanol intoxication. Ann Intern Med 1985;102(2):274.
109. Shook TL, Kirshenbaum JM, Hundley RF, Shorey JM, Lamas GA. Ethanol intoxication complicating intravenous nitroglycerin therapy. Ann Intern Med 1984;101(4):498–9.
110. Andrien P, Lemberg L. An unusual complication of intravenous nitroglycerin. Heart Lung 1986;15(5):534–6.
111. Shorey J, Bhardwaj N, Loscalzo J. Acute Wernicke's encephalopathy after intravenous infusion of high-dose nitroglycerin. Ann Intern Med 1984;101(4):500.
112. Shergy WJ, Gilkeson GS, German DC. Acute gouty arthritis and intravenous nitroglycerin. Arch Intern Med 1988;148(11):2505–6.
113. Habbab MA, Haft JI. Intravenous nitroglycerin and heparin resistance. Ann Intern Med 1986;105(2):305.
114. Pye M, Oldroyd KG, Conkie JA, Hutton I, Cobbe SM. A clinical and in vitro study on the possible interaction of intravenous nitrates with heparin anticoagulation. Clin Cardiol 1994;17(12):658–61.
115. Bechtold H, Kleist P, Landgraf K, Moser K. Einfluss einer niedrigdosierten intravenosen Nitrattherapie auf die antikoagulatorische Wirkung von Heparin. [Effect of low-dosage intravenous nitrate therapy on the anticoagulant effect of heparin.] Med Klin (Munich) 1994;89(7):360–6.
116. Lancaster L, Fenster PE. Complete heart block after sublingual nitroglycerin. Chest 1983;84(1):111–2.
117. Antonelli D, Barzilay J. Complete atrioventricular block after sublingual isosorbide dinitrate. Int J Cardiol 1986; 10(1):71–3.
118. Hyldstrup L, Mogensen NB, Nielsen PE. Orthostatic response before and after nitroglycerin in metoprolol- and verapamil-treated angina pectoris. Acta Med Scand 1983;214(2):131–4.
119. Silfvast T, Kinnunen A, Varpula T. Laryngeal oedema after isosorbide dinitrate spray and sublingual nifedipine. BMJ 1995;311(6999):232.
120. Cheitlin MD, Hutter AM Jr, Brindis RG, Ganz P, Kaul S, Russell RO Jr, Zusman RM. Use of sildenafil (Viagra) in patients with cardiovascular disease. Technology and Practice Executive Committee. Circulation 1999; 99(1):168–77.
121. Kloner RA, Hutter AM, Emmick JT, Mitchell MI, Denne J, Jackson G. Time course of the interaction between tadalafil and nitrates. J Am Coll Cardiol 2003;42:1855–60.

122. Webb DJ, Freestone S, Allen MJ, Muirhead GJ. Sildenafil citrate and blood-pressure-lowering drugs: results of drug interaction studies with an organic nitrate and a calcium antagonist. Am J Cardiol 1999;83(5A):C21–8.
123. Bhalerao S. A new suicide. J Fam Pract 2001;50(6):551.
124. Kloner RA, Jarow JP. Erectile dysfunction and sildenafil citrate and cardiologists. Am J Cardiol 1999;83(4):576–82.
125. Cheitlin MD, Hutter AM Jr, Brindis RG, Ganz P, Kaul S, Russell RO Jr, Zusman RM. Use of sildenafil (Viagra) in patients with cardiovascular disease. Technology and Practice Executive Committee. Circulation 1999; 99(1):168–77.
126. Webb DJ, Freestone S, Allen MJ, Muirhead GJ. Sildenafil citrate and blood-pressure-lowering drugs: results of drug interaction studies with an organic nitrate and a calcium antagonist. Am J Cardiol 1999;83(5A):C21–8.
127. Kelly EA, Ahmed RM, Horowitz JD. Withdrawal of intravenous glyceryl trinitrate: absence of rebound phenomena with transition to oral isosorbide dinitrate. Clin Exp Pharmacol Physiol 2005;32(4):269–72.

Nitrendipine

See also Calcium channel blockers

General Information

Nitrendipine is a dihydropyridine calcium channel blocker.

Skin

Cutaneous lupus erythematosus has been attributed to nitrendipine (1).

- A 66-year-old man developed widespread subacute cutaneous lupus erythematosus after taking nitrendipine 40 mg/day for hypertension for about 3 weeks. The skin lesions abated 2 weeks after drug withdrawal, and the immunological effects normalized within a few months. Nitrendipine was the only drug that the patient was taking at the time of the event.

Drug–Drug Interactions

Ciclosporin

Nitrendipine had a small but significant nephroprotective effect independent of its antihypertensive action in patients treated for 2 years with ciclosporin and no adverse effect on renal allografts (2). Nitrendipine did not affect blood concentrations of ciclosporin. Thus, nitrendipine, in contrast to most other calcium channel blockers, does not affect ciclosporin blood concentrations and can be safely added in transplant recipients.

References

1. Marzano AV, Borghi A, Mercogliano M, Facchetti M, Caputo R. Nitrendipine-induced subacute cutaneous lupus erythematosus. Eur J Dermatol 2003;13:213–6.
2. Rahn KH, Barenbrock M, Fritschka E, Heinecke A, Lippert J, Schroeder K, Hauser I, Wagner K, Neumayer HH. Effect of nitrendipine on renal function in renal-transplant patients treated with cyclosporin: a randomised trial. Lancet 1999;354(9188):1415–20.

Perindopril

See also Angiotensin converting enzyme inhibitors

General Information

Perindopril is a prodrug ester of perindoprilat, an ACE inhibitor that has been used in patients with hypertension and heart failure. An updated review of its use in hypertension has appeared (1).

A large French open postmarketing study in 47 351 hypertensive patients treated with perindopril for 12 months did not show unexpected adverse effects. The safety profile was similar to that of ACE inhibitors as a class (2).

Organs and Systems

Cardiovascular

The effects of perindopril on the renin–angiotensin system and its adverse effects have been reviewed (3).

Respiratory

Perindopril has previously been reported to cause pulmonary eosinophilia. Various types of drug-induced eosinophilic lung diseases have been described, varying from mild, simple pulmonary eosinophilia-like syndrome to fulminant, acute eosinophilic pneumonia-like syndromes. A further case of pulmonary eosinophilia-like syndrome has been reported, with symptoms 5 weeks after the introduction of perindopril for hypertension (4). No rechallenge was performed, but multiple investigations to exclude other causes and the temporal association were fairly convincing for drug-induced pulmonary eosinophilia.

Two cases of pneumonitis have been reported with perindopril (5). The second case was more convincing, because of the result of involuntary rechallenge, but the two cases had typical features of drug-induced pneumonitis. The authors referred to two previous reports of less typical but very likely cases of pneumonitis with captopril.

Ear, nose, and throat

Dysphonia has been associated with perindopril (6).

- An 84-year-old man, who had been taking nifedipine for hypertension for 10 years, was switched to lisinopril 4 mg/day because of inadequate control. One week later the patient became hoarse. Because of persistent hoarseness, he came to hospital 3 weeks later. The blood pressure was normal and apart from dysphonia there were no abnormalities of the ear, nose, and throat. The presence or absence of cough was not reported. Lisinopril was withdrawn and amlodipine started. Within 72 hours her voice was normal. Two months later she ran out of amlodipine and started taking lisinopril tablets she still had at home. Four days later the dysphonia recurred. Laryngoscopy was normal and 3 days after stopping lisinopril her voice returned to normal.

The author pointed out that although ACE inhibitors can cause dry cough, dysphonia has not been reported before.

Pancreas

ACE inhibitors have often been implicated in pancreatitis, although few cases have occurred in association with perindopril. A case of probable perindopril-induced pancreatitis has been reported (7).

Skin

UVA photosensitivity induced by perindopril has been reported (8).

References

1. Hurst M, Jarvis B. Perindopril: an updated review of its use in hypertension. Drugs 2001;61(6):867–96.
2. Speirs C, Wagniart F, Poggi L. Perindopril postmarketing surveillance: a 12 month study in 47,351 hypertensive patients. Br J Clin Pharmacol 1998;46(1):63–70.
3. Ferrari R, Pasanisi G, Notarstefano P, Campo G, Gardini E, Ceconi C. Specific properties and effect of perindopril in controlling the renin–angiotensin system. Am J Hypertens 2005;18(9 Pt 2):142S–154S.
4. Rochford AP, Smith PR, Khan SJ, Pearson AJ. Perindopril and pulmonary eosinophilic syndrome. J R Soc Med 2005;98(4):163–5.
5. Benard A, Melloni B, Gosselin B, Bonnaud F, Wallaert B. Perindopril-associated pneumonitis. Eur Respir J 1996;9(6):1314–6.
6. Vázquez JF. Dysphonia secondary to perindopril treatment. Am J Hypertens 2003;16:329–30.
7. Famularo G, Minisola G, Nicotra GC, De SC. Idiosyncratic pancreatitis associated with perindopril. JOP 2005;6(6):605–7.
8. Le Borgne G, Leonard F, Cambie MP, Serpier J, Germain ML, Kalis B. UVA photosensitivity induced by perindopril (Coversyl): first reported case. Nouv Dermatol 1996;15:378–80.

Phosphodiesterase type III inhibitors

General Information

There is a range of bipyridines that are selective inhibitors of a specific isoenzyme of phosphodiesterase, F-III. These include amrinone (inamrinone), enoximone (fenoximone), milrinone, olprinone, pimobendan, sulmazole, and vesnarinone. Their clinical pharmacology has been reviewed (1). The pharmacology, clinical pharmacology, uses, therapeutic value and adverse effects of positive inotropic drugs other than digitalis have been reviewed (2,3,4,5,6,7), as has the suggestion that their long-term use may be deleterious (6). The use of intravenous vasodilators, including amrinone and milrinone, in treating congestive heart failure has been reviewed (8).

The under-reporting of the results of clinical trials in patients with heart failure has been reviewed (9). Some trials that have been unpublished or published only in abstract or preliminary form have involved drugs with positive inotropic effects, such as the phosphodiesterase inhibitor vesnarinone (SEDA-23, 195), the beta-adrenoceptor partial agonist xamoterol, and the dopamine receptor agonist ibopamine.

Although the phosphodiesterase inhibitors are effective in the treatment of acute cardiac failure in various settings, overall mortality during long-term treatment of heart failure is increased, and these drugs should not be used for that purpose (10).

Amrinone

The adverse effects of amrinone (11,12,13,14,15) include thrombocytopenia (10%), hypotension, tachydysrhythmias (sometimes resulting in syncope and death) (9%), worsening cardiac ischemia (7%), worsening heart failure (15%), gastrointestinal disturbances (39%), neurological complications (17%), liver damage (7%), fever (6%), nephrogenic diabetes insipidus, hyperuricemia, flaking of the skin, brown discoloration of the nails, and reduced tear secretions. The figures in parentheses are taken from a study of the use of amrinone in 173 patients with chronic ischemic heart disease or idiopathic cardiomyopathies (13).

Other reported adverse effects include acute pleuropericardial effusions, perforated duodenal ulcer, acute myositis and pulmonary infiltrates, vasculitis with pulmonary infiltrates and jaundice, influenza-like illnesses, chest pain, headache, dizziness, anxiety, maculopapular rash, and night sweats (16).

Enoximone

The most common adverse effects of are enoximone gastrointestinal and cardiac. The gastrointestinal effects include anorexia, nausea, vomiting, and diarrhea (17,18). Other rarely reported adverse effects include thrombocytopenia, leukocytosis, increased appetite, increased serum activity of alanine transaminase, hyperglycemia, headache, lethargy, anxiety, dyspnea, and skin rashes (19,20,21,22).

Milrinone

Although the long-term oral use of inhibitors of phosphodiesterase type III is associated with increased mortality (SEDA-17, 217), milrinone has been used intravenously in patients with heart failure, both for short-term and longer-term therapy (SEDA-21, 196; SEDA-22, 203; SEDA-23, 109). It has been suggested to be particularly useful in tiding patients over while they are waiting for definitive treatment.

- A 48-year-old man with an inflammatory aneurysm of the ascending aorta and severe heart failure due to massive aortic regurgitation was given a continuous intravenous infusion of milrinone 0.5 micrograms/kg/minute (23). His pulmonary arterial pressure fell and his symptoms improved over 7 weeks while he was taking corticosteroids. The diseased tissue was successfully replaced at operation and the milrinone was tapered uneventfully.

The authors suggested that continuous milrinone infusion may be suitable for patients with surgically correctable inflammatory cardiovascular diseases complicating severe heart failure in whom maintenance of optimal hemodynamics is necessary for several weeks before operation.

The disappointing results with positive inotropic drugs in treating acute and chronic heart failure may be due to the fact that they increase both systolic and diastolic calcium concentrations in the myocardium (24).

Olprinone

Olprinone is given intravenously and is mostly eliminated by the kidneys. Its pharmacological effects have been reviewed (25). Its major adverse effects are cardiac dysrhythmias and thrombocytopenia, the latter with a reported incidence of 0.43%.

Pimobendan

Pimobendan derives its inotropic effect from a combination of phosphodiesterase III inhibition and sensitisation of myocardial contractile proteins to calcium (26).

Sulmazole

Sulmazole commonly causes adverse gastrointestinal effects; dose-related anorexia, nausea, and vomiting have been reported in about 50% of patients given an intravenous infusion (27,28) and also after single oral doses (29). Cardiac dysrhythmias, mostly ventricular, have been reported occasionally with sulmazole (30). Other reported adverse effects in small numbers of patients include headache (27), temporary visual disturbances (28,31), discoloration of the urine (attributed to a metabolite) (28), and a small reduction in platelet count (27).

Vesnarinone

Vesnarinone increases the risk of sudden cardiac death in patients with congestive heart failure and can cause neutropenia. It has also been used to treat cancer, and its effects have been reviewed (32). It has immunomodulatory effects, including suppression of tumor necrosis factor-induced activation of NF-κB, NK cell activity, and endotoxin-induced production of inflammatory cytokines, and altered expression of E selectin in endothelial cells, and it induces apoptosis and suppresses growth in a variety of in vitro cell lines. The precautions for using vesnarinone and its adverse effects have been reviewed (33). The most important adverse effects are an increased risk of sudden death and neutropenia. They are dose-related in the therapeutic range of doses (i.e. collateral adverse effects). Patients with renal impairment should not receive vesnarinone and care should be taken in patients who are also taking drugs that inhibit CYP3A.

Observational studies

In a retrospective study of the effects of long-term combination therapy with intravenous milrinone and oral beta-blockers in 65 patients with severe congestive heart failure (New York Heart Association class IV and ejection fraction less than 25%) refractory to oral therapy, 51 successfully began beta-blocker therapy while receiving intravenous milrinone (34). The mean duration of milrinone treatment was 269 (range 14–1026) days and functional class improved from IV to II–III. In 24 patients beta-blocker up-titration was well tolerated and milrinone was stopped. During combination therapy 16 patients died; one died of sudden cardiac death (on day 116) and the other 15 died of progressive heart failure or other complications.

There has been a study of the use of intermittent intravenous milrinone in 10 patients with end-stage congestive heart failure (35). Some hemodynamic benefit was obtained, but this was an open uncontrolled study and firm conclusions are impossible. There was also some improvement in the quality of life. The drug was well tolerated and only one patient had an increase in dysrhythmias (an increased frequency of ventricular extra beats).

On the premise that phosphodiesterase inhibitors also inhibit the production of cytokines, milrinone has been used in the treatment of nine patients with the systemic inflammatory response syndrome and compared with seven patients with congestive heart failure (36). In both groups milrinone significantly altered cardiac index, pulmonary capillary wedge pressure, and left ventricular stroke work index. In the patients with cardiac failure it also reduced systemic vascular resistance index, and the dose of adrenaline had to be increased substantially during milrinone infusion to counteract vasodilatation.

Comparative studies

In 12 patients with mild stable asthma, intravenous olprinone 30 micrograms/minute was compared with aminophylline 2.25 mg/minute and the combination over 150 minutes (37). The mean maximal increases in FEV_1 were 7.8% (95%CI = 2.4, 13), 17% (10, 24), 17% (11, 22), and 1% (−1.1, 3.1) during infusion of aminophylline, olprinone, aminophylline + olprinone, and saline respectively. Olprinone alone and in combination with aminophylline lowered diastolic blood pressure and increased heart rate.

In a randomized, open, parallel-group study of the hemodynamic effects of milrinone and glyceryl trinitrate in 119 patients with advanced decompensated heart failure, milrinone was significantly more effective than glyceryl trinitrate (38). Adverse effects caused the withdrawal of milrinone in three of the 58 patients who took it; one had ventricular extra beats, one had renal insufficiency, and one had hypokalemia. Headache was the most common adverse effect in both groups, but was less common in those who took milrinone (12% versus 29%). In a randomized, open study of milrinone and dobutamine in 120 patients with low cardiac output after cardiac surgery, the two drugs were roughly comparable in efficacy (39). Adverse events were more frequent with dobutamine (77%) than with milrinone (58%). With milrinone cardiovascular adverse events were the most common, including hypertension (13%), hypotension (20%), bradycardia (13%), new atrial fibrillation (5%), and tachycardia (8%). Other adverse events included hemorrhage and oliguria. Adverse events required withdrawal from the study in 20 patients who took milrinone.

Placebo-controlled studies

In a double-blind, placebo-controlled study in 238 children after cardiac surgery, milrinone was used in a low dose (25 micrograms/kg bolus over 60 minutes followed by 0.25 micrograms/kg/minute by infusion for 35 hours) and a high dose (75 micrograms/kg bolus followed by 0.75 micrograms/kg/minute by infusion for 35 hours) (40). Low cardiac output syndrome developed in the first 36 hours after surgery in 26%, 18%, and 12% in those given placebo, low-dose milrinone, and high-dose milrinone respectively. High-dose milrinone significantly reduced the risk of low cardiac output syndrome compared with placebo, with a relative risk reduction of 55%. There were two deaths, both after infusion of milrinone.

OPTIME-CHF (41) was a randomized placebo-controlled study in which 951 patients (mean age 65 years; 92% with baseline NYHA class III or IV; mean left ventricular ejection fraction 23%) with acute exacerbations of chronic heart failure in 78 community and tertiary care hospitals in the USA were randomly assigned to a 48-hour infusion of either milrinone (0.5 μg/kg/minute initially for 24 hours) or saline (42). The median number of days in hospital for cardiovascular causes within 60 days after randomization did not differ significantly between patients given milrinone (6 days) or placebo (7 days). Sustained hypotension requiring intervention (11% versus 3.2%) and new atrial dysrhythmias (4.6% versus 1.5%) were more common in the patients who received milrinone. There was no difference in hospital mortality (3.8% versus 2.3%), 60-day mortality (10% versus 8.9%), or the composite incidence of death or readmission (35% versus 35%). The authors concluded that these results do not support the routine use of intravenous milrinone as an adjunct to standard therapy in patients with an exacerbation of chronic heart failure.

Systematic reviews The role of intravenous inotropic agents with vasodilator properties (so-called inodilators) in the management of acute heart failure syndromes has been reviewed (43). Randomized controlled trials of the currently available medications, dobutamine, dopamine, and milrinone, have failed to show benefits and the results suggest that acute, intermittent, or continuous intravenous use of such drugs may increase morbidity and mortality. Their use should be restricted to patients who are hypotensive as a result of low cardiac output despite a high left ventricular filling pressure.

Organs and Systems

Cardiovascular

The inhibitors of phosphodiesterase type III have been associated with increased mortality in patients with chronic congestive heart failure [Leier 120G,Noni 121,44), although pimobendan may be an exception (45,46,47). Acute exacerbation of chronic heart failure has also been reported with milrinone (48).

In 3833 patients with heart failure of Class III or IV and a left ventricular ejection fraction of 30% or less despite optimal treatment, followed for 286 days, vesnarinone 60 mg/day was associated with significantly more deaths and a shorter duration of survival (64). The increased death rate was due to an increase in sudden deaths, presumably because of cardiac dysrhythmias. There were similar trends in patients who took 30 mg/day, but the changes were not significant. In contrast to the increased mortality with vesnarinone there was a short-term increase in quality of life, which occurred during the first 16 weeks but was not maintained at 6 months. Probably the increased mortality due to vesnarinone outweighs the improvement in quality of life.

Dysrhythmias

In the study called OPTIME, the Outcomes of a Prospective Trial of Intravenous Milrinone for Exacerbations, in which 949 patients with decompensated heart failure were randomized to receive intravenous milrinone or placebo, there were 59 new dysrhythmic events in 6% of the population; a post-hoc analysis has now shown that there was a higher risk among those who received milrinone (49).

The cardiac effects of enoximone include hypotension, transient atrial fibrillation, and bradycardia (19). Ventricular tachydysrhythmias have been reported in about 4% of patients (18) and myocardial ischemia can also occur (20,50)

In a retrospective view of 63 patients who received intravenous milrinone for more than 24 hours for advanced cardiac failure, the mean dose was 0.43 micrograms/kg/minute and the mean duration of therapy 12 (range 1–70) days (51). After 24 hours of therapy there was significant improvement in pulmonary artery pressure, pulmonary capillary wedge pressure, and cardiac index. Because of the nature of the study, which was not placebo- controlled, it is impossible to be sure what events could have been attributed to the milrinone. However, the authors reported five cases of asymptomatic non-sustained ventricular tachycardia, six of symptomatic ventricular tachycardia, and three deaths, one in ventricular tachycardia and two in heart failure. There was no difference in the incidence of these adverse events in patients who received milrinone for more than 7 days compared with the others.

Milrinone can cause a tachycardia, partly because of its vasodilatory effects and partly perhaps by a direct effect on the heart.

- A 74-year-old man had a tachycardia of 145/minute during infusion of milrinone after an operation for repair of an abdominal aortic aneurysm (52). The tachycardia was controlled by esmolol on one occasion and more impressively by metoprolol on a second occasion. However, the hemodynamic effects of milrinone were not altered by beta-blockade.

Presumably the beneficial effects of beta-blockade in this case were non-specific, since milrinone does not affect beta-adrenoceptors.

Of 19 children (12 infants and seven children aged 1–13 years) who were given either two boluses of 25 micrograms/kg followed by an infusion of 0.5 micrograms/kg/min, or a bolus of 50 micrograms/kg followed by a bolus of 25 micrograms/kg followed by an infusion of 0.75 micrograms/kg/minute, two infants developed junctional ectopic tachycardia during infusion of milrinone (53). In a nested case-cohort analysis of 33 patients (aged 1 day to 10.5 years; median 1.8 months) the postoperative use of dopamine or milrinone, longer cardiopulmonary bypass times, and younger age were associated with junctional ectopic tachycardia (54).

There has been one report of paroxysmal supraventricular tachycardia in one of 16 patients with cardiogenic shock given amrinone (55).

Hypotension

The main limitations of the short-term use of milrinone is the risk of hypotension and ventricular dysrhythmias (56,57,58,59) and sudden death has occurred (57).

Vasopressin has been used to treat hypotension due to milrinone. In seven patients with congestive heart failure who developed hypotension (systolic arterial pressure below 90 mmHg), vasopressin 0.03–0.07 units/minute increased the systolic arterial pressure to 127 mmHg (60). This effect was due to peripheral vasoconstriction, since the systemic vascular resistance increased from 1112 to 1460 dyne.s/cm5 with no change in cardiac index. Urine output also improved significantly. In three patients in whom milrinone caused hypotension, vasopressin (0.03-0.07 units/minute) increased the systolic arterial pressure from 90 to 130 mmHg and reduced the dosages of catecholamines that were being used (61). The authors hypothesized that vasopressin may have inhibited the milrinone-induced accumulation of cyclic AMP in vascular smooth muscle.

Metabolism

Hypoglycemia has been attributed to pimobendan (62).

- A 74-year-old man with severe congestive heart failure due to aortic insufficiency with severe left ventricular dysfunction was given enalapril 2.5 mg/day, furosemide 30 mg/day, and digoxin 0.125 mg/day. The heart failure worsened and pimobendan 2.5 mg/day was added and the dose increased to 5 mg/day 4 weeks later. After a further 5 weeks he became unconscious with a plasma glucose concentration of 0.5 mmol/l. He recovered after intravenous infusion of glucose and had no further episodes of hypoglycemia after pimobendan was withdrawn.

Pimobendan increases insulin release from pancreatic beta cells by directly sensitizing the calcium-sensitive exocytotic mechanism (63).

Fluid balance

In one series fluid retention was very common with milrinone (57).

Hematologic

Vesnarinone causes reversible neutropenia in up to 2.5% of patients (44). In 3833 patients with heart failure of Class III or IV and a left ventricular ejection fraction of 30% or less despite optimal treatment, followed for 286 days, there was dose-dependent agranulocytosis, in 0.2% of those taking vesnarinone 30 mg/day and 1.2% of those taking 60 mg/day (64).

Amrinone and milrinone can cause thrombocytopenia (65).

Whole blood platelet aggregation has been examined in 45 cardiac surgery patients who were randomly given milrinone 50 micrograms /kg plus 0.5 micrograms /kg/minute for 10 hours, amrinone 1.5 mg/kg plus 10 micrograms/kg/minute for 10 hours, or placebo after release of aortic cross-clamp (66). The mean postoperative platelet counts 3 days in those given milrinone and amrinone did not differ significantly from those given placebo. Although there were changes in platelet aggregation after surgery there was no significant differences in platelet aggregation or other hematological values among the three groups.

In a historically controlled trial in 24 children with severe pulmonary edema induced by enterovirus 71 (EV71) in southern Taiwan the mortality was lower in those who were given milrinone (36% versus 92%) (67). There were significant reductions in white blood cell count (10.8 versus 19.5 x 10^9/l) and platelet count (257 versus 400 x 10^9/l).

Falls in platelet counts have been seen with milrinone (68) and it can cause thrombocytopenia (SEDA-20, 174). Of 19 children (12 infants and seven children aged 1–13 years) who were given either two boluses of 25 micrograms/kg followed by an infusion of 0.5 micrograms/kg/min, or a bolus of 50 micrograms/kg followed by a bolus of 25 micrograms/kg followed by an infusion of 0.75 micrograms/kg/minute, 11 developed thrombocytopenia, defined as a platelet count below 100 x 109/l, during milrinone infusion (69). Of these, two infants required a platelet transfusion. By comparison, in 128 patients who did not receive milrinone the incidence of thrombocytopenia was 25%, significantly lower than in those given milrinone. The occurrence of thrombocytopenia increased with increasing duration of infusion.

Gastrointestinal

In 3833 patients with heart failure of Class III or IV and a left ventricular ejection fraction of 30% or less despite optimal treatment, followed for 286 days, there was a slight increase in the risk of diarrhea, which occurred in 17% of those taking 60 mg/day compared with 12% of those taking placebo and 14.5% of those taking 30 mg/day (64).

Musculoskeletal

Enoximone has been reported to cause increased contraction in skeletal muscle in vitro in patients with a predisposition to malignant hyperthermia (70,71).

- Vastus lateralis muscle from a 48-year-old man with rhabdomyolysis had an increased in vitro response to increasing concentrations of enoximone above 0.6 mmol/l (72). He also had a heterozygous mutation in his ryanodine receptor gene, with substitution of arginine for glycine in position 2433, a mutation that is associated with malignant hyperthermia.

Phosphodiesterase inhibitors increase the release of calcium from the sarcoplasmic reticulum by activating ryanodine receptors, and the authors suggested that this may have been the mechanism whereby enoximone precipitated rhabdomyolysis in this case.

Death

Long-term treatment with inhibitors of phosphodiesterase type III is associated with increased mortality in congestive heart failure (SEDA-17, 217).

In 102 patients taking digoxin and diuretics the addition of enoximone did not confer any therapeutic advantage and there was a significantly higher drop-out rate and a significantly higher mortality rate in those taking enoximone (73).

The use of a lower dose of enoximone in 105 patients with heart failure (New York Heart Association classes 2 or 3) has been studied over 12 weeks (74). Enoximone 25–50 mg tds improved exercise capacity and reduced dyspnea. There was no evidence of a dysrhythmic effect. The rates of adverse events were similar with enoximone and placebo, and there were fewer cases of dizziness, vertigo, or hypotension in those who took enoximone. There were two deaths in the 70 patients who took enoximone compared with four of the 35 who took placebo. However, this small short-term study does not rule out the possibility that even this small dose of enoximone can cause increased mortality during long-term administration or in patients with more severe cardiac failure.

Susceptibility Factors

Renal disease

In six patients with severe congestive heart failure and renal insufficiency being treated with continuous venovenous hemofiltration, the pharmacokinetics of milrinone 0.25 mg/kg/minute by continuous intravenous infusion were different from those that have been previously reported in patients with normal renal function, with a prolonged half-life and a raised mean steady-state concentration, suggestive of reduced clearance (75). The half-life of milrinone was 20 hours, compared with reported half-lives of around 3 hours.

Drug administration

Drug administration route

Inhaled milrinone has been used to treat patients with pulmonary hypertension after cardiac surgery (76). Milrinone produced a significant dose-related reduction in pulmonary vascular resistance in nine patients. There were no systemic adverse effects.

Drug-drug interactions

Gemcitabine

In 26 patients who received combinations of vesnarinone and gemcitabine there was no pharmacokinetic interaction between the two drugs (77). Although there were cases of neutropenia and thrombocytopenia, those effects could have occurred with gemcitabine alone and could not necessarily be attributed to vesnarinone. Other adverse effects of the combination included nausea and vomiting, anorexia and fatigue, diarrhea, headache, and fever. There were no cases of QT interval prolongation or ventricular dysrhythmias. There were rises in hepatic transaminase activities in up to 25% of cases. Again it was not possible to distinguish between adverse effects of gemcitabine and vesnarinone.

Vecuronium

In 30 adults randomly assigned to intravenous saline or milrinone (5 micrograms/kg/minute for 10 minutes followed by 0.5 micrograms/kg/minute) 30 minutes before anesthesia, the onset of neuromuscular block produced by vecuronium 0.1 mg/kg was significantly delayed by milrinone and recovery was significantly quicker (78). The mechanism of this effect is not known, but the authors suggested that milrinone blocks adenosine A_1 receptors, increasing the release of acetylcholine in motor nerve endings, and might also make the neuromuscular junction more sensitive to the effects of acetylcholine.

References

1. Frielingsdorf J, Kiowski W: Pharmacology and clinical use of newer inotropic agents. Anesthetic Pharmacol Rev 1994;2:332.
2. Colucci WS, Wright RF, Braunwald E. New positive inotropic agents in the treatment of congestive heart failure. Mechanisms of action and recent clinical developments. 1. N Engl J Med 1986;314(5):290–9.
3. Colucci WS, Wright RF, Braunwald E. New positive inotropic agents in the treatment of congestive heart failure.

Mechanisms of action and recent clinical developments. 2. N Engl J Med. 1986;314(6):349–58.
4. Webster MW, Sharpe DN. Adverse effects associated with the newer inotropic agents. Med Toxicol 1986;1(5):335–42.
5. Rocci ML Jr, Wilson H. The pharmacokinetics and pharmacodynamics of newer inotropic agents. Clin Pharmacokinet 1987;13(2):91–109.
6. Leier CV. Current status of non-digitalis positive inotropic drugs. Am J Cardiol 1992;69(18):120G–128G.
7. Sasayama S. What do the newer inotropic drugs have to offer? Cardiovasc Drugs Ther 1992;6(1):15–18.
8. Moazemi K, Chana JS, Willard AM, Kocheril AG. Intravenous vasodilator therapy in congestive heart failure. Drugs Aging 2003;20:485–508.
9. van Veldhuisen DJ, Poole-Wilson PA. The underreporting of results and possible mechanisms of 'negative' drug trials in patients with chronic heart failure. Int J Cardiol 2001; 80(1):19–27.
10. Nony P, Boissel JP, Lievre M, Leizorovicz A, Haugh MC, Fareh S, de Breyne B. Evaluation of the effect of phosphodiesterase inhibitors on mortality in chronic heart failure patients. A meta-analysis. Eur J Clin Pharmacol 1994; 46(3):191–6.
11. Wilmshurst PT, Webb-Peploe MM. Side effects of amrinone therapy. Br Heart J 1983;49(5):447–51.
12. DiBianco R, Shabetai R, Silverman BD, Leier CV, Benotti JR. Oral amrinone for the treatment of chronic congestive heart failure: results of a multicenter randomized double-blind and placebo-controlled withdrawal study. J Am Coll Cardiol 1984;4(5):855–66.
13. Johnston DL, Humen DP, Kostuk WJ. Amrinone therapy in patients with heart failure. Lack of improvement in functional capacity and left ventricular function at rest and during exercise. Chest 1984;86(3):394–400.
14. Packer M, Medina N, Yushak M. Hemodynamic and clinical limitations of long-term inotropic therapy with amrinone in patients with severe chronic heart failure. Circulation 1984;70(6):1038–47.
15. Klepzig M, Kleinhans E, Büll U, Strauer BE. Amrinone in Akut- und Langzeittherapie. [Amrinone in acute and long-term therapy.] Z Kardiol 1985;74(2):85–90.
16. Leier CV, Dalpiaz K, Huss P, Hermiller JB, Magorien RD, Bashore TM, Unverferth DV. Amrinone therapy for congestive heart failure in outpatients with idiopathic dilated cardiomyopathy. Am J Cardiol 1983;52(3):304–8.
17. Shah PK, Amin DK, Hulse S, Shellock F, Swan HJ. Inotropic therapy for refractory congestive heart failure with oral fenoximone (MDL-17,043): poor long-term results despite early hemodynamic and clinical improvement. Circulation 1985;71(2):326–31.
18. Kereiakes D, Chatterjee K, Parmley WW, Atherton B, Curran D, Kereiakes A, Spangenberg R. Intravenous and oral MDL 17043 (a new inotrope-vasodilator agent) in congestive heart failure: hemodynamic and clinical evaluation in 38 patients. J Am Coll Cardiol 1984;4(5):884–9.
19. Leeman M, Lejeune P, Melot C, Naeije R. Reduction in pulmonary hypertension and in airway resistances by enoximone (MDL 17,043) in decompensated COPD. Chest 1987;91(5):662–6.
20. Martin JL, Likoff MJ, Janicki JS, Laskey WK, Hirshfeld JW Jr, Weber KT. Myocardial energetics and clinical response to the cardiotonic agent MDL 17043 in advanced heart failure. J Am Coll Cardiol 1984;4(5):875–83.
21. Rubin SA, Tabak L. MDL 17,043: short- and long-term cardiopulmonary and clinical effects in patients with heart failure.J Am Coll Cardiol 1985;5(6):1422–7.
22. Uretsky BF, Generalovich T, Verbalis JG, Valdes AM, Reddy PS. MDL 17,043 therapy in severe congestive heart failure: characterization of the early and late hemodynamic, pharmacokinetic, hormonal and clinical response. J Am Coll Cardiol 1985;5(6):1414–21.
23. Takeno M, Takagi S, Sakuragi S, Suzuki S, Tsutsumi Y, Nonogi H, Goto Y. Continuous milrinone infusion during preoperative anti-inflammatory therapy in inflammatory aortic aneurysm complicating severe congestive heart failure. Heart Vessels 2002;17(1):42–4.
24. Poole-Wilson PA. Treatment of acute heart failure: out with the old, in with the new. JAMA 2002;287(12):1578–80.
25. Mizushige K, Ueda T, Yukiiri K, Suzuki H. Olprinone: a phosphodiesterase III inhibitor with positive inotropic and vasodilator effects. Cardiovasc Drug Rev 2002;20(3): 163–74.
26. Fitton A, Brogden RN. Pimobendan. A review of its pharmacology and therapeutic potential in congestive heart failure. Drugs Aging 1994;4(5):417–41.
27. Renard M, Jacobs P, Dechamps P, Dresse A, Bernard R. Hemodynamic and clinical response to three-day infusion of sulmazol (AR-L 115 BS) in severe congestive heart failure. Chest 1983;84(4):408–13.
28. Renard M, Jacobs P, Melot C, Dresse A, Bernard R. Le sulmazol: un nouvel agent inotrope positif. [Sulmazole: a new positive inotropic agent.] Ann Cardiol Angeiol (Paris) 1984;33(4):219–22.
29. Berkenboom GM, Sobolski JC, Depelchin PE, Contu E, Dieudonné PM, Degré SG. Clinical and hemodynamic observations on orally administered sulmazol (ARL115BS) in refractory heart failure. Cardiology 1984; 71(6):323–30.
30. Hagemeijer F, Segers A, Schelling A. Cardiovascular effects of sulmazol administered intravenously to patients with severe heart failure. Eur Heart J 1984 Feb;5(2):158–67.
31. Thormann J, Schlepper M, Kramer W, Gottwik M, Kindler M. Effects of AR-L 115 BS (Sulmazol), a new cardiotonic agent, in coronary artery disease: improved ventricular wall motion, increased pump function and abolition of pacing-induced ischemia. J Am Coll Cardiol 1983;2(2):332–7.
32. Kawamata H, Omotehara F, Nakashiro K.-I, Uchida D, Hino S, Fujimori T. Vesnarinone: a differentiation-inducing anti-cancer drug. Anti-Cancer Drugs 2003;14:391–5.
33. Bertolet BD. Precautions for use and adverse effects of vesnarinone: potential mechanisms and future therapies. Drug Saf 2004;27 Suppl 1:11–18.
34. Zewail AM, Nawar M, Vrtovec B, Eastwood C, Kar B, Delgado III RM. Intravenous milrinone in treatment of advanced congestive heart failure. Texas Heart Inst J 2003;30:109–13.
35. Cesario D, Clark J, Maisel A. Beneficial effects of intermittent home administration of the inotrope/vasodilator milrinone in patients with end-stage congestive heart failure: a preliminary study. Am Heart J 1998;135(1):121–9.
36. Heinz G, Geppert A, Delle Karth G, Reinelt P, Gschwandtner ME, Neunteufl T, Zauner C, Frossard M, Siostrzonek P. IV milrinone for cardiac output increase and maintenance: comparison in nonhyperdynamic SIRS/sepsis and congestive heart failure. Intensive Care Med 1999;25(6):620–4.
37. Myou S, Fujimura M, Kamio Y, Hirose T, Kita T, Tachibana H, Ishiura Y, Watanabe K, Hashimoto T, Nakao S. Bronchodilator effects of intravenous olprinone, a phosphodiesterase 3 inhibitor, with and without aminophylline in asthmatic patients. 2003;55:341–6.
38. Loh E, Elkayam U, Cody R, Bristow M, Jaski B, Colucci WS. A randomized multicenter study comparing the

efficacy and safety of intravenous milrinone and intravenous nitroglycerin in patients with advanced heart failure. J Card Fail 2001;7(2):114–21.
39. Feneck RO, Sherry KM, Withington PS, Oduro-Dominah A; European Milrinone Multicenter Trial Group. Comparison of the hemodynamic effects of milrinone with dobutamine in patients after cardiac surgery. J Cardiothorac Vasc Anesth 2001;15(3):306–15.
40. Hoffman TM, Wemovsky G, Atz AM, Kulik TJ, Nelson DP, Chang AC, Bailey JM, Akbary A, Kocsis JF, Kaczmarek R, Spray TL, Wessel DL. Circulation 2003; 107:996–1002.
41. Thackray S, Witte K, Clark AL, Cleland JG. Clinical trials update: OPTIME-CHF, PRAISE-2, ALL-HAT. Eur J Heart Fail 2000;2(2):209–12.
42. Cuffe MS, Califf RM, Adams KF Jr, Benza R, Bourge R, Colucci WS, Massie BM, O'Connor CM, Pina I, Quigg R, Silver MA, Gheorghiade M; Outcomes of a Prospective Trial of Intravenous Milrinone for Exacerbations of Chronic Heart Failure (OPTIME-CHF) Investigators. Short-term intravenous milrinone for acute exacerbation of chronic heart failure: a randomized controlled trial. JAMA 2002;287(12):1541–7.
43. Bayram M, De Luca L, Massie MB, Gheorghiade M. Reassessment of dobutamine, dopamine, and milrinone in the management of acute heart failure syndromes. Am J Cardiol 2005;96(6A):47G–58G.
44. Feldman AM, Bristow MR, Parmley WW, Carson PE, Pepine CJ, Gilbert EM, Strobeck JE, Hendrix GH, Powers ER, Bain RP, et al. Effects of vesnarinone on morbidity and mortality in patients with heart failure. Vesnarinone Study Group. N Engl J Med 1993;329(3):149–55.
45. Lubsen J, Just H, Hjalmarsson AC, La Framboise D, Remme WJ, Heinrich-Nols J, Dumont JM, Seed P. Effect of pimobendan on exercise capacity in patients with heart failure: main results from the Pimobendan in Congestive Heart Failure (PICO) trial. Heart 1996;76(3):223–31.
46. Effects of Pimobendan on Chronic Heart Failure Study (EPOCH Study). Effects of pimobendan on adverse cardiac events and physical activities in patients with mild to moderate chronic heart failure: the effects of pimobendan on chronic heart failure study (EPOCH study). Circ J 2002;66(2):149–57.
47. Kato K. Clinical efficacy and safety of pimobendan in treatment of heart failure–experience in Japan. Cardiology 1997;88 Suppl 2:28–36.
48. Takei K, Haruta S. Three cases of acute exacerbation of chronic heart failure treated with milrinone. Ther Res 1999; 20(8):2433–9.
49. Benza RL, Tallaj JA, Felker GM, Zabel KM, Kao W, Bourge RC, Pearce D, Leimberger JD, Borzak S, O'connor CM, Gheorghiade M;OPTIME-CHF Investigators. The impact of arrhythmias in acute heart failure. J Card Fail 2004;10(4):279–84.
50. Amin DK, Shah PK, Hulse S, Shellock FG, Swan HJ. Myocardial metabolic and hemodynamic effects of intravenous MDL-17,043, a new cardiotonic drug, in patients with chronic severe heart failure. Am Heart J 1984;108(5): 1285–92.
51. Milfred-LaForest SK, Shubert J, Mendoza B, Flores I, Eisen HJ, Piña IL. Tolerability of extended duration intravenous milrinone in patients hospitalized for advanced heart failure and the usefulness of uptitration of oral angiotensin-converting enzyme inhibitors. Am J Cardiol 1999;84(8):894–9.
52. Alhashemi JA, Hooper J. Treatment of milrinone-associated tachycardia with beta-blockers. Can J Anaesth 1998;45(1):67–70.
53. Ramamoorthy C, Anderson G, Williams G, Lynn A. Pharmacokinetics and side effects of milrinone in infants and children after open heart surgery. Anesth Analg 1998;86(2):283–9.
54. Hoffman TM, Bush DM, Wernovsky G, Cohen MI, Wieand TS, Gaynor JW, Spray TL, Rhodes LA. Postoperative junctional ectopic tachycardia in children: incidence, risk factors, and treatment. Ann Thorac Surg 2002;74(5):1607–11.
55. Bichel T, Steinbach G, Olry L, Lambert H. Utilisation de l'amrinone dans le traitement du choc cardiogenique. [Use of intravenous amrinone in the treatment of cardiogenic shock.] Agressologie 1988;29(3):187–92.
56. Tsuneyoshi Y, Minami K, Nakayama S, Yoshikawa E. Comparative efficacy of milrinone and amrinone in patients undergoing coronary artery bypass grafting. Ther Res 1999; 20(6):1860–4.
57. Simonton CA, Chatterjee K, Cody RJ, Kubo SH, Leonard D, Daly P, Rutman H. Milrinone in congestive heart failure: acute and chronic hemodynamic and clinical evaluation. J Am Coll Cardiol 1985;6(2):453–9.
58. Jaski BE, Fifer MA, Wright RF, Braunwald E, Colucci WS. Positive inotropic and vasodilator actions of milrinone in patients with severe congestive heart failure. Dose-response relationships and comparison to nitroprusside. J Clin Invest 1985;75(2):643–9.
59. Anderson JL, Askins JC, Gilbert EM, Menlove RL, Lutz JR. Occurrence of ventricular arrhythmias in patients receiving acute and chronic infusions of milrinone. Am Heart J 1986;111(3):466–74.
60. Gold J, Cullinane S, Chen J, Seo S, Oz MC, Oliver JA, Landry DW. Vasopressin in the treatment of milrinone-induced hypotension in severe heart failure. Am J Cardiol 2000;85(4):506–8, A11.
61. Gold JA, Cullinane S, Chen J, Oz MC, Oliver JA, Landry DW.Vasopressin as an alternative to norepinephrine in the treatment of milrinone-induced hypotension. Crit Care Med 2000;28(1):249–52.
62. Ako J, Eto M, Watanabe T, Ouchi Y. A case of severe hypoglycemia due to pimobendan. Int J Cardiol 2001;80(1):83–4.
63. Fujimoto S, Ishida H, Kato S, The novel insulinotropic mechanism of pimobendan: direct enhancement of the exocytotic process of insulin secretory granules by increased Ca^{2+} sensitivity in beta-cells. Endocrinology 1998; 139:1133–40.
64. Cohn JN, Goldstein SO, Greenberg BH, Lorell BH, Bourge RC, Jaski BE, Gottlieb SO, McGrew F 3rd, DeMets DL, White BG. A dose-dependent increase in mortality with vesnarinone among patients with severe heart failure. Vesnarinone Trial Investigators. N Engl J Med 1998; 339(25):1810–6.
65. Patnode NM, Gandhi PJ. Drug-induced thrombocytopenia in the coronary care unit. J Thromb Thrombolysis 2000;10(2):155–67.
66. Kikura M, Sato S. Effects of preemptive therapy with milrinone or amrinone on perioperative platelet function and haemostasis in patients undergoing coronary bypass grafting. Platelets 2003;14:277–82.
67. Wang SM, Lei HY, Huang MC, Wu JM, Chen CT, Wang JN, Wang JR, Liu CC. Therapeutic efficacy of milrinone in the management of enterovirus 71-induced pulmonary edema. Pediatr Pulmonol 2005;39(3):219–23.
68. Timmis AD, Smyth P, Jewitt DE. Milrinone in heart failure. Effects on exercise haemodynamics during short term treatment. Br Heart J 1985;54(1):42–7.
69. Ramamoorthy C, Anderson GD, Williams GD, Lynn AM. Pharmacokinetics and side effects of milrinone in infants

and children after open heart surgery. Anesth Analg 1998;86(2):283–9.
70. Fiege M, Wappler F, Scholz J, von Richthofen V, Brinken B, Schulte am Esch J. Diagnostik der Disposition zur malignen Hyperthermie durch einen in vitro Kontrakturtest mit dem Phosphodiesterase-III-Hemmstoff Enoximon. [Diagnosis of susceptibility to malignant hyperthermia with an in vitro contracture test with the phosphodiesterase iii inhibitor enoximone.] Anasthesiol Intensivmed Notfallmed Schmerzther 1998;33(9):557–63.
71. Fiege M, Wappler F, Scholz J, Weisshorn R, von Richthofen V, Schulte am Esch J. Effects of the phosphodiesterase-III inhibitor enoximone on skeletal muscle specimens from malignant hyperthermia susceptible patients. J Clin Anesth 2000;12(2):123–8.
72. Riess FC, Fiege M, Moshar S, Bergmann H, Bleese N, Kormann J, Weisshorn R, Wappler F. Rhabdomyolysis following cardiopulmonary bypass and treatment with enoximone in a patient susceptible to malignant hyperthermia. Anesthesiology 2001;94(2):355–7.
73. Uretsky BF, Jessup M, Konstam MA, Dec GW, Leier CV, Benotti J, Murali S, Herrmann HC, Sandberg JA. Multicenter trial of oral enoximone in patients with moderate to moderately severe congestive heart failure. Lack of benefit compared with placebo. Enoximone Multicenter Trial Group. Circulation 1990;82(3):774–80.
74. Lowes BD, Higginbotham M, Petrovich L, DeWood MA, Greenberg MA, Rahko PS, Dec GW, LeJemtel TH, Roden RL, Schleman MM, Robertson AD, Gorczynski RJ, Bristow MR. Low-dose enoximone improves exercise capacity in chronic heart failure. Enoximone Study Group. J Am Coll Cardiol 2000;36(2):501–8.
75. Taniguchi T, Shibata K, Saito S, Matsumoto H, Okeie K. Pharmacokinetics of milrinone in patients with congestive heart failure during continuous venovenous hemofiltration. Intensive Care Med 2000;26(8):1089–93.
76. Haraldsson Å, Kieler-Jensen N, Ricksten S-E. The additive pulmonary vasodilatory effects of inhaled prostacyclin and inhaled milrinone in postcardiac surgical patients with pulmonary hypertension. Anesth Analg 2001;93(6):1439–45.
77. Patnaik A, Rowinsky EK, Tammara BK, Hidalgo M, Drengler RL, Garner AM, Siu LL, Hammond LA, Felton SA, Mallikaarjun S, Von Hoff DD, Eckhardt SG. Phase I and pharmacokinetic study of the differentiating agent vesnarinone in combination with gemcitabine in patients with advanced cancer. J Clin Oncol 2000;18(23):3974–85.
78. Nakajima H, Hattori H, Aoki K, Katayama T, Saitoh Y, Murakawa M. Effect of milrinone on vecuronium-induced neuromuscular block. Anaesthesia 2003;58:643–6.

Pindolol

See also Beta-adrenoceptor antagonists

General Information

Pindolol is a beta-adrenoceptor antagonist with beta$_2$-adrenoceptor agonist action and some membrane-stabilizing activity (1).

Organs and Systems

Nervous system

Pindolol can cause resting tremor in some patients (2).

References

1. Aellig WH, Clark BJ. Is the ISA of pindolol beta$_2$-adrenoceptor selective? Br J Clin Pharmacol 1987;24(Suppl 1):S21–8.
2. Hod H, Har-Zahav J, Kaplinsky N, Frankl O. Pindolol-induced tremor. Postgrad Med J 1980;56(655):346–7.

Practolol

See also Beta-adrenoceptor antagonists

General Information

Practolol is a highly cardioselective beta-adrenoceptor antagonist with partial agonist activity.

Although practolol has long been withdrawn from general oral use (because of the oculomucocutaneous syndrome and sclerosing peritonitis) (1), it is still available for intravenous administration in some countries. The practolol syndrome included a psoriasiform rash, xerophthalmia due to lacrimal gland fibrosis, secretory otitis media, fibrinous peritonitis, and a lupus-like syndrome (SED-8, 444) (SEDA-3, 161) (SEDA-2, 170). The pathogenesis of this adverse effect is unknown, but it appears to be unique to practolol.

Reference

1. Mann RD. A ranked presentation of the MHRA/CSM (Medicines & Health Care Regulatory Agency/Committee on Safety of Medicines) Drug Analysis Print (DAP) data on practolol. Pharmacoepidemiol Drug Saf 2005;14(10):705–10.

Prazosin

See also Alpha-adrenoceptor antagonists

General Information

Prazosin is an alpha$_1$-adrenoceptor antagonist.

Organs and Systems

Cardiovascular

Postural hypotension and reflex tachycardia, particularly on standing, are features of the first-dose response to prazosin, but adaptive receptor responses lead to re-setting of the reflex mechanisms within the first few days of treatment, and there are therefore generally no significant changes in heart rate during long-term treatment. An orthostatic component to the hypotensive response persists during long-term treatment, and this may be significant and symptomatic if high dosages are used. It is less frequent with modified-release prazosin.

Metabolism

Prazosin and other quinazolines are associated with small but significant changes in plasma lipid profiles. Generally these are potentially beneficial changes, with reductions in LDL cholesterol, total cholesterol, and triglycerides, and increases in HDL cholesterol.

Urinary tract

Blockade of $alpha_1$-adrenoceptors in the urinary tract leads to smooth muscle relaxation and improvement in urinary flow, and this pharmacological action has been used to ameliorate the urinary symptoms of benign prostatic hyperplasia. Prazosin can occasionally lead to urinary incontinence, particularly stress incontinence in women (1).

Sexual function

There is some evidence that male sexual dysfunction occurs less often with alpha-blockers than with other types of antihypertensive drugs (2). For example, in a comparison of the effects of prazosin and hydrochlorothiazide on sexual function in 12 hypertensive men, plethysmographic measurements and subjective assessments showed less dysfunction with prazosin than with hydrochlorothiazide (3). There is no evidence that this effect can be used therapeutically, but cases of priapism have been reported: for example, a 55-year-old man presented with priapism having taken prazosin 7.5 mg tds for 4 months. After a further 3 months, prazosin was discontinued and erectile function became normal (4).

Drug–Drug Interactions

Alcohol

In a well-controlled study in 10 Japanese patients with mild hypertension, the blood pressure reduction caused by alcohol 1 ml/kg was significantly increased by concurrent treatment with prazosin 1 mg tds (5). At 2–4 hours after ingestion the blood pressure fell by 18/12 mmHg without prazosin and by 24/18 mmHg with prazosin. These results raise the possibility that heavy drinking may cause symptomatic hypotension in patients taking prazosin.

Calcium channel blockers

An interaction of prazosin with nifedipine or verapamil resulted in acute hypotension(6,7). The mechanism appears to be partly kinetic (the systemic availability of prazosin increasing by 60%) and partly dynamic.

References

1. Wall LL, Addison WA. Prazosin-induced stress incontinence. Obstet Gynecol 1990;75(3 Pt 2):558–60.
2. Neaton JD, Grimm RH Jr, Prineas RJ, Stamler J, Grandits GA, Elmer PJ, Cutler JA, Flack JM, Schoenberger JA, McDonald R, Lewis CE, Liebson PR. Treatment of Mild Hypertension Study. Final results. Treatment of Mild Hypertension Study Research Group. JAMA 1993;270(6):713–24.
3. Scharf MB, Mayleben DW. Comparative effects of prazosin and hydrochlorothiazide on sexual function in hypertensive men. Am J Med 1989;86(1B):110–2.
4. Bullock N. Prazosin-induced priapism. Br J Urol 1988;62(5):487–8.
5. Kawano Y, Abe H, Kojima S, Takishita S, Omae T. Interaction of alcohol and an alpha1-blocker on ambulatory blood pressure in patients with essential hypertension. Am J Hypertens 2000;13(3):307–12.
6. Jee LD, Opie LH. Acute hypotensive response to nifedipine added to prazosin in treatment of hypertension. BMJ (Clin Res Ed) 1983;287(6404):1514.
7. Pasanisi F, Meredith PA, Elliott HL, Reld JL. Verapamil and prazosin: pharmacodynamic and pharmacokinetic interactions in normal man. Br J Clin Pharmacol 1984;18:290.

Prenylamine

See also Calcium channel blockers

General Information

Prenylamine is a coronary vasodilator that depletes myocardial catecholamine stores and has some calcium-channel blocking activity. It has been used in the treatment of angina pectoris, but it often causes ventricular dysrhythmias and has been superseded by less toxic drugs.

Organs and Systems

Cardiovascular

Ventricular tachycardia associated with QT prolongation (torsade de pointes) has often been described with prenylamine (1,2).

Drug–Drug Interactions

Iothalamate

Ventricular tachycardia has been reported after the intravenous administration of sodium iothalamate in a patient taking prenylamine (3). Both drugs can prolong the QT interval.

References

1. Riccioni N, Bartolomei C, Soldani S. Prenylamine-induced ventricular arrhythmias and syncopal attacks with Q-T prolongation. Report of a case and comment on therapeutic use of lignocaine. Cardiology 1980;66(4):199–203.

2. Burri C, Ajdacic K, Michot F. Syndrom der verlangerten QT-Zeit und Kammer-Tachykardie "en torsade de pointe" nach Behandlung mit Prenylamin (Segontin). [Prolonged QT interval and "torsades de pointes" after therapy with prenylamin (Segontin).] Schweiz Rundsch Med Prax 1981;70(16):717–20.
3. Duncan JS, Ramsay LE. Ventricular tachycardia precipitated by sodium iothalamate (Conray 420) injection during prenylamine treatment: a predictable adverse drug interaction. Postgrad Med J 1985;61(715):415–7.

Propranolol

See also Beta-adrenoceptor antagonists

General Information

Propranolol is a lipophilic, non-selective, pure antagonist at beta-adrenoceptors, with membrane-stabilizing action.

Organs and Systems

Cardiovascular

An unusual dysrhythmia has been attributed to propranolol: alternating sinus rhythm with intermittent sinoatrial block. The authors suggested that this was accounted for by the existence of sinoatrial conduction via two pathways, the first with 2:1 block and the second with a slightly longer conduction time and intermittent 2:1 block (1).

Gastrointestinal

Propranolol can rarely cause mesenteric ischemia (2).

- A 59-year-old white man with hyperthyroidism was given propylthiouracil 100 mg tds and propranolol 20 mg bd. The next day, he developed increased abdominal pain and bloody diarrhea. Angiography showed superior mesenteric artery occlusion. Antegrade aorta-mesenteric bypass surgery was performed for revascularization.

Propranolol can reduce splanchnic blood flow by reducing cardiac output and selectively inhibiting vasodilatory receptors in the splanchnic circulation.

Immunologic

Propranolol has been implicated in hypersensitivity pneumonitis (3,4), although other beta-blockers have also been associated with this complication.

Drug administration

Drug overdose

In one case an overdose of propranolol unmasked Brugada syndrome (5).

- A 24-year-old healthy man took propranolol 2.28 g. After gastric lavage, electrolytes, cardiac enzymes, chest X-ray, and echocardiography were normal, but an electrocardiogram showed the typical coved pattern of Brugada syndrome. An ajmaline test confirmed Brugada syndrome.

The electrocardiographic effects in this case may have been due to the membrane stabilizing effect of high concentration of propranolol and/or inhibition of calcium channels.

References

1. Ozturk M, Demiroglu C. Alternating sinus rhythm and intermittent sinoatrial block induced by propranolol. Eur Heart J 1984;5(11):890–5.
2. Köksal AS, Usküdar O, Köklü S, Yüksel O, Beyazit Y, Sahin B. Propranolol-exacerbated mesenteric ischemia in a patient with hyperthyroidism. Ann Pharmacother 2005;39(3):559–62.
3. Aellig WH, Clark BJ. Is the ISA of pindolol beta 2-adrenoceptor selective? Br J Clin Pharmacol 1987;24(Suppl 1):S21–8.
4. Gauthier-Rahman S, Akoun GM, Milleron BJ, Mayaud CM. Leukocyte migration inhibition in propranolol-induced pneumonitis. Evidence for an immunologic cell-mediated mechanism. Chest 1990;97(1):238–41.
5. Aouate P, Clerc J, Viard P, Seoud J. Propranolol intoxication revealing a Brugada syndrome. J Cardiovasc Electrophysiol 2005;16(3):348–51.

Quinapril

See also Angiotensin converting enzyme inhibitors

General Information

The pharmacology, therapeutic uses, and safety of the ACE inhibitor quinapril have been reviewed (1).

Organs and Systems

Psychological, psychiatric

Depression has been attributed to quinapril (2).

- A 90-year-old man with a history of peripheral arterial disease and mild heart failure presented with reduced appetite, insomnia, anhedonia, reduced energy, and suicidal ideation. His symptoms had started a month before, when he had begun to take quinapril 10 mg/day. He was also taking furosemide 20 mg/day and digoxin. He was alert, coherent, and cognitively intact, but depressed with a flat affect. He had no psychotic symptoms. Quinapril was withdrawn. He improved within 48 hours and recovered fully in 5 days.

The authors cited several reports of depression, mania, and psychosis with various ACE inhibitors.

Acute psychosis has been attributed to quinapril (3).

- A 93-year-old woman with heart failure was given quinapril 2.5 mg bd. Two hours after the first dose she became confused, disoriented, and anxious. During the night, she made many frantic telephone calls to her daughter and other family members complaining that she was being assaulted. Her anxiety, disorientation, and visual hallucinations continued for 5 days. She had had an episode of hallucinations 2 years before, while taking a beta-blocker. Quinapril was withdrawn. She recovered over the next day.

The authors commented on three previous cases of visual hallucination with other ACE inhibitors.

Pancreas

A possible association between quinapril and pancreatitis has been reported in an 85-year-old woman, very similar to other previously reported cases with other ACE inhibitors (4).

Skin

Photosensitivity has been described in two patients taking quinapril; subsequent photoprovocation tests were positive (5).

Pemphigus vulgaris and pemphigus foliaceus have both been attributed to ACE inhibitors, including quinapril.

- A 54-year-old man developed a skin reaction 2 months after starting to take quinapril, with skin biopsy findings of IgG and complement fragment C3 deposits, consistent with drug-induced pemphigus (6).

References

1. Plosker GL, Sorkin EM. Quinapril. A reappraisal of its pharmacology and therapeutic efficacy in cardiovascular disorders. Drugs 1994;48(2):227–52.
2. Gunduz H, Georges JL, Fleishman S. Quinapril and depression. Am J Psychiatry 1999;156(7):1114–5.
3. Tarlow MM, Sakaris A, Scoyni R, Wolf-Klein G. Quinapril-associated acute psychosis in an older woman. J Am Geriatr Soc 2000;48(11):1533.
4. Arjomand H, Kemp DG. Quinapril and pancreatitis. Am J Gastroenterol 1999;94(1):290–1.
5. Rodriguez Granados MT, Abalde T, Garcia Doval I, De la Torre C. Systemic photosensitivity to quinapril. J Eur Acad Dermatol Venereol 2004;18(3):389–90.
6. Centre de Pharmacovigilance. Angiotensin-converting enzyme inhibitors and pemphigus. Folia Pharmacother 2005;32(1):9.

Ramipril

See also Angiotensin converting enzyme inhibitors

General Information

Ramipril is an ACE inhibitor, a prodrug that is rapidly hydrolysed after absorption to its active metabolite ramiprilat. It has been used in patients with hypertension, heart failure, and myocardial infraction.

Organs and Systems

Metabolism

Hypoglycemia has been attributed to several drugs in a complicated sequence of effects (1).

- A 64-year-old man with type II diabetes, hypertension, and bilateral renal artery stenosis presented with confusion and dysarthria related to profound hypoglycemia (2.2 mmol/l). He was taking naproxen 500 mg bd, ramipril 2.5 mg/day, glibenclamide 2.5 mg bd, metformin 850 mg bd, a thiazide diuretic, terazosin, ranitidine, paracetamol, and codeine. His plasma creatinine concentration, previously 185 μmol/l, was 362 μmol/l and it fell to 210 μmol/l after the withdrawal of ramipril and naproxen.

The authors discussed the possible role of renal insufficiency, resulting from co-prescription of naproxen and ramipril in the presence of volume depletion, which may have increased the risk of hypoglycemia related to glibenclamide plus metformin.

Electrolyte balance

Hyperkalemia is a concern when ACE inhibitors are co-administered with potassium-sparing diuretics, such as triamterene or amiloride, or aldosterone antagonists, such as spironolactone. Both classes of agents are increasingly being used, in both heart failure and hypertension. A 59-year-old patient with both of these diagnoses was given ramipril 5 mg/day, hydrochlorothiazide 12.5 mg/day, and spironolactone 25 mg/day and developed acute renal impairment and severe hyperkalemia (9.3 mmol/l) (2).

Hematologic

Neutropenia and agranulocytosis have both rarely been attributed to ACE inhibitors.

- Ramipril-induced agranulocytosis has been reported in a 55-year-old man with hypertensive chronic renal insufficiency (3). He was also taking metoprolol, clonidine, furosemide, simvastatin, aspirin, and amlodipine, but was given ramipril 4 days before developing weakness and a neutropenic fever. A bone marrow biopsy showed moderate hypocellularity. An in vitro lymphocyte cytotoxicity assay was performed with furosemide and ramipril; there was a cytotoxic response to ramipril.

Salivary glands

Sialadenitis has been reported with ACE inhibitors, although the mechanism is unclear and probably multifactorial. The primary pathogenesis is reduced salivary volume and/or flow rate. Disruption in salivary secretion at the cellular level can affect volumes, while flow rates can be affected by spasm or edema in the salivary ducts.

- An adult woman developed recurrent, painless, unilateral parotid gland swelling, which was initially thought to be recurrent sialadenitis (4). However, her symptoms resolved on withdrawal of ramipril. The clue to the diagnosis came from the fact that after multiple episodes of localized, pre-auricular gland swelling, she developed swelling that extended anteriorly to her nasal-labial crease, which became indistinct. This was considered to be a sign of angioedema and the ACE inhibitor was withdrawn. There was no recurrence.

It was presumed that the angioedema was the cause of the gland swelling, due to obstruction of the salivary gland ducts.

Gastrointestinal

Two cases of severe vomiting, dyspepsia, and headache, with falls in body weight and plasma albumin, have been reported in patients on chronic peritoneal dialysis (5). Both occurred a few days after they started to take ramipril (dose not reported) and totally resolved after withdrawal. Both patients subsequently took losartan, which was well tolerated. This led the authors to suggest that the mechanism was mediated by bradykinin and/or prostaglandins, through an interaction with gastrointestinal motility, which may also be affected by peritoneal dialysis.

Diarrhea is a recognized infrequent adverse effect of ACE inhibitors. A convincing case of diarrhea associated with ramipril has been reported (6).

Pancreas

Pancreatitis has been reported in patients taking ramipril (7,8). Pancreatitis, a rare adverse effect of ACE inhibitors, has also been reported with benazepril, captopril, enalapril, lisinopril, and quinapril.

Skin

Recurrent angioedema and urticaria associated with ramipril 2.5 mg/day has been reported (9). The patient had recurrent episodes, including three emergency admissions for respiratory distress, over 5 years. After she stopped taking ramipril her signs and symptoms cleared and she had no relapses during 6 months of follow-up. The authors pointed out that the rarity of this adverse effect can result in delayed diagnosis and they advocated greater awareness of ACE inhibitor-induced allergy.

Serosae

Polyserositis is an unusual adverse drug reaction.

- A 63-year-old man developed bilateral pleural effusions and a substantial pericardial effusion (10). He had taken nitrendipine, ramipril, hydrochlorothiazide, and carvedilol for 10 years. After withdrawal of these drugs and treatment of his condition, ramipril alone was restarted. Within 1 week the symptoms and effusions recurred.

A subsequent positive lymphocyte transformation test facilitated the probable diagnosis of an adverse drug reaction to ramipril.

References

1. Collin M, Mucklow JC. Drug interactions, renal impairment and hypoglycaemia in a patient with type II diabetes. Br J Clin Pharmacol 1999;48(2):134–7.
2. Nurnberger J, Daul A, Philipp T. Patient mit schwerer Hyperkaliämie—ein Notfall nach RALES. Dtsch Med Wochenschr 2005;130(36):2008–11.
3. Horowitz N, Molnar M, Levy Y, Pollack S. Ramipril-induced agranulocytosis confirmed by a lymphocyte cytotoxicity test. Am J Med Sci 2005;329(1):52–3.
4. Moss JR, Zanation AM, Shores CG. ACE inhibitor associated recurrent intermittent parotid gland swelling. Otolaryngol Head Neck Surg 2005;133(6):992–4.
5. Riley S, Rutherford PA. Gastrointestinal side effects of ramipril in peritoneal dialysis patients. Perit Dial Int 1998;18(1):83–4.
6. Tosetti C. Angiotensin-converting enzyme inhibitors and diarrhea. J Clin Gastroenterol 2002;35(1):105–6.
7. Kanbay M, Korkmaz M, Yilmaz U, Gur G, Boyacioglu S. Acute pancreatitis due to ramipril therapy. Postgrad Med J 2004;80(948):617–8.
8. Anagnostopoulus GK, Kostopoulus P, Tsiakos Í, Margantinis G, Arvantidis A. Fulminant pancreatitis associated with ramipril therapy. Pancreas 2003;3:278.
9. Bhalla M, Thami GP. Delayed diagnosis of angiotensin-converting enzyme (ACE) inhibitor induced angioedema and urticaria. Clin Exp Dermatol 2003;28:333–4.
10. Brunkhorst FM, Bloos F, Klein R. Ramipril induced polyserositis with pericardial tamponade and pleural effusion. Int J Cardiol 2005;102(2):355–6.

Reserpine

General Information

The *Rauwolfia* alkaloids, including reserpine, are now little used. They act by depleting neurotransmitter stores of catecholamines and reducing sympathetic nervous activity, but their effects are non-specific, and nervous system adverse effects (depression, drowsiness, tiredness, confusion) are prominent. There is also a troublesome incidence of diarrhea, hyperprolactinemia, gynecomastia, and a possible withdrawal syndrome.

Organs and Systems

Psychological, psychiatric

Depression is a very common adverse effect of reserpine (SED-9, 328) (1).

- A 66-year-old woman was admitted to hospital suffering from an agitated depressive psychosis. This settled with standard antipsychotic therapy. It was subsequently found that she had been taking reserpine in an over-the-counter formulation and had stopped this a week before her admission (2). Her doctor felt that the syndrome had occurred as a result of nervous system hypersensitivity after reserpine withdrawal.

Endocrine

In women reserpine causes a small increase in circulating concentrations of prolactin (3), which could be related to the small increase in the risk of breast cancer. In 27 hypertensive men reserpine 0.25 mg/day for 3 months had no effect on testosterone, dihydrotestosterone, estradiol, luteinizing hormone, or prolactin (4).

Long-Term Effects

Drug withdrawal

Acute psychosis has been reported after the withdrawal of reserpine (2,5). During long-term therapy in animals reserpine causes denervation sensitivity to dopamine in the basal ganglia and chemotactic trigger zone in man and to catecholaminergic agents in the basal ganglia and mesolimbic system; withdrawal could lead to rebound supersensitivity of these systems to endogenous catecholamines, causing the reported symptoms.

Tumorigenicity

Early retrospective studies suggested that reserpine was associated with breast cancer, but prospective studies and meta-analyses of case-control studies have shown only a weak association (6). In vitro studies have shown that rauwolfia alkaloids are not genotoxic or mutagenic (7).

References

1. Riddiough MA. Preventing, detecting and managing adverse reactions of antihypertensive agents in the ambulant patient with essential hypertension. Am J Hosp Pharm 1977;34(5):465–79.
2. Samuels AH, Taylor AJ. Reserpine withdrawal psychosis. Aust N Z J Psychiatry 1989;23(1):129–30.
3. Ross RK, Paganini-Hill A, Krailo MD, Gerkins VR, Henderson BE, Pike MC. Effects of reserpine on prolactin levels and incidence of breast cancer in postmenopausal women. Cancer Res 1984;44(7):3106–8.
4. Boyden TW, Nugent CA, Ogihara T, Maeda T. Reserpine, hydrochlorothiazide and pituitary–gonadal hormones in hypertensive patients. Eur J Clin Pharmacol 1980;17(5):329–32.
5. Kent TA, Wilber RD. Reserpine withdrawal psychosis: the possible role of denervation supersensitivity of receptors. J Nerv Ment Dis 1982;170(8):502–4.
6. Grossman E, Messerli FH, Goldbourt U. Carcinogenicity of antihypertensive therapy. Curr Hypertens Rep 2002;4(3):195–201.
7. von Poser G, Andrade HH, da Silva KV, Henriques AT, Henriques JA. Genotoxic, mutagenic and recombinogenic effects of rauwolfia alkaloids. Mutat Res 1990;232(1):37–43.

Rilmenidine

General Information

Rilmenidine is an imidazoline derivative that lowers blood pressure by an interaction with imidazoline (II) receptors in the brainstem and kidneys (1).

In a number of trials in hypertension the adverse effects profile of rilmenidine was qualitatively similar to that of clonidine, although the overall incidence of adverse events appeared to be lower. The main complaints were of dry mouth, drowsiness, and constipation (2–4). In one of these reviews, in which the impact on vigilance was particularly examined, it was concluded that although there were dose-related effects there was no statistically significant difference between rilmenidine and placebo in relation to sedation and drowsiness. However, these results consistently showed a rank order among placebo, rilmenidine, and clonidine, leading to progressively greater degrees of sedation (5).

Organs and Systems

Skin

A photosensitivity reaction has been attributed to rilmenidine (6).

- A 51-year-old woman developed erythema and swelling on sun-exposed areas and complained of a local burning sensation and pruritus 10 days after she started to take rilmenidine 1 mg/day for mild hypertension. She recovered fully 1 week after rilmenidine withdrawal and treatment with prednisolone. The chronology and the results of patch and photopatch tests suggested a phototoxic reaction to rilmenidine.

The authors suggested that this may have been related to the double bond in the oxazoline ring of the drug.

References

1. Reid JL. Update on rilmenidine: clinical benefits. Am J Hypertens 2001;14(11 Pt 2):S322–4.
2. Ostermann G, Brisgand B, Schmitt J, Fillastre JP. Efficacy and acceptability of rilmenidine for mild to moderate systemic hypertension. Am J Cardiol 1988;61(7):D76–80.
3. Beau B, Mahieux F, Paraire M, Laurin S, Brisgand B, Vitou P. Efficacy and safety of rilmenidine for arterial hypertension. Am J Cardiol 1988;61(7):D95–D102.
4. Fillastre JP, Letac B, Galinier F, Le Bihan G, Schwartz J. A multicenter double-blind comparative study of rilmenidine and clonidine in 333 hypertensive patients. Am J Cardiol 1988;61(7):D81–5.
5. Mahieux F. Rilmenidine and vigilance. Review of clinical studies. Am J Med 1989;87(3C):S67–72.
6. Mota AV, Vasconcelos C, Correia TM, Barros MA, Mesquita-Guimaraes J. Rilmenidine-induced photosensitivity reaction. Photodermatol Photoimmunol Photomed 1998;14(3–4):132–3.

Sitaxsentan

General Information

Sitaxsentan is an endothelin receptor antagonist that is reported to be more selective at endothelin A (ET_A) receptors (1).

Placebo-controlled studies

In a randomized, placebo-controlled trial in patients with symptomatic pulmonary arterial hypertension sitaxsentan 100 mg/day or 300 mg/day had similar adverse effects to those of bosentan, namely headache, peripheral edema, nausea, nasal congestion, and dizziness (2). There were also dose-dependent liver function abnormalities: transaminase activities three times the upper limit of normal were reached, with a cumulative risk at 9 months of 8% in the 100 mg/day group and 32% in the 300 mg/day group. Up to 80% of the patients were taking warfarin, and the INR was often increased because of inhibition of CYP2C9 by sitaxsentan. In a 1-year open extension phase of this trial, 11 patients were studied and one died from progressive pulmonary artery hypertension (3). Of the remaining 10 patients who were followed (four taking placebo, three sitaxsentan 100 mg/day, and three sitaxsentan 300 mg/day), none had liver function abnormalities or complications related to interaction of sitaxsentan with warfarin.

References

1. Widlitz AC, Barst RJ, Horn EM. Sitaxsentan: a novel endothelin-A receptor antagonist for pulmonary arterial hypertension. Exp Rev Cardiovasc Ther 2005;3(6):985–91.
2. Barst RJ, Langleben D, Frost A, Horn EM, Oudiz R, Shapiro S, McLaughlin V, Hill N, Tapson VF, Robbins IM, Zwicke D, Duncan B, Dixon RA, Frumkin LR; STRIDE-1 Study Group. Sitaxsentan therapy for pulmonary arterial hypertension. Am J Respir Crit Care Med 2004;169(4):441–7.
3. Langleben D, Hirsch AM, Shalit E, Lesenko L, Barst RJ. Sustained symptomatic, functional, and hemodynamic benefit with the selective endothelin-A receptor antagonist, sitaxsentan, in patients with pulmonary arterial hypertension: a 1-year follow-up study. Chest 2004;126(4):1377–81.

Sotalol

General Information

Sotalol is a non-selective beta-adrenoceptor antagonist with no partial agonist or membrane-stabilizing activity. It is also a class III antidysrhythmic drug. The beta-adrenoceptor antagonist properties of D,L-sotalol reside in the L-isomer, the D-isomer having class III antidysrhythmic activity and no clinically significant beta-blocking action (1).

Quinidine and sotalol have been compared in a prospective multicenter trial of 121 patients after conversion of atrial fibrillation (2). Patients with low left ventricular ejection fractions (below 0.4) or high left atrial diameters (over 5.2 cm) were excluded. After 6 months the percentages of patients remaining in sinus rhythm were similar in the two groups (around 70%), but when the dysrhythmia recurred, it occurred later with sotalol than with quinidine (69 versus 10 days). There was also a difference between patients who had been converted from recent onset atrial fibrillation, in whom sotalol was more effective than quinidine, and in patients who had had chronic atrial fibrillation for more than 3 days, in whom quinidine was more effective. There were significant adverse effects requiring withdrawal of therapy in 17 patients, of whom nine were taking quinidine; three patients had gastrointestinal symptoms, two had central nervous symptoms, two had allergic reactions, one had undefined palpitation, and one had QT interval prolongation. One patient taking quinidine, a 65-year-old man, had frequent episodes of non-sustained ventricular tachycardia.

Organs and Systems

Cardiovascular

Sotalol prolongs the QT interval and can predispose to torsade de pointes (3), sustained ventricular tachycardia, and cardiac arrest, especially in those taking 160 mg/day or more (SEDA-16, 191). Torsade de pointes has been reported in 1.2% of those exposed (data on file, Bristol-Myers Squibb, Princeton, New Jersey). In a review it was concluded that torsade de pointes is also more likely in patients with depressed left ventricular function and a history of sustained ventricular tachydysrhythmias (4).

Four cases of torsade de pointes have been reported in patients taking sotalol, prescribed for preventing atrial fibrillation episodes (5). All four were women and all had normal left ventricular function and QT intervals before treatment. The patients were taking no other drugs and the dosages were relatively low (80–240 mg/day). These cases confirm that there is a risk of torsade de pointes, even with low dosages of sotalol, normal QT intervals, and normal renal function before treatment.

Several publications involving 962 patients treated with an intravenous infusion of racemic sotalol have been reviewed, with the aim of describing the risk of torsade de pointes (6). Torsade de pointes occurred in only one case (0.1%; 95% CI = 0.003, 0.6), which is less often than with oral sotalol (2–4%). This difference can be explained by the shorter duration of treatment. The other reported complications were hypotension (0.3%), severe bradycardia (0.2%), and atrioventricular block necessitating drug withdrawal (0.03%).

In a survey of 1288 patients taking sotalol, dysrhythmias occurred in 56, in 24 cases torsade de pointes (3). There was no relation between these dysrhythmias and previously associated factors, such as bradycardia, a long QT interval, and hypokalemia, but in patients on hemodialysis even a low dosage (40 mg bd) can be associated with torsade de pointes (7). Similar prodysrhythmic effects have been reported in children (8).

Coronary artery spasm can be induced by beta-adrenoceptor antagonists not only in patients with variant angina but also in those without documented coronary artery disease.

- A middle-aged man with severe left ventricular dysfunction due to dilated cardiomyopathy developed a symptomatic sustained ventricular tachycardia (9). Sotalol prevented it. After 1 month of treatment, there were brief episodes of ST segment elevation during electrocardiographic monitoring for almost 3 consecutive days. Emergency coronary angiography showed no significant stenosis of the coronary tree. The ST segment elevation completely disappeared after sotalol was withdrawn and replaced by long-acting diltiazem.

The authors concluded that sotalol even in standard doses can cause coronary artery spasm through beta-blockade.

Death

A trial of the efficacy of D-sotalol in patients with left ventricular dysfunction (10) was discontinued early when an interim analysis of 2762 patients showed an overall mortality of 3.9% in the D-sotalol group, compared with 2% in the placebo group (11).

Susceptibility Factors

Some patients with thyrotoxicosis have occult cardiac dysfunction. However, the use of beta-blockers in the treatment of thyrotoxicosis can have severe consequences in terms of severe cardiac dysfunction (12).

- A 52-year-old woman developed atrial fibrillation with a ventricular rate of 220/minute. She had had thyrotoxicosis 20 years before, but was taking no medications. She had an 8-month history of weight loss with a normal appetite. During the previous month she had had excessive sweating, palpitation, and exercise intolerance. The diagnosis was thyroid storm. She was given intravenous sotalol 1 mg/kg over 15 minutes and within 5 minutes sinus rhythm was restored, with improvement of symptoms. However, immediately after the infusion she had an episode of ventricular tachycardia followed by sinus bradycardia, with a resultant fall in blood pressure, followed by an asystolic cardiac arrest. She underwent endotracheal intubation and was treated with inotropes. Right-heart catheterization showed severe cardiac dysfunction (cardiac index 1.2 l/minute). She made an uneventful recovery.

Patients with severe hyperthyroidism can have an occult cardiomyopathy that makes them extremely sensitive to beta-blockers. The long duration of action of sotalol in this case necessitated prolonged inotropic and vasopressor support. A shorter-acting beta-blocker, such as esmolol, could theoretically be safer in such cases.

References

1. Yasuda SU, Barbey JT, Funck-Brentano C, Wellstein A, Woosley RL. d-Sotalol reduces heart rate in vivo through a beta-adrenergic receptor-independent mechanism. Clin Pharmacol Ther 1993;53(4):436–42.
2. de Paola AA, Veloso HH. Efficacy and safety of sotalol versus quinidine for the maintenance of sinus rhythm after conversion of atrial fibrillation. SOCESP Investigators. The Cardiology Society of Sao Paulo. Am J Cardiol 1999;84(9):1033–7.
3. Soyka LF, Wirtz C, Spangenberg RB. Clinical safety profile of sotalol in patients with arrhythmias. Am J Cardiol 1990;65(2):A74–81.
4. Hohnloser SH, Woosley RL. Sotalol. N Engl J Med 1994;331(1):31–8.
5. Tan HH, Hsu LF, Kam RML, Chua T, Teo WS. A case series of sotalol-induced torsade de pointes in patients with atrial fibrillation. A tale with a twist. Ann Acad Med Singapore 2003;32:403–7.
6. Marill KA, Runge T. Meta-analysis of the risk of torsades de pointes in patients treated with intravenous racemic sotalol. Acad Emerg Med 2001;8(2):117–24.
7. Huynh-Do U, Wahl C, Sulzer M, Buhler H, Keusch G. Torsades de pointes during low-dosage sotalol therapy in haemodialysis patients. Nephrol Dial Transplant 1996; 11(6):1153–4.
8. Pfammatter JP, Paul T, Lehmann C, Kallfelz HC. Efficacy and proarrhythmia of oral sotalol in pediatric patients. J Am Coll Cardiol 1995;26(4):1002–7.
9. Muto S, Ashizawa N, Arakawa S, Tanaka K, Komiya N, Toda G, Seto S, Yano K. Sotalol-induced coronary spasm in a patient with dilated cardiomyopathy associated with sustained ventricular tachycardia. Intern Med 2004; 43(11):1051–5.
10. Waldo AL, Camm AJ, deRuyter H, Freidman PL, MacNeil DJ, Pitt B, Pratt CM, Rodda BE, Schwartz PJ. Survival with oral d-sotalol in patients with left ventricular dysfunction after myocardial infarction: rationale, design, and methods (the SWORD trial). Am J Cardiol 1995; 75(15):1023–7.
11. Choo V. SWORD slashed. Lancet 1994;344:1358.
12. Fraser T, Green D. Weathering the storm: beta-blockade and the potential for disaster in severe hyperthyroidism. Emerg Med (Fremantle) 2001;13(3):376–80.

Tamsulosin

See also Alpha-adrenoceptor antagonists

General Information

Tamsulosin is an $alpha_1$-adrenoceptor antagonist that was specially designed for the treatment of benign prostatic hyperplasia, since it is highly selective for the urinary tract $alpha_{1A}$-adrenoceptors (1). Indeed, it produces little or no cardiovascular effects, no first-dose effect, and much less dizziness. In clinical trials, adverse effects included dizziness, weakness, headache, and nasal congestion. Abnormal ejaculation was the most frequent adverse effect, in 8% of the patients at 0.4 mg/day, and in 18% at 0.8 mg/day (2,3).

Sensory systems

The intraoperative floppy iris syndrome (IFIS) has been reported in current or recent users of α-adrenoceptor antagonists undergoing phacoemulsification cataract surgery, as reported by the manufacturers of Flomax®, a formulation of tamsulosin hydrochloride (4). The company stated that no definitive causal relation has been established, but it has changed its product information to warn patients to tell their surgeons if they are taking tamsulosin, who may need to alter their surgical technique to minimize the risk of IFIS. A retrospective and prospective study has substantiated this association (5).

Gastrointestinal

Life-threatening esophagitis has been reported after ingestion of a single tamsulosin capsule (6).

Drug Administration

Drug overdose

Overdosage of tamsulosin has been reported (7).

- A 78-year-old woman took an unintentional overdose of tamsulosin 2 mg, mistaking it for herbal tablets. On the next day she developed headache and dizziness and had hypotension, orthostatic hypotension, and bradycardia. She recovered rapidly after supportive care.

Overdose with alpha-adrenoceptor antagonists usually causes hypotension, often accompanied by reflex tachycardia or bradycardia, as in this case. The mechanism of the bradycardia is unclear.

References

1. Michel MC, de la Rosette JJ. Efficacy and safety of tamsulosin in the treatment of urological diseases. Expert Opin Pharmacother 2004;5(1):151–60.
2. Narayan P, Tewari A. Overview of alpha-blocker therapy for benign prostatic hyperplasia Urology 1998;51(Suppl 4A): 38–45.
3. de Mey C. Cardiovascular effects of alpha-blockers used for the treatment of symptomatic BPH: impact on safety and well-being. Eur Urol 1998;34(Suppl 2):18–28.
4. Boehringer Ingelheim. Important safety information on Intraoperative Floppy Iris Syndrome (IFIS): 14 Oct 2005. http: //hc-sc.gc.ca/dhp-mps/medeff/advisories-avis/prof/2005/flomax_hpc-cps_e.html.
5. Chang DF, Campbell JR. Intraoperative floppy iris syndrome associated with tamsulosin. J Cataract Refract Surg 2005;31(4):664–73.
6. D'Agostino L, Manguso F, Bennato R, Scaramuzzo A. A life-threatening case of stenosing pill hypopharynx-oesophagitis caused by a tamsulosin capsule. Dig Liver Dis 2004;36(9):632–4.
7. Anand JS, Chodorowski Z, Wisniewski M. Acute intoxication with tamsulosin hydrochloride. Clin Toxicol (Phila) 2005;43(4):311.

Telmisartan

See also Angiotensin II receptor antagonists

General Information

Telmisartan is a highly selective antagonist at type 1 angiotensin II (AT_1) receptors.

Organs and Systems

Respiratory

The incidence of cough with telmisartan has been assessed in a multicenter, randomized, parallel-group, double-blind, placebo-controlled study, for 8 weeks using a visual analogue scale in 88 patients with hypertension who had previously had ACE inhibitor-related cough (1). Cough was reported in 60% of the patients taking lisinopril 20 mg/day, 16% of those taking telmisartan 80 mg/day, and 9.7% with placebo.

Gastrointestinal

Previous case reports have described visceral angioedema as a rare complication of ACE inhibition (2) and it has also been described with an angiotensin receptor antagonist (3).

- A 65-year-old woman who had had recurrent intermittent attacks of severe abdominal pain while taking enalapril, thought to be due to intestinal angioedema, took losartan and developed the same symptoms.

No further details were given about the likelihood that losartan was responsible, but this is the first reported case of possible intestinal angioedema with an angiotensin receptor antagonist.

Nails

Fingernail clubbing and discoloration (chromonychia) have been reported in association with angiotensin receptor antagonists (4).

- A 76-year-old Caucasian man developed clubbing of the fingernails and discoloration of both the fingernails and toenails after taking losartan 50 mg/day for 27 days. Even though treatment was changed to valsartan, the nail changes persisted for another 6 months. He was then given captopril, and the changes gradually abated over 17 months.

An extensive literature search revealed no reports of this effect in association with angiotensin receptor antagonists. However, one manufacturer had received spontaneous reports. Despite careful consideration of other possible causes of the patient's symptoms, the temporal association suggested a possible drug-related adverse event.

Urinary tract

Losartan has been previously been implicated in acute renal insufficiency in a patient with renal artery stenosis, and a further case has been reported (5).

- A 52-year-old woman with a 5-year history of hypertension developed accelerated hypertension and was given losartan and a loop diuretic. She developed anuria after a diarrheal illness and was found to have a right-sided atrophic kidney. Her treatment was continued and she subsequently was readmitted with a further anuric episode and acute renal insufficiency. There was severe stenosis in the solitary renal artery, and after percutaneous transluminal renal balloon angioplasty to the stenotic segment her urine flow increased and renal function normalized.

Immunologic

Severe angioedema has been attributed to telmisartan (6).

Second-Generation Effects

Teratogenicity
Intra-uterine exposure to angiotensin II receptor antagonists can be associated with neonatal abnormalities.

- A neonate exposed to telmisartan during the first trimester of pregnancy developed acute renal insufficiency, presenting with oligohydramnios, and survived (7).

In reviewing the literature the authors found reports of five fetal deaths and one neonatal death associated with first-trimester exposure to angiotensin II receptor antagonists. All of these cases presented with severe oligohydramnios. In three fetuses there were abnormalities of the feet and face with hypoplastic skull bones. The kidneys were enlarged with tubular dysgenesis.

Susceptibility factors

Renal disease

Telmisartan has been studied in patients with varying severity of chronic kidney disease in a non-randomized study (8). It was an effective antihypertensive with few adverse effects.

Drug Administration

Drug overdose

Pancreatitis has previously been described with angiotensin receptor antagonists. Biochemical alterations suggesting acute pancreatitis were reported in a patient after a suicide attempt with high doses of telmisartan and oxazepam (9). The authors suggested that the occurrence of mild pancreatitis after an overdose of telmisartan is a rare class effect and, although the patient had taken other drugs, the concomitant overdose of oxazepam was not considered responsible, because it has never been associated with pancreatic injury.

Drug–Drug Interactions

Cardiac glycosides

Digoxin

Multiple-dose telmisartan 120 mg od administered with digoxin 0.25 mg od resulted in higher serum digoxin concentrations (10). Digoxin AUC and C_{max} rose by 22% and 50% respectively; the rise in C_{min} (13%) was not significant. These results suggest that telmisartan reduces the clearance of digoxin. The magnitude of this effect is comparable to that observed with calcium channel blockers, carvedilol, captopril, amiodarone, quinidine, and propafenone. Monitoring serum digoxin concentrations should be considered when patients first receive telmisartan and in the event of any changes in the dosage of telmisartan.

Paracetamol or ibuprofen

The single-dose pharmacokinetics of telmisartan 120 mg were not affected by a concurrent single dose of paracetamol 1 g or ibuprofen 400 mg tds for 7 days (11).

References

1. Lacourciere Y Telmisartan Cough Study Group. The incidence of cough: a comparison of lisinopril, placebo and telmisartan, a novel angiotensin II antagonist. Int J Clin Pract 1999;53(2):99–103.
2. Schmidt TD, McGrath KM. Angiotensin-converting enzyme inhibitor angioedema of the intestine: a case report and review of the literature. Am J Med Sci 2002;324(2):106–8.
3. Bucca C. Take the side-effects of drugs into account. Lancet 2004;364(9441):1285.
4. Packard KA, Arouni AJ, Hilleman DE, Gannon JM. Fingernail clubbing and chromonychia associated with the use of angiotensin II receptor blockers. Pharmacotherapy 2004;24(4):546–50.
5. Kiykim AA, Boz M, Ozer C, Camsari A, Yildiz A. Two episodes of anuria and acute pulmonary edema in a losartan-treated patient with solitary kidney. Heart Vessels 2004;19(1):52–4.
6. Borazan A, Üstin H, Yilmaz A. Angioedema induced by angiotensin II blocker telmisartan. Eur J Allergy Clin Immunol 2003;58:454.
7. Pieterement C, Malot L, Santerne B, Roussel B, Motte J, Morville P. Neonatal acute renal failure secondary to maternal exposure to telmisartan, angiotensin II receptor antagonist. J Perinatol 2003;23:254–5.
8. Sharma AM, Hollander A, Koster J. Telmisartan in patients with mild/moderate hypertension and chronic kidney disease. Clin Nephrol 2005;63(4):250–7.
9. Baffoni L, Durante V, Grossi M. Acute pancreatitis induced by telmisartan overdose. Ann Pharmacother 2004;38(6):1088.
10. Stangier J, Su CA, Hendriks MG, van Lier JJ, Sollie FA, Oosterhuis B, Jonkman JH. The effect of telmisartan on the steady-state pharmacokinetics of digoxin in healthy male volunteers. J Clin Pharmacol 2000;40(12 Pt 1):1373–9.
11. Stangier J, Su CA, Fraunhofer A, Tetzloff W. Pharmacokinetics of acetaminophen and ibuprofen when coadministered with telmisartan in healthy volunteers. J Clin Pharmacol 2000;40(12 Pt 1):1338–46.

Temocapril

See also Angiotensin converting enzyme inhibitors

General Information

Temocapril is an ACE inhibitor, a prodrug that is converted to the active metabolite temocaprilat after absorption.

Pharmacological and clinical studies of temocapril including its toxicology, pharmacokinetics, and adverse effects have been reviewed (1).

Cardiovascular

A 72-year-old woman developed exercise-related syncope during treatment with amlodipine and temocapril for hypertension (2).

Drug–Drug Interactions

Warfarin

Temocapril had no effect on the pharmacokinetics or pharmacodynamics of multiple-dose warfarin in 24 healthy subjects (3). There have been similar negative interaction studies with other ACE inhibitors and oral anticoagulants.

References

1. Yasunari K, Maeda K, Nakamura M, Watanabe T, Yoshikawa J, Asada A. Pharmacological and clinical studies with temocapril, an angiotensin converting enzyme inhibitor that is excreted in the bile. Cardiovasc Drug Rev 2004;22(3):189–98.
2. Ogimoto A, Mizobuchi T, Shigematsu Y, Hara Y, Ohtsuka T, Fukuoka T, Okura T, Higaki J. Exercise-related syncope induced by vasodilator therapy in an elderly hypertensive patient. J Am Geriatr Soc 2005;53(2):351–2.
3. Lankhaar G, Eckenberger P, Ouwerkerk MJA, Dingemanse J. Pharmacokinetic–pharmacodynamic investigation of a possible interaction between steady-state temocapril and warfarin in healthy subjects. Clin Drug Invest 1999;17:399–405.

Terazosin

See also Alpha-adrenoceptor antagonists

General Information

The alpha$_1$-adrenoceptor antagonist terazosin has similar adverse effects to those of prazosin. The long-term efficacy and safety of terazosin have been reviewed, in both monotherapy and combination therapy in whites and blacks (1,2).

In order to investigate the mechanisms of adverse events associated with alpha$_1$-adrenoceptor antagonists, the Veterans Affairs Cooperative Study database was analysed with respect to the relation between adverse events and hypotension in 1229 men with benign prostate hyperplasia. Treatment with terazosin produced the following rates of adverse events: dizziness 19%, weakness 6%, postural hypotension 6%, and syncope 1%. Of these adverse events only postural hypotension was associated with orthostatic blood pressure changes. Weakness, dizziness, and postural hypotension occurred to the same extent in patients with falls in systolic blood pressure of 5 mmHg or more or less than 5 mmHg. Thus, dizziness and weakness do not seem to be associated with changes in blood pressure, suggesting that these adverse events associated with alpha-blockers are not related to vascular effects. Designing a subtype selective alpha$_1$-antagonist that has less effect on blood pressure may not result in marked improvement in tolerability over currently available alpha$_1$-antagonists (3).

Organs and Systems

Skin

Terazosin has reportedly caused a rash (4).

- A 59-year-old man developed a generalized rash 3 days after starting to take terazosin 2 mg/day for benign prostatic hyperplasia. He had mild fever, weakness, intense pruritus, and a widespread eruption of scaling erythematous plaques with a violaceous hue on the trunk and extremities. The clinical examination was otherwise unremarkable and laboratory and serological test were negative. A skin biopsy was suggestive of a drug reaction. Terazosin was stopped, and treatment with oral prednisolone and emollients resulted in complete recovery in 2 weeks.

Sexual function

Priapism is rare with alpha-blockers (5).

- A 20-year-old man with a cervical spinal cord injury and a neuropathic bladder was given terazosin 1 mg at night increasing 5 days later to 2 mg. He developed a full erection of his penis, which lasted 5 hours and subsided spontaneously. Terazosin was stopped and the patient experienced no further priapism.

References

1. Saunders E. The safety and efficacy of terazosin in the treatment of essential hypertension in blacks. Am Heart J 1991;122(3 Pt 2):936–42.
2. Cohen JD. Long-term efficacy and safety of terazosin alone and in combination with other antihypertensive agents. Am Heart J 1991;122(3 Pt 2):919–25.
3. Lepor H, Jones K, Williford W. The mechanism of adverse events associated with terazosin: an analysis of the Veterans Affairs Cooperative Study. J Urol 2000;163(4):1134–7.
4. Hernandez-Cano N, Herranz P, Lazaro TE, Mayor M, Casado M. Severe cutaneous reaction due to terazosin. Lancet 1998;352(9123):202–3.
5. Vaidyanathan S, Soni BM, Singh G, Sett P, Krishnan KR. Prolonged penile erection association with terazosin in a cervical spinal cord injury patient. Spinal Cord 1998;36(11):805.

Tezosentan

See also Endothelin receptor antagonists

General Information

Tezosentan is a dual endothelin-A and endothelin-B receptor antagonist. Its common adverse effects include headaches, nausea, and hypotension compared with placebo (1).

Organs and Systems

Nervous system

Tezosentan has been studied in single intravenous doses of 5, 20, 50, 100, 200, 400, and 600 mg for 1 hour in sequential groups of six men in a randomized, placebo-controlled, double-blind design (2). Headache was the most frequently reported adverse event and it was more common than with placebo at doses of 100 mg and over. There were no clinically important changes in vital signs, electrocardiography, or clinical laboratory tests. The volume of distribution at steady state, about 16 liters, and the clearance, 30 l/hour, were independent of dose. Endothelin-1 concentrations increased dose- and concentration-dependently and returned slowly to baseline after the end of the infusion.

During chronic double-blind, placebo-controlled infusion in healthy men, tezosentan 100 mg/hour for 6 hours in six subjects (total dose 600 mg) and 5 mg/hour for 72 hours in eight subjects (total dose 360 mg), headache was the most common adverse event (75–100% with tezosentan and 50% with placebo) (3).

References

1. Tovar JM, Gums JG. Tezosentan in the treatment of acute heart failure. Ann Pharmacother 2003;37(12):1877–83.
2. Dingemanse J, Clozel M, van Giersbergen PL. Entry-into-humans study with tezosentan, an intravenous dual endothelin receptor antagonist. J Cardiovasc Pharmacol 2002;39(6):795–802.
3. Dingemanse J, Clozel M, van Giersbergen PL. Pharmacokinetics and pharmacodynamics of tezosentan, an intravenous dual endothelin receptor antagonist, following chronic infusion in healthy subjects. Br J Clin Pharmacol 2002;53(4):355–62.

Timolol

See also Beta-adrenoceptor antagonists

General Information

Timolol is a hydrophilic, non-selective beta-blocker with little partial agonist and membrane-stabilizing activity.

Organs and Systems

Cardiovascular

Of 153 consecutive patients treated with timolol eye-drops, three complained of unexplained falls and two of them had dizziness and blackouts (1). Two had a cardioinhibitory carotid sinus syndrome (a period of asystole greater than 3 seconds after carotid sinus massage) and the third a vasodepressor carotid sinus syndrome (a reduction in systolic blood pressure higher than 50 mmHg after carotid sinus massage). In all three cases, timolol eye-drops were withdrawn. Follow-up carotid sinus massage in the two patients with a cardioinhibitory carotid sinus syndrome was negative, and all reported complete remission of symptoms.

Significant cardiovascular adverse effects have been reported after topical administration of timolol maleate. Bradycardia with frank syncope can occur, especially in elderly patients (2). After topical administration its action begins in 20 minutes, peaks in 4 hours, and lasts 24 hours. Episodes of dizziness and occasional falls can occur 1–2 hours after instillation of timolol, as has been described in an otherwise healthy elderly patient (3).

- An 80-year-old woman with gastrointestinal bleeding had a sinus bradycardia (52/minute) despite acute blood loss. The only drug she had used that was an AV nodal depressant was timolol maleate 0.5%, one drop to both eyes every day. Continuous electrocardiography showed transient complete AV block without ventricular escape for nearly 6 seconds about 1 hour after instillation of timolol eye drops. She also reported having previously had episodes of dizziness and occasional falls 1–2 hours after instillation of her eye drops. Timolol was withheld and a temporary pacemaker was inserted. Rechallenge with timolol was associated with recurrence of third-degree AV block. She subsequently had a permanent dual chamber pacemaker implanted.

Respiratory

In a randomized controlled clinical trial in 21 otherwise healthy individuals with high-pressure primary open-angle glaucoma, six of 11 who used topical timolol 0.5% for 3 years had subclinical increases in bronchial reactivity not completely reversible in three cases on withdrawal of the beta-blocker (4).

Psychological, psychiatric

Depressive symptoms were reported in 17 of 165 patients after the administration of timolol over two decades (5). Depression accounted for 17% of 369 central nervous system reactions to timolol reported to a National Registry of Drug-induced Ocular Side Effects during 7 years: of these, 20 cases were of acute suicidal depression.

Skin

Toxic epidermal necrolysis occurred 1 week after the combined use of dorzolamide, timolol, and latanoprost eyedrops in an otherwise healthy 60-year-old woman;

the authors attributed it to the combination of timolol and dorzolamide (6).

Sexual function

Peyronie's disease has been attributed to long-term therapy with timolol eye drops 0.25% bd in a 74-year-old man who had used it for 22 years; the conditon imporoved when he stopped using timolol (7). This seems to have been the first report of this condition in a patient using beta-blocker eye drops.

Long-Term Effects

Tumorigenicity

The development of liver tumors in mice and an increased incidence of mammary fibroadenomata in female rats at one time caused concern in the USA, although animal studies elsewhere and clinical studies have shown no evidence of tumor-forming potential.

Drug Administration

Drug formulations

Timolol hydrogel eyedrops 0.1% have been compared with timolol aqueous solution 0.5% in a randomized, double-blind, crossover study in 24 healthy subjects (8). The aqueous solution had about 10 times more systemic availability than the hydrogel. There was no difference between the two formulations in efficacy in reducing intraocular pressure, but the mean peak heart rate during exercise was reduced by 19/minute by the aqueous solution and by only 5/minute by the hydrogel.

Drug–Drug Interactions

Brimonidine

The effects of topical brimonidine and timolol have been compared in two trials in 926 subjects with glaucoma or ocular hypertension already using systemic beta-blockers (9). Concurrent systemic beta-blocker therapy had no deleterious effects on ocular hypotensive efficacy and no impact on safety with topical brimonidine, but the combination of timolol and brimonidine significantly reduced systolic and diastolic blood pressures and heart rate compared with brimonidine alone. This observation suggests that ocular hypotensive agents other than beta-blockers, such as brimonidine, may be appropriate as a first-choice therapy for glaucoma in patients concurrently taking systemic beta-blockers.

Quinidine

The combination of quinidine with timolol causes an increased risk of bradycardia (10).

References

1. Mulcahy R, Allcock L, O'Shea D. Timolol, carotid sinus hypersensitivity, and elderly patients. Lancet 1998;352(9134):1147–8.
2. Frishman WH, Kowalski M, Nagnur S, Warshafsky S, Sica D. Cardiovascular considerations in using topical, oral, and intravenous drugs for the treatment of glaucoma and ocular hypertension: focus on beta-adrenergic blockade. Heart Dis 2001;3(6):386–97.
3. Sharifi M, Koch JM, Steele RJ, Adler D, Pompili VJ, Sopko J. Third degree AV block due to ophthalmic timilol solution. Int J Cardiol 2001;80(2–3):257–9.
4. Gandolfi SA, Chetta A, Cimino L, Mora P, Sangermani C, Tardini MG. Bronchial reactivity in healthy individuals undergoing long-term topical treatment with beta-blockers. Arch Ophthalmol 2005;123(1):35–8.
5. Schweitzer I, Maguire K, Tuckwell V. Antiglaucoma medication and clinical depression. Aust NZ J Psychiatry 2001;35(5):569–71.
6. Flórez A, Rosón E, Conde A, González B, García-Doval I, de la Torre C, Cruces M. Toxic epidermal necrolysis secondary to timolol, dorzolamide, and latanoprost eyedrops. J Am Acad Dermatol 2005;53(5):909–11.
7. Ross JJ, Rahman I, Walters RF. Peyronie's disease following long-term use of topical timolol. Eye 2006;20(8):974–6.
8. Uusitalo H, Nino J, Tahvanainen K, Turjanmaa V, Ropo A, Tuominen J, Kahonen M. Efficacy and systemic side-effects of topical 0.5% timolol aqueous solution and 0.1% timolol hydrogel. Acta Ophthalmol Scand. 2005;83(6):723–8.
9. Schuman JSBrimonidine Study Groups 1 and 2. Effects of systemic beta-blocker therapy on the efficacy and safety of topical brimonidine and timolol. Ophthalmology 2000;107(6):1171–217.
10. Hartshorn EA. Interactions of cardiac drugs. Drug Intell Clin Pharm 1970;4:272.

Trandolapril

See also Angiotensin converting enzyme inhibitors

General Information

Trandolapril is a non-sulfhydryl ACE inhibitor that has been used in patients with hypertension, congestive heart failure, and myocardial infarction. It is a prodrug that is hydrolyzed to the active diacid trandolaprilat.

Comprehensive reviews and the results of a large trial have shown that trandolapril has a pattern of adverse effects reflecting those usually documented with other ACE inhibitors (1).

Organs and Systems

Respiratory

In 3402 patients with hypertension taking trandolapril, cough, assessed by visual analogue scale, was less common in smokers than in non-smokers (2).

References

1. Kober L, Torp-Pedersen C, Carlsen JE, Bagger H, Eliasen P, Lyngborg K, Videbaek J, Cole DS, Auclert L, Pauly NCTrandolapril Cardiac Evaluation (TRACE) Study Group. A clinical trial of the angiotensin-converting-enzyme inhibitor trandolapril in patients with left ventricular dysfunction after myocardial infarction. N Engl J Med 1995;333(25):1670–6.
2. Genes N, Vaur L, Etienne S, Clerson P. Evaluation de l'influence du tabac sur la tolérance du trandolapril. [Evaluation of the effect of tobacco on trandolapril tolerance.] Therapie 1999;54(6):693–7.

Valsartan

See also Angiotensin II receptor antagonists

General Information

Valsartan is a highly selective antagonist at angiotensin II type 1 (AT_1) receptors (1). It has a half-life of about 8 hours, and is metabolized to a negligible extent. Most of its clearance is via the feces.

In 1378 patients with mild to moderate hypertension valsartan 80 mg/day has been compared with losartan 50 mg/day and placebo for 8 weeks (2). This study confirmed the excellent tolerance profile, not different from placebo, of both angiotensin II receptor antagonists.

Organs and Systems

Respiratory

Pneumonitis has been reported in association with valsartan (3).

- A 70-year-old man with dyspnea had diffuse pulmonary infiltrates on chest X-ray and ground-glass changes and non-segmental consolidation on CT imaging. He had been taking the Chinese herbal drug saiko-ka-ryukotsu-borei-to and valsartan for about 3 months, and drug-induced pneumonitis was suspected. The drugs were withdrawn and oral prednisolone given. He improved symptomatically.

Psychological

Two different reports have pointed to nightmares and depression with valsartan.

- Nightmares from valsartan for hypertension improved when it was withdrawn and recurred when it was restarted in a 64-year-old woman (4).
- A 43-year-old woman reported depression and attempted suicide after taking valsartan in combination with hydrochlorothiazide for hypertension (5). The patient was also taking atenolol, but the time course and complete resolution after withdrawal of valsartan/hydrochlorothiazide suggested that this was a drug-induced episode of depression.

Skin

A lymphocytic dermatitis has been attributed to valsartan (6).

- A 40-year-old man, who had previously had chronic dermatitis while taking enalapril for hypertension, developed recurrent dermatological adverse effects after starting to take valsartan instead. He developed a skin reaction within a few weeks of changing therapy, and biopsy showed a lymphocytic infiltrate of Jessner–Kanof pattern. The dermatitis resolved completely without recurrence after withdrawal of valsartan.

Second-Generation Effects

Pregnancy

Anhydramnios has been attributed to valsartan (7).

- A 43-year-old patient developed complete anhydramnios by week 20 of her pregnancy. She was taking valsartan, which was stopped at 20 weeks in view of a possible adverse effect. The anhydramnios resolved within a further 4 weeks and a healthy infant was born near term.

This is the second reported case of anhydramnios related to the use of valsartan in pregnancy, and the evidence suggests that angiotensin receptor antagonists are likely to have similar fetotoxic effects to ACE inhibitors and should therefore be avoided in pregnancy.

Drug Administration

Drug overdose

The effects of valsartan overdose have been described (8).

- A 25-year-old woman developed hypotension, tachycardia, and painful muscle cramps after taking a significant dose of valsartan in a suicide attempt.

References

1. Thurmann PA. Valsartan: a novel angiotensin type 1 receptor antagonist. Expert Opin Pharmacother 2000;1(2):337–50.
2. Hedner T, Oparil S, Rasmussen K, Rapelli A, Gatlin M, Kobi P, Sullivan J, Oddou-Stock P. A comparison of the angiotensin II antagonists valsartan and losartan in the treatment of essential hypertension. Am J Hypertens 1999;12(4 Pt 1):414–7.
3. Tokunaga T. A case of drug-induced pneumonitis associated with Chinese herbal drugs and valsartan. Nihon Kokyuki Gakkai Zasshi 2005;43(7):406–11.
4. Kastalli K, Aïdli S, Klouz A, Sraïri S. Nightmares induced by valsartan. Pharmacoepidemiol Drug Saf 2003;12(Suppl 2): 236.
5. Ullrich H, Passenberg P, Agelink MW. Episodes of depression with attempted suicide after taking valsartan with hydrochlorothiazide. Dtsch Med Wochenschr 2003;128(48):2534–6.
6. Schepis C, Lentini M, Siragusa M, Batolo D. ACE-inhibitor-induced drug eruption resembling lymphocytic infiltration

(of Jessner–Kanof) and lupus erythematosus tumidus. Dermatology 2004;208(4):354–5.
7. Berkane N, Carlier P, Verstraete L, Mathieu E, Heim N, Uzan S. Fetal toxicity of valsartan and possible reversible adverse side effects. Birth Defects Res A Clin Mol Teratol 2004;70(8):547–9.
8. Kumbasar B, Atlibatur Akbas F, Serez K, Ger E, Uzunoglu S, Ergen K, Ayer M. High-dose exposure to valsartan with suicidal intention. Int J Clin Pharmacol Ther 2004;42(6):328–9.

Verapamil

See also Calcium channel blockers

General Information

Verapamil is a phenylalkylamine calcium channel blocker, with actions mainly on the heart.

Comparative studies

In a double-blind, randomized study in 1033 patients the combination of verapamil plus quinidine was not inferior to sotalol in preventing symptomatic episodes of paroxysmal atrial fibrillation and both were superior to placebo (1). However, recurrence rates were relatively high in all groups, including placebo, in the presence of a low but definite risk of severe and potentially life-threatening adverse effects. For this reason the authors suggested that antidysrhythmic drug therapy should be restricted to highly symptomatic patients.

Placebo-controlled studies

In a double-blind, randomized study in 848 patients successfully cardioverted for persistent atrial fibrillation the combination of verapamil plus quinidine was not inferior to sotalol in preventing recurrence of atrial fibrillation or death, but it was superior to placebo (2). There were adverse events in 75% of those taking either sotalol or verapamil plus quinidine and in 61% of those taking placebo. Serious adverse events had the same incidences in the three groups, but there were five deaths in those who took quinidine plus verapamil, six in those who took sotalol, and none in those who took placebo. There were 10 episodes of torsade de pointes in those taking sotalol, and none in those taking verapamil plus quinidine. The authors suggested that verapamil might have prevented the prodysrhythmic effects of quinidine.

Organs and Systems

Cardiovascular

Heart failure

Verapamil can cause cardiac failure, because it has a potent negative inotropic effect and causes increased capillary filtration pressure by vasodilatation (3). There are also reports of drug-induced acute right heart failure in patients with pulmonary hypertension due to valvular heart and chronic pulmonary obstructive airways diseases (4).

- A 16-year-old boy who took long-term verapamil after a Mustard operation for transposition of the great arteries developed severe congestive heart failure, which did not respond to diuretics. Systemic vascular resistance was increased by 75% and pulmonary vascular resistance by 150%; the cardiac index was reduced from 3.0 to 1.8 l/minute/m^2. Ejection fraction and atrial pressure were unchanged and neurohormonal causes were excluded. The heart failure resolved after withdrawal of verapamil.

Cardiogenic shock after the ingestion of verapamil has been reported, in one case with a small dose (5).

- A 78-year-old woman with a history of biventricular heart failure developed cardiogenic shock after she took a single tablet of verapamil 80 mg. She was resuscitated with artificial ventilation, dobutamine, noradrenaline, and calcium gluconate. Toxicological analysis showed an unexpectedly high plasma verapamil concentration, which was attributed to liver failure.
- A 37-year-old-man with rheumatoid disease developed rapid atrial fibrillation 4.5 years after aortic and mitral valve replacement and was given amiodarone 400 mg/day and verapamil 40 mg tds (6). Several months later amiodarone was withdrawn because of thyrotoxicosis and the dosage of verapamil was increased to 80 mg tds in order to slow his heart rate. About 3 hours after the third dose of verapamil, he developed severe bradycardia and cardiogenic shock, complicated by fulminant hepatic failure. His bradycardia was reversed by calcium and atropine and his blood pressure was maintained satisfactorily. However, he died from intracranial bleeding.

The authors suggested that the liver damage was probably caused by verapamil-induced cardiogenic shock and induced a vicious cycle of raised blood verapamil concentrations, which in turn aggravated the cardiogenic shock by cardiodepression, causing further liver damage. Thyrotoxicosis made his heart more susceptible to the effects of verapamil.

Cardiac dysrhythmias and heart block

Relatively harmless and asymptomatic disturbances of heart rhythm are common during treatment with verapamil. For example, Wenckebach phenomenon (Mobitz type I second-degree atrioventricular block), sinus bradycardia and junctional escape, and accelerated junctional rhythm have been seen during the treatment of angina pectoris (7). In the DAVIT II study (8), 23 of 878 patients who took verapamil and seven of 897 who took placebo in a survival study after myocardial infarction developed second- or third-degree AV block. It has been suggested that the extent of PR prolongation may be proportional to the pharmacological effect of verapamil in patients with sinus rhythm (9). Patients with diseased conduction systems can develop atrioventricular block, sinoatrial block, sinus arrest, or sinus bradycardia with verapamil (10).

Severe cardiac conduction disturbances have been reported in patients with chronic renal insufficiency taking verapamil, mainly related to modified-release formulations.

- A serious bradydysrhythmia with complete atrioventricular block occurred after some months of conventional verapamil at normal doses and dose-adjusted digoxin in a 72-year-old woman with chronic renal insufficiency of unknown cause; the atrioventricular block resolved after 2 hours of hemodialysis (11).

Problems can also occur in patients with aberrant conduction pathways; for example, verapamil caused an increased ventricular response in patients with Wolff–Parkinson–White syndrome associated with atrial fibrillation (12–14). The danger lies in provocation of ventricular fibrillation from advanced anterograde conduction in accessory pathways (15).

Of 169 consecutive patients, mainly elderly with structural heart disease, with second- or third-degree atrioventricular block not related to other causes (such as acute myocardial infarction, digitalis toxicity, or vasovagal syncope), 92 (54%) who were taking a beta-blocker and/or verapamil or diltiazem had similar clinical and electrocardiographic characteristics to patients who had atrioventricular block in the absence of drugs (16). Drug withdrawal was followed by resolution of atrioventricular block in 41% of cases, whereas there was spontaneous improvement in atrioventricular conduction in 23% of patients who had atrioventricular block in the absence of drugs. However, 56% of the patients in whom drug withdrawal led to resolution of atrioventricular block had a recurrence in the absence of therapy. There was atrioventricular block that was "truly caused by drugs" in only 15% of patients who had second- or third-degree block during therapy with beta-blockers, verapamil, or diltiazem. The authors concluded that atrioventricular block is commonly related to drugs but rarely caused by drugs.

Nervous system

Myoclonic dystonia has been described in a 70-year-old man who took verapamil for supraventricular tachycardia for 10 months; the abnormal movements disappeared within 3 weeks of substituting diltiazem, but rechallenge was not attempted (17).

Myoclonic seizures have been attributed to verapamil (18).

- An 18-month-old girl with supraventricular tachycardia was given intravenous verapamil 0.2 mg/kg, which was discontinued after half the dose had been given, because she developed irregular, repetitive, jerky movements in both upper and lower limbs which lasted for 2 minutes. As the supraventricular tachycardia had not responded, second and third doses of 0.2 mg/kg were given intravenously under diazepam cover, but similar myoclonic seizures occurred again. There were no predisposing factors.

Fasciculation occurred when a patient with background peripheral neuropathy was given verapamil (SEDA-16, 197).

Feelings of "painful coldness and numbness" in the legs, in the absence of a neurological deficit, have been reported with verapamil (19); the symptoms disappeared with drug withdrawal and recurred with rechallenge.

Endocrine

There was a significant rise in serum alkaline phosphatase of skeletal origin in a group of patients taking verapamil for hypertension. It was associated with a slight increase in parathyroid hormone, indicating involvement of bone metabolism, although there was no change in the urinary excretion of calcium, phosphate, or potassium. It has yet to be shown whether verapamil causes osteopenia in man, as it does in animals (SEDA-16, 197).

Verapamil-induced hyperprolactinemia has been reported in a 74-year-old man, who presented with impotence and who was subsequently found to have a benign 6 mm pituitary microadenoma (20). The withdrawal of verapamil was associated with the normalization of serum prolactin concentrations.

Marked hyperprolactinemia has been attributed to verapamil in a 42-year-old woman who may have been hypersusceptible to this action of verapamil (21).

The mechanism for this adverse effect is unclear and has not been reported with other calcium channel blockers.

Mouth

Four patients with cluster headache developed gingival enlargement after taking verapamil. In two it was treated by optimizing oral hygiene and dental plaque control (22). In the other one it was also necessary to reduce the dose of verapamil, and in one the verapamil had to be withdrawn to reverse the gingival enlargement.

Skin

Subacute cutaneous lupus erythematosus has been attributed in a single case to verapamil (23).

A skin rash consistent with lymphomatoid papulosis has been attributed to verapamil (24).

- A 71 year old man with chronic cluster headaches took verapamil, starting in December 2000 at a dose of 240 mg/day and then 120–480 mg/day. He stopped taking it in September 2001 and restarted in November because his headaches recurred. Within 1 week of restarting he developed a distinctive rash that took the form of 0.5 cm pruritic, erythematous papules on his neck, armpits, and groin. They persisted for about 2 weeks and resolved when the dose of verapamil was reduced from 240 to 200 mg/day. A skin biopsy showed histological features consistent with lymphomatoid papulosis.

By modulating the cellular immune system, verapamil may promote the expansion of an atypical T cell subpopulation, with subsequent progression to lymphomatoid papulosis, which is important to recognize because it can evolve into malignant lymphoma.

Hair

Increased hair growth has been associated with modified-release verapamil (SEDA-17, 238).

Musculoskeletal

Arthralgia has been associated with verapamil (25).

- A 28-year-old man with migraine and vascular pain of the face was given verapamil 480 mg/day when sumatriptan did not control the pain. Verapamil was withdrawn after 1 month and reintroduced in a dosage of 960 mg/day after a new episode of pain. He then developed arthralgia in the right hand and arm, which disappeared after withdrawal and recurred after rechallenge.

Susceptibility Factors

In therapeutic doses, verapamil has a negative dromotropic effect, reflected in plasma concentration-dependent prolongation of the PR interval and AV nodal block. PR interval prolongation is detectable even after small single doses. The effects of rheumatoid arthritis on the pharmacokinetics and pharmacodynamics of verapamil have been studied in eight patients and eight age- and sex-matched healthy volunteers (26). Verapamil and norverapamil concentrations were substantially increased in the patients with rheumatoid arthritis, accompanied by significantly less dromotropic activity. Regardless of the mechanism involved in the altered pharmacokinetics of verapamil, the rise in drug concentrations was accompanied by a significant reduction in dromotropic activity. Therefore, administration of verapamil to patients with rheumatoid arthritis may require close attention to prevent therapeutic failure.

Drug Administration

Drug overdose

Verapamil poisoning can cause cardiac toxicity with cardiogenic shock, conduction abnormalities (including life-threatening dysrhythmias), hypotension, and death.

- A 15-year-old woman survived a total of 65 minutes cardiac arrest after taking 7200 mg of verapamil and 240 mg of paroxetine, which potentiates verapamil toxicity (27). Despite the length of cardiopulmonary resuscitation, with evidence of subsequent myocardial damage and renal impairment, she was discharged 17 days after admission, having made a full recovery, without evidence of neurological impairment.

The authors considered that the lack of neurological damage may have been related to the possible neuroprotective effect of a large dose of verapamil.

- A 28-year-old man developed rhabdomyolysis and acute renal insufficiency 10 hours after taking four capsules containing verapamil 180 mg plus trandolapril 2 mg. He survived with gastric lavage and activated charcoal 70 g. By 4 hours after admission he was awake and complained of diffuse muscle cramps and myalgia. Creatinine and creatine kinase reached 565 μmol/l (6.7 mg/dl) and 10 700 U/l respectively. After 6 days the laboratory tests were all within the reference ranges.

The authors suggested that rhabdomyolysis should be considered in patients with myalgia and muscle cramps taking verapamil plus trandolapril, and that routine serum creatinine kinase should be checked (28).

Non-cardiogenic pulmonary edema has been reported after the ingestion of a large quantity of verapamil (29).

- Two 19-year-old-girls developed pulmonary edema after taking massive overdoses of verapamil (6000 mg and 7200 mg). In each case a chest X-ray showed diffuse bilateral patchy infiltration. Left ventricular size and function was normal on transthoracic echocardiography. They were both treated successfully with mechanical ventilatory support.

Several mechanisms may be involved in non-cardiogenic pulmonary edema subsequent to verapamil intoxication, including leaky capillary syndrome attributable to inhibition of prostacyclin, a cellular membrane protector. Prolonged hypotension and a shock-like state may also contribute. The authors recommended pressor/inotropic therapy and mechanical ventilation as therapy.

- A 51-year old man with a cardiomyopathy died after taking 7200 mg of sustained-release (SR) verapamil (30). Plasma verapamil and norverapamil concentrations on admission were 3-4 times higher than the highest therapeutic concentrations previously reported.

Different therapeutic approaches have been described in suicide attempts with verapamil.

- Enoximone has been used to treat cardiogenic shock in a case of self-poisoning with verapamil 2.4 g in a 40-year-old woman; treatment with calcium and noradrenaline had been unsuccessful (31).
- A 41-year-old man who had taken 4800–6400 mg of verapamil presented 5 hours later and had a cardiac arrest; cardiopulmonary resuscitation was started and he underwent percutaneous cardiopulmonary by-pass, to maintain adequate tissue perfusion and sufficient cerebral oxygen supply until the drug concentration was reduced and restoration of spontaneous circulation was achieved (32).
- Severe poisoning with modified-release verapamil was associated with respiratory failure in a 27-year-old man, who presented 2 hours after taking 24 g, one of the largest reported overdoses (33). He survived after receiving 4 days of partial liquid ventilation as a part of his medical management.
- A 61-year-old woman developed profound hypotension, bradycardia, oliguria, and multifocal myoclonus after verapamil overdose. Her serum verapamil concentration was 800 ng/ml (1760 nmol/l) (usual target range 50–200 ng/ml, 110–440 nmol/l). She was treated with supportive care, vasopressin, dopamine, and calcium gluconate. After about 60 hours the myoclonus and hemodynamic compromise resolved completely (34).

- Two patients took 2400 mg and 9600 mg of verapamil, resulting in life-threatening hypotension and bradycardia, needing cardiac pacing and resuscitation (35). Gastric lavage was not carried out in the first patient, owing to clouded consciousness. Both patients underwent plasmapheresis within 4 hours of ingestion, with dramatic improvement. However, the first patient died 38 hours after admission with multiorgan failure, probably because of the long period of cardiogenic shock. The second patient survived without infirmity, despite the higher verapamil dose.

The authors of the last report emphasized the role of vigorous gastrointestinal lavage with repeated administration of activated charcoal to reduce gastrointestinal absorption of the drug.

The management of verapamil overdose has been reviewed in the context of a case report (36).

- A 29-year-old woman took an overdose of sotalol 4.8 g and verapamil 3.6 g and had cardiovascular collapse (37). After 4 hours of normothermic cardiopulmonary resuscitation extracorporeal heart lung assist was established. Vasoactive drugs were withheld after 2 days. She recovered after 5 days, after experiencing several complications (intestinal bleeding, transient nerve paralysis, and renal insufficiency due to rhabdomyolysis).

Drug–Drug Interactions

ACE inhibitors

The addition of low doses of verapamil to ACE inhibitor therapy reversed ACE inhibitor-induced increases in creatinine concentrations in eight elderly hypertensive patients (38). During an average of 25 weeks, ACE inhibitors significantly reduced blood pressure, but serum creatinine concentrations rose. During an average of 10 weeks, the addition of verapamil did not reduce the blood pressure further, but the serum creatinine concentrations were normalized. Verapamil appears to have a beneficial effect, through dilatation of constricted afferent and efferent arterioles and reduction of the mesangial cell contraction induced by endothelin-1, factors that have been implicated in the increase in intraglomerular pressure and proteinuria due to ACE inhibitors.

Beta-blockers

Verapamil has a marked negative inotropic effect and can cause refractory cardiogenic shock, particularly when it is used in combination with a beta-blocker, as combination that is generally to be avoided (39).

Calcium adipinate and calciferol

Treatment of osteoporosis with calcium adipinate and calciferol counteracts the antidysrhythmic effect of verapamil (40).

Cardiac glycosides

Verapamil increases plasma digoxin concentrations at steady state by inhibiting the active tubular secretion and non-renal clearance of digoxin (41,42). There is only anecdotal evidence that this can result in digitalis toxicity (42,43).

Verapamil reversed digoxin-induced splanchnic vasoconstriction in healthy men (44), but this has no direct effect on systemic hemodynamics.

Cephalosporins

Competitive albumin binding of drugs with high serum protein affinity can increase pharmacologically active unbound concentrations and enhance the metabolism of low clearance drugs. Acute verapamil toxicity with ceftriaxone and clindamycin may be explained by this mechanism (45).

Chloroquine and hydroxychloroquine

Verapamil completely reversed pre-existing in vitro resistance to chloroquine to below the cut-off point of 70 nmol/l (46).

Ciclosporin

Calcium channel blockers are given to transplant patients for their protective effect against ciclosporin-induced nephrotoxicity. Verapamil has been particularly preferred, as it causes a significant increase in plasma ciclosporin concentrations and also seems to have a direct immunosuppressive action. However, in a study in 152 kidney transplant recipients, verapamil increased the incidence of postoperative infections (47). The patients, all of whom were taking ciclosporin, were assigned either to verapamil 240 mg/day or to no verapamil; during a postoperative period of 2–14 months, the incidence of infections was 22% (17/77) of those given verapamil compared with 5% (4/75) of the others. However, since the study was not randomized, it is not possible to draw reliable conclusions.

Dofetilide

By increasing its rate of absorption, verapamil produced transient rises in peak plasma concentrations of dofetilide in 12 young healthy male volunteers (48). During combination treatment at steady state there was a significant, albeit modest, increase in the mean C_{max} and AUC of dofetilide. These changes were associated with corresponding short-lived increases in its pharmacodynamic effect, as measured by changes in the QT_c interval. These two drugs should not be administered concurrently.

Pharmacokinetic and pharmacodynamic interactions between dofetilide 0.5 mg bd and verapamil 80 mg tds have been studied in 12 healthy men (49). At steady state verapamil increased the peak plasma concentration of dofetilide from 2.40 to 3.43 ng/ml, without other pharmacokinetic effects. This was accompanied by a small increase in the prolongation of the QT_c interval produced by dofetilide alone, from 20 to 26 ms. Although this small effect is unlikely to be of clinical significance, it would be wise to avoid verapamil in patients taking dofetilide.

Flecainide

Although verapamil reduces the clearance of flecainide, this is probably not clinically important (50). However, there is also a pharmacodynamic interaction, since both drugs increase the PR interval and have additive effects on myocardial contractility and atrioventricular conduction (51).

Grapefruit juice

The interaction with grapefruit juice observed with several calcium channel blockers, including felodipine, nifedipine, and nisoldipine, has not been confirmed with verapamil. A single dose of grapefruit juice had no effect on the pharmacokinetics of verapamil in 10 hypertensive patients taking chronic verapamil (52).

HMG Co-A reductase inhibitors

Rhabdomyolysis has been associated with the co-administration of diltiazem with simvastatin or atorvastatin.

- A 63-year-old white man taking ciclosporin after cardiac transplantation developed a fever and diffuse muscle pain (53). He had been taking several medications, including simvastatin, for more than 5 years, and verapamil had been started 4 weeks before for hypertension. A diagnosis of rhabdomyolysis was made and his medications were withdrawn. His creatine kinase activity reached a maximum of 24 028 units/l on day 4 and fell to normal in 2 weeks.

While rhabdomyolysis from statins is rare, the risk is increased when they are used in combination with agents that share similar metabolic pathways. Statins are metabolized by CYP3A4, which is inhibited by verapamil and ciclosporin; competitive interference with CYP3A4 can increase statin concentrations, and increase the risk of adverse events.

Macrolide antibiotics

Verapamil is both a substrate and an inhibitor of CYP3A4, which is inhibited by clarithromycin and erythromycin. Giving these macrolide antibiotics during verapamil therapy is likely to reduce the first-pass metabolism of verapamil, increase its systemic availability, and impair its elimination. In patients taking this combination, verapamil should be started in a low dosage and its hemodynamic effects should be monitored closely.

- A 77-year-old woman taking verapamil and propranolol for hypertrophic cardiomyopathy and paroxysmal atrial fibrillation developed symptomatic bradycardia on two separate occasions within days of taking either erythromycin or clarithromycin. The proposed mechanism of the interaction was inhibition of cytochrome P450, since verapamil is a substrate of an isoenzyme that is inhibited by some macrolides (54,55).

Severe hypotension and bradycardia after combined therapy with verapamil and clarithromycin has been reported in another case (56).

Clarithromycin

- A 53-year-old woman had periods of dizziness and episodes of fainting when she stood up 24 hours after having been given clarithromycin for an acute exacerbation of chronic obstructive pulmonary disease and verapamil for atrial fibrillation (57). One day later she developed severe hypotension and bradycardia (57). Since her symptoms matched those of severe verapamil overdosage, the drug was withdrawn and her condition improved within two days.
- A 77-year-old hypertensive woman receiving both verapamil and propranolol for hypertrophic cardiomyopathy and paroxysmal atrial fibrillation developed symptomatic bradycardia within 2–4 days of initiation of both erythromycin or clarithromycin on two separate occasions, for pneumonia and sinusitis respectively (58).

Erythromycin

- A 79-year-old white woman developed extreme fatigue and dizziness (59). Her heart rate was 40/minute and her blood pressure 80/40 mmHg. An electrocardiogram showed complete atrioventricular block, an escape rhythm at 50/minute, and QT_c interval prolongation to 583 milliseconds. This event was attributed to concomitant treatment with verapamil 480 mg/day and erythromycin 2000 mg/day, which had been prescribed 1 week before admission.

This is the first report of complete AV block and prolongation of the QT interval after co-administration of erythromycin and verapamil, both of which are principally metabolized by CYP3A4. Both drugs are potent inhibitors of CYP3A4 and P-glycoprotein, which may be the basis of this interaction.

The effects of a combination of erythromycin and verapamil on the pharmacokinetics of a single dose of simvastatin have been studied in a randomized, double-blind, cross-over study in 12 healthy volunteers simultaneously taking the three drugs. Both erythromycin and verapamil interacted with simvastatin, producing significant increases in the serum concentrations of simvastatin and its active metabolite simvastatin acid. The mean C_{max} of active simvastatin acid was increased about five-fold and the AUC_{0-24} four-fold by erythromycin; verapamil increased the C_{max} of simvastatin acid 3.4-fold and the AUC_{0-24} 2.8-fold. There was a substantial interindividual variation in the extent of these interactions. Concomitant use of erythromycin, verapamil, and simvastatin should be avoided (60).

Telithromycin

Verapamil toxicity has been attributed to inhibition of verapamil metabolism by telithromycin (61).

- A 76-year-old white woman taking verapamil 180 mg/day for hypertension took telithromycin 800 mg/day for 2 days for acute sinusitis and became short of breath, weak, profoundly hypotensive and bradycardic, with a systolic blood pressure of 50–60 mmHg and a heart rate of 30/minute. She recovered after drug withdrawal and supportive measures.

Telithromycin is a substrate of CYP3A4 and presumably inhibited the metabolism of verapamil.

Rifampicin

Rifampicin reduces the systemic availability of verapamil, presumably by enzyme induction in the liver (62).

References

1. Patten M, Maas R, Bauer P, Lüderitz B, Sonntag F, Dluzniewski M, Hatala R, Opolski G, Müller HW, Meinertz T, for the SOPAT Investigators. Suppression of paroxysmal atrial tachyarrhythmias – results of the SOPAT trial. Eur Heart J 2004;25:1395–404.
2. Fetsch T, Bauer P, Engberding R, Koch HP, Lukl J, Meinertz T, Oeff M, Seipel L, Trappe HJ, Treese N, Breithardt G, for the Prevention of Atrial Fibrillation after Cardioversion Investigators. Prevention of atrial fibrillation after cardioversion: results of the PAFAC trial. Eur Heart J 2004;25:1385–94.
3. Buchhorn R, Motz R, Bursch J. Hemodynamic and neurohormonal causes of a severe verapamil induced cardiac decompensation in a child after mustard operation. Herz Kreisl 2000;32:74–7.
4. Shimoni A, Maor-Kendler Y, Neuman Y. Verapamil-induced acute right heart failure. Am Heart J 1996;132(1 Pt 1):193–4.
5. Stajer D, Bervar M, Horvat M. Cardiogenic shock following a single therapeutic oral dose of verapamil. Int J Clin Pract 2001;55(1):69–70.
6. Margolin L. Fatal cardiogenic shock and liver failure induced by verapamil in a thyrotoxic patient. Clin Drug Invest 2003;23:285–6.
7. Pine MB, Citron PD, Bailly DJ, Butman S, Plasencia GO, Landa DW, Wong RK. Verapamil versus placebo in relieving stable angina pectoris. Circulation 1982;65(1):17–22.
8. Held PH, Yusuf S, Furberg CD. Calcium channel blockers in acute myocardial infarction and unstable angina: an overview. BMJ 1989;299(6709):1187–92.
9. Dominic JA, Bourne DW, Tan TG, Kirsten EB, McAllister RG Jr. The pharmacology of verapamil. III. Pharmacokinetics in normal subjects after intravenous drug administration. J Cardiovasc Pharmacol 1981;3(1):25–38.
10. Opie LH. Drugs and the heart. III. Calcium antagonists. Lancet 1980;1(8172):806–10.
11. Martin-Gago J, Pascual J, Rodriguez-Palomares JR, Marcen R, Teruel JL, Liano F, Ortuno J. Complete atrioventricular blockade secondary to conventional-release verapamil in a patient on hemodialysis. Nephron 1999; 83(1):89–90.
12. Harper R, Whitford E, Middlebrook K, et al. Verapamil in patients with Wolff–Parkinson–White syndrome—a potential hazard? NZ J Med 1981;11:456.
13. Strasberg B, Sagie A, Rechavia E, Katz A, Ovsyscher IA, Sclarovsky S, Agmon J. Deleterious effects of intravenous verapamil in Wolff–Parkinson–White patients and atrial fibrillation. Cardiovasc Drugs Ther 1989;2(6):801–6.
14. Garratt C, Antoniou A, Ward D, Camm AJ. Misuse of verapamil in pre-excited atrial fibrillation. Lancet 1989;1(8634):367–9.
15. Jacob AS, Nielsen DH, Gianelly RE. Fatal ventricular fibrillation following verapamil in Wolff–Parkinson–White syndrome with atrial fibrillation. Ann Emerg Med 1985;14(2):159–60.
16. Zeltser D, Justo D, Halkin A, Rosso R, Ish-Shalom M, Hochenberg M, Viskin S. Drug-induced atrioventricular block: prognosis after discontinuation of the culprit drug. JACC 2004;44:105–8.
17. Hicks CB, Abraham K. Verapamil and myoclonic dystonia. Ann Intern Med 1985;103(1):154.
18. Maiteh M, Daoud AS. Myoclonic seizure following intravenous verapamil injection: case report and review of the literature. Ann Trop Paediatr 2001;21(3):271–2.
19. Kumana CR, Mahon WA. Bizarre perceptual disorder of extremities in patients taking verapamil. Lancet 1981;1(8233):1324–5.
20. Dombrowski RC, Romeo JH, Aron DC. Verapamil-induced hyperprolactinemia complicated by a pituitary incidentaloma. Ann Pharmacother 1995;29(10):999–1001.
21. Krysiak R, Okopieh B, Herman ZS. Hiperprolaktynemia spowodowana przez werapamil. Opis przypadku. Arch Med Wewn 2005;113(2):155–8.
22. Matharu MS, van Vliet JA, Ferrari MD, Goadsby PJ. Verapamil induced gingival enlargement in cluster headache. J Neurol Neurosurg Psychiatry 2005;76(1):124–7.
23. Kurtis B, Larson MJ, Hoang MP, Cohen JB. Case report: verapamil-induced subacute cutaneous lupus erythematosus. J Drugs Dermatol 2005;4(4):506–8.
24. Afridi S, Bacon CM, Bowling J, Goadsby PJ. Verapamil and lymphomatoid papulosis in chronic cluster headache. J Neurol 2004;251:473–5.
25. Nicolas X, Bellard S, Zagnoli F. Arthralgies induites par de fortes doses de vérapamil. [Arthralgia induced by large doses of verapamil.] Presse Méd 2001;30(25 Pt 1): 1256–7.
26. Mayo PR, Skeith K, Russell AS, Jamali F. Decreased dromotropic response to verapamil despite pronounced increased drug concentration in rheumatoid arthritis. Br J Clin Pharmacol 2000;50(6):605–13.
27. Evans JS, Oram MP. Neurological recovery after prolonged verapamil-induced cardiac arrest. Anaesth Intensive Care 1999;27(6):653–5.
28. Gokel Y, Paydas S, Duru M. High-dose verapamil–trandolapril induced rhabdomyolysis and acute renal failure. Am J Emerg Med 2000;18(6):738–9.
29. Sami Karti S, Ulusoy H, Yandi M, Gunduz A, Kosucu M, Erol K, Ratip S. Non-cardiogenic pulmonary oedema in the course of verapamil intoxication. Emerg Med J 2002;19(5):458–9.
30. Tracqui A, Tournoud C, Kintz P, Villain M, Kummerlen C, Sauder P, Ludes B. HPLC/MS findings in a fatality involving sustained-release verapamil. Hum Exp Toxicol 2003;22:515–21.
31. Link A, Hammer B, Weisgerber K, Bohm M. Therapie der Verapamil-Intoxikation mit Noradrenalin und dem Phosphodiesterasehemmer Enoximon. [Therapy of verapamil poisoning with noradrenaline and the phosphodiesterase inhibitor enoximone.] Dtsch Med Wochenschr 2002;127(39):2006–8.
32. Holzer M, Sterz F, Schoerkhuber W, Behringer W, Domanovits H, Weinmar D, Weinstabl C, Stimpfl T. Successful resuscitation of a verapamil-intoxicated patient with percutaneous cardiopulmonary bypass. Crit Care Med 1999;27(12):2818–23.
33. Szekely LA, Thompson BT, Woolf A. Use of partial liquid ventilation to manage pulmonary complications of acute verapamil-sustained release poisoning. J Toxicol Clin Toxicol 1999;37(4):475–9.

34. Vadlamudi L, Wijdicks EF. Multifocal myoclonus due to verapamil overdose. Neurology 2002;58(6):984.
35. Kuhlmann U, Schoenemann H, Muller TF, Keuchel M, Lange H. Plasmapheresis in fatal overdose with verapamil. Intensive Care Med 1999;25(12):1473.
36. Morales MG, Guerrero SG, Garcia GR, Villalobos SJ, Camarena AG, Aguirre SJ, Martinez SJ. Intoxicacion grave con verapamilo. Arch Cardiol Mex 2005;75 Suppl 3:S3-100–5.
37. Rygnestad T, Moen S, Wahba A, Lien S, Ingul CB, Schrader H, Knapstad SE. Severe poisoning with sotalol and verapamil. Recovery after 4 h of normothermic CPR followed by extra corporeal heart lung assist. Acta Anaesthesiol Scand 2005;49(9):1378–80.
38. Bitar R, Flores O, Reverte M, Lopez-Novoa JM, Macias JF. Beneficial effect of verapamil added to chronic ACE inhibitor treatment on renal function in hypertensive elderly patients. Int Urol Nephrol 2000;32(2):165–9.
39. Nanda U, Ashish A, Why HJ. Modified release verapamil induced cardiogenic shock. Emerg Med J 2005;22(11):832–3.
40. Bar-Or D, Gasiel Y. Calcium and calciferol antagonise effect of verapamil in atrial fibrillation. BMJ (Clin Res Ed) 1981;282(6276):1585–6.
41. Pedersen KE, Dorph-Pedersen A, Hvidt S, Klitgaard NA, Nielsen-Kudsk F. Digoxin–verapamil interaction. Clin Pharmacol Ther 1981;30(3):311–6.
42. Klein HO, Lang R, Weiss E, Di Segni E, Libhaber C, Guerrero J, Kaplinsky E. The influence of verapamil on serum digoxin concentration. Circulation 1982;65(5):998–1003.
43. Zatuchni J. Verapamil–digoxin interaction. Am Heart J 1984;108(2):412–3.
44. Gasic S, Eichler HG, Korn A. Effect of calcium antagonists on basal and digitalis-dependent changes in splanchnic and systemic hemodynamics. Clin Pharmacol Ther 1987;41(4):460–6.
45. Kishore K, Raina A, Misra V, Jonas E. Acute verapamil toxicity in a patient with chronic toxicity: possible interaction with ceftriaxone and clindamycin. Ann Pharmacother 1993;27(7–8):877–80.
46. Ndifor AM, Howells RE, Bray PG, Ngu JL, Ward SA. Enhancement of drug susceptibility in *Plasmodium falciparum* in vitro and *Plasmodium berghei* in vivo by mixed-function oxidase inhibitors. Antimicrob Agents Chemother 1993;37(6):1318–23.
47. Nanni G, Panocchia N, Tacchino R, Foco M, Piccioni E, Castagneto M. Increased incidence of infection in verapamil-treated kidney transplant recipients. Transplant Proc 2000;32(3):551–3.
48. Johnson BF, Cheng SL, Venitz J. Transient kinetic and dynamic interactions between verapamil and dofetilide, a class III antiarrhythmic. J Clin Pharmacol 2001;41(11): 1248–56.
49. Spodick DH. Well concealed atrial tachycardia with Wenckebach (Mobitz I) atrioventricular block: digitalis toxicity. Am J Geriatr Cardiol 2001;10(1):59.
50. Holtzman JL, Finley D, Mottonen L, Berry DA, Ekholm BP, Kvam DC, McQuinn RL, Miller AM. The pharmacodynamic and pharmacokinetic interaction between single doses of flecainide acetate and verapamil: effects on cardiac function and drug clearance. Clin Pharmacol Ther 1989;46(1):26–32.
51. Buss J, Lasserre JJ, Heene DL. Asystole and cardiogenic shock due to combined treatment with verapamil and flecainide. Lancet 1992;340(8818):546.
52. Zaidenstein R, Dishi V, Gips M, Soback S, Cohen N, Weissgarten J, Blatt A, Golik A. The effect of grapefruit juice on the pharmacokinetics of orally administered verapamil. Eur J Clin Pharmacol 1998;54(4):337–40.
53. Chiffoleau A, Trochu JN, Veyrac G, Petit T, Abadie P, Bourin M, Jolliet P. Rhabdomyolysis in a cardiac transplant recipient due to verapamil interaction with simvastatin and cyclosporine treatment. Thérapie 2003;58:168–70.
54. Arola O, Peltonen R, Rossi T. Arthritis, uveitis, and Stevens-Johnson syndrome induced by trimethoprim. Lancet 1998;351(9109):1102.
55. Steenbergen JA, Stauffer VL. Potential macrolide interaction with verapamil. Ann Pharmacother 1998;32(3):387–8.
56. Kaeser YA, Brunner F, Drewe J, Haefeli WE. Severe hypotension and bradycardia associated with verapamil and clarithromycin. Am J Health Syst Pharm 1998;55(22): 2417–8.
57. Kaeser YA, Brunner F, Drewe J, Haefeli WE. Severe hypotension and bradycardia associated with verapamil and clarithromycin. Am J Health Syst Pharm 1998;55(22):2417–8.
58. Steenbergen JA, Stauffer VL. Potential macrolide interaction with verapamil. Ann Pharmacother 1998;32(3):387–8.
59. Goldschmidt N, Azaz-Livshits T, Gotsman, Nir-Paz R, Ben-Yehuda A, Muszkat M. Compound cardiac toxicity of oral erythromycin and verapamil. Ann Pharmacother 2001;35(11):1396–9.
60. Kantola T, Kivisto KT, Neuvonen PJ. Erythromycin and verapamil considerably increase serum simvastatin and simvastatin acid concentrations. Clin Pharmacol Ther 1998;64(2):177–82.
61. Reed M, Wall GC, Shah NP, Heun JM, Hicklin GA. Verapamil toxicity resulting from a probable interaction with telithromycin. Ann Pharmacother 2005;39(2):357–60.
62. Barbarash RA, Bauman JL, Fischer JH, Kondos GT, Batenhorst RL. Near-total reduction in verapamil bioavailability by rifampin. Electrocardiographic correlates. Chest 1988;94(5):954–9.

Xamoterol

See also Beta-adrenoceptor antagonists

General Information

Xamoterol is a beta-adrenoceptor antagonist/partial agonist that was developed for use in mild cases of cardiac failure and to treat atrial fibrillation (1). In more severe cases, however, it can actually worsen heart failure and increase mortality (2), because when sympathetic nervous system activity is high its beta-adrenoceptor antagonist properties predominate. It has therefore been withdrawn from the market (SEDA-18, 159).

References

1. Lawson-Matthew PJ, McLean KA, Dent M, Austin CA, Channer KS. Xamoterol improves the control of chronic atrial fibrillation in elderly patients. Age Ageing 1995; 24(4):321–5.
2. Anonymous. New evidence on xamoterol. Lancet 1990; 336(8706):24.

Zofenopril

See also Angiotensin converting enzyme inhibitors

General Information

Zofenopril is a prodrug that, once absorbed, undergoes rapid and complete hydrolysis to the sulfhydryl-containing active ACE inhibitor zofenoprilat. The use of zofenopril in hypertension and in acute myocardial infarction has been extensively reviewed (1).

Reference

1. Borghi C, Ambrosioni E. Zofenopril: a review of the evidence of its benefits in hypertension and acute myocardial infarction. Clin Drug Invest 2000;20:371–84.

DIURETICS

General Information

Diuretics are among the most widely used drugs, particularly for the treatment of hypertension and of various conditions associated with sodium retention.

Considering the widespread use of diuretics over a long period (chlorothiazide was introduced in 1957) their safety record is remarkable, and reports of adverse effects of any significance with the best-known drugs of this type are uncommon.

When problems do arise they usually reflect either interactions, which with caution could have been avoided, or relative overdosage. In the course of time the recommended antihypertensive doses of diuretics have been reduced, and some adverse effects that were noted in the early years are now of less significance; these include hypotension, dehydration, reduction of the glomerular filtration rate, and severe hypokalemia. Continued use of thiazides in excessive doses may reflect ignorance of their very flat dose–response curve (1). At currently recommended low doses, diuretics improve overall quality of life, even in asymptomatic patients with mild hypertension (2). The large HANE study (3) provided no evidence of superior efficacy or tolerability of new classes of antihypertensive drugs.

Uses

Hypertension

The popularity of thiazide diuretics in the management of hypertension reflects three major factors (4):

- recognition of the effectiveness of much lower dosages than those used previously, thereby providing good antihypertensive efficacy with fewer adverse effects (SED-13, 558);
- the excellent reductions in morbidity and mortality that have been achieved by low-dosage diuretic-based therapy in multiple randomized, controlled trials (SED-13, 558);
- the increasing awareness that some diuretic-induced shrinkage of effective blood volume is essential for adequate treatment of most patients with hypertension.

The Seventh Joint Committee Report on Detection, Evaluation, and Treatment of High Blood Pressure (5) concluded that a diuretic should be the first-step drug of choice, unless there are specific indications for other drugs. Although there may be theoretical advantages of certain newer types of drugs, the data thus far have not consistently shown that these drugs are more effective in reducing morbidity and mortality than therapy based on diuretics or beta-blockers (6). Emphasis is correctly placed on the important role of ACE inhibitors in retarding progression of renal insufficiency in diabetic and other nephropathies (7). However, in these circumstances, ACE inhibitors are added to a background of other antihypertensive therapies, commonly including a diuretic. Therefore, the renal protective action of ACE inhibitors is in the context of combination regimens. The ability of low doses of diuretics to enhance efficacy has been demonstrated for all other classes of drugs (8). Moreover, the tendency for increased retention of sodium by the hypertensive kidney when non-diuretic drugs cause the blood pressure to fall has long been recognized to contribute to loss of antihypertensive efficacy, which can be restored immediately by the addition of a diuretic.

Clinical trials

The LIVE (Left ventricular regression, Indapamide Versus Enalapril) study was a 1-year, prospective, randomized, double-blind comparison of modified-release indapamide 1.5 mg and enalapril 20 mg in reducing left ventricular mass in 411 hypertensive patients with left ventricular hypertrophy (9). For equivalent reductions in blood pressure, indapamide was significantly more effective than enalapril in reducing left ventricular mass index.

The effectiveness of specific first-line antihypertensive drugs in lowering blood pressure and preventing adverse outcomes has been systematically quantified in a meta-analysis of randomized controlled trials that lasted at least 1 year, and compared one of six possible first-line antihypertensive therapies either with another of the six drug therapies or with no treatment (10). Of 38 trials identified, 23 in 50 853 patients met the inclusion criteria. Four drug classes were evaluated: thiazides (21 trials), beta-blockers (5 trials), calcium antagonists (4 trials), and ACE inhibitors (1 trial). In five comparisons of thiazides with beta-blockers, thiazides were associated with a significantly lower rate of withdrawal because of adverse effects (RR = 0.69; 95% CI = 0.63, 0.76). In the trials that had an untreated control group, low-dose thiazide therapy was associated with a significant reduction in the risk of death (RR = 0.89; CI = 0.81, 0.99), stroke (RR = 0.66; CI = 0.56, 0.79), coronary heart disease (RR = 0.71, CI = 0.60, 0.84), and cardiovascular events (RR = 0.68; CI = 0.62, 0.75). Low-dose thiazide therapy reduced the absolute risk of cardiovascular events by 5.7% (CI = 4.2, 7.2); the number needed to treat (NNT) for approximately 5 years to prevent one event was 18. High-dose thiazide, beta-blocker, and calcium antagonist therapy did not significantly reduce the risk of death or coronary heart disease. Thiazides were significantly better than the other drugs in reducing systolic pressure, but antihypertensive efficacy did not differ between the high- and low-dose thiazide trials.

In the UK Medical Research Council (MRC) trial, the outcome of antihypertensive treatment based on diuretics was compared with placebo in a very large number of hypertensive subjects (11). Treatment based on a thiazide did not increase the incidence of coronary events or sudden death; indeed, thiazide-based treatment reduced the incidence of strokes by 67% and of all cardiovascular complications by 20%. It should be noted that the dose of bendroflumethiazide used in the MRC trial (10 mg/day) is now known to be unnecessarily high and that it was used without prophylaxis against hypokalemia. Even so, a subgroup analysis of data from the MRC Trial provided no evidence that the association between major electrocardiographic abnormalities and an increased likelihood of a clinical event was strengthened by bendroflumethiazide treatment (12).

A series of trials in elderly hypertensive subjects has shown a very pronounced reduction in cardiac events as a result of treatment based on thiazide diuretics. In the European Working Party on Hypertension in the Elderly (EWPHE) trial (13), total cardiovascular deaths were reduced by 38%, all cardiac deaths by 43%, and deaths due to myocardial infarction by 60%. Benefits in the Systolic Hypertension in the Elderly Program (SHEP) included a reduction in fatal and non-fatal myocardial infarction of 25% and major cardiovascular events of 32% (14) and were seen in those with and without electrocardiographic abnormalities at entry. The risk of heart failure was also reduced in patients taking chlortalidone-based therapy (15). Relative risk was similar in patients with and without non-insulin dependent diabetes mellitus; absolute risk reduction was twice as great in the diabetic subjects (16). The Swedish Trial of Old Patients with Hypertension (STOP-Hypertension) reported a significant reduction in myocardial infarction and all-cause mortality (17). In the MRC Trial in elderly adults (18), diuretic treatment reduced coronary events by 44% and fatal cardiovascular events by 35%.

In the MRC trials (11,18), the IPPPSH trial (19), and the HAPPHY trial (20), antihypertensive treatment based on a thiazide diuretic was compared with treatment based on a beta-blocker. The results with diuretic treatment were no less favorable as regards cardiac events than those when using a "cardioprotective" beta-blocker (20). In the IPPPSH trial, the group using no beta-blockers (but with a higher incidence of diuretic use and of hypokalemia) showed no excess of cardiac events, even in patients who had an abnormal electrocardiogram when they entered the study. The MRC trial in elderly adults (18) showed a significantly lower risk of cardiovascular events with the diuretic compared with the beta-blocker, raising the possibility that diuretics confer benefit through a mechanism other than the reduction of blood pressure.

In the MAPHY study (21), total mortality was significantly lower for metoprolol than for thiazide, because of fewer deaths from coronary heart disease and stroke. However, the MAPHY study population comprised a subgroup of about half of the patients from the HAPPHY trial followed for an extended period. The difference in mortality between metoprolol and diuretics did not emerge during this extended follow-up, but was present during the first period of observation (that is during the HAPPHY trial), when there was no overall difference between beta-blockers and thiazides. Therefore, patients treated with atenolol in the HAPPHY trial must have fared worse than those treated with thiazides and much worse than those treated with metoprolol. Since there was no prior hypothesis for a difference between atenolol and metoprolol (and no plausible explanation for it), it seems reasonable to conclude that the apparent advantage of metoprolol was a chance finding produced by post-hoc subgroup analysis. The MAPHY study should be interpreted with extreme caution.

A review of the large trials has shown that the reduction in the incidence of events with usual thiazide-based treatment is 16% (95% CI = 8, 23%) against the prediction from epidemiological studies of 20–25% (22). This shortfall in benefit could easily be due to chance.

In the Antihypertensive and Lipid-Lowering Treatment to Prevent Heart Attack Trial (ALLHAT), over 40 000 participants aged 55 years or older with hypertension and at least one other risk factor for coronary heart disease were randomized to chlortalidone, amlodipine, doxazosin, or lisinopril (23,24). Doxazosin was discontinued prematurely because chlortalidone was clearly superior in preventing cardiovascular events, particularly heart failure (24). Otherwise, mean follow-up was 4.9 years. There were no differences between chlortalidone, amlodipine, and lisinopril in the primary combined outcome or all-cause mortality. Compared with chlortalidone heart failure was more common with amlodipine and lisinopril, and chlortalidone was better than lisinopril at preventing stroke. The authors concluded that diuretics are superior to other antihypertensive drugs in preventing one or more major form of cardiovascular disease and are less expensive. They should therefore be used as the preferred first-step treatment of hypertension. Although the design, conduct, and analyses of ALLHAT can be criticized (25), it was a huge, prospective, randomized study and its results suggest that diuretic-based therapy is unsurpassed in the management of hypertension.

Adverse effects

The frequencies and the profile of adverse effects of five major classes of antihypertensive agents have been assessed in an unselected group of 2586 chronically drug-treated hypertensive patients (26). This was accompanied by a questionnaire-based survey among patients attending a general practitioner. The percentages of patients who reported adverse effects spontaneously, on general inquiry, and on specific questioning were 16, 24, and 62% respectively. The percentage of patients in whom discontinuation was due to adverse effects with diuretics was 2.8%. The authors did not find a significant effect of age on the pattern of adverse effects. Women reported more effects and effects that were less related to the pharmacological treatment.

Heart failure

Loop and thiazide diuretics

Loop and thiazide diuretics have been the mainstay of treatment for symptomatic heart failure (27), relieving symptoms and improving cardiovascular hemodynamics. However, despite their widespread use, they have not been shown to improve survival in patients with heart failure. As it is not feasible to conduct such a trial in patients with pulmonary edema due to heart failure, the place of diuretic therapy in the management of heart failure appears secure.

Aldosterone receptor antagonists

Although they are widely used in the management of heart failure, loop and thiazide diuretics have not been shown to prolong survival. However, spironolactone and eplerenone must be added to the list of medications that offer improved survival for patients with heart failure.

In the Randomized Aldactone Evaluation Study (RALES) in 1663 patients with New York Heart Association (NYHA) class III (70%) or IV (30%) symptoms and an ejection fraction less than 35%, the addition of spironolactone 25 mg/day to conventional treatment (an ACE inhibitor, a loop diuretic, in most cases digoxin, and in 11% a beta-blocker) for an average of 24 months lowered the risk of all-cause mortality by 30% (from 46% to 35%), death from progressive heart failure, and sudden death (28). There were similar reductions in hospital admissions for worsening heart failure and for all cardiac causes. The magnitude of the overall effect was similar and additional to the proven benefit from ACE inhibition in severe heart failure.

In the Eplerenone Post-Acute Myocardial Infarction Heart Failure Efficacy and Survival Study (EPHESUS), in 6632 patients with an acute myocardial infarction complicated by left ventricular dysfunction and heart failure, the addition of eplerenone 25–50 mg/day to optimal medical therapy significantly reduced all-cause mortality by 15% and cardiovascular mortality by 17% over a mean follow-up period of 16 months; hospitalization rates were also reduced (29).

Mechanisms

Several mechanisms have been postulated to underlie the benefits of aldosterone receptor antagonists in heart failure (30). Aldosterone-induced cardiac fibrosis may reduce systolic function, impair diastolic function, and promote intracardiac conduction defects, with the potential for serious dysrhythmias. Aldosterone may also increase vulnerability to serious dysrhythmias by other mechanisms. The diuretic and hemodynamic effects of spironolactone in RALES and EPHESUS were subtle, and there were no significant changes in body weight, sodium retention, or systemic blood pressure.

Clinical use

The safe and effective dose of spironolactone remains uncertain (30). Pilot data from RALES showed that the frequency of hyperkalemia and uremia increased with doses of spironolactone above 50 mg/day (SEDA-20, 202). Doses up to 50 mg/day are appropriate, with adequate monitoring of serum electrolytes and renal function. The optimum strategy in the face of hyperkalemia, uremia, or symptomatic hypotension (reduction in frequency of spironolactone to alternate-day dosing, reduction in dose of ACE inhibitor, and/or increased dose of loop diuretics) is unclear, and how frequent such dose adjustments were necessary in RALES was not stated (28).

Adverse effects

The only frequent adverse effects were gynecomastia, breast pain, or both in 10% of men. The rate of discontinuation because of these events was 2%. The risk of gynecomastia should not be an argument against the use of spironolactone in men with severe heart failure, since it reduces both morbidity and death.

At the dose of spironolactone used in RALES (28), there was serious hyperkalemia, defined as a serum potassium concentration over 6.0 mmol/l, in 2% (compared with 1% of controls) and uremia was rare. However, a serum potassium concentration over 5 mmol/l and a serum creatinine concentration over 220 μmol/l were exclusion criteria. Although 29% of patients in the spironolactone group used potassium supplements, the benefit of spironolactone in these patients was similar to that in patients who did not.

In EPHESUS, as in RALES, the exclusion criteria included a serum potassium concentration over 5 mmol/l and a serum creatinine concentration over 220 μmol/l. There was serious hyperkalemia (a serum potassium concentration of 6.0 mmol/l or over) in 5.5% of those who took eplerenone and in 3.9% of those who took placebo. In each treatment group the incidence of hyperkalemia was higher among patients with the lowest baseline creatinine clearances.

The use of diuretics in patients with renal insufficiency

About 84% of the patients in the Reduction of Endpoints in NIDDM with the Angiotensin II Antagonist Losartan Study (RENAAL) required diuretic therapy to effect blood pressure control (31). Although not specifically reported, a similarly high proportion of patients were likely to have required diuretic therapy to reach the target blood pressure in the Irbesartan Diabetic Nephropathy Trial (IDNT) (32). These studies remind us once again of the importance of targeting volume control in order to reduce blood pressure in patients with chronic renal insufficiency.

In a cohort study of 552 patients with acute renal insufficiency studied from 1989 to 1995 diuretic use was associated with a significant increase in the risk of death or non-recovery of renal function (33). This increased risk was largely borne by patients who were relatively unresponsive to diuretics. Although this study was observational, which prohibits causal inference, it is unlikely that diuretics afford any material benefit in the setting of acute renal insufficiency.

Thiazide and loop diuretics

Various factors influence the choice of a diuretic in patients with renal insufficiency. First, it is widely believed that thiazide diuretics are ineffective once renal function falls below a creatinine clearance of 40–50 ml/minute. Diuretics need to enter the renal tubule to reach their luminal sites of action. In renal insufficiency, higher diuretic doses are required to overcome disease-related impediments to drug delivery to their sites of action. In the case of the loop diuretics the common practice is to titrate the dose of the diuretic until the desired response occurs (34). No such titration generally occurs with thiazide-type diuretics; consequently they are likely to fail, although for no reason other than their not being titrated to a truly effective dose. The true basis for "failure" of thiazide-type diuretics in patients with renal insufficiency resides in the fact that these diuretics are not of sufficient maximal efficacy to produce adequate volume control in these typically volume-expanded patients (35).

However, resistance to loop diuretics can occur by various mechanisms (36). These include poor adherence to therapy, poor absorption, progressive worsening of heart failure, excess volume loss, renal insufficiency, secondary hyperaldosteronism, and hypertrophy of the tubular cells of the distal nephron. Resistance due to inadequate drug absorption—either its speed or extent—is common with furosemide, which is poorly absorbed (34). Once recognized, this hurdle to response can be overcome by using loop diuretics that are predictably well absorbed, such as bumetanide and torasemide or by giving intravenous furosemide (37).

If loop diuretics fail to produce the desired diuretic response, combination diuretic therapy can be considered by adding a thiazide or a thiazide-like diuretic, such as metolazone. Such combinations are generally quite successful in both advanced congestive heart failure and late-stage chronic renal insufficiency, although excessive diuresis is a constant risk with such combinations. An excess diuresis with combination loop plus thiazide diuretic therapy is best managed by temporarily withdrawing both diuretics. Generally, diuretic doses are reduced when therapy is resumed (38).

An alternative in the diuretic-resistant patient is the use of continuous infusions of loop diuretics rather than bolus diuretic therapy. Such infusions can also be given with a small volume of hypertonic saline, with good effect (39). The reasons why continuous infusions of loop diuretics work when bolus doses have failed may relate to a more efficient time-course of diuretic delivery and/or less activation of the renin–angiotensin system (40). Furosemide and torasemide may be the safest loop diuretics to be given as infusions, in that infusion of bumetanide has been associated with severe musculoskeletal symptoms (41).

The importance of volume control in patients with renal insufficiency extends beyond its effect on blood pressure. Accordingly, the addition of hydrochlorothiazide can overcome the blunting by a high sodium intake of the therapeutic efficacy of ACE inhibition on proteinuria (42). This presumably relates to volume-related activation of the renin–angiotensin system.

Finally, loop diuretics reduce the metabolic demand of tubular cells, reducing oxygen requirements and thereby, in theory, increasing resistance to ischemic insults and perhaps other toxic circumstances. This property has been advanced as the basis for using diuretics in acute renal insufficiency. Although an attractive hypothesis to date, there is no compelling evidence to suggest any benefit from loop diuretics in established acute renal insufficiency. Alternatively, loop diuretics can convert oliguric to non-oliguric acute renal insufficiency, thereby easing the fluid restriction that would otherwise be necessary in such patients (43).

Aldosterone receptor antagonists

It is no longer appropriate to consider the endocrine or paracrine properties of aldosterone as being restricted to the classical target cells. Hemodynamic and humoral actions of aldosterone have important clinical implications for the pathogenesis of progressive renal disease, and may therefore affect future antihypertensive strategies. Initially, one might anticipate that the adverse effects of aldosterone could be attenuated merely by blocking aldosterone release with either an ACE inhibitor or angiotensin-II receptor antagonists. However, this appears not to be the case. Several investigators have now shown that ACE inhibitors acutely reduce aldosterone concentrations, but that with continued use this suppression fades. Thus, the presumption that ACE inhibitors would suppress the production of both aldosterone and angiotensin-II was incorrect. So, although ACE inhibitors and angiotensin-II receptor antagonists are individually very effective in retarding disease progression, additional benefit may be realized with a concurrent aldosterone receptor antagonist. As observed in clinical studies of congestive heart failure, as well as in animals with renal disease, antagonism of aldosterone receptors protects against end-organ damage through a combination of both hemodynamic and direct cellular actions (44).

An important consideration regarding the feasibility of aldosterone receptor antagonist therapy in chronic renal insufficiency is the risk of provoking hyperkalemia. Many patients with chronic renal insufficiency are already taking an ACE inhibitor or an angiotensin-II receptor antagonist, with the attendant risk of hyperkalemia. Despite such concerns, the results of the RALES and EPHESUS trials have been reassuring (28). In those studies, patients taking an ACE inhibitor who were randomized to spironolactone or eplerenone 25–50 mg/day had only small increases in potassium concentrations. Although the differences between those who took aldosterone receptor antagonists and those who took placebo were statistically significant, the mean increases were not clinically important, and serious hyperkalemia was uncommon. However, in clinical practice the risk of hyperkalemia may be greater (45), and close laboratory monitoring and judicious use of these drugs is necessary to minimize the risk.

Although it is an effective aldosterone receptor antagonist, spironolactone is limited by its tendency to cause undesirable sexual adverse effects. At standard doses, impotence and gynecomastia can occur in men, and premenopausal women can have menstrual disturbances. These adverse effects, caused by the binding of spironolactone to progesterone and androgen receptors, are substantial causes of drug withdrawal. In the RALES study there was a 10% incidence of gynecomastia or breast pain in men, compared with 1% with placebo, and significantly more patients discontinued treatment (2 versus 0.2%). Although troublesome, these adverse effects are reversible and dose-related. The advent of selective aldosterone receptor antagonists, such as eplerenone, should reduce these adverse effects and thereby improve patient compliance. In EPHESUS there was no increase in the incidence of gynecomastia, breast pain, or impotence in men or menstrual irregularities in women who took eplerenone.

Organs and Systems

Cardiovascular

Cardiac dysrhythmias

Changes in potassium metabolism supposedly cause electrical instability in the heart, cardiac dysrhythmias, and increased mortality; replacement of potassium has been said to eliminate the risk of dysrhythmias (46–48). Mild hypokalemia might be expected to cause dysrhythmias in patients with serious organic heart disease (cardiomegaly, an abnormal electrocardiogram, frequent ventricular extra beats before treatment). However, the evidence suggests that hypokalemia after myocardial infarction is not the cause of dysrhythmias, but that both are the result of excess catecholamines. Furthermore, although hypertensive patients with left ventricular hypertrophy have an increased frequency of ventricular dysrhythmias, extra beats do not increase in frequency during diuretic treatment, even in the face of profound hypokalemia (49).

Early evidence linking thiazide-induced hypokalemia with dysrhythmias and sudden death was indirect and tenuous at best (50,51). One study suggested that diuretics are not responsible for the relation between hypokalemia and ventricular fibrillation in acute myocardial infarction (50). Chronic preoperative hypokalemia due to diuretics was not a risk factor for intraoperative dysrhythmias (52). Two large studies using 24-hour electrocardiographic monitoring failed to show a relation between diuretic-induced hypokalemia and ventricular dysrhythmias (53,54).

However, a retrospective analysis of 6797 patients with ejection fractions below 0.36 enrolled in the Studies Of Left Ventricular Dysfunction (SOLVD) was conducted to assess the relation between diuretic use at baseline and the subsequent risk of dysrhythmic death (54). Patients who were taking a diuretic at baseline (n = 2901) were significantly more likely to have such an event than those not taking a diuretic (n = 2896): 3.1 versus 1.7 per 1000 person-years. On univariate analysis and after controlling for important co-variates, the relation remained significant (relative risks 1.85 and 1.37 respectively). However, the association was seen only with non-potassium-sparing diuretics (n = 2495; relative risk 1.33); for potassium-sparing diuretics, alone or in combination with a non-potassium-sparing diuretic (n = 406), the relative risk was 0.90.

These data suggest that diuretic-induced potassium disturbances can cause fatal dysrhythmias in patients with left ventricular systolic dysfunction. SOLVD were not randomized trials of the risk of dysrhythmic death caused by diuretics. On average, patients retaking diuretics not only had lower serum potassium concentrations, but were also older, had more severe heart failure and were more likely to be taking antidysrhythmic drugs at baseline, although they had fewer indicators of ischemic heart disease. Even controlling for bias in multivariate analysis does not exclude the influence of unrecognized confounders. It is unknown whether diuretics were continued or changed during the 3 years of the trial. Thus, it remains uncertain that diuretic therapy is related to a risk of sudden dysrhythmic death in patients with heart failure.

Diuretic-induced hypokalemia is undoubtedly associated with a risk of serious ventricular dysrhythmias if diuretics are co-administered with drugs that prolong the QT interval (SED-13, 563). Diuretics increase the risk of torsade de pointes during antidysrhythmic drug therapy, independent of serum potassium concentration (54). The list of cardiac and non-cardiac drugs that prolong the QT interval continues to lengthen (54). It includes various antidysrhythmic drugs, including ibutilide, almokalant, and dofetilide; antimicrobial drugs, including clarithromycin, clindamycin, co-trimoxazole, pentamidine, imidazoles (such as ketoconazole), some fluoroquinolones, and antimalarial drugs (quinine, halofantrine); histamine H_1 receptor antagonists (terfenadine and astemizole); the antidepressant zimeldine; antipsychotic drugs (pimozide and sertindole); tricyclic/tetracyclic antidepressants; and cisapride. Particular care should be taken to avoid diuretic-induced hypokalemia when any of these agents is co-prescribed.

Coronary heart disease

There is no valid evidence that diuretics contribute to myocardial infarction, sudden death, or a failure of antihypertensive treatment or other risk factor interventions to prevent coronary deaths (50). An association between diuretics and sudden death has been suggested only in selected subset analyses, which allow no valid conclusions. Even in subjects with electrocardiographic abnormalities before treatment, there is no sound or consistent evidence to support the suggestion that diuretics predispose to sudden death.

Two retrospective case-control studies have reported an increased risk of cardiac arrest in hypertensive patients treated with thiazide-type diuretics (55,56). The risk was less among those treated with low-dose thiazides (equivalent to hydrochlorothiazide 25 mg/day) or with thiazides plus potassium-sparing drugs. The case-control design lacks one of the major advantages of randomized clinical trials, a tendency to equalize unknown, but important, differences between the comparison groups. Underadjustment is usual in case-control studies; for example, patients with more severe hypertension were more likely to have been treated with higher doses and less likely to have received potassium-sparing drugs because of renal dysfunction. Failure to randomize treatment tends to exaggerate differences between groups. Data from the prospective observational Gothenburg study suggested that metabolic changes during long-term treatment with antihypertensive drugs (predominantly beta-blockers and thiazides) are not associated with increased risk of coronary heart disease (57).

Sensory systems

Eyes

In 1077 patients with intraocular pressures of 22–29 mmHg, normal visual fields, and normal optic discs, who were followed every 6 months for 5 years in the European Glaucoma Prevention Study, a multivariate analysis showed that the use of diuretics (HR = 2.41, 95% CI = 1.12, 5.19) was one of several factors associated with the development of open-angle glaucoma (58). Systemic hypertension was not associated with the conversion to open-angle glaucoma. Information concerning the use of any medications was obtained at baseline and every 6 months throughout the study. However, particular classes of diuretics were not separately cited. This contributory effect of diuretics may be explained by the play of chance alone or by as yet unknown detrimental effects of diuretics on the retinal ganglion cells. A possible reduction in ocular perfusion pressure induced by a diuretic-related reduction in systemic pressure may be an additional mechanism. In this study, diuretics were more often used in combination with other antihypertensive drugs, particularly in the patients who developed open-angle glaucoma. Blood pressure readings were not obtained during the study, and so it could not be determined whether there was hypotension.

Ears

Sensorineural hearing loss *is* an important adverse outcome in survivors of neonatal intensive care illnesses, particularly those with persistent pulmonary hypertension of the newborn. The relations between ototoxic drugs and 4-year sensorineural hearing loss have been assessed in a prospective, longitudinal outcome study in near-term and term survivors of severe neonatal respiratory failure who were enrolled in the Canadian arm of the Neonatal Inhaled Nitric Oxide Study (59). A combination of loop diuretic use for more than 14 days and an average dose of neuromuscular blocker greater than 0.96 mg/kg/day contributed to sensorineural hearing loss among survivors (OR = 5.2; 95% CI = 1.6, 17). Cumulative dose and duration of diuretic use and overlap of diuretic use with neuromuscular blockers, aminoglycosides, and vancomycin were individually linked to sensorineural hearing loss. These studies implicate loop diuretics and neuromuscular blockers individually and possibly synergistically in sensorineural hearing loss.

Metabolism

Despite their safety record, speculation persists that the metabolic effects of long-term diuretic treatment predispose to myocardial infarction or sudden death, and that diuretic treatment may therefore be hazardous. It is worth noting that much of this speculation is found outside the columns of the legitimate medical press (SEDA-10, 185). The supposed risks of diuretics are broadcast in countless symposium proceedings, monographs, and such like, sponsored by pharmaceutical companies with a vested interest in diverting prescriptions from diuretics to other drugs. Needless to say, these publications do not present a balanced view. Studies of dubious quality are published repeatedly without ever appearing in refereed journals, and eventually come to be cited in independent reviews and articles. There can be no doubt that these publications have a large impact on prescribing practices.

It is relatively easy to foster these concerns, particularly since antihypertensive therapy would be expected to prevent myocardial infarction and sudden death, but in selected studies does not appear to do so. To explain this, it is suggested that the beneficial effects of lowering blood pressure are offset in part by adverse effects related to thiazide-induced biochemical disturbances. Hypokalemia and hypomagnesemia might, for example, be dysrhythmogenic and cause sudden death; or hyperlipidemia and impaired glucose tolerance might be atherogenic and promote myocardial infarction. Some authors have suggested that thiazides should certainly be avoided in left ventricular hypertrophy and coronary heart disease, because of an increased risk of ventricular extra beats (60,61), and that diuretics are not appropriate options in those who already have hyperglycemia, hyperuricemia, or hyperlipidemia (62). The argument has been taken ad absurdum by a suggestion that the use of diuretics as first-line treatment of hypertension is illogical (27). Others have argued effectively that we should not be impressed by such speculations and that treatment should be based on long-term experience (63), a conclusion similar to that reached in volumes of SEDA, from SEDA-9 to SEDA-19.

Impaired glucose tolerance and insulin resistance

There is little doubt that high dosages of diuretics carry an appreciable risk of impairing diabetic control in patients with established diabetes mellitus. However, their role in causing de novo glucose intolerance is not clear (64). Long-acting diuretics are more likely to alter glucose metabolism. Impaired glucose tolerance is a relatively rare complication with loop diuretics, although isolated cases of non-ketotic hyperglycemia in diabetics have been described (SED-9, 350).

The effects of thiazide-type diuretics on carbohydrate tolerance cannot be ignored (50). There is a definite relation between diuretic treatment, impaired glucose tolerance, and biochemical diabetes, and a possible relation with insulin resistance (65). It is well established that the effect of thiazides on blood glucose is dose-related, probably linearly, while the antihypertensive effect has little relation to dose (66–68). There is relatively little information on the time-course; numerous short-term studies have shown that the blood glucose concentration increases in 4–8 weeks (69). The evidence that current low dosages impair glucose tolerance in the long term is not entirely consistent, perhaps because of differences between studies in dosages, diuretics used, durations of treatment, and types of patient (69,70). In SHEP, low-dosage chlortalidone in elderly patients for 3 years resulted in a non-significant excess of diabetes (8.6 versus 7.5%) compared with placebo (71). The apparent differences between diuretics may be due to comparisons of dosages that are not equivalent (72). Important differences between individual diuretics will be established only when their complete dose–response relations for metabolic variables and blood pressure have been defined (69).

In contrast to the wealth of evidence on impaired glucose tolerance in diabetics, sound clinical trials of the effect on insulin resistance are difficult to find, considering the amount of comment and speculation on the topic (SEDA-15, 216). It is not known whether insulin resistance is completely or even partly responsible for the changes in glucose tolerance that occur during long-term thiazide treatment; impaired insulin secretion may also have a role (73). Hypokalemia or potassium depletion may contribute to impaired glucose tolerance, by inhibiting insulin secretion rather than by causing insulin resistance, but is not the only or even the main cause of impaired glucose tolerance during long-term diuretic treatment (69,70). The routine use of potassium-sparing diuretics with relatively low dosages of thiazides does not prevent impaired glucose tolerance.

Diuretics worsen metabolic control in established diabetes, but it is not known whether this adversely affects prognosis (69). Disturbances of carbohydrate homeostasis have been detected by detailed biochemical testing, but their clinical importance is uncertain (65). The major clinical trials have not shown a major risk of diabetes mellitus. The incidence of diabetes mellitus in diuretic-treated subjects is only about 1%, even when large dosages are used (74). In ALLHAT, among patients classified as non-diabetic at baseline, the incidence of diabetes after more than 4 years was 12% with chlortalidone compared with 9.8% (amlodipine) and 8.1% (lisinopril). Despite these trends, there was no excess of cardiovascular events or mortality from chlortalidone in the entire population or among patients with diabetes. Although these data are reassuring, observational data suggest that diuretic-induced new-onset diabetes carries an increased risk of cardiovascular morbidity and mortality, but that this may take 10–15 years to become fully apparent (75).

Since changes in glucose balance after diuretics tend to be reversible on withdrawal, measures of carbohydrate homeostasis should be assessed after several months of thiazide treatment to detect those few patients who experience significant glucose intolerance (76). With this approach, the small risk of diabetes mellitus secondary to diuretic therapy can be minimized.

Hyperuricemia

Most diuretics cause hyperuricemia. Increased reabsorption of uric acid (along with other solutes) in the proximal tubule as a consequence of volume depletion is one reason; however, diuretics also compete with uric acid for excretory transport mechanisms. There is a small increased risk of acute gout in susceptible subjects (74). In the large outcome trials, about 3–5% of subjects treated with diuretics for hypertension developed clinical gout (77). In those with acute gout during diuretic treatment, attacks were more strongly related to loop diuretics than to thiazides (78). Gout was significantly associated with obesity and a high alcohol intake in the subgroup taking only a thiazide diuretic. About 40% of cases of acute gout may have been prevented by avoiding thiazides in those 20% of men who weighed over 90 kg and/or who consumed more than 56 units of alcohol per week.

Well-conducted studies have shown that diuretic-induced changes in serum uric acid are dose-related (67,79). In low-dosage regimens, as currently recommended, alterations are minor, and other than the risk of gout the long-term consequences of an increased serum uric acid are unknown.

The issues of whether hyperuricemia is an independent risk factor for cardiovascular disease and the clinical relevance of the rise in serum uric acid caused by diuretic treatment are controversial (SED-14, 660) (50). In the Systolic Hypertension in the Elderly Program (SHEP), diuretic-based treatment in 4327 men and women, aged 60 years or more, with isolated systolic hypertension was associated with significant reduction in cardiovascular events (SED-14, 657). Serum uric acid independently predicted cardiovascular events in these patients (80). The benefit of active treatment was not affected by baseline serum uric acid. After randomization, however, an increase in serum uric acid of less than 0.06 mmol/l (median change) in the active treatment group was associated with a hazard ratio (HR) of 0.58 (CI = 0.37, 0.92) for coronary heart disease compared with those whose serum uric acid rose by 0.6 mmol/l or more. This was despite a slight but significantly greater reduction in both systolic and diastolic blood pressures in the latter. Those with serum uric acid increases of 0.6 mmol/l or more in the active group had a similar risk of coronary events as those in the placebo group.

This analysis of the SHEP database confirms the findings of a systematic worksite hypertension program (81). In 7978 treated patients with mild to moderate hypertension, cardiovascular disease was significantly associated with serum uric acid (HR = 1.22; CI = 1.11, 1.35), controlling for known cardiovascular risk factors. The cardioprotective effect of diuretics increased from 31% to 38% after adjustment for serum uric acid.

These observations suggest that persistent elevation of serum uric acid during diuretic-based antihypertensive therapy may detract from the benefit of blood pressure reduction. However, the relation between serum uric acid and cardiovascular disease was independent of the effects of diuretics. Furthermore, low-dose thiazide regimens have a smaller impact on serum uric acid (SED-14, 660).

Lipid metabolism

Reviews of the influence of diuretics on serum lipids (82–85) are in broad agreement as regards short-term effects. Thiazide and loop diuretics increase low-density lipoprotein (LDL) cholesterol, very-low-density lipoprotein (VLDL) cholesterol, total cholesterol, and triglycerides. The effect on high-density lipoprotein (HDL) cholesterol has been variable. The ratio of LDL/HDL or total cholesterol/HDL is generally increased, but not in all studies. Spironolactone 50 mg bd caused modest falls in HDL cholesterol and triglycerides (86). The effects of other potassium-sparing drugs on lipid metabolism have not been well documented. The possible mechanisms of these various short-term effects have been discussed (85).

Diuretic-induced effects on lipid metabolism are dose-related (67,68): at low dosages of thiazides, changes are very slight while antihypertensive efficacy is well maintained. Diuretic-induced lipid changes have not been prominent in studies lasting one year or longer (11,20,82,85,87). An association between thiazide use or antihypertensive treatment and changes in serum lipids has been shown in some population surveys (88,89) but not in others (90).

Most studies of the effects of diuretics on serum lipids have lacked a placebo control to allow identification of time-dependent or environmental changes, or have been confounded intentionally or unknowingly by life-style interventions, including weight loss, diet, and exercise (70). The argument that the effects of diuretics on lipids may limit the beneficial effect of blood pressure reduction is difficult to sustain (50). The effect of thiazides is largely transient, and in the long term total cholesterol and LDL cholesterol are raised only slightly and HDL cholesterol is unchanged. It is not known whether diuretic-induced changes in cholesterol carry the same prognostic significance as naturally occurring hyperlipidemia (50,91). Attempts to calculate a potential impact of diuretic-induced increases in total cholesterol and LDL cholesterol on coronary prognosis are premature. Observations have so far been limited largely to serum concentrations. However, binding of lipids to vascular cells, rather than in the bloodstream, is decisive for atherogenesis, and the effect at the cellular level remains to be investigated.

It has been suggested that even small changes in lipids might be clinically significant (92). However, the study was underpowered to show statistical significance of minor effects. Since under 50% of the patients in each group were followed for 1 year, selection bias may also have been introduced. It should be remembered that lipid changes are subsidiary to mortality. In this respect, diuretics are the best established of the antihypertensive drug classes.

In a community-based sample of 585 adults with essential hypertension who took monotherapy with hydrochlorothiazide 25 mg/day for 4 weeks the mean changes in response to hydrochlorothiazide were 0.16 mmol/l (6.13 mg/dl) for total cholesterol, 0.19 mmol/l (17.2 mg/dl) for triglycerides, and 0.19 mmol/l (3.5 mg/dl) for plasma glucose (93). A range of demographic, environmental, and genetic variables taken together only accounted for 13%, 17%, and 11% of the variations in total cholesterol, triglyceride, and glucose, and less than half of this predicted variation in response was explained by measured genotypes. This suggests that there are no predictors of the adverse metabolic effects of thiazide-type diuretics.

There is little or no evidence that thiazides should be avoided in patients with hyperlipidemia (76), although some physicians continue to make this recommendation. Serum lipids should be checked within 3–6 months of starting thiazides to detect the very few patients who have an increase in total cholesterol or LDL cholesterol. This should not add to the cost of care, since serum chemistry need only be obtained once or twice a year and is no reason to avoid the use of these drugs as initial monotherapy.

Electrolyte balance

Potassium-wasting diuretics can cause sodium and potassium depletion with hyponatremia and hypokalemia. Potassium-retaining diuretics can cause hyperkalemia.

Hyponatremia

DoTS classification (BMJ 2003;327:1222-5):
Dose-relation: Collateral
Time course: Time independent, but typically occurs within weeks of starting therapy
Susceptibility factors: Age, female sex, reduced solute intake

Hyponatremia is a serious complication of diuretic therapy (94,95,96). Thiazide diuretics are more likely than loop diuretics to cause hyponatremia. Loop diuretics inhibit sodium transport in the loop of Henle in the renal medulla and thereby preclude the generation of a maximal osmotic gradient. Thus, they impair urinary concentrating ability. Conversely, thiazide-type diuretics increase sodium excretion in the distal convoluted tubule in the renal cortex and so prevent maximal urine dilution while preserving the kidney's innate concentrating capacity. Patients can present with variable hypovolemia or apparent euvolemia, depending on the magnitude of sodium loss and water retention.

Time course

When diuretic-related hyponatremia occurs it is usually shortly after therapy starts (within the first 2 weeks) (95), although it can occur even after several years of therapy, particularly when loop diuretics are the cause.

Susceptibility factors

Diuretic-related hyponatremia is more common in elderly women (95). There are multiple factors that select women, including age, reduced body mass, an exaggerated natriuretic response to thiazide diuretics, a reduced capacity to excrete free water, and low solute intake. However, it has been suggested that the excess of thiazide-induced hyponatremia in women may relate to overrepresentation of women in thiazide-treated cohorts rather than an inherent propensity for women to develop this electrolyte disturbance (95).

The use of fixed combination of a thiazide and a potassium-sparing drug, often Moduretic (hydrochlorothiazide 50 mg with amiloride 5 mg), has been consistently implicated in diuretic-induced hyponatremia. Treatment with chlorpropamide (200–800 mg/day) along with Moduretic has precipitated hyponatremia in several cases (97). Simultaneous use of Moduretic with trimethoprim has also been reported to increase the risk (98). The mechanism appears to be impairment of the clearance of free water, resulting in dilutional hyponatremia. Whether data such as these point to a special risk of Moduretic as a product or merely reflect its extraordinarily widespread use in old people is not clear. In a survey of electrolyte disturbances in 1000 geriatric admissions, the incidence of hyponatremia with the combination of hydrochlorothiazide and amiloride was twice that with other diuretics, although the difference failed to reach statistical significance (99).

Presentation

The presenting symptoms of diuretic-related hyponatremia can be vague and include anorexia, nausea, lethargy, and apathy; there is often evidence of volume depletion. More advanced symptoms include disorientation, agitation, seizures, reduced reflexes, focal neurological deficits and eventually Cheyne–Stokes respiration. Coma and seizures usually occur only with acute reduction of the serum sodium concentration to less than 120 mmol/l.

Two major surveys have reviewed the features of diuretic-induced hyponatremia (100,101). Thiazide-like diuretics alone or in combination with potassium-sparing diuretics are responsible for more than 90% of cases. Hyponatremia occurs mainly in elderly women, although the relation with age probably merely reflects the widespread use of diuretics in older subjects. In most cases, the interval between starting thiazide and clinical presentation is less than 2 weeks; serum sodium may fall by 5 mmol/l or more in 24 hours or less in patients who develop severe hyponatremia. In contrast, when loop diuretics cause hyponatremia, the lag period is usually several months. Hypertension is the indication for diuretics in over 80% of cases. The patient is clinically euvolumic, but in the majority excess antidiuretic hormone (ADH) activity, hypokalemia, excess water intake, and increased free water clearance singly or together appear to contribute to the development of hyponatremia. The urine is inappropriately concentrated and contains moderately large amounts of sodium. The findings may be clinically indistinguishable from those of the syndrome of inappropriate ADH secretion. Severe neurological complications are seen in about 60%; seizures (30%), stupor or coma (30%), and death (5–10%). The extent of irreversible loss of intellectual function is unknown. Serum sodium concentration is usually in the range 105–120 mmol/l. If diuretics are withdrawn, the serum sodium concentration returns to normal, but hyponatremia is reproducible on rechallenge.

Management

There is no consensus about the optimal treatment of symptomatic hyponatremia. The authors of a comprehensive review recommended a targeted rate of correction that does not exceed 8 mmol/l on any day of treatment (94). Remaining within this target, the initial rate of correction can still be 1–2 mmol/l/hour for several hours in patients with severe symptoms (SED-13, 562). They suggested the following formulae for calculating the effect of giving 1 liter of infusate on serum sodium.

- Solutions containing sodium only:

Change in serum sodium concentration (mmol/l) = (infusate Na^+ – serum Na^+)/(total body water + 1)

- Solutions containing sodium and potassium:

Change in serum sodium concentration (mmol/l) = (infusate Na^+ + infusate K^+ – serum Na^+)/(total body water + 1)

Estimated total body water (in liters) is calculated as a function of body weight: 0.6 l/kg in children; 0.6 and 0.5 l/kg in non-elderly men and women respectively; 0.5 and 0.45 l/kg in elderly men and women respectively.

Comprehensive accounts of the pathophysiology, etiology, and management of hyponatremia have stressed the care required in correction of symptomatic hyponatremia (102). Neurological signs are an indication for active sodium replacement using hypertonic saline. Correction should be planned over 24–48 hours to increase serum sodium by 1 mmol/l each hour with a target increase of 20–25 mmol/l, a serum sodium concentration of 130 mmol/l or abolition of symptoms. If onset is within 24 hours of starting diuretics, correction should be rapid but within a total increase of 20 mmol/l in the first 24 hours. If the onset is over several days or longer, correction should be slower, aiming for 12–15 mmol/l in 24 hours. There is concern that rapid correction of hyponatremia and a relatively high total correction (more than 20 mmol/l in the first 24 hours) can be associated with higher morbidity or a demyelinating syndrome (101), but the rate of correction does not appear to be important to the outcome if the absolute increase is limited to 25 mmol/l over 48 hours (102). An extensive review of the literature suggested that rate of correction is not a factor in the genesis of hyponatremic brain syndrome (102).

Loop diuretics can cause hypernatremia by increasing free water clearance (net water loss in the form of hypotonic fluid) (103). Over-rapid correction should be avoided.

Recently vasopressin receptor antagonists have become available. These compounds act on vasopressin V_2 receptors. The several compounds in this class, including conivaptan, lixivaptan, and tolvaptan, all improve renal water handling and correct hyponatremia in conditions associated with fluid retention. Optimal doses and dosing frequencies have still not been established for these compounds. The availability of vasopressin receptor antagonists will simplify the management of hyponatremic states, particularly when excess amounts of vasopressin are involved (104).

Hypokalemia

The loss of potassium caused by diuretics is their most intensively debated adverse effect, and the extent and significance of the problem has long been disputed. The effect of diuretics on potassium balance and their clinical consequences have been reviewed extensively (50–52,66,105–107). The risks of diuretic-induced hypokalemia have been greatly exaggerated (50,66,107). A fall in plasma potassium is common, but sound studies have consistently showed that diuretics do not deplete body potassium or cause potassium deficiency during long-term therapy in hypertensive patients (50).

Susceptibility factors

In most patients, routine monitoring of serum potassium and routine administration of potassium-sparing diuretics is unnecessary (66,107), although patients at special risk require particular attention. Routine administration of potassium-sparing diuretics may have some justification in such patients. The following postulates seem to summarize the present position:

- When there is severe hypokalemia, it should not be attributed immediately to diuretic treatment. It may well be due to primary hyperaldosteronism, occult chronic liver disease, or abuse of licorice or laxatives.
- Relatively mild degrees of hypokalemia can be dangerous in patients taking cardiac glycosides, because their effects on the myocardium are potentiated in the absence of potassium (51).

- Diuretic-induced hypokalemia should also be avoided in patients taking drugs that prolong the QT interval, for example some antidysrhythmic drugs (quinidine, procainamide, disopyramide, encainide, sotalol, amiodarone), some psychotropic drugs (thioridazine, imipramine, phenothiazines), the lipid lowering drug probucol, the antibiotic erythromycin, and the 5-HT_2 receptor antagonist ketanserin.
- The risk of a clinically important degree of hypokalemia is increased in patients with liver cirrhosis and those with severe cardiac failure complicated by secondary hyperaldosteronism.
- In heart failure, depletion of potassium can provoke fatigue and lethargy and can cause ventricular dysrhythmias in the failing heart (107). Potassium depletion can occur in cardiac failure when neurohumoral systems are stimulated by diuretics and can be especially profound when skeletal muscle wasting is advanced (106). However, since heart failure itself, independent of diuretic treatment, is associated with loss of total body potassium, it is difficult to assess the independent contribution of diuretic treatment to this potassium deficit.

Critical reviews have concluded that in most patients (that is in the absence of the occasional risk factors listed above) diuretics can be prescribed alone, and that there is no need to take precautions against hypokalemia (64,108). Precautions, including the monitoring of serum potassium, have to be taken when there are risk factors or when symptoms (for example muscle pain) occur, or in patients who require aggressive treatment with both a thiazide and a loop diuretic; such a combination is indeed likely to provoke considerable falls in serum potassium, even to concentrations below 2.5 mmol/l (109). An observational study in 43 738 American men aged 40–75 years and without cardiovascular disease or diabetes showed that stroke was less likely in those with high potassium intake and that use of potassium supplements was related inversely to stroke in men taking diuretics (110). Diuretic-based treatment was associated with a significant mean reduction in serum potassium in SHEP (71). Although there were notable reductions in stroke in SHEP (14), potassium supplements were used if the serum potassium fell below 3.5 mmol/l. Therefore, avoidance of hypokalemia may be important in ensuring the cardiovascular benefits of thiazides in the management of hypertension.

Well-conducted studies (67,79) have underlined the fact that the dose–response curves for lowering blood pressure and lowering plasma potassium during diuretic treatment are dissociated. There is little if any loss of antihypertensive effect with low doses of diuretics, whereas hypokalemia is much less prominent. Although low dosages of diuretics are to be preferred in uncomplicated hypertension, very low dosages of thiazides may not be as effective as higher dosages (111).

Management

In patients in whom hypokalemia needs to be prevented or corrected, potassium-sparing diuretics should be preferred to potassium supplements, which are relatively ineffective, even in high dosages (SEDA-14, 180) (SEDA-15, 214).

The National Council on Potassium in Clinical Practice has provided guidelines on potassium replacement (112). In hypertensive patients with drug-induced hypokalemia, an effort should be made to achieve and maintain the serum potassium concentration at 4.0 mmol/l or over. Potassium replacement should be considered routinely in patients with congestive heart failure, even if the initial potassium determination is normal. Regular monitoring of serum potassium is essential in these patients, because of the risk of hyperkalemia in patients given potassium (or potassium-sparing diuretics) and ACE inhibitors or angiotensin receptor antagonists. Maintenance of optimal potassium concentrations (4.0 mmol/l or over) is critical in patients with cardiac dysrhythmias, and again routine potassium monitoring is obligatory.

Angiotensin-converting enzyme inhibitors are widely believed to attenuate diuretic-induced falls in plasma potassium. However, captopril 25 mg bd had no effect on plasma potassium in hypertensive patients treated with bendroflumethiazide 5 mg/day (113).

Hyperkalemia

All potassium-sparing diuretics can cause hyperkalemia. Several groups of patients are at particularly high risk of developing hyperkalemia if they take potassium-sparing diuretics (114). These include patients with moderate to severe chronic renal insufficiency, hypoaldosteronism, and diseases associated with impaired response to the potassium secretory effects of aldosterone. Hypoaldosteronism is often seen in elderly patients, those with chronic renal impairment or diabetic nephropathy, those with AIDS, or those with primary adrenal disease. In addition, patients with sickle cell disease, obstructive uropathy, systemic lupus erythematosus with nephropathy, and renal transplants may have tubular resistance to the effects of mineralocorticoids. Spironolactone 300 mg can significantly increase the serum potassium concentration in hemodialysis patients, suggesting that aldosterone may affect the cellular handling and gastrointestinal excretion of potassium. Commonly forgotten is the patient for whom these diuretics are prescribed as a precautionary measure and who for many years has been flavoring a low-salt diet with "salt substitutes," which are often based on potassium chloride (115).

Life-threatening hyperkalemia has been observed in patients receiving potassium-sparing diuretics (116). The most distressing finding in this study was that these agents were far more often prescribed in patients with abnormal kidney function than were other diuretics. Combining potassium-sparing diuretics with potassium supplements borders on malpractice; regular monitoring of serum potassium is mandatory in all patients treated with potassium-sparing diuretics, particularly if they are elderly or have abnormal renal or hepatic function.

Mineral balance

Thiazide and thiazide-like diuretics reduce the renal clearance of calcium by inhibiting the tubular secretion of calcium ions. In contrast, furosemide (which does not promote reabsorption of calcium in the distal tubule) causes transient hypercalciuria, an effect that has been exploited on occasion in the treatment of hypercalcemia (SED-9, 346; 118).

Hypercalcemia associated with thiazide diuretics is probably due to reduced urine calcium excretion. However, the incidence of thiazide diuretic-associated hypercalcemia has not been reported. In a study that included 72 patients with thiazide diuretic-associated hypercalcemia (68 women and 4 men; mean age 64 years), the overall age and sex-adjusted incidence was 7.7 per 100 000 (95% CI = 5.9, 9.5), with an increase in incidence after 1996, peaking at 16 per 100 000 (95% CI = 8.3, 24) in 1998 (118). The highest rate was in women aged 70–79 years (55/100 000). Thiazide diuretic-associated hypercalcemia was typically discovered several years (mean 6.0) after the start of therapy. The mean highest serum calcium was 2.7 (range 2.54–2.88) mmol/l and the serum parathyroid hormone concentration (obtained in 53 patients) was 4.8 pmol/l (reference range 1.0–5.2). In 21 of 33 patients who stopped taking the thiazide diuretic there was persistent hypercalcemia, and 18 were subsequently found to have primary hyperparathyroidism. The observed sex discrepancy in these studies suggests that postmenopausal women are predisposed to hypercalcemia with thiazide diuretics and/or that women more often take thiazides for the treatment of hypertension. This phenomenon tends to be mild, asymptomatic, and non-progressive; it calls for establishing whether primary hyperparathyroidism is present.

Loop diuretics can affect bone by increasing urine calcium excretion and modifying the diurnal rhythm of parathyroid hormone. In a double-blind, randomized study 87 otherwise healthy postmenopausal women with osteopenia were randomized to bumetanide 2 mg/day or placebo for 1 year (119). Calcium (800 mg/day) and vitamin D (10 micrograms/day) were administered throughout the study. Compared with placebo, urine calcium and plasma parathyroid hormone concentrations rose by 17% and 9% respectively. Bone mineral density fell by 2% at the hip (95% CI = 0.7, 3.2%) and ultradistal forearm. Six months after completion of bumetanide therapy between-group differences in bone mineral density were no longer statistically significant. The negative effect of bumetanide on bone metabolism and density was probably mitigated in these studies by calcium and vitamin D.

Metal metabolism

Diuretic-induced magnesium loss has been heavily stressed both by some independent investigators and by those seeking for commercial reasons to discredit the diuretics; there are far more reviews on the subject than sound original studies. It has been declared (120) that diuretic-induced magnesium deficiency "positively contributes" to myocardial infarction, delayed infarct healing, coronary and cerebral arterial spasm, hyperlipidemia, and cardiac dysrhythmias, with the addition of a horrific catalog of clinical manifestations, including ataxia, delirium, convulsions, coma, and ventricular fibrillation. Ramsay has remarked that this review, and others like it, belongs to the category of science fiction rather than that of serious medical writing (SEDA-10, 187).

Most reviews of the effects of diuretics on magnesium metabolism are uncritical and extrapolate wildly from the little sound evidence that is available (50). Several uncontrolled studies and very few controlled studies have suggested that long-term thiazide treatment causes a small fall in serum magnesium within the reference range. The complex relation between intracellular free and total magnesium content remains to be clarified, and there is absolutely no evidence that diuretic treatment reduces intracellular magnesium, and the few competent investigations have suggested that it does not. Diuretic-induced disturbances of magnesium balance do not cause depletion of intracellular potassium. The clinical significance of the small diuretic-induced alterations in magnesium balance, if any, is obscure. There is no satisfactory evidence that diuretic-induced magnesium disturbances cause or predispose to cardiac dysrhythmias, either in general or specifically after myocardial infarction (121).

A balanced account (122) concluded that the controlled trials, of which there are few, do not substantiate a role of diuretics in causing magnesium deficiency. Consequently, the vast majority of patients taking conventional doses of thiazides do not need magnesium supplements. On balance, potassium-sparing diuretics tend to increase serum and intracellular magnesium content, but this should not be taken as evidence of prior magnesium deficiency. It remains theoretically possible that large doses of loop diuretics given more than once-daily for long periods could induce negative magnesium balance and magnesium deficiency. However, it is difficult to conduct appropriately controlled trials in heart failure, in which such treatment is needed, and until more reliable information becomes available no absolute recommendation can be made. Magnesium depletion should be regarded as no more than a possible, as yet unproven, risk factor for cardiovascular morbidity and mortality (123).

In ordinary practice, serum magnesium need not be monitored in patients taking diuretics, and potassium-sparing drugs should not be used to prevent non-existent magnesium problems (SEDA-10, 190).

Gastrointestinal

Potassium, given separately or in combination with a diuretic, introduces the risk of esophageal injury; attempts to provide potassium in modified-release formulations have simply transposed the site of injury from the esophagus to the small bowel (SEDA-10, 187).

Biliary tract

There is experimental evidence in humans suggesting that thiazide diuretics increase biliary cholesterol saturation, the main determinant of cholesterol gallstone development; however, epidemiological data on the association between thiazide diuretics and gallbladder disease is sparse. In a prospective study in 81 351 US women aged 30–55 in 1980 and followed to 2000, 8607 reported undergoing a cholecystectomy (124). There was a modest positive relation between the use of thiazide diuretics and cholecystectomy. Compared with never users and current users of thiazide diuretics, the multivariate relative risks of cholecystectomy were 1.16 (95%CI = 1.08, 1.24) and 1.39 (95%CI = 1.29, 1.50) respectively. These findings are compatible with the possibility that thiazide diuretics increase the risk of symptomatic cholecystitis; however, measurement error is a concern in these studies, since information was lacking on the validity of self-reported thiazide diuretic use.

Musculoskeletal

In a nested case-control study in patients with type 2 diabetes there were 26 cases of a first leg amputation among 12 140 cases. Subjects who used thiazide diuretics alone or together with other antihypertensive medications had a higher risk of leg amputation (crude odds ratio = 6.11) compared with subjects taking ACE inhibitor monotherapy. Thiazide diuretics were also associated with an increased risk of leg amputation than non-thiazide antihypertensive drugs (adjusted odds ratio = 7.04). These findings suggest that thiazide-type diuretics be used cautiously in patients with type 2 diabetes particularly when there is significant lower limb peripheral vascular disease (125).

Although loop diuretics increase renal calcium excretion, there have been variable results in studies of their effects on the risk of fractures. In a case–control study of the risk of fracture in 44 001 patients who had taken a loop diuretic in the preceding 5 years and 194 111 age- and sex-matched controls, "ever" use of a loop diuretic was associated with a crude 51% increased risk of any fracture (OR = 1.51; 95% CI = 1.48, 155) and a 72% increased risk of hip fracture (OR = 1.72; 95% CI = 1.64, 181) (126). After adjustment for potential confounders, the risk reduction was only slightly increased for any fracture (OR = 1.04) and for hip fracture (OR = 1.16). With an increased average daily dose the estimates increased in former users but fell in current users. Furosemide was associated with higher risk estimates than bumetanide. This study does not provide a definitive answer to the risk of fractures with loop diuretics but suggests that particular caution should be taken.

Urinary tract

Since diuretics act primarily on the kidney, it is hardly surprising that renal damage can occur including interstitial nephritis (127,128). However, toxic damage is very rare in relation to the number of patients treated. Renal damage due to interstitial nephritis is more likely to occur in patients with pre-existing glomerular disease. It tends to present 4–10 weeks after starting treatment, with non-oliguric acute renal insufficiency and in some cases pyrexia and eosinophilia. This complication probably always recovers spontaneously and completely, although some consider it prudent to give high doses of glucocorticoids.

It has been hypothesized that population-wide use of diuretics might be associated with acceleration of the incidence of end-stage renal disease. Using data fusion techniques, pooled data trends in disease incidence and the use of antihypertensive medications were examined to determine whether changes in drug use patterns predict disease state emergence. There was a statistically significant inverse relation between all-cause cardiovascular disease mortality and the incidence of end-stage renal disease for the period from 1980 to 1998. There was a statistically significant time-lagged relation between both annual changes in diuretic distribution and total diuretic expenditure (measured by diuretic sales) to annual changes in the growth rate of end-stage renal disease. The methods used in these particular studies were clearly limited, as there was no patient-specific information, nor was there any indication of the specific agents used, the doses, or the exposure times in affected individuals. Moreover, the analysis did not distinguish between diuretic monotherapy and combination therapy. Thus, although the findings from these studies present an interesting hypothesis (129,130), considerably more supportive scientific evidence is needed before firm practice recommendations can be made about the propensity for end-stage renal disease and the use of diuretics.

Sexual function

The adverse effects of thiazide and thiazide-like diuretics on male sexual function include reduced libido, erectile dysfunction, and difficulty in ejaculating. The exact incidence of sexual dysfunction in patients taking diuretics is poorly documented, perhaps because of the personal nature of the problem and the reluctance of patients and/or physicians to discuss it. However, these abnormalities have been reported with incidence rates of 3–32%. The true incidence of sexual dysfunction probably lies closer to the lower end of this range (131). In a meta-analysis of 13 randomized, placebo-controlled trials conducted over a mean of 4 years the NNH (number needed to harm) for erectile impotence with thiazide diuretics in hypertension was 20 and the relative risk was 5.0 (132).

The mechanisms by which thiazides affect erectile dysfunction or libido are unclear, but it has been suggested that they have a direct effect on vascular smooth muscle cells or reduce the response to catecholamines. Sexual dysfunction does not appear to be mediated by either a low serum potassium concentration or a low blood pressure. Since sexual dysfunction can adversely affect the quality of life of hypertensive patients, physicians or health-care providers should take an accurate baseline sexual history and monitor sexual status for changes during therapy. If there are significant changes in sexual function, diuretic therapy can be withdrawn and an alternative drug class substituted. However, not uncommonly sexual dysfunction will persist despite withdrawal of the diuretic, suggesting that elements of the hypertensive state itself contribute to the process.

Analysis of the risk of erectile impotence has been hampered by the fact that impotence is common in men beyond middle age and is likely to be even more common in hypertension, treated or otherwise. The UK MRC trial (11) was the first to provide quantified evidence of a link with diuretics; in men taking bendroflumethiazide 10 mg/day, the withdrawal rate for impotence was 19.6 cases per 1000 treatment years, compared with 0.9 cases with placebo.

However, these figures may exaggerate the effect, since the trial was single-blind. A questionnaire pointed to a less striking difference; after 3 months impotence was noted in 16% of diuretic-treated patients and 9% of those taking placebo, and the roughly 2:1 ratio was again noted after 2 years. Later studies (16,87,133–135), pointed to a similar effect, even when low-dosage regimens were used (16,87). Some studies (135) but not others (87) have suggested that weight reduction ameliorates diuretic-induced sexual dysfunction. Erectile dysfunction occurs early and is often tolerable; onset after 2 years is unlikely (87).

Erectile impotence is particularly common in men with diabetes (16), who are likely to have difficulties because of autonomic dysfunction. It is unclear whether younger men and women are similarly affected and whether normotensive men have fewer such problems (136). Most investigations of the effects of diuretics on sexual function have been characterized by poor study design (137); the majority had no placebo control and relied on comparisons with baseline. The best studies have suggested an increase in erectile dysfunction in thiazides users compared with placebo. Bearing in mind all the confounding factors, it can be concluded that diuretics will sometimes cause impotence, but that in the population as a whole the effect is slight compared with other causes (SEDA-11, 197; SEDA-11, 198).

Sexual problems due to diuretics are very rare in women (16,87).

Immunologic

Hypersensitivity reactions to sulfonamide derivatives

In patients who are allergic to antimicrobial sulfonamides a decision has to be made as to whether other sulfonamide derivatives can be given safely.

Sulfonamide structures

Sulfonamides are characterized by a sulfur dioxide (SO_2) moiety and a nitrogen (N) moiety directly linked to a benzene ring (Figure 1). Many classes of drugs contain this chemical structure, including antimicrobial sulfonamides (for example sulfamethoxazole), carbonic anhydrase inhibitors (for example acetazolamide), thiazide, thiazide-like, and loop diuretics (for example hydrochlorothiazide, chlortalidone, furosemide, bumetanide), sulfonylureas (for example glibenclamide, glipizide), uricosuric drugs (for example probenecid), sulfasalazine, selective serotonin receptor agonists (for example sumatriptan), and the selective cyclo-oxygenase (COX-2) inhibitor celecoxib.

However, there are several structural differences among these classes. A major difference between sulfonamide antimicrobial drugs and other sulfonamide-containing compounds is the presence of an arylamine group (NH_2) at the N4 position. It has been hypothesized that only sulfonamides that contain the N4 primary amine are implicated in allergic reactions. This primary arylamine is oxidized to a highly reactive hydroxylamine intermediate, which is further oxidized to a highly immunogenic nitroso metabolite with a high affinity for cysteine residues on proteins. Non-antimicrobial sulfonamides are thought to be less immunogenic because they lack the N4 primary amine.

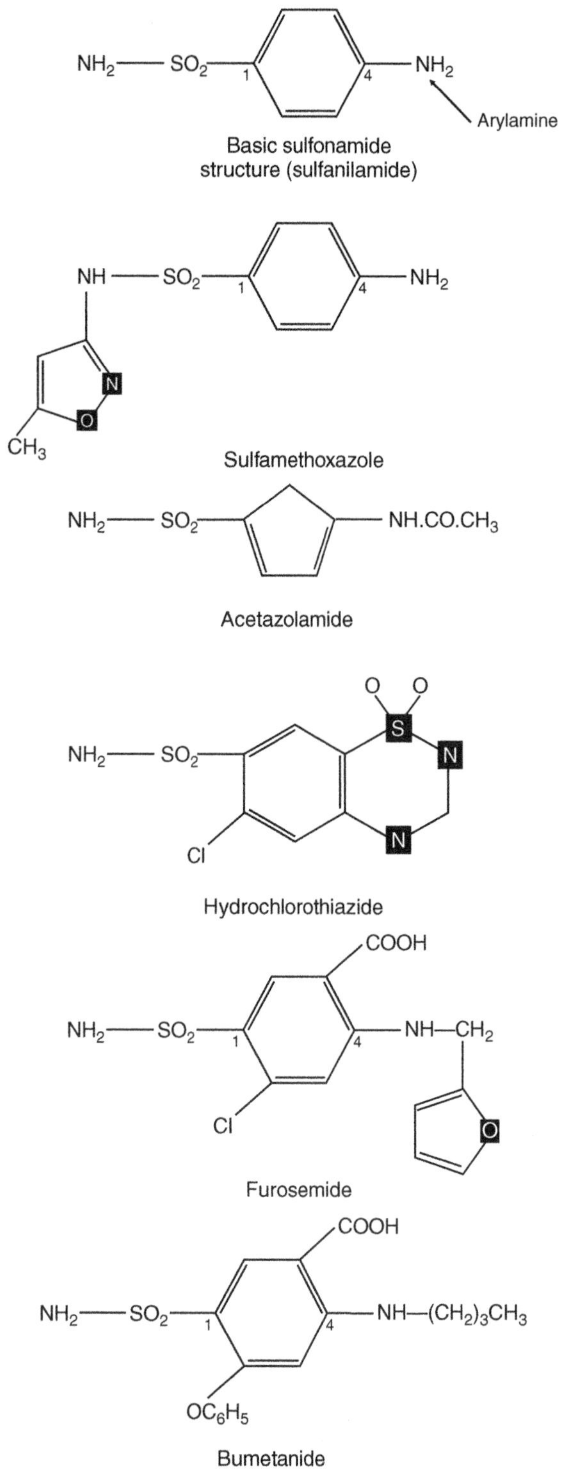

NH2—SO2—[1 benzene 4]—NH2 (Arylamine)

Basic sulfonamide structure (sulfanilamide)

Sulfamethoxazole

Acetazolamide

Hydrochlorothiazide

Furosemide

Bumetanide

Figure 1

Case reports

- A 57-year-old woman with decompensated heart failure and a history of allergy to co-trimoxazole (rash and pancreatitis) was given etacrynic acid instead of a sulfonamide diuretic (138). However, owing to a shortage of oral etacrynic acid she went without diuretic therapy and developed acute decompensated heart failure. She was given intravenous furosemide 40 mg bd and 5 weeks later developed nausea, vomiting, and abdominal pain and raised serum amylase and lipase activities. Intravenous furosemide was withdrawn and 2 days later the abdominal symptoms resolved. Because of worsening pulmonary edema, she was given intravenous bumetanide and within 24 hours developed myalgias. After 3 days bumetanide was withdrawn and intravenous torsemide begun. Four days later she developed severe abdominal pain similar to prior episodes, and laboratory and imaging studies were consistent with a diagnosis of pancreatitis. Torsemide was withdrawn and her symptoms resolved quickly. Intravenous etacrynic acid was started with no further complications.

- An 82-year-old woman, who had a history of angioedema with co-trimoxazole, developed angioedema, severe dysphagia, dyspnea, and a rash after taking valsartan and hydrochlorothiazide for 4 months (139). She was given subcutaneous adrenaline and intravenous antihistamines and glucocorticoids. The angioedema resolved in under 24 hours, and she was discharged with instructions to continue her regular medications, with the exception of valsartan. After 12 weeks she had another episode of angioedema, dyspnea, dysphagia, and rash. Hydrochlorothiazide was withdrawn but was later restarted. Within 2 days she developed mild swelling of the face and lips. Hydrochlorothiazide was permanently withdrawn.

In the second patient the similarity between the allergic reactions that occurred during treatment with hydrochlorothiazide and co-trimoxazole suggested true cross-reactivity.

Formal studies

There have been very few studies of the incidence of cross-reactivity between antimicrobial sulfonamides and other sulfonamide derivatives. In a retrospective cohort study, of 969 patients who had an allergic reaction to an antimicrobial sulfonamide, 96 (9.9%; adjusted OR = 2.8; 95% CI = 2.1, 3.7) had an allergic reaction to a non-antimicrobial sulfonamide (140). Of 19 257 who had no allergic reaction to an antimicrobial sulfonamide, 315 (1.6 %) had an allergic reaction to a non-antimicrobial sulfonamide (adjusted OR = 2.8; 95 % CI = 2.1, 3.7). However, the risk of an allergic reaction was even greater with a penicillin among patients with a prior hypersensitivity reaction to an antimicrobial sulfonamide, compared with patients with no such history (adjusted OR = 3.9; 95 % CI = 3.5, 4.3). Finally, the risk of an allergic reaction with a non-antimicrobial sulfonamide was lower among patients with a history of hypersensitivity to an antimicrobial sulfonamide than among those with a history of hypersensitivity to a penicillin (adjusted OR = 0.6; 95 % CI = 0.5, 0.8).

Thus, although a history of an allergic reaction to antimicrobial sulfonamides marks a heightened risk on exposure to non-antimicrobial sulfonamides, this risk is not exclusive to antimicrobial sulfonamides. In point of fact, patients with a history of hypersensitivity to an antimicrobial sulfonamide are at even greater risk of a reaction to a penicillin than to a non-antimicrobial sulfonamide.

Prescribers should appreciate that patients with a history of allergic reactions to antimicrobial sulfonamides may be at increased risk of all drug-induced adverse events of a seeming allergic nature.

Death

In 1354 patients (76% men, mean age 53 years) with advanced systolic heart failure (mean ejection fraction 24%) four groups were identified according to daily furosemide equivalent dosages: 0–40, 41–80, 81–160, and over 160 mg/day (141). The four groups were comparable as regards sex, body mass index, a history of hypertension, ischemic etiology of heart failure, and use of spironolactone. The highest dose was associated with lower hemoglobin and serum sodium concentrations and higher blood urea nitrogen and serum creatinine concentrations. There was reduced survival with increased diuretic dosage at 2 years: 83, 81, 68, and 53%. Diuretic dosage was an independent predictor of mortality at 1 year and 2 years, even with co-variate adjustment for multiple potentially confounding variables. Kaplan–Meier survival estimate curves showed persistence of the prognostic value of diuretic dosage in men/women, patients with and without coronary artery disease, and those with a serum creatinine above or below 133 μmol/l. The association of larger doses of loop diuretics with poorer outcomes has been demonstrated in a number of retrospective analyses. However, this study differed, in that it showed a stepwise, dose-dependent effect of loop diuretics on mortality. Propensity matching was not performed in these studies and so it is not known whether the relation between loop diuretic dosage and increased mortality is causative.

Long-Term Effects

Drug abuse

When diuretics are abused it is mostly in the course of a misguided attempt to lose weight; in the past, various "slimming remedies" offered for sale outside normal trading channels have been found to contain diuretics, sometimes with components such as thyroid extract. The unnecessary use of diuretics by a healthy individual, perhaps in excessive doses, can lead to dehydration, hypokalemia, and hypotension; when furosemide is abused, even tetany can occur because of hypocalcemia (142). The weight loss achieved by using diuretics in this way is purely due to dehydration and will soon be annulled by extra fluid intake.

One complication of long-term diuretic therapy in otherwise healthy individuals is edema, and it has been suggested that surreptitious use of diuretics can explain some otherwise paradoxical cases of idiopathic edema; presumably the diuretic induces a persistent increase in plasma renin activity and secondary hyperaldosteronism, and attempts to stop the diuretic intake can at first actually aggravate the condition (143). However, three studies have furnished strong evidence that diuretic abuse is not an important cause of idiopathic edema (144–146).

In some patients the covert use of diuretics leads to a complete replica of Bartter's syndrome, characterized by hypokalemia and hyper-reninemia without hypertension (147,148). In a patient who denies having taken diuretics, the true diagnosis may only be made by finding the drug or its metabolites in the urine, although challenge with diuretic may provoke the syndrome, while withdrawal leads to weight gain and the resolution of metabolic alkalosis.

The dangers of diuretic abuse among athletes have been discussed (SEDA-21, 228). Diuretics are used illegitimately by many athletes and amount for 6% of all drugs abused. They are taken by body builders to dehydrate the tissues, giving better definition to muscle shape, as well as to offset the fluid retaining effect of concomitant anabolic steroids and growth hormone, used to increase muscle mass. Diuretics are used in other sports to reduce weight, allowing competition in lower body weight classes, and to dilute the urine, reducing its specific gravity and making difficult the detection of other banned drugs. This practice is not only unethical but also hazardous.

Tumorigenicity

There is an association between diuretic use and renal cell carcinoma (149). Some of the studies that support this association can be dismissed, since the epidemiological data on which they were based were not suitably adjusted for confounding variables, including obesity, hypertension, age, and cigarette smoking. However, other case-control studies have shown a small risk of renal cell carcinoma in patients taking long-term diuretics after adjustment of the data for potentially confounding variables.

The carcinogenic mechanism of diuretics is not known, but could be related to a carcinogenic action of *N*-nitroso metabolic derivatives of thiazide and loop diuretics or structural changes in the transporting tubular epithelia, which provoke different stages of apoptosis. Rats and mice treated with diuretics have been reported to develop nephropathies and renal adenomas. Renal cell carcinoma arises in renal tubular cells, which are the principal site of action of diuretics. Contact over years or decades may have a low-grade carcinogenic effect. However, most prospective randomized trials provide too short a period of observation to assess the potential for carcinogenicity unequivocally. Furthermore, the findings may have been confounded by other risk factors for renal cell carcinoma. Adjustment for confounders greatly attenuated the risk (to non-significance) in one study, and in another the association with diuretic use disappeared completely. Therefore, the findings from these observational studies may have resulted from uncontrolled confounding by known or unrecognized risk factors.

The relation between diuretic therapy and the risk of malignancies has been examined in a review of pertinent publications between 1966 and 1998 (150). In nine case control studies (4185 cases), the odds ratio for renal cell carcinoma in patients treated with diuretics was 1.55 (95% CI = 1.42, 1.71) compared with non-users of diuretics. In three cohort studies of 1 226 229 patients (802 cases), patients taking diuretics had a more than two-fold risk of renal cell carcinoma compared with patients not taking diuretics. Women had an odds ratio of 2.01 (CI = 1.56, 1.67) compared with 1.96 (CI = 1.34, 2.13) in men. Thus, the cumulative evidence suggests that long-term use of diuretics may be associated with renal cell carcinoma.

The findings linking diuretic therapy with renal cell carcinoma need careful scrutiny. The strength of evidence provided by observational studies is limited, and such studies have yielded contradictory and controversial results in the past. An accompanying editorial (151) pointed out that some of the studies reviewed appear to have been designed to evaluate predictors of renal cell carcinoma without an a priori hypothesis that diuretics might be implicated. Statistical significance (set at $P < 0.05$) may have emerged merely by chance if 20 risk factors were examined.

Another commentary (152) emphasized the potential bias of observational studies and also publication bias in meta-analysis. The contemporary relevance of the findings is further reduced, since many of the studies included patients taking very high doses of thiazides. It is difficult to disentangle a drug-related effect from the association between hypertension and renal cell carcinoma.

Since renal cell carcinoma is rare, the practical importance of these observations is small: one extra case of renal cell carcinoma in 1500 patients treated for 20 years. If the hypothesis is correct, antihypertensive therapy with diuretics will prevent 20–40 strokes, 3–28 heart attacks, 3–10 cardiovascular deaths, and 4–14 deaths overall for every extra case of renal cell carcinoma. Even middle-aged women would be spared six strokes for each potential case of renal cell carcinoma. If a low grade carcinogen is involved, most patients will not live long enough for its effect to be expressed. The available information does not support a change in current prescription practices for diuretics in the treatment of hypertension and cardiac failure. Physicians should be more concerned about controlling blood pressure rather than concerning themselves with what at best might be a small risk of renal cell carcinoma.

Other cancer types have been evaluated for their association with diuretic therapy. The development of colon cancer has been studied in 14 166 patients aged 45–74 years with a previous myocardial infarction and/or stable angina, screened for participation in the Bezafibrate Infarction Prevention Study (153). Of these, 2153 used diuretics and 12 013 did not. Multivariate analysis identified diuretics as an independent predictor of an increased incidence of colon cancer (hazard ratio 2.0) and colon cancer mortality (hazard ratio 3.7). However, the association between diuretic therapy and a higher incidence of colon cancer was observed only among non-users of aspirin. There was a relatively lower incidence of colon cancer in furosemide users and a higher incidence in the small combined subgroup of those who took amiloride and/or hydrochlorothiazide. Further studies to test the association between diuretics and colon cancer, as well as the potential protective effects of aspirin, are needed.

Until these data become available, physicians should be aware of the potential effects of diuretics, especially when choosing long-term treatment for young patients with mild hypertension.

Second-Generation Effects

Pregnancy

Diuretics have been used for the treatment of both hypertension and edema in pregnancy. While regimens including thiazides have improved the outcome in hypertension developing during the first 20 weeks of pregnancy, the situation in later pregnancy seems to be different. Some controlled studies have shown that the thiazides do not influence the development of pre-eclampsia (154) and can even aggravate it.

Edema is a relatively frequent finding in normal pregnancy. It appears to be benign and can even be associated with improved obstetric performance (155). Edema, even in the presence of hypertension or proteinuria, is not a useful predictor of obstetric complications (156). These observations are important, in view of the finding that treatment with thiazide diuretics is associated with a reduced birth weight in normotensive pregnant subjects (157).

In a review of eight large-scale trials, four well-controlled studies did not show any benefit, whereas four others, which showed some benefit, were judged to be totally unsatisfactory, either because control was not strict enough or because inclusion and exclusion criteria were either not sufficiently well-defined or were scientifically unacceptable (158). The general consensus in earlier publications was that diuretics are useless, because they reduce the already lowered cardiac output in pre-eclamptic toxemia and so further reduce placental and uterine perfusion (159,160); diuretics do not alter the incidence of pre-eclamptic toxemia and eclampsia and they do not improve perinatal mortality or birth weight. Both the maternal adverse effects (hypokalemia, metabolic acidosis, hyperglycemia, hyperuricemia, hemorrhagic pancreatitis, and even death due to abuse) and the fetal and perinatal adverse effects (hyponatremia, fetal dysrhythmias induced by hypokalemia, thrombocytopenia, and jaundice) preclude the use of diuretics in pre-eclamptic toxemia.

Nine randomized, controlled studies of the use of diuretics in pre-eclampsia have been analysed (161). There were almost 7000 subjects in whom thiazides had been compared with placebo or no treatment. In the pooled analysis, diuretic treatment was associated with a non-significant reduction in perinatal mortality (about 10%) and stillbirths (about 33%). There was no excess of serious adverse reactions, either in the mothers or the neonates. These findings thus give no support to the belief that diuretics are harmful in pregnancy. It should not be concluded that diuretics ought to be used in pre-eclampsia in preference to other well-established agents, but they certainly can be used with confidence when they are really needed, for example in patients with severe hypertension that predates pregnancy and in heart failure.

Nevertheless, substantial numbers of pregnant women use these agents during pregnancy. In a Swedish study (162), 11% of 341 women took diuretics in the third trimester, while in a Tennessee Medical study (163) diuretics were prescribed as frequently as hypnotics and minor tranquilizers (7% for each category of drug). On the other hand, a study based on a University obstetrics programme in Florida (164) found that diuretics had been prescribed for only 2.4% of the patients.

Obviously treatment with any drug during pregnancy will affect both mother and fetus, and two aspects of such therapy are important: the adverse effects of the drug and the potential for teratogenicity.

Chlorothiazide readily crosses the placenta (165), but there have been few studies of the effects of diuretic treatment on the fetus. There are case reports of abnormalities of glucose handling (166) and severe electrolyte disturbances (167) in pregnant women taking diuretics, but these reports generally do not mention the metabolic values in the neonate. There has been one report of chronic fetal bradycardia associated with a low serum potassium in the diuretic-treated mother (168). Infusion of potassium restored maternal electrolytes and fetal heart rate. Normal delivery followed later, and electrolyte values in the child were normal. With regard to glucose handling, Senior and others (169) have raised the possibility that thiazide treatment of the mother can lead to neonatal hypoglycemia. In the absence of other predisposing factors, thiazides had been given to the mothers of 57% of their affected children, and, in the general population, the risk of neonatal hyperglycaemia was increased five-fold by thiazide treatment of the mother.

Hematological effects in the neonate have also been ascribed to maternal diuretic treatment. Seven cases of neonatal thrombocytopenia attributed to chlorothiazide have been described (170) and one case in which hydrochlorothiazide may have played a role (171). None of the mothers had thrombocytopenia. There was one death among the affected infants. Neonatal hemolysis occurred in two infants and was attributed to maternal thiazide therapy (172).

There is no evidence of a teratogenic effect of diuretics in humans, although one case of teratoma has been reported in the offspring of a woman treated for the first 19 weeks of pregnancy with acetazolamide (173), which is teratogenic in experimental animals. The lack of evidence of teratogenicity is not surprising. Assuming a basic malformation rate of 0.1% and that a drug that is used by 2% of pregnant women increases the malformation rate six times, it would require more than 100 000 women to be sure of detecting such an effect (162).

Susceptibility Factors

Age

There has been much discussion about the safety or risks of diuretics in elderly people. Although diuretic use has been implicated as a cause of urinary incontinence, no evidence has been presented to confirm this. An epidemiological survey of 1956 respondents aged at least 60 years showed no significant difference between diuretic-users and non-users in the prevalence of incontinence (174).

Orthostatic hypotension, although often loosely referred to in older literature (SED-10, 370) (175), is in fact only likely to become a problem in very old subjects, aged 90 years or more, even if a potent loop diuretic is used; this was the clear conclusion of a good prospective study published in 1978 (175). In 843 independent living men and women aged 60–87 years, postural fall in systolic blood pressure was not related to treatment with diuretics, after correction for initial blood pressure (176). At currently recommended doses, older subjects do not generally experience particular problems from hypokalemia and do not appear to be at special risk of cardiac dysrhythmias in the face of diuretic-induced hypokalemia (SEDA-15, 218).

Reviews of the age-related effects of diuretics in hypertensive subjects (177,178) have concluded that symptomatic adverse reactions are not more frequent in older patients and some trials have suggested a lower frequency than in younger subjects. Although the elderly may be susceptible to some metabolic effects of diuretics, the evidence for cardiotoxic potential is not conclusive.

The best overall evidence of the safety of diuretics in old people comes from the large-scale outcome trials in hypertensive patients (11,13,15,17,18). These studies in over 10 000 subjects aged over 60 years showed clearly that thiazide-based treatment reduces the risk of stroke, coronary heart disease events, and cardiovascular events in older hypertensive patients. A meta-analysis (179) of randomized trials lasting at least 1 year and involving 16 164 individuals aged at least 60 years showed that diuretics were superior to beta-blockers with regard to all endpoints (stroke, coronary heart disease events, cardiovascular mortality, and all-cause mortality). The beneficial effects noted in these trials should dispel any doubts about the safety and efficacy of diuretics in old people.

Drug Administration

Drug formulations

Fixed combinations of thiazides and loop diuretics with potassium and of thiazides with beta-blockers serve little useful purpose and can in fact do harm. Combinations of thiazides and loop diuretics with potassium-sparing diuretics serve the needs of the small minority of patients who develop clinically significant hypokalemia when given diuretics alone, or in whom hypokalemia is particularly risky. In fact, these combinations are much too widely used, and since individual needs vary so much there is a spectrum of risk, ranging from hypokalemia to hyperkalemia (SEDA-10, 370) (SEDA-10, 371).

Drug–Drug Interactions

Antidysrhythmic drugs

Hypokalemia due to diuretics potentiates the dysrhythmogenic actions of antidysrhythmic drugs that prolong the QT interval, such as Class I and Class III antidysrhythmic drugs, increasing the risk of torsade de pointes (180). This can also happen with other drugs that prolong the QT interval, such as phenothiazines (181).

Beta-blockers

Interaction of thiazides with beta-blockers can cause a greater risk of hyperglycemia than if either component is given separately (182,183).

Cardiac glycosides

Hypokalemia due to diuretics potentiates the actions of cardiac glycosides and can result in digitalis toxicity. In 67 patients taking a maintenance dose of digoxin, 42 with digoxin toxicity and 25 without, the mean serum digoxin concentration was significantly higher in the former and the mean serum potassium concentration was significantly lower (184). Among the patients with toxicity, 24% had hypokalemia due to diuretics. The mean serum digoxin concentration in those with hypokalemia and toxicity was significantly lower than in those with normokalemia and toxicity.

References

1. Cranston WI, Juel-Jensen BE, Semmence AM, Jones RP, Forbes JA, Mutch LM. Effects of oral diuretics on raised arterial pressure. Lancet 1963;186:966–70.
2. Grimm RH Jr, Grandits GA, Cutler JA, Stewart AL, McDonald RH, Svendsen K, Prineas RJ, Liebson PR. Relationships of quality-of-life measures to long-term lifestyle and drug treatment in the Treatment of Mild Hypertension Study. Arch Intern Med 1997;157(6):638–48.
3. Philipp T, Anlauf M, Distler A, Holzgreve H, Michaelis J, Wellek S. Randomised, double blind, multicentre comparison of hydrochlorothiazide, atenolol, nitrendipine, and enalapril in antihypertensive treatment: results of the HANE study. HANE Trial Research Group. BMJ 1997;315(7101):154–9.
4. Kaplan NM. Diuretics as a basis of antihypertensive therapy. An overview. Drugs 2000;59(Suppl 2):21–5.
5. Chobanian AV, Bakris GL, Black HR, Cushman WC, Green LA, Izzo JL J Jr, Jones DW, Materson BJ, Oparil S, Wright JT Jr, Roccella EJ. Joint National Committee on Prevention, Detection, Evaluation, and Treatment of High Blood Pressure. National Heart, Lung, and Blood Institute; National High Blood Pressure Education Program Coordinating Committee. Seventh report of the Joint National Committee on Prevention, Detection, Evaluation, and Treatment of High Blood Pressure. Hypertension 2003;42(6):1206–52.
6. Moser M. National recommendations for the pharmacological treatment of hypertension: should they be revised? Arch Intern Med 1999;159(13):1403–6.
7. Zanchetti A. Contribution of fixed low-dose combinations to initial therapy in hypertension. Eur Heart J 1999;1(Suppl L):L5–9.
8. Kaplan N. Low dose combinations in the treatment of hypertension: theory and practice. J Hum Hypertens 1999;13(10):707–10.
9. Gosse P, Sheridan DJ, Zannad F, Dubourg O, Gueret P, Karpov Y, de Leeuw PW, Palma-Gamiz JL, Pessina A, Motz W, Degaute JP, Chastang C. Regression of left ventricular hypertrophy in hypertensive patients treated with

indapamide SR 1.5 mg versus enalapril 20 mg: the LIVE study J Hypertens 2000;18(10):1465–75.

10. Wright JM, Lee CH, Chambers GK. Systematic review of antihypertensive therapies: does the evidence assist in choosing a first-line drug? CMAJ 1999;161(1):25–32.
11. Medical Research Council Working Party. MRC trial of treatment of mild hypertension: principal results. BMJ (Clin Res Ed) 1985;291(6488):97–104.
12. Medical Research Council Working Party on Mild Hypertension. Coronary heart disease in the Medical Research Council trial of treatment of mild hypertension. Br Heart J 1988;59(3):364–78.
13. authors>Amery A, Birkenhager W, Brixko P, Bulpitt C, Clement D, Deruyttere M, De Schaepdryver A, Dollery C, Fagard R, Forette F, . Mortality and morbidity results from the European Working Party on High Blood Pressure in the Elderly trialLancet 1985;1(8442):1349–54.
14. SHEP Cooperative Research Group. Prevention of stroke by antihypertensive drug treatment in older persons with isolated systolic hypertension. Final results of the Systolic Hypertension in the Elderly Program (SHEP). JAMA 1991;265(24):3255–64.
15. Kostis JB, Davis BR, Cutler J, Grimm RH Jr, Berge KG, Cohen JD, Lacy CR, Perry HM Jr, Blaufox MD, Wassertheil-Smoller S, Black HR, Schron E, Berkson DM, Curb JD, Smith WM, McDonald R, Applegate WB. Prevention of heart failure by antihypertensive drug treatment in older persons with isolated systolic hypertension. SHEP Cooperative Research Group. JAMA 1997;278(3):212–6.
16. Curb JD, Pressel SL, Cutler JA, Savage PJ, Applegate WB, Black H, Camel G, Davis BR, Frost PH, Gonzalez N, Guthrie G, Oberman A, Rutan GH, Stamler J. Effect of diuretic-based antihypertensive treatment on cardiovascular disease risk in older diabetic patients with isolated systolic hypertension. Systolic Hypertension in the Elderly Program Cooperative Research Group. JAMA 1996;276(23):1886–92.
17. Dahlof B, Lindholm LH, Hansson L, Schersten B, Ekbom T, Wester PO. Morbidity and mortality in the Swedish Trial in Old Patients with Hypertension (STOP-Hypertension). Lancet 1991;338(8778):1281–5.
18. MRC Working Party. Medical Research Council trial of treatment of hypertension in older adults: principal results. BMJ 1992;304(6824):405–12.
19. The IPPPSH Collaborative Group. Cardiovascular risk and risk factors in a randomized trial of treatment based on the beta-blocker oxprenolol: the International Prospective Primary Prevention Study in Hypertension (IPPPSH). J Hypertens 1985;3(4):379–92.
20. Wilhelmsen L, Berglund G, Elmfeldt D, Fitzsimons T, Holzgreve H, Hosie J, Hornkvist PE, Pennert K, Tuomilehto J, Wedel H. Beta-blockers versus diuretics in hypertensive men: main results from the HAPPHY trial. J Hypertens 1987;5(5):561–72.
21. Wikstrand J, Warnold I, Olsson G, Tuomilehto J, Elmfeldt D, Berglund G. Primary prevention with metoprolol in patients with hypertension. Mortality results from the MAPHY study. JAMA 1988;259(13):1976–82.
22. Collins R, MacMahon S. Blood pressure, antihypertensive drug treatment and the risks of stroke and of coronary heart disease. Br Med Bull 1994;50(2):272–98.
23. ALLHAT Collaborative Research Group. Major cardiovascular events in hypertensive patients randomized to doxazosin vs chlorthalidone: the antihypertensive and lipid-lowering treatment to prevent heart attack trial (ALLHAT). JAMA 2000;283(15):1967–75.
24. ALLHAT Officers and Coordinators for the ALLHAT Collaborative Research Group. The Antihypertensive and Lipid-Lowering Treatment to Prevent Heart Attack Trial. Major outcomes in high-risk hypertensive patients randomized to angiotensin-converting enzyme inhibitor or calcium channel blocker vs diuretic: The Antihypertensive and Lipid-Lowering Treatment to Prevent Heart Attack Trial (ALLHAT). JAMA 2002;288(23):2981–97.
25. McInnes GT. Size isn't everything—ALLHAT in perspective. J Hypertens 2003;21(3):459–61.
26. Olsen H, Klemetsrud T, Stokke HP, Tretli S, Westheim A. Adverse drug reactions in current antihypertensive therapy: a general practice survey of 2586 patients in Norway. Blood Press 1999;8(2):94–101.
27. Kramer BK, Schweda F, Riegger GA. Diuretic treatment and diuretic resistance in heart failure. Am J Med 1999;106(1):90–6.
28. Pitt B, Zannad F, Remme WJ, Cody R, Castaigne A, Perez A, Palensky J, Wittes J. The effect of spironolactone on morbidity and mortality in patients with severe heart failure. Randomized Aldactone Evaluation Study Investigators. N Engl J Med 1999;341(10):709–17.
29. Pitt B, Remme W, Zannad F, Neaton J, Martinez F, Roniker B, Bittman R, Hurley S, Kleiman J, Gatlin M. Eplerenone Post-Acute Myocardial Infarction Heart Failure Efficacy and Survival Study Investigators. Eplerenone, a selective aldosterone blocker, in patients with left ventricular dysfunction after myocardial infarction. N Engl J Med 2003;348(14):1309–21.
30. Richards AM, Nicholls MG. Aldosterone antagonism in heart failure. Lancet 1999;354(9181):789–90.
31. Brenner BM, Cooper ME, de Zeeuw D, Keane WF, Mitch WE, Parving HH, Remuzzi G, Snapinn SM, Zhang Z, Shahinfar S. RENAAL Study Investigators. Effects of losartan on renal and cardiovascular outcomes in patients with type 2 diabetes and nephropathy. N Engl J Med 2001;345(12):861–9.
32. Lewis EJ, Hunsicker LG, Clarke WR, Berl T, Pohl MA, Lewis JB, Ritz E, Atkins RC, Rohde R, Raz ICollaborative Study Group. Renoprotective effect of the angiotensin-receptor antagonist irbesartan in patients with nephropathy due to type 2 diabetes. N Engl J Med 2001;345(12):851–60.
33. Mehta RL, Pascual MT, Soroko S, Chertow GMPICARD Study Group. Diuretics, mortality, and nonrecovery of renal function in acute renal failure. JAMA 2002;288(20):2547–53.
34. Brater DC. Diuretic therapy. N Engl J Med 1998;339(6):387–95.
35. Schwenger V, Zeier M, Ritz E. Antihypertensive therapy in renal patients—benefits and difficulties. Nephron 1999;83(3):202–13.
36. Grahame-Smith DG. The Lilly Prize Lecture. 1996 "Keep on taking the tablets": pharmacological adaptation during long-term drug therapy. Br J Clin Pharmacol 1997;44(3):227–38.
37. Knauf H, Mutschler E. Clinical pharmacokinetics and pharmacodynamics of torasemide. Clin Pharmacokinet 1998;34(1):1–24.
38. Sica DA, Gehr TW. Diuretic combinations in refractory oedema states: pharmacokinetic–pharmacodynamic relationships. Clin Pharmacokinet 1996;30(3):229–49.
39. Paterna S, Di Pasquale P, Parrinello G, Amato P, Cardinale A, Follone G, Giubilato A, Licata G. Effects of high-dose furosemide and small-volume hypertonic saline solution infusion in comparison with a high dose of furosemide as a bolus, in refractory congestive heart failure. Eur J Heart Fail 2000;2(3):305–13.

40. Ravnan SL, Ravnan MC. Management of adult heart failure: bolus versus continuous infusion loop diuretics, a review of the literature. Hosp Pharm 2000;35:832–6.
41. Howard PA, Dunn MI. Severe musculoskeletal symptoms during continuous infusion of bumetanide. Chest 1997;111(2):359–64.
42. Buter H, Hemmelder MH, Navis G, de Jong PE, de Zeeuw D. The blunting of the antiproteinuric efficacy of ACE inhibition by high sodium intake can be restored by hydrochlorothiazide. Nephrol Dial Transplant 1998;13(7): 1682–5.
43. Dishart MK, Kellum JA. An evaluation of pharmacological strategies for the prevention and treatment of acute renal failure. Drugs 2000;59(1):79–91.
44. Epstein M. Aldosterone as a mediator of progressive renal disease: pathogenetic and clinical implications. Am J Kidney Dis 2001;37(4):677–88.
45. Juurlink DN, Mamdani MM, Lee DS, Kopp A, Austin PC, Laupacis A, Redelmeier DA. Rates of hyperkalemia after publication of the Randomized Aldactone Evaluation Study. N Engl J Med 2004;351(6):543–51.
46. Andersson OK, Gudbrandsson T, Jamerson K. Metabolic adverse effects of thiazide diuretics: the importance of normokalaemia. J Intern Med Suppl 1991;735: 89–96.
47. Dyckner T. Relation of cardiovascular disease to potassium and magnesium deficiencies. Am J Cardiol 1990;65(23):K44–6.
48. Schulman M, Narins RG. Hypokalemia and cardiovascular disease. Am J Cardiol 1990;65(10):E4–9.
49. Papademetriou V, Burris JF, Notargiacomo A, Fletcher RD, Freis ED. Thiazide therapy is not a cause of arrhythmia in patients with systemic hypertension. Arch Intern Med 1988;148(6):1272–6.
50. McInnes GT, Yeo WW, Ramsay LE, Moser M. Cardiotoxicity and diuretics: much speculation—little substance. J Hypertens 1992;10(4):317–35.
51. Papademetriou V. Diuretics in hypertension: clinical experiences. Eur Heart J 1992;13(Suppl G):92–5.
52. Restrick LJ, Huddy N, Hoffbrand BI. Diuretic-induced hypokalaemia and surgery: much ado about nothing? Postgrad Med J 1992;68(799):318–20.
53. Kostis JB, Lacy CR, Hall WD, Wilson AC, Borhani NO, Krieger SD, Cosgrove NMThe SHEP Study Group. The effect of chlorthalidone on ventricular ectopic activity in patients with isolated systolic hypertension. Am J Cardiol 1994;74(5):464–7.
54. Neaton JD, Grimm RH Jr, Prineas RJ, Stamler J, Grandits GA, Elmer PJ, Cutler JA, Flack JM, Schoenberger JA, McDonald R, Lewis CE, Liebson PR. Treatment of Mild Hypertension Study. Final results. Treatment of Mild Hypertension Study Research Group. JAMA 1993;270(6):713–24.
55. Cooper HA, Dries DL, Davis CE, Shen YL, Domanski MJ. Diuretics and risk of arrhythmic death in patients with left ventricular dysfunction. Circulation 1999;100(12):1311–5.
56. Viskin S. Long QT syndromes and torsade de pointes. Lancet 1999;354(9190):1625–33.
57. Yap YG, Camm J. Risk of torsades de pointes with noncardiac drugs. Doctors need to be aware that many drugs can cause QT prolongation. BMJ 2000; 320(7243):1158–9.
58. Miglior S, Torri V, Zeyen T, Pfeiffer N, Vaz JC, Adamsons I; EGPS Group. Intercurrent factors associated with the development of open-angle glaucoma in the European Glaucoma Prevention Study. Am J Ophthalmol 2007;144:266–75.
59. Robertson CM, Tyebkhan JM, Peliowski A, Etches PC, Cheung PY. Ototoxic drugs and sensorineural hearing loss following severe neonatal respiratory failure. Acta Paediatrica 2006;95:214–23.
60. Hoes AW, Grobbee DE, Lubsen J, Man in't Veld AJ, van der Does E, Hofman A. Diuretics, beta-blockers, and the risk for sudden cardiac death in hypertensive patients. Ann Intern Med 1995;123(7):481–7.
61. Siscovick DS, Raghunathan TE, Psaty BM, Koepsell TD, Wicklund KG, Lin X, Cobb L, Rautaharju PM, Copass MK, Wagner EH. Diuretic therapy for hypertension and the risk of primary cardiac arrest. N Engl J Med 1994;330(26):1852–7.
62. Samuelsson O, Pennert K, Andersson O, Berglund G, Hedner T, Persson B, Wedel H, Wilhelmsen L. Diabetes mellitus and raised serum triglyceride concentration in treated hypertension—are they of prognostic importance? Observational study. BMJ 1996;313(7058):660–3.
63. Kaplan NM. How bad are diuretic-induced hypokalemia and hypercholesterolemia? Arch Intern Med 1989;149(12):2649.
64. Weinberger MH. Selection of drugs for initial treatment of hypertension. Pract Cardiol 1989;15:81.
65. Moser M. In defense of traditional antihypertensive therapy. Hypertension 1988;12(3):324–6.
66. Thompson WG. An assault on old friends: thiazide diuretics under siege. Am J Med Sci 1990;300(3):152–8.
67. Anonymous. Potassium-sparing diuretics—when are they really needed. Drug Ther Bull 1985;23(5):17–20.
68. Ramsay LE, Yeo WW, Jackson PR. Diabetes, impaired glucose tolerance and insulin resistance with diuretics. Eur Heart J 1992;13(Suppl G):68–71.
69. Saunders A, Wilson SM. Do diuretics differ in degree of hypokalaemia, and does it matter? Aust J Hosp Pharm 1991;21:120–1.
70. Carlsen JE, Kober L, Torp-Pedersen C, Johansen P. Relation between dose of bendrofluazide, antihypertensive effect, and adverse biochemical effects. BMJ 1990;300(6730):975–8.
71. McVeigh GE, Dulie EB, Ravenscroft A, Galloway DB, Johnston GD. Low and conventional dose cyclopenthiazide on glucose and lipid metabolism in mild hypertension. Br J Clin Pharmacol 1989;27(4):523–6.
72. Ramsay LE, Yeo WW, Jackson PR. Influence of diuretics, calcium antagonists, and alpha-blockers on insulin sensitivity and glucose tolerance in hypertensive patients. J Cardiovasc Pharmacol 1992;20(Suppl 11):S49–54.
73. Weinberger MH. Mechanisms of diuretic effects on carbohydrate tolerance, insulin sensitivity and lipid levels. Eur Heart J 1992;13(Suppl G):5–9.
74. Savage PJ, Pressel SL, Curb JD, Schron EB, Applegate WB, Black HR, Cohen J, Davis BR, Frost P, Smith W, Gonzalez N, Guthrie GP, Oberman A, Rutan G, Probstfield JL, Stamler J. Influence of long-term, low-dose, diuretic-based, antihypertensive therapy on glucose, lipid, uric acid, and potassium levels in older men and women with isolated systolic hypertension: The Systolic Hypertension in the Elderly Program. SHEP Cooperative Research Group. Arch Intern Med 1998;158(7):741–51.
75. Green TP, Johnson DE, Bass JL, Landrum BG, Ferrara TB, Thompson TR. Prophylactic furosemide in severe respiratory distress syndrome: blinded prospective study. J Pediatr 1988;112(4):605–12.
76. Johnston GD. Treatment of hypertension in older adults. BMJ 1992;304:639.

77. Moser M. Diuretics and cardiovascular risk factors. Eur Heart J 1992;13(Suppl G):72–80.
78. Verdecchia P, Reboldi G, Angeli F, Borgioni C, Gattobigio R, Filippucci L, Norgiolini S, Bracco C, Porcellati C. Adverse prognostic significance of new diabetes in treated hypertensive subjects. Hypertension 2004; 43(5):963–9.
79. Moser M. Diuretics should continue to be recommended as initial therapy in the treatment of hypertension. In: Puschett JB, Greenberg A, editors. Diuretics IV: Chemistry, Pharmacology and Clinical Applications. Amsterdam: Elsevier, 1993:465–76.
80. Moser M. Do different hemodynamic effects of antihypertensive drugs translate into different safety profiles? Eur J Clin Pharmacol 1990;38(Suppl 2):S134–8.
81. Waller PC, Ramsay LE. Predicting acute gout in diuretic-treated hypertensive patients. J Hum Hypertens 1989;3(6):457–61.
82. McVeigh G, Galloway D, Johnston D. The case for low dose diuretics in hypertension: comparison of low and conventional doses of cyclopenthiazide. BMJ 1988; 297(6641):95–8.
83. Franse LV, Pahor M, Di Bari M, Shorr RI, Wan JY, Somes GW, Applegate WB. Serum uric acid, diuretic treatment and risk of cardiovascular events in the Systolic Hypertension in the Elderly Program (SHEP). J Hypertens 2000;18(8):1149–54.
84. Alderman MH, Cohen H, Madhavan S, Kivlighn S. Serum uric acid and cardiovascular events in successfully treated hypertensive patients. Hypertension 1999; 34(1):144–50.
85. Freis ED, Papademetriou V. How dangerous are diuretics? Drugs 1985;30(6):469–74.
86. Spence JD. Effects of antihypertensive drugs on atherogenic factors: possible importance of drug selection in prevention of atherosclerosis. J Cardiovasc Pharmacol 1985;7(Suppl 2):S121–5.
87. Weinberger MH. Antihypertensive therapy and lipids. Evidence, mechanisms, and implications. Arch Intern Med 1985;145(6):1102–5.
88. Weidmann P, Ferrier C, Saxenhofer H, Uehlinger DE, Trost BN. Serum lipoproteins during treatment with antihypertensive drugs. Drugs 1988;35(Suppl 6):118–34.
89. Falch DK, Schreiner A. The effect of spironolactone on lipid, glucose and uric acid levels in blood during long-term administration to hypertensives. Acta Med Scand 1983;213(1):27–30.
90. Grimm RH Jr, Grandits GA, Prineas RJ, McDonald RH, Lewis CE, Flack JM, Yunis C, Svendsen K, Liebson PR, Elmer PJ, Stamler J. Long-term effects on sexual function of five antihypertensive drugs and nutritional hygienic treatment in hypertensive men and women. Treatment of Mild Hypertension Study (TOMHS). Hypertension 1997; 29(1 Pt 1):8–14.
91. MacMahon SW, Macdonald GJ, Blacket RB. Plasma lipoprotein levels in treated and untreated hypertensive men and women. The National Heart Foundation of Australia Risk Factor Prevalence Study. Arteriosclerosis 1985; 5(4):391–6.
92. Wallace RB, Hunninghake DB, Chambless LE, Heiss G, Wahl P, Barrett-Connor E. A screening survey of dyslipoproteinemias associated with prescription drug use. The Lipid Research Clinics Program Prevalence Study. Circulation 1986;73(1 Pt 2):I70–9.
93. Maitland-van der Zee AH, Turner ST, Schwartz GL, Chapman AB, Klungel OH, Boerwinkle E. Demographic, environmental, and genetic predictors of metabolic side effects of hydrochlorothiazide treatment in hypertensive subjects. Am J Hypertens 2005;18:1077–83.
94. Tuomilehto J, Salonen JT, Nissinen A. Factors associated with changes in serum cholesterol during a community-based hypertension programme. Acta Med Scand 1985;217(3):243–52.
95. Chow KM, Szeto CC, Wong TY, Leung CB, Li PK. Risk factors for thiazide-induced hyponatraemia. Q J Med 2003;96:911–7.
96. Sica DA. Diuretic-related side effects: development and treatment. J Clin Hypertens (Greenwich) 2004;6:532–40.
97. Fernandez P, Choi M. Thiazide-induced hyponatraemia. In: Puschett JB, Greenberg A, editors. Diuretics IV: Chemistry, Pharmacology and Clinical Applications. Amsterdam: Elsevier, 1993:199–209.
98. Sonnenblick M, Friedlander Y, Rosin AJ. Diuretic-induced severe hyponatremia. Review and analysis of 129 reported patients. Chest 1993;103(2):601–6.
99. Arieff AI. Management of hyponatraemia. BMJ 1993;307(6899):305–8.
100. Weidmann P, de Courten M, Ferrari P. Effect of diuretics on the plasma lipid profile. Eur Heart J 1992;13(Suppl G): 61–7.
101. Golomb BA, Criqui MH. Antihypertensives: much ado about lipids. Arch Intern Med 1999;159(6):535–7.
102. Adrogué HJ, Madias NE. Hyponatremia. N Engl J Med 2000;342(21):1581–9.
103. Zalin AM, Hutchinson CE, Jong M, Matthews K. Hyponatraemia during treatment with chlorpropamide and Moduretic (amiloride plus hydrochlorothiazide). BMJ (Clin Res Ed) 1984;289(6446):659.
104. Goldsmith SR, Gheorghiade M. Vasopressin antagonism in heart failure. J Am Coll Cardiol 2005;46:1785–91.
105. Hart TJ, Johnston LJ, Edmonds MW, Brownscombe L. Hyponatraemia secondary to thiazide–trimethoprim interaction. Can J Hosp Pharm 1989;42:243–6.
106. Byatt CM, Millard PH, Levin GE. Diuretics and electrolyte disturbances in 1000 consecutive geriatric admissions. J R Soc Med 1990;83(11):704–8.
107. Adrogue HJ, Madias NE. Hypernatremia. N Engl J Med 2000;342(20):1493–9.
108. Frohlich ED. Current issues in hypertension. Old questions with new answers and new questions. Med Clin North Am 1992;76(5):1043–56.
109. Nicholls MG. Interaction of diuretics and electrolytes in congestive heart failure. Am J Cardiol 1990;65(10):E17–21.
110. McInnes GT. Potassium and diuretics—when does it matter? Med Resource 1992;6:21–4.
111. Kassirer JP. Does the benefit of aggressive potassium replacement in diuretic-treated patients outweigh the risk? J Cardiovasc Pharmacol 1984;6(Suppl 3):S488–92.
112. Shintani S, Shiigai T, Tsukagoshi H. Marked hypokalemic rhabdomyolysis with myoglobinuria due to diuretic treatment. Eur Neurol 1991;31(6):396–8.
113. Ascherio A, Rimm EB, Hernan MA, Giovannucci EL, Kawachi I, Stampfer MJ, Willett WC. Intake of potassium, magnesium, calcium, and fiber and risk of stroke among US men. Circulation 1998;98(12):1198–204.
114. Harper R, Ennis CN, Sheridan B, Atkinson AB, Johnston GD, Bell PM. Effects of low dose versus conventional dose thiazide diuretic on insulin action in essential hypertension. BMJ 1994;309(6949):226–30.
115. Cohn JN, Kowey PR, Whelton PK, Prisant LM. New guidelines for potassium replacement in clinical practice: a contemporary review by the National Council on

Potassium in Clinical Practice. Arch Intern Med 2000; 160(16):2429–36.

116. Murdoch DL, Gillen GJ, Morton JJ, Leckie B, Murray GD, Davies DL, McInnes GT. Twice-daily low-dose captopril in diuretic-treated hypertensives. J Hum Hypertens 1989;3(1):29–33.
117. Perazella MA. Drug-induced hyperkalemia: old culprits and new offenders. Am J Med 2000;109(4):307–14.
118. Wermers RA, Kearns AE, Jenkins GD, Melton LJ. Incidence and clinical spectrum of thiazide-associated hypercalcemia. Am J Med 2007;120:e9–15.
119. Rejnmark L, Vestergaard P, Heickendorff L, Andreasen F, Mosekilde L. Loop diuretics increase bone turnover and decrease BMD in osteopenic postmenopausal women: results from a randomized controlled study with bumetanide. J Bone Min Res 2006;21:163–70.
120. McCaughan D. Hazards of non-prescription potassium supplements. Lancet 1984;1(8375):513–4.
121. Lawson DH, O'Connor PC, Jick H. Drug attributed alterations in potassium handling in congestive cardiac failure. Eur J Clin Pharmacol 1982;23(1):21–5.
122. Suki WN, Yium JJ, Von Minden M, Saller-Hebert C, Eknoyan G, Martinez-Maldonado M. Actue treatment of hypercalcemia with furosemide. N Engl J Med 1970;283(16):836–40.
123. Reyes AJ, Leary WP. Cardiovascular toxicity of diuretics related to magnesium depletion. Hum Toxicol 1984; 3(5):351–71.
124. Leitzmann MF, Tsai CJ, Stampfer MJ, Willett WC, Giovannucci E. Thiazide diuretics and the risk of gallbladder disease requiring surgery in women. Arch Intern Med 2005;165:567–73.
125. Erkens JA, Klungel OH, Stolk RP, Spoelstra JA, Grobbee DE, Leufkens HG. Antihypertensive drug therapy and the risk of lower extremity amputations in pharmacologically treated type 2 diabetes patients. Pharmacoepidemiol Drug Saf 2004;13:139–46.
126. Rejnmark L, Vestergaard P, Mosekilde L. Fracture risk in patients treated with loop diuretics. J Intern Med 2006;259:117–24.
127. Gettes LS. Electrolyte abnormalities underlying lethal and ventricular arrhythmias. Circulation 1992;85(Suppl 1):I70–6.
128. Davies DL, Fraser R. Do diuretics cause magnesium deficiency? Br J Clin Pharmacol 1993;36(1):1–10.
129. Hawkins RG, Houston MC. Is population-wide diuretic use directly associated with the incidence of end-stage renal disease in the United States? A hypothesis. Am J Hypertens 2005;18:744–9.
130. Akram M, Reddan D. Diuretic use and ESRD, precipitating factor or epiphenomenon. Am J Hypertens. 2005; 18:739–40.
131. Leary WP. Diuretics and increase in urinary magnesium excretion: possible clinical relevance. In: Puschett JB, Greenberg A, editors. Diuretics IV: Chemistry, Pharmacology and Clinical Applications. Amsterdam: Elsevier, 1993:261–5.
132. Jennings M, Shortland JR, Maddocks JL. Interstitial nephritis associated with frusemide. J R Soc Med 1986; 79(4):239–40.
133. Magil AB. Drug-induced acute interstitial nephritis with granulomas. Hum Pathol 1983;14(1):36–41.
134. Fogari R, Zoppi A. Effects of antihypertensive therapy on sexual activity in hypertensive men. Curr Hypertens Rep 2002;4(3):202–10.
135. Loke Y. A systematic review of the benefits and harms of antihypertensive drug therapy. European Council for Blood Pressure and Cardiovascular Research Meeting, Holland 9–11 Oct 1999. Hypertension 1999;34(4):710.
136. Grimm RH Jr, Cohen JD, Smith WM, Falvo-Gerard L, Neaton JD. Hypertension management in the Multiple Risk Factor Intervention Trial (MRFIT). Six-year intervention results for men in special intervention and usual care groups. Arch Intern Med 1985;145(7):1191–9.
137. Helgeland A, Strommen R, Hagelund CH, Tretli S. Enalapril, atenolol, and hydrochlorothiazide in mild to moderate hypertension. A comparative multicentre study in general practice in Norway. Lancet 1986;1(8486):872–5.
138. Juang P, Page RL II, Zolty R. Probable loop diuretic-induced pancreatitis in a sulfonamide-allergic patient. Ann Pharmacother 2006;40:128–34.
139. Ruscin JM, Page RL, Scott J. Hydrochlorothiazide-induced angioedema in a patient allergic to sulfonamide antibiotics: evidence from a case report and a review of the literature. Am J Geriatr Pharmacother 2006;4:325–9.
140. Strom BL, Schinnar R, Apter AJ, Margolis DJ, Lautenbach E, Hennessy S, Bilker WB, Pettitt D. Absence of cross-reactivity between sulfonamide antibiotics and sulfonamide nonantibiotics. N Engl J Med 2003;349:1628–35.
141. Eshaghian S, Horwich TB, Fonarow GC. Relation of loop diuretic dose to mortality in advanced heart failure. Am J Cardiol 2006;97:1759–64.
142. Wassertheil-Smoller S, Blaufox MD, Oberman A, Davis BR, Swencionis C, Knerr MO, Hawkins CM, Langford HG. Effect of antihypertensives on sexual function and quality of life: the TAIM Study. Ann Intern Med 1991;114(8):613–20.
143. Smith PJ, Talbert RL. Sexual dysfunction with antihypertensive and antipsychotic agents. Clin Pharm 1986;5(5):373–84.
144. Prisant LM, Carr AA, Bottini PB, Solursh DS, Solursh LP. Sexual dysfunction with antihypertensive drugs. Arch Intern Med 1994;154(7):730–6.
145. Kaufmann H, Elijovich F, Yahr MD. An unusual cause of tetany: surreptitious use of furosemide. Mt Sinai J Med 1984;51(5):625–8.
146. De Wardener HE. Idiopathic edema: role of diuretic abuse. Kidney Int 1981;19(6):881–91.
147. Dunnigan MG, Denning DW, Henry JA, de Wolff FA. Idiopathic oedema and diuretics. Postgrad Med J 1987;63(735):25–6.
148. Pelosi AJ, Sykes RA, Lough JR, Muir WJ, Dunnigan MG. A psychiatric study of idiopathic oedema. Lancet 1986; 2(8514):999–1002.
149. Young JB, Brownjohn AM, Lee MR. Diuretics and idiopathic oedema. Nephron 1986;43(4):311–2.
150. Marty H. Pseudo-Bartter Syndrom bei Diuretika-Abusus. [Pseudo-Bartter syndrome in diuretics abuse.] Schweiz Med Wochenschr 1985;115(7):250–2.
151. Lopez Jimenez M, Barbado FJ, Mateos F, Pena JM, Gil A, Arnalich F, Tovar I, Alonso FG, Vazquez Rodriguez JJ. Sindrome de Bartter factitio inducido por la ingestion subrepticia de diureticos. [Factitious Bartter's syndrome induced by the surreptitious ingestion of diuretics.] Med Clin (Barc) 1985;84(1):23–6.
152. Schmieder RE, Delles C, Messerli FH. Diuretic therapy and the risk for renal cell carcinoma. J Nephrol 2000; 13(5):343–6.
153. Grossman E, Messerli FH, Goldbourt U. Does diuretic therapy increase the risk of renal cell carcinoma? Am J Cardiol 1999;83(7):1090–3.
154. Lee IM, Hennekens CH. Diuretics and renal cell carcinoma. Am J Cardiol 1999;83(7):1094.

155. Lip GY, Ferner RE. Diuretic therapy for hypertension: a cancer risk? J Hum Hypertens 1999;13(7):421–3.
156. Tenenbaum A, Grossman E, Fisman EZ, Adler Y, Boyko V, Jonas M, Behar S, Motro M, Reicher-Reiss H. Long-term diuretic therapy in patients with coronary disease: increased colon cancer-related mortality over a 5-year follow-up. J Hum Hypertens 2001;15(6):373–9.
157. Chesley LC. Preeclampsia: Early signs and development. In: Hypertensive Disorders in Pregnancy. New York: Appleton-Century-Crofts, 1978:302–6.
158. Thomson AM, Hytten FE, Billewicz WZ. The epidemiology of oedema during pregnancy. J Obstet Gynaecol Br Commonw 1967;74(1):1–10.
159. Friedman EA, Neff RK. Pregnancy outcome as related to hypertension, edema and proteinuria. Hypertension in PregnancyNew York: John Wiley and Sons;. 1976.
160. Campbell DM, MacGillivray I. The effect of a low calorie diet or a thiazide diuretic on the incidence of pre-eclampsia and on birth weight. Br J Obstet Gynaecol 1975;82(7):572–7.
161. Beilin LJ, Redman CWG. The use of antihypertensive drugs in pregnancy. In: Lewis PJ, editor. Therapeutic Problems in Pregnancy. Lancaster: MTP Press, 1977:1.
162. Gant NF, Madden JD, Siteri PK, MacDonald PC. The metabolic clearance rate of dehydroisoandrosterone sulfate. III. The effect of thiazide diuretics in normal and future pre-eclamptic pregnancies. Am J Obstet Gynecol 1975;123(2):159–63.
163. Zuspan FP, Zuspan KJ, Wilson AL. Acute and chronic hypertension in pregnancy. In: Rayburn WF, Zuspan FP, editors. Drug Therapy in Obstetrics and Gynecology. Norwalk, CT: Appleton-Century-Crofts, 1982:65.
164. Collins R, Yusuf S, Peto R. Overview of randomised trials of diuretics in pregnancy. BMJ (Clin Res Ed) 1985;290(6461):17–23.
165. Boethius G. Recording of drug prescriptions in the county of Jamtland, Sweden. II. Drug exposure of pregnant women in relation to course and outcome of pregnancy. Eur J Clin Pharmacol 1977;12(1):37–43.
166. Brocklebank JC, Ray WA, Federspiel CF, Schaffner W. Drug prescribing during pregnancy. A controlled study of Tennessee Medicaid recipients. Am J Obstet Gynecol 1978;132(3):235–44.
167. Doering PL, Stewart RB. The extent and character of drug consumption during pregnancy. JAMA 1978;239(9):843–6.
168. Garnet JD. Placental transfer of chlorothiazide. Obstet Gynecol 1963;21:123–5.
169. Goldman JA, Neri A, Ovadia J, Eckerling B, de Vries A. Effect of chlorothiazide on intravenous glucose tolerance in pregnancy. Am J Obstet Gynecol 1969;105(4):556–60.
170. Pritchard JA, Walley PJ. Severe hypokalemia due to prolonged administration of chlorothiazide during pregnancy. Am J Obstet Gynecol 1961;81:1241–4.
171. Anderson GG, Hanson TM. Chronic fetal bradycardia: possible association with hypokalemia. Obstet Gynecol 1974;44(6):896–8.
172. Senior B, Slone D, Shapiro S, Mitchell AA, Heinonen OP. Letter: Benzothiadiazides and neonatal hypoglycaemia. Lancet 1976;2(7981):377.
173. Rodriguez SU, Leikin SL, Hiller MC. Neonatal thrombocytopenia associated with ante-partum administration of thiazide drugs. N Engl J Med 1964;270:881–4.
174. Combe R. Perturbations sanguines chez un premature apres administration de guanéthidiine et hydrochloride au cours de la grossesse. [Blood disorders in a premature infant after guanethidine and hydrochlorothiazide administration during pregnancy.] Pédiatrie 1978; 33(6):599–601.
175. Harley JD, Robin H, Robertson SE. Thiazide-induced neonatal haemolysis? BMJ 1964;5384:696–7.
176. Worsham F Jr, Beckman EN, Mitchell EH. Sacrococcygeal teratoma in a neonate. Association with maternal use of acetazolamide. JAMA 1978;240(3):251–2.
177. Diokno AC, Brown MB, Herzog AR. Relationship between use of diuretics and continence status in the elderly. Urology 1991;38(1):39–42.
178. Myers MG, Kearns PM, Kennedy DS, Fisher RH. Postural hypotension and diuretic therapy in the elderly. Can Med Assoc J 1978;119(6):581–4.
179. Burke V, Beilin LJ, German R, Grosskopf S, Ritchie J, Puddey IB, Rogers P. Postural fall in blood pressure in the elderly in relation to drug treatment and other lifestyle factors. Q J Med 1992;84(304):583–91.
180. Applegate WB. Hypertension in elderly patients. Ann Intern Med 1989;110(11):901–15.
181. Nicholls MG. Age-related effects of diuretics in hypertensive subjects. J Cardiovasc Pharmacol 1988;12(Suppl 8):S51–9.
182. Messerli FH, Grossman E, Goldbourt U. Are beta-blockers efficacious as first-line therapy for hypertension in the elderly? A systematic review. JAMA 1998;279(23):1903–7.
183. Moro C, Romero J, Corres Peiretti MA. Amiodarone and hypokalemia. A dangerous combination. Int J Cardiol 1986;13(3):365–8.
184. Jarchovsky J, Zamir D, Plavnik L. [Torsade de pointes and use of phenothiazine and diuretic.]Harefuah 1991;121(11):435–6.

Amiloride

General Information

Amiloride is a potassium-sparing diuretic that acts in the distal convoluted tubule independently of the action of aldosterone, inhibiting sodium channels. It is a relatively safe drug with few reported adverse effects.

Patients with congenital nephrogenic diabetes insipidus are often treated with a combination of a thiazide and a potassium-sparing diuretic, without consensus on the preferred potassium-sparing diuretic. A Japanese adult was systematically studied to determine the renal effects of hydrochlorothiazide plus amiloride and hydrochlorothiazide plus triamterene (1). The combination with amiloride was superior to that with triamterene in preventing excessive urinary potassium loss, hypokalemia, and metabolic alkalosis. These results suggest that amiloride is the preferred add-on therapy to hydrochlorothiazide in nephrogenic diabetes insipidus.

Organs and Systems

Cardiovascular

An isolated report (2) suggests that amiloride may have prodysrhythmic potential in a small proportion of patients with inducible sustained ventricular tachycardia.

Electrolyte balance

Amiloride has a specific effect on sodium flux in the renal tubules; severe hyponatremia has been reported with the combination of a thiazide diuretic and amiloride.

As a potassium-sparing diuretic, amiloride can cause hyperkalemia (3), even in patients who are taking a potassium-wasting diuretic (4). This effect can be enhanced by concomitant therapy with ACE inhibitors or angiotensin-II receptor antagonists. In five patients with diabetes mellitus over 50 years of age who were taking an ACE inhibitor the serum potassium rose markedly 8–18 days after the addition of amiloride (5). All but one had some degree of renal impairment. In four cases potassium concentrations were between 9.4 and 11 mmol/l.

Skin

Rashes with diarrhea and eosinophilia have been reported (6).

Immunologic

Sweden's National Adverse Reaction Monitoring System called the attention of physicians to a case in which a 73-year-old woman developed anaphylactic shock after taking only a single tablet of Moduretic (7). The National System had received three other reports on anaphylactic reactions to this combination, and by 1988 the WHO center in Uppsala had received eight others from other countries. Except for mild skin reactions, hypersensitivity reactions to hydrochlorothiazide alone are highly unusual and not of this type.

Drug–Drug Interactions

ACE inhibitors

Co-administration of potassium-sparing diuretics with ACE inhibitors can cause severe hyperkalemia (SED-14, 674). In a retrospective study, five patients developed extreme hyperkalemia (9.4–11 mmol/l) within 8–18 days of starting combination therapy with co-amilozide (amiloride + hydrochlorothiazide) and an ACE inhibitor (5).

Amphotericin

Amiloride is a therapeutic option in reducing potassium losses in patients receiving amphotericin. When it was given to 19 oncology patients with marked amphotericin-induced potassium depletion mean serum potassium concentrations increased in the 5 days before and after administration (from 3.4 to 3.9 mmol/l) (8). There was also a trend toward reduced potassium supplementation (48 versus 29 mmol/day). Adverse reactions were limited to hyperkalemia in two patients who took amiloride 20 mg/day and a high potassium intake.

Carbenoxolone

Amiloride inhibits the ulcer-healing properties of carbenoxolone (9).

Lithium

Although amiloride may reduce the renal clearance of lithium, it appears to be free of the troublesome interaction with lithium that complicates the use of thiazides and loop diuretics.

Pentamidine

Pentamidine is structurally similar to amiloride and can cause severe hyperkalemia if co-prescribed with potassium-sparing diuretics (10). This is a particularly important interaction in patients with AIDS.

Quinidine

Quinidine has a pharmacodynamic interaction with amiloride, further prolonging the duration of the QRS complex, but not the QT interval (11).

Trimethoprim

Trimethoprim (SED-14, 675) is structurally similar to amiloride and can cause severe hyperkalemia if co-prescribed with potassium-sparing diuretics (10). This is a particularly important interaction in patients with AIDS.

Interference with Diagnostic Tests

Creatinine clearance

Amiloride does not alter renal function (12), but it does block the tubular secretion of creatinine, leading to falsely high measurements of creatinine clearance; inulin clearance is not affected similarly (6).

References

1. Konoshita T, Kuroda M, Kawane T, Koni I, Miyamori I, Tofuku Y, Mabuchi H, Takeda R. Treatment of congenital nephrogenic diabetes insipidus with hydrochlorothiazide and amiloride in an adult patient. Horm Res 2004; 61(2):63–7.
2. Duff HJ, Mitchell LB, Kavanagh KM, Manyari DE, Gillis AM, Wyse DG. Amiloride. Antiarrhythmic and electrophysiologic actions in patients with inducible sustained ventricular tachycardia. Circulation 1989;79(6):1257–63.
3. Maddox RW, Arnold WS, Dewell WM Jr. Extreme hyperkalemia associated with amiloride. South Med J 1985; 78(3):365.
4. Jaffey L, Martin A. Malignant hyperkalaemia after amiloride/hydrochlorothiazide treatment. Lancet 1981;1(8232): 1272.
5. Chiu TF, Bullard MJ, Chen JC, Liaw SJ, Ng CJ. Rapid life-threatening hyperkalemia after addition of amiloride HCl/ hydrochlorothiazide to angiotensin-converting enzyme inhibitor therapy. Ann Emerg Med 1997;30(5):612–5.
6. Vidt DG. Mechanism of action, pharmacokinetics, adverse effects, and therapeutic uses of amiloride hydrochloride, a new potassium-sparing diuretic. Pharmacotherapy 1981; 1(3):179–87.
7. Anonymous. Moduretic-anaphylactic shock. Uppsala: WHO Collaborating Centre for Adverse Drug Reaction Monitoring. Adv React Newslett 1988;(3–4):.

8. Bearden DT, Muncey LA. The effect of amiloride on amphotericin B-induced hypokalaemia. J Antimicrob Chemother 2001;48(1):109–11.
9. Reed PI, Lewis SI, Vincent-Brown A, Holdstock DJ, Gribble RJ, Murgatroyd RE, Baron JH. The influence of amiloride on the therapeutic and metabolic effects of carbenoxolone in patients with gastric ulcer. A double-blind controlled trial. Scand J Gastroenterol Suppl 1980;65:51–7.
10. Perazella MA. Drug-induced hyperkalemia: old culprits and new offenders. Am J Med 2000;109(4):307–14.
11. Wang L, Sheldon RS, Mitchell LB, Wyse DG, Gillis AM, Chiamvimonvat N, Duff HJ. Amiloride–quinidine interaction: adverse outcomes. Clin Pharmacol Ther 1994;56(6 Pt 1):659–67.
12. Maronde RF, Milgrom M, Vlachakis ND, Chan L. Response of thiazide-induced hypokalemia to amiloride. JAMA 1983;249(2):237–41.

Bemetizide

General Information

Bemetizide is chemically unrelated to the thiazides, but it shares many of their actions and adverse effects.

Susceptibility Factors

Age

The pharmacokinetics and pharmacodynamics of a fixed combination of bemetizide 25 mg and triamterene 50 mg have been evaluated in 15 elderly patients (aged 70–84 years) and 10 young volunteers (aged 18–30 years) after single doses (on day 1) and multiple doses (at steady state on day 8) (1). Mean plasma concentrations of bemetizide, triamterene, and the active metabolite of triamterene, hydroxytriamterene, were significantly higher in the elderly subjects after single and multiple doses, and urine flow and sodium excretion rates fell in tandem with the accumulation of these drugs. The glomerular filtration rate, which is reduced in elderly people, was further reduced at higher concentrations of bemetizide and triamterene, which may explain why there were limited diuretic and saliuretic effects after multiple doses. This study clearly points to a modulating effect of the degree of renal function on the diuretic actions of these compounds in elderly people.

Drug–Drug Interactions

Triamterene

In 1988 the Federal German Health Authorities issued a warning about fixed combinations of the thiazide bemetizide and triamterene that they could cause allergic vasculitis (2). Thiazides can occasionally cause allergic vasculitis and it is not clear on what basis the combination might be more likely to cause the same problem.

References

1. Muhlberg W, Mutschler E, Hofner A, Spahn-Langguth H, Arnold O. The influence of age on the pharmacokinetics and pharmacodynamics of bemetizide and triamterene: a single and multiple dose study. Arch Gerontol Geriatr 2001;32(3):265–73.
2. Anonymous. Bemetizide/triamterene: warning of allergic vasculitis. WHO Drug Inform 1988;2:148.

Bumetanide

General Information

Bumetanide is very similar in most respects to furosemide, although it is more potent mole for mole. An oral dose of 1–4 mg often suffices for ordinary use compared with 40–160 mg of furosemide.

Most of the adverse effects of bumetanide are the same as those of furosemide. However, it is less ototoxic and can cause muscle cramps, which furosemide does not.

Organs and Systems

Respiratory

Bumetanide has been implicated in the development of pulmonary fibrosis (SEDA-16, 223), but causality could not be determined with certainty.

Sensory systems

One advantage of bumetanide is that it is less ototoxic than furosemide (1–3). It is sensible to prefer bumetanide to furosemide in patients with hearing problems or who concurrently need ototoxic drugs, such as an aminoglycoside antibiotic.

Metabolism

It has been suggested that bumetanide has a smaller effect on blood glucose than furosemide does, but that is not at all clear.

Gastrointestinal

Very high doses of bumetanide used for renal insufficiency have sometimes produced colic (SED-10, 376).

Pancreas

A patient in whom pancreatitis was induced by both furosemide and bumetanide has been documented (SEDA-14, 184).

Skin

Bullous pemphigoid induced by bumetanide has been reported (4).

- A 67-year-old man presented with an acute bullous eruption 6 weeks after starting bumetanide. He had numerous large tense bullae on erythematous skin, with superficial ulceration on the thighs, arms, and anterior trunk. Pruritus was severe. Routine laboratory tests were normal, except for blood eosinophilia. Biopsy of a blister showed subepidermal bullae associated with dermal infiltrates of neutrophils and eosinophils. Direct immunofluorescence showed continuous linear deposits of C3 and IgG at the basement membrane zone, confirmed by immunoelectron microscopy. Circulating IgG antibasement membrane antibodies were localized in the roof of the blister. Compete clinical healing and normalization of immunology occurred within 2 months of withdrawal of bumetanide.

Pseudoporphyria (SEDA-16, 222) and Stevens–Johnson syndrome have also been reported (SED-10, 376).

Musculoskeletal

Various musculoskeletal symptoms have been reported with bumetanide, including pain, cramping, and weakness (SEDA-22, 236). The symptoms are usually mild and self-limiting, with an incidence of less than 2%, but disabling reactions have occurred after oral administration and intravenous injection, and severity correlates with dose. The onset is usually 2–4 hours after dosing; the onset may be slower during infusion.

References

1. Tuzel IH. Comparison of adverse reactions to bumetanide and furosemide. J Clin Pharmacol 1981;21(11–12 Pt 2):615–9.
2. Halstenson CE, Matzke GR. Bumetanide: a new loop diuretic (Bumex, Roche Laboratories). Drug Intell Clin Pharm 1983;17(11):786–97.
3. Ward A, Heel RC. Bumetanide. A review of its pharmacodynamic and pharmacokinetic properties and therapeutic use. Drugs 1984;28(5):426–64.
4. Boulinguez S, Bernard P, Bedane C, le Brun V, Bonnetblanc JM. Bullous pemphigoid induced by bumetanide. Br J Dermatol 1998;138(3):548–9.

Carbonic anhydrase inhibitors

General Information

The carbonic anhydrase inhibitors, of which acetazolamide (rINN), a non-competitive inhibitor, is the prototype, are not suitable for normal diuretic use, because tolerance soon develops. However, they are well suited to brief intermittent use, particularly in the relief of glaucoma and in the prevention of acute mountain sickness. Acetazolamide and methazolamide (rINN) should be used with caution in the long-term control of glaucoma because of its serious systemic adverse effects. However, brinzolamide (rINN) and dorzolamide (rINN) are available for long-term topical administration.

The safety profile and efficacy of 2% dorzolamide hydrochloride (Trusopt) eye-drops have been evaluated. It was as effective as pilocarpine 2% and its ocular hypotensive efficacy was comparable with that of betaxolol 0.5%. The patients reported less interference with quality of life with dorzolamide than pilocarpine, particularly in regard to limitations in their ability to drive, read, and perform moderate activities. Long-term use was not associated with important electrolyte disturbances or the systemic effects commonly observed with oral carbonic anhydrase inhibitors (1–3).

In a 3-month prospective study of the adverse effects and efficacy of topical dorzolamide in 39 patients intolerant of systemic carbonic anhydrase inhibitors, the effect on mean intraocular pressure was similar to that of acetazolamide, and health assessment scores improved significantly in seven of the eight categories of the SF-36 health assessment questionnaire used to evaluate changes in well-being and quality of life (4). There were no adverse effects with the switch in medication.

General adverse effects

The incidence and severity of many adverse reactions to carbonic anhydrase inhibitors are dose-related and the problems usually abate when the dose is reduced or the drug is withdrawn. Symptoms of depression, confusion, fatigue, impotence, irritability, malaise, nervousness, and weight loss are often present to some extent in patients on long-term acetazolamide or methazolamide therapy. These symptoms may be related to systemic metabolic acidosis (due to renal excretion of bicarbonate), which is often accompanied by a reduction in serum potassium concentration (5). Gastrointestinal intolerance, manifested as abdominal cramping, dyspepsia, and nausea, with or without diarrhea, is another common problem. Carbonic anhydrase inhibitors reduce the urinary excretion of citrate and uric acid, which may lead to renal calculi and gouty arthritis. Pulmonary edema, taste disorders, and alopecia have also been reported. The most serious adverse effect of carbonic anhydrase inhibitors is bone-marrow depression.

Organs and Systems

Cardiovascular

Anaphylactic shock has been reported after a single oral dose of acetazolamide (6).

- A 70-year-old man was given acetazolamide 250 mg to control postoperative intravascular pressure 5 hours after cataract removal under local anesthetic. Thirty minutes later he complained of nausea, became cyanotic, and had an acute respiratory arrest. His systolic blood pressure was 70 mmHg, his heart rate was 180/minute, and there was tachypnea (40 breaths/minute). Arterial gases confirmed hypoxemia (PaO_2 6.34 kPa, 47 mmHg). Pulmonary embolism and high-pressure pulmonary edema were excluded by perfusion lung scanning andsided heart catheterization. Management was with ventilatory support, vasopressors, intravenous hydrocortisone, and diphenhydramine. Clinical improvement occurred over 12 hours. After stabilization, sulfonamide hypersensitivity was confirmed by skin testing, suggesting cross-sensitivity with a sulfonamide derivative (acetazolamide).

Physicians should be aware of the risk of anaphylaxis to acetazolamide, particularly in patients with a history of allergy to sulfonamides.

Nervous system

A case of possible Gerstmann syndrome has been attributed to acetazolamide (7).

- A 60-year-old woman became acutely confused 2 days after the removal of a cataract. She had long-standing diabetes mellitus, hypertension, and ischemic heart disease. There had been a minor stroke with complete recovery 2 years before. Her medication included aspirin, indapamide, enalapril, and oral hypoglycaemic agents. Acetazolamide 500 mg bd was added shortly after her eye operation. Neurological examination showed finger agnosia, nominal and receptive dysphasia, acalculia, astereognosis, and left–right disorientation. Other systems were normal. Withdrawal of acetazolamide resulted in rapid improvement with no residual neurological signs two days after admission.

Acetazolamide toxicity was suspected, because of the temporal association between drug treatment and the onset of the neurological symptoms, together with metabolic acidosis. Gerstmann syndrome is usually due to an acute stroke. Although a brain CT scan was negative, such an event was likely in this patient, who had a history of cerebrovascular disease and multiple risk factors, and a causal relation to acetazolamide must be considered tenuous.

Acetazolamide typically causes increased cerebral blood flow as a result of vasodilatation. In patients with stenosis or occlusion of cerebral vessels, perfusion pressure drops and cerebral autoregulatory mechanisms are called into play to maintain cerebral blood flow. In such patients, cerebral blood flow may not increase appropriately after acetazolamide, and it may redistribute, potentially stealing blood from areas that are liable to underperfusion injury (stroke).

- A 62-year-old woman with symptoms compatible with vertebral artery disease underwent a CT perfusion scan in anticipation of possible vertebral artery bypass (8). She underwent a routine unenhanced CT scan, followed by dynamic enhanced perfusion CT before and after intravenous administration of acetazolamide 1000 mg. Almost immediately she noted dizziness and perioral numbness and shortly thereafter slurred speech and ataxia. Her symptoms resolved entirely within 24 hours without any therapy. It was thought that she had had a transient ischemic attack possibly related to a steal phenomenon.

In this case, the shift in regional cerebral blood flow produced by acetazolamide resulted in significant underperfusion. Patients with significant cerebrovascular disease should be monitored carefully when they are given acetazolamide.

Acetazolamide is used to reduce the frequency of attacks of ataxia in patients with episodic ataxia type 2. However, the metabolic acidosis that acetazolamide causes can result in nervous system complications (9).

- A 49-year-old man with episodic ataxia type 2 responded to acetazolamide 250 mg qds. However, during an attack of ataxia he developed gaze-evoked nystagmus, positional nystagmus, dysarthria, and aggravated gait ataxia; he was also excessively drowsy and unable to speak unless subjected to painful stimuli. Imaging studies and laboratory data were normal, except for hyperammonemia (1.58 mg/l, reference range 0.08–0.48 mg/l) and a compensated metabolic acidosis. Acetazolamide was continued, but 3 days later the ammonia concentration increased to 6.6 mg/l, with worsening of the ataxia and dysarthria. Acetazolamide was withdrawn, and within 10 days his symptoms had almost completely resolved and the serum ammonia concentration had fallen to 1.22 mg/l.

This is the first report of ataxia secondary to acetazolamide-induced hyperammonemia in a patient with episodic ataxia type 2. Acetazolamide increases the renal production of NH_3, and the consequent metabolic acidosis can cause hyperammonemia. Acetazolamide should therefore be used cautiously in patients with a history of hepatic or renal disease (conditions that predispose to hyperammonemia).

Metabolic acidosis due to acetazolamide causes increased minute ventilation, which can cause increased intracranial pressure and result in neurological complications (10).

- A 19-year-old woman with postoperative bilateral raised intracranial pressure was given intravenous acetazolamide 500 mg followed by 250 mg every 6 hours. After 3 days she developed a metabolic acidosis (serum HCO_3 18 mmol/l), which was attributed to acetazolamide. The metabolic acidosis progressed over the next 3 days (HCO_3 15 mmol/l) with appropriate hypocapnia (P_aCO_2 3.2–3.6 kPa). After 6 days she became agitated, and a propofol infusion was begun, but her mechanical respiratory rate was not increased to maintain her prior hypocapnia. After 5 hours her P_aCO_2 was 4.7 kPa and her arterial pH was 7.26. She developed extreme hypertension and a tachycardia. A CT scan of the brain showed cerebral edema, brain stem herniation, and bilateral watershed infarcts. At post mortem there were multiple fat emboli in the brain and lungs.

The use of acetazolamide in the presence of unrecognized cerebral edema due to fat embolism, with sudden normalization of brain CO_2, as occurred in this patient when her previous state of hypocapnia was no longer sustained by ventilatory effort, resulted in cerebral acidosis, vasodilatation, and a further increase in intracranial pressure. This proved catastrophic and led to brainstem herniation and brain death. Acetazolamide should be avoided if at all possible in patients with bony and traumatic brain injuries, particularly during weaning from mechanical ventilation, since it can precipitate coning in patients with raised intracranial pressure.

Sensory systems

Eyes

In a multicenter, double-blind, prospective, parallel-group comparison of brinzolamide 1.0% bd or tds, dorzolamide

2.0% tds, and timolol 0.5% bd in 572 patients with primary open-angle glaucoma or ocular hypertension, the three drugs were equally effective (11). Brinzolamide 1.0% caused less ocular discomfort (burning and stinging) (bd 1.8%; tds 3.0%) than dorzolamide (16%).

- A 68-year-old woman developed bilateral marginal keratitis 2 weeks after starting to use dorzolamide eye-drops (12). One week after withdrawal she was asymptomatic, with complete resolution of her corneal infiltrates.

In this case the allergic reaction was caused by dorzolamide hydrochloride and not the preservative, benzalkonium chloride, since therapy was uneventfully continued with timolol maleate, which also contains benzalkonium chloride as a preservative. This is the first report of this phenomenon with a carbonic anhydrase inhibitor.

Acetazolamide-induced transient myopia and angle closure glaucoma can occur in patients without glaucoma (13).

- A 66-year-old man with chronic open-angle glaucoma underwent routine left cataract surgery and intraocular lens implantation. He was given oral acetazolamide 250 mg the evening before and immediately after surgery and 3 hours later developed severe left eye pain. He had corneal edema, a uniformly shallow anterior chamber, and an intraocular pressure of 52 mmHg. The right eye showed circumcorneal congestion, corneal edema, and an intraocular pressure of 40 mmHg. He was treated with intravenous mannitol, oral acetazolamide, intraocular timolol, ofloxacin, and dexamethasone. After 24 hours the corneal edema and shallow anterior chamber persisted, with an intraocular pressure of 32 mmHg in each eye. Acetazolamide was withdrawn. Ultrasound showed bilateral choroidal effusions and intravenous dexamethasone was begun. On day 5, all ocular signs had resolved and the intraocular pressure was 18 mmHg bilaterally.

Angle closure glaucoma has been reported as an adverse effect of acetazolamide. It has been suggested that this relates to an induced forward shift of the crystalline lens in addition to ciliary body edema. Although glucocorticoids have been reported to cause choroidal effusions, an exaggeration of this phenomenon by the combination of a glucocorticoid and acetazolamide has not been previously reported.

Topical brinzolamide has been associated with corneal edema (14,15).

- A 57-year-old man with glaucoma used a 1% solution of brinzolamide in both eyes for 24 months and developed bilateral corneal edema, which resolved 1 week after withdrawal of brinzolamide and the addition of topical prednisolone acetate.
- A 77-year-old man with glaucoma used a 1% solution of brinzolamide in his right eye for 15 months and developed right corneal edema. Brinzolamide was withdrawn and the corneal edema gradually resolved over 3 months.

There has been a limited number of previous case reports of resolution of corneal edema with brinzolamide. When corneal edema occurs with brinzolamide immediate withdrawal of therapy, with or without the use of topical glucocorticoids, can lead to complete resolution.

Eyes that have not undergone surgery have not been reported to undergo choroidal detachment with topical hypotensive agents.

- A 76-year-old woman with a 7-year history of open-angle glaucoma presented with distorted visual acuity after applying two doses of dorzolamide eye-drops 2% bd to both eyes (16). Other systemic medications included hydrochlorothiazide and verapamil. There was a peripheral choroidal detachment in the left eye, which completely resolved 1 week after withdrawal of dorzolamide together with topical glucocorticoid therapy.

In this case, the systemic hydrochlorothiazide may have sensitized the choroidal epithelium sufficiently to result in aqueous fluid shutdown.

Smell

Anosmia has been attributed to topical dorzolamide (17)

- A 49-year-old man with glaucoma was given 2% dorzolamide in addition to timolol. After 1 month he developed reduced smell and after 2 months anosmia. Dorzolamide was withdrawn and latanoprost substituted. His sense of smell returned to normal within 20 days. Rechallenge with several different glaucoma medications resulted in recurrence of the anosmia whenever dorzolamide was used.

Carbonic anhydrase exists in multiple forms in the nasal mucosa, where its inhibition may have resulted in anosmia. However, the absence of previous reports of anosmia with dorzolamide suggests a patient-specific isoenzyme variability/susceptibility to carbonic anhydrase inhibition.

Taste

Acetazolamide can cause altered taste perception of carbonated drinks. The mechanism is unknown, but carbonic drinks may be a source of high concentrations of carbonic acid, which are reduced by carbonic anhydrase activity (SEDA-15, 219).

Endocrine

Thyrotoxic periodic paralysis occurs only occasionally in Caucasians, but for unknown reasons it can be worsened by acetazolamide (SEDA-15, 219).

Electrolyte balance

The carbonic anhydrase inhibitors are the most kaliuretic of all diuretics and can cause severe hypokalemia during the first few days of administration. However, hypokalemia is not a problem during long-term administration, because of compensatory potassium retention secondary to acidosis (18).

- A 46-year-old man with hypokalemic periodic paralysis and diabetes mellitus had worse muscle weakness after taking acetazolamide, possibly because of reduced muscle uptake of potassium (19).

Children with heart disease often require high-dose diuretic therapy, which can lead to hypochloremic metabolic alkalosis. There are limited data on the safety of acetazolamide in the treatment of hypochloremic metabolic alkalosis in children. In 28 patients, median age 2 (range 0.3–20) months who took acetazolamide 5 mg/kg for 3 days, there were no adverse events (20). There was no significant difference in any electrolyte concentration, except for serum HCO_3, which fell from 36 to 31 mmol/l, and serum chloride, which rose from 91 to 95 mmol/l. There was no change in urine output. Acetazolamide appears to be safe in very young patients when given for 3 consecutive days.

Acid–base balance

Acetazolamide can produce severe lactic acidosis, with an increased lactate:pyruvate ratio, ketosis with a low beta-hydroxybutyrate:acetoacetate ratio, and a urinary organic acid profile consistent with pyruvate carboxylase deficiency. The acquired enzymatic injury that results from inhibition of mitochondrial carbonic anhydrase V, which provides bicarbonate to pyruvate carboxylase, can damage the tricarboxylic acid cycle.

Four preterm neonates with posthemorrhagic ventricular dilatation developed severe metabolic acidosis after being given acetazolamide (21). The acidosis suddenly disappeared after a transfusion of packed erythrocytes, which was attributed to the citrate contained in the blood.

Acetazolamide can cause a metabolic acidosis in 50% of elderly patients (SEDA-11, 199); occasionally (particularly if salicylates are being given or renal function is poor) the acidosis can be severe. It does this by inhibiting renal bicarbonate reabsorption. This effect is of particular use in treating patients with chronic respiratory acidosis with superimposed metabolic alkalosis. Life-threatening metabolic acidosis is rarely observed in the absence of renal insufficiency and/or diabetes mellitus. In three patients with central nervous system pathology alone conventional doses of acetazolamide resulted in severe metabolic acidosis (22). After withdrawal it took up to 48 hours for the metabolic acidosis and accompanying hyperventilation to resolve.

Metabolic acidosis has also been described with the topical carbonic anhydrase inhibitor dorzolamide.

- A 5-day-old boy, weighing 2.3 kg, developed a metabolic acidosis after receiving topical dorzolamide for 1 week for bilateral Peter's anomaly (a congenital corneal disorder characterized by a central leukoma and adhesions at the periphery of the corneal opacity) (23). The maximum base deficit was 20.2 mmol/l. On withdrawal of topical dorzolamide his acidosis resolved within 1 day.

Factors that may have contributed to this metabolic acidosis included low birth weight, renal tubular immaturity, and impaired renal function, which may have resulted in systemic accumulation with repetitive dosing. This case stresses the fact that topical medications can cause systemic effects if a sufficient amount of drug is absorbed in a susceptible subject.

Topical brinzolamide can be systemically absorbed and cause systemic adverse effects. In one case it was reported to have caused a metabolic acidosis (24).

- A 66-year-old man with glaucoma was given brinzolamide twice daily to only one eye in addition to topical latanoprost. After 3 months he reported having stopped using brinzolamide because of lethargy and a bad taste in the mouth. He also reported that a silver chain worn around his neck had turned black within 2 days of starting brinzolamide and again soon after it had been cleaned. Electrolyte and acid-base status were not assessed.

Although a metabolic acidosis could occur with topical brinzolamide if sufficient drug were systemically absorbed, a change in the color of a silver chain worn around the neck is hardly an adequate basis for a diagnosis of a metabolic acidosis.

Fluid balance

Acetazolamide can cause rapid volume changes.

- A 47-year-old woman with diabetes who took acetazolamide 250 mg bd for 6 days) for left cystoid macular edema developed profound hyperosmolar non-ketotic hyperglycemia (25). This occurred as the result of marked diuresis and an associated fall in glomerular filtration rate.

Hematologic

Aplastic anemia occurs in about one in 18 000 patient-years of exposure to oral carbonic anhydrase inhibitors. Most cases occur within the first 6 months, and peak at 2–3 months.

Eleven cases of acetazolamide-associated aplastic anemia were reported in Sweden during a 17-year period (26). The median dose was 500 mg/day and the median duration of therapy was 3 (2–71) months. Ten of the eleven patients died within 8 weeks of diagnosis. The relative risk of aplastic anemia with acetazolamide was 13.3 (95% CI = 6.8, 25) and the estimated frequency was 1 in 18 000. These findings suggest that acetazolamide is associated with a substantially increased risk of aplastic anemia.

However, the National Registry of Drug-Induced Ocular Side Effects has not received a report of blood dyscrasias in patients who have taken oral carbonic anhydrase inhibitors for less than 2 weeks (27). Since there is generally some time before abnormal blood counts progress to bone-marrow failure, seeking a symptomless "early window" is precisely the point of routine hematological screening; thus, patients taking long-term oral carbonic anhydrase inhibitors should have erythrocyte, leukocyte, and platelet counts bimonthly during the first 6 months of therapy and then every 6 months thereafter. Routine hematological surveillance should always be accompanied by patient education about warning signs and symptoms of progressive marrow failure.

Acetazolamide can cause rapid hematological changes, including thrombocytopenia.

- A 67-year-old man developed isolated thrombocytopenia (platelet count 31×10^9/l) after taking acetazolamide 250/mg day for 2 days for raised intraocular pressure (28). Several months later acetazolamide 375 mg/day was prescribed again and 2 weeks later he developed extensive purpura and a platelet count of 3×10^9/l. Acetazolamide was withdrawn and the platelet count rose spontaneously to 20, 73, and 246×10^9/l after 1, 3, and 10 days respectively.

Urinary tract

In a French analysis of 22 510 urinary calculi performed by infrared spectroscopy, drug-induced urolithiasis was divided into two categories: first, stones with drugs physically embedded (*n* = 238; 1.0%), notably indinavir monohydrate (*n* = 126; 53%), followed by triamterene (*n* = 43; 18%), sulfonamides (*n* = 29; 12%), and amorphous silica (*n* = 24; 10%); secondly, metabolic nephrolithiasis induced by drugs (*n* = 140; 0.6%), involving mainly calcium/vitamin D supplementation (*n* = 56; 40%) and carbonic anhydrase inhibitors (*n* = 33; 24%) (29). Drug-induced stones are responsible for about 1.6% of all calculi in France. Physical analysis and a thorough drug history are important elements in the diagnosis.

Skin

Two Japanese-American women developed clinical features that satisfied the criteria for Stevens–Johnson syndrome during methazolamide treatment (30). Methazolamide is a sulfonamide, the most common group of drugs associated with Stevens–Johnson syndrome.

- A 62-year-old man with cerebrovascular disease underwent regional cerebrovascular reactivity to intravenous acetazolamide 1000 mg using single-photon emission computed tomography with [^{123}I]-N-isopropyl-para-iodoamfetamine 0.45 mg (31). Three days later he developed erythematous eruptions of varying sizes on his back, which then spread over his entire body. The presumptive diagnosis was Stevens–Johnson syndrome and he was given a glucocorticoid. The skin and mucous lesions became bullous or erosive, ruptured spontaneously, and dried with crusting. He gradually improved over 21 days. Lymphocyte transformation tests were positive with acetazolamide.

All previous cases have occurred in patients taking oral acetazolamide for glaucoma and have been limited to Japanese or Indian patients or patients of Japanese descent, as was the case here.

Topical dorzolamide caused severe periorbital dermatitis after an average exposure time of 20 weeks in 14 patients (32). Although dermatitis due to dorzolamide can resolve when it is withdrawn, this does not always occur, and in some patients all topical medications containing benzalkonium chloride must be withdrawn. Allergic contact blepharoconjunctivitis has also been reported with dorzolamide in a 72-year-old man (33).

Dorzolamide has been associated with contact dermatitis (34).

- Two 71-year-old men with open-angle glaucoma developed eyelid dermatitis while using long-term topical dorzolamide. Concomitant glaucoma medications included pilocarpine and timolol. In each case the lesions disappeared within 1 month of withdrawal.

This is the first report of contact dermatitis due to dorzolamide in which stripping (tape-stripping of the upper stratum corneum by 10-fold application and removal of conventional adhesive tape) and the scarified patch test (scratching the epidermis with a lancet) were used in diagnosis.

- An 80-year-old woman developed conjunctival inflammation and a severe eczematous rash affecting the eyelids and cheeks within 1 week of switching from latanoprost + benzalkonium chloride to Cosopt® (dorzolamide hydrochloride 2% + timolol maleate 0.5% + benzalkonium chloride) (35). Cosopt® was withdrawn and the dermatitis was treated with topical glucocorticoids and emollients, with rapid resolution. Subsequent patch testing was positive with pure dorzolamide 40% and timolol 10%, with no reactions to product excipients or benzalkonium chloride.

Contact dermatitis to two separate topical agents, timolol and dorzolamide, is decidedly uncommon. Patch testing, albeit at higher drug concentrations, showed sensitivity to both compounds.

Musculoskeletal

Chronic acetazolamide therapy is associated with greater spinal bone mineral density. This is probably the result of metabolic acidosis; urine calcium is increased and serum phosphate reduced. Osteomalacia has been reported during long-term therapy in combination with barbiturates in two patients.

Immunologic

Fatal anaphylactic shock with massive pulmonary edema has been reported in a 66-year-old woman who was taking acetazolamide for glaucoma (36). She had a history of sulfonamide allergy, and acetazolamide is a sulfonamide derivative. Sulfonamide allergy should be regarded as a contraindication to acetazolamide.

- Non-fatal anaphylactic shock with acute pulmonary edema has been reported in a 79-year-old woman after a first dose of acetazolamide (37). There was no history of sulfonamide allergy and she had been taking hydrochlorothiazide for some time.

Anaphylactic shock with acetazolamide should be recognized to occur as a first-dose phenomenon with no prior demonstrable sulfonamide allergy.

Long-Term Effects

Drug tolerance

Acetazolamide alkalinizes the urine and increases its volume. Urinary sodium, potassium, and bicarbonate concentrations rise, and chloride concentration falls. These effects are due to inhibition of carbonic anhydrase in the nephron, mainly in the proximal convoluted tubule, which reduces the number of protons available for Na/H antiporter exchange. Consequently, the concentration of bicarbonate in extracellular fluids falls, causing a metabolic acidosis, which itself limits the diuretic response to acetazolamide. Thus, the initial kaliuretic effect is lost within a few weeks.

Second-Generation Effects

Teratogenicity

Possible teratogenicity due to acetazolamide resulted in one case of congenital glaucoma, microphthalmia, and patent ductus arteriosus and one case of sacrococcygeal teratoma. One child was born with acidosis, hypercalcemia, and hypomagnesemia; these features resolved rapidly, but at 8 months there was mild hypotonicity of the legs. It has been suggested that acetazolamide should be avoided in the first trimester of pregnancy (SEDA-21, 228).

Fetotoxicity

The use of acetazolamide in pregnancy is ill-advised.

- Renal tubular acidosis occurred in a preterm boy shortly after birth (38). His mother had taken oral acetazolamide during pregnancy for glaucoma. When renal tubular acidosis developed, acetazolamide was detected in his serum, demonstrating transplacental passage of acetazolamide.

Susceptibility Factors

Age

Dorzolamide has been poorly studied in children with glaucoma. In a prospective study of 2% dorzolamide tds in children aged 1 week to under 6 years the most common drug-related adverse effects were ocular hyperemia (5.4%) in those aged 2 years or under and ocular burning/stinging (12%) in those from aged 2–6 years (39). Two patients treated with dorzolamide discontinued therapy because of eye pain, ocular hyperemia, ocular burning/stinging, or ocular itching. The higher incidence of ocular burning and stinging in the youngest stage may reflect the preverbal stage. Two other subjects had a reduction in total bicarbonate; however, concurrent illnesses and sampling considerations complicated interpretation of changes in total bicarbonate.

For the most part, the safety profile of dorzolamide in children is analogous to that in adults. Systemic dorzolamide-related adverse effects are infrequent in both populations.

Other features of the patient

Carbonic anhydrase inhibitors should be used with caution in patients with respiratory acidosis or those with severe loss of respiratory capacity, and in patients with diabetes mellitus. They are contraindicated in patients with hepatic disease or insufficiency, reduced serum concentrations of sodium or potassium, adrenocortical insufficiency, hyperchloremic acidosis, or severe renal disease or dysfunction. They should also be avoided in patients taking salicylates.

Drug Administration

Drug administration route

Brinzolamide eye-drops can cause transient blurred vision (40).

Dorzolamide eye-drops can cause irreversible corneal edema in glaucoma patients with endothelial compromise (41).

Drug overdose

Overdosage with acetazolamide can destroy the gastric mucosal barrier, by interfering with prostaglandin and bicarbonate release, and can cause thrombocytopenia by bone marrow suppression, eventually leading to hemorrhagic gastritis. Although acetazolamide binds strongly to plasma proteins, and clearance by hemodialysis is generally poor, this approach and hemoperfusion (without heparin) appears to be effective (SEDA-22, 236).

Acetazolamide overdose has been reported in a child.

- A 12-month-old girl, weighing 10 kg, developed metabolic acidosis after taking 500–1250 mg of acetazolamide (42). The maximum base deficit recorded was 11.6. She was treated with sodium bicarbonate and recovered completely.

Accidental poisoning with acetazolamide should be included in the differential diagnosis of metabolic acidosis.

Drug–Drug Interactions

Acetylsalicylic acid

In two children, aspirin potentiated the slight metabolic acidosis induced by carbonic anhydrase inhibitors (SEDA-9, 79; 43).

Two elderly patients developed lethargy, confusion, and incontinence, attributed to an interaction between acetazolamide and salicylate (44). Salicylate may reduce acetazolamide clearance by inhibiting its renal tubular secretion. Analgesic doses of salicylates should probably be avoided in elderly patients taking acetazolamide.

- A 50-year-old woman with chronic renal insufficiency treated with acetazolamide for simple glaucoma developed confusion, cerebellar ataxia, and metabolic acidosis 2 weeks after starting to take aspirin for acute pericarditis (45). A diagnosis of salicylism was made despite low serum salicylate concentrations.

The authors suggested that acetazolamide-induced acidosis increases the concentration of the unionized form of salicylate, which crosses membranes more rapidly, and hence explains the cerebral toxicity associated with low serum concentrations of salicylate. However, an increase in the plasma or erythrocyte concentrations of acetazolamide (not measured) is a much more convincing explanation for this patient's symptoms.

Barbiturates

The intracarotid amobarbital procedure (IAP) is used to determine language laterality and as a predictor of postoperative memory deficits in candidates for temporal lobectomy. A valid test requires that the patient demonstrate sufficient anesthetization in the injected hemisphere, marked by loss of contralateral muscle tone and strength, electroencephalographic slowing, and hemisphere-specific behavioral change. In 11 of 56 patients undergoing an intracarotid amobarbital procedure there was either too rapid recovery (no more than 1 minute) or anesthetic failure. Of these, 10 were taking a medication with some carbonic anhydrase activity: topiramate (n = 7), zonisamide (n = 2), hydrochlorothiazide (n = 1), and furosemide (n = 1). Procedures performed on patients weaned from topiramate showed a correlation between the duration of drug withdrawal and the duration of the anesthetic effect. This apparent drug interaction can last for 4 weeks after the last dose of a carbonic anhydrase inhibitor. Its mechanism is speculative. Carbonic anhydrase inhibitors, including topiramate and zonisamide, should be tapered at least 8 weeks before an intracarotid amobarbital procedure (46,47).

Furosemide

Acetazolamide inhibits cerebrospinal fluid production, and its effects are augmented by furosemide, suggesting a role for this combination in infants with ventricular dilatation resulting from severe periventricular hemorrhage. The International Posthaemorrhagic Ventricular Drug Trial Group has reported the results of a randomized, controlled trial of standard therapy alone or with the addition of acetazolamide plus furosemide in the first 151 of 177 randomized patients (48). The addition of acetazolamide plus furosemide was not only ineffective but also worsened the already poor outcome in these infants. At 1 year, death and the need for shunt placement were substantially commoner in the diuretic-treated group. There were adverse events (for example acidosis, nausea, anorexia, and diarrhea) in 37% of those given acetazolamide plus furosemide; in addition, 27% developed nephrocalcinosis (SEDA-21, 228). Altered cerebral blood flow due to acetazolamide may have caused additional brain injury.

In a commentary on this trial (49) it was pointed out that the use of acetazolamide alone or with furosemide in the treatment of posthemorrhagic hydrocephalus is another example of the use of untried therapy in neonatal care. The safety and efficacy of acetazolamide in children has never been established.

Other Environmental Interactions

Ketogenic diet

A ketogenic diet is sometimes used to control intractable seizures. Acetazolamide should be discontinued before starting the diet, because of the potential risk of severe secondary metabolic acidosis (50). Acetazolamide can be reintroduced once the acid–base status of the patient has stabilized.

Interference with Diagnostic Tests

HPLC assays for theophyline

Acetazolamide interferes with certain HPLC assays for theophylline, leading to spuriously increased theophylline concentrations (SEDA-21, 228).

References

1. Palmberg P. A topical carbonic anhydrase inhibitor finally arrives. Arch Ophthalmol 1995;113(8):985–6.
2. Strahlman E, Tipping R, Vogel RInternational Dorzolamide Study Group. A double-masked, randomized 1-year study comparing dorzolamide (Trusopt), timolol, and betaxolol. Arch Ophthalmol 1995;113(8):1009–16.
3. Laibovitz R, Strahlman ER, Barber BL, Strohmaier KM. Comparison of quality of life and patient preference of dorzolamide and pilocarpine as adjunctive therapy to timolol in the treatment of glaucoma. J Glaucoma 1995;4:306–13.
4. Nesher R, Ticho U. Switching from systemic to the topical carbonic anhydrase inhibitor dorzolamide: effect on the quality of life of glaucoma patients with drug-related side effects. Isr Med Assoc J 2003;5(4):260–3.
5. Fraunfelder FT, Meyer SM. Systemic adverse reactions to glaucoma medications. Int Ophthalmol Clin 1989;29(3):143–6.
6. Tzanakis N, Metzidaki G, Thermos K, Spyraki CH, Bouros D. Anaphylactic shock after a single oral intake of acetazolamide. Br J Ophthalmol 1998;82(5):588.
7. Lee YT, Wu JC, Chan FK. Acetazolamide-induced Gerstmann syndrome. Int J Clin Pract 1999;53(7):560–1.
8. Choksi V, Hughes M, Selwa L, Hoeffner E. Transient neurologic deficit after acetazolamide challenge for computed tomography perfusion imaging. J Comput Assist Tomogr 2005;29:278–80.
9. Kim JM, Ryu WS, Hwang YH, Kim JS. Aggravation of ataxia due to acetazolamide induced hyperammonaemia in episodic ataxia. J Neurol Neurosurg Psychiatry 2007;78:771–2.
10. Walshe CM, Cooper JD, Kossmann T, Hayes I, Iles L. Cerebral fat embolism syndrome causing brain death after long-bone fractures and acetazolamdie therapy. Crit Care Resusc 2007;9:184–6.
11. Silver LHBrinzolamide Primary Therapy Study Group. Clinical efficacy and safety of brinzolamide (Azopt), a new topical carbonic anhydrase inhibitor for primary open-angle glaucoma and ocular hypertension. Am J Ophthalmol 1998;126(3):400–8.
12. Taguri AH, Khan MA, Sanders R. Marginal keratitis: an uncommon form of topical dorzolamide allergy. Am J Ophthalmol 2000;130(1):120–2.
13. Parthasarathi S, Myint K, Singh G, Mon S, Sadasivam P, Dhillon B. Bilateral acetazolamide-induced choroidal effusion following cataract surgery. Eye 2007;21:870–2.

14. Zhao JC, Chen T. Brinzolamide induced reversible corneal decompensation. Br J Ophthalmol 2005;89:389–90.
15. Tanimura H, Minamoto A, Narai A, Hirayama T, Suzuki M, Mishima HK. Corneal edema in glaucoma patients after the addition of brinzolamide 1% ophthalmic suspension. Jpn J Ophthalmol 2005;49:332–3.
16. Goldberg S, Gallily R, Bishara S, Blumenthal EZ. Dorzolamide-induced choroidal detachment in a surgically untreated eye. Am J Ophthalmol 2004;138:285–6.
17. Turgut B, Türkçüoğlu P, Güler M, Akyol N, Celiker U, Demir T. Anosmia as an adverse effect of dorzolamide. Acta Ophthalmol Scand 2007;85:228–9.
18. Critchlow AS, Freeborn SN, Roddie RA. Potassium supplements during treatment of glaucoma with acetazolamide. BMJ (Clin Res Ed) 1984;289(6436):21.
19. Ikeda K, Iwasaki Y, Kinoshita M, Yabuki D, Igarashi O, Ichikawa Y, Satoyoshi E. Acetazolamide-induced muscle weakness in hypokalemic periodic paralysis. Intern Med 2002;41(9):743–5.
20. Moffett BS, Moffett TI, Dickerson HA. Acetazolamide therapy for hypochloremic metabolic alkalosis in pediatric patients with heart disease. Am J Ther 2007;14:331–5.
21. Filippi L, Bagnoli F, Margollicci M, Zammarchi E, Tronchin M, Rubaltelli FF. Pathogenic mechanism, prophylaxis, and therapy of symptomatic acidosis induced by acetazolamide. J Investig Med 2002;50(2):125–32.
22. Venkatesha SL, Umamaheswara Rao GS. Metabolic acidosis and hyperventilation induced by acetazolamide in patients with central nervous system pathology. Anesthesiology 2000;93(6):1546–8.
23. Morris S, Geh V, Nischal KK, Sahi S, Ahmed MA. Topical dorzolamide and metabolic acidosis in a neonate. Br J Ophthalmol 2003;87(8):1052–3.
24. Menon GJ, Vernon SA. Topical brinzolamide and metabolic acidosis. Br J Ophthalmol 2006;90:247–8.
25. Zaidi FH, Kinnear PE. Acetazolamide, alternate carbonic anhydrase inhibitors and hypoglycaemic agents: comparing enzymatic with diuresis induced metabolic acidosis following intraocular surgery in diabetes. Br J Ophthalmol 2004;88:714–15.
26. Keisu M, Wiholm BE, Ost A, Mortimer O. Acetazolamide-associated aplastic anaemia. J Intern Med 1990;228(6):627–32.
27. Fraunfelder FT, Bagby GC. Monitoring patients taking oral carbonic anhydrase inhibitors. Am J Ophthalmol 2000;130(2):221–3.
28. Kodjikian L, Durand B, Burillon C, Rouberol F, Grange J-D, Renaudier P. Acetazolamide-induced thrombocytopenia. Arch Ophthalmol 2004;122:1543–4.
29. Cohen-Solal F, Abdelmoula J, Hoarau MP, Jungers P, Lacour B, Daudon M. Les lithiases urinaires d'origine medicamenteuse. [Urinary lithiasis of medical origin.] Therapie 2001;56(6):743–50.
30. Flach AJ, Smith RE, Fraunfelder FT. Stevens–Johnson syndrome associated with methazolamide treatment reported in two Japanese–American women. Ophthalmology 1995; 102(11):1677–80.
31. Ogasawara K, Tomitsuka N, Kobayashi M, Komoribayashi N, Fukuda T, Saitoh H, Inoue T, Ogawa A. Stevens–Johnson syndrome associated with intravenous acetazolamide administration for evaluation of cerebrovascular reactivity. Neurol Med Chir 2006;46:161–3.
32. Delaney YM, Salmon JF, Mossa F, Gee B, Beehne K, Powell S. Periorbital dermatitis as a side effect of topical dorzolamide. Br J Ophthalmol 2002;86(4):378–80.
33. Mancuso G, Berdondini RM. Allergic contact blepharoconjunctivitis from dorzolamide. Contact Dermatitis 2001;45(4):243.
34. Linares Mata T, Pardo Sanchez J, de la Cuadra Oyanguren J. Contact dermatitis caused by allergy to dorzolamide. Contact Dermatitis 2005;52:111–2.
35. Kalavala M, Statham BN. Allergic contact dermatitis from timolol and dorzolamide eye drops. Contact Dermatitis 2006;54:345.
36. Gerhards LJ, van Arnhem AC, Holman ND, Nossent GD. Fatale anafylactische reactie na inname van acetazolamide (Diamox) wegens glaucoom. [Fatal anaphylactic reaction after oral acetazolamide (Diamox) for glaucoma.] Ned Tijdschr Geneeskd 2000;144(25):1228–30.
37. Gallerani M, Manzoli N, Fellin R, Simonato M, Orzincolo C. Anaphylactic shock and acute pulmonary edema after a single oral dose of acetazolamide. Am J Emerg Med 2002;20(4):371–2.
38. Ozawa H, Azuma E, Shindo K, Higashigawa M, Mukouhara R, Komada Y. Transient renal tubular acidosis in a neonate following transplacental acetazolamide. Eur J Pediatr 2001;160(5):321–2.
39. Ott EZ, Mills MD, Arango S, Getson AJ, Assaid CA, Adamsons IA. A randomized trial assessing dorzolamide in patients with glaucoma who are younger than 6 years. Arch Ophthalmol 2005;123:1177–86.
40. Doyle JW, Smith MF. New aqueous inflow inhibitors. Semin Ophthalmol 1999;14(3):159–63.
41. Konowal A, Morrison JC, Brown SV, Cooke DL, Maguire LJ, Verdier DV, Fraunfelder FT, Dennis RF, Epstein RJ. Irreversible corneal decompensation in patients treated with topical dorzolamide. Am J Ophthalmol 1999;127(4):403–6.
42. Baer E, Reith DM. Acetazolamide poisoning in a toddler. J Paediatr Child Health 2001;37(4):411–2.
43. Cowan RA, Hartnell GG, Lowdell CP, Baird IM, Leak AM. Metabolic acidosis induced by carbonic anhydrase inhibitors and salicylates in patients with normal renal function. BMJ (Clin Res Ed) 1984;289(6441):347–8.
44. Sweeney KR, Chapron DJ, Brandt JL, Gomolin IH, Feig PU, Kramer PA. Toxic interaction between acetazolamide and salicylate: case reports and a pharmacokinetic explanation. Clin Pharmacol Ther 1986; 40(5):518–24.
45. Hazouard E, Grimbert M, Jonville-Berra AP, De Toffol MC, Legras A. Salicylisme et glaucome: augmentation reciproque de la toxicité de l'acétazolamide et de l'acide acetyl salicylique. [Salicylism and glaucoma: reciprocal augmentation of the toxicity of acetazolamide and acetylsalicylic acid.] J Fr Ophtalmol 1999;22(1):73–5.
46. Bookheimer S, Schrader LM, Rausch R, Sankar R, Engel J Jr. Reduced anesthetization during the intracarotid amobarbital (Wada) test in patients taking carbonic anhydrase-inhibiting medications. Epilepsia 2005;46:236–43.
47. Ringman JM, Grant AC. Carbonic anhydrase inhibitors and amobarbital resistance. Epilepsia 2005;46:1333.
48. International PHVD Drug Trial Group. International randomised controlled trial of acetazolamide and furosemide in posthaemorrhagic ventricular dilatation in infancy. Lancet 1998;352(9126):433–40.
49. Hack M, Cohen AR. Acetazolamide plus furosemide for periventricular dilatation: lessons for drug therapy in children. Lancet 1998;352(9126):418–9.
50. Tallian KB, Nahata MC, Tsao CY. Role of the ketogenic diet in children with intractable seizures. Ann Pharmacother 1998;32(3):349–61.

Chlortalidone

General Information

Chlortalidone (chlorthalidone) is chemically unrelated to the thiazide diuretics but shares many of their actions and adverse effects.

It is so closely similar to the thiazide diuretics in most respects that data have generally been regarded as interchangeable.

Comparative studies

In the Antihypertensive and Lipid-Lowering Treatment to Prevent Heart Attack Trial (ALLHAT), over 40 000 participants aged 55 years or older with hypertension and at least one other risk factor for coronary heart disease were randomized to chlortalidone, amlodipine, doxazosin, or lisinopril (1,2). Doxazosin was discontinued prematurely because chlortalidone was clearly superior in preventing cardiovascular events, particularly heart failure (2). Otherwise, mean follow-up was 4.9 years. There were no differences between chlortalidone, amlodipine, and lisinopril in the primary combined outcome or all-cause mortality. Compared with chlortalidone, heart failure was more common with amlodipine and lisinopril, and chlortalidone was better than lisinopril at preventing stroke.

Organs and Systems

Chlortalidone

Sensory systems

Chlortalidone has been associated with acute myopia (3).

- A 30-year-old man with hypertension developed visual blurring after taking chlortalidone 12.5 mg/day for 4 days in addition to atenolol. Visual acuity was 20/120 in both eyes and slit-lamp biomicroscopy showed retinal striae radiating from the fovea. B-scan ultrasonography showed shallow peripheral serous choroidal detachment. The blood pressure was 140/100, hematological and renal function tests and serum electrolytes were within the reference ranges, and a CT scan of the head was normal. Atenolol and chlortalidone were withdrawn and 5 days later his visual acuity had returned to 20/20. Fundoscopy showed disappearance of the retinal striae at the macula and resolution of the peripheral choroidal effusion.

The exact mechanism whereby chlortalidone caused acute myopia in this case was unclear, but the authors suggested that it could have been due to ciliary body effusion, peripheral uveal effusion, ciliary spasm, and lens swelling.

Electrolyte balance

In the past chlortalidone was thought to carry a greater risk of hypokalemia and hyponatremia (SED-8, 488), but this probably reflected relative overdosage with accumulation, because of its long half-life. With the much lower dosages now in use there appears to be no clear difference from the thiazides.

Skin

Pseudoporphyria has been attributed to chlortalidone (4).

Body temperature

Drug fever has been attributed to chlortalidone (5).

- A 58-year-old woman presented with intermittent nocturnal fever (up to 39.5°C) for 4 weeks beginning 2 weeks after the introduction of chlortalidone for hypertension. Otherwise, physical examination was normal. Investigations showed a normochromic normocytic anemia and a raised erythrocyte sedimentation rate and C-reactive protein. Other biochemical tests, blood cultures, and serological and immunological tests were negative. Chest radiography, lung scintigraphy, and abdominal ultrasound were unremarkable. There was no further fever after chlortalidone was withdrawn, and all biochemical tests became normal. The lymphocyte transformation test showed stimulation by chlortalidone.

References

1. ALLHAT Collaborative Research Group. Major cardiovascular events in hypertensive patients randomized to doxazosin vs chlorthalidone: the antihypertensive and lipid-lowering treatment to prevent heart attack trial (ALLHAT). JAMA 2000;283(15):1967–75.
2. ALLHAT Officers and Coordinators for the ALLHAT Collaborative Research Group. The Antihypertensive and Lipid-Lowering Treatment to Prevent Heart Attack Trial. Major outcomes in high-risk hypertensive patients randomized to angiotensin-converting enzyme inhibitor or calcium channel blocker vs diuretic: The Antihypertensive and Lipid-Lowering Treatment to Prevent Heart Attack Trial (ALLHAT). JAMA 2002;288(23):2981–97.
3. Mahesh G, Giridhar A, Saikumar SJ, Fegde S. Drug-induced myopia following chlorthalidone treatment. Indian J Ophthalmol 2007;55:386–8.
4. Baker EJ, Reed KD, Dixon SL. Chlorthalidone-induced pseudoporphyria: clinical and microscopic findings of a case. J Am Acad Dermatol 1989;21(5 Pt 1):1026–9.
5. Osterwalder P, Koch J, Wuthrich B, Pichler WJ, Vetter W. Unklarer status febrilis. [Intermittent fever of unknown origin.] Dtsch Med Wochenschr 1998;123(24):761–5.

Eplerenone

General Information

Eplerenone is a potassium-sparing diuretic. It is similar to spironolactone as an aldosterone antagonist, but has less affinity for androgen and progesterone receptors and may therefore have fewer adverse effects (1).

Eplerenone is metabolized by CYP3A4 (2) and should not be given with inhibitors of CYP3A4 because of the risk of hyperkalemia (3).

Eplerenone has been compared with spironolactone in patients with heart failure (NYHA classes II–IV). In a dose-finding study, 321 patients maintained on ACE inhibitors and diuretics, with or without digoxin, were randomized to receive eplerenone 25–100 mg/day, spironolactone 25 mg/day, or placebo (4). After 12 weeks there was no improvement in heart failure in patients taking either eplerenone or spironolactone. Increases in plasma testosterone were significantly greater in men taking spironolactone than in those taking eplerenone. However, hyperkalemia was more frequent in patients taking eplerenone 100 mg/day (12%) than in those taking spironolactone 25 mg/day (8.7%). The potency of eplerenone relative to spironolactone remains to be established in patients with heart failure.

Organs and Systems

Electrolyte balance

Hyperkalemia is always a consideration with spironolactone and eplerenone, but the relation between a beneficial response to eplerenone and the change in serum potassium concentration has been poorly characterized. In 397 hypertensive patients (responders and non-responders) taking eplerenone 50–200 mg/day without other susceptibility factors for hyperkalemia, eplerenone caused an increase in serum potassium concentration of 0.2 mmol/l (5). The blood pressure lowering effect of eplerenone did not correlate with the change in serum potassium concentration. This suggests that in patients with uncomplicated hypertension, eplerenone can be used without a significant risk of hyperkalemia.

In the Eplerenone Post-Acute Myocardial Infarction Heart Failure Efficacy and Survival Study (EPHESUS), the addition of eplerenone 25–50 mg/day to optimal medical therapy in patients with an acute myocardial infarction complicated by left ventricular dysfunction and heart failure significantly reduced all-cause and cardiovascular mortality (6). Patients with a serum potassium concentration over 5.0 mmol/l at baseline were excluded. The incidence of hypokalemia was significantly lower with eplerenone. However, there was serious hyperkalemia (serum potassium 6.0 mmol/l or more) in 5.5% of those who took eplerenone compared with 3.9% of those who took placebo.

Reproductive system

Eplerenone has been reported to produce less gynecomastia than spironolactone (7,8). In patients with gynecomastia from spironolactone, eplerenone can be substituted and gynecomastia status reassessed.

Susceptibility factors

Renal disease

In 64 subjects mild, moderate, or severe renal impairment had no effect on the pharmacokinetics of a single dose of eplerenone 100 mg followed by 100 mg/day on days 3–8 (9). Hemodialysis removed about 10% of the dose.

Drug–Drug Interactions

ACE inhibitors

Co-administration of potassium-sparing diuretics with ACE inhibitors can cause severe hyperkalemia (SED-14, 674).

NSAIDs

The interaction between potassium-sparing diuretics and NSAIDs is well documented (SED-14, 674). The major complications are deterioration of renal function and hyperkalemia. The risk associated with the non-selective COX-2 inhibitors is unknown. However, three patients had hyperkalemia (8.5, 5.4, and 5.1 mmol/l) after developing acute renal insufficiency while taking these drugs (10).

References

1. Zillich AJ, Carter BL. Eplerenone—a novel selective aldosterone blocker. Ann Pharmacother 2002;36(10):1567–76.
2. Cook CS, Berry LM, Kim DH, Burton EG, Hribar JD, Zhang L. Involvement of CYP3A in the metabolism of eplerenone in humans and dogs: differential metabolism by CYP3A4 and CYP3A5. Drug Metab Dispos 2002;30(12):1344–51.
3. Moore TD, Nawarskas JJ, Anderson JR. Eplerenone: a selective aldosterone receptor antagonist for hypertension and heart failure. Heart Dis 2003;5(5):354–63.
4. Pitt B, Roniker B. Eplerenone a novel selective aldosterone receptor antagonist (SARA): dose finding study in patients with heart failure. J Am Coll Cardiol 1999;33(Suppl A):A188–9.
5. Levy DG, Rocha R, Funder JW. Distinguishing the antihypertensive and electrolyte effects of eplerenone. J Clin Endocrinol Metab 2004;89:2736–40.
6. Keating GM, Plosker GL. Eplerenone : a review of its use in left ventricular systolic dysfunction and heart failure after acute myocardial infarction. Drugs 2004;64(23):2689–707.
7. Burgess ED, Lacourciere Y, Ruilope-Urioste LM, Oparil S, Kleiman JH, Krause S, Roniker B, Maurath C. Long-term safety and efficacy of the selective aldosterone blocker eplerenone in patients with essential hypertension. Clin Ther 2003;25(9):2388–404.
8. Hollenberg NK, Williams GH, Anderson R, Akhras KS, Bittman RM, Krause SL. Symptoms and the distress they cause: comparison of an aldosterone antagonist and a calcium channel blocking agent in patients with systolic hypertension. Arch Intern Med 2003;163(13):1543–8.
9. Ravis WR, Reid S, Sica DA, Tolbert DS. Pharmacokinetics of eplerenone after single and multiple dosing in subjects with and without renal impairment. J Clin Pharmacol 2005;45(7):810–21.
10. Perazella MA, Eras J. Are selective COX-2 inhibitors nephrotoxic? Am J Kidney Dis 2000;35(5):937–40.

Etacrynic acid

General Information

Etacrynic acid (ethacrynic acid) is a loop diuretic with similar actions to furosemide and bumetanide. However, it has more adverse effects than the other loop diuretics and offers no clear advantages.

Organs and Systems

Sensory systems

Etacrynic acid is ototoxic after intravenous administration (1–3), an effect that is dose-related. It can sometimes be permanent (3,4). An association with nystagmus has also been reported (3,5). The risk has been estimated to be about seven per 1000 injections (6). The risk is increased in renal insufficiency and in patients who are also being given aminoglycoside antibiotics. Animal data suggest that etacrynic acid is more ototoxic than furosemide (7).

Metabolism

Like other diuretics etacrynic acid can impair glucose tolerance in patients with type 2 diabetes mellitus (8). Non-ketotic hyperglycemia has also been reported (9). However, hypoglycemia has also been reported in two patients with uremia (10).

There has been an anecdotal report of hyperuricemia and acute gout in a patient taking etacrynic acid (11).

Electrolyte balance

Etacrynic acid can cause excessive diuresis, natriuresis, and kaliuresis, leading respectively to dehydration and sodium and potassium depletion (12–17).

Mineral balance

Etacrynic acid can cause hypocalcemia and hypomagnesemia (18).

Hematologic

There have been anecdotal reports of agranulocytosis (12,19), hemolytic anemia (20), and thrombocytopenia (16,21).

Gastrointestinal

Nausea, abdominal pain (22), and diarrhea are relatively common with etacrynic acid (23,24), more so than with other loop diuretics. Data from the Boston Drug Surveillance Program showed that intravenous etacrynic acid was associated with an increased incidence of gastrointestinal bleeding (25,26). Overall 20% of the patients treated with etacrynic acid bled, compared with 5% of those who received furosemide. With intravenous etacrynic acid the risk of gastrointestinal hemorrhage was 26%, compared with 10% after oral administration. However, 12 of 28 patients who bled did so before etacrynic acid had been given (27), and it is not clear whether there is a true difference in risk with the different routes of administration.

Liver

Like other diuretics etacrynic acid can precipitate hepatic encephalopathy in the treatment of ascites in patients with hepatic cirrhosis (28). There has been an anecdotal report of focal hepatic necrosis in a patient taking etacrynic acid (29).

Pancreas

There has been an anecdotal report of necrotizing pancreatitis in a patient taking etacrynic acid (30).

Immunologic

Two patients developed a Henoch-Schönlein type of necrotic hemorrhagic rash of the legs and lower part of the body accompanied by histological evidence of vasculitis (31). In both cases the lesions appeared about 2–3 weeks after the start of treatment; in another case a hemorrhagic rash was accompanied by acute gastric and duodenal ulceration (32).

Drug Administration

Drug administration route

Intravenous administration of etacrynic acid causes burning at the site of injection (18).

Drug–Drug Interactions

Aminoglycoside antibiotics

Etacrynic acid potentiates aminoglycoside ototoxicity by facilitating the entry of the antibiotics from the systemic circulation into the endolymph (33). Animal evidence suggests that this effect may be potentiated by glutathione depletion (34). Conversely, neomycin can enhance the penetration of etacrynic acid into the inner ear (35).

Cardiac glycosides

Etacrynic acid potentiates the actions of cardiac glycosides if it causes potassium depletion.

- A patient developed digitalis intoxication and quinidine-induced cardiac dysrhythmias when hypokalemia occurred during etacrynic acid therapy (36).

In rats etacrynic acid inhibited the absorption of digitoxin from the small intestine (37), but an interaction of this sort has not been described in man.

Cefaloridine

Etacrynic potentiated the nephrotoxic effects of cefaloridine in mice, causing renal tubular necrosis (38).

Glucocorticoids

In a survey of the effects of etacrynic acid in 16 646 patients with gastrointestinal bleeding, the risk of major bleeding was increased in patients taking etacrynic acid (4.5% of 111 patients versus 0.2% of 10 637 patients taking no drugs). The risk was further increased in patients taking etacrynic acid plus glucocorticoids (9.1% of 22 patients) (25).

Glyceryl trinitrate

Etacrynic acid inhibited relaxation of large coronary microvessels in response to glyceryl trinitrate in pigs (39). The relevance of this to the clinical use of these drugs is not known.

Lithium

The loop diuretics increase the renal excretion of lithium after single-dose intravenous administration in both animals (40) and man (41). Furosemide has been used to treat lithium intoxication (42). The effect of etacrynic acid is larger than those of furosemide and bumetanide (41). However, long-term treatment with furosemide and bumetanide can cause lithium intoxication in some patients (43,44), perhaps by causing sodium depletion and a secondary increase in lithium reabsorption. An adverse interaction of lithium during long-term therapy with etacrynic acid is therefore theoretically likely.

Methotrexate

Etacrynic acid interacts with human serum albumin and modifies its binding properties (45). Since it binds to two binding sites on albumin, the benzodiazepine binding site and the warfarin binding site, it can displace drugs that bind at those sites (46). It competitively displaced 7-hydroxymethotrexate from its binding proteins in vitro (47). The clinical significance of this effect is not known.

Warfarin

Etacrynic acid interacts with human serum albumin and modifies its binding properties (45). Since it binds to two binding sites on albumin, the benzodiazepine binding site and the warfarin binding site, it can displace drugs that bind at those sites (46,48). This has been described for warfarin, whose anticoagulant effect it potentiated (49).

References

1. Beauchamp GD, Crouch TC. Deafness. Review of intravenous ethacrynic acid. J Kans Med Soc 1975;76(7):166–8180.
2. Schneider WJ, Becker EL. Acute transient hearing loss after ethacrynic acid therapy. Arch Intern Med 1966;117(5):715–7.
3. Schwartz FD, Pillay VK, Kark RM. Ethacrynic acid: its usefulness and untoward effects. Am Heart J 1970;79(3):427–8.
4. Meriwether WD, Mangi RJ, Serpick AA. Deafness following standard intravenous dose of ethacrynic acid. JAMA 1971;216(5):795–8.
5. Gomolin IH, Garschick E. Ethacrynic acid-induced deafness accompanied by nystagmus. N Engl J Med 1980;303(12):702.
6. Boston Collaborative Drug Surveillance Program. A cooperative study. Drug-induced deafness. JAMA 1973;224(4):515–6.
7. Brown RD. Comparison of the cochlear toxicity of sodium ethacrynate, furosemide, and the cysteine adduct of sodium ethacrynate in cats. Toxicol Appl Pharmacol 1975;31(2):270–82.
8. Russell RP, Lindeman RD, Prescott LF. Metabolic and hypotensive effects of ethacrynic acid. Comparative study with hydrochlorothiazide. JAMA 1968;205(1):81–5.
9. Cowley AJ, Elkeles RS. Diabetes and therapy with potent diuretics. Lancet 1978;1(8056):154.
10. Maher JF, Schreiner GE. Studies on ethacrynic acid in patients with refractory edema. Ann Intern Med 1965;62:15–29.
11. Melvin KE, Farrelly RO, North JD. Ethacrynic acid: a new oral diuretic. BMJ 1963;5344:1521–4.
12. Ledingham JG, Bayliss RI. Ethacrynic acid: two years' experience with a new diuretic. BMJ 1965;5464:732–5.
13. Daley D, Evans B. Diuretic action of ethacrynic acid in congestive heart failure. BMJ 1963;5366:1169–71.
14. DeRubertis FR, Michelis MF, Beck N, Davis BB. Complications of diuretic therapy: severe alkalosis and syndrome resembling inappropriate secretion of antidiuretic hormone. Metabolism 1970;19(9):709–19.
15. Fort AT, Morrison JC, Fish SA. Iatrogenic hypokalemia of pregnancy by furosemide and ethacrynic acid. J Reprod Med 1971;6(5):207–8.
16. Schroeder G, Sannerstedt R, Werkoe L. Clinical experiences with ethacrynic acid, a new non-thiazide saluretic agent (MK-595). Acta Med Scand 1964;175:781–6.
17. Sullivan RC, Freemon FR, Caranasos GJ. Complications from diuretic therapy with ethacrynic acid and furosemide. South Med J 1971;64(7):869–72.
18. Lacreta FP, Brennan JM, Nash SL, Comis RL, Tew KD, O'Dwyer PJ. Pharmacokinetics and bioavailability study of ethacrynic acid as a modulator of drug resistance in patients with cancer. J Pharmacol Exp Ther 1994;270(3):1186–91.
19. Walker JG. Fatal agranulocytosis complicating treatment with ethacrynic acid. Report of a case. Ann Intern Med 1966;64(6):1303–5.
20. Hanna M. Ethacrynic acid (MK595) as a diuretic—some early observations. Med J Aust 1966;1(13):534–7.
21. O'Dwyer PJ, LaCreta F, Nash S, Tinsley PW, Schilder R, Clapper ML, Tew KD, Panting L, Litwin S, Comis RL, et al. Phase I study of thiotepa in combination with the glutathione transferase inhibitor ethacrynic acid. Cancer Res 1991;51(22):6059–65.
22. Hagedorn CW, Kaplan AA, Hulet WH. Prolonged administration of ethacrynic acid in patients with chronic renal disease. N Engl J Med 1965;272:1152–5.
23. Dollery CT, Parry EH, Young DS. Diuretic and hypotensive properties of ethacrynic acid: a comparison with hydrochlorothiazide. Lancet 1964;41:947–52.
24. Lauwers L. Clinical experience with a new diuretic: ethacrinic acid. Acta Cardiol 1966;21(1):79–89.
25. Jick H, Porter J. Drug-induced gastrointestinal bleeding. Report from the Boston Collaborative Drug Surveillance Program, Boston University Medical Center. Lancet 1978;2(8080):87–9.
26. Slone D, Jick H, Lewis GP, Shapiro S, Miettinen OS. Intravenously given ethacrynic acid and gastrointestinal bleeding. A finding resulting from comprehensive drug surveillance. JAMA 1969;209(11):1668–71.
27. Wilkinson WH, Ciminera JL, Simpkins GT. Intravenously given ethacrynic acid and gastrointestinal bleeding. JAMA 1969;210:347.
28. Sherlock S, Senewiratne B, Scott A, Walker JG. Complications of diuretic therapy in hepatic cirrhosis. Lancet 1966;1(7446):1049–52.
29. Datey KK, Deshmukh SN, Dalvi CP, Purandare NM. Hepatocellular damage with ethacrynic acid. BMJ 1967; 3(558):152–3.
30. Schmidt P, Friedman IS. Adverse effects of ethacrynic acid. NY State J Med 1967;67(11):1438–42.
31. Bar-On H, Eisenberg S, Eliakim M. Clinical experience with ethacrynic acid with reference to a possible complication of the Schoenlein–Henoch type. Isr J Med Sci 1967;3(1):113–8.
32. Pain AK. Acute gastric ulceration associated with drug therapy. BMJ 1967;1(540):634.
33. Conlon BJ, McSwain SD, Smith DW. Topical gentamicin and ethacrynic acid: effects on cochlear function. Laryngoscope 1998;108(7):1087–9.

34. Hoffman DW, Whitworth CA, Jones-King KL, Rybak LP. Potentiation of ototoxicity by glutathione depletion. Ann Otol Rhinol Laryngol 1988;97(1):36–41.
35. Orsulakova A, Schacht J. A biochemical mechanism of the ototoxic interaction between neomycin and ethacrynic acid. Acta Otolaryngol 1982;93(1–2):43–8.
36. Oravetz J, Slodki SJ. Recurrent ventricular fibrillation precipitated by quinidine. Report of a patient with recovery after 28 paroxysms. Arch Intern Med 1968;122(1):63–5.
37. Braun W, Damm KH. Drug interactions in intestinal absorption of ^{3}H-digitoxin in rats. Experientia 1976; 32(5):613–4.
38. Dodds MG, Foord RD. Enhancement by potent diuretics of renal tubular necrosis induced by cephaloridine. Br J Pharmacol 1970;40(2):227–36.
39. Sellke FW, Tomanek RJ, Harrison DG. L-cysteine selectively potentiates nitroglycerin-induced dilation of small coronary microvessels. J Pharmacol Exp Ther 1991; 258(1):365–9.
40. Stokke ES, Ostensen J, Hartmann A, Kiil F. Loop diuretics reduce lithium reabsorption without affecting bicarbonate and phosphate reabsorption. Acta Physiol Scand 1990; 140(1):111–8.
41. Beutler JJ, Boer WH, Koomans HA, Dorhout Mees EJ. Comparative study of the effects of furosemide, ethacrynic acid and bumetanide on the lithium clearance and diluting segment reabsorption in humans. J Pharmacol Exp Ther 1992;260(2):768–72.
42. Hansen HE, Amdisen A. Lithium intoxication. (Report of 23 cases and review of 100 cases from the literature). Q J Med 1978;47(186):123–44.
43. Jefferson JW, Kalin NH. Serum lithium levels and long-term diuretic use. JAMA 1979;241(11):1134–6.
44. Huang LG. Lithium intoxication with coadministration of a loop-diuretic. J Clin Psychopharmacol 1990;10(3):228.
45. Bertucci C, Wainer IW. Improved chromatographic performance of a modified human albumin based stationary phase. Chirality 1997;9(4):335–40.
46. Fehske KJ, Muller WE. High-affinity binding of ethacrynic acid is mediated by the two most important drug binding sites of human serum albumin. Pharmacology 1986; 32(4):208–13.
47. Slordal L, Sager G, Jaeger R, Aarbakke J. Interactions with the protein binding of 7–hydroxy-methotrexate in human serum in vitro. Biochem Pharmacol 1988;37(4):607–11.
48. Sellers EM, Koch-Weser J. Displacement of warfarin from human albumin by diazoxide and ethacrynic, mefenamic, and nalidixic acids. Clin Pharmacol Ther 1970;11(4):524–9.
49. Petrick RJ, Kronacher N, Alcena V. Interaction between warfarin and ethacrynic acid. JAMA 1975;231(8):843–4.

Furosemide

General Information

Furosemide is the classic member of the group of so-called high-ceiling or loop diuretics, which can achieve a much greater peak diuresis than the thiazides. It is widely and frequently used both orally and parenterally over a wider dosage range than the thiazide diuretics, because its concentration-effect curve is steeper and because it is effective in patients with moderate renal insufficiency (creatinine clearance 5–25 ml/minute), in whom the thiazide diuretics and most related compounds are ineffective.

Furosemide is quickly and almost completely excreted by kidney unmetabolized. The usual oral dose is 20–120 mg, but much larger doses (for example 1000 mg) have been used in renal insufficiency. It is very effective after intravenous injection, and doses of 500 mg and more can be used in emergencies (renal insufficiency, pulmonary edema). The majority of adverse effects occur with the use of high doses, 95% of the reactions being dose dependent (1).

General adverse effects

Disturbances of fluid and electrolyte balance, such as hyponatremia, hypokalemia, and dehydration with circulatory disturbances (such as dizziness, postural hypotension, and syncope), have been reported. Rarely gastrointestinal symptoms are problems with high dosages and in the elderly. Pancreatitis and jaundice seem to occur more often than with the thiazide diuretics, but deterioration of glucose tolerance seems to be less common. At serum concentrations over 50 μg/ml, tinnitus, vertigo, and deafness, sometimes permanent, have been reported. Hematological disorders, particularly thrombocytopenia, and serious skin disorders occur occasionally, as do hypersensitivity reactions. Neither tumor-inducing effects nor second-generation effects have been documented.

Organs and Systems

Cardiovascular

Because furosemide, apart from its diuretic effect, has transient but pronounced vasodilator properties (changes in both venous capacitance and peripheral arteriolar resistance have been described in anephric patients), it can cause postural hypotension and syncope, particularly if given together with other blood pressure-lowering drugs (SED-8, 458) (SED-9, 349) (SED-9, 350). Ischemic complications have been reported in elderly patients. Reactions due to extracellular volume depletion accounted for 9% of all adverse effects observed in 535 patients treated with furosemide (1). Neuroendocrine activation and resultant increased peripheral resistance (afterload) after furosemide reduce cardiac output and stroke volume and increase cardiac work, with the possibility of worsening myocardial and tissue ischemia. Since many patients with heart failure have underlying myocardial ischemia or infarction, initial symptomatic benefit from furosemide can be followed by detrimental effects on myocardial perfusion, with extension or completion of myocardial necrosis.

When it is used in cardiac failure, furosemide acts in two ways: besides its diuretic effect it produces an immediate fall in left ventricular filling pressure, which is independent of and precedes diuresis. If furosemide is given intravenously in stable chronic heart failure (which it normally is not), this can be an unwanted effect, causing deterioration (SEDA-11, 199), particularly in patients with pure right ventricular failure.

Nervous system

Visual disturbances and drowsiness have been described, but it is not clear whether these were caused by reduced cerebral perfusion or by a direct effect of the drug itself.

Sensory systems

At high doses, and especially if serum concentrations are over 50 μg/ml, furosemide can cause ototoxic reactions, such as tinnitus, vertigo, and even deafness, sometimes permanent (SED-9, 351). Subclinical, audiometrically determined, high-tone deafness has been reported to occur in 6.4% of furosemide-treated patients (2). It is generally considered advisable to use another diuretic in patients whose hearing is already impaired, and to avoid using furosemide along with other ototoxic drugs, such as the aminoglycosides.

Sensorineural hearing loss occurs in a small proportion of very premature babies who are given furosemide. Various causative mechanisms have been suggested, including bilirubin, drugs, infection, and/or hypoxic brainstem injury. In a case-control study of 15 children and 30 controls born before 33 weeks of gestation, renal insufficiency and/or aminoglycoside use in conjunction with furosemide was associated with sensorineural hearing loss (3).

A retrospective chart review (July 2000 to January 2002) of all survivors in a neonatal intensive care unit was undertaken to evaluate the effect of furosemide on hearing loss (4). Of the 57 neonates who had received furosemide nine had a subsequent abnormal hearing screen, and of the 207 neonates who had not received furosemide 33 also had an abnormal hearing screen. This suggests that hearing loss in these neonates is not directly related to the use of furosemide.

Endocrine

Furosemide rarely causes the syndrome of inappropriate antidiuretic hormone secretion (SIADH) (although it has been found useful in treating some patients with SIADH who cannot tolerate water restriction (5)). In furosemide-induced cases (SEDA-7, 246), serum ADH concentrations were raised, total body sodium was normal, total body potassium greatly reduced, and intracellular water raised at the expense of extracellular fluid volume. However, such cases are rare, and no new cases have been published since this complication was reported in SEDA-7.

Nutrition

Patients with congestive heart failure taking high doses of furosemide can develop thiamine deficiency, which is improved by thiamine supplementation. There is whole-blood thiamine phosphate deficiency, but no reduction in the storage form of thiamine, thiamine diphosphate. These observations suggest that thiamine supplementation may not be necessary in elderly patients taking furosemide for congestive heart failure (6).

Erythrocyte transketolase activity suggested severe thiamine deficiency in 24 of 25 patients with heart failure who were taking at least 80 mg/day of furosemide and in four of seven patients who were taking 40 mg/day (OR = 19; CI = 1.1, 601) (7). Thiamine status was not associated with any other clinical variables. These findings suggest that thiamine deficiency occurs in a substantial proportion of patients with heart failure who are taking furosemide; however, this is of unclear clinical significance.

Hematologic

Agranulocytosis, thrombocytopenia, and hemolytic anemia have been reported occasionally (SED-8, 484) (SED-9, 350) (SEDA-1, 180). The commonest complication is thrombocytopenia (8), although it is often mild and asymptomatic (9).

Salivary glands

Diuretics cause altered salivary flow rate and composition and have been associated with both subjective and objective evidence of xerostomia. In a randomized trial in 12 healthy women randomly assignment to placebo, bendroflumethiazide (2.5 mg/day for 7 days) and furosemide (40 mg/day for 7 days), xerostomia increased with furosemide in conjunction with a reduction in submandibular-sublingual salivary secretion (10). This was particularly so at lunchtime. This suggests that diuretics may potentiate dryness of the mouth and should be used carefully in patients with abnormal salivary flow.

Gastrointestinal

With normal doses, nausea, vomiting, and diarrhea are very uncommon, accounting for less than 1% of all adverse reactions (SED-9, 350). The incidence rises with higher doses and in the presence of uremia.

Liver

Although furosemide is very hepatotoxic in experimental animals, only a few cases of jaundice have been reported (SED-8, 484) (SED-8, 485), and no fully documented cases have so far been published. However, in patients with cirrhosis furosemide readily precipitates hepatic encephalopathy (SED-8, 485), even when low doses are used (1).

Biliary tract

Biliary colic has been attributed to furosemide (SED-11, 427).

Pancreas

Increases in serum isoamylase (SEDA-6, 213) and pancreatitis have been reported in patients taking furosemide.

Urinary tract

Excessive diuresis and dehydration often cause a transient reduction in glomerular filtration rate and a rise in serum urea (about 8% of all adverse reactions) (SED-8, 350). The sudden diuresis can cause loin pain, particularly in elderly patients, and acute urinary retention and overflow incontinence in elderly men with prostatic hyperplasia.

Although it is often described in children, medullary nephrocalcinosis with furosemide has been rarely described in adults.

- A 40-year-old woman who had taken furosemide (40–160 mg/day) for 15 years developed medullary nephrocalcinosis (11).

Chronic tubulointerstitial nephritis has been reported with furosemide.

- A 25-year-old woman developed biopsy-proven chronic tubulointerstitial nephritis with accompanying distal renal tubular acidosis in association with furosemide abuse (up to 1.2 g/day for several months) (12).

Four children with the nephrotic syndrome developed transient hypercalciuria and intraluminal calcification in renal histopathological specimens without radiological evidence of renal calcification. These children were resistant to corticosteroids and were receiving furosemide plus albumin for the management of edema (13). This result stresses the pervasive effect of furosemide, and probably all loop diuretics, in increasing urinary calcium excretion, with resultant nephrocalcinosis. Whenever possible, steps should be taken to limit the hypercalciuric effect of loop diuretics. Such maneuvers could include limiting the sodium content of the diet and/or combining the loop diuretic with a thiazide diuretic.

- A 900-gram girl born before term with bronchopulmonary dysplasia developed ureteral obstruction, urinoma, and acute renal insufficiency as a result of furosemide-related hypercalciuria (cumulative dose 27.5 mg) and nephrolithiasis (14). Percutaneous drainage of the urinoma plus conversion to hydrochlorothiazide resolved the urinoma and hydronephrosis.

Acute renal insufficiency carries a high mortality and morbidity. Diuretics may increase mortality in patients with acute renal insufficiency, but this has not been studied prospectively. In a prospective, multicenter, multinational, epidemiological study of 1743 consecutive patients, who were either treated with renal replacement therapy or who fulfilled predefined criteria for acute renal insufficiency, about 70% were taking diuretics at enrolment (15). Severe sepsis/septic shock (48%), major surgery (39%), low cardiac output (30%), and hypovolemia (28%) were the most common conditions associated with the development of acute renal insufficiency. Furosemide was the most common diuretic used (98%). In all three multivariate models, diuretic use was not associated with a significantly increased risk of mortality. The use of diuretics in patients with acute renal insufficiency should continue, but only according to a specific need to control volume excess.

Skin

Rashes seem to be just as common with furosemide (SED-8, 484) as with other oral diuretics, but severe skin reactions (exfoliative dermatitis, erythema multiforme, acquired epidermolysis bullosa), which are rare with other diuretics, have been occasionally reported with high doses of furosemide in renal insufficiency (SED-12, 503) (SEDA-18, 234).

Cases of furosemide-induced lupus-like syndrome (16), bullous pemphigoid (17,18), and lichenoid drug eruptions (19) have been reported.

A review of furosemide-induced skin reactions included a description of an 88-year-old man who developed an eruption that clinically and histologically simulated Sweet's syndrome (acute febrile neutrophilic vasculitis) after 6 weeks (20). Atypical features and rapid resolution suggested a drug eruption rather than true Sweet's syndrome. However, a similar mechanism may have been implicated a hypersensitivity reaction involving immune complexes.

- A 46-year-old woman with congestive heart failure was given intravenous furosemide and 3 days later developed a low-grade fever followed by tender, papular, erythematous, non-pruritic skin eruptions bilaterally on the wrists, forearms, arms, and thighs (21). There was associated redness in both eyes and photophobia, consistent with episcleritis and iritis. A skin biopsy showed a superficial dermal nodular neutrophilic infiltrate, associated with nuclear dust and rare eosinophils, findings consistent with Sweet syndrome. Furosemide was withdrawn and the skin lesions and eye symptoms gradually subsided.

Sweet syndrome is a disorder of poorly understood pathogenesis that has several features suggesting a hypersensitivity reaction. It has been associated with autoimmune diseases, malignancies, and drugs. This patient was different from the previous one in having symptoms and signs suggestive of iritis.

Linear IgA bullous dermatosis associated with the administration of furosemide has been reported (22).

- An 86-year-old woman, with a history of stable schizophrenia, chronic obstructive pulmonary disease, ischemic cardiomyopathy, and type 2 diabetes, was admitted with cardiac insufficiency, which was treated by introduction of enalapril. A chest infection was treated with co-amoxiclav with gradual alleviation of symptoms over 10 days. At this point, furosemide was begun, because of persistent signs of heart failure. After 3 days, erythema and bullae were noted on her palms and soles, and later on the trunk, extremities, hard palate, and buccal mucosa. Biopsy showed the characteristic features of linear IgA bullous dermatosis, with linear deposition of IgA along the basement membrane. Co-amoxiclav and furosemide were withdrawn; no new lesions were noted thereafter.

Furosemide has previously been related to other bullous dermatoses, particularly bullous pemphigoid (SED-14, 671). In this case only a temporal relation was demonstrated, as rechallenge was judged to be unethical. The other putative offending drug, co-amoxiclav, was regarded as unlikely to be causative, because it had been given many times before without noticeable skin lesions.

Furosemide has been associated with disseminated superficial porokeratosis, a heritable disorder of cornification (23).

A man developed acute generalized exanthematous pustulosis while taking furosemide (24). A positive lymphocyte transformation test suggested an immunological mechanism.

Musculoskeletal

Although loop diuretics increase renal calcium excretion, there have been variable results in studies of their effects on the risk of fractures. In a case–control study of the risk of fracture in 44 001 patients who had taken a loop diuretic in the preceding 5 years and 194 111 age- and sex-matched controls, "ever" use of a loop diuretic was associated with a crude 51% increased risk of any fracture (OR = 1.51; 95% CI = 1.48, 155) and a 72% increased risk of hip fracture (OR = 1.72; 95% CI = 1.64, 181) (25). After adjustment for potential confounders, the risk reduction was only slightly increased for any fracture (OR = 1.04) and for hip fracture (OR = 1.16). With an increased average daily dose the estimates increased in former users but fell in current users. Furosemide was associated with higher risk estimates than bumetanide. This study does not provide a definitive answer to the risk of fractures with loop diuretics but suggests that particular caution should be taken.

Immunologic

It has long been thought that loop and thiazide diuretics pose a theoretical risk of cross-sensitivity in patients with sulfonamide allergy because of their common structures. However, the available literature does not provide sufficient numbers of well-documented cases to support this impression (26). It seems that careful administration of diuretics is permissible in patients with documented sulfonamide allergy, but as always such a drug challenge should not be attempted without careful follow-up. A furosemide rechallenge protocol, based on a method that has been used to rechallenge with a sulfa-containing antimicrobial agent, safely allowed the long-term reinstitution of loop diuretic therapy with furosemide (27).

Furosemide rarely causes type 1 allergic reactions.

- A 24-year-old woman took one tablet of furosemide 40 mg and 10 minutes later developed oral itching, generalized urticaria, facial angioedema, dyspnea, and hypotension (28). She recovered after the administration of parenteral adrenaline, methylprednisolone, and diphenhydramine. A furosemide skin prick test 10 mg/ml was negative. An intradermal skin test was positive for furosemide 1 % and sulfamethoxazole 0.03 mg/ml.

IgE-mediated reactions to furosemide are infrequent, but can be life-threatening. The positive intradermal test to sulfamethoxazole in this case raises the question of cross-reactivity between non-aromatic and antimicrobial sulfonamides.

There is little clinical or pharmacological evidence that a self-reported sulfa allergy is likely to be associated with a life-threatening cross-reaction with acetazolamide or furosemide. (29,30). Most patients who report sulfa allergy have actually had an adverse reaction to a sulfonamide antimicrobial drug. There are significant structural differences between sulfonamide antibiotics and other sulfonamide non-antimicrobial drugs such as furosemide and acetazolamide, and generally cross-reactivity would not be expected.

In a retrospective cohort study using the General Practice Research Base in the UK, 969 cases with a so-called allergic reaction to an antimicrobial sulfonamide were reviewed (30). Only 96 (9.9%) had an allergic reaction after receiving a non-antimicrobial sulfonamide. Of 19 257 who had no allergic reaction after a antimicrobial sulfonamide, 315 (1.6%) had an allergic reaction However, the risk of allergic reactions was even greater after the use of a penicillin among patients with a prior hypersensitivity reaction to a antimicrobial sulfonamide, compared with patients with no such history (adjusted odds ratio = 3.9) after receiving a non-antimicrobial sulfonamide (adjusted odds ratio = 2.8). In addition, among those with a prior hypersensitivity reaction after an antimicrobial sulfonamide, the risk of an allergic reaction after the subsequent receipt of a non-antimicrobial sulfonamide was lower than the risk of an allergic reaction with penicillin (adjusted odds ratio = 0.7). Finally, the risk of an allergic reaction after an antimicrobial sulfonamide was lower among patients with a history of hypersensitivity to an antimicrobial sulfonamide than among patients with a history of hypersensitivity to penicillins (adjusted odds ratio = 0.6).

These results suggest that there is an association between hypersensitivity after an antimicrobial sulfonamide and a subsequent allergic reaction after a non-antimicrobial sulfonamide, but this association appears to be due to a predisposition to allergic reactions rather than to cross-reactivity with sulfonamide-based drugs.

Body temperature

Several cases of furosemide-associated fever have been reported (SEDA-20, 204) (SEDA-21, 229) (SEDA-22, 238). These had certain features in common: (1) the affected infants were in their first year of life; (2) there was congestive heart failure as a result of congenital abnormalities or cardiomyopathy; (3) the temperature was raised to 38.5–40°C and there was a resemblance to septic fever; (4) concomitant treatment with digoxin; (5) negative physical examination and investigations for a source of infection; (6) withdrawal of furosemide was followed by disappearance of fever within 1–2 days. In some cases, the use of furosemide in lower dosages or on alternate days avoided fever. The mechanism of this adverse effect is unclear, but it appears to be dose-related. Consideration of furosemide as a cause of fever in such patients may save unnecessary laboratory studies and lead to early resolution.

Long-Term Effects

Drug withdrawal

The more intensive diuretic treatment is the greater the risk from sudden withdrawal. In one series of 38 patients, taking 20–40 mg furosemide or equivalent, generally for heart failure, withdrawal was attempted when all had been free of heart failure or hypertension for at least 3 months; it was followed by clinical or radiological relapse in 29% and one of the patients died (31). In another five patients, withdrawal in three led to severe symptoms necessitating admission to hospital (SEDA-10, 192) (SEDA-10, 193). The best advice is to withdraw intensive diuretic treatment only with the utmost caution and not to attempt withdrawal at all if there is any radiological evidence of heart failure. In patients with congestive heart failure, independent predictors of the need for continued diuretic were furosemide dosage greater than 40 mg/day, a left ventricular ejection fraction less than 0.27, and hypertension (32).

The effects of furosemide withdrawal on postprandial blood pressure have been assessed in 20 elderly patients (mean age 73 years) with heart failure and preserved left ventricular systolic function (ejection fraction 61%) (33). In 13 who were able to discontinue furosemide (mean dose 32 mg/day), maximum systolic blood pressure fell significantly from 25 mmHg to 11 mmHg and diastolic blood pressure from 18 to 9 mmHg over 3 months. In the continuation group (mean furosemide dose 21 mg/day), there was no change in the postprandial fall.

However, there are sometimes opportunities for withdrawal of furosemide (SEDA-22, 237). If diuretics are withdrawn suddenly in patients with a normal sodium intake, there will be rebound retention of sodium and water, because compensatory mechanisms that maintain sodium balance in the face of diuretics continue to act for several days after the diuresis has worn off. There are two methods of mitigating rebound retention of sodium and water: gradual reduction of the dosage or institution of a low sodium diet so that only a small amount of sodium can be retained when the diuretic is withdrawn. Rebound retention of sodium and water, with consequent edema, may convince the doctor that continued diuretic therapy is necessary, and the patient is then committed to life-time exposure. Even in patients with heart failure, reduced intake of salt may remove the need for diuretics or allow the use of lower dosages.

Second-Generation Effects

Pregnancy

Although furosemide has embryotoxic properties in some animal species, it has been widely used in pregnant women without any adverse effects. Nevertheless, it should be used with great caution, since hypovolemia can lead to reduced uterine and placental blood flow. Careful monitoring of fetal heart action is necessary. Furosemide passes the placenta and increases fetal urine production. It can also increase acid concentrations in maternal serum, fetal serum, and amniotic fluid, thus masking a useful index for the development of pre-eclampsia (34). Its use in pregnant women should therefore be restricted to the treatment of cardiac failure.

Lactation

Furosemide passes into the breast milk in a concentration 80% of that in serum at pH 7.0, but the total amount of drug ingested is probably too low to affect the neonate (35).

Susceptibility Factors

Age

In premature infants furosemide can cause persistent patent ductus arteriosus (SEDA-9, 208). The use of furosemide in this age group needs to be re-evaluated (SEDA-14, 182) (SEDA-15, 217).

Renal disease

High doses of furosemide in uremia are potentially ototoxic (36).

Hepatic disease

Although high-dose furosemide can be very useful in incipient acute renal insufficiency, it should be used with extreme caution in patients with severe hepatic insufficiency (SEDA-14, 183) and not at all in hepatic coma.

Drug–Drug Interactions

ACE inhibitors

Acute renal insufficiency with severe hyponatremia has been attributed to vigorous diuretic treatment (metolazone, furosemide, spironolactone) with an ACE inhibitor (37). Because ACE inhibition impairs renal protection against reduced perfusion, the combination of an ACE inhibitor with high-dose furosemide causes a reduction in glomerular filtration rate linearly related to the change in blood pressure.

In congestive heart failure, single conventional doses of captopril (25–75 mg) attenuated the natriuretic response to furosemide, while a low dose (1 mg) significantly enhanced furosemide-induced natriuresis (SEDA-17, 268) (SEDA-18, 236). The mechanism of this interaction is uncertain, but captopril did not affect delivery of furosemide to its site of action (38). In the long-term, intensive treatment with captopril enhanced the natriuretic response to furosemide (SEDA-17, 268).

Allopurinol

Interstitial nephritis with granulomatous hepatitis has been attributed to an interaction of furosemide with allopurinol (39). As in previous reports (SEDA-11, 198) the evidence for this interaction is not convincing. The role of allopurinol in causing the illness is credible, but the role of furosemide is doubtful.

Aminoglycoside antibiotics

Furosemide increases the ototoxic risks of aminoglycoside antibiotics (40,41) by reducing their clearance by about 35% (42); permanent deafness has resulted from the use of this combination.

- A 60-year-old woman developed moderately severe sensorineural hearing loss bilaterally after receiving five doses of gentamicin and one of furosemide 20 mg (43).

The cumulative dose and duration of aminoglycoside therapy are more important than serum concentrations in the development of gentamicin ototoxicity, except when interacting medications such as furosemide are co-administered.

Aspirin

Furosemide inhibits the absorption of aspirin (44), and aspirin 75 mg/day and 300 mg/day inhibits the venodilator effect of furosemide (45). These data raise the issue of whether aspirin should be routinely used in patients with congestive heart failure using furosemide.

Carbonic anhydrase inhibitors

Acetazolamide inhibits cerebrospinal fluid production, and its effects are augmented by furosemide, suggesting a role for this combination in infants with ventricular dilatation resulting from severe periventricular hemorrhage. The International Posthaemorrhagic Ventricular Drug Trial Group has reported the results of a randomized, controlled trial of standard therapy alone or with the addition of acetazolamide plus furosemide in the first 151 of 177 randomized patients (46). The addition of acetazolamide plus furosemide was not only ineffective but also worsened the already poor outcome in these infants. At 1 year, death and the need for shunt placement were substantially commoner in the diuretic-treated group. There were adverse events (for example acidosis, nausea, anorexia, and diarrhea) in 37% of those given acetazolamide plus furosemide; in addition, 27% developed nephrocalcinosis (SEDA-21, 228). Altered cerebral blood flow due to acetazolamide may have caused additional brain injury.

In a commentary on this trial (47) it was pointed out that the use of acetazolamide alone or with furosemide in the treatment of posthemorrhagic hydrocephalus is another example of the use of untried therapy in neonatal care. The safety and efficacy of acetazolamide in children has never been established.

Cephaloridine

The nephrotoxic effects of cefaloridine are potentiated by concurrent administration of furosemide (48,49), perhaps by a direct interaction and probably also because furosemide lowers the clearance of the antibiotic (50). Such combinations are better avoided.

Digoxin

It is advisable to monitor serum potassium concentrations closely if furosemide is combined with any cardiac glycoside.

Fluoroquinolones

The renal excretion of lomefloxacin and ofloxacin was reduced and plasma concentrations increased by furosemide, resulting in higher concentrations (51).

Flurbiprofen

Flurbiprofen reduces the diuretic action of furosemide (52).

Lithium

Furosemide can cause lithium toxicity by inhibiting the renal tubular excretion of lithium ions (53).

Of 10 615 elderly patients continuously taking lithium for over 10 years, 413 (3.9%) were admitted to the hospital at least once with lithium toxicity. After adjustment for potential confounders, there was a 5.5-fold increase in the relative risk of lithium toxicity within 1 month of starting therapy with a loop diuretic (54). Thiazide diuretics were not independently associated with an increased risk of hospitalization for lithium toxicity. This population-based nested case-control study stresses the importance of monitoring for lithium toxicity whenever a loop diuretic is started.

Mannitol

The combination of furosemide with mannitol can rapidly cause acute renal insufficiency (55) and potentiate the effect of curare (56).

NSAIDs

Furosemide inhibits the absorption of indometacin (57), while the diuretic and hypotensive effects of most diuretics are blunted by indometacin and probably also other NSAIDs (58). Intravenous furosemide is commonly given to patients with acute heart failure to relieve pulmonary congestion. Symptomatic relief occurs before the onset of diuresis, and the beneficial effect is believed to result from a venodilator action of furosemide, which precedes its diuretic effect. This venodilator response is inhibited by indometacin, suggesting that it occurs through local prostaglandin release.

Furosemide inhibits the absorption of meloxicam (59), an NSAID that is relatively selective for COX-2, sparing the physiologically important isoform that mediates vasodilator prostaglandins, which maintain renal function.

Pancuronium bromide

Furosemide (1 mg/kg) shortened the recovery time from pancuronium blockade in neurosurgical patients with normal renal function (60). Phosphodiesterase inhibition and increased pancuronium excretion were suggested as possible explanations.

Phenytoin

The diuretic response to furosemide is reduced by approximately 50% during concurrent administration of phenytoin (61), probably through reduction of furosemide absorption (62).

Potassium-wasting drugs

The hypokalemic effects of furosemide can be potentiated by drugs such as licorice (63) and carbenoxolone (64).

Piroxicam

Furosemide natriuresis and kaliuresis can be reduced by short-term treatment with piroxicam in hypertensive patients with impaired renal function (SEDA-16, 113).

Sulindac

Data on sulindac inhibition of the diuretic effect of furosemide are contradictory (65).

Suramin

In 26 patients treated with suramin for hormone-refractory prostate cancer, furosemide reduced the total body clearance of suramin by 36% (66). In view of the increased risk of severe adverse effects after treatment with higher plasma concentrations of suramin, it would be prudent to alter dosage schemes in patients treated with both suramin and furosemide.

Theophylline

Furosemide increases steady-state theophylline concentrations (SED-11, 428) (SED-14, 673).

A randomized, controlled study has shown an enhanced diuretic response to furosemide in infants taking theophylline during extracorporeal membrane oxygenation (67). The underlying mechanism was uncertain, but may have been an increase in glomerular filtration rate.

Thiazide diuretics

The use of thiazide and loop diuretics in combination to treat resistant hypertension often causes severe deterioration in renal function (68,69). It is not clear whether this is the result of excess diuresis or excessive blood pressure reduction.

Vancomycin

Ten preterm infants receiving regular theophylline for apnea of prematurity, who subsequently received vancomycin and furosemide, have been studied (70). When vancomycin was introduced in the infants who were established on furosemide and theophylline, there was a consistent failure to achieve therapeutic concentrations. Starting furosemide in infants who were already receiving vancomycin resulted in falls in serum vancomycin to subtherapeutic concentrations in all but one case. Serum concentrations fell by a mean of 24% (range 12–43%), in the 24 hours after the start of furosemide treatment. Two of the 10 infants had persistence of coagulase-negative staphylococcal sepsis while vancomycin concentrations were suboptimal. While the mechanism of this interaction is uncertain, the changes in serum vancomycin concentration may indicate acute changes in glomerular filtration rate. Preterm infants receiving theophylline and furosemide need shorter vancomycin dosage intervals to avoid therapeutic failure.

Preterm infants often receive theophylline for apnea of prematurity and furosemide to improve dynamic lung compliance and to reduce total pulmonary resistance. In 10 preterm infants receiving furosemide plus theophylline, vancomycin was introduced according to the generally recommended dosing schedules but failed to achieve anticipated therapeutic concentrations (71). In nine infants the addition of furosemide to vancomycin plus theophylline resulted in a fall in serum vancomycin concentrations to subtherapeutic within 24 hours, and only in one did vancomycin concentrations remain within the target range. The addition of furosemide reduced the serum creatinine concentration by 24% (range 12–43). These data suggest that furosemide enhances the excretion of vancomycin, and an acute and transient increase in glomerular filtration rate and creatinine clearance may be the explanation.

Warfarin

In patients taking warfarin, furosemide, like other diuretics, reduces the prothrombin time without changing the plasma warfarin concentration (SEDA-22, 238). The reduced anticoagulant effect is probably secondary to diuretic-induced volume depletion, mediating increases in clotting factors. Any effect is likely to be short-lived.

References

1. Naranjo CA, Busto U, Cassis L. Furosemide-induced adverse reactions during hospitalization. Am J Hosp Pharm 1978;35(7):794–8.
2. Tuzel IH. Comparison of adverse reactions to bumetanide and furosemide. J Clin Pharmacol 1981;21(11–12 Pt 2):615–9.
3. Marlow ES, Hunt LP, Marlow N. Sensorineural hearing loss and prematurity. Arch Dis Child Fetal Neonatal Ed 2000;82(2):F141–4.
4. Rais-Bahrami K, Majd M, Veszelovszky E, Short BL. Use of furosemide and hearing loss in neonatal intensive care survivors. Am J Perinatol 2004;21:329–32.
5. Decaux G, Waterlot Y, Genette F, Hallemans R, Demanet JC. Inappropriate secretion of antidiuretic hormone treated with frusemide. BMJ (Clin Res Ed) 1982; 285(6335):89–90.
6. Hardig L, Daae C, Dellborg M, Kontny F, Bohmer T. Reduced thiamine phosphate, but not thiamine diphosphate, in erythrocytes in elderly patients with congestive heart failure treated with furosemide. J Intern Med 2000; 247(5):597–600.
7. Zenuk C, Healey J, Donnelly J, Vaillancourt R, Almalki Y, Smith S. Thiamine deficiency in congestive heart failure patients receiving long term furosemide therapy. Can J Clin Pharmacol 2003;10:184–8.
8. Bottiger LE, Westerholm B. Thrombocytopenia. II. Drug-induced thrombocytopenia. Acta Med Scand 1972; 191(6):541–8.

9. De Gruchy GC. Thrombocytopenia. In: Drug-Induced Blood-Disorders. Oxford: Blackwell, 1975:118.
10. Nederfors T, Nauntofte B, Twetman S. Effects of furosemide and bendroflumethiazide on saliva flow rate and composition. Arch Oral Biol 2004;49:507–13.
11. Simoes A, Domingos F, Prata MM. Nephrocalcinosis induced by furosemide in an adult patient with incomplete renal tubular acidosis. Nephrol Dial Transplant 2001;16(5):1073–4.
12. Park CW, You HY, Kim YK, Chang YS, Shin YS, Hong CK, Kim YC, Bang BK. Chronic tubulointerstitial nephritis and distal renal tubular acidosis in a patient with frusemide abuse. Nephrol Dial Transplant 2001;16(4): 867–9.
13. Mocan H, Yildiran A, Camlibel T, Kuzey GM. Microscopic nephrocalcinosis and hypercalciuria in nephrotic syndrome. Hum Pathol 2000;31(11):1363–7.
14. Alpert SA, Noe HN. Furosemide nephrolithiasis causing ureteral obstruction and urinoma in a preterm neonate. Urology 2004;64:589.
15. Uchino S, Doig GS, Bellomo R, Morimatsu H, Morgera S, Schetz M, Tan I, Bouman C, Nacedo E, Gibney N, Tolwani A, Ronco C, Kellum JA. Diuretics and mortality in acute renal failure. Crit Care Med 2004;32:1669–77.
16. Lin RY. Unusual autoimmune manifestations in furosemide-associated hypersensitivity angiitis. NY State J Med 1988;88(8):439–40.
17. Ihu H, Shimozuma H. Bullous pemphigoid induced by furosemide. Nishinihon J Dermatol 1993;55:890–3.
18. Guerrera V, Carbone RL. Bullous pemphigoid induced by furosemide. G Ital Dermatol Venereol 1994;129:239–41.
19. Eom SC, Chae YS, Suh KS, Kim ST. The clinical features of lichenoid drug eruption and the histopathological differentiation between lichenoid drug eruption and lichen planus. Korean J Dermatol 1994;32:1019–25.
20. Cobb MW. Furosemide-induced eruption simulating Sweet's syndrome. J Am Acad Dermatol 1989;21(2 Pt 2):339–43.
21. Govindarajan G, Bashir Q, Kuppuswamy S, Brooks C. Sweet syndrome associated with furosemide. South Med J 2005;98:570–2.
22. Cerottini JP, Ricci C, Guggisberg D, Panizzon RG. Drug-induced linear IgA bullous dermatosis probably induced by furosemide. J Am Acad Dermatol 1999;41(1):103–5.
23. Kroiss MM, Stolz W, Hohenleutner U, Landthaler M. Disseminated superficial porokeratosis induced by furosemide. Acta Derm Venereol 2000;80(1):52–3.
24. Noce R, Paredes BE, Pichler WJ, Krahenbuhl S. Acute generalized exanthematic pustulosis (AGEP) in a patient treated with furosemide. Am J Med Sci 2000;320(5):331–3.
25. Rejnmark L, Vestergaard P, Mosekilde L. Fracture risk in patients treated with loop diuretics. J Intern Med 2006;259:117–24.
26. Phipatanakul W, Adkinson NF Jr. Cross-reactivity between sulfonamides and loop or thiazide diuretics: a theoretical or actual risk? Allergy Clin Immunol Int 2000;12:26–8.
27. Earl G, Davenport J, Narula J. Furosemide challenge in patients with heart failure and adverse reactions to sulfa-containing diuretics. Ann Intern Med 2003; 138(4):358–9.
28. Dominguez-Ortega J, Martinez-Alonso JC, Dominguez-Ortega C, Fuentes MJ, Frades A, Fernandez-Colino T. Anaphylaxis to oral furosemide. Allergol Immunopathol (Madr) 2003;31:345–7.
29. Lee AG, Anderson R, Kardon RH, Wall M. Presumed sulfa allergy in patients with intracranial hypertension treated with acetazolamide or furosemide: cross-reactivity, myth or reality? Am J Ophthalmol 2004;138:114–18.
30. Strom BL, Schinnar R, Apter AJ, Margolis DJ, Lautenbach E, Hennessey S, Bilker WB. Absence of cross-reactivity between sulfonamide antibiotics and sulfonamide nonantibiotics. New Engl J Med 2003;349:1628–35.
31. Taggart AJ, McDevitt DG. Diuretic withdrawal—a need for caution. Curr Med Res Opin 1983;8(7):501–8.
32. Grinstead WC, Francis MJ, Marks GF, Tawa CB, Zoghbi WA, Young JB. Discontinuation of chronic diuretic therapy in stable congestive heart failure secondary to coronary artery disease or to idiopathic dilated cardiomyopathy. Am J Cardiol 1994;73(12):881–6.
33. van Kraaij DJ, Jansen RW, Bouwels LH, Hoefnagels WH. Furosemide withdrawal improves postprandial hypotension in elderly patients with heart failure and preserved left ventricular systolic function. Arch Intern Med 1999;159(14):1599–605.
34. Berkowitz RL, Coustan DR, Mochizuki TK. Furosemide (Lasix). In: Handbook for Prescribing Medications during Pregnancy. Boston, MA: Little, Brown and Co, 1981:95.
35. Dailey JW. Anticoagulant and cardiovascular drugs. In: Wilson JT, editor. Drugs in Breast Milk. Lancaster: MTP Press, 1981:61.
36. Wigand ME, Heidland A. Ototoxic side-effects of high doses of frusemide in patients with uraemia. Postgrad Med J 1971;47(Suppl):54–6.
37. Hogg KJ, Hillis WS. Captopril/metolazone induced renal failure. Lancet 1986;1(8479):501–2.
38. Reed S, Greene P, Ryan T, Cerimele B, Schwertschlag U, Weinberger M, Voelker J. The renin angiotensin aldosterone system and frusemide response in congestive heart failure. Br J Clin Pharmacol 1995;39(1):51–7.
39. Mousson C, Justrabo E, Tanter Y, Chalopin JM, Rifle G. Néphrite interstitielle et hépatite aiguës granulomateuses d'origine médicamenteuse: rôle possible de l'association allopurinol–furosémide. [Acute granulomatous interstitial nephritis and hepatitis caused by drugs. Possible role of an allopurinol–furosemide combination.] Nephrologie 1986; 7(5):199–203.
40. Brown CB, Ogg CS, Cameron JS, Bewick M. High dose frusemide in acute reversible intrinsic renal failure. A preliminary communication. Scott Med J 1974;19(Suppl 1): 35–9.
41. Thomsen J, Bech P, Szpirt W. Otologic symptoms in chronic renal failure. The possible role of aminoglycoside–furosemide interaction. Arch Otorhinolaryngol 1976;214(1):71–9.
42. Lawson DH, Tilstone WJ, Gray JM, Srivastava PK. Effect of furosemide on the pharmacokinetics of gentamicin in patients. J Clin Pharmacol 1982;22(5–6):254–8.
43. Bates DE, Beaumont SJ, Baylis BW. Ototoxicity induced by gentamicin and furosemide. Ann Pharmacother 2002; 36(3):446–51.
44. Bartoli E, Arras S, Faedda R, Soggia G, Satta A, Olmeo NA. Blunting of furosemide diuresis by aspirin in man. J Clin Pharmacol 1980;20(7):452–8.
45. Jhund PS, Davie AP, McMurray JJ. Aspirin inhibits the acute venodilator response to furosemide in patients with chronic heart failure. J Am Coll Cardiol 2001;37(5):1234–8.
46. International PHVD Drug Trial Group. International randomised controlled trial of acetazolamide and furosemide in posthaemorrhagic ventricular dilatation in infancy. Lancet 1998;352(9126):433–40.
47. Hack M, Cohen AR. Acetazolamide plus furosemide for periventricular dilatation: lessons for drug therapy in children. Lancet 1998;352(9126):418–9.

48. Dodds MG, Foord RD. Enhancement by potent diuretics of renal tubular necrosis induced by cephaloridine. Br J Pharmacol 1970;40(2):227–36.
49. Simpson IJ. Nephrotoxicity and acute renal failure associated with cephalothin and cephaloridine. NZ Med J 1971;74(474):312–5.
50. Norrby R, Stenqvist K, Elgefors B. Interaction between cephaloridine and furosemide in man. Scand J Infect Dis 1976;8(3):209–12.
51. Sudoh T, Fujimura A, Shiga T, Sasaki M, Harada K, Tateishi T, Ohashi K, Ebihara A. Renal clearance of lomefloxacin is decreased by furosemide. Eur J Clin Pharmacol 1994;46(3):267–9.
52. Rawles JM. Antagonism between non-steroidal anti-inflammatory drugs and diuretics. Scott Med J 1982;27(1):37–40.
53. Hurtig HI, Dyson WL. Letter: Lithium toxicity enhanced by diuresis. N Engl J Med 1974;290(13):748–9.
54. Juurlink DN, Mamdani MM, Kopp A, Rochon PA, Shulman KI, Redelmeier DA. Drug-induced lithium toxicity in the elderly: a population-based study. J Am Geriatr Soc 2004;52:794–8.
55. Plouvier B, Baclet JL, de Coninck P. Une association néphrotoxique: mannitol et furosémide. [A nephrotoxic combination: mannitol and furosemide.] Nouv Presse Méd 1981;10(21):1744–5.
56. Miller RD, Sohn YJ, Matteo RS. Enhancement of d-tubocurarine neuromuscular blockade by diuretics in man. Anesthesiology 1976;45(4):442–5.
57. Brooks PM, Bell P, Lee P, et al. The effect of frusemide on indomethacin plasma levels. Br J Clin Pharmacol 1974;1:485.
58. Benet LZ. Pharmacokinetics/pharmacodynamics of furosemide in man: a review. J Pharmacokinet Biopharm 1979;7(1):1–27.
59. Muller FO, Middle MV, Schall R, Terblanche J, Hundt HK, Groenewoud G. An evaluation of the interaction of meloxicam with frusemide in patients with compensated chronic cardiac failure. Br J Clin Pharmacol 1997;44(4):393–8.
60. Azar I, Cottrell J, Gupta B, Turndorf H. Furosemide facilitates recovery of evoked twitch response after pancuronium. Anesth Analg 1980;59(1):55–7.
61. Ahmad S. Renal insensitivity to frusemide caused by chronic anticonvulsant therapy. BMJ 1974;3(5932):657–9.
62. Fine A, Henderson IS, Morgan DR, Tilstone WJ. Malabsorption of frusemide caused by phenytoin. BMJ 1977;2(6094):1061–2.
63. Famularo G, Corsi FM, Giacanelli M. Iatrogenic worsening of hypokalemia and neuromuscular paralysis associated with the use of glucose solutions for potassium replacement in a young woman with licorice intoxication and furosemide abuse. Acad Emerg Med 1999;6(9):960–4.
64. Sarkar SK. Stridor due to drug-induced hypokalaemic alkalosis. J Laryngol Otol 1987;101(2):197–8.
65. Brater DC, Anderson S, Baird B, Campbell WB. Effects of ibuprofen, naproxen, and sulindac on prostaglandins in men. Kidney Int 1985;27(1):66–73.
66. Piscitelli SC, Forrest A, Lush RM, Ryan N, Whitfield LR, Figg WD. Pharmacometric analysis of the effect of furosemide on suramin pharmacokinetics. Pharmacotherapy 1997;17(3):431–7.
67. Lochan SR, Adeniyi-Jones S, Assadi FK, Frey BM, Marcus S, Baumgart S. Coadministration of theophylline enhances diuretic response to furosemide in infants during extracorporeal membrane oxygenation: a randomized controlled pilot study. J Pediatr 1998;133(1):86–9.
68. Freestone S, Ramsay LE. Frusemide and spironolactone in resistant hypertension: a controlled trial. J Hypertens 1983;1(Suppl 2):326.
69. Wollam GL, Tarazi RC, Bravo EL, Dustan HP. Diuretic potency of combined hydrochlorothiazide and furosemide therapy in patients with azotemia. Am J Med 1982;72(6):929–38.
70. Yeung MY, Smyth JP. Concurrent frusemide–theophylline dosing reduces serum vancomycin concentrations in preterm infants. Aust J Hosp Pharm 1999;29:269–72.
71. Yeung MY, Smyth JP. Concurrent frusemide–theophylline dosing reduces serum vancomycin concentrations in preterm infants. Aust J Hosp Pharm 1999;29:269–72.

Hydrochlorothiazide

Hydrochlorothiazide is a thiazide diuretic, the adverse effects of all of which are similar, as discussed in the general monograph at the start of this section.

Respiratory

There have been several reports of non-cardiogenic pulmonary edema as a paradoxical adverse effect of hydrochlorothiazide (1,2,3,4,5,6,7,8,9,10). This rare reaction has been suggested to be allergic.

Sensory systems

Bilateral angle closure glaucoma has been reported with hydrochlorothiazide (11).

- A 70-year-old woman with hypertension developed bilateral worsening vision over 1 day. She had started to take hydrochlorothiazide 1 month before. Visual acuity was substantially impaired and intraocular pressures were about 40 mmHg bilaterally. Slit lamp examination showed bilateral shallow anterior chambers with anterior bowing of the iris and a centered posterior chamber intraocular lens. She was given dorzolamide, timolol, brimonidine, and prednisolone acetate and by the next day the intraocular pressures were ≈ 15 mmHg, although the anterior chambers remained shallow. Anterior choroidal effusions were suspected and the hydrochlorothiazide was withdrawn. After 3 days later the best-corrected vision was substantially improved, deep anterior chambers were present bilaterally, and the intraocular pressures were 12 mmHg.

Hydrochlorothiazide has only been associated with bilateral secondary angle closure once before. The proposed mechanism involves an idiosyncratic reaction in the uvea, with expansion of the extracellular tissues of the ciliary body and choroid. Presumably in cases in which there is mild expansion of the ciliary body and choroid there might only be a period of transient myopia. If uveal expansion is unrelenting there is appositional closure of the angle to such a degree that the intraocular pressure cannot equilibrate to normal. It is unclear whether there is cross-reactivity amongst various sulfonamide derivatives for this ocular complication.

Pancreas

Fatal necrotizing pancreatitis has been reported in a patient taking hydrochlorothiazide (12).

- A 49-year-old man with hypertension, who had taken hydrochlorothiazide 12.5 mg/day and lisinopril 10 mg/day for 1 year, developed severe diffuse abdominal pain and vomiting. The diagnosis was pancreatic inflammation and necrosis. Other causes of pancreatitis, including hypertriglyceridemia, hypercalcemia, HIV infection, alcohol, and trauma, were excluded. Despite aggressive treatment he developed multi-organ failure and refractory shock and died shortly thereafter.

Thiazide-induced necrotizing pancreatitis is very rare, only three cases having previously been reported. The exact mechanism is incompletely understood and the susceptibility factors are not known. The intensity varies and can range from mild to life-threatening.

References

1. Goetschalckx K, Ceuppens J, Van Mieghem W. Hydrochlorothiazide-associated noncardiogenic pulmonary oedema and shock: a case report and review of the literature. Acta Cardiol 2007;62(2):215–20.
2. Eisenhut M. Hydrochlorothiazide induced pulmonary oedema and alveolar epithelial ion transport. Int J Cardiol 2007;118(1):118–9.
3. Gamboa PM, Achotegui V, Irigoyen J, Pérez-Asenjo J, Merino J, Sanz ML. Hydrochlorothiazide-induced acute non-cardiogenic pulmonary edema. J Investig Allergol Clin Immunol.2005;15(4):299–301.
4. Solar M, Ceral J, Kvasnicka J. Hydrochlorothiazide induced pulmonary edema—a rare side effect of common diuretic drug. Int J Cardiol 2006;112(2):251–2.
5. Knowles SR, Wong GA, Rahim SA, Binkley K, Phillips EJ, Shear NH. Hydrochlorothiazide-induced noncardiogenic pulmonary edema: an underrecognized yet serious adverse drug reaction. Pharmacotherapy 2005;25(9):1258–65.
6. Jara Chinarro B, de Miguel Díez J, García Satue JL, Juretschke Moragues MA, Serrano Iglesias JA. [Acute non-cardiogenic pulmonary edema secondary to hydrochlorothiazide therapy.] Arch Bronconeumol 2003; 39(2):91–3.
7. d'Aloia A, Fiorina C, Vizzardi E, Faggiano P, Dei Cas L. [Recurrent episodes of acute non-cardiogenic pulmonary edema caused by ingesting hydrochlorothiazide.] Ital Heart J Suppl 2001;2(8):904–7.
8. Torné Cachot J, Peñarrubia María MT, Moner Coromina L. [Non-cardiogenic acute pulmonary edema and recurrent leukopenia caused by hydrochlorothiazide.] Rev Clin Esp 2001;201(2):107–8.
9. Almoosa KF. Hydrochlorothiazide-induced pulmonary edema. South Med J 1999;92(11):1100–2.
10. Bernal C, Patarca R. Hydrochlorothiazide-induced pulmonary edema and associated immunologic changes. Ann Pharmacother 1999;33(2):172–4.
11. Lee GC, Tam CP, Danesh-Meyer HV, Myers JS, Katz LJ. Bilateral angle closure glaucoma induced by sulphonamide-derived medications. Clin Experiment Ophthalmol 2007;35:55–8.
12. Bedrossian S, Vahid B. A case of fatal necrotizing pancreatitis: complication of hydrochlorothiazide and lisinopril therapy. Dig Dis Sci 2007;52:558–60.

Indapamide

General Information

Indapamide is a thiazide-like diuretic. Although it was introduced as a specific antihypertensive drug without appreciable diuretic action, the effects of indapamide are no different from those of bendroflumethiazide.

Organs and Systems

Cardiovascular

Indapamide has been associated with QT interval prolongation and torsade de pointes independent of electrolyte changes (1).

- A 42-year-old woman with systemic lupus erythematosus and hypertension developed palpitation and near syncope. Her medications included prednisolone 5 mg/day and indapamide 2.5 mg/day. A baseline electrocardiogram showed sinus rhythm, several episodes of non-sustained ventricular tachycardia, and a QT_c interval of 510 ms. Serum potassium, magnesium, and free calcium concentrations were all normal. Despite withdrawal of indapamide and prednisolone, torsade de pointes occurred and degenerated into ventricular fibrillation. Over the next two days the QT_c interval gradually shortened to 430 ms. Coronary angiography did not show any significant stenotic lesions.

Several non-antidysrhythmic drugs, including diuretics, prolong cardiac repolarization, predisposing to torsade de pointes. Diuretic-induced hypokalemia can exacerbate this effect. However, indapamide blocks the slow component of the delayed inward rectifier potassium current, leading to excessive lengthening of cardiac repolarization and thereby a predisposition to torsade de pointes. Indapamide should therefore be used with caution in patients who are at risk of torsade de pointes, especially when the patient is also taking drugs that block the rapid component of the delayed inward rectifier potassium current. Such drugs include class III antidysrhythmic drugs, macrolide antibiotics, antipsychotic drugs, and antihistamines.

Sensory systems

Transient myopia associated with diffuse choroidal thickening has been described in a 38-year-old white man who had taken indapamide for hypertension; it resolved after withdrawal (2).

Metabolism

The metabolic effects of indapamide appear to be as common as those of thiazides (SEDA-14, 185) (SEDA-15, 216). The metabolic effects of hydrochlorothiazide 25 mg/day and indapamide 2.5 mg/day for 6 months have been compared in a randomized, double-blind study in 44 patients with mild to moderate hypertension (3). There was little difference between the effects of the drugs on a wide range of lipid parameters, glucose, and potassium. The purported metabolic differences with indapamide are unlikely to be of sufficient magnitude to warrant its preferential use in hyperlipidemia.

Electrolyte balance

Cases of severe indapamide-induced hypokalemia have been reported (SED-11, 424) (SEDA-15, 213).

Liver

Severe acute hepatitis associated with indapamide has been reported (4). Serum bilirubin and liver enzymes were greatly raised. All normalized over 6 months after withdrawal of indapamide.

Urinary tract

Interstitial nephritis leading to acute renal insufficiency has been reported with indapamide (5).

Skin

Hypersensitivity to indapamide can provoke serious adverse skin reactions (SEDA-17, 260) (SEDA-18, 234).

One patient had several episodes of fixed drug eruption during treatment with indapamide (6). The diagnosis was confirmed by positive controlled oral challenge.

Pemphigus foliaceus has been described in relation to indapamide (7).

Drug–Drug Interactions

Sulfamethoxazole

The possibility of cross-reactivity with other sulfonamide derivatives was investigated by controlled oral challenge tests with sulfamethoxazole, sulfadiazine, and furosemide. The test with sulfamethoxazole was positive.

References

1. Letsas KP, Alexanian IP, Pappas LK, Kounas SP, Efremidis M, Sideris A, Kardaras F. QT interval prolongation and torsade de pointes associated with indapamide. Int J Cardiol 2006;112:373–4.
2. Blain P, Paques M, Massin P, Erginay A, Santiago P, Gaudric A. Acute transient myopia induced by indapamide. Am J Ophthalmol 2000;129(4):538–40.
3. Spence JD, Huff M, Barnett PA. Effects of indapamide versus hydrochlorothiazide on plasma lipids and lipoproteins in hypertensive patients: a direct comparison. Can J Clin Pharmacol 2000;7(1):32–7.
4. Safer L, Ben Mimoun H, Brahem A, Hanza J, Harzallah S, Abdellati S, Bdioui F, Saffar H. Severe acute hepatitis induced by indapamide. Report of a case. Sem Hop Paris 1998;74:1274.
5. Newstead CG, Moore RH, Barnes AJ. Interstitial nephritis associated with indapamide. BMJ 1990;300(6735):1344.
6. De Barrio M, Tornero P, Zubeldia JM, Sierra Z, Matheu V, Herrero T. Fixed drug eruption induced by indapamide. Cross-reactivity with sulfonamides. J Investig Allergol Clin Immunol 1998;8(4):253–5.
7. Bayramgurler D, Ercin C, Apaydin R, Unal G. Indapamide-induced pemphigus foliaceus. J Dermatol Treat 2001; 12(3):175–7.

Mannitol

General Information

Mannitol is an osmotic diuretic that has been used in acute oliguric renal insufficiency, acute cerebral edema, and the short-term management of glaucoma, especially to reduce intraocular pressure before ophthalmic surgery. Other indications include promotion of the excretion of toxic substances by forced diuresis, bladder irrigation during transurethral resection of the prostate, and oral administration as an osmotic laxative for bowel preparation. Mannitol is used as a diluent and excipient in pharmaceutical formulations and as a bulk sweetener.

Large doses of mannitol used in treating cerebral edema can alter extracellular fluid volume, osmolality, and composition and can lead under some circumstances to acute renal insufficiency, cardiac decompensation, and other complications (1). The patient's body habitus, age, total body water content relative to body weight, pretreatment plasma sodium concentration and plasma osmolality, and the presence of edema or ascites can influence the degree of extracellular fluid change and the rate of mannitol excretion to a significant degree.

Mannitol is excreted unchanged through the kidneys, and when renal function is impaired it accumulates and the movement of water into the intravascular space results in cellular dehydration. Two patients have been reported who suffered reversible acute oliguric renal insufficiency after mannitol infusion given as treatment for intracranial hypertension (2). Both had nausea and vomiting and became increasingly lethargic with the development of generalized edema. Congestive cardiac failure occurred. Laboratory tests showed severe dilutional hyponatremia with hyperosmolality.

The commonly recognized complications of osmotic agents used in patients with acute closed-angle glaucoma are mild headache, neck pain, nausea, and vomiting.

Uses of mannitol

The number of uses of mannitol continues to increase.

Reducing raised intracranial pressure

The use of mannitol in the treatment of raised intracranial pressure has been reviewed, in the light of disagreements about the appropriate timing of administration, the optimal fluid management protocol, and the mechanisms of action of osmotic diuretics (3,4,5).

The effects of four methods of infusion of mannitol and glycerol on raised intracranial pressure, as monitored by epidural pressure recordings, have been studied in 65 patients (6).

A. mannitol 0.5 g/kg was infused over 15, 30, or 60 minutes;
B. mannitol 1.0 g/kg was infused over 30, 60, or 90 minutes;
C. glycerol 0.5 g/kg in 5% fructose was infused over 30, 60, or 90 minutes;
D. glycerol 1.0 g/kg was infused over 60, 120, or 180 minutes.

In group A, there were no differences in the reduction in intracranial pressure across the three infusion rates. In group B, the degree of reduction in intracranial pressure increased with shorter times of infusion. In groups C and D the reduction in intracranial pressure was inversely related to the rate of infusion. In each group, the slower the infusion rate of the same dosage, the longer the reduction in intracranial pressure lasted. There was a rebound increase in intracranial pressure in 12% of those given mannitol and 34% of those given glycerol. The dose and the rate of mannitol infusion did not affect the rebound.

In 22 patients with meningoencephalitis and hypertensive cranial syndrome from cerebral edema, mannitol was given to 13 and dexamethasone to nine (7). There were three therapeutic failures in those given mannitol and none in those given dexamethasone, although the two drugs had similar effects on the duration of the hypertensive cranial syndrome (39–44 hours). The patients who were treated with mannitol had hyponatremia after 48 hours.

In 43 patients, osmotherapy with mannitol 20% and sorbitol 40% increased the serum concentration of lactate but not pyruvate, causing an increased ratio of lactate to pyruvate (8). Sorbitol had the greater effect, with a maximum at 1 hour. Mannitol had its maximum effect at 4 hours. The author concluded that acidosis, shock, diabetes mellitus, and hepatic dysfunctions increase the risk of osmotherapy, especially with sorbitol.

Randomized trials of mannitol in patients with acute traumatic brain injury of any severity have been reviewed (9). In the pre-operative management of patients with acute intracranial hemorrhage high-dose mannitol reduced mortality (RR = 0.55; 95%CI = 0.36, 0.84) and reduced death and severe disability (RR = 0.58; 95%CI = 0.45, 0.74) compared with conventional-dose mannitol. In one trial treatment intended to lower intracranial pressure was compared with "standard care" (RR for death = 0.83; 95%CI = 0.47, 1.46). In one trial mannitol and pentobarbital were compared (RR for death = 0.85; 95% CI = 0.52, 1.38). In one trial pre-hospital mannitol was compared with placebo (RR for death = 1.75; 95% CI = 0.48, 6.38). The reviewers concluded that high-dose mannitol is preferable to conventional-dose mannitol in the pre-operative management of patients with acute intracranial hematomas. However, there is little evidence about the use of mannitol as a continuous infusion in patients with raised intracranial pressure who do not have an operable intracranial hematoma.

In 20 patients with head trauma and persistent coma who required infusions of an osmotic agent to treat episodes of intracranial hypertension resistant to standard modes of therapy, isovolumic infusions of either 7.5% hypertonic saline reduced the number of episodes of intracranial hypertension per day (6.9 versus 13.3) and the daily duration of episodes of intracranial hypertension (67 versus 131 minutes) compared with 20% mannitol (10).

Stroke

In 805 patients who were given intravenous mannitol (mean dose, 47 g/day; mean duration, 6 days) or no treatment within 72 hours of the onset of a stroke, the case fatality was 25% versus 16% at 30 days and 38% versus 25% at 1 year (11). The prognostic scores on the Scandinavian Neurological Stroke Scale were similar in treated and untreated patients, both in ischemic and hemorrhagic strokes. However, the patient groups differed in several factors that might have affected survival. Thus, this uncontrolled study was inconclusive.

Serum and cerebrospinal fluid osmolarity were measured in 30 patients with severe head injuries or subarachnoid hemorrhage, 10 of whom received mannitol for at least 72 hours, 10 of whom received it for 24–48 hours, and 10 of whom were controls (12). Serum osmolarity increased quickly in all those who received mannitol and was unchanged in controls. Average cerebrospinal fluid osmolarity increased slowly in all those who received mannitol and was unchanged in controls. This is a potentially dangerous effect and the authors recommended that cerebrospinal fluid osmolarity should be measured regularly in all patients who receive mannitol for longer than 24 hours.

Disrupting the blood-brain barrier

Mannitol has been used to disrupt the blood-brain barrier temporarily in order to allow better penetration of chemotherapeutic drugs.

In eight patients with gliomas and one with a primary lymphoma of the central nervous system the blood-brain or blood-tumor barrier was reversibly opened by intra-arterial injection of hyperosmolar mannitol 25% (13). There was tumor regression or a tumor progression-free interval in five patients.

In 10 patients with malignant gliomas intra-arterial chemotherapy with 5-fluorouracil, nitrosourea, or interferon beta was given after osmotic blood-brain barrier disruption with intra-arterial 20% mannitol (14). In nine evaluable cases there were one complete and three partial responses; in five there was no change and no progressive disease on CT. The most untoward effect was myelosuppression: platelet and leukocyte counts fell below 20×10^9/l and 2×10^9/l respectively in three patients, of whom two died of severe infections. The other complications were eye pain during mannitol infusion in all cases in which selective catheterization of the internal carotid artery failed to pass the origin of the ophthalmic artery. There was reduced activity in 70%, nausea and vomiting in 50%, swelling of the external decompression area in 33%, and increased neurological deficits in 20%. However, all these adverse effects were transient.

Of 21 patients with malignant brain tumors, 16 were treated by operation, irradiation, and two or more courses of intracarotid infusion of nitrosourea 100 mg after 20% mannitol 200 ml, and five were treated similarly but without mannitol (15). The 2-year survival rate in those who received mannitol was 79% (11 of 14 cases followed for longer than 2 years) and the 3-year survival rate was 67%. Five of seven patients with grade 4 astrocytomas survived for more than 18 months, whereas four of five patients with grade 4 astrocytomas who did not receive mannitol died within 18 months.

Over 4 years, 37 patients with high-grade malignant gliomas underwent 246 treatment procedures with a combination of methotrexate, cyclophosphamide, and procarbazine given together with hyperosmolar mannitol-induced transient breakdown of the blood-brain barrier (16). There were complete remissions in 16% and 24 patients (65%) had partial or temporary remissions. Progression-free intervals were 1–47 (mean 15) months and median survival was 22 months. Neurotoxicity was minimal with one periprocedural death and five instances of worsened neurological deficits after a procedure.

The delivery of chemotherapeutic agents in the treatment of malignant brain tumors is improved by osmotic opening of the blood-brain barrier by prior infusion of mannitol into the internal carotid or vertebral artery. Over 4200 blood-brain barrier disruption procedures have been performed in over 400 patients with primary central nervous system lymphomas, gliomas, primitive neuroectodermal tumors, germ cell and metastatic cancers in the National Blood-Brain Barrier Program (17). In patients with primary nervous system lymphomas, long-lasting responses have been obtained without loss of cognitive function and without the use of radiotherapy. The results in patients with primitive neuroectodermal tumors and germ cell tumors are also said to be very encouraging.

The efficacy of mannitol in augmenting the tumoricidal effect of etoposide has been studied in 99 children aged 1–21 years with recurrent brain tumors (18). They were randomly assigned to intravenous etoposide 150 mg/m^2 with or without mannitol 15 g/m^2, daily for 5 days every 3 weeks for 1 year or until disease progression or death. CT or MRI scans, obtained after three cycles of therapy, were compared with pre-therapy scans. Of 87 evaluable patients, 12 had an objective response according to the radiologist and of 66 patients reviewed centrally, seven responded (two of 12 low grade astrocytomas, four of 26 medulloblastomas or primitive neuroectodermal tumors, one of 13 high-grade astrocytomas, and one of 15 brain stem gliomas). Survival at 1 year was 53% for low grade astrocytomas, 38% for medulloblastomas or primitive neuroectodermal tumors, 28% for high-grade astrocytomas and 9% for brain stem gliomas. Mannitol had no beneficial effect.

Use in bowel cleansing

Mannitol has been used for preoperative bowel cleansing before radiological investigations (19), diagnostic and operative endoscopy (20), and bowel surgery.

In whole gut irrigation mannitol is badly tolerated and leaves a bowel full of gas and fluid although it causes only small changes in serum electrolytes (21). In a study of the effect of an intravenous infusion of saline on the volume of rectal effluent and quality of bowel preparation produced by a smaller oral dose of mannitol, 19 patients drank 2–3 l of 5% mannitol, supplemented by an intravenous infusion of isotonic saline and 19 patients drank 4–5 l of 5% mannitol (22). The volume of rectal effluent and the quality of bowel preparation was the same in both groups. Loss of sodium in the oral group was corrected by the intravenous infusion, but the infusion resulted in greater water retention. There was no difference in the incidence of vomiting between the two groups.

Polyethylene glycol electrolyte lavage solution has been compared with 10% mannitol for preoperative colonic cleansing in 80 patients (23). Colonic cleansing was better with polyethylene glycol (90% optimal cleansing versus 75%). mannitol caused subclinical dehydration according to hematological, biochemical, and weight changes before and after bowel preparation and caused more nausea, cramps, and abdominal pain. Two patients given mannitol had combustible amounts of hydrogen gas in the colon.

Three formulations, two based on magnesium citrate and one an optimized oral mannitol regimen, have been compared for their effectiveness in clearing the large bowel before double-contrast barium enema and for effects on barium mucosal coating (24). The formulations based on magnesium citrate were equally good and caused significantly less nausea and vomiting than mannitol. The authors concluded that mannitol should not be used for preparing the bowel for barium enema.

Two hypotonic non-hemolysing irrigating solutions, sorbitol + mannitol (2% + 1%) and glycine (1.5%), have been compared in patients undergoing transurethral resection of the prostate (25). Ethanol (1%) was added to the irrigating fluid as a marker to allow early detection of fluid absorption by breath analysis. There was very little absorption (less than 1 liter).

However, in other cases large volumes of fluid have been absorbed. In 39 patients having transurethral resection of the prostate for benign prostatic hyperplasia, large quantities of mannitol, which was used as the irrigating fluid, entered the circulation (26). There was a corresponding fall in serum sodium concentration. Patients who had serum mannitol concentrations over 4 mg/ml had hypotension and bradycardia; because they were nearly all hypovolemic, the bradycardia was thought to be inappropriate.

Irrigating fluid bags containing mannitol 3% or glycine 1.5%, both with added ethanol 1% as an indicator of fluid absorption, were used to investigate adverse effects in a randomized, double-blind study during 394 transurethral prostatic resections (27). The incidence of 13 symptoms was studied in 52 patients (13%) who absorbed more than 500 ml of fluid. The incidence of circulatory symptoms did not differ between the fluids, but the risk of neurological symptoms, such as nausea, was 4.8 times higher with glycine 1.5%. An increase of 1000 ml. in the volume of irrigant absorbed increased the overall risk of circulatory symptoms by a factor of 3.4 and the risk of neurological symptoms by a factor of 4.4. The authors concluded that absorption of mannitol 3% during transurethral prostatic resection is associated with fewer neurological symptoms than glycine 1.5%.

In 80 patients randomized for precolonoscopic cleansing with either 10% mannitol 750 ml or sodium phosphate 180 ml there were statistically significant differences in serum sodium, phosphorus, potassium, and calcium between the two groups, but no clinical symptoms and no significant differences in the frequencies of adverse effects (28). Six of eight patients who were treated with sodium phosphate and who had had mannitol for a previous colonoscopy preferred sodium phosphate. The endoscopists, who were blinded to the treatment, reported excellent or good bowel preparation in 85% of those prepared with sodium phosphate versus 83% for mannitol. The authors concluded that although the quality of preparation and the frequencies of adverse effects were similar with the two solutions, retention of sodium and phosphate ions contraindicates the use of sodium phosphate in patients with renal insufficiency, cirrhosis, ascites, and heart failure.

In a retrospective study of patients who underwent elective surgery for colorectal carcinomas, traditional bowel preparation was performed the day before the operation either with oral castor oil 30 ml and three soap enemas (n = 154) or with mannitol 500 ml (n = 36) (29). There were infectious wound complications in 26 patients (17%) pretreated with castor oil compared with 13 patients (36%) treated with mannitol. There were no differences in the incidence of anastomotic leaks or mortality rate.

Systemic antimicrobial prophylaxis with metronidazole and gentamicin has been compared with metronidazole alone in elective colorectal surgery in a prospective randomized trial, in which all the patients received 10% mannitol solution before surgery (30). Although there were no serious infections in either group, the incidence of superficial wound infections was relatively high: 19% in those given metronidazole and gentamicin prophylaxis and 25% in those given metronidazole alone. Escherichia coli was isolated from all these wounds, and no obligate anerobic bacteria were cultured. The high rate of wound infection was probably caused by overgrowth after irrigation, due to residues of mannitol in the colon, which serve as a nutrient for Escherichia coli.

Both sets of authors concluded that mannitol should not be used for preoperative mechanical preparation of the large bowel before elective colorectal surgery.

Hydrogen gas can accumulate in the colon after the administration of mannitol (23). This can cause a risk of explosion.

- Colonic explosion during colonoscopic polypectomy occurred after mannitol had been used for bowel preparation and the colon was completely clean (31). In spite of emergency surgery, with transfusion of 45 units of blood, uncontrollable hemorrhage persisted from multiple bleeding points and the patient died.

Use as a radiocontrast medium

In 56 patients undergoing abdominal CT the gastrointestinal tract was defined by negative contrast with 2.5% mannitol instead of the conventional positive contrast from an iodine-containing contrast medium (32). The number of artifacts due to high-contrast boundaries was slightly greater with negative contrast than it would have been with positive contrast, but differentiation of the gastrointestinal tract from other abdominal organs was equally good. Negative contrast was poor for diagnosing cystic tumors but much better than positive contrast for evaluating the wall of the gastrointestinal tract.

The effect of oral mannitol in an aqueous solution in enhancing pelvic MRI has been reported in a retrospective study in 72 patients with suspected or proven pelvic abnormalities: In 36 patients bowel marking was not carried out and in 36 patients the bowel was contrast-enhanced by oral mannitol 1000 ml (33). Mannitol significantly improved delineation of the intestinal structures and pelvic organs or pathological lesions, but eight patients had diarrhea, nausea, or meteorism.

Diagnosis of diarrhea

In chronic diarrhea intestinal permeability to sugars, such as raffinose, lactose, lactulose, sucrose and mannitol (34), can detect intestinal damage. The absorption of a combined dose of lactulose and mannitol has been studied in 261 consecutive patients with three or more bowel movements daily for at least 3 weeks; 120 (46%) were found to have an organic cause for chronic diarrhea, whereas in 141 (54%) a functional condition was diagnosed (35). The lactulose/mannitol test and C-reactive protein were independent predictors for the final diagnosis of an organic cause of chronic diarrhea, with odds ratios of 1.5 (95% CI = 1.29, 1.78) and 5.2 (95% CI = 1.90, 14.12) respectively.

Cardiopulmonary bypass

The effects of mannitol and dopamine, alone and in combination, on beta$_2$-microglobulin excretion rates in 100 patients undergoing coronary artery bypass graft surgery with cardiopulmonary bypass have been studied in a double-blind, randomized, placebo-controlled study (36). Mannitol 1 g/kg was added to the cardiopulmonary bypass prime and dopamine 2 micrograms/kg/minute was given from the time of induction of anesthesia to 1 hour after bypass. Dopamine significantly increased the excretion rate of beta$_2$-microglobulin compared with placebo. Thus, rather than being protective during cardiopulmonary bypass, dopamine reduces renal tubular dysfunction. This effect was not ameliorated by the addition of mannitol.

Treatment of cisplatin nephrotoxicity

Saline alone, saline + furosemide, and saline + mannitol have been used to prevent nephrotoxicity in 49 women who received cisplatin 75 mg/m^2 every 3 weeks (37). Hydration with saline or saline + furosemide was associated with less cisplatin nephrotoxicity than hydration with saline + mannitol.

Comparative studies

Equiosmolar loads of mannitol 20% and hypertonic saline 7.5% over 10 minutes have been compared with isotonic saline in 30 ASA I and II patients undergoing non-hemorrhagic surgery under general anesthesia (38). The serum sodium concentration was lowest after mannitol (129 mmol/l) at the end of the infusion and highest after hypertonic saline (151 mmol/l), with normalization at 60 minutes. The hemodynamic effects were similar in the three groups. Temperature was lower after isotonic saline, because of the volume infused.

Organs and Systems

Cardiovascular

Glycine (1% or 1.5%) 15 ml/kg, or 3% mannitol (all containing 1% ethanol), or sorbitol 2% + mannitol 1% (with no ethanol) were infused intravenously over 20 minutes into 10 healthy men to determine their hemodynamic effects using Doppler ultrasonography (39). All reduced cardiac output 30 minutes after infusion, and glycine reduced the heart rate and cardiac output and raised mean arterial pressure, indicating an increase in systemic resistance. Almost identical breath-ethanol curves were obtained with the three fluids containing ethanol and all of them caused slight hypoglycemia. There was no evidence of ethanol-induced tachycardia.

Respiratory

Inhaled mannitol increases airway responsiveness in asthmatic subjects (40) and may act through mast cell activation (41). It has been used to investigate the effects of therapeutic drugs or to predict responsiveness to their effects and to determine susceptibility to exercise-induced bronchoconstriction (42). In 18 asthmatic subjects the inhaled glucocorticoid budesonide caused a reduction in airway sensitivity and reactivity to inhaled mannitol and this was associated with expected improvements in lung function and symptoms (43).

Nervous system

Rebound intracranial hypertension can develop as a result of mannitol administration in neurological patients and is associated with an increase in CSF mannitol and osmolality after intravenous administration (44,45). The reported incidence of rebound intracranial hypertension with mannitol varies widely. There is no agreement that osmotic gradients and their reversal are the respective mechanisms of intracranial pressure reduction and rebound after mannitol. Another explanation is that mannitol reduces intracranial pressure by reducing viscosity and compensatory vasoconstriction, thus reducing cerebral blood volume and thereby reducing intracranial pressure. Rebound can then be explained by the delayed increase in viscosity after mannitol-induced diuresis and elimination of mannitol. It is unlikely that this provides a comprehensive explanation for the effects of mannitol. It may be that passage of mannitol across the blood–brain barrier is significantly influenced by the presence and nature of neurological disease. There may be patient subsets, such as those with diffusely increased blood–brain barrier permeability or those with focally increased blood–brain barrier permeability but diffuse distribution of mannitol throughout the cerebrospinal fluid. Such patients might be especially susceptible. These observations need to be regarded as speculative for the time being, but the idea of identifying subsets of patients who are at special risk is challenging.

Sensory systems

The effects of intravenous mannitol on aqueous fluid protein concentration have been evaluated in healthy young adults (average age 20 years) and older adults (average age 61 years), and in patients with diabetes mellitus, hypertension, or pseudoexfoliation syndrome who were about to undergo intraocular surgery (average age 66 years) (46). Mannitol increased aqueous humor protein concentration in all subjects, with a maximum effects at around 1 hour. The magnitude and the duration of the effect were significantly greater in the healthy older subjects than in the young subjects, but the same in older adults with and without diseases. The effect was reversed within 6 hours.

Psychological, psychiatric

Acute mania has been reported in a patient who was given intravenous mannitol (47).

- A previously well 75-year-old woman without a personal or family history of mental disorders developed severe major depression and was given nortriptyline 50 mg/day. After 10 days a diagnosis of bilateral acute angle-closure glaucoma was made and she was given a continuous intravenous infusion of mannitol 20%, oral acetazolamide 500 mg, and topical pilocarpine 2%, timolol 0.5%, and dexamethasone 0.1% (each 1 drop every 15 minutes). She became euphoric 30 minutes later and remained overactive, overly affectionate, and talkative, telling jokes, with pressured speech and flight of ideas. Her manic state remitted 1 hour after the end of the mannitol infusion and her severe depression recurred dramatically.

Mannitol can cause both acute expansion of extracellular fluid volume and a rapid reduction of intracellular fluid volume with retention of brain electrolytes; these may have been the mechanisms in this case.

Electrolyte balance

Perioperative hyperkalemia has been reported with rapid infusion of mannitol after craniotomy (48).

- A 34-year-old man received 300 ml of a 20% solution of mannitol intravenously over 20 minutes immediately after a craniotomy for clipping of an anterior communicating artery aneurysm. Ten minutes after the end of the infusion there was electrocardiographic evidence of hyperkalemia (serum potassium concentration 5.4 mmol/l). After treatment with calcium gluconate 425 mg and 250 ml of 5% dextrose containing insulin 10 U, the serum potassium fell to 3.6 mmol/l and the electrocardiographic changes resolved.
- A 68-year-old man received 500 ml of a 20% solution of mannitol intravenously over 45 minutes immediately after a craniotomy for evacuation of an intracerebral hematoma, and 60 minutes after the end of the infusion there was electrocardiographic evidence of hyperkalemia (serum potassium concentration 6.1 mmol/l, the baseline value having been 4.1 mmol/l). After 40ml of a 7% bicarbonate solution the serum potassium fell to 5.4 mmol/l and the electrocardiographic changes resolved.
- A 31-year old woman received two 40 g doses of intravenous mannitol over about 45 minutes during a right frontal craniotomy and debulking of a recurrent astrocytoma (49). About 15 minutes later she developed peaked T waves on the electrocardiogram in conjunction with a plasma potassium concentration of 6.1 mmol/l. Serum potassium was lowered and the electrocardiographic changes resolved.
- A 41-year old man received 200 g of 20% intravenous mannitol during elective craniotomy for clipping of a left middle cerebral artery aneurysm (50). After about 40 minutes, about 120 g having been infused, his electrocardiogram showed peaked T waves and a prolonged QRS duration. Despite withdrawal of mannitol the electrocardiogram deteriorated into a sine-wave ventricular tachycardia and soon afterwards coarse ventricular fibrillation. The serum potassium concentration was 7.5 mmol/l and the serum sodium 116 mmol/l. The serum potassium was reduced to 3.9 mmol/l and sinus rhythm was eventually restored.

These cases illustrate that rapidly infused hypertonic mannitol can trigger an extracellular shift of potassium, with the potential for cardiotoxicity, despite there being only modest increases in serum potassium concentration. The rate of change of serum potassium concentration as well as the absolute concentration reached has a bearing on the development of hyperkalemic cardiotoxicity. Accordingly, plasma potassium concentration should be monitored whenever significant quantities of mannitol (80 grams or more) are given rapidly (in under 1 hour).

Gastrointestinal

Oral mannitol can raise intracolonic hydrogen concentrations sufficiently to cause explosions at diathermy (51).

Urinary tract

Four cases of acute renal insufficiency have been described in men aged 20–42 years who received mannitol 1172 (sd 439) g over 58 (sd 28) hours (52). The onset of acute renal insufficiency was detected 48 (sd 22) hours after the start of infusion. All the patients had dilutional hyponatremia (average 120 mmol/l) and serum hyperosmolarity (osmolar gap 70 mosm/kg water). In the three anuric cases, in which hemodialysis was performed, there was immediate recovery of diuresis. This emphasizes the risk of renal insufficiency with mannitol and stresses the importance of early hemodialysis. Mannitol is dialysable and once its suppressive effect on renal perfusion is eliminated functional recovery is prompt.

Acute renal insufficiency after infusion of mannitol has most commonly been seen after neurosurgery when mannitol has been used to reduce intracranial pressure. Treatment guidelines recommend that mannitol be withheld if the serum osmolality exceeds 320 mOsm/kg, lest renal insufficiency develops. In a retrospective analysis of 95 patients treated with mannitol in a neurology–neurosurgery intensive care unit, 11 met criteria for acute renal insufficiency (serum creatinine increase by 44 μmol/l (0.5 mg/dl) or 88 μmol/l if the baseline value was 176 μmol/l or more) (53). There was no relation between the onset of acute renal insufficiency and total mannitol dose, maximum daily dose of mannitol, serum osmolality, or osmotic gap. The serum osmolality and the osmotic gap measured just before the onset of acute renal insufficiency did not differ from values observed in patients who did not develop acute renal insufficiency. The severity of illness and the presence of co-morbid diseases more often predicted the development of acute renal insufficiency than serum osmolality did. The concept of a specific osmolality barrier of 320 mOsm/kg for acute renal insufficiency, beyond which mannitol should be withheld, needs to be reconsidered.

Musculoskeletal

Mannitol can cause a compartment syndrome if it extravasates into soft tissues (54,55).

- A 17-year-old woman was treated for carbosulfan poisoning with atropine, isotonic saline, and intravenous mannitol (20%) (54). Her right forearm became edematous, cyanotic, and tender at the site of the mannitol infusion and compartment pressures were raised. A fasciotomy was required and recovery occurred over several days.

References

1. Oken DE. Renal and extrarenal considerations in high-dose mannitol therapy. Ren Fail 1994;16(1):147–59.
2. Suzuki K, Miki M, Ono Y, Saito Y, Yamanaka H. [Acute renal failure following mannitol infusion.]Hinyokika Kiyo 1993;39(8):721–4.
3. Paczynski R-P. Osmotherapy. Basic concepts and controversies. Crit Care Clin 1997;13:105–29.
4. Better OS, Rubinstein I, Winaver JM, Knochel JP. Mannitol therapy revisited (1940–1997). Kidney Int 1997;52:886–94.
5. Hansen P-H, Rosenorn J, Westergaard L. Mannitolbehandling ved forhojet intrakranielt tryk. Ugeskr Laeger 1983;145:1125–7.

6. Node Y, Nakazawa S. Clinical study of mannitol and glycerol on raised intracranial pressure and on their rebound phenomenon. Adv Neurol 1990;52:359–63.
7. Sanchez R, Brindis LC, Fierro H, Strecker C, Munoz O. Uso de manitol y dexametasona en el manejo del edema cerebral agudo de origen infeccioso. Bol Med Hosp Infant Mex 1977;34:283–90.
8. Spring A. Veranderungen des Laktat-Pyruvat-Spiegels im Blut auf eine Osmotherapie mit Mannit und Sorbit. Neurochirurgia (Stuttg) 1980;23:176–81.
9. Roberts I, Schierhout G, Wakai A. Mannitol for acute traumatic brain injury. Cochrane Database Syst Rev 2003;(2):CD001049.
10. Vialet R, Albanese J, Thomachot L, Antonini F, Bourgouin A, Alliez B, Martin C. Isovolume hypertonic solutes (sodium chloride or mannitol) in the treatment of refractory posttraumatic intracranial hypertension: 2 ml/kg 7.5% saline is more effective than 2 ml/kg 20% mannitol. Crit Care Med 2003;31:1683–7.
11. Bereczki D, Mihalka L, Szatmari S, Fekete K, Di Cesar D, Fulesdi B, Csiba L, Fekete I. Mannitol use in acute stroke: case fatality at 30 days and 1 year. Stroke 2003;34:1730–5.
12. Polderman KH, van de Kraats G, Dixon JM, Vandertop WP, Girbes AR. Increases in spinal fluid osmolarity induced by mannitol. Crit Care Med 2003;31:584–90.
13. Heimberger K, Samec P, Binder H, Podreka I, Reisner T, Deecke L, Horaczek A, Dittrich C, Steger G, Zimpfer M, et al. Blood brain barrier modification and chemotherapy. Interventional neuroradiology in the treatment of malignant gliomas. Acta Radiol Suppl 1986;369:223–6.
14. Yamada K, Takahama H, Nakai O, Takanashi T, Hosoya T. Intra-arterial chemotherapy of malignant glioma after osmotic blood-brain barrier disruption [in Japanese].Gan To Kagaku Ryoho 1989;16:2692–6.
15. Miyagami M, Tsubokawa T, Tazoe M, Kagawa Y. Intra-arterial ACNU chemotherapy employing 20% mannitol osmotic blood-brain barrier disruption for malignant brain tumors. Neurol Med Chir (Tokyo) 1990;30:582–90.
16. Gumerlock MK, Belshe BD, Madsen R, Watts C. Osmotic blood-brain barrier disruption and chemotherapy in the treatment of high grade malignant glioma: patient series and literature review. J Neurooncol 1992;12:33–46.
17. Doolittle ND, Petrillo A, Bell S, Cummings P, Eriksen S. Blood-brain barrier disruption for the treatment of malignant brain tumors: The National Program. J Neurosci Nurs 1998;30:81–90.
18. Kobrinsky NL, Packer RJ, Boyett JM, Stanley P, Shiminski-Maher T, Allen JC, Garvin JH, Stewart DJ, Finlay JL. Etoposide with or without mannitol for the treatment of recurrent or primarily unresponsive brain tumors: a Children's Cancer Group Study. J Neurooncol. 1999;45:47–54.
19. Lou-Moller P, Olsen L, Schierbeck J, Hansen H, Christau B, Bonnevie O. Bisakodyl (Toilax) og peroral mannitol som udrensningsmidler for rontgenundersogelse af colon. Ugeskr Laeger 1983;145:3093–6.
20. Noya G, Dettori G, Muscas AG, Delogu L, Antona C, Marongiu G, Frassetto A, Biglioli P. Il mannitolo nella preparazione del colon alla endoscopia diagnostica ed operativa. Minerva Dietol Gastroenterol 1984;30:397–9.
21. Kujat R, Pichlmayr R. Nebenwirkungen verschiedener Spullosungen bei der orthograden Darmspulung. Chirurg Z Geb Operat Med 1983;54:669–72.
22. Hares MM, Nevah E, Minervini S, Bentley S, Keighley M, Alexander-Williams J. An attempt to reduce the side effects of mannitol bowel preparation by intravenous infusion. Dis Colon Rectum 1982;25:289–91.
23. Beck DE, Fazio VW, Jagelman DG. Comparison of oral lavage methods for preoperative colonic cleansing. Dis Colon Rectum 1986;29:699–703.
24. Foord KD, Morcos SK, Ward P. A comparison of mannitol and magnesium citrate preparations for double-contrast barium enema. Clin Radiol 1983;34:309–12.
25. Dimberg M, Norlen H, Allgen LG, Allgen T, Wallin M. A comparison between two hypotonic irrigating solutions used in transurethral resections of the prostate: sorbitol (2%)-mannitol (1%) and 1.5% glycine solutions. Scand J Urol Nephrol 1992;26:241–7.
26. Logie JR, Keenan RA, Whiting PH, Steyn JH. Fluid absorption during transurethral prostatectomy. Br J Urol 1980;52:526–8.
27. Hahn RG, Sandfeldt L, Nyman CR. Double-blind randomized study of symptoms associated with absorption of glycine 1.5% or mannitol 3% during transurethral resection of the prostate. J Urol 1998;160:397–401.
28. Habr-Gama A, Bringel RW, Nahas SC, Araujo SE, Souza-Junior AH, Calache JE, Alves PA. Bowel preparation for colonoscopy: comparison of mannitol and sodium phosphate. Results of a prospective randomized study. Rev Hosp Clin Fac Med Sao Paulo 1999;54:187–92.
29. Todorov AT, Mantchev ID, Atanasov TB. Traditional bowel preparation versus osmotic agent mannitol for preoperative colonic cleansing in elective colorectal surgery. Folia Med (Plovdiv) 2002;44:36–9.
30. Weidema WF, Van den Boogaard AE, Wesdorp RI, Van Boven CP, Greep JM. 24-hour systemic antimicrobial prophylaxis with gentamicin and metronidazole, or metronidazole alone, in elective colorectal surgery after mechanical bowel preparation with mannitol and whole gut irrigation. Acta Chir Belg 1985;85:349–53.
31. Bigard MA, Gaucher P, Lassalle C. Fatal colonic explosion during colonoscopic polypectomy. Gastroenterology 1979; 77:1307–10.
32. Schunk K, Wiessner J, Schadmand S, Kaltenborn H, Duber C, Brunier A. Zur Frage der Darmkontrastierung in der abdominellen Computertomographie. Rofo Fortschr Geb Rontgenstr Neuen Bildgeb Verfahr 1992;156:443–7.
33. Schunk K, Kersjes W, Schadmand-Fischer S, Grebe P, Kauczor HU, Thelen M. Eine Mannitollosung als orales Kontrastmittel in der pelvinen MRT. Rofo Fortschr Geb Rontgenstr Neuen Bildgeb Verfahr 1995;163:60–6.
34. Hessels J, Eidhof HH, Steggink J, Roeloffzen WW, Wu K, Tan G, Van de Stadt J, Van Bergeijk L. Assessment of hypolactasia and site-specific intestinal permeability by differential sugar absorption of raffinose, lactose, sucrose and mannitol. Clin Chem Lab Med 2003;41:1056–63.
35. Di Leo V, D'Inca R, Diaz-Granado N, Fries W, Venturi C, D'Odorico A, Martines D, Sturniolo GC. Lactulose/mannitol test has high efficacy for excluding organic causes of chronic diarrhea. Am J Gastroenterol 2003;98: 2245–52.
36. Carcoana OV, Mathew JP, Davis E, Byrne DW, Hayslett JP, Hines RL, Garwood S. Mannitol and dopamine in patients undergoing cardiopulmonary bypass: a randomized clinical trial. Anesth Analg 2003;97:1222–9.
37. Santoso JT, Lucci JA 3rd, Coleman RL, Schafer I, Hannigan EV. Saline, mannitol, and furosemide hydration in acute cisplatin nephrotoxicity: a randomized trial.Cancer Chemother Pharmacol 2003;52:13–18.
38. Erard AC, Walder B, Ravussin P. Effêts de charges equiosmolaires de mannitol 20%, de NaCl 7.5% et de NaCl 0.9% sur l'osmolarité, l'hémodynamique et les electrolytes plasmatiques. Ann Fr Anesth Réanim 2003;22:18–24.

39. Nilsson A, Randmaa I, Hahn RG. Haemodynamic effects of irrigating fluids studied by Doppler ultrasonography in volunteers. Br J Urol 1996;77:541–6.
40. Barben J, Roberts M, Chew N, Carlin JB, Robertson CF. Repeatability of bronchial responsiveness to mannitol dry powder in children with asthma. Pediatr Pulmonol 2003;36:490–4.
41. Brannan JD, Gulliksson M, Anderson SD, Chew N, Kumlin M. Evidence of mast cell activation and leukotriene release after mannitol inhalation. Eur Respir J 2003;22:491–6.
42. Holzer K, Anderson SD, Chan HK, Douglass J. Mannitol as a challenge test to identify exercise-induced bronchoconstriction in elite athletes. Am J Respir Crit Care Med 2003;167:534–7.
43. Brannan JD, Koskela H, Anderson SD, Chan HK. Budesonide reduces sensitivity and reactivity to inhaled mannitol in asthmatic subjects. Respirology 2002;7:37–44.
44. Kofke WA. Mannitol: potential for rebound intracranial hypertension? J Neurosurg Anesthesiol 1993;5(1):1–3.
45. Rudehill A, Gordon E, Ohman G, Lindqvist C, Andersson P. Pharmacokinetics and effects of mannitol on hemodynamics, blood and cerebrospinal fluid electrolytes, and osmolality during intracranial surgery. J Neurosurg Anesthesiol 1993;5(1):4–12.
46. Miyake Y, Miyake K, Maekubo K, Kayazawa F. Increase in aqueous flare by a therapeutic dose of mannitol in humans [in Japanese]. Nippon Ganka Gakkai Zasshi 1989;93:1149–53.
47. Navarro V, Vieta E, Gasto C. Mannitol-induced acute manic state. J Clin Psychiatry 2001;62(2):126.
48. Hirota K, Hara T, Hosoi S, Sasaki Y, Hara Y, Adachi T. Two cases of hyperkalemia after administration of hypertonic mannitol during craniotomy. J Anesth 2005;19:75–7.
49. Hassan ZU, Kruer JJ, Fuhrman TM. Electrolyte changes during craniotomy caused by administration of hypertonic mannitol. J Clin Anesth 2007;19:307–9.
50. Flynn BC. Hyperkalemia cardiac arrest with hypertonic mannitol infusion: the strong ion difference revisited. Anesth Analg 2007;104:225–6.
51. La Brooy SJ, Avgerinos A, Fendick CL, Williams CB, Misiewicz JJ. Potentially explosive colonic concentrations of hydrogen after bowel preparation with mannitol. Lancet 1981;1(8221):634–6.
52. Perez-Perez AJ, Pazos B, Sobrado J, Gonzalez L, Gandara A. Acute renal failure following massive mannitol infusion. Am J Nephrol 2002;22(5–6):573–5.
53. Gondim Fde A, Aiyagari V, Shackleford A, Diringer MN. Osmolality not predictive of mannitol-induced acute renal insufficiency. J Neurosurg 2005;103:444–7.
54. Eroghu A, Uzunlar H. Forearm compartment syndrome after intravenous mannitol extravasation in a carbosulfan poisoning patient. J Toxicol Clin Toxicol 2004;42:649–652.
55. Edwards JJ, Samuels D, Fu ES. Forearm compartment syndrome from intravenous mannitol extravasation during general anesthesia. Anesth Analg 2003;96:245–6.

Mefruside

General Information

Mefruside is a diuretic that is structurally related to furosemide, but with pharmacological actions more similar to those of the thiazide diuretics (SED-8, 488) (1). Its effects on serum lipids (SEDA-11, 196) and serum magnesium (SEDA-10, 188) are also similar to those of the thiazides.

Reference

1. Brogden RN, Speight TM, Avery GS. Mefruside: a preliminary report of its pharmacological properties and therapeutic efficacy in oedema and hypertension. Drugs 1974;7(6):419–25.

Metolazone

General Information

Metolazone occupies an intermediate position between the thiazide diuretics and the more potent loop diuretics (1). It is more effective than the thiazides in moderate to advanced renal insufficiency (SED-8, 488) (SED-9, 355). In patients with normal renal function its antihypertensive effect compares favorably with that of bendroflumethiazide. When severe heart failure is refractory to conventional triple therapy (high-dose loop diuretic, digoxin, and angiotensin-converting enzyme inhibitor) metolazone can restore diuresis, with weight loss and clinical improvement.

Organs and Systems

Electrolyte balance

Metolazone causes a greater degree of potassium depletion than thiazides (SED-11, 424). In order to avoid serious electrolyte disturbances when metolazone is introduced in patients taking loop diuretics, the metolazone should be given in low doses to start with in hospital, and at the same time the dosage of the loop diuretic should be reduced under careful biochemical monitoring (SEDA-16, 225).

Long-Term Effects

Drug abuse

Neutropenia is a rare complication of metolazone (2).

References

1. Sica DA. Metolazone and its role in edema management. Congest Heart Fail 2003;9(2):100–5.
2. Donovan KL. Neutropenia and metolazone. BMJ 1989;299(6705):981.

Muzolimine

General Information

Muzolimine is a diuretic with a long duration of action, slightly more effective than furosemide. In Germany muzolimine was withdrawn 2 years after its introduction.

Organs and Systems

Nervous system

Of 29 patients with chronic renal insufficiency treated with high doses of muzolimine, five developed a neurological syndrome very similar to multiple sclerosis (1). Other reports of severe neurotoxicity with muzolimine have appeared (SEDA-16, 225).

Reference

1. Gilli M, Papurello D, Chiado Cutin I, Bradac GB, Delsedime M, Dettoni E, Giangrandi C, Fidelio T, Rocci E, Riccio A. Azione neurotossica della muzolimina ad alte dosi in pazienti uremica. Osservazione su un gruppo di 29 soggetti. [Neurotoxic action of muzolimine at high doses in uremic patients. Observation of a group of 29 subjects.] Minerva Urol Nefrol 1989;41(3):215–8.

Piretanide

General Information

Piretanide is a close relative of furosemide, with diuretic and kinetic properties very similar to those of furosemide and bumetanide (1,2). Its potency lies somewhere between the two. As with some other diuretics, attempts have been made to show specific advantages for piretanide, for example that it is a "potassium-stable" diuretic. However, there is no good evidence that it has such advantages (SEDA-10, 188) (SEDA-15, 213).

References

1. Clissold SP, Brogden RN. Piretanide. A preliminary review of its pharmacodynamic and pharmacokinetic properties, and therapeutic efficacy. Drugs 1985;29(6):489–530.
2. Knauf H, Mutschler E. Das Wirkprofil von Diuretika. Piretanid im Vergleich zu Thiaziden und Autikaliuretika. Internist (Berl) 1992;33(suppl 1):S16–22.

Quinethazone

General Information

Quinethazone is chemically related to the thiazides; early work suggested that it was in all essential respects identical (SED-8, 488), and there is no newer evidence that it has any special characteristics.

Spironolactone

General Information

Spironolactone is a competitive antagonist at aldosterone receptors. It acts through its active metabolite, canrenone. Canrenone itself has also been used as a potassium-sparing diuretic for intravenous use and its potassium salt has been used orally, in the hope of avoiding the hormonal adverse effects of spironolactone.

Spironolactone has been used as a potassium-sparing diuretic in cardiac failure and in the management of ascites and edema associated with hepatic cirrhosis with secondary hyperaldosteronism. It is also used to treat hyperaldosteronism due to adrenal tumors or adrenal hyperplasia. It has a weak positive inotropic effect and a modest antihypertensive effect, in keeping with its natriuretic action.

Although spironolactone has been available for more than 30 years, its efficacy and safety in patients with heart failure have only recently been recognized in the Randomized Aldosterone Evaluation Study (RALES), in which it reduced mortality (1). Based on this and numerous smaller trials, the use of spironolactone, in conjunction with ACE inhibitors, other diuretics, and possibly beta-blockers or digoxin, represents a promising strategy for patients with severe heart failure. Its main adverse effects are hyperkalemia and antiadrenergic complications (SED-14, 675).

General adverse effects

The major limitation to the use of spironolactone is its liability to cause (sometimes lethal) hyperkalemia, particularly in the elderly, in patients with reduced renal function, and in patients who simultaneously take potassium supplements or ACE inhibitors. As with other diuretics, hyponatremia and dehydration can occur. Other less frequent adverse effects are gastrointestinal intolerance, neurological symptoms, and skin rashes. Hypersensitivity rashes and a lupus-like syndrome have been reported rarely. A few cases of mammary carcinoma have been reported and potential human metabolic products of spironolactone are carcinogenic in rodents. Second-generation effects have not been reported.

Organs and Systems

Nervous system

Central nervous system effects, such as weakness, drowsiness, and confusion, have been reported in patients taking spironolactone. Because most patients with such adverse effects were taking spironolactone for edema and ascites in hepatic cirrhosis, it is not yet clear whether they were caused by the drug itself or by hepatic encephalopathy. The incidence of these complaints in such patients is quite high (9.8%) (SED-9, 357).

Of 35 consecutive patients with acne, mean age 21 years, who took spironolactone 100 mg/day on 16 days each month for 3 months, only two withdrew from the study, because of severe vertigo and dizziness (2). The possibility of spironolactone-related adverse effects at a dose as high as 100 mg/day should not be an a priori deterrent to using it.

Endocrine

In an open study in 35 women with acne, mean age 21 years, spironolactone 100 mg/day on 16 days each month for 3 months had no effect on serum total testosterone concentrations but reduced serum dehydroepiandrosterone sulfate concentrations (2)

Electrolyte balance

The beneficial effects of spironolactone in congestive cardiac failure and hypertension are additive to those of ACE inhibitors. In the RALES, patients taking an ACE inhibitor who were randomized to spironolactone 25–50 mg/day had only a 0.3 mmol/l increase in median potassium concentration (1). Although the difference between the spironolactone and placebo groups was statistically significant, the mean increase was not clinically important, and the incidence of serious hyperkalemia was minimal in both groups of patients.

However, since the publication of RALES the incidence of hyperkalemia attributable to spironolactone has increased. Trends in the rate of spironolactone prescriptions and the rate of hospitalization for hyperkalemia in ambulatory patients before and after the publication of RALES have been examined in a population time-based series analysis (3). Among patients treated with ACE inhibitors who had recently been hospitalized for heart failure, the spironolactone prescription rate increased immediately after the publication of RALES, from 3.4% in 1994 to 15% in late 2001. The rate of hospitalization for hyperkalemia rose from 0.2% in 1994 to 11% and the associated mortality rose from 0.3 to 2.0 per 1000 patients. These increases were particularly marked in elderly people.

In a retrospective analysis of 840 patients being seen in a large cardiology clinic and given a new prescription for spironolactone, 91% had baseline laboratory values and 34% did not have any serum potassium or creatinine determinations within 3 months of starting therapy (4). Of 551 patients with follow-up laboratory values, 15% developed hyperkalemia (≥5.5 mmol/l) and 6% developed severe hyperkalemia (≥6.0 mmol/l). There was renal dysfunction in 51 patients (9%), of whom 25 developed hyperkalemia within 3 months. Hyperkalemia developed in 48 of 138 patients (35%) with baseline serum creatinine values of at least 130 μmol/l (1.5 mg/dl) and 12 of 19 (63%) with baseline serum creatinine values of at least 220 μmol/l (2.5 mg/dl). Hyperkalemia is a significant issue in spironolactone-treated patients with heart failure, who require routine biochemical follow-up if the actual incidence is to be appreciated (4,5).

In 3995 patients for whom spironolactone was prescribed in a single teaching hospital and two community hospitals, hyperkalemia was identified in 419 (10.5%); the median age of all the patients was 65 years and the median daily dose was 25 mg (6).

One would expect the risk of hyperkalemia in patients taking spironolactone to be increased by concomitant therapy with an ACE inhibitor or an angiotensin receptor antagonist (7,8). However, this effect may be mitigated by the addition of a potassium-wasting diuretic, as exemplified by a retrospective study of 259 patients with chronic heart failure who were taking spironolactone (25 mg/day, n = 210, or 50 mg/day, n = 49) plus furosemide 40 mg/day, plus either an ACE inhibitor or an angiotensin receptor antagonist, and 251 who were taking the same drugs without spironolactone (9). The serum potassium concentration was raised only in those who were taking spironolactone, regardless of other potassium-sparing therapy (enalapril maleate 5 mg/day, losartan potassium 50 mg/day, or candesartan cilexetil 8 mg/day).

It is essential to identify patients who are likely to develop serious hyperkalemia during combined treatment and to evaluate the associated morbidity and mortality. Spironolactone is used in patients with end-stage renal disease undergoing either peritoneal (10) or maintenance hemodialysis (11), although there is scant experimental evidence to support such use in advanced dialysis-dependent renal disease. Carefully selected dialysis patients exposed to doses of spironolactone ranging from 25 mg three times a week to 25 mg/day appeared not to have an increased frequency of hyperkalemia. However, these studies were preliminary and the results should not be extrapolated to the general dialysis population. If spironolactone is contemplated in patients with end-stage renal disease, periodic surveillance is necessary and other causes of hyperkalemia, such as ACE inhibitor therapy and/or a high potassium intake, should be controlled.

Close laboratory monitoring and judicious use of spironolactone is needed to reduce the risk of hyperkalemia. The excess of hyperkalemia in clinical practice compared with RALES can be explained largely by the use of higher doses of spironolactone and the inclusion of patients with lower glomerular filtration rates and whose aldosterone-mediated compensatory distal tubular potassium excretion is already attenuated (12). Such patients include elderly people, people with diabetes, and those taking beta-blockers, non-steroidal anti-inflammatory drugs, potassium salts, potassium-sparing diuretics, or trimethoprim.

Hematologic

Rare cases of agranulocytosis have been reported in patients taking spironolactone (SEDA-22, 239).

- Agranulocytosis occurred in an 87-year-old man with congestive heart failure who took spironolactone 25 mg/day for 3 weeks (13). The agranulocytosis rapidly reversed after withdrawal of spironolactone and a short course of granulocyte-colony-stimulating-factor.
- A 43-year-old woman with cirrhosis and hepatocellular carcinoma took spironolactone for 7 days before leukopenia was detected (14). The leukocyte count returned to normal 8 days after withdrawal of spironolactone.

In general, the onset of agranulocytosis with spironolactone ranges from 4 days to 5 weeks and it takes 5–7 days to resolve without the aid of a growth factor.

Gastrointestinal

Nausea and vomiting are more common with spironolactone than with other diuretics (SED-9, 358) and were reported in 11% of patients in the Boston Collaborative Drug Surveillance Program (15). These symptoms can be reduced in patients taking high doses by spreading the dose out during the day.

Liver

Two cases of spironolactone-induced parenchymal hepatitis have been reported. The patients recovered uneventfully (SEDA-20, 205).

- A 50-year-old woman taking spironolactone for androgenic alopecia developed hepatitis with minimal cholestasis 6 weeks after starting therapy (16). After withdrawal of spironolactone, her symptoms resolved and liver function tests improved. She was not rechallenged.

Urinary tract

Of 226 patients with heart failure in a retrospective analysis, 25 stopped therapy because of renal dysfunction, 13 primarily because of hyperkalemia and 12 primarily because of a rising creatinine (17). The mean baseline creatinine in the latter was 140 μmol/l, rising to a mean of 197 μmol/l at the time of stopping therapy. Eight of 11 patients had a serum creatinine over 200 μmol/l. Spironolactone should be considered as one of the several pharmacological factors that can cause deterioration of renal function in heart failure.

Skin

Rashes (sometimes with eosinophilia), lichen planus (SED-9, 358), and a lupus–like syndrome with a positive rechallenge (SEDA-5, 230) have been reported, but on the whole skin reactions seem to be rare.

Drug rash with eosinophilia and systemic symptoms (DRESS) has been attributed to spironolactone.

- A 58-year-old man developed erythroderma, fever, anorexia, peripheral edema, eosinophilia, and multi-organ failure while taking several drugs, including spironolactone (18). All the medications were stopped and the condition completely remitted over 4 months. Patch testing was positive for spironolactone.

Cutaneous vasculitis has been attributed to spironolactone (SED-11, 430). It is unclear whether a patient presenting with erythema annulare centrifugum while taking spironolactone (19) had a similar reaction.

Other reported adverse effects of spironolactone include urticaria, alopecia, and chloasma (SEDA-14, 185).

- A 76-year-old patient developed eczema-like lesions and severe pruritus; histological and immunological investigations showed pemphigoid (20). The skin lesions regressed spontaneously within 15 days of spironolactone withdrawal and no relapse was noted over the next 30 months.

Sexual function

Because of its antiandrogenic action, spironolactone causes gynecomastia, reduced libido, and erectile failure in 4–30% of men. These effects seem to be both dose- and time-dependent.

Reproductive system

Very high doses of spironolactone (over 450 mg/day) can cause infertility, but such doses are rarely used. As with other diuretics, studies with spironolactone have mostly been small, not placebo-controlled, and often anecdotal (SEDA-18, 235).

Breasts

Spironolactone causes breast tenderness and enlargement, mastodynia, infertility, chloasma, altered vaginal lubrication, and reduced libido in women, probably because of estrogenic effects on target tissue. Menstrual irregularities were experienced by almost all women taking spironolactone 400 mg/day and most developed amenorrhea at doses of 100–200 mg/day. Normal menstruation was resumed within 2 months of withdrawal.

In patients with hepatic cirrhosis, gynecomastia was twice as common during treatment with spironolactone than with potassium canrenoate (42 versus 20%) at equiactive doses (21). This difference may be related to structural characteristics unique to metabolites of spironolactone, namely a thiol group at the 7-alpha position (SEDA-11, 200).

In a man with chronic heart failure and gynecomastia attributed to spironolactone there was bilaterally increased uptake of ^{67}Ga citrate and ^{18}F FDG in the breast tissues (22).

Long-Term Effects

Tumorigenicity

Spironolactone is antiandrogenic and increases the peripheral metabolism of testosterone to estradiol (23). It often causes gynecomastia in men and breast enlargement and soreness in women. Five cases of mammary carcinoma have been reported. Potential human metabolic products of spironolactone are carcinogenic in rodents, and the UK Committee on Safety of Medicines in 1988 restricted the approved indications for the drug, removing the indications of essential hypertension and idiopathic edema (24).

Susceptibility Factors

Age

Children

In 100 children (average age 21 months, weight 9.5 kg), 62 of whom had heart disease, 29 had chronic lung disease, and 9 had other conditions, spironolactone 1.7–2.0 mg/kg/day plus a potassium-wasting diuretic was associated with hyperkalemia initially, but hypokalemia was more frequent during long-term use (25).

Elderly patients

It has been argued that potassium-sparing diuretics present a real risk of renal insufficiency when they are used in elderly people (26). In large-scale studies in elderly hypertensive patients there is indeed some slight increase in the incidence of renal insufficiency when combinations including potassium-sparing diuretics are used. Although the overall incidence of nephrotoxicity is quite low, elderly patients and those with prior renal dysfunction are at particular risk. Special care is necessary in these circumstances.

In a retrospective survey of 64 patients aged over 75 years, there was hyperkalemia (over 5.5 mmol/l) in 36% and severe hyperkalemia (over 6.0 mmol/l) in 11% (27). The incidence of severe renal insufficiency was 11% and 38% of the patients had a more than 50% rise in creatinine concentration. Three deaths were thought to be drug related. Only severe intercurrent illness predicted the adverse outcomes. In contrast to the results of other studies, age, baseline creatinine, creatinine clearance, ACE inhibitor dose, NYHA class, diabetes mellitus, intensity of monitoring, co-morbidity, and number of co-medications were not predictive. These results suggest that elderly patients are more susceptible to hyperkalemia and renal insufficiency when taking spironolactone.

Renal disease

In patients with renal insufficiency there is an increased risk of hyperkalemia with spironolactone; serum potassium should be monitored regularly.

Hepatic disease

Hepatic failure carries an unwarranted risk unless serum potassium is regularly monitored (28).

Drug–Drug Interactions

Angiotensin-converting enzyme (ACE) inhibitors

Co-administration of potassium-sparing diuretics with ACE inhibitors can cause severe hyperkalemia (SED-14, 674).

The effects of ACE inhibitors plus spironolactone have been evaluated in 25 patients (11 men, 14 women, mean age 74 years, five with diabetes mellitus) with a mean serum potassium concentration of 7.7 mmol/l (29). The mean serum creatinine was 336 μmol/l, the mean arterial pH 7.3, and the mean plasma bicarbonate 18 mmol/l. The main causes of acute renal insufficiency were dehydration (n = 12) and worsening heart failure (n = 9). The mean dose of spironolactone was 57 mg/day and 12 patients were also taking other drugs that can cause hyperkalemia. Two patients died and two were resuscitated and survived. Hemodialysis was necessary in 17 patients. The mean duration of hospitalization was 12 days. The combination of ACE inhibitors and spironolactone should be used cautiously in patients with chronic renal insufficiency, diabetes, older age, worsening cardiac failure, a risk of dehydration (for example diarrhea), and in those who are taking other drugs that can cause hyperkalemia (30). A dose of spironolactone of 25 mg/day should be exceeded only with caution. In four similar elderly patients with underlying renal insufficiency taking spironolactone there was an increased risk of hyperkalemia associated with diarrhea (31).

In a population-based, nested, case-control study, all patients aged 66 years or over who were taking an ACE inhibitor and who had been hospitalized for hyperkalemia were evaluated for interacting medications (32). Compared with controls taking ACE inhibitors who did not have hyperkalemia, before adjusting for confounding variables, the patients were 27 times more likely to have received a prescription for a potassium-sparing diuretic (type not specified) in the week before hospitalization. There was no such association between hospital admission for hyperkalemia and the use of indapamide in patients taking ACE inhibitors.

Angiotensin receptor antagonists

In eight healthy subjects, treatment with spironolactone and losartan increased mean plasma potassium concentration by 0.8 mmol/l (up to 5.0 mmol/l) and reduced mean urinary potassium excretion from 108 to 87 mmol/l (33). Until more data are available, it is prudent to consider angiotensin II receptor antagonists similar to ACE inhibitors as risk factors for hyperkalemia in patients taking potassium-sparing diuretics.

Antipyrine

Spironolactone has a weak enzyme-inducing effect and enhances the metabolic breakdown of phenazone (antipyrine) (SED-9, 358).

Acetylsalicylic acid

The renal tubular secretion of canrenone, the main metabolite of spironolactone, is blocked by aspirin (34), abolishing the diuretic response, although the antihypertensive effect is apparently not affected (35).

Carbenoxolone

Spironolactone abolishes the ulcer-healing properties of carbenoxolone (36).

Cardiac glycosides

Spironolactone inhibits the active tubular secretion of digoxin by about 25% and in some cases digoxin dosages may have to be reduced (37). There may also be a pharmacodynamic interaction with digoxin. The clinical importance of these observations is uncertain (SEDA-9, 209).

Colestyramine

Colestyramine in combination with spironolactone has been reported to cause severe hyperkalemia (38), presumably because it causes exchange of chloride for bicarbonate in the small bowel, predisposing to hyperchloremic acidosis.

Digitoxin

Spironolactone has a weak enzyme-inducing effect and enhances the metabolic breakdown of digitoxin (SED-9, 358) (39). However, this effect may not be significant, because in six healthy subjects spironolactone prolonged the half-life of digitoxin from 142 to 192 hours (40).

NSAIDs

Some non-steroidal anti-inflammatory drugs, notably indometacin and mefenamic acid, inhibit the excretion of canrenone (41).

Warfarin

It has been suggested that spironolactone may reduce the anticoagulant effect of warfarin (SED-11, 430).

Interference with Diagnostic Tests

Radioimmunoassay of digoxin

Overestimation of the digoxin concentration

Spironolactone can alter the results of some digoxin radioimmunoassays, because it and its metabolites, such as canrenone and 7-alpha-thiomethylspironolactone, are immunoreactive with some forms of antidigoxin antibody (42–44). This results in an overestimate of the true digoxin concentration, because the assay reads the interfering substances as digoxin.

In a study of the use of the TDx II assay in 80 children and adults, there was apparent digoxin immunoreactivity in 3.7% of healthy subjects (n = 80), 3.6% of pregnant women (n = 28), 10% of patients with renal transplants (n = 31), and 23% of immature infants (n = 40) (45).

Underestimation of the digoxin concentration

In contrast to this, it has also been reported that spironolactone and its metabolite canrenone caused falsely low readings in a common assay for digoxin (AXSym MEIA) because of negative cross-reactivity (46). Misleading subtarget concentrations were repeatedly reported, and falsely guided increases in drug doses, resulting in digoxin intoxication.

In another study canrenoate interfered with the immunoassay of digoxin (46), reducing the apparent digoxin concentration. The effect was largest with the AxSym MEIA II assay, and there was also some interference with the EMIT assay. However, the TDx assay was not affected. Spironolactone also caused some interference in the AxSym assay but less than canrenoate. In this case the failure to measure high digoxin concentrations resulted in clinical toxicity in a 71-year-old man who was given 3.8 mg over 11 days.

In nine assays (AxSYM, IMx, TDx, Emit, Dimension, aca, TinaQuant, Elecsys, and Vitros) interference by spironolactone, canrenone, and three metabolites was sought in vitro, and all routine digoxin measurements using the AxSYM system over 16.5 months (n = 3089) were reviewed (47). There was a reduction in the expected concentrations by canrenone (3125 ng/ml) in the following assays: AxSYM (42% of expected), IMx (51%), and Dimension (78%). There was positive bias in aca (0.7 ng/ml), TDx (0.62 ng/ml), and Elecsys (>0.58 ng/ml). Of 669 routinely monitored patients, 25 had falsely low results and 19 of them actually had potentially toxic digoxin concentrations; this was attributable to concurrent therapy with spironolactone, canrenone, hydrocortisone, or prednisolone. However, standard doses of spironolactone (up to 50 mg/day) in patients with heart failure produced less than 11% inhibition.

Although several modern digoxin assays suffer from potentially negative interference from spironolactone and canrenone (47), there are assays that are less susceptible to interference by spironolactone (48). Laboratories should determine the false negative potential of the analytical method that they use.

References

1. Pitt B, Zannad F, Remme WJ, Cody R, Castaigne A, Perez A, Palensky J, Wittes J. The effect of spironolactone on morbidity and mortality in patients with severe heart failure. Randomized Aldactone Evaluation Study Investigators. N Engl J Med 1999;341(10):709–17.
2. Yemisci A, Gorgulu A, Piskin S. Effects and side-effects of spironolactone therapy in women with acne. J Eur Acad Dermatol Venereol 2005;19:163–6.
3. Juurlink DN, Mamdani MM, Lee DS, Kopp A, Austin PC, Laupacis A, Redelmeier DA. Rates of hyperkalemia after publication of the Randomized Aldactone Evaluation Study. N Engl J Med 2004;351(6):543–51.
4. Shah KB, Rao K, Sawyer R, Gottlieb SS. The adequacy of laboratory monitoring in patients treated with spironolactone for congestive heart failure. J Am Coll Cardiol 2005;46:845–9.
5. Tamirisa KP, Aaronson KD, Koelling TM. Spironolactone-induced renal insufficiency and hyperkalemia in patients with heart failure. Am Heart J 2004;148:971–8.
6. Huang C, Noirot LA, Reichley RM, Bouselli DA, Dunagan WC, Bailey TC. Automatic detection of spironolactone-related adverse drug events. AMIA Annu Symp Proc 2005:989.
7. Fujii H, Nakahama H, Yoshihara F, Nakamura S, Inenaga T, Kawano Y. Life-threatening hyperkalemia during a combined therapy with the angiotensin receptor blocker candesartan and spironolactone. Kobe J Med Sci 2005; 51(1–2):1–6.
8. Kauffmann R, Orozco R, Venegas JC. [Severe hyperkalemia associated to the use of losartan and spironolactone: case report] Rev Med Chil 2005;133(8):947–52.
9. Saito M, Takada M, Hirooka K, Isobe F, Yasumura Y. Serum concentration of potassium in chronic heart failure patients administered spironolactone plus furosemide and either enalapril maleate, losartan potassium or candesartan cilexetil. J Clin Pharm Ther 2005;30(6):603–10.
10. Hausmann MJ, Liel-Cohen N. Aldactone therapy in a peritoneal dialysis patient with decreased left ventricular function. Nephrol Dial Transplant 2002;17(11):2035–6.

11. Hussain S, Dreyfus DE, Marcus RJ, Biederman RW, McGill RL. Is spironolactone safe for dialysis patients? Nephrol Dial Transplant 2003;18(11):2364–8.
12. McMurray JJ, O'Meara E. Treatment of heart failure with spironolactone—trial and tribulations. N Engl J Med 2004;351(6):526–8.
13. Hui CH, Das PK, Horvath N. Spironolactone and agranulocytosis. Aust NZ J Med 2000;30(4):515.
14. Hsiao SH, Lin YJ, Hsu MY, Wu TJ. Spironolactone-induced agranulocytosis: a case report. Kaohsiung J Med Sci 2003;19(11):574–8.
15. Ochs HR, Greenblatt DJ, Bodem G, Smith TW. Spironolactone. Am Heart J 1978;96(3):389–400.
16. Thai KE, Sinclair RD. Spironolactone-induced hepatitis. Australas J Dermatol 2001;42(3):180–2.
17. Witham MD, Gillespie ND, Struthers AD. Tolerability of spironolactone in patients with chronic heart failure – a cautionary message. Br J Clin Pharmacol 2004; 58:554–7.
18. Ghislain PD, Bodarwe AD, Vanderdonckt O, Tennstedt D, Marot L, Lachapelle JM. Drug-induced eosinophilia and multisystem failure with positive patch-test reaction to spironolactone: DRESS syndrome. Acta Dermatol Venereol 2004;84:65–8.
19. Carsuzaa F, Pierre C, Dubegny M. Erythème annulaire centrifugé à l'Aldactone. [Erythema annulare centrifugum caused by Aldactone.] Ann Dermatol Venereol 1987;114(3):375–6.
20. Modeste AB, Cordel N, Courville P, Gilbert D, Lauret P, Joly P. Pemphigoide regressive après arrêt d'un diurétique contenant de l'Aldactone. [Bullous pemphigoid induced by spironolactone.] Ann Dermatol Venereol 2002;129(1 Pt 1):56–8.
21. Emili M, Cuppone R, Ricci GL. Comparative clinical study of spironolactone and potassium canrenoate. A randomized evaluation with double cross-over. Arzneimittelforschung 1988;38(10):1492–5.
22. Fukuchi K, Sasaki H, Yokoya T, Noguchi T, Goto Y, Hayashida K, Ishida Y. Ga-67 citrate and F-18 FDG uptake in spironolactone-induced gynecomastia. Clin Nucl Med 2005;30(2):105–6.
23. Rose LI, Underwood RH, Newmark SR, Kisch ES, Williams GH. Pathophysiology of spironolactone-induced gynecomastia. Ann Intern Med 1977;87(4):398–403.
24. Committee on Safety of Medicines. Spironolactone. Curr Probl 1988;21.
25. Buck ML. Clinical experience with spironolactone in pediatrics. Ann Pharmacother 2005;39(5):823–8.
26. Bailey RR. Adverse renal reactions to non-steroidal anti-inflammatory drugs and potassium-sparing diuretics. Adv Drug React Bull 1988;131:492.
27. Dinsdale C, Wani M, Steward J, O'Mahony MS. Tolerability of spironolactone as adjunctive treatment for heart failure in patients over 75 years of age. Age Ageing 2005;34(4):395–8.
28. Rado JP, Marosi J, Szende L, Tako J. Hyperkalemic changes during spironolactone therapy for cirrhosis and ascites, with special reference to hyperkalemic intermittent paralysis. J Am Geriatr Soc 1968;16(8):874–86.
29. Schepkens H, Vanholder R, Billiouw JM, Lameire N. Life-threatening hyperkalemia during combined therapy with angiotensin-converting enzyme inhibitors and spironolactone: an analysis of 25 cases. Am J Med 2001;110(6): 438–41.
30. Vanpee D, Swine CH. Elderly heart failure patients with drug-induced serious hyperkalemia. Aging (Milano) 2000;12(4):315–9.
31. Berry C, McMurray JJ. Serious adverse events experienced by patients with chronic heart failure taking spironolactone. Heart 2001;85(4):E8.
32. Juurlink DN, Mamdani M, Kopp A, Laupacis A, Redelmeier DA. Drug–drug interactions among elderly patients hospitalized for drug toxicity. JAMA 2003; 289(13):1652–8.
33. Henger A, Tutt P, Hulter HM, Krapf R. Acid-base effects of inhibition of aldosterone and angiotensin II action in chronic metabolic acidosis in humans. J Am Soc Nephrol 1999;10:121A.
34. McInnes GT, Shelton JR, Ramsay LE. Evaluation of aldosterone antagonists in healthy man. Methods Find Exp Clin Pharmacol 1982;4(1):49–71.
35. Hollifield JW. Failure of aspirin to antagonize the antihypertensive effect of spironolactone in low-renin hypertension. South Med J 1976;69(8):1034–6.
36. Doll R, Langman MJ, Shawdon HH. Treatment of gastric ulcer with carbenoxolone: antagonistic effect of spironolactone. Gut 1968;9(1):42–5.
37. Waldorff S, Andersen JD, Heeboll-Nielsen N, Nielsen OG, Moltke E, Sorensen U, Steiness E. Spironolactone-induced changes in digoxin kinetics. Clin Pharmacol Ther 1978; 24(2):162–7.
38. Zapater P, Alba D. Acidosis and extreme hyperkalemia associated with cholestyramine and spironolactone. Ann Pharmacother 1995;29(2):199–200.
39. Taylor SA, Rawlins MD, Smith SE. Spironolactone—a weak enzyme inducer in man. J Pharm Pharmacol 1972;24(7):578–9.
40. Carruthers SG, Dujovne CA. Cholestyramine and spironolactone and their combination in digitoxin elimination. Clin Pharmacol Ther 1980;27(2):184–7.
41. Tweeddale MG. Antagonism between antipyretic analgesic drugs and spironolactone in man. Clin Res 1974;22:727A.
42. Huffman DH. The effect of spironolactone and canrenone on the digoxin radioimmunoassay. Res Commun Chem Pathol Pharmacol 1974;9(4):787–90.
43. Pleasants RA, Williams DM, Porter RS, Gadsden RH Sr. Reassessment of cross-reactivity of spironolactone metabolites with four digoxin immunoassays. Ther Drug Monit 1989;11(2):200–4.
44. Valdes R Jr, Jortani SA. Unexpected suppression of immunoassay results by cross-reactivity: now a demonstrated cause for concern. Clin Chem 2002;48(3):405–6.
45. Capone D, Gentile A, Basile V. Possible interference of digoxin-like immunoreactive substances using the digoxin fluorescence polarization immunoassay. J Appl Ther Res 1999;2:305–8.
46. Steimer W, Muller C, Eber B, Emmanuilidis K. Intoxication due to negative canrenone interference in digoxin drug monitoring. Lancet 1999;354(9185):1176–7.
47. Steimer W, Muller C, Eber B. Digoxin assays: frequent, substantial, and potentially dangerous interference by spironolactone, canrenone, and other steroids. Clin Chem 2002;48(3):507–16.
48. Datta P, Dasgupta A. A new turbidometric digoxin immunoassay on the ADVIA 1650 analyzer is free from interference by spironolactone, potassium canrenoate, and their common metabolite canrenone. Ther Drug Monit 2003; 25(4):478–82.

Tienilic acid

General Information

Tienilic acid, a uricosuric diuretic (1), had to be withdrawn soon after its release in 1980, after it had caused several hundreds of cases of severe liver damage, many of them fatal (SEDA-5, 229). In France almost 500 reports of liver injury were received. Tienilic acid had no advantage over other diuretics; it certainly had the worst benefit:harm balance (SEDA-17, 269). For a time it was considered that the hepatic complication might be geographically limited (for example because of differences in manufacturing in different countries), but there is now no doubt that the drug itself was responsible. The hepatic complication was thoroughly studied in retrospect (SED-11, 424). Other adverse effects, including acute renal insufficiency, urate stones, and interactions with other drugs, such as oral anticoagulants and phenytoin, helped tienilic acid to its early demise.

Reference

1. Maass AR, Snow B, Beg M, Stote RM. Pharmacokinetics and mode of action of tienilic acid. Clin Exp Hypertens A 1982;4(1–2):139–60.

Torasemide

General Information

Torasemide is a long-acting loop diuretic promoted for use in hypertension. Like piretanide, it is claimed to be potassium neutral, but the assertion is premature. There is no evidence that torasemide has metabolic advantages over thiazides (SEDA-16, 226) (SEDA-17, 264) (SEDA-18, 237).

Organs and Systems

Skin

Torasemide has been associated with various rashes, including non-specific erythematous lesions, pruritus, and photoallergic lichenoid lesions (SEDA-22, 239).

Immunologic

Two possible cases of vasculitis with renal insufficiency have been reported in patients taking torasemide (1,2). This adverse effect is not surprising, since torasemide is structurally similar to sulfa drugs, which can cause vasculitis.

- A 70-year-old man developed heart failure secondary to ischemic heart disease and severe aortic stenosis (1). Furosemide 20 mg/day was replaced by torasemide 5 mg/day. After the second dose he developed oliguria and an erythematous morbilliform rash with palpable violet petechial lesions on the legs. Chest X-ray showed bilateral alveolar infiltrates. Serum creatinine and potassium were raised (212 μmol/l and 6.7 mmol/l respectively). Skin biopsy showed leukocytoclastic vasculitis. After withdrawal of torasemide, his renal function improved (serum creatinine 97 μmol/l) and the skin lesions resolved (leaving residual pigmented areas) within 8 days.
- An 84-year-old man with ischemic heart disease and hypertension took torasemide 10 mg/day for persistent edema (1). About 24 hours after the first dose of torasemide, he developed painless, non-palpable, petechial lesions on the limbs and trunk, with oliguria. His serum creatinine was 256 μmol/l and his serum potassium 6.2 mmol/l. Skin biopsy showed non-leukocytoclastic vasculitis with a mixed inflammatory infiltrate including eosinophils. He was symptom free 15 days after withdrawal of torasemide.

Neither patient had a previous history of drug hypersensitivity. Both patients had previously tolerated furosemide, another sulfonamide derivative. The temporal correlation with torasemide administration suggested a causal relation, but the mechanism was unclear.

Drug–Drug Interactions

NSAIDs

NSAIDs block the natriuretic effect of torasemide (SEDA-21, 229).

Probenecid

Probenecid blocks the natriuretic effect of torasemide (SEDA-21, 229).

References

1. Palop-Larrea V, Sancho-Calabuig A, Gorriz-Teruel JL, Martinez-Mir I, Pallardo-Mateu LM. Vasculitis with acute kidney failure and torasemide. Lancet 1998;352(9144):1909–10.
2. Sanfelix Genoves J, Benlloch Nieto H, Verdu Tarraga R, Costa Alcaraz AM. Erupcion purpurica compatible con vasculitis y torasemida. [Eruption of purpura compatible with vasculitis and torasemide.] Aten Primaria 1998;21(4):252–3.

Triamterene

General Information

Triamterene is a potassium-sparing diuretic that acts in the distal convoluted tubule independently of the action of aldosterone, inhibiting sodium channels (1).

Patients with congenital nephrogenic diabetes insipidus are often treated with a combination of a thiazide and a potassium-sparing diuretic, without consensus on the preferred potassium-sparing diuretic. A Japanese adult was systematically studied to determine the renal effects of hydrochlorothiazide plus amiloride and hydrochlorothiazide plus triamterene (2). The combination with amiloride was superior to that with triamterene in preventing excessive urinary potassium loss, hypokalemia, and metabolic alkalosis. These results suggest that amiloride is the preferred add-on therapy to hydrochlorothiazide in nephrogenic diabetes insipidus.

Organs and Systems

Hematologic

Triamterene blocks dihydrofolate reductase and can cause folate deficiency with megaloblastic anemia and pancytopenia, particularly in patients with hepatic cirrhosis, who have reduced clearance of the drug (SED-11, 431). When this has been reported, all patients were taking doses of 150–600 mg/day for ascites and all had hepatic cirrhosis, often due to alcohol abuse (SEDA-17, 269). It is advisable to use spironolactone rather than triamterene in patients with cirrhosis.

Gastrointestinal

Of the milder adverse effects of triamterene, nausea, vomiting, and diarrhea are fairly common (3).

Urinary tract

Like other diuretics, triamterene occasionally causes interstitial nephritis (4), but it has also been responsible for other renal problems, notably reversible non-oliguric renal insufficiency when it is given along with indometacin (or presumably any other inhibitor of prostaglandin synthesis) (5).

Triamterene can cause transient asymptomatic crystalluria in acidic urine (6). In most cases, the crystals are associated with brown casts, also due to triamterene. Triamterene crystals should be regarded as a potential cause of severe tubular injury.

Triamterene can cause nephrolithiasis. Microcrystals of the parahydroxy metabolite form a particularly suitable nucleus for crystalline calcium oxalate deposition (7). However, a cause-and-effect relation has been questioned on epidemiological grounds and in the light of a controlled study that found no link (8), and triamterene could be an innocent bystander in stone formation (SEDA-11, 200). Nevertheless, triamterene is sometimes detected in urinary calculi. In a French analysis of 22 510 urinary calculi performed by infrared spectroscopy, drug-induced urolithiasis was divided into two categories: first, stones with drugs physically embedded ($n = 238$; 1.0%), notably indinavir monohydrate ($n = 126$; 53%), followed by triamterene ($n = 43$; 18%), sulfonamides ($n = 29$; 12%), and amorphous silica ($n = 24$; 10%); secondly, metabolic nephrolithiasis induced by drugs ($n = 140$; 0.6%), involving mainly calcium/vitamin D supplementation ($n = 56$; 40%) and carbonic anhydrase inhibitors ($n = 33$; 24%) (9). Drug-induced stones are responsible for about 1.6% of all calculi in France. Physical analysis and a thorough drug history are important elements in the diagnosis.

Skin

Rashes, including photodermatitis (10), sometimes occur with triamterene.

Pseudoporphyria has been attributed to co-triamterzide (hydrochlorothiazide plus triamterene, Dyazide) in a patient with vitiligo (11). It is uncertain which constituent was responsible.

Body temperature

Drug fever has been attributed to triamterene (12).

Second-Generation Effects

Pregnancy

Because of the potential risk of folate deficiency it is sensible to avoid triamterene in pregnancy (13).

Teratogenicity

Neural tube defects, characterized by a failure of the neural tube to close properly after conception, affect about one in 1000 live births in the USA. Periconceptional folic acid supplementation reduces the risk. To determine whether periconceptional exposure to folic acid antagonists might therefore increase the risk of neural tube defects, data from a case-control study of birth defects (1979–98) in the USA and Canada have been examined (14). Data on 1242 infants with neural tube defects (spina bifida, anencephaly, and encephalocele) were compared with data from a control group of 6660 infants with malformations not related to vitamin supplementation. Triamterene is a folic acid antagonist and in this series was associated with the development of neural tube defects, but there were too few cases to estimate an odds ratio.

Susceptibility Factors

Age

Triamterene is often given in combination with the thiazide diuretic bemetizide. The pharmacokinetics and the pharmacodynamics of a fixed combination of bemetizide 25 mg and triamterene 50 mg have been evaluated in 15 elderly patients (aged 70–84 years) and 10 young volunteers (aged 18–30 years) after single doses (on day 1) and multiple doses (at steady state on day 8) (15). Mean plasma concentrations of bemetizide, triamterene, and the active metabolite of triamterene, hydroxytriamterene, were significantly higher in the elderly subjects after single and multiple doses and urine flow and sodium excretion rates fell in tandem with the accumulation of these drugs. The glomerular filtration rate, known to be reduced in elderly people, was further reduced at higher concentrations of bemetizide and triamterene, which may explain why there were limited diuretic and saliuretic effects after multiple doses. This study clearly points to a modulating effect of the degree of renal function on the diuretic actions of these compounds in the elderly.

Drug–Drug Interactions

Bemetizide

In 1988 the Federal German Health Authorities issued a warning about fixed combinations of the thiazide bemetizide and triamterene that they could cause allergic vasculitis (17). Thiazides can occasionally cause allergic vasculitis and it is not clear on what basis the combination might be more likely to cause the same problem.

Diclofenac

The interaction between diclofenac and triamterene causes renal impairment (19).

Methotrexate

An interaction between triamterene and methotrexate (also an inhibitor of dihydrofolate reductase), leading to pancytopenia, has been reported (16). Dehydration due to diuretic treatment may have contributed to renal impairment and reduced clearance of methotrexate, further increasing the risk of bone marrow suppression.

Drugs that inhibit folate metabolism increase the likelihood of serious adverse reactions to methotrexate, particularly hematological toxicity. Bone marrow suppression and reduced plasma folate concentrations resulted from the concomitant administration of triamterene with methotrexate(18).

References

1. Sica DA, Gehr TW. Triamterene and the kidney. Nephron 1989;51(4):454–61.
2. Konoshita T, Kuroda M, Kawane T, Koni I, Miyamori I, Tofuku Y, Mabuchi H, Takeda R. Treatment of congenital nephrogenic diabetes insipidus with hydrochlorothiazide and amiloride in an adult patient. Horm Res 2004; 61(2):63–7.
3. Bender AD, Carter CL, Hansen KB. Use of a diuretic combination of triamterene and hydrochlorothiazide in elderly patients. J Am Geriatr Soc 1967;15(2):166–73.
4. Bailey RR, Lynn KL, Drennan CJ, Turner GA. Triamterene-induced acute interstitial nephritis. Lancet 1982;1(8265):226.
5. Favre L, Glasson P, Vallotton MB. Reversible acute renal failure from combined triamterene and indomethacin: a study in healthy subjects. Ann Intern Med 1982;96(3): 317–20.
6. Fogazzi GB. Crystalluria: a neglected aspect of urinary sediment analysis. Nephrol Dial Transplant 1996;11(2):379–87.
7. White DJ, Nancollas GH. Triamterene and renal stone formation. J Urol 1982;127(3):593–7.
8. Jick H, Dinan BJ, Hunter JR. Triamterene and renal stones. J Urol 1982;127(2):224–5.
9. Cohen-Solal F, Abdelmoula J, Hoarau MP, Jungers P, Lacour B, Daudon M. Les lithiases urinaires d'origine médicamenteuse. [Urinary lithiasis of medical origin.] Therapie 2001;56(6):743–50.
10. Fernandez de Corres L, Bernaola G, Fernandez E, Leanizbarrutia I, Munoz D. Photodermatitis from triamterene. Contact Dermatitis 1987;17(2):114–5.
11. Motley RJ. Pseudoporphyria due to Dyazide in a patient with vitiligo. BMJ 1990;300(6737):1468.
12. Safdi MA. Fever secondary to triamterene therapy. N Engl J Med 1980;303(12):701.
13. Corcino J, Waxman S, Herbert V. Mechanism of triamterene-induced megaloblastosis. Ann Intern Med 1970; 73(3):419–24.
14. Hernandez-Diaz S, Werler MM, Walker AM, Mitchell AA. Neural tube defects in relation to use of folic acid antagonists during pregnancy. Am J Epidemiol 2001;153(10):961–8.
15. Muhlberg W, Mutschler E, Hofner A, Spahn-Langguth H, Arnold O. The influence of age on the pharmacokinetics and pharmacodynamics of bemetizide and triamterene: a single and multiple dose study. Arch Gerontol Geriatr 2001;32(3):265–73.
16. Richmond R, McRorie ER, Ogden DA, Lambert CM. Methotrexate and triamterene—a potentially fatal combination? Ann Rheum Dis 1997;56(3):209–10.
17. Anonymous. Bemetizide/triamterene: warning of allergic vasculitis. WHO Drug Inform 1988;2:148.
18. Richmond R, McRorie ER, Ogden DA, Lambert CM. Methotrexate and triamterene—a potentially fatal combination? Ann Rheum Dis 1997;56(3):209–10.
19. Weinblatt ME. Drug interactions with non steroidal anti-inflammatory drugs (NSAIDs). Scand J Rheumatol Suppl 1989;83:7–10.

Xipamide

General Information

Xipamide is a non-thiazide diuretic that acts mainly on the distal tubule (1). Its maximal diuretic effect is as great as that of furosemide, but its duration of action is longer and similar to that of the thiazides. Thus, like metolazone, it occupies an intermediate position between the two main groups of diuretics and can be used in renal insufficiency. However, xipamide does appear to present some risks, and it is not clear that these are outweighed by any advantages.

Organs and Systems

Metabolism

At equivalent therapeutic doses, the metabolic effects of xipamide are greater than those of the thiazides or furosemide (SED-11, 200).

Skin

A photoallergic skin reaction has been described in a patient taking xipamide (SEDA-16, 222).

Reference

1. Prichard BN, Brogden RN. Xipamide. A review of its pharmacodynamic and pharmacokinetic properties and therapeutic efficacy. Drugs 1985;30(4):313–32.

ANTIDYSRHYTHMIC DRUGS

General Information

Classification

Drugs used in dysrhythmias can be classified in different ways, the usual classification being according to their effects on the cardiac action potential (1), as shown in Table 1.

Antidysrhythmic drugs with Class I activity reduce the rate of the fast inward sodium current during Phase I of the action potential and increase the duration of the effective refractory period expressed as a proportion of the total action potential duration. The action potential duration is itself affected in different ways by subgroups of the Class I drugs:

- class Ia drugs, of which quinidine is the prototype, prolong the action potential;
- class Ib drugs, of which lidocaine is the prototype, shorten the action potential;
- class Ic drugs, of which flecainide is the prototype, do not alter action potential duration.

The beta-adrenoceptor antagonists (class II) and bretylium inhibit the effect of catecholamines on the action potential.

Antidysrhythmic drugs with class III activity prolong the total action potential duration. These drugs act by effects on potassium channels, altering the rate of repolarization.

Antidysrhythmic drugs with class IV activity prolong total action potential duration by prolonging the plateau phase (phase III) of the action potential via calcium channel blockade.

Other classifications of antidysrhythmic drugs have been proposed, but the most useful clinical classification relates to the sites of action of the antidysrhythmic drugs on the various cardiac tissues, as shown in Table 2.

General adverse effects

There have been many reviews of the pharmacology, clinical pharmacology, pharmacokinetics, and adverse effects and interactions of antidysrhythmic drugs (2–13).

The patterns of adverse effects of the antidysrhythmic drugs depend on three features:

1. All antidysrhythmic drugs have effects on the cardiac conducting tissues and can all therefore cause cardiac dysrhythmias.
2. All antidysrhythmic drugs have a negative inotropic effect on the heart, and can result in heart failure. However, the degree of negative inotropy varies from drug to drug; for example, it is less marked with drugs such as lidocaine and phenytoin and very marked with the beta-adrenoceptor antagonists, verapamil, and class 1a drugs.
3. Each antidysrhythmic drug has its own non-cardiac effects, which can result in adverse effects. These are summarized in Table 3.

Clinical studies

Management of atrial fibrillation

The efficacy of a large range of antidysrhythmic drugs in converting atrial fibrillation to sinus rhythm acutely and in maintaining it during long-term treatment has been the subject of a systematic review (14). Adverse effects were too sporadically reported to be suitable for proper review. The efficacy results are summarized in Table 4.

There is some doubt about whether conversion to sinus rhythm produces a better long-term outcome than rate control. Five randomized controlled comparisons of rhythm control versus rate control, mostly in patients with persistent atrial fibrillation ($n = 5175$ in all), have all suggested that there are no major differences in beneficial outcomes between the two strategies (15,16), although there were fewer adverse drug reactions in patients randomized to rate control in three of the studies and in all the studies rate control was associated with fewer hospital admissions. Furthermore, in an analysis of cost-effectiveness, rate

Table 1 Electrophysiological classification of antidysrhythmic drugs

Class I	Ia	Ib	Ic
	Quinidine	Lidocaine	Flecainide
	Procainamide	Aprindine	Encainide
	Disopyramide	Mexiletine	Lorcainide
		Phenytoin	Propafenone (also has class II activity)
		Tocainide	
Class II	Beta-adrenoceptor antagonists		
	Bretylium		
Class III	Amiodarone		
	Sotalol (d-sotalol; l-sotalol has class II activity)		
Class IV	Verapamil		

Table 2 Classification of antidysrhythmic drugs by their actions in different parts of the heart

Sinus node	*Anomalous pathways*	*Atria*	*Ventricles*	*Atrioventricular node*
Class Ic	Class Ia	Class Ia	Class Ia	Class Ic
Class II	Class Ic	Class Ic	Class Ib	Class II
Class IV	Class III	Class III	Class Ic	Class III
				Class IV

Table 3 Non-cardiac adverse effects of some antidysrhythmic drugs

Drug	*Common non-cardiac adverse effects*
Acecainide	Gastrointestinal and nervous system effects
Adenosine	Flushing, dyspnea
Ajmaline derivatives	Liver damage; agranulocytosis; nervous system effects
Amiodarone	Corneal microdeposits; altered thyroid function; lipofuscin deposition in skin, lungs, liver, nerves, muscles
Aprindine	Agranulocytosis; nervous system effects; liver damage
Cibenzoline	Gastrointestinal and nervous system effects; hypoglycemia
Disopyramide	Anticholinergic effects
Dofetilide	Nervous system effects
Encainide	Nervous system effects
Flecainide	Nervous system effects
Lidocaine	Nervous system effects
Lorcainide	Nervous system effects
Mexiletine	Nervous system effects
Moracizine	Nervous system effects
Procainamide	Lupus-like syndrome; neutropenia
Propafenone	Nervous system effects
Quinidine	Anticholinergic effects; hypersusceptibility reactions
Tocainide	Nervous system effects

Table 4 The results of a systematic review of the efficacy of antidysrhythmic drugs in converting atrial fibrillation to sinus rhythm and maintaining it

Drug	*Number of subjects*	*Efficacy in converting AF to sinus rhythm (odds ratio versus other drugs[1])*	*Efficacy in maintaining sinus rhythm (odds ratio versus other drugs[1])*	*Ventricular dysrhythmias[2] (%)*	*Other dysrhythmias[3] (%)*	*Drug withdrawal or dosage reduction (%)*
Amiodarone	108	5.7		0–15	0–9	
Disopyramide	30	7.0	3.4	0	0	0–55
Dofetilide/ ibutilide	530	29.0		3–9		
Flecainide	169	25.0	3.1	0–2	0–12	0–20
Propafenone	1168	4.6	3.7	0–3	0–17	0–55
Quinidine	200	2.9	4.1	0–12	0–28	0–58
Sotalol	34	0.4	7.1	0–1	2–44	4–44

[1]Digoxin, diltiazem, or verapamil
[2]Ventricular fibrillation, polymorphic ventricular tachycardia, torsade de pointes
[3]Symptomatic bradycardia, junctional rhythm, non-sustained and/or monomorphic ventricular tachycardia

control plus warfarin was much cheaper than rhythm control in preventing thromboembolism, largely because of the use of expensive modern antidysrhythmic drugs for the latter (17). Although these results suggest that rate control might be preferable to rhythm control, they do not give any information about patients in whom sinus rhythm is established permanently after pharmacological or physical conversion, since many of the patients in whom rhythm control is used as a strategy will actually have paroxysmal atrial fibrillation.

In a meta-analysis of 91 randomized controlled trials of the effectiveness of antidysrhythmic drugs in promoting sinus rhythm in patients with atrial fibrillation followed for a median of 1 day (range 0.04–1096 days), the median proportion of patients in sinus rhythm at follow up was 55% (range 0–100%) of those who took active treatment and 32% (range 0–90%) of those who took placebo (18). Median survival was 99% (range 55–100%) and 99% (range 55–100%). Compared with placebo, the following drugs were associated with increased frequencies of sinus rhythm at follow-up:

- class IA (disopyramide, procainamide, and quinidine; treatment difference 22%, 95%CI = 16, 27);
- class IC (flecainide, pilsicainide, and propafenone; treatment difference 33%, 95%CI = 23, 43);
- class III (amiodarone, dofetilide, and ibutilide; treatment difference 17%, 95%CI = 12, 23).

Class IC drugs were associated with a higher frequency of sinus rhythm at follow-up than class IV drugs (treatment difference 43%; 95%CI = 12, 75). Adverse effects were not consistently reported in these studies and could not be analysed, but there was no significant difference in mortality between any drug classes.

However, there is some doubt about whether conversion to sinus rhythm produces a better long-term outcome than rate control. Five randomized controlled comparisons of rhythm control versus rate control, mostly in patients with persistent atrial fibrillation (n = 5175 in all), have all suggested that there are no major differences in beneficial outcomes between the two strategies (19,20,21,22), although there were fewer adverse drug reactions in patients randomized to rate control in three of the studies and in all the studies rate control was associated with fewer hospital admissions. The results are summarized in Table 5 (23,24,25,26,27). Furthermore, in an analysis of cost-effectiveness, rate control plus warfarin was much cheaper than rhythm control in preventing thromboembolism, largely because of the use of expensive modern antidysrhythmic drugs for rhythm control (28). Although these results suggest that rate control might be preferable to rhythm control, they do not give any information about patients in whom sinus rhythm is established permanently after pharmacological or physical conversion, since many of the patients in whom rhythm control is used as a strategy will actually have paroxysmal atrial fibrillation.

In a substudy of the AFFIRM study (25) different antidysrhythmic drugs were compared, by randomly assigning the first drug treatment to amiodarone, sotalol, or a class I drug (29). At one year, in 222 patients randomized between amiodarone and class I agents, 62% were successfully treated with amiodarone, compared with 23% taking class I agents. In 256 patients randomized between amiodarone and sotalol, 60% versus 38% were successfully treated. In 183 patients randomized between sotalol and class I agents, 34% versus 23% were successfully treated, although this portion of the substudy was stopped early when amiodarone was shown to be better than class I agents. Sinus rhythm was achieved in nearly 80% of patients at 1 year. There was only one case of torsade de pointes in this substudy (in a patient who had taken quinidine for more than 1 year). There were no cases of agranulocytosis or lupus syndrome induced by procainamide. However, adverse effects that caused discontinuation of the antidysrhythmic drugs during the first year were frequent (Table 6), and occurred in 12% of patients taking amiodarone, 11% of those taking sotalol, and 28% of those taking class I agents. Among those who were randomized to amiodarone, *pulmonary toxicity* was diagnosed in two by 1 year, three by 2 years, and no additional patients by 3 years. Gastrointestinal adverse events were a common reason for stopping class I drugs.

Comparative studies with other treatments

The Australian Intervention Randomized Control of Rate in Atrial Fibrillation Trial (AIRCRAFT) was a multicenter randomized trial of atrioventricular junction ablation and pacing compared with pharmacological ventricular rate control in 99 patients, mean age 68 years, with mildly to moderately symptomatic permanent atrial fibrillation (30). At 12 months follow-up there was no significant difference in left ventricular ejection fraction or exercise duration on treadmill testing; however, the peak ventricular rate was lower in the ablation group during exercise (112 versus 153) as was a score of activities of daily life. The CAST quality-of-life questionnaire showed that patients who had ablation had fewer symptoms at 6 and 12 months, with a relative risk reduction in symptoms at 12 months of 18%. Global subjective semiquantitative measurement of quality of life using the "ladder of life" showed that ablation produced a 6% better quality of life at 6 months. There were no differences in adverse events between the two treatments.

Organs and Systems

Cardiovascular

Cardiac dysrhythmias

Antidysrhythmic drugs can themselves cause cardiac dysrhythmias, their major adverse effect. The risk of antidysrhythmic-induced cardiac dysrhythmias (prodysrhythmic effects) has been estimated at about 11–13% in non-invasive studies (31,32) and at up to 20% in invasive electrophysiological studies. However, the risk varies from drug to drug and is particularly low with class III drugs. In one study the quoted risks of dysrhythmias were: flecainide 30%, quinidine 18%, propafenone 7%, sotalol 6%, and amiodarone 0% (33). However, amiodarone does cause dysrhythmias, especially when the QT_c interval is over 600 ms.

The prodysrhythmic effects of antidysrhythmic drugs have been extensively reviewed (34–49), as have drugs that prolong the QT interval (50,51,52).

Dysrhythmias secondary to antidysrhythmic drugs are arbitrarily defined as either early (within 30 days of starting treatment) or late (35,36). A lack of early dysrhythmias in response to antidysrhythmic drugs does not predict the risk of late dysrhythmias (37).

Ventricular dysrhythmias due to drugs may be either monomorphic or polymorphic. The class Ia drugs are particularly likely to cause polymorphic dysrhythmias, as is amiodarone (although to a lesser extent). In contrast, the class Ic drugs are more likely to cause monomorphic dysrhythmias (38).

Class Ic antidysrhythmic drugs have been reported to cause the characteristic electrocardiographic changes of Brugada syndrome, which consists of right bundle branch block, persistent ST segment elevation, and sudden cardiac death, in two patients (53). Class Ia drugs did not cause the same effect.

The prodysrhythmic effects of antidysrhythmic drugs have been reviewed in discussions of the pharmacological conversion of atrial fibrillation (54) and the relative benefits of rate control in atrial fibrillation or maintaining sinus rhythm after cardioversion (55).

The major drugs that have been implicated in prolonging the QT interval in one way or another, including cardiac and non-cardiac drugs, are listed in Table 7.

Mechanisms

There are four major mechanisms whereby antidysrhythmic drugs cause dysrhythmias (34):

1. Worsening of a pre-existing dysrhythmia. For example, ventricular extra beats can be converted to ventricular tachycardia or the ventricular rate in atrial flutter can be accelerated when slowing of the atrial rate results in

5 Outcomes of five randomized comparisons of rhythm control and rate control in atrial fibrillation (20)

	PIAF	*PAF2*	*AFFIRM*	*RACE*	*STAF*
er	252	141	4060	522	200
duration ars)	1.0	1.3	3.5	2.3	1.7
of AF	Persistent	Paroxysmal with severe symptoms	Persistent or paroxysmal (2:1)	Persistent	Persistent
m control	Amiodarone, cardioversion	Amiodarone, flecainide, propafenone, sotalol	Amiodarone, propafenone, sotalol, cardioversion	Amiodarone, flecainide, propafenone, sotalol, cardioversion	Amiodarone, propafenon cardioversic
control	Digitalis, diltiazem, beta-blockers, ablation	Ablation	Digitalis, diltiazem, beta-blockers, verapamil, ablation	Digitalis, diltiazem, beta-blockers, verapamil, ablation	Digitalis, diltia beta-blocke verapamil,
rin	Continued	Discontinued in sinus rhythm	Discontinued in sinus rhythm	Discontinued in sinus rhythm	Discontinued rhythm
me	No difference in symptoms or quality of life Rhythm control: better functional capacity Rate control: fewer adverse drug reactions	Rhythm control: less permanent AF Rate control less worsening heart failure	No difference in mortality, quality of life, or functional capacityRate control: fewer adverse drug reactions	No overall difference or in quality of life Rate control: fewer adverse drug reactions	No overall diff

= Pharmacological Intervention in Atrial Fibrillation;
= Paroxysmal Atrial Fibrillation 2;
RM = Atrial Fibrillation Follow-up Investigation of Rhythm Management;
= Rate Control versus Electrical Cardioversion for Persistent Atrial Fibrillation;
= Strategies of Treatment of Atrial Fibrillation.

Table 6 Adverse effects causing discontinuation of first antidysrhythmic drug within the first year (25)

Adverse event	*Amiodarone (n = 154)*	*Sotalol (n = 135)*	*Class I (n = 121)*
Congestive heart failure	0	3	2
Pulmonary events	4	1	1
Gastrointestinal events	4	6	14
Symptomatic bradycardia	0	3	4
Prolonged QT_c (>520 ms)	0	0	5
Syncope	0	1	3
Ocular effects	1	0	1
Other	11	7	17

Table 7 Drugs that prolong the QT interval and/or might cause torsade de pointes (50–52)

Class	*Drug*
Class IA antidysrhythmic drugs	Ajmaline, aprindine, cibenzoline, disopyramide, pirmenol, procainamide, propafenone, quinidine
Class IB antidysrhythmic drugs	Bretylium
Class IC antidysrhythmic drugs	Flecainide
Class III antidysrhythmic drugs	Amiodarone, dofetilide, ibutilide, nifekalant, sotalol
Class IV antidysrhythmic drugs	Bepridil, lidoflazine, prenylamine
Calcium channel blockers	Isradipine, nicardipine
Antibacterial drugs	Ciprofloxacin, clarithromycin, clindamycin, co-trimoxazole, erythromycin, grepafloxacin, levofloxacin, moxifloxacin, sparfloxacin, spiramycin, troleandomycin
Antidepressants	Amitriptyline, citalopram, clomipramine, desipramine, doxepin, fluoxetine, imipramine, maprotiline, nortriptyline, venlafaxine, zimeldine
Antiepileptic drugs	Felbamate, fosphenytoin
Antifungal drugs	Amphotericin, fluconazole, itraconazole, ketoconazole, miconazole
Antihistamines	Astemizole, azelastine, clemastine, diphenhydramine, ebastine, hydroxyzine, oxatomide, terfenadine
Antihypertensive drugs	Ketanserin
Antimalarial drugs	Chloroquine, halofantrine, mefloquine, quinine
Antiprotozoal drugs	Pentamidine
Antipsychotic drugs	Chlorpromazine, droperidol, fluphenazine, haloperidol, lithium, mesoridazine, pimozide, prochlorperazine, quetiapine, risperidone, sertindole, sultopride, thioridazine, timiperone, trifluoperazine, ziprasidone
Antiviral drugs	Foscarnet
Cytotoxic and immunosuppressant drugs	Arsenic trioxide, amsacrine, doxorubicin, tacrolimus, zorubicin
Diuretics	Indapamide, triamterene
Histamine H_2 receptor antagonists	Cimetidine, famotidine, ranitidine
Hormones	Octreotide, vasopressin
Miscellaneous drugs	Amantadine, aminophylline, budipine, chloral hydrate, cisapride, fenoxidil, ketanserin, prednisone, probucol, salbutamol, salmeterol, suxamethonium, terodiline, vincamine

the conduction of an increased number of atrial impulses through the AV node.

2. The induction of heart block or suppression of an escape mechanism. For example, slowing of conduction through the AV node can impair a mechanism that allows the conducting system to escape a re-entry mechanism.
3. The uncovering of a hidden mechanism of dysrhythmia. For example, antidysrhythmic drugs can cause early or delayed afterdepolarizations, which can result in dysrhythmias.
4. The induction of a new mechanism of dysrhythmia. For example, a patient in whom myocardial ischemia has predisposed to dysrhythmias may be more at risk when an antidysrhythmic drug alters conduction.

Combinations of these different mechanisms are also possible.

The prodysrhythmic effects of antidysrhythmic drugs have been reviewed, with regard to mechanisms at the cellular level (56) and molecular level (57). As far as the cellular mechanisms are concerned, the antidysrhythmic drugs have been divided into three classes (which do not overlap with the classes specified in the electrophysiological classification).

1. Group 1 drugs have fast-onset kinetics and the block saturates at rapid rates (about 300 beats/minute).
2. Group 2 drugs have slow-onset kinetics and the block saturates at rapid rates.

3. Group 3 drugs have slow-onset kinetics and there is saturation of frequency-dependent block at slow heart rates (about 100 beats/minute).

The fast-onset kinetics of the Group 1 drugs makes them the least likely to cause dysrhythmias. Group 2 drugs, which include encainide, flecainide, procainamide, and quinidine, are the most likely to cause dysrhythmias, because of their slow-onset kinetics. Although this also applies to the Group 3 drugs, which include propafenone and disopyramide, block is less likely to occur during faster heart rates and serious dysrhythmias are therefore less likely during exercise.

The most common mechanism of dysrhythmias at the molecular level is by inhibition of the potassium channels known as IK_r, which are encoded by the human ether-a-go-go-related gene (HERG). The antidysrhythmic drugs that affect these channels include almokalant, amiodarone, azimilide, bretylium, dofetilide, ibutilide, sematilide, D-sotalol, and tedisamil (all drugs with Class III actions) and bepridil, disopyramide, prenylamine, procainamide, propafenone, quinidine, and terodiline (all drugs with Class I actions). Other drugs that affect these channels but are not used to treat cardiac dysrhythmias include astemizole and terfenadine (antihistamines), cisapride, erythromycin, haloperidol, sertindole, and thioridazine.

Prolongation of the QT_c interval, resulting from inhibition of the human ether-a-go-go related gene (HERG) potassium channels by antidysrhythmic drugs, can cause serious ventricular dysrhythmias and sudden death. In 284 426 patients with suspected adverse reactions to drugs that are known to inhibit HERG channels reported to the International Drug Monitoring Program of the World Health Organization (WHO-UMC) up to the first quarter of 2003, 5591 cases (cardiac arrest, sudden death, torsade de pointes, ventricular fibrillation, and ventricular tachycardia) were compared with 278 835 non-cases (58). HERG inhibitory activity was defined as the effective therapeutic unbound plasma concentration divided by the HERG IC_{50} value of the suspected drug. There was a significant association between HERG inhibitory activity and the risk of serious ventricular dysrhythmias and sudden death (Table 8). The antidysrhythmic drugs that least followed the predicted pattern were amiodarone, bepridil, flecainide, ibutilide, and sotalol, for which the odds were higher than expected, and aprindine, for which the odds were lower than expected.

The mechanism of action of class I antidysrhythmic drugs has been studied in 14 patients with accessory pathways and orthodromic atrioventricular re-entrant tachycardia (59). The drugs were cibenzoline (n = 7), pilsicainide (n = 2), disopyramide (n = 2), and procainamide (n = 3). In four of six patients with a manifest accessory pathway, class I drugs induced unidirectional conduction block of the accessory pathway (anterograde conduction block associated with preserved retrograde conduction) and enhanced the induction of atrioventricular re-entrant tachycardia with atrial extrastimulation. In eight patients with a concealed accessory pathway, there was outward or inward expansion of the tachycardia induction zone in patients who had greater prolongation of the conduction time than the refractory period of the retrograde accessory pathway after class I drugs. During ventricular extrastimulation, induction of bundle branch re-entry after class I drugs initiated atrioventricular re-entrant tachycardia in all the patients. The authors concluded that the adverse effects of all class I drugs in patients with accessory pathways are mainly due to induction of unidirectional retrograde conduction in manifest accessory pathways and greater prolongation of retrograde conduction time in concealed accessory pathways than the refractory period, regardless of the subtype of drug.

Susceptibility factors

There are no good predictors of the occurrence of dysrhythmias, but there are several susceptibility factors (39,40), including a history of sustained tachydysrhythmias, poor left ventricular function, and myocardial ischemia. Potassium depletion and prolongation of the QT interval are particularly important, and these particularly predispose to polymorphous ventricular dysrhythmias (for example torsade de pointes). Altered metabolism of antidysrhythmic drugs (for example liver disease, polymorphic acetylation or hydroxylation, and drug interactions) can also contribute.

The prodysrhythmic effects of antidysrhythmic drugs have been reviewed in the context of whether patients who are to be given class I or class III antidysrhythmic

Table 8 HERG inhibitory activities of antidysrhythmic drugs and the frequencies of dysrhythmias

Drug	*Log HERG inhibitory activity*	*Cases*	*Non-cases*	*Cases/total (%)*
Amiodarone	−3.3	271	10 467	2.52
Cibenzoline	−1.4	13	214	5.73
Bepridil	−1.3	59	125	32.07
Procainamide	−0.8	101	2652	3.67
Flecainide	−0.7	332	1894	14.92
Disopyramide	−0.4	110	1843	5.63
Dofetilide	−0.4	68	676	9.14
Propafenone	−0.3	97	1146	7.80
Aprindine	0.0	1	164	0.61
Quinidine	1.0	181	3399	5.06
Ibutilide	1.1	154	27	85.08

drugs should first be admitted to hospital for observation in the hope of identifying those who are most likely to develop dysrhythmias (33). The risk of sudden death in patients taking amiodarone was significantly increased in those who had had a prior bout of torsade de pointes. The risk of sotalol-induced torsade de pointes was higher in patients with pre-existing heart failure. Women are at a greater risk of prodysrhythmic drug effects (SEDA-18, 199). The highest risk was in women with heart failure who took more than 320 mg/day (22%); the corresponding figure in men was 8%. The authors delineated certain subgroups that they considered to be at specific risk of dysrhythmias, listing drugs that should be avoided in those subjects. They recommended avoiding drugs of classes Ia and III in women without coronary artery disease, drugs of class Ic in men with coronary artery disease, and drugs of classes Ia, Ic, and III in men with congestive heart failure and women with coronary artery disease.

Factors that predict atrial flutter with 1:1 conduction as a prodysrhythmic effect of class I antidysrhythmic drugs (cibenzoline, disopyramide, flecainide, propafenone, and quinidine) have been studied in 24 patients (aged 46–78 years) with 1:1 atrial flutter and in 100 controls (60). Underlying heart disease was present in nine patients. There was a short PR interval (PR < 0.13 ms) with normal P wave duration in leads V5 and V6 in nine of the 26 patients and only seven of the 100 controls. Signal-averaged electrocardiography showed pseudofusion between the P wave and QRS complex in 19 of the 26 patients and only 11 of the 100 controls. There was rapid atrioventricular nodal conduction (a short AH interval or second-degree atrioventricular block during atrial pacing at over 200 minute) in 19 of the 23 patients. Pseudofusion of the P wave and QRS complex had a sensitivity of 100% and a specificity of 89% for the prediction of an atrial prodysrhythmic effect of class I antidysrhythmic drugs.

The pharmacogenetic aspects of drug-induced torsade de pointes have been reviewed (61). Major mutations and functional polymorphisms in the congenital long QT syndrome (cLQTS) genes, KCNE1, KCNE2, KCNH2, KCNQ1, and SCN5A, have been associated with an increased risk of torsade de pointes in patients taking antidysrhythmic drugs (62).

Reducing the risk

The methods for minimizing the risks of prodysrhythmic effects of antidysrhythmic drugs (63) are as follows:

- Care in choosing those who are likely to benefit from antidysrhythmic drug therapy.
- Identification and correction, if possible, of impaired pump function and ischemic damage.
- Correction of electrolyte abnormalities.
- Exercise testing before and during the early stages of drug therapy: widening of the QRS complex during exercise predicts a high risk of ventricular tachycardia as does prolongation of the QT interval.
- Instruction of patients about the signs and symptoms that can occur with dysrhythmias.
- Monitoring renal and hepatic function in order to predict reduced drug elimination.
- Avoiding drug interactions or changing the dosage of the antidysrhythmic drug in anticipation of a change in its disposition secondary to an interaction.

Measurement of the concentrations of antidysrhythmic drugs and their metabolites in the plasma can be useful in recognizing the need for changing dosage requirements when cardiac, hepatic, or renal dysfunction occurs, in maintaining serum drug or metabolite concentrations within optimal ranges, and for predicting dosage changes required when interacting drugs are added (41). However, in most hospitals plasma drug concentration measurement is not routinely available for these drugs.

Another strategy for reducing the risk of prodysrhythmias is to use combinations of different classes of antidysrhythmic drugs in lower dosages than those used in monotherapy.

Torsade de pointes can be prevented by withholding antidysrhythmic drug therapy from patients who have pre-existing prolongation of the QT interval, and by correction of low serum potassium and magnesium concentrations before therapy. During therapy patients at risk should have frequent monitoring of the electrocardiogram and serum electrolytes.

The prodysrhythmic risks of using antidysrhythmic drugs have been mentioned in the context of a set of guidelines on the management of patients with atrial fibrillation (64,65). The recommended drugs for maintaining sinus rhythm after cardioversion vary depending on the presence of different risk factors for dysrhythmias:

- heart failure: amiodarone and dofetilide;
- coronary artery disease: sotalol and amiodarone;
- hypertensive heart disease: propafenone and flecainide.

Management

The management of drug-induced cardiac dysrhythmias includes withdrawal of the drug and the administration of potassium if necessary to maintain the serum potassium concentration at over 4.5 mmol/l and magnesium sulfate (SEDA-23, 196). Magnesium sulfate is given intravenously on a dose of 2 g over 2–3 minutes, followed by continuous intravenous infusion at a rate of 2–4 mg/minute; if the dysrhythmia recurs, another bolus of 2 g should be given and the infusion rate increased to 6–8 mg/minute; rarely, a third bolus of 2 g may be required (66). If magnesium is ineffective, cardiac pacing should be tried.

There is some anecdotal evidence that atrioventricular nodal blockade with verapamil or a beta-blocker can also be effective. However, in two cases the addition of a beta-blocker (either atenolol or metoprolol) to treatment with class I antidysrhythmic drugs (cibenzoline in one case and flecainide in the other) did not prevent the occurrence of atrial flutter with a 1:1 response (67). However, the author suggested that in these cases, although the beta-blockers had not suppressed the dysrhythmia, they had at least improved the patient's tolerance of it. In both cases the uses of class I antidysrhythmic drugs was contraindicated by virtue of structural damage, in the first case due to mitral valvular disease and in the second due to an ischemic cardiomyopathy.

Adverse hemodynamic effects of antidysrhythmic drugs

Many antidysrhythmic drugs have negative inotropic effects (68–70). This means that such drugs should be avoided in patients with a history of heart failure, a low left ventricular ejection fraction, or a cardiomyopathy. The general risk of induction or a worsening of heart failure is up to about 5%, but those who have risk factors have a risk of up to 10%. The negative inotropic effects are most marked with drugs of classes Ia, Ic, II, and IV. For drugs with class I activity there is a strong relation between their negative inotropic effect and the extent to which they block the inward sodium current (70). Thus, class Ib drugs that are associated with a short recovery time of sodium channels have a smaller negative inotropic effect than class Ia drugs, which in turn have less of an effect than class Ic drugs. However, the overall hemodynamic effects of antidysrhythmic drugs depend not only on their negative inotropic effects on the heart, but also on their effects on the peripheral circulation (71). Thus, although all drugs with class I activity have similar negative inotropic effects on the heart, disopyramide has large hemodynamic effects (because it increases peripheral resistance) and its hemodynamic effect is therefore greater than that of mexiletine, for example. Similarly the adverse hemodynamic effects of encainide and tocainide are greater than those of procainamide (72).

Death

Sudden death due to antidysrhythmic drugs has been reported in several trials in patients who have had ventricular dysrhythmias after myocardial infarction. The drugs that have been incriminated include disopyramide, encainide, flecainide, mexiletine, moracizine, procainamide, and quinidine (73–79). The class III drug d-sotalol has also been associated with an increased risk of mortality in such patients (80). This increase in mortality is thought to be due to an increased risk of cardiac dysrhythmias, perhaps as a consequence of rate-dependent conduction block and preferential slowing of conduction in the ischemic areas. Cardiac dysrhythmias of this sort may also occur through slowing of the rate of conduction around non-conducting ischemic or infracted areas in the heart.

Second-Generation Effects

Pregnancy

The use of antidysrhythmic drugs in pregnancy and breast-feeding has been reviewed (81). The FDA system that is used to classify the risks does not clearly distinguish between teratogenicity (occurring in the first trimester), which would be expected to be irreversible, and fetotoxicity (occurring after the first trimester), some effects of which will be reversible. The FDA classification of antidysrhythmic drugs (82) is shown in Table 9 with some of the reported teratogenic and fetotoxic effects, most of which have been reported only anecdotally.

Susceptibility Factors

Age

The safety of antidysrhythmic drugs in children has not been thoroughly studied. However, the risk of prolongation of the QT interval seems to be considerably less than that in adults (83), although it has been reported with quinidine, disopyramide, amiodarone, sotalol, and diphemanil.

Drug Administration

Drug overdose

The use of techniques of circulatory support (extracorporeal oxygenation and intra-aortic balloon pump) in seven cases of overdose with antidysrhythmic drugs (disopyramide, flecainide, prajmaline, and quinidine) has been reviewed (84).

Table 9 The FDA teratogenicity classes of antidysrhythmic drugs

Drug	*Category*[*]	*Reported teratogenic or fetotoxic effects of therapeutic maternal doses*
Adenosine	C	None
Amiodarone	D	Sinus bradycardia, QT interval prolongation; congenital nystagmus; impaired language skills; altered thyroid function
Digoxin	C	None
Flecainide	C	Electrocardiographic changes; respiratory distress
Lidocaine	B	Bradycardia; acidosis
Mexiletine	C	None
Procainamide	C	None
Quinidine	C	Thrombocytopenia; eighth nerve damage

[*]FDA categories:

A: Controlled studies show no risk to the fetus. Adequate, well-controlled studies in pregnant women have failed to demonstrate a risk to the fetus.

B: No evidence of risk in humans. Either animal studies show risk but human findings do not, or if no adequate human studies have been done, animal findings are negative.

C: Risk cannot be ruled out. Human studies are lacking and animal studies are either positive for fetal risk or lacking. However, potential benefits may justify potential harm.

D: Positive evidence of risk. Investigational or post-marketing data show risk of harm to the fetus. Nevertheless, potential benefits may outweigh the potential harm.

X: Contraindicated in pregnancy. Studies in animals or humans or investigational or post-marketing reports have shown risk of fetal harm, which clearly outweighs any possible benefit to the patient.

Table 10 Some important drug–drug interactions with antidysrhythmic drugs

Object drug(s)	*Precipitant drug(s)*	*Result of interaction*
Adenosine	Dipyridamole	Increased effect
Adenosine	Theophylline	Reduced effect
Anticholinergic drugs	Disopyramide, quinidine	Potentiation
Antihypertensive drugs	Bretylium	Severe hypotension
Beta-adrenoceptor antagonists	Propafenone	Potentiation
Class I drugs	Beta-adrenoceptor antagonists	Negative inotropy
Class I drugs	Class I drugs	Potentiation
Class I drugs	Drugs that cause potassium depletion	Prodysrhythmic effects
Digoxin	Amiodarone, quinidine, verapamil	Digoxin toxicity
Disopyramide	Enzyme-inducing drugs	Increased metabolism
Neuromuscular blockers	Quinidine	Potentiation
Procainamide	Cimetidine, trimethoprim	Reduced metabolism
Quinidine	Enzyme-inducing drugs	Increased metabolism
Theophylline	Mexiletine	Cardiac dysrhythmias
Verapamil	Beta-adrenoceptor antagonists	Negative inotropy/bradycardia/asystole
Warfarin	Amiodarone, quinidine	Warfarin toxicity

Drug–Drug Interactions

Amiodarone

Amiodarone can potentiate the dysrhythmogenic actions of some Class I antidysrhythmic drugs (SEDA-18, 203), particularly because of the risk of QT interval prolongation (SEDA-19, 194). In 26 patients taking mexiletine plus amiodarone for 1 month and 155 taking mexiletine alone, there was no significant difference in the apparent oral clearance of mexiletine (85). However, the lack of a pharmacokinetic interaction does not reduce the risk that dangerous QT interval prolongation may occur with a combination such as this.

Bepridil

Because it prolongs the QT interval, bepridil can potentiate the effects of other drugs with the same effect (for example other Class I antidysrhythmic drugs and amiodarone).

Cardiac glycosides

Digoxin does not interact with a variety of antidysrhythmic drugs, including ajmaline, aprindine, lidocaine, lidoflazine (86), and moracizine (87). Other drugs that may have minor and clinically unimportant interactions include captopril, carvedilol, disopyramide, and flosequinan.

Diuretics

Hypokalemia due to diuretics potentiates the dysrhythmogenic actions of antidysrhythmic drugs that prolong the QT interval, such as Class I and Class III antidysrhythmic drugs, increasing the risk of torsade de pointes (88). This can also happen with other drugs that prolong the QT interval, such as phenothiazines (89).

Disopyramide

There is an increased risk of dysrhythmias if disopyramide is used in conjunction with other drugs that prolong the QT interval, for example class I or class III antidysrhythmic drugs (90).

General

Some important drug–drug interactions with antidysrhythmic drugs are summarized in Table 10.

Mexiletine

Interactions of mexiletine with other cardioactive drugs have been reviewed (91). The most important are beneficial interactions with beta-adrenoceptor antagonists, quinidine, and amiodarone in the suppression of ventricular tachydysrhythmias. During these interactions the adverse effects of mexiletine may also be less common, although this effect is inconsistent (92).

Procainamide

The effects of procainamide on the QT interval can be potentiated by other drugs with this action, for example other class I antidysrhythmic drugs (93).

Suxamethonium

Quinidine potentiates not only non-depolarizing muscle relaxants but also depolarizing drugs (94).

Verapamil can potentiate the block produced by both types of neuromuscular blocking agent (95).

Beta-blockers can prolong and possibly exaggerate the rise in serum potassium resulting from the injection of suxamethonium (SEDA-10, 108) (SEDA-11, 122) (96,97).

Tricyclic antidepressants

Because of their similar lipophilic and surfactant properties, tricyclic antidepressants interact with antidysrhythmic drugs of the quinidine type, interfering with the voltage-dependent stimulus and producing dose-related synergy (98). Cardiac glycosides and beta-blockers are free of this interaction, although animal studies have suggested increased lethality of digoxin in rats pretreated with tricyclic antidepressants (99), while propranolol may potentiate direct depression of myocardial contractility due to tricyclic antidepressants. For all of these reasons, the preferred treatment for tricyclic-induced dysrhythmias is lidocaine, but even this is reported to be only variably effective and possibly to potentiate the hypotensive effects of tricyclic drugs (100).

Tubocurarine

Class I antidysrhythmic drugs, such as procainamide, lidocaine, propranolol, diphenylhydantoin (101), quinidine (102), and lidocaine (103) have all been claimed to enhance neuromuscular blockade by D-tubocurarine and other non-depolarizing agents. Bretylium (104) and disopyramide (105) are also reported to have their neuromuscular blocking activities potentiated by low concentrations of D-tubocurarine in animal experiments; neostigmine failed to reverse disopyramide-induced blockade (SEDA-13, 102) (106). The greatest hazard from these agents is that they can cause "recurarization" when given postoperatively. With bretylium this can occur several hours after its administration, as a result of its slow kinetics (104). Effects in man have still to be documented for bretylium, but "recurarization" 15 minutes after adequate reversal of vecuronium blockade with neostigmine has been described in a patient given disopyramide intravenously (SEDA-14, 116) (107).

Monitoring therapy

Of 36 patients receiving antidysrhythmic drugs for supraventricular or ventricular dysrhythmias, 12 were treated with flecainide, 12 with pilsicainide, and 12 with pirmenol (108). Signal-averaged electrocardiograms were recorded before starting therapy, 1 month later, and twice during subsequent therapy. All three drugs, but especially flecainide and pilsicainide, prolonged the filtered QRS and the duration of low-amplitude signals at the terminal portion of the QRS complex. Differences in the duration of the filtered QRS between recordings correlated significantly with differences in serum drug concentrations ($r = 0.91$ for flecainide, $r = 0.70$ for pilsicainide, and $r = 0.61$ for pirmenol). There were no significant correlation between drug concentration and other parameters. The authors suggested that changes in the serum concentrations of flecainide, pilsicainide, and pirmenol can be estimated from changes in the duration of the filtered QRS on signal-averaged electrocardiograms and that periodic electrocardiographic monitoring in this way could substitute for drug concentration measurement.

References

1. Vaughan Williams EM. A classification of antiarrhythmic actions reassessed after a decade of new drugs. J Clin Pharmacol 1984;24(4):129–47.
2. Mason DT, DeMaria AN, Amsterdam EA, Zelis R, Massumi RA. Antiarrhythmic agents. Drugs 1973; 5(4):261–317.
3. Winkle RA, Glantz SA, Harrison DC. Pharmacologic therapy of ventricular arrhythmias. Am J Cardiol 1975; 36(5):629–50.
4. Singh BN. Side effects of antiarrhythmic drugs. Pharmacol Ther 1977;2:151.
5. Harrison DC, Meffin PJ, Winkle RA. Clinical pharmacokinetics of antiarrhythmic drugs. Prog Cardiovasc Dis 1977;20(3):217–42.
6. Anderson JL, Harrison DC, Meffin PJ, Winkle RA. Antiarrhythmic drugs: clinical pharmacology and therapeutic uses. Drugs 1978;15(4):271–309.
7. Zipes DP, Troup PJ. New antiarrhythmic agents: amiodarone, aprindine, disopyramide, ethmozin, mexiletine, tocainide, verapamil. Am J Cardiol 1978;41(6):1005–24.
8. Nattel S, Zipes DP. Clinical pharmacology of old and new antiarrhythmic drugs. Cardiovasc Clin 1980;11(1):221–48.
9. Schwartz JB, Keefe D, Harrison DC. Adverse effects of antiarrhythmic drugs. Drugs 1981;21(1):23–45.
10. Keefe DL, Kates RE, Harrison DC. New antiarrhythmic drugs: their place in therapy. Drugs 1981;22(5):363–400.
11. Kowey PR, Marinchak RA, Rials SJ, Bharucha DB. Intravenous antiarrhythmic therapy in the acute control of in-hospital destabilizing ventricular tachycardia and fibrillation. Am J Cardiol 1999;84(9A):R46–51.
12. Lip GYH, Kamath S. Adverse reactions of drugs used to treat arrhythmia. Adverse Drug React Bull 2000;201:767–70.
13. Wooten JM, Earnest J, Reyes J. Review of common adverse effects of selected antiarrhythmic drugs. Crit Care Nurs Q 2000;22(4):23–38.
14. Miller MR, McNamara RL, Segal JB, Kim N, Robinson KA, Goodman SN, Powe NR, Bass EB. Efficacy of agents for pharmacologic conversion of atrial fibrillation and subsequent maintenance of sinus rhythm: a meta-analysis of clinical trials. J Fam Pract 2000;49(11):1033–46.
15. Gronefeld G, Hohnloser SH. Rhythm or rate control in atrial fibrillation: insights from the randomized controlled trials. J Cardiovasc Pharmacol Ther 2003;8(Suppl 1):S39–44.
16. Wyse DG. Rhythm versus rate control trials in atrial fibrillation. J Cardiovasc Electrophysiol 2003;14(Suppl 9):S35–9.
17. The Research Group for Antiarrhythmic Drug Therapy. Cost-Effectiveness of antiarrhythmic drugs for prevention of thromboembolism in patients with paroxysmal atrial fibrillation. Jpn Circ J 2001;65(9):765–8.
18. Nichol G, McAlister F, Pham B, Laupacis A, Shea B, Green M, Tang A, Wells G. Meta-analysis of randomised controlled trials of the effectiveness of antiarrhythmic agents at promoting sinus rhythm in patients with atrial fibrillation. Heart 2002;87:535–43.
19. Gronefeld G, Hohnloser SH. Rhythm or rate control in atrial fibrillation: insights from the randomized controlled trials. J Cardiovasc Pharmacol Ther 2003;8 Suppl 1:S39–44.
20. Wyse DG. Rhythm versus rate control trials in atrial fibrillation. J Cardiovasc Electrophysiol 2003;14 Suppl:S35–9.
21. Wyse DG. Rhythm management in atrial fibrillation: less is more. J Am Coll Cardiol 2003;41:1703–6.
22. Freestone B, Lip GYH. Editorial review. Managing acute and recent-onset atrial fibrillation. Evidence-based Cardiovasc Med 2003;7:111–6.
23. Hohnloser SH, Kuck KH, Lilienthal J. Rhythm or rate control in atrial fibrillation—Pharmacological Intervention in Atrial Fibrillation (PIAF): a randomised trial. Lancet 2000;356:1789–94.
24. Brignole M, Menozzi C, Gasparini M, Bongiorni MG, Botto GL, Ometto R, Alboni P, Bruna C, Vincenti A, Verlato R;PAF 2 Study Investigators. An evaluation of the strategy of maintenance of sinus rhythm by antiarrhythmic drug therapy after ablation and pacing therapy in patients with paroxysmal atrial fibrillation. Eur Heart J 2002;23:892–900.

25. Wyse DG, Waldo AL, DiMarco JP, Domanski MJ, Rosenberg Y, Schron EB, Kellen JC, Greene HL, Mickel MC, Dalquist JE, Corley SD; Atrial Fibrillation Follow-up Investigation of Rhythm Management (AFFIRM) Investigators. A comparison of rate control and rhythm control in patients with atrial fibrillation. New Engl J Med 2002;347:1825–33.
26. Van Gelder IC, Hagens VE, Bosker HA, Kingma JH, Kamp O, Kingma T, Said SA, Darmanata JI, Timmermans AJ, Tijssen JG, Crijns HJ;Rate Control versus Electrical Cardioversion for Persistent Atrial Fibrillation Study Group. A comparison of rate control and rhythm control in patients with recurrent persistent atrial fibrillation. New Engl J Med 2002;347:1834–40.
27. Carlsson J, Miketic S, Windeler J, Cuneo A, Haun S, Micus S, Walter S, Tebbe U;STAF Investigators. Randomized trial of rate-control versus rhythm-control in persistent atrial fibrillation: the Strategies of Treatment of Atrial Fibrillation (STAF) study. J Am Coll Cardiol 2003;41:1690–6.
28. The Research Group for Antiarrhythmic Drug Therapy. Cost-effectiveness of antiarrhythmic drugs for prevention of thromboembolism in patients with paroxysmal atrial fibrillation. Jpn Circ J 2001;65:765–8.
29. AFFIRM First Antiarrhythmic Drug Substudy Investigators. Maintenance of sinus rhythm in patients with atrial fibrillation. An AFFIRM substudy of the first antiarrhythmic drug. J Am Coll Cardiol 2003;42:20–9.
30. Weerasooriya R, Davis M, Powell A, Szili-Torok T, Shah C, Whalley D, Kanagaratnam L, Heddle W, Leitch J, Perks A, Ferguson L, Bulsara M. The Australian Intervention Randomized Control of Rate in Atrial Fibrillation Trial (AIRCRAFT). J Am Coll Cardiol 2003;41:1697–702.
31. Rinkenberger RL, Prystowsky EN, Jackman WM, Naccarelli GV, Heger JJ, Zipes DP. Drug conversion of nonsustained ventricular tachycardia to sustained ventricular tachycardia during serial electrophysiologic studies: identification of drugs that exacerbate tachycardia and potential mechanisms. Am Heart J 1982;103(2):177–84.
32. Velebit V, Podrid P, Lown B, Cohen BH, Graboys TB. Aggravation and provocation of ventricular arrhythmias by antiarrhythmic drugs. Circulation 1982;65(5):886–94.
33. Thibault B, Nattel S. Optimal management with Class I and Class III antiarrhythmic drugs should be done in the outpatient setting: protagonist. J Cardiovasc Electrophysiol 1999;10(3):472–81.
34. Wellens HJ, Smeets JL, Vos M, Gorgels AP. Antiarrhythmic drug treatment: need for continuous vigilance. Br Heart J 1992;67(1):25–33.
35. Morganroth J. Early and late proarrhythmia from antiarrhythmic drug therapy. Cardiovasc Drugs Ther 1992;6(1):11–4.
36. Morganroth J. Proarrhythmic effects of antiarrhythmic drugs: evolving concepts. Am Heart J 1992;123(4 Pt 2):1137–9.
37. Hilleman DE, Mohiuddin SM, Gannon JM. Adverse reactions during acute and chronic class I antiarrhythmic therapy. Curr Ther Res 1992;51:730–8.
38. Hilleman DE, Larsen KE. Proarrhythmic effects of antiarrhythmic drugs. PT 1991;520–4June.
39. Libersa C, Caron J, Guedon-Moreau L, Adamantidis M, Nisse C. Adverse cardiovascular effects of anti-arrhythmia drugs. Part I: Proarrhythmic effects. Therapie 1992; 47(3):193–8.
40. Podrid PJ, Fogel RI. Aggravation of arrhythmia by antiarrhythmic drugs, and the important role of underlying ischemia. Am J Cardiol 1992;70(1):100–2.
41. Follath F. Clinical pharmacology of antiarrhythmic drugs: variability of metabolism and dose requirements. J Cardiovasc Pharmacol 1991;17(Suppl 6):S74–6.
42. Cowan JC, Coulshed DS, Zaman AG. Antiarrhythmic therapy and survival following myocardial infarction. J Cardiovasc Pharmacol 1991;18(Suppl 2):S92–8.
43. Friedman L, Schron E, Yusuf S. Risk-benefit assessment of antiarrhythmic drugs. An epidemiological perspective. Drug Saf 1991;6(5):323–31.
44. Furberg CD, Yusuf S. Antiarrhythmics and VPD suppression. Circulation 1991;84(2):928–30.
45. Luderitz B. Möglichkeiten und Grenzen der Arrhythmiebehandlung. [Possibilities and limitations of treatment for arrhythmia.] Z Gesamte Inn Med 1991;46(12):425–30.
46. Podrid PJ. Safety and toxicity of antiarrhythmic drug therapy: benefit versus risk. J Cardiovasc Pharmacol 1991; 17(Suppl 6):S65–73.
47. Zimmermann M. Antiarrhythmic therapy for ventricular arrhythmias. J Cardiovasc Pharmacol 1991;17(Suppl 6): S59–64.
48. Fauchier JP, Babuty D, Fauchier L, Rouesnel P, Cosnay P. Les effets proarythmiques des antiarythmiques. [Proarrhythmic effects of antiarrhythmic drugs.] Arch Mal Coeur Vaiss 1992;85(6):891–7.
49. Leenhardt A, Coumel P, Slama R. Torsade de pointes. J Cardiovasc Electrophysiol 1992;3:281–92.
50. Walker BD, Krahn AD, Klein GJ, Skanes AC, Wang J, Hegele RA, Yee R. Congenital and acquired long QT syndromes. Can J Cardiol 2003;19:76–87.
51. Fermini B, Fossa AA. The impact of drug-induced QT interval prolongation on drug discovery and development. Nature Rev Drug Disc 2003;2:439–47.
52. Horie M. Genetic background predisposing the drug-induced long QT syndrome. Folia Pharmacol Japon 2003;121:401–7.
53. Fujiki A, Usui M, Nagasawa H, Mizumaki K, Hayashi H, Inoue H. ST segment elevation in the right precordial leads induced with class IC antiarrhythmic drugs: insight into the mechanism of Brugada syndrome. J Cardiovasc Electrophysiol 1999;10(2):214–8.
54. Boriani G. New options for pharmacological conversion of atrial fibrillation. Card Electrophysiol Rev 2001;5:195–200.
55. Donahue TP, Conti JB. Atrial fibrillation: rate control versus maintenance of sinus rhythm. Curr Opin Cardiol 2001;16(1):46–53.
56. Chaudhry GM, Haffajee CI. Antiarrhythmic agents and proarrhythmia. Crit Care Med 2000;28(Suppl 10):N158–64.
57. Witchel HJ, Hancox JC. Familial and acquired long QT syndrome and the cardiac rapid delayed rectifier potassium current. Clin Exp Pharmacol Physiol 2000;27(10):753–66.
58. De Bruin ML, Pettersson M, Meyboom RH, Hoes AW, Leufkens HG. Anti-HERG activity and the risk of drug-induced arrhythmias and sudden death. Eur Heart J 2005;26(6):590–7.
59. Fujiki A, Tani M, Yoshida S, Inoue H. Electrophysiologic mechanisms of adverse effects of class I antiarrhythmic drugs (cibenzoline, pilsicainide, disopyramide, procainamide) in induction of atrioventricular re-entrant tachycardia. Cardiovasc Drugs Ther 1996;10:159–66.
60. Brembilla-Perrot B, Houriez P, Beurrier D, Claudon O, Terrier de la Chaise A, Louis P. Predictors of atrial flutter with 1:1 conduction in patients treated with class I antiarrhythmic drugs for atrial tachyarrhythmias. Int J Cardiol 2001;80(1):7–15.
61. Shah RR. Pharmacogenetic aspects of drug-induced torsade de pointes: potential tool for improving clinical drug development and prescribing. Drug Saf 2004;27(3):145–72.

62. Paulussen AD, Gilissen RA, Armstrong M, Doevendans PA, Verhasselt P, Smeets HJ, Schulze-Bahr E, Haverkamp W, Breithardt G, Cohen N, Aerssens J. Genetic variations of KCNQ1, KCNH2, SCN5A, KCNE1, and KCNE2 in drug-induced long QT syndrome patients. J Mol Med 2004;82(3):182–8.
63. Feldman AM, Bristow MR, Parmley WW, Carson PE, Pepine CJ, Gilbert EM, Strobeck JE, Hendrix GH, Powers ER, Bain RP, et alVesnarinone Study Group. Effects of vesnarinone on morbidity and mortality in patients with heart failure. N Engl J Med 1993; 329(3):149–55.
64. Fuster V, Rydèn LE, Asinger RW, Cannom DS, Crijns HJ, Frye RL, Halperin JL, Kay GN, Klein WW, Levy S, McNamara RL, Prystowsky EN, Wann LS, Wyse DG, Gibbons RJ, Antman EM, Alpert JS, Faxon DP, Fuster V, Gregoratos G, Hiratzka LF, Jacobs AK, Russell RO, Smith SC Jr, Klein WW, Alonso-Garcia A, Blomstrom-Lundqvist C, de Backer G, Flather M, Hradec J, Oto A, Parkhomenko A, Silber S, Torbicki A. American College of Cardiology/American Heart Association Task Force on Practice Guidelines; European Society of Cardiology Committee for Practice Guidelines and Policy Conferences (Committee to Develop Guidelines for the Management of Patients with Atrial Fibrillation); North American Society of Pacing and Electrophysiology. ACC/AHA/ESC Guidelines for the Management of Patients with Atrial Fibrillation: Executive Summary. A Report of the American College of Cardiology/American Heart Association Task Force on Practice Guidelines and the European Society of Cardiology Committee for Practice Guidelines and Policy Conferences (Committee to Develop Guidelines for the Management of Patients with Atrial Fibrillation) Developed in Collaboration with the North American Society of Pacing and Electrophysiology. Circulation 2001;104(17):2118–50.
65. Fuster V, Ryden LE, Asinger RW, Cannom DS, Crijns HJ, Frye RL, Halperin JL, Kay GN, Klein WW, Levy S, McNamara RL, Prystowsky EN, Wann LS, Wyse DG. American College of Cardiology; American Heart Association; European Society of Cardiology; North American Society of Pacing and Electrophysiology. ACC/AHA/ESC Guidelines for the Management of Patients with Atrial Fibrillation. A report of the American College of Cardiology/American Heart Association Task Force on Practice Guidelines and the European Society of Cardiology Committee for Practice Guidelines and Policy Conferences (Committee to Develop Guidelines for the Management of Patients with Atrial Fibrillation) developed in collaboration with the North American Society of Pacing and Electrophysiology. Eur Heart J 2001; 22(20):1852–923.
66. Banai S, Tzivoni D. Drug therapy for torsade de pointes. J Cardiovasc Electrophysiol 1993;4(2):206–10.
67. Brembilla-Perrot B, Houriez P, Claudon O, Yassine M, Suty-Selton C, Vancon AC, Abo el Makarem Y, Makarem E, Courtelour JM. Les effets proarythmiques supraventricularires des antiarythmiques de classe IC sont-ils prévenus par l'association avec des bétabloquants?. [Can the supraventricular proarrhythmic effects of class 1C antiarrhythmic drugs be prevented with the association of beta blockers?.] Ann Cardiol Angeiol (Paris) 2000;49(8):439–42.
68. Scholz H. Antiarrhythmischer und Kardiodepressive Wirkungen antiarrhythmischer Substanzen. [Anti-arrhythmic and cardiodepressive effects of anti-arrhythmia agents.] Z Kardiol 1988;77(Suppl 5):113–9.
69. Luderitz B, Manz M. Hämodynamic bei ventrikularen Rhythmusstörungen und bei ihrer Behandlung. [Hemodynamics in ventricular arrhythmias and in their treatment.] Z Kardiol 1988;77(Suppl 5):143–9.
70. Schlepper M. Cardiodepressive effects of antiarrhythmic drugs. Eur Heart J 1989;10(Suppl E):73–80.
71. Seipel L, Hoffmeister HM. Hemodynamic effects of anti-arrhythmic drugs: negative inotropy versus influence on peripheral circulation. Am J Cardiol 1989;64(20):J37–40.
72. Hammermeister KE. Adverse hemodynamic effects of antiarrhythmic drugs in congestive heart failure. Circulation 1990;81(3):1151–3.
73. The Cardiac Arrhythmia Suppression Trial (CAST) Investigators. Preliminary report: effect of encainide and flecainide on mortality in a randomized trial of arrhythmia suppression after myocardial infarction. N Engl J Med 1989;321(6):406–12.
74. The Cardiac Arrhythmia Suppression Trial II Investigators. Effect of the antiarrhythmic agent moricizine on survival after myocardial infarction. N Engl J Med 1992;327(4):227–33.
75. Impact Research Group. International mexiletine and placebo antiarrhythmic coronary trial: I. Report on arrhythmia and other findings. J Am Coll Cardiol 1984; 4(6):1148–63.
76. Coplen SE, Antman EM, Berlin JA, Hewitt P, Chalmers TC. Efficacy and safety of quinidine therapy for maintenance of sinus rhythm after cardioversion. A meta-analysis of randomized control trials. Circulation 1990;82(4):1106–16Erratum in: Circulation 1991;83(2):714.
77. Flaker GC, Blackshear JL, McBride R, Kronmal RA, Halperin JL, Hart RG. Antiarrhythmic drug therapy and cardiac mortality in atrial fibrillation. The Stroke Prevention in Atrial Fibrillation Investigators. J Am Coll Cardiol 1992;20(3):527–32.
78. Nattel S, Hadjis T, Talajic M. The treatment of atrial fibrillation. An evaluation of drug therapy, electrical modalities and therapeutic considerations. Drugs 1994;48(3):345–71.
79. Moosvi AR, Goldstein S, VanderBrug Medendorp S, Landis JR, Wolfe RA, Leighton R, Ritter G, Vasu CM, Acheson A. Effect of empiric antiarrhythmic therapy in resuscitated out-of-hospital cardiac arrest victims with coronary artery disease. Am J Cardiol 1990;65(18):1192–7.
80. Waldo AL, Camm AJ, deRuyter H, Friedman PL, MacNeil DJ, Pauls JF, Pitt B, Pratt CM, Schwartz PJ, Veltri EP. Effect of d-sotalol on mortality in patients with left ventricular dysfunction after recent and remote myocardial infarction. The SWORD Investigators. Survival With Oral d-Sotalol. Lancet 1996;348(9019):7–12Erratum in: Lancet 1996;348(9024):416.
81. Lee JCR, Wetzel G, Shannon K. Maternal arrhythmia management during pregnancy in patients with structural heart disease. Progr Pediatr Cardiol 2004;19:71–82.
82. Doering PL, Boothby LA, Cheok M. Review of pregnancy labeling of prescription drugs: is the current system adequate to inform of risks? Am J Obstet Gynecol 2002;187(2):333–9.
83. Villain E. Les syndromes de QT long chez l'enfant. [Long QT syndromes in children.] Arch Fr Pediatr 1993;50(3):241–7.
84. Bosquet C, Jaeger A. Exceptional treatments in toxic circulatory and respiratory failures. Reanim Urgences 2001;10:402–11.

85. Yonezawa E, Matsumoto K, Ueno K, Tachibana M, Hashimoto H, Komamura K, Kamakura S, Miyatake K, Tanaka K. Lack of interaction between amiodarone and mexiletine in cardiac arrhythmia patients. J Clin Pharmacol 2002;42(3):342–6.
86. Doering W. Quinidine-digoxin interaction: Pharmacokinetics, underlying mechanism and clinical implications. N Engl J Med 1979;301(8):400–4.
87. Antman EM, Arnold M, Friedman PL, White H, Bosak M, Smith TW. Drug interactions with cardiac glycosides: evaluation of a possible digoxin–ethmozine pharmacokinetic interaction. J Cardiovasc Pharmacol 1987;9(5): 622–7.
88. Applegate WB. Hypertension in elderly patients. Ann Intern Med 1989;110(11):901–15.
89. Nicholls MG. Age-related effects of diuretics in hypertensive subjects. J Cardiovasc Pharmacol 1988;12(Suppl 8): S51–9.
90. Ellrodt G, Singh BN. Adverse effects of disopyramide (Norpace): toxic interactions with other antiarrhythmic agents. Heart Lung 1980;9(3):469–74.
91. Bigger JT Jr. The interaction of mexiletine with other cardiovascular drugs. Am Heart J 1984;107(5 Pt 2): 1079–1085.
92. Poole JE, Werner JA, Bardy GH, Graham EL, Pulaski WP, Fahrenbruch CE, Greene HL. Intolerance and ineffectiveness of mexiletine in patients with serious ventricular arrhythmias. Am Heart J 1986;112(2):322–6.
93. Windle J, Prystowsky EN, Miles WM, Heger JJ. Pharmacokinetic and electrophysiologic interactions of amiodarone and procainamide. Clin Pharmacol Ther 1987;41(6):603–10.
94. Miller RD, Way WL, Katzung BG. The potentiation of neuromuscular blocking agents by quinidine. Anesthesiology 1967;28(6):1036–41.
95. Durant NN, Nguyen N, Katz RL. Potentiation of neuromuscular blockade by verapamil. Anesthesiology 1984;60(4):298–303.
96. O'Brien DJ, Moriarty DC, Hope CE. The effect of preexisting beta blockade on potassium flux in patients receiving succinylcholine. Can Anaesth Soc J 1986;3:S89.
97. McCammon RL, Stoelting RK. Exaggerated increase in serum potassium following succinylcholine in dogs with beta blockade. Anesthesiology 1984;61(6):723–5.
98. Cocco G, Ague C. Interactions between cardioactive drugs and antidepressants. Eur J Clin Pharmacol 1977;11(5):389–93.
99. Attree T, Sawyer P, Turnbull MJ. Interaction between digoxin and tricyclic antidepressants in the rat. Eur J Pharmacol 1972;19(2):294–6.
100. Hoffman JR, McElroy CR. Bicarbonate therapy for dysrhythmia hypotension in tricyclic antidepressant overdose. West J Med 1981;134(1):60–4.
101. Harrah MD, Way WL, Katzung BG. The interaction of d-tubocurarine with antiarrhythmic drugs. Anesthesiology 1970;33(4):406–10.
102. Miller RD, Way WL, Katzung BG. The potentiation of neuromuscular blocking agents by quinidine. Anesthesiology 1967;28(6):1036–41.
103. Katz RL, Gissen AJ. Effects of intravenous and intra-arterial procaine and lidocaine on neuromuscular transmission in man. Acta Anaesthesiol Scand Suppl 1969;36:103–13.
104. Welch GW, Waud BE. Effect of bretylium on neuromuscular transmission. Anesth Analg 1982;61(5):442–4.
105. Healy TE, O'Shea M, Massey J. Disopyramide and neuromuscular transmission. Br J Anaesth 1981;53(5):495–8.
106. Jones SV, Marshall IG. Non-competitive effects of disopyramide at the neuromuscular junction: evidence for end-plate ion channel block. Br J Anaesth 1987;59(6):776–83.
107. Baurain M, Barvais L, d'Hollander A, Hennart D. Impairment of the antagonism of vecuronium-induced paralysis and intra-operative disopyramide administration. Anaesthesia 1989;44(1):34–6.
108. Sutovsky I, Katoh T, Takayama H, Ono T, Takano T. Therapeutic monitoring of class I antiarrhythmic agents using high-resolution electrocardiography instead of blood samples. Circ J 2003;67:195–8.

Acecainide

General Information

Acecainide (*N*-acetylprocainamide) is the main metabolite of procainamide, and it has antidysrhythmic activity (1). However, in contrast to procainamide, which has Class Ib activity, the main action of acecainide is that of Class III.

Apart from the lupus-like syndrome, the adverse effects of acecainide are as common as those of procainamide. The commonest affect the gastrointestinal tract and the central nervous system. Anorexia, nausea, vomiting, diarrhea, and abdominal pain are common, as are insomnia, dizziness, light-headedness, tingling sensations, and blurred vision. Other reported unwanted effects include skin rashes, constipation, and reduced sexual function (2–5).

Organs and Systems

Cardiovascular

Acecainide prolongs the QT interval and can therefore cause ventricular dysrhythmias (6). The risk is increased in renal insufficiency, since acecainide is mainly eliminated unchanged via the kidneys.

Immunologic

The main advantage of acecainide over procainamide is the lower incidence of the lupus-like syndrome. Many fewer patients develop antinuclear antibodies during long-term treatment with acecainide than during long-term treatment with procainamide (7).

There are also reports of remission of lupus-like syndrome without recurrence in patients in whom acecainide has been used as a replacement for procainamide (8–10). Furthermore, patients in whom procainamide has previously caused a lupus-like syndrome have been reported not to suffer from the syndrome on subsequent long-term treatment with acecainide (8). However, one patient suffered mild arthralgia while taking acecainide, having had a more severe arthropathy while taking procainamide (8).

Susceptibility Factors

Renal disease

Because acecainide is eliminated mostly unchanged by renal excretion, with a half-life of about 7 hours, its clearance is reduced in patients with renal impairment, who are at increased risk of adverse effects. This means that elderly people, who generally have a degree of renal impairment, are also at increased risk.

Monitoring Drug Therapy

The target plasma concentration range of acecainide is 15–25 μg/ml. The adverse effects of acecainide increase in frequency at concentrations above 30 μg/ml (11).

References

1. Atkinson AJ Jr, Ruo TI, Piergies AA. Comparison of the pharmacokinetic and pharmacodynamic properties of procainamide and N-acetylprocainamide. Angiology 1988;39(7 Pt 2):655–67.
2. Roden DM, Reele SB, Higgins SB, Wilkinson GR, Smith RF, Oates JA, Woosley RL. Antiarrhythmic efficacy, pharmacokinetics and safety of N-acetylprocainamide in human subjects: comparison with procainamide. Am J Cardiol 1980;46(3):463–8.
3. Winkle RA, Jaillon P, Kates RE, Peters F. Clinical pharmacology and antiarrhythmic efficacy of N-acetylprocainamide. Am J Cardiol 1981;47(1):123–30.
4. Atkinson AJ Jr, Lertora JJ, Kushner W, Chao GC, Nevin MJ. Efficacy and safety of N-acetylprocainamide in long-term treatment of ventricular arrhythmias. Clin Pharmacol Ther 1983;33(5):565–76.
5. Domoto DT, Brown WW, Bruggensmith P. Removal of toxic levels of N-acetylprocainamide with continuous arteriovenous hemofiltration or continuous arteriovenous hemodiafiltration. Ann Intern Med 1987;106(4):550–2.
6. Piergies AA, Ruo TI, Jansyn EM, Belknap SM, Atkinson AJ Jr. Effect kinetics of N-acetylprocainamide-induced QT interval prolongation. Clin Pharmacol Ther 1987;42(1):107–12.
7. Lahita R, Kluger J, Drayer DE, Koffler D, Reidenberg MM. Antibodies to nuclear antigens in patients treated with procainamide or acetylprocainamide. N Engl J Med 1979;301(25):1382–5.
8. Kluger J, Leech S, Reidenberg MM, Lloyd V, Drayer DE. Long-term antiarrhythmic therapy with acetylprocainamide. Am J Cardiol 1981;48(6):1124–32.
9. Kluger J, Drayer DE, Reidenberg MM, Lahita R. Acetylprocainamide therapy in patients with previous procainamide-induced lupus syndrome. Ann Intern Med 1981;95(1):18–23.
10. Stec GP, Lertora JJ, Atkinson AJ Jr, Nevin MJ, Kushner W, Jones C, Schmid FR, Askenazi J. Remission of procainamide-induced lupus erythematosus with N-acetylprocainamide therapy. Ann Intern Med 1979;90(5):799–801.
11. Connolly SJ, Kates RE. Clinical pharmacokinetics of N-acetylprocainamide. Clin Pharmacokinet 1982;7(3):206–20.

Adenosine and adenosine triphosphate (ATP)

General Information

Adenosine and adenosine triphosphate (ATP), its phosphorylated derivative, have been used to treat acute paroxysmal supraventricular tachycardias and adenosine has also been used in the diagnosis of narrow-and broad-complex tachycardias (SEDA-16, 176).

Several reviews of the clinical pharmacology, actions, therapeutic uses, and adverse reactions and interactions of adenosine and ATP have appeared (1–4). After intravenous administration adenosine enters cells, disappearing from the blood with a half-life of less than 10 seconds; intracellularly it is phosphorylated to cyclic AMP. Its mechanism of action as an antidysrhythmic drug is not known, but it may act by an effect at adenosine receptors on the cell membrane. Its electrophysiological effects are to prolong AV nodal conduction time by prolonging the AH interval, without an effect on the HV interval. The pharmacological and adverse effects of adenosine triphosphate are similar to those of adenosine.

Although adenosine and ATP very commonly cause adverse effects, they are generally mild and usually transient, because adenosine is rapidly eliminated from the blood (with a half-life of less than 10 seconds). Adverse effects have been reported in 81% of patients given adenosine and 94% of patients given ATP (5). Exercise reduces the non-cardiac adverse effects and the incidence of major dysrhythmias (6). Reducing the duration of adenosine infusion from 6 to 4 minutes reduced the incidence of chest discomfort and ischemic ST segment changes, but had no impact on non-cardiac effects (7).

Several studies have reported the efficacy and safety of adenosine and ATP in the treatment of tachycardias in children (8–11).

In 18 children with aortic valve disease or Kawasaki disease, adenosine stress myocardial perfusion imaging was associated with the usual adverse effects, most commonly flushing and dyspnea (12).

Exercise reduces both non-cardiac adverse effects and dysrhythmias in patients who are given adenosine for diagnostic purposes in myocardial perfusion imaging (SEDA-21, 197). This has been confirmed in two studies. In the first of these, 793 patients were given an intravenous infusion of adenosine 140 micrograms/kg/minute while exercising for 6 minutes or for a similar time without exercise (13). The rate of hypotension and dysrhythmias was significantly less in those who exercised (14 of 507) than in those who did not exercise (16 of 286). Overall reactions were more common in women than in men (5.7 versus 1.8%). All the adverse effects were transient and no specific therapy was required. The authors attributed the difference to the increase in sympathetic tone during exercise, which would have partly counteracted the hypotension and the negative chronotropic and negative dromotropic effects of adenosine. However, there was a major difference between the two groups, in that those who did not take exercise were considered unfit for exercise, which may have been associated with an increased risk of adverse effects. Nevertheless, the authors discarded that possibility, because the frequency of adverse reactions in those who did not take exercise was similar to frequencies that have previously been reported.

In the second study 19 patients received an intravenous infusion of adenosine 140 micrograms/kg/minute for 4 minutes during exercise or for 6 minutes without exercise; the patients undertook both protocols (14). Again, there were fewer adverse effects in those who took exercise, but only hypotension, chest pain, and headache were significantly different; there was a reduction in the frequency of flushing, which was almost significant. In addition, adverse effects were experienced for longer and the severity was greater in those who did not take exercise.

Observational studies

In 44 patients with paroxysmal supraventricular tachycardia, adenosine terminated the tachycardia in 16 and revealed the type of tachycardia in 21; in three of six patients it contributed to the diagnosis of broad-complex tachycardia; latent ventricular pre-excitation was induced in two of 12 patients (15). Subjective complaints after adenosine were common (at least one symptom in 50 of 62 patients), but all were transient.

Studies of the use of intravenous adenosine with ^{201}Tl in myocardial scintigraphy continue to be published, showing adverse effects that have been previously reported. In a phase II study in 44 patients given adenosine 120 or 140 micrograms/kg/minute for 6 minutes there was chest pain or discomfort in 23 and flushing or a feeling of warmth in 12 (16). Adenosine reversibly lowered blood pressure and increased heart rate slightly; the fall in systolic blood pressure was more than 20 mmHg from baseline in 26% of patients who received 120 micrograms/kg/minute and in 52% of those who received 140 micrograms/kg/minute. The same authors reported similar adverse effects in a phase III study in 207 patients given 120 micrograms/kg/minute (17) and in a clinical trial in 31 patients (18).

Comparative studies

The perioperative antinociceptive and analgesic effects of intraoperative adenosine 50–500 micrograms/kg/minute have been compared with those of remifentanil 0.05–0.5 micrograms/kg/minute in 62 patients undergoing major surgical procedures in a randomized, double-blind study (19). Intraoperative inhibition of the cardiovascular responses to surgical stimulation was similar after by adenosine and remifentanil, and both maintained excellent hemodynamic stability. However, there were striking postoperative differences:

1. initial pain score was significantly reduced by 60% by adenosine compared with remifentanil and it remained lower throughout the 48-hour recovery period;
2. postoperative morphine requirements during the first 0.25, 2, and 48 hours were consistently lower after adenosine than after remifentanil;
3. patients who received adenosine were significantly less sedated;
4. postoperative end-tidal and arterial carbon dioxide pressures were significantly higher after remifentanil.

In a comparison of intracoronary adenosine 24–288 micrograms with and without nitroprusside in 53 patients with no reflow despite coronary artery reperfusion, one patient had advanced atrioventricular block, which responded to atropine, and another had bradycardia, which resolved spontaneously (20).

In 50 patients in whom intravenous adenosine 140 micrograms/kg/minute was compared with intracoronary adenosine 60–150 micrograms, intravenous adenosine caused *angina* (26%), dyspnea (16%), and nausea (2%), while intracoronary adenosine caused dose-related atrioventricular block (21).

The effects of abciximab or intracoronary adenosine distal to the occlusion on immediate angiographic results and 6-month left ventricular remodelling have been studied in 90 patients undergoing primary angioplasty with coronary stenting (22). Abciximab enhanced myocardial reperfusion, with a reduced incidence of 6-month left ventricular remodelling. In contrast, adenosine improved angiographic results but did not prevent left ventricular remodelling. Adverse events were not reported.

Placebo-controlled studies

In 608 patients with ST-elevation acute myocardial infarction randomized to receive infusions of saline or adenosine 10 micrograms/kg/minute for 6 hours after the start of thrombolysis, there was a trend to reduced cardiovascular mortality with adenosine after 12 months, 9% versus 12% with placebo among all patients (OR = 0.71, 95%CI = 0.4, 1.2) and 8% versus 15% among patients with anterior myocardial infarction (OR = 0.53, 95% CI = 0.23, 1.24) (23). There were no adverse effects of adenosine. A much larger trial would be needed to confirm the trend to a beneficial effect on mortality after myocardial infarction.

Organs and Systems

Cardiovascular

The most common cardiac effects are atrioventricular block, sinus bradycardia, and ventricular extra beats. Occasionally serious dysrhythmias occur (SEDA-17, 219), including ventricular fibrillation (24). ATP can cause transient atrial fibrillation (25). Chest pain occurs in 30–50% of patients and dyspnea and chest discomfort in 35–55%. Chest pain can occur in patients with and without coronary artery disease, and the symptoms are not always typical of cardiac pain.

Myocardial ischemia

Adenosine can cause cardiac ischemia by activating adenosine A1 receptors in the heart. However, in a double-blind, placebo-controlled, crossover study in eight healthy volunteers, adenosine 100 μg/kg/minute did not alter ischemic pain in an exercising arm (26). Otherwise, the usual adverse effects were noted, including facial flushing and mild chest tightness.

ST segment depression can occur during adenosine myocardial perfusion imaging and is an independent predictor of subsequent cardiac events and worse outcome, particularly in association with ischemic defects. In a retrospective analysis of 3231 patients undergoing adenosine myocardial perfusion imaging, 228 (7%) had ischemic electrocardiographic changes during adenosine infusion (27). Of these, 66 (29%, 2% of all patients) had normal imaging. An age- and sex-matched group of 200 patients with normal imaging without electrocardiographic changes served as controls. During a mean follow-up of 29 months, those who had had electrocardiographic changes during imaging had significantly more adverse cardiac events than those in the control group (non-fatal myocardial infarction, 7.6% versus 0.5%; subsequent revascularization, 14% versus 2.5%). Although cardiac death alone did not differ between the two groups (3.0% versus 1.0%), cumulative survival free from cardiac death and non-fatal myocardial infarction was worse in patients with ST segment depression during adenosine infusion and normal imaging (11% versus 1.5%). The authors concluded that patients with normal myocardial perfusion images in whom ST segment depression occurs during adenosine administration are at higher risk of future cardiac events than similar patients without electrocardiographic evidence of ischemia.

In 75 patients with aortic stenosis, intravenous adenosine caused the usual adverse effects (flushing, chest pain, dyspnea, dizziness, headache, and nausea) in 8–41% of patients, second-degree heart block in seven, and third-degree heart block in two; there was transient ST segment depression greater than 1 mm in six patients (28).

Despite this, acute myocardial infarction after adenosine is rare, but has been reported in a 71-year-old man who was given an intravenous infusion of adenosine 140 micrograms/kg/minute for 3 minutes (29). The authors suggested that coronary vasodilatation had led to reduced perfusion pressure in collaterals, with a further contribution from reduced flow secondary to aortic stenosis.

Hypotension

When adenosine (70 micrograms/kg/minute) was given by intravenous infusion to 45 patients with acute myocardial infarction preceding balloon angioplasty, one patient developed persisting hypotension in conjunction with a large inferolateral myocardial infarction (30). Transient hypotension in three other patients resolved with a reduction in dosage. There were no cases of atrioventricular block. Symptomatic hypotension has occasionally been reported in patients with myocardial infarction who have been given adenosine (SEDA-20, 174).

An infusion of adenosine, 25 micrograms/kg, in 15 women undergoing anesthesia for major gynecological procedures was effective in maintaining hemodynamic stability during operation in addition to conventional anesthesia (31). It caused a significantly greater fall in systolic blood pressure and increase in heart rate than remifentanil in a comparable group. In four cases ephedrine was required for hypotension that was refractory to intravenous fluids or a temporary reduction in the infusion rate of adenosine. Two patients also required atropine for prolonged bradycardia.

Cardiac dysrhythmias

In patients with ischemic heart disease adenosine can prolong the QT_c interval and can increase the frequency of ventricular extra beats when there is myocardial scarring. It also causes increased release of catecholamines, and this may be the mechanism whereby it causes dysrhythmias in susceptible patients. If a dysrhythmia occurs, theophylline or one of its derivatives may be beneficial (32).

Of 100 patients who received intravenous adenosine in hospital (mean dose 7.8 mg) two had a dysrhythmia other than that for which they were being treated (33).

- A 53-year-old man with a dilated cardiomyopathy was given adenosine 6 mg for a regular broad-complex tachycardia; the dysrhythmia resolved but was followed by prolonged asystole and cyanosis for about 15 seconds.
- A 64-year-old woman with atrial fibrillation was given adenosine 12 mg; she developed a non-sustained polymorphous ventricular tachycardia followed by sustained ventricular fibrillation requiring DC shock.

In the whole series, about 40% of the patients received adenosine unnecessarily, having atrial fibrillation or atrial flutter, and the authors suggested that misuse of this sort resulted in unnecessary expense and increased risks of adverse effects. Most of this misuse was attributed to misdiagnosis by house officers who thought that rapid atrial fibrillation was a paroxysmal supraventricular tachycardia. Very few thought that adenosine would be likely to terminate atrial fibrillation.

Adenosine is contraindicated in patients with aberrant conduction pathways, because it can cause cardiac dysrhythmias. Supraventricular dysrhythmias occurred in three children with Wolff–Parkinson–White syndrome who were given intravenous adenosine (34).

There have been reports of cardiac dysrhythmias in patients given either an intravenous infusion of adenosine or a single bolus dose.

- A 38-year-old man was given intravenous adenosine 6 mg for a narrow-complex tachycardia (32). Within about 1 minute his heart rate fell from 230/minute to bradycardia and then asystole. Cardiopulmonary resuscitation was ineffective. At autopsy there was a 75% occlusion of one of the coronary arteries (unspecified).

The cause of the dysrhythmia in response to adenosine was not clear. He was not known to be taking other drugs (for example dipyridamole) that might have potentiated the action of adenosine.

- A 56-year-old man was given adenosine 12 mg for a narrow-complex tachycardia on four occasions, and on each occasion developed transient atrial fibrillation for a few minutes thereafter. He had a concealed left-sided accessory pathway, which was successfully ablated (35).
- An 86-year-old woman was given adenosine 12 mg intravenously for sustained supraventricular tachycardia, which terminated but was followed by atrial fibrillation and paroxysmal ventricular tachycardia (36). Cardioversion was unsuccessful, but normal sinus rhythm was obtained with procainamide. This followed an anteroseptal myocardial infarction.
- A 75-year-old man who had had coronary bypass surgery was given an intravenous infusion of adenosine for stress testing (37). After 1 minute he developed a three-beat run of wide-complex tachycardia, followed by a 20-second run of a regular wide-complex tachycardia at a rate of 115/minute. There was left bundle branch block, and the tachycardia ended spontaneously. Adenosine infusion was continued and some ventricular extra beats with the same configuration occurred. In this case there was impaired perfusion of the left ventricle.
- In a 60-year-old woman with atrial flutter with 2:1 block and a ventricular rate of 130/minute, the ventricular rate increased paradoxically to 260/minute with 1:1 conduction after intravenous administration of adenosine 6 mg; it responded to intravenous amiodarone 300 mg (38).
- A 52-year-old woman with a wide-complex tachycardia was given adenosine 6, 12, and another 12 mg as intravenous bolus doses; immediately after the third dose she developed ventricular fibrillation (39). She recovered with cardioversion.

In the last case the authors did not discuss the possibility that the presence of digoxin (serum concentration 1.8 ng/ml) may have contributed; the risk of cardiac dysrhythmias after electrical cardioversion is increased in the presence of digoxin (SEDA-8, 174), and the same might be true of chemical cardioversion.

In a prospective study of 187 episodes of tachycardia in 127 unselected patients adenosine was given in an average dose of 9.7 mg (40). In 108 cases, adenosine induced transient ventricular extra beats or non-sustained ventricular tachycardia after successful termination of supraventricular tachycardia; more than half had a right bundle branch block morphology that suggested that the dysrhythmias had originated from the inferior left ventricular septum.

Heart block

The frequency of atrioventricular block has been studied in 600 patients who underwent stress testing with intravenous adenosine 140 micrograms/kg/minute for 6 minutes (41). The patients were young (under 49 years old; $n = 75$), middle-aged (50–65 years; $n = 214$), old (66–75 years; $n = 195$), or very old (over 75 years; $n = 116$). The respective frequencies of first-degree atrioventricular block were 15, 9.3, 14, and 17% (overall average 13%), of second-degree block 15, 7.0, 8.7, and 16% (overall average 10%), and of third-degree block 2.7, 2.3, 1.0, and 2.6% (overall average 2.0%). The differences with age were not statistically significant. All types of atrioventricular block were of short duration, were well tolerated, and did not require withdrawal of adenosine or specific treatment.

Atrioventricular block is usually transient after adenosine, even after intracoronary administration (42), but it can be sustained, as in the case of a 2-year-old child with a supraventricular tachycardia (43).

In four out of nine patients with heart transplants second-degree or third-degree atrioventricular block occurred during the administration of adenosine 140 micrograms/kg/minute over 6 minutes (44). In two patients the infusion had to be interrupted because of severe discomfort and chest pain.

The incidence of atrioventricular block has been reported in 600 consecutive patients who underwent stress myocardial perfusion imaging with adenosine (140 micrograms/kg/minute for 6 minutes), and of whom 43 had first-degree heart block before adenosine and 557 had a baseline PR interval less than 200 ms (Table 1) (45). The heart block in all cases was of short duration, was not associated with any specific symptoms, and in no case required specific treatment. The risk of atrioventricular block during adenosine infusion was not increased by the presence of other drugs that might have caused atrioventricular block (digitalis, beta-blockers, diltiazem, verapamil).

Accessory conduction pathways

Accessory conduction pathways contraindicate adenosine.

- A 39-year-old woman with bouts of palpitation and a narrow-complex tachycardia was given intravenous adenosine 6 mg and developed a broad-complex tachycardia, due to supraventricular tachycardia, which spontaneously converted to sinus rhythm after 10 seconds (46). She was then found to have Wolff–Parkinson–White syndrome with bilateral accessory pathways.

However, it is not always possible to detect an accessory conduction pathway.

Table 1 The incidence of atrioventricular block with adenosine

Type of block	*Baseline PR interval over 200 ms (n = 43)*	*Baseline PR interval under 200 ms (n = 557)*
Further prolongation of PR interval	49%	10%
Second-degree block	37%	8%
Third-degree block	14%	1%

- A previously healthy 35-year-old man was given adenosine 6 mg for a narrow-complex tachycardia at 217/minute (47). The rhythm suddenly changed to wide QRS atrial fibrillation at an average ventricular rate of 250/minute and transient higher rates over 300/minute, degenerating into ventricular fibrillation after a few minutes. DC shock 200 J resulted in sinus rhythm.

An accessory pathway was subsequently diagnosed.

Sinus arrest during adenosine stress testing has been reviewed in the context of three cases in liver transplant recipients with graft failure (48). The authors concluded that patients with orthotopic liver transplants have an increased risk of sinus arrest from adenosine. They attributed this to stimulation of adenosine A1 receptors in the sinoatrial node, activating an outward potassium current, and resulting in a direct negative chronotropic effect.

Respiratory

Adenosine can cause bronchoconstriction with asthma (49), and a history of bronchoconstriction is a contraindication to intravenous adenosine.

In 94 patients with chronic obstructive pulmonary disease who were given adenosine in an initial dosage of 50 micrograms/kg/minute, increasing to 140 micrograms/kg/minute if adverse effects did not occur, there was only a slight and insignificant fall in FEV_1 at the highest dose of adenosine (50). However, four patients had a fall in FEV_1 of 20% or more, although without shortness of breath or evidence of bronchospasm; in these the dosage of adenosine was reduced to 100 micrograms/kg/minute. Two other patients had shortness of breath with no fall in FEV_1 or bronchospasm, and the dosage was reduced to 100 micrograms/kg/minute. There was no difference in the fall in FEV_1 between patients who had a history of asthma and those who did not. Other adverse effects included light-headedness ($n = 26$), dyspnea ($n = 17$), headache ($n = 14$), flushing ($n = 8$), hypotension ($n = 7$), chest pain ($n = 6$), and nausea ($n = 2$). In a subsequent study in 117 patients, two had symptomatic bronchospasm during adenosine infusion. In two other patients in whom bronchospasm was present before treatment, bronchospasm did not develop when adenosine was infused at the highest dosage.

In another study, 63 of 122 patients had breathlessness during cardiac stress testing with adenosine but none had associated bronchospasm (51). Pre-test lung function did not predict the risk of breathlessness and neither chronic obstructive airways disease nor smoking increased the risk. The authors concluded that breathlessness during adenosine stress testing is not due to bronchospasm.

Nervous system

Adenosine has been used intrathecally to treat pain, but can itself cause backache (SEDA-23, 197; 52). In a placebo-controlled study in 40 healthy volunteers, who were given intrathecal adenosine 2 mg in 2 ml of saline, 13 had a mild headache, nine had mild to moderate backache, and one had mild aching in the thigh, compared with none of those who were given saline alone (53). No headaches or leg aches occurred later than 6 hours after the injection, but the backaches occurred at 6–24 hours; there were no later symptoms.

In a randomized, double-blind study of two doses of intrathecal adenosine in 35 volunteers with experimental hypersensitivity induced by capsaicin, intrathecal adenosine 0.5 or 2 mg in 2 ml of saline, but not saline alone, equally reduced areas of allodynia and hyperalgesia from capsaicin (54). There were adverse effects in 1, 2, and 6 of the volunteers who received saline, 0.5 mg, and 2.0 mg of adenosine respectively. The adverse effects were headache, backache, and leg or groin ache. Intravenous aminophylline 5 mg/kg, given 2 hours after the adenosine, did not reverse the effects of adenosine.

Of 12 healthy volunteers given an intrathecal injection of adenosine (500–2000 micrograms) one volunteer had transient lumbar pain lasting 30 minutes after an injection of 2000 micrograms (55). There were no adverse effects at lower doses.

Adenosine can cause increased intracranial pressure (56).

A tonic-clonic seizure lasting 5–10 seconds has been attributed to adenosine (57).

- A 41-year-old man with a history of chronic alcohol abuse, bronchitis, hypertension, and an episode of tachydysrhythmia 10 years before, was given adenosine 6 mg for a supraventricular tachycardia, and 3–5 minutes later another 9 mg, which caused only a warm feeling. After a further 3–5 minutes he was given 12 mg and just over 8 minutes later had a tonic-clonic seizure with flushing of the skin. In addition, the supraventricular tachycardia degenerated into ventricular tachycardia.

The authors thought that the adenosine had precipitated the seizure and that alcohol abuse had been a contributory factor. A seizure has been associated with adenosine in a previous case. However, in this case the association with adenosine was not clear, because the seizure did not occur until many minutes after the last injection and adenosine has a half-life of a few seconds. Furthermore, adenosine has been reported to be antiepileptic in animals (58).

Gastrointestinal

Adenosine can cause transient epigastric pain mimicking that of peptic ulceration (59).

Immunologic

- An anaphylactic reaction has been reported in a 75-year-old woman who was given adenosine 12 mg for a supraventricular tachycardia. She developed bronchospasm and profound inspiratory stridor, her arterial blood pressure fell to 50/30 mmHg from an arterial systolic pressure of 70 mmHg, and she recovered with appropriate treatment (60).

Death

Two cases of sudden death have been reported soon after the administration of adenosine for presumed supraventricular tachycardia, which turned out to be atrial fibrillation (61). The authors thought that both patients may have been unable to cope with the sudden momentary loss of cardiac function that would have occurred immediately after the administration of adenosine; in one case, a patient with chronic lung disease, bronchospasm may have contributed.

Drug Administration

Drug administration route

The standard regimen for stress testing with intravenous adenosine is 140 micrograms/kg/minute for 6 minutes. However, in 599 patients a 3-minute infusion was associated with a lower frequency of some adverse effects (specifically flushing, headache, neck pain, and atrioventricular block) and had similar sensitivity in the diagnosis of coronary artery disease (62).

Intracoronary adenosine has been compared with intravenous adenosine for the measure of fractional flow reserve in 52 patients with coronary artery lesions (63). The intravenous dose was 140 micrograms/kg/minute and the intracoronary bolus dose was 15–20 micrograms to the right coronary artery and 18–24 micrograms to the left coronary artery. The two routes of administration were equally effective in measuring hyperemic flow, and adverse effects were limited to two patients who received intravenous adenosine; one patient had severe nausea and one patient with asthma had an episode of bronchospasm.

The use of intrathecal adenosine in patients with chronic neuropathic pain (52,64) has been briefly reviewed (65).

Drug–Drug Interactions

General

Adenosine does not interact with digoxin, disopyramide, flecainide, or quinidine.

Ciclosporin

Endogenous plasma adenosine concentrations were measured in 14 kidney transplant recipients taking ciclosporin and compared with five transplant recipients not taking ciclosporin, two taking sirolimus (FK506), six patients with chronic renal insufficiency, and ten controls (66). Plasma adenosine concentrations were significantly higher in those taking ciclosporin and sirolimus and in the patients taking ciclosporin the plasma adenosine concentrations correlated with serum ciclosporin concentrations. An in vitro study showed that ciclosporin inhibited the uptake of adenosine by erythrocytes. The authors concluded that since adenosine is immunosuppressant, the raised concentrations of adenosine in patients taking ciclosporin might contribute to the immunosuppressive action of ciclosporin. A further mechanism of the increase in adenosine concentration was possibly increased tissue release secondary to ciclosporin-induced vasoconstriction. The relevance of these results to the use of therapeutic intravenous adenosine in patients already taking ciclosporin is not clear.

Dipyridamole

Dipyridamole inhibits the uptake of adenosine by cells and so increases its effects; this causes a large reduction in the effective dose of adenosine (67).

Sirolimus

In two kidney transplant recipients taking sirolimus (FK506), plasma adenosine concentrations were significantly increased (66). The relevance of these results to the use of therapeutic intravenous adenosine in patients already taking sirolimus is not clear.

Xanthines

Antagonists at adenosine receptors should inhibit the action of adenosine, and indeed theophylline increases the dose of adenosine needed for conversion of supraventricular tachycardia (68).

Diagnosis of Adverse Drug Reactions

In 34 patients given midazolam or placebo in a double-blind study, midazolam significantly reduced patients' experiences of palpitation and chest pain but had no effects on other adverse events (69). These effects were probably due to amnesia rather than a true reduction in the incidence of adverse events, and it is uncertain that the benefit to harm ratio is worth while. However, the authors suggested that midazolam might be useful in patients who have previously had unpleasant adverse reactions to adenosine.

References

1. Camm AJ, Garratt CJ. Adenosine and supraventricular tachycardia. N Engl J Med 1991;325(23):1621–9.
2. Harper KJ. Adenosine in the acute treatment of PSVT. Drug Ther 1992;53–72March.
3. Rankin AC, Brooks R, Ruskin JN, McGovern BA. Adenosine and the treatment of supraventricular tachycardia. Am J Med 1992;92(6):655–64.
4. Hori M, Kitakaze M. Adenosine, the heart, and coronary circulation. Hypertension 1991;18(5):565–74.
5. Rankin AC, Oldroyd KG, Chong E, Dow JW, Rae AP, Cobbe SM. Adenosine or adenosine triphosphate for supraventricular tachycardias? Comparative double-blind randomized study in patients with spontaneous or inducible arrhythmias. Am Heart J 1990;119(2 Pt 1):316–23.
6. Pennell DJ, Mavrogeni SI, Forbat SM, Karwatowski SP, Underwood SR. Adenosine combined with dynamic exercise for myocardial perfusion imaging. J Am Coll Cardiol 1995;25(6):1300–9.
7. O'Keefe JH Jr, Bateman TM, Handlin LR, Barnhart CS. Four- versus 6-minute infusion protocol for adenosine thallium-201 single photon emission computed tomography imaging. Am Heart J 1995;129(3):482–7.
8. Dimitriu AG, Nistor N, Russu G, Cristogel F, Streanga V, Varlam L. Value of intravenous ATP in the diagnosis and treatment of tachyarrhythmias in children. Rev Med Chir Soc Med Nat Iasi 1998;102(3–4):100–2.

9. Pfammatter JP, Bauersfeld U. Safety issues in the treatment of paediatric supraventricular tachycardias. Drug Saf 1998;18(5):345–56.
10. Sherwood MC, Lau KC, Sholler GF. Adenosine in the management of supraventricular tachycardia in children. J Paediatr Child Health 1998;34(1):53–6.
11. Bakshi F, Barzilay Z, Paret G. Adenosine in the diagnosis and treatment of narrow complex tachycardia in the pediatric intensive care unit. Heart Lung 1998;27(1):47–50.
12. Prabhu AS, Singh TP, Morrow WR, Muzik O, Di Carli MF. Safety and efficacy of intravenous adenosine for pharmacologic stress testing in children with aortic valve disease or Kawasaki disease. Am J Cardiol 1999;83(2):284–6.
13. Thomas GS, Prill NV, Majmundar H, Fabrizi RR, Thomas JJ, Hayashida C, Kothapalli S, Payne JL, Payne MM, Miyamoto MI. Treadmill exercise during adenosine infusion is safe, results in fewer adverse reactions, and improves myocardial perfusion image quality. J Nucl Cardiol 2000;7(5):439–46.
14. Elliott MD, Holly TA, Leonard SM, Hendel RC. Impact of an abbreviated adenosine protocol incorporating adjunctive treadmill exercise on adverse effects and image quality in patients undergoing stress myocardial perfusion imaging. J Nucl Cardiol 2000;7(6):584–9.
15. Dúbrava J, Jurkovičová O. Účinnost' a bezpečnost' adenozínu v terapii a diagnostike arytmií. Vnitrni Lekarstvi 2003;49:267–72.
16. Sakata Y, Nishimura T, Yamazaki J, Nishimura S, Kaivyas T, Kodama K, Kato K. [Diagnosis of coronary artery disease by thallium-201 myocardial scintigraphy with intravenous infusion of SUNY4001 (adenosine) in effort angina pectoris—the clinical trial report at multi-center: phase II.] Kaku Igaku 2004;41(2):123–32.
17. Yamazaki J, Nishimura T, Nishimura S, Kajiya T, Kodama K, Kato K. [The diagnostic value for ischemic heart disease of thallium-201 myocardial scintigraphy by intravenous infusion of SUNY4001 (adenosine)—the report of clinical trial at multi-center: phase III.] Kaku Igaku 2004;41(2):133–42.
18. Nishimura S, Nishimura T, Yamazaki J, Doi O, Konishi T, Iwasaki T, Kajiya T, Fukuyama T, Akaishi M, Kato K, Nakashima M. [Comparison of myocardial perfusion imaging by thallium-201 single-photon emission computed tomography with SUNY4001 (adenosine) and exercise–crossover clinical trial at multi-center.] Kaku Igaku 2004;41(2):143–54.
19. Fukunaga AF, Alexander GE, Stark CW. Characterization of the analgesic actions of adenosine: comparison of adenosine and remifentanil infusions in patients undergoing major surgical procedures. Pain 2003;101:129–38.
20. Barcin C, Denktas AE, Lennon RJ, Hammes L, Higano ST, Holmes DR Jr, Garratt KN, Lerman A. Comparison of combination therapy of adenosine and nitroprusside with adenosine alone in the treatment of angiographic no-reflow phenomenon. Catheter Cardiovasc Interv 2004;61(4):484–91.
21. Casella G, Leibig M, Schiele TM, Schrepf R, Seelig V, Stempfle HU, Erdin P, Rieber J, Konig A, Siebert U, Klauss V. Are high doses of intracoronary adenosine an alternative to standard intravenous adenosine for the assessment of fractional flow reserve? Am Heart J 2004;148(4):590–5.
22. Petronio AS, De Carlo M, Ciabatti N, Amoroso G, Limbruno U, Palagi C, Di Bello V, Romano MF, Mariani M. Left ventricular remodeling after primary coronary angioplasty in patients treated with abciximab or intracoronary adenosine. Am Heart J 2005;150(5):1015.
23. Quintana M, Hjemdahl P, Sollevi A, Kahan T, Edner M, Rehnqvist N, Swahn E, Kjerr A-C, Nasman P. Left ventricular function and cardiovascular events following adjuvant therapy with adenosine in acute myocardial infarction treated with thrombolysis. Results of the ATTenuation by Adenosine of Cardiac Complications (ATTACC) study. Eur J Clin Pharmacol 2003;59:1–9.
24. Mulla N, Karpawich PP. Ventricular fibrillation following adenosine therapy for supraventricular tachycardia in a neonate with concealed Wolff–Parkinson–White syndrome treated with digoxin. Pediatr Emerg Care 1995;11(4):238–9.
25. Strickberger SA, Man KC, Daoud EG, Goyal R, Brinkman K, Knight BP, Weiss R, Bahu M, Morady F. Adenosine-induced atrial arrhythmia: a prospective analysis. Ann Intern Med 1997;127(6):417–22.
26. Rae CP, Mansfield MD, Dryden C, Kinsella J. Analgesic effect of adenosine on ischaemic pain in human volunteers. Br J Anaesth 1999;82(3):427–8.
27. Abbott BG, Afshar M, Berger AK, Wackers FJTh. Prognostic significance of ischemic electrocardiographic changes during adenosine infusion in patients with normal myocardial perfusion imaging. J Nucl Cardiol 2003;10: 9–16.
28. Patsilinakos SP, Spanodimos S, Rontoyanni F, Kranidis A, Antonelis IP, Sotirellos K, Antonatos D, Tsaglis E, Nikolaou N, Tsigas D. Adenosine stress myocardial perfusion tomographic imaging in patients with significant aortic stenosis. J Nucl Cardiol 2004;11(1):20–5.
29. Reyes E, Wechalekar K, Loong CY, Underwood SR. Acute myocardial infarction during adenosine myocardial perfusion imaging. J Nucl Cardiol 2004;11(1):97–9.
30. Garratt KN, Holmes DR Jr, Molina-Viamonte V, Reeder GS, Hodge DO, Bailey KR, Lobl JK, Laudon DA, Gibbons RJ. Intravenous adenosine and lidocaine in patients with acute mycocardial infarction. Am Heart J 1998;136(2):196–204.
31. Zarate E, Sa Rego MM, White PF, Duffy L, Shearer VE, Griffin JD, Whitten CW. Comparison of adenosine and remifentanil infusions as adjuvants to desflurane anesthesia. Anesthesiology 1999;90(4):956–63.
32. Christopher M, Key CB, Persse DE. Refractory asystole and death following the prehospital administration of adenosine. Prehosp Emerg Care 2000;4(2):196–8.
33. Knight BP, Zivin A, Souza J, Goyal R, Man KC, Strickberger A, Morady F. Use of adenosine in patients hospitalized in a university medical center. Am J Med 1998;105(4):275–80.
34. Jaeggi E, Chiu C, Hamilton R, Gilljam T, Gow R. Adenosine-induced atrial pro-arrhythmia in children. Can J Cardiol 1999;15(2):169–72.
35. Israel C, Klingenheben T, Gronefeld G, Hohnloser SH. Adenosine-induced atrial fibrillation. J Cardiovasc Electrophysiol 2000;11(7):825.
36. Kaplan IV, Kaplan AV, Fisher JD. Adenosine induced atrial fibrillation precipitating polymorphic ventricular tachycardia. Pacing Clin Electrophysiol 2000;23(1):140–1.
37. Misra D, Van Tosh A, Schweitzer P. Adenosine induced monomorphic ventricular tachycardia. Pacing Clin Electrophysiol 2000;23(6):1044–6.
38. Ruiz Ruiz MJ, Rivero Guerrero JA, Barrera Cordero A, de Teresa E. Empeoramiento de la taquicardia Supraventricular tras la administracion de adenosina: un efecto paradojico. [Worsening of supraventricular tachycardia after intravenous after adenosine administration: a paradoxical effect.] Med Clin (Barc) 2001;117(7):276.
39. Parham WA, Mehdirad AA, Biermann KM, Fredman CS. Case report: adenosine induced ventricular fibrillation in a patient with stable ventricular tachycardia. J Interv Card Electrophysiol 2001;5(1):71–4.

40. Tan HL, Spekhorst HH, Peters RJ, Wilde AA. Adenosine induced ventricular arrhythmias in the emergency room. Pacing Clin Electrophysiol 2001;24(4 Pt 1):450–5.
41. Alkoutami GS, Reeves WC, Movahed A. The frequency of atrioventricular block during adenosine stress testing in young, middle-aged, young-old, and old-old adults. Am J Geriatr Cardiol 2001;10(3):159–61.
42. Lopez-Palop R, Saura D, Pinar E, Lozano I, Perez-Lorente F, Pico F, Valdez M. Adequate intracoronary adenosine doses to achieve maximum hyperaemia in coronary functional studies by pressure derived fractional flow reserve: a dose response study. Heart 2004;90(1):95–6.
43. Soult Rubio JA, Muñoz Sáez M, López Castilla JD, Sánchez Ganfornina I, Navas López VM, Romero Parreño A. Efecto secundario grave tras la administración de adenosina en un niño. Rev Esp Pediatr 2004;60(2):160–1.
44. Toft J, Mortensen J, Hesse B. Risk of atrioventricular block during adenosine pharmacologic stress testing in heart transplant recipients. Am J Cardiol 1998;82(5): 696–7.
45. Alkoutami GS, Reeves WC, Movahed A. The safety of adenosine pharmacologic stress testing in patients with first-degree atrioventricular block in the presence and absence of atrioventricular blocking medications. J Nucl Cardiol 1999;6(5):495–7.
46. Tsai CL, Chang WT. A wide QRS complex tachycardia following intravenous adenosine. Resuscitation 2004;61(2):240–1.
47. Copetti R, Proclemer A, Paolo Pillinini P, Chizzola G. Life-threatening proarrhythmia in a patient with orthodromic atrioventricular tachycardia treated with low-dose adenosine. J Cardiovasc Electrophysiol 2005;16(1):106.
48. Giedd KN, Bokhari S, Daniele TP, Johnson LL. Sinus arrest during adenosine stress testing in liver transplant recipients with graft failure: three case reports and a review of the literature. J Nucl Cardiol 2005;12(6):696–702.
49. Ng WH, Polosa R, Church MK. Adenosine bronchoconstriction in asthma: investigations into its possible mechanism of action. Br J Clin Pharmacol 1990;30(Suppl 1):S89–98.
50. Johnston DL, Scanlon PD, Hodge DO, Glynn RB, Hung JC, Gibbons RJ. Pulmonary function monitoring during adenosine myocardial perfusion scintigraphy in patients with chronic obstructive pulmonary disease. Mayo Clin Proc 1999;74(4):339–46.
51. Balan KK, Critchley M. Is the dyspnea during adenosine cardiac stress test caused by bronchospasm? Am Heart J 2001;142(1):142–5.
52. Belfrage M, Segerdahl M, Arner S, Sollevi A. The safety and efficacy of intrathecal adenosine in patients with chronic neuropathic pain. Anesth Analg 1999;89(1): 136–42.
53. Eisenach JC, Hood DD, Curry R. Phase I safety assessment of intrathecal injection of an American formulation of adenosine in humans. Anesthesiology 2002;96(1):24–8.
54. Eisenach JC, Curry R, Hood DD. Dose response of intrathecal adenosine in experimental pain and allodynia. Anesthesiology 2002;97(4):938–42.
55. Rane K, Segerdahl M, Goiny M, Sollevi A. Intrathecal adenosine administration: a phase 1 clinical safety study in healthy volunteers, with additional evaluation of its influence on sensory thresholds and experimental pain. Anesthesiology 1998;89(5):1108–15.
56. Clarke KW, Brear SG, Hanley SP. Rise in intracranial pressure with intravenous adenosine. Lancet 1992; 339(8786):188–9.
57. Hempe S, Hof H, Bornemann J, Lierz P. Akuter grand malanfall unter antiarrhythmischer therapie mit adenosin. Intensivmed Notfallmed 2003;40:233–6.
58. Anschel DJ, Ortega EL, Kraus AC, Fisher RS. Focally injected adenosine prevents seizures in the rat. Exp Neurol 2004;190:544–7.
59. Watt AH, Lewis DJ, Horne JJ, Smith PM. Reproduction of epigastric pain of duodenal ulceration by adenosine. BMJ (Clin Res Ed) 1987;294(6563):10–2.
60. Shaw AD, Boscoe MJ. Anaphylactic reaction following intravenous adenosine. Anaesthesia 1999;54(6):608.
61. Haynes BE. Two deaths after prehospital use of adenosine. J Emerg Med 2001;21(2):151–4.
62. Treuth MG, Reyes GA, He ZX, Cwajg E, Mahmarian JJ, Verani MS. Tolerance and diagnostic accuracy of an abbreviated adenosine infusion for myocardial scintigraphy: a randomized, prospective study. J Nucl Cardiol 2001;8(5):548–54.
63. Jeremias A, Whitbourn RJ, Filardo SD, Fitzgerald PJ, Cohen DJ, Tuzcu EM, Anderson WD, Abizaid AA, Mintz GS, Yeung AC, Kern MJ, Yock PG. Adequacy of intracoronary versus intravenous adenosine-induced maximal coronary hyperemia for fractional flow reserve measurements. Am Heart J 2000;140(4):651–7.
64. Sjolund KF, Segerdahl M, Sollevi A. Adenosine reduces secondary hyperalgesia in two human models of cutaneous inflammatory pain. Anesth Analg 1999;88(3):605–10.
65. Kopf A, Ruf W. Novel drugs for neuropathic pain. Curr Opin Anaesthesiol 2000;13:577–83.
66. Guieu R, Dussol B, Devaux C, Sampol J, Brunet P, Rochat H, Bechis G, Berland YF. Interactions between cyclosporine A and adenosine in kidney transplant recipients. Kidney Int 1998;53(1):200–4.
67. Watt AH, Bernard MS, Webster J, Passani SL, Stephens MR, Routledge PA. Intravenous adenosine in the treatment of supraventricular tachycardia: a dose-ranging study and interaction with dipyridamole. Br J Clin Pharmacol 1986;21(2):227–30.
68. diMarco JP, Sellers TD, Lerman BB, Greenberg ML, Berne RM, Belardinelli L. Diagnostic and therapeutic use of adenosine in patients with supraventricular tachyarrhythmias. J Am Coll Cardiol 1985;6(2):417–25.
69. Hourigan C, Safih S, Rogers I, Jacobs I, Lockney A. Randomized controlled trial of midazolam premedication to reduce the subjective adverse effects of adenosine. Emerg Med (Fremantle) 2001;13(1):51–6.

Adenosine receptor agonists

General Information

To date, four subtypes of adenosine receptor have been described: A_1, A_{2A}, A_{2B}, and A_3 (Giedd 696). Stimulation of specific cell-surface A_1 receptors shortens the duration, depresses the amplitude, and reduces the rate of rise of the action potential in the atrioventricular node, slowing conduction; this is the mechanism by which adenosine terminates re-entrant supraventricular tachycardias. In myocardial perfusion imaging adenosine causes coronary vasodilatation by stimulating A_{2A} receptors in vascular endothelium and smooth muscle cells. Stimulation of A_{2B} receptors in mast cells causes bronchospasm in susceptible individuals. Stimulation of A_3 receptors reduces the degree of apoptosis resulting from ischemia reperfusion injury in the heart, although most of the cardioprotective effects of adenosine are thought to be mediated by A_1 receptors.

A high proportion of patients experience transient adverse effects after the administration of adenosine; selective agonists developed for clinical use might be safer and would be longer acting. Over a dozen selective agonists are now in clinical trials: A_1 agonists for cardiac dysrhythmias and neuropathic pain; A_{2A} agonists for myocardial perfusion imaging and as anti-inflammatory agents (1); A_{2B} agonists for treatment of cardiac ischemia; and A_3 receptor agonists for rheumatoid arthritis and colorectal cancer (2). Apadenoson, binodenoson, and regadenoson are new selective adenosine A_{2A} receptor agonists (3).

Observational studies

Regadenoson has been studied in 36 patients as a pharmacological stress agent for detecting reversible myocardial hypoperfusion combined with single-photon emission computed tomography (SPECT) (4). The reported adverse events (frequencies in parentheses) were chest pain (33), flushing (31), dyspnea (31), headache (25), dizziness (19), abdominal pain (11), hypesthesia (8), and palpitation (6).

Comparative studies

Binodenoson and adenosine have been compared in a multicenter, randomized, single-blind, two-arm crossover study in 226 patients who underwent two single photon emission computed tomographic (SPECT) imaging studies (5). There were fewer adverse events, specifically chest pain, dyspnea, and flushing, with any dose of binodenoson than with adenosine, and the adverse effects were less severe. There were no cases of atrioventricular block with binodenoson compared with seven cases among 226 patients who were given adenosine.

References

1. Lappas CM, Sullivan GW, Linden J. Adenosine A2A agonists in development for the treatment of inflammation. Expert Opin Investig Drugs 2005;14(7):797–806.
2. Gao ZG, Jacobson KA. Emerging adenosine receptor agonists. Expert Opin Emerg Drugs 2007;12(3):479–92.
3. Miller DD. Impact of selective adenosine A2A receptor agonists on cardiac imaging: feeling the lightning, waiting on the thunder. J Am Coll Cardiol 2005;46(11):2076–8.
4. Hendel RC, Bateman TM, Cerqueira MD, Iskandrian AE, Leppo JA, Blackburn B, Mahmarian JJ. Initial clinical experience with regadenoson, a novel selective A_{2A} agonist for pharmacologic stress single-photon emission computed tomography myocardial perfusion imaging. J Am Coll Cardiol 2005;46(11):2069–75.
5. Udelson JE, Heller GV, Wackers FJ, Chai A, Hinchman D, Coleman PS, Dilsizian V, DiCarli M, Hachamovitch R, Johnson JR, Barrett RJ, Gibbons RJ. Randomized, controlled dose-ranging study of the selective adenosine A2A receptor agonist binodenoson for pharmacological stress as an adjunct to myocardial perfusion imaging. Circulation 2004;109(4):457–64.

Ajmaline and its derivatives

General Information

Ajmaline and its derivatives, prajmalium bitartrate (rINN; *N*-propylajmaline), lorajmine (rINN; chloroacetylajmaline), detajmium bitartrate (rINN), and diethylaminohydroxypropylajmaline, are *Rauwolfia* alkaloids. Their use is restricted by serious adverse effects, such as neutropenia and cardiac dysrhythmias, which have been reviewed (1). Other adverse effects include dizziness, headache, and a sensation of warmth after intravenous injection.

Organs and Systems

Cardiovascular

Ajmaline occasionally causes cardiac dysrhythmias (SEDA-17, 219). Of 1995 patients who were given ajmaline 1 mg/kg intravenously during an electrophysiological study, 63 developed a supraventricular tachydysrhythmia (atrial flutter, fibrillation, or tachycardia), and seven an atrioventricular re-entrant tachycardia (2). Those most at risk were older patients, those with underlying cardiac disease, and those with a history of dysrhythmias or sinus node dysfunction.

Two cases of torsade de pointes have been reported in association with prolongation of the QT interval (3). Polymorphous ventricular tachycardia has been reported in three cases (4–6).

- A 13-year-old boy with Brugada syndrome (right bundle branch block with persistent ST segment elevation) was given an injection of ajmaline 1 mg/kg and developed greater ST segment elevation and more marked right bundle branch block morphology (7). This was followed by short runs of non-sustained polymorphic ventricular tachycardia, gradually increasing until monomorphic ventricular tachycardia occurred. The dysrhythmia eventually resolved without further treatment.

It is unwise to give antidysrhythmic drugs to patients with Brugada syndrome.

The diagnostic electrocardiographic pattern in Brugada syndrome can be absent and can be unmasked by sodium channel blockers, such as ajmaline. The risks of a standardized ajmaline challenge test have been studied in 158 patients, who were given intravenous ajmaline 10 mg every 2 minutes up to a target dose of 1 mg/kg (8). In 37 patients (23%) the typical coved ST pattern of the Brugada syndrome was unmasked. During the test, symptomatic ventricular tachycardia occurred in two patients. In all other patients, the drug challenge did not induce ventricular tachycardia if the endpoints of the test were the administration of the target dose, QRS prolongation over 30%, the presence of the typical electrocardiogram, or the occurrence of ventricular extra beats. There was a positive response to ajmaline in two of 94 patients with a normal baseline electrocardiogram, who underwent evaluation solely for syncope of unknown origin.

Ajmaline challenge was positive in 48 of 103 patients with Brugada syndrome (9). J wave elevation in -2V2 and a reduced T wave amplitude in V3 at baseline were independent predictors of a positive response.

Nervous system

Neurological effects have occasionally been reported in patients taking ajmaline derivatives; they include confusion and cranial nerve palsies (10,11).

Hematologic

Neutropenia is a relatively common and important adverse effect of ajmaline (12). Of the three main mechanisms that cause neutropenia (immune, toxic, and autoimmune) two have been associated with ajmaline: immune and autoimmune neutropenia.

Liver

Ajmaline can cause hepatitis or cholestasis. Cholestasis has been reported in association with neutropenia (13) and with fever and eosinophilia (14). Although acute liver damage due to ajmaline is usually reversible, there has been a report of persistent jaundice due to long-lasting cholestasis (15).

Immunologic

Hypersensitivity to ajmaline is rare, but there has been a report of an immune interstitial nephritis in association with fever (16).

Drug Administration

Drug overdose

In overdosage ajmaline can cause heart block and dysrhythmias, hypotension, malaise, vertigo, respiratory depression, and coma (17). In one series of 38 cases there were nine deaths (24%) (18). Treatment of overdosage includes the intravenous administration of molar sodium lactate for dysrhythmias, conduction disturbances, and circulatory failure; a pacemaker may be required.

- After an overdose of detajmium bitartrate in a dose of 18 mg/kg, a 36-year-old woman developed ventricular flutter, which responded to treatment with lidocaine, defibrillation, glucagon, noradrenaline, and sodium chloride (19). Hypokalemia responded to intravenous potassium chloride.
- A 57-year-old man took ajmaline 1000 mg with suicidal intent (20). He was unconscious and hypotensive and had serious disturbances in cardiac conduction. His serum and urine ajmaline concentrations were high. Although only 4% of the ingested dose was excreted following forced diuresis, all evidence of toxicity disappeared within 21 hours.

Various types of cardiac dysrhythmia have previously been reported after overdosage of ajmaline (SEDA-2, 162).

References

1. Schwartz JB, Keefe D, Harrison DC. Adverse effects of antiarrhythmic drugs. Drugs 1981;21(1):23–45.
2. Brembilla-Perrot B, Terrier de la Chaise A. Provocation of supraventricular tachycardias by an intravenous class I antiarrhythmic drug. Int J Cardiol 1992;34(2):189–98.
3. Haverkamp W, Monnig G, Kirchhof P, Eckardt L, Borggrefe M, Breithardt G. Torsade de pointes induced by ajmaline. Z Kardiol 2001;90(8):586–90.
4. Kaul U, Mohan JC, Narula J, Nath CS, Bhatia ML. Ajmaline-induced torsade de pointes. Cardiology 1985;72(3):140–3.
5. Kolar J, Humhal J, Karetova D, Novak M. "Torsade de pointes" po mezokainu a ajmalinu u nemocne s intermitertni sinokomorovou blokadou. Priznivy recebny vliv vysokych davek izoprenalinu. [Torsade de pointes after mesocaine and ajmaline in a patient with intermittent atrioventricular block. Favorable therapeutic effect of high doses of isoprenaline.] Cas Lek Cesk 1987;126(48):1503–7.
6. Schmitt C, Brachmann J, Schols W, Beyer T, Kubler W. Proarrhythmischer Effekt von Ajmalin bei idiopathischer ventrikulärer Tachykardie. [Proarrhythmic effect of ajmaline in idiopathic ventricular tachycardia.] Dtsch Med Wochenschr 1989;114(3):99–102.
7. Pinar Bermudez E, Garcia-Alberola A, Martinez Sanchez J, Sanchez Munoz JJ, Valdes Chavarri M. Spontaneous sustained monomorphic ventricular tachycardia after administration of ajmaline in a patient with Brugada syndrome. Pacing Clin Electrophysiol 2000;23(3):407–9.
8. Rolf S, Bruns H-J, Wichter T, Kirchhof P, Ribbing M, Wasmer K, Paul M, Breithardt G, Haverkamp W, Eckardt L. The ajmaline challenge in Brugada syndrome: diagnostic impact, safety, and recommended protocol. Eur Heart J 2003;24:1104–12.
9. Hermida JS, Denjoy I, Jarry G, Jandaud S, Bertrand C, Delonca J. Electrocardiographic predictors of Brugada type response during Na channel blockade challenge. Europace 2005;7(5):447–53.
10. Aquaro G, Marra S, Paolillo V, Pavia M. Complicanze neurologiche in corso di terapia con 17-MDCAA. [Neurological complications during therapy with 17 MDCAA.] G Ital Cardiol 1977;7(3):304–8.
11. Lessing JB, Copperman IJ. Severe cerebral confusion produced by prajmalium bitartrate. BMJ 1977;2(6088):675.
12. Brna TG Jr. Agranulocytosis from antiarrhythmic agents. What to watch for when a medication is first prescribed. Postgrad Med 1991;89(1):181–8.
13. Offenstadt G, Boisante L, Onimus R, Amstutz P. Agranulocytose et hepatite cholestatique au cours d'un traitement par l'ajmaline. [Agranulocytosis and cholestatic hepatitis during treatment with ajmaline.] Ann Med Interne (Paris) 1976;127(8–9):622–7.

14. Buscher HP, Talke H, Rademacher HP, Gessner U, Oehlert W, Gerok W. Intrahepatische Cholestase durch N-Propyl-Ajmalin. [Intrahepatic cholestasis due to N-propyl ajmaline.] Dtsch Med Wochenschr 1976;101(18):699–703.
15. Chammartin F, Levillain P, Silvain C, Chauvin C, Beauchant M. Hepatite prolongée a l'ajmaline—description d'un cas et revue de la litterature. [Prolonged hepatitis due to ajmaline—description of a case and review of the literature.] Schweiz Rundsch Med Prax 1989;78(20):582–4.
16. Dupond JL, Herve P, Saint-Hillier Y, Guyon B, Colas JM, Perol C, Leconte des Floris R. Anurie recidivant a 3 reprises; complication exceptionelle d'un traitement antiarythmique. J Med Besancon 1975;11:231.
17. Tempe JD, Jaeger A, Beissel J, Burg E, Mantz JM. Intoxications aigués par trois drogues cardiotropes: l'ajmaline, la chloroquine, la digitaline. J Med Strasbourg (Eur Med) 1976;7:569.
18. Conso F, Bismuth C, Riboulet G, Efthymiou ML. Intoxication aiguë par l'ajmaline. [Acute poisoning by ajmaline.] Therapie 1979;34(4):529–30.
19. Mobis A, Minz DH. Suizidale Tachmalcor-Intoxikation–Ein Fallbericht. [Suicidal Tachmalcor poisoning—a case report.] Anaesthesiol Reanim 1999;24(4):109–10.
20. Almog C, Maidan A, Pik A, Schlesinger Z. Acute intoxication with ajmaline. Isr J Med Sci 1979;15(7):570–2.

Amiodarone

General Information

Amiodarone is highly effective in treating both ventricular and supraventricular dysrhythmias (1). Its pharmacology, therapeutic uses, and adverse effects and interactions have been extensively reviewed (2–14).

General adverse effects

Amiodarone prolongs the QT interval and can therefore cause dysrhythmias; there have also been reports of conduction disturbances. Abnormalities of thyroid function tests can occur without thyroid dysfunction, typically increases in serum T4 and reverse T3 and a reduction in serum T3. However, in up to 6% of patients frank thyroid dysfunction can occur (either hypothyroidism or hyperthyroidism). Several of the adverse effects of amiodarone are attributable to deposition of phospholipids in the tissues. These include its effects on the eyes, nerves, liver, skin, and lungs. Almost all patients develop reversible corneal microdeposits, which can occasionally interfere with vision. There are reports of peripheral neuropathy and other neurological effects. Changes in serum activities of aspartate transaminase and lactate dehydrogenase can occur without other evidence of liver disease, but liver damage can occur in the absence of biochemical evidence. Skin sensitivity to light occurs commonly, possibly due to phototoxicity. There may also be a bluish pigmentation of the skin. Interstitial pneumonitis and alveolitis have been reported and may be fatal. Lung damage due to amiodarone may be partly due to hypersensitivity. Tumor-inducing effects have not been reported.

Studies in the prevention of dysrhythmias

In a study of the use of implantable defibrillators or antidysrhythmic drugs (amiodarone or metoprolol) in 288 patients resuscitated from cardiac arrest, the defibrillator was associated with a slightly lower rate of all-cause mortality than the antidysrhythmic drugs (15). However, the small difference was not statistically significant. There was hyperthyroidism in three of those given amiodarone. Drug withdrawal was required in nine of those given amiodarone and 10 of those given metoprolol. There were deaths in five patients fitted with a defibrillator and two patients given amiodarone. There was crossover to the other therapy in 6% in each group, usually because of recurrence of the dysrhythmia. When sudden cardiac death was analysed, the reduction in mortality with defibrillation was much larger (61%). There were no differences in all-cause mortality and sudden death rates between those given amiodarone and those given metoprolol.

Amiodarone and carvedilol have been used in combination in 109 patients with severe heart failure and left ventricular ejection fractions of 0.25 (16). They were given amiodarone 1000 mg/week plus carvedilol titrated to a target dose of 50 mg/day. A dual-chamber pacemaker was inserted and programmed in back-up mode at a basal rate of 40. Significantly more patients were in sinus rhythm after 1 year, and in 47 patients who were studied for at least 1 year the resting heart rate fell from 90 to 59. Ventricular extra beats were suppressed from 1 to 0.1/day and the number of bouts of tachycardia over 167 per minute was reduced from 1.2 to 0.3 episodes per patient per 3 months. The left ventricular ejection fraction increased from 0.26 to 0.39 and New York Heart Association Classification improved from 3.2 to 1.8. The probability of sudden death was significantly reduced by amiodarone plus carvedilol compared with 154 patients treated with amiodarone alone and even more so compared with 283 patients who received no treatment at all. However, the study was not randomized, and this vitiates the results. The main adverse effect was symptomatic bradycardia, which occurred in seven patients; two of those developed atrioventricular block and four had sinoatrial block and/or sinus bradycardia; one patient developed slow atrial fibrillation.

A meta-analysis of 13 randomized trials has shown that both total mortality and sudden death or dysrhythmic death was less common over 24 months after randomization to amiodarone than in control subjects (17).

Studies in atrial fibrillation

Cardiac glycosides such as digoxin are commonly used to treat uncomplicated atrial fibrillation. In those in whom digitalis is not completely effective or in whom symptoms (for example bouts of palpitation) persist despite adequate digitalization, a calcium antagonist, such as verapamil or diltiazem, can be added, or amiodarone used as an alternative.

The use of oral amiodarone in preventing recurrence of atrial fibrillation, for preventing recurrence after cardioversion or for pharmacological cardioversion of atrial fibrillation, has been reviewed (18). There is insufficient evidence to support its use as a first-line drug for preventing recurrence of atrial fibrillation or in preventing paroxysmal atrial fibrillation.

In 186 patients randomized equally to amiodarone 200 mg/day, sotalol 160–480 mg/day, or placebo, the incidence of atrial fibrillation after 6 months was higher in those taking placebo compared with amiodarone and sotalol and higher in those taking sotalol compared with amiodarone (19). Of the 65 patients who took amiodarone, 15 had significant adverse effects after an average of 16 months. There were eight cases of hypothyroidism, four of hyperthyroidism, two of symptomatic bradycardia, and one of ataxia. There were minor adverse effects in 9% of the patients, including gastrointestinal discomfort, nausea, photosensitivity, and eye problems. These patients had recurrent symptomatic atrial fibrillation. In contrast, only two patients using sotalol developed symptomatic bradycardia and one had severe dizziness.

In 208 patients with atrial fibrillation of various duration, including 50 with chronic atrial fibrillation, randomized to amiodarone or placebo, 80% converted to sinus rhythm after amiodarone compared with 40% of those given placebo (20). Amiodarone was given as an intravenous loading dose of 300 mg for 1 hour and 20 mg/kg for 24 hours, followed by 600 mg/day orally for 1 week and 400 mg/day for 3 weeks. Those who converted to sinus rhythm had had atrial fibrillation for a shorter duration and had smaller atria than those who did not convert. The shorter the duration of fibrillation and the smaller the atria the sooner conversion occurred. There was significant hypotension in 12 of the 118 patients who received amiodarone during the first hour of intravenous administration, but in all cases this responded to intravenous fluids alone. There was phlebitis at the site of infusion in 17 patients, and the peripheral catheter was replaced by a central catheter. There were no dysrhythmic effects.

In 40 patients with atrial fibrillation, some with severe heart disease (including cardiogenic shock in eight and pulmonary edema in 12), amiodarone 450 mg was given through a peripheral vein within 1 minute, followed by 10 ml of saline; 21 patients converted to sinus rhythm, 13 within 30 minutes and another 8 within 24 hours (21). There were two cases of hypotension, but in those that converted to sinus rhythm there was a slight increase in systolic blood pressure. There were no cases of thrombophlebitis. Efficacy is hard to judge from this study, because it was not placebo-controlled.

In 72 patients with paroxysmal atrial fibrillation randomized to either amiodarone 30 mg/kg or placebo, those who received amiodarone converted to sinus rhythm more often than those given placebo (22). The respective conversion rates were about 50 and 20% at 8 hours, and 87 and 35% after 24 hours. The time to conversion in patients who converted did not differ. One patient developed slow atrial fibrillation (35/minute) with a blood pressure of 75/55 mmHg. Three other patients who received amiodarone had diarrhea and one had nausea. In the control group two patients had headache, one had diarrhea, one had nausea, and two had episodes of sinus arrest associated with syncope during conversion to sinus rhythm; the last of these was thought to have sick sinus syndrome.

In a single-blind study 150 patients with acute atrial fibrillation were randomized to intravenous flecainide, propafenone, or amiodarone (23). At 12 hours there was conversion to sinus rhythm in 45 of 50 patients given flecainide, 36 of the 50 given propafenone, and 32 of the 50 given amiodarone. Thus, flecainide and propafenone were both more effective than amiodarone. There were no differences between the groups in the incidences of adverse effects; there was one withdrawal in each group, due to cerebral embolism in a patient given amiodarone, heart failure in a patient given propafenone, and atrial flutter in a patient given flecainide. There were no ventricular dysrhythmias during the study.

Amiodarone and magnesium have been compared in a placebo-controlled study to reduce the occurrence of atrial fibrillation in 147 patients after coronary artery bypass graft surgery (24). Amiodarone was given as a infusion of 900 mg/day for 3 days and magnesium by infusion of 4 g/day for 3 days. The cumulative occurrences of atrial fibrillation with placebo, amiodarone, and magnesium were 27, 14, and 23% respectively. These differences were not significant. Amiodarone delayed the onset of the first episode of dysrhythmia significantly, but the slight benefit was associated with a longer period of invasive monitoring and was not considered worthwhile. Patients who were more likely to develop atrial fibrillation were older and had a plasma magnesium concentration at 24 hours of under 0.95 mmol/l. Patients who were given amiodarone had a slightly higher rate of adverse events, including hypotension, atrioventricular block, and bradycardia; adverse events led to withdrawal in four cases.

Amiodarone, sotalol, and propafenone have been compared for the prevention of atrial fibrillation in 403 patients who had had at least one episode of atrial fibrillation within the previous 6 months; the study was not placebo-controlled (25). The rate of recurrence of atrial fibrillation was significantly higher in those given sotalol or propafenone than in those given amiodarone. During the study nine patients given amiodarone died, compared with eight given sotalol or propafenone. Four deaths were thought to be dysrhythmic, three in patients given amiodarone. There were major non-fatal adverse events in 36 of the 201 patients given amiodarone and in 35 of the 202 patients given propafenone or sotalol. These included one case of torsade de pointes in a patient who received propafenone, and congestive heart failure in 11 patients given amiodarone and nine given sotalol or propafenone. There were strokes and intracranial hemorrhages in one patient given amiodarone and nine patients given sotalol or propafenone, of whom most were taking warfarin at the time. In all, 68 of the patients who were given amiodarone and 93 of those given sotalol or propafenone withdrew from the study; 17 of those taking amiodarone withdrew because of lack of efficacy compared with 56 of those taking sotalol or propafenone; 36 of those who took amiodarone withdrew because of adverse events compared with 23 of those who took sotalol or propafenone, and this was almost statistically significant.

Amiodarone, propafenone, and sotalol have also been compared in the prevention of atrial fibrillation in 214 patients with recurrent symptomatic atrial fibrillation. They were randomized to amiodarone 200 mg/day, propafenone 450 mg/day, or sotalol 320 mg/day. There was recurrence of atrial fibrillation in 25 of the 75 patients who took amiodarone compared with the 51 of 75 who took sotalol and 24 of the 64 who took propafenone. There were adverse effects requiring withdrawal of treatment in 14 patients who took amiodarone, five who took sotalol and one who took propafenone while they were in sinus rhythm. These effects included symptomatic bradycardia in three patients, hyperthyroidism in six, hypothyroidism in four, and ataxia in one patient who took amiodarone. In those taking sotalol the adverse effects were bradycardia in three and severe dizziness in two. In the one patient in whom propafenone was withheld the reason was symptomatic bradycardia. Thus, amiodarone and propafenone were both more effective than sotalol, but amiodarone also caused more adverse effects requiring withdrawal (26).

In a meta-analysis of five randomized, placebo-controlled trials of amiodarone 200–1200 mg/day for 2–7 days in the treatment of postoperative atrial fibrillation and flutter in 764 patients, the incidence of adverse events with amiodarone was no greater than with placebo (27).

In a meta-analysis of five randomized, placebo-controlled trials of intravenous amiodarone about 500–2200 mg over 24 hours in the treatment of recent-onset atrial fibrillation in 410 patients, the incidence of adverse events was 27% with amiodarone and 11% with placebo (28). Intravenous amiodarone was significantly more effective than placebo in producing cardioversion. The most common adverse effects of intravenous amiodarone were phlebitis, bradycardia, and hypotension; most of these effects were not considered to be dose-limiting.

Of 85 patients with persistent atrial fibrillation after balloon mitral valvotomy given amiodarone (600 mg/day for 2 weeks and 200 mg/day thereafter), 33 converted to sinus rhythm (29). Of the other 52 patients, who underwent DC cardioversion at 6 weeks, 41 converted to sinus rhythm. Six patients had adverse effects attributable to amiodarone. Five had mild gastrointestinal symptoms, such as abdominal discomfort and nausea. One developed hypothyroidism after 3 months, which resolved when the dosage of amiodarone was reduced to 100 mg/day.

In 83 patients (27 women, 56 men; mean age 61 years) disopyramide, propafenone, or sotalol were used to prevent recurrence after elective electrical cardioversion for persistent atrial fibrillation (30). If there was recurrence cardioversion was repeated and the patient was given one of the other antidysrhythmic drugs. If there was further recurrence, amiodarone was used, a third cardioversion was performed, and, if sinus rhythm was restored, amiodarone 100–200 mg/day was continued. Patients in whom the initial cardioversion was not successful were given amiodarone and underwent repeated cardioversion. The follow-up duration was 12 months. The first electrical cardioversion was effective in 44 (53%) patients, and after 1 year 23 (52%) of them were still in sinus rhythm. None of the patients who underwent a second cardioversion and received a second antidysrhythmic drug stayed in sinus rhythm. Amiodarone as a third antidysrhythmic agent was effective in 10 (48%) patients. After 12 months of antidysrhythmic drug therapy sinus rhythm was maintained in 75% of patients in whom the first cardioversion had been effective, accounting for 40% of all the patients selected for cardioversion. In the 83 patients, sequential antidysrhythmic treatment effectively maintained sinus rhythm in 54 (65%), of whom 31 (57%) took amiodarone. The authors concluded that repeated electrical cardioversion and antidysrhythmic drug therapy enabled maintenance of sinus rhythm in 68% of patients for 1 year, that there was limited efficacy of the first antidysrhythmic drug given after a first effective electrical cardioversion, regardless of the drug used, excluding amiodarone, and that when atrial fibrillation recurred, a second antidysrhythmic drug, other than amiodarone, was completely ineffective. There were very few adverse events in this study. One patient taking amiodarone developed hyperthyroidism and two had symptomatic bradycardia.

Amiodarone 30 mg/kg orally for the first 24 hours plus, if necessary, 15 mg/kg over 24 hours has been compared with propafenone 600 mg in the first 24 hours plus, if necessary, 300 mg in the next 24 hours in 86 patients with recent onset atrial fibrillation (31). Conversion to sinus rhythm occurred faster with propafenone (2.4 hours) than amiodarone (6.9 hours). However, by 24 hours and 48 hours the same proportions of patients were in sinus rhythm; one patient given amiodarone had a supraventricular tachycardia and one a non-sustained ventricular tachycardia.

The effects of additional intravenous amiodarone (300 mg in 1 hour followed by 15 mg/kg over 24 hours) have been studied in 45 patients with acute atrial fibrillation who were already taking oral amiodarone for maintenance of sinus rhythm (32). In 20 of 23 patients given amiodarone there was conversion to sinus rhythm, compared with 13 of 22 who were given placebo. There were no prodysrhythmic effects and the only adverse effect of intravenous amiodarone was thrombophlebitis in two patients.

In 44 patients who underwent percutaneous balloon mitral commissurotomy for chronic persistent atrial fibrillation, with a procedural success rate of 100% and no immediate morbidity or mortality, amiodarone maintained sinus rhythm in eight patients compared with none in the control group (33). The adverse effects of amiodarone included bradycardia in two patients and shortness of breath in one; the last required drug withdrawal. Another patient developed long sinus pauses at 15 months and was treated with a permanent pacemaker without withdrawing amiodarone. Otherwise, there were no serious adverse effects or electrocardiographic abnormalities.

In a double-blind, placebo-controlled trial 665 patients who were taking anticoagulants and had persistent atrial fibrillation were randomized to amiodarone (n = 267), sotalol (n = 261), or placebo (n = 137) for 1.0–4.5 years (34). Amiodarone and sotalol were equally effective in producing cardioversion to sinus rhythm (27% and 24% versus placebo 0.8%), but the effect of amiodarone lasted significantly longer (487, 74, and 6 days according to intention to treat, and 809, 209, and 13 days according to treatment received). There were no significant differences in the rates of adverse events, except *minor bleeding*, which was significantly more common with amiodarone than sotalol or placebo (8.33 versus 6.37 and 6.71 per 100 patient-years). The rates of major bleeding were 2.07, 3.10, and 3.97 per 100 patient-years; of minor strokes 1.19, 0.68, and 0.96; and of major strokes 0.87, 2.03, and 0.95. There were two cases of non-fatal adverse pulmonary effects with amiodarone and one with placebo. There was one case of non-fatal torsade de pointes with sotalol.

In a systematic review of the efficacy and safety of amiodarone for pharmacological cardioversion of recent-onset atrial fibrillation in 21 studies, amiodarone was efficacious in 34–69% with bolus-only regimens, and 55–95% with a bolus followed by an infusion (35). The highest 24-hour conversion rates occur with an intravenous regimen of 125 mg/hour until conversion or a maximum of 3 g and an oral regimen of 25–30 mg/kg given as a single loading-dose (over 90% and over 85% respectively). Most conversions occur after 6–8 hours of the start of therapy. Predictors of successful conversion are shorter duration of atrial fibrillation, smaller left atrial size, and higher amiodarone dose. Amiodarone is not superior to other antidysrhythmic drugs but is relatively safe in patients with structural heart disease and in those with depressed left ventricular function. No major prodysrhythmic events, such as sustained ventricular tachycardia, ventricular fibrillation, or torsade de pointes, were reported in these studies. There were minor cardiac effects: first-degree atrioventricular block, self-limited sinus bradycardia, hypotension, and non-sustained ventricular tachycardia. These adverse effects were more common after intravenous administration. Asymptomatic sinus bradycardia was reported in up to 10% of patients and hypotension in up to 18% of patients who received intravenous amiodarone. All the episodes of hypotension were transient and responded to saline volume expansion or inotropic support. Hypotension with intravenous amiodarone is reportedly due to the vehicle. Phlebitis at the amiodarone infusion site occurs up to 16% of patients. Other rare adverse effects were nausea, diarrhea, blurred vision, and allergic reactions. Gastrointestinal adverse effects were predominantly reported after oral administration.

In another systematic review of the studies of the use of amiodarone in the treatment of atrial fibrillation, 21 studies met the eligibility criteria, including 10 of those covered in the systematic review mentioned above (36). Bradydysrhythmias and hypotension were the most commonly reported adverse effects. Death rates were reported in 18 studies; there were five deaths among 816 patients given amiodarone and five among 696 in comparison groups.

However, information about adverse events in these randomized trials was inconsistently reported and too scanty to allow proper analysis. This stresses yet again the need for standard methods of reporting adverse events in clinical trials.

Studies in atrial flutter

Antidysrhythmic drugs have been compared with radiofrequency ablation in 61 patients with atrial flutter (37). Drug treatment was with at least two drugs, one of which was amiodarone. Of the 30 patients who took drug therapy, 19 needed to come into hospital one or more times, whereas after radiofrequency ablation that happened in only seven of 31 cases. In those who took the antidysrhythmic drugs the mean number of drugs was 3.4 and the range of drugs used was very wide. Quality-of-life and symptoms scores improved significantly in those in whom radiofrequency ablation was used, but not in those who took the antidysrhythmic drugs, apart from the symptom of palpitation, which improved in both groups, but to a greater extent in the non-drug group. Adverse effects were not discussed in this study, but it is clear that it suggests that radiofrequency ablation is to be preferred in these patients.

Studies in ventricular dysrhythmias

The effects of amiodarone in 55 patients with sustained ventricular tachycardia after myocardial infarction have been assessed in a long-term follow-up study (38). The patients underwent programmed ventricular stimulation after having been loaded with amiodarone. They were divided into those in whom ventricular tachydysrhythmias could be induced or not, and all were then given amiodarone 200 mg/day. In 11 cases a cardioverter defibrillator was implanted, because the first episode of ventricular tachycardia had been poorly tolerated or had caused hemodynamic instability. A defibrillator was also implanted in five other cases during follow-up, because of recurrence of dysrhythmias. There was a non-significant trend to a difference between the cumulative rates of dysrhythmias during long-term follow-up, with more events in those in whom a dysrhythmia had been inducible after loading. However, mortality rates in the two groups did not differ, and was around 25% at a mean follow-up of 42 months. Survival was significantly higher in patients with a left ventricular ejection fraction over 0.4, and the lower the left ventricular ejection fractions the higher the mortality. Amiodarone was withdrawn in six patients after a mean of 34 months because of neuropathy ($n = 1$), hypothyroidism ($n = 1$), prodysrhythmia with incessant ventricular tachycardia ($n = 2$), and non-specific adverse effects ($n = 2$). There was no pulmonary toxicity and no cases of torsade de pointes. In two patients there was evidence of hypothyroidism, mild neuropathy, and skin discoloration, but these events did not lead to withdrawal. In two patients the doses of amiodarone was reduced to 100 mg/day because of sinus bradycardia.

In a comparison of amiodarone ($n = 23$) with sotalol ($n = 22$) in patients with spontaneous sustained ventricular tachydysrhythmias secondary to myocardial infarction, sotalol was much more effective, 75% of those taking it remaining free of dysrhythmias compared with 38% of those taking amiodarone (39). Adverse effects requiring withdrawal occurred in 17% of those taking amiodarone at a median time of 3.5 months. The adverse effects included malaise, rash, headaches, flushing, and dyspnea due to pulmonary fibrosis.

Studies in myocardial infarction and heart failure

There have been reviews of the results of major trials of amiodarone after myocardial infarction (40) and in chronic heart failure (41).

- In the Basel Antiarrhythmic Study of Infarct Survival (BASIS) amiodarone significantly reduced all-cause mortality from 13 to 5%, compared with no antidysrhythmic drug therapy (42).
- In the Polish Arrhythmia Trial (PAT) amiodarone reduced all-cause mortality from 10.7 to 6.9% compared with placebo and cardiac mortality from 10.7 to 6.2% (43).
- In the Spanish Study of Sudden Death (SSD) amiodarone reduced all-cause mortality from 15.4 to 3.5% compared with metoprolol; however, the mortality in those receiving no antidysrhythmic drugs at all was only 7.7%, and in those the effect of amiodarone was not significant (44).
- In the European Myocardial Infarction Arrhythmia Trial (EMIAT) amiodarone reduced the risk of dysrhythmic deaths from 8.5 to 4.1% compared with placebo (45).
- In the Canadian Amiodarone Myocardial Infarction Arrhythmia Trial (CAMIAT) amiodarone reduced dysrhythmic deaths from 6.0 to 3.3% compared with placebo; non-dysrhythmic deaths were not affected (46).

In a meta-analysis of 10 studies of the use of amiodarone in patients with heart failure, the overall odds ratio for mortality with amiodarone compared with placebo was 0.79 (95% CI = 0.68, 0.92). The corresponding odds ratio for adverse effects was 2.29 (1.97, 2.66) (41). The benefit to risk ratio of the use of amiodarone in these patients is not yet clear. The dosage of amiodarone in these studies varied from 50 to 400 mg/day, with an average of around 250 mg/day.

Organs and Systems

Cardiovascular

The incidence of cardiac dysrhythmias with amiodarone is under 3% (47), lower than with many other antidysrhythmic drugs, and several randomized controlled trials have failed to show any prodysrhythmic effect (48).

Other cardiac effects that have been reported include sinus bradycardia, atrioventricular block, infra-His block, asystole, and refractoriness to DC cardioversion (SEDA-10, 147).

There is a risk of hypotension and atrioventricular block when amiodarone is given intravenously.

Cardiogenic shock has been reported in 73-year-old woman with a dilated cardiomyopathy who had digitalis and amiodarone toxicity (49).

Ventricular dysrhythmias

Amiodarone can prolong the QT interval, and this can be associated with torsade de pointes (50), although this is uncommon. This effect is potentiated by hypokalemia (51).

Ventricular dysrhythmias due to drugs can be either monomorphic or polymorphic. The class Ia drugs are particularly likely to cause polymorphic dysrhythmias, as is amiodarone (although to a lesser extent). In contrast, the class Ic drugs are more likely to cause monomorphic dysrhythmias (52).

- A 40-year-old woman developed torsade de pointes within the first 24 hours of intravenous administration of amiodarone 150 mg followed by 35 mg/hour (53). The association with amiodarone was confirmed by subsequent rechallenge.
- Three boys with congenital cardiac defects developed polymorphous ventricular tachycardia after having been given intravenous amiodarone; two died (54).
- An 8-day-old boy was given intravenous amiodarone 5 mg/kg over 60 minutes followed by 10 mg/kg/day for a postoperative junctional ectopic tachycardia after a cardiac operation. He developed ventricular fibrillation 12 hours later, but recovered with defibrillation and internal cardiac massage. His serum amiodarone concentration was 1–2.5 mg/l, within the usual target range.
- A 3-month-old boy underwent a cardiac operation and 6 hours later developed a junctional ectopic tachycardia. He was given amiodarone as a continuous intravenous infusion of 10 mg/kg/day for 3 hours and developed ventricular fibrillation, from which he was not resuscitated. The serum amiodarone concentration was 0.3 mg/l.
- A 3-month-old boy developed a postoperative junctional ectopic tachycardia 48 hours after operation and was given a continuous intravenous infusion of amiodarone 10 mg/kg/day. After 2 hours he developed ventricular fibrillation and was not resuscitated. His serum amiodarone concentration was in the target range.

It is not clear that the dysrhythmias in these cases were due to amiodarone, particularly since the doses had been very low and the serum concentrations no higher than the usual target range; QT intervals were not reported.

- A 79-year-old woman took amiodarone 4800 mg over 6 days and developed a polymorphous ventricular tachycardia; the associated precipitating factors were a prolonged QT interval and hypokalemia (55).
- A 71-year-old Japanese man with bouts of sustained monomorphic ventricular tachycardia, in whom non-sustained polymorphic ventricular tachycardia was induced by rapid pacing during electrophysiological studies, was given amiodarone and developed three different types of sustained monomorphic ventricular tachycardia, with slightly different cycle lengths, induced and terminated by rapid pacing (56).

The authors proposed that amiodarone had modulated the threshold of induction and/or termination of ventricular tachycardia.

- In an 84-year-old woman torsade de pointes occurred after oral amiodarone therapy for 4 days in the presence of multiple exacerbating factors, including hypokalemia and digoxin toxicity (57). Transient prolongation of the QT interval during bladder irrigation prompted the episode. When amiodarone was withdrawn, bladder irrigation did not induce torsade de pointes, despite hypokalemia and hypomagnesemia.
- A 69-year-old woman with a history of coronary heart disease, myocardial infarction, and paroxysmal atrial fibrillation had an occipital stroke (58). She was given amiodarone 600 mg/day, beta-acetyldigoxin 0.1 mg/day, and

bisoprolol 1.25 mg/day, and developed significant QT interval prolongation (maximum 700 ms; QT_c 614 ms) and repetitive short-lasting torsade de pointes, which terminated spontaneously. Her serum electrolytes were normal and plasma concentrations of digoxin (1.8 ng/ml) and amiodarone (1.9 μg/ml) were within the usual target ranges.

In the first case the authors speculated that increased vagal tone during bladder irrigation was responsible for QT interval prolongation associated with bradycardia in the presence of amiodarone. In the second case the authors suggested that the dysrhythmia was due to the triple combination of amiodarone with a beta-blocker and digitalis in a patient with atrial fibrillation and structural heart disease; again it is possible that bradycardia played a part.

Of five patients with torsade de pointes due to amiodarone, three had hypokalemia and those with negative T waves were at greater risk of ventricular fibrillation than those with positive T waves (59). Of six patients with torsade de pointes taking chronic amiodarone, five were women, three were taking drugs that inhibit CYP3A4 (loratadine or trazodone), three had hypokalemia, and four had reduced left ventricular function (60).

Of 189 patients, five had torsade de pointes and all five had prolonged QT intervals (61). Two of the five, all women, also had raised blood glucose concentrations, and the authors suggested that hyperglycemia is a risk factor for torsade de pointes. However, the number of cases reported in this series was too small to justify such a conclusion.

It has been suggested that women are more likely to develop torsade de pointes than men in response to antidysrhythmic drugs (62), and this has been confirmed in the case of amiodarone in a study of 189 patients given intravenous amiodarone (61). This is also reminiscent of the finding that prolongation of the QT interval due to quinidine is greater in women than in men at equivalent serum concentrations (63).

T wave alternans is an occasional presentation, in association with ventricular dysrhythmias (SEDA-18, 201).

- A 65-year-old man with atrial fibrillation was given intravenous amiodarone 450 mg over 30 minutes followed by 900 mg over 24 hours (64). He reverted to sinus rhythm, but the electrocardiogram showed giant T wave alternans with a variable QT interval (0.52–0.84 seconds). He had a short bout of torsade de pointes and was given magnesium. Two days later the electrocardiogram was normal.
- In a 62-year-old man with dilated cardiomyopathy and an implantable cardioverter defibrillator for ventricular tachycardia, microvolt T wave alternans differed when amiodarone was added (65). The onset heart rate with T wave alternans was lower and the alternans voltage higher with amiodarone than without it.

The effects of amiodarone appeared to be related to exacerbations of ventricular tachycardia and an increased defibrillation threshold.

Atrial dysrhythmias

Amiodarone has been reported to cause atrial flutter in 10 patients who had been given it for paroxysmal atrial fibrillation (66). In nine of those the atrial flutter was successfully treated by catheter ablation. However, during a mean follow-up period of 8 months after ablation, atrial fibrillation occurred in two patients who had continued to take amiodarone; this was a lower rate of recurrence than in patients in whom atrial flutter was not associated with amiodarone. The authors therefore suggested that in patients with atrial flutter secondary to amiodarone given for atrial fibrillation, catheter ablation allows continuation of amiodarone therapy.

Amiodarone can sometimes cause atrial flutter, even though it is also used to treat it (SEDA-25, 180). There has been a report of seven cases (six men and one woman, aged 34–75 years) of 1:1 atrial flutter with oral amiodarone (67). Four of them had underlying cardiac disease; none had hyperthyroidism. The initial dysrhythmia was 2:1 atrial flutter (n = 4), 1:1 atrial flutter (n = 2), or atrial fibrillation (n = 1). One patient was taking amiodarone 200 mg/day and one was taking 400 mg/day plus carvedilol. The other five all received loading doses of 9200 (sd 2400) mg over 10 (sd 4) days. There was an adrenergic trigger factor (exertion, fever, esophageal stimulation, or a beta-adrenoceptor agonist aerosol) in five patients. One required emergency cardioversion.

In another case there was prolongation of the flutter cycle and infra-Hissian block (68).

Of 136 patients with atrial fibrillation treated with either amiodarone (n = 96) or propafenone (n = 40), 15 developed subsequent persistent atrial flutter, nine of those taking amiodarone and six of those taking propafenone (69). In all cases radiofrequency ablation was effective. It is not clear to what extent these cases of atrial flutter were due to the drugs, although the frequency of atrial flutter in previous studies with propafenone has been similar. Atrial enlargement was significantly related to the occurrence of persistent atrial flutter in these patients.

Bradycardia

Bradycardia has been reported to occur in about 5% of patients taking amiodarone (SEDA-20, 176).

Of 2559 patients admitted to an intensive cardiac care unit over 3 years, 64 with major cardiac iatrogenic problems were reviewed (70). Of those, 58 had dysrhythmias, mainly bradydysrhythmias, secondary to amiodarone, beta-blockers, calcium channel blockers, electrolyte imbalance, or a combination of those. Amiodarone was implicated in 19 cases, compared with 44 cases attributed to beta-blockers and 28 to calcium channel blockers. Of the 56 patients with sinus bradycardia, 10 were taking a combination of amiodarone and a beta-blocker, six were taking amiodarone alone, and three were taking amiodarone plus a calcium channel blocker.

Amiodarone is superior to placebo for cardioversion of recent onset atrial fibrillation, and even though the onset of conversion is delayed compared with class Ic drugs, efficacy is similar at 24 hours (71). However, among 8770 patients aged over 65 years with a new diagnosis of atrial fibrillation who had had a previous myocardial infarction there were 477 cases of bradydysrhythmias requiring a permanent pacemaker and they were matched 1: 4 to 1908 controls (Essebag, 249[C]); the use of amiodarone was associated with an increased risk of pacemaker insertion (OR = 2.14; 95%CI = 1.30, 3.54). This effect was modified by sex, with a greater risk in women (OR = 3.86; 95% CI = 1.70, 8.75) than in men (OR = 1.52; 95% CI = 0.80, 2.89).

During 409 trials of antidysrhythmic drugs to maintain sinus rhythm in patients with previous atrial fibrillation or atrial flutter amiodarone was used in 212 patients (52%), type 1C drugs in 127 (31%), sotalol in 37 (9.0%), and a type 1A drug in 33 (8.1%) (72). There were adverse events in 17 patients: three died, three had bradycardia that required permanent pacemaker implantation, and 11 had bradycardia requiring a reduction in drug dosage. Most of the events were due to bradycardia in patients who received amiodarone. There was a significant association between amiodarone-associated bradycardia and female sex. The only event that occurred during the first 48 hours was an episode of bradycardia in a patient who received amiodarone and was managed as an out-patient.

Heart block

- In a 66-year-old woman taking amiodarone 1200 mg/week there was marked prolongation of the QT interval, to 680 ms; the succeeding P waves fell within the refractory period of the preceding beat and were unable to institute conduction (73). This resulted in 2:1 atrioventricular block. Amiodarone was withdrawn and the QT interval normalized with a time-course consistent with the long half-life of amiodarone. A subsequent rechallenge with intravenous amiodarone caused further prolongation of the QT interval.

The authors hypothesized that this patient had a silent mutation in one of the genes coding for the two major potassium channel proteins (IK_r or IK_s) that are involved in the mode of action of amiodarone. However, they did not present any genetic studies to support this hypothesis.

Pacemaker requirements

Of 8770 patients aged 65 years or over with a new diagnosis of atrial fibrillation, 477 had bradydysrhythmias requiring a permanent pacemaker and were matched with 1908 controls (74). The use of amiodarone was associated with an increased risk of pacemaker insertion (OR = 2.14; 95%CI = 1.30, 3.54). Women had a greater risk than men (OR = 3.86 versus 1.52).

In a retrospective study of 82 patients with an implanted pacemaker cardioverter defibrillator, those who were also taking amiodarone (for 24 consecutive months without interruption) had a significant three-fold increase in episodes of defibrillation compared with those who did not take amiodarone (75). This is an unexpected finding, for which the authors had no explanation. However, the finding was vitiated by the retrospective nature of the study.

Hypotension

Intravenous amiodarone can cause hypotension in anesthetized patients undergoing cardiac surgery. In a prospective double-blind study, 30 patients undergoing coronary artery bypass graft surgery were randomly assigned to receive intravenous amiodarone or placebo (76). At 6 minutes, amiodarone reduced mean arterial pressure by 14 mmHg and placebo reduced it by 4 mmHg. The changes in mean arterial pressure and systolic and diastolic blood pressures between groups were statistically different for the first 15 minutes after drug administration. Hypotension required intervention in three of 15 patients given amiodarone and none of the 15 given placebo. The mean heart rate was 12/minute less after amiodarone, but pulmonary artery pressure, central venous pressure, mixed venous oxygen saturation, and fractional left ventricular area change were not different between the groups. The authors concluded that the hypotension that amiodarone caused during the first 15 minutes after administration was not accompanied by altered left ventricular function, suggesting that selective arterial vasodilatation was the primary cause.

Hypotension has been attributed to solvents in the intravenous formulation of amiodarone, and this hypothesis was supported by the observation that in four trials an aqueous formulation caused hypotension in only three of 278 patients and only during bouts of ventricular tachycardia (77). In addition, six patients had cardiac dysrhythmias or heart block, two had erythema and/or pain at the site of injection, and two had thrombophlebitis.

Respiratory

There have been reviews of the lung complications of amiodarone toxicity (78,79,80) and of its mechanisms (81).

Frequency

The risk of lung toxicity is about 5–6% (82) and is greatest during the first 12 months of treatment and among patients over 40 years of age. The mortality rate in those who develop respiratory involvement is about 9% (about 0.5% of the total).

The number of reports to the FDA of serious adverse events in patients taking amiodarone increased from under 50 in each year from 1986 to 1992 to nearly 250 in 2001 and 2002 (83). The total number of such reports from 1986 to 2002 was about 2000, of which the most common were dyspnea (n = 264), pneumonia (230), unspecified lung disorders (224), and pulmonary fibrosis (210). Reports of parenchymal lung damage represented about 14% of all serious adverse events. Lung damage can occur within days or weeks of the start of therapy and death can occur. The prognosis is worse in those with pre-existing lung damage and the incidence can be reduced by using lower loading and maintenance doses.

Time-course

Lung damage due to amiodarone can occur within days or weeks of the start of therapy, and death can occur. The prognosis is worse in those with pre-existing lung damage and the incidence can be reduced by using lower loading and maintenance doses. The speed with which amiodarone-induced lung damage can occur has been illustrated by the case of a 75-year-old man who received a total dose of amiodarone of only 1500 mg and developed dyspnea, tachypnea, and hypoxemia, with diffuse crackles over both lungs, multiple bilateral acinar pattern infiltrates without Kerley B lines or peribronchial cuffing on the chest X ray, and diffuse ground glass opacities associated with smooth interlobular thickening, more prominent in the lower lung zones, and intralobular interstitial thickening in subpleural regions on a high-resolution chest CT scan; there were foamy macrophages in the bronchoalveolar lavage fluid (84).

Of 613 Chinese patients taking amiodarone 200 mg/day 12 (1.9%) had amiodarone-induced lung damage; nine were men (85). Their mean age was 77 years. The average duration of therapy was 14 (range 1–27) months. Three patients developed the complication within 4 months of starting the medication. Eight developed their complications at 12–24 months. Two died of respiratory failure.

Lung damage from amiodarone can occur quite quickly after lung resection.

- A 73-year-old man underwent resection of the right middle lobe of lung for a squamous cell carcinoma (86). Postoperatively he developed atrial fibrillation and was given amiodarone 450 mg intravenously followed by 800 mg intravenously for 2 days, when sinus rhythm was restored. However, he then developed cough and fever, and a CT scan showed bilateral patchy infiltrates. After 20 days (total dose of amiodarone 9000 mg) he developed adult respiratory distress syndrome. Bronchoalveolar lavage showed eosinophils, mast cells, and foamy macrophages. Amiodarone was withdrawn and he was given intravenous methylprednisolone 200 mg/day followed by oral prednisolone. He recovered over 8 weeks.

The authors suggested that this man had amiodarone-induced pneumonitis, which occurred early because of pre-existing lung damage.

The speed with which amiodarone-induced lung damage can occur has also been illustrated by the case of a 53-year-old man who developed dyspnea and bilateral pulmonary infiltrates and pleural effusions within 9 days (87).

Mechanism

Amiodarone causes lung damage either by direct deposition of phospholipids in the lung tissue or by some immunologically mediated reaction. Other mechanisms have also been proposed, including oxidant-mediated damage, a direct detergent effect, and a direct toxic effect of iodide (SEDA-15, 168).

It has been suggested that the serum activity of lactate dehydrogenase (LDH) may be related to the occurrence of amiodarone-induced pneumonitis, as occurred in a 72-year-old woman in whom the serum LDH activity rose from a baseline of around 750 U/l to around 1500 U/l during acute pneumonitis and resolved with resolution of a clinical condition after withdrawal of amiodarone (88). The LDH activity in bronchoalveolar lavage fluid was also increased. The proposed mechanism was leakage of lactate dehydrogenase from the pulmonary interstitial cells into the blood. Of course, a rise in the serum LDH activity is highly non-specific, and it is not clear whether it might also rise in bronchoalveolar lavage fluid in other conditions.

Presentation

The commonest form of lung damage is an interstitial alveolitis, although pneumonitis and bronchiolitis obliterans have also been reported, as have solitary localized fibrotic lesions, non-cardiac pulmonary edema, pleural effusions, acute respiratory failure, acute pleuritic chest pain, and adult respiratory distress syndrome (SEDA-17, 220; SEDA-18, 201; 89–91). Amiodarone has also been reported to cause impairment of lung function, even in patients who do not develop pneumonitis (92), and pre-existing impairment of lung function may constitute a contraindication to amiodarone.

Lung damage due to amiodarone usually develops slowly, but it can occasionally have a rapid onset, particularly in patients who are given high concentrations of inspired oxygen, and there is experimental evidence that amiodarone enhances the toxic effects of oxygen on the lungs (93).

- Adult respiratory distress syndrome occurred very rapidly in a 66-year-old man who took amiodarone 200 mg/day for a few weeks only (94).
- Pulmonary infiltrates occurred in a 72-year-old man after treatment with amiodarone (total dose 6800 mg) for only 7 days (95).
- Two patients with dilated cardiomyopathy developed pneumonitis after 6 weeks and 8 months while taking amiodarone 400 and 200 mg/day respectively (96).

Some cases of amiodarone-induced lung toxicity, some in patients who took very large doses, illustrate the wide variety of possible presentations.

- A 77-year-old man without a history of lung disease was given amiodarone 7 days after bypass surgery because of supraventricular dysrhythmias and non-sustained ventricular tachycardia (97). He had taken 1600 mg/day for a week followed by a maintenance dosage of 400 mg/day, and 15 days later became pale, sweaty, febrile, and tachypneic. His blood pressure was 100/60 and his heart rate 100/minute. There were reduced breath sounds and crackles throughout the lung fields. A chest X-ray showed diffuse interstitial and alveolar infiltrates and small bilateral pleural effusions. A high-resolution CT scan of the chest showed diffuse ground-glass attenuation and patchy peripheral opacities, consistent with an acute hypersensitivity pneumonitis, and other diagnoses were ruled out. He responded to gluco-corticoids.
- A 72-year-old man developed hypoxemic respiratory failure while taking amiodarone 300 mg/day (98). He had no history of lung disease. His CT scan was similar to that of the first patient. He responded to treatment with corticosteroids.
- In a 79-year-old man with emphysema taking amiodarone 200 mg/day, the diagnosis of amiodarone-induced lung

toxicity was complicated by the fact that emphysema has the opposite effect on lung volumes and spirometry from interstitial lung disease (99). His FEV_1, which had been reduced, became normal and then increased. However, the combination of emphysema with amiodarone-induced lung disease led to worsening dyspnea, and a chest X-ray showed patchy mixed interstitial and airspace disease, most marked in the mid to upper lung zones bilaterally, and ground-glass opacification in the left lower lobe, suggesting an acute alveolitis. He responded to prednisone after withdrawal of amiodarone. His carbon monoxide diffusing capacity, which had fallen, returned to normal. A CT scan showed marked bullous emphysema and ground-glass interstitial changes. The FEV_1 almost doubled, from being severely reduced to within the reference range.

- A 77-year-old man who had taken amiodarone 400 mg/day for 11 months developed crackles at the lung bases and scattered respiratory wheeze (100). His leukocyte count was raised at $13.5 \times 109/l$ and he had progressive reduction in carbon monoxide diffusing capacity, serially measured. A chest X-ray showed bilateral opacities in the upper zones, peripheral in distribution, and a CT scan showed dense bilateral lung parenchymal opacities. The symptoms of dyspnea on exertion, cough with minimal sputum, pleuritic chest pain, and low-grade fever abated after withdrawal, and the upper lobe densities resolved.
- A 62-year-old man took amiodarone 400 mg bd and developed several adverse effects, including bilateral apical opacities with left hilar lymphadenopathy (101). Amiodarone was withdrawn and he was given glucocorticoids, with good effect; there was dramatic radiographic resolution within 3 weeks and he was no longer breathless with 1 week. The lung biopsy showed typical foamy macrophages. He had fibrosis of the bronchioles and interstitium, foci of obliterative bronchiolitis, and thickening of the alveolar walls. He had an accompanying peripheral neuropathy, which improved after withdrawal, and impaired visual acuity, about which no further information was given. Biopsy of the right vastus lateralis muscle showed type II atrophy with vacuolization, which the authors suggested supported the suspicion of amiodarone toxicity.

Amiodarone can occasionally cause isolated lung masses (SEDA-15, 168). One case was associated with a vasculitis; the lesions resolved completely 4 months after amiodarone withdrawal (102). In another case an isolated mass was associated with multiple small nodules in both lungs; the lesions resolved completely 6 months after amiodarone withdrawal (103).

- A 73-year-old man who had taken amiodarone 200 mg/day, 5 days a week, for 15 years was given a glucocorticoid for suspected giant cell arteritis (104). The glucocorticoid was suddenly withdrawn 2 weeks later, and 10 days later he developed dyspnea and fever and rapidly developed acute respiratory failure. Post-mortem findings were consistent with amiodarone-induced acute interstitial pneumonitis, with mild fibrosis and numerous intra-alveolar foamy macrophages.

The authors hypothesized that the glucocorticoid had masked amiodarone-associated lung damage.

Diagnosis

Diagnosis of amiodarone-induced lung damage can be difficult. The clinical symptoms and signs, the changes on chest radiography, and abnormalities of lung function tests are all non-specific. The presence of lymphocytes and foamy macrophages in bronchial lavage fluids and of phospholipidosis in lung biopsies are all suggestive. Measurement of the diffusing capacity of carbon monoxide has been used, but is unreliable.

The sialylated carbohydrate antigen Krebs von den Lungen-6 (KL-6) has been reported to be a serum marker of the activity of interstitial pneumonitis in seven patients with amiodarone-induced pulmonary toxicity (105,106). The dosages of amiodarone were 200–800 mg as an oral loading dose followed by 75–200 mg/day. Pulmonary complications occurred at 17 days to 48 months of treatment. In two patients with severe dyspnea and interstitial shadows on chest X-ray the KL-6 concentrations were very high (2100 and 3000 U/ml). In one of these the concentration increased from 695 to 2100 U/ml at a time when the interstitial changes on the CT scan worsened. In contrast, in two patients in whom pneumonia resolved with antibiotic treatment and without withdrawal of amiodarone, the serum KL-6 concentrations were lower (120 and 330 U/ml). In a patient in whom congestion of the lungs due to congestive cardiac failure had been confused with interstitial shadows the KL-6 concentration was only 190 U/ml. In two patients with lung cancers the concentrations were 260 and 360 U/ml. The authors proposed that a KL-6 concentration above the reference range (more than 520 U/ml) might be useful in differentiating patients with amiodarone-induced pneumonitis from patients with similar features not associated with amiodarone.

In 25 patients, three had proven interstitial pneumonitis and KL-6 serum concentrations of 414, 848, and 1217 U/ml; in contrast, all of the other 22 patients had normal CT scans and normal KL-6 concentrations (under 500 U/ml) (107). In the same study the limitations of carbon monoxide diffusing capacity in the diagnosis of amiodarone-induced lung disease (SEDA-15, 168) were again demonstrated.

A 69-year-old woman with lung damage due to amiodarone had increased blood concentrations of KL-6 (108).

Several scanning techniques have been used in the diagnosis of amiodarone-induced lung damage.

Computed tomography

Computed tomography may show a typical pattern of basal peripheral high-density pleuroparenchymal linear opacities, although these may be absent (109). It has also been suggested that high-resolution CT scanning may be able to detect iodine deposition from the drug (110). Of 16 patients taking long-term amiodarone, eight had severe respiratory and other symptoms and eight either had no symptoms or had only mild or chronic respiratory symptoms. All eight controls had negative high-resolution CT scans with no areas of high attenuation, while all eight cases had a least one high-attenuation lesion.

^{67}Gallium scintigraphy

^{67}Gallium scintigraphy has been used to diagnose amiodarone-induced lung damage (SEDA-15, 168).

- A 75-year-old man, who had taken amiodarone 200 mg/day for 4 years, developed acute dyspnea, chest pain, fever, and sweats (111). The chest X-ray showed diffuse alveolar and interstitial infiltrates, particularly at the lung bases. No pathogenic organisms were isolated and antibiotics had no effect. There was no evidence of sarcoidosis. Pulmonary ^{67}gallium scintigraphy showed extensive uptake of tracer throughout both lungs, consistent with amiodarone pneumonitis on a background of asbestosis with interstitial fibrosis. Treatment with corticosteroids after withdrawal of amiodarone resulted in marked clinical improvement.

The authors said that the extensive changes on gallium scanning, not present on the chest X-ray, had helped them to make the diagnosis, although a high-resolution CT scan had also shown widespread changes.

^{99m}Technetium-diethylene triamine penta-acetic acid (DPTA) aerosol scintigraphy

Another scanning technique, ^{99m}Tc-diethylene triamine penta-acetic acid (DPTA) aerosol scintigraphy, has been compared with ^{67}Ga scanning in 26 patients, seven with amiodarone-induced lung damage, eight taking amiodarone without lung damage, and 11 healthy controls (112). ^{67}Ga scintigraphy was positive in four of the seven patients with lung damage but normal in the others. There was a positive correlation between ^{99m}Tc-DTPA clearance and the cumulative dose of amiodarone. The mean clearance values were 2%/minute in those with amiodarone-induced lung damage, 1.3%/minute in those without lung damage, and 0.9%/minute in the controls. The authors concluded that ^{67}Ga lung scintigraphy is useful for detecting amiodarone-induced lung damage but that ^{99m}Tc-DTPA aerosol scintigraphy is better.

Management

Although early reports suggested that glucocorticoids might be beneficial in management, this has not been subsequently confirmed (113).

Nervous system

The most common forms of neurological damage attributed to amiodarone are tremor, peripheral neuropathy, and ataxia (114). Other effects that have been reported include delirium (115), Parkinsonian tremor (116), and pseudotumor cerebri (117). Acute myolysis has recently been described at high dose (118). The peripheral neuropathy is probably due to intracellular lipidosis (119).

- Periodic ataxia has been attributed to amiodarone in a 67-year-old man taking amiodarone 200 mg/day (120). The ataxia responded to acetazolamide and eventually to withdrawal of amiodarone. It recurred with rechallenge.
- An 84-year-old woman with hypertrophic obstructive cardiomyopathy and paroxysmal atrial fibrillation developed a progressively debilitating ataxia, which abated over 4 months after withdrawal of amiodarone (121). Despite the long half-life of amiodarone, her symptoms began to improve after several days, and she was walking without assistance within 1 week.

Amiodarone-induced neuromyopathy has been studied in three patients by a review of their records, electromyography, and histopathology of muscle and nerve (122). Two patients had a slightly asymmetric, mixed, but primarily demyelinating sensorimotor polyneuropathy and the third had an acute neuropathy resembling Guillain–Barré syndrome. Creatine kinase activity did not correlate with clinical or electromyographic evidence of myopathy. In the peripheral nerves there was demyelination, some axon loss, and a variable number of characteristic lysosomal inclusions. Muscle specimens from two patients showed evidence of a vacuolar myopathy. After withdrawal of amiodarone, two patients improved and one died with a cardiac dysrhythmia.

Benign orgasmic headache has been associated with amiodarone (123).

- A 52-year-old man, who had taken amiodarone 800 mg/day for 7 months, developed acute, severe, throbbing headaches precipitated by coitus and occasionally other forms of exertion. An MRI scan of the brain was normal. When the dose of amiodarone was reduced to 200 mg/day the headaches diminished in frequency and severity. When the dose was increased again to 400 mg/day they increased in frequency and severity. The amiodarone was withdrawn and the headaches resolved.

Amiodarone was originally developed as a vasodilator, and that may have been the cause of headaches in this case.

Sensory systems

The adverse effects of amiodarone on the eyes have been reviewed (124). The most common effect is corneal microdeposits. In some cases chronic blepharitis and conjunctivitis have been reported (125), but the relation of these to amiodarone is not clear.

Keratopathy

In almost all patients (126,127) corneal microdeposits of lipofuscin occur secondary to the deposition of amiodarone. These are generally of no clinical significance, but occasionally patients complain of haloes around lights, particularly at night, photophobia, blurring of vision, dryness of the eyes, or lid irritation. Occasionally amiodarone can cause anterior subcapsular deposits, which are usually asymptomatic. In 22 patients taking long-term amiodarone there were corneal drug deposits in all of the eyes, slight anterior subcapsular lens opacities in 22%, and dry eyes in 9% (128).

Verticillate epithelial keratopathy due to amiodarone, in which there is whorl-shaped pigmentation of the cornea, has been proposed to be worsened by soft contact lenses (SEDA-15, 171). Two patients with hard contact lenses and amiodarone-associated keratopathy both complained of increased sensitivity to sunlight and were fitted with ultraviolet light-blocking lenses instead, as a precaution against further corneal damage; however, the authors did not think that the contact lenses had contributed to the damage (129).

The eyes of 11 patients (eight men and three women) taking amiodarone have been compared with those of 10 healthy sex- and age-matched controls by confocal microscopy (130). All those taking amiodarone had bright, highly reflective intracellular inclusions in the epithelial layers, particularly in the basal cell layers. In eyes with advanced keratopathy there were bright microdots in the anterior and posterior stroma and on the endothelial cell layer. Keratocyte density in the anterior stroma was lower in the treated subjects than in the controls, and there was marked irregularity of the stromal nerve fibers. The authors concluded that in some patients taking long-term amiodarone corneal damage may penetrate deeper than has previously been suspected.

Morphological changes in the cornea caused by amiodarone have been evaluated by in vivo slit scanning confocal microscopy in 49 eyes of 25 patients taking amiodarone and 26 eyes of 13 age- and sex-matched healthy controls (131). The mean dosage of amiodarone was 224 mg/day and the mean duration of treatment was 21 months. There were deposits in all but eight eyes of the patients who took amiodarone, and they were detected as early as 2 months after the start of treatment. Deposition correlated significantly with the duration of treatment and therefore the cumulative dose. The deposits were seen in the basal lamina in all eyes and in the superficial epithelium, anterior stroma, mid-stroma, and subepithelial nerves in eyes with grades 2–4 keratopathy. There were also abnormalities in anterior stromal keratocytes, subepithelial and stromal nerves, and endothelium. The authors suggested that confocal microscopy will prove to be useful in early diagnosis and in understanding the pathophysiology of amiodarone keratopathy.

Unilateral amiodarone vortex keratopathy in an 87-year-old woman was explained by the presence of corneal dysplasia in the unaffected eye, which did not allow amiodarone to bind to corneal lipid (132).

Color vision disturbances

Amiodarone can cause impaired color vision associated with keratopathy (SEDA-12, 153) (133). Of 22 patients taking long-term amiodarone, two who had otherwise healthy eyes had abnormal blue color vision (134). Otherwise, color vision, contrast sensitivity, and visual fields were normal or could be explained by eye diseases such as cataract.

Optic neuropathy

A more serious effect of amiodarone on the eye is an optic neuropathy (SEDA-15, 171; 135). Although this resolves on withdrawal there can be residual field defects (136,137). Blindness has been attributed to bilateral optic neuropathy in a patient taking amiodarone (138). The incidence of optic neuropathy with amiodarone (SEDA-23, 199) has been estimated at 1.8% (SEDA-13, 141; 139). There has been a recent review of 73 cases of optic neuropathy associated with the use of amiodarone, including 16 published case reports and 57 other reports from the National Registry of Drug-Induced Ocular Side Effects, the US FDA, and the WHO (140). Amiodarone-induced optic neuropathy is of insidious onset, with slow progression, bilateral visual loss, and protracted disc swelling, which tends to stabilize within several months of withdrawal. These features all distinguish it from non-arteritic ischemic optic neuropathy. The pathology of amiodarone-induced optic neuropathy is associated with lipid deposition, as with other forms of adverse effects of amiodarone.

- A 51-year-old man developed blurred vision after having taken amiodarone 600 mg/day for 3 months and 400 mg/day for 5 months (141). There was mild optic disc palor and edema on the right side, with a nearby flame-shaped hemorrhage; the optic disc on the left side was normal. There were accompanying corneal opacities in both eyes. Amiodarone was withdrawn and the optic neuropathy and corneal opacities improved.
- A 48-year-old man developed bilateral blurred vision and visual field changes after having taken amiodarone 400 mg/day for 2 months; 3 weeks after withdrawal of amiodarone his symptoms improved (142). There was no optic disc edema.

An unusual case has been reported in which bilateral inferior field loss progressed to upper and lower field loss bilaterally despite withdrawal of the amiodarone (143).

In a retrospective study, three patients with amiodarone-induced optic neuropathy had mildly impaired vision, visual field defects, and bilateral optic disc swelling; on withdrawal of amiodarone, visual function and optic disc swelling slowly improved in all three (144).

Other effects

The absence of optic disc edema in the last case is unusual; most cases are accompanied by some form of swelling of the optic disc.

Rare effects include raised intracranial pressure with papilledema (SEDA-12, 153) (145), retinal maculopathy (146), and retinopathy (124).

Multiple chalazia have been reported on the eyelids, due to lipogranulomata that contained a lot of amiodarone (147).

Sicca syndrome has occasionally been reported (9,148–150).

Brown discoloration of implanted lenses has been attributed to amiodarone (151).

- A 66-year-old woman, who had had two silicone intraocular lenses inserted because of cataract, developed progressive brown discoloration of the lens while taking amiodarone (dosage not stated). The discoloration progressed markedly after vitrectomy, suggesting that it was due to leakage of the drug into the eye. She also had an amiodarone-induced keratopathy.

In a case-control study in 14 patients there were significant changes in visual evoked responses (152). There was no relation to the duration of therapy. Intraocular pressure was unaffected and fundoscopy was normal.

Psychological, psychiatric

Delirium has rarely been reported with amiodarone (SEDA-13, 140).

- A 54-year-old man with no previous psychiatric history took amiodarone 400 mg bd (153). After a few days he became depressed and paranoid, suffered from insomnia, and had rambling speech. The dosage of amiodarone was reduced to 200 mg bd and he improved. However, 3 days later he became confused, with tangential thinking, labile effect, and a macular rash on the limbs. His serum sodium was reduced at 127 mmol/l and his blood urea nitrogen was raised. A CT scan of the head was normal. Amiodarone was withdrawn and 4 days later he was alert and oriented. About a week later he started taking amiodarone again and within 4 days became increasingly agitated, confused, and paranoid. He once more recovered after withdrawal of amiodarone.
- Depression has been attributed to amiodarone in a 65-year-old woman who was taking amiodarone (dosage not stated) (154). Because the mode of presentation was atypical in onset, course, duration, and its response to antidepressant drugs, amiodarone was withdrawn, and she improved rapidly. There was no evidence of thyroid disease.

Endocrine

Testes

Amiodarone can cause endocrine testicular dysfunction, as judged by increases in serum concentrations of FSH and LH and hyper-responsiveness to GnRH (155).

Syndrome of inappropriate ADH secretion

Amiodarone-induced hyponatremia, due to the syndrome of inappropriate secretion of antidiuretic hormone, is rare (SEDA-21, 199; 156, 157). The mechanism is unknown. Unlike other adverse effects of amiodarone, it seems to occur rapidly and to resolve rapidly after withdrawal.

- A 63-year-old man reduced his dietary sodium intake to combat fluid retention and was taking furosemide 40 mg/day, spironolactone 50 mg/day, and enalapril 2.5 mg/day (158). He then took amiodarone 800 mg/day for 7 days and his serum sodium concentration fell to 119 mmol/l; his plasma vasopressin concentration was raised at 2.6 pmol/l. The dose of amiodarone was reduced to 100 mg/day, with fluid restriction; his sodium rose to 130 mmol/l and his vasopressin fell to 1.4 pmol/l.
- An 87-year-old man reduced his dietary sodium intake to combat fluid retention and was taking furosemide 40 mg/day and spironolactone 25 mg/day (158). He then took amiodarone 200 mg/day for 7 days and 100 mg/day for 8 days and his serum sodium concentration fell to 121 mmol/l; his plasma vasopressin concentration was raised at 11 pmol/l. Amiodarone was continued, with fluid restriction; his sodium rose to 133 mmol/l and his vasopressin fell to 2.4 pmol/l.
- A 67-year-old man, who had taken amiodarone 200 mg/day for 3 months, developed hyponatremia (serum sodium concentration 117 mmol/l) (159). He was also taking furosemide 20 mg/day, spironolactone 25 mg/day, and lisinopril 40 mg/day. His urine osmolality was 740 mosmol/kg with a normal serum osmolality. Fluid restriction was ineffective, but when amiodarone was withdrawn the sodium rose to 136 mmol/l.
- A 62-year-old woman with paroxysmal atrial fibrillation who had taken amiodarone 300 mg/day had a serum sodium concentration of 120 mmol/l with a normal serum potassium and a reduced serum osmolality (240 mmol/kg); the urinary sodium concentration was 141 mmol/l and the urine osmolality 422 mmol/kg (156). There was no evident cause of inappropriate secretion of ADH and within 5 days of withdrawal of amiodarone the serum sodium concentration had risen to 133 mmol/l and rose further to 143 mmol/l 14 days later. There was no rechallenge and no recurrence of hyponatremia during the next 6 months.

In some of these cases other factors may have contributed to the hyponatremia that amiodarone seems to have caused.

Thyroid gland

The effects of amiodarone on thyroid function tests and in causing thyroid disease, both hyperthyroidism and hypothyroidism, have been reviewed in the context of the use of perchlorate, which acts by inhibiting iodine uptake by the thyroid gland (160), and there have been several other reviews (161–165).

Effects on thyroid function tests

Amiodarone causes altered thyroid function tests, with rises in serum concentrations of T4 and reverse T3 and a fall in serum T3 concentration. This is due to inhibition of the peripheral conversion of T4 to T3, causing preferential conversion to reverse T3. These changes can occur in the absence of symptomatic abnormalities of thyroid function.

Hyperthyroidism

> DoTS classification (BMJ 2003;327:1222-5)
> Dose-relation: collateral effect
> Time-course: delayed
> Susceptibility factors: genetic (unoperated or palliated cyanotic congenital heart disease; beta-thalassemia major); sex (conflicting results); altered physiology (iodine intake, conflicting results)

Frequency

Apart from its effects on thyroid function tests, amiodarone is also associated with both functional hyperthyroidism and hypothyroidism, in up to 6% of patients. The frequency of thyroid disease in patients taking amiodarone has been retrospectively studied in 90 patients taking amiodarone 200 mg/day for a mean duration of 33 months (166). Hypothyroidism occurred in five patients and hyperthyroidism in 11. Hyperthyroidism became more frequent with time and was associated with recurrent supraventricular dysrhythmias in four of the 11 patients.

In a nested case-control analysis of 5522 patients with a first prescription for an antidysrhythmic drug and no previous use of thyroid drugs, cases were defined as all patients who had started a thyroid-mimetic or antithyroid drug no sooner than 3 months after the start of an antidysrhythmic drug and controls were patients with a comparable follow-up period who had not taken any thyroid drugs during the observation period (167). There were 123 patients who had started antithyroid drugs and 96 who had started a thyroid-mimetic drug. In users of amiodarone there was an adjusted odds ratios of 6.3 (95% CI = 3.9, 10) for hyperthyroidism compared with users of other antidysrhythmic drugs. Patients who were exposed to a cumulative dose of amiodarone over 144 g had an adjusted odds ratio of 13 (6, 27) for hyperthyroidism.

Mechanisms

Amiodarone causes two different varieties of hyperthyroidism (SEDA-23, 199), one by the effects of excess iodine in those with latent disease (so-called type 1 hyperthyroidism), the other through a destructive thyroiditis in a previously normal gland (so-called type 2 hyperthyroidism). The two varieties can be distinguished by differences in radio-iodine uptake by the gland: in type 1 hyperthyroidism radio-iodine uptake is normal or increased, whereas in type 2 it is reduced. In type 1 hyperthyroidism, thyroid ultrasound shows a nodular, hypoechoic gland of increased volume, whereas in type 2 the gland is normal.

This distinction may be important, because type 1 typically responds to thionamides and perchlorate while type 2 responds to high-dose glucocorticoids. Color-flow Doppler sonography can be of use in distinguishing the two types, because type 1 is associated with increased vascularity and type 2 is not. In a retrospective study of 24 patients with amiodarone-induced hyperthyroidism in an iodine-replete environment, 13 had little or no vascularity, of whom seven were prednisolone-responsive; of 11 patients with increased vascularity, four responded to antithyroid drugs alone and only one of seven responded to prednisolone (168). Euthyroidism was achieved twice as rapidly in patients with low vascularity than in those with increased vascularity. Thus, responsiveness to prednisolone was not consistently predicted by lack of vascularity, but the presence of flow appeared to correlate with non-responsiveness to prednisolone.

Thyroid hormone-producing thyroid carcinoma is an uncommon cause of thyrotoxicosis. Precipitation of thyrotoxicosis by iodine-containing compounds in patients with thyroid carcinoma is rare, but has been attributed to amiodarone in a 77-year-old man with extensive hepatic metastases from a well-differentiated thyroid carcinoma (169).

Iodine intake may be important in determining the type of amiodarone-induced thyroid disease. In 229 patients taking long-term amiodarone hyperthyroidism was more common (9.6 versus 2%) in West Tuscany, where dietary iodine intake is low, and hypothyroidism more common (22 versus 5%) in Massachusetts, where iodine intake is adequate (SEDA-10, 148) (170). However, other factors may play a part. In a retrospective inter-regional study in France there was a greater incidence of amiodarone-induced hyperthyroidism in the maritime areas Aquitaine and Languedoc–Roussillon, and a greater incidence of amiodarone-induced hypothyroidism in Midi–Pyrenees, a non-maritime area, in which iodine intake is lower than in Languedoc–Roussillon (171).

There have also been reports of painful thyroiditis associated with amiodarone (SEDA-15, 170).

Time-course

Thyroid function tests were measured before and after treatment of amiodarone-induced hyperthyroidism (n = 12) and the response to combined antithyroid and glucocorticoid treatment (n = 11) was recorded (172). One patient had type 1 hyperthyroidism, nine had type 2, and two probably had a mixed form. Six patients had diffuse hypoechoic goiters. The median time to euthyroidism (defined as a normal free T3 concentration) with a thionamide + prednisolone (starting dose 20–75 mg/day) was 2 (interquartile range 1.0–2.7) months. Thionamide treatment was stopped after a median duration of 5.7 (4.2–8.7) months and glucocorticoids were completely withdrawn after 6.7 (5.5–8.7) months.

Susceptibility factors

There has been a retrospective study of the frequency of amiodarone-associated thyroid dysfunction in adults with congenital heart disease (173). Of 92 patients who had taken amiodarone for at least 6 months (mean age 35, range 18–60 years), 36% developed thyroid dysfunction—19 became hyperthyroid and 14 hypothyroid. The mean dosage was 194 (100–300) mg/day, and the median duration of therapy was 3 (0.5–15) years. Female sex (OR = 3) and unoperated or palliated cyanotic congenital heart disease (OR = 7) were significant susceptibility factors for thyroid dysfunction. The risk was also dose-related. Although the authors conceded that they may have overestimated the risk of thyroid dysfunction, because of the selected nature of the population they studied, the risks were markedly higher than in previous studies of older patients with acquired heart disease, despite a lower maintenance dosage of amiodarone.

In contrast, it has been suggested that men are more susceptible to hyperthyroidism due to amiodarone (174). Of 122 600 patients in 12 practices in the West Midlands in the UK, 142 men and 74 women were taking amiodarone and 27 (12.5%) had thyroid disease. Of those, 11 men (7.7%) and 4 women (5.4%) had hypothyroidism, a non-significant difference; however, 12 men (8.5%) had hyperthyroidism compared with no women. This difference is particularly striking because hyperthyroidism is usually more common in women.

Patients with beta-thalassemia major have an increased risk of primary hypothyroidism. In 23 patients with beta-thalassemia amiodarone was associated with a high risk of overt hypothyroidism (33 versus 3% in controls) (175). This occurred at up to 3 months after starting amiodarone. The risk of subclinical hypothyroidism was similar in the two groups. In one case overt hypothyroidism resolved spontaneously after withdrawal, but the other patients were given thyroxine. After 21–47 months of treatment three patients developed thyrotoxicosis, with remission after withdrawal. There were no cases of hyperthyroidism in the controls. The authors proposed that patients with beta-thalassemia may be more susceptible to iodine-induced hypothyroidism, related to an underlying defect in iodine in the thyroid, perhaps associated with an effect of iron overload.

Of 26 fetuses with hydrops fetalis and supraventricular tachycardias, 25 received transplacental drug therapy; prenatal conversion occurred in 15 (176). Nine fetuses were converted to sinus rhythm using either flecainide (n = 7) or amiodarone (n = 2) as first-line therapy, while digoxin either alone or in association with sotalol failed to restore sinus rhythm in all cases. After first-line therapy, supraventricular tachycardia persisted in 10 fetuses, nine of whom received amiodarone alone or in association with digoxin as second-line therapy, and five of whom converted to sinus rhythm. Of 11 neonates who received amiodarone in utero, two developed raised thyroid stimulating hormone concentrations on postnatal days 3–4; they received thyroid hormone and had normal outcomes.

Presentation

Many examples of hyperthyroidism due to amiodarone have been published.

- A 72-year-old woman with dilated cardiomyopathy was given amiodarone for fast atrial flutter and 6 months later developed abnormal thyroid function tests, with a suppressed TSH and a raised serum thyroxine. The autoantibody profile was negative and a thyroid uptake scan showed reduced uptake (177).

Despite the fact that she was clinically euthyroid, the authors suggested that this patient had amiodarone-induced hyperthyroidism. However, amiodarone inhibits the peripheral conversion of thyroxine to triiodothyronine; it can therefore increase the serum thyroxine and suppress the serum TSH, as in this case. On the other hand, the reduced uptake by the thyroid gland is consistent with type 2 amiodarone-induced hyperthyroidism. The authors did not report the serum concentrations of free thyroxine and triiodothyronine.

- A 67-year-old man took amiodarone 200 mg/day for 20 months, after which it was withdrawn; 8 months later his serum TSH was suppressed and the free thyroxine and free triiodothyronine were both raised; there were no thyroid antibodies and an ultrasound scan showed a diffuse goiter with a nodule in the right lobe and reduced iodine uptake (178). Histological examination of the nodule showed a papillary cancer.

The authors attributed these changes to an effect of amiodarone, but it is not clear that amiodarone-induced changes would have taken so long to become manifest after withdrawal. However, the diagnosis of type 2 amiodarone-induced hyperthyroidism was supported by a poor response to prednisone, potassium perchlorate, and methimazole. Lithium produced temporary benefit, but thyroidectomy was required.

- In five patients who presented in Tasmania during 1 year, all of whom were taking amiodarone 200 mg/day, serum TSH was undetectable and the free thyroxine and triiodothyronine concentrations were raised (179). In one case there was a low titer of TSH receptor antibodies and in another a high titer of antithyroid peroxidase antibodies. In all cases the hyperthyroidism was severe and occurred after at least 2 years of treatment with amiodarone. In one of two patients in whom it was measured the serum concentration of interleukin-6 was raised, as has been previously shown (SEDA-19, 193). In two cases the hyperthyroidism was refractory to treatment with propylthiouracil, lithium, and dexamethasone; in these cases thyroidectomy was required. Two patients responded to propylthiouracil, lithium, and dexamethasone, and one responded to carbimazole.

Amiodarone-induced hyperthyroidism can occasionally be fatal (180).

- A 62-year-old man took amiodarone for 2 years and developed hyperthyroidism; carbimazole 40 mg/day, prednisolone, lithium, and colestyramine were ineffective and he died with hepatic encephalopathy and multiorgan failure.
- A 55-year-old man took amiodarone for 4 years and developed hyperthyroidism; carbimazole 60 mg/day, prednisolone, and lithium were ineffective and he died with septicemia and multiorgan failure.

In three other cases reported in the same paper, severe hyperthyroidism responded severally to treatment with carbimazole, carbimazole plus lithium, or propylthiouracil. In one case amiodarone therapy was restarted after prophylactic subtotal thyroidectomy.

Diagnosis

The diagnosis of amiodarone-induced thyroid disorders can be difficult, because amiodarone often alters thyroid function tests without disturbing clinical thyroid function. Although radio-iodine uptake by the thyroid gland is not helpful in making a diagnosis, the discharge of iodine from the thyroid gland in response to perchlorate is reduced in patients with hypothyroidism (181). The test is not abnormal in patients with hyperthyroidism and it is not clear how helpful it is in hypothyroidism.

Since the measurement of serum T3 and T4 concentrations may not be helpful, an alternative would be to measure metabolic status. Measurement of the serum concentration of co-enzyme Q10 may distinguish patients with clinical thyroid dysfunction from those who simply have abnormalities of thyroid function tests (182), but the value of this test remains to be established.

Color-flow Doppler sonography of the thyroid and measurement of serum interleukin-6 (IL-6) have been studied as diagnostic tools in a retrospective case-note study of patients with amiodarone-associated hyperthyroidism (183). There were 37 patients with amiodarone-associated hyperthyroidism (mean age 65, range 20–86 years), and 25 underwent color-flow Doppler sonography. Of those, 10 were classified as type 1 (based on increased vascularity) and 10 as type 2 (based on patchy or reduced vascularity); 5 were indeterminate. In those with type 1 hyperthyroidism, free serum thyroxine tended to be lower (52 versus 75 pmol/l), free serum triiodothyronine was lower (8.8 versus 16 pmol/l), the cumulative amiodarone dose was lower (66 versus 186 g), and less prednisolone was used (because the diagnosis of type 1 disease encouraged steroid withdrawal); however, carbimazole doses were not different and the time to euthyroidism was the same in the two groups (81 versus 88 days). IL-6 was raised in two patients with type 1 and in one patient with type 2 hyperthyroidism. The authors proposed that color-flow Doppler sonography could be used to distinguish the two subtypes, confirming an earlier report (184), but that IL-6 measurement was unhelpful.

Management

The treatment of amiodarone-induced hyperthyroidism is difficult. It often does not respond to conventional therapy with carbimazole, methimazole, or radio-iodine. However, corticosteroids and the combination of methimazole with potassium perchlorate have been reported to be effective (185), even if amiodarone is continued (186). Other regimens that have been used include combinations of corticosteroids with carbimazole (187), corticosteroids and benzylthiouracil (188), or propylthiouracil (SEDA-15, 170). Potassium perchlorate has also been used (SEDA-21, 199). Other forms of treatment that have been successful have been plasma exchange and in very severe cases subtotal thyroidectomy (189) or total thyroidectomy (SEDA-15, 170) (SEDA-17, 220) (190).

It has been suggested that potassium perchlorate should be used in the treatment of type 1 hyperthyroidism and glucocorticoids in the treatment of type 2 (SEDA-21, 199). Since hypothyroidism due to amiodarone tends to occur in areas in which there is sufficient iodine in the diet, it has been hypothesized that an iodinated organic inhibitor of hormone synthesis is formed and that the formation of this inhibitor is inhibited by perchlorate to a greater extent than thyroid hormone iodination is inhibited, since the iodinated lipids that are thought to be inhibitors require about 10 times more iodide than the hormone. However, there is a high risk of recurrence after treatment with potassium perchlorate, and it can cause serious adverse effects (SED-13, 1281).

When five patients with type 2 amiodarone-induced hyperthyroidism were treated with a combination of an oral cholecystographic agent (sodium ipodate or sodium iopanoate, which are rich in iodine and potent inhibitors of 5′-deiodinase) plus a thionamide (propylthiouracil or methimazole) after amiodarone withdrawal, all improved substantially within a few days and became euthyroid or hypothyroid in 15–31 weeks (191). Four of the five became hypothyroid and required long-term treatment with levothyroxine.

In another study, three patients with type 1 disease, two of whom had not responded to methimazole plus perchlorate, were successfully treated with a short course of iopanoic acid 1 g/day, resulting in a marked reduction in the peripheral conversion of T4 to T3 (192). Euthyroidism was restored in 7–12 days, allowing uneventful thyroidectomy. The patients were then treated with levothyroxine for hypothyroidism and amiodarone was safely restarted. The authors suggested that iopanoic acid is the drug of choice for rapid restoration of normal thyroid function before thyroidectomy in patients with drug-resistant type 1 amiodarone-induced hyperthyroidism.

However, others have suggested that the differentiation of amiodarone-induced hyperthyroidism into two types is not helpful in determining suitable therapy (193). Of 28 consecutive patients there was spontaneous resolution of hyperthyroidism in 5 and 23 received carbimazole alone as first-line therapy. Long-term euthyroidism was achieved in 11, 5 became hypothyroid and required long-term thyroxine, and 5 relapsed after withdrawal of carbimazole and became euthyroid with either long-term carbimazole ($n = 3$) or radioiodine ($n = 2$). Four were intolerant of carbimazole and received propylthiouracil, with good effect in three. One was resistant to thionamides and responded to corticosteroids. There was no difference in presentation or outcome between those in whom amiodarone was continued or stopped or between possible type 1 or type 2 disease (defined clinically and by serum IL-6 measurement). The authors concluded that continuing amiodarone has no adverse effect on the response to treatment of hyperthyroidism and that first-line therapy with a thionamide alone, whatever the type of disease, is appropriate in iodine-replete areas, thus avoiding potential complications of other drugs. However, it is not clear how good their differentiation of types 1 and 2 disease was. A previous prospective study in 24 patients showed that differentiation predicted response to treatment (194).

It is generally recommended that amiodarone should be withdrawn (195,196) worsening of thyrotoxic symptoms and heart function has been reported after withdrawal of amiodarone. When withdrawal of amiodarone is not an option, near-total thyroidectomy may be preferred. If surgery is not possible plasmapheresis can be helpful.

The use of local anesthesia for total thyroidectomy in patients with amiodarone-induced hyperthyroidism and cardiac impairment has been reviewed in the context of six patients (197).

The management of hyperthyroidism due to amiodarone has been reviewed in the light of the practices of 101 European endocrinologists (198). Most (82%) treat type I amiodarone-induced hyperthyroidism with thionamides, either alone (51%) or in combination with potassium perchlorate (31%); the preferred treatment for type II hyperthyroidism is a glucocorticoid (46%). Some initially treat all cases, before the type has been established, with a combination of thionamides and glucocorticoids. After restoration of normal thyroid function, 34% recommend ablative therapy in type I hyperthyroidism and only 8% in type II. If amiodarone therapy needs to be restarted, 65% recommend prophylactic thyroid ablation in type I hyperthyroidism and 70% recommend a wait-and-see strategy in type II.

Two patients with cardiomyopathy and resistant dysrhythmias developed thyrotoxicosis while taking amiodarone (199). Despite medical therapy, they failed to improve. Both underwent total thyroidectomy without difficulty or complications. Most reported cases of amiodarone-induced thyrotoxicosis that have been treated surgically have been of type II, i.e. with no underlying thyroid disease.

- A 40 year-old patient with severe amiodarone-induced hyperthyroidism after heart transplantation did not respond to high doses of antithyroid drugs combined with glucocorticoids (200). A low dose of lithium carbonate resulted in normalization of thyroid function.

Plasmapheresis, to remove iodine and thyroid hormones, was reportedly successful in treating amiodarone-induced hyperthyroidism in two of three patients, and was followed by thyroidectomy (201). It has been suggested that this would be ineffective in type II hyperthyroidism (202).

Prevention of recurrence of amiodarone-induced hyperthyroidism has been successfully attempted with ^{131}I in 18 patients, in 16 of whom amiodarone was reintroduced (203); the same authors reported the first 15 of these patients in two separate papers (204,205). The problem of whether to restart amiodarone therapy after hyperthyroidism has resolved has been discussed in the light of a case (206).

Hypothyroidism

Amiodarone-induced hypothyroidism has been reviewed in the light of a case of 74-year-old woman (187).

The clinical, biochemical, and therapeutic aspects of amiodarone-induced hypothyroidism have been reviewed in the light of 18 elderly patients (207). Free thyroxine (T4) concentrations were reduced only in those with severe hypothyroidism and free triiodothyronine (T3) concentrations were always normal. Withdrawal of amiodarone in five patients led to improvement in four and worsening in one.

In a nested case-control analysis of 5522 patients with a first prescription for an antidysrhythmic drug and no previous use of thyroid drugs, cases were defined as all patients who had started a thyroid-mimetic or antithyroid drug no sooner than 3 months after the start of an antidysrhythmic drug and controls were patients with a comparable follow-up period who had not taken any thyroid drugs during the observation period (167). There were 123 patients who had started antithyroid drugs and 96 who had started a thyroid-mimetic drug. In users of amiodarone there was an adjusted odds ratios of 6.6 (3.9, 11) for hypothyroidism compared with users of other antidysrhythmic drugs.

The risk of amiodarone-induced hypothyroidism may be greater in patients who have pre-existing thyroid autoimmune disease (208). There is some evidence that the risk of hypothyroidism due to amiodarone is increased in elderly patients (209), but the data are not conclusive.

In amiodarone-induced hypothyroidism the simplest method of treatment is to continue with amiodarone and to add thyroxine as required.

Metabolism

Amiodarone can cause altered serum lipid concentrations (210). Serum cholesterol rises, as can blood glucose and serum triglyceride concentrations. The mechanisms of these effects are not known; nor is it known to what extent they are due to changes in thyroid function.

Hematologic

Although phospholipid inclusion bodies commonly occur in the neutrophils of patients taking amiodarone (211), adverse hematological effects have rarely been attributed to amiodarone. However, there have been reports of thrombocytopenia (212) and of impaired platelet aggregation, associated with gingival bleeding and ecchymoses of the legs (213). Coombs'-positive hemolytic anemia has also been reported (214).

Bone marrow granulomata have rarely been reported in patients taking amiodarone (215).

- A 53-year-old woman developed leukoerythroblastosis with giant thrombocytes in the peripheral blood and was subsequently given amiodarone. The bone marrow became hypocellular with atypical megakaryocytes and several granulomata.
- A 78-year-old woman with a raised erythrocyte sedimentation rate, a mild anemia, and a polyclonal gammopathy on serum immunoelectrophoresis. The bone marrow became hypocellular with atypical megakaryocytes and several granulomata.

In the first case amiodarone was given after the onset of the peripheral blood film abnormalities and the only change in the bone marrow was the occurrence of the granulomata. The authors proposed that the granulomata had occurred because of phospholipid accumulation.

- Bone marrow biopsy in a patient taking amiodarone 100 mg/day showed multiple non-caseating epithelioid granulomata, which resolved 3 months after the withdrawal of amiodarone in a 67-year-old man (216). Similar granulomas were found in a 77-year-old woman who had taken amiodarone 100 mg/day for many years and had thrombocytopenia; the bone marrow contained an increased number of megakaryocytes. Her platelet count normalized 1 month after amiodarone had been withdrawn, and after 3 months there were fewer granulomata in the bone marrow.
- A 76-year-old man, who had taken amiodarone for an unspecified time, developed a monoclonal gammopathy with bone marrow granulomata (217). After another 2 years he developed hepatic granulomata and the amiodarone was withdrawn. The bone marrow granulomata resolved within a few months. Infections were excluded and there was no evidence of sarcoidosis.

Other cases have been reported (218). The mechanism of this effect is unknown.

Liver

Amiodarone often causes rises in the serum activities of aspartate transaminase and lactate dehydrogenase to about twice normal, without changes in alkaline phosphatase or bilirubin, and without clinical evidence of liver dysfunction (219). Changes of this kind were originally reported to be transient and dose-related, returning to normal when the dose was reduced (220).

In 125 patients without clinical liver damage there was a weak correlation between alanine transaminase activity and serum amiodarone concentration (221). An effect compartment model predicted that 6% of patients will have a rise in alanine transaminase activity to more than three times the upper limit of the reference range if serum amiodarone concentrations are maintained at below 2.5 mg/l; concentrations below 1.5.mg/l were associated with no predicted change in alanine transaminase. The authors suggested that amiodarone-induced hepatotoxicity could be efficiently detected by measuring the alanine transaminase activity at baseline, at 1, 3, and 6 months, and every 6 months thereafter.

However, amiodarone can also cause liver damage, which usually takes the form of a hepatitis associated with phospholipid deposition, and there can be changes similar to those of alcoholic hepatitis (222–224). In some cases there is progression of cirrhosis (225). The risk of hepatic impairment in patients taking amiodarone is not known, but relatively severe liver damage can occur even in the absence of symptoms and with only minor associated changes in liver function tests.

- A 40-year-old man who had taken amiodarone 400 mg/day for 6 weeks developed an acute hepatitis accompanied by clusters of light brown granular cells, which were identified as macrophages (226). There were phospholipid inclusions in the macrophages and hepatocytes.

The authors proposed that the granular macrophages represented an early marker of amiodarone-induced hepatotoxicity. Their unusual color was attributed to the deposition of a combination of phospholipid, lipofuscin, and bile breakdown products.

A monitoring strategy has been devised for predicting liver damage in patients taking amiodarone (227). Liver function tests were monitored for over 2 years in 50 patients who were given a loading dose of amiodarone followed by an average maintenance dose of 300 mg/day. Only the serum transaminase activities changed significantly, and alanine transaminase activity showed the greatest discrimination between those taking a low dose of amiodarone and those taking a high dose. The authors proposed the following monitoring strategy:

- establish baseline activities of transaminases, alkaline phosphatase, and lactate dehydrogenase before starting therapy, for future reference;
- measure alanine transaminase activity after 1 month, to rule out hypersusceptibility reactions;
- measure alanine transaminase activity at 3 and 6 months, to determine the maximum response to accumulated amiodarone;
- measure alanine transaminase activity every 6 months thereafter, to screen for late effects;
- further investigation of liver function should only be necessary if the alanine transaminase activity rises to above three times to upper limit of the reference range or if other evidence of liver damage occurs.

Other forms of liver damage that have been reported include a syndrome resembling Reye's syndrome (228). Severe hepatitis at low dosage and thought to be immunologically based has been described (229).

Chronic liver damage with amiodarone is much more common than acute hepatitis, but cholestatic jaundice is one of the relatively rare presentations (SEDA-19, 193) (230,231).

- An 84-year-old woman, who had taken amiodarone 400 mg/day for 4–5 years, developed weakness, fatigue, anorexia, and abnormal liver function tests, with an aspartate transaminase activity of 234 U/l, alanine transaminase 154 U/l, and alkaline phosphatase 316 U/l (232). She had a normal serum bilirubin and the serum concentrations of amiodarone and desethylamiodarone were both within the usual target ranges. Apart from gallstones, endoscopic retrograde cholangiography and abdominal ultrasound showed a normal biliary tree. After withdrawal of amiodarone her liver function tests improved, but 4 months later she developed a rapidly rising serum bilirubin concentration (142 mmol/l), her serum albumin concentration fell to 30 g/l, and her serum cholesterol concentration was high (11 mmol/l). She had bilirubin and urobilinogen in the urine and there was no evidence of viral or immunological hepatitis. Abdominal ultrasonography was normal, as was a liver scan, but a CT scan of the abdomen showed a diffusely hyperdense liver consistent with the effects of amiodarone. Liver biopsy also showed findings consistent with amiodarone-induced cholestatic liver damage, with distorted architecture, portal fibrosis, pericellular sinusoidal fibrosis, and focally irregular lobular sinusoidal fibrosis, but without bridging fibrosis or cirrhosis. There was ductal proliferation and a mild lymphocytic infiltrate but no cholangitis. Electron microscopy showed lamella inclusions compatible with phospholipidosis.

If amiodarone was responsible in this case, it is hard to reconcile the improvement after amiodarone withdrawal with the cholestasis that occurred 4 months later. The patient was also taking felodipine, furosemide, potassium chloride, aspirin (500 mg/day), and cisapride, but the authors argued that the changes were not consistent with liver damage due to any of those drugs. In particular they thought that the felodipine was unlikely to have caused cholestasis, although that has been previously reported, because of the presence of Mallory bodies in the liver biopsy; however, Mallory bodies have previously been reported with felodipine, and although cholestatic liver injury caused by felodipine has not been reported before, that seems as likely a candidate in this case as amiodarone.

Another case has been reported in a 78-year-old man taking 200 mg/day; it resolved after withdrawal (233).

Amiodarone can occasionally cause cirrhosis that mimics alcohol damage, and another case has been reported (234).

- A 79-year-old man who had taken amiodarone 200 mg/day for 33 months developed chronic liver disease. A liver biopsy showed established cirrhosis with extensive fibrosis, with polymorphonuclear leukocyte infiltration, reduplicating bile ducts in nodules, and degenerating hepatocytes. Numerous investigations ruled out other causes of cirrhosis. Liver function deteriorated despite amiodarone withdrawal and he died 3 months later.

However, the authors did not mention intracellular deposition of phospholipids as a feature of this case and the attribution to amiodarone is not clear.

- A 63-year-old man developed ascites after taking amiodarone 200 mg/day for 23 months (235). A liver biopsy showed grade 3 chronic hepatitis and micronodular cirrhosis. The presence of striking microvesicular steatosis on light microscopy and lysosomal inclusion bodies on electron microscopy suggested amiodarone-induced hepatotoxicity.
- An 85-year-old man took amiodarone for 7 years (total dose 528 g) (236). He had hepatomegaly and mild elevations of serum transaminases. Liver biopsy showed cirrhosis, and electron microscopy showed numerous lysosomes with electron-dense, whorled, lamellar inclusions characteristic of a secondary phospholipidosis. Initially, withdrawal of amiodarone led to a slight improvement, but his general condition deteriorated and he died from complications of pneumonia and renal insufficiency.

The rare reports of hepatic cirrhosis attributed to amiodarone have been briefly reviewed (237).

Acute hepatitis with liver failure can occur after intravenous administration and may be due to the solvent, polysorbate (SEDA-14, 149; SEDA-16, 178; SEDA-18, 202; SEDA-22, 206; 238).

- A 69-year-old man was given amiodarone intravenously 1500 mg for multiple coupled ventricular extra beats and 24 hours later developed acute hepatitis, with a 50-fold increase in serum transaminase activities and simultaneous increases in lactate dehydrogenase, gamma-glutamyl transferase, bilirubin, and prothrombin time; there was a moderate leukocytosis and mild renal insufficiency (239). No further amiodarone was given and there was full recovery within 2 weeks. Other causes of acute hepatitis were excluded.

Acute hepatic damage after intravenous amiodarone can be fatal. Three cases of acute hepatocellular injury after intravenous amiodarone in critically ill patients have been described and another 25 published cases and six cases reported to the Swiss Pharmacovigilance Center (Swissmedic) discussed (240). The authors suggested that acute liver damage after intravenous amiodarone may have been caused by the solubilizing excipient polysorbate 80. However, ischemic hepatitis due to hemodynamic changes could not be ruled out (241).

Liver damage can occur very quickly after the start of amiodarone therapy (242).

- A 54-year-old man with 70% burns developed atrial fibrillation and was given intravenous amiodarone 150 mg followed by 1 mg/minute and then 0.5 mg/minute overlapped with oral doses of 400 mg tds (total dose 6.2 g). Liver function tests had been normal, but after 5 days the aspartate transaminase and alanine transaminase activities rose to 739 and 1303 units/l respectively. Amiodarone was withdrawn and the transaminase activities fell immediately.

The mechanism of this effect is not known, but the histology includes Mallory bodies, steatosis, intralobular inflammatory infiltrates, and fibrosis; electron microscopy suggests phospholipidosis.

It has been suggested that liver injury due to amiodarone is either due to a direct biochemical action or perhaps metabolic idiosyncrasy. Because there have been cases in which oral administration has not led to a recurrence, it has also been suggested that the vehicle in which amiodarone is usually dissolved, polysorbate (Tween) 80, is responsible rather than the amiodarone itself (SEDA-18, 202).

Pancreas

There have been two reported cases of pancreatitis and in one the patient died of progressive liver failure (243,244). Whether this was a direct effect of amiodarone is unclear.

Urinary tract

Increases in serum creatinine concentrations, correlated with serum amiodarone concentrations, have been reported (245).

Skin

Amiodarone commonly causes phototoxicity reactions (246,247). The risk of phototoxicity increases with the duration of the exposure. Window glass and sun screens do not give protection, although zinc or titanium oxide formulations and narrow band UVB photo therapy can help (248–250). For most patients this adverse effect will be no more than a nuisance, and the benefit of therapy may be worthwhile. However, in a few cases treatment may have to be withdrawn. Histological examination of skin biopsies shows intracytoplasmic inclusions of phospholipids (251). There has been a single report of a severe case of photosensitivity in conjunction with a syndrome resembling porphyria cutanea tarda, resulting in bullous lesions (252).

Amiodarone can cause a cosmetically annoying bluish pigmentation of the skin (SEDA-15, 171; SEDA-22, 207; SEDA-23, 199; 253).

- A 76-year-old woman who had taken amiodarone 200 mg/day for 4 years developed blue-gray discoloration of the skin of the face resembling cyanosis; amiodarone was withdrawn and substantial improvement occurred within 4 months (254)

- A 54-year-old man who took the drug for 1 year developed bluish-gray discoloration of the face (255). The discoloration almost completely resolved within 9 months of withdrawal.
- A 55-year-old woman who had taken amiodarone 250 mg/day for about 10 years developed bluish-gray discoloration of the face (256). The pigmentation responded to treatment with a Q-switched ruby laser at an energy of 8 J/cm^2 and a wave length of 694 mm.

The authors suggested that the ruby laser had damaged pigment-containing cells. However, the amiodarone was continued, so presumably the laser also destroyed lipofuscin in situ.

- A 69-year-old white man who had taken amiodarone 400 mg/day for 3 years developed blue–gray discoloration of the face and other exposed areas (257). Areas that had been protected from the sun (the forehead by a broad-brimmed hat and the skin under his wrist watch) were not affected.
- A 70-year-old man developed grey–blue pigmentation on sun-exposed areas of his skin after taking amiodarone for 10 years and minocycline for 4 years (258).

Minocycline can also cause skin pigmentation and it is not clear in this last case what the interaction of amiodarone with minocycline was.

Phototoxicity affecting the peri-oral skin has been reported in a 70-year-old white man who was taking amiodarone 100 mg/day, but also losartan 50 mg/day, co-amilofruse, aspirin 150 mg/day, and diclofenac 50 mg/day; the rash resolved after withdrawal of amiodarone (259).

Grey–blue discoloration of the skin during amiodarone therapy has been presumed to be due to lipofucsin deposition. However, it has also been suggested that amiodarone may block the maturation of melanosomes, in view of a case of discoloration associated with a reduced number of mature melanosomes and an increased number of premelanosomes in sun-exposed areas of the skin, but normal numbers in non-exposed areas (260).

It has been suggested that the skin and mucosal toxicity of amiodarone may be enhanced by radiotherapy (SEDA-16, 178; SEDA-17, 221). However, in a retrospective review of 10 patients who took amiodarone when having external beam radiation therapy there were no unexpected acute sequelae (261).

Other skin reactions that have been reported include iododerma (262), erythema nodosum, psoriasis (263), and exfoliative dermatitis (264).

Amiodarone can increase the risk of mucosal and skin toxicity due to radiotherapy and rarely causes hair loss (SEDA-17, 221) and vasculitis (265).

The severe form of erythema multiforme known as toxic epidermal necrolysis has rarely been attributed to amiodarone (266).

- A 71-year-old woman, who had taken amiodarone 200 mg/day for 3 months and diltiazem for 8 months, developed extensive erythema, blistering, and erosions affecting 50% of the body surface area, with a maculopapular rash on the limbs (267). She developed bilateral pneumonia and septicemia and died after 7 days.

The adverse effects of amiodarone on the skin usually resolve within 2 years of drug withdrawal. However, in a 67-year-old woman who developed both phototoxicity and a slate-grey discoloration while taking amiodarone, the dyspigmentation gradually resolved after withdrawal but the phototoxicity persisted for more than 17 years (268). The authors offered four possible explanations: continuing phototoxicity due to persistence of amiodarone or its metabolites in the skin, which seems unlikely; persistent postinflammatory cutaneous hypersensitivity; transfer to a photoallergic mechanism; conversion to light-exacerbated seborrheic dermatitis.

Musculoskeletal

A proximal myopathy has occasionally been reported in patients taking amiodarone (92) and there has been a report of an acute necrotizing myopathy (269).

Two patients with atrial fibrillation had acute devastating low back pain a few minutes after the start of intravenous loading with amiodarone (270).

Sexual function

Epididymitis has been reported in patients taking high dosages of amiodarone, resolving with dosage reduction or withdrawal (SEDA-18, 203).

Reproductive system

Non-infective epididymitis has occasionally been reported in patients taking amiodarone (SEDA-10, 148) (SEDA-18, 203).

- A 25-year-old man who had taken amiodarone 200 mg bd for 1 year developed epididymitis, which resolved within 3 months of withdrawal of amiodarone and recurred within 2 months of its reintroduction (271).

This case was unusual in that it involved both testes. The mechanism of this effect is not known, but it has been reported to be dose-dependent, although anti-amiodarone antibodies have also been reported (272). The incidence is not known, but has been reported to be as high as 11%. The author of one very brief report (273) claimed to have seen 20 cases of epididymitis, some of which were bilateral, since the late 1980s. He claimed that withdrawal produced dramatic resolution of symptoms within 10–20 days, and that amiodarone in a dose of 200 mg/day did not usually cause symptoms, even in patients who had had epididymitis at higher dosages.

Immunologic

Lupus-like syndrome has rarely been attributed to amiodarone.

- A 71-year-old woman, who had been taking amiodarone 200 mg bd for 2 years, developed malaise, intermittent fever, arthralgia, and weight loss (274). She had a malar rash and hypoventilation at both lung bases. Her erythrocyte sedimentation rate was markedly raised (90 mm/hour), there was a mild normochromic normocytic anemia (10 g/dl), a slight lymphopenia, and otherwise normal routine tests. Her rheumatoid factor was raised in a titer of 1:320, and circulating complexes of IgG-C1q were positive. Antinuclear antibody was positive (1:640), but all other antibodies were negative. There was progressive improvement on withdrawal of amiodarone and all the biochemical tests returned to normal.

- A 59-year-old man, who had taken amiodarone 200 mg/day for 2 years, developed fever, pleuritic chest pain, dyspnea at rest, a non-productive cough, malaise, and joint pains (275). He had a verrucous endocarditis and a pleuropericardial effusion. He had raised titers of antinuclear antibodies (1:320) with anti-Ro specificity. Serum complement was normal and there were no circulating immune complexes, no cryoglobulins, and no anti-dsDNA, anti-La, anti-U1 ribonucleoprotein, anti-Sm, anti-Scl, 70, anti-Jo 1, antihistone, antiphospholipid, anticentromere, anticardiolipin, or anticytoplasmic antibodies. Within 7 days of withdrawal of amiodarone the signs and symptoms started to resolve, and he recovered fully with the addition of prednisolone.

Angioedema has been reported in a 70-year-old woman who had taken amiodarone 200 mg/day for 8 years (276). The amiodarone was withdrawn and the symptoms disappeared. Rechallenge produced facial flush and facial angioedema within 20 minutes of a 200 mg dose.

Two other cases have been reported (277).

Oral formulations of amiodarone contain iodine, about 10% of which is released into the circulatory system and may increase the risks of hypersensitivity reactions in iodine-sensitive patients. Reactions to iodinated contrast media are usually due to their high osmolar or ionic content. Although amiodarone has a high content of iodine, there is no known association between amiodarone and reactions to contrast media. Three patients who were allergic to iodine were given amiodarone for chemical cardioversion of dysrhythmias (278). There were no anaphylactic or anaphylactoid reactions. However, the authors suggested that in patients with true iodine hypersensitivity there is a possibility of such reactions.

Death

The effect of intravenous and oral amiodarone on morbidity and mortality has been studied in 1073 patients during the first hours after the onset of acute myocardial infarction (279). The patients were randomized to receive amiodarone or placebo for 6 months. The interim analysis showed an increased mortality, albeit not significant, with high-dose amiodarone (16 versus 10%) and the dose was therefore reduced from 400 to 200 mg/day. Low-dose amiodarone was associated with a reduced death rate (6.6 versus 9.9%). There were non-fatal adverse events in 108 patients taking amiodarone and 73 taking placebo. The only non-fatal adverse effect that occurred significantly more often with amiodarone was hypotension during the initial intravenous loading phase, a well-known effect. In the context of this study, it should be remembered that in several previous studies amiodarone has been shown to reduce mortality after myocardial infarction (SEDA-23, 198; SEDA-24, 206; 280).

In a retrospective study of 20 patients who had taken amiodarone within the 3 months before heart transplantation, survival was lower than in 65 patients who had not taken amiodarone (50 versus 85% at 1 year) (281). The patients who had taken amiodarone also had a greater risk of adult respiratory distress syndrome (ARDS) after transplant and had more bleeding complications. The risks were greater the longer the duration of amiodarone use before transplantation.

Long-Term Effects

Tumorigenicity

Basal cell carcinoma has been rarely reported in patients taking amiodarone (282,283), and another case has been reported (284). The rareness of the reports and the commonness of the tumor make this association hard to substantiate.

Second-Generation Effects

Teratogenicity

There are no major teratogenic effects of amiodarone (285).

Fetotoxicity

Exposure to amiodarone in utero has only occasionally been described.

Cardiovascular

There have been reports of sinus bradycardia (SEDA-8, 179; SEDA-20, 176; 286,287) and prolongation of the QT interval (285).

Nervous system

Congenital nystagmus has been described (SEDA-20, 176).

Psychological

In one child there was evidence of mental delay, hypotonia, hypertelorism, and micrognathia (SEDA-20, 176), although the authors thought that the link between amiodarone and neurotoxicity was speculative. However, there has also been a retrospective study of 10 children who were exposed to amiodarone during pregnancy, compared with matched controls (288). There was no change in IQ score, but the children who had been exposed to amiodarone had impaired expressive language skills and one child had global developmental delay. However, most of the mothers were not concerned about their children's development, and so any effect of amiodarone on neurological development was probably small. One child had transient neonatal hypothyroidism, which responded to a short course of thyroxine; another had mild transient neonatal hyperthyroidism; in neither case was there any difference in development from the other children who had been exposed to amiodarone. One child was born with a congenital jerk nystagmus and had relatively poor reading and comprehension skills for both words and passages, low scores on several of the verbal subtests of the WISC-R test for information, arithmetic, and vocabulary, and below average spelling.

Endocrine

The major adverse effect on the fetus is altered thyroid function (SEDA-13, 141; SEDA-14, 149; SEDA-19, 194; SEDA-20, 176). There have been individual reports of neonatal hyperthyroxinemia (289), goiter (286), and hypothyroidism (290). In the patient with goiter there was associated hypotonia, bradycardia, large fontanelles, and macroglossia (286).

Two neonates who had been given intravenous amiodarone as fetuses at 26 and 29 weeks and whose mothers had also taken it orally developed hypothyroidism (291). The authors suggested that low dietary iodine intake by the mothers may have contributed, by enhancing the Wolff–Chaikoff effect.

Transient hypothyroidism has been reported in five infants born to 26 mothers who were given amiodarone for treatment of fetal tachycardias (292). Two similar cases have been reported elsewhere (293).

Susceptibility Factors

Age

The safety and efficacy of amiodarone for supraventricular tachycardia have been studied in 50 infants (mean age 1.0 month, 35 boys) (294). They had congenital heart disease (24%), congestive heart failure (36%), or ventricular dysfunction (44%). Six, who were critically ill, received a loading dose of intravenous amiodarone 5 mg/kg over 1 hour, and all took 20 mg/kg/day orally for 7–10 days, followed by 100 mg/day; if this failed to control the dysrhythmia, oral propranolol (2 mg/kg/day) was added. Follow-up was for an average of 16 months. Rhythm control was achieved in all patients. Growth and development were normal. The higher dose of amiodarone was associated with an increase in the QT_c interval to over 0.44 seconds, but there were no dysrhythmias. Two infants had hypotension during intravenous loading, as has previously been reported in infants (SEDA-19, 194). Aspartate transaminase and alanine transaminase activities and thyrotropin (TSH) concentrations all increased, but remained within their reference ranges. There were no adverse effects that necessitated drug withdrawal.

Young patients are more likely to develop adverse effects in the skin (295).

In children under 10 years of age the risk of adverse effects is less than in adults (296,297). It is not clear whether older children are at greater or lesser risks of adverse effects than are adults.

The safety of antidysrhythmic drugs in children has not been thoroughly studied. However, the risk of prolongation of the QT interval seems to be considerably less than that in adults (298), although it has been reported with quinidine, disopyramide, amiodarone, sotalol, and diphemanil.

Other features of the patient

Amiodarone-induced liver damage is more common in those with reduced left ventricular function (295).

Anesthesia and surgery

It has been reported that there is an increased risk of adverse reactions to amiodarone in patients undergoing anesthesia (SEDA-15, 171). However, in a retrospective survey of 12 patients who underwent anesthesia for urgent thyroidectomy due to amiodarone there were no anesthetic complications or deaths (299).

There is an increased risk of some of the adverse effects of amiodarone (including dysfunction of the liver and lungs) in patients who have had or who are having surgery (300). In addition the perioperative mortality in these patients is higher than in controls (301). The factors that increase the risks of amiodarone-associated adverse cardiovascular effects during surgery (302) include pre-existing ventricular dysfunction, too rapid a rate of intravenous infusion, hypocalcemia, and an interaction between amiodarone and both the general anesthetics used and other drugs with negative inotropic or chronotropic effects. It has therefore been recommended (302) that serum concentrations of calcium, amiodarone, and digoxin should be within the reference or target ranges before operation, and that other drugs with negative inotropic or chronotropic effects should be withdrawn before surgery.

The use of amiodarone in the prevention of atrial fibrillation after cardiac surgery has been reviewed (303). When an intravenous loading dose of amiodarone was used, bradycardia was a common adverse effect but was rarely severe enough to warrant withdrawal. When only oral amiodarone was used there were no serious adverse reactions.

Lung disease

In the Atrial Fibrillation Follow-up Investigation of Rhythm Management (AFFIRM) study, pre-existing lung disease was present in 591 of 4060 patients and was associated with a higher risk of pulmonary death and a higher risk of amiodarone-associated pulmonary toxicity (304). At 4 years amiodarone-induced pulmonary toxicity had occurred in 52 of 1468 patients (3.5%) and was present in more patients with pre-existing pulmonary disease (14 of 238 patients, 5.9%) than in those without (38 of 1230 patients, 3.1%). However, the use of amiodarone in the presence of pre-existing pulmonary disease did not increase the rates of pulmonary deaths or all-cause mortality. The authors concluded that cautious use of amiodarone to treat atrial fibrillation is acceptable in elderly patients, even if there is pre-existing pulmonary disease.

Drug Administration

Drug formulations

The steady-state plasma concentrations of amiodarone and desethylamiodarone in 77 patients taking two different formulations, a new generic formulation and Cordarone, were comparable (305).

Drug administration route

In contrast to its effects during oral administration, the therapeutic and short-term unwanted effects of amiodarone during intravenous administration arise within minutes or hours (2). The reason for this is not clear; plasma concentrations after single oral and intravenous doses of 400 mg are very similar (2), but that does not rule out a pharmacokinetic explanation for the paradox. The possible role of the solvent used in the intravenous formulation, polysorbate (Tween) 80, has not been fully elucidated. Amiodarone certainly has different electrophysiological effects when it is given intravenously. For example, intravenous amiodarone prolongs the AH interval, while oral amiodarone prolongs atrial and ventricular refractory periods and the HV interval (306). Furthermore, the blocking effects of amiodarone on sodium and calcium channels and its beta-adrenoceptor blocking action occur earlier than its Class III action (307). An anecdotal report of torsade de pointes after both intravenous and oral administration of amiodarone on different occasions has underlined this difference.

- A 70-year-old woman with dilated cardiomyopathy, ventricular tachydysrhythmia, and a QT_c interval of 0.49 seconds was given intravenous amiodarone 240 mg over 15 minutes, and 30 minutes later developed a junctional escape rhythm (48/minute) with QT_c prolongation to 0.68 seconds; 8 hours later she developed torsade de pointes (308). A few years later she was given oral amiodarone 100 mg/day and 7 weeks later presented with congestive heart failure. Her QT_c interval was prolonged (0.50 seconds) and increased further to 0.64 seconds after the addition of dopamine 3 micrograms/kg/minute; torsade de pointes again developed. Amiodarone was withdrawn and the QT_c interval shortened, but she continued to have recurrent episodes of sustained ventricular tachycardia.

The authors suggested that torsade de pointes induced by intravenous amiodarone depended on heart rate during a bout of bradycardia, while that after oral amiodarone depended on increased sympathetic nervous system activity, and that therefore different electrophysiological mechanisms had been at play. However, it is by no means clear from their description of this case that that was so. They did not report plasma concentrations of amiodarone or desethylamiodarone, its active metabolite.

After rapid intravenous administration hypotension, shock, and atrioventricular block can occur and can be fatal (2). The rate of infusion should not exceed 5 mg/minute. Other adverse effects reported during intravenous infusion include sinus bradycardia (309), facial flushing, and thrombophlebitis (309–312). The risk of this last complication can be reduced by infusing the drug into as large a vein as possible and preferably via a central venous catheter, or perhaps by using a very dilute solution of the drug (313).

The use of intravenous amiodarone for atrial fibrillation has been reviewed (314). The most commonly reported adverse effects in all studies have been hypotension and bradycardia. Other effects include worsening of heart failure, thrombophlebitis at the site of infusion, non-sustained ventricular tachycardia, facial rash, and nightmares.

A few studies have also reported the effects and adverse effects of intravenous amiodarone in patients with atrial fibrillation. Of 67 patients with atrial fibrillation, of whom 33 received amiodarone and 34 received placebo, conversion to sinus rhythm occurred in 16 of the patients who received amiodarone and in none of those who received placebo (315). In five patients the systolic blood pressure fell significantly during the first trial of intravenous drug administration. There were no cardiac dysrhythmias. Thrombophlebitis occurred in 12 patients who received amiodarone.

In a randomized, placebo-controlled trial of 100 patients with paroxysmal atrial fibrillation, intravenous amiodarone 125 mg/hour was compared with placebo (316). There were no serious adverse effects; five patients given amiodarone developed significant sinus bradycardia, in all cases after conversion to sinus rhythm. In this series there were no significant episodes of hypotension. Thrombophlebitis occurred in eight patients who received amiodarone.

Drug overdose

The features of amiodarone overdose and its management have been reviewed (317).

- There has been a report of acute self-poisoning with 8 g of amiodarone orally (318). Initially the only abnormal physical sign was profuse sweating; the electrocardiogram showed sinus rhythm with a normal QT_c interval and the blood pressure was normal. No active measures were taken. The QT_c interval subsequently lengthened on the third and fourth days after overdosage, and there was sinus bradycardia between the second and fifth days. Over 3 months of follow-up there were no effects on thyroid or liver function and no evidence of lung, skin, or corneal involvement.

Drug–Drug Interactions

Anesthetics

The risks of cardiovascular adverse effects in patients undergoing surgery may be partly related to an interaction of amiodarone with anesthetics, either directly or via some interaction with the catecholamines that are released during anesthesia (319). The risk of hypotension during cardiopulmonary bypass in patients taking amiodarone can be increased by the concurrent administration of an ACE inhibitor. There is a high incidence of lung complications when patients treated with amiodarone are ventilated with 100% oxygen, including acute adult respiratory distress syndrome.

Antimony and antimonials

A pharmacodynamic interaction has been described between amiodarone and meglumine antimoniate, both of which prolong the QT interval; the interaction resulted in torsade de pointes (320).

- A 73-year-old man with visceral leishmaniasis was given meglumine antimoniate intramuscularly 75 mg/kg/day. At that time his QT_c interval was normal at 0.42 seconds. Three weeks later his QT_c interval was prolonged to 0.64 seconds and he was given metildigoxin 0.4 mg and amiodarone 450 mg intravenously over 8.25 hours; 12 hours later he had a cardiac arrest with torsade de pointes, which was cardioverted by two direct shocks of 300 J and lidocaine 100 mg in two bolus injections. Because he had frequent episodes of paroxysmal atrial fibrillation, he was given amiodarone 100 mg over the next 40 hours, and developed recurrent self-limiting episodes of torsade de pointes associated with QT_c interval prolongation, which responded to intravenous magnesium 1500 mg. After withdrawal of amiodarone there was no recurrence and a week later the QT_c interval was 0.48 seconds. The plasma potassium concentration was not abnormal in this case.

In view of this report it is probably wise to avoid co-administration of antimonials and amiodarone.

Beta-blockers

The combination of amiodarone with beta-adrenoceptor antagonists can be beneficial in the treatment of refractory ventricular tachycardia, especially when low doses of the beta-blockers are used. However, there have also been reports of adverse interactions in these circumstances, and various correspondents have commented on the possibility that beta-blockers may enhance the effects of amiodarone in reducing mortality in patients who have had a myocardial infarction or are in heart failure (250,321,322).

In a systematic review, the beneficial interaction has been confirmed (323). Four groups of patients who had been studied in EMIAT and CAMIAT (SEDA-21, 198) were defined: patients who had taken amiodarone plus beta-blockers, patients who had taken beta-blockers or amiodarone alone, and patients who had taken neither. The relative risks for all-cause mortality and all forms of cardiac death or resuscitated cardiac arrest were lower in the patients who had taken amiodarone plus beta-blockers than in the other three groups. The results of this post hoc analysis should be regarded with caution, but in view of previous similar reports they are suggestive of a beneficial interaction of amiodarone with beta-blockers in patients who have had a previous myocardial infarction. The interaction was statistically significant for cardiac deaths and for dysrhythmic deaths or resuscitated cardiac arrest. In all other cases the relative risk was reduced, although not significantly. The risk was not affected by heart rate. This interaction has been reviewed (324).

Budesonide

The corticosteroid budesonide undergoes a high degree of first-pass elimination in the liver after oral administration, and therefore causes few systemic adverse effects. It was therefore surprising that Cushing's syndrome occurred in an 81-year-old man taking oral budesonide 9 mg/day and amiodarone 100 mg/day (325). When amiodarone was withdrawn the clinical effects of Cushing's syndrome disappeared. The authors suggested that amiodarone had inhibited the metabolism of budesonide by hepatic CYP3A.

Cardiac glycosides

Amiodarone inhibits the renal tubular secretion of digoxin (SEDA-22, 201) and it has also been suggested that it increases its absorption (SEDA-10, 144; SEDA-12, 150). This interaction has also been reported with acetyldigoxin (SEDA-18, 198; 326).

The time courses of the interactions of amiodarone with digoxin and warfarin have been compared (327). In 79 patients who had been taking fixed maintenance doses of warfarin (n = 77) and/or digoxin (n = 54), amiodarone reduced the clearance of S-warfarin within about the first 2 weeks of co-administration after which the interaction stabilized; there was only a small reduction in the clearance of R-warfarin. In contrast, the clearance of digoxin fell gradually with time, and did not become significantly reduced until about 6 weeks, during which time amiodarone and desethylamiodarone concentrations rose towards steady state; there was a good inverse correlation between amiodarone and desethylamiodarone concentrations and digoxin clearance. The authors concluded that relatively short-term monitoring of the effect of warfarin is required when amiodarone is co-administered, compared with long-term monitoring of digoxin.

Amiodarone also interacts with digitoxin (328). In two cases the half-life of digitoxin was prolonged, but there was no other information that suggested a mechanism. The author suggested that amiodarone might displace digitoxin from tissue sites, but that would have led to a shortening of the half-life rather than a prolongation. It seems more likely that amiodarone inhibits the clearance of digitoxin by inhibiting renal and gut P glycoprotein and perhaps by inhibiting its metabolism.

Ciclosporin

Amiodarone can increase the blood concentrations of ciclosporin and thus impair renal function.

- A 66-year-old developed a ventricular tachycardia after kidney transplantation and was given amiodarone. Maintenance immunosuppression included prednisone, azathioprine, and ciclosporin. Ciclosporin concentrations before amiodarone initiation were stable (range 100–150 ng/ml). During amiodarone therapy, the ciclosporin concentration increased more than two-fold.

The authors proposed that changes in protein binding or metabolism might have explained this interaction.

Class I antidysrhythmic drugs

Amiodarone can potentiate the dysrhythmogenic actions of some Class I antidysrhythmic drugs (SEDA-18, 203), particularly because of the risk of QT interval prolongation (SEDA-19, 194).

Cyclophosphamide

Both amiodarone and cyclophosphamide can cause lung damage.

- Interstitial pneumonitis has been reported in a 59-year-old man, who had taken amiodarone for 18 months, 18 days after a single dose of cyclophosphamide; 1 year before he had also received six cycles of chemotherapy containing cyclophosphamide, vincristine, and prednisone, followed by four cycles of cisplatin, cytarabine, and dexamethasone (329).

The authors suggested that the lung damage had been due to the cyclophosphamide, enhanced by the presence of amiodarone, but in view of the fact that previous similar exposure on six occasions had not resulted in the same effect, it is perhaps more likely that this was a long-term adverse effect of amiodarone alone. The presence of foamy histiocytes in the lung biopsy was consistent with this interpretation (SEDA-15, 168). It is true, however, that lung damage due to amiodarone is usually of a more insidious onset than was reported in this case, although a more rapid onset can occur in patients who are given high concentrations of inspired oxygen. On the other hand, lung damage has occasionally been reported to occur rapidly (94).

Diltiazem

Sinus arrest with hypotension has been reported in a patient with a congestive cardiomyopathy when diltiazem was added to amiodarone (330).

Flecainide

The combination of flecainide with amiodarone can result in reduced conduction, predisposing to bundle branch block and dysrhythmias (331,332).

Indinavir

Indinavir inhibits CYP3A4, which is responsible for the de-ethylation of amiodarone to desethylamiodarone.

- A 38-year-old man, who had taken amiodarone 200 mg/day for more than 6 months, was given postexposure prophylaxis for HIV infection after a needle injury; this included zidovudine, lamivudine, and indinavir (333). During the 4 weeks of therapy his serum amiodarone concentration rose from 0.9 to 1.3 mg/l, with only a small rise in the serum concentration of desethylamiodarone from 0.4 to 0.5 mg/l. After withdrawal of the prophylactic therapy the plasma amiodarone concentration gradually fell to the pretreatment value, and there was no further change in the concentration of desethylamiodarone.

No adverse effects of this interaction were reported.

Loratadine

In therapeutic doses loratadine does not prolong the QT interval, but it can do so if its metabolism is inhibited.

- A 73-year-old woman with hypertension and hyperlipidemia, who was taking amiodarone, cilazapril, pravastatin, and warfarin, was given loratadine 10 mg/day for an allergic reaction (334). She had a bout of syncope in association with a QT_c interval of 688 ms. Rhythm monitoring showed episodes of long-short QT cycles preceded by short self-terminating bouts of torsade de pointes. Amiodarone and loratadine were withdrawn and over the next 4 days the QT interval returned to the reference range and she became asymptomatic.

The authors attributed QT interval prolongation in this case to a toxic effect of loratadine after inhibition by amiodarone of its metabolism by CYP3A4. They did not place much emphasis on a possible pharmacodynamic interaction between the two drugs, although that could also have contributed.

Meglumine antimoniate

A pharmacodynamic interaction has been described between amiodarone and meglumine antimoniate, both of which prolong the QT interval; the interaction resulted in torsade de pointes (335).

- A 73-year-old man with visceral leishmaniasis was given meglumine antimoniate intramuscularly 75 mg/kg/day. At that time his QT_c interval was normal at 0.42 seconds. Three weeks later his QT_c interval was prolonged to 0.64 seconds and he was given methyldigoxin 0.4 mg and amiodarone 450 mg intravenously over 8.25 hours; 12 hours later he had a cardiac arrest with torsade de pointes, which was cardioverted by two direct shocks of 300 J and lidocaine 100 mg in two bolus injections. Because he had frequent episodes of paroxysmal atrial fibrillation, he was given amiodarone 100 mg over the next 40 hours, and developed recurrent self-limiting episodes of torsade de pointes associated with QT_c interval prolongation, which responded to intravenous magnesium 1500 mg. After withdrawal of amiodarone there was no recurrence and a week later the QT_c interval was 0.48 seconds. The plasma potassium concentration was not abnormal in this case.

Metoprolol

Amiodarone increased mean metoprolol plasma concentrations twofold after a loading dose of 1.2 g/day for 6 days in 10 patients (336). The extent of the effect depended on the CYP2D6 genotype.

Metronidazole

Amiodarone toxicity has been attributed to impaired metabolism by metronidazole (337).

- A 71-year-old Caucasian woman was given metronidazole 500 mg tds for antibiotic-associated pseudomembranous colitis. The QT_c interval was 440 ms. After 3 days she developed atrial fibrillation and was given intravenous amiodarone as a 450 mg bolus followed by 900 mg/day. Conversion to sinus rhythm occurred 2 days later and the QT_c interval was prolonged to 625 ms. Later she developed sustained polymorphic torsade de pointes. Metronidazole and amiodarone were immediately withdrawn, and over the next 6 days the QT_c interval gradually normalized.

Mexiletine

In view of this report it is probably wise to avoid co-administration of these two drugs.

In 26 patients taking mexiletine plus amiodarone for 1 month and 155 taking mexiletine alone, there was no significant difference in the apparent oral clearance of mexiletine (338). However, the lack of a pharmacokinetic interaction does not reduce the risk that dangerous QT interval prolongation may occur with a combination such as this.

Orlistat

One might expect the absorption of lipophilic drugs to be reduced by the lipase inhibitor orlistat. In a double-blind, placebo-controlled, randomized study in 32 healthy volunteers aged 18–65 years, body mass index 18–30 kg/m^2, orlistat significantly reduced the C_{max} and AUC of amiodarone by about 25%; the C_{max} and AUC of desethylamiodarone were also significantly reduced (339). However, orlistat did not affect the t_{max} or half-life of amiodarone. These results suggest that orlistat reduces the extent of absorption of amiodarone but not its rate of absorption. In parallel studies orlistat did not affect the pharmacokinetics of fluoxetine or simvastatin.

Phenazone (antipyrine)

The clearance of phenazone (antipyrine) is reduced by amiodarone (340).

Phenytoin

Amiodarone increases the plasma concentrations of phenytoin, probably by inhibiting its metabolism (341,342), while phenytoin increases the metabolism of amiodarone and perhaps also of its metabolite desethylamiodarone (343).

Procainamide

The effects of procainamide on the QT interval may be potentiated by other drugs with this action, for example amiodarone (344).

The pharmacokinetics of procainamide are altered by amiodarone, with a reduction in clearance of about 25% due to changes in both renal and non-renal clearances (345).

Quinidine

Quinidine prolongs the QT interval and will therefore potentiate the effects of other antidysrhythmic drugs that have the same effect (for example amiodarone) (345).

Rifampicin

- In a 33-year-old woman taking amiodarone 400 mg/day the addition of rifampicin 600 mg/day resulted in paroxysms of atrial fibrillation and atrial flutter, with a very low serum amiodarone concentration and an undetectable concentration of desethylamiodarone (346). This was attributed to induction of the metabolism of amiodarone and desethylamiodarone, and after withdrawal of rifampicin the concentrations of the two compounds rose to within the target ranges.

This interaction has been demonstrated in human liver microsomes (347).

Rifampicin has been reported to reduce the effects of amiodarone (348).

- A 33-year-old woman was given rifampicin to suppress an MRSA infection of a pacing system that could not be removed. She was already taking amiodarone which, with the pacing system, was intended to manage her complex dysrhythmias. The introduction of antibiotic therapy was followed by an increase in bouts of palpitation and in shocks from her defibrillator. Her amiodarone concentrations had fallen and returned to the target range, with disappearance of her symptoms, when the rifampicin was withdrawn.

The authors discussed the possible reasons for this interaction, including a reduction in systemic availability of amiodarone or induction of metabolism by rifampicin.

Simvastatin

Rhabdomyolysis has been reported in a 63-year-old man who was taking amiodarone 1 g/day and simvastatin 40 mg/day (349). The authors hypothesized that amiodarone had inhibited the metabolism of simvastatin via CYP3A4 but did not have plasma concentration measurements to back up their assumption.

Trazodone

A patient taking both trazodone and amiodarone developed prolongation of the QT interval and a polymorphous ventricular tachycardia, perhaps by mutual potentiation (350).

Warfarin

That amiodarone can potentiate the action of warfarin by inhibiting its metabolism is well known (SEDA-11, 156). However, potentiation of the action of warfarin has been attributed to amiodarone-induced thyrotoxicosis (344). A metabolic interaction in this case was unlikely, because the patient had taken both drugs together for 2 years before the increase in response to warfarin, coincident with the emergence of thyrotoxicosis.

These interactions have been briefly reviewed in the context of three cases (351).

In a study of this interaction in 43 patients who took both amiodarone and warfarin for at least 1 year, the interaction peaked at 7 weeks and the mean dosage of warfarin fell by 44% from 5.2 to 2.9 mg/day (352). The dosage of warfarin correlated inversely with the maintenance dose of amiodarone. There were minor bleeding episodes in five patients. The authors recommended reducing the daily warfarin dose by about 25, 30, 35, and 40% in patients taking amiodarone 100, 200, 300, and 400 mg/day respectively.

Interference with Diagnostic Tests

Serum creatinine

Amiodarone has been reported to cause a small and reversible increase in serum creatinine concentration (353). It is not clear whether this effect is due to true renal impairment, or to some effect on either the kinetics of creatinine or its measurement in the blood.

Thyroid function tests

Amiodarone causes altered thyroid function tests, with rises in serum concentrations of T4 and reverse T3 and a fall in serum T3 concentration (354). This is due to inhibition of the peripheral conversion of T4 to T3, causing preferential conversion to reverse T3. These changes can occur in the absence of symptomatic abnormalities of thyroid function.

Blood urea nitrogen

Amiodarone has also been reported to cause a small and reversible increase in blood urea nitrogen concentration (353). It is not clear whether this effect is due to true renal impairment, or to some effect on either the kinetics of urea or its measurement in the blood.

Double potential interval

Ablation of the cavotricuspid isthmus generates a corridor of double potentials along the ablation line, and the double potential interval is shortened by isoprenaline. However, in 32 patients amiodarone prolonged the double potential interval, in both the presence and absence of isoprenaline, and this effect should be taken into account when assessing the completeness of this ablation procedure (354).

Diagnosis of Adverse Drug Reactions

The differences in the rates of onset of effects of amiodarone after oral and intravenous administration in the face of similar plasma concentrations suggest that there can be no simple relation between the plasma concentrations of amiodarone and its therapeutic effects. Matters are further complicated by metabolism of amiodarone to desethylamiodarone, which has pharmacological activity. However, evidence (2,355) suggests that a plasma amiodarone concentration of around 1.0–2.5 mg/ml is associated with a high likelihood of therapeutic efficacy in patients with dysrhythmias. However, adverse effects can still occur when the plasma concentration is within this range, and there is no clear limit to the concentration above which toxicity starts to become important. Similarly, the therapeutic range of concentrations for desethylamiodarone is unclear, although it has been suggested to be around 0.5–1.0 mg/ml (SEDA-10, 146).

In one careful study EC_{50} values for certain effects of amiodarone were calculated (356). The respective concentrations of amiodarone and desethylamiodarone that were associated with effects were as follows: reduction in heart rate 1.2 and 0.5 mg/ml; QT_c prolongation 2.6 and 1.4 mg/ml; corneal microdeposits 2.2 and 1.1 mg/ml.

Because amiodarone prolongs the QT_c interval, it has been suggested that this might be a useful measure of its efficacy. The percentage prolongation of the QT_c interval correlates well with both daily dose and the plasma and myocardial concentrations of amiodarone (357), although this is not a universal finding during long-term administration (310,358), and the QT_c interval is not prolonged after short-term intravenous use (2).

Since amiodarone inhibits the peripheral conversion of thyroxine (T4) to tri-iodothyronine (T3), there is an increase in serum concentrations of reverse tri-iodothyronine (rT3). However, there have been conflicting results in studies of the relation between serum concentrations of rT3 and the therapeutic and adverse effects of amiodarone (SEDA-15, 172).

Monitoring therapy

In many areas of drug therapy evidence to support monitoring recommendation is scant, and this is true of amiodarone, as a systematic review has shown (359). The authors found 43 articles that provided specific monitoring recommendations, but none that compared the outcomes of patients managed with different monitoring regimens. In a study of 99 patients, 52 received minimum baseline evaluations, 22 underwent continuing surveillance, 75 had appropriate responses to abnormal surveillance results, and 71 had timely follow-up visits. They concluded that current standards for monitoring amiodarone toxicity are based on expert opinion with limited evidence to support most recommendations, that monitoring practices vary significantly, and that few patients receive all of the recommended monitoring.

References

1. Rosenbaum MB, Chiale PA, Halpern MS, Nau GJ, Przybylski J, Levi RJ, Lazzari JO, Elizari MV. Clinical efficacy of amiodarone as an antiarrhythmic agent. Am J Cardiol 1976;38(7):934–44.
2. McGovern B, Garan H, Ruskin JN. Serious adverse effects of amiodarone. Clin Cardiol 1984;7(3):131–7.
3. Latini R, Tognoni G, Kates RE. Clinical pharmacokinetics of amiodarone. Clin Pharmacokinet 1984;9(2):136–56.
4. Heger JJ, Prystowsky EN, Miles WM, Zipes DP. Clinical use and pharmacology of amiodarone. Med Clin North Am 1984;68(5):1339–66.

5. Cetnarowski AB, Rihn TL. A review of adverse reactions to amiodarone. Cardiovasc Rev Rep 1985;6:1206–22.
6. Kadish A, Morady F. The use of intravenous amiodarone in the acute therapy of life-threatening tachyarrhythmias. Prog Cardiovasc Dis 1989;31(4):281–94.
7. Kopelman HA, Horowitz LN. Efficacy and toxicity of amiodarone for the treatment of supraventricular tachyarrhythmias. Prog Cardiovasc Dis 1989;31(5):355–66.
8. Heger JJ. Monitoring and treating side effects of amiodarone therapy. Cardiovasc Rev Rep 1988;9:47.
9. Kerin NZ, Aragon E, Faitel K, Frumin H, Rubenfire M. Long-term efficacy and toxicity of high- and low-dose amiodarone regimens. J Clin Pharmacol 1989;29(5):418–23.
10. Somani P. Basic and clinical pharmacology of amiodarone: relationship of antiarrhythmic effects, dose and drug concentrations to intracellular inclusion bodies. J Clin Pharmacol 1989;29(5):405–12.
11. Vorperian VR, Havighurst TC, Miller S, January CT. Adverse effects of low dose amiodarone: a meta-analysis. J Am Coll Cardiol 1997;30(3):791–8.
12. Marcus FI. Drug combinations and interactions with class III agents. J Cardiovasc Pharmacol 1992;20(Suppl 2):S70–4.
13. Tsikouris JP, Cox CD. A review of class III antiarrhythmic agents for atrial fibrillation: maintenance of normal sinus rhythm. Pharmacotherapy 2001;21(12):1514–29.
14. Trappe HJ. Amiodarone. Intensivmed Notf Med 2001; 38:169–78.
15. Kuck KH, Cappato R, Siebels J, Ruppel R. Randomized comparison of antiarrhythmic drug therapy with implantable defibrillators in patients resuscitated from cardiac arrest: the Cardiac Arrest Study Hamburg (CASH). Circulation 2000;102(7):748–54.
16. Nagele H, Bohlmann M, Eck U, Petersen B, Rodiger W. Combination therapy with carvedilol and amiodarone in patients with severe heart failure. Eur J Heart Fail 2000;2(1):71–9.
17. Connolly SJ. Meta-analysis of antiarrhythmic drug trials. Am J Cardiol 1999;84(9A):R90–3.
18. Levy S. Amiodarone in atrial fibrillation. Int J Clin Pract 1998;52(6):429–31.
19. Kochiadakis GE, Igoumenidis NE, Marketou ME, Kaleboubas MD, Simantirakis EN, Vardas PE. Low dose amiodarone and sotalol in the treatment of recurrent, symptomatic atrial fibrillation: a comparative, placebo controlled study. Heart 2000;84(3):251–7.
20. Vardas PE, Kochiadakis GE, Igoumenidis NE, Tsatsakis AM, Simantirakis EN, Chlouverakis GI. Amiodarone as a first-choice drug for restoring sinus rhythm in patients with atrial fibrillation: a randomized, controlled study. Chest 2000;117(6):1538–45.
21. Hofmann R, Wimmer G, Leisch F. Intravenous amiodarone bolus immediately controls heart rate in patients with atrial fibrillation accompanied by severe congestive heart failure. Heart 2000;84(6):635.
22. Peuhkurinen K, Niemela M, Ylitalo A, Linnaluoto M, Lilja M, Juvonen J. Effectiveness of amiodarone as a single oral dose for recent-onset atrial fibrillation. Am J Cardiol 2000;85(4):462–5.
23. Martinez-Marcos FJ, Garcia-Garmendia JL, Ortega-Carpio A, Fernandez-Gomez JM, Santos JM, Camacho C. Comparison of intravenous flecainide, propafenone, and amiodarone for conversion of acute atrial fibrillation to sinus rhythm. Am J Cardiol 2000;86(9):950–3.
24. Treggiari-Venzi MM, Waeber JL, Perneger TV, Suter PM, Adamec R, Romand JA. Intravenous amiodarone or magnesium sulphate is not cost-beneficial prophylaxis for atrial fibrillation after coronary artery bypass surgery. Br J Anaesth 2000;85(5):690–5.
25. Roy D, Talajic M, Dorian P, Connolly S, Eisenberg MJ, Green M, Kus T, Lambert J, Dubuc M, Gagne P, Nattel S, Thibault B. Amiodarone to prevent recurrence of atrial fibrillation. Canadian Trial of Atrial Fibrillation Investigators. N Engl J Med 2000;342(13):913–20.
26. Kochiadakis GE, Marketou ME, Igoumenidis NE, Chrysostomakis SI, Mavrakis HE, Kaleboubas MD, Vardas PE. Amiodarone, sotalol, or propafenone in atrial fibrillation: which is preferred to maintain normal sinus rhythm? Pacing Clin Electrophysiol 2000;23(11 Pt 2):1883–7.
27. Wurdeman RL, Mooss AN, Mohiuddin SM, Lenz TL. Amiodarone vs. sotalol as prophylaxis against atrial fibrillation/flutter after heart surgery: a meta-analysis. Chest 2002;121(4):1203–10.
28. Hilleman DE, Spinler SA. Conversion of recent-onset atrial fibrillation with intravenous amiodarone: a meta-analysis of randomized controlled trials. Pharmacotherapy 2002;22(1):66–74.
29. Kapoor A, Kumar S, Singh RK, Pandey CM, Sinha N. Management of persistent atrial fibrillation following balloon mitral valvotomy: safety and efficacy of low-dose amiodarone. J Heart Valve Dis 2002;11(6):802–9.
30. Kosior D, Karpinski G, Wretowski D, Stolarz P, Stawicki S, Rabczenko D, Torbicki A, Opolski G. Sequential prophylactic antiarrhythmic therapy for maintenance of sinus rhythm after cardioversion of persistent atrial fibrillation—one year follow-up. Kardiol Pol 2002; 56:361–7.
31. Blanc JJ, Voinov C, Maarek MPARSIFAL Study Group. Comparison of oral loading dose of propafenone and amiodarone for converting recent-onset atrial fibrillation. Am J Cardiol 1999;84(9):1029–32.
32. Kanoupakis EM, Kochiadakis GE, Manios EG, Igoumenidis NE, Mavrakis HE, Vardas PE. Pharmacological cardioversion of recent onset atrial fibrillation with intravenous amiodarone in patients receiving long-term amiodarone therapy: is it reasonable? J Intervent Cardiac Electrophysiol 2003;8:19–26.
33. Liu T-J, Hsueh C-W, Lee W-L, Lai H-C, Wang K-Y, Ting C-T. Conversion of rheumatic atrial fibrillation by amiodarone after percutaneous balloon mitral commissurotomy. Am J Cardiol 2003;92:1244–6.
34. Singh BN, Singh SN, Reda DJ, Tang XC, Lopez B, Harris CL, Fletcher RD, Sharma SC, Atwood JE, Jacobson AK, Lewis HD Jr, Raisch DW, Ezekowitz MD;Sotalol Amiodarone Atrial Fibrillation Efficacy Trial (SAFE-T) Investigators. Amiodarone versus sotalol for atrial fibrillation. N Engl J Med 2005;352(18):1861–72.
35. Khan IA, Mehta NJ, Gowda RM. Amiodarone for pharmacological cardioversion of recent-onset atrial fibrillation. Int J Cardiol 2003;89:239–48.
36. Letelier LM, Udol K, Ena J, Weaver B, Guyatt GH. Effectiveness of amiodarone for conversion of atrial fibrillation to sinus rhythm. A meta-analysis. Arch Intern Med 2003;163:777–85.
37. Natale A, Newby KH, Pisano E, Leonelli F, Fanelli R, Potenza D, Beheiry S, Tomassoni G. Prospective randomized comparison of antiarrhythmic therapy versus first-line radiofrequency ablation in patients with atrial flutter. J Am Coll Cardiol 2000;35(7):1898–904.
38. Maury P, Zimmermann M, Metzger J, Reynard C, Dorsaz P, Adamec R. Amiodarone therapy for sustained ventricular tachycardia after myocardial infarction: long-term follow-up, risk assessment and predictive value of

programmed ventricular stimulation. Int J Cardiol 2000;76(2–3):199–210.
39. Kovoor P, Eipper V, Byth K, Cooper MJ, Uther JB, Ross DL. Comparison of sotalol with amiodarone for long-term treatment of spontaneous sustained ventricular tachyarrhythmia based on coronary artery disease. Eur Heart J 1999;20(5):364–74.
40. Cairns JA. Antiarrhythmic therapy in the post-infarction setting: update from major amiodarone studies. Int J Clin Pract 1998;52(6):422–4.
41. Piepoli M, Villani GQ, Ponikowski P, Wright A, Flather MD, Coats AJ. Overview and meta-analysis of randomised trials of amiodarone in chronic heart failure. Int J Cardiol 1998;66(1):1–10.
42. Pfisterer ME, Kiowski W, Brunner H, Burckhardt D, Burkart F. Long-term benefit of 1-year amiodarone treatment for persistent complex ventricular arrhythmias after myocardial infarction. Circulation 1993;87(2):309–11.
43. Ceremuzynski L, Kleczar E, Krzeminska-Pakula M, Kuch J, Nartowicz E, Smielak-Korombel J, Dyduszynski A, Maciejewicz J, Zaleska T, Lazarczyk-Kedzia E, et al. Effect of amiodarone on mortality after myocardial infarction: a double-blind, placebo-controlled, pilot study. J Am Coll Cardiol 1992;20(5):1056–62.
44. Navarro-Lopez F, Cosin J, Marrugat J, Guindo J, Bayes de Luna A. Comparison of the effects of amiodarone versus metoprolol on the frequency of ventricular arrhythmias and on mortality after acute myocardial infarction. SSSD Investigators. Spanish Study on Sudden Death. Am J Cardiol 1993;72(17):1243–8.
45. Julian DG, Camm AJ, Frangin G, Janse MJ, Munoz A, Schwartz PJ, Simon P. Randomised trial of effect of amiodarone on mortality in patients with left-ventricular dysfunction after recent myocardial infarction: EMIAT. European Myocardial Infarct Amiodarone Trial Investigators. Lancet 1997;349(9053):667–74.
46. Cairns JA, Connolly SJ, Roberts R, Gent M. Randomised trial of outcome after myocardial infarction in patients with frequent or repetitive ventricular premature depolarisations: CAMIAT. Canadian Amiodarone Myocardial Infarction Arrhythmia Trial Investigators. Lancet 1997;349(9053):675–82.
47. Kerin NZ, Blevins RD, Kerner N, Faitel K, Frumin H, Maciejko JJ, Rubenfire M. A low incidence of proarrhythmia using low-dose amiodarone. J Electrophysiol 1988;2:289–95.
48. Hohnloser SH. Proarrhythmia with class III antiarrhythmic drugs: types, risks, and management. Am J Cardiol 1997;80(8A):G82–9.
49. Crocco F, Severino M, Scrivano P, Calcaterra R, Cristiano G. Cardiogenic shock in a case of amiodarone intoxication. Gazz Med Ital Arch Sci Med 1999;158:159–63.
50. Lin S-L, Hsieh P-L, Liu C-P, Chiang H-T, Tak T. Ventricular tachycardia after amiodarone: report of an unusual case. J Appl Res 2003;3:159–62.
51. In: Krikler DM, McKenna WJ, Chamberlain DA, editors. Amiodarone and Arrhythmias. Oxford: Pergamon Press, 1983:473.
52. Hilleman DE, Larsen KE. Proarrhythmic effects of antiarrhythmic drugs. PT 1991;520–4June.
53. Tomcsanyi J, Merkely B, Tenczer J, Papp L, Karlocai K. Early proarrhythmia during intravenous amiodarone treatment. Pacing Clin Electrophysiol 1999;22(6 Pt 1):968–70.
54. Yap SC, Hoomtje T, Sreeram N. Polymorphic ventricular tachycardia after use of intravenous amiodarone for postoperative junctional ectopic tachycardia. Int J Cardiol 2000;76(2–3):245–7.
55. Nkomo VT, Shen WK. Amiodarone-induced long QT and polymorphic ventricular tachycardia. Am J Emerg Med 2001;19(3):246–8.
56. Shimoshige S, Uno K, Miyamoto K, Nakahara N, Wakabayashi T, Tsuchihashi K, Shimamoto K, Murakami H. Amiodarone modulates thresholds of induction and/or termination of ventricular tachycardia and ventricular fibrillation—a case of VT with previous myocardial infarction. Ther Res 2001;22:861–6.
57. Voigt L, Coromilas J, Saul BI, Kassotis J. Amiodarone-induced torsade de pointes during bladder irrigation: an unusual presentation. A case report. Angiology 2003; 54:229–31.
58. Schrickel J, Bielik H, Yang A, Schwab JO, Shlevkov N, Schimpf R, Luderitz B, Lewalter T. Amiodarone-associated 'torsade de pointes'. Relevance of concomitant cardiovascular medication in a patient with atrial fibrillation and structural heart disease. Zeitschr Kardiol 2003; 92:889–92.
59. Kukla P, Slowiak-Lewinska T. Torsade de pointes jako proarytmiczny efekt dziaania amiodaronu. Analiza 5 przypadków. [Amiodarone-induced torsade de pointes—five case reports.] Kardiol Pol 2004;60(4):365–70.
60. Antonelli D, Atar S, Freedberg NA, Rosenfeld T. Torsade de pointes in patients on chronic amiodarone treatment: contributing factors and drug interactions. Isr Med Assoc J 2005;7(3):163–5.
61. Psirropoulos D, Lefkos N, Boudonas G, Efthimiadis A, Eklissiarhos D, Tsapas G. Incidence of and predicting factors for torsades de pointes during intravenous administration of amiodarone. Heart Drug 2001;1:186–91.
62. Makkar RR, Fromm BS, Steinman RT, Meissner MD, Lehmann MH. Female gender as a risk factor for torsades de pointes associated with cardiovascular drugs. JAMA 1993;270(21):2590–7.
63. Benton RE, Sale M, Flockhart DA, Woosley RL. Greater quinidine-induced QTc interval prolongation in women. Clin Pharmacol Ther 2000;67(4):413–8.
64. Tomcsanyi J, Somloi M, Horvath L. Amiodarone-induced giant T wave alternans hastens proarrhythmic response. J Cardiovasc Electrophysiol 2002;13(6):629.
65. Matsuyama T, Tanno K, Kobayashi Y, Obara C, Ryu S, Adachi T, Ezumi H, Asano T, Miyata A, Koba S, Baba T, Katagiri T. T wave alternans for predicting adverse effects of amiodarone in a patient with dilated cardiomyopathy. Jpn Circ J 2001;65(5):468–70.
66. Reithmann C, Hoffmann E, Spitzlberger G, Dorwarth U, Gerth A, Remp T, Steinbeck G. Catheter ablation of atrial flutter due to amiodarone therapy for paroxysmal atrial fibrillation. Eur Heart J 2000;21(7):565–72.
67. Aouate P, Elbaz N, Klug D, Lacotte J, Raguin D, Frank R, Lelouche D, Dubois-Rande JL, Tonet J, Fontaine G. Flutter atrial à conduction nodo-ventricular 1/1 sous amiodarone. De la physiopathologie au dépistage. [Atrial flutter with 1/1 nodo-ventricular conduction with amiodarone. From physiopathology to diagnosis.] Arch Mal Coeur Vaiss 2002;95(12):1181–7.
68. Perdrix-Andujar L, Paziaud O, Ricard G, Diebold B, Le Heuzey JY. Flutter atrial a conduction nodovenhiculaire 1/1 sous amiodarone. [1/1 nodo-ventricular conduction atrial flutter with amiodarone] Arch Mal Coeur Vaiss 2005;98(3):259–62.
69. Tai CT, Chiang CE, Lee SH, Chen YJ, Yu WC, Feng AN, Ding YA, Chang MS, Chen SA. Persistent atrial flutter in patients treated for atrial fibrillation with amiodarone and propafenone: electrophysiologic characteristics, radiofrequency catheter ablation, and risk prediction. J Cardiovasc Electrophysiol 1999;10(9):1180–7.

70. Hammerman H, Kapeliovich M. Drug-related cardiac iatrogenic illness as the cause for admission to the intensive cardiac care unit. Isr Med Assoc J 2000;2(8):577–9.
71. Chevalier P, Durand-Dubief A, Burri H, Cucherat M, Kirkorian G, Touboul P. Amiodarone versus placebo and class IC drugs for cardioversion of recent-onset atrial fibrillation: a meta-analysis. J Am Coll Cardiol 2003;41:255–62.
72. Hauser TH, Pinto DS, Josephson ME, Zimetbaum P. Safety and feasibility of a clinical pathway for the outpatient initiation of antiarrhythmic medications in patients with atrial fibrillation or atrial flutter. Am J Cardiol 2003;91:1437–41.
73. Ravina T, Gutierrez J. Amiodarone-induced AV block and ventricular standstill. A forme fruste of an idiopathic long QT syndrome. Int J Cardiol 2000;75(1):105–8.
74. Essebag V, Hadjis T, Platt RW, Pilote L. Amiodarone and the risk of bradyarrhythmia requiring permanent pacemaker in elderly patients with atrial fibrillation and prior myocardial infarction. J Am Coll Cardiol 2003;41:249–54.
75. Shinde AA, Juneman EB, Mitchell B, Pierce MK, Gaballa MA, Goldman S, Thai H. Shocks from pacemaker cardioverter defibrillators increase with amiodarone in patients at high risk for sudden cardiac death. Cardiology 2003;100:143–8.
76. Cheung AT, Weiss SJ, Savino JS, Levy WJ, Augoustides JG, Harrington A, Gardner TJ. Acute circulatory actions of intravenous amiodarone loading in cardiac surgical patients. Annals of Thoracic Surgery 2003;76:535–41.
77. Somberg JC, Timar S, Bailin SJ, Lakatos F, Haffajee CI, Tarjan J, Paladino WP, Sarosi I, Kerin NZ, Borbola J, Bridges DE, Molnar J;Amio-Aqueous Investigators. Lack of a hypotensive effect with rapid administration of a new aqueous formulation of intravenous amiodarone. Am J Cardiol 2004;93(5):576–81.
78. Dunn M, Glassroth J. Pulmonary complications of amiodarone toxicity. Prog Cardiovasc Dis 1989;31(6):447–53.
79. Kennedy JI Jr. Clinical aspects of amiodarone pulmonary toxicity. Clin Chest Med 1990;11(1):119–29.
80. Camus P, Martin WJ 2nd, Rosenow EC 3rd. Amiodarone pulmonary toxicity. Clin Chest Med 2004;25(1):65–75.
81. Martin WJ 2nd. Mechanisms of amiodarone pulmonary toxicity. Clin Chest Med 1990;11(1):131–8.
82. Dusman RE, Stanton MS, Miles WM, Klein LS, Zipes DP, Fineberg NS, Heger JJ. Clinical features of amiodarone-induced pulmonary toxicity. Circulation 1990;82(1):51–9.
83. Brinker A, Johnston M. Acute pulmonary injury in association with amiodarone. Chest 2004;125(4):1591–2.
84. Skroubis G, Galiatsou E, Metafratzi Z, Karahaliou A, Kitsakos A, Nakos G. Amiodarone-induced acute lung toxicity in an ICU setting. Acta Anaesthesiol Scand 2005;49(4):569–71 [erratum 2005;49(6):886].
85. Fung RC, Chan WK, Chu CM, Yue CS. Low dose amiodarone-induced lung injury. Int J Cardiol 2006;113(1):144–5.
86. Handschin AE, Lardinois D, Schneiter D, Bloch K, Weder W. Acute amiodarone-induced pulmonary toxicity following lung resection. Respiration 2003;70:310–12.
87. Dittmann C, Lutz H, Lehmann H. Pulmonale, amiodaroninduzierte, reversible Vershattungen nach akuten Herzinfarkt und Kammerflimmern. [Reversible, amiodarone-induced pulmonary opacities after acute myocardial infarction and ventricular fibrillation.] Internist Prax 2004;44:243–50.
88. Drent M, Cobben NA, Van Dieijen-Visser MP, Braat SH, Wouters EF. Serum lactate dehydrogenase activity: indicator of the development of pneumonitis induced by amiodarone. Eur Heart J 1998;19(6):969–70.
89. Gonzalez-Rothi RJ, Hannan SE, Hood CI, Franzini DA. Amiodarone pulmonary toxicity presenting as bilateral exudative pleural effusions. Chest 1987;92(1):179–82.
90. Carmichael LC, Newman JH. Lymphocytic pleural exudate in a patient receiving amiodarone. Br J Clin Pract 1996;50(4):228–30.
91. Valle JM, Alvarez D, Antunez J, Valdes L. Bronchiolitis obliterans organizing pneumonia secondary to amiodarone: a rare aetiology. Eur Respir J 1995;8(3):470–1.
92. Kudenchuk PJ, Pierson DJ, Greene HL, Graham EL, Sears GK, Trobaugh GB. Prospective evaluation of amiodarone pulmonary toxicity. Chest 1984;86(4):541–8.
93. Donica SK, Paulsen AW, Simpson BR, Ramsay MA, Saunders CT, Swygert TH, Tappe J. Danger of amiodarone therapy and elevated inspired oxygen concentrations in mice. Am J Cardiol 1996;77(1):109–10.
94. Liverani E, Armuzzi A, Mormile F, Anti M, Gasbarrini G, Gentiloni N. Amiodarone-induced adult respiratory distress syndrome after nonthoracotomy subcutaneous defibrillator implantation. J Intern Med 2001;249(6):565–6.
95. Kaushik S, Hussain A, Clarke P, Lazar HL. Acute pulmonary toxicity after low-dose amiodarone therapy. Ann Thorac Surg 2001;72(5):1760–1.
96. Alter P, Grimm W, Maisch B. Amiodaron-induzierte Pneumonitis bei dilatativer Kardiomyopathie. [Amiodarone induced pulmonary toxicity.] Pneumologie 2002;56(1):31–5.
97. Kanji Z, Sunderji R, Gin K. Amiodarone-induced pulmonary toxicity. Pharmacotherapy 1999;19(12):1463–6.
98. Cockcroft DW, Fisher KL. Near normalization of spirometry in a subject with severe emphysema complicated by amiodarone lung. Respir Med 1999;93(8):597–600.
99. Kagawa FT, Kirsch CM, Jensen WA, Wehner JH. A 77-year-old man with bilateral pulmonary infiltrates and shortness of breath. Semin Respir Infect 2000;15(1):90–2.
100. Burns KE, Piliotis E, Garcia BM, Ferguson KA. Amiodarone pulmonary, neuromuscular and ophthalmological toxicity. Can Respir J 2000;7(2):193–7.
101. Scharf C, Oechslin EN, Salomon F, Kiowski W. Clinical picture: Amiodarone-induced pulmonary mass and cutaneous vasculitis. Lancet 2001;358(9298):2045.
102. Rodriguez-Garcia JL, Garcia-Nieto JC, Ballesta F, Prieto E, Villanueva MA, Gallardo J. Pulmonary mass and multiple lung nodules mimicking a lung neoplasm as amiodarone-induced pulmonary toxicity. Eur J Intern Med 2001;12(4):372–6.
103. Endoh Y, Hanai R, Uto K, Uno M, Nagashima H, Takizawa T, Narimatsu A, Ohnishi S, Kasanuki H. [Diagnostic usefulness of KL-6 measurements in patients with pulmonary complications after administration of amiodarone.]J Cardiol 2000;35(2):121–7.
104. Charles PE, Doise JM, Quenot JP, Muller G, Aube H, Baudouin N, Piard F, Besancenot JF, Blettery B. Amiodarone-related acute respiratory distress syndrome following sudden withdrawal of steroids. Respiration 2006;73(2):248–9.
105. Endoh Y, Hanai R, Uto K, Uno M, Nagashima H, Narimatsu A, Takizawa T, Onishi S, Kasanuki H. KL-6 as a potential new marker for amiodarone-induced pulmonary toxicity. Am J Cardiol 2000;86(2):229–31.
106. Esato M, Sakurada H, Okazaki H, Kimura T, Nomizo A, Endou M, Tamura T, Hiyoshi Y, Mishizaki M, Teshima T, Yanase O, Hiraoka M. Evaluation of pulmonary toxicity by CT, pulmonary function tests, and KL-6 measurements in amiodarone-treated patients. Ther Res 2001;22:867–73.

107. Nicholson AA, Hayward C. The value of computed tomography in the diagnosis of amiodarone-induced pulmonary toxicity. Clin Radiol 1989;40(6):564–7.
108. Bernal Morell E, Hernández Madrid A, Marín Marín I, Rodríguez Pena R, González Gordaliza MC, Moro C. Nodulos pulmonares multiples y amiodarona. KL-6 como nueva herramienta diagnostica. [Multiple pulmonary nodules and amiodarone. KL-6 as a new diagnostic tool.] Rev Esp Cardiol 2005;58(4):447–9.
109. Siniakowicz RM, Narula D, Suster B, Steinberg JS. Diagnosis of amiodarone pulmonary toxicity with high-resolution computerized tomographic scan. J Cardiovasc Electrophysiol 2001;12(4):431–6.
110. Lim KK, Radford DJ. Amiodarone pneumonitis diagnosed by gallium-67 scintigraphy. Heart Lung Circ 2002;11:59–62.
111. Dirlik A, Erinc R, Ozcan Z, Atasever A, Bacakoglu F, Nalbantgil S, Ozhan M, Burak Z. Technetium-99m-DTPA aerosol scintigraphy in amiodarone induced pulmonary toxicity in comparison with Ga-67 scintigraphy. Ann Nucl Med 2002;16(7):477–81.
112. Biour M, Hugues FC, Hamel JD, Cheymol G. Les effets indésirables pulmonaires de l'amiodarone: analyse de 162 observations. [Adverse pulmonary effects of amiodarone. Analysis of 162 cases.] Therapie 1985;40(5):343–8.
113. Palakurthy PR, Iyer V, Meckler RJ. Unusual neurotoxicity associated with amiodarone therapy. Arch Intern Med 1987;147(5):881–4.
114. Trohman RG, Castellanos D, Castellanos A, Kessler KM. Amiodarone-induced delirium. Ann Intern Med 1988;108(1):68–9.
115. Werner EG, Olanow CW. Parkinsonism and amiodarone therapy. Ann Neurol 1989;25(6):630–2.
116. Borruat FX, Regli F. Pseudotumor cerebri as a complication of amiodarone therapy. Am J Ophthalmol 1993;116(6):776–7.
117. Itoh K, Kato R, Hotta N. A case report of myolysis during high-dose amiodarone therapy for uncontrolled ventricular tachycardia. Jpn Circ J 1998;62(4):305–8.
118. Jacobs JM, Costa-Jussa FR. The pathology of amiodarone neurotoxicity. II. Peripheral neuropathy in man. Brain 1985;108(Pt 3):753–69.
119. Onofrj M, Thomas A. Acetazolamide-responsive periodic ataxia induced by amiodarone. Mov Disord 1999; 14(2):379–81.
120. Pulipaka U, Lacomis D, Omalu B. Amiodarone-induced neuromyopathy: three cases and a review of the literature. J Clin Neuromuscular Dis 2002;3:97–105.
121. Krauser DG, Segal AZ, Kligfield P. Severe ataxia caused by amiodarone. Am J Cardiol 2005;96(10):1463–4.
122. Biran I, Steiner I. Coital headaches induced by amiodarone. Neurology 2002;58(3):501–2.
123. Mantyjarvi M, Tuppurainen K, Ikaheimo K. Ocular side effects of amiodarone. Surv Ophthalmol 1998;42(4):360–6.
124. Duff GR, Fraser AG. Impairment of colour vision associated with amiodarone keratopathy. Acta Ophthalmol (Copenh) 1987;65(1):48–52.
125. Ingram DV, Jaggarao NS, Chamberlain DA. Ocular changes resulting from therapy with amiodarone. Br J Ophthalmol 1982;66(10):676–9.
126. Ingram DV. Ocular effects in long-term amiodarone therapy. Am Heart J 1983;106(4 Pt 2):902–5.
127. Ikaheimo K, Kettunen R, Mantyjarvi M. Visual functions and adverse ocular effects in patients with amiodarone medication. Acta Ophthalmol Scand 2002;80(1):59–63.
128. Astin CLK. Amiodarone keratopathy and rigid contact lens wear. Contact Lens Anterior Eye 2001;24:80–2.
129. Ciancaglini M, Carpineto P, Zuppardi E, Nubile M, Doronzo E, Mastropasqua L. In vivo confocal microscopy of patients with amiodarone-induced keratopathy. Cornea 2001;20(4):368–73.
130. Feiner LA, Younge BR, Kazmier FJ, Stricker BH, Fraunfelder FT. Optic neuropathy and amiodarone therapy. Mayo Clin Proc 1987;62(8):702–17.
131. In vivo confocal microscopy of megalocornea with central mosaic dystrophy. Uçakhan OO, Kanpolat A, Yilmaz N. Clin Experiment Ophthalmol 2005;33(1):102–5.
132. Chilov MN, Moshegov CN, Booth F. Unilateral amiodarone keratopathy. Clin Experiment Ophthalmol 2005;33(6):666–8.
133. Mansour AM, Puklin JE, O'Grady R. Optic nerve ultrastructure following amiodarone therapy. J Clin Neuroophthalmol 1988;8(4):231–7.
134. Dewachter A, Lievens H. Amiodarone and optic neuropathy. Bull Soc Belge Ophtalmol 1988;227:47–50.
135. Garret SN, Kearney JJ, Schiffman JS. Amiodarone optic neuropathy. J Clin Neuro-ophthalmol 1988;8:105.
136. Mindel JM. Amiodarone and optic neuropathy—a medicolegal issue. Surv Ophthalmol 1998;42(4):358–9.
137. Macaluso DC, Shults WT, Fraunfelder FT. Features of amiodarone-induced optic neuropathy. Am J Ophthalmol 1999;127(5):610–2.
138. Eryilmaz T, Atilla H, Batioglu F, Gunalp I. Amiodarone-related optic neuropathy. Jpn J Ophthalmol 2000; 44(5):565–8.
139. Murphy MA, Murphy JF. Amiodarone and optic neuropathy: the heart of the matter. J Neuroophthalmol 2005;25(3):232–6.
140. Speicher MA, Goldman MH, Chrousos GA. Amiodarone optic neuropathy without disc edema. J Neuroophthalmol 2000;20(3):171–2.
141. Fikkers BG, Bogousslavsky J, Regli F, Glasson S. Pseudotumor cerebri with amiodarone. J Neurol Neurosurg Psychiatry 1986;49(5):606.
142. Thystrup JD, Fledelius HC. Retinal maculopathy possibly associated with amiodarone medication. Acta Ophthalmol (Copenh) 1994;72(5):639–41.
143. Clement CI, Myers P, Tan KP. Bilateral optic neuropathy due to amiodarone with recurrence. Clin Experiment Ophthalmol 2005;33(2):222–5.
144. Nagra PK, Foroozan R, Savino PJ, Castillo I, Sergott RC. Amiodarone induced optic neuropathy. 2003;87:420–2.
145. Reifler DM, Verdier DD, Davy CL, Mostow ND, Wendt VE. Multiple chalazia and rosacea in a patient treated with amiodarone. Am J Ophthalmol 1987; 103(4):594–5.
146. Dickinson EJ, Wolman RL. Sicca syndrome associated with amiodarone therapy. BMJ (Clin Res Ed) 1986;293:510.
147. Vrobel TR, Miller PE, Mostow ND, Rakita L. A general overview of amiodarone toxicity: its prevention, detection, and management. Prog Cardiovasc Dis 1989;31(6):393–426.
148. Greene HL, Graham EL, Werner JA, Sears GK, Gross BW, Gorham JP, Kudenchuk PJ, Trobaugh GB. Toxic and therapeutic effects of amiodarone in the treatment of cardiac arrhythmias. J Am Coll Cardiol 1983;2(6):1114–28.
149. Katai N, Yokoyama R, Yoshimura N. Progressive brown discoloration of silicone intraocular lenses after vitrectomy in a patient on amiodarone. J Cataract Refract Surg 1999;25(3):451–2.

150. Barry JJ, Franklin K. Amiodarone-induced delirium. Am J Psychiatry 1999;156(7):1119.
151. Ambrose A, Salib E. Amiodarone-induced depression. Br J Psychiatry 1999;174:366–7.
152. Domingues MF, Barros H, Falcao-Reis FM. Amiodarone and optic neuropathy. Acta Ophthalmol Scand 2004;82(3 Pt 1):277–82.
153. Dobs AS, Sarma PS, Guarnieri T, Griffith L. Testicular dysfunction with amiodarone use. J Am Coll Cardiol 1991;18(5):1328–32.
154. Odeh M, Schiff E, Oliven A. Hyponatremia during therapy with amiodarone. Arch Intern Med 1999;159(21):2599–600.
155. Ikegami H, Shiga T, Tsushima T, Nirei T, Kasanuki H. Syndrome of inappropriate antidiuretic hormone secretion (SIADH) induced by amiodarone: a report on two cases. J Cardiovasc Pharmacol Ther 2002;7(1):25–8.
156. Patel GP, Kasiar JB. Syndrome of inappropriate antidiuretic hormone-induced hyponatremia associated with amiodarone. Pharmacotherapy 2002;22(5):649–51.
157. Aslam MK, Gnaim C, Kutnick J, Kowal RC, McGuire DK. Syndrome of inappropriate antidiuretic hormone secretion induced by amiodarone therapy. Pacing Clin Electrophysiol 2004;27(6 Pt 1):831–2.
158. Wolff J. Perchlorate and the thyroid gland. Pharmacol Rev 1998;50(1):89–105.
159. Wiersinga WM, Trip MD. Amiodarone and thyroid hormone metabolism. Postgrad Med J 1986;62(732):909–14.
160. Mason JW. Amiodarone. N Engl J Med 1987;316(8):455–66.
161. Tajiri J, Higashi K, Morita M, Umeda T, Sato T. Studies of hypothyroidism in patients with high iodine intake. J Clin Endocrinol Metab 1986;63(2):412–7.
162. Nademanee K, Piwonka RW, Singh BN, Hershman JM. Amiodarone and thyroid function. Prog Cardiovasc Dis 1989;31(6):427–37.
163. Newman CM, Price A, Davies DW, Gray TA, Weetman AP. Amiodarone and the thyroid: a practical guide to the management of thyroid dysfunction induced by amiodarone therapy. Heart 1998;79(2):121–7.
164. Rouleau F, Baudusseau O, Dupuis JM, Victor J, Geslin P. Incidence et delai d'apparition des dysthyroïdies sours traitement chronique par amiodarone. [Incidence and timing of thyroid dysfunction with long-term amiodarone therapy.] Arch Mal Coeur Vaiss 2001;94(1):39–43.
165. Bouvy ML, Heerdink ER, Hoes AW, Leufkens HG. Amiodarone-induced thyroid dysfunction associated with cumulative dose. Pharmacoepidemiol Drug Saf 2002; 11(7):601–6.
166. Martino E, Safran M, Aghini-Lombardi F, Rajatanavin R, Lenziardi M, Fay M, Pacchiarotti A, Aronin N, Macchia E, Haffajee C, et al. Environmental iodine intake and thyroid dysfunction during chronic amiodarone therapy. Ann Intern Med 1984;101(1):28–34.
167. Bagheri H, Lapeyre-Mestre M, Levy C, Haramburu F, Hillaire-Buys D, Blayac JP, Montastruc JL. Dysthyroidism due to amiodarone: comparison of spontaneous reporting in Aquitaine, Midi–Pyrenees and Languedoc–Roussillon. [Inter-regional differences in dysthyroidism due to amiodarone: comparison of spontaneous notifications in Aquitaine, Midi–Pyrenees and Languedoc–Roussillon.] Therapie 2001;56(3):301–6.
168. Wong R, Cheung W, Stockigt JR, Topliss DJ. Heterogeneity of amiodarone-induced thyrotoxicosis: evaluation of colour-flow Doppler sonography in predicting therapeutic response. Intern Med J 2003;33:420–6.
169. Mackie GC, Shulkin BL. Amiodarone-induced hyperthyroidism in a patient with functioning papillary carcinoma of the thyroid and extensive hepatic metastases. Thyroid 2005;15(12):1337–40.
170. Thorne SA, Barnes I, Cullinan P, Somerville J. Amiodarone-associated thyroid dysfunction: risk factors in adults with congenital heart disease. Circulation 1999;100(2):149–54.
171. Sidhu J, Jenkins D. Men are at increased risk of amiodarone-associated thyrotoxicosis in the UK. QJM 2003;96(12):949–50.
172. Dietlein M, Schicha H. Amiodarone-induced thyrotoxicosis due to destructive thyroiditis: therapeutic recommendations. Exp Clin Endocrinol Diabetes 2005;113(3):145–51.
173. Mariotti S, Loviselli A, Murenu S, Sau F, Valentino L, Mandas A, Vacquer S, Martino E, Balestrieri A, Lai ME. High prevalence of thyroid dysfunction in adult patients with beta-thalassemia major submitted to amiodarone treatment. J Endocrinol Invest 1999;22(1):55–63.
174. Findlay PF, Seymour DG. Hyperthyroidism in an elderly patient. Postgrad Med J 2000;76(893):173–5.
175. Cattaneo F. Type II amiodarone-induced thyrotoxicosis and concomitant papillary cancer of the thyroid. Eur J Endocrinol 2000;143(6):823–4.
176. Jouannic J-M, Delahaye S, Fermont L, Le Bidois J, Villain E, Dumez Y, Dommergues M. Fetal supraventricular tachycardia: a role for amiodarone as second-line therapy? Prenatal Diagn 2003;23:152–6.
177. Claxton S, Sinha SN, Donovan S, Greenaway TM, Hoffman L, Loughhead M, Burgess JR. Refractory amiodarone-associated thyrotoxicosis: an indication for thyroidectomy. Aust NZ J Surg 2000;70(3):174–8.
178. Leung PM, Quinn ND, Belchetz PE. Amiodarone-induced thyrotoxicosis: not a benign condition. Int J Clin Pract 2002;56(1):44–6.
179. Martino E, Bartalena L, Mariotti S, Aghini-Lombardi F, Ceccarelli C, Lippi F, Piga M, Loviselli A, Braverman L, Safran M, et al. Radioactive iodine thyroid uptake in patients with amiodarone-iodine-induced thyroid dysfunction. Acta Endocrinol (Copenh) 1988;119(2):167–73.
180. Mancini A, De Marinis L, Calabro F, Sciuto R, Oradei A, Lippa S, Sandric S, Littarru GP, Barbarino A. Evaluation of metabolic status in amiodarone-induced thyroid disorders: plasma coenzyme Q10 determination. J Endocrinol Invest 1989;12(8):511–6.
181. Eaton SE, Euinton HA, Newman CM, Weetman AP, Bennet WM. Clinical experience of amiodarone-induced thyrotoxicosis over a 3-year period: role of colour-flow Doppler sonography. Clin Endocrinol (Oxf) 2002; 56(1):33–8.
182. Bogazzi F, Bartalena L, Brogioni S, Mazzeo S, Vitti P, Burelli A, Bartolozzi C, Martino E. Color flow Doppler sonography rapidly differentiates type I and type II amiodarone-induced thyrotoxicosis. Thyroid 1997;7(4):541–5.
183. Martino E, Aghini-Lombardi F, Mariotti S, Lenziardi M, Baschieri L, Braverman LE, Pinchera A. Treatment of amiodarone associated thyrotoxicosis by simultaneous administration of potassium perchlorate and methimazole. J Endocrinol Invest 1986;9(3):201–7.
184. Reichert LJ, de Rooy HA. Treatment of amiodarone induced hyperthyroidism with potassium perchlorate and methimazole during amiodarone treatment. BMJ 1989;298(6687):1547–8.
185. Stephens JW, Baynes C, Hurel SJ. Amiodarone and thyroid dysfunction. A case-illustrated guide to management. Br J Cardiol 2001;8:499–506.
186. Broussolle C, Ducottet X, Martin C, Barbier Y, Bornet H, Noel G, Orgiazzi J. Rapid effectiveness of prednisone and thionamides combined therapy in severe amiodarone

iodine-induced thyrotoxicosis. Comparison of two groups of patients with apparently normal thyroid glands. J Endocrinol Invest 1989;12(1):37–42.
187. Marketou ME, Simantirakis EN, Manios EG, Vardas PE. Electrical storm due to amiodarone induced thyrotoxicosis in a young adult with dilated cardiomyopathy: thyroidectomy as the treatment of choice. Pacing Clin Electrophysiol 2001;24(12):1827–8.
188. Daniels GH. Amiodarone-induced thyrotoxicosis. J Clin Endocrinol Metab 2001;86(1):3–8.
189. Chopra IJ, Baber K. Use of oral cholecystographic agents in the treatment of amiodarone-induced hyperthyroidism. J Clin Endocrinol Metab 2001;86(10):4707–10.
190. Bogazzi F, Aghini-Lombardi F, Cosci C, Lupi I, Santini F, Tanda ML, Miccoli P, Basolo F, Pinchera A, Bartalena L, Braverman LE, Martino E. Lopanoic acid rapidly controls type I amiodarone-induced thyrotoxicosis prior to thyroidectomy. J Endocrinol Invest 2002;25(2):176–80.
191. Osman F, Franklyn JA, Sheppard MC, Gammage MD. Successful treatment of amiodarone-induced thyrotoxicosis. Circulation 2002;105(11):1275–7.
192. Bartalena L, Brogioni S, Grasso L, Bogazzi F, Burelli A, Martino E. Treatment of amiodarone-induced thyrotoxicosis, a difficult challenge: results of a prospective study. J Clin Endocrinol Metab 1996;81(8):2930–3.
193. Leger AF, Massin JP, Laurent MF, Vincens M, Auriol M, Helal OB, Chomette G, Savoie JC. Iodine-induced thyrotoxicosis: analysis of eighty-five consecutive cases. Eur J Clin Invest 1984;14(6):449–55.
194. Brennan MD, van Heerden JA, Carney JA. Amiodarone-associated thyrotoxicosis (AAT): experience with surgical management. Surgery 1987;102(6):1062–7.
195. Williams M, Lo Gerfo P. Thyroidectomy using local anesthesia in critically ill patients with amiodarone-induced thyrotoxicosis: a review and description of the technique. Thyroid 2002;12(6):523–5.
196. Martino E, Aghini-Lombardi F, Bartalena L, Grasso L, Loviselli A, Velluzzi F, Pinchera A, Braverman LE. Enhanced susceptibility to amiodarone-induced hypothyroidism in patients with thyroid autoimmune disease. Arch Intern Med 1994;154(23):2722–6.
197. Hyatt RH, Sinha B, Vallon A, Bailey RJ, Martin A. Noncardiac side-effects of long-term oral amiodarone in the elderly. Age Ageing 1988;17(2):116–22.
198. Bartalena L, Wiersinga WM, Tanda ML, Bogazzi F, Piantanida E, Lai A, Martino E. Diagnosis and management of amiodarone-induced thyrotoxicosis in Europe: results of an international survey among members of the European Thyroid Association. Clin Endocrinol (Oxf) 2004;61(4):494–502.
199. Franzese CB, Fan CY, Stack BC. Surgical management of amiodarone-induced thyrotoxicosis. Otolaryngol Head Neck Surg 2003;129:565–70.
200. Boeving A, Cubas ER, Santos CM, Carvalho GA, Graf H. O uso de carbonato de litio no tratamento da tireotoxicose induzida por amiodarona. [Use of lithium carbonate for the treatment of amiodarone-induced thyrotoxicosis.] Arq Bras Endocrinol Metabol 2005;49(6):991–5.
201. Diamond TH, Rajagopal R, Ganda K, Manoharan A, Luk A. Plasmapheresis as a potential treatment option for amiodarone-induced thyrotoxicosis. Intern Med J 2004;34(6):369–70.
202. Topliss DJ, Wong R, Stockigt JR A. Plasmapheresis as a potential treatment option for amiodarone-induced thyrotoxicosis. Reply. Intern Med J 2004;34(6):370–71.
203. Hermida JS, Jarry G, Tcheng E, Moullart V, Arlot S, Rey JL, Schvartz C. Prévention des récidives d'hyperthyroïdie a l'amiodarone par l'iode131. [Prevention of recurrent amiodarone-induced hyperthyroidism by iodine-131.] Arch Mal Coeur Vaiss 2004;97(3):207–13.
204. Hermida JS, Tcheng E, Jarry G, Moullart V, Arlot S, Rey JL, Delonca J, Schvartz C. Radioiodine ablation of the thyroid to prevent recurrence of amiodarone-induced thyrotoxicosis in patients with resistant tachyarrhythmias. Europace 2004;6(2):169–74.
205. Hermida JS, Jarry G, Tcheng E, Moullart V, Arlot S, Rey JL, Delonca J, Schvartz C. Radioiodine ablation of the thyroid to allow the reintroduction of amiodarone treatment in patients with a prior history of amiodarone-induced thyrotoxicosis. Am J Med 2004;116(5):345–8.
206. Ryan LE, Braverman LE, Cooper DS, Ladenson PW, Kloos RT. Can amiodarone be restarted after amiodarone-induced thyrotoxicosis? Thyroid 2004;14(2):149–53.
207. Gheri RG, Pucci P, Falsetti C, Luisi ML, Cerisano GP, Gheri CF, Petruzzi I, Pinzani P, Salvadori B, Petruzzi E. Clinical, biochemical and therapeutical aspects of amiodarone-induced hypothyroidism (AIH) in geriatric patients with cardiac arrhythmias. Arch Gerontol Geriatr 2004;38(1):27–36.
208. Pollak PT, Sharma AD, Carruthers SG. Elevation of serum total cholesterol and triglyceride levels during amiodarone therapy. Am J Cardiol 1988;62(9):562–5.
209. Adams PC, Sloan P, Morley AR, Holt DW. Peripheral neutrophil inclusions in amiodarone treated patients. Br J Clin Pharmacol 1986;22(6):736–8.
210. Weinberger I, Rotenberg Z, Fuchs J, Ben-Sasson E, Agmon J. Amiodarone-induced thrombocytopenia. Arch Intern Med 1987;147(4):735–6.
211. Berrebi A, Shtalrid M, Vorst EJ. Amiodarone-induced thrombocytopathy. Acta Haematol 1983;70(1):68–9.
212. Arpin MP, Alt M, Kheiralla JC, Chabrier G, Welsch M, Imbs JL, Imler M. Hyperthyroïdie et anémie hémolytique immune après traitement par amiodarone. [Hyperthyroidism and immune hemolytic anemia following amiodarone therapy.] Rev Med Interne 1991; 12(4):309–11.
213. Rosenbaum H, Ben-Arie Y, Azzam ZS, Krivoy N. Amiodarone-associated granuloma in bone marrow. Ann Pharmacother 1998;32(1):60–2.
214. Boutros NY, Dilly S, Bevan DH. Amiodarone-induced bone marrow granulomas. Clin Lab Haematol 2000;22(3):167–70.
215. Moran SK, Manoharan A. Amiodarone-induced bone marrow granulomas. Pathology 2002;34(3):267–9.
216. Heger JJ, Prystowsky EN, Jackman WM, Naccarelli GV, Warfel KA, Rinkenberger RL, Zipes DP. Amiodarone. N Engl J Med 1982;305:539–45.
217. Bexton RS, Camm AJ. Drugs with a class III antiarrhythmic action. Pharmacol Ther 1982;17(3):315–55.
218. Mukhopadhyay S, Mukhopadhyay S, Abraham NZ Jr, Jones LA, Howard L, Gajra A. Unexplained bone marrow granulomas: is amiodarone the culprit? A report of 2 cases. Am J Hematol 2004;75(2):110–2.
219. Freneaux E, Larrey D, Pessayre D. Phospholipidose et lesions pseudoalcooliques hepatiques medicamenteuses. Rev Fr Gastroenterol 1988;24:879–84.
220. Adams PC, Bennett MK, Holt DW. Hepatic effects of amiodarone. Br J Clin Pract Suppl 1986;44:81–95.
221. Pollak PT, Shafer SL. Use of population modeling to define rational monitoring of amiodarone hepatic effects. Clin Pharmacol Ther 2004;75(4):342–51.
222. Geneve J, Zafrani ES, Dhumeaux D. Amiodarone-induced liver disease. J Hepatol 1989;9(1):130–3.

223. Harrison RF, Elias E. Amiodarone-associated cirrhosis with hepatic and lymph node granulomas. Histopathology 1993;22(1):80–2.
224. Jain D, Bowlus CL, Anderson JM, Robert ME. Granular cells as a marker of early amiodarone hepatotoxicity. J Clin Gastroenterol 2000;31(3):241–3.
225. Jones DB, Mullick FG, Hoofnagle JH, Baranski B. Reye's syndrome-like illness in a patient receiving amiodarone. Am J Gastroenterol 1988;83(9):967–9.
226. Breuer HW, Bossek W, Haferland C, Schmidt M, Neumann H, Gruszka J. Amiodarone-induced severe hepatitis mediated by immunological mechanisms. Int J Clin Pharmacol Ther 1998;36(6):350–2.
227. Pollak PT, You YD. Monitoring of hepatic function during amiodarone therapy. Am J Cardiol 2003;91:613–16.
228. Tilz GP, Liebig E, Pristautz H. Cholestase bei Amiodaron-eine seltene Komplikation der antiarrhythmischen Therapie. Med Welt 1989;40:985.
229. Salti Z, Cloche P, Weber P, Houssemand G, Vollmer F. A propos d'un cas d'hepatite choléstatique à l'amiodarone. [A case of cholestatic hepatitis caused by amiodarone.] Ann Cardiol Angeiol (Paris) 1989;38(1):13–6.
230. Chang CC, Petrelli M, Tomashefski JF Jr, McCullough AJ. Severe intrahepatic cholestasis caused by amiodarone toxicity after withdrawal of the drug: a case report and review of the literature. Arch Pathol Lab Med 1999; 123(3):251–6.
231. Rhodes A, Eastwood JB, Smith SA. Early acute hepatitis with parenteral amiodarone: a toxic effect of the vehicle? Gut 1993;34(4):565–6.
232. Iliopoulou A, Giannakopoulos G, Mayrikakis M, Zafiris E, Stamatelopoulos S. Reversible fulminant hepatitis following intravenous amiodarone loading. Amiodarone hepatotoxicity. Int J Clin Pharmacol Ther 1999;37(6):312–3.
233. Assy N, Khair G, Schlesinger S, Hussein O. Severe cholestatic jaundice in the elderly induced by low-dose amiodarone. Dig Dis Sci 2004;49(3):450–2.
234. Singhal A, Ghosh P, Khan SA. Low dose amiodarone causing pseudo-alcoholic cirrhosis. Age Ageing 2003;32:224–5.
235. Puli SR, Fraley MA, Puli V, Kuperman AB, Alpert MA. Hepatic cirrhosis caused by low-dose oral amiodarone therapy. Am J Med Sci 2005;330(5):257–61.
236. Oikawa H, Maesawa C, Sato R, Oikawa K, Yamada H, Oriso S, Ono S, Yashima-Abo A, Kotani K, Suzuki K, Masuda T. Liver cirrhosis induced by long-term administration of a daily low dose of amiodarone: a case report. World J Gastroenterol 2005;11(34):5394–7.
237. Chow KM, Liu ZC. Amiodarone and cirrhosis. Age Ageing 2004;33(2):207–8.
238. Sastri SV, Diaz-Arias AA, Marshall JB. Can pancreatitis be associated with amiodarone hepatotoxicity? J Clin Gastroenterol 1990;12(1):70–3.
239. Bosch X, Bernadich O. Acute pancreatitis during treatment with amiodarone. Lancet 1997;350(9087):1300.
240. Rätz Bravo AE, Drewe J, Schlienger RG, Krähenbühl S, Pargger H, Ummenhofer W. Hepatotoxicity during rapid intravenous loading with amiodarone: description of three cases and review of the literature. Crit Care Med 2005;33(1):128–34.
241. Guglin M. Intravenous amiodarone: offender or bystander? Crit Care Med 2005;33(1):245–6.
242. Maker AV, Orgill DP. Rapid acute amiodarone-induced hepatotoxicity in a burn patient. J Burn Care Rehabil 2005;26(4):341–3.
243. Pollak PT, Sharma AD, Carruthers SG. Creatinine elevation in patients receiving amiodarone correlates with serum amiodarone concentration. Br J Clin Pharmacol 1993;36(2):125–7.
244. Anonymous. Amiodarone—a new type of antiarrhythmic drug. Drug Ther Bull 1981;19(22):86–8.
245. Chalmers RJ, Muston HL, Srinivas V, Bennett DH. High incidence of amiodarone-induced photosensitivity in North-west England. BMJ (Clin Res Ed) 1982; 285(6338):341.
246. Ferguson J, de Vane PJ, Wirth M. Prevention of amiodarone-induced photosensitivity. Lancet 1984;2(8399):414.
247. Ferguson J, Addo HA, Jones S, Johnson BE, Frain-Bell W. A study of cutaneous photosensitivity induced by amiodarone. Br J Dermatol 1985;113(5):537–49.
248. Collins P, Ferguson J. Narrow-band UVB (TL-01) phototherapy: an effective preventative treatment for the photodermatoses. Br J Dermatol 1995;132(6):956–63.
249. Waitzer S, Butany J, From L, Hanna W, Ramsay C, Downar E. Cutaneous ultrastructural changes and photosensitivity associated with amiodarone therapy. J Am Acad Dermatol 1987;16(4):779–87.
250. Parodi A, Guarrera M, Rebora A. Amiodarone-induced pseudoporphyria. Photodermatol 1988;5(3):146–7.
251. Beukema WP, Graboys TB. Spontaneous disappearance of blue-gray facial pigmentation during amiodarone therapy (out of the blue). Am J Cardiol 1988;62(16): 1146–7.
252. Sra J, Bremner S. Images in cardiovascular medicine: amiodarone skin toxicity. Circulation 1998;97(11):1105.
253. Karrer S, Hohenleutner U, Szeimies RM, Landthaler M, Hruza GJ. Amiodarone-induced pigmentation resolves after treatment with the Q-switched ruby laser. Arch Dermatol 1999;135(3):251–3.
254. Ioannides MA, Moutiris JA, Zambartas C. A case of pseudocyanotic coloring of skin after prolonged use of amiodarone. Int J Cardiol 2003;90:345–6.
255. Rogers KC, Wolfe DA. Amiodarone-induced blue-gray syndrome. Ann Pharmacother 2000;34(9):1075.
256. Erdmann SM, Poblete P. Hyperpigmentation in a patient under treatment with amiodarone and minocycline. H G Z Hautkr 2001;76:746–8.
257. Haas N, Schadendorf D, Hermes B, Henz BM. Hypomelanosis due to block of melanosomal maturation in amiodarone-induced hyperpigmentation. Arch Dermatol 2001;137(4):513–4.
258. Wilkinson CM, Weidner GJ, Paulino AC. Amiodarone and radiation therapy sequelae. Am J Clin Oncol 2001;24(4):379–81.
259. Shah N, Warnakulasuriya S. Amiodarone-induced peri-oral photosensitivity. J Oral Pathol Med 2004;33(1):56–8.
260. Zantkuyl CF, Weemers M. Iododerma caused by amiodarone (Cordarone). Dermatologia (Basel) 1975;151:311.
261. Muir AD, Wilson M. Amiodarone and psoriasis. NZ Med J 1982;95(717):711.
262. Moots RJ, Banerjee A. Exfoliative dermatitis after amiodarone treatment. BMJ (Clin Res Ed) 1988; 296(6632):1332–3.
263. Dootson G, Byatt C. Amiodarone-induced vasculitis and a review of the cutaneous side-effects of amiodarone. Clin Exp Dermatol 1994;19(5):422–4.
264. Bencini PL, Crosti C, Sala F, Bertani E, Nobili M. Toxic epidermal necrolysis and amiodarone treatment. Arch Dermatol 1985;121(7):838.
265. Yung A, Agnew K, Snow J, Oliver F. Two unusual cases of toxic epidermal necrolysis. Australas J Dermatol 2002;43(1):35–8.
266. Clouston PD, Donnelly PE. Acute necrotising myopathy associated with amiodarone therapy. Aust NZ J Med 1989;19(5):483–5.

267. Gabal-Shehab LL, Monga M. Recurrent bilateral amiodarone induced epididymitis. J Urol 1999;161(3):921.
268. Yones SS, O'Donoghue NB, Palmer RA, Menagé Hdu P, Hawk JL. Persistent severe amiodarone-induced photosensitivity. Clin Exp Dermatol 2005;30(5):500–2.
269. Sadek I, Biron P, Kus T. Amiodarone-induced epididymitis: report of a new case and literature review of 12 cases. Can J Cardiol 1993;9(9):833–6.
270. Korantzopoulos P, Pappa E, Karanikis P, Kountouris E, Dimitroula V, Siogas K. Acute low back pain during intravenous administration of amiodarone: a report of two cases. Int J Cardiol 2005;98(2):355–7.
271. Kirkali Z. Re: Recurrent bilateral amiodarone induced epididymitis. J Urol 1999;162(3 Pt 1):808–9.
272. Susano R, Caminal L, Ramos D, Diaz B. Amiodarone induced lupus. Ann Rheum Dis 1999;58(10):655–6.
273. Sheikhzadeh A, Schafer U, Schnabel A. Drug-induced lupus erythematosus by amiodarone. Arch Intern Med 2002;162(7):834–6.
274. Burches E, Garcia-Verdegay F, Ferrer M, Pelaez A. Amiodarone-induced angioedema. Allergy 2000;55(12):1199–200.
275. Elizari MV, Martinez JM, Belziti C, Ciruzzi M, Perez de la Hoz R, Sinisi A, Carbajales J, Scapin O, Garguichevich J, Girotti L, Cagide A. Morbidity and mortality following early administration of amiodarone in acute myocardial infarction. GEMICA study investigators, GEMA Group, Buenos Aires, Argentina. Grupo de Estudios Multicentricos en Argentina. Eur Heart J 2000;21(3):198–205.
276. Scheinman MM. Amiodarone after acute myocardial infarction. Eur Heart J 2000;21(3):177–8.
277. Lahiri K, Malakar S, Sarma N. Amiodarone-induced angioedema: report of two cases. Indian J Dermatol Venereol Leprol 2005;71(1):46–7.
278. Brouse SD, Phillips SM. Amiodarone use in patients with documented allergy to iodine-containing compounds. Pharmacotherapy 2005;25(3):429–34.
279. Foster CJ, Love HG. Amiodarone in pregnancy. Case report and review of the literature. Int J Cardiol 1988;20(3):307–16.
280. De Wolf D, De Schepper J, Verhaaren H, Deneyer M, Smitz J, Sacre-Smits L. Congenital hypothyroid goiter and amiodarone. Acta Paediatr Scand 1988;77(4):616–8.
281. Blomberg PJ, Feingold AD, Denofrio D, Rand W, Konstam MA, Estes NA 3rd, Link MS. Comparison of survival and other complications after heart transplantation in patients taking amiodarone before surgery versus those not taking amiodarone. Am J Cardiol 2004;93(3):379–81.
282. Monk B. Amiodarone-induced photosensitivity and basal-cell carcinoma. Clin Exp Dermatol 1990;15(4):319–20.
283. Monk BE. Basal cell carcinoma following amiodarone therapy. Br J Dermatol 1995;133(1):148–9.
284. Hall MA, Annas A, Nyman K, Talme T, Emtestam L. Basalioma after amiodarone therapy—not only in Britain. Br J Dermatol 2004;151(4):932–3.
285. Hofmann R, Leisch E. Symptomatische Bradykardien unter Amiodaron bei Patienten mit praexistenter Reizleitungs storung. [Symptomatic bradycardia with amiodarone in patients with pre-existing conduction disorders.] Wien Klin Wochenschr 1995;107(21):640–4.
286. Magee LA, Nulman I, Rovet JF, Koren G. Neurodevelopment after in utero amiodarone exposure. Neurotoxicol Teratol 1999;21(3):261–5.
287. Tubman R, Jenkins J, Lim J. Neonatal hyperthyroxinaemia associated with maternal amiodarone therapy: case report. Ir J Med Sci 1988;157(7):243.
288. De Catte L, De Wolf D, Smitz J, Bougatef A, De Schepper J, Foulon W. Fetal hypothyroidism as a complication of amiodarone treatment for persistent fetal supraventricular tachycardia. Prenat Diagn 1994;14(8):762–5.
289. Vanbesien J, Casteels A, Bougatef A, De Catte L, Foulon W, De Bock S, Smitz J, De Schepper J. Transient fetal hypothyroidism due to direct fetal administration of amiodarone for drug resistant fetal tachycardia. Am J Perinatol 2001;18(2):113–6.
290. Etheridge SP, Craig JE, Compton SJ. Amiodarone is safe and highly effective therapy for supraventricular tachycardia in infants. Am Heart J 2001;141(1):105–10.
291. Tisdale JE, Follin SL, Ordelova A, Webb CR. Risk factors for the development of specific noncardiovascular adverse effects associated with amiodarone. J Clin Pharmacol 1995;35(4):351–6.
292. Strasburger JF, Cuneo BF, Michon MM, Gotteiner NL, Deal BJ, McGregor SN, Oudijk MA, Meijboom EJ, Feinkind L, Hussey M, Parilla BV. Amiodarone therapy for drug-refractory fetal tachycardia. Circulation 2004;109(3):375–9.
293. Lomenick JP, Jackson WA, Backeljauw PF. Amiodarone-induced neonatal hypothyroidism: a unique form of transient early-onset hypothyroidism. J Perinatol 2004;24(6):397–9.
294. Garson A Jr, Gillette PC, McVey P, Hesslein PS, Porter CJ, Angell LK, Kaldis LC, Hittner HM. Amiodarone treatment of critical arrhythmias in children and young adults. J Am Coll Cardiol 1984;4(4):749–55.
295. Guccione P, Paul T, Garson A Jr. Long-term follow-up of amiodarone therapy in the young: continued efficacy, unimpaired growth, moderate side effects. J Am Coll Cardiol 1990;15(5):1118–24.
296. Villain E. Les syndromes de QT long chez l'enfant. [Long QT syndromes in children.] Arch Fr Pediatr 1993;50(3):241–7.
297. Sutherland J, Robinson B, Delbridge L. Anaesthesia for amiodarone-induced thyrotoxicosis: a case review. Anaesth Intensive Care 2001;29(1):24–9.
298. Kupferschmid JP, Rosengart TK, McIntosh CL, Leon MB, Clark RE. Amiodarone-induced complications after cardiac operation for obstructive hypertrophic cardiomyopathy. Ann Thorac Surg 1989;48(3):359–64.
299. Andersen HR, Bjorn-Hansen LS, Kimose HH, et al. Amiodaronebehandling og arytmikirurgi. Ugeskr Laeger 1989;151:2264.
300. Perkins MW, Dasta JF, Reilley TE, Halpern P. Intraoperative complications in patients receiving amiodarone: characteristics and risk factors. DICP 1989;23(10):757–63.
301. Morady F. Prevention of atrial fibrillation in the post-operative cardiac patient: significance of oral class III antiarrhythmic agents. Am J Cardiol 1999;84(9A):R156–60.
302. Sauro SC, DeCarolis DD, Pierpont GL, Gornick CC. Comparison of plasma concentrations for two amiodarone products. Ann Pharmacother 2002;36(11):1682–5.
303. Wellens HJ, Brugada P, Abdollah H, Dassen WR. A comparison of the electrophysiologic effects of intravenous and oral amiodarone in the same patient. Circulation 1984;69(1):120–4.
304. Olshansky B, Sami M, Rubin A, Kostis J, Shorofsky S, Slee A, Greene HL;NHLBI AFFIRM Investigators. Use of amiodarone for atrial fibrillation in patients with preexisting pulmonary disease in the AFFIRM study. Am J Cardiol 2005;95(3):404–5.
305. Mitchell LB, Wyse DG, Gillis AM, Duff HJ. Electropharmacology of amiodarone therapy initiation.

Time courses of onset of electrophysiologic and antiarrhythmic effects. Circulation 1989;80(1):34–42.
306. Yamada S, Kuga K, Yamaguchi I. Torsade de pointes induced by intravenous and long-term oral amiodarone therapy in a patient with dilated cardiomyopathy. Jpn Circ J 2001;65(3):236–8.
307. Morady F, Scheinman MM, Shen E, Shapiro W, Sung RJ, DiCarlo L. Intravenous amiodarone in the acute treatment of recurrent symptomatic ventricular tachycardia. Am J Cardiol 1983;51(1):156–9.
308. Holt P, Curry PVL, Way B, Storey G, Holt DW. Intravenous amiodarone in the management of tachyarrhythmias. In: Breithardt H, Loogen F, editors. New Aspects in the Medical Treatment of Tachyarrhythmas. Urban and Schwartzenberg: Munich, 1983:136–41.
309. Faniel R, Schoenfeld P. Efficacy of i.v. amiodarone in converting rapid atrial fibrillation and flutter to sinus rhythm in intensive care patients Eur Heart J 1983;4(3):180–5.
310. Antonelli D, Barzilay J. Acute thrombophlebitis following IV amiodarone administration. Chest 1983;84(1):120.
311. Kerin NZ, Blevins R, Rubenfire M, Faital K, Householder S. Acute thrombophlebitis following IV amiodarone administration. Chest 1983;84(1):120.
312. Reiffel JA. Intravenous amiodarone in the management of atrial fibrillation. J Cardiovasc Pharmacol Ther 1999;4(4):199–204.
313. Kochiadakis GE, Igoumenidis NE, Solomou MC, Kaleboubas MD, Chlouverakis GI, Vardas PE. Efficacy of amiodarone for the termination of persistent atrial fibrillation. Am J Cardiol 1999;83(1):58–61.
314. Cotter G, Blatt A, Kaluski E, Metzkor-Cotter E, Koren M, Litinski I, Simantov R, Moshkovitz Y, Zaidenstein R, Peleg E, Vered Z, Golik A. Conversion of recent onset paroxysmal atrial fibrillation to normal sinus rhythm: the effect of no treatment and high-dose amiodarone. A randomized, placebo-controlled study. Eur Heart J 1999;20(24):1833–42.
315. Leatham EW, Holt DW, McKenna WJ. Class III antiarrhythmics in overdose. Presenting features and management principles. Drug Saf 1993;9(6):450–62.
316. Bonati M, D'Aranno V, Galletti F, Fortunati MT, Tognoni G. Acute overdosage of amiodarone in a suicide attempt. J Toxicol Clin Toxicol 1983;20(2):181–6.
317. Liberman BA, Teasdale SJ. Anaesthesia and amiodarone. Can Anaesth Soc J 1985;32(6):629–38.
318. Woeber KA, Warner I. Potentiation of warfarin sodium by amiodarone-induced thyrotoxicosis. West J Med 1999;170(1):49–51.
319. Davies PH, Franklyn JA. The effects of drugs on tests of thyroid function. Eur J Clin Pharmacol 1991;40(5):439–51.
320. Segura I, Garcia-Bolao I. Meglumine antimoniate, amiodarone and torsades de pointes: a case report. Resuscitation 1999;42(1):65–8.
321. Landray MJ, Kendall MJ. Effect of amiodarone on mortality. Lancet 1998;351(9101):523.
322. McCullough PA, Redle JD, Zaman AG, Archbold A, Alamgir F, Ulahannan TJ, Daoud EG, Morady F. Amiodarone prophylaxis for atrial fibrillation after cardiac surgery. N Engl J Med 1998;338(19):1383–4.
323. Boutitie F, Boissel JP, Connolly SJ, Camm AJ, Cairns JA, Julian DG, Gent M, Janse MJ, Dorian P, Frangin G. Amiodarone interaction with beta-blockers: analysis of the merged EMIAT (European Myocardial Infarct Amiodarone Trial) and CAMIAT (Canadian Amiodarone Myocardial Infarction Trial) databases. The EMIAT and CAMIAT Investigators. Circulation 1999;99(17):2268–75.
324. Ogunyankin KO, Singh BN. Mortality reduction by antiadrenergic modulation of arrhythmogenic substrate: significance of combining beta blockers and amiodarone. Am J Cardiol 1999;84(9A):R76–82.
325. Ahle GB, Blum AL, Martinek J, Oneta CM, Dorta G. Cushing's syndrome in an 81-year-old patient treated with budesonide and amiodarone. Eur J Gastroenterol Hepatol 2000;12(9):1041–2.
326. Lelarge P, Bauer P, Royer-Morrot MJ, Meregnani JL, Larcan A, Lambert H. Intoxication digitalique après administration conjointe d'acétyl digitoxine et d'amiodarone. Ann Med Nancy Est 1993;32:307.
327. Matsumoto K, Ueno K, Nakabayashi T, Komamura K, Kamakura S, Miyatake K. Amiodarone interaction time differences with warfarin and digoxin. J Pharm Technol 2003;19:83–90.
328. Laer S, Scholz H, Buschmann I, Thoenes M, Meinertz T. Digitoxin intoxication during concomitant use of amiodarone. Eur J Clin Pharmacol 1998;54(1):95–6.
329. Bhagat R, Sporn TA, Long GD, Folz RJ. Amiodarone and cyclophosphamide: potential for enhanced lung toxicity. Bone Marrow Transplant 2001;27(10):1109–11.
330. Lee TH, Friedman PL, Goldman L, Stone PH, Antman EM. Sinus arrest and hypotension with combined amiodarone–diltiazem therapy. Am Heart J 1985; 109(1): 163–4.
331. Chouty F, Coumel P. Oral flecainide for prophylaxis of paroxysmal atrial fibrillation. Am J Cardiol 1988; 62(6):D35–7.
332. Saoudi N, Galtier M, Hidden F, Gerber L, Letac B. Bundle-branch reentrant ventricular tachycardia: a possible mechanism of flecainide proarrhythmic effect. J Electrophysiol 1988;2:365–71.
333. Lohman JJ, Reichert LJ, Degen LP. Antiretroviral therapy increases serum concentrations of amiodarone. Ann Pharmacother 1999;33(5):645–6.
334. Atar S, Freedberg NA, Antonelli D, Rosenfeld T. Torsades de pointes and QT prolongation due to a combination of loratadine and amiodarone. PACE - Pacing Clin Electrophysiol 2003;26:785–6.
335. Segura I, Garcia-Bolao I. Meglumine antimoniate, amiodarone and torsades de pointes: a case report. Resuscitation 1999;42(1):65–8.
336. Werner D, Wuttke H, Fromm MF, Schaefer S, Eschenhagen T, Brune K, Daniel WG, Werner U. Effect of amiodarone on the plasma levels of metoprolol. Am J Cardiol 2004;94(10):1319–21.
337. Kounas SP, Letsas KP, Sideris A, Efraimidis M, Kardaras F. QT interval prolongation and torsades de pointes due to a coadministration of metronidazole and amiodarone. Pacing Clin Electrophysiol 2005;28(5):472–3.
338. Yonezawa E, Matsumoto K, Ueno K, Tachibana M, Hashimoto H, Komamura K, Kamakura S, Miyatake K, Tanaka K. Lack of interaction between amiodarone and mexiletine in cardiac arrhythmia patients. J Clin Pharmacol 2002;42(3):342–6.
339. Zhi J, Moore R, Kanitra L, Mulligan TE. Effects of orlistat, a lipase inhibitor, on the pharmacokinetics of three highly lipophilic drugs (amiodarone, fluoxetine, and simvastatin) in healthy volunteers. J Clin Pharmacol 2003; 43:428–35.
340. Staiger C, Jauernig R, de Vries J, Weber E. Influence of amiodarone on antipyrine pharmacokinetics in three patients with ventricular tachycardia. Br J Clin Pharmacol 1984;18(2):263–4.

341. McGovern B, Geer VR, LaRaia PJ, Garan H, Ruskin JN. Possible interaction between amiodarone and phenytoin. Ann Intern Med 1984;101(5):650–1.
342. Lesko LJ. Pharmacokinetic drug interactions with amiodarone. Clin Pharmacokinet 1989;17(2):130–40.
343. Nolan PE Jr, Marcus FI, Karol MD, Hoyer GL, Gear K. Effect of phenytoin on the clinical pharmacokinetics of amiodarone. J Clin Pharmacol 1990;30(12):1112–9.
344. Windle J, Prystowsky EN, Miles WM, Heger JJ. Pharmacokinetic and electrophysiologic interactions of amiodarone and procainamide. Clin Pharmacol Ther 1987;41(6):603–10.
345. Tartini R, Kappenberger L, Steinbrunn W. Gefahrliche Interaktionen zwischen Amiodaron und Antiarrhythmika der Klasse I. [Harmful interactions of amiodarone and class I anti-arrhythmia agents.] Schweiz Med Wochenschr 1982;112(45):1585–7.
346. Zarembski DG, Fischer SA, Santucci PA, Porter MT, Costanzo MR, Trohman RG. Impact of rifampin on serum amiodarone concentrations in a patient with congenital heart disease. Pharmacotherapy 1999;19(2):249–51.
347. Fabre G, Julian B, Saint-Aubert B, Joyeux H, Berger Y. Evidence for CYP3A-mediated N-deethylation of amiodarone in human liver microsomal fractions. Drug Metab Dispos 1993;21(6):978–85.
348. Zarembski DG, Fischer SA, Santucci PA, Porter MT, Costanzo MR, Trohman RG. Impact of rifampin on serum amiodarone concentrations in a patient with congenital heart disease. Pharmacotherapy 1999; 19(2):249–51.
349. Roten L, Schoenenberger RA, Krahenbuhl S, Schlienger RG. Rhabdomyolysis in association with simvastatin and amiodarone. Ann Pharmacother 2004;38(6):978–81.
350. Mazur A, Strasberg B, Kusniec J, Sclarovsky S. QT prolongation and polymorphous ventricular tachycardia associated with trasodone-amiodarone combination. Int J Cardiol 1995;52(1):27–9.
351. Kurnik D, Loebstein R, Farfel Z, Ezra D, Halkin H, Olchovsky D. Complex drug-drug-disease interactions between amiodarone, warfarin, and the thyroid gland. Medicine (Baltimore) 2004;83(2):107–13.
352. Sanoski CA, Bauman JL. Clinical observations with the amiodarone/warfarin interaction: dosing relationships with long-term therapy. Chest 2002;121(1):19–23.
353. Jacobs MB. Serum creatinine increase associated with amiodarone therapy. NY State J Med 1987;87(6):358–9.
354. Tada H, Ozaydin M, Chugh A, Scharf C, Oral H, Pelosi F Jr, Knight BP, Strickberger SA, Morady F. Effects of isoproterenol and amiodarone on the double potential interval after ablation of the cavotricuspid isthmus. J Cardiovasc Electrophysiol 2003;14:935–9.
355. Rotmensch HH, Belhassen B, Swanson BN, Shoshani D, Spielman SR, Greenspon AJ, Greenspan AM, Vlasses PH, Horowitz LN. Steady-state serum amiodarone concentrations: relationships with antiarrhythmic efficacy and toxicity. Ann Intern Med 1984;101(4):462–9.
356. Pollak PT, Sharma AD, Carruthers SG. Correlation of amiodarone dosage, heart rate, QT interval and corneal microdeposits with serum amiodarone and desethylamiodarone concentrations. Am J Cardiol 1989; 64(18):1138–43.
357. Debbas NM, du Cailar C, Bexton RS, Demaille JG, Camm AJ, Puech P. The QT interval: a predictor of the plasma and myocardial concentrations of amiodarone. Br Heart J 1984;51(3):316–20.
358. Willems J, De Geest H. Digitalis intoxicatie. Ned Tijdschr Geneeskd 1968;24:617.
359. Stelfox HT, Ahmed SB, Fiskio J, Bates DW. Monitoring amiodarone's toxicities: recommendations, evidence, and clinical practice. Clin Pharmacol Ther 2004;75(1): 110–22.

Aprindine

General Information

The pharmacology, clinical pharmacology, clinical uses, efficacy, and adverse effects of aprindine have been extensively reviewed (1–5).

The adverse effects of aprindine most commonly affect the central nervous system. However, other less common but serious and potentially fatal adverse effects (neutropenia and liver damage) occur, and these limit its usefulness.

Aprindine is metabolized by CYP2D6, and one would therefore expect interactions with drugs that inhibit this isozyme or are metabolized by it.

Comparative studies

In a comparison of oral aprindine and propafenone in 32 patients (25 men and 7 women, aged 43–82) with paroxysmal or persistent atrial fibrillation, aprindine was effective in five of 29 and propafenone in six of 28; adverse effects were not reported (6).

There has been a multicenter, randomized, placebo-controlled, double-blind comparison of aprindine and digoxin in the prevention of atrial fibrillation and its recurrence in 141 patients with symptomatic paroxysmal or persistent atrial fibrillation who had converted to sinus rhythm (7). They were randomized in equal numbers to aprindine 40 mg/day, digoxin 0.25 mg/day, or placebo and followed every 2 weeks for 6 months. After 6 months the Kaplan–Meier estimates of the numbers of patients who had no recurrences with aprindine, digoxin, and placebo were 33, 29, and 22% respectively. The rates of adverse events were similar in the three groups. This suggests that aprindine has a very small beneficial effect in preventing relapse of symptomatic atrial fibrillation after conversion to sinus rhythm. Furthermore, recurrence occurred later with aprindine than with placebo or digoxin (about 60% recurrence at 115 days compared with 30 days).

Organs and Systems

Respiratory

Pneumonitis has been attributed to aprindine (SEDA-16, 179).

Nervous system

Aprindine can cause adverse effects in the central nervous system (SEDA-1, 156).

Hematologic

The incidence of leukopenia has been estimated at about 2 cases per 1000 patient-years (8).

In a case-control study, 177 cases of agranulocytosis were compared with 586 sex-, age-, and hospital-matched control subjects with regard to previous use of medicines (9). The annual incidence of community-acquired agranulocytosis was 3.46 per million, and it increased with age. The fatality rate was 7.0% and the mortality rate was 0.24 per million. The following drugs were most strongly associated with a risk of agranulocytosis:

- ticlopidine hydrochloride (OR = 103; 95% CI = 13, 837);
- calcium dobesilate (78; 4.5, 1346);
- antithyroid drugs (53; 5.8, 478);
- dipyrone (metamizole sodium and metamizole magnesium) (26; 8.4, 179);
- spironolactone (20; 2.3, 176).

Aprindine was among other drugs associated with a significant risk. However, the size of the risk was not stated.

Liver

Aprindine can cause liver damage (10).

References

1. Zipes DP, Troup PJ. New antiarrhythmic agents: amiodarone, aprindine, disopyramide, ethmozin, mexiletine, tocainide, verapamil. Am J Cardiol 1978;41(6):1005–24.
2. Schwartz JB, Keefe D, Harrison DC. Adverse effects of antiarrhythmic drugs. Drugs 1981;21(1):23–45.
3. Danilo P Jr. Aprindine. Am Heart J 1979;97(1):119–24.
4. Zipes DP, Elharrar V, Gilmour RF Jr, Heger JJ, Prystowsky EN. Studies with aprindine. Am Heart J 1980;100(6 Pt 2):1055–62.
5. Kodama I, Ogawa S, Inoue H, Kasanuki H, Kato T, Mitamura H, Hiraoka M, Sugimoto T. Profiles of aprindine, cibenzoline, pilsicainide and pirmenol in the framework of the Sicilian Gambit. The Guideline Committee for Clinical Use of Antiarrhythmic Drugs in Japan (Working Group of Arrhythmias of the Japanese Society of Electrocardiology). Jpn Circ J 1999;63(1):1–12.
6. Shibata N, Shirato K, Manaka M, Sugimoto C. Comparison of aprindine and propafenone for the treatment of atrial fibrillation. Ther Res 2001;22:794–6.
7. Atarashi H, Inoue H, Fukunami M, Sugi K, Hamada C, Origasa H. Sinus Rhythm Maintenance in Atrial Fibrillation Randomized Trial (SMART) Investigators. Double-blind placebo-controlled trial of aprindine and digoxin for the prevention of symptomatic atrial fibrillation. Circ J 2002;66(6):553–6.
8. Ibanez L, Juan J, Perez E, Carne X, Laporte JR. Agranulocytosis associated with aprindine and other antiarrhythmic drugs: an epidemiological approach. Eur Heart J 1991;12(5):639–41.
9. Ibáñez L, Vidal X, Ballarín E, Laporte JR. Population-based drug-induced agranulocytosis. Arch Intern Med 2005;165(8):869–74.
10. Elewaut A, Van Durme JP, Goethals L, Kauffman JM, Mussche M, Elinck W, Roels H, Bogaert M, Barbier F. Aprindine-induced liver injury. Acta Gastroenterol Belg 1977;40(5–6):236–43.

Bepridil

General Information

Bepridil is an antidysrhythmic drug with unusual pharmacological properties in that it belongs to both class I and class IV. In other words, it blocks both the fast inward sodium current and the slow outward calcium current in excitable cardiac cells (SEDA-13, 141). It was withdrawn because of its serious prodysrhythmic effects.

Bepridil has been the subject of a brief general review (1) and its pharmacokinetics have been specifically reviewed (2). Although it is highly protein-bound, bepridil does not take part in protein-binding displacement interactions (2).

The main adverse effect of bepridil is torsade de pointes due to QT interval prolongation. After intravenous infusion bepridil can cause local reactions (3) and phlebothrombosis (4). Other minor adverse effects that have been reported include urticaria (5), gastrointestinal disturbances (especially diarrhea) (6,7), and dizziness (6–8). Hepatic enzymes can rise (9,10).

Comparative studies

In a randomized study in 61 patients with symptomatic paroxysmal atrial fibrillation, bepridil 200 mg/day (n = 23) was compared with flecainide 100–200 mg/day or pilsicainide 75–150 mg/day (n = 38) (11). Both bepridil and the class Ic drugs effectively prevented paroxysmal atrial fibrillation (15/23 versus 24/38). In those who took the class Ic drugs, the f-f interval on the surface electrocardiogram during atrial fibrillation before treatment was significantly longer in responders (114 ms) than in non-responders (68 ms). In contrast, in those who took bepridil the f-f interval was significantly shorter in responders (85 ms) than in non-responders (152 ms). In non-responders the class Ic drugs prolonged the f-f interval from 78 ms to 128 ms whereas bepridil had no significant effect (109 versus 135 ms). Although bepridil has been primarily classified as a drug with class I and class IV properties, the authors suggested that these results marked it as acting primarily by a class III mechanism in paroxysmal atrial fibrillation. This is consistent with reports that bepridil inhibits a slow component of the cardiac delayed rectifier potassium current IK_s in HEK293 cells (12).

Organs and Systems

Cardiovascular

Bepridil can cause hypotension after rapid intravenous injection (6,13), but not during long-term oral therapy (14,15).

Bepridil prolongs the QT interval (3,7,8,16), an effect that is dose-related (16). It can therefore cause dysrhythmias, including polymorphous ventricular tachycardia, the risk of which is greater in patients with potassium depletion, those with pre-existing prolongation of the QT interval, those with a history of serious ventricular dysrhythmias, and those who are also taking other drugs that prolong the QT interval (17).

Of 75 elderly patients who took bepridil 200 mg/day, 23 had prolongation of the QT interval. The factors that were associated with this were hypokalemia, bradycardia, renal insufficiency, and an increased plasma bepridil concentration (18).

Respiratory

Interstitial pneumonitis has been attributed to bepridil (19).

- A 65-year-old man with paroxysmal atrial fibrillation took bepridil 150 mg/day and 2 weeks later developed a cough and fever which did not respond to antimicrobial drugs. He had fine crackles at the lung bases and severe hypoxia. An X-ray and a CT scan showed bilateral reticular shadows and microfibrosis, mainly in the lower lungs. Bepridil was withdrawn and he was given prednisolone, to which he responded.

Drug–Drug Interactions

Antidysrhythmic drugs

Because it prolongs the QT interval, bepridil can potentiate the effects of other drugs with the same effect (for example other Class I antidysrhythmic drugs and amiodarone).

Beta-adrenoceptor antagonists

The effect of a beta-blocker (metoprolol 30–40 mg/day or bisoprolol 2.5–5.0 mg/day for 1 month) on the change in QT interval, QT dispersion, and transmural dispersion of repolarization caused by bepridil has been studied in 10 patients with paroxysmal atrial fibrillation resistant to various antidysrhythmic drugs (20). Bepridil significantly prolonged the QTc interval from 0.42 to 0.50 seconds, QT dispersion from 0.07 to 0.14 seconds, and transmural dispersion of repolarization from 0.10 to 0.16 seconds. The addition of a beta-blocker shortened the QTc interval from 0.50 to 0.47 seconds, QTc dispersion from 0.14 to 0.06 seconds, and transmural dispersion of repolarization from 0.16 to 0.11 seconds. The authors therefore suggested that combined therapy with bepridil and a beta-blocker might be useful for intractable atrial fibrillation.

Bepridil does not interact with propranolol (8).

Digoxin

Bepridil does not interact with digoxin (21).

Phenazone (antipyrine)

Bepridil increases the rate of clearance of phenazone (antipyrine) and might therefore be expected to enhance the rate of clearance of other drugs that are metabolized (22).

References

1. Anonymous. Bepridil. Lancet 1988;1(8580):278–9.
2. Benet LZ. Pharmacokinetics and metabolism of bepridil. Am J Cardiol 1985;55(7):C8–C13.
3. Ponsonnaille J, Citron B, Threil F, Heiligenstein D, Gras H. Etude des effets electrophysiologiques du bepridil utilisé par voie veineuse. [The electrophysiologic effects of intravenously administered bepridil.] Arch Mal Coeur Vaiss 1982;75(12):1415–23.
4. Rowland E, McKenna WJ, Krikler DM. Electrophysiologic and antiarrhythmic actions of bepridil. Comparison with verapamil and ajmaline for atrioventricular reentrant tachycardia. Am J Cardiol 1985;55(13 Pt 1):1513–9.
5. Brembilla-Perrot B, Aliot E, Clementy J, Cosnay P, Djiane P, Fauchier JP, Kacet S, Lellouche D, Mabo P, Richard M, Victor J. Evaluation of bepridil efficacy by electrophysiologic testing in patients with recurrent ventricular tachycardia: comparison of two regimens. Cardiovasc Drugs Ther 1992;6(2):187–93.
6. Fauchier JP, Cosnay P, Neel C, Rouesnel P, Bonnet P, Quilliet L. Traitement des tachycardies supraventriculaires et ventriculaires paroxystiques par le bépridil. [Treatment of supraventricular and paroxysmal ventricular tachycardia with bepridil.] Arch Mal Coeur Vaiss 1985;78(4):612–9.
7. Roy D, Montigny M, Klein GJ, Sharma AD, Cassidy D. Electrophysiologic effects and long-term efficacy of bepridil for recurrent supraventricular tachycardias. Am J Cardiol 1987;59(1):89–92.
8. Frishman WH, Charlap S, Farnham DJ, Sawin HS, Michelson EL, Crawford MH, DiBianco R, Kostis JB, Zellner SR, Michie DD, et al. Combination propranolol and bepridil therapy in stable angina pectoris. Am J Cardiol 1985;55(7):C43–9.
9. DiBianco R, Alpert J, Katz RJ, Spann J, Chesler E, Ferri DP, Larca LJ, Costello RB, Gore JM, Eisenman MJ. Bepridil for chronic stable angina pectoris: results of a prospective multicenter, placebo-controlled, dose-ranging study in 77 patients. Am J Cardiol 1984;53(1):35–41.
10. Hill JA, O'Brien JT, Alpert JS, Gore JM, Zusman RM, Christensen D, Boucher CA, Vetrovec G, Borer JS, Friedman C, et al. Effect of bepridil in patients with chronic stable angina: results of a multicenter trial. Circulation 1985;71(1):98–103.
11. Yoshida T, Niwano S, Inuo K, Saito J, Kojima J, Ikeda-Murakami K, Hara H, Izumi T. Evaluation of the effect of bepridil on paroxysmal atrial fibrillation: relationship between efficacy and the f-f interval in surface ECG recordings. Circ J 2003;67:11–15.
12. Yumoto Y, Horie M, Kubota T, Ninomiya T, Kobori A, Takenaka K, Takano M, Niwano S, Izumi T. Bepridil block of recombinant human cardiac IKs current shows a time-dependent unblock. J Cardiovasc Pharmacol 2004; 43:178–82.
13. Flammang D, Waynberger M, Jansen FH, Paillet R, Coumel P. Electrophysiological profile of bepridil, a new

anti-anginal drug with calcium blocking properties. Eur Heart J 1983;4(9):647–54.
14. Canicave JC, Deu J, Jacq J, Paillet R. Un nouvel antiangoreux, le bépridil: appreciation de son efficacité par l'epreuve d'effort au cours d'un essai a double insu contre placébo. [A new antianginal drug, bepridil: efficacy estimation by exertion test during a double blind test against a placebo.] Therapie 1980;35(5):607–12.
15. Upward JW, Daly K, Campbell S, Bergman G, Jewitt DE. Electrophysiologic, hemodynamic and metabolic effects of intravenous bepridil hydrochloride. Am J Cardiol 1985;55(13 Pt 1):1589–95.
16. Perelman MS, McKenna WJ, Rowland E, Krikler DM. A comparison of bepridil with amiodarone in the treatment of established atrial fibrillation. Br Heart J 1987;58(4):339–44.
17. Singh BN. Bepridil therapy: guidelines for patient selection and monitoring of therapy. Am J Cardiol 1992; 69(11):D79–85.
18. Viallon A, Laporte-Simitsidis S, Pouzet V, Venet C, Tardy B, Zeni F, Bertrand JC. Bépridil: intérêt du dosage sérique dans la surveillance du traitement. [Bepridil: importance of serum level in treatment surveillance.] Presse Méd 2000;29(12):645–7.
19. Gaku S, Naoshi K, Teruhiko A. A case of bepridil induced interstitial pneumonitis. Heart 2003;89:1415.
20. Yoshiga Y, Shimizu A, Yamagata T, Hayano T, Ueyama T, Ohmura M, Itagaki K, Kimura M, Matsuzaki M. Beta-blocker decreases the increase in QT dispersion and transmural dispersion of repolarization induced by bepridil. Circ J 2002;66(11):1024–8.
21. Stern H, Aust P, Belz GG, Schneider HT. Interaction entre bépridil et digoxine. Rev Med 1983;24:1279.
22. Funck-Brentano C, Chaffin PL, Wilkinson GR, McAllister B, Woosley RL. Effect of oral administration of a new calcium channel blocking agent, bepridil on antipyrine clearance in man. Br J Clin Pharmacol 1987; 24(4):559–60.

Bretylium

General Information

Bretylium, originally introduced as a hypotensive agent but no longer used as such, has not been extensively used as an antidysrhythmic drug. Its clinical pharmacology, uses, efficacy, and adverse effects have been reviewed (1–4). Adverse effects have reportedly caused the need for withdrawal in about 7% of patients (3).

Organs and Systems

Cardiovascular

Postural hypotension is common, and a significant fall in arterial pressure can occur, even in patients who are supine (5).

Occasionally bretylium causes a transient rise in blood pressure after intravenous administration, due to release of noradrenaline from sympathetic nerve endings.

Ventricular tachydysrhythmias have been reported occasionally (6).

Mouth and teeth

Long-term treatment with bretylium can cause parotid pain and swelling and pain in the tongue (7).

Body temperature

Intravenous bretylium can cause severe hyperthermia (8,9).

Drug Administration

Drug administration route

Bretylium is poorly absorbed and is therefore given parenterally. Despite its poor systemic availability, bretylium was available for some years and was often combined with a tricyclic antidepressant. After intravenous administration bretylium can cause hypotension, nausea, vomiting, and diarrhea. Bretylium should be infused over 30–60 minutes to minimize these effects.

References

1. Schwartz JB, Keefe D, Harrison DC. Adverse effects of antiarrhythmic drugs. Drugs 1981;21(1):23–45.
2. Cooper JA, Frieden J. Bretylium tosylate. Am Heart J 1971;82(5):703–6.
3. Koch-Weser J. Drug therapy: bretylium. N Engl J Med 1979;300(9):473–7.
4. Rapeport WG. Clinical pharmacokinetics of bretylium. Clin Pharmacokinet 1985;10(3):248–56.
5. Taylor SH, Saxton C, Davies PS, Stoker JB. Bretylium tosylate in prevention of cardiac dysrhythmias after myocardial infarction. Br Heart J 1970;32(3):326–9.
6. Anderson JL, Popat KD, Pitt B. Paradoxical ventricular tachycardia and fibrillation after intravenous bretylium therapy. Report of two cases. Arch Intern Med 1981;141(6):801–2.
7. Heinrich KW, Effert S. Bretylium-Tosylat zur Behandlung maligner Arrhythmien: erste Resultate. [Bretylium tosylate in the therapy of malignant arrhythmia. First results.] Med Welt 1973;24(24):1000–2.
8. Thibault J. Hyperthermia associated with bretylium tosylate injection. Clin Pharm 1989;8(2):145–6.
9. Perlman PE, Adams WG Jr, Ridgeway NA. Extreme pyrexia during bretylium administration. Postgrad Med 1989;85(1):111–4.

Cardiac glycosides

General Information

Many aspects of the pharmacology, clinical pharmacology, and adverse effects and interactions of cardiac glycosides have been reviewed (1–12).

Table 1 Cardiac glycosides that have been used therapeutically

Drug	*Source*	*Main route of elimination*
Acetyldigoxin	Semisynthetic derivative of digoxin	Renal
Digitoxin	*Digitalis purpurea*	Hepatic
Digoxin	*Digitalis lanata*	Renal
Gitoformate	*Digitalis purpurea*	Hepatic
Gitoxin	*Digitalis purpurea*	Hepatic
Lanatoside C	*Digitalis lanata*	Renal
Metildigoxin (betamethyldigoxin)	Semisynthetic derivative of digoxin	Hepatic/renal
Ouabain(strophanthin-g)	*Strophanthus gratus, Acokanthera schimperi, Acokanthera ouabaio*	Renal
Peruvoside	*Nerium peruviana*	Hepatic
Proscillaridin	*Drimia maritima*	Hepatic
Strophanthin-k	*Strophanthus kombe*	Renal

Note on nomenclature

The most commonly used cardiac glycosides, digoxin and digitoxin, are derived from foxgloves, respectively *Digitalis lanata* and *Digitalis purpurea*. For this reason they are generally known as "digitalis." Most other cardiac glycosides, such as ouabain and proscillaridin, do not come from foxgloves but are nevertheless also commonly called "digitalis." Thus, the terms "cardiac glycoside" and "digitalis" are used interchangeably.

The cardiac glycosides that have been used therapeutically are listed in Table 1.

Mechanisms of digitalis toxicity

There is a large amount of evidence that the mechanisms of action of cardiac glycosides are mediated directly or indirectly by inhibition of the sodium/potassium pump enzyme, Na/K-ATPase (13). Their toxic effects on the myocardium may be due to excessive inhibition of cardiac Na/K-ATPase, although there is also evidence that effects on the nervous input to the heart may be involved (14), and it is not clear to what extent such an effect is mediated by inhibition of Na/K-ATPase. However, color vision disturbances associated with cardiac glycosides are due to inhibition of Na/K-ATPase (15).

Epidemiology of digitalis toxicity

Digitalis toxicity is common, since all cardiac glycosides have a low therapeutic index. Estimates vary widely from study to study, but in large prospective studies of hospital inpatients the frequency of digitalis toxicity has been as high as 29% (16). In outpatients the figure may be as high as 16% (7). The lower frequency in outpatients may be due partly to poor compliance and partly to digitalis toxicity being a reason for admission to hospital, thus increasing the numbers of toxic inpatients. The risk of toxicity may be lower with digitoxin than with digoxin (4), but when toxicity occurs it lasts longer, because of the very long half-life of digitoxin.

The overall mortality from digitalis toxicity also varies widely, having been reported as low as 4% and as high as 36% (7). However, it varies with dysrhythmias, and for paroxysmal supraventricular tachycardia with block may be as high as 50% (7).

In a study of serum digoxin concentrations in 1433 patients admitted to hospital, 115 had a raised concentration (17). Of the 82 in whom the blood sample had been taken at an appropriate time, 59 had electrocardiographic or clinical features of digoxin toxicity. The patients whose serum digoxin concentrations were over 3.2 nmol/l (2.5 ng/ml) were slightly older (78 versus 73 years) and had higher serum creatinine concentrations (273 versus 123 μmol/l) than those whose plasma concentrations were below 3.1 nmol/l. Of 47 patients with raised digoxin concentrations on admission, 21 were admitted because of digoxin toxicity, and impaired or worsening renal function contributed to high concentrations in 37 patients. A drug interaction was a contributory factor in 10 cases. These results suggest that digoxin toxicity is still very common and confirms the increased risk in elderly patients, patients with renal impairment, and patients taking drugs that may interact with digoxin. Serum potassium concentrations were not reported in this study.

In another study of this sort, serum digoxin concentrations were measured in 2009 patients (18). The concentration was over 2.6 nmol/l in 320 cases (9.3%) but in 51 of those the sample had been drawn too soon after the dose. When other results were omitted in cases in which the sampling time was not known, there were 138 evaluable patients, of whom 83 had clinical evidence of digoxin toxicity, an overall incidence of 4.1%. The authors concluded that digoxin toxicity was less common in their series than has previously been reported. There were no differences between the groups in serum potassium, calcium, or magnesium concentrations, but the serum creatinine concentration was significantly higher in those who had definite and possible toxicity. The mean age of the patients was 69 years. It is likely that the differences across studies of this sort are largely due to differences in renal function and age in the population being studied.

In a multicenter survey, conducted between 1988 and 1997, of 28 411 patients aged 70 (s.d. 16) years admitted to 81 hospitals throughout Italy, 1704 had adverse drug reactions (19). In 964 cases (3.4% of all admissions), adverse reactions were considered to be the cause of admission. Of these, 187 were regarded as severe. Gastrointestinal complaints (19%) were the most common, followed by metabolic and hemorrhagic complications (9%). The drugs most often responsible were diuretics, calcium channel blockers, non-steroidal anti-inflammatory drugs, and digoxin. Female sex (OR = 1.30; 95% CI = 1.10, 1.54), alcohol use (OR = 1.39; 95% CI = 1.20, 1.60), and number of drugs (OR = 1.24; 95% CI = 1.20, 1.27) were independent predictors of admission for adverse reactions. For severe adverse reactions, age (for age 65–79, OR = 1.50; 95% CI = 1.01, 2.23; for age 80 and over, OR = 1.53; 95% CI = 1.00, 2.33), co-morbidity (OR = 1.12; 95% CI = 1.05, 1.20 for each point on the Charlson Comorbidity Index), and number of drugs (OR = 1.18; 95% CI = 1.11, 1.25) were predisposing factors. Of the 28 411 patients, about 6700 were taking digoxin, and they suffered 82 adverse effects, either gastrointestinal (n = 28) or unspecified dysrhythmias (n = 44), or presumably both (data not given); of those, 11 were graded as severe (two gastrointestinal and nine dysrhythmias).

Of 603 adults aged 79 years, of whom 59% were women and 18% African-American, 376 patients (62%) were discharged taking digoxin, and 223 (37%) had no indication for its use, based on the absence of left ventricular systolic dysfunction or atrial fibrillation (20). After adjustment for various factors, prior digoxin use (OR = 11; 95% CI = 5.7, 23) and pulse over 100/minute (OR = 2.33; 95% CI = 1.1, 4.9) were associated with inappropriate digoxin use. Unfortunately, the authors did not report the frequency of adverse effects, and it is not therefore clear whether patients in whom digoxin is used inappropriately are more or less likely to suffer adverse reactions.

General adverse effects

The adverse effects of cardiac glycosides can be cardiac or non-cardiac. They mostly occur through toxicity and are time-independent (DoTS classification); susceptibility factors include electrolyte abnormalities (particularly hypokalemia), renal insufficiency, and age.

Frequent non-cardiac reactions include gastrointestinal effects (anorexia, nausea, vomiting, and diarrhea), central nervous system effects (drowsiness, dizziness, confusion, delirium), and less commonly visual effects (color vision abnormalities, photophobia, and blurred vision). Hypersensitivity reactions are rare and include thrombocytopenia and skin rashes. Tumor-inducing effects have not been reported.

Frequent cardiac adverse effects include heart block and ectopic dysrhythmias (ventricular extra beats, other ventricular tachydysrhythmias, and paroxysmal supraventricular tachycardia). The combination of heart block with an ectopic dysrhythmia, for example paroxysmal supraventricular tachycardia with block, is particularly suggestive of toxicity due to cardiac glycosides. Any other dysrhythmia can occasionally be caused by cardiac glycosides.

Of 332 residents of a nursing home, 52 had to be admitted to hospital because of adverse drug reactions (21). The drugs most commonly associated with adverse effects were non-steroidal anti-inflammatory drugs (n = 30), psychotropic drugs (n = 14), and digoxin (n = 5).

Individual cardiac glycosides

There are major pharmacokinetic differences among the different cardiac glycosides, the principal difference being between those that are mainly excreted via the kidneys (for example digoxin, metildigoxin, beta-acetyldigoxin, ouabain, and k-strophanthin) and those that are mainly excreted via hepatic metabolism (including digitoxin, gitoxin, pengitoxin (16-acetyldigoxin), and gitoformate).

It has also been suggested that there may be some pharmacodynamic differences among different cardiac glycosides (22), but these may at least partly be determined by differences in tissue distribution.

It is debatable whether any of these differences makes any particular cardiac glycoside preferable to another. The most strongly argued case is that digitoxin is preferable to digoxin in patients with renal insufficiency, since digitoxin is metabolized and digoxin is excreted by the kidneys. However, digitoxin has a much longer duration

Table 2 Some plants that contain cardiac glycosides

Plant	*Common name(s)*	*Cardiac glycoside(s)*	*Comments*
Adonis vernalis	False hellebore, pheasant's eye	Adonitoxin, strophanthidin	
Antiaris toxicaria	Upas tree	Antiarin	A Javanese tree of the mulberry family, used as an arrow poison
Convallaria majalis	Lily of the valley	Convallamarin	
Erysimum helveticum	Wallflower	Helveticoside	
Helleborus niger	Black hellebore, Christmas rose	Helleborcin	Also called melampodium after the Greek physician Melampus, who used it as a purgative
Nerium oleander	Pink oleander	Neriifolin	
Periploca graeca	Silk vine	Periplocin	
Tanghinia venenifera	Ordeal tree	Tanghinin	At one time used in Madagascar to test the guilt of someone suspected of a crime
Thevetia peruviana	Yellow oleander	Peruvoside, thevetins	Seeds widely used for self-poisoning in Southern India and Sri Lanka
Urginea (Scilla) maritima	Squill	Proscillaridin	Squill was a common remedy for dropsy in ancient times and up to the 19th century

of action, and if toxicity occurs it will take longer to resolve. Furthermore, determining the effective dose of digitoxin is much more difficult, since there is great interindividual variability in the extent to which digitoxin is metabolized, and hepatic metabolic function cannot be directly measured. Although digoxin excretion also varies from patient to patient, it can at least be gauged by measurement of creatinine clearance. The arguments for and against these preferences have been outlined (4,23) and it is probably best to choose a particular drug according to individual patient requirements.

Herbal formulations

Numerous plants worldwide contain cardiac glycosides that have been used both therapeutically as herbal formulations and for the purposes of self-poisoning. Details are given in Table 2.

- A 59-year-old man developed third-degree atrioventricular block after using an extract of *Nerium oleander* transdermally to treat psoriasis (24). A fatality due to drinking a herbal tea prepared from *N. oleander* leaves, erroneously believed to be eucalyptus leaves, has been reported (25).

Poisoning from ingestion of the seeds of *Thevetia peruviana* (yellow oleander) can be treated with oral multiple-dose activated charcoal, which reduce mortality (26). The lack of efficacy of activated charcoal in one trial (27) was probably due to failure to select the patients that were most likely to benefit (28). Fab fragments of antidigoxin antibody have also been used to treat oleander intoxication, for example in a 7-year-old child (29) and a 44-year-old man (30), but in a randomized controlled trial did not affect mortality (31). A retrospective study suggested that the antibody fragments might reduce mortality (32), but was confounded by the simultaneous introduction of activated charcoal.

In a Turkish case, the ingestion of two bulbs or *Urginea maritima* as a folk remedy for arthritic pains was sufficient to result in fatal poisoning (33).

Organs and Systems

Cardiovascular

Cardiac dysrhythmias and heart block

Percentage incidence figures for digitalis-induced dysrhythmias were given by Chung in his review of 726 patients (5). The commonest dysrhythmias are ventricular extra beats (54% of all dysrhythmias), coupled ventricular extra beats (25%), and supraventricular tachycardia (33%). Sinus tachycardia was not common (3.4%). Atrial fibrillation (1.7%) or atrial flutter (1.8%) can cause difficulty in diagnosis, since digitalis is often used to treat those dysrhythmias.

Atrioventricular block was common (42%): first-degree, 14%; second-degree, 17%; and complete, 11%). However, first-degree heart block (that is prolongation of the PR interval) without higher degrees of atrioventricular nodal block can occur in the absence of digitalis intoxication.

Digitalis-induced dysrhythmias can be classified according to their sites of origin in the sinus node, the atria and atrioventricular node, and the ventricles.

Sinoatrial node

Digitalis can cause sinus bradycardia as a toxic effect, although patients with sinus bradycardia at rest often have no other evidence of digitalis toxicity, and this effect may simply represent increased vagal tone (34). Digitalis inhibits conduction through the sinoatrial node and has been reported to cause a syndrome mimicking that of the sick sinus syndrome (35–37); however, it is not clear whether or not it can impair sinus node function in patients who have previously normal sinoatrial nodes (38,39). Digitalis can certainly worsen sinus node function that has been otherwise impaired, for example by hyperthyroidism (40) or endotracheal suction (41).

Among 8770 patients aged over 65 years with a new diagnosis of atrial fibrillation who had had a previous myocardial infarction, there were 477 cases of bradydysrhythmias requiring a permanent pacemaker, and they were matched 1: 4 to 1908 controls; the use of digoxin was associated with an increased risk of pacemaker insertion (OR = 1.78; 95% CI = 1.37, 2.31) (42).

Atria and atrioventricular node

Digitalis can cause supraventricular extra beats or tachycardia. The combination of such dysrhythmias with atrioventricular block is particularly suggestive of digitalis toxicity and carries a high mortality rate (3,43). Rarely atrial fibrillation (44) and atrial flutter (45) may be attributed to digitalis toxicity. The frequency of atrioventricular nodal block is mentioned above.

In two patients with chronic atrial fibrillation taking digoxin, the administration of Fab fragments of antidigoxin antibodies for digoxin toxicity caused conversion to sinus rhythm, in one case maintained for 6 months before atrial fibrillation recurred and in the other case maintained for at least 15 days (46). In these cases digoxin may have caused atrial fibrillation, the adverse effect being reversed by the antibody.

Paroxysmal atrial tachycardia with Wenckebach (Mobitz type I) atrioventricular block has been reported in a patient with a serum digoxin concentration of 3.2 ng/ml (47) and in a patient who in error took three times the recommended dose (48).

Ventricles

Ventricular extra beats, including coupled beats (that is ventricular bigeminy), are the most common cardiac effects of digitalis toxicity, although they are not specific. In more severe cases ventricular tachycardia, bidirectional tachycardia, and ventricular fibrillation can occur. There have also been reports of accelerated idioventricular rhythm (49,50).

Digoxin can cause ventricular fibrillation in children with Wolff–Parkinson–White syndrome (51,52).

- A male infant, whose narrow-complex tachycardia at birth had responded to adenosine, was treated with digoxin and 1 week later, during transesophageal electrophysiology with isoprenaline, developed coarse ventricular fibrillation after the induction of a supraventricular tachycardia (53). The serum digoxin concentration was not measured. The isoprenaline was withdrawn and the dysrhythmia resolved spontaneously at 160 seconds.

The effects of digoxin, isoprenaline, and transesophageal stimulation may have combined in this case to cause ventricular fibrillation.

- Bidirectional ventricular tachycardia occurred in an 86-year-old woman with renal insufficiency whose serum digoxin concentration was 13 ng/ml (digoxin dose 250 micrograms/day) (54); the tachycardia resolved after 6 days, when the digoxin concentration fell to 1.6 ng/ml.
- A broad complex tachycardia due to hyperkalemia and mild digoxin toxicity occurred in a 78-year-old woman (55).
- Asystolic cardiac arrest occurred 1 week after surgical repair of a congenital heart anomaly in a 12-week-old girl (56).

Effects of digitalis on the electrocardiogram

Digitalis can prolong the PR interval and cause shortening of the QT interval, depression of the ST segment, and asymmetrical T wave inversion. These effects are nonspecific and can occur in the absence of toxicity. However, there is evidence that the effects on the ST segment and T wave may be more common in patients with co-existing ischemic heart disease (57). Digitalis can also rarely cause both left (58) and right (59) bundle branch block.

The electrocardiographic effects of cardiac glycoside toxicity in 688 patients have been reviewed in the context of three cases of digoxin toxicity (60). The three cases featured bidirectional tachycardia in a 50-year-old man with a plasma digoxin concentration of 3.7 ng/ml, junctional tachycardia in a 59-year-old man with a plasma digoxin concentration of 4.3 ng/ml, and complete heart block in a 90-year-old woman whose postmortem digoxin concentration was 5.0 ng/ml.

Heart failure

In toxic doses digitalis impairs myocardial contractility and can cause or worsen heart failure. In one series of 148 patients with digitalis intoxication, worsening heart failure was diagnosed in 7.5% (61). In some cases worsening heart failure may be attributable to a cardiac dysrhythmia (62).

Vasoconstrictor and hypertensive effects

Giving a cardiac glycoside rapidly intravenously causes a transient increase in blood pressure, which has been attributed to an increase in peripheral resistance (63). However, digitalis does not seem to increase blood pressure during long-term treatment.

Myocardial ischemia

Subacute digitalis intoxication in dogs causes myocardial damage (64), and after intravenous administration there is increased creatine kinase activity in the plasma in man (65), suggesting ischemic damage.

There has been a report of coronary vasoconstriction in patients who were given acetyldigoxin 0.8 mg intravenously at angiography (66). Pretreatment with nisoldipine 10 mg, 2 hours before angiography, prevented the digoxin-induced vasoconstriction. These patients all had pre-existing coronary artery disease, but the vasoconstrictor effect occurred in both normal and abnormal coronary segments. However, the effect on high-grade stenoses was more pronounced. There is other evidence that ischemic damage can occur in patients who have been given digoxin intravenously, including an increase in the activity of serum creatinine kinase (65) and impaired left ventricular function after acute myocardial infarction (67).

- A 26-year-old woman who had taken a herbal supplement for stress relief which contained *Scutellaria lateriflora*, *Pedicularis canadensis*, *Cimifuga racemosa*, *Humulus lupulus*, *Valeriana officinalis*, and *Capsicum annuum* developed chest pain of 7 hours duration (68). Her medical history was otherwise unremarkable. Examination of her heart showed no abnormality, but during monitoring her heart rate fell to 39/minute and her blood pressure to 59/36 mmHg. Her serum digoxin concentration was 0.9 ng/ml. The authors therefore concluded that the herbal remedy contained digoxin-like factors that had caused digitalis toxicity.

Long-term use and cardiovascular adverse effects of cardiac glycosides

There have been many studies of the long-term efficacy of digitalis in patients in heart failure in sinus rhythm and also in patients with atrial fibrillation. These have been reviewed (SEDA-4, 123) (SEDA-14, 145) (SEDA-18, 196). The following is a brief resumé.

Atrial fibrillation is not necessarily an indication for long-term therapy with digitalis. In patients with controlled atrial fibrillation whose plasma digitalis concentration is below the lower limit of the target range (0.8 ng/ml for digoxin and 10 ng/ml for digitoxin) withdrawal rarely if ever results in deterioration. However, in those who have plasma digitalis concentrations within the target range withdrawal should not be attempted, since the risk of worsening atrial fibrillation outweighs the risk of toxicity, if there is careful monitoring of the plasma concentration.

In patients with heart failure in sinus rhythm there is no way of predicting which patients will benefit from long-term therapy, but the following recommendations can be made:

1. If the plasma digitalis concentration is below the therapeutic range (0.8 ng/ml for digoxin and 10 ng/ml for digitoxin) withdrawal is very unlikely to produce deterioration.
2. If a patient's condition is stable, and the plasma digitalis concentration is in the therapeutic range, with

little risk of toxicity, withdrawal is probably not worthwhile because of the risk of deterioration.
3. If there is an increased risk of toxicity (for example because of renal impairment or if potassium balance is difficult to maintain) careful withdrawal of digitalis may be worth attempting.
4. In patients who have evidence of poor left ventricular function it may be better to continue therapy, even if there is an increased risk of toxicity, since these patients are very likely to deteriorate following withdrawal. In these cases careful monitoring of therapy will help to reduce the risk of toxicity.

The long-term adverse cardiovascular effects of digitalis have been reviewed (SEDA-10, 142) (SEDA-11, 153) (SEDA-15, 165). Briefly, in a number of retrospective studies, although mortality in the digitalis-treated patients was generally higher than in those not treated with digitalis, the difference was reduced when allowance was made for other confounding factors, such as the degree of heart failure, a history of dysrhythmias, and the use of other drugs. There may also be a higher mortality rate in patients who take long-term digitalis therapy after coronary artery bypass graft surgery (69), in patients who have had a cardiac arrest (70), and in those without evidence of heart failure or atrial fibrillation (71). In the first two studies the risk was increased further among those who were taking digitalis with diuretics, and it may be that these effects are due to digitalis toxicity secondary to potassium depletion, although it may simply indicate a greater prevalence of hypertension or heart failure among those treated with digitalis and diuretics.

Despite these earlier results, in the Digitalis Investigation Group (DIG) study (72) digoxin had no overall impact on mortality in patients with heart failure in sinus rhythm and with left ventricular ejection fractions equal to or less than 0.45.

It produced a small reduction in hospitalizations due to heart failure (nine per 1000 patients-years) balanced by a significant increase in deaths from presumed dysrhythmias. Digitalis is therefore indicated for a small number of patients who have severe heart failure associated with sinus rhythm after treatment with diuretics, vasodilators, beta-blockers, and spironolactone. It remains the drug of first choice in patients with heart failure accompanied by fast atrial fibrillation, especially if due to myocardial or mitral valve disease. A trial of withdrawal of digitalis therapy can be considered in some cases (as noted in point 3 above).

Of course, there are alternatives to digitalis in the long-term treatment of heart failure in sinus rhythm. It is not clear that any of these offers any particular advantage over digitalis in terms of therapeutic efficacy, although there may be fewer problems with toxicity. The comparative studies have been reviewed (SEDA-14, 141).

Sinus rhythm

The question of whether digoxin should be used to treat patients with mild to moderate heart failure in sinus rhythm, in the wake of randomized, controlled trials of its efficacy, including PROVED, RADIANCE, and DIG (SEDA-18, 196) (SEDA-20, 173), has been reviewed (73). The authors concluded that digoxin is effective in producing symptomatic improvement in patients with mild or moderate heart failure, but that because of concerns about its safety careful consideration must be taken in each case before using it.

Other positive inotropic drugs carry no extra benefit, and can increase mortality during long-term administration. There is no evidence that the combination of two drugs with positive inotropic actions is beneficial in chronic congestive heart failure. Vasodilators are as efficacious as digitalis, but there is a rationale for combining digitalis and a vasodilator, since by doing so it is possible to affect simultaneously the three important factors determining cardiac output (contractility, pre-load, and after-load). Furthermore, a vasodilator will oppose the small effect that digitalis has in increasing peripheral resistance, and which may reduce the beneficial effect of digitalis on cardiac output.

Of 2254 elderly patients, 724 were being treated with digoxin, of whom 187 had congestive heart failure, 90 had atrial fibrillation, and 447 were both free from heart failure and in sinus rhythm (71). Among those who did not have heart failure or atrial fibrillation, cardiovascular and total mortality were significantly higher among those taking digoxin. Digoxin was a predictor of mortality in those subjects. In addition, the incidence of non-fatal heart failure was higher among those taking digoxin. This is yet another non-randomized study purporting to show deleterious effects of digoxin during long-term use, in this case in patients in whom it was not indicated in the first place. Since similar non-randomized studies in patients with heart failure, which also showed deleterious effects (SEDA-20, 173), have since been contradicted by proper prospective randomized studies, this result should be ignored.

Atrial fibrillation

In uncomplicated atrial fibrillation a cardiac glycoside such as digoxin remains the drug of first choice. However, in those in whom digitalis is not completely effective or in whom symptoms (for example bouts of palpitation) persist despite adequate digitalization, a calcium antagonist, such as verapamil or diltiazem, can be added, or amiodarone used as an alternative. In patients with atrial fibrillation due to hyperthyroidism, a beta-adrenoceptor antagonist should be used in preference to digitalis, but digitalis can be added if there is an incomplete effect. In patients with atrial fibrillation secondary to an anomalous conduction pathway (for example Wolff–Parkinson–White syndrome), in most of whom digitalis is contraindicated, a calcium antagonist would be the treatment of choice. Paroxysmal atrial fibrillation generally does not respond to digitalis, and digitalis may in fact prolong the duration of a paroxysmal attack when it occurs. The treatment of paroxysmal atrial fibrillation is problematic, but many use amiodarone. Sotalol, propafenone, and flecainide are options, but there are doubts about the long-term safety of flecainide and sotalol, particularly in those who have had an acute myocardial infarction.

There has been a multicenter, randomized, placebo-controlled, double-blind comparison of aprindine and digoxin in the prevention of atrial fibrillation and its recurrence in 141 patients with symptomatic paroxysmal or persistent atrial fibrillation who had converted to sinus rhythm (74). They were randomized in equal numbers to aprindine 40 mg/day, digoxin 0.25 mg/day, or placebo, and were followed every 2 weeks for 6 months. After 6 months the Kaplan-Meier estimates of the numbers of patients who had no recurrences while taking aprindine, digoxin, and placebo were 33%, 29%, and 22% respectively. The rates of adverse events were similar in the three groups. This confirms that digoxin does not prevent relapse of symptomatic atrial fibrillation after conversion to sinus rhythm.

Cardioversion and digitalis

The presence of digitalis increases the risk of serious dysrhythmias after electrical cardioversion, even in the absence of frank toxicity (75). In order to minimize the risk of dysrhythmias in these circumstances digitalis should be withdrawn if possible a day or two before cardioversion and potassium depletion should be corrected. If cardioversion is required acutely, it has been recommended that low energies (for example 10 J) should be used initially (76).

Nervous system

Toxic effects of digitalis on the nervous system occur relatively often. Although in severe toxicity the incidence may be as high as 65% (77), in most series it has been below 25% (78). In 8220 patients aged at least 65 years, the crude relative risk of nervous system dysfunction of unspecified types was about 2.0 (95% CI = 1.7, 2.5) among the 16.5% who were taking digoxin; co-morbidity and the number of out-patient medical services did not affect the risk (79).

Anorexia, nausea, and vomiting are mediated by the central nervous system. Other common nervous system effects of digitalis include confusion, dizziness, drowsiness, bad dreams, restlessness, nervousness, agitation, and amnesia.

Epilepsy occurs rarely and can be accompanied by electroencephalographic changes (80–82).

Other reported effects include transient global amnesia (83), trigeminal neuralgia (84,85), nightmares (86), organic brain syndrome (including impairment of long-term and short-term memory) (87), impairment of learning and memory (23), and a clinical syndrome resembling herpes encephalitis (88).

- Progressive stupor has been reported in an 85-year-old woman with mild renal insufficiency who was given digoxin 0.25 mg/day (89). The plasma digoxin concentration was 7.8 nmol/l. She recovered within 2 weeks after digoxin withdrawal, consistent with the likely half-life of digoxin. A 24-hour electrocardiogram showed one period of asystole for 4 seconds, but that is unlikely to have explained her symptoms.

Chorea has occasionally been reported in adults taking digitalis (SEDA-13, 138; 90), and also in a child (91).

- A 7-year-old girl with severe congenital heart disease who was given digoxin 0.125 mg bd developed chorea and had a serum digoxin concentration of 3.8 ng/ml. When digoxin was withheld and the serum concentration fell to 1.5 ng/ml her symptoms resolved. They recurred 4 days after rechallenge when her digoxin concentration was 2.5 mg/ml and again resolved after it had fallen to 1.3 mg/ml.

The authors hypothesized that digoxin caused chorea by virtue of an estrogenic effect in the basal ganglia, similar to the effect that is occasionally produced by oral contraceptives.

Sensory systems

Color vision abnormality is a well-known adverse effect of digitalis (SEDA-20, 173), and particularly occurs in patients with digitalis toxicity.

There have been two cases of digoxin-related visual disturbances in patients whose blood concentrations were in the usual target range (92).

- A 68-year-old woman had shimmering lights in her field of vision in both eyes when in sunlight, and a 63-year-old woman complained of blurring of vision in both eyes. The serum digoxin concentrations were 2.2 and 1.3 nmol/l (1.7 and 1.0 ng/ml) respectively. Withdrawal of digoxin caused resolution of their symptoms within 1–2 weeks.

Unfortunately the authors did not report serum electrolyte concentrations, and it is not clear in these cases whether the effect of digoxin was potentiated by potassium depletion.

In 30 patients (mean age 81 years) taking digoxin and an age-matched control group there was no correlation between color vision impairment and serum digoxin concentration (93). There was slight to moderate red-green impairment in 20–30% of those taking digoxin, depending on the test used; about 20% had a severe tritan defect. The authors suggested that color vision testing in elderly patients would have limited value in the detection of digitalis toxicity. However, this conclusion was based on using the digoxin concentration as a standard, while the point of pharmacodynamic tests, such as color vision measurement, is that they are supposed to reflect the effect of the drug better than the serum concentration.

Psychological, psychiatric

Acute psychosis and delirium can occur in digitalis toxicity, particularly in elderly people (94–96), and can be accompanied by visual or auditory hallucinations (97,98).

- Acute delirium occurred in a 61-year-old man whose serum digitoxin concentration was 44 ng/ml (99).

Digitalis toxicity can occasionally cause depression (100).

- A 77-year-old woman developed extreme fatigue, anorexia, psychomotor retardation, and social withdrawal 1 month after starting to take digoxin 0.5 mg/day for congestive heart failure (101). She did not respond to intravenous clomipramine 25 mg/day for 7 months. Her serum digoxin concentration was 3.2 ng/ml. Digoxin was withdrawn, and 12 days later, when her serum digoxin concentration was 0.5 ng/ml, she had improved, but was left with a memory disturbance, which was attributed to background dementia.

Endocrine

Digitalis has effects on sex hormones. It causes increased serum concentrations of follicle-stimulating hormone (FSH) and estrogen and reduced concentrations of luteinizing hormone (LH) and testosterone (102–105). These effects are probably not related to any direct estrogen-like structure of digitalis (despite structural similarities), but rather to an effect involving the synthesis or release of sex hormones. There are three possible clinical outcomes of these effects.

Gynecomastia in men and breast enlargement in women

Effects of cardiac glycosides in the breasts can be associated with demonstrable histological changes (106,107).

Stratification of the vaginal squamous epithelium in postmenopausal women

This can cause difficulty in the pathological interpretation of vaginal smears for cancer diagnosis (108).

A possible modifying effect on breast cancer

Digitalis can reduce the heterogeneity of breast cancer cell populations and reduce the rate of distant metastases (109). There is also evidence that the 5-year recurrence rate after mastectomy is lower in women who have been treated with digitalis (110). Early studies suggested that when breast tumors occurred in women with congestive heart failure taking cardiac glycosides, tumor size was significantly smaller and the tumor cells more homogeneous (SEDA-7, 194). It was originally thought that this action was due to an estrogen-like effect of cardiac glycosides, but more recent evidence suggests that it occurs because inhibition of the Na/K pump is involved in inhibiting proliferation and inducing apoptosis in various cell lines (111–114). Cardiac glycosides have different potencies in their effects on cell lines such as those of ovarian carcinoma and breast carcinoma (order of potency: proscillaridin A > digitoxin > digoxin > ouabain > lanatoside C) (115).

Metabolism

In three patients with diabetes mellitus, withdrawal of digoxin improved blood glucose control, implying that digoxin had impaired glucose tolerance (116). The authors conceded that the effect might have occurred coincidentally, but in one case glucose tolerance deteriorated again after rechallenge. Insulin increases the cellular uptake of glucose and stimulates the sodium/potassium pump, and it may be that inhibition of the sodium/potassium pump by digoxin has the opposite effect.

In 14 patients with morbid obesity, who were being given digoxin in the hope that reduced production of cerebrospinal fluid, with the consequent reduction in pressure, might be associated with weight reduction, the dosage of digoxin (Lanacrist 0.13 mg, equivalent to 0.065 mg of digoxin) was titrated to produce a minimum serum digoxin concentration of 1.0 nmol/l (117). One patient was already diabetic, and five developed fasting blood glucose concentrations greater than 5.0 mmol/l on three consecutive occasions, with accompanying glycosuria. Another had fasting blood glucose concentrations of 6.0–8.5 mmol/l. There was a significant relation between the dose of digoxin and the risk of impaired glucose tolerance. However, the diabetes mellitus did not abate after digoxin withdrawal, and since all these patients were obese, the occurrence of diabetes was probably coincidental.

Hematologic

Thrombocytopenia has been reported in patients taking digitoxin, acetyldigoxin, and digoxin (118–121). Cardiac glycosides inhibit Na/K-ATPase, causing changes in intracellular calcium concentration. Increased intracellular calcium is a key event in platelet activation, and there is evidence that cardiac glycosides activate platelets in vitro, albeit in high concentrations (122,123). Platelet and endothelial functions have been studied in 30 patients with non-valvular atrial fibrillation, 16 of whom were taking digoxin (mean plasma digoxin concentration 1.2 nmol/l) (124). Digoxin significantly increased platelet CD62P expression and platelet–leukocyte conjugates and markedly increased EMP62E and EMP31, markers of endothelial activation. After adjusting for potential confounders (including age, congestive heart failure, coronary artery disease, ejection fraction, antiplatelet drugs, beta-blockers, and calcium channel blockers), the differences persisted. The authors concluded that if digitalis activates endothelial cells and platelets it could predispose to thrombosis and vascular events. However, there is no evidence that that happens clinically.

There have been rare reports of eosinophilia in patients taking cardiac glycosides (125).

Gastrointestinal

Gastrointestinal symptoms are common in digitalis toxicity. These include anorexia, nausea, and vomiting (126), probably as a result of stimulation of the chemoreceptor trigger zone in the brain.

Diarrhea occurs occasionally (127).

Dysphagia has been rarely reported (128,129).

Other rare events include intestinal ischemia (SEDA-17, 215) (130), and hemorrhagic intestinal necrosis (131).

- A 79-year old woman with a serum digoxin concentration of 4.9 ng/ml had a mesenteric infarction (132). At postmortem no other causes were discovered.
- An 84-year-old woman developed abdominal pain in association with symptoms of digitalis toxicity while

taking digitoxin 0.07 mg/day and other drugs, including furosemide (133). Her serum potassium concentration was 2.9 mmol/l and the serum digitoxin concentration was 32 ng/ml (usual target range 13–25). She had first-degree heart block, incomplete left bundle branch block, and typical ST segment changes. All medications were withdrawn and the hypokalemia was corrected with intravenous potassium. Abdominal X-ray and ultrasonography showed paralytic ileus and she died 48 hours later. At autopsy there was hemorrhagic congestion of the heart, lungs, and other organs, and the intestines were edematous and hemorrhagic, with submucosal edema, necrotic ulceration, and intramural bleeding. There was no thromboembolism.

The authors attributed this effect to digitoxin toxicity. Verapamil, diltiazem, and antidigoxin antibody fragments have all been reported to be beneficial in mesenteric ischemia induced by cardiac glycosides. Intestinal ischemia responds to verapamil and to antidigoxin antibody (131).

Necrotic enterocolitis might have been related to digoxin toxicity in a neonate (134). However, the authors did not highlight this in their report, but instead emphasized that digitalis intoxication in neonates may present with vomiting and no cardiac signs of toxicity.

Death

In a retrospective, non-randomized study of 484 patients, 90 of whom were taking digoxin, there was an increased death rate (RR = 2.12, CI = 1.21, 3.74) in those taking digoxin (135). In another non-randomized, retrospective analysis of the effects of digoxin in patients with acute myocardial infarction there was a higher rate of mortality in the 243 patients taking digoxin compared with the 1743 patients who were not (136). The results of these studies are reminiscent of the results of previous similar retrospective analyses. However, the prospective DIG study clearly showed no increase in mortality (SEDA-20, 173), and the results of these later non-randomized retrospective studies should be ignored.

A post hoc re-analysis of the data from the DIG study suggested that mortality might be increased in patients with higher plasma digoxin concentrations. If this were true, it would suggest that lower digoxin concentrations (0.5–0.8 ng/ml) would be associated with a reduced death rate and this has again been hypothesized (137).

Evidence that that is in fact so has come from another retrospective analysis of the Digitalis Investigation Group (DIG) study, in which continuous multivariable analysis showed a significant linear relation between serum digoxin concentration and mortality, with no effect of sex (138). The hazard ratios (HR) for mortality varied with serum concentration, with reduced mortality at serum concentrations of 0.5–0.9 ng/ml (HR = 0.81; 95% CI = 0.71, 0.92) and increased mortality at concentrations of 1.2–2.0 ng/ml (HR = 1.21; CI = 1.05, 1.40). However, digoxin reduced the risk of hospitalization at all serum concentrations.

In 345 patients with heart failure randomized to either digoxin (*n* = 175) or captopril (*n* = 170) and followed for a median of 4.5 years the death rate at 48 months was lower with captopril (21%) than with digoxin (32%), although this did not reach conventional significance (139). Since there was no placebo group for comparison it is not clear whether digoxin altered mortality in this study. Of the numerous adverse effects that were reported, the only one that differed between the two treatments was cough, which was significantly more frequent with captopril. In the absence of a placebo comparison it is impossible to say whether any of the other adverse effects were drug-related.

Despite the fact that the prospective study called DIG clearly showed that there was no increase in mortality in patients in taking long-term digoxin therapy (SEDA-20, 173), retrospective, non-randomized studies continue to be reported (SEDA-24, 201). In 180 patients with idiopathic dilated cardiomyopathy the overall mortality was 19% in those taking digoxin and 10% in those not taking digoxin (140). However, when the use of digoxin was adjusted for several predictive variables it no longer predicted cardiac death. This finding is reassuring, but results of studies like this, whatever their results, should be ignored, in view of the evidence that is currently available from the one large prospective, randomized study.

When interpreting the evidence presented in other accounts of the association between drug therapy and death it is important to remember that the current evidence suggests that digoxin does not cause excess mortality. For example, digoxin was the second most commonly encountered medication in an investigation of 2233 deaths reported to an American County Medical Examiner's office, with a medication history available in 775 cases (141). Furosemide was mentioned 181 times, digoxin 131 times, and glyceryl trinitrate 103 times. All other drugs were mentioned less than 100 times each. The authors suggested that the presence of digoxin at a death scene should suggest heart failure or a cardiac dysrhythmia, but they did not go further and stress that in such a case digoxin need not necessarily be implicated in the death. Postmortem diagnosis of digoxin toxicity is exceptionally difficult, but measurement of digoxin in the vitreous fluid can be helpful.

In a survey of 2 312 203 deaths in the USA in 1995, 206 (0.009%) were attributed to adverse drug reactions on death certificates (142). At the same time in the MedWatch program, 6894 deaths were reportedly attributed to adverse drug reactions, representing 6.3% of the 108 735 reports of adverse drug reactions. In the death certificate study 18 deaths were attributed to cardiac glycosides and in the MedWatch survey 15 deaths. This compares with figures of 289 and 782 from antimicrobial drugs, 449 and 280 from hormones, and 947 and 477 from drugs that affect the constituents of the blood (for example anticoagulants).

Sex-based differences in the effect of digoxin have been explored in a post-hoc analysis of the data from the DIG) study (SEDA-20, 173; 143). There was an absolute difference of 5.8% (95 CI = 0.5, 11) between men and women in the effect of digoxin on the case fatality rate from any cause. Women who were randomly assigned to digoxin had a significantly higher fatality rate than women who were randomly assigned to placebo (33 versus 29%), while the fatality rate was similar among men randomly assigned to digoxin or placebo (35% versus 37%). However, serum digoxin concentrations were higher in the women at 1 month, and this may have contributed to the increased risk of death (144).

In another post-hoc analysis of the data from the DIG study the patients who had been randomized to digoxin were divided into three groups, according to serum digoxin concentration, 0.5–0.8 ng/ml, 0.9–1.1 ng/ml, and 1.2 ng/ml and over (145). Higher concentrations were associated with higher all-cause fatality rates: 30%, 39%, and 48% respectively.

Both of these studies suggest that lower serum concentrations of digoxin (0.5–1.0 ng/ml) may be beneficial for routine therapy of heart failure than have traditionally been recommended (0.8–2.0 ng/ml), and this has been discussed in a brief review (146).

Second-Generation Effects

Pregnancy

Digoxin has been used to cause fetal death before termination of pregnancy (147). However, in a double-blind study in 126 women who had terminations by dilatation and evacuation at 20–23 weeks gestation intra-amniotic injection of digoxin 1 mg did not alter blood loss or pain; nor did it reduce difficulties with or the complications of the procedure (148). Significantly more women vomited after intra-amniotic digoxin. Digoxin given by this route is slowly absorbed into the systemic circulation, with a peak plasma concentration of 0.8 ng/ml at 11 hours (149).

Fetotoxicity

Digoxin crosses the placenta and enters the neonatal circulation (150). It has therefore been used, for example, to improve fetal cardiac function (151). In normal circumstances there seem to be no adverse effects on the neonate, and neonatal plasma concentrations are below those generally considered to be therapeutic. There has been one report of fatal toxicity in the fetus of a woman who took an overdose of digitoxin (152).

Susceptibility Factors

Age

Elderly people

The risk of digoxin toxicity is increased in old people, partly because they have poor renal function and lower body weight, factors that tend to increase the concentration of drug at the active site during steady-state therapy, and partly because they are liable to electrolyte imbalances, such as hypokalemia, which tend to increase the response of the tissues to a given concentration. Other factors, such as altered Na/K pump activity, may also contribute to increased tissue sensitivity. This means that the serum digoxin concentration that is associated with an increased risk of toxicity is slightly lower in elderly people than in younger people, and this has been confirmed in a recent study of 899 patients taking digoxin for heart failure or atrial fibrillation (153). No patients with serum digoxin concentrations below 1.4 ng/ml had evidence of digoxin toxicity. All patients who had a concentration of 3.0 ng/ml or more had severe toxicity. However in the range 1.4–2.9 ng/ml there were patients with and without evidence of toxicity, and the overlap was age-dependent. In patients aged 51–60 there was more evidence of toxicity with concentrations of 2.4–2.9 ng/ml; in patients aged 61–70 the range was 1.8–2.9 ng/ml, in patients aged 71–80 it was 1.4–2.7 ng/ml, and in those aged over 80 it was 1.4–2.6 ng/ml. The authors therefore suggested that serum digoxin concentrations should be no greater than 1.4 ng/ml during routine steady-state therapy. The incidences of toxicity were 16% in patients over 70 years of age and 7.3% in the whole group. The risk of toxicity was increased in the presence of renal insufficiency.

Because digitoxin is metabolized rather than being renally eliminated, the effects of renal impairment in elderly patients may not be so important in precipitating digitoxin toxicity. In 80 patients hospitalized 147 times, toxicity with digitoxin occurred in 7.6% of 92 admissions and digoxin toxicity occurred in 18.3% of 55 admissions (154). On the basis of these results the authors suggested that digitoxin is safer than digoxin in elderly patients. This is an old debate, and there are arguments in favor of both digoxin and digitoxin (155). However, there is currently no information on the long-term toxicity of digitoxin, and in particular its effects on mortality in patients with heart failure. Neither the severity of toxicity nor its duration was reported in this study.

A retrospective Bayesian analysis in 60 patients confirmed that age is a major factor in digoxin toxicity (156). However, an analysis of the data from the DIG study (SEDA-20, 173) has shown that while mortality in heart failure increases with age, the actions of digoxin are independent of age (157).

Children

Matters are also more complicated in young people. The pharmacokinetics of cardiac glycosides are different (158): the apparent volume of distribution of digoxin is higher in neonates, infants, and older children than in adults, and renal digoxin clearance is lower in children under 4–6 months. However, there may also be increased resistance to the effects of digoxin in infants because of changes in digitalis tissue receptors (159). Seriously ill children of low birth weight may be particularly at risk, even when low dosages of digitalis are used (160).

The risk of digitalis toxicity during the therapeutic use of cardiac glycosides is similar in children to that in adults, ranging in 12 separate published series from 12 to 50% (median 21%) (161). The most common non-cardiac effects are vomiting and feeding problems, and the most common cardiac effects are conduction defects, particularly atrioventricular block and ectopic rhythms, although (as in adults) any dysrhythmia can occur.

The pharmacokinetics of digoxin have been studied in 181 neonates and children with and without congestive heart failure (162). The clearance rate was lower in premature neonates than in neonates born at full term. Children with congestive heart failure also had lower digoxin clearance.

A population pharmacokinetic study in 172 neonates and infants showed that the clearance of digoxin is affected significantly by total body weight, age, renal function, and congestive heart failure (163).

Sex

Post-hoc analysis of the results of the DIG study suggested that digoxin may adversely affect survival in women but not in men. Among patients with left ventricular dysfunction enrolled in the Studies of Left Ventricular Dysfunction (SOLVD) with left ventricular ejection fractions of 0.35 and below, there was no interaction between sex and digitalis treatment in respect of survival, and there was no significant difference in the hazard ratios for men and women taking digitalis with respect to all-cause mortality, cardiovascular mortality, heart failure mortality, or dysrhythmic death with worsening heart failure (164). The authors concluded that there was no evidence of a difference between men and women in the effect of digitalis on survival. This has been confirmed by a reanalysis of the data from the Digitalis Investigation Group (DIG) study (42).

Evidence of sex differences in the response to digoxin has been sought in two studies in Sweden (165), following the observation that in the DIG study women who were randomly assigned to digoxin had a significantly higher fatality rate than women who were randomly assigned to placebo (33% versus 29%), while the fatality rate was similar among men randomly assigned to digoxin or placebo (35% versus 37%). In the first study, in which 363 women and 257 men were compared, the women were taking significantly smaller doses of digoxin (0.16 versus 0.18 mg/day), but had significantly higher trough steady-state serum digoxin concentrations, both unadjusted (1.48 versus 1.26 nmol/l) and adjusted for age, dose, and serum creatinine (1.54 versus 1.20 nmol/l); furthermore, significantly more women had serum digoxin concentrations above 2.5 nmol/l (OR = 4; 9%% CI = 1.6, 10). In the second study the authors searched the Swedish national register of adverse drug reactions and found that there were significantly more reports of adverse reactions to digoxin in women (165 versus 112) and significantly more serious reactions (30 versus 9), despite similar numbers of prescriptions. These data support the suggestion that women may be more susceptible to the adverse effects of digoxin than men.

Other features of the patient

Several factors increase patient susceptibility to digitalis intoxication. They can be considered in two groups (166).

Factors that alter the amount of digitalis that accumulates in the body or the plasma concentration at a fixed dose (pharmacokinetic factors)

Pharmacokinetic factors affect different cardiac glycosides differently.

Altered tissue distribution

The apparent volume of distribution of digoxin is reduced in hypothyroidism (167) and in renal insufficiency (168). This leads to increased plasma concentrations after a loading dose and hence an increased risk of toxicity, but does not affect the plasma concentration at steady state. The opposite occurs in hyperthyroidism.

Altered renal elimination

The effects of renal insufficiency on the pharmacokinetics of cardiac glycosides have been reviewed (168). The most important effect of renal failure is a reduced rate of elimination of digoxin, leading to increased accumulation during steady-state treatment. The same applies to some other glycosides, including beta-methyldigoxin, beta-acetyldigoxin, ouabain, and k-strophanthin, but not to glycosides that are mostly metabolized, such as digitoxin, the proscillaridins, and peruvoside (4). Drug interactions can also lead to reduced digoxin renal elimination.

A retrospective Bayesian analysis in 60 patients confirmed that renal impairment is a major factor in digoxin toxicity (156). In a 64-year-old with infective endocarditis digoxin toxicity was the mode of presentation (169).

A population pharmacokinetic study in 172 neonates and infants showed that the clearance of digoxin is affected significantly by renal function (163).

The risk of primary cardiac arrest associated with digoxin therapy at three levels of renal function has been investigated in a case-control study (170). The results are shown in Table 3. After adjustment for other clinical characteristics, digoxin therapy for congestive heart failure was not associated with an increased risk of cardiac arrest among patients with normal renal function but was associated with a modest increase in risk among patients with mild renal impairment and a twofold increase in risk among patients with moderate renal impairment.

Table 3 The odds ratio of cardiac arrest according to renal function and digoxin therapy in patients with congestive heart failure (170)

	Digoxin				*No digoxin*			
Creatinine (μmol/l)	*Cases (n)*	*Controls (n)*	*OR*	*95% CI*	*Cases (n)*	*Controls (n)*	*OR*	*95%CI*
<100	111	95	0.98	0.59, 1.64	66	58	1.00	–
110–130	120	71	1.60	0.94, 2.74	54	41	1.02	0.55, 1.87
140–320	161	53	2.70	1.52, 4.82	61	44	1.12	0.61, 2.08

Altered non-renal elimination
See Drug–Drug Interactions.

Factors that alter the clinical response to digitalis at a fixed amount of digitalis in the body and a fixed plasma concentration (pharmacodynamic factors)
Pharmacodynamic factors affect all cardiac glycosides in the same way.

Electrolyte disturbances
Of electrolyte disturbances that alter the response to a cardiac glycoside, hypokalemia is the most important. It has been estimated that a fall in plasma potassium concentration from 3.5 to 3.0 mmol/l is associated with a 50% increase in sensitivity to digoxin (166). Total body potassium depletion, even in the absence of hypokalemia, has a similar effect (171).

There is evidence that hypomagnesemia has the same effect as hypokalemia (172).

Hypercalcemia has the same effect as hypokalemia; hypocalcemia has the opposite effect, that is it causes resistance to the effects of digitalis.

Hypoxia and acidosis increase the risk of digitalis intoxication.

Renal insufficiency
In addition to its effect in reducing the elimination of digoxin, renal insufficiency may be associated with an increased sensitivity to the actions of digitalis (173).

Thyroid disease
Apart from the pharmacokinetic differences in thyroid disease (mentioned above), there may also be changes in tissue responsiveness, with reduced sensitivity in hyperthyroidism and the reverse in hypothyroidism (167). The reasons for these changes are not known, but they may be related to differences in tissue Na/K-ATPase activity.

Cardiac disease
All cardiac glycosides are best avoided in patients with acute myocardial infarction, since they increase oxygen demand in ischemic tissue, increase peripheral vascular resistance, and carry an increased risk of dysrhythmias, especially in the presence of tissue hypoxia and acidosis. Furthermore, there is evidence that digitalis is of little value in patients with acute myocardial infarction and either left ventricular failure or cardiogenic shock (174). The evidence that mortality in patients who take digitalis after an acute myocardial infarction is increased is discussed in the section Death in this monograph.

Cardiac glycosides are contraindicated in conditions in which there is obstruction to ventricular outflow, for example hypertrophic obstructive cardiomyopathy, constrictive pericarditis, and cardiac tamponade. Acute myocarditis may also increase the risk of toxicity.

Direct current cardioversion increases the risk of digitalis-induced dysrhythmias, but digitalis treatment is not a contraindication to cardioversion (175).

Hypercalcemia
The effects of digitalis are enhanced in the presence of hypercalcemia.

- An 81-year-old woman with congestive heart failure and hypercalcemia secondary to squamous cell carcinoma of the bronchus developed first-degree heart block and symptomatic sinus pauses when her serum digoxin concentration was only 1.5 mg/ml (176).

Drug Administration

Drug contamination

Digoxin toxicity has been reported in two patients who took herbal remedies (177). *D. lanata* was found as a contaminant.

Drug administration route

Intramuscular injection of digitalis can be painful and can cause local muscle necrosis, sometimes with pyrexia (178). The systemic availability of digitalis after intramuscular injection is poor (179), and this route of administration should be avoided if possible.

Digoxin has a rapid onset of action after oral administration, and there is rarely any justification for giving digoxin intravenously. This route of administration should be restricted to those who have severe atrial fibrillation that requires rapid treatment and in whom cardioversion or other drug therapy is not possible or indicated. If digoxin is given intravenously it should be infused over at least half-an-hour, since there is a risk of hypertension if it is infused faster.

There has been a randomized, double-blind comparison of intravenous diltiazem and digoxin in 40 patients with atrial fibrillation and a ventricular rate of over 100/minute (180). One patient given intravenous digoxin had a burning sensation at the site of injection.

Drug overdose

The most important complication of overdosage of digitalis in all age groups is disturbance of cardiac conduction, but in addition any dysrhythmia can occur. Death can result from asystole or ventricular fibrillation. Hyperkalemia is common, and the higher the plasma potassium concentration the poorer the prognosis (181).

Other common effects of overdosage are nausea, vomiting, and central nervous system and visual disturbances (77).

- Accidental overdose of digoxin in a 22-month-old boy caused vomiting, lethargy, and dehydration. The plasma digoxin concentration was 12 mg/ml. There was a relative bradycardia of 90/minute and Mobitz type 1 second-degree heart block on the electrocardiogram (182).

The pharmacokinetics of digoxin are altered after overdosage, the half-life being reportedly rapid, but there is too little information to define the kinetics precisely (181).

- Plasma glycoside concentrations have been documented after an overdose with purple foxglove in a 36-year-old woman (183). Apart from gitaloxin, which peaked on the fifth day at 113 ng/ml, all the glycosides detected peaked on the first day (gitoxin 13 ng/ml, digitoxin 113 ng/ml, digitoxigenin 3.3 ng/ml, and digitoxigenin monodigitoxoside 8.9 ng/ml). There was a second peak of digitoxin at about 70 hours, and this is consistent with the known enterohepatic recirculation of digitoxin.

The risk of death after overdose probably increases with the number of different drugs taken. Combined toxicity with digoxin, metoprolol, and verapamil has been reported (184).

- A 39-year-old man was found dead in his room, with a lot of empty packets of prescribed drugs nearby. The blood concentrations of digoxin, metoprolol, and verapamil were 3.2 ng/ml, 3.6 μg/ml, and 9.2 μg/ml respectively. The cause of death was given as cardiac failure, hypotension, and bradycardia, due to a mixed drug overdose of digoxin, metoprolol, and verapamil.

The concentrations of the individual drugs were not high enough for any one drug to have caused death, and the authors speculated that the toxicity of verapamil is potentiated by interaction with metoprolol and digoxin.

Fatal poisoning involving verapamil, metoprolol, and digoxin has been reported (185).

- A 39-year-old man was found dead in his room with empty packets of prescribed drugs nearby. The blood concentrations of verapamil, norverapamil, metoprolol and digoxin were 9.2 μg/ml, 3.0 μg/ml, 3.6 μg/ml, and 3.2 ng/ml respectively.

This was a complicated case of overdose. There is a pharmacodynamic interaction of beta-blockers with verapamil, and verapamil may have inhibited the metabolism of metoprolol via CYP2D6 (186); verapamil also inhibits the clearance of digoxin. Based on the results of an autopsy and toxicological examination, death was attributed to cardiac failure, hypotension, and bradycardia due to combined overdose of verapamil, metoprolol, and digoxin.

Treatment of toxicity and overdosage

The treatment of digitalis toxicity and overdose have been reviewed (SEDA-5, 172; SEDA-12, 149). In summary, the following measures should be taken.

Remove digitalis from the stomach

If the patient is seen within 1 hour of overdosage try to remove whatever drug still remains in the stomach by gastric lavage, although the risk of dysrhythmias may be increased. There is no evidence that emesis is beneficial.

Give activated charcoal

The use of activated charcoal in the treatment of digitalis overdose has been reviewed (187). The rationale for its use is that after absorption into the systemic circulation digitoxin is secreted into the bile and digoxin (188) (and probably other cardiac glycosides) is secreted into the gut lumen by the action of the P glycoprotein. Activated charcoal in the gut binds this secreted digoxin and encourages further secretion. To be fully effective charcoal should be given at regular intervals (for example 50 g 4-hourly).

In pigs, repeated doses of activated charcoal reduced the half-life of intravenous digoxin significantly from 65 to 17 hours and increased the clearance from 2.3 to 7.1 ml/minute/kg (189).

There is evidence of the efficacy of charcoal in healthy volunteers. In six adult volunteers repeated doses of activated charcoal significantly reduced the half-life of digoxin from 23 to 17 hours without a significant increase in clearance; the half-life of digitoxin was also reduced from 110 to 51 hours, and this was accompanied by a significant increase in clearance from 0.24 to 0.47 l/hour (190). In a volunteer with chronic renal insufficiency charcoal reduced the digoxin half-life from 93 to 29 hours and increased the clearance from 3.6 to 10 l/hour (190). Similarly, during maintenance therapy in six individuals, daily activated charcoal significantly reduced the mean plasma digoxin concentration by 31% and serum digitoxin concentration by 18% (191). In 10 healthy volunteers given intravenous digoxin 10 μg/kg repeated doses of activated charcoal significantly increased the total body clearance from 12 to 18 l/hour and reduced the half-life from 37 to 22 hours (188).

There have also been reports of the value of multiple doses of activated charcoal in patients with digoxin and digitoxin poisoning. In 23 patients with plasma digoxin concentrations over 2.5 ng/ml multiple doses of activated charcoal increased the mean clearance of digoxin to 98 ml/minute compared with 55 ml/minute in 16 patients who were not treated, and reduced the half-life from 68 to 36 hours (192). Anecdotal reports have also appeared.

- In a 69-year-old man the plasma digoxin concentration of 8.3 ng/ml fell with a half-life of 14 hours (193).
- In a 71-year-old woman with chronic renal insufficiency and a plasma digoxin concentration of 9 ng/ml charcoal shortened the half-life from 7.3 to 1.4 days (194).
- In a 66-year-old man with chronic renal insufficiency whose serum digoxin concentration did not respond to daily hemodialysis, multiple doses of activated charcoal caused a rapid reduction in the serum concentration (195).
- In a patient with a peak plasma digoxin concentration of 264 ng/ml multiple doses of activated charcoal reduced the half-life of digitoxin from 162 hours to 18 hours (196).
- In a patient on hemodialysis, activated charcoal caused a fall in the plasma digoxin concentration with a half-life of 29 hours (197).
- In another patient there was convincing evidence of a change in the digoxin half-life after repeated doses of activated charcoal (193).
- In a 73-year-old woman who took 12.5 mg of digoxin, gastric lavage and activated charcoal tided the patient over until antibodies became available (198).

However, the best evidence of the usefulness of repeated oral doses of activated charcoal in cardiac glycoside poisoning comes from the results of a randomized, placebo-controlled study in 402 individuals who took overdoses of the seeds of the yellow oleander tree in

Sri Lanka. Repeated doses of activated charcoal reduced mortality from 8.0% to 2.5% (26).

Thus, the use of repeated doses of activated charcoal in patients with digitalis toxicity is a cheap way of increasing the rate of cardiac glycoside clearance.

Colestyramine has been used for the same purpose. In two elderly patients with congestive heart failure and raised serum digoxin concentrations it enhanced the elimination of digoxin (199). In one case the half-life of digoxin was 20 hours and in the other 24 hours.

Correct electrolyte disturbances

Hypokalemia should be treated with potassium chloride.

Hyperkalemia carries a poor prognosis and is usually an indication for antidigoxin antibody. The suggestion that intravenous calcium should be used to treat the hyperkalemia that can occur in digitalis intoxication (200) has been challenged, on the grounds that it can increase the risk of cardiac dysrhythmias in such cases (201).

Give antidigoxin antibody

Antidigoxin antibody (Fab fragments) is the treatment of choice in patients with severe digitalis toxicity due to any cardiac glycoside. It is effective after self-poisoning with digitoxin, lanatoside C, and acylated forms of digoxin, as well as digoxin itself (202). It should be used when there are life-threatening dysrhythmias or heart block, and when the plasma potassium concentration is above 5.0 mmol/l or is rising, since hyperkalemia is evidence of serious toxicity and carries a poor prognosis. Evidence of the efficacy of antidigoxin Fab antibody fragments in reducing the risk of cardiac dysrhythmias in the treatment of intoxication due to oleander poisoning, previously only anecdotally reported (203,204), has come from a prospective study (205), although the study was too small to assess the effect on mortality. The role of antidigoxin antibody fragments in treating milder forms of digitalis toxicity has not been fully assessed.

In a systematic review of 250 publications no controlled, randomized trials were found, and the authors concluded that there was little or no scientific evidence of efficacy, because there are no randomized, controlled trials (206). However, this is one case in which the cumulative anecdotal evidence is overwhelmingly convincing, and there can be no doubt that antidigoxin antibodies are highly effective in the treatment of digoxin intoxication and of intoxication with other cardiac glycosides. The important question is whether there are cases in which the antibodies need not be used, and guidelines have yet to be developed.

The rapidity with which antidigoxin antibody Fab fragments can act in the treatment of digitalis toxicity has been illustrated in the case of a neonate who developed severe hyperkalemia (7.8 mmol/l) and an atrioventricular junctional tachycardia after the administration of several doses of digoxin for cardiac dysrhythmias (serum digoxin concentration 9.2 nmol/l) (207). Within 1 minute of the administration of Digibind® the tachycardia converted to sinus rhythm and within 20 minutes the serum potassium concentration fell to within the reference range. The rate of reversal in this case was unusually fast.

In another case an 89-year-old woman with bradycardia and hyperkalemia was treated with atropine, glucagon, insulin, and intravenous calcium chloride before it was realized that she had digoxin toxicity; antidigoxin antibody caused rapid resolution (208). Although calcium can potentiate the actions of cardiac glycosides, this patient did not seem to have suffered any adverse consequences as a result of its administration.

- Overdose of digoxin in a neonate caused complete atrioventricular block and cardiogenic shock, which were completely reversed within 4 hours after administration of the first dose of antidigoxin antibody; a second dose was given 48 hours later, when first-degree atrioventricular block occurred (209).

However, it is not known whether antidigoxin antibody reduces mortality after digitalis poisoning; one randomized, controlled study (205) was too small to detect an effect on mortality after self-poisoning with seeds of the yellow oleander, which contains cardiac glycosides such as peruvoside. Another study that suggested reduced mortality (210) was confounded by being retrospective and inferential, based on changes in death rates as antibody was introduced and reintroduced; furthermore, the change in mortality that was attributed to the use of antibody occurred at the same time as the introduction of multiple-dose activated charcoal, which reduces mortality.

The dose of antidigoxin antibody should be based on the dose of digitalis taken and, when possible, the plasma digitalis concentration. The recommended doses for cases of poisoning with digoxin and digitoxin are given in Table 4. Although the clearance of the fragments is reduced in patients who also have severe renal impairment (211,212), there is no need to reduce the dose of antibody in such patients (213–216).

Different formulations of antidigoxin antibody are available. Digibind is an ovine antibody to digoxin, whose production is stimulated by the administration of digoxin conjugated to human albumin. In contrast, DigiFab, also ovine, is stimulated by injecting a conjugate of digoxin to keyhole-limpet hemocyanin (217). In 15 adults with digoxin toxicity who were given DigiFab, electrocardiographic abnormalities resolved within 4 hours in 10 patients and the signs of toxicity completely resolved within 4 hours in seven patients and within 20 hours in 14 patients. In one patient loss of the effect of digoxin resulted in pulmonary edema, pleural effusions, and renal insufficiency. The half-life of DigiFab is slightly shorter than that of Digibind (15 versus 23 hours), but their pharmacodynamic properties are similar. It is said that DigiFab may be preferred in patients who are allergic to sheep proteins, papain, chymopapain, or bromelains.

Antidigoxin antibody fragments cause adverse events in about 7% of cases. These include allergic responses, possible recurrence of digitalis toxicity after treatment, and some effects attributable to the withdrawal of digitalis, such as worsening of heart failure. In one series allergic reactions occurred in only six of 717 patients reviewed,

Table 4 Methods for calculating the required dose of antidigoxin antibody fragments in cases of digoxin or digitoxin intoxication

Digoxin

(1) When the ingested dose is known:

(a) Tablets	dose in mg × 40
(b) Elixir	dose in mg × 48
(c) Capsules	dose in mg × 55
(d) Intravenous	dose in mg × 60

(2) When the plasma or serum concentration is known:
ng/ml × lean body weight × 0.34
or
nmol/l × lean body weight × 0.26

Example: Plasma digoxin concentration = 24 ng/ml in a 75 kgpatient
Dose of antibody = 24 × 75 × 0.34 = 612 mg
Give 640 mg (for example 16 ampoules of Digibind)

Digitoxin

(1) When the ingested dose is known (all formulations):
Dose in mg × 60
(2) When the plasma or serum concentration is known:
ng/ml × lean body weight × 0.034
or
nmol/l × lean body weight × 0.026

Example: Plasma digitoxin concentration = 280 ng/ml in a 70 kgpatient
Dose of antibody = 280 × 70 × 0.034 = 666 mg
Give 680 mg (for example 17 ampoules of Digibind)

and consisted of pruritic rash and flushing or facial swelling (218). The risk of an allergic reaction was increased in patients who had a history of previous allergy or asthma. Recurrence of digitalis toxicity after treatment was usually due to inadequate treatment with the antibody.

Plasma exchange has been used to enhance the rate of removal of antidigoxin antibody Fab fragments in a 46-year-old man with renal insufficiency (219). Removal of the digoxin-Fab complexes in this case prevented their subsequent dissociation and a further increase in the unbound concentration of digoxin. The authors proposed that plasma exchange is best used in these cases within the first 3 hours after the administration of antidigoxin antibodies.

Treat dysrhythmias

Cardiac dysrhythmias in digitalis overdose should be treated only if they are life-threatening. Phenytoin is probably the treatment of choice for ventricular tachydysrhythmias, but lidocaine or a beta-adrenoceptor antagonist, such as propranolol, are options. After an overdose of 300 tablets of digoxin (plasma digoxin concentration 50 ng/ml), recurrent ventricular fibrillation was successfully treated with bretylium tosylate (220). Sinus bradycardia may respond to atropine.

Treat or anticipate heart block

Heart block is the most serious consequence of digitalis poisoning and should be anticipated by the insertion of a temporary pacemaker. If this is postponed until heart block or dysrhythmias occur, there may be difficulty in inserting the pacemaker (because of ventricular excitability), and delay in treatment can be deleterious.

The use of temporary cardiac pacing has been studied retrospectively in 70 patients (mean age 74 years, 30 men) with digoxin toxicity not due to self-poisoning (221). A transvenous pacemaker was used in 24 patients with sinus arrest and junctional bradydysrhythmias (n = 9), atrial fibrillation with a slow ventricular rate (n = 11), and high-degree atrioventricular block (n = 4). The mean duration of pacemaker implantation was 5.8 (2–12) days. There were no major dysrhythmic events or deaths. Two of the 46 patients who did not have a transvenous pacemaker inserted died of ventricular tachydysrhythmias. The authors concluded that temporary cardiac pacing is safe for patients with digoxin overdose complicated by symptomatic bradycardia.

Other measures

Because digoxin has a large apparent volume of distribution, plasma exchange, hemodialysis, and hemoperfusion are not effective methods of removing digoxin from the body.

Hemoperfusion

Hemoperfusion has been used to treat digitalis overdose (SEDA-5, 174) but once a cardiac glycoside has been distributed to the body tissues hemoperfusion is unlikely to be of benefit. For digitoxin, which has a lower apparent volume of distribution than digoxin, charcoal hemoperfusion may be more valuable.

- In an 88-year-old woman whose serum digoxin concentration was 6.4 ng/ml, hemoperfusion with an adsorption column containing beta$_2$-microglobulin caused a fall in serum concentration from 6 ng/ml to 2.3 ng/ml, with improvement in gastrointestinal symptoms (222). The serum digoxin concentration then rose again to 3.5 ng/ml over the next 3 days, and a further hemoperfusion treatment reduced it to 1.7 ng/ml, after which the serum concentration gradually fell and the patient improved. Because of chronic renal insufficiency, she also had repeated hemodialyses on alternate days, which also contributed to the reduction in serum digoxin concentrations.

The clearance of digoxin was measured in eight patients receiving hemodialysis during the use of a beta$_2$-microglobulin column (223). After 240 minutes of hemoperfusion the serum digoxin concentration fell from 1.11 to 0.57 ng/ml and digoxin clearance was about 145 ml/minute. However, this clearance rate cannot have been the true total body clearance, since it merely reflected the change in plasma concentration during hemoperfusion, which would have been almost entirely due to removal of digoxin from the plasma only and not from the tissues. This method cannot be recommended as a substitute for the use of antidigoxin antibodies in the treatment of digitalis toxicity.

Hemofiltration

Hemofiltration has been used to treat digitalis poisoning (SEDA-12, 149), but there is no convincing evidence of its efficacy.

Exchange transfusion

There have been reports of the use of exchange transfusion to remove Fab antibody fragment-digoxin complexes in patients with acute anuric renal insufficiency (224).

- Plasma exchange was used to enhance the rate of removal of antidigoxin antibody Fab fragments in a 46-year-old man with renal insufficiency (219). Removal of the digoxin-Fab complexes in this case prevented their subsequent dissociation and a further increase in the unbound concentration of digoxin.

The authors proposed that plasma exchange is best used in these cases within the first 3 hours after the administration of antidigoxin antibodies.

- A 70-year-old man with alcoholic cirrhosis was given amiodarone and digoxin for atrial fibrillation after a hemicolectomy for adenocarcinoma (225). He developed acute renal insufficiency and digoxin toxicity, with a serum concentration of 4.4 ng/ml. A dose of Fab antidigoxin antibody fragments was followed 16 hours later by exchange transfusion and another dose was followed by two exchanges. He recovered slowly over the next few days. The total digoxin concentration (antibody bound and unbound) rose after the first dose of Fab fragments but did not fall until after the second plasma exchange (after the second dose of Fab fragments).

In this case digoxin was recovered from the plasma collection bags, but the total amount recovered seems to have been less than 100 μg, so the efficacy of plasma exchange was not clear.

Drug–Drug Interactions

General

Drug interactions with cardiac glycosides can be subdivided into six types, according to mechanism. Interactions with digoxin have been reviewed (SEDA-6, 173) (226–228). Drug interactions with digitoxin have been briefly reviewed in the context of its use in the treatment of congestive heart failure (229).

Absorption

Absorption interactions with digitalis are probably not of great clinical importance, since (a) the dosages of drugs that reduce the absorption of digitalis are usually larger than those used clinically, and (b) the major effect on absorption probably occurs only if the two drugs are taken together.

The effect of activated charcoal in reducing digitalis absorption is mentioned under the treatment of poisoning, and binding resins such as colestyramine and colestipol have similar actions (230,231).

Certain combinations of cytotoxic drugs (cyclophosphamide, vincristine, and prednisone, with and without procarbazine) reduced plasma digoxin concentrations by about 50% during treatment with beta-acetyldigoxin, perhaps through impaired absorption of beta-acetyldigoxin (232); digitoxin was not affected (233).

Absorption interactions involving altered gastrointestinal motility have been described. These include interactions of digoxin with propantheline (234), metoclopramide (234), and cisapride (235). However, the clinical relevance of these interactions is unclear and they are probably unimportant.

Some antacids, such as magnesium trisilicate, reduce the absorption of digoxin slightly, but these interactions are probably of no clinical importance (236).

Protein-binding displacement

Interactions of this kind are of no importance for digoxin, which is only about 20% bound and has a high apparent volume of distribution.

An interaction of heparin with digitoxin has been described and ascribed to altered protein binding, secondary to changes in fatty acid concentrations, but the clinical relevance, for example in patients undergoing hemodialysis, is unclear (237).

Renal clearance

Digoxin is cleared by the kidneys by glomerular filtration and active secretion, and retained by passive reabsorption, the last two roughly balancing each other, so that clearance is usually proportional to creatinine clearance. A major mechanism for drug interactions with digoxin is inhibition of its renal tubular secretion by inhibition of P glycoprotein. This mechanism has been reviewed in relation to an in vitro tissue culture model, consisting of confluent polarized renal tubular cell monolayers (238).

This model has confirmed the action of several drugs that can inhibit the renal tubular secretion of digoxin in this way, including amiodarone, ciclosporin, itraconazole and ketoconazole, mifepristone, propafenone, quinidine, spironolactone, verapamil, and vinblastine and vincristine.

In three large studies using either population pharmacokinetic analysis or Bayesian techniques, drugs that inhibit the transport of digoxin by inhibiting P glycoprotein significantly increased the serum digoxin concentration (163,239,240). These drugs included quinidine, spironolactone, and the calcium channel blockers diltiazem, nicardipine, nifedipine, and verapamil. The effects varied from about 22% to about 36%.

Vasodilators may increase the active secretion of digoxin, which is a high clearance process and is therefore affected by renal blood flow.

Uptake by the end-organ

The interaction of quinidine with digoxin involves displacement of digoxin from tissues.

Hyperkalemia, due to potassium chloride, potassium-retaining diuretics, ACE inhibitors, or angiotensin-receptor antagonists, reduces the apparent affinity of digitalis for Na/K-ATPase and thereby reduces its tissue binding.

Response of the end-organ

Interactions involving changes in the response of the end-organ to digitalis are the most common of all interactions with cardiac glycosides.

Potassium depletion, for example due to diuretics or corticosteroids, potentiates the effects of cardiac glycosides on the myocardium and may also have a small effect in reducing the renal tubular secretion of digoxin (241,242).

Magnesium depletion may have a similar effect (243), but the data are not as clear-cut as those for potassium.

Anecdotal reports suggest that intravenous infusion of calcium salts in patients taking digitalis can result in dangerous cardiac dysrhythmias. This may also be the basis of reports that edrophonium and suxamethonium enhance the actions of digitalis, since both of these drugs might cause altered disposition of calcium. Conversely, there is good anecdotal evidence that hypocalcemia causes reduced plasma responsiveness to digoxin (244).

Alpha-glucosidase inhibitors

Acarbose

In patients taking steady-state digoxin therapy the addition of acarbose reduced the plasma digoxin concentration (245,246). The association was confirmed in both cases by withdrawal of acarbose. The likely mechanism of this interaction is inhibition of the absorption of digoxin by acarbose, although the authors also suggested that acarbose might interfere with the hydrolysis of digoxin before absorption, thus altering the pattern of metabolites in the blood; however, this is a much less likely mechanism. The authors recommended that if acarbose be used in conjunction with digoxin, the doses should be separated by about 6 hours. This absorption interaction has previously been highlighted (247,248).

However, in a formal study of the pharmacokinetics of a single dose of digoxin 0.75 mg before and after the administration of acarbose 50 mg tds for 12 days in healthy volunteers, apart from a small increase in C_{max}, the pharmacokinetics of digoxin were unaffected by acarbose (249). It is not uncommon for anecdotal reports of a possible interaction to be unconfirmed by formal kinetic studies, and it is possible in such cases that there is a subset of patients who are susceptible to the interaction who have not been included in the formal study. In this case, for example, it may be that the interaction occurs in people with diabetes and not in healthy subjects. There may also be a difference in the effect of acarbose on a single dose of digoxin, compared with steady-state therapy. Advice that acarbose and digoxin should be administered 6 hours apart is still reasonable.

Voglibose

In eight healthy men voglibose, another inhibitor of alpha-glucosidase, had no effect on the pharmacokinetics of digoxin (250). This result suggests that inhibition of alpha-glucosidase is not the mechanism whereby acarbose alters the absorption of digoxin; however, acarbose also inhibits alpha-amylase and such inhibition cannot be ruled out as a mechanism of the interaction with acarbose, perhaps by altered gastrointestinal motility.

Amiodarone

Amiodarone inhibits the renal tubular secretion of digoxin (SEDA-22, 201) and toxicity can occur as a result (251). It has also been suggested that it increases its absorption (SEDA-10, 144; SEDA-12, 150). This interaction has also been reported with acetyldigoxin (SEDA-18, 198; 252).

- A 69-year-old man with heart failure was given digoxin 1 mg over 12 hours followed by amiodarone 1 mg/kg for 6 hours and then 0.5 mg/hour. He developed sustained monomorphic ventricular tachycardia and later bidirectional ventricular tachycardia. The serum potassium concentration was 3.5 mmol/l. The serum digoxin concentration was 4.3 ng/ml (usual target range 0.8–2.0). Amiodarone was withdrawn and he was given intravenous lidocaine and potassium. Sinus rhythm was restored after 5 days as the serum digoxin concentration fell to 1.5 ng/ml.

Amiodarone also interacts with digitoxin (253). In two cases the half-life of digitoxin was prolonged, but there was no other information that suggested a mechanism. The author suggested that amiodarone might displace digitoxin from tissue sites, but that would have led to a shortening of the half-life rather than a prolongation. It seems more likely that amiodarone inhibits the clearance of digitoxin by inhibiting renal and gut P glycoprotein and perhaps by inhibiting its metabolism.

Antacids

Although digoxin is not extensively metabolized in most patients, it is metabolized before its absorption from the gut by two mechanisms, hydrolysis by gastric acid and hydrogenation by intestinal bacteria. In patients who have hypochlorhydria presystemic metabolism is reduced and increasing concentrations of digoxin are achieved systemically. This means that drugs that reduce gastric acid secretion, such as cimetidine, ranitidine, other histamine H_2 receptor antagonists, and omeprazole would be expected to increase the systemic availability of digoxin, and there is some supporting evidence (254,255), although this is not conclusive (SEDA-17, 216).

Antidysrhythmic drugs

Digoxin does not interact with a variety of antidysrhythmic drugs, including ajmaline, aprindine, lidocaine, lidoflazine (256), and moracizine (257). Other drugs that may have minor and clinically unimportant interactions include captopril, carvedilol, disopyramide, and flosequinan.

Antifungal imidazoles

Itraconazole

Itraconazole increases steady-state serum digoxin concentrations, perhaps by inhibiting the renal tubular secretion of digoxin (SEDA-22, 202; 258–260). An alternative proposed mechanism is inhibition of CYP3A (SEDA-21, 196), and this has been reported in rats with ketoconazole (261), although an effect on P glycoprotein was also possible. Whatever the mechanism, ketoconazole increased the systemic availability of digoxin from 0.68 to 0.84 and reduced the mean absorption time from 1.1 hours to 0.3 hours. The increased systemic availability could have been explained by inhibition of CYP3A or P glycoprotein in the gut, but the increased rate of absorption could only be explained by inhibition of the P glycoprotein. Since the t_{max} was unaffected, the authors hypothesized that inhibition of P glycoprotein increased the absorption rate, which would have tended to reduce the t_{max}, while inhibition of CYP3A, which would have reduced the elimination rate of digoxin, would have tended to increase the t_{max}. Thus a combination of these two effects would have had no effect on t_{max}. It should be noted that CYP3A is an important route of metabolism of digoxin in rats, but not in man.

Digoxin toxicity sometimes accompanies this interaction.

- A 62-year-old woman who was taking digoxin took itraconazole 400 mg/day for 3 days developed nausea, anorexia, and lethargy; the symptoms improved within 48 hours after withdrawal of itraconazole (262). The serum digoxin concentrations were not reported.
- In a 75-year-old man who took itraconazole in a low dose (200 mg/day) the steady-state serum digoxin concentration only rose from 0.8 to 1.1 ng/ml after 8 days (263).
- Two renal transplant patients developed digoxin toxicity when they also took itraconazole (264).

Itraconazole increases the digoxin AUC_{0-72} by about 50%, and reduces its renal clearance by about 20% (265). Apart from inhibition of the renal secretion of digoxin, which is probably mediated by inhibition of P glycoprotein, a study in guinea pigs also showed significantly reduced biliary excretion of digoxin by itraconazole, suggesting that the interaction between itraconazole and digoxin may be due not only to a reduction in renal clearance, but also to a reduction in the metabolic clearance of digoxin by itraconazole (266).

Voriconazole

The effect of multiple-dose voriconazole on the steady-state pharmacokinetics of digoxin in healthy men has been studied in a double-blind, randomized, placebo-controlled study (267). All the subjects took oral digoxin for 22 days (0.5 mg bd on day 1, 0.25 mg bd on day 2, and 0.25 mg/day on days 3–22). On days 11–22 they were randomized to either voriconazole 200 mg bd or placebo. Voriconazole did not significantly alter the C_{max}, C_{min}, AUC, t_{max}, or clearance of digoxin at steady state. There were no significant differences in adverse events, all of which were classified as mild and transient.

Aprepitant

Aprepitant, a selective neurokinin-1 receptor antagonist is a substrate and a weak inhibitor of P glycoprotein. The effect of aprepitant on the pharmacokinetics of digoxin has been studied in a double-blind, randomized, placebo-controlled, crossover study in 12 healthy subjects (268). Each took oral digoxin 0.25 mg/day for 13 days and aprepitant 125 mg (or matching placebo) on day 7 and 80 mg (or matching placebo) on days 8–11. Aprepitant did not affect the pharmacokinetics of digoxin.

Argatroban

The thrombin inhibitor argatroban had no effect on the steady-state pharmacokinetics of oral digoxin 0.375 mg/day in 12 healthy volunteers; the argatroban was given as an intravenous infusion of 2 micrograms/kg/minute on days 11–15 (269).

Bosentan

In a randomized study in 18 young men, bosentan caused a small (12%) reduction in the AUC of digoxin, without changing either the C_{max} or the C_{min} (270). This was a steady-state study, with plasma concentration measurements for only 24 hours, and so the half-life of digoxin was not measured. Bosentan is an antagonist at endothelin type 1 receptors and could therefore have increased renal blood flow by pre- and postglomerular vasoconstriction, increasing the elimination of digoxin. Furthermore, bosentan is a substrate for P glycoprotein, which mediates the renal tubular secretion of digoxin, and may induce its expression, increasing digoxin renal clearance. However, whatever the mechanism of this putative interaction, it is clearly unlikely to be of clinical significance.

Calcium channel blockers

Calcium channel blockers have varying effects on the disposition of digoxin. The calcium channel blockers for which varying amounts of information are available include cinnarizine, diltiazem, felodipine, fendiline, gallopamil, isradipine, lidoflazine, mibefradil, nicardipine, nifedipine, nitrendipine, tiapamil, and verapamil (271). Interactions of digoxin with nitrendipine (272) and bepridil (273) have also been described.

The main mechanism is inhibition of digoxin renal tubular secretion by inhibition of P glycoprotein. In a review of the interactions of calcium channel blockers with digoxin, in which their clinical relevance was assessed, it was concluded that serious consequences can be prevented by careful monitoring, especially in patients whose serum digoxin concentration is already near the upper end of the therapeutic range (274).

Diltiazem

Studies of the effects of diltiazem on the pharmacokinetics of digoxin have yielded variable results. In one study, diltiazem reduced the steady-state total body clearance of beta-acetyldigoxin in 12 healthy men, perhaps because of reduced renal and non-renal clearances (SEDA-10, 145). In some studies diltiazem 120–240 mg/day increased steady-state plasma digoxin concentrations by about 20–50% (SEDA-14, 146; 275,276,277), although not in other studies (SEDA-14, 146), and reduced the total body clearance of digoxin, with changes in both renal and non-renal clearances (SEDA-11, 154), although others did not find this (278). In at least one case digoxin toxicity was attributed to this interaction (SEDA-17, 216). In eight patients with chronic heart failure taking digoxin 0.25 mg/day, diltiazem 180 mg/day increased the AUC and mean steady-state serum concentrations of digoxin by 50% and reduced its total clearance (279).

Mibefradil

In 40 healthy subjects mibefradil 50 or 100 mg/day for 6 days had no significant effects on the steady-state pharmacokinetics of digoxin, apart from a very small increase in the C_{max} (280).

Nifedipine

The interaction of digoxin with nifedipine increases plasma digoxin concentrations by only about 15% (281,282).

Tiapamil

Tiapamil reversed digoxin-induced splanchnic vasoconstriction in healthy men (283), but this has no direct effect on systemic hemodynamics.

Verapamil

Verapamil increases plasma digoxin concentrations at steady state by inhibiting the active tubular secretion and non-renal clearance of digoxin (284,285). There is only anecdotal evidence that this can result in digitalis toxicity (285,286).

Verapamil reversed digoxin-induced splanchnic vasoconstriction in healthy men (283), but this has no direct effect on systemic hemodynamics.

Verapamil suppresses renal digoxin elimination acutely, but this suppression disappears over a few weeks (287). However, inhibition of the extrarenal clearance of digoxin persists, and the result of this complex interaction is an increase in steady-state plasma digoxin concentrations of less than 100%. Patients taking both drugs should be carefully monitored. However, the pharmacodynamic effects of digoxin are apparently reduced by verapamil (288), so that dosage adjustment may be unnecessary. Cardiovascular collapse and/or asystole has followed the use of intravenous verapamil in patients taking oral digoxin alone (289) or in combination with quinidine, propranolol, or disopyramide (290).

Verapamil and diltiazem, but not nifedipine, increase steady-state plasma digitoxin concentrations (291).

Captopril

There have been contradictory results in studies of the effects of captopril on digoxin pharmacokinetics. In some cases, captopril increased steady-state plasma digoxin concentrations (292), while in others there was no evidence of an interaction (293–296). In eight patients with NHYA Class IV congestive heart failure, captopril 12.5 mg tds for 1 week increased the peak digoxin concentration, reduced the time to peak, and increased the AUC during a single dosage interval at steady state; trough digoxin concentrations did not change (297). There are two possible explanations for these findings: that captopril reduced the clearance of digoxin or that it increased both the rate and extent of absorption of digoxin; unfortunately, the authors did not measure either the mean steady-state concentration or the half-life, which would have clarified this. On the whole, however, it is unlikely that this interaction is of any clinical significance, since in no study has there been evidence of digoxin toxicity during concomitant captopril therapy, and in one formal study (296) there was evidence of no pharmacodynamic interaction.

Carvedilol

In eight children aged 2 weeks to 8 years) carvedilol reduced the oral clearance of digoxin by half; two of the children had digoxin toxicity (298).

- A 12-year-old boy with dilated cardiomyopathy was treated with digoxin 250 micrograms in the morning and 125 micrograms in the evening. Carvedilol 0.07 mg/kg/day was begun, in an attempt to enhance ventricular performance. Within several days, he developed vomiting and anorexia; the serum digoxin concentration, which was usually 2–3 nmol/l, was 5.4 nmol/l. On withdrawal of digoxin, the serum concentration fell with a half-life of 72 hours (normal 12–36 hours); the serum creatinine was normal (42 μmol/l). Subsequently, digoxin was restarted at half the original dose.

The authors proposed that digoxin doses be reduced by at least 25% in children who also start to take carvedilol, with further adjustments as required after a new steady-state occurs, which can take up to 2 weeks, because of the reduced clearance.

Cilomilast

In 12 healthy young adults the phosphodiesterase inhibitor cilomilast 15 mg bd for 5 days had no effect on the steady-state pharmacokinetics of digoxin, apart from a small reduction in the maximal concentration and a small increase in the time to peak (299). This was consistent with a small effect on the rate of digoxin absorption, which is unlikely to be of clinical significance. Digoxin had no effect on the disposition of cilomilast.

Cisapride

- In a 90-year-old woman the addition of cisapride 5 mg bd reduced the mean steady-state digoxin concentration from 0.9 ng/ml to 0.6 ng/ml and 5 mg tds reduced it to 0.4 ng/ml, with recurrence of her severe biventricular failure (300). When the doses of digoxin and cisapride were separated, the serum digoxin concentration rose again. The mechanism of this effect is thought to be increased gastrointestinal motility, although in this case the effect of separating the doses suggests that there might be a direct chemical interaction between the two drugs.

Clopidogrel

In 12 healthy men steady-state clopidogrel 75 mg/day had no effect on the steady-state plasma concentrations of digoxin (301).

Crataegus oxyacantha (hawthorn)

Crataegus oxyacantha (hawthorn) contains flavonoids, and some flavonoids are substrates of P glycoprotein (302). The interaction of St John's wort with digoxin might be due to flavonoids, although St John's wort contains many other types of compound that could also be responsible. The interaction between hawthorn (*Crataegus* extract 450 mg bd for 21 days) and digoxin 0.25 mg has been studied in a randomized crossover trial in eight healthy volunteers (303). Hawthorn extract had no effect on the pharmacokinetics of digoxin.

Dihydroergocriptine

The effect of dihydroergocriptine on the pharmacokinetics of a single oral dose of digoxin has been studied in 12 healthy men aged 23–39 years (304). There was no interaction.

Dipyridamole

In a placebo-controlled study in 12 healthy volunteers dipyridamole 300 mg/day for 3 days altered the pharmacokinetics of a single oral dose of digoxin 0.5 mg, increasing its AUC by about 13% (305). This may have been due to reduced clearance of digoxin, since dipyridamole inhibits P glycoprotein. There was no difference in the effect of dipyridamole between subjects with the different MDR1 genotypes, TT and CC.

Diuretics

Hypokalemia due to diuretics potentiates the actions of cardiac glycosides and can result in digitalis toxicity. In 67 patients taking a maintenance dose of digoxin, 42 with digoxin toxicity and 25 without, the mean serum digoxin concentration was significantly higher in the former and the mean serum potassium concentration was significantly lower (306). Among the patients with toxicity, 24% had hypokalemia due to diuretics. The mean serum digoxin concentration in those with hypokalemia and toxicity was significantly lower than in those with normokalemia and toxicity.

Dofetilide

In 14 healthy men dofetilide 250 μg bd for 5 days had no effect on the pharmacokinetics of digoxin at a steady-state trough concentration of 1.0 ng/ml (307). However, in a placebo-controlled study in patients with atrial fibrillation or atrial flutter, conversion to sinus rhythm in patients given dofetilide was more likely if they were also given digoxin (308), so there may be a pharmacodynamic interaction.

Etacrynic acid

Etacrynic acid potentiates the actions of cardiac glycosides if it causes potassium depletion.

- A patient developed digitalis intoxication and quinidine-induced cardiac dysrhythmias when hypokalemia occurred during etacrynic acid therapy (309).

In rats etacrynic acid inhibited the absorption of digitoxin from the small intestine (310), but an interaction of this sort has not been described in man.

Etanercept

The interaction of digoxin with etanercept has been studied at steady state in an open, non-randomized, crossover, 3-period study in 12 healthy men, who received loading oral doses of digoxin 0.5 mg every 12 hours on day 1 and 0.25 mg every 12 hours on day 2, followed by a daily maintenance dose of 0.25 mg for a total of 27 days (311). Etanercept was given subcutaneously 25 mg twice weekly starting on day 9 and continuing for a total of 9 doses. There were no significant pharmacokinetic or pharmacodynamic interactions.

Exenatide

In an open study exenatide had little effect on the steady-state pharmacokinetics of digoxin in 21 healthy men (312). A small fall in $C_{ss.max}$ and a small prolongation of $t_{ss.max}$ were thought to be of no clinical importance.

Fondaparinux

Fondaparinux sodium, a selective inhibitor of coagulation factor Xa, is eliminated by the kidneys. In a randomized, crossover study in 24 healthy volunteers the pharmacokinetics and pharmacodynamics of digoxin 0.25 mg/day orally for 7 days were unaffected by fondaparinux sodium 10 mg/day subcutaneously for 7 days (313).

Gentamicin

Gentamicin sulfate 80 mg bd intramuscularly for 7 days increased the serum digoxin concentration from 0.8 to 1.7 ng/ml in 12 patients with congestive heart failure and from 0.8 to 2.6 ng/ml in 12 with diabetes mellitus (314). This effect was probably due to impaired renal function, since the serum creatinine concentration also rose. Other mechanisms that the authors invoked were unlikely. However, since the rise in serum digoxin concentration was not accompanied by evidence of digoxin toxicity, it is also possible that gentamicin in some way interfered with the measurement of serum digoxin.

Hydralazine

The vasodilator hydralazine increases the renal clearance of digoxin, perhaps by increasing renal blood flow and therefore renal tubular secretion (315). This causes a small fall in plasma digoxin concentration the clinical significance of which is unclear.

Hypericum perforatum (St John's wort)

There are conflicting reports about a possible interaction of *Hypericum perforatum* with digoxin.

The effect of an extract of *Hypericum perforatum* (St John's wort) on the pharmacokinetics of digoxin have been investigated in 25 subjects in a single-blind, parallel, placebo-controlled study (316). St John's wort had no effect on the pharmacokinetics of digoxin after a single dose but after 10 days there was a 25% reduction in the steady-state AUC of digoxin and reductions in $C_{ss.max}$ (33%) and $C_{ss.min}$ (26%). The authors suggested that the mechanism might be induction of P glycoprotein.

In another randomized, placebo-controlled, parallel-group study in 96 healthy volunteers a 7-day loading phase with digoxin was followed by 14 days of co-medication with placebo or one of 10 St John's wort products varying in dose and type of formulation (317). The high-dose hyperforin-rich extract LI 160 reduced the steady-state AUC of digoxin by 25% (95% CI=21, 28), the $C_{ss.max}$ by 37% (32, 42), and the $C_{ss.min}$ by 19% (11, 27). Co-medication with hypericum powder 4 g with comparable hyperforin content resulted in reductions in digoxin AUC by 27% (16, 37), $C_{ss.max}$ by 38% (18, 48), and $C_{ss.min}$ by 19% (10, 27). Hypericum powder 2 g with half the hyperforin content reduced AUC by 18% (14, 22), $C_{ss.max}$ by 21% (2, 40), and $C_{ss.min}$ by 13% (5, 21). The authors concluded that the interaction of St John's wort with digoxin correlates with the dose of hyperforin.

In contrast, in a randomized, placebo-controlled study of a low-dose extract of *Hypericum*, hyperforin had no effect on the steady-state AUC of digoxin (318).

- An 80-year-old man taking long-term digoxin started to take St John's wort herbal tea 2 l/day and later developed digoxin poisoning, with a nodal bradycardia at 36/minute and bigeminy (319).

This was not a convincing report.

It is likely that St John's wort has a small effect on the pharmacokinetics of digoxin, since the transport of digoxin in the gut and kidneys is partly mediated by P glycoprotein, which is induced by St John's wort. However, different formulations of St John's wort are likely to have different effects and it is not clear whether any interaction has clinical significance.

In Japan enough patients take St John's wort with a cardiac glycoside to make this interaction potentially important; of 741 outpatients taking St John's wort, 171 had been given a prescription for either digoxin or metildigoxin (320)

Indometacin

Indometacin increased plasma digoxin concentrations in premature neonates with a patent ductus arteriosus (321), but a formal study in healthy adults showed no interaction (322). It may be that pre-existing impairment of renal function is required for this interaction, but this remains to be elucidated.

Ion exchange resins

Colestyramine and colestipol bind digoxin and digitoxin and can affect their absorption. When the ion exchange resins and digoxin are co-administered the absorption of digoxin is reduced (323). However, even when they are not administered together, the resins can bind cardiac glycosides that have re-entered the gut after absorption by virtue of enterohepatic recycling and enteral secretion, increasing their rate of elimination. This action has been used to treat digoxin toxicity (324–326), digitoxin toxicity (327), and toxicity from derivatives of digoxin (328).

Lasofoxifene

In 12 healthy postmenopausal women lasofoxifene had no effect on the steady-state pharmacokinetics of digoxin (329).

Levetiracetam

In 11 healthy adults in a double-blind, placebo-controlled study levetiracetam had no effect on the steady-state pharmacokinetics of digoxin or the actions of digoxin on the electrocardiogram (330).

Macrogol

Macrogol 20 g/day for 8 days caused a 30% reduction in the AUC and a 40% reduction in the C_{max} of a single dose of digoxin in 18 healthy volunteers; the t_{max} and half-life were not affected (331). This interaction was probably due to reduced absorption of digoxin, perhaps by a physicochemical interaction between the two compounds.

Macrolide antibiotics

Although digoxin is not extensively metabolized in the majority of patients, in some it is metabolized before its absorption from the gut, by two mechanisms: hydrolysis by gastric acid and hydrogenation by intestinal bacteria, mainly *Eubacterium lentum*. The macrolide antibiotics and tetracycline increase the systemic availability of digoxin, by inhibiting its breakdown by intestinal bacteria (332,333).

Macrolides have also been suggested to interact with digoxin by inhibiting P glycoprotein (SEDA-26, 200). However, in a study of this mechanism in nine healthy Japanese men, clarithromycin 200 mg bd and erythromycin 200 mg qds did not alter the plasma concentration versus time curve of a single intravenous dose of digoxin 0.5 mg, but increased its renal clearance (334). This contrasts with an observation of reduced renal clearance of digoxin in two patients taking clarithromycin (SEDA-23, 194). That inhibition of renal P glycoprotein may not reduce the renal clearance of digoxin has also been suggested by studies with talinolol (SEDA-25, 172) and atorvastatin (335). Other transport mechanisms for digoxin, including the organic anion-transporting polypeptides, have not been well studied and may play a role in digoxin disposition and hence drug interactions.

Azithromycin

- In a 31-month-old boy with Down's syndrome and Fallot's tetralogy during a 5-day course of azithromycin 5 mg/kg/day the serum digoxin concentration rose and the child had anorexia, diarrhea, and second-degree atrioventricular block with junctional extra beats (336).

The mechanism was not investigated.

Clarithromycin

- In a 72-year-old woman taking digoxin 0.25 mg/day, the addition of clarithromycin caused a rise in the serum digoxin concentration to 4.6 ng/ml (337).

In two other cases in which clarithromycin increased serum digoxin concentrations there was an associated reduction in the rate of renal digoxin clearance, which may be another mechanism for this interaction (338). The authors hypothesized that clarithromycin inhibited P glycoprotein. This was supported by the observation of a concentration-dependent effect of clarithromycin on in vitro transcellular transport of digoxin.

There is anecdotal evidence that this interaction may be clinically important. Digoxin toxicity occurred in patients taking clarithromycin (339,340).

There have been conflicting reports that clarithromycin can either reduce or increase the renal clearance of digoxin. In a double-blind, randomized, placebo-controlled, crossover study 12 healthy men took single oral doses of digoxin 0.75 mg with either placebo or clarithromycin 250 mg bd for 3 days (341). Three of the subjects also received single intravenous doses of digoxin 0.01 mg/kg with oral placebo or clarithromycin. Clarithromycin increase the AUC of oral digoxin 1.7-fold and reduced its non-glomerular renal clearance. The ratios of mean digoxin plasma concentrations with and without clarithromycin were highest during the absorption phase of clarithromycin. The pharmacokinetics of intravenous digoxin were not affected by clarithromycin. The authors concluded that increased oral systemic availability and reduced non-glomerular renal clearance of digoxin both contribute to the interaction between digoxin and clarithromycin, probably due to inhibition of intestinal and renal P glycoprotein.

Table 5 Percentage changes in steady-state serum digoxin concentrations during co-administration of various inhibitors of P glycoprotein (342)

P glycoprotein inhibitor	*Increase in steady-state serum digoxin concentration (%)*
Amiodarone	70–100
Clarithromycin	100–150
Ciclosporin	10–80
Itraconazole	35–80
Propafenone	35–60
Quinidine	100–200
Spironolactone	0–20
Verapamil	40–80

This interpretation has been supported by observations of increased steady-state plasma digoxin concentrations in patients taking clarithromycin 400 mg/day (1.4 versus 0.8 ng/ml) (Table 5).

In six men with end-stage renal disease clarithromycin increased serum digoxin concentrations by 1.8–4.0 times (343). In three cases the increase occurred within 12 days and in the other three at 53–190 days. The authors attributed the increase in serum digoxin to inhibition by clarithromycin of P glycoprotein in the intestine and/or bile capillaries rather than the kidneys, since renal function was dramatically impaired and four of the patients were anuric.

Erythromycin

- Digoxin toxicity occurred in a neonate who was also given erythromycin (344). She had bradycardia and coupled extra beats. Digoxin and erythromycin were withdrawn and she was given antidigoxin antibodies. Her plasma digoxin concentration, which had previously been 1.8 ng/ml, had risen to 8.0 ng/ml.

Meglitinides

In 12 healthy volunteers aged 19–36 years nateglinide had no effects on the pharmacokinetics of a single dose of digoxin (345). Similarly, in 14 healthy adults, repaglinide 2 mg three times had no effect on the steady-state pharmacokinetics of digoxin (346). These results suggest that the meglitinides do not affect P glycoprotein.

Nitrates

The effects of nitrates on the pharmacokinetics of digoxin are probably small.

In eight patients with chronic heart failure taking digoxin 0.25 mg/day, isosorbide dinitrate 10 mg tds caused only a small increase in C_{max} (15%) and had no effect on the mean steady-state serum concentration or AUC (279).

Nitroprusside increased the renal clearance of digoxin, perhaps by increasing renal blood flow and therefore renal tubular secretion (315). This caused a small fall in plasma digoxin concentration, the clinical significance of which is unclear.

Phenobarbital

The metabolism of digitoxin can be increased, with resulting increasing dosage requirements, by enzyme-inducing drugs (230). For example, phenobarbital 100 mg daily reduces steady-state serum digitoxin concentrations by 50%.

Polyethoxylated castor oil

Polyethoxylated castor oil (Cremophor EL) inhibits P glycoprotein in vitro and in vivo. In a double-blind, randomized, placebo-controlled, crossover study in 12 healthy individuals a single oral dose of digoxin 0.5 mg in a hard gelatin capsule was given in combination with multiple doses of oral polyethoxylated castor oil (Cremophor RH40) 600 mg tds or placebo (347). Cremophor RH40 delayed and enhanced the absorption of digoxin in the first 5 hours after dosing: the lag time was prolonged from 0.36 to 0.53 hours; the C_{max} rose by 22%, from 2.21 to 2.69 ng/ml; and the $AUC_{0\rightarrow5}$ increased by 22%. The authors concluded that Cremophor RH40 alters the pharmacokinetics of digoxin by inhibiting P glycoprotein and delaying the dissolution time of digoxin tablets.

Propafenone

Propafenone causes a small increase in plasma digoxin concentrations by an unknown mechanism (348,349). Although this effect is perhaps not clinically significant, in patients with dysrhythmias it was accompanied by an increase in PR interval (244).

Quinidine

Digoxin

The quinidine–digoxin interaction has been reviewed (SEDA-6, 173) (SEDA-7, 195) (SEDA-9, 159) (SEDA-10, 145) (SEDA-15, 166) (SEDA-18, 198). Although the major mechanism of the interaction is probably inhibition of the active tubular secretion of digoxin by quinidine, other mechanisms are involved, including reduced non-renal clearance and displacement of digoxin from the tissues. The reduction in non-renal clearance is at least partly due to a reduction in biliary clearance. This interaction affects most patients and on average causes a two-fold increase in steady-state plasma digoxin concentrations. Because both clearance and apparent volume of distribution are reduced, the half-life of digoxin is either unaffected or perhaps slightly prolonged.

In addition to the pharmacokinetic interaction, there may be a pharmacodynamic interaction, since there is some evidence that quinidine reduces the positive inotropic effect of digoxin on the heart, in addition to having a negative inotropic effect of its own. Thus, the outcome of the interaction is a 24-fold increased risk of digoxin toxicity and a reduction in its beneficial effect on the heart, at least in sinus rhythm. If the two drugs are used together, the initial digoxin dosage should be halved and adjusted subsequently on the basis of the patient's clinical condition and the plasma digoxin concentration.

Lanatoside C and metildigoxin are theoretically likely to be similarly affected by quinidine, and there is anecdotal evidence of this (350).

The pharmacokinetic interaction of quinidine with digoxin also occurs with quinine (351,352) and hydroxychloroquine (353). However, the effects of these drugs are smaller than those with quinidine. Quinine reduces the extrarenal clearance of digoxin, perhaps by altering its biliary secretion (354).

Digitoxin

In five healthy adults quinidine prolonged the half-life of a single dose of digitoxin from 174 to 261 hours and reduced its total body clearance from 1.54 to 1.09 ml/hour/kg and renal clearance from 0.65 to 0.46 ml/hour/kg. Digitoxin volume of distribution and protein binding were unaffected. Quinidine caused a rise in serum digitoxin concentrations (355).

In eight healthy subjects steady-state digitoxin plasma concentrations and renal excretion increased from 13.6 ng/ml and 16.1 μg/day before dosing to 19.7 ng/ml and 23.4 μg/day respectively during quinidine dosing for 32 days (356). Renal digitoxin clearance was not noticeably changed by quinidine, but total and extrarenal digitoxin clearances fell by 32% and 41% respectively. The half-life of digitoxin was prolonged from 150 to 203 hours. There were corresponding pharmacodynamic effects, as assessed by electrocardiography and systolic time intervals.

Quinolone antibiotics

The quinolone antimicrobial drugs gemifloxacin, levofloxacin, and sparfloxacin do not interact with digoxin.

In a crossover study in 14 healthy elderly individuals gemifloxacin 320 mg/day had no effect on the steady-state pharmacokinetics of digoxin (357).

Levofloxacin 500 mg bd for 6 days in 12 healthy volunteers did not interact with digoxin (358).

The interaction of sparfloxacin with digoxin, both at steady state, has been studied in a double-blind, placebo-controlled, crossover study in 24 healthy men aged 20–49 years. Sparfloxacin had no effect on the steady-state mean plasma concentration, C_{min}, t_{max}, or AUC of digoxin (359). Conversely digoxin did not alter the steady-state pharmacokinetics of sparfloxacin.

Rifampicin

The metabolism of digitoxin can be increased, with resulting increasing dosage requirements, by rifampicin (360).

Rifampicin also induces P glycoprotein and should therefore increase the clearance of digoxin by that mechanism. Some reports (361,362) have suggested that rifampicin reduces the steady-state concentration of digoxin. Another study in eight healthy men has shown that steady-state rifampicin significantly reduced the AUC of oral digoxin but had less of an effect on intravenous digoxin (363). The authors attributed this to induction of P glycoprotein in the intestine, with increased secretion of the drug into the gut lumen. Rifampicin has previously been reported to reduce steady-state plasma digitoxin concentrations, an effect that was attributed to induction of the metabolism of digitoxin to digoxin (360); however, in that study plasma digoxin concentrations were not measured separately, and it may be that rifampicin also increases the secretion of digitoxin into the gut lumen.

Ritonavir

Ritonavir is an inhibitor of P glycoprotein and might therefore be expected to interact with digoxin.

- A 61-year-old woman taking digoxin 0.25 mg/day, coumadin, indinavir 800 mg tds, lamivudine 150 mg bd, and stavudine 40 mg bd developed nausea and vomiting 3 days after she started to take ritonavir 200 mg bd (364). Her serum digoxin concentration 5 hours after the last dose was 7.2 nmol/l.

Ropinirole

In 10 patients with Parkinson's disease steady-state ropinirole treatment caused small reductions in the AUC and C_{max} of digoxin at steady state (10% and 25% respectively) (365). However, the C_{min} was unaffected and overall there was probably no significant effect of ropinirole on digoxin disposition.

Selective serotonin reuptake inhibitors (SSRIs)

There is in vitro evidence that some serotonin-selective reuptake inhibitors (SSRIs) inhibit P glycoprotein, which is responsible for active transport of digoxin in the gut and kidneys. In a nested case–control study of patients aged 66 years or older, hospital admissions for digoxin toxicity were related to SSRI therapy in the previous 30 days (366). Among 245 305 patients taking digoxin there were 3144 cases of digoxin toxicity. After adjusting for potential confounders, there was an increased risk of digoxin toxicity after the start of therapy with paroxetine (OR = 2.8; 95% CI = 1.6, 4.7), fluoxetine (OR = 2.9; CI = 1.5, 5.4), sertraline (OR = 3.0; CI = 1.9, 4.7), and fluvoxamine (OR = 3.0; CI = 1.5, 5.7). However, there was also an increased risk with tricyclic antidepressants (OR = 1.5; 95% CI = 1.0, 2.4) and benzodiazepines (OR = 2.1; 95% CI = 1.7, 2.5), drug classes that have no known pharmacokinetic interaction with digoxin and do not inhibit P glycoprotein. There were no statistical differences in the risks of digoxin toxicity among any of the agents tested. The authors therefore suggested that SSRIs do not increase the risk of digoxin toxicity and that inhibition of P glycoprotein by sertraline and paroxetine is unlikely to be of major clinical significance.

In 11 healthy adults citalopram 40 mg/day had no effect on the pharmacokinetics of a single oral dose of digoxin 1 mg (367). Digoxin did not affect citalopram pharmacokinetics.

Sevelamer

Sevelamer, a non-absorbed phosphate-binding polymer, in a dose of 2.4 g, had no effect on the pharmacokinetics of single doses of digoxin in 19 healthy volunteers (368).

Spironolactone

Spironolactone inhibits the active tubular secretion of digoxin by about 25% and in some cases digoxin dosages may have to be reduced (369).

Statins

The effects of statins on the pharmacokinetics of digoxin have been variable. There is no simple explanation that reconciles the disparate findings.

- A 52-year-old man developed rhabdomyolysis while taking simvastatin, digoxin, ciclosporin, and verapamil (370). The authors proposed that this had been due in part to inhibition of the biliary secretion of simvastatin by digoxin; however, it is likely that the major mechanism of the interaction was inhibition of CYP3A4 by ciclosporin.

Atorvastatin

Atorvastatin 80 mg/day increased the AUC and C_{max} of digoxin 0.25 mg/day by 15% and 20% respectively during steady-state therapy, without affecting renal digoxin clearance (335).

Cerivastatin

In 20 healthy men, cerivastatin had no effect on the steady-state pharmacokinetics of digoxin (371). Digoxin had no significant effect on the pharmacokinetics of cerivastatin.

Fluvastatin

A single dose of fluvastatin increased the steady-state C_{max} of digoxin 0.125–0.5 mg/day by 11% and renal clearance by 15%, without changing AUC or t_{max} (372).

Rosuvastatin

Rosuvastatin 40 mg/day had no effect on the pharmacokinetics of a single oral dose of digoxin 0.5 mg in 18 healthy men (373).

Suxamethonium

Cardiac glycosides and suxamethonium can interact, resulting in an increased risk of dysrhythmias (374), perhaps through alterations in intracellular calcium (375). In 24 patients with ischemic heart disease taking digoxin who underwent abdominal surgery ventricular extra beats with bigemini or severe bradycardia were recorded in two patients and episodes of torsade de pointes occurred in two others during endotracheal intubation (376). The authors suggested that endotracheal intubation in digitalized patients should be performed without suxamethonium. However, considering the frequency with which digitalized patients receive suxamethonium and the paucity of reports of clinical problems, this interaction is probably of minor importance.

Talinolol

The effects of talinolol on the pharmacokinetics of digoxin have been studied in 10 healthy volunteers aged 23–30 years in a crossover study (377). Oral talinolol 100 mg increased the AUC of digoxin significantly, but the renal clearance and half-life of digoxin were unchanged. Intravenous talinolol 30 mg had no effect on the pharmacokinetics of oral digoxin. The authors concluded that the change in AUC after oral talinolol was due to increased systemic availability of digoxin, through inhibition of intestinal P glycoprotein. They did not discuss the possibility that talinolol had also reduced the non-renal clearance of digoxin by inhibiting its biliary secretion, and indeed there was a small, albeit non-significant reduction in non-renal clearance of digoxin after both oral and intravenous talinolol. In contrast, digoxin did not affect the kinetics of talinolol.

Tamsulosin

The alpha$_1$-adrenoceptor antagonist tamsulosin 0.4 and 0.8 mg/day had no effect on the pharmacokinetics of a single intravenous dose of digoxin 0.5 mg in 10 healthy men (378).

Tegaserod

In 12 healthy subjects the 5-HT$_4$ receptor partial agonist tegaserod 6 mg bd for 3 days had a small effect on the pharmacokinetics of a single oral dose of digoxin 1 mg, reducing the mean AUC by 12% and the C_{max} by 15% (379). There was a small delay in the time to peak, which was not significant. These results suggest that tegaserod slightly reduces the systemic availability of digoxin, perhaps because it increases gastrointestinal motility, but that the effect is of no clinical significance.

Telmisartan

Multiple-dose telmisartan 120 mg/day administered with digoxin 0.25 mg/day resulted in higher serum digoxin concentrations (380). Digoxin AUC rose by 22% and C_{max} by 50%; the rise in C_{min} (13%) was not significant. These results suggest that telmisartan reduces the clearance of digoxin. The magnitude of this effect is comparable to that observed with calcium channel blockers, carvedilol, captopril, amiodarone, quinidine, and propafenone. Monitoring serum digoxin concentrations should be considered when patients first use telmisartan and when the dosage of telmisartan is changed.

Teriparatide

Teriparatide (recombinant human parathyroid hormone (1–34)) causes small transient increases in serum calcium concentration, and might therefore interact with digoxin. However, in 15 healthy subjects, teriparatide 20 micrograms on days 14 and 15 of steady-state digoxin therapy had no effects on the action of digoxin on systolic time intervals or heart rate compared with placebo (381).

Tetracycline

Although digoxin is not extensively metabolized in the majority of patients, in some it is metabolized before its absorption from the gut, by two mechanisms: hydrolysis by gastric acid and hydrogenation by intestinal bacteria, mainly *E. lentum*. Tetracycline increases the systemic availability of digoxin, by inhibiting its breakdown by intestinal bacteria (297).

Tiagabine

Tiagabine, which is principally metabolized by CYP3A, had no effect on the steady-state pharmacokinetics of digoxin in 13 healthy volunteers (382). This is evidence that CYP3A is not important in the metabolism of digoxin.

Warfarin

It has been claimed that digoxin potentiated the effects of warfarin in a 66-year-old man (383). However, the discussion of the possible mechanisms of this observation was flawed, and there is no reason to expect such an interaction, which probably does not occur (384).

Ximelagatran

The interaction of digoxin with ximelagatran has been investigated in a randomized, double-blind, two-way, crossover study in 16 healthy men and women, who took ximelagatran 36 mg or placebo bd for 8 days and a single oral dose of digoxin 0.5 mg on day 4; there was no pharmacokinetic interaction (385).

Interference with diagnostic tests Although spironolactone can interfere with digoxin radioimmunoassay, this was not found in a study of the AxSYM Digoxin II assay (386). However, the conclusion was based on a comparison with an Emit assay, and so systematic interference by both assays could have been missed. Furthermore, the dose of spironolactone was low (25 mg/day).

In one case *Eleutheroccus senticosus* (Siberian ginseng) increased the serum concentration of digoxin (387), probably because ginseng can give false readings in digoxin radioimmunoassay rather than by an in vivo interaction. This effect varies from formulation to formulation and from assay to assay (388,389).

Therapeutic doses of cardiac glycosides increase serum concentrations of A and B natriuretic peptides (ANP and BNP) (390,391). There has now been a report of a large increase in BNP in a patient with digitalis toxicity (392).

- A 78-year-old woman with congestive heart failure developed nausea, increasing dyspnea on exertion, and leg edema. Her heart rate was 50/minute and her respiratory rate 30/minute. An electrocardiogram showed sinus bradycardia with left bundle-branch block. The serum potassium concentration was 7.0 mmol/l, the serum creatinine was 221 mol/l, and she had a compensated metabolic acidosis. The serum digoxin concentration was 4.2 ng/ml. The serum concentration of B natriuretic peptide (BNP) was greater than 1300 pg/ml (the upper limit of the assay).

Zaleplon

The effects of zaleplon on the pharmacokinetics and pharmacodynamics of steady-state digoxin have been studied in 20 healthy men aged 18–45 years (393). There was no interaction.

Interference with Diagnostic Tests

Digoxin radioimmunoassays

Spironolactone interferes with some digoxin radioimmunoassays, because it and its metabolites, such as canrenone and 7-alpha-thiomethylspironolactone, are immunoreactive with some forms of antidigoxin antibody (394–396). However, in contrast to this there has been a recent report that canrenoate, the main metabolite of spironolactone, interfered with the immunoassay of digoxin (397). The effect was largest with the AxSym MEIA II assay, and there was also some interference with the EMIT assay. However, the TDx assay did not show any interference. Spironolactone also caused some interference in the AxSym assay but less than canrenoate. In this case the failure to measure high digoxin concentrations resulted in clinical toxicity in a 71-year-old man who was given 3.8 mg over 11 days.

In a study of the use of the TDx II assay in 80 children and adults, there was apparent digoxin immunoreactivity in 3.7% of healthy subjects (n = 80), 3.6% of pregnant women (n = 28), 10% of patients with renal transplants (n = 31), and 23% of immature infants (n = 40) (398).

In nine assays (AxSYM, IMx, TDx, Emit, Dimension, aca, TinaQuant, Elecsys, and Vitros) interference by spironolactone, canrenone, and three metabolites was sought in vitro, and all routine digoxin measurements using the AxSYM system over 16.5 months (n = 3089) were reviewed (399). There was a reduction in the expected concentrations by canrenone (3125 ng/ml) in the following assays: AxSYM (42% of expected), IMx (51%), and Dimension (78%). There was positive bias in aca (0.7 ng/ml), TDx (0.62 ng/ml), and Elecsys (>0.58 ng/ml). Of 669 routinely monitored patients, 25 had falsely low results and 19 of them actually had potentially toxic digoxin concentrations; this was attributable to concurrent therapy with spironolactone, canrenone, hydrocortisone, or prednisolone. However, standard doses of spironolactone (up to 50 mg/day) in patients with heart failure produced less than 11% inhibition.

In a patient with digoxin toxicity the plasma digoxin concentration measured with a microparticle enzyme immunoassay (MEIA/AxSYM) gave consistently lower results than other methods, enzyme multiplied immunoassay (EMIT/Cobas), fluorescence polarization immunoassay (FPIA/FLx), and liquid chromatography–electrospray–mass spectrometry (400). The source of the negative interference was not discovered.

Monitoring Therapy

The use of plasma digitalis concentrations in monitoring therapy has been reviewed (153,401).

In a retrospective analysis of 210 randomly selected digoxin plasma concentration determinations in inpatients, the indications were considered to have been inappropriate in 67, appropriate in 81, and unevaluable in 4 (402). Timing of the blood sample was wrong in 17 cases (samples should be taken at least 6 hours and preferably about 11 hours after the dose). Of the measurements whose indications were considered to have been inappropriate, most (63) were performed as part of "routine" monitoring.

The following are the main uses of plasma digitalis concentration measurements.

Individualizing therapy

In the absence of factors that alter the response to digitalis it may be worth measuring the plasma concentration during the initial stages of therapy to ensure that a reasonable concentration has been achieved (1.0–1.5 ng/ml for digoxin, 10–15 ng/ml for digitoxin). In cases where there is still a poor response to treatment it is justifiable to increase digitalis dosages cautiously, but the risk of toxicity starts to rise markedly at plasma concentrations above 2.0 and 20 ng/ml respectively. If there are subsequent changes in the patient's condition, for example renal impairment, then plasma concentration measurement may help in readjusting dosages.

Monitoring adherence

The commonest cause of a low plasma digitalis concentration in a patient taking a cardiac glycoside is poor adherence to therapy.

Making decisions about long-term therapy

If the plasma digitalis concentration is below the target range in a patient whose condition is stable (for example atrial fibrillation with a ventricular rate of 80/minute) digitalis can usually be withdrawn safely.

Diagnosis of toxicity

The underlying principles in using plasma digitalis concentrations to diagnose toxicity are as follows:

1. The plasma digitalis concentration must be considered in conjunction with other clinical information, that is symptoms, the signs of possible intoxication, the stability of the underlying condition, age, renal function, the dosage, and biochemical measurements such as the plasma potassium concentration.
2. At plasma digitalis concentrations above 3.0 ng/ml (digoxin) or 30 ng/ml (digitoxin) toxicity is highly likely. At concentrations below 1.5 or 15 ng/ml, respectively, toxicity is unlikely. However, toxicity can occur even with low concentrations and should be suspected particularly if there is hypokalemia.
3. Certain factors increase the risk of digitalis toxicity at a given plasma concentration (see Susceptibility Factors). These factors will alter the interpretation of the plasma digitalis concentration and should lower the threshold for suspicion.

4. When in doubt it is far better to withhold digitalis and monitor progress than to continue treatment, thereby running the risk of perpetuating toxicity.

In patients whose condition is satisfactory and stable and whose plasma digoxin concentration is low (below 0.8 ng/ml), withdrawal of digoxin is recommended and is highly unlikely to affect the patient's condition (discussed in the section Making decisions about long-term therapy). Probably the same applies for digitoxin at concentrations below 8.0 ng/ml, although that has not been demonstrated.

It has yet again been confirmed that the serum digoxin concentration distinguishes between patients with and without digoxin toxicity, but with considerable overlap (403). Of 99 patients, 41 with toxicity had mean serum digoxin concentrations of 3.1 ng/ml compared with 1.6 ng/ml in 58 non-toxic patients. However the digoxin concentration was below 2 ng/ml in 10 patients with toxicity and higher than 2 ng/ml in 16 patients without. There were no significant differences in serum electrolyte concentrations between the toxic and non-toxic patients, and the authors therefore concluded that such abnormalities are less important than they have usually been considered to be. However, this study does not demonstrate that at all; rather it shows that even if serum electrolyte concentrations are well controlled it may not be possible to avoid digitalis toxicity for other reasons. Indeed, in this study the patients with toxicity had significantly worse renal function, which would have explained their increased risk.

References

1. Various authors. Digoxin symposium. Br Heart J 1985;54:227.
2. William Withering: An account of the foxglove and some of its medical uses 1785–1985. Fisch, C editor. J Am Coll Cardiol 1985;5(Suppl A):.
3. Cardiac Glycosides 1785–1984. Biochemistry, Pharmacology, Clinical Relevance. In: Erdmann E, Greeff K, Skou JC, editors. Darmstadt: Steinkopff Verlag 1985:473.
4. Rietbrock N, Woodcock BG. Handbook of Renal-Independent Cardiac Glycosides: Pharmacology and Clinical PharmacologyChichester: Ellis Horwood;. 1989.
5. Chung EK. Digitalis IntoxicationAmsterdam: Excerpta Medica;. 1969.
6. Smith TW, Antman EM, Friedman PL, Blatt CM, Marsh JD. Digitalis glycosides: mechanisms and manifestations of toxicity. Prog Cardiovasc Dis 1984;26(6):495–5401984;27(1):21–56.
7. Aronson JK. Digitalis intoxication. Clin Sci (Lond) 1983;64(3):253–8.
8. Buchanan JF, Olson KR. Current management of digitalis toxicity. Part I: Clinical manifestations. Pract Cardiol 1988;14:75–9.
9. Buchanan JF, Olson KR. Current management of digitalis toxicity. Part II: Treatment of digitalis intoxication. Pract Cardiol 1988;14:92–5.
10. Hauptman PJ, Kelly RA. Digitalis. Circulation 1999;99(9):1265–70.
11. Haji SA, Movahed A. Update on digoxin therapy in congestive heart failure. Am Fam Physician 2000;62(2):409–16.
12. Gibbs CR, Davies MK, Lip GY. ABC of heart failure. Management: digoxin and other inotropes, beta blockers, and antiarrhythmic and antithrombotic treatment. BMJ 2000;320(7233):495–8.
13. Schwartz A, Lindenmayer GE, Allen JC. The sodium-potassium adenosine triphosphatase: pharmacological, physiological and biochemical aspects. Pharmacol Rev 1975;27(01):3–134.
14. Levitt B, Cagin N, Kleid J, Somberg J, Gillis R. Role of the nervous system in the genesis of cardiac rhythm disorders. Am J Cardiol 1976;37(7):1111–3.
15. Aronson JK, Ford AR. The use of colour vision measurement in the diagnosis of digoxin toxicity. Q J Med 1980;49(195):273–82.
16. Beller GA, Smith TW, Abelmann WH, Haber E, Hood WB Jr. Digitalis intoxication. A prospective clinical study with serum level correlations. N Engl J Med 1971;284(18):989–97.
17. Marik PE, Fromm L. A case series of hospitalized patients with elevated digoxin levels. Am J Med 1998;105(2):110–5.
18. Williamson KM, Thrasher KA, Fulton KB, LaPointe NM, Dunham GD, Cooper AA, Barrett PS, Patterson JH. Digoxin toxicity: an evaluation in current clinical practice. Arch Intern Med 1998;158(22):2444–9.
19. Onder G, Pedone C, Landi F, Cesari M, Della Vedova C, Bernabei R, Gambassi G. Adverse drug reactions as cause of hospital admissions: results from the Italian Group of Pharmacoepidemiology in the Elderly (GIFA). J Am Geriatr Soc 2002;50(12):1962–8.
20. Ahmed A, Allman RM, DeLong JF. Inappropriate use of digoxin in older hospitalized heart failure patients. J Gerontol A Biol Sci Med Sci 2002;57(2):M138–43.
21. Cooper JW. Adverse drug reaction-related hospitalizations of nursing facility patients: a 4-year study. South Med J 1999;92(5):485–90.
22. Joubert PH. Are all cardiac glycosides pharmacodynamically similar? Eur J Clin Pharmacol 1990;39(4):317–20.
23. Tucker AR, Ng KT. Digoxin-related impairment of learning and memory in cardiac patients. Psychopharmacology (Berl) 1983;81(1):86–8.
24. Wojtyna W, Enseleit F. A rare cause of complete heart block after transdermal botanical treatment for psoriasis. Pacing Clin Electrophysiol 2004;27(12):1686–8.
25. Haynes BE, Bessen HA, Wightman WD. Oleander tea: herbal draught of death. Ann Emerg Med 1985;14(4):350–3.
26. de Silva HA, Fonseka MM, Pathmeswaran A, Alahakone DG, Ratnatilake GA, Gunatilake SB, Ranasinha CD, Lalloo DG, Aronson JK, de Silva HJ. Multiple-dose activated charcoal for treatment of yellow oleander poisoning: a single-blind, randomised, placebo-controlled trial. Lancet 2003;361(9373):1935–8.
27. Eddleston M, Juszczak E, Buckley NA, Senarathna L, Mohamed F, Dissanayake W, Hittarage A, Azher S, Jeganathan K, Jayamanne S, Sheriff MR, Warrell DA; Ox-Col Poisoning Study collaborators. Multiple-dose activated charcoal in acute self-poisoning: a randomised controlled trial. Lancet 2008;371(9612):579–87.
28. de Silva HA, Pathmeswaran A, Lalloo DG, de Silva HJ, Aronson JK. Multiple-dose activated charcoal in yellow oleander poisoning. Lancet 2008;371(9631):2171.
29. Camphausen C, Haas NA, Mattke AC. Successful treatment of oleander intoxication (cardiac glycosides) with

digoxin-specific Fab antibody fragments in a 7-year-old child: case report and review of literature. Z Kardiol 2005;94(12):817–23.
30. Bourgeois B, Incagnoli P, Hanna J, Tirard V. Traitement par anticorps antidigitalique d'une intoxication volontaire par laurier rose. Ann Fr Anesth Reanim 2005;24(6):640–2.
31. Eddleston M, Rajapakse S, Rajakanthan, Jayalath S, Sjöström L, Santharaj W, Thenabadu PN, Sheriff MH, Warrell DA. Anti-digoxin Fab fragments in cardiotoxicity induced by ingestion of yellow oleander: a randomised controlled trial. Lancet 2000;355(9208):967–72.
32. Eddleston M, Senarathna L, Mohamed F, Buckley N, Juszczak E, Sheriff MH, Ariaratnam A, Rajapakse S, Warrell D, Rajakanthan K. Deaths due to absence of an affordable antitoxin for plant poisoning. Lancet 2003;362(9389):1041–4.
33. Tuncok Y, Kozan O, Cavdar C, Guven H, Fowler J. Urginea maritima (squill) toxicity. J Toxicol Clin Toxicol 1995;33(1):83–6.
34. Williams P, Aronson J, Sleight P. Is a slow pulse-rate a reliable sign of digitalis toxicity? Lancet 1978;2(8104–5):1340–2.
35. Hamer SS, Lemberg L. Digitalis excess mimicking the sick sinus syndrome. Heart Lung 1976;5(4):652–6.
36. Di Giacomo V, Carmenini G, Sciacca A. Su un caso di malattia del nodo del seno insorto in corso di trattamento digilatico. Progr Med 1977;33:775.
37. Margolis JR, Strauss HC, Miller HC, Gilbert M, Wallace AG. Digitalis and the sick sinus syndrome. Clinical and electrophysiologic documentation of severe toxic effect on sinus node function. Circulation 1975;52(1):162–9.
38. Engel TR, Schaal SF. Digitalis in the sick sinus syndrome. The effects of digitalis on sinoatrial automaticity and atrioventricular conduction. Circulation 1973;48(6): 1201–7.
39. Vera Z, Miller RR, McMillin D, Mason DT. Effects of digitalis on sinus nodal function in patients with sick sinus syndrome. Am J Cardiol 1978;41(2):318–23.
40. Talley JD, Wathen MS, Hurst JW. Hyperthyroid-induced atrial flutter-fibrillation with profound sinoatrial nodal pauses due to small doses of digoxin, verapamil, and propranolol. Clin Cardiol 1989;12(1):45–7.
41. McCauley CS, Boller LR. Bradycardic responses to endotracheal suctioning. Crit Care Med 1988;16(11):1165–6.
42. Essebag V, Hadjis T, Platt RW, Pilote L. Amiodarone and the risk of bradyarrhythmia requiring permanent pacemaker in elderly patients with atrial fibrillation and prior myocardial infarction. J Am Coll Cardiol 2003;41:249–54.
43. Lown B, Wyatt NF, Levine HD. Paroxysmal atrial tachycardia with block. Circulation 1960;21:129–43.
44. Tawakkol AA, Nutter DO, Massumi RA. A prospective study of digitalis toxicity in a large city hospital. Med Ann Dist Columbia 1967;36(7):402–9.
45. Agarwal BL, Agrawal BV, Agarwal RK, Kansal SC. Atrial flutter. A rare manifestation of digitalis intoxication. Br Heart J 1972;34(4):392–5.
46. Siniorakis E, Arvanitakis S, Ralli D, Ciubotariou-Petsa I, Barbis C, Bonoris P. Digoxin intoxication: arrhythmogenic or antiarrhythmic? Int J Cardiol 2003;91:111–12.
47. Spodick DH. Well concealed atrial tachycardia with Wenckebach (Mobitz I) atrioventricular block: digitalis toxicity. Am J Geriatr Cardiol 2001;10(1):59.
48. Barold SS, Hayes DL. Non-paroxysmal junctional tachycardia with type I exit block. Heart 2002;88(3):288.
49. Pellegrino L. Ritmo idioventriculare accelerato da intossicazione digitalica. Studio clinico ed elettrocardiografico su due casi. [Accelerated idioventricular rhythm in patients with digitalic intoxication. Clinical and electrocardiographic study of two cases.] G Ital Cardiol 1976;6(3):527–31.
50. Castellanos A, Shin EK, Luceri RM, Myerburg RJ. Parasystolic accelerated idioventricular rhythms producing bidirectional tachycardia patterns. J Electrophysiol 1988;2:296.
51. Deal BJ, Keane JF, Gillette PC, Garson A Jr. Wolff-Parkinson–White syndrome and supraventricular tachycardia during infancy: management and follow-up. J Am Coll Cardiol 1985;5(1):130–5.
52. Pfammatter JP, Stocker FP. Re-entrant supraventricular tachycardia in infancy: current role of prophylactic digoxin treatment. Eur J Pediatr 1998;157(2):101–6.
53. Sanatani S, Saul JP, Walsh EP, Gross GJ. Spontaneously terminating apparent ventricular fibrillation during transesophageal electrophysiological testing in infants with Wolff-Parkinson-White syndrome. Pacing Clin Electrophysiol 2001;24(12):1816–8.
54. Grimard C, De Labriolle A, Charbonnier B, Babuty D. Bidirectional ventricular tachycardia resulting from digoxin toxicity. J Cardiovasc Electrophysiol 2005;16(7):807–8.
55. Goranitou G, Stavrianaki D, Babalis D. Wide QRS tachycardia caused by severe hyperkalaemia and digoxin intoxication. Acta Cardiol 2005;60(4):437–41.
56. Eyal D, Molczan KA, Carroll LS. Digoxin toxicity: pediatric survival after asystolic arrest. Clin Toxicol (Phila) 2005;43(1):51–4.
57. Lehmann HU, Witt E, Hochrein H. Zunahme von Angina pectoris und ST-Strecken-Senkung im EKG durch Digitalis (Koronare Funktionsuntersuchungen bei Gesunden und Koranarkranken ohne Herzinsuffizienz mittels rechtsatrialer Frequenzstimulation). [Digitalis-induced increase in angina pectoris and segment depression on electrocardiograms (Investigations of coronary function of healthy subjects and of coronary patients without cardiac insufficiency by means of atrial pacing).] Z Kardiol 1978;67(1):57–66.
58. Singh RB, Agrawal BV, Somani PN. Left bundle branch block: a rare manifestation of digitalis intoxication. Acta Cardiol 1976;31(2):175–9.
59. Gould L, Patel C, Betzu R, Judge D, Lee J. Right bundle branch block: a rare manifestation of digitalis toxicity—case report. Angiology 1986;37(7):543–6.
60. Ma G, Brady WJ, Pollack M, Chan TC. Electrocardiographic manifestations: digitalis toxicity. J Emerg Med 2001;20(2):145–52.
61. Von Capeller D, Copeland GD, Stern TN. Digitalis intoxication: a clinical report of 148 cases. Ann Intern Med 1959;50(4):869–78.
62. Somlyo AP. The toxicology of digitalis. Am J Cardiol 1960;5:523–33.
63. Braunwald E, Bloodwell RD, Goldberg LI, Morrow AG. Studies on digitalis. IV. Observations in man on the effects of digitalis preparations on the contractility of the non-failing heart and on total vascular resistance. J Clin Invest 1961;40:52–9.
64. Teske RH, Bishop SP, Righter HF, Detweiler DK. Subacute digoxin toxicosis in the beagle dog. Toxicol Appl Pharmacol 1976;35(2):283–301.
65. Varonkov Y, Shell WE, Smirnov V, Gukovsky D, Chazov EI. Augmentation of serum CPK activity by

digitalis in patients with acute myocardial infarction. Circulation 1977;55(5):719–27.
66. Nolte CW, Jost S, Mugge A, Daniel WG. Protection from digoxin-induced coronary vasoconstriction in patients with coronary artery disease by calcium antagonists. Am J Cardiol 1999;83(3):440–2.
67. Balcon R, Hoy J, Sowton E. Haemodynamic effects of rapid digitalization following acute myocardial infarction. Br Heart J 1968;30(3):373–6.
68. Scheinost ME. Digoxin toxicity in a 26-year-old woman taking a herbal dietary supplement. J Am Osteopath Assoc 2001;101(8):444–6.
69. Eaker ED, Kronmal R, Kennedy JW, Davis K. Comparison of the long-term, postsurgical survival of women and men in the Coronary Artery Surgery Study (CASS). Am Heart J 1989;117(1):71–81.
70. Ross DL, Davis KB, Pettinger MB, Alderman EL, Killip T, Mason JW. Features of cardiac arrest episodes with and without acute myocardial infarction in the Coronary Artery Surgery Study (CASS). Am J Cardiol 1987;60(16):1219–24.
71. Casiglia E, Tikhonoff V, Pizziol A, Onesto C, Ginocchio G, Mazza A, Pessina AC. Should digoxin be proscribed in elderly subjects in sinus rhythm free from heart failure? A population-based study. Jpn Heart J 1998;39(5):639–51.
72. The Digitalis Investigation Group. The effect of digoxin on mortality and morbidity in patients with heart failure. N Engl J Med 1997;336(8):525–33.
73. Soler-Soler J, Permanyer-Miralda G. Should we still prescribe digoxin in mild-to-moderate heart failure? Is quality of life the issue rather than quantity? Eur Heart J 1998;19(Suppl P):P26–31.
74. Atarashi H, Inoue H, Fukunami M, Sugi K, Hamada C, Origasa H. Sinus Rhythm Maintenance in Atrial Fibrillation Randomized Trial (SMART) Investigators. Double-blind placebo-controlled trial of aprindine and digoxin for the prevention of symptomatic atrial fibrillation. Circ J 2002;66(6):553–6.
75. Deglin S, Deglin J, Chung EK. Direct current shock and digitalis therapy. Drug Intell Clin Pharm 1977;11:76.
76. Ali N, Dais K, Banks T, Sheikh M. Titrated electrical cardioversion in patients on digoxin. Clin Cardiol 1982;5(7):417–9.
77. Lely AH, van Enter CH. Large-scale digitoxin intoxication. BMJ 1970;3(725):737–40.
78. Lely AH, van Enter CH. Non-cardiac symptoms of digitalis intoxication. Am Heart J 1972;83(2):149–52.
79. Wang PS, Schneeweiss S, Glynn RJ, Mogun H, Avorn J. Use of the case-crossover design to study prolonged drug exposures and insidious outcomes. Ann Epidemiol 2004;14:296–303.
80. Miller S, Forker AD. Digitalis toxicity. Neurologic manifestations. J Kans Med Soc 1974;75(8):263–4.
81. Kerr DJ, Elliott HL, Hillis WS. Epileptiform seizures and electroencephalographic abnormalities as manifestations of digoxin toxicity. BMJ (Clin Res Ed) 1982;284(6310):162–3.
82. Douglas EF, White PT, Nelson JW. Three per second spike-wave in digitalis toxicity. Report of a case. Arch Neurol 1971;25(4):373–5.
83. Greenlee JE, Crampton RS, Miller JQ. Transient global amnesia associated with cardiac arrhythmia and digitalis intoxication. Stroke 1975;6(5):513–6.
84. Bernat JL, Sullivan JK. Trigeminal neuralgia from digitalis intoxication. JAMA 1979;241(2):164.
85. Batterman RC, Guter LB. Hitherto undescribed neurological manifestations of digitalis toxicity. Am Heart J 1984;36:582–6.
86. Brezis M, Michaeli J, Hamburger R. Nightmares from digoxin. Ann Intern Med 1980;93(4):639–40.
87. Eisendrath SJ, Gershengorn KN, Unger R. Digoxin-induced organic brain syndrome. Am Heart J 1983;106(2):419–20.
88. Greenaway JR, Abuaisha B, Bramble MG. Digoxin toxicity presenting as encephalopathy. Postgrad Med J 1996;72(848):367–8.
89. Eberhard SM, Woolley S, Zellweger U. Adynamie und Apathie bei Digitalisintoxikation. [Powerlessness and apathy in digitalis intoxication.] Schweiz Rundsch Med Prax 1999;88(17):772–4.
90. Mulder LJ, van der Mast RC, Meerwaldt JD. Generalised chorea due to digoxin toxicity. BMJ (Clin Res Ed) 1988;296(6631):1262Erratum in: BMJ (Clin Res Ed) 1988;297(6647):562.
91. Sekul EA, Kaminer S, Sethi KD. Digoxin-induced chorea in a child. Mov Disord 1999;14(5):877–9.
92. Wolin MJ. Digoxin visual toxicity with therapeutic blood levels of digoxin. Am J Ophthalmol 1998;125(3):406–7.
93. Lawrenson JG, Kelly C, Lawrenson AL, Birch J. Acquired colour vision deficiency in patients receiving digoxin maintenance therapy. Br J Ophthalmol 2002;86(11):1259–61.
94. Singh RB, Singh VP, Somani PN. Psychosis: a rare manifestation of digoxin intoxication. J Indian Med Assoc 1977;69(3):62–3.
95. Shear MK, Sacks MH. Digitalis delirium: report of two cases. Am J Psychiatry 1978;135(1):109–10Digitalis delirium: psychiatric considerations. Int J Psychiatry Med 1977–78;8(4):371–81.
96. Portnoi VA. Digitalis delirium in elderly patients. J Clin Pharmacol 1979;19(11–12):747–50.
97. Gorelick DA, Kussin SZ, Kahn I. Paranoid delusions and auditory hallucinations associated with digoxin intoxication. J Nerv Ment Dis 1978;166(11):817–9.
98. Volpe BT, Soave R. Formed visual hallucinations as digitalis toxicity. Ann Intern Med 1979;91(6):865–6.
99. Kardels B, Beine KH. Acute delirium as a result of digitalis intoxication. Notf Med 2001;27:542–5.
100. Wamboldt FS, Jefferson JW, Wamboldt MZ. Digitalis intoxication misdiagnosed as depression by primary care physicians. Am J Psychiatry 1986;143(2):219–21.
101. Song YH, Terao T, Shiraishi Y, Nakamura J. Digitalis intoxication misdiagnosed as depression—revisited. Psychosomatics 2001;42(4):369–70.
102. Donat J, Jirkalova V, Havel V, Mikulecka D. Kotazce estrogenniho ucinku digitalisu u zen po menopauze. [On the question of the estrogenic effect of digitalis in women after menopause.] Cesk Gynekol 1980;45(1):19–23.
103. Burckhardt D, Vera CA, LaDue JS. Effect of digitalis on urinary pituitary gonadotrophine excretion. A study in postmenopausal women. Ann Intern Med 1968;68(5):1069–71.
104. Stoffer SS, Hynes KM, Jiang NS, Ryan RJ. Digoxin and abnormal serum hormone levels. JAMA 1973;225(13):1643–4.
105. Neri A, Aygen M, Zukerman Z, Bahary C. Subjective assessment of sexual dysfunction of patients on long-term administration of digoxin. Arch Sex Behav 1980;9(4): 343–347.
106. LeWinn EB. Gynecomastia during digitalis therapy; report of eight additional cases with liver-function studies. N Engl J Med 1953;248(8):316–20.

107. Calov WL, Whyte MH. Oedema and mammary hypertrophy: a toxic effect of digitalis leaf. Med J Aust 1954;41(1:15):556–7.
108. Navab A, Koss LG, LaDue JS. Estrogen-like activity of digitalis: its effect on the squamous epithelium of the female genital tract. JAMA 1965;194(1):30–2.
109. Stenkvist B, Bengtsson E, Eklund G, Eriksson O, Holmquist J, Nordin B, Westman-Naeser S. Evidence of a modifying influence of heart glucosides on the development of breast cancer. Anal Quant Cytol 1980;2(1):49–54.
110. Stenkvist B, Pengtsson E, Dahlqvist B, Eriksson O, Jarkrans T, Nordin B. Cardiac glycosides and breast cancer, revisited. N Engl J Med 1982;306(8):484.
111. Haux J, Lam M, Marthinsen ABL, Strickert T, Lundgren S. Digitoxin, in non toxic concentrations, induces apoptotic cell death in Jurkat cells in vitro. Z Onkol 1999;31:14–20.
112. Haux J. Digitoxin is a potential anticancer agent for several types of cancer. Med Hypotheses 1999;53(6):543–8.
113. Haux J, Solheim O, Isaksen T, Anglesen A. Digitoxin, in non toxic concentrations, inhibits proliferation and induces cell death in prostate cancer cell line. Z Onkol 2000;32:11–6.
114. Nobel CS, Aronson JK, van den Dobbelsteen DJ, Slater AF. Inhibition of Na+/K(+)-ATPase may be one mechanism contributing to potassium efflux and cell shrinkage in CD95-induced apoptosis. Apoptosis 2000;5(2):153–63.
115. Johansson S, Lindholm P, Gullbo J, Larsson R, Bohlin L, Claeson P. Cytotoxicity of digitoxin and related cardiac glycosides in human tumor cells. Anticancer Drugs 2001;12(5):475–83.
116. Spigset O, Mjorndal T. Increased glucose intolerance related to digoxin treatment in patients with type 2 diabetes mellitus. J Intern Med 1999;246(4):419–22.
117. Hannerz J. Decrease of intracranial pressure and weight with digoxin in obesity. J Clin Pharmacol 2001;41(4):465–8.
118. Karpatkin S. Drug-induced thrombocytopenia. Am J Med Sci 1971;262(2):68–78.
119. Schneider AW, Gilfrich HJ, Fechler L. Thrombozytopenie bei Digitoxin-Intoxikation. [Thrombocytopenia in digitoxin poisoning.] Dtsch Med Wochenschr 1992;117(9):337–40.
120. Pirovino M, Ohnhaus EE, von Felten A. Digoxin-associated thrombocytopaenia. Eur J Clin Pharmacol 1981;19(3):205–7.
121. Forzy P, Joram F. Thrombopénie sévère en rapport avec une intoxication a l'acétyl digitoxine. Sem Hop Paris 1989;65:235–6.
122. Prodouz KN, Poindexter BJ, Fratantoni JC. Ouabain affects platelet reactivity as measured in vitro. Thromb Res 1987;46:337–46.
123. Andersson TL, Vinge E. Effects of ouabain on 86Rb-uptake, 3H-5-HT-uptake and aggregation by 5-HT and ADP in human platelets. Pharmacol Toxicol 1988;62:172–6.
124. Chirinos JA, Castrellon A, Zambrano JP, Jimenez JJ, Jy W, Horstman LL, Willens HJ, Castellanos A, Myerburg RJ, Ahn YS. Digoxin use is associated with increased platelet and endothelial cell activation in patients with nonvalvular atrial fibrillation. Heart Rhythm 2005;2(5):525–9.
125. Almeyda J, Levantine A. Cutaneous reactions to cardiovascular drugs. Br J Dermatol 1973;88(3):313–9.
126. Holt DW, Volans GN. Gastrointestinal symptoms of digoxin toxicity. BMJ 1977;2(6088):704.
127. Willems J, De Geest H. Digitalis intoxicatie. Ned Tijdschr Geneeskd 1968;24:617.
128. Kelton JG, Scullin DC. Digitalis toxicity manifested by dysphagia. JAMA 1978;239(7):613–4.
129. Cordeiro MF, Arnold KG. Digoxin toxicity presenting as dysphagia and dysphonia. BMJ 1991;302(6783):1025.
130. Adar R, Salzman EW. Letter: Intestinal ischemia and digitalis. JAMA 1974;229(12):1577.
131. Bourhis F, Riard P, Danel V, Hostein J, Fournet J. Intoxication digitalique avec ischémie colique grave: évolution favorable après traitement par anticorps spécifiques. [Digitalis poisoning with severe ischemic colitis: a favorable course after treatment with specific antibodies.] Gastroenterol Clin Biol 1990;14(1):95.
132. Guglielminotti J, Tremey B, Maury E, Alzieu M, Offenstadt G. Fatal non-occlusive mesenteric infarction following digoxin intoxication. Intensive Care Med 2000;26(6):829.
133. Weil J, Sen Gupta R, Herfarth H. Nonocclusive mesenteric ischemia induced by digitalis. Int J Colorectal Dis 2004;19(3):277–80.
134. Nybo M, Damkier P. Gastrointestinal symptoms as an important sign in premature newborns with severely increased S-digoxin. Basic Clin Pharmacol Toxicol 2005;96(6):465–8.
135. Lindsay SJ, Kearney MT, Prescott RJ, Fox KA, Nolan J. Digoxin and mortality in chronic heart failure. UK Heart Investigation. Lancet 1999;354(9183):1003.
136. Spargias KS, Hall AS, Ball SG. Safety concerns about digoxin after acute myocardial infarction. Lancet 1999;354(9176):391–2.
137. Wang L, Song S. Digoxin may reduce the mortality rates in patients with congestive heart failure. Med Hypotheses 2005;64(1):124–6.
138. Adams KF Jr, Patterson JH, Gattis WA, O'Connor CM, Lee CR, Schwartz TA, Gheorghiade M. Relationship of serum digoxin concentration to mortality and morbidity in women in the Digitalis Investigation Group trial a retrospective analysis, J Am Coll Cardiol 2005;46:497–504.
139. Cosin-Aguilar J, Marrugat J, Sanz G, Masso J, Gil M, Vargas R, Perez-Casar F, Simarro E, De Armas D, Garcia-Garcia J, Azpitarte J, Diago JL, Rodrigo-Trallero G, Lekuona I, Domingo E, Marin-Huerta E. Long-term results of the Spanish trial on treatment and survival of patients with predominantly mild heart failure. J Cardiovasc Pharmacol 1999;33(5):733–40.
140. Fauchier L, Babuty D, Cosnay P, Fauchier JP. Digoxin and mortality in idiopathic dilated cardiomyopathy. Eur Heart J 2000;21(10):858–9.
141. Heninger MM. Commonly encountered prescription medications in medical-legal death investigation: a guide for death investigators and medical examiners. Am J Forensic Med Pathol 2000;21(3):287–99.
142. Chyka PA. How many deaths occur annually from adverse drug reactions in the United States? Am J Med 2000;109(2):122–30.
143. Rathore SS, Wang Y, Krumholz HM. Sex-based differences in the effect of digoxin for the treatment of heart failure. N Engl J Med 2002;347(18):1403–11.
144. Eichhorn EJ, Gheorghiade M. Digoxin—new perspective on an old drug. N Engl J Med 2002;347(18):1394–5.
145. Rathore SS, Curtis JP, Wang Y, Bristow MR, Krumholz HM. Association of serum digoxin concentration and outcomes in patients with heart failure. JAMA 2003;289(7):871–8.
146. Sameri RM, Soberman JE, Finch CK, Self TH. Lower serum digoxin concentrations in heart failure and

reassessment of laboratory report forms. Am J Med Sci 2002;324(1):10–3.

147. Hern WM, Zen C, Ferguson KA, Hart V, Haseman MV. Outpatient abortion for fetal anomaly and fetal death from 15–34 menstrual weeks' gestation: techniques and clinical management. Obstet Gynecol 1993;81(2):301–6.
148. Jackson RA, Teplin VL, Drey EA, Thomas LJ, Darney PD. Digoxin to facilitate late second-trimester abortion: a randomized, masked, placebo-controlled trial. Obstet Gynecol 2001;97(3):471–6.
149. Drey EA, Thomas LJ, Benowitz NL, Goldschlager N, Darney PD. Safety of intra-amniotic digoxin administration before late second-trimester abortion by dilation and evacuation. Am J Obstet Gynecol 2000;182(5):1063–6.
150. Aronson JK. Clinical pharmacokinetics of digoxin. Clin Pharmacokinet 1980;5(2):137–49.
151. Rotmensch HH, Rotmensch S, Elkayam U. Management of cardiac arrhythmias during pregnancy. Current concepts. Drugs 1987;33(6):623–33.
152. Nishimura H, Tanimura T. Clinical Aspects of the Teratogenicity of DrugsAmsterdam: Excerpta Medica;. 1976.
153. Miura T, Kojima R, Sugiura Y, Mizutani M, Takatsu F, Suzuki Y. Effect of aging on the incidence of digoxin toxicity. Ann Pharmacother 2000;34(4):427–32.
154. Roever C, Ferrante J, Gonzalez EC, Pal N, Roetzheim RG. Comparing the toxicity of digoxin and digitoxin in a geriatric population: should an old drug be rediscovered? South Med J 2000;93(2):199–202.
155. Aronson JK. Book review: Handbook of renal-independent cardiac glycosides: pharmacology and clinical pharmacology, by N Rietbrock and BG Woodcock. Lancet 1989;2:1130–1.
156. Lecointre K, Pisante L, Fauvelle F, Mazouz S. Digoxin toxicity evaluation in clinical practice with pharmacokinetic correlations. Clin Drug Invest 2001;21:225–32.
157. Rich MW, McSherry F, Williford WO, Yusuf S. Digitalis Investigation Group. Effect of age on mortality, hospitalizations and response to digoxin in patients with heart failure: the DIG study. J Am Coll Cardiol 2001;38(3):806–13.
158. Steinberg C, Notterman DA. Pharmacokinetics of cardiovascular drugs in children. Inotropes and vasopressors. Clin Pharmacokinet 1994;27(5):345–67.
159. Kearin M, Kelly JG, O'Malley K. Digoxin "receptors" in neonates: an explanation of less sensitivity to digoxin than in adults. Clin Pharmacol Ther 1980;28(3):346–9.
160. Johnson GL, Desai NS, Pauly TH, Cunningham MD. Complications associated with digoxin therapy in low-birth weight infants. Pediatrics 1982;69(4):463–5.
161. Hastreiter AR, van der Horst RL, Chow-Tung E. Digitalis toxicity in infants and children. Pediatr Cardiol 1984;5(2):131–48.
162. Suematsu F, Minemoto M, Yukawa E, Higuchi S. Population analysis for the optimization of digoxin treatment in Japanese paediatric patients. J Clin Pharm Ther 1999;24(3):203–8.
163. Suematsu F, Yukawa E, Yukawa M, Minemoto M, Ohdo S, Higuchi S, Goto Y. Population-based investigation of relative clearance of digoxin in Japanese neonates and infants by multiple-trough screen analysis. Eur J Clin Pharmacol 2001;57(1):19–24.
164. Domanski M, Fleg J, Bristow M, Knox S. The effect of gender on outcome in digitalis-treated heart failure patients. J Card Fail 2005;11(2):83–6.
165. Hallberg P, Michaëlsson K, Melhus H. Digoxin for the treatment of heart failure. New Engl J Med 2003;348:661–2.
166. Aronson JK. Digoxin: clinical aspects. In: Richens A, Marks V, editors. Therapeutic Drug Monitoring. London, Edinburgh: Churchill-Livingstone, 1981:404.
167. Shenfield GM. Influence of thyroid dysfunction on drug pharmacokinetics. Clin Pharmacokinet 1981;6(4):275–97.
168. Aronson JK. Clinical pharmacokinetics of cardiac glycosides in patients with renal dysfunction. Clin Pharmacokinet 1983;8(2):155–78.
169. Soran H, Murray L, Younis N, Wong SPY, Currie P, Jones IR. Digoxin toxicity: an unusual presentation of infective endocarditis. Br J Cardiol 2003;10:308–9.
170. Rea TD, Siscovick DS, Psaty BM, Pearce RM, Raghunathan TE, Whitsel EA, Cobb LA, Weinmann S, Anderson GD, Arbogast P, Lin D. Digoxin therapy and the risk of primary cardiac arrest in patients with congestive heart failure: effect of mild-moderate renal impairment. J Clin Epidemiol 2003;56:646–50.
171. Brater DC, Morrelli HF. Digoxin toxicity in patients with normokalemic potassium depletion. Clin Pharmacol Ther 1977;22(1):21–33.
172. Young IS, Goh EM, McKillop UH, Stanford CF, Nicholls DP, Trimble ER. Magnesium status and digoxin toxicity. Br J Clin Pharmacol 1991;32(6):717–21.
173. Piergies AA, Worwag EM, Atkinson AJ Jr. A concurrent audit of high digoxin plasma levels. Clin Pharmacol Ther 1994;55(3):353–8.
174. Hamer J. The paradox of the lack of the efficacy of digitalis in congestive heart failure with sinus rhythm. Br J Clin Pharmacol 1979;8(2):109–13.
175. Hagemeijer F, Van Houwe E. Titrated energy cardioversion of patients on digitalis. Br Heart J 1975;37(12):1303–7.
176. Vella A, Gerber TC, Hayes DL, Reeder GS. Digoxin, hypercalcaemia, and cardiac conduction. Postgrad Med J 1999;75(887):554–6.
177. Slifman NR, Obermeyer WR, Aloi BK, Musser SM, Correll WA Jr, Cichowicz SM, Betz JM, Love LA. Contamination of botanical dietary supplements by Digitalis lanata. N Engl J Med 1998;339(12):806–11.
178. Andersen KE, Damsgaard T. The effect on serum enzymes of intramuscular injections of digoxin, bumetanide, pentazocine and isotonic sodium chloride. Acta Med Scand 1976;199(4):317–9.
179. Lewis WS, Doherty JE. Another disadvantage of intramuscular digoxin. N Engl J Med 1973;288(20):1077.
180. Tisdale JE, Padhi ID, Goldberg AD, Silverman NA, Webb CR, Higgins RS, Paone G, Frank DM, Borzak S. A randomized, double-blind comparison of intravenous diltiazem and digoxin for atrial fibrillation after coronary artery bypass surgery. Am Heart J 1998;135(5 Pt 1):739–47.
181. Bismuth C, Gaultier M, Conso F, Efthymiou ML. Hyperkalemia in acute digitalis poisoning: prognostic significance and therapeutic implications. Clin Toxicol 1973;6(2):153–62.
182. Gittelman MA, Stephan M, Perry H. Acute pediatric digoxin ingestion. Pediatr Emerg Care 1999;15(5):359–62.
183. Lacassie E, Marquet P, Martin-Dupont S, Gaulier JM, Lachatre G. A non-fatal case of intoxication with foxglove, documented by means of liquid chromatography–electrospray-mass spectrometry. J Forensic Sci 2000;45(5):1154–8.
184. Kinoshita H, Taniguchi T, Nishiguchi M, Ouchi H, Minami T, Utsumi T, Motomura H, Tsuda T, Ohta T, Aoki S, Komeda M, Kamamoto T, Kubota A, Fuke C, Arao T, Miyazaki T, Hishida S. An autopsy case of combined drug intoxication involving verapamil, metoprolol and digoxin. Forensic Sci Int 2003;133:107–12.

185. Kinoshita H, Taniguchi T, Nishiguchi M, Ouchi H, Minami T, Utsumi T, Motomura H, Tsuda T, Ohta T, Aoki S, Komeda M, Kamamoto T, Kubota A, Fuke C, Arao T, Miyazaki T, Hishida S. An autopsy case of combined drug intoxication involving verapamil, metoprolol and digoxin. Forensic Sci Int 2003;133:107–12.
186. Ma B, Prueksaritanont T, Lin JH. Drug interactions with calcium channel blockers: possible involvement of metabolite-intermediate complexation with CYP3A. Drug Metab Dispos 2000;28(2):125–30.
187. American Academy of Clinical Toxicology. European Association of Poisons Centres and Clinical Toxicologists. Position statement and practice guidelines on the use of multi-dose activated charcoal in the treatment of acute poisoning. J Toxicol Clin Toxicol 1999;37(6):731–51.
188. Lalonde RL, Deshpande R, Hamilton PP, McLean WM, Greenway DC. Acceleration of digoxin clearance by activated charcoal. Clin Pharmacol Ther 1985;37(4):367–71.
189. Chyka PA, Holley JE, Mandrell TD, Sugathan P. Correlation of drug pharmacokinetics and effectiveness of multiple-dose activated charcoal therapy. Ann Emerg Med 1995;25(3):356–62.
190. Park GD, Goldberg MJ, Spector R, Johnson GF, Feldman RD, Quee CK, Roberts P. The effects of activated charcoal on digoxin and digitoxin clearance. Drug Intell Clin Pharm 1985;19(12):937–41.
191. Reissell P, Manninen V. Effect of administration of activated charcoal and fibre on absorption, excretion and steady state blood levels of digoxin and digitoxin. Evidence for intestinal secretion of the glycosides. Acta Med Scand Suppl 1982;668:88–90.
192. Ibanez C, Carcas AJ, Frias J, Abad F. Activated charcoal increases digoxin elimination in patients. Int J Cardiol 1995;48(1):27–30.
193. Boldy DA, Smart V, Vale JA. Multiple doses of charcoal in digoxin poisoning. Lancet 1985;2(8463):1076–7.
194. Lake KD, Brown DC, Peterson CD. Digoxin toxicity: enhanced systemic elimination during oral activated charcoal therapy. Pharmacotherapy 1984;4(3):161–3.
195. Critchley JA, Critchley LA. Digoxin toxicity in chronic renal failure: treatment by multiple dose activated charcoal intestinal dialysis. Hum Exp Toxicol 1997;16(12):733–5.
196. Pond S, Jacobs M, Marks J, Garner J, Goldschlager N, Hansen D. Treatment of digitoxin overdose with oral activated charcoal. Lancet 1981;2(8256):1177–8.
197. Papadakis MA, Wexman MP, Fraser C, Sedlacek SM. Hyperkalemia complicating digoxin toxicity in a patient with renal failure. Am J Kidney Dis 1985;5(1):64–6.
198. Lopez-Gomez D, Valdovinos P, Comin-Colet J, Esteve F, Sabate X, Esplugas E. Intoxicación grave por digoxina. Utilización exitosa del tratamiento clásico. [Severe digoxin poisoning. The successful use of the classic treatment.] Rev Esp Cardiol 2000;53(3):471–2.
199. Roberge RJ, Sorensen T. Congestive heart failure and toxic digoxin levels: role of cholestyramine. Vet Hum Toxicol 2000;42(3):172–3.
200. Ahee P, Crowe AV. The management of hyperkalaemia in the emergency department. J Accid Emerg Med 2000;17(3):188–91.
201. Davey M. Calcium for hyperkalaemia in digoxin toxicity. Emerg Med J 2002;19(2):183.
202. Lapostolle F, Adnet F, Baud F, Lapandry C. Intoxications digitaliques: l'antidote existe. Rev Prat Med Gen 2000;14:345–7.
203. Safadi R, Levy I, Amitai Y, Caraco Y. Beneficial effect of digoxin-specific Fab antibody fragments in oleander intoxication. Arch Intern Med 1995;155(19):2121–5.
204. Shumaik GM, Wu AW, Ping AC. Oleander poisoning: treatment with digoxin-specific Fab antibody fragments. Ann Emerg Med 1988;17(7):732–5.
205. Eddleston M, Rajapakse S, Rajakanthan, Jayalath S, Sjostrom L, Santharaj W, Thenabadu PN, Sheriff MH, Warrell DA. Anti-digoxin Fab fragments in cardiotoxicity induced by ingestion of yellow oleander: a randomised controlled trial. Lancet 2000;355(9208):967–72.
206. Gonzalez Andres VL. Revisión sistemática sobre la efectividad e indicaciones de los anticuerpos antidigoxina en la intoxicación digitálica. [Systematic review of the effectiveness and indications of antidigoxin antibodies in the treatment of digitalis intoxication.] Rev Esp Cardiol 2000;53(1):49–58.
207. Husby P, Farstad M, Brock-Utne JG, Koller ME, Segadal L, Lund T, Ohm OJ. Immediate control of life-threatening digoxin intoxication in a child by use of digoxin-specific antibody fragments (Fab). Paediatr Anaesth 2003;13:541–6.
208. Van Deusen SK, Birkhahn RH, Gaeta TJ. Treatment of hyperkalemia in a patient with unrecognized digitalis toxicity. J Toxicol Clin Toxicol 2003;41:373–6.
209. Laurent G, Poulet B, Falcon-Eicher S, Petit A, Ballout J, Iovescu D, Gouyon JB, Louis P. Anticorps antidigoxine au cours d'une intoxication digitalique sévère chez un nouveau-né de 11 jours. Revue de la litterature. [Anti-digoxin antibodies in severe digitalis poisoning in an 11-day old infant. Review of the literature.] Ann Cardiol Angeiol (Paris) 2001;50(5):274–84.
210. Eddleston M, Senarathna L, Mohamed F, Buckley N, Juszczak E, Sheriff MH, Ariaratnam A, Rajapakse S, Warrell D, Rajakanthan K. Deaths due to absence of an affordable antitoxin for plant poisoning. Lancet 2003;362(9389):1041–4.
211. Erdmann E, Mair W, Knedel M, Schaumann W. Digitalis intoxication and treatment with digoxin antibody fragments in renal failure. Klin Wochenschr 1989;67(1):16–9.
212. Clifton GD, McIntyre WJ, Zannikos PN, Harrison MR, Chandler MH. Free and total serum digoxin concentrations in a renal failure patient after treatment with digoxin immune Fab. Clin Pharm 1989;8(6):441–5.
213. Proudfoot AT. A star treatment for digoxin overdose? BMJ (Clin Res Ed) 1986;293(6548):642–3.
214. Butler VP Jr, Smith TW. Immunologic treatment of digitalis toxicity: a tale of two prophecies. Ann Intern Med 1986;105(4):613–4.
215. Robinson CP. Digoxin immune Fab (ovine). Drugs Future 1986;11:922–6.
216. Aronson J. Digitalis intoxication and its treatment. Top Circ 1987;2:9–12.
217. Thompson CA. FDA approves digoxin-toxicity remedy. Am J Health Syst Pharm 2001;58(21):2021.
218. Kelly RA, Smith TW. Recognition and management of digitalis toxicity. Am J Cardiol 1992;69(18):G108–18.
219. Zdunek M, Mitra A, Mokrzycki MH. Plasma exchange for the removal of digoxin-specific antibody fragments in renal failure: timing is important for maximizing clearance. Am J Kidney Dis 2000;36(1):177–83.
220. Krynicki R, Szumski B, Perkowska J, Dyduszynski A. Digitalis intoxication complicated by recurrent ventricular fibrillation and successfully treated with bretylium tosylate—a case report. Kardiol Pol 1999;50:230–4.

221. Chen JY, Liu PY, Chen JH, Lin LJ. Safety of transvenous temporary cardiac pacing in patients with accidental digoxin overdose and symptomatic bradycardia. Cardiology 2004;102(3):152–5.
222. Kaneko T, Kudo M, Okumura T, Kasiwagi T, Turuoka S, Simizu M, Iino Y, Katayama Y. Successful treatment of digoxin intoxication by haemoperfusion with specific columns for beta$_2$-microgloblin-adsorption (Lixelle) in a maintenance haemodialysis patient. Nephrol Dial Transplant 2001;16(1):195–6.
223. Tsuruoka S, Osono E, Nishiki K, Kawaguchi A, Arai T, Furuyoshi S, Saito T, Takata S, Sugimoto K, Kurihara S, Fujimura A. Removal of digoxin by column for specific adsorption of beta(2)-microglobulin: a potential use for digoxin intoxication. Clin Pharmacol Ther 2001;69(6):422–30.
224. Rabetoy GM, Price CA, Findlay JW, Sailstad JM. Treatment of digoxin intoxication in a renal failure patient with digoxin-specific antibody fragments and plasmapheresis. Am J Nephrol 1990;10(6):518–21.
225. Chillet P, Korach JM, Petitpas D, Vincent N, Poiron L, Barbier B, Boazis M, Berger PH. Digoxin poisoning and anuric acute renal failure: efficiency of the treatment associating digoxin-specific antibodies (Fab) and plasma exchanges. Int J Artif Organs 2002;25(6):538–41.
226. Binnion PF. Drug interactions with digitalis glycosides. Drugs 1978;15(5):369–80.
227. Brown DD, Spector R, Juhl RP. Drug interactions with digoxin. Drugs 1980;20(3):198–206.
228. Lampe D, Lampe H, Banaschak H. Arzneimittelwechselwirkungen mit Herzglykosiden. Dtsch Gesundheitsw 1980;35:1081–7.
229. Belz GG, Breithaupt-Grogler K, Osowski U. Treatment of congestive heart failure—current status of use of digitoxin. Eur J Clin Invest 2001;31(Suppl 2):10–7.
230. Bazzano G, Bazzano GS. Digitalis intoxication. Treatment with a new steroid-binding resin. JAMA 1972;220(6):828–30.
231. Fresard F, Balant L, Noble J, Garcia B, Muller AF. Choléstyramine et intoxication a la digoxine: éfficacité therapeutique?. [Cholestyramine and digoxin intoxication: therapeutic efficacy?.] Schweiz Med Wochenschr 1979;109(12):431–6.
232. Kuhlmann J, Zilly W, Wilke J. Effects of cytostatic drugs on plasma level and renal excretion of beta-acetyldigoxin. Clin Pharmacol Ther 1981;30(4):518–27.
233. Kuhlmann J, Wilke J, Rietbrock N. Cytostatic drugs are without significant effect on digitoxin plasma level and renal excretion. Clin Pharmacol Ther 1982;32(5):646–51.
234. Manninen V, Apajalahti A, Melin J, Karesoja M. Altered absorption of digoxin in patients given propantheline and metoclopramide. Lancet 1973;1(7800):398–400.
235. Kirch W, Janisch HD, Santos SR, Duhrsen U, Dylewicz P, Ohnhaus EE. Effect of cisapride and metoclopramide on digoxin bioavailability. Eur J Drug Metab Pharmacokinet 1986;11(4):249–50.
236. D'Arcy PF, McElnay JC. Drug–antacid interactions: assessment of clinical importance. Drug Intell Clin Pharm 1987;21(7–8):607–17.
237. Lohman JJ, Merkus FW. Plasma protein binding of digitoxin and some other drugs in renal disease. Pharm Weekbl Sci 1987;9(2):75–8.
238. Woodland C, Ito S, Koren G. A model for the prediction of digoxin-drug interactions at the renal tubular cell level. Ther Drug Monit 1998;20(2):134–8.
239. Yukawa E, Suematu F, Yukawa M, Minemoto M, Ohdo S, Higuchi S, Goto Y, Aoyama T. Population pharmacokinetics of digoxin in Japanese patients: a 2-compartment pharmacokinetic model. Clin Pharmacokinet 2001;40(10):773–81.
240. Nakamura T, Kakumoto M, Yamashita K, Takara K, Tanigawara Y, Sakaeda T, Okumura K. Factors influencing the prediction of steady state concentrations of digoxin. Biol Pharm Bull 2001;24(4):403–8.
241. Shapiro W. Correlative studies of serum digitalis levels and the arrhythmias of digitalis intoxication. Am J Cardiol 1978;41(5):852–9.
242. Steiness E. Suppression of renal excretion of digoxin in hypokalemic patients. Clin Pharmacol Ther 1978;23(5):511–4.
243. Storstein O, Hansteen V, Hatle L, Hillestad L, Storstein L. Studies on digitalis. XIII. A prospective study of 649 patients on maintenance treatment with digitoxin. Am Heart J 1977;93(4):434–43.
244. Chopra D, Janson P, Sawin CT. Insensitivity to digoxin associated with hypocalcemia. N Engl J Med 1977;296(16):917–8.
245. Ben-Ami H, Krivoy N, Nagachandran P, Roguin A, Edoute Y. An interaction between digoxin and acarbose. Diabetes Care 1999;22(5):860–1.
246. Nagai Y, Hayakawa T, Abe T, Nomura G. Are there different effects of acarbose and voglibose on serum levels of digoxin in a diabetic patient with congestive heart failure? Diabetes Care 2000;23(11):1703.
247. Serrano JS, Jimenez CM, Serrano MI, Balboa B. A possible interaction of potential clinical interest between digoxin and acarbose. Clin Pharmacol Ther 1996;60(5):589–92.
248. Miura T, Ueno K, Tanaka K, Sugiura Y, Mizutani M, Takatsu F, Takano Y, Shibakawa M. Impairment of absorption of digoxin by acarbose. J Clin Pharmacol 1998;38(7):654–7.
249. Cohen E, Almog S, Staruvin D, Garty M. Do therapeutic doses of acarbose alter the pharmacokinetics of digoxin? Isr Med Assoc J 2002;4(10):772–5.
250. Kusumoto M, Ueno K, Fujimura Y, Kameda T, Mashimo K, Takeda K, Tatami R, Shibakawa M. Lack of kinetic interaction between digoxin and voglibose. Eur J Clin Pharmacol 1999;55(1):79–80.
251. Lien W-C, Huang C-H, Chen W-J.Bidirectional ventricular tachycardia resulting from digoxin and amiodarone treatment of rapid atrial fibrillation. Am J Emerg Med 2004;22:235–6.
252. Lelarge P, Bauer P, Royer-Morrot MJ, Meregnani JL, Larcan A, Lambert H. Intoxication digitalique après administration conjointe d'acétyl digitoxine et d'amiodarone. Ann Med Nancy Est 1993;32:307.
253. Laer S, Scholz H, Buschmann I, Thoenes M, Meinertz T. Digitoxin intoxication during concomitant use of amiodarone. Eur J Clin Pharmacol 1998;54(1):95–6.
254. Fraley DS, Britton HL, Schwinghammer TL, Kalla R. Effect of cimetidine on steady-state serum digoxin concentrations. Clin Pharm 1983;2(2):163–5.
255. Oosterhuis B, Jonkman JH, Andersson T, Zuiderwijk PB, Jedema JN. Minor effect of multiple dose omeprazole on the pharmacokinetics of digoxin after a single oral dose. Br J Clin Pharmacol 1991;32(5):569–72.
256. Doering W. Quinidine-digoxin interaction: Pharmacokinetics, underlying mechanism and clinical implications. N Engl J Med 1979;301(8):400–4.

257. Antman EM, Arnold M, Friedman PL, White H, Bosak M, Smith TW. Drug interactions with cardiac glycosides: evaluation of a possible digoxin–ethmozine pharmacokinetic interaction. J Cardiovasc Pharmacol 1987;9(5):622–7.
258. Alderman CP, Jersmann HP. Digoxin–itraconazole interaction. Med J Aust 1993;159(11–12):838–9.
259. Sachs MK, Blanchard LM, Green PJ. Interaction of itraconazole and digoxin. Clin Infect Dis 1993;16(3):400–3.
260. Lomaestro BM, Piatek MA. Update on drug interactions with azole antifungal agents. Ann Pharmacother 1998;32(9):915–28.
261. Salphati L, Benet LZ. Effects of ketoconazole on digoxin absorption and disposition in rat. Pharmacology 1998;56(6):308–13.
262. Brodell RT, Elewski B. Antifungal drug interactions. Avoidance requires more than memorization. Postgrad Med 2000;107(1):41–3.
263. Mochizuki M, Murase S, Takahashi K, Shimada S, Kume H, Iizuka T, Fukuda M. Serum itraconazole and hydroxyitraconazole concentrations and interaction with digoxin in a case of chronic hypertrophic pachymenigitis caused by Aspergillus flavus. Nippon Ishinkin Gakkai Zasshi 2000;41(1):33–9.
264. Mathis AS, Friedman GS. Coadministration of digoxin with itraconazole in renal transplant recipients. Am J Kidney Dis 2001;37(2):E18.
265. Jalava KM, Partanen J, Neuvonen PJ. Itraconazole decreases renal clearance of digoxin. Ther Drug Monit 1997;19(6):609–13.
266. Nishihara K, Hibino J, Kotaki H, Sawada Y, Iga T. Effect of itraconazole on the pharmacokinetics of digoxin in guinea pigs. Biopharm Drug Dispos 1999;20(3):145–9.
267. Purkins L, Wood N, Kleinermans D, Nichols D. Voriconazole does not affect the steady-state pharmacokinetics of digoxin. Br J Clin Pharmacol 2003;56 Suppl 1:45–50.
268. Feuring M, Lee Y, Orlowski LH, Michiels N, De Smet M, Majumdar AK, Petty KJ, Goldberg MR, Murphy MG, Gottesdiener KM, Hesney M, Brackett LE, Wehling M. Lack of effect of aprepitant on digoxin pharmacokinetics in healthy subjects. J Clin Pharmacol 2003;43:912–17.
269. Inglis AM, Sheth SB, Hursting MJ, Tenero DM, Graham AM, DiCicco RA. Investigation of the interaction between argatroban and acetaminophen, lidocaine, or digoxin. Am J Health Syst Pharm 2002;59(13):1258–66.
270. Weber C, Banken L, Birnboeck H, Nave S, Schulz R. The effect of bosentan on the pharmacokinetics of digoxin in healthy male subjects. Br J Clin Pharmacol 1999;47(6):701–6.
271. Pliakos ChC, Papadopoulos K, Parcharidis G, Styliadis J, Tourkantonis A. Effects of calcium channel blockers on serum concentrations of digoxin. Epitheorese Klin Farmakol Farmakokinetikes 1991;9:118–25.
272. Kirch W, Hutt HJ, Heidemann H, Ramsch K, Janisch HD, Ohnhaus EE. Drug interactions with nitrendipine. J Cardiovasc Pharmacol 1984;6(Suppl 7):S982–5.
273. Belz GG, Wistuba S, Matthews JH. Digoxin and bepridil: pharmacokinetic and pharmacodynamic interactions. Clin Pharmacol Ther 1986;39(1):65–71.
274. De Vito JM, Friedman B. Evaluation of the pharmacodynamic and pharmacokinetic interaction between calcium antagonists and digoxin. Pharmacotherapy 1986;6(2):73–82.
275. North DS, Mattern AL, Hiser WW. The influence of diltiazem hydrochloride on trough serum digoxin concentrations. Drug Intell Clin Pharm 1986;20(6):500–3.
276. Oyama Y, Fujii S, Kanda K, Akino E, Kawasaki H, Nagata M, Goto K. Digoxin–diltiazem interaction. Am J Cardiol 1984;53(10):1480–1.
277. D'Arcy PF. Diltiazem–digoxin interactions. Pharm Int 1985;6:148.
278. Halawa B, Mazurek W. Interakcje digoksyny z nifedypina i diltiazemem. [Interactions of digoxin with nifedipine and diltiazem.] Pol Tyg Lek 1990;45(23–24):467–9.
279. Mahgoub AA, El-Medany AH, Abdulatif AS. A comparison between the effects of diltiazem and isosorbide dinitrate on digoxin pharmacodynamics and kinetics in the treatment of patients with chronic ischemic heart failure. Saudi Med J 2002;23(6):725–31.
280. Peters J, Welker HA, Bullingham R. Pharmacokinetic and pharmacodynamic aspects of concomitant mibefradil–digoxin therapy at therapeutic doses. Eur J Drug Metab Pharmacokinet 1999;24(2):133–40.
281. Kleinbloesem CH, van Brummelen P, Hillers J, Moolenaar AJ, Breimer DD. Interaction between digoxin and nifedipine at steady state in patients with atrial fibrillation. Ther Drug Monit 1985;7(4):372–6.
282. Kirch W, Hutt HJ, Dylewicz P, Graf KJ, Ohnhaus EE. Dose-dependence of the nifedipine–digoxin interaction? Clin Pharmacol Ther 1986;39(1):35–9.
283. Gasic S, Eichler HG, Korn A. Effect of calcium antagonists on basal and digitalis-dependent changes in splanchnic and systemic hemodynamics. Clin Pharmacol Ther 1987;41(4):460–6.
284. Pedersen KE, Dorph-Pedersen A, Hvidt S, Klitgaard NA, Nielsen-Kudsk F. Digoxin–verapamil interaction. Clin Pharmacol Ther 1981;30(3):311–6.
285. Klein HO, Lang R, Weiss E, Di Segni E, Libhaber C, Guerrero J, Kaplinsky E. The influence of verapamil on serum digoxin concentration. Circulation 1982;65(5):998–1003.
286. Zatuchni J. Verapamil–digoxin interaction. Am Heart J 1984;108(2):412–3.
287. Pedersen KE, Dorph-Pedersen A, Hvidt S, Klitgaard NA, Pedersen KK. The long-term effect of verapamil on plasma digoxin concentration and renal digoxin clearance in healthy subjects. Eur J Clin Pharmacol 1982;22(2):123–7.
288. Schwartz JB, Keefe D, Kates RE, Kirsten E, Harrison DC. Acute and chronic pharmacodynamic interaction of verapamil and digoxin in atrial fibrillation. Circulation 1982;65(6):1163–70.
289. Kounis NG. Asystole after verapamil and digoxin. Br J Clin Pract 1980;34(2):57–8.
290. Saini RK, Fulmor IE, Antonaccio MJ. Effect of tiapamil and nifedepine during critical coronary stenosis and in the presence of adrenergic beta-receptor blockade in anesthetized dogs. J Cardiovasc Pharmacol 1982;4(5):770–6.
291. Kuhlmann J. Effects of verapamil, diltiazem, and nifedipine on plasma levels and renal excretion of digitoxin. Clin Pharmacol Ther 1985;38(6):667–73.
292. Cleland JG, Dargie HJ, Pettigrew A, Gillen G, Robertson JI. The effects of captopril on serum digoxin and urinary urea and digoxin clearances in patients with congestive heart failure. Am Heart J 1986;112(1):130–5.
293. Douste-Blazy P, Blanc M, Montastruc JL, Conte D, Cotonat J, Galinier F. Is there any interaction between digoxin and enalapril? Br J Clin Pharmacol 1986;22(6):752–3.
294. Magelli C, Bassein L, Ribani MA, Liberatore S, Ambrosioni E, Magnani B. Lack of effect of captopril on

serum digoxin in congestive heart failure. Eur J Clin Pharmacol 1989;36(1):99–100.
295. Miyakawa T, Shionoiri H, Takasaki I, Kobayashi K, Ishii M. The effect of captopril on pharmacokinetics of digoxin in patients with mild congestive heart failure. J Cardiovasc Pharmacol 1991;17(4):576–80.
296. de Mey C, Elich D, Schroeter V, Butzer R, Belz GG. Captopril does not interact with the pharmacodynamics and pharmacokinetics of digitoxin in healthy man. Eur J Clin Pharmacol 1992;43(4):445–7.
297. Kirimli O, Kalkan S, Guneri S, Tuncok Y, Akdeniz B, Ozdamar M, Guven H. The effects of captopril on serum digoxin levels in patients with severe congestive heart failure. Int J Clin Pharmacol Ther 2001;39(7):311–4.
298. Ratnapalan S, Griffiths K, Costei AM, Benson L, Koren G. Digoxin-carvedilol interactions in children. J Pediatr 2003;142:572–4.
299. Zussman BD, Kelly J, Murdoch RD, Clark DJ, Schubert C, Collie H. Cilomilast: pharmacokinetic and pharmacodynamic interactions with digoxin. Clin Ther 2001;23(6):921–31.
300. Kubler PA, Pillans PI, McKay JR. Possible interaction between cisapride and digoxin. Ann Pharmacother 2001;35(1):127–8.
301. Peeters PA, Crijns HJ, Tamminga WJ, Jonkman JH, Dickinson JP, Necciari J. Clopidogrel, a novel antiplatelet agent, and digoxin: absence of pharmacodynamic and pharmacokinetic interaction. Semin Thromb Hemost 1999;25(Suppl 2):51–4.
302. Conseil G, Baubichon-Cortay H, Dayan G, Jault JM, Barron D, Di Pietro A. Flavonoids: a class of modulators with bifunctional interactions at vicinal ATP- and steroid-binding sites on mouse P-glycoprotein. Proc Natl Acad Sci USA 1998;95:9831–6.
303. Tankanow R, Tamer HR, Streetman DS, Smith SG, Welton JL, Annesley T, Aaronson KD, Bleske BE. Interaction study between digoxin and a preparation of hawthorn (Crataegus oxyacantha). J Clin Pharmacol 2003;43:637–42.
304. Retzow A, Althaus M, de Mey C, Mazur D, Vens-Cappell B. Study on the interaction of the dopamine agonist alpha-dihydroergocryptine with the pharmacokinetics of digoxin. Arzneimittelforschung 2000;50(7):591–6.
305. Verstuyft C, Strabach S, El-Morabet H, Kerb R, Brinkmann U, Dubert L, Jaillon P, Funck-Brentano C, Trugnan G, Becquemont L. Dipyridamole enhances digoxin bioavailability via P-glycoprotein inhibition. Clin Pharmacol Ther 2003;73:51–60.
306. Jarchovsky J, Zamir D, Plavnik L. [Torsade de pointes and use of phenothiazine and diuretic.]Harefuah 1991;121(11):435–6.
307. Kleinermans D, Nichols DJ, Dalrymple I. Effect of dofetillide on the pharmacokinetics of digoxin. Am J Cardiol 2001;87(2):248–50.
308. Norgaard BL, Wachtell K, Christensen PD, Madsen B, Johansen JB, Christiansen EH, Graff O, Simonsen EH. Efficacy and safety of intravenously administered dofetilide in acute termination of atrial fibrillation and flutter: a multicenter, randomized, double-blind, placebo-controlled trial. Danish Dofetilide in Atrial Fibrillation and Flutter Study Group. Am Heart J 1999;137(6):1062–9.
309. Oravetz J, Slodki SJ. Recurrent ventricular fibrillation precipitated by quinidine. Report of a patient with recovery after 28 paroxysms. Arch Intern Med 1968;122(1):63–5.
310. Braun W, Damm KH. Drug interactions in intestinal absorption of ^{3}H-digitoxin in rats. Experientia 1976;32(5):613–4.
311. Zhou H, Parks V, Patat A, Le Coz F, Simcoe D, Korth-Bradley J. Absence of a clinically relevant interaction between etanercept and digoxin. J Clin Pharmacol 2004;44(11):1244–51.
312. Kothare PA, Soon DK, Linnebjerg H, Park S, Chan C, Yeo A, Lim M, Mace KF, Wise SD. Effect of exenatide on the steady-state pharmacokinetics of digoxin. J Clin Pharmacol 2005;45(9):1032–7.
313. Mant T, Fournie P, Ollier C, Donat F, Necciari J. Absence of interaction of fondaparinux sodium with digoxin in healthy volunteers. Clin Pharmacokinet 2002;41(Suppl 2):39–45.
314. Alkadi HO, Nooman MA, Raja'a YA. Effect of gentamicin on serum digoxin level in patients with congestive heart failure. Pharm World Sci 2004;26(2):107–9.
315. Cogan JJ, Humphreys MH, Carlson CJ, Benowitz NL, Rapaport E. Acute vasodilator therapy increases renal clearance of digoxin in patients with congestive heart failure. Circulation 1981;64(5):973–6.
316. Johne A, Brockmoller J, Bauer S, Maurer A, Langheinrich M, Roots I. Pharmacokinetic interaction of digoxin with an herbal extract from St. John's wort (Hypericum perforatum). Clin Pharmacol Ther 1999;66(4):338–45.
317. Mueller SC, Uehleke B, Woehling H, Petzsch M, Majcher-Peszynska J, Hehl EM, Sievers H, Frank B, Riethling AK, Drewelow B. Effect of St John's wort dose and preparations on the pharmacokinetics of digoxin. Clin Pharmacol Ther 2004;75(6):546–57.
318. Arold G, Donath F, Maurer A, Diefenbach K, Bauer S, Henneicke-von Zepelin HH, Friede M, Roots I. No relevant interaction with alprazolam, caffeine, tolbutamide, and digoxin by treatment with a low-hyperforin St John's wort extract. Planta Med 2005;71(4):331–7.
319. Anđelić S. Bigeminija—rezultat interakcije digoksina i kantariona. [Bigeminy—the result of interaction between digoxin and St. John's wort.] Vojnosanit Pregl 2003;60(3):361–4.
320. Homma M, Takeda M, Yamamoto Y, Suga H, Horiuchi M, Satoh S, Kohda Y. [Consultation and survey for drug interaction in outpatients taking the medicines potentially interact with St. John's Wort.]Yakugaku Zasshi 2000;120(12):1435–40.
321. Schimmel MS, Inwood RJ, Eidelman AI, Eylath U. Toxic digitalis levels associated with indomethacin therapy in a neonate. Clin Pediatr (Phila) 1980;19(11):768–9.
322. Finch MB, Kelly JG, Johnston GD, McDevitt DG. Evidence against a digoxin–indomethacin interaction. Br J Clin Pharmacol 1983;16:P212–3.
323. Brown DD, Schmid J, Long RA, Hull JH. A steady-state evaluation of the effects of propantheline bromide and cholestyramine on the bioavailability of digoxin when administered as tablets or capsules. J Clin Pharmacol 1985;25(5):360–4.
324. Roberge RJ, Sorensen T. Congestive heart failure and toxic digoxin levels: role of cholestyramine. Vet Hum Toxicol 2000;42(3):172–3.
325. Krivoy N, Eisenman A. [Cholestyramine for digoxin intoxication.]Harefuah 1995;128(3):145–7199.
326. Payne VW, Secter RA, Noback RK. Use of colestipol in a patient with digoxin intoxication. Drug Intell Clin Pharm 1981;15(11):902–3.

327. Hantson P, Vandenplas O, Mahieu P, Wallemacq P, Hassoun A. Repeated doses of activated charcoal and cholestyramine for digitoxin overdose: pharmacokinetic data and urinary elimination. J Toxicol Clin Exp 1991;11(7–8):401–5.
328. Kuhlmann J. Use of cholestyramine in three patients with beta-acetyldigoxin, beta-methyldigoxin and digitoxin intoxication. Int J Clin Pharmacol Ther Toxicol 1984;22(10):543–8.
329. Roman D, Bramson C, Ouellet D, Randinitis E, Gardner M. Effect of lasofoxifene on the pharmacokinetics of digoxin in healthy postmenopausal women. J Clin Pharmacol 2005;45(12):1407–12.
330. Levy RH, Ragueneau-Majlessi I, Baltes E. Repeated administration of the novel antiepileptic agent levetiracetam does not alter digoxin pharmacokinetics and pharmacodynamics in healthy volunteers. Epilepsy Res 2001;46(2):93–9.
331. Ragueneau I, Poirier JM, Radembino N, Sao AB, Funck-Brentano C, Jaillon P. Pharmacokinetic and pharmacodynamic drug interactions between digoxin and macrogol 4000, a laxative polymer, in healthy volunteers. Br J Clin Pharmacol 1999;48(3):453–6.
332. Lindenbaum J, Rund DG, Butler VP Jr, Tse-Eng D, Saha JR. Inactivation of digoxin by the gut flora: reversal by antibiotic therapy. N Engl J Med 1981;305(14):789–94.
333. Ford A, Smith LC, Baltch AL, Smith RP. Clarithromycin-induced digoxin toxicity in a patient with AIDS. Clin Infect Dis 1995;21(4):1051–2.
334. Tsutsumi K, Kotegawa T, Kuranari M, Otani Y, Morimoto T, Matsuki S, Nakano S. The effect of erythromycin and clarithromycin on the pharmacokinetics of intravenous digoxin in healthy volunteers. J Clin Pharmacol 2002;42(10):1159–64.
335. Boyd RA, Stern RH, Stewart BH, Wu X, Reyner EL, Zegarac EA, Randinitis EJ, Whitfield L. Atorvastatin coadministration may increase digoxin concentrations by inhibition of intestinal P-glycoprotein-mediated secretion. J Clin Pharmacol 2000;40(1):91–8.
336. Ten Eick AP, Sallee D, Preminger T, Weiss A, Reed MD. Possible drug interaction between digoxin and azithromycin in a young child. Clin Drug Invest 2000;20:61–4.
337. Gooderham MJ, Bolli P, Fernandez PG. Concomitant digoxin toxicity and warfarin interaction in a patient receiving clarithromycin. Ann Pharmacother 1999;33(7–8):796–9.
338. Wakasugi H, Yano I, Ito T, Hashida T, Futami T, Nohara R, Sasayama S, Inui K. Effect of clarithromycin on renal excretion of digoxin: interaction with P-glycoprotein. Clin Pharmacol Ther 1998;64(1):123–8.
339. Trivedi S, Hyman J, Lichstein E. Clarithromycin and digoxin toxicity. Ann Intern Med 1998;128(7):604.
340. Kiran N, Azam S, Dhakam S. Clarithromycin induced digoxin toxicity: case report and review. J Pak Med Assoc 2004;54(8):440–1.
341. Rengelshausen J, Goggelmann C, Burhenne J, Riedel K-D, Ludwig J, Weiss J, Mikus G, Walter-Sack I, Haefeli WE. Contribution of increased oral bioavailability and reduced nonglomerular renal clearance of digoxin to the digoxin-clarithromycin interaction. Br J Clin Pharmacol 2003;56:32–8.
342. Tanaka H, Matsumoto K, Ueno K, Kodama M, Yoneda K, Katayama Y, Miyatake K, Nazario M, Laplante S. Effect of clarithromycin on steady-state digoxin concentrations. Ann Pharmacother 2003;37:178–81.
343. Hirata S, Izumi S, Furukubo T, Ota M, Fujita M, Yamakawa T, Hasegawa I, Ohtani H, Sawada Y. Interactions between clarithromycin and digoxin in patients with end-stage renal disease. Int J Clin Pharmacol Ther 2005;43(1):30–6.
344. Coudray S, Janoly A, Belkacem-Kahlouli A, Bourhis Y, Bleyzac N, Bourgeois J, Putet G, Aulagner G. Erythromycin-induced digoxin toxicity in a neonatal intensive care unit. J Pharm Clin 2001;20:129–31.
345. Zhou H, Walter YH, Smith H, Devineni D, McLeod JF. Nateglinide, a new mealtime glucose regulator. Lack of pharmacokinetic interaction with digoxin in healthy volunteers. Clin Drug Invest 2000;19:465–71.
346. Hatorp V, Thomsen MS. Drug interaction studies with repaglinide: repaglinide on digoxin or theophylline pharmacokinetics and cimetidine on repaglinide pharmacokinetics. J Clin Pharmacol 2000;40(2):184–92.
347. Tayrouz Y, Ding R, Burhenne J, Riedel K-D, Weiss J, Hoppe-Tichy T, Haefeli WE, Mikus G. Pharmacokinetic and pharmaceutic interaction between digoxin and Cremophor RH40. Clin Pharmacol Ther 2003;73:397–405.
348. Cardaioli P, Compostella L, De Domenico R, Papalia D, Zeppellini R, Libardoni M, Pulido E, Cucchini F. Influenza del propafenone sulla farmacocinetica della digossina somministrata per via orale: studio su volontari sani. [Effect of propafenone on the pharmacokinetics of digoxin administered orally: a study in healthy volunteers.] G Ital Cardiol 1986;16(3):237–40.
349. Palumbo E, Svetoni N, Casini M, Spargi T, Biagi G, Martelli F, Lanzetta T. Interazione digoxina–propafenone: valori e limiti del dosaggio plasmatico dei due farmaci. Efficacia antiaritmica del propafenone. [Digoxin–propafenone interaction: values and limitations of plasma determination of the 2 drugs. Anti-arrhythmia effectiveness of propafenone.] G Ital Cardiol 1986;16(10):855–62.
350. Malek I, Gebauerova M, Stanek V. Riziko soucasného podávání chinidinu a digitalisu. [Risk of simultaneous administration of quinidine and digitalis.] Vnitr Lek 1980;26(3):358–61.
351. Pedersen KE, Lysgaard Madsen J, Klitgaard NA, Kjaer K, Hvidt S. Effect of quinine on plasma digoxin concentration and renal digoxin clearance. Acta Med Scand 1985;218(2):229–32.
352. Aronson JK, Carver JG. Interaction of digoxin with quinine. Lancet 1981;1(8235):1418.
353. Leden I. Digoxin–hydroxychloroquine interaction? Acta Med Scand 1982;211(5):411–2.
354. Hedman A. Inhibition by basic drugs of digoxin secretion into human bile. Eur J Clin Pharmacol 1992;42(4):457–9.
355. Fenster PE, Powell JR, Graves PE, Conrad KA, Hager WD, Goldman S, Marcus FI. Digitoxin–quinidine interaction: pharmacokinetic evaluation. Ann Intern Med 1980;93(5):698–701.
356. Kuhlmann J, Dohrmann M, Marcin S. Effects of quinidine on pharmacokinetics and pharmacodynamics of digitoxin achieving steady-state conditions. Clin Pharmacol Ther 1986;39(3):288–94.
357. Vousden M, Allen A, Lewis A, Ehren N. Lack of pharmacokinetic interaction between gemifloxacin and digoxin in healthy elderly volunteers. Chemotherapy 1999;45(6):485–90.
358. Chien SC, Rogge MC, Williams RR, Natarajan J, Wong F, Chow AT. Absence of a pharmacokinetic interaction between digoxin and levofloxacin. J Clin Pharm Ther 2002;27(1):7–12.

359. Johnson RD, Dorr MB, Hunt TL, Conway S, Talbot GH. Pharmacokinetic interaction of sparfloxacin and digoxin. Clin Ther 1999;21(2):368–79.
360. Boman G, Eliasson K, Odar-Cederlof I. Acute cardiac failure during treatment with digitoxin—an interaction with rifampicin. Br J Clin Pharmacol 1980;10(1):89–90.
361. Gault H, Longerich L, Dawe M, Fine A. Digoxin–rifampin interaction. Clin Pharmacol Ther 1984;35(6):750–4.
362. Novi C, Bissoli F, Simonati V, Volpini T, Baroli A, Vignati G. Rifampin and digoxin: possible drug interaction in a dialysis patient. JAMA 1980;244(22):2521–2.
363. Greiner B, Eichelbaum M, Fritz P, Kreichgauer HP, von Richter O, Zundler J, Kroemer HK. The role of intestinal P-glycoprotein in the interaction of digoxin and rifampin. J Clin Invest 1999;104(2):147–53.
364. Phillips EJ, Rachlis AR, Ito S. Digoxin toxicity and ritonavir: a drug interaction mediated through p-glycoprotein? AIDS 2003;17:1577–8.
365. Taylor A, Beerahee A, Citerone D, Davy M, Fitzpatrick K, Lopez-Gil A, Stocchi F. The effect of steady-state ropinirole on plasma concentrations of digoxin in patients with Parkinson's disease. Br J Clin Pharmacol 1999;47(2):219–22.
366. Juurlink DN, Mamdani MM, Kopp A, Herrmann N, Laupacis A. A population-based assessment of the potential interaction between serotonin-specific reuptake inhibitors and digoxin. Br J Clin Pharmacol 2005;59(1):102–7.
367. Larsen F, Priskorn M, Overo KF. Lack of citalopram effect on oral digoxin pharmacokinetics. J Clin Pharmacol 2001;41(3):340–6.
368. Burke S, Amin N, Incerti C, Plone M, Watson N. Sevelamer hydrochloride (Renagel), a nonabsorbed phosphate-binding polymer, does not interfere with digoxin or warfarin pharmacokinetics. J Clin Pharmacol 2001;41(2):193–8.
369. Waldorff S, Andersen JD, Heeboll-Nielsen N, Nielsen OG, Moltke E, Sorensen U, Steiness E. Spironolactone-induced changes in digoxin kinetics. Clin Pharmacol Ther 1978;24(2):162–7.
370. Kusus M, Stapleton DD, Lertora JJ, Simon EE, Dreisbach AW. Rhabdomyolysis and acute renal failure in a cardiac transplant recipient due to multiple drug interactions. Am J Med Sci 2000;320(6):394–7.
371. Weber P, Lettieri JT, Kaiser L, Mazzu AL. Lack of mutual pharmacokinetic interaction between cerivastatin, a new HMG-CoA reductase inhibitor, and digoxin in healthy normocholesterolemic volunteers. Clin Ther 1999;21(9):1563–75.
372. Garnett WR, Venitz J, Wilkens RC, Dimenna G. Pharmacokinetic effects of fluvastatin in patients chronically receiving digoxin. Am J Med 1994;96(6A):S84–6.
373. Martin PD, Kemp J, Dane AL, Warwick MJ, Schneck DW. No effect of rosuvastatin on the pharmacokinetics of digoxin in healthy volunteers. J Clin Pharmacol 2002;42(12):1352–7.
374. Avery GS. Check-list to potential clinically important interactions. Drugs 1973;5(3):187–211.
375. Smith RB, Petruscak J. Succinylcholine, digitalis, and hypercalcemia: a case report. Anesth Analg 1972;51(2):202–5.
376. Blanloeil Y, Pinaud M, Nicolas F. Arythmies per-operatoires chez le coronarien digitalise. Vingt-quatre cas. [Perioperative cardiac arrhythmias in digitalized patients with ischemic heart disease.] Anesth Analg (Paris) 1980;37(11–12):669–74.
377. Westphal K, Weinbrenner A, Giessmann T, Stuhr M, Franke G, Zschiesche M, Oertel R, Terhaag B, Kroemer HK, Siegmund W. Oral bioavailability of digoxin is enhanced by talinolol: evidence for involvement of intestinal P-glycoprotein. Clin Pharmacol Ther 2000;68(1):6–12.
378. Miyazawa Y, Paul Starkey L, Forrest A, Schentag JJ, Kamimura H, Swarz H, Ito Y. Effects of the concomitant administration of tamsulosin (0.8 mg) on the pharmacokinetic and safety profile of intravenous digoxin (Lanoxin) in normal healthy subjects: a placebo-controlled evaluation J Clin Pharm Ther 2002;27(1):13–9.
379. Zhou H, Horowitz A, Ledford PC, Hubert M, Appel-Dingemanse S, Osborne S, McLeod JF. The effects of tegaserod (HTF 919) on the pharmacokinetics and pharmacodynamics of digoxin in healthy subjects. J Clin Pharmacol 2001;41(10):113.
380. Stangier J, Su CA, Hendriks MG, van Lier JJ, Sollie FA, Oosterhuis B, Jonkman JH. The effect of telmisartan on the steady-state pharmacokinetics of digoxin in healthy male volunteers. J Clin Pharmacol 2000;40(12 Pt 1):1373–9.
381. Benson CT, Voelker JR. Teriparatide has no effect on the calcium-mediated pharmacodynamics of digoxin. Clin Pharmacol Ther 2003;73:87–94.
382. Snel S, Jansen JA, Pedersen PC, Jonkman JH, van Heiningen PN. Tiagabine, a novel antiepileptic agent: lack of pharmacokinetic interaction with digoxin. Eur J Clin Pharmacol 1998;54(4):355–7.
383. Bhattacharyya A, Bhavnani M, Tymms DJ. Serious interaction between digoxin and warfarin. Br J Cardiol 2002;9:356–7.
384. Richards D, Aronson JK. Serious interaction between digoxin and warfarin. Br J Cardiol 2002;9:446.
385. Sarich TC, Schutzer KM, Wollbratt M, Wall U, Kessler E, Eriksson UG. No pharmacokinetic or pharmacodynamic interaction between digoxin and the oral direct thrombin inhibitor ximelagatran in healthy volunteers. J Clin Pharmacol 2004;44(8):935–41.
386. Howard G, Barclay M, Florkowski C, Moore G, Roche A. Lack of clinically significant interference by spironolactone with the AxSym Digoxin II assay. Ther Drug Monit 2003;25:112–13.
387. McRae S. Elevated serum digoxin levels in a patient taking digoxin and Siberian ginseng. CMAJ 1996;155(3):293–5.
388. Dasgupta A, Wu S, Actor J, Olsen M, Wells A, Datta P. Effect of Asian and Siberian ginseng on serum digoxin measurement by five digoxin immunoassays. Significant variation in digoxin-like immunoreactivity among commercial ginsengs. Am J Clin Pathol 2003;119(2):298–303.
389. Dasgupta A, Reyes MA. Effect of Brazilian, Indian, Siberian, Asian, and North American ginseng on serum digoxin measurement by immunoassays and binding of digoxin-like immunoreactive components of ginseng with Fab fragment of antidigoxin antibody (Digibind). Am J Clin Pathol 2005;124(2):229–36.
390. Kobusiak-Prokopowicz M, Swidnicka-Szuszkowska B, Mysiak A. Effect of digoxin on ANP, BNP, and cGMP in patients with chronic congestive heart failure. Pol Arch Med Wew 2001;105(6):475–82.
391. Tsutamoto T, Wada A, Maeda K, Hisanaga T, Fukai D, Maeda Y, Ohnishi M, Mabuchi N, Kinoshita M. Digitalis increases brain natriuretic peptide in patients with severe congestive heart failure. Am Heart J 1997;134(5 Pt 1):910–6.
392. Heinrich K, Prendergast HM, Erickson T. Chronic digoxin toxicity and significantly elevated BNP levels in the presence of mild heart failure. Am J Emerg Med 2005;23(4):561–2.
393. Sanchez Garcia P, Paty I, Leister CA, Guerra P, Frias J, Garcia Perez LE, Darwish M. Effect of zaleplon on

digoxin pharmacokinetics and pharmacodynamics. Am J Health Syst Pharm 2000;57(24):2267–70.
394. Huffman DH. The effect of spironolactone and canrenone on the digoxin radioimmunoassay. Res Commun Chem Pathol Pharmacol 1974;9(4):787–90.
395. Pleasants RA, Williams DM, Porter RS, Gadsden RH Sr. Reassessment of cross-reactivity of spironolactone metabolites with four digoxin immunoassays. Ther Drug Monit 1989;11(2):200–4.
396. Valdes R Jr, Jortani SA. Unexpected suppression of immunoassay results by cross-reactivity: now a demonstrated cause for concern. Clin Chem 2002;48(3):405–6.
397. Steimer W, Muller C, Eber B, Emmanuilidis K. Intoxication due to negative canrenone interference in digoxin drug monitoring. Lancet 1999;354(9185):1176–7.
398. Capone D, Gentile A, Basile V. Possible interference of digoxin-like immunoreactive substances using the digoxin fluorescence polarization immunoassay. J Appl Ther Res 1999;2:305–8.
399. Steimer W, Muller C, Eber B. Digoxin assays: frequent, substantial, and potentially dangerous interference by spironolactone, canrenone, and other steroids. Clin Chem 2002;48(3):507–16.
400. Tribut O, Gaulier JM, Allain H, Bentué-Ferrer D. Major discrepancy between digoxin immunoassay results in a context of acute overdose: a case report. Clin Chim Acta 2005;354(1-2):201–3.
401. Aronson JK. Digoxin. In: Widdop B, editor. Contemporary Issues in Biochemistry. Edinburgh: Churchill Livingstone, 1985:3.
402. Mordasini MR, Krahenbuhl S, Schlienger RG. Appropriateness of digoxin level monitoring. Swiss Med Wkly 2002;132(35–36):506–12.
403. Abad-Santos F, Carcas AJ, Ibanez C, Frias J. Digoxin level and clinical manifestations as determinants in the diagnosis of digoxin toxicity. Ther Drug Monit 2000;22(2):163–8.

Cibenzoline

General Information

Cibenzoline is an antidysrhythmic drug of Class Ia, with some additional properties of drugs of Class III and Class IV. Its pharmacology, electrophysiological effects, therapeutic effects and indications, pharmacokinetics, and adverse effects have been reviewed (1,2). Prolongation of the QT interval, leading to cardiac dysrhythmias, is the major adverse effect. Other effects include gastrointestinal disturbances, effects on the central nervous system, and hypoglycemia, perhaps related to inhibition of ATP-dependent potassium channels in the pancreas.

Comparative studies

Cibenzoline and flecainide have been compared in the prevention of recurrence of atrial tachydysrhythmias in 139 patients (3). During the study, 27 patients withdrew, in 13 cases with adverse effects, seven of which were due to cibenzoline. Overall there were 26 adverse effects in 23% of the patients taking cibenzoline; these included one case of ventricular dysrhythmia, four minor cardiac events, four cases of nausea or epigastric pain, eight cases of weakness, four cases of depression or insomnia, one skin rash, and one case of hypoglycemia. The QRS complex was prolonged by more than 13% in 14 patients, but the QT interval was not prolonged. Although this was not a placebo-controlled study, the incidence of adverse effects with cibenzoline was probably as one would expect in such a population.

Organs and Systems

Cardiovascular

Cibenzoline prolongs the PR interval, the QRS interval, and the QT_c interval (4–7). It also prolongs the AH and HV intervals (5,8) and shortens the sinus cycle length (5–8). Because of these effects it can cause dysrhythmias (4,6,7,9).

- Cardiac dysrhythmias have been attributed to cibenzoline in a 60-year-old man with hypertrophic cardiomyopathy (10).
- In a 72-year-old woman cibenzoline was associated with left bundle branch block and heart failure (11). Excess cibenzoline accumulation was suspected, because of reduced renal function, but plasma cibenzoline concentrations were not reported.

Right bundle branch block has also been reported (6).

In three patients in whom cibenzoline had caused sinus node dysfunction, normal sinus node recovery time was restored by cilostazol (12).

Cibenzoline has a negative inotropic effect and can therefore cause hypotension (6,7,13,14) and worsening heart failure (4).

Cibenzoline has been reported to have unmasked Brugada syndrome in a 61-year-old woman (15). Ajmaline reproduced the effect. The authors concluded that cibenzoline should probably be avoided in patients with Brugada syndrome.

Nervous system

Various nervous system complaints have been reported in occasional patients, including headache (16), disturbances of visual accommodation (16), and tremulousness (17). Dizziness and light-headedness have more commonly been reported, sometimes in association with hypertension (7,18).

There have been frequent reports of anticholinergic adverse effects, including dry mouth (7,9,13), blurred vision (4,9), and difficulty in micturition (19).

- Choreiform movements associated with persistent orofacial dystonia have been attributed to cibenzoline in a 77-year-old woman who took 260 mg/day for 1 week (20). When cibenzoline was eventually withdrawn the effects resolved within 1 month.

The authors proposed that the effect was due to inhibition of potassium channels.

A myasthenia-like syndrome has occasionally been attributed to cibenzoline (SEDA-21, 199) (21).

- A 57-year-old man with chronic renal insufficiency treated by continuous and ambulatory peritoneal dialysis took cibenzoline 150 mg/day for a ventricular dysrhythmia. Four days later he developed proximal muscle weakness, progressing to generalized muscle weakness, with dysphagia and dysarthria. Hemodiafiltration on six occasions caused complete improvement and cibenzoline was withdrawn. There was no further recurrence, even when other drugs that he had been taking were restarted.

The authors suggested that cibenzoline may have inhibited ATP-dependent potassium channels in skeletal muscle. The plasma cibenzoline concentration at the height of this patient's symptoms was very high at 1890 μg/ml (usual target range 300–600) and the authors counselled caution in patients with renal insufficiency (SEDA-15, 174).

Metabolism

Several cases of hypoglycemia have been attributed to cibenzoline (22–24) (SEDA-18, 204). In a case-control study of 14 156 outpatients, 91 had hypoglycemia, and each was matched with five controls (25). Eight of those with hypoglycemia were taking cibenzoline and three were taking disopyramide. In contrast, only seven of the controls were taking cibenzoline, a significant difference. However, 20 of the controls were taking disopyramide, which was not significant from the patients with hypoglycemia, although disopyramide is known to cause hypoglycemia. Insulin was also associated with hypoglycemia, but sulfonylureas were not. Furthermore, there was a positive association with what were termed "thyroid agents." All of these features cast some doubt on the validity of these results in relation to cibenzoline.

- A 65-year-old woman developed hypoglycemia while taking cibenzoline and alacepril (26).

It is possible that hypoglycemia due to cibenzoline can be enhanced by ACE inhibitors, which can increase insulin sensitivity (SED-14, 640).

Gastrointestinal

Various gastrointestinal adverse effects have been reported not infrequently, including nausea (16), sometimes in association with abdominal pain ((7,9,13), vomiting, and diarrhea (7,8,14,17).

Liver

Hepatotoxicity with cibenzoline has been rarely reported. There has been a report of slight rises in the activities of serum transaminases (17), and one of ischemic hepatitis (27).

- A 67-year-old woman, who also had mild thrombocytopenia developed markedly abnormal liver function tests, which normalized within 3 months of withdrawal (28).

Susceptibility Factors

Renal disease

Cibenzoline is 60% eliminated by the kidneys (14,29,30). Its renal clearance falls with age (31); this is attributable to renal impairment.

- Three patients with severe renal insufficiency (creatinine clearance 10–16 ml/minute) had increased plasma cibenzoline concentrations during treatment with 300 mg/day (32). They developed prolonged QTc intervals, widened QRS complexes, dysrhythmias, hypotension, and hypoglycemia. Their plasma cibenzoline concentrations were 1944–2580 μg/l, 5–10 times higher than the usual target range. The half-lives of cibenzoline immediately after withdrawal were 69, 116, and 198 hours, 3–10 times longer than reported in patients with end-stage renal insufficiency (about 20 hours).

Drug Administration

Drug overdose

Overdose of cibenzoline has been reported.

- An 80-year-old woman who took an unknown number of 300 mg tablets had impaired consciousness, a low blood pressure (58/30 mmHg), coarse crackles in the lung, and a prolonged QT_c interval (0.64 seconds) (33). There was mild left ventricular global hypokinesia with a 50% left ventricular ejection fraction, severe mitral regurgitation, and left atrial dilatation. There was a mild partially compensated metabolic acidosis and hypoglycemia (2.9 mmol/l). The plasma concentration of cibenzoline was 2580 ng/ml. She was treated with sodium bicarbonate, dopamine, and noradrenaline. A pre-existing right ventricular pacemaker was not functioning, and the amplitude was increased from 2.5 to 7.5 volts at a rate of 90 per minute. Charcoal perfusion for 5 hours caused a dramatic fall in the plasma cibenzoline concentration in association with a reduction in the QT_c interval to 0.48 seconds and improvement in the pacing threshold.
- Fatal intoxication with cibenzoline has been reported in an 83-year-old man (34).

Diagnosis of Adverse Drug Reactions

It has been suggested that plasma concentrations of 100–200 mg/ml are necessary for efficacy and that at concentrations over 400 mg/ml there is an increased risk of adverse reactions (7). In most studies plasma cibenzoline concentrations have been around 200–400 mg/ml (4,6,8,18). Plasma concentrations correlate well with electrophysiological effects (6), its hemodynamic effects (34), and reduction in ventricular extra beats (35). In one study, patients who had adverse effects had a mean concentration of 913 mg/ml compared with 312 mg/ml in those who did not (8).

References

1. Miura DS, Keren G, Torres V, Butler B, Aogaichi K, Somberg JC. Antiarrhythmic effects of cibenzoline. Am Heart J 1985;109(4):827–33.
2. Kodama I, Ogawa S, Inoue H, Kasanuki H, Kato T, Mitamura H, Hiraoka M, Sugimoto T. Profiles of aprindine, cibenzoline, pilsicainide and pirmenol in the framework of the Sicilian Gambit. The Guideline Committee for Clinical Use of Antiarrhythmic Drugs in Japan (Working Group of Arrhythmias of the Japanese Society of Electrocardiology). Jpn Circ J 1999;63(1):1–12.
3. Babuty D, Maison-Blanche P, Fauchier L, Brembilla-Perrot B, Medvedowsky JL, Bine-Scheck F. Double-blind comparison of cibenzoline versus flecainide in the prevention of recurrence of atrial tachyarrhythmias in 139 patients. Ann Noninvasive Electrocardiol 1999;4:53–9.
4. Miura DS, Keren G, Torres V, Butler B, Aogaichi K, Somberg JC. Antiarrhythmic effects of cibenzoline. Am Heart J 1985;109(4):827–33.
5. Thizy JF, Jandot V, Andre-Fouet X, Viallet M, Pout M. Etude électrophysiologique de l'UP 339–01 chez l'homme. Lyon Med 1981;245:119–22.
6. Hoffmann E, Mattke S, Haberl R, Steinbeck G. Randomized crossover comparison of the electrophysiologic and antiarrhythmic efficacy of oral cibenzoline and sotalol for sustained ventricular tachycardia. J Cardiovasc Pharmacol 1993;21(1):95–100.
7. Kostis JB, Davis D, Kluger J, Aogaichi K, Smith M. Cifenline in the short-term treatment of patients with ventricular premature complexes: a double-blind placebo-controlled study. J Cardiovasc Pharmacol 1989;14(1):88–95.
8. Kushner M, Magiros E, Peters R, Carliner N, Plotnick G, Fisher M. The electrophysiologic effects of oral cibenzoline. J Electrocardiol 1984;17(1):15–23.
9. Browne KF, Prystowsky EN, Zipes DP, Chilson DA, Heger JJ. Clinical efficacy and electrophysiologic effects of cibenzoline therapy in patients with ventricular arrhythmias. J Am Coll Cardiol 1984;3(3):857–64.
10. Nishida K, Fujiki A, Mizumaki K, Nagasawa H, Sakabe M, Sakurai K, Inoue H. Exercise-induced ventricular fibrillation during treatment with cibenzoline in a patient with hypertrophic cardiomyopathy. Ther Res 2001;22:832–6.
11. Paelinck BP, De Raedt H, Conraads V. Blurred vision, left bundle-branch block and cardiac failure. Acta Cardiol 2001;56(1):39–40.
12. Yamaji S, Imai S, Watanabe T, Takahashi N, Uenishi T, Matsudaira K, Sugino K, Yagi H, Kanmatsuse K. Cilostazol improved sinus nodal dysfunction induced by cibenzoline which was used for hybrid therapy in patients with paroxysmal atrial fibrillation. Ther Res 2002;23:882–6.
13. Humen DP, Lesoway R, Kostuk WJ. Acute, single, intravenous doses of cibenzoline: an evaluation of safety, tolerance, and hemodynamic effects. Clin Pharmacol Ther 1987;41(5):537–45.
14. Katoh T, Ishihara S, Tanaka T, Kobagasi Y, Takada K, Shimai S, Seino Y, Tanaka K, Takano T, Hayakawa H. Hemodynamic effects of intravenous cibenzoline, a new antiarrhythmic agent. Jpn J Clin Pharmacol Ther 1988;19:707–16.
15. Sarkozy A, Caenepeel A, Geelen P, Peytchev P, de Zutter M, Brugada P. Cibenzoline induced Brugada ECG pattern. Europace 2005;7(6):537–9.
16. Cocco G, Strozzi C, Pansini R, Rochat N, Bulgarelli R, Padula A, Sfrisi C, Kamal Al Yassini A. Antiarrhythmic use of cibenzoline, a new class 1 antiarrhythmic agent with class 3 and 4 properties, in patients with recurrent ventricular tachycardia. Eur Heart J 1984;5(2):108–14.
17. Klein RC, House M, Rushforth N. Efficacy and safety of oral cibenzoline in treatment of ventricular ectopy. Clin Res 1984;32:9A.
18. Lee MA, Fenster PE, Garcia ZM, Kipps JE, Huang SK. Cibenzoline for symptomatic ventricular arrhythmias: a prospective, randomized, double-blind, placebo controlled trial and a long term open label study. Can J Cardiol 1989;5(6):295–8.
19. Miura D, Torres V, Butler B, Gottlieb S, Aogaicki K, Somberg J. Effects of cibenzoline in patients with ventricular tachycardia. J Clin Pharmacol 1984;24:413.
20. Devos D, Defebvre L, Destee A, Caron J. Choreic movements induced by cibenzoline: an Ic class antiarrhythmic effect? Mov Disord 2000;15(5):1030–1.
21. Wakutani Y, Matsushima E, Son A, Shimizu Y, Goto Y, Ishida H. Myasthenialike syndrome due to adverse effects of cibenzoline in a patient with chronic renal failure. Muscle Nerve 1998;21(3):416–7.
22. Lefort G, Haissaguerre M, Floro J, Beauffigeau P, Warin JF, Latapie JL. Hypoglycémies au cours de surdosages par un nouvel anti-arythmique: la cibenzoline; trois observations. [Hypoglycemia caused by overdose of a new anti-arrhythmia agent: cibenzoline. 3 cases.] Presse Méd 1988;17(14):687–91.
23. Jeandel C, Preiss MA, Pierson H, Penin F, Cuny G, Bannwarth B, Netter P. Hypoglycaemia induced by cibenzoline. Lancet 1988;1(8596):1232–3.
24. Gachot BA, Bezier M, Cherrier JF, Daubeze J. Cibenzoline and hypoglycaemia. Lancet 1988;2(8605):280.
25. Takada M, Fujita S, Katayama Y, Harano Y, Shibakawa M. The relationship between risk of hypoglycemia and use of cibenzoline and disopyramide. Eur J Clin Pharmacol 2000;56(4):335–42.
26. Ogimoto A, Hamada M, Saeki H, Hiasa G, Ohtsuka T, Hashida H, Hara Y, Okura T, Shigematsu Y, Hiwada K. Hypoglycemic syncope induced by a combination of cibenzoline and angiotensin converting enzyme inhibitor. Jpn Heart J 2001;42(2):255–9.
27. Gutknecht J, Larrey D, Ychou M, Fedkovic Y, Janbon C. Ischémie hépatique grave après prise de cibenzoline. [Severe ischemic hepatitis after taking cibenzoline.] Ann Gastroenterol Hepatol (Paris) 1991;27(6):269–70.
28. Binois F, Guiserix J, Kilian D. Hépatite aiguë au cours d'un traitement par la cibenzoline. [Acute hepatitis during cibenzoline therapy.] Presse Méd 2000;29(13):703.
29. Canal M, Flouvat B, Tremblay D, Dufour A. Pharmacokinetics in man of a new antiarrhythmic drug, cibenzoline. Eur J Clin Pharmacol 1983;24(4):509–15.
30. Brazzell RK, Rees MM, Khoo KC, Szuna AJ, Sandor D, Hannigan J. Age and cibenzoline disposition. Clin Pharmacol Ther 1984;36(5):613–9.
31. Takahashi M, Echizen H, Takahashi K, Shimada S, Aoyama N, Izumi T. Extremely prolonged elimination of cibenzoline at toxic plasma concentrations in patients with renal impairments. Ther Drug Monit 2002;24(4):492–6.
32. Aoyama N, Sasaki T, Yoshida M, Suzuki K, Matsuyama K, Aizaki T, Izumi T, Kondo R, Kamijo Y, Soma K, Ohwada T. Effect of charcoal hemoperfusion on clearance of cibenzoline succinate (cifenline) poisoning. J Toxicol Clin Toxicol 1999;37(4):505–8.
33. Sadeg N, Richecoeur J, Dumontet M. Intoxication mortelle a la cibenzoline. [Fatal poisoning by cibenzoline.] Therapie 2001;56(2):188–9.

34. van den Brand M, Serruys P, de Roon Y, Aymard MF, Dufour A. Haemodynamic effects of intravenous cibenzoline in patients with coronary heart disease. Eur J Clin Pharmacol 1984;26(3):297–302.
35. Khoo KC, Szuna AJ, Colburn WA, Aogaichi K, Morganroth J, Brazzell RK. Single-dose pharmacokinetics and dose proportionality of oral cibenzoline. J Clin Pharmacol 1984;24(7):283–8.

Disopyramide

General Information

The use, clinical pharmacology, and adverse effects of disopyramide have been reviewed thoroughly (1,2).

Disopyramide

Observational studies

Because of its negative inotropic effect, disopyramide reduces the left ventricular outflow gradient and improves symptoms in patients with hypertrophic obstructive cardiomyopathy. However, its long-term effects have not been well studied. In 118 patients who took disopyramide (mean dose 432 mg/day for a mean of 3.1 years) and 373 who did not, all-cause annual cardiac death rates (1.4% versus 2.6% per year) and sudden death rates (1.0% versus 1.8% per year) did not differ significantly, although there was a tendency for disopyramide to reduce mortality (3). The authors concluded that disopyramide is not prodysrhythmic in hypertrophic obstructive cardiomyopathy. The main adverse effects of disopyramide were attributable to its anticholinergic effects—dry mouth and prostatism, which required drug withdrawal in 7% of the patients.

Comparative studies

Disopyramide (by intravenous infusion of 2 mg/kg/minute up to a maximum total dose of 100 mg) has been compared with pilsicainide (in a single oral dose of 100–150 mg) in the treatment of paroxysmal atrial fibrillation in 72 patients (4). Conversion to sinus rhythm occurred in 29 of the 40 patients given pilsicainide and 18 of 32 patients given disopyramide, a non-significant difference. However, the mean time to conversion was faster with disopyramide (23 versus 60 minutes). No adverse effects were observed with either drug.

General adverse reactions

The adverse effects of disopyramide are mostly mediated by its effects on the cardiovascular system and by its anticholinergic effects. Disopyramide has a strong negative inotropic effect on the myocardium and can cause heart failure and hypotension. It prolongs the QT interval and can cause serious ventricular tachydysrhythmias. Anticholinergic effects can cause dry mouth, blurred vision, urinary retention, glaucoma, and erectile impotence. Hypoglycemia can also occur. Disopyramide can cause uterine contractions and should not be used during pregnancy. Angioedema has been reported rarely. Tumor-inducing effects have not been reported.

Organs and Systems

Cardiovascular

Disopyramide has three effects that can lead to cardiovascular complications (5).

1. *Anticholinergic* The anticholinergic effects of disopyramide on the vagus have been reported to cause tachycardia with bundle branch block or conversion to 1:1 conduction of a supraventricular tachycardia with block.
2. QT_c *interval prolongation* There have been several reports of ventricular dysrhythmias (for example polymorphous ventricular tachycardia, ventricular fibrillation, ventricular tachycardia) in association with a prolonged QT_c interval (SEDA-5, 180).
3. *Negative inotropic effect* Disopyramide can worsen cardiac failure and occasionally causes hypotension.

The risk of adverse cardiac effects of disopyramide during intravenous administration relates to the speed of its administration rather than to the total dose given (SEDA-10, 149).

Torsade de pointes due to disopyramide is well described (SEDA-4, 180). This effect is associated with prolongation of the QT interval. There has been a study of the effects of disopyramide on the QT interval in patients with pre-existing QT interval prolongation (6). In eight patients with QT interval prolongation during bradycardia and five patients without QT interval prolongation, disopyramide significantly prolonged the QT interval; however, the change was more pronounced in those with pre-existing bradycardia (78 versus 35 ms). The authors proposed that this difference might be due to an underlying abnormality of potassium channels in those with pre-existing bradycardia. Thus, those who are genetically predisposed to cardiac dysrhythmias may be at greater risk of the prodysrhythmic effects of antidysrhythmic drugs.

The risk of myocardial depression with consequent hypotension is greatest when disopyramide is infused rapidly intravenously (SEDA-10, 149). Loading doses of disopyramide should therefore be infused slowly (over 30–60 minutes).

Respiratory

- Pneumonitis has been attributed to disopyramide in a 72-year-old man; the symptoms began soon after the first dose (7). Bronchoalveolar lavage fluid contained a high percentage of lymphocytes (65%) and a high CD4:CD8 ratio (69:1).

The results of a lymphocyte stimulation test suggested that disopyramide had been responsible.

Nervous system

Through its anticholinergic effects disopyramide causes dry mouth and blurred vision and can occasionally cause serious adverse effects, including glaucoma and acute urinary retention (1).

Neuropathy has rarely been attributed to disopyramide (8).

- A 71-year-old woman, who had taken disopyramide 500 mg/day for 4 years, developed fatigue, paresthesia, pain, and cramps in her legs (9). She had proximal weakness in all four limbs and an unsteady gait. Electrophysiology showed a sensorimotor polyneuropathy, with reduced motor conduction velocity and muscle denervation. All antibodies were negative. The symptoms did not respond to prednisone but improved in the months after disopyramide withdrawal.

Psychological, psychiatric

Acute psychosis has been attributed to disopyramide (10,11).

Metabolism

Disopyramide can cause hypoglycemia (SEDA-6, 180; SEDA-17, 222) (12), perhaps due to increased secretion of insulin, and can also potentiate the effects of conventional hypoglycemic drugs (13). This effect may be due to its chief metabolite mono-*N*-dealkyldisopyramide, since many of the reported cases of hypoglycemia have been in patients with renal impairment, in which the metabolite accumulates. In six subjects who were being considered for treatment with disopyramide, serum glucose concentrations were measured at 13, 15, 17, and 19 hours after supper, with no further food, with and without the added administration of two modified-released tablets of disopyramide 150 mg with supper and 12 hours later (14). Disopyramide significantly reduced the serum glucose concentration at all measurement times by an average of 0.54 mmol/l. The fall in serum glucose concentration was not related to the serum concentration of disopyramide or the serum creatinine concentration; it was greater in older patients and in underweight patients.

- Hypoglycemia has also been reported in a 70-year-old woman with type 2 diabetes mellitus taking disopyramide (15).

Hematologic

Disopyramide has caused neutropenia (16) and a coagulopathy (17).

Liver

Liver damage was reported in 22 (0.35%) of 6294 patients given disopyramide, with jaundice in 6 (0.09%) (18). Liver damage due to disopyramide can be associated with direct hepatocellular damage (19) and intrahepatic cholestasis (20,21). However, it can also occur indirectly, because of heart failure and hepatic congestion (22). Thus, the incidence of direct liver damage quoted above may be an overestimate.

Sexual function

Erectile impotence has been attributed to disopyramide (23,24).

Immunologic

Angioedema has been attributed to disopyramide (25).

Second-Generation Effects

Pregnancy

Disopyramide can cause uterine contractions (SEDA-3, 156) (26), and has been reported to have caused the onset of uterine contractions in eight of 10 patients at term (26).

- A 26-year-old woman with Wolff–Parkinson–White syndrome was given two doses of disopyramide at 36 weeks and shortly afterwards went into active labor with prepartum hemorrhage (27). The child was delivered by cesarean section and the woman made a full recovery.

In view of these reports disopyramide should be avoided in pregnancy.

Lactation

Although disopyramide and its *N*-monodesalkyl metabolite are both excreted in breast milk, the amounts are probably too small to be of importance (28).

Susceptibility Factors

Renal disease

In renal insufficiency there are complex changes in the pharmacokinetics of disopyramide, but the overall effect is accumulation of it and its active metabolite, due to reduced renal clearance (29).

Other features of the patient

Because of the anticholinergic effects of disopyramide, care should be taken both in patients with symptoms of prostatic hyperplasia (because of the risk of urinary retention) and in patients with glaucoma.

Disopyramide is highly bound to plasma proteins and this binding is saturable within the therapeutic range. Thus, at high dosages there may be an increase in the unbound fraction of drug in the plasma with proportionately greater effects. However, this is probably of no clinical relevance.

Drug Administration

Drug overdose

Overdosage of disopyramide is associated with apnea, loss of consciousness, loss of spontaneous respiration, hypotension, and cardiac dysrhythmias (30,31). Suggested treatment (32) includes arterial blood pressure monitoring, correction of acidosis and hypokalemia, and the intravenous infusion of a pressor agent for severe hypotension. Cardiac depressant drugs (for example class I antidysrhythmic drugs) should not be used to treat dysrhythmias, and the use of pyridostigmine to reverse the anticholinergic effects of disopyramide (33) is not recommended (SEDA-10, 149).

Drug–Drug Interactions

Class I antidysrhythmic drugs

There is an increased risk of dysrhythmias if disopyramide is used in conjunction with other drugs that prolong the QT interval, for example class I or class III antidysrhythmic drugs (34).

Macrolide antibiotics

Some of the macrolide antibiotics have been reported to inhibit the clearance of disopyramide (SEDA-21, 200; SEDA-22, 207), resulting in serious dysrhythmias or hypoglycemia. The mechanism of this interaction is presumed to be inhibition of dealkylation of disopyramide to its major metabolite, mono-*N*-dealkyldisopyramide. For example, in human liver microsomes the macrolide antibiotic troleandomycin significantly inhibited the mono-*N*-dealkylation of disopyramide enantiomers by inhibition of CYP3A4 (35). This interaction can result in serious dysrhythmias or other adverse effects of disopyramide.

- A 76-year-old woman developed torsade de pointes 5 days after starting to take clarithromycin 200 mg bd in addition to disopyramide 100 mg tds (36). Her serum potassium concentration was 2.8 mmol/l and the QT_c interval was prolonged to 0.71 seconds. The plasma disopyramide concentration was in the usual target range (3.2 μg/ml). The disopyramide and clarithromycin were withheld and potassium was given; 14 hours later the serum potassium concentration was 4.3 mmol/l and there was no further dysrhythmia, despite prolongation of the QT_c interval to 0.67 seconds, falling to 0.45 seconds 10 days later.
- A 35-year-old woman taking disopyramide phosphate modified-release capsules 150 mg qds was given azithromycin 500 mg initially and 250 mg/day thereafter (37). In 11 days she developed malaise, light-headedness, and urinary retention. After the insertion of a urinary catheter she developed a monomorphic ventricular tachycardia with left bundle branch block. She was successfully cardioverted and the electrocardiogram showed a markedly prolonged QT interval of 560 ms and T wave inversion in the anterolateral leads. Her serum disopyramide concentration, which had previously been 2.6 μg/ml, was 11 μg/ml.
- In a 59-year-old man taking disopyramide 50 mg/day the addition of clarithromycin 600 mg/day caused hypoglycemia, and the serum disopyramide concentration rose from 1.5 to 8.0 μg/ml (38). The ratio of plasma insulin concentration to blood glucose concentration was greatly increased, suggesting that hypersecretion of insulin was responsible, confirming the likelihood that the hypoglycemia was due to disopyramide intoxication secondary to inhibition of its metabolism by clarithromycin. There was also slight prolongation of the QT_c interval, but no cardiac dysrhythmias. After withdrawal of clarithromycin and disopyramide both the blood glucose concentration and the QT_c interval returned to normal.
- An 86-year-old woman presented with severe hypoglycemia after clarithromycin 500 mg/day had been added for 3 days to her other therapy, which included disopyramide 500 mg/day (39). The hypoglycemia resolved completely after withdrawal of disopyramide.

Severe cardiac dysrhythmias and major hypoglycemia have occurred in patients taking disopyramide with some macrolide antibiotics, especially erythromycin and clarithromycin (40).

Clarithromycin

- In a 76-year old woman, severe prolongation of the QT_c interval and self-terminating torsade de pointes were induced by the combined use of clarithromycin and disopyramide; concomitant hypokalemia may have contributed to the cardiac dysrhythmia (41).
- Symptomatic hypoglycemia developed in a 59-year-old man treated with a combination of disopyramide (50 mg/day) and clarithromycin (600 mg/day) (42).

Additional investigations in the second case suggested enhanced insulin secretion induced by toxic disopyramide concentrations as the probable mechanism.

Erythromycin

Disopyramide alters the protein binding of erythromycin, and this results in increased plasma concentrations in vitro (43). The interaction between erythromycin and disopyramide was potentially fatal in two cases (44).

Roxithromycin

Disopyramide alters the protein binding of roxithromycin, and this results in increased plasma concentrations in vitro (45). However, this effect has not been observed with roxithromycin in vivo.

Nitrates, organic

Disopyramide, by producing xerostomia, can prevent the dissolution of sublingual isosorbide tablets (45).

Potassium-sparing drugs

Hyperkalemia has been reported to increase the risk of dysrhythmias in patients taking disopyramide (46), and disopyramide should therefore be used with caution in patients who are taking drugs that can increase body potassium, such as potassium-sparing diuretics and ACE inhibitors.

Practolol

The combination of disopyramide and practolol can cause profound sinus bradycardia and asystole (47,48).

Vecuronium bromide

Disopyramide has also been associated with impairment of neostigmine antagonism of vecuronium-induced neuromuscular blockade (49).

- A 63-year-old man was given vecuronium 70 μg/kg followed by increments of 20 μg/kg, and three intravenous doses of disopyramide 10 mg for supraventricular extra beats, followed by an infusion of 25 mg/hour. Paralysis was reversed using atropine 0.75 mg and neostigmine 2.5 mg. The twitch height returned to normal and the train-of-four was above 85%, but the responses to tetanic stimulation at 100 and 50 Hz remained severely depressed (10 and 45% respectively). The plasma concentration of disopyramide was 5.1 μg/ml.

Warfarin

Disopyramide can potentiate the effects of warfarin (50), although it is not known whether this is of any importance (51).

Monitoring therapy

In 20 patients taking disopyramide 100–600 mg/day, there was no correlation between the dose and total or unbound plasma disopyramide concentrations (52). The unbound fraction of disopyramide in plasma was 0.25–0.57 and did not correlate with the plasma albumin concentration, but did correlate with the concentration of alpha$_1$-acid glycoprotein. In 16 of the 20 patients (four had adverse effects) there was a significant difference in plasma disopyramide concentrations in responders and non-responders, but the overlap was large. However, the unbound concentration of disopyramide was above 0.8 μg/ml in responders and below 0.8 μg/ ml in non-responders. There was no correlation between the unbound concentration of mono-N-dealkyldisopyramide and effectiveness. The authors suggested that plasma concentration of unbound disopyramide could be used as an index of efficacy.

References

1. Heel RC, Brogden RN, Speight TM, Avery GS. Disopyramide: a review of its pharmacological properties and therapeutic use in treating cardiac arrhythmias. Drugs 1978;15(5):331–68.
2. Koch-Weser J. Disopyramide. N Engl J Med 1979;300(17):957–62.
3. Sherrid MV, Barac I, McKenna WJ, Elliott PM, Dickie S, Chojnowska L, Casey S, Maron BJ. Multicenter study of the efficacy and safety of disopyramide in obstructive hypertrophic cardiomyopathy. J Am Coll Cardiol 2005;45(8):1251–8.
4. Kumagai K, Abe H, Hiraki T, Nakashima H, Oginosawa Y, Ikeda H, Nakashima Y, Imaizumi T, Saku K. Single oral administration of pilsicainide versus infusion of disopyramide for termination of paroxysmal atrial fibrillation: a multicenter trial. Pacing Clin Electrophysiol 2000;23(11 Pt 2):1880–2.
5. Warrington SJ, Hamer J. Some cardiovascular problems with disopyramide. Postgrad Med J 1980;56(654):229–33.
6. Furushima H, Niwano S, Chinushi M, Ohhira K, Abe A, Aizawa Y. Relation between bradycardia dependent long QT syndrome and QT prolongation by disopyramide in humans. Heart 1998;79(1):56–8.
7. Yamamoto Y, Narasaki F, Futsuki Y, Fukushima K, Tomono K, Kadota J, Kohno S. Disopyramide-induced pneumonitis, diagnosed by lymphocyte stimulation test using bronchoalveolar lavage fluid. Intern Med 2001;40(8):775–8.
8. Dawkins KD, Gibson J. Peripheral neuropathy with disopyramide. Lancet 1978;1(8059):329.
9. Briani C, Zara G, Negrin P. Disopyramide-induced neuropathy. Neurology 2002;58(4):663.
10. Falk RH, Nisbet PA, Gray TJ. Mental distress in patient on disopyramide. Lancet 1977;1(8016):858–9.
11. Padfield PL, Smith DA, Fitzsimons EJ, McCruden DC. Disopyramide and acute psychosis. Lancet 1977;1(8022):1152.
12. Otsu T, Ito T, Inagaki Y, Amano I, Masamoto S, Niwa M. [Accumulation of a disopyramide metabolite in renal failure.]Nippon Jinzo Gakkai Shi 1993;35(9):1065–71Asaio J 1993;39:M609–13.
13. Series C. Hypoglycémie induite ou favorisée par le disopyramide. [Hypoglycemia induced or facilitated by disopyramide.] Rev Med Interne 1988;9(5):528–9.
14. Hasegawa J, Mori A, Yamamoto R, Kinugawa T, Morisawa T, Kishimoto Y. Disopyramide decreases the fasting serum glucose level in man. Cardiovasc Drugs Ther 1999;13(4):325–7.
15. Reynolds RM, Walker JD. Hypoglycaemia induced by disopyramide in a patient with Type 2 diabetes mellitus. Diabet Med 2001;18(12):1009–10.
16. Conrad ME, Cumbie WG, Thrasher DR, Carpenter JT. Agranulocytosis associated with disopyramide therapy. JAMA 1978;240(17):1857–8.
17. Handa SP. Disopyramide-induced toxic cutaneous blisters and coagulopathy. Dialysis Transplant 1982;11:706–7.
18. Anonymous. Hepatic damage due to disopyramide. Jpn Med Gaz 1981;June 20;11.
19. Doody PT. Disopyramide hepatotoxicity and disseminated intravascular coagulation. South Med J 1982;75(4):496–8.
20. Meinertz T, Langer KH, Kasper W, Just H. Disopyramide-induced intrahepatic cholestasis Lancet 1977;2(8042): 828–9.
21. Riccioni N, Bozzi L, Susini N, Roni P. Disopyramide-induced intrahepatic cholestasis. Lancet 1977;2(8052–8053):1362–3.
22. Scheinman SJ, Poll DS, Wolfson S. Acute cardiac failure and hepatic ischemia induced by disopyramide phosphate. Yale J Biol Med 1980;53(5):361–6.
23. McHaffie DJ, Guz A, Johnston A. Impotence in patient on disopyramide. Lancet 1977;1(8016):859.
24. Hasegawa J, Mashiba H. Transient sexual dysfunction observed during antiarrhythmic therapy by long-acting disopyramide in a male Wolff–Parkinson–White patient. Cardiovasc Drugs Ther 1994;8(2):277.
25. Porterfield JG, Antman EM, Lown B. Respiratory difficulty after use of disopyramide. N Engl J Med 1980;303(10):584.
26. Tadmor OP, Keren A, Rosenak D, Gal M, Shaia M, Hornstein E, Yaffe H, Graff E, Stern S, Diamant YZ. The effect of disopyramide on uterine contractions during pregnancy. Am J Obstet Gynecol 1990;162(2):482–6.

27. Abbi M, Kriplani A, Singh B. Preterm labor and accidental hemorrhage after disopyramide therapy in pregnancy. A case report. J Reprod Med 1999;44(7):653–5.
28. Barnett DB, Hudson SA, McBurney A. Disopyramide and its N-monodesalkyl metabolite in breast milk. Br J Clin Pharmacol 1982;14(2):310–2.
29. Perlman PE, Adams WG Jr, Ridgeway NA. Extreme pyrexia during bretylium administration. Postgrad Med 1989;85(1):111–4.
30. Hayler AM, Holt DW, Volans GN. Fatal overdosage with disopyramide. Lancet 1978;1(8071):968–9.
31. Larcan A, Lambert H, Laprevote-Heully MC, Delorme N, Royer MJ, Guillet J. Les intoxications aiguës volontaires au disopyramide: à propos de 20 observations. Ann Med Nancy 1981;20:901–17.
32. Hayler AM, Medd RK, Holt DW, O'Keefe BD. Treatment of disopyramide overdosage. Vet Hum Toxicol 1979;21(Suppl):93–5.
33. Teichman SL, Fisher JD, Matos JA, Kim SG. Disopyramide–pyridostigmine: report of a beneficial drug interaction. J Cardiovasc Pharmacol 1985;7(1):108–13.
34. Ellrodt G, Singh BN. Adverse effects of disopyramide (Norpace): toxic interactions with other antiarrhythmic agents. Heart Lung 1980;9(3):469–74.
35. Echizen H, Tanizaki M, Tatsuno J, Chiba K, Berwick T, Tani M, Gonzalez FJ, Ishizaki T. Identification of CYP3A4 as the enzyme involved in the mono-N-dealkylation of disopyramide enantiomers in humans. Drug Metab Dispos 2000;28(8):937–44.
36. Hayashi Y, Ikeda U, Hashimoto T, Watanabe T, Mitsuhashi T, Shimada K. Torsades de pointes ventricular tachycardia induced by clarithromycin and disopyramide in the presence of hypokalemia. Pacing Clin Electrophysiol 1999;22(4 Pt 1):672–4.
37. Granowitz EV, Tabor KJ, Kirchhoffer JB. Potentially fatal interaction between azithromycin and disopyramide. Pacing Clin Electrophysiol 2000;23(9):1433–5.
38. Iida H, Morita T, Suzuki E, Iwasawa K, Toyo-oka T, Nakajima T. Hypoglycemia induced by interaction between clarithromycin and disopyramide. Jpn Heart J 1999;40(1):91–6.
39. Morlet-Barla N, Narbonne H, Vialettes B. Hypoglycémie grave et récidivante secondaire à l'interaction disopyramide–clarithromicine. [Severe hypoglycemia and recurrence caused by disopyramide–clarithromycin interaction.] Presse Méd 2000;29(24):1351.
40. Anonymous. Disopyramide: interactions with marcolide antibiotics. Prescrire Int 2001;10(55):151.
41. Hayashi Y, Ikeda U, Hashimoto T, Watanabe T, Mitsuhashi T, Shimada K. Torsades de pointes ventricular tachycardia induced by clarithromycin and disopyramide in the presence of hypokalemia. Pacing Clin Electrophysiol 1999;22(4 Pt 1):672–4.
42. Iida H, Morita T, Suzuki E, Iwasawa K, Toyo-oka T, Nakajima T. Hypoglycemia induced by interaction between clarithromycin and disopyramide. Jpn Heart J 1999;40(1):91–6.
43. Zini R, Fournet MP, Barre J, Tremblay D, Tillement JP. In vitro study of roxithromycin binding to serum proteins and erythrocytes in man. Br J Clin Pract 1987;42(Suppl 5):54.
44. Ragosta M, Weihl AC, Rosenfeld LE. Potentially fatal interaction between erythromycin and disopyramide. Am J Med 1989;86(4):465–6.
45. Barletta MA, Eisen H. Isosorbide dinitrate–disopyramide phosphate interaction. Drug Intell Clin Pharm 1985;19(10):764.
46. Maddux BD, Whiting RB. Toxic synergism of disopyramide and hyperkalemia. Chest 1980;78(4):654–6.
47. Cumming AD, Robertson C. Interaction between disopyramide and practolol. BMJ 1979;2(6200):1264.
48. Gelipter D, Hazell M. Interaction between disopyramide and practolol. BMJ 1980;280:52.
49. Baurain M, Barvais L, d'Hollander A, Hennart D. Impairment of the antagonism of vecuronium-induced paralysis and intra-operative disopyramide administration. Anaesthesia 1989;44(1):34–6.
50. Haworth E, Burroughs AK. Disopyramide and warfarin interaction. BMJ 1977;2(6091):866–7.
51. Sylven C, Anderson P. Evidence that disopyramide does not interact with warfarin. BMJ (Clin Res Ed) 1983;286(6372):1181.
52. Ohkawa H, Watanabe M, Saito Y, Uchiwa H, Fujii S, Ino H, Yokogawa K, Miyamoto K-I. Evidence for the usefulness of unbound plasma concentration of disopyramide in individual anti-arrhythmic therapy. Jpn J Clin Pharmacol Ther 2003;34:1–6.

Dofetilide

General Information

Dofetilide is a pure Class III antidysrhythmic drug, without actions of any other class. It was developed following the observation that bis(arylalkyl)amines with methanesulfonamido moieties on both aryl groups prolong the cardiac action potential without significantly altering the maximum rate of depolarization (1). The pharmacology, clinical pharmacology, uses, adverse effects, and interactions of dofetilide have been reviewed (2–8). Cardiovascular adverse effects of dofetilide are the most troublesome. Other common effects have included mild headache, dizziness, dyspepsia, nausea, and vomiting (9).

Pharmacology

Dofetilide is a highly selective blocker of the rapidly activating component of the inward rectifier potassium channel, IK_r (10–16). It therefore delays ventricular repolarization, which becomes less heterogeneous (17), and prolongs the action potential duration and effective refractory period (14,18,19); it has the same effects in ventricular muscle in dilated cardiomyopathy and chronic ischemic cardiomyopathy (10). It has greater affinity for atrial than ventricular tissues in animals (20), but probably not in man (14,18). It preferentially blocks open channels and has Group 3 actions (SEDA-25, 209), with slow-onset kinetics (21) and an increased likelihood of being prodysrhythmic at slower heart rates (22).

Dofetilide causes a dose-related and plasma concentration-related prolongation of the QT_c interval (14,18,19,23–27), and either reduces QT_c dispersion (26) or has no effect on it (28,29). The effects on the QT_c interval are rate-dependent, being greater at slower heart rates (30). Dofetilide does not usually broaden the QRS complex, but this was reported (and published twice) in a single patient with atrial fibrillation, in whom it was attributed to aberrant conduction (31,32).

During repeated oral administration of dofetilide 1.0–2.5 mg/day for 5 days the effect on the QT interval was slightly greater on day 1 than on day 5 at a range of plasma concentrations of about 1–4 ng/ml; this observation suggests the occurrence of tolerance, but in that case one would have expected a clockwise hysteresis loop in the effect concentration curve, and no hysteresis was seen either after a single dose or at steady state (25).

Dofetilide has a small positive inotropic effect in animal hearts (15,33). In a double-blind, placebo-controlled study of oral dofetilide 125, 250, or 500 mg bd for the maintenance of sinus rhythm after cardioversion of sustained atrial fibrillation or flutter in 201 patients, there were small changes in echocardiographic measures of atrial contractility, but no changes in stroke volume or cardiac output (34).

Clinical pharmacology

The pharmacokinetics of dofetilide are linear after single oral doses of 2–10 micrograms/kg (24,35) and repeated doses of 1.0–2.5 mg/day (25). Dofetilide is well absorbed (about 90%) after oral administration (24,35,36). Its absorption is relatively slow and peak concentrations are not reached for 1–2.5 hours; absorption is slower after food. It is a low clearance drug, with a clearance rate of about 6 ml/minute/kg, and has a volume of distribution of about 3 l/kg (23,36). It is mostly excreted unchanged by the kidneys, with a half-life of about 8 hours. Its clearance is therefore roughly proportional to creatinine clearance, particularly at high rates of clearance. A small proportion is metabolized in the liver by CYP3A4 to inactive metabolites (37).

In a pharmacokinetic–pharmacodynamic study in 10 healthy volunteers intravenous dofetilide 0.5 mg caused a mean maximum prolongation of the QT_c interval of 121 ms and the mean plasma concentration associated with half-maximal effect was 2.2 ng/ml (36).

Uses

Dofetilide has been used to convert atrial fibrillation and atrial flutter to sinus rhythm, in maintaining sinus rhythm thereafter, in suppressing paroxysmal supraventricular tachycardia, inducible atrioventricular nodal re-entry tachycardia, and inducible sustained ventricular tachycardia, in suppressing the dysrhythmias of the Wolff–Parkinson–White syndrome, and in facilitating conversion of ventricular fibrillation.

Open clinical studies

In 19 patients with atrial fibrillation and five with atrial flutter, dofetilide 2.5–8.0 micrograms/kg caused conversion to sinus rhythm in 14 (10 with atrial fibrillation and four with atrial flutter) (38).

In patients with sustained monomorphic ventricular tachycardia inducible by programmed electrical stimulation, who had previously been unsuccessfully treated with 0–7 other drugs, intravenous dofetilide 3–15 micrograms/kg suppressed or slowed inducible ventricular tachycardia in 17 of 41 patients, compared with none of nine patients who received only 1.5 micrograms/kg (19).

Intravenous dofetilide 2.5–5.0 micrograms/kg produced sinus rhythm in seven patients with paroxysmal atrial fibrillation of recent onset (under 7 days) and terminated paroxysmal supraventricular tachycardia in four of six patients (39).

In patients with electrically inducible atrioventricular re-entrant tachycardia intravenous dofetilide 1.5–15 micrograms/kg had no effect on tachycardia inducibility at two lower doses but prevented the re-induction of tachycardia at three higher doses in 11 of 31 patients (40).

Placebo-controlled studies

Conversion of atrial fibrillation and flutter

In a crossover, placebo-controlled study in 16 patients with recent onset atrial fibrillation, cardioversion was achieved in two of six patients who received dofetilide 8 micrograms/kg and in two of nine who received 12 micrograms/kg (41). None cardioverted with placebo. However, the average duration of atrial fibrillation was 35 days in those who cardioverted with dofetilide and 83 days in those who did not. The authors concluded that dofetilide had only limited effect in cardioverting atrial fibrillation of moderate duration.

In a double-blind, placebo-controlled study 98 patients, who developed atrial fibrillation/flutter within 1–6 days after coronary artery bypass graft surgery, were given dofetilide 4 or 8 micrograms/kg intravenously over 15 minutes (42). Eight of 33 patients converted to sinus rhythm after placebo, 12 of 33 after dofetilide 4 micrograms/kg, and 14 of 32 after dofetilide 8 micrograms/kg.

In a double-blind, placebo-controlled study in patients with sustained atrial fibrillation (n = 75) or atrial flutter (n = 16), dofetilide 8 micrograms/kg terminated the dysrhythmia in nine of 29 patients, compared with only four of 32 who received 4 micrograms/kg and none of 30 who received placebo (43). Patients with atrial flutter had a greater response to dofetilide (six of 11) than those with atrial fibrillation (five of 49).

In a placebo-controlled study in patients with atrial fibrillation or atrial flutter with a median dysrhythmia duration of 62 (range 1–180) days, there was conversion to sinus rhythm in 20 of 66 patients given dofetilide, compared with one of 30 patients given placebo (44). The conversion rate was higher in atrial flutter (seven of 11 patients) than in atrial fibrillation (13 of 55).

In a double-blind, placebo-controlled study in 325 patients with atrial fibrillation or flutter cardioversion, rates for dofetilide 125, 250, and 500 micrograms bd were 6.1, 9.8, and 30% respectively, compared with 1.2% with placebo (45). The probabilities of remaining in sinus rhythm at 1 year with dofetilide 125, 250, and 500 micrograms bd were 0.40, 0.37, and 0.58 respectively, and 0.25 for placebo.

In a double-blind, placebo-controlled study in 69 patients with atrial fibrillation or flutter, intravenous dofetilide 2–8 micrograms/kg caused conversion to sinus rhythm in 16 of 51 patients, compared with one of 18 who were given placebo; conversion of atrial flutter occurred in five of seven who were given dofetilide compared with none of three who were given placebo (46).

In a randomized, placebo-controlled, crossover study in 15 men, mean age 34 (range 18–63) years, with Wolff–Parkinson–White syndrome and atrial fibrillation or atrioventricular re-entrant tachycardia induced electrophysiologically, six of ten patients who were given dofetilide converted to sinus rhythm, compared with one of five who were given placebo (47). There were no dysrhythmias.

Ventricular tachydysrhythmias

In a placebo-controlled study, sustained ventricular tachycardia or fibrillation, reproducibly inducible electrophysiologically, was no longer inducible in eight of 18 patients who were given intravenous dofetilide 0.1–8.0 ng/ml, compared with one of six patients who received placebo (48).

In a randomized, double-blind, placebo-controlled study in 32 patients with ventricular extra beats (more than 30/hour on two consecutive 24-hour Holter recordings while drug free and more than 50/hour during 2-hour telemetric electrocardiography), dofetilide 7.5 micrograms/kg produced an 83% and placebo a 2.9% median reduction in ventricular extra beats (49).

Sudden death

The Danish Investigations of Arrhythmia and Mortality ON Dofetilide (DIAMOND) study comprised two studies in patients at high risk of sudden death: one in patients with congestive heart failure and one in patients with acute myocardial infarction within the previous 7 days (50).

In the congestive heart failure study, 1518 patients with symptomatic congestive heart failure and severe left ventricular dysfunction were recruited at 34 Danish hospitals; 762 were randomized double-blind to dofetilide and 756 to placebo (51). After 1 month, 22 of 190 patients with atrial fibrillation at baseline had sinus rhythm restored by dofetilide, compared with only three of 201 who took placebo. Dofetilide was also significantly more effective than placebo in maintaining sinus rhythm (hazard ratio for the recurrence of atrial fibrillation = 0.35; CI = 0.22, 0.57). Dofetilide significantly reduced the risk of hospitalization for worsening congestive heart failure (risk ratio = 0.75; CI = 0.63, 0.89). During a median follow-up of 18 months, 311 patients taking dofetilide and 317 patients taking placebo died (hazard ratio = 0.95; CI = 0.81, 1.11).

In the corresponding myocardial infarction study, 1510 patients with severe left ventricular dysfunction after myocardial infarction were recruited in 37 Danish coronary-care units; 749 were randomized double-blind to dofetilide and 761 to placebo (52). There were no significant differences between dofetilide and placebo in all-cause mortality or total dysrhythmic deaths. Dofetilide was significantly better than placebo at restoring sinus rhythm in patients with atrial fibrillation or flutter.

Comparative studies with other antidysrhythmic drugs

Amiodarone

In a comparison of intravenous dofetilide (8 micrograms/kg; $n = 48$), amiodarone (5 mg/kg; $n = 50$), or placebo ($n = 52$) in converting atrial fibrillation or flutter to sinus rhythm in 150 patients, sinus rhythm was restored in 35, 4, and 4% respectively (53).

Flecainide

In a non-randomized comparison of flecainide (2 mg/kg; $n = 11$) and dofetilide (8 micrograms/kg; $n = 10$) in patients with atrial flutter, only one patient given flecainide converted to sinus rhythm compared with seven of the 10 patients given dofetilide (54).

Propafenone

In a randomized, placebo-controlled, parallel-group comparison of oral dofetilide 500 micrograms bd, propafenone 150 mg tds, or placebo in preventing the recurrence of paroxysmal supraventricular tachycardia in 122 symptomatic patients, the respective probabilities of remaining free of episodes of paroxysmal supraventricular tachycardia were 50, 54, and 6%; both dofetilide and propafenone also reduced the frequency of episodes (median numbers 1, 0.5, and 5 respectively) (55).

Sotalol

In a double-blind, randomized, crossover comparison of oral dofetilide 500 micrograms bd with sotalol 160 mg bd in 128 patients with ischemic heart disease and inducible sustained ventricular tachycardia, 46 patients responded to dofetilide and 43 to sotalol; however, only 23 patients responded to both dofetilide and sotalol (9).

The hemodynamic effects of dofetilide 500 micrograms bd and sotalol 160 mg bd for 3–5 days have been studied in 12 patients with ischemic heart disease and sustained ventricular tachycardia (56). There were significant reductions in heart rate, mean systemic pressure, and cardiac index (–13%) with sotalol, but cardiac index increased significantly with dofetilide (11%) with no effect on heart rate or systemic blood pressure. The authors suggested that oral dofetilide could be useful in patients with ventricular tachydysrhythmias associated with impaired left ventricular function. One patient taking dofetilide reported mild dizziness and there were no cardiac dysrhythmias.

Organs and Systems

Cardiovascular

Because dofetilide prolongs the QT interval, there is a risk of ventricular tachydysrhythmias, which have often been reported, after both intravenous and oral administration.

In 154 patients with implantable cardioverter-defibrillators randomly assigned to dofetilide or placebo, there were pause-dependent runs of polymorphic ventricular tachycardia in 15 of the 87 patients who received dofetilide and in only five of the 87 who received placebo (57). There were five early events (at less than 3 days of therapy), all torsade de pointes in patients taking dofetilide. There were 15 late events, 10 with dofetilide and five with placebo. The median time to a late event was 22 (range 6–107) days for dofetilide and 99 (34–207) days for placebo.

In the DIAMOND study in congestive heart failure, there were 25 cases of torsade de pointes in the dofetilide group (3.3%) compared with none in the placebo group (51). In the DIAMOND myocardial infarction study there were seven cases of torsade de pointes (0.93%), all in those who were given dofetilide (52).

In a double-blind, placebo-controlled study in 325 patients with atrial fibrillation or atrial flutter randomized to dofetilide 125, 250, or 500 micrograms bd or placebo, there were two cases of torsade de pointes, one on day 2 and the other on day 3 (0.8% of all patients given the active drug); there was one sudden cardiac death, classified as prodysrhythmic, on day 8 (0.4% of all patients given the active drug). The authors recommended that dosage adjustment based on QT_c interval and renal function would minimize the small but not negligible prodysrhythmic risk of dofetilide.

In 128 patients who received dofetilide and sotalol in a crossover study, there were treatment-related adverse events in 2.3% of the patients who received dofetilide and 8.6% of those who received sotalol (9). Three patients who took dofetilide had torsade de pointes.

In a comparison of intravenous dofetilide (8 micrograms/kg; $n = 48$), amiodarone (5 mg/kg; $n = 50$), or placebo ($n = 52$) in converting atrial fibrillation or flutter to sinus rhythm in 150 patients, two patients given dofetilide had non-sustained ventricular tachycardias; four had torsade de pointes, in one case requiring electrical cardioversion (53).

There was more than 15% prolongation of the QT interval in 19 of 107 patients after the first dose of dofetilide and in 28 after subsequent doses; there were no cases of torsade de pointes (58).

Torsade de pointes occurs in 0.8–3.3% of patients taking dofetilide, and most episodes occur within 3 days of the start of therapy (59). Two unusual variants of ventricular dysrhythmias have been reported in patients taking dofetilide—non-sustained runs of monomorphic ventricular tachycardia shortly after the first dose of dofetilide, confirmed by rechallenge, and torsade de pointes that followed isolated ventricular extra beats during an exercise test in a patient without baseline QT interval prolongation but with significant QT interval prolongation after the postectopic pause (60).

Polymorphisms in potassium channels increase the risk of prolongation of the QT interval, and the two polymorphisms designated KCNQ1 and KCNH2 account for 80% of cases of congenital long QT syndrome. Of 105 patients from the DIAMOND study, seven had torsade de pointes, of whom two had a polymorphism (R1047L) in the hERG potassium channel; only 5 of the 98 patients without torsade de pointes carried the polymorphism (61). The only patient who was homozygous for the polymorphism was one with torsade de pointes. The affected hERG channel, transfected into HEK-293 cells, had significantly slower activation and inactivation kinetics than wild-type channels.

- In one of 10 healthy men given intravenous dofetilide 0.5 micrograms/kg there was prolongation of the QT_c interval from 451 to 808 ms 5 minutes after the end of the infusion; this was associated with five beats of polymorphic ventricular tachycardia, several multifocal ventricular extra beats, and ventricular couplets and triplets, all within 10 minutes after the end of the infusion (24).
- In a 37-year-old woman with atrial flutter with 1:1 conduction and partial right bundle branch block, intravenous dofetilide 5 micrograms/kg given over 5 minutes not only suppressed the atrioventricular nodal block to 2:1 or 3:1 but also caused complete right bundle branch block and QT interval prolongation (62).
- Self-limiting torsade de pointes developed in a 67-year-old man who was given 12.8 micrograms/kg (plasma concentration 7.1 ng/ml) in an open study; the QT_c interval was prolonged to over 600 ms (19).

This patient had stopped taking amiodarone 1 month before the administration of dofetilide, and that may have contributed to the prolongation of the QT_c interval.

- A woman with atrial fibrillation developed torsade de pointes after receiving intravenous dofetilide 6 micrograms/kg (plasma concentration 26 ng/ml) (41).
- A patient with atrial fibrillation received intravenous dofetilide 8 micrograms/kg and developed torsade de pointes before reverting to sinus rhythm (41).
- One of 18 patients with sustained ventricular tachycardia or fibrillation, reproducibly inducible electrophysiologically, developed torsade de pointes after receiving intravenous dofetilide 8 micrograms/kg (plasma concentration 5.3 ng/ml) (48).
- Short episodes of aberrant ventricular conduction and ventricular tachycardia occurred in three of 32 patients with atrial fibrillation who were given dofetilide 8 micrograms/kg (42).
- Torsade de pointes occurred in two of 62 patients with atrial tachydysrhythmias who received dofetilide 4 or 8 micrograms/kg; two other patients had ventricular extra beats associated with prolongation of the QT_c interval (43).
- In a placebo-controlled study of the effect of dofetilide 8 micrograms/kg in converting atrial fibrillation or flutter, transient torsade de pointes occurred in two men, aged 57 and 67, with prolongation of the QT_c interval from 370 and 420 ms to 450 and 510 ms respectively (44).
- A 58-year-old woman developed torsade de pointes with prolongation of the QT_c interval to 490 ms after receiving intravenous dofetilide 4.3 micrograms/kg; she responded to intravenous magnesium sulfate plus isoprenaline (46).

Other cardiac dysrhythmias that have been reported have included episodes of junctional rhythm with bundle branch block, spontaneous atrioventricular re-entrant tachycardia, and sustained supraventricular tachycardia (40).

Hypotension occasionally occurs after intravenous dofetilide (40).

Death

The risk of death in patients with supraventricular dysrhythmias taking dofetilide has been studied in a systematic review of randomized controlled trials (63). After adjusting for the effects of dysrhythmia diagnosis, age, sex, and structural heart disease, the hazard ratio was 1.1 (CI = 0.3, 4.3).

Drug–Drug Interactions

Cardiac glycosides

In 14 healthy men dofetilide 250 micrograms bd for 5 days had no effect on the pharmacokinetics of digoxin at a steady-state trough concentration of 1.0 ng/ml (66). However, in a placebo-controlled study in patients with atrial fibrillation or atrial flutter, conversion to sinus rhythm in patients given dofetilide was more likely if they were also given digoxin (44), suggesting that there may be a pharmacodynamic interaction.

Histamine (H_2) receptor antagonists

In a randomized, placebo-controlled study of the effects of cimetidine and ranitidine on the pharmacokinetics and pharmacodynamics of a single dose of dofetilide 500 micrograms in 20 healthy men, ranitidine 150 mg bd did not affect the pharmacokinetics or pharmacodynamics of dofetilide, but there was a dose-dependent increase in exposure to dofetilide with cimetidine (64). With cimetidine 100 and 400 mg bd the AUC of dofetilide increased by 11 and 48%, the maximum plasma dofetilide concentration increased by 11 and 29%, renal clearance fell by 13 and 33%, and non-renal clearance by 5 and 21%; dofetilide-induced prolongation of the QT_c interval was increased by 22 and 33%. The authors suggested that cimetidine inhibited renal tubular dofetilide secretion, an effect that is specific to cimetidine in its class. Cimetidine should be avoided in patients taking dofetilide.

Verapamil

Pharmacokinetic and pharmacodynamic interactions between dofetilide 0.5 mg bd and verapamil 80 mg tds have been studied in 12 healthy men (65). At steady state verapamil increased the peak plasma concentration of dofetilide from 2.40 to 3.43 ng/ml, without other pharmacokinetic effects. This was accompanied by a small increase in the prolongation of the QT_c interval produced by dofetilide alone, from 20 to 26 ms. Although this small effect is unlikely to be of clinical significance, it would be wise to avoid verapamil in patients taking dofetilide.

By increasing its rate of absorption, verapamil produced transient rises in peak plasma concentrations of dofetilide in 12 young healthy male volunteers (67). During combination treatment at steady state there was a significant, albeit modest, increase in the mean C_{max} and AUC of dofetilide. These changes were associated with corresponding short-lived increases in its pharmacodynamic effect, as measured by changes in the QT_c interval. These two drugs should not be administered concurrently.

Management of adverse drug reactions In the USA there is a risk-management program for dofetilide, which restricts its distribution and requires education of prescribers. Charts for 47 patients taking dofetilide and 117 patients taking sotalol were reviewed (68). The recommended starting dose was prescribed significantly more often in the dofetilide group than in the sotalol group (79% versus 35%). Significantly more patients given dofetilide had baseline tests for serum potassium (100% versus 82%), magnesium (89% versus 38%), and creatinine (100% versus 82%), and electrocardiography (94% versus 67%). Significantly more patients given dofetilide had electrocardiography after the first dose (94% versus 43%) and subsequent doses (80% versus 3.5%). The authors concluded that there was better adherence to dosing and monitoring recommendations in those given dofetilide, which may have been attributable to the risk-management program. However, there was low usage of dofetilide during the study period, perhaps an unintended, negative consequence of the program.

References

1. Cross PE, Arrowsmith JE, Thomas GN, Gwilt M, Burges RA, Higgins AJ. Selective class III antiarrhythmic agents. 1 Bis(arylalkyl)amines. J Med Chem 1990;33(4):1151–5.
2. Anonymous. Dofetilide for atrial fibrillation. Med Lett Drugs Ther 2000;42(1078):41–2.
3. Torp-Pedersen C, Moller M, Kober L, Camm AJ. Dofetilide for the treatment of atrial fibrillation in patients with congestive heart failure. Eur Heart J 2000;21(15):1204–6.
4. Al-Dashti R, Sami M. Dofetilide: a new class III antiarrhythmic agent. Can J Cardiol 2001;17(1):63–7.
5. Ansani NT. Dofetilide: a new treatment of arrhythmias. P&T 2001;26:372–8.
6. Lauer MR. Dofetilide: is the treatment worse than the disease? J Am Coll Cardiol 2001;37(4):1106–10.
7. Saoudi N, Rinaldi JP, Yaici K, Bergonzi M. Dofetilide: what role in the treatment of ventricular tachyarrhythmias? Eur Heart J 2001;22(23):2141–3.
8. Tran A, Vichiendilokkul A, Racine E, Milad A. Practical approach to the use and monitoring of dofetilide therapy. Am J Health Syst Pharm 2001;58(21):2050–9.
9. Boriani G, Lubinski A, Capucci A, Niederle R, Kornacewicz-Jack Z, Wnuk-Wojnar AM, Borggrefe M, Brachmann J, Biffi M, Butrous GSVentricular Arrhythmias Dofetilide Investigators. A multicentre, double-blind randomized crossover comparative study on the efficacy and safety of dofetilide vs sotalol in patients with inducible sustained ventricular tachycardia and ischaemic heart disease. Eur Heart J 2001;22(23):2180–91.
10. Sanguinetti MC, Jurkiewicz NK. Two components of cardiac delayed rectifier K^+ current. Differential sensitivity to block by class III antiarrhythmic agents. J Gen Physiol 1990;96(1):195–215.
11. Tande PM, Bjornstad H, Yang T, Refsum H. Rate-dependent class III antiarrhythmic action, negative chronotropy, and positive inotropy of a novel Ik blocking drug, UK-68,798: potent in guinea pig but no effect in rat myocardium. J Cardiovasc Pharmacol 1990;16(3):401–10.
12. Gwilt M, Arrowsmith JE, Blackburn KJ, Burges RA, Cross PE, Dalrymple HW, Higgins AJ. UK-68,798: a novel, potent and highly selective class III antiarrhythmic

agent which blocks potassium channels in cardiac cells. J Pharmacol Exp Ther 1991;256(1):318–24.

13. Gwilt M, Blackburn KJ, Burges RA, Higgins AJ, Milne AA, Solca AM. Electropharmacology of dofetilide, a new class III agent, in anaesthetised dogs. Eur J Pharmacol 1992;215(2–3):137–44.
14. Sedgwick ML, Rasmussen HS, Cobbe SM. Clinical and electrophysiologic effects of intravenous dofetilide (UK-68,798), a new class III antiarrhythmic drug, in patients with angina pectoris. Am J Cardiol 1992;69(5):513–7.
15. Abrahamsson C, Duker G, Lundberg C, Carlsson L. Electrophysiological and inotropic effects of H 234/09 (almokalant) in vitro: a comparison with two other novel IK blocking drugs, UK-68,798 (dofetilide) and E-4031. Cardiovasc Res 1993;27(5):861–7.
16. Montero M, Schmitt C. Recording of transmembrane action potentials in chronic ischemic heart disease and dilated cardiomyopathy and the effects of the new class III antiarrhythmic agents D-sotalol and dofetilide. J Cardiovasc Pharmacol 1996;27(4):571–7.
17. Gwilt M, King RC, Milne AA, Solca AM. Dofetilide, a new class III antiarrhythmic agent, reduces pacing induced heterogeneity of repolarisation in vivo. Cardiovasc Res 1992;26(11):1102–8.
18. Sedgwick ML, Dalrymple I, Rae AP, Cobbe SM. Effects of the new class III antiarrhythmic drug dofetilide on the atrial and ventricular intracardiac monophasic action potential in patients with angina pectoris. Eur Heart J 1995;16(11):1641–6.
19. Yuan S, Wohlfart B, Rasmussen HS, Olsson S, Blomstrom-Lundqvist C. Effect of dofetilide on cardiac repolarization in patients with ventricular tachycardia. A study using simultaneous monophasic action potential recordings from two sites in the right ventricle. Eur Heart J 1994;15(4):514–22.
20. Baskin EP, Lynch JJ Jr. Differential atrial versus ventricular activities of class III potassium channel blockers. J Pharmacol Exp Ther 1998;285(1):135–42.
21. Snyders DJ, Chaudhary A. High affinity open channel block by dofetilide of HERG expressed in a human cell line. Mol Pharmacol 1996;49(6):949–55.
22. Sager PT. Frequency-dependent electrophysiologic effects of dofetilide in humans. Circulation 1995;92:1774.
23. Sedgwick M, Rasmussen HS, Walker D, Cobbe SM. Pharmacokinetic and pharmacodynamic effects of UK-68,798, a new potential class III antiarrhythmic drug. Br J Clin Pharmacol 1991;31(5):515–9.
24. Tham TC, MacLennan BA, Burke MT, Harron DW. Pharmacodynamics and pharmacokinetics of the class III antiarrhythmic agent dofetilide (UK-68,798) in humans. J Cardiovasc Pharmacol 1993;21(3):507–12.
25. Allen MJ, Nichols DJ, Oliver SD. The pharmacokinetics and pharmacodynamics of oral dofetilide after twice daily and three times daily dosing. Br J Clin Pharmacol 2000;50(3):247–53.
26. Boriani G, Biffi M, De Simone N, Bacchi L, Martignani C, Bitonti F, Zannoli R, Butrous G, Branzi A. Repolarization changes in a double-blind crossover study of dofetilide versus sotalol in the treatment of ventricular tachycardia. Pacing Clin Electrophysiol 2000;23(11 Pt 2):1935–8.
27. Bashir Y, Thomsen PE, Kingma JH, Moller M, Wong C, Cobbe SM, Jordaens L, Campbell RW, Rasmussen HS, Camm AJDofetilide Arrhythmia Study Group. Electrophysiologic profile and efficacy of intravenous dofetilide (UK-68,798), a new class III antiarrhythmic drug, in patients with sustained monomorphic ventricular tachycardia. Am J Cardiol 1995;76(14):1040–4.
28. Sedgwick ML, Rasmussen HS, Cobbe SM. Effects of the class III antiarrhythmic drug dofetilide on ventricular monophasic action potential duration and QT interval dispersion in stable angina pectoris. Am J Cardiol 1992;70(18):1432–7.
29. Demolis JL, Funck-Brentano C, Ropers J, Ghadanfar M, Nichols DJ, Jaillon P. Influence of dofetilide on QT-interval duration and dispersion at various heart rates during exercise in humans. Circulation 1996;94(7):1592–9.
30. Lande G, Maison-Blanche P, Fayn J, Ghadanfar M, Coumel P, Funck-Brentano C. Dynamic analysis of dofetilide-induced changes in ventricular repolarization. Clin Pharmacol Ther 1998;64(3):312–21.
31. Crijns HJ, Kingma JH, Gosselink AT, Lie K. Comparison in the same patient of aberrant conduction and bundle branch reentry after dofetilide, a new selective class III antiarrhythmic agent. Pacing Clin Electrophysiol 1993;16(5 Pt 1):1006–16.
32. Crijns HJ, Kingma JH, Gosselink AT, Dalrymple HW, De Langen CD, Lie K. Sequential bilateral bundle branch block during dofetilide, a new class III antiarrhythmic agent, in a patient with atrial fibrillation. J Cardiovasc Electrophysiol 1993;4(4):459–66.
33. Doggrell SA, Nand V. Effects of dofetilide on cardiovascular tissues from normo- and hypertensive rats. J Pharm Pharmacol 2002;54(5):707–15.
34. DeCara JM, Pollak A, Dubrey S, Falk RH. Positive atrial inotropic effect of dofetilide after cardioversion of atrial fibrillation or flutter. Am J Cardiol 2000;86(6):685–8.
35. Gemmill JD, Howie CA, Meredith PA, Kelman AW, Rasmussen HS, Hillis WS, Elliott HL. A dose-ranging study of UK-68,798, a novel class III anti-arrhythmic agent, in normal volunteers. Br J Clin Pharmacol 1991;32(4):429–32.
36. Le Coz F, Funck-Brentano C, Morell T, Ghadanfar MM, Jaillon P. Pharmacokinetic and pharmacodynamic modeling of the effects of oral and intravenous administrations of dofetilide on ventricular repolarization. Clin Pharmacol Ther 1995;57(5):533–42.
37. Walker DK, Alabaster CT, Congrave GS, Hargreaves MB, Hyland R, Jones BC, Reed LJ, Smith DA. Significance of metabolism in the disposition and action of the antidysrhythmic drug, dofetilide. In vitro studies and correlation with in vivo data. Drug Metab Dispos 1996;24(4):447–55.
38. Suttorp MJ, Polak PE, van't Hof A, Rasmussen HS, Dunselman PH, Kingma JH. Efficacy and safety of a new selective class III antiarrhythmic agent dofetilide in paroxysmal atrial fibrillation or atrial flutter. Am J Cardiol 1992;69(4):417–9.
39. Kobayashi Y, Atarashi H, Ino T, Kuruma A, Nomura A, Saitoh H, Hayakawa H. Clinical and electrophysiologic effects of dofetilide in patients with supraventricular tachyarrhythmias. J Cardiovasc Pharmacol 1997;30(3):367–73.
40. Cobbe SM, Campbell RW, Camm AJ, Nathan AW, Rowland E, Bloch-Thomsen PE, Moller M, Jordaens L. Effects of intravenous dofetilide on induction of atrioventricular re-entrant tachycardia. Heart 2001;86(5):522–6.
41. Sedgwick ML, Lip G, Rae AP, Cobbe SM. Chemical cardioversion of atrial fibrillation with intravenous dofetilide. Int J Cardiol 1995;49(2):159–66.
42. Frost L, Mortensen PE, Tingleff J, Platou ES, Christiansen EH, Christiansen NDofetilide Post-CABG Study Group. Efficacy and safety of dofetilide, a new class

III antiarrhythmic agent, in acute termination of atrial fibrillation or flutter after coronary artery bypass surgery. Int J Cardiol 1997;58(2):135–40.
43. Falk RH, Pollak A, Singh SN, Friedrich T. Intravenous dofetilide, a class III antiarrhythmic agent, for the termination of sustained atrial fibrillation or flutter. Intravenous Dofetilide Investigators. J Am Coll Cardiol 1997;29(2):385–90.
44. Norgaard BL, Wachtell K, Christensen PD, Madsen B, Johansen JB, Christiansen EH, Graff O, Simonsen EHDanish Dofetilide in Atrial Fibrillation and Flutter Study Group. Efficacy and safety of intravenously administered dofetilide in acute termination of artrial fibrillation and flutter: a multicenter, randomized, double-blind, placebo-controlled trial. Am Heart J 1999;137(6):1062–9.
45. Singh S, Zoble RG, Yellen L, Brodsky MA, Feld GK, Berk M, Billing CB Jr. Efficacy and safety of oral dofetilide in converting to and maintaining sinus rhythm in patients with chronic atrial fibrillation or atrial flutter: the symptomatic atrial fibrillation investigative research on dofetilide (SAFIRE-D) study. Circulation 2000;102(19):2385–90.
46. Lindeboom JE, Kingma JH, Crijns HJ, Dunselman PH. Efficacy and safety of intravenous dofetilide for rapid termination of atrial fibrillation and atrial flutter. Am J Cardiol 2000;85(8):1031–3.
47. Krahn AD, Klein GJ, Yee R. A randomized, double-blind, placebo-controlled evaluation of the efficacy and safety of intravenously administered dofetilide in patients with Wolff–Parkinson–White syndrome. Pacing Clin Electrophysiol 2001;24(8 Pt 1):1258–60.
48. Echt DS, Lee JT, Murray KT, Vorperian V, Borganelli SM, Crawford DM, Friedrich T, Roden DM. A randomized, double-blind, placebo-controlled, dose-ranging study of dofetilide in patients with inducible sustained ventricular tachyarrhythmias. J Cardiovasc Electrophysiol 1995;6(9):687–99.
49. Pool PE, Singh SN, Friedrich T. Effects of intravenous dofetilide in patients with frequent premature ventricular contractions: a clinical trial. Clin Cardiol 2000;23(6):415–6.
50. Danish Investigations of Arrhythmia and Mortality ON Dofetilide. Dofetilide in patients with left ventricular dysfunction and either heart failure or acute myocardial infarction: rationale, design, and patient characteristics of the DIAMOND studies. Clin Cardiol 1997;20(8):704–10.
51. Torp-Pedersen C, Moller M, Bloch-Thomsen PE, Kober L, Sandoe E, Egstrup K, Agner E, Carlsen J, Videbaek J, Marchant B, Camm AJDanish Investigations of Arrhythmia and Mortality on Dofetilide Study Group. Dofetilide in patients with congestive heart failure and left ventricular dysfunction. N Engl J Med 1999;341(12):857–65.
52. Kober L, Bloch Thomsen PE, Moller M, Torp-Pedersen C, Carlsen J, Sandoe E, Egstrup K, Agner E, Videbaek J, Marchant B, Camm AJDanish Investigations of Arrhythmia and Mortality on Dofetilide (DIAMOND) Study Group. Effect of dofetilide in patients with recent myocardial infarction and left-ventricular dysfunction: a randomised trial. Lancet 2000;356(9247):2052–8.
53. Bianconi L, Castro A, Dinelli M, Alboni P, Pappalardo A, Richiardi E, Santini M. Comparison of intravenously administered dofetilide versus amiodarone in the acute termination of atrial fibrillation and flutter. A multicentre, randomized, double-blind, placebo-controlled study. Eur Heart J 2000;21(15):1265–73.
54. Crijns HJ, Van Gelder IC, Kingma JH, Dunselman PH, Gosselink AT, Lie KI. Atrial flutter can be terminated by a class III antiarrhythmic drug but not by a class IC drug. Eur Heart J 1994;15(10):1403–8.
55. Tendera M, Wnuk-Wojnar AM, Kulakowski P, Malolepszy J, Kozlowski JW, Krzeminska-Pakula M, Szechinski J, Droszcz W, Kawecka-Jaszcz K, Swiatecka G, Ruzyllo W, Graff O. Efficacy and safety of dofetilide in the prevention of symptomatic episodes of paroxysmal supraventricular tachycardia: a 6-month double-blind comparison with propafenone and placebo. Am Heart J 2001;142(1):93–8.
56. Boriani G, Biffi M, Bacchi L, Martignani C, Zannoli R, Butrous GS, Branzi A. A randomised cross-over study on the haemodynamic effects of oral dofetilide compared with oral sotalol in patients with ischaemic heart disease and sustained ventricular tachycardia. Eur J Clin Pharmacol 2002;58(3):165–9.
57. Mazur A, Anderson ME, Bonney S, Roden DM. Pause-dependent polymorphic ventricular tachycardia during long-term treatment with dofetilide: a placebo-controlled, implantable cardioverter–defibrillator-based evaluation. J Am Coll Cardiol 2001;37(4):1100–5.
58. Guanzon AV, Crouch MA. Phase IV trial evaluating the effectiveness and safety of dofetilide. Ann Pharmacother 2004;38(7–8):1142–7.
59. Nagra BS, Ledley GS, Kantharia BK. Marked QT prolongation and torsades de pointes secondary to acute ischemia in an elderly man taking dofetilide for atrial fibrillation: a cautionary tale. J Cardiovasc Pharmacol Ther 2005;10(3):191–5.
60. Reiffel JA. Atypical proarrhythmia with dofetilide: monomorphic VT and exercise-induced torsade de pointes. Pacing Clin Electrophysiol 2005;28(8):877–9.
61. Sun Z, Milos PM, Thompson JF, Lloyd DB, Mank-Seymour A, Richmond J, Cordes JS, Zhou J. Role of a KCNH2 polymorphism (R1047 L) in dofetilide-induced torsades de pointes. J Mol Cell Cardiol 2004;37(5):1031–9.
62. Deal BJ, Keane JF, Gillette PC, Garson A Jr. Wolff–Parkinson–White syndrome and supraventricular tachycardia during infancy: management and follow-up. J Am Coll Cardiol 1985;5(1):130–5.
63. Pfammatter JP, Stocker FP. Re-entrant supraventricular tachycardia in infancy: current role of prophylactic digoxin treatment. Eur J Pediatr 1998;157(2):101–6.
64. Sanatani S, Saul JP, Walsh EP, Gross GJ. Spontaneously terminating apparent ventricular fibrillation during transesophageal electrophysiological testing in infants with Wolff-Parkinson-White syndrome. Pacing Clin Electrophysiol 2001;24(12):1816–8.
65. Spodick DH. Well concealed atrial tachycardia with Wenckebach (Mobitz I) atrioventricular block: digitalis toxicity. Am J Geriatr Cardiol 2001;10(1):59.
66. Kleinermans D, Nichols DJ, Dalrymple I. Effect of dofetillide on the pharmacokinetics of digoxin. Am J Cardiol 2001;87(2):248–50.
67. Johnson BF, Cheng SL, Venitz J. Transient kinetic and dynamic interactions between verapamil and dofetilide, a class III antiarrhythmic. J Clin Pharmacol 2001;41(11):1248–56.
68. Allen LaPointe NM, Chen A, Hammill B, DeLong E, Kramer JM, Califf RM. Evaluation of the dofetilide risk-management program. Am Heart J 2003;146: 894–901.

Encainide

General Information

Encainide is a class I antidysrhythmic drug. Reviews of its clinical pharmacology, clinical use, efficacy, and adverse effects have appeared (1–6).

Organs and Systems

Cardiovascular

In the wake of the preliminary and final reports of the Cardiac Arrhythmia Suppression Trial (CAST) (7,8), which showed that there was an increased risk of death among patients who took encainide and flecainide after myocardial infarction, there have been many publications in which the implications of these findings have been thoroughly discussed (9–13). The relative risk of death or cardiac arrest due to dysrhythmias in the treated patients was 2.6 and the relative risk due to all causes was 2.38. The risk of non-fatal cardiac adverse effects was no different in treated patients from that in those taking placebo and there was no difference between the groups in the use of other drugs.

Although there is a consensus that encainide and flecainide were associated with an increase in the rate of mortality in CAST, there are still some open questions. First, all the patients recruited to CAST had asymptomatic ventricular dysrhythmias after myocardial infarction, and it is not clear whether the results can be extrapolated to other patients. Secondly, the reasons for the increased mortality in the treated patients are not clear: ventricular dysrhythmias and worsening of left ventricular function are both possible. Thirdly, it is not clear whether the results of CAST in patients with asymptomatic ventricular dysrhythmias after myocardial infarction can also be applied to other Class I antidysrhythmic drugs.

Separate studies have confirmed the prodysrhythmic actions of encainide (14–17). The incidence of prodysrhythmias is much higher in patients being treated for ventricular dysrhythmias than in those being treated for supraventricular dysrhythmias (18).

Encainide has been reported to cause sinus node arrest in association with prolonged sinus node recovery time (19). It also raises the pacing threshold in patients with chronic implanted pacemakers (20), although this has not been reported to increase the failure rate of pacemakers.

Encainide has a negative inotropic effect on the heart and can cause hypotension (21) or worsen heart failure (22).

Nervous system

The most common non-cardiac effects of encainide are on the central nervous system, and include abnormal or blurred vision (11%), dizziness (7.3%), headaches (6.0%), nausea (4.3%), vertigo (2.3%), insomnia, and fatigue. The figures in parentheses are taken from a review of 349 patients with supraventricular dysrhythmias treated with encainide (14). These effects are common during long-term therapy, but may also occur transiently during intravenous administration and appear to be dose-related (23,24).

A case of encephalopathy has been attributed to encainide (25).

Sensory systems

There have been reports of a metallic taste in the mouth after intravenous administration of encainide (24).

Endocrine

Encainide can cause hyperglycemia (26), perhaps due to insulin resistance.

Gastrointestinal

Nausea can occur during long-term therapy with encainide, probably due to a central nervous effect (27).

Skin

Skin rashes have been reported occasionally during long-term therapy with encainide (27).

Musculoskeletal

There have been reports of leg cramps after intravenous administration of encainide (24).

Susceptibility Factors

Genetic factors

Encainide is metabolized, at least partly, by oxidation to *O*-desmethylencainide, 3-methoxy-*O*-desmethylencainide, *N*-desmethylencainide, and *N,O*-didesmethylencainide. Of 112 healthy Caucasians 9 (8%) were defective in their ability to 4-hydroxylate debrisoquine (28). The cumulative frequency distribution of the 8-hour urinary recovery ratio of encainide/*O*-desmethylencainide indicated two distinct populations, an extensive metabolizer (EM) phenotype and a poor metabolizer (PM) phenotype. There was no 3-methoxy-*O*-desmethylencainide in the urine of the poor metabolizers. As *O*-desmethylencainide is a more potent antidysrhythmic drug than encainide and 3-methoxy-*O*-desmethylencainide is at least equipotent, these metabolites contribute significantly to the overall antidysrhythmic effect in extensive metabolizers. The low plasma concentrations of *O*-desmethylencainide and 3-methoxy-*O*-desmethylencainide in poor metabolizers would be expected to result in ineffective therapy when usual doses of encainide are given. However, in such individuals, chronic oral therapy results in accumulation of unmetabolized encainide to far higher concentrations than in extensive metabolizers, and as encainide itself has antidysrhythmic activity at these concentrations, this generally results in the desired response. It is possible that poor hydroxylators are at greater risk of adverse effects than extensive hydroxylators.

In 110 healthy subjects the changes in atrioventricular (PR) and intraventricular (QRS) conduction times produced by encainide were different in extensive and poor metabolizers and correlated with CYP2D6 activity, although the electrocardiographic response was never 100% specific and sensitive for the identification of either phenotype (29). Moreover, genotypic identification of heterozygous and homozygous extensive metabolizer subjects did not predict CYP2D6 activity, as determined by encainide metabolic ratios or encainide responses, as determined by intraventricular and atrioventricular changes.

Age

Children under the age of 6 months may be more liable to the prodysrhythmic effects of encainide than older children (30).

References

1. Keefe DL, Kates RE, Harrison DC. New antiarrhythmic drugs: their place in therapy. Drugs 1981;22(5):363–400.
2. Harrison DC, Winkle R, Sami M, Mason J. Encainide: a new and potent antiarrhythmic agent. Am Heart J 1980;100(6 Pt 2):1046–54.
3. Lynch JJ, Lucchesi BR. New antiarrhythmic agents. II. The pharmacology and clinical use of encainide. Pract Cardiol 1984;10:109–32.
4. Roden DM, Woosley RL. Clinical pharmacology of the new antiarrhythmic encainide. Clin Progr Pacing Electrophysiol 1984;2:112–9.
5. Rinkenberger RL, Naccarelli GV, Dougherty AH. New antiarrhythmic agents. X. Safety and efficacy of encainide in the treatment of ventricular arrhythmias. Pract Cardiol 1987;13:110–32.
6. Naccarelli GV, et al. A symposium: The use of encainide in supraventricular tachycardias. Am J Cardiol 1988;62(19):1L–84LJune 3 and 4, 1988, Bermuda. Proceedings.
7. The Cardiac Arrhythmia Suppression Trial (CAST) Investigators. Preliminary report: effect of encainide and flecainide on mortality in a randomized trial of arrhythmia suppression after myocardial infarction. N Engl J Med 1989;321(6):406–12.
8. Echt DS, Liebson PR, Mitchell LB, Peters RW, Obias-Manno D, Barker AH, Arensberg D, Baker A, Friedman L, Greene HL, et al. Mortality and morbidity in patients receiving encainide, flecainide, or placebo. The Cardiac Arrhythmia Suppression Trial. N Engl J Med 1991;324(12):781–8.
9. Gottlieb SS. The use of antiarrhythmic agents in heart failure: implications of CAST. Am Heart J 1989;118(5 Pt 1):1074–7.
10. Podrid PJ, Marcus FI. Lessons to be learned from the Cardiac Arrhythmia Suppression Trial. Am J Cardiol 1989;64(18):1189–91.
11. Bigger JT Jr. The events surrounding the removal of encainide and flecainide from the Cardiac Arrhythmia Suppression Trial (CAST) and why CAST is continuing with moricizine. J Am Coll Cardiol 1990;15(1):243–5.
12. Akhtar M, Breithardt G, Camm AJ, Coumel P, Janse MJ, Lazzara R, Myerburg RJ, Schwartz PJ, Waldo AL, Wellens HJ, et al. CAST and beyond. Implications of the Cardiac Arrhythmia Suppression Trial. Task Force of the Working Group on Arrhythmias of the European Society of Cardiology. Circulation 1990;81(3):1123–7.
13. Thomis JA. Encainide—an updated safety profile. Cardiovasc Drugs Ther 1990;4(Suppl 3):585–94.
14. Soyka LF. Safety considerations and dosing guidelines for encainide in supraventricular arrhythmias. Am J Cardiol 1988;62(19):L63–8.
15. Miles WM, Zipes DP, Rinkenberger RL, Markel ML, Prystowsky EN, Dougherty AH, Heger JJ, Naccarelli GV. Encainide for treatment of atrioventricular reciprocating tachycardia in the Wolff–Parkinson–White syndrome. Am J Cardiol 1988;62(19):L20–5.
16. Rinkenberger RL, Naccarelli GV, Miles WM, Markel ML, Dougherty AH, Prystowsky EN, Heger JJ, Zipes DP. Encainide for atrial fibrillation associated with Wolff–Parkinson–White syndrome. Am J Cardiol 1988;62(19):L26–30.
17. Naccarelli GV, Jackman WM, Akhtar M, Rinkenberger RL, Friday KJ, Dougherty AH, Tchou P, Yeung-Lai-Wah JA. Efficacy and electrophysiologic effects of encainide for atrioventricular nodal reentrant tachycardia. Am J Cardiol 1988;62(19):L31–6.
18. The Encainide-Ventricular Tachycardia Study Group. Treatment of life-threatening ventricular tachycardia with encainide hydrochloride in patients with left ventricular dysfunction. Am J Cardiol 1988;62(9):571–5.
19. Lemery R, Talajic M, Nattel S, Theroux P, Roy D. Sinus node dysfunction and sudden cardiac death following treatment with encainide. Pacing Clin Electrophysiol 1989;12(10):1607–12.
20. Salel AF, Seagren SC, Pool PE. Effects of encainide on the function of implanted pacemakers. Pacing Clin Electrophysiol 1989;12(9):1439–44.
21. Rinkenberger RL, Prystowsky EN, Jackman WM, Naccarelli GV, Heger JJ, Zipes DP. Drug conversion of nonsustained ventricular tachycardia to sustained ventricular tachycardia during serial electrophysiologic studies: identification of drugs that exacerbate tachycardia and potential mechanisms. Am Heart J 1982;103(2):177–84.
22. DiBianco R, Fletcher RD, Cohen AI, Gottdiener JS, Singh SN, Katz RJ, Bates HR, Sauerbrunn B. Treatment of frequent ventricular arrhythmia with encainide: assessment using serial ambulatory electrocardiograms, intracardiac electrophysiologic studies, treadmill exercise tests, and radionuclide cineangiographic studies. Circulation 1982;65(6):1134–47.
23. Kesteloot H, Stroobandt R. Clinical experience of encainide (MJ 9067): a new anti-arrhythmic drug. Eur J Clin Pharmacol 1979;16(5):323–6.
24. Sami M, Mason JW, Peters F, Harrison DC. Clinical electrophysiologic effects of encainide, a newly developed antiarrhythmic agent. Am J Cardiol 1979;44(3):526–32.
25. Tartini A, Kesselbrenner M. Encainide-induced encephalopathy in a patient with chronic renal failure. Am J Kidney Dis 1990;15(2):178–9.
26. Winter WE, Funahashi M, Koons J. Encainide-induced diabetes: analysis of islet cell function. Res Commun Chem Pathol Pharmacol 1992;76(3):259–68.

27. Sami M, Harrison DC, Kraemer H, Houston N, Shimasaki C, DeBusk RF. Antiarrhythmic efficacy of encainide and quinidine: validation of a model for drug assessment. Am J Cardiol 1981;48(1):147–56.
28. McAllister CB, Wolfenden HT, Aslanian WS, Woosley RL, Wilkinson GR. Oxidative metabolism of encainide: polymorphism, pharmacokinetics and clinical considerations. Xenobiotica 1986;16(5):483–90.
29. Funck-Brentano C, Thomas G, Jacqz-Aigrain E, Poirier JM, Simon T, Bereziat G, Jaillon P. Polymorphism of dextromethorphan metabolism: relationships between phenotype, genotype and response to the administration of encainide in humans. J Pharmacol Exp Ther 1992;263(2):780–6.
30. Strasburger JF, Smith RT Jr, Moak JP, Gothing C, Garson A Jr. Encainide for resistant supraventricular tachycardia in children: follow-up report. Am J Cardiol 1988;62(19):L50–4.

Flecainide

General Information

Flecainide is a class Ic antidysrhythmic drug. Its clinical pharmacology, clinical use, and adverse effects have been reviewed (1–3).

Comparative studies

In some patients treatment with a class I antidysrhythmic drug converts atrial fibrillation to atrial flutter. Of 187 patients with paroxysmal atrial fibrillation who were treated with flecainide or propafenone, 24 developed atrial flutter, which was typical in 20 cases (4). These patients underwent radiofrequency ablation, which failed in only one case. All the patients continued to take their pre-existing drugs, and during a mean follow-up period of 11 months the incidence of atrial fibrillation was higher in patients who were taking combined therapy than in those taking monotherapy. The authors suggested that in patients with atrial fibrillation who developed typical atrial flutter due to class Ic antidysrhythmic drugs, combined catheter ablation and continued drug treatment is highly effective in reducing the occurrence and duration of atrial tachydysrhythmias. They did not report adverse effects.

In 33 patients with symptomatic and inducible supraventricular tachycardias single doses of placebo, flecainide 3 mg/kg, or diltiazem 120 mg plus propranolol 80 mg were used to terminate the dysrhythmia (5). Conversion to sinus rhythm was achieved within 2 hours in 17 patients with placebo, in 20 with flecainide, and in 31 with diltiazem plus propranolol. Time to conversion was shorter with diltiazem plus propranolol (32 minutes) than with flecainide (74 minutes) or placebo (77 minutes). Of those who were given flecainide, two had hypotension and one had sinus bradycardia.

Systematic reviews

A systematic review of 22 studies of the effects of flecainide used for at least 3 months in the treatment of supraventricular dysrhythmias suggested that flecainide is associated with a variety of adverse reactions, many of which are well tolerated, but carries a small risk of serious cardiac events (2%), which can lead to death (0.13%) (SEDA-21, 200).

In a systematic review of trials of the use of a single oral loading dose of flecainide for cardioversion of recent-onset atrial fibrillation most of the trials used a single oral dose of 300 mg for loading (6). The success rate was 57–68% at 2–4 hours and 75–91% at 8 hours. Adverse effects were mild non-cardiac adverse effects, reversible QRS complex widening, transient dysrhythmias, and left ventricular decompensation. The transient dysrhythmias occurred chiefly at the time of conversion and included atrial flutter and sinus pauses; there were no life-threatening ventricular dysrhythmias or deaths.

In a meta-analysis of 122 prospective studies of the use of flecainide in 4811 patients with supraventricular dysrhythmias, 21 were placebo-controlled and 37 were comparative studies with other antidysrhythmic drugs (7). The total exposure time was 2015 patient-years, with a mean oral flecainide dose of 216 mg/day. There were eight deaths (total mortality 0.17%, fatality rate per 100 patient-years 0.40; 95% CI = 0.17, 0.78), confirming the earlier finding. Three deaths were non-cardiac (cancer, suicide, urinary sepsis). Of the cardiac deaths, all but two occurred in patients with coronary heart disease. In controls, there was one death. There were prodysrhythmic events in 120 patients taking flecainide (2.7%) and 88 controls (4.8%), 58 (7.4%) of which occurred in patients taking placebo. Non-cardiac adverse effects are listed in Table 1. Thus, flecainide is safe in patients with supraventricular dysrhythmias with no cardiac damage, in contrast to patients with ventricular dysrhythmias after myocardial infarction.

Organs and Systems

Cardiovascular

In a review of 60 original articles detailing 1835 courses of intravenous and/or oral flecainide in both placebo-controlled and comparative studies as well as a large number of uncontrolled studies, unwanted cardiac events occurred in 8% of patients (8). The cardiac events were hypotension (1.3%), heart failure (0.4%), sinus node dysfunction (1.6%), bundle branch block (1.0%), atrial dysrhythmias (1.6%), and ventricular dysrhythmias (1.3%). However, in 8505 patients, 5507 of whom were administered flecainide for more than 4 weeks and most of whom took dosages of 100–300 mg/day, cardiac adverse effects occurred in only about 2% and non-cardiac effects in about 10% (9). The most common cardiac adverse effects were angina pectoris, dysrhythmias, worsening of heart failure, and hemodynamic changes. Of the long-term non-cardiac adverse effects the most common were nausea,

Table 1 Numbers (%) of cardiac and non-cardiac adverse effects of flecainide in a meta-analysis of 4375 treatment courses compared with 1818 treatment courses in controls

System	*Adverse effect*	*Flecainide*	*Controls*
Cardiovascular	Angina pectoris	43 (1.0)	25 (1.3)
	Palpitation	17 (0.4)	6 (0.3)
	Hypotension	33 (0.8)	24 (1.3)*
	Syncope	5 (0.1)	3 (0.2)
	Heart failure/dyspnea	40 (0.9)	13 (0.7)
	Sinus node dysfunction	52 (1.2)	22 (1.2)
	Bundle branch block	29 (0.7)	7 (0.4)
	Atrioventricular block	24 (0.5)	7 (0.4)
Nervous system	Total	412 (9.4)	65 (3.4)*
	Headache	88 (2.0)	53 (2.9)*
	Dizziness	148 (3.4)	45 (2.5)
	Vertigo	137 (3.1)	42 (2.3)
Sensory systems	Visual disturbances	175 (4.0)	16 (0.9)*
Gastrointestinal	Total	144 (3.3)	121 (6.7)*
	Diarrhea	29 (0.7)	50 (2.8)*
	Nausea	71 (1.6)	33 (1.8)

*Significantly different.

vomiting, dizziness, bowel disturbances, headache, and visual disturbances.

Cardiac dysrhythmias

In the wake of the preliminary and final reports of the Cardiac Arrhythmia Suppression Trial (CAST) (10,11), which showed that there was an increased risk of death among patients who took encainide and flecainide after myocardial infarction, there have been many publications in which the implications of these findings have been thoroughly discussed (12–16). The relative risk of death or cardiac arrest due to dysrhythmias in the treated patients was 2.6 and the relative risk due to all causes was 2.4. The risk of non-fatal cardiac adverse effects was no different in treated patients from that in those taking placebo and there was no difference between the groups in the use of other drugs.

Although there is a consensus that encainide and flecainide were associated with an increase in the rate of mortality in CAST, there are still some open questions. First, all the patients recruited to CAST had asymptomatic ventricular dysrhythmias after myocardial infarction, and it is not clear whether the results can be extrapolated to other patients. Secondly, the reasons for the increased mortality in the treated patients are not clear: ventricular dysrhythmias and worsening of left ventricular function are both possible. Thirdly, it is not clear whether the results of CAST in patients with asymptomatic ventricular dysrhythmias after myocardial infarction can also be applied to other Class I antidysrhythmic drugs.

The prodysrhythmic effects of flecainide, by prolongation of the QT_c interval, have been widely discussed (SEDA-15, 175) (17–21). However, overall, cardiac dysrhythmias are less common with flecainide than with other antidysrhythmic drugs of Class I, and the dysrhythmogenic effects of flecainide in patients with supraventricular dysrhythmias may not be great (SEDA-22, 207). When dysrhythmias occur, prolongation of the QT interval is an important mechanism, but in a recent case it was suggested that tachycardia was due to re-entry within the His-Purkinje system (22). In another case flecainide reportedly caused a wide-complex tachycardia due to atypical atrial flutter with 1:1 conduction and aberrant QRS complexes (23). Although drugs of Class Ic, such as flecainide, can slow atrial and atrioventricular nodal conduction in patients with atrial fibrillation or atrial flutter, they do not alter the refractoriness of the atrioventricular node, and this allows 1:1 atrioventricular conduction as the atrial rate slows. This happens despite prolongation of the PR interval.

- Syncope occurred in a patient whose QRS complex duration was prolonged (24).
- In a 67-year-old woman taking flecainide 150 mg bd, widening of the QRS complex occurred during exercise; the effect did not occur at rest or with a dose of 50 mg bd (25).
- A 68-year-old woman developed flecainide-induced syncope due to torsade de pointes, before the onset of which her QT_c interval reached 680 ms without a change in the QRS duration (26). None of the usual triggers were found and she was taking no other drugs.
- A baby developed a supraventricular tachycardia in utero at 38 weeks and was delivered by cesarian section (27). Flecainide 2 mg/kg was given postnatally and 48 hours later, after four doses had been given, a broad-complex tachycardia developed. Flecainide was withdrawn, but 4 hours later ventricular fibrillation developed. After resuscitation a re-entrant supraventricular tachycardia was treated with digoxin and amiodarone.

In the second case the serum flecainide concentration 24 hours after the last dose was 630 ng/ml, and since flecainide has a half-life of about 12 hours in children (28) this suggest that the flecainide concentration at the time of the ventricular fibrillation was probably quite high.

Sodium channel blockers have been used to improve muscle strength and relaxation in myotonic dystrophy. However, myotonic dystrophy is associated with cardiac abnormalities and 30% of deaths are attributable to cardiac causes, mainly dysrhythmias.

- A 41-year-old woman with myotonic dystrophy developed a ventricular tachycardia while talking flecainide (29). When flecainide was withdrawn ventricular tachycardia could not be induced.

The authors recommended careful cardiac assessment, risk stratification, and consideration of high-risk patients for electrophysiological studies, especially if considering use of a class I antidsysrhythmic drug.

Hypokalemia can increase the risk of torsade de pointes with flecainide, as with other antidysrhythmic drugs.

- Torsade pointes has been attributed to mosapride (which is related to cisapride) and flecainide in a 68-year-old man with a plasma potassium concentration of 3.2 mmol/l and prolongation of the QT_c interval from 0.48 to 0.56 seconds (30). His plasma flecainide concentration was just above the target range at 1013 ng/ml, but the mosapride concentration was not reported.

The authors speculated that mosapride may have inhibited the metabolism of flecainide by CYP2D6.

In a retrospective analysis of 24 patients who developed atrial flutter while taking flecainide (n = 12) or propafenone (n = 12), the electrocardiogram was classified as typical atrial flutter in 13 cases, atypical atrial flutter in eight, or coarse atrial fibrillation in three (31). Counterclockwise atrial flutter was the predominant dysrhythmia. The acute results of ablation suggested that the flutter circuit was located in the right atrium and that the isthmus was involved in the re-entry mechanism. There was better long-term control of recurrent atrial fibrillation in patients with typical atrial flutter (85%) compared with atypical atrial flutter (50%). The authors suggested that patients who develop coarse drug-induced atrial fibrillation may not be candidates for ablation.

Flecainide can cause changes of the Brugada syndrome on the electrocardiogram, and another case has been reported in a 70-year-old man (32).

Cardioversion

In 24 patients with atrial fibrillation who underwent elective transvenous cardioversion for atrial fibrillation, flecainide reduced the energy requirements for further defibrillation after induction of atrial fibrillation by atrial pacing (33). There were no ventricular dysrhythmias, but transient bradycardia requiring ventricular pacing occurred in two patients. Two patients had transient asymptomatic hypotension after flecainide and one reported transient dizziness and some light-headedness.

Antidysrhythmic drugs increase the pacing threshold, but failure to capture is a rare consequence. It has, however, been reported twice with flecainide (34,35).

Electrocardiographic changes

ST segment elevation in leads II, III, and aVf, resembling an acute inferior myocardial infarction, have been reported with oral flecainide (36). Brugada syndrome is partial right bundle branch block with ST segment elevation in the right precordial leads of the electrocardiogram; it is due to an abnormality of sodium channels and occurs in 0.05–0.1% of the population; some cases are inherited (37). Flecainide can bring out Brugada-type changes on the electrocardiogram (38–40).

Hypotension

Flecainide has been reported to cause acute hypotension after intravenous administration (41).

Respiratory

Interstitial pneumonitis has only rarely been attributed to flecainide (SEDA-16, 181).

- Interstitial pneumonitis with acute respiratory failure was attributed to flecainide in a 59-year-old man with congenital heart disease related to the LEOPARD syndrome, in which there are multiple freckles (Lentigines), Electrocardiographic abnormalities, Ocular hypertelorism, Pulmonic stenosis, Abnormalities of the genitalia, Retarded growth, and sensorineural Deafness (42). A CT scan showed diffuse interstitial injury characterized by thickening of the intralobular septa, with areas of ground-glass pattern. Flecainide was withdrawn and within 2 weeks the changes on CT scan had almost completely disappeared.
- A 75-year-old man, who had taken flecainide 100 mg/day for 22 months, developed fever, headache, and a dry cough (43). A CT scan of the lungs was normal and he responded to prednisone. His symptoms disappeared, but when prednisone was withdrawn they returned, with breathlessness, a dry cough, and weight loss. A chest X-ray showed bilateral patchy opacities and a CT scan subpleural ground-glass opacities and septal thickening. He had impaired lung function, including a reduced diffusion capacity. Biopsy showed diffuse interstitial thickening with lymphocytic and eosinophilic infiltrates. Flecainide was withdrawn and prednisone given, and he made a full recovery within 1 month.
- A 73-year-old man, who had taken flecainide 100 mg/day for 4 months, developed fever, weight loss, breathlessness, and a dry cough (43). A chest X-ray showed patchy infiltrates and a CT scan ground-glass opacities and subpleural septal thickening. He had normal lung function, apart from a reduced diffusion capacity. Flecainide was withdrawn and prednisone given, and he made a full recovery within a few months.

Nervous system

The most common non-cardiac adverse effects of flecainide are on the central nervous system and include dizziness, drowsiness, visual disturbances, headache, nausea, paresthesia, nervousness, and tremor. The incidence of these adverse effects has varied widely (44,45).

Other reported effects of flecainide include dysarthria and visual hallucinations (46), abnormal taste sensations, flushing, a glove-and-stocking type of peripheral neuropathy (47), and dystonia (SEDA-17, 223).

Sensory systems

Ocular adverse effects of flecainide have included corneal deposits in two patients, due to deposition of flecainide (48). In 38 patients taking flecainide 100–300 mg/day there were brown corneal epithelial deposits in 11 eyes, dryness in eight eyes, and slight blurring of vision on lateral gaze in four patients (49). Four patients had local symptoms, including tearing, itching, and burning. Color vision, contrast sensitivity, and visual fields were all unaffected.

Hematologic

Flecainide can cause neutropenia (50).

Liver

Flecainide can cause increases in the serum activities of transaminases (51). In one case cholestasis and jaundice occurred (52).

Urinary tract

Flecainide can cause acute urinary retention, perhaps due to a local anesthetic effect on the bladder mucosa (53).

Skin

Flecainide reportedly caused a psoriasiform eruption (54).

Sexual function

Flecainide inhibits sperm motility in vitro (55), but this has not been reported to be of clinical relevance.

Flecainide can cause erectile impotence (56).

Second-Generation Effects

Pregnancy

Flecainide is occasionally used to treat fetal cardiac dysrhythmias by administration to the mother (57,58), although occasionally it can cause adverse effects in the child (SEDA-25, 184) and in the mother (59).

- At 30 weeks of gestation in a 41-year-old woman the fetus had hydrops, ascites, a pericardial effusion, and bilateral hydroceles. A supraventricular tachycardia with 1:1 conduction was treated by giving the mother oral flecainide 150 mg bd. However, during the next few weeks the mother developed evidence of hepatic cholestasis. The dosage of flecainide was reduced to 50 mg bd and the liver damage resolved. The child was born healthy but later required sotalol for a re-entry tachycardia.

Fetotoxicity

The serum flecainide concentration at time of birth in a neonate whose mother had been given flecainide during pregnancy was 1030 ng/ml and there was a broad QRS complex, which resolved after 3 days (60).

- A fetus was treated for a supraventricular tachycardia at 27 weeks of gestation by giving the mother first flecainide then amiodarone plus flecainide (61). The girl was born at 33 weeks by cesarian section and had poor cardiac contractility, a prolonged PR interval, a broad QRS complex, and a long QT_c interval (532 ms).

The authors attributed the cardiac abnormalities to flecainide toxicity.

- A pregnant woman was given digoxin and flecainide at 29 weeks of gestation for fetal tachycardia and hydrops fetalis (62). The child was delivered spontaneously at 33 weeks and had mild respiratory distress. His electrocardiogram showed bifid P waves, a prolonged PR interval, deep wide Q waves, and raised ST segments. The QT interval was not prolonged. The serum digoxin concentration in the neonate was 1.2 mg/ml. The abnormalities resolved within 3 weeks of birth, despite continued digoxin therapy.

The authors attributed the electrocardiographic abnormalities to maternal use of flecainide.

Susceptibility Factors

Genetic

Among 40 patients taking flecainide or propafenone adverse effects were more frequent in poor CYP2D6 metabolizers (21%) than in extensive metabolizers (4.8%) (63). Only the poor metabolizers had adverse effects that required drug withdrawal.

Age

Reports of the use of flecainide in children suggest that the risk of adverse effects is low, although these studies have been very small (64–66).

Renal disease

Flecainide is cleared partly by dose-dependent hydroxylation and partly unchanged via the kidneys. Therefore, severe renal impairment and hepatic impairment both cause a reduction in its rate of clearance (67,68). Dosages should be reduced in these circumstances.

Hepatic disease

Flecainide is cleared partly by dose-dependent hydroxylation and partly unchanged via the kidneys. Therefore, severe renal impairment and hepatic impairment both cause a reduction in its rate of clearance (67,68). Dosages should be reduced in these circumstances.

Other features of the patient

Flecainide half-life is prolonged in poor hydroxylators (69) and poor metabolizers may be at an increased risk of adverse effects.

Drug Administration

Drug overdose

There have been a few reports of the effects of overdosage of flecainide (70–75). Various treatments have been used in these circumstances; none is specific.

- A 20-year-old woman took 3–4 g of flecainide and developed circulatory failure unresponsive to pacing, inotropic drugs, and sodium bicarbonate (76). She was then successfully treated with cardiopulmonary bypass for 30 hours. At peak, the plasma flecainide concentration was in excess of 4000 μg/ml, and although this fell during bypass to below 3.5 μg/ml, clinical recovery preceded this fall. Other complications included a coagulopathy with intravascular hemolysis, requiring the use of blood products, and renal insufficiency requiring hemodiafiltration.

The authors proposed that extracorporeal circulatory support had allowed increased perfusion of the liver and therefore more effective metabolism of the flecainide.

- Death has been reported in two patients who took flecainide (77). The postmortem femoral blood flecainide concentrations were 5.4 and 1.2 μg/ml (target range 0.2–1.0).
- Fatal flecainide overdose has been reported in a 65-year-old man who probably took 20 tablets of 100 mg each (78). However, no clinical details were available, because he was found dead. Flecainide was detected in his blood, gastric contents, and liver. *O*-dealkylated flecainide was found in his urine.

Of other fatal cases reviewed in the paper, in only one was there a slightly lower blood concentration of flecainide (7.3 compared with 7.7 mg/kg). The authors proposed that pre-existing cardiac damage could have predisposed this man to a dysrhythmic death.

- A 15-year-old girl was found dead in bed a few days a suicide attempt with benzodiazepines (79). At autopsy, no specific cause of death was identified. However, toxicological analysis of post-mortem blood and urine showed the presence of high concentrations of flecainide and its two major metabolites. The flecainide concentrations in blood and urine were 19 and 28 μg/ml respectively, and the metabolites were detected only in urine at the following concentrations: meta-O-dealkylated flecainide 9.4 μg/ml and meta-O-dealkylated flecainide lactam 8.6 μg/ml.

Drug–Drug Interactions

Amiodarone

The combination of flecainide with amiodarone can result in reduced conduction, predisposing to bundle branch block and dysrhythmias (80,81).

Beta-adrenoceptor antagonists

The combination of flecainide with propranolol results in additive hypotensive and negative inotropic effects (82).

Lipophilic beta-adrenoceptor antagonists are metabolized to varying degrees by oxidation by liver microsomal cytochrome P450 (for example propranolol by CYP1A2 and CYP2D6 and metoprolol by CYP2D6). They can therefore reduce the clearance and increase the steady-state plasma concentrations of other drugs that undergo similar metabolism, potentiating their effects. Drugs that interact in this way include flecainide (83).

Digoxin

Despite an early report that flecainide might alter the pharmacokinetics of digoxin, this action is minimal and probably of no clinical significance (83). The combination causes a significant increase in the PR interval; but the clinical significance of this is unclear, it may be important for patients with impaired sinus node function (84).

Dofetilide

In a non-randomized comparison of flecainide (2 mg/kg; $n = 11$) and dofetilide (8 micrograms/kg; $n = 10$) in patients with atrial flutter, only one patient given flecainide converted to sinus rhythm compared with seven of the 10 patients given dofetilide (85).

Quinidine

Flecainide is metabolized by CYP2D6, and is subject to polymorphic metabolism. In extensive metabolizers its clearance is reduced by quinine (86). Quinidine reduces the clearance of *R*-flecainide but not that of *S*-flecainide (87).

Quinine

Flecainide is metabolized by CYP2D6, and is subject to polymorphic metabolism. In extensive metabolizers its clearance is reduced by quinine (88).

Verapamil

Although verapamil reduces the clearance of flecainide, this is probably not clinically important (89). However, there is also a pharmacodynamic interaction, since both drugs increase the PR interval and have additive effects on myocardial contractility and atrioventricular conduction (90).

Monitoring therapy

The usual target range for plasma flecainide concentrations is 200–1000 ng/ml (91). The relation between plasma flecainide concentration and its effect on atrial fibrillatory rate has been studied in 10 patients during acute and maintenance therapy of persistent lone atrial fibrillation (92). Flecainide was given as a single oral bolus of 300 mg followed by 200–400 mg/day for 5 days. The initial bolus resulted in plasma concentrations of 288–629 ng/ml; day 5 plasma concentrations were lower in those taking a dosage of 200 mg/day than in those taking higher doses (mean 508 versus 974 ng/ml). The fibrillatory rate was reduced after the initial bolus but remained stable thereafter and was independent of flecainide plasma concentration. There was significant prolongation of the QRS duration but no change in RR or QT_c interval and no correlation between plasma flecainide concentration and QRS duration. Presumably plasma flecainide concentrations do not well reflect concentrations in the heart.

In a double-blind, randomized, placebo-controlled study of the effects of flecainide in six men with congenital long QT syndrome with the DeltaKPQ gene deletion, the lowest dose of flecainide associated with at least a 40 ms reduction in the QT_c interval was determined in an initial open, dose-ranging investigation using one-quarter or half of the recommended maximal antidysrhythmic dose; QT_c interval reduction was achieved with a dose of 1.5 mg/kg/day in four subjects and with 3.0 mg/kg/day in two (93). They were then randomized to four 6-month alternating periods of flecainide and placebo. Average QT_c intervals during placebo and flecainide were 534 ms and 503 ms respectively, with an adjusted reduction in QT_c interval of –27 ms (95% CI = –37, –17 at a mean flecainide blood concentration of 110 ng/ml.

References

1. International Symposium on Supraventricular Arrhythmias: Focus on Flecainide. October 23–26, 1987, Paradise Island, Nassau, Bahamas. Proceedings. Am J Cardiol 1988;62(6):D1–D67.
2. Schneeweiss A. New antiarrhythmic drugs. II Flecainide. Pediatr Cardiol 1990;11(3):143–6.
3. Falk RH, Fogel RI. Flecainide. J Cardiovasc Electrophysiol 1994;5(11):964–81.
4. Schumacher B, Jung W, Lewalter T, Vahlhaus C, Wolpert C, Luderitz B. Radiofrequency ablation of atrial flutter due to administration of class IC antiarrhythmic drugs for atrial fibrillation. Am J Cardiol 1999;83(5):710–3.
5. Alboni P, Menozzi C. Episodic drug therapy for paroxysmal supraventricular tachycardia. Cardiol Rev 2002;19:44–6.
6. Khan IA. Oral loading single dose flecainide for pharmacological cardioversion of recent-onset atrial fibrillation. Int J Cardiol 2003;87:121–28.
7. Wehling M. Meta-analysis of flecainide safety in patients with supraventricular arrhythmias. Arzneimittelforschung 2002;52(7):507–14.
8. Hohnloser SH, Zabel M. Short- and long-term efficacy and safety of flecainide acetate for supraventricular arrhythmias. Am J Cardiol 1992;70(5):A3–A10.
9. Schulze JJ, Inhester B. Arrhythmiebehandlung unter Praxisbedingungen. Therapiewoche 1985;35:5898.
10. The Cardiac Arrhythmia Suppression Trial (CAST) Investigators. Preliminary report: effect of encainide and flecainide on mortality in a randomized trial of arrhythmia suppression after myocardial infarction. N Engl J Med 1989;321(6):406–12.
11. Echt DS, Liebson PR, Mitchell LB, Peters RW, Obias-Manno D, Barker AH, Arensberg D, Baker A, Friedman L, Greene HL, Huther ML, Richardson DW. Mortality and morbidity in patients receiving encainide, flecainide, or placebo. The Cardiac Arrhythmia Suppression Trial. N Engl J Med 1991;324(12):781–8.
12. Gottlieb SS. The use of antiarrhythmic agents in heart failure: implications of CAST. Am Heart J 1989;118(5 Pt 1):1074–7.
13. Podrid PJ, Marcus FI. Lessons to be learned from the Cardiac Arrhythmia Suppression Trial. Am J Cardiol 1989;64(18):1189–91.
14. Bigger JT Jr. The events surrounding the removal of encainide and flecainide from the Cardiac Arrhythmia Suppression Trial (CAST) and why CAST is continuing with moricizine. J Am Coll Cardiol 1990;15(1):243–5.
15. Akhtar M, Breithardt G, Camm AJ, Coumel P, Janse MJ, Lazzara R, Myerburg RJ, Schwartz PJ, Waldo AL, Wellens HJ, et al. CAST and beyond. Implications of the Cardiac Arrhythmia Suppression Trial. Task Force of the Working Group on Arrhythmias of the European Society of Cardiology. Circulation 1990;81(3):1123–7.
16. Thomis JA. Encainide—an updated safety profile. Cardiovasc Drugs Ther 1990;4(Suppl 3):585–94.
17. Anderson JL, Jolivette DM, Fredell PA. Summary of efficacy and safety of flecainide for supraventricular arrhythmias. Am J Cardiol 1988;62(6):D62–6.
18. Morganroth J, Horowitz LN. Flecainide: its proarrhythmic effect and expected changes on the surface electrocardiogram. Am J Cardiol 1984;53(5):B89–94.
19. Nathan AW, Hellestrand KJ, Bexton RS, Spurrell RA, Camm AJ. The proarrhythmic effects of flecainide. Drugs 1985;29(Suppl 4):45–53.
20. Podrid PJ, Morganroth J. Aggravation of arrhythmia during drug therapy: experience with flecainide acetate. Pract Cardiol 1985;11:55–70.
21. Wehr M, Noll B, Krappe J. Flecainide-induced aggravation of ventricular arrhythmias. Am J Cardiol 1985;55(13 Pt 1):1643–4.
22. Chalvidan T, Cellarier G, Deharo JC, Colin R, Savon N, Barra N, Peyre JP, Djiane P. His–Purkinje system reentry as a proarrhythmic effect of flecainide. Pacing Clin Electrophysiol 2000;23(4 Pt 1):530–3.
23. Mackstaller LL, Marcus FI. Rapid ventricular response due to treatment of atrial flutter or fibrillation with Class I antiarrhythmic drugs. Ann Noninvasive Electrocardiol 2000;5:101–4.
24. Kawabata M, Hirao K, Horikawa T, Suzuki K, Motokawa K, Suzuki F, Azegami K, Hiejima K. Syncope in patients with atrial flutter during treatment with class Ic antiarrhythmic drugs. J Electrocardiol 2001;34(1):65–72.
25. Turner N, Thwaites BC. Exercise induced widening of the QRS complex in a patient on flecainide. Heart 2001;85(4):423.
26. Thevenin J, Da Costa A, Roche F, Romeyer C, Messier M, Isaaz K. Flecainide induced ventricular tachycardia (torsades de pointes). PACE - Pacing Clin Electrophysiol 2003;26:1907–8.
27. Ackland F, Singh R, Thayyil S. Flecainide induced ventricular fibrillation in a neonate. Heart 2003;89:1261.
28. Perry JC, Garson A Jr. Flecainide acetate for treatment of tachyarrhythmias in children: review of world literature on efficacy, safety, and dosing. Am Heart J 1992;124:1614–21.

29. Gorog DA, Russell G, Casian A, Peters NS. A cautionary tale. The risks of flecainide treatment for myotonic dystrophy. J Clin Neuromuscular Dis 2005;7:25–8.
30. Ohki R, Takahashi M, Mizuno O, Fujikawa H, Mitsuhashi T, Katsuki T, Ikeda U, Shimada K. Torsades de pointes ventricular tachycardia induced by mosapride and flecainide in the presence of hypokalemia. Pacing Clin Electrophysiol 2001;24(1):119–21.
31. Nabar A, Rodriguez LM, Timmermans C, van Mechelen R, Wellens HJ. Class IC antiarrhythmic drug induced atrial flutter: electrocardiographic and electrophysiological findings and their importance for long term outcome after right atrial isthmus ablation. Heart 2001;85(4):424–9.
32. Hudson CJ, Whitner TE, Rinaldi MJ, Littmann L. Brugada electrocardiographic pattern elicited by inadvertent flecainide overdose. Pacing Clin Electrophysiol 2004;27(9):1311–3.
33. Boriani G, Biffi M, Capucci A, Bronzetti G, Ayers GM, Zannoli R, Branzi A, Magnani B. Favorable effects of flecainide in transvenous internal cardioversion of atrial fibrillation. J Am Coll Cardiol 1999;33(2):333–41.
34. Walker PR, Papouchado M, James MA, Clarke LM. Pacing failure due to flecainide acetate. Pacing Clin Electrophysiol 1985;8(6):900–2.
35. Antonelli D, Freedberg NA, Rosenfeld T. Acute loss of capture due to flecainide acetate. Pacing Clin Electrophysiol 2001;24(7):1170.
36. Nakamura W, Segawa K, Ito H, Tanaka S, Yoshimoto N. Class IC antiarrhythmic drugs, flecainide and pilsicainide, produce ST segment elevation simulating inferior myocardial ischemia. J Cardiovasc Electrophysiol 1998;9(8):855–8.
37. Chandrasekaran B, Kurbaan AS. Brugada syndrome: a review. Br J Cardiol 2002;9:406–10.
38. Priori SG, Napolitano C, Terrence L, et al. Incomplete penetrance and variable response to sodium channel blockade in Brugada's syndrome. Eur Heart J 1999;20(Suppl):465A.
39. Brugada R, Brugada J, Antzelevitch C, Kirsch GE, Potenza D, Towbin JA, Brugada P. Sodium channel blockers identify risk for sudden death in patients with ST-segment elevation and right bundle branch block but structurally normal hearts. Circulation 2000;101(5):510–5.
40. Priori SG, Napolitano C, Schwartz PJ, Bloise R, Crotti L, Ronchetti E. The elusive link between LQT3 and Brugada syndrome: the role of flecainide challenge. Circulation 2000;102(9):945–7.
41. Saishu T, Iwatsuki N, Tajima T, Hashimoto Y. Flecainide is effective against premature supraventricular and ventricular contractions during general anesthesia. J Anesth 1994;8:284–7.
42. Robain A, Perchet H, Fuhrman C. Flecainide-associated pneumonitis with acute respiratory failure in a patient with the LEOPARD syndrome. Acta Cardiol 2000;55(1):45–7.
43. Pesenti S, Lauque D, Daste G, Boulay V, Pujazon MC, Carles P. Diffuse infiltrative lung disease associated with flecainide. Report of two cases. Respiration 2002;69(2):182–5.
44. Gentzkow GD, Sullivan JY. Extracardiac adverse effects of flecainide. Am J Cardiol 1984;53(5):B101–5.
45. Epstein M, Jardine RM, Obel IW. Flecainide acetate in the treatment of resistant supraventricular arrhythmias. S Afr Med J 1988;74(11):559–62.
46. Ramhamadany E, Mackenzie S, Ramsdale DR. Dysarthria and visual hallucinations due to flecainide toxicity. Postgrad Med J 1986;62(723):61–2.
47. Malesker MA, Sojka SG, Fagan NL. Flecainide-induced neuropathy. Ann Pharmacother 2005;39(9):1580.
48. Moller HU, Thygesen K, Kruit PJ. Corneal deposits associated with flecainide. BMJ 1991;302(6775):506–7.
49. Ikaheimo K, Kettunen R, Mantyjarvi M. Adverse ocular effects of flecainide. Acta Ophthalmol Scand 2001;79(2):175–6.
50. Samlowski WE, Frame RN, Logue GL. Flecanide-induced immune neutropenia. Documentation of a hapten-mediated mechanism of cell destruction. Arch Intern Med 1987;147(2):383–4.
51. Kuhlkamp V, Haasis R, Seipel L. Flecainidinduzierte Hepatitis. [Flecainide-induced hepatitis.] Z Kardiol 1988;77(10):678–80.
52. Mikloweit P, Bienmuller H. Medikamentös induzierte intrahepatische Cholestase durch Flecainidacetat und Enalapril. [Drug-induced intrahepatic cholestasis caused by flecainide acetate and enalapril.] Internist (Berl) 1987;28(3):193–5.
53. Ziegelbaum M, Lever H. Acute urinary retention associated with flecainide. Cleve Clin J Med 1990;57(1):86–7.
54. Mancuso G, Tampieri E, Berdondini RM. Eruzione psoriasiforme da flecainide. [Psoriasis-like eruption caused by flecainide.] G Ital Dermatol Venereol 1988;123(4):171–2.
55. Penhall RK, Hong CY, Muhiddin KA. The effect of flecainide on human sperm motility. Br J Clin Pharmacol 1982;14:147P.
56. Zehender M, Treese N, Kasper W, Pop T, Meinertz T. Effectiveness and tolerance in long-term treatment with flecainide. Circulation 1982;66(Suppl II):144.
57. Allan LD, Chita SK, Sharland GK, Maxwell D, Priestley K. Flecainide in the treatment of fetal tachycardias. Br Heart J 1991;65(1):46–8.
58. Edwards A, Peek MJ, Curren J. Transplacental flecainide therapy for fetal supraventricular tachycardia in a twin pregnancy. Aust NZ J Obstet Gynaecol 1999;39(1):110–2.
59. D'Souza D, MacKenzie WE, Martin WL. Transplacental flecainide therapy in the treatment of fetal supraventricular tachycardia. J Obstet Gynaecol 2002;22(3):320–2.
60. Rasheed A, Simpson J, Rosenthal E. Neonatal ECG changes caused by supratherapeutic flecainide following treatment for fetal supraventricular tachycardia. Heart 2003;89:470.
61. Hall CM, Platt MPW. Neonatal flecainide toxicity following supraventricular tachycardia treatment. Ann Pharmacother 2003;37:1343–4.
62. Trotter A, Kaestner M, Pohlandt F, Lang D. Unusual electrocardiogram findings in a preterm infant after fetal tachycardia with hydrops fetalis treated with flecainide. Pediatr Cardiol 2000;21(3):259–62.
63. Martínez-Sellés M, Castillo I, Montenegro P, Martín ML, Almendral J, Sanjurjo M. Estudio farmacogenetico de la respuesta a flecainida y propafenona en pacientes con fibrilacion auricular. [Pharmacogenetic study of the response to flecainide and propafenone in patients with atrial fibrillation.] Rev Esp Cardiol 2005;58(6):745–8.
64. Musto B, D'Onofrio A, Cavallaro C, Musto A, Greco R. Electrophysiologic effects and clinical efficacy of flecainide in children with recurrent paroxysmal supraventricular tachycardia. Am J Cardiol 1988;62(4):229–33.
65. Priestley KA, Ladusans EJ, Rosenthal E, Holt DW, Tynan MJ, Jones OD, Curry PV. Experience with flecainide for the treatment of cardiac arrhythmias in children. Eur Heart J 1988;9(12):1284–90.
66. Perry JC, McQuinn RL, Smith RT Jr, Gothing C, Fredell P, Garson A Jr. Flecainide acetate for resistant arrhythmias in

the young: efficacy and pharmacokinetics. J Am Coll Cardiol 1989;14(1):185–91.
67. Williams AJ, McQuinn RL, Walls J. Pharmacokinetics of flecainide acetate in patients with severe renal impairment. Clin Pharmacol Ther 1988;43(4):449–55.
68. McQuinn RL, Pentikainen PJ, Chang SF, Conard GJ. Pharmacokinetics of flecainide in patients with cirrhosis of the liver. Clin Pharmacol Ther 1988;44(5):566–72.
69. Beckmann J, Hertrampf R, Gundert-Remy U, Mikus G, Gross AS, Eichelbaum M. Is there a genetic factor in flecainide toxicity? BMJ 1988;297(6659):1316.
70. Rodin SM, Johnson BF, Wilson J, Ritchie P, Johnson J. Comparative effects of verapamil and isradipine on steady-state digoxin kinetics. Clin Pharmacol Ther 1988;43(6):668–72.
71. Kirch W, Logemann C, Heidemann H, Santos SR, Ohnhaus EE. Nitrendipine/digoxin interaction. J Cardiovasc Pharmacol 1987;10(Suppl 10):S74–5.
72. Dunselman PH, Scaf AH, Kuntze CE, Lie KI, Wesseling H. Digoxin–felodipine interaction in patients with congestive heart failure. Eur J Clin Pharmacol 1988;35(5):461–5.
73. Bruserud O, Skadberg BT, Ohm OJ. Combined intoxication with digitoxin and verapamil. The possible inhibition of sensitisation to digitalis-specific antiserum by toxic drug concentrations. J Clin Lab Immunol 1988;25(4):167–71.
74. Ferrari E, Fournier JP, Gibelin P, Drici MD, Morand P. Le traitement par le lactate molaire de l'intoxication par le flécainide est-il sans danger?. [Is treatment with molar lactate in flecainide poisoning safe?.] Presse Méd 1989;18(28):1395.
75. Yang XS, Sun JP, Zhi GN. Acute flecainide toxicity. Chin Med J (Engl) 1990;103(7):606–7.
76. Corkeron MA, van Heerden PV, Newman SM, Dusci L. Extracorporeal circulatory support in near-fatal flecainide overdose. Anaesth Intensive Care 1999;27(4):405–8.
77. Lynch MJ, Gerostamoulos J. Flecainide toxicity: cause and contribution to death. Leg Med (Tokyo) 2001;3(4):233–6.
78. Romain N, Giroud C, Michaud K, Augsburger M, Mangin P. Fatal flecainide intoxication. Forensic Sci Int 1999;106(2):115–23.
79. Benijts T, Borreyl D, Lambert WE, De Letter EA, Piette MHA, Van Peteghem C, De Leenheer AP. Analysis of flecainide and two metabolites in biological specimens by HPLC: Application to a fatal intoxication. J Anal Toxicol 2003;27:47–52.
80. Chouty F, Coumel P. Oral flecainide for prophylaxis of paroxysmal atrial fibrillation. Am J Cardiol 1988;62(6):D35–7.
81. Saoudi N, Galtier M, Hidden F, Gerber L, Letac B. Bundle-branch reentrant ventricular tachycardia: a possible mechanism of flecainide proarrhythmic effect. J Electrophysiol 1988;2:365–71.
82. Almeyda J, Levantine A. Cutaneous reactions to cardiovascular drugs. Br J Dermatol 1973;88(3):313–9.
83. Lewis GP, Holtzman JL. Interaction of flecainide with digoxin and propranolol. Am J Cardiol 1984;53(5):B52–7.
84. Hellestrand KJ, Nathan AW, Bexton RS, Camm AJ. Response of an abnormal sinus node to intravenous flecainide acetate. Pacing Clin Electrophysiol 1984;7(3 Pt 1):436–9.
85. Crijns HJ, Van Gelder IC, Kingma JH, Dunselman PH, Gosselink AT, Lie KI. Atrial flutter can be terminated by a class III antiarrhythmic drug but not by a class IC drug. Eur Heart J 1994;15(10):1403–8.
86. Munafo A, Reymond-Michel G, Biollaz J. Altered flecainide disposition in healthy volunteers taking quinine. Eur J Clin Pharmacol 1990;38(3):269–73.
87. Birgersdotter UM, Wong W, Turgeon J, Roden DM. Stereoselective genetically-determined interaction between chronic flecainide and quinidine in patients with arrhythmias. Br J Clin Pharmacol 1992;33(3):275–80.
88. Munafo A, Reymond-Michel G, Biollaz J. Altered flecainide disposition in healthy volunteers taking quinine. Eur J Clin Pharmacol 1990;38(3):269–73.
89. Holtzman JL, Finley D, Mottonen L, Berry DA, Ekholm BP, Kvam DC, McQuinn RL, Miller AM. The pharmacodynamic and pharmacokinetic interaction between single doses of flecainide acetate and verapamil: effects on cardiac function and drug clearance. Clin Pharmacol Ther 1989;46(1):26–32.
90. Buss J, Lasserre JJ, Heene DL. Asystole and cardiogenic shock due to combined treatment with verapamil and flecainide. Lancet 1992;340(8818):546.
91. Breindahl T. Therapeutic drug monitoring of flecainide in serum using high-performance liquid chromatography and electrospray mass spectrometry. J Chromatogr B Biomed Sci Appl 2000;746(2):249–54.
92. Husser D, Binias KH, Stridh M, Sornmo L, Olsson SB, Molling J, Geller C, Klein HU, Bollmann A. Pilot study: noninvasive monitoring of oral flecainide's effects on atrial electrophysiology during persistent human atrial fibrillation using the surface electrocardiogram. Ann Noninvasive Electrocardiol 2005;10(2):206–10.
93. Moss AJ, Windle JR, Hall WJ, Zareba W, Robinson JL, McNitt S, Severski P, Rosero S, Daubert JP, Qi M, Cieciorka M, Manalan AS. Safety and efficacy of flecainide in subjects with long QT-3 syndrome (DeltaKPQ mutation): a randomized, double-blind, placebo-controlled clinical trial. Ann Noninvasive Electrocardiol 2005;10(4 Suppl):59–66.

Lidocaine (Lignocaine)

General Information

Lidocaine is the most widely used aminoamide local anesthetic agent, with a low toxic potential; its effects are mostly typical for this class of drug. It can be given by injection or topically and is also combined with prilocaine in Emla for topical administration. It is also used as an antidysrhythmic drug and has occasionally been used in other conditions, such as multiple sclerosis, chronic daily headache, migraine and cluster headaches, and neuropathic pain, such as postherpetic neuralgia.

Local anesthetic gels and creams used liberally on traumatized epithelium can be rapidly absorbed, resulting in systemic effects, such as convulsions, particularly if excessive quantities are used. This has been highlighted in the case of a 40-year-old woman who developed seizures after lidocaine gel 40 ml was injected into the ureter during an attempt to remove a stone (1). Site of administration is also important, as local conditions, particularly vascularity, affect the rate of absorption. Adverse effects of lidocaine when it is used as a local anesthetic can also occur after inadvertent intravascular injection.

The incidence of adverse effects to lidocaine in antidysrhythmic dosages is low. In one series of 750 patients given lidocaine intravenously for cardiac dysrhythmias, adverse reactions occurred in only 47 (6.3%) and were thought to have been life-threatening in 12 (1.6%) (2). However, the risk of adverse effects is dose-related and increases at intravenous infusion rates of around 3 mg/minute (3). Most of the adverse effects are on the cardiovascular and central nervous systems. Nervous system toxicity is directly related to blood concentrations, with symptoms that include light-headedness, headache, dizziness, tremor, confusion, tinnitus, dysarthria, paresthesia, alterations in the level of consciousness from drowsiness to coma, respiratory depression, and convulsions. Cardiovascular effects, including dysrhythmias and very rarely worsening of cardiac function, only occur at very high blood concentrations. The intravenous dose of lidocaine required to produce cardiovascular collapse is seven times that which causes seizures. Risks of serious systemic effects do not increase with age. Deaths have occurred with voluntary intoxication, primarily because of the cardiac effects.

The active metabolites of lidocaine, glycinexylidide and monoethylglycinexylidide, are toxic and intravenous infusion should not continue for more than 24–48 hours.

Hypersensitivity reactions are rare, and not all reports are clear, but cases do occur and are usually mild (SED-12, 255) (4). Some patients are highly sensitive to lidocaine, yet insensitive to other aminoamide local anesthetics (5), and the reverse has also been found (SEDA-14, 109). True anaphylaxis with rechallenge has been documented (6). A few cases of contact dermatitis have been reported.

Even topical administration of lidocaine continues to generate reports with tragic outcomes, as absorption from mucosal surfaces is underestimated.

- A patient due to have a bronchoscopy was given an overdose of lidocaine to anesthetize the airway by an inexperienced health worker. He was then left unobserved and subsequently developed convulsions and cardiopulmonary arrest (7). He survived with severe cerebral damage.

His lidocaine concentration was 24 μg/ml about 1 hour after initial administration (a blood concentration over 6 μg/ml is considered to be toxic).

Observational studies

Lidocaine has been used to treat some of the symptoms of multiple sclerosis in 30 patients with painful tonic seizures, attacks of neuralgia, paroxysmal itching, and Lhermitte's sign (8). Lidocaine was given by intravenous infusion for 5.5 hours in a maintenance dose of 2.0–2.8 mg/kg/hour after a loading dose, and the mean steady-state concentration was 2.4 μg/ml. Lidocaine almost completely abolished the paroxysmal symptoms and markedly alleviated the persistent symptoms of multiple sclerosis. Adverse effects were not specifically mentioned, but in one case, when the plasma concentration of lidocaine rose above 3.5 μg/ml, weakness of the left leg became marked and was associated with an extensor plantar response; this disappeared when the lidocaine was replaced by saline single-blind, but subsequently the positive symptoms recurred.

Intravenous lidocaine has been used to treat severe chronic daily headache in 19 patients (three men, median age 37 years) (9). There were adverse effects during four infusions of lidocaine: hyperkalemia (6.4 mmol/l), which did not resolve after withdrawal of lidocaine; transient hypotension (75/50 mmHg), which was attributed to concomitant droperidol; an unspecified abnormality of cardiac rhythm and on another occasion a transient bradycardia; and chest pain with a normal electrocardiogram, fever, and intractable nausea. The study was neither randomized nor placebo-controlled, and in no case was the adverse event strongly associated with the administration of lidocaine.

Placebo-controlled studies

In a double-blind, placebo-controlled study of the use of intravenous lidocaine for neuropathic pain, 16 patients were given 5 ml/kg intravenously over 30 minutes (10). Lidocaine was better than placebo in relieving pain. The major adverse effect was light-headedness, which occurred in seven patients given lidocaine and none given saline. Other adverse effects included somnolence, nausea and vomiting, dysarthria or garbled speech, blurred vision, and malaise. In two patients the rate of infusion had to be reduced because of adverse effects.

Organs and Systems

Cardiovascular

Lidocaine can cause dysrhythmias and hypotension. The dysrhythmias that have been reported include sinus bradycardia, supraventricular tachycardia (11), and rarely torsade de pointes (12). There have also been rare reports of cardiac arrest (2) and worsening heart failure (13). Lidocaine can also cause an increased risk of asystole after repeated attempts at defibrillation (14). Lidocaine may increase mortality after acute myocardial infarction, and it should be used only in patients with specific so-called warning dysrhythmias (that is frequent or multifocal ventricular extra beats, or salvos) (15).

Sinus bradycardia has been seen after a bolus injection of 50 mg, atrioventricular block after a dose of 800 mg given over 12 hours, and left bundle branch block after a mere subconjunctival injection of 2% lidocaine.

- High-grade atrioventricular block has been reported in a 14-day-old infant who was given lidocaine 2 mg/kg intravenously (SED-12, 255) (16).

A death due to ventricular fibrillation after 50 mg and another due to sinus arrest after 100 mg have been reported (SED-12, 255) (17). Two cases of ventricular fibrillation and cardiopulmonary arrest occurred after local infiltration of lidocaine for cardiac catheterization (SEDA-21, 136).

Lidocaine does not usually cause conduction disturbances, but two cases have been reported in the presence of hyperkalemia (18).

- A 57-year-old man with a wide-complex tachycardia was given lidocaine 100 mg intravenously and immediately became asystolic. Resuscitation was unsuccessful.
- A 31-year-old woman had a cardiac arrest and was resuscitated to a wide-complex tachycardia, which was treated with intravenous lidocaine 100 mg. She immediately became asystolic but responded to calcium chloride.

In both cases there was severe hyperkalemia, and the authors suggested that hyperkalemia-induced resting membrane depolarization had increased the number of inactivated sodium channels, thus increasing the binding of lidocaine and potentiating its effects.

The degree of hypotension occurring after epidural anesthesia with alkalinized lidocaine (with adrenaline) was greater than with a standard commercial solution (SED-12, 255) (19).

In 23 patients there was a significant dose-dependent reduction in blood pressure following submucosal infiltration of lidocaine plus adrenaline compared with saline plus adrenaline for orthognathic surgery (20). The study was randomized but small; larger studies are needed to confirm effects that could easily have been due to multifactorial causes in patients undergoing general anesthesia.

Respiratory

Topical anesthesia of the airways is commonly used to facilitate endoscopy and sometimes manipulation of the airways. This can result in an increase in airway flow resistance, possibly due to laryngeal dysfunction (21). Lidocaine spray 10%, used for upper airways anesthesia for fiberoptic intubation in a grossly obese patient, caused acute airway obstruction. The patient went on to have a percutaneous tracheotomy, and it was postulated that the local anesthetic had abolished laryngeal receptors responsible for airway maintenance, or that laryngospasm and reduced muscle tone due to the lidocaine might have been the cause (SEDA-22, 140).

Life-threatening bronchospasm can occur after either spinal or topical use of lidocaine. In one series of patients being treated with lidocaine spray 40 mg for persistent cough, there was an increase of airway resistance (SED-12, 255; (22).

Ear, nose, throat

Local anesthesia to the larynx, for example with 4% lidocaine, is generally safe. Laryngeal edema has been reported in a few cases and could be due to the propellant rather than to lidocaine itself (23).

Intranasal 4% lidocaine has been used for migraine and cluster headaches with success and few serious adverse effects: a bitter taste was common and some patients complained of nasal burning and oropharyngeal numbness (SEDA-20, 127).

Lidocaine gel is not recommended for lubrication of laryngeal masks. It confers no benefits and increases the incidence of adverse effects such as intraoperative hiccups, postoperative hoarseness, nausea, vomiting, and tongue paresthesia (24).

Nervous system

Nervous system toxicity is most often seen with rapid intravenous infusion (3,25,26). The effects include headache, dizziness, tremor, confusion, tinnitus, dysarthria, paresthesia, respiratory depression, altered level of consciousness (from drowsiness to coma), and convulsions.

Two cases have illustrated the effects of lidocaine in precipitating partial seizures in patients with a previous history of epilepsy (27).

- A 36-year-old woman developed chest pain and ventricular tachycardia. She had a 14-year history ofsided focal motor seizures controlled with phenytoin. After receiving intravenous lidocaine 100 mg to treat the dysrhythmias, she developed a typical seizure involving the right side of her face and arm. She was given a loading dose of phenytoin and the seizure abated. However, the ventricular tachycardia persisted and was treated with additional lidocaine 50 mg followed by an infusion of 3.3 mg/minute; 6 hours later she had a generalized seizure with a venous blood lidocaine concentration of 21 μg/ml. The infusion was stopped and the seizure was treated with intravenous diazepam 10 mg.
- A 41-year-old woman with a long-standing history of focal and secondarily generalized seizures controlled with carbamazepine underwent cerebral arteriography, during which she was inadvertently given lidocaine 20 mg via an intra-arterial catheter in the right internal carotid artery; within 20 seconds she had a focal seizure.

These two patients had their typical partial seizures triggered by high doses of lidocaine. In both cases the serum concentrations of their usual anticonvulsants were initially low. The first patient received a loading dose of phenytoin after the partial seizure, was then given a second bolus of lidocaine and an infusion, and then had a second seizure, which was generalized. There was no evidence that this second seizure evolved from the left seizure focus. The authors concluded that lidocaine can activate seizure foci in patients with a history of partial seizures and that this may be more likely if the serum concentrations of anticonvulsants are low. However, therapeutic concentrations of antiepileptic drugs may not prevent generalized seizures that result from the widespread lowering of seizure threshold caused by high concentrations of lidocaine.

A tonic-clonic seizure occurred after the application of 400 mg of lidocaine jelly to traumatized ureteric mucosa (SEDA-22, 142).

- A 54-year-old woman who was given lidocaine, 200 mg intravenously, for ventricular fibrillation during cardiopulmonary bypass, had a tonic-clonic seizure (28). The seizure occurred immediately after the administration of lidocaine and was relieved by the intravenous administration of thiopental and midazolam. Her ventricular fibrillation responded to procainamide 1 g intravenously over 10 minutes.

The pharmacokinetics of lidocaine are altered by cardiopulmonary bypass, because of hemodilution, changed protein binding, the exclusion of the lungs as an organ for first-pass elimination, altered acid-base balance, and sometimes drug interactions. In particular, reduced protein binding may have contributed in this case to the risk of seizure, but plasma lidocaine concentrations were not measured.

- A 30-year-old woman received two 5 g applications of 40% lidocaine cream with occlusion by plastic wrap during and after laser therapy to areas of her skin (29). She developed dizziness and headache postoperatively, followed 45 minutes later by light-headedness, increasing dizziness, and confusion. The dressings were removed. The lidocaine concentration was 2.7 μg/ml 7 hours later.

It is recommended that repeat applications of lidocaine, especially in high-concentration formulations, be avoided and the area of application limited.

- A 16-year-old woman had had an adverse reaction after administration of an unknown local anesthetic agent for a dental procedure. Patch testing had elicited similar symptoms with lidocaine only, and 20 minutes after subcutaneous lidocaine 0.05 mg she developed perioral paresthesia, nausea, vomiting, vertigo, dizziness, mild agitation, drowsiness, and euphoria. Hemodynamic parameters remained stable but her symptoms were thought to be part of a genuine non-allergic, neuropsychiatric reaction, as the patch testing was double-blind and placebo-controlled (30).

Transient and permanent nerve damage can occur after regional anesthesia, particularly neuraxial anesthesia. The mechanism of this nerve damage is unclear. Some studies have shown an indirect effect. However, in crayfish giant axon, lidocaine had a dose- and time-dependent effect on isolated nerve function in vitro (31). At high concentrations lidocaine caused irreversible conduction block and total loss of resting membrane potential. These results in an isolated nerve suggest a direct neurotoxic effect of lidocaine.

Sensory systems

Tinnitus and visual disturbances are early components of a systemic toxic reaction to lidocaine.

Eyes

A potentially beneficial effect of lidocaine has been studied in a randomized, double-blind, placebo-controlled study of the effects of preinstillation of lidocaine on tropicamide-induced mydriasis (32). Pupillary diameter was significantly increased by the instillation of lidocaine before tropicamide. It was thought that lidocaine can enhance intraocular penetration and hence potentiate the effect of tropicamide.

Double vision and difficulty in focusing have been attributed to lidocaine applied to the tongue (33).

- A 22-year-old man developed double vision and difficulty in focusing after using 2% viscous lidocaine for a painful tongue ulcer. He used viscous 2% lidocaine 10 ml hourly and developed symptoms when the daily dose exceeded 240 ml (4800 mg of lidocaine hydrochloride) after 10 days of use. At that time his serum lidocaine concentration was 6.7 μg/ml. His symptoms persisted when the serum concentration of lidocaine fell to below toxic concentrations, implying that metabolites of lidocaine had contributed.

Temporary blindness, an unusual feature of lidocaine toxicity, has been reported in an otherwise healthy young woman (34).

- A 21-year-old 50 kg woman, previously fit, was to have an open reduction and fixation of a fractured proximal phalanx with intravenous regional anesthesia. As a result of misreading the vial label, 30 ml (600 mg) of 2% lidocaine was injected, and this inadvertent error was immediately recognized. The decision was made to continue with the procedure, which was uneventful, with a tourniquet time of 45 minutes. At this point the patient complained of severe tourniquet pain, and without the anesthesiologist's knowledge the cuff was deflated. Immediately she developed a tachycardia, complained of visual disturbances, and became unconscious. She had a seizure, which lasted 30 seconds and resolved with midazolam. She became more alert, but complained of reduced vision. Neurological examination was normal, apart from temporary blindness; this fully resolved within 10 minutes. There were no long-term neurological or visual sequelae.

The authors suggested that the visual symptoms could have occurred as a result of occipital lobe seizure activity or subcortical stimulation, due to the acute high cerebral concentration of lidocaine. The speed of spontaneous resolution was consistent with the pharmacokinetics of lidocaine.

Pupillary mydriasis occurred in a neonate who was given intravenous lidocaine 3 mg/kg/hour as an anticonvulsant (35).

Ears

The efficacy of intravenous lidocaine 100 mg over 5 minutes in treatment-resistant pruritus in chronic cholestatic liver disease has been studied in a placebo-controlled study in 18 patients (36). There were no severe adverse events. Five patients who were given lidocaine had mild tinnitus, associated in two cases with lingual paresthesia during infusion.

Taste

Taste disturbance has been reported with lidocaine (37).

- A 73-year-old woman was given a Nadbath Rehman block behind the left pinna to provide motor blockade of cranial nerve VII, before retrobulbar block for cataract surgery. Several minutes later she complained of a metallic taste in her mouth. After surgery she had altered taste sensation on the anterior left side of the tongue, with recovery a day later.

The author postulated this to be due to block of the chorda tympani, which runs with cranial nerve VII close to the site of the Nadbath Rehman block.

Metabolism

High systemic doses of lidocaine can cause transient hypoglycemia (SED-12, 255) (38).

Electrolyte balance

There has been one report of hypokalemia (2.2 mmol/l), probably due to potassium channel blockade, after administration of high-dose intravenous lidocaine (8 mg/l) for raised intracranial pressure (SEDA-21, 136).

Hematologic

Severe thrombocytopenic purpura with a lidocaine-mediated antiplatelet IgM antibody has been reported (SED-12, 255) (39).

Three cases of lidocaine-induced methemoglobinemia have been reported in patients undergoing topical anesthesia of the airway and oropharynx (40).

- A 26-year-old woman undergoing bronchoscopy received lidocaine jelly 2% to each nostril, lidocaine solution 2% sprayed on the throat, and 10 ml of lidocaine solution 2% into the trachea. She was also given intravenous diazepam 5 mg and pethidine 75 mg and intramuscular atropine 0.6 mg. She developed dyspnea and cyanosis after the procedure and despite 100% oxygen, her SpO_2 was 85%. Her methemoglobin concentration was 14%.
- A 61-year-old woman was given 15 ml of lidocaine solution 2% and lidocaine spray 4% for topical anesthesia of the throat and oropharynx before upper gastrointestinal endoscopy. She was also sedated with intravenous midazolam 2 mg and pethidine 75 mg. She became cyanosed and desaturated (SpO_2 78%) immediately after the procedure. Her SpO_2 did not recover, despite 100% oxygen. Her methemoglobin concentration was 37%.
- In preparation for transesophageal echocardiogram, a 73-year-old woman was given 15 ml of lidocaine solution 2% and lidocaine spray 4% to anesthetize the oropharynx, plus intravenous midazolam 1 mg and pethidine 12.5 mg. She very rapidly became cyanosed, but remained asymptomatic. Her SpO_2 was 85% on oxygen 2 l/minute and her methemoglobin concentration was 25%.

Liver

Liver damage due to lidocaine has rarely been reported. However, severe liver damage has been reported shortly after the withdrawal of mexiletine 300 mg/day and the introduction of lidocaine 1000 mg/day, although lidocaine in the same dose had been used during the previous week (41). The lidocaine was withdrawn and the liver enzymes normalized after treatment with prednisolone.

Skin

Topical 5% lidocaine to 33 patients with postherpetic neuralgia in a crossover trial provided significantly more pain relief than a vehicle patch placebo (42). There was no difference in reported adverse effects: skin redness or rash was reported by 9 in the lidocaine patch phase and 11 in the placebo phase. One patient stopped using the placebo patch owing to red irritated skin, which resolved after the application of lidocaine patches.

Treatment of 27 HIV-infected patients with distal sensory polyneuropathy (the most common neurological disorder associated with HIV) with 5% lidocaine gel resulted in effective analgesia in 75% of patients; three had dry skin and one had blisters (43).

In a phase IV trial, 66% patients with postherpetic neuralgia gained relief from a 5% lidocaine patch applied to the most painful area of the body (44). The lidocaine patch was well tolerated, a rash being the most common adverse effect, in 14% of patients.

Several cases of contact dermatitis have been reported with lidocaine. Generalized exfoliative dermatitis has also been noted once. Local inflammation and necrosis, possibly due to mechanical pressure, are both complications at the injection site.

- A 60-year-old woman was given infiltration anesthesia with lidocaine hydrochloride for removal of a melanoma (45). She developed an itchy dermatitis over the area 36 hours later. Conventional patch testing was negative at 48 and 72 hours to lidocaine and mepivacaine (both amides), as was intracutaneous testing with lidocaine 2%, mepivacaine 2%, and bupivacaine 0.5%. However, intradermal testing at 1/100 dilutions was positive, with itching and erythema at 48 hours with lidocaine and mepivacaine, suggesting delayed hypersensitivity to these drugs, but not with bupivacaine.

It has previously been reported that lidocaine and mepivacaine have a high degree of cross-reactivity not seen with bupivacaine.

Sexual function

Two cases of impotence after anesthesia for elective circumcision in adults have been described (SED-12, 256) (46), but it is very doubtful whether this was a pharmacological and not merely a psychological effect.

Immunologic

There have been 62 reports of allergic contact dermatitis to lidocaine worldwide between 1972 and 1996; 49 were in Australia and several showed cross-reactivity with other amide local anesthetics, such as bupivacaine, mepivacaine, and prilocaine (47).

Body temperature

There is no reliable evidence to support reports of malignant hyperthermia due to local anesthetic agents. In 307 dental patients susceptible to malignant hyperthermia who received local anesthesia, only one had ever developed symptoms suggestive of malignant hyperthermia, after mepivacaine and on another occasion lidocaine (48). Both reactions resolved without specific therapy. There has been one case report of cyanosis, muscle rigidity, tachycardia, tachypnea, a temperature of 41.5 °C, and loss of consciousness in a patient who received epidural lidocaine and bupivacaine (49). However, perioperative stress may itself be a potential trigger of malignant hyperthermia.

Death

Previous meta-analysis suggested that the use of lidocaine in patients with an acute myocardial infarction is associated with excess mortality, and this led to 1996 guidelines that lidocaine should be regarded as a drug for which there is "evidence and/or general agreement that a procedure/treatment is not useful/effective and in some cases may be harmful" (50). This conclusion has been challenged in a description of a 32-year follow-up observational study of 4254 patients with acute myocardial infarction, of whom 4150 received prophylactic lidocaine and 104 (treated after 1996) did not (51). The incidence of primary ventricular fibrillation was 0.5% among those who received prophylactic lidocaine and 10% among those who did not. Mortality rates were 11% in patients without primary ventricular fibrillation and 25% in patients with. This impressive result was vitiated by the fact that the study was neither prospective nor randomized.

In New York City, five of 50 000 deaths over a 5-year period were associated with tumescent liposuction; all had received lidocaine in doses of 10–40 mg/kg in association with general anesthesia and/or intravenous sedation and analgesia (52). Three patients died as a result of severe acute intraoperative hypotension and bradycardia with no identified cause, one died of fluid overload, and another died of pulmonary embolism. The authors speculated that lidocaine toxicity or lidocaine-related drug interactions could have contributed to some of the deaths, but other causes could not be ruled out.

In California, six cases of cardiac arrest or severe hypoxemia associated with outpatient liposuction resulted in four deaths over a 3.5-year period, all in women aged 38–62 years; one had a cardiac arrest after sedation and the administration of local anesthetic but before liposuction was started, four had respiratory difficulties and cardiac arrest after liposuction, and one had respiratory difficulties during liposuction (53). Whether the cause of morbidity and mortality in any of these cases was related to local anesthetic toxicity was not mentioned.

A weak solution of lidocaine has sometimes been injected into excess fat before liposuction, so that the procedure can be carried out without general anesthesia. The technique is generally regarded as safe (54). However, deaths are increasingly reported, associated with local anesthetic toxicity or drug interactions (52).

A 19-year-old healthy volunteer undergoing bronchoscopy was given about 1200 mg of lidocaine to anesthetize the airway and was sent home after the procedure, despite complaining of chest pain. Shortly afterwards she had a tonic-clonic seizure and cardiopulmonary arrest and died 2 days later. The research protocol had failed to specify an upper dose limit for lidocaine (55).

Second-Generation Effects

Fetotoxicity

Because of rapid transfer across the placenta and the prolonged half-life of lidocaine in neonates, lidocaine can cause fetal acidosis (SEDA-8, 127). Fetal bradycardia is usually observed only in those fetuses with pre-existing heart rate deceleration. Despite massive intoxication at birth, one child had normal behavioral development at 7 months of age (SED-12, 256) (56).

Lactation

Low concentrations of lidocaine and its metabolite monoethylglycinexylidide (MEGX) have been found in breast milk after a dental procedure, but no risk seems to be involved (57).

Susceptibility Factors

Age

Children

Two reports have illustrated the need for particular care when using local anesthetics in neonates and small children. A 2-year-old child died from the combined effects of chloral hydrate, lidocaine, and nitrous oxide for a dental procedure (58). The doses used were not clarified, but in postmortem blood the plasma concentration of lidocaine was 12 μg/ml. The level and adequacy of perioperative monitoring was also not clear.

A neonate who needed a tracheostomy 10 days after a tracheoesophageal fistula repair was given intravenous lidocaine, 1 mg/kg followed 15–20 minutes later by 0.7 mg/kg. Immediately after, tonic-clonic seizures developed. The child recovered, with no observable ill effects at 6 months.

The authors pointed out that the dose of lidocaine used was well within recommended dosage limits. However, they stressed that a more appropriate dosing schedule should be worked out for neonates.

Lidocaine pharmacokinetics tend to follow a single compartment model in neonates, with an increased half-life, and substantially reduced protein binding, leading to a much larger volume of distribution than in adults, but an increased proportion of unbound drug (59).

Elderly people

In the elderly, some local anesthetics (including lidocaine and bupivacaine) have longer durations of action (60).

Sex

That sex differences can affect lidocaine pharmacokinetics is suggested by a report of higher blood concentrations in men than in women after administration of the same dose (SED-12, 256) (61).

Hepatic disease

In patients with heart and liver disease, the dosage requirement of lidocaine is reduced; the half-life of lidocaine is substantially longer in patients with liver disease (62).

The clearance of lidocaine was 12 ml/minute/kg in 10 healthy individuals, 9.8 ml/minute/kg in 10 patients with mild liver impairment, and 4.2 ml/minute/kg in 10 patients with severe liver impairment; the half-life was proportionately prolonged (63). CYP1A2 is the major enzyme responsible for the metabolism of lidocaine in people with normal liver function, and when fluvoxamine was used to inhibit CYP1A2 the effect was much greater in the healthy subjects than in those with liver disease.

Other features of the patient

The adverse effects of lidocaine are dose-related, and are more common in people of light weight and in patients with acute myocardial infarction or congestive cardiac failure. There is also an increased risk of central nervous system effects during cardiopulmonary bypass (64). In cardiac failure, shock, and postoperatively, there are reductions in both the metabolism and the apparent volume of distribution of lidocaine; dosages should be altered accordingly (65).

Drug Administration

Drug additives

The addition of dextran to a lidocaine + adrenaline solution used for infiltration reduced the absorption of both (66).

Alkalinization of local anesthetic solutions should theoretically lead to a faster onset of effect and prolonged anesthesia. However, raising the pH of the solution can cause the local anesthetic to precipitate out of solution, and one study with 2% lidocaine has shown no difference in quality or onset of anesthesia (SEDA-20, 129).

Adrenaline 1:100 000, added to lidocaine 2%, has caused full-thickness skin necrosis when used for ambulatory phlebectomy for varicose veins (SEDA-21, 136).

Drug administration route

Creams and gels

Cutaneous absorption of lidocaine is negligible through normal skin after short-term application. However, when applied to erosive lesions over large body areas, significant absorption may occur. When the drug is applied to mucous membranes, blood levels simulate those resulting from intravenous injection. Local anesthetic creams and gels used liberally on traumatized epithelium can be rapidly absorbed, resulting in systemic effects, such as convulsions, particularly if excessive quantities are used. This has been highlighted in the case of a 40-year-old woman who developed seizures after lidocaine gel 40 ml was injected into the ureter during an attempt to remove a stone (1).

Topical administration of lidocaine to the nasal mucosa occasionally causes severe methemoglobinemia in patients who have the heterozygous form of NADH methemoglobin reductase deficiency (67).

Subcutaneously in liposuction

Some have suggested that lidocaine is unnecessary and potentially toxic in liposuction, and that it provides no postoperative pain relief (68). Others think that lidocaine toxicity is not a major cause of death during liposuction, stating that all reported deaths after liposuction have been associated with general anesthesia or sedation, including the five in New York, and that doses of lidocaine higher than those used in these cases (10–40 mg/kg) are routinely used in tumescent liposuction, no deaths having been reported (53,69). It is possible that adrenaline, high pressure injection, removal of lidocaine by liposuction, and the development of tolerance all contribute to delay in absorption and lack of toxic symptoms at higher than expected plasma concentrations (70).

Patches

Lidocaine is available as a topical analgesic in an adhesive patch formulation for the pain of postherpetic neuralgia. The pharmacokinetics and safety of the 5% lidocaine patches have been studied in 20 healthy volunteers, who applied four patches to the skin either every 24 hours or every 12 hours for 3 days (71). Mean steady-state plasma concentrations were 186 and 225 ng/ml respectively, well below those required for an antidysrhythmic effect (1500 ng/ml) or a risk of toxicity (5000 ng/ml). The patches were well tolerated, with no major cutaneous adverse effects. This is in line with data from postmarketing surveillance studies, which have shown that since the availability of lidocaine patches in 1999, no adverse cardiac or other serious adverse events have been reported (72).

The pharmacokinetics of lidocaine in patches have been investigated in two studies. In 20 healthy volunteers, 5% lidocaine patches were applied for 18 hours/day on 3 consecutive days (73). The mean peak concentrations on days 1, 2, and 3 were 145, 153, and 154 ng/ml respectively; the median values of t_{max} were 18.0, 16.5, and 16.5 hours; and the mean trough concentrations were 83, 86, and 77 ng/ml. The patches were well tolerated; local skin reactions were generally minimal and self-limiting. In 20 healthy volunteers, 4 lidocaine patches were applied every 12 or 24 hours on 3 consecutive days (71). The mean maximum-plasma lidocaine concentrations at steady state were 225 and 186 ng/ml respectively. There was no loss of sensation at the site of application. No patient had edema and most cases of erythema were very slight. No systemic adverse events were judged to be related to the patches.

Nebulization

Nebulized 4% lidocaine 100 mg qds has been used to treat 50 patients with mild-to-moderate asthma in a randomized, saline-controlled study for 8 weeks (74). Lidocaine improved FEV_1 and reduced night-time awakenings, symptoms, bronchodilator use, and blood eosinophil counts; there was no effect on the use of inhaled glucocorticoids. There were no serious adverse effects. Four patients taking lidocaine dropped out, one with a cold feeling in the throat, one with a feeling of claustrophobia, one with a cough attributed to the medication, and one with wheezing and a reduced FEV_1 after inhalation.

Drug overdose

Inadvertent intravenous injection of lidocaine 1 g resulted in asystole, apnea, and tonic-clonic seizures, with full recovery after 6 hours of intensive resuscitation (SED-12, 256) (75).

Fatal accidental overdose has been reported in a child (76).

- An 18-month-old infant died after swallowing an unknown amount of 2% viscous lidocaine. He rapidly became unwell at home, with convulsions, followed by an asystolic cardiorespiratory arrest. He was intubated and resuscitated by paramedics, but continued to have seizures. He was given anticonvulsants and cardiorespiratory resuscitation was unsuccessful. Toxicological tests identified high concentrations of lidocaine and its metabolites.

Owing to the rare but serious poisonings reported to date, 2% viscous lidocaine should not be prescribed for children under 6 years of age.

An unusual case of homicide using an overdose of intravenous lidocaine has been described (77).

- A 32-year-old man, who had been in hospital for several months because of acute intermittent porphyria and chronic pancreatitis, had a seizure and an asystolic cardiac arrest. Resuscitation was unsuccessful. There was a suspicion of patient mistreatment by one of the attending nurses, and toxicological analyses showed high blood concentrations of lidocaine, diazepam, phenytoin, and promethazine. Diazepam and phenytoin had been administered during resuscitation but lidocaine had not.

The cause of death was given as a ventricular dysrhythmia caused by a lidocaine overdose (total dose about 1500 mg); a nurse was later arrested and tried for murder.

Drug–Drug Interactions

Argatroban

The thrombin inhibitor argatroban had no effect on the pharmacokinetics of intravenous lidocaine 1.5 mg/kg for 10 minutes followed by 2 mg/kg/hour for 16 hours in 12 healthy volunteers; the argatroban was given as an intravenous infusion of 2 μg/kg/minute for 16 hours (78).

Beta-adrenoceptor antagonists

The combination of lidocaine with beta-adrenoceptor antagonists is associated with a slightly increased risk of some minor non-cardiac adverse events (dizziness, numbness, somnolence, confusion, slurred speech, and nausea and vomiting) (79). The combination is not associated with an increased risk of dysrhythmias.

Some beta-blockers reduce hepatic blood flow and inhibit microsomal enzymes, reducing the clearance of lidocaine; there is a clinically significant increase in the plasma concentration of lidocaine during concomitant propranolol therapy (80).

Cimetidine

Cimetidine inhibits the metabolism of lidocaine (81,82) and reduces protein binding, increasing toxicity.

Erythromycin

The effects of erythromycin, an inhibitor of CYP3A4, on the pharmacokinetics of lidocaine have been studied in nine healthy volunteers. Steady-state oral erythromycin had no effect on the plasma concentration versus time curve of lidocaine after intravenous administration, but erythromycin increased the plasma concentrations of the major metabolite of lidocaine, MEGX (83). It is not clear what the interpretation of these results is, particularly since the authors did not study enough subjects to detect what might have been small but significant changes in various disposition parameters of lidocaine and did not report unbound concentrations of lidocaine or its metabolites. However, whatever the pharmacokinetic explanation, the clinical relevance is that one would expect that erythromycin would potentiate the toxic effects of lidocaine that are mediated by MEGX.

The effect of erythromycin on the pharmacokinetics of intravenous lidocaine and its two pharmacologically active metabolites, monoethylglycinexylidide and glycinexylidide, has been studied in 10 healthy volunteers, 10 patients with hepatic cirrhosis Child's class A, and 10 with hepatic cirrhosis class C, in a double-blind, randomized, two-way, crossover study (84). Erythromycin caused statistically significant but small changes in the pharmacokinetics of lidocaine and monoethylglycinexylidide. In healthy subjects, lidocaine clearance fell from 9.93 to 8.15 ml/kg/minute (82%; 95%CI = 65, 98) and the half-life was prolonged from 2.23 to 02.80 hours (130%; 95%CI = 109, 151); the AUC of monoethylglycinexylidide rose to 129% (95%CI = 102, 156). There were quantitatively similar modifications in the two groups of patients with cirrhosis, but only in the patients with Child's grade C liver cirrhosis were lidocaine pharmacokinetics significantly different than in the healthy subjects; clearance was approximately halved, steady-state volume of distribution was increased, and terminal half-life was more than doubled. The authors concluded that no dosage adjustment is needed in patients with moderate liver cirrhosis, but that the dose of lidocaine should be halved in patients with severe cirrhosis.

Fluvoxamine

In a double-blind, placebo-controlled, randomized, three-way, crossover study, fluvoxamine (a CYP1A2 inhibitor) 100 mg/day reduced the clearance of intravenous lidocaine by 41% and prolonged its half-life from 2.6 to 3.5 hours (85). During co-administration of fluvoxamine + erythromycin (a CYP3A4 inhibitor) 500 mg tds, lidocaine clearance was reduced by 53% compared with placebo and 21% compared with fluvoxamine alone and the half-life was further prolonged to 4.3 hours. The apparent volume of distribution of lidocaine was not affected. The half-life of monoethylglycinexylidide, an active metabolite of lidocaine, was significantly prolonged by both fluvoxamine alone and the combination of fluvoxamine + erythromycin. The authors concluded that the minor effect of inhibitors of CYP3A4 on the pharmacokinetics of lidocaine can be explained by compensatory metabolism by CYP1A2.

Itraconazole

The effects of itraconazole, an inhibitor of CYP3A4, on the pharmacokinetics of lidocaine have been studied in nine healthy volunteers. Steady-state oral itraconazole had no effect on the plasma concentration versus time curve of lidocaine after intravenous administration nor on the plasma concentrations of the major metabolite of lidocaine, MEGX (83).

Itraconazole (200 mg/day for 4 days) had no effect on the pharmacokinetics of inhaled lidocaine or its major metabolite, monoethylglycinexylidide, in 10 healthy volunteers in a randomized, placebo-controlled, crossover study (86).

Mexiletine

An interaction of lidocaine with mexiletine, which resulted in toxic concentrations of lidocaine, has been reported (87).

- An 80-year-old man with a dilated cardiomyopathy was given a lidocaine infusion started at 90 mg/hour for a ventricular tachycardia. He was already taking mexiletine 400 mg/day, and the plasma concentration was within the usual target range; however, the dose was reduced to 200 mg/day to avoid possible adverse effects. Intermittent ventricular tachycardia persisted, and so the lidocaine infusion was increased to 120 mg/day, but adverse effects (involuntary movements, muscle rigidity) were observed. The lidocaine infusion was stopped and within 20 minutes the adverse effects abated; the lidocaine concentration was 6.84 μg/ml. The ventricular tachycardia persisted, lidocaine was restarted at a lower rate, and the oral dose of mexiletine was increased to 450 mg/day. This resulted in an unexpectedly high concentration of lidocaine and the lidocaine concentration was significantly higher while the mexiletine dose was high.

Further studies suggested that mexiletine had displaced lidocaine from tissue binding sites. The authors suggested that this finding has implications for loading doses and acute effects of lidocaine in the concurrent therapy of lidocaine and mexiletine and highlighted the importance of close monitoring of lidocaine concentrations in this setting.

Opioid analgesics

A synergistic interaction of intrathecal fentanyl 100 μg and morphine 0.5 mg, given before induction, with systemically administered lidocaine 200 mg 4 hours later for ventricular tachycardia, resulted in potentiation of opioid effects in a 74-year-old man with major heart disease after coronary artery bypass grafting; during the 5 minutes after lidocaine he had a respiratory arrest with loss of consciousness and miotic pupils, all reversed by naloxone (88). The proposed mechanism was thought to be a reduction in calcium ion concentrations in opioid-sensitive CNS sites.

Propafenone

The CNS toxicity of lidocaine was increased in 11 healthy volunteers who simultaneously received propafenone, which reduced the metabolism of lidocaine (89).

Propofol

Propofol dose-dependently reduced the threshold for lidocaine-induced convulsions in rats (90). Higher doses of propofol completely abolished convulsions. However, there was no difference in the dose of lidocaine that caused cardiac arrest and death, when it was given with three different propofol infusions and placebo.

Ranitidine

Ranitidine inhibits the clearance of lidocaine (82).

Suxamethonium

Procaine and cocaine are esters that are hydrolysed by plasma cholinesterase and may therefore competitively enhance the action of suxamethonium (91). Chloroprocaine may have a similar action. Lidocaine also interacts, although the mechanism is not clear unless very high doses are used (92).

References

1. Pantuck AJ, Goldsmith JW, Kuriyan JB, Weiss RE. Seizures after ureteral stone manipulation with lidocaine. J Urol 1997;157(6):2248.
2. Pfeifer HJ, Greenblatt DJ, Koch-Weser J. Clinical use and toxicity of intravenous lidocaine. A report from the Boston Collaborative Drug Surveillance Program. Am Heart J 1976;92(2):168–73.
3. Greenspon AJ, Mohiuddin S, Saksena S, Lengerich R, Snapinn S, Holmes G, Irvin J, Sappington E, et al. Comparison of intravenous tocainide with intravenous lidocaine for treating ventricular arrhythmias. Cardiovasc Rev Rep 1989;10:55–9.
4. Adriani J, Coffman VD, Naraghi M. The allergenicity of lidocaine and other amide and related local anesthetics. Anesthesiol Rev 1986;13:30–6.
5. Bonnet MC, du Cailar G, Deschodt J. Anaphylaxie à la lidocaine. [Anaphylaxis caused by lidocaine.] Ann Fr Anesth Reanim 1989;8(2):127–9.
6. Kennedy KS, Cave RH. Anaphylactic reaction to lidocaine. Arch Otolaryngol Head Neck Surg 1986;112(6):671–3.

7. Avery JK. Routine procedure—bad outcome. Tenn Med 1998;91(7):280–1.
8. Sakurai M, Kanazawa I. Positive symptoms in multiple sclerosis: their treatment with sodium channel blockers, lidocaine and mexiletine. J Neurol Sci 1999;162(2):162–8.
9. Hand PJ, Stark RJ. Intravenous lignocaine infusions for severe chronic daily headache. Med J Aust 2000;172(4):157–9.
10. Attal N, Gaude V, Brasseur L, Dupuy M, Guirimand F, Parker F, Bouhassira D. Intravenous lidocaine in central pain: a double-blind, placebo-controlled, psychophysical study. Neurology 2000;54(3):564–74.
11. Ziegelbaum M, Lever H. Acute urinary retention associated with flecainide. Cleve Clin J Med 1990;57(1):86–7.
12. Krikler DM, Curry PV. Torsade de pointes, an atypical ventricular tachycardia. Br Heart J 1976;38(2):117–20.
13. Gottlieb SS, Packer M. Deleterious hemodynamic effects of lidocaine in severe congestive heart failure. Am Heart J 1989;118(3):611–2.
14. Weaver WD, Fahrenbruch CE, Johnson DD, Hallstrom AP, Cobb LA, Copass MK. Effect of epinephrine and lidocaine therapy on outcome after cardiac arrest due to ventricular fibrillation. Circulation 1990;82(6):2027–34.
15. Tisdale JE. Lidocaine prophylaxis in acute myocardial infarction. Henry Ford Hosp Med J 1991;39(3–4):217–25.
16. Garner L, Stirt JA, Finholt DA. Heart block after intravenous lidocaine in an infant. Can Anaesth Soc J 1985;32(4):425–8.
17. Hansoti RC, Ashar PN. Atrioventicular block and ventricular fibrillation due to lidocaine therapy. Bombay Hosp J 1975;17:26.
18. McLean SA, Paul ID, Spector PS. Lidocaine-induced conduction disturbance in patients with systemic hyperkalemia. Ann Emerg Med 2000;36(6):615–8.
19. Parnass SM, Curran MJ, Becker GL. Incidence of hypotension associated with epidural anesthesia using alkalinized and nonalkalinized lidocaine for cesarean section. Anesth Analg 1987;66(11):1148–50.
20. Enlund M, Mentell O, Krekmanov L. Unintentional hypotension from lidocaine infiltration during orthognathic surgery and general anaesthesia. Acta Anaesthesiol Scand 2001;45(3):294–7.
21. Beydon L, Lorino AM, Verra F, Labroue M, Catoire P, Lofaso F, Bonnet F. Topical upper airway anaesthesia with lidocaine increases airway resistance by impairing glottic function. Intensive Care Med 1995;21(11):920–6.
22. Howard P, Cayton RM, Brennan SR, Anderson PB. Lignocaine aerosol and persistent cough. Br J Dis Chest 1977;71(1):19–24.
23. Ryder W. "Two cautionary tales". Anaesthesia 1994;49(2):180–1.
24. Keller C, Sparr HJ, Brimacombe JR. Laryngeal mask lubrication. A comparative study of saline versus 2% lignocaine gel with cuff pressure control. Anaesthesia 1997;52(6):592–7.
25. Stargel WW, Shand DG, Routledge PA, Barchowsky A, Wagner GS. Clinical comparison of rapid infusion and multiple injection methods for lidocaine loading. Am Heart J 1981;102(5):872–6.
26. Olthoff D, Vetter B, Deutrich C, Burkhardt U. Pharmakokinetische Untersuchungen zu den Ursachen der erhohten Neurotoxizität des Lidokains während kardiochirurgischer Operationen. [Pharmacokinetic studies on the causes of increased neurotoxicity of lidocaine during heart surgery.] Anaesthesiol Reanim 1989;14(4):207–14.
27. DeToledo JC, Minagar A, Lowe MR. Lidocaine-induced seizures in patients with history of epilepsy: effect of antiepileptic drugs. Anesthesiology 2002;97(3):737–9.
28. Lee DL, Ayoub C, Shaw RK, Fontes ML. Grand mal seizure during cardiopulmonary bypass: probable lidocaine toxicity. J Cardiothorac Vasc Anesth 1999;13(2):200–2.
29. Goodwin DP, McMeekin TO. A case of lidocaine absorption from topical administration of 40% lidocaine cream. J Am Acad Dermatol 1999;41(2 Pt 1):280–1.
30. Anibarro B, Seoane FJ. Adverse reaction to lidocaine. Allergy 1998;53(7):717–8.
31. Kanai Y, Katsuki H, Takasaki M. Graded, irreversible changes in crayfish giant axon as manifestations of lidocaine neurotoxicity in vitro. Anesth Analg 1998;86(3):569–73.
32. Ghose S, Garodia VK, Sachdev MS, Kumar H, Biswas NR, Pandey RM. Evaluation of potentiating effect of a drop of lignocaine on tropicamide-induced mydriasis. Invest Ophthalmol Vis Sci 2001;42(7):1581–5.
33. Yamashita S, Sato S, Kakiuchi Y, Miyabe M, Yamaguchi H. Lidocaine toxicity during frequent viscous lidocaine use for painful tongue ulcer. J Pain Symptom Manage 2002;24(5):543–5.
34. Sawyer RJ, von Schroeder H. Temporary bilateral blindness after acute lidocaine toxicity. Anesth Analg 2002;95(1):224–6.
35. Berger I, Steinberg A, Schlesinger Y, Seelenfreund M, Schimmel MS. Neonatal mydriasis: intravenous lidocaine adverse reaction. J Child Neurol 2002;17(5):400–1.
36. Villamil AG, Bandi JC, Galdame OA, Gerona S, Gadano AC. Efficacy of lidocaine in the treatment of pruritus in patients with chronic cholestatic liver diseases. Am J Med 2005;118(10):1160–3.
37. Bigeleisen PE. An unusual presentation of metallic taste after lidocaine injections. Anesth Analg 1999;89(5):1239–40.
38. Janda A, Salem C. Hypoglykämie durch Lidocain-Überdosierung. [Hypoglycemia caused by lidocaine overdosage.] Reg Anaesth 1986;9(3):88–90.
39. Stefanini M, Hoffman MN. Studies on platelets: XXVIII: acute thrombocytopenic purpura due to lidocaine (Xylocaine)-mediated antibody. Report of a case. Am J Med Sci 1978;275(3):365–71.
40. Karim A, Ahmed S, Siddiqui R, Mattana J. Methemoglobinemia complicating topical lidocaine used during endoscopic procedures. Am J Med 2001;111(2):150–3.
41. Kakinoki K, Tachibana Y, Yonejima H, Ogino H, Satomura Y, Unoura M. A case of mexiletine and lidocaine induced severe liver injury. Acta Hepatol Jpn 2000;41:812–6.
42. Galer BS, Rowbotham MC, Perander J, Friedman E. Topical lidocaine patch relieves postherpetic neuralgia more effectively than a vehicle topical patch: results of an enriched enrollment study. Pain 1999;80(3):533–8.
43. Dorfman D, Dalton A, Khan A, Markarian Y, Scarano A, Cansino M, Wulff E, Simpson D. Treatment of painful distal sensory polyneuropathy in HIV-infected patients with a topical agent: results of an open-label trial of 5% lidocaine gel. AIDS 1999;13(12):1589–90.
44. Anonymous. Lidocaine patch shown to relieve postherpetic neuralgia. J Pharm Technol 2001;17:154.
45. Scala E, Giani M, Pirrotta L, Guerra EC, Girardelli CR, De Pita O, Puddu P. Simultaneous allergy to ampicillin and local anesthetics. Allergy 2001;56(5):454–5.
46. Palmer JM, Link D. Impotence following anesthesia for elective circumcision. JAMA 1979;241(24):2635–6.

47. Weightman W, Turner T. Allergic contact dermatitis from lignocaine: report of 29 cases and review of the literature. Contact Dermatitis 1998;39(5):265–6.
48. Minasian A, Yagiela JA. The use of amide local anesthetics in patients susceptible to malignant hyperthermia. Oral Surg Oral Med Oral Pathol 1988;66(4):405–15.
49. Klimanek J, Majewski W, Walencik K. A case of malignant hyperthermia during epidural analgesia. Anaesth Resusc Intensive Ther 1976;4(2):143–5.
50. Ryan TJ, Anderson JL, Antman EM, Braniff BA, Brooks NH, Califf RM, Hillis LD, Hiratzka LF, Rapaport E, Riegel BJ, Russell RO, Smith EE Jr, Weaver WD. ACC/AHA guidelines for the management of patients with acute myocardial infarction. A report of the American College of Cardiology/American Heart Association Task Force on Practice Guidelines (Committee on Management of Acute Myocardial Infarction). J Am Coll Cardiol 1996;28(5):1328–428.
51. Wyman MG, Wyman RM, Cannom DS, Criley JM. Prevention of primary ventricular fibrillation in acute myocardial infarction with prophylactic lidocaine. Am J Cardiol 2004;94(5):545–51.
52. Rao RB, Ely SF, Hoffman RS. Deaths related to liposuction. N Engl J Med 1999;340(19):1471–5.
53. Ginsberg MM, Gresham L, Vermeulen C, Serra M, Roujeau JC, Talmor M, Barie PS, Klein JA, Rigel DS, Wheeland RG, Schnur P, Penn J, Fodor PB. Deaths related to liposuction. N Engl J Med 1999;341(13):1000–3.
54. Klein JA. Tumescent technique for local anesthesia improves safety in large-volume liposuction. Plast Reconstr Surg 1993;92(6):1085–100.
55. Day RO, Chalmers DR, Williams KM, Campbell TJ. The death of a healthy volunteer in a human research project: implications for Australian clinical research. Med J Aust 1998;168(9):449–51.
56. Kim WY, Pomerance JJ, Miller AA. Lidocaine intoxication in a newborn following local anesthesia for episiotomy. Pediatrics 1979;64(5):643–5.
57. Lebedevs TH, Wojnar-Horton RE, Yapp P, Roberts MJ, Dusci LJ, Hackett LP, Ilett K. Excretion of lignocaine and its metabolite monoethylglycinexylidide in breast milk following its use in a dental procedure. A case report. J Clin Periodontol 1993;20(8):606–8.
58. Engelhart DA, Lavins ES, Hazenstab CB, Sutheimer CA. Unusual death attributed to the combined effects of chloral hydrate, lidocaine, and nitrous oxide. J Anal Toxicol 1998;22(3):246–7.
59. Resar LM, Helfaer MA. Recurrent seizures in a neonate after lidocaine administration. J Perinatol 1998;18(3):193–5.
60. Chauvin M. Toxicité aiguë des anesthésiques locaux en fonction du terrain. [Acute toxicity of local anesthetics as a function of the patient's condition.] Ann Fr Anesth Reanim 1988;7(3):216–23.
61. Bruguerolle B, Isnardon R, Valli M, Vadot G. Influence du sexe sur les taux plasmatiques de lidocaine en anesthésie dentaire. Thérapie (Paris) 1982;37:593.
62. Thomson PD, Rowland M, Melmon KL. The influence of heart failure, liver disease, and renal failure on the disposition of lidocaine in man. Am Heart J 1971;82(3):417–21.
63. Orlando R, Piccoli P, De Martin S, Padrini R, Floreani M, Palatini P. Cytochrome P450 1A2 is a major determinant of lidocaine metabolism in vivo: effects of liver function. Clin Pharmacol Ther 2004;75(1):80–8.
64. Bauer LA, Brown T, Gibaldi M, Hudson L, Nelson S, Raisys V, Shea JP. Influence of long-term infusions on lidocaine kinetics. Clin Pharmacol Ther 1982;31(4):433–7.
65. Kumana CR. Therapeutic drug monitoring—antidysrhythmic drugs. In: Richens A, Marks V, editors. Therapeutic Drug Monitoring. Ch 16A. London, Edinburgh: Churchill-Livingstone, 1981:370 .
66. Adams HA, Biscoping J, Kafurke H, Muller H, Hoffmann B, Boerner U, Hempelmann G. Influence of dextran on the absorption of adrenaline-containing lignocaine solutions: a protective mechanism in local anaesthesia. Br J Anaesth 1988;60(6):645–50.
67. Kotler RL, Hansen-Flaschen J, Casey MP. Severe methaemoglobinaemia after flexible fibreoptic bronchoscopy. Thorax 1989;44(3):234–5.
68. Perry AW, Petti C, Rankin M. Lidocaine is not necessary in liposuction. Plast Reconstr Surg 1999;104(6):1900–2.
69. Klein JA. Lidocaine is not necessary in liposuction: discussion. Plast Reconstr Surg 1999;104:1903–6.
70. Rubin JP, Bierman C, Rosow CE, Arthur GR, Chang Y, Courtiss EH, May JW Jr. The tumescent technique: the effect of high tissue pressure and dilute epinephrine on absorption of lidocaine. Plast Reconstr Surg 1999;103(3):990–1002.
71. Gammaitoni AR, Alvarez NA, Galer BS. Pharmacokinetics and safety of continuously applied lidocaine patches 5%. Am J Health Syst Pharm 2002;59(22):2215–20.
72. Galer BS. Effectiveness and safety of lidocaine patch 5%. J Fam Pract 2002;51(10):867–8.
73. Gammaitoni AR, Davis MW. Pharmacokinetics and tolerability of lidocaine patch 5% with extended dosing. Ann Pharmacother 2002;36(2):236–40.
74. Hunt LW, Frigas E, Butterfield JH, Kita H, Blomgren J, Dunnette SL, Offord KP, Gleich GJ. Treatment of asthma with nebulized lidocaine: a randomized, placebo-controlled study. J Allergy Clin Immunol 2004;113(5):853–9.
75. Finkelstein F, Kreeft J. Massive lidocaine poisoning. N Engl J Med 1979;301(1):50.
76. Nisse P, Lhermitte M, Dherbecourt V, Fourier C, Leclerc F, Houdret N, Mathieu-Nolf M. Intoxication mortelle après ingestion accidentelle de Xylocaine visqueuse a 2% chez une jeune enfant. [Fatal intoxication after accidental ingestion of viscous 2% lidocaine in a young child.] Acta Clin Belg Suppl 2002;(1):51–3.
77. Kalin JR, Brissie RM. A case of homicide by lethal injection with lidocaine. J Forensic Sci 2002;47(5):1135–8.
78. Inglis AM, Sheth SB, Hursting MJ, Tenero DM, Graham AM, DiCicco RA. Investigation of the interaction between argatroban and acetaminophen, lidocaine, or digoxin. Am J Health Syst Pharm 2002;59(13):1258–66.
79. Wyse DG, Kellen J, Tam Y, Rademaker AW. Increased efficacy and toxicity of lidocaine in patients on beta-blockers. Int J Cardiol 1988;21(1):59–70.
80. Naguib M, Magboul MM, Samarkandi AH, Attia M. Adverse effects and drug interactions associated with local and regional anaesthesia. Drug Saf 1998;18(4):221–50.
81. Jackson JE, Bentley JB, Glass SJ, Fukui T, Gandolfi AJ, Plachetka JR. Effects of histamine-2 receptor blockade on lidocaine kinetics. Clin Pharmacol Ther 1985;37(5):544–8.
82. Kowalsky SF. Lidocaine interaction with cimetidine and ranitidine: a critical analysis of the literature. Adv Ther 1988;5:229–44.
83. Isohanni MH, Neuvonen PJ, Palkama VJ, Olkkola KT. Effect of erythromycin and itraconazole on the pharmacokinetics of intravenous lignocaine. Eur J Clin Pharmacol 1998;54(7):561–5.
84. Orlando R, Piccoli P, De Martin S, Padrini R, Palatini P. Effect of the CYP3A4 inhibitor erythromycin on the pharmacokinetics of lignocaine and its pharmacologically active

metabolites in subjects with normal and impaired liver function. Br J Clin Pharmacol 2003;55:86–93.
85. Olkkola KT, Isohanni MH, Hamunen K, Neuvonen PJ. The effect of erythromycin and fluvoxamine on the pharmacokinetics of intravenous lidocaine. Anesth Analg 2005;100(5):1352–6.
86. Isohanni MH, Neuvonen PJ, Olkkola KT. Effect of itraconazole on the pharmacokinetics of inhaled lidocaine. Basic Clin Pharmacol Toxicol 2004;95(3):120–3.
87. Maeda Y, Funakoshi S, Nakamura M, Fukuzawa M, Kugaya Y, Yamasaki M, Tsukiai S, Murakami T, Takano M. Possible mechanism for pharmacokinetic interaction between lidocaine and mexiletine. Clin Pharmacol Ther 2002;71(5):389–97.
88. Jensen E, Nader ND. Potentiation of narcosis after intravenous lidocaine in a patient given spinal opioids. Anesth Analg 1999;89(3):758–9.
89. Ujhelyi MR, O'Rangers EA, Fan C, Kluger J, Pharand C, Chow MS. The pharmacokinetic and pharmacodynamic interaction between propafenone and lidocaine. Clin Pharmacol Ther 1993;53(1):38–48.
90. Lee VC, Moscicki JC, DiFazio CA. Propofol sedation produces dose-dependent suppression of lidocaine-induced seizures in rats. Anesth Analg 1998;86(3):652–7.
91. Matsuo S, Rao DB, Chaudry I, Foldes FF. Interaction of muscle relaxants and local anesthetics at the neuromuscular junction. Anesth Analg 1978;57(5):580–7.
92. Usubiaga JE, Wikinski JA, Morales RL, Usubiaga LE. Interaction of intravenously administered procaine, lidocaine and succinylcholine in anesthetized subjects. Anesth Analg 1967;46(1):39–45.

Lorcainide

General Information

The clinical pharmacology, clinical use, efficacy, and adverse effects of lorcainide have been reviewed (1–3).

Organs and Systems

Cardiovascular

Cardiovascular effects of lorcainide are reportedly uncommon (under 1% of cases) and have mostly been associated with intravenous administration. They include hypotension (1) and heart block. Pre-existing sinoatrial disease and heart block are contraindications to the use of lorcainide (4).

Nervous system

Nervous system effects account for most adverse reactions to lorcainide after both intravenous and oral administration. During intravenous treatment patients may complain of vertigo, feeling hot, numbness of the feet (5), dizziness, blurred vision, muscle tremor (6), and tingling sensations in the fingers or tongue (7). All of these effects are transient.

During long-term oral therapy the most frequent adverse effect is sleep disturbance of one kind or another, the frequency being up to 45% (6,8–13). The main types of disturbance are difficulty in falling asleep, nightmares, and lively dreams. In about half of the patients affected there is also excessive sweating (10,11). These effects become less severe after a week or two of therapy, and during that time benzodiazepines may give symptomatic relief (14). In contrast, these patients do not suffer drowsiness or other central nervous system effects during the day.

Seizures and hallucinations have been reported rarely (14).

Endocrine

Lorcainide has been reported to cause hyponatremia, attributed to inappropriate secretion of ADH (15).

Gastrointestinal

Nausea occurs in about 5% of patients during long-term oral therapy. It is probably of central nervous origin (14).

Skin

There have been a few reports of mild allergic skin reactions (14).

Susceptibility Factors

The adverse effects of lorcainide may be partly caused by its active metabolite, norlorcainide, which accumulates during treatment (15).

Drug Administration

Drug overdose

- Lorcainide 2500 mg produced bradycardia, convulsions, and coma in a 15-year-old girl who took a deliberate overdose and died rapidly (16).

References

1. Amery WK, Heykants JJ, Xhonneux R, Towse G, Oettel P, Gough DA, Janssen PA. Lorcainide (R 15 889), a first review. Acta Cardiol 1981;36(3):207–34.
2. Anonymous. Lorcainide hydrochloride. Med Actual 1982;18:12–22.
3. Eiriksson C, Brogden RN. Lorcainide. A preliminary review of its pharmacodynamic properties and therapeutic efficacy. Drugs 1984;27(4):279–300.
4. Kasper W, Meinertz T, Kersting F, Lollgen H, Lang K, Just H. Electrophysiological actions of lorcainide in patients with cardiac disease. J Cardiovasc Pharmacol 1979;1(3):343–52.
5. Shita A, Bernard R, Mostinckx R, Debacker M. Haemodynamic reactions after intravenous injection of lorcainide hydrochloride in acute myocardial infarction. Eur J Cardiol 1981;12(5):237–42.
6. Kesteloot H, Stroobandt R. Clinical experience with lorcainide (R 15 889), a new anti-arrhythmic drug. Arch Int Pharmacodyn Ther 1977;230(2):225–34.
7. Somani P. Pharmacokinetics of lorcainide, a new antiarrhythmic drug, in patients with cardiac rhythm disorders. Am J Cardiol 1981;48(1):157–63.

8. Klotz U, Muller-Seydlitz P, Heimburg P. Disposition and antiarrhythmic effect of lorcainide. Int J Clin Pharmacol Biopharm 1979;17(4):152–8.
9. Cocco G, Strozzi C. Initial clinical experience of lorcainide (Ro 13-1042), a new antiarrhythmic agent. Eur J Clin Pharmacol 1978;14(2):105–9.
10. Klotz U, Muller-Seydlitz PM, Heimburg P. Lorcainide infusion in the treatment of ventricular premature beats (VPB). Eur J Clin Pharmacol 1979;16(1):1–6.
11. Meinertz T, Kasper W, Kersting F, Bechtold H, Just H, Jahnchen E. Antiarrhythmic effect of lorcainide during chronic treatment. Arzneimittelforschung 1980;30(9):1593–5.
12. Myburgh DP, Goldman AP, Schamroth JM. Lorcainide—an anti-arrhythmic agent for ventricular arrhythmias. S Afr Med J 1980;57(7):236–9.
13. Keefe DL, Peters F, Winkle RA. Randomized double-blind placebo controlled crossover trial documenting oral lorcainide efficacy in suppression of symptomatic ventricular tachyarrhythmias. Am Heart J 1982;103(4 Pt 1):511–8.
14. Somani P, Temesy-Armos PN, Leighton RF, Goodenday LS, Fraker TD Jr. Hyponatremia in patients treated with lorcainide, a new antiarrhythmic drug. Am Heart J 1984;108(6):1443–8.
15. Meinertz T, Kasper W, Kersting F, Just H, Bechtold H, Jahnchen E. Lorcainide. II. Plasma concentration-effect relationship. Clin Pharmacol Ther 1979;26(2):196–204.
16. Evers J, Buttner-Belz U. Fatal lorcainide poisoning. J Toxicol Clin Toxicol 1995;33(2):157–9.

Mexiletine

General Information

Mexiletine is a class Ib antidysrhythmic drug, similar in action to lidocaine, but it can be given orally. Its adverse effects occur in up to 50% of patients (1) and withdrawal is often necessary (2). The most common adverse effects are on the cardiovascular and central nervous systems. The pharmacokinetics, clinical use, and adverse effects and interactions of mexiletine have been reviewed widely (3–8).

In addition to its use as an antidysrhythmic drug, mexiletine has also been used in the treatment of various types of neuropathic pain and dystonias (9,10). The adverse effects in these circumstances have been reported (11–13) and reviewed (14).

Observational studies

In an open study of the antidystonic effect of mexiletine (200 mg/day increasing to a maximum of 800 mg/day) in spasmodic torticollis in six patients, mexiletine produced significant improvement and there were no adverse effects in five of the six patients; in the other patient dizziness occurred at the highest dose and required a reduction in dosage (11).

In studies in patients with diabetic peripheral neuropathy, adverse effects included nausea, hiccups, tremor, headache, weakness and dizziness, tachycardia, and allergic reactions (14).

Mexiletine has been used to treat painful peripheral neuropathy in patients with HIV infection, without any evidence of efficacy (15,16). In one study of 22 patients, nine had adverse effects probably related to mexiletine, including nausea in five, vomiting in four, and abdominal pain, diarrhea, dizziness, insomnia, rises in liver enzymes, and skin rash in one patient each (15). Adverse effects in seven patients required dosage reduction in four cases and withdrawal in three (because of a rash in one case and gastrointestinal effects in two).

In a study in which 48 patients with painful peripheral neuropathy due to HIV infection were treated with mexiletine, 10 had nausea and vomiting that required dosage modification; dosage modification was also occasioned by dizziness in one case and urinary retention in three cases (16).

Comparative studies

In 51 patients with serious ventricular tachydysrhythmias mexiletine alone or in combination with other class Ia drugs (quinidine, procainamide, disopyramide) was ineffective in 41% and there was no difference between those who had mexiletine alone and those who had the combinations (17). The incidence of adverse effects was also the same in both groups (about 70%), despite a lower mean dose of mexiletine in the latter.

The adverse effects of mexiletine have been compared with those of quinidine in a randomized, double-blind comparison in 491 patients (18). The commonest adverse effects in the 246 patients given mexiletine were upper gastrointestinal distress (38%), light-headedness (20%), tremor (13%), and co-ordination difficulties (11%). These all occurred more frequently than with quinidine. However, quinidine more often caused diarrhea (35 versus 7%) and dysrhythmias (19 versus 12 patients). The other reported adverse effects, associated in equal frequency with the two drugs, were fatigue, changes in sleep habit, weakness, headache, angina-like pain, and rash. Visual problems (7 versus 3%) and nervousness (6 versus 2%) occurred slightly more often with mexiletine.

Placebo-controlled studies

In a double-blind, placebo-controlled, crossover study of the use of mexiletine in 20 patients with neuropathic pain with prominent allodynia the dosage was titrated to maximum of 900 mg/day or until dose-limiting adverse effects occurred. Mexiletine produced little beneficial effect and the two most common adverse effects were nausea and sedation (12). Other adverse effects that occurred in one or two patients each included insomnia, trismus, headache, agitation, nightmares, and tremor.

In a double-blind, placebo-controlled, crossover study in 12 healthy volunteers mexiletine, in a dosage that was titrated to a maximum of 1350 mg/day or until there were dose-limiting adverse effects, was used to alleviate capsaicin-induced allodynia and hyperalgesia (13). Mexiletine had no significant effect on any of the major measures of pain or neurosensory thresholds after intradermal capsaicin; however, it did reduce the flare response. All 12 subjects had dose-limiting adverse effects and the mean maximum tolerable daily dose was 850 mg. The adverse effects included nausea, light-headedness, muscle twitching and weakness, blurred vision, headache, tremor, difficulty in concentrating, dysphoria, sedation, pruritus, and rash. These adverse effects occurred at an average daily dose of 993 mg. The three most common adverse effects were nausea, dizziness, and tremor in ten, nine, and four of the subjects respectively.

The analgesic effects of mexiletine 600 mg/day and gabapentin 1200 mg/day have been investigated in 75 patients with acute and chronic pain associated with breast surgery in a double-blind, randomized, placebo-controlled study for 10 days (19). Pain at rest and after movement was reduced by both drugs on the third postoperative day and pain after movement was reduced by gabapentin at 2–5 days postoperatively. Two women given mexiletine withdrew from the study because of adverse events, one with an axillary vein thrombosis and one with nausea and vomiting.

Organs and Systems

Cardiovascular

Prodysrhythmic effects of mexiletine have been reported in up to 29% of patients, although it has been suggested that on average the incidence is lower than has been reported with other antidysrhythmic drugs (20).

Mexiletine can have major hemodynamic effects in patients with pre-existing impairment of left ventricular dysfunction (21). Circulatory depression with bradycardia and hypotension have been reported (17,22,23).

Other cardiovascular effects include widening of the QRS complex, atrioventricular dissociation, heart block, sinus arrest, and cardiac arrest (23,24).

Respiratory

Mexiletine has been associated with pulmonary fibrosis and eventual respiratory failure (25).

Nervous system

The commonest adverse effects of mexiletine are on the nervous system, and include centrally mediated gastrointestinal distress (38%), light-headedness (20%), tremor (12%), and coordination difficulties (11%). The figures in parentheses are quoted from a study in which mexiletine was compared with quinidine (26). Other reported adverse effects include changes in sleep habit, weakness, headache, visual problems, and nervousness. Slurred speech, dysarthria, diplopia, and ataxia have also been reported (18).

Mexiletine 300–400 mg/day has been used to treat some of the symptoms of multiple sclerosis in 30 patients with painful tonic seizures, attacks of neuralgia, paroxysmal itching, and Lhermitte's sign (27). Mexiletine produced similar therapeutic effects to lidocaine. In one patient, weakness worsened during the administration of mexiletine. In two patients when mexiletine was replaced by placebo there was a suggestion of some rebound in painful tonic seizures.

In three patients with neuropathic pain, mexiletine added to the analgesic regimen caused some improvement in visual analogue scores for pain (10). All three had some symptoms of nausea, and two became depressed. One patient also reported feeling "trembly and shaky" on occasion, but there was no objective evidence of tremor.

Sensory systems

In two patients with neuropathic pain and pre-existing ocular disease, mexiletine caused persistent ophthalmic changes (28).

- A 39-year-old woman took mexiletine 300 mg tds for 3 days and developed transient blindness with residual reduced visual acuity due to an acute pigmentary retinopathy. Her vision improved markedly after withdrawal of mexiletine, but when she restarted it she developed clouding of the vision, which resolved again on withdrawal. The pigmentary changes persisted.
- A man with a history of glaucoma took mexiletine 300 mg tds and developed worsening visual acuity. Mexiletine was withdrawn and he took carbamazepine 200 mg/day instead. However, he started to see red and green spots. The serum carbamazepine concentration was below that usually associated with visual disturbances. No structural abnormalities were detected.

Hematologic

Mexiletine can cause a spuriously low platelet count as a result of clumping of platelets due to antibodies (29); however, it can also cause a true thrombocytopenia (30,31).

Eosinophilia and atypical lymphocytosis associated with liver dysfunction have also been described (32).

Gastrointestinal

Nausea and vomiting are common and probably central in origin, since they occur with intravenous as well as oral therapy (17).

Mexiletine affects the esophagus, and can cause heartburn (1) and esophageal spasm and ulceration (1,33).

Liver

Changes in liver function tests, resolving on withdrawal, have been reported (34). In one case there was histological evidence of cholestasis.

Mexiletine caused increases in the serum activities of aspartate transaminase, alanine transaminase, and alkaline phosphatase in a few patients (8).

Skin

Mexiletine can occasionally cause generalized rashes (SEDA-17, 224) (26,32), pseudolymphomatous change with erythroderma (35), and contact urticaria (36).

- Exfoliative dermatitis has been reported in a 68-year-old man who had taken mexiletine and diltiazem for 3 weeks (37). Patch tests with 1, 10, and 30% mexiletine and diltiazem in petrolatum were positive, but a lymphocyte stimulation test was negative.

- Drug eruptions occurred in a 56-year-old woman, a 50-year-old man, and a 66-year-old woman, who developed disseminated maculopapular eruptions with high fever after oral mexiletine (38). In all cases the liver transaminase activities were raised and there was an eosinophilia with atypical lymphocytes; in two cases there was a lymphadenopathy. In all cases patch tests were positive.
- An acute exanthematous pustular eruption has been reported in a 56-year-old man who had taken mexiletine 300 mg/day for 1 month (39). There was mild liver dysfunction. Patch tests with mexiletine 10 and 20% were subsequently positive, but a lymphocyte stimulation test was negative.

In addition to these cases, 37 cases of drug eruption due to mexiletine have been reported in Japan, with several common clinical features (37). The interval between initial drug therapy and the start of the eruption was relatively long (48–88 days); there was a high proportion of positive patch tests (86–97%) but a low incidence of positive lymphocyte stimulation tests (23–27%); there were frequent systemic symptoms, such as fever (93–94%) and liver dysfunction (43–78%); finally, some patients had multiple drug eruptions.

Immunologic

Mexiletine caused an increased incidence of positive antinuclear antibody (ANA) titers in some studies (8,40), but not in others (41,42). The clinical significance of this effect is not clear. For example, there have been no reports of a lupus-like syndrome attributable to mexiletine.

Susceptibility Factors

In a case of drug-induced hypersensitivity syndrome, in which there was a leukocytosis, eosinophilia, and liver damage, there was reactivation of human herpesvirus-6 (HHV-6) IgM (43). This patient was one of 17 in whom hypersensitivity reactions, toxic epidermal necrolysis, or Stevens–Johnson syndrome had occurred in response to a wide range of drugs. More than 3 weeks after the onset of drug-induced hypersensitivity syndrome, HHV-6 serological tests showed a rise in IgG antibodies in six patients, including one treated without glucocorticoids. HHV-6 DNA was detected in blood from three patients. In one patient with drug-induced hypersensitivity syndrome, there was reactivation of cytomegalovirus without reactivation of HHV-6, whereas in three patients anti-cytomegalovirus IgG antibodies rose after the rise in anti-HHV-6 IgG. Anti-HHV-7 IgG did not show change. The authors concluded that reactivation of human herpesvirus-6 in patients with drug-induced hypersensitivity syndrome is not due to non-specific reactivation induced by glucocorticoids, but to events specific to drug-induced hypersensitivity syndrome, and they hypothesized that drug-induced hypersensitivity syndrome may occur as a result of reactivation of human herpesvirus infection, especially HHV-6, accompanied by an allergic reaction to drugs, followed by a marked immune response to the virus, which is probably responsible for visceral involvement.

Acute severe diabetes mellitus has been reported as part of the spectrum of presentation of drug-induced hypersensitivity syndrome (44).

- A 46-year-old man with type 2 diabetes mellitus was given mexiletine 300 mg/day for diabetic peripheral neuropathy. After 41 days he developed pruritus, a diffuse macropapular rash with facial edema and erythema, and bilateral inguinal lymph node enlargement. His temperature was 38.1°C, heart rate 110/minute, and blood pressure 125/69 mmHg. His white blood cell count was 19 x 10^9/l (eosinophils 7%), aspartate transaminase activity 0.23 μkat/l (0.20–0.53), alanine transaminase 1.10 μkat/l (0.11–0.55), C-reactive protein 89 mg/l, and serum amylase 10.4 μkat/l (0.8–2.4). The titer of anti-human herpesvirus-6 IgG was over 1:320. He was given oral prednisolone 10 mg/day for 3 days, followed by daily intravenous betamethasone. Insulin was required to control the blood glucose concentration and he needed up to 62 U/day despite gradually reduced doses of betamethasone and normalization of the serum amylase activity to 1.33 μkat/l.

The authors proposed that fulminant type 1 diabetes had been associated with hypersensitivity to mexiletine and that reactivation of HHV-6 had played a part. The patient's HLA status also included DQA10303 and DQB10401, which have been associated with fulminant type 1 diabetes (45).

A drug-induced hypersensitivity syndrome occurred in a 66-year-old man taking mexiletine (46). Withdrawal of the mexiletine and glucocorticoid treatment led to temporary improvement, but tapering the glucocorticoid dose twice led to recrudescence, during which there were raised antibody titers against HHV-6 and cytomegalovirus and viral DNA in the blood, suggesting that these two viruses may have been involved in the recrudescence.

Renal disease

Mexiletine is mainly cleared polymorphically by CYP2D6 in the liver, and poor metabolizers and the slower among the extensive metabolizers have a higher incidence of mild adverse effects (nausea and light-headedness) (47). However, renal insufficiency can also be associated with an increase in plasma mexiletine concentrations (48).

Hepatic disease

Mexiletine is mainly cleared polymorphically by CYP2D6 in the liver, and poor metabolizers and the slower among the extensive metabolizers have a higher incidence of mild adverse effects (nausea and light-headedness) (47).

Drug Administration

Drug overdose

Overdosage with mexiletine resulted in death in two cases due to cardiovascular effects (49,50).

- Status epilepticus was the chief presenting feature in a 17-year-old boy who took an unspecified amount of mexiletine (51). The seizures responded to intravenous diazepam and phenytoin, but he also had agitation and hallucinations, which took 24 hours to abate. A urine specimen was positive for both benzodiazepines and amphetamines, but this was subsequently found to be a false positive result, because of the presence of large amounts of mexiletine, confirmed by thin-layer chromatography. The serum mexiletine concentration was 44 μmol/l, the usual target range being about 4–11 μmol/l.

Drug–Drug Interactions

Antacids

Mexiletine interacts with antacids, but the interaction is unlikely to be clinically important (52).

Antidysrhythmic drugs

Interactions of mexiletine with other cardioactive drugs have been reviewed (52). The most important are beneficial interactions with beta-adrenoceptor antagonists, quinidine, and amiodarone in the suppression of ventricular tachydysrhythmias. During these interactions the adverse effects of mexiletine may also be less common, although this effect is inconsistent (17).

Fluvoxamine

Mexiletine is metabolized by CYP2D6, CYP1A2, and CYP3A4; fluvoxamine inhibits CYP1A2. It is not surprising therefore that fluvoxamine 50 mg bd for 7 days increased the C_{max} and AUC of a single oral dose of mexiletine 200 mg in six healthy Japanese men (53).

Lidocaine

An interaction of lidocaine with mexiletine, which resulted in toxic concentrations of lidocaine, has been reported (54).

- An 80-year-old man with a dilated cardiomyopathy was given a lidocaine infusion started at 90 mg/hour for a ventricular tachycardia. He was already taking mexiletine 400 mg/day, and the plasma concentration was within the usual target range; however, the dose was reduced to 200 mg/day to avoid possible adverse effects. Intermittent ventricular tachycardia persisted, and so the lidocaine infusion was increased to 120 mg/day, but adverse effects (involuntary movements, muscle rigidity) were observed. The lidocaine infusion was stopped and within 20 minutes the adverse effects abated; the lidocaine concentration was 6.84 μg/ml. The ventricular tachycardia persisted, lidocaine was restarted at a lower rate, and the oral dose of mexiletine was increased to 450 mg/day. This resulted in an unexpectedly high concentration of lidocaine and the lidocaine concentration was significantly higher while the mexiletine dose was high.

Further studies suggested that mexiletine had displaced lidocaine from tissue binding sites. The authors suggested that this finding has implications for loading doses and acute effects of lidocaine in the concurrent therapy of lidocaine and mexiletine and highlighted the importance of close monitoring of lidocaine concentrations in this setting.

Omeprazole

Mexiletine is metabolized mainly by CYP2D6 and CYP1A2. Omeprazole is an inducer of CYP1A2 and might therefore be expected to interact with mexiletine. However, in a study in nine healthy men there was no evidence of an effect of steady-state omeprazole 40 mg/day on the single-dose kinetics of mexiletine 200 mg (55).

Phenytoin

Mexiletine interacts with phenytoin, but the interaction is unlikely to be clinically important (52).

Propafenone

Mexiletine and propafenone are metabolized by the same enzymes, CYP2D6, CYP1A2, and CYP3A4. In 15 healthy volunteers, eight of whom were extensive metabolizers of CYP2D6, administration of oral mexiletine 100 mg bd on days 1–8 and oral propafenone 1 mg bd on days 5–12 significantly reduced the clearance of R(–) mexiletine from 41 to 28 l/hour and of S(+) mexiletine from 43 to 29 l/hour in the extensive metabolizers (56). The new values were no different from the clearance values in the poor metabolizers. Propafenone also reduced the partial metabolic clearances of mexiletine to hydroxymethylmexiletine, parahydroxymexiletine, and metahydroxymexiletine by about 70% in the extensive metabolizers. Propafenone had no effect on the kinetics of mexiletine in the poor metabolizers. There were no electrocardiographic changes during this interaction. Smokers had higher clearance rates than non-smokers, but the effects of propafenone were similar in the two groups. In contrast, mexiletine had little effect on the disposition of propafenone. The authors proposed that these effects could explain at least in part the increased efficacy that sometimes occurs when mexiletine and propafenone are combined in patients in whom a single drug was not effective. They also recommended that the dosages of the drugs should be titrated slowly when they are used together, in order to reduce the risk of adverse effects.

Rifamycins

The interaction of rifampicin with mexiletine (52) has been reviewed (57). The mean half-life of a single dose of mexiletine 400 mg fell from 8.5 to 5 hours in eight healthy subjects who took rifampicin 600 mg/day for 10 days (58). This interaction is unlikely to be clinically important (52).

Sevoflurane

- A 79-year-old woman was given mexiletine 125 mg intravenously over 10 minutes during anesthesia after having been given lidocaine 100 mg intravenously, and had a marked drop in blood pressure 1 hour later (59). The blood pressure rose when sevoflurane was withdrawn.

The authors proposed that the effect had been brought about by the combination of mexiletine and sevoflurane, although it is more likely that the effect was due to the combination of mexiletine with lidocaine.

Xanthines

Mexiletine reduces the clearance of theophylline, and this combination has been reported to cause ventricular tachycardia (60). A similar interaction with caffeine has been reported (61).

References

1. Kerin NZ, Aragon E, Marinescu G, Faitel K, Frumin H, Rubenfire M. Mexiletine. Long-term efficacy and side effects in patients with chronic drug-resistant potentially lethal ventricular arrhythmias. Arch Intern Med 1990;150(2):381–4.
2. Murray KT, Barbey JT, Kopelman HA, Siddoway LA, Echt DS, Woosley RL, Roden DM. Mexiletine and tocainide: a comparison of antiarrhythmic efficacy, adverse effects, and predictive value of lidocaine testing. Clin Pharmacol Ther 1989;45(5):553–61.
3. Schwartz JB, Keefe D, Harrison DC. Adverse effects of antiarrhythmic drugs. Drugs 1981;21(1):23–45.
4. Grech-Belanger O. Clinical pharmacokinetics of mexiletine. Clin Progr Electrophysiol Pacing 1986;4:553.
5. Roden DM. Use of mexiletine in combination with other antiarrhythmic drugs. Clin Progr Electrophysiol Pacing 1986;4:561–7.
6. Halinen MO. Mexiletine for the management of ventricular arrhythmias in ischemic heart disease. Clin Progr Electrophysiol Pacing 1986;4:580–1.
7. Sami MH. Mexiletine: its role in the management of chronic ventricular arrhythmias. Clin Progr Electrophysiol Pacing 1986;4:582–8.
8. Flaker GC, Beach CL, Chapman D. Adverse side effects associated with mexiletine. Clin Progr Electrophysiol Pacing 1986;4:602–7.
9. Kopf A, Ruf W. Novel drugs for neuropathic pain. Curr Opin Anaesthesiol 2000;13:577–83.
10. Sloan P, Basta M, Storey P, von Gunten C. Mexiletine as an adjuvant analgesic for the management of neuropathic cancer pain. Anesth Analg 1999;89(3):760–1.
11. Lucetti C, Nuti A, Gambaccini G, Bernardini S, Brotini S, Manca ML, Bonuccelli U. Mexiletine in the treatment of torticollis and generalized dystonia. Clin Neuropharmacol 2000;23(4):186–9.
12. Wallace MS, Magnuson S, Ridgeway B. Efficacy of oral mexiletine for neuropathic pain with allodynia: a double-blind, placebo-controlled, crossover study. Reg Anesth Pain Med 2000;25(5):459–67.
13. Ando K, Wallace MS, Braun J, Schulteis G. Effect of oral mexiletine on capsaicin-induced allodynia and hyperalgesia: a double-blind, placebo-controlled, crossover study. Reg Anesth Pain Med 2000;25(5):468–74.
14. Nabulsi LH, McLendon BM, Vondracek TG. Mexiletine for diabetic peripheral neuropathy. J Pharm Technol 2000;16:8–11.
15. Kemper CA, Kent G, Burton S, Deresinski SC. Mexiletine for HIV-infected patients with painful peripheral neuropathy: a double-blind, placebo-controlled, crossover treatment trial. J Acquir Immune Defic Syndr Hum Retrovirol 1998;19(4):367–72.
16. Kieburtz K, Simpson D, Yiannoutsos C, Max MB, Hall CD, Ellis RJ, Marra CM, McKendall R, Singer E, Dal Pan GJ, Clifford DB, Tucker T, Cohen B, Jatlow P, Kasdan P, Shriver S, Martinez A, Millar L, Colquhoun D, Zaborski L, Dias V, Jubelt B, Noseworthy J, Barton B, Sharer L, Kerza A, Sperber K, Chusid E, Gerits P, et alAIDS Clinical Trial Group 242 Protocol Team. A randomized trial of amitriptyline and mexiletine for painful neuropathy in HIV infection. Neurology 1998;51(6):1682–8.
17. Poole JE, Werner JA, Bardy GH, Graham EL, Pulaski WP, Fahrenbruch CE, Greene HL. Intolerance and ineffectiveness of mexiletine in patients with serious ventricular arrhythmias. Am Heart J 1986;112(2):322–6.
18. Morganroth J. Comparative efficacy and safety of oral mexiletine and quinidine in benign or potentially lethal ventricular arrhythmias. Am J Cardiol 1987;60(16):1276–81.
19. Fassoulaki A, Patris K, Sarantopoulos C, Hogan Q. The analgesic effect of gabapentin and mexiletine after breast surgery for cancer. Anesth Analg 2002;95(4):985–91.
20. Manolis AS, Deering TF, Cameron J, Estes NA 3rd. Mexiletine: pharmacology and therapeutic use. Clin Cardiol 1990;13(5):349–59.
21. Gottlieb SS, Weinberg M. Cardiodepressant effects of mexiletine in patients with severe left ventricular dysfunction. Eur Heart J 1992;13(1):22–7.
22. Talbot RG, Nimmo J, Julian DG, Clark RA, Neilson JM, Prescott LF. Treatment of ventricular arrhythmias with mexiletine (Ko 1173). Lancet 1973;2(7826):399–404.
23. Campbell NP, Kelly JG, Shanks RG, Chaturvedi NC, Strong JE, Pantridge JF. Mexiletine (Ko 1173) in the management of ventricular dysrhythmias. Lancet 1973;2(7826):404–7.
24. Campbell NP, Chaturvedi NC, Shanks RG, Kelly JG, Strong JE, Adgey AA. The development of mexiletine in the management of ventricular dysrhythmias. Postgrad Med J 1977;53(Suppl 1):114–9.
25. Bero CJ, Rihn TL. Possible association of pulmonary fibrosis with mexiletine. DICP 1991;25(12):1329–31.
26. Roos JC, Paalman DC, Dunning AJ. Electrophysiological effects of mexiletine in man. Postgrad Med J 1977;53(Suppl 1):92–6.
27. Sakurai M, Kanazawa I. Positive symptoms in multiple sclerosis: their treatment with sodium channel blockers, lidocaine and mexiletine. J Neurol Sci 1999;162(2):162–8.
28. Leong MS, Isolani F, Gaeta RR. Mexiletine and persistent ophthalmic changes. Pain Med 2001;2(3):228–9.
29. Girmann G, Pees H, Scheurlen PG. Pseudothrombocytopenia and mexiletine. Ann Intern Med 1984;100(5):767.
30. Campbell NP, Pantridge JF, Adgey AA. Long-term oral antiarrhythmic therapy with mexiletine. Br Heart J 1978;40(7):796–801.
31. Fasola GP, D'Osualdo F, de Pangher V, Barducci E. Thrombocytopenia and mexiletine. Ann Intern Med 1984;100(1):162.
32. Higa K, Hirata K, Dan K. Mexiletine-induced severe skin eruption, fever, eosinophilia, atypical lymphocytosis, and liver dysfunction. Pain 1997;73(1):97–9.
33. Rudolph R, Seggewiss H, Seckfort H. Ösophagus-Ulcus durch Mexiletin. [Esophageal ulcer caused by mexiletine.] Dtsch Med Wochenschr 1983;108(26):1018–20.
34. Pernot C, Marcon F, Weber JL, Nether P, Trechot P. Effets indésirables hépatiques de la méxiletine. Therapie 1983;38:695–700.

35. Sigal M, Pulik M. Pseudo-lymphomes médicamenteux à expression cutanée prédominante. [Drug-induced pseudolymphoma with predominantly cutaneous manifestation.] Ann Dermatol Venereol 1993;120(2):175–80.
36. Yamazaki S, Katayama I, Kurumaji Y, Yokozeki H, Nishioka K. Contact urticaria induced by mexiletine hydrochloride in a patient receiving iontophoresis. Br J Dermatol 1994;130(4):538–40.
37. Umebayashi Y. Drug eruption due to mexiletine and diltiazem. Nishinihon J Dermatol 2000;62:80–2.
38. Kayaba M, Tanaka T, Misago N, Narisawa Y. Three cases of hypersensitivity syndrome due to mexiletine hydrochloride. Nishinihon J Dermatol 2000;62:338–42.
39. Sasaki K, Yamamoto T, Kishi M, Yokozeki H, Nishioka K. Acute exanthematous pustular drug eruption induced by mexiletine. Eur J Dermatol 2001;11(5):469–71.
40. Stein J, Podrid PJ, Lampert S, Hirsowitz G, Lown B. Long-term mexiletine for ventricular arrhythmia. Am Heart J 1984;107(5 Pt 2):1091–8.
41. Johansson BW, Stavenow L. Long-term clinical effects and side effects of mexiletine in patients with ventricular arrhythmias. Clin Progr Electrophysiol Pacing 1986;4:589–94.
42. Johansson BW, Stavenow L, Hanson A. Long-term clinical experience with mexiletine. Am Heart J 1984;107(5 Pt 2):1099–102.
43. Aihara M, Mitani N, Kakemizu N, Yamakawa Y, Inomata N, Ito N, Komatsu H, Aihara Y, Ikezawa Z. Human herpesvirus infection in drug-induced hypersensitivity syndrome, toxic epidermal necrolysis and Stevens–Johnson syndrome. Allergol Int 2004;53:23–9.
44. Seino Y, Yamauchi M, Hirai C, Okumura A, Kondo K, Yamamoto M, Okazaki Y. A case of fulminant Type 1 diabetes associated with mexiletine hypersensitivity syndrome. Diabet Med 2004;21(10):1156–7.
45. Tanaka S, Kobayashi T, Nakanishi K, Koyama R, Okubo M, Murase T, Odawara M, Inoko H. Association of HLA-DQ genotype in autoantibody-negative and rapid-onset type 1 diabetes. Diabetes Care 2002;25:2302–7.
46. Sekiguchi A, Kashiwagi T, Ishida-Yamamoto A, Takahashi H, Hashimoto Y, Kimura H, Tohyama M, Hashimoto K, Iizuka H. Drug-induced hypersensitivity syndrome due to mexiletine associated with human herpes virus 6 and cytomegalovirus reactivation. J Dermatol 2005;32(4):278–81.
47. Lledo P, Abrams SM, Johnston A, Patel M, Pearson RM, Turner P. Influence of debrisoquine hydroxylation phenotype on the pharmacokinetics of mexiletine. Eur J Clin Pharmacol 1993;44(1):63–7.
48. Nora MO, Chandrasekaran K, Hammill SC, Reeder GS. Prolongation of ventricular depolarization. ECG manifestation of mexiletine toxicity. Chest 1989;95(4):925–8.
49. Jequier P, Jones R, Mackintosh A. Fatal mexiletine overdose. Lancet 1976;1(7956):429.
50. Mackintosh AF, Jequier P. Fatal mexiletine overdose. Postgrad Med J 1977;53(Suppl 1):134.
51. Kozer E, Verjee Z, Koren G. Misdiagnosis of a mexiletine overdose because of a nonspecific result of urinary toxicologic screening. N Engl J Med 2000;343(26):1971–2.
52. Bigger JT Jr. The interaction of mexiletine with other cardiovascular drugs. Am Heart J 1984;107(5 Pt 2):1079–85.
53. Kusumoto M, Ueno K, Oda A, Takeda K, Mashimo K, Takaya K, Fujimura Y, Nishihori T, Tanaka K. Effect of fluvoxamine on the pharmacokinetics of mexiletine in healthy Japanese men. Clin Pharmacol Ther 2001;69(3):104–7.
54. Maeda Y, Funakoshi S, Nakamura M, Fukuzawa M, Kugaya Y, Yamasaki M, Tsukiai S, Murakami T, Takano M. Possible mechanism for pharmacokinetic interaction between lidocaine and mexiletine. Clin Pharmacol Ther 2002;71(5):389–97.
55. Kusumoto M, Ueno K, Tanaka K, Takeda K, Mashimo K, Kameda T, Fujimura Y, Shibakawa M, Guzman WM, Laplante S. Lack of pharmacokinetic interaction between mexiletine and omeprazole. Ann Pharmacother 1998;32(2):182–4.
56. Labbe L, O'Hara G, Lefebvre M, Lessard E, Gilbert M, Adedoyin A, Champagne J, Hamelin B, Turgeon J. Pharmacokinetic and pharmacodynamic interaction between mexiletine and propafenone in human beings. Clin Pharmacol Ther 2000;68(1):44–57.
57. Baciewicz AM, Self TH. Rifampin drug interactions. Arch Intern Med 1984;144(8):1667–71.
58. Pentikainen PJ, Koivula IH, Hiltunen HA. Effect of rifampicin treatment on the kinetics of mexiletine. Eur J Clin Pharmacol 1982;23(3):261–6.
59. Kudo M, Ohke H, Kawai T, Kato M, Kokubu M, Shinya N. Geriatric patient who suffered transitory cardiovascular collapse under sevoflurane anesthesia due to continuous medication with mexiletine hydrochloride. J Jpn Dent Soc Anesthesiol 1999;27:614–8.
60. Kessler KM, Interian A Jr, Cox M, Topaz O, De Marchena EJ, Myerburg RJ. Proarrhythmia related to a kinetic and dynamic interaction of mexiletine and theophylline. Am Heart J 1989;117(4):964–6.
61. Joeres R, Richter E. Mexiletine and caffeine elimination. N Engl J Med 1987;317(2):117.

Moracizine

General Information

Moracizine (moricizine) has Class I antidysrhythmic actions that cannot be easily further subclassified. Its pharmacology, clinical pharmacology, clinical uses, adverse effects and interactions have been reviewed (1–3).

In a retrospective study of 85 patients with recurrent atrial fibrillation (mean left atrial size 46 mm, mean left ventricular ejection fraction 0.51), 69 of whom had structural heart disease, moricizine (mean dose 609 mg/day) was withdrawn because of unacceptable adverse effects in six patients: frequent ventricular extra beats and short runs of non-sustained ventricular tachycardia ($n = 2$); significant widening of the QRS complex ($n = 1$); first-degree heart block ($n = 1$); a significant rise in liver enzymes ($n = 1$); and a severe rash ($n = 1$) (4). Six patients developed transient adverse effects that resolved spontaneously without withdrawal: brief self-limiting episodes of atrial flutter ($n = 2$); a small increase in blood pressure ($n = 1$); generalized weakness and tremor ($n = 2$); and reduced appetite ($n = 1$).

Organs and Systems

Cardiovascular

The overall risk of dysrhythmias with moracizine seems to be similar to that of other antidysrhythmic drugs, at around 10% (5), although lower rates have been reported. In

CAST-II (the Cardiac Arrhythmia Suppression Trial) moracizine increased mortality in patients with asymptomatic ventricular dysrhythmias after myocardial infarction (6).

There have been anecdotal reports of conduction defects (SEDA-17, 225).

Moracizine has little or no effect on hemodynamics, but it sometimes has adverse hemodynamic effects and can exacerbate heart failure (SEDA-17, 225).

Hematologic

Moracizine has an anti-aggregatory effect on platelets, and thrombocytopenia has been reported (7).

Liver

Moracizine can cause increased serum bilirubin concentrations and serum transaminase activities (7).

Susceptibility Factors

Hepatic disease

Moracizine is cleared by the liver, and dosages should be reduced in liver disease (8).

Drug–Drug Interactions

Cimetidine

Cimetidine inhibits the metabolism of moracizine, but the effect is probably clinically unimportant (9).

Diltiazem

Moracizine is an enzyme inducer; it increases the rate of clearance of diltiazem (10). Conversely, diltiazem caused a doubling of the AUC of moracizine (10) in healthy volunteers.

Ethacizine

The combination of moracizine with another antidysrhythmic drug ethacizine, in the weight ratio of 6:1, has been marketed in Russia under the name of metacizine (Ethmocor®). The pharmacokinetics of the two drugs when given separately and together in single doses have been studied in eight healthy subjects (11). Ethacizine prolonged the half-life of moracizine by increasing its volume of distribution without a change in clearance. Moracizine prolonged the half-life of ethacizine by reducing its clearance despite a parallel reduction in volume. Exposure to both drugs was increased when they were given in combination.

Phenazone (antipyrine)

Moracizine is an enzyme inducer; it increases the rate of clearance of phenazone (antipyrine) (12).

Theophylline

Moracizine is an enzyme inducer; it increases the rate of clearance of theophylline (13).

References

1. Carnes CA, Coyle JD. Moricizine: a novel antiarrhythmic agent. DICP 1990;24(7–8):745–53.
2. Podrid PJ. Moricizine (ethmozine HCl)—a new antiarrhythmic drug: is it unique? Am J Cardiol 1991;68(15):1521–5.
3. Clyne CA, Estes NA 3rd, Wang PJ. Moricizine. N Engl J Med 1992;327(4):255–60.
4. Geller JC, Geller M, Carlson MD, Waldo AL. Efficacy and safety of moricizine in the maintenance of sinus rhythm in patients with recurrent atrial fibrillation. Am J Cardiol 2001;87(2):172–7.
5. Podrid PJ, Lampert S, Graboys TB, Blatt CM, Lown B. Aggravation of arrhythmia by antiarrhythmic drugs—incidence and predictors. Am J Cardiol 1987;59(11):E38–44.
6. The Cardiac Arrhythmia Suppression Trial II Investigators. Effect of the antiarrhythmic agent moricizine on survival after myocardial infarction. N Engl J Med 1992;327(4):227–33.
7. Kennedy HL. Noncardiac adverse effects and organ toxicity of moricizine during short- and long-term studies. Am J Cardiol 1990;65(8):D47–50.
8. Kurapov AP, Nekrasova OV, Gneushev ET, Ryzhenkova AP, Kukes VG. Farmokokinetika etmozina pri nedostatochnosti funkstii pecheni. [Ethmozine pharmacokinetics in liver insufficiency.] Sov Med 1990;(5):34–6.
9. Biollaz J, Shaheen O, Wood AJ. Cimetidine inhibition of ethmozine metabolism. Clin Pharmacol Ther 1985;37(6):665–8.
10. Shum L, Pieniaszek HJ Jr, Robinson CA, Davidson AF, Widner PJ, Benedek IH, Flamenbaum W. Pharmacokinetic interactions of moricizine and diltiazem in healthy volunteers. J Clin Pharmacol 1996;36(12):1161–8.
11. Beloborodov VL, Bugrii EM, Zalesskaya MA, Tyukavkina NA, Kaverina NN. Clinical pharmacokinetics of ethmozine and ethacizine in the course of combined administration. Pharm Chem J 2004;38(2):59–62.
12. Benedek IH, Davidson AF, Pieniaszek HJ Jr. Enzyme induction by moricizine: time course and extent in healthy subjects. J Clin Pharmacol 1994;34(2):167–75.
13. Pieniaszek HJ Jr, Davidson AF, Benedek IH. Effect of moricizine on the pharmacokinetics of single-dose theophylline in healthy subjects. Ther Drug Monit 1993;15(3):199–203.

Procainamide

See also Acecainide

General Information

In a prospective study of 488 inpatients there were adverse reactions in 45 cases (9.2%), thought to have been life-threatening in 7 (1.4%), none of whom died (1). The seven patients all had cardiovascular effects. Common adverse effects included gastrointestinal upsets (19 cases) and fever (8 cases). Reactions were more common at daily dosages of 3 g and more.

Comparative studies

The adverse effects of intravenous procainamide (400 mg up to three times infused over 10 minutes) have been reported in 60 adults with atrial flutter or fibrillation in a comparison with ibutilide (2). The adverse effects were headache in 11%, hypotension in 11%, flushing in 3.1%, dizziness in 3.1%, and hypesthesia in 3.1%. The mean fall in systolic blood pressure was about 20 mmHg and occurred at 30–35 minutes after infusion; the corresponding fall in diastolic blood pressure was 10 mmHg. However, in seven patients there was severe hypotension, with a fall in diastolic blood pressure of up to 67 mmHg; in three cases withdrawal of the infusion was required and these patients were treated with intravenous fluids, dopamine, or both. In the severe cases the hypotension occurred during or immediately after the infusion of procainamide.

A comparison between procainamide and propafenone in 62 patients, who had undergone coronary artery bypass grafting or valvular surgery within 3 weeks and developed sustained atrial fibrillation, showed that both drugs converted the dysrhythmia to sinus rhythm in up to 76% of cases, but that propafenone did it more quickly (3). Symptomatic arterial hypotension occurred more frequently with procainamide (nine of 33 patients) than with propafenone (two of 29 patients). Other adverse effects of procainamide were nausea (n = 2) and junctional escape rhythm (n = 2).

General adverse effects

The adverse effects of procainamide are predominantly on the heart. It causes reduced myocardial contractility and hypotension, and prolongs the QT interval, with consequent dysrhythmias and conduction defects. The lupus-like syndrome most commonly causes polyarthralgia, myalgia, fever, and pleurisy. Neutropenia has been relatively commonly reported in patients taking modified-release formulations. Other adverse effects are uncommon; these include muscle weakness, ataxia, mental confusion, cholestasis, and skin rashes. Hypersensitivity reactions include fever and hematological reactions, including neutropenia, pancytopenia, and pure red cell aplasia. Tumor-inducing effects have not been reported.

Organs and Systems

Cardiovascular

Procainamide has a negative inotropic effect and can cause hypotension after both intravenous and oral administration (4,5). When given intravenously it should therefore be infused slowly, at no more than 20 mg/minute. In patients with poor cardiac function procainamide can worsen heart failure, and it may reduce survival after myocardial infarction (6).

Procainamide prolongs the QT interval (7) and can cause dysrhythmias. It can also impair cardiac conduction and can cause bradycardia and heart block (1). In the sick sinus syndrome it can alter sinus node recovery time (8), although the clinical significance of this is not clear.

Pericarditis and tamponade have been reported as rare complications of procainamide-induced lupus-like syndrome (9).

Respiratory

Procainamide can cause lung damage in the context of a lupus-like syndrome (SEDA-17, 226).

Nervous system

Procainamide rarely causes nervous system effects. Acute confusion (10), cerebellar ataxia (11), tremor (12), and muscle weakness (13–16) have all been occasionally reported. In high dosages procainamide has anticholinergic effects (17).

Procainamide has been reported to cause a chronic inflammatory demyelinating polyradiculoneuropathy (18).

- A 68-year-old man took procainamide 500 mg qds for 3 years and developed distal paresthesia and dysesthesia in the legs, followed by progressive muscle weakness, mainly affecting the legs. His gait became unsteady and was wide-based. He had antinuclear antibodies directed against histones in a titer of 1:320, but no antibodies to double-stranded DNA. He had a circulating lupus anticoagulant. The serum procainamide concentration was 3.3 μg/ml (target range 4–8). Nerve conduction studies showed a reduction in sensory nerve action potential amplitudes, a mild reduction in sensory nerve conduction velocity, prolongation of distal motor latencies, and reduced conduction velocities, but no conduction block or temporal dispersions. Electromyography was normal. A left sural nerve biopsy showed perivascular inflammation around a single vessel, without evidence of vasculitis. Myelinated nerve fibers were reduced, and scattered nerve fibers showed thin myelin sheaths. About 30% of the fibers showed randomly distributed demyelinated or remyelinated segments. Procainamide was withdrawn and prednisone was given in combination with six plasma exchanges over 2 weeks; after 1 month there was clinical improvement.

This case of polyneuropathy was attributed to a lupus-like effect of procainamide.

Sensory systems

Scleritis has been reported as part of a procainamide-induced, lupus-like syndrome (19).

Psychological, psychiatric

Acute psychosis has been attributed to procainamide (20).

- A 45-year-old woman developed an acute psychosis within 72 hours of starting to take procainamide 75 mg intravenously, followed by a continuous infusion of 2 mg/minute for atrial fibrillation. The plasma procainamide concentration was 8.2 μg/ml and the plasma concentration of the main acetylated metabolite, acecainide, was 4.6 μg/ml. She was then given oral procainamide 500 mg qds, and 2 days later her trough concentrations of procainamide and acecainide were 4.5 and 4.9 μg/ml respectively. The following day she had visual hallucinations and was later found wandering the hospital asking about the babies under her bed. She had no previous history of psychiatric illness and she recovered completely 24 hours after withdrawal of procainamide.

There have been a few previous reports of similar adverse effects with procainamide in therapeutic dosages, and in most cases the plasma concentrations of procainamide and acecainide have been within the usual target ranges, as in this case.

Hematologic

Hematological abnormalities can occur in the absence of a lupus-like syndrome. Hemolytic anemia (21), a circulating anticoagulant (22), thrombocytopenia (23), granulocytopenia (24), and pancytopenia (25) have all been reported occasionally. Thrombocytopenia may be more common in patients taking modified-release formulations (23). Pure red cell aplasia has also been reported (26,27,28), but it was not clear whether or not there was an associated lupus-like syndrome.

The most common adverse hematological effect of procainamide is neutropenia, which has often been reported, particularly in patients taking modified-release formulations. In one case-control study 4.4% of 114 patients taking modified-release formulations had neutropenia, compared with none in a control group of 509 patients (29). However, a larger subsequent case-control study failed to confirm this association (30). Nevertheless, reports of neutropenia attributed to procainamide continue to appear (31).

There was a positive direct antiglobulin (Coombs') test in about 20% of elderly patients taking procainamide (32). Although a positive Coombs' test can occur in association with a lupus-like syndrome, there was no relation in these cases between positive tests and the presence of antinuclear antibodies. Three of the patients had an autoimmune hemolytic anemia.

Gastrointestinal

Nausea, vomiting, and diarrhea in response to procainamide are common with dosages of 4 g/day or more (33). Pseudo-obstruction has been attributed to the use of a modified-release formulation of procainamide, perhaps due to its anticholinergic effects (34).

Liver

Procainamide can cause intrahepatic cholestasis, perhaps as part of a hypersensitivity reaction (35).

Skin

Lichen planus has been attributed to procainamide as part of a lupus-like syndrome (36).

Procainamide has been reported to cause urticarial vasculitis, although it was not clear whether or not this was part of a lupus-like syndrome (37).

- A 70-year-old man developed a maculopapular skin rash on the trunk 10 days after starting to take procainamide 1.5 g/day (38). Resolution occurred soon after the withdrawal of procainamide and a similar skin rash occurred on rechallenge. In addition to the maculopapular rash, there was swelling and erythema around the eyes and a purpuric rash on the legs. At the time of rechallenge he also had an eosinophilia.

Musculoskeletal

Procainamide can cause an arthropathy as part of a lupus-like syndrome, and the histological findings are indistinguishable from those in idiopathic SLE (39).

Procainamide has occasionally been reported to cause muscle weakness (SEDA-14, 152) (SEDA-16, 183), and it can also cause or exacerbate myasthenia gravis (40).

Necrotizing myopathy of the diaphragm has been reported, perhaps as part of a lupus-like syndrome (41).

Immunologic

Lupus-like syndrome

DoTS classification (BMJ 2003;327:1222–5)
Dose-relation: collateral effect
Time-course: delayed
Susceptibility factors: genetic (slow acetylators)

Procainamide is one of the common causes of drug-induced lupus-like syndrome (42), which is contrasted with idiopathic lupus erythematosus in Table 1.

Frequency

About 29–35% of patients taking procainamide for at least a year are affected and the effect is dose-related. The average age of onset is 59–68 years and 35–58% of the subjects are women. The syndrome can come on within a few weeks, but has been reported as late as 9 years after starting treatment.

Table 1 The contrast between drug-induced lupus-like syndrome and idiopathic lupus erythematosus

Feature	*Idiopathic lupus erythematosus*	*Drug-induced lupus-like syndrome*
Age and sex	Typically young women	Any (depends on use)
Acetylator status	Any	More likely in slow acetylators
Organs involved	Any	Kidneys usually spared
Antinuclear antibody	Usually present	Usually present
Complement	Can be reduced	Usually normal
Anti-DNA antibodies	Usually present (native DNA)	Only to single-stranded DNA

Mechanism

The mechanisms whereby procainamide causes this lupus-like syndrome are not clear. Procainamide is associated with the production of many antibodies, including antihistone, antiguanosine, anti-DNA, and antiphospholipid antibodies (43). The production of autoantibodies may be due to one of two major mechanisms: first, procainamide may act as a hapten, binding to DNA, nuclear protein, or some membrane constituent, the hapten-protein complex stimulating the production of antibodies; secondly, it may alter suppressor cell function (44). There is also evidence that procainamide can combine with ribonucleoprotein from damaged myocardium after myocardial infarction, thus precipitating the production of antibodies to ribonucleoprotein (45).

Thymus function in 10 patients with symptomatic procainamide-induced lupus has been compared with that in 13 asymptomatic patients who only developed drug-induced autoantibodies (46). Newly generated T cells were detected in all the subjects. Although there was no overall quantitative difference between the symptomatic and asymptomatic patients, there was a correlation between the level of T cell receptor rearrangement excision circles in peripheral lymphocytes and serum IgG antichromatin antibody activity in patients with drug-induced lupus. These results support the hypothesis that the thymus is important in the genesis of drug-induced lupus-like syndrome and that the production of autoreactive T cells starts in the thymus when procainamide hydroxylamine alters T cell tolerance.

In procainamide-induced lupus there is an increase in the number of B cells in both blood and pleural fluid to about 80% (normal 10–25%). Concentrations of IL-6 and soluble IL-2R are also increased (47).

IgG antibodies to the (H2A–H2B)-DNA complex are common in patients with immunological reactions to procainamide (SEDA-20, 178). In a prospective study of 62 patients who had taken procainamide for a mean duration of 23 months (range 1 month to 16 years), and excluding patients with pre-existing systemic lupus erythematosus or who were taking other drugs that have been associated with the lupus-like syndrome, nine developed evidence of lupus-like syndrome and were compared with the other 53 (48). The mean dosage in the patients with lupus-like syndrome was 3.7 g/day compared with 3 g/day in the others, and the durations of administration were 8.8 and 38 months respectively. All four patients who had polyarticular arthritis and/or pleural effusions were positive for the IgG antibodies and four of the other five, who had significantly lower IgG concentrations, presented with the more typical manifestations of vasculitic rashes, pericarditis, or agranulocytosis. In all cases the symptoms and signs either improved markedly or resolved within 3 weeks of withdrawal of procainamide. Seven asymptomatic patients had IgG concentrations comparable with those seen in the patients with lupus-like syndrome, and high concentrations of antibody, although not as high as the two highest in the latter group. Anti-double-stranded DNA antibodies were not detected in any patients. In 33 control subjects there were no detectable antibodies of any kind. The authors therefore suggested that the (H2A-H2B)-DNA IgG antibody is not a useful marker for the occurrence of lupus-like syndrome in patients taking procainamide, although they conceded that the presence of IgG antibody supported the diagnosis of lupus-like syndrome when the main clinical manifestations were arthritis or pleurisy. The significance of the presence of IgG antibodies in asymptomatic patients is not clear, but the authors suggested that it meant that those patients merited careful observation.

The lupus-like syndrome in patients taking procainamide is thought to be due to the procainamide itself, since it has rarely been reported when the main metabolite of procainamide, acecainide, has been administered itself (SEDA-18, 206). Alternatively, the syndrome may be due to a different metabolite of procainamide, and a hydroxylated derivative has been implicated (SEDA-15, 178) (49). CYP2D6 is the major isoform of cytochrome P450 that is involved in the hydroxylation of procainamide, and in the production of some other metabolites (50). It remains to be seen whether extensive CYP2D6 metabolizers are more likely to develop lupus-like syndrome than poor metabolizers.

Susceptibility factors

The lupus-like syndrome is more likely to occur in slow acetylators than in fast acetylators (51), and the rate of development of antinuclear antibody depends on acetylator status (52).

Presentation

The most common feature is arthralgia, in about 77% of cases, with pleural or lung involvement in about 75%. Other common features include myalgia, fever, hepatomegaly, pericarditis, arthritis, and splenomegaly. Skin rashes, adenopathy, and Raynaud's phenomenon occur in 5–10% of cases, and neuropsychiatric and renal involvement are rare. Thrombotic problems can occur because of the properties of anti-DNA and antiphospholipid antibodies (discussed in the next case report). The so-called lupus anticoagulant can be detected in people taking procainamide, even without clinical evidence of lupus (53). Angioedema has been reported in a patient who had no history of hereditary angioedema (54).

A lupus-like syndrome with an antiphospholipid syndrome has been attributed to procainamide in a patient with pre-existing systemic sclerosis (55).

- A 51-year-old Korean man with systemic sclerosis was given procainamide 2–3 g/day for suppression of ventricular dysrhythmias. About 2 years later he noticed a new skin ulcer on one of his toes. His dorsalis pedis arteries were not palpable and there was tenderness over the proximal interphalangeal and metacarpophalangeal joints. He had a pancytopenia, prolonged coagulation (with prolongation of the activated partial thromboplastin time and prothrombin time and reduced concentrations of factors XI and XII), an increase in the plasma concentration of von Willebrand factor, a raised serum creatinine concentration, a raised serum C-reactive protein concentration, hypergammaglobulinemia, and hypocomplementemia. He had positive circulating immune complexes and mixed-type cryoglobulinemia, antinuclear antibodies, anti-DNA topoisomerase I antibodies, and a positive LE cell preparation. He had anti-DNA antibodies with a high titer of antibodies to single-stranded DNA and a slightly raised titer of antibodies to double-stranded DNA. Anti-U1 ribonuclear protein, anti-Sm, and anticentromere antibodies were negative. There were high titers of beta$_2$-glycoprotein-I-dependent IgG anticardiolipin antibodies and the lupus anticoagulant test was positive. There were antihistone antibodies. Procainamide was withdrawn and prednisolone and azathioprine given. The pancytopenia, coagulopathy, and renal dysfunction resolved, and his general condition improved; serum concentrations of several of the antibodies returned to normal.

The authors suggested that the pre-existence of systemic sclerosis in this case and the presence of allele HIA-DQBP1* 0303 had increased the patient's susceptibility to the lupus-like syndrome and antiphospholipid syndrome.

Diagnosis

The antinuclear antibody is positive in virtually all cases and the ESR is often raised. Antihistone antibodies are also present in most cases. The prevalence of serum autoantibodies to high-mobility group (HMG) proteins in the serum of patients with drug-induced lupus-like syndrome varies from protein to protein: 67% for HMG-14 and/or HMG-17 compared with 21% for HMG-1 and/or HMG-2. Procainamide-induced lupus is also associated with antibodies to the H2A–H2B dimer (56,57).

Antinuclear antibody is usually present (in 83% of cases), but antibodies to native DNA are not found, although there may be antibodies to single-stranded DNA. Patients taking procainamide can have antinuclear antibodies without developing a lupus-like syndrome.

Management

The syndrome usually regresses rapidly after withdrawal of procainamide, but in a few patients recovery may be delayed; if there are serious effects oral glucocorticoid therapy may be required.

Antiphospholipid antibody syndrome

The antiphospholipid antibody syndrome can cause widespread cutaneous necrosis, which has been reported in a case attributed to procainamide (58).

- A 51-year-old man developed multiple diffuse painful nodules, which coalesced, increased in size and number, ulcerated, and turned reddish-black. He had taken procainamide (Procan-SR) 500 mg qds for atrial tachydysrhythmias for 6 months. His other medications included fosinopril and glipizide. He had a normochromic, normocytic anemia, mild leukopenia, and thrombocytopenia. Cultures of the blood, urine, and skin were negative. Serology for auto-antibodies was negative, except for a low-titer of antinuclear antibody. The prothrombin time was 13 seconds and the activated partial thromboplastin time was prolonged to 57 seconds. A lupus anticoagulant was identified. Serology for anticardiolipin antibody was negative. A biopsy from the edge of the ulcer showed microthrombi in the dermal microvasculature with very minimal mononuclear inflammatory infiltration in the perivascular area. There was no leukocytoclastic vasculitis. Procainamide was withdrawn and high-dose intravenous methylprednisolone, heparin, and aspirin started. No further skin lesions developed and the ulcers healed.

Body temperature

Fever occasionally occurs in patients taking procainamide (SEDA-1, 155) and has been attributed to an allergic reaction (59).

Susceptibility Factors

Renal disease

Maintenance dosages should be reduced in the presence of renal or hepatic impairment, including hepatic congestion due to cardiac failure (60).

Hepatic disease

Maintenance dosages should also be reduced in the presence of renal or hepatic impairment, including hepatic congestion due to cardiac failure (60).

Other features of the patient

Care must be taken in patients with pre-existing connective tissue disease or heart failure. Loading dosages should be reduced in cardiac failure, because of a lowered apparent volume of distribution (60).

Drug Administration

Drug overdose

Procainamide overdose has been reported.

- A 14-year-old boy took about 21 g of procainamide and developed abdominal pain, weakness, blurred vision, dry mouth, pain on swallowing, and headache (61). His pupils were dilated, his skin dry and pale, and his mucous membranes dry. His blood pressure was 106/49 mmHg, his heart rate 91/minute. Following a tonic-clonic seizure his blood pressure was 125/57 mmHg and his heart rate 136/minute in sinus tachycardia. He became lethargic with slurred speech. He was given repeated doses of activated charcoal and made a full recovery. The serum procainamide and acecainide (*N*-acetylprocainamide) concentrations were 63 and 80 μg/ml respectively.
- A 79-year-old man took about 19 g of procainamide and developed lethargy, vomiting, a wide-complex tachycardia, hypotension, and coma (62). His serum procainamide concentration was 77 μg/ml at 3 hours. He was treated with vasopressors and peritoneal dialysis.

- A 67-year-old woman took about 7 g of procainamide and developed nausea, vomiting, lethargy, a junctional tachycardia, hypotension, and oliguria (63). She was treated with hemodialysis.

Drug–Drug Interactions

Amiodarone

The effects of procainamide on the QT interval may be potentiated by other drugs with this action, for example amiodarone (64).

The pharmacokinetics of procainamide are altered by amiodarone, with a reduction in clearance of about 25% due to changes in both renal and non-renal clearances (64).

Cimetidine

The renal clearance of procainamide is inhibited by cimetidine (65,66).

Class I antidysrhythmic drugs

The effects of procainamide on the QT interval can be potentiated by other drugs with this action, for example other class I antidysrhythmic drugs (64).

Glucose

Procainamide interacts with glucose in vitro to form glucosylamines (65). The reaction was pH-dependent, with a maximum rate of association at a pH of 3.0 and a maximum rate of dissociation at a pH of 1.5. The authors suggested that the loss of procainamide in an intravenous solution of glucose could be marked.

Levofloxacin

In a randomized, crossover drug interaction study in 10 healthy adults, levofloxacin reduced the renal clearances and prolonged the half-lives of both procainamide and N-acetylprocainamide (acecainide); ciprofloxacin reduced the renal clearances of procainamide and acecainide only slightly (67).

Ofloxacin

The renal clearance of procainamide is inhibited by ofloxacin (68).

Trimethoprim

The renal tubular clearance of procainamide is inhibited by trimethoprim (69,70).

Monitoring Therapy

The use of serum procainamide concentration measurements in monitoring therapy has been reviewed (60,71). Serum concentrations of 4–10 μg/ml are associated with therapeutic benefit in over 90% of patients with ventricular tachydysrhythmias, and toxicity becomes highly likely over 12 μg/ml. However, the main metabolite of procainamide, acecainide, has antidysrhythmic activity of its own; thus, because metabolism varies widely between individuals, and because acecainide is eliminated by the kidneys, serum concentration measurement of procainamide alone has limited usefulness, particularly in renal insufficiency. There is currently little information on the interpretation of combined measurement of the two compounds.

References

1. Lawson DH, Jick H. Adverse reactions to procainamide. Br J Clin Pharmacol 1977;4(5):507–11.
2. Volgman AS, Carberry PA, Stambler B, Lewis WR, Dunn GH, Perry KT, Vanderlugt JT, Kowey PR. Conversion efficacy and safety of intravenous ibutilide compared with intravenous procainamide in patients with atrial flutter or fibrillation. J Am Coll Cardiol 1998;31(6):1414–9.
3. Geelen P, O'Hara GE, Roy N, Talajic M, Roy D, Plante S, Turgeon J. Comparison of propafenone versus procainamide for the acute treatment of atrial fibrillation after cardiac surgery. Am J Cardiol 1999;84(3):345–7.
4. Koch-Weser J, Klein SW, Foo-Canto LL, Kastor JA, DeSanctis RW. Antiarrhythmic prophylaxis with procainamide in acute myocardial infarction. N Engl J Med 1969;281(23):1253–60.
5. Kosowsky BD, Taylor J, Lown B, Ritchie RF. Long-term use of procaine amide following acute myocardial infarction. Circulation 1973;47(6):1204–10.
6. Hallstrom AP, Cobb LA, Yu BH, Weaver WD, Fahrenbruch CE. An antiarrhythmic drug experience in 941 patients resuscitated from an initial cardiac arrest between 1970 and 1985. Am J Cardiol 1991;68(10):1025–31.
7. Miller RR, Hilliard G, Lies JE, Massumi RA, Zelis R, Mason DT, Amsterdam EA. Hemodynamic effects of procainamide in patients with acute myocardial infarction and comparison with lidocaine. Am J Med 1973;55(2):161–8.
8. Goldberg D, Reiffel JA, Davis JC, Gang E, Livelli F, Bigger JT Jr. Electrophysiologic effects of procainamide on sinus function in patients with and without sinus node disease. Am Heart J 1982;103(1):75–9.
9. Mohindra SK, Udeani GO, Abrahamson D. Cardiac tamponade associated with drug-induced systemic lupus erythematosus. Crit Care Med 1989;17(9):961–2.
10. McCrum ID, Guidry JR. Procainamide-induced psychosis. JAMA 1978;240(12):1265–6.
11. Schwartz AB, Klausner SC, Yee S, Turchyn M. Cerebellar ataxia due to procainamide toxicity. Arch Intern Med 1984;144(11):2260–1.
12. Rubinstein A, Cabili S. Tremor induced by procainamide. Am J Cardiol 1986;57(4):340–1.
13. Miller B, Skupin A, Rubenfire M, Bigman O. Respiratory failure produced by severe procainamide intoxication in a patient with pre-existing peripheral neuropathy caused by amiodarone. Chest 1988;94(3):663–5.
14. Godley PJ, Morton TA, Karboski JA, Tami JA. Procainamide-induced myasthenic crisis. Ther Drug Monit 1990;12(4):411–4.
15. Putnam JB Jr, Bolling SF, Kirsh MM. Procainamide-induced respiratory insufficiency after cardiopulmonary bypass. Ann Thorac Surg 1991;51(3):482–3.
16. Sayler DJ, DeJong DJ. Possible procainamide-induced myopathy. DICP 1991;25(4):436.
17. Prendergast MD, Nasca TJ. Anticholinergic syndrome with procainamide toxicity. JAMA 1984;251(22):2926–7.

18. Erdem S, Freimer ML, O'Dorisio T, Mendell JR. Procainamide-induced chronic inflammatory demyelinating polyradiculoneuropathy. Neurology 1998;50(3):824–5.
19. Turgeon PW, Slamovits TL. Scleritis as the presenting manifestation of procainamide-induced lupus. Ophthalmology 1989;96(1):68–71.
20. Bizjak ED, Nolan PE Jr, Brody EA, Galloway JM. Procainamide-induced psychosis: a case report and review of the literature. Ann Pharmacother 1999;33(9):948–51.
21. Kleinman S, Nelson R, Smith L, Goldfinger D. Positive direct antiglobulin tests and immune hemolytic anemia in patients receiving procainamide. N Engl J Med 1984;311(13):809–12.
22. Galanakis DK, Newman J, Summers D. Circulating thrombin time anticoagulant in a procainamide-induced syndrome. JAMA 1978;239(18):1873–4.
23. Meisner DJ, Carlson RJ, Gottlieb AJ. Thrombocytopenia following sustained-release procainamide. Arch Intern Med 1985;145(4):700–2.
24. Abe H, Suzuka H, Tasaki H, Kuroiwa A. Sustained-release procainamide-induced reversible granulocytopenia after myocardial infarction. Jpn Heart J 1995;36(4):483–7.
25. Bluming AZ, Plotkin D, Rosen P, Thiessen AR. Severe transient pancytopenia associated with procainamide ingestion. JAMA 1976;236(22):2520–1.
26. Giannone L, Kugler JW, Krantz SB. Pure red cell aplasia associated with administration of sustained-release procainamide. Arch Intern Med 1987;147(6):1179–80.
27. Agudelo CA, Wise CM, Lyles MF. Pure red cell aplasia in procainamide induced systemic lupus erythematosus. Report and review of the literature. J Rheumatol 1988;15(9):1431–2.
28. Pasha SF, Pruthi RK. Procainamide-induced pure red cell aplasia. Int J Cardiol 2006;110(1):125–6.
29. Ellrodt AG, Murata GH, Riedinger MS, Stewart ME, Mochizuki C, Gray R. Severe neutropenia associated with sustained-release procainamide. Ann Intern Med 1984;100(2):197–201.
30. Meyers DG, Gonzalez ER, Peters LL, Davis RB, Feagler JR, Egan JD, Nair CK. Severe neutropenia associated with procainamide: comparison of sustained release and conventional preparations. Am Heart J 1985;109(6):1393–5.
31. Hoffman HS. Severe neutropenia with procainamide therapy. Conn Med 1990;54(2):59–61.
32. Kleinman S, Nelson R, Smith L, Goldfinger D. Positive direct antiglobulin tests and immune hemolytic anemia in patients receiving procainamide. N Engl J Med 1984;311(13):809–12.
33. Bigger JT Jr, Heissenbuttel RH. The use of procaine amide and lidocaine in the treatment of cardiac arrhythmias. Prog Cardiovasc Dis 1969;11(6):515–34.
34. Peterson AM, Conrad SD, Bell JM. Procainamide-induced pseudo-obstruction in a diabetic patient. DICP 1991;25(12):1334–5.
35. Chuang LC, Tunier AP, Akhtar N, Levine SM. Possible case of procainamide-induced intrahepatic cholestatic jaundice. Ann Pharmacother 1993;27(4):434–7.
36. Sherertz EF. Lichen planus following procainamide-induced lupus erythematosus. Cutis 1988;42(1):51–3.
37. Knox JP, Welykyj SE, Gradini R, Massa MC. Procainamide-induced urticarial vasculitis. Cutis 1988;42(5):469–72.
38. Numata T, Abe H, Nakashima Y, Yamamoto O, Kohshi K. [Procainamide-induced skin eruption associated with disseminated intravascular coagulation in a patient with sustained ventricular tachycardia.]J UOEH 1999;21(3):235–40.
39. Vivino FB, Schumacher HR Jr. Synovial fluid characteristics and the lupus erythematosus cell phenomenon in drug-induced lupus. Findings in three patients and review of pertinent literature. Arthritis Rheum 1989;32(5):560–8.
40. Miller CD, Oleshansky MA, Gibson KF, Cantilena LR. Procainamide-induced myasthenia-like weakness and dysphagia. Ther Drug Monit 1993;15(3):251–4.
41. Venkayya RV, Poole RM, Pentz WH. Respiratory failure from procainamide-induced myopathy. Ann Intern Med 1993;119(4):345–6.
42. Yung RL, Richardson BC. Drug-induced lupus. Rheum Dis Clin North Am 1994;20(1):61–86.
43. Smiley JD, Moore SE Jr. Molecular mechanisms of autoimmunity. Am J Med Sci 1988;295(5):478–96.
44. Green BJ, Wyse DG, Duff HJ, Mitchell LB, Matheson DS. Procainamide in vivo modulates suppressor T lymphocyte activity. Clin Invest Med 1988;11(6):425–9.
45. Burlingame RW, Rubin RL. Drug-induced anti-histone autoantibodies display two patterns of reactivity with substructures of chromatin. J Clin Invest 1991;88(2):680–90.
46. Rubin RL, Salomon DR, Guerrero RS. Thymus function in drug-induced lupus. Lupus 2001;10(11):795–801.
47. Winfield JB, Koffler D, Kunkel HG. Development of antibodies to ribonucleoprotein following short-term therapy with procainamide. Arthritis Rheum 1975;18(6):531–4.
48. Lau CC, Clos TD. Anti-[(H2A/2B)-DNA] IgG supports the diagnosis of procainamide-induced arthritis or pleuritis. Arthritis Rheum 1999;42(6):1300–1.
49. Sim E. Drug-induced immune-complex disease. Complement Inflamm 1989;6(2):119–26.
50. Lessard E, Hamelin BA, Labbe L, O'Hara G, Belanger PM, Turgeon J. Involvement of CYP2D6 activity in the N-oxidation of procainamide in man. Pharmacogenetics 1999;9(6):683–96.
51. Uetrecht JP, Woosley RL. Acetylator phenotype and lupus erythematosus. Clin Pharmacokinet 1981;6(2):118–34.
52. Woosley RL, Drayer DE, Reidenberg MM, Nies AS, Carr K, Oates JA. Effect of acetylator phenotype on the rate at which procainamide induces antinuclear antibodies and the lupus syndrome. N Engl J Med 1978;298(21):1157–9.
53. Heyman MR, Flores RH, Edelman BB, Carliner NH. Procainamide-induced lupus anticoagulant. South Med J 1988;81(7):934–6.
54. Ponte CD, Horner P. Suspected procainamide-induced angioedema. Drug Intell Clin Pharm 1985;19(2):139–40.
55. Kameda H, Mimori T, Kaburaki J, Fujii T, Takahashi T, Akaishi M, Ikeda Y. Systemic sclerosis complicated by procainamide-induced lupus and antiphospholipid syndrome. Br J Rheumatol 1998;37(11):1236–9.
56. Klimas NG, Patarca R, Perez G, Garcia-Morales R, Schultz D, Schabel J, Fletcher MA. Case report: distinctive immune abnormalities in a patient with procainamide-induced lupus and serositis. Am J Med Sci 1992;303(2):99–104.
57. Rubin RL, Burlingame RW, Arnott JE, Totoritis MC, McNally EM, Johnson AD. IgG but not other classes of anti-[(H2A-H2B)-DNA] is an early sign of procainamide-induced lupus. J Immunol 1995;154(5):2483–93.
58. El-Rayes BF, Edelstein M. Unusual case of antiphospholipid antibody syndrome presenting with extensive cutaneous infarcts in a patient on long-term procainamide therapy. Am J Hematol 2003;72:154.

59. Murray KD, Vlasnik JJ. Procainamide-induced postoperative pyrexia. Ann Thorac Surg 1999;68(3):1072–4.
60. Kumana CR. Therapeutic drug monitoring-antidysrhythmic drugs. In: Richens A, Marks V, editors. Therapeutic Drug Monitoring. London, Edinburgh: Churchill-Livingstone, 1981:370.
61. White SR, Dy G, Wilson JM. The case of the slandered Halloween cupcake: survival after massive pediatric procainamide overdose. Pediatr Emerg Care 2002;18(3):185–8.
62. Villalba-Pimentel L, Epstein LM, Sellers EM, Foster JR, Bennion LJ, Nadler LM, Bough EW, Koch-Weser J. Survival after massive procainamide ingestion. Am J Cardiol 1973;32(5):727–30.
63. Atkinson AJ Jr, Krumlovsky FA, Huang CM, del Greco F. Hemodialysis for severe procainamide toxicity: clinical and pharmacokinetic observations. Clin Pharmacol Ther 1976;20(5):585–92.
64. Windle J, Prystowsky EN, Miles WM, Heger JJ. Pharmacokinetic and electrophysiologic interactions of amiodarone and procainamide. Clin Pharmacol Ther 1987;41(6):603–10.
65. Somogyi A, McLean A, Heinzow B. Cimetidine–procainamide pharmacokinetic interaction in man: evidence of competition for tubular secretion of basic drugs. Eur J Clin Pharmacol 1983;25(3):339–45.
66. Bauer LA, Black D, Gensler A. Procainamide–cimetidine drug interaction in elderly male patients. J Am Geriatr Soc 1990;38(4):467–9.
67. Bauer LA, Black DJ, Lill JS, Garrison J, Raisys VA, Hooton TM. Levofloxacin and ciprofloxacin decrease procainamide and N-acetylprocainamide renal clearances. Antimicrob Agents Chemother 2005;49(4):1649–51.
68. Martin DE, Shen J, Griener J, Raasch R, Patterson JH, Cascio W. Effects of ofloxacin on the pharmacokinetics and pharmacodynamics of procainamide. J Clin Pharmacol 1996;36(1):85–91.
69. Kosoglou T, Rocci ML Jr, Vlasses PH. Trimethoprim alters the disposition of procainamide and N-acetylprocainamide. Clin Pharmacol Ther 1988;44(4):467–77.
70. Trujillo TC, Nolan PE. Antiarrhythmic agents: drug interactions of clinical significance. Drug Saf 2000;23(6):509–32.
71. Koch-Weser J. Serum procainamide levels as therapeutic guides. Clin Pharmacokinet 1977;2(6):389–402.

Propafenone

General Information

Propafenone is both a class I antidysrhythmic drug and a beta-adrenoceptor antagonist. Its pharmacological effects, clinical pharmacology, therapeutic uses, adverse effects, and interactions have been reviewed (1–5).

The main adverse effects of propafenone are cardiovascular (27%), central nervous (21%), and gastrointestinal (20%) (6). Other adverse effects occur in under 6% of cases. The overall risk of non-cardiac effects is around 14%. These adverse effects are dose-related: the incidence is 11% at 300 mg/day, 22% at 450 mg/day, 33% at 600 mg/day, and 48% at 900 mg/day (6).

Observational studies

In 87 patients with atrial fibrillation who were given propafenone 2 mg/kg intravenously over 10 minutes, four had hypotension at 8–45 minutes after the start of infusion (7). In two cases this was accompanied by sinus bradycardia, nausea, and slight malaise. In all cases the hypotension resolved rapidly with saline infusion; the drug was withdrawn in only one case. In two cases atrial fibrillation was transformed to asymptomatic atrial flutter with 2:1 atrioventricular conduction.

Quinidine has been added to propafenone with the intention of inhibiting propafenone metabolism via CYP2D6 in the hope of improving outcome (8). Of 60 patients with paroxysmal atrial fibrillation given propafenone 300–450 mg/day for 8 weeks there were 19 refractory cases, who were then randomized double-blind to receive either a higher dose of propafenone (450–675 mg/day) or the standard dose of propafenone with extra low-dose quinidine (150 mg/day), each for 8 weeks, with subsequent crossover to the alternative. Patients who even then were not adequately controlled were given the standard dose of propafenone plus a standard dose of quinidine (600 mg/day) for a further 8 weeks. The plasma propafenone concentrations during the four phases were as follows:

1. standard-dose propafenone alone 128 ng/ml;
2. standard-dose propafenone plus low-dose quinidine 259 ng/ml;
3. high-dose propafenone alone 336 ng/ml;
4. standard-dose propafenone plus standard-dose quinidine 490 ng/ml.

The beneficial effects were related to these plasma concentrations, as were the time to the first bout of atrial fibrillation, the frequency of bouts of atrial fibrillation, and the time between episodes. However, when atrial fibrillation occurred there was no difference in the ventricular rate in the different groups. Adverse effects necessitated drug withdrawal in four patients; one had heart failure and two had gastrointestinal symptoms. These effects were not dose-related, although there were too few occurrences for a definitive conclusion. The authors suggested that this stepwise approach, with increasing doses of propafenone and increasing doses of quinidine could be beneficial in the treatment of paroxysmal atrial fibrillation.

Comparative studies

Amiodarone

Amiodarone 30 mg/kg orally for the first 24 hours plus, if necessary, 15 mg/kg over 24 hours has been compared with propafenone 600 mg in the first 24 hours plus, if necessary, 300 mg in the next 24 hours in 86 patients with recent onset atrial fibrillation (9). Conversion to sinus rhythm occurred faster with propafenone (2.4 hours) than amiodarone (6.9 hours). However, by 24 hours and 48 hours the same proportions of patients were in sinus rhythm; one patient given amiodarone had a supraventricular tachycardia and one a non-sustained ventricular tachycardia.

Of 136 patients with atrial fibrillation treated with either amiodarone (n = 96) or propafenone (n = 40), 15 developed subsequent persistent atrial flutter, nine of those taking amiodarone and six of those taking propafenone (10). In all cases radiofrequency ablation was effective. It is not clear to what extent these cases of atrial flutter were due to the drugs, although the frequencies of atrial flutter in previous studies with propafenone have been similar. Atrial enlargement was significantly related to the occurrence of persistent atrial flutter in these patients.

Procainamide

A comparison between procainamide and propafenone in 62 patients, who had undergone coronary artery bypass grafting or valvular surgery within 3 weeks and developed sustained atrial fibrillation, showed that both drugs converted the dysrhythmia to sinus rhythm in up to 76% of cases, but that propafenone did it more quickly (11). Symptomatic arterial hypotension occurred more frequently with procainamide (nine of 33 patients) than propafenone (two of 29 patients). Other adverse effects of procainamide were nausea (n = 2) and junctional escape rhythm (n = 2). Other adverse effects of propafenone were hot flushes (n = 1), nausea (n = 3), bronchospasm (n = 1), and junctional escape rhythm (n = 2).

Quinidine

A placebo-controlled study of the use of propafenone 450–600 mg orally, either alone or in combination with digoxin, has been carried out in 176 patients with atrial fibrillation; a further 70 patients were given digitalis plus quinidine (12). There were no significant differences across the groups in terms of percentage conversion to sinus rhythm, although conversion occurred more quickly in those given digoxin plus propafenone; this catch-up of the other treatments was attributed to spontaneous conversion in those groups. There were no serious adverse effects in this study. The QT_c interval was slightly prolonged by digitalis plus quinidine and not by the other treatments. In six patients taking propafenone alone there were mild non-cardiac effects, including sickness in two, headache in one, gastrointestinal disturbances in two, and paresthesia in one. Four patients taking propafenone plus digitalis had either sickness or dizziness, and nine patients taking digitalis plus quinidine had gastrointestinal disturbances, sickness, dizziness, or headache. There were no major dysrhythmias; four of the patients who took digitalis plus propafenone had asymptomatic ventricular extra beats, as did one patient who took digitalis plus quinidine. In nine patients who took digitalis plus quinidine there were asymptomatic short-lasting episodes of atrial flutter with atrial ventricular conduction of at least 2:1 immediately before the restoration of sinus rhythm; this happened in 13 patients who took propafenone, 12 patients who took digitalis plus propafenone, and three patients who took placebo. There were two cases of complete left bundle branch block in patients who took digitalis plus quinidine, in three patients who took propafenone, and in two who took digitalis plus propafenone. In two patients who took digitalis plus quinidine and two who took propafenone there was reversible asymptomatic sinoatrial block of Wenckebach type II. Transient mild hypotension occurred in one patient taking digitalis plus quinidine, five taking propafenone, one taking digitalis plus propafenone, and one taking placebo, but the hypotension was transient and not severe. The authors concluded that the addition of digitalis to propafenone hastened cardioversion from atrial fibrillation, although they conceded that the balance of other evidence suggests that digitalis is not effective in restoring sinus rhythm and were unable to explain the efficacy of the combination of digitalis with propafenone.

Sotalol

In a randomized, double-blind, placebo-controlled comparison of propafenone (mean dose 13 mg/kg/day; n = 102) and sotalol (mean dose 3 mg/kg/day; n = 106) in maintaining sinus rhythm after conversion of recurrent symptomatic atrial fibrillation in 300 patients, efficacy was comparable (13). Tolerable adverse effects in those who took propafenone were gastrointestinal discomfort (n = 15), neurological disturbances (n = 9), a metallic taste (n = 4), and generalized weakness (n = 1); nine patients withdrew owing to adverse effects, four with gastrointestinal disorders, three with dizziness, and two with headache; there were no prodysrhythmias.

Propafenone 450 mg/day and sotalol 240 mg/day have been compared in a placebo-controlled study of 300 patients with atrial fibrillation (14). The two drugs had similar efficacy. There were adverse events in 38 of the patients who took propafenone, compared with 12 of those who took placebo. These included gastrointestinal discomfort, neurological disturbances, asymptomatic bradycardia, a metallic taste, and general weakness. In nine patients the adverse effects were sufficient to cause withdrawal of propafenone.

Placebo-controlled studies

In controlled trials in patients with recent-onset atrial fibrillation without heart failure, oral propafenone (450–600 mg as a single dose) had a relatively quick effect (within 3–4 hours) and a high rate of efficacy (72–78% within 8 hours) (15).

The adverse effects of propafenone in placebo-controlled trials in patients with atrial tachydysrhythmias have been reviewed (16). The following effects were reported after single intravenous oral doses to produce conversion of atrial fibrillation to sinus rhythm. Non-cardiac adverse effects included mild dizziness. Mild hypotension was also noted, but only required withdrawal of propafenone in one of 29 patients in one study. There have been prodysrhythmic effects in several studies, including atrial flutter with a broad QRS complex, which can occur in up to 5% of cases; in some cases atrial flutter can have a rapid ventricular response due to 1:1 atrioventricular conduction, which has been attributed to slowing of atrial conduction and reduced refractoriness of the atrioventricular node. Other prodysrhythmic effects in a few patients included sinus bradycardia with sinus pauses and effects on atrioventricular conduction.

In patients taking long-term propafenone for supraventricular dysrhythmias adverse effects were more common and have been reported in 14–60% of cases. Cardiac adverse effects were more common in patients with structural heart disease. The non-cardiac effects were either gastrointestinal (nausea, vomiting, taste disturbances) or neurological (dizziness). Adverse effects are dose-related. In one large study there was no difference between propafenone and placebo in the risk of death.

The use of propafenone in atrial fibrillation (SEDA-23, 202) has been studied in a randomized, double-blind, placebo-controlled trial in 55 patients (17). The dose of propafenone was chosen according to body weight: 450, 600, and 750 mg for those weighing 50–64, 65–80, and over 80 kg respectively. Propafenone converted atrial fibrillation to sinus rhythm significantly more quickly than placebo, and most patients given propafenone had converted by 6 hours. However, by 24 hours there was no significant difference between the two groups. Four patients had hypotension after propafenone, in three cases transiently. The patient with sustained hypotension had poor left ventricular systolic function, but it responded promptly to the administration of fluids and electrical cardioversion. In one patient with transient hypotension there was a brief episode of sinus bradycardia and in another an isolated sinus pause.

In a randomized, double-blind, placebo-controlled study of a modified-release formulation of propafenone (propafenone SR) patients with a history of symptomatic atrial fibrillation who were in sinus rhythm were randomized to placebo or propafenone SR 225, 325, 425 or mg, all twice daily (18). In the primary efficacy analysis, propafenone SR significantly prolonged the time to first symptomatic recurrence of atrial dysrhythmia at all three doses compared with placebo. The median time to recurrence was 41 days with placebo, 112 days with propafenone 225 mg, 291 days with 325 mg, and over 300 days with 425 mg. The numbers of patients who reported at least one adverse event were 91 (72%) with placebo, 97 (77%) with propafenone 225 mg, 113 (84%) with propafenone 325 mg, and 113 (83%) with propafenone 425 mg. Adverse events that led to withdrawal occurred in 17 (14%), 16 (13%), 19 (14%), and 34 (25%) patients in each of the four groups respectively. In all of the propafenone treatment groups, the most commonly reported adverse events that also exceeded the percent reported by patients taking placebo by at least 5% were dizziness, dyspnea, taste disturbances, fatigue, and constipation. Propafenone increased the PR interval and QRS duration dose-relatedly but did not change the QT_c interval. The average changes from baseline in PR interval were 1.0 ms with placebo, 9.1 ms with propafenone 225 mg, 12 ms with propafenone 325 mg, and 21 ms with propafenone 425 mg. The changes in QRS duration were ?1.6 ms with placebo, 4.0 ms with propafenone 225 mg, 6.3 ms with propafenone 325 mg, and 6.3 ms with propafenone 425 mg. There were no deaths during treatment or within 30 days of withdrawal. There were no cases of ventricular tachycardia.

Organs and Systems

Cardiovascular

Cardiovascular adverse effects have been reported in 13–27% of patients taking propafenone and ventricular dysrhythmias in 8–19% in small studies. However, in large studies the risk has been reported to be about 5%.

Conduction disturbances are common with propafenone and can result in sinus bradycardia, sinoatrial block, sinus arrest, any degree of atrioventricular block, and right or left bundle-branch block (SEDA-10, 151; SEDA-15, 179).

The adverse effects of a single oral dose of propafenone for cardioversion of recent-onset atrial fibrillation have been evaluated in a systematic review (19). The adverse effects were transient dysrhythmias (atrial flutter, bradycardia, pauses, and junctional rhythm), reversible widening of the QRS complex, transient hypotension, and mild non-cardiac effects (nausea, headache, gastrointestinal disturbances, dizziness, and paresthesia).

Dysrhythmias can occur; these include ventricular tachycardia, ventricular flutter, and atrial fibrillation (20–22). Hypotension and worsening of heart failure have occasionally been reported (SEDA-10, 151) (23).

- Wide-complex tachycardias occurred in two elderly patients (a 74-year-old man and an 80-year-old woman) who had taken propafenone for atrial fibrillation (24). In the first case the dysrhythmia was due to atrial flutter with 1:1 conduction.

Although drugs of class Ic, such as propafenone, can slow atrial and atrioventricular nodal conduction in patients with atrial fibrillation or atrial flutter, they do not alter the refractoriness of the atrioventricular node, and this allows 1:1 atrioventricular conduction as the atrial rate slows. This happens despite prolongation of the PR interval.

Class Ic drugs can also convert atrial fibrillation to atrial flutter, reportedly in 3.5–5% of patients. Of 187 patients with paroxysmal atrial fibrillation who were treated with flecainide or propafenone, 24 developed atrial flutter, which was typical in 20 cases (25). These patients underwent radiofrequency ablation, which failed in only one case. All the patients continued to take their preexisting drugs, and during a mean follow-up period of 11 months, the incidence of atrial fibrillation was higher in patients who were taking combined therapy than in those taking monotherapy. The authors suggested that in patients with atrial fibrillation who developed typical atrial flutter due to class Ic antidysrhythmic drugs, combined catheter ablation and continued drug treatment is highly effective in reducing the occurrence and duration of atrial tachydysrhythmias. They did not report adverse effects.

In controlled trials of oral propafenone (450–600 mg as a single dose) in patients with recent-onset atrial fibrillation without heart failure, atrial flutter with 1:1 atrioventricular conduction occurred in only two of 709 patients (0.3%) who received propafenone (15).

Propafenone and ibutilide have been compared in 40 patients with atrial flutter (26). Ibutilide was superior to propafenone for treating atrial flutter (90% versus 30%) and the respective mean conversion times were 11 and 35 minutes. Bradycardia (2/20) and hypotension (4/20) were more common adverse effects with propafenone.

Respiratory

Since propafenone is a beta-adrenoceptor antagonist it can cause shortness of breath or worsening of asthma (27).

Nervous system

Adverse effects on the nervous system are common with propafenone and include somnolence, weakness and disorientation, global amnesia (28), dizziness and vertigo, tremor, visual disturbances, and convulsions (SEDA-10, 151).

- Peripheral neuropathy has been reported in a 41-year-old man who took propafenone 450 mg/day for about a year (29).

Ataxia has been reported in patients taking propafenone (30).

- An 80-year-old man taking propafenone 150 mg tds for paroxysmal atrial fibrillation developed progressive generalized ataxia and weakness 4 days after starting treatment. He had a bilateral symmetrical ataxia, unclear speech, impairment of gait, altered hand coordination, and tremor. The ataxia resolved completely within 3 days of withdrawal.
- A 73-year-old woman taking propafenone 150 mg tds for paroxysmal atrial tachycardia underwent cardioversion during an attack, and the dose of propafenone was increased to 300 mg tds. After 5 days she developed severe ataxia and progressive weakness. The ataxia was symmetrical and there was severe impairment of gait, altered hand coordination, and tremor. The dose of propafenone was reduced to 600 mg/day and the ataxia resolved completely within 6 days. A year later, when the dose of propafenone was increased to 900 mg/day, progressive ataxia again developed after 2 days and became severe within 1 week. Propafenone was withdrawn and the ataxia resolved within a few days.
- An 85-year-old woman took propafenone 150 mg tds for paroxysmal atrial fibrillation and 2 months later developed a progressive ataxia and recurrent falls. The ataxia was symmetrical and there was altered hand coordination, impairment of gait, and tremor. The propafenone was withdrawn and the ataxia resolved completely within 4 days.

Endocrine

Propafenone can cause hyponatremia due to inappropriate secretion of ADH (31).

Hematologic

Propafenone can occasionally cause neutropenia, with a calculated incidence of about one in 10 000 prescriptions per year (32).

Gastrointestinal

Unwanted gastrointestinal effects are the most common adverse effects of propafenone, occurring in up to 30% of cases. They include anorexia, nausea and vomiting, dry mouth, a metallic or bitter taste in the mouth, abdominal discomfort, and constipation (SEDA-10, 151).

Liver

Propafenone can increase the serum activities of transaminases and other enzymes associated with liver function (33). There have also been reports of cholestatic jaundice (34).

- A 67-year-old woman who had taken glibenclamide 5 mg/day and enalapril 20 mg/day for 5 years became jaundiced 6 weeks after starting to take propafenone 450 mg/day (35). The liver function tests suggested an obstructive jaundice. Viral serology and autoantibodies were negative and gallstones and other obstructive lesions were ruled out by ultrasound. She recovered 15 days after withdrawal of propafenone.
- A 69-year-old woman who had taken digitoxin 0.25 mg/day and enalapril 5 mg/day for 6 years became jaundiced 7 months after starting to take propafenone 300 mg/day (35). Viral serology was negative and obstruction was ruled out by ultrasound. All autoantibodies were negative except for an antinuclear factor titer of 1/160. A liver biopsy showed expanded portal tracts with an infiltrate of lymphocytes, monocytes, and macrophages and proliferation of bile ductules, sinusoidal dilatation, ballooning of hepatocytes, and bile thrombi. There were also some eosinophils among the hepatocytes. She recovered gradually after withdrawal of propafenone and her antinuclear factor titer returned to normal.

The occurrence of eosinophilia in most of the reported cases of cholestatic jaundice in patients taking propafenone suggests that it is a hypersensitivity reaction.

Skin

Skin rashes have been reported occasionally (31); these include an acneiform rash and urticaria.

Acute generalized exanthematous pustulosis has been attributed to propafenone in an 85-year-old man (36). The mechanism of this type of eruption is thought to be via the preferential production of IL-8 by T cells after the drug binds to the cells and elicits a CD4 and CD8 immune reaction.

Sexual function

Impotence has occasionally been reported, in one case with a reduced sperm count (37,38).

Immunologic

Propafenone can cause a rise in antinuclear antibody titers (38) and has once been reported to have caused a lupus-like syndrome (39).

Body temperature

Drug fever without agranulocytosis has been attributed to propafenone (40).

Second-Generation Effects

Pregnancy

When a pregnant woman was given propafenone from the fifth month to term in a dosage of 300 mg tds her dysrhythmias responded satisfactorily and the neonate was healthy (41).

Susceptibility Factors

Genetic factors

Poor and extensive oxidation phenotypes for CYP2D6, which metabolizes propafenone, have been studied in 42 patients, aged 36–75 years, with paroxysmal atrial fibrillation (42). Efficacy was 100% in poor metabolizers, 61% in extensive metabolizers, and 0% in very extensive metabolizers. There was a significant correlation between oxidation phenotype and the ability to maintain sinus rhythm.

Age

The incidence of adverse effects of propafenone in children has varied from study to study, but has sometimes been as high as 25%, requiring withdrawal in 6% of cases (43). Elderly people are at increased risk of adverse effects.

The safety of oral propafenone in the treatment of dysrhythmias has been studied retrospectively in infants and children (44). There were significant electrophysiological adverse effects and prodysrhythmia in 15 of 772 patients (1.9%). These included sinus node dysfunction in four, complete atrioventricular block in two, aggravation of supraventricular tachycardia in two, acceleration of ventricular rate during atrial flutter in one, ventricular prodysrhythmia in five, and unexplained syncope in one. Cardiac arrest or sudden death occurred in five patients (0.6%); two had a supraventricular tachycardia due to Wolff–Parkinson–White syndrome; the other three had structural heart disease. Adverse cardiac events were more common in the presence of structural heart disease and there was no difference between patients with supraventricular and ventricular dysrhythmias.

Drug Administration

Drug dosage regimens

Three different regimens of oral propafenone have been compared in patients with paroxysmal atrial fibrillation (45). In 48 patients who took 600 mg followed 8 hours later by 150 mg there was a higher rate of early successful cardioversion with a lower incidence of adverse effects than in two other groups who took either 300 mg three times over 8 hours ($n = 82$) or four doses of 150 mg over 9 hours ($n = 58$). The rates of conversion were around 80%, similar to those found in other studies. There was QRS prolongation in all three groups, and four of those who took a total of 900 mg developed a broad complex tachycardia.

Drug overdose

About 60 cases of propafenone poisoning have been reported. Even low doses can lead to serious poisoning. Some symptoms, such as gastrointestinal or neurological symptoms, are misleading. Cardiovascular abnormalities include bradycardia, atrioventricular block, abnormal intraventricular conduction, shock, and electromechanical dissociation. The simultaneous presence of neurological symptoms (especially seizures) and abnormal cardiac conduction suggests serious poisoning.

The mortality from propafenone overdose is 23%, according to a retrospective series of 120 patients in Germany over 14 years (46). Severity was related to cardiac conduction disorders and cardiac hyperexcitability.

Seizures have been reported in cases of overdose.

- A 24-year-old woman who took an overdose of propafenone 2.7 g had a convulsion and complete heart block (47). The plasma propafenone concentration 4 hours after ingestion was 2930 ng/ml (target range 400–1600 ng/ml). She was given activated charcoal 50 g, mannitol 200 ml, and clonazepam 1 mg intravenously. By 19 hours after ingestion the electrocardiogram was normal and the plasma propafenone concentration was 858 ng/ml.
- A 22-year-old woman took an overdose of propafenone (amount unknown) and developed tetany and then generalized convulsions requiring intravenous clonazepam (48). She had a low blood pressure and first-degree atrioventricular block associated with prolonged intraventricular conduction. She was intubated and given intravenous fluids, equimolar sodium lactate, dopamine, and adrenaline. Her cardiac conduction returned to normal.
- A 16-year-old girl was discovered in her bedroom having tonic-clonic seizures and later had a cardiorespiratory arrest (49). She was thought to have taken an overdose of up to 9 g of propafenone. Despite vigorous resuscitation she died after 2.5 hours. Her plasma concentration of propafenone was 5.9 μg/ml, of 5-hydroxypropafenone 0.25 μg/ml, and of N-depropylpropafenone 0.38 μg/ml.

- A 13-year-old boy took about 3 g of propafenone and had a seizure, extreme bradycardia, cyanosis, and hypotension, followed by a cardiorespiratory arrest (50). He had hypokalemia, hypercalcemia, hyperphosphatemia, and hypermagnesemia. His electrocardiogram showed first-degree heart block with right bundle branch block and a prolonged QT interval. His plasma propafenone concentration on admission was 4.6 μg/ml and his 5-hydroxypropafenone concentration was 17 μg/ml. He was resuscitated with cardiac massage, adrenaline, and sodium lactate and was given multiple doses of activated charcoal 25 g 4-hourly for 24 hours.
- A 3-year-old boy unintentionally took one 300 mg tablet of propafenone (15 mg/kg); 3 hours later he had a seizure and became cyanosed and hypotensive followed by a cardiac arrest (51). His electrocardiogram showed irregular bradycardia with bundle branch block and a prolonged QT interval. His serum propafenone concentration was 1.8 μg/ml. He was resuscitated with cardiac massage, atropine, adrenaline, dobutamine, midazolam, and sodium bicarbonate and recovered within 15 hours.

Treatment is symptomatic. Sodium lactate may be required for abnormal intraventricular conduction. Plasma exchange has been successfully used to treat propafenone overdose (52).

- An 18-year-old woman took 35 tablets of propafenone, 300 mg each. She had dilated, non-reactive pupils and greatly increased activity of neuron-specific enolase. She had atrioventricular and intraventricular conduction disorders, and repeated resuscitation was necessary. Propafenone was eliminated by plasma exchange, and the conduction disturbances disappeared rapidly during treatment.

Drug–Drug Interactions

Cardiac glycosides

Propafenone causes a small increase in plasma digoxin concentrations by an unknown mechanism (53,54). Although this effect is perhaps not clinically significant, in patients with dysrhythmias it was accompanied by an increase in PR interval (55).

Lidocaine

The CNS toxicity of lidocaine was increased in 11 healthy volunteers who simultaneously received propafenone, which reduced the metabolism of lidocaine (56).

Metoprolol

Propafenone reduces the apparent oral clearance of metoprolol, but it is not clear whether the mechanism is by reduction of true clearance or an increase in systemic availability (57). In addition the beta-blocking action of metoprolol is increased by propafenone; this could be due either to reduced clearance of metoprolol or to a true pharmacodynamic interaction between the two drugs, since propafenone has beta-blocking activity.

Mexiletine

Mexiletine and propafenone are metabolized by the same enzymes, CYP2D6, CYP1A2, and CYP3A4. In 15 healthy volunteers, eight of whom were extensive metabolizers of CYP2D6, administration of oral mexiletine 100 mg bd on days 1–8 and oral propafenone 1 mg bd on days 5–12 significantly reduced the clearance of R(–) mexiletine from 41 to 28 l/hour and of S(+) mexiletine from 43 to 29 l/hour in the extensive metabolizers (58). The new values were no different from the clearance values in the poor metabolizers. Propafenone also reduced the partial metabolic clearances of mexiletine to hydroxymethylmexiletine, parahydroxymexiletine, and metahydroxymexiletine by about 70% in the extensive metabolizers. Propafenone had no effect on the kinetics of mexiletine in the poor metabolizers. There were no electrocardiographic changes during this interaction. Smokers had higher clearance rates than non-smokers, but the effects of propafenone were similar in the two groups. In contrast, mexiletine had little effect on the disposition of propafenone. The authors proposed that these effects could explain at least in part the increased efficacy that sometimes occurs when mexiletine and propafenone are combined in patients in whom a single drug was not effective. They also recommended that the dosages of the drugs should be titrated slowly when they are used together, in order to reduce the risk of adverse effects.

Monitoring Therapy

Therapeutic benefit is most likely when the plasma propafenone concentration is in the range 0.5–2.0 mg/ml, although the correlation is poor (59), and there is a large overlap between therapeutic and toxic concentrations (34). The therapeutic effect of propafenone correlates better with prolongation of the PR and QRS intervals (60).

References

1. Birgersdotter-Green U. Propafenone for cardiac arrhythmias. Am J Med Sci 1992;303(2):123–8.
2. Bryson HM, Palmer KJ, Langtry HD, Fitton A. Propafenone. A reappraisal of its pharmacology, pharmacokinetics and therapeutic use in cardiac arrhythmias. Drugs 1993;45(1):85–130.
3. Cobbe SM. Drug therapy of supraventricular tachyarrhythmias—based on efficacy or futility? Eur Heart J 1994;15(Suppl A):22–6.
4. Paul T, Janousek J. New antiarrhythmic drugs in pediatric use: propafenone. Pediatr Cardiol 1994;15(4):190–7.
5. Valderrabano M, Singh BN. Electrophysiologic and Antiarrhythmic Effects of Propafenone: Focus on Atrial Fibrillation. J Cardiovasc Pharmacol Ther 1999;4(3):183–98.
6. Ravid S, Podrid PJ, Novrit B. Safety of long-term propafenone therapy for cardiac arrhythmia—experience with 774 patients. J Electrophysiol 1987;1:580–90.
7. Bianconi L, Mennuni M. Comparison between propafenone and digoxin administered intravenously to patients with acute atrial fibrillation. PAFIT-3 Investigators. The

Propafenone in Atrial Fibrillation Italian Trial. Am J Cardiol 1998;82(5):584–8.
8. Lau CP, Chow MS, Tse HF, Tang MO, Fan C. Control of paroxysmal atrial fibrillation recurrence using combined administration of propafenone and quinidine. Am J Cardiol 2000;86(12):1327–32.
9. Blanc JJ, Voinov C, Maarek MPARSIFAL Study Group. Comparison of oral loading dose of propafenone and amiodarone for converting recent-onset atrial fibrillation. Am J Cardiol 1999;84(9):1029–32.
10. Tai CT, Chiang CE, Lee SH, Chen YJ, Yu WC, Feng AN, Ding YA, Chang MS, Chen SA. Persistent atrial flutter in patients treated for atrial fibrillation with amiodarone and propafenone: electrophysiologic characteristics, radiofrequency catheter ablation, and risk prediction. J Cardiovasc Electrophysiol 1999;10(9):1180–7.
11. Geelen P, O'Hara GE, Roy N, Talajic M, Roy D, Plante S, Turgeon J. Comparison of propafenone versus procainamide for the acute treatment of atrial fibrillation after cardiac surgery. Am J Cardiol 1999;84(3):345–7.
12. Capucci A, Villani GQ, Aschieri D, Piepoli M. Safety of oral propafenone in the conversion of recent onset atrial fibrillation to sinus rhythm: a prospective parallel placebo-controlled multicentre study. Int J Cardiol 1999;68(2):187–96.
13. Bellandi F, Simonetti I, Leoncini M, Frascarelli F, Giovannini T, Maioli M, Dabizzi RP. Long-term efficacy and safety of propafenone and sotalol for the maintenance of sinus rhythm after conversion of recurrent symptomatic atrial fibrillation. Am J Cardiol 2001;88(6):640–5.
14. Bellandi F, Leoncini M, Maioli M, Gallopin M, Dabizzi RP. Comparing agents for prevention of atrial fibrillation recurrence. Cardiol Rev 2002;19:18–21.
15. Chopra IJ, Baber K. Use of oral cholecystographic agents in the treatment of amiodarone-induced hyperthyroidism. J Clin Endocrinol Metab 2001;86(10):4707–10.
16. Rae AP, Camm J, Winters S, Page R. Placebo-controlled evaluations of propafenone for atrial tachyarrhythmias. Am J Cardiol 1998;82(8A):N59–65.
17. Azpitarte J, Alvarez M, Baun O, Garcia R, Moreno E, Navarrete A, Fernandez R. Using propafenone to convert recent-onset atrial fibrillation. Cardiol Rev 2000;17:37–43.
18. Pritchett ELC, Page RL, Carlson M, Undesser K, Fava G. Efficacy and safety of sustained-release propafenone (propafenone SR) for patients with atrial fibrillation. Am J Cardiol 2003;92:941–6.
19. Khan IA. Single oral loading dose of propafenone for pharmacological cardioversion of recent-onset atrial fibrillation. J Am Coll Cardiol 2001;37(2):542–7.
20. Antman EM, Beamer AD, Cantillon C, McGowan N, Friedman PL. Therapy of refractory symptomatic atrial fibrillation and atrial flutter: a staged care approach with new antiarrhythmic drugs. J Am Coll Cardiol 1990;15(3):698–707.
21. Escande M, Diadema B, Maarek-Charbit M. Étude a long terme de la propafénone dans l'extrasystolie ventriculaire grave du sujet agé. [Long-term study of propafenone in severe ventricular extrasystole in elderly subjects.] Ann Cardiol Angeiol (Paris) 1989;38(9):555–60.
22. Colas A, Maarek-Charbit M. Propafenone per os dans les troubles du rythme ventriculaire. [Propafenone per os in ventricular arrhythmia.] Cah Anesthesiol 1989;37(4):241–4.
23. Cobbe SM, Rae AP, Poloniecki JDUK Propafenone PSVT Study Group. A randomized, placebo-controlled trial of propafenone in the prophylaxis of paroxysmal supraventricular tachycardia and paroxysmal atrial fibrillation. Circulation 1995;92(9):2550–7.
24. Mackstaller LL, Marcus FI. Rapid ventricular response due to treatment of atrial flutter or fibrillation with Class I antiarrhythmic drugs. Ann Noninvasive Electrocardiol 2000;5:101–4.
25. Schumacher B, Jung W, Lewalter T, Vahlhaus C, Wolpert C, Luderitz B. Radiofrequency ablation of atrial flutter due to administration of class IC antiarrhythmic drugs for atrial fibrillation. Am J Cardiol 1999;83(5):710–3.
26. Sun JL, Guo JH, Zhang N, Zhang HC, Zhang P. Clinical comparison of ibutilide and propafenone for converting atrial flutter. Cardiovasc Drugs Ther 2005;19(1):57–64.
27. Veale D, McComb JM, Gibson GJ. Propafenone. Lancet 1990;335(8695):979.
28. Jones RJ, Brace SR, Vander Tuin EL. Probable propafenone-induced transient global amnesia. Ann Pharmacother 1995;29(6):586–90.
29. Galasso PJ, Stanton MS, Vogel H. Propafenone-induced peripheral neuropathy. Mayo Clin Proc 1995;70(5):469–72.
30. Odeh M, Seligmann H, Oliven A. Propafenone-induced ataxia: report of three cases. Am J Med Sci 2000;320(2):151–3.
31. Hammill SC, Sorenson PB, Wood DL, Sugrue DD, Osborn MJ, Gersh BJ, Holmes DR Jr. Propafenone for the treatment of refractory complex ventricular ectopic activity. Mayo Clin Proc 1986;61(2):98–103.
32. Miwa LJ, Jolson HM. Propafenone associated agranulocytosis. Pacing Clin Electrophysiol 1992;15(4 Pt 1):387–90.
33. Connolly SJ, Kates RE, Lebsack CS, Harrison DC, Winkle RA. Clinical pharmacology of propafenone. Circulation 1983;68(3):589–96.
34. Mondardini A, Pasquino P, Bernardi P, Aluffi E, Tartaglino B, Mazzucco G, Bonino F, Verme G, Negro F. Propafenone-induced liver injury: report of a case and review of the literature. Gastroenterology 1993;104(5):1524–6.
35. Cocozzella D, Curciarello J, Corallini O, Olivera A, Alburquerque MM, Fraquelli E, Zamagna L, Olenchuck A, Cremona A. Propafenone hepatotoxicity. Report of two new cases. Dig Dis Sci 2003;48:354–7.
36. Huang YM, Lee WR, Hu CH, Cheng KL. Propafenone-induced acute generalized exanthematous pustulosis. Int J Dermatol 2005;44(3):256–7.
37. Korst HA, Brandes JW, Littmann KP. Potenz- und Spermiogenesestorungen durch Propafenon. [Disturbances of potency and spermiogenesis due to propafenon.] Dtsch Med Wochenschr 1980;105(34):1187–9.
38. Gaita F, Richiardi E, Bocchiardo M, Asteggiano R, Pinnavaia A, Di Leo M, Rosettani E, Brusca A. Short- and long-term effects of propafenone in ventricular arrhythmias. Int J Cardiol 1986;13(2):163–70.
39. Guindo J, Rodriguez de la Serna A, Borja J, Oter R, Jane F, Bayes de Luna A. Propafenone and a syndrome of the lupus erythematosus type. Ann Intern Med 1986;104(4):589.
40. O'Rourke DJ, Palac RT, Holzberger PT, Gerling BR, Greenberg ML. Propafenone-induced drug fever in the absence of agranulocytosis. Clin Cardiol 1997;20(7):662–4.
41. Brunozzi LT, Meniconi L, Chiocchi P, Liberati R, Zuanetti G, Latini R. Propafenone in the treatment of chronic ventricular arrhythmias in a pregnant patient. Br J Clin Pharmacol 1988;26(4):489–90.
42. Jazwinska-Tarnawska E, Orzechowska-Juzwenko K, Niewinski P, Rzemislawska Z, Loboz-Grudzien K, Dmochowska-Perz M, Slawin J. The influence of CYP2D6 polymorphism on the antiarrhythmic efficacy of propafenone in patients with paroxysmal atrial fibrillation during 3 months propafenone prophylactic treatment. Int J Clin Pharmacol Ther 2001;39(7):288–92.

43. Vignati G, Mauri L, Figini A. The use of propafenone in the treatment of tachyarrhythmias in children. Eur Heart J 1993;14(4):546–50.
44. Janousek J, Paul TWorking Group on Pediatric Arrhythmias and Electrophysiology of the Association of European Pediatric Cardiologists. Safety of oral propafenone in the treatment of arrhythmias in infants and children (European Retrospective Multicenter Study). Am J Cardiol 1998;81(9):1121–4.
45. Antonelli D, Darawsha A, Rimbrot S, Freedberg NA, Rosenfeld T. [Propafenone dose for emergency room conversion of paroxysmal atrial fibrillation.]Harefuah 1999;136(11):857–9915.
46. Koppel C, Oberdisse U, Heinemeyer G. Clinical course and outcome in class IC antiarrhythmic overdose. J Toxicol Clin Toxicol 1990;28(4):433–44.
47. Rambourg-Schepens MO, Grossenbacher F, Buffet M, Lamiable D. Recurrent convulsions and cardiac conduction disturbances after propafenone overdose. Vet Hum Toxicol 1999;41(3):153–4.
48. Genty A, De Brabant F, Pibarot N, Busseuil C, Dubien PY, Ducluze R. Seizure disclosing acute propafenone poisoning. Jeur 2001;14:248–54.
49. Palette C, Maroun N, Gaulier J-M, Priolet B, Lachatre G, Bedos J-P, Advenier C, Therond P. Intoxication fatale par la propafénone: à propos d'un cas documenté par des dosages danguins. Thérapie 2003;58:384–6.
50. Sadeg N, Richecoeur J, Dumontet M. Intoxication à la propafénone. Thérapie 2003;58:381–3.
51. Molia AC, Tholon J-P, Lamiable DL, Trenque TC. Unintentional pediatric overdose of propafenone. Ann Pharmacother 2003;37:1147–8.
52. Schwenger V. Plasma separation in severe propafenone intoxication. Intensivmed Notf Med 2001;38:124–7.
53. Cardaioli P, Compostella L, De Domenico R, Papalia D, Zeppellini R, Libardoni M, Pulido E, Cucchini F. Influenza del propafenone sulla farmacocinetica della digossina somministrata per via orale: studio su volontari sani. [Effect of propafenone on the pharmacokinetics of digoxin administered orally: a study in healthy volunteers.] G Ital Cardiol 1986;16(3):237–40.
54. Palumbo E, Svetoni N, Casini M, Spargi T, Biagi G, Martelli F, Lanzetta T. Interazione digoxina–propafenone: valori e limiti del dosaggio plasmatico dei due farmaci. Efficacia antiaritmica del propafenone. [Digoxin–propafenone interaction: values and limitations of plasma determination of the 2 drugs. Anti-arrhythmia effectiveness of propafenone.] G Ital Cardiol 1986;16(10):855–62.
55. Chopra D, Janson P, Sawin CT. Insensitivity to digoxin associated with hypocalcemia. N Engl J Med 1977;296(16):917–8.
56. Ujhelyi MR, O'Rangers EA, Fan C, Kluger J, Pharand C, Chow MS. The pharmacokinetic and pharmacodynamic interaction between propafenone and lidocaine. Clin Pharmacol Ther 1993;53(1):38–48.
57. Wagner F, Kalusche D, Trenk D, Jahnchen E, Roskamm H. Drug interaction between propafenone and metoprolol. Br J Clin Pharmacol 1987;24(2):213–20.
58. Labbe L, O'Hara G, Lefebvre M, Lessard E, Gilbert M, Adedoyin A, Champagne J, Hamelin B, Turgeon J. Pharmacokinetic and pharmacodynamic interaction between mexiletine and propafenone in human beings. Clin Pharmacol Ther 2000;68(1):44–57.
59. Dinh H, Murphy ML, Baker BJ, De Soyza N. Propafenone:a new antiarrhythmic for treatment of chronic ventricular arrhythmias. Clin Progr Electrophysiol Pacing 1986;4:535–45.
60. De Soyza N, Murphy M, Sakhaii M, Treat L. The safety and efficacy of propafenone in suppressing ventricular ectopy. In: Shlepper M, Olsen B, editors. Cardiac Arrhythmias. Berlin: Springer Verlag, 1983:221.

Quinidine

General Information

Quinidine is a class I antidysrhythmic drug. Its actions, clinical use, interactions, and adverse effects have been reviewed (1).

In 652 consecutive inpatients 91 adverse reactions to quinidine occurred in as many patients (14%); of these, 51 were gastrointestinal, 16 dysrhythmic, 11 febrile, six dermatological, and one hematological; there were six cases of cinchonism (2). Although there were four cases of potentially fatal dysrhythmias, there were no deaths.

In a study of 245 patients, the most common adverse effects were diarrhea (35%), upper gastrointestinal distress (22%), and light-headedness (15%). Other common adverse effects included fatigue (7%), palpitation (7%), headache (7%), angina-like pain (6%), weakness (5%), and rash (5%) (3).

Observational studies

Of 35 patients given hydroquinidine for Brugada syndrome, seven had diarrhea, five during the first month, and two withdrew as a result (4). One patient had hepatitis during the first month, which reversed after withdrawal. Two patients had syncope during long-term treatment, one associated with prolongation of the QT interval.

Comparative studies

Quinidine and sotalol have been compared in a prospective multicenter trial of 121 patients after conversion of atrial fibrillation (5). Patients with low left ventricular ejection fractions (below 0.4) or high left atrial diameters (over 5.2 cm) were excluded. After 6 months the percentages of patients remaining in sinus rhythm were similar in the two groups (around 70%), but when the dysrhythmia recurred it occurred later with sotalol than with quinidine (69 versus 10 days). There was also a difference between patients who had been converted from recent-onset atrial fibrillation, in whom sotalol was more effective than quinidine, and in patients who had had chronic atrial fibrillation for more than 3 days, in whom quinidine was more effective. There were significant adverse effects requiring withdrawal of therapy in 17 patients, of whom nine were taking quinidine; three patients had gastrointestinal symptoms, two had central nervous symptoms, two had allergic reactions, one had undefined palpitation, and one had QT interval prolongation. One patient taking quinidine, a 65-year-old man, had frequent episodes of nonsustained ventricular tachycardia.

Placebo-controlled studies

In a prospective multicenter study in Germany, Poland, and the Slovak Republic, 1033 patients (mean age 60 years, 62% men) with frequent episodes of symptomatic paroxysmal atrial fibrillation were randomized to quinidine + verapamil 480/240 mg/day (n = 263), quinidine + verapamil 320/160 mg/day (n = 255), sotalol 320 mg/day (n = 264), or placebo (n = 251) (6). The three treatments were equally effective and the incidences of adverse effects were the same. There were four deaths, 13 cases of syncope, and one case of ventricular tachycardia; one death and the ventricular tachycardia were related to quinidine + verapamil in the higher doses.

In a prospective, multicenter, double-blind, placebo-controlled, randomized study in 1182 patients with persistent atrial fibrillation, 848 were successfully cardioverted and then randomized to sotalol (n = 383), quinidine plus verapamil (n = 377), or placebo (n = 88) (7). Quinidine plus verapamil was significantly superior to both placebo and sotalol in preventing recurrence. Adverse events were comparable (about 24% in each group), except that all nine cases of torsade de pointes occurred in patients taking sotalol.

Quinidine has previously been found to have relatively poor efficacy in maintaining sinus rhythm after cardioversion of atrial fibrillation (8), and can have serious adverse effects, including cardiac dysrhythmias ("quinidine syncope"), thrombocytopenia, and liver damage. These studies do not rehabilitate it in the management of atrial fibrillation, despite arguments to the contrary (9), since apparent efficacy may have been largely due to the verapamil with which quinidine was combined in both these studies.

General adverse reactions

Gastrointestinal symptoms, including anorexia, nausea, and vomiting, are common. Quinidine has a negative inotropic effect on the heart and can cause heart failure and hypotension. It prolongs the QRS complex and QT interval and can cause cardiac dysrhythmias, which can result in syncope (so-called "quinidine syncope"). "Cinchonism" is the term used to describe a cluster of adverse effects that occur at high dosages, including nausea, vomiting, diarrhea, tinnitus, dizziness, and blurred vision; in severe cases there may be deafness and toxic amblyopia in addition to cardiac abnormalities. Other adverse effects that have been described include dementia, psychosis, esophagitis, sometimes resulting in stricture, and exacerbation of myasthenia.

Various hypersusceptibility reactions have been attributed to quinidine, including fever, skin rashes, various types of hematological abnormalities (particularly thrombocytopenia), hepatitis, asthma, and anaphylactic shock. Quinidine has rarely been reported to cause a lupus-like syndrome. Tumor inducing effects have not been reported.

Organs and Systems

Cardiovascular

Quinidine prolongs the QRS complex and QT interval, the effect being related to plasma quinidine concentrations (10) and being greater in women (11). As a result, torsade de pointes or other ventricular tachydysrhythmias can occur and may lead to syncope ("quinidine syncope"). The minimum risk of torsade de pointes has been estimated at 1.5% per year (12).

The relation between serum quinidine concentrations and QT interval dispersion has been studied in 11 patients with atrial dysrhythmias and subtherapeutic or therapeutic serum quinidine concentrations (1.48 and 3.78 μg/ml respectively) (13). The baseline QT_c interval was 430 ms. At subtherapeutic and therapeutic serum quinidine concentrations, mean QT_c intervals were 451 and 472 ms respectively. Mean QT dispersion was 47 ms at baseline, 98 ms at subtherapeutic concentrations, and 71 ms at therapeutic concentrations. Despite QT interval lengthening with increasing serum quinidine concentrations, QT dispersion was greatest at subtherapeutic concentrations.

Quinidine has also been incriminated in cases of sinoatrial block and sinus arrest, but it was not clearly established that quinidine was responsible (14,15). The anticholinergic effects of quinidine can increase the risk of dysrhythmias (16).

Treatment of tachydysrhythmias secondary to quinidine is problematic. Other class I antidysrhythmic drugs are theoretically contraindicated. Some success has been reported with bretylium and with a combination of a beta-adrenoceptor antagonist and phenytoin (17,18). Overdrive pacing has also been reported to be of value (19).

- Quinidine syncope in a 46-year-old woman, associated with bradycardia and torsade de pointes due to prolongation of the QT interval (660 ms), responded to intravenous sodium bicarbonate and pacing (20).

Intravenous quinidine has been used to study the susceptibility to QT interval prolongation in 14 relatives of patients who had safely tolerated chronic therapy with a QT interval-prolonging drug (controls) and 12 relatives of patients who had developed acquired long QT syndrome (21). The interval from the peak to the end of the T wave, an index of transmural dispersion of repolarization, was significantly prolonged (from 63 to 83 ms) by quinidine in the relatives of those with acquired long QT syndrome but not in control relatives (from 66 to 71 ms). The time from the peak to the end of the T wave as a fraction of the QT interval was similar in the two groups at baseline but was longer in the relatives of those with acquired long QT syndrome after quinidine. The authors concluded that first-degree relatives of patients with acquired long QT syndrome have greater drug-induced prolongation of terminal repolarization than control relatives, supporting a genetic predisposition to acquired long QT syndrome.

Hydroquinidine has been used to treat the short QT syndrome in six patients and has been compared with other antidysrhythmic drugs (22). Hydroquinidine prolonged the QTc interval from 290 to 405 ms and ventricular fibrillation was no longer inducible. Diarrhea was the main adverse effect.

Quinidine has a negative inotropic effect on the heart and causes peripheral vasodilatation. Hypotension can occur secondary to these effects (23).

Nervous system

Quinidine rarely causes effects on the nervous system when used in therapeutic dosages, although there have been occasional reports of dementia (24–26). However, in toxicity it can cause vertigo and tinnitus (27).

Sensory systems

Quinidine rarely affects vision, but can cause scotomata, impaired color vision, and toxic amblyopia (28), altered vision (29), sicca syndrome (30), keratopathy (31), and granulomatous uveitis (32).

Hematologic

Hypersusceptibility reactions to quinidine include thrombocytopenia, hemolytic anemia, and neutropenia.

The overall annual incidence of acute thrombocytopenia has been estimated at 18 cases per million (33), and two possible mechanisms have been invoked:

1. direct combination of quinidine with platelets, causing the production of antibodies, which then cause platelet lysis;
2. formation of quinidine-antibody complexes, which are then deposited on the platelet (the so-called "innocent bystander" mechanism).

There is evidence in favor of both of these mechanisms (34).

- A 72-year-old woman developed repeated bouts of thrombocytopenia; quinidine-dependent antibodies were found after it transpired that she had been taking her husband's tablets intermittently to treat bouts of palpitation (35).

Sensitization may have occurred because of the intermittent nature of exposure in this case.

Other hematological effects are much less common, and include hemolytic anemia (34,36), neutropenia (37,38), and eosinophilia (39).

Withdrawal of quinidine and the administration of corticosteroids is the usual treatment for thrombocytopenia (40), although intravenous immune globulin has also been used (41).

In a review of all English-language reports on drug-induced thrombocytopenia, excluding heparin, 561 case articles reporting on 774 patients were analysed (42). A definite or probable causal role for the drug used was attributed in 247 case reports, and of the 98 that were implicated quinidine was mentioned in 38 cases. The next most common drugs involved were gold salts (11 cases) and co-trimoxazole (10 cases).

Mouth and teeth

A blue–black pigmentation of the oral mucosa has been reported with quinidine (43).

Gastrointestinal

Anorexia, nausea, vomiting, and diarrhea are common adverse effects of quinidine and occur in up to 30% of patients (2,44). These effects can be minimized by the use of modified-release formulations (45).

In one series diarrhea was reported commonly (46), and it can occur late in treatment (47).

Quinidine sometimes causes esophagitis (48,49), especially when there is some abnormality of the esophagus or cardiomegaly. This sometimes results in esophageal stricture (50).

The quinidine derivative hydroquinidine had some beneficial effects in 10 patients with myotonic dystrophy with slow saccadic eye movements, apathy, and hypersomnia (51,52). However, two patients had nausea and epigastric pain and withdrew while taking the active treatment. Although there were no cases of cardiac abnormalities, the authors raised the concern that in patients with myotonic dystrophy, who have a high frequency of cardiac disturbances, the risk of cardiac dysrhythmias with quinidine derivatives may be too high to take.

Liver

Hypersusceptibility reactions to quinidine include granulomatous hepatitis. In one retrospective series of 487 patients, 32 had evidence of hypersusceptibility, 10 of whom had hepatotoxicity (53). In another series of 1500 patients, quinidine-induced hepatitis was identified in 33 (2.2%) (54); these represented one-third of all cases of drug-induced hepatitis in those patients. In all cases the liver damage resolved on withdrawal.

Urinary tract

Glomerulonephritis has been reported in association with Henoch–Schönlein purpura (55) and nephrotic syndrome, possibly as part of a lupus-like syndrome (56).

Skin

Skin rashes are uncommon (57), but can occur as part of a hypersusceptibility reaction. In a review of drug-induced skin disorders, from a list of 26 drugs or groups of drugs, only quinidine was mentioned of all antidysrhythmic drugs (58). The reaction rate was quoted as 12 per 1000 recipients. Several different kinds of rash have been reported, including photosensitivity (59), which can result in a variety of types of rash, contact dermatitis (60), pigmentation (61), urticaria (62), exfoliative dermatitis (63,64), granuloma annulare (65), and exacerbation of psoriasis.

Musculoskeletal

Quinidine is a neuromuscular blocking drug and can exacerbate myasthenia gravis (66). It can also cause an increase in the serum activity of the muscle-specific isozyme of creatine phosphokinase (67–69).

- A 56-year-old man who took quinidine 324 mg five times daily for prevention of paroxysmal atrial fibrillation developed a syndrome similar to polymyalgia rheumatica which settled on drug withdrawal (69).

Quinidine can cause polyarthropathy (70), both in association with and independent of a lupus-like syndrome.

Immunologic

A variety of immune syndromes have occasionally been reported with quinidine, including a lupus-like syndrome, polymyalgia rheumatica, and vasculitis (SEDA-20, 179) (SEDA-23, 202).

Life-threatening vasculitis has been attributed to quinidine in a healthy volunteer taking part in a clinical trial (71).

- A 58-year-old man took quinidine 200 mg tds for 7 days as part of an interaction study with a new alpha-blocker. He developed widespread maculopapular purpuric lesions on the limbs, trunk, and ears. His temperature rose to 38.4°C and some of the lesions on his fingers, toes, ears, and nose became necrotic. He had peripheral edema with a bluish purpuric discoloration of the hands and feet. There was mucous membrane involvement with purpuric, partially necrotic lesions on the tongue and palate. A skin biopsy showed necrotizing vasculitis with focal leukocytoclasia. Direct immunofluorescence showed microgranular deposits of IgA, IgM, and C3 around the superficial skin vessels. Quinidine was withdrawn and he was given intravenous methyl prednisolone followed by oral prednisone for one month. He recovered completely within 3 weeks.

There has been a report of a dermatomyositis-like illness in a man taking quinidine (72).

- A previously healthy 63-year-old man, who had taken quinidine gluconate 972 mg/day for 9 months, developed diffuse edematous erythema on the extensive surfaces of the hands, arms, and face, with marked accentuation over the joints. His nail-fold capillaries were dilated and the shoulder abductors were slightly weak. His erythrocyte sedimentation rate was slightly raised (29 mm/hour) and there was a positive ANA titer (1:640) with a speckled pattern. There were no antibodies to Sm, ribonucleoprotein, SSA or SSB antigens, or histones. There was no evidence of inflammatory myopathy on electromyography, and a skin biopsy showed a mild, superficial, perivascular, lymphocytic inflammation with positive direct immunofluorescence for IgG and IgM at the dermoepidermal junction. There was no evidence of malignancy. All these abnormalities resolved rapidly after quinidine withdrawal.

Lupus-like syndrome

A lupus-like syndrome has occasionally been reported in patients taking quinidine (69,70,73). It usually presents with polyarthralgia, a raised erythrocyte sedimentation rate, and a raised antinuclear antibody titer. It can occasionally be associated with antihistone antibodies and a circulating coagulant. In two cases (74) the syndrome was associated with quinidine and not with procainamide. Lupus anticoagulant has been reported with the use of quinine and quinidine, and an associated antiphospholipid syndrome has been described (75).

Susceptibility Factors

Age

Maintenance dosages should be reduced in elderly people (76).

Sex

The effect of quinidine on the QT interval is greater in women than in men at equivalent serum concentrations (6). In 12 men and 12 women who received a single intravenous dose of quinidine (4 mg/kg) in a randomized, single-blind, placebo-controlled, crossover study, total and unbound serum concentrations of quinidine and 3-hydroxyquinidine were measured and QT intervals were corrected for differences in heart rate using Bazett's method. The QT interval at baseline was longer in women than in men (407 versus 395 ms). The slope of the relation between the serum concentration of quinidine and the change in the QT_c interval from baseline was 44% greater in the women than in the men. However, there were no significant differences between the men and the women in the disposition of quinidine, apart from a small reduction in the unbound fraction of 3-hydroxyquinidine in the men (0.47 versus 0.53). The authors proposed that estrogens and androgens differentially affect the expression and activity of potassium channels in the heart, and that a lower density of potassium channels could contribute to a larger effect of quinidine in the women. They suggested that women are at greater risk of quinidine-induced cardiac dysrhythmias and that altering dosages according to weight would not correct for this difference. They also pointed to the fact that in the SWORD study the risk of excess mortality in those taking D-sotalol was greater in the women than in the men.

Hepatic disease

Maintenance dosages should be reduced if hepatic metabolism is reduced, for example in congestive cardiac failure and hepatic disease and in elderly people (76).

Other features of the patient

Care must be taken in patients with pre-existing conduction tissue disease or heart failure. Loading doses should be reduced in cardiac failure, because of a lowered apparent volume of distribution (76).

Drug Administration

Drug overdose

In acute overdosage the chief effect of quinidine is profound hypotension, due to the combination of peripheral vasodilatation, and a negative inotropic effect on the heart (77,78). Charcoal hemoperfusion has been successfully used in the treatment of quinidine overdose (79).

Drug–Drug Interactions

Amiloride

Quinidine has a pharmacodynamic interaction with amiloride, further prolonging the duration of the QRS complex, but not the QT interval (80).

Amiodarone

Quinidine prolongs the QT interval and will therefore potentiate the effects of other antidysrhythmic drugs that have the same effect (for example amiodarone) (81).

Antacids

The interactions of quinidine with antacids have been reviewed (82). Although the data are inconclusive, it has been suggested that at least 2 hours should elapse between quinidine and antacid doses.

Bupivacaine

Quinidine displaces bupivacaine from plasma proteins (83).

Cardiac glycosides

Digoxin
The quinidine–digoxin interaction has been reviewed (SEDA-6, 173) (SEDA-7, 195) (SEDA-9, 159) (SEDA-10, 145) (SEDA-15, 166) (SEDA-18, 198). Although the major mechanism of the interaction is probably inhibition of the active tubular secretion of digoxin by quinidine, other mechanisms are involved, including reduced non-renal clearance and displacement of digoxin from the tissues. The reduction in non-renal clearance is at least partly due to a reduction in biliary clearance. This interaction affects most patients and on average causes a two-fold increase in steady-state plasma digoxin concentrations. Because both clearance and apparent volume of distribution are reduced, the half-life of digoxin is either unaffected or perhaps slightly prolonged.

In addition to the pharmacokinetic interaction, there may be a pharmacodynamic interaction, since there is some evidence that quinidine reduces the positive inotropic effect of digoxin on the heart, in addition to having a negative inotropic effect of its own. Thus, the outcome of the interaction is a 24-fold increased risk of digoxin toxicity and a reduction in its beneficial effect on the heart, at least in sinus rhythm. If the two drugs are used together, the initial digoxin dosage should be halved and adjusted subsequently on the basis of the patient's clinical condition and the plasma digoxin concentration.

Lanatoside C and metildigoxin are theoretically likely to be similarly affected by quinidine, and there is anecdotal evidence of this (84).

The pharmacokinetic interaction of quinidine with digoxin also occurs with quinine (85,86) and hydroxychloroquine (87). However, the effects of these drugs are smaller than those with quinidine. Quinine reduces the extrarenal clearance of digoxin, perhaps by altering its biliary secretion (88).

Digitoxin
In five healthy adults quinidine prolonged the half-life of a single dose of digitoxin from 174 to 261 hours and reduced its total body clearance from 1.54 to 1.09 ml/hour/kg and renal clearance from 0.65 to 0.46 ml/hour/kg. Digitoxin volume of distribution and protein binding were unaffected. Quinidine caused a rise in serum digitoxin concentrations (89).

In eight healthy subjects steady-state digitoxin plasma concentrations and renal excretion increased from 13.6 ng/ml and 16.1 μg/day before dosing to 19.7 ng/ml and 23.4 μg/day respectively during quinidine dosing for 32 days (90). Renal digitoxin clearance was not noticeably changed by quinidine, but total and extrarenal digitoxin clearances fell by 32% and 41% respectively. The half-life of digitoxin was prolonged from 150 to 203 hours. There were corresponding pharmacodynamic effects, as assessed by electrocardiography and systolic time intervals.

Ciprofloxacin

Serum quinidine concentrations rose during concomitant administration of ciprofloxacin (91). The authors speculated that the mechanism was inhibition of cytochrome P450 by ciprofloxacin.

Class I antidysrhythmic drugs

Quinidine prolongs the QT interval and will therefore potentiate the effects of other antidysrhythmic drugs that have the same effect (for example Class I antidysrhythmic drugs). Specific interactions of this kind have been reported with tocainide (92) and mexiletine (93); in these cases the interactions were reported to be beneficial in terms of antidysrhythmic effects.

Codeine

Quinidine inhibits the hepatic metabolism of codeine to morphine. Whether this diminishes or abolishes the analgesic effect of codeine is uncertain (94,95).

Dextromethorphan

The effects of quinidine sulfate, 50 mg orally, an inhibitor of cytochrome CYP2D6, on the metabolism of dextromethorphan 50 mg have been studied in seven healthy volunteers in a randomized, double-blind, crossover, placebo-controlled study (96). Quinidine suppressed the conversion of dextromethorphan to dextrorphan in extensive metabolizers to the extent seen in poor metabolizers. The increased concentrations of dextromethorphan increased subjective and objective pain thresholds by 35 and 45% respectively. This result suggests that debrisoquine/sparteine-type polymorphisms account for important differences in the effect of dextromethorphan and the balance between the analgesic effect of dextromethorphan and the hallucinogenic effect of dextrorphan. Concomitant use of quinidine or other inhibitors of CYP2D6 could further affect this balance, increasing the risk of serotonergic- and narcotic-related adverse effects.

In a double-blind, randomized, crossover study 22 subjects took placebo, dextromethorphan hydrobromide 30 mg, dextromethorphan hydrobromide 60 mg, and dextromethorphan hydrobromide 30 mg preceded at 1 hour by quinidine hydrochloride 50 mg (97). Cough was elicited using citric acid. The intrinsic clearance of dextromethorphan estimated from a pharmacokinetic model was 59–1536 1/h, which overlapped with that extrapolated from in vitro data (12–261 1/h). Quinidine reduced the clearance of dextromethorphan with an estimated average K_i of 0.017 µmol/l, and prolonged its half-life more than three-fold to 58 hours; it also increased its rate of absorption, which the authors thought might be due to an increased rate of gastric emptying.

In two multiple-dose studies the lowest oral dose of quinidine that could be used in a fixed combination with three doses of dextromethorphan, in order to suppress its O-demethylation maximally, was determined to be 25–30 mg (98). This dose was then used in a randomized study of the use of the combination of quinidine + dextromethorphan, compared with either alone, in the treatment of pseudobulbar affect in 129 patients with amyotrophic lateral sclerosis (99). There were significantly higher incidences of nausea, dizziness, and somnolence in those who took the combination, all attributed to reduced clearance of dextromethorphan.

Diclofenac

In human liver microsomes, diclofenac inhibited testosterone 6-beta-hydroxylation with characteristics that suggested that it inactivated CYP3A4 (100). Quinidine, which stimulates CYP3A4-mediated diclofenac 5-hydroxylation, did not affect the inactivation of CYP3A4 assessed by testosterone 6-beta-hydroxylation activity but accelerated the inactivation assessed by diazepam 3-hydroxylation activity.

In 30 healthy young men the pharmacokinetics of a single oral dose of quinidine 200 mg were studied before and during the daily administration of diclofenac 100 mg (a substrate of CYP2C9) (72). The clearance of quinidine by *N*-oxidation was reduced by diclofenac, but only by 27%. This small effect of diclofenac suggests a minor role for CYP2C9 in the metabolism of quinidine.

Erythromycin

Erythromycin reduced the total clearance of quinidine, reduced its partial clearance by 3-hydroxylation, and increased its maximal serum concentration in an open study in 30 healthy young volunteers (101).

Enzyme-inducing drugs

The metabolism of quinidine is enhanced by drugs that induce hepatic microsomal drug-metabolizing enzymes (76), such as rifampicin (102) and phenytoin and phenobarbital (103). This leads to an increase in the first-pass metabolism of quinidine, and thus increased requirements of oral quinidine but little change in intravenous dosages.

Quinidine inhibits P glycoprotein, and has been used to study the pharmacokinetics of fentanyl (104). Quinidine increased oral fentanyl plasma C_{max} and AUC and prolonged its half-life. However, it did not alter the magnitude or time to maximum miosis, time-specific pupil diameter, or subjective self-assessments after intravenous fentanyl.

Flecainide

Flecainide is metabolized by CYP2D6, and is subject to polymorphic metabolism. In extensive metabolizers its clearance is reduced by quinine (105). Quinidine reduces the clearance of *R*-flecainide but not that of *S*-flecainide (106).

Inhibitors of CYP3A4

There is in vitro evidence that the oxidation of quinidine to 3-hydroxyquinidine is catalysed by CYP3A4. However, the extent to which other cytochrome P450 isoforms, such as CYP2C9 and CYP2E1, are involved in the oxidation of quinidine is not clear. In 30 healthy young men the pharmacokinetics of a single oral dose of quinidine 200 mg were studied before and during the daily administration of diclofenac 100 mg (a substrate of CYP2C9), disulfiram 200 mg (an inhibitor of CYP2E1), three inhibitors of CYP3A4 (grapefruit juice, itraconazole, and erythromycin), and probes of other enzyme activities, namely caffeine (CYPA2), sparteine (CYP2D6), mephenytoin (CYP2C19), and tolbutamide (CTP2C9) (72). The clearance of quinidine by *N*-oxidation was reduced by 27% by diclofenac. Itraconazole, grapefruit juice, and erythromycin all reduced the clearance of quinidine, including its partial clearance by 3-hydroxylation and *N*-oxidation, by up to 84%, confirming the involvement of CYP3A4 in the in vivo oxidation of quinidine. Specifically, itraconazole reduced quinidine total clearance, partial clearance by 3-hydroxylation, and partial clearance by *N*-oxidation by 61, 84, and 73% respectively. Thus, it is likely that inhibitors of CYP3A4 will cause significant drug interactions with quinidine.

Itraconazole

In vitro studies have suggested that the oxidation of quinidine to 3-hydroxyquinidine is a specific marker reaction for CYP3A4 activity. In six healthy young men the pharmacokinetics of a single oral dose of quinidine 200 mg were studied before and during daily administration of itraconazole 100 mg (107). Itraconazole reduced quinidine total clearance, partial clearance by 3-hydroxylation, and partial clearance by *N*-oxidation by 61, 84, and 73% respectively.

Methadone

Quinidine inhibits P glycoprotein, and has been used to study the pharmacokinetics of methadone (108). Quinidine altered the systemic availability of methadone without altering its pharmacodynamic effects after intravenous administration. The authors therefore suggested that P glycoprotein has less of an effect on the brain access of fentanyl and methadone than on their systemic availability.

Metoprolol

Quinidine inhibits drug hydroxylation and can thus convert extensive hydroxylators to poor hydroxylators, as in the case of metoprolol (109). Inhibition of hydroxylation does not occur in poor hydroxylators.

Metronidazole

Serum quinidine concentrations rose during concomitant administration of metronidazole (110). The authors speculated that the mechanism was inhibition of cytochrome P_{450}.

Neuromuscular blocking drugs

Quinidine is a non-depolarizing muscle blocker and potentiates the effects of neuromuscular blocking drugs (111).

Nifedipine

Nifedipine may enhance the elimination of quinidine (112) and cause hypotension and a loss of antidysrhythmic effect (113).

Rifamycins

Quinidine plasma concentrations can be lowered by rifampicin because of enzyme induction (114).

Thioridazine

Concurrent administration of quinidine with neuroleptic drugs, particularly thioridazine, can cause myocardial depression (115).

Timolol

The combination of quinidine with timolol causes an increased risk of bradycardia (116).

Tramadol

Inhibition of the hepatic metabolism of tramadol to morphine by quinidine may reduce its opioid effects (SEDA-21, 90).

Venlafaxine

Venlafaxine is metabolized by CYP2D6. In healthy volunteers the oral clearance of venlafaxine (37.5 mg/day for 2 days) was fourfold less in poor metabolizers ($n = 6$) than extensive metabolizers ($n = 8$) (117). Administration of the CYP2D6 inhibitor quinidine, 200 mg/day for 2 days, to the extensive metabolizers reduced the oral clearance of venlafaxine to the level seen in poor metabolizers. Quinidine had no effect on venlafaxine clearance in subjects who were poor metabolizers before treatment. The authors suggested that poor metabolizers may be at particular risk of venlafaxine toxicity, as could subjects who take inhibitors of CYP2D6.

Warfarin

Quinidine potentiates the effects of warfarin by a pharmacodynamic effect (118).

However, it has also been reported that quinidine increased the 4′-hydroxylation of *S*-warfarin and the 10-hydroxylation of *R*-warfarin in human liver microsomes and intact hepatocytes by stimulation of CYP3A4 (119). The increases were concentration-dependent and respectively maximized at about three and five times control values. In contrast, warfarin did not affect the 3-hydroxylation of quinidine. These results are consistent with previous findings suggesting that there is more than one binding site on CYP3A4 through which interactions can occur.

References

1. Malcolm AD, David GK. Quinidine in cardiology. Acta Leiden 1987;55:87–98.
2. Cohen IS, Jick H, Cohen SI. Adverse reactions to quinidine in hospitalized patients: findings based on data from the Boston Collaborative Drug Surveillance Program. Prog Cardiovasc Dis 1977;20(2):151–63.
3. Roos JC, Paalman DC, Dunning AJ. Electrophysiological effects of mexiletine in man. Postgrad Med J 1977;53(Suppl 1):92–6.
4. Hermida JS, Denjoy I, Clerc J, Extramiana F, Jarry G, Milliez P, Guicheney P, Di Fusco S, Rey JL, Cauchemez B, Leenhardt A. Hydroquinidine therapy in Brugada syndrome. J Am Coll Cardiol 2004;43(10):1853–60.
5. de Paola AA, Veloso HH. Efficacy and safety of sotalol versus quinidine for the maintenance of sinus rhythm after conversion of atrial fibrillation. SOCESP Investigators. The Cardiology Society of Sao Paulo. Am J Cardiol 1999;84(9):1033–7.
6. Patten M, Maas R, Bauer P, Luderitz B, Sonntag F, Dluzniewski M, Hatala R, Opolski G, Muller HW, Meinertz T;SOPAT Investigators. Suppression of paroxysmal atrial tachyarrhythmias–results of the SOPAT trial. Eur Heart J 2004;25(16):1395–404.
7. Fetsch T, Bauer P, Engberding R, Koch HP, Lukl J, Meinertz T, Oeff M, Seipel L, Trappe HJ, Treese N, Breithardt G;Prevention of Atrial Fibrillation after Cardioversion Investigators. Prevention of atrial fibrillation after cardioversion: results of the PAFAC trial. Eur Heart J 2004;25(16):1385–94.
8. McNamara RL, Tamariz LJ, Segal JB, Bass EB. Management of atrial fibrillation: review of the evidence for the role of pharmacologic therapy, electrical cardioversion, and echocardiography. Ann Intern Med 2003;139(12):1018–33.
9. van Hemel N. Quinidine rehabilitated and more lessons from the PAFAC and SOPAT anti-arrhythmic drug trials for the prevention of paroxysmal atrial fibrillation. Eur Heart J 2004;25(16):1371–3.
10. White NJ, Looareesuwan S, Warrell DA, Chongsuphajaisiddhi T, Bunnag D, Harinasuta T.

Quinidine in falciparum malaria. Lancet 1981;2(8255):1069–71.
11. Benton RE, Sale M, Flockhart DA, Woosley RL. Greater quinidine-induced QTc interval prolongation in women. Clin Pharmacol Ther 2000;67(4):413–8.
12. Roden DM, Woosley RL, Primm RK. Incidence and clinical features of the quinidine-associated long QT syndrome: implications for patient care. Am Heart J 1986;111(6):1088–93.
13. Mathis AS, Gandhi AJ. Serum quinidine concentrations and effect on QT dispersion and interval. Ann Pharmacother 2002;36(7–8):1156–61.
14. Grayzel J, Angeles J. Sino-atrial block in man provoked by quinidine. J Electrocardiol 1972;5(3):289–94.
15. Jeresaty RM, Kahn AH, Landry AB Jr. Sinoatrial arrest due to lidocaine in a patient receiving guinidine. Chest 1972;61(7):683–5.
16. Cappato R, Alboni P, Codeca L, Guardigli G, Toselli T, Antonioli GE. Direct and autonomically mediated effects of oral quinidine on RR/QT relation after an abrupt increase in heart rate. J Am Coll Cardiol 1993;22(1):99–105.
17. VanderArk CR, Reynolds EW, Kahn DR, Tullett G. Quinidine syncope. A report of successful treatment with bretylium tosylate. J Thorac Cardiovasc Surg 1976;72(3):464–7.
18. Koster RW, Wellens HJ. Quinidine-induced ventricular flutter and fibrillation without digitalis therapy. Am J Cardiol 1976;38(4):519–23.
19. DiSegni E, Klein HO, David D, Libhaber C, Kaplinsky E. Overdrive pacing in quinidine syncope and other long QT-interval syndromes. Arch Intern Med 1980;140(8):1036–40.
20. Tsai CL. Quinidine cardiotoxicity. J Emerg Med 2005;28(4):463–5.
21. Kannankeril PJ, Roden DM, Norris KJ, Whalen SP, George AL Jr, Murray KT. Genetic susceptibility to acquired long QT syndrome: pharmacologic challenge in first-degree relatives. Heart Rhythm 2005;2(2):134–40.
22. Gaita F, Giustetto C, Bianchi F, Schimpf R, Haissaguerre M, Calo L, Brugada R, Antzelevitch C, Borggrefe M, Wolpert C. Short QT syndrome: pharmacological treatment. J Am Coll Cardiol 2004;43(8):1494–9.
23. Luchi RJ, Helwig J Jr, Conn HL Jr. Quinidine toxicity and its treatment. An Experimental study. Am Heart J 1963;65:340–8.
24. Gilbert GJ. Quinidine dementia. JAMA 1977;237(19):2093–4.
25. Billig N, Buongiorno P. Quinidine-induced organic mental disorders. J Am Geriatr Soc 1985;33(7):504–6.
26. Deleu D, Schmedding E. Acute psychosis as idiosyncratic reaction to quinidine: report of two cases. BMJ (Clin Res Ed) 1987;294(6578):1001–2.
27. Abrams J. Quinidine toxicity: a review. Rocky Mt Med J 1973;70(5):31–4.
28. Bolton FG. Thrombocytopenic purpura due to quinidine. II. Serologic mechanisms. Blood 1956;11(6):547–64.
29. Fisher CM. Visual disturbances associated with quinidine and quinine. Neurology 1981;31(12):1569–71.
30. Naschitz JE, Yeshurun D. Quinidine induced sicca syndrome. J Toxicol Clin Toxicol 1983;20(4):367–71.
31. Zaidman GW. Quinidine keratopathy. Am J Ophthalmol 1984;97(2):247–9.
32. Hustead JD. Granulomatous uveitis and quinidine hypersensitivity. Am J Ophthalmol 1991;112(4):461–2.
33. Kaufman DW, Kelly JP, Johannes CB, Sandler A, Harmon D, Stolley PD, Shapiro S. Acute thrombocytopenic purpura in relation to the use of drugs. Blood 1993;82(9):2714–8.
34. Garratty G. Review: Immune hemolytic anemia and/or positive direct antiglobulin tests caused by drugs. Immunohematol 1994;10(2):41–50.
35. Reddy JC, Shuman MA, Aster RH. Quinine/quinidine-induced thrombocytopenia: a great imitator. Arch Intern Med 2004;164(2):218–20.
36. Barzel US. Quinidine-sulfate-induced hypoplastic anemia and agranulocytosis. JAMA 1967;201(5):325–7.
37. Castro O, Nash I. Quinidine leukopenia and thrombocytopenia with a drug-dependent leukoagglutinin. N Engl J Med 1977;296(10):572.
38. Eisner EV, Carr RM, MacKinney AR. Quinidine-induced agranulocytosis. JAMA 1977;238(8):884–6.
39. Religa H, Roznіecki J, Szmidt M. Eozynofilia w przebiegu leczenia lanatozydem C i siarczanem chinidyny jako jedyny przejaw nadwrazliwoski polekowej. [Eosinophilia in the course of lanatoside C and quinidine treatment as the sole sign of drug hypersensitivity.] Pol Tyg Lek 1972;27(44):1727–9.
40. Saleh MN, Dhodaphar N, Allen K, LoBuglio AF. Quinidine-induced immune thrombocytopenia. Henry Ford Hosp Med J 1989;37(1):28–32.
41. Redell MA, Moore BR, Fass L. Use of i.v. immune globulin for presumed quinidine-induced thrombocytopenia Clin Pharm 1989;8(2):89.
42. George JN, Raskob GE, Shah SR, Rizvi MA, Hamilton SA, Osborne S, Vondracek T. Drug-induced thrombocytopenia: a systematic review of published case reports. Ann Intern Med 1998;129(11):886–90.
43. Birek C, Main JH. Two cases of oral pigmentation associated with quinidine therapy. Oral Surg Oral Med Oral Pathol 1988;66(1):59–61.
44. Rokseth R, Storstein O. Quinidine therapy of chronic auricular fibrillation. The occurrence and mechanism of syncope. Arch Intern Med 1963;111:184–9.
45. Mahon WA, Mayersohn M, Inaba T. Disposition kinetics of two oral forms of quinidine. Clin Pharmacol Ther 1976;19(5 Pt 1):566–75.
46. Kennedy HL, DeMaria AN, Sprague MK, Wiens RD, Redd RM, Janosik DL, Buckingham TA. Comparative efficacy of moricizine and quinidine for benign and potentially lethal ventricular arrhythmias. Am J Noninvas Cardiol 1988;2:98–105.
47. Zahger D, Gilon D, Gotsman MS. Delayed quinidine-induced diarrhea after five years of treatment. Chest 1992;101(1):296.
48. Bott SJ, McCallum RW. Medication-induced oesophageal injury. Survey of the literature. Med Toxicol 1986;1(6):449–57.
49. Wong RK, Kikendall JW, Dachman AH. Quinaglute-induced esophagitis mimicking an esophageal mass. Ann Intern Med 1986;105(1):62–3.
50. Bonavina L, DeMeester TR, McChesney L, Schwizer W, Albertucci M, Bailey RT. Drug-induced esophageal strictures. Ann Surg 1987;206(2):173–83.
51. Di Costanzo A, Mottola A, Toriello A, Di Iorio G, Tedeschi G, Bonavita V. Does abnormal neuronal excitability exist in myotonic dystrophy? I. Effects of the antiarrhythmic drug hydroquinidine on slow saccadic eye movements. Neurol Sci 2000;21(2):73–80.
52. Di Costanzo A, Mottola A, Toriello A, Di Iorio G, Tedeschi G, Bonavita V. Does abnormal neuronal excitability exist in myotonic dystrophy? II. Effects of the

antiarrhythmic drug hydroquinidine on apathy and hypersomnia. Neurol Sci 2000;21(2):81–6.
53. Geltner D, Chajek T, Rubinger D, Levij IS. Quinidine hypersensitivity and liver involvement. A survey of 32 patients. Gastroenterology 1976;70(5 Pt 1):650–2.
54. Knobler H, Levij IS, Gavish D, Chajek-Shaul T. Quinidine-induced hepatitis. A common and reversible hypersensitivity reaction. Arch Intern Med 1986;146(3):526–8.
55. Aviram A. Henoch-Schonlein syndrome associated with quinidine. JAMA 1980;243(5):432–3.
56. Chisholm JC Jr. Quinidine-induced nephrotic syndrome. J Natl Med Assoc 1985;77(11):920–2.
57. Pariser DM, Taylor JR. Quinidine photosensitivity. Arch Dermatol 1975;111(11):1440–3.
58. Duncan KO. Severe cutaneous adverse reactions to medications. Prim Care Case Rev 2001;4:171–85.
59. Lang PG Jr. Quinidine-induced photodermatitis confirmed by photopatch testing. J Am Acad Dermatol 1983;9(1):124–8.
60. Fowler JF. Allergic contact dermatitis to quinidine. Contact Dermatitis 1985;13(4):280–1.
61. Mahler R, Sissons W, Watters K. Pigmentation induced by quinidine therapy. Arch Dermatol 1986;122(9):1062–4.
62. Shaftel N, Halpern A. The quinidine problem. Angiology 1958;9(1):34–46.
63. Taylor DR, Potashnick R. Quinidine-induced exfoliative dermatitis; with a brief review of quinidine idiosyncrasies. JAMA 1951;145(9):641–2.
64. Gouffault J, Pawlotsky Y, Morel H, Bourel M. Erythrodermie d'origine quinidinique. [Erythrodermia of quinidine origin.] Sem Hop 1965;41(22):1350–3.
65. Ross V, Cobb M. Generalized granuloma annulare associated with quinidine therapy. J Assoc Military Dermatol 1991;17:16–7.
66. Stoffer SS, Chandler JH. Quinidine-induced exacerbation of myasthenia gravis in patient with Graves' disease. Arch Intern Med 1980;140(2):283–4.
67. Weiss M, Hassin D, Eisenstein Z, Bank H. Elevated skeletal muscle enzymes during quinidine therapy. N Engl J Med 1979;300(21):1218.
68. Ramsey R, Higbee M, Wood JS. Quinidine-induced creatine phosphokinase elevations:case report and prospective case survey in the elderly. J Geriatr Drug Ther 1989;3:97.
69. Alloway JA, Salata MP. Quinidine-induced rheumatic syndromes. Semin Arthritis Rheum 1995;24(5):315–22.
70. Yagiela JA, Benoit PW. Skeletal-muscle damage from quinidine. N Engl J Med 1979;301(8):437.
71. Lipsker D, Walther S, Schulz R, Nave S, Cribier B. Life-threatening vasculitis related to quinidine occurring in a healthy volunteer during a clinical trial. Eur J Clin Pharmacol 1998;54(9–10):815.
72. Gilliland WR. Quinidine-induced dermatomyositis-like illness. J Clin Rheumatol 1999;5:39.
73. Tebas P, Lozano I, de la Fuente J, Ortigosa J, Perez Maestu R, Masa C, de Letona JM. Lupus inducido por quinidina. [Lupus induced by quinidine.] Rev Clin Esp 1991;189(3):123–4.
74. Amadio P Jr, Cummings DM, Dashow L. Procainamide, quinidine, and lupus erythematosus. Ann Intern Med 1985;102(3):419.
75. Bird MR, O'Neill AI, Buchanan RR, Ibrahim KM, Des Parkin J. Lupus anticoagulant in the elderly may be associated with both quinine and quinidine usage. Pathology 1995;27(2):136–9.
76. Kumana CR. Therapeutic drug monitoring—antidysrhythmic drugs. In: Richens A, Marks V, editors. Therapeutic Drug Monitoring. London, Edinburgh: Churchill-Livingstone, 1981:370.
77. Kerr F, Kenoyer G, Bilitch M. Quinidine overdose. Neurological and cardiovascular toxicity in a normal person. Br Heart J 1971;33(4):629–31.
78. Woie L, Oyri A. Quinidine intoxication treated with hemodialysis. Acta Med Scand 1974;195(3):237–9.
79. Haapanen EJ, Pellinen TJ. Hemoperfusion in quinidine intoxication. Acta Med Scand 1981;210(6):515–6.
80. Wang L, Sheldon RS, Mitchell LB, Wyse DG, Gillis AM, Chiamvimonvat N, Duff HJ. Amiloride–quinidine interaction: adverse outcomes. Clin Pharmacol Ther 1994;56(6 Pt 1):659–67.
81. Tartini R, Kappenberger L, Steinbrunn W. Gefahrliche Interaktionen zwischen Amiodaron und Antiarrhythmika der Klasse I. [Harmful interactions of amiodarone and class I anti-arrhythmia agents.] Schweiz Med Wochenschr 1982;112(45):1585–7.
82. Sadowski DC. Drug interactions with antacids. Mechanisms and clinical significance. Drug Saf 1994;11(6):395–407.
83. Ghoneim MM, Pandya H. Plasma protein binding of bupivacaine and its interaction with other drugs in man. Br J Anaesth 1974;46(6):435–8.
84. Malek I, Gebauerova M, Stanek V. Riziko soucasného podávání chinidinu a digitalisu. [Risk of simultaneous administration of quinidine and digitalis.] Vnitr Lek 1980;26(3):358–61.
85. Pedersen KE, Lysgaard Madsen J, Klitgaard NA, Kjaer K, Hvidt S. Effect of quinine on plasma digoxin concentration and renal digoxin clearance. Acta Med Scand 1985;218(2):229–32.
86. Aronson JK, Carver JG. Interaction of digoxin with quinine. Lancet 1981;1(8235):1418.
87. Leden I. Digoxin–hydroxychloroquine interaction? Acta Med Scand 1982;211(5):411–2.
88. Hedman A. Inhibition by basic drugs of digoxin secretion into human bile. Eur J Clin Pharmacol 1992;42(4):457–9.
89. Fenster PE, Powell JR, Graves PE, Conrad KA, Hager WD, Goldman S, Marcus FI. Digitoxin–quinidine interaction: pharmacokinetic evaluation. Ann Intern Med 1980;93(5):698–701.
90. Kuhlmann J, Dohrmann M, Marcin S. Effects of quinidine on pharmacokinetics and pharmacodynamics of digitoxin achieving steady-state conditions. Clin Pharmacol Ther 1986;39(3):288–94.
91. Teppo AM, Haltia K, Wager O. Immunoelectrophoretic "tailing" of albumin line due to albumin-IgG antibody complexes: a side effect of nitrofurantoin treatment? Scand J Immunol 1976;5(3):249–61.
92. Barbey JT, Thompson KA, Echt DS, Woosley RL, Roden DM. Tocainide plus quinidine for treatment of ventricular arrhythmias. Am J Cardiol 1988;61(8):570–3.
93. Giardina EG, Wechsler ME. Low dose quinidine–mexiletine combination therapy versus quinidine monotherapy for treatment of ventricular arrhythmias. J Am Coll Cardiol 1990;15(5):1138–45.
94. Desmeules J, Gascon MP, Dayer P, Magistris M. Impact of environmental and genetic factors on codeine analgesia. Eur J Clin Pharmacol 1991;41(1):23–6.
95. Sindrup SH, Arendt-Nielsen L, Brosen K, Bjerring P, Angelo HR, Eriksen B, Gram LF. The effect of quinidine on the analgesic effect of codeine. Eur J Clin Pharmacol 1992;42(6):587–91.

96. Desmeules JA, Oestreicher MK, Piguet V, Allaz AF, Dayer P. Contribution of cytochrome P-450 2D6 phenotype to the neuromodulatory effects of dextromethorphan. J Pharmacol Exp Ther 1999;288(2):607–12.
97. Moghadamnia AA, Rostami-Hodjegan A, Abdul-Manap R, Wright CE, Morice AH, Tucker GT. Physiologically based modelling of inhibition of metabolism and assessment of the relative potency of drug and metabolite: dextromethorphan vs. dextrorphan using quinidine inhibition. Br J Clin Pharmacol 2003;56:57–67.
98. Pope LE, Khalil MH, Berg JE, Stiles M, Yakatan GJ, Sellers EM. Pharmacokinetics of dextromethorphan after single or multiple dosing in combination with quinidine in extensive and poor metabolizers. J Clin Pharmacol 2004;44(10):1132–42.
99. Brooks BR, Thisted RA, Appel SH, Bradley WG, Olney RK, Berg JE, Pope LE, Smith RA;AVP-923 ALS Study Group. Treatment of pseudobulbar affect in ALS with dextromethorphan/quinidine: a randomized trial. Neurology 2004;63(8):1364–70.
100. Masubuchi Y, Ose A, Horie T. Diclofenac-induced inactivation of CYP3A4 and its stimulation by quinidine. Drug Metab Dispos 2002;30(10):1143–8.
101. Damkier P, Hansen LL, Brosen K. Effect of diclofenac, disulfiram, itraconazole, grapefruit juice and erythromycin on the pharmacokinetics of quinidine. Br J Clin Pharmacol 1999;48(6):829–38.
102. Twum-Barima Y, Carruthers SG. Quinidine–rifampin interaction. N Engl J Med 1981;304(24):1466–9.
103. Data JL, Wilkinson GR, Nies AS. Interaction of quinidine with anticonvulsant drugs. N Engl J Med 1976;294(13):699–702.
104. Kharasch ED, Hoffer C, Altuntas TG, Whittington D. Quinidine as a probe for the role of p-glycoprotein in the intestinal absorption and clinical effects of fentanyl. J Clin Pharmacol 2004;44(3):224–33.
105. Munafo A, Reymond-Michel G, Biollaz J. Altered flecainide disposition in healthy volunteers taking quinine. Eur J Clin Pharmacol 1990;38(3):269–73.
106. Birgersdotter UM, Wong W, Turgeon J, Roden DM. Stereoselective genetically-determined interaction between chronic flecainide and quinidine in patients with arrhythmias. Br J Clin Pharmacol 1992;33(3):275–80.
107. Damkier P, Hansen LL, Brosen K. Effect of diclofenac, disulfiram, itraconazole, grapefruit juice and erythromycin on the pharmacokinetics of quinidine. Br J Clin Pharmacol 1999;48(6):829–38.
108. Kharasch ED, Hoffer C, Whittington D. The effect of quinidine, used as a probe for the involvement of P-glycoprotein, on the intestinal absorption and pharmacodynamics of methadone. Br J Clin Pharmacol 2004;57(5):600–10.
109. Leemann T, Dayer P, Meyer UA. Single-dose quinidine treatment inhibits metoprolol oxidation in extensive metabolizers. Eur J Clin Pharmacol 1986;29(6):739–41.
110. Cooke CE, Sklar GE, Nappi JM. Possible pharmacokinetic interaction with quinidine: ciprofloxacin or metronidazole? Ann-Pharmacother 1996;30(4):364–6.
111. Hartshorn EA. Interactions of cardiac drugs. Drug Intell Clin Pharm 1970;4:272.
112. Green JA, Clementi WA, Porter C, Stigelman W. Nifedipine–quinidine interaction. Clin Pharm 1983;2(5):461–5.
113. Farringer JA, Green JA, O'Rourke RA, Linn WA, Clementi WA. Nifedipine-induced alterations in serum quinidine concentrations. Am Heart J 1984;108(6):1570–2.
114. Damkier P, Hansen LL, Brosen K. Rifampicin treatment greatly increases the apparent oral clearance of quinidine. Pharmacol Toxicol 1999;85(6):257–62.
115. Risch SC, Groom GP, Janowsky DS. The effects of psychotropic drugs on the cardiovascular system. J Clin Psychiatry 1982;43(5 Pt 2):16–31.
116. Dinai Y, Sharir M, Naveh N, Halkin H. Bradycardia induced by interaction between quinidine and ophthalmic timolol. Ann Intern Med 1985;103(6 Pt 1):890–1.
117. Lessard E, Yessine MA, Hamelin BA, O'Hara G, LeBlanc J, Turgeon J. Influence of CYP2D6 activity on the disposition and cardiovascular toxicity of the antidepressant agent venlafaxine in humans. Pharmacogenetics 1999;9(4):435–43.
118. Koch-Weser J. Quinidine-induced hypoprothrombinemic hemorrhage in patients on chronic warfarin therapy. Ann Intern Med 1968;68(3):511–7.
119. Ngui JS, Chen Q, Shou M, Wang RW, Stearns RA, Baillie TA, Tang W. In vitro stimulation of warfarin metabolism by quinidine: increases in the formation of 4′- and 10-hydroxywarfarin. Drug Metab Dispos 2001;29(6):877–86.

Tocainide

General Information

The clinical pharmacology, uses, efficacy, and adverse effects of tocainide have been extensively reviewed (1–9).

The adverse effects of tocainide occur with increasing frequency above plasma concentrations of 10 μg/ml. They are mostly related to the nervous system. In two large series withdrawal of tocainide was required in 11–16% of patients because of adverse effects (10,11). The high incidence of blood dyscrasias severely limits the use of tocainide.

Organs and Systems

Cardiovascular

Adverse cardiovascular effects have been reported in 6–55% of cases. After a single dose of tocainide the most common effect is hypotension with bradycardia (12). Angina pectoris has also been reported (13).

During repeated administration cardiovascular adverse effects are relatively uncommon. Increasing heart failure (14,15), worsening dysrhythmias (15,16), pericarditis (15,17,18), and sinus arrest with sinoatrial block (19) have all been reported. Tocainide can worsen ventricular tachycardia (20).

Respiratory

There have been several reports of interstitial pneumonitis attributable to tocainide (21,22), and it may also cause severe pulmonary fibrosis (23,24).

Nervous system

A wide variety of central nervous system symptoms has been reported in almost every study, varying in incidence from 10 to 100%. Common reactions include dizziness, tremor and tremulousness, dysesthesia and paresthesia, light-headedness, and blurred vision. Various mental changes have been described, including paranoid psychosis (25–27). Nausea is common and probably central in origin, since it occurs after intravenous as well as oral administration. These adverse effects can be reduced in frequency during oral administration by taking the tablets with food.

Sensory systems

There have been not infrequent reports of odd taste sensations (peppermint and menthol) and of coolness of the throat, hands, and feet.

Hematologic

Tocainide can cause blood dyscrasias (neutropenia, thrombocytopenia, pancytopenia, and aplastic anemia) in one in 300 patients (28–31). There have been isolated reports of eosinophilia with an allergic rash and anemia with pericarditis (17,32).

Gastrointestinal

Nausea is very common, particularly during the initial stages of treatment, after which it tends to disappear (14). There have been occasional reports of constipation (17) and of anorexia and vomiting (14).

Liver

Tocainide can increase the activities of serum transaminases (33) and can cause fatty change (33) and granulomatous hepatitis (34).

Urinary tract

Non-membranous glomerulonephritis has been reported (35).

Skin

There have been a few reports of skin rashes attributed to tocainide (17,36).

Sweat glands

Night sweats have been reported in three patients taking tocainide (17).

Immunologic

Arthralgia has been reported in two cases with positive antinuclear antibody titers, suggesting the possibility of a lupus-like syndrome (37). In another case tocainide treatment was associated with both a lupus-like syndrome and neutropenia (38). Cross-reactivity of tocainide with lidocaine has been reported (39).

Drug Administration

Drug overdose

Convulsions, complete heart block, and asystole developed in a case of fatal self-poisoning with 400 mg of tocainide (40).

Drug–Drug Interactions

Theophylline

Tocainide reduces the clearance of theophylline and increases its half-life, but to an extent that is probably clinically insignificant (41).

References

1. Nattel S, Zipes DP. Clinical pharmacology of old and new antiarrhythmic drugs. Cardiovasc Clin 1980;11(1):221–48.
2. Danilo P Jr. Tocainide. Am Heart J 1979;97(2):259–62.
3. Holmes B, Brogden RN, Heel RC, Speight TM, Avery GS. Tocainide. A review of its pharmacological properties and therapeutic efficacy. Drugs 1983;26(2):93–123.
4. ADIS Editors and Consultants. Tocainide (Tonocard): a review of its pharmacological properties and therapeutic efficacy. Curr Ther 1984;July:17.
5. Schweyen DH. Tocainide: a new antiarrhythmic. Hosp Pharm 1984;19:558–65.
6. Keefe DL, Somberg JC. New therapy focus: tocainide. Cardiovasc Rev Rep 1984;5:1023–30.
7. Lynch JJ, Lucchesi BR. New antiarrhythmic agents. IV. The pharmacology and clinical use of tocainide. Pract Cardiol 1985;2:108.
8. Hasegawa GR. Tocainide: a new oral antiarrhythmic. Drug Intell Clin Pharm 1985;19(7–8):514–7.
9. Kutalek SP, Morganroth J, Horowitz LN. Tocainide: a new oral antiarrhythmic agent. Ann Intern Med 1985;103(3):387–91.
10. Horn HR, Hadidian Z, Johnson JL, Vassallo HG, Williams JH, Young MD. Safety evaluation of tocainide in the American Emergency Use Program. Am Heart J 1980;100(6 Pt 2):1037–40.
11. Young MD, Hadidian Z, Horn HR, Johnson JL, Vassallo HG. Treatment of ventricular arrhythmias with oral tocainide. Am Heart J 1980;100(6 Pt 2):1041–5.
12. Greenspon AJ, Mohiuddin S, Saksena S, Lengerich R, Snapinn S, Holmes G, Irvin J, Sappington E, et al. Comparison of intravenous tocainide with intravenous lidocaine for treating ventricular arrhythmias. Cardiovasc Rev Rep 1989;10:55–9.
13. Winkle RA, Anderson JL, Peters F, Meffin PJ, Fowles RE, Harrison DC. The hemodynamic effects of intravenous tocainide in patients with heart disease. Circulation 1978;57(4):787–92.
14. Maloney JD, Nissen RG, McColgan JM. Open clinical studies at a referral center: chronic maintenance tocainide therapy in patients with recurrent sustained ventricular tachycardia refractory to conventional antiarrhythmic agents. Am Heart J 1980;100(6 Pt 2):1023–30.
15. Cheesman M, Ward DE. Exacerbation of ventricular tachycardia by tocainide. Clin Cardiol 1985;8(1):47–50.
16. Winkle RA, Meffin PJ, Harrison DC. Long-term tocainide therapy for ventricular arrhythmias. Circulation 1978;57(5):1008–16.

17. Roden DM, Reele SB, Higgins SB, Carr RK, Smith RF, Oates JA, Woosley RL. Tocainide therapy for refractory ventricular arrhythmias. Am Heart J 1980;100(1):15–22.
18. Gould LA, Betzu R, Vacek T, Muller R, Pradeep V, Downs L. Sinoatrial block due to tocainide. Am Heart J 1989;118(4):851–3.
19. Van Natta B, Lazarus M, Li C. Irreversible interstitial pneumonitis associated with tocainide therapy. West J Med 1988;149(1):91–2.
20. Perlow GM, Jain BP, Pauker SG, Zarren HS, Wistran DC, Epstein RL. Tocainide-associated interstitial pneumonitis. Ann Intern Med 1981;94(4 Pt 1):489–90.
21. Braude AC, Downar E, Chamberlain DW, Rebuck AS. Tocainide-associated interstitial pneumonitis. Thorax 1982;37(4):309–10.
22. Anonymous. Tocainide hydrochloride pulmonary fibrosis. Swed Adv Drug React Adv Comm Bull 1988;54:.
23. Feinberg L, Travis WD, Ferrans V, Sato N, Bernton HF. Pulmonary fibrosis associated with tocainide: report of a case with literature review. Am Rev Respir Dis 1990;141(2):505–8.
24. Currie P, Ramsdale DR. Paranoid psychosis induced by tocainide. BMJ (Clin Res Ed) 1984;288(6417):606–7.
25. Harrison DJ, Wathen CG. Paranoid psychoses induced by tocainide. BMJ (Clin Res Ed) 1984;288(6422):1010–1.
26. Clarke CW, el-Mahdi EO. Confusion and paranoia associated with oral tocainide. Postgrad Med J 1985;61(711):79–81.
27. Woosley RL, McDevitt DG, Nies AS, Smith RF, Wilkinson GR, Oates JA. Suppression of ventricular ectopic depolarizations by tocainide. Circulation 1977;56(6):980–4.
28. Anonymous. Tocainide and blood disorders. Aust Adv Drug React Bull 1986;April.
29. Drost RA. Voorzorgen bij gebruik van Tonocard (tocainide). Meded Coll Beoord Geneesmidd 1986;121:167.
30. Soff GA, Kadin ME. Tocainide-induced reversible agranulocytosis and anemia. Arch Intern Med 1987;147(3):598–9.
31. Morrill GB, Gibson SM. Tocainide-induced aplastic anemia. DICP 1989;23(1):90–1.
32. Engler R, Ryan W, LeWinter M, Bluestein H, Karliner JS. Assessment of long-term antiarrhythmic therapy: studies on the long-term efficacy and toxicity of tocainide. Am J Cardiol 1979;43(3):612–8.
33. Farquhar DL, Davidson NM. Possible hepatoxicity of tocainide. Scott Med J 1984;29(4):238.
34. Tucker LE. Tocainide-induced granulomatous hepatitis. JAMA 1986;255(24):3362.
35. Winkle RA, Mason JW, Harrison DC. Tocainide for drug-resistant ventricular arrhythmias: efficacy, side effects, and lidocaine responsiveness for predicting tocainide success. Am Heart J 1980;100(6 Pt 2):1031–6.
36. Nyquist O, Forssell G, Nordlander R, Schenck-Gustafsson K. Hemodynamic and antiarrhythmic effects of tocainide in patients with acute myocardial infarction. Am Heart J 1980;100(6 Pt 2):1000–5.
37. Mohiuddin SM, Esterbrooks D, Mooss AN, Dahl JM, Hilleman DE. Efficacy and tolerance of tocainide during long-term treatment of malignant ventricular arrhythmias. Clin Cardiol 1987;10(8):457–62.
38. Oliphant LD, Goddard M. Tocainide-associated neutropenia and lupus-like syndrome. Chest 1988;94(2):427–8.
39. Duff HJ, Roden DM, Marney S, Colley DG, Maffucci R, Primm RK, Oates JA, Woosley RL. Molecular basis for the antigenicity of lidocaine analogues: tocainide and mexiletine. Am Heart J 1984;107(3):585–9.
40. Barnfield C, Kemmenoe AV. A sudden death due to tocainide overdose. Hum Toxicol 1986;5(5):337–40.
41. Loi CM, Wei X, Parker BM, Korrapati MR, Vestal RE. The effect of tocainide on theophylline metabolism. Br J Clin Pharmacol 1993;35(4):437–40.

DRUGS THAT ACT ON THE CEREBRAL AND PERIPHERAL ARTERIAL AND VENOUS CIRCULATIONS

Buflomedil

General Information

Buflomedil hydrochloride is a vasoactive drug with a variety of actions. It is an alpha-adrenoceptor antagonist and a weak calcium channel blocker. It inhibits platelet aggregation and improves erythrocyte deformability. However, its mechanism of action in peripheral vascular disease is not known.

Buflomedil has generally been well tolerated by most patients in clinical trials (1). The most frequently reported adverse effects include flushing, headache, vertigo, gastrointestinal discomfort, and dizziness. These rarely require drug withdrawal. In controlled trials, adverse effects have occurred in 20% of patients assigned to buflomedil and 18% of those assigned to placebo; only gastrointestinal discomfort occurred more frequently in buflomedil-treated patients (3.4 versus 2%) (2).

Organs and Systems

Nervous system

Extrapyramidal symptoms have been reported in a few frail elderly women (SEDA-13, 169). A few cases of myoclonic encephalopathy have been observed at therapeutic dosages; old people, patients of low body weight, and patients with renal insufficiency due to dehydration appear to be especially vulnerable (SEDA-9, 189) (SEDA-10, 172) (SEDA-11, 179).

Psychological, psychiatric

Depression has been reported in a few frail elderly women taking buflomedil (SEDA-13, 169).

Immunologic

An anaphylactic reaction to buflomedil has been reported (3).

- A 53-year-old woman with Raynaud's phenomenon developed an urticarial rash, pruritus, and hypotension 10 minutes after the parenteral administration of buflomedil. She received corticosteroids and recovered within 6 hours. When she later underwent skin tests with buflomedil, there was an immediate positive reaction, suggesting a type I hypersensitivity mechanism.

Drug Administration

Drug overdose

Buflomedil is generally considered to be innocuous at therapeutic dosages. Acute toxicity is due to accidental or intentional overdosage. Overdosage causes generalized seizures and cardiac conduction abnormalities, eventually leading to cardiac arrest (SEDA-21, 215).

Three cases of self-poisoning with buflomedil have been reported (4–6); two were fatal.

- A 15-year-old girl took 24 tablets of buflomedil (300 mg) and had a plasma concentration of 93.7 μg/ml. She was deeply comatose and in a convulsive state. Shortly after, she developed cardiac arrest from which she did not recover despite initially successful resuscitation.
- A young girl took a single dose of buflomedil 7.5 g and a young man 18 g. Coma with convulsions occurred in both, soon followed by rhythm and conduction disturbances. Intensive resuscitation attempts could not prevent a fatal outcome in the girl.

In two other fatal cases there were high concentrations of buflomedil in body fluids; a man was found dead and a young woman died in the hospital after presumed overdoses (7,8). Chronic overdose can also lead to convulsions (9).

- A 75-year-old woman had fever and convulsions. She had diabetes mellitus and angina pectoris and took buflomedil for peripheral arterial disease. No cause for her symptoms was found, but she had a high plasma concentration of buflomedil (6.3 μg/ml, usual target range 4–4.5 μg/ml). The drug was withdrawn and the symptoms did not recur. On questioning, it appeared that she had mistakenly forgotten to abandon her old commercial formulation when her pharmacist proposed a cheaper, new, generic brand, but took both at the correctly prescribed dose.

References

1. Clissold SP, Lynch S, Sorkin EM. Buflomedil. A review of its pharmacodynamic and pharmacokinetic properties, and therapeutic efficacy in peripheral and cerebral vascular diseases. Drugs 1987;33(5):430–60.
2. Bachand RT, Dubourg AY. A review of long-term safety data with buflomedil. J Int Med Res 1990;18(3):245–52.
3. Scala E, Guerra EC, Pirrotta L, Giani M, De Pita O, Puddu P. Anaphylactic reactions to buflomedil. Allergy 1999;54(3):288–9.
4. Tomassini E, Poussel JF, Guiot P. Intoxication mortelle au buflomédil. [Fatal poisoning caused by buflomedil.] Ann Fr Anesth Reanim 1999;18(10):1091–2.
5. Perrot C, Rifler JP, Freysz M. Intoxication volontaire fatale au buflomedil: à propos d'un cas. JEUR 2000;13:135–8.
6. Vandemergel X, Biston P, Lenearts L, Marecaux G, Daune M. Buflomedil poisoning: a potentially life-threatening intoxication. Intensive Care Med 2000; 26(11):1713.
7. Neri C, Barbareschi M, Turrina S, De Leo D. Suicide by buflomedil HCl: a case report. J Clin Forensic Med 2004:11:15–16.
8. Babel B, Tatschner T, Patzelt D. Suicidal buflomedil intoxication. Arch Kriminol 2004;213:108–13.
9. Chiffoleau A, Yatim D, Garrec F, Veyrac G, Raoult P, Larousse C, Bourin M. Warning! One buflomedil may hide another one! Therapie 2000;55(1):221–3

Calcium dobesilate

General Information

Calcium dobesilate is an antioxidant that has been used to treat diabetic retinopathy, in which it slows progression of the disease during long-term oral treatment by reducing microvascular permeability, leading to improved visual acuity (1). It not only acts as an antioxidant but also stimulates endothelial production of nitric oxide.

In early studies the most common adverse effects of calcium dobesilate after oral administration were gastrointestinal disturbances and, occasionally, nervousness and fever (SEDA-3, 181; SEDA-16, 204).

In a systematic review of the published literature from 1970 to 2003, a postmarketing surveillance report covering the period 1974–1998, and periodic safety update reports covering the period 1995–2003 from the French regulatory authorities pharmacovigilance database, the following adverse effects were reported: fever (26%), gastrointestinal disorders (12.5%), skin reactions (8.2%), arthralgia (4.3%), and agranulocytosis (4.3%) (2).

Safety data from post-marketing surveillance over 25 years has been extensively summarized (3). The data suggest that most adverse events are rare and frequently unrelated to the drug. Fever, gastrointestinal intolerance, skin reactions, and arthralgia are the most frequent adverse effects ascribed to calcium dobesilate. There have been a few reports of agranulocytosis, but its frequency is lower than in the general population and methodological bias is therefore suspected.

Organs and Systems

Hematologic

Agranulocytosis associated with calcium dobesilate has occasionally been reported (4). Although the risk has been calculated at 121 cases per million per year from case-control and case-population studies (5), there have been many fewer spontaneous reports than are consistent with this relatively high figure, and the true incidence is unknown (6).

During the period 1978–2000, nine cases of agranulocytosis associated with the use of calcium dobesilate were reported in Spain. This is substantially less than expected on the basis of calculations derived from case-control and case-population strategies (7). Factors that could explain the disagreement are spontaneous under-reporting, the duration of use, and the age of the patients.

In an overview calcium dobesilate ranked second in a list of drugs associated with agranulocytosis (8). Nevertheless, in a previous review it was estimated that the incidence is very low (less than one in a million) and lower than in the general population.

Drug Administration

Drug administration route

Intramuscular calcium dobesilate can be painful (SEDA-3, 181).

References

1. Berthet P, Farine JC, Barras JP. Calcium dobesilate: pharmacological profile related to its use in diabetic retinopathy. Int J Clin Pract 1999;53(8):631–6.
2. Allain H, Ramelet AA, Polard E, Bentue-Ferrer D. Safety of calcium dobesilate in chronic venous disease, diabetic retinopathy and haemorrhoids. Drug Saf 2004;27(9):649–60.
3. Allain H, Ramelet AA, Polard E, Bentue-Ferrer D. Safety of calcium dobesilate in chronic venous disease, diabetic retinopathy and haemorrhoids. Drug Saf 2004;27:649–60.
4. Garcia Benayas E, Garcia Diaz B, Perez G. Calcium dobesilate-induced agranulocytosis. Pharm World Sci 1997;19(5):251–2.
5. Ibanez L, Ballarin E, Vidal X, Laporte JR. Agranulocytosis associated with calcium dobesilate clinical course and risk estimation with the case-control and the case-population approaches. Eur J Clin Pharmacol 2000;56(9–10):763–7.
6. Zapater P, Horga JF, Garcia A. Risk of drug-induced agranulocytosis: the case of calcium dobesilate. Eur J Clin Pharmacol 2003;58(11):767–72.
7. Zapater P, Horga JF, Garcia A. Risk of drug-induced agranulocytosis: the case of calcium dobesilate. Eur J Clin Pharmacol 2003;58:767–72.
8. Ibanez L, Vidal X, Ballarin E, Laporte JR. Population-based drug-induced agranulocytosis. Arch Intern Med 2005;165:869–74.

Cilostazol

General Information

Cilostazol is a phosphodiesterase inhibitor that suppresses platelet aggregation and also acts as a direct arterial vasodilator. Small studies in Japan suggested that it might be useful for treating chronic arterial disease and symptoms of intermittent claudication. A trial in 81 patients with claudication substantiated this claim: claudication distance was improved by 35% for initial and 41% for absolute claudication distance.

Placebo-controlled studies

The efficacy and safety data of cilostazol in placebo-controlled clinical trials have been repeatedly subjected to meta-analysis, with the same conclusion (1,2). Cilostazol is well tolerated; headache, bowel complaints, and palpitation are the most common but mild adverse effects.

Six multicenter placebo-controlled trials have been conducted in the USA (3,4). They involved more than 2000 patients with intermittent claudication and established the efficacy of cilostazol in improving walking distance in these patients.

Cilostazol was approved by the FDA in January 1999 for the treatment of symptoms of intermittent claudication; from 1984 to 1999 pentoxifylline was the only drug approved in the USA for this indication. The two drugs have been compared with placebo in a large, randomized, double-blind, placebo-controlled trial (5). After 24 weeks of treatment the mean increase in maximal walking distance was 54% with cilostazol and only 30% with pentoxifylline and 34% with placebo. Headache, diarrhea, abnormal stools, and bouts of palpitation were significantly more common with cilostazol. They were reported as generally mild to moderate and self-limiting and have been previously recognized as being related to cilostazol.

The efficacy of antithrombotic prophylaxis with cilostazol for the secondary prevention of cerebral infarction has been studied in a Japanese, placebo-controlled trial in 1095 patients (6). There was a 42% relative risk reduction compared with placebo. As in the trials in patients with peripheral arterial disease, mild to moderate headache and palpitation were the most commonly observed symptomatic adverse events attributed to cilostazol; they respectively occurred in 13 and 5.3%. Headache with cilostazol is attributed to cerebral vasodilatation induced by relaxation of vascular smooth muscle.

In another placebo-controlled trial, there were gastrointestinal complaints in 44% of the cilostazol-treated patients and in 15% of the placebo group. The most commonly reported adverse effects included diarrhea, loose stools, flatulence, and nausea; they were usually mild and transient but persisted in some patients. Headache occurred in 20% of cilostazol-treated patients but also in 15% of those given placebo (7).

Safety data relating to the use of cilostazol in 2702 patients who participated in eight USA–UK placebo-controlled trials have been re-analysed (8). The most frequently recorded adverse events were headache (32%), diarrhea (17%), and abnormal stools (14%). Palpitation, tachycardia, and dizziness were additional events that occurred more often in cilostazol-treated patients and were considered to be probably related to treatment. Headache led to withdrawal of cilostazol in 3.5% of patients, and palpitation and diarrhea led to withdrawal in another 1%. All adverse events quickly resolved after withdrawal. Cardiovascular and all-cause mortality were similar with cilostazol and placebo.

Organs and Systems

Cardiovascular

Ventricular tachycardia has been reported during cilostazol therapy (9).

- A 92-year-old woman developed sudden runs of ventricular tachycardia, a few days after she started to take cilostazol because of subacute leg ischemia. She was known to have atrial fibrillation and intraventricular conduction delay. The runs did not respond to empirical magnesium therapy but subsided shortly after withdrawal of cilostazol.

The phosphodiesterase III inhibitors have no known direct dysrhythmogenic effects. However, the authors speculated that raised concentrations of cAMP may have contributed to the ventricular tachycardia, mainly because ventricular dysrhythmias have been mentioned with other agents of the same family in patients with heart failure. Whether some populations are particularly vulnerable is unknown.

Long-Term Effects

Drug withdrawal

Withdrawal of cilostazol led to sinus node dysfunction in a patient taking concurrent atenolol (10).

- A 72-year-old patient with diabetes was taking insulin, cilostazol, atenolol, and amlodipine. A few days before surgery for vascular repair the cilostazol was withdrawn to reduce blood loss. His heart rate gradually fell (from 75 to 60/minute). Anesthesia led to a further fall, and shortly after declamping of the operated artery he had an asystolic cardiac arrest. Closed-chest cardiac massage and injections of atropine and adrenaline restored sinus rhythm.

The authors argued that withdrawal of cilostazol had removed a rhythm accelerating effect and left the bradycardic action of atenolol unopposed.

Drug-drug Interactions

Antiplatelet agents

Cilostazol has antiplatelet properties in addition to an action on vascular smooth muscle and was therefore initially thought to be unsafe in combination with other commonly used antiplatelet drugs, such as aspirin or clopidogrel. However, in a crossover study in 21 patients with peripheral disease cilostazol did not significantly increase bleeding time, nor further prolong it when it was added to any regimen containing aspirin or clopidogrel (11). Thus, cilostazol is not expected to increase the risk of adverse bleeding effects when combined with aspirin or clopidogrel.

References

1. Thompson PD, Zimet R, Forbes WP, Zhang P. Meta-analysis of results from eight randomized, placebo-controlled trials on the effect of cilostazol on patients with intermittent claudication. Am J Cardiol 2002;90(12):1314–9.
2. Regensteiner JG, Ware JE Jr, McCarthy WJ, Zhang P, Forbes WP, Heckman J, Hiatt WR. Effect of cilostazol on treadmill walking, community-based walking ability, and health-related quality of life in patients with intermittent claudication due to peripheral arterial disease: meta-analysis of six randomized controlled trials. J Am Geriatr Soc 2002;50(12):1939–46.
3. Comp PC. Treatment of intermittent claudication in peripheral arterial disease. Recent clinical experience with cilostazol. Today's Ther Trends 1999;17:99–112.
4. Sorkin EM, Markham A. Cilostazol. Drugs Aging 1999;14(1):63–71.

5. Dawson DL, Cutler BS, Hiatt WR, Hobson RW 2nd, Martin JD, Bortey EB, Forbes WP, Strandness DE Jr. A comparison of cilostazol and pentoxifylline for treating intermittent claudication. Am J Med 2000;109(7):523–30.
6. Gotoh F, Tohgi H, Hirai S, Terashi A, Fukuuchi Y, Otomo E, Shinohara Y, Itoh E, Matsuda T, Sawada T, Yamaguchi T, Nishimaru K, Ohashi Y. Cilostazol stroke prevention study: a placebo-controlled double-blind trial for secondary prevention of cerebral infarction. J Stroke Cerebrovasc Dis 2000;9:147–57.
7. Dawson DL, Cutler BS, Meissner MH, Strandness DE Jr. Cilostazol has beneficial effects in treatment of intermittent claudication: results from a multicenter, randomized, prospective, double-blind trial. Circulation 1998;98(7):678–86.
8. Cariski AT. Cilostazol: a novel treatment option in intermittent claudication. Int J Clin Pract Suppl 2001;(119):11–8.
9. Gamssari F, Mahmood H, Ho JS, Villareal RP, Liu B, Rasekh A, Garcia E, Massumi A. Rapid ventricular tachycardias associated with cilostazol use. Tex Heart Inst J 2002;29(2):140–2.
10. Ishiyama T, Oguchi T, Yamaguchi T, Kumazawa T. Sinus node dysfunction associated with discontinuation of cilostazol in a patient taking atenolol. Br J Anaesth 2004;93:472.
11. Comerota AJ. Effect on platelet function of cilostazol, clopidogrel, and aspirin, each alone or in combination. Atheroscler Suppl 2005;6:13–19.

Cyclandelate

General Information

The spasmolytic action of cyclandelate, an ester of mandelic acid, was described as early as 1959, but only in later years have its properties been more fully investigated. It appears to act as a calcium channel blocker in smooth muscle and platelets, this effect being partly due to inhibition of phosphodiesterases. It also produces increased deformability of erythrocytes, the mechanism of which is so far unknown, although phosphodiesterase inhibition may again be responsible. Cyclandelate also reduces the activity of the rate-limiting enzyme in the biosynthesis of cholesterol (HMG-CoA), and its "antidiabetic" properties may be due to inhibition of aldose reductase.

Cyclandelate is mainly given to elderly patients with mild to moderate cognitive impairment; it is also used in the prophylaxis of migraine. Its efficacy is not impressive (SEDA-21, 215) (1).

Major adverse effects have not been reported with dosages of 1.6 g/day. Gastrointestinal upset, flushing, and tingling are rare and minor complaints. Even elderly patients apparently tolerate 3.2 g/day without problems (SEDA-9, 189).

Reference

1. Diener HC, Krupp P, Schmitt T, Steitz G, Milde K, Freytag S. Cyclandelate in the prophylaxis of migraine: a placebo-controlled study. Cephalalgia 2001;21(1):66–70.

Defibrotide

General Information

Defibrotide is a polydeoxyribonucleotide extracted from mammalian organ. Its antithrombotic activity is partly ascribed to enhancement of eicosanoid metabolism, in particular increased release of prostacyclin, with ensuing vasodilatation and inhibition of platelet aggregation. An additional mechanism is activation of the fibrinolytic system, primarily increased activation of tissue plasminogen in the vessel wall.

The antithrombotic potential of defibrotide has been reported in patients with hepatic veno-occlusive disease after stem cell transplantation. In a randomized, placebo-controlled trial in 310 patients with claudication there was significant improvement in walking distance with defibrotide, but no difference in efficacy between the two doses of the drug tested (800 and 1200 mg/day) (1). Twenty patients stopped taking the drug because of cardiovascular events, the preset endpoints of the study. Seven others stopped because of adverse drug reactions, mainly gastrointestinal intolerance and skin reactions. They were equally distributed among the placebo and the two defibrotide groups.

Reference

1. Violi F, Marubini E, Coccheri S, Nenci GG. Improvement of walking distance by defibrotide in patients with intermittent claudication—results of a randomized, placebo-controlled study (the DICLIS study). Defibrotide Intermittent CLaudication Italian Study. Thromb Haemost 2000;83(5):672–7.

Isoxsuprine

See also Adrenoceptor agonists

General Information

Isoxsuprine is a beta$_2$-adrenoceptor agonist that also has antagonist action at alpha-adrenoceptors. It has been variously presented as a beta-agonist, a specific vasorelaxant, a uterine relaxant, and an agent that reduces blood viscosity. In high dosages it also inhibits platelet aggregation. There is slim evidence that isoxsuprine improves cognitive function and mental performance in a limited number of patients, but the practical benefit is minor. There is no convincing proof of its efficacy in patients with claudication. It has been widely used in horses for treating navicular syndrome and laminitis, with little evidence of efficacy (1).

Since its efficacy has been doubted with respect to both its vascular and its obstetric use, one must suspect that underdosage of the drug is the best explanation for the low reported incidence of adverse reactions, which are generally of the type regarded as anecdotal (nausea, vomiting, palpitation, dizziness, weakness). Minor facial flushing and tremor are common. As patients tend to increase the dosage gradually, diarrhea, vomiting, headache, vertigo, and rash can occur (SEDA-3, 177) (SEDA-7, 228). The important adverse reactions of isoxsuprine are tachycardia and orthostatic hypotension, but they occur only at high dosages.

Second-Generation Effects

Fetotoxicity

The beta-adrenoceptor agonist effect is sufficient to cause fetal tachycardia when isoxsuprine is given intravenously in late pregnancy, although one large survey suggested that it may be better tolerated by the mother (as regards cardiac and stimulant effects) than salbutamol (SED-12, 313). The fact that maternal blood pressure can fall may reflect a direct relaxant effect on the blood vessels.

Reference

1. Erkert RS, Macallister CG. Isoxsuprine hydrochloride in the horse: a review. J Vet Pharmacol Ther 2002;25(2):81–7.

Ketanserin

General Information

Ketanserin is a selective 5-HT_2 receptor antagonist with antihypertensive action. Because it antagonizes the vasoconstriction and platelet aggregation induced by 5-HT and increases erythrocyte deformability, its potential usefulness in peripheral vascular disease has been extensively investigated, although never convincingly demonstrated. It has been reported to be beneficial in pre-eclampsia (1).

Most of the adverse effects attributable to ketanserin seem, not surprisingly, to be on the central nervous system. They include drowsiness, fatigue, headache, sleep disturbances, and dry mouth. Other complaints include dizziness, light-headedness, lack of concentration, and dyspepsia. A gradual increase in dosage is therefore recommended. These adverse effects occur in about 10% of patients and lead to withdrawal of the drug in 3–4% (SEDA-17, 243).

Organs and Systems

Cardiovascular

Dose-related prolongation of the QT interval on the electrocardiogram occurs in roughly one-third of patients taking ketanserin (2). Several cases of ventricular dysrhythmias with QT prolongation, leading to syncope, have been reported (SED-11, 392) (SED-12, 473), but the exact incidence of significant ventricular dysrhythmias on the basis of QT prolongation during ketanserin therapy is unknown.

Drug–Drug Interactions

Ecstasy

In 14 healthy subjects, ketanserin attenuated the perceptual changes, emotional excitation, and acute adverse responses induced by ecstasy, but had little effect on positive mood, well-being, extroversion, and short-term sequelae (3). Body temperature was lower with ecstasy plus ketanserin than with ecstasy alone.

Potassium-wasting diuretics

Prolongation of the QT interval by ketanserin is more pronounced when it used in combination with potassium-wasting diuretics (4). This combination must therefore be avoided.

References

1. Bolte AC, van Eyck J, Gaffar SF, van Geijn HP, Dekker GA. Ketanserin for the treatment of preeclampsia. J Perinat Med 2001;29(1):14–22.
2. Zehender M, Meinertz T, Hohnloser S, Geibel A, Hartung J, Seiler KU, Just H. Incidence and clinical relevance of QT prolongation caused by the new selective serotonin antagonist ketanserin. Am J Cardiol 1989;63(12):826–32.
3. Liechti ME, Saur MR, Gamma A, Hell D, Vollenweider FX. Psychological and physiological effects of MDMA ("Ecstasy") after pretreatment with the 5-HT(2) antagonist ketanserin in healthy humans. Neuropsychopharmacology 2000;23(4):396–404.
4. Prevention of Atherosclerotic Complications with Ketanserin Trial Group. Prevention of atherosclerotic complications: controlled trial of ketanserin. BMJ 1989;298(6671):424–30. Erratum in: BMJ 1989;298(6674):644.

Naftidrofuryl

General Information

Naftidrofuryl is a complex acid ester of diethylaminoethanol, with direct vasodilatory properties and antagonistic effects on 5-HT (via 5-HT_2 receptors) and bradykinin. It also causes an intracellular increase in ATP concentrations, improves cellular oxidative metabolism (by activating succinate dehydrogenase), and reduces blood and plasma viscosity and fibrinogen concentrations.

Naftidrofuryl has been marketed in Europe for over 20 years for the treatment of peripheral and cerebrovascular diseases and for senile dementia (1). A few placebo-controlled studies have reported increases in walking distance in patients with claudication, but no substantial hemodynamic improvement. Subjective improvement with naftidrofuryl was also noted in patients with Raynaud's phenomenon, but it had variable effects on digital blood flow and pressure during cold provocation. In patients with an acute ischemic stroke there was better neurological progress, leading to a shorter stay in hospital in some but not all studies. Several double-blind trials have reported improvements in the deficits associated with mild senile organic brain syndrome.

Oral naftidrofuryl (300–600 mg/day) was generally well tolerated in controlled trials: gastrointestinal complaints were as frequent with placebo as with the drug. Esophageal ulceration has been ascribed to naftidrofuryl in a few patients. Rare cases of reversible liver dysfunction have been reported.

Parenteral naftidrofuryl was withdrawn from the market following reports of severe adverse effects with intravenous or intra-arterial bolus injections, including intracardiac conduction defects, epileptic seizures, severe anaphylactic reactions, and acute renal insufficiency secondary to deposition of oxalate crystals in the tubules (SED-12, 473; SEDA-17, 244).

Liver

Very few cases of reversible liver dysfunction have been reported with naftidrofuryl. Biopsy-proven naftidrofuryl-induced liver injury has been reported (2).

- A 44-year-old woman developed deteriorating jaundice, despite a recent cholecystectomy for presumed cholecystitis/cholangitis, which was not confirmed at operation. There was no pain, but fatigue and loss of appetite, and she had dark urine. Her liver enzymes were raised. The only drug she admitted to having taken was naftidrofuryl 100 mg bd for intermittent positional dizziness. Autoantibodies and detailed laboratory tests for other causes of liver damage were negative. Liver biopsy showed moderate portal infiltration of lymphocytes and eosinophils with mild extension to adjacent liver parenchyma, compatible with drug-induced liver damage.

The authors concluded that the rarity of hepatic injury as an adverse effect of naftidrofuryl reflects a hypersensitivity reaction rather than a toxic effect on the liver in this case

References

1. Barradell LB, Brogden RN. Oral naftidrofuryl: A review of its pharmacology and therapeutic use in the management of peripheral occlusive arterial disease. Drugs Aging 1996;8(4):299–322.
2. Cholongitas E, Papatheodoridis GV, Mavrogiannaki A, Manesis E. Naftidrofuryl-induced liver injury. Am J Gastroenterol 2003;98:1448–50.

Pentoxifylline

General Information

Pentoxifylline (oxipentifylline) is a methylxanthine that antagonizes the vasoconstrictor effects of catecholamines and increases cyclic AMP concentrations, causing smooth muscle to relax. It has also been claimed to correct impaired microcirculation, by improving various factors that disturb blood rheology, and to reduce the generation of toxic free radicals from leukocytes during ischemic leg exercise in patients with intermittent claudication. Pentoxifylline has been used to suppress overproduction of tumor necrosis factor alfa in conditions such as falciparum malaria and rheumatoid arthritis and in transplant recipients, with varied success.

Comparative studies

In 52 adults with cerebral malaria who were randomized to either quinine dihydrochloride alone ($n = 32$) or a combination of quinine and pentoxifylline ($n = 20$), the addition of pentoxifylline significantly improved coma resolution time from 64 to 22 hours and reduced mortality from 25% to 10% (1). Three days after therapy, serum tumor necrosis factor alfa concentrations fell significantly in those who received pentoxifylline. Pentoxifylline caused no serious adverse effects that necessitated withdrawal.

Placebo-controlled studies

The efficacy of pentoxifylline in treating claudication has been evaluated in double-blind, controlled trials in Europe and the USA. Several of these trials have shown that pentoxifylline 400 mg tds increases the walking distance significantly more than placebo in patients with claudication. However, critics remain skeptical about the real value of pentoxifylline, because of a negative correlation between the trial sample sizes and its effects, reflecting an overestimate of drug effect in highly selected trial populations (2). In a trial in almost 300 patients with acute ischemic stroke, the neurological deficit improved more rapidly with pentoxifylline in the initial phase, but the difference was not significant at the end of 1 week (3).

In a placebo-controlled trial in 114 patients with critical limb ischemia, twice-daily intravenous pentoxifylline 600 mg produced unimpressive results (4).

Unwanted effects of pentoxifylline recognized in the double-blind studies were gastrointestinal symptoms (chiefly nausea, vomiting, and bloating) and dizziness. Although common, they required drug withdrawal in only about 3% of patients.

Organs and Systems

Nervous system

Aseptic meningitis has been reported in a patient taking pentoxifylline (5).

- A 37-year-old woman with mixed connective tissue disease took pentoxifylline 400 mg/day for Raynaud's phenomenon. After 12 days she complained of headache, myalgia, and neck pain and developed a fever. She recovered promptly after withdrawal of the drug. A few weeks later, she took a single tablet of pentoxifylline and within 1 hour developed the same symptoms together with chills, vomiting, and diarrhea. Neurological examination was normal, but the cerebrospinal fluid contained a large number of leukocytes. Bacteriological and virological tests were negative.

A diagnosis of aseptic meningitis was made and pentoxifylline was thought to have played a causative role because of the suggestive symptoms after first exposure, the close temporal relation on rechallenge, and the quick recovery after withdrawal. The authors argued that the underlying disease may have had a predisposing role, since aseptic meningitis secondary to pentoxifylline has never been reported in the thousands of patients who have used it for peripheral arterial disease. On the other hand, aseptic meningitis occurs with NSAIDs in patients with connective tissue diseases.

Psychological, psychiatric

Rare cases of hallucinations in elderly people have been ascribed to a stimulant effect of pentoxifylline on the central nervous system (SEDA-18, 219).

Hematologic

A single case of bleeding duodenal ulcer after a single dose of pentoxifylline was reported as being possibly secondary to disturbed platelet function induced by the drug (6). However, the effects of pentoxifylline on platelet function have not been very consistent, and the use of pentoxifylline may have been purely coincidental.

Skin

Acute urticaria triggered by pentoxifylline and confirmed by skin tests and oral rechallenge has been reported in a 60-year-old man (7).

Immunologic

In an open, randomized, controlled trial in 56 children with cerebral malaria, the 26 children who received pentoxifylline 10 mg/kg/day by continuous infusion had significantly shorter periods of coma than the controls. The pentoxifylline recipients showed a trend toward a lower mortality. Pentoxifylline has an inhibitory effect on the synthesis of tumor necrosis factor alfa. The better outcome in the treated group was associated with a fall in tumor necrosis factor alfa serum concentrations on the third day of treatment in a few subjects; this was not seen in the controls (8).

However, in a later, randomized, placebo-controlled trial pentoxifylline neither reduced tumor necrosis factor alfa serum concentrations nor affected the clinical course in 51 patients who received it as adjunctive treatment to standard antimalarial therapy in a dosage of 20 mg/kg/day over 5 days (9).

References

1. Das BK, Mishra S, Padhi PK, Manish R, Tripathy R, Sahoo PK, Ravindran B. Pentoxifylline adjunct improves prognosis of human cerebral malaria in adults. Trop Med Int Health 2003;8(8):680–4.
2. Cameron HA, Waller PC, Ramsay LE. Drug treatment of intermittent claudication: a critical analysis of the methods and findings of published clinical trials, 1965–1985. Br J Clin Pharmacol 1988;26(5):569–76.
3. Hsu CY, Norris JW, Hogan EL, Bladin P, Dinsdale HB, Yatsu FM, Earnest MP, Scheinberg P, Caplan LR, Karp HR. Pentoxifylline in acute nonhemorrhagic stroke. A randomized, placebo-controlled double-blind trial. Stroke 1988;19(6):716–22.
4. Norwegian Pentoxifylline Multicenter Trial Group. Efficacy and clinical tolerance of parenteral pentoxifylline in the treatment of critical lower limb ischemia. A placebo controlled multicenter study. Int Angiol 1996;15(1):75–80.
5. Mathian A, Amoura Z, Piette JC. Pentoxifylline-induced aseptic meningitis in a patient with mixed connective tissue disease. Neurology 2002;59(9):1468–9.
6. Oren R, Yishar U, Lysy J, Livshitz T, Ligumsky M. Pentoxifylline-induced gastrointestinal bleeding. DICP 1991;25(3):315–6.
7. Gonzales-Mahave I, Del Pozo MD, Blasco A, Lobera T, Venturini M. Urticaria due to pentoxyfylline. Allergy, 2005;60:705.
8. Di Perri G, Di Perri IG, Monteiro GB, Bonora S, Hennig C, Cassatella M, Micciolo R, Vento S, Dusi S, Bassetti D, et al. Pentoxifylline as a supportive agent in the treatment of cerebral malaria in children. J Infect Dis 1995;171(5):1317–22.
9. Hemmer CJ, Hort G, Chiwakata CB, Seitz R, Egbring R, Gaus W, Hogel J, Hassemer M, Nawroth PP, Kern P, Dietrich M. Supportive pentoxifylline in falciparum malaria: no effect on tumor necrosis factor alpha levels or clinical outcome: a prospective, randomized, placebo-controlled study. Am J Trop Med Hyg 1997;56(4):397–403.

Phosphodiesterase type V inhibitors

General Information

Sildenafil is a potent inhibitor of phosphodiesterase type V, and therefore of the breakdown of cyclic guanosine monophosphate (cGMP) in the corpora cavernosa. The increased concentration of cGMP leads to nitric oxide-mediated relaxation of the smooth muscle cells and vasodilatation in the corpus cavernosum, which is essential for normal erection. Thus, sildenafil increases the penile response to sexual stimulation and is effective in erectile dysfunction (1,2). Tadalafil and vardenafil are phosphodiesterase type V inhibitors that have been launched to compete with sildenafil in a highly lucrative market. The adverse effects and drug interactions of these compounds appear to be similar to those of sildenafil.

Reviews of inhibitors of phosphodiesterase type V have stressed their overall safety in men with erectile dysfunction (3,4,6), cardiovascular disease (6,7), and pulmonary hypertension (8,9,10,11,12,13). The reviewed data have mainly been obtained from clinical trials.

The most common complaints are headache, facial flushing, nasal congestion, and dyspepsia. Prolonged erection and priapism are rare.

The main adverse effects reported in clinical studies were flushing, headache, dyspepsia, visual disturbances, and rhinitis, some of which show that vasodilatation is not confined to the corpora cavernosa. They were mild, and only 1–2% of the patients discontinued sildenafil because of adverse effects.

The adverse effects of a single dose of sildenafil 50 mg have been evaluated in a placebo-controlled study in 40 young healthy volunteers (14). The most commonly reported adverse effects with sildenafil and placebo respectively were flushing (75 and 0%), headache (50 and 5%), and dyspepsia (15 and 5%). This adverse effects profile was similar to that observed in clinical trials. Heart rate changed significantly, but blood pressure did not.

There have been several case reports of the use of sildenafil to ameliorate rebound pulmonary hypertension (15).

- A six-week-old 3.1 kg girl developed severe pulmonary hypertension and systemic hypotension after the removal of a bilateral pulmonary vein obstruction due to a left atrial membrane. Nitric oxide 20 ppm reduced the pulmonary arterial pressure from 57 to 33 mmHg, and plasma cGMP concentrations increased from 12 nmol/l at baseline to 28 nmol/l with nitric oxide. After three unsuccessful weaning attempts, due to rebound, sildenafil 1 mg was given via a nasogastric tube and nitric oxide was withdrawn 90 minutes later, with minimal increase in pulmonary artery pressure and a rise in cGMP concentration to 45 nmol/l.
- A 3.5 kg newborn girl, who underwent corrective surgery for an infradiaphragmatic totally anomalous pulmonary venous connection, had postoperative systemic hypotension, low cardiac output, and pulmonary hypertension. Inhaled nitric oxide 20 ppm improved her pulmonary hemodynamics but could not be withdrawn by the third postoperative day, owing to reflex pulmonary hypertension. Sildenafil 1 mg via nasogastric tube again allowed withdrawal of nitric oxide, with preservation of plasma cGMP concentrations above baseline.
- A 4-month-old 4.1 kg boy with severe bilateral pulmonary vein stenosis developed moderate pulmonary hypertension after surgical revision. Sildenafil 1.1 mg via nasogastric tube did not increase the plasma cGMP concentration or reduce the rebound effect.

The authors suggested that in this case the effect of sildenafil may have been reduced by impaired gastrointestinal absorption. They speculated that exogenous nitric oxide inhibits nitric oxide synthase activity, with a consequent reduction in pulmonary vascular smooth muscle cGMP concentration. Phosphodiesterase type V inhibitors, such as sildenafil, increase cGMP concentration and ameliorate the rebound effect.

Organs and Systems

Cardiovascular

In several reviews, in which the same data have been analysed, sildenafil has been rated as being well tolerated (16,17) and extremely safe (18). Concerns about its cardiovascular safety profile have stemmed primarily from sporadic reports of myocardial infarction and stroke.

- One patient without a history of previous chest pain or risk factors for cardiovascular disease, developed a well-documented myocardial infarction 30 minutes after he took sildenafil 50 mg and before any attempt at sexual intercourse (19).

The interpretation of these sporadic cases is controversial, although some have argued that the reported cardiovascular adverse effects occur more often with sildenafil than with other pharmacological treatments of erectile dysfunction. It is at present unclear whether there is an increased risk with sildenafil. For example, in placebo-controlled trials there have been no differences in the incidences of myocardial infarction, angina, or coronary artery disorders between sildenafil and placebo (20). Exclusion criteria in clinical trials may have prevented the inclusion of patients who are at increased risk of adverse events. On the other hand, sexual activity itself increases cardiac workload and the risk of myocardial infarction. Patients with cardiovascular disease should be cautious in their use of sildenafil.

Prolongation of the QT_c interval has been reported but usually without serious dysrhythmias, although ventricular tachycardia has been reported after sildenafil.

- A healthy 50-year-old man took vardenafil 10 mg and 15 minutes later developed persistent palpitation lasting 2 hours (21). He had atrial fibrillation and no structural heart disease. The dysrhythmia converted spontaneously to sinus rhythm 4 hours later.

The authors suggested that hypotension induced by vardenafil had led to a reflex tachycardia.

- A 54-year-old man with hypertrophic cardiomyopathy, for which he took verapamil, felt unwell after a single tablet of sildenafil (22). Holter monitoring after repeat sildenafil showed an increase in ventricular extra beats and episodes of non-sustained ventricular tachycardia. On echocardiography, left ventricular dimensions were reduced and the subaortic gradient was markedly increased.
- Two other patients with severely depressed left ventricular function developed ventricular tachycardia (23).

Ear, nose, throat

Two men developed prolonged epistaxis a few hours after taking sildenafil 50 mg to enhance their sexual performance (before the nose bleeding started); both had well-controlled hypertension (24). Epistaxis is not an unusual problem in elderly people with hypertension, and venous engorgement is thought to be the main causative factor. Whether this is amplified by sildenafil (and/or by sexual activity) is an open question.

Nervous system

Headache is the one of the most common adverse effects of inhibitors of phosphodiesterase type V, being reported by 15–30% of patients. There is debate about whether migraine attacks are provoked by these drugs (25). A typical migraine attack occurred after ingestion of sildenafil in 10 of 12 women with a history of migraine (26).

Clinical trials of sildenafil have not shown increased risks of stroke or myocardial infarction. However, postmarketing drug surveillance programs have mentioned strokes associated with sildenafil, and case reports have been published.

- A 50-year-old man took sildenafil 50 mg, and 2 hours later developed asided hemiparesis and altered hemibody sensation, afacial paresis, and slurred speech (27). The symptoms gradually disappeared 4 hours later, but recurred the next week when he took sildenafil 100 mg. On the second occasion the symptoms did not resolve, and an MRI scan showed a recent infarct in the left internal capsule and lateral thalamus. No other cause of the stroke was found by evaluation of the heart and extracranial vessels. The symptoms gradually improved over 6 months.
- A 44-year-old man developed a severe headache and vomiting after taking four tablets of sildenafil (of unknown strength) followed by sexual intercourse (28). A CT scan showed a left-sided temporal intracranial hemorrhage. He died of cerebral edema and pneumonia a few days later. Autopsy showed no vascular abnormality.
- A 67-year-old man took two tablets of sildenafil 25 mg 1 hour apart (29). He complained of headache, confusion, and nervousness after the first tablet, his symptoms increased, and he developed language difficulty after the second tablet. He did not have sexual intercourse. A few days later, an MRI scan showed a large left temporal subcortical hemorrhage. The symptoms resolved partially over a few days.
- A 62-year-old hypertensive man developed left-sided hemiballismus secondary to a small hemorrhage in the right subthalamic–thalamic region (30). He had recently taken sildenafil 50 mg before having sexual intercourse.

Whereas the suspected mechanism for ischemic stroke is analogous to that leading to myocardial infarction (hypoperfusion distal to a critical lesion), intracerebral bleeding (28,29) may be more difficult to explain. The authors considered the likelihood of sildenafil-induced spontaneous intracerebral hemorrhage due to the vasodilatory effects of the drug on the cerebral vasculature (as evidenced by headache, flushing, and nasal congestion).

- A third-nerve palsy occurred 36 hours after a second dose of sildenafil in a 56-year-old man.

The authors suggested that sildenafil had caused systemic hypotension sufficient to cause neurological dysfunction, but 36 hours is a long lag time for a drug with a half-life of only a few hours (31).

Unexpected functional disturbances, which occur shortly after the use of sildenafil, are likely to be attributed to the drug.

- A 51-year-old man had transient global amnesia 30 minutes after taking sildenafil 25 mg (32).
- A 79-year-old man had acute vertigo, vomiting, and tinnitus resembling vestibular neuronitis 2 hours after first taking 50 mg; the symptoms lasted for 24 hours (33).
- Generalized tonic-clonic seizures have been reported after a first 50 mg tablet in two men aged 54 and 63, neither of whom had organic brain lesions on imaging; one had tonic-clonic seizures again on rechallenge with the drug 3 months later (34).

Sensory systems

Sildenafil has weak inhibitory effects on phosphodiesterase type V in the retina, leading to temporary changes in the perception of color hue and brightness. The importance of reversible changes in the electroretinogram observed in volunteers after sildenafil 100 mg, ascribed to inhibition of phosphodiesterase type 5 in the retina, is unclear (35–37).

There have been reports of a temporal association between vascular events in the eye and sildenafil administration.

- A fit, healthy, 69-year-old man presented with sudden painless loss of vision in the left eye a few hours after taking sildenafil 100 mg (38). Fundus examination showed occlusion of a branch of the retinal artery. No cardiovascular abnormality was detected.
- A 52-year-old man developed sweating, headache, and blurred vision in his left eye 1 hour after a first dose of sildenafil 50 mg (39). The same symptoms recurred on the next night, after a second dose of sildenafil. Fundoscopy a few days later showed an ischemic optic neuropathy.
- A 42-year-old man presented with anterior ischemic optic neuropathy, leading to a visual field defect in that eye within 24 hours of having taken sildenafil (40).

Non-arteritic anterior ischemic optic neuropathy (NAION) is a disorder whose pathophysiology is poorly understood. It is the most common acquired optic neuropathy after the age of 50 and has repeatedly been connected to inhibitors of phosphodiesterase type V, most often sildenafil. Case reports are being published in ever increasing numbers, but they do not clarify the problem. In a review the weak points in the possible connection between inhibitors of phosphodiesterase type V and NAION have been emphasized (41). The difference between the intraocular pressure and the perfusion pressure in the posterior ciliary arteries determines the circulation in the optic disc, and a reduction in this difference may contribute to ischemia.

- A 51-year-old man with poorly controlled hypertension had sudden superior hemifield loss in the left eye during sexual activity 4 hours after taking sildenafil 100 mg (42). He had used sildenafil repeatedly over the previous few weeks without unwanted effects. Fundoscopy and fluorescein angiography confirmed an embolic occlusion of the inferior hemiretinal artery.

The authors thought that debris from an atherosclerotic plaque at the carotid bifurcation had been dislodged as a result of increased cardiac workload during sexual activity, rather than a direct effect of sildenafil itself.

Several other cases of non-arteritic ischemic optic neuropathy have been reported in men taking sildenafil (43,44).

Serous chorioretinopathy has been attributed to sildenafil (45).

- A 37-year-old man took regular sildenafil and developed acute visual loss and a pressure sensation in his right eye. Fundoscopy showed an area of subretinal fluid, pointing to central serous chorioretinopathy. He was encouraged to refrain from sildenafil, but continued to use it and the symptoms worsened. When he later stopped using it, the disorder completely resolved within 3 weeks.Palpebral edema has been attributed to tadalafil (46).
- A 56-year-old patient with diabetes and no history of allergy noticed bilateral eyelid edema the morning after a first pill of tadalafil. The symptoms regressed spontaneously within 72 hours but recurred 1 week later after a second dose of the drug. The absence of any other known cause of palpebral edema, the time course, and recurrence on rechallenge were strong arguments for a causative link with tadalafil.

Liver

Acute hepatitis has been reported in a patient using sildenafil (47).

- A 65-year-old man with diabetes and hypertension had taken 50 mg of sildenafil about once every 2 weeks for 1 year, when he suddenly felt generally unwell. He had a tender liver, and blood tests showed mild thrombocytopenia, a lymphocytosis, and markedly raised transaminases, which had been normal shortly before and returned to normal over a few weeks after withdrawal of sildenafil, while he continued to take his antidiabetic and antihypertensive drugs. Other causes of hepatitis were ruled out by appropriate tests.

Definite proof of liver toxicity of a drug is difficult to provide. The authors invoke an ischemic rather than an immunoallergic pathogenesis to explain the hepatotoxic effect of sildenafil in this patient. This could also have explained his subsequent occasional use of sildenafil without recurrence of liver toxicity. The concomitant use of antihypertensive drugs may have facilitated the single episode of hepatitis.

There have been further reports of acute hepatitis in two men, aged 49 and 56 years, without apparent risk factors for liver disease (48,49). The evidence in favor of a link between acute liver disease and sildenafil is largely circumstantial but corresponds to accepted criteria for drug-induced hepatitis.

- Fatal variceal rupture occurred in a 41-year-old man with alcoholic liver cirrhosis and portal hypertension 3–4 hours after he took an unknown dose of sildenafil (50).

The authors hypothesized that sildenafil caused vasodilatation and increased splanchnic blood flow, thereby augmenting intravariceal pressure. He also had a high blood concentration of ethanol, another vasodilator.

- A 68-year-old man with alcoholic cirrhosis and small esophageal varices (degree of severity classified as Child I) started bleeding after taking sildenafil 25 mg for the first time (51).

Sudden overload of the portal venous system related to splanchnic vasodilatation was a possible provoking factor in this case. Gastroesophageal reflux secondary to a lower esophageal sphincter tone and causing mucosal erosion was an alternative explanation.

Skin

Rashes have been attributed to sildenafil.

- A 57-year-old man developed lichenoid lesions on the upper half of the body (52). Biopsy showed degeneration of keratinocytes and a dense lymphohistioid infiltrate arranged in a lichenoid pattern. He had been taken sildenafil irregularly. The eruption resolved within 3 weeks of withdrawal of sildenafil and recurred after rechallenge a few weeks later.
- A 67-year-old man developed toxic epidermal necrolysis, having during the previous 48 hours taken sildenafil between 300 and 400 mg (partly as a commercially available drug and partly in a Chinese aphrodisiac herbal medicine) (53). He had also been taking eight different medicines for diabetes, hypertension, hyperlipidemia, heart failure, gout, and osteoarthritis. The serum concentration of tumor necrosis factor was initially raised. He was given infliximab and recovered.

Acute widespread urticaria was reported after vardenafil consumption in a 48-year-old man with no other identifiable causative factors (54).

Sexual function

Whereas priapism has not been reported with sildenafil in controlled clinical trials, it is being mentioned in postmarketing drug surveillance programs, and two case reports have appeared in a healthy young man and a patient with sickle cell trait (55,56).

Reproductive system

Edema of the male breast secondary to strong vasodilatation after the use of inhibitors of phosphodiesterase type V can occur (57).

- A 32-year-old man complained of breast tenderness after taking sildenafil 50 mg/day for 3 weeks. There was tenderness confined to the nipples and areolae of both breasts, without signs of infection or inflammation. The symptoms subsided within 48 hours after withdrawal of the drug and recurred within 72 hours after its reintroduction.

Death

There has been debate on the risk of death after sildenafil: can it be entirely attributed to co-existing cardiovascular disease with an inherent high risk of mortality, particularly during sexual activity, or does the drug contribute (58,59).

- A 44-year-old man developed an acute myocardial infarction after taking sildenafil but before sexual intercourse (60).

This case appears to point to the drug as a potential trigger in people with unknown critical coronary lesions

A postmarketing survey over the first 6 months of sales of sildenafil in the USA (April to the middle of November 1998) has reported details on 130 patients who died after having been given sildenafil (over 3.5 million having received a prescription for it). As expected, deaths were mostly cardiovascular. There was a close time relation to the administration of sildenafil (within 4–5 hours) in 44 patients, 27 of whom died during or immediately after sexual intercourse. The disturbing finding was that 16 men also either took or were given glyceryl trinitrate, contrary to product labeling and to several warnings (61). A survey conducted in the UK on recreational use of sildenafil among night-club customers detected combined use of sildenafil with amyl nitrate. As both drugs are vasodilators, combined use may expose users to the same risk as the combination with glyceryl trinitrate (62).

Long-Term Effects

Drug tolerance

Potential tachyphylaxis with long-term use of sildenafil has been discussed following a survey in which 200 patients were questioned twice, 2 years apart, about the effects of sildenafil; on the second occasion 20% of them reported needing an increased dose, and 17% had discontinued therapy because of lack of effect (63). However, others have commented that more likely reasons for the loss of effect over time included worsening of the disease (64–68), reducing testosterone concentrations with age (65,66), or lack of proper arousal (69). Although the mRNA for phosphodiesterase type 5 is down-regulated in the retina by long-term administration of sildenafil in rats (70), it is not known whether that happens in human corpus cavernosum, nor whether long-term sildenafil use is associated with tachyphylaxis.

Drug Administration

Drug overdose

A case of suspected overdose of sildenafil has been reported (71).

- A 56-year-old man with a history of diabetes mellitus and hypertension was found dead at home, with an empty package of sildenafil (12 tablets of 50 mg) near the body. The concentration of sildenafil in postmortem blood was high. Autopsy showed a dilated cardiomyopathy and diffuse coronary atherosclerosis, but an overdose of sildenafil was suspected to have provoked a fatal ventricular dysrhythmia.

Drug–Drug Interactions

Alcohol

Lovers of red wine can be reassured: sildenafil 100 mg together with a full bottle of Australian Cabernet Sauvignon did not disturb the hemodynamics of young volunteers more than the wine did on its own (72). Alcohol stimulates sexual desire but impairs potency, but the authors did not report if sildenafil helped to maintain potency after rapid drinking of the wine.

Amlodipine

The effect of sildenafil on arterial pressure has been tested in 16 hypertensive men taking amlodipine 5–10 mg/day (73). Sildenafil did not affect amlodipine pharmacokinetics, but caused a further additive fall in blood pressure. Adverse events with the combination of sildenafil and amlodipine, headache, dyspepsia, and nausea, did not require drug withdrawal.

Calcium channel blockers

Retrospective analysis of clinical trials has suggested that the concomitant use of antihypertensive drugs did not lead to an increase in adverse events in patients also taking sildenafil (74).

Hypertensive patients taking amlodipine, in contrast to glyceryl trinitrate, had only a minor supplementary fall in blood pressure when challenged with a single dose of sildenafil, and a few had a mild to moderate headache (75).

Diltiazem is metabolized by CYP3A4 and was held responsible for unanticipated prolonged hypotension after sublingual glyceryl trinitrate in a patient who underwent coronary angiography 2 days after last using sildenafil (76).

- A 72-year-old man, who regularly took aspirin, metoprolol, diltiazem, and sublingual glyceryl trinitrate for stable angina, reported chest pain during elective prognostic coronary angiography, which resolved with half of a sublingual tablet of glyceryl trinitrate. Within 2 minutes he developed severe hypotension, with an unchanged electrocardiogram and no evidence of anaphylaxis. He had taken sildenafil 50 mg 48 hours before angiography.

The interval after which even short-acting nitrates can be safely given after the use of sildenafil is likely to be substantially longer than 24 hours when elderly patients are concurrently taking a CYP3A4 inhibitor, such as diltiazem.

Cannabinoids

Myocardial infarction has been attributed to the combination of cannabis and sildenafil.

- A 41-year-old man developed chest tightness radiating down both arms (77). He had taken sildenafil and cannabis recreationally the night before. His vital signs were normal and he had no signs of heart failure. However, electrocardiography showed an inferior evolving non-Q-wave myocardial infarct and his creatine kinase activity was raised (431 U/l).

Cannabis inhibits CYP3A4, which is primarily responsible for the metabolism of sildenafil, increased concentrations of which may have caused this cardiac event.

Indinavir

A drug interaction between sildenafil and HIV antiretroviral drugs is suspected, since both are metabolized by and act as inhibitors of the same cytochrome P450 isoforms (78). A pharmacokinetic study of sildenafil in HIV-positive patients taking indinavir has shown that the AUC of sildenafil was 4.4 times higher than data from historical controls (79). The magnitude of the interaction suggested that a lower starting dose of sildenafil may be more appropriate in patients taking indinavir.

There is good evidence that indinavir can substantially increase plasma concentrations of sildenafil (80). Since HIV infection commonly leads to erectile dysfunction, the drugs may well be used together and it will then be prudent to use a lower dose of sildenafil.

Nitrates, organic

The concomitant use of nitrates and sildenafil can precipitate a hypotensive reaction, and this combination should therefore be avoided (81,82,83,84,85). More than 3.6 million prescriptions for sildenafil were issued as of August 1998; of 69 deaths, 12 were attributable to the interaction with nitrates, as reported to the FDA.

Potentiation by sildenafil of the hypotensive effects of glyceryl trinitrate and amlodipine has been investigated. The fall in systolic blood pressure with glyceryl trinitrate was amplified four-fold by sildenafil in healthy subjects.

The pressure-lowering effect of a step-wise intravenous infusion of glyceryl trinitrate was significantly increased in 12 healthy volunteers treated for 4 days with sildenafil 25 mg tds; episodes of symptomatic hypotension were also more frequent with sildenafil (86). When sublingual glyceryl trinitrate was used, the reduction in systolic blood pressure was fourfold greater in association with sildenafil.

An elderly man with anginal pain died after treatment with glyceryl trinitrate (87). The family physician discovered that the man had also taken sildenafil, and he suggested suicidal intent, since he claimed that the patient was aware of the adverse cardiac events of the association of these two drugs.

Tacrolimus

The effects of sildenafil can be potentiated by drugs that are metabolized by CYP3A4. Tacrolimus is an example. When sildenafil was given to patients with kidney transplants taking regular tacrolimus, peak concentrations were much higher and the half-life much longer than expected from data in healthy volunteers (88). However, an effect of the underlying disease and other concomitant drugs obviously could not be excluded.

References

1. Boolell M, Gepi-Attee S, Gingell JC, Allen MJ. Sildenafil, a novel effective oral therapy for male erectile dysfunction. Br J Urol 1996;78(2):257–61.
2. Goldstein I, Lue TF, Padma-Nathan H, Rosen RC, Steers WD, Wicker PASildenafil Study Group. Oral sildenafil in the treatment of erectile dysfunction. N Engl J Med 1998;338(20):1397–404.
3. Rashid A. The efficacy and safety of PDE5 inhibitors. Clin Cornerstone 2005;7:47–56.
4. Reffelmann T, Kloner RA. Pharmacotherapy of erectile dysfunction: focus on cardiovascular safety. Expert Opin Drug Saf. 2005;4:531–40.
5. Carson CC 3rd. Cardiac safety in clinical trials of phosphodiesterase 5 inhibitors. Am J Cardiol 2005;96:37M–41M.
6. Tran D, Howes LG. Cardiovascular safety of sildenafil. Drug Saf 2003;26:453–60.
7. Salonia A, Rigatti P, Montorsi F. Sildenafil in erectile dysfunction: a critical review. Curr Med Res Opin 2003;19:241–62.
8. Hatzichristou D, Montorsi F, Buvat J, Laferriere N, Bandel TJ, Porst H; European Vardenafil Study Group. The efficacy and safety of flexible-dose vardenafil (Levitra) in a broad population of European men. Eur Urol 2004;45:634–41.
9. Montorsi F, Verheyden B, Meuleman E, Junemann KP, Moncada I, Valiquette L, Casabe A, Pacheco C, Denne J, Knight J, Segal S, Watkins VS. Long-term safety and tolerability of tadalafil in the treatment of erectile dysfunction. Eur Urol 2004;45:339–44.
10. Kloner RA. Cardiovascular effects of the 3 phosphodiesterase-5 inhibitors approved for the treatment of erectile dysfunction. Circulation 2004;110:3149–55.
11. Alaeddini J, Uber PA, Park MH, Scott RL, Ventura HO, Mehra MR. Efficacy and safety of sildenafil in the evaluation of pulmonary hypertension in severe heart failure Am J Cardiol 2004;94:1475–7.
12. Eardley I,Gentile V, Austoni E, Hackett G, Lembo D, Wang C, Beardswoth A. Efficacy and safety of tadalafil in a Western European population of men with erectile dysfunction. BJU Int 2004;94:871–7.
13. Seftel AD, Wilson SK, Knapp PM, Shin J, Wang WC, Ahuja S. The efficacy and safety of tadalafil in United States and Puerto Rican men with erectile dysfunction. J Urol 2004;172:652–7.
14. Dundar M, Kocak I, Dundar SO, Erol H. Evaluation of side effects of sildenafil in group of young healthy volunteers. Int Urol Nephrol 2001;32(4):705–8.
15. Atz AM, Wessel DL. Sildenafil ameliorates effects of inhaled nitric oxide withdrawal. Anesthesiology 1999;91(1):307–10.

16. Padma-nathan H, Eardley I, Kloner RA, Laties AM, Montorsi F. A 4-year update on the safety of sildenafil citrate (Viagra). Urology 2002;60(2 Suppl 2):67–90.
17. Fink HA, Mac Donald R, Rutks IR, Nelson DB, Wilt TJ. Sildenafil for male erectile dysfunction: a systematic review and meta-analysis. Arch Intern Med 2002;162(12):1349–60.
18. Lim PH, Moorthy P, Benton KG. The clinical safety of Viagra. Ann NY Acad Sci 2002;962:378–88.
19. Feenstra J, van Drie-Pierik RJ, Lacle CF, Stricker BH. Acute myocardial infarction associated with sildenafil. Lancet 1998;352(9132):957–8.
20. Morales A, Gingell C, Collins M, Wicker PA, Osterloh IH. Clinical safety of oral sildenafil citrate (Viagra) in the treatment of erectile dysfunction. Int J Impot Res 1998;10(2):69–73.
21. Veloso HH, de Paola AAV. Atrial fibrillation after vardenafil therapy. Emerg Med J 2005;22:823.
22. Stauffer JC, Ruiz V, Morard JD. subaortic obstruction after sildenafil in a patient with hypertrophic cardiomyopathy. N Engl J Med 1999;341(9):700–1.
23. Shah PK. Sildenafil in the treatment of erectile dysfunction. N Engl J Med 1998;339(10):699.
24. Hicklin LA, Ryan C, Wong DK, Hinton AE. Nose-bleeds after sildenafil (Viagra). J R Soc Med 2002;95(8):402–3.
25. Evans RW, Kruuse C. Phosphodiesterase-5 inhibitors and migraine. Headache 2004;44:925–6.
26. Kruuse C, Thomsen LL, Birk S, Olesen J. Migraine can be induced by sildenafil without changes in middle cerebral artery diameter. Brain 2003;126:241–7.
27. Morgan JC, Alhatou M, Oberlies J, Johnston KC. Transient ischemic attack and stroke associated with sildenafil (Viagra) use. Neurology 2001;57(9):1730–1.
28. Buxton N, Flannery T, Wild D, Bassi S. Sildenafil (Viagra)-induced spontaneous intracerebral haemorrhage. Br J Neurosurg 2001;15(4):347–9.
29. Monastero R, Pipia C, Camarda LK, Camarda R. Intracerebral haemorrhage associated with sildenafil citrate. J Neurol 2001;248(2):141–2.
30. Marti I, Marti Masso JF. Hemiballism due to sildenafil use. Neurology 2004;63:534.
31. Wigley FM, Korn JH, Csuka ME, Medsger TA Jr, Rothfield NF, Ellman M, Martin R, Collier DH, Weinstein A, Furst DE, Jimenez SA, Mayes MD, Merkel PA, Gruber B, Kaufman L, Varga J, Bell P, Kern J, Marrott P, White B, Simms RW, Phillips AC, Seibold JR. Oral iloprost treatment in patients with Raynaud's phenomenon secondary to systemic sclerosis: a multicenter, placebo-controlled, double-blind study. Arthritis Rheum 1998;41(4):670–7.
32. Savitz SA, Caplan LR. Transient global amnesia after sildenafil (Viagra) use. Neurology 2002;59(5):778.
33. Hamzavi J, Schmetterer L, Formanek M. Vestibular symptoms as a complication of sildenafil: a case report. Wien Klin Wochenschr 2002;114(1–2):54–5.
34. Gilad R, Lampl Y, Eshel Y, Sadeh M. Tonic–clonic seizures in patients taking sildenafil. BMJ 2002;325(7369):869.
35. Vobig MA, Klotz T, Staak M, Bartz-Schmidt KU, Engelmann U, Walter P. Retinal side-effects of sildenafil. Lancet 1999;353(9150):375.
36. Zrenner E. No cause for alarm over retinal side-effects of sildenafil. Lancet 1999;353(9150):340–1.
37. Vobig MA. Retinal side-effects of sildenafil. Lancet 1999;353(9162):1442.
38. Tripathi A, O'Donnell NP. Branch retinal artery occlusion; another complication of sildenafil. Br J Ophthalmol 2000;84(8):934–5.
39. Egan R, Pomeranz H. Sildenafil (Viagra) associated anterior ischemic optic neuropathy. Arch Ophthalmol 2000;118(2):291–2.
40. Payne B, Sasse B, Franzen D, Hailemariam S, Gemsenjager E. Manifold manifestations of ergotism. Schweiz Med Wochenschr 2000;130(33):1152–6.
41. Tomsak R. PDE5 inhibitors and permanent visual loss. Int J Impot Res 2005;17:547–9.
42. Bertolucci A, Latkany RA, Gentile RC, Rosen RB. Hemiretinal artery occlusion associated with sexual activity and sildenafil citrate (Viagra). Acta Ophthalmol Scand 2003;81:198–200.
43. Pomeranz HD, Smith KH, Hart WM Jr, Egan RA. Sildenafil-associated nonarteritic anterior ischemic optic neuropathy. Ophthalmology 2002;109(3):584–7.
44. Boshier A, Pambakian N, Shakir SA. A case of nonarteritic ischemic optic neuropathy (NAION) in a male patient taking sildenafil. Int J Clin Pharmacol Ther 2002;40(9):422–3.
45. Allibhai ZA, Gale JS, Sheidow TS. Central serous chorioretinopathy in a patient taking sildenafil citrate. Ophtalmic Surg Lasers Imaging 2004;35:165–7.
46. Chandeclerc ML, Martin S, Petitpain N, Barbaud A, Schmutz JL. Tadalafil and palpebral edema. South Med J 2004;97:1142–3.
47. Maroy B. Hépatite aiguë cytolitique probablement due à la prise de sildénafil (Viagra). [Cytolytic acute hepatitis probably due to sildenafil (Viagra).] Gastroenterol Clin Biol 2003;27:564–5.
48. Balian A, Touati F, Huguenin B, Prevot S, Perlemuter G, Naveau S, Chaput JC. Probable sildenafil induced acute hepatitis in a patient with no other risk factors. Gastroenterol Biol Clin 2005;29:89.
49. Daghfous R, El Aidli S, Zaiem A, Loueslati MH, Belkahia C Sildenafil-associated hepatotoxicity. Am J Gastroenterol 2005;100(8):1895–6.
50. Finley DS, Lugo B, Ridgway J, Teng W, Imagawa D. Fatal variceal rupture after sildenafil use: report of a case. Curr Surg 2005;62:55–6.
51. Tzathas C, Christidou A, Ladas SD. Sildenafil (Viagra) is a risk factor for acute variceal bleeding. Am J Gastroenterol 2002;97(7):1856.
52. Antiga E, Melani L, Cardinali C, Giomi B, Caproni M, Francalanci S, Fabbri P. A case of lichenoid drug eruption associated with sildenafil citratus. J Dermatol 2005;32:972–5.
53. Al-Shouli S, Abouchala N, Bogusz MJ, Al Tufail M, Thestrup-Pedersen K. Toxic epidermal necrolysis associated with high intake of sildenafil and is response to infliximab. Acta Derm Venereol 2005;85:534–5.
54. Minciullo PL, Saija A, Patafi M, Giannetto L, Marotta G, Ferlazzo B, Gangemi S. Vardenafil-induced generalized urticaria. J Clin Pharmacol Ther 2004;29:483–4.
55. Sur RL, Kane CJ. Sildenafil citrate-associated priapism. Urology 2000;55(6):950.
56. Kassim AA, Fabry ME, Nagel RL. Acute priapism associated with the use of sildenafil in a patient with sickle cell trait. Blood 2000;95(5):1878–9.
57. Chattopadhyay S, Dhar S. Mastalgia: an adverse effect of sildenafil. Dermatology 2004;209:346.
58. Kloner RA. Cardiovascular risk and sildenafil. Am J Cardiol 2000;86(2A):F57–61.
59. Mitka M. Some men who take Viagra die—why? JAMA 2000;283(5):590593.
60. Muniz AE, Holstege CP. Acute myocardial infarction associated with sildenafil (Viagra) ingestion. Am J Emerg Med 2000;18(3):353–5.

61. Jackson G, Sweeney M, Osterloh IH. Sildenafil citrate (Viagra): a cardiovascular overview. Br J Cardiol 1999;6:325–33.
62. Aldridge J, Measham F. Sildenafil (Viagra) is used as a recreational drug in England. BMJ 1999;318(7184):669.
63. El-Galley R, Rutland H, Talic R, Keane T, Clark H. Long-term efficacy of sildenafil and tachyphylaxis effect. J Urol 2001;166(3):927–31.
64. Billups KL. Re: Long-term efficacy of sildenafil and tachyphylaxis effect. J Urol 2002;168(1):204–5.
65. Carson CC. Re: Long-term efficacy of sildenafil and tachyphylaxis effect. J Urol 2002;168(1):205.
66. Guay AT. Re: Long-term efficacy of sildenafil and tachyphylaxis effect. J Urol 2002;168(1):205–6.
67. Mumtaz FH, Khan MA, Mikhailidis DP, Morgan RJ. Long-term efficacy of sildenafil and tachyphylaxis effect. J Urol 2002;168(1):206.
68. Tomera K. Re: Long-term efficacy of sildenafil and tachyphylaxis effect. J Urol 2002;168(1):206.
69. Basson R, Robinow O. Re: Long-term efficacy of sildenafil and tachyphylaxis effect. J Urol 2002;168(1):204.
70. Steers WD. Tachyphylaxis and phosphodiesterase type 5 inhibitors. J Urol 2002;168(1):207.
71. Tracqui A, Miras A, Tabib A, Raul JS, Ludes B, Malicier D. Fatal overdosage with sildenafil citrate (Viagra): first report and review of the literature. Hum Exp Toxicol 2002;21(11):623–9.
72. Leslie SJ, Atkins G, Oliver JJ, Webb DJ. No adverse hemodynamic interaction between sildenafil and red wine. Clin Pharmacol Ther 2004;76:365–70.
73. Knowles S, Gupta AK, Shear NH. The spectrum of cutaneous reactions associated with diltiazem: three cases and a review of the literature. J Am Acad Dermatol 1998;38(2 Pt 1):201–6.
74. Stauffer JC, Ruiz V, Morard JD. Subaortic obstruction after sildenafil in a patient with hypertrophic cardiomyopathy. N Engl J Med 1999;341(9):700–1.
75. Spencer CM, Gunasekara NS, Hills C. Zolmitriptan: a review of its use in migraine. Drugs 1999;58(2):347–74.
76. Khoury V, Kritharides L. Diltiazem-mediated inhibition of sildenafil metabolism may promote nitrate-induced hypotension. Aust NZ J Med 2000;30(5):641–2.
77. McLeod AL, McKenna CJ, Northridge DB. Myocardial infarction following the combined recreational use of Viagra and cannabis. Clin Cardiol 2002;25(3):133–4.
78. Granier I, Garcia E, Geissler A, Boespflug MD, Durand-Gasselin J. Postpartum cerebral angiopathy associated with the administration of sumatriptan and dihydroergotamine—a case report. Intensive Care Med 1999;25(5):532–4.
79. Liston H, Bennett L, Usher B Jr, Nappi J. The association of the combination of sumatriptan and methysergide in myocardial infarction in a premenopausal woman. Arch Intern Med 1999;159(5):511–3.
80. Merry C, Barry MG, Ryan M, Tjia JF, Hennessy M, Eagling VA, Mulcahy F, Back DJ. Interaction of sildenafil and indinavir when co-administered to HIV-positive patients. AIDS 1999;13(15):F101–7.
81. Kloner RA, Jarow JP. Erectile dysfunction and sildenafil citrate and cardiologists. Am J Cardiol 1999;83(4):576–82.
82. Cheitlin MD, Hutter AM Jr, Brindis RG, Ganz P, Kaul S, Russell RO Jr, Zusman RM. Use of sildenafil (Viagra) in patients with cardiovascular disease. Technology and Practice Executive Committee. Circulation 1999;99(1):168–77.
83. Curran M, Keating G. Tadalafil. Drugs 2003;63:2203–12.
84. Meuleman EJ. Review of tadalafil in the treatment of erectile dysfunction. Expert Opin Pharmacother 2003;4:2049–56.
85. Hellstrom WJ. Vardenafil: a new approach to the treatment of erectile dysfunction. Curr Urol Rep 2003;4:479–87.
86. Webb DJ, Freestone S, Allen MJ, Muirhead GJ. Sildenafil citrate and blood-pressure-lowering drugs: results of drug interaction studies with an organic nitrate and a calcium antagonist. Am J Cardiol 1999;83(5A):C21–8.
87. Bhalerao S. A new suicide. J Fam Pract 2001;50(6):551.
88. Christ B, Brockmeier D, Hauck EW, Friemann S. Interactions of sildenafil and tacrolimus in men with erectile dysfunction after kidney transplantation. Urology 2001;58(4):589–93.

Piridoxilate

General Information

Piridoxilate, an equimolar mixture of glyoxylic acid and pyridoxine, is marketed in a few countries (for example France) for peripheral arterial occlusive disease and functional venous disorders.

Organs and Systems

Urinary tract

A few reports have dealt with the occurrence of calcium oxalate renal calculi associated with long-term administration of piridoxilate.

- A 23-year-old man developed acute renal insufficiency due to hyperoxaluria and intratubular deposits of oxalate crystals after attempting suicide with an overdose of piridoxilate (SEDA-11, 180).
- An old woman developed end-stage chronic renal insufficiency and histological evidence of renal oxalosis ascribed to 10 years of piridoxilate treatment; renal function did not improve after withdrawal of the drug, and chronic hemodialysis was required (1).

Reference

1. Mousson C, Justrabo E, Rifle G, Sgro C, Chalopin JM, Gerard C. Piridoxilate-induced oxalate nephropathy can lead to end-stage renal failure. Nephron 1993;63(1):104–6.

Ruscus aculeatus (Liliaceae)

General Information

Ruscus aculeatus (butcher's broom, knee holy, knee holly, knee holm. Jew's myrtle, sweet broom, pettigree) has been used topically for vasoconstrictor treatment of varicose veins and hemorrhoids (1), and for chronic venous insufficiency, both alone (2,3) and in the combination known as Cyclo-3 fort, marketed in France, which contains an extract of *Ruscus aculeatus* 150 mg, hesperidin methyl chalcone 150 mg, ascorbic acid 100 mg, and metesculetol.

Systematic reviews

In a meta-analysis of the efficacy of Cyclo-3-fort in patients with chronic venous insufficiency 20 double-blind, randomized, placebo-controlled studies and five randomized comparison studies in 10 246 subjects were included (4). Cyclo 3 fort significantly reduced the severity of pain, cramps, heaviness, and paresthesia compared with placebo. There were also significant reductions in venous capacity and severity of edema.

Organs and Systems

Gastrointestinal

Cyclo 3 fort occasionally causes chronic diarrhea, which is thought to be secondary to altered gastrointestinal motility (SEDA-16, 205; SEDA-17, 244). or rarely to lymphocytic colitis (5). The difference between the two may be a question of the extent of investigation of the symptoms.

- A 55-year-old woman took Cyclo 3 fort three tablets a day for 48 hours and started to have loose watery stools at least four times a day (6). She lost 3 kg in weight over 5 weeks. She remembered that she had had a similar problem when taking Cyclo 3 fort a few months before. Clinical examination, laboratory tests for inflammatory disease, stool culture, and extensive endoscopy (stomach, terminal ileum, and colon) were all normal. Several biopsies taken from the colon all showed lymphocytic infiltration of the superficial layers of the epithelium. The diarrhea disappeared shortly after withdrawal of the drug.

Without biopsy proof of lymphocytic colitis, this case would probably have been classified as "functional" diarrhea.

Skin

Ruscus aculeatus can cause allergic contact dermatitis (7).

References

1. MacKay D. Hemorrhoids and varicose veins: a review of treatment options. Altern Med Rev 2001;6(2):126–40.
2. Vanscheidt W, Jost V, Wolna P, Lücker PW, Müller A, Theurer C, Patz B, Grützner KI. Efficacy and safety of a Butcher's broom preparation (*Ruscus aculeatus L. extract*) compared to placebo in patients suffering from chronic venous insufficiency. Arzneimittelforschung 2002;52(4):243–50.
3. Beltramino R, Penenory A, Buceta AM. An open-label, randomized multicenter study comparing the efficacy and safety of Cyclo 3 Fort versus hydroxyethyl rutoside in chronic venous lymphatic insufficiency. Angiology 2000;51(7):535–44.
4. Boyle P, Diehm C, Robertson C. Meta-analysis of clinical trials of Cyclo 3 Fort in the treatment of chronic venous insufficiency. Int Angiol 2003;22(3):250–62.
5. Tysk C. Lakemedelsutlost enterokolit. Viktig differentialdiagnos vid utredning av diarre och tarmblodning. [Drug-induced enterocolitis. Important differential diagnosis in the investigation of diarrhea and intestinal hemorrhage.] Lakartidningen 2000;97(21):2606–10.
6. Thiolet C, Bredin C, Rimlinger H, Nizou C, Mennecier D, Farret O. Colite lymphocytaire secondaire à la prise de Cyclo 3 fort. Presse Med 2003;32:1323–4.
7. Landa N, Aguirre A, Goday J, Ratón JA, Díaz-Pérez JL. Allergic contact dermatitis from a vasoconstrictor cream. Contact Dermatitis 1990;22(5):290–1.

Tolazoline

General Information

Tolazoline is an alpha$_2$-adrenoceptor antagonist that increases skin blood flow in healthy subjects and has been used to relieve acute vasospasm. However, convincing evidence of its therapeutic value in patients with chronic conditions, such as claudication or cerebrovascular impairment, is not available. It has a pulmonary vasodilator action and seems to be a useful adjunct in the management of neonates with persistent fetal circulation.

Tolazoline has a histamine-like action. Its most common adverse effects are palpitation, tachycardia, flushing, sweating, headache, and paresthesia of the skin with piloerection. Unwanted gastrointestinal effects include nausea, vomiting, and diarrhea. Postural hypotension and failure to ejaculate have also been reported.

Organs and Systems

Gastrointestinal

An infant with persistent fetal circulation treated with tolazoline developed gastric ulceration and perforation (1). The first symptoms occurred 14 hours after the start of therapy, and there was free air in the abdomen 34 hours later. Duodenal perforation also occurred in a 31-hour-old infant with meconium aspiration treated with tolazoline for 6 hours; the defect required surgical correction (2).

Drug–Drug Interactions

Clonidine

Tolazoline can antagonize the effects of intravenous clonidine and prevent the brief hypertensive reaction that it often causes (SEDA-2, 192; SEDA-3, 180).

References

1. von Muhlendahl KE. Perforating gastric ulceration during tolazoline therapy for persistent foetal circulation. Z Kinderchir 1988;43(1):48–9.

2. Matsuo M, Aida M, Yamada T, Takemine H, Tsugawa C, Kimura K, Matsumoto Y. Duodenal perforation with tolazoline therapy. J Pediatr 1982;100(6):1005–6.

Troxerutin

General Information

Troxerutin, a flavonoid derivative, is claimed to reduce capillary fragility. It is generally well tolerated (SEDA-19, 207), but ineffective (1).

Organs and Systems

Skin

Troxerutin is a yellow substance (SEDA-13, 170), and a few cases of yellow discoloration of the skin, but not of the sclerae, have been observed. The discoloration, which can be mistaken for jaundice, vanishes when the drug is withdrawn.

Reference

1. Anonymous. Paroven: not much effect in trials. Drug Ther Bull 1992;30(2):7–8.

Vincamine

General Information

Vincamine is an alkaloid extracted from the plant *Vinca minor*. Ethyl apovincaminate is a related synthetic ethyl ester of vincaminic acid. These drugs have spasmolytic effects similar to those of reserpine, but also have metabolic effects, including, in high doses, inhibition of phosphodiesterase. Although increased cerebral blood flow has been reported after the intravenous administration of vincamine, there have been no reliable studies of blood flow after oral medication. Improvement in scores on some psychometric tests have been obtained in some patients with cerebrovascular disease, but no clear-cut practical benefit has been demonstrated.

Organs and Systems

Cardiovascular

Commercial formulations obtained from the alkaloids in *Vinca minor* may not be free from adverse effects. Ventricular dysrhythmias have been observed, but only after intramuscular or intravenous administration, and they seem to reflect a direct effect on myocardial cells. Hypokalemia and a prolonged QT interval seem to be predisposing factors. Vincamine should therefore be avoided in patients with a prolonged QT interval (SEDA-2, 183) (SEDA-5, 206) (SEDA-5, 207) (SEDA-7, 228).

Psychological, psychiatric

Vincamine has minor psychoactive effects (1).

Reference

1. Crispi G, Di Lorenzo RS, Gentile A, Florino A, Pannone B, Sciorio G. Azione psicoattiva della vincamina in un gruppo di soggetti affetti da "sindrome depressiva recidivante". Nota preventiva. [Psychoactive effect of vincamine in a group of subjects affected by recurrent depressive syndrome. Preliminary note.] Minerva Med 1975;66(70): 3683–5.

ANTICOAGULANTS, THROMBOLYTIC AGENTS, AND ANTI-PLATELET DRUGS

ANTICOAGULANTS AND THEIR ANTIDOTES

Coumarin anticoagulants

General Information

The coumarins were first discovered in Wisconsin, when bleeding in cattle was found to be due to the consumption of bruised sweet clover in the 1920s (1). The causative agent, dicoumarol, was isolated in 1940, and a range of related compounds was then synthesized, the most popular of which proved to be warfarin (named after the Wisconsin Alumni Research Foundation). Other coumarins that have been used are acenocoumarol (nicoumalone), bishydroxycoumarin, dicoumarol, ethyl biscoumacetate, and phenprocoumon.

The coumarins act as competitive inhibitors of vitamin K epoxide reductase, which is responsible for regenerating reduced vitamin K from vitamin K epoxide after it has been consumed as a co-factor in the synthesis of coagulation factors II, VII, IX, and X.

Uses

The main use of the coumarins is in the treatment and prevention of thromboembolic disease, including deep vein thrombosis, pulmonary embolism, and cerebral embolism from cardiac and other sources.

Protein C, another vitamin K-dependent serine protease zymogen in plasma, is a regulatory protein that, when activated, limits the activity of two activated procoagulant co-factors, factors Va and VIIIa. Heterozygotes for hereditary isolated protein C deficiency tend to develop a thrombotic disease which has been successfully treated with long-term coumarins (2,3). Apparently, the balance between the activities of protein C and the procoagulant factors (II, VII, IX, and X), which is disturbed in protein C-deficient patients, is restored during long-term treatment with coumarins.

Non-anticoagulant uses

Warfarin reduces calcium deposition in spontaneously degenerated bioprosthetic valves (4).

Both direct and indirect antitumor actions of anticoagulants have been postulated on the basis of experimental findings in animals (5). Warfarin given alone or in combination with cytostatic drugs reduces the size of fibrosarcomas in animals (5,6) and of osteosarcomas in man (7), and consequently prolongs survival times in both. The Cooperative Studies Program of the Veterans Administration Medical Research Service suggests that, among patients with various tumors, those with small-cell carcinoma of the lung had a significantly longer survival (about 4 years instead of 2) when warfarin was added to standard treatment (8). There were no differences in survival between warfarin-treated and control groups for advanced non-small-cell lung cancers, colorectal, head and neck, and prostate cancers (9). However, the contention that cancer morbidity and/or mortality is reduced in anticoagulated patients is not substantiated by the results of a retrospective study of 378 patients who had been taking anticoagulants for about 10 years on average (10).

It has been suggested that warfarin may have a specific effect on tumor-cell growth via the inhibition of protein synthesis (11). Another explanation of the effect of the drug is a reduction in the co-adherence of tumor cells, which renders them more vulnerable to the actions of defence mechanisms. The reduction in co-adherence is thought to be caused by an inhibitory action of anticoagulants on the fibrin network that is vital for tumor cell growth. This hypothesis, which is supported by dose dependence of the inhibitory effect of both heparin and warfarin (12), has been studied in a controlled randomized trial of the use of streptokinase after surgery for tumors of the large bowel (6).

Supercoumarins

Anticoagulant pesticides are used widely in agricultural and urban rodent control. The emergence of warfarin-resistant strains of rats led to the introduction of a group of anticoagulant rodenticides variously referred to as "supercoumarins", "superwarfarins", "single dose" rodenticides, or "long-acting" rodenticides (13). This group includes the second generation 4-hydroxycoumarins brodifacoum, bromadiolone, difenacoum, and flocoumafen and the indanedione derivatives chlorophacinone and diphacinone; these drugs typically have longer half-lives than warfarin.

The greater potency and duration of action of long-acting anticoagulant rodenticides is attributed to: (i) their greater affinity for vitamin K epoxide reductase; (ii) their ability to disrupt the vitamin K epoxide cycle at more than one point; (iii) hepatic accumulation; and (iv) unusually long half-lives, due to high lipid solubility and enterohepatic recirculation.

There have been several published case reports of accidental or intentional poisoning due to these drugs (14,15), which can result in prolonged coagulopathy (16,17).

Most cases of anticoagulant rodenticide exposure involve young children, and so the amounts ingested are almost invariably small. In contrast, intentional ingestion of large quantities of long-acting anticoagulant rodenticides can cause anticoagulation for several weeks or months (18). Occupational exposure has also been reported.

Substantial ingestion produces epistaxis, gingival bleeding, widespread bruising, hematomas, hematuria with flank pain, menorrhagia, gastrointestinal bleeding, rectal bleeding, and hemorrhage into any internal organ; anaemia can result. Spontaneous hemoperitoneum has been described. Severe blood loss can result in hypovolemic shock, coma, and death. The first clinical signs of bleeding can be delayed and patients can remain anticoagulated for several days (warfarin) or days, weeks, or months (long-acting anticoagulants) after ingesting large amounts.

In 10 762 children aged 6 years and under who unintentionally took single doses of brodifacoum, there were no deaths or major adverse effects, although 67 reported evidence of coagulopathy (19). There were minor and moderate adverse effects in 38 and 54 children, respectively. About half of the children received some form of gastrointestinal decontamination, which had no effect on the distribution of outcomes but caused adverse effects in 42 patients.

Ingestion of a small amount of a superwarfarin does not require specific therapy (20,21). However, if the coagulopathy due to a superwarfarin is severe, large doses of vitamin K may be required (16). Recombinant activated factor VII has been successfully used to treat superwarfarin poisoning (22).

There are now sufficient data in young children exposed to anticoagulant rodenticides to conclude that routine measurement of the international normalized ratio (INR) is unnecessary. In all other cases, the INR should be measured 36–48 hours after exposure. If the INR is normal at this time, even in the case of long-acting formulations, no further action is required. If active bleeding occurs, prothrombin complex concentrate (which contains factors II, VII, IX, and X) 50 units/kg, or recombinant activated factor VII 1.2–4.8 mg, or fresh frozen plasma 15 ml/kg (if no concentrate is available) and phytomenadione 10 mg intravenously (100 micrograms/kg for a child) should be given. If there is no active bleeding and the INR is under 4.0, no treatment is required; if the INR is 4.0 or higher phytomenadione 10 mg should be given intravenously. When cases of unexplained acquired coagulopathy and selective deficiency of vitamin K-dependent clotting factors occur in patients in the absence of liver disease or inhibitors, physicians should consider the possibility of superwarfarin poisoning as a cause.

General adverse effects

The major adverse effect of the coumarins is hemorrhage and, exceptionally, hemorrhagic skin necrosis. Administration during pregnancy can cause an embryopathy. Allergic reactions are extremely rare. Tumor-inducing effects have not been reported.

In the common database of the German spontaneous reporting system, 1164 reports of adverse drug reactions were registered that had been attributed to therapy with vitamin K antagonists during the period from 1990 to 2002 (phenprocoumon: 91%; warfarin: 8.3%; acenocoumarol: 0.9%) (23). Among these reactions a reduction in prothrombin time was the most common (15%), followed by gastrointestinal hemorrhage (13%), cerebral hemorrhage (9.1%), melena (7.4%), and increased hepatic enzymes (7.3%). Unspecified hemorrhage, intracranial hemorrhage, and hematomas accounted for 6.0% each, hepatitis 5.7%, and hematuria 4.9%. There were 42 reports (3.0%) of skin necrosis and seven of hepatic necrosis. Altogether, there were 609 cases of drug-induced hemorrhage. On average 47 cases of hemorrhage were attributed to phenprocoumon or warfarin each year from 1990, with spikes in the numbers of cases in 1997 (n = 107) and 2002 (n = 110). During the entire period the amount of prescribing increased continuously. Total sales reached 132.2 million defined daily doses (DDDs) in 1997 and 190.0 million DDDs in 2001. It was therefore not surprising that the number of adverse reactions reports increased.

Organs and Systems

Cardiovascular

Vasodilatory effects on the coronary arteries, peripheral veins, and capillaries, with purple toes as one of the most obvious consequences (24,25), have been reported. Sensations of cold may be due to increased loss of body heat caused by peripheral vasodilatation (26).

Cholesterol embolization, which promptly improves after the drug is withdrawn (27), may explain the purple-toe phenomenon.

Nervous system

Apart from cerebral hemorrhage, coumarins have no direct adverse effects on the nervous system.

Endocrine

An antithyroid effect of bishydroxycoumarin has been suspected (28).

Metabolism

Dicoumarol has been reported to have a uricosuric effect (29).

Hematologic

Hemorrhage

Bleeding is the major complication of coumarin anticoagulants. The annual incidence of major bleeding among 4060 patients in the AFFIRM trial, who were followed for an average of 3.5 years, was about 2% per year (30).

Dose-dependency

The intensity and stability of treatment, in addition to the beneficial effect of the coumarins, determine the rate and severity of bleeding complications. The average annual frequencies of fatal, major, and major and minor bleeding during warfarin therapy were 0.6, 3.0, and 9.6% respectively (31). The relationship between the intensity of anticoagulant therapy and the risk of bleeding is very strong, both in patients with deep vein thrombosis tissue and in those with mechanical heart valves. In randomized trials for these indications, the incidence of major bleeding in patients randomly assessed to less intense warfarin therapy (targeted INR 2.0–3.0) has been less than half the incidence found in patients randomly assigned to more intense anticoagulation (INR more than 3) (32). The bleeding risk increases dramatically when the INR is higher than 4.0 (33), especially the risk of intracranial hemorrhage. Less intense warfarin therapy has been used in patients with non-rheumatic atrial fibrillation, with only a very slight increase in major bleeding compared with placebo. In such patients, an INR of 2.5 (range 2.0–3.0) minimizes both hemorrhage and thromboembolism (34).

Time-dependency

Although bleeding can occur at any time, if the effective dose of drug increases, whether because of a change in dose or an interaction, most studies have reported higher frequencies of bleeding soon after the start of treatment. In one study the incidence of major bleeding fell from 3.0% per month during the first month of outpatient warfarin therapy to 0.8% per month during the rest of the first year of therapy and to 0.3% per month thereafter (35). Increased variation in anticoagulant effect, as reflected by time-dependent variation in the INR is associated with an increased frequency of hemorrhage independent of the mean INR (36,37).

Susceptibility factors

Important susceptibility factors include age, endogenous coagulation defects, thrombocytopenia, hypertension, cerebrovascular disease, thyroid disease, renal insufficiency, liver disease, tumors, cerebrovascular disease, alcoholism, a history of gastrointestinal bleeding (peptic ulcer disease alone without past bleeding is not associated with an increased risk of bleeding), and an inability to adhere to the regimen.

In a case-control study 170 patients with atrial fibrillation who developed intracranial hemorrhage while taking warfarin were compared with 1020 matched controls who did not (38). The cases were significantly older than the controls (median age 78 versus 75 years) and had higher median INRs (2.7 versus 2.3). The risk of intracranial hemorrhage increased at 85 years of age or older (adjusted OR = 2.5; 95% CI = 1.3, 4.7; referent age 70–74 years) and at an INR range of 3.5–3.9 (adjusted OR = 4.6; CI = 2.3, 9.4; referent INR 2.0–3.0). The risk of intracranial hemorrhage at INRs below 2.0 did not differ statistically from the risk at INRs of 2.0–3.0 (adjusted OR = 1.3; CI = 0.8, 2.2). In contrast, an analysis of patients with atrial fibrillation in the Framingham Heart Study did not show age as a risk factor for bleeding (39).

Management

The management of coumarin toxicity depends on the INR and whether there is bleeding.

- INR more than 0.5 over the target but under 5.0: reduce the dose
- INR 5.0–8.0: withhold the drug and restart once the INR is below 5.0
- INR over 8.0 and no bleeding: withhold the drug and restart once the INR is below 5.0
- INR over 8.0 with bleeding: withhold the drug and give vitamin K_1 (phytomenadione) 0.5 mg intravenously or 5 mg orally
- INR over 8.0 with life-threatening hemorrhage: withhold the drug and give a prothrombin complex concentrate (such as Beriplex-P/N or Prothromplex-T) 50 U/kg (40); if this is not available, give fresh frozen plasma 15 ml/kg.

If rapid and complete reversal is required (for example before a procedure such as a liver biopsy), vitamin K_1 (phytomenadione) 5–10 mg can be given slowly (1 mg/minute).

Other effects

Blood and plasma viscosity fall by 5–10% during the administration of coumarins in healthy volunteers and in patients with coronary artery disease (41). This may also explain, at least partly, the antianginal effect of coumarins. The mechanism might be related to changes in the protein composition of the plasma.

A reversible increase in white cell count has been reported with long-term use of acenocoumarol (42).

Hemolytic anemia thought to be related to warfarin has been reported (43).

Gastrointestinal

Gastrointestinal complications due to coumarins are limited to hemorrhage, such as small bowel obstruction due to bleeding (44).

- A 53-year-old woman developed abdominal pain and vomiting while taking warfarin after aortic and mitral valve surgery. There was jejunal narrowing consistent with a stricture, probably as a result of submucosal bleeding. Warfarin was withdrawn and she was given heparin, with complete resolution of symptoms.

Liver

Only a few cases of hepatic injury have been documented in patients taking coumarins. In some of them, rechallenge caused a relapse (45). The usual presentation was with a cholestatic illness, beginning about 10 days after the coumarin anticoagulant was started, sometimes associated with eosinophilia. In Switzerland, oral anticoagulants were involved in only 11 of 674 reports of drug-induced liver disease collected during 1981–95. Seven cases were due to phenprocoumon and three to acenocoumarol. The severity ranged from asymptomatic rises in liver enzymes to cholestatic hepatitis. The interval between administration and onset of symptoms varied from 2 days to several weeks or months (46).

Subacute liver failure necessitating orthotopic liver transplantation has been reported with phenprocoumon (47).

- A 39-year-old woman developed idiopathic thrombosis of the posterior tibial vein. Oral contraceptives and resistance to activated protein C were identified as risk factors. After initial treatment with intravenous heparin, she was given phenprocoumon and the oral contraceptive was withdrawn. After 4 months she developed subacute liver failure and phenprocoumon was withdrawn immediately. Autoimmune disease, viral hepatitis, toxic causes, and Budd–Chiari syndrome were excluded. Despite symptomatic treatment, she deteriorated further and orthotopic liver transplantation was performed. Histopathology of the explanted liver further excluded ischemic liver cell necrosis and Budd–Chiari syndrome.

Urinary tract

In one case, multisystem abnormalities, including renal insufficiency, were caused by cholesterol embolization. Withdrawal of the anticoagulant resulted in dramatically improved renal function (27).

Skin

Maculopapular rashes with cross-sensitivity between coumarin derivates have been reported (48). Non-pruritic purpuric skin eruptions, histologically presenting as vasculitis and reappearing on rechallenge with warfarin or acenocoumarol, have been described (49).

Skin necrosis

DoTS classification (BMJ 2003:327:1222-5):
Dose-relation: Collateral reaction
Time-course: Early
Susceptibility factors: Genetic (thrombophilic defects); diseases (obesity, heparin-induced thrombocytopenia)

Vitamin K antagonists, such as warfarin, prevent gamma-carboxylation of the vitamin K- dependent procoagulant factors II, VII, IX, and X and the natural anticoagulants protein C and protein S. When therapy is begun these factors fall at rates that depend on their individual half-lives. As the natural anticoagulant protein C has a short half-life and some of the procoagulants (X and particularly II) have longer half-lives, coumarins can be procoagulant until equilibrium is reached. During this induction period thrombosis in the dermal vessels can occur, resulting in skin necrosis.

Skin necrosis is a rare but serious complication of oral anticoagulation, seen typically during induction of therapy and occurring in about 0.01–0.1% of patients (50). It is conceivable that this condition is more common than is generally recognized, because many formes frustes can occur, presenting as painful cellulitis without hemorrhagic necrosis. Severe skin necrosis in a patient taking an oral anticoagulant was first described in 1943, but not documented in a series of patients until 11 years later (51). The morbidity of this complication is high and, despite treatment, about 50% of patients ultimately require surgical intervention and in some cases skin grafting.

There is convincing evidence that skin necrosis occurs exclusively in patients with excessively severe initial coumarin-induced hypocoagulability (52). As a rule, excessive doses have been given, resulting in severe and rapid reductions in the concentrations of factor VII and protein C (53,54).

Time-course

Familiarity with the clinical and histological pictures is essential, because the earliest signs and symptoms must be recognized if necrosis is to be prevented. The lesion does not usually occur before the second day of treatment or after the second week of treatment has started; most commonly it appears between the third and fifth day (55,56). There are occasionally exceptions to this rule; warfarin-induced skin necrosis can occur several days after discontinuation of warfarin (57) and some cases have been reported during long-term treatment (58,59). However, as a general rule, when skin necrosis occurs after 10 days of warfarin therapy, another cause must be sought.

Presentation

The lesion often appears symmetrically or at a pair of unrelated sites, with a predilection for parts of the body rich in fatty tissue, such as the breasts, abdomen, buttocks, thighs, and calves (60). Feet and toes are seldom affected, male genitalia only rarely (51,61,62), and the vagina and uterus very exceptionally (63).

The lesion begins with an evanescent, painful, slightly raised, more or less clearly demarcated erythematous patch. Histological examination at this stage reveals slight round-cell perivascular infiltration of the corium, edema, and swelling of the capillary endothelium, particularly at the cutis/subcutis boundary (the dermovascular loop), with fibrin thrombi in the small venules. Patches of necrotic fatty tissue, slight polymorphonuclear perivascular infiltration, and patchy interstitial edema as well as bleeding are present at this stage. Very soon, petechiae appear and become confluent within 24 hours, forming purple ecchymotic lesions surrounded by a sharply defined zone of hyperemia. During the next 24 hours, thrombotic occlusion of the veins causes infarction with necrosis of the skin, subcutaneous fat, and sometimes also of deeper anatomical structures. Hemorrhagic blisters characterize the onset of irreversible necrosis of the skin. Laboratory investigation may reveal diffuse intravascular coagulation and even hemolytic anemia (64).

- A 50-year-old woman with a left leg deep venous thrombosis and subsequent pulmonary embolism was first anticoagulated with low-molecular-weight heparin and subsequently warfarin (65). Within 4 days she developed abdominal skin necrosis. She had protein S deficiency and a mutation in the methylenetetrahydrofolate reductase gene (MTHFR).
- A 38-year-old obese woman was given heparin and warfarin for a presumed pulmonary embolism (66).

On day 5 she developed a tender mass in the left breast and was thought to have an inflammatory carcinoma. On the next day irregular greyish blue areas of the skin, ecchymoses, and hemorrhagic bullae were noted overlying the mass. Urgent surgical debridement showed extensive necrosis of the skin and breast substance.

- A 29-year-old woman with an acute pulmonary embolism was given heparin and warfarin and on day 5 developed severe pain, purple discoloration, and swelling in her right breast (67). A diagnosis of warfarin-induced skin necrosis was made. Thrombophilia testing showed heterozygosity for factor V Leiden. The concentrations of protein C and S were normal.
- A 75-year-old man who had taken acenocoumarol for 7 years was given diclofenac for a painful knee (68). Two days later, his renal function deteriorated and skin necrosis became evident. Biopsy showed histological changes consistent with coumarin-induced necrosis. Protein C and S concentrations were normal. The authors concluded that acute renal insufficiency could have precipitated a transient defect in the protein C pathway.
- Six patients with heparin-induced thrombocytopenia (HIT), who developed frank or impending venous limb gangrene (n = 2) or central skin necrosis (n = 5) (one had both) temporally related to warfarin therapy, developed these complications after either taking warfarin alone for 2–7 days (n = 4) or while a direct thrombin inhibitor was being withdrawn (n = 2) (69). All had supratherapeutic international normalized ratios. One patient required leg and breast amputations, and another died.
- A 54-year-old woman had an above knee amputation for limb ischemia due to arteriosclerosis and postoperatively developed subclavian vein thrombosis. She was given enoxaparin and warfarin and after a single dose of 10 mg her INR was 6.7 (70). Three days later she developed skin necrosis over her right upper thigh. She was subsequently found to be protein S deficient.
- An obese woman developed extensive cutaneous necrosis while taking acenocoumarol for a deep venous thrombosis (71). She had a heterozygous deficit for protein C. The histopathological findings of vessel thrombi and erythrocyte extravasation were consistent with the clinical picture.
- A 61-year-old woman with a 12-month history of Raynaud's phenomenon developed multiple digital necrosis following aortic valve replacement with a mechanical prosthesis for aortic insufficiency caused by non-bacterial thrombotic endocarditis (72). Postoperatively she had daily episodes of ischemia of the fingers and toes, which improved with local warming. However, coincident with the occurrence of immune heparin-induced thrombocytopenia, and while undergoing routine warfarin anticoagulation because of the mechanical valve prosthesis, she abruptly developed progression of the digital ischemia to multiple digital necrosis on postoperative day 8, when the international normalized ratio reached its peak value of 4.3. Subsequently, she was found to have metastatic breast adenocarcinoma.

The authors of the last report suggested that multiple digital gangrene can result from the interaction of various localizing and systemic factors, including compromised microvascular blood flow (Raynaud's phenomenon), increased thrombin generation (heparin-induced thrombocytopenia, adenocarcinoma), and warfarin-induced failure of the protein C natural anticoagulant pathway.

These reports re-enforce and add to our knowledge of coumarin-induced skin necrosis. In two cases the patient was described as obese [[Byrne 356,Valdivielso 211), and in another three this was apparent from the clinical photographs [[Byrne 356,Moll 628,Tai 237). Defects of the protein C pathway are a well recognized susceptibility factor, and in these cases two were protein S deficient [[Byrne 356,Tai 237) and one was protein C deficient [[Valdivielso 211). Two patients had factor V Leiden [[Byrne 356,Moll 628), so other thrombophilic defects may also predispose to this adverse effect. There is no other evidence that a mutation in methylene tetrahydrofolate reductase [[Byrne 356) or renal impairment [[Muniesa 502) play a role. Inadequate heparin while starting coumarin therapy is an important factor [[Moll 628), but adequate heparin does not offer complete protection [[Byrne 356,Khalid 268). Finally, there is a risk of severe thrombosis when starting warfarin in patients with heparin-induced thrombocytopenia [[Srinivasan 66,Warkentin 56).

Mechanism

The pathogenesis is still not completely elucidated, but data suggest that transient protein C deficiency may be causative.

Hemorrhagic skin necrosis has been described with all coumarins and indanediones. During the initial phase of oral anticoagulant treatment with these agents, the plasma concentration of protein C falls rapidly in parallel with factor VII. The half-lives of protein C and factor VII are much shorter than those of factors II, IX, and X, and during this initial phase of oral anticoagulation, there is therefore a striking imbalance between procoagulant factors (factors IX, X, and II) and anticoagulant vitamin K-dependent factors (protein C).

Susceptibility factors

Probably because of the sites of predilection, women account for about 80% of cases (55).

Patients with hereditary protein C deficiency (73,74) or acquired functional protein C deficiency (75) are particularly susceptible to the development of hemorrhagic skin necrosis. However, skin necrosis has also been reported in patients with deficiency of protein S (a co-factor for protein C), in patients at high thrombogenic risk linked to a constitutional antithrombin III deficit, and in patients with antiphospholipid antibodies associated with systemic lupus erythematosus (76,77).

Prevention

Prevention of coumarin-induced skin necrosis can be achieved by avoiding initial overshooting of the coumarin effect. A primary cautious initial dosage is mandatory

(74), especially in elderly people, who require a smaller dose than younger age groups (one-third less on average) (78). Secondly, adequate patient surveillance is essential. The development of full-blown skin necrosis can usually be prevented by the administration, at the first signs of a developing lesion, of vitamin K_1 (52). If the lesions progress, the oral anticoagulant should preferably be withdrawn.

Prevention of recurrence of coumarin necrosis in patients with protein C deficiency, if treatment is necessary, could consist of transient simultaneous infusion of fresh frozen plasma (leading to a constant concentration of protein C) and heparin both before and at the first time of administration of an oral anticoagulant, associated or not with protein C concentrate (79,80).

Musculoskeletal

Preliminary results that suggested coumarin-associated reduction in bone density have not been confirmed. In a study from the Osteoporotic Fractures Research Group, 6201 postmenopausal women who were either users ($n = 149$) or non-users of warfarin were assessed for fractures and bone mineral density. Over 2 years, the two groups had similar age-adjusted heel and hip bone mineral density measurements. During an average of 3.5 years, non-traumatic, non-vertebral fractures occurred in 10% of warfarin users and 9.3% of non-users (81).

However, bones that fracture during oral anticoagulant therapy require more time to form adequate amounts of callus. This may be explained by an anticoagulant-induced increase in the size of fracture hematoma. However, the observation that warfarin inhibits calcification inside artificial hearts implanted in calves adds to the body of evidence indicating that coumarin depresses not only coagulation factors, but also other (Gla)-containing proteins, for example osteocalcin, a shortage of which may also delay skeletal calcification (82). Osteocalcin is a non-collagenous bone matrix protein containing gamma-carboxyglutamic acid, the synthesis of which is vitamin K-dependent. During oral anticoagulation with phenprocoumon, osteocalcin concentrations were lower than in control subjects, whereas the proportion of non-carboxylated osteocalcin was significantly higher than in healthy subjects (83). Since osteocalcin concentrations reflect bone formation, but not bone resorption activity, the reduced serum total osteocalcin concentrations during oral anticoagulation do not necessarily imply that bone loss occurs in these patients. However, reduced bone formation and impaired gamma-carboxylation of osteocalcin in patients treated with phenprocoumon can be clinically important in circumstances such as fracture healing or when there is pre-existing bone disease. It has been suggested that vitamin D regulates the synthesis of vitamin K-dependent bone protein, but no significant effect of the duration of phenprocoumon therapy on parathormone and vitamin D concentrations has been observed (83).

Immunologic

Skin-test reactivity, that is induration and tissue factor generation by monocytes, is reduced by therapeutic doses of oral anticoagulants, but lymphocyte transformation activity is not. This constitutes the rationale for the use of oral anticoagulants in the treatment of immune diseases characterized by fibrin deposition, such as allograft rejection and lupus nephritis (84).

Second-Generation Effects

Pregnancy

Antithrombotic treatment during pregnancy carries a well-established and substantial risk for both mother and fetus (85). The mother has an increased chance of abortion and of perinatal bleeding complications.

Contraceptive counselling must be given to all women who need anticoagulants. Recommendations for the use of anticoagulants during pregnancy have been reassessed in various publications (86–89). In cases of previous venous thrombosis and/or pulmonary embolism, two reasonable approaches are possible. One can either use low-dose heparin throughout pregnancy followed by an oral anticoagulant postpartum for 4–6 weeks, or one can choose to initiate clinical surveillance combined with periodic venous non-invasive tests followed by an oral anticoagulant postpartum for 4–6 weeks. If venous thrombosis occurs, heparin should be used until term, withdrawn immediately before delivery, and then both heparin and an oral anticoagulant can be started postpartum. When pregnancy is planned in patients who are taking long-term oral anticoagulants, the physician should either replace the oral anticoagulant with heparin before conception is attempted, or perform frequent pregnancy tests and substitute oral anticoagulant for heparin when pregnancy is achieved.

In patients with artificial heart valves, management in pregnancy is problematic, because the efficacy of heparin has not been established. Two approaches have been recommended. One can use heparin in a therapeutic dosage throughout pregnancy or use heparin until the 13th week, followed by an oral anticoagulant until the middle of the third trimester, and then heparin until delivery.

Teratogenicity

Vitamin K antagonists, including hydroxycoumarin derivatives and indan-1,3-dione-derived drugs, can be teratogenic and can also induce bleeding in the fetus (85,86,90–93). Adverse fetal outcomes occur in about one-third of pregnancies after either oral anticoagulants or heparin (91). This surprisingly high figure, which is often quoted, should be tempered by the consideration that these data are largely (if not exclusively) derived from women with heart valves taking relatively large doses of warfarin in order to maintain the INR at around 4. Some have suggested that the risk of fetal embryopathy is lower in women taking lower doses for prophylaxis or treatment of deep venous thrombosis (94).

The pattern of teratogenicity of congeners of the vitamin K antagonist group is generally known as warfarin embryopathy, also referred to as the fetal-warfarin syndrome and Conradi–Hünermann syndrome, although the latter term also covers an identical hereditary disorder. The pattern is now considered to represent a specific group of malformations occurring in some fetuses exposed to a vitamin K antagonist during the first trimester, with a critical period during the sixth to twelfth weeks of gestation. The minimal criteria for the diagnosis include either nasal hypoplasia or stippled epiphyses. In severe cases the forehead is bossed and the nose is sunken, with deep grooves between the alae nasi and the tip of the nose; other skeletal deformities can also be present. Half the affected children have upper airway obstruction, secondary to underdeveloped cartilage. Radiography shows stippling (caused by abnormal focal calcification of the epiphyseal regions), preferentially in the axial skeleton, for example the proximal femur, and in the calcanei. Children with milder defects can show catch-up growth, and stippling can disappear after the first year of life; however, in severe cases the nose remains small and sunken. It is questionable whether mental retardation is part of the syndrome, since it has only been observed in cases of exposure for at least two trimesters.

The belief that the incidence of the overt syndrome is of the order of 5% seems to have been confirmed by a review of 186 studies: among 1325 pregnant women who took anticoagulant drugs, 970 were allocated to warfarin and the incidence of warfarin embryopathy was 4.6% (95). However, in a prospective survey of 72 pregnant women with cardiac valve prostheses, the incidence of coumarin embryopathy was 25% in the 12 pregnancies in which heparin was not substituted for acenocoumarol until after the 7th week, and 30% in the 37 pregnancies in which acenocoumarol was given throughout pregnancy. There were no signs of coumarin embryopathy when acenocoumarol was withdrawn from the sixth to the twelfth weeks of gestation and the women were treated instead with heparin (96).

Oral anticoagulation may cause warfarin embryopathy by inhibiting post-translational carboxylation of proteins needed in the normal ossification process. Intensity of treatment appears to be of importance, since there were no cases of warfarin embryopathy in 44 consecutive children of 42 mothers exposed in the first trimester, but who had prothrombin times prolonged by 40–60% (97,98). Experiments in rats in which highly intensive long-term anticoagulant therapy produced excessive mineralization disorders favor this causal relation (99).

Another possible adverse reaction is the occurrence of central nervous system anomalies in fetuses exposed to vitamin K antagonists at any time during pregnancy. The anomalies may result from fetal intracerebral hemorrhage with scarring. There are two patterns: one consists of dorsal midline dysplasia, expressed as agenesis of the corpus callosum, Dandy-Walker malformation, midline cerebral atrophy, or possible encephalocele; the other consists of ventral midline dysplasia, characterized by optic atrophy. The patients are never completely normal on follow-up, and the resulting personal and social burdens are considerable. According to a review (95), there were central nervous system anomalies in 26 (2.7%) of 970 pregnancies in which warfarin was used. Neonates can also have hemorrhagic complications if a pregnant mother takes an oral anticoagulant near term. The neonatal liver is immature, and concentrations of vitamin K-dependent coagulation factors are low. Although maternal warfarin concentrations are in the therapeutic range, bleeding can occur in neonates. Warfarin in particular should be avoided beyond 36 weeks of gestation (100).

Other consequences of oral anticoagulation during pregnancy are spontaneous abortion, stillbirth, and premature birth (101).

Lactation

In contrast to heparin, coumarins are secreted into the breast milk, but it has long been known that prothrombin activity in the plasma of neonates whose mothers take coumarins is not significantly reduced (102–104) and that warfarin does not have anticoagulant effects in breast-fed infants when given to nursing mothers (103,105). These conclusions are subject to the reservation that in some of the studies the dose of anticoagulant was low (103). Acenocoumarol-treated breastfeeding mothers can as a rule safely breastfeed their infants (106,107); nevertheless, it is prudent to check the infant's prothrombin time in such cases.

There are differences between infants in their sensitivity to coumarins. Some experts therefore recommended weekly oral administration of 1 mg of vitamin K_1 to the child if the mother is taking a coumarin and breastfeeding (108).

Susceptibility Factors

Genetic factors

There is a strong association between CYP2C9 variant alleles and warfarin dosage requirements (109), and the CYP2C9*2 and CYP2C9*3 variant alleles, which are associated with reduced enzyme activity, have been associated with significant reductions in mean warfarin dosage requirements (110). The possession of a variant allele may also be associated with an increased risk of adverse effects (111). The CYP2C9 genotype is also relevant to acenocoumarol (112). Other polymorphisms, such as those that affect CYP3A4 or CYP1A2, may also be important (110). The molecular basis of warfarin resistance is unclear but could be due to unusually high CYP2C9 activity (pharmacokinetic resistance) or to abnormal activity of vitamin K epoxide reductase (pharmacodynamic resistance) (110).

There is also a rare ala-10 mutation in the propeptide of factor IX, which leads to an increased risk of bleeding when starting anticoagulation with coumarin anticoagulants (113,114).

An effect on acenocoumarol dose requirements appears to be absent for the CYP2C9*2 allele and the consequences for phenprocoumon metabolism have not yet been established. In 1124 patients from the Rotterdam Study who took acenocoumarol or phenprocoumon there was a statistically significant difference in the first INR between patients with variant genotypes and those with the wild type (115). Almost all acenocoumarol-treated patients with a variant genotype had a significantly higher mean INR and a higher risk of an INR of 6.0 or over during the first 6 weeks of treatment and there was a clear genotype–dose relation. Individuals with one or more CYP2C9*2 or CYP2C9*3 alleles required a significantly lower dose of acenocoumarol than wild- type patients. In patients taking phenprocoumon there were no significant differences between variant genotypes and the wild type genotype. The authors concluded that phenprocoumon is a clinically useful alternative in patients carrying the CYP2C9*2 and CYP2C9*3 alleles.

Age

Age is generally considered as an important susceptibility factor for bleeding during oral anticoagulant treatment (116). The frequency of bleeding was higher in elderly patients in four of five cohort studies of warfarin-related bleeding, published over a 10-year period (32,36). In one study (33), age was the only significant independent risk factor for subdural hemorrhage, apart from the intensity of anticoagulation, whereas age was only of borderline significance for intracerebral hemorrhage. Various explanations could account for the effect of age: co-medication, co-morbidity, increased sensitivity to warfarin, and impaired vascular integrity. Elderly patients, especially those with moderate hypertension, cerebral thrombosis, or a latent gastrointestinal ulcer, should be supervised closely (117).

Heart failure

In a cohort study of all 1077 patients in an outpatient anticoagulation clinic taking acenocoumarol or phenprocoumon between 1 January 1990 and 1 January 2000, 396 developed an INR of 6.0 or over, and the risk of over-anticoagulation was increased in patients with heart failure (118).

Other features of the patient

Susceptibility factors for bleeding with oral anticoagulants, apart from age, include (32):

- endogenous coagulation defects
- thrombocytopenia
- hypertension
- cerebrovascular disease
- thyroid disease
- renal insufficiency
- liver disease
- tumors
- cerebrovascular disease
- alcoholism
- a history of gastrointestinal bleeding (peptic ulcer disease alone without past bleeding is not associated with an increased risk of bleeding)
- an inability to adhere to the regimen.

Patients with hypothyroidism and hyperthyroidism should all be carefully monitored because the response is greatly reduced in hypothyroidism and greatly enhanced in hyperthyroidism (119). The increased warfarin sensitivity is said to be related to increased degradation of clotting factors. When the thyroid disease is treated, the susceptibility of these patients to coumarins gradually normalizes.

Since vitamin K is one of the vitamins whose absorption is significantly affected by diarrhea, a raised INR and bleeding can occur during episodes of diarrhea (120). Therefore, patients who have diarrhea or reduced food intake should have their INR evaluated more often and their anticoagulant doses adjusted appropriately.

For oral anticoagulant prophylaxis, the overall number of bleeding patients with organic lesions exceeds 50%. Most have prothrombin times within the usual target range. It is therefore clear that all cases of major (or recurrent mild) internal hemorrhagic complications should be investigated for possible underlying organic disease. In one study, patients with hemorrhagic complications, despite a prothrombin time within the target range, were more likely to have an underlying pathological lesion as a direct cause of bleeding than patients whose bleeding occurred when the prothrombin time was above the target range (121).

Drug–Drug Interactions

General

Commonly prescribed drugs can potentiate or antagonize the anticoagulant effect of coumarin drugs in several ways:

- by interfering with the absorption of the drug
- by inhibiting intestinal bacterial production and absorption of vitamin K_2
- by inhibiting the absorption of vitamin K_1 present in food
- by altering the binding of the coumarins to plasma proteins
- by altering the metabolism of the drug by liver microsomes.

Competition for metabolic pathways (for example cytochrome P_{450} isozymes) could lead to an increase in the anticoagulant effect, and/or the pharmacological effects of a drug whose active metabolites are produced by the same metabolic pathway. Hence, any patient taking an oral anticoagulant who has another drug added to or withdrawn from the regimen must have prothrombin time carefully monitored to avoid important changes in the intensity of treatment.

The information on interactions given in Table 1 is mainly based on well-documented prospective studies and, in a few instances, on anecdotal reports. A given interaction may hold for one but not for another coumarin congener.

Alcohol

Alcoholics can drink ethanol in amounts large enough (200 g) to inhibit warfarin clearance as an acute effect.

Table 1 Important drug–drug interactions with coumarin anticoagulants

Drug	*Effect on anticoagulant response*	*Mechanism*	*Comment*
Acetylsalicylic acid and other salicylates	Potentiation (122,123) (at doses over 1.5 g/day) even with topical application (124)	Unknown	Avoid concurrent use: can also impair platelet aggregation and cause peptic erosion and ulceration
Allopurinol	Potentiation (125)	Inhibition of microsomal enzymes (126)	Adjust dosage
Aminoglutethimide	Reduction possible (127)	Unknown	Adjust dosage
Aminoglycoside antibiotics	Potentiation possible (128)	Reduced availability of vitamin K (129)	Monitor INR
Amiodarone	Potentiation (130–133)	Reduced warfarin and acenocoumarol clearances (130,133)	Adjust dosage
Anabolic steroids and androgens (C17-alkylated)	Potentiation (134)	Unknown	Adjust dosage
Antacids containing magnesium	Potentiation possible (135) (only documented for dicoumarol)	Increased absorption (135)	Of no practical importance
Azapropazone	Potentiation (136)	Displacement from binding sites on plasma proteins (137); inhibition of metabolism causes peptic ulceration	Avoid concurrent use
Azathioprine	Reduction (138,139)	Unknown	
Azithromycin	Potentiation (140)	Inhibition of microsomal enzymes possible	Monitor INR or avoid concurrent use
Barbiturates	Reduction (141)	Induction of microsomal enzymes (142)	Adjust dosage
Benziodarone	Potentiation (143)	Unknown	Adjust dosage
Bezafibrate	Potentiation (144,145)	Unknown	Adjust dosage
Carbamazepine	Reduction (146,147)	Induction of microsomal enzymes (147)	Adjust dosage
Carbimazole	Reduction (148)	Decreased catabolism of vitamin K-dependent clotting	Adjust dosage
Cetirizine	Potentiation (149)	Unknown	Anecdotal
Cephalosporins	Potentiation possible (150,151)	Inhibition of hepatic vitamin K metabolism (152)	Monitor INR
Chloral hydrate and related compounds	Minor potentiation during initial phase of anticoagulation therapy, not during maintenance therapy (124,153)	Displacement from binding sites on plasma proteins	Usually of no practical importance
Chloramphenicol	Potentiation (154,155) even with ocular administration (156)	Unknown	Adjust dosage
			Monitor INR
Chlortalidone	Pseudoreduction	Hemoconcentration (157)	No practical importance
Colestyramine	Reduction (158,159)	Reduced absorption and interruption of enterohepatic circulation (158,159)	Adjust dosage: separate dosage of these two drugs by a long time interval
Cimetidine	Potentiation (160); not documented for phenprocoumon (161)	Stereoselective inhibition of warfarin metabolism (162)	Adjust dosage
Ciprofloxacin	Potentiation possible (163,164)	Unknown	Monitor INR
Cisapride	Potentiation (165)	Inhibition of microsomal enzymes possible	Monitor INR
Clarithromycin	Potentiation (166–168)	Inhibition of microsomal enzymes possible	Monitor INR or avoid concurrent use
Clofibrate	Potentiation (169)	Unknown	Adjust dosage
Co-trimoxazole	Potentiation (170,171)	Unknown	Adjust dosage
Cyclophosphamide	Reduction possible (172)	Unknown	Anecdotal
Ciclosporin	Reduction possible (173)	Unknown	Adjust dosage
Danazol	Potentiation (174)		
Danshen	Potentiation (175)	Unknown	Anecdotal
Dicloxacillin	Reduction (176)	Unknown	Monitor INR
Diflunisal	Potentiation possible (177)	Can impair platelet aggregation and cause peptic ulceration	Avoid concurrent use
Disulfiram	Potentiation (178,179)	Inhibition of microsomal enzymes (178)	Adjust dosage
Erythromycin	Potentiation possible (180,181)	Reduced warfarin clearance (182)	Monitor INR

(Continued)

Table 1 (Continued)

Drug	*Effect on anticoagulant response*	*Mechanism*	*Comment*
Etacrynic acid	Potentiation possible (183)	Unknown	Anecdotal
Fenofibrate	Potentiation (184)	Displacement from binding sites on plasma proteins	Adjust dosageMonitor INR
Floctafenine	Minor potentiation (185)	Unknown	Monitor INR
Fluconazole	Potentiation possible (186–188)	Unknown	Monitor INR
Fluorouracil	Potentiation possible (189)	Unknown	Monitor INR
Fluoxetine	Potentiation (190)	Unknown	Adjust dosage or avoid concurrent use
Flurbiprofen	Potentiation possible (191,192)	Unknown	Monitor INR
Fluvastatin	Potentiation possible (193)	Unknown	Monitor INR
Furosemide	Reduction (194)	Hemoconcentration	Monitor INR
Ginseng	Reduction (195)	Unknown	Anecdotal
Glucagon	Potentiation (196)	Unknown	Avoid concurrent use
Glucocorticoids	Ambiguous (197,198)		No practical importance
Glutethimide	Reduction (199)	Induction of microsomal enzymes (199)	Adjust dosage
Griseofulvin	Reduction (200)	Unknown	Adjust dosage
Haloperidol	Reduction possible (201)	Unknown	Anecdotal
Indometacin	Potentiation possible (202)	Unknown	Monitor INR
Isoniazid	Potentiation possible (203)	Unknown	Anecdotal
Itraconazole	Potentiation possible (204)	Unknown	Monitor INR
Ketoconazole	Potentiation possible (205)	Unknown	Monitor INR
Lovastatin	Potentiation possible (206)	Unknown	Monitor INR
Mefenamic acid	Potentiation possible (207)	Fibrinolysis	No clinical importance
Mercaptopurine	Reduction (208)	Unknown	Adjust dosage
Mesalazine	Reduction (209,210)	Unknown	Monitor INR
Metronidazole	Potentiation (211)	Stereoselective inhibition of metabolism (warfarin) (211)	Adjust dosage
Miconazole	Potentiation possible (212–214) even with topical application (213,215,216)	Inhibition of metabolism (217)	Avoid concurrent use or contraindicated
Nafcillin	Reduction possible (218,219)	Unknown	Anecdotal
Nalidixic acid	Potentiation possible (220,221)	Unknown	Anecdotal
Norfloxacin	Potentiation (163)	Unknown	Monitor INR
Nutritional formulations	Reduction (222)	Increased availability of vitamin K (222)	Adjust dosage
Ofloxacin	Potentiation (223)	Unknown	Monitor INR
Omeprazole	Potentiation (224)	Unknown	Monitor INR
Oral contraceptives	Ambiguous (225,226)	Unknown	No practical importance
Paracetamol	Potentiation during long-term administration (227,228)	Unknown	Use paracetamol for limited periodsMonitor INR
Phenylbutazone	Potentiation (229)	Displacement from binding sites on plasma proteins; stereoselective inhibition of metabolism (warfarin); causes peptic ulceration (230,231)	Avoid concurrent use
Phenytoin	Reduction (232); warfarin may give potentiation (233,234)	Unknown	Adjust dosage
Piracetam	Potentiation possible (235)	Unknown	Anecdotal
Piroxicam	Potentiation (236)	Unknown	Monitor INR
Pravastatin	Potentiation (237)	Unknown	Monitor INR
Proguanil	Potentiation (238)	Unknown	Monitor INR
Propafenone	Potentiation (239)	Unknown	Monitor INR
Propranolol	Potentiation (240)	Inhibition of metabolism	Monitor INR
Quinidine	Potentiation possible (241,242)	Pharmacodynamic interaction	Monitor INR
Ranitidine	Possible potentiation at high dosages (243)	Unknown	Monitor INR if high dosage
Rifampicin	Reduction (244)	Induction of microsomal enzymes (245,246)	Adjust dosage
Ritonavir	Reduction (247)	Unknown	Monitor INR
Saquinavir	Potentiation (248)	Inhibition of microsomal enzymes	Monitor INR
Simvastatin	Potentiation (249,250)	Unknown	Adjust dosage Monitor INR

Table 1 (Continued)

Drug	*Effect on anticoagulant response*	*Mechanism*	*Comment*
Spironolactone	Pseudoreduction (251)	Hemoconcentration	No practical importance
Sucralfate	Reduction possible (252,253)		Anecdotal
Sulfinpyrazone	Potentiation possible (254–257) (not documented with phenprocoumon) (256)	Inhibition of microsomal enzymes (258)	Adjust dosage
Sulfonamides	Potentiation possible (259,260)	Unknown	Anecdotal
Sulindac	Potentiation possible (261,262)	Unknown	Monitor INR
Tamoxifen	Potentiation possible (263,264)	Unknown	Adjust dosage
Tetracyclines	Potentiation possible (155,170)	Reduced availability of vitamin K (129)	Monitor INR
Thiouracils	Reduction	Reduced catabolism of vitamin K-dependent clotting factors	Adjust dosage
Thyroid hormones	Potentiation (164,265)	Increased catabolism of vitamin K-dependent clotting factors (148)	Adjust dosage
Tramadol	Potentiation possible (266,267)	Inhibition of microsomal enzymes	Monitor INR
Ubidecarone	Reduction (268)	Unknown	Adjust dosage
Vitamin E	Potentiation possible (269)	Unknown	Anecdotal
Zafirlukast	Potentiation possible (270)	Unknown	Monitor INR
Zileuton (271)	Potentiation	Unknown	Adjust dosageMonitor INR

However, daily consumption of ethanol for a period of 3 weeks in the form of wine taken with meals in moderate amounts (28 g/day) or even in liberal amounts (56 g/day) had no effect on the hypoprothrombinemia induced by warfarin (272).

Although long-term alcohol can induce some cytochrome P_{450} isozymes, such as CYP1A1, CYP1A2 (273), CYP2B1, and CYP2E1 (274), these effects are not relevant to the coumarins, which are mostly metabolized by CYP2C9.

A report of ethanol potentiation of aspirin-induced prolongation of bleeding time (275) is of importance, in view of a putative increase, due to alcohol intake, of the hemorrhagic risk accompanying outpatient oral anticoagulation.

Antiplatelet drugs

The impact on major bleeding rates of antiplatelet drug therapy among warfarin users has been examined in a retrospective cohort analysis of 10 093 patients with atrial fibrillation (mean age 77 years), of whom 19% took antiplatelet drugs (276). Antiplatelet drugs significantly increased the rates of major bleeding from 1.3% to 1.9%. In a multivariate analysis, the factors associated with bleeding events included anemia (OR = 2.52; 95% CI = 1.64, 3.88), a history of bleeding (OR = 2.40; 95% CI = 1.71, 3.38), and concurrent antiplatelet drug therapy (OR = 1.53; 95% CI = 1.05, 2.22). Although concerns about increased bleeding risk when warfarin and antiplatelet drugs are combined are not unfounded, the risk of bleeding is only increased by 50% and the decision to use concurrent antiplatelet drugs will be tempered by cardiac and bleeding risk factors.

Carnitine

There was a rise in INR when a 33-year-old man taking acenocoumarol took carnitine 100 mg/day in an over-the-counter product recommended to complement body-building (277). No mechanism was suggested.

Co-amoxiclav

In a case-control study in 300 previously stable outpatients taking phenprocoumon or acenocoumarol with INR values of 6.0 or more and 302 matched controls with INR values within the target range, potentially interacting drugs were predominantly antibacterial drugs (278). A course of co-amoxiclav was associated with a small increase in risk (OR = 2.4; 95% CI = 1.0, 5.5). Of 87 potentially interacting drugs, 45 were not used during the 4-week study and only 15 drugs were used by at least 10 patients. The authors concluded that unless no therapeutic alternative is available, co-amoxiclav should be avoided in patients taking coumarins. If no therapeutic alternative is available, increased monitoring of INR values is warranted to prevent overanticoagulation and potential bleeding complications.

Co-trimoxazole

In a case-control study in 300 previously stable outpatients taking phenprocoumon or acenocoumarol, with INR values of 6 or more, and 302 matched controls with INR values within the target range, potentially interacting drugs were predominantly antibacterial drugs (278). A course of co-trimoxazole was associated with a large increase in the risk of excess anticoagulation (adjusted OR = 24; 95% CI = 2.8, 209). Of 87 potentially interacting drugs 45 were not used during the 4-week study and only 15 drugs were used by at least 10 patients. The authors concluded that unless no therapeutic alternative is available, co-trimoxazole should be avoided in patients taking coumarins. If no therapeutic alternative is available, increased monitoring of INR values is warranted to prevent overanticoagulation and potential bleeding complications.

Dicloxacillin

An interaction with dicloxacillin resulted in increased warfarin dosage requirements for 2 weeks after dicloxacillin was withdrawn (279). The mechanism is probably non-stereoselective induction of warfarin metabolism by dicloxacillin (280,281).

Fluoroquinolones

A series of cases of coagulopathy due to an interaction of warfarin with ciprofloxacin has been extracted from the FDA's Spontaneous Reporting System database, including all cases reported from 1987 to 1997, combined with two of the author's own cases (282). Soon after the introduction of ciprofloxacin in 1987, hemorrhagic events from hypoprothrombinemia were reported in patients taking warfarin. However, several prospective studies of combining ciprofloxacin and other fluoroquinolones with warfarin failed to show a significant change in INR (Table 2). Although it is impossible to estimate the frequency of the ciprofloxacin/warfarin interaction from these or other descriptive data, it can be concluded that potentiation of anticoagulant effect by ciprofloxacin occurs most often in older patients and those taking multiple medications.

Herbal medicines

In a meta-analysis of interactions of warfarin with other drugs, herbal medicines, Chinese herbal drugs, and foods 642 citations were retrieved, of which 181 eligible articles contained original reports on 120 drugs or foods (290). Of all the reports, 72% described potentiation of the effect of warfarin, and the authors considered that 84% were of poor quality, 86% of which were single case reports. The 31 incidents of clinically significant bleeding were all single case reports. Relatively few anecdotal reports of adverse event–drug associations are followed up with formal studies (291), and reports of interactions of warfarin with herbal medicines are no exception—most are based on anecdotal reports.

Drug interactions of warfarin with herbal preparations have been reviewed (292,293). In a systematic review, warfarin was the most common cardiovascular drug involved in interactions with herbal medicines (294). Medicines that resulted in increased anticoagulation include *Allium sativum* (garlic), *Angelica sinensis* (dong quai), *Carica papaya* (papaya), curbicin (from *Cucurbita pepo* seed and *Serenoa repens* fruit), *Ginkgo biloba* (maidenhair), *Harpagophytum procumbens* (devil's claw), *Lycium barbarum* (Chinese wolfberry), *Mangifera indica* (mango) (295), *Peumus boldus* (boldo) (296), *Salvia miltiorrhiza* (danshen), *Trigonella foenum graecum* (fenugreek), and PC-SPES (a patented combination of eight herbs). Medicines that resulted in reduced anticoagulation include *Camellia sinensis* (green tea), milk prepared from *Hypericum perforatum* (St. John's wort), and *Panax ginseng* (ginseng).

In a retrospective analysis of the pharmaceutical care plans of 631 patients, 170 (27%) were taking some form of complementary or alternative medicine and 99 were using a medicine that could interact with warfarin, the commonest being cod-liver oil and garlic (297).

Allium sativum (garlic)

The interaction of garlic with warfarin has been reviewed (298). Certain organosulfur components inhibit human platelet aggregation in vitro and in vivo; some garlic components have an anticoagulant effect and might thus enhance the effect of warfarin. However, there is only anecdotal evidence that this occurs. Two case reports have suggested that the combination of warfarin with garlic extract prolonged the clotting time and increased the international normalized ratio (INR). There have also been reports that garlic can cause postoperative bleeding and spontaneous spinal epidural hematoma. Garlic should be withdrawn 4–8 weeks before an operation or in those taking long-term warfarin.

Angelica sinesis (dong quai)

Although Angelica sinesis is a commonly used herbal medicine, there are no clinical data on drug interactions except for one report of a 46-year-old African–American woman with atrial fibrillation stabilized on warfarin who

Table 2 Prospective interaction studies of fluoroquinolones with warfarin

Study design	*N*	*Source of subjects*	*Drug*	*Coagulation effect*	*Reference*
Single arm	9	Warfarin clinic	Ciprofloxacin	None	(283)
Single arm	16	Warfarin clinic	Ciprofloxacin	*R*-warfarin	(284)
Placebo-controlled[a]	36	Warfarin clinic	Ciprofloxacin	*R*-warfarin	(285)
Randomized crossover[b]	6	Healthy volunteers	Enoxacin	*R*-warfarin	(286)
Randomized crossover[c]	10	Healthy volunteers	Ofloxacin	None	(287)
Single arm	7	Healthy volunteers	Ofloxacin	None	(288)
Single arm	10	Healthy volunteers	Temafloxacin	None	(289)

[a]Randomized, double-blind, placebo-controlled study
[b]Randomized, double-blind, placebo-controlled, multicenter study
[c]Randomized, two-way, crossover study

had a greater than two-fold increase in prothrombin time and INR after taking Angelica sinesis for 4 weeks (299). Angelica sinesis extract and its active ingredient, ferulic acid, inhibit rat platelet aggregation in vivo. In rabbits oral administration Angelica sinesis root extract (2 g/kg bd) significantly reduced the prothrombin time when combined with warfarin (2 mg/kg), while the pharmacokinetics of warfarin were not altered (300). However, in rats an aqueous extract of Angelica sinesis increased the activities of CYP2D6 and CYP3A (301), and in in vitro studies components from Angelica sinesis root altered CYP3A4 and CYP1A activity, indicating a potential for drug interactions with CYP substrates. For example, a decoction or infusion of Angelica sinesis root inhibited CYP3A4-catalysed testosterone 6-beta-hydroxylation in human liver microsomes, whereas ferulic acid (0.5 μmol/l) from Angelica sinesis root significantly inhibited ethoxyresorufin O-methylase (CYP1A) activity. All of these findings suggest that precautionary advice should be given to patients who self-medicate with Angelica sinensis root preparations while taking long-term warfarin. Well-designed case-control studies are needed to evaluate these effects of Angelica sinesis root.

Camellia sinensis (green tea)

Camellia sinensis has been anecdotally reported to reduce the effect of warfarin (302).

- A 44-year-old white man taking warfarin had an INR of 3.8. He then drank green tea 0.5–1 gallon/day (4.5 l/day) for about 1 week and the INR fell to 1.37. He stopped drinking green tea and the INR rose to 2.55.

Green tea is a source of vitamin K. Dry green leaves contain 1428 micrograms of vitamin K per 100 g of leaves compared with only 262 micrograms per 100 g of dry black tea leaves (303). The amount of vitamin K ingested will obviously depend on the dilution and amount of tea leaves used to brew the tea and the quantity of tea consumed.

Cucurbita pepo

Curbicin has been anecdotally reported to cause altered coagulation in the absence of anticoagulant therapy and to enhance the anticoagulant action of warfarin; however, the authors attributed this effect to the vitamin E that was also present in the curbicin tablets (304).

Ginkgo biloba

There are anecdotal reports of possible interactions of ginkgo with warfarin (305). However, formal, albeit small, studies in patients and healthy volunteers have not confirmed this. In an open, crossover, randomized study, 12 healthy men took a single dose of warfarin 25 mg either alone or after pretreatment with Ginkgo biloba for 7 days; ginkgo did not significantly affect clotting or the pharmacokinetics or pharmacodynamics of warfarin (306). In a randomized, double-blind, placebo-controlled, crossover study, oral ginkgo extract 100 mg/day for 4 weeks did not alter the INR in 24 Danish outpatients (14 women and 10 men) taking stable, long-term warfarin, and the geometric mean dosage of warfarin did not change (307).

The mechanism for this interaction, if it occurs, is unknown, but both pharmacokinetic and pharmacodynamic mechanisms may be involved, given that ginkgo extracts can modulate various CYP isoenzymes and exert antiplatelet activity. Ginkgolides are also potent inhibitors of platelet-activating factor (308). There are reports of postoperative bleeding and spontaneous hemorrhage attributed to consumption of gingko (309,310) and interactions have been described with antiplatelet drugs. For example, spontaneous hyphema occurred when ginkgo extract was combined with aspirin (acetylsalicylic acid) (311) and fatal intracerebral bleeding was associated with the combined use of ginkgo extract and ibuprofen (312). Ginkgo extract also enhanced the antiplatelet and antithrombotic effects of ticlopidine in rats, resulting in prolongation of the bleeding time by 150% (313). However, in a double-blind, randomized, placebo-controlled study in 32 young healthy men oral ginkgo extract 120, 240, or 480 mg/day for 14 days did not alter platelet function or coagulation (314). Bleeding attributed to ginkgo often occurs in elderly or postoperative patients who may have had impaired platelet function before the use of ginkgo.

Hypericxum perforatum (St John's wort)

An interaction of St John's wort with warfarin has been reported anecdotally, including 22 spontaneous reports of reduced warfarin effect after treatment with St John's wort submitted to regulatory authorities in Europe between 1998 and 2000 (315). These interactions all resulted in unstable INR values, a reduction in INR being the most common effect. Although no thromboembolic episodes occurred, the reduction in anticoagulant activity was considered clinically significant. Anticoagulant activity was restored when St John's wort was withdrawn or the warfarin dose was increased.

In a crossover study, healthy volunteers who took hypericum extract LI 160, 900 mg/day for 11 days before a single dose of phenprocoumon had a lower AUC of the unbound fraction than when they took placebo (316).

In an open, three-way, crossover, randomized study in 12 healthy men who took a single dose of warfarin 25 mg alone or after pretreatment for 14 days with St John's wort, the apparent clearance of S-warfarin was 3.3 ml/minute before St John's wort was added and 3.7 ml/minute after (317). The respective apparent clearances of R-warfarin were 1.8 and 2.4 ml/minute. The mean ratios of the apparent clearances were 1.29 (95%CI = 1.16, 1.46) for S-warfarin and 1.23 (1.11, 1.37) for R-warfarin. St John's wort did not affect the apparent volume of distribution or protein binding of either enantiomer of warfarin. The authors concluded that St John's wort induces the clearance of both enantiomers of warfarin. INR was slightly reduced as a result, but platelet aggregation was not altered.

These observations suggest that St John's wort increases the clearance of both warfarin and phenprocoumon, possibly because of induction of CYP isozymes, particularly CYP2C9 and CYP3A4.

***Lycium barbarum* (Chinese wolfberry)**

There has been a single anecdotal report of a possible interaction of *Lycium barbarum* with warfarin (318).

- A 61-year-old Chinese woman, previously stabilized on warfarin (INR 2–3), drank a concentrated Chinese herbal tea made from *Lycium barbarum* fruits (3–4 glasses/day) for 4 days; her INR rose to 4.1. Warfarin was withheld for 1 day and then restarted at a lower dose. She stopped drinking the tea, and 7 days later her INR was 2.4.

In vitro studies showed that *Lycium barbarum* tea inhibited S-warfarin metabolism by CYP2C9; however, the inhibition was weak, with a dissociation constant of 3.4 g/l, suggesting that the observed interaction may have been caused by other mechanisms.

Panax ginseng

Ginseng can reduce the effect of warfarin (319), and there have been anecdotal reports of such an interaction (320,321). There have also been several formal studies of the pharmacokinetic and pharmacodynamic effects of ginseng on warfarin.

The effects of American ginseng (*Panax quinquefolium*) have been studied in a double-blind, randomized, placebo-controlled trial in 20 young healthy subjects (322). Warfarin was given for 3 days during weeks 1 and 4, and starting in week 2 the subjects were assigned to either ginseng or placebo. The peak INR fell significantly after 2 weeks of ginseng administration compared with placebo; the difference between ginseng and placebo was –0.19 (95% CI = –0.36, –0.07).

However, in an open, three-way, crossover, randomized study in 12 healthy men who took a single dose of warfarin 25 mg alone or after pretreatment for 7 days with ginseng, there was no change in the pharmacokinetics or pharmacodynamics of either S-warfarin or R-warfarin (323). If the mechanism of this interaction is induction of warfarin metabolism by ginseng, the discrepancy between these two studies could be explained by the difference in duration.

In 20 healthy volunteers who took 100 mg of an extract of *Panax ginseng* standardized to 4% ginsenosides twice daily for 14 days there were no effects on CYP3A and this result was confirmed in in vitro studies (324).

The discrepancies between these studies could be explained by differing susceptibilities of different populations or by different effects of ginseng from different sources.

Both pharmacokinetic and pharmacodynamic components could play a role in an interaction of ginseng with warfarin. Ginseng extracts have an antiplatelet effect. Ginsenosides Rg3 and protopanaxadiol-type saponins were platelet-activating factor antagonists with IC_{50} values of 49–92 μmol/l (325). Modulation of various CYP isoenzymes could also be a mechanism. In rats the pharmacokinetics and pharmacodynamics of warfarin after a single dose and at steady state were not altered by co-administered ginseng (326). However, extensive in vitro and in vivo animal studies have shown that constituents of ginseng can modulate various CYP isoenzymes that metabolize warfarin. Ginsenoside Rd was weakly inhibitory against recombinant CYP3A4, CYP2D6, CYP2C19, and CYP2C9, whereas ginsenoside Re and ginsenoside Rf (200 μmol/l) increased the activity of CYP2C9 and CYP3A4 (327). In rats, the standardized saponin of red ginseng was inhibitory on p-nitrophenol hydroxylase (CYP2E1) activity in a dose-related manner (328).

***Salvia miltiorrhiza* (danshen)**

There have been anecdotal reports of enhanced anticoagulation and bleeding when patients taking long-term warfarin therapy consumed *Salvia miltiorrhiza* root (159,329,330). As these patients were also taking other medications, the contribution of *Salvia miltiorrhiza* to the interaction was difficult to determine. However, the author of a systematic review concluded that danshen should be avoided in patients taking warfarin (331).

The direct anticoagulant activity of *Salvia miltiorrhiza* root itself may provide a partial explanation for the interactions. However, pharmacokinetic interactions may also play a role. Warfarin is mainly metabolized by CYP2C9 and to a smaller extent by CYP1A2 and CYP3A4. In mice oral, administration of an ethyl acetate extract of danshen caused a dose-related increase in liver microsomal 7-methoxyresorufin O-demethylation activity, with a three-fold increase in warfarin 7-hydroxylation (332). However, the aqueous extract had no effects. Immunoblot analysis of microsomal proteins showed that ethyl acetate extraction increased the proteins associated with CYP1A and CYP3A. At a dose corresponding to its content in the ethyl acetate extract, tanshinone IIA, the main diterpene quinone in *Salvia miltiorrhiza*, increased mouse liver microsomal 7-methoxyresorufin O-demethylation activity. These results suggest that there are inducing agents for mouse CYP1A, CYP2C, and CYP3A in ethyl acetate extracts but not in aqueous extracts of *Salvia miltiorrhiza*.

In rats treatment with *Salvia miltiorrhiza* root extract 5 g/kg bd for 3 days followed by a single oral dose of racemic warfarin increased the absorption rate constants, AUC, C_{max}, and half-life of warfarin but reduced the clearances and apparent volumes of distribution of both (R)-warfarin and (S)-warfarin (333). A similar effect was observed during steady-state warfarin administration. The anticoagulant effect of warfarin was also potentiated. *Salvia miltiorrhiza* root extract itself had no effect on prothrombin time at this dose, suggesting that altered warfarin metabolism was a possible mechanism.

After a single oral dose of racemic warfarin 2 mg/kg in rats, an oral extract of *Salvia miltiorrhiza* 5 g/kg bd for 3 days significantly altered the pharmacokinetics of both R-warfarin and S-warfarin and increased the plasma concentrations of both enantiomers over a period of 24 hours and the prothrombin time over 2 days (334). Steady-state concentrations of racemic warfarin during administration of 0.2 mg/kg/day for 5 days with extract of *Salvia miltiorrhiza* 5 g/kg bd for 3 days not only prolonged the prothrombin time but also increased the steady-state plasma concentrations of R-warfarin and S-warfarin. These results suggested that *Salvia miltiorrhiza* increases the absorption rate, exposure, and half-lives of both R-warfarin and S-warfarin, but reduces their clearances and apparent volumes of distribution.

In addition, *Salvia miltiorrhiza* root extract might change the plasma protein binding of warfarin. Both (R)-warfarin and (S)-warfarin bind to the so-called site I of albumin with high affinity. *Salvia miltiorrhiza* root extract was 50–70% bound by albumin and in vitro Salvia miltiorrhiza root extract displaced salicylate from protein binding, thereby increasing the unbound salicylate concentration (335). However, kangen-karyu, a mixture of six herbs (peony root, *Cnidium* rhizome, safflower, *Cyperus* rhizome, *Saussurea* root, and root of *Salvia miltiorrhiza*), significantly increased the plasma warfarin concentration and prothrombin time in rats, but did not alter the serum protein binding of warfarin (336). Further studies are required to explore the effects of Salvia miltiorrhiza root extract on the metabolism and plasma protein binding of drugs such as warfarin in humans.

Zingiber officinale (ginger)

Despite anecdotal reports of a possible interaction (337,338), several studies in rats and humans have shown no effect of ginger on warfarin pharmacokinetics or pharmacodynamics (Vaes 1478,Engelson 1868,339,340).

Herbal mixtures

When mixtures of herbs are used, as is common practice in the far East, it is not possible to be sure which component was responsible for a reported interaction.

PC-Spes

PC-Spes is a mixture of eight herbs: *Chrysanthemum morifolium*, *Isatis indigotica*, *Glycyrrhiza glabra* (licorice), *Ganoderma lucidum*, *Panax pseudoginseng*, *Robdosia rubescens*, *Serenoa repens* (saw palmetto), and *Scutellaria baicalensis* (skullcap). It has been reported to increase the INR in a 79-year-old man with prostate cancer taking warfarin, an effect that was attributed to inhibition of warfarin metabolism (341). However, warfarin has also been found in formulations of PC-Spes (342).

Quilinggao

Quilinggao, a popular Chinese mixture that contains a multitude of herbal ingredients (including *Fritillaria cirrhosa* and other *Fritillaria* species, *Paeoniae rubra*, *Lonicera japonica*, and *Poncirus trifoliata*, in many different brands), has been anecdotally reported to enhance the actions of warfarin (343).

- A 61-year-old man taking stable warfarin therapy developed gum bleeding, epistaxis, and skin bruising 5 days after taking quilinggao. His international normalized ratio was above 6. His warfarin was withdrawn and the international normalized ratio normalized. Days later he tried taking quilinggao again, with a similar result.

The authors pointed out that several herbs in this mixture have anticoagulant activity.

Ketolide antibiotics

There has been a report that telithromycin increases the effect of warfarin (344). The mechanism is unknown, but the authors suggested that it resulted from inhibition of the metabolism of the R-isomer of warfarin, which is metabolized predominantly by CYP1A2 and less by CYP3A4.

Laxatives

In a large population-based cohort study in 1124 patients, 351 developed an International Normalized Ratio of 6.0 or more (345). The only laxative with a moderate but significantly increased relative risk of overanticoagulation was lactulose (relative risk 3.4; 95% CI = 2.2, 5.3).

Macrolide antibiotics

There has been a report that azithromycin increases the effect of warfarin (346). Azithromycin, unlike erythromycin and clarithromycin, is not known to inhibit cytochrome P450 isozymes and is presumed to be the macrolide of choice in patients already taking warfarin. However, reports of azithromycin–warfarin interactions support the possibility that azithromycin does interact with warfarin, although the exact mechanism is not understood.

Methylprednisolone

Potentiation of the effects of vitamin K antagonists by high-dose intravenous methylprednisolone has been prospectively studied in 10 consecutive patients and 5 controls after the observation of a sharp increase in the International Normalized Ratio (INR) in a patient taking oral anticoagulation after concomitant administration of methylprednisolone (1 g/day for 3 days) (347). The mean INR was 2.8 (range 2.0–3.8) at baseline and increased to 8.0 (5.3–20). The maximum increase in INR occurred after a mean of 93 (29–156) hours. The coumarins taken by these patients were fluindione in eight and acenocoumarol in two. The prothrombin time in the controls remained stable. A similar rise in INR was described in two patients with multiple sclerosis taking warfarin (348).

The mechanism of the interaction of methylprednisolone with oral anticoagulants has not been elucidated, but it might be due to inhibition of anticoagulant catabolism. In the three patients in whom the authors assayed fluindione concentrations, these increased in parallel with the INR. The administration of high-dose methylprednisolone to patients taking oral anticoagulants is not rare; daily monitoring of INR or even reducing the anticoagulant dose before administering methylprednisolone is advised.

Miconazole

Interactions between coumarins and miconazole have been reported, although the mechanism of the interaction is not known. There have been two separate reports of potentiation of the anticoagulant effect of acenocoumarol by non-systemic administration of miconazole. Three patients taking long-term acenocoumarol who used miconazole gel for oral candidiasis had significant rises in INR (213). In two other patients vaginal miconazole potentiated the anticoagulant effect of acenocoumarol (214).

Non-steroidal anti-inflammatory drugs (NSAIDs)

NSAIDs are reported to increase the risk of bleeding in coumarin users. The mechanism underlying this risk is inhibition of platelet aggregation, but in some cases a pharmacokinetic mechanism, resulting in an increased International Normalized Ratio (INR), has been proposed. In a retrospective cohort study in 112 out-patients stabilized on acenocoumarol 52 had an increase in INR above the target value after they started to take diclofenac, naproxen, or ibuprofen (349). In 12 patients the INR increased above 6.0. The INR in the other 60 patients remained constant. There were no statistically significant differences between patients with an increased INR and patients without with regard to age, sex, target range, and average dose of acenocoumarol. Genotyping was performed in 80 patients, of whom 36 had an increased INR as a result of an interaction with an NSAID; there no association between the CYP2C9 genotype and an increased INR.

Over 1 year all patients in an anticoagulant clinic who reported bleeding were sent a questionnaire about NSAID use (350). The local pharmacists detected patients with concomitant coumarin and NSAID prescriptions (but no bleeding). In 681 coumarin users there were 738 hemorrhages, in 12% of which an NSAID had been involved. In contrast, in the whole population of coumarin users, 2.5% were taking an NSAID. Therefore, the relative risk of NSAID use with regard to bleeding complications was 5.8 (95% CI = 2.3, 13.6). Only 7% of bleeding was gastrointestinal; 35% were hematomas and 29% epistaxis. The concomitant use of NSAIDs and coumarin derivatives should be avoided if possible. Choice of long-term analgesia in patients taking warfarin is restricted, and theoretically COX2-selective NSAIDs might be preferred if a NSAID is required.

The COX-2 selective non-steroidal anti-inflammatory drug celecoxib did not alter the pharmacokinetics or the hypoprothrombinemic effect of warfarin in 24 healthy subjects (33). However, there have been at least two reports of increased INR in patients taking stable warfarin therapy and celecoxib (34,116).

Another COX-2-selective non-steroidal anti-inflammatory drug, rofecoxib, increased plasma concentrations of the biologically less active isomer, *R*(+) warfarin, which accounted for an approximately 8% increase in INR at steady state in healthy volunteers (36).

Standard monitoring of INR in patients taking warfarin should be performed when therapy with celecoxib or rofecoxib is begun or changed.

Paracetamol

In a prospective, case-control study, designed to determine causes of INRs over 6.0 in an outpatient anticoagulant unit, there was a clear dose-dependent association between the use of paracetamol (acetaminophen) and having an INR greater than 6.0 (228). The authors studied 93 patients with INRs over 6.0 (cases) and 196 patients with INRs of 1.7–3.3 (controls) during warfarin therapy. The likelihood of an INR greater than 6.0 increased from an odds ratio of 3.5 for doses of 2275–4549 mg per week, to 6.9 for doses of 4550–9099 mg per week, to a 10-fold increase at a dose of over 9100 mg per week.

Paracetamol also potentiates the effects of acenocoumarol (227) and although a study with phenprocoumon showed no interaction (351), the study was too small and not well designed to detect an effect of the sort that has been shown in case-control studies.

Whether paracetamol increases the effect of warfarin has been controversial. The authors of a literature review concluded that this was still an unanswered question and recommended that the INR be closely monitored if paracetamol is used (352).

Ribavirin

Inhibition of the effect of warfarin by ribavirin has been described.

- In a 61-year-old man, who was taking long-term warfarin after a heart valve replacement, the dose of warfarin had to be increased by 40% after the introduction of ribavirin for chronic hepatitis C; the inhibition of the effect of warfarin was reproduced on rechallenge (353).

The mechanism of this effect was not clear.

Ritonavir

Inhibition of the effect of acenocoumarol by ritonavir has been described.

- A 46-year-old man with two prosthetic heart valves taking acenocoumarol started to take ritonavir 600 mg/day. Despite a progressive increase in the dose of acenocoumarol to three times the original dose, it was impossible to achieve a therapeutic INR, and the ritonavir was withdrawn (354).

The authors proposed that ritonavir had induced CYP1A2, CYP1A4, and CYP2C9/19 activity, leading to increased metabolism, at least of acenocoumarol. However, this effect was the opposite of what was expected, since ritonavir is a potent inhibitor of most hepatic isoenzymes.

Salicylates

Salicylates can interact with warfarin in different ways. They displace warfarin from albumin binding sites, and this can result in a transient increase in anticoagulant activity, which is usually clinically unimportant. They can also cause peptic ulceration and reduce platelet aggregation, making bleeding more likely and slower to stop when it occurs.

There was a rise in INR from 2.8 to 12 when a topical pain-relieving gel containing methylsalicylate was applied to the knees of a 22-year-old white woman taking stable warfarin anticoagulation (355).

Other similar cases have been reported (356).

Food–Drug Interactions

Cranberry juice

Cranberry juice has been reported to enhance the action of warfarin (357).

- After a chest infection, a man in his 70s had a poor appetite for 2 weeks and ate next to nothing, taking only cranberry juice as well as his regular drugs (digoxin, phenytoin, and warfarin) (358). Six weeks later his international normalised ratio was over 50, having previously been stable. He died of gastrointestinal and pericardial haemorrhage. He had not taken any over-the-counter preparations or herbal medicines, and he had been taking his drugs correctly.

The Committee on Safety of Medicines has received several other reports through the yellow card reporting scheme about a possible interaction between warfarin and cranberry juice. In one case, this effect was suggested to have been due to contamination with salicylic acid, which displaces warfarin from protein binding sites (359).

Smoking

Cigarette smoking altered the clearance and apparent volume of distribution of warfarin, although the net effect on anticoagulant activity was negligible both in volunteers (360) and patients (361).

Monitoring Therapy

There are no clear guidelines about how often the INR should be measured in patients taking coumarins. However, it should be done quite frequently at the start of therapy when the risk of bleeding is higher (35). Monitoring anticoagulant therapy is greatly improved by the use of anticoagulation clinics or anticoagulation management services in which the management is conducted by registered nurses, pharmacists, or physicians using dosage adjustment protocols developed by experts in the field (362).

References

1. Duxbury BM, Poller L. The oral anticoagulant saga: past, present, and future. Clin Appl Thromb Hemost 2001;7(4):269–75.
2. Griffin JH, Evatt B, Zimmerman TS, Kleiss AJ, Wideman C. Deficiency of protein C in congenital thrombotic disease. J Clin Invest 1981;68(5):1370–3.
3. Marlar RA, Kleiss AJ, Griffin JH. Mechanism of action of human activated protein C, a thrombin-dependent anticoagulant enzyme. Blood 1982;59(5):1067–72.
4. Stein PD, Riddle JM, Kemp SR, Lee MW, Lewis JW, Magilligan DJ Jr. Effect of warfarin on calcification of spontaneously degenerated porcine bioprosthetic valves. J Thorac Cardiovasc Surg 1985;90(1):119–25.
5. Hilgard P, Schulte H, Wetzig G, Schmitt G, Schmidt CG. Oral anticoagulation in the treatment of a spontaneously metastasising murine tumour (3LL). Br J Cancer 1977;35(1):78–86.
6. Thornes RD. Adjuvant therapy of cancer via the cellular immune mechanism or fibrin by induced fibrinolysis and oral anticoagulants. Cancer 1975;35(1):91–7.
7. Hoover HC Jr, Ketcham AS, Millar RC, Gralnick HR. Osteosarcoma: improved survival with anticoagulation and amputation. Cancer 1978;41(6):2475–80.
8. Zacharski LR, Henderson WG, Rickles FR, Forman WB, Cornell CJ Jr, Forcier RJ, Edwards R, Headley E, Kim SH, O'Donnell JR, O'Dell R, Tornyos K, Kwaan HC. Effect of warfarin on survival in small cell carcinoma of the lung. Veterans Administration Study No. 75. JAMA 1981;245(8):831–5.
9. Zacharski LR, Henderson WG, Rickles FR, Forman WB, Cornell CJ Jr, Forcier RJ, Edwards RL, Headley E, Kim SH, O'Donnell JF, et al. Effect of warfarin anticoagulation on survival in carcinoma of the lung, colon, head and neck, and prostate. Final report of VA Cooperative Study #75. Cancer 1984;53(10):2046–52.
10. Annegers JF, Zacharski LR. Cancer morbidity and mortality in previously anticoagulated patients. Thromb Res 1980;18(3–4):399–403.
11. Hilgard P, Maat B. Mechanism of lung tumour colony reduction caused by coumarin anticoagulation. Eur J Cancer 1979;15(2):183–7.
12. Lione A, Bosmann HB. The inhibitory effect of heparin and warfarin treatments on the intravascular survival of B16 melanoma cells in syngeneic C57 mice. Cell Biol Int Rep 1978;2(1):81–6.
13. Watt BE, Proudfoot AT, Bradberry SM, Vale JA. Anticoagulant rodenticides. Toxicol Rev 2005;24(4): 259–69.
14. Poovalingam V, Kenoyer DG, Mahomed R, Rapiti N, Bassa F, Govender P. Superwarfarin poisoning—a report of 4 cases. S Afr Med J 2002;92(11):874–6.
15. Sharma P, Bentley P. Of rats and men: superwarfarin toxicity. Lancet 2005;365(9459):552–3.
16. Tsutaoka BT, Miller M, Fung SM, Patel MM, Olson KR. Superwarfarin and glass ingestion with prolonged coagulopathy requiring high-dose vitamin K1 therapy. Pharmacotherapy 2003;23(9):1186–9.
17. Sarin S, Mukhtar H, Mirza MA. Prolonged coagulopathy related to superwarfarin overdose. Ann Intern Med 2005;142(2):156.
18. Rodrigo Casanova P, Rodríguez Fernández V, García Peña JM, Aguilera Celorrio L. Intento autolítico con superwarfarinas. [Attempted suicide with superwarfarin.] Rev Esp Anestesiol Reanim 2005;52(8):506–7.
19. Shepherd G, Klein-Schwartz W, Anderson BD. Acute, unintentional pediatric brodifacoum ingestions. Pediatr Emerg Care 2002;18(3):174–8.
20. Mullins ME, Brands CL, Daya MR. Unintentional pediatric superwarfarin exposures: do we really need a prothrombin time? Pediatrics 2000;105(2):402–4.
21. Ingels M, Lai C, Tai W, Manning BH, Rangan C, Williams SR, Manoguerra AS, Albertson T, Clark RF. A

prospective study of acute, unintentional, pediatric superwarfarin ingestions managed without decontamination. Ann Emerg Med 2002;40(1):73–8.
22. Zupancic-Salek S, Kovacevic-Metelko J, Radman I. Successful reversal of anticoagulant effect of superwarfarin poisoning with recombinant activated factor VII. Blood Coagul Fibrinolysis 2005;16(4):239–44.
23. Tiaden JD, Wenzel E, Berthold HK, Muller-Oerlinghausen B. Adverse reactions to anticoagulants and to antiplatelet drugs recorded by the German spontaneous reporting system. Semin Thromb Hemost 2005;31(4):371–80.
24. Akle CA, Joiner CL. Purple toe syndrome. J R Soc Med 1981;74(3):219.
25. Feder W, Auerbach R. "Purple toes": an uncommon sequela of oral coumarin drug therapy. Ann Intern Med 1961;55:911–7.
26. Burton JL, Pennock P. Anticoagulants and "feeling cold". Lancet 1979;1(8116):608.
27. Bruns FJ, Segel DP, Adler S. Control of cholesterol embolization by discontinuation of anticoagulant therapy. Am J Med Sci 1978;275(1):105–8.
28. Walters MB. The relationship between thyroid function and anticoagulant thrapy. Am J Cardiol 1963;11:112–4.
29. Christensen F. Uricosuric effect of dicoumarol. Acta Med Scand 1964;175:461–8.
30. DiMarco JP, Flaker G, Waldo AL, Corley SD, Greene HL, Safford RE, Rosenfeld LE, Mitrani G, Nemeth M. AFFIRM Investigators. Factors affecting bleeding risk during anticoagulant therapy in patients with atrial fibrillation: observations from the Atrial Fibrillation Follow-up Investigation of Rhythm Management (AFFIRM) study. Am Heart J 2005;149(4):650–6.
31. Landefeld CS, Beyth RJ. Anticoagulant-related bleeding: clinical epidemiology, prediction, and prevention. Am J Med 1993;95(3):315–28.
32. Levine MN, Raskob G, Landefeld S, Kearon C. Hemorrhagic complications of anticoagulant treatment. Chest 1998;114(5 Suppl):S511–23.
33. Hylek EM, Singer DE. Risk factors for intracranial hemorrhage in outpatients taking warfarin. Ann Intern Med 1994;120(11):897–902.
34. Hylek EM, Skates SJ, Sheehan MA, Singer DE. An analysis of the lowest effective intensity of prophylactic anticoagulation for patients with nonrheumatic atrial fibrillation. N Engl J Med 1996;335(8):540–6.
35. Landefeld CS, Goldman L. Major bleeding in outpatients treated with warfarin: incidence and prediction by factors known at the start of outpatient therapy. Am J Med 1989;87:153.
36. Fihn SD, McDonell M, Martin D, Henikoff J, Vermes D, Kent D, White RH. Risk factors for complications of chronic anticoagulation. A multicenter study. Warfarin Optimized Outpatient Follow-up Study Group. Ann Intern Med 1993;118(7):511–20.
37. The Stroke Prevention in Atrial Fibrillation Investigators. Bleeding during antithrombotic therapy in patients with atrial fibrillation. Arch Intern Med 1996;156(4):409–16.
38. Fang MC, Chang Y, Hylek EM, Rosand J, Greenberg SM, Go AS, Singer DE. Advanced age, anticoagulation intensity, and risk for intracranial hemorrhage among patients taking warfarin for atrial fibrillation. Ann Intern Med 2004;141(10):745–52.
39. Sam C, Massaro JM, D'Agostino RB Sr, Levy D, Lambert JW, Wolf PA, Benjamin EJ; Framingham Heart Study. Warfarin and aspirin use and the predictors of major bleeding complications in atrial fibrillation (the Framingham Heart Study). Am J Cardiol 2004;94(7):947–51.
40. Makris M, Greaves M, Phillips WS, Kitchen S, Rosendaal FR, Preston EF. Emergency oral anticoagulant reversal: the relative efficacy of infusions of fresh frozen plasma and clotting factor concentrate on correction of the coagulopathy. Thromb Haemost 1997;77(3):477–80.
41. Mayer GA. Blood viscosity and oral anticoagulant therapy. Am J Clin Pathol 1976;65(3):402–6.
42. Herrmann KS, Kreuzer H. Beobachtung einer Acenocoumarol induzierten Granulocytose. [Observation of acenocoumarol-induced granulocytosis.] Klin Wochenschr 1988;66(14):639–42.
43. Dybedal I, Lamvik J. Warfarin as a probable cause of haemolytic anaemia. Thromb Haemost 1990;63(1):143.
44. Manu N, Martin L. Warfarin-induced small bowel obstruction. Clin Lab Haematol 2005;27(5):350–2.
45. Adler E, Benjamin SB, Zimmerman HJ. Cholestatic hepatic injury related to warfarin exposure. Arch Intern Med 1986;146(9):1837–9.
46. Ciorciaro C, Hartmann K, Stoller R, Kuhn M. Leberschäden durch Coumarin-antikoagulantien: Erfahrungen der IKS und der SANZ. [Liver injury caused by coumarin anticoagulants: experience of the IKS (Intercanton Monitoring Station) and the SANZ (Swiss Center for Drug Monitoring).] Schweiz Med Wochenschr 1996;126(49):2109–13.
47. Mix H, Wagner S, Boker K, Gloger S, Oldhafer KJ, Behrend M, Flemming P, Manns MP. Subacute liver failure induced by phenprocoumon treatment. Digestion 1999;60(6):579–82.
48. Kruis-de Vries MH, Stricker BH, Coenraads PJ, Nater JP. Maculopapular rash due to coumarin derivatives. Dermatologica 1989;178(2):109–11.
49. Susano R, Garcia A, Altadill A, Ferro J. Hypersensitivity vasculitis related to nicoumalone. BMJ 1993;306(6883):973.
50. Cole MS, Minifee PK, Wolma FJ. Coumarin necrosis–a review of the literature. Surgery 1988;103(3):271–7.
51. Verhagen H. Local haemorrhage and necrosis of the skin and underlying tissues, during anti-coagulant therapy with dicumarol or dicumacyl. Acta Med Scand 1954;148(6):453–67.
52. van Amstel WJ, Boekhout-Mussert MJ, Loeliger EA. Successful prevention of coumarin-induced hemorrhagic skin necrosis by timely administration of vitamin K_1. Blut 1978;36(2):89–93.
53. Loeliger EA, Paul LC, Van Brummelen P, Bieger R. Is coumarin-induced haemorrhagic necrosis of the skin the result of hypoproconvertinaemic bleeding in non-specifically inflamed skin patches? Neth J Med 1975;18(1):12–6.
54. Horn JR, Danziger LH, Davis RJ. Warfarin-induced skin necrosis: report of four cases. Am J Hosp Pharm 1981;38(11):1763–8.
55. Eby CS. Warfarin-induced skin necrosis. Hematol Oncol Clin North Am 1993;7(6):1291–300.
56. Gelwix TJ, Beeson MS. Warfarin-induced skin necrosis. Am J Emerg Med 1998;16(5):541–3.
57. Wynn SS, Jin DK, Essex DW. Warfarin-induced skin necrosis occurring four days after discontinuation of warfarin. Haemostasis 1997;27(5):246–50.
58. Goldberg SL, Orthner CL, Yalisove BL, Elgart ML, Kessler CM. Skin necrosis following prolonged administration of coumarin in a patient with inherited protein S deficiency. Am J Hematol 1991;38(1):64–6.

59. Sternberg ML, Pettyjohn FS. Warfarin sodium-induced skin necrosis. Ann Emerg Med 1995;26(1):94–7.
60. Torngren S, Somell A. Warfarin skin necrosis of the breast. Acta Chir Scand 1982;148(5):471–2.
61. Barkley C, Badalament RA, Metz EN, Nesbitt J, Drago JR. Coumarin necrosis of the penis. J Urol 1989;141(4):946–8.
62. Kandrotas RJ, Deterding J. Genital necrosis secondary to warfarin therapy. Pharmacotherapy 1988;8(6):351–4.
63. Haefeli H. Uterusnekrose bei Cumarin-Therapie. Fortschr Geburtslilfe Gynekol 1969;39:49.
64. DiCato M-A, Ellman L. Letter: Coumadin-induced necrosis of breast, disseminated intravascular coagulation, and hemolytic anemia. Ann Intern Med 1975;83(2):233–4.
65. Byrne JS, Abdul Razak AR, Patchett S, Murphy GM. Warfarin skin necrosis associated with protein S deficiency and a mutation in the methylenetetrahydrofolate reductase gene. Clin Exp Dermatol 2004;29(1):35–6.
66. Khalid K. Warfarin-induced necrosis of the breast: case report. J Postgrad Med 2004;50(4):268–9.
67. Moll S. Warfarin-induced skin necrosis. Br J Haematol 2004;126(5):628.
68. Muniesa C, Marcoval J, Moreno A, Gimenez S, Sanchez J, Ferreres JR, Peyri J. Coumarin necrosis induced by renal insufficiency. Br J Dermatol 2004;151(2):502–4.
69. Srinivasan AF, Rice L, Bartholomew JR, Rangaswamy C, La Perna L, Thompson JE, Murphy S, Baker KR. Warfarin-induced skin necrosis and venous limb gangrene in the setting of heparin-induced thrombocytopenia. Arch Intern Med 2004;164(1):66–70.
70. Tai CY, Ierardi R, Alexander JB. A case of warfarin skin necrosis despite enoxaparin anticoagulation in a patient with protein S deficiency. Ann Vasc Surg 2004;18(2): 237–42.
71. Valdivielso M, Longo I, Lecona M, Lazaro P. Cutaneous necrosis induced by acenocoumarol. J Eur Acad Dermatol Venereol 2004;18(2):211–5.
72. Warkentin TE, Whitlock RP, Teoh KH. Warfarin-associated multiple digital necrosis complicating heparin-induced thrombocytopenia and Raynaud's phenomenon after aortic valve replacement for adenocarcinoma-associated thrombotic endocarditis. Am J Hematol 2004;75(1):56–62.
73. Broekmans AW, Bertina RM, Loeliger EA, Hofmann V, Klingemann HG. Protein C and the development of skin necrosis during anticoagulant therapy. Thromb Haemost 1983;49(3):251.
74. Samama M, Horellou MH, Soria J, Conard J, Nicolas G. Successful progressive anticoagulation in a severe protein C deficiency and previous skin necrosis at the initiation of oral anticoagulant treatment. Thromb Haemost 1984;51(1):132–3.
75. Teepe RG, Broekmans AW, Vermeer BJ, Nienhuis AM, Loeliger EA. Recurrent coumarin-induced skin necrosis in a patient with an acquired functional protein C deficiency. Arch Dermatol 1986;122(12):1408–12.
76. Comp PC, Elrod JP, Karzenski S. Warfarin-induced skin necrosis. Semin Thromb Hemost 1990;16(4):293–8.
77. Wattiaux MJ, Herve R, Robert A, Cabane J, Housset B. Imbert JC. Coumarin-induced skin necrosis associated with acquired protein S deficiency and antiphospholipid antibody syndrome. Arthritis Rheum 1994;37(7):1096–100.
78. Shepherd AM, Hewick DS, Moreland TA, Stevenson IH. Age as a determinant of sensitivity to warfarin. Br J Clin Pharmacol 1977;4(3):315–20.
79. Lewandowski K, Zawilska K. Protein C concentrate in the treatment of warfarin-induced skin necrosis in the protein C deficiency. Thromb Haemost 1994;71(3):395.
80. Zauber NP, Stark MW. Successful warfarin anticoagulation despite protein C deficiency and a history of warfarin necrosis. Ann Intern Med 1986;104(5):659–60.
81. Jamal SA, Browner WS, Bauer DC, Cummings SR. Warfarin use and risk for osteoporosis in elderly women. Study of Osteoporotic Fractures Research Group. Ann Intern Med 1998;128(10):829–32.
82. Pierce WS, Donachy JH, Rosenberg G, Baier RE. Calcification inside artificial hearts: inhibition by warfarin-sodium. Science 1980;208(4444):601–3.
83. Pietschmann P, Woloszczuk W, Panzer S, Kyrle P, Smolen J. Decreased serum osteocalcin levels in phenprocoumon-treated patients. J Clin Endocrinol Metab 1988;66(5):1071–4.
84. Edwards RL, Rickles FR. Delayed hypersensitivity in man: effects of systemic anticoagulation. Science 1978;200(4341):541–3.
85. Hirsh J, Cade JF, Gallus AS. Anticoagulants in pregnancy: a review of indications and complications. Am Heart J 1972;83(3):301–5.
86. Ginsberg JS, Hirsh J. Use of antithrombotic agents during pregnancy. Chest 1998;114(Suppl 5):S524–30.
87. Ginsberg JS, Hirsh J. Use of antithrombotic agents during pregnancy. Chest 1992;102(Suppl 4):S385–90.
88. Ginsberg JS, Hirsh J. Use of antithrombotic agents during pregnancy. Chest 1995;108(Suppl 4):S305–11.
89. Greer IA. Thrombosis in pregnancy: maternal and fetal issues. Lancet 1999;353(9160):1258–65.
90. Ginsberg JS, Hirsh J. Optimum use of anticoagulants in pregnancy. Drugs 1988;36(4):505–12.
91. Hall JG, Pauli RM, Wilson KM. Maternal and fetal sequelae of anticoagulation during pregnancy. Am J Med 1980;68(1):122–40.
92. Stevenson RE, Burton OM, Ferlauto GJ, Taylor HA. Hazards of oral anticoagulants during pregnancy. JAMA 1980;243(15):1549–51.
93. Wellesley D, Moore I, Heard M, Keeton B. Two cases of warfarin embryopathy: a re-emergence of this condition? Br J Obstet Gynaecol 1998;105(7):805–6.
94. Vitale N, De Feo M, De Santo LS, Pollice A, Tedesco N, Cotrufo M. Dose-dependent fetal complications of warfarin in pregnant women with mechanical heart valves. J Am Coll Cardiol 1999;33(6):1637–41.
95. Ginsberg JS, Hirsh J. Use of anticoagulants during pregnancy. Chest 1989;95(Suppl 2):S156–60.
96. Iturbe-Alessio I, Fonseca MC, Mutchinik O, Santos MA, Zajarias A, Salazar E. Risks of anticoagulant therapy in pregnant women with artificial heart valves. N Engl J Med 1986;315(22):1390–3.
97. Kort HI, Cassel GA. An appraisal of warfarin therapy during pregnancy. S Afr Med J 1981;60(15):578–9.
98. Olwin JH, Koppel JL. Anticoagulant therapy during pregnancy. A new approach. Obstet Gynecol 1969;34(6):847–52.
99. Price PA, Williamson MK, Haba T, Dell RB, Jee WS. Excessive mineralization with growth plate closure in rats on chronic warfarin treatment. Proc Natl Acad Sci USA 1982;79(24):7734–8.
100. Bates SM, Ginsberg JS. In: Anticoagulants in pregnancy: fetal defects. London: Bailliere Tindall, 1997:479.
101. Chong MK, Harvey D, de Swiet M. Follow-up study of children whose mothers were treated with warfarin during pregnancy. Br J Obstet Gynaecol 1984;91(11):1070–3.

102. Ludwig H. Antikoagulantien in der Schwangerschaft und im Wochenbett. [Anticoagulants in pregnancy and puerperium.] Geburtshilfe Frauenheilkd 1970;30(4):337–47.
103. Orme ML, Lewis PJ, de Swiet M, Serlin MJ, Sibeon R, Baty JD, Breckenridge AM. May mothers given warfarin breast-feed their infants? BMJ 1977;1(6076):1564–5.
104. De Swiet M, Lewis PJ. Excretion of anticoagulants in human milk. N Engl J Med 1977;297(26):1471.
105. McKenna R, Cole ER, Vasan U. Is warfarin sodium contraindicated in the lactating mother? J Pediatr 1983;103(2):325–7.
106. Houwert-de Jong M, Gerards LJ, Tetteroo-Tempelman CA, de Wolff FA. May mothers taking acenocoumarol breast feed their infants? Eur J Clin Pharmacol 1981;21(1):61–4.
107. Fondevila CG, Meschengieser S, Blanco A, Penalva L, Lazzari MA. Effect of acenocoumarine on the breast-fed infant. Thromb Res 1989;56(1):29–36.
108. Eckstein HB, Jack B. Breast-feeding and anticoagulant therapy. Lancet 1970;1(7648):672–3.
109. Aithal GP, Day CP, Kesteven PJ, Daly AK. Association of polymorphisms in the cytochrome P450 CYP2C9 with warfarin dose requirement and risk of bleeding complications. Lancet 1999;353(9154):717–9.
110. Daly AK, Aithal GP. Genetic regulation of warfarin metabolism and response. Semin Vasc Med 2003;3(3):231–8.
111. Oldenburg J, Kriz K, Wuillemin WA, Maly FE, von Felten A, Siegemund A, Keeling DM, Baker P, Chu K, Konkle BA, Lammle B, Albert T. Study Group on Hereditary Warfarin Sensitivity. Genetic predisposition to bleeding during oral anticoagulant therapy: evidence for common founder mutations (FIXVal-10 and FIXThr-10) and an independent CpG hotspot mutation (FIXThr-10). Thromb Haemost 2001;85(3):454–7.
112. Visser LE, van Vliet M, van Schaik RH, Kasbergen AA, De Smet PA, Vulto AG, Hofman A, van Duijn CM, Stricker BH. The risk of overanticoagulation in patients with cytochrome P450 CYP2C9*2 or CYP2C9*3 alleles on acenocoumarol or phenprocoumon. Pharmacogenetics 2004;14(1):27–33.
113. Chu K, Wu SM, Stanley T, Stafford DW, High KA. A mutation in the propeptide of Factor IX leads to warfarin sensitivity by a novel mechanism. J Clin Invest 1996;98(7):1619–25.
114. Baker P, Clarke K, Giangrande P, Keeling D. Ala-10 mutations in the factor IX propeptide and haemorrhage in a patient treated with warfarin. Br J Haematol 2000;108(3):663.
115. Visser LE, van Vliet M, van Schaik RH, Kasbergen AA, De Smet PA, Vulto AG, Hofman A, van Duijn CM, Stricker BH. The risk of overanticoagulation in patients with cytochrome P450 CYP2C9*2 or CYP2C9*3 alleles on acenocoumarol or phenprocoumon. Pharmacogenetics 2004;14(1):27–33.
116. Beyth RJ, Shorr RI. Epidemiology of adverse drug reactions in the elderly by drug class. Drugs Aging 1999;14(3):231–9.
117. Launbjerg J, Egeblad H, Heaf J, Nielsen NH, Fugleholm AM, Ladefoged K. Bleeding complications to oral anticoagulant therapy: multivariate analysis of 1010 treatment years in 551 outpatients. J Intern Med 1991;229(4):351–5.
118. Visser LE, Bleumink GS, Trienekens PH, Vulto AG, Hofman A, Stricker BH. The risk of overanticoagulation in patients with heart failure on coumarin anticoagulants. Br J Haematol 2004;127(1):85–9.
119. Self TH, Straughn AB, Weisburst MR. Effect of hyperthyroidism on hypoprothrombinemic response to warfarin. Am J Hosp Pharm 1976;33(4):387–9.
120. Smith JK, Aljazairi A, Fuller SH. INR elevation associated with diarrhea in a patient receiving warfarin. Ann Pharmacother 1999;33(3):301–4.
121. Landefeld CS, Rosenblatt MW, Goldman L. Bleeding in outpatients treated with warfarin: relation to the prothrombin time and important remediable lesions. Am J Med 1989;87(2):153–9.
122. Watson RM, Pierson RN Jr. Effect of anticoagulant therapy upon aspirin-induced gastrointestinal bleeding. Circulation 1961;24:613–6.
123. Chesebro JH, Fuster V, Elveback LR, McGoon DC, Pluth JR, Puga FJ, Wallace RB, Danielson GK, Orszulak TA, Piehler JM, Schaff HV. Trial of combined warfarin plus dipyridamole or aspirin therapy in prosthetic heart valve replacement: danger of aspirin compared with dipyridamole. Am J Cardiol 1983;51(9):1537–41.
124. Boston Collaborative Drug Surveillance Program, Boston Universtiy Medical Center. Interaction between chloral hydrate and warfarin. N Engl J Med 1972;286(2):53–5.
125. Jahnchen E, Meinertz I, Gilfrich HJ. Interaction of allopurinol with phenprocoumon in man. Klin Wochenschr 1977;55(15):759–61.
126. Vesell ES, Passananti GT, Greene FE. Impairment of drug metabolism in man by allopurinol and nortriptyline. N Engl J Med 1970;283(27):1484–8.
127. Bruning PF. Personal communication, Antoni van Leeuwenhoekhuis. Het Nederlands Kanker Instituut, Amsterdam 1983;.
128. Udall JA. Human sources and absorption of vitamin K in relation to anticoagulation stability. JAMA 1965;194(2):127–9.
129. O'Reilly RA, Aggeler PM. Determinants of the response to oral anticoagulant drugs in man. Pharmacol Rev 1970;22(1):35–96.
130. Caraco Y, Chajek-Shaul T. The incidence and clinical significance of amiodarone and acenocoumarol interaction. Thromb Haemost 1989;62(3):906–8.
131. Hamer A, Peter T, Mandel WJ, Scheinman MM, Weiss D. The potentiation of warfarin anticoagulation by amiodarone. Circulation 1982;65(5):1025–9.
132. Martinowitz U, Rabinovich J, Goldfarb D, Many A, Bank H. Interaction between warfarin sodium and amiodarone. N Engl J Med 1981;304(11):671–2.
133. Watt AH, Stephens MR, Buss DC, Routledge PA. Amiodarone reduces plasma warfarin clearance in man. Br J Clin Pharmacol 1985;20(6):707–9.
134. Pyerele K, Kekki M. Decreased anticoagulant tolerance during methandrostenolone therapy. Scand J Clin Lab Invest 1963;15:367–74.
135. Ambre JJ, Fischer LJ. Effect of coadministration of aluminum and magnesium hydroxides on absorption of anticoagulants in man. Clin Pharmacol Ther 1973;14(2):231–7.
136. Green AE, Hort JF, Korn HE, Leach H. Potentiation of warfarin by azapropazone. BMJ 1977;1(6075):1532.
137. McElnay JC, D'Arcy PF. Interaction between azapropazone and warfarin. Experientia 1978;34(10):1320–1.
138. Rivier G, Khamashta MA, Hughes GR. Warfarin and azathioprine: a drug interaction does exist. Am J Med 1993;95(3):342.
139. Singleton JD, Conyers L. Warfarin and azathioprine: an important drug interaction. Am J Med 1992;92(2):217.
140. Woldtvedt BR, Cahoon CL, Bradley LA, Miller SJ. Possible increased anticoagulation effect of warfarin

induced by azithromycin. Ann Pharmacother 1998;32(2):269–70.
141. Robinson DS, MacDonald MG. The effect of phenobarbital administration on the control of coagulation achieved during warfarin therapy in man. J Pharmacol Exp Ther 1966;153:250.
142. Levy G, O'Reilly RA, Aggeler PM, Keech GM. Parmacokinetic analysis of the effect of barbiturate on the anticoagulant action of warfarin in man. Clin Pharmacol Ther 1970;11(3):372–7.
143. Pyorala K, Ikkala E, Siltanen P. Benziodarone (Amplivix) and anticoagulant therapy. Acta Med Scand 1963;173:385–9.
144. Blum A, Seligmann H, Livneh A, Ezra D. Severe gastrointestinal bleeding induced by a probable hydroxycoumarin-bezafibrate interaction. Isr J Med Sci 1992;28(1):47–9.
145. Beringer TR. Warfarin potentiation with bezafibrate. Postgrad Med J 1997;73(864):657–8.
146. Denbow CE, Fraser HS. Clinically significant hemorrhage due to warfarin-carbamazepine interaction. South Med J 1990;83(8):981.
147. Hansen JM, Siersbaek-Nielsen K, Skovsted L. Carbamazepine-induced acceleration of diphenylhydantoin and warfarin metabolism in man. Clin Pharmacol Ther 1971;12(3):539–43.
148. Loeliger EA, Van der Esch B, Mattern MJ, Hemker HC. The biological disappearance rate of prothrombin, factors VII, IX and X from plasma in hypothyroidism, hyperthuroidism, and during fever. Thromb Diath Haemorrh 1964;10:267–77.
149. Berod T, Mathiot I. Probable interaction between cetirizine and acenocoumarol. Ann Pharmacother 1997;31(1):122.
150. Rymer W, Greenlaw CW. Hypoprothrombinemia associated with cefamandole. Drug Intell Clin Pharm 1980;14:780.
151. Parker SW, Baxter J, Beam TR Jr. Cefoperazone-induced coagulopathy. Lancet 1984;1(8384):1016.
152. Bechtold H, Andrassy K, Jahnchen E, Koderisch J, Koderisch H, Weilemann LS, Sonntag HG, Ritz E. Evidence for impaired hepatic vitamin K_1 metabolism in patients treated with N-methyl-thiotetrazole cephalosporins. Thromb Haemost 1984;51(3):358–61.
153. Udall JA. Warfarin interactions with chloral hydrate and glutethimide. Curr Ther Res Clin Exp 1975;17(1):67–74.
154. Christensen LK, Skovsted L. Inhibition of drug metabolism by chloramphenicol. Lancet 1969;2(7635):1397–9.
155. Magid E. Tolerance to anticoagulants during antibiotic therapy. Scand J Clin Lab Invest 1962;14:565.
156. Leone R, Ghiotto E, Conforti A, Velo G. Potential interaction between warfarin and ocular chloramphenicol. Ann Pharmacother 1999;33(1):114.
157. O'Reilly RA, Sahud MA, Aggeler PM. Impact of aspirin and chlorthalidone on the pharmacodynamics of oral anticoagulant drugs in man. Ann NY Acad Sci 1971;179:173–86.
158. Robinson DS, Benjamin DM, McCormack JJ. Interaction of warfarin and nonsystemic gastrointestinal drugs. Clin Pharmacol Ther 1971;12(3):491–5.
159. Meinertz T, Gilfrich HJ, Groth U, Jonen HG, Jahnchen E. Interruption of the enterohepatic circulation of phenprocoumon by cholestyramine. Clin Pharmacol Ther 1977;21(6):731–5.
160. Serlin MJ, Sibeon RG, Mossman S, Breckenridge AM, Williams JR, Atwood JL, Willoughby JM. Cimetidine: interaction with oral anticoagulants in man. Lancet 1979;2(8138):317–9.
161. Harenberg J, Zimmermann R, Staiger C, de Vries JX, Walter E, Weber E. Lack of effect of cimetidine on action of phenprocoumon. Eur J Clin Pharmacol 1982;23(4):365–7.
162. Toon S, Hopkins KJ, Garstang FM, Rowland M. Comparative effects of ranitidine and cimetidine on warfarin in man. Br J Clin Pharmacol 1986;21:P565.
163. Jolson HM, Tanner LA, Green L, Grasela TH Jr. Adverse reaction reporting of interaction between warfarin and fluoroquinolones. Arch Intern Med 1991;151(5):1003–4.
164. Costigan DC, Freedman MH, Ehrlich RM. Potentiation of oral anticoagulant effect by L-thyroxine. Clin Pediatr (Phila) 1984;23(3):172–4.
165. Raburn M. Hypoprothrombinemia induced by warfarin sodium and cisapride. Am J Health Syst Pharm 1997;54(3):320–1.
166. Grau E, Real E, Pastor E. Interaction between clarithromycin and oral anticoagulants. Ann Pharmacother 1996;30(12):1495–6.
167. Sanchez B, Muruzabal MJ, Peralta G, et al. Clarithromycin oral anticoagulants interaction: report of five cases. Clin Drug Invest 1997;13:220–2.
168. Oberg KC. Delayed elevation of international normalized ratio with concurrent clarithromycin and warfarin therapy. Pharmacotherapy 1998;18(2):386–91.
169. O'Reilly RA, Sahud MA, Robinson AJ. Studies on the interaction of warfarin and clofibrate in man. Thromb Diath Haemorrh 1972;27(2):309–18.
170. O'Donnell D. Antibiotic-induced potentiation of oral anticoagulant agents. Med J Aust 1989;150(3):163–4.
171. O'Reilly RA, Motley CH. Racemic warfarin and trimethoprim–sulfamethoxazole interaction in humans. Ann Intern Med 1979;91(1):34–6.
172. Tashima CK. Cyclophosphamide effect on coumarin anticoagulation. South Med J 1979;72(5):633–4.
173. Snyder DS. Interaction between cyclosporine and warfarin. Ann Intern Med 1988;108(2):311.
174. Meeks ML, Mahaffey KW, Katz MD. Danazol increases the anticoagulant effect of warfarin. Ann Pharmacother 1992;26(5):641–2.
175. Izzat MB, Yim AP, El-Zufari MH. A taste of Chinese medicine!. Ann Thorac Surg 1998;66(3):941–2.
176. Mailloux AT, Gidal BE, Sorkness CA. Potential interaction between warfarin and dicloxacillin. Ann Pharmacother 1996;30(12):1402–7.
177. Serlin MJ, Mossman S, Sibeon RG, Tempero KF, Breckenridge AM. Interaction between diflunisal and warfarin. Clin Pharmacol Ther 1980;28(4):493–8.
178. O'Reilly RA. Interaction of sodium warfarin and disulfiram (antabuse) in man. Ann Intern Med 1973;78(1):73–6.
179. O'Reilly RA. Dynamic interaction between disulfiram and separated enantiomorphs of racemic warfarin. Clin Pharmacol Ther 1981;29(3):332–6.
180. Bartle WR. Possible warfarin–erythromycin interaction. Arch Intern Med 1980;140(7):985–7.
181. Grau E, Fontcuberta J, Felez J. Erythromycin-oral anticoagulants interaction. Arch Intern Med 1986;146(8):1639.
182. Bachmann K, Schwartz JI, Forney R Jr, Frogameni A, Jauregui LE. The effect of erythromycin on the disposition kinetics of warfarin. Pharmacology 1984;28(3):171–6.
183. Petrick RJ, Kronacher N, Alcena V. Interaction between warfarin and ethacrynic acid. JAMA 1975;231(8):843–4.

184. Ascah KJ, Rock GA, Wells PS. Interaction between fenofibrate and warfarin. Ann Pharmacother 1998;32(7–8):765–8.
185. Boeijinga JK, van de Broeke RN, Jochemsen R, Breimer DD, Hoogslag MA, Jeletich-Bastiaanse A. De invloed van floctafenine (Idalon) op antistollingsbehandeling met coumarinderivaten. [The effect of floctafenine (Idalon) on anticoagulant treatment with coumarin derivatives.] Ned Tijdschr Geneeskd 1981;125(47):1931–5.
186. Seaton TL, Celum CL, Black DJ. Possible potentiation of warfarin by fluconazole. DICP 1990;24(12):1177–8.
187. Baciewicz AM, Menke JJ, Bokar JA, Baud EB. Fluconazole–warfarin interaction. Ann Pharmacother 1994;28(9):1111.
188. Gericke KR. Possible interaction between warfarin and fluconazole. Pharmacotherapy 1993;13(5):508–9.
189. Brown MC. Multisite mucous membrane bleeding due to a possible interaction between warfarin and 5-fluorouracil. Pharmacotherapy 1997;17(3):631–3.
190. Dent LA, Orrock MW. Warfarin–fluoxetine and diazepam–fluoxetine interaction. Pharmacotherapy 1997;17(1):170–2.
191. Marbet GA, Duckert F, Walter M, Six P, Airenne H. Interaction study between phenprocoumon and flurbiprofen. Curr Med Res Opin 1977;5(1):26–31.
192. Stricker BH, Delhez JL. Interactions between flurbiprofen and coumarins. BMJ (Clin Res Ed) 1982;285(6344):812–3.
193. Trilli LE, Kelley CL, Aspinall SL, Kroner BA. Potential interaction between warfarin and fluvastatin. Ann Pharmacother 1996;30(12):1399–402.
194. Laizure SC, Madlock L, Cyr M, Self T. Decreased hypoprothrombinemic effect of warfarin associated with furosemide. Ther Drug Monit 1997;19(3):361–3.
195. Janetzky K, Morreale AP. Probable interaction between warfarin and ginseng. Am J Health Syst Pharm 1997;54(6):692–3.
196. Koch-Weser J. Potentiation by glucagon of the hypoprothrombinemic action of warfarin. Ann Intern Med 1970;72(3):331–5.
197. Chatterjea JB, Salomon L. Antagonistic effect of A.C.T.H. and cortisone on the anticoagulant activity of ethyl biscoumacetate BMJ 1954;4891:790–2.
198. Hellem AJ, Solem JH. The influence of ACTH on prothrombin-proconvertin values in blood during treatment with dicumarol and phenylindanedione. Acta Med Scand 1954;150(5):389–93.
199. MacDonald MG, Robinson DS, Sylwester D, Jaffe JJ. The effects of phenobarbital, chloral betaine, and glutethimide administration on warfarin plasma levels and hypoprothrombinemic responese in man. Clin Pharmacol Ther 1969;10(1):80–4.
200. Cullen SI, Catalano PM. Griseofulvin–warfarin antagonism. JAMA 1967;199(8):582–3.
201. Oakley DP, Lautch H. Haloperidol and anticoagulant treatment. Lancet 1963;41:1231.
202. Chan TY, Lui SF, Chung SY, Luk S, Critchley JA. Adverse interaction between warfarin and indomethacin. Drug Saf 1994;10(3):267–9.
203. Rosenthal AR, Self TH, Baker ED, Linden RA. Interaction of isoniazid and warfarin. JAMA 1977;238(20):2177.
204. Yeh J, Soo SC, Summerton C, Richardson C. Potentiation of action of warfarin by itraconazole. BMJ 1990;301(6753):669.
205. Smith AG. Potentiation of oral anticoagulants by ketoconazole. BMJ (Clin Res Ed) 1984;288(6412):188–9.
206. Ahmad S. Lovastatin. Warfarin interaction. Arch Intern Med 1990;150(11):2407.
207. Holmes EL. Experimental observations on flufenamic, mefenamic, and meclofenamic acids. IV. Toleration by normal human subjects. Ann Phys Med 1966;(Suppl):36–49.
208. Spiers AS, Mibashan RS. Letter: Increased warfarin requirement during mercaptopurine therapy: a new drug interaction. Lancet 1974;2(7874):221–2.
209. Self TH. Interaction of warfarin and aminosalicylic acid. JAMA 1973;223(11):1285.
210. Marinella MA. Mesalamine and warfarin therapy resulting in decreased warfarin effect. Ann Pharmacother 1998;32(7–8):841–2.
211. O'Reilly RA. The stereoselective interaction of warfarin and metronidazole in man. N Engl J Med 1976;295(7):354–7.
212. Watson PG, Lochan RG, Redding VJ. Drug interactions with coumarin derivative anticoagulants. BMJ (Clin Res Ed) 1982;285(6347):1045–6.
213. Ortin M, Olalla JI, Muruzabal MJ, Peralta FG, Gutierrez MA. Miconazole oral gel enhances acenocoumarol anticoagulant activity: a report of three cases. Ann Pharmacother 1999;33(2):175–7.
214. Lansdorp D, Bressers HP, Dekens-Konter JA, Meyboom RH. Potentiation of acenocoumarol during vaginal administration of miconazole. Br J Clin Pharmacol 1999;47(2):225–6.
215. Colquhoun MC, Daly M, Stewart P, Beeley L. Interaction between warfarin and miconazole oral gel. Lancet 1987;1(8534):695–6.
216. Marotel C, Cerisay D, Vasseur P, Rouvier B, Chabanne JP. Potentialisation des effets de l'acenocoumarol par le gel buccal de miconazole. [Potentiation of the effects of acenocoumarol by a buccal gel of miconazole.] Presse Méd 1986;15(33):1684–5.
217. O'Reilly RA, Goulart DA, Kunze KL, Neal J, Gibaldi M, Eddy AC, Trager WF. Mechanisms of the stereoselective interaction between miconazole and racemic warfarin in human subjects. Clin Pharmacol Ther 1992;51(6):656–67.
218. Qureshi GD, Reinders TP, Somori GJ, Evans HJ. Warfarin resistance with nafcillin therapy. Ann Intern Med 1984;100(4):527–9.
219. Taylor AT, Pritchard DC, Goldstein AO, Fletcher JL Jr. Continuation of warfarin–nafcillin interaction during dicloxacillin therapy. J Fam Pract 1994;39(2):182–5.
220. Hoffbrand BI. Interaction of nalidixic acid and warfarin. BMJ 1974;2(920):666.
221. Leor J, Levartowsky D, Sharon C. Interaction between nalidixic acid and warfarin. Ann Intern Med 1987;107(4):601.
222. Lee M, Schwartz RN, Sharifi R. Warfarin resistance and vitamin K. Ann Intern Med 1981;94(1):140–1.
223. Baciewicz AM, Ashar BH, Locke TW. Interaction of ofloxacin and warfarin. Ann Intern Med 1993;119(12):1223.
224. Ahmad S. Omeprazole–warfarin interaction. South Med J 1991;84(5):674–5.
225. de Teresa E, Vera A, Ortigosa J, Pulpon LA, Arus AP, de Artaza M. Interaction between anticoagulants and contraceptives: an unsuspected finding. BMJ 1979;2(6200):1260–1261.
226. Monig H, Baese C, Heidemann HT, Ohnhaus EE, Schulte HM. Effect of oral contraceptive steroids on the pharmacokinetics of phenprocoumon. Br J Clin Pharmacol 1990;30(1):115–8.

227. Bagheri H, Bernhard NB, Montastruc JL. Potentiation of the acenocoumarol anticoagulant effect by acetaminophen. Ann Pharmacother 1999;33(4):506.
228. Hylek EM, Heiman H, Skates SJ, Sheehan MA, Singer DE. Acetaminophen and other risk factors for excessive warfarin anticoagulation. JAMA 1998;279(9):657–62.
229. Aggeler PM, O'Reilly RA, Leong L, Kowitz PE. Potentiation of anticoagulant effect of warfarin by phenylbutazone. N Engl J Med 1967;276(9):496–501.
230. Lewis RJ, Trager WF, Chan KK, Breckenridge A, Orme M, Roland M, Schary W. Warfarin. Stereochemical aspects of its metabolism and the interaction with phenylbutazone. J Clin Invest 1974;53(6):1607–17.
231. O'Reilly RA, Goulart DA. Comparative interaction of sulfinpyrazone and phenylbutazone with racemic warfarin: alteration in vivo of free fraction of plasma warfarin. J Pharmacol Exp Ther 1981;219(3):691–4.
232. Hansen JM, Siersbaek-Nielsen K, Kristensen M, Skovsted L, Christensen LK. Effect of diphenylhydantoin on the metabolism of dicoumarol in man. Acta Med Scand 1971;189(1–2):15–9.
233. Nappi JM. Warfarin and phenytoin interaction. Ann Intern Med 1979;90(5):852.
234. Levine M, Sheppard I. Biphasic interaction of phenytoin warfarin interaction. Postgrad Med J 1991;67:98.
235. Pan HY, Ng RP. The effect of Nootropil in a patient on warfarin. Eur J Clin Pharmacol 1983;24(5):711.
236. Rhodes RS, Rhodes PJ, Klein C, Sintek CD. A warfarin–piroxicam drug interaction. Drug Intell Clin Pharm 1985;19(7–8):556–8.
237. Trenque T, Choisy H, Germain ML. Pravastatin: interaction with oral anticoagulant? BMJ 1996;312(7035):886.
238. Jassal SV. Warfarin potentiated by proguanil. BMJ 1991;303(6805):789.
239. Kates RE, Yee YG, Kirsten EB. Interaction between warfarin and propafenone in healthy volunteer subjects. Clin Pharmacol Ther 1987;42(3):305–11.
240. Scott AK, Park BK, Breckenridge AM. Interaction between warfarin and propranolol. Br J Clin Pharmacol 1984;17(5):559–64.
241. Koch-Weser J. Quinidine-induced hypoprothrombinemic hemorrhage in patients on chronic warfarin therapy. Ann Intern Med 1968;68(3):511–7.
242. Trenk D, Mohrke W, Warth L, Jahnchen E. Determination of the interaction of 3S-hydroxy-10,11-dihydroquinidine on the pharmacokinetics and pharmacodynamics of warfarin. Arzneimittelforschung 1993;43(8):836–41.
243. Baciewicz AM, Morgan PJ. Ranitidine-warfarin interaction. Ann Intern Med 1990;112(1):76–7.
244. O'Reilly RA. Interaction of chronic daily warfarin therapy and rifampin. Ann Intern Med 1975;83(4):506–8.
245. O'Reilly RA. Interaction of sodium warfarin and rifampin. Studies in man. Ann Intern Med 1974;81(3):337–40.
246. Heimark LD, Gibaldi M, Trager WF, O'Reilly RA, Goulart DA. The mechanism of the warfarin–rifampin drug interaction in humans. Clin Pharmacol Ther 1987;42(4):388–94.
247. Knoell KR, Young TM, Cousins ES. Potential interaction involving warfarin and ritonavir. Ann Pharmacother 1998;32(12):1299–302.
248. Darlington MR. Hypoprothrombinemia during concomitant therapy with warfarin and saquinavir. Ann Pharmacother 1997;31(5):647.
249. Grau E, Perella M, Pastor E. Simvastatin–oral anticoagulant interaction. Lancet 1996;347(8998):405–6.
250. Risc IR. Bilateral subdural hematoma caused by simvastatin during warfarin treatment. Case report. Acta Neurol Scandinavica 1997;96:339.
251. O'Reilly RA. Spironolactone and warfarin interaction. Clin Pharmacol Ther 1980;27(2):198–201.
252. Braverman SE, Marino MT. Sucralfate—warfarin interaction. Drug Intell Clin Pharm 1988;22(11):913.
253. Mungall D, Talbert RL, Phillips C, Jaffe D, Ludden TM. Sucralfate and warfarin. Ann Intern Med 1983;98(4):557.
254. Michot F, Holt NF, Fontanilles F. Uber die Beeinflussung der gerinnungshemmenden Wirkung von Acenocoumarol durch Sulfinpyrazon. [The effect of sulfinpyrazone on the coagulation-inhibiting action of acenocoumarol.] Schweiz Med Wochenschr 1981;111(8):255–60.
255. Nenci GG, Agnelli G, Berrettini M. Biphasic sulphinpyrazone–warfarin interaction. BMJ (Clin Res Ed) 1981;282(6273):1361–2.
256. O'Reilly RA. Phenylbutazone and sulfinpyrazone interaction with oral anticoagulant phenprocoumon. Arch Intern Med 1982;142(9):1634–7.
257. Toon S, Low LK, Gibaldi M, Trager WF, O'Reilly RA, Motley CH, Goulart DA. The warfarin–sulfinpyrazone interaction: stereochemical considerations. Clin Pharmacol Ther 1986;39(1):15–24.
258. Walter E, Staiger C, de Vries J, Zimmermann R, Weber E. Induction of drug metabolizing enzymes by sulfinpyrazone. Eur J Clin Pharmacol 1981;19(5):353–8.
259. Self TH, Evans W, Ferguson T. Interaction of sulfisoxazole and warfarin. Circulation 1975;52(3):528.
260. Sioris LJ, Weibert RT, Pentel PR. Potentiation of warfarin anticoagulation by sulfisoxazole. Arch Intern Med 1980;140(4):546–7.
261. Carter SA. Potential effect of sulindac on response of prothrombin-time to oral anticoagulants. Lancet 1979;2(8144):698–9.
262. Loftin JP, Vesell ES. Interaction between sulindac and warfarin: different results in normal subjects and in an unusual patient with a potassium-losing renal tubular defect. J Clin Pharmacol 1979;19(11–12):733–42.
263. Lodwick R, McConkey B, Brown AM. Life threatening interaction between tamoxifen and warfarin. BMJ (Clin Res Ed) 1987;295(6606):1141.
264. Gustovic P, Baldin B, Tricoire MJ, Chichmanian RM. Interaction tamoxifene–acenocoumarol. Une interaction potentiellement dangereuse. [Tamoxifen–acenocoumarol interaction. A potentially dangerous interaction.] Therapie 1994;49(1):55–6.
265. Owens JC, Neely WB, Owen WR. Effect of sodium dextrothyroxine in patients receiving anticoagulants. N Engl J Med 1962;266:76–9.
266. Scher ML, Huntington NH, Vitillo JA. Potential interaction between tramadol and warfarin. Ann Pharmacother 1997;31(5):646–7.
267. Sabbe JR, Sims PJ, Sims MH. Tramadol-warfarin interaction. Pharmacotherapy 1998;18(4):871–3.
268. Spigset O. Reduced effect of warfarin caused by ubidecarenone. Lancet 1994;344(8933):1372–3.
269. Corrigan JJ Jr, Marcus FI. Coagulopathy associated with vitamin E ingestion. JAMA 1974;230(9):1300–1.
270. Morkunas A, Graeme K. Zafirlukast warfarin drug interaction with gastrointestinal bleeding. J Toxicol 1997;35:501.
271. Awni WM, Hussein Z, Granneman GR, Patterson KJ, Dube LM, Cavanaugh JH. Pharmacodynamic and stereoselective pharmacokinetic interactions between zileuton and warfarin in humans. Clin Pharmacokinet 1995;29(Suppl 2):67–76.

272. O'Reilly RA. Lack of effect of mealtime wine on the hypoprothrombinemia of oral anticoagulants. Am J Med Sci 1979;277(2):189–94.
273. Kukongviriyapan V, Senggunprai L, Prawan A, Gaysornsiri D, Kukongviriyapan U, Aiemsa-Ard J. Salivary caffeine metabolic ratio in alcohol-dependent subjects. Eur J Clin Pharmacol 2004;60(2):103–7.
274. Badger TM, Huang J, Ronis M, Lumpkin CK. Induction of cytochrome P450 2E1 during chronic ethanol exposure occurs via transcription of the CYP 2E1 gene when blood alcohol concentrations are high. Biochem Biophys Res Commun 1993;190(3):780–5.
275. Deykin D, Janson P, McMahon L. Ethanol potentiation of aspirin-induced prolongation of the bleeding time. N Engl J Med 1982;306(14):852–4.
276. Shireman TI, Howard PA, Kresowik TF, Ellerbeck EF. Combined anticoagulant-antiplatelet use and major bleeding events in elderly atrial fibrillation patients. Stroke 2004;35(10):2362–7.
277. Bachmann HU, Hoffmann A. Interaction of food supplement L-carnitine with oral anticoagulant acenocoumarol. Swiss Med Wkly 2004;134(25–26):385.
278. Penning-van Beest FJ, van Meegen E, Rosendaal FR, Stricker BH. Drug interactions as a cause of overanticoagulation on phenprocoumon or acenocoumarol predominantly concern antibacterial drugs. Clin Pharmacol Ther 2001;69(6):451–7.
279. Lacey CS. Interaction of dicloxacillin with warfarin. Ann Pharmacother 2004;38:898.
280. Mailloux AT, Gidal BE, Sorkness CA. Potential interaction between warfarin and dicloxacillin. Ann Pharmacother 1996;30(12):1402–7.
281. Cropp JS, Bussey HI. A review of enzyme induction of warfarin metabolism with recommendations for patient management. Pharmacotherapy 1997;17(5):917–28.
282. Ellis RJ, Mayo MS, Bodensteiner DM. Ciprofloxacin–warfarin coagulopathy: a case series. Am J Hematol 2000;63(1):28–31.
283. Rindone JP, Kelley CL, Jones WN, Garewal HS. Hypoprothrombinemic effect of warfarin not influenced by ciprofloxacin. Clin Pharm 1991;10(2):136–8.
284. Bianco TM, Bussey HI, Farnett LE, Linn WD, Roush MK, Wong YW. Potential warfarin–ciprofloxacin interaction in patients receiving long-term anticoagulation. Pharmacotherapy 1992;12(6):435–9.
285. Israel DS, Stotka J, Rock W, Sintek CD, Kamada AK, Klein C, Swaim WR, Pluhar RE, Toscano JP, Lettieri JT, Heller AH, Polk RE. Effect of ciprofloxacin on the pharmacokinetics and pharmacodynamics of warfarin. Clin Infect Dis 1996;22(2):251–6.
286. Toon S, Hopkins KJ, Garstang FM, Aarons L, Sedman A, Rowland M. Enoxacin-warfarin interaction: pharmacokinetic and stereochemical aspects. Clin Pharmacol Ther 1987;42(1):33–41.
287. Rocci ML Jr, Vlasses PH, Distlerath LM, Gregg MH, Wheeler SC, Zing W, Bjornsson TD. Norfloxacin does not alter warfarin's disposition or anticoagulant effect. J Clin Pharmacol 1990;30(8):728–32.
288. Verho M, Malerczyk V, Rosenkranz B, Grotsch H. Absence of interaction between ofloxacin and phenprocoumon. Curr Med Res Opin 1987;10(7):474–9.
289. Wyld P, Nimmo W, Millar W, Coles S, Abbott S. The lack of potentiation of the anticoagulant effect of warfarin when administered concurrently with temafloxacin 1991;.
290. Holbrook AM, Pereira JA, Labiris R, McDonald H, Douketis JD, Crowther M, Wells PS. Systematic overview of warfarin and its drug and food interactions. Arch Intern Med 2005;165(10):1095–106.
291. Loke YK, Price D, Derry S, Aronson JK. Case reports of suspected adverse drug reactions–systematic literature survey of follow-up. BMJ 2006;332(7537):335–9.
292. Hu Z, Yang X, Ho PC, Chan SY, Heng PW, Chan E, Duan W, Koh HL, Zhou S. Herb-drug interactions: a literature review. Drugs 2005;65(9):1239–82.
293. Williamson EM. Interactions between herbal and conventional medicines. Expert Opin Drug Saf 2005;4(2):355–78.
294. Izzo AA, Di Carlo G, Borrelli F, Ernst E. Cardiovascular pharmacotherapy and herbal medicines: the risk of drug interaction. Int J Cardiol 2005;98(1):1–14.
295. Monterrey-Rodríguez J. Interaction between warfarin and mango fruit. Ann Pharmacother 2002;36(5):940–1.
296. Lambert JP, Cormier J. Potential interaction between warfarin and boldo–fenugreek. Pharmacotherapy 2001;21(4):509–12.
297. Ramsay NA, Kenny MW, Davies G, Patel JP. Complimentary and alternative medicine use among patients starting warfarin. Br J Haematol 2005;130(5):777–80.
298. Vaes LP, Chyka PA. Interactions of warfarin with garlic, ginger, ginkgo, or ginseng: nature of the evidence. Ann Pharmacother 2000;34(12):1478–82.
299. Page RL 2nd, Lawrence JD. Potentiation of warfarin by dong quai. Pharmacotherapy 1999;19(7):870–6.
300. Lo AC, Chan K, Yeung JH, Woo KS. Danggui (*Angelica sinensis*) affects the pharmacodynamics but not the pharmacokinetics of warfarin in rabbits. Eur J Drug Metab Pharmacokinet 1995;20(1):55–60.
301. Tang JC, Zhang JN, Wu YT, Li ZX. Effect of the water extract and ethanol extract from traditional Chinese medicines *Angelica sinensis (Oliv.) Diels, Ligusticum chuanxiong Hort. and Rheum palmatum L.* on rat liver cytochrome P450 activity. Phytother Res 2006;20(12):1046–51.
302. Taylor JR, Wilt VM. Probable antagonism of warfarin by green tea. Ann Pharmacother 1999;33(4):426–8.
303. Cheng TO. Green tea may inhibit warfarin. Int J Cardiol 2007;115(2):236.
304. Yue QY, Jansson K. Herbal drug curbicin and anticoagulant effect with and without warfarin: possibly related to the vitamin E component. J Am Geriatr Soc 2001;49(6):838.
305. Matthews MK. Association of *Ginkgo biloba* with intracerebral haemorrhage. Neurology 1998;5:1933.
306. Jiang X, Williams KM, Liauw WS, Ammit AJ, Roufogalis BD, Duke CC, Day RO, McLachlan AJ. Effect of ginkgo and ginger on the pharmacokinetics and pharmacodynamics of warfarin in healthy subjects. Br J Clin Pharmacol 2005;59:425–32.
307. Engelsen J, Nielsen JD, Hansen KF. Effekten af coenzym Q10 og *Ginkgo biloba* pa warfarindosis hos patienter i laengerevarende warfarinbehandling. Et randomiseret, dobbeltblindt, placebokontrolleret overkrydsningsforsog. Ugeskr Laeger 2003;165(18):1868–71.
308. Koch E. Inhibition of platelet activating factor (PAF)-induced aggregation of human thrombocytes by ginkgolides: considerations on possible bleeding complications after oral intake of *Ginkgo biloba* extracts. Phytomedicine 2005;12(1–2):10–16.
309. Destro MW, Speranzini MB, Cavalheiro Filho C, Destro T, Destro C. Bilateral haematoma after rhytidoplasty and blepharoplasty following chronic use of *Ginkgo biloba.* Br J Plast Surg 2005;58(1):100–1.

310. Bebbington A, Kulkarni R, Roberts P. *Ginkgo biloba:* persistent bleeding after total hip arthroplasty caused by herbal self-medication. J Arthroplasty 2005;20(1):125–6.
311. Rosenblatt M, Mindel J. Spontaneous hyphema associated with ingestion of *Ginkgo biloba extract.* N Engl J Med 1997;336(15):1108.
312. Meisel C, Johne A, Roots I. Fatal intracerebral mass bleeding associated with *Ginkgo biloba* and ibuprofen. Atherosclerosis 2003;167(2):367.
313. Kim YS, Pyo MK, Park KM, Park PH, Hahn BS, Wu SJ, Yun-Choi HS. Antiplatelet and antithrombotic effects of a combination of ticlopidine and *Ginkgo biloba ext* (EGb 761). Thromb Res 1998;91(1):33–8.
314. Bal Dit Sollier C, Caplain H, Drouet L. No alteration in platelet function or coagulation induced by EGb761 in a controlled study. Clin Lab Haematol 2003;25(4):251–3.
315. Henderson L, Yue QY, Bergquist C, Gerden B, Arlett P. St John's wort (*Hypericum perforatum*): drug interactions and clinical outcomes. Br J Clin Pharmacol 2002;54(4):349–56.
316. Maurer A, Johne A, Bauer S. Interaction of St. John's wort extract with phenprocoumon. Eur J Clin Pharmacol 1999;55:A22.
317. Jiang X, Williams KM, Liauw WS, Ammit AJ, Roufogalis BD, Duke CC, Day RO, McLachlan AJ. Effect of St John's wort and ginseng on the pharmacokinetics and pharmacodynamics of warfarin in healthy subjects. Br J Clin Pharmacol 2004;57(5):592–9. Erratum 2004;58(1):102.
318. Lam AY, Elmer GW, Mohutsky MA. Possible interaction between warfarin and *Lycium barbarum* L. Ann Pharmacother 2001;35(10):1199–201.
319. Coon JT, Ernst E. *Panax ginseng:* a systematic review of adverse effects and drug interactions. Drug Saf 2002;25(5):323–44.
320. Janetzky K, Morreale AP. Probable interaction between warfarin and ginseng. Am J Health Syst Pharm 1997;54(6):692–3.
321. Rosado MF. Thrombosis of a prosthetic aortic valve disclosing a hazardous interaction between warfarin and a commercial ginseng product. Cardiology 2003;99(2):111.
322. Yuan CS, Wei G, Dey L, Karrison T, Nahlik L, Maleckar S, Kasza K, Ang-Lee M, Moss J. American ginseng reduces warfarin's effect in healthy patients: a randomized, controlled trial. Ann Intern Med 2004;141(1):23–7.
323. Jiang X, Williams KM, Liauw WS, Ammit AJ, Roufogalis BD, Duke CC, Day RO, McLachlan AJ. Effect of St John's wort and ginseng on the pharmacokinetics and pharmacodynamics of warfarin in healthy subjects. Br J Clin Pharmacol 2004;57(5):592–9. Erratum 2004;58(1):102.
324. Anderson GD, Rosito G, Mohustsy MA, Elmer GW. Drug interaction potential of soy extract and *Panax ginseng.* J Clin Pharmacol 2003;43:643–8.
325. Jung KY, Kim DS, Oh SR, Lee IS, Lee JJ, Park JD, Kim SI, Lee HK. Platelet activating factor antagonist activity of ginsenosides. Biol Pharm Bull 1998;21(1):79–80.
326. Zhu M, Chan KW, Ng LS, Chang Q, Chang S, Li RC. Possible influences of ginseng on the pharmacokinetics and pharmacodynamics of warfarin in rats. J Pharm Pharmacol 1999;51(2):175–80.
327. Henderson GL, Harkey MR, Gershwin ME, Hackman RM, Stern JS, Stresser DM. Effects of ginseng components on c-DNA-expressed cytochrome P450 enzyme catalytic activity. Life Sci 1999;65(15):PL209–14.
328. Kim HJ, Chun YJ, Park JD, Kim SI, Roh JK, Jeong TC. Protection of rat liver microsomes against carbon tetrachloride-induced lipid peroxidation by red ginseng saponin through cytochrome P450 inhibition. Planta Med 1997;63(5):415–8.
329. Tam LS, Chan TYK, Leung WK, Critchley JAJH. Warfarin interactions with Chinese traditional medicines: danshen and methyl salicylate medicated oil. Aust NZ J Med 1995;25:258.
330. Yu CM, Chan JCN, Sanderson JE. Chinese herbs and warfarin potentiation by "danshen". J Int Med 1997;241:337–9.
331. Chan TY. Interaction between warfarin and danshen (*Salvia miltiorrhiza*). Ann Pharmacother 2001;35(4):501–4.
332. Kuo YH, Lin YL, Don MJ, Chen RM, Ueng YF. Induction of cytochrome P450-dependent monooxygenase by extracts of the medicinal herb *Salvia miltiorrhiza.* J Pharm Pharmacol 2006;58(4):521–7.
333. Lo AC, Chan K, Yeung JH, Woo KS. The effects of danshen (*Salvia miltiorrhiza*) on pharmacokinetics and pharmacodynamics of warfarin in rats. Eur J Drug Metab Pharmacokinet 1992;17(4):257–62.
334. Chan K, Lo AC, Yeung JH, Woo KS. The effects of danshen (*Salvia miltiorrhiza*) on warfarin pharmacodynamics and pharmacokinetics of warfarin enantiomers in rats. J Pharm Pharmacol 1995;47(5):402–6.
335. Gupta D, Jalali M, Wells A, Dasgupta A. Drug–herb interactions: unexpected suppression of free danshen concentrations by salicylate. J Clin Lab Anal 2002;16(6):290–4.
336. Makino T, Wakushima H, Okamoto T, Okukubo Y, Deguchi Y, Kano Y. Pharmacokinetic interactions between warfarin and kangen-karyu, a Chinese traditional herbal medicine, and their synergistic action. J Ethnopharmacol 2002;82(1):35–40.
337. Lesho EP, Saullo L, Udvari-Nagy S. A 76-year-old woman with erratic anticoagulation. Cleve Clin J Med 2004;71(8):651–6.
338. Anonymous. Medical mystery. A woman with too-thin blood. Why was the patient bleeding? What's her case mean to you? Heart Advis 2004;7(12):4–5.
339. Weidner MS, Sigwart K. The safety of a ginger extract in the rat. J Ethnopharmacol 2000;73(3):513–20.
340. Jiang X, Blair EY, McLachlan AJ. Investigation of the effects of herbal medicines on warfarin response in healthy subjects: a population pharmacokinetic–pharmacodynamic modeling approach. J Clin Pharmacol 2006;46(11):1370–8.
341. Davis NB, Nahlik L, Vogelzang NJ. Does PC-Spes interact with warfarin? J Urol 2002;167(4):1793.
342. Duncan GG. Re: Does PC-Spes interact with warfarin? J Urol 2003;169(1):294–5.
343. Wong ALN, Chan TYK. Interaction between warfarin and the herbal product quilinggao. Ann Pharmacother 2003;37:836–8.
344. Kolilekas L, Anagnostopoulos GK, Lampaditis I, Eleftheriadis I. Potential interaction between telithromycin and warfarin. Ann Pharmacother 2004;38(9):1424–7.
345. Visser LE, Penning-van Beest FJ, Wilson JH, Vulto AG, Kasbergen AA, De Smet PA, Hofman A, Stricker BH. Overanticoagulation associated with combined use of lactulose and coumarin anticoagulants. Br J Clin Pharmacol 2004;57(4):522–4.
346. Rao KB, Pallaki M, Tolbert SR, Hornick TR. Hypoprothrombinemia with warfarin due to azithromycin. Ann Pharmacother 2004;38:982–5.
347. Leizorovicz A, Haugh MC, Chapuis FR, Samama MM, Boissel JP. Low molecular weight heparin in prevention of perioperative thrombosis. BMJ 1992;305(6859):913–20.
348. Nurmohamed MT, Rosendaal FR, Buller HR, Dekker E, Hommes DW, Vandenbroucke JP, Briet E. Low-

molecular-weight heparin versus standard heparin in general and orthopaedic surgery: a meta-analysis. Lancet 1992;340(8812):152–6.

349. van Dijk KN, Plat AW, van Dijk AA, Piersma-Wichers M, de Vries-Bots AM, Slomp J, de Jong-van den Berg LT, Brouwers JR. Potential interaction between acenocoumarol and diclofenac, naproxen and ibuprofen and role of CYP2C9 genotype. Thromb Haemost 2004;91(1): 95–101.
350. Knijff-Dutmer EA, Schut GA, Van de Laar MA. Concomitant coumarin–NSAID therapy and risk for bleeding. Ann Pharmacother 2003;37:12–16.
351. Gadisseur AP, Van Der Meer FJ, Rosendaal FR. Sustained intake of paracetamol (acetaminophen) during oral anticoagulant therapy with coumarins does not cause clinically important INR changes: a randomized double-blind clinical trial. J Thromb Haemost 2003;1(4):714–7.
352. Mahe I, Caulin C, Bergmann JF. Does paracetamol potentiate the effects of oral anticoagulants? A literature review. Drug Saf 2004;27(5):325–33.
353. Schulman S. Inhibition of warfarin activity by ribavirin. Ann Pharmacother 2002;36(1):72–4.
354. Llibre JM, Romeu J, Lopez E, Sirera G. Severe interaction between ritonavir and acenocoumarol. Ann Pharmacother 2002;36(4):621–3.
355. Leizorovicz A, Simonneau G, Decousus H, Boissel JP. Comparison of efficacy and safety of low molecular weight heparins and unfractionated heparin in initial treatment of deep venous thrombosis: a meta-analysis. BMJ 1994;309(6950):299–304.
356. Ramanathan M. Warfarin—topical salicylate interactions: case reports. Med J Malaysia 1995;50(3):278–9.
357. Grant P. Warfarin and cranberry juice: an interaction? J Heart Valve Dis 2004;13(1):25–6.
358. Suvarna R, Pirmohamed M, Henderson L. Possible interaction between warfarin and cranberry juice. BMJ 2003;327(7429):1454.
359. Isele H. Todliche Blutung unter Warfarin plus Preiselbeersaft. Liegt's an der Salizylsaure?. [Fatal bleeding under warfarin plus cranberry juice. Is it due to salicylic acid?.] MMW Fortschr Med 2004;146(11):13.
360. Bachmann K, Shapiro R, Fulton R, Carroll FT, Sullivan TJ. Smoking and warfarin disposition. Clin Pharmacol Ther 1979;25(3):309–15.
361. Weiner B, Faraci PA, Fayad R, Swanson L. Warfarin dosage following prosthetic valve replacement: effect of smoking history. Drug Intell Clin Pharm 1984;18(11):904–6.
362. Ansell JE, Buttaro ML, Thomas OV, Knowlton CH. Consensus guidelines for coordinated outpatient oral anticoagulation therapy management. Anticoagulation Guidelines Task Force. Ann Pharmacother 1997;31(5):604–15.

Danaparoid sodium

General Information

Danaparoid is a low molecular heparinoid consisting of a mixture of sulfated glycosaminoglycans (heparan, dermatan, and chondroitin sulfates). It has an antithrombotic effect via antithrombin III-mediated inhibition of factor Xa and to a lesser extent through inhibition of factor IIa. The ratio of antifactor Xa to antifactor IIa is over 20.

Danaparoid is as effective as heparins in inhibiting the formation of thrombi. Its major advantage over low molecular weight heparins is its low rate of cross-reactivity with heparin-associated antibodies from patients with heparin-induced thrombocytopenia (1). It is therefore indicated in patients with heparin-associated thrombocytopenia who require further anticoagulation after withdrawal of heparin. However, it should not be used if there is in vitro cross-reactivity between heparin and danaparoid. Cross-reactivity is uncommon but can result in thrombotic complications, as reported in a case with a fatal outcome (2).

Organs and Systems

Hematologic

Severe bleeding is uncommon with danaparoid but was 3.1% in one series (3).

Immunologic

Delayed hypersensitivity reactions have been reported in patients given danaparoid (4).

References

1. Wilde MI, Markham A. Danaparoid. A review of its pharmacology and clinical use in the management of heparin-induced thrombocytopenia. Drugs 1997;54(6):903–24.
2. Tardy B, Tardy-Poncet B, Viallon A, Piot M, Mazet E. Fatal danaparoid-sodium induced thrombocytopenia and arterial thromboses. Thromb Haemost 1998;80(3):530.
3. Magnani HN. Orgaran (danaparoid sodium) use in the syndrome of heparin-induced thrombocytopenia. Platelets 1997;8:74.
4. Koch P, Munssinger T, Rupp-John C, Uhl K. Delayed-type hypersensitivity skin reactions caused by subcutaneous unfractionated and low-molecular-weight heparins: tolerance of a new recombinant hirudin. J Am Acad Dermatol 2000;42(4):612–9.

Direct factor Xa inhibitors

General Introduction

The direct inhibitors of Factor Xa are short polysaccharides. They include BAY 59-7939 (5-chloro-N-({(5S)-2-oxo-3- (4-(3-oxomorpholin-4-yl)phenyl)-1,3-oxazolidin-5-yl}methyl)thiophene-2-carboxamide), DX-9065a ((+)-2S-2-[4-[[(3S)-1-acetimidoyl-3-pyrrolidinyl] oxy]phenyl]-3-[7-amidino-2-naphthyllpropanoic acid hydrochloride pentahydrate), and otamixaban (1,2). Others have been synthesized (3).

Placebo-controlled studies

BAY 59-7939 has been investigated in a single-center, placebo-controlled, single-blind, parallel-group, multiple-dose escalation study in healthy men aged 20–45 years, body mass index 19–31 kg/m^2, who took BAY 59-7939 (n = 8 per dosage regimen) or placebo (n = 4 per dosage regimen) on days 0 and 3–7 (4). Dosage regimens were 5 mg once, twice, or three times a day, and 10 mg, 20 mg, or 30 mg twice a day. There were no clinically relevant changes in bleeding time or other safety variables across all doses and regimens. There was no dose-related increase in the frequency or severity of adverse events. Maximum inhibition of factor Xa activity occurred after about 3 hours and inhibition was maintained for at least 12 hours at all doses. Prothrombin time, activated partial thromboplastin time, and HepTest were prolonged to a similar extent to inhibition of factor Xa activity. The half-life of BAY 59-7939 was 5.7–9.2 hours at steady state.

The effects of otamixaban have been studied in 10 consecutive parallel groups of healthy men, of whom eight received escalating intravenous doses of otamixaban as 6-hour infusions (1.7–183 micrograms/kg/hour) and two received a bolus dose (30 or 120 micrograms/kg) with a 6-hour infusion (60 or 140 micrograms/ kg/hour) (5). Otamixaban was said to be "well tolerated". Plasma concentrations increased with increasing dose, were maximal at the end-of-infusion, and fell rapidly as the infusion was stopped. Anti-factor Xa activity coincided with otamixaban plasma concentrations and clotting time measurements followed the same pattern.

Organs and Systems

Hematologic

In a multicenter, parallel-group, double-blind, double-dummy study, 621 patients undergoing elective total knee replacement were randomly assigned to oral BAY 59-7939 (2.5, 5, 10, 20, and 30 mg bd), starting 6–8 hours after surgery, or subcutaneous enoxaparin (30 mg bd, starting 12–24 hours after surgery) (6). Treatment was continued for 5–9 days. The frequency of major postoperative bleeding increased with increasing doses of BAY 59-7939. Bleeding end points were lower for the 2.5–10 mg bd dosages compared with higher dosages.

References

1. Nutescu EA, Pater K. Drug evaluation: the directly activated Factor Xa inhibitor otamixaban. IDrugs 2006;9(12):854–65.
2. Guertin KR, Choi YM. The discovery of the Factor Xa inhibitor otamixaban: from lead identification to clinical development. Curr Med Chem 2007;14(23):2471–81.
3. Stürzebecher A, Dönnecke D, Schweinitz A, Schuster O, Steinmetzer P, Stürzebecher U, Kotthaus J, Clement B, Stürzebecher J, Steinmetzer T. Highly potent and selective substrate analogue factor Xa inhibitors containing D-homophenylalanine analogues as P3 residue: part 2. ChemMedChem 2007;2(7):1043–53.
4. Kubitza D, Becka M, Wensing G, Voith B, Zuehlsdorf M. Safety, pharmacodynamics, and pharmacokinetics of BAY 59-7939—an oral, direct factor Xa inhibitor—after multiple dosing in healthy male subjects. Eur J Clin Pharmacol 2005;61(12):873–80.
5. Paccaly A, Ozoux ML, Chu V, Simcox K, Marks V, Freyburger G, Sibille M, Shukla U. Pharmacodynamic markers in the early clinical assessment of otamixaban, a direct factor Xa inhibitor. Thromb Haemost 2005;94(6):1156–63.
6. Turpie AG, Fisher WD, Bauer KA, Kwong LM, Irwin MW, Kälebo P, Misselwitz F, Gent M; OdiXa-Knee Study Group. BAY 59-7939: an oral, direct factor Xa inhibitor for the prevention of venous thromboembolism in patients after total knee replacement. A phase II dose-ranging study. J Thromb Haemost 2005;3(11):2479–86.

Direct thrombin inhibitors

General Information

The class of direct thrombin inhibitors includes hirudin, desirudin, lepirudin, and bivalirudin, and the tripeptide or peptidomimetic compounds argatroban, efegatran, inogatran, napsagatran, melagatran, and ximelagatran. They act by binding to the active site on thrombin and inhibiting its enzymatic activity. They thus inhibit fibrin formation, activation of anticoagulant factors V, VIII, and XIII, and protein C, and platelet aggregation. The antithrombin action occurs rapidly and is quickly reversible.

In a meta-analysis of individual patients' data from 11 randomized comparisons of direct thrombin inhibitors (hirudin, bivalirudin, argatroban, efegatran, or inogatran) with heparin, 35 970 patients were treated for up to 7 days and followed for at least 30 days (1). Compared with heparin, the direct thrombin inhibitors were associated with a lower risk of death or myocardial infarction at the end of treatment (OR = 0.85; 95% CI = 0.77, 0.94) and at 30 days (OR = 0.91; CI = 0.84, 0.99). There was no excess of intracranial hemorrhages with the direct thrombin inhibitors.

Argatroban

Argatroban has been used for the prophylaxis and treatment of thrombosis in patients with heparin-induced thrombocytopenia (2,3,4,5), heparin allergy (6), and various thrombotic disorders. Its effects can be monitored using the activated partial thromboplastin time for low doses and the activated clotting time for high doses. Its pharmacology, clinical pharmacology, and uses have been reviewed (7–14).

Dabigatran

Dabigatran etexilate is a prodrug of dabigatran, a specific, competitive, reversible inhibitor of thrombin. Dabigatran etexilate is rapidly absorbed after oral administration and converted to dabigatran. Its half-life is about 8 hours after a single dose and 14–17 hours after multiple doses. It is cleared renally.

Desirudin

Desirudin, similarly produced by recombinant techniques, has been used to prevent venous thromboembolism. It has been investigated in studies in patients with total hip replacements. In a large comparison of desirudin and low molecular weight heparin (enoxaparin), desirudin produced better prophylaxis against proximal deep vein thrombosis after total hip replacement (15) with no differences between the two groups with respect to perioperative, postoperative, and total blood loss. There were no cases of thrombocytopenia associated with desirudin.

Hirudin

Hirudin is the principal anticoagulant of the medicinal leech (*Hirudo medicinalis*). Though long known in its natural form, this is now produced by recombinant DNA techniques and characterized by a high affinity for thrombin, whether it is free in the plasma or absorbed onto the fibrin clot. The high incidence of hemorrhagic strokes in patients treated with both thrombolytic drugs and hirudin led to the redesign of three major trials (GUSTO-IIA, TIMI-9A, HIT III) of the effect of these antithrombotic agents in myocardial infarction (16–18). Hirudin is now regarded as the treatment of choice for heparin-induced thrombocytopenia. It is not recommended for use in pregnancy as it can cross the placenta; however, its safe use has been reported (19).

There is no specific antidote for the direct thrombin inhibitors: they are not neutralized by protamine sulfate.

Lepirudin

Lepirudin is a direct thrombin inhibitor obtained by recombinant technology from the medicinal leech and used for treatment of heparin-induced thrombocytopenia.

Ximelagatran

Ximelagatran is a prodrug that is converted to the active compound melagatran. With a fixed dose (which nevertheless needed to be altered in renal insufficiency) and no monitoring it was thought to be easier to use than warfarin and it was thought that it might replace it as the anticoagulant of choice for the prevention of stroke in high-risk patients with atrial fibrillation (20,21). However, it was associated with mainly asymptomatic rises in alanine transaminase activity during long-term use (>35 days) in a mean of 7.9% of patients in long-term clinical trials. Nearly all of the cases occurred within the first 6 months of therapy. Rare symptomatic cases have occurred. Owing to the unexplained relation between raised alanine transaminase, and in some cases a raised bilirubin, with a fatal outcome in three patients, the US Food and Drug Administration did not license ximelagatran. Consequently, the manufacturers withdrew ximelagatran from the European market.

Comparative studies

Ximelagatran versus dalteparin

In a randomized, controlled comparison of ximelagatran 24, 36, 48, or 60 mg bd and dalteparin followed by warfarin in 350 patients with deep vein thrombosis of the leg, there was no difference in the rate of regression of the thrombus; there was progression in 8% and 3% of patients respectively (22). Treatment was withdrawn because of bleeding in two patients taking ximelagatran (24 mg and 36 mg) and in two patients taking dalteparin and warfarin.

Ximelagatran versus enoxaparin

In a randomized, multicenter, double-blind study in patients undergoing total hip replacement ximelagatran 24 mg bd or subcutaneous enoxaparin 30 mg bd were compared with placebo for 7–12 days (23). Overall rates of total venous thromboembolism were 7.9% (62/782) with ximelagatran and 4.6% (36/775) with enoxaparin (absolute difference = 3.3%; 95%CI = 0.9, 5.7%). There were major bleeding events in 0.8% (7/906) patients who used ximelagatran and in 0.9% (8/910) of those who used enoxaparin.

In a double-blind study in 2788 patients undergoing total hip or knee replacement subcutaneous melagatran 3 mg followed by oral ximelagatran 24 mg bd was compared with subcutaneous enoxaparin 40 mg/day (24). Venous thromboembolism occurred in 355/1146 (31%) and 306/1122 (27%) patients in the ximelagatran and enoxaparin groups respectively, a difference in risk of 3.7% in favor of enoxaparin. Bleeding was comparable in the two groups.

Ximelagatran versus enoxaparin + warfarin

Ximelagatran 36 mg bd has been compared with standard enoxaparin + warfarin for prevention of recurrent venous thromboembolism in a 6-month, double-blind, randomized, non-inferiority study, the Thrombin Inhibitor in Venous Thromboembolism (THRIVE) Treatment Study in 2489 patients with acute deep vein thrombosis, of whom about one-third had a concomitant pulmonary embolism (25). Major bleeding occurred in 1.3% and 2.2% of those who used ximelagatran and enoxaparin + warfarin respectively, and the deaths rates were 2.3% and 3.4%. Alanine transaminase activity rose to more than three times the upper limit of normal in 119 patients (9.6%) and 25 patients (2.0%) respectively. The increased enzyme activity was mainly asymptomatic. Retrospective analysis of locally reported adverse events showed a higher rate of serious coronary events with ximelagatran (10/1240 patients) compared with enoxaparin + warfarin (1/1249 patients).

Ximelagatran versus warfarin

In a randomized, double-blind comparison of oral ximelagatran 24 or 36 mg bd for 7–12 days or warfarin in 1851 patients after total knee replacement, ximelagatran 36 mg bd was better than warfarin in preventing venous thromboembolism and death from all causes (20% versus 28%) (26). There were no significant differences in major bleeding (0.8% and 0.7% respectively), perioperative indicators of bleeding, wound characteristics, or the composite secondary end point of proximal deep-vein thrombosis, pulmonary embolism, and death (2.7% and 4.1%).

Placebo-controlled studies

Ximelagatran plus aspirin versus aspirin alone
In a placebo-controlled, double-blind, multicenter study, 1883 patients who had had a recent myocardial infarction were randomized to oral ximelagatran 24 mg, 36 mg, 48 mg, or 60 mg bd, or placebo for 6 months; all took aspirin 160 mg/day (27). Ximelagatran significantly reduced the risks of all-cause death, non-fatal myocardial infarction, and severe recurrent ischemia compared with placebo from 16% (102/638) to 13% (154/1245) (hazard ratio = 0.76; 95% CI = 0.59, 0.98). Major bleeding events were rare: 1.8% (23/1245) and 0.9% (6/638) (hazard ratio = 1.97; 95% CI = 0.80, 4.84) in the combined ximelagatran and placebo groups respectively.

Organs and Systems

Hematologic

Argatroban
In some patients overanticoagulation has occurred with argatroban and care with dosage is clearly required (28).

Dabigatran
In a multicenter, parallel-group, double-blind study, 1973 patients undergoing total hip or knee replacement were randomized to oral dabigatran etexilate for 6–10 days starting 1–4 hours after surgery or to subcutaneous enoxaparin starting 12 hours before surgery (29). Major bleeding with dabigatran was dose-related; it was significantly lower with dabigatran 50 mg bd than with enoxaparin (0.3% versus 2.0%) but higher at higher doses, nearly reaching statistical significance at a dose of 300 mg/day (4.7%).

Lepirudin
In 25 patients with a history of heparin-induced thrombocytopenia, who underwent a total of 36 percutaneous interventions and were given lepirudin, there was one procedure-related death from a retroperitoneal bleed and three patients had minor bleeding (30).

- A 50-year-old patient with heart failure and a dilated left ventricle had biventricular floating thrombi, for which lepirudin was given; the thrombi dissolved within 17 days, but the patient developed petechial bleeding, hemoptysis, and gross hematuria, and died of a subarachnoid hemorrhage (31).

Ximelagatran
In the SPORTIF III trial 1704 patients were treated with ximelagatran and 1703 with warfarin (32). The rates of major bleeding in the two groups were 1.3 and 1.8 per 100 patient years respectively (difference –0.5, 95% CI = –1.2, 0.2). This suggests that ximelagatran is at least as safe as warfarin in this regard. In 6% of patients taking ximelagatran serum alanine transaminase activity was over three times the upper limit of normal, consistent with the results of previous studies.

In a double-blind study, 2835 consecutive patients undergoing total hip or knee replacement were randomized to either melagatran/ximelagatran or enoxaparin for 8–11 days (33). The rates of major and total venous thromboembolism were significantly lower with melagatran/ximelagatran and fatal bleeding, critical site bleeding, and bleeding requiring re-operation did not differ between the two groups. However, "excessive bleeding as judged by the investigator", which was based on an observational assessment of bleeding from the wound, either during the procedure or during drainage, was more frequent with melagatran/ximelagatran than with enoxaparin.

In a meta-analysis of 12 randomized controlled trials of ximelagatran there was an absolute risk of major venous thromboembolism of 4.04% and 1.69% and of major bleeding episodes of 1.68% and 1.03% in prophylaxis and treatment trials respectively (34). In prophylaxis trials, there was significant excess mortality (OR = 2.5; 95% CI = 1.02, 6.13) and an excess of major bleeding episodes (OR = 1.41; 95% CI = 0.93, 2.14) in the whole ximelagatran group. There was an increase in the absolute risk of bleeding (from 1.04% to 3.03%) between postoperative and preoperative administration of ximelagatran.

Liver

DoTS classification (BMJ 2003;327:1222-5):
Reaction: Liver damage due to ximelagatran
Dose-relation: Collateral
Time-course: Intermediate, with tolerance
Susceptibility factors: Other diseases (simultaneous acute illnesses)

In a double-blind, placebo-controlled study, THRIVE III, patients with venous thromboembolism who had taken an anticoagulant for 6 months were randomized to extended secondary prevention with ximelagatran 24 mg bd or placebo for 18 months without monitoring (35). Death from any cause occurred in six patients who took ximelagatran and seven who took placebo; bleeding occurred in 134 and 111 patients respectively (HR = 1.19; 95% CI = 0.93, 1.53). The cumulative risk of a transient rise in the alanine transaminase activity to more than three times the upper limit of the reference range was 6.4% with ximelagatran compared with 1.2% with placebo. The main reason for withdrawal from the study was a rise in serum alanine transaminase activity (36).

In a double-blind, randomized study in 254 patients with non-valvular atrial fibrillation, ximelagatran (n = 187) 20, 40, or 60 mg bd was compared with warfarin (n = 67) (37). Alanine transaminase increased in eight patients taking ximelagatran, but normalized with continuous treatment or withdrawal.

Adjusted-dose warfarin and fixed-dose oral ximelagatran 36 mg bd have been compared in a double-blind, randomized, multicenter study in 3922 patients with non-valvular atrial fibrillation and additional risk factors for stroke (38). There was no difference between the groups in the rates of major bleeding, but there was more total bleeding (major and minor) with warfarin (37% versus 47% per year; 95% CI for the difference = 6, 14). Serum alanine transaminase activities rose to greater than three times the upper limit of the reference range in 6% of the patients who took ximelagatran, usually within 6 months, and typically improved whether or not treatment continued; however, there was one clear case of fatal liver disease and one other suggestive case.

In a prospective analysis of 6948 patients randomized to ximelagatran and 6230 patients randomized to a comparator (warfarin, low-molecular-weight heparin followed by warfarin, or placebo), the alanine transaminase activity rose to more than three times the upper limit of the reference range in 7.9% of the patients who received ximelagatran and 1.2% in the comparator group (39). The increase in alanine transaminase occurred at 1–6 months after the start of therapy, and there was recovery to less than twice the upper limit of the reference range in 96% of patients, whether they continued to take ximelagatran or not. A raised alanine transaminase activity was more common in those with simultaneous acute illnesses (acute myocardial infarction or venous thromboembolism). Combined rises in alanine transaminase activity (to three times the upper limit of normal) and total bilirubin concentration (to twice the upper limit of normal within 1 month of the rise in alanine transaminase), regardless of cause, were infrequent, occurring in 37 patients (0.5%) taking ximelagatran, of whom one had a severe hepatic illness that appeared to be resolving when the patient died from a gastrointestinal hemorrhage. No deaths were directly related to hepatic failure caused by ximelagatran.

Immunologic

Desirudin

Desirudin has a very low immunogenic potential. During repeated administration to 263 healthy volunteers, there were no signs or symptoms directly attributable to desirudin and only three volunteers exposed to a second course had allergic reactions with pruritic erythema attributable to desirudin in one case (40). In this study, specific antibodies directed against desirudin were detected in only one subject.

Lepirudin

The effects on activated partial thromboplastin time and the incidence and clinical relevance of antihirudin antibodies in patients treated with lepirudin have been studied using data from two prospective multicenter studies, in which patients with heparin-induced thrombocytopenia received one of four intravenous lepirudin dosage regimens (41). Of 196 evaluable patients, 87 (44%) had IgG antihirudin antibodies. The development of antihirudin antibodies depended on the duration of treatment (antibody-positive patients 18.6 days versus antibody-negative patients 11.6 days). Antihirudin antibodies were not associated with increases in clinical endpoints (limb amputation, new thromboembolic complications, or major bleeding). In 23 of 51 evaluable patients in whom antihirudin antibodies developed during treatment with lepirudin, the antibodies enhanced the anticoagulatory effect of lepirudin. During prolonged treatment with lepirudin, anticoagulatory activity should be monitored daily.

About 40% of patients who receive lepirudin for 5-10 days develop antihirudin antibodies, which lead to reduced renal elimination of lepirudin, increasing its effects. They also recognize epitopes on bivalirudin (42). A few patients with antihirudin antibodies develop hypersensitivity reactions to lepirudin on re-exposure.

- A 45-year-old African-American woman received lepirudin for anticoagulation after having type II heparin-induced thrombocytopenia (43). She developed supratherapeutic anticoagulation after 10 days, and lepirudin was withheld for 6 days. After it was restarted, she had an anaphylactic reaction. After the reaction had resolved, she was rechallenged with lepirudin, and the anaphylactic reaction recurred.

Using databases of lepirudin studies, 26 possible cases of anaphylaxis/severe allergy were found from 1994 to 2002 (44). Nine patients were judged to have had severe anaphylaxis within minutes of intravenous lepirudin, and four died. In these four cases, there had been a previous uneventful treatment course with lepirudin 1–12 weeks earlier. About 35 000 patients have received lepirudin, and so the authors estimated the risk of anaphylaxis as being about 0.015% on first exposure and 0.16% on re-exposure.

Lepirudin rarely causes allergic reactions after re-exposure. In a retrospective analysis of the medical records of 43 adults who had received at least two courses of lepirudin there were no cases of anaphylaxis or allergic reactions (45). On the first day of lepirudin therapy 10 patients had lower systolic blood pressures (by at least 20 mmHg) and four had systolic blood pressures of less than 100 mmHg. However, isolated asymptomatic falls in blood pressure after re-exposure to lepirudin most probably do not reflect anaphylactic reactions. Isolated and uncommon cases of anaphylaxis temporally related to lepirudin exposure should not preclude its use in patients with heparin-induced thrombocytopenia and past lepirudin exposure.

In two patients with a history of heparin-induced thrombocytopenia and anti-lepirudin antibodies who received argatroban and lepirudin intravenously, IgG reacting against lepirudin was not generated, in contrast to two patients taking lepirudin, in whom anti-lepirudin antibodies developed (46).

Susceptibility factors

Genetic

In 36 young healthy men of African, Asian, or Caucasian origin, ethnicity had no effect on the pharmacokinetics and pharmacodynamics of melagatran after oral ximelagatran 50 mg (47).

Age

The pharmacokinetics of oral ximelagatran and intravenous melagatran in elderly patients with non-valvular atrial fibrillation were similar to with those in matched healthy controls and dosage adjustment is therefore not necessary in these patients (48). In six young and six older subjects there were no age-dependent differences in the absorption and biotransformation of ximelagatran and differences in exposure to melagatran were explicable by differences in renal function (49).

Obesity

There were no differences in the pharmacokinetics or pharmacodynamics of melagatran between 12 obese and 12 non-obese subjects after oral administration of ximelagatran24 mg, suggesting that dose adjustment of ximelagatran in obesity (BMI up to 39 kg/m^2) is not necessary (50).

Hepatic impairment

The pharmacokinetics of ximelagatran were similar in 12 subjects with mild-to-moderate hepatic impairment (classified as Child-Pugh A or B) and 12 age-, weight-, and sex-matched controls with normal hepatic function, although the renal clearance of melagatran was 13% higher in hepatic impairment (51). Baseline prothrombin time was slightly longer in hepatic impairment, but when concentrations of melagatran were at their peak, the increase in prothrombin time from baseline values was the same in the two groups. Thus, dosage adjustment is not necessary in patients with mild-to-moderate impairment of hepatic function.

Renal impairment

After administration of subcutaneous melagatran and oral ximelagatran, 12 subjects with severe renal impairment had significantly higher melagatran exposure and longer half-lives than 12 controls with normal renal function, because of lower renal clearance of melagatran (52). These results suggest that in patients with severe renal impairment the dosage of ximelagatran should be reduced.

Other features of the patient

Little is known about the factors that increase the risk of adverse effects of argatroban.

- A 54-year-old white woman with an artificial mitral valve developed anasarca secondary to acute renal insufficiency and was given prophylactic argatroban (53). Despite normal hepatic function, she had a raised activated partial thromboplastin time for a prolonged period of time and required a significant dosage reduction. This prolonged effect persisted despite hemodialysis.

These observations suggest that in patients who are fluid-overloaded the anticoagulant effects of argatroban may be prolonged and that argatroban may not be removed by hemodialysis.

Drug Administration

Drug dosage regimens

Nine patients received lepirudin for thromboembolic disease according to the dosage recommendations approved by the European Agency for the Evaluation of Medicinal Products (EMEA): a 0.4 mg/kg bolus followed by 0.15 mg/kg/hour by intravenous infusion, adjusted to the activated partial thromboplastin time, aPTT, in order to maintain a patient:mean normal aPTT ratio of 1.5–2.5 (54). However, this dosage regimen turned out to be excessive. There were episodes of overdosage in eight cases, usually within the first 4 hours, after which the infusion was stopped for 2 hours and restarted at 50% of the previous dose. The dosage was then gradually reduced until equilibrium was achieved in the target range. The minimal maintenance infusion rate could be as low as 0.01 mg/kg/hour and the median rate was 0.04 mg/kg/hour. There were neither hemorrhagic nor thrombotic events. The authors' suggested that a bolus dose of lepirudin should be omitted in patients who do not have massive, life-threatening thrombosis, especially in elderly patients, and that therapy should start with an initial infusion rate of 0.10 mg/kg/hour only.

Drug–Drug Interactions

Acetylsalicylic acid

In a double-blind, randomized, two-way, crossover study in 12 healthy subjects aspirin, 450 mg on the day 1 and 150 mg just before intravenous administration of melagatran 4.12 mg on day 2, had no effect on the pharmacokinetics or pharmacodynamics of melagatran (55). However, the authors' conclusion that the two can be safely coadministered needs to be tempered by caution until evidence is available in patients who use these drugs.

Alcohol

In a randomized, open, two-way, crossover study in 26 men and women, alcohol 0.5 g/kg (women) or 0.6 g/kg (men) did not alter melagatran-induced prolongation of the activated partial thromboplastin time after the administration of a single oral dose of ximelagatran 36 mg (56).

Atorvastatin

In a randomized, two-way, crossover study in 16 healthy men and women, there were no reciprocal pharmacokinetic or pharmacodynamic interactions of a single oral dose of atorvastatin 40 mg and ximelagatran 36 mg bd for 5 days (57).

Digoxin

The pharmacokinetics of intravenous argatroban 1.5–2.0 μg/kg/minute were unaffected by co-administration of oral digoxin in healthy volunteers (58).

In a randomized, double-blind, placebo-controlled, two-way, crossover study in 16 healthy men and women, ximelagatran 36 mg for 8 days had no pharmacokinetic or pharmacodynamic effects on a single dose of digoxin 0.5 mg (59).

Erythromycin

Argatroban is metabolized by CYP3A4/5, and its pharmacokinetics might therefore be expected to be altered by inhibitors of CYP3A. However, in 14 healthy men erythromycin 500 mg qds had no effects on the pharmacokinetics of argatroban 1 μg/kg/minute infused over 5 hours (60).

Inhibitors of CYP isozymes

In vitro studies have shown no evidence of involvement of CYP isozymes in either the production of melagatran from ximelagatran or its elimination. Diclofenac (an inhibitor of CYP2C9) 50 mg orally, diazepam (an inhibitor of CYP2C19) 0.1 mg/kg intravenously, and nifedipine (an inhibitor of CYP3A4) 60 mg orally were coadministered with oral ximelagatran 24 mg in healthy volunteers; none of these drugs altered the pharmacokinetics of ximelagatran and ximelagatran did not alter their pharmacokinetics (61).

Lidocaine

The pharmacokinetics of intravenous argatroban 1.5–2.0 μg/kg/minute were unaffected by co-administration of intravenous lidocaine in healthy volunteers (58).

Paracetamol

The pharmacokinetics of intravenous argatroban 1.5–2.0 μg/kg/minute were unaffected by co-administration of oral paracetamol in healthy volunteers (58).

Monitoring therapy

The robustness and sensitivity of the different methods of monitoring therapy with direct thrombin inhibitors have been assessed in an international collaborative study using a panel of plasma samples spiked with lepirudin and argatroban (62). Activated partial thromboplastin time and the TAS analyser with ecarin clotting time cards gave the most reproducible results.

The ecarin clotting time (ECT) specifically reflects inhibition of meizothrombin by direct thrombin inhibitors (63) and is prolonged by vitamin K antagonists. Concomitant use of vitamin K antagonists with direct thrombin inhibitors may affect the two published ecarin clotting time methods differently. In 12 samples of normal plasma and 12 samples of plasma from patients taking stable warfarin, to which lepirudin (100–3000 ng/ml), argatroban (300–3000 ng/ml), and melagatran (30–1000 ng/ml) were added, two different assay methods produced different results. Use of the ecarin clotting time ratio improved but did not abolish the differences between the methods.

References

1. Direct Thrombin Inhibitor Trialists' Collaborative Group. Direct thrombin inhibitors in acute coronary syndromes: principal results of a meta-analysis based on individual patients' data. Lancet 2002;359(9303):294–302.
2. Verme-Gibboney CN, Hursting MJ. Argatroban dosing in patients with heparin-induced thrombocytopenia. Ann Pharmacother 2003;37:970–5.
3. Sakai K, Oda H, Honsako A, Takahashi K, Miida T, Higuma N. Obstinate thrombosis during percutaneous coronary intervention in a case with heparin-induced thrombocytopenia with thrombosis syndrome successfully treated by argatroban anticoagulant therapy. Catheter Cardiovasc Interv 2003;59:351–4.
4. Edwards JT, Hamby JK, Worrall NK. Successful use of Argatroban as a heparin substitute during cardiopulmonary bypass: heparin-induced thrombocytopenia in a high-risk cardiac surgical patient. Ann Thorac Surg 2003;75:1622–4.
5. Kieta DR, McCammon AT, Holman WL, Nielsen VG. Hemostatic analysis of a patient undergoing off-pump coronary artery bypass surgery with argatroban anticoagulation. Anesth Analg 2003;96:956–8.
6. Ohno H, Higashidate M, Yokosuka T. Argatroban as an alternative anticoagulant for patients with heparin allergy during coronary bypass surgery. Heart Vessels 2003;18:40–2.
7. Hursting MJ, Alford KL, Becker JC, Brooks RL, Joffrion JL, Knappenberger GD, Kogan PW, Kogan TP, McKinney AA, Schwarz RP Jr. Novastan (brand of argatroban): a small-molecule, direct thrombin inhibitor. Semin Thromb Hemost 1997;23(6):503–16.
8. Walenga JM. An overview of the direct thrombin inhibitor argatroban. Pathophysiol Haemost Thromb 2002;32(Suppl 3):9–14.
9. Ikoma H. Development of argatroban as an anticoagulant and antithrombin agent in Japan. Pathophysiol Haemost Thromb 2002;32(Suppl 3):23–8.
10. Fareed J, Hoppensteadt D, Iqbal O, Tobu M, Lewis BE. Practical issues in the development of argatroban: a perspective. Pathophysiol Haemost Thromb 2002;32(Suppl 3):56–65.
11. Hauptmann J. Pharmacokinetics of an emerging new class of anticoagulant/antithrombotic drugs. A review of small-molecule thrombin inhibitors. Eur J Clin Pharmacol 2002;57(11):751–8.
12. Kathiresan S, Shiomura J, Jang IK. Argatroban. J Thromb Thrombolysis 2002;13(1):41–7.
13. Kaplan KL, Francis CW. Direct thrombin inhibitors. Semin Hematol 2002;39(3):187–96.
14. Breddin HK. Experimentelle und klinische Befunde mit dem Thrombinhemmer Argatroban. [Experimental and clinical results with the thrombin inhibitor Argatroban.] Hämostaseologie 2002;22(3):55–9.
15. Eriksson BI, Wille-Jorgensen P, Kalebo P, Mouret P, Rosencher N, Bosch P, Baur M, Ekman S, Bach D, Lindbratt S, Close P. A comparison of recombinant hirudin with a low-molecular-weight heparin to prevent thromboembolic complications after total hip replacement. N Engl J Med 1997;337(19):1329–35.
16. Antman EM. Hirudin in acute myocardial infarction. Safety report from the Thrombolysis and Thrombin Inhibition in

Myocardial Infarction (TIMI) 9A Trial. Circulation 1994;90(4):1624–30.
17. Neuhaus KL, von Essen R, Tebbe U, Jessel A, Heinrichs H, Maurer W, Doring W, Harmjanz D, Kotter V, Kalhammer E, et al. Safety observations from the pilot phase of the randomized r-Hirudin for Improvement of Thrombolysis (HIT-III) study. A study of the Arbeitsgemeinschaft Leitender Kardiologischer Krankenhausarzte (ALKK) Circulation 1994;90(4):1638–42.
18. The Global Use of Strategies to Open Occluded Coronary Arteries (GUSTO) IIa Investigators. Randomized trial of intravenous heparin versus recombinant hirudin for acute coronary syndromes. Circulation 1994;90(4):1631–7.
19. Huhle G, Geberth M, Hoffmann U, Heene DL, Harenberg J. Management of heparin-associated thrombocytopenia in pregnancy with subcutaneous r-hirudin. Gynecol Obstet Invest 2000;49(1):67–9.
20. Bergsrud EA, Gandhi PJ. A review of the clinical uses of ximelagatran in thrombosis syndromes. J Thromb Thrombolysis 2003;16:175–88.
21. Hrebickova L, Nawarskas JJ, Anderson JR. Ximelagatran: a new oral anticoagulant. Heart Dis 2003;5:397–408.
22. Eriksson H, Wahlander K, Gustafsson D, Welin LT, Frison L, Schulman S; Thrive Investigators. A randomized, controlled, dose-guiding study of the oral direct thrombin inhibitor ximelagatran compared with standard therapy for the treatment of acute deep vein thrombosis: THRIVE I. J Thromb Haemost 2003;1:41–7.
23. Colwell CW Jr, Berkowitz SD, Davidson BL, Lotke PA, Ginsberg JS, Lieberman JR, Neubauer J, McElhattan JL, Peters GR, Francis CW. Comparison of ximelagatran, an oral direct thrombin inhibitor, with enoxaparin for the prevention of venous thromboembolism following total hip replacement. A randomized, double-blind study. J Thromb Haemost 2003;1:2119–30.
24. Eriksson BI, Agnelli G, Cohen AT, Dahl OE, Mouret P, Rosencher N, Eskilson C, Nylander I, Frison L, Ogren M; METHRO III Study Group. Direct thrombin inhibitor melagatran followed by oral ximelagatran in comparison with enoxaparin for prevention of venous thromboembolism after total hip or knee replacement. Thromb Haemost 2003;89:288–96.
25. Fiessinger JN, Huisman MV, Davidson BL, Bounameaux H, Francis CW, Eriksson H, Lundstrom T, Berkowitz SD, Nystrom P, Thorsen M, Ginsberg JS; THRIVE Treatment Study Investigators. Ximelagatran vs low-molecular-weight heparin and warfarin for the treatment of deep vein thrombosis: a randomized trial. JAMA 2005;293(6):681–9.
26. Francis CW, Berkowitz SD, Comp PC, Lieberman JR, Ginsberg JS, Paiement G, Peters GR, Roth AW, McElhattan J, Colwell CW Jr; EXULT A Study Group. Comparison of ximelagatran with warfarin for the prevention of venous thromboembolism after total knee replacement. New Engl J Med 2003;349:1703–12.
27. Wallentin L, Wilcox RG, Weaver WD, Emanuelsson H, Goodvin A, Nystrom P, Bylock A; ESTEEM Investigators. Oral ximelagatran for secondary prophylaxis after myocardial infarction: the ESTEEM randomised controlled trial. Lancet. 2003;362:789–97.
28. Reichert MG, MacGregor DA, Kincaid EH, Dolinski SY. Excessive argatroban anticoagulation for heparin-induced thrombocytopenia. Ann Pharmacother 2003;37:652–4.
29. Eriksson BI, Dahl OE, Buller HR, Hettiarachchi R, Rosencher N, Bravo ML, Ahnfelt L, Piovella F, Stangier J, Kalebo P, Reilly P; BISTRO II Study Group. A new oral direct thrombin inhibitor, dabigatran etexilate, compared with enoxaparin for prevention of thromboembolic events following total hip or knee replacement: the BISTRO II randomized trial. J Thromb Haemost 2005;3(1):103–11.
30. Cochran K, DeMartini TJ, Lewis BE, O Brien J, Steen LH, Grassman ED, Leya F. Use of lepirudin during percutaneous vascular interventions in patients with heparin-induced thrombocytopenia. J Invasive Cardiol 2003;15:617–21.
31. Skowasch D, Potzsch B, Kuntz-Hehner S, Gampert T, Rox J, Omran H, Bauriedel G, Luderitz B. Biventrikulare Thrombenauflosung und Antikorper-Bildung unter Lepirudin-Therapie. Dtsch Med Wochenschr 2003;128:1531–4.
32. Olsson SB. Stroke prevention with the oral direct thrombin inhibitor ximelagatran compared with warfarin in patients with non-valvular atrial fibrillation (SPORTIF III): randomised controlled trial. Lancet 2003;362:1691–8.
33. Eriksson BI, Agnelli G, Cohen AT, Dahl OE, Lassen MR, Mouret P, Rosencher N, Kalebo P, Panfilov S, Eskilson C, Andersson M, Freij A; EXPRESS Study Group. The direct thrombin inhibitor melagatran followed by oral ximelagatran compared with enoxaparin for the prevention of venous thromboembolism after total hip or knee replacement: the EXPRESS study. J Thromb Haemost 2003;1:2490–6.
34. Iorio A, Guercini F, Ferrante F, Nenci GG. Safety and efficacy of ximelagatran: meta-analysis of the controlled randomized trials for the prophylaxis or treatment of venous thromboembolism. Curr Pharm Des 2005;11(30):3893–918.
35. Schulman S, Wahlander K, Lundstrom T, Clason SB, Eriksson H; THRIVE III Investigators. Secondary prevention of venous thromboembolism with the oral direct thrombin inhibitor ximelagatran. N Engl J Med 2003;349:1713–21.
36. Schulman S, Lundstrom T, Walander K, Billing Clason S, Eriksson H. Ximelagatran for the secondary prevention of venous thromboembolism: a complementary follow-up analysis of the THRIVE III study. Thromb Haemost 2005;94(4):820–4.
37. Petersen P, Grind M, Adler J; SPORTIF II Investigators. Ximelagatran versus warfarin for stroke prevention in patients with nonvalvular atrial fibrillation. SPORTIF II: a dose-guiding, tolerability, and safety study. J Am Coll Cardiol 2003;41:1445–51.
38. Albers GW, Diener HC, Frison L, Grind M, Nevinson M, Partridge S, Halperin JL, Horrow J, Olsson SB, Petersen P, Vahanian A; SPORTIF Executive Steering Committee for the SPORTIF V Investigators. Ximelagatran vs warfarin for stroke prevention in patients with nonvalvular atrial fibrillation: a randomized trial. JAMA 2005;293(6):690–8.
39. Lee WM, Larrey D, Olsson R, Lewis JH, Keisu M, Auclert L, Sheth S. Hepatic findings in long-term clinical trials of ximelagatran. Drug Saf 2005;28(4):351–70.
40. Close P, Bichler J, Kerry R, Ekman S, Bueller HR, Kienast J, Marbet GA, Schramm W, Verstraete M. Weak allergenicity of recombinant hirudin CGP 39393 (REVASC) in immunocompetent volunteers. The European Hirudin in Thrombosis Group (HIT Group). Coron Artery Dis 1994;5(11):943–9.
41. Eichler P, Friesen HJ, Lubenow N, Jaeger B, Greinacher A. Antihirudin antibodies in patients with heparin-induced thrombocytopenia treated with lepirudin: incidence, effects on aPTT, and clinical relevance. Blood 2000;96(7):2373–8.
42. Eichler P, Lubenow N, Strobel U, Greinacher A. Antibodies against lepirudin are polyspecific and recognize epitopes on bivalirudin. Blood 2004;103(2):613–6.

43. Badger NO, Butler K, Hallman LC. Excessive anticoagulation and anaphylactic reaction after rechallenge with lepirudin in a patient with heparin-induced thrombocytopenia. Pharmacotherapy 2004;24(12):1800–3.
44. Greinacher A, Lubenow N, P Eichler. Anaphylactic and anaphylactoid reactions associated with lepirudin in patients with heparin-induced thrombocytopenia. Circulation 2003;108:2062–5.
45. Cardenas GA, Deitcher SR. Risk of anaphylaxis after reexposure to intravenous lepirudin in patients with current or past heparin-induced thrombocytopenia. Mayo Clin Proc 2005;80(4):491–3.
46. Harenberg J, Jorg I, Fenyvesi T, Piazolo L. Treatment of patients with a history of heparin-induced thrombocytopenia and anti-lepirudin antibodies with argatroban. J Thromb Thrombolysis 2005;19(1):65–9.
47. Johansson LC, Andersson M, Fager G, Gustafsson D, Eriksson UG. No influence of ethnic origin on the pharmacokinetics and pharmacodynamics of melagatran following oral administration of ximelagatran, a novel oral direct thrombin inhibitor, to healthy male volunteers. Clin Pharmacokinet 2003;42:475–84.
48. Wolzt M, Wollbratt M, Svensson M, Wahlander K, Grind M, Eriksson UG. Consistent pharmacokinetics of the oral direct thrombin inhibitor ximelagatran in patients with nonvalvular atrial fibrillation and in healthy subjects. Eur J Clin Pharmacol 2003;59:537–43.
49. Johansson LC, Frison L, Logren U, Fager G, Gustafsson D, Eriksson UG. Influence of age on the pharmacokinetics and pharmacodynamics of ximelagatran, an oral direct thrombin inhibitor. Clin Pharmacokinet 2003;42:381–92.
50. Sarich TC, Teng R, Peters GR, Wollbratt M, Homolka R, Svensson M, Eriksson UG. No influence of obesity on the pharmacokinetics and pharmacodynamics of melagatran, the active form of the oral direct thrombin inhibitor ximelagatran. Clin Pharmacokinet 2003;42:485–92.
51. Wahlander K, Eriksson-Lepkowska M, Frison L, Fager G, Eriksson UG. No influence of mild-to-moderate hepatic impairment on the pharmacokinetics and pharmacodynamics of ximelagatran, an oral direct thrombin inhibitor. Clin Pharmacokinet 2003;42:755–64.
52. Eriksson UG, Johansson S, Attman PO, Mulec H, Frison L, Fager G, Samuelsson O. Influence of severe renal impairment on the pharmacokinetics and pharmacodynamics of oral ximelagatran and subcutaneous melagatran. Clin Pharmacokinet 2003;42:743–53.
53. De Denus S, Spinler SA. Decreased argatroban clearance unaffected by hemodialysis in anasarca. Ann Pharmacother 2003;37:1237–40.
54. Hacquard M, de Maistre E, Lecompte T. Lepirudin: is the approved dosing schedule too high? J Thromb Haemost 2005;3(11):2593–6.
55. Fager G, Cullberg M, Eriksson-Lepkowska M, Frison L, Eriksson UG. Pharmacokinetics and pharmacodynamics of melagatran, the active form of the oral direct thrombin inhibitor ximelagatran, are not influenced by acetylsalicylic acid. Eur J Clin Pharmacol 2003;59:283–9.
56. Sarich TC, Johansson S, Schutzer KM, Wall U, Kessler E, Teng R, Eriksson UG. The pharmacokinetics and pharmacodynamics of ximelagatran, an oral direct thrombin inhibitor, are unaffected by a single dose of alcohol. J Clin Pharmacol 2004;44(4):388–93.
57. Sarich TC, Schutzer KM, Dorani H, Wall U, Kalies I, Ohlsson L, Eriksson UG. No pharmacokinetic or pharmacodynamic interaction between atorvastatin and the oral direct thrombin inhibitor ximelagatran. J Clin Pharmacol 2004;44(8):928–34.
58. Inglis AM, Sheth SB, Hursting MJ, Tenero DM, Graham AM, DiCicco RA. Investigation of the interaction between argatroban and acetaminophen, lidocaine, or digoxin. Am J Health Syst Pharm 2002;59(13):1258–66.
59. Sarich TC, Schutzer KM, Wollbratt M, Wall U, Kessler E, Eriksson UG. No pharmacokinetic or pharmacodynamic interaction between digoxin and the oral direct thrombin inhibitor ximelagatran in healthy volunteers. J Clin Pharmacol 2004;44(8):935–41.
60. Tran JQ, Di Cicco RA, Sheth SB, Tucci M, Peng L, Jorkasky DK, Hursting MJ, Benincosa LJ. Assessment of the potential pharmacokinetic and pharmacodynamic interactions between erythromycin and argatroban. J Clin Pharmacol 1999;39(5):513–9.
61. Bredberg E, Andersson TB, Frison L, Thuresson A, Johansson S, Eriksson-Lepkowska M, Larsson M, Eriksson UG. Ximelagatran, an oral direct thrombin inhibitor, has a low potential for cytochrome P450-mediated drug-drug interactions. Clin Pharmacokinet 2003;42:765–77.
62. Gray E, Harenberg J; ISTH Control of Anticoagulation SSC Working Group on Thrombin Inhibitors. Collaborative study on monitoring methods to determine direct thrombin inhibitors lepirudin and argatroban. J Thromb Haemost 2005;3(9):2096–7.
63. Fenyvesi T, Harenberg J, Weiss C, Jorg I. Comparison of two different ecarin clotting time methods. J Thromb Thrombolysis 2005;20(1):51–6.

Fondaparinux and Idraparinux

General Information

Fondaparinux is a synthetic pentasaccharide that mimics the site of heparin that binds to antithrombin III and inhibits factor Xa activity, which in turn inhibits thrombin generation. It does not release tissue factor pathway inhibitor. It is nearly completely absorbed after subcutaneous administration, has a rapid onset of action and a long half-life (14–20 hours), and is excreted by the kidneys. Its pharmacology, clinical pharmacology, and uses have been reviewed (1–16).

Idraparinux has a similar structure to that of fondaparinux but is modified, mostly by multiple methylation, to give it a higher affinity for antithrombin III and therefore a longer duration of action.

A review of clinical trials has shown that fondaparinux has a similar safety profile to that of enoxaparin with respect to clinically relevant major bleeding, including fatal bleeding, non-fatal bleeding, and bleeding requiring repeat surgery (17).

Fondaparinux has been approved in some countries for the prophylaxis of venous thrombosis after orthopedic surgery in a fixed dose of 2.5 mg/day without monitoring. A structural analogue, idraparinux sodium, has additional methyl groups, a long half-life, and once-weekly administration. Both drugs are being developed as antithrombotic drugs for venous and arterial thrombosis, acute coronary syndrome, and stroke, and as adjuncts to thrombolytic therapy.

There is currently no specific antidote for fondaparinux: it is not neutralized by protamine sulfate. Fondaparinux shows no cross-reactivity with antibodies associated with heparin-induced thrombocytopenia.

Comparative studies

There have been two double-blind, randomized comparisons of fondaparinux 2.5 mg/day or enoxaparin 40 mg/day after hip fracture surgery (18), or elective major knee surgery (19). Fondaparinux reduced the risk of thromboembolism by 56% in the first study (n = 1250) and by 55% in the second (n = 724). There were no significant differences between the two groups in the first study in the incidence of death or clinically important bleeding. In the second study, major bleeding occurred more often with fondaparinux, but there were no significant differences between the two groups in the incidence of bleeding leading to death or reoperation, or occurring in a critical organ.

In a double-blind, randomized study of 2309 consecutive patients undergoing elective hip replacement, postoperative subcutaneous fondaparinux 2.5 mg/day was compared with preoperative enoxaparin 40 mg/day (20). By day 11, venous thromboembolism had occurred in 85 (9%) of 919 patients assigned to enoxaparin and in 37 (4%) of 908 patients assigned to fondaparinux, a relative risk reduction of 56% (95% CI = 33, 73). In a similar comparison of postoperative fondaparinux 2.5 mg/day and enoxaparin 30 mg bd in 2275 consecutive patients undergoing elective hip replacement, by day 11 venous thromboembolism had occurred in 48 (6%) of 787 patients assigned to fondaparinux and in 66 (8%) of 797 patients assigned to enoxaparin, a relative risk reduction of 26% (95% CI = 11, 53) (21).

In a meta-analysis of these studies, bleeding was slightly, but not significantly, more frequent with fondaparinux than with enoxaparin (22,23).

Hematologic

Of 1103 patients randomly assigned to receive fondaparinux 42 (3.8%) had recurrent thromboembolic events, compared with 56 of 1110 patients randomly assigned to receive unfractionated heparin (5.0%), an absolute difference of –1.2% in favor of fondaparinux (95%CI = –3.0, 0.5) (24). There was major bleeding in 1.3% of those treated with fondaparinux and 1.1% of those treated with unfractionated heparin. Mortality rates at 3 months were similar in the two groups.

In four phase III multicenter, randomized, parallel-group, double-blind trials in 375 centers in Argentina, Australia and New Zealand, Europe, North America, and South Africa, 7344 patients (mean age 68 years, 60% women) undergoing elective total hip arthroplasty (two studies), elective knee arthroplasty (one study), or hip fracture surgery (one study) were given either subcutaneous fondaparinux (n = 3668) 2.5 mg once daily, beginning six hours after surgery or subcutaneous enoxaparin (n = 3676) 30 mg bd beginning 12 hours before surgery; 7237 received at least one dose of the study drug and 5385 completed the study (25). The incidence of venous thromboembolism was significantly lower with fondaparinux (182/2682) than enoxaparin (371/2703). However, major bleeding events were significantly more frequent with fondaparinux (96/3616) than enoxaparin (63/3621). The incidences of death, minor bleeding, and other adverse events did not differ between the two groups.

In a double-blind multicenter trial, 656 patients undergoing hip fracture surgery were randomly assigned to receive prophylaxis with once-daily subcutaneous fondaparinux sodium 2.5 mg of placebo for 19–23 days (26). Before randomization, all had received fondaparinux for 6–8 days. Fondaparinux reduced the incidence of venous thromboembolism from 35% (77/220) to 1.4% (3/208), with a relative risk reduction of 96% (95%CI = 87, 100%). Although there was a trend towards more major bleeding with fondaparinux than with placebo, there were no differences between the two groups in the incidence of clinically relevant bleeding (leading to death, re-operation, or critical organ bleeding).

Liver

The effects of idraparinux on plasma liver enzyme activities (gamma-glutamyl-transferase, aspartate transaminase, and alanine transaminase) have been studied in 37 patients with deep vein thrombosis in the PERSIST trial (27). Patients were first treated with weight-adjusted enoxaparin for 4–7 days and then randomized to either idraparinux (2.5, 5, 7.5 or 10 mg) or warfarin. Gamma-glutamyl-transferase was significantly increased after enoxaparin at the baseline visit and at week 2, but returned to screening values at week 3 for the remainder of the study. Transaminases were significantly increased at the baseline visit and returned to screening values at week 2 for the remainder of the study. There was no significant difference between the mean values of plasma liver enzymes in the four idraparinux groups and the warfarin group. The authors concluded that idraparinux, in contrast to enoxaparin, does not alter plasma liver enzyme activities significantly.

Immunologic

Delayed-type hypersensitivity reactions at heparin injection sites, in the form of eczema-like infiltrated plaques, with occasional generalized reactions, are common adverse effects; cross-reactivity with fondaparinux has been reported [[Hohenstein 149).

- An 85-year-old woman developed localized infiltrated plaques followed by a disseminated exanthematous reaction 10 days after the start of dalteparin therapy. Intradermal tests with a comprehensive series of undiluted commercial drugs yielded positive reactions on day 7 to all forms of heparin and the heparinoid danaparoid, whereas fondaparinux and lepirudin were negative. Subcutaneous provocation with lepirudin was tolerated, but there was a positive reaction to fondaparinux at each site of three therapeutic subcutaneous injections. Later, a second intradermal test with fondaparinux was positive.

Cross-reactivity between high- and low-molecular-weight heparins is well known and the pentasaccharide fondaparinux has been recommended as a safe alternative. However, the authors concluded that in patients with delayed-type hypersensitivity to high- and low-molecular-weight heparins and danaparoid, fondaparinux cannot reliably be recommended as a therapeutic alternative.

In another series of seven women with delayed-type allergy to heparins and semisynthetic heparinoids one also had a delayed-type allergic reaction to fondaparinux (28). The authors suggested that the rare cross-reaction between fondaparinux and heparins may be due to differences in the response to haptens.

In patients with delayed-type hypersensitivity to heparin, cross-reactions often occur. Ultra-low-molecular-weight heparins may be a therapeutic alternative in some cases, but not in all patients with delayed-type hypersensitivity skin reactions at subcutaneous heparin injection sites (29).

- A 59-year-old woman developed localized pruritic skin lesions after receiving nadroparin for 3 days. Patch tests and prick and intracutaneous tests with heparin, low molecular weight heparins, fondaparinux, heparinoid, and xylanolpolyhydrogensulfates were positive to all except pentosanpolysulfate, a recombinant form of desirudin. Fondaparinux produced a progressive itchy erythematous infiltrated patch at the injection site. Hirudin was well tolerated.

Second-Generation Effects

Pregnancy

Five pregnant patients in whom anticoagulant therapy was indicated and who had severe cutaneous allergic reactions to low molecular weight heparin were treated with fondaparinux 2.5 mg/day (30). There were no allergic reactions at injection sites, thromboembolic events, or abnormal bleeding and no adverse effects in the neonates. In four of the patients, anti-factor Xa activity in umbilical cord blood was increased. The concentration of fondaparinux detected in umbilical cord plasma by the chromogenic assay was about one-tenth the concentration in maternal plasma. The authors concluded that the use of fondaparinux in pregnant women might best be limited to those for whom there are no obvious therapeutic alternatives, such as patients with heparin-induced thrombocytopenia or severe allergic reactions to heparin.

Drug–Drug Interactions

Aspirin

In healthy volunteers, steady-state fondaparinux did not alter the pharmacodynamic effects of a single dose of aspirin 975 mg (31). Aspirin did not alter the pharmacokinetics of fondaparinux.

Digoxin

In a randomized, crossover study in 24 healthy volunteers, the pharmacokinetics of fondaparinux sodium 10 mg/day subcutaneously for 7 days were unaffected by digoxin 0.25 mg/day orally for 7 days (32).

Piroxicam

In healthy volunteers, steady-state fondaparinux did not alter the pharmacodynamic effects of a single dose of piroxicam 20 mg (31). Piroxicam did not alter the pharmacokinetics of fondaparinux.

Warfarin

There was no pharmacodynamic interaction of subcutaneous fondaparinux 4 mg with warfarin in 12 healthy men (33).

References

1. Keam SJ, Goa KL. Fondaparinux sodium. Drugs 2002;62(11):1673–85.
2. Walenga JM, Jeske WP, Samama MM, Frapaise FX, Bick RL, Fareed J. Fondaparinux: a synthetic heparin pentasaccharide as a new antithrombotic agent. Expert Opin Investig Drugs 2002;11(3):397–407.
3. Regazzoni S, de Moerloose P. Deux nouveaux agents antithrombotiques très prometteurs: le pentasaccharide et le ximelagatran. [Two new very promising antithrombotic agents: pentasaccharide and ximelagatran.] Rev Med Suisse Romande 2002;122(1):29–33.
4. Bounameaux H, Perneger T. Fondaparinux: a new synthetic pentasaccharide for thrombosis prevention. Lancet 2002;359(9319):1710–1.
5. Bauer KA, Hawkins DW, Peters PC, Petitou M, Herbert JM, van Boeckel CA, Meuleman DG. Fondaparinux, a synthetic pentasaccharide: the first in a new class of antithrombotic agents—the selective factor Xa inhibitors. Cardiovasc Drug Rev 2002;20(1):37–52.
6. Bauer KA. Selective inhibition of coagulation factors: advances in antithrombotic therapy. Semin Thromb Hemost 2002;28(Suppl 2):15–24.
7. Turpie AG. Optimizing prophylaxis of venous thromboembolism. Semin Thromb Hemost 2002;28(Suppl 2):25–32.
8. Buller HR. Treatment of symptomatic venous thromboembolism: improving outcomes. Semin Thromb Hemost 2002;28(Suppl 2):41–8.
9. Turpie AG. Pentasaccharides. Semin Hematol 2002;39(3):158–71.
10. Gallus AS, Coghlan DW. Heparin pentasaccharide. Curr Opin Hematol 2002;9(5):422–9.
11. Bauer KA, Eriksson BI, Lassen MR, Turpie AG. Factor Xa inhibition in the prevention of venous thromboembolism and treatment of patients with venous thromboembolism. Curr Opin Pulm Med 2002;8(5):398–404.
12. Garces K. Fondaparinux for post-operative venous thrombosis prophylaxis. Issues Emerg Health Technol 2002;(37):1–4.
13. Petitou M, Duchaussoy P, Herbert JM, Duc G, El Hajji M, Branellec JF, Donat F, Necciari J, Cariou R, Bouthier J, Garrigou E. The synthetic pentasaccharide fondaparinux: first in the class of antithrombotic agents that selectively inhibit coagulation factor Xa. Semin Thromb Hemost 2002;28(4):393–402.
14. Cheng JW. Fondaparinux: a new antithrombotic agent. Clin Ther 2002;24(11):1757–69.
15. Reverter JC. Fondaparinux sodium. Drugs Today (Barc) 2002;38(3):185–94.

16. Samama MM, Gerotziafas GT, Elalamy I, Horellou MH, Conard J. Biochemistry and clinical pharmacology of new anticoagulant agents. Pathophysiol Haemost Thromb 2002;32(5–6):218–24.
17. Tran AH, Lee G. Fondaparinux for prevention of venous thromboembolism in major orthopedic surgery. Ann Pharmacother 2003;37:1632–43.
18. Eriksson BI, Bauer KA, Lassen MR, Turpie AG. Steering Committee of the Pentasaccharide in Hip-Fracture Surgery Study. Fondaparinux compared with enoxaparin for the prevention of venous thromboembolism after hip-fracture surgery. N Engl J Med 2001;345(18):1298–304.
19. Bauer KA, Eriksson BI, Lassen MR, Turpie AG. Steering Committee of the Pentasaccharide in Major Knee Surgery Study. Fondaparinux compared with enoxaparin for the prevention of venous thromboembolism after elective major knee surgery. N Engl J Med 2001;345(18):1305–10.
20. Lassen MR, Bauer KA, Eriksson BI, Turpie AG. European Pentasaccharide Elective Surgery Study (EPHESUS) Steering Committee. Postoperative fondaparinux versus preoperative enoxaparin for prevention of venous thromboembolism in elective hip-replacement surgery: a randomised double-blind comparison. Lancet 2002;359(9319):1715–20.
21. Turpie AG, Bauer KA, Eriksson BI, Lassen MR. PENTATHALON 2000 Study Steering Committee. Postoperative fondaparinux versus postoperative enoxaparin for prevention of venous thromboembolism after elective hip-replacement surgery: a randomised double-blind trial. Lancet 2002;359(9319):1721–6.
22. Turpie AG, Bauer KA, Eriksson BI, Lassen MR. Fondaparinux vs enoxaparin for the prevention of venous thromboembolism in major orthopedic surgery: a meta-analysis of 4 randomized double-blind studies. Arch Intern Med 2002;162(16):1833–40.
23. Turpie AG, Eriksson BI, Lassen MR, Bauer KA. A meta-analysis of fondaparinux versus enoxaparin in the prevention of venous thromboembolism after major orthopaedic surgery. J South Orthop Assoc 2002;11(4):182–8.
24. Buller HR, Davidson BL, Decousus H, Gallus A, Gent M, Piovella F, Prins MH, Raskob G, Van den Berg-Segers AE, Cariou R, Leeuwenkamp O, Lensing AW; Matisse Investigators. Subcutaneous fondaparinux versus intravenous unfractionated heparin in the initial treatment of pulmonary embolism. New Engl J Med 2003;349:1695–702; erratum 2004;350:423.
25. Hardy JF. Best evidence in anesthetic practice: prevention: fondaparinux is better than enoxaparin for prevention of major venous thromboembolism after orthopedic surgery. Can J Anaesth 2003;50:764–6.
26. Eriksson BI, Lassen MR; PENTasaccharide in HIp-FRActure Surgery Plus Investigators. Duration of prophylaxis against venous thromboembolism with fondaparinux after hip fracture surgery: a multicenter, randomized, placebo-controlled, double-blind study. Arch Intern Med 2003;163:1337–42.
27. Reiter M, Bucek RA, Koca N, Heger J, Minar E; PERSIST. Idraparinux and liver enzymes: observations from the PERSIST trial. Blood Coagul Fibrinolysis 2003;14:61–5.
28. Jappe U, Juschka U, Kuner N, Hausen BM, Krohn K. Fondaparinux: a suitable alternative in cases of delayed-type allergy to heparins and semisynthetic heparinoids? A study of 7 cases. Contact Dermatitis 2004;51(2):67–72.
29. Maetzke J, Hinrichs R, Schneider L-A, Scharffetter-Kochanek K. Unexpected delayed-type hypersensitivity skin reactions to the ultra-low-molecular-weight heparin fondaparinux. Allergy 2005;60:413–5.
30. Dempfle CE. Minor transplacental passage of fondaparinux in vivo. N Engl J Med 2004;350(18):1914–5.
31. Ollier C, Faaij RA, Santoni A, Duvauchelle T, van Haard PM, Schoemaker RC, Cohen AF, de Greef R, Burggraaf J. Absence of interaction of fondaparinux sodium with aspirin and piroxicam in healthy male volunteers. Clin Pharmacokinet 2002;41(Suppl 2):31–7.
32. Mant T, Fournie P, Ollier C, Donat F, Necciari J. Absence of interaction of fondaparinux sodium with digoxin in healthy volunteers. Clin Pharmacokinet 2002;41(Suppl 2):39–45.
33. Faaij RA, Burggraaf J, Schoemaker RC, Van Amsterdam RG, Cohen AF. Absence of an interaction between the synthetic pentasaccharide fondaparinux and oral warfarin. Br J Clin Pharmacol 2002;54(3):304–8.

Heparins

General Information

Heparins are mucopolysaccharides whose molecules are of varying lengths. Unfractionated (standard) heparin contains molecules of average molecular weight of 12 000–15 000 Da. Low molecular weight heparins contain molecules whose average molecular weight is below 5000 Da. Each formulation of a low molecular weight heparin contains a different range of sizes of molecules, but they have in common more activity against factor X than thrombin compared with standard heparin, a longer duration of action, and a lower risk of thrombocytopenia. They can be given subcutaneously once a day without the monitoring of activated partial thromboplastin time that is necessary for standard heparin. Low molecular weight heparins include bemiparin, certoparin, dalteparin, enoxaparin, nadroparin, reviparin, and tinzaparin.

Non-anticoagulant uses

A therapeutic role for heparin has been proposed in inflammatory bowel disease, particularly in ulcerative colitis (1). This beneficial response may result from mechanisms other than anticoagulation, including the restoration of high-affinity receptor binding by antiulcerogenic growth factors, such as basic fibroblast growth factor, that normally rely on the presence of heparian sulfate proteoglycans (2).

Observational studies

A new entity of autoimmune sensorineural hearing loss has been proposed and treated with subcutaneous enoxaparin 2000 IU bd for 10 days in a placebo-controlled study in 30 patients (3). All those who received enoxaparin had both subjective and objective improvement and there were no adverse effects.

General adverse effects

The major adverse effect of heparin is bleeding. Thrombocytopenia occurs in some 5% of patients receiving standard heparin but is uncommon in those receiving low molecular weight heparins; bleeding as a result is rare. Long-term use can lead to osteoporosis. General vasospastic reactions have been reported and, exceptionally, skin necrosis can occur. Allergic reactions to heparin are well-known but rare. Tumor-inducing effects have not been described and the possibility of an anti-tumor effect has been raised.

Organs and Systems

Cardiovascular

Bolus administration of heparin causes vasodilatation and a fall in arterial blood pressure of 5–10 mmHg (4). Some convincing data have been reported concerning the role in these reactions of chlorbutol which has been used as a bactericidal and fungicidal ingredient in some heparin formulations (5).

Cardiogenic shock can occur in parallel with disseminated intravascular coagulation (6). In these circumstances heparin is thought to act as a hapten in a heparin–protein interaction that stimulates antibody production and an antigen–antibody reaction associated with release of platelet and vasoactive compounds.

General vasospastic reactions have been described in patients receiving heparin, exceptionally complicated by skin necrosis (7). Vasospastic reactions are probably part of the syndrome of thrombohemorrhagic complications considered above.

Cholesterol crystal embolism is a rare complication of anticoagulant treatment of ulcerative atheroma of the great arteries and has been attributed to low-molecular-weight heparins in three cases (8).

Nervous system

Effects of heparin on the nervous system, other than those due to bleeding, have not been reported.

Endocrine

Heparin-induced hypoaldosteronism is well documented, both in patients treated with standard heparin, even at low doses, and in patients treated with low molecular weight heparin (9,10). The most important mechanism of aldosterone inhibition appears to be a reduction in both the number and affinity of angiotensin II receptors in the zona glomerulosa (9). A direct effect of heparin on aldosterone synthesis, with inhibition of conversion of corticosterone to 18-hydroxycorticosterone, has also been suggested. This effect is believed to be responsible for the hyperkalemia that can occur in heparin-treated patients with impaired renal function and particularly in patients on chronic hemodialysis (11), or with diabetes mellitus, or who are taking other potentially hyperkalemic drugs.

Metabolism

Heparin has a strong clearing action on postprandial lipidemia by activating lipoprotein lipase. This has been thought to be associated with an increase in free fatty acid-induced dysrhythmias and death in patients with myocardial infarction.

- Substantial hypertriglyceridemia occurred in a pregnant woman who received long-term subcutaneous heparin treatment (12).

However, the risk of hyperlipidemia seems to have been exaggerated, since the extent of lipolysis is usually small (13).

Electrolyte balance

Hyperkalemia is an occasional complication of heparin therapy, and is often forgotten until life-threatening dysrhythmias have occurred (14). It has been attributed to hypoaldosteronism, and fludrocortisone has been used to treat it (15). It has been suggested that marked hyperkalemia is only likely to occur in the presence of other factors that alter potassium balance (16).

Low molecular weight heparin is less likely to cause hyperkalemia than standard heparin. In 28 men, mean age 70 years, given low molecular weight heparin (40 mg subcutaneously every 12 hours) for deep venous thrombosis prophylaxis after an operation, the serum potassium concentration did not change significantly after 4 days of therapy (4.25 mmol/l before therapy and 4.35 mmol/l) (17).

However, in 85 patients enoxaparin therapy was associated with an increase in mean potassium concentration from 4.26 mmol/l at baseline to 4.43 mmol/l on the third day; potassium concentrations exceeded 5.0 mmol/l in 9% (18). There was no life-threatening or symptomatic hyperkalemia. Neither plasma renin activity nor aldosterone concentrations changed significantly and there was no correlation between the increase in potassium concentrations and the presence of diabetes mellitus or treatment with angiotensin converting enzymes inhibitors, angiotensin receptor blockers, beta-blockers, or potassium-wasting diuretics.

Hematologic

Hemorrhage

The major adverse effect of heparin is bleeding, and it can occur with low molecular weight heparins as well as with standard heparin (19).

> DoTS classification (BMJ 2003;327:1222–5)
> Dose-relation: toxic effect
> Time-course: early
> Susceptibility factors: treatment with antiplatelet drugs, sex, age, renal function

Frequency

An authoritative review of the relevant literature was published in 1993 and is still valid (20). Its authors concluded that the use of heparin in therapeutic dosages (over 15 000 IU/day) is typically marked by an average daily frequency of fatal, major, and major and minor bleeding episodes of 0.05, 0.8, and 2.0% respectively; these frequencies are about twice those expected without heparin therapy. In 2656 medical patients studied by the Boston Collaborative Drug Surveillance Program, the crude risk of bleeding from heparin (route unspecified) was 9% but varied with dose: 4.9% for doses below 50 IU/kg per dose, 8.1% for doses of 50–99 IU/kg per dose, and 17% for doses of 100 IU/kg per dose or more (21). The 7-day cumulative risk for any kind of bleeding was 9.1%. Melena, hematomas, and macrohematuria were the most frequent manifestations; intracranial bleeding and pulmonary bleeding were rare.

Presentation

In the Boston Collaborative Drug Surveillance Program study mentioned above (21), the localization of bleeding, expressed in percentages and broadly classified as major/minor, was as follows:

- gastrointestinal hemorrhage 49/11
- vaginal hemorrhage 11/8 (1/3 postpartum)
- bleeding from wounds and accidental soft-tissue trauma 11/28
- retroperitoneal bleeding 6/0
- genitourinary bleeding other than vaginal 6/35
- intracranial hemorrhage 5/0
- epistaxis 4/10
- other forms of blood loss 8/8.

An FDA Health Advisory warning was issued in December 1997 after 30 cases of spinal hematoma had been reported in patients undergoing spinal or epidural anesthesia while receiving low molecular weight heparin perioperatively. In the European literature, the risk of spinal hematoma in patients receiving low molecular weight heparin was not considered clinically significant. In Europe, low molecular weight heparin is given once a day and at a smaller total daily dose; this enables the placement (and removal) of needles and catheters during periods of reduced low molecular weight heparin activity. Identification of further risk factors is difficult, owing to the rarity of spinal hematoma. However, several possible risk factors have been suggested: about 75% of the patients were elderly women, in 22 patients an epidural catheter was used, and in 17 patients the first dose of low molecular weight heparin was administered while the catheter was indwelling; 12 patients received antiplatelet drugs and/or warfarin in addition to low molecular weight heparin (22). The authors formulated recommendations for minimizing the risk of spinal hematoma, including using the smallest effective dose of low molecular weight heparin perioperatively, delaying heparin therapy as long as possible postoperatively, and removing catheters when anticoagulant activity is low. Furthermore, they warned against combining low molecular weight heparin with antiplatelet drugs or oral anticoagulants.

Relation to dose

Subgroup analyses of randomized trials and prospective cohort studies point to an association, as one would expect, between the incidence of bleeding and the anticoagulant response as measured by a test of blood coagulation, for example the activated partial thromboplastin time (APTT). For instance, there were five major bleeding episodes in 10 patients whose APTT was prolonged to more than twice the upper limit of their therapeutic range but only one episode in 40 patients whose APTT remained in the therapeutic range (relative risk 20) (23). Furthermore, there is an increased rate of major bleeding with intermittent intravenous use compared with continuous intravenous heparin infusion. No differences in major bleeding are generally detected between continuous intravenous and subcutaneous heparin (24).

Time-course

In the Boston Collaborative Drug Surveillance Program study mentioned above, the peak incidence of bleeding was on the third day after the start of heparin therapy, which suggests a relatively safe initial 48-hour period in systemic heparinization (21). However, bleeding can be expected at any time during therapy, if the dosage becomes excessive or if susceptibility factors, such as trauma, intervene.

Susceptibility factors

In the Boston Collaborative Drug Surveillance Program study mentioned above, bleeding correlated with aspirin treatment, with sex, and probably with age and renal function (21). A meta-analysis of randomized trials reported that those aged over 70 are associated with the risk of major bleeding (25). Recent surgery or trauma also increase the risk.

Low molecular weight heparin

Because of its reduced activity on overall clotting, low molecular weight heparin is expected to cause less bleeding than standard heparin. However, several meta-analyses of randomized comparisons of low molecular weight heparin with standard heparin in the prevention of postoperative deep venous thrombosis showed no difference in the incidence of major bleeding (26,27). For low molecular weight heparin, the rates of major bleeding ranged from 0 to 3% and fatal bleeding from 0 to 0.8% (24). A later meta-analysis of randomized comparisons in the initial curative treatment of deep venous thrombosis showed a 35% reduction in major bleeding in patients treated with low molecular weight heparin, but this result was not significant (28).

Thrombocytopenia

Thrombocytopenia (100×10^9/l or lower) occurs in about 3–7% of patients, independent of the mode of administration or the dose and type of heparin (29,30). The incidence varies largely among published series, but the overall average may be some 5% (31). There was a higher incidence in some older studies (32), but this may have been due to the presence of impurities in the heparin then available, and some of the studies clearly included cases of thrombocytopenia not attributable to heparin. Even the current incidence is to some extent uncertain, because of differences in methods of diagnosis and in the definitions used (33,34). Low molecular weight heparin is significantly less likely to cause thrombocytopenia than unfractionated heparin (30).

Heparin-induced thrombocytopenia (35) has been recognized in adults for some time, but only recently in neonates and children (36). There are two types. Type I is non-immunogenic, mild, and self-limiting. Type II is a less common but severe immune reaction that leads to thrombocytopenia and often thromboembolic complications; it is generally seen after a week or more of treatment.

A third variety, so-called "delayed-onset heparin-induced thrombocytopenia" has also been described in several reports. In 12 patients, recruited from secondary and tertiary care hospitals, thrombocytopenia and associated thrombosis occurred at a mean of 9.2 (range 5–19) days after the withdrawal of heparin; nine received additional heparin, with further falls in platelet counts (37). In a retrospective case series, 14 patients, seen over a 3-year period, developed thromboembolic complications a median of 14 days after treatment with heparin (38). The emboli were venous ($n = 10$), or arterial ($n = 2$), or both ($n = 2$); of the 12 patients with venous embolism, 7 had pulmonary embolism. Platelet counts were mildly reduced in all but two patients at the time of the second presentation. On readmission, 11 patients received therapeutic heparin, which worsened their clinical condition and further reduced the platelet count.

Heparin can rarely cause acute cardiorespiratory reactions, and some reports relate this to underlying heparin-induced thrombocytopenia. Over 2 years four cardiovascular surgery patients were identified who had eight episodes of cardiorespiratory collapse immediately after heparin administration (39). All had underlying heparin-induced thrombocytopenia. They received intravenous boluses of unfractionated heparin. Two had severe respiratory distress within 15 minutes, for which they required endotracheal intubation. Two others had cardiac arrest or a dysrhythmia within minutes of receiving intravenous heparin. Serological tests for heparin-induced antibodies were positive in all cases. In three cases, the platelet count was normal or near normal, but fell dramatically immediately after the heparin bolus. Three patients had prior diagnoses of heparin-induced thrombocytopenia, but their prescribers gave them heparin either unaware of the diagnosis or ignorant of its significance. The authors noted that heparin administration to patients with heparin-induced antibodies can result in life-threatening pulmonary or cardiac events.

A spinal-epidural hematoma occurred after combined spinal–epidural anesthesia in a woman who had been taking clopidogrel and had received perioperative dalteparin for thromboprophylaxis (40). This occurred despite adherence to standard guidelines on the administration of low-molecular-weight heparin perioperatively and withdrawal of clopidogrel 7 days before the anesthetic.

- A 44-year-old woman developed delayed-onset thrombocytopenia and cerebral thrombosis 7 days after a single dose of standard heparin 5000 units (41).

Type I heparin-induced thrombocytopenia

DoTS classification (BMJ 2003;327:1222–5)
Dose-relation: toxic effect
Time-course: early
Susceptibility factors: not known

Type I heparin-induced thrombocytopenia is common and is characterized by a mild transient thrombocytopenia (with platelet counts that usually do not fall below 50×10^9/l); the thrombocytopenia occurs on the first few days of heparin administration (usually 1–5 days) and requires careful monitoring but not usually withdrawal of heparin. Type I thrombocytopenia is generally harmless and very probably results from direct heparin-induced platelet aggregation. Thrombocytopenia is most common when large doses of heparin are used, or in some particular circumstances, such as after thrombolytic therapy (42) or in the early orthopedic postoperative period (43); it can abate in spite of continued therapy. Type I thrombocytopenia is a non-immune reaction, probably due to a direct activating effect of heparin on platelets.

Type II heparin-induced thrombocytopenia

DoTS classification (BMJ 2003;327:1222–5)
Dose-relation: hypersusceptibility effect
Time-course: early persistent
Susceptibility factors: previous heparin therapy, renal disease co-administration of antiplatelet drugs or oral anticoagulants

Type II heparin-induced thrombocytopenia is less common than Type I but is often associated with severe thrombocytopenia. It is generally accepted that heparin-induced thrombocytopenia refers to platelet counts of less than 150×10^9/l or a reduction in platelet count of 30–50% to a previous count. In some cases, platelet counts that have fallen to 50% from a high normal previous count would remain apparently normal but should be considered as potentially heparin-induced thrombocytopenia.

The literature on heparin-induced thrombocytopenia has been reviewed in the context of a case in a neonate after heart surgery (44).

Incidence

The incidence of type II thrombocytopenia is 2–5% in adults and may be equally high in neonates and children. The mortality rate in adults is 7–30% and is unknown but potentially high in neonates.

The most frequently suspected drug registered by the German spontaneous reporting system in cases of thrombocytopenia was unfractionated heparin (45). Of 3291 adverse reactions reports 78% were associated with unfractionated heparin, 13% with enoxaparin, 11% with certoparin, 5.5% with dalteparin, 2.8% with heparin fractions, 2.5% with reviparin, and 1.2% with tinzaparin.

Heparin-induced thrombocytopenia was the most common adverse effect (38%), followed by pulmonary embolism (11%), hematomas (6.8%), erythematous rashes (4.8%), and unspecified bleeding (4.5%). Injection site reactions were common (15%) and included skin necrosis and injection site necrosis in 1.8% of cases. Antibodies to heparin–platelet factor (PF)4 complex can be demonstrated in almost all patients with type II heparin-induced thrombocytopenia. There was a positive specific antibody result in 736 (59%) of 1245 cases.

The incidences of heparin-induced thrombocytopenia in surgical and medical patients receiving thromboprophylaxis with either unfractionated or low-molecular-weight heparin have been studied in a systematic review of all relevant randomized and non-randomized studies identified in MEDLINE (1984–2004), not limited by language, and from reference lists of key articles (46). Heparin-induced thrombocytopenia was defined as a fall in platelet count to less than 50% or less than 100 x 10^9/l and a positive laboratory diagnostic assay, including enzyme-linked immunosorbent assay (ELISA), [^{14}C]serotonin release assay, or adenosine triphosphate lumi-aggregometry. There were 15 eligible studies (7287 patients).

The odds ratios were as follows:

- two randomized controlled trials (n = 1014): OR = 0.10 (95% CI = 0.01, 0.2);
- three prospective studies with non-randomized comparison groups (n = 1464): OR = 0.10 (95% CI = 0.03, 0.33);
- all 15 studies (including ten in which only thrombocytopenia was measured): OR = 0.47 (95% CI = 0.22, 1.02).

The absolute risks were 0.2% with low-molecular-weight heparin and 2.6% with unfractionated heparin.

In a retrospective study of 389 consecutive patients with subarachnoid hemorrhage, 59 (15%) met the clinical diagnostic criteria for heparin-induced thrombocytopenia type II (47). The average platelet count nadir was 69 x 10^9/l. Women and patients with Fisher Grade 3 were at higher risk. There were systemic thrombotic complications in 37% compared with 7% of those without thrombocytopenia. There were more new hypodensities on CT scan in those with thrombocytopenia (66% versus 40%) and more deaths (29% versus 12%).

Dose relation

Heparin-induced thrombocytopenia can occur after minimal heparin exposure, including heparin flushes (48).

Time-course

The fall in platelet count usually occurs 5–10 days after the first exposure to heparin.

Case reports

Thrombocytopenia has been reported in a patient who received intraperitoneal heparin (49).

- A 52-year-old man with end-stage renal disease and peritonitis associated with CAPD was given intraperitoneal heparin 1000 U/day for 7 days. His platelet count 13 days before this had been 260 x 10^9/l, but 14 days after the last dose of heparin he developed epistaxis and petechiae on his trunk and lower legs and the platelet count was 25 x 10^9/l. His platelet count spontaneously normalized over the next 7 days.

Heparin-induced thrombocytopenia was confirmed by detection of antibodies against the heparin–PF4 complex using a serotonin release assay.

Thrombosis can occur when heparin causes a fall in the platelet count within the reference range if there are associated antibodies (50).

- A 45-year-old man with pulmonary embolism was given heparin and developed massive thrombosis after insertion of a filter on day 3; the platelet count was 221 x 10^9/l. Heparin was replaced by argatroban on day 13 and the platelet count rose to 355 x 10^9/l on day 15. There were antibodies against complexes of heparin and platelet factor 4.

Since 1992, miniaturized pulsatile air-driven ventricular assist devices, the so-called "Berlin Heart", have been used in children at many institutions (36 cases in North America in 19 different institutions). Heparin-induced thrombocytopenia can cause thrombosis in such devices (51).

- A 13-month-old girl, weight 8.1 kg, who required support with a left ventricular assist device for cardiogenic shock of unclear cause, developed a persistent low-grade fever, heparin-induced thrombocytopenia, and impaired renal function. On post-implant day 10, the pump required replacement because of concerns about an inlet valve thrombus; the explanted device contained a nearly occlusive clot.

Mechanism

Type II thrombocytopenia is probably an immune-mediated phenomenon, a fact that has been the subject of much specific investigation (30,52–57). It has been proposed that the diagnosis should depend on two criteria: the association of one or more clinical events and laboratory evidence of a heparin-dependent immunoglobulin (43).

Diagnosis

Laboratory diagnosis of heparin-induced thrombocytopenia can be made either using functional tests (demonstration of platelet activation of normal donor platelets in vitro by the patient's serum in the presence of heparin) or by screening for antibodies (30,58). The duration of the antibody response can vary from weeks to months up to 1 year. The two types of tests are complementary; both should be used when the reported results of either are inconsistent with the clinical problem.

Susceptibility factors

The risk of thrombocytopenia is increased in patients with a history of previous heparin therapy (59). Thrombocytopenia and/or thromboembolic complications can occur sooner in patients with a history of previous exposure to heparin, suggesting an anamnestic response (31,58).

The results of some prospective studies, including mainly transient benign thrombocytopenia (type I heparin-induced thrombocytopenia), have suggested that thrombocytopenia is more common in patients who have received the bovine lung type of heparin than in those who have received the porcine intestinal mucosa type (60–62). Type II thrombocytopenia is certainly more common in patients treated with standard heparin than in those treated with low molecular weight heparin (43), but there is usually cross-reactivity between standard heparin and low molecular weight heparin.

Heparin-coated catheters can sustain the thrombocytopenia in patients with heparin-associated antiplatelet antibodies (63,64) and therefore have to be removed from such patients.

There is less published experience with low-molecular-weight heparins in children than in adults, but the low frequency of significant bleeding appears to be similar. A child who received therapeutic doses of a low-molecular-weight heparin for a deep vein thrombosis spontaneously developed an intramural hemorrhage in the small bowel, leading to infarction, which required partial bowel resection (65).

Unlike unfractionated heparin, dalteparin is mainly cleared through the kidney and can therefore accumulate if renal function is impaired, increasing the risk of hemorrhage.

- An 84-year-old woman with chronic renal insufficiency had angioplasty for a stenosis in a femorofibular bypass, developed a deep vein thrombosis, and was given dalteparin (66). After 4 days she developed a pronounced hematoma on her flank and her hemoglobin fell to 5.5 g/dl. Dalteparin was withdrawn and she was given protamine 2500 U and packed red blood cells. She had no further bleeding during treatment with unfractionated heparin and an oral anticoagulant.

Dalteparin should be avoided in patients with severe renal impairment or used only with close monitoring of antifactor Xa activity. As an alternative, unfractionated heparin can be used, since renal impairment does not affect its short half-life.

Two patients with chronic kidney disease had retroperitoneal hematomas requiring blood transfusion after the administration of enoxaparin (67). Enoxaparin should be administered with great caution in patients with chronic kidney disease, especially if antiplatelet agents or other anticoagulants are administered concomitantly.

Presentation

Heparin-induced thrombocytopenia usually occurs with a delayed onset (a week or more) and is often complicated by paradoxical recurrence of thromboembolic events, leading to life-threatening complications. The general term "heparin-induced thrombocytopenia" is as a rule used to designate this phenomenon.

Thrombosis can occur before the onset of overt thrombocytopenia (43). In some cases dramatic thrombotic events occur in association with antibodies without significant fall in platelet count (58). Thrombotic complications include new or recurrent arterial thromboembolism, often localized in the distal aorta and legs or presenting as myocardial infarction and hemiplegia (68), but also in the brachial artery (69), often requiring surgical treatment. In older reports there was a higher incidence of arterial than of venous thrombosis, but later reports documented that venous thrombosis is four times more common than arterial thrombosis (70). Heparin-induced thrombocytopenia can be associated with deep venous thrombosis and pulmonary embolism, but unusual forms of thrombosis, such as adrenal hemorrhagic infarction due to adrenal vein thrombosis, a complication that must be considered in patients who develop abdominal pain or unexplained hypotension during heparin therapy, are not uncommon.

Venous limb gangrene due to heparin-induced thrombocytopenia is characterized by distal tissue losses, with extensive venous thrombosis involving large veins and small venules (71). This reaction seems to be related to acquired deficiency of protein C induced by concomitant oral anticoagulants.

Patients with heparin-induced thrombocytopenia have a reported mortality of 25–30% and amputation rates of up to 25% (72). The development of the syndrome is not related to the dose of heparin in the therapeutic range (that is it is a hypersusceptibility reaction). This has been confirmed by the fact that thrombocytopenia with thromboembolic complications sometimes occurs after the limited exposure that is involved in "flushing" with heparin and saline to maintain the patency of venous catheters (73).

Heparin-induced skin necrosis is an immune-complex phenomenon associated with heparin-induced thrombocytopenia (HIT); it can rarely occur in the presence of HIT IgG alone (serological HIT). It is thought to be caused by an antibody-mediated local prothrombotic condition associated with platelet activation and increased thrombin production. If skin necrosis occurs, treatment with an alternative thromboprophylactic agent should be considered.

Skin necrosis has been reported in patients receiving tinzaparin (74), enoxaparin (75), and unfractionated heparin (76).

- A 76-year-old man with polycythemia vera, hypertension, diabetes mellitus, hyperuricemia, atrial fibrillation, and chronic bronchitis, taking hydroxyurea, digoxin, allopurinol, and enalapril, was given prophylactic subcutaneous enoxaparin 60 mg bd (75). After 5 days he developed two symmetrical erythematous patches, 5 cm in diameter, on the abdominal wall at injection sites. The lesions enlarged over 24 hours and formed purplish-blue necrotic plaques 15 cm x 5 cm. The hemoglobin was 7.4 g/dl, the white blood cell count, 29 x 10^9/l, and the platelet count 1025 x 10^9/l; there was no significant change in the platelet count throughout the admission. The prothrombin time ratio was 1.17 and the thromboplastin time ratio 1.36. Protein C was normal but protein S was reduced to 56% (reference range 71–142%), with a low free protein S concentration (63%; reference range 72–139%) and normal total protein S (protein S deficiency type III). There were no IgG antiphospholipid antibodies, but IgM was raised. Heparin–platelet factor 4 (PF4) antibodies were also demonstrated. A skin biopsy showed multiple fibrin thrombi in the dermal microvasculature with ischemic necrosis of the overlying epidermis. Enoxaparin was withdrawn, the lesions were treated locally, and there was complete healing after about 1 month.

- A 69-year-old woman with severe bronchopneumonia was given subcutaneous prophylactic unfractionated sodium heparin (5000 IU bd) (76). By day 7 she had developed blistering skin lesions with central necrosis and surrounding erythema at the heparin injection sites. The platelet count was stable at 275 x 10^9/l (range 196–338). There was a circulating IgG antibody against heparin-platelet factor 4 and anticardiolipin IgG antibodies. A skin biopsy showed extensive focal epidermal necrosis with marked neutrophil infiltration and extensive fibrin deposition within the small vessels of the dermis. All the lesions resolved within 5 day of withdrawal of heparin.

The association of heparin-induced skin necrosis with antibodies directed against heparin–PF4 is well-established, but the participation of other procoagulant factors has received little attention. The observation of heparin-induced skin necrosis should motivate a systematic search for the presence of anti-PF4 antibodies, but also for additional genetic or acquired procoagulant factors. Heparin-induced skin necrosis may be a marker of an increased risk of systemic arterial or venous thromboembolism.

Catastrophic antiphospholipid syndrome is a medical emergency characterized by thrombosis of multiple small vessels of the internal organs and the brain (77). In one case hepatic, renal, and splenic artery thromboses, as well as cerebral venous thrombosis, were complicated by severe thrombocytopenia and hemolytic anemia. However, there were no anti-platelet-factor-4 antibodies, making heparin-induced thrombocytopenia unlikely.

Management

In view of the severity of the Type II syndrome, it has been recommended that heparin therapy be monitored by twice-weekly platelet counts (78).

Heparin-induced thrombocytopenia demands immediate withdrawal of heparin (58), after which platelet counts usually return to normal within 2–3 days, that is much faster than after early-onset thrombocytopenia. Rechallenge with heparin leads to an abrupt fall in the number of platelets and to reappearance of the clinical signs and symptoms, sometimes leading to sudden death (79).

From the therapeutic point of view, prophylaxis of thrombosis must be continued after withdrawal of heparin, since even when there is no evidence of thrombosis in association with heparin-induced thrombocytopenia, thrombosis can follow after some days (70). Because of cross-reactivity, low molecular weight heparin should not be used when heparin has been withdrawn because of heparin-induced thrombocytopenia; nor should warfarin be used, because of the risk of venous gangrene, at least until the thrombocytopenia has resolved. Patients with life-threatening or limb-threatening thrombosis can be treated with thrombolytic drugs. Current views are that two antithrombotic drugs should be used, for example danaparoid plus lepirudin (80).

Lepirudin has been used in patients with heparin-induced thrombocytopenia in a prospective study in 205 patients with 120 historical controls (HAT-3) and in a combined analysis of all HAT study data (81). Patients with laboratory-confirmed thrombocytopenia were treated with lepirudin in three different aPTT-adjusted dose regimens and during cardiopulmonary bypass. Mean lepirudin maintenance doses were 0.07–0.11 mg/kg/hour. End points were new thromboembolic complications, limb amputations, and death and major bleeding. The combined end point occurred in 43 (21%) of those treated with lepirudin; 30 died, 10 underwent limb amputation, and 11 had new thromboembolic complications. There was major bleeding in 40 patients, seven during cardiopulmonary bypass. Combining all the prospective HAT trials (n = 403), after the start of lepirudin treatment, the combined end point occurred in 82 patients (20%), with 47 deaths, 22 limb amputations, 30 new thromboembolic complications, and 71 episodes of major bleeding. Compared with the historical controls, the combined end point after the start of treatment was significantly reduced (30% versus 52%), primarily because of a reduction in new episodes of thrombosis (12% versus 32%). Major bleeding was more frequent in the lepirudin-treated patients (29% versus 9.1%). Thus, the rate of new thromboembolic complications in patients with heparin-induced thrombocytopenia is low after lepirudin treatment. The rate of major bleeding of 18% might be reduced by reducing the starting dose to 0.1 mg/kg/hour.

Argatroban has been used in 13 patients who developed heparin-induced thrombocytopenia after exposure to heparin 10–13 000 U from an intravascular catheter or filter flush, with a mean exposure of 8 days (82). They were compared with 10 historical controls who had received no direct thrombin inhibitors. The platelet count recovered to a mean of 207 x 10^9/l (n = 12) after 5.5 days of argatroban therapy and to a mean of 127 x 10^9/l (n = 8) 5 days after baseline in the control group. A composite end point of death, amputation, or new thrombosis within 37 days occurred in five argatroban-treated patients and four controls. Death was the most common untoward outcome (about 30% in each group). No argatroban-treated patient and two control patients had new episodes of thrombosis. Major bleeding was comparable.

Eosinophilia

Eosinophilia has rarely been attributed to heparin, with positive rechallenge (70,83).

Liver

Minor increases in serum transaminases without evidence of liver dysfunction are common in patients receiving standard heparin or low molecular weight heparin given therapeutically or prophylactically (84,85). This rise is more pronounced for alanine transaminase than for aspartate transaminase and occurs after 5–10 days of heparin treatment (85). The source of heparin has no relation to the development of raised transaminases. After withdrawal of heparin and sometimes even in spite of continued treatment (84,85), the transaminases return to normal (86,87). The mechanism of these increases has not been elucidated. A concomitant increase in gamma-glutamyl transpeptidase activity has been described in some patients (88).

In one study of patients receiving heparin, the isoenzyme pattern of lactate dehydrogenase was studied; all had rises in the hepatic form of the enzyme, suggesting hepatocellular damage as the most likely source (84).

Heparin-associated hepatotoxicity has not been reported.

Skin

Immediate-type reactions to heparins are rare and delayed (type IV) reactions more common. The lesions usually develop 2–4 days after injection and are erythematous, infiltrated, and sometimes eczematous; they can become generalized if therapy is continued (89).

In four patients with such reactions, patch, intradermal, and subcutaneous tests were performed with a panel of unfractionated heparins, low-molecular-weight heparins, heparinoids, recombinant hirudins, and fondaparinux sodium (90). Three were sensitized to all the unfractionated heparins and low-molecular-weight heparins. Tinzaparin sodium was a possible substitute in one patient and the heparinoid pentosan polysulfate in another. The recombinant hirudins and fondaparinux sodium were tolerated without any adverse effects.

- A 68-year-old woman had repeated localized skin reactions at injection sites after administration of nadroparin, certoparin, and heparin sodium (89). The lesions were itchy and eczematous, developed soon after starting the anticoagulants, and persisted for several days. Patch testing with multiple heparin preparations (heparin sodium, heparin calcium, certoparin, dalteparin, nadroparin, the xylanopolyhydrogen sulfate pentosanpolysulfate, and the synthetic heparin pentasaccharide fondaparinux) showed type IV sensitization to certoparin. Prick testing was negative, but intracutaneous tests were positive to nadroparin, dalteparin, heparin calcium, and heparin sodium. Subcutaneous pentosanpolysulfate and fondaparinux had no effects.
- A 62-year-old woman presented had localized skin reactions at injection sites after nadroparin (89). Patch tests showed type IV reactions to nadroparin and dalteparin. Intracutaneous tests caused delayed reactions after 6 days to heparin sodium and heparin calcium. Subcutaneous challenge with pentosanpolysulfate was positive but negative with fondaparinux.
- A 57-year-old immobile obese Caucasian woman with a history of delayed hypersensitivity reactions to several heparins was given danaparoid for prophylactic anticoagulation while being treated for venous ulcers (91). Two days later she developed erythematous plaques and disseminating papulovesicles at the injection sites. She was then given fondaparinux instead, which was well tolerated over 2 weeks. Skin allergy tests 6 weeks later were positive to heparin at days 4–8, with erythematous plaques at the injection sites. Pentosan polysulfate was also injected subcutaneously, and produced a similar lesion. In contrast, all skin tests with fondaparinux were negative, and re-exposure was uneventful.
- A 67-year-old man with unstable angina pectoris developed an urticarial rash 10 hours after the second exposure to heparin during coronary artery bypass (92). Lymphocyte stimulation tests were positive with porcine heparin, coumadin, and nicorandil. Off-pump coronary bypass surgery was successfully performed with argatroban as an alternative to heparin.

Erythematous nodules or infiltrated and sometimes eczema-like plaques at the site of injection are common adverse effects of subcutaneous standard heparin 3–21 days after starting heparin treatment. They are probably delayed-type hypersensitivity reactions and are also seen with low molecular weight heparin (93–97). There can be cross-reactivity between standard heparin and low molecular weight heparin (97).

Erythromelalgia has been described with enoxaparin (98).

In cases of local reactions, skin tests should not be performed, since they are rarely helpful in detecting potential cross-reactivity between low molecular weight heparin and heparin (93,99).

Skin necrosis

Heparin-induced skin necrosis was first described in 1973 (100) and was later observed in patients with the thrombohemorrhagic syndrome (101,102). The skin pathology develops 6–9 days after the start of subcutaneous heparin treatment and usually develops at the site of subcutaneous injection. However, it can also occur at sites distant from the site of injection (103) or after intravenous therapy (104). Vasculitis is likely to be present, perhaps with fever (105). There is an increased risk of thrombocytopenia and thrombotic complications in patients who have had previous heparin-associated skin necrosis episodes (106).

Various physiopathological mechanisms have been suggested (107), particularly heparin-induced thrombocytopenia, vasculitis caused by a type III hypersensitivity reaction, local trauma at the site of injection, and poor vascularization of adipose tissue resulting in reduced absorption of heparin, as seen in diabetic lipodystrophy. This form of skin necrosis must be distinguished from vasospastic skin necrosis and from skin necrosis induced by cholesterol embolization. Ergotism will be a differential diagnosis in patients who are receiving prophylaxis with the combination of low dose heparin and dihydroergotamine (108). In evaluating any necrotic skin reaction, one must always consider its infectious origin (particularly *Escherichia coli* or *Pseudomonas*) related to unsterile injections (109).

Several patients with heparin-induced skin necrosis were positive for HIT-IgG using the platelet ^{14}C-serotonin release assay, even though they did not develop thrombocytopenia (110). Skin necrosis has been particularly reported in patients receiving standard heparin but also occasionally with low molecular weight heparins, such as dalteparin (111) or enoxaparin (106).

The clinical and histological pictures of heparin-induced skin necrosis are similar to those found in so-called "coumarin necrosis." Laboratory examinations show inflammatory changes, with anemia, leukocytosis, eosinophilia, a raised erythrocyte sedimentation rate, and a positive capillary fragility test. The vasculitis can lead to organ involvement, such as glomerulonephritis (112).

Skin necrosis due to low-molecular-weight heparin has been reviewed in a systematic review of 20 articles (21 cases) in which skin necrosis occurred locally and distant from the injection site (113). Heparin-induced antibodies were common (positive in 9/11 cases, negative in 2/11). However, severe thrombocytopenia (platelet count below 100 x 10^9/l) occurred in only four cases, while the platelet count was normal in half of the cases. Recovery after withdrawal was usually benign, but two patients needed reconstructive surgery. The authors concluded that skin necrosis due to low-molecular-weight heparin can occur as part of heparin-induced thrombocytopenia, but that there can be other mechanisms, including allergic reactions and local trauma. When heparin-induced thrombocytopenia is excluded it is safe to switch to unfractionated heparin; otherwise, drugs such as hirudin or fondaparinux should be preferred.

As variation in the molecular weights of different heparin formulations has repeatedly been implicated in determining the frequency of sensitization, it has been suggested that the pentasaccharide fondaparinux may provide a practicable and safe alternative, because of its low molecular weight (114). Patients with cutaneous reactions after subcutaneous anticoagulant treatment (n = 12) underwent a series of in vivo skin allergy and challenge tests with unfractionated heparin, low-molecular-weight heparins (certoparin, dalteparin, enoxaparin, nadroparin, and tinzaparin), danaparoid, and fondaparinux. There was a high degree of cross-reactivity among heparins and heparinoids. In contrast, there was rarely cross-sensitization with fondaparinux. Molecular weight was a key determinant of sensitization to heparins and other oligosaccharides.

Heparin-induced skin necrosis can be associated with high morbidity and occasional mortality. Heparin should be withdrawn when it occurs. The risk of recurrence if heparin is given again later is not known (115).

Fat necrosis

Subcutaneous fat necrosis has been attributed to heparin (116).

- A 91-year-old woman with diabetes, hypertension, and unstable angina was given subcutaneous enoxaparin. After 5 days she developed extensive induration of the skin and subcutaneous fat of the upper part of the left breast and bruising of the overlying skin. There were a few patches of ecchymosis were over other parts of her body, but none at injection sites on the abdomen. Coagulation screen and platelet count were normal. Mammography showed asymmetrical nodular densities over the upper inner quadrant of the left breast and ultrasound of the area was consistent with fat necrosis.

Subcutaneous calcinosis

Accumulation of calcium in the skin is usually classified as a group of disorders referred to as calcinosis cutis. Calcinosis has been attributed to subcutaneous nadroparin (117).

- A 92-year-old man developed multiple cutaneous plaques and nodules on the abdomen and right thigh after daily subcutaneous administration of nadroparin calcium for 5 months. The lesions were asymptomatic, firm, and roundish and measured up to 3 cm. Some were isolated and others were confluent, forming annular shapes. Most were ulcerated. Some developed an erythematous, yellow, annular border, with a darker ring inside. Histopathology showed marked hyperkeratosis, mild epidermal acanthosis, and central ulceration. Within the superficial and mid-dermis there were marked regressive-degenerative changes of the collagen fibers, leading to artefactual dermo-epidermal and intradermal clefts, marked fragmentation of the elastic fibers, focal liponecrosis, and deposition of calcium salts.

Calcifying panniculitis is a rare form of calcinosis cutis that belongs to the spectrum of calciphylaxis that has almost invariably been described in patients with severe renal impairment. It has been reported in a patient with hyperparathyroidism and normal renal function who received subcutaneous nadroparin calcium and resolved after withdrawal (118). The authors suggested that low-molecular-weight calcium-containing heparins should probably be used with caution in patients with hyperparathyroidism.

Two renal transplant recipients developed calcinosis cutis (tender erythematous subcutaneous nodules with induration, ulceration, and necrosis), confirmed by bone scan and biopsies, at the site of subcutaneous administration of nadroparin (119). Both had secondary hyperparathyroidism and a raised calcium-phosphate product at the time of administration. The authors suggested that the high concentration of calcium in nadroparin (220 mmol/l) had been an important factor, since in the same patients subcutaneous injections of nicomorphine or dalteparin did not cause subcutaneous nodules. Three more cases were found after retrospective examination of all 51 adult patients who underwent a renal transplantation in the same hospital during 7 months, and there were no other cases of calcinosis cutis at the injection sites of low molecular weight heparins after nadroparin was replaced by dalteparin.

Musculoskeletal

Heparin-associated osteoporosis is observed with both standard heparin and low molecular weight heparin, although it is far less pronounced with the latter. The incidence of heparin-associated osteoporosis is not known, but it is probably less than 5% with standard heparin. The highest estimates are derived from studies of pregnant patients and should be interpreted cautiously, since pregnancy, even in the absence of heparin therapy, is sometimes itself associated with osteoporosis. Overall, it appears that osteoporosis can occur in patients given heparin for longer than 10 weeks and usually for a minimum of 4 months. The effect is rarely complicated by fractures. It has been suggested that during long-term heparin therapy, calcium supplements be given (120).

The processes involved are bone resorption and inhibition of bone deposition. Standard heparin may affect both phenomena, whereas low molecular weight heparin seems to affect only the rate of bone formation. Heparin-associated osteoporosis has been attributed to enhancement of collagenolysis or to enzyme inhibition by heparin. Other authors have stressed the resemblance to hyperparathyroid osteopathy. Further possibilities include deficiency of vitamin D analogues (121) and ascorbic acid deprivation at the site of osteoblast activity. Heparin-induced osteolysis may explain the metastatic calcification that can occur in patients with end-stage renal insufficiency treated by hemodialysis, which is sometimes associated with calcified subcutaneous nodules (122).

In osteoblast cultures unfractionated heparin, dalteparin, and enoxaparin significantly reduced matrix collagen type II content and calcification in concentrations equal to or higher than those that are generally associated with a therapeutic effect; in contrast, fondaparinux had no inhibitory in vitro effects on human osteoblasts at concentrations of 0.01–100 μg/ml (123). Mitochondrial activity and protein synthesis in osteoblasts treated with fondaparinux were significantly higher than in cells treated with heparins.

Sexual function

Heparin has been associated with priapism (124). The frequency and severity of iatrogenic priapism as a result of heparin therapy seems to be greater than with any other type of medical treatment (125). However, it is uncertain whether it is heparin or the underlying thrombotic condition that causes the complication.

Immunologic

A wide range of allergic reactions have been described in patients receiving heparin, including urticaria, conjunctivitis, rhinitis, asthma, cyanosis, tachypnea, a feeling of oppression, fever, chills, angioedema, and anaphylactic shock.

- A patient with end-stage renal disease developed recurrent anaphylaxis after receiving heparin during hemodialysis (126). There were raised concentrations of total and mature tryptase at 1 hour, but although the latter returned to normal by 24 hours the former did not. Prick tests were negative with heparin, enoxaparin, and danaparoid, but intradermal skin tests were positive with heparin and enoxaparin. Danaparoid was used as an anticoagulant during dialysis for the next 3 years without any adverse effects.

In some cases of allergy to a heparin formulation, the precipitating agent will prove to be a preservative, such as chlorocresol (127) or chlorbutol (128), rather than heparin itself.

Heparin has immunomodulatory effects, such as enhancement of cytokine production in monocytes induced by endotoxin (lipopolysaccharide). In whole blood from five healthy volunteers heparin 20 IU/ml or more significantly increased lipopolysaccharide-induced production of interleukin-8; fondaparinux did not (129). The authors suggested that fondaparinux is not an immunomodulator like heparin, and may therefore lack adverse effects in patients with endotoxemia.

However, in another case there was cross-reactivity with fondaparinux after a delayed-type hypersensitivity reaction to heparin (130); this is distinct from heparin-induced thrombocytopenia due to IgG antibody against PF4.

Lepirudin can be used safely for prophylaxis of recurrent venous thromboembolism throughout pregnancy in patients with hypersensitivity reactions to heparins (131).

- A 26-year-old woman with recurrent venous thromboembolism, lupus pernio with antiphospholipid antibodies, and local intolerance to heparin, low-molecular-weight heparin, and danaparoid, was given phenprocoumon during pregnancy and was later switched to subcutaneous lepirudin and acetylsalicylic acid 100 mg/day. At week 16 she developed lupus pernio, with painful lesions on the toes and fingers and later also on the cheeks and nose. Aspirin was withdrawn and she was given cortisone and morphine. Fetal examination was normal. At week 29, she was given intravenous lepirudin. At week 30 a healthy child was delivered by cesarean section. After delivery the lupus pernio resolved within 2 weeks and subcutaneous lepirudin was restarted, followed by phenprocoumon.

Infection risk

Following a cluster of cases of unexpected hospital-acquired suspected to be related to intravenous heparin infusion, all cases of hospital-acquired primary bacteremia in low-risk patients were analysed over 4 years (132). Of 1618 episodes of hospital-acquired bacteremia a peripheral intravenous line was the only risk factor in 96 (6%). These patients were divided into two groups: 60 patients with phlebitis and 36 without local signs of inflammation. The baseline features in the two groups were comparable, but there was a significant association between intravenous heparin use, the predominance of Gram-negative organisms (especially *Klebsiella*, *Serratia*, and *Enterobacter* species), and the absence of phlebitis. However, in spite of a clear statistical association, the mechanism whereby the heparin solution became contaminated with Gram-negative organisms was unknown. Following implementation of infection control methods in handling heparin, there were no more cases.

Second-Generation Effects

Pregnancy

The maternal rate of bleeding complications during heparin treatment is about 2%. This is consistent with the reported rates of bleeding associated with heparin therapy in non-pregnant women, and in warfarin therapy when used for the treatment of venous thrombosis (133). Subcutaneous heparin given just before labor can also cause a persistent anticoagulant effect at the time of delivery; the mechanism of this prolonged effect is unclear (134).

Another pregnancy-associated hazard of long-term heparin administration is maternal osteoporosis, as clearly illustrated by several reports of pregnant women who developed severe osteoporosis after having received heparin, in some cases complicated by multiple vertebral fractures (135,136). Long-term heparin therapy during pregnancy results in changes in calcium homeostasis, which may be dose-related in the therapeutic range of doses (137). The incidence of osteoporotic fractures was 2.2% (four patients) in 184 women receiving long-term subcutaneous prophylaxis with twice-daily heparin during pregnancy (138). In another study, the incidence of osteopenia in 70 pregnant women on long-term heparin was 17%; there was also a regression in the degree of osteopenia postpartum (139). It is difficult to distinguish between the mean bone loss that occurs in normal pregnancy and the additive effect on bone demineralization that results from heparin therapy; further prospective studies on bone demineralization during pregnancy are needed to clarify this point.

There has been a systematic review of studies on the use of low-molecular-weight heparins for thromboprophylaxis and treatment of venous thromboembolism in pregnancy (140). Data on recurrence of venous thromboembolism and adverse effects were extracted and cumulative incidences calculated. Of 81 reports, 64 reporting 2777 pregnancies were included. In 15 studies (174 patients) the indication for low-molecular-weight heparin was treatment of acute venous thromboembolism, and in 61 studies (2603 pregnancies) it was thromboprophylaxis or an adverse pregnancy outcome. There were no maternal deaths. Venous thromboembolism and arterial thrombosis (associated with antiphospholipid syndrome) were reported in 0.86% of pregnancies (95% CI = 0.55, 1.28) and 0.50% of pregnancies (95% CI = 0.28, 0.84) respectively. There was significant bleeding, generally associated with primary obstetric causes, in 1.98% of pregnancies (95% CI = 1.50, 2.57), allergic skin reactions in 1.80% (95% CI = 1.34, 2.37), heparin-induced thrombocytopenia in 0%, thrombocytopenia (unrelated to heparin) in 0.11% (95% CI = 0.02, 0.32), and osteoporotic fractures in 0.04% (95% CI <0.01, 0.20). Overall, live births were reported in 94.7% of pregnancies, including 85% in those who received low-molecular-weight heparin for recurrent pregnancy loss.

Fetotoxicity

An adverse fetal outcome occurs in about one-third of pregnancies after either oral anticoagulant or heparin treatment (141). However, the high rate of adverse fetal outcomes associated with heparin during pregnancy is in part a reflection of the severity of pre-existing maternal diseases (142). Moreover, some studies have suggested that unfractioned heparin is relatively safe for the fetus and is the anticoagulant therapy of choice during pregnancy, whereas oral anticoagulants may not be, particularly during the first trimester (133,143). Heparin does not cross the placenta, and therefore does not have the potential to cause fetal bleeding or teratogenicity (144). Low molecular weight heparins and heparinoids similarly do not cross the placenta (145,146).

Lactation

Heparin is not secreted into breast milk and can be given safely to nursing mothers (147).

Drug–Drug Interactions

Benzodiazepines

In healthy non-fasting subjects, 100–1000 IU of heparin given intravenously caused a rapid increase in the unbound fractions of diazepam, chlordiazepoxide, and oxazepam (148,149), but no change in the case of lorazepam (148). The clinical implications of this finding are not known.

Cardiac glycosides

Reduced plasma protein binding of digitoxin has been reported after the administration of heparin (150). In 10 hemodialysed patients taking maintenance digitoxin therapy, there was reduced binding in vitro because of heparin-induced lipolysis, and not as a consequence of in vivo binding of digitoxin to plasma proteins.

Dobutamine hydrochloride

Dobutamine hydrochloride and heparin should not be mixed or infused through the same intravenous line, as this causes precipitation (151).

Glyceryl trinitrate

Resistance to heparin has been observed in patients in coronary care units receiving intravenous glyceryl trinitrate. Although this has been imputed to propylene glycol in glyceryl trinitrate formulations (152), resistance has also been described with propylene glycol-free formulations. The underlying mechanism might be a qualitative abnormality of antithrombin III induced by glyceryl trinitrate (153). It has also been reported that heparin concentrations can be markedly reduced by glyceryl trinitrate (154).

Interference with Diagnostic Tests

Acid phosphatase activity

Concentrations of heparin as low as 1 IU/ml give 80% inhibition of acid phosphatase activity in leukocytes (155).

Aminoglycosides

Heparin can interfere with the determination of aminoglycosides by enzyme-multiplied immunoassays, resulting in lower values than with the use of other assays (156,157). In the case of gentamicin, the lower values may be due to direct binding of gentamicin to heparin, as well as a more complex interaction involving heparin, gentamicin, and proteins (156). Blood samples for the determination of aminoglycosides by enzyme immunoassays should not be collected in heparinized tubes or from indwelling lines (157).

Arterial blood gases

Heparin can affect the analysis of arterial blood gases (158). Addition of excess heparin and acid mucopolysaccharide to the syringe can alter the Henderson–Hasselbalch equation, simulating metabolic acidosis. It is advisable to coat the syringe with heparin and remove all excess heparin before blood sampling.

Calcium

Pseudo-hypocalcemia can occur in hemodialysis patients, particularly in those who have chylomicronemia before administration of heparin (159). This spurious hypocalcemia is thought to result from lipolytic activity in vitro, sufficient to produce calcium soaps of fatty acid. This can be detected and eliminated by the analysis of blood samples immediately after venepuncture. During hemodialysis, there is a significant fall in plasma ionized calcium after intravenous administration of heparin (an average reduction of 0.03 mmol/l after 10 000 IU) (160).

Chromogenic lysate assay for endotoxin

Heparin has a dose-related inhibitory effect on the chromogenic lysate assay for endotoxin (161). There was a 90% reduction in detectable endotoxin at concentrations of heparin as low as 30 U/ml.

Ciclosporin

Measured ciclosporin concentrations differ according to the use of heparin or EDTA as anticoagulant (162).

Propranolol

Interference by heparin with the measurement of propranolol when the blood is collected from a heparinized cannula is well documented (163). There is no in vivo interaction, since the beta-blockade is unchanged (164).

Thyroid function tests

Artefactual increases of as much as 50% in total thyroxine, estimated by a competitive protein-binding assay, and of as much as 30% in triiodothyronine resin uptake are probably due to rapid and continuing lipolytic hydrolysis of triglycerides after blood has been drawn (165). Thyroid function tests should therefore always be performed on blood samples taken before (or a sufficient time after) heparin treatment (166). An increase in serum-free thyroxine concentrations has also been reported after low molecular weight heparin, by up to 171% in specimens taken 2–6 hours after injection. When specimens were obtained 10 hours after injection, the effects were smaller, but with concentrations still up to 40% above normal the results can still cause errors of interpretation (167).

References

1. Brazier F, Yzet T, Boruchowicz A, Columbel JC, Duchmann JC, Dupas JL. Treatment of ulcerative colitis with heparin. Gastroenterology 1996;110:A872.
2. Day R, Forbes A. Heparin, cell adhesion, and pathogenesis of inflammatory bowel disease. Lancet 1999;354(9172):62–5.
3. Mora R, Jankowska B, Passali GC, Mora F, Passali FM, Crippa B, Quaranta N, Barbieri M. Sodium enoxaparin treatment of sensorineural hearing loss: an immune-mediated response• Int Tinnitus J 2005;11(1):38–42.
4. Bjoraker DG, Ketcham TR. Hemodynamic and platelet response to the bolus intravenous administration of porcine heparin. Thromb Haemost 1983;49(1):1–4.
5. Bowler GM, Galloway DW, Meiklejohn BH, Macintyre CC. Sharp fall in blood pressure after injection of heparin containing chlorbutol. Lancet 1986;1(8485): 848–9.
6. Schimpf K, Barth P. Heparinshock mit Verbrauchsreaktion des Blutgerinnungssystems?. [Heparin shock with consumption reaction of the blood coagulation system?.] Klin Wochenschr 1966;44(10):544–7.
7. Gayer W. Seltene Beobachtungen von Heparin-Unvertreglichkeit. [Unusual case of heparin intolerance.] Gynaecologia 1968;166(1):25.
8. Calota F, Vilcea D, Intorcaciu M, Pairvanescu H, Enache D, Comanescu V, Vasile I, Scurtu S. Emboliile cu cristale de colesterol in cursul tratamentului cu heparine cu greutate molecular mic. [Cholesterol crystal embolisation in the course of treatment with low-molecular-weight heparins.] Chirurgia (Bucur) 2005;100(6):605–8.
9. Oster JR, Singer I, Fishman LM. Heparin-induced aldosterone suppression and hyperkalemia. Am J Med 1995;98(6):575–86.
10. Levesque H, Cailleux N, Noblet C, Gancel A, Moore N, Courtois H. Hypoaldosteronisme induit par les héparines de bas poids moleculaire. [Hypoaldosteronism induced by low molecular weight heparins.] Presse Méd 1991;20(1):35.
11. Hottelart C, Achard JM, Moriniere P, Zoghbi F, Dieval J, Fournier A. Heparin-induced hyperkalemia in chronic hemodialysis patients: comparison of low molecular weight and unfractionated heparin. Artif Organs 1998;22(7):614–7.
12. Watts GF, Cameron J, Henderson A, Richmond W. Lipoprotein lipase deficiency due to long-term heparinization presenting as severe hypertriglyceridaemia in pregnancy. Postgrad Med J 1991;67(794):1062–4.
13. Wolf R, Beck OA, Hochrein H. Der Einfluss von Heparin auf die Haufigkeit von Rhythmusstorungen beim akuten Myokardinfarkt. [The effect of heparin on the incidence of arrhythmias after acute myocardial infarction.] Dtsch Med Wochenschr 1974;99(30):1549–53.
14. Su HM, Voon WC, Chu CS, Lin TH, Lai WT, Sheu SH. Heparin-induced cardiac tamponade and life-threatening hyperkalema in a patient with chronic hemodialysis. Kaohsiung J Med Sci 2005;21(3):128–33.

15. Sherman DS, Kass CL, Fish DN. Fludrocortisone for the treatment of heparin-induced hyperkalemia. Ann Pharmacother 2000;34(5):606–10.
16. Canova CR, Fischler MP, Reinhart WH. Effect of low-molecular-weight heparin on serum potassium. Lancet 1997;349(9063):1447–8.
17. Abdel-Raheem MM, Potti A, Tadros S, Koka V, Hanekom D, Fraiman G, Danielson BD. Effect of low-molecular-weight heparin on potassium homeostasis. Pathophysiol Haemost Thromb 2002;32(3):107–10.
18. Koren-Michowitz M, Avni B, Michowitz Y, Moravski G, Efrati S, Golik A. Early onset of hyperkalemia in patients treated with low molecular weight heparin: a prospective study. Pharmacoepidemiol Drug Saf 2004;13(5):299–302.
19. Pachter HL, Riles TS. Low dose heparin: bleeding and wound complications in the surgical patient. A prospective randomized study. Ann Surg 1977;186(6):669–74.
20. Landefeld CS, Beyth RJ. Anticoagulant-related bleeding: clinical epidemiology, prediction, and prevention. Am J Med 1993;95(3):315–28.
21. Walker AM, Jick H. Predictors of bleeding during heparin therapy. JAMA 1980;244(11):1209–12.
22. Horlocker TT, Wedel DJ. Spinal and epidural blockade and perioperative low molecular weight heparin: smooth sailing on the Titanic. Anesth Analg 1998;86(6):1153–6.
23. Norman CS, Provan JL. Control and complications of intermittent heparin therapy. Surg Gynecol Obstet 1977;145(3):338–42.
24. Levine MN, Raskob G, Landefeld S, Kearon C. Hemorrhagic complications of anticoagulant treatment. Chest 2001;119(1 Suppl):S108–21.
25. Campbell NR, Hull RD, Brant R, Hogan DB, Pineo GF, Raskob GE. Aging and heparin-related bleeding. Arch Intern Med 1996;156(8):857–60.
26. Leizorovicz A, Haugh MC, Chapuis FR, Samama MM, Boissel JP. Low molecular weight heparin in prevention of perioperative thrombosis. BMJ 1992;305(6859):913–20.
27. Nurmohamed MT, Rosendaal FR, Buller HR, Dekker E, Hommes DW, Vandenbroucke JP, Briet E. Low-molecular-weight heparin versus standard heparin in general and orthopaedic surgery: a meta-analysis. Lancet 1992;340(8812):152–6.
28. Leizorovicz A, Simonneau G, Decousus H, Boissel JP. Comparison of efficacy and safety of low molecular weight heparins and unfractionated heparin in initial treatment of deep venous thrombosis: a meta-analysis. BMJ 1994;309(6950):299–304.
29. Scott BD. Heparin-induced thrombocytopenia. A common but controllable condition. Postgrad Med 1989;86(5):153–5158.
30. Warkentin TE, Chong BH, Greinacher A. Heparin-induced thrombocytopenia: towards consensus. Thromb Haemost 1998;79(1):1–7.
31. Kelton JG. Heparin-induced thrombocytopenia. Haemostasis 1986;16(2):173–86.
32. Bell WR, Tomasulo PA, Alving BM, Duffy TP. Thrombocytopenia occurring during the administration of heparin. A prospective study in 52 patients. Ann Intern Med 1976;85(2):155–60.
33. Kahl K, Heidrich H. The incidence of heparin-induced thrombocytopenias. Int J Angiol 1998;7(3):255–7.
34. Yamamoto S, Koide M, Matsuo M, Suzuki S, Ohtaka M, Saika S, Matsuo T. Heparin-induced thrombocytopenia in hemodialysis patients. Am J Kidney Dis 1996;28(1):82–5.
35. Arnold DM, Kelton JG. Heparin-induced thrombocytopenia: an iceberg rising. Mayo Clin Proc 2005;80(8):988–90.
36. McNulty I, Katz E, Kim KY. Shah PS, Ng E, Sinha AK. Heparin for prolonging peripheral intravenous catheter use in neonates. Cochrane Database Syst Rev 2005;(4):CD002774.
37. Warkentin TE, Kelton JG. Delayed-onset heparin-induced thrombocytopenia and thrombosis. Ann Intern Med 2001;135(7):502–6.
38. Rice L, Attisha WK, Drexler A, Francis JL. Delayed-onset heparin-induced thrombocytopenia. Ann Intern Med 2002;136(3):210–5.
39. Mims MP, Manian P, Rice L. Acute cardiorespiratory collapse from heparin: a consequence of heparin-induced thrombocytopenia. Eur J Haematol 2004;72(5):366–9.
40. Tam NL, Pac-Soo C, Pretorius PM. Epidural haematoma after a combined spinal–epidural anaesthetic in a patient treated with clopidogrel and dalteparin. Br J Anaesth 2006;96(2):262–5.
41. Warkentin TE, Bernstein RA. Delayed-onset heparin-induced thrombocytopenia and cerebral thrombosis after a single administration of unfractionated heparin. N Engl J Med 2003;348(11):1067–9.
42. Balduini CL, Noris P, Bertolino G, Previtali M. Heparin modifies platelet count and function in patients who have undergone thrombolytic therapy for acute myocardial infarction. Thromb Haemost 1993;69(5):522–3.
43. Warkentin TE, Levine MN, Hirsh J, Horsewood P, Roberts RS, Gent M, Kelton JG. Heparin-induced thrombocytopenia in patients treated with low-molecular-weight heparin or unfractionated heparin. N Engl J Med 1995;332(20):1330–5.
44. Martchenke J, Boshkov L. Heparin-induced thrombocytopenia in neonates. Neonatal Netw 2005;24(5):33–7.
45. Tiaden JD, Wenzel E, Berthold HK, Muller-Oerlinghausen B. Adverse reactions to anticoagulants and to antiplatelet drugs recorded by the German spontaneous reporting system. Semin Thromb Hemost 2005; 31(4): 371–80.
46. Martel N, Lee J, Wells PS. Risk for heparin-induced thrombocytopenia with unfractionated and low-molecular-weight heparin thromboprophylaxis: a meta-analysis. Blood 2005;106(8):2710–5.
47. Hoh BL, Aghi M, Pryor JC, Ogilvy CS. Heparin-induced thrombocytopenia type II in subarachnoid hemorrhage patients: incidence and complications. Neurosurgery 2005;57(2):243–8.
48. Frost J, Mureebe L, Russo P, Russo J, Tobias JD. Heparin-induced thrombocytopenia in the pediatric intensive care unit population. Pediatr Crit Care Med 2005;6(2):216–9.
49. Kaplan GG, Manns B, McLaughlin K. Heparin induced thrombocytopaenia secondary to intraperitoneal heparin exposure. Nephrol Dial Transplant 2005;20(11):2561–2.
50. Ishibashi H, Takashi O, Hosaka M, Sugimoto I, Takahashi M, Nihei T, Kawanishi J, Ishiguchi T. Heparin-induced thrombocytopenia complicated with massive thrombosis of the inferior vena cava after filter placement. Int Angiol 2005;24(4):387–90.
51. Eghtesady P, Nelson D, Schwartz SM, Wheeler D, Pearl JM, Cripe LH, Manning PB. Heparin-induced thrombocytopenia complicating support by the Berlin Heart. ASAIO J 2005;51(6):820–5.
52. Chong BH, Fawaz I, Chesterman CN, Berndt MC. Heparin-induced thrombocytopenia: mechanism of interaction of the heparin-dependent antibody with platelets. Br J Haematol 1989;73(2):235–40.

53. Burgess JK, Lindeman R, Chesterman CN, Chong BH. Single amino acid mutation of Fc gamma receptor is associated with the development of heparin-induced thrombocytopenia. Br J Haematol 1995;91(3):761–6.
54. Brandt JT, Isenhart CE, Osborne JM, Ahmed A, Anderson CL. On the role of platelet Fc gamma RIIa phenotype in heparin-induced thrombocytopenia. Thromb Haemost 1995;74(6):1564–72.
55. Carlsson LE, Santoso S, Baurichter G, Kroll H, Papenberg S, Eichler P, Westerdaal NA, Kiefel V, van de Winkel JG, Greinacher A. Heparin-induced thrombocytopenia: new insights into the impact of the FcgammaRIIa-R-H131 polymorphism. Blood 1998;92(5):1526–31.
56. Barrowcliffe TW, Johnson EA, Thomas D. Antithrombin III and heparin. Br Med Bull 1978;34(2):143–50.
57. Boshkov LK, Warkentin TE, Hayward CP, Andrew M, Kelton JG. Heparin-induced thrombocytopenia and thrombosis: clinical and laboratory studies. Br J Haematol 1993;84(2):322–8.
58. Warkentin TE. Heparin-induced thrombocytopenia. Pathogenesis, frequency, avoidance and management. Drug Saf 1997;17(5):325–41.
59. Kakkasseril JS, Cranley JJ, Panke T, Grannan K. Heparin-induced thrombocytopenia: a prospective study of 142 patients. J Vasc Surg 1985;2(3):382–4.
60. Bell WR, Royall RM. Heparin-associated thrombocytopenia: a comparison of three heparin preparations. N Engl J Med 1980;303(16):902–7.
61. Green D, Martin GJ, Shoichet SH, DeBacker N, Bomalaski JS, Lind RN. Thrombocytopenia in a prospective, randomized, double-blind trial of bovine and porcine heparin. Am J Med Sci 1984;288(2):60–4.
62. Rao AK, White GC, Sherman L, Colman R, Lan G, Ball AP. Low incidence of thrombocytopenia with porcine mucosal heparin. A prospective multicenter study. Arch Intern Med 1989;149(6):1285–8.
63. Laster J, Silver D. Heparin-coated catheters and heparin-induced thrombocytopenia. J Vasc Surg 1988;7(5):667–72.
64. Pelouze GA, Coste B, Rassam T, Valat JD. Thrombopénie immunoallergique déclenche par le revêtement héparine d'un catheter. [Immunoallergic thrombopenia caused by heparin coating of a catheter.] Presse Méd 1989;18(30):1481.
65. Shaw PH, Ranganathan S, Gaines B. A spontaneous intramural hematoma of the bowel presenting as obstruction in a child receiving low-molecular-weight heparin. J Pediatr Hematol Oncol 2005;27(10):558–60.
66. Egger SS, Sawatzki MG, Drewe J, Krahenbuhl S. Life-threatening hemorrhage after dalteparin therapy in a patient with impaired renal function. Pharmacotherapy 2005;25(6):881–5.
67. Malik A, Capling R, Bastani B. Enoxaparin-associated retroperitoneal bleeding in two patients with renal insufficiency. Pharmacotherapy 2005;25(5):769–72.
68. Cimo PL, Moake JL, Weinger RS, Ben-Menachem YB, Khalil KG. Heparin-induced thrombocytopenia: association with a platelet aggregating factor and arterial thromboses. Am J Hematol 1979;6(2):125–33.
69. Moore JR, Weiland AJ. Heparin-induced thromboembolism: a case report. J Hand Surg [Am] 1979;4(4):382–5.
70. Warkentin TE, Kelton JG. A 14-year study of heparin-induced thrombocytopenia. Am J Med 1996;101(5):502–7.
71. Warkentin TE, Elavathil LJ, Hayward CP, Johnston MA, Russett JI, Kelton JG. The pathogenesis of venous limb gangrene associated with heparin-induced thrombocytopenia. Ann Intern Med 1997;127(9):804–12.
72. Wallis DE, Lewis BE, Pifarre R, Scanlon PJ. Active surveillance for heparin induced thrombocytopenia or thromboembolism. Chest 1994;106:120.
73. Rizzoni WE, Miller K, Rick M, Lotze MT. Heparin-induced thrombocytopenia and thromboembolism in the postoperative period. Surgery 1988;103(4):470–6.
74. Patel GK, Knight AG. Generalised cutaneous necrosis: a complication of low-molecular-weight heparin. Int Wound J 2005;2(3):267–70.
75. Toll A, Gallardo F, Abella ME, Fontcuberta J, Barranco C, Pujol RM. Low-molecular-weight heparin-induced skin necrosis: a potential association with pre-existent hypercoagulable states. Int J Dermatol 2005;44(11):964–6.
76. Patel R, Lim Z, Dawe S, Salisbury J, Arya R. Heparin-induced skin necrosis. Br J Haematol 2005;129(6):712.
77. Zeller L, Almog Y, Tomer A, Sukenik S, Abu-Shakra M. Catastrophic thromboses and severe thrombocytopenia during heparin therapy in a patient with anti-phospholipid syndrome. Clin Rheumatol. 2006;25(3):426–9.
78. Godal HC. Report of the International Committee on Thrombosis and Haemostasis. Thrombocytopenia and heparin. Thromb Haemost 1980;43(3):222–4.
79. Becker PS, Miller VT. Heparin-induced thrombocytopenia. Stroke 1989;20(11):1449–59.
80. Hirsh J, Warkentin TE, Raschke R, Granger C, Ohman EM, Dalen JE. Heparin and low-molecular-weight heparin: mechanisms of action, pharmacokinetics, dosing considerations, monitoring, efficacy, and safety. Chest 1998;114(Suppl 5):S489–510.
81. Lubenow N, Eichler P, Lietz T, Greinacher A; HIT Investigators Group. Lepirudin in patients with heparin-induced thrombocytopenia—results of the third prospective study (HAT-3) and a combined analysis of HAT-1, HAT-2, and HAT-3. J Thromb Haemost 2005;3(11):2428–36.
82. McNulty I, Katz E, Kim KY. Thrombocytopenia following heparin flush. Prog Cardiovasc Nurs 2005;20(4):143–7.
83. Bircher AJ, Itin PH, Buchner SA. Skin lesions, hypereosinophilia, and subcutaneous heparin. Lancet 1994;343(8901):861.
84. Dukes GE Jr, Sanders SW, Russo J Jr, Swenson E, Burnakis TG, Saffle JR, Warden GD. Transaminase elevations in patients receiving bovine or porcine heparin. Ann Intern Med 1984;100(5):646–50.
85. Toulemonde F, Kher A. Heparine et transaminases: une énigme sans importance en 1994? Therapie 1994;49:355.
86. Carlson MK, Gleason PP, Sen S. Elevation of hepatic transaminases after enoxaparin use: case report and review of unfractionated and low-molecular-weight heparin-induced hepatotoxicity. Pharmacotherapy 2001;21(1):108–13.
87. Hui CK, Yuen MF, Ng IO, Tsang KW, Fong GC, Lai CL. Low molecular weight heparin-induced liver toxicity. J Clin Pharmacol 2001;41(6):691–4.
88. Lambert M, Laterre PF, Leroy C, Lavenne E, Coche E, Moriau M. Modifications of liver enzymes during heparin therapy. Acta Clin Belg 1986;41(5):307–10.
89. Sacher C, Hunzelmann N. Tolerance to the synthetic pentasaccharide fondaparinux in heparin sensitization. Allergy 2003;58:1318–19.
90. Koch P. Delayed-type hypersensitivity skin reactions due to heparins and heparinoids. Tolerance of recombinant hirudins and of the new synthetic anticoagulant fondaparinux. Contact Dermatitis 2003;49:276–80.
91. Ludwig RJ, Beier C, Lindhoff-Last E, Kaufmann R, Boehncke WH. Tolerance of fondaparinux in a patient

allergic to heparins and other glycosaminoglycans. Contact Dermatitis 2003;49:158–9.
92. Ohno H, Higashidate M, Yokosuka T. Argatroban as an alternative anticoagulant for patients with heparin allergy during coronary bypass surgery. Heart Vessels 2003;18:40–2.
93. Phillips JK, Majumdar G, Hunt BJ, Savidge GF. Heparin-induced skin reaction due to two different preparations of low molecular weight heparin (LMWH). Br J Haematol 1993;84(2):349–50.
94. Moreau A, Dompmartin A, Esnault P, Michel M, Leroy D. Delayed hypersensitivity at the injection sites of a low-molecular-weight heparin. Contact Dermatitis 1996;34(1):31–4.
95. Valdes F, Vidal C, Fernandez-Redondo V, Peteiro C, Toribio J. Eczema-like plaques to enoxaparin. Allergy 1998;53(6):625–6.
96. Mendez J, Sanchis ME, de la Fuente R, Stolle R, Vega JM, Martinez C, Armentia A, Sanchez P, Fernandez A. Delayed-type hypersensitivity to subcutaneous enoxaparin. Allergy 1998;53(10):999–1003.
97. Koch P, Hindi S, Landwehr D. Delayed allergic skin reactions due to subcutaneous heparin-calcium, enoxaparin-sodium, pentosan polysulfate and acute skin lesions from systemic sodium-heparin. Contact Dermatitis 1996;34(2):156–8.
98. Conri CL, Azoulai P, Constans J, Sebban A, Le Mouroux A, Midy D, Baste JC. Erythromélalgie et héparine de bas poids moléculaire. [Erythromelalgia and low molecular weight heparin.] Therapie 1994;49(6):518–9.
99. Wutschert R, Piletta P, Bounameaux H. Adverse skin reactions to low molecular weight heparins: frequency, management and prevention Drug Saf 1999;20(6): 515–525.
100. O'Toole RD. Heparin: adverse reaction. Ann Intern Med 1973;79(5):759.
101. Fried M, Kahanovich S, Dagan R. Enoxaparin-induced skin necrosis. Ann Intern Med 1996;125(6):521–2.
102. White PW, Sadd JR, Nensel RE. Thrombotic complications of heparin therapy: including six cases of heparin-induced skin necrosis. Ann Surg 1979;190(5):595–608.
103. Levine LE, Bernstein JE, Soltani K, Medenica MM, Yung CW. Heparin-induced cutaneous necrosis unrelated to injection sites. A sign of potentially lethal complications. Arch Dermatol 1983;119(5):400–3.
104. Kelly RA, Gelfand JA, Pincus SH. Cutaneous necrosis caused by systemically administered heparin. JAMA 1981;246(14):1582–3.
105. Stavorovsky M, Lichtenstein D, Nissim F. Skin petechiae and ecchymoses (vasculitis) due to anticoagulant therapy. Dermatologica 1979;158(6):451–61.
106. Fowlie J, Stanton PD, Anderson JR. Heparin-associated skin necrosis. Postgrad Med J 1990;66(777):573–5.
107. Vanderweidt P. Necroses cutanées étendues et multiples sous HBPM. Nouv Dermatol 1996;15:450.
108. Schlag G, Poigenfurst J, Gaudernak T. Risk/benefit of heparin–dihydroergotamine thromboembolic prophylaxis. Lancet 1986;2(8521–22):1465.
109. Kumar PD. Heparin-induced skin necrosis. N Engl J Med 1997;336(8):588–9.
110. Warkentin TE. Heparin-induced skin lesions. Br J Haematol 1996;92(2):494–7.
111. Santamaria A, Romani J, Souto JC, Lopez A, Mateo J, Fontcuberta J. Skin necrosis at the injection site induced by low-molecular-weight heparin: case report and review. Dermatology 1998;196(2):264–5.
112. Jones BF, Epstein MT. Cutaneous heparin necrosis associated with glomerulonephritis. Australas J Dermatol 1987;28(3):117–8.
113. Handschin AE, Trentz O, Kock HJ, Wanner GA. Low molecular weight heparin-induced skin necrosis-a systematic review. Langenbecks Arch Surg 2005;390(3):249–54.
114. Ludwig RJ, Schindewolf M, Alban S, Kaufmann R, Lindhoff-Last E, Boehncke WH. Molecular weight determines the frequency of delayed type hypersensitivity reactions to heparin and synthetic oligosaccharides. Thromb Haemost 2005;94(6):1265–9.
115. Sallah S, Thomas DP, Roberts HR. Warfarin and heparin-induced skin necrosis and the purple toe syndrome: infrequent complications of anticoagulant treatment. Thromb Haemost 1997;78(2):785–90.
116. Das AK. Low-molecular-weight heparin-associated fat necrosis of the breast. Age Ageing 2005;34(2):193–4.
117. Giorgini S, Martinelli C, Massi D, Lumini A, Mannucci M, Giglioli L. Iatrogenic calcinosis cutis following nadroparin injection. Int J Dermatol 2005;44(10):855–7.
118. Campanelli A, Kaya G, Masouye I, Borradori L. Calcifying panniculitis following subcutaneous injections of nadroparin–calcium in a patient with osteomalacia. Br J Dermatol 2005;153(3):657–60.
119. van Haren FM, Ruiter DJ, Hilbrands LB. Nadroparin-induced calcinosis cutis in renal transplant recipients. Nephron 2001;87(3):279–82.
120. Chigot P, De Gennes C, Samama MM. Ostéoporose induite soit par l'héparine non fractionée soit par l'héparine de bas poids moleculaire. [Osteoporosis induced either by unfractionated heparin or by low molecular weight heparin.] J Mal Vasc 1996;21(3):121–5.
121. Dahlman T, Lindvall N, Hellgren M. Osteopenia in pregnancy during long-term heparin treatment: a radiological study post partum. Br J Obstet Gynaecol 1990;97(3):221–8.
122. Fox JG, Walli RK, Jaffray B, Simpson HK. Calcified subcutaneous nodules due to calcium heparin injections in a patient with chronic renal failure. Nephrol Dial Transplant 1994;9(2):187–8.
123. Matziolis G, Perka C, Disch A, Zippel H. Effects of fondaparinux compared with dalteparin, enoxaparin and unfractionated heparin on human osteoblasts. Calcif Tissue Int 2003;73:370–9.
124. Clark SK, Tremann JA, Sennewald FR, Donaldson JA. Priapism: an unusual complication of heparin therapy for sudden deafness. Am J Otolaryngol 1981;2(1):69–72.
125. Adjiman S, Fava P, Bitker MO, Chatelain C. Priapisme induit par l'héparine: un pronostic plus sombre?. [Priapism induced by heparin. A more serious prognosis?.] Ann Urol (Paris) 1988;22(2):125–8.
126. Berkun Y, Haviv YS, Schwartz LB, Shalit M. Heparin-induced recurrent anaphylaxis. Clin Exp Allergy 2004;34(12):1916–8.
127. Ainley EJ, Mackie IG, Macarthur D. Adverse reaction to chlorocresol-preserved heparin. Lancet 1977;1(8013):705.
128. Dux S, Pitlik S, Perry G, Rosenfeld JB. Hypersensitivity reaction to chlorbutol-preserved heparin. Lancet 1981;1(8212):149.
129. Heinzelmann M, Bosshart H. Fondaparinux sodium lacks immunomodulatory effects of heparin. Am J Surg 2004;187(1):111–3.
130. Hohenstein E, Tsakiris D, Bircher AJ. Delayed-type hypersensitivity to the ultra-low-molecular-weight heparin fondaparinux. Contact Dermatitis 2004;51(3): 149–51.

131. Harenberg J, Jorg I, Bayerl C, Fiehn C. Treatment of a woman with lupus pernio, thrombosis and cutaneous intolerance to heparins using lepirudin during pregnancy. Lupus 2005;14(5):411–2.
132. Siegman-Igra Y, Jacobi E, Lang R, Schwartz D, Carmeli Y. Unexpected hospital-acquired bacteraemia in patients at low risk of bloodstream infection: the role of a heparin drip. J Hosp Infect 2005;60(2):122–8.
133. Ginsberg JS, Hirsh J. Use of antithrombotic agents during pregnancy. Chest 1992;102(Suppl 4):S385–90.
134. Anderson DR, Ginsberg JS, Burrows R, Brill-Edwards P. Subcutaneous heparin therapy during pregnancy: a need for concern at the time of delivery. Thromb Haemost 1991;65(3):248–50.
135. de Swiet M, Ward PD, Fidler J, Horsman A, Katz D, Letsky E, Peacock M, Wise PH. Prolonged heparin therapy in pregnancy causes bone demineralization. Br J Obstet Gynaecol 1983;90(12):1129–34.
136. Douketis JD, Ginsberg JS, Burrows RF, Duku EK, Webber CE, Brill-Edwards P. The effects of long-term heparin therapy during pregnancy on bone density. A prospective matched cohort study. Thromb Haemost 1996;75(2):254–7.
137. Dahlman T, Sjoberg HE, Hellgren M, Bucht E. Calcium homeostasis in pregnancy during long-term heparin treatment. Br J Obstet Gynaecol 1992;99(5):412–6.
138. Dahlman TC. Osteoporotic fractures and the recurrence of thromboembolism during pregnancy and the puerperium in 184 women undergoing thromboprophylaxis with heparin. Am J Obstet Gynecol 1993;168(4):1265–70.
139. Nelson-Piercy C, Letsky EA, de Swiet M. Low-molecular-weight heparin for obstetric thromboprophylaxis: experience of sixty-nine pregnancies in sixty-one women at high risk. Am J Obstet Gynecol 1997;176(5):1062–8.
140. Greer IA, Nelson-Piercy C. Low-molecular-weight heparins for thromboprophylaxis and treatment of venous thromboembolism in pregnancy: a systematic review of safety and efficacy. Blood 2005;106(2):401–7.
141. Hall JG, Pauli RM, Wilson KM. Maternal and fetal sequelae of anticoagulation during pregnancy. Am J Med 1980;68(1):122–40.
142. Ginsberg JS, Kowalchuk G, Hirsh J, Brill-Edwards P, Burrows R. Heparin therapy during pregnancy. Risks to the fetus and mother. Arch Intern Med 1989;149(10):2233–6.
143. Ginsberg JS, Hirsh J. Use of antithrombotic agents during pregnancy. Chest 1995;108(Suppl 4):S305–11.
144. Flessa HC, Kapstrom AB, Glueck HI, Will JJ. Placental transport of heparin. Am J Obstet Gynecol 1965;93(4):570–3.
145. Forestier F, Daffos F, Capella-Pavlovsky M. Low molecular weight heparin (PK 10169) does not cross the placenta during the second trimester of pregnancy study by direct fetal blood sampling under ultrasound. Thromb Res 1984;34(6):557–60.
146. Omri A, Delaloye JF, Andersen H, Bachmann F. Low molecular weight heparin Novo (LHN-1) does not cross the placenta during the second trimester of pregnancy. Thromb Haemost 1989;61(1):55–6.
147. O'Reilly R. In: Anticoagulant, Antithrombotic and Thrombolytic Drugs. New York: MacMillan, 1980:1347.
148. Desmond PV, Roberts RK, Wood AJ, Dunn GD, Wilkinson GR, Schenker S. Effect of heparin administration on plasma binding of benzodiazepines. Br J Clin Pharmacol 1980;9(2):171–5.
149. Routledge PA, Kitchell BB, Bjornsson TD, Skinner T, Linnoila M, Shand DG. Diazepam and N-desmethyldiazepam redistribution after heparin. Clin Pharmacol Ther 1980;27(4):528–32.
150. Lohman JJ, Hooymans PM, Koten ML, Verhey MT, Merkus FW. Effect of heparin on digitoxin protein binding. Clin Pharmacol Ther 1985;37(1):55–60.
151. Hasegawa GR, Eder JF. Dobutamine–heparin mixture inadvisable. Am J Hosp Pharm 1984;41(12):25882590.
152. Col J, Col-Debeys C, Lavenne-Pardonge E, Meert P, Hericks L, Broze MC, Moriau M. Propylene glycol-induced heparin resistance during nitroglycerin infusion. Am Heart J 1985;110(1 Pt 1):171–3.
153. Becker RC, Corrao JM, Bovill EG, Gore JM, Baker SP, Miller ML, Lucas FV, Alpert JA. Intravenous nitroglycerin-induced heparin resistance: a qualitative antithrombin III abnormality. Am Heart J 1990;119(6):1254–61.
154. Brack MJ, Gershlick AM. Nitrate infusion even in lower dose decreases the anticoagulant effect of heparin through a direct effect on heparin levels. Eur Heart J 1991;12:80.
155. DeChatelet LR, McCall CE, Cooper MR, Shirley PS. Inhibition of leukocyte acid phosphatase by heparin. Clin Chem 1972;18(12):1532–4.
156. Walters MI, Roberts WH. Gentamicin/heparin interactions: effects on two immunoassays and on protein binding. Ther Drug Monit 1984;6(2):199–202.
157. Krogstad DJ, Granich GG, Murray PR, Pfaller MA, Valdes R. Heparin interferes with the radioenzymatic and homogeneous enzyme immunoassays for aminoglycosides. Clin Chem 1982;28(7):1517–21.
158. Ordog GJ, Wasserberger J, Balasubramaniam S. Effect of heparin on arterial blood gases. Ann Emerg Med 1985;14(3):233–8.
159. Godolphin W, Cameron EC, Frohlich J, Price JD. Spurious hypocalcemia in hemodialysis patients after heparinization. In-vitro formation of calcium soaps. Am J Clin Pathol 1979;71(2):215–8.
160. Biswas CK, Ramos JM, Kerr DN. Heparin effect on ionised calcium concentration. Clin Chim Acta 1981;116(3):343–7.
161. McConnell JS, Cohen J. Effect of anticoagulants on the chromogenic *Limulus* lysate assay for endotoxin. J Clin Pathol 1985;38(4):430–2.
162. Prasad R, Maddux MS, Mozes MF, Biskup NS, Maturen A. A significant difference in cyclosporine blood and plasma concentrations with heparin or EDTA anticoagulant. Transplantation 1985;39(6):667–9.
163. Wood M, Shand DG, Wood AJ. Altered drug binding due to the use of indwelling heparinized cannulas (heparin lock) for sampling. Clin Pharmacol Ther 1979;25(1):103–7.
164. De Leveld, Piafasky KM. Lack of heparin effect on propranolol-induced beta adrenoreceptor blockade. Clin Pharmacol Ther 1982;31:216.
165. Thompson JE, Baird SG, Thomson JA. Effect of i.v. heparin on serum free triiodothyronine levels Br J Clin Pharmacol 1977;4:701.
166. Wilkins TA, Midgley JE, Giles AF. Treatment with heparin and results for free thyroxin: an in vivo or an in vitro effect? Clin Chem 1982;28(12):2441–3.
167. Stevenson HP, Archbold GP, Johnston P, Young IS, Sheridan B. Misleading serum free thyroxine results during low molecular weight heparin treatment. Clin Chem 1998;44(5):1002–7.

Indanediones

General Information

Indanediones have the same anti-vitamin K action as the coumarins. However they are generally more toxic than the coumarins, and can cause allergic reactions capable of involving many organs and sometimes resulting in death (1). These reactions are mainly observed with phenindione and are very rare with other indanediones, such as fluindione.

In an MRC trial (2), a 2.25% incidence of reactions necessitated withdrawal of phenindione.

Organs and Systems

Cardiovascular

As with the coumarins, the main complication of the indanediones is bleeding. The incidence of hemorrhage varies, depending on the age of the patient, the intensity of treatment, and the indication for using an anticoagulant. Hematuria, bruising, and gastrointestinal bleeding are the commonest signs. The most common site of fatal hemorrhage is intracranial (3).

Hematologic

Agranulocytosis can occur during the first month of treatment with phenindione (4). Anemia, thrombocytopenia (5), and leukemoid reactions (6) have also been described.

Liver

Hepatitis can rarely occur during indanedione treatment, most often with phenindione, although occasionally with fluindione (7,8). The mechanism is probably allergic. Resistance to treatment may be associated with the appearance of this reaction (8).

Urinary tract

Phenindione can cause renal insufficiency and the nephrotic syndrome, usually preceded by fever and skin reactions. In such cases, renal biopsy shows interstitial edema with infiltration by eosinophils and plasma cells. Tubular necrosis has also been observed(9).

Skin

Skin reactions are the commonest manifestations of allergic reactions. They generally occur during the first month of treatment and include erythematous, scarlatiniform, papular, and urticarial rashes, sometimes accompanied by fever. Severe exfoliative dermatitis has been reported (10,11).

Skin necrosis can occur during therapy with phenindione (12), as it does with the coumarin congeners.

Second-Generation Effects

Lactation

Massive neonatal scrotal hematoma occurred while the breast-feeding mother was taking phenindione (13).

References

1. Perkins J. Phenindione sensitivity. Lancet 1962;1:127–30.
2. Report of the Working Party on Anticoagulant Therapy in Coronary Thrombosis to the Medical Research Council. Assessment of short-anticoagulant administration after cardiac infarction. BMJ 1969;1(640):335–42.
3. Lacroix P, Portefaix O, Boucher M, Ramiandrisoa H, Dumas M, Ravon R, Christides C, Laskar M. Condition de survenue des accidents hémorragiques intracraniens des antivitamines K. [The causes of intracranial hemorrhagic complications induced by antivitamins K.] Arch Mal Coeur Vaiss 1994;87(12):1715–9.
4. Ager JA, Ingram GI. Agranulocytosis during phenindione therapy. BMJ 1957;(5027):1102–3.
5. Farwell C. Thrombocytopenia due to phenindione (Hedulin) sensitivity; report of a case. Med Ann Dist Columbia 1959;28(2):82.
6. Wright JS. Phenindione sensitivity with leukaemoid reaction and hépato-renal damage. Postgrad Med J 1970;46(537):452–5.
7. Biour M, Davy JM, Poynard T, Levy VG. La fluindione est-elle hepatotoxique?. [Is fluindione hepatotoxic?.] Gastroenterol Clin Biol 1990;14(10):782.
8. Penot JP, Fontenelle P, Dorleac D. Hépatite aiguë cytolytique asymptomatique à la fluindione s'accompagnant d'une resistance au traitement. A propos d'un cas. Revue de la litterature. [Asymptomatic acute cytolytic hepatitis due to fluindione and associated with resistance to treatment: a case report and review of the literature.] Arch Mal Coeur Vaiss 1998;91(2):267–70.
9. Smith K. Acute renal failure in phenindione sensitivity. BMJ 1965;5452:24–6.
10. Hollman A, Wong HO. Phenindione sensitivity. BMJ 1964;5411:730–2.
11. Copeman PW. Phenindione toxicity. BMJ 1965;5456:305.
12. Eldon K, Lindahl F. Hudnekroser efter behandling med fenindion (Dindevan). [Skin necrosis after treatment with phenindione (Dindevan).] Ugeskr Laeger 1980;142(15):965.
13. Eckstein HB, Jack B. Breast-feeding and anticoagulant therapy. Lancet 1970;1(7648):672–3.

Protamine

General Information

Protamine, derived from salmon sperm, combines with and neutralizes heparin through an acid-base interaction. It is most commonly used as the sulfate, but chloride and hydrochloride salts are also occasionally used.

The relative insolubility of protamine–insulin suspension in water slows the subcutaneous resorption of the hormone, and consequently prolongs its biological activity.

In a prospective study of patients undergoing cardiopulmonary bypass, 10.7% of patients receiving protamine sulfate had adverse reactions of varying types (1).

Protamine sulfate regularly induces hypotension, bradycardia, transient flushing, and a feeling of warmth when administered too rapidly. These non-allergic effects are not severe, provided the injection is given slowly (50 mg, the maximum recommended dose, over 10 minutes).

Comparative studies

In a double-blind, randomized trial, 167 patients undergoing aortocoronary bypass graft surgery received either heparinase I (maximum 35 μg/kg) or protamine (maximum 650 mg) for heparin reversal (2). Although the two treatments had similar efficacy, protamine had a better safety profile. Those given heparinase I had longer hospital stays, were more likely to have a serious adverse event, and were less likely to avoid transfusion. A composite morbidity score was not different, and there were similar rates of hemodynamic instability.

Organs and Systems

Cardiovascular

Severe acute pulmonary vasoconstriction with cardiovascular collapse was identified in 1983 and subsequent studies showed that the incidence of protamine-induced pulmonary artery vasoconstriction was about 1.5% after cardiac surgery (3). Protamine-induced pulmonary artery vasoconstriction is accompanied by generation of large quantities of thromboxane A_2, a potent vasoconstrictor (3).

The precise mechanisms that explain protamine-mediated systemic hypotension are unknown, but there is evidence that it may be mediated by the endothelium and dependent on nitric oxide/cyclic guanosine monophosphate; it has been suggested that methylthioninium chloride (methylene blue) may be of use in treating hemodynamic complications caused by the use of protamine after cardiopulmonary bypass (4).

Immunologic

Although protamine itself was for a long time not generally considered allergenic, allergic reactions can occur in susceptible individuals. Attributed to residual fish antigens that remained after purification, they are characterized by flushing, urticaria, wheezing, angioedema, and hypotension, and they can occur even after slow intravenous administration. True anaphylaxis with bronchospasm and/or anaphylactic shock is very rare (5).

It now seems to be widely believed that, exceptionally, protamine itself can act as an allergen. There were positive skin tests with protamine in patients who had anaphylaxis and who had all previously received protamine (5). Cases of anaphylactic shock after slow intravenous administration of protamine sulfate to patients with diabetes mellitus suggested cross-allergy to protamine present in protamine zinc insulin (6–8), especially in patients with serum antiprotamine IgE or IgG antibodies (9). Insulin-dependent patients with diabetes who use protamine insulin may be at greater risk of adverse effects when they receive protamine sulfate. A retrospective study in patients who received protamine sulfate to reverse the effects of heparin during catheterization procedures or cardiac surgery showed a relative risk of anaphylaxis four times higher in patients with diabetes who had used protamine insulin than in non-diabetic controls (10). In these patients, the risk of anaphylaxis on administration of protamine sulfate is about 1%.

References

1. Weiler JM, Gellhaus MA, Carter JG, Meng RL, Benson PM, Hottel RA, Schillig KB, Vegh AB, Clarke WR. A prospective study of the risk of an immediate adverse reaction to protamine sulfate during cardiopulmonary bypass surgery. J Allergy Clin Immunol 1990;85(4):713–719.
2. Stafford-Smith M, Lefrak EA, Qazi AG, Welsby IJ, Barber L, Hoeft A, Dorenbaum A, Mathias J, Rochon JJ, Newman MF. Members of the Global Perioperative Research Organization. Efficacy and safety of heparinase I versus protamine in patients undergoing coronary artery bypass grafting with and without cardiopulmonary bypass. Anesthesiology 2005;103(2):229–40.
3. Lowenstein E, Zapol WM. Protamine reactions, explosive mediator release, and pulmonary vasoconstriction. Anesthesiology 1990;73(3):373–5.
4. Viaro F, Dalio MB, Evora PR. Catastrophic cardiovascular adverse reactions to protamine are nitric oxide/cyclic guanosine monophosphate dependent and endothelium mediated: should methylene blue be the treatment of choice? Chest 2002;122(3):1061–6.
5. Doolan L, McKenzie I, Krafchek J, Parsons B, Buxton B. Protamine sulphate hypersensitivity. Anaesth Intensive Care 1981;9(2):147–9.
6. Gottschlich GM, Gravlee GP, Georgitis JW. Adverse reactions to protamine sulfate during cardiac surgery in diabetic and non-diabetic patients. Ann Allergy 1988;61(4):277–81.
7. Gupta SK, Veith FJ, Ascer E, Wengerter KR, Franco C, Amar D, el-Gaweet ES, Gupta A. Anaphylactoid reactions to protamine: an often lethal complication in insulin-dependent diabetic patients undergoing vascular surgery. J Vasc Surg 1989;9(2):342–50.
8. Stewart WJ, McSweeney SM, Kellett MA, Faxon DP, Ryan TJ. Increased risk of severe protamine reactions in NPH insulin-dependent diabetics undergoing cardiac catheterization. Circulation 1984;70(5):788–92.
9. Weiss ME, Nyhan D, Peng ZK, Horrow JC, Lowenstein E, Hirshman C, Adkinson NF Jr. Association of protamine IgE and IgG antibodies with life-threatening reactions to intravenous protamine. N Engl J Med 1989;320(14):886–92.
10. Vincent GM, Janowski M, Menlove R. Protamine allergy reactions during cardiac catheterization and cardiac surgery: risk in patients taking protamine–insulin preparations. Cathet Cardiovasc Diagn 1991;23(3):164–8.

Vitamin K analogues

General Information

Vitamin K is the name of a group of compounds, all of which contain the 2-methyl-1,4-naphthoquinone moiety. Common nomenclature is:

- vitamin K_1 (phytomenadione; rINN)
- vitamin K_2 (phytonadione; rINN)
- vitamin K_3 (menadione; rINN).

Vitamin K is routinely administered in many countries to newborn babies in varying doses and by various routes, most commonly either orally or by intramuscular injection, to prevent hemorrhagic disease of the newborn (see Table 1).

Phytomenadione given intravenously is generally tolerated well. However, intravenous administration of small amounts (2–5 mg) can be followed by severe short-lasting cyanosis, dyspnea, tachycardia, and low blood pressure in patients with cardiac failure. Flushing and sweating can occur. Anaphylactoid reactions may be due to the excipient Cremophor EL. The most important adverse effects are jaundice and kernicterus, which can occur in small and premature babies, even after small doses, probably because of immature liver function.

Organs and Systems

Cardiovascular

Results of intravenous administration of 100 sequential doses of vitamin K in 45 patients (34 men, 11 women) have been retrospectively reviewed (1). Vitamin K_1 was administered as Aquamephyton. The dose averaged 9.7 (range 1–20) mg. The mean age was 44 (range 18–66) years. Many of the patients were seriously ill and had significant underlying surgical and medical problems, as evidenced by the fact that 16 died. However, none of the deaths was attributable to vitamin K. In the 11 instances in which the duration of administration was noted, the mean time was 49 (range 12.5–60) minutes. Only one patient had an episode of transient hypotension.

- A 37-year-old man with a history of ethanol abuse presented with hepatic failure and non-cardiogenic pulmonary edema after an overdose of paracetamol, codeine, ibuprofen, and diazepam. He received two doses of Aquamephyton (phytonadione) over 2 days and 10 minutes after the second dose his blood pressure fell to 79/40 mmHg.

In this study the incidence of adverse effects was 1%, contrasting with the impression that one obtains from reading the medical literature on vitamin K, which suggests an overall incidence of less than 1%.

Nervous system

Kernicterus can occur in small and premature babies, even after small doses of vitamin K, probably because of immature liver function. In its 1997 clinical practice guidelines, the Canadian Paediatric Society recommended that, in order to prevent haemorrhagic disease of the newborn, phytomenadione be given as a single intramuscular injection to all newborns within 6 hours of birth, in a dose of 1 mg for infants with a birth weight of more than 1500 g and half this dose for smaller infants. The dose is crucial, since overdosage carries a serious risk of kernicterus, especially in premature infants. There is only a remote possibility that this treatment might increase the risk of childhood cancer.

Cerebral vein thrombosis developed after the administration of 10 mg of phytomenadione in two patients with chronic intestinal inflammation assumed to be part of an autoimmune disease (2).

Hematologic

High doses of vitamin K can occasionally cause hemolytic anemia (3).

Severe hemolysis has been attributed to vitamin K in patients with glucose-6-phosphate dehydrogenase deficiency, particularly if infection is present (4,5).

In severe hepatocellular disease the prothrombin concentration can be further depressed by high doses of vitamin K (6,7).

Skin

Skin reactions have been observed after intramuscular injection of phytomenadione.

Table 1 Classification of hemorrhagic disease of the newborn

Syndrome	*Time of presentation*	*Common bleeding sites*	*Comments*
Early disease	0–24 hours	Cephalohematoma, intracranial, intrathoracic, intra-abdominal	Maternal drugs a frequent cause (for example warfarin, anticonvulsants)
Classic disease	1–7 days	Gastrointestinal, skin, nasal, circumcision	Mainly idiopathic, maternal drugs
Late disease	2–12 weeks	Intracranial, skin, gastrointestinal	Mainly idiopathic, can be the presenting feature of an underlying disease (for example cystic fibrosis, α_1-antitrypsin deficiency, biliary atresia; there is often some degree of cholestasis

Presentation

The English-language literature on adverse skin reactions associated with intramuscular or subcutaneous phytomenadione has been reviewed (8). Vitamin K is generally well tolerated subcutaneously or intramuscularly. However, erythematous eczematous plaques have been well documented. Of 39 skin reactions due to phytomenadione, 32 were eczematous and 91% of the patients were women, average age 39 (range 9–64) years; in four cases there were small vesicles within the plaques (9–12).

As a rule, only a rash occurs, generally after high doses of the oil-soluble analogue. However, intramuscular injection of phytomenadione can also cause lumbar scleroderma around the injection site. The scleroderma is reported to develop in four stages; erythematous, erythematous pigmented, established scleroderma, and resolvent scleroderma. Localized pseudoscleroderma (Texier's syndrome) has also been reported (13). This particular adverse reaction has occurred mainly in patients with hepatic insufficiency (SEDA-12, 330), but skin reactions have also been seen in patients with a normal hepatic condition (14,15). Brown pigmentation in the trochanteric region has been reported in a patient with cirrhosis.

Localized eczema at the site of subcutaneous injection has been reported (8).

- A 50-year-old woman taking warfarin had an INR of 8 and was given phytomenadione 10 mg subcutaneously (Sabex, containing propylene glycol 2% and polyethylene glycol 10%) and 12 hours later another 5 mg. One week later she developed two red, pruritic, warm, indurated areas, measuring 2 × 4 and 8 × 10 cm, at the two separate injection sites. A skin biopsy showed minimal spongiosis of the epidermis and an edematous dermis with a dense perivascular lymphocytic infiltrate and numerous eosinophils.

A severe localized and subsequently generalized dermatitis occurred after intramuscular injection of phytomenadione (16).

- A 40-year-old woman with no pre-existing hepatic disease was given intramuscular phytomenadione (Konakion) 1 ml/day (10 mg/ml) on 4 days before open cholecystectomy for gallstones. She then noticed a pruritic erythematous patch on her anterior left thigh, and over the next 4 weeks four patches appeared on her thighs and became larger (maximum diameter 15 cm), vesicular, weeping, hot, indurated, and intensely itchy. Topical calamine lotion, 1% hydrocortisone cream, and Paxyl cream (lidocaine and benzalkonium chloride) and oral cefalexin had no effect. The eruption then spread to the operation wound site, the face, neck, and arms. Oral prednisolone resulted in gradual improvement over the next 4 weeks, but residual erythema persisted for nearly 6 months.

Other presentations include erythematous pruritic plaques associated with epidermal keratosis, focal spongiosis, and perivascular infiltrates of lymphocytes and eosinophils or scleroderma-like hypodermic indurations (17). In such cases, intradermal tests may be positive for phytonadione.

Timing

The lesions usually appear 4–16 days after intramuscular injection, and persist for up to 2 months (18,19), sometimes leaving a scar lasting for years (20). The median number of days between the administration of phytomenadione and the appearance of the eruption was 13 days, but eruptions have appeared as early as 30 minutes and as late as 4 weeks after injection. In 13 of 32 cases it took more than 2 months for the reaction to resolve.

Frequency

Skin reactions to vitamin K are rare, only about 40 cases having been reported.

Relation to dose

In some early reports it was postulated that a minimum dose was necessary to cause eruptions, but there have been later reports of adverse skin reactions after small doses (range 10–440 mg).

Histopathology

Histopathological examination typically shows epidermal changes, including spongiosis with or without intraepidermal vesicles. In the dermis there is a perivascular mononuclear cell infiltrate, which may also be interstitial, often containing eosinophils.

Mechanism

Various formulations of phytomenadione contain different inactive ingredients, including polysorbate 80, propylene glycol, sodium acetate, glacial acetic acid, polyethoxylated castor oil (Cremophor EL), dextrose, and benzyl alcohol. Based on negative results with these ingredients, it appears that phytomenadione itself is the antigen that leads to adverse skin reactions. Of 10 patients with liver disease and prior exposure to phytomenadione (Konakion), 4 had positive results to patch-testing with vitamin K (21). Intracutaneous tests have supported the generally accepted hypothesis that the phenomenon is due to type IV hypersensitivity.

Treatment

No particular therapy is effective. It is not known whether the minute quantities of phytomenadione that are present in some foods, such as parsley, kale, brussels sprouts, spinach, cucumber, soy bean oil, and green and black tea leaves, preclude effective dietary therapy. Since the mechanism of this reaction is thought to be delayed hypersensitivity, another potential therapeutic approach is topical application of tacrolimus (FK-506), a potent inhibitor of interleukin 2 and T cell activation. Tacrolimus up to now has only been shown to suppress allergic contact dermatitis to dinitrophenol.

Immunologic

Allergic reactions have been attributed to phytomenadione, menadione, and vitamin K_4. In mice, menadione caused marked hypodynamia and hypothermia. This effect was potentiated by riboflavin. Allergic reactions to the systemic administration of vitamin K are immunologically mediated, and generally arise in patients with coagulation or liver problems.

Intradermal tests with phytomenadione and menadione caused an allergic skin reaction 7–22 days after injection in 13 of 145 healthy subjects. The results suggested that the index of cutaneous sensitivity lies somewhere between 5.5 and 8.9%. On the other hand, the absence of adverse effects with oral phytomenadione is striking. Continuation of treatment orally can in some cases prevent dermatitis (10). No cross-sensitivity has been seen between phytomenadione and menadione (22). Anaphylactoid reactions, some fatal, to phytomenadione have been reported (SEDA-15, 416) mainly from intravenous use (23).

The recommendation that prolongation of the international normalized ratio (INR) to over 6.0 should be corrected with parenteral phytomenadione (24,25) is not accompanied by the caveat that the intravenous route entails the risk of life-threatening, non-IgE-mediated anaphylactic reactions and even death, due to the use of polyethoxylated castor oil (Cremophor EL) as a solvent (26). A severe reaction to intravenous phytomenadione has been reported (27).

- A 74-year-old woman, taking warfarin, presented with an INR of 6.2. She was given a slow intravenous injection of phytomenadione (Konakion, Roche Products Ltd) 0.25 ml (500 μg) over about 60 seconds, and soon after felt profoundly unwell and complained of severe backache. She was given chlorphenamine 10 mg by rapid intravenous injection and adrenaline 1 mg in 10 ml of water over 30 minutes, with a good result.

An unusual case of allergy to vitamin K, with a relapsing and remitting eczematous reaction, has been described after an intramuscular injection of vitamin K_1 (28).

- A 27-year-old woman with cystic fibrosis and pancreatic insufficiency was given intramuscular vitamin K_1 into her thigh. The next day transient erythema occurred over the injection site and 6 weeks later there was localized pain, erythema, and edema. She was given intravenous cefuroxime for presumed cellulitis, but over the next few days the features became more consistent with localized eczema; she was given oral prednisolone 30 mg/day, super-potent topical corticosteroids, corticosteroid injections, and corticosteroids under occlusion with Duoderm, all of which failed to result in improvement. She then also developed an eczematous reaction to the Duoderm dressing. Patch tests were positive to vitamin K_1 and cross-reacted with vitamin K_4. She was also positive to colophonium and ester gum rosin, the dressing adhesive. Recurrent angio-edema persisted for several months and 2 years later she still had symptoms at the injection sites.

In all reported cases, only the whole formulation of vitamin K_1 (in its vehicle) or vitamin K_1 alone elicited positive patch tests. When individual additives were tested the results were negative. No previous exposure to vitamin K_1 was required for the development of type IV hypersensitivity, and primary sensitization occurred within 1–2 weeks or after a longer time period, as in the patient described here.

Long-Term Effects

Tumorigenicity

The relation between neonatal vitamin K administration and childhood cancer has been investigated in three case-control studies.

In a retrospective study 685 children who developed cancer before their 15th birthday were compared with 3442 controls matched for date and hospital of birth (29). There was no association between the administration of vitamin K and the development of all childhood cancers (unadjusted odds ratio 0.89; 95% CI = 0.69, 1.15) or of all cases of acute lymphoblastic leukemia (1.20; 0.75, 1.92), but there was a raised odds ratio for acute lymphoblastic leukemia 1–6 years after birth (1.79; 1.02, 3.15).

However, no such association was seen in a separate cohort-based study not dependent on case-note retrieval, in which the rates of acute lymphoblastic leukemia in children born in hospital units in which all babies received vitamin K were compared with those born in units in which less than a third received prophylaxis. It was concluded that on the basis of currently published evidence neonatal intramuscular vitamin K administration does not increase the risk of early childhood leukemia.

In another study children aged 0–14 years with leukemia (n = 150), lymphomas (n = 46), central nervous system tumors (n = 79), a range of other solid tumours (n = 142), and a subset of acute lymphoblastic leukemia (n = 129) were compared with 777 children matched for age and sex (30). Odds ratios showed no significant positive association for leukemias (1.30; 0.83, 2.03), acute lymphoblastic leukemia (1.21; 0.74, 1.97), lymphomas (1.06; 0.46, 2.42), central nervous system tumors (0.74; 0.40, 1.34), and other solid tumors (0.59; 0.37, 0.96). There was no association with acute lymphoblastic leukemia in children aged 1–6 years. The authors concluded that they had not confirmed the observation of an increased risk of childhood leukemia and cancer associated with intramuscular vitamin K.

In another study 597 cases and matched controls were compared. The association between cancer generally and intramuscular vitamin K was of borderline significance (odds ratio 1.44); the association was strongest for acute lymphoblastic leukemia (1.73) (31). However, there was also an effect of abnormal delivery. The authors suggested from the lack of consistency between the various studies so far published, including their own, and the low relative risks found in most of them, that the risk, if any, attributable to the use of vitamin K cannot be large, but that the possibility that there is some risk cannot be excluded. They recommended that prophylaxis using the commonly used intramuscular dose of 1 mg should be restricted to babies at particularly high risk of vitamin K deficiency; alternatively, a lower dose might be given to a larger proportion of those at risk.

Ecological studies of the relation between hospital policies on neonatal vitamin K administration and subsequent occurrences of childhood cancer have been analysed using data from selected large maternity units in Scotland, England, and Wales (32). The study covered 94 hospitals, with a total of 2.3 million births during periods when intramuscular vitamin K was routinely used and 1.4 million births when a selective policy was in operation. An increased risk was occasionally associated with vitamin K (highest odds ratio 1.25 for acute lymphoblastic leukemia in one hospital), but the overall results were not significant, and there was no evidence to support the previously suggested doubling of the risk of childhood cancer.

On the basis of all these results it is unlikely that there is a greatly increased risk of childhood cancer attributable to intramuscular vitamin K given to newborns, if indeed there is any risk at all.

Susceptibility Factors

High doses of vitamin K can cause relapses in patients with thrombosis and myocardial infarction treated with dicoumarol (33).

Drug Administration

Drug formulations

In a worldwide post-marketing surveillance program a conventional formulation of phytomenadione (Konakion), which contains Cremophor EL (polyethoxylated castor oil), was compared with a new mixed micellar formulation (Konakion MM) (34). During 1974–95 an estimated 635 million adults and 728 million children were given Konakion or Konakion MM. Of the 404 adverse events reported in 286 subjects (see Table 2) 387 (96%) were associated with Konakion, which had 95% of sales. Konakion MM accounted for 4% ($n = 17$) of the reported adverse effects and 5% of sales. Of the 17 adverse events 13 reported with Konakion MM were minor injection site reactions. Overall, 120 of the adverse events were serious, of which 117 (98%) were associated with Konakion. There were 85 probable anaphylactoid reactions (of which six were fatal) with conventional Konakion, compared with one non-fatal anaphylactoid reaction with Konakion MM. During the last 12 months of postmarketing surveillance, there were 14 serious adverse events reported in an estimated 21 million individuals treated with Konakion, but none in the 13 million who received Konakion MM. These results suggest that formulations of phytomenadione that are solubilized with Cremophor EL have a higher profile of adverse events,

Table 2 The distribution of 404 adverse events associated with phytomenadione in 286 patients, by system; the figures in parentheses refer to serious adverse events

System	*Konakion (Cremophor EL)*	*Konakion (mixed micelles)*	*Total*	*% oftotal (n = 404)*
General disorders	83 (37)	2 (1)	85 (38)	21.0
Cardiovascular				
Heart rate and rhythm disorders	31 (10)	1 (0)	32 (10)	7.9
Vascular	18 (11)		18 (11)	4.5
Extracardiac	8 (3)		8 (3)	2.0
Myocardial, endocardial, pericardial and valve disorders	1 (1)		1 (1)	0.2
Respiratory	31 (15)	1 (1)	32 (16)	7.9
Nervous system				
Central nervous and peripheral nervous systems	10 (4)		10 (4)	2.5
Autonomic nervous system	3 (0)		3 (0)	0.7
Sensory systems				
Hearing and vestibular disorders	2 (0)		2 (0)	0.5
Endocrine, metabolic, nutritional	1 (0)		1 (0)	0.2
Hematologic				
Erythrocyte disorders	3 (1)		3 (1)	0.7
Leukocyte and reticuloendothelial system disorders	2 (2)		2 (2)	0.5
Platelet, bleeding, and clotting disorders	5 (2)		5 (2)	1.2
Liver, biliary	14 (3)		14 (3)	3.4
Gastrointestinal	14 (2)		14 (2)	3.4
Urinary system	3 (3)		3 (3)	0.7
Skin	90 (9)	5 (0)	95 (9)	23.5
Musculoskeletal	2 (1)		2 (1)	0.5
Neonates and infants	13 (7)		13 (7)	3.2
Application site disorders	53 (6)	8 (1)	61 (7)	15.1
Total	387 (117)	17 (3)	404 (120)	100%

including anaphylactoid reactions, than the newer mixed micellar formulation Konakion MM.

Drug administration route

A case-control study in 1990 suggested a causal relation between intramuscular administration of vitamin K to neonates and the subsequent development of leukemia (35). Later studies have not been able to confirm this finding (30,36–38), although the results are not sufficiently conclusive to rule out the possibility that there is some risk (29,31).

Rapid intravenous administration of vitamin K can cause facial flushing, sweating, fever, a rise in blood pressure, a feeling of constriction in the chest, and cyanosis. Cases of cardiovascular collapse after intravenous injection of phytomenadione and phytonadione have been reported (39). Intravenous injection of phytomenadione is recommended to be performed slowly at a rate not exceeding 5 mg/minute.

The safety and efficacy of intravenously administered phytonadione have been retrospectively studied in patients taking long-term oral anticoagulants (40). Of 105 patients, 85 initially received intravenous phytonadione 0.5–1 mg, and 29 received a second dose. Only two of the 105 patients had suspected adverse reactions to phytonadione. Both were men with prior lung disease (one with an adenocarcinoma and one with chronic obstructive lung disease) and were to receive a 0.5 mg dose of phytonadione. During the phytonadione infusion both patients developed dyspnea and chest tightness, which resolved within 15 minutes of stopping the infusion; neither patient had hypotension.

Drug–Drug Interactions

Warfarin

An interaction of warfarin with a nutritional supplement containing vitamin K has been reported (41).

- A 72-year-old man taking warfarin (42.5 mg/week) to prevent thromboembolism related to atrial fibrillation/flutter wanted to take an over-the-counter nutritional supplement containing vitamin K (Nature's Life Greens, a blend of more than 20 herbs). He was instructed to withhold two doses of warfarin. Two weeks later he had an INR of 1.68; however, he had missed three doses of warfarin within the previous 10 days because of cataract surgery. Two weeks later his INR was 3.34. The warfarin dosage was reduced from 42.5 to 40.5 mg/week and the INR was in the target range 2 weeks later.

This report underlines the need for monitoring the use of nutritional and herbal products in patients taking warfarin.

Ubidecarenone (coenzyme Q10; ubiquinone-10) is chemically closely related to phytonadione. It is also widely available as a non-orthodox over-the-counter product, and several patients have been described in whom a reduced effect of warfarin was observed after the addition of non-orthodox ubidecarenone (42,43).

References

1. Bosse GM, Mallory MN, Malone GJ. The safety of intravenously administered vitamin K. Vet Hum Toxicol 2002;44(3):174–6.
2. Florholmen J, Waldum H, Nordoy A. Cerebral thrombosis in two patients with malabsorption syndrome treated with vitamin K. BMJ 1980;281(6239):541.
3. Jablonska A, Gadamska T. Niedokrwistosc hemolityczna u nowrodka po przedawkowaniu witaminy K. Med Wiejska 1973;1:79.
4. Ludin H. Blut- und Knochenmarkschädigungen durch Medikamente. [Blood and bone marrow damage caused by drugs.] Schweiz Med Wochenschr 1965;95(31):1027–32.
5. Garrett JV, Hallum J, Scott P. Urinary antiseptics causing haemolytic anaemia in pregnancy in a West Indian woman with red cell enzyme deficiency. J Obstet Gynaecol Br Commonw 1963;70:1073–5.
6. Roe DA. Nutrient toxicity with excessive intake. NY St J Med 1966;66:869.
7. Gupta KD, Banerji A. Paradoxical effect of vitamin K therapy in aggravating hypoprothrombinaemia. J Indian Med Assoc 1967;49(10):482–4.
8. Wilkins K, DeKoven J, Assaad D. Cutaneous reactions associated with vitamin K1. J Cutan Med Surg 2000;4(3):164–8.
9. Bruynzeel I, Hebeda CL, Folkers E, Bruynzeel DP. Cutaneous hypersensitivity reactions to vitamin K: 2 case reports and a review of the literature. Contact Dermatitis 1995;32(2):78–82.
10. Pigatto PD, Bigardi A, Fumagalli M, Altomare GF, Riboldi A. Allergic dermatitis from parenteral vitamin K. Contact Dermatitis 1990;22(5):307–8.
11. Joyce JP, Hood AF, Weiss MM. Persistent cutaneous reaction to intramuscular vitamin K injection. Arch Dermatol 1988;124(1):27–8.
12. Keough GC, English JC 3rd, Meffert JJ. Eczematous hypersensitivity from aqueous vitamin K injection. Cutis 1998;61(2):81–3.
13. Brunskill NJ, Berth-Jones J, Graham-Brown RA. Pseudosclerodermatous reaction to phytomenadione injection (Texier's syndrome). Clin Exp Dermatol 1988;13(4):276–8.
14. Sanders MN, Winkelmann RK. Cutaneous reactions to vitamin K. J Am Acad Dermatol 1988;19(4):699–704.
15. Moreau-Cabarrot A, Giordano-Labadie F, Bazex J. Hypersensibilité cutanée au point d'injection de vitamin K1. [Cutaneous hypersensitivity at the site of injection of vitamin K1.] Ann Dermatol Venereol 1996;123(3):177–9.
16. Wong DA, Freeman S. Cutaneous allergic reaction to intramuscular vitamin K1. Australas J Dermatol 1999;40(3):147–52.
17. Balato N, Cuccurullo FM, Patruno C, Ayala F. Adverse skin reactions to vitamin K1: report of 2 cases. Contact Dermatitis 1998;38(6):341–2.
18. Barnes HM, Sarkany I. Adverse skin reaction from vitamin K1. Br J Dermatol 1976;95(6):653–6.
19. Baer RL. Cutaneous skin changes probably due to pyridoxine abuse. J Am Acad Dermatol 1984;10(3):527–8.
20. Chung JY, Ramos-Caro FA, Beers B, Ford MJ, Flowers FP. Hypersensitivity reactions to parenteral vitamin K. Cutis 1999;63(1):33–4.

21. Bullen AW, Miller JP, Cunliffe WJ, Losowsky MS. Skin reactions caused by vitamin K in patients with liver disease. Br J Dermatol 1978;98(5):561–5.
22. Hwang SW, Kim YP, Chung BS, Kim HK. Vitamin K1 dermatitis. Korean J Dermatol 1983;21:91.
23. ADRAC Slow down on parenteral vitamin K. Aust Adverse Drug React Bull 1991;10:3.
24. Hirsh J, Dalen JE, Deykin D, Poller L, Bussey H. Oral anticoagulants. Mechanism of action, clinical effectiveness, and optimal therapeutic range. Chest 1995;108(Suppl 4):S231–46.
25. Routledge PA. Practical prescribing: warfarin. Prescr J 1997;37:173–9.
26. Martin JC. Anaphylactoid reactions and vitamin K. Med J Aust 1991;155(11–12):851.
27. Jolobe OM, Penny E. Severe reaction to i.v. vitamin K Pharm J 1999;262:112.
28. Sommer S, Wilkinson SM, Peckham D, Wilson C. Type IV hypersensitivity to vitamin K. Contact Dermatitis 2002;46(2):94–6.
29. Parker L, Cole M, Craft AW, Hey EN. Neonatal vitamin K administration and childhood cancer in the north of England: retrospective case-control study. BMJ 1998;316(7126):189–93.
30. McKinney PA, Juszczak E, Findlay E, Smith K. Case-control study of childhood leukaemia and cancer in Scotland: findings for neonatal intramuscular vitamin K. BMJ 1998;316(7126):173–7.
31. Passmore SJ, Draper G, Brownbill P, Kroll M. Case-control studies of relation between childhood cancer and neonatal vitamin K administration. BMJ 1998;316(7126):178–84.
32. Passmore SJ, Draper G, Brownbill P, Kroll M. Ecological studies of relation between hospital policies on neonatal vitamin K administration and subsequent occurrence of childhood cancer. BMJ 1998;316(7126):184–9.
33. Reuter H. Vitamine, Chemie und KlinikStuttgart: Hippokrates Verlag;. 1970.
34. Pereira SP, Williams R. Adverse events associated with vitamin K1: results of a worldwide postmarketing surveillance programme. Pharmacoepidemiol Drug Saf 1998;7(3):173–82.
35. Golding J, Paterson M, Kinlen LJ. Factors associated with childhood cancer in a national cohort study. Br J Cancer 1990;62(2):304–8.
36. Klebanoff MA, Read JS, Mills JL, Shiono PH. The risk of childhood cancer after neonatal exposure to vitamin K. N Engl J Med 1993;329(13):905–8.
37. Ekelund H, Finnstrom O, Gunnarskog J, Kallen B, Larsson Y. Administration of vitamin K to newborn infants and childhood cancer. BMJ 1993;307(6896):89–91.
38. McWhirter WR. Vitamin K and childhood cancer. Med J Aust 1993;159(8):499.
39. Pelletier G, Attali P, Ink O. Arrêt cardiorespiratoire après injection intraveineuse de vitamine K1. [Cardiopulmonary arrest after intravenous injection of vitamin K1.] Gastroenterol Clin Biol 1986;10(8–9):615.
40. Shields RC, McBane RD, Kuiper JD, Li H, Heit JA. Efficacy and safety of intravenous phytonadione (vitamin K1) in patients on long-term oral anticoagulant therapy. Mayo Clin Proc 2001;76(3):260–6.
41. Bransgrove LL. Interaction between warfarin and a vitamin K-containing nutritional supplement: a case report. J Herbal Pharmacother 2001;1:85–9.
42. Spigset O. Reduced effect of warfarin caused by ubidecarenone. Lancet 1994;344(8933):1372–3.
43. Anonymous. Coenzym Q10 (Qumin Q10 u.a.) stört orale Antikoagulation Arznei-Telegramm 1994;12:120.

THROMBOLYTIC AGENTS

General Information

The major thrombolytic agents are:

- streptokinase
- urokinase
- anistreplase (anisoylated plasminogen streptokinase activator complex or APSAC)
- pro-urokinase (single-chain urokinase-type plasminogen activator or scu-PA)
- alteplase (recombinant tissue plasminogen activator or rt-PA)
- reteplase (r-PA: a deletion mutant of rt-PA)
- tenecteplase (a triple combination mutant variant of alteplase).

Although their pharmacodynamic properties do not differ significantly, some characteristics, as a result of their origin or mode of action, explain specific differences in their adverse effects and pharmacokinetics.

General adverse effects

Hemorrhage is the major risk of thrombolytic drugs; there are some differences in risks between the various agents, and certain susceptibility factors can be identified. Transient hypotensive reactions have been described with all thrombolytic agents, but they are in principle reversible. Hypersensitivity reactions are most often seen in patients who have been treated with compounds derived from cultures of streptococci (streptokinase and anistreplase). Tumor-inducing effects have not been reported.

Organs and Systems

Cardiovascular

Transient hypotension can occur after thrombolysis, and the incidence after infusion of streptokinase in myocardial infarction is high. Hypotension has been related to plasminemia, which leads to bradykinin generation from kallikrein, activation of complement, and potentially endothelial prostacyclin release (1). During treatment of myocardial infarction, some other mechanisms may explain hypotension: vagal reflexes precipitated by posterior wall reperfusion (2) and left ventricular dysfunction associated with the infarct process. Hypotension occurs shortly after the start of infusion and can be accompanied by bradycardia, flushing, and anxiety. It is reversible if fluids are given and the infusion temporarily stopped (1); vasopressors may be needed. The magnitude of this hypotensive reaction is directly related to the rate of infusion of streptokinase. The incidence of such reactions does not seem to be affected by premedication with glucocorticoids. Like streptokinase, anistreplase can also cause transient hypotension (3). Alteplase seems to be associated with a lower risk of hypotension than streptokinase or anistreplase. In the ISIS-3 trial, the incidences of profound hypotension necessitating drug therapy were: 6.8% with streptokinase, 7.2% with anistreplase, and 4.3% with alteplase (4).

Embolic detachment of components of venous or mural thrombi can sometimes be involved in the development of thromboembolism or cholesterol embolization. There is a 10% incidence of pulmonary embolism during thrombolysis, lethal in 0–5% (5), pointing to the risk of detachment of white components of venous thrombi, especially if large veins, such as those in the pelvis, are involved (6). However, the risk has not been proven to exceed that reported in patients treated with heparin and/or oral anticoagulants.

Cholesterol embolization is thought to occur after removal of mural thrombi covering atherosclerotic plaques leading to direct exposure of the soft lipid-laden core of these plaques to the arterial circulation. The contents of the soft lipid core, including crystallized cholesterol, shower the downstream circulation. Cholesterol crystals are impervious to dissolution or lysis and they lodge in small arterioles, causing obstruction. Cholesterol embolization shortly after thrombolytic treatment is characterized by livedo reticularis or multiple necrotic lesions of the skin of both legs (7) but can also be associated with lower limb ischemia, gangrene, visceral ischemia, or pseudovasculitis (8). Cholesterol embolization is difficult to diagnose, and may therefore be more common than suspected. It can cause acute renal insufficiency (9), and renal biopsy shows acute tubular necrosis, in some instances with arteriolar clefts representing cholesterol crystals (10).

Reperfusion dysrhythmias can occur when thrombolytic drugs are used to treat myocardial infarction. The most common is transient ventricular tachycardia. However, controlled trials have failed to show an increase in serious ventricular dysrhythmias, the incidence of ventricular fibrillation being actually reduced by thrombolytic therapy in myocardial infarction (11).

The prognostic value of ventricular late potentials and the character of rhythm disturbances have been studied in 64 patients with acute coronary syndrome (12). The overall rate of ventricular late potentials increased 8 hours after admission from 69% to 86%. In all cases of reperfusion dysrhythmias there was deterioration of signal-averaged electrocardiographic parameters and ventricular late potentials. However, in those who received thrombolysis the rate of ventricular late potentials was lower than in those who did not: 31% versus 48% on day 10 and in 11% versus 41% by the end of hospital treatment. Beta-blockers improved signal-averaged electrocardiography and cardiac rhythm variability in both groups.

- A coronary artery aneurysm occurred in a 49-year-old man 1 month after successful percutaneous transluminal coronary recanalization with streptokinase (13).

Phlebitis at the site of infusion of streptokinase has been reported.

Respiratory

Streptokinase can cause acute bronchospasm, sometimes fatal (14), or dyspnea.

Nervous system

Guillain–Barré syndrome has been reported in patients who were treated with streptokinase and anistreplase (15–19). As streptokinase and anistreplase are derived from streptococci, an immunological reaction is thought to have been responsible.

Sensory systems

Iritis and uveitis have been reported with streptokinase (20,21).

Hematologic

Hemorrhage is the major risk of the use of thrombolytic agents and can result from the lysis of thrombi at other sites where hemostasis is required. Hemostatic incompetence related to thrombolysis is also related to reductions in circulating fibrinogen and factors V and VIII and to the generation of fibrin and fibrinogen degradation products with anticoagulant properties. The most serious complication is intracranial hemorrhage, which has a high mortality (60% in the GUSTO-1 trial) and a high rate of severe disability (up to one-third of patients) (22).

A systematic review identified three studies (GISSI-2/ISG, GUSTO-1, and ISIS-3) in which patients had been randomized between the two treatments and recorded the risk of hemorrhagic stroke (23). Hemorrhagic stroke was more common with alteplase (0.57%) than streptokinase (0.36%) (OR = 1.8; CI = 1.1, 2.9). If GUSTO-1, in which alteplase was given in an accelerated fashion, was excluded, the hemorrhagic stroke rates were 0.50% and 0.23% (OR = 2.1, CI = 1.0, 4.4). There was no difference in 35-day mortality between the drugs.

Hemorrhagic stroke due to streptokinase is not uncommon and has again been reported (24).

In trials of intravenous streptokinase without angiography, hemorrhagic complications, apart from local venepuncture sites, were reported in about 4% of patients, with an absolute excess over controls of about 3% (11). Major bleeding, defined by transfusion requirements, occurred in only 0.5% of patients, and cerebral hemorrhage was seen in only 0.1%. In the AIMS trial, all hemorrhagic complications during hospitalization were 14% with anistreplase and 4.1% with placebo (25). Transfusion was required in 0.8% in each group. The absolute excess of cerebral hemorrhage rates was 0.3% with anistreplase compared with placebo. In the ASSET trial, the absolute excess risks of minor and major hemorrhagic complications with alteplase were respectively 6 and 0.9%. The excess risk of hemorrhagic strokes was 0.2% (26).

Several large clinical trials have compared alteplase and anistreplase with streptokinase. No clear-cut differences in severe or life-threatening bleeding were observed in these studies.

However, in interpreting these results, the role of adjunctive antithrombotic therapy must also be taken into account, since it can increase the risk of hemorrhage (27). In most of the trials, heparin-treated patients had significantly more major bleeding, and in the ISIS-3 trial, the incidence of cerebral hemorrhage was significantly increased by heparin (4). In the TIMI II B trial (28) in patients with unstable angina, the addition of a low dose of alteplase (0.8 mg/kg) to aspirin and heparin was associated with a 2.4% incidence of major hemorrhagic events (versus 0.6% with placebo) and a 0.55% incidence of intracranial hemorrhage (no such event occurring with placebo).

A study in general practice has suggested that intracranial hemorrhage during thrombolytic therapy for myocardial infarction may be more frequent than in clinical trials (10). The authors analysed the data accumulated in a US large registry called NRMI 2 (1484 hospitals, 71 073 patients having received alteplase in the initial treatment of acute myocardial infarction). The overall incidence of intracranial hemorrhage was 0.95%. This is a higher incidence than has been reported in clinical trials, and may have been due to the stringent enrolment criteria.

The susceptibility factors for intracranial hemorrhage have also been examined (10). The main factors were:

- advanced age; compared with patients under 65, the odds ratio was 2.71 (CI = 2.18, 3.37) for individuals aged 65–74 years and as high as 4.34 (3.45, 5.45) for patients aged 75 years or more
- female sex
- black ethnicity
- a history of stroke
- hypertension (an increased risk in patients with a systolic blood pressure of 140 mmHg or more and/or a diastolic of 100 mmHg or more)
- the dose of alteplase.

Because of the significant correlation with age and a previous history of stroke, it is now widely considered that in patients over 75 years with a history of stroke other therapies, such as primary coronary angioplasty, must be considered. As reported in the GUSTO-1 trial, patients with very high systolic blood pressures have an increased risk of death or disabling stroke (29), and the pulse pressure has been reported to be the clearest predictor of the blood pressure as regards the risk of intracranial hemorrhage (30).

The risk of major bleeding during thrombolytic therapy is undoubtedly a real problem, but it has to be considered against the background of the underlying mortality risk and the absolute benefit likely to be achieved in patients with acute myocardial infarction (31). This equation is fundamental in older patients, who have the highest mortality risk after myocardial infarction and thus the most to gain from thrombolysis (32).

Some reports of serious adverse consequences in patients with aortic dissection or pericarditis that mimic clinical or electrocardiographic myocardial infarction have emphasized the necessity for accurate diagnosis before thrombolysis is instituted in patients with chest pain or with atypical electrocardiographic changes (33).

The incidence of hemorrhagic complications is particularly high when high doses of streptokinase are used in the treatment of deep venous thrombosis (34). High-dose streptokinase thrombolysis for 2–3 days with an initial dose of 500 000 units followed by a maintenance dose of 3 600 000 U/day led to a 10% rate of major spontaneous bleeding complications, with a fatal outcome in four older subjects out of the total of 98 patients. The fatality rate of bleeding caused by streptokinase amounts to 7% in patients with peripheral arterial occlusion, but is much lower in younger patients with venous thromboembolism.

Secondary intraventricular hemorrhage has been described in two patients who received intraventricular tissue-type plasminogen activator to hasten the lysis of intraventricular hemorrhage (35).

Various other bleeding complications have been reported after thrombolysis, including atraumatic compartment syndrome in the thigh (36), bleeding into the neck (37), and a duodenal hematoma (38).

Hemolysis after intravenous streptokinase is rare (39).

Transient lymphopenia, possibly related to immunological destruction of T-helper cells by streptokinase, has been described (40).

Platelet aggregation due to an antibody-mediated reaction has been suggested as a cause of streptokinase-enhanced coronary thrombosis in patients with a specific type of antistreptokinase antibodies (41).

Thrombocytopenia has been described after thrombolytic therapy but it is most probably multifactorial, since patients generally also received intravenous heparin (42).

Disseminated intravascular coagulation has been reported with urokinase (43).

Liver

There were no differences in the activities of serum transaminases, lactate dehydrogenase, or creatine kinase in patients with myocardial infarction who received an 18-hour infusion of urokinase compared with patients who received glucose alone (44), but subacute alterations of liver function tests have been described with streptokinase and anistreplase (45). Unexplained increases in transaminase activities have been reported in almost 25% of patients treated with streptokinase (46). In view of the greater prominence of liver dysfunction with streptokinase than with alteplase it could be wiser to choose alteplase rather than streptokinase in patients with previous impaired hepatic function (47).

A few cases of liver dysfunction with or without jaundice have been reported with streptokinase and urokinase (48–51). Jaundice with hepatic cytolysis can occur after allergic reactions to streptokinase (52).

Urinary tract

Hematuria and proteinuria can occur without alteration of immunohistochemical features, suggesting a non-immune cause, such as a direct effect on the glomeruli or a reflection of the hypocoagulable state.

- Acute interstitial nephritis associated with cholesterol embolization during streptokinase therapy has been reported (53). In this case, cholesterol clefts were found in the preglomerular and interlobular arterioles.

Cholesterol embolization can cause acute renal insufficiency (9), and renal biopsy shows acute tubular necrosis with, in some instances, arteriolar clefts representing cholesterol crystals (10).

Immunologic

Since streptokinase is a natural product of cultures of streptococci and therefore has similar antigenic properties, most of the population have anti-streptokinase antibodies. These antibodies may explain both the allergic reactions that can occur and resistance to the drug, which occurs in some cases. Within 3 or 4 days of streptokinase administration, the titer of neutralizing antibodies has become sufficient to inactivate the usual doses. Persistence of these antibodies is observed in up to 80% of patients 1 year after treatment and in about 50% of patients after 2–4 years (54,55). If streptokinase has to be re-administered within 8 months of previous exposure, the neutralizing effects of plasma should be taken into account and the dose of streptokinase should be adjusted to overcome these effects (56). However, the extent to which streptokinase antibodies decline after earlier exposure to streptokinase is controversial, and persistently raised titers are found in a large proportion of patients (57). For this reason, an alternative thrombolytic agent should be recommended in patients who have already received streptokinase. In addition, it appears that allergic reactions occur more often when streptokinase is reused (58).

Antistreptokinase antibody titers have been determined in 47 consecutive streptokinase-naive patients with the acute coronary syndrome from Australian communities with endemic group A streptococcal infection, because of the implications for streptokinase thrombolysis (59). Antistreptolysin O and anti-DNAse B titers were also determined. Indigenous patients were significantly more likely to have anti-streptokinase antibodies than the non-indigenous patients. Anti-streptokinase antibody titers also correlated well with antistreptolysin O and anti-DNAse B titers. The authors concluded that streptokinase should not be used for thrombolysis in populations with endemic group A streptococcal infection.

In a double-blind, randomized study in 50 consecutive patients with acute myocardial infarction who were given streptokinase 1.5 or 2.5 million units and underwent angiography within 24 hours, the presence of antistreptokinase antibodies or the administration of an increased dose of streptokinase had no effect on improving the patency rate of the infarct-related artery (60). A larger study is needed to confirm these observations.

The use of streptokinase as a preflush in non-heart-beating kidney donors did not cause the production of anti-streptokinase antibodies in 18 recipients of renal transplants (61).

Angioedema is a rare acute and potentially fatal reaction to streptokinase, which should be diagnosed and treated quickly to guarantee the best prognosis (62).

- A 65-year-old white man with a myocardial infarction was given streptokinase 1 500 000 U intravenously. After infusion of 1/6th of the dose, he developed progressive hoarseness and nasal obstruction, secondary to angioedema. The infusion was stopped immediately and he was given adrenaline and hydrocortisone. He rapidly developed respiratory difficulty and wheezing, requiring intubation He also had edema of the eyebrows, lips, and tongue and later developed hemolytic anemia, which was attributed to activation of complement as part of the anaphylactic reaction.

Anistreplase, a compound consisting of streptokinase and anisoylated plasminogen, can also cause allergic reactions, but has a longer half-life, allowing intravenous administration in a relatively short interval of time.

Urokinase is extracted from human urine or prepared from cultures of fetal kidney cells and does not seem to cause allergic reactions.

Pro-urokinase, alteplase, reteplase, and tenecteplase, which are recombinant products, also appear to be free from allergic reactions. Pro-urokinase and alteplase have short half-lives (3–8 minutes) and require continuous infusion administration, which may in some cases be an advantage as it allows rapid surgical intervention when necessary (63). Reteplase and tenecteplase have substantially longer half-lives, allowing bolus administration.

The incidence of acute generalized allergic reactions in patients treated with streptokinase or anistreplase was originally reported to be high, but with the introduction of more highly purified forms of streptokinase the figure has fallen to 1–5%. The most common manifestations of non-anaphylactic reactions to streptokinase or anistreplase are skin rashes and pyrexia. However, anaphylactic shock with streptokinase is also well known, although rare; it occurred in 0.1% of the patients in the GISSI trial (64). Life-threatening angioedema has been described with streptokinase (65,66) and with alteplase (67–69). Skin rashes are seen especially with streptokinase, but rarely, if ever, with alteplase.

Several cases of low back pain associated with streptokinase or anistreplase injections for acute myocardial infarction have been described, with rapid resolution once the streptokinase was stopped (70,71). It is presumed that the mechanism is allergic, since no such report has been published with alteplase (72).

Streptokinase is regularly reported to have induced an immune complex syndrome, characterized by plasmacytosis, often severe, and accompanied by fever and the development of hemolytic anemia, occurring as early as the first week after the start of treatment; in some cases, temporary alterations in renal function also occur (73).

Vasculitis has been rarely described after streptokinase or anistreplase (74–77), but not after urokinase or alteplase. It is characterized by lymphocyte infiltration and deposition of immune complexes, fibrin, and complement in the skin microvasculature.

Other patients have the typical picture of serum sickness, sometimes associated with acute renal insufficiency (78–80) or of Henoch–Schönlein purpura (81) with a purpuric rash, joint and abdominal pains, and sometimes hematuria (82).

Adult respiratory distress syndromes and multisystem organ failure have been reported with streptokinase and anistreplase (83–86); the timing of the onset of symptoms and the antibody profile in one case suggested an immunological response (84).

The clinical efficacy of thrombolytic drugs such as streptokinase does not appear to be compromised by the occurrence of allergic reactions, according to data from GUSTO-1 (87).

There have been three reports of anaphylactoid reactions, mostly orolingual angioedema, after therapy of acute ischemic stroke with alteplase.

- In one patient there was marked edema of the lip about 45 minutes after a bolus dose of alteplase; it subsided within 2 hours without any intervention (88).

Of 105 consecutive patients treated with alteplase for acute ischemic stroke, 2 developed anaphylactoid reactions (89). The first had a rash and extensive bilateral swelling of the tongue, epiglottis, and uvula, requiring intubation. The second developed unilateral swelling of the tongue and lips without a rash or hypotension.

In 230 patients treated with alteplase for acute ischemic stroke, there were two cases of orolingual angioedema (90). Both presented with localized symptoms only, symmetrical in one and asymmetrical in the other.

IgE-associated anaphylactic reactions can also occur.

- A 70-year-old woman was treated with intravenous alteplase for thrombolysis in acute ischemic stroke and 30 minutes later had acute sinus tachycardia and hypotension, followed by cyanosis and loss of consciousness (91). Serum samples analysed by ELISA were positive for IgE antibodies to alteplase.

Such reactions are very rare; there have been four reported cases in over 1 million administrations.

Body temperature

Pyrexia can occur after the initial dose of streptokinase (92). Concurrent administration of glucocorticoids does not prevent it completely. A marked rise in body temperature with chills can be accompanied by hypotension, abdominal pain (particularly low back pain), and very occasionally mild psychotic reactions.

Second-Generation Effects

Pregnancy

Pregnancy is not considered to be incompatible with treatment with thrombolytic agents. No adverse effects on the fetus have been reported. Nevertheless, the possibility cannot be excluded that placental separation and uterine bleeding will occur. Only minimal amounts of streptokinase cross the placenta, and these are not sufficient to cause fibrinolysis in the fetus (93). Although the passage of streptokinase is blocked by the placenta, streptokinase antibodies do cross to the fetus (94). This passive sensitization would have clinical importance only if the neonate required streptokinase therapy. Septicemia during treatment with streptokinase has been also described (95). These different hazards may preclude the use of thrombolytic drugs during pregnancy but in acute life-threatening problems, efficacy has to be weighed up against the risk of adverse effects.

Susceptibility factors

There have been two studies of the susceptibility factors for intracranial hemorrhage in patients with acute coronary syndromes undergoing fibrinolytic therapy (96,97). FASTRAK II is a prospective registry of acute coronary syndromes involving 111 Canadian hospitals, in which trained medical personnel record admission, treatment, and discharge data on patients admitted with acute coronary syndromes (96). From 1 January 1998 to 31 December 2000, 12 739 patients received fibrinolytic therapy for acute myocardial infarction. Of these, 146 patients (1.15%) sustained strokes and 82 (0.65%) had an intracranial hemorrhage. Advanced age, female sex, a history of stroke, and systolic hypertension (over 160 mmHg) on arrival were important independent risks factors for intracranial hemorrhage. Patients who received streptokinase had a lower risk of intracranial hemorrhage.

Of 592 patients in the Global Utilization of Streptokinase and tPA for Occluded Arteries-I (GUSTO-I) trial who had a stroke during initial hospitalization, the risk of intracranial hemorrhage was significantly greater in those with recent facial or head trauma (OR = 13, 95% CI = 3.4, 86); dementia was also associated with an increased risk for intracranial hemorrhage (OR = 3.4, 95% CI = 1.2, 10) (97). Because facial or head trauma can greatly influence treatment decisions, it was suggested that this risk factor should be incorporated into models designed to estimate the benefit to harm balance of fibrinolytic therapy.

Drug Administration

Drug administration route

Pleural space infection is an important cause of sepsis, and traditional treatment comprises drainage of the pleural space, either through chest tubes or surgically, combined with antibiotics. Attempts at closed drainage with a chest tube often fail, owing to the development of loculations within the fluid and the high viscosity of the infected fluid. This problem has led to the use of intrapleural fibrinolytic drugs to improve drainage, hasten resolution of infection, and reduce the need for surgical intervention. At first, the use of intrapleural streptokinase and streptodornase was complicated by frequent allergic reactions, but with the availability of purified forms of streptokinase and urokinase these agents have been widely used for intrapleural fibrinolysis and this has been reviewed.

Streptokinase

Several reports, predominantly in the form of case series, have described the use of intrapleural streptokinase for complicated pleural effusions (98–104). The dose of streptokinase used varies, but is usually around 250 000 IU instilled once or twice daily, with tube clamping for 2–4 hours, repeated over several days. Local adverse effects are rare, but transient chest pain at the time of instillation occurs occasionally (105).

There are occasional reports of fever attributed to the intrapleural instillation of streptokinase (98,100), but the most important adverse event due to intrapleural streptokinase is systemic fibrinolysis. In many of the case series cited above, simple tests of clotting activity were performed before and after treatment and there were no significant changes. However, in 1984, a case report described major hemorrhage in one patient after the intrapleural use of streptokinase 500 000 IU, with tube clamping for 6 hours (106). Within 12 hours this patient developed a generalized coagulopathy, with features of disseminated intravascular coagulation. Following this report, in a carefully designed study, the systemic fibrinolytic effects of one dose (250 000 IU) and multiple doses of intrapleural streptokinase (250 000 IU bd for 3 days) were examined in healthy subjects (107). There were no physiological or statistical changes in any coagulation indices after intrapleural streptokinase. This suggests that the hemorrhage seen in the 1984 case report was probably not related to streptokinase, but was perhaps a complication of underlying sepsis.

Although intrapleural streptokinase does not cause systemic fibrinolytic effects, there can be local fibrinolytic effects. In a case series describing the use of intrapleural streptokinase or urokinase in 26 patients, one developed major "oozing" from rib fractures sustained 1 month before therapy (101). This local bleeding required two thoracotomies. It is not clear from the report if streptokinase or urokinase was used in this patient, but streptokinase was used in most of patients in this series. Furthermore, the dose used was also not clear, with streptokinase doses of 100 000–750 000 IU.

In 1998, major local hemorrhage after the use of intrapleural streptokinase was described in two patients (108). One patient had undergone mitral valve replacement 6 weeks before and collapsed after 2 days of standard intrapleural streptokinase, with hemorrhage into the chest. The second had undergone a mitral valve replacement 9 months before intrapleural streptokinase and collapsed with bleeding into the chest after 3 days. Both patients recovered, but the authors felt that recent cardiac surgery presented a contraindication to the use of intrapleural fibrinolytic drugs. It is not clear from this report if these two patients were taking oral anticoagulants, such as aspirin or warfarin, which may have confounded these findings. Another report of local hemorrhage has recently been published (109). The patient died after the instillation of intrapleural streptokinase for presumed empyema;

autopsy showed an unsuspected abdominal aortic dissection with extension of blood clot into the thoracic cavity. These reports suggest that intrapleural streptokinase should be used with caution when there has been prior surgery or trauma involving the thorax.

Another potential adverse event associated with the use of intrapleural streptokinase is the development of antistreptokinase antibodies. These antibodies have been documented after intravenous streptokinase and can cause serious adverse events with re-exposure to streptokinase or can limit the efficacy of streptokinase in myocardial revascularization.

Urokinase

Urokinase has also been widely used intrapleurally for fibrinolysis, in doses of 40 000–250 000 IU and is as effective as streptokinase (110–113). Furthermore, urokinase is not antigenic and does not produce febrile reactions. As with streptokinase, transient chest pain at the time of instillation has occasionally been reported. These considerations have led some authors to prefer urokinase to streptokinase for intrapleural therapy, but urokinase is about twice as expensive and there is also more published experience with streptokinase.

Drug–Drug Interactions

Diltiazem

Clinical and experimental data suggested a possible increase of hemorrhage risk during thrombolytic therapy with alteplase in combination with diltiazem (114).

Heparin

The combination of thrombolytic agents with an anticoagulant and/or aspirin has been said to be life-threatening. An excess of major bleeding episodes with combined subcutaneous heparin and streptokinase or alteplase treatments (1.0% with heparin versus 0.5% without heparin) has been reported in the International Study Group Trial (115) in patients with suspected acute myocardial infarction.

Interference with Diagnostic Tests

Coagulation tests

Thrombolytic agents interfere with the measurement of plasma fibrinogen and other coagulation tests by promoting continuing fibrinolysis after the specimen is taken; thus, fibrinogen cannot be measured in a sample that is taken less than an hour after the administration of a thrombolytic agent.

References

1. Lew AS, Laramee P, Cercek B, Shah PK, Ganz W. The hypotensive effect of intravenous streptokinase in patients with acute myocardial infarction. Circulation 1985;72(6):1321–6.
2. Wei JY, Markis JE, Malagold M, Braunwald E. Cardiovascular reflexes stimulated by reperfusion of ischemic myocardium in acute myocardial infarction. Circulation 1983;67(4):796–801.
3. Been M, de Bono DP, Muir AL, Boulton FE, Fears R, Standring R, Ferres H. Clinical effects and kinetic properties of intravenous APSAC—anisoylated plasminogen-streptokinase activator complex (BRL 26921) in acute myocardial infarction. Int J Cardiol 1986;11(1):53–61.
4. ISIS-3 (Third International Study of Infarct Survival) Collaborative Group. ISIS-3: a randomised comparison of streptokinase vs tissue plasminogen activator vs anistreplase and of aspirin plus heparin vs aspirin alone among 41,299 cases of suspected acute myocardial infarction. Lancet 1992;339(8796):753–70.
5. Meissner AJ, Misiak A, Ziemski JM, Scharf R, Rudowski W, Huszcza S, Kucharski W, Wislawski S. Hazards of thrombolytic therapy in deep vein thrombosis. Br J Surg 1987;74(11):991–3.
6. Grimm W, Schwieder G, Wagner T. Todliche Lungenembolie bei Bein-Beckenvenenthrombose unter Lysetherapie. [Fatal pulmonary embolism in venous thrombosis of the leg and pelvis during lysis therapy.] Dtsch Med Wochenschr 1990;115(31–32):1183–7.
7. Queen M, Biem HJ, Moe GW, Sugar L. Development of cholesterol embolization syndrome after intravenous streptokinase for acute myocardial infarction. Am J Cardiol 1990;65(15):1042–3.
8. Blankenship JC. Cholesterol embolisation after thrombolytic therapy. Drug Saf 1996;14(2):78–84.
9. Gupta BK, Spinowitz BS, Charytan C, Wahl SJ. Cholesterol crystal embolization-associated renal failure after therapy with recombinant tissue-type plasminogen activator. Am J Kidney Dis 1993;21(6):659–62.
10. Gurwitz JH, Gore JM, Goldberg RJ, Barron HV, Breen T, Rundle AC, Sloan MA, French W, Rogers WJ. Risk for intracranial hemorrhage after tissue plasminogen activator treatment for acute myocardial infarction. Participants in the National Registry of Myocardial Infarction 2. Ann Intern Med 1998;129(8):597–604.
11. Cairns JA, Kennedy JW, Fuster V. Coronary thrombolysis. Chest 1998;114(5 Suppl):S634–57.
12. Tatarchenko IP, Pozdniakova NV, Petranin AIu, Morozova OI. [Ventricular arrhythmias and cardiac late potentials inpatients with acute coronary syndrome after reperfusion therapy.] Klin Med (Mosk) 2005;83(5):19–22.
13. Chen MF, Liau CS, Lee YT. Coronary arterial aneurysm after percutaneous transluminal coronary recanalization with streptokinase. Int J Cardiol 1990;28(1):117–9.
14. Shaw CE, Easthope RN. Fatal bronchospasm following streptokinase. NZ Med J 1993;106(956):207.
15. Ancillo P, Duarte J, Cortina JJ, Sempere AP, Claveria LE. Guillain–Barré syndrome after acute myocardial infarction treated with anistreplase. Chest 1994;105(4):1301–2.
16. Cicale MJ. Guillain–Barré syndrome after streptokinase therapy. South Med J 1987;80(8):1068.
17. Eden KV. Possible association of Guillain–Barré syndrome with thrombolytic therapy. JAMA 1983;249(15):2020–1.
18. Leaf DA, MacDonald I, Kliks B, Wilson R, Jones SR. Streptokinase and the Guillain–Barré syndrome. Ann Intern Med 1984;100(4):617.
19. Taylor BV, Mastaglia FL, Stell R. Guillain–Barré syndrome complicating treatment with streptokinase. Med J Aust 1995;162(4):214–5.
20. Gray MY, Lazarus JH. Iritis after treatment with streptokinase. BMJ 1994;309(6947):97.

21. Kinshuck D. Bilateral hypopyon and streptokinase. BMJ 1992;305(6865):1332.
22. Gore JM, Sloan M, Price TR, Randall AM, Bovill E, Collen D, Forman S, Knatterud GL, Sopko G, Terrin ML. Intracerebral hemorrhage, cerebral infarction, and subdural hematoma after acute myocardial infarction and thrombolytic therapy in the Thrombolysis in Myocardial Infarction Study. Thrombolysis in Myocardial Infarction, phase II, pilot and clinical trial. Circulation 1991;83(2):448–59.
23. Dundar Y, Hill R, Dickson R, Walley T. Comparative efficacy of thrombolytics in acute myocardial infarction: a systematic review. Q J Med 2003;96:103–13.
24. Nandhagopal R, Vengamma B, Krishnamoorthy SG, Rajasekhar D, Subramanyam G. Thalamic disequilibrium syndrome after thrombolytic therapy for acute myocardial infarction. Neurology 2005;65(3):494.
25. AIMS Trial Study Group. Effect of intravenous APSAC on mortality after acute myocardial infarction: preliminary report of a placebo-controlled clinical trial. Lancet 1988;1(8585):545–9.
26. Wilcox RG, von der Lippe G, Olsson CG, Jensen G, Skene AM, Hampton JR. Trial of tissue plasminogen activator for mortality reduction in acute myocardial infarction. Anglo-Scandinavian Study of Early Thrombolysis (ASSET). Lancet 1988;2(8610):525–30.
27. Habib GB. Current status of thrombolysis in acute myocardial infarction. Part III. Optimalization of adjunctive therapy after thrombolytic therapy. Ches 1995;107(3):809–16.
28. Bovill EG, Tracy RP, Knatterud GL, Stone PH, Nasmith J, Gore JM, Thompson BW, Tofler GH, Kleiman NS, Cannon C, Braunwald E. Hemorrhagic events during therapy with recombinant tissue plasminogen activator, heparin, and aspirin for unstable angina (Thrombolysis in Myocardial Ischemia, phase IIIB trial). Am J Cardiol 1997;79(4):391–6.
29. Aylward PE, Wilcox RG, Horgan JH, White HD, Granger CB, Califf RM, Topol EJ. Relation of increased arterial blood pressure to mortality and stroke in the context of contemporary thrombolytic therapy for acute myocardial infarction. A randomized trial. GUSTO-I Investigators. Ann Intern Med 1996;125(11):891–900.
30. Selker HP, Beshansky JR, Schmid CH, Griffith JL, Longstreth WT Jr, O'Connor CM, Caplan LR, Massey EW, D'Agostino RB, Laks MM, et al. Presenting pulse pressure predicts thrombolytic therapy-related intracranial hemorrhage. Thrombolytic Predictive Instrument (TPI) Project results. Circulation 1994;90(4):1657–61.
31. Simoons ML, Maggioni AP, Knatterud G, Leimberger JD, de Jaegere P, van Domburg R, Boersma E, Franzosi MG, Califf R, Schroder R, et al. Individual risk assessment for intracranial haemorrhage during thrombolytic therapy. Lancet 1993;342(8886–8887):1523–8.
32. Woods KL, Ketley D. Utilisation of thrombolytic therapy in older patients with myocardial infarction. Drugs Aging 1998;13(6):435–41.
33. Blankenship JC, Almquist AK. Cardiovascular complications of thrombolytic therapy in patients with a mistaken diagnosis of acute myocardial infarction. J Am Coll Cardiol 1989;14(6):1579–82.
34. Conard J, Samama M, Milochevitch R, Horellou MH, Chabrun B, Prestat J. Complications hémorragiques au cours de 98 traitements par la stréptokinase. Place de la surveillance biologique. [Haemorrhagic complications using streptokinase during 98 treatments. Place of the biological surveillance.] Nouv Presse Méd 1979;8(16):1319–25.
35. Jenkins LA, Lau S, Crawford M, Keung YK. Delayed profound thrombocytopenia after c7E3 Fab (abciximab) therapy. Circulation 1998;97(12):1214–5.
36. Reuben A, Clouting E. Compartment syndrome after thrombolysis for acute myocardial infarction. Emerg Med J 2005;22(1):77.
37. Ahmed J, Philpott J, Lew-Gor S, Blunt D. Airway obstruction: a rare complication of thrombolytic therapy. J Laryngol Otol 2005;119(10):819–21.
38. Cahill RA, Siddique S, O'Connor J. Vomiting in the recently anticoagulated patient. Gut 2005;54(1):90, 102.
39. Mathiesen O, Grunnet N. Haemolysis after intravenous streptokinase. Lancet 1989;1(8645):1016–7.
40. Blum A, Shohat B. CD-4 lymphopenia induced by streptokinase. Circulation 1995;91(6):1899.
41. Vaughan DE, Kirshenbaum JM, Loscalzo J. Streptokinase-induced, antibody-mediated platelet aggregation: a potential cause of clot propagation in vivo. J Am Coll Cardiol 1988;11(6):1343–8.
42. Harrington RA, Sane DC, Califf RM, Sigmon KN, Abbottsmith CW, Candela RJ, Lee KL, Topol EJ. Clinical importance of thrombocytopenia occurring in the hospital phase after administration of thrombolytic therapy for acute myocardial infarction. The Thrombolysis and Angioplasty in Myocardial Infarction Study Group. J Am Coll Cardiol 1994;23(4):891–8.
43. Oyama H, Iwakoshi T, Niwa M, Kida Y, Tanaka T, Kitamura R, Maezawa S, Kobayashi T. Coagulation and fibrinolysis study after local thrombolysis of a cerebral artery with urokinase. Neurol Med Chir (Tokyo) 1996;36(5):300–4.
44. A European Collaborative Study. Controlled trial of urokinase in myocardial infarction. Lancet 1975;2(7936): 624–6.
45. Sallen MK, Efrusy ME, Kniaz JL, Wolfson PM. Streptokinase-induced hepatic dysfunction. Am J Gastroenterol 1983;78(8):523–4.
46. Maclennan AC, Ahmad N, Lawrence JR. Activities of aminotransferases after treatment with streptokinase for acute myocardial infarction. BMJ 1990;301(6747): 321–2.
47. Freimark D, Leor R, Hod H, Elian D, Kaplinsky E, Rabinowitz B. Impaired hepatic function tests after thrombolysis for acute myocardial infarction. Am J Cardiol 1991;67(6):535–7.
48. Mager A, Birnbaum Y, Zlotikamien B, Strasberg B, Rechavia E, Sagie A, Sclarovsky S. Streptokinase-induced jaundice in patients with acute myocardial infarction. Am Heart J 1991;121(5):1543–4.
49. Phillips E, Woolfrey S, Cameron E. Streptokinase-induced jaundice. Postgrad Med J 1994;70(819):55.
50. Polkey MI, Oliver RM, Walker JM. Hepatic dysfunction induced by streptokinase. Am J Gastroenterol 1992;87(8):1062.
51. Pavlou H, Panagiotopoulos A, Graham A, Alexopoulos D. Urokinase-induced cyto-hepatolysis in a patient with acute myocardial infarction. Eur Heart J 1995;16(2):291–2.
52. Gilutz H, Cohn G, Battler A. Jaundice induced by streptokinase. Angiology 1996;47(3):281–4.
53. Adorati M, Pizzolitto S, Franzon R, Vallone C, Artero M, Moro A. Cholesterol embolism and acute interstitial nephritis: two adverse effects of streptokinase thrombolytic therapy in the same patient. Nephrol Dial Transplant 1998;13(5):1262–4.

54. Elliot JM, Cross DB, Cederholm-Williams S, et al. Streptokinase titers 1 to 4 years after intravenous streptokinase. Circulation 1991;84(Suppl 2):116.
55. Massel D, Turpie AGG, Oberhardt BJ, et al. Estimation of resistance to streptokinase: a preliminary report of a rapid bedside test. Can J Cardiol 1993;9:E134.
56. Jalihal S, Morris GK. Antistreptokinase titres after intravenous streptokinase. Lancet 1990;335(8683):184–5.
57. Cross DB. Should streptokinase be readministered? Insights from recent studies of antistreptokinase antibodies. Med J Aust 1994;161(2):100–1.
58. Cross DB, White HD. Allergic reactions to streptokinase: does antibody formation prevent reuse in a second myocardial infarction? Clin Immunother 1994;2:415.
59. Blackwell N, Hollins A, Gilmore G, Norton R. Antistreptokinase antibodies: implications for thrombolysis in a region with endemic streptococcal infection. J Clin Pathol 2005;58(9):1005–7.
60. Abuosa AM, Akhras F, Sorour K, El-Said G, El-Tobgy S, Kinsara AJ. Effect of pretreatment antistreptokinase antibody and streptococcal infection on the efficacy and dosage of streptokinase in acute myocardial infarction. Saudi Med J 2005;26(6):934–6.
61. Mi H, Gupta A, Gok MA, Asher J, Shenton BK, Stamp S, Carter V, Del Rio Martin J, Soomro NA, Jaques BC, Manas DM, Talbot D. Do recipients of kidneys from donors treated with streptokinase develop anti-streptokinase antibodies? Transplant Proc 2005;37(8):3272–3.
62. Oliveira DC, Coelho OR, Paraschin K, Ferraroni NR, Zolner Rde L. Angioedema related to the use of streptokinase. Arq Bras Cardiol 2005;85(2):131–4.
63. Jolliet P, Magnin C, Unger PF. Pulmonary embolectomy after intravenous thrombolysis with alteplase. Lancet 1990;335(8684):290–1.
64. Gruppo Italiano per lo Studio della Streptocochinasi nell'Infarto miocardico (GSSI). Effectiveness of intravenous thrombolytic treatment in acute myocardial infarction. Lancet 1987;1:397.
65. Cooper JP, Quarry DP, Beale DJ, Chappell AG. Life-threatening, localized angio-oedema associated with streptokinase. Postgrad Med J 1994;70(826):592–3.
66. Stephens MB, Pepper PV. Streptokinase therapy. Recognizing and treating allergic reactions. Postgrad Med 1998;103(3):89–90.
67. Francis CW, Brenner B, Leddy JP, Marder VJ. Angioedema during therapy with recombinant tissue plasminogen activator. Br J Haematol 1991;77(4):562–3.
68. Purvis JA, Booth NA, Wilson CM, Adgey AA, McCluskey DR. Anaphylactoid reaction after injection of alteplase. Lancet 1993;341(8850):966–7.
69. Pancioli A, Brott T, Donaldson V, Miller R. Asymmetric angioneurotic edema associated with thrombolysis for acute stroke. Ann Emerg Med 1997;30(2):227–9.
70. Dickinson RJ, Rosser A. Low back pain associated with streptokinase. BMJ 1991;302(6768):111–2.
71. Hannaford P, Kay CR. Back pain and thrombolysis. BMJ 1992;304(6831):915.
72. Lear J, Rajapakse R, Pohl J. Low back pain associated with streptokinase. Lancet 1992;340(8823):851.
73. Chan NS, White H, Maslowski A, Cleland J. Plasmacytosis and renal failure after readministration of streptokinase for threatened myocardial reinfarction. BMJ 1988;297(6650):717–8.
74. Bucknall C, Darley C, Flax J, Vincent R, Chamberlain D. Vasculitis complicating treatment with intravenous anisoylated plasminogen streptokinase activator complex in acute myocardial infarction. Br Heart J 1988;59(1):9–11.
75. Gemmill JD, Sandler M, Hillis WS, Tillman J, Wakeel R. Vasculitis complicating treatment with intravenous anisoylated plasminogen streptokinase activator complex in acute myocardial infarction. Br Heart J 1988;60(4):361.
76. Ong AC, Handler CE, Walker JM. Hypersensitivity vasculitis complicating intravenous streptokinase therapy in acute myocardial infarction. Int J Cardiol 1988;21(1): 71–3.
77. Sorber WA, Herbst V. Lymphocytic angiitis following streptokinase therapy. Cutis 1988;42(1):57–8.
78. Albert F, Dubourg O, Steg G, Delorme G, Bourdarias JP. Maladie sérique après fibrinoyse par streptokinase intraveineuse au cours d'un infarctus du myocarde. [Serum sickness after fibrinolysis using intravenous streptokinase in myocardial infarction.] Arch Mal Coeur Vaiss 1988;81(8):1013–5.
79. Davies KA, Mathieson P, Winearls CG, Rees AJ, Walport MJ. Serum sickness and acute renal failure after streptokinase therapy for myocardial infarction. Clin Exp Immunol 1990;80(1):83–8.
80. Noel J, Rosenbaum LH, Gangadharan V, Stewart J, Galens G. Serum sickness-like illness and leukocytoclastic vasculitis following intracoronary arterial streptokinase. Am Heart J 1987;113(2 Pt 1):395–7.
81. Verstraete M, Vermylen J, Donati MB. The effect of streptokinase infusion on chronic arterial occlusions and stenoses. Ann Intern Med 1971;74(3):377–82.
82. Argent N, Adam PC. Proteinuria and thrombolytic agents. Lancet 1990;335(8681):106–7.
83. Le SP, Chatterjee K, Wolfe CL. Adult respiratory distress syndrome following thrombolytic therapy with APSAC for acute myocardial infarction. Am Heart J 1992;123(5):1368–9.
84. Tio RA, Voorbij RH, Enthoven R. Adult respiratory distress syndrome after streptokinase. Am J Cardiol 1992;70(20):1632–3.
85. Montserrat I, Altimiras J, Dominguez M, Lamich R, Olle A, Fontcuberta J. Adverse reaction to streptokinase with multiple systemic manifestations. Pharm World Sci 1995;17(5):168–71.
86. Montgomery HE, McIntyre CW, Almond MK, Davies K, Pumphrey CW, Bennett D. Rhabdomyolysis and multiple system organ failure with streptokinase. BMJ 1995;311(7018):1472.
87. Tsang TS, Califf RM, Stebbins AL, Lee KL, Cho S, Ross AM, Armstrong PW. Incidence and impact on outcome of streptokinase allergy in the GUSTO-I trial. Global Utilization of Streptokinase and t-PA in Occluded Coronary Arteries. Am J Cardiol 1997;79(9):1232–5.
88. Papamitsakis NI, Kuyl J, Lutsep HL, Clark WM. Benign angioedema after thrombolysis for acute stroke. J Stroke Cerebrovasc Dis 2000;9:79–81.
89. Hill MD, Barber PA, Takahashi J, Demchuk AM, Feasby TE, Buchan AM. Anaphylactoid reactions and angioedema during alteplase treatment of acute ischemic stroke. CMAJ 2000;162(9):1281–4.
90. Rudolf J, Grond M, Schmulling S, Neveling M, Heiss W. Orolingual angioneurotic edema following therapy of acute ischemic stroke with alteplase. Neurology 2000;55(4):599–600.
91. Rudolf J, Grond M, Prince WS, Schmulling S, Heiss WD. Evidence of anaphylaxy after alteplase infusion. Stroke 1999;30(5):1142–3.

92. Marder VJ, Soulen RL, Atichartakarn V, Budzynski AZ, Parulekar S, Kim JR, Edward N, Zahavi J, Algazy KM. Quantitative venographic assessment of deep vein thrombosis in the evaluation of streptokinase and heparin therapy. J Lab Clin Med 1977;89(5):1018–29.
93. Pfeifer GW. Distribution and placental transfer of 131-I streptokinase. Australas Ann Med 1970;19(Suppl 1):17–8.
94. Ludwig H. Results of streptokinase therapy in deep venous thrombosis during pregnancy. Postgrad Med J 1973;49(Suppl 5).
95. Goring G. Kasuistischer Beitrag zur Streptokinase-behandlung in der Schwangerschaft. [Case report on streptokinase therapy during pregnancy.] Geburtshilfe Frauenheilkd 1971;31(4):348–53.
96. Huynh T, Cox JL, Massel D, Davies C, Hilbe J, Warnica W, Daly PA; FASTRAK II Network. Predictors of intracranial hemorrhage with fibrinolytic therapy in unselected community patients: a report from the FASTRAK II project. Am Heart J 2004;148(1):86–91.
97. Kandzari DE, Granger CB, Simoons ML, White HD, Simes J, Mahaffey KW, Gore J, Weaver WD, Longstreth WT Jr, Stebbins A, Lee KL, Califf RM, Topol EJ; Global Utilization of Streptokinase and tPA for Occluded Arteries-I Investigators. Risk factors for intracranial hemorrhage and nonhemorrhagic stroke after fibrinolytic therapy (from the GUSTO-I trial). Am J Cardiol 2004;93(4):458–61.
98. Robinson LA, Moulton AL, Fleming WH, Alonso A, Galbraith TA. Intrapleural fibrinolytic treatment of multiloculated thoracic empyemas. Ann Thorac Surg 1994;57(4):803–13discussion 813–14.
99. Kennedy L, Rusch VW, Strange C, Ginsberg RJ, Sahn SA. Pleurodesis using talc slurry. Chest 1994;106(2):342–6.
100. Bouros D, Schiza S, Panagou P, Drositis J, Siafakas N. Role of streptokinase in the treatment of acute loculated parapneumonic pleural effusions and empyema. Thorax 1994;49(9):852–5.
101. Temes RT, Follis F, Kessler RM, Pett SB Jr, Wernly JA. Intrapleural fibrinolytics in management of empyema thoracis. Chest 1996;110(1):102–6.
102. Davies CW, Traill ZC, Gleeson FV, Davies RJ. Intrapleural streptokinase in the management of malignant multiloculated pleural effusions. Chest 1999;115(3):729–33.
103. Sasse SA. Parapneumonic effusions and empyema. Curr Opin Pulm Med 1996;2(4):320–6.
104. Sahn SA. Use of fibrinolytic agents in the management of complicated parapneumonic effusions and empyemas. Thorax 1998;53(Suppl 2):S65–72.
105. Taylor RF, Rubens MB, Pearson MC, Barnes NC. Intrapleural streptokinase in the management of empyema. Thorax 1994;49(9):856–9.
106. Godley PJ, Bell RC. Major hemorrhage following administration of intrapleural streptokinase. Chest 1984;86(3):486–7.
107. Davies CW, Lok S, Davies RJ. The systemic fibrinolytic activity of intrapleural streptokinase. Am J Respir Crit Care Med 1998;157(1):328–30.
108. Porter J, Banning AP. Intrapleural streptokinase. Thorax 1998;53(8):720.
109. Srivastava P, Godden DJ, Kerr KM, Legge JS. Fatal haemorrhage from aortic dissection following instillation of intrapleural streptokinase Scott Med J 2000;45(3): 86–7.
110. Moulton JS, Benkert RE, Weisiger KH, Chambers JA. Treatment of complicated pleural fluid collections with image-guided drainage and intracavitary urokinase. Chest 1995;108(5):1252–9.
111. Bouros D, Schiza S, Tzanakis N, Drositis J, Siafakas N. Intrapleural urokinase in the treatment of complicated parapneumonic pleural effusions and empyema. Eur Respir J 1996;9(8):1656–9.
112. Bouros D, Schiza S, Patsourakis G, Chalkiadakis G, Panagou P, Siafakas NM. Intrapleural streptokinase versus urokinase in the treatment of complicated parapneumonic effusions: a prospective, double-blind study. Am J Respir Crit Care Med 1997;155(1):291–5.
113. Krishnan S, Amin N, Dozor AJ, Stringel G. Urokinase in the management of complicated parapneumonic effusions in children. Chest 1997;112(6):1579–83.
114. Becker RC, Caputo R, Ball S, Corrao JM, Baker S, Gore JM. Hemorrhagic potential of combined diltiazem and recombinant tissue-type plasminogen activator administration. Am Heart J 1993;126(1):11–4.
115. The International Study Group. In-hospital mortality and clinical course of 20,891 patients with suspected acute myocardial infarction randomised between alteplase and streptokinase with or without heparin. Lancet 1990;336(8707):71–5.

ANTI-PLATELET DRUGS

Abciximab

General Information

Abciximab is a Fab fragment of the chimeric human-murine monoclonal antibody 7E3, which binds to the platelet glycoprotein IIb/IIIa receptor and inhibits platelet aggregation (1).

Abciximab is used for prevention of cardiac ischemic events in patients undergoing percutaneous coronary intervention and to prevent myocardial infarction in patients with unstable angina who do not respond to conventional treatment. It has also been used for thrombolysis in patients with peripheral arterial occlusive disease and arterial thrombosis (2).

It has been studied in percutaneous coronary intervention, ST-elevation myocardial infarction, and non-ST-elevation acute coronary syndromes (3). The use of glycoprotein IIb/IIIa inhibitors, such as abciximab, is associated with a mortality benefit, but does not alter the risk of restenosis (4).

Besides bleeding, other adverse reactions that have been associated with abciximab include back pain, hypotension, nausea, and chest pain (but with an incidence not significantly different from that observed with placebo).

Comparative studies

In the RAPPORT trial, the addition of abciximab was evaluated in 483 patients undergoing angioplasty. Abciximab was associated with an increase in *major bleeding complications* compared with heparin alone, 17% versus 9.5%. The safety profile of abciximab can be improved by weight-adjusted heparin dosing and reduced thrombolytic therapy (3).

Abciximab (an intravenous bolus dose of 0.25 mg/kg or a 12-hour infusion of 0.125 micrograms/kg/minute) was evaluated in combination with intravenous reteplase (0.25–1.0 U/hour) and intravenous heparin (100–400 U/hour) in 50 adults with arterial occlusive disease (5). Eight developed a *major hematoma* that necessitated blood transfusion (1–11 U). Two had hematuria and one had hemoptysis. There was *thrombocytopenia* in two patients, including one with zero platelets. Four developed distal embolization after thrombolysis.

Organs and Systems

Respiratory

Lung hemorrhage is a rare but potentially lethal complication of antithrombotic and antiplatelet therapy. The incidence of spontaneous pulmonary hemorrhage after the use of platelet glycoprotein IIb/IIIa inhibitors has been analysed from the medical records of 1020 consecutive patients who underwent coronary interventions (6). Diffuse pulmonary hemorrhage developed in seven patients, two of whom died and five of whom had activated clotting times greater than 250 seconds during the procedure. Activated partial thromboplastin time measured at the time of lung hemorrhage was raised in all cases (mean 85, range 69–95 seconds). All had a history of congestive heart failure, and had raised pulmonary capillary wedge pressures and/or left ventricular end-diastolic pressures at the time of the procedure. Six patients also had evidence of baseline radiographic abnormalities.

Nervous system

Seven patients undergoing neurointerventional procedures who received abciximab developed fatal intracerebral hemorrhages (7). The procedures included angioplasty and stent placement in the cervical internal carotid artery ($n = 4$), angioplasty of the intracranial carotid artery ($n = 1$), and angioplasty of the middle cerebral artery ($n = 2$). Aggressive antithrombotic treatment is used as adjuvant to angioplasty and/or stent placement to reduce the rate of ischemic and thrombotic complications associated with these procedures. Intravenous abciximab has a short life (10 minutes), but its inhibitory effect on platelets lasts for 48 hours. The exact cause of abciximab-associated intracerebral hemorrhage is unclear.

In a database review of 7244 consecutive interventions, 6190 had been performed with abciximab, including 515 in patients with a history of stroke, either recent (n = 101) or remote (> 2 years; n = 414) (8). The rate of stroke after intervention was significantly higher in those with a prior stroke (2.06% versus 0.35% for all stroke; 0.38% versus 0.03% for intracerebral hemorrhage). However, the incidence of intracerebral hemorrhage among the abciximab-treated patients was 0.065%, and a history of prior stroke did not increase the incidence.

Distal embolization is the main potential risk of carotid stenting, and techniques to minimize the risk are evolving. Between July 1998 and March 2002, 305 consecutive patients who underwent elective or urgent percutaneous carotid intervention at The Cleveland Clinic were followed. During this period, the practice of carotid stenting evolved from the routine use of glycoprotein IIb-IIIa inhibitors to the routine of an emboli-prevention device (9). In all, 199 patients received adjunctive glycoprotein IIb-IIIa inhibitors (91% abciximab), and 106 patients had an emboli-prevention device inserted (85% filter design, 15% occlusive balloon). At 30 days, the composite end point of neurological death, non-fatal stroke, and major bleeding, including intracranial hemorrhage, was significantly less among patients treated with emboli-prevention device than in those treated with glycoprotein IIb-IIIa inhibitors (0% versus 5.1%).

Hematologic

Bleeding

The primary risk associated with abciximab is bleeding. In the EPIC trial in high-risk angioplasty, 14% of patients who received a bolus of abciximab followed by an infusion had a major bleeding complication rate, versus 7% in the placebo group (10). The most marked excess of major bleeding episodes occurred at the site of vascular puncture, but there were also a substantial number of gastrointestinal haemorrhages. However, the therapeutic regimen used was not adjusted for body weight, and the risk of major bleeding was also related to the heparin dose per kg and not only to the use of abciximab (11).

In 7800 patients with chest pain and either ST segment depression or a positive troponin test, the addition of abciximab to unfractionated heparin or low molecular weight heparin in the treatment of acute coronary syndrome was not associated with any significant reduction in cardiac events, but a doubled risk of bleeding (12).

An analysis of data from the EPIC trial identified a series of factors that predicted vascular access site bleeding or the need for vascular access site surgery in abciximab-treated patients (13). They comprised larger vascular access sheath size, the presence of acute myocardial infarction at enrolment, female sex, higher baseline hematocrits, lower body weight, and a longer time spent in the catheterization laboratory.

It must be emphasized that patients in the EPIC trial received high-dose heparin and that vascular access site sheaths were left in place for 12–16 hours. In subsequent studies, the risk of vascular site bleeding was probably reduced by using lower doses of heparin and removing sheaths sooner. This was the case in the EPILOG trial in which heparin was withdrawn immediately after the coronary procedure and vascular sheaths were removed as soon as possible (14). The incidence of major bleeding in this study was not significantly higher with abciximab than with placebo. Nevertheless, the incidence of minor bleeding complications was significantly higher in the abciximab plus standard dose heparin group (but not in the abciximab plus low dose heparin group) compared with placebo. In the EPISTENT trial, all patients received low dose, body weight-adjusted heparin: Here the incidence of both major and minor bleeding complications was low and not significantly different between treatment groups (15).

It would therefore seem possible to reduce the incidence of bleeding complications when using abciximab during prophylactic coronary revascularization procedures. This is unfortunately not the case so far in the setting of primary angioplasty for myocardial infarction after intense anticoagulation (17% of major hemorrhagic complications versus 9.5 in placebo recipients) (16). The risk of serious bleeding complications is also increased in rescue situations when high doses of heparin have been used (17), but here it can be reduced by giving protamine to reverse heparin anticoagulation before abciximab therapy (18). There is also a high incidence of major bleeding in patients who receive abciximab during percutaneous coronary revascularization after unsuccessful thrombolytic therapy. It has been suggested that abciximab should not be administered within 18 hours after thrombolytic therapy (19).

It must be emphasized that very few episodes of abciximab-related bleeding are life-threatening and that in none of the trials with abciximab as well as with other glycoprotein IIb/IIIa antagonists has there been an excess of intracranial hemorrhage (20).

However, the bleeding risk in patients enrolled in trials may not be representative of the population actually being given abciximab. To clarify this, a review of adverse events in patients receiving glycoprotein IIb/IIIa inhibitors reported to the FDA has been undertaken (21,22). The FDA received 450 reports of deaths related to treatment with glycoprotein IIb/IIIa inhibitors between November 1, 1997 and December 31, 2000; these were reviewed and a standard rating system for assessing causation was applied to each event. Of the 450 deaths, 44% were considered to be definitely or probably attributable to glycoprotein IIb/IIIa inhibitors. The mean age of patients who died was 69 years and 47% of the deaths were in women. All of the deaths that were deemed to be definitely or probably associated with glycoprotein IIb/IIIa inhibitors were associated with excessive bleeding, most often in the nervous system.

Thrombocytopenia

DoTS classification (BMJ 2003;327:1222–5):
Dose-relation: Hypersusceptibility
Time-course: First dose or early persistent
Susceptibility factors: Female sex; drug-drug interactions (heparin, acetylsalicylic acid, ticlopidine); renal disease for some forms of heparin

Abciximab is associated with higher incidence of thrombocytopenia (1–5%) than other glycoprotein IIb/IIIa inhibitors, such as eptifibatide and tirofiban (23). It can occur within hours, and fatal outcomes have been described (24). However, it can be delayed and is potentially prothrombotic (25,26). Profound thrombocytopenia occurring 1 week after administration of abciximab is uncommon, self-limiting, and mostly uneventful (27). However, a quick diagnosis is essential, since other forms of thrombocytopenia associated with concomitant antithrombotic therapies may be much more severe and require prompt treatment. Awareness of this reaction may avoid unnecessary and risky withdrawal of other antiplatelet drugs in the critical phase after coronary stenting.

In consecutive patients undergoing percutaneous coronary interventions who were given either eptifibatide (n = 342) or abciximab (n = 300) during the procedure, thrombocytopenia was more frequent in those given abciximab (6%, versus 0%), including five patients who developed severe thrombocytopenia (under 20×10^9/l) (28).

In the CADILLAC study, 2082 patients who had an acute myocardial infarction within 12 hours without shock were prospectively randomized to receive balloon angioplasty with or without abciximab versus stenting with or without abciximab (29). Acquired thrombocytopenia, defined as a nadir platelet count below 100 x 10^9/l in patients who did not have baseline thrombocytopenia, developed in 50 of 1975 patients (2.5%). The independent predictors of acquired thrombocytopenia were a platelet count below 200 x 10^9/l on admission (OR = 5.35; 95% CI = 2.91, 9.81), non-insulin-requiring diabetes mellitus (4.42; 2.19, 8.91), previous statin administration (OR = 3.19; 1.55, 6.57), and use of abciximab (2.06; 1.11, 3.83). Thrombocytopenia was *less* likely in those with a greater body mass index (0.90; 0.84, 0.97) and previous aspirin use (0.27; 0.11, 0.63). Patients who developed thrombocytopenia had significantly higher rates of major hemorrhagic complications than those who did not (10% versus 2.7%), greater requirements for blood transfusions (10% versus 3.9%), and longer hospital stays (median 4.8 versus 3.6 days), and they incurred higher costs (median $14 466 versus $11 629). All-cause mortality was markedly increased at 30 days in patients who developed thrombocytopenia (8.0% versus 1.6%) and at 1 year (10% versus 3.9%). Data pooled from three major trials showed that thrombocytopenia (under 100 × 10^9/l) was significantly more frequent in those who received a bolus dose of abciximab followed by an infusion than in placebo recipients (3.7 versus 2%). Severe thrombocytopenia (under 50 × 10^9/l) was also more frequent with abciximab (1.1 versus 0.5%) (30). Very acute and profound thrombocytopenia (under 20 × 10^9/l) within 24 hours after administration has been observed in 0.3–0.7% of patients treated with abciximab for the first time (20,30–32).

During postmarketing surveillance of the first 4000 patients treated with abciximab in France, 25 cases of thrombocytopenia (0.6%) were reported, with five severe cases (0.15%) and three acute profound forms (0.08%). In all cases reported, the role of heparin must be taken into account. The thrombocytopenia associated with abciximab differs with that associated with heparin by its rapid onset (within 24 hours), its reversal after platelet transfusion, and its possible association with hemorrhage but not with thrombosis.

Positive human anti-chimeric antibodies have been detected in 6% of patients (generally in low titers) but were not associated with hypersensitivity or allergic reactions. Preliminary data indicate that abciximab can be safety readministered, although a greater incidence of thrombocytopenia after administration has been reported with a lesser efficacy of platelet transfusion (17).

The clinical aspects of thrombocytopenia that results from sensitivity to glycoprotein IIb-IIIa inhibitors and the evidence that the platelet destruction is antibody-mediated have been reviewed (33).

Time-course

In contrast to other types of drug-induced thrombocytopenia, this complication can occur within a few hours of first exposure, and usually occurs within 12–96 hours. However, there has been a report of acute profound thrombocytopenia after 7 days (34).

- A 65-year-old woman with type 2 diabetes mellitus and coronary artery disease received a 0.25 mg/kg bolus of abciximab at the time of intervention followed by an infusion of 10 micrograms/minute for 12 hours. Her baseline platelet counts were 286 × 10^9/l before use, 385 × 10^9/l at 2 hours, and 296 × 10^9/l at 18 hours. On day 7 she developed petechiae over her legs and her platelet count was 1 × 10^9/l. Coagulation tests were normal and there was no evidence of heparin-induced thrombocytopenia. She received 10 units of single-donor platelets and recovered slowly over the next 4 days. The platelet count was 114 × 10^9/l on day 12.

In another case of profound thrombocytopenia after abciximab there was a delayed onset (6 days after therapy) (35). The authors speculated that preceding treatment with methylprednisolone may have delayed the onset of thrombocytopenia.The mechanism of severe thrombocytopenia associated with abciximab is unclear. Further administration should be avoided, but other glycoprotein IIb/IIIa inhibitors (eptifibatide and tirofiban) have been successfully used in patients with history of abciximab-induced thrombocytopenia.

Thrombocytopenia after a second exposure to abciximab in nine patients showed that each had a strong immunoglobulin IgG antibody that recognized platelets sensitized with abciximab (36). Five patients also had IgM antibodies. Thrombocytopenia occurred four times as often as after the first exposure. The mechanism is not understood, but these findings suggest that it may be antibody-mediated. These antibodies were also found in 77 of 104 healthy patients, but in the patients the antibodies were specific for murine sequences in abciximab, causing the life-threatening thrombocytopenia.

Nine patients who developed profound thrombocytopenia after a second exposure to abciximab had an IgG antibody that recognized platelets sensitized with abciximab. In contrast, in 104 healthy subjects, in whom IgG antibodies reactive with abciximab-coated platelets were found in 77, the antibodies were specific for murine sequences in abciximab and were capable of causing life-threatening thrombocytopenia (36).

Frequency

There have been 2780 international reports of adverse events in patients taking abciximab, including 1046 cases of thrombocytopenia (38%). Of 250 adverse events reports of patients treated with abciximab in Germany, there were 76 (30.4%) cases of thrombocytopenia and 93 (37%) of hemorrhage; 44 patients (18%) had both thrombocytopenia and hemorrhage (Tiaden 371). Abciximab is often used in combination with heparin, acetylsalicylic acid, or ticlopidine, and combination therapy leads to a greater risk of severe and lethal bleeding. Thrombocytopenia due to abciximab usually resolves 10 days after withdrawal. However, persistent thrombocytopenia has been reported for up to 21 days and in one case was resistant to platelet transfusions and two courses of intravenous gammaglobulin (37).

Mechanism

Five patients with thrombocytopenia that developed 7–12 days after the start of abciximab infusion had IgG antibodies against abciximab-coated platelets, and antibody-containing plasma from three patients induced abciximab-dependent activation and aggregation of normal platelets (25). Abciximab-induced thrombocytopenia refractory to platelet transfusions has also been observed in a patient with idiopathic thrombocytopenic purpura (38).

Ethylenediaminetetra-acetate can cause pseudothrombocytopenia by activating platelet agglutination, resulting in a spuriously low platelet count (SEDA-21, 250). Of 66 patients who received abciximab after coronary revascularization, 17 developed thrombocytopenia and 9 developed severe thrombocytopenia (39). However, of these 26 patients, 18 had pseudothrombocytopenia. True thrombocytopenia occurred at 4 hours after infusion whereas pseudothrombocytopenia occurred within the first 24 hours. The mechanism of pseudothrombocytopenia may be the effect of EDTA on the calcium-dependant glycoprotein IIb/IIIa complex, which frees the antigenic binding site on glycoprotein IIb available to IgM antibody. This increased antibody binding may cause platelet clumping and lead to false thrombocytopenia. True thrombocytopenia did not lead to hemorrhagic complications, but the patients required platelet transfusion.

Bleeding

Bleeding complications were prospectively recorded in 344 consecutive patients who underwent percutaneous coronary intervention with adjunctive use of abciximab (40). There was major bleeding in six patients (1.7%) one of whom had pulmonary hemorrhage and one intracranial hemorrhage. There was minor bleeding in 20 (5.8%). There was thrombocytopenia in 13 (3.9%), mild in four cases, severe in four, and profound in five. Female sex and the "bail-out" use of abciximab were susceptibility factors for bleeding complications.

In a retrospective chart analysis of 348 patients, 79 who received abciximab (n = 30), eptifibatide (n = 24), or tirofiban (n = 25) were included (41). There were 21 bleeding events not related to coronary artery bypass grafting, one major bleed (on tirofiban), and 20 minor bleeds. Bleeding rates were not significantly different between the groups: nine with abciximab, five with eptifibatide, and seven with tirofiban. Significant risk factors for bleeding included weight, infusion duration, and baseline platelet count.

In a meta-analysis of randomized comparisons of abciximab and placebo or control therapy in primary percutaneous coronary intervention for acute myocardial infarction in 3266 patients, treatment with abciximab resulted in an increased likelihood of major bleeding (OR = 1.74; 95% CI = 1.11, 2.72) (42).

In a meta-analysis of randomized controlled trials of abciximab as adjunctive therapy to percutaneous coronary interventions for acute myocardial infarction abciximab was associated with an increased risk of major bleeding (OR = 1.39, 95% CI = 1.03, 1.87), but bleeding was observed only when the heparin bolus was 100 U/kg followed by a maintenance infusion (OR = 1.89, 95% CI = 1.10, 3.28) and not with a bolus of 70 U/kg (OR = 1.22, 95% CI = 0.85, 1.73) (43).

In the Global Use of Strategies to Open Occluded Arteries in Acute Coronary Syndromes (GUSTO IV-ACS) trial, abciximab (either a 24-hour or a 48-hour infusion) was compared with placebo in 7800 patients with an acute coronary syndrome (44). There were major bleeds in 98 patients (1.2%), including eight with intracerebral hemorrhage, and 215 patients (2.8%) had a minor bleed. The most significant predictors of spontaneous bleeding were: use of low-molecular weight heparin (OR = 6.6, CI = 5.3-8.4), duration of abciximab infusion (24 hours OR = 3.2, CI = 2.6, 4.1; 48 hours OR = 4.8 CI = 3.8, 6.0), advanced age (over 70 years OR = 1.7, CI = 1.4, 2.0), and female sex (OR = 1.5, CI = 1.3, 1.8). The authors concluded that treatment with abciximab in patients with non-ST-elevation acute coronary syndromes is safe, because major bleeding and stroke are rare and most events are clinically manageable or have few clinical consequences.

The ReoPro Readministration Registry found clinically significant bleeding in 31 (2.3%) of 1342 patients given abciximab, including one with intracerebral hemorrhage (45). There was thrombocytopenia (less than 100 x 10^9/l) in 5% and profound thrombocytopenia (less than 20 x 10^9/l) in 2%. In patients who received abciximab within 1 month of a previous treatment (n = 115), the incidences of thrombocytopenia and profound thrombocytopenia were 17% and 12% respectively.

Splenic rupture

Splenic rupture with massive intra-abdominal hemorrhage, as a consequence of secondary bleeding into multiple pre-existing splenic infarctions, has been reported in a patient taking a glycoprotein IIb-IIIa antagonist (46).

Immunologic

Human antichimeric antibodies, specific to the murine epitope of Fab antibody fragments, have been observed in patients treated with abciximab. These antibodies are IgG antibodies and have so far not correlated with any adverse effects (17).

Because of its antigenic potential, there are theoretical concerns about the readministration of abciximab, and this has been studied in 1342 patients, who underwent percutaneous coronary interventions and received abciximab at least twice (47). There were no cases of anaphylaxis, and there were only five minor allergic reactions, none of which required termination of the infusion. There was clinically significant bleeding in 31 patients, including one with intracranial hemorrhage. There was thrombocytopenia (platelet count below 100 × 10^9/l) in 5% and profound thrombocytopenia (platelet count below 20 × 10^9/l) in 2%. In patients who received abciximab within 1 month of a previous treatment (n = 115), the risks of thrombocytopenia and profound thrombocytopenia were 17 and 12% respectively. Human chimeric antibody titers before readministration did not correlate with adverse outcomes or bleeding, but were associated with thrombocytopenia and profound thrombocytopenia.

An anaphylactic reaction to abciximab has been reported (48).

- An obese 46-year-old woman with prolonged angina pectoris underwent coronary angiography. She had no known drug allergies, but on administration of an iodinated contrast media she developed anaphylactic shock. After successful resuscitation angiography was completed and she was given aspirin, ticlopidine for a month, and metoprolol. Five months later she developed chest pain again, and angiography was repeated after pretreatment with prednisone and diphenhydramine and she was given abciximab. Within 5 minutes she had an anaphylactic reaction, requiring resuscitation.

This case shows that anaphylactic reactions to abciximab can occur even after pretreatment with prednisone and diphenhydramine for a known allergy to iodine.

Susceptibility Factors

Renal disease

The available data do not suggest an increased risk of bleeding with abciximab among patients with mild to moderate renal insufficiency (31), even though there is reduced platelet aggregation in renal insufficiency.

Drug-drug interactions

Heparin

In 30 patients with acute myocardial infarction undergoing percutaneous coronary intervention, abciximab significantly increased the concentration of the platelet activation marker P selectin, and reduced the dosage of unfractionated heparin that was required to prolong the aPTT by over 60 seconds (49). When abciximab was withdrawn the heparin dosage requirement increased to a greater extent and reached the level found in the untreated patients, even when platelet aggregation was still inhibited. The authors concluded that the increased platelet activation found at the end of abciximab treatment points to a procoagulant condition that should be carefully monitored and treated by altering the doses of anticoagulants and antiplatelet drugs.

References

1. Ibbotson T, McGavin JK, Goa KL. Abciximab: an updated review of its therapeutic use in patients with ischaemic heart disease undergoing percutaneous coronary revascularisation. Drugs 2003;63(11):1121–63.
2. Schweizer J, Kirch W, Koch R, Muller A, Hellner G, Forkmann L. Use of abciximab and tirofiban in patients with peripheral arterial occlusive disease and arterial thrombosis. Angiology 2003;54(2):155–61.
3. Barbaro G, Grisorio B, Fruttaldo L, Bacca D, Babudieri S, Torre D, Francavilla R, Rizzo G, Belloni G, Lucchini A, Annese M, Matarazzo F, Hazra C, Barbarini G. Good safety profile and efficacy of leucocyte interferon-alfa in combination with oral ribavirin in treatment-naive patients with chronic hepatitis C: a multicentre, randomised, controlled study. BioDrugs 2003;17:433–9.
4. Rebeiz AG, Adams J, Harrington RA. Interventional cardiovascular pharmacotherapy: current issues. Am J Cardiovasc Drugs 2005;5(2):93–102.
5. Tripi S, Soresi M, Di Gaetano G, Carroccio A, Giannitrapani L, Vuturo O, Di Giovanni G, Montalto G. Leucocyte interferon-alfa for patients with chronic hepatitis C intolerant to other alfa-interferons. BioDrugs 2003;17:201–5.
6. Ali A, Hashem M, Rosman HS, Kazmouz G, Gardin JM, Schrieber TL. Use of platelet glycoprotein IIb/IIIa inhibitors and spontaneous pulmonary hemorrhage. J Invasive Cardiol 2003;15(4):186–8.
7. Qureshi AI, Saad M, Zaidat OO, Suarez JI, Alexander MJ, Fareed M, Suri K, Ali Z, Hopkins LN. Intracerebral hemorrhages associated with neurointerventional procedures using a combination of antithrombotic agents including abciximab. Stroke 2002;33(7):1916–9.
8. Deliargyris EN, Upadhya B, Applegate RJ, Kontos JL, Kutcher MA, Riesmeyer JS, Sane DC. Safety of abciximab administration during PCI of patients with previous stroke. J Thromb Thrombolysis 2005;19(3):147–53.
9. Chan AW, Yadav JS, Bhatt DL, Bajzer CT, Gum PA, Roffi M, Cho L, Agah R, Topol EJ. Comparison of the safety and efficacy of emboli prevention devices versus platelet glycoprotein IIb/IIIa inhibition during carotid stenting. Am J Cardiol 2005;95(6):791–5.
10. The EPIC Investigation. Use of a monoclonal antibody directed against the platelet glycoprotein IIb/IIIa receptor in high-risk coronary angioplasty. N Engl J Med 1994;330(14):956–61.
11. Aguirre FV, Topol EJ, Ferguson JJ, Anderson K, Blankenship JC, Heuser RR, Sigmon K, Taylor M, Gottlieb R, Hanovich G, et al. Bleeding complications with the chimeric antibody to platelet glycoprotein IIb/IIIa integrin in patients undergoing percutaneous coronary intervention. EPIC Investigators. Circulation 1995;91(12):2882–90.
12. James S, Armstrong P, Califf R, Husted S, Kontny F, Niemminen M, Pfisterer M, Simoons ML, Wallentin L. Safety and efficacy of abciximab combined with dalteparin in treatment of acute coronary syndromes. Eur Heart J 2002;23(19):1538–45.
13. Blankenship JC, Hellkamp AS, Aguirre FV, Demko SL, Topol EJ, Califf RM. Vascular access site complications after percutaneous coronary intervention with abciximab in the Evaluation of c7E3 for the Prevention of Ischemic Complications (EPIC) trial. Am J Cardiol 1998;81(1):36–40.
14. The EPILOG Investigators. Platelet glycoprotein IIb/IIIa receptor blockade and low-dose heparin during percutaneous coronary revascularization. N Engl J Med 1997;336(24):1689–96.
15. The EPISTENT Investigators. Evaluation of Platelet IIb/IIIa Inhibitor for Stenting. Randomised placebo-controlled and balloon-angioplasty-controlled trial to assess safety of coronary stenting with use of platelet glycoprotein-IIb/IIIa blockade. Lancet 1998;352(9122):87–92.
16. Brener SJ, Barr LA, Burchenal JE, Katz S, George BS, Jones AA, Cohen ED, Gainey PC, White HJ, Cheek HB, Moses JW, Moliterno DJ, Effron MB, Topol EJ. Randomized, placebo-controlled trial of platelet glycoprotein IIb/IIIa blockade with primary angioplasty for acute myocardial infarction. ReoPro and Primary PTCA Organization and Randomized Trial (RAPPORT) Investigators. Circulation 1998;98(8):734–41.

17. Ferguson JJ, Kereiakes DJ, Adgey AA, Fox KA, Hillegass WB Jr, Pfisterer M, Vassanelli C. Safe use of platelet GP IIb/IIIa inhibitors. Am Heart J 1998;135(4):S77–89.
18. Kereiakes DJ, Broderick TM, Whang DD, Anderson L, Fye D. Partial reversal of heparin anticoagulation by intravenous protamine in abciximab-treated patients undergoing percutaneous intervention. Am J Cardiol 1997;80(5):633–4.
19. Kleiman NS. A risk-benefit assessment of abciximab in angioplasty. Drug Saf 1999;20(1):43–57.
20. Pinton P. Thrombopénies sous abciximab dans le traitement des syndromes coronariens aigus par angioplastie. [Abciximab-induced thrombopenia during treatment of acute coronary syndromes by angioplasty.] Ann Cardiol Angeiol (Paris) 1998;47(5):351–8.
21. Brown DL. Deaths associated with platelet glycoprotein IIb/IIIa inhibitor treatment. Heart 2003;89(5):535–7.
22. McLenachan JM. Who would I not give IIb/IIIa inhibitors to during percutaneous coronary intervention? Heart 2003;89(5):477–8.
23. Eryonucu B, Tuncer M, Erkoc R. Repetitive profound thrombocytopenia after treatment with tirofiban: a case report. Cardiovasc Drugs Ther 2004;18(6):503–5.
24. Aster RH, Curtis BR, Bougie DW. Thrombocytopenia resulting from sensitivity to GPIIb-IIIa inhibitors. Semin Thromb Hemost 2004;30(5):569–77.
25. Nurden P, Clofent-Sanchez G, Jais C, Bermejo E, Leroux L, Coste P, Nurden AT. Delayed immunologic thrombocytopenia induced by abciximab. Thromb Haemost 2004;92(4):820–8.
26. Curtis BR, Divgi A, Garritty M, Aster RH. Delayed thrombocytopenia after treatment with abciximab: a distinct clinical entity associated with the immune response to the drug. J Thromb Haemost 2004;2(6):985–92.
27. Trapolin G, Savonitto S, Merlini PA, Caimi MT, Klugmann S. Delayed profound thrombocytopenia after abciximab administration for coronary stenting in acute coronary syndrome. Case reports and review of the literature. Ital Heart J 2005;6(8):647–51.
28. Suleiman M, Gruberg L, Hammerman H, Aronson D, Halabi M, Goldberg A, Grenadier E, Boulus M, Markiewicz W, Beyar R. Comparison of two platelet glycoprotein IIb/IIIa inhibitors, eptifibatide and abciximab: outcomes, complications and thrombocytopenia during percutaneous coronary intervention. J Invasive Cardiol 2003;15:319–23.
29. Nikolsky E, Sadeghi HM, Effron MB, Mehran R, Lansky AJ, Na Y, Cox DA, Garcia E, Tcheng JE, Griffin JJ, Stuckey TD, Turco M, Carroll JD, Grines CL, Stone GW. Impact of in-hospital acquired thrombocytopenia in patients undergoing primary angioplasty for acute myocardial infarction. Am J Cardiol 2005 15;96(4):474–81.
30. Berkowitz SD, Harrington RA, Rund MM, Tcheng JE. Acute profound thrombocytopenia after C7E3 Fab (abciximab) therapy. Circulation 1997;95(4):809–13.
31. Foster RH, Wiseman LR. Abciximab. An updated review of its use in ischaemic heart disease. Drugs 1998;56(4):629–65.
32. Joseph T, Marco J, Gregorini L. Acute profound thrombocytopenia after abciximab therapy during coronary angioplasty. Clin Cardiol 1998;21(11):851–2.
33. Aster RH. Immune thrombocytopenia caused by glycoprotein IIb/IIIa inhibitors. Chest 2005;127(2 Suppl):53S–59S.
34. Sharma S, Bhambi B, Nyitray W, Sharma G, Shambaugh S, Antonescu A, Shukla P, Denny E. Delayed profound thrombocytopenia presenting 7 days after use of abciximab (ReoPro). J Cardiovasc Pharmacol Ther 2002;7(1):21–4.
35. Schwarz S, Schwab S, Steiner HH, Hacke W. Secondary hemorrhage after intraventricular fibrinolysis: a cautionary note: a report of two cases. Neurosurgery 1998;42(3): 659–663.
36. Curtis BR, Swyers J, Divgi A, McFarland JG, Aster RH. Thrombocytopenia after second exposure to abciximab is caused by antibodies that recognize abciximab-coated platelets. Blood 2002;99(6):2054–9.
37. Lown JA, Hughes AS, Cannell P. Prolonged profound abciximab associated immune thrombocytopenia complicated by transient multispecific platelet antibodies. Heart 2004;90(9):e55.
38. Mendez TC, Diaz O, Enriquez L, Baz JA, Fernandez F, Goicolea J. Severe thrombocytopenia refractory to platelet transfusions, secondary to abciximab readministration, in a patient previously diagnosed with idiopathic thrombocytopenic purpura. A possible etiopathogenic link. Rev Esp Cardiol 2004;57(8):789–91.
39. Schell DA, Ganti AK, Levitt R, Potti A. Thrombocytopenia associated with c7E3 Fab (abciximab). Ann Hematol 2002;81(2):76–9.
40. Razakjr OA, Tan HC, Yip WL, Lim YT. Predictors of bleeding complications and thrombocytopenia with the use of abciximab during percutaneous coronary intervention. J Interv Cardiol 2005;18(1):33–7.
41. Brouse SD, Wiesehan VG. Evaluation of bleeding complications associated with glycoprotein IIb/IIIa inhibitors. Ann Pharmacother 2004;38(11):1783–8.
42. Kandzari DE, Hasselblad V, Tcheng JE, Stone GW, Califf RM, Kastrati A, Neumann FJ, Brener SJ, Montalescot G, Kong DF, Harrington RA. Improved clinical outcomes with abciximab therapy in acute myocardial infarction: a systematic overview of randomized clinical trials. Am Heart J 2004;147(3):457–62.
43. de Queiroz Fernandes Araujo JO, Veloso HH, Braga De Paiva JM, Filho MW, Vincenzo De Paola AA. Efficacy and safety of abciximab on acute myocardial infarction treated with percutaneous coronary interventions: a meta-analysis of randomized, controlled trials. Am Heart J 2004;148(6):937–43.
44. Lenderink T, Boersma E, Ruzyllo W, Widimsky P, Ohman EM, Armstrong PW, Wallentin L, Simoons ML; GUSTO IV-ACS Investigators. Bleeding events with abciximab in acute coronary syndromes without early revascularization: an analysis of GUSTO IV-ACS. Am Heart J 2004;147(5):865–73.
45. Dery JP, Braden GA, Lincoff AM, Kereiakes DJ, Browne K, Little T, George BS, Sane DC, Cines DB, Effron MB, Mascelli MA, Langrall MA, Damaraju L, Barnathan ES, Tcheng JE; ReoPro Readministration Registry Investigators. Final results of the ReoPro readministration registry. Am J Cardiol 2004;93(8):979–84.
46. Friedrich EB, Kindermann M, Link A, Bohm M. Splenic rupture complicating periinterventional glycoprotein IIb/IIIa antagonist therapy for myocardial infarction in polycythemia vera. Z Kardiol 2005;94(3):200–4.
47. Dery JP, Braden GA, Lincoff AM, Kereiakes DJ, Browne K, Little T, George BS, Sane DC, Cines DB, Effron MB, Mascelli MA, Langrall MA, Damaraju L, Barnathan ES, Tcheng JEReoPro Readministration Registry Investigators. Final results of the ReoPro readministration registry. Am J Cardiol 2004;93(8):979–84.

48. Pharand C, Palisaitis DA, Hamel D. Potential anaphylactic shock with abciximab readministration. Pharmacotherapy 2002;22(3):380–3.
49. Piorkowski M, Priess J, Weikert U, Jaster M, Schwimmbeck PL, Schultheiss HP, Rauch U. Abciximab therapy is associated with increased platelet activation and decreased heparin dosage in patients with acute myocardial infarction. Thromb Haemost 2005;94(2):422–6.

Acetylsalicylic acid and related compound

General Information

A century after its introduction, acetylsalicylic acid (aspirin) is by far the most commonly used analgesic, sharing its leading position with the relative newcomer paracetamol (acetaminophen), and notwithstanding the fact that other widely used compounds of their class, like ibuprofen and naproxen, have in recent years been introduced in over-the-counter versions. Both are also still being prescribed by physicians and are generally used for mild to moderate pain, fever associated with common everyday illnesses, and disorders ranging from head colds and influenza to toothache and headache. Their greatest use is by consumers who obtain them directly at the pharmacy, and in many countries outside pharmacies as well. Perhaps this wide availability and advertising via mass media lead to a lack of appreciation by the lay public that these are medicines with associated adverse effects. Both have at any rate been subject to misuse and excessive use, leading to such problems as chronic salicylate intoxication with aspirin, and severe hepatic damage after overdose with paracetamol. Both aspirin and paracetamol have featured in accidental overdosage (particularly in children) as well as intentional overdosage.

In an investigation of Canadian donors who had not admitted to drug intake, 6–7% of the blood samples taken were found to have detectable concentrations of acetylsalicylic acid and paracetamol (1). Such drugs would be potentially capable of causing untoward reactions in the recipients.

To offer some protection against misuse of analgesics, many countries have insisted on the use of packs containing total quantities less than the minimum toxic dose (albeit usually the one obtained for healthy young volunteers and thus disregarding the majority of the population), and supplied in child-resistant packaging. Most important, however, is the need to provide education for the lay public to respect such medicines in general for the good they can do, but more especially for the harm that can arise but which can be avoided. There is a definite role for the prescribing physician, as informing the patient seems to prevent adverse events (2).

The sale of paracetamol or aspirin in dosage forms in which they are combined with other active ingredients offers considerable risk to the consumer, since the product as sold may not be clearly identified as containing either of these two analgesics. Brand names sometimes obscure the actual composition of older formulations that contain one or both of these analgesics in combination with, for example, a pyrazolone derivative and/or a potentially addictive substance. For instance, in Germany, with the EC harmonization of the Drug Law of 1990, the manufacturers of drugs already marketed before 1978 had the opportunity of exchanging even the active principles without being obliged to undergo a new approval procedure or to abandon their brand name. Combination formulations are still being promoted and sold, and not exclusively in developing countries. Consequently, the patient who is so anxious to allay all his symptoms that he takes several medications concurrently may without knowing it take several doses of aspirin or paracetamol at the same time, perhaps sufficient to cause toxicity. It is essential that product labels clearly state their active ingredients by approved name together with the quantity per dosage form (3).

The antipyretic analgesics, with the non-steroidal anti-inflammatory drugs (NSAIDs), share a common mechanism of action, namely the inhibition of prostaglandin synthesis from arachidonic acid and their release. More precisely their mode of action is thought to result from inhibition of both the constitutive and the inducible isoenzymes (COX-1 and COX-2) of the cyclo-oxygenase pathway (4). However, aspirin and paracetamol are distinguishable from most of the NSAIDs by their ability to inhibit prostaglandin synthesis in the nervous system, and thus the hypothalamic center for body temperature regulation, rather than acting mainly in the periphery.

Endogenous pyrogens (and exogenous pyrogens that have their effects through the endogenous group) induce the hypothalamic vascular endothelium to produce prostaglandins, which activate the thermoregulatory neurons by increasing AMP concentrations. The capacity of the antipyretic analgesics to inhibit hypothalamic prostaglandin synthesis appears to be the basis of their antipyretic action. Neither aspirin nor paracetamol affects the synthesis or release of endogenous pyrogens and neither will lower body temperature if it is normal.

While aspirin significantly inhibits peripheral prostaglandin and thromboxane synthesis, paracetamol is less potent as a synthetase inhibitor than the NSAIDs, except in the brain, and paracetamol has only a weak anti-inflammatory action. It is simple to ascribe the analgesic activity of aspirin to its capacity to inhibit prostaglandin synthesis, with a consequent reduction in inflammatory edema and vasodilatation, since aspirin is most effective in the pain associated with inflammation or injury. However, such a peripheral effect cannot account for the analgesic activity of paracetamol, which is less well understood.

As a prostaglandin synthesis inhibitor, aspirin, like other NSAIDs, is associated with irritation of and damage to the gastrointestinal mucosa. In low doses it can also increase bleeding by inhibiting platelet aggregation; in high doses, prolongation of the prothrombin time will contribute to the bleeding tendency. Intensive treatment can also produce unwanted nervous system effects (salicylism).

Depending on the criteria used, the incidence of aspirin hypersensitivity is variously estimated as being as low as 1% or as high as 50%, the highest frequency being found in asthmatics. The condition is characterized by bronchospasm (asthma), urticaria, angioedema, and vasomotor rhinitis, each occurring alone or in combination, often leading to severe and even life-threatening reactions. There is no clear evidence of an association with tumors, apart from the possible peripheral contribution of aspirin to the development of urinary tract neoplasms in patients with analgesic nephropathy. Indeed, some authors have suggested a role for salicylates in reducing the incidence of colorectal tumors and breast tumors.

The following are absolute contraindications to the use of aspirin:

- children under 16;
- people with hypersensitivity to salicylates, NSAIDs, or tartrazine;
- people with peptic ulceration;
- people with known coagulopathies, including those induced as part of medical therapy.

The following are relative contraindications to the use of long-term analgesic doses of aspirin:

- gout, since normal analgesic doses impede the excretion of uric acid (high doses have a uricosuric effect); an additional problem in gout is that salicylates reduce the uricosuric effects of sulfinpyrazone and probenecid;
- variant angina; a daily dose of 4 g has been found to provoke attacks both at night-time and during the day (5,6), perhaps owing to direct triggering of coronary arterial spasm; blockade of the synthesis of PGI_2, which normally protects against vasoconstriction, could be involved;
- diabetes mellitus, in which aspirin can in theory interfere with the actions of insulin and glucagon sufficiently to derange control;
- some days before elective surgery (even in coronary artery bypass grafting) or delivery, especially if extradural anesthesia is used (7), although recent data seem reassuring (8); aspirin increases bleeding at dental extraction or perioperatively;
- in elderly people, who may develop gastrointestinal bleeding;
- anorectal inflammation (suppositories);
- pre-existing gastrointestinal disease, liver disease, hypoalbuminemia, hypovolemia, in the third trimester of pregnancy, perioperatively, or in patients with threatening abortion.

Assessing the benefit-to-harm balance of low-dose aspirin in preventing strokes and heart attacks

Although there is clear evidence of benefit of acetylsalicylic acid (aspirin) in secondary prevention of strokes and heart attacks, the question of whether aspirin should also be prescribed for primary prevention in asymptomatic people is still debatable. Trials in primary prevention have given contrasting results (9,10), and aspirin can cause major harms (for example severe gastrointestinal bleeding and hemorrhagic stroke).

Furthermore, despite evidence of the efficacy of aspirin in secondary prevention, its use in patients at high risk of strokes and heart attacks remains suboptimal (11). A possible explanation for this underuse may be concern about the relative benefit in relation to the potential risk for serious hemorrhagic events. Accurate evaluation of the benefits and harms of aspirin is therefore warranted.

Two meta-analyses have provided some information. The first examined the benefit and harms of aspirin in subjects without known cardiovascular or cerebrovascular disease (primary prevention) (12). The authors selected articles published between 1966 and 2000—five large controlled studies of primary prevention that lasted at least 1 year and nine studies of the effects of aspirin on gastrointestinal bleeding and hemorrhagic stroke. The five randomized, placebo-controlled trials included more than 50 000 patients and the meta-analysis showed that aspirin significantly reduced the risk of the combined outcome (confirmed non-fatal myocardial infarction or death from coronary heart disease) (OR = 0.72; 95% CI = 0.60, 0.87). However, aspirin increased the risk of major gastrointestinal bleeding (OR = 1.7; CI = 1.4, 2.1) significantly, while the small increase found for hemorrhagic stroke (OR = 1.4; CI = 0.9, 2.0) was not statistically significant. All-cause mortality was not significantly affected (OR = 0.93; CI = 0.84; 1.02). Most important was the finding that the net effect of aspirin improved with increasing risk of coronary heart disease. The meta-analysis showed that for 1000 patients with a 5% risk of coronary heart disease events over 5 years, aspirin would prevent 6–20 myocardial infarctions but would cause also 0–2 hemorrhagic strokes and 2–4 major gastrointestinal bleeds. For patients at lower risk (1% over 5 years), aspirin would prevent 1–4 myocardial infarctions but would still cause 0–2 hemorrhagic strokes and 2-4 major gastrointestinal bleeds.

Therefore when deciding to use aspirin in primary prophylaxis, one should take account of the relative utility of the different outcomes that are prevented or caused by aspirin.

The other meta-analysis (13) compared the benefits of aspirin in secondary prevention with the risk of gastrointestinal bleeding. An earlier analysis of this problem included patients at various levels of risk and doses of aspirin that would currently be regarded as too high (14), and may therefore have either under-represented the benefit or exaggerated the risk. In another analysis there was no difference in the risk of gastrointestinal bleeding across the whole range of doses used (15).

The meta-analysis reviewed all randomized, placebo-controlled, secondary prevention trials of at least 3-months duration published from 1970 to 2000. The dosage of aspirin was 50–325 mg/day. Six studies contributed 6300 patients to the analysis (3127 on aspirin and 3173 on placebo). Aspirin reduced all-cause mortality by 18%, the number of strokes by 20%, myocardial infarctions by 30%, and other vascular events by 30%. On the other hand, patients who took aspirin were 2.5 times more likely than those who took placebo to have gastrointestinal tract bleeds. The number of patients needed to be treated (NNT) to prevent one death from any cause was 67 and the NNT to cause one gastrointestinal bleeding event was 100. In other words 1.5 lives can be saved for every gastrointestinal bleed attributed to aspirin. Although the risk of gastrointestinal bleeding was increased by aspirin, the hemorrhagic events were manageable and led to no deaths. On the basis of these data we can conclude that the benefits–harm balance for low-dose aspirin in the secondary prevention of cardiovascular and cerebrovascular events is highly favorable. The same conclusions have been drawn from the systematic overview published by the Antithrombotic Trialists Collaboration Group, which analysed data from 287 studies involving 135 000 patients (16).

As far as primary prevention of cardiovascular events is concerned, it appears that aspirin can reduce heart attacks and strokes but increases gastrointestinal and intracranial bleeding. The decision to use aspirin in primary prevention should therefore take into account the fact that the net effect of aspirin improves with increasing risk of coronary heart disease as well as the values that patients attach to the main favorable and unfavorable outcomes.

Organs and Systems

Cardiovascular

Apart from rare reports of variant angina pectoris and vasculitis theoretically related to thromboxane, aspirin is not associated with adverse effects on the cardiovascular system (17,18), except an increase in circulating plasma volume after large doses.

The effects of aspirin on blood pressure have been investigated in 100 untreated patients with mild hypertension who took aspirin on awakening or before bedtime (19). There was no change in blood pressure after dietary recommendations alone or when aspirin was given on awakening. However, there was a highly significant reduction in blood pressure in those who took aspirin before bedtime (reductions of 6 and 4 mmHg in systolic and diastolic blood pressures respectively). As aspirin is given once a day for its cardioprotective effect, giving it in the evening could be of greater benefit if it also results in a reduction in blood pressure.

Respiratory

The effect of aspirin on bronchial musculature is discussed in the section on Immunologic in this monograph.

Salicylates can cause pulmonary edema, particularly in the elderly, especially if they are or have been heavy smokers (20).

Chronic salicylate toxicity can cause pulmonary injury, leading to respiratory distress. Lung biopsy may show diffuse alveolar damage and fibrosis (21).

Nervous system

Salicylism is a reaction to very high circulating concentrations of salicylate, characterized by tinnitus, dizziness, confusion, and headache.

Encephalopathy secondary to hyperammonemia has been reported in those rare cases of liver failure that are associated with high doses of aspirin, and this also forms a major feature of Reye's syndrome (see the section on Liver in this monograph).

One case-control study showed no increased risk of intracerebral hemorrhage in patients using aspirin or other NSAIDs in low dosages as prophylaxis against thrombosis (22). However, intracerebral hemorrhage has been reported with aspirin, even in low doses, and in the SALT study (23) and the Physicians Health Study of 1989 (24) hemorrhagic stroke and associated deaths occurred with aspirin.

Sensory systems

Eyes

Well-documented acute myopia and increased ocular pressure attributed to aspirin has been described (25).

Ears

With the high concentrations achieved in attempted suicide, tinnitus and hearing loss, leading to deafness, develop within about 5 hours, usually with regression within 48 hours, but permanent damage can occur. Disturbed balance, often with vertigo, can develop, as well as nausea, usually with maintenance of consciousness, even without treatment. It has been postulated that in this state depolarization of the cochlear hair cells occurs, similar to the changes induced by pressure. Tinnitus is also a symptom of salicylism.

Aspirin has been reported to cause damage to the semicircular canals.

- A 61-year-old man with a monoclonal gammopathy developed severe persistent bilateral vestibular dysfunction after taking a high dose of aspirin (5–6 g/day for 3 days) (26). His symptoms (unsteadiness, a broad-based gait, blurred vision, and apparent visual motion when he moved his head and when he walked) persisted for 9 months. Investigations showed a bilateral dynamic deficit of his horizontal semicircular canal.

Metabolism

Aspirin lowers plasma glucose concentrations in C-peptide-positive diabetic subjects and in normoglycemic persons (27). This is of no clinical significance.

Fluid balance

NSAIDs can cause fluid retention, but this has rarely been reported with aspirin.

- Severe fluid retention, possibly due to impaired renal tubular secretion, has been reported in a 29-year-old woman taking aspirin (1.5 g/day for several days) for persistent headache (28). During rechallenge with aspirin (0.5 g tds for 3 days) a dynamic renal scintigram showed a substantial fall in tubular filtration. Withdrawal was followed by complete uneventful recovery.

Pulmonary edema is a feature of salicylate intoxication, but this patient was taking a therapeutic dosage.

Hematologic

Thrombocytopenia, agranulocytosis, neutropenia, aplastic anemia, and even pancytopenia have been reported in association with aspirin. The prospect for recovery from the latter is poor, mortality approaching 50%.

Hemolytic anemia can occur in patients with glucose-6-phosphate dehydrogenase deficiency or erythrocyte glutathione peroxidase deficiency (SED-9, 128) (29–31). Whether these reports have anything more than anecdotal value (SEDA-17, 97) is not known.

Simple iron deficiency caused by occult blood loss occurs with a frequency of 1%, and upper gastrointestinal bleeding resulting from regular aspirin ingestion is the reason for hospitalization in about 15 patients per 100 000 aspirin users per year. Aspirin causes bleeding of sufficient severity to lead to iron deficiency anemia in 10–15% of patients taking it continuously for chronic arthritis. Some individuals are particularly at risk because of pregnancy, age, inadequate diet, menorrhagia, gastrectomy, or malabsorption syndromes.

Macrocytic anemia associated with folate deficiency has been described in patients with rheumatoid arthritis (32) and also in patients who abuse analgesic mixtures containing aspirin (32).

Effects on coagulation

Aspirin in high doses for several days can reduce prothrombin concentrations and prolong the prothrombin time. This will contribute to bleeding problems initiated by other factors, including aspirin's local irritant effects on epithelial cells. It is therefore very risky to use aspirin in patients with bleeding disorders. The effect will contribute to increased blood loss at parturition, spontaneous abortion, or menorrhagia, and may be linked to persistent ocular hemorrhage, particularly in older people, with or without associated surgical intervention (33,34).

By virtue of its effects on both cyclo-oxygenase isoenzymes, aspirin inhibits platelet thromboxane A_2 formation. This effect in the platelet is irreversible and will persist for the lifetime of the platelet (that is up to 10 days), since the platelet cannot synthesize new cyclo-oxygenase. It is of clinical significance that the dose of aspirin necessary to inhibit platelet thromboxane A_2 (around 40 mg/day) is much lower than that needed to inactivate the subendothelial prostacyclin (PGI_2). Hence, platelet aggregation is inhibited, with some associated dilatation of coronary and cerebral arterioles, at doses that do not interfere with prostacyclin inhibition. It is important, in considering the dosage of aspirin for prophylaxis (see below), to appreciate that prostacyclin is a general inhibitor of platelet aggregation, while aspirin, as a cyclo-oxygenase inhibitor, affects aggregation from a limited number of stimuli, for example ADP, adrenaline, thromboxane A_2. It is also worth recalling that the vascular endothelium can synthesize new cyclo-oxygenase, so that any effect on prostacyclin synthesis is of limited duration only (SEDA-12, 74) (35).

Several long-term studies have been carried out since the 1980s to determine the prophylactic usefulness of these effects on clotting. It is now clear that aspirin in dosages of around 300 mg/day can be used successfully for secondary prophylaxis in patients with coronary artery disease, in order to reduce the incidence of severe myocardial infarction, and in patients with cerebrovascular disease to reduce the incidence of transient ischemic attacks and strokes. There is some suggestion that higher doses of aspirin may be required in women. A major drawback has been the high incidence of gastrointestinal adverse effects and particularly bleeding in aspirin-treated groups (5,6,10,36). In view of the age group involved, bleeding can have serious implications. In an attempt to avoid this high proportion of ill-effects and yet retain the benefits of prophylactic antithrombotic treatment, a few trials have been conducted using aspirin in a dose of 162 mg (ISIS-2) (37) and 75 mg (RISC) (38) in symptomatic coronary heart disease, with good evidence of efficacy. Two studies have been reported in patients with cerebrovascular events, namely the Dutch TIA trial with aspirin 30 versus 283 mg (22) and the SALT study with aspirin 75 mg (23). The former did not show any difference in efficacy between the 30 and 283 mg dose groups, but there was no placebo control. The latter study showed a significant reduction in thrombotic stroke. However, intracerebral hemorrhage has been reported with aspirin, even in low doses, and in this as well as in the Physicians Health Study of 1989 (24), hemorrhagic stroke and associated deaths occurred with aspirin. On the other hand, the incidence of serious gastrointestinal events was much lower than previously described.

As nearly all of the risks seem to be dose-related (SEDA-21, 96), there is a good prospect that an even lower daily dose of aspirin may offer advantages in antithrombotic prophylaxis without an increased risk of bleeding, but the results of further such studies are still awaited (9).

Relatively few patients developed a prolonged bleeding time while taking aspirin or other NSAIDs and only few had significant intraoperative blood loss. There is variation in the response of patients for unknown reasons and so the recommendation that NSAIDs should be withdrawn before elective surgery awaits confirmation (SEDA-19, 96).

Gastrointestinal

Gastric ulceration and hemorrhage

DoTS classification
Dose-relation: collateral effect
Time-course: intermediate
Susceptibility factors: age (over 65); sex (women); disease (peptic ulceration)

The gastrointestinal adverse effects of aspirin and the other NSAIDs are the most common. While some argue against a causative relation between aspirin ingestion and chronic gastric ulceration, the current consensus favors such a relation, while admitting that other factors, such as *Helicobacter pylori*, are likely to play a part. Patients aged over 65 years and women are more at risk, as are those who take aspirin over prolonged periods in a daily dose of about 2 g or more.

However, there is no ambiguity about the association of aspirin with gastritis, gastric erosions, or extensions of existing peptic ulcers, all of which are demonstrable by endoscopy. Even after one or two doses, superficial erosions have been described in over 50% of healthy subjects. This association is now almost universally accepted as the standard basis for comparative testing of NSAIDs and other drugs (22,39–41). Whether it is of benefit to use other drugs concomitantly to prevent the effect of gastric acid on the mucosa, and thus reduce the risk of gastric ulceration, is discussed further in this monograph.

Dyspepsia, nausea, and vomiting occur in 2–6% of patients after aspirin ingestion. Patients with rheumatoid arthritis seem to be more sensitive, and the frequency of aspirin-induced dyspepsia in this group is 10–30% (SEDA-9, 129). However, these symptoms are generally poor predictors of the incidence of mucosal damage (SEDA-18, 90).

The bleeding that occurs is usually triggered by erosions and aggravated by the antithrombotic action of aspirin. While it is reported to occur in up to 100% of regular aspirin takers, bleeding tends to be asymptomatic in young adults, unless it is associated with peptic ulceration, but it is readily detectable by endoscopy and the presence of occult blood in the feces. Hematemesis and melena are less often seen, the odds ratio being 1.5–2.0 in an overview of 21 low-dose aspirin prevention studies (42). A degree of resultant iron deficiency anemia is common. Such events are more commonly seen in older people in whom there is a significant proportion of serious bleeding and even deaths. Major gastrointestinal bleeding has an incidence of 15 per 100 000 so-called heavy aspirin users. However, the interpretation of "heavy" and of quantities of aspirin actually taken is to a large extent subjective and very dependent on the questionable accuracy of patient reporting. The risk appears to be greater in women, smokers, and patients concurrently taking other NSAIDs, and is possibly affected by other factors not yet established (43). Gastrointestinal perforation can occur without prodromes. Aspirin increases the risk of major upper gastrointestinal bleeding and perforation two- to three-fold in a dose-related manner, but deaths are rare.

Incidence

Of the estimated annual 65 000 upper gastrointestinal emergency admissions in the UK, nearly 20% (including deaths in 3.4%) are attributable to the use of prostaglandin synthesis inhibitors (44). As might be expected with an inhibitor of prostaglandin synthesis, the cytoprotective effects of prostaglandin E and prostacyclin (PGI_2) are reduced by aspirin, as is the inhibitory action on gastric acid secretion. This effect may be both direct, as is the case with aspirin released in the stomach (or the lower rectum in the case of aspirin suppositories), and indirect following absorption and distribution via the systemic circulation; attempts to reduce the problems by coating and buffering can therefore have only limited success. The indirect type of effect is shown by the fact that these adverse gastric effects can also be exerted by parenteral lysine acetylsalicylate (SEDA-10, 72). The local effects depend in part on the tablet particle size, solubility, and rate of gastric absorption, while the most important variable appears to be gastric pH. On the other hand, within-day changes in the pharmacokinetics of the analgesic compounds may be involved in the prevalence of gastrointestinal adverse effects.

The estimates of gastrointestinal complication rates from aspirin are generally derived from clinical trials (SEDA-21, 100). However, the applicability of the results of such trials to the general population may be debatable, as protocols for these studies often are designed precisely to avoid enrollment of patients who are at risk of complications. Indeed differences in benefit-to-harm balance have been found in trials using the same dose of aspirin (45,46). For this reason, a population-based historical cohort study on frequency of major complications of aspirin used for secondary stroke prevention may be of interest (47). The study identified 588 patients who had a first ischemic stroke, transient ischemic attack, or amaurosis fugax during the study period. Of these, 339 patients had taken aspirin for an average of 1.7 years. The mean age of patients who had taken aspirin was 74 years. Complications occurred within 30 days of initiation of treatment in one patient, between 30 days and 6 months in 10 patients, between 6 months and 1 year in seven patients, and between 1 year and 2 years in two patients. Estimated standardized morbidity ratio of gastrointestinal hemorrhage (determined on the basis of 10 observed events and 0.661 expected events, during 576 person-years of observation) was 15 (95% CI = 7, 28). The estimated standardized morbidity ratio of intracerebral hemorrhage (determined on the basis of only one event and 0.59 expected events) was 1.7 (CI = 0.04, 9.4). One patient had a fatal gastrointestinal hemorrhage. Unfortunately these complication rates must be considered estimates, because aspirin therapy was not consistently recorded. However, the rates of complications were similar to those observed in some randomized clinical trials. On the basis of these data and of those of a meta-analysis of 16 trials involving more than 95 000 patients (47), the overall benefits of aspirin, measured in terms of preventing myocardial infarction and ischemic stroke, clearly outweigh the risks.

Dose-relatedness

The question of whether the risk of gastrointestinal hemorrhage with long-term aspirin is related to dose within the usual therapeutic dosage range (SEDA-12, 100; 15,48) merits attention. In a meta-analysis of the incidence of gastrointestinal hemorrhage associated with long-term aspirin and the effect of dose in 24 randomized, controlled clinical trials including almost 66 000 patients exposed for an average duration of 28 months to a wide range of different doses of aspirin (50–1500 mg/day), gastrointestinal hemorrhage occurred in 2.47% of patients taking aspirin compared with 1.42% taking placebo (OR = 1.68; 95% CI = 1.51, 1.88) (15). In patients taking low doses of aspirin (50–162.5 mg/day; n = 49 927), gastrointestinal hemorrhage occurred in 2.3% compared with 1.45% taking placebo (OR = 1.56; 95% CI = 1.40, 1.81). The pooled OR for gastrointestinal hemorrhage with low-dose aspirin was 1.59 (95% CI = 1.4, 1.81). A meta-regression to test for a linear relation between the daily dose of aspirin and the risk of gastrointestinal hemorrhage gave a pooled OR of 1.015 (95% CI = 0.998, 1.047) per 100 mg dose reduction. The reduction in the incidence of gastrointestinal hemorrhage was estimated to be 1.5% per 100 mg dose reduction, but this was not significant. In other studies the incidence of upper gastrointestinal haemorrhage has been reported to be similar in patients taking a range of doses of aspirin from 75 mg to 325 mg/day (49,50).

These data are in apparent contrast with others previously reported (SEDA-21, 100) (14), which showed that gastrointestinal hemorrhage was related to dose in the usual dosage range. Many reasons may explain these contrasting results, the most important being differences in the definition of the hemorrhagic events, in study design, in the population studied, and in the presence of accessory risk factors (51–53).

The recent trends toward the use of lower doses of aspirin have been driven by the belief that these offer a better safety profile while retaining equivalent therapeutic efficacy. Despite the large number of patients enrolled in randomized clinical trials and included in meta-analyses, there is no firm evidence that dose reduction significantly lowers the risk of gastrointestinal bleeding. Patients and doctors therefore need to consider the trade-off between the benefits and harms of long-term treatment with aspirin. Meanwhile, it seems wise to use the lowest dose of proven efficacy.

A systematic review of 17 epidemiological studies conducted between 1990 and 2001 has provided further data on this topic (54). The effect of aspirin dosage was investigated in five studies. There was a greater risk of gastrointestinal complications with aspirin in dosages over 300 mg/day than in dosages of 300 mg/day or less. However, users of low-dose aspirin still had a two-fold increased risk of such complications compared with non-users, with no clear evidence of a dose-response relation at dosages under 300 mg/day, confirming previous findings (15). The study also addressed the question of whether the aspirin formulation affects gastrotoxicity. The pooled relative risks of gastrointestinal complications in four studies were 2.4 (95%; CI = 1.9, 2.9) for enteric-coated aspirin, 5.3 (3.0, 9.2) for buffered formulations, and 2.6 (2.3, 2.9) for plain aspirin, compared with non-use. These data confirm those from previous studies (SEDA-21, 100) (15), which negate any protective effect of the most frequently used aspirin formulations. Furthermore, there were higher relative risks, compared with non-use, for gastrointestinal complications in patients who used aspirin regularly (RR = 3.2; CI = 2.6, 5.9) than in patients who used it occasionally (2.1; 1.7, 2.6), and during the first month of use (4.4; 3.2, 6.1) compared with subsequent months (2.6; 2.1, 3.1).

Comparative studies

A comparative study of gastrointestinal blood loss after aspirin 972 mg qds for 4 days versus different doses of piroxicam (20 mg od, 5 mg qds, and 10 mg qds) showed that piroxicam did not increase fecal blood loss, whereas aspirin did. Gastroscopic evidence of irritation was also greater with aspirin (55).

In a randomized trial comparing ticlopidine (500 mg/day) with aspirin (1300 mg/day) for the prevention of stroke in high-risk patients, the incidence of bleeding was similar in both groups, although more patients treated with aspirin developed peptic ulceration or gastrointestinal hemorrhage (56).

Risk factors

A study of the risk factors for gastrointestinal perforation, a much less frequent event than bleeding, has confirmed that aspirin and other NSAIDs increase the risk of both upper and lower gastrointestinal perforation (OR 6.7, CI 3.1–14.5 for NSAIDs) (57). Gastrointestinal perforation has been associated with other factors, such as coffee consumption, a history of peptic ulcer, and smoking. The combination of NSAIDs, smoking, and alcohol increased the risk of gastrointestinal perforation (OR 10.7, CI 3.8–30) (SEDA-21, 97).

Associated effects

Aspirin can also play a role in esophageal bleeding, ulceration, or benign stricture, and it should be considered as a possible cause in patients, particularly the elderly, who present with any of these features. There have also been reports of rectal stricture in the elderly, associated with the use of aspirin suppositories. Effects on both these strictures emphasize the significance of a direct local action of aspirin as well as a systemic action and underlines the relevance of the involvement of oxygen-derived free radicals in the pathogenesis of mucosal lesions in the gastrointestinal tract (58–60).

A gastrocolic fistula developed in a 47-year-old woman taking aspirin and prednisone for rheumatoid arthritis (61). Other similar case reports have been published (62,63).

Long-term effects

The effects on the stomach of continued exposure to aspirin remain controversial. While in short-term use, gastric mucosal erosions may often be recurrent but transient and comparatively trivial lesions, with longer administration there seems to be an increased risk of progression to ulceration.

Prophylaxis

Intravenous administration, or the use of enteric-coated formulations or modified-release products all appear to reduce the risk both of bleeding and more particularly of erosions/ulceration. However, because of the indirect effect noted above, such formulations do not eliminate the risk, although they may reduce the incidence of gastric or duodenal ulcer, as may buffered aspirin (64,65).

Enteric-coated aspirin has been associated with gastroduodenal ulcer formation; the enteric coating has been shown to be toxic to the bowel and it is postulated that it is also toxic to the stomach (66).

Considerable attention in recent years has been directed toward the efficacy of using synthetic forms of PGE_2, histamine H_2 receptor antagonists, proton pump inhibitors, or antacids, either to heal peptic ulcers associated with use of prostaglandin inhibitors or more significantly to act prophylactically to protect against ulceration or bleeding associated with aspirin or the NSAIDs. With the exception of PGE_2, there is no convincing evidence to justify their prophylactic use, as they do not reduce the risk of significant gastrointestinal events. In contrast, their soothing effect on gastrointestinal symptoms may ultimately result in more severe complications (67). Since all these agents carry their own potential risks, it is more than questionable whether administration to a patient with normal gastrointestinal mucosa is justified. Generally, use of prostaglandin inhibitors should be limited to the shortest possible duration, thereby minimizing, but not eliminating, the risk of gastrointestinal damage. Only high-risk patients should be eligible for prophylactic drug therapy. Well-known risk factors for the development of mucosal lesions of the gastrointestinal tract are age (over 75 years), a history of peptic ulcer, or gastrointestinal bleeding, and concomitant cardiac disease.

Liver

Aspirin can cause dose-related focal hepatic necrosis that is usually asymptomatic or anicteric. Much of the evidence for hepatotoxicity of aspirin and the salicylates has been shown in children (68,69), usually in patients with connective tissue disorders, taking relatively high long-term dosages for Still's disease, rheumatoid arthritis, or occasionally systemic lupus erythematosus. Rises in serum transaminases seem to be the most common feature (in up to 50% of patients) and are usually reversible on withdrawal, but they occasionally lead to fatal hepatic necrosis. Severe and even fatal metabolic encephalopathy can also occur, as in Reye's syndrome (see the section on Reye's syndrome in this monograph). One can easily overload the young patient's individual metabolic capacity. The co-existence of hypoalbuminemia may be a particular risk factor; in patients with hypoalbuminemia of 35 g/l or less, close monitoring of the aspartate transaminase is advisable, especially if the concentration of total serum salicylate is 1.1 mmol/l or higher (70). Plasma salicylate concentrations in serious cases have usually been in excess of 1.4 mmol/l and liver function tests return rapidly to normal when the drug is withdrawn. Finally, a very small number of cases of chronic active hepatitis have been attributed to aspirin (71).

Reye's syndrome

First defined as a distinct syndrome in 1963, Reye's syndrome came to be regarded some years later as an adverse effect of aspirin. In fact, the position is more complex, and the syndrome still cannot be assigned a specific cause. There is general agreement that the disorder presents a few days after the prodrome of a viral illness. Well over a dozen different viruses have so far been implicated, including influenza A and B, adenovirus, *Varicella*, and reovirus. Various other factors have also been incriminated, including aflatoxins, certain pesticides, and such antioxidants as butylated hydroxytoluene. Only in the case of aspirin have some epidemiological studies been conducted, and these appeared to show a close correlation with cases of Reye's syndrome. It was these studies that led to regulatory action against the promotion of salicylate use in children. However, doubt has been thrown on the clarity of the link, and it now seems increasingly likely that while there is some association with aspirin, the etiology is in fact multifactorial, including some genetic predisposition. Studies in Japan did not support the US findings, while studies in Thailand and Canada invoked other factors.

Two characteristic phenomena are present in Reye's syndrome.

1. Damage to mitochondrial structures, with pleomorphism, disorganization of matrix, proliferation of smooth endoplasmic reticulum, and an increase in peroxisomes; mitochondrial enzyme activity is severely reduced, but cytoplasmic enzymes are unaffected. The changes first appear in single cells, but may spread to all hepatocytes. Recovery may be complete by 5–7 days. While these changes are most evident in liver cells, similar effects have been seen in cerebral neurons and skeletal muscle. There appears to be a block in beta-oxidation of fatty acids (inhibition of oxidation of NAB-linked substrates). In vitro aspirin selectively inhibits mitochondrial oxidation of medium- and long-chain fatty acids.
2. An acute catabolic state with hypoglycemia, hyperammonemia, raised activities of serum aspartate transaminase and creatine phosphokinase, and increased urinary nitrogen and serum long chain dicarboxylic acid.

Despite our lack of understanding of the syndrome, the decision taken in many countries to advise against the use of salicylates in children under 12 made an impact, in terms of a falling incidence of Reye's syndrome (SEDA-16, 96; SEDA-17, 97).

Over the last 25 years, in the USA, the incidence of Reye's syndrome has fallen significantly—from the time that the advice was introduced up to 1999 there were 25 reported cases, but 15 were in adolescents aged 12–17 years, and 8% of cases occurred in patients aged 15 years or over (72). In the UK, in view of these findings, the Commission on Safety of Medicines (CSM) amended its original statement and advised that aspirin should be avoided in febrile illnesses or viral infections in patients aged under 16 years. However, the appropriateness of this decision has been challenged (73). This is because the incidence of Reye's syndrome is already low and is falling; furthermore, restricting the use of aspirin leaves paracetamol and ibuprofen as the only available therapeutic alternatives, and their safety is not absolutely guaranteed and might be even worse than that of aspirin.

Pancreas

There are conflicting findings in the literature regarding the possibility that long-term use of aspirin is associated with an increased risk of pancreatic cancer (74). New data from a recent study have suggested that extended periods of regular aspirin use appear to be associated with a statistically significant increased risk of pancreatic cancer among women (75). However, the results of this study were inconsistent and require confirmation.

Urinary tract

Aspirin is associated with a small but significant risk of hospitalization for acute renal insufficiency (SEDA-19, 95).

Studies of the association between the long-term use of aspirin or other NSAIDs and end-stage renal disease have given conflicting results. In order to examine this association, a case-control study was carried out in 583 patients with end-stage renal disease and 1190 controls (76). Long-term use of any analgesic was associated with an overall non-significant odds ratio of 1.22 (CI = 0.89, 1.66). For specific groups of drugs the risks were:

- aspirin 1.56 (1.05, 2.30);
- paracetamol 0.80 (0.39, 1.63);
- pyrazolones 1.03 (0.60, 1.76);
- other NSAIDs 0.94 (0.57, 1.56).

There was thus a small increased risk of end-stage renal disease associated with aspirin, which was related to the cumulative dose and duration of use; it was particularly high among the subset of patients with vascular nephropathy as underlying disease.

These results suggest that long-term use of non-aspirin analgesics and NSAIDs is not associated with an increased risk of end-stage renal disease but that long-term use of aspirin is associated with a small increase in the risk of end-stage renal disease. However, these results should be taken with caution since they have arisen from a subgroup analysis, and should be confirmed by other studies.

When aspirin is used by patients on sodium restriction or with congestive heart failure, there tends to be a reduction in the glomerular filtration rate, with preservation of normal renal plasma flow. Some renal tubular epithelial shedding can also occur.

Severe systemic disease involving the heart, liver, or kidneys seems to predispose the patient to the effects of aspirin and other NSAIDs on renal function (77).

Chronic renal disease

Renal papillary necrosis has been reported after long-term intake or abuse of aspirin and other NSAIDs (SEDA-11, 85) (SEDA-12, 79). The relation between long-term heavy exposure to analgesics and the risk of chronic renal disease has been the object of intensive toxicological and epidemiological research for many years (SEDA-24, 120) (78). Most of the earlier reports suggested that phenacetin-containing analgesics probably cause renal papillary necrosis and interstitial nephritis. In contrast, there was no convincing epidemiological evidence that non-phenacetin-containing analgesics (including paracetamol, aspirin, mixtures of the two, and NSAIDs) cause chronic renal disease. Moreover, findings from epidemiological studies should be interpreted with caution, because of a number of inherent limitations and potential biases in study design (79). Two methodologically sound studies have provided information on this topic.

The first was the largest cohort study conducted thus far to assess the risk of renal dysfunction associated with analgesic use (80). Details of analgesic use were obtained from 11 032 men without previous renal dysfunction participating in the Physicians' Health Study (PHS), which lasted 14 years. The main outcome measure was a raised creatinine concentration defined as 1.5 mg/dl (133 μmol/l) or higher and a reduced creatinine clearance of 55 ml/minute or less. In all, 460 men (4.2%) had a raised creatinine concentration and 1258 (11%) had a reduced creatinine clearance. Mean creatinine concentrations and creatinine clearances were similar among men who did not use analgesics and those who did. This was true for all categories of analgesics (paracetamol and paracetamol-containing mixtures, aspirin and aspirin-containing mixtures, and other NSAIDs) and for higher-risk groups, such as those aged 60 years or over or those with hypertension or diabetes.

These data are convincing, as the large size of the PHS cohort should make it possible to examine and detect even modest associations between analgesic use and a risk of renal disease. Furthermore, this study included more individuals who reported extensive use of analgesics than any prior case-control study. However, the study had some limitations, the most important being the fact that the cohort was composed of relatively healthy men, most of whom were white. These results cannot therefore be generalized to the entire population. However, the study clearly showed that there is not a strong association between chronic analgesic use and chronic renal dysfunction among a large cohort of men without a history of renal impairment.

The second study was a Swedish nationwide, population-based, case-control study of early-stage chronic renal insufficiency in men whose serum creatinine concentration exceeded 3.4 mg/dl (300 μmol/l) or women whose serum creatinine exceeded 2.8 mg/dl (250 μmol/l) (81). In all, 918 patients with newly diagnosed renal insufficiency and 980 controls were interviewed and completed questionnaires about their lifetime consumption of analgesics. Compared with controls, more patients with chronic renal insufficiency were regular users of aspirin (37 versus 19%) or paracetamol (25 versus 12%). Among subjects who did not use aspirin regularly, the regular use of paracetamol was associated with a risk of chronic renal insufficiency that was 2.5 times as high as that for non-users of paracetamol. The risk increased with increasing cumulative lifetime dose. Patients who took 500 g or more over a year (1.4 g/day) during periods of regular use had an increased odds ratio for chronic renal insufficiency (OR = 5.3; 95% CI = 1.8, 15). Among subjects who did not use paracetamol regularly, the regular use of aspirin was associated with a risk of chronic renal insufficiency that was 2.5 times as high as that for non-users of aspirin. The risk increased significantly with an increasing cumulative lifetime dose of aspirin. Among the patients with an average intake of 500 g or more of aspirin per year during periods of regular use, the risk of chronic renal insufficiency was increased about three-fold (OR = 3.3; CI = 1.4, 8.0). Among patients who used paracetamol in addition to aspirin, the risk of chronic renal insufficiency was increased about two-fold when regular aspirin users served as the reference group (OR = 2.2; CI = 1.4, 3.5) and non-significantly when regular paracetamol users were used as controls (OR = 1.6; CI = 0.9, 2.7). There was no relation between the use of other analgesics (propoxyphene, NSAIDs, codeine, and pyrazolones) and the risk of chronic renal insufficiency. Thus, the regular use of paracetamol, or aspirin, or both was associated dose-dependently with an increased risk of chronic renal insufficiency. The OR among regular users exceeded 1.0 for all types of chronic renal insufficiency, albeit not always significantly. These results are consistent with exacerbating effects of paracetamol and aspirin on chronic renal insufficiency, regardless of accompanying disease.

How can we explain the contrasting results of these two studies? A possible explanation lies in the different populations studied. In the PHS study, relatively healthy individuals were enrolled while in the Swedish study all the patients had pre-existing severe renal or systemic disease, suggesting that such disease has an important role in causing analgesic-associated chronic renal insufficiency. People without pre-existing disease who use analgesics may have only a small risk of end-stage renal disease.

Skin

Hypersensitivity reactions, such as urticaria and angioedema, are relatively common in subjects with aspirin hypersensitivity. Purpura, hemorrhagic vasculitis, erythema multiforme, Stevens–Johnson syndrome, and Lyell's syndrome have also been reported, but much less often. Fixed drug eruptions, probably hypersensitive in origin, are periodically described. In some patients they do not recur on rechallenge, that is the sensitivity disappears (82).

Musculoskeletal

There is evidence that salicylates together with at least some NSAIDs suppress proteoglycan biosynthesis independently of effects on prostaglandin synthesis (83). Thus, prolonged use of these agents can accentuate deterioration of articular cartilage in weight-bearing arthritic joints. If this is proved, the problem will be of greatest relevance to elderly people with osteoarthritis, a condition in which this use of prostaglandin inhibitors is questionable.

Immunologic

Aspirin hypersensitivity

Of adult asthmatics 2–20% have aspirin hypersensitivity (9). The mechanism is related to a deficiency in bronchodilator prostaglandins; prostaglandin inhibition may make arachidonic acid produce more leukotrienes with bronchoconstrictor activity. Oral challenge in asthmatic patients is an effective but potentially dangerous method for establishing the presence of aspirin hypersensitivity (68).

The term "aspirin allergy" is better avoided, in the absence of identification of a definite antigen–antibody reaction. This topic has been reviewed (SEDA-17, 94) (SEDA-18, 90).

Epidemiology

Aspirin hypersensitivity is relatively common in adults (about 20%). Estimates of the prevalence of aspirin-induced asthma vary from 3.3 to 44% in different reports (SEDA-5, 169), although it is often only demonstrable by challenge tests with spirometry, and only 4% have problems in practice. Patients with existing asthma and nasal polyps or chronic urticaria have a greater frequency of hypersensitivity (84), and women appear to be more susceptible than men, perhaps particularly during the child-bearing period of life (85). Acute intolerance to aspirin can develop even in patients who have taken the drug for some years without problems.

There is considerable cross-reactivity with other NSAIDs and the now widely banned food colorant tartrazine (86). Cross-sensitization between aspirin and tartrazine is common; for example, in one series 24% of aspirin-sensitive patients also reacted to tartrazine (SEDA-9, 76).

Mechanism

The current theory of the mechanism relates to the inhibition of cyclo-oxygenases (87) and a greater degree of interference with PGE_2 synthesis, allowing the bronchoconstrictor PGF_2 to predominate in susceptible individuals.

PGE_2 inhibition in macrophages may also unleash bronchial cytotoxic lymphocytes, generated by chronic viral infection, leading to destruction of virus-infected cells in the respiratory tract (88). When urticaria occurs, it may result from increased release of leukotrienes LTC_4, D_4, and E_4, which also induces bronchoconstriction, with a shunt of arachidonic acid toward lipoxygenation in aspirin-sensitive asthmatics (SEDA-18, 93). Aspirin-induced asthma patients show hyper-reactivity to inhaled metacholine and sulpyrine.

Two studies have examined possible biochemical pathways in aspirin-induced asthma. In a study of the generation of 15-hydroxyeicosatetraenoic acid (15-HETE) and other eicosanoids by peripheral blood leukocytes from aspirin-sensitive and aspirin-tolerant asthmatics incubation with aspirin 2, 20, or 200 μmol/l resulted in a dose-dependent increase in 15-HETE generation (mean change +85%, +189%, and +284% at each aspirin concentration respectively) only in aspirin-sensitive patients (89). In a study of the cyclo-oxygenase pathways in airway fibroblasts from patients with aspirin-tolerant asthma (n = 9), and patients with aspirin-intolerant asthma (n = 7), patients with asthma had a low capacity for

PGE2 production after stimulation (90). In non-asthmatic patients mean PGE2 production was 32 ng/ml (35 times basal production), in the patients with aspirin-tolerant asthma it was 16 ng/ml (16 times basal), and in the patients with aspirin-intolerant asthma it was only 5.3 ng/ml (4 times basal). These studies show biochemical differences in the effects of aspirin in patents with aspirin-induced asthma. That this is mediated by inhibition of cyclo-oxygenase type 1 is suggested by a study in 33 subjects with a typical history of aspirin-induced asthma, who tolerated the cyclo-oxygenase-2 selective celecoxib; there were no changes in lung function or in urinary excretion of leukotriene E4 (91).

Features

The features of aspirin hypersensitivity include bronchospasm, acute and usually generalized urticaria, angioedema, severe rhinitis, and shock. These reactions can occur alone or in various combinations, developing within minutes or a few hours of aspirin ingestion, and lasting until elimination is complete. They can be life-threatening. The bronchospastic type of reaction predominates in adults, only the urticarial type being found in children. The frequency of recurrent urticaria is significantly greater in adults (3.8 versus 0.3%).

People with asthma may be particularly sensitive to acetylsalicylic acid, which may be given alone or as a constituent of a combination medicine. The association between aspirin sensitivity, nasal polyps, and rhinitis in asthma is well known.

Henoch–Schönlein purpura has been reported (92).

Life-threatening respiratory distress, facial edema, and lethargy occurred in a woman with a history of severe asthma and aspirin hypersensitivity (SEDA-22, 118).

Aspirin-sensitive subjects may have attacks induced by other NSAIDs (93).

Fish oil can also cause exacerbation of asthma in aspirin-sensitive patients (94).

Prophylaxis and treatment

Asthma induced by aspirin is often severe and resistant to treatment. Avoidance of aspirin and substances to which there is cross-sensitivity is the only satisfactory solution. Desensitization is not usually successful and repeated treatments are needed to maintain any effect (95,96).

Long-Term Effects

Tumorigenicity

Studies on the tumor-inducing effects of heavy use of analgesics, especially those that contain phenacetin, have given contrasting results (SEDA-21, 100) (97,98). There has been a case-control study of the role of habitual intake of aspirin on the occurrence of urothelial cancer and renal cell carcinoma (99). In previous studies there was a consistent association between phenacetin and renal cell carcinoma, but inconclusive results with respect to non-phenacetin analgesics. In 1024 patients with renal cell carcinoma and an equal number of matched controls, regular use of analgesics was a significant risk factor for renal cell carcinoma (OR = 1.6; CI = 1.4, 1.9). The risk was significantly increased by aspirin, NSAIDs, paracetamol, and phenacetin, and within each class of analgesic the risk increased with increasing exposure. Individuals in the highest exposure categories had about a 2.5-fold increase in risk relative to non-users or irregular users of analgesics. However, exclusive users of aspirin who took aspirin 325 mg/day or less for cardiovascular problems were not at an increased risk of renal cell carcinoma (OR = 0.9; CI = 0.6, 1.4).

Second-Generation Effects

Teratogenicity

It is perhaps surprising that aspirin, which is teratogenic in rodents, and which by virtue of its capacity to inhibit prostaglandin synthesis would be expected to affect the development of the renal and cardiovascular systems, has shown no evidence of teratogenesis in humans, despite very widespread use in pregnant women. Perhaps increased production of prostaglandins during pregnancy overrides the effects of aspirin in the usual dosages, and the intervention of placental metabolism protects the human fetus from exposure to aspirin. Whatever the explanation, there are very few reports in which aspirin can be implicated as a human teratogen and a few studies (100,101) have provided positive reassurance.

Fetotoxicity

Because aspirin is an antithrombotic agent and can promote bleeding, it should be avoided in the third trimester of pregnancy and at parturition (102). At parturition there is a second reason for avoiding aspirin, since its prostaglandin-inhibiting capacity could mean that it will delay parturition and induce early closure of the ductus arteriosus in the near-term fetus, as other NSAIDs do (103). However, its use in low doses in pregnancy may prevent retardation of fetal growth (104).

Susceptibility Factors

Age

In view of the association with Reye's syndrome, aspirin should be avoided in children aged under 16.

Drug Administration

Drug formulations

Although the use of enteric-coated aspirin can reduce its direct adverse effect on the stomach (SEDA-10, 72), it could in principle transfer these to some extent to the intestine; modified-release NSAIDs have sometimes caused intestinal perforation.

Enteric coating reduces the rate of absorption of aspirin. In cases of severe overdosage this can cause difficulties in diagnosis and treatment, since early plasma salicylate measurements are unreliable, maximum blood concentrations sometimes not being reached until 60 or 70 hours after overdose (105,106). Another complication of the use of enteric-coated aspirin is the risk of gastric outlet obstruction and the resulting accumulation of tablets because of subclinical pyloric stenosis.

Drug overdose

Acute poisoning

Acute salicylate poisoning is a major clinical hazard (107), although it is associated with low major morbidity and mortality, in contrast to chronic intoxication (SEDA-17, 98). It can cause alkalemia or acidemia, alkaluria or aciduria, hyperglycemia or hypoglycemia, and water and electrolyte imbalances. However, the usual picture is one of hypokalemia with metabolic acidosis and respiratory alkalosis. Effects on hearing have been referred to in the section on Sensory systems in this monograph. Nausea, vomiting, tinnitus, hyperpnea, hyperpyrexia, confusion, disorientation, dizziness, coma, and/or convulsions are common. They are expressions of the nervous system effects of the salicylates. Gastrointestinal hemorrhage is frequent.

Serum salicylate concentrations above 3.6 mmol/l are likely to be toxic, and concentrations of 5.4 mmol/l can easily prove fatal.

Aspirin overdose in children can be particularly serious.

- A 5-year-old girl died after taking an aspirin overdose. Autopsy showed a pattern of necrosis resembling acute toxic myocarditis (108).

After ingestion, drug absorption can be prevented by induction of emesis, gastric lavage, and the administration of active charcoal; drug excretion is enhanced by administering intravenous alkalinizing solutions, hemoperfusion, and hemodialysis (109). Forced diuresis is dangerous and unnecessary.

Fluid and electrolyte management is the mainstay of therapy. The immediate aim must be to correct acidosis, hyperpyrexia, hypokalemia, and dehydration. In severe cases vitamin K_1 should be given to counteract hypoprothrombinemia.

Chronic poisoning

Chronic salicylate intoxication is commonly associated with chronic daily headaches, lethargy, confusion, or coma. Since headache is a feature, it can easily be misdiagnosed if the physician is not aware that aspirin has been over-used. Depression of mental status is usually present at the time of diagnosis, when the serum salicylate concentration is at a peak. The explanation of depression, manifested by irritability, lethargy, and unresponsiveness, occurring 1–3 days after the start of therapy for aspirin intoxication, lies in a persistently high concentration of salicylate in the central nervous system, while the serum salicylate concentration falls to non-toxic values. The delayed unresponsiveness associated with salicylate intoxication appears to be closely associated with the development of cerebral edema of uncertain cause. The encephalopathy that ensues appears to be directly related to increased intracranial pressure, a known effect of prostaglandin synthesis inhibitors; it responds to mannitol (110).

Drug–Drug Interactions

Alcohol

Although ethanol itself has no effect on bleeding time, it enhances the effect of aspirin when given simultaneously or up to at least 36 hours after aspirin ingestion (111). Ethanol also promotes gastric bleeding.

The FDA has announced its intention to require alcohol warnings on all over-the-counter pain medications that contain acetylsalicylic acid, salicylates, paracetamol, ibuprofen, ketoprofen, or naproxen. The proposed warnings are aimed at alerting consumers to the specific risks incurred from heavy alcohol consumption and its interaction with analgesics. For products that contain paracetamol, the warning indicates the risks of liver damage in those who drink more than three alcoholic beverages a day. For formulations that contain salicylates or the mentioned NSAIDs, three or more alcoholic beverages will increase the risk of stomach bleeding (112).

Angiotensin-converting enzyme (ACE) inhibitors

Many large, prospective, randomized studies have shown that aspirin and ACE inhibitors reduce the risk of death and major adverse cardiovascular events in patients who have left ventricular dysfunction with or without congestive heart failure. Thus, both types of drugs are often taken concomitantly.

There is a controversy arose about whether there is a risk of a negative interaction between ACE inhibitors and COX inhibitors, in particular aspirin.

It is important to understand the theoretic basis for this potential interaction. ACE not only converts angiotensin I to angiotensin II, but it is also responsible for the degradations of kinins; thus, ACE inhibitors can increase bradykinin concentrations. Bradykinin, a potent vasodilator, activates endothelial β_2-kinin receptors, which promote the formation of vasodilatory prostaglandins through the action of phospholipase A2 and cyclo-oxygenase (COX). ACE inhibitors reduce arterial blood pressure by reducing angiotensin II production and increasing the vasodilators bradykinin, PGI_2, and PGE_3. Some investigators have suggested that aspirin (and other NSAIDs) blunt the blood pressure lowering effects of ACE inhibitors by inhibiting the production of vasodilatory prostaglandins. Others have suggested that aspirin causes reduced synthesis of renal PGE_2, which might augment unwanted ACE inhibitor-induced impairment of renal function, resulting in increased retention of sodium and water. For example, in 31 sodium-restricted patients with essential hypertension both aspirin and indomethacin blocked the increased in concentration of a PGE metabolite in response to captopril and blunted the depressor response (113). Consequently, it has been postulated that the beneficial effects of ACE inhibitors might be reduced in patients taking concomitant aspirin.

All of the studies of the clinical relevance of this possible interaction were post-hoc analyses or retrospective cohort studies of large trials of ACE inhibitors, and these studies have given different results. Some of them have shown possible interactions (114, 115), while others have given conflicting results (116,117) or have not supported the hypothesis that aspirin has a negative effect on survival in patients taking ACE inhibitors (118,119,120,121).

A systematic review and two retrospective studies have provided more information on this topic.

The systematic review assessed the effects of ACE inhibitors in patients with or without aspirin use at baseline (122). Individual patient data were collected on 22 060 patients from six long-term, randomized, placebo-controlled studies of ACE inhibitors (123,124,125,126,127,128) each of which included more than 1000 patients. The results from all of the trials, except SOLVD, did not suggest any significant differences between the proportional reductions in risk with ACE inhibitors in the presence or absence of aspirin for the major clinical outcomes (death; myocardial infarction and reinfarction; stroke; hospital admission for congestive heart failure; revascularization; and a combination of major vascular events) or in the risk of any of its individual components, except myocardial infarction. Overall ACE inhibitors significantly reduced the risk of the major clinical outcomes by 22% with clear reductions in risk among those taking aspirin at baseline (OR 0.80; 99% CI = 0.73, 0.88) and those who were not (OR 0.71; 99% CI = 0.62, 0.81).

Considering the totality of evidence on all major vascular outcomes in these studies, there is only weak evidence of any reduction in the benefit of ACE inhibitor therapy when added to aspirin. On the other hand, there is strong evidence of clinically important benefits with respect to these major clinical outcomes with ACE inhibitors, irrespective of whether aspirin is used concomitantly.

The authors of this meta-analysis concluded that at least some of the differences in the effects of ACE inhibitors on outcomes in SOLVD (129) among patients taking aspirin, compared with those who were not, might have suggested differences in the effects of ACE inhibitors in different types of patients rather than an interaction between ACE inhibitor and aspirin.

Evidence that aspirin does not interact with ACE inhibitors has come from two retrospective studies.

The first was a retrospective analysis of 755 stable patients with left ventricular systolic dysfunction and congestive heart failure, 92% of whom were taking ACE inhibitors (130). Compared with previous retrospective trials this study had some specific favorable features. It was a single-center study with the same kind of management used for all patients (including diagnostic procedures), all the patients had congestive heart failure related to left ventricular systolic dysfunction, and treatment (including aspirin and its dosage) was precisely recorded at entry. The mean dose of aspirin at entry was 183 mg/day and 74% of the patients took under 200 mg/day. Using a Cox regression model there were no interactions among aspirin, ACE inhibitors, and survival in the overall population or in subgroups of patient with ischemic or non-ischemic cardiomyopathies. Therefore, small doses of aspirin did not affect survival in patients with stable congestive heart failure taking ACE inhibitors.

The importance of the dose of aspirin was confirmed in the second study, a retrospective analysis of 344 patients taking ACE inhibitors admitted to hospital for congestive heart failure, in whom information was available about aspirin therapy during a follow-up period of 37 months (131). Cox proportional hazards regression analysis showed that the combination of high dose aspirin (325 mg/day and over) with an ACE inhibitor was independently associated with the risk of death, but that the combination with low-dose aspirin (under 160 mg/day) was not.

The results of these two studies must be interpreted with caution. Not only do they have the limitations common to cohort studies, including their retrospective nature and lack of randomization, but they were also small and biased by potential confounders related to patient characteristics.

However, taken together, the evidence for a significant interaction between low-dose aspirin and ACE inhibitors in patient with congestive heart failure is probably negligible and all patients should receive low-dose aspirin together with full-dose ACE inhibition if both are needed.

Anticoagulants

The effects on coagulation are additive if aspirin is used concurrently with anticoagulants. There are also other interaction mechanisms: the effect of the coumarins is temporarily increased by protein binding displacement, and if aspirin causes gastric hemorrhages, the latter may well be more severe when anticoagulants are being given.

Aspirin should therefore generally be avoided in patients adequately treated with anticoagulants. The most relevant information on hemorrhagic complications occurring during prophylaxis with antiplatelet drugs, whether used singly or in combination, has been provided by well-controlled prospective trials with aspirin (17), aspirin combined with dipyridamole (132), or aspirin compared with oral anticoagulants (133).

Antihypertensive drugs

An increase in mean supine blood pressure has been reported with aspirin (SEDA-19, 92). Aspirin may therefore interfere with antihypertensive pharmacotherapy, warranting caution, especially in the elderly.

Carbonic anhydrase inhibitors

In two children, aspirin potentiated the slight metabolic acidosis induced by carbonic anhydrase inhibitors (SEDA-9, 79) (134).

Clopidogrel

The combination of aspirin with clopidogrel can increase the risk of bleeding (135).

- A 76-year-old man with a history of myocardial infarction and unstable angina developed spontaneous hemarthrosis in his knee 2 weeks after starting to take clopidogrel 75 mg/day and aspirin 100 mg/day. He suddenly developed pain in the right knee while resting in bed. There was massive swelling, tenderness, and an intra-articular effusion; an X-ray showed osteoarthritis. Hemorrhagic fluid was aspirated. His coagulation status was normal. Treatment was withdrawn and recovery was uneventful.

In the CAPRIE study clopidogrel was superior to aspirin in patients with previous manifestations of atherothrombotic disease, and its benefit was amplified in some high-risk subgroups of patients (136). To assess whether the addition of aspirin to clopidogrel could have a greater benefit than clopidogrel alone in preventing vascular events with a potentially higher bleeding risk, patients who had recently had an ischemic stroke or a transient ischemic attack and were taking clopidogrel 75 mg/day, were randomized to receive additional aspirin 75 mg/day (n = 3797) or placebo (n = 3802) for 18 months (137). The primary end-point was a composite of ischemic stroke, myocardial infarction, vascular death, or rehospitalization for acute ischemia. Aspirin was associated with a small non-significant reduction in the risk of the primary end-point (relative risk reduction of 6.4%; CI = –4.6, 16); absolute risk reduction 1.1%; (CI = –0.6, 2.7). However, the incidence of life-threatening bleeding was higher with aspirin than placebo: 96 (2.6%) versus 49 (1.3%). The absolute increase in risk was 1.3% (CI = –0.6, 1.9) as was the incidence of major bleeding. These possibly increased risks might therefore offset any beneficial effect of adding aspirin to clopidogrel treatment in these patients.

Case reports documenting an increased risk of bleeding with the concomitant use of clopidogrel with aspirin in perioperative setting have been published (138).

Glucocorticoids

The effects of aspirin on gastrointestinal mucosa will lead to additive effects if it is used concurrently with other drugs that have an irritant effect on the stomach, notably other NSAIDs or glucocorticoids (139,140).

Heparin

Risk factors for heparin-induced bleeding include concomitant use of aspirin (141).

Intrauterine contraceptive devices

The supposed mechanisms of action of intrauterine contraceptive devices (IUCDs) include a local inflammatory response and increased local production of prostaglandins that prevent sperm from fertilizing ova (142,143). As aspirin has both anti-inflammatory and antiprostaglandin properties, the contraceptive effectiveness of an IUCD can be reduced by the drug, although the effect on periodic bleeding may prevail.

Methotrexate

Aspirin displaces methotrexate from its binding sites and also inhibits its renal tubular elimination, so that the dosage of concurrently used methotrexate should be reduced (except once-a-week low-dose treatment in rheumatoid arthritis) (144).

Nitrates

Aspirin in low dosages (under 300 mg/day) is widely used in cardiovascular prophylaxis, but its use is accompanied by an increased risk of gastrointestinal bleeding (SEDA-21, 100). Of particular interest therefore are data from a retrospective case-control study showing that nitrate therapy may reduce the risk of aspirin-induced gastrointestinal bleeding (145). As nitrates are often used in the same population of patients, such data merit further confirmation from larger prospective studies.

NSAIDs

The effects of aspirin on gastrointestinal mucosa will lead to additive effects if it is used concurrently with other drugs that have an irritant effect on the stomach, notably other NSAIDs or glucocorticoids (139,140). Salicylates can be displaced from binding sites by some NSAIDs such as naproxen, or in turn displace others such as piroxicam.

Sodium valproate

Aspirin displaces sodium valproate from protein binding sites (146) and reduces its hepatic metabolism (147).

Streptokinase

Major hemorrhagic complications, including cerebral hemorrhage, can occur with aspirin (SEDA-23, 116) and the same is also true for thrombolytic therapy of acute ischemic stroke (148). A post hoc analysis of the Multicenter Acute Stroke Trial in Italy showed a negative interaction of aspirin and streptokinase in acute ischemic stroke (149). In 156 patients who received streptokinase plus aspirin and 157 patients treated with streptokinase alone, the combined regimen significantly increased early case fatality at days 3–10 (53 versus 30; OR = 2.5; CI = 1.2, 3.6). The excess in deaths was solely due to treatment and was not explained by the main prognostic predictors. Deaths in the combination group were mainly cerebral (42 versus 24; OR = 2.0; CI = 1.3, 3.7) and associated with hemorrhagic transformation (22 versus 11; OR = 2.2; CI = 1.0, 5.0). The data suggest that aspirin should be avoided when thrombolytic agents are used for acute ischemic stroke.

Uricosuric drugs

In low dosages (up to 2 g/day), aspirin reduces urate excretion and blocks the effects of probenecid and other uricosuric agents (150). However, in 11 patients with gout, aspirin 325 mg/day had no effect on the uricosuric action of probenecid (151). In higher dosages (over 5 g/day), salicylates increase urate excretion and inhibit the effects of spironolactone, but it is not clear that these phenomena are of importance.

Food-Drug Interactions

Food allergens

Aspirin seems to potentiate the effects of food allergens, but this is uncertain (SEDA-10, 72).

Interference with Diagnostic Tests

Thyroid function tests

Through competitive binding to thyroid-binding globulin, salicylates in high concentrations can displace thyroxine and triiodothyronine, thus interfering with the results of diagnostic thyroid function tests (152).

Management of adverse drug reactions

Desensitization was attempted in 16 patients with acute coronary artery disease and a history of aspirin hypersensitivity (of whom three had a history of angioedema) in a protocol that lasted a few hours (153). None received pretreatment with antihistamines or glucocorticoids, and beta-blockers were withheld. The first seven received eight oral doses of aspirin, starting at 1 mg and doubling each 30 minutes; the next nine patients underwent a shorter version using five doses (5, 10, 20, 40, and 75 mg). The patients were monitored in the coronary care unit; blood pressure, pulse, and peak expiratory flow were measured every 30 minutes, and cutaneous, naso-ocular, and pulmonary reactions were monitored closely until 3 hours after the procedure. Immediate tolerance was obtained in 14 patients, all of whom continued treatment uneventfully. One patient developed angioedema 3 hours after the procedure, which resolved immediately with a glucocorticoid and adrenaline. The patient was rechallenged successfully 2 days later and continued to take aspirin. Another patient, who had had a severe recent attack of asthma, developed nasal swelling and shortness of breath 1 hour after the last dose; although the symptoms resolved rapidly with inhaled salbutamol, rechallenge was not attempted. In 11 patients who then underwent coronary stenting aspirin + clopidogrel was given for 9–12 months; four were treated with aspirin alone. There were no major adverse cardiac events or new revascularization during a median follow-up of 14 months (range 1–35).

References

1. MacIntyre A, Gray JD, Gorelick M, Renton K. Salicylate and acetaminophen in donated blood. CMAJ 1986;135(3):215–6.
2. Wynne HA, Long A. Patient awareness of the adverse effects of non-steroidal anti-inflammatory drugs (NSAIDs). Br J Clin Pharmacol 1996;42(2):253–6.
3. National Drugs Advisory Board. Availability of aspirin and paracetamol. Annual Report 1987;24.
4. Mitchell JA, Akarasereenont P, Thiemermann C, Flower RJ, Vane JR. Selectivity of nonsteroidal antiinflammatory drugs as inhibitors of constitutive and inducible cyclooxygenase. Proc Natl Acad Sci USA 1993;90(24):11693–7.
5. Antiplatelet Trialists' Collaboration. Secondary prevention of vascular disease by prolonged antiplatelet treatment. BMJ (Clin Res Ed) 1988;296(6618):320–31.
6. Hennekens CH, Buring JE, Sandercock P, Collins R, Peto R. Aspirin and other antiplatelet agents in the secondary and primary prevention of cardiovascular disease. Circulation 1989;80(4):749–56.
7. Macdonald R. Aspirin and extradural blocks. Br J Anaesth 1991;66(1):1–3.
8. de Swiet M, Redman CW. Aspirin, extradural anaesthesia and the MRC Collaborative Low-dose Aspirin Study in Pregnancy (CLASP). Br J Anaesth 1992;69(1):109–10.
9. Steering Committee of the Physicians' Health Study Research Group. Final report on the aspirin component of the ongoing Physicians' Health Study. N Engl J Med 1989;321(3):129–35.
10. Peto R, Gray R, Collins R, Wheatley K, Hennekens C, Jamrozik K, Warlow C, Hafner B, Thompson E, Norton S, et al. Randomised trial of prophylactic daily aspirin in British male doctors. BMJ (Clin Res Ed) 1988;296(6618):313–6.
11. Stafford RS. Aspirin use is low among United States outpatients with coronary artery disease. Circulation 2000;101(10):1097–101.
12. Hayden M, Pignone M, Phillips C, Mulrow C. Aspirin for the primary prevention of cardiovascular events: a summary of the evidence for the U.S. Preventive Services Task Force Ann Intern Med 2002;136(2):161–72.
13. Weisman SM, Graham DY. Evaluation of the benefits and risks of low-dose aspirin in the secondary prevention of cardiovascular and cerebrovascular events. Arch Intern Med 2002;162(19):2197–202.
14. Roderick PJ, Wilkes HC, Meade TW. The gastrointestinal toxicity of aspirin: an overview of randomised controlled trials. Br J Clin Pharmacol 1993;35(3):219–26.
15. Derry S, Loke YK. Risk of gastrointestinal haemorrhage with long term use of aspirin: meta-analysis. BMJ 2000;321(7270):1183–7.
16. Antithrombotic Trialists' Collaboration. Collaborative meta-analysis of randomised trials of antiplatelet therapy for prevention of death, myocardial infarction, and stroke in high risk patients. BMJ 2002;324(7329):71–86.
17. Aspirin Myocardial Infarction Study Research Group. A randomized, controlled trial of aspirin in persons recovered from myocardial infarction. JAMA 1980;243(7):661–9.
18. Habbab MA, Szwed SA, Haft JI. Is coronary arterial spasm part of the aspirin-induced asthma syndrome? Chest 1986;90(1):141–3.
19. Hermida RC, Ayala DE, Calvo C, Lopez JE, Fernandez JR, Mojon A, Dominguez MJ, Covelo M. Administration

time-dependent effects of aspirin on blood pressure in untreated hypertensive patients. Hypertension 2003;41:1259–67.

20. Heffner JE, Sahn SA. Salicylate-induced pulmonary edema. Clinical features and prognosis. Ann Intern Med 1981;95(4):405–9.
21. Grabe DW, Manley HJ, Kim JS, McGoldrick MD, Bailie GR. Respiratory distress caused by salicylism confirmed by lung biopsy. Clin Drug Invest 1999;17:79–81.
22. The Dutch TIA Trial Study Group. A comparison of two doses of aspirin (30 mg vs. 283 mg a day) in patients after a transient ischemic attack or minor ischemic stroke N Engl J Med 1991;325(18):1261–6.
23. The SALT Collaborative Group. Swedish Aspirin Low-Dose Trial (SALT) of 75 mg aspirin as secondary prophylaxis after cerebrovascular ischaemic events Lancet 1991;338(8779):1345–9.
24. Steering Committee of the Physicians' Health Study Research Group. Final report on the aspirin component of the ongoing Physicians' Health Study. N Engl J Med 1989;321(3):129–35.
25. Rohr WD. Transitorische Myopisierung und Drucksteigerung als Medikamentennebenwirkung. [Transitory myopia and increased ocular pressure as side effects of drugs.] Fortschr Ophthalmol 1984;81(2):199–200.
26. Strupp M, Jahn K, Brandt T. Another adverse effect of aspirin: bilateral vestibulopathy. J Neurol Neurosurg Psychiatry 2003;74:691.
27. Prince RL, Larkins RG, Alford FP. The effect of acetylsalicylic acid on plasma glucose and the response of glucose regulatory hormones to intravenous glucose and arginine in insulin treated diabetics and normal subjects. Metabolism 1981;30(3):293–8.
28. Manfredini R, Ricci L, Giganti M, La Cecilia O, Kuwornu Afi H, Chierici F, Gallerani M. An uncommon case of fluid retention simulating a congestive heart failure after aspirin consumption. Am J Med Sci 2000;320(1):72–4.
29. Necheles TF, Steinberg MH, Cameron D. Erythrocyte glutathione-peroxidase deficiency. Br J Haematol 1970;19(5):605–12.
30. Meloni T, Forteleoni G, Ogana A, Franca V. Aspirin-induced acute haemolytic anaemia in glucose-6-phosphate dehydrogenase-deficient children with systemic arthritis. Acta Haematol 1989;81(4):208–9.
31. Levy M, Heyman A. Hematological adverse effects of analgesic anti-inflammatory drugs. Hematol Rev 1990;4:177.
32. Williams JO, Mengel CE, Sullivan LW, Haq AS. Megaloblastic anemia associated with chronic ingestion of an analgesic. N Engl J Med 1969;280(6):312–3.
33. Kingham JD, Chen MC, Levy MH. Macular hemorrhage in the aging eye: the effects of anticoagulants. N Engl J Med 1988;318(17):1126–7.
34. Werblin TP, Peiffer RL. Persistent hemorrhage after extracapsular surgery associated with excessive aspirin ingestion. Am J Ophthalmol 1987;104(4):426.
35. Hanley SP, Bevan J, Cockbill SR, Heptinstall S. Differential inhibition by low-dose aspirin of human venous prostacyclin synthesis and platelet thromboxane synthesis. Lancet 1981;1(8227):969–71.
36. The Canadian Cooperative Study Group. A randomized trial of aspirin and sulfinpyrazone in threatened stroke. N Engl J Med 1978;299(2):53–9.
37. ISIS-2 (Second International Study of Infarct Survival) Collaborative Group. Randomised trial of intravenous streptokinase, oral aspirin, both, or neither among 17,187 cases of suspected acute myocardial infarction: ISIS-2. Lancet 1988;2(8607):349–60.
38. The RISC Group. Risk of myocardial infarction and death during treatment with low dose aspirin and intravenous heparin in men with unstable coronary artery disease. Lancet 1990;336(8719):827–30.
39. Blower AL, Brooks A, Fenn GC, Hill A, Pearce MY, Morant S, Bardhan KD. Emergency admissions for upper gastrointestinal disease and their relation to NSAID use. Aliment Pharmacol Ther 1997;11(2):283–91.
40. Piper DW, McIntosh JH, Ariotti DE, Fenton BH, MacLennan R. Analgesic ingestion and chronic peptic ulcer. Gastroenterology 1981;80(3):427–32.
41. Petroski D. Endoscopic comparison of various aspirin preparations-gastric mucosal adaptability to aspirin restudied. Curr Ther Res 1989;45:945.
42. Szabo S. Pathogenesis of gastric mucosal injury. S Afr Med J 1988;74(Suppl):35.
43. Faulkner G, Prichard P, Somerville K, Langman MJ. Aspirin and bleeding peptic ulcers in the elderly. BMJ 1988;297(6659):1311–3.
44. Freeland GR, Northington RS, Hedrich DA, Walker BR. Hepatic safety of two analgesics used over the counter: ibuprofen and aspirin. Clin Pharmacol Ther 1988;43(5):473–9.
45. Hansson L, Zanchetti A, Carruthers SG, Dahlof B, Elmfeldt D, Julius S, Menard J, Rahn KH, Wedel H, Westerling S. Effects of intensive blood-pressure lowering and low-dose aspirin in patients with hypertension: principal results of the Hypertension Optimal Treatment (HOT) randomised trial. HOT Study Group. Lancet 1998;351(9118):1755–62.
46. Meade TW, Brennan PJ, Wilkes HC, Zuhrie SR. Thrombosis prevention trial: randomised trial of low-intensity oral anticoagulation with warfarin and low-dose aspirin in the primary prevention of ischaemic heart disease in men at increased risk. The Medical Research Council's General Practice Research Framework. Lancet 1998;351(9098):233–41.
47. Petty GW, Brown RD Jr, Whisnant JP, Sicks JD, O'Fallon WM, Wiebers DO. Frequency of major complications of aspirin, warfarin, and intravenous heparin for secondary stroke prevention. A population-based study. Ann Intern Med 1999;130(1):14–22.
48. Sorensen HT, Mellemkjaer L, Blot WJ, Nielsen GL, Steffensen FH, McLaughlin JK, Olsen JH. Risk of upper gastrointestinal bleeding associated with use of low-dose aspirin. Am J Gastroenterol 2000;95(9):2218–24.
49. Fisher M, Knappertz V. Comments in response to "Analysis of risk of bleeding complications after different doses of aspirin in 192,036 patients enrolled in 31 randomised controlled trials". Am J Cardiol 2005;96(10):1467.
50. Laine L, McQuaid K. Bleeding complications related to aspirin dose. Am J Cardiol 2005;96(7):1035–6.
51. Tramèr MR, Moore RA, Reynolds DJ, McQuay HJ. Quantitative estimation of rare adverse events which follow a biological progression: a new model applied to chronic NSAID use. Pain 2000;85(1–2):169–82.
52. Weil J, Langman MJ, Wainwright P, Lawson DH, Rawlins M, Logan RF, Brown TP, Vessey MP, Murphy M, Colin-Jones DG. Peptic ulcer bleeding: accessory risk factors and interactions with non-steroidal anti-inflammatory drugs. Gut 2000;46(1):27–31.
53. Tramèr MR. Aspirin, like all other drugs, is a poison. BMJ 2000;321(7270):1170–1.

54. Garcia Rodriguez LA, Hernandez-Diaz S, de Abajo FJ. Association between aspirin and upper gastrointestinal complications: systematic review of epidemiologic studies. Br J Clin Pharmacol 2001;52(5):563–71.
55. Bianchine JR, Procter RR, Thomas FB. Piroxicam, aspirin, and gastrointestinal blood loss. Clin Pharmacol Ther 1982;32(2):247–52.
56. Hass WK, Easton JD, Adams HP Jr, Pryse-Phillips W, Molony BA, Anderson S, Kamm B. A randomized trial comparing ticlopidine hydrochloride with aspirin for the prevention of stroke in high-risk patients. Ticlopidine Aspirin Stroke Study Group. N Engl J Med 1989;321(8):501–7.
57. Lanas A, Serrano P, Bajador E, Esteva F, Benito R, Sainz R. Evidence of aspirin use in both upper and lower gastrointestinal perforation. Gastroenterology 1997;112(3):683–9.
58. Bonavina L, DeMeester TR, McChesney L, Schwizer W, Albertucci M, Bailey RT. Drug-induced esophageal strictures. Ann Surg 1987;206(2):173–83.
59. Schreiber JB, Covington JA. Aspirin-induced esophageal hemorrhage. JAMA 1988;259(11):1647–8.
60. Barrier CH, Hirschowitz BI. Controversies in the detection and management of nonsteroidal antiinflammatory drug-induced side effects of the upper gastrointestinal tract. Arthritis Rheum 1989;32(7):926–32.
61. Suazo-Barahona J, Gallegos J, Carmona-Sanchez R, Martinez R, Robles-Diaz G. Nonsteroidal anti-inflammatory drugs and gastrocolic fistula. J Clin Gastroenterol 1998;26(4):343–5.
62. Gutnik SH, Willmott D, Ziebarth J. Gastrocolic fistula-secondary to aspirin abuse. S D J Med 1993;46(10):358–60.
63. Levine MS, Kelly MR, Laufer I, Rubesin SE, Herlinger H. Gastrocolic fistulas: the increasing role of aspirin. Radiology 1993;187(2):359–61.
64. Mielants H, Verbruggen G, Schelstraete K, Veys EM. Salicylate-induced gastrointestinal bleeding: comparison between soluble buffered, enteric-coated, and intravenous administration. J Rheumatol 1979;6(2):210–8.
65. Malfertheiner P, Stanescu A, Rogatti W, Ditschuneit H. Effects of microencapsulated vs. enteric-coated acetylsalicylic acid on gastric and duodenal mucosa: an endoscopic study. J Clin Gastroenterol 1988;10(3):269–72.
66. Graham D, Chan F. Endoscopic ulcers with low-dose aspirin and reality testing. Gastroenterology 2005;128(3):807.
67. Singh G, Ramey DR, Morfeld D, Shi H, Hatoum HT, Fries JF. Gastrointestinal tract complications of nonsteroidal anti-inflammatory drug treatment in rheumatoid arthritis. A prospective observational cohort study. Arch Intern Med 1996;156(14):1530–6.
68. Ward MR. Reye's syndrome: an update. Nurse Pract 1997;22(12):45–649–50, 52–3.
69. Food and Drug Administration, HHS. Labeling for oral and rectal over-the-couter drug products containing aspirin and nonaspirin salicylates; Reye's Syndrome warning. Final rule. Fed Regist 2003;68(74):18861–9.
70. Zimmerman HJ. Effects of aspirin and acetaminophen on the liver. Arch Intern Med 1981;141(3 Spec No):333–42.
71. Gitlin N. Salicylate hepatotoxicity: the potential role of hypoalbuminemia. J Clin Gastroenterol 1980;2(3):281–5.
72. Belay ED, Bresee JS, Holman RC, Khan AS, Shahriari A, Schonberger LB. Reye's syndrome in the United States from 1981 through 1997. N Engl J Med 1999;340(18):1377–82.
73. Langford NJ. Aspirin and Reye's syndrome: is the response appropriate? J Clin Pharm Ther 2002;27(3): 157–60.
74. Baron JA. What now for aspirin and cancer prevention? J Natl Cancer Inst 2004;96:22–8.
75. Schernhammer ES, Kang JH, Chan AT, Michaud DS, Skinner HG, Giovannucci E, Colditz GA, Fuchs CS. A prospective study of aspirin use and the risk of pancreatic cancer in women. J Natl Cancer Inst 2004;96:22–8.
76. Ibáñez L, Morlans M, Vidal X, Martínez MJ, Laporte J-R. Case-control study of regular analgesic and nonsteroidal anti-inflammatory use and end-stage renal disease. Kidney Int 2005;67:2393–8.
77. Plotz PH, Kimberly RP. Acute effects of aspirin and acetaminophen on renal function. Arch Intern Med 1981;141(3 Spec No):343–8.
78. Delzell E, Shapiro S. A review of epidemiologic studies of nonnarcotic analgesics and chronic renal disease. Medicine (Baltimore) 1998;77(2):102–21.
79. McLaughlin JK, Lipworth L, Chow WH, Blot WJ. Analgesic use and chronic renal failure: a critical review of the epidemiologic literature. Kidney Int 1998;54(3): 679–86.
80. Rexrode KM, Buring JE, Glynn RJ, Stampfer MJ, Youngman LD, Gaziano JM. Analgesic use and renal function in men. JAMA 2001;286(3):315–21.
81. Fored CM, Ejerblad E, Lindblad P, Fryzek JP, Dickman PW, Signorello LB, Lipworth L, Elinder CG, Blot WJ, McLaughlin JK, Zack MM, Nyren O. Acetaminophen, aspirin, and chronic renal failure. N Engl J Med 2001;345(25):1801–8.
82. Kanwar AJ, Belhaj MS, Bharija SC, Mohammed M. Drugs causing fixed eruptions. J Dermatol 1984;11(4):383–5.
83. Brandt KD, Palmoski MJ. Effects of salicylates and other nonsteroidal anti-inflammatory drugs on articular cartilage. Am J Med 1984;77(1A):65–9.
84. Oates JA, FitzGerald GA, Branch RA, Jackson EK, Knapp HR, Roberts LJ 2nd. Clinical implications of prostaglandin and thromboxane A2 formation (1). N Engl J Med 1988;319(11):689–98.
85. Settipane RA, Constantine HP, Settipane GA. Aspirin intolerance and recurrent urticaria in normal adults and children. Epidemiology and review. Allergy 1980;35(2):149–54.
86. Farr RS, Spector SL, Wangaard CH. Evaluation of aspirin and tartrazine idiosyncrasy. J Allergy Clin Immunol 1979;64(6 pt 2):667–8.
87. Szczeklik A. The cyclooxygenase theory of aspirin-induced asthma. Eur Respir J 1990;3(5):588–93.
88. Szczeklik A. Aspirin-induced asthma: pathogenesis and clinical presentation. Allergy Proc 1992;13(4):163–73.
89. Kowalski ML, Ptasinska A, Bienkiewicz B, Pawliczak R, DuBuske L, Differential effects of aspirin and misoprostol on 15-hydroxyeicosatetraenoic acid generation by leukocytes from aspirin-sensitive asthmatic patients. J Allergy Clin Immunol 2003;112:505–12.
90. Pierzchalska M, Szabo Z, Sanak M, Soja J, Szczeklik A. Deficient prostaglandin E2 production by bronchial fibroblasts of asthmatic patients, with special reference to aspirin-induced asthma. J Allergy Clin Immunol 2003;111:1041–8.
91. Gyllfors P, Bochenek G, Overholt J, Drupka D, Kumlin M, Sheller J, Nizankowska E, Isakson PC, Mejza F, Lefkowith JB, Dahlen SE, Szczeklik A, Murray JJ, Dahlen B. Biochemical and clinical evidence that aspirin-intolerant asthmatic subjects tolerate the cyclooxygenase

2-selective analgetic drug celecoxib. J Allergy Clin Immunol 2003;111:1116–21.

92. Sola Alberich R, Jammoul A, Masana L. Henoch-Schonlein purpura associated with acetylsalicylic acid. Ann Intern Med 1997;126(8):665.
93. Martelli NA. Bronchial and intravenous provocation tests with indomethacin in aspirin-sensitive asthmatics. Am Rev Respir Dis 1979;120(5):1073–9.
94. Ritter JM, Taylor GW. Fish oil in asthma. Thorax 1988;43(2):81–3.
95. Anonymous. Aspirin sensitivity in asthmatics. BMJ 1980;281(6246):958–9.
96. Pleskow WW, Stevenson DD, Mathison DA, Simon RA, Schatz M, Zeiger RS. Aspirin desensitization in aspirin-sensitive asthmatic patients: clinical manifestations and characterization of the refractory period. J Allergy Clin Immunol 1982;69(1 Pt 1):11–9.
97. Dubach UC, Rosner B, Pfister E. Epidemiologic study of abuse of analgesics containing phenacetin. Renal morbidity and mortality (1968-1979). N Engl J Med 1983;308(7):357–62.
98. Dubach UC, Rosner B, Sturmer T. An epidemiologic study of abuse of analgesic drugs. Effects of phenacetin and salicylate on mortality and cardiovascular morbidity (1968 to 1987). N Engl J Med 1991;324(3):155–60.
99. Gago-Dominguez M, Yuan JM, Castelao JE, Ross RK, Yu MC. Regular use of analgesics is a risk factor for renal cell carcinoma. Br J Cancer 1999;81(3):542–8.
100. Slone D, Siskind V, Heinonen OP, Monson RR, Kaufman DW, Shapiro S. Aspirin and congenital malformations. Lancet 1976;1(7974):1373–5.
101. Werler MM, Mitchell AA, Shapiro S. The relation of aspirin use during the first trimester of pregnancy to congenital cardiac defects. N Engl J Med 1989;321(24): 1639–1642.
102. Rumack CM, Guggenheim MA, Rumack BH, Peterson RG, Johnson ML, Braithwaite WR. Neonatal intracranial hemorrhage and maternal use of aspirin. Obstet Gynecol 1981;58(Suppl 5):S52–6.
103. Shapiro S, Siskind V, Monson RR, Heinonen OP, Kaufman DW, Slone D. Perinatal mortality and birthweight in relation to aspirin taken during pregnancy. Lancet 1976;1(7974):1375–6.
104. Uzan S, Beaufils M, Breart G, Bazin B, Capitant C, Paris J. Prevention of fetal growth retardation with low-dose aspirin: findings of the EPREDA trial. Lancet 1991;337(8755):1427–31.
105. Anonymous. Poisoning with enteric-coated aspirin. Lancet 1981;2(8238):130.
106. Pierce RP, Gazewood J, Blake RL Jr. Salicylate poisoning from enteric-coated aspirin. Delayed absorption may complicate management. Postgrad Med 1991;89(5):61–4.
107. Temple AR. Acute and chronic effects of aspirin toxicity and their treatment. Arch Intern Med 1981;141(3 Spec No):364–9.
108. Pena-Alonso YR, Montoya-Cabrera MA, Bustos-Cordoba E, Marroquin-Yanez L, Olivar-Lopez V. Aspirin intoxication in a child associated with myocardial necrosis: is a drug-related lesion? Pediatr Dev Pathol 2003;6:342–7.
109. Meredith TJ, Vale JA. Non-narcotic analgesics. Problems of overdosage. Drugs 1986;32(Suppl 4):177–205.
110. Dove DJ, Jones T. Delayed coma associated with salicylate intoxication. J Pediatr 1982;100(3):493–6.
111. Deykin D, Janson P, McMahon L. Ethanol potentiation of aspirin-induced prolongation of the bleeding time. N Engl J Med 1982;306(14):852–4.
112. Anonymous. Alcohol warning on over-the-counter pain medications. WHO Drug Inf 1998;12:16.
113. Moore TJ, Crantz FR, Hollenberg NK, Koletsky RJ, Leboff MS, Swartz SL, Levine L, Podolsky S, Dluhy RG, Williams GH. Contribution of prostaglandins to the antihypertensive action of captopril in essential hypertension. Hypertension 1981;3(2):168–73.
114. Al-Khadra AS, Salem DN, Rand WM, Udelson JE, Smith JJ, Konstam MA. Antiplatelet agents and survival: a cohort analysis from the Studies on Ventricular Dysfunction (SOLVD) trial. J Am Coll Cardiol 1998; 31: 419–25.
115. Nguyen KN, Aursnes I, Kjekshus J. Interaction between enalapril and aspirin on mortality after acute myocardial infarction: subgroup analysis of the Cooperative New Scandinavian Enalapril Survival Study II (CONSENSUS II). Am J Cardiol 1997;79:115–9.
116. Baur LH, Schipperheyn JJ, Van den Laarse A, Souverijin JH, Frolich M, De Groot Voogd PJ, Vroom TF, Cats VM, Keirse MJ, Bruschke AVG. Combining salicylate and enalapril in patients with coronary artery disease and heart failure. Br Heart J. 1995;73:227–36.
117. Van Wijngaarden J, Smit AJ, De Graeff PA, Van Glist WH, Van der Broek SA, Van Veldhuisen DJ, Lie KI, Wesseling H. Effects of acetylsalicylic acid on peripheral hemodynamics in patients chronic heart failure treated with angiotensin-converting enzyme inhibitors. J Cardiovasc Pharmacol 1994;23:240–5.
118. Oosterga M, Anthonio RL, de Kam PJ, Kingma JH, Crijns HJ, Van Gilst WH. Effects of aspirin on angiotensin-converting enzyme inhibition and left ventricular dilatation one year after acute myocardial infarction. Am J Cardiol 1998;81:1178–81.
119. Leor J, Reicher-Reiss H, Goldbourt U, Boyko V, Gottlieb S, Battler A, Behar S. Aspirin and mortality in patients treated with angiotensin-converting enzyme inhibitors: a cohort study of 11,575 patients with coronary artery disease. J Am Coll Cardiol 2000;35:817–9.
120. Latini R, Tognoni G, Maggioni AP, Baigent C, Braunwald E, Chen ZM, Collins R, Flather M, Franzosi MG, Kjekashus J, Kober L, Liu LS, Peto R, Pfeffer M, Pizzetti F, Santoro E, Sleight P, Swedberg K, Tavazzi L, Wang W, Yusuf S. Clinical effects of early angiotensin-converting enzyme inhibitor treatment for acute myocardial infarction are similar in the presence and absence of aspirin: systematic overview of individual data from 96,712 randomized patients. Angiotensin-converting Enzyme Inhibitor Myocardial Infarct Collaborative Group. J Am Coll Cardiol 2000;35:1801–7.
121. Nawarskas JJ, Spinler SA. Does aspirin interfere with the therapeutic efficacy of angiotensin-converting enzyme inhibitors in hypertension or congestive heart failure. Pharmacotherapy 1998;18:1041–52.
122. Teo KK, Yusuf S, Pfeffer M, Torp-Pedersen C, Kober L, Hall A, Pogue J, Latini R, Collins R. ACE Inhibitors Collaborative Group. Effects of long-term treatment with angiotensin-converting-enzyme inhibitors in the presence or absence of aspirin: a systematic review. Lancet 2002;360:1037–43.
123. The SOLVD investigators. Effect of enalapril on survival in patients with reduced left ventricular ejection fractions and congestive heart failure. New Engl J Med 1991;325:293–302.
124. The SOLVD investigators. Effects of enalapril on mortality and the development of heart failure in asymptomatic patients with reduced left ventricular ejection fractions. New Engl J Med 1991;327:685–91.

125. Pfeffer MA, Braunwald E, Moye LA, Basta L, Brown EJ Jr, Cuddy TE, Davis BR, Geltman EM, Goldman S, Flaker GC, Klein M, Lamas GA, Packer M, Rouleau J, Rouleau JL, Rutherford J, Wertheimer JH, Hawkins CM. Effect of captopril on mortality and morbidity in patients with left ventricular dysfunction after myocardial infarction. Results of the survival and ventricular enlargement trial. The SAVE Investigators. New Engl J Med 1992;327:669–77.
126. The Acute Infarction Ramipril Efficacy (AIRE) Study Investigators. Effect of ramipril on mortality and morbidity of survivors of acute myocardial infarction with clinical evidence of heart failure. Lancet 1993;342:821–8.
127. Kober L, Torp-Pederson C, Carlsen JE, Bagger H, Eliasen P, Lyngborg K, Videbaek J, Cole DS, Aucler L, Pauly NC. A clinical trial of the angiotensin-converting-enzyme inhibitor trandolapril in patients with left ventricular dysfunction after myocardial infarction. Trandolapril Cardiac Evaluation (TRACE) Study Group. New Engl J Med 1995;333:1670–6.
128. The Heart Outcomes Prevention Evaluation Investigators. Effect of an angiotensin converting enzyme inhibitors ramipril, on cardiovascular events in high risk patients. New Engl J Med 2000;324:145–53.
129. Langman M, Kong SX, Zhang Q, Kahler KH, Finch E. Safety and patient tolerance of standard and slow-release formulation. NSAIDs. Pharmacoepidemiol Drug Saf 2003;12:61–6.
130. Aumègeat V, Lamblin N, De Groote P, Mc Fadden EP, Millaire A, Bauters C, Lablanche JM. Aspirin does not adversely affect survival in patients with stable congestive heart failure treated with angiotensin-converting enzyme inhibitors. Chest 2003;124:1250–8.
131. Guazzi M, Brambilla R, Reina G, Tuminello G, Guazzi MD. Aspirin-angiotensin-converting enzyme inhibitor coadministration and mortality in patients with heart failure a dose-related adverse effect of aspirin. Arch Intern Med 2003;163:1574–9.
132. Diener HC, Cunha L, Forbes C, Sivenius J, Smets P, Lowenthal AEuropean Stroke Prevention Study. 2. Dipyridamole and acetylsalicylic acid in the secondary prevention of stroke. J Neurol Sci 1996;143(1-2):1–13.
133. Enquete de prevention secondaire de l'infarctus du Myocarde' Research Group. A controlled comparison of aspirin and oral anticoagulants in prevention of death after myocardial infarction. N Engl J Med 1982;307(12):701–8.
134. Cowan RA, Hartnell GG, Lowdell CP, Baird IM, Leak AM. Metabolic acidosis induced by carbonic anhydrase inhibitors and salicylates in patients with normal renal function. BMJ (Clin Res Ed) 1984;289(6441):347–8.
135. Gille J, Bernotat J, Bohm S, Behrens P, Lohr JF. Spontaneous hemarthrosis of the knee associated with clopidogrel and aspirin treatment. Z Rheumatol 2003;62:80–1.
136. CAPRIE Steering Committee. A randomised, blinded, trial of clopidogrel versus aspirin in patients at risk of ischaemic events (CAPRIE). Lancet 1996;348(9038):1329–39.
137. Diener HC, Bogousslavsky J, Brass LM, Cimminiello C, Csiba L, Kaste M, Leys D, Matias-Guiu J, Rupprecht HJ; Match investigators. Aspirin and clopidogrel compared with clopidogrel alone after recent ischaemic stroke or transient ischaemic attack in high-risk patients (MATCH): randomised, double-blind, placebo-controlled trial. Lancet 2004;364:331–7.
138. Moore M, Power M. Perioperative hemorrhage and combined clopidogrel and aspirin therapy. Anesthesiology 2004;101:792–4.
139. Brooks PM, Day RO. Nonsteroidal antiinflammatory drugs—differences and similarities. N Engl J Med 1991;324(24):1716–25.
140. McInnes GT, Brodie MJ. Drug interactions that matter. A critical reappraisal. Drugs 1988;36(1):83–110.
141. Levine MN, Raskob G, Landefeld S, Kearon C. Hemorrhagic complications of anticoagulant treatment. Chest 1998;114(Suppl 5):S511–23.
142. World Health Organization (WHO). Mechanism of action, safety and efficacy of intrauterine devices. In: Technical Report Series 753. Geneva: WHO, 1987:91.
143. Croxatto HB, Ortiz ME, Valdez E. IUD mechanisms of action. In: Bardin CW, Mishell DR, editors. Proceedings from the 4th International Conference on IUDs. Boston: Butterworth-Heinemann, 1994:44.
144. Offerhaus L. Drug interactions at excretory mechanisms. Pharmacol Ther 1981;15(1):69–78.
145. Lanas A, Bajador E, Serrano P, Arroyo M, Fuentes J, Santolaria S. Effects of nitrate and prophylactic aspirin on upper gastrointestinal bleeding: a retrospective case-control study. J Int Med Res 1998;26(3):120–8.
146. Orr JM, Abbott FS, Farrell K, Ferguson S, Sheppard I, Godolphin W. Interaction between valproic acid and aspirin in epileptic children: serum protein binding and metabolic effects. Clin Pharmacol Ther 1982;31(5):642–9.
147. Abbott FS, Kassam J, Orr JM, Farrell K. The effect of aspirin on valporic acid metabolism. Clin Pharmacol Ther 1986;40(1):94–100.
148. Multicentre Acute Stroke Trial—Italy (MAST-I) Group. Randomised controlled trial of streptokinase, aspirin, and combination of both in treatment of acute ischaemic stroke. Lancet 1995;346(8989):1509–14.
149. Ciccone A, Motto C, Aritzu E, Piana A, Candelise L. Negative interaction of aspirin and streptokinase in acute ischemic stroke: further analysis of the Multicenter Acute Stroke Trial—Italy. Cerebrovasc Dis 2000;10(1):61–4.
150. Akyol SM, Thompson M, Kerr DN. Renal function after prolonged consumption of aspirin. BMJ (Clin Res Ed) 1982;284(6316):631–2.
151. Harris M, Bryant LR, Danaher P, Alloway J. Effect of low dose daily aspirin on serum urate levels and urinary excretion in patients receiving probenecid for gouty arthritis. J Rheumatol 2000;27(12):2873–6.
152. Samuels MH, Pillote K, Ashex D, Nelson JC. Variable effects of nonsteroidal antiinflammatory agents on thyroid test results. J Clin Endocrinol Metab 2003;88(12):5710–6.
153. Silberman S, Neukirch-Stoop C, Steg PG. Rapid desensitization procedure for patients with aspirin hypersensitivity undergoing coronary stenting. Am J Cardiol 2005;95(4):509–10.

Anagrelide

General Information

Anagrelide was developed as an inhibitor of platelet aggregation and acts by inhibiting phosphodiesterase. It was was later found to reduce the platelet count, in doses lower than those required to inhibit platelet aggregation, by interfering with megakaryocyte differentiation and proliferation. It is used for the treatment of essential thrombocythemia.

Observational studies

In a long-term study of 39 young patients with essential thrombocythemia treated with anagrelide, 20 had adverse effects: tachycardia (n = 9), gastric distress (n = 6), anemia (n = 4), headache (n = 2), capillary leak syndrome (n = 2), acute fluid retention (n = 1), alopecia (n = 1), and a rash (n = 1) (1).

In a prospective study of 97 patients the most frequent adverse effects after 1 month were headache (n = 24), diarrhea (n = 8), and bouts of palpitation (n = 8) (2).

In a prospective study in 120 patients with myeloproliferative disease the adverse effects were bouts of palpitation (n = 84), headache (n = 62), nausea (n = 42), diarrhea or flatulence (n = 38), edema (n = 260), and fatigue (n = 280) (3).

Organs and Systems

Cardiovascular

High-output heart failure has been attributed to anagrelide in a patient with essential thrombocytosis; there was dramatic improvement after withdrawal of anagrelide (4). Of 577 patients taking anagrelide, 14 developed CHF; 2 died suddenly (5). In another study of 942 patients taking anagrelide for thrombocytosis, 15 died of cardiac causes (6). The authors suggested that increased cardiac output was due to the positive inotropic activity through phosphodiesterase inhibition.

Psychiatric

There has been a single case report of visual hallucinations in a patient taking anagrelide, with recurrence on re-challenge (7).

Urinary tract

Renal tubular damage has been attributed to anagrelide in a 60-year-old man with Crohn's disease and essential thrombocytosis (8).

Sexual function

In a retrospective study of 52 patients with chronic myeloproliferative diseases, 42 had adverse effects and in 15 the adverse effects necessitated withdrawal (9). Two patients had erectile dysfunction, which has been described only once before in association with anagrelide.

References

1. Mazzucconi MG, Redi R, Bernasconi S, Bizzoni L, Dragoni F, Latagliata R, Santoro C, Mandelli F. A long-term study of young patients with essential thrombocythemia treated with anagrelide. Haematologica 2004;89(11):1306–13.
2. Steurer M, Gastl G, Jedrzejczak WW, Pytlik R, Lin W, Schlogl E, Gisslinger H. Anagrelide for thrombocytosis in myeloproliferative disorders: a prospective study to assess efficacy and adverse event profile. Cancer 2004;101(10):2239–46.
3. Birgegard G, Bjorkholm M, Kutti J, Larfars G, Lofvenberg E, Markevarn B, Merup M, Palmblad J, Mauritzson N, Westin J, Samuelsson J. Adverse effects and benefits of two years of anagrelide treatment for thrombocythemia in chronic myeloproliferative disorders. Haematologica 2004;89(5):520–7.
4. Engel PJ, Johnson H, Baughman RP, Richards AI. High-output heart failure associated with anagrelide therapy for essential thrombocytosis. Ann Intern Med 2005;143(4):311–3.
5. Anagrelide Study Group. Anagrelide, a therapy for thrombocythemic states: experience in 577 patients. Am J Med 1992;92:69–76.
6. Petitt RM, Silverstein MN, Petrone ME. Anagrelide for control of thrombocythemia in polycythemia and other myeloproliferative disorders. Semin Hematol 1997;34:51–4.
7. Swords R, Fay M, O'Donnell R, Murphy PT. Anagrelide-induced visual hallucinations in a patient with essential thrombocythemia. Eur J Haematol 2004;73(3):223–4.
8. Rodwell GE, Troxell ML, Lafayette RA. Renal tubular injury associated with anagrelide use. Nephrol Dial Transplant 2005;20(5):988–90.
9. Penninga E, Jensen BA, Hansen PB, Clausen NT, Mourits-Andersen T, Nielsen OJ, Hasselbalch HC. Anagrelide treatment in 52 patients with chronic myeloproliferative diseases. Clin Lab Haematol 2004;26(5):335–40.

Clopidogrel

General Information

Clopidogrel is a thienopyridine compound, structurally related to ticlopidine, which inhibits ADP-induced platelet aggregation (1). Its efficacy in stroke, myocardial infarction, and other vascular causes of death has been demonstrated in the CAPRIE Study (2), a large trial in which 9599 patients were treated with clopidogrel. The most frequent adverse effects were gastrointestinal symptoms, although they were not significantly more frequent than with aspirin. Bleeding disorders were as common with clopidogrel as with aspirin (9.3% with each). More patients withdrew from treatment because of rashes from clopidogrel (0.9%) than from aspirin (0.4%). Neutropenia and thrombocytopenia occurred very rarely and with similar rates in the two groups.

The adverse effects of clopidogrel have been reviewed (3). From 1990 to 2002, 475 reports of adverse reactions due to clopidogrel and 691 due to ticlopidine were registered in the database of the German spontaneous reporting system (4). The breakdowns are reported under separate headings below.

Organs and Systems

Hematologic

Thrombocytopenia

In the German pharmacovigilance database thrombocytopenia was the most frequent adverse event attributed to clopidogrel (60 cases; 13%), followed by gastrointestinal hemorrhage (56 cases; 12%), anemia (38 cases; 8.0%), hematoma (29 cases; 6.1%), and melena (27 cases; 5.7%) (21). White cell disorders were reported in only 8.9% of cases compared with 42% with ticlopidine: There were three reports of clopidogrel-associated thrombotic thrombocytopenic purpura (0.6%).

During active surveillance by medical directors of blood banks, hematologists, and the manufacturers of clopidogrel, 11 patients were identified in whom thrombotic thrombocytopenic purpura developed during or soon after treatment with clopidogrel (5). Of these 11 patients, 10 had taken clopidogrel for 14 days or less before the onset. From this study it is not possible to calculate the frequency of thrombotic thrombocytopenic purpura, since the size of the population from which the 11 cases were drawn is unknown.

Leukopenia

There have been reports of clopidogrel-associated leukopenia (6).

Aplastic anemia

- Fatal aplastic anemia has been reported in two patients taking clopidogrel (7,8). Aplastic anemia was diagnosed 5 months after starting clopidogrel in the first patient and after 3 months in the second. Both died from infection (sepsis and pneumonia). Except for allopurinol in the first case, these patients did not take any medications associated with aplastic anemia.

Hemorrhage

Intraoperative and postoperative outcomes have been studied in 505 consecutive patients who underwent isolated coronary artery bypass grafting; some had taken clopidogrel until 72 hours before surgery (n = 136) and others had not taken clopidogrel (n = 369) (9). Chest tube fluid drainage was significantly higher during the first 24 hours after bypass in those who had taken clopidogrel (1485 versus 780 ml). These patients also required more transfusion of platelets and fresh frozen plasma. The re-exploration rate because of bleeding was significantly higher in those who had taken clopidogrel (5.9% versus 1.2%).

The use of antiplatelet agents in elderly patients with trauma significantly increases the risk of death when head injury involves intracranial hemorrhage. In a retrospective analysis, patients older than 50 years who had had a traumatic intracranial hemorrhage over the previous 4 years associated with the use of aspirin, clopidogrel, or a combination of the two were compared with a control group of patients who had had a hemorrhage but were not taking antiplatelet drugs (10). There were no significant differences between the 90 patients and the 89 controls in terms of demographics, mechanism of injury, Injury Severity Score, Glasgow Coma Score, or hospital length of stay. The patients who were taking antiplatelet drugs had significantly more co-morbid conditions (71% versus 35%); 21 patients and eight controls died (23% versus 8.9%) and age over 76 years and a Glasgow Coma Score lower than 12 correlated significantly with increased mortality.

An epidural hematoma occurred after combined spinal–epidural anesthesia in an 80-year-old woman who was given clopidogrel and dalteparin (11).

Major bleeding is uncommon with clopidogrel. In a randomized, double-blind, placebo-controlled, 18-month study of the effects of aspirin 75 mg/day in 7599 high-risk patients with recent ischemic stroke or transient ischemic attacks and at least one additional vascular risk factor who were already taking clopidogrel 75 mg/day, life-threatening bleeding was more common in the group who took aspirin and clopidogrel compared with clopidogrel alone: 96 (2.6%) versus 49 (1.3%); absolute risk increase 1.3% (95% CI 0.6, 1.9) (12). Episodes of major bleeding were also more common in those taking aspirin plus clopidogrel, but there was no difference in mortality.

Hemophilia

Acquired autoimmune hemophilia A has been attributed to clopidogrel (13).

- Two women aged 70 and 67 years developed excessive bruising and soft tissue bleeding 2–3 months after starting clopidogrel therapy for peripheral vascular disease (14). Their drug therapy had not changed recently. Both had acquired hemophilia A. They had no clinical symptoms or signs of malignancy, antiphospholipid syndrome, or collagen vascular disease. Both were given prednisolone and the inhibitor became undetectable within 8 weeks.

Liver

Hepatotoxicity has been attributed to clopidogrel.

- A 74-year-old man with acute coronary syndrome was given clopidogrel 75 mg/day and pantoprazole 40 mg/day and 4 weeks later developed tea-coloured urine and jaundice (15). The serum bilirubin concentration was 91 μmol/l, alkaline phosphatase 172 IU/l, and alanine transaminase 212 IU/l; serum albumin concentration and coagulation tests were normal, but there was mild thrombocytopenia (135 x 10^9/l) and a low white cell count (3.8 x 10^9/l). Serological tests for acute viral causes of hepatitis were negative, as were autoantibody tests. Clopidogrel was withdrawn. After 5 days the liver biochemistry started to improve gradually and became normal after 4 weeks.
- A 59-year-old man with acute coronary syndrome was given aspirin, clopidogrel, enoxaparin, transdermal nitrates, omeprazole, carvedilol, and captopril (16). Four days later his alanine transaminase activity rose to 318 U/l (5–41). Drug-induced liver disease was suspected, and clopidogrel, omeprazole, carvedilol, and captopril were withdrawn and 11 days later the transaminase activity returned to normal. Clopidogrel was reintroduced because of imminent coronary angiography and 4 days later the alanine transaminase activity rose again to 119 UI/l. Clopidogrel was again withdrawn, and the transaminases returned to normal after 7 days. Bilirubin and alkaline phosphatase were normal throughout.

Skin

Clopidogrel can cause pruritus and urticaria (17).

Immunologic

Allergic rashes are common in patients taking clopidogrel and can require withdrawal.

A severe hypersensitivity syndrome in a patient taking clopidogrel resolved on dechallenge and recurred on rechallenge (18). The reaction included neutropenia, rash, fever, tachycardia, nausea, and vomiting. This presentation is very similar to the frequently reported hypersensitivity reactions to ticlopidine and very similar to a case involving clopidogrel reported previously.

In another case hypersensitivity to clopidogrel in a 62-year-old man was associated with fever, severe pruritus, raised liver enzymes, pancytopenia, and a rash, with erythematous macules and papules symmetrically distributed on the face, trunk, and limbs, but without mucosal lesions (19).

- A 57-year-old man took clopidogrel after a myocardial infarction and after 5 days developed a fever, rash, pruritus, and abdominal pain (20). Three days later he developed shock. He had thrombocytopenia, lymphopenia, aseptic leukocyturia, and raised serum activities of transaminases, amylase, and gamma-glutamyl transpeptidase. Blood cultures were negative. Clopidogrel was withdrawn and within 1 week he had completely recovered and all blood tests had returned to normal. One month later, he took clopidogrel again; 4 hours later the same symptoms reappeared, with aseptic leukocyturia and raised transaminases and gamma-glutamyl transpeptidase. Drug allergy was suspected and clopidogrel was withdrawn. All the symptoms disappeared within a few days and did not recur during the following year.

It is highly probable that this reaction was provoked by clopidogrel because of the positive rechallenge and because the patient did not take any other drug.

Long-Term Effects

Drug resistance

Resistance to aspirin and clopidogrel, which can have serious consequences, such as recurrent myocardial infarction, stroke, or death, has been reviewed (21). In its broadest sense, resistance refers to the continued occurrence of ischemic events despite adequate antiplatelet therapy and adherence to therapy. However, the lack of a standard definition of resistance and the lack of standard diagnostic methods has hampered identification and treatment. Attempts have been made to develop a more useful definition, with the goal of correlating laboratory tests with clinical outcomes, but there is no current definition that unites biochemical and clinical expressions of failed treatment. Rates of aspirin resistance are reported at 5–45%, depending on the study and the method of determining therapeutic failure. However, rather than characterizing patients as resistant or sensitive to a medication, resistance is probably better regarded as a continuous variable, similar to blood pressure. The mechanisms of resistance to antiplatelet drugs are incompletely defined, but there are specific clinical, cellular, and genetic factors that affect therapeutic failure, including failure to prescribe these medications despite appropriate indications and polymorphisms of platelet membrane glycoproteins. As new bedside tests are developed for the rapid and accurate measurement of the response to antiplatelet drugs, it may become easier to individualize antiplatelet drug therapy.

Drug-drug interactions

HMG Co-A reductase inhibitors

Clopidogrel is a pro-drug that is converted to its active form by CYP3A4. The active drug irreversibly blocks one specific platelet adenosine 5′-diphosphate (ADP) receptor (P2Y12). As certain lipophilic statins (atorvastatin, lovastatin, simvastatin) are substrates of CYP3A4, drug interactions are possible, as suggested in two studies. Platelet aggregation was measured in 44 patients undergoing coronary artery stent implantation taking clopidogrel, clopidogrel plus pravastatin, or clopidogrel plus atorvastatin (22). Atorvastatin, but not pravastatin, attenuated the antiplatelet activity of clopidogrel; in the presence of clopidogrel plus atorvastatin 0, 10, 20, and 40 mg, platelet aggregation was respectively 34%, 58%, 74%, and 89% of normal.

In 47 patients with coronary artery disease blood samples were taken before and 5 and 48 hours after oral clopidogrel (loading dose 300 mg followed by 75 mg/day) (23). ADP-stimulated expression of P-selectin (CD62P) on platelets was measured by flow cytometry and used as a marker for the antiplatelet effect of clopidogrel. Pre-treatment with statins (atorvastatin, simvastatin) reduced the inhibitory effects of clopidogrel during the loading phase (relative reduction after 5 hours 29%) and to a lesser extent during the maintenance phase (relative reduction after 48 hours 17%).

Both sets of authors suggested that if clopidogrel and a statin are given together one should either use a statin that is not metabolized by CYP3A4 or monitor platelet function.

This interaction was also studied in the Clopidogrel for the Reduction of Events During Observation (CREDO) trial (24). This was a double-blind, placebo-controlled, randomized comparison of two regimens after percutaneous coronary interventions: pretreatment with clopidogrel 300 mg followed by 75 mg/day for 1 year, or clopidogrel 75 mg/day for 1 month without pretreatment. All the patients took aspirin. The 1-year primary end-point was a composite of death, myocardial infarction, and stroke. Of the 2116 patients enrolled, 1001 took a statin that was metabolized by CYP3A4 and 158 took one that was not. In the overall study population, the primary end-point was significantly reduced by clopidogrel (8.5% versus 11.5%, RRR = 27%). This benefit was similar in those who also took a statin, irrespective of whether the statin inhibited CYP3A4 (7.6% versus 11.8%; RRR = 36%; 95% CI = 3.9, 58) or not (5.4% versus 14%; RRR = 61%, 95% CI = 24, 87). The authors concluded that although ex vivo testing has suggested a potential negative interaction when a CYP3A4-metabolized statin is co-administered with clopidogrel, this was not observed clinically in a post-hoc analysis of a placebo-controlled study. Thus, there is currently insufficient evidence to recommend avoiding a CYP3A4-metabolized statin or to recommend platelet function testing in patients taking clopidogrel.

The Interaction of Atorvastatin and Clopidogrel Study included 75 patients undergoing coronary stenting (25). All took aspirin 325 mg/day for at least 1 week and clopidogrel 300 mg immediately before stent implantation. They had been taking atorvastatin (n = 25), any other statin (n = 25), or no statin (n = 25) for at least 30 days beforehand. At baseline, the patients in both statin groups had reduced platelet aggregation and reduced platelet expression of G-protein-coupled protease-activated thrombin receptor (PAR)-1. There were no significant differences in measured platelet characteristics among the study groups at 4 and 24 hours after clopidogrel, with the exception of lower collagen-induced aggregation at 24 hours and a constantly reduced expression of PAR-1 in patients taking any statin. The authors concluded that statins in general, and atorvastatin in particular, do not affect the ability of clopidogrel to inhibit platelet function in patients undergoing coronary stenting. These prospective data also suggest that statins may inhibit platelets directly via yet unknown mechanisms, possibly related to the regulation of PAR-1 thrombin receptors.

Management of adverse reactions

Type I allergic reactions to drugs may be amenable to desensitization. A protocol for clopidogrel desensitization over 8 hours has been described, using 15 doubling doses of oral clopidogrel to achieve a maintenance dose of 75 mg/day (Table 1) (26).

Table 1 Clopidogrel desensitization protocol

Desensitizing dose (mg)	*Concentration (mg/ml)*	*Volume (ml)*
0.005	0.5	0.01
0.010		0.02
0.020		0.04
0.040		0.08
0.080		0.16
0.160		0.32
0.300		0.60
0.600		1.20
1.200	5	0.24
2.5		0.5
5		1
10		2
20		4
40		8
75	75 mg tablet	1 tablet

Notes: Doses are given orally every 30 minutes if no adverse reaction occurs. The patient is monitored closely during the procedure, with equipment available to treat anaphylaxis. If any adverse reactions occur the protocol is adjusted. The patient is kept under observation for at least 1 hour after the last dose.

- A 71-year-old man with a history of allergy to penicillin was given clopidogrel after intracoronary stenting. After 7 days he developed generalized pruritic erythematous macules, without dyspnea, hypotension, or mucosal or vesiculobullous lesions. Clopidogrel was withdrawn and he was given ticlopidine instead. However, 2 weeks later he developed severe neutropenia, fever, and sinusitis, and ticlopidine was withdrawn. IgE-mediated drug allergy to clopidogrel was diagnosed and he was deemed suitable for desensitization, which was successfully completed without adverse effects. He continued to take clopidogrel and had no recurrence.
- A 68-year-old woman with a history of allergy to penicillin was given clopidogrel after intracoronary stenting and after 3 days developed a pruritic maculopapular rash on her lower abdomen. Clopidogrel was withdrawn and the rash resolved with a short course of prednisone. Later, after repeat stenting, she was given ticlopidine and after 2 days developed a similar rash. The ticlopidine was withdrawn and the rash resolved with prednisone. She underwent out-patient oral desensitization to clopidogrel. After the 12th dose she felt flushing of her forearms without objective changes, and the procedure was successfully completed without further incident. She then took clopidogrel 75 mg/day with no recurrence.
- A 71-year-old woman with no drug allergies was given clopidogrel after abdominal aneurysm stenting and 3 weeks later developed an erythematous macular rash on her trunk, arms, and upper thighs. Clopidogrel was withdrawn and the rash resolved with a short course of prednisone. She was given ticlopidine and 1 week later developed a similar pruritic rash on her face and limbs with associated vomiting. Ticlopidine was withdrawn and the rash resolved spontaneously within 1 week. She underwent clopidogrel desensitization, but 3 weeks later she had mild generalized pruritus, easy bruising, and melena. Clopidogrel was withdrawn. She later underwent redesensitization to clopidogrel as an outpatient, but developed a pruritic erythematous rash on her lower abdomen 10 minutes after taking a dose of 5 mg. She was given oral cetirizine and after 30 minutes, when the pruritus and erythema had subsided, the desensitization protocol was restarted with a dose of 2.5 mg and was completed at the 40 mg dose (total cumulative dose 87 mg) with no further complications. She then took clopidogrel 75 mg/day without recurrence.

This case series suggests that patients who have had a type I allergic reaction to clopidogrel can be rapidly desensitized. Further studies in a larger number of patients are needed to confirm the safety and efficacy of this regimen.

References

1. Coukell AJ, Markham A. Clopidogrel. Drugs 1997;54(5):745–50.
2. CAPRIE Steering Committee. A randomised, blinded, trial of clopidogrel versus aspirin in patients at risk of ischaemic events (CAPRIE). Lancet 1996;348(9038):1329–39.
3. Sharis PJ, Cannon CP, Loscalzo J. The antiplatelet effects of ticlopidine and clopidogrel. Ann Intern Med 1998;129(5):394–405.
4. Tiaden JD, Wenzel E, Berthold HK, Muller-Oerlinghausen B. Adverse reactions to anticoagulants and to antiplatelet drugs recorded by the German spontaneous reporting system. Semin Thromb Hemost 2005;31(4):371–80.
5. Bennett CL, Connors JM, Carwile JM, Moake JL, Bell WR, Tarantolo SR, McCarthy LJ, Sarode R, Hatfield AJ, Feldman MD, Davidson CJ, Tsai HM. Thrombotic thrombocytopenic purpura associated with clopidogrel. N Engl J Med 2000;342(24):1773–7.
6. McCarthy MW, Kockler DR. Clopidogrel-associated leukopenia. Ann Pharmacother 2003;37:216–9.
7. Trivier JM, Caron J, Mahieu M, Cambier N, Rose C. Fatal aplastic anaemia associated with clopidogrel. Lancet 2001;357(9254):446.
8. Meyer B, Staudinger T, Lechner K. Clopidogrel and aplastic anaemia. Lancet 2001;357(9266):1446–7.
9. Englberger L, Faeh B, Berdat PA, Eberli F, Meier B, Carrel T. Impact of clopidogrel in coronary artery bypass grafting. Eur J Cardiothorac Surg 2004;26(1):96–101.
10. Ohm C, Mina A, Howells G, Bair H, Bendick P. Effects of antiplatelet agents on outcomes for elderly patients with traumatic intracranial hemorrhage. J Trauma 2005;58(3):518–22.
11. Tam NL, Pac-Soo C, Pretorius PM. Epidural haematoma after a combined spinal–epidural anaesthetic in a patient treated with clopidogrel and dalteparin. Br J Anaesth 2006; 96(2): 262–5.
12. Diener HC, Bogousslavsky J, Brass LM, Cimminiello C, Csiba L, Kaste M, Leys D, Matias-Guiu J, Rupprecht HJMATCH investigators. Aspirin and clopidogrel compared with clopidogrel alone after recent ischaemic stroke or transient ischaemic attack in high-risk patients (MATCH): randomised, double-blind, placebo-controlled trial. Lancet 2004;364(9431):331–7.
13. Haj M, Dasani H, Kundu S, Mohite U, Collins PW. Acquired haemophilia A may be associated with clopidogrel. BMJ 2004;329(7461):323.
14. Chau TN, Yim KF, Mok NS, Chan WK, Leung VK, Leung MF, Lai ST. Clopidogrel-induced hepatotoxicity after percutaneous coronary stenting. Hong Kong Med J 2005;11(5):414–6.
15. Beltran-Robles M, Marquez Saavedra E, Sanchez-Munoz D, Romero-Gomez M. Hepatotoxicity induced by clopidogrel. J Hepatol 2004;40(3):560–2.
16. Haj M, Dasani H, Kundu S, Mohite U, Collins PW. Acquired haemophilia A may be associated with clopidogrel. BMJ 2004;329(7461):323.
17. Khambekar SK, Kovac J, Gershlick AH. Clopidogrel induced urticarial rash in a patient with left main stem percutaneous coronary intervention: management issues. Heart 2004;90(3):e14.
18. Doogue MP, Begg EJ, Bridgman P. Clopidogrel hypersensitivity syndrome with rash, fever, and neutropenia. Mayo Clin Proc 2005;80(10):1368–70.
19. Comert A, Akgun S, Civelek A, Kavala M, Sarigul S, Yildirim T, Arsan S. Clopidogrel-induced hypersensitivity syndrome associated with febrile pancytopenia. Int J Dermatol 2005;44(10):882–4.
20. Sarrot-Reynauld F, Bouillet L, Bourrain JL. Severe hypersensitivity associated with clopidogrel. Ann Intern Med 2001;135(4):305–6.
21. Wang TH, Bhatt DL, Topol EJ. Aspirin and clopidogrel resistance: an emerging clinical entity. Eur Heart J 2006;27(6):647–54.
22. Lau WC, Waskell LA, Watkins PB, Neer CJ, Horowitz K, Hopp AS, Tait AR, Carville DG, KE Guyer, Bates ER. Atorvastatin reduces the ability of clopidogrel to inhibit platelet aggregation: a new drug-drug interaction. Circulation 2003;107:32–7.
23. Neubauer H, Gunesdogan B, Hanefeld C, Spiecker M, Mugge A. Lipophilic statins interfere with the inhibitory effects of clopidogrel on platelet function—a flow cytometry study. Eur Heart J 2003;24:1744–9.
24. Saw J, Steinhubl SR, Berger PB, Kereiakes DJ, Serebruany VL, Brennan D, Topol EJ. Lack of adverse clopidogrel–atorvastatin clinical interaction from secondary analysis of a randomized, placebo-controlled clopidogrel trial. Circulation 2003;108:921–4.
25. Serebruany VL, Midei MG, Malinin AI, Oshrine BR, Lowry DR, Sane DC, Tanguay JF, Steinhubl SR, Berger PB, O'Connor CM, Hennekens CH. Absence of interaction between atorvastatin or other statins and clopidogrel: results from the interaction study. Arch Intern Med 2004;164(18):2051–7.
26. Camara MG, Almeda FQ. Clopidogrel (Plavix) desensitization: a case series. Catheter Cardiovasc Interv 2005;65(4):525–7.

Dipyridamole

General Information

Dipyridamole inhibits platelet aggregation by inhibiting platelet cyclic AMP phosphodiesterase, potentiating adenosine inhibition of platelet function by blocking adenosine reuptake by vascular and blood cells and breakdown of adenosine, and by potentiating prostaglandin I_2 anti-aggregatory activity and enhancement of its synthesis. These independent processes inhibit platelet aggregation by increasing platelet cyclic AMP through a reduction in enzymatic breakdown of cyclic AMP and stimulation of cyclic AMP formation by activation of adenyl cyclase by adenosine and possibly prostaglandin I_2 (1).

Dipyridamole is also a potent coronary arteriolar vasodilator, perhaps by opening of vascular K_{ATP} channels (2). However, the Food and Drug Administration withdrew conditional approval for certain drug products containing dipyridamole, because of a lack of sufficient evidence of effectiveness in the long-term therapy of angina pectoris (3).

Observational studies

The incidence of major adverse reactions to dipyridamole was determined in a multicenter retrospective study, involving 73 806 patients who underwent intravenous dipyridamole stress imaging in 59 hospitals and 19 countries (4). The main conclusion was that the risk of serious dipyridamole-induced adverse effects is very low, a conclusion that is in line with other reports (5), and comparable to that reported for exercise testing in a similar patient population. Combined major adverse events among the entire patient population included 7 cardiac deaths (0.95 per 10 000), 13 non-fatal myocardial infarctions (1.76 per 10 000), 6 non-fatal sustained ventricular dysrhythmias (0.81 per 10 000) (ventricular tachycardia in 2 and ventricular fibrillation in 4), 9 transient cerebral ischemic attacks (1.22 per 10 000), 1 stroke, and 9 severe cases of bronchospasm (1.22 per 10 000). Minor non-cardiac adverse effects were less frequent among the elderly and more frequent in women and patients taking maintenance aspirin.

Placebo-controlled studies

The efficacy and safety of dipyridamole have been assessed in patients with chronic stable angina in a large-scale, international, randomized, placebo-controlled, parallel-group study, in which 400 patients with chronic stable angina pectoris and a positive treadmill exercise test were randomized to receive either modified-release dipyridamole (200 mg bd orally; $n = 198$) or placebo ($n = 202$) for 24 weeks as an add-on to conventional anti-anginal therapy and for 4 additional weeks as monotherapy, the latter after withdrawal of standard treatment with calcium channel blockers and/or beta-blockers and/or long-acting (prophylactic) nitrates (6). Of the 198 patients randomized to dipyridamole, 134 completed the add-on phase but only 12 completed the monotherapy phase. Of the 202 patients randomized to placebo, 162 reached the add-on phase but only 12 reached the monotherapy phase. There were serious adverse events in 15 patients who took dipyridamole and 12 who took placebo (7.6 versus 6.0%). These included chest pain, angina pectoris, and non-cardiac adverse effects, such as diarrhea, nausea, and headache.

Organs and Systems

Cardiovascular

Dipyridamole is used with ^{201}thallium imaging in the detection of coronary artery disease, but can cause dysrhythmias.

- A 41-year-old man with hypertension was investigated for chest tightness by dipyridamole–thallium single-photon emission computed tomography (7). A standard dose of dipyridamole (0.56 mg/kg) was infused intravenously over 4 minutes, during which his heart rate increased from 68 to 88/minute and his blood pressure fell slightly (from 160/80 to 140/76 mmHg). He had no subjective symptoms, such as palpitation, dizziness, or chest tightness, but had ventricular extra beats 40 seconds after completion of the dipyridamole infusion, followed 1 minute later by a sustained ventricular tachycardia. His blood pressure fell to 80/50 mmHg and he complained of dizziness. Intravenous aminophylline 125 mg was given immediately. About 30 seconds later, the ventricular tachycardia terminated and his hemodynamics stabilized. The ventricular extra beats persisted for another 30 seconds.
- Two cases of severe bradydysrhythmias (one complete heart block, one sinus bradycardia) occurred after intravenous dipyridamole (8).

Dipyridamole can also cause myocardial ischemia.

- A 43-year-old man had an acute myocardial infarction immediately after exercise and pharmacological stress echocardiography with dipyridamole + atropine 1 month after successful stent implantation (9).
- A 59-year-old man developed unstable angina 1 month after coronary artery bypass surgery (10). During dipyridamole scintigraphy, 2 minutes after the beginning of dipyridamole infusion, ST segment elevation occurred in the inferior electrocardiographic leads and there were two marked anteroseptal and inferior defects on myocardial scintigraphy.

A patient with aorto-iliac occlusive vascular disease and hypertension suffered a stroke 6.5 minutes after administration of intravenous dipyridamole during a ^{201}thallium myocardial study (11). Aminophylline did not reverse its progression.

Respiratory

The practicability of dipyridamole ^{13}N-ammonia myocardial positron emission tomography for perioperative risk assessment of coronary artery disease in patients with severe chronic obstructive pulmonary disease undergoing lung volume reduction surgery has been studied in 13 men and 7 women (mean age 57 years) without symptoms of coronary artery disease (12). Nine patients had intolerable dyspnea due to bronchoconstriction and required intravenous aminophylline. Dipyridamole cannot be recommended as a pharmacological stress in this setting.

- A 69-year-old non-smoking woman with stable asthma developed sudden bronchospasm within minutes of receiving intravenous dipyridamole during a thallium stress test; it responded to intravenous aminophylline 150 mg (13).

The authors proposed that dipyridamole had increased circulating concentrations of adenosine, a bronchoconstrictor.

Fatal respiratory insufficiency after dipyridamole–thallium imaging has been described in patients with a history of chronic obstructive lung disease (14,15). The authors of these reports concluded that patients with a history of chronic obstructive lung disease may have an increased risk of bronchospasm after dipyridamole infusion, and caution is advised in such patients.

Nervous system

Dipyridamole in combination with aspirin was more effective in preventing secondary stroke than low-dose aspirin or dipyridamole alone in only one of several studies (16). There is some evidence that dipyridamole can sometimes cause transient ischemic attacks (17).

- A 74-year-old woman had a 3-year history of mild dysarthria, dizziness, and gait ataxia, accompanied by two transient ischemic attacks with involuntary ballistic movements of her left arm lasting several seconds each, and another transient ischemic attack with a right homonymous hemianopia lasting 30 minutes. About 45 minutes after her first-ever oral administration of dipyridamole plus aspirin she developed a transient cerebellar deficit that reproduced features of previous vertebrobasilar ischemic events, as well as severe headache, flushing, and diarrhea.

The acute onset, the pattern of the cerebellar deficit, and the absence of features of epilepsy suggested that the episode was a transient ischemic attack. Aspirin is not known to cause transient ischemic attacks, and only rarely causes headache, flushing, and diarrhea. Since headache, flushing, and diarrhea, which can be caused by dipyridamole, occurred at the same time as the transient ischemic attacks and did not recur after withdrawal, dipyridamole may have caused the transient ischemic attacks. However, it was not clear whether the attacks occurred despite treatment rather than because of it.

Headache

Dipyridamole, used together with aspirin, has been evaluated in secondary prevention after ischemic stroke in several studies and in a large number of patients. The largest study, the Second European Stroke Prevention Study (18), showed that a combination of aspirin and modified-release dipyridamole reduced the risk of stroke or death by 24% (significantly more than the 13% observed with aspirin alone). However, one problem was that patients using dipyridamole alone or a combination of dipyridamole and aspirin discontinued treatment because of headache significantly more often than patients taking placebo. More than one-third of the patients who used dipyridamole reported headache as an adverse effect. The headache mostly occurred early, during the initial phase of dipyridamole treatment, and was a reason for early withdrawal.

In the Second European Stroke Prevention Study, headaches associated with dipyridamole (in 8% of patients taking dipyridamole or dipyridamole + aspirin versus 2% of patients taking aspirin alone or placebo) often led to withdrawal of therapy.

The mechanisms of headache associated with dipyridamole are unknown, although it is a vasodilator and there are similarities with vascular headaches in migraine or in patients who use nitrates. Several different regimens have been suggested to lower the prevalence of headache associated with dipyridamole, for example concurrent treatment with analgesics and an initial titration phase with a low daily dose of dipyridamole (19). In healthy individuals, headache is more common during the first days of treatment with dipyridamole and then declines in frequency (20). This supports the use of dose titration during the initial period of treatment as a means of reducing the frequency of headache, and could lead to increased adherence to treatment.

The predictive factors for headaches were explored in a study of the bioequivalence of two formulations of dipyridamole 200 mg in a modified-release combination with aspirin 25 mg (21). The conclusion was that the rapid fall in the incidence of headaches over time implied that most patients quickly develop tolerance to dipyridamole-associated headaches. However, in the European Stroke Prevention Study 2, headache was the most common adverse event, and it occurred more often in dipyridamole-treated patients (22).

Biliary tract

Recurrent drug-containing gallstones, 18 months after a previous stone had been removed endoscopically, were attributed to dipyridamole (23).

Immunologic

Infusion of dipyridamole caused an acute allergic reaction during myocardial scintigraphy (24).

- A 56-year old man with a history of allergy to aspirin, tetracycline, and penicillin, including angioedema and dyspnea, was given a dipyridamole stress test. About 1 minute after the infusion was started he reported periorbital pruritus. The infusion was completed uneventfully with the administration of ^{99m}technetium sestamibi at 7 minutes. Twenty minutes later he had tightness in the neck, dyspnea, and generalized facial swelling. He was given oxygen and intravenous promethazine hydrochloride and hydrocortisone. He improved over the next 2 hours, with residual periorbital edema but complete recovery from the respiratory symptoms. The cardiac study was completed without further events. The result was normal.

Drug overdose

An unusual adverse effect has been attributed to dipyridamole overdose, yellow skin discoloration (25).

- A 66-year-old man who took an overdose dipyridamole (Persantin Depot) of about 34 g developed intense neon-yellow coloring of the skin and urine. He also had an acute myocardial infarction and renal insufficiency. He gradually recovered but hemodialysis was needed for 2 weeks because of acute tubular necrosis after a prolonged period of hypotension.

Drug-drug interactions

Beta-blockers

Dipyridamole, in combination with [^{99m}Tc]-sestamibi, is used to image myocardial perfusion, but the presence and severity of coronary artery disease may be underestimated in patients who are also taking beta-blockers (26).

Cardiac glycosides

In a placebo-controlled study in 12 healthy volunteers, dipyridamole 300 mg/day for 3 days altered on the pharmacokinetics of a single oral dose of digoxin 0.5 mg, increasing its AUC by about 13% (27). This may have been due to reduced clearance of digoxin, since dipyridamole inhibits the P glycoprotein that secretes digoxin into the gut lumen after its primary absorption and into the urine via the renal tubules. However, the effects were very small and probably clinically unimportant. There was no difference in the effect of dipyridamole between subjects with the different MDR1 genotypes, TT and CC.

Cytarabine

Cytarabine (cytosine arabinoside) is eliminated by deoxycytidine monophosphate deaminase and cytidine deaminase. Severe hepatic failure has been previously reported after the use of cytosine arabinoside but the mechanism is not clear. In reporting the case of a middle-aged man who developed multi-organ failure when treated with cytarabine and dipyridamole concurrently, the authors suggested that liver toxicity had been caused by inhibition of the extracellular transport of cytarabine by dipyridamole, thereby increasing retention of cytarabine in hepatocytes (28).

Glibenclamide

Dipyridamole, in combination with [^{99m}Tc]-sestamibi, is used to image myocardial perfusion, but the presence and severity of coronary artery disease may be enhanced in patients who are taking glibenclamide (29).

References

1. Harker LA, Kadatz RA. Mechanism of action of dipyridamole. Thromb Res Suppl 1983;4:39–46.
2. Bijlstra P, van Ginneken EE, Huls M, van Dijk R, Smits P, Rongen GA. Glyburide inhibits dipyridamole-induced forearm vasodilation but not adenosine-induced forearm vasodilation. Clin Pharmacol Ther 2004;75(3):147–56.
3. Anonymous. Dipyridamole—withdrawn. WHO Pharm Newslett 1999;5/6:2.
4. Lette J, Tatum JL, Fraser S, Miller DD, Waters DD, Heller G, Stanton EB, Bom HS, Leppo J, Nattel S. Safety of dipyridamole testing in 73,806 patients: the Multicenter Dipyridamole Safety Study. J Nucl Cardiol 1995;2(1):3–17.
5. Beller GA. Pharmacologic stress imaging. JAMA 1991;265(5):633–8.
6. Picano E. PISA (Persantin In Stable Angina) study group. Dipyridamole in chronic stable angina pectoris; a randomized, double blind, placebo-controlled, parallel group study. Eur Heart J 2001;22(19):1785–93.
7. Chang WT, Lin LC, Yen RF, Huang PJ. Persistent myocardial ischemia after termination of dipyridamole-induced ventricular tachycardia by intravenous aminophylline: scintigraphic demonstration. J Formos Med Assoc 2000;99(3):264–6.
8. Bielen M, Karsera D, Melon P, Kulbertus H. Bradyarythmies sevères au cours d'ure scintigraphie myocardique de perfusion avec injection de dipyridamole (persantine). [Severe bradyarrhythmia during myocardial perfusion scintigraphy with injection of dipyridamole (Persantine).] Rev Med Liege 1999;54(2):105–8.
9. Nedeljkovic MA, Ostojic M, Beleslin B, Nedeljkovic IP, Stankovic G, Stojkovic S, Saponjski J, Babic R, Vukcevic V, Ristic AD, Orlic D. Dipyridamole–atropine-induced myocardial infarction in a patient with patent epicardial coronary arteries. Herz 2001;26(7):485–8.
10. Wartski M, Caussin C, Lancelin B. Spasme coronaire plurifocal declenche par l'injection de dipyridamole. Med Nucl 2001;25:153–9.
11. Whiting JH Jr, Datz FL, Gabor FV, Jones SR, Morton KA. Cerebrovascular accident associated with dipyridamole thallium-201 myocardial imaging: case report. J Nucl Med 1993;34(1):128–30.
12. Thurnheer R, Laube I, Kaufmann PA, Stumpe KD, Stammberger U, Bloch KE, Weder W, Russi EW. Practicability and safety of dipyridamole cardiac imaging in patients with severe chronic obstructive pulmonary disease. Eur J Nucl Med 1999;26(8):812–7.
13. Cogen F, Zweiman B. Dipyridamole (Persantin)-induced asthma during thallium stress testing. J Allergy Clin Immunol 2005;115(1):203–4.
14. Ottervanger JP, Haan D, Gans SJ, Hoorntje JC, Stricker BH. Bronchospasme, apnoe en hartstilstand na een dipyridamol-perfusiescintigrafie. [Bronchospasm, apnea and heart arrest following dipyridamole perfusion scintigraphy.] Ned Tijdschr Geneeskd 1993;137(3):142–3.
15. Hillis GS, al-Mohammad A, Jennings KP. Respiratory arrest during dipyridamole stress testing. Postgrad Med J 1997;73(859):301–2.
16. Diener HC, Cunha L, Forbes C, Sivenius J, Smets P, Lowenthal AEuropean Stroke Prevention Study. Dipyridamole and acetylsalicylic acid in the secondary prevention of stroke. J Neurol Sci 1996;143(1–2):1–132.
17. Siegel AM, Sandor P, Kollias SS, Baumgartner RW. Transient ischemic attacks after dipyridamole–aspirin therapy. J Neurol 2000;247(10):807–8.
18. Diener HC, Cunha L, Forbes C, European Stroke Prevention Study 2. Dipyridamole and acetylsalicylic acid in the secondary prevention of stroke. J Neurol Sci 1996;143:1–13.
19. Lindgren A. Management of vasodilation headache with dipyridamole. In: Wahlgren NG, editor. Update on Stroke Therapy 2000/2001. Stockholm: Karolinska Stroke Update;2001:261–70.
20. Theis JGW, Deichsel G. Marshall S. Rapid development of tolerance to dipyridamole-associated headaches. Br J Clin Pharmacol 1999;48:750–5.
21. Theis JG, Deichsel G, Marshall S. Rapid development of tolerance to dipyridamole-associated headaches. Br J Clin Pharmacol 1999;48(5):750–5.
22. Diener H, Cunha L, Forbes C, Sirenius J, Smets P, Lowenthal A. European stroke prevention study 2. Nervenheilkunde 1999;18:380–90.
23. Sautereau D, Moesch C, Letard JC, Cessot F, Gainant A, Pillegand B. Recurrence of biliary drug lithiasis due to dipyridamole. Endoscopy 1997;29(5):421–3.
24. Angelides S, Van der Wall H, Freedman SB. Acute reaction to dipyridamole during myocardial scintigraphy. N Engl J Med 1999;340(5):394.
25. Personne M, Mansten A, Svensson JO. Medvetslös patient med neongul hudvar intoxikerad med dipyridamol. Lakartidningen 2003;100:4194–5.
26. Taillefer R, Ahlberg AW, Masood Y, White CM, Lamargese I, Mather JF, McGill CC, Heller GV. Acute beta-blockade reduces the extent and severity of myocardial

perfusion defects with dipyridamole Tc-99m sestamibi SPECT imaging. J Am Coll Cardiol 2003;42:1475–83.
27. Verstuyft C, Strabach S, El-Morabet H, Kerb R, Brinkmann U, Dubert L, Jaillon P, Funck-Brentano C, Trugnan G, Becquemont L. Dipyridamole enhances digoxin bioavailability via P-glycoprotein inhibition. Clin Pharmacol Ther 2003;73:51–60.
28. Babaoglu MO, Karadag O, Saikawa Y, Altundag K, Elkiran T, Yasar U, Bozkurt A. Hepatotoxicity due to a possible interaction between cytosine arabinoside and dipyridamole: a case report. Eur J Clin Pharmacol 2004;60(6):455–6.
29. Mortensen UM, Nielsen-Kudsk JE, Jakobsen P, Nielsen TT. Glibenclamide blunts coronary flow reserve induced by adenosine and dipyridamole. Scand Cardiovasc J 2003;37:247–52.

Eptifibatide

General Information

Eptifibatide is a cyclic heptapeptide inhibitor of glycoprotein IIb/IIIa. It is derived from barbourin from the venom of the South-Eastern pigmy rattlesnake (1). It has low affinity for other integrins and is a potent inhibitor of platelet aggregation. It has a rapid onset of action and its effects are rapidly reversible.

Organs and Systems

Hematologic

Eptifibatide can cause thrombocytopenia (2,3,4,5) in an estimated 0.2% of patients (6). In 305 patients taking eptifibatide there were four cases of acute thrombocytopenia (platelet count less than 100 x 10^9/l) (7). One had been previously exposed to eptifibatide. The platelet counts fell within 6 hours of the first dose and recovered within 6–30 hours after withdrawal. There were no adverse outcomes from thrombocytopenia. Proposed mechanisms include sequestration of platelets in the liver and an interaction of eptifibatide with naturally-occurring antibodies to ligand-induced binding sites on glycoprotein IIb/IIIa (Morel 2685).

- A 61-year-old woman with acute coronary syndrome was given heparin and eptifibatide and developed thrombocytopenia (platelet count 2.0 x 10^9/l) and severe refractory hypotension (8). There were no heparin-induced antibodies, but eptifibatide-dependent antibodies specific for platelets were detected.

The authors attributed the thrombocytopenia and the hypotension to the eptifibatide antibodies.

The risk of hemorrhagic complications after the use of eptifibatide in patients undergoing rescue angiography with or without stenting after failed thrombolysis has been studied in 43 consecutive patients (9). There were bleeding complications in 13 patients; four had major bleeding. The predictors of major bleeding complications were older age, female sex, lower baseline platelet count, and time to administration of eptifibatide after failed thrombolysis.

In one patient thrombocytopenia occurred 2 hours after eptifibatide and resolved within 12 hours (10).

Susceptibility factors

Renal disease

Eptifibatide is cleared by both renal and non-renal mechanisms, and renal clearance accounts for about 40% of total body clearance (11). Severe renal dysfunction prolongs the duration of action of eptifibatide, as has been shown in three patients (12). One patient with acute renal insufficiency prolonged inhibition of platelet aggregation; one with acute on chronic renal insufficiency had an intracerebral hemorrhage and normal platelet aggregation was restored by hemodialysis; in a third, with end-stage renal disease, platelet function returned to normal after hemodialysis.

The pharmacokinetics of eptifibatide during steady-state infusion have been evaluated in an open study in 31 patients with mild, moderate, or severe renal impairment (13). There was a strong correlation between eptifibatide clearance and creatinine clearance, and the break point for the purpose of dosage adjustment was a creatinine clearance of 56 ml/minute. In patients with moderate or severe renal impairment, clearance rates and steady-state concentrations of eptifibatide were respectively about 50% lower and almost two-fold higher than in patients with normal renal function or mild renal impairment. Inhibition of platelet aggregation exceeded the clinically significant threshold of 80% in all the groups. Five subjects had a mild bleeding event and four had mild to moderate non-bleeding events; none of the events required intervention. The authors concluded that in patients with renal impairment infusion doses should be adjusted accordingly, but that bolus doses need no adjustment.

References

1. Phillips DR, Scarborough RM. Clinical pharmacology of eptifibatide. Am J Cardiol 1997;80(4A):11B–20B.
2. Tanaka KA, Vega JD, Kelly AB, Hanson SR, Levy JH. Eptifibatide-induced thrombocytopenia and coronary bypass operation. J Thromb Haemost 2003;1:392–4.
3. Nagge J, Jackevicius C, Dzavik V, Ross JR, Seidelin P. Acute profound thrombocytopenia associated with eptifibatide therapy. Pharmacotherapy 2003;23:374–9.
4. Salengro E, Mulvihill NT, Farah B. Acute profound thrombocytopenia after use of eptifibatide for coronary stenting. Catheter Cardiovasc Interv 2003;58:73–5.
5. Coons JC, Barcelona RA, Freedy T, Hagerty MF. Eptifibatide-associated acute, profound thrombocytopenia. Ann Pharmacother 2005;39(2):368–72.
6. Morel O, Jesel L, Chauvin M, Freyssinet JM, Toti F. Eptifibatide-induced thrombocytopenia and circulating procoagulant platelet-derived microparticles in a patient with acute coronary syndrome. J Thromb Haemost 2003;1:2685–7.

7. Khaykin Y, Paradiso-Hardy FL, Madan M. Acute thrombocytopenia associated with eptifibatide therapy. Can J Cardiol 2003;19:797–801.
8. Rezkalla SH, Hayes JJ, Curtis BR, Aster RH. Eptifibatide-induced acute profound thrombocytopenia presenting as refractory hypotension. Catheter Cardiovasc Interv 2003;58:76–9.
9. Ali A, Rehan A, Ganji J, Munir A, Moser L, Davis T, Khan S, Lalonde T, Schreiber T. Eptifibatide and risk of bleeding after failed thrombolysis. J Invasive Cardiol 2004;16(1):20–2.
10. Gupta N, Kapoor R, Bhandari S. Eptifibatide-induced profound thrombocytopenia. Indian Heart J 2004;56(3):250–1.
11. Alton KB, Kosoglou T, Baker S, Affrime MB, Cayen MN, Patrick JE. Disposition of 14C-eptifibatide after intravenous administration to healthy men. Clin Ther 1998;20:307–23.
12. Sperling RT, Pinto DS, Ho KK, Carrozza JP Jr. Platelet glycoprotein IIb/IIIa inhibition with eptifibatide: prolongation of inhibition of aggregation in acute renal failure and reversal with hemodialysis. Catheter Cardiovasc Interv 2003;59:459–62.
13. Gretler DD, Guerciolini R, Williams PJ. Pharmacokinetic and pharmacodynamic properties of eptifibatide in subjects with normal or impaired renal function. Clin Ther 2004;26(3):390–8.

Ticlopidine

General Information

Ticlopidine is a thienopyridine derivative with potent antiplatelet activity associated with inhibition of ADP-induced platelet aggregation. It was first used in Europe in 1978 in the secondary prevention of stroke and coronary events, the treatment of peripheral vascular disease, and after vascular stent placement. However, the use of ticlopidine has been progressively restricted in some countries because of its serious adverse effects. It has largely been superseded by clopidogrel.

General adverse effects

There is a risk of hemorrhage from drugs that inhibit platelet aggregation. The adverse effects most commonly reported with ticlopidine are skin rashes and gastrointestinal effects, including nausea, dyspepsia, vomiting, anorexia, epigastric pain, and diarrhea. Hypersensitivity reactions underly some of the hematological complications of ticlopidine, such as thrombotic thrombocytopenic purpura.

From 1990 to 2002, 691 reports of adverse reactions due to ticlopidine were registered in the database of the German spontaneous reporting system (38; see Hematologic and Liver below).

Organs and Systems

Respiratory

Bronchiolitis obliterans organizing pneumonia has been attributed to ticlopidine (1).

- A 76-year-old non-smoking woman with giant-cell arteritis who had a normal chest X-ray was taking prednisone 45 mg/day and ticlopidine 250 mg bd for persistence of cloudy vision. After 1 month of ticlopidine therapy, she developed increasing dyspnea and a pruritic rash. Chest radiography showed diffuse interstitial infiltrates, predominantly affecting the peripheries of both lungs. Transbronchial biopsy showed widening of the alveolar fields, with a mixed inflammatory infiltrate. Ticlopidine was withdrawn and prednisone was continued in the same dosage. Her symptoms completely resolved within 3 months and her chest X-ray normalized within 5 months.

Sensory systems

Two patients developed retinal vasculitis 3 and 4 weeks after starting to take ticlopidine (2). The symptoms and signs of retinitis resolved within 4–6 weeks after withdrawal of ticlopidine, with no reactivation within the next 2 or 3 years.

Hematologic

Ticlopidine can prolong bleeding time by up to 30 minutes. In well-controlled studies the incidence of minor bleeding was 10% (menorrhagia, bruises, and epistaxis) (3). Since the effect on platelet function is irreversible, the bleeding time does not return to normal sooner than 5–10 days after withdrawal, although it can be corrected immediately by platelet transfusion. In a retrospective comparison of the hemorrhagic risk in patients who had received platelet antiaggregants preoperatively for coronary surgery revascularization, there was a significant increase in postoperative hemorrhage and a higher incidence of re-operations in the patients who received ticlopidine compared with other antiaggregants (4).

Various other hematological complications have been observed during ticlopidine treatment (5). They include neutropenia, agranulocytosis, thrombocytopenia, aplastic anemia, pancytopenia, and thrombotic thrombocytopenic purpura (6). These reactions are potentially severe and some are fatal. Major bleeding is rare. Among 188 cases of hematological complications associated with ticlopidine and reported to the US Food and Drug Administration, 36 (19%) resulted in death (7).

In the German spontaneous reporting system there were 107 cases (16%) of agranulocytosis, 79 of leukopenia (111%) and 48 of thrombocytopenia (6.9%) (8). Other important hematological adverse effects were granulocytopenia (43 cases; 6.2%), anemia (29 cases; 4.2%), pancytopenia (18 cases; 2.6%), thrombotic thrombocytopenic purpura (14 cases; 2.0%), and bone marrow depression (14 cases; 2.0%).

Neutropenia

Neutropenia was the most common hematological complication in some of the major trials with ticlopidine. Severe neutropenia occurs in about 1% of patients. Neutropenia typically occurs in the first few months after the start of therapy, but is seen infrequently in the first 2–3 weeks. The mechanism has been reported to be inhibition of myeloid colony growth. On the basis of major randomized, controlled studies with ticlopidine, there is a 2.4% overall incidence of neutropenia and a 0.85% incidence of severe neutropenia and agranulocytosis (6). The neutropenia generally occurs at 2–5 weeks after the start of treatment (9). In some cases, neutropenia is delayed (10) or can develop after drug withdrawal. It has therefore been recommended that one should perform a complete blood count at baseline and every 2 weeks for the first 3 months of treatment, and that ticlopidine should be withdrawn if the neutrophil count falls below 1.2×10^9/l. In some cases, granulocyte colony-stimulating factors have been used, but their efficacy needs to be evaluated further (11). Although some fatal cases have been reported, neutropenia induced by ticlopidine is usually reversible.

Thrombocytopenia

Isolated thrombocytopenia has been reported at anything from 4 weeks to 5 years after the start of drug therapy (12,13). In these cases, platelet antibodies were also detected.

Thrombotic thrombocytopenic purpura has been repeatedly reported (6). The incidence has been estimated at 1 in 1600 (based on the observation of 5 cases out of a total of 7842 patients studied after coronary stent placements) (14). This rare adverse reaction is generally observed after a short exposure period of about 3–8 weeks. The clinical features are generally indistinguishable from those of idiopathic thrombotic thrombocytopenic purpura (15). The pathogenesis is unknown. Platelet antibodies were found in one patient, suggesting an immunologic mechanism (6). The outcome appears to be favorable in comparison to that of idiopathic thrombotic thrombocytopenic purpura, but several authors have reported cases in which death resulted (16,17). Early diagnosis and the rapid institution of plasmapheresis provide the best hope of complete recovery.

Sixty cases of thrombotic thrombocytopenic purpura have been reviewed (18). Ticlopidine had been prescribed for under 1 month in 80% of the patients, and platelet counts were normal within 2 weeks of the onset of thrombotic thrombocytopenic purpura in most patients. Mortality rates were higher among patients who were not treated with plasmapheresis than among those who underwent plasmapheresis (50% compared with 24%). The authors concluded that the onset of ticlopidine-associated thrombotic thrombocytopenic purpura is difficult to predict, despite close monitoring of platelet counts.

In 43 322 patients treated with coronary stents and ticlopidine for 1 year there were 9 cases of thrombotic thrombocytopenic purpura (0.02%) (19). The risk of thrombotic thrombocytopenic purpura during the use of ticlopidine after coronary stenting was 50-fold higher than in the general population. Ten other cases of thrombotic thrombocytopenic purpura related to ticlopidine were identified from the participating centers. Four of the 19 patients died, and all four deaths occurred in patients who were not treated with plasmapheresis. The authors stressed that early recognition and treatment is crucial for minimizing mortality.

Deficiency of, or auto-antibodies to, a von Willebrand-cleaving metalloproteinase are pathogenic in idiopathic thrombotic thrombocytopenic purpura (20). Seven consecutive patients, who developed thrombotic thrombocytopenic purpura 2–7 weeks after starting to take ticlopidine, had markedly reduced concentrations of von Willebrand factor metalloproteinase. In six cases initial samples were available and were positive for immunoglobulin G inhibitors to von Willebrand factor metalloproteinase.

Aplastic anemia

Rare cases of aplastic anemia have been reported (21). Some authors have suggested that the incidence of this complication may have been underestimated (22). The reaction is characterized by pancytopenia and the bone marrow shows profound hypocellularity, absence of precursor cells, and fatty replacement. Some studies have suggested a direct toxic effect of ticlopidine on bone marrow cells, particularly the myeloid line. In some cases, agranulocytosis or thrombocytopenia have been reported to precede the development of aplastic anemia. One should be alert to the fact that this reaction may still occur after the recommended 3 months of hematological monitoring have elapsed (6).

Gastrointestinal

Diarrhea is among the most common adverse effects of ticlopidine. It can occur in up to 20% of those who take it, in contrast to an incidence of 10% among placebo- and aspirin-treated patients (23) and usually occurs early in the course of treatment. It can be prevented by taking ticlopidine with food or by reducing the dosage. In rare cases, severe chronic diarrhea with marked body weight loss and anorexia can occur, sometimes long after the start of therapy (24).

In a few reports, lymphocytic colitis has been attributed to ticlopidine; in some cases the pathological features in the colon biopsy resolved after withdrawal of the drug (25–27).

Liver

Several cases of ticlopidine-induced hepatotoxicity have been reported (28). Between 10 days and 12 weeks after the start of treatment, patients develop jaundice, usually without fever, eosinophilia, or pain. Laboratory tests show a cholestatic or mixed cholestatic-hepatocellular pattern of injury. There is usually clinical and biochemical recovery within 1–11 months. Frank ticlopidine-associated liver injury is uncommon, but in one study, 44% of patients had abnormal liver function tests and about one-half of them had to stop taking the drug (23).

In the German spontaneous reporting system there were 283 cases (41%) of liver and biliary system disorders (8). These included increases in hepatic enzymes (123 cases; 18%), hepatitis and cholestatic hepatitis (65 cases; 9.4%), and jaundice (23 cases; 3.3%). There have been other reports of ticlopidine-induced cholestasis (29,30), in one case associated with pure red cell aplasia (31).

Skin

In a 1-year prospective study of 136 patients taking ticlopidine to prevent thrombosis after coronary stenting, 16 had adverse skin reactions (32). The most common were urticaria, pruritus, and maculopapular eruptions. Three patients had previously unreported reactions: a fixed drug eruption, an erythromelalgia-like eruption, and an erythema multiforme-like eruption.

Immunologic

A lupus-like illness (fever, rash, arthritis, renal involvement, and positive antinuclear and antihistone antibodies) developed in three patients 2–8 weeks after they started to take ticlopidine (33). After withdrawal, there was slow but complete resolution in all patients.

Susceptibility Factors

If it is considered necessary to withdraw ticlopidine before elective surgery, it should be done 1 week in advance (34). In emergencies, desmopressin or fresh platelet concentrates reduce the risk of bleeding.

Drug–Drug Interactions

Carbamazepine

In one patient taking ticlopidine and carbamazepine, serum carbamazepine concentrations rose in association with nervous system toxicity (35).

Omeprazole

In six Japanese CYP2C19 extensive metabolizers, ticlopidine 300 mg/day for 6 days significantly increased the C_{max} of a single oral dose of omeprazole 40 mg and reduced its oral clearance of omeprazole (36). The authors concluded that ticlopidine inhibits CYP2C19 but not, or to a lesser extent, CYP3A4, and that the magnitude of inhibition by ticlopidine is related to the in vivo activity of CYP2C19 before inhibition.

Phenytoin

Several reports have described ticlopidine-induced phenytoin toxicity with increased serum phenytoin concentrations (37–40). Data obtained in vitro using hepatic microsomal preparations have shown that ticlopidine inhibits the activity of CYP2C19 (41).

References

1. Alonso-Martinez JL, Elejalde-Guerra JI, Larrinaga-Linero D. Bronchiolitis obliterans–organizing pneumonia caused by ticlopidine. Ann Intern Med 1998;129(1):71–2.
2. Barak A, Morse LS, Schwab IR. Atypical reginal vasculitis associated with ticlopidine hydrochloridine use. Am J Ophthalmol 2000;129(5):684–5.
3. Editorial. Ticlopidine. Lancet 1991;23:459.
4. Criado A, Juffe A, Carmona J, Otero C, Avello F. Ticlopidine as a hemorrhagic risk factor in coronary surgery. Drug Intell Clin Pharm 1985;19(9):673–6.
5. Sharis PJ, Cannon CP, Loscalzo J. The antiplatelet effects of ticlopidine and clopidogrel. Ann Intern Med 1998;129(5):394–405.
6. Love BB, Biller J, Gent M. Adverse haematological effects of ticlopidine. Prevention, recognition and management. Drug Saf 1998;19(2):89–98.
7. Wysowski DK, Bacsanyi J. Blood dyscrasias and hematologic reactions in ticlopidine users. JAMA 1996;276:952.
8. Tiaden JD, Wenzel E, Berthold HK, Muller-Oerlinghausen B. Adverse reactions to anticoagulants and to antiplatelet drugs recorded by the German spontaneous reporting system. Semin Thromb Hemost 2005;31(4):371–80.
9. Haushofer A, Halbmayer WM, Prachar H. Neutropenia with ticlopidine plus aspirin. Lancet 1997;349(9050):474–5.
10. Farver DK, Hansen LA. Delayed neutropenia with ticlopidine. Ann Pharmacother 1994;28(12):1344–6.
11. Gur H, Wartenfeld R, Tanne D, Solomon F, Sidi Y. Ticlopidine-induced severe neutropenia. Postgrad Med J 1998;74(868):126–7.
12. Claas FH, de Fraiture WH, Meyboom RH. Thrombopénie causée par des anticorps induits par la ticlopidine. [Thrombocytopenia due to antibodies induced by ticlopidine.] Nouv Rev Fr Hematol 1984;26(5):323–4.
13. Takishita S, Kawazoe N, Yoshida T, Fukiyama K. Ticlopidine and thrombocytopenia. N Engl J Med 1990;323(21):1487–8.
14. Bennett CL, Kiss JE, Weinberg PD, Pinevich AJ, Green D, Kwaan HC, Feldman MD. Thrombotic thrombocytopenic purpura after stenting and ticlopidine. Lancet 1998;352(9133):1036–7.
15. Muszkat M, Shapira MY, Sviri S, Linton DM, Caraco Y. Ticlopidine-induced thrombotic thrombocytopenic purpura. Pharmacotherapy 1998;18(6):1352–5.
16. Ellie E, Durrieu C, Besse P, Julien J, Gbipki-Benissan G. Thrombotic thrombocytopenic purpura associated with ticlopidine. Stroke 1992;23(6):922–3.
17. Kupfer Y, Tessler S. Ticlopidine and thrombotic thrombocytopenic purpura. N Engl J Med 1997;337(17):1245.
18. Bennett CL, Weinberg PD, Rozenberg-Ben-Dror K, Yarnold PR, Kwaan HC, Green D. Thrombotic thrombocytopenic purpura associated with ticlopidine. A review of 60 cases. Ann Intern Med 1998;128(7):541–4.
19. Steinhubl SR, Tan WA, Foody JM, Topol EJ. Incidence and clinical course of thrombotic thrombocytopenic purpura due to ticlopidine following coronary stenting. EPISTENT Investigators. Evaluation of Platelet IIb/IIIa Inhibitor for Stenting. JAMA 1999;281(9):806–10.
20. Tsai HM, Rice L, Sarode R, Chow TW, Moake JL. Antibody inhibitors to von Willebrand factor metalloproteinase and increased binding of von Willebrand factor to platelets in ticlopidine-associated thrombotic thrombocytopenic purpura. Ann Intern Med 2000;132(10):794–9.
21. Elias M, Reichman N, Flatau E. Bone marrow aplasia associated with ticlopidine therapy. Am J Hematol 1993;44(4):289–90.
22. Lesesve JF, Callat MP, Lenormand B, Monconduit M, Noblet C, Moore N, Caron F, Humbert G, Stamatoullas A, Tilly H. Hematological toxicity of ticlopidine. Am J Hematol 1994;47(2):149–50.
23. Gent M, Blakely JA, Easton JD, Ellis DJ, Hachinski VC, Harbison JW, Panak E, Roberts RS, Sicurella J, Turpie AG. The Canadian American Ticlopidine Study (CATS) in thromboembolic stroke. Lancet 1989;1(8649):1215–20.

24. Mansoor GA, Aziz K. Delayed chronic diarrhea and weight loss possibly due to ticlopidine therapy. Ann Pharmacother 1997;31(7–8):870–2.
25. Martinez Aviles P, Gisbert Moya C, Berbegal Serra J, Lopez Benito I. Colitis linfocitaria inducida por ticlopidina. [Ticlopidine-induced lymphocytic colitis.] Med Clin (Barc) 1996;106(8):317.
26. Swine C, Cornette P, Van Pee D, Delos M, Melange M. Ticlopidine, diarrhée et colite lymphocytaire. [Ticlopidine, diarrhea and lymphocytic colitis.] Gastroenterol Clin Biol 1998;22(4):475–6.
27. Brigot C, Courillon-Mallet A, Roucayrol AM, Cattan D. Colite lymphocytaire et ticlopidine. [Lymphocytic colitis and ticlopidine.] Gastroenterol Clin Biol 1998;22(3): 361–2.
28. Martinez Perez-Balsa A, De Arce A, Castiella A, Lopez P, Ruibal M, Ruiz-Martinez J, Lopez De Munain A, Marti Masso JF. Hepatotoxicity due to ticlopidine. Ann Pharmacother 1998;32(11):1250–1.
29. Gandolfi A, Mengoli M, Rota E, Tolomelli S, Zanghieri G, Bernini MV, Lusetti L. Epatite acuta colestatica da ticlopidina. [Ticlopidine-induced acute cholestatic hepatitis. A case report]. Recenti Prog Med 2004;95(2):96–9.
30. Leone N, Giordanino C, Baronio MB, Morgando A, David E, Rizzetto M. Ticlopidine-induced cholestatic hepatitis successfully treated with corticosteroids: a case report. Hepatol Res 2004;28:109–12.
31. Yamamoto N, Shiraki K, Saitou Y, Kawakita T, Okano H, Sugimoto K, Murata K, Nakano T. Ticlopidine induced acute cholestatic hepatitis complicated with pure red cell aplasia. J Clin Gastroenterol 2004;38(1):84.
32. Yosipovitch G, Rechavia E, Feinmesser M, David M. Adverse cutaneous reactions to ticlopidine in patients with coronary stents. J Am Acad Dermatol 1999;41(3 Pt 1):473–6.
33. Braun-Moscovici Y, Schapira D, Balbir-Gurman A, Sevilia R, Menachem Nahir A. Ticlopidine-induced lupus. J Clin Rheumatol 2001;7:102–5.
34. Harder S, Klinkhardt U, Alvarez JM. Avoidance of bleeding during surgery in patients receiving anticoagulant and/or antiplatelet therapy: pharmacokinetic and pharmacodynamic considerations. Clin Pharmacokinet 2004;43(14):963–81.
35. Brown RI, Cooper TG. Ticlopidine–carbamazepine interaction in a coronary stent patient. Can J Cardiol 1997;13(9):853–4.
36. Tateishi T, Kumai T, Watanabe M, Nakura H, Tanaka M, Kobayashi S. Ticlopidine decreases the in vivo activity of CYP2C19 as measured by omeprazole metabolism. Br J Clin Pharmacol 1999;47(4):454–7.
37. Riva R, Cerullo A, Albani F, Baruzzi A. Ticlopidine impairs phenytoin clearance: a case report. Neurology 1996;46(4):1172–3.
38. Privitera M, Welty TE. Acute phenytoin toxicity followed by seizure breakthrough from a ticlopidine–phenytoin interaction. Arch Neurol 1996;53(11):1191–2.
39. Rindone JP, Bryan G 2nd. Phenytoin toxicity associated with ticlopidine administration. Arch Intern Med 1996;156(10):1113.
40. Klaassen SL. Ticlopidine-induced phenytoin toxicity. Ann Pharmacother 1998;32(12):1295–8.
41. Donahue SR, Flockhart DA, Abernethy DR, Ko JW. Ticlopidine inhibition of phenytoin metabolism mediated by potent inhibition of CYP2C19. Clin Pharmacol Ther 1997;62(5):572–7.

Tirofiban

General Information

Tirofiban is a synthetic, non-peptide inhibitor of platelet glycoprotein IIb/IIIa receptors. It has a rapid onset and short duration of action after intravenous administration, and aggregation returns to normal 4–8 hours after it is withdrawn. It is eliminated by the kidneys and dosage requirements are reduced in renal disease (1).

Organs and Systems

Respiratory

Fatal diffuse alveolar hemorrhage has been reported after the use of tirofiban in a patient with acute coronary syndrome (2).

Hematologic

Tirofiban causes thrombocytopenia less commonly than other glycoprotein IIb/IIIa inhibitors. In the Do Tirofiban and Reopro Give Similar Efficacy Outcomes (TARGET) study, the frequencies of thrombocytopenia (nadir platelet count under 100×10^9/l) were 2.4% with abciximab and 0.5% with tirofiban (3). In one case a patient developed acute profound thrombocytopenia on two occasions (4). Monitoring of platelet counts at 2–6 hours and 24 hours after administration has been advised (5). Intravenous immunoglobulin has been used in treatment (6).

References

1. Kondo K, Umemura K. Clinical pharmacokinetics of tirofiban, a nonpeptide glycoprotein IIb/IIIa receptor antagonist: comparison with the monoclonal antibody abciximab. Clin Pharmacokinet 2002;41(3):187–95.
2. Yilmaz MB, Akin Y, Biyikoglu SF, Guray U, Korkmaz S. Diffuse alveolar hemorrhage following administration of tirofiban in a patient with acute coronary syndrome: a fatal complication. Int J Cardiol 2004;93(1):81–2.
3. Merlini PA, Rossi M, Menozzi A, Buratti S, Brennan DM, Moliterno DJ, Topol EJ, Ardissino D. Thrombocytopenia caused by abciximab or tirofiban and its association with clinical outcome in patients undergoing coronary stenting. Circulation 2004;109(18):2203–6.
4. Eryonucu B, Tuncer M, Erkoc R. Repetitive profound thrombocytopenia after treatment with tirofiban: a case report. Cardiovasc Drugs Ther 2004;18(6):503–5.
5. Patel S, Patel M, Din I, Reddy CV, Kassotis J. Profound thrombocytopenia associated with tirofiban: case report and review of literature. Angiology 2005;56(3):351–5.
6. Clofent-Sanchez G, Harizi H, Nurden A, Coste P, Jais C, Nurden P. A case of profound and prolonged tirofiban-induced thrombocytopenia and its correction by intravenous immunoglobulin G. J Thromb Haemost 2007;5(5):1068–70.

ADVERSE CARDIOVASCULAR EFFECTS OF NON-CARDIOVASCULAR DRUGS

5-HT$_3$ receptor antagonists

Hypertension or disturbances of cardiac rhythm can occur with 5HT$_3$ receptor antagonists, especially in elderly patients (SEDA-20, 316).

Some patients have electrocardiographic changes, such as prolongation of the QT interval. In a double-blind, placebo-controlled, randomized study of the effects of dolasetron mesylate 0.6–5.0 mg/kg in 80 subjects, there were transient and asymptomatic electrocardiographic changes (small mean increases in PR interval and QRS complex duration versus baseline) in several subjects at 1–2 hours after infusion at doses of 3.0 mg/kg and over (1).

In one trial a patient who took ondansetron had syncope, presumably because of a change in blood pressure (SEDA-18, 370).

Acetylcholinesterase inhibitors

With any acetylcholinesterase inhibitor, bradycardia can, with excessive dosage, proceed to dysrhythmias (2) and even asystole.

- A 67-year-old man underwent left upper lobectomy for a presumed malignancy 11 years after cardiac transplantation (3). He had had no cardiac symptoms since his transplant. Suxamethonium was used as a muscle relaxant and was reversed with glycopyrrolate 0.8 mg and neostigmine 4 mg. Within a few minutes, he developed asystole, which lasted for about 45 seconds. He subsequently made a full recovery.

The authors speculated that some degree of cardiac reinnervation may have occurred; they recommended that this type of response should be anticipated in future anesthesia in such patients and that therapeutic measures, such as a beta-adrenoceptor agonist, should be available.

Another case of asystole has been reported with the very short-acting cholinesterase inhibitor edrophonium (4).

- A 49-year-old woman was given intravenous edrophonium chloride 2 mg as part of the investigation of an acute myopathy following gastrointestinal surgery. She had also received 60 mg of intravenous labetalol in the 14 hours before the edrophonium was given: presumably this was for a raised blood pressure, but that was not specified. Labetalol caused transient but severe bradycardia (heart rate about 20/minute). Immediately after the injection of edrophonium, she developed asystole, which was treated immediately with atropine and recovered in 10 seconds.

Such reactions are extremely rare, but in this case the risk was undoubtedly enhanced by previous beta-blockade.

With physostigmine, hypertension has been both demonstrated in animal experiments and observed in a series of patients after intravenous use in relatively high doses; it has also occurred during use of low doses of oral physostigmine in an elderly patient with Alzheimer's disease (5).

Acrylic bone cement

Cardiovascular reactions to acrylic bone cement are a common complication in bone surgery. It is believed that cementation activates an adrenocortical response, increasing the blood pressure during general anesthesia (6,7); during spinal anesthesia this response is suppressed and the blood pressure falls. The mechanism is thought to be by a direct effect on the blood pressure through the kallikrein–kinin system, since aprotinin (Trasylol), an inhibitor of kallikrein, prevents the fall in arterial pressure if it is given during the application of acrylic bone cement (8).

Some investigators suggested that implantation of acrylic bone cement into the femur increases plasma histamine, which, especially in elderly patients with pre-existing cardiac diseases and/or hypovolemia, can cause serious, sometimes fatal, cardiovascular complications (9).

Adrenaline

When the limits of tolerance are approached, there may be palpitation, extra beats, and a rise in blood pressure. In sensitive individuals or at high doses, ventricular fibrillation, subarachnoid hemorrhage, and even hemiplegia have been known to occur. Adrenaline can occasionally cause pulmonary edema (10,11). It is possible that in at least some of these cases the drug has been inadvertently injected intravenously.

In two cases in which adrenaline was used in noncritical circumstances there were severe adverse effects, including myocardial ischemia and dysrhythmias (12).

- A 64-year-old man was given an EpiPen to treat episodes of benign angioedema affecting the face and tongue, although he never considered the reaction to be life-threatening, and on two occasions was treated in an accident and emergency department with intramuscular adrenaline (doses not specified). On one occasion he developed chest pain and electrocardiographic signs of ischemia, which had not occurred before. The EpiPen was withdrawn and his angioedema was prevented by regular antihistamine administration.
- A 40-year-old woman developed angioedema after taking pseudoephedrine and diphenhydramine for sinusitis. Her breathing was not compromised and she was not hypotensive but she was given 1 ml of 1 in 1000 adrenaline intravenously, with subsequent ventricular tachycardia requiring cardiopulmonary resuscitation. Despite this an EpiPen was dispensed, though later withdrawn.

The authors commented that adrenaline could and should have been avoided in both cases. In response to this report immunologists from Melbourne commented that in the first case it was difficult to be sure that angioedema of the tongue was really benign, while in the second adrenaline should not have been given intravenously and certainly not in such a high dose (13). They therefore counselled against excessive caution in the use of adrenaline, provided the dose and route of administration are correct.

Ventricular dysrhythmias have been reported in a case of adrenaline overdose (14).

- A 5-year-old boy was given subcutaneous adrenaline 1:1000 after a severe allergic reaction to a bee sting. Inadvertently, 10 times the correct dose was given. He developed extra beats and two brief runs of ventricular tachycardia, but recovered fully after about 20 minutes. Creatine kinase activity, both total and the MB fraction, was slightly raised in this patient (total 603 IU/l, MB fraction 161 IU/l; upper limits of the local reference range 243 and 15 IU/l), suggesting cardiac damage.

Life-threatening torsade de pointes has been observed when an epidural anesthetic was given using 20 ml of bupivacaine containing only 1:200 000 adrenaline (15).

Long QT syndrome has been described as the mechanism of adrenaline-induced cardiac arrest (16).

- A 9-year-old boy with multiple congenital abnormalities was given epidural anesthesia containing 0.25% bupivacaine and 1:200 000 adrenaline before closure of a colostomy. This led within a few minutes to ventricular tachycardia and then two episodes of ventricular fibrillation. There was full recovery after cardiopulmonary resuscitation, without the need for cardioversion. It was then found that the epidural catheter had in fact entered a vein. Subsequent electrocardiograms showed consistent prolongation of the QT_c interval and the patient was therefore clearly at increased risk.

The main point of this case is the potential danger of accidental intravenous adrenaline injection even at very high dilutions.

In two Japanese patients, one man one woman, in whom adrenaline was used for this purpose during mandibular surgery, atrioventricular junctional rhythm occurred (17). In the man 11 ml of 2% lidocaine with 1:80 000 adrenaline was injected into the gums, while in the woman the dose was 9 ml of the same formulation injected sequentially into each side of the jaw. Both were anesthetized with sevoflurane. The episodes of dysrhythmia were self-limiting, lasting only 3.5 minutes in the man and about 10 and 3 minutes in the woman after each injection. The authors suggested that lidocaine may facilitate the absorption of adrenaline.

In a prospective double-blind study of the hemodynamic effects of infiltration with lidocaine + 1:200,000 adrenaline 76 patients were randomly allocated to three groups (18). Group I received 2% lidocaine 2 ml with adrenaline, group II received saline 2 ml with adrenaline, and group III received saline 2 ml without adrenaline. Adrenaline, with and without lidocaine, caused significant hemodynamic changes compared with saline. The changes lasted no more than 4 minutes. The authors concluded that the changes were due to the effects of adrenaline on β_2 adrenoceptors.

When adrenaline is used as an adjunct to interscalene block during shoulder surgery, which is often performed with the patient in a sitting position, up to 28% of patients develop hypotension, bradycardia, or both. In a prospective study intended to determine the role of adrenaline in these reactions in non-arthroscopic shoulder surgery 55 patients were given a scalene block with 0.5% bupivacaine and 2.5% lidocaine, together with 1:200 000 adrenaline, while a matched group of 55 patients had the same local anesthetics without the adrenaline (19). The incidence of hypotension or bradycardia was 28% in the adrenaline group but only 11% in the control group. Overall, however, the adrenaline-treated patients had higher intraoperative heart rates and blood pressures. Understandably, the authors argued against the use adrenaline in this setting.

It is somewhat surprising to see cellular phones cited as a cause of acute poisoning, but just that has been suggested in a report from New York (20).

- An 18-year-old man with septic shock was given an infusion of adrenaline at an initial rate of 0.15 micrograms/kg/minute. During the following 9 hours his systolic blood pressure varied from 92 to 105 mmHg and the infusion rate was 0.1–0.2 micrograms/kg/minute. He then complained of a sudden headache and chest and abdominal pain. His blood pressure was 250/188 mm Hg with a pulse of 198/minute. He developed pulmonary edema and raised cardiac enzymes. It was noted that the infusion bag was much emptier than it should have been—he had acutely received 10.5 mg of adrenaline more than had been intended.

Just before this episode a member of the family had received a call on a cellular phone. Subsequent testing by hospital engineers showed that the proximity of the phone during a call could have triggered the pump to run at nearly 1 litre/hour. Most hospitals, certainly in the UK, ask visitors to switch off their phones when entering clinical areas, although the need for this has been questioned (21,22). It is possible that this problem varies with the type and make of equipment, but this report suggests that it needs careful examination.

When adrenaline 0.4 ml of a 1 mg/ml solution was inadvertently injected into the penile skin of a 12-hour-old neonate the skin blanched and the error was immediately understood (23). After repeated doses of phentolamine (total 0.65 mg) the skin regained its normal color. There were no sequelae.

Adrenaline is occasionally used as a hemostatic agent, with rare complications. However, they do occur, as noted in a report from Lyon (24).

- A 64-year-old man with diabetes and hypertension bled from a site in the lower rectum. A local injection of adrenaline 0.2 mg successfully stopped the hemorrhage, but very soon after he became hypotensive, with rapid atrial fibrillation (ventricular rate not given), the first time he had experienced this. He reverted spontaneously to sinus rhythm within 24 hours.

The authors suggested that if this type of procedure is contemplated in elderly patients with cardiovascular disease an anesthetist should be present to monitor cardiovascular status; it may in any case be wiser to avoid adrenaline altogether in favor of other means of hemostasis.

Hypertension

DoTS classification (BMJ 2003; 327: 1222-5):
Dose-relation: Toxic
Time-course: Time-independent
Susceptibility factors: ?Age

Adrenaline is widely used in topical anesthesia, when of course systemic administration is not intended.

- A 19-year-old woman was given subcutaneous 0.25% bupivacaine with 1:100 000 adrenaline during arthroscopic knee surgery (25). Dilute adrenaline (calculated to be 0.3 μg/ml) was also used to irrigate the joint space during the procedure. There was sudden onset of hypertension and tachycardia (240/150 mmHg, heart rate 150/minute) and esmolol was immediately administered, but this was followed by acute pulmonary edema. She was treated with furosemide and hydrocortisone, but required mechanical ventilation. She recovered within 24 hours, with no evidence of ischemic myocardial damage.

The authors noted that in over 50 patients in whom dilute adrenaline irrigation was used for orthopedic procedures there were no cardiopulmonary complications (26). The authors could not be sure why this almost fatal complication occurred on this occasion, but they warned others of the possibility and emphasized the importance of accurate calculation of adrenaline concentrations (although there is no evidence of error in this case).

- A 73-year-old man in North Carolina developed acute confusion, fluent aphasia, and vomiting. He had no lateralizing motor signs other than slight right arm weakness, although both plantar responses were extensor. The blood pressure was 160/81 mmHg. A CT scan of the brain showed a left thalamic hemorrhage. It emerged that he was using large doses of inhaled adrenaline (0.22 mg/puff) as self-medication for chronic obstructive pulmonary disease. The maximum recommended dose was 2 puffs every 3 hours on not more than 2 days of the week, and this had clearly been exceeded by a large margin. He made an almost complete recovery, with slight residual weakness, and stopped using inhaled adrenaline.

Presumably the cause of the cerebral hemorrhage in this case was an acute episode of hypertension.

Inhaled adrenaline is available as an over-the-counter medication in at least some parts of the USA. In one case the use of inhaled adrenaline resulted in a stroke secondary to presumed hypertension (27).

In susceptible individuals, an attack of angina pectoris can be precipitated by adrenaline, and in any form of cardiac disease caution is indicated; at one time an attempt was made to use high doses of adrenaline for the early treatment of ventricular fibrillation, but its pharmacological effects swing the balance against its use, the immediate survival rate actually being reduced.

- A 79-year-old woman developed pituitary apoplexy in an adenomatous gland and was being prepared for trans-sphenoidal hypophysectomy (28). Topical adrenaline (1:1000) was applied to both nostrils and then 1.5 ml of 1% lidocaine containing 1:100 000 adrenaline was injected into the nasal mucosa. The blood pressure immediately rose from 100/50 to 230/148 mmHg and the pulse rate from 48 to 140/minute. Although she was treated immediately with esmolol and intravenous glyceryl trinitrate, resulting in normalization of her blood pressure, subsequent investigations showed that she had had a painless myocardial infarction. She made a full recovery after pituitary surgery.

The authors suggested that if adrenaline is to be used in such cases, even lower concentrations might be advisable. This is reasonable, although one also wonders in this case whether her blood pressure may have been lowered too rapidly.

Adrenoceptor agonists

See also Beta$_2$-adrenoceptor agonists

Adrenoceptor agonists evoke physiological responses similar to those produced by stimulation of adrenergic nerves or the physiological release of adrenaline (see Table 1). For many of these responses it is currently possible to conclude that only an alpha-adrenoceptor or a beta-adrenoceptor is involved, and in some cases one can distinguish a beta$_1$ from a beta$_2$ response. In some cases, however, the distinction is not clear: most adrenoceptor agonists, however specific to a particular receptor type they are claimed to be, will for example on occasion stimulate central nervous functions, resulting in nervousness, insomnia, tremors, dizziness, or headache. In some organ systems both alpha-adrenoceptors and beta-adrenoceptors are present; thus, the nature of the response produced will depend either on the concentrations achieved or on other factors; whether, for example, the uterus contracts or relaxes in response to an adrenergic drug depends in part on the hormonal balance in the system at that moment.

Alpha-adrenoceptor agonists, such as clonidine, are little used nowadays in the treatment of hypertension or migraine. Clonidine is used epidurally, in combination with opioids, neostigmine, and anesthetic and analgesic agents, to produce segmental analgesia, particularly for postoperative relief of pain after obstetrical and surgical procedures. Apraclonidine is available for the short-term reduction of intraocular pressure.

The drugs that were developed some 40 years ago as general beta-adrenoceptor agonists have largely fallen into disuse with the development of more selective beta$_1$-adrenoceptor agonists (for use in cardiac failure) and beta$_2$-adrenoceptor agonists (for use in airways disease and threatened premature labor).

Alatrofloxacin and trovafloxacin

Phlebitis can occur during parenteral administration of trovafloxacin. High concentrations of trovafloxacin

Table 1 Adrenoceptors and the effects of agonists

Organs and systems	*Receptor*	*Response to an agonist*
Cardiovascular		
Heart		
Sinoatrial node	β_1	Increased heart rate
Atria	β_1	Increased contractility and conduction velocity
Atrioventricular node and conduction system	β_1	Increased conduction velocity and automaticity
Ventricles	β_1	Increased contractility, conduction velocity, automaticity, rate of idiopathic pacemakers
Blood vessels		
Coronary	α, β_2	Constriction
Skin, mucosa	α	Constriction
Skeletal muscle	α or β_2	Constriction or dilatation
Cerebral	α	Slight constriction
Pulmonary	α or β_2	Constriction or dilatation
Abdominal viscera	α or β_2	Constriction or dilatation
Salivary glands	α	Constriction
Respiratory		
Bronchial muscle	β_2	Relaxation
Bronchial glands	α_1, β_2	Decreased or increased secretion
Nervous system		
Cerebral function	Various	Stimulation
Eyes		
Radial muscle, iris	α	Contraction (mydriasis)
Ciliary muscle	β	Relaxation for far vision (slight)
Hematologic		
Spleen capsule	α	Contraction
Salivary glands	α_1	Potassium and water secretion
	β	Amylase secretion
Gastrointestinal		
Motility and tone	α_1, β_1, β_2	Decrease (usually)
Sphincters	α	Contraction (usually)
Secretion of various substances	Various	Inhibition
Liver		
Glycogenolysis and gluconeogenesis	α_1, β_2	Stimulation
Gallbladder		
Bile ducts	β_2	Relaxation
Urinary tract		
Ureter; tone, motility	β_2	Relaxation (usually)
Bladder; detrusor	β	Relaxation (usually)
Trigone, sphincter	α	Contraction
Renal vessels	α_1, β_1, β_2	Primary contraction
Skin		
Pilomotor muscles	α	Contraction
Sweat glands	α	Slight local secretion
Musculoskeletal		
Muscle glycogenolysis	β	Stimulation
Sexual function		
Uterus	α, β_2	Variable effect[a]
Male sex function	α_1	Ejaculation

[a]Response depends inter alia on hormonal status.

(2 mg/ml) significantly reduced intracellular ATP content in cultured endothelial cells and reduced concentrations of ADP, GTP, and GDP (29). These in vitro data suggest that high doses of trovafloxacin are not compatible with maintenance of endothelial cell function and may explain the occurrence of phlebitis. Commercial formulations should be diluted and given into large veins.

Albumin

Albumin infusions can increase capillary permeability (30) and albumin infusion during resuscitation can result in hypervolemia, due to overfilling of the circulation (31).

It has been postulated that cardiac pump function can be reduced by binding of free calcium to albumin (31).

Alcuronium

Tachycardia, hypotension, and a fall in total peripheral resistance all occur to an extent similar to that seen with D-tubocurarine, according to most studies (32–35). Others have reported that these effects are short-lived (36). Doses of 0.2 mg/kg or more may be associated with the more extreme cardiovascular effects. Blockade of cardiac muscarinic receptors (37), histamine release, and, possibly, some ganglionic blockade (although it has a very low ganglion-blocking activity in animals) (37) may all play a role in the production of the cardiovascular effects of alcuronium.

Aldesleukin

Hemodynamic and cardiac complications are the major limitations of high-dose aldesleukin and have been described in both adults (38,39) and children (40). Significant hypotension requiring meticulous maintenance therapy with intravenous fluids or low-dose vasopressors was observed in most patients (41). The clinical findings were very similar to the hemodynamic pattern seen in early septic shock. Aldesleukin-induced increases in plasma nitrate and nitrite concentrations correlated with the severity of hypotension (42).

Among other cardiovascular complications, cardiac dysrhythmias were reported in 6–10% of patients, angina pectoris or documented myocardial infarction in 3–4%, and mortality due to myocardial infarction in 1–2% (43). Severe myocardial dysfunction, myocarditis, and cardiomyopathy have been seldom reported (SED-13, 1103; SEDA-20, 334; SEDA-22, 406).

The cardiopulmonary toxicity of high-dose intravenous bolus aldesleukin has been analysed in 199 metastatic melanoma or renal cell carcinoma patients without underlying cardiac disease (44). Cardiovascular events occurred within hours after starting infusion, persisted throughout aldesleukin therapy, and normalized within 1–3 days after treatment withdrawal. Hypotension was the most frequent adverse effect (53% of treatment courses) and resolved promptly with vasopressor treatment. Unexpectedly, the response to treatment was significantly better in patients with melanoma who had hypotension. There were cardiac dysrhythmias in 9% of patients; they mostly consisted of easily manageable atrial fibrillation or supraventricular tachycardia. Further courses of aldesleukin in 11 of these patients produced recurrent dysrhythmias in only two, and long-term treatment of dysrhythmias was never required. High-degree atrioventricular block and repetitive episodes of ventricular tachycardia were each observed once. Although 11% of patients had raised creatine kinase activity before or during treatment, only 2.5% had a documented rise in the MB isoenzyme fraction.

At-risk patients include those with pre-existing cardiac disease, whereas age, performance status, and sex are not significantly associated with cardiopulmonary toxicity. In view of this risk, it is reasonable to monitor cardiac function and creatine kinase activity closely in all patients, or to exclude those with significant underlying coronary or cardiorespiratory disease. Pretreatment cardiac screening has greatly reduced the incidence of myocardial infarction, ischemia, and related dysrhythmias, and two-dimensional and Doppler echocardiography was suggested to be helpful to anticipate cardiovascular toxicity (45). A reduction in systemic vascular resistance, stroke work index, and left ventricular ejection fraction are usually involved in the pathophysiology of cardiac dysfunction. Clinical, electrocardiographic, and radionuclide ventriculography monitoring in 22 patients undergoing a 5-day continuous intravenous infusion of aldesleukin for various cancers showed that reversible left ventricular dysfunction accounted for most of the observed hemodynamic changes (46). Indeed, significant coronary disease was usually not observed in patients undergoing cardiac catheterization, which argues for direct myocardial damage (44). In isolated reports, clinical and histological findings of eosinophilic, lymphocytic, or mixed lymphocytic–eosinophilic myocarditis also suggested an immune-mediated drug reaction (SED-13, 1103) (SEDA-22, 406).

Aldesleukin-induced cardiac eosinophilic infiltration has been reported (47).

- After 25 days of treatment with continuous aldesleukin infusion (up to 150 000 units/kg/day) for stage IV Hodgkin's disease, a 26-year-old woman had increased fatigue, tachycardia, hypotension, and hypothermia. Echocardiography showed bilateral intraventricular masses. Her maximal absolute eosinophil count was $11.4 \times 10^9/l$ and the platelet count was $17 \times 10^9/l$. Despite aldesleukin withdrawal, her condition deteriorated and she died. Postmortem examination showed biventricular thrombi and prominent eosinophilic infiltration of the endomyocardium.

Of 10 subsequent patients who received prolonged infusions of aldesleukin and were monitored by echocardiography, one developed asymptomatic changes in cardiac function, with features suggestive of early thrombus formation and a reduced ejection fraction during weeks 6–8. The maximal absolute eosinophil count was $5 \times 10^9/l$. These abnormalities resolved on aldesleukin withdrawal.

Conjugates of aldesleukin with polyethylene glycol produce less cardiovascular toxicity (SEDA-20, 333).

Alemtuzumab

Alemtuzumab caused cardiotoxicity in four of eight patients with mycosis fungoides/Sézary syndrome and no prior history of cardiac problems. The effects included congestive heart failure or dysrhythmias that improved after alemtuzumab withdrawal (48).

Cardiac ischemia has been attributed to alemtuzumab in a man with chronic lymphocytic leukemia (49).

- A 58 year-old man with B cell chronic lymphocytic leukemia underwent two coronary angioplasties and was asymptomatic thereafter. Chemotherapy before alemtuzumab consisted of fludarabine, melphalan, prednisone, chlorambucil, cyclophosphamide, vincristine, and cladribine. After the second dose of alemtuzumab

10 mg he developed severe chills and a fever, followed by shortness of breath, typical chest pain, hypotension, and hypoxemia. An electrocardiogram showed depressed ST segments laterally. A chest X-ray showed pulmonary edema. The MB fraction of creatine kinase activity was raised. Echocardiography showed segmental hypokinesia but a normal ejection fraction and moderate aortic stenosis. He improved and was given two more doses of alemtuzumab 10 mg subcutaneously, followed a week later by 30 mg; 7 hours after the last infusion he developed chest pain and dyspnea with electrocardiographic changes. He was admitted to the coronary unit where he recovered without complications. He was given further doses of alemtuzumab 10 mg subcutaneously and a glyceryl trinitrate patch with each injection. He had no further symptoms and received six more doses.

Alfentanil

When alfentanil 30 μg/kg was given to six healthy volunteers there were no clinical changes in respiratory or cardiovascular function (SEDA-16, 78).

Bradycardia often occurs with the combination of a potent short-acting opioid with suxamethonium during induction of anesthesia, and alfentanil has been reported to have caused sinus arrest in three patients (SEDA-17, 79) (50).

In one study alfentanil was particularly likely to cause hemodynamic instability and myocardial ischemia; however, drug interactions or the dosage regimen may have been responsible (51).

Aliphatic alcohols

Skin disinfection before insertion of peripheral infusion catheters is standard practice. Ethanol 70% has been compared with 2% iodine dissolved in 70% ethanol in a prospective, randomized trial in 109 patients who were given infusions of prednisone and theophylline (52). Phlebitis occurred six times in the ethanol group and 12 times in the iodine group. The relative risk reduction of 53% failed to reach significance, but the power of the study was only 0.55, so there was a 45% chance of missing a true difference. As vast numbers of catheters are inserted each year, a small difference in phlebitis rate could save many patients discomfort.

$Alpha_1$-antitrypsin

Chest pain and cyanosis occurred in 14 patients but the symptoms seemed to be related to the presence of sucrose in the product (53).

Alpha-glucosidase inhibitors

The effects of miglitol 25 or 50 mg tds have been compared with placebo and various doses of glibenclamide in 411 patients aged over 60 years (mean 68 years) with mild type 2 diabetes insufficiently controlled by diet for 56 weeks (54). HbA_{1c} fell after 1 year by 0.92% (glibenclamide), 0.49% (miglitol 25 mg), and 0.40% (miglitol 50 mg). Gastrointestinal events were most common in patients taking miglitol. Most (88%) of the hypoglycemic events were minor and occurred in the patients taking glibenclamide. Mean body weight increased continuously and significantly in those taking glibenclamide. In the other groups weight fell by more than 1 kg. The rate of withdrawal was the same in all groups; withdrawal was mostly occasioned by cardiovascular effects in those taking glibenclamide and by hyperglycemia, flatulence, or diarrhea in those taking miglitol.

When patients with inadequately controlled type 2 diabetes used glibenclamide plus metformin, miglitol, or placebo for 24 weeks in addition to their earlier therapy, fasting blood glucose concentrations improved with miglitol (55). Flatulence and diarrhea were significantly more common with miglitol. No patient stopped taking miglitol because of adverse effects.

A review of miglitol included data on adverse effects in 3585 patients in well-designed clinical trials (56). Only the adverse effects in the gastrointestinal tract occurred with a significantly greater incidence with miglitol 50 or 100 mg tds. The adverse effects were the same as with other drugs in this class: flatulence, diarrhea, dyspepsia, and abdominal pain. There were no differences with monotherapy or combination therapy or in relation to age or ethnicity. There were more episodes of hypoglycemia when miglitol was combined with insulin but not with oral agents. The incidence of cardiovascular events was the same as with placebo.

Long-term acarbose had a good effect on late dumping syndrome in six patients with type 2 diabetes; one patient complained of increased flatulence (57).

Although acarbose often causes abdominal complaints, dietary manipulation has not been used to reduce the complaints (58).

In 120 patients with type 1 diabetes, acarbose lowered postprandial glucose but did not reduce HbA_{1c} (59). Four patients taking acarbose withdrew because of gastrointestinal effects, which improved after withdrawal. One of the placebo group withdrew because of gastrointestinal problems and one other patient taking acarbose withdrew with a Bell's palsy, which was not considered to be related to acarbose.

Ileus has also been reported with acarbose (60–63).

Paralytic ileus with intestinal pneumatosis cystoides has been reported (62).

- An 87-year-old woman, who took acarbose, glibenclamide, and mannitol (for constipation), developed abdominal distention and loss of appetite. An X-ray showed distention of the small intestine, with pockets of small gas bubbles in the submucosal When her drugs were withdrawn, her symptoms subsided and the radiological evidence of ileus disappeared by 5 days. Although she had an atonic bladder, there were no signs of neuropathy. She was also hypothyroid, which could have contributed.

Acarbose may also have caused pneumatosis cystoides intestinalis in a 55-year-old woman with pemphigus vulgaris (64).

Most cases of ileus with acarbose have been reported in Japan.

- A 73-year-old man with diabetic gangrene who had used insulin and acarbose 300 mg/day for 15 months developed ileus with abdominal pain and vomiting after he took PL granules (containing salicylamide, paracetamol, anhydrous caffeine, and promethazine methylene disalicylate) for a common cold (63). The ileus subsided after acarbose and the other drugs were withdrawn.

Although the ileus in this case was not clearly related to the use of acarbose, the combination of acarbose, which can cause ileus, with the other drugs that the patient was taking, may have caused it. The anticholinergic effect of promethazine methylene disalicylate may have contributed.

Lymphocytic colitis activated by acarbose has been reported (65).

- A 52-year-old man developed watery diarrhea 6–8 times a day 2 weeks after he had started to take acarbose 100 mg. In 3 weeks he lost 3 kg. Duodenal biopsies were normal; colon biopsies showed a large increase in intraepithelial lymphocytes. The mononuclear cells expressed CD-25, and HLA-DR antigen was increased in epithelial cells. Within 4 days of acarbose withdrawal the diarrhea had disappeared, and biopsies 4 months later showed that CD-25 expression in the cells of the lamina propria was improved and HLA-DR was no longer expressed by the epithelial cells. On rechallenge the diarrhea recurred within 3 days. Biopsies showed pronounced HLA-DR in the epithelial cells and CD-25 expression in some mononuclear cells in the lamina propria.

Non-digestable sugar substitutes and alpha-glucosidase inhibitors should probably not be used in combination.

Alprazolam

Alprazolam has been associated with hypotension (66).

- A 76-year-old woman, who had a history of hypertension, valvular heart disease (mitral regurgitation) with chronic atrial fibrillation, chronic obstructive airways disease, diverticular disease of the sigmoid colon, and generalized anxiety disorder, developed severe hypotension with a tachycardia after taking alprazolam for 7 days. She also had severe weakness, depressed mood, and impaired gait and balance, without clinical features of neuromuscular disease.

Alprostadil

Moderate or severe phlebitis can occur at the site of venepuncture in some patients who receive alprostadil by infusion. It is sometimes severe enough to necessitate withdrawal of therapy. The frequency and severity of phlebitis has been investigated in 18 men, mean age 63 (range 47–78) years, with peripheral vascular disease who received a 2-hour infusion twice daily (67). Although it is usual to dissolve 60 micrograms of alprostadil in 500 ml of fluid to avoid phlebitis, in this study 200 ml was used to prevent volume overload. The solution was neutralized to pH 7.4 with 4 ml of 7% sodium bicarbonate. Two patients had grade 0, four grade 1, 11 grade 2, and one grade 3 phlebitis (by Dinley's criteria (68)). Age correlated negatively with the severity of phlebitis. Usually, alprostadil infusion therapy is stopped when phlebitis reaches grade 4 or more, but there were no such cases in this study.

Amantadine

Reversible congestive cardiac failure has been attributed to amantadine (69).

Amfebutamone (bupropion)

The UK Committee on Safety of Medicines has received over 200 reports of chest pain in patients taking amfebutamone, and a case of myocardial infarction has been reported.

- A 43-year-old man who had smoked up to 20 cigarettes daily for several years and had a family history of heart disease was given amfebutamone and stopped smoking after reaching the recommended dose (300 mg/day) (70). Three days later he developed central chest and arm pain and 1 day later stopped taking amfebutamone. Three days after this he developed classical symptoms of acute inferoposterolateral myocardial infarction. He was treated with thrombolytic therapy and discharged taking secondary prevention therapy.

It is difficult to know how far amfebutamone might have contributed to this acute cardiac event. However, continuing vigilance and case-control studies are warranted.

Aminocaproic acid

Aminocaproic acid has been reported to cause acute right heart failure (71).

Aminoglycoside antibiotics

Anecdotal reports refer to tachycardia, electrocardiographic changes, hypotension, and even cardiac arrest (72). In practice, effects on the cardiovascular system are unlikely to be of any significance.

Aminosalicylates

A single Scandinavian case report has reliably attributed a fatal myocarditis to mesalazine (73).

The time between the start of treatment and the onset of mesalazine-associated pericarditis usually ranges from a few days to 7 months. However, the delay can be longer.

- A 53-year-old man with Crohn's disease, who had taken mesalazine 500 mg/day for the past 8 years, developed pericarditis with an effusion (74). The pericarditis resolved rapidly on drug withdrawal. Investigations excluded other common causes of pericarditis.

- Pericarditis in a 16-year-old boy with inflammatory bowel disease taking mesalazine resolved on withdrawal, but recurred after starting sulfasalazine 500 mg tds (75).
- Acute pericarditis has been reported in a 17-year-old man with severe ulcerative colitis who had taken mesalazine 1.5 g/day for 2 weeks (76). The pericarditis resolved on withdrawal and recurred on rechallenge with a low dose (62.5 mg) of mesalazine 3 weeks later.
- Constrictive pericarditis has been reported in a 37-year-old woman with chronic ulcerative colitis who had taken mesalazine 2 g/day for 2 weeks (77). She recovered after radical pericardiectomy.

Bradycardia has been attributed to mesalazine.

- Severe symptomatic bradycardia has been reported in a 29-year-old woman with ulcerative colitis who was taking mesalazine (78). The bradycardia resolved on withdrawal. Six weeks later mesalazine was restarted for a relapse of her colitis, and symptomatic bradycardia recurred. Again this resolved on withdrawal.

Chest pain has been attributed to mesalazine (79).

- A 37-year-old man with ulcerative colitis developed severe retrosternal chest pain with non-specific ST–T wave changes in the inferolateral leads of the electrocardiogram after taking mesalazine 800 mg qds for 1 week. Cardiac enzymes, coronary angiography, left ventricular function, and pulmonary angiography were normal. Mesalazine was omitted and glucocorticoids were tapered. He recovered completely and his electrocardiogram normalized. Two weeks later he was given mesalazine 800 mg qds and again developed retrosternal chest pain with T wave inversion in the lateral leads. Mesalazine was withdrawn and his symptoms resolvedwithin 24 hours. His chest pain did not recur over an 18 month follow-up period while not taking mesalazine.

Raynaud's phenomenon has been attributed to sulfasalazine (80,81).

Amisulpride

QT interval prolongation has been attributed to amisulpride.

- Sinus bradycardia and QT interval prolongation occurred in a 25-year-old man taking amisulpride 800 mg/day (82). The dosage of amisulpride was reduced to 600 mg/day and the electrocardiogram normalized within a few days.

Amodiaquine

Prolongation of the QT interval is a recognized effect of 4-aminoquinolines. In 20 adult Cameroonian patients with non-severe falciparum malaria treated with amodiaquine (total dose 30 mg/kg or 35 mg/kg over 3 days) there was asymptomatic sinus bradycardia ($n = 16$) and prolongation of the PQ, QRS, and QT intervals at the time of maximum cumulative concentration of drug (day 2 of treatment) (83).

Amphetamines

See also Methylenedioxymetamfetamine

Tachycardia, dysrhythmias, and a rise in blood pressure have been described after the administration of centrally acting sympathomimetic amines. Amfetamine acutely administered to men with a history of amfetamine abuse enhanced the pressor effects of tyramine and noradrenaline, while continuous amfetamine led to tolerance of the pressor response to tyramine. As with intravenous amphetamines, cardiomyopathy, cardiomegaly, and pulmonary edema have been reported with smoking of crystal metamfetamine (84–86).

During short-term treatment with a modified-release formulation of mixed amfetamine salts in children with ADHD, changes in blood pressure, pulse, and QT_c interval were not statistically significantly different from the changes that were seen in children with ADHD taking placebo (87). Short-term cardiovascular effects were assessed during a 4-week, double-blind, randomized, placebo-controlled, forced-dose titration study with once-daily mixed amfetamine salts 10, 20, and 30 mg (n = 580). Long-term cardiovascular effects were assessed in 568 subjects during a 2-year, open extension study of mixed amfetamine salts 10–30 mg/day. The mean increases in blood pressure after 2 years of treatment (systolic 3.5 mmHg, diastolic 2.6 mmHg) and pulse (3.4/minute) were clinically insignificant. These findings differ from previously reported linear dose-response relations with blood pressure and pulse with immediate-release methylphenidate during short-term treatment (88). These differences may be attributable to differences in timing between dosing and cardiovascular measurements or to differences in formulations. Both amphetamine and methylphenidate have sympathomimetic effects that can lead to increases in systolic blood pressure and diastolic blood pressure at therapeutic doses, although the sizes of the effects on blood pressure may differ (89).

Amfetamine

Coronary artery rupture has been associated with amfetamine abuse (90).

- A 31-year-old woman suddenly developed central chest pain, with a normal electrocardiogram. Changes in troponin and creatine kinase MB were consistent with acute myocardial infarction. Drug screening was positive for amphetamines and barbiturates. Coronary angiography showed an aneurysm with 99% occlusion of the proximal left circumflex coronary artery and extravasation of contrast material. A stent was inserted percutaneously and antegrade flow was achieved without residual stenosis.

An uncommon presentation of amfetamine-related acute myocardial infarction due to coronary artery spasm has been reported (91).

- A 24-year-old man developed an acute myocardial infarction involving the anterior and inferior walls within 3 hours of taking intravenous amfetamine. A coronary angiogram showed plaques in the mid-portion of the left anterior descending artery, which developed spasm after the administration of intracoronary ergonovine. He was discharged after treatment with verapamil, isosorbide mononitrate, and aspirin. He subsequently developed early morning chest tightness 2 weeks, 1 month, 2 months, and 9 months after discharge. On each occasion he left against medical advice.

These findings suggest that coronary artery plaques played a role in endothelial dysfunction resulting from amfetamine use, and that induction of coronary artery spasm, a finding not reported before, was the likely mechanism of amfetamine-related acute myocardial infarction.

The cardiovascular response to an oral dose of *d*-amfetamine 0.5 mg/kg has been determined in 81 subjects with schizophrenia, 8 healthy controls who took amfetamine, and 7 subjects with schizophrenia who took a placebo (92). Blood pressure increased in both amphetamine groups, whereas placebo had no effect. However, pulse rate did not change in the schizophrenic group and only increased after 3 hours in the controls. Intramuscular haloperidol 5 mg produced a more rapid fall in systolic blood pressure in six subjects, compared with 12 subjects who did not receive haloperidol. The authors concluded that increased blood pressure due to amfetamine may have a dopaminergic component. They also suggested that haloperidol may be beneficial in the treatment of hypertensive crises caused by high doses of amfetamine or metamfetamine.

Two cases of myocardial infarction after the use of amfetamine have been reported (93,94).

- A 34-year-old man who smoked a pack of cigarettes a day took amfetamine for mild obesity. He developed an acute myocardial infarction 1 week later. Echocardiography showed inferior left ventricular hypokinesia and a left ventricular ejection fraction of 50%. Coronary cineangiography showed normal coronary arteries but confirmed the inferior left ventricular hypokinesia. Blood and urine toxicology were positive only for amfetamine.
- A 31-year-old man developed generalized discomfort after injecting four doses of amfetamine and metamfetamine over 48 hours, but no chest pain or tightness or shortness of breath. Electrocardiography showed inverted T-waves and left bundle branch block. Echocardiography showed reduced anterior wall motion.

The authors reviewed other reported cases of myocardial infarction associated with amphetamines. The patients were in their mid-thirties and most were men. The interval from the use of amphetamines to the onset of symptoms varied from a few minutes to years. No specific myocardial site was implicated. Coronary angiography in most cases showed non-occlusion. The cause of myocardial ischemia in these cases was uncertain, even though coronary artery spasm followed by thrombus formation was considered the most likely underlying mechanism. Some have suggested that electrocardiographic and biochemical cardiac marker testing should be considered in every patient, with or without symptoms suggesting acute coronary syndrome, after the use of amphetamines. Others have suggested that calcium channel blockers may play an important role in the treatment of myocardial infarction due to amfetamine use or abuse. In one patient, administration of beta-blockers caused anginal pain, suggesting that they should be avoided. All the patients except one had a good outcome.

Vertebral artery dissection has been described in a previously healthy man with a 3-year history of daily oral amfetamine abuse (95).

- A healthy 40-year-oldhanded man presented with a 3-day history of an occipital headache and imbalance. He had a 3-year history of daily oral amfetamine abuse with escalating quantities, the last occasion being 12 hours before the onset of the symptoms. He had a history of "speed" abuse and a 20-pack-year history of tobacco use. He had mild right arm dysmetria without ataxia. His brain CT scan without contrast was normal. He then developed nausea, vomiting, visual loss, and progressive obtundation. He had hypertension (160/90 mmHg), bilateral complete visual loss, right lower facial weakness, mild dysarthria without tongue deviation, divergent gaze attenuated by arousal, bilateral truncal and appendicular dysmetria with inability to stand and walk, and generalized symmetrical hyperreflexia with extensor plantar reflexes. His urine screen was positive for metamfetamine. A brain MRI scan showed infarction of both medial temporal lobes, the left posteromedial thalamus, and the right superior and left inferior cerebellum. Magnetic resonance angiography and fat saturation MRI showed reduced flow in the left vertebral artery and a ring of increased signal within its lumen, consistent with hematoma and dissection. He was treated with anticoagulants and made a partial recovery.

Since this patient had no known risk factors for vertebral artery dissection and had abused amfetamine daily for 3 years with escalating amounts, an association between metamfetamine and vertebral artery dissection cannot be excluded. The local and systemic vascular impacts of amfetamine could have contributed to initial changes (along with smoking), resulting in dissection.

Metamfetamine

The frequency of acute coronary syndrome in patients who developed chest pain after using metamfetamine has been described. In 33 patients (25 men, 8 women, mean age 40 years) metamfetamine abuse was confirmed by urine screening. Acute coronary syndrome was diagnosed in nine patients. Three patients (two of whom had acute coronary syndrome) had cardiac dysrhythmias. The authors concluded that acute coronary syndrome is common in patients with chest pain after metamfetamine use, and that the frequency of other potentially life-threatening cardiac complications is not negligible.

A normal electrocardiogram reduces the likelihood of acute coronary syndrome, but an abnormal electrocardiogram is not helpful in distinguishing patients with or without acute coronary syndrome (96). These conclusions were based on a retrospective chart review conducted to explore metamfetamine-associated acute coronary syndromes in patients who presented to the emergency room at a University Center between 1994 and 1996. There were 36 admissions, three of which were repeat patients. Nine of these patients had acute coronary syndrome. Of these, one had an acute anterior Q wave myocardial infarction with cardiac arrest, seven had non-Q wave myocardial infarctions, and one had unstable angina. There were potentially life-threatening cardiac complications in three subjects (8%). The authors suggested that acute coronary syndromes and life-threatening complications associated with the use of metamfetamine are not uncommon, as evidenced by their experience in this study. The generalizability of these findings is limited.

- A 44-year-old female metamfetamine abuser died unexpectedly due to right-sided infective endocarditis (97).
- A 27-year-old man who had used intravenous amfetamine had an acute myocardial infarction (98).

The authors suggested that coronary angiography with an ergometrine provocation test might be necessary in patients who use amfetamine and who develop acute coronary syndrome.

The cardiovascular effects of metamfetamine in 11 patients with Parkinson's disease have been described and compared with six healthy controls (99). All were tested twice, once with intravenous saline and once with intravenous metamfetamine 0.3 mg/kg. Cardiovascular measurements were taken for 15 minutes before drug administration and for 103 minutes after. Both groups had significant increases in blood pressure after metamfetamine, but the patients with Parkinson's disease had a shorter duration of increased blood pressure and a lower increase from baseline than the controls. The authors proposed that these results suggested cardiac and vascular hyposensitivity to metamfetamine in Parkinson's disease due to impaired metamfetamine-induced catecholamine release.

Others

Of the other central stimulants, aminorex, doxapram, fenfluramine, and fenfluramine plus phentermine can cause chronic pulmonary hypertension, as can chlorphentermine, phentermine, phenmetrazine, and D-norpseudoephedrine (SED-9, 8). A genetic predisposition may be involved (SED-9, 8). Pulmonary hypertension may develop or be diagnosed a long time after the drug has been withdrawn.

Amphotericin

Electrolyte disturbances (hyperkalemia, hypomagnesemia, renal tubular acidosis) due to renal toxicity can be additional factors that precipitate cardiac reactions.

Effects on blood pressure

Changes in blood pressure (hypotension as well as hypertension) have been reported (100,101).

- A 67-year-old man with multiple intraperitoneal and urinary fungal pathogens and a history of well-controlled chronic hypertension developed severe hypertension associated with an infusion of ABLC (102). He received a 5 mg test dose, which was tolerated without incident. About 60 minutes into the infusion (5 mg/kg), his blood pressure rapidly increased to 262/110 mmHg from a baseline of 150/80 mmHg. His temperature increased to 39.8°C, and tachycardia developed (up to 121/minute). The infusion was stopped, and he was given morphine, propranolol, and paracetamol. His blood pressure returned to baseline over the next 2 hours. Rechallenge with ABLC on the next day resulted in an identical reaction despite premedication with pethidine, diphenhydramine, and morphine. ABLC was permanently withdrawn, and the infection was managed with high dosages of fluconazole.

The etiology of amphotericin-associated hypertension has not been elucidated, but it may be related to vasoconstriction. Of note, the traditional test dose appears not to identify individuals predisposed to hypertensive reactions; four of six cases of amphotericin-associated hypertension received test doses without incident.

Cardiac dysrhythmias

In 6 of 90 children given intravenous amphotericin there was a significant fall in heart rate, and monitoring of heart rate was recommended in children with underlying heart disease (103). These immediate reactions follow intravenous administration and occur particularly with excessively rapid infusion of DAMB.

Ventricular dysrhythmias have been reported after rapid infusion of large doses of DAMB (104) in patients with hyperkalemia and renal insufficiency, but not in patients with normal serum creatinine and potassium concentrations, even if they have received the drug over a period of 1 hour. Slower infusion rates and infusion during hemodialysis have been advocated in patients with terminal kidney insufficiency, in order to avoid hyperkalemia.

Chest pain

Three cases of chest discomfort associated with infusion of L-AmB at a dosage of 3 mg/kg/hour for 1 hour have been reported (105).

- The first patient had chest tightness and difficulty in breathing and the second had dyspnea and acute hypoxia (PaO_2 55 mmHg; 7.3 kPa), both within 10 minutes of the start of the infusion. The third complained of chest pain 5 minutes after the start of two infusions.

In all cases the symptoms resolved on terminating therapy. Two patients were later rechallenged with slower infusions and tolerated the drug well.

A review of the literature showed that similar reactions had been reported anecdotally in several clinical trials of L-AmB, with all other formulations, and with liposomal daunorubicin and doxorubicin. While the pathophysiology of such reactions is yet unclear, the authors recommended infusing L-AmB over at least 2 hours with careful monitoring of adverse events.

Myocardial ischemia

Rare instances of cardiac arrest have been reported (106).

Cardiomyopathy

Reversible dilated cardiomyopathy has been attributed to amphotericin B deoxycholate (DAMB).

- A 20-year old man with fluconazole-refractory disseminated coccidioidomycosis without evidence of cardiac involvement developed dilated cardiomyopathy and clinical congestive heart failure after 2 months of therapy with amphotericin B (0.7 mg/kg/day of amphotericin B deoxycholate, switched after 1 month of treatment, because of rising serum creatinine concentrations, to amphotericin B lipid complex 5mg/kg/day. His echocardiographic abnormalities and heart failure resolved within 6 weeks, posaconazole having been substituted for amphotericin B after about 90 days (107).

Vascular effects

Phlebitis occurs in over 5% of patients receiving amphotericin deoxycholate through peripheral veins, which limits the concentration advisable for this route of administration.

Raynaud's syndrome has been attributed to DAMB phenomenon (108).

Extravasation can cause severe local reactions, including tissue necrosis. Safe venous access, preferably via a central line is advisable. The recommendation to use sodium heparin or buffered dextrose is not supported by clinical data.

Amrinone in amrinone

There has been one report of paroxysmal supraventricular tachycardia in one of 16 patients with cardiogenic shock given amrinone (109).

Androgens and anabolic steroids

Particularly when androgens/anabolics are misused to promote extreme muscular development, there is a risk of cardiomegaly and ultimate cardiac failure. Androgen-induced hypertension may be due to a hypertensive shift in the pressure-natriuresis relation, either by an increase in proximal tubular reabsorption or by activation of the renin–angiotensin system (110). This effect is not related to higher doses or longer treatment and can develop after a few months but can also be delayed for many years.

Antacids

The sodium content of antacids varies greatly; a daily dose of some products may contain sodium equivalent to more than 1 g of salt. This may not be clear from the labeling or the name of the formulation. However, the amount can be sufficient to precipitate heart failure in predisposed individuals (111).

Anthracyclines and related compounds

Cardiomyopathy

Anthracyclines can cause the late complication of a cardiomyopathy, which can be irreversible and can proceed to congestive cardiac failure, ventricular dysfunction, conduction disturbances, or dysrhythmias several months or years after the end of treatment (112,113). Doxorubicin can cause abnormalities of right ventricular wall motion (114). A significant number of patients receiving anthracyclines develop cardiac autonomic dysfunction (115).

Dose-relatedness

The development of anthracycline-induced cardiomyopathy is closely related to the cumulative lifetime dose of the anthracycline. The recommended maximum cumulative lifetime dose of doxorubicin is 450–550 mg/m^2 (116) and of daunorubicin 400–550 mg/m^2 intravenously in adults (117,118). About 5% of doxorubicin-treated patients develop congestive cardiac failure at this dose; however, the incidence approaches 50% at cumulative doses of 1000 mg/m^2 (116,119–120). These figures are derived from experience with doxorubicin administered as a bolus or by infusion of very short duration (under 30 minutes). The incidence of clinical cardiotoxicity falls dramatically with other schedules of administration (that is weekly doses or continuous infusion for more than 24 hours).

In a randomized study of adjuvant chemotherapy comparing bolus against continuous intravenous infusion of doxorubicin 60 mg/m^2, cardiotoxicity, defined as a 10% or greater reduction in left ventricular ejection fraction, occurred in 61% of patients on a bolus median dose equal to 420 mg/m^2 compared with 42% on the continuous infusion schedule with a median dose of 540 mg/m^2; the rate of cardiotoxicity as a function of the cumulative dose of doxorubicin was significantly higher in the bolus treatment arm (121).

In 11 patients with anthracycline cardiotoxicity studied by heart catheterization and endomyocardial biopsy, myocytic damage correlated linearly with cumulative dose (122). There was a non-linear relation between electron microscopic changes and the extent of hemodynamic impairment. There was pronounced fibrous thickening of the endocardium in most patients, especially in the left ventricle. Endocardial fibrosis may be the first morphological sign of cardiotoxicity.

In 630 patients who received cumulative anthracycline doses of 550 mg/m^2, 164 (26%) had congestive heart failure compared with the literature estimate of 7%. Age was a critical risk factor after a cumulative dose of 400 mg /m^2, and patients over 65 years of age had a greater incidence rate (123).

In 184 children (101 with acute lymphoblastic leukemia and 83 with Wilms' tumors) the cumulative anthracycline dose was reduced to under 250 mg/m^2, followed by serial echocardiograms over 10 years. In children who were given less than this dose there was no deterioration of left ventricular function (124).

Although relatively modest cumulative doses of anthracyclines were given to 111 children with acute lymphoblastic leukemia, there were subclinical abnormalities of left ventricular performance. Continuous drug administration by 6-hourly infusion did not appear to be advantageous compared with bolus injection in respect to the development of late cardiotoxicity (125).

Cardiac failure and severe cardiac abnormalities (mainly conduction abnormalities) were of major clinical importance in 229 childhood survivors who had received doxorubicin, even after follow-up of 15 years or more (126). The cardiotoxic risk increases with the dose of doxorubicin and the amount of radiation received by the heart, without evidence of a threshold.

Susceptibility factors

The risk of cardiotoxicity is greater in children and patients with pre-existing cardiac disease or concomitant or prior mediastinal or chest wall irradiation (127,128).

Of 682 patients, 144 who were over 65 years of age all had doses up to but not exceeding the usual cumulative dose for doxorubicin (129). The authors concluded that older patients without cardiovascular co-morbidity are at no greater risk of congestive heart failure.

The use of doxorubicin in childhood impairs myocardial growth, resulting in a progressive increase in left ventricular afterload, sometimes associated with impaired myocardial contractility (130). Of 201 children who received doxorubicin and/or daunorubicin 200–1275 mg/m^2, 23% had abnormal cardiac function 4–20 years afterwards. Of those who were followed for more than 10 years, 38% had abnormal cardiac function compared with 18% in those who were followed for less than 10 years (131,132). In another study, more than half of the children studied by serial echocardiography after doxorubicin therapy for acute lymphoblastic leukemia developed increased left ventricular wall stress due to reduced wall thickness. This stress progressed with time (133).

Predisposing factors to mitoxantrone cardiotoxicity include increasing age, prior anthracycline therapy, previous cardiovascular disease, mediastinal radiotherapy, and a cumulative dose of the drug exceeding 120 mg/m^2. In 801 patients treated with mitoxantrone, prior treatment with doxorubicin and mitoxantrone was significantly associated with risk of cardiotoxicity; however, age, sex, and prior mediastinal radiotherapy were not useful predictors (134).

Anesthesia is difficult in patients with cumulative anthracycline-induced cardiotoxicity, and it has proved fatal on occasions (135).

Pretreatment with anthracyclines is a risk factor for prolongation of the QT_c interval during isoflurane anesthesia (136). In 40 women undergoing surgery for breast cancer, who were pretreated with anthracyclines (n = 20) or who were chemotherapy naive, there was significantly greater prolongation of the QT_c interval in those who had previously received an anthracycline (137).

Comparative studies of anthracyclines

All anthracyclines have cardiotoxic potential. However, because only a few cycles of treatment are administered in most regimens, few patients reach the cardiotoxic threshold of cumulative anthracycline dose. There is therefore limited information about the comparative cardiotoxic potential of these agents.

Epirubicin is considered to cause substantially less cardiotoxicity than doxorubicin on a molar basis (113,138). This has been attributed to its more rapid clearance rather than a different action (139). In a randomized, double-blind comparison of epirubicin and doxorubicin, there was a significant reduction in left ventricle ejection fraction with doxorubicin but not with epirubicin (140). However, data from large clinical series and from morphological examination of endomyocardial biopsies in smaller series of patients suggest that the incidence and severity of cumulative cardiac toxicity associated with epirubicin 900 mg/m^2 is similar to that associated with doxorubicin 450–550 mg/m^2 (141). In 29 patients treated with epirubicin in cumulative doses ranging from 147 to 888 mg/m^2 the ultrastructural myocardial lesions were similar to those produced by doxorubicin (partial and total myofibrillar loss in individual myocytes) (142). With both drugs, severe lesions were associated with replacement fibrosis. None of the patients who received epirubicin in the study developed congestive cardiac failure.

Both mitoxantrone and the oral formulation of idarubicin have been thought to be less cardiotoxic than doxorubicin (143,144). The South West Oncology Group reported on 801 patients treated with mitoxantrone; 1.5% developed congestive cardiac failure, an additional 1.5% had a reduced left ventricular ejection fraction (LVEF), and 0.25% developed acute myocardial infarction (134). Idarubicin has been reported to cause short-term cardiac toxicity when used in high doses in leukemia, and there is no doubt that it causes cumulative dose-related toxicity as well (145). Electrocardiographic changes occurred in 7% of adults with acute leukemia receiving aclarubicin (146).

Presentation

The main effects of anthracycline-induced cardiotoxicity are reduced left ventricular function and chronic congestive heart failure. Other cardiotoxic events occur only rarely. Occasionally, acute transient electrocardiographic changes (ST–T wave changes, prolongation of the QT interval) and dysrhythmias can occur. Acute conduction disturbances, acute myopericarditis, and acute cardiac failure are also rare. In a study of the effects of anthracyclines on myocardial function in 50 long-term survivors of childhood cancer, there was cardiac failure in one patient and electrocardiographic abnormalities (non-specific ST segment and T wave changes) in two (128). In one patient with a VVI pacemaker, who received the combination of vincristine, doxorubicin, and dexamethasone, the pacemaker had to be reset after each cycle of treatment, as the pacing threshold had increased, resulting in bradycardia (147).

Hypokinetic heart wall motion abnormalities and early signs of chronic cardiomyopathy have been identified as a significant toxic effect of mitoxantrone in patients who received cumulative doses of 32–174 mg (148). Electrocardiographic T wave inversion and cardiac complications have been described from intensive therapy with mitoxantrone 40 mg/m^2 over 5 days and cyclophosphamide 1550 mg/m^2 for 4 days, given before bone marrow transplantation for metastatic breast cancer. All the patients had had previous exposure to doxorubicin in cumulative doses that did not exceed 442 mg/m^2 (134).

The authors of a study of the use of MRI scans to assess the subclinical effects of the anthracyclines concluded that increased MRI enhancement equal to or greater than 5 on day 3 compared with the baseline predicted significant reduction in ejection fraction at day 28 (149). In 1000 patients given doxorubicin chemotherapy and irradiation there were six cases of congestive heart failure and three cases of myocardial infarction; there was a cumulative cardiac mortality of 0.4% in all anthracycline-exposed patients (150).

Diagnosis

The diagnosis of anthracycline cardiomyopathy is based on the clinical presentation and investigations such as radionuclide cardiac angiography, which can show a reduced ejection fraction (151), and echocardiography, which can show reduced or abnormal ventricular function (152,153). Dysrhythmias can be detected by electrocardiography, and QT_c interval prolongation may offer an easy, non-invasive test to predict patients who are at special risk of late cardiac decompensation after anthracycline treatment for childhood cancer (154). Radioimmunoscintigraphy can be used to highlight damaged myocytes, and changes such as myocardial fibrosis are characteristic on endomyocardial biopsy (128,155,156).

The subtle chronic abnormalities in myocardial function that occur 10–20 years after anthracycline exposure in childhood are best detected by exercise echocardiography, since these patients may have normal resting cardiac function (157).

It has been suggested that monitoring B type natriuretic peptide concentrations after anthracycline administration can reflect cardiac tolerance, and through serial monitoring allow a picture of the degree of left ventricular dysfunction to be established (158).

Mechanisms

Several mechanisms contribute to anthracycline cardiotoxicity. The principal mechanism is thought to be oxidative stresses placed on cardiac myocytes by reactive oxygen species. Amelioration of this toxicity is possible using dexrazoxane, an intracellular metal-chelating agent of the dioxopiperazine class (112). Dexrazoxane acts by depleting intracellular iron, thus reducing the formation of cardiotoxic hydroxyl anions and radicals. In patients without heart failure, in-vivo measurements of myocardial oxidative metabolism and blood flow did not change in patients with cancer receiving doxorubicin (159).

Anthracyclines have the ability inherent in their quinone structure to form free-radical semiquinones which result in very reactive oxygen species, causing peroxidation of the lipid membranes of the heart. However, this reaction has not been demonstrated with mitoxantrone, and the mechanism of its cardiotoxicity is unknown.

Abnormalities of left ventricular ejection fraction have been described in 46% of patients (n = 14) treated with mitoxantrone (14 mg/m^2) and with vincristine and prednisolone (160). A history of cardiac disease or of previous anthracycline exposure was excluded. Only one patient developed clinically overt congestive cardiac failure. Other reports have described less cardiotoxicity compared with the parent compound, doxorubicin (113,161).

Management

Anthracycline cardiomyopathy, although reportedly difficult to treat, often responds to current methods used to manage congestive cardiac failure.

Severe anthracycline-induced cardiotoxicity is generally considered irreversible, and it is associated with a poor prognosis and high mortality. However, in four cases the advanced cardiac dysfunction associated with doxorubicin recovered completely after withdrawal (162). Of 19 patients with anthracycline-induced congestive cardiac failure, 12 recovered after withdrawal, although reversal was modest (163).

The prolongation of the QT interval that occurs in patients who have recently finished doxorubicin therapy is slowly reversible over at least 3 years and the degree of prolongation is related to the cumulative dose (164).

Heart transplantation has been successful in patients with late, progressive cardiomyopathy without recurrence of the underlying malignant disease (165).

Cardiac dysrhythmias

Cardiac dysrhythmias have been reported after amsacrine therapy in association with hypokalemia. Pre-existing supraventricular dysrhythmias or ventricular extra beats are not absolute contraindications to its use (166). Of 5430 patients treated with amsacrine, 65 developed cardiotoxicity, including prolongation of the QT interval, non-specific ST–T wave changes, ventricular tachycardia, and ventricular fibrillation (167). There were serious ventricular dysrhythmias resulting in cardiopulmonary arrest in 31 patients; 14 died as a result. The dysrhythmias occurred within minutes to several hours after drug administration. The cardiotoxicity was not related to total cumulative dose, and hypokalemia was possibly a risk factor for dysrhythmias.

Anthracyclines—liposomal formulations

The incidence of cardiotoxicity in anthracycline-treated patients has been related to the peak plasma drug concentration (168,169). One of the aims in developing pegylated liposomal doxorubicin was to reduce plasma concentrations of free doxorubicin and restrict myocardial penetration, to minimize cardiotoxicity. Preclinical data suggested that the liposomal formulation was indeed less cardiotoxic than the free drug: about 50% more pegylated liposomal doxorubicin than free doxorubicin can be given to rabbits without producing the same frequency of cardiotoxicity (170).

Cardiac adverse events that have been considered probably or possibly related to pegylated liposomal doxorubicin have been reported in 3–9% of patients (171–173). These include hypotension, pericardial effusion, thrombophlebitis, heart failure, and tachycardia (171,172).

Left ventricular failure has been reported in a few patients, particularly those who received high cumulative lifetime doses of pegylated liposomal doxorubicin (over 550 mg/m^2) (171,172). However, cumulative doses of 450 mg/m^2 or more and 550 mg/m^2 have been administered without significant reduction in ejection fraction or the development of cardiac failure (174,175). To date, no or minimal cardiotoxicity has been observed in patients with AIDS-related Kaposi's sarcoma who received pegylated liposomal doxorubicin in high cumulative doses (176).

Both peak and overall concentrations of doxorubicin in myocardial tissue are reduced by 30–40% after Myocet relative to conventional doxorubicin (177). This reduced myocardial exposure resulted in a significant reduction in cardiotoxicity, assessed both functionally and histologically (178,179). Compared with free doxorubicin 75 mg/m^2 given 3-weekly, Myocet 75 mg/m^2 caused significantly less congestive cardiac failure (1 versus 6%) (180). However, a high dose of Myocet (135 mg/m^2, median cumulative dose 405 mg/m^2) caused a significant increase in cardiac toxicity: 38% of patients had a protocol-defined cardiac event, including 13% who developed congestive heart failure (181).

In one study there was a significant (over 20%) reduction in the shortening fraction with liposomal daunorubicin measured by echocardiography (182). In contrast, in another study there was no significant fall in cardiac function, even after cumulative doses of liposomal daunorubicin over 1000 mg/m^2 (183).

Women with metastatic breast cancer were randomized to receive either liposomal doxorubicin (Myocet) 75 mg/m^2 (n = 108) or conventional doxorubicin 75 mg/m^2 (n = 116) (184). The liposomal formulation was less cardiotoxic than the conventional one, and the cumulative doses before the onset of cardiotoxicity were 780 versus 570 mg/m^2 respectively; the liposomal formulation provided comparable antitumor activity. In another study the authors tried to define the cumulative toxic intravenous dose of daunorubicin (DaunoXome) and concluded that it may be 750–900 mg/m^2 or even higher (exceeding 1000 mg/m^2) (185).

Anti-CD40 antibody (BG9588)

In an open, multiple-dose study, 28 patients with active proliferative lupus nephritis received intravenous BG9588 20 mg/kg at biweekly intervals for the first three doses and at a monthly interval for four additional doses (186). The study was prematurely terminated because of thromboembolic events in this and other BG9588 protocols. One or more adverse event was reported in 27/28 patients including headache (32%), fatigue (25%), chest pain (21%), and pharyngeal pain (18%). Severe or moderately severe adverse events occurred in 17/28 (61%), including myocardial infarction (n = 2), death (n = 1), progression to end-stage renal failure (n = 1), tracheobronchitis (n = 1), and fever/chills (n = 1). Serum C3 concentrations rose significantly at 1 month after the last dose. The authors discussed recent data suggesting that CD40L stabilizes arterial thrombi by a β3 integrin-dependent mechanism. Inhibition by anti-CD40L antibody may render platelet plugs unstable and thus ready to embolize.

Anticholinergic drugs

The potential dysrhythmogenic effect of anticholinergic drugs has been examined retrospectively in nearly 4000 patients taking flavoxate, oxybutynin, and hyoscyamine between 1991 and 1995, compared with over 10 000 patients who had no exposure to these drugs (187). All the patients were over 65 years old and they were reasonably matched for co-morbidities, although not surprisingly more women than men were taking the antispasmodic drugs (75% women in the treated group versus 67% in the control population). The encouraging conclusion was that there was no evidence that these drugs promote ventricular dysrhythmias or sudden death. However, the authors noted that more detailed analysis of these data is needed and that it may also be advisable to consider data from newer drugs.

Anticoagulant proteins

In a phase III trial of drotrecogin alfa, hypertension (2.6 versus 0.6% with placebo) was one of the most frequent adverse effects (188).

Antidepressants, second-generation

The newer antidepressants that have followed the monoamine oxidase inhibitors and tricyclic antidepressants are listed in Table 2. Most of them are covered in separate monographs. They have a wide variety of chemical structures and pharmacological profiles, and are categorized as "second generation" antidepressants purely for convenience. Although these drugs are widely considered to be as effective as each other and as any of the older compounds, they have different adverse effects profiles. No new antidepressant has proven to be sufficiently free of adverse effects to establish itself as a routine first line compound; some share similar adverse effects profiles with the tricyclic compounds, while others have novel or unexpected adverse effects. Complete categorization of each compound will rest on wide-scale general use beyond the artificial confines of clinical trials. This also includes the experience that accumulates from cases of overdosage, which cannot be anticipated before a new

Table 2 Second-generation antidepressants (rINNs except where stated)

Compound	*Structure*	*Comments*
Amfebutamone (bupropion)	Aminoketone	Modulates dopaminergic function Increased risk of seizures in high doses
Maprotiline	Tetracyclic	Strong inhibitory effect on noradrenaline uptake Skin rashes (3%) Increased incidence of seizures in overdose Similar adverse effects profile to tricyclic compounds
Mianserin (pINN)	Tetracyclic	Sedative profile Increased incidence of agranulocytosis Possibly safer in overdose Fewer cardiac effects
Milnacipran	Tetracyclic	Inhibitor of serotonin and noradrenaline reuptake
Mirtazapine	Piperazinoazepine	Noradrenergic and specific serotonergic antidepressant (NaSSA); similar to mianserin
Nefazodone	Phenylpiperazine	Weak serotonin reuptake inhibitor Blocks 5-HT_2 receptors Chemically related to trazodone
Reboxetine	Morpholine	Selective noradrenaline reuptake inhibitor (NRI or NARI)
Trazodone	Triazolopyridine	Weak effect on serotonin uptake Blocks 5-HT_2 receptors Fewer peripheral anticholinergic properties Sedative profile
Tryptophan	Amino acid	Precursor of serotonin Eosinophilia-myalgia syndrome
Venlafaxine	Bicyclic; cyclohexanol	Serotonin and noradrenaline uptake inhibitor Nausea, sexual dysfunction, and cardiovascular adverse effects
Viloxazine	Bicyclic	Fewer anticholinergic or sedative effects and weight gain Causes nausea, vomiting, and weight loss Can precipitate migraine

drug is released. The selective serotonin reuptake inhibitors are dealt with as a separate group, since they have many class-specific adverse effects.

Antidysrhythmic drugs

Cardiac dysrhythmias

Antidysrhythmic drugs can themselves cause cardiac dysrhythmias, their major adverse effect. The risk of antidysrhythmic-induced cardiac dysrhythmias (prodysrhythmic effects) has been estimated at about 11–13% in non-invasive studies (189,190) and at up to 20% in invasive electrophysiological studies. However, the risk varies from drug to drug and is particularly low with class III drugs. In one study the quoted risks of dysrhythmias were: flecainide 30%, quinidine 18%, propafenone 7%, sotalol 6%, and amiodarone 0% (191). However, amiodarone does cause dysrhythmias, especially when the QT_c interval is over 600 ms.

The prodysrhythmic effects of antidysrhythmic drugs have been extensively reviewed (192–207).

Dysrhythmias secondary to antidysrhythmic drugs are arbitrarily defined as either early (within 30 days of starting treatment) or late (193,194). A lack of early dysrhythmias in response to antidysrhythmic drugs does not predict the risk of late dysrhythmias (195).

Ventricular dysrhythmias due to drugs may be either monomorphic or polymorphic. The class Ia drugs are particularly likely to cause polymorphic dysrhythmias, as is amiodarone (although to a lesser extent). In contrast, the class Ic drugs are more likely to cause monomorphic dysrhythmias (196).

Class Ic antidysrhythmic drugs have been reported to cause the characteristic electrocardiographic changes of Brugada syndrome, which consists of right bundle branch block, persistent ST segment elevation, and sudden cardiac death, in two patients (208). Class Ia drugs did not cause the same effect.

The prodysrhythmic effects of antidysrhythmic drugs have been reviewed in discussions of the pharmacological conversion of atrial fibrillation (209) and the relative benefits of rate control in atrial fibrillation or maintaining sinus rhythm after cardioversion (210).

Mechanisms

There are four major mechanisms whereby antidysrhythmic drugs cause dysrhythmias (192):

1. Worsening of a pre-existing dysrhythmia. For example, ventricular extra beats can be converted to ventricular tachycardia or the ventricular rate in atrial flutter can be accelerated when slowing of the atrial rate results in the conduction of an increased number of atrial impulses through the AV node.
2. The induction of heart block or suppression of an escape mechanism. For example, slowing of conduction through the AV node can impair a mechanism that

allows the conducting system to escape a re-entry mechanism.
3. The uncovering of a hidden mechanism of dysrhythmia. For example, antidysrhythmic drugs can cause early or delayed afterdepolarizations, which can result in dysrhythmias.
4. The induction of a new mechanism of dysrhythmia. For example, a patient in whom myocardial ischemia has predisposed to dysrhythmias may be more at risk when an antidysrhythmic drug alters conduction.

Combinations of these different mechanisms are also possible.

The prodysrhythmic effects of antidysrhythmic drugs have been reviewed, with regard to mechanisms at the cellular level (211) and molecular level (212). As far as the cellular mechanisms are concerned, the antidysrhythmic drugs have been divided into three classes (which do not overlap with the classes specified in the electrophysiological classification).

1. Group 1 drugs have fast-onset kinetics and the block saturates at rapid rates (about 300 beats/minute).
2. Group 2 drugs have slow-onset kinetics and the block saturates at rapid rates.
3. Group 3 drugs have slow-onset kinetics and there is saturation of frequency-dependent block at slow heart rates (about 100 beats/minute).

The fast-onset kinetics of the Group 1 drugs makes them the least likely to cause dysrhythmias. Group 2 drugs, which include encainide, flecainide, procainamide, and quinidine, are the most likely to cause dysrhythmias, because of their slow-onset kinetics. Although this also applies to the Group 3 drugs, which include propafenone and disopyramide, block is less likely to occur during faster heart rates and serious dysrhythmias are therefore less likely during exercise.

The most common mechanism of dysrhythmias at the molecular level is by inhibition of the potassium channels known as IK_r, which are encoded by the human ether-a-go-go-related gene (HERG). The antidysrhythmic drugs that affect these channels include almokalant, amiodarone, azimilide, bretylium, dofetilide, ibutilide, sematilide, D-sotalol, and tedisamil (all drugs with Class III actions) and bepridil, disopyramide, prenylamine, procainamide, propafenone, quinidine, and terodiline (all drugs with Class I actions). Other drugs that affect these channels but are not used to treat cardiac dysrhythmias include astemizole and terfenadine (antihistamines), cisapride, erythromycin, haloperidol, sertindole, and thioridazine.

Susceptibility factors

There are no good predictors of the occurrence of dysrhythmias, but there are several susceptibility factors (197,198), including a history of sustained tachydysrhythmias, poor left ventricular function, and myocardial ischemia. Potassium depletion and prolongation of the QT interval are particularly important, and these particularly predispose to polymorphous ventricular dysrhythmias (for example torsade de pointes). Altered metabolism of antidysrhythmic drugs (for example liver disease, polymorphic acetylation or hydroxylation, and drug interactions) can also contribute.

The prodysrhythmic effects of antidysrhythmic drugs have been reviewed in the context of whether patients who are to be given class I or class III antidysrhythmic drugs should first be admitted to hospital for observation in the hope of identifying those who are most likely to develop dysrhythmias (191). The risk of sudden death in patients taking amiodarone was significantly increased in those who had had a prior bout of torsade de pointes. The risk of sotalol-induced torsade de pointes was higher in patients with pre-existing heart failure. Women are at a greater risk of prodysrhythmic drug effects (SEDA-18, 199). The highest risk was in women with heart failure who took more than 320 mg/day (22%); the corresponding figure in men was 8%. The authors delineated certain subgroups that they considered to be at specific risk of dysrhythmias, listing drugs that should be avoided in those subjects. They recommended avoiding drugs of classes Ia and III in women without coronary artery disease, drugs of class Ic in men with coronary artery disease, and drugs of classes Ia, Ic, and III in men with congestive heart failure and women with coronary artery disease.

Factors that predict atrial flutter with 1:1 conduction as a prodysrhythmic effect of class I antidysrhythmic drugs (cibenzoline, disopyramide, flecainide, propafenone, and quinidine) have been studied in 24 patients (aged 46–78 years) with 1:1 atrial flutter and in 100 controls (213). Underlying heart disease was present in nine patients. There was a short PR interval (PR < 0.13 ms) with normal P wave duration in leads V5 and V6 in nine of the 26 patients and only seven of the 100 controls. Signal-averaged electrocardiography showed pseudofusion between the P wave and QRS complex in 19 of the 26 patients and only 11 of the 100 controls. There was rapid atrioventricular nodal conduction (a short AH interval or second-degree atrioventricular block during atrial pacing at over 200 minute) in 19 of the 23 patients. Pseudofusion of the P wave and QRS complex had a sensitivity of 100% and a specificity of 89% for the prediction of an atrial prodysrhythmic effect of class I antidysrhythmic drugs.

Reducing the risk

The methods for minimizing the risks of prodysrhythmic effects of antidysrhythmic drugs (214) are as follows:

- Care in choosing those who are likely to benefit from antidysrhythmic drug therapy.
- Identification and correction, if possible, of impaired pump function and ischemic damage.
- Correction of electrolyte abnormalities.
- Exercise testing before and during the early stages of drug therapy: widening of the QRS complex during exercise predicts a high risk of ventricular tachycardia as does prolongation of the QT interval.
- Instruction of patients about the signs and symptoms that can occur with dysrhythmias.
- Monitoring renal and hepatic function in order to predict reduced drug elimination.

- Avoiding drug interactions or changing the dosage of the antidysrhythmic drug in anticipation of a change in its disposition secondary to an interaction.

Measurement of the concentrations of antidysrhythmic drugs and their metabolites in the plasma can be useful in recognizing the need for changing dosage requirements when cardiac, hepatic, or renal dysfunction occurs, in maintaining serum drug or metabolite concentrations within optimal ranges, and for predicting dosage changes required when interacting drugs are added (199). However, in most hospitals plasma drug concentration measurement is not routinely available for these drugs.

Another strategy for reducing the risk of prodysrhythmias is to use combinations of different classes of antidysrhythmic drugs in lower dosages than those used in monotherapy.

Torsade de pointes can be prevented by withholding antidysrhythmic drug therapy from patients who have pre-existing prolongation of the QT interval, and by correction of low serum potassium and magnesium concentrations before therapy. During therapy patients at risk should have frequent monitoring of the electrocardiogram and serum electrolytes.

The prodysrhythmic risks of using antidysrhythmic drugs have been mentioned in the context of a set of guidelines on the management of patients with atrial fibrillation (215,216). The recommended drugs for maintaining sinus rhythm after cardioversion vary depending on the presence of different risk factors for dysrhythmias:

- heart failure: amiodarone and dofetilide;
- coronary artery disease: sotalol and amiodarone;
- hypertensive heart disease: propafenone and flecainide.

Management

The management of drug-induced cardiac dysrhythmias includes withdrawal of the drug and the administration of potassium if necessary to maintain the serum potassium concentration at over 4.5 mmol/l and magnesium sulfate (SEDA-23, 196). Magnesium sulfate is given intravenously on a dose of 2 g over 2–3 minutes, followed by continuous intravenous infusion at a rate of 2–4 mg/minute; if the dysrhythmia recurs, another bolus of 2 g should be given and the infusion rate increased to 6–8 mg/minute; rarely, a third bolus of 2 g may be required (217). If magnesium is ineffective, cardiac pacing should be tried.

There is some anecdotal evidence that atrioventricular nodal blockade with verapamil or a beta-blocker can also be effective. However, in two cases the addition of a beta-blocker (either atenolol or metoprolol) to treatment with class I antidysrhythmic drugs (cibenzoline in one case and flecainide in the other) did not prevent the occurrence of atrial flutter with a 1:1 response (218). However, the author suggested that in these cases, although the beta-blockers had not suppressed the dysrhythmia, they had at least improved the patient's tolerance of it. In both cases the uses of class I antidysrhythmic drugs was contraindicated by virtue of structural damage, in the first case due to mitral valvular disease and in the second due to an ischemic cardiomyopathy.

Adverse hemodynamic effects of antidysrhythmic drugs

Many antidysrhythmic drugs have negative inotropic effects (219–221). This means that such drugs should be avoided in patients with a history of heart failure, a low left ventricular ejection fraction, or a cardiomyopathy. The general risk of induction or a worsening of heart failure is up to about 5%, but those who have risk factors have a risk of up to 10%. The negative inotropic effects are most marked with drugs of classes Ia, Ic, II, and IV. For drugs with class I activity there is a strong relation between their negative inotropic effect and the extent to which they block the inward sodium current (221). Thus, class Ib drugs that are associated with a short recovery time of sodium channels have a smaller negative inotropic effect than class Ia drugs, which in turn have less of an effect than class Ic drugs. However, the overall hemodynamic effects of antidysrhythmic drugs depend not only on their negative inotropic effects on the heart, but also on their effects on the peripheral circulation (222). Thus, although all drugs with class I activity have similar negative inotropic effects on the heart, disopyramide has large hemodynamic effects (because it increases peripheral resistance) and its hemodynamic effect is therefore greater than that of mexiletine, for example. Similarly the adverse hemodynamic effects of encainide and tocainide are greater than those of procainamide (223).

Antiepileptic drugs

See also individual compounds

Cardiac dysrhythmias induced by anticonvulsants are rare and occur mainly in patients other than those known to be at high risk of sudden death (23). Phenytoin has been rarely associated with bradydysrhythmias, almost exclusively after intravenous dosing, and some of these have been fatal. Hypotension can also complicate intravenous phenytoin. Carbamazepine can depress cardiac conduction, mostly in elderly or otherwise predisposed patients. Third-degree atrioventricular block occurred in one patient with pre-existing right bundle branch block treated with topiramate, but a cause-and-effect relation was uncertain (SEDA-21, 76).

Antiestrogens

Evidence of the wanted or unwanted effects of antiestrogens on the cardiovascular system has been sought in an extensive literature study (224). Earlier papers had suggested that selective estrogen receptor modulators might have cardioprotective effects and could lower the risk of atherosclerotic disease, but both tamoxifen and raloxifene had also been shown to increase the risk of thromboembolic events. The situation is still unclear, but raloxifene is currently being studied in a prospective, randomized, controlled trial (RUTH = Raloxifene Use for the Heart) to assess its effects on coronary heart disease and to set these alongside whatever adverse effects are observed.

There is much evidence that when tamoxifen is used to prevent and treat breast cancer it significantly increases the risk of venous thromboembolism, but there has been doubt as to whether the risk is greater than with other treatments for this condition. An extensive literature survey published in 2004 and covering a 7-year period sought to resolve these doubts, taking careful account of other susceptibility factors that might distort the picture (225). Accurate determination of the rate of thromboembolism was impaired by the lack in most studies of routine assessments to detect asymptomatic cases. However, based on symptomatic cases the risk of thromboembolism was increased two- to three-fold during use of either tamoxifen or raloxifene to prevent breast carcinoma. It is not known whether the risk is increased further in women with inherited hypercoagulable states. In the case of early-stage breast carcinoma, the risk of thromboembolism is increased with both tamoxifen and anastrozole, although the problem appeared to be somewhat less when using anastrozole.

Antihistamines

See also individual compounds

Tachycardia and hypertension have long been known as problems arising incidentally reported with various classic antihistamines (SEDA-22, 176).

- A 9-year old girl taking hydroxyzine for nocturnal itch developed a supraventricular tachycardia (226). She had a history of three episodes of palpitation with chest tightness while taking hydroxyzine for 5 months. Her clinical course suggested that the supraventricular episodes occurred in association with hydroxyzine. Hydroxyzine was withdrawn and after treatment her heart rate and rhythm returned to normal.

Cardiac toxicity is very rarely reported with hydroxyzine; before this case there had been no reported cases of supraventricular tachycardia due to hydroxyzine or its active metabolite, cetirizine. Cetirizine itself is reportedly free of such adverse effects (227).

Prolonged QT interval and ventricular dysrhythmias

DoTS classification
Dose-relation: toxic effect
Time-course: time-independent
Susceptibility factors: genetic (long QT syndrome); altered physiology (hypokalemia); drug interactions (metabolism inhibitors; drugs that prolong the QT interval); diseases (liver disease; cardiac disease with prolongation of QT interval)

Several antihistamines can cause ventricular dysrhythmias of the torsade de pointes type (228), first reported with astemizole (229) and later with terfenadine (230). Astemizole and terfenadine both have a dose-dependent effect on cardiac repolarization and cause prolongation of the QT interval, which can lead to ventricular dysrhythmias (such as torsade de pointes), syncope, and cardiac arrest. Reported cases relate preponderantly to overdosage, especially in children (SEDA-12, 142; SEDA-14, 135; SEDA-14, 137; SEDA-17, 196) (229–231). Terfenadine and astemizole have been described as having dysrhythmogenic actions, and deaths have been described (232,233). The effects of some antihistamines on the QT interval are listed in Table 3.

With a few exceptions, antihistamines are rapidly and completely absorbed after oral administration; peak plasma concentrations are reached after 1–4 hours and are highly variable, owing to differences in tissue distribution and metabolism (234). Many of the second-generation antihistamines (for example astemizole, ebastine, loratadine, and terfenadine) undergo extensive first-pass metabolism to pharmacologically active metabolites; as a common feature, the reaction is primarily supported by CYP3A4. Under normal circumstances this extensive metabolism leads to low or undetectable plasma concentrations of the parent drug. However, sometimes metabolism of the parent compound can be compromised. Accumulation of unmetabolized astemizole or terfenadine can result in blockade of cardiac potassium channels in the ventricular myocytes that regulate the duration of the action potential; consequent prolongation of the QT interval can result in potentially life-threatening ventricular tachycardia (235).

Dysrhythmias can also occur with therapeutic doses of these and other antihistamines, if certain other susceptibility factors are present:

- impaired hepatic metabolism due to liver disease;
- simultaneous treatment with drugs that are inhibitors of the cytochrome P450 enzyme CYP3A4 (for example macrolide antibiotics, antifungal azoles, or grapefruit juice), leading to increased plasma concentrations thereby raising the risk of cardiotoxic effects (SEDA-17, 196) (236);
- pre-existing QT prolongation caused by congenital long QT syndrome, other heart disease, or treatment with antidysrhythmic drugs, such as class I antidysrhythmic drugs, amiodarone, or sotalol;
- electrolyte imbalance; in particular, hypokalemia predisposes to dysrhythmias.

Terfenadine is especially likely to cause torsade de pointes in patients in whom these risk factors are present (SEDA-19, 176; SEDA-21, 176). Ventricular dysrhythmias can also occur after overdosage of antihistamines that prolong the QT interval.

The mechanism responsible for dysrhythmias has been identified as blockade of HERG potassium channels (237). The dysrhythmogenic potential of antihistamines has been evaluated in vitro using cloned human potassium channels or guinea-pig heart muscle cells, and using an in vivo guinea-pig model. Studies in humans, including the assessment of drug interactions, are considered more reliable. Investigations in human volunteers have shown that there are no significant electrocardiographic changes with azelastine, cetirizine, fexofenadine, and loratadine even at several times the therapeutic doses, which shows that cardiotoxicity is not a class effect (238) (SEDA-19, 172). Mizolastine also appears to cause no cardiac problems in

Table 3 The sedative, anticholinergic, and QT prolonging effects of antihistamines (all rINNs, except where stated)

Drug	*Sedative effect*	*Anticholinergic effect*	*QT interval prolongation*
Acrivastine	+	–	–
Alimemazine (trifluomeprazine, trimeprazine)	++++	+++	
Antazoline			
Astemizole	–		+++
Azelastine			–
Betahistine	+	++	
Brompheniramine	+	++	
Carebastine			–
Cetirizine	±	±	–
Chlorphenamine (chlorpheniramine)	++	±	–
Cinnarizine	+	+	–
Clemastine	+	+	±
Cyclizine	++	++	
Cyproheptadine	++	++	
Desloratadine	–	+	–
Dexbrompheniramine			
Dexchlorpheniramine			
Dimenhydrinate	+++	+++	
Dimetindene			
Diphenhydramine	+++	++	+
Diphenylpyraline			
Doxylamine	+		
Ebastine	–		–
Emedastine			
Fexofenadine	–	–	–
Flunarizine			
Hydroxyzine	+	–	±
Ketotifen	+		–
Levocabastine			
Levocetirizine	–		–
Loratadine	–	±	–
Mebhydrolin			
Meclozine (pINN)			
Mepyramine	++	+	
Mequitazine	+		+
Methapyrilene			
Mizolastine	–	–	–
Oxatomide			
Phenindamine			
Pheniramine			
Promethazine	++++	+++	
Terfenadine	±	+	+++
Thiazinamium			
Tripelennamine			
Triprolidine	++		

humans (239). Large doses of ebastine have shown cardiac effects in guinea pigs, but QT prolongation has not occurred in human studies with up to three times therapeutic doses (240). Slight QT prolongation was seen on further increased doses to 100 mg/day and when subjects were given erythromycin or ketoconazole, but the effect was less than the effect of terfenadine and was not considered clinically relevant (240). The active metabolite of ebastine, carebastine, had no effect on the QT interval, even in large doses.

The absolute risk of antihistamine-induced dysrhythmias is low in the general population. In an epidemiological study using a general practice database, the crude incidence of ventricular dysrhythmias was 1.9 per 10 000 person-years, corresponding to a relative risk of 4.2 for all antihistamines compared with non-use. Astemizole presented the highest relative risk, whereas terfenadine was in the range of other non-sedating antihistamines. Older age was associated with greater risk. The absolute risk in this study was one case per 5300 person-years of use (241).

In the USA, terfenadine was withdrawn from the market in 1998, and in other countries terfenadine has been moved from over-the-counter to prescription-only, with only 60 mg tablets available. The active metabolite of terfenadine, fexofenadine, is marketed as an alternative. For astemizole this option was not available, since the main metabolite (desmethylastemizole) is also cardiotoxic and has a half-life of 10 days; astemizole was therefore withdrawn from the market worldwide in June 1999.

Although it is widely believed that cardiotoxicity of antihistamines is limited to second-generation compounds, both hydroxyzine and diphenhydramine can block potassium channels. Caution should therefore be exercised in prescribing first-generation antihistamines for patients with a predisposition to cardiac dysrhythmias. For example, therapeutic doses of diphenhydramine caused prolongation of the QT interval in healthy volunteers and in patients undergoing angioplasty (242), and one cannot exclude the possibility that first-generation drugs that modulate potassium channels may in some circumstances cause dysrhythmias (243). All antihistamines should be screened for cardiotoxicity, as some patients may be poor metabolizers or may be susceptible to plasma concentrations near to the usual therapeutic range. Useful information may be obtained from pharmacokinetic studies using potential inhibitors (see under Drug–Drug Interactions).

The single- and multiple-dose pharmacokinetics of ebastine (10 mg) have been determined in elderly and young healthy subjects using 24-hour Holter monitoring (244). There were no clinically relevant effects.

The incidence of ventricular dysrhythmias associated with non-sedating antihistamines (including cetirizine) has also been assessed using the UK-based General Practice Research Database (241). There were 18 cases over the period 1992–96. Astemizole was associated with the highest relative risk. The risk associated with terfenadine was no different from that with other non-sedating antihistamines, and there was no single case of ventricular dysrhythmia with the concomitant use of P450 inhibitors and terfenadine.

In a comparison of the dysrhythmogenic potential of a series of second-generation antihistamines, the antihistamines were given intravenously and electrocardiographic and cardiovascular parameters (blood pressure and heart rate) were measured. The lowest dose that produced significant prolongation of the QT_c interval was compared with the dose required to inhibit by 50% the peripheral bronchospasm elicited by histamine 10 micrograms/kg intravenously. Astemizole, ebastine, and terfenadine produced pronounced dose-dependent QT_c interval prolongation. In contrast, terfenadine carboxylate, norastemizole, and carebastine, the major metabolites of terfenadine, astemizole, and ebastine, and cetirizine had no effects (245).

Antimony and antimonials

Cardiotoxicity of pentavalent antimonials is not new but continues to be reported (246). In April and May 2000, an outbreak of fatal cardiotoxicity occurred in Nepal amongst patients with visceral leishmaniasis who were treated with a recently introduced batch of generic sodium stibogluconate. Eight of 23 patients died and in five cases death was attributed to cardiotoxicity. This contrasted with the low total death rate (3.2%) and death rate due to cardiotoxicity (0.8%) observed among 252 patients treated between August 1999 and December 2001 with generic sodium stibogluconate from another company.

- A 22-year-old man presented with an ulcer in the left leg diagnosed as cutaneous leishmaniasis and was treated with sodium stibogluconate (Shandong Xinhua) equivalent to a dose of antimony of 10 mg/kg/day for 20 days (247). After dose 3 he developed arthralgia, myalgia, nausea, and weakness. During continued therapy his symptoms worsened, with abdominal pain and irradiation into the thorax. After dose 7 he developed mild dyspnea and thoracic pain. After dose 9 there was further deterioration, but therapy was continued up to dose 11, when he deteriorated to such an extent that he was hospitalized in the intensive care unit. He died from cardiorespiratory insufficiency.
- A 12-year-old-girl was referred to hospital with recurrent syncopal attacks for 4 days, each lasting 10–15 minutes (248). She had seven such episodes. She had completed a course of sodium stibogluconate 20 mg/kg/day for 30 days for visceral leishmaniasis 5 days before. She had frequent runs of sustained polymorphous ventricular tachycardia (torsade de pointes) with a prolonged QT_c interval (460 ms). Lidocaine was temporarily beneficial and hypokalemia (3.3 mmol/l) was managed by intravenous potassium. Recurrent ventricular tachycardia with giddiness was aborted by synchronized cardioversion (0.5 J/kg), as was a later episode of ventricular fibrillation. She recovered slowly, with episodes of sustained ventricular tachycardia, each requiring cardioversion, over the next 2 days. Her QT_c interval eventually shortened to 350 ms.

Electrocardiographic changes are common in patients taking antimony salts; in one group there was an incidence of 7%. The most common changes are ST segment changes, T wave inversion, and a prolonged QT interval. The role of conduction disturbances in cases of cardiac failure and sudden death is not known. Cases of sudden death have been seen early in treatment after a second injection (SEDA-13, 838; SEDA-16, 311).

Changes in the electrocardiogram depend on the cumulative dose of antimony, and sudden death can occur rarely (249).

- A 4-year-old boy with visceral leishmaniasis was given intravenous sodium stibogluconate 20 mg/kg/day (1200 mg/day) and oral allopurinol 16 mg/kg/day (100 mg tds). On day 3 he reported chest pain and a persistent cough. Electrocardiography was unremarkable. The drugs were withdrawn and 3 days later he developed a petechial rash on the legs. Sepsis and other causes of petechial rashes were ruled out. Three days after treatment was discontinued he developed ventricular fibrillation and died.

The authors suggested that patients taking antimony compounds should be observed cautiously for signs of cardiological and hematological changes.

Myocarditis with electrocardiographic changes has been well described, but the risk of dysrhythmias is usually small. There have been reports of severe cardiotoxicity, leading in some cases to death (250,251). This may largely be due to changes in physicochemical properties of the drug; one cluster of cases was associated with a high-osmolarity lot of sodium stibogluconate (250).

Because of concerns regarding the cardiac adverse effects of antimonials, it is good practice to admit patients for the duration of therapy whenever practicable. This may mean admitting otherwise fit young patients for several weeks for treatment of a non-healing ulcer. To address the safety of outpatient management, a recent small study of 13 marines in the UK showed that they could be safely managed as outpatients with daily stibogluconate injections, provided there was close monitoring of electrocardiograms and blood tests to provide early warning of bone marrow toxicity (252). Three patients developed minor electrocardiographic changes and one developed thrombocytopenia. All these adverse effects resolved when treatment was withdrawn. Patients with a predisposition to dysrhythmias (such as some with ischemic heart disease) are best treated with pentavalent antimonials as inpatients to identify and manage adverse effects early when resources allow.

The cardiac toxicity of antimony has been explored in cultured myocytes (253,253). Potassium antimony tartrate disrupted calcium handling, leading to a progressive increase in the resting or diastolic internal calcium concentration and eventual cessation of beating activity and cell death. An interaction with thiol homeostasis is also involved. Reduced cellular ATP concentrations paralleled toxicity but appeared to be secondary to other cellular changes initiated by exposure to antimony.

Even the normal dose of sodium stibogluconate can lead to both cardiotoxicity and hemotoxicity, because of its cumulative effects.

- Fatal accumulation of sodium stibogluconate occurred in a 4-year-old boy with visceral leishmaniasis treated with intravenous sodium stibogluconate 20 mg/kg (1200 mg/day) and oral allopurinol 16 mg/kg/day (100 mg tds) (249). On day 3 he reported chest pain and persistent cough, and the drugs were withdrawn. Three days later he developed a petechial rash on the legs and died with ventricular fibrillation.

Antipsychotic drugs

See also individual drugs

Since the early 1960s, sudden cardiac death has been reported with antipsychotic drugs. In a population-based retrospective case-control study performed in the Integrated Primary Care Information project, a longitudinal observational database, the use of antipsychotic drugs, particularly haloperidol, was associated with a significant increase in the risk of sudden death (adjusted OR = 5.0; 95% CI = 1.6, 15 for antipsychotic drugs as a whole; adjusted OR = 5.6; 95% CI = 1.6, 19 for haloperidol; number of cases and controls 775 and 6297 respectively; mean ages 71 and 69 years respectively) (255). Sudden death was defined as a natural death due to cardiac causes heralded by abrupt loss of consciousness within 1 hour after the onset of acute symptoms or unwitnessed, or an unexpected death of someone seen in a stable medical condition under 24 hours before with no evidence of a non-cardiac cause. For each case of sudden death, up to 10 controls were randomly drawn from the source population matched for age, sex, and practice.

The susceptibility factors for drug-induced torsade de pointes are: female sex, hypokalemia, bradycardia, recent conversion from atrial fibrillation (especially with a QT interval-prolonging drug), congestive heart failure, digitalis therapy, a high drug concentration, baseline QT interval prolongation, subclinical long QT syndrome, ion channel polymorphisms, and severe hypomagnesemia (256). The authors of a review identified 45 reports containing 70 cases of torsade de pointes, most associated with antipsychotic drugs (267). Female sex, heart disease, hypokalemia, high doses of the offending agent, concomitant use of a QT interval prolonging agent, and a history of long QT syndrome were identified as susceptibility factors. A thorough review of antipsychotic drugs and QT prolongation included some practical observations and suggestions (258):

- there are almost no antipsychotic drugs available that do not prolong the QT interval;
- thioridazine, mesoridazine, and pimozide should not be prescribed for patients with known heart disease, a personal history of syncope, a family history of sudden death under age 40 years, or congenital long QT syndrome;
- a baseline electrocardiogram should be obtained in all patients to determine the QT_c interval as well as the presence of other abnormalities suggesting a cardiac disorder.

Brugada-like electrocardiographic abnormalities have occurred after thioridazine overdose (259).

- A previously healthy 58-year-old woman whose blood concentration of thioridazine was 1480 μg/l (usual therapeutic concentrations are up to 200 μg/l) became comatose and had muscular rigidity. An electrocardiogram showed sinus rhythm with significant QT prolongation and 1 day later evolved into a Brugada-like pattern. Over the next 72 hours, both the electrocardiogram and the clinical abnormalities resolved.

Information on the strength of evidence linking some neuroleptic drugs to torsade de pointes has been reviewed (260).

- A 34-year-old alcoholic man with acute pancreatitis was given continuous intravenous infusion of haloperidol (2 mg/hour) for agitation; after 7 hours he received a bolus dose of haloperidol 10 mg for worsening agitation and 20 minutes later, QT interval was 560 ms (420 ms before treatment) (261). He developed torsade de pointes and ventricular fibrillation, which resolved with electric defibrillation. He was a smoker and was also taking tiapride and alprazolam for depression, in addition to pantoprazole, piperacillin + tazobactam, paracetamol, and vitamins B1, B6, and B12.

Critically ill patients are particularly susceptible to torsade de pointes, owing to various co-morbidities, electrolyte disturbance, and many drugs.

- A 58-year-old woman with pneumonia and multiple co-morbidities developed disorientation, hypoxia, and respiratory failure (262). She was given intravenous haloperidol 70, 30, 300, 270, and 340 mg on days 1, 2, 3, 4, and 5 respectively; concomitant drugs were intravenous levofloxacin 500 mg/day for 14 days, piperacillin + tazobactam, doxycycline, midazolam, morphine, diltiazem, enoxaparin, famotidine, metoclopramide, hydroxychloroquine, and transdermal nicotine. On day 5, she developed non- sustained runs of ventricular tachycardia and torsade de pointes (QT interval 533 ms), which resolved after haloperidol was withdrawn and magnesium sulfate 4 g was given.

Of 86 439 patients who had been exposed to neuroleptic drugs, 59 developed a cardiovascular adverse effect (263). Among the commonly used neuroleptic drugs, the highest rate of cardiovascular adverse effects was found for clozapine (4.5 cases per 10 000 patients) including a case of myocarditis. The study was supported by a pharmaceutical company, and since no exposure times were provided, comparative estimates cannot be calculated.

Antithymocyte globulin

Antithymocyte globulin is a polyclonal antibody directed against T lymphocyte surface antigens. It is used to induce immunosuppression and to treat acute rejection after transplantation. Rabbit antithymocyte globulin (RATG) is recommended for central venous administration, and there is no evidence to justify the common practice of adding heparin to the infusion bag. However, evaluation of thrombocytopenia may be complicated by co-administration. A total of 330 central or peripheral courses of RATG in 288 patients resulted in nine cases of deep vein thrombosis (264). In five there were prior infusion-related reactions and heparin was not used. In most cases the venous thrombosis occurred near the site of the infusion. These results provide justification for adding heparin when RATG is infused, especially peripherally.

Aprotinin

The issue of whether the use of aprotinin is associated with an increased risk of vein graft thrombosis in cardiac bypass surgery has not been resolved (265,266). The use of aprotinin was not associated with an increased rate of early occlusion of saphenous vein or internal mammary artery grafts in controlled studies with coronary angiography (267–270).

In a randomized, placebo-controlled, multicenter study of aprotinin in coronary artery bypass surgery, there was no increase in mortality or the incidence of myocardial infarction (271).

There was no evidence of an increased risk of venous thromboembolism in patients receiving aprotinin in a small study after hip replacement (272).

Araliaceae

The WHO database contains seven cases from five countries of ginseng intake followed by arterial hypertension (273). In five of them no other medication was noted. In four cases the outcome was mentioned, which was invariably full recovery without sequelae after withdrawal of ginseng.

Arsenic

Atrioventricular block is very rare after arsenic trioxide treatment for refractory acute promyelocytic leukemia (274).

- A 34-year-old woman with acute promyelocytic leukemia was given arsenic trioxide solution 0.1%, 10 ml/day for 7 days. A generalized skin rash appeared, and her serum transaminases rose. An electrocardiogram was normal. A second course of arsenic trioxide was used about 3 months later. She felt palpitation and mild dyspnea and had complete atrioventricular block. Echocardiography showed a normal left ventricle. A ^{201}thallium myocardial perfusion scan did not show a perfusion defect. Arsenic trioxide was withheld. Sinus rhythm returned 3 days later. Complete atrioventricular block recurred later when arsenic trioxide was re-administered, albeit in a lower dosage for a shorter period of time.

The heart block due to arsenic trioxide in this case was reversible and did not correlate with the patient's leukemic status.

In 19 patients with hematological malignancies given arsenic trioxide 10–20 mg in 500 ml of 5% dextrose/ isotonic saline over 3 hours daily for up to 60 days, there were three cases of torsade de pointes (275).

Artemisinin derivatives

Sinus bradycardia and a reversible prolongation of the QT interval have been reported (SEDA-21, 352).

A combination of artemether + lumefantrine (co-artemether, six doses over 3 days) followed by quinine (a 2-hour intravenous infusion of 10 mg/kg, not exceeding 600 mg in total, 2 hours after the last dose of co-artemether) was given to 42 healthy volunteers in a double-blind, parallel, three-group study (14 subjects per group) to examine the electrocardiographic effects of these drugs (276). Co-artemether had no effect on the QT_c interval. The infusion of quinine alone caused transient prolongation of the QT_c interval, and this effect was slightly but significantly greater when quinine was infused after co-artemether. Thus, the inherent risk of QT_c prolongation by intravenous quinine was enhanced by prior administration of co-artemether. Overlapping therapy with co-artemether and intravenous quinine in the treatment of patients with complicated or multidrug-resistant *Plasmodium falciparum* malaria may result in a modest increased risk of QT_c prolongation, but this is far outweighed by the potential therapeutic benefit.

The effects on the QT_c interval of single oral doses of halofantrine 500 mg and artemether 80 mg + lumefantrine 480 mg have been studied in 13 healthy men in a double-blind, randomized, crossover study (277). The length of the QT_c interval correlated positively with halofantrine exposure but was unchanged by co-artemether.

Articaine

The incidence of hypotension and headache after spinal anesthesia was similar to that encountered with lidocaine (SED-12, 256; 278).

Artificial sweeteners

Two patients with Raynaud's phenomenon (279) and one man with fibromyalgia had all used aspartame 6–15 g/day (0.12–0.16 mg/kg/day) as well as a dietary drink containing aspartame, and no other risk factors were identified. All three described regular keyboarding to the extent of 30 hours per week but had used wrist rests, "stretch breaks," and other steps to optimize their work practices. Complete resolution of symptoms occurred over 2 weeks after they had eliminated aspartame from the diet, despite no changes in the intensity of keyboarding or other work practices. Nerve conduction velocities had been within normal limits before the withdrawal of aspartame and were not repeated.

Asparaginase

The risk of venous thromboembolism in patients with cancer is increased by the use of asparaginase (280).

Astemizole

Astemizole is metabolized by CYP3A4 to desmethylastemizole and norastemizole, although these metabolites may not be free of the potential to prolong the QT interval.

- A 77-year-old woman with QT interval prolongation and torsade de pointes had been taking astemizole 10 mg/day for 6 months (281). She had markedly raised plasma concentrations of astemizole and was also taking cimetidine.

However, cardiac dysrhythmias in patients taking antihistamines may be related to other factors, and in this case the patient was also taking another antihistamine and had a history of hepatitis.

Atracurium dibesilate

There have been reports of hypotension (SEDA-15, 125; 280–284), attributed to histamine release by atracurium. A large prospective surveillance study involving more than 1800 patients given atracurium showed a 10% incidence of adverse reactions, with moderate hypotension (20–50% decrease) in 3.5% of patients (285). In one study cardiovascular stability was maintained with atracurium up to doses of 0.4 mg/kg (286). However, at higher doses (0.5 and 0.6 mg/kg) arterial pressure fell by 13 and 20% and heart rate increased by 5 and 8% respectively. These effects were maximal at 1–1.5 minutes. Since these cardiovascular effects were associated with facial flushing, it was suggested that they might have resulted from histamine release. In a subsequent study the same investigators linked significant cardiovascular changes to increased plasma histamine concentrations at a dose of atracurium of 0.6 mg/kg (287). Injecting this dose slowly over 75 seconds caused less histamine release and adverse hemodynamic effects (288). However, other investigators found no correlation between histamine plasma concentrations and hemodynamic reactions after atracurium administration (289).

Cardiovascular effects, apart from those resulting from histamine release, appear to be almost entirely limited to bradycardia. From animal studies, vagolytic (290) and ganglion-blocking (291) effects are very unlikely to occur at neuromuscular blocking doses, and these predictions appear to be borne out by investigations in man, cardiovascular effects being reported only at high dosages associated with signs suggestive of histamine release (286,287,292). The bradycardia (293–295) is occasionally severe, but, as with vecuronium, the explanation seems to be that the bradycardic effects of other agents used during anesthesia are not attenuated by atracurium as they are by alcuronium, gallamine, or pancuronium, which have vagolytic (or sympathomimetic) effects. The possibility that bradycardia can be caused by one of the metabolites, such as laudanosine (SEDA-12, 115), which is structurally similar to apomorphine, has yet to be excluded. An animal study has suggested that noradrenaline release from sympathetic nerve terminals can be increased by very large doses of atracurium, probably because of high concentrations of laudanosine (SEDA-14, 117; 296). Clinically, cardiovascular effects from this source would only be expected in circumstances that produced much higher than usual laudanosine concentrations.

Hypoxemia has been incidentally reported (SEDA-15, 125; 297), and most probably resulted from an increase ino-left cardiac shunting (in a patient with a ventricular septal defect and pulmonary atresia). Atracurium (0.2 mg/kg) may have produced a fall in systemic vascular resistance, perhaps from histamine release; pancuronium was subsequently given without incident.

Atropine

See also Anticholinergic drugs

In a classic study, more than a generation ago, of patients given atropine sulfate intravenously as premedication in a total dose of 1 mg, dysrhythmias occurred in over one-third of the subjects, and in over half of those younger than 20 years.

In adults, atrioventricular dissociation was common and in children atrial rhythm disturbances (298). In volunteers, atropine in doses of 1.6 mg/70 kg/minute causes episodes of nodal rhythm with absent P waves on the electrocardiogram (299); the episodes occurred before the heart rate had increased under the influence of the drug. In healthy men being anesthetized for dental surgery a dose of only 0.4 mg atropine intravenously 5 minutes before induction caused reductions in mean arterial pressure, stroke volume, and total peripheral resistance (300).

Second- or third-degree heart block occurred in three of 23 male heart transplant recipients given intravenous atropine (SEDA-22, 187). The mechanism is unknown but it appears that particular caution is needed when atropine is used in this group of patients.

The use of atropine in myocardial infarction to increase the heart rate succeeds as a rule, but in some patients with second-degree heart block the ventricular rate is slowed by atropine, resulting in bradycardia. In contrast, other patients can have tachycardia, and even ventricular fibrillation has been seen, occasionally even in doses as low as 0.5 mg (301).

Possible precipitation of acute myocardial infarction has been discussed by two American emergency medicine specialists (302).

- A 62-year-old woman developed chest pain and sinus bradycardia (41/minute). She had third-degree heart block and was given atropine 1 mg intravenously. Three minutes later, her chest pain increased and the electrocardiogram now showed an acute inferior myocardial infarction, confirmed by serum markers. Angioplasty recanalized the right coronary artery.

The authors discussed the possibility, suggested by others, that atropine can precipitate acute myocardial infarction in an ischemic setting. They concluded that while this may be true, on the whole the advantages of successfully correcting bradycardia outweigh the risks of this rare complication.

Azathioprine and mercaptopurine

In a follow-up study of 157 patients receiving azathioprine or mercaptopurine for Crohn's disease, the long-term risks (mainly hematological toxicity and malignancies) over 4 years of treatment were deemed to outweigh the therapeutic benefit (303). In contrast to these findings, both drugs were considered efficacious and reasonably safe in patients with inflammatory bowel disease, provided that patients are carefully selected and regularly investigated for bone marrow toxicity (304). Similar opinions were expressed regarding renal transplant patients. Conversion from ciclosporin to azathioprine in selected and carefully monitored patients had beneficial effects, by improving renal function, reducing cardiovascular risk factors, and reducing financial costs, without increasing the incidence of chronic rejection and graft loss (305).

Experience in children with juvenile chronic arthritis or chronic inflammatory bowel disease has also accumulated, and the toxicity profile of azathioprine or mercaptopurine appears to be very similar to that previously found in the adult population (SEDA-21, 381; SEDA-22, 410).

Azithromycin

Torsade de pointes and cardiorespiratory arrest have been reported in a patient with congenital long QT syndrome who took azithromycin (306). In a prospective study of 47 previously healthy people, there was a modest statistically insignificant prolongation of the QT_c interval without clinical consequences after the end of a course of azithromycin 3 g/day for 5 days (307).

Bacille Calmette–Guérin (BCG) vaccine

A mycotic aneurysm, a rare complication of intravesical BCG therapy, has been reported (308).

- A 71-year-old man with bladder carcinoma in situ received six instillations of BCG at weekly intervals followed 3 months later by three booster instillations at weekly intervals. Four months later an inflammatory aortic aneurysm, which had ruptured into a pseudoaneurysm, was diagnosed and excised. *Mycobacterium bovis* was found. After treatment with isoniazid and rifampicin he recovered. There was no sign of tumor in the bladder at cystocopy 8 months after the last BCG instillation.

Bambuterol

A retrospective study of prescriptions from three cohort studies suggested a possible adverse effect of oral bambuterol: cardiac failure. A cohort of 12 294 patients who received at least one prescription for nedocromil acted as the control group and was compared with 15 407 patients given inhaled salmeterol and 8098 patients given oral bambuterol. Questionnaires were sent to each prescriber asking for details of significant medical events after the first prescription (prescription event monitoring). From this information, rates and relative risks of non-fatal cardiac failure and ischemic heart disease were calculated. The age- and sex-adjusted relative risk of non-fatal cardiac failure associated with bambuterol was 3.41 (CI = 1.99, 5.86) compared with nedocromil. When salmeterol was compared with nedocromil the relative risk for developing non-fatal cardiac failure was 1.1 (CI = 0.63, 1.91). The adjusted relative risks of non-fatal ischemic heart disease with bambuterol and salmeterol compared with nedocromil were 1.23 (CI = 0.73, 2.08) and 1.07 (CI = 0.69, 1.66) respectively. In the month after the first prescription, the relative risk of non-fatal ischemic heart disease was 3.95 (CI = 1.38, 11.31), when bambuterol was compared with nedocromil. The authors concluded that care should be exercised when prescribing long-acting oral $beta_2$-adrenoceptor agonists for patients at risk of cardiac failure.

This study at best generates the hypothesis that the oral use of a particular $beta_2$-agonist is linked to an increased risk of cardiac disease. However, the study had several flaws. The three cohorts of patients were not strictly comparable: they were not treated concurrently nor were they sufficiently matched for age and diagnosis before prescription of the study drugs. It took nearly 3 years to recruit the patients, who preferred or needed an oral agent. A smaller proportion of patients in this cohort were treated for "asthmatic wheeze"—57 versus 70% in the salmeterol cohort. More of the bambuterol patients were given the drug for other indications, such as dyspnea, bronchitis, cough, chest infection, emphysema, and bronchitis—15 versus 2.8%. The salmeterol cohort consisted mostly of patients who were changing to a longer-acting agent and those using nedocromil were mostly changing from cromoglicate. The bambuterol group was more heterogeneous. Some patients with impending or undiagnosed heart failure may have presented with dyspnea, cough, or wheeze and received bambuterol. During the first month the correct diagnosis would become evident.

A review of the preclinical studies, clinical studies, and postmarketing surveillance data has given no support to the proposed association between bambuterol and cardiac failure. The UK Committee on Safety of Medicines has received no spontaneous reports of cardiac failure due to bambuterol. Data from the WHO database, INTDIS, show no reports of cardiac failure with bambuterol, in contrast to ten reports for salmeterol.

So, there is no evidence that oral $beta_2$-adrenoceptor agonists cause cardiac failure. The epidemiological study proposing the link is inadequate, and the authors themselves have emphasized the need for prospective, randomized trials.

Barium sulfate

Electrocardiographic changes have been recorded during administration of barium enemas and could represent a hazard in cases of cardiac disease (309–311).

Batanopride

In a double-blind, randomized, crossover comparison of batanopride and metoclopramide in 21 chemotherapy-naive patients who received cisplatin at least 70 mg/m^2, the study was terminated when hypotension was observed after infusion of batanopride at other institutions testing similar drug schedules, although the authors themselves saw no cases of hypotension after treatment with batanopride (312). However, they did note asymptomatic prolongation of the QTc interval, PR interval, and QRS complex.

Benzatropine and etybenzatropine

See also Anticholinergic drugs

A paradoxical sinus bradycardia in a psychotic patient was attributed to benzatropine since it abated when benzatropine (but not others) was withdrawn (SEDA-17, 174).

Benfluorex

Valvular heart disease has been attributed to the anorectic agent benfluorex (313).

- A 50-year-old woman who had been taking benfluorex intermittently for 1 year developed severe fibrosis and regurgitation of the mitral, aortic, and tricuspid valves. Clinical, echocardiographic, and histopathological findings were analogous to those reported with fenfluramines.

The similarity between the histopathological lesion documented in patients treated with appetite suppressants and the valvular diseases associated with ergot-related drugs suggests a common pathophysiological mechanism and a central role for serotonin in the development of the disease. Further support for this hypothesis has emerged from a report of two patients who took pergolide in doses of more than 5 mg/day and who developed symptomatic heart failure due to restrictive valvular disease (314). An earlier report described strikingly similar features of pergolide-induced valvular heart disease (315). Fibrotic valvular heart disease has also been described with another ergot derivative, bromocriptine (316).

Benzodiazepines

See also individual compounds

Hypotension follows the intravenous injection of benzodiazepines, but is usually mild and transient (SED-11, 92; 317), except in neonates who are particularly sensitive to this effect (318). Local reactions to injected diazepam are quite common and can progress to compartment syndrome (SEDA-17, 44). In one study (319), two-thirds of the patients had some problem, and most eventually progressed to thrombophlebitis. Flunitrazepam is similar to diazepam in this regard (320). Altering the formulation by changing the solvent or using an emulsion did not greatly affect the outcome (321). Midazolam, being water-soluble, might be expected to produce fewer problems; in five separate studies there were no cases of thrombophlebitis, and in two others the incidence was 8–10%, less than with diazepam but similar to thiopental and saline (322).

The effect of hypnotics during spaceflight on the high incidence of post-flight orthostatic hypotension has been studied in astronauts who took no treatment (n = 20), temazepam 15 or 30 mg (n = 9), or zolpidem 5 or 10 mg (n = 8) (323). Temazepam and zolpidem were only taken the night before landing. On the day of landing, systolic pressure fell significantly and heart rate increased significantly in the temazepam group, but not in the control group or in the zolpidem group. Temazepam may aggravate orthostatic hypotension after spaceflight, when astronauts are hemodynamically compromised. It should not be the initial choice as a sleeping aid for astronauts; zolpidem may be a better choice.

Midazolam is often used for conscious sedation during transesophageal echocardiography. In a prospective study of the effects of midazolam or no sedation in addition to pharyngeal local anesthesia with lidocaine on the cardiorespiratory effects of transesophageal echocardiography in patients in sinus rhythm midazolam (median dose 3.3, range 1-5 mg) caused significantly higher heart rates and significantly lower blood pressures and oxygen saturations (324).

Benzyl alcohol

A gasping syndrome in small premature infants who had been exposed to intravenous formulations containing benzyl alcohol 0.9% as a preservative has been described (SEDA-10, 421; SEDA-11, 475; 325–327). The affected infants presented with a metabolic acidosis, seizures, neurological deterioration, hepatic and renal dysfunction, and cardiovascular collapse. Death was reported in 16 children who received a minimum of 99 mg/kg/day of benzyl alcohol. This metabolic acidosis is caused by accumulation of the metabolite benzoic acid and is mainly related to an excessive body burden relative to body weight, so that the load of the metabolite may exceed the capacity of the immature liver and kidney for detoxification. The FDA has recommended that neither intramuscular flushing solutions containing benzyl alcohol nor dilutions with this preservative should be used in newborn infants.

In a review of the hospital and autopsy records of infants admitted to a nursery during the previous 18 months, 218 patients had been given fluids containing benzyl alcohol as flush solutions and they were compared with 218 neonates admitted during the following 18 months (328). Withdrawal of benzyl alcohol as a preservative had no demonstrable effect on mortality, but the development of kernicterus was significantly associated with benzyl alcohol in 15 of 49 exposed patients, and no cases occurred after withdrawal of the preservative. However, this apparent association was not confirmed in a 5-year study of the use of benzyl alcohol as a preservative in intravenous medications in a neonatal intensive care unit (329). In 129 neonates who died between the ages of 2 and 28 days, there was no difference in the rate of kernicterus and the exposure to benzyl alcohol between neonates who developed kernicterus and the control group of unaffected infants who were born during the same period and who were of the same birth weight and gestation age. In this study, only estimates of the extent of exposure to benzyl alcohol were given, rather than exact doses and serum concentrations.

Beraprost

Beraprost is a stable, orally active analogue of PGI_2. It has been tested in patients with intermittent claudication in a randomized, placebo-controlled trial (330). Beraprost improved walking distance more often than placebo. It also reduced the incidence of critical cardiovascular events, but the trial was not powered for statistical validation of this effect. As with iloprost, headache and flushing were the most common adverse effects.

Beta$_2$-adrenoceptor agonists

See also Adrenoceptor agonists and individual compounds

Beta$_1$- and beta$_2$-adrenoceptors co-exist in the heart, in a ratio of about 3:1 (331), and the cardiac effects of beta$_2$-adrenoceptor agonists are a consequence of direct activation of cardiac beta$_2$-adrenoceptors. There has been an extended meta-analysis of the cardiovascular effects of various beta$_2$-adrenoceptor agonists in randomized, placebo-controlled trials in patients with asthma or chronic obstructive pulmonary disease (COPD), including 13 single-dose trials (n = 232) and 20 trials of longer durations (n = 6623) (332). At the start of treatment heart rate rose and potassium concentrations fell compared with placebo. In trials lasting from 3 days to 1 year, beta$_2$-adrenoceptor agonists significantly increased the risk of a cardiovascular event (RR = 2.54; 95% CI = 1.59, 4.05) compared with placebo. In most cases the event was sinus tachycardia. The relative risk of a cardiovascular event other than sinus tachycardia (ventricular tachycardia, atrial fibrillation, syncope, congestive heart failure, myocardial infarction, cardiac arrest, and sudden death) was not significantly increased (RR = 1.66; 95% CI = 0.76, 3.6).

A standard dose of nebulized salbutamol 5 mg significantly enhanced atrioventricular nodal conduction and reduced atrioventricular nodal, atrial, and ventricular refractoriness, in addition to its positive chronotropic effect, in adults patients with asthma or COPD (333). Although dysrhythmias were not reported in this study, these effects could contribute to the generation of spontaneous dysrhythmias.

Tolerance to the hemodynamic effects of regularly inhaled salbutamol has been studied in 10 asthmatic patients and 10 matched healthy controls in a randomized, placebo-controlled, double-blind, crossover study (334). Asthmatic adults using intermittent beta$_2$-adrenoceptor agonists had a greater cardiovascular response to sympathetic stimulation by isometric exercise, but also an attenuated hemodynamic response (tachycardia, hypertonia, increased cardiac index) to acute inhalation of salbutamol 200 micrograms, suggesting hemodynamic tolerance. It is unclear if these findings represent a group effect of beta$_2$-adrenoceptor agonists or apply to salbutamol only.

Eight men with mild asthma underwent measurement of forearm blood flow, a surrogate marker for peripheral vasodilatation (335). All received in sequential order the following: normoxia plus placebo, normoxia plus inhaled salbutamol 800 micrograms, hypoxia (SpO_2 82%) plus placebo, and hypoxia plus inhaled salbutamol 800 micrograms. The period of mask breathing was 60 minutes and inhalation of salbutamol/placebo started after 30 minutes. While there were non-significant differences in blood pressure and potassium concentrations between the different treatments, forearm blood flow increased significantly by 45% in hypoxic patients inhaling salbutamol versus normoxic patients inhaling placebo. The authors concluded that the combination of hypoxia and inhalation of beta$_2$-agonists has serious systemic vascular adverse effects, potentially leading to pulmonary shunting and reduced venous return, which may be associated with sudden death. Furthermore, asthmatic patients in respiratory distress should be given $beta_2$-agonists and oxygen concomitantly whenever possible.

Intravenous and intracoronary salbutamol (10–30 micrograms/minute and 1–10 micrograms/minute respectively), and intravenous isoprenaline (1–5 micrograms/minute), a mixed $beta_1$/$beta_2$-adrenoceptor agonist, were infused in 85 patients with coronary artery disease and 22 healthy controls during fixed atrial pacing (336). Both salbutamol and isoprenaline produced large increases in QT dispersion (QT_{onset}, QT_{peak}, and QT_{end}), more pronouncedly in patients with coronary artery disease. Dispersion of the QT interval is thought to be a surrogate marker for cardiac dysrhythmia (337). The authors concluded that $beta_2$-adrenoceptors mediate important electrophysiological effects in human ventricular myocardium and can trigger dysrhythmias in susceptible patients.

In a blind, randomized study, 29 children aged under 2 years, with moderate to severe acute exacerbations of hyper-reactive airways disease, were treated with either a standard dose of nebulized salbutamol (0.15 mg/kg) or a low dose of nebulized salbutamol (0.075 mg/kg) plus nebulized ipratropium bromide 250 micrograms (338). Standard and low-dose nebulized salbutamol was given three times at intervals of 20 minutes and nebulized ipratropium bromide was given once. Clinical improvement, measured as O_2 saturation and relief of respiratory distress, was similar in both groups. QT dispersion was measured at baseline and after treatment and was significantly increased only by the standard dose of nebulized salbutamol.

Beta-lactam antibiotics

See also individual compounds

The Jarisch–Herxheimer reaction is a systemic reaction that occurs hours after initial treatment of spirochete infections, such as syphilis, leptospirosis, Lyme disease, and relapsing fever, and presents with fever, rigors, hypotension, and flushing (339,340). In patients with syphilis the reaction is more frequent in secondary syphilis and can cause additional manifestations, such as flare-up of cutaneous lesions, sudden aneurysmal dilatation of the aortic arch (341), and angina pectoris or acute coronary occlusion (SED-8, 559). It can easily be mistaken for a drug-induced hypersensitivity reaction. The underlying mechanism is initiated by antibiotic-induced release of spirochete-derived pyrogens. Transient rises in TNF, IL-6, and IL-8 have been detected (342). The role of TNF-alpha in the pathogenesis of the Jarisch–Herxheimer reaction is further underscored by the observation that in patients undergoing penicillin treatment for louse-borne relapsing fever, pretreatment with anti-TNF antibody Fab fragments partially protected against the reaction (343). The reaction lasts 12–24 hours and can be alleviated by aspirin. Alternatively, prednisone can be used and is recommended as adjunctive treatment of symptomatic cardiovascular syphilis or neurosyphilis.

Bevacizumab

Bevacizumab 5 mg/kg intravenously every 2 weeks can cause mild hypertension (344).

Biguanides

The cardiovascular effects of metformin have been reviewed (345). Metformin reduces blood pressure and has a beneficial effect on blood lipid concentrations.

In a retrospective study, cardiovascular deaths in patients using a sulfonylurea only ($n = 741$) were compared with deaths in patients taking a sulfonylurea + metformin ($n = 169$) (346). In patients taking the combination the adjusted odds ratios (95% CI) were:

- overall mortality 1.63 (1.27, 2.09);
- mortality from ischemic heart disease 1.73 (1.17, 2.55);
- stroke 2.33 (1.17, 4.63).

The patients taking the combination were younger, had had diabetes for longer, were more obese, and had higher blood glucose concentrations.

Biotin

Eosinophilic pleural effusion with eosinophilic pericardial tamponade has been attributed to concomitant use of pantothenic acid and biotin (347).

- A 76-year-old woman developed chest pain and difficulty in breathing. She had no history of allergy and had been taking biotin 10 mg/day and pantothenic acid 300 mg/day for 2 months for alopecia. Chest X-rays showed pleural effusions and cardiac enlargement. Blood tests showed an inflammatory syndrome, with an erythrocyte sedimentation rate of 51 mm/hour and an eosinophil count of $1.2–1.5 \times 10^9/l$. Pericardiotomy showed an eosinophilic infiltrate. There was no evidence of vasculitis. Serological studies were negative for antinuclear antibodies, rheumatoid factor, viruses, bacteria, and Lyme disease. Stool examination and parasitological serologies were negative. A malignant tumor was excluded by mammography, thoracoscopy, and a CT scan. Myelography, a biopsy specimen of the iliac crest bone, and the concentrations of IgE, lysozyme, and vitamin B_{12} were also normal. A week after withdrawal of pantothenic acid and biotin she improved dramatically and her eosinophilia resolved.

Bismuth

From 1973 onward many cases of an encephalopathy were reported among bismuth users. By 1979, 945 cases had been recorded in France alone, 72 of them fatal; the worldwide total exceeded 1000 cases. Bismuth encephalopathy is characterized by ataxia, confusion, speech disorder, and myoclonus. The subgallate and oxychloride have been implicated, as has the subcitrate when used in a patient with impaired bismuth clearance. The chelate tripotassium dicitratobismuthate, which contains very small amounts of bismuth, appears to be safe in this respect for normal use, as does the occasional use of more traditional bismuth products for conditions such as travelers' diarrhea. Use of bismuth subnitrate ("Roter" tablets) does not seem to lead to metal absorption (348), but such absorption does occur from bismuth salicylate (349).

- A 49-year-old woman with chronic gastric ulcers and 5 years of bismuth abuse developed bismuth encephalopathy, with progressive dementia, dysarthria, and myoclonic jerks 1 week after increasing the dosage of bismuth (350). Electroencephalography showed generalized spike-wave complexes, suggesting that the myoclonus was epileptic in nature. Bismuth was withdrawn and she was given valproate, which reduced the frequency of the myoclonic jerks. Administration of the chelator dimercaptopropane sulfonic acid (DMPS) enhanced bismuth elimination but aggravated the clinical symptoms and it was therefore withdrawn. She slowly recovered.
- An 86-year-old woman underwent partial maxillectomy for squamous cell carcinoma of the right alveolus and hard palate, and the maxillary antrum was packed with a length of bismuth iodoform paraffin paste (351). Postoperatively she developed delirium and by day 5 was exhausted, light-headed, and unsteady. She became increasingly aggressive and by day 11 was eating very little and having fainting episodes. Worsening confusion and some paranoid ideation was apparent and tremor was pronounced, but there were no focal neurological signs. The bismuth iodoform paraffin pack was removed on day 14 and 7 days later she had improved and was very cooperative and alert when she was discharged 5 days later. The bismuth serum concentration on day 14 was 146 nmol/l (reference range 0–4 nmol/l) and on day 22 was 81 nmol/l. The original peak bismuth concentration was extrapolated to over 300 nmol/l.

Three phases of the encephalopathy that bismuth can cause can be distinguished (SEDA-4 167; 352):

1. *Prodromal phase* This is characterized by such vague complaints as weakness, mental slowness, short memory, reduced working capacity, insomnia, headache, and anxiety. These symptoms can easily be dismissed as non-specific, and can persist for more than 2 years before further signs develop. In general, however, the course is gradually progressive toward the symptoms of the next phase.
2. *Acute phase* Sudden deterioration to an acute encephalopathy occurs over a period ranging from several hours to 1–2 days. Patients develop dysarthria, severe locomotor disturbances such as ataxia, difficulty in walking, intention tremor, myoclonic jerks (especially of the upper limbs, face, and trunk), incontinence due to loss of sphincter control, hyper-reflexia, and sometimes generalized convulsions. The patient may be confused, disoriented, and agitated. Excitation, hallucinations, delirium, and fluctuations of mental alertness, ranging from dizziness to loss of consciousness and coma, can occur. In about 7–9% of cases the outcome is fatal, owing to bronchopulmonary, cardiovascular, thromboembolic, or infectious complications.
3. *Recovery phase* After withdrawal of bismuth, the toxic symptoms usually abate rapidly, although physical and psychological weakness, depressive mood, memory impairment, intellectual deterioration, sleep disturbances, headache, and other symptoms occasionally persist for several months, and in isolated cases for more than a year. Exceptionally, psychic and intellectual capacity remain permanently impaired.

The reason for the suddenness of the outbreak of bismuth encephalopathy in the 1970s after the drug had been used for so many years is uncertain. A major French epidemiological study undertaken at the time (SEDA-6, 217) failed to detect any clear link to particular dosage habits, topographical factors, or drug interactions. Unknown environmental factors may have altered either the effects of bismuth itself or the sensitivity of the brain. It was later suggested that a change in intestinal flora converted bismuth salts into more soluble neurotoxic compounds (353), although it is not clear why this should have happened during the 1970s. Since measures were taken to counter the traditional use of bismuth, only very occasional new cases of encephalopathy have been reported since then. It is currently unresolved whether the lower doses of bismuth in gastric powders, antacids, astringents, demulcents, purgatives, and antidiarrheal formulations that are still in use are in the long run innocuous; if an unknown environmental factor did indeed lead to toxic manifestations with the older products, it might similarly modify the effects of others; watchfulness is therefore essential. In fact, one suspects something of a renaissance of bismuth encephalopathy in the 1990s; in 1993 cases were reported from several countries, associated inter alia with abuse of bismuth subsalicylate (Pepto-Bismol) (354) and the use of bismuth subgallate (355).

As late as 1994 a toxic encephalopathy, relapsing but reversible, was reported in a woman who had received a large dose of bismuth when bismuth iodoform paraffin paste was inserted extradurally (356). Elsewhere neurotoxicity has been reported (SEDA-19, 446).

Bleomycin

A syndrome of acute chest pain occurring during bleomycin infusion has been described in 10 patients with features that could not be ascribed to pulmonary fibrosis, hypersensitivity pneumonitis, or cardiovascular toxicity (357). The pain was sudden in onset and occurred during the first or second course, usually on the second or third day. There was retrosternal pressure or pleuritic pain, in some cases severe enough to require narcotic analgesics. Stopping or slowing the infusion produced marked improvement. In two of seven patients who received a subsequent course of treatment, the pain recurred. One patient had dyspnea and two developed an erythematous rash. One had a pleural friction rub and one a pericardial friction rub. There was fever during five episodes. There were no other physical abnormalities. There were electrocardiographic changes suggestive of pericarditis in two patients; one patient developed transient blunting of a costophrenic angle on a chest X-ray, and another a transient and small retrocardiac pulmonary infiltrate. The pain resolved spontaneously with analgesics, and there were no long-term pulmonary or cardiac sequelae. Possible underlying mechanisms include pleuropericarditis due to serosal inflammation or vascular pathology.

Severe morbidity has been reported after the use of regimens containing bleomycin: 10% of patients developed adult respiratory distress syndrome and a further 9% needed prolonged ventilation (358). The authors thought that these rates were higher than expected and attributed this to a combination of the toxic effects of bleomycin on the lung and a large retroperitoneal and/or pulmonary tumor burden.

Blood cell transfusion and bone marrow transplantation

Bone marrow for transplantation is usually kept cryopreserved in dimethylsulfoxide (DMSO), and residual DMSO may be responsible for toxic reactions: in one report, all of a series of 10 patients had falls in heart rate and blood pressure (359). However, other factors, such as cell lysis or rapid intravenous infusion of large volumes, may be responsible.

Bruising in the antecubital fossa is one of the commonest adverse effects of blood donation. The common practice of flexing the arm over a cotton wool ball or swab can aggravate bleeding, and direct compression over the puncture site and elevation of the extended arm is recommended (360).

Brofaromine

A randomized comparison of brofaromine with imipramine in inpatients with major depression showed that brofaromine was as effective as imipramine in the treatment of major depression but had a different adverse effects profile (361). Brofaromine was more likely to cause sleep disturbances but lacked the anticholinergic and certain cardiovascular adverse effects of imipramine.

Bromocriptine

Occasionally, patients taking low doses have postural hypotension, leading to dizziness or syncope. Raynaud's syndrome, with blanching of the extremities in response to cold, is rare. At high doses some 30% suffer a peripheral cold reaction of this type, which may not become manifest for a time, while leg cramps occur in some 10% of cases. Under 10% of patients have flushing or erythromelalgia, and hypotension occurs in some 5%. Occasional cases of bradycardia or acute left ventricular failure have been described. Patients with angina pectoris can experience aggravation of their symptoms (SED-12, 376). Even acute myocardial infarction after coronary spasm has been reported (SEDA-12, 123; SED-12, 376).

Bromocriptine has been associated with myocardial infarction, presumably because of coronary artery spasm.

- A 29-year-old woman took bromocriptine 5 mg/day postpartum to suppress lactation (362). Four days later she developed an acute anterior myocardial infarction. Angiography showed dissection of the left main and anterior descending arteries, with occlusion of the latter. She recovered after emergency arterial grafting.
- A 21-year-old woman had an inferior myocardial infarction, in the absence of cardiovascular risks and with normal coronary arteries on angiography (363). She made a good recovery, but with some persistent posterior-wall akinesia.

In another similar case the myocardial infarction proved fatal (364).

- A 30-year-old woman collapsed and died after a first dose of bromocriptine 2.5 mg. She had severe atheroma narrowing the right coronary artery proximal to the site of thrombosis. The only obvious risk factor was heavy smoking, 30 cigarettes per day.
- A 38-year-old woman had an acute occlusion of the right popliteal artery, which gradually resolved without surgical intervention after treatment with vasodilators, anticoagulants, and antiplatelet drugs (none of which were specified) (363). Her serum cholesterol concentration was 8 mmol/l, but she had no other risk factors.

Myocardial infarction occurred postpartum in two women taking bromocriptine (365).

- A 33-year-old woman taking bromocriptine 5 mg/day for suppression of lactation was given ergotamine 2.25 mg for acute migraine, having taken ergotamine intermittently for over 20 years. She had a myocardial infarction involving the left anterior descending coronary artery, without apparent pre-existing atherosclerosis. She made a good recovery following thrombolysis.
- A 29-year-old woman taking bromocriptine 5 mg/day postpartum had a dissection of the left anterior descending coronary artery and needed emergency bypass grafting. She made a good recovery.

These cases emphasize the potential danger of these drugs, even in young and apparently healthy individuals, although in the first case the use of two ergot derivatives simultaneously may have been ill-advised.

Edema can occur with this and other dopamine receptor agonists, the association easily being overlooked (SEDA-18, 190).

Fibrotic complications can extend to the heart, resulting in constrictive pericarditis.

Bucricaine

See also Local anesthetics

Nausea, vomiting, bradycardia, backache, shivering, and hypotension can occur with a similar incidence to that of lidocaine (SED-12, 256) (366). It has been suggested that bucricaine is more potent than lidocaine and has few cardiovascular and nervous system adverse effects in animals, at high doses. One study in humans suggested that bucricaine may be associated with fewer cardiovascular adverse effects than lidocaine, but there are insufficient data to confirm this impression (367).

Bupivacaine

See also Local anesthetics.

Bupivacaine-induced cardiotoxicity, notably after epidural use, is a matter of concern and controversy (368–370). The risk can be greatly reduced or eliminated by careful dosage and/or the use of lower concentrations (SEDA-12, 108; 368).

All studies of the cardiotoxic effects of local anesthetics on the isolated heart published from 1981 to 2001 have been reviewed (371). Thirteen studies were identified, all of which studied bupivacaine, either alone or compared with other local anesthetics. The general conclusions were:

- Highly lipid-soluble, extensively protein-bound, highly potent local anesthetics, such as tetracaine, bupivacaine, and etidocaine, are much more cardiotoxic than less lipid-soluble, protein-bound, and potent local anesthetics, such as lidocaine and prilocaine.
- Bupivacaine has a potent depressant effect on electrical conduction in the heart, primarily via an action on voltage-gated sodium channels that govern the initial rapid depolarization of the cardiac action potential.
- The $S(-)$ isomer of bupivacaine is less cardiotoxic than the $R(+)$ form.
- Bupivacaine predisposes the heart to re-entrant dysrhythmias.
- The actions of bupivacaine on channels other than voltage-gated sodium channels probably contribute to the dose-dependent cardiotoxic effects of bupivacaine.

The recommended safe upper dose limit for bupivacaine is commonly 2–2.5 mg/kg. However, some authors recommend a lower dose of 1.25 mg/kg as the safe upper limit in dental practice (SEDA-20, 128; 372).

Hyperkalemia, acidosis, severe hypoxia, and myocardial ischemia increase the cardiovascular depressive effects of bupivacaine.

There has been a report of T wave changes on the electrocardiogram during caudal administration of local anesthetics (373).

- A 4.2 kg 2-month-old baby was given a caudal injection under general anesthesia for an inguinal hernia repair. A mixture of 1% lidocaine 2 ml and 0.25% bupivacaine 2 ml was injected. Every 1 ml was preceded by an aspiration test and followed by observation for 20 seconds for electrocardiographic changes. On administration of the third 1 ml dose, there was a significant increase in T wave amplitude. The aspiration test was repeated and was positive for blood. The caudal injection was stopped and the electrocardiogram returned to normal after 35 seconds. The patient remained cardiovascularly stable with no postoperative sequelae.

Previous reports have suggested that an increase in T wave amplitude could result from inadvertent intravascular administration of adrenaline-containing local anesthetics. This is the first case report of local anesthetics alone causing significant T wave changes.

Bradycardia has been reported very occasionally (374).

Bupivacaine can cause ventricular extra beats (375). Ventricular dysrhythmias and seizures were reported in a patient who received 0.5% bupivacaine 30 ml with adrenaline 5 micrograms/ml for lumbar plexus block, after a negative aspiration test (376). The patient developed ventricular fibrillation and required advanced cardiac life support for 1 hour, including 15 defibrillations, and adrenaline 40 mg before sinus rhythm could be restored. There were no neurological sequelae.

Infusions of 0.25% bupivacaine into pig coronary arteries caused ventricular fibrillation at lower rates of infusion than 0.25% bupivacaine with 1% lidocaine (377). The lidocaine/bupivacaine mixture did not have a greater myocardial depressant effect than bupivacaine alone. The authors suggested that when regional anesthesia requires high doses of local anesthetics, bupivacaine should not be used alone but in a mixture with lidocaine, and that lidocaine should be useful in the management of bupivacaine-induced ventricular fibrillation.

An animal study of the mechanism of bupivacaine-induced dysrhythmias has shown that bupivacaine facilitates early after-depolarization in rabbit sinoatrial nodal cells by blocking the delayed rectifier potassium current (378).

Inadvertent administration of bupivacaine can lead to fatal cardiovascular collapse that may be refractory to conventional resuscitation. A study in rats has suggested that in addition to its direct cardiotoxic effect, bupivacaine may have a toxic action on the brainstem, and that cardiovascular collapse may result from dysfunction of vital cardiorespiratory control systems (379).

Buprenorphine

- A 27-year-old man injected a 2 mg suspension of crushed oral buprenorphine into his left ulnar artery, leading to acute ischemia of the hand; he was successfully treated with iloprost and dextran-40 (380).
- A 22-year-old man snorted an 8 mg crushed tablet of buprenorphine and 2 hours later had crushing chest pain, which resolved within a few minutes (381). The symptom recurred 3 weeks later after another inhalation of buprenorphine. An electrocardiogram suggested an acute anterior myocardial infarction caused by buprenorphine-induced coronary artery spasm.

Busulfan

Pericardial fibrosis has been reported after busulfan treatment in a man with chronic myeloid leukemia (382). Endocardial fibrosis has also been reported (383).

Butorphanol

Although cardiovascular toxicity with butorphanol is slight, raised pulmonary wedge pressure has occurred at cardiac catheterization (384).

Caffeine

There is debate over the association between caffeine intake and cardiovascular disease. Increases in mean blood pressure, blood glucose and free fatty acid concentrations, and urinary catecholamine excretion have been found after acute ingestion of 150 mg of caffeine (SEDA-4, 4). While not particularly potent, caffeine appears to produce cardiac dysrhythmias, including ventricular tachycardia. Furthermore, in non-smokers, daily consumption of five cups of percolated coffee per day (about 680 mg of caffeine) was associated with a modestly increased risk of having a cardiac arrest without a prior history of cardiovascular disease (385). In patients with pre-existing heart disease, caffeine lowers the effective and functional refractory period of the atrioventricular node (SEDA-4, 5). A two-fold increased risk of myocardial infarction has been suggested for women who drink six or more cups of coffee a day (386), and in two studies, men who drank five or more cups of coffee a day had an approximately two-fold increase in the risk of myocardial infarction (387) or coronary artery disease (388). However, prospective studies of the relation between caffeine consumption and an increased risk of coronary artery disease (or stroke) have been negative (SEDA-15, 1) or inconclusive (389,390), as has been the association between the consumption of coffee and serum cholesterol concentration assessed in 24 cross-sectional epidemiological studies (SEDA-15, 1). However, a Finnish study did show a positive-dose relation with serum cholesterol after adjustment for confounding variables, in men but not in women. It has also been suggested that the relation is seen only with boiled and not instant or filter coffee.

Guaraná is produced from the guaraná plant (*Paullina cupana*), the seeds of which contain 3.6–5.8% caffeine.

- A 25-year-old woman, who had pre-existing mitral valve prolapse and a history of having had bouts of palpitation with caffeine, developed intractable ventricular fibrillation after consuming a "natural energy" guaraná health drink containing a high concentration of caffeine (391). At autopsy, she was found to have sclerosis and myxoid changes in the mitral valve leaflets. The caffeine concentration in her aortic blood was 19 μg/ml.

This case highlights the need for more careful regulation of "natural" products, including warning patients with underlying health problems, and clear labeling to document the presence of any constituents with potentially adverse effects. It also shows the need for medical practitioners to be familiar with the more widely used "natural remedies" and their toxicological profiles. Following the death of this patient, the Western Australian Coroner recommended that Race 2005 Energy Blast should be removed from the local market, and the product was recalled nationally in August 1999.

Calcitonin

Flushing occurs soon after administration of calcitonin in up to 20% of patients and usually settles within several minutes (392).

Cannabinoids

Marijuana has several effects on the cardiovascular system, and can increase resting heart rate and supine blood pressure and cause postural hypotension. It is associated with an increase in myocardial oxygen demand and a decrease in oxygen supply. Peripheral vasodilatation, with increased blood flow, orthostatic hypotension, and tachycardia, can occur with normal recreational doses of cannabis. High doses of THC taken intravenously have often been associated with ventricular extra beats, a shortened PR interval, and reduced T wave amplitude, to which tolerance readily develops and which are reversible on withdrawal. While the other cardiovascular effects tend to decrease in chronic smokers, the degree of tachycardia continues to be exaggerated with exercise, as shown by bicycle ergometry.

Marijuana use is most popular among young adults (18–25 years old). However, with a generation of post-1960s smokers growing older, the use of marijuana in the age group that is prone to coronary artery disease has increased. The cardiovascular effects may present a risk to those with cardiovascular disorders, but in adults with normal cardiovascular function there is no evidence of permanent damage associated with marijuana (393,394,395), and it is not known whether marijuana can precipitate myocardial infarction, although mixed use of tobacco and cannabis make the evaluation of the effects of cannabis very difficult.

Ischemic heart disease

Investigators in the Determinants of Myocardial Infarction Onset Study recently reported that smoking marijuana is a rare trigger of acute myocardial infarction (396). Interviews of 3882 patients (1258 women) were conducted on an average of 4 days after infarction. Reported use of marijuana in the hour preceding the first symptoms of myocardial infarction was compared with use in matched controls. Among the patients, 124 reported smoking marijuana in the previous year, 37 within 24 hours, and 9 within 1 hour of cardiac symptoms. The risk of myocardial infarction was increased 4.8 times over baseline in the 60 minutes after marijuana use and then fell rapidly. The authors emphasized that in a majority of cases, the mechanism that triggered the onset of myocardial infarction involved a ruptured atherosclerotic plaque secondary to hemodynamic stress. It was not clear whether marijuana has direct or indirect hemodynamic effects sufficient to cause plaque rupture.

- Two young men, aged 18 and 30 years, developed retrosternal pain with shortness of breath, attributed to acute coronary syndrome (397). Each had smoked marijuana and tobacco and admitted to intravenous drug use. Urine toxicology was positive for tetrahydrocannabinol. Aspartate transaminase and creatine kinase activities and troponin-I and C-reactive protein concentrations were raised. Echocardiography in the first patient showed hypokinesia of the posterior and inferior walls and in the second hypokinesia of the basal segment of the anterolateral wall. Coronary angiography showed normal coronary anatomy with coronary artery spasm. Genetic testing for three common genetic polymorphisms predisposing to acute coronary syndrome was negative.

The authors suggested that marijuana had increased the blood carboxyhemoglobin concentration, leading to reduced oxygen transport capacity, increased oxygen demand, and reduced oxygen supply.

Two other cases have been reported (398).

- A 48-year-old man, a chronic user of cannabis who had had coronary artery bypass grafting 10 years before and recurrent angina over the past 18 months, developed chest pain. An electrocardiogram showed intermittent resting ST segment changes and coronary angiography showed that of the three previous grafts, only one was still patent. There was also sub-total occlusion of a stent in the left main stem. After 24 hours he had a cardiac arrest while smoking cannabis and had multiple episodes of ventricular fibrillation, requiring both electrical and pharmacological cardioversion. He then underwent urgent percutaneous coronary intervention which involved stenting of his left main stem. He eventually stabilized and recovered for discharge 11 weeks later.
- A 22-year-old man had two episodes of tight central chest pain with shortness of breath after smoking cannabis. He had been a regular marijuana smoker since his mid-teens and had used more potent and larger amounts during the previous 2 weeks. An electrocardiogram showed ST segment elevation in leads V1-5, with reciprocal ST segment depression in the inferior limb leads. A provisional diagnosis of acute myocardial infarction was made. Thrombolysis was performed, but the electrocardiographic changes continued to evolve. Angiography showed an atheromatous plaque in the left anterior descending artery which was dilated and stented. There was early diffuse disease in the cardiac vessels.

The authors suggested that in the first case ventricular fibrillation had been caused by increased myocardial oxygen demand in the presence of long-standing coronary artery disease. In the second case, they speculated that chronic cannabis use may have contributed to the unexpectedly severe coronary artery disease in a young patient with few risk factors.

Coronary no-flow and ventricular tachycardia after habitual marijuana use has been reported (399).

- A 34-year-old man developed palpitation, shortness of breath, and chest pain. He had smoked a quarter to a half an ounce of marijuana per week and had taken it 3 hours before the incident. He had ventricular tachycardia at a rate of 200/minute with a right bundle branch block pattern. Electrical cardioversion restored sinus rhythm. Angiography showed a significant reduction in left anterior descending coronary artery flow rate, which was normalized by intra-arterial verapamil 200 micrograms.

The authors thought that marijuana may have enhanced triggered activity in the Purkinje fibers along with a reduction in coronary blood flow, perhaps through coronary spasm.

Cardiac dysrhythmias

In terms of its potential for inducing cardiac dysrhythmias, cannabis is most likely to cause palpitation due to a dose-related sinus tachycardia. Other reported dysrhythmias include sinus bradycardia, second-degree atrioventricular block, and atrial fibrillation. Also reported are ventricular extra beats and other reversible electrocardiographic changes. Supraventricular tachycardia after the use of cannabis has been reported (400).

- A 35-year-old woman with a 1-month history of headaches was found to be hypertensive, with a blood pressure of 179/119 mmHg. She smoked 20 cigarettes a day and used cannabis infrequently. Her family history included hypertension. Electrocardiography suggested left ventricular hypertrophy but echocardiography was unremarkable. She was given amlodipine 10 mg/day and the blood pressure improved. While in the hospital, she smoked marijuana and about 30 minutes later developed palpitation, chest pain, and shortness of breath. The blood pressure was 233/120 mmHg and the pulse rate 150/minute. Electrocardiography showed atrial flutter with 2:1 atrioventricular block. Cardiac troponin was normal at 12 hours. Urine toxicology was positive for cannabis only. Two weeks later, while she was taking amlodipine 10 mg/day and atenolol 25 mg/day, her blood pressure was 117/85 mmHg.

The authors reviewed the biphasic effect of marijuana on the autonomic nervous system. At low to moderate doses it causes increased sympathetic activity, producing a tachycardia and increase in cardiac output; blood pressure therefore increases. At high doses it causes increased parasympathetic activity, leading to bradycardia and hypotension. They thought that this patient most probably had adrenergic atrial flutter.

Paroxysmal atrial fibrillation has been reported in two cases after marijuana use (401).

- A healthy 32-year-old doctor, who smoked marijuana 1–2 times a month, had paroxysmal tachycardia for several months. An electrocardiogram was normal and a Holter recording showed sinus rhythm with isolated supraventricular extra beats. He was treated with propranolol. He later secretly smoked marijuana while undergoing another Holter recording, which showed numerous episodes of paroxysmal atrial tachycardia and atrial fibrillation lasting up to 2 minutes. He abstained from marijuana for 12 months and maintained stable sinus rhythm.

- A 24-year-old woman briefly lost consciousness and had nausea and vomiting several minutes after smoking marijuana. She had hyporeflexia, atrial fibrillation (maximum 140/minute with a pulse deficit), and a blood pressure of 130/80 mmHg. Echocardiography was unremarkable. Within 12 hours, after metoprolol, propafenone, and intravenous hydration with electrolytes, sinus rhythm was restored.

The authors discussed the possibility that Δ9-THC, the active ingredient of marijuana, can cause intra-atrial re-entry by several mechanisms and thereby precipitate atrial fibrillation.

Sustained atrial fibrillation has also been attributed to marijuana (402).

- A 14-year-old African-American man with no cardiac history had palpitation and dizziness, resulting in a fall, within 1 hour of smoking marijuana. After vomiting several times he had a new sensation of skipped heartbeats. The only remarkable finding was a flow murmur. The electrocardiogram showed atrial fibrillation. Echocardiography was normal. Serum and urine toxicology showed cannabis. He was given digoxin, and about 12 hours later his cardiac rhythm converted to sinus rhythm. Digoxin was withdrawn. He abstained from marijuana over the next year and was symptom free.

The authors noted that marijuana's catecholaminergic properties can affect autonomic control, vasomotor reflexes, and conduction-enhancement of perinodal fibers in cardiac muscle, and thus lead to an event such as this.

Cerebral blood flow

Postural syncope after marijuana use has been studied in 29 marijuana-experienced volunteers, using transcranial Doppler to measure cerebral blood velocity in the middle cerebral artery in response to postural changes (403). They were required to abstain from marijuana and other drugs for 2 weeks before the assessment, as confirmed by urine drug screening. They were then given marijuana, tetrahydrocannabinol, or placebo and lying and standing measurements were made. When marijuana or tetrahydrocannabinol was administered, 48% reported a dizziness rating of three or four and had significant falls in standing cerebral blood velocity, mean arterial blood pressure, and systolic blood pressure. Eight subjects were so dizzy that they had to be supported. The authors suggested that marijuana interferes with the protective mechanisms that maintain standing blood pressure and cerebral blood velocity. All but one of the subjects who took marijuana or tetrahydrocannabinol reported some degree of dizziness. Women tended to be dizzier. As the postural dizziness was significant and unrelated to plasma concentrations of tetrahydrocannabinol or other indices, the authors raised concerns about marijuana use in those who are medically compromised or elderly.

Arteritis

A case of progressive arteritis associated with cannabis use has been reported (404).

- A 38-year-old Afro-Caribbean man was admitted after 3 months of severe constant ischemic pain and numbness affecting the right foot. The pain was worse at night. He also had intermittent claudication after walking 100 yards. He had a chronic history of smoking cannabis about 1 ounce/day, mixed with tobacco in the early years of usage. However, at the time of admission, he had not used tobacco in any form for over 10 years. He had patchy necrosis and ulceration of the toes and impalpable pulses in the right foot. The serum cotinine concentrations were consistent with those found in non-smokers of tobacco. Angiography of his leg was highly suggestive of Buerger's disease (thromboangiitis obliterans).

Remarkably, this patient, despite having abstained from tobacco for more than 10 years, developed a progressive arteritis leading to ischemic changes. While arterial pathology with cannabis has been reported before, it has been difficult to dissociate the effects of other drugs.

Popliteal artery entrapment occurred in a patient with distal necrosis and cannabis-related arteritis, two rare or exceptional disorders that have never been described in association (405).

- A 19-year-old man developed necrosis in the distal third right toe, with loss of the popliteal and foot pulses. Arteriography showed posterior popliteal artery compression in the right leg and unusually poor distal vascularization in both legs. An MRI scan did not show a cyst and failed to identify the type of compression and the causal agent. Surgery showed that the patient had type III entrapment. Surprisingly, the pain failed to regress and the loss of distal pulses persisted despite a perfect result on the postoperative MRA scan. The patient then admitted consuming cannabis 10 times a day for 4 years, which suggested a Buerger-type arteritis related to cannabis consumption. A 21-day course of intravenous vasodilators caused the leg pain to disappear and the toe necrosis to regress. An MRA scan confirmed permanent occlusion of three arteries on the right side of the leg and the peroneal artery on the left side. Capillaroscopy excluded Buerger's disease.

The authors suggested that popliteal artery entrapment in a young patient with non-specific symptoms should raise the suspicion of a cannabis-related lesion. Their review of literature suggested that this condition affects young patient and that complications secondary to popliteal artery entrapment did not occur in those who were under 38 years age.

Capecitabine

The incidence of symptomatic cardiotoxicity with capecitabine has been estimated to average 3%, based on two phase III trials, which is similar to intravenous fluorouracil (406).

Carbamazepine

Carbamazepine can depress the cardiac conducting system (SEDA-16, 85). There have been a few reports of reversible atrioventricular block (SED-12, 162; 407–409), and asystole has been described in a patient with Guillain–Barré syndrome (SEDA-18, 76).

In one patient carbamazepine caused pacemaker failure until the pacemaker was adjusted, no doubt because of its effects on cardiac conduction (SEDA-18, 76).

Three cases of carbamazepine-induced Stokes–Adams attacks caused by intermittent total atrioventricular block, sinoatrial block with functional escape rhythm, and intermittent asystole have been described; it was suggested that cardiac conduction should be assessed if syncope or changes in seizure type occur in patients taking carbamazepine (410).

Bradydysrhythmias of different types and severity have also been reported, especially in the elderly, but they do not seem to be common.

There have been reports of carbamazepine-induced hypertension.

- A 33-year-old man with complex partial seizures was switched from phenytoin to carbamazepine (411). His blood pressure was 118/70, but 4 months later, while he was taking carbamazepine 600 mg bd, his blood pressure was 150/112 and 1 month later 142/110. He was taking no medications beside carbamazepine. Secondary causes of hypertension were ruled out. Carbamazepine was withdrawn and gabapentin prescribed. One month later his blood pressure was normal, and it remained so at subsequent follow-up appointments over the next 2 years.
- A 68-year-old man with epilepsy and well controlled hypertension had a marked increase in arterial pressure and a transient neurological deficit after starting to take carbamazepine (412). His blood pressure normalized when the dose of carbamazepine was reduced or discontinued, but because this was the only antiepileptic drug that controlled his seizures it had to be used again, resulting in hypertension each time the dose was increased.

Although uncontrolled hypertension in the second patient could have been a direct adverse effect of carbamazepine, it is more likely that carbamazepine increased the clearance of antihypertensive drugs (atenolol, felodipine, ramipril, and irbesartan) by inducing oxidative enzymes.

A case of congestive heart failure possibly caused by carbamazepine has been described (SED-12, 162) (413), but there have been no subsequent reports.

Potentially fatal eosinophilic myocarditis may be a manifestation of carbamazepine hypersensitivity (SEDA-22, 110).

Four additional cases of carbamazepine-induced sinus node dysfunction ($n = 3$) and atrioventricular block ($n = 1$) were described in elderly Japanese women taking 200–600 mg/day. In two of the three patients rechallenged, sinus arrest recurred within 48 hours (414).

In 12 patients in whom carbamazepine was withdrawn, power-spectrum analysis of RR interval variability was used to investigate changes in sympathetic/parasympathetic autonomic equilibrium (415). Abrupt withdrawal of carbamazepine altered the sympathovagal balance during non-REM sleep, shifting the sympathovagal balance toward sympathetic predominance. However, analysis of the before and after withdrawal cardiac Holter recordings showed no serious cardiac dysrhythmias in any patient.

Autopsy in an 11-year-old boy who had been taking carbamazepine for almost 7 years showed diffuse coronary artery disease and a reduced coronary artery lumen diameter by as much as 40% (416). The authors considered that coronary artery disease could have been caused by carbamazepine-induced dyslipidemia; however, serum lipid testing had not been performed, and there had been two cases of sudden death in young maternal age relatives, suggesting other mechanisms in the pathogenesis of occlusive arterial disease in this patient.

Catheters

A major complication of intravenous infusion is thrombophlebitis, which is a principle limitation of peripheral parenteral nutrition. Its precise pathogenesis is unclear, but venospasm has been proposed as the most likely cause. However, in a study with ultrasound techniques to monitor vein caliber, there was no evidence to support this hypothesis, although thrombophlebitis was observed (417). The author suggested that the initiating event may be venous endothelial trauma, caused by the venepuncture itself, abrasion at the catheter tip, or the delivery of the feeding solution.

Venous reactions could also theoretically be influenced by the composition of the fat emulsion, because long-chain triglycerides, in particular, generate prostaglandin synthesis which can in turn effect vein tolerance. This potentially important issue has been assessed in a randomized, comparative trial of peripheral parenteral nutrition regimens with fat emulsions containing either long-chain triglycerides alone or in equal proportions with medium-chain triglycerides (418). All other factors were standardized. Long-chain triglyceride-based fat emulsions significantly prolonged the life of the peripheral vein, compared with mixtures of medium-chain and long-chain triglycerides. The authors hypothesized that this effect was due to a reduced reaction of the venous epithelium to the irritating nutritional mixture.

Superior vena cava thrombosis has been described after frequent central venous catheterization and total parenteral nutrition, with eventual partial recovery (419). The possible etiological factors included the catheter material, catheter-related sepsis, endothelial trauma, osmotic injury, and hypercoagulability. Although thrombosis of the great veins of the thorax is rare, it is life-threatening, characterized by swelling of the head, upper limbs, and torso, and on chest X-ray by mediastinal widening. Confirmation of thrombosis is best achieved by contrast venography or contrast-enhanced CT scan.

Cardiac tamponade is a serious complication of central venous catheterization. A classical case history has been described with detailed discussion of prevention and management (420). Most serious complications, including air embolism, pneumothorax, hemothorax, chylothorax, chylopericardium, rupture of the right atrium, ventricular dysrhythmias, and cardiac tamponade, are essentially mechanical injuries relating to catheter insertion. Cardiac tamponade can be caused by acute perforation of the superior vena cava during insertion. Alternatively, a delayed event may be due to catheter-related erosion of the vascular wall, either in the vena cava or in the ventricular wall. The consequences are impairment of diastolic filling and a dramatic decrease in cardiac output, with a very high death rate (about 70%).

- A 63-year-old man with cancer of the esophagus developed severe dysphagia. A central venous catheter was introduced for presurgical parenteral nutrition and 3 hours later he reported severe epigastric and retrosternal pain. His condition deteriorated rapidly, with loss of consciousness, a weak pulse, hypotension, distant heart sounds, and jugular venous distension. A chest X-ray showed an enlarged mediastinal shadow and an electrocardiogram showed reduced voltage. The catheter was promptly removed. An emergency laparotomy showed only hepatic engorgement and about 100 ml of ascites, but at thoracotomy the pericardial sac was distended by about 500 ml of clear fluid. There was no apparent injury to the right subclavian artery or evidence of pleural hemorrhage.

The authors concluded that the right ventricle had been perforated by the catheter and they pointed out that these events can be insidious, and can take several months before symptoms suddenly start, requiring quick diagnosis and immediate intervention. This includes immediate removal of the catheter and often also emergency surgical intervention.

Celastraceae

When 80 healthy volunteers chewed fresh khat leaves for 3 hours there were significant progressive rises in systolic and diastolic blood pressures and heart rate, without return to baseline 1 hour after chewing had ceased (421).

Of 247 chronic khat chewers 169 (62%) had hemorrhoids and 124 (45%) underwent hemorrhoidectomy; by comparison, of 200 non-khat chewers 8 (4%) had hemorrhoids and one underwent hemorrhoidectomy (422).

Cephalosporins

In isolated cases, cefalotin (423) and cefaclor (424) have been suspected to cause hypersensitivity myocarditis.

Cephalosporins

Angina and myocardial infarction can occur in the absence of angiographically stenosed coronary arteries, because of arterial vasospasm during a drug-induced allergic reaction. This rare condition is called the Kounis syndrome, and it has been reported in a 70-year-old woman after intravenous cefuroxime (425). The authors suggested that individuals in whom there is increased mast cell degranulation may be more susceptible to this effect.

Cetirizine

In a prospective, double-blind, parallel-group study for 18 months in 817 children with atopic dermatitis aged 12–24 months, cetirizine 0.25 mg/kg bd had no effect on the QT_c interval (426).

In a double-blind, randomized, placebo-controlled study in 85 infants aged 6–11 months there were no differences in the frequencies of adverse events between those who were given cetirizine or placebo, and no electrocardiographic changes, particularly in the QT interval (427).

Chloral hydrate

A single dose (1.2 g) in a 4-year-old girl gave rise to a reversible ventricular dysrhythmia (SEDA-21, 39).

Chloramphenicol

The "gray syndrome" is the term given to the vasomotor collapse that occurs in neonates who are given excessive parenteral doses of chloramphenicol. The syndrome is characterized by an ashen gray, cyanotic color of the skin, a fall in body temperature, vomiting, a protuberant abdomen, refusal to suck, irregular and rapid respiration, and lethargy. It is mainly seen in newborn infants, particularly when premature. It usually begins 2–9 days after the start of treatment.

Inadequate glucuronyl transferase activity combined with reduced glomerular filtration in the neonatal period is responsible for a longer half-life and accumulation of the drug. In addition, the potency of chloramphenicol to inhibit protein synthesis is higher in proliferating cells and tissues. The most important abnormality seems to be respiratory deficiency of mitochondria, due, for example to suppressed synthesis of cytochrome oxidase. The dosage should be adjusted according to the age of the neonate, and blood concentrations should be monitored. In most cases of gray syndrome, the daily dose of chloramphenicol has been higher than 25 mg/kg (428,429). Occasionally, treatment of older children and teenagers with large doses of chloramphenicol (about 100 mg/kg) has resulted in a similar form of vasomotor collapse (430).

Chlorhexidine

However, although randomized studies have failed to show an association between hypersensitivity reactions and chlorhexidine-impregnated central venous catheters, there have been reports of anaphylaxis after insertion of these catheters (SEDA-22, 262) (431,432), and two life-threatening episodes of anaphylaxis in the same patient were attributed to a central venous catheter that had been impregnated with chlorhexidine and sulfadiazine (433).

- In a 51-year-old man, two episodes of pronounced, refractory cardiovascular collapse accompanied the insertion of a chlorhexidine-coated central venous catheter (434). Sensitivity to chlorhexidine was not at first suspected, but 5 months later, a skin prick test with

> chlorhexidine resulted in a characteristic sustained wheal and flare response, strongly suggesting IgE-mediated sensitivity. The patient subsequently underwent uneventful surgery following strict avoidance of chlorhexidine exposure.

Bronchospasm was not a feature in any of these cases.

The FDA has issued a public health notice to inform health-care professionals about the potential for serious hypersensitivity reactions to medical devices impregnated with chlorhexidine. The Agency is also seeking information and reports to better evaluate the potential health hazard these products might pose, and to decide on what action, if any, should be taken. Devices that incorporate chlorhexidine that the FDA has cleared for marketing include intravenous catheters, topical antimicrobial skin dressings, and implanted antimicrobial surgical mesh. The notice describes non-US reports of systemic reactions to chlorhexidine-impregnated gels or lubricants used during urological procedures and similarly impregnated central venous catheters. It also describes other types of reactions that have been reported in the USA, including localized reactions to impregnated patches in neonates and occupational asthma in nurses exposed to chlorhexidine and alcohol aerosols (435).

A meta-analysis of the clinical and economic effects of chlorhexidine and silver sulfadiazine antiseptic-impregnated catheters has been undertaken (436). The costs of hypersensitivity reactions were considered as part of the analysis, and the use of catheters impregnated with antiseptics resulted in reduced costs. The analysis used the higher estimated incidence of hypersensitivity reactions occurring in Japan, where the use of chlorhexidine-impregnated catheters is still banned (437).

Chloroprocaine

Of 25 patients who received epidural chloroprocaine for various day procedures, 23 had a fall in arterial blood pressure of 15%, and in two it fell by 25% (438).

Chloroquine and hydroxychloroquine

Electrocardiographic changes, comprising altered T waves and prolongation of the QT interval, are not uncommon during high-dose treatment with chloroquine. The clinical significance of this is uncertain. With chronic intoxication, a varying degree of atrioventricular block can be seen; first-degree right bundle branch block and total atrioventricular block have been described. Symptoms depend on the severity of the effects: syncope, Stokes–Adams attacks, and signs of cardiac failure can occur. Acute intoxication can cause cardiovascular collapse and/or respiratory failure. Cardiac complications can prove fatal in both chronic and acute intoxication.

Third-degree atrioventricular conduction defects have been reported in two patients with rheumatoid arthritis after prolonged administration of chloroquine (439,440).

Intravenous administration can result in dysrhythmias and cardiac arrest; the speed of administration is relevant, but also the concentration reached: deaths have been recorded with blood concentrations of 1 μg/ml; concentrations after a 300 mg dose are usually 50–100 μg/ml (SEDA-13, 803).

Long-term chloroquine can cause cardiac complications, such as conduction disorders and cardiomyopathy (restrictive or hypertrophic), by structural alteration of the interventricular septum (441). Thirteen cases of cardiac toxicity associated with long-term chloroquine and hydroxychloroquine have been reported in patients with systemic autoimmune diseases. The cumulative doses were 600–2281 g for chloroquine and 292–4380 g for hydroxychloroquine.

- A 64-year-old woman with systemic lupus erythematosus took chloroquine for 7 years (cumulative dose 1000 g). She developed syncope, and the electrocardiogram showed complete heart block; a permanent pacemaker was inserted. The next year she presented with biventricular cardiac failure, skin hyperpigmentation, proximal muscle weakness, and chloroquine retinopathy. Coronary angiography was normal. An echocardiogram showed a restrictive cardiomyopathy. A skeletal muscle biopsy was characteristic of chloroquine myopathy. Chloroquine was withdrawn and she improved rapidly with diuretic therapy.
- Chloroquine cardiomyopathy occurred during long-term (7 years) treatment for rheumatoid polyarthritis in a 42-year-old woman, who had an isolated acute severe conduction defect, confirmed by histological study with electron microscopy (442).
- A 50-year-old woman took chloroquine for 6 years for rheumatoid arthritis and developed a restrictive cardiomyopathy, which required heart transplantation (443).

Regular cardiac evaluation should be considered for those who have taken a cumulative chloroquine dose of 1000 g, particularly elderly patients.

More than one mechanism may underlie the cardiac adverse effects of chloroquine. Severe hypokalemia after a single large dose of chloroquine has been documented, and some studies show a correlation between plasma potassium concentrations and the severity of the cardiac effects (444).

Light and electron microscopic abnormalities were found on endomyocardial biopsy in two patients with cardiac failure. The first had taken hydroxychloroquine 200 mg/day for 10 years, then 400 mg/day for a further 6 years; the second had taken hydroxychloroquine 400 mg/day for 2 years (SEDA-13, 239). A similar case was reported after the use of 250 mg/day for 25 years (SEDA-18, 286).

Chlorpromazine

The possibility that some of the cardiac effects of chlorpromazine may be related to metabolites as well as the parent compound has been explored (445,446).

Some cases of sudden death in apparently young healthy individuals may be directly attributable to cardiac dysrhythmias after treatment with thioridazine or chlorpromazine (447).

Ciclosporin

See also Sirolimus

Calcineurin inhibitors potentially contribute to the risk of cardiovascular events through the development of new-onset diabetes mellitus, hypertension, and hyperlipidemia. Trials have consistently shown a higher incidence of new-onset diabetes mellitus with tacrolimus, which has been borne out in large-scale registry analyses. However, the risk of hypertension is about 5% higher with ciclosporin than tacrolimus, as is the risk of hyperlipidemia (448).

Major cardiovascular risk factors have been analysed and the risk of coronary artery disease estimated in a comparative 6-month study of microemulsified ciclosporin (n = 271) versus tacrolimus (n = 286) concomitant with azathioprine and glucocorticoids (449). The primary endpoints were the incidence of and time to acute rejection. Blood pressure, serum cholesterol, HDL cholesterol, triglycerides, and blood glucose were measured at baseline and at months 1, 3, and 6. The 10-year risk of coronary heart disease was estimated according to the Framingham risk algorithm. Tacrolimus lowered serum cholesterol and mean arterial blood pressure, but in a higher summary measure of blood glucose than ciclosporin. Serum triglycerides were not different between tacrolimus and ciclosporin. The mean 10-year coronary artery disease risk estimate was significantly lowered in men who took tacrolimus, but was unchanged in women.

Hypertension

Compared with azathioprine, hypertension has been considered one of the main long-term risks in patients taking ciclosporin, with major concerns about the post-transplantion increase in cardiovascular morbidity and mortality. However, there are many susceptibility factors for cardiovascular disease in transplant patients (450), and it is difficult to take into account their complex interplay. Ciclosporin-associated hypertension appears to be dose-related, and higher whole-blood ciclosporin concentrations were found during the preceding months in patients who had thromboembolic complications compared with patients who did not (451).

De novo or aggravated hypertension is very common in patients taking ciclosporin, with the highest incidence in cases of heart transplant (71–100%) and the lowest incidence in bone marrow transplant recipients (33–60%) (452). In addition, 30–45% of patients with psoriasis, rheumatoid arthritis, or uveitis had hypertension, suggesting that ciclosporin is a significant cause of hypertension in organ transplantation. Ciclosporin-associated hypertension can cause acute vascular injury, with microangiopathic hemolysis, encephalopathy, seizures, and intracranial hemorrhage.

The incidence, clinical features, consequences, and management of ciclosporin-induced hypertension have been reviewed (453). The prevalence was 29–54% in non-transplant patients and 65–100% in heart and liver transplant patients also taking glucocorticoids. Disturbed circadian rhythm with a loss of nocturnal blood pressure fall was the main characteristic, and patients therefore had higher risks of left ventricular hypertrophy, cerebrovascular damage, microalbuminuria, and other target organ damage.

The pathophysiology of ciclosporin-induced hypertension is complex and not yet fully elucidated. Increased systemic vascular resistance subsequent to altered vascular endothelium function, renal vasoconstriction with reduced glomerular filtration and sodium-water retention, and/or increased activity of the sympathetic nervous system were suggested, while only a minor role or none was attributed to the renin–angiotensin system (454). However, hypertension often occurs before changes in renal function or sodium balance can be demonstrated, and ciclosporin nephrotoxicity alone does not explain ciclosporin-associated hypertension (452,455).

The effects of antihypertensive agents have been evaluated in patients taking ciclosporin. Collectively, dihydropyridine calcium channel blockers that do not affect ciclosporin blood concentrations substantially or at all (felodipine, isradipine, and nifedipine) are usually considered to be the drugs of choice. However, the risk of gingival hyperplasia with nifedipine, which ciclosporin also causes, should be borne in mind. Combination therapy with angiotensin-converting enzyme inhibitors or beta-blockers, or the use of other calcium channel blockers (verapamil or diltiazem) should also be considered, but careful monitoring of ciclosporin blood concentrations is recommended with the latter because they inhibit ciclosporin metabolism.

Arterial hypertension is common after kidney transplantation. In 3365 adults with a functioning graft after the first year, the prevalence of hypertension increased progressively and significantly during follow-up (456). The presence of arterial hypertension at 1 year was significantly associated with recipient sex (male), donor age (under 60 years), immunosuppressive therapy (ciclosporin), serum creatinine, and year of transplantation. Arterial hypertension was not associated with graft survival or cardiovascular mortality. The prevalence and severity of hypertension was significantly lower in patients treated with tacrolimus than with ciclosporin.

Other effects

Erythromelalgia is a symptom complex of painful inflammatory vasodilatation of the extremities, often regarded as the inverse of Raynaud's phenomenon. It is usually idiopathic or due to thrombocythemia, and rarely caused by calcium channel blockers.

- A 37 year old man took ciclosporin 150 mg/day for psoriasis and after 4 weeks developed marked erythema, edema, and tenderness over the fingers and toes (457). His symptoms increased with warmth and were partly relieved by cold compresses. His full blood count, serum biochemistry, urine analysis, and collagen profile were normal. Ciclosporin was withdrawn and the lesions regressed within 1 week but recurred when ciclosporin was restarted.

A possible role of ciclosporin in the exacerbation or development of Raynaud's disease has been suggested on one occasion; such an effect could be linked to endothelial damage or changes in platelet function (458).

A capillary leak syndrome with subsequent pulmonary edema has also been reported after intravenous ciclosporin (SEDA-21, 455).

Infusion phlebitis has been attributed to intravenous ciclosporin (459), as has recurrent infusion phlebitis during ciclosporin treatment (459).

- A 28-year-old man with ulcerative colitis had acute recurrent infusion phlebitis during administration of intravenous ciclosporin following intravenous hydrocortisone. The intravenous catheter and its site needed to be replaced repeatedly during treatment, which eventually led to complete remission of the ulcerative colitis. After 8 months, he was still in remission, with no permanent signs of damage to the phlebitic veins.

The association between the risk of thrombotic microangiopathy and the use of the combinations ciclosporin + mycophenolate mofetil, ciclosporin + sirolimus, tacrolimus + mycophenolate mofetil, and tacrolimus + sirolimus has been studied in 368 kidney or kidney-pancreas transplant recipients (460). Biopsy-proven thrombotic microangiopathy was detected in 13 patients in the absence of vascular rejection. The incidence of thrombotic microangiopathy was highest with ciclosporin + sirolimus (21%). The relative risk of thrombotic microangiopathy was 16 (95% CI = 4.3, 61) for ciclosporin + sirolimus compared with tacrolimus + mycophenolate mofetil. Ciclosporin + sirolimus was the only regimen that had concomitant pro-necrotic and anti-angiogenic effects on arterial endothelial cells. This suggests that ciclosporin + sirolimus causes thrombotic microangiopathy through dual effects on endothelial cell death and repair.

Cimetidine

Rapid infusion of cimetidine causes an increase in plasma histamine concentration, and this could be one reason for its cardiac effects. A marked degree of bradycardia (for example a 30% reduction in heart rate) is uncommon although well recognized (SEDA-17, 417); sinoatrial and atrioventricular conduction effects and dysrhythmias of every possible type have occasionally been noted, particularly after infusion but also with oral therapy. It has been suggested that if a patient is particularly at risk (for example because of poor renal function) the electrocardiogram should be monitored (SEDA-15, 394).

Symptoms of vasodilatation have been attributed to cimetidine (461).

- A 31-year-old white man taking maintenance hemodialysis had frequent episodes of hot flushes, sweating, palpitation, and dizziness, which started after about 1 month of treatment with cimetidine 400 mg/day. When the drug was temporarily discontinued and later when it was totally withdrawn, he noted marked improvement in 2–5 days.

Ciprofloxacin

See also Fluoroquinolones

Ciprofloxacin causes prolongation of the QT interval (462,463).

Cisapride

The FDA reported that between September 1993 and April 1996 it received reports of prolonged QT intervals in 23 patients and torsade de pointes or ventricular fibrillation in 34. Some proved fatal (SEDA-21, 430).

Several later studies confirmed the finding of QT interval prolongation during cisapride therapy in children (464). The effects of cisapride on the QT_c interval, heart rate, and cardiac rhythm were reported in a controlled study of 83 infants aged 2–54 months who received cisapride for a minimum of 4 days for gastro-esophageal reflux and 77 controls, using continuous bipolar limb lead electrocardiography for 8 hours (465). The QT_c interval was significantly prolonged by cisapride in infants under 3 months old. There was no significant difference in heart rates and there were no dysrhythmias. None of the infants was receiving drugs that inhibit the hepatic metabolism of cisapride via CYP3A4.

The effect of cisapride 0.8 mg/kg/day for 14 days on the QT_c interval has been studied prospectively in 50 infants with feeding intolerance, apnea, and bradycardia episodes secondary to gastro-esophageal reflux and gastrointestinal dysmotility (466). In 15 infants there was prolongation of the QT_c interval at some time during the 14 days. Infants with a QT_c interval on day 3 at least two standard deviations above the mean baseline QT_c interval were more likely to develop a prolonged QT_c interval.

However, electrocardiography in 175 children aged 1.5 months to 17 years showed no changes in QT_c interval after 15 days of treatment with cisapride 0.2 mg/kg tds or qds (467).

Over 30 drugs (for example clarithromycin, erythromycin, and troleandomycin; nefazodone; fluconazole, itraconazole, and ketoconazole; indinavir and ritonavir) and other substances (for example grapefruit juice) can interact with cisapride and enhance its dysrhythmogenic effect. Despite regulatory action, including strengthened warnings in the product information, reports of dysrhythmias have repeatedly appeared. For example, in the UK the Medicines Control Agency received 60 reports of serious cardiac adverse events between 1988 and 2000; five were fatal (468). The corresponding worldwide figures were 386 serious ventricular dysrhythmias with 125 deaths and 50 unexplained deaths.

- An 81-year-old woman, who had a permanent pacemaker for complete heart block with symptomatic bradycardia-dependent torsade de pointes, had breakthrough torsade de pointes during therapy with cisapride 10 mg tds for 22 days and paroxetine for 9 days (469). She made a good recovery on withdrawal of the drugs.

In some countries therefore the product licence was suspended (468) and cisapride was subsequently withdrawn from the market by Janssen Cilag in 2003 (470).

Before its withdrawal in the UK cisapride was specifically contraindicated in premature babies for up to 3 months after birth, because of the risk of QT interval prolongation (468). Between 1988 and 2000 the Medicines Control Agency received 64 reports of suspected adverse effects of cisapride in children under 13 years, of which two were cases of QT prolongation and two were sudden unexplained deaths. Another 106 cardiovascular events were reported from other countries, including 30 cases of QT prolongation, six cases of ventricular fibrillation or tachycardia, and four sudden unexplained deaths.

Electrocardiographic changes and predisposition to cardiac dysrhythmias were investigated in 63 children (mean age 29 months) with gastro-esophageal reflux who had taken cisapride 0.2 mg/kg tds for at least 15 days and 57 control children (mean age 27 months) who were hospitalized for other reasons and were not given cisapride or other oral treatment (471). All the children had an electrocardiogram performed at inclusion, and 24-hour Holter recording was performed in all children with prolonged QT intervals. When a prolonged QT interval was detected cisapride was withdrawn and a new electrocardiogram was recorded. Five children in the treatment group and six controls had prolonged QT intervals, which normalized in three of the five children after cisapride was withdrawn. Holter recording was normal in all children.

The effect of cisapride 0.2 mg/kg tds for 8 weeks on cardiac rhythm was studied in a placebo-controlled, double-blind trial in 49 children aged 0.5–4 years with gastro-esophageal reflux resistant to other medical therapy (472). None had underlying cardiac disease or electrolyte imbalance. Cisapride had no effect on cardiac electrical function. However, in a prospective study cisapride 2 mg/kg qds given for 72 hours to 10 premature infants caused a significant increase in the QT_c interval compared with pretreatment values (473).

The relation between cisapride plasma concentrations, QT_c interval, and cardiac rhythm was evaluated in a controlled study in 211 infants undergoing routine 8-hour polysomnography (474). Cisapride was given for at least 4 days. At comparable doses of cisapride and comparable plasma concentrations, the QT_c was significantly higher in infants below 3 months of age.

In a postmarketing study of the safety of cisapride during 1993–9, 341 patients had cardiac effects, of whom 80 (23%) died, the deaths being directly or indirectly associated with a dysrhythmic event (475). The cardiac effects included QT interval prolongation, torsade de pointes, polymorphous ventricular tachycardia, ventricular fibrillation, ventricular tachycardia, cardiac arrest, unspecified "serious" dysrhythmias, and sudden death. In most instances the dysrhythmia occurred in the presence of risk factors such as other drugs or medical conditions.

Cisatracurium besilate

With doses up to eight times the ED_{95} no cardiovascular adverse effects were observed (476) and in other studies cisatracurium had only minor cardiovascular adverse effects (477,478,479). Patients with coronary artery disease undergoing myocardial revascularization tolerated cisatracurium doses up to several fold the ED_{95} well; hemodynamic changes from pre- to postinjection were minimal (480,481).

Citalopram and escitalopram

There has been some concern about the cardiovascular safety of citalopram, mainly because of animal studies showing effects on cardiac conduction. These most commonly occur in large overdoses, in which a variety of cardiac abnormalities, including QT_c prolongation, have been noted. However, this can occur with therapeutic doses too.

- Bradycardia (34/minute) with a prolonged QT_c interval of 463 ms occurred in a patient taking citalopram 40 mg/day (482). The bradycardia resolved when citalopram was withdrawn. The patient also had alcohol dependence and evidence of cardiomyopathy; presumably this may have potentiated the effect of citalopram on cardiac conduction.
- A 21-year-old woman developed QT_c prolongation (457 ms) after taking a fairly modest overdose (400 mg) of citalopram (usual daily dose 20–60 mg) (483). The QT_c prolongation resolved uneventfully over the next 30 hours.

This suggests that even modest overdoses of citalopram can cause QT_c prolongation and that cardiac monitoring should be considered. Based on the pharmacokinetic profile of citalopram and the temporal pattern of QT_c change, the authors suggested that the effect of citalopram on the QT_c interval was mediated by one of its metabolites, dimethylcitalopram.

Prolongation of the QT_c interval has been reported in five patients who made non-fatal suicide attempts by taking large amounts of citalopram. Their electrocardiograms showed other conduction disorders, including sinus tachycardia and inferolateral repolarization disturbances (SEDA-21, 12).

Citric acid and citrates

Citric acid toxicity has been reported previously, but only after intravenous administration. It was originally seen with massive transfusion of blood products with citrate as the anticoagulant. Two case reports have described accidental intravenous administration of citrate or citric acid; at a maximum serum concentration of citrate (4.1 mmol/l) there were profound alterations in blood pressure and QT interval; these were reversed by calcium infusion (484).

Clarithromycin

See also Macrolide antibiotics

- QT interval prolongation and a ventricular dysrhythmia occurred in an HIV-positive 30-year-old man at the start of intravenous clarithromycin therapy 500 mg 12-hourly (485).

Intravenous clarithromycin caused thrombophlebitis in four patients when it was given inappropriately as a rapid bolus injection instead of a short infusion; the manufacturers have received other reports of similar reactions, even with infusions, but the incidence seems to be considerably lower than with erythromycin (486). In a prospective, non-randomized study, phlebitis occurred in 15 of 19 patients treated with intravenous erythromycin (incidence rate of 0.40 episodes/patient-day) and in 19 of 25 patients treated with intravenous clarithromycin (0.35 episodes/patient-day) (487).

Clenbuterol

Induction of physiological cardiac hypertrophy has important implications in various clinical settings, particularly in training of the left ventricle in operations for certain types of congenital heart disease. Positive effects of beta$_2$-adrenoceptor overexpression on cardiac function have been shown in experiments in mice (488).

The hypothesis that clenbuterol improves right ventricular systolic function in large mammalian species when given at the time of induction of pressure-overload cardiac hypertrophy has been tested in 15 open-chest operated sheep before and after 6 weeks of pulmonary artery banding (489). The animals were randomly assigned to either saline solution or clenbuterol. There was a highly significant improvement in the slope of the end-systolic pressure–volume loops in the clenbuterol-treated animals, without detrimental hemodynamic effects. No predictions can be made as to the lasting effects of clenbuterol's ability to augment systolic function in a chronically pressure-overloaded, thin-walled ventricle beyond 6 weeks, or the potential for tachyphylaxis.

Clomipramine

Venous thrombosis is a recognized complication of tricyclic antidepressants. Thrombosis of the cerebral veins occurred in a 61-year-old woman after intravenous clomipramine, and the authors suggested that the risk may be greater when the intravenous route is used (490).

Intravenous administration invariably produced electrocardiographic changes, sometimes slow to reverse, in elderly patients (491).

At therapeutic doses, tricyclic antidepressants can cause postural hypotension, but they are regarded as being safe in patients who require general anesthesia. However, hypotension during surgery has been associated with clomipramine (492).

- A 57-year-old man due to undergo mitral valve surgery took clomipramine (150 mg at night) up to the night before surgery. His blood pressure before induction with thiopental (250 mg) and fentanyl (250 μg) was 105/65 mmHg, with a heart rate of 70 beats/minute. Anesthesia was maintained with isofluorothane, and 45 minutes after induction, his systolic blood pressure fell to 90 mmHg. Ephedrine (30 mg total), phenylephrine (500 μg total), or dopamine (10 μg/kg/minute) did not increase the blood pressure. After sternotomy, his systolic blood pressure fell to 55 mmHg and his pulse rate to 60 beats/minute, and he had third-degree atrioventricular block. Further ephedrine, phenylephrine, and adrenaline were without effect. During cardiopulmonary bypass, a noradrenaline infusion was started (0.2 μg/kg/minute) and isofluorothane was withdrawn. After he had been weaned from bypass the noradrenaline infusion was continued at a dose of 0.2–0.8 μg/kg/minute, sufficient to maintain the systolic blood pressure at 90–100 mmHg. After the operation, clomipramine was withheld and the noradrenaline infusion tapered off, and 3 days later the hypotension had resolved.

The hypotension in this case was severe and refractory to noradrenergic stimulation, perhaps because of the alpha$_1$-adrenoceptor antagonist properties of clomipramine. The fall in systolic blood pressure was accompanied by a paradoxical fall in heart rate, perhaps because the anticholinergic effect of clomipramine removed the effect of vagal tone on the resting heart rate. It seems likely that the hypotensive effect of clomipramine was potentiated by general anesthesia; however, such a reaction is rare and the underlying cardiac problem may have contributed to this severe adverse reaction. This case reinforces current advice that tricyclic antidepressants are best avoided in patients with significant cardiac disease.

Clozapine

Clozapine has been associated with cardiomyopathy (SED-14, 142; SEDA-21, 52; 493), changes in blood pressure (SEDA-21, 52; SEDA-21, 52; SEDA-22, 57; 494,495), electrocardiographic changes (SEDA-22, 57; 496–498), and venous thromboembolism (SEDA-20, 47).

Since selenium is an essential antioxidant, and its deficiency has been implicated in myocarditis and cardiomyopathy, the aim of an observational study was to measure plasma and erythrocyte selenium concentrations in random venous blood samples from four groups: patients with mood disorders (n = 36), patients with schizophrenia taking clozapine (n = 54), patients with schizophrenia not taking clozapine (n = 41), and healthy controls (n = 56) (499). Selenium concentrations in plasma and erythrocytes were significantly lower in the patients taking clozapine compared with all the others. Thus, low selenium concentrations in patients taking clozapine may be important in the pathogenesis of life-threatening cardiac adverse effects associated with clozapine.

Hypertension

Several cases of hypertension have been associated with clozapine (SEDA-22, 57), and alpha$_2$-adrenoceptor blockade has been proposed as a possible mechanism (496). Four patients developed pseudopheochromocytoma syndrome associated with clozapine (500); all had hypertension, profuse sweating, and obesity. The authors suggested that clozapine could increase plasma noradrenaline concentrations by inhibiting presynaptic reuptake mediated by alpha$_2$-adrenoceptors.

Hypotension

Hypotension is the most commonly observed cardiovascular adverse effect of neuroleptic drugs, particularly after administration of those that are also potent alpha-adrenoceptor antagonists, such as chlorpromazine, thioridazine, and clozapine (501). A central mechanism involving the vasomotor regulatory center may also contribute to the lowering of blood pressure.

- A 51-year-old man taking maintenance clozapine developed profound hypotension after cardiopulmonary bypass (498).

Cardiac dysrhythmias

A substantial portion of patients taking clozapine develop electrocardiographic abnormalities; the prevalence was originally estimated at 10% (SEDA-20, 47; SEDA-22, 57). However, although the prevalence may be higher, most of the effects are benign and do not need treatment. In 61 patients with schizophrenia taking clozapine, in whom a retrospective chart review was conducted to identify electrocardiographic abnormalities, the prevalence of electrocardiographic abnormalities in those who used neuroleptic drugs other than clozapine was 14% (6/44), while in the neuroleptic drug-free patients it was 12% (2/17); when treatment was switched to clozapine, the prevalence of electrocardiographic abnormalities rose to 31% (19/61) (502).

The correlation between plasma clozapine concentration and heart rate variability has been studied in 40 patients with schizophrenia treated with clozapine 50–600 mg/day (497). The patients had reduced heart rate variability parameters, which correlated negatively with plasma clozapine concentration.

Clozapine can cause prolongation of the QT interval (503).

- In a 30-year-old man taking clozapine there were minor electrocardiographic abnormalities, including a prolonged QT interval. A power spectrum analysis of heart rate variability showed marked abnormalities in autonomic nervous system activity. When olanzapine was substituted, power spectrum analysis studies showed that his heart rate had improved significantly and that his cardiovascular parameters had returned to normal. Serial electrocardiograms showed minimal prolongation of the QT interval.

Tachycardia is the most common cardiovascular adverse effect of clozapine, and atrial fibrillation has also been reported (SEDA-22, 57; 496).

The reports of sudden death associated with clozapine and the possibility that it may have direct prodysrhythmic properties have been reviewed (504).

A patient developed ventricular fibrillation and atrial fibrillation after taking clozapine for 2 weeks (498).

- A 44-year-old man with no significant cardiac history was given clozapine and 12 days later had bibasal crackles in the chest and ST segment elevation in leads V2 and V3 of the electrocardiogram. He then developed ventricular tachycardia and needed resuscitation. He also developed atrial fibrillation for 24 hours, which subsequently resolved.

Cardiomyopathy

Cardiomyopathy has been associated with neuroleptic drugs, including clozapine (505–507), and partial data initially suggested an incidence of 1 in 500 in the first month.

A thorough study of the risk of myocarditis or cardiomyopathy in Australia detected 23 cases (mean age 36 years; 20 men) out of 8000 patients treated with clozapine from January 1993 to March 1999 (absolute risk 0.29%; relative risk about 1000–2000) (493). All the accumulated data on previous reports of sudden death, myocarditis, or cardiac disease noted in connection with clozapine treatment were requested from the Adverse Drug Reactions Advisory Committee (ADRAC); there were 15 cases of myocarditis (five fatal) and 8 of cardiomyopathy (one fatal) associated with clozapine. All cases of myocarditis occurred within 3 weeks of starting clozapine. Cardiomyopathy was diagnosed up to 36 months after clozapine had been started. There were no confounding factors to account for cardiac illness. Necropsy results showed mainly eosinophilic infiltrates with myocytolysis, consistent with an acute drug reaction.

The manufacturers analysed 125 reports of myocarditis with clozapine and found 35 cases with fatal outcomes (508). A total of 53% occurred in the first month of therapy, and a small number (4.8%) occurred more than 2 years after the start of treatment. In this series, 70% of the patients were men.

Taking into account the results from an epidemiological study of deaths in users and former users of clozapine (509), the cardiovascular mortality risk related to clozapine may be outweighed by the overall lower mortality risk associated with its beneficial effects, since the death rate was lower among current users (322 per 100 000 person years) than among past users (696 per 100 000 person years). The reduction in death rate during current use was largely accounted for by a reduction in the suicide rate compared with past use (RR = 0.25; CI = 0.10, 0.30).

In a review of articles on adverse cardiac effects associated with clozapine, it has been estimated that clozapine is associated with a low risk of potentially fatal myocarditis or cardiomyopathy (0.01–0.19%), and that this low risk of serious adverse cardiac events should be outweighed by a reduction in suicide risk in most patients (510).

Since cardiomyopathy is potentially fatal, some precautions must be taken. If patients taking clozapine present with flu-like symptoms, fever, myalgia, dizziness or faintness, chest pain, dyspnea, tachycardia or palpitation, and other signs or symptoms of heart failure, consideration should always be given to a diagnosis of myocarditis. Suspicion should be heightened if the symptoms develop during the first 6–8 weeks of therapy. It should be noted, however, that flu-like symptoms can also occur during the titration period, supposedly as a result of alpha-adrenoceptor antagonism by clozapine. Patients in whom myocarditis is suspected should be referred immediately to a cardiac unit for evaluation.

Clozapine rechallenge after myocarditis has been described (511).

- A 23-year-old man with no history of cardiac disease was given clozapine 12.5 mg/day, increasing to 200 mg/day over 3 weeks; 5 weeks later he complained of shortness of breath and non-specific aches and pains in his legs and body. There was marked ST-segment depression and T wave inversion in the lateral and inferior leads of the electrocardiogram. There was no eosinophilia, and creatine kinase activity was not raised. An echocardiogram showed a hyperdynamic heart and left ventricular size was at the upper limit of normal. The heart valves were normal. Clozapine was withdrawn, but his mental state and quality of life deteriorated, and 2 years later clozapine was restarted because other drugs had not produced improvement. The dose of clozapine was built up to 225 mg at night and he remained well and free from cardiac adverse effects.

In this case a consultant cardiologist diagnosed myocarditis secondary to clozapine, as no other confounding co-morbidity was identified. However, the negative rechallenge suggests either that the clozapine was not responsible or that there was tolerance to the effect.

Pericarditis and pericardial effusion

Serositis (pericarditis and pericardial effusion, with or without pleural effusion) has been reported in patients taking clozapine (512–514).

- A 43-year-old man developed a pericardial effusion after taking clozapine for 7 years. The condition resolved when the drug was withdrawn.
- A 16-year-old girl developed pericarditis associated with clozapine. There were electrocardiographic changes and serial rises in serum troponin I, a highly sensitive and specific marker of myocardial injury.

The latter is said to be the first reported case of pericarditis due to clozapine demonstrating rises in troponin I, which resolved despite continuation of therapy. The authors suggested that troponin I is the preferred marker for monitoring the cardiac adverse effects of clozapine.

Pericardial effusion associated with clozapine can be accompanied by pleural effusions (515).

- A 21-year-old man with paranoid schizophrenia was treated with zuclopentixol, which was withdrawn because of extrapyramidal adverse effects, He was given clozapine 300 mg/day, and from day 43 developed breathlessness and complained of pain in his shoulders on deep inspiration. A chest X-ray showed an enlarged cardiac silhouette and bilateral pleural effusions. An echocardiogram showed pericardial and pleural effusions with no compromise of cardiac function. Clozapine was withdrawn and all the symptoms resolved within 2 weeks.

Venous thromboembolism

Typical neuroleptic drugs have been associated with an increased risk of venous thromboembolism (516). Data from the Swedish Reactions Advisory Committee suggested that clozapine is also associated with venous thromboembolic complications (517). Between 1 April 1989 and 1 March 2000, 12 cases of venous thromboembolism were collected; in 5 the outcome was fatal. Symptoms occurred in the first 3 months of treatment in eight patients; the mean clozapine dose was 277 mg/day (75–500). Although during the study total neuroleptic drug sales, excluding clozapine, accounted for 96% of all neuroleptic drug sales, only three cases of thromboembolism associated with those neuroleptic drugs were reported. The reported risk of thromboembolism associated with clozapine is estimated to be 1 per 2000–6000 treated patients, the true risk being higher owing to under-reporting. These conclusions were consistent with those from an observational study (509).

Between February 1990, when clozapine was first marketed in the USA, and December 1999 the FDA received 99 reports of venous thromboembolism (83 mentioned pulmonary embolism with or without deep vein thrombosis and 16 mentioned deep vein thrombosis alone) (518). In 63 cases death had resulted from pulmonary embolism; 32 were confirmed by necropsy. Of 36 non-fatal cases, only 7 had been documented objectively by such diagnostic techniques as perfusion-ventilation lung scanning and venography. Thus, in 39 of the 99 reports there was objective evidence of pulmonary embolism or deep vein thrombosis. The median age of the 39 individuals was 38 (range 17–70) years and 20 were women. The median daily dose was 400 (range 125–900) mg. The median duration of clozapine exposure before diagnosis was 3 months (range 2 days to 6 years). Information on risk factors for pulmonary embolism and deep vein thrombosis varied; however, 18 of the 39 patients were obese. The frequency of fatal pulmonary embolism in this study is consistent with that described in the labelling for clozapine in the USA.

As of 31 December 1993, there were 18 cases of fatal pulmonary embolism in association with clozapine therapy in users aged 10–54 years. Based on the extent of use recorded in the Clozapine National Registry, the mortality rate associated with pulmonary embolism was 1 death per 3450 person years of use. This rate was about 28 times higher than that in the general population of a similar age and sex (95% CI = 17, 42). Whether pulmonary embolism can be attributed to clozapine or some characteristic(s) of its users is not clear (519).

Fatal pulmonary embolism occurred in a 29-year-old man who was not obese, did not smoke, and had not had recent surgery, after he had taken clozapine 300 mg/day for 6 weeks (520).

The mortality rate associated with pulmonary embolism in patients taking clozapine has been estimated to be about 28 times higher than in the general population of similar age and sex; it is not clear whether pulmonary embolism can be attributed to clozapine or some characteristics of its users.

- A 58-year-old white deaf man, with a history of pulmonary embolism, two first-degree relatives with a history of stroke and myocardial infarction, and one first-degree relative who died suddenly, developed a new episode of pulmonary embolism shortly after clozapine was begun (521).

The authors pointed out that this case suggests that susceptibility factors, such as a previous history of pulmonary embolism or venous thrombosis or a strong family history, could be viewed as relative contraindications to treatment with clozapine; however, in an invited comment it was pointed out that in this case it was not mentioned whether the patient was immobile or not, because a crucial risk factor for thrombosis in psychiatric patients is reduced physical activity (522).

In another case venous thromboembolism occurred on two occasions in a 22-year-old man taking clozapine (523). Resolution of the first episode was most probably due to treatment with anticoagulants, and the authors thought that withdrawal of anticoagulants could be regarded as equivalent to reintroduction of clozapine. They suggested that, although the interval between the withdrawal of anticoagulant therapy and the second episode was long (26 months), the sequence of events suggested a causal relation between clozapine and venous thromboembolism.

Clusiaceae

A 41-year-old man who had taken St. John's wort had a hypertensive crisis after taking cheese with red wine (524). St. John's wort is a monoamine oxidase inhibitor, and the authors believed that this explained how the concomitant use of a tyramine-rich food with St. John's wort had caused this problem.

Coagulation factors

Two factor XI products, both containing antithrombin III and heparin, have been associated with evidence of coagulation activation and thrombotic events in patients with pre-existing vascular disease (525). It has been recommended that doses of more than 30 U/kg and factor XI peak concentrations in severely deficient patients of more than 500–700 U/l should be avoided. In addition, the concurrent use of tranexamic acid or other antifibrinolytic drugs should be avoided.

Problems such as thrombosis may be related to the use of a Port-A-Cath system, but it has been suggested that high dosages of coagulation factors may increase the risk (526).

To arrest bleeding in hemophilia A with inhibitors or in acquired hemophilia due to anti-factor VIII autoantibodies, the options are treatment with high-dose factor VIII, porcine factor VIII, activated prothrombin complex concentrate, or recombinant factor VIIa, all of which carry a risk of thrombosis (527,528). Recombinant factor VIIa is considered to be relatively safe, because it acts locally at the bleeding site. Since the approval of recombinant factor VIIa in 1996, more than 500 000 doses have been used and 24 cases of thrombosis have been reported (529). In most of these cases there were other risk factors, such as advanced age. Three of 40 patients treated with recombinant factor VIIa for uncontrollable bleeding developed thrombosis unrelated to dose; all three had susceptibility factors (530,531).

Whereas low-purity factor IX products have been associated with thromboembolic events, high-purity factor IX is said to be rarely associated with such events (532).

Treatment of patients with factor VII or factor XI deficiency with fresh-frozen plasma can cause circulatory overload (533,534).

Coagulation factors, especially in young children with hemophilia who often require central venous access, can cause thrombus formation or infections near central venous access devices (535).

Coronary artery disease is less frequent in patients with hemophilia than non-hemophiliacs, but it can occur.

- A 62-year-old man with mild hemophilia A received 4000 IU of recombinant factor VIII before coronary angiography. He had an acute myocardial infarction (536).
- A 50-year-old man with von Willebrand disease type 2A with angiodysplasia and gastrointestinal hemorrhage had a large myocardial infarction when he was given recombinant activated factor VII (537).

The second patient had several pre-existing risk factors for thrombosis (hypertension, insulin-dependent diabetes, heavy smoking, and a family history of atherosclerosis). Although acute myocardial infarction is very unusual in patients with hemophilia, it has been associated with the use of intermediate purity factor IX concentrates, owing to the presence of the activated form of factor IX.

- A 65-year-old man with severe hemophilia B treated with cryosupernatant had an acute myocardial infarction (538).

Recombinant factor VIIa is not approved except in patients with hemophilia with inhibitors, but is also used in cases of excessive bleeding. It is associated with a low incidence of thrombotic complications (539,540,541,542,543). Activated prothrombin complex concentrate (also called FEIBA), used to bypass factor VIII inhibitors in patients with hemophilia A or acquired hemophilia, can do the same (544,545,546,547).

Prothrombin complex concentrate also carries the risk of thrombosis, particularly when repeated doses are required. The risk of thrombosis from prothrombin complex concentrate is suggested to be higher in the treatment of factor VII deficiencies, because the other coagulation factors with longer half-lives can accumulate (533). Recombinant factor VIIa, prothrombin complex concentrate, and activated prothrombin complex concentrate should be used with caution in patients with a history of thrombotic complications, in patients with established thrombotic disorders, or when there is excessive bleeding (539,540).

Thrombophlebitis at the site of venous access is a common adverse effect of continuous infusion of factor VIII (548).

Von Willebrand factor concentrates vary in factor VIII content. Although concentrates with a high concentration of factor VIII are more effective in acute surgery, they can produce a high concentration of factor VIII with increased risk of thrombosis (549).

Serious complications of recombinant factor VIIa, such as thromboembolic events, occur but are rare (550). The binding of recombinant factor VIIa to exposed tissue factor at the site of endothelial injury forms a complex that activates factor X, thereby producing thrombin. Atherosclerotic plaques express tissue factor, so that pharmacological doses of recombinant factor VIIa may cause acute thrombosis. If tissue factor is expressed on the surface of monocytes, as is the case in sepsis, widespread coagulation can occur and cause disseminated intravascular coagulation. Case series and dose-ranging studies have reported events of thrombosis, disseminated intravascular coagulation, and anaphylaxis at doses of recombinant factor VIIa of 5–120 micrograms/kg. In hemophiliac patients with inhibitors, for whom the product is indicated, it is estimated that one thrombotic event occurs for every 11 300 doses (551). The frequency of thromboembolic events after the administration of recombinant factor VIIa 40 micrograms/kg to 108 patients with intracerebral hemorrhage was 7% compared with 2% with placebo (n = 96) (552).

Cocaine

Cocaine abuse is a risk factor for myocardial ischemia, infarction, and dysrhythmias, as well as pulmonary edema, ruptured aortic aneurysm, infectious endocarditis, vascular thrombosis, myocarditis, and dilated cardiomyopathy (553). Acutely, cocaine suppresses myocardial contractility, reduces coronary caliber and coronary blood flow, induces electrical abnormalities in the heart, and increases heart rate and blood pressure. These effects can lead to myocardial ischemia (554,555). However, intranasal cocaine in doses used medicinally or recreationally does not have a deleterious effect on intracardiac pressures or left ventricular performance (556).

Tachycardia and vasoconstriction from cocaine can exacerbate coronary insufficiency, complicated by dysrhythmias and hypertensive and vascular hemorrhage (557). Sudden deaths have been reported in patients with angina (558). Chronic dosing includes cardiomyopathy and cardiomegaly; other chronic conditions include endocarditis and thrombophlebitis. Crack smoking has led to pneumopericardium (559).

The cardiac effects of intracoronary infusion of cocaine have been studied in dogs and humans (560). The procedure can be performed safely and does not alter coronary arterial blood flow. The effects of direct intracoronary infusion of cocaine on left ventricle systolic and diastolic performance have been studied in 20 patients referred for cardiac catheterization for evaluation of chest pain. They were given saline or cocaine hydrochloride (1 mg/minute) in 15-minute intracoronary infusions, and cardiac measurements were made during the final 2–3 minutes of each infusion. The blood cocaine concentration obtained from the coronary sinus was 3.0 μg/ml, which is similar in magnitude to the blood–cocaine concentration reported in abusers who die of cocaine intoxication. Minimal systemic effects were produced. The overall results were that cocaine caused measurable deterioration of left ventricular systolic and diastolic performance.

Possible predictors of cardiovascular responses to smoked cocaine have been studied in 62 crack cocaine users (24 women and 38 men, aged 20–45 years) who used a single dose of smoked cocaine 0.4 mg/kg (561). Physiological responses to smoked cocaine, such as changes in heart rate and blood pressure, were monitored. The findings suggested that higher baseline blood pressure and heart rate, a greater amount and frequency of current cocaine use, and current cocaine snorting predicted a reduced cardiovascular response to cocaine. By contrast, factors such as male sex, African-American race, higher body weight, and current marijuana use were associated with a greater cardiovascular response.

Fatal pulmonary edema developed in a 36-year-old man shortly after injecting free-base cocaine intravenously (562).

Myocardial ischemia

Resentation

Myocardial infarction

> DoTS classification (BMJ 2003; 327: 1222-5):
> Dose-relation: Toxic
> Time-course: Time independent
> Susceptibility factors: Not known

Cocaine users often present with complaints suggestive of acute cardiac ischemia (chest pain, dyspnea, syncope, dizziness, and palpitation). Two studies have shown that the risk of actual acute cardiac ischemia among cocaine users with such symptoms was low. The first study reviewed the clinical database from the Acute Cardiac Ischemia–Time Insensitive Predictive Instrument Clinical Trial, a multicenter prospective clinical trial conducted in the USA in 1993 (563). Among 10 689 enrolled patients, 293 (2.7%) had cocaine-associated complaints. This rate varied from 0.3% to 8.4% in the 10 participating hospitals. Only six of these patients had a diagnosis of acute cardiac ischemia (2.0%), four with unstable angina and two with acute myocardial infarction. The cocaine users were admitted to the coronary care unit as often as other study participants (14 versus 18%), but were much less likely to have confirmed unstable angina (1.4 versus 9.3%). A second study also suggested that cocaine users who present with chest pain have a very low risk of adverse cardiac events (564). Emergency departments have instituted centers for the evaluation and treatment of patients with chest pain who are at low to moderate risk of acute coronary syndromes. In this particular study, patients with a history of coronary artery disease or presentations that included hemodynamic instability, electrocardiographic changes consistent with ischemia, or clinically unstable angina were directly admitted to hospital. In a retrospective study of 179 patients with reliable 30-day follow-up in chest pain centers, there was one cardiac complication due to cocaine use.

Case reports

- A 26-year-old man reported smoking 10 cigarettes/day and using cocaine by inhalation at weekends (565). His electrocardiogram showed raised anterolateral ST segments. He had raised creatine kinase MB activity and troponin I concentration. He was normotensive with signs of pulmonary congestion. Ventriculography showed anterolateral and apical hypokinesia and an ejection fraction of 21%. Angiography showed massive thrombosis of the left anterior descending coronary artery. He was given recombinant tissue plasminogen activator by intravenous infusion. Angiography 4 days later showed thrombus resolution, and ventriculography showed an improved ejection fraction. He was symptom-free 6 months later.
- A 50-year-old man had 12 hours of chest pain and shortness of breath after a cocaine binge. His history included hyperlipidemia, cigarette smoking, and cocaine use (566). The troponin concentration was increased, at 25 ng/ml. An electrocardiogram showed anteroseptal ST segment elevation and inferolateral depression with T wave inversion. An angiogram showed multi-vessel occlusion of the left anterior descending artery, right coronary artery, and left circumflex artery. Ventriculography showed severe anteroapical and inferior wall hypokinesis with an ejection fraction of 25%. Angioplasty was performed urgently and final angiography showed no residual stenosis in the treated vessels.
- In a cocaine abuser, resolution of intracoronary thrombosis with direct thrombin has been successfully attempted (567). Medical treatment was started with tirofiban and low molecular weight heparin, and 48 hours later they were replaced with bivalirudin, a direct thrombin inhibitor, in an initial bolus dose of 0.1 mg/kg followed by an infusion of 0.25 mg/kg/hour. Repeat angiography 48 hours after bivalirudin showed near total thrombus dissolution with resolution of the electrocardiographic abnormalities.

Accelerated extensive atherosclerosis secondary to chronic cocaine abuse has also been reported (568).

- A 32-year-old man, who was a cigarette smoker (10 cigarettes/day) and had been a frequent cocaine user for 16 years, developed acute chest pain 2 hours after heavy cocaine use. He had electrocardiographic abnormalities consistent with an acute myocardial infarction, confirmed by serial enzyme measurements. There was no family history of atherosclerosis. Echocardiography showed a large akinetic anteroapical segment. Serum lipoprotein concentrations were normal. Despite management with aspirin, glyceryl trinitrate, heparin, and morphine, he required emergency cardiac catheterization because of prolonged chest pain. Coronary angiography showed severe atherosclerosis of the middle and distal segments of the left anterior descending coronary artery and a large diagonal branch. Intracoronary glyceryl trinitrate and nitroprusside ruled out coronary spasm. Intravascular ultrasound localized the lesion to the middle of the left anterior descending artery and showed diffuse plaques along the entire artery, with variable composition, including fibrocalcific changes consistent with a chronic process. Two overlapping stents were inserted.

The authors acknowledged that the mechanisms for accelerated atherosclerotic process are not known. However, they cited research that suggests that cocaine has pro-atherogenic effects in blood vessels.

Fiberoptic bronchoscopy is often done after intratracheal injection of 2.5% cocaine solution and lidocaine spray. Acute myocardial infarction after fiberoptic bronchoscopy with intratracheal cocaine has been reported (569).

- A 73-year-old man with a history of breathlessness, cough, and weight loss had some ill-defined peripheral shadow in the upper zones of a chest X-ray. He had fiberoptic bronchoscopy with cocaine and lidocaine and 5 minutes later became distressed, with dyspnea, chest pain, and tachycardia. Electrocardiography showed an evolving anterior myocardial infarction. Coronary angiography showed a stenosis of less than 25% in the proximal left anterior descending artery with coronary artery spasm. He made an uneventful recovery.

The authors suggested that the principal cardiac effects of cocaine can be attributed to or are mediated by the following mechanisms: increased myocardial oxygen demand due to an acute rise in systemic blood pressure and heart rate; coronary vasoconstriction caused by alpha-adrenergic effects and calcium-dependent direct vasoconstriction; and promotion of arteriosclerosis and endothelial dysfunction, which predisposes to vasoconstriction and thrombosis.

Spontaneous acute coronary dissection after cocaine abuse has been reported (570).

- A 34-year-old woman developed chest pain suggestive of acute coronary syndrome, having inhaled cocaine 30 minutes before. Her blood pressure was 180/100 mmHg and an electrocardiogram showed sinus rhythm with anterolateral ischemia, ST segment depression, and T wave inversion. An echocardiogram showed a large hypokinetic area, including the middle and apical segments of the anterior septum and the anterior and lateral walls, and mild reduction in the left ventricular ejection fraction (45%). Troponin I and creatine kinase were slightly raised but the MB fraction was normal. Unstable angina was diagnosed but despite full medical therapy, the chest pain and ischemic changes did not resolve. Immediate catheterization showed a dissection flap within the left main trunk extending to the proximal portion of the descending anterior and circumflex arteries. There was no atherosclerosis in the coronary vessels. The flap resulted in a 90% stenosis of the proximal left anterior descending artery. Urgent coronary artery bypass surgery was successful.

Spontaneous coronary artery dissection is an unusual cause of acute coronary syndrome. Only three other cases secondary to cocaine use have been described.

Aortic thrombus and renal infarction has been reported in a patient who used nasal cocaine (571).

- A 52-year-old woman with a history of hypertension for 15 years developed acute left flank pain, nausea, and vomiting. On a previous similar occasion 2 weeks before she had had a trace of proteinuria and microscopic hematuria. A contrast-enhanced CT scan of the abdomen had not shown stones, hydronephrosis, or morphological abnormalities. She had had no rash. Her urine contained cocaine. Creatine kinase and lactate dehydrogenase activities were raised and there was a leukocytosis. A second abdominal CT scan with contrast showed a segmental infarct of the left kidney. A transesophageal echocardiogram showed a 2x2 cm mobile mass, consistent with a thrombus, attached to the aortic arch, distal to the left subclavian artery. There was no evidence of atherosclerosis. She was given anticoagulants and aggressive fluid therapy for rhabdomyolysis.

The authors speculated that cocaine may have caused aortic inflammation by assuming that cocaine-related increased sympathetic tone along with possible cocaine-related aortic inflammation (similar to reported cases of cocaine-related vascular injury) may have led to enhance aggregation of platelets at the inflamed area, which would act as a nidus to form a thrombus. The thrombus resolved with anticoagulation within 13 days and similar results have been reported before (572). The patient had very high creatine kinase activity, suggesting rhabdomyolysis, which could have been caused by intense cocaine-related vasoconstriction.

Frequency

By one estimate, since the first report in 1982, over 250 cases of myocardial infarction due to cocaine have been reported, mostly in the USA. The first report from the UK was published in 1999 (573).

The incidence of acute myocardial infarction in cocaine-associated chest pain is small but significant (574), but as cocaine use has become more widespread, the number of cocaine-related cardiovascular events has increased (575). The electrocardiogram has a higher false-positive rate in these patients. A normal electrocardiogram reduces the likelihood of myocardial injury but does not exclude it.

During the hour after cocaine is used, the risk of myocardial infarction is 24 times the baseline risk (576). Cocaine users have a lifetime risk of non-fatal myocardial infarction that is seven times the risk in non-users, and cocaine use accounts for up to 25% of cases of acute myocardial infarction in patients aged 18-45 years (577). In 2000, there were 175 000 cocaine-related visits to emergency departments in the USA (578), with chest discomfort in 40% of the patients (579), 57% of whom were admitted to the hospital and had an admission lasting an average of 3 days (580), involving huge costs (581).

Cocaine use may account for up to 25% of acute myocardial infarctions among patients aged 18–45 years. The safety of a 12-hour observation period in a chest pain unit followed by discharge in individuals with cocaine-associated chest discomfort who are at low risk of cardiovascular events has been evaluated in 302 consecutive patients aged 18 years or older (66% men, 70% black, 84% tobacco users) who developed chest pain within 1 week of cocaine use or who tested positive for cocaine (582). Cocaine use was self-reported by 247 of the 302 subjects and rest had urine positive for cocaine; 203 had used crack cocaine, 51 reported snorting, and 10 had used it intravenously. Of the 247 who reported cocaine use, 237 (96%) said they had used it in the week before presentation and 169 (68%) within 24 hours before presentation. Follow-up information was obtained for 300 subjects. There were no deaths from cardiovascular causes. Four patients had a non-fatal myocardial infarction during the 30-day period; all four had continued to use cocaine. Of the 42 who were directly admitted to hospital, 20 had acute coronary syndrome. The authors suggested that in this group of subjects, observation for 9–12 hours with follow-up is appropriate.

In a review of the literature, 91 patients with cocaine-induced myocardial infarction were identified (583). Myocardial infarction occurred in 44 patients after intranasal use, in 27 after smoking, and in 19 after intravenous use. Almost half had a prior episode of chest pain. Two-thirds had their myocardial infarction within 3 hours of use. There were acute complications related to the myocardial infarction in 18 patients. Of 24 patients followed up, 58% had subsequent cardiac complications. Two-phase myocardial imaging with ^{99m}Tc-sestamibi can be helpful in the definitive diagnosis of cocaine-induced myocardial infarction in patients with a history of cocaine use, chest pain, and a non-diagnostic electrocardiogram (584). Damage to the myocardium associated with cocaine can be unrecognized by the abuser (585). Major electrocardiographic findings (including myocardial infarction, myocardial ischemia, and bundle branch block) were recorded during a review of 99 electrocardiograms of known cocaine abusers. None of the 11 patients with major electrocardiographic changes had a past history of cardiac disease or had complained of chest pain. The mechanism by which cocaine causes acute myocardial damage is unclear. Of 20 healthy cocaine abusers given intravenous cocaine, in doses commonly self-administered, or placebo, none developed myocardial ischemia or ventricular dysfunction on two-dimensional echocardiography during the test (586).

Myocardial infarction has been documented in 6% of patients who present to emergency departments with cocaine-associated chest pain (587,588). Treatment of cocaine-associated myocardial infarction has previously generally been conservative, using benzodiazepines, aspirin, glyceryl trinitrate, calcium channel blockers, and thrombolytic drugs. In the context of 10 patients with cocaine-associated myocardial infarction, who were treated with percutaneous interventions, including angioplasty, stenting, and AngioJet mechanical extraction of thrombus, the authors suggested that percutaneous intervention can be performed in such patients safely and with a high degree of procedural success (589). Patients with cocaine-associated chest pain and electrocardiographic ST-segment elevation should first undergo coronary angiography, if available, followed by percutaneous intervention. Alternatively, thrombolytic drugs can be used. However, the relative safety and efficacy of thrombolytic drugs compared with percutaneous intervention is undefined in patients with cocaine-associated myocardial infarction.

Mechanisms

The hypothesis that cocaine users have increased coronary microvascular resistance, even in the absence of recent myocardial infarction, coronary artery disease, or spasm, has been assessed in 59 consecutive cocaine users without acute or recent myocardial infarction or angiographically significant epicardial stenosis or spasm (590). Microvascular resistance was significantly increased by 26-54% in cocaine users. There was an abnormally high resistance in the left anterior descending artery in 61% of the patients, in the left circumflex artery in 69%, and in the right coronary artery in 47%. Increased microvascular resistance may explain many important cardiovascular effects of cocaine and has therapeutic implications. For example, slow coronary filling in diagnostic tests may suggest the possibility of cocaine use in patients in whom it was not otherwise suspected. There was increased microvascular resistance in the coronary bed even after the acute effects of a dose of cocaine would have worn off, suggesting that cocaine may have long-lasting effects on coronary microvasculature. This implies that medical therapy of vasoconstriction should be continued for extended periods. Moreover, heightened microvascular resistance in cocaine users may explain the development of chest pain and myocardial ischemia in patients who do not have epicardial stenosis due to coronary artery disease or spasm. The authors suggested that in the absence of coronary artery disease, the small vessel effects of cocaine may be more important. As the process is diffuse rather than confined to one vascular territory, electrocardiographic findings may not be localizing.

In a review of 114 cases, coronary anatomy, defined either by angiography or autopsy, was normal in 38% of chronic cocaine users who had had a myocardial infarction (591). The authors of another review concluded that "the vast majority of patients dying with cocaine toxicity, either have no pathological changes in the heart, or only minimal changes" (592). There can be a delay between the use of cocaine and the development of chest pain (593). The results of a study of 101 consecutive patients admitted with acute chest pain related to cocaine suggested that it commonly causes chest pain that may not be secondary to myocardial ischemia (594). The use of intranasal cocaine for therapeutic purposes (to treat epistaxis) was associated with myocardial infarction in a 57-year-old man with hypertension and stable angina (595).

Susceptibility factors

Myocardial ischemia and infarction associated with cocaine are unrelated to the route of administration, the amount taken, and the frequency of use. The risk of acute myocardial infarction is increased after acute cocaine use and it can occur in individuals with normal coronary arteries at angiography. The patients are typically young men and smokers and do not have other risk factors for atherosclerosis.

In a report of cocaine-associated chest pain, the authors studied the incidence and predictors of underlying significant coronary disease in 90 patients with and without myocardial infarction (596). Patients with 50% or more stenosis of coronary arteries or major branches or bypass graft were included and 50% of them had significant disease: one-vessel disease in 32%, two-vessel disease in 10%, three-vessel disease in 6%, and significant graft stenosis in 3%. There was significant disease in 77% of patients with myocardial infarction or a raised troponin I concentration, compared with 35% of patients without myonecrosis. Predictors of significant coronary disease included myocardial damage, a prior history of coronary disease, and a raised cholesterol. Only seven of the 39 patients without myonecrosis or a history of coronary disease had significant angiographic disease. The authors concluded that significant disease is found in most patients with cocaine-associated myocardial damage. In contrast, only a minority of those without myonecrosis have significant coronary disease.

Pregnancy increases the incidence of dysrhythmias in patients with Wolff–Parkinson–White syndrome. This association may relate to an effect of estrogen, increased plasma volume, or increased maternal stress or anxiety. A case of cocaine-induced myocardial ischemia in pregnancy mistaken for Wolff–Parkinson–White syndrome has been reported (597).

- A 22-year-old pregnant woman at 38 weeks' gestation developed chest pain, palpitation, and shortness of breath. Her blood pressure was 148/97 mmHg, heart rate 105/minute, and respiratory rate 22/minute. An electrocardiogram showed a short PR interval and a broad QRS complex with slurred upstrokes of the R waves in leads V2 and V3. Other laboratory studies were normal. The fetal heart rate was 160/minute. The differential diagnosis included a new onset dysrhythmia, Wolff–Parkinson–White syndrome, and myocardial ischemia/infarction. With supportive treatment and monitoring the dysrhythmia resolved. Serial creatine kinase measurements were normal. However, there were cocaine metabolites in the urine. The patient admitted to having used cocaine 3-4 hours before the episode.

The use of cocaine among women of reproductive age is increasing and pregnancy also enhances the potential cardiovascular toxicity of cocaine.

Management

Based on a retrospective study of 344 patients with cocaine-associated chest pain, it has been suggested that patients who do not have evidence of ischemia or cardiovascular complications over 9–12 hours in a chest-pain observation unit have a very low risk of death or myocardial infarction during the 30 days after discharge (598). Nevertheless, patients with cocaine-associated chest pain should be evaluated for potential acute coronary syndromes; those who do not have recurrent symptoms, increased concentrations of markers of myocardial necrosis, or dysrhythmias can be safely discharged after 9–12 hours of observation. A protocol of this sort should incorporate strategies for treating substance abuse, since there is an increased likelihood of non-fatal myocardial infarction in patients who continue to use cocaine.

In a randomized, controlled trial in 36 patients with cocaine-associated chest pain, the early use of lorazepam together with glyceryl trinitrate was more efficacious than glyceryl trinitrate alone in relieving the chest pain (599). These findings contrast with those of an earlier study in which there was no evidence of additional benefit from diazepam in managing cocaine-related chest pain (600). However, the Advanced Cardiovascular Life Support (ACLS) guidelines state that in cocaine-associated acute coronary syndromes a nitrate should be first-line therapy together with a benzodiazepine (601).

Although percutaneous revascularization for cocaine-associated myocardial infarction is the preferred method of treatment (602), the feasibility and safety of multivessel primary angioplasty has been demonstrated. In patients with persistent myocardial ischemia despite medical therapy, or evidence of cardiogenic shock, aggressive early intervention is particularly beneficial. Beta-blockers should be avoided in patients who have recently used cocaine; they fail to control the heart rate, enhance cocaine-induced vasoconstriction, increase the likelihood of seizures, and reduce survival (603).

Other vascular disease

Dissection of the aorta has been reported during cocaine use (604,605). The authors of these two reports noted that all six cases of this rare complication reported in the past 5 years were in men with pre-existing essential hypertension. In a review of emergency visits to a hospital during a 20-year period, 14 of 38 cases of acute aortic dissection involved cocaine use; 6 were of type A and 8 of type B (606). Crack cocaine had been smoked in 13 cases and powder cocaine had been snorted in one case. The mean time of onset of chest pain was 12 hours after cocaine use. The chronicity of cocaine use was not known in most of the cases. The cocaine users were typically younger than the non-cocaine users. Chronic untreated hypertension and cigarette smoking were often present.

- A 43-year-old man with untreated hypertension developed transient mild chest pressure followed by shortness of breath for 4 hours (607). He had long used tobacco, alcohol, and cocaine and admitted to having used cocaine within the last 12 hours. He had a tachycardia with a pansystolic murmur suggesting mitral regurgitation. Urine drug screen was positive for cocaine metabolites. A chest X-ray showed mild cardiomegaly and prominent upper lobe vasculature. An electrocardiogram showed atrial flutter at a rate of 130/minute and non-specific T wave changes. The diagnoses were myocardial infarction due to cocaine, with mild congestive heart failure, mitral regurgitation and atrial flutter. However, transesophageal echocardiography showed severe aortic insufficiency and a dissection flap in the ascending aorta. He underwent emergency repair of the aortic root and resuspension of the aortic valve.

Intramural hematoma of the ascending aorta has been reported in a cocaine user (608).

- A healthy 39-year-old man developed retrosternal chest pain radiating to the back with nausea and sweating. About 10–15 minutes before, he had inhaled cocaine for 2 hours and then smoked crack cocaine. He had an aortic dissection, which was repaired surgically.

The authors identified hypertension secondary to the use of cocaine as the risk factor for this complication.

Coronary artery dissection associated with cocaine is rare. The first case was reported in 1994 (609) and two other cases have been reported (610,611).

- A healthy 33-year-old man with prior cocaine use had a small myocardial infarction and, 36 hours later, having inhaled cocaine, developed a dissection of the left main coronary artery, extending distally to the left anterior descending and circumflex arteries. There was marked anterolateral and apical hypokinesis.
- A 23-year-old man with a history of intravenous drug abuse and hepatitis C was found unconscious, hypoxic, and hypotensive. A urine drug screen was positive for cocaine metabolites, benzodiazepines, and opiates. An electrocardiogram suggested a myocardial infarction, verified by raised troponin I and the MB fraction of creatine kinase. He had severe hypokinesia with a left ventricular ejection fraction of 10%, falling to less than 5%. He became septic, developed multiorgan system failure, and died. The postmortem findings included dissection of the left anterior descending artery with complete occlusion of the true lumen and thrombosis of the false lumen. The left ventricle showed extensive transmural myocardial necrosis with adjacent contraction band necrosis. He also had deep vein thromboses in veins in the neck and abdomen and multiple pulmonary infarctions.

Peripheral vascular disease in the fingers has been attributed to cocaine.

- A 48-year-old man who smoked cigarettes and used cocaine developed ischemia of the right index finger due to occlusion of the distal ulnar artery (612). He had a history of recurrent deep vein thrombosis. A venous bypass graft was performed. Two years later he had non-healing gangrene of the left index finger. His blood pressure was normal in both arms. Urine toxicology was positive for cocaine. Angiography of the left arm showed small-vessel vasculitis.
- A healthy 36-year-old man, who had used intranasal crack cocaine daily in increasing doses for 2 weeks, developed pain, numbness, swelling, and cyanosis of

the fingers and toes aggravated by cold and an ulcer on one finger (613). Ultrasound Doppler of the hand confirmed ischemic finger necrosis. He was treated unsuccessfully with aspirin, diltiazem, and heparin, but responded to intravenous infusions of iloprost for 5 days.

Another less common complication of cocaine use is cerebral vasculitis (614), and benign cocaine-induced cerebral angiopathy has been reported (615). (Stroke is discussed under the section Nervous system in this monograph).

Cardiac dysrhythmias

Dysrhythmias seem to be the most likely cause of sudden death from cocaine, but cardiac conduction disorders are more common in patients with acute cocaine toxicity. Severe cocaine toxicity also causes acidemia and cardiac dysfunction (616). Four patients developed seizures, psychomotor agitation, and cardiopulmonary arrest; two of these are briefly summarized here.

- A 43-year-old man injected a large dose of cocaine in a suicide attempt and had a seizure and cardiopulmonary arrest, from which he was resuscitated. His arterial blood pH was 6.72 and his electrocardiogram showed a wide complex tachycardia. An infusion of sodium bicarbonate maintained the blood pH at 7.50 and the electrocardiogram became normal. The bicarbonate infusion was discontinued after 12 hours.
- A 25-year-old man had a cardiac arrest after taking one "knot" or sealed bag of crack cocaine (2.5 g) and was resuscitated. His arterial blood pH was 6.92 and an electrocardiogram showed sinus rhythm, QRS axis 300°, and terminal 40 msec of the QRS axis 285°. After an infusion of sodium bicarbonate, his blood pH was 7.30, his QRS axis 15°, and the terminal 40 msec QRS axis 30°. He passed the bag of cocaine rectally within 12 hours of admission.

These patients' initial laboratory values showed acidosis, prolongation of the QRS complex and QT_c interval, and right axis deviation. Appropriate treatment included hyperventilation, sedation, active cooling, and sodium bicarbonate, which led to correction of the blood pH and of the cardiac conduction disorders. The authors suggested that when intracellular pH is lowered, myocardial contractility is depressed as a result of reduced calcium availability. During acidosis, there are abnormalities of repolarization and depolarization, which potentiate dysrhythmias.

- A 37-year-old man developed severe chest pain and shortness of breath after smoking 200+ rocks of crack cocaine in 72 hours while attempting to walk several miles in order to get more drug money (617). His symptoms resolved before the ambulance arrived. His medical history included a prior episode of chest pain after a 3-day crack cocaine binge. He also occasionally used alcohol and marijuana. He was anxious but had a normal heart rate and blood pressure. His drug screen showed cocaine. Cardiac enzymes were within the reference ranges. An initial electrocardiogram showed a QT_c interval of 621 ms, which shortened to 605 ms after 2 hours, 530 ms after 7 hours, and 543 ms after 15 hours. One month later the QT_c interval was 453 ms.

Brugada syndrome has been attributed to cocaine.

- A 36-year old man became comatose 14 hours after inhaling an unspecified amount of heroin for an unspecified duration (618). He had been taking lithium, chlorpromazine, and diazepam for a chronic psychosis. His family reported recreational use of drugs of abuse. His Glasgow coma score was 4 without focal deficits. A toxicology screen was positive for cocaine and remained positive for 4 days after admission. His electrocardiogram showed prominent coved ST elevation and J wave amplitude of at least 2 mm in leads V1-V 3, followed by negative T waves with no isoelectric separation, associated with right incomplete bundle branch block indicative of type 1 Brugada syndrome. His serum potassium concentration was 5.9 mmol/l. He was immediately treated with 42% sodium bicarbonate intravenously. The Brugada pattern completely resolved in 24 hours and his creatinine improved significantly in 48 hours. Transthoracic echocardiography was normal.

The authors of this report thought that the Brugada syndrome was probably not due to chlorpromazine or lithium in this patient, and it has not been previously described with heroin. It may have been due to hyperkalemia (as the Brugada pattern normalized when the serum potassium concentration normalized), perhaps facilitated by cocaine.

Brugada syndrome has also been described in a case of cocaine overdose.

- A 28-year-old man aboard a flight reported feeling ill, became confused, and had a generalized seizure (619). After landing he had two further generalized seizures and a cardiac arrest. With medical management, his dysrhythmia, a wide complex tachycardia with features of right bundle branch block (Brugada syndrome) and ST segment elevation in the anteroseptal leads, was converted to sinus rhythm. He had raised troponin and creatine kinase MB. In the next 24 hours, he developed acute renal insufficiency and was treated with intermittent hemodialysis. A contrast CT showed generalized edema of the brain and features of coning, subarachnoid hemorrhage, and acute hydrocephalus. A plain abdominal x-ray showed multiple radio-opaque shadows. A CT scan of the abdomen confirmed the presence of multiple radio-opaque densities consistent with foreign bodies. There was free gas in the abdomen, consistent with bowel perforation. He died 96 hours after admission. Autopsy showed 120 condoms containing cocaine in the gastrointestinal tract; one had ruptured.

The electrocardiographic PR interval increased with increasing abstinence from crack cocaine in a study of 441 chronic cocaine users who had smoked at least 10 g of cocaine in the 3 months before enrollment (620). The authors suggested that this may have reflected the normalization of a depolarization defect. Chronic cocaine users have shortened PR intervals, indicative of rapid cardiac depolarization.

Cardiovascular effects of cocaine as a local anesthetic

Cardiovascular effects due to enhanced sympathetic activity include tachycardia, increased cardiac output, vasoconstriction, and increased arterial pressure. Myocardial infarction is the most common adverse cardiac effect (561), and there is an increased risk of myocardial depression when amide-type local anesthetics, such as bupivacaine, levobupivacaine, lidocaine, or ropivacaine are administered with antidysrhythmic drugs.

- A woman who inappropriately used cocaine on the nasal mucosa to treat epistaxis had a myocardial infarction (585).
- A patient who was treated with intranasal cocaine and phenylephrine during a general anesthetic had a myocardial infarction and a cardiac arrest due to ventricular fibrillation (SEDA-20, 128).
- Myocardial ischemia was reported in a fit 29-year-old patient after the nasal application of cocaine for surgery. No relief was gained from vasodilators or intracoronary verapamil, and there were no other signs of cocaine toxicity. Although coronary vasoconstriction and platelet activation are systemic effects of cocaine, pre-existing thrombus may also have played a part (SEDA-22, 142).

Previous cocaine abuse has also been implicated in increasing the risk of myocardial ischemia when other local anesthetics are used.

Cardiac dysrhythmias have also been described in patients after the use of topical cocaine for nasal surgery (SEDA-20, 128).

- A patient who was treated with intranasal cocaine and submucosal lidocaine during general anesthesia developed ventricular fibrillation (SEDA-17, 142).

These events do not appear to have been related to the concomitant use of a vasoconstrictor, but more to excessive doses of cocaine.

Substantial systemic absorption of cocaine can cause severe cardiovascular complications (621).

- An 18-year-old man had both nasal cavities prepared with a pack soaked in 3–5 ml of Brompton solution (3% cocaine, about 3 mg/kg, plus adrenaline 1:4000) 2 hours preoperatively. In the anesthetic room he was anxious and withdrawn, with a mild tachycardia. Ten minutes later the nasal pack was removed and polypectomy was begun, with immediate sinus tachycardia and marked ST depression on lead II of the electrocardiogram. Increasing the depth of anesthesia and giving fentanyl had little effect, and the procedure was terminated. After extubation a further electrocardiogram showed T wave flattening in leads II, III, aVF, and aVL. Further cardiac investigations ruled out a myocardial infarction, an anatomical defect, or other pathological or metabolic processes. On day 4 a stress electrocardiogram showed no ischemic changes.

Absorption of cocaine from the nasal mucosa in eight patients using cotton pledglets soaked in 4 ml of 4% cocaine and applied for 10 or 20 minutes resulted in an absorption rate four times higher than expected, but was not associated with any cardiovascular disturbance; however, one of four patients who received 4 ml of 10% cocaine for 20 minutes developed intraoperative hypertension and another transient ventricular tachycardia (622). The authors advised against topical use of 10% cocaine.

Collagen and gelatin

Hypotension has been described in a patient who received 500 ml of gelofusine during surgery for an infrarenal abdominal aneurysm (623). Within 5 minutes there was a significant fall in blood pressure, in the absence of surgical bleeding and ischemic changes on the electrocardiogram. There was no change in oxygenation. A further 500 ml of gelofusine was rapidly infused, with an additional fall in blood pressure to 50/30. Subsequently, after recovery, a small test dose (100 ml) of gelofusine reproduced the transient hypotension.

A patient undergoing anesthesia for coronary artery bypass surgery developed a probable anaphylactic reaction to Gelofusine (624). The principal feature of the reaction was cardiovascular depression. An infusion of the angiotensin analogue angiotensinamide was effective in treating the reaction.

Colony-stimulating factors

In a randomized non-placebo-controlled study in 93 patients with myelodysplastic syndrome, the addition of GM-CSF (molgramostim) to standard induction chemotherapy in 46 patients did not improve the remission rate and was associated with significantly higher rates of non-hematological adverse effects (625). In particular, there were fluid retention, pulmonary edema, or weight gain in 14 patients receiving GM-CSF compared with only one in those who were not given GM-CSF. All six patients who had acute myocardial infarctions were receiving GM-CSF. According to experimental evidence that endogenous GM-CSF might contribute to the development of acute coronary syndrome, the authors suggested caution in elderly patients with advanced coronary atherosclerosis.

Aortitis has been attributed to molgramostim in a previously healthy patient (626).

- A 55-year-old woman was given a 5-day course of G-CSF 5 micrograms/kg/day for peripheral blood stem cell mobilization. Two days after the completion of treatment she developed severe abdominal pain, fever, and vomiting. Acute aortitis of the descending aorta was seen on MRI. Infectious causes, giant cell arteritis, and several systemic diseases were ruled out. She recovered without aneurysm after glucocorticoid treatment.

As this patient also had a neutrophilia and major leukocyturia with hematuria during the acute episode, a neutrophilic-mediated aortitis was the suggested mechanism.

The use of G-CSF, either alone or before intracoronary cell infusion of mobilized peripheral blood stem cells to improve cardiac function in patients with myocardial infarction, has been associated with in-stent restenosis in seven of the first 10 enrolled patients (627). This unexpectedly high rate of restenosis led to premature discontinuation of the trial.

Capillary leak syndrome is caused by damage to endothelial cells, resulting in extravasation of plasma proteins and fluid from the capillaries into the extravascular space. This results in large amounts of extravascular lung water and pulmonary vascular permeability, necessitating mechanical ventilation. Capillary leak syndrome in which serial extravascular lung water measurements were performed has been reported in a patient receiving granulocyte colony-stimulating factor (628).

- A 68-year-old woman with a stage IIIB diffuse large B cell lymphoma developed severe capillary leak syndrome during treatment with G-CSF after autologous hematological stem cell transplantation, having received chemotherapy before transplantation. She was given subcutaneous G-CSF 480 micrograms/day and treatment was withdrawn on day 11 when the absolute neutrophil count exceeded 500 x 10^6/l. She developed renal impairment from day 2, became somnolent, and developed leg edema. Her blood pressure was 98/45 mmHg, pulse 137/minute, central venous pressure 14 mmHg, respiratory rate 20/minute, and temperature 38.4°C. A chest X-ray showed marked pulmonary congestion. She was given slow extended daily dialysis, noradrenaline, and dobutamine, but became more hypoxic and required mechanical ventilation. Her liver function deteriorated and she became icteric without hepatomegaly or ascites. Liver biopsy showed drug-induced hepatitis without veno-occlusive disease. Her extravascular lung water volume was increased. Continuous venovenous hemofiltration and hydrocortisone 100 mg tds were ineffective. Hypoxia, respiratory and metabolic acidosis, and hemodynamic instability worsened and she died.

Five other cases have been reported before. The white blood cell count at the onset was 0–90 × 10^9/l. The symptoms started on days 5–9 after the start of G-CSF treatment, and all the patients had fever. Mechanical ventilation was required in three and renal replacement therapy in four; two died.

Complementary and alternative medicine

Acupuncture has been associated with hemopericardium due to ventricular puncture (629).

- An 83-year-old Austrian woman developed syncope and cardiogenic shock shortly after acupuncture over the sternum. Echocardiography showed cardiac tamponade and pericardiocentesis revealed hemopericardium. At operation a small bleeding perforation of the right ventricle was found and closed. The acupuncture at the point "Ren 17" was above a sternal foramen, which allowed the needle to penetrate the heart.

A case in which acupuncture was apparently responsible for cardiac dysrhythmias has been reported (630).

Corticosteroids—glucocorticoids

The considerable body of evidence that glucocorticoids can cause increased rates of vascular mortality and the underlying mechanisms (increased blood pressure, impaired glucose tolerance, dyslipidemia, hypercoagulability, and increased fibrinogen production) have been reviewed (631). In view of their adverse cardiovascular effects, the therapeutic options should be carefully considered before long-term glucocorticoids are begun; although they can be life-saving, dosages should be regularly reviewed during long-term therapy, in order to minimize complications.

A study has been undertaken to clarify whether glucocorticoid excess affects endothelium-dependent vascular relaxation in glucocorticoid treated patients and whether dexamethasone alters the production of hydrogen peroxide and the formation of peroxynitrite, a reactive molecule between nitric oxide and superoxide, in cultured human umbilical endothelial cells (632). Glucocorticoid excess impaired endothelium-dependent vascular relaxation in vivo and enhanced the production of reactive oxygen species to cause increased production of peroxynitrite in vitro. Glucocorticoid-induced reduction in nitric oxide availability may cause vascular endothelial dysfunction, leading to hypertension and atherosclerosis.

Hypertension

The secondary mineralocorticoid activity of glucocorticoids can lead to salt and water retention, which can cause hypertension. Although the detailed mechanisms are as yet uncertain, glucocorticoid-induced hypertension often occurs in elderly patients and is more common in patients with total serum calcium concentrations below the reference range and/or in those with a family history of essential hypertension (SEDA-20, 368; 633).

Hemangioma is the most common tumor of infancy, with a natural history of spontaneous involution. Some hemangiomas, however, as a result of their proximity to vital structures, destruction of facial anatomy, or excessive bleeding, can be successfully treated with systemic glucocorticoids between other therapies. The risk of hypertension is poorly documented in this setting. In one prospective study of 37 infants (7 boys, 17 girls; mean age 3.5 months, range 1.5–10) with rapidly growing complicated hemangiomas treated with oral prednisone 1–5 mg/kg/day, blood pressure increased in seven cases (634). Cardiac ultrasound examination in five showed two cases of myocardial hypertrophy, which was unrelated to the hypertension and which regressed after withdrawal of the prednisone.

Myocardial ischemia

Cortisone-induced cardiac lesions are sometimes reported and electrocardiographic changes have been seen in patients taking glucocorticoids (635). Whereas abnormal myocardial hypertrophy in children has perhaps been associated more readily with corticotropin, it has been seen on occasion during treatment with high dosages of glucocorticoids, with normalization after dosage reduction and withdrawal.

Fatal myocardial infarction occurred after intravenous methylprednisolone for an episode of ulcerative colitis (636).

- A day after a dose of intravenous methylprednisolone 60 mg a 79-year-old woman developed acute thoracic pain and collapsed. An electrocardiogram showed signs of a myocardial infarction and her cardiac enzyme activities were raised. She died within several hours. Autopsy showed an anterior transmural myocardial infarction and mild atheromatous lesions in the coronary arteries.

This report highlights the risk of cardiovascular adverse effects with short courses of glucocorticoid therapy in elderly patients with inflammatory bowel disease, even with rather low-dosage regimens. Acute myocardial infarction occurred in an old man with coronary insufficiency and giant cell arteritis after treatment with prednisolone (SEDA-10, 409) but could well have been coincidental.

Myocardial ischemia has been reportedly precipitated by intramuscular administration of betamethasone (SEDA-21, 413; 637). It has been suggested that long-term glucocorticoid therapy accelerates atherosclerosis and the formation of aortic aneurysms, with a high risk of rupture (SEDA-20, 369; 638).

Patients with seropositive rheumatoid arthritis taking long-term systemic glucocorticoids are at risk of accelerated cardiac rupture in the setting of transmural acute myocardial infarction treated with thrombolytic drugs (639).

- Two women and one man, aged 53–74 years, died after they received thrombolytic therapy for acute myocardial infarction. All three had a long history of seropositive rheumatoid arthritis treated with prednisone 5–20 mg/day for many years.

Long-term systemic administration of glucocorticoids might be expected, because of their effects on vascular fragility and wound healing, to increase the risk of vascular complications during percutaneous coronary intervention. To assess the potential risk of long-term glucocorticoid use in the setting of coronary angioplasty, 114 of 12 883 consecutively treated patients who were taking long-term glucocorticoids were compared with those who were not. Glucocorticoid use was not associated with an increased risk of composite events of major ischemia but was associated with a threefold risk of major vascular complications and a three- to fourfold risk of coronary perforation (640).

Cardiomyopathy

Postnatal exposure to glucocorticoids has been associated with hypertrophic cardiomyopathy in neonates. Such an effect has not previously been described in infants born to mothers who received antenatal glucocorticoids. Three neonates (gestational ages 36, 29, and 34 weeks), whose mothers had been treated with betamethasone prenatally in doses of 12 mg twice weekly for 16 doses, 8 doses, and 5 doses respectively, developed various degrees of hypertrophic cardiomyopathy diagnosed by echocardiography (641). There was no maternal evidence of diabetes, except for one infant whose mother had a normal fasting and postprandial blood glucose before glucocorticoid therapy, but an abnormal 1-hour postprandial glucose after 8 weeks of betamethasone therapy, with a normal HbA_{1C} concentration. There was no family history of hypertrophic cardiomyopathy, no history of maternal intake of other relevant medications, no hypertension, and none of the infants received glucocorticoids postnatally. Follow-up echocardiography showed complete resolution in all infants. The authors suggested that repeated antenatal maternal glucocorticoids might cause hypertrophic cardiomyopathy in neonates. These changes appear to be dose- and duration-related and are mostly reversible.

Transient hypertrophic cardiomyopathy is a rare sequel of the concurrent administration of glucocorticoid and insulin excess (SEDA-21, 412; 642). The heart is also almost certainly a site for myopathic changes analogous to those that affect other muscles.

Transient hypertrophic cardiomyopathy has been attributed to systemic glucocorticoid administration for a craniofacial hemangioma (643).

- A 69-day-old white child presented with a rapidly growing 2.5 × 1.5 cm hemangioma of the external left nasal side wall. He was normotensive and there was no family history of cardiomyopathy or maternal gestational diabetes. Because of nasal obstruction and possible visual obstruction, he was given prednisolone 3 mg/kg/day. After 10 weeks his weight had fallen from 7.6 to 7.1 kg and 2 weeks later he became tachypneic with a respiratory rate of 40/minutes. A chest X-ray showed cardiomegaly and pulmonary venous congestion. An echocardiogram showed hypertrophic cardiomyopathy. The left ventricular posterior wall thickness was 10 mm (normal under 4 mm), and the peak left ventricular outflow gradient was 64 mmHg. He was given a beta-blocker and a diuretic and the glucocorticoid dose was tapered. The cardiomyopathy eventually resolved.

Dilated cardiomyopathy caused by occult pheochromocytoma has been described infrequently.

- A 34-year-old woman had acute congestive heart failure 12 hours after administration of dexamethasone 16 mg for an atypical migraine (644). The authors postulated that the acute episode had been induced by the dexamethasone, which increased the production of adrenaline, causing beta$_2$-adrenoceptor stimulation, peripheral vasodilatation, and congestive heart failure.

In an addendum the authors reported another similar case.

Obstructive cardiomyopathy has been attributed to a glucocorticoid in a child with subglottal stenosis (645).

- A 4-month-old boy (weight 4 kg) developed fever, nasal secretions, and stridor due to a subglottal granuloma. Dexamethasone 1 mg/kg/day was started and tapered over 1 week. The mass shrank to 25% of its original size but the symptoms recurred 2 weeks later. The granuloma was excised and dexamethasone 1 mg/kg/day was restarted. After 5 days he developed a tachycardia (140/minute) and a new systolic murmur. Echocardiography showed severe ventricular hypertrophy with dynamic left ventricular outflow tract obstruction. The dexamethasone was weaned over several days. Over the next 3 weeks several echocardiograms showed rapid resolution of the outflow tract obstruction and gradual improvement of the cardiac hypertrophy. After 8 months there was no further problem.

Cardiac dysrhythmias

Serious cardiac dysrhythmias and sudden death have been reported with pulsed methylprednisolone. Oral methylprednisolone has been implicated in a case of sinus bradycardia (646).

- A 14-year-old boy received an intravenous dose of methylprednisolone 30 mg/kg for progressive glomerulonephritis. After 5 hours, his heart rate had fallen to 50/minute and an electrocardiogram showed sinus bradycardia. His heart rate then fell to 40/minutes and a temporary transvenous pacemaker was inserted and methylprednisolone was withdrawn. His heart rate increased to 80/minutes over 3 days. After a further 3 days, he was treated with oral methylprednisolone 60 mg/m^2/day and his heart rate fell to 40/minutes in 5 days. Oral methylprednisolone was stopped on day 8 of treatment and his heart rate normalized.

Hypokalemia, secondary to mineralocorticoid effects, can cause cardiac dysrhythmias and cardiac arrest.

Recurrent cardiocirculatory arrest has been reported (647).

- A 60-year-old white man was admitted for kidney transplantation. Immediately after reperfusion and intravenous methylprednisolone 500 mg, he developed severe bradycardia with hypotension and then cardiac arrest. After resuscitation, his clinical state improved quickly, but on the morning of the first postoperative day directly after the intravenous administration of methylprednisolone 250 mg, he had another episode of severe bradycardia, hypotension, and successful cardiopulmonary resuscitation. A third episode occurred 24 hours later after intravenous methylprednisolone 100 mg, again followed by rapid recovery after resuscitation. Two weeks later, during a bout of acute rejection, he was given intravenous methylprednisolone 500 mg, after which he collapsed and no heartbeat or breathing was detectable; after cardiopulmonary resuscitation he was transferred to the intensive care unit, where he died a few hours later.

If patients at risk are identified, glucocorticoid bolus therapy should be avoided or, if that is not possible, should only be done under close monitoring.

Thromboembolism

A case report with a review of 27 cases of thromboembolic events after the administration of intravenous globulin with or without glucocorticoids has been published (648). The authors suggested that this combined therapy should be administered with caution because of its potential synergistic thrombotic risk.

Pericarditis

- Disseminated *Varicella* and staphylococcal pericarditis developed in a previously healthy girl after a single application of triamcinolone cream 0.1% to relieve pruritus associated with *Varicella* skin lesions (SEDA-22, 443; 649).

Vasculitis

Long-term treatment with glucocorticoids can cause arteritis, but patients with rheumatoid arthritis have a special susceptibility to vascular reactions, and cases of periarteritis nodosa after withdrawal of long-term glucocorticoids have been reported (650).

Corticosteroids—mineralocorticoids

Edema, hypertension, and cardiac hypertrophy can occur with fludrocortisone (651). Hypertension is the most common reason for reducing the dosage.

Fluid retention due to mineralocorticoid effects can cause cardiac failure (651).

- Congestive heart failure occurred in a 47-year-old woman after she had taken fludrocortisone 100 micrograms/day for 2 weeks for Addison's disease (652). Ten months later, fludrocortisone 25 micrograms/day was restarted, and the dosage was increased to 100 micrograms/day over 2 months. At follow-up after 4 months she was well, without fluid retention or electrolyte abnormalities.

Corticotrophins (corticotropin and tetracosactide)

Corticotropin has been reported to cause enlargement of cardiac tumors in tuberous sclerosis (653).

- A female infant with tuberous sclerosis had multiple large cardiac tumors in the left and right ventricles. Corticotropin was given (dose not stated; once a day for 2 weeks, tapering over 3 months) at 4 months for infantile spasms. At 6 months a heart murmur was detected. Echocardiography showed pronounced enlargement of the tumors in both ventricles and a small tumor extending from the upper portion of the interventricular septum into the left ventricular outflow tract. An electrocardiogram showed 2–3 mm ST segment depression in leads I, aVL, and V4-6. Gated single photon emission CT showed low perfusion at the lateral and inferior regions of the left ventricle, indicating myocardial ischemia. Corticotropin was withdrawn and 3 months later the patient was asymptomatic. An echocardiogram showed that the tumors had reduced in size, and there was concomitant improvement in the electrocardiogram.

There is a risk of myocardial hypertrophy in children on prolonged treatment with ACTH (654), an effect that could reflect increased androgen secretion and thus be more likely to occur than with glucocorticoids.

Hypertension, with or without simultaneous hypertrophic myopathy, is a common feature of adrenal stimulation that seems to be common with depot tetracosactide but not simple tetracosactide (655). During treatment for infantile spasms, hypertension occurred more often in those treated with high doses (SEDA-19, 374; 656), and changes in cardiac function, such as left ventricular shortening fraction, can occur early and sometimes before systolic hypertension (SEDA-19, 374; 657).

Coumarin anticoagulants

Vasodilatory effects on the coronary arteries, peripheral veins, and capillaries, with purple toes as one of the most obvious consequences (658,659), have been reported. Sensations of cold may be due to increased loss of body heat caused by peripheral vasodilatation (660).

Cholesterol embolization, which promptly improves after the drug is withdrawn (661), may explain the purple-toe phenomenon.

COX-2 inhibitors

See also Non-steroidal anti-inflammatory drugs

Prothrombotic effects and cardiovascular disease

Selective inhibitors of the cyclo-oxygenase type 2 isoenzyme (COX-2) were developed with the expectation that their use would be accompanied by a reduction in the severe gastrointestinal and renal adverse effects associated with conventional non-steroidal anti-inflammatory drugs (NSAIDs), adverse effects that were thought to be largely a result of inhibition of the COX-1 isoenzyme (662). However, evidence has since accumulated that COX-2 selective inhibitors (coxibs) not only cause gastrointestinal and renal toxicity but can also contribute to an increased risk of adverse cardiovascular events.

The possible explanation of this phenomenon came from pharmacological studies that identified COX-2 inhibition as a plausible, albeit not the only, pharmacological mechanism for induction of thrombotic events. In fact, suppression of COX-2 dependent formation of PGI2 by coxibs, without significant concomitant inhibition of thromboxane A2 biosynthesis, can predispose patients to acute thromboembolic events. The pharmacological evidence that supports the mechanism for prothrombotic effects of coxibs has been reviewed (663,664,665).There is evidence from experimental and clinical studies that COX-2 may be atherogenic and thrombogenic and that selective COX-2 inhibitors may actually reduce, rather than cause, atherothrombotic vascular events (666). Some evidence supporting this hypothesis comes from laboratory and clinical studies. A randomized trial suggested that simvastatin reduces inflammation and suppresses the expression of cyclo-oxygenase-2 and prostaglandin E synthase in plaque macrophages, and this effect may in turn contribute to plaque stabilization by inhibition of metalloproteinase-induced plaque rupture (667). Inhibition of COX-2 by celecoxib compared with placebo improved endothelium-dependent vasodilatation and reduced oxidative stress in men with severe coronary artery disease (668). Figure 1 shows the physiological effects of COX-2 inhibition and the potential clinical implications (669).

Information about this topic comes from the two large studies, CLASS and VIGOR, from a pooled cardiovascular safety analysis using individual patient data derived from all rofecoxib phase IIb to V trials conducted by the manufacturer that lasted at least 2 weeks, and from the adverse events reporting system in the USA (670). As a result of these and other studies the COX-2 inhibitors, rofecoxib and valdecoxib, have been voluntarily withdrawn from the market by their manufacturers, and other COX-2 inhibitors are under scrutiny.

Rofecoxib

The first clinically-based demonstration of an increase in cardiovascular risk with the use of coxibs stemmed from the findings of the VIGOR (Vioxx Gastrointestinal Outcome Research) study, in which 8076 patients with rheumatoid arthritis were randomized to rofecoxib (50 mg/day, twice the usual dose) or naproxen (500 mg bd) for a median follow up of 9 months (671). The aim of the study was to discover if upper gastrointestinal toxicity was less with rofecoxib than naproxen. The rates of confirmed ulcer complications were indeed lower with rofecoxib (0.6 versus 1.0 per 100 patient-years; RR = 0.4; 95% CI = 0.2, 0.8; NNT_H = 125), but an important unexpected finding was a significant increase in the incidence of myocardial infarctions in patients taking rofecoxib compared with naproxen (0.4% versus 0.1%; RR = 5.0; 95% CI = 1.7, 14). These results sparked a debate about whether they represented a harmful effect of rofecoxib or a protective effect of naproxen or were simply due to chance. Concern about the potential for adverse cardiovascular effects with coxibs in general called for a separate analysis of adjudicated cardiovascular events from the study (665).

The vascular events referred for adjudication included coronary events (myocardial infarction, unstable angina, cardiac thrombus, resuscitated cardiac arrest, and sudden or unexplained death), cerebrovascular events (stroke and transient ischemic attacks), and venous thrombosis and pulmonary embolism. Most of the thrombotic cardiac events were myocardial infarction: rofecoxib 20 (0.5%) versus naproxen 4 (0.1%). Of the serious vascular adverse events that met the criteria for adjudication, more than twice were reported with rofecoxib (65 versus 33), but owing to incomplete documentation the number of adjudicated events was lower (45 versus 19 for rofecoxib and naproxen respectively). Low-dose aspirin for cardiovascular prophylaxis was not allowed in the study, but stratification according to whether aspirin was indicated or not indicated (according the criteria of the Antiplatelet Trialists Collaborative Group, ATCG) showed a significantly lower rate of adjudicated events in those who took naproxen. There was no difference in overall mortality (0.5% versus 0.4%) nor in the incidence of cardiovascular deaths (both 1%). There were lower risks of myocardial infarction and stroke in those taking naproxen compared with those taking rofecoxib in whom aspirin would have been indicated (RR = 0.26: CI = 0.07, 0.91). The risk of serious thrombotic events was also significantly lower with naproxen than with rofecoxib (RR = 0.42; CI = 0.25, 0.72).

In summary the adjudication analysis confirmed the first impression of an increased risk of severe cardiovascular events in patients taking rofecoxib. Consistent with this result were the time to event curves, which showed a clear divergence in survival early after starting treatment (within 2 months), which persisted for the remainder of the study.

The results of VIGOR might be explained by a significant prothrombotic effect of rofecoxib and/or an antithrombotic effect of naproxen, which may have a significant antiplatelet effect. To clarify this, the annualized myocardial infarction rates in both VIGOR and CLASS (see below, under Celecoxib) were compared with those found in placebo-treated patients with similar cardiac risk factors enrolled in three meta-analyses of four aspirin primary prevention trials (670,672). The analysis showed 0.24 and 0.30% increases over placebo in cardiovascular events for rofecoxib and celecoxib respectively, suggesting a prothrombotic potential of both COX-2 inhibitors. However, these results have been heavily criticized (673–679), because of many potential pitfalls in the comparisons of patient populations in different trials. The results from a recently reported pooled analysis of individual patient data combined across rofecoxib phase IIb to phase V trials seem to be more reliable (680). Cardiovascular events were assessed across 23 studies. Comparisons were made between patients taking rofecoxib and those taking placebo, naproxen, or other non-selective NSAIDs (diclofenac, ibuprofen, and nabumetone). The major outcome measure was the combined end-point used by the Anti-platelet Trialist Collaboration, which includes cardiovascular death, hemorrhagic death, death of unknown cause, non-fatal myocardial infarction, and non-fatal stroke. More than 28 000 patients, representing more than 14 000 patient-years at risk, were analysed. The relative risk for an end-point was 0.84 (95% CI = 0.51, 1.38) when comparing rofecoxib with placebo; 0.79 (95% CI = 0.40, 1.55) when comparing rofecoxib with non-selective NSAIDs; and 1.69 (95% CI = 1.07, 2.69) when comparing rofecoxib with naproxen. These data provide no evidence for an excess of cardiovascular events for rofecoxib relative to either placebo or the non-selective NSAIDs that were studied. Instead, the difference between rofecoxib and naproxen showed that naproxen was associated with a reduced risk of cardiovascular events.

The cardiovascular results in VIGOR may have occurred simply by chance, given the low number of events, or because naproxen may have a cardioprotective effect similar to that of aspirin, or because rofecoxib 50 mg/day could have prothrombotic effects, especially in the absence of concomitant COX-1 inhibition in patients at increased risk of cardiovascular thromboembolic events. Because there was no untreated group in VIGOR, we do not know whether this finding suggests a protective effect of naproxen or a harmful effect of rofecoxib. All three explanations are plausible, and they are not mutually exclusive.

It has been suggested that the increase in thrombotic cardiovascular events in rofecoxib-treated patients probably represents the antiplatelet effect of naproxen (670,681,682). Naproxen has a long pharmacodynamic half-life and inhibits platelet aggregation by 88% for up to 8 hours (683).

In VIGOR (671,684), the incidence of confirmed thrombotic cardiovascular events was 0.6% higher with rofecoxib than with naproxen (RR = 2.4; 95% CI = 3.9, 4.0).

The debate about whether the results of VIGOR represented a harmful effect of rofecoxib rather than a cardioprotective effect of naproxen stimulated further research, which yielded contrasting results. In one meta-analysis (680) of randomized trials of rofecoxib and in three case-control studies on naproxen and myocardial infarction the interpretation of a cardioprotective effect of naproxen was reiterated (685,686,687), while other studies (688,689,690) suggested an increased risk of acute cardiac events, at least in patients taking rofecoxib in doses above 25 mg/day. Although contrasting data have been published (691), the hypothesis that naproxen has a cardioprotective effect has gained wider support (692). Only one of the four most recently published studies negated a potential cardioprotective effect of naproxen.

However, an analysis of FDA data suggested an increase in serious cardiac events with celecoxib (691). The incidence of serious cardiac adverse events (myocardial infarction, combined anginal events, and atrial dysrhythmias) was 0.6% higher with celecoxib than with other NSAIDs (RR = 1.55; CI = 1.04, 2.30). The reasons for these inconsistencies are not clear.

In an 11-year observational study in new users of non-selective, non-aspirin NSAIDs (n = 181 441) and an equal number of non-users there was no evidence of a protective effect of naproxen (693). During 532 634 person-years of follow-up there were 6382 cases of serious coronary heart disease (11.9 per 1000 person-years). Multivariate-adjusted rate ratios for current and former use of non-aspirin NSAIDs were 1.05 (95% CI = 0.97, 1.14) and 1.02 (0.97, 1.08) respectively. Rate ratios for ibuprofen, naproxen, and other NSAIDs were 1.15 (1.02, 1.28), 0.95 (0.82, 1.09), and 1.03 (0.92, 1.16) respectively. There was no protection in long-term users with uninterrupted use; the rate ratio among current users with more than 60 days of continuous use was 1.05 (0.91, 1.21). When naproxen was directly compared with ibuprofen, the rate ratio in current users was 0.83 (0.69, 0.98). This study therefore seems to have shown no cardioprotective effect of naproxen. However, the study had a number of important limitations, including: lack of information about some important confounders (smoking, obesity), possible exposure misclassification, and lack of information about over-the-counter use of aspirin.

Opposite evidence has emerged from three case-control studies from the USA, Canada, and the UK, which showed that the rates of myocardial infarction in patients taking naproxen were lower than in patients not taking any NSAIDs (685,686) and those taking other NSAIDs (687).

In the first study, 4425 patients hospitalized with acute myocardial infarction who used NSAIDs were compared with 17 700 controls in a large health-care database in the USA (685). Multivariate models were constructed to control for potential confounders. A quarter of the cases and controls had also filled a prescription for an NSAID in the 6 months before the study. Overall, the NSAID users had the same risk of acute myocardial infarction as non-users, but naproxen users had a significantly lower risk of acute myocardial infarction compared with those who were not taking NSAIDs (adjusted OR = 0.84; 95% CI = 0.72, 0.98). The cardioprotective effect of naproxen was very modest compared with aspirin (a 44% reduction in the risk of acute myocardial infarction in the Physician Health Study (694)).

The second study was a case-control study sponsored by Merck & Co (the manufacturers of celecoxib), in which the risk of acute thromboembolic cardiovascular events among 16 937 patients aged 40–75 years with rheumatoid arthritis using naproxen was examined using the British General Practice Research Database (686). Each patient with a first thromboembolic event (n = 809: 435 myocardial infarctions, 347 strokes, 27 sudden deaths) was matched with four controls. The results suggested that patients with rheumatoid arthritis who currently use naproxen have a significantly lower risk of thromboembolic events relative to those who have not used naproxen in the past year (RR = 0.61; 0.39, 0.94). However, the risk was not lower with previous use of naproxen, suggesting that any effect of naproxen is likely to be short-lived. Moreover, the significantly lower risk of myocardial infarction with current naproxen was not found when myocardial infractions were analysed separately (RR = 0.57; 0.3, 1.06). There was no protective effect for thromboembolic events with current use of non-naproxen NSAIDs.

The third study was also sponsored by Merck & Co and was designed to examine the association between the use of naproxen and other non-selective NSAIDs and hospitalization for acute myocardial infarction (687). In a database of 14 163 patients aged 65 years or older who were hospitalized for acute myocardial infarction and an equal number of age-matched controls, concurrent exposure to naproxen had a protective effect against myocardial infarction compared with the other non-selective non-aspirin NSAIDs (RR = 0.79; CI = 0.63, 0.99). This effect was present only with concurrent naproxen exposure and was stronger in long-term users. However, this study also had several limitations: some important risk factors, such as smoking and obesity, could not be assessed, patients who died of myocardial infarction before reaching the hospital were not included, and there was uncertainty about concurrent use of over-the-counter drugs, especially aspirin.

Unexpectedly, in September 2004, Merck Sharp & Dohme announced the voluntary worldwide withdrawal of rofecoxib. This followed the release of data from a long-term, randomized, placebo-controlled trial, the Adenomatous Polyp Prevention on Vioxx the APPROVe trial, which showed that the use of rofecoxib was associated with an increased risk of thrombotic events (695,696). The study was designed to evaluate the hypothesis that treatment with rofecoxib for 3 years would reduce the risk of recurrent adenomatous polyps among patients with a history of colorectal adenomas. A total of 2586 patients underwent randomization (1287 were assigned to rofecoxib 25 mg/day and 1299 to placebo). All investigator-reported serious adverse events that represented potential thrombotic cardiovascular events were adjudicated in a blinded fashion by an independent committee, and all the safety data were monitored by an external safety monitoring committee. The mean duration of treatment was 2.4 years with rofecoxib and 2.6 years with placebo. In the rofecoxib group 46 patients had confirmed thrombotic events during 3059 patient-years of follow-up (1.50 events per 100 patient years) compared with 26 patients who took placebo during 3327 patient-years of follow up (0.78 events per 100 patient years). The relative risk was 1.92 (95% CI = 1.19, 3.11). However, the increased relative risk became apparent only after 18 months of treatment; during the first 18 months the events rates were similar in the two groups. The difference between the two groups was mainly due to an increased number of myocardial infarctions and strokes in those taking rofecoxib.

Those who took rofecoxib also had higher risks of non-adjudicated cardiovascular events (hypertension and edema-related events) compared with those who took placebo. The Kaplan–Meier curves for the cumulative incidence of congestive heart failure, pulmonary edema, and cardiac failure showed an early separation of the two groups, at about 5 months.

Further data consistent with the results of APPROVe were found in a standard and cumulative random effects meta-analysis (697). This study aimed at establishing whether robust evidence on the adverse cardiovascular effects of rofecoxib had been available before September 2004, the date on which rofecoxib was withdrawn. The meta-analysis identified 18 randomized controlled trials, including 25 273 patients with chronic musculoskeletal disorders, in which rofecoxib was compared with other NSAIDs or placebo, and 11 observational studies on naproxen and cardiovascular risk. Myocardial infarction was the primary end-point. By the end of 2000 (52 myocardial infarctions, 20 742 patients) the relative risk in randomized controlled trials was 2.30

(95% CI = 1.22, 4.33), and 1 year later (64 events, 21 432 patients) it was 2.24 (1.24, 4.02). There was no evidence that the relative risk differed depending on the control group (placebo, a non-naproxen NSAID, or naproxen) or the duration of the trial. In observational studies, the cardioprotective effect of naproxen was small (combined estimate 0.86; CI = 0.75, 0.99) and could not have explained the findings of VIGOR. These findings suggest that evidence of the adverse cardiovascular effects was available before September 2004 and that rofecoxib should have been withdrawn several years earlier. The reasons why the manufacturers and regulatory authorities did not continuously monitor and summarize the accumulating evidence need to be clarified (695).

Besides the VIGOR and APPROVe studies, a number of other clinical and epidemiological studies have signalled an increased risk of harmful cardiovascular effects with rofecoxib. Among them three epidemiological studies merit attention.

The first study (688) was a retrospective cohort study of individuals on the expanded Tennessee Medicaid programme, in which the occurrence of serious coronary heart disease was assessed both in non-users (n = 202 916) and users of rofecoxib (n = 24 132) and other NSAIDs (n = 15 728). The patients were aged 50–84 years, lived in the community, and had no life-threatening non-cardiovascular illnesses. The incidence of acute myocardial infarction or cardiac death in new users of rofecoxib at doses above 25 mg/day was almost twice that in non-users (24 versus 13 events per 1000 patient years; RR = 1.93; 95% CI = 1.00, 3.43). In contrast, there was no evidence of an increased risk of coronary heart disease among users of rofecoxib at doses of 25 mg/day or less or among users of other NSAIDs.

The second study was a matched case-control study of the relation between coxibs, non-selective NSAIDs, and hospitalization for acute myocardial infarction in a large population of patients aged 65 years or over (689). The study database contained information on more than 50 000 patients and included 10 895 cases of acute myocardial infarction. Current use of rofecoxib was associated with an increased relative risk of acute myocardial infarction compared with celecoxib (OR = 1.24; CI = 1.05, 1.46) and with no NSAIDs (OR = 1.14; CI = 1.00, 1.31). A dosage of rofecoxib of over 25 mg/day was associated with a higher risk than 25 mg/day or less. The risk was increased in the first 90 days but not thereafter.

The timing of cardiovascular risks associated with the use of COX-2 inhibitors is unclear. The APPROVe trial reported a two-fold increase in cardiovascular toxicity after 18 months of use, while in the VIGOR study a similar increase in risk was reported after only 9 months of treatment. However, the risk curves in the VIGOR study started to diverge after the first month of therapy (698). In the APC trial celecoxib was associated with a dose-dependent increase in risk after 3 years of use, but the combination of intravenous parecoxib plus oral valdecoxib resulted in an increased risk after only 10 days of exposure. A more recent study has provided new insights into the timing of the cardiovascular risks associated with the use of rofecoxib and celecoxib (699). The study, which was a time-matched, nested, case-control study, showed that among elderly users of rofecoxib and celecoxib the cardiovascular risks associated with the use of rofecoxib were more acute than has previously been recognized. The risk was highest in the first 6–13 days (median 9 days) after starting treatment and did not increase with further treatment. Indeed, the risk of myocardial infarction appeared to reduce over time despite continued exposure, presumably owing to attrition of susceptible individuals, i.e. the healthy survivor effect that is seen in an adverse effect of intermediate time course (700).

Further evidence that rofecoxib increases the risk of serious coronary heart disease comes from a third case-control study commissioned by the US Food and Drug Administration (690). Cases of serious coronary heart disease (acute myocardial infarction and sudden cardiac death) in a cohort of NSAID-treated patients were risk-matched with four controls. Current exposure to coxibs (rofecoxib and celecoxib) and standard NSAIDs was compared with remote exposure to any NSAID, and rofecoxib was compared with celecoxib. During 2 302 029 person-years of follow up, there were 8143 cases of serious coronary heart disease, of which 2210 were fatal. Multivariate adjusted odds ratios for rofecoxib versus celecoxib were: 1.59 (all doses; CI = 1.10, 2.32); 1.47 (rofecoxib 25 mg/day or less; CI = 0.99, 2.17); and 3.58 (rofecoxib over 25 mg/day; CI = 1.00, 4.30). For naproxen versus remote NSAIDs the adjusted OR was 1.14 (1.00, 1.30). The interpretation of these results was that rofecoxib increases the risk of acute myocardial infarction and sudden death compared with celecoxib and that naproxen does not protect against serious coronary heart disease. This study also provided data relevant to some other controversies about the cardiovascular safety of coxibs. In particular there was a substantially higher risk with high-dose rofecoxib, in accordance with the results of other studies (688,689). Moreover the mean duration of use before the occurrence of an event was identical with high doses and standard doses of rofecoxib (about 110 days), consistent with the idea that the risk of cardiovascular toxicity begins early in treatment. That is in accord with the analysis by the FDA of data from VIGOR, which showed that the survival curve for acute myocardial infarction with high-dose rofecoxib began to diverge from the naproxen curve after 1 month. There was no evidence that the relative risk differed depending on the control group (placebo, a non-naproxen NSAID, or naproxen) or the duration of treatment. This suggests that patients are at risk of myocardial infarction even if rofecoxib is taken for a few months only. Therefore, the reassuring statement by Merck that there is no excess risk in the first 18 months is not supported by this meta-analysis.

In contrast to these findings, two earlier meta-analyses from Merck laboratories showed no evidence for an excess of cardiovascular thrombotic events for rofecoxib relative to either placebo or non-naproxen NSAIDs (ibuprofen, nabumetone, diclofenac) or an increased risk in trials in which rofecoxib and naproxen were compared (701,702).

In conclusion these data provide further convincing evidence that naproxen does not have a cardioprotective effect, in accordance with the results of some studies (693,703) although not others (704,705,706). Indeed, the current data show even a possibility of a small increased risk of coronary heart disease.

The demonstration of a lack of protective effect of naproxen is important, because it is often used as a comparator in clinical trials of new coxibs. Thus, a result that shows that a new drug has an increased risk of cardiovascular disease relative to naproxen should alert prescribers to potential cardiotoxic effects.

There are two further crucial questions:

- Is the apparent treatment-associated increase in severe cardiovascular events unique to rofecoxib or does it apply to other COX-2 selective drugs, suggesting a COX-2 class effect, and even to NSAIDs in general?
- Does it apply to all patients or only to a subset, such as those with known susceptibility factors for vascular disease?

Two later reports help in answering these questions. The first is a meta-analysis of randomized controlled trials of selective COX-2 inhibitors versus placebo or traditional NSAIDs or both (707). Eligible studies were randomized trials that lasted at least 4 weeks, with information on serious vascular events (defined as myocardial infarction, stroke, and vascular death). Data were available from 138 randomized trials involving a total of 143 373 participants. NSAIDs were subdivided into naproxen and others. In placebo comparisons, allocation to a COX-2 selective inhibitor was associated with a 42% relative increase in the incidence of serious vascular events: 1.2% per year versus 0.9% per year (RR = 1.42; 95% CI = 1.13, 1.78) with no significant heterogeneity among the different coxibs. This increase was chiefly attributable to an increased risk of myocardial infarction (0.6% per year versus 0.3% per year; RR = 1.86; 95% CI = 1.93, 2.59), with little apparent difference in other vascular outcomes. Among trials of at least 1 year duration (mean 2.7 years) the rate ratio for vascular events was 1.45 (CI = 1.12, 1.89). There were too few vascular events to allow an assessment of the dose-response relation in placebo-controlled trials of all coxibs, with the exception of celecoxib, for which there was a significant trend towards an increased incidence of serious vascular events with higher daily doses. The size of the risk of vascular events in placebo-controlled trials that allowed concomitant use of aspirin was similar among aspirin users and non-users.

Comparisons of selective COX-2 inhibitors versus traditional NSAIDs showed a similar incidence of serious vascular events (1.0% per year versus 0.9% per year; RR = 1.16; CI = 0.97, 138). However, there was marked heterogeneity across the rate ratios for vascular events in trials that compared a coxib with naproxen and trials that compared a selective COX-2 inhibitor with a non-naproxen NSAID. Overall, compared with naproxen, allocation to a selective COX-2 inhibitor was associated with a highly significant increase in the incidence of a serious vascular event (RR = 1.57; CI = 1.21, 2.03) and a two-fold increased risk of a myocardial infarction (RR = 2.04; CI = 1.41, 2.96), but no difference in the incidences of stroke or vascular death. In contrast the comparison of COX-2 selective inhibitors with non-naproxen NSAIDs showed non-significant differences in the incidences of vascular events, myocardial infarctions, and vascular deaths, but selective COX-2 inhibitors were associated with a significantly lower incidence of stroke than any non-naproxen NSAIDs (RR = 0.62; CI = 0.41, 0.95).

When the rate ratio of serious vascular events with traditional NSAIDs was compared with that of placebo, high-dose ibuprofen and high-dose diclofenac were both associated with an increased risk of cardiovascular events (ibuprofen RR = 1.51; CI = 0.96, 2.37; diclofenac RR = 1.63; CI = 1.12, 237), while naproxen was not.

Thus, it appears that allocation to a selective COX-2 inhibitor is associated with about three extra vascular events per 1000 patients per year. Most of this excess is attributable to myocardial infarction.

The other study that has contributed to better evaluation of the data on cardiovascular risk due to inhibition of cyclo-oxygenase is a systematic review of the available controlled pharmacoepidemiological studies (708). There were 23 eligible studies that reported on cardiovascular events (predominantly myocardial infarction) with COX-2 inhibitors, NSAIDs, or both, with non-use/remote use of the drug as the reference exposure for calculation of the relative risk. The studies were 17 case-control studies involving 86 193 cases with cardiovascular events and about 52 800 controls and six cohort studies involving 75 520 users of selective COX-2 inhibitors, 375 619 users of non-selective NSAIDs, and 594 720 unexposed participants. The results confirmed that there is an increased risk of cardiovascular events with rofecoxib. The summary relative risk with doses in excess of 25 mg/day was 2.19 (CI = 1.64, 2.91) compared with 1.33 (CI = 1.00, 1.79) with 25 mg/day or less. The risk was increased during the first month of treatment. In contrast to the evidence from the meta-analysis of randomized trials, this study showed no increase in risk with celecoxib 200 mg/day. However, one must remember that the randomized clinical trials only showed an increased risk with doses of celecoxib of 400 mg/day and above. The study also provided more data on the cardiovascular toxicity of traditional NSAIDs. Naproxen was not associated with a reduction in risk, as was suggested in previous comparisons with rofecoxib.

Of more concern is evidence, from both randomized and non-randomized trials, that diclofenac increases the risk of cardiovascular events (RR = 1.63; CI = 1.12, 2.37). Less convincing is the evidence of potential cardiovascular risk with other NSAIDs, such as ibuprofen, meloxicam, and indometacin.

Celecoxib

In December 2004 the results of the Adenoma Prevention with Celecoxib trial (the APC trial) were presented. This was a randomized, double-blind, multicenter study of the effects of two doses of celecoxib (200 mg or 400 mg bd) or placebo in the prevention of colorectal adenomas (709). All potentially serious cardiovascular events among 2035 patients with a history of colorectal neoplasia were identified, with a follow up of 2.8–3.1 years. Celecoxib was associated with a dose-related increase in the composite end-point of death from cardiovascular causes, myocardial infarction, stroke, or heart failure. The composite end-point was reached in seven of 679 patients in the placebo group (1%) compared with 16 of 685 patients who took celecoxib 200 mg bd (2.3%; hazard ratio = 2.3; CI = 0.9, 5.5) and 23 of 671 patients who took celecoxib 400 mg bd (3.4%; hazard ratio = 3.4; CI = 1.4, 7.8). The annualized incidence of death due to cardiovascular causes per 1000 patient years was 3.4 events in the placebo group; there were 7.5 events in patients who took celecoxib 200 mg bd and 11.4 in those who took 400 mg bd. The hazard ratio associated with celecoxib was not significantly affected by any baseline characteristics, including use of aspirin. Moreover, the results were also consistent among the individual components of the composite end-point. Based on these statistically significant findings, the sponsor of the study, the USA National Cancer Institute suspended the study.

However, to underscore the uncertainty of the available data on cardiovascular risk, we should consider the results of another large long-term trial with celecoxib 400 mg/day, designed to investigate the role of NSAIDs in preventing Alzheimer's disease, the Alzheimer Disease Anti-inflammatory Prevention Trial (ADAPT), in which celecoxib was compared with naproxen (440 mg/day). The unpublished results of the study showed that celecoxib did not increase cardiovascular risk, while naproxen was associated with an increased risk (710). The results from the APC and ADAPT studies are disturbing, because they are inconsistent with the results of most previous studies. The following data summarize the currently available information on the cardiovascular safety of celecoxib.

In two retrospective re-analyses of randomized controlled trial data for celecoxib (711,712) the cardiovascular events in patients enrolled in the CLASS study (713) were examined, as were those reported across the entire controlled arthritis clinical trial database for celecoxib. The results of the reanalysis of CLASS showed no evidence of an increase in investigator-reported serious cardiovascular adverse events in patients taking celecoxib. In the analysis of the celecoxib comparative trials database the incidence of cardiovascular events was not significantly different between celecoxib and placebo or between celecoxib and naproxen, regardless of aspirin use. Thus, these comparative analyses failed to show an increased risk of thrombotic events associated with celecoxib compared with conventional NSAIDs, naproxen specifically, or placebo.

Most information on the potential cardiovascular toxicity of celecoxib has come from a number of observational studies (688,689,690,712,714,715,716). Most of them were population-based case-control studies in which the relative risk of acute myocardial infarction was assessed in patients who took celecoxib, rofecoxib, conventional NSAIDs, or no NSAIDs at all. None of these studies identified a significantly increased risk of severe cardiovascular events with celecoxib. However, the results of these observational studies must be interpreted with caution. In fact, although findings from observational studies are more generalizable, as they are larger and include less carefully selected patients than randomized controlled trials, they suffer from potential bias and confounding, which could have contributed to the failure to confirm the hypothesis of a consistent difference in severe cardiovascular risk between all the coxibs and other NSAIDs. Whether this safety concern represents a class effect requires additional information from other randomized controlled trials, not yet available (716).

Valdecoxib and parecoxib

Further support of the hypothesis that the increase in cardiovascular risk is a COX-2 class effect has come from an analysis of clinical trials of valdecoxib and its pro-drug parecoxib.

The cardiovascular safety of valdecoxib was initially assessed in a study that pooled results from 10 clinical trials that included nearly 8000 patients with osteoarthritis and rheumatoid arthritis and compared the incidence of cardiovascular events in patients taking valdecoxib (10–80 mg/day) with those of controls taking diclofenac, ibuprofen, naproxen, or placebo (717). The incidences of cardiovascular thrombotic events (cardiac, cerebrovascular, and peripheral vascular or arterial thrombotic) were similar with valdecoxib, the conventional NSAIDs, and placebo. Short-term and intermediate-term treatment with valdecoxib in therapeutic doses (10 or 20 mg/day) and supratherapeutic doses (40–80 mg/day) was not associated with an increased incidence of thrombotic events. In contrast, recent studies have raise serious doubts about the safety of valdecoxib in patients who are at high risk of thrombotic complications, such those undergoing coronary artery bypass surgery.

The cardiovascular toxicity of valdecoxib and its pro-drug parecoxib has been studied in two randomized placebo-controlled trial and one meta-analysis. The first study involved 462 patients and evaluated the safety and efficacy of intravenous parecoxib (parecoxib 40 mg intravenously every 12 hours for 3 days postoperatively), followed by oral valdecoxib (40 mg every 12 hours for a total of 14 days) (718). In the second study, which involved more than 1600 patients, a similar schedule of administration was used, but the dosage and duration of treatment were reduced (20 mg bd for 10 days) (719). In both studies there were clusters of cardiovascular adverse events, including myocardial infarction, stroke, deep vein thrombosis, and pulmonary embolism, which were more frequent in those who were given parecoxib than in the controls, although the difference was not statistically significant. However, when the coronary and cerebrovascular events were combined in a meta-analysis parecoxib/valdecoxib was associated with a three-fold risk of cardiovascular events compared with placebo: 31/1399 events versus 5/699 (RR = 3.08; CI = 1.20, 7.87) (720).

Following this important information a new warning contraindicating the use of parecoxib/valdecoxib in patients undergoing coronary artery bypass grafting was added to the label by many drug control agencies. Meanwhile the FDA received reports of 87 cases of severe skin reactions associated with valdecoxib, including Stevens–Johnson syndrome and toxic epidermal necrolysis, with four deaths; 20 of the 87 cases involved patients with a known allergy to sulfa-containing drugs, of which valdecoxib is one; thus, patients allergic to sulfa drugs may be at greater risk of severe skin reactions with valdecoxib. These data have raised concerns not only about the cardiovascular safety of valdecoxib in the general population but also about its overall benefit to harm profile. The drug manufacturers and regulatory agencies in the USA and Europe agreed to suspend valdecoxib in April 2005 (721).

Despite claims from the results of a meta-analysis conducted by its manufacturer that valdecoxib appears to be associated with significantly fewer adverse gastrointestinal events that non-selective NSAIDs (722), its marketing authorization has been suspended in many countries because of an increased risk of serious cardiovascular outcomes.

Lumiracoxib

Lumiracoxib is a COX-2 selective inhibitor, a phenylacetic acid derivative with a short half-life and a higher selectivity than any other coxib. Clinical data on its efficacy and safety are limited.

The cardiovascular safety of lumiracoxib has been assessed in the Therapeutic Arthritis and Gastrointestinal Event Trial (TARGET) in 18 325 patients aged 50 years or older with osteoarthritis, who were randomized to lumiracoxib (400 mg/day; n = 9156), ibuprofen (800 mg tds; n = 4415), or naproxen (500 mg bd; n = 4754) for 52 weeks in two substudies of identical design (lumiracoxib versus ibuprofen and lumiracoxib versus naproxen) (723). Randomization was stratified for low-dose aspirin and age. The primary cardiovascular end points were the ATC end-points of non-fatal and silent myocardial infarction, stroke, or cardiovascular death. At 1 year of follow up, the incidence of the primary end-point was low and did not differ between treatment groups: lumiracoxib 59 events (0.65%), NSAIDs 50 events (0.55%; hazard ratio = 1.14; CI = 0.28, 1.66). The incidence of myocardial infarction in the overall population in the individual substudies was 0.38% with lumiracoxib (18 events) versus 0.21% with naproxen (10 events), and 0.11% with lumiracoxib (5 events) versus 0.16% (7 events) with ibuprofen. In both substudies the rates of myocardial infarction did not differ significantly between lumiracoxib and NSAIDs, irrespective of aspirin use.

These TARGET trial data suggest that lumiracoxib does not have the potential to precipitate adverse cardiovascular events more often than NSAIDs. However, these results must be interpreted with caution (724) for several reasons:

- patients with known and significant pre-existing coronary artery disease were excluded, which explains the overall low frequency of cardiovascular events and introduced bias into the generalizability of the TARGET results;
- the statistical power was inadequate to detect significant differences in the rates of myocardial infarction, again raising concern about an excess, albeit not a statistically significant one, of myocardial infarctions with lumiracoxib compared with naproxen: 18 events (0.38%) versus 10 events (0.21%); hazard ratio 1.77 (0.82, 3.84).
- lumiracoxib significantly reduced the frequency of upper gastrointestinal ulcer complications (NNT to prevent one ulcer complication 139) but only in patients not taking low-dose aspirin (725), confirming the results of other coxib trials (CLASS).
- the data raised a concern about possible hepatotoxicity of lumiracoxib; the proportion of patients with transaminase activities more than three times the upper limit of the reference range differed significantly between lumiracoxib (2.57%, n = 230) and NSAIDs (0.63%, n = 56; hazard ratio 3.97; CI = 2.96, 5.32).

Etoricoxib

Data on the efficacy and safety of etoricoxib, a highly selective COX-2 inhibitor, are limited. A combined analysis of all randomized, double-blind trials of long-term treatment showed a significantly lower incidence of peptic ulcer bleeding with etoricoxib (n = 9226) than with conventional NSAIDs (n = 2215), but there was no information about potential cardiovascular toxicity (726).

There has been a systematic review and meta-analysis of double-blind, randomized, placebo-controlled trials of etoricoxib of at least 6 weeks duration (five studies with a total of 2919 patients); the main outcome measure was cardiovascular thromboembolic events (727). There were seven cardiovascular thromboembolic events in 1441 patients (0.5%) taking etoricoxib, and one event in 906 patients (0.1%) taking placebo. The pooled fixed-effects estimate of the absolute risk difference was 0.5% (CI = 0.1, 1.0). The odds ratio for the risk of cardiovascular events was 1.49 (0.42, 5.31). These limited data provide weak evidence of an increased cardiovascular risk with etoricoxib.

A large sufficiently powered comparison of etoricoxib and diclofenac is currently under way and should clarify this issue.

Meloxicam

The NSAIDs in Unstable Angina Treatment-2 (NUT-2) pilot study, which compared meloxicam (15 mg/day until 30 days after discharge) with aspirin plus heparin in 120 patients who had a non-ST segment elevation acute coronary syndrome, showed that patients who took meloxicam had a significant reduction in the primary composite outcome of recurrent angina, myocardial infarction, or death during their stay in the coronary care unit (728): the relative risk reduction was 61% (95% CI = 23, 80). Larger trials are required to confirm the findings of this pilot study.

Conclusions

At present at least three important questions about the adverse cardiovascular effects of COX-2 selective inhibitors remain unanswered.

First, the mechanism whereby COX-2 inhibitors facilitate ischemic cardiovascular events is unclear. The major unanswered question is whether unopposed COX-2 inhibition or other drug-specific mechanisms cause the increased cardiovascular risk. It is worth noting that in addition to ischemic cardiovascular disease, COX-2 inhibitors are associated with other adverse cardiovascular effects, such as heart failure (729,730), hypertension (731,732), and edema (733).

Secondly, it is still unknown whether the cardiovascular risk is or is not a class effect. Since at least three separate drugs in the class (rofecoxib, valdecoxib, and celecoxib) have now been associated with increased cardiovascular morbidity, the burden of proof has been shifted to those who deny a class effect. We must remember that absence of evidence is not evidence of absence (663).

Thirdly, there is uncertainty about what physicians should do if they decide to prescribe an NSAID. In light of the current uncertainty about whether cardiotoxicity is a class effect, coxibs, in particular at high dosages and for long-term use, should not be prescribed, particularly in populations at high risk, such as elderly patients and those with established cardiovascular disease. This recommendation is fully justified in light of the possible consequences of the current heavy unjustified promotion of these compounds to both patients and prescribers. A recent epidemiological study in Ontario, Canada showed that the use of NSAIDs in patients aged 66 years or older increased by 41% after the introduction of coxibs (734). This rise was entirely attributable to the use of coxibs and was accompanied by a 10% increase in hospitalization for upper gastrointestinal bleeding.

The continued commercial availability of coxibs is therefore troubling, and probably unjustified, in the light of their marginal efficacy, heightened risk, and higher costs compared with traditional NSAIDs.

Hypertension

The shift in hemostatic balance toward a prothrombotic state might not be the only mechanism by which COX-2 inhibitors could increase the risk of cardiovascular adverse effects. In fact, non-selective NSAIDs can raise blood pressure and antagonize the hypotensive effect of antihypertensive medications to an extent that may increase hypertension-related morbidity (735,736). The problem is clinically relevant, as arthritis and hypertension are common co-morbid conditions in elderly people, requiring concurrent therapy.

Information on the effect of COX-2-selective inhibitors on arterial blood pressure is scanty. In VIGOR, more patients developed hypertension with rofecoxib than naproxen. For rofecoxib, the mean increase in blood pressure was 4.6/1.7 mmHg compared with a 1.0/0.1 mmHg increase with naproxen (670). Previous work has shown that a 2 mmHg reduction in diastolic blood pressure can result in about a 40% reduction in the rate of stroke and a 25% reduction in the rate of myocardial infarction (737). The effect of celecoxib on blood pressure was evaluated in a post hoc analysis using the safety database generated during the celecoxib clinical development program in more than 13 000 subjects (738). The incidence of hypertension after celecoxib was greater than that after placebo but similar to that after non-selective NSAIDs. Hypertension and exacerbation of pre-existing hypertension occurred, respectively, in 0.8 and 0.6% of patients. Furthermore, there was no evidence of interactions between celecoxib and other antihypertensive drugs.

The safety profiles of celecoxib (200 mg/day) and rofecoxib (25 mg/day) have recently been compared in a 6-week, randomized, parallel-group, double-blind trial in 810 patients with osteoarthritis aged 65 years or over taking antihypertensive drugs (739). The primary endpoints were edema and changes in systolic and diastolic blood pressures, measured at baseline and after 1, 2, and 6 weeks. Systolic blood pressure rose significantly in 17% of rofecoxib-treated patients (n = 399) compared with 11% of celecoxib-treated patients (n = 411), a statistically significant difference, at any time in the study. Diastolic blood pressure rose in 2.3% of rofecoxib-treated patients compared with 1.5% of celecoxib-treated patients, a non-significant difference. At week 6, the change from baseline in mean systolic blood pressure was +2.6 mmHg for rofecoxib compared with –0.5 mmHg for celecoxib, a highly significant difference. Nearly twice as many rofecoxib-treated than celecoxib-treated patients had edema. Despite some limitations, this study provides some evidence that COX-2-selective inhibitors may differ in their ability to alter arterial blood pressure.

Hypertensive patients taking rofecoxib are more likely to require an increase in their antihypertensive drug dosages (740) than those taking celecoxib: OR = 1.68 (95% CI = 1.09, 2.6).

Cardiac dysrhythmias

There have been reports of celecoxib-associated torsade de pointes in three patients who had never had any complaints before celecoxib administration; the dysrhythmia did not recur after the drug was withdrawn (741). However, all three patients had cardiac abnormalities that might have predisposed them to the development of torsade de pointes, and the follow-up period was too short to evaluate possible spontaneous recurrence. The hypothesis that celecoxib is dysrhythmogenic requires confirmation.

Cromoglicate sodium

- A woman developed peripheral eosinophilia and pericarditis with cardiac tamponade after using cromoglicate (SEDA-4, 120). Cellular and humoral sensitivity to cromoglicate were demonstrated and she recovered following pericardiocentesis.

Cyclobenzaprine

Cyclobenzaprine can occasionally cause marked arteriolar spasm due to increased adrenergic tone, precipitating Raynaud's syndrome (SEDA-6, 132).

Cyclophosphamide

Cardiac toxicity can be observed at high doses of cyclophosphamide (usually over 1.5 g/m^2/day), and acute myocardial necrosis or severe cardiac failure have been anecdotally reported after smaller dosages (SEDA-21, 386).

High-dose cyclophosphamide (120–200 mg/kg) can cause lethal cardiotoxicity, and severe congestive heart failure can develop 1–10 days after the first dose. Severe congestive heart failure is accompanied by electrocardiographic findings of diffuse voltage loss, cardiomegaly, pulmonary vascular congestion, and pleural and pericardial effusions. Pathological findings include hemorrhagic myocardial necrosis, thickening of the left ventricular wall, and fibrinous pericarditis.

Of 80 patients who received cyclophosphamide 50 mg/kg/day for 4 days in preparation for bone marrow grafting 17% had symptoms consistent with cyclophosphamide cardiotoxicity (742). Six died from congestive heart failure. Older patients were at greatest risk of developing cardiotoxicity.

In six patients who developed heart failure after high-dose conditioning therapy before stem cell transplantation, cyclophosphamide was suspected, despite the possible involvement of four drugs (743). The authors suggested monitoring high-risk patients.

Corrected QT dispersion was a predictor of acute heart failure after high-dose cyclophosphamide chemotherapy (5.6 g/m^2 over 4 days) in 19 patients (744).

Cyclopropane

Cardiac dysrhythmias, which can be ventricular, can complicate the use of cyclopropane (745), and the risk is increased in patients who have also been given catecholamines (746).

Cyproterone acetate

Cyproterone acetate in combination with ethinylestradiol has been associated with a greater risk of venous thromboembolism than oral contraceptives. However, in a rigorous case-control study the risk of venous thromboembolism with cyproterone acetate + ethinylestradiol was not significantly greater than the risk in women who took conventional oral contraceptives (747).

Cytarabine

Acute pericarditis has been described during high-dose cytarabine treatment. Interruption of high-dose therapy and the use of glucocorticoids were effective in resolving the pericardial effusion. Besides direct toxic effects on the pericardium, mechanisms mediated by the immune system may be involved (748).

- A 37-year-old patient with acute myeloblastic leukemia and nervous system involvement received a first consolidation course of chemotherapy containing mitoxantrone and cytarabine 3 g/m^2 (once daily on days 1–5) after remission with daunorubicin and cytarabine (749). On day 3, after the fifth dose of subsequent consolidation treatment containing high-dose cytarabine, the patient suddenly complained of dyspnea and sharp anterior chest pain worsened by inspiration. His temperature was 37.8°C, blood pressure 100/60 mm Hg, pulse 120/minute. Echocardiography showed a small circumferential pericardial effusion. On day 4, fever was still present, and a pericardial friction rub was heard. The pericardial effusion had increased in size and there was paradoxical septal movement. Cytarabine was withdrawn and oral methyprednisolone 0.5 mg/kg/day was given together with morphine. The fever and chest pain resolved.

Danazol

A 42-year-old woman, with no history of smoking, had an acute myocardial infarct 2 months after she had started to take danazol 100 mg/day (750). There have been occasional reports of thrombosis during danazol treatment, possibly because of effects on low-density lipoprotein cholesterol, raised glucose concentrations, or an increase in the platelet count. It would at all events seem wise to avoid danazol in patients at particular risk of coronary artery disease.

Dantrolene

The combination of dantrolene with calcium channel blockers, such as verapamil, can result in severe cardiovascular depression and hyperkalemia (SEDA-12, 113; 751,752), so that extreme care is required.

Dapsone and analogues

Signs of heart failure with edema, ascites, and severe hypoalbuminemia have been described in the treatment of dermatitis herpetiformis with dapsone (753).

Decamethonium

Cardiovascular effects are less frequent than with suxamethonium; a reduction in heart rate sometimes occurs after a second dose.

Deferiprone

Since circulating deferiprone-iron complexes can easily dissociate, deferiprone can redistribute iron in the body (754). Although this might theoretically lead to the precipitation or aggravation of heart failure (754), no such cases have as yet been reported.

Deferoxamine (desferrioxamine)

Although rarely described, hypotension can be a significant problem in patients receiving deferoxamine, especially when it is given too rapidly by intravenous injection (755); it is possibly due to histamine release (756). Dose reduction alleviates the hypotension. Anaphylactic shock has only rarely been reported (SEDA-7, 262).

In a single report, soft-tissue swelling around the elbow and localized mild pitting edema were thought to have been induced by deferoxamine (757). Although the clinical features suggested a deep-vein thrombosis, this was ruled out by a phlebogram.

In iron storage disease, ascorbic acid should be given only after adequate serum concentrations of deferoxamine have been attained, in order to prevent serious cardiac arrhythmia (758). Opportunistic fungal infections associated with deferoxamine may also involve the heart muscle and usually have a fatal outcome (759–761).

There was severe phlebitis in cancer patients receiving deferoxamine (50 mg/kg/day by intravenous infusion over 72 hours) and iron sorbitol citrate in an attempt to enhance doxorubicin activity (762). Dilution of the drug in large volumes of saline did not prevent this adverse effect.

Desflurane

Desflurane increases the heart rate and reduces both mean arterial pressure and systemic vascular resistance while maintaining cardiac output (763,764). In high concentrations it can cause transient activation of the sympathetic nervous system, predisposing to hypertension and dysrhythmias (765).

Despite some coronary vasodilatation in dogs, there is no evidence of coronary steal in man. Desflurane may benefit elderly patients by allowing more rapid recovery from anesthesia (766).

Cardiac arrest due to desflurane toxicity has been attributed to accidental delivery of a high concentration of desflurane due to vaporizer malfunction (767).

- A healthy 36-year-old woman underwent anesthesia maintained with desflurane, which was delivered at 3.5% using a Tec 6 Plus Vaporizer (Datex Ohmeda, Steeton, England) via a partially closed circuit with a low flow of fresh gases (1 l/minute). Five minutes after induction, she developed hypoxia and bradycardia, rapidly followed by cardiac arrest with asystole. She was resuscitated and a chest x-ray showed pulmonary edema.

Examination of the memory of the halogenated anesthetic monitor (Viridia 24 C; Hewlett Packard, Boeblingen, Germany) showed a progressive increase in end-expiratory desflurane concentration up to 23%. There was an internal crack in the control dial, which normally regulates the control valve, but the damage did not limit the rotation of the control valve, which remained uncontrolled. The authors thought that this defect had been responsible for massive administration of desflurane in the inhalation circuit. Cardiac arrest was probably due to the negative inotropic effect of desflurane.

Duchenne's muscular dystrophy can be associated with cardiac arrest during anesthesia, and this has been reported in a 16-year-old boy who was anesthetized with desflurane (768).

Desloratadine

In a large, multicenter, double-blind, placebo-controlled, parallel-group study of the efficacy and tolerability of desloratadine in 346 patients with seasonal allergic rhinitis, the symptoms improved significantly and there was no significant effect on the QT_c interval (769).

In a multicenter, randomized, double-blind, placebo-controlled study in 190 patients, desloratadine was effective in the treatment of moderate to severe chronic idiopathic urticaria, with no adverse electrocardiographic effects (770).

In healthy volunteers, 12 men and 12 women, there was no prolongation of the QT_c interval after co-administration of desloratadine with erythromycin (771).

Desmopressin

Facial flushing occurred in two of 25 children with either hemophilia or von Willebrand disease given high-dose intranasal desmopressin (150 micrograms) in a single-dose open study (772).

Marked hypotension with circulatory collapse has occasionally been reported with desmopressin (773,774), although both of these reports related to patients with pre-existing cardiac conditions.

Thrombotic disorders

Desmopressin stimulates the release from endothelial cells of all the multimeric forms of von Willebrand factor found in normal plasma, including large forms that are not normally present (775). These abnormal multimers can aggregate platelets, particularly at the high levels of fluid shear stress that occur at sites of arterial stenosis.

Myocardial thrombosis

There have been several reports of arterial thrombosis associated with the use of desmopressin, including myocardial infarction (776–780). One of these reports concerned a case of fatal myocardial infarction in a blood donor in excellent health, with no risk factors and no signs of vascular disease (778).

- A 59-year-old woman with hemolytic–uremic syndrome and a recent history of atypical chest pain was given prophylactic desmopressin 0.4 micrograms/kg immediately before a renal biopsy (781). Within 30 minutes she developed chest pain and bradycardia due to myocardial infarction.

Three other cases of myocardial infarction in the absence of desmopressin have been reported in patients with hemolytic–uremic syndrome, who already have an increased risk of thrombosis.

A meta-analysis of placebo-controlled trials of desmopressin in 702 cardiac surgery patients showed a significantly increased risk of myocardial infarction in treated patients (RR = 2.39, CI = 1.02, 5.60) (782). Overall mortality was not different from placebo. Desmopressin was less efficacious in reducing perioperative blood loss than either aprotinin or lysine analogues.

Cerebral thrombosis

Cerebral infarction has also been reported in association with the use of desmopressin in children (783,784). One of these cases involved a 7-month-old child with congenital nephrotic syndrome who developed a cerebral infarction after surgery (785). One child developed cerebral ischemia after *Varicella* infection and desmopressin for enuresis (784).

Thromboembolism

There are isolated reports of thromboembolic complications in recipients of desmopressin; most occurred in patients with pre-existing vascular disease. However, in nine trials of the hemostatic efficacy of desmopressin in reducing blood and transfusion requirements in 763 patients, there were no significant differences between the frequencies of thromboembolism in subjects treated with desmopressin and controls (785). An analysis of 31 clinical trials of desmopressin in patients undergoing cardiac, vascular, orthopedic, or other major surgery showed that desmopressin did not increase the incidence of thrombosis (786).

Most of the reported thromboembolic complications occurred in elderly patients and desmopressin should not be used in patients with documented arterial disease or even in elderly patients, in whom some degree of latent arterial disease may be assumed to be present (786). Concomitant use of antifibrinolytic agents, such as tranexamic acid, should also be avoided.

Dextrans

High output left ventricular failure has been described after hysteroscopic lysis of adhesions using dextran as a distension medium. Prolonged surgical dissection of the uterine wall (the precise duration of the operation was not stated in the report) and the large volume of dextran and fluid (2 liters of 5% dextrose and an additional 800 ml of dextran) probably caused the dextran to enter into the systemic circulation, inducing a significant shift of fluid into the intravascular compartment (787).

Dextropropoxyphene

Dextropropoxyphene-induced cardiogenic shock has been described (788).

- A 32-year-old man became deeply comatose, with intraventricular conduction disturbances, after taking dextropropoxyphene 4.6 g. Treatment-resistant seizures lasted for hours. He was treated with an intra-aortic balloon pump and a continuous infusion of milrinone for 7 days and recovered fully.

The mechanism of cardiotoxicity of dextropropoxyphene is unknown, but the membrane-stabilizing effect of its major metabolite, norpropoxyphene, seems to play a central part. The cardiac effects are not reversed by naloxone (789), but dopamine may be effective.

Dezocine

Dezocine is structurally related to pentazocine (790). It reacts primarily with OP_3 (μ) receptors, but also has some affinity for OP_1 (δ) and OP_2 (κ) receptors. It is slightly more potent than morphine, but with similar adverse effects at effective doses (2.5–10 mg). The most common adverse effects (3–9%) are nausea and vomiting, sedation, or local injection site reactions; dizziness/vertigo have also been reported (1–3%) (SEDA-16, 88). However, in some trials nausea and/or vomiting were reported in 5–22%, while headache was the most common CNS complaint (16–35%). Other adverse effects reported in 1% of patients involve the cardiovascular system, respiratory system, urogenital system, CNS, gastrointestinal system, and visual senses.

Diazepam

Cases of inadvertent intra-arterial injection of diazepam have been reported.

- A 51-year-old woman with an acute claustrophobic anxiety attack developed gangrene of the fingers after she was inadvertently given diazepam 10 mg intra-arterially (791).
- Inadvertent intra-arterial injection of diazepam (2.5 mg in 0.5 ml) has been reported in an 8-year-old girl (792). Gangrene resulted and amputation of the 4th and 5th fingers was required.

Gangrene has been previously reported with intra-arterial injection of diazepam and is also well known with other classes of drugs, such as barbiturates and phenothiazines. It appears to be caused by the drug rather than the solvent used in the intravenous formulations.

Diethylstilbestrol

In a randomized study of men treated hormonally for prostatic cancer (793), cardiovascular adverse effects were reported more often in patients treated with diethylstilbestrol than in those treated with cyproterone acetate. The risk was highest during the first 6 months of treatment.

Diphenhydramine

CYP2D6 is the major enzyme involved in the metabolism of venlafaxine, and diphenhydramine alters the disposition of venlafaxine, increasing plasma concentrations and predisposing to cardiovascular adverse effects (794).

There have been several reports of overdose with diphenhydramine, the features of which include cardiotoxicity. In addition to inhibition of cardiac fast sodium channels, higher concentrations of diphenhydramine also inhibit potassium channels, which can result in QT interval prolongation.

- A 40-year-old woman took 25 tablets of Tylenol-PM (McNeil Pharmaceutical, Raritan, NJ, USA), an over-the-counter combination of diphenhydramine 25 mg and paracetamol 500 mg (795). She was intubated and a 12-lead electrocardiogram showed a sinus tachycardia with a markedly prolonged QT interval (QT_c 588 ms). Ventricular repolarization on the electrocardiogram was abnormal, with broad biphasic T waves, which were more apparent in the mid-precordial leads. Torsade de pointes was absent. By day 2, the QT interval and T waves were normal. She recovered fully.

Paracetamol does not prolong the QT interval or affect cardiac repolarization. The authors concluded that the tachycardia caused by the anticholinergic and hypotensive effects of diphenhydramine may have protected against torsade de pointes. They further suggested that it may be practical to avoid bradycardia in the acute phase of diphenhydramine toxicity.

- A 17-year-old man was found unconscious next to a bottle of diphenhydramine, which had contained up to 40 tablets of 50 mg each (796). He had a tachycardia of 150/minute, a tachypnea of 32/minute, a metabolic acidosis, with a pH of 7.21, and an unspecified number of seizures. Electrocardiography showed right bundle branch block and QT interval prolongation at 522 ms. He was treated with intravenous benzodiazepines to suppress seizures and with physostigmine 0.5 ml (concentration not stated). He was then able to speak and the metabolic and cardiovascular parameters all improved and reverted to normal, though the time scale was unclear.

The authors noted that although the clinical features of diphenhydramine toxicity are well described, electrocardiographic changes have not been reported extensively. They emphasized that diphenhydramine overdose can occasionally cause prolongation of the QT interval. Although torsade de pointes has been described with diphenhydramine overdose it did not occur here. Presumably the cardiovascular toxicity of diphenhydramine is partly related to its anticholinergic activity, similar to that seen with tricyclic antidepressants. However, other antihistamines without anticholinergic effects have been implicated in QT prolongation and torsade de pointes. Furthermore, the dramatic response to physostigmine, a cholinesterase inhibitor, in this case is very interesting, in view of the controversy associated with its use in poisoning.

- A 39-year-old man took an overdose of diphenhydramine and became unconscious and hypotensive. His electrocardiogram showed changes suggestive of the Brugada syndrome (797). However, normalization of the electrocardiogram as he recovered and a negative flecainide test ruled out Brugada syndrome.
- A 26 year-old man took 50 tablets each containing 50 mg of diphenhydramine in a suicide attempt (798). Despite supportive therapy he developed significant torsade de pointes and nearly died.

Diphtheria vaccine

Myopericarditis has been attributed to Td-IPV vaccine (799).

- A 31-year-old man developed arthralgia and chest pain 2 days after Td-IPV immunization and had an acute myopericarditis. He recovered within a few days with high-dose aspirin.

The authors discussed two possible causal mechanisms, natural infection or an immune complex-mediated mechanism. Infection was excluded by negative bacterial and viral serology and the favorable outcome within a few days without antimicrobial treatment.

Dipyridamole

Dipyridamole is used with ^{201}thallium imaging in the detection of coronary artery disease, but can cause dysrhythmias.

- A 41-year-old man with hypertension was investigated for chest tightness by dipyridamole–thallium single-photon emission computed tomography (800). A standard dose of dipyridamole (0.56 mg/kg) was infused intravenously over 4 minutes, during which his heart rate increased from 68 to 88/minute and his blood pressure fell slightly (from 160/80 to 140/76 mmHg). He had no subjective symptoms, such as palpitation, dizziness, or chest tightness, but had ventricular extra beats 40 seconds after completion of the dipyridamole infusion, followed 1 minute later by a sustained ventricular tachycardia. His blood pressure fell to 80/50 mmHg and he complained of dizziness. Intravenous aminophylline 125 mg was given immediately. About 30 seconds later, the ventricular tachycardia terminated and his hemodynamics stabilized. The ventricular extra beats persisted for another 30 seconds.
- Two cases of severe bradydysrhythmias (one complete heart block, one sinus bradycardia) occurred after intravenous dipyridamole (801).

Dipyridamole can also cause myocardial ischemia.

- A 43-year-old man had an acute myocardial infarction immediately after exercise and pharmacological stress echocardiography with dipyridamole + atropine 1 month after successful stent implantation (802).
- A 59-year-old man developed unstable angina 1 month after coronary artery bypass surgery (803). During dipyridamole scintigraphy, 2 minutes after the beginning of dipyridamole infusion, ST segment elevation occurred in the inferior electrocardiographic leads and there were two marked anteroseptal and inferior defects on myocardial scintigraphy.

A patient with aorto-iliac occlusive vascular disease and hypertension suffered a stroke 6.5 minutes after administration of intravenous dipyridamole during a ^{201}thallium myocardial study (804). Aminophylline did not reverse its progression.

Disulfiram

Disulfiram in a dose of 250–300 mg/day does not affect pulse rate, blood pressure, or plasma noradrenaline concentrations, but 500 mg/day causes an increase in plasma noradrenaline, increased systolic blood pressure both recumbent and erect, and an increased erect pulse rate. The raised blood pressure does not reach hypertensive values, but the results suggest increased sympathetic nervous system activity in patients who take disulfiram. Caution should therefore be exercised in using disulfiram in hypertensive patients. Close monitoring of blood pressure is advised, and the dose of disulfiram should preferably be reduced to 250 mg/day (805).

Cardiac dysrhythmias occurred during a disulfiram + alcohol test in a 48-year-old man who had been an alcoholic for 5 years (806). After drinking a test amount of alcohol, he developed flushing, nausea, vomiting, sweating, dyspnea, and hyperventilation, palpitation, tremor, confusion, and syncope. The electrocardiogram showed atrial fibrillation and non-sustained bouts of ventricular tachycardia of 7–8/minute. He also had severe hypotension.

Dobutamine

Dobutamine stress testing, particularly combined with echocardiography, is a very widely used tool in cardiological investigation, and its safety continues to be examined in great detail.

In a review of 37 publications, each reporting on 100 or more patients, with a total of over 26 000 tests, 79 life-threatening complications were described, including acute myocardial infarction, a variety of cardiac dysrhythmias, and severe hypotension (807). The authors also referred to 29 isolated case reports of severe complications, including two deaths. They concluded that there must be a clear indication for the procedure, informed consent must be obtained, a physician should be present during the test, and patients should be carefully followed as outpatients in case of delayed problems.

An Israeli group has reviewed the hemodynamic and adverse effects of dobutamine in 400 patients, of whom 187 were aged 65–79 years and 49 were 80 years or older (808). They found a very low incidence (1.5%) of serious adverse effects, even at high doses of up to 50 micrograms/kg/minute, and noted that most problems occur transiently at a lower dose and then resolve. Hypotension and dysrhythmias were the most common adverse effects, and their incidence was not related to age.

Acute subaortic left ventricular outflow tract obstruction has been described during dobutamine infusion in a patient who had no evidence of this at rest but developed severe obstruction when his pulse rate exceeded 105/minute (809). The subaortic gradient eventually reached the very high figure of 182 mmHg, even though the patient remained asymptomatic and there was no clear reason why this occurred in this particular patient.

- A 73-year-old woman with chest pain had ST segment elevation on the electrocardiogram and raised creatine kinase activity and troponin concentrations (810). However, coronary angiography showed no significant obstructive lesions. She was given intravenous glyceryl trinitrate, with resolution of both the chest pain and the electrocardiographic changes. Stress echocardiography with dobutamine 10 micrograms/kg/minute was carried out 5 days later. This led to severe midventricular obstruction (a pressure gradient of 150 mmHg) and mitral regurgitation, with marked pulmonary hypertension. There was no pulmonary edema or cardiogenic shock, and she recovered fully after being given propranolol.

The authors commented that sympathetic stimulation may play a crucial role in this syndrome, whether it occurs spontaneously or after sympathomimetic drug administration.

Cardiac rupture has been reported after dobutamine stress echocardiography (811).

- A 57-year-old man with a history of hypertension had an inferior myocardial infarction and was referred for dobutamine stress echocardiography. An unnamed intravenous ultrasound contrast agent was used to improve endocardial visualization. Resting echocardiographic images showed akinesia of the inferior and posterior basal segments without a pericardial effusion. At the peak dobutamine dose (40 μg/kg/minute), he had chest pain and ST segment elevation in the inferior leads. Simultaneously, extravasation of the contrast material into the pericardial cavity was detected by echocardiography. A few seconds later, electromechanical dissociation with cardiopulmonary arrest occurred. All attempts at cardiopulmonary resuscitation failed and he died.

This seems to have been the first report of acute cardiac rupture during dobutamine stress testing, diagnosed by the use of intravenous contrast echocardiography.

In 47 consecutive patients (mean age 64 years, 46 men) with three or more cardiovascular risk factors, intravenous dobutamine was given at a rate of 40 micrograms/kg/minute until the target heart rate was achieved, which took a mean of 11.6 minutes (812). Subjective sensations occurred in 49% of the patients (palpitation 21%, chest pain 6%, nausea 6%, headache 6%, dizziness 13%), while half the patients had abnormal cardiac rhythms (ventricular extra beats 38%, supraventricular tachycardia 10%, and non-sustained ventricular tachycardia 2%). The authors concluded that the safety and tolerability of this procedure is comparable to that of standard dobutamine stress testing, although its specificity and selectivity are still uncertain.

The value and safety of dobutamine stress echocardiography have been studied in 135 patients aged 70 years or older (mean age 74 years; 58% men) soon after myocardial infarction (813). A diagnostic end-point was reached in 83% of patients, a figure comparable to younger populations. No major complications were reported. The procedure was terminated prematurely because of symptomatic hypotension ($n = 8$), non-sustained ventricular tachycardia ($n = 2$), systolic hypertension ($n = 2$), atrial fibrillation ($n = 1$), or bladder discomfort ($n = 1$). All of these problems resolved rapidly after the procedure was stopped. At the time of discontinuation, seven of these 16 patients had objective evidence of myocardial ischemia.

Cardiac dysrhythmias

Dobutamine increases cardiac output after acute myocardial infarction without exacerbating myocardial infarction or ventricular dysrhythmias. However, a mild increase in heart rate often occurs, and occasionally a more marked tachycardia.

Cardiac dysrhythmias of various types (including occasional ventricular dysrhythmias) when dobutamine is used as a stress-inducing agent during echocardiography have been described (SEDA-18, 158).

- A 56-year-old woman with dilated cardiomyopathy developed QT interval prolongation and torsade de pointes during infusion of dobutamine in a low dose (2.5 micrograms/kg/minute). She had no previous documented dysrhythmias (814). She was hypokalemic and when her plasma potassium concentration was restored to normal the dysrhythmia disappeared and the QT interval normalized. The authors observed that this was the first case of its kind and emphasized the importance of normokalemia in dobutamine-treated patients.
- A 55-year-old man with stable angina was admitted for dobutamine stress testing (815). He had been taking metoprolol, but this was withdrawn 2 days before the test. During the lowest dose of the dobutamine infusion (5 micrograms/kg/minute) his heart rate rose to 143/minute and he developed chest pain and ST segment depression in the lateral chest leads of the electrocardiogram. The drug was stopped and metoprolol and glyceryl trinitrate were given immediately, but a few minutes later he developed torsade de pointes followed by ventricular fibrillation. Resuscitation was unsuccessful. There was no evidence of acute myocardial infarction at autopsy.
- A 74-year-old man with idiopathic dilated cardiomyopathy was given dobutamine (5 micrograms/kg/minute) to determine whether it would produce a positive inotropic effect (816). After 14 minutes he developed asymptomatic pulsus alternans, which resolved within 20 minutes of withdrawal.

The prognostic implications of non-sustained ventricular tachycardia during dobutamine stress testing have been considered in 1266 consecutive patients, of whom 65 (5.1%) had this dysrhythmia (817). After 3 years of follow-up there was no significant difference in all-cause mortality between patients who had ventricular tachycardia and those who did not (22% versus 17%). However, further analysis showed that patients who had (i) non-sustained ventricular tachycardia, (ii) no evidence of inducible ischemia, and (iii) a moderately reduced ejection fraction (0.35–0.45) did have significantly reduced survival. On the other hand, as the authors pointed out, this was a retrospective study with post hoc subgroup analysis, and the results should be confirmed by a prospective study. Even allowing for this, these results support earlier findings that this type of dysrhythmia when induced by dobutamine does not in itself indicate significantly increased risk.

A much more serious complication of dobutamine therapy is ventricular dysrhythmias. Of 305 patients with acutely decompensated congestive heart failure, 58 were given dobutamine (although it is difficult to ascertain the dose), 44 were given other standard inotropic drugs such as milrinone, and 203 were treated with brain natriuretic peptide (nesiritide, 0.015 or 0.03 micrograms/kg/minute) (818). Of those given dobutamine 7% had sustained ventricular tachycardia, 17% had non-sustained ventricular tachycardia, and 5% had a cardiac arrest. In contrast, the figures for nesiritide were 1, 11, and 0% respectively. There was no analysis of other outcomes but these results certainly do not encourage the use of dobutamine in these very vulnerable patients.

A study from Rotterdam, has examined the consequences of adding atropine to dobutamine in 200 patients with impaired left ventricular function (ejection fraction less than 35%) (819). There were cardiac dysrhythmias in 6% of patients and significant hypotension in 11%, figures comparable to those in other published studies. In the 36 patients who required atropine to achieve target heart rates the incidence of adverse effects was not increased. The same group has studied over 1000 consecutive patients undergoing dobutamine stress scintigraphic imaging of the myocardium (820). In these patients the incidence of dysrhythmias was about 8%, of whom about half had transient ventricular tachycardia, but only about 3% had significant hypotension (defined as a fall in systolic blood pressure of 40 mmHg). Atropine was required in nearly 40% of the patients in order to achieve target heart rates, but it is difficult to determine whether this had any influence on the occurrence of adverse effects.

A report from the Mayo Clinic has described 27 elderly patients (mean age 71 years) with aortic stenosis in whom dobutamine stress hemodynamic testing was used to assess the severity of the stenosis (821). There were no severe adverse effects, but relatively minor problems occurred in 16 patients, including chest pain and ventricular extra beats ($n = 9$ each) and atrial dysrhythmias ($n = 4$). The authors concluded that the procedure appears to be safe in these high-risk patients, although its diagnostic value may be limited.

There are variants of the standard procedure that do not appear to be associated with increased risk. An accelerated high-dose protocol, in which a constant infusion of 50 micrograms/kg/minute was given for up to 10 minutes in 100 patients, has been compared with the standard stepwise procedure in a similar number (822). The cumulative dose was somewhat lower in the accelerated protocol, while the duration of the test was halved. Dysrhythmic adverse effects occurred in 28 patients on the accelerated protocol and in 39 of those tested by the standard method: this difference was said to be non-significant. In another study, transesophageal was compared with transthoracic dobutamine stress echocardiography in 63 and 100 patients respectively (823). Baseline pulse and blood pressure were higher in the transesophageal group. The authors noted that there were no cases of ventricular tachycardia or fibrillation in either group, or of myocardial infarction. They also stated that the incidence of less serious dysrhythmias was similar in the two groups; however, no figures were quoted.

In the PRECEDENT study, dobutamine was compared with nesiritide (B natriuretic peptide) in patients with acutely decompensated congestive heart failure (824). The primary objective of the study was to assess the risk of ventricular dysrhythmias with the two therapies. Altogether 255 patients (mean age 61 years, 67% men) were randomized to receive dobutamine 5 micrograms/kg/hour or one of two doses of nesiritide 0.015 or 0.03 micrograms/kg/hour. Dobutamine significantly increased the number of episodes of ventricular tachycardia by 48/day, ventricular extra beats by 69/hour, and overall heart rate by 5/minute. These parameters were unchanged or improved by nesiritide. Since the two drugs had similar effects on the hemodynamic features of heart failure and its symptoms, the authors concluded that nesiritide should be considered as an alternative to dobutamine in this type of patient.

Myocardial ischemia

Angina and coronary artery spasm (825) have also occurred under these and other conditions. Some 10% of patients require treatment for angina occurring during stress testing, but up to a quarter have ischemic changes on the electrocardiogram (SEDA-17, 163), while others have headache, palpitation, anxiety, nausea, tingling, and flushing.

Myocardial ischemia has also been reported in susceptible patients. A Japanese group carried out dobutamine stress echocardiography in 51 patients with a presumptive diagnosis of variant angina (826). All had coronary vasospasm in response to intracoronary acetylcholine and seven also had chest pain and reversible ST segment elevation. One must incidentally wonder whether this procedure was entirely advisable.

In 47 patients in Mannheim (mean age 61 years, 34 men) dobutamine echocardiography was carried out, with blood sampling immediately before and after the procedure and then at 1, 2, 4, 6, and 12 hours (827). Assays were carried out for creatine kinase-MB, troponins I and T, myoglobin, and fibrin monomer antigen. There were no significant increases in these markers of myocardial damage and coagulation, regardless of the outcome of the stress test. These findings have confirmed those of an earlier study, although the data do not absolutely exclude abnormal findings in a minority of individuals (828).

Hypotension

In one study 38% of patients undergoing dobutamine stress echocardiography developed hypotension. Increases in blood pressure are more in line with what one would expect. Although dobutamine does not as a rule cause a marked increase in systolic blood pressure in normotensive patients, hypertensive patients can develop marked systolic hypertension during an infusion of the drug. When stress echocardiography with dobutamine is performed in subjects who prove to be entirely healthy, an audible Still's-like vibratory systolic ejection murmur is nevertheless produced.

Severe peripheral ischemia (even leading to dermal necrosis) can occur, as with dopamine (829).

Domperidone

Intravenous infusion of domperidone, withdrawn some years ago, has caused convulsions and cardiac arrest (830–832).

Donepezil

Symptomatic sinus bradycardia is a possible adverse effect of treatment with donepezil in Alzheimer's disease (833).

- An 84-year-old patient with hypertensive cardiomyopathy developed bradycardia, fainting, and left-sided heart failure 3 weeks after starting treatment with donepezil. When donepezil was withdrawn, the sinus bradycardia disappeared; 24-hour electrocardiography showed no signs of sinus node disease, and no episodes of this type recurred during the next 6 months.

One well-recognized complication of donepezil is non-sustained ventricular tachycardia. The prognostic implications have been considered in 1266 consecutive patients, of whom 65 (5.1%) had this dysrhythmia (817). After 3 years of follow-up there was no significant difference in all-cause mortality between patients who had ventricular tachycardia and those who did not (22% versus 17%). However, further analysis showed that patients who had (i) non-sustained ventricular tachycardia, (ii) no evidence of inducible ischemia, and (iii) a moderately reduced ejection fraction (0.35–0.45) did have significantly reduced survival. On the other hand, as the authors pointed out, this was a retrospective study with post hoc subgroup analysis, and the results should be confirmed by a prospective study. Even allowing for this, these results support earlier findings that this type of dysrhythmia when induced by dobutamine does not in itself indicate significantly increased risk.

Soon after the start of donepezil treatment three patients with Alzheimer's disease developed cardiac syncope (834). In two cases, a bradydysrhythmia was documented and pacemaker implantation was considered justified rather than donepezil withdrawal.

Exaggeration of hypotension during donepezil treatment, due to interference with autonomic control, has been described (835).

The causes of syncope in patients with Alzheimer's disease treated with donepezil have been reported in 16 consecutive patients (12 women, 4 men) with Alzheimer's disease, mean age 80 years, who underwent staged evaluation, ranging from physical examination to electrophysiological testing (836). The mean dose of donepezil was 7.8 mg/day and the mean duration of donepezil treatment at the time of syncope was 12 months. Among the causes of syncope, carotid sinus syndrome (n = 3), complete atrioventricular block (n = 2), sinus node dysfunction (n = 2), and paroxysmal atrial fibrillation (n = 1) were diagnosed. No cause of syncope was found in six patients. Non-invasive evaluation is recommended before withdrawing cholinesterase inhibitors in patients with Alzheimer's disease and unexplained syncope.

It is important to emphasize that disorders of cardiac rhythm associated with the use of donepezil are extremely unusual.

Dopamine

The major risk during dopamine treatment is that of severe peripheral ischemia, particularly in patients in whom the peripheral circulation is already impaired, since dopamine is converted to noradrenaline; gangrene has repeatedly resulted. In some of the reported cases the error lay in extravasation of dopamine from a peripheral venous infusion site; in others the dosage had been high and prolonged, or ergometrine had also been given. In cases of pre-existing vascular damage from arteriosclerosis, diabetes, Raynaud's disease, or frostbite, particular care must be taken. If discoloration appears, the infusion should be stopped and phentolamine 5–10 mg given intravenously. Nitroprusside may fail to prevent the onset of gangrene.

Dosulepin

Dosulepin (dothiepin) is a tricyclic antidepressant that has been available in Europe for over 30 years and is particularly popular in the UK. Its animal and clinical pharmacology has been described in an extensive review (837). It appears to be equivalent to amitriptyline, although few studies have reported dosages above 225 mg/day. Dosulepin is as effective and sedative as amitriptyline, with somewhat fewer anticholinergic adverse effects in several studies. However, it does not appear to have been compared with other less sedative or anticholinergic tricyclic compounds or second-generation drugs. Although it is claimed to have fewer cardiovascular effects, this has not been well substantiated in controlled comparative studies, and the cardiovascular and other effects of overdosage appear to be identical for all tricyclic compounds. Fetal tachydysrhythmias were believed to be caused by maternal ingestion of dosulepin (838).

Doxacurium chloride

No rise in plasma histamine concentration was found with bolus doses up to 0.08 mg/kg (839), but in one case there was transient hypotension 1 minute after a bolus dose of 0.05 mg/kg via a pulmonary artery cannula, with cutaneous flushing at 2 minutes, suggesting that histamine release can occur on occasion (840). There was no tachycardia or bronchospasm, but the mean arterial pressure fell from 88 to 40 mmHg and recovered with therapy within 3 minutes, by which time the skin flushing was fading.

In a study of 54 patients, plasma histamine concentrations increased by 200% following doxacurium in two patients, but there were no changes in heart rate or blood pressure (841). Indeed, cardiovascular stability has been reported in several studies (839,841) and only minor clinically insignificant changes have been seen with doses up to 0.08 mg/kg, even in cardiac patients (AHA classes III–IV) (842,843).

Bradycardia is occasionally seen (844), but this may be due to co-administration of vagotonic drugs with atracurium, vecuronium, and pipecuronium.

Doxapram

Dysrhythmias consisting of self-limiting or single ventricular extra beats were observed after doxapram was given by injection; it was suggested that electrocardiographic dysrhythmias occurring in anesthesia could be associated with analeptic drug administration (845).

Doxapram is used to treat idiopathic apnea in premature infants. Second-degree atrioventricular heart block developed after its administration to three neonates (846).

- Thirty-six hours after an infusion of doxapram was started, the first infant developed second-degree AV block, with QT interval prolongation and an increase in the QRS interval. There was no hypotension. Sinus rhythm returned 92 hours after stopping the infusion.
- Doxapram was given orally, every 6 hours, to the second infant. After 43 hours second-degree AV block with a prolonged QT interval was noted. Echocardiography showed a normal heart. Doxapram was discontinued and 8 hours later sinus rhythm returned.
- The third infant was given aminophylline orally and doxapram by intravenous infusion; 5 days later cisapride was added to treat suspected gastro-esophageal reflux. The next day the infant had developed second-degree AV block, with a prolonged QT interval. Doxapram was withdrawn and sinus rhythm returned 36 hours later.

Heart block did not occur in the third case until cisapride, which can prolong the QT interval, was added.

There has been another report of three cases of second-degree atrioventricular block and QT interval prolongation associated with doxapram; sinus rhythm returned after withdrawal (847).

The mean pulmonary arterial pressure was significantly, but not severely, increased in 10 patients receiving intravenous infusions of doxapram (848).

There have been two studies of doxapram in very low birth weight infants before extubation. In one it was concluded that doxapram did not increase the likelihood of successful extubation. In the other there was an increase in systolic blood pressure along with much higher plasma doxapram concentrations than expected (849,850).

Doxorubicin

Repeated cycles of doxorubicin 75 mg/m^2 intravenously followed by bevacizumab 15 mg/kg intravenously every 3 weeks have been studied in 17 patients with metastatic soft-tissue sarcomas (851). Dexrazoxane was also given when the total dose of doxorubicin was over 300 mg/m^2. In all, 85 cycles of doxorubicin + bevacizumab were administered; the median number of cycles was four. Six patients developed cardiac toxicity of grade 2 or worse: four had grade 2 (cumulative doxorubicin dose 75, 150, 300, 300 mg/m^2), one had grade 3 (total doxorubicin dose 591 mg/m^2), and one had grade 4 (total doxorubicin dose 420 mg/m^2). One patient with extensive lung disease died of recurrent bilateral pneumothorax, possibly related to treatment. The 12% response rate for these patients was no greater than that observed for single-agent doxorubicin.

Droperidol

Droperidol has been associated with QT interval prolongation (SED-14, 141; 852–854) and torsade de pointes has been reported (855).

- A 59-year-old woman with no history of cardiac problems, except for hypertension, who was taking amlodipine 5 mg qds, cyclobenzaprine 10 mg qds, and co-triamterzide 37.5 + 25 mg qds, and who had a QT_c interval of 497 ms, was given intravenous droperidol 0.625 mg and metoclopramide 10 mg 45 minutes before surgery. About 1.75 hours after surgery she developed a polymorphic ventricular tachycardia with findings consistent with torsade de pointes, which resolved with defibrillation.

In late 2001 the FDA decided to introduce a "black box" warning regarding the use of droperidol because of its potential cardiac effects (QT prolongation leading to torsade de pointes and death). This decision, based mainly on post-marketing surveillance (MedWatch program and other relevant sources), has produced several reactions. The authors of a review (856) of three clinical studies, one published abstract, and seven case reports, as well as MedWatch reports, stressed the following points: there are several risk factors for dysrhythmias, including underlying illnesses, other drug exposures, different clinical settings (including surgery, emergencies, and psychiatry), and high doses of the drug (up to 100 mg intramuscularly). Long and widespread clinical use is also mentioned. They found no convincing evidence of a causal relation between droperidol and serious cardiac events. Others reached the same conclusion after reviewing 10 cases, reported in the FDA database, that were possibly related to the administration of droperidol in doses of 1.25 or less (857).

A review of the cases in which serious cardiovascular events were probably related to droperidol at doses of 1.25 mg/day or less showed, according to the authors, that there are many confounding factors, such as the concomitant use of drugs that can cause QT prolongation and other risk factors (858,859). It has also been suggested that using the more expensive 5-HT_3 receptor antagonists as first-line agents for antiemetic prophylaxis has significant cost implications, with no evidence that they are any safer than droperidol (860). On the other hand, several issues have been raised in defence of the FDA decision (861); it is assumed, for instance, that cardiovascular events would be more likely to occur in those with other risk factors, since such factors would be expected co-variates of droperidol-induced dysrhythmias.

Duloxetine

Early descriptions of duloxetine suggested that it might be less likely than venlafaxine to cause increased blood pressure. The cardiovascular profile of duloxetine has been reviewed from a database of eight double-blind randomized trials in depression, in which 1139 patients took duloxetine (40–120 mg/day) and 777 took placebo (862). Relative to placebo, duloxetine produced a small but significant increase in heart rate (about 2/minute). Duloxetine produced a greater rate of sustained increase in systolic blood pressure than placebo (1% versus 0.4%, relative risk 2.1) but no difference in diastolic blood pressure. There were also significantly more instances of significantly increased systolic blood pressure in patients taking duloxetine than placebo at any clinic visit (19% versus 13%). These findings suggest that duloxetine can produce increases in blood pressure in some people, presumably through its potentiation of noradrenaline function. How the effect of duloxetine compares with that of venlafaxine remains to be established.

Ebastine

Cardiac effects were reported from early experimental work using very high doses of ebastine, but they are not believed to be clinically relevant in normal use. In one study serial electrocardiograms showed no changes with doses up to the maximum used (30 mg). Ebastine in doses up to five times the recommended therapeutic dose did not cause clinically relevant changes in QT_c interval in healthy subjects (863). Co-administration of ebastine with ketoconazole or erythromycin did not lead to significant changes in the QT_c interval (864,865).

Enflurane

Despite conflicting results, enflurane is generally considered to have little effect on the cardiovascular system. Cardiac output was mildly influenced in healthy men and the negative inotropic effects of enflurane (866) were more pronounced in patients with congestive heart failure (867). Myocardial damage was suggested to be an unlikely complication of enflurane anesthesia, even in patients with ischemic heart disease (868).

Cardiac dysrhythmias are generally considered to be less frequent, or at least less severe, with enflurane than with halothane (869,870). However, caution in the use of adrenaline is advisable, especially in patients with cardiac disease or hyperthyroidism. Isorhythmic atrioventricular dissociation was seen in 16 of 105 patients after the use of 1.0–1.5% enflurane (871).

Enoximone

The cardiac effects of enoximone include hypotension, transient atrial fibrillation, and bradycardia (872). Ventricular tachydysrhythmias have been reported in about 4% of patients (873) and myocardial ischemia can also occur (874,875).

Enprofylline

The most severe adverse effects of theophylline, namely seizures, are claimed to be absent even at high doses of enprofylline, although there are other disadvantages, such as cardiovascular effects, nausea, vomiting, and headache. Much more information is needed before claims of greater safety can be taken seriously.

The possibility of obtaining a steady-state plasma enprofylline concentration of 5 μg/ml by two constant rate infusions was examined in six asthmatic patients (876). The resulting adverse effects and bronchodilatation were compared with those obtained with theophylline at a steady-state concentration of 15 μg/ml. Headache, nausea, and vomiting became pronounced in two patients in whom the plasma enprofylline concentration was about 6 μg/ml. The authors concluded that by varying the infusion rate the plasma concentration of enprofylline can be controlled like that of theophylline, but they stressed the need for further studies of efficacy versus adverse effects.

Ephedra, ephedrine, and pseudoephedrine

Cardiologists and pharmacologists from Boston have reviewed cases of possible cardiovascular toxicity associated with the use of Ma huang, which contains ephedrine. Among over 900 cases of possible toxicity reported to the Food and Drug Administration, they identified 37 patients (23 women) in whom stroke ($n = 16$), myocardial infarction ($n = 10$), and sudden death ($n = 11$) appeared to be temporally related to consumption of Ma huang. The authors noted that in all but one of these cases the doses used were within the range recommended by the manufacturers. In the seven patients on whom autopsies were performed, coronary artery disease was found in three and cardiomyopathy in another three, with features suggesting sympathomimetic toxicity. The authors emphasized the toxicity of Ma huang in relatively young individuals—the mean age of the 37 patients was about 43 years—even though the relevant pathogenic mechanisms remain to be fully defined.

Hypertension

The cardiovascular effects, subjective effects, and abuse potential of single intranasal doses of ephedrine 5 and 10 mg have been compared with oral doses of (−)ephedrine 50 mg in 16 healthy Caucasian men with no drug/alcohol/nicotine abuse or dependence (877). Intranasal ephedrine caused an increase in blood pressure but associated orthostatic hypotension.

Pseudoephedrine is a component of some non-prescription basal decongestants given by mouth; even quite ordinary doses of such products (for example 120 mg) can cause a hypertensive reaction in sensitive subjects, such as those with a pheochromocytoma and those with at least a family history of hypertension (SEDA-17, 162).

- A 21-year-old man presented with hypertension (blood pressure 220/110 mmHg) and ventricular dysrhythmias after taking four capsules of herbal ecstasy (878). He was treated with lidocaine and sodium nitroprusside and his symptoms resolved within 9 hours.

Severe hypertension has been attributed to pseudoephedrine abuse (879).

- A 36-year-old man with hypertension taking no less than seven antihypertensive drugs had outpatient systolic pressures of over 190 mmHg. Investigations for primary causes of hypertension were negative and there was increasing suspicion of treatment non-compliance or factitious hypertension. Urine screening showed the presence of pseudoephedrine, which the patient could not explain. When he was given his normal antihypertensive drugs under close supervision his systolic blood pressure fell to 70 mmHg and his serum creatinine doubled. His blood pressure became normal when his medication was briefly suspended but he continued to deny any deliberate attempt to alter his blood pressure and discharged himself soon afterwards.

The authors concluded that this represented factitious hypertension due to pseudoephedrine, the first such case reported and a very unusual example of Munchausen's syndrome.

Cardiac dysrhythmias

Cardiac dysrhythmias have been attributed to ephedrine both in therapeutic doses (880) and when used as a drug of abuse (878).

- A 25-year-old woman became hypotensive after the administration of epidural anesthesia for an elective cesarean section. She was given intravenous ephedrine 9 mg, after which she complained of nausea. For an unexplained reason this was taken as an indication to give a further 9 mg of ephedrine. She immediately developed sinus tachycardia with atrial and multifocal ventricular extra beats, followed by short runs of ventricular tachycardia. She remained asymptomatic and recovered after about 5 minutes.

One must have some concerns about the high dose used here for a relatively modest level of hypotension with a systolic blood pressure of 100 mmHg.

- A 21-year-old Canadian man presented with severe headache, nausea and vomiting, hypertension (220/120 mmHg), and a sinus tachycardia of about 120/minute with frequent multifocal ventricular extra beats. He was treated with intravenous sodium nitroprusside and lidocaine and recovered fully within 24 hours.

The authors concluded that ephedrine was very largely responsible for the cardiovascular toxicity.

Metabolife 356, a dietary supplement that contains *Ephedra* and caffeine, increased the mean maximal QT_c interval and systolic blood pressure in a double-blind, crossover, randomized, placebo-controlled study in 15 young healthy volunteers with normal BMI (881). Those who took Metabolife were more likely to have a shortening of the QT_c interval by at least 30 milliseconds compared with placebo (RR 2.67; CI=1.40, 5.10). All those who took Metabolife reported non-specific symptoms (882). There were no adverse effects while patients were taking placebo. One woman developed a sinus tachycardia of 120/minute with palpitation, 1 hour after taking Metabolife. A hand tremor developed in one subject and subsided after 5 hours.

In a meta-analysis of the effects of *Ephedra* or ephedrine- containing products compared with control, the odds ratio of palpitation was 2.29 (95% CI = 1.27, 4.32) (883).

Myocardial ischemia

Myocardial infarction has been attributed to *Ephedra* (884).

- A previously healthy 19-year-old man took tablets containing a total of 24 mg of Ephedra alkaloids and 100 mg of caffeine, and 15 minutes later developed severe chest pain radiating down the left arm. An electrocardiogram showed an inferolateral myocardial infarct, confirmed by creatine kinase and troponin I measurements. He made a full recovery, and coronary angiography showed only minimal atherosclerotic disease of the left anterior descending artery.

The authors emphasized the dangers of *Ephedra*-containing over-the-counter formulations, even in fit young people.

Ephedrine is sometimes used to treat vasovagal episodes and has been reported to cause coronary artery spasm and myocardial infarction in these circumstances (885).

- Two apparently healthy women, aged 26 and 34 years, who were given spinal anesthesia for pelvic or hip surgery, both developed hypotension and bradycardia and were given intravenous ephedrine in divided doses, in one case a total of 20 mg and in the other 30 mg. In both cases this resulted in ventricular tachycardia and raised troponin I concentrations and creatine kinase activity, but Q waves did not follow. Coronary angiography showed normal arteries in both cases and the women recovered, apparently fully.

The authors rightly pointed out that atropine is a safer and more appropriate intervention in circumstances such as these.

Arteritis

Arteritis has been attributed to ephedrine (886).

- A 44-year-old French woman was given intravenous ephedrine during anesthesia and developed hypertension (systolic pressure 180 mmHg) with nausea, vomiting, and progressive drowsiness. She had a history of migraine, antiphospholipid antibodies, mitral valve prolapse, and atrial fibrillation treated with propranolol. Immediately after anesthesia she developed headache, vomiting, and drowsiness. CT scanning showed frontal and parietal hemorrhagic infarcts, and cerebral angiography showed multiple segments of constriction and dilatation consistent with arteritis. She was left with residual visual and memory deficits.

This is the first case of this kind associated with ephedrine, and it seems plausible that the patient's previous history indicated a genetic predisposition to vascular disease.

Others

A form of toxic shock syndrome on more than one occasion occurred in one patient who used pseudoephedrine (SEDA-18, 158).

Two young men took *Ephedra* supplements and developed severe cardiomyopathies and global cardiac hypokinesis (887). Both were treated with standard treatments for heart failure but one died nevertheless.

Ephedra and coronary dissection have been linked (888).

- A 50-year-old African American woman developed a myocardial infarction 2 days after taking a supplement containing *Ephedra* standardized to ephedrine 20 mg/day. Subsequent investigations showed a dissection in the mid-distal segment of the left anterior descending artery and a thrombosis occluding a large first obtuse marginal branch. She had bypass surgery, but later developed refractory heart failure.

Ergot derivatives

Ergot derivatives are powerful vasoconstrictors. The extremities become pale and cold, and arterial spasm in the arms and legs has been demonstrated; even the face can be affected. The condition can develop acutely even after brief use of the drug, and there is real risk of gangrene; if given early, intra-arterial infusion of prostaglandin E_1 can reverse the spasms. Protracted coronary spasm has also been reported in some cases (SEDA-14, 122). Renal arterial spasm has occurred after a dose of ergotamine of 10 mg in the form of suppositories given over 60 hours (SED-8, 308; 889), and bilateral papillitis with ischemia of the periaxial fibers has resulted from 2 weeks of maximum-dose treatment. High doses have also led on occasion to mesenteric vascular constriction, ischemic bowel disease, and partial necrosis of the tongue. Arterial stenosis can even result in aneurysm formation (890). The absence of symptoms does not mean that there is no adverse effect; in 30 patients who had taken ergotamine 1–5 mg/day for a year all had lowered systolic blood pressures in the foot.

Arterial spasm induced by ergotamine can be due to overdose, but some patients have an exaggerated response to a therapeutic dose. Interaction with other drugs, leading to potentiation, is a third mechanism. Four patients developed extreme limb ischemia after a therapeutic dose of ergotamine: one even took only a single dose of 1 mg. The symptoms responded to vasodilator therapy after withdrawal of ergotamine, but one needed a minor amputation. All patients were taking antiviral treatment for HIV infection, and ritonavir and indinavir may have potentiated the effect of ergotamine by inhibiting its metabolism (891–894).

Treatment of migraine can lead to subclinical ergotism for a prolonged period, and thence to occlusive peripheral vascular disease; peripheral systolic pressure (and liver function tests) should be monitored in patients taking regular ergotamine (SEDA-15, 135). Tolerance to these vasoconstrictive effects varies widely among individuals; such symptoms as cyanosis of the limbs, syncope, hypotension, and paresthesia have been seen in sensitive subjects after doses of up to 8 mg taken over as little as 10 days. The Swedish drug authorities have recommended that treatment should not be continued for more than 7 days (SEDA-13, 113).

- A 78-year-old woman developed gangrene in three fingertips on the right hand and two on the left after being given dihydroergotamine 10 mg/day as migraine prophylaxis (895). She had had Raynaud's syndrome for at least the previous 5 years and was thought to have a relatively mild form of systemic sclerosis.

- A 44-year-old woman developed claudication and rest pain after gross overuse of ergotamine 100 mg suppositories (up to 6 times a day) for chronic headaches over a period of several years (896). Angiography showed occlusion of both femoral arteries. Intra-arterial prostaglandin E (presumably E_1) followed by chemical sympathectomy normalized the circulation in both the legs.
- A 63-year-old Canadian woman who had been taking ergotamine (dosage unspecified) for migraine for 20 years developed acute ischemia of the right arm, with no palpable pulses below the axilla (897). Angiography showed multiple filling defects in upper limb arteries, partially reversible with intravenous phentolamine. She was given prazosin 2 mg/day and became pain-free after 5 days. Treatment was continued for 3 months and during this time attacks of migraine were treated with sumatriptan without adverse effects. Her pulses were entirely normal after 3 months, as was angiography.

Myocardial infarction has been reported.

- A 27-year-old woman with familial hypercholesterolemia already treated with lipid-lowering drugs developed acute chest pain after a prophylactic intramuscular injection of 0.5 mg ergometrine, given during the late stages of labor (998). Angiography showed three-vessel atherosclerotic disease and occlusion of the left anterior descending coronary artery. Angioplasty with stenting was successful and she made an excellent recovery.
- A 34-year-old woman had a myocardial infarction after being given ergonovine for an atonic uterus after cesarean section (999). Within minutes she became unresponsive, with bradycardia and then asystole followed by ventricular fibrillation during cardiopulmonary resuscitation. An electrocardiogram showed an acute anterior infarct and coronary angiography showed diffuse spasm of the circumflex and left anterior descending arteries, with subtotal occlusion of the latter. The spasm was reversed with intracoronary glyceryl trinitrate but she required ventilation for another 2 days and was eventually discharged 11 days after the infarct, with a borderline left ventricular ejection fraction of 45%.

The authors commented that the latter patient was of Asian origin and that such individuals are thought to have increased susceptibility to the vasoconstrictor effects of ergot derivatives.

The role of arterial spasm as a cause of angina after angioplasty has been studied in two men, aged 45 and 58 years, who had emergency angioplasties for acute coronary thrombosis (900). Although the primary procedures were successful in both cases, ischemic chest pain returned after 4 and 6 months respectively. Perhaps a little surprisingly, in both cases intravenous ergonovine (0.4 mg) was given during coronary angiography, causing severe arterial spasm, which resolved with intravenous isosorbide dinitrate. There was no evidence of restenosis. The authors noted that recurrence of angina does not necessarily imply restenosis, although it is not clear how many cardiologists will repeat this procedure with their own patients.

- A 48-year-old woman developed a cold pulseless right leg and no measurable blood pressure at the right ankle (901). She was a migraine sufferer and had been taking over-the-counter medications, some of which contained ergot derivatives, although the nature and quantity were not specified. Arteriography showed severe stenosis of the superficial femoral artery, with no identifiable tibial vessels. There was an initial improvement with intra-arterial glyceryl trinitrate infusion, and sustained normalization of the circulation in the leg after administration of sodium nitroprusside, nifedipine, prazosin, and heparin. She made a full recovery.

The authors reviewed the pharmacology of the ergot alkaloids and the acute and subclinical ischemic syndromes that they can produce. They pointed out that in some countries, notably in Latin America, ergot-containing formulations are freely available without a prescription.

Acute hypertensive encephalopathy has occurred in a patient given methylergotamine (SEDA-3, 121).

Diverticulum formation in the internal carotid artery occurred in a patient who had taken ergotamine for 4 years (SEDA-10, 119). Rupture of a splenic artery aneurysm in a 46-year-old woman was ascribed to excessive ergotamine ingestion for migraine (902). The authors thought that significant vasospasm may have led to damage and weakening of the vessel wall and consequently to the development of a false aneurysm. It should be noted that splanchnic aneurysms are occasionally discovered in otherwise healthy people and carry a 5% risk of spontaneous rupture.

Long-term abuse of ergotamine can occasionally cause fibrosis of the cardiac valves (SEDA-18, 160).

Erythromycin

Erythromycin has antidysrhythmic properties similar to those of Class IA antidysrhythmic drugs, and causes an increase in atrial and ventricular refractory periods. This is only likely to be a problem in patients with heart disease or in those who are receiving drugs that delay ventricular repolarization (903). High-doses intravenously have caused ventricular fibrillation and torsade de pointes (904). Each episode of dysrhythmia, QT interval prolongation, and myocardial dysfunction occurred 1–1.5 hours after erythromycin infusion and resolved after withdrawal.

In an FDA database analysis, 346 cases of cardiac dysrhythmias associated with erythromycin were identified. There was a preponderance of women, as there was among those with life-threatening ventricular dysrhythmias and deaths after intravenous erythromycin lactobionate. A sex difference in cardiac repolarization response to erythromycin is a potential contributing factor, since in an in vitro experiment on rabbit hearts, erythromycin caused significantly greater QT prolongation in female than in male hearts (905).

In 35 women and 28 men erythromycin caused QT interval prolongation after the first few doses of erythromycin (906). Similarly, in a prospective, comparative study in 19 patients with uncomplicated community-acquired pneumonia, a single dose of intravenous erythromycin 500 mg increased the heart rate and prolonged the QT interval. These effects were seen after 15 minutes of infusion and disappeared 5 minutes after the infusion had been stopped (907).

Owing to prolongation of the QT interval, a newborn with congenital AV block developed ventricular extra beats and non-sustained ventricular tachycardia after intravenous erythromycin; the QT interval normalized after withdrawal (908).

- Intravenous erythromycin (1 g 6-hourly by intravenous infusion over 30 minutes) resulted in QT interval prolongation, ventricular fibrillation, and torsade de pointes in a 32-year-old woman (904).

Intravenous administration of erythromycin into peripheral veins relatively commonly causes thrombophlebitis, although the lactobionate form of erythromycin may be less irritating to veins than other parenteral forms (909,910). In a prospective study of 550 patients with 1386 peripheral venous catheters, the incidence of phlebitis was 19% with antibiotics and 8.8% without; erythromycin was associated with an increased risk (911).

Erythropoietin, epoetin alfa, epoetin beta, epoetin gamma, and darbepoetin

In a randomized study of 180 patients with anemia due to hormone-refractory prostate cancer, who were treated with epoetin beta 1000 IU or 5000 IU subcutaneously 3 times a week for 12 weeks, cardiovascular events were more frequent with the higher dosage. Four patients had deep vein thrombosis and two had myocardial infarctions; all were taking the higher dosage. However, only one of the patients with deep vein thrombosis had a high hemoglobin concentration (912).

Hypertension

Increases in blood pressure are not uncommon after treatment with epoetin or darbepoetin alfa and often require antihypertensive drugs for correction (913,914,915). It causes or aggravates hypertension in about 20–35% of dialysis patients (916–920). It can be accompanied by encephalopathy or seizures (921). In 44 children with chronic renal insufficiency treated with epoetin 150 U/kg/week, hypertension was mostly observed in patients on hemodialysis (66%) compared with peritoneal dialysis (33%) and predialysis patients (16%) (922).

In 341 dialysis patients treated with darbepoetin alfa for anemia, there was one death due to a stroke in an 87-year-old man with a history of cardiac disease (923). This event was judged as being possibly related to darbepoetin alfa, as the hemoglobin concentration was increased at the time of death.

Several factors contribute to the development of hypertension. One is the loss of the hypoxic vasodilatory response, leading to an increase in peripheral vascular resistance (924), but more important is the rise in blood viscosity, which increases with the hematocrit in both normotensive and hypertensive individuals (925). It is still being debated whether hypertension occurs only in patients with pre-existing hypertension or in normotensive patients as well, but about 30% of all patients require increased or de novo antihypertensive therapy as they respond to erythropoietin treatment (926).

Hypertension after epoetin has mostly been seen in uremic patients (927). However, in two of 44 cancer patients treated with epoetin during cisplatin-containing chemotherapy, epoetin was withdrawn owing to hypertension (diastolic pressure over 100 mmHg) (928).

Hypertensive encephalopathy can arise in connection with the sudden and extreme rises in blood pressure that occur in some patients given epoetin (929,930).

- A 14-year-old child developed hypertensive encephalopathy, a known rare adverse effect of erythropoietin, after 2 months (931).

During an open, uncontrolled study in 22 patients with end-stage renal disease treated with epoetin omega there was one case of hypertensive encephalopathy (932).

Susceptibility factors

It is not clear why certain patients develop hypertension and hypertensive encephalopathy and others do not, but transfusion-dependent anemic patients with a low hematocrit (<20%) are particularly susceptible, as are those with previous hypertension and seizures. Careful control of blood pressure at the start of epoetin treatment and the use of low doses are therefore advised in patients at high risk.

If epoetin is given preoperatively without autologous predonation, there is an increased risk of hypertension, an increased risk of graft thrombosis and myocardial infarction in cardiac surgery, and an increased risk of venous thromboembolism in orthopedic patients (933).

Because of an increase in cardiac-related deaths, it has been recommended that in patients with congestive heart failure or ischemic heart disease the hematocrit should not be raised above 42% (934).

Erythropoietin in patients with cancer was associated with a risk of hypertension (over 19% higher) and of thromboembolic events (over 58% more common) compared with controls, but the increases were not significant (935,936). Likewise, in patients with chronic renal anemia undergoing peritoneal dialysis, one of the most common adverse event was hypertension (937).

Mechanisms

Several mechanisms have been postulated, such as increased blood viscosity, attenuation of hypoxic vasodilatation, direct vasoconstriction, activation of neurohumoral systems, including catecholamines, activation of the renin-angiotensin-aldosterone system, or altered concentrations of vasoactive substances, such as endothelin and nitric oxide. Involvement of the kidneys was excluded in a reported case of erythropoietin-induced hypertension in a nephrectomized patient (938). However in other cases of hypertension the kidneys may play a role.

One of the possible mechanisms of epoetin-induced hypertension is an imbalance of local endothelial factors, such as endothelium-derived relaxation factor and endothelin (921), a potent vasoactive peptide produced by endothelial cells (939), increased concentrations of which have been observed in adults with an increase in mean blood pressure of more than 10 mmHg. In contrast, in preterm infants receiving epoetin there were no acute effects of epoetin on endothelin-1 concentrations or mean blood pressure (939). When intravenous and subcutaneous epoetin were compared, only intravenous epoetin was accompanied by hypertension (917,921). Plasma concentrations of proendothelin-1 and endothelin-1 increased after infusion of epoetin only in patients with hypertension. In addition the molar ratio of endothelin-1 to proendothelin-1 was significantly higher in patients with hypertension than in patients without (922). The authors suggested that endothelin-1 converting enzyme may play a role in the pathogenesis of epoetin-induced hypertension (921). Studies in rats have confirmed that epoetin stimulates endothelin release and synthesis of vascular tissue (940). There is a relation between the dose of epoetin and postdialysis blood pressure, but not predialysis blood pressure (917). Risk factors for the development or worsening of hypertension are pre-existing hypertension, the presence of native kidney, a rapid increase in hematocrit, a low baseline hematocrit, high dosages of epoetin, and intravenous administration (917).

Epoetin has vascular effects that can cause an imbalance between vasoconstrictor-proproliferative-proatherogenic factors (angiotensin, endothelin, thromboxane) and vasodilator-antiproliferative-antiatherogenic factors (nitric oxide, prostacyclin). These changes may be related to the occurrence or aggravation of pre-existing hypertension in humans and can cause vascular hypertrophy and potentially accelerate the development of atherosclerosis (927).

Adrenomedullin, an endocrine peptide with vasodilatory and natriuretic actions, is increased in patients with hypertension and chronic renal insufficiency. In 54 patients with renal anemia, treated with epoetin 6000 IU once a week, there was a correlation between the progression of renal disease and circulating adrenomedullin; however, there was no relation between adrenomedullin and epoetin-induced hypertension (941).

The induction of hypertension by epoetin has been attributed to increased blood viscosity after a rapid increase in hematocrit and a loss of hypoxia-induced vasodilatation (919). Another pathogenic mechanism for the induction of hypertension by epoetin is enhanced endothelin production by endothelial cells (919).

In 10 normotensive hemodialysis patients with severe anemia treated with epoetin there was increased adrenergic activity (942).

Epoetin-induced hypertension may be associated with an angiotensinogen gene polymorphism; the incidence of hypertension was increased in patients carrying a homozygous T on position 235 of the angiotensinogen gene (943). The authors speculated that erythropoietin causes a rise in blood pressure via the T allele, which influences components of the renin–angiotensin system, thereby stimulating production of angiotensinogen, which leads to hypertension.

Management

Blood pressure should be monitored during treatment (944).

Erythropoietin-induced hypertension can easily be treated by initiating or increasing antihypertensive medication (945). For example, two of 26 pregnant women treated with epoetin for iron deficiency anemia had worsening of mild hypertension, which was managed effectively with methyldopa 250 mg tds (946).

Antiplatelet therapy reduces the incidence of epoetin-induced hypertension in predialysis patients (920).

Vascular disease

There is an increased incidence of peripheral vascular diseases in diabetic patients who receive peritoneal dialysis and epoetin (947). In these patients the time to a first vascular incident is shorter, the number of vascular events is increased, and more hospital days associated with vascular disease have been reported compared with patients receiving peritoneal dialysis without epoetin (947). Significant risk factors for the development of peripheral vascular disease are epoetin therapy, epoetin dose, and smoking (947). Peripheral vascular disease may be related to increased blood viscosity or other changes in blood rheology (947).

Thromboembolism

Thrombotic events related to epoetin include vascular access thrombosis, renal and temporal vein thrombosis, transient ischemic attacks, and myocardial infarction (948).

Vascular access thrombosis has been reported in up to 26% of patients treated with epoetin alfa (949,950,951). Most of the failures occurred in polytetrafluoroethylene grafts. There was no comparison with patients not treated with epoetin. It has been suggested that the increased risk of extracorporeal circuit clotting and the higher heparin requirements during hemodialysis may not be due to a hypercoagulable state, but rather to an increase in erythrocyte mass and consequently in whole blood viscosity (948).

- A patient with end-stage renal disease treated with epoetin-alfa developed a dural sinus thrombosis (952).

It was postulated that it was caused by polycythemia, because the hematocrit more than doubled in under 2 months, reaching 0.55 (952).

In 312 patients with breast, lung, or gynecological malignancies and chemotherapy-induced anemia who were treated with 2-weekly darbepoetin or weekly epoetin alfa in a randomized study, there were two cases of deep venous thrombosis, one in each treatment group, and one case of pulmonary embolism in those who used epoetin alfa (953).

There was a high incidence of deep vein thrombosis, related to the use of recombinant human erythropoietin, in seven of 53 anemic women with local advanced cervical cancer (954).

A randomized, controlled trial in 27 patients with metastatic breast cancer and mild anemia was terminated because of four thrombotic events in 14 patients who were treated with erythropoietin compared with no events in 13 patients who were not treated with erythropoietin (955).

In 1265 hemodialysis patients with cardiac disease randomized to erythropoietin to maintain a hematocrit of 35% or 42%, there was a higher mortality (35%) in the higher hematocrit group compared with 29% in the lower hematocrit group, probably because of a higher incidence of thrombotic events (956).

A meta-analysis of 12 randomized controlled studies showed a 1.55 times increased risk of thromboembolic events and a 1.25 times increased risk of hypertension during recombinant erythropoietin treatment in anemic patients with cancers (957).

Estrogens

See also Hormonal contraceptives (various), Hormone replacement therapy

Estrogens have both wanted and unwanted effects on the cardiovascular system, depending on the manner in which they are used. Hormone replacement therapy is used in the hope of reducing the risk of ischemic heart disease after the menopause. The reduction in risk may be as much as 50% and is attributed variously to vasodilatation mediated by the endothelial production of prostaglandin I_2 (prostacyclin), effects on coagulation factors and endothelial function, and improvements in serum lipids (increased concentrations of HDL cholesterol and reduced concentrations of LDL and total cholesterol) (958), but variable effects on triglycerides. However, estrogens (especially as used in contraception but also postmenopausally) can have a marked effect on clotting factors and renin substrate, increasing the risk of thromboembolism.

The Coronary Drug Project in men taking different doses of estrogens showed a dose-related increase in myocardial infarction and thromboembolic diseases (959).

Because any possible effect of estrogens, favorable or unfavorable, on atherosclerosis is difficult to detect without very long-term experience, some workers have sought to use indirect measures that might be relevant. A Greek group set out to determine in a randomized, double-blind study the effect of hormonal or antihormonal therapy on serum VE-cadherin in 28 healthy postmenopausal women, who received either 17-beta-estradiol (2 mg/day) with norethisterone acetate (1 mg/day) or alternatively raloxifene HCl alone (60 mg/day) for 6 months (960). Serum VE-cadherin, which was estimated at baseline and at month 6, fell significantly in both groups. These findings suggest that these drugs may preserve interendothelial junction integrity and control vascular permeability. Although this effect may favorably influence the progress of an atheromatous lesion, its clinical impact, for example on coronary artery disease, remains uncertain.

Etamsylate

Modest but transient hypotension has occasionally been observed after intravenous injection, and elderly subjects appear to be more susceptible (961).

Ethanol

Alcohol is one of many drugs that cause or aggravate systemic hypertension. Acute alcohol exposure has an inconsistent effect on blood pressure, but cross-sectional population studies have shown a relation between chronic alcohol consumption and blood pressure, and the prevalence of hypertension up to three times higher in heavy drinkers (962). Although the mechanism of hypertension caused by chronic alcohol consumption is not known, it is suspected that it is partly related to repeated episodes of acute withdrawal, causing increased sympathoadrenomedullary activity, an increase in plasma renin activity, and increased ACTH secretion, which may be sufficient to have a mineralocorticoid effect (963).

Etherified starches

In a comprehensive comparison of the pharmacokinetics and pharmacodynamics of dextran and etherified starch (964), the effects of etherified starch on the cardiovascular system have been delineated. The mean arterial pressure, central venous pressure, wedge pressure, cardiac index, left ventricular stroke work index, and stroke output all rise, whereas the pulmonary vascular resistance falls. Oxygen availability to the tissues is improved. The effects of etherified starch on blood viscosity and erythrocyte aggregation, in particular, are more pronounced than with dextran.

Etomidate

The cardiorespiratory tolerance of etomidate is usually excellent (965), but cardiovascular instability has been described after a bolus dose (966).

Eye-drops and ointments

In the very young and the very old patients atropine eye-drops carry a risk of cardiovascular collapse and neuropsychiatric disturbances (SEDA-16, 543).

Systemic problems can also develop with antimicrobial eye-drops (967) and contact lens products (968,969).

Fabaceae

In healthy volunteers who took licorice corresponding to glycyrrhizinic acid 75–540 mg/day for periods of 2–4 weeks, there was an average increase in systolic blood pressure of 3.1–14.4 mmHg (970). The increase in blood pressure was dose-related and the authors concluded that as little as 50 g/day of licorice for 2 weeks would have caused a significant rise in blood pressure.

Factor VII

Thrombophlebitis at the infusion site is a common complication of continuous infusion of various clotting factor concentrates and has been noted after infusion of factor VIIa (971,972). Thrombophlebitis occurred in one of eight hemophiliacs with inhibitors who received continuous infusion of recombinant factor VIIa to allow elective surgery (973). In 25 hemophilia patients with inhibitors, who received recombinant factor VIIa for surgical procedures or spontaneous bleeding, there was one case of thrombophlebitis in 35 continuous infusion courses (974). In most instances, thrombophlebitis can be prevented by parallel infusion of saline or heparin.

- A 38-year-old patient with hemophilia A with factor VIII inhibitors was treated with recombinant factor VIIa for about 1 month and 18 days after the last infusion developed a distal deep venous thrombosis. An effect of the factor VIIa could not be ruled out, but long-term immobilization and severe infection could have contributed (975).

Angina pectoris and tachycardia have been reported after the use of recombinant factor VIIa (976). Among patients who had more than 2400 treatment episodes with recombinant factor VIIa, there were two cases of acute myocardial infarction (977). One of these patients had a history of cardiovascular disease and the other was very overweight and received massive transfusions, human factor VIII, activated prothrombin complex, and finally recombinant factor VIIa to treat severe intra-abdominal bleeding and shock.

Factor VIII

Thromboembolic complications developed in two of 81 patients with Von Willebrand disease treated with a high-purity factor VIII/Von Willebrand factor (VWF) concentrate (978). Four cases of venous thrombosis were reported in patients with Von Willebrand disease treated with an intermediate-purity factor VIII/VWF concentrate. Use of pure VWF concentrate without increased factor VIII:C is preferable (979).

Factor IX

Thrombophlebitis at the infusion site is a common complication of continuous infusion of various clotting factor concentrates. Continuous infusion of recombinant factor IX in six patients with hemophilia B undergoing surgery or suffering bleeding was complicated in two cases by thrombophlebitis (980). In most instances, thrombophlebitis can be prevented by parallel infusion of saline or heparin.

Fazadinium

Cardiovascular effects account for the relative unpopularity of fazadinium. It has some ganglion-blocking activity (981) and blocks cardiac muscarinic receptors in the therapeutic dose range (982). Its vagolytic potency is about the same as that of gallamine. Fazadinium, like pancuronium, also blocks the reuptake of noradrenaline into sympathetic nerve endings. These actions explain its major cardiac adverse effect, namely significant tachycardia (983), which occurs even with small doses and is persistent. It is dose-related (984), the increase in heart rate varying between 30 and 100%, and is associated with a rise in cardiac output and falls in stroke volume and peripheral resistance. Hypertension or hypotension can occur (985,986). If fazadinium is used injudiciously, extreme and dangerous cardiovascular changes can ensue (987).

Fenfluramines

Spontaneous rupture of a retroperitoneal aneurysm occurred in a 70-year-old woman who had been taking phentermine hydrochloride, 30 mg/day, for about 1 month (988). Other long-term medications included fluoxetine and amitriptyline, and she had no history of coronary artery disease, hypertension, diabetes, or complications of pregnancy. Although it is plausible that phentermine could have contributed to the ruptured aneurysm, other possibilities should be considered, particularly rupture of an anomalous retroperitoneal blood vessel.

Cardiomyopathy

Restrictive cardiomyopathy due to endocardial fibrosis occurred in a 35-year-old woman 5 months after she had started to take fenfluramine 10 mg tds and phentermine 15 mg/day (989). The endocardial findings strongly resembled the valvular lesions associated with the use of fenfluramine–phentermine. Endocardial and valvular fibrosis associated with anorectic drugs is strikingly similar to the plaque material found in patients with carcinoid syndrome and those exposed to methysergide, and all possibly arise from a common mechanism.

Valvulopathy

The fenfluramines and phentermine can cause valvular heart disease (990–992), and this has been reviewed (993). Fenfluramine was voluntarily withdrawn by the manufacturers on 15 September 1997, and the US Department of Health and Human Services issued interim recommendations for people previously exposed to fenfluramine or dexfenfluramine with cardiac valvulopathies (SEDA-22, 3).

The use of fenfluramine or dexfenfluramine alone or in combination with phentermine, in 2524 adult participants in the population-based Hypertension Genetic Epidemiology Network Study, was associated with aortic regurgitation independent of aortic dilatation or fibrocalcification (994). The association between the use of fenfluramine or dexfenfluramine (alone or with phentermine) and aortic regurgitation adjusted for potential confounders was analysed. Nineteen participants, all of whom had hypertension, were being treated with fenfluramine or dexfenfluramine (5 on these agents alone, 14 also with phentermine). Aortic regurgitation was present in 32% ($n = 6$) of those taking fenfluramine/dexfenfluramine versus 6% (162/2505) of the remaining subjects. In multivariate analyses, after adjusting for important confounders, in particular aortic root structure, treatment with fenfluramine or dexfenfluramine was associated with aortic regurgitation (OR, 5.2; 95% CI, 1.7–14) and fibrocalcification (OR, 5.2; 95% CI, 1.9–15).

The autopsy findings in the heart and lungs of a patient with pulmonary hypertension associated with fenfluramine and phentermine have been described (995).

- A 36-year-old woman with a body mass index of 47.5 kg/m^2, took fen–phen for 7 months and developed pulmonary hypertension. Her pulmonary arterial pressure was 56 mmHg and echocardiography showed right ventricular dilatation and hypokinesia. She had a cardiopulmonary arrest duringheart catheterization and died 3 days later. At autopsy, there was right ventricular dilatation with a fibroproliferative tricuspid valve. The pulmonary arteries had fibroproliferative plaques which were more severe and prominent in the upper lobes than in the lower lobes.

More autopsy cases of patients with a history of fen–phen use are warranted to document the frequency of combined cardiac valvular disease and pulmonary hypertension.

Progressive pulmonary hypertension occurred in two patients who took fenfluramine for only 8 months (SEDA-6, 9). The symptoms abated on withdrawal but returned in one patient when rechallenged.

In 1996, in a case-control study, 95 patients from 35 centers in France, Belgium, the UK, and the Netherlands were compared with 355 age- and sex-matched controls (996). The use of anorexic drugs (mainly derivatives of fenfluramine) was associated with an increased risk of primary pulmonary hypertension. Association with recognized risk factors such as a family history of primary pulmonary hypertension, infection with HIV, or the use of intravenous drugs was also confirmed. The absolute risk for obese patients who took anorexic agents for more than 3 months was 30 times higher than in non-users.

Echocardiography with color Doppler in 22 patients aged 25–69 years (19 women and three men) who had taken fen–phen for more than 3 months showed that one patient with newly discovered aortic insufficiency was asymptomatic. Some were taking several other drugs, none of which is known to precipitate valvular heart disease. Echocardiography was normal in 12 cases and abnormal in 10 including significant aortic insufficiency and significant mitral regurgitation. Ten of the patients had significant aortic insufficiency and nine had at least mild mitral insufficiency. The author inferred that fenfluramine was the likely offending agent, because (a) while it is known to cause release of serotonin, phentermine does not; (b) carcinoid tumors, which secrete serotonin and ergotamine, a serotonergic drug, are known to cause valvular heart disease; *and* (c) none of the obese patients who took phentermine and fluoxetine for more than 2 years developed pulmonary hypertension (997) and the author found no valvular heart disease in this cohort either (998). The recommendation that phentermine should be combined with fluoxetine, sertraline, or fluvoxamine as safer alternatives (998) requires prospective studies.

It was suggested that in patients who met the FDA criteria for cardiac valvular abnormalities on echocardiography performed soon after the withdrawal of appetite suppressants, there was a possibility (ranging from as low as 5% to as high as 67%) that the abnormality was a naturally occurring phenomenon and not a consequence of drug use (999). However, various studies have supported earlier reports of an association between fenfluramine or its d-isomer and cardiac valvular regurgitation, although they have differed with regard to the strength and clinical significance of the association (999). Differences in design, including a lack of baseline cardiac evaluation in echocardiographic assessment (991,992,1000), have precluded comparisons. Additional evidence linking the use of fenfluramine or dexfenfluramine to cardiac valvular regurgitation has reaffirmed the wisdom of the FDA's decision to withdraw them from the market.

Why was this type of valvulopathy not recognized sooner? Changes in medical practice seem to have played a role as long-term and widespread use of these drugs evolved in the 1990s. Furthermore, cardiac murmurs can be more difficult to detect in obese patients (1001).

Prevalence

Although initially a prevalence of up to 30% was estimated, subsequent reports have suggested much lower rates. There are several reasons for this disparity including uncontrolled data and the limitations of echocardiography (1002–1004). The FDA surveys and the University of Minnesota study reported point prevalences and inherently overestimated the association of appetite suppressants with valvulopathy because a certain percentage of patients have pre-existing valvular lesions. The method of detection also plays a crucial role. Echocardiography is far more sensitive than clinical examination in detecting valvular regurgitation. The issue may also be confounded by lesion regression after withdrawal of therapy (1005). Moreover, case-control studies are no substitute for objective evidence of the status of cardiac valves before drug exposure (990). Also the duration of exposure has varied widely in different reports.

Pulmonary hypertension and valvular heart disease associated with anorexigens have been described predominantly in women, which raises important questions about biological and psychological risk factors and ethical practice. Do women respond differently to these drugs because of genetic or physiological factors or are these drugs being prescribed almost exclusively for women? Was it realistic for the regulatory authorities to believe that these drugs would be used only to treat morbid obesity? Most important, what view of the benefit to harm balance of using anorexigenic drugs has allowed women and their physicians to justify the use of potentially lethal drugs to deal with concerns about body image and weight? These questions (1006) are pertinent to the current scenario of appetite-suppressant drug-related concerns.

Studies in which baseline echocardiography was carried out before the drug was used showed that the risk of new or prospective valvular heart disease was much lower than implied by previous prevalence studies (990). In 46 patients who used fenfluramine or dexfenfluramine for 14 days or more, the primary outcome was new or worsening valvulopathy, defined as progression of either aortic regurgitation or mitral regurgitation by at least one degree of severity and disease that met FDA criteria. Two patients taking fen–phen developed valvular heart disease. One had mild aortic regurgitation that progressed to moderate regurgitation and the second developed new moderate aortic insufficiency. The authors argued that the referral bias in their study, which required an echocardiogram for inclusion, would have tended to result in a higher incidence of valvular disease.

In an amended randomized double-blind placebo-controlled comparison of dexfenfluramine with an investigational modified-release formulation of dexfenfluramine, the study medication was discontinued and echocardiographic examinations were performed on 1072 overweight patients within a median of 1 month after withdrawal of treatment (1000). These patients, 80% of whom were women, had been randomly assigned to receive dexfenfluramine (366 patients), modified-release dexfenfluramine (352 patients), or placebo (354 patients). The average duration of treatment was 71–72 days in each group. Echocardiograms were assessed blind. Pooling the fenfluramine groups, there was a higher prevalence of any degree of aortic regurgitation (17 versus 12%) and mitral regurgitation (61 versus 54%) with fenfluramine. Analyses carried out using the criteria set by the FDA showed that aortic regurgitation of mild or greater severity occurred in 5% of the patients taking dexfenfluramine, 5.4% of those in the two fenfluramine groups combined, and 3.6% of those in the placebo group. Moderate or severe mitral regurgitation occurred in 1.7 and 1.8% of those taking fenfluramine and 1.2% of those taking placebo. Aortic regurgitation of mild or greater severity, mitral regurgitation of moderate or greater severity, or both occurred in 6.5, 6.9, and 4.5% respectively.

This was an unusual study because patients enrolled for a different purpose were analysed mid-way through the study in response to withdrawal of fenfluramine. Exposure to fenfluramine in this study was relatively short (2–3 months) and the prevalences of mitral regurgitation and aortic regurgitation in this study were much lower than previously described (1007).

Although the findings of this study may be reassuring for patients who have taken dexfenfluramine for 2–3 months, they should not preclude the appropriate investigation of a new murmur or new symptoms in any patient with a history of exposure to dexfenfluramine as specified in the American College of Cardiology Guidelines (1008).

In 24 women who were evaluated an average of 12 months after starting to take fen–phen, echocardiography showed that all had unusual valvular morphology and regurgitation affecting valves on the right and left sides (1009). Eight women also had newly documented pulmonary hypertension. Histopathological findings included plaque-like encasement of the leaflets and chordal structures with intact valve architecture. The histopathological features were like those seen in carcinoid or ergotamine-induced valve disease. As of the end of September 1997, the FDA had received 144 individual spontaneous reports (including the 24 cases reported earlier) involving fenfluramine or dexfenfluramine with or without phentermine in association with valvulopathy, 113 with complete information (1010). Of these, 98% occurred in women of whom 2% used fenfluramine alone, 14% used dexfenfluramine alone, 79% used fenfluramine with phentermine, and 5% used a combination of all three; none had used phentermine alone. The median duration of drug use was 9 months (range 1–39 months). Cardiac valve replacement surgery was required in 24% and there was an 11% mortality.

Based on a prospective study carried out in 226 obese subjects (183 women and 43 men) with a mean body mass index of 40 kg/m^2, therapy with fen–phen was associated with low prevalence of significant valvular regurgitation (1011). The authors suggested that valvular regurgitation in these subjects may have reflected age-related degenerative changes. However, several limitations of this study have been pointed out: (a) there was no control group and not all the subjects had echocardiography; (b) multiple readers interpreted the echocardiograms, rendering the comparison less accurate; (c) there was inherent inaccuracy in differentiating mild degrees of valvular regurgitation especially using qualitative scoring systems; (d) there was selection bias; and (e) neither direct inspection nor histopathological confirmation of valvular lesions was performed on any patient.

The risk of a subsequent clinical diagnosis of a valvular disorder of uncertain origin has been assessed in a population-based follow-up study using nested case-control analysis of 6532 subjects who took dexfenfluramine, 2371 who took fenfluramine, and 862 who took phentermine (992). The control group comprised 9281 obese subjects who did not take appetite suppressants matched with the treated subjects for age, sex, and weight. No subject had cardiovascular disease at the start of the follow-up for an average duration of 5 years. There were 11 cases of newly diagnosed idiopathic valvular disorders, five with dexfenfluramine and six with fenfluramine. There were six cases of aortic regurgitation, two of mitral regurgitation, and three of combined aortic and mitral regurgitation. There were no cases of idiopathic cardiac valve abnormalities among the controls or those who took phentermine. The 5-year cumulative incidence of idiopathic cardiac valve disorders was 0 per 10 000 among both those who had not taken appetite suppressants (95 CI = 0, 15) and those who took phentermine alone (CI = 0, 77), 7.1 per 10 000 among those who took either fenfluramine or dexfenfluramine for less than 4 months (CI = 3.6, 18), and 35 per 10 000 among those who took either of these medications for 4 months or more (CI = 16, 76). The authors concluded that the use of fenfluramine or dexfenfluramine, particularly when used for

4 months or longer, is associated with an increased risk of newly diagnosed cardiac valve disorders, particularly aortic regurgitation.

The above study was based on information derived from the General Practice Research Database in the UK. Subjects who had been given at least one prescription for dexfenfluramine, fenfluramine, or phentermine after 1 January 1988, and who were 70 years or younger at the time of their first prescription were included. Subjects were considered to have a new cardiac abnormality if they had no history, on the basis of clinical records, of cardiac valvular abnormalities and if there was evidence of a new valvular disorder on the basis of echocardiography or clinical examination after exposure to appetite suppressants. All the data had been recorded before the publication of recent reports of an association between appetite suppressants and cardiac valve disorders (1007,1009,1012–1014) or primary pulmonary hypertension (996). Hence, it was possible to exclude the possibility that enhanced awareness of possible serious adverse effects of appetite suppressants had led to closer surveillance of patients who were taking these drugs. Nevertheless, the study did not provide information on the frequency of idiopathic cardiac valve disorders that are asymptomatic or otherwise not clinically diagnosed.

Using the FDA case definition of appetite-suppressant related valvulopathy, the prevalence was 31% (60/191) in a selected group of Mayo Clinic patients at Rochester (1015). The most common finding was mild aortic regurgitation. Of asymptomatic patients 28% had abnormal echocardiographic findings. This study emphasized the spectrum of diet- or drug-related cardiac disease and the potential for valvulopathy in asymptomatic patients.

In patients who had taken dexfenfluramine ($n = 479$) or fen–phen ($n = 455$) continuously for 30 days or more in the previous 14 months, there was an increase in the prevalence of aortic regurgitation compared with 539 control subjects (1016). There was no increase in the prevalence of moderate or severe aortic regurgitation in treated patients, and no difference in the prevalence of mitral regurgitation between the untreated and treated groups, irrespective of duration of therapy. All evaluations were carried out using the FDA criteria. The authors were careful to point out that their study was not specifically designed or adequately powered to evaluate specific categories of anorexigen therapy duration.

Further evidence that the prevalence of significant valvular regurgitation is low in patients who take fen–phen has been reported (1017). Transthoracic echocardiography was performed in 343 obese patients in a 3-year prospective study that began within 4 months from the withdrawal of fenfluramine and dexfenfluramine from the market. There were 281 women and 62 men, mean age 47 years, and mean body mass index 40 kg/m^2. Using the FDA's criteria, only 21 subjects (6.1%) had significant valvular lesions. Aortic regurgitation was detected in 18 subjects, mitral regurgitation in 3, and both aortic and mitral regurgitation in 1. Significant valvular disease did not correlate with age, sex, initial or final body mass index, drug dose, or the duration of therapy.

Mechanisms and risk factors

The determinants of valvulopathy in patients treated with dexfenfluramine have been investigated: age and blood pressure can also affect the prevalence of regurgitation (1018,1019), as can duration of exposure (1018). Others have found no correlation between valvular disease associated with appetite suppressants and either dose or duration of drug exposure (1011).

Cases of severe diffuse multivalvular disease associated with fen–phen have been described (1020,1021).

- A 52-year-old woman had a transesophageal echocardiogram 1 year before starting to take fen–phen, and had no significant valvular disease. She presented a year later with a new heart murmur and eventually required isolated aortic valve replacement. Pathological evaluation of the excised aortic valve was consistent with that described with fen–phen use.
- A 44-year-old woman who had previously taken appetite suppressants, developed valvular disease consistent with the effects of fen–phen. She had an identical twin who, despite having been treated with the same medication, remained symptom-free and without abnormal echocardiography. Both the patient and her sister took fen–phen for 2 years. However, the patient took a daily dose of fenfluramine of 60–120 mg (and often as much as 240 mg) and phentermine 90 mg (at times 180 mg), whereas her twin sister adhered to the daily amount prescribed (fenfluramine 60 mg and phentermine 24 mg).

The latter case suggests that dosage is important in the production of the valvular pathology. Mitral and tricuspid insufficiency developed in a 36-year-old woman who had taken fen–phen for 24 months (1024). Transmission electron microscopy of the mitral and tricuspid valves showed many areas that appeared to contain intracellular, virus-like particles clustered in the cytoplasm, with a mean diameter of 32 nm. Whether this finding was incidental or related to the underlying pathology was uncertain.

There is further evidence, from an uncontrolled observational study in 85 patients, that the dose and duration of administration of fen–phen affects the risk of significant valvular disease (1023). The authors suggested that it would be prudent to consider diagnostic echocardiography in patients who have used fen–phen either in a dosage of at least 60 mg/day or for at least 9 months. They also raised concerns that for patients with mild obesity, the prolonged use of larger cumulative amounts may lead to a higher risk of valve regurgitation.

There is further evidence of the relation between the duration of treatment with fen–phen and the prevalence of valvular abnormalities (1024). In 1163 patients who had taken anorexigens within the previous 5 years and 672 control patients who had not, valvular abnormalities primarily involved those who had taken anorexigens for more than 6 months, and predominantly resulted in mild aortic regurgitation. The study had some noteworthy limitations: since fenfluramine has been withdrawn from use, a randomized trial was impossible; also the lack of baseline echocardiograms before treatment implies that one cannot be certain that the valvular regurgitation developed subsequent to drug treatment.

Diagnosis

It has been proposed that valvular disease can be attributable to appetite suppressants only if the following criteria are satisfied:

- the macroscopic and microscopic features are consistent with fenfluramine-related valvulopathy;
- clinical, echocardiographic, and intraoperative findings support the diagnosis;
- the history of drug exposure predates the development or exacerbation of valvular dysfunction (1025).

It is obvious that these criteria can be applicable only in cases in which cardiac valves are explanted and are available for histopathological studies.

The prevalence and diagnostic value of cardiac murmurs for valvular regurgitation has been determined in 223 patients taking dexfenfluramine for 6.9 months and 189 matched controls. Experienced physicians, non-cardiologists, who were unaware of the echocardiographic findings, took a history and performed cardiac auscultation. Based on their findings the authors recommended that cardiac auscultation should be the screening method of choice for detecting valvular regurgitation in users of anorexigens (1026). In this study, the absence of cardiac murmurs predicted the absence of clinically important valvular regurgitation in 93% of dexfenfluramine users. These results support the recommendation by the American Heart Association and American College of Cardiology (1027) that asymptomatic users of anorexigens without a cardiac murmur do not warrant echocardiography. The data also suggest that in users of anorexigens with a 10% prevalence of valvular regurgitation and a 10–15% prevalence of cardiac murmurs, cardiac auscultation will prevent 85–90% of patients from undergoing unnecessary echocardiography. These implications may apply to all users of anorexigens because the prevalence of valvular regurgitation in recent large series is similar to that in this study (SEDA-24, 4). There are therefore large potential cost savings of cardiac auscultation, by preventing a large proportion of the more than 6 million Americans who are exposed to anorexigens from undergoing initial and follow-up echocardiography or from receiving empiric antibiotic prophylaxis for emergency procedures that preclude further cardiac evaluation.

Effects of withdrawal of therapy on valvulopathy

In a patient who was followed-up for 2 years after withdrawal, multivalvular regurgitation associated with fenfluramine and phentermine may have regressed (1005).

- A 44-year-old woman with morbid obesity but no history of cardiac disease developed atypical chest pain. Myocardial infarction was ruled out, and an echocardiogram showed normal chamber sizes and mildly reduced global systolic function. However, moderate to moderately severe aortic regurgitation, mild mitral regurgitation, and moderate tricuspid regurgitation were present. The estimated pulmonary artery pressure was slightly raised. Her only medications were fenfluramine 60 mg/day and phentermine 30 mg/day, which she had taken for the previous 50 weeks, during which time she had lost 40 kg. These drugs were withdrawn and 6 months later an echocardiogram showed improved left ventricular function and a reduction in the severity of all her valvular lesions with no clinically significant change in the estimated pulmonary artery pressure. An echocardiogram obtained 2 years after the initial study showed only trace aortic and tricuspid regurgitation without mitral regurgitation.

In this case, serial echocardiography over 2 years documented regression of multivalvular regurgitation, first discovered while the patient was taking fenfluramine and phentermine. The authors argued that although she was also given lisinopril, the marked degree of improvement in all the valvular lesions after withdrawal of the appetite suppressants was unlikely to be attributable to this alone (1028).

The small increase in prevalence of minor degrees of aortic regurgitation and mitral regurgitation in 941 patients treated with dexfenfluramine for 2–3 months was no longer present 3–5 months (median 137 days) after withdrawal (1029). Echocardiograms were acquired using a standardized protocol and were assessed blindly.

In 50 patients with fenfluramine-associated valvular heart disease followed by serial echocardiography for 6–24 months after withdrawal of therapy (1030), in most cases valvular heart disease either did not change or improved at least by one grade. Mitral and aortic regurgitation improved in some patients, and tricuspid and pulmonic regurgitation improved in most patients after withdrawal. When improvement did occur, regression of regurgitation often involved multiple valves on both the left and right sides of the heart, rather than affecting one valve in isolation. Although most of the patients stabilized or improved, a few had worsening of valvular regurgitation despite withdrawal.

Comparable results were also reported in a larger series in another study (1031). Sequential echocardiographic evaluation 1 year after withdrawal of dexfenfluramine showed a significant reduction in aortic regurgitation. There were no significant changes in mitral regurgitation or any other valvular variables. Although these results can be applied only to the population studied (predominantly middle-aged, obese, white women who took dexfenfluramine for 2–3 months), the implications are considerable. Because valvular regurgitation remained stable or improved in most of the patients, surgical referral for patients with severe regurgitation may be delayed. Improvement in valvular regurgitation often occurred within months after drug withdrawal. Watchful waiting with serial echocardiography, prophylaxis against endocarditis, and medical therapy may be a reasonable management strategy in patients with severe regurgitation, minimal symptoms, and no evidence of left ventricular dysfunction (1030).

However, reversibility may not occur in all cases (1032).

- Cardiac allograft transplantation was carried out from a 35-year-old hypertensive donor with prolonged exposure to fenfluramine and phentermine. There was non-specific mitral valve thickening, with trivial mitral regurgitation and poor approximation of the mitral valve leaflets, due to reduced posterior leaflet mobility. There was no evidence of any other valvular lesion. Examination of the donor heart during cardiac implantation showed three discontinuous lesions along the left atrial surface of the mitral valve annulus and another firm nodular lesion of the annular endocardium. Transplantation of the heart was uneventful and intraoperative transesophageal echocardiography, performed after weaning from cardiopulmonary bypass, showed trivial mitral regurgitation with excellent allograft contractility. The postoperative course was uneventful and the patient was discharged on the eighth postoperative day. Histological examination of the specimen showed a glistening appearance with proliferating myofibroblasts and associated fibrinous vegetations. There was no evidence of acute or chronic inflammation, and Gram staining did not show bacterial or fungal elements. A transthoracic echocardiogram 6 weeks after transplantation showed only trivial mitral regurgitation with improved mobility of the posterior leaflet. Hemodynamic data showed normal allograft function. However, after 6 months of follow-up, Doppler echocardiography showed worsening of mitral regurgitation to moderate severity, but no adverse effects of this hemodynamic load were noted and the patient remained stable.

Although conclusions based on single cases have limitations, they can often provide useful insights and act as catalysts for further studies. This report illustrates a few important features of cardiac valvulopathy associated with anorexigen use. There was no involvement of chordal apparatus and so the pathological changes within the valve leaflets and annulus represented the earliest site of an anorexigen-induced valvulopathy. This opportunity to observe a case of "early" valvulopathy visually and histopathologically offered insight into pathogenesis, and may have been helpful in staging the lesions temporally.

Of 120 patients who had follow-up echocardiography at least twice after stopping fen–phen, 99 met FDA criteria for valvulopathy (1033). On second echocardiography, 57 of these 99 had no change in valvulopathy, 33 had improved, and 9 had deteriorated; nine patients no longer met FDA criteria for valvulopathy. The authors suggested that physicians must continue to be vigilant with patients who develop valvulopathy after taking fen–phen.

Obesity as a confounding factor in cardiac valvulopathy

It is not clear whether valvular insufficiency is related to the use of appetite suppressants or is simply a consequence of obesity. Obese patients who took dexfenfluramine alone, dexfenfluramine in combination with phentermine, or fenfluramine in combination with phentermine have been compared with a matched group of obese-control subjects who had not taken these medications (991). A total of 1.3% of the controls (3 of 233) and 23% of the patients (53 of 233) met the case definition for cardiac valve abnormalities (OR = 23). The odds ratios for such cardiac valve abnormalities were 13 with dexfenfluramine alone, 25 with dexfenfluramine and phentermine, and 26 with fenfluramine and phentermine. This study showed that the prevalence of valvular insufficiency is significantly higher among obese patients who have taken appetite suppressants than among subjects matched for age, sex, and body mass index who did not take such drugs. Since a higher percentage of patients than controls had trace aortic valve insufficiency, the authors questioned whether the case definition threshold for cardiac valve abnormalities in association with appetite suppressants set by the FDA and Centers for Disease Control and Prevention is perhaps too high. In this study the factors that predisposed patients to valvular insufficiency were (a) age at the start of therapy, (b) use of dexfenfluramine, (c) combination of dexfenfluramine with phentermine, and (d) combination of fenfluramine with phentermine. Hence, neither the clinical significance nor the natural history of this type of valvular disease has been defined.

Other epidemiological studies have ruled out the possibility that obesity itself causes a high prevalence of cardiac valvular regurgitation (1034–1035).

Pulmonary hypertension

Pulmonary hypertension associated with fenfluramine was first reported in the early 1980s. A retrospective study further established the link (1037). Subsequently a multicenter case-control study in 95 patients showed a high incidence of pulmonary hypertension in patients who had used fenfluramine or dexfenfluramine (996). Moreover, there was a strong suggestion of a dose–response effect, longer periods of use being associated with a progressive increase in the relative risk of pulmonary hypertension. In 1997, the first case of pulmonary hypertension in association with fen–phen was reported (1038). Eight of the 24 patients with valvular disease had newly diagnosed pulmonary hypertension, although in most cases it was attributable to valvular abnormalities (1039). It is not clear whether a combination of these agents poses a higher risk in predisposed individuals.

The results of a Belgian study in 35 patients with pulmonary hypertension and 85 matched controls have been published (1040). The data were collected when there was no restriction on prescribing of appetite suppressants. Of the patients, 23 had previously taken appetite suppressants, mainly fenfluramines, compared with 5 controls. Moreover, the patients who had been exposed to appetite suppressants tended to be on an average more severely ill and to have a shorter median delay between the onset of symptoms and diagnosis.

Pulmonary artery pressure and cardiac valvular status were determined in a series of 156 mostly asymptomatic patients taking fenfluramine and phentermine (1041). The anorexigen was withdrawn when abnormalities were noted. Pulmonary artery pressure was estimated and valvular examination was performed using Doppler echocardiography. There was borderline or mildly elevated pulmonary artery pressure in 21 patients and 31 patients had notable valvular abnormalities. It has therefore been established that asymptomatic patients may have significant echocardiographic abnormalities, representing early lesions.

- A 30-year-old woman who had taken dexfenfluramine for 7 months developed pulmonary hypertension and right heart failure during late pregnancy. She died of septicemia with multiorgan failure 4 days after a cesarean section (1042).
- Pulmonary hypertension and multivalvular damage after prolonged use of fenfluramine with phentermine have been reported in a 70-year-old Israeli woman (1043).
- Fatal pulmonary hypertension occurred in a 32-year-old man who had been taking phentermine in unknown doses for 4 months (1044).

Incidence

The epidemiological association of pulmonary hypertension with aminorex and dexfenfluramine, both with respect to the strength of the association (estimate of relative risk) and its impact on public health, has been investigated (1045). Control rates of exposure were used to estimate population exposure prevalences. The estimated odds ratio for the association between pulmonary hypertension and any exposure to aminorex was 98 and for dexfenfluramine 3.7. The strong association between aminorex and pulmonary hypertension projected a fivefold increase in the incidence of pulmonary hypertension, and thus a very noticeable epidemic. In contrast, the association with dexfenfluramine is expected to result in an incidence of only 20% and thus a repeat epidemic seems unlikely.

In a prospective surveillance study of 579 patients with pulmonary hypertension at 12 large referral centers in North America, 205 had primary pulmonary hypertension and 374 had secondary pulmonary hypertension (1046). Among the drugs surveyed, only fenfluramine had a significant association with primary pulmonary hypertension compared with secondary pulmonary hypertension (adjusted odds ratio for use for more than 6 months = 7.5; 95% CI = 1.7, 32). The association was stronger with longer duration of use compared with shorter duration of use and was more pronounced in recent users than in remote users. An unexpectedly high (11%) number of patients with secondary pulmonary hypertension had used anorexigens. The high prevalence of anorexigen use in patients with secondary pulmonary hypertension also raised the possibility that these drugs precipitate pulmonary hypertension in patients with underlying conditions associated with secondary pulmonary hypertension.

The age-adjusted mortality rates from primary pulmonary hypertension in the years immediately preceding the use of fen–phen were not different from those reported during the years of widespread use among patients aged 20–54 years. This analysis failed to support the hypothesis that the widespread use of fen–phen in the years 1992–1997 increased the incidence of primary pulmonary hypertension. "If the use [of fen–phen] during these years created an epidemic of primary pulmonary hypertension, as some have declared, such an epidemic is not reflected in the mortality database maintained by CDC" (1047).

Prognosis

Of 62 patients (61 women) exposed to fenfluramine compared with 125 sex-matched patients with primary pulmonary hypertension, 33 had used dexfenfluramine alone, 7 had used fenfluramine alone, and 5 had used both (1048). In 17 cases fenfluramines were taken with amphetamines. Most of the patients (81%) had taken fenfluramines for at least 3 months. The interval between the start of therapy and the onset of dyspnea was 49 months (range 27 days to 23 years). The two groups differed significantly in terms of age and body mass index. Both groups had similar severe baseline hemodynamics, but the percentage of responders to an acute vasodilator was higher in patients with primary pulmonary hypertension. Hence, more patients with primary pulmonary hypertension were treated with oral vasodilators, and long-term epoprostenol infusion was more often used in fenfluramine users. Overall survival was similar in the two groups, with a 3-year survival rate of 50%.

Mechanism and pathophysiology

The mechanism of fenfluramine-associated pulmonary hypertension has been reviewed (SEDA-21, 3). Since only a minority of patients exposed to fenfluramines develop pulmonary hypertension, it has been postulated that a subset may be genetically susceptible. Whether there is a related genetic abnormality in the familial PPH gene located on chromosome 2q (1049) or an abnormality of the angiotensin-converting enzyme gene (1050) has yet to be explored.

Anorexigenic drugs accumulate in the lung and other tissues, especially in cellular organelles with an internal acid pH, such as lysosomes, where they bind to acidic enzymes. Lipid enzyme inhibition, lysosomal lipidosis, and associated myeloidosis are key events in the pathological cascade (1051). Since several stimulants and other psychotropic drugs are cationic amphiphilic compounds that accumulate in the lung, brain, and other tissues, the variety of pathological mechanisms involved in the effects of this group of drugs should be kept in mind during long-term use (1052,1053).

The hypothesis that nitric oxide deficiency predisposes affected individuals to anorexigen-associated pulmonary hypertension has been tested in a prospective case-control comparison with two sex-matched sets of controls: patients with primary pulmonary hypertension (n = 8) and healthy volunteers (n = 12) (1054). Lung production of nitric oxide and systemic plasma oxidation products of nitric oxide were measured at rest and during exercise, and were lower in patients with anorexigen-associated pulmonary hypertension than in patients with primary pulmonary hypertension. This deficiency may have resulted from increased oxidative inactivation of nitric oxide, as the concentrations of their oxidative products were raised in inverse proportion to nitric oxide. These findings, and earlier evidence from animal studies (1055), have given support to the hypothesis incriminating nitric oxide as a determinant of individual susceptibility to anorexigen-associated pulmonary hypertension.

The pressure response to endothelin-1 in the canine circulation has been investigated in isolated perfused dog lung (1056). Acute treatment of the isolated lobes with fenfluramine increased pulmonary arterial pressure. Chronic treatment with fenfluramine potentiated the pulmonary vasoconstrictor response to endothelin-1. Based on these findings, the authors proposed that the pulmonary vasculature becomes hyper-reactive to vasoactive substances, such as serotonin and endothelin-1, possibly leading to pulmonary hypertension.

Anorexigen-associated severe pulmonary hypertension is clinically and histopathologically indistinguishable from idiopathic or primary pulmonary hypertension. Analysis of clonality in microdissected endothelial cells of plexiform lesions in two patients with anorexigen-associated pulmonary hypertension showed a monoclonal expansion of pulmonary endothelial cells. Accelerated growth of pulmonary endothelial cells in response to anorexigens in patients with predisposition to primary pulmonary hypertension has been speculated (1057).

Hypertension

In a few patients, hypertension was induced or aggravated by fenfluramine. The hypertension disappeared on withdrawal, but could not in all instances be reinduced by rechallenge (1058).

Fenoterol

The cardiovascular safety of high doses of inhaled fenoterol and salbutamol has been compared in acute severe asthma (SEDA-21, 183). It was concluded that in adequately oxygenated patients a total dose of 3.2 mg of fenoterol or 1.6 mg of salbutamol given over 60 minutes was safe in terms of cardiovascular effects in acute severe asthma.

Fentanyl

A hypertensive crisis occurred in a patient with a previously unknown pheochromocytoma (1059).

Fexofenadine

Fexofenadine is said to have little cardiotoxicity (1060). In one study fexofenadine was well tolerated, and there were no statistically significant changes in PR interval, QT interval, QRS complex, or heart rate (1061).

There has been a report of ventricular fibrillation during fexofenadine administration in a man with a pre-existing long QT interval (1062). However, causality between fexofenadine and the cardiac effects was unclear.

The safety of fexofenadine in children aged 6–11 years with seasonal allergic rhinitis has been assessed in a large double-blind, randomized, placebo-controlled, parallel study (1063). There were no statistically significant electrocardiographic effects, suggesting that fexofenadine is both efficacious and well tolerated in children with allergic disease.

Fibrin glue

The primary purpose of using fibrin glue is to reduce blood loss and hence the need for transfusion. It is sprayed on to a surgical field in aerosolized form with a double-barrelled syringe, using either compressed air or nitrogen. Its hemostatic and adhesive properties can be used in any surgical specialty, for example to control bleeding after organ injury (1064). Its usefulness is particularly well documented in the fields of cardiovascular surgery (1065,1066), ENT surgery (1067), neurosurgery (1068), and thoracic surgery (1069).

Fibrin glue has been widely used to treat anal fistulae. In a systematic review of 19 studies the reported success rates ranged from 0 to 100%, which may have been due to differences in patient selection (including fistula aetiology and type), treatment protocols, and follow-up duration (1070).

The hemostatic efficacy of fibrin glue in a nasal spray has been studied in 24 patients with hereditary hemorrhagic telangiectasia and epistaxis (1071). Fibrin glue produced immediate hemostasis and good healing of bleeding sites, no secondary bleeding, and no inflammation. Adverse events, including local swelling, pain, and slow healing of the bleeding site with atrophy of the nasal mucosa, were more frequent in those who were given foam nasal packing rather than fibrin glue spray.

Fibrin glue has also been used experimentally to deliver a high concentration of drug to a local site, as in the example of the use of losartan to prevent neointimal hyperplasia in pig saphenous artery (1072).

In three patients who underwent cardiovascular surgery subsequent abnormalities in hemostasis, characterized by increased activated partial thromboplastin time, prothrombin time, and bovine thrombin time, and by a markedly reduced concentration of factor V, developed between the seventh and eighth postoperative days after exposure to fibrin glue containing bovine thrombin (1073). It was suggested that the glue also contains small amounts of factor V and that this may have caused the abnormalities.

Finasteride

The long-term effects and adverse effects of finasteride have been studied in a multicenter study of 3270 men (1074). There was a background history of cardiovascular disease in 40% of the patients at baseline, and myocardial infarction was reported in 1.5% of those who took finasteride and 0.5% of those who took placebo, a significant difference.

Fish oils

Some of the beneficial effects of fish oils after acute myocardial infarction have been attributed to an antidysrhythmic effect on the heart (1075). However, the results of a randomized trial in 200 patients with implantable cardioverter defibrillators are at variance with this: the rate of cardioversion was higher in those taking fish oils 1.8 g/day than in a control group who took olive oil

(1076). The lack of benefit and the suggestion that fish oil supplementation may increase the risk of ventricular tachycardia or ventricular fibrillation in some patients with implantable cardioverter defibrillators can reasonably be interpreted as evidence that the routine use of fish oil supplementation in patients with implantable cardioverter defibrillators and recurrent ventricular dysrhythmias should be avoided.

Fluconazole

Prolongation of the QT interval is a class effect of the antifungal azoles and has occasionally been reported with fluconazole, with a risk of torsade de pointes.

- A 68-year-old woman with *Candida glabrata* isolated from a presacral abscess developed torsade de pointes after 8 days treatment with oral fluconazole 150 mg/day (1077). She had no other risk factors for torsade de pointes, including coronary artery disease, cardiomyopathy, congestive heart failure, or electrolyte abnormalities. The dysrhythmia resolved when fluconazole was withdrawn, but she continued to have ventricular extra beats and non-sustained ventricular tachycardia for 6 days.
- A 59-year-old woman with liver cirrhosis and *Candida* peritonitis developed long QT syndrome and torsade de pointes after intravenous therapy with 400–800 mg/day of fluconazole for 65 weeks, followed by intraperitoneal administration (150 mg/day) (1078). One day after the second intraperitoneal administration, she developed palpitation, multifocal ventricular extra beats, and syncope. In contrast to a normal electrocardiogram on admission, electrocardiography showed polymorphic ventricular extra beats, T wave inversion, alternating T wave amplitude, and a prolonged QT_c interval of 606 ms. Torsade de pointes required cardiopulmonary resuscitation. The fluconazole plasma concentration was 216 μg/ml (usual target range at 400–800 mg/day: 18–28 μg/ml). Fluconazole was withdrawn and all conduction abnormalities reversed fully within 3 weeks.

These patients were not taking any concomitant drugs that prolong the QT interval, suggesting that fluconazole was to blame.

- A 25-year-old woman with worsening endocarditis had a prolonged QT interval at baseline and developed monomorphic ventricular dysrhythmias, which were managed successfully with pacing and antidysrhythmic therapy, including amiodarone (1079). Several days later, she was given high-dose fluconazole (800 mg/day) for fungemia and after 3 days had episodes of torsade de pointes.

In this case torsade de pointes developed in the presence of known risk factors—hypokalemia, hypomagnesemia, female sex, baseline QT interval prolongation, and ventricular dysrhythmias.

Flucytosine

Life-threatening fluorouracil-like cardiotoxicity has been attributed to flucytosine (1080).

- A 34-year-old woman took flucytosine, 500 mg 12 times a day for 2 days, for vaginal candidiasis. After the last dose she complained of chest pain, which persisted for a week and was associated with ST segment elevation during exercise. Coronary angiography showed normal coronary arteries. One month later she was rechallenged with 500 mg 12 times a day for 2 days. The day after completion of this regimen, she developed severe chest pain. Electrocardiography showed widespread ST segment elevation and echocardiography showed apicolateral septal hypokinesia with a left ventricular ejection fraction of less than 15%. Her flucytosine plasma concentration 48 hours after the last dose was not high, but the fluorouracil concentration was similar to that found during a 5-day continuous infusion of 5-fluorouracil. Her lymphocytes showed no abnormalities of intracellular flucytosine clearance, and cytosine deaminase, the enzyme that converts flucytosine to fluorouracil, was not detectable.

Similar cardiotoxicity has been reported with 5-fluorouracil. The reported events were generally consistent with a drug- or metabolite-induced increase in coronary vasomotor tone and spasm, leading to myocardial ischemia. The authors concluded that more attention should be given to the conversion of flucytosine to fluorouracil; however, it is not clear whether flucytosine should be contraindicated in patients with vasospastic or exertional angina.

Fludarabine

Conditioning with fludarabine (25 mg/m^2/day on 5 consecutive days) and melphalan (70 mg/m^2 on 2 consecutive days) has been associated with cardiotoxicity as a unique complication (1081). However, analysis of the benefit to harm balance showed that the regimen led to durable remissions in patients with hematological malignancies.

Fluoroquinolone antibiotics

See also individual compounds

Fluoroquinolones cause prolongation of the QT interval and can cause torsade de pointes (1082). In an in vitro study in isolated canine cardiac Purkinje fibers the rank order of potency in prolonging action potential duration was sparfloxacin > grepafloxacin = moxifloxacin > ciprofloxacin (1083). In guinea-pig ventricular myocardium sparfloxacin prolonged the action potential duration by about 8% at 10 μmol/l and 41% at 100 μmol/l (1084). Gatifloxacin, grepafloxacin, and moxifloxacin were less potent, but prolonged the action potential duration at 100 μmol/l by about 13%, 24%, and 25% respectively. In contrast, ciprofloxacin, gemifloxacin, levofloxacin, sitafloxacin, tosufloxacin, and trovafloxacin had little or no effect on the action potential at concentrations as high as 100 μmol/l.

Preclinical and clinical trial data and data from phase IV studies have shown that levofloxacin, moxifloxacin, and gatifloxacin cause prolongation of the QT interval, but that the potential for torsade de pointes is rare and is influenced by several independent variables (for example concurrent administration of class Ia and III antidysrhythmic agents)

(1085). There is a moderate increase in the QT interval associated with sparfloxacin, averaging 3%, and the few serious adverse cardiovascular events that have been reported during postmarketing surveillance all occurred in patients with underlying heart disease (1086).

In patients taking quinolones (ciprofloxacin 11 477, enoxacin 2790, ofloxacin 11 033, and norfloxacin 11 110; mean ages 49–57 years) there was no evidence of drug-induced dysrhythmias associated with enoxacin within 42 days of drug administration (1087). Of the other quinolones, atrial fibrillation was reported most often within 42 days of ciprofloxacin administration, with no change in event rate over that time. The crude rate of palpitation did not change significantly with ciprofloxacin, norfloxacin, or ofloxacin. Syncope and tachycardia were also reported with ciprofloxacin and ofloxacin. There was no evidence of drug-induced hepatic dysfunction within 42 days of drug administration with any of the quinolones used.

In a retrospective database analysis 25 cases of torsade de pointes associated with ciprofloxacin ($n = 2$), ofloxacin ($n = 2$), levofloxacin ($n = 13$), and gatifloxacin ($n = 8$) were identified in the USA (1088). Ciprofloxacin was associated with a significantly lower rate of torsade de pointes (0.3 cases/10 million prescriptions) than levofloxacin (5.4/10 million) or gatifloxacin (27/10 million). When the analysis was limited to the first 16 months after initial approval of the drug, the rates for levofloxacin (16/10 million) and gatifloxacin (27/10 million) were similar.

In 16 trials worldwide, gemifloxacin has been reported to produce small, non-significant QT interval prolongation (1089).

Grepafloxacin has been removed from the US market because of deaths as a result of torsade de pointes (1090).

In healthy volunteers who took levofloxacin 1000 mg or moxifloxacin the QT_c interval was significantly prolonged compared with placebo (1091). However, torsade de pointes has only been reported in one case associated with moxifloxacin (1092).

Sparfloxacin can cause prolongation of the QT interval (1093).

In vitro and in dogs prulifloxacin did not prolong the QT_c interval (1094). In rabbits intravenous mexiletine 3 mg/kg reduced the electrical vulnerability of the heart during sparfloxacin overdose and may be a pharmacological strategy against the drug-induced long QT syndrome (1095). Pazufloxacin 3–30 mg/kg intravenously had a low potential for QT interval prolongation in an animal model (1096). In vitro and in vivo studies have suggested that the QT interval is unlikely to be prolonged by prulifloxacin (1097).

- Cardiac arrest temporally related to ciprofloxacin occurred in two women (aged 44 and 67 years) when they developed marked QT_c interval prolongation (590 and 680 ms) within 24 hours of ciprofloxacin administration, with recurrent syncope and documented torsade de pointes requiring defibrillation (1098). The QT_c interval normalized after withdrawal of ciprofloxacin.
- A 76-year-old man with acute on chronic renal insufficiency taking ciprofloxacin developed torsade de pointes in combination with hypocalcemia, triggered by hemodialysis (1099). The QT interval prolongation was corrected by treating the hypocalcemia.
- A 95-year-old woman took gatifloxacin and developed recurrent episodes of torsade de pointes on the fourth day of treatment and 1 hour after infusion (1100).
- A 65-year-old woman had torsade de pointes after receiving levofloxacin 250 mg/day intravenously for 3 days (1090).

Fluorouracil

Fluorouracil can cause anginal chest pain, with non-specific ST–T electrocardiographic changes, during infusion (1101). The outcome is favourable if the drug is withdrawn. Re-introduction of the drug has been associated with occasional fatal outcomes and is not recommended (1102). The cardiotoxicity of 5-fluorouracil in 135 reported cases has been reviewed (1103).

Presentation

More frequent use of fluorouracil by continuous infusion, increased awareness of the problems, and more sophisticated monitoring have increased the reported incidence. By 1990, more than 67 clinical cases had been described (1104) and an incidence ranging up to 68% of silent ischemic electrocardiographic changes was identified in patients monitored by continuous 24-hour ambulatory electrocardiography during fluorouracil infusion (1105). The clinical features include the following:

- Precordial pain (both non-specific and anginal) (1104).
- Electrocardiographic ST–T wave changes (non-specific and ischemic) (1104,1105).
- Acute myocardial infarction (rare) (1106,1107).
- Atrial dysrhythmias (including atrial fibrillation) and less often, ventricular extra beats (including refractory ventricular tachycardia and fibrillation) (1105–1107).
- Ventricular dysfunction (usually global, less frequently segmental).
- Cardiac failure, pulmonary edema, and cardiogenic shock (with and without ischemic symptoms) (1105–1110).
- Sudden death, presumed to be caused by ventricular fibrillation (1107,1111,1112).

In most patients with chest pain, with or without electrocardiographic changes, the creatinine kinase MB fraction remained normal (1104,1107,1109).

Acute dilated cardiomyopathy with left ventricular dysfunction related temporally to fluorouracil and cisplatin infusion, with subsequent complete recovery, has been tentatively linked to fluorouracil (1113). Other similar events have been reported (1114,1115). The association is more striking in patients who receive a continuous infusion of fluorouracil and in patients who receive concomitant cisplatin (1116,1117). For example, myocardial ischemia and infarction occur in about 10% of patients who receive fluorouracil by infusion and sudden death has occurred (1118).

Five cases of paroxysmal atrial fibrillation and sinus bradycardia attributed to fluorouracil have been reported (1119).

Acute pulmonary edema leading to lethal cardiogenic shock has been reported with fluorouracil. This occurred despite the fact that the patient had received eight infusions of leucovorin 100 mg/m^2 at weekly intervals (1120).

Most often, cardiotoxicity develops during the second or later course of treatment, but some patients have problems during the first course (1101). Those who develop cardiac toxicity and recover usually have symptoms again when re-challenged with another infusion (1104,1107).

Fluorouracil has also been associated with a number of vascular effects, particularly thromboembolic or circulatory in nature (1121). Although Raynaud's phenomenon has been reported after cisplatin-based chemotherapy, the first case of digital ischemia and Raynaud's phenomenon has been reported with fluorouracil given in a De Gramond type schedule (1122).

Mechanisms and pathophysiology

The mechanisms of fluorouracil cardiotoxicity are not known. Those that have been suggested include:

- direct uncoupling of electromechanical myocardial function at the level of ATP generation (1114);
- an immunoallergic reaction following sensitization by a complex of fluorouracil and cardiac cells;
- vasospasm secondary either to fluorouracil or to released products;
- a direct toxic effect of the drug on the myocardium.

Most reports have attributed chest pain to vasospasm (1123). Certainly, the ischemic-like pains and electrocardiographic findings, lack of changes in creatine kinase, and frequent responses to nitrates and at times to calcium antagonists in the setting of anatomically normal coronary angiography, plus reversible contractility defects suggest coronary vasospasm as a mechanism of fluorouracil cardiotoxicity. However, global dysfunction possibly due to stunned myocardium and the lack of universal response to coronary vasodilators leaves some questions about this hypothesis. Some investigators have postulated myocarditis or myocardiopathy (1124–1126). In 43 patients it did not interfere with the electrical properties of myocardial fibers (1127).

Findings on autopsy and endomyocardial biopsy have shown diffuse, interstitial edema, intracytoplasmic vacuolization of myocytes, and no inflammatory infiltrate (1128). Acute myocardial infarction has been demonstrated pathologically in some, but not all, patients with clinical infarction (1104).

In patients with fluorouracil cardiotoxicity endothelin plasma concentrations were raised (1129).

Susceptibility factors

With regard to susceptibility factors for cardiotoxicity with fluorouracil, there was no effect of age or sex on incidence (1104). Symptoms have been reported in a 38-year-old man (1110) and in several women in their forties (1104,1108) with no prior cardiac history. Cardiac findings have occurred when fluorouracil was given by infusion or bolus as a single agent or with cisplatin and other drugs (1104,1115). Although some felt that cardiac irradiation and pre-existing heart disease were susceptibility factors (1105,1130), others did not (1104,1131). Several investigators have documented normal coronary arteries in patients with severe symptoms (1107,1108).

Patients with pre-existing ischemic heart disease are at a higher risk of severe complications, including sudden death or cardiogenic shock. Higher fluorouracil dosages confer a higher risk of cardiotoxicity (1132,1133). Among several hypotheses, including formulation impurities (1134,1135), fluorouracil-induced arterial vasoconstriction has been identified as an important first step in cardiotoxicity (1136,1137).

Frequency

The cardiotoxicity of fluorouracil was first identified in 1975 (1138). Of 140 patients treated with intravenous 5-fluorouracil, 4 developed ischemic chest pain within 18 hours of either the second or third dose. In three of these patients the pain recurred after subsequent doses. Predose electrocardiograms in two cases were normal. None of the four patients had a history of ischemic heart disease, although all had received left ventricular irradiation (1139).

A 5% incidence of cardiotoxicity-complicating high-dose infusion of fluorouracil 1000 mg/m^2/day for 4 days has been reported and correlated with plasma fluorouracil concentrations in excess of 450 mg/ml (1140).

In 910 patients toxicity was life-threatening in 0.55% (1141). A combination of cisplatin, fluorouracil, and etoposide given for advanced non-small cell cancer of the lung caused only the expected amount of hematological toxicity, but was associated with a higher than expected incidence of cardiac, pulmonary, and cerebrovascular toxicity, including two myocardial infarctions, two cases of congestive heart failure, one pulmonary embolus, and one cerebrovascular accident in a study of 35 patients (1142).

In 1083 patients there was cardiotoxicity in 1.1% of all patients and in 4.6% of patients with prior evidence of heart disease (1130).

Management

Some investigators have reported success in preventing cardiotoxicity with calcium antagonists, such as nifedipine and diltiazem (1123), while others had less success (1109,1143). Two patients with proven fluorouracil cardiotoxicity did not have cardiotoxicity when treated with the specific thymidylate synthase inhibitor raltitrexed 3 mg/m^2 every 3 weeks (1144). The authors commented that fluorouracil cardiotoxicity is therefore not mediated via thymidylate synthase.

In most cases, fluorouracil-induced dysrhythmias were treatable and the ischemic-like symptoms and electrocardiographic changes disappeared if the infusion was discontinued or responded to nitrates, allowing the infusion to continue. The abnormalities of segmental and global ventricular function reverted to normal within days to weeks of withdrawal. In some patients intravenous inotropic and vasodilator support was needed during the initial period (848–850,853). Transdermal glyceryl trinitrate has been recommended as prophylaxis in patients with fluorouracil-related angina-like symptoms (1137).

However, in one case, both oral nitrates and calcium antagonists failed to prevent chest pain associated with 5-fluorouracil (1145).

- About 48 hours after starting her first course of 5-fluorouracil (1000 $mg/m^2/day$) a woman developed anginal chest pain and electrocardiographic changes that eventually normalized. She was readmitted for her second cycle whilst taking amlodipine 10 mg/day and isosorbide dinitrate 40 mg/day, and after 42 hours into the second cycle had the same chest pain and electrocardiographic changes. These were only controlled by withdrawal of the 5-fluorouracil and the intravenous administration of glyceryl trinitrate.

Fluoxetine

Fluoxetine appears not to have the cardiovascular effects associated with tricyclic compounds, but 10 patients did discontinue treatment because of tachycardia, palpitation, and dyspnea (1146). Two older women each had a myocardial infarction and subsequently died, although these events may not have been drug-related.

In general, SSRIs are assumed to be safe in patients with cardiovascular disease, although there have been few systematic investigations in these patients. In a prospective study of 27 depressed patients with established cardiac disease, fluoxetine (up to 60 mg/day for 7 weeks) produced a statistically significant reduction in heart rate (6%) and an increase in supine systolic blood pressure (2%) (1147). One patient had worsening of a pre-existing dysrhythmia and this persisted after fluoxetine withdrawal. These findings suggest that, relative to tricyclic antidepressants, fluoxetine may have a relatively benign profile in patients with cardiovascular disease. However, the authors cautioned that in view of the small number of patients studied, these findings cannot be widely generalized.

The effects of fluoxetine (20 mg/day for 12 weeks) on sitting and standing blood pressures have been reported (1148). Fluoxetine modestly but significantly lowered sitting and standing systolic and diastolic blood pressures by about 2 mmHg. Patients with pre-existing cardiovascular disease showed no change. This study confirms that fluoxetine has little effect on blood pressure in physically healthy depressed patients and in those with moderate cardiovascular disease.

Fluoxetine-induced remission of Raynaud's phenomenon has been reported (SEDA-18, 20) (1149).

Cardiac dysrhythmias

Fluoxetine has reportedly caused prolongation of the QT_c interval (1150).

- A 52-year-old man had an abnormally prolonged QT_c interval of 560 ms, with broad-based T-waves. He had taken fluoxetine 40 mg/day over the previous 3 months, before which an electrocardiogram had shown a normal QT_c interval (380 ms). The fluoxetine was withdrawn, and 10 days later the QT_c interval was 380 ms. His only other medication was verapamil which he had taken for 3 years for hypertension.

Systematic studies of fluoxetine as monotherapy have not shown evidence of QT_c prolongation. It is possible in this case that fluoxetine interacted with verapamil to produce a conduction disorder.

An elderly man developed atrial fibrillation and bradycardia shortly after starting fluoxetine, and again on rechallenge (SEDA-16, 9). Dose-dependent bradycardia with dizziness and syncope has also been reported in a few patients taking fluoxetine (SEDA-16, 9) and in a presenile patient (1151).

Fluvoxamine

A slight, clinically unimportant reduction in heart rate has been reported with fluvoxamine (1152,1153). There has been one report of supraventricular tachycardia in a woman with no previous cardiovascular disease, but the association with fluvoxamine was unclear since there was no rechallenge (SEDA-16, 9).

Formoterol

The cardiac effects of formoterol and salmeterol have been studied in 12 patients with COPD, hypoxemia (PaO_2 below 60 mmHg), and cardiac disease. Holter monitoring showed that the heart rate was higher after formoterol 24 micrograms than after either formoterol 12 micrograms or salmeterol 50 micrograms. Supraventricular or ventricular extra beats occurred more often after formoterol 24 micrograms. Formoterol 24 micrograms caused a significant reduction in the plasma potassium concentration for 9 hours after administration. The authors suggested that in patients with COPD, hypoxemia, and pre-existing cardiac dysrhythmias, long-acting $beta_2$-adrenoceptor agonists can cause adverse cardiac effects. However, the recommended single dose of salmeterol or formoterol allows a higher safety margin than inhaled formoterol 24 micrograms (1154).

The non-pulmonary effects of formoterol have been carefully studied in an uncontrolled observational trial in 10 patients with asthma who were already taking regular inhaled budesonide (400 micrograms bd) (1155). A 24-hour Holter recording was taken at baseline, and after a 2-week treatment period with formoterol 12 micrograms bd, there were no significant changes in blood pressure, heart rate, cardiac morphology, or the circadian rhythm of autonomic regulation as assessed by measurements taken from the Holter monitor after the treatment period.

FTY720

FTY 720 was associated with a mean 10% reduction in supine heart rate in patients with psoriasis (1156). Heart rate changes were asymptomatic in all cases. One subject had asymptomatic second-degree type 1 atrioventricular (Wenckebach) block. A mild reduction in heart rate was reported in a phase 2a multicenter study in kidney recipients (1157).

Gadolinium

The classical macromolecular blood-pool contrast agents are based on gadolinium DTPA or gadobenate dimeglumine, which are linked to albumin, dextran, or polylysine (1158). Albumin-based agents are not considered optimal for clinical development, as it is difficult to obtain highly consistent synthetic products. There are also problems with cardiovascular toxicity and retention of gadolinium in bones and liver. Dextran-based agents appear to be safer, and an agent called CMD-A2-gadobenate dimeglumine (CMD = carboxymethyldextran) has a favorable toxicity profile.

Gadolinium chelates include ionic high-osmolar agents, such as gadolinium pentetate dimeglumine (Gd-DTPA) and low osmolar non-ionic agents, such as gadodiamide. They generally have similar MR relaxivities, pharmacokinetics, and biodistribution, behaving as non-specific extracellular fluid space agents analogues to iodinated contrast media. The similar safety and efficacy of ionic and non-ionic gadolinium-based extracellular contrast agents suggests that the choice of contrast medium will in future be determined by economic rather than clinical considerations (1159).

The use of gadolinium-based contrast agents as an alternative to iodinated contrast agents has been reported in a patient with a history of allergy to the latter (1160).

- A 77-year-old woman had a gadolinium-enhanced MRI scan followed by gadolinium-enhanced spiral CT pulmonary angiography for suspected pulmonary embolism. Gadodiamide 0.4 mmol/kg (60 ml) was injected intravenously at a rate of 2 ml/second. There were no adverse reactions.

Gallamine triethiodide

Tachycardia invariably accompanies the use of gallamine. It is seen after doses as low as 20 mg and reaches a maximum at around 100 mg in adults (1161). It is often extreme, rates above 120 per minute being not uncommon. The increase in heart rate outlasts the neuromuscular blocking effect (1161). Usual clinical doses also result in a slight increase in mean arterial pressure, a slight fall in systemic vascular resistance, and a marked rise in cardiac index (1162,1163). These cardiovascular effects are principally accounted for by the strong vagolytic action of gallamine, the cardiac muscarinic receptors being almost as sensitive to its blocking action as the acetylcholine receptors of the neuromuscular junction (1164). Blockade of noradrenaline reuptake and an increased release of noradrenaline from cardiac adrenergic nerve endings (1165,1166) may contribute, although an inotropic effect in man is disputed (1167). Ganglion-blocking activity is slight and is not seen in the usual dose range. The possible mechanisms have been reviewed (1168–1170). Gallamine should therefore not be used when tachycardia has to be avoided.

Gatifloxacin

See also Fluoroquinolone antibiotics.

Although early studies suggested that gatifloxacin has little effect on the QT interval of the electrocardiogram (1171,1172), clinical trial data and data from phase IV studies have shown that it prolongs the QT interval (1173). Four cases of gatifloxacin-associated cardiac toxicity have been reported in patients with known risk factors for this adverse event (1174).

Gemcitabine

In a retrospective chart review of patients with gemcitabine-associated thrombotic microangiopathy diagnosed between January 1997 and February 2002, the cumulative incidence was 0.31% (8 cases among 2586 patients), higher than previously reported (0.015%) (1175). The median age was 53 years, the median time to development was 8 (range 3–18) months, and the cumulative dose was 9–56 g/m^2. New or exacerbated hypertension was a prominent feature in seven of nine patients and preceded the diagnosis by 0.5–10 weeks. Treatment included withdrawal of gemcitabine, antihypertensive therapy, plasma exchange, and dialysis. Six patients survived and three died of disease progression; none died as a direct result of thrombotic microangiopathy, but two developed renal insufficiency requiring dialysis, and one developed chronic renal insufficiency.

Gemcitabine can occasionally cause systemic capillary leak syndrome (1176,1177).

Gemeprost

Two women developed myocardial ischemia during treatment with gemeprost for termination of pregnancy (1178).

- A 29-year-old woman, a smoker with a history of renal insufficiency, obesity, hypertension, hypercholesterolemia, and cardiac dysrhythmias, underwent termination of pregnancy at 10 weeks with a pessary of gemeprost 1 mg and 5 hours later dilatation and evacuation, followed by tubal ligation. After surgery, her blood pressure became unmeasurable, her heart rate dropped to 40/minute, and she developed ventricular fibrillation. She was given streptokinase and intravenous heparin for suspected pulmonary embolism; her blood pressure rose and was maintained with adrenaline and noradrenaline. Angiography showed an 80% stenosis of her right coronary artery and complete occlusion of the anterior interventricular branch. Blood flow was re-established by coronary angioplasty.
- A 32-year-old woman, a smoker, had an evacuation after the death of her fetus at 18 weeks. Two pessaries of gemeprost 1 mg were inserted 7.25 hours apart, and about 90 minutes later she became unconscious, apneic, and cyanotic, and had dilated pupils and no detectable blood pressure or pulse. She was given 100% oxygen, intravenous adrenaline and dobutamine, and a crystalloid infusion. Her systolic pressure rose to 100 mmHg. Coronary angiography showed left and circumflex coronary artery spasm.

The author commented that the myocardial ischemia experienced by both of these patients was thought to be due to prostaglandin-induced coronary spasm. It would be prudent to monitor every woman treated with gemeprost during the course of an abortion.

Gemtuzumab

Gemtuzumab ozogamicin has been associated with veno-occlusive disease. Of 62 patients with previously treated acute myelogenous leukemia or myelodysplastic syndrome who underwent allogeneic stem cell transplantation and who were studied retrospectively, 14 received gemtuzumab before stem cell transplantation (1179). Veno-occlusive disease developed in 13 patients, including nine of 14 who had had prior exposure to gemtuzumab, compared with four of 48 without prior exposure. Nine of 10 patients who underwent stem cell transplantation up to 3.5 months after gemtuzumab developed veno-occlusive disease compared with none of four patients who underwent stem cell transplantation more than 3.5 months after gemtuzumab administration. Three of 14 patients who received gemtuzumab before stem cell transplantation died of veno-occlusive disease.

In another study, children with acute myelogenous leukemia were treated with intravenous gemtuzumab monotherapy 4-9 mg/m^2 for up to three courses (1180). One of 15 children developed veno-occlusive disease and one grade 3 hypotension.

Gemtuzumab ozogamicin was used as initial treatment in 12 patients with acute myeloid leukemia over the age of 65; the response rate was 27% (1181). The adverse effects were acceptable, although five patients developed cardiac toxicity, three of whom had grade 3 and/or 4. One of these patients had underlying coronary artery disease and required coronary stent placement. Another developed hypoxia related to pulmonary edema several hours after completing the initial infusion and eventually died of respiratory and ventilatory complications. Another had a myocardial infarction after the second infusion, and one had chest pains half an hour after the initial infusion. It is therefore recommended that elderly patients be screened for cardiac diseases. Patients with coronary artery disease or cardiac dysfunction should not receive gemtuzumab ozogamicin.

General anesthetics

See also individual compounds

Volatile anesthetic agents depress cardiac output, especially in the elderly. A study of 80 patients aged over 60 years compared the effects of halothane and isoflurane with and without nitrous oxide 50% (1182). Doses were carefully adjusted to be equipotent in all four groups. Isoflurane caused a 30% reduction in systolic and diastolic arterial pressures compared with a 17% reduction with halothane. The reductions in cardiac index were similar with the two agents, about 17%. The addition of nitrous oxide attenuated the reductions in arterial pressure. In the case of the combination of isoflurane with nitrous oxide, there was a small increase in cardiac index and a small reduction in the halothane/nitrous oxide group. Systemic vascular resistance was reduced by a greater extent with isoflurane compared with halothane and little altered by the addition of nitrous oxide. The result suggests that nitrous oxide supplementation may be advantageous in the elderly, but interpretation is limited by the fact that it does not include the effects of surgery on these important cardiac parameters.

The long QT syndrome is associated with potentially fatal ventricular dysrhythmias under anesthesia. The effect of halothane and isoflurane on the QT interval was studied in 51 healthy children (1183). Isoflurane 2.3–3.0% increased the average QT interval from 425 to 475 milliseconds at the time of induction. Halothane reduced the average QT interval from 428 to 407 milliseconds. The result suggested that halothane may be the more desirable agent in children with a prolonged QT interval.

The frequencies of cardiac dysrhythmias during halothane and sevoflurane inhalation have been compared in 150 children aged 3–15 years undergoing outpatient general anesthesia for dental extraction (1184). They were randomized into three groups and received either halothane or sevoflurane in 66% nitrous oxide whilst breathing spontaneously. One group received 0.75% increments of halothane every two to three breaths to a maximum of 3.0% for induction, and then 1.5% for maintenance of anesthesia. One group received sevoflurane in 2% increments to a maximum of 8% and then a maintenance dose of 4%. The final group received 8% sevoflurane for induction and then a maintenance dose of 4%. The children who received halothane had a 48% incidence of dysrhythmias, significantly higher than the 16% incidence in the sevoflurane group and 8% in the incremental sevoflurane group. The halothane-associated dysrhythmias mainly occurred during dental extraction or emergence from anesthesia, and were usually ventricular. Six children in the halothane group had ventricular tachycardia. The longest run of ventricular tachycardia lasted 5.5 seconds, and one child had 13 separate episodes. Sevoflurane-associated dysrhythmias were mainly single supraventricular extra beats, and did not differ between the two administration methods. Although there was insufficient evidence to suggest that transient dysrhythmias associated with halothane in dental anesthesia can lead to cardiac arrest, sustained ectopic ventricular activity, including ventricular tachycardia, even if self-limiting, results in reduced cardiac output and cannot be ignored. These results imply that sevoflurane may be the preferable agent in this setting.

The hemodynamic responses to induction and maintenance of anesthesia with halothane have been compared with those of sevoflurane in 68 unpremedicated children aged 1–3 years undergoing adenoidectomy (1185). The children received either sevoflurane 8% or halothane 5% + nitrous oxide 66% for induction of anesthesia and tracheal intubation, without neuromuscular blocking drugs. Anesthesia was maintained by adjusting the inspired concentration of the volatile anesthetic to maintain arterial blood pressure within 20% of baseline values, and the electrocardiogram was continuously recorded. The incidence of cardiac dysrhythmias was 23% with halothane and 6% with sevoflurane. Most of the dysrhythmias were short-lasting/self-limiting supraventricular extra beats or ventricular extra beats. Although the overall incidence of dysrhythmias was low in both groups, the result again shows that sevoflurane causes fewer dysrhythmias in children and may be the preferable agent.

QT dispersion, defined as the difference between QT_{max} and QT_{min} in the 12-lead electrocardiogram, is a measure of regional variation in ventricular repolarization (1186). It is greater in patients with dysrhythmias. The effects of halothane and isoflurane on QT dispersion have been studied in 46 adult patients undergoing general anesthesia. QT dispersion was increased in both groups both with and without correction for heart rate. The increase was significantly greater with halothane than with isoflurane. The clinical significance of this finding is not known. In isolation, QT dispersion reflects an abnormality in ventricular repolarization and correlates with dysrhythmic events. Although there were no overt dysrhythmias in this study, the effect suggests a reason for the variable results of past studies of the QT interval: most studies showed prolongation of the QT interval, but some showed no change, or shortening. Larger studies are needed to elaborate on the possible clinical importance of this phenomenon, but it may be a significant cause of dysrhythmias with volatile anesthetics.

Glafenine

A rise in blood pressure coupled with renal adverse effects has been reported. Coronary artery spasm leading to myocardial infarction was described as part of an allergic reaction with Quincke's edema (SED-11, 190) (1187).

Glucagon

Glucagon has been reported to have caused myocardial ischemia (1188).

Glutaral

Nine of 184 implantations of glutaral-preserved mitral valves became incompetent (1189). This incidence of dehiscence substantially exceeded that previously noted with synthetic valves, and the authors suggested that incomplete removal of glutaral from the prosthesis might have contributed to failure of healing. Collagen ultrastructure investigations of glutaral-treated porcine aortic valve tissue showed that the long-term mechanical durability of treated aortic valves can be substantially increased if careful consideration is given to the pressure at which initial fixation of glutaral is carried out.

Glycyrrhiza species (*Fabaceae*)

Liquorice can cause hyperaldosteronism and hence hypertension.

- A 52-year-old woman had high blood pressure that had previously been resistant to antihypertensive drugs (1190). Her history was unremarkable and her blood pressure readings varied between 140/70 and 200/80 mmHg without apparent reason. On detailed questioning she admitted to eating two bars of an unnamed liquorice sweet daily. Discontinuation of this habit normalized her blood pressure within 1 month without the need for any other medical intervention.

Gold and gold salts

Acute vasodilatory (nitritoid) reactions occur in a minority of patients receiving parenteral gold, especially sodium aurothiomalate (SEDA-22, 245). A few minutes after injection the patient experiences weakness, flushing, hypotension, tachycardia, palpitation, sweating, and sometimes syncope (1191). Very rarely myocardial infarction and stroke follow (1191,1192). The mechanism is unknown, but it has been suggested that the vehicle might be responsible, and that aurothioglucose might therefore be preferable to sodium aurothiomalate in elderly patients or in those with a history of cardiovascular disease.

Gonadorelin

Gonadorelin inhibits nitric oxide-mediated arterial relaxation, which disappears within 3 months after stopping treatment. This effect was abolished with "add-back" hormone replacement in a prospective, randomized study of 50 women treated for 6 months (1193).

Granulocyte colony-stimulating factor (G-CSF)

Cardiovascular events have seldom been described in patients given colony-stimulating factors. However, possible excesses of cardiovascular events and unexpected deaths have been suggested (SED-13, 1115; 1194), although the actual risk was not fully evaluated. In three isolated reports, acute arterial thrombosis or angina pectoris were deemed to have resulted from hypercoagulability with extreme leukocytosis and G-CSF-induced abnormalities in platelet aggregation (SEDA-20, 337; SEDA-21, 377; 1195). Increased platelet aggregation has also been found in healthy volunteers (1196). Although the relevance of these findings is unclear, caution is warranted in patients predisposed to thromboembolic events.

- A 46-year-old donor denied pre-existing cardiac symptoms, but smoking and a family history of coronary artery disease were noted as possible risk factors (1197). The pre-treatment electrocardiogram was normal. Six hours after the second and the third doses of G-CSF 10 micrograms/kg before peripheral blood progenitor cell collection, he developed symptoms and signs of cardiac ischemia, including palpitation, chest discomfort, trigeminy, and T wave inversion. However, troponin was unchanged. Cardiac catheterization showed severe coronary artery occlusion and he underwent percutaneous transluminal coronary angioplasty. He finally admitted mild exertional chest discomfort 2 weeks before the first dose of G-CSF.

Although not described during clinical trials, typical capillary leak syndrome has been anecdotally observed after G-CSF administration, illustrating the possible consequences of accelerated release of activated granulocytes (SEDA-22, 407; 1198).

Microthrombotic necrotizing panniculitis has been reported (1199).

- A 49-year-old woman received subcutaneous filgrastim 300 micrograms/day into the upper thighs for neutropenia prophylaxis after treatment of relapsing Hodgkin's disease with mitoguazone, etoposide, vinorelbine, and ifosfamide. After 3 days she suddenly developed fever, painful livedo, deeply infiltrated edema on the legs and thighs, and inflamed livedoid erythema on both soles. Deep biopsy specimens showed small vessel thrombosis with subcutaneous necrosis and hemorrhage. She recovered over the next 4 weeks after filgrastim withdrawal and prednisone treatment.

Although a causal relation was difficult to ascertain in the context of malignancy and cytotoxic chemotherapy, the short time to occurrence after G-CSF favored a causative role.

Granulocyte–macrophage colony-stimulating factor (GM-CSF)

Mild local phlebitis sometimes occurs at intravenous sites of administration of GM-CSF. Central venous catheter site thrombosis, inferior vena cava thrombosis, and possible pulmonary embolism have sometimes been observed (1200,1201). Although chemotherapy for breast cancer is associated with a higher risk of developing vascular thrombosis, iliac artery thrombosis was attributed to GM-CSF in two patients (1202).

Raynaud's phenomenon has been reported, but confounding factors including the use of high-dose antineoplastic drugs are possible (1200).

A rapidly reversible first-dose syndrome (dyspnea, hypoxia, tachycardia, and hypotension) can occur within the first hour after the first continuous infusion in 15–30% of patients (1200). A dose-limiting vascular leak syndrome was consistently described in patients receiving GM-CSF 30 micrograms/kg/day or more, but lower doses were also reported to induce a clinically relevant capillary leak syndrome (SEDA-22, 408; 1203,1204). Continuation of GM-CSF treatment at the same dose or lower and careful management was possible in some patients. Endothelial cell damage with an increase in the transcapillary escape rate of albumin and the possible role of IL-1 and TNF production by GM-CSF-activated monocytes were suggested as possible mechanisms. This was consistent with the observation of marked hypoalbuminemia in some instances associated with edema and ascites after GM-CSF in four of nine patients treated for myelodysplastic syndrome or aplastic anemia (1205).

Grepafloxacin

See also Fluoroquinolone antibiotics.

Grepafloxacin is a synthetic fluoroquinolone antibiotic with extensive tissue distribution and strong antibacterial activity in vivo (1206,1207). However, it was withdrawn in 1999 because of its adverse cardiovascular effects, which included dysrhythmias (1208–1210).

Growth hormone (human growth hormone, hGH, somatotropin)

The use of growth hormone other than in growth hormone-deficient people continues to be investigated. In 80 post-menopausal women with osteoporosis aged 50-70 years, two doses of growth hormone, 1.0 U/day (n = 28) or 2.5 U/day (n = 27) for 3 years, were compared with placebo (1211). Plasma fibrinogen increased in both growth hormone groups at 4 years only; the significance of this finding is uncertain. Fibrinogen concentrations are raised in acromegaly and since fibrinogen is a risk factor for cardiovascular disease, especially stroke, people with acromegaly have an increased risk of cardiovascular disease.

Of 23 adolescent patients with growth hormone deficiency in childhood who were reassessed when they had reached adult bone age and completed puberty, eight were no longer thought to be growth hormone deficient and therapy was withdrawn (1212). The other 15 had a 6-month break from growth hormone therapy and then restarted. Compared with a control group at the time of withdrawal, the eight patients without growth hormone deficiency had increased thickness of the intima media, which fell to normal values by 12 months. These results support the recommendation that children with idiopathic growth hormone deficiency should be retested after completion of growth to assess the need for continued administration of growth hormone.

Halofantrine

In 1993, the sudden cardiac death of a 37-year-old woman after her ninth dose of halofantrine was reported. A subsequent prospective study showed that halofantrine was associated with a dose-related lengthening of the QT interval by more than 25% (SEDA-17, 328). Mefloquine did not cause such changes, but the combination of mefloquine with halofantrine had a more pronounced effect on the electrocardiogram. However, in the region where this investigation was carried out (on the Thai–Burmese border area) thiamine deficiency is common, and patients in this area have longer baseline QT intervals than are usually reported (SEDA-17, 328). Two patients, mother and son, both with congenital prolongation of the QT interval, suffered sustained episodes of torsade de pointes after a total dose of 1000 mg of halofantrine, and there have been other reports of dysrhythmias, including death in a patient who took mefloquine and halofantrine (SEDA-20, 260; SEDA-21, 295). Dysrhythmias due to halofantrine may respond to propranolol (1213).

African children who received halofantrine (three doses of 8 mg/kg 6-hourly) for uncomplicated *P. falciparum* malaria had increases in both the PR interval and the QT_c interval; out of 42 children in the study, two children developed first-degree heart block and one child second-degree heart block; the QT_c interval either increased by more than 125% of baseline value or by more than 0.44 seconds (an effect that persisted for at least 48 hours) (1214).

There have been recent reports in the French medical press of cases of significant QT_c prolongation in children returning to France, and caution in its use has been urged (1215). A small trial in non-immune adults in the Netherlands and France also showed increased QT_c dispersion with halofantrine but not artemether + lumefantrine (1216).

- Death due to a dysrhythmia was reported in a woman who had taken halofantrine for malaria (1217). She had a normal electrocardiogram before treatment and no family history of heart disease.

Prolongation of the QT interval occurred in 10 of 25 children treated with halofantrine (24 mg/kg oral suspension in three divided doses) for acute falciparum malaria (1218).

Electrocardiographic monitoring is recommended for children and adults taking halofantrine.

Haloperidol

Intravenous haloperidol is often prescribed to treat agitation, and torsade de pointes has on occasions occurred (SEDA-20, 36). In a cross-sectional cohort study QT_c intervals were measured before the intravenous administration of haloperidol plus flunitrazepam, and continuous electrocardiographic monitoring was performed for at least 8 hours after (n = 34) (1219); patients who received only flunitrazepam served as controls. The mean QT_c interval after 8 hours in those who were given haloperidol was longer than in those who were given flunitrazepam alone; four patients given haloperidol had a QT_c interval of more than 500 ms after 8 hours. However, none developed ventricular tachydysrhythmias.

In a case-control study, haloperidol-induced QT_c prolongation was associated with torsade de pointes (1220). The odds ratio of developing torsade de pointes in a patient with QT_c prolongation to over 550 ms compared with those with QT_c intervals shorter than 550 ms was 33 (95% CI = 6, 195). The sample consisted of all critically ill adult patients in medical, cardiac, and surgical intensive care units at a tertiary hospital who received intravenous haloperidol and had no metabolic, pharmacological, or neurological risk factors known to cause torsade de pointes, or if the dysrhythmia developed more than 24 hours after intravenous haloperidol. Of 223 patients who fulfilled the inclusion criteria, eight developed torsade de pointes. A group of 41 patients, randomly selected from the 215 without torsade de pointes, served as controls. The length of hospital stay after the development of haloperidol-associated torsade de pointes was significantly longer than that after the maximum dose of intravenous haloperidol in the control group. The overall incidence of torsade de pointes was 3.6% and 11% in patients who received intravenous haloperidol 35 mg or more over 24 hours.

Conflicting results have been found in two crossover studies with regard to haloperidol-induced QT interval prolongation. In the first study, QT interval prolongation was associated with sulpiride but not haloperidol (1221). Eight schizophrenic patients who had been free of medication for at least 2 weeks took sulpiride 15 mg/kg for 2 weeks and then haloperidol 0.25 mg/kg for another 2 weeks. QT_c intervals during sulpiride treatment were significantly prolonged by 5.1% and 8.5% compared with haloperidol and no treatment. Conversely, in the second study there was a statistically longer mean QT_c interval with haloperidol (422 ms) than placebo (408 ms) 10 hours after haloperidol or placebo administration (1222). The subjects of this study were 16 healthy volunteers who randomly took haloperidol (a single dose of 10 mg) or placebo during the first study period (4 days) and the alternative during the second period (4 days). Despite a statistically significant longer mean half-life of haloperidol (19 versus 13 hours) in poor metabolizers of CYP2D6 than in extensive metabolizers, this exposure change did not translate into marked changes in the QT_c interval.

Several cases of torsade de pointes have been reported with intravenous haloperidol used with low-dose oral haloperidol (1223).

The effects of haloperidol dose and plasma concentration and CYP2D6 activity on the QT_c interval have been studied in 27 Caucasian patients taking oral haloperidol (aged 23–77 years, dosages 1.5–30 mg/day) (1224). Three patients had a QT_c interval longer than 456 ms, which can be considered as the cut-off value for a risk of cardiac dysrhythmias. There was no correlation between QT_c interval and haloperidol dosage or plasma concentrations or CYP2D6 activity.

Asystolic cardiac arrest has been reported after intravenous haloperidol (1225).

In one case, the use of carbamazepine and haloperidol led to prolongation of the QT_c interval and cardiac complications (1226).

- A 75-year-old man developed ventricular fibrillation and cardiac arrest after intravenous haloperidol (1227). His past history included coronary bypass surgery and coronary angioplasty. As he continued to have severe chest pain, emergency angioplasty was performed. On day 3 he received haloperidol by infusion 2 mg/hour, with 2 mg increments every 10 minutes (up to 20 mg in 6 hours) as needed for relief of agitation. Before haloperidol, his QT_c interval was normal; after haloperidol it increased to 570 ms. The next day he developed ventricular fibrillation. Subsequent electrocardiograms showed prolonged QT_c intervals of 579 and 615 ms, and haloperidol was withdrawn; the QT_c returned to normal.
- A 76-year-old man developed torsade de pointes while taking tiapride 300 mg/day; the QT_c interval 1 day after starting treatment was 600 ms; the dysrhythmia resolved when tiapride was withdrawn (1228).
- A 39-year-old man died suddenly 1 hour after taking a single oral dose of haloperidol 5 mg (1229). He had myasthenia, alcoholic hepatitis, and electrolyte abnormalities due to inadequate nutritional state. His electrocardiogram showed prolongation of the QT_c interval (460 ms). Autopsy showed a cardiomyopathy but no explanation for sudden death.

However, malignant dysrhythmias can occur without changes in the QT interval (SEDA-24, 54).

- A 64-year-old woman underwent coronary artery bypass surgery and was given intravenous haloperidol for agitation and to avoid postoperative delirium; she developed torsade de pointes (1230).
- Asystolic cardiac arrest occurred in a 49-year-old woman after she had received haloperidol 10 mg

intramuscularly for 2 days; no previous QT_c prolongation had been observed (1231).

Halothane

Halothane, isoflurane, and sevoflurane are potent coronary vasodilators, able to produce some degree of coronary steal in ischemic regions. Despite this, halothane may preferentially dilate large coronary arteries and/or interfere with platelet aggregation. If these experimental effects are confirmed, halothane may be the anesthetic of choice in the non-failing ischemic heart (1232).

Halothane has a mild depressive effect on cardiac performance (1233). In human ventricular myocardium, halothane interacted with L-type calcium channels by interfering with the dihydropyridine binding site; this may, at least in part, explain its negative inotropic effect (1234).

Halothane depressed cardiovascular function significantly more than isoflurane in younger adults, but the falls in systolic and diastolic blood pressures in elderly patients were significantly greater with isoflurane (1235).

Cardiac dysrhythmias

Halothane produces bradycardia, but dysrhythmias, most often ventricular in origin, also occur during maintenance of anesthesia. They were noted in 53% of 679 patients (1236). Concomitant administration of catecholamines increases the risk of dysrhythmias.

Bundle branch block and aberrant conduction were noted in children during halothane anesthesia (1237).

In a double-blind, randomized, controlled study of 77 children undergoing halothane anesthesia for adenoidectomy, the effects of atropine 0.02 mg/kg, glycopyrrolate 0.04 mg/kg, and physiological saline were compared (1238). There was no difference in the incidence of ventricular dysrhythmias. Atropine prevented bradycardia but was associated with sinus tachycardia in most patients. The bradycardias that occurred in the groups that received glycopyrrolate or placebo were short-lived and resolved spontaneously.

Pulsus alternans in association with hypercapnia occurred in a study of 120 patients who breathed spontaneously during halothane anesthesia (1239). End-tidal carbon dioxide concentration was allowed to rise freely until pulsus alternans or other cardiac dysrhythmias occurred. Ten of the patients developed pulsus alternans, which was promptly relieved on institution of positive pressure ventilation and the return of end-tidal carbon dioxide concentration to normal. The mechanism and the significance of this phenomenon are not well understood.

Hemoglobins

Current hemoglobin-based oxygen carriers can cause vasoconstriction, which is probably due to scavenging of nitric oxide or the production of the endogenous vasoconstrictor endothelin-1 (1240,1241). However liposome-encapsulated hemoglobin does probably not scavenge nitric oxide, because these relatively large particles are less able to penetrate the endothelium to bind nitric oxide (1242).

There was a mild reduction in cardiac output in a study in which 10 patients received diaspirin cross-linked hemoglobin in a dose of 50 mg/kg or 35 ml for a 70 kg patient after repair of an abdominal aortic aneurysm (1241).

Heparins

Bolus administration of heparin causes vasodilatation and a fall in arterial blood pressure of 5–10 mmHg (1243). Some convincing data have been reported concerning the role in these reactions of chlorbutol which has been used as a bactericidal and fungicidal ingredient in some heparin formulations (1244).

Cardiogenic shock can occur in parallel with disseminated intravascular coagulation (1245). In these circumstances heparin is thought to act as a hapten in a heparin–protein interaction that stimulates antibody production and an antigen–antibody reaction associated with release of platelet and vasoactive compounds.

General vasospastic reactions have been described in patients receiving heparin, exceptionally complicated by skin necrosis (1246). Vasospastic reactions are probably part of the syndrome of thrombohemorrhagic complications considered above.

Herbal medicines

Several reviews have focused on herbal medicines and have covered:

- the toxicity of medicinal plants (1247–1249);
- the safety of herbal products in general (1250–1261);
- adverse effects in specific countries, for example the USA (1262) and Malaysia (1263);
- adverse effects on specific organs (1264), such as the cardiovascular system (1265), the liver (1266,1267), and the skin (1268,1269);
- the safety of herbal medicines in vulnerable populations: elderly patients (1270), pregnant women (1271), and surgical patients (1272,1273);
- carcinogenicity (1274);
- the adverse effects of herbal antidepressants (1275);
- the adverse effects of Chinese herbal medicaments (1276,1277);
- the adverse effects of Ayurvedic medicines (1278);
- herb–drug interactions (1279–1290);
- pharmacovigilance of herbal medicines (1291).

Direct effects associated with herbal medicines can occur in several ways:

- hypersusceptibility reactions
- collateral reactions
- toxic reactions
- drug interactions
- contamination
- false authentication
- lack of quality control.

Some of these effects relate to product quality. While there are some data on certain of these aspects, information on other aspects is almost entirely lacking. For example, there are isolated case reports of interactions between conventional medicines and complementary (usually herbal) remedies (1292,1293,1294), although further information is largely theoretical (1297).

Even a perfectly safe remedy (mainstream or unorthodox) can become unsafe when used incompetently. Medical competence can be defined as doing everything in the best interest of the patient according to the best available evidence. There are numerous circumstances, both in orthodox and complementary medicine, when competence is jeopardized:

- missed diagnosis
- misdiagnosis
- disregarding contraindications
- preventing/delaying more effective treatments (for example misinformation about effective therapies; loss of herd immunity through a negative attitude toward immunization)
- clinical deterioration not diagnosed
- adverse reaction not diagnosed
- discontinuation of prescribed drugs
- self-medication.

The attitude of consumers toward herbal medicines can also constitute a risk. When 515 users of herbal remedies were interviewed about their behavior vis a vis adverse effects of herbal versus synthetic over-the-counter drugs, a clear difference emerged. While 26% would consult their doctor for a serious adverse effect of a synthetic medication, only 0.8% would do the same in relation to herbal remedies (1296).

The only way to minimize incompetence is by proper education and training, combined with responsible regulatory control. While training and control are self-evident features of mainstream medicine they are often not fully incorporated in complementary medicine. Thus the issue of indirect health risk is particularly pertinent to complementary medicine. Whenever complementary practitioners take full responsibility for a patient, this should be matched with full medical competence; if on the other hand, competence is not demonstrably complete, the practitioner in question should not assume full responsibility (1297).

Severe hypotension has been attributed to a Chinese herbal mixture (1298).

- A 57-year-old man developed nausea, epigastric pain, dizziness, and diarrhea 4 hours after taking a decoction made of 14 Chinese herbs. On admission his blood pressure was 77/46 and his pulse 6 per minute. He was given intravenous fluids and the hypotension normalized within hours.

The authors pointed out that seven of the 14 herbal constituents are known to have vasodilatory effects. They therefore believed that this herbal mixture synergistically caused the hypotensive crisis.

Histamine H_2 receptor antagonists

See also individual compounds

Rapid intravenous administration of ranitidine, cimetidine, or famotidine can precipitate cardiovascular complications, notably bradycardia, hypotension, and dysrhythmias (SEDA-20, 317).

HMG coenzyme-A reductase inhibitors

The term "transaminitis" has been coined to describe a rise in the activities of serum transaminases without clinical symptoms. One author has suggested that in such cases one should switch from one statin to another, thereby preventing unnecessary withdrawal of statin treatment in dyslipidemic patients at high cardiovascular risk (1299).

A return to normal or only slightly increased values of transaminases is often seen after a short period. The overall probability of having an increase in transaminase activity more than three times the top of the reference range is 0.7% (1300). The probability may be increased in patients with pre-existing minor hepatic changes, as has been seen in one patient with systemic lupus erythematosus (1301). There seems to be no difference between the various drugs in this respect (1302), but when simvastatin and atorvastatin, each at a dose of 80 mg/day, were compared in 826 hypercholesterolemic patients there were fewer drug-related gastrointestinal symptoms and clinically significant transaminase rises with simvastatin (1303). Frank hepatitis is rare. A cholestatic picture has also been reported (1304). The mechanisms of these reactions are not known.

In one series, the frequency of liver toxicity was similar in patients taking pravastatin or simvastatin (1305), while in another study there was a difference 6 months after the start of the study when the simvastatin group showed increases in liver enzymes (SEDA-13, 1327) (1306).

In a comparison of atorvastatin with pravastatin, of 224 patients taking atorvastatin, two had clinically significant increases in alanine transaminase activity (1307). They recovered during the next 4 months, one after withdrawal of atorvastatin and the other after a dosage reduction. Withdrawals due to adverse effects were similar in the two groups. One patient developed hepatitis while taking atorvastatin, but was able to tolerate simvastatin (1308). The authors concluded that this adverse effect was not a class effect. Eosinophils in a liver-biopsy specimen pointed to an immunological mechanism.

Hormonal contraceptives—oral

See also Estrogens, Hormone replacement therapy

The cardiovascular complications of oral contraceptives include venous thrombosis and thromboembolism, arterial damage, and hypertension.

Venous thromboembolism

A central issue almost from the beginning of the oral contraceptive era has been the undoubted ability of these products to increase the risk of thromboembolic and allied complications. It was the dominant reason for the progressive reduction in hormonal content of these products during their first 20 years; it led at one point to a precipitate and poorly motivated replacement of mestranol by ethinylestradiol as the estrogenic component; and rightly or wrongly it has played a central role in the recent debate concerning modified products based on newer progestogens.

History

The fact that some women could develop thromboembolic complications as a result of taking oral contraceptives first emerged in 1961, although at that time the evidence was anecdotal and poorly quantified. The first reasonably quantified investigation conducted on a sufficient scale to merit conclusions was published in 1967 by the UK Medical Research Council (1309). This and other large studies conducted during the early years (and considered in older volumes in this series) concluded that women using oral contraceptives ran a greater risk than non-users of developing deep venous thrombosis, pulmonary embolism, cerebral thrombosis, myocardial infarction, and retinal thrombosis. Later papers and case reports described deep venous thrombosis, portal venous thrombosis, and pulmonary embolism (1310–1312). The Boston Collaborative Drug Surveillance Program follow-up study of more than 65 000 healthy women in 1980–82 found a positive association between current oral contraceptive use and venous thromboembolism (rate ratio 2.8); there was also a positive association between current oral contraceptive use and stroke or myocardial infarction (1313). A UK study using data from 1978, by which time lower-dose products were increasing in use, pointed to an approximate doubling of the risk of thromboembolism compared with controls (SEDA-7, 387). The early 1980s were nevertheless marked by a series of critical papers that sought to question the entire concept of there being a link.

Much of the work on both sides of the argument was less than watertight. Some studies failed to consider the confounding effects of other risk factors (notably smoking) or the likelihood of detection bias (particularly for venous thromboembolism, which is much more common in young women than myocardial infarction or stroke). The results of some studies were also confounded by uncertainties in the history of drug exposure. A landmark paper to resolve the issue concluded that the link with venous thromboembolism in subjects without predisposition had been consistently observed in case-control and cohort studies (1314). However, the evidence regarding myocardial infarction, various types of stroke, and cardiovascular mortality was less consistent. By 1990, an international Consensus Development Meeting reached agreement on the following statement regarding the relation between oral contraceptive use and cardiovascular disease (1315):

> The majority of epidemiological studies strongly suggest an association between current oral contraceptive use and certain cardiovascular deaths. Although the relative risk is increased, the absolute risk is small. Because the risk of myocardial infarction is apparent in current users, disappears on cessation of use, and is not associated with duration of use, there is no epidemiologic support for the hypothesis that risk of cardiovascular diseases is of atherogenic origin ⋯ . Whether particular formulations or progestogens have qualitative advantages or disadvantages merits further study. Estrogens and progestogens interact at many levels, and in epidemiologic studies of users of combined oral contraceptives it is difficult to assign a risk to either component separately. Moreover, it is physiologically unsound to do so ⋯ . Alterations in plasma lipid, carbohydrate, and hemostasis variables are of major importance for the development of cardiovascular diseases, and their concentrations can be influenced by sex steroids, including artificial steroids contained in oral contraceptives. The pharmacodynamic responses ⋯ are dependent on not only the type and dosage of sex steroids, but also on intra- and interindividual variability in pharmacokinetics.

Frequency

It has been confirmed that the incidence and mortality rates of thrombotic diseases among young women are low (1316). However, the risk is increased by oral contraceptives and there is variation in the degree of risk, depending on the accompanying progestogen. The spontaneous incidence of venous thromboembolism in healthy non-pregnant women (not taking any oral contraceptive) is about five cases per 100 000 women per year of use. The incidence in users of third-generation formulations is about 15 per 100 000 women per year. The incidence in users of third-generation formulations is about 25 per 100 000 women per year: this excess incidence has not been satisfactorily explained by bias or confounding. The risks of venous thromboembolism increase with age and is likely to be increased in women with other known factors for venous thromboembolism, such as obesity. The risk in pregnancy has been estimated at 60 cases per 100 000 pregnancies.

The Medicines Commission of the UK has reviewed all currently available relevant data and has confirmed that the incidence of venous thromboembolism is about 25 per 100 000 women per year of use (1317). The incidence of venous thrombembolism in users of second-generation combined oral contraceptives is about 15 per 100 000 women per year of use. This indicates a small excess risk of about 100 000 women-years for women using third-generation combined oral contraceptives containing desogestrel or gestodene, which has not been satisfactorily explained by bias or confounding. However, the absolute risk of venous thromboembolism in women taking combined oral contraceptives containing desogestrel or gestodene is very small and is much less than the risk of venous thromboembolism in pregnancy.

The incidence of venous thromboembolic disease in about 540 000 women born between 1941 and 1981 and taking oral contraceptives was 4.1–4.2 cases per 10 000 woman-years (1318).

In another study, the figures ranged from 1895 events per 100 000 women-years when norgestimate was used to 3969 per 100 000 women-years when desogestrel was used (1319). Although the authors did not find the difference statistically significant, it runs parallel to findings from other work regarding a higher risk when third-generation progestogens are used.

In women aged 15–29 years who used oral contraceptives containing third-generation progestogens, venous thromboembolism was twice as common as arterial complications. In women aged 30–44 years of age the number of arterial complications exceeded the number of venous complications by about 50%. However, in women under 30 years, deaths from arterial complications were 3.5 times more common than deaths from venous complications and in women aged 30–44 years 8.5 times more common. Women over 30 years of age who take oral contraceptives containing third-generation progestogens may have a lower risk of thrombotic morbidity, disability, and mortality than users of second-generation progestogens. However, a weighted analysis such as this does not result in any consistent recommendation of a particular progestogen type.

Nevertheless, some groups continue to produce data from their own systems that fail to confirm this. Some of these studies, including unpublished data circulated to experts for purposes of special pleading, have used selected material, and one can only consider them flawed. On the other hand, Jick et al. may be entirely right in their finding that, insofar as the special risk of idiopathic cerebral hemorrhage is concerned, no material difference in risk has been demonstrated between products of the second generation and those of the third generation (1320).

And in another study it was found that users of oral contraceptives with second-generation progestogens have 30% greater increased risk of thrombotic diseases, a 260% greater increased risk of thrombotic deaths, and a 220% greater increased risk of post-thrombotic disability than users of oral contraceptives with third-generation progestogens (1316).

Effects of dosage and formulation

As noted above, the early recognition of thromboembolic complications had repercussions for the formulation of the oral contraceptives; the progestogen and estrogen contents were both progressively reduced, and in 1969/70 mestranol was replaced by ethinylestradiol, on the grounds that the dose could thereby be halved (although it is not at all certain that this reduced the estrogenic contribution to thromboembolic events). By the late 1980s, when a major cohort study based in Oxford examined the problem (1321), products containing estrogen 50 micrograms accounted for about 70% of the woman-years of tablet use, most of the remainder being accounted for by lower doses.

By the mid-1990s it was reasonably well proven that progressive dose reduction had reduced the risk of thrombotic complications (1322). Epidemiological studies showed that users of low-dose combined products had small, and often statistically non-significant, rises in the risks of myocardial infarction (SEDA-16, 465; 1323,1324), thrombotic stroke (1325–1327), venous thromboembolism (1328), and subarachnoid hemorrhage (1325,1327,1329). Several of these studies compared the risks presented by different doses and found somewhat higher rates for products containing more than 50 micrograms of estrogen but somewhat lower rates among women currently using the lower-dose formulations (1323,1325,1326,1330).

A group in The Netherlands has stressed the fact that even though the risk of venous thrombosis is small in absolute terms, oral contraceptives form the major cause of thrombotic disease in young women. The risk is higher during the first year of use (up to one per 1000 per year), among women with a prothrombotic predisposition, and with third-generation progestogens (1331).

Presentation

The commonest presentation is deep venous thrombosis in the leg, which can lead to pulmonary embolism. Fatal pulmonary embolism has even been reported after intravenous injections of conjugated estrogens (1332). Despite the improvement noted with reduced doses, incidental case reports of severe cardiovascular events during the use of low-dose products have continued to appear. They include incidents of cerebral venous thrombosis and subarachnoid hemorrhage, fatal central angiitis, sinus thrombosis, and cerebral ischemia. In one series of 22 cases of cerebral infarction involving either arteries or veins, all the oral contraceptives that had been used contained a low dose of ethinylestradiol (1333). Thromboembolism in other veins, such as the hepatic vein, that is Budd–Chiari syndrome, has occasionally been reported (1334); the first 10 such cases were reported as long ago as 1972 and the increasing number since then has at times raised some concern (SEDA-6, 344; SEDA-7, 386). A case of renal vein thrombosis has also occurred (1335). Incidental reports continue to appear of thrombotic incidents in relatively unusual forms, including a further case of mesenteric thrombosis leading to intestinal necrosis (1336) and a report of fatal pulmonary embolism following intravenous injections of conjugated estrogens (1332).

The incidence of hepatic veno-occlusive disease in 249 consecutive women treated with norethisterone who underwent allogenic hemopoietic stem cell transplantation was 27% compared with 3% in women without this treatment (1337). One-year survival rates were 17% and 73% in patients with ($n = 24$) or without veno-occlusive disease ($n = 225$) respectively. Because of this adverse effect, norethisterone should not be used in patients undergoing bone-marrow transplantation. Heparin prophylaxis does not affect the risk of death from veno-occlusive disease.

Susceptibility factors

By 1980 it was considered clear that the risk of thromboembolic events was further increased under particular conditions. It was higher in smokers, in older women, and in the obese, and appropriate warnings were issued. The fact that these warnings to a large extent eliminated the high-risk individuals who had formed part of the early population of oral contraceptive users means that data from the early period cannot be used to provide a valid historical comparison with later findings (1338,1339).

Thrombotic diathesis

Early epidemiological data on the recurrence of thrombosis (1340) indicated something of an inherited predisposition, and others found a low content of fibrinolytic activators in the vessel wall of women who 6–12 months earlier had experienced a thrombotic complication while using oral contraceptives (1341); high doses of estrogen affected the concentrations of such activators (1342). However, such lesions are apparently not exclusive to users of oral contraceptives (SEDA-8, 360) and examination of the vessel wall is not of predictive value in determining risk.

The risk of venous thrombosis among carriers of the factor V Leiden mutation is increased eight-fold overall and 30-fold among carriers who take an oral contraceptive (1343,1344). This mutation results in resistance to activated protein C and thereby potentiates the prothrombotic effect of oral contraceptives. Early work suggested a greater risk after major surgery (1345,1346). However, epidemiological studies of postoperative venous thromboembolism are limited and disputed (1347). There is no documented excess risk of postoperative thrombosis associated with low-dose combined oral contraceptives among women without other risk factors (1348). If it is correct that some effects on coagulation persist for several weeks, it is wise to withdraw these products a month before surgery (1349). From a practical point of view, any decision regarding possible discontinuation of combined oral contraceptives before surgery should take into consideration the need for alternative and adequate contraception during the interim. If the woman chooses to stop taking combined oral contraceptives, progestogen-only formulations (as well as barrier methods) are deemed suitable (1347). A recommended alternative to discontinuation of a combined oral contraceptive is heparin in low doses (1348).

Age

The increase in risk with age is clear (1349), although the underlying risk of cardiovascular disease also rises as age progresses. The US Food and Drug Administration has concluded that the benefits may outweigh the risks in healthy non-smoking women over age 40, and it has in most countries been common for over two decades to advise reticence in the use of oral contraceptives after the age of 35.

Obesity

Obesity has repeatedly been shown to play a role, and its relevance to particular types of complication has been demonstrated. The Oxford Family Planning Association's 1987 data showed that the risk of myocardial infarction or angina increased significantly with weight (1350).

Smoking

Smoking has been very clearly incriminated as a susceptibility factor for thromboembolism and arterial thrombosis in women taking the oral contraceptive, and its apparently synergistic role has been well defined and quantified (1351,1352). The 1989 case-control analysis of the RCGP cohort study estimated the relative risk of myocardial infarct during current oral contraceptive use at only 0.9 for non-smokers, but at 3.5 for women smoking under 15 cigarettes per day, and as much as 21 for users of more than 15 cigarettes per day (1353). Smoking increases not only the risk of myocardial infarction among oral contraceptive uses (1323), but also the risk of angina pectoris (1350), thrombotic stroke (1324,1325), and subarachnoid hemorrhage, and it can double or treble mortality (1356–1356). There has also been further confirmation that the effect is dose-related, light smokers having twice the risk of coronary heart disease and heavier smokers having up to four times the risk, compared with non-smokers; cessation of smoking is accompanied by a reduction in risk of coronary heart disease to the level prevalent among non-smokers within 3–5 years (1352).

Smoking contributes to effects on the procoagulation process in young women (1357). The effects of oral contraceptives on the coagulation system are much greater in smokers than non-smokers (1341). Oral contraceptive users who were smokers generally have significantly lower fibrinolytic activity than non-smokers (1358), but not consistently (1359).

Ethnicity

Much of the evidence on the occurrence of thromboembolic complications with oral contraceptives or hormone replacement therapy has been gathered from European or American populations, and it can be helpful to identify data from other parts of the world, where factors such as body weight, climate, or diet could affect the incidence. When a Japanese group sought to obtain national data by mailing questionnaires to a large number of institutes monitoring these forms of treatment, 771 (71%) of 1083 institutes responded (1360). Follow-up questionnaires were sent to 39 institutions that reported having experienced in all 53 cases of thromboembolism during hormone therapy; 29 cases related to oral contraceptives and 13 to hormone replacement therapy, while 11 had taken other forms of hormone treatment. Of the 29 patients taking contraceptives, eight had developed arterial thromboembolism (including two with myocardial infarctions).

Others

Women of blood group O have less of a risk of thromboembolism (1311). The risk of thromboembolic complications may be greater where there is a history of diabetes, hypertension, and pre-eclamptic toxemia. In some studies there has been an association with type II hyperlipoproteinemia, hypercholesterolemia, and atheroma (1361–1365). Hypertension may be an additional risk factor when considered in relation to oral contraceptive use.

Mechanisms

Study of the mechanisms that might underlie the link between oral contraceptives and thromboembolic events is of importance in developing safer formulations, but also in identifying, if possible, individuals at particular risk who should be advised to change to alternative contraceptive methods.

The 1990 international consensus statement cited above noted that: "Oral contraceptives induce alterations in hemostasis variables. There are changes in the concentrations of a large number of specific plasma components of the coagulation and fibrinolytic systems, although usually within the normal range ⋯ . It is conceivable that these effects are estrogen-mediated because they have not been demonstrated in progestogen-only preparations. There is a dose-dependent relationship in the case of estrogen, although in combination tablets, the progestogens might exert a modifying effect ··· . Further attention should be given to changes in factor VIIc and fibrinogen induced by oral contraceptives and also to the association between carbohydrate metabolism and fibrinolysis."

Quantification of coagulation factors is notoriously difficult, because of the interrelations among the various components of the coagulation cascade, the broad range of normal values, and considerable inter-laboratory variability (1348). This variability is illustrated by a WHO study of users of combined oral contraceptives, conducted on several continents, which showed statistically significant differences among clinical centers in prothrombin time, fibrin plate lysis, plasminogen, and activated partial thromboplastin time (SEDA-16, 464). Effects also vary between different populations, users of different doses, users of different products, and tests performed at different periods of the medication cycle (1359,1366).

The term "hypercoagulability" has been used to describe a supposed pre-thrombotic state, identifiable by certain changes in the hemostatic system, but to date there is no broad-spectrum laboratory test for assessing the risk of thrombosis in a given individual, although coagulation changes in vitro have sometimes been regarded as proof of a thrombotic state. Deviations in laboratory data from patients with thromboses have often been interpreted as demonstrating the cause of the thrombosis, whereas they may simply be a consequence.

Despite the variations that are found, the overall conclusion is that oral contraceptives cause an increase in coagulation factors I (fibrinogen), II, VII, IX, X, and XII, and a reduction in antithrombin III concentrations, which would be expected to predispose to venous thromboembolism, especially if not counterbalanced by an increase either in fibrinolytic activity or of other inhibitory proteins of the coagulation, such as protein C (1367).

There is also fairly strong evidence that immunological mechanisms play a role in thrombotic episodes associated with oral contraceptives, especially when they occur in the absence of risk factors for vascular disease (1368), although this has been contested (1369). In one series of reports on cerebral infarction, circulating immune complexes and/or specific antihormone antibodies were found in 15 of 20 patients (1333). In a large series of women with venous or arterial thrombosis, anti-ethinylestradiol antibodies were absent in non-users but present in 72% of users; they were also present in 33% of healthy oral contraceptive users without thrombosis (SEDA-16, 465). In half of the cases there were both anti-ethinylestradiol antibodies and a history of smoking were found jointly in half of the cases.

There is a significant rise in fibrinogen concentrations during the early months of oral contraceptive use, and concentrations return to baseline after withdrawal (1370). Prolonged use of oral contraception also seems to lower concentrations of antiaggregatory prostacyclin (1371).

Some work that was considered to show severe acquired plasma resistance to activated protein C among users of third-generation (as opposed to second-generation) products has been re-examined by a French group (1372). In their view the technical measures used to demonstrate the effect of activated protein C introduced a bias of interpretation and hence false results; they have further argued that such a test cannot demonstrate the presence of a raised thromboembolic risk in asymptomatic women taking these contraceptives, since it is non-specific and subject to changes in the plasma concentrations of many coagulation factors that are themselves increased or decreased by estrogens and progestogens. They point, for example, to protein S (1373), changes in which account for the differential effect of oral contraceptives on Rosing's assay (1374), but which are in their view irrelevant to issues of thromboembolic risk with oral contraceptives; the androgenic potential of the progestogen may further counteract the effect of estrogens in the test. More generally, in such a complex situation in which there is a "modification of the modification," there is no hemostasis-related test that provides a risk indicator for thrombosis. This argument is sound, but it naturally remains theoretical; the question of thromboembolism with the third-generation products must, as pointed out above, be resolved on the basis of epidemiological data, and certainly those data now strongly point to an increased risk.

Relative roles of estrogens and progestogens

The risk of thrombosis is closely associated with the estrogen component (1375) for both arterial and venous events and with the progestogen for arterial events; however, if a particular progestogen is metabolized to estrogen or raises estrogen concentrations, it will make a contribution to venous complications. Estrogen alone has after all been incriminated as a cause of thromboembolism when given to men (1376); the risk of puerperal thromboembolism after estrogen inhibition of lactation has been shown in several studies (1377); non-contraceptive estrogens clearly increase the risk of acute myocardial infarction in women under 46 years of age (1378). Changes in coagulation factors appear to be related to the estrogen dose (1341,1342,1379,1380). Progestogen-only formulations do not have any significant effect on the coagulation system.

Third-generation oral contraceptives and thromboembolism

There are now reasons to doubt whether the third-generation oral contraceptives are indeed safer than their predecessors in respect to thromboembolism and substantial grounds for believing that they present greater risks.

The first reason is theoretical. The demonstrated effects of the new substances and combinations on lipids and carbohydrates do not have any major relevance to the thromboembolic process. The latter is linked primarily to changes in the hemostatic system and blood coagulation, involving platelet aggregation, coagulation factors, fibrinogen concentrations, and blood viscosity.

The second reason is kinetic. It is true that the dose of estrogen (probably the main instrument in inducing thromboembolism) has been kept to a minimum, but the new progestogen gestodene tends to accumulate in the system with continued use, and the concentrations of ethinylestradiol increase simultaneously; this increase is due to the ability of gestodene to inhibit cytochrome P450 and therefore to inhibit the breakdown of estrogen, as well as its own metabolism (1381). Similar findings emerged with desogestrel, although they were somewhat less marked (1382).

The third reason is hematological. The third-generation contraceptives have greater adverse effects on the clotting system than those of the second generation. In particular, women using the third-generation products have a greater resistance to activated protein C (1383), a shift that is associated with a higher risk of thrombosis.

The fourth reason is epidemiological. During the period 1987–88, when the third-generation products were relatively new, anecdotal reports of thromboembolic events appeared, including at least one death, and partly for this reason a series of large controlled studies were set up. The findings of three such studies (British, European, and global) became available to the drug control authorities late in 1995 and were subsequently published. The UK Committee on Safety of Medicines, considering all three, concluded that the risk of venous thromboembolic events in these third-generation oral contraceptives was about double that in users of the previous generation of products using the older progestogens (30 as opposed to 15 per 100 000 woman-years, the risk in healthy women being only five per 100 000). Despite the different populations studied, the individual studies produced broadly consistent results. The global study on four continents found a relative risk of 2.6 when comparing the desogestrel/gestodene products with the older variety, while the European study found a relative risk of 1.5–1.6. Various later papers pointed in the same direction. In Denmark, there was an increase in hospital admissions for primary venous thromboembolism in young women coinciding with the introduction of the third-generation products (1384). Papers from The Netherlands have confirmed the main trend (1385,1386).

Authoritative reviews and editorials have further confirmed the correctness of the above findings. There has been some criticism of the individual studies on various points of detail, but it is difficult to see that this in any way undermines what is now very consistent evidence that the third-generation oral contraceptives increase the risk of thromboembolic events to a substantially greater degree than previous products. Some work that has been advanced as pointing to the safety of third-generation products (1387) proves to relate primarily to the second-generation combinations, with only a few late entrants using the more recent oral contraceptives, and other work was performed on a very small scale.

As a rule, the study of adverse reactions must relate to current and emergent issues. However, now and again it can be instructive to look back into recent history. When a drug problem has been fairly clearly defined, and particularly when it has for a time been the subject of debate and even frank controversy, one can learn something from the processes involved. How did the facts become known? Why did the controversy emerge? And could the risk have been detected and eliminated earlier?

Since their appearance in the late 1950s, oral contraceptives have gone through several stages of development. What are now in retrospect referred to as first-generation oral contraceptives were high-dose combinations of progestogens (more particularly norethynodrel, norethisterone, and lynestrenol in doses of 2.5 mg or more) and the estrogen mestranol 75 micrograms. A decade later a second generation emerged, with substantially lower doses, commonly half of those used earlier and some new progestogens, notably the more potent levonorgestrel. Finally, in the early 1980s some manufacturers introduced so-called third-generation products, a particular characteristic of which was the use of entirely new, very potent progestogens, among them desogestrel and gestodene. Clinical studies of contraceptives that contained gestodene and desogestrel suggested that they are very similar to one another, although differences in dosage and potency could account for reports that products that contain gestodene provide better cycle control (1388).

Almost from the earlier years, the risk of thromboembolic complications among users of "the pill" was recognized, and by the mid-1960s it was well documented (1389,1390). Progressive reductions in dosage, in particular that of the estrogenic component, during the period that first- and second-generation products held sway were widely regarded as having reduced this risk to manageable proportions, although it was not eliminated. The relative risk with first-generation products was highly variable (2–11), but the best work in the UK and the USA fairly consistently reached an estimate of 4–6 (1391–1393). With the second-generation products the relative risk of thromboembolic complications was again variously estimated, but a large cohort study published in 1991 set it at 1.5 with products containing the lowest doses of estrogen, and 1.7 with products containing intermediate doses of estrogen (1330).

The fact that both prescribers and users of medicines are likely to anticipate that new drugs will be in some way better than those that have gone before means that both groups are in principle receptive to potentially spurious claims and suggestions. By the time the third-generation oral contraceptives were marketed, this type of contraception had been around for a quarter of a century; the risk of thromboembolism, the most widely publicized problem in the field, seemed by that time to have receded with progressive reductions in dosage. There was every reason to hope that it would recede further with the newest generation of products. That expectation was further nurtured by the even lower doses latterly attainable. It also seems to have been fostered by some of the suggestive promotion that appeared, although that in fact related as a rule merely to an improved lipid spectrum, which in turn raised the theoretical possibility, also discussed but not documented by some clinical investigators (1394), that arterial and cardiac risks might be less.

What in fact happened was that by 1989 alarm bells began to ring in Germany, where the regulatory authorities were alerted to the submission of an unusually high number of spontaneous reports of thromboembolic complications thought to be associated with the new products. Cases continued to accumulate, long-term studies already begun were completed, and in 1995 Britain's Committee on Safety of Medicines made a public statement to the effect that the risk of thromboembolic complications among hitherto healthy users of third-generation products was approximately twice than that seen with second-generation products (SEDA-19, xix). The studies in question, including work by the World Health Organization and others (SED-14, 1410), were subsequently published and confirmed that conclusion, as did later work (1395). It was further reinforced by others (1396), who worked on a smaller scale but provided well-documented evidence that while a factor V Leiden mutation or a biased family history could increase the risk in individual cases, they did not explain the higher thrombosis risk seen with a product based on desogestrel than with contraceptives that incorporated levonorgestrel, norethisterone, or lynestrenol.

Currently one must ask why the particular risk of the third-generation contraceptives was identified so late. These third-generation products had been in development since the late 1970s and the first had been marketed in 1981–82, some 14 years before the Committee on Safety of Medicines issued its statement. Could society not have done better and thereby reduced the risks to which women were exposed? There are two principal answers, both of them at least partly in the affirmative.

The first is that products of this type could well have been entered at an earlier date into large studies of oral contraception and their effects. A series of university centers around the world, as well as bodies such as Britain's Royal College of Physicians and Royal College of General Practitioners, have throughout the oral contraceptive era either sponsored or participated in prolonged cohort and case-control studies of these products. Experience with data on thromboembolism suggests that significant data are likely to be obtainable in a cohort study of manageable size within some 5–7 years. The use of third-generation products may have been small in the early years, but they were aggressively promoted in major oral contraceptive markets to ensure rapid growth, in all probability sufficient to provide adequate recruitment. One would hesitate to argue that such studies should be a universal condition of the marketing of drugs, but when the products concerned have immense social significance and considerable potential for good and harm, as the oral contraceptives do, and when the compounds involved are entirely new, there is at least a sound medical reason for such work in every case. That work was performed with successive forms of the earlier oral contraceptive products, in which dosages were progressively reduced, and there was particular reason to set it in motion on the introduction of products that contained new chemical components with some significant structural and pharmacological differences from the older progestogens. A straightforward statistical calculation shows that an early cohort study involving some 30 000–50 000 women taking a third-generation product could within 2 years have shown the degree of increase in the thrombotic risk, which was actually not elicited until much later.

The second answer with respect to the earlier acquisition of risk data must come from the laboratory. Not from animal studies, which in this field are of very restricted value, but from biochemical and particularly hematological work. When during the 1990s various groups began to examine in detail the effects of the third-generation contraceptives on processes related to the clotting system, they identified a series of properties that could very well explain an increased incidence of thrombosis.

The first of these was an increase in circulating concentrations of factor VII produced by the desogestrel plus estrogen combination, which was some 20–30% higher than that seen with a second-generation product based on levonorgestrel (1397). The methods used to carry out this work were available before 1988 (1398), and it is not at all clear from the published material whether there was a failure to compare the two generations in this respect at an early date, or whether such work was performed and either overlooked or misinterpreted.

A second finding related to the effects of activated protein C on thrombin generation in low-platelet plasma via the intrinsic or extrinsic clotting pathways. Using a method developed on the basis of work first published in 1997 (1399), a Dutch group in Maastricht found that all types of combined oral contraceptives induced acquired resistance to activated protein C. With the third-generation contraceptives, however, the effect was significantly more marked than with those of the second generation: in other words, these drugs significantly reduced the ability of activated protein C to down-regulate the formation of thrombin (1383). However, this work only became feasible in the late 1990s.

A third underlying mechanism seems to involve a reduction in concentrations of free protein S, again more pronounced with third-generation products. When protein S falls, the antifibrinolytic effect of the so-called thrombin-activated fibrinolysis inhibitor is increased; in other words, fibrinolysis is impeded, with an increased risk of clotting problems (1400). Again, however, these are recent methods, which were not available when the third-generation products were launched.

The laboratory findings therefore suggest that a greater thrombosis-inducing effect of the third-generation oral contraceptives can be explained and even anticipated on the basis of known mechanisms. Not all the relevant methods were available in the early years, but that relating to factor VII most certainly was. It is unfortunate, to say the least, that such work was either not performed or not properly interpreted.

All in all, had a combination of hematological methods and field studies been initiated sufficiently soon, the increased risk of thromboembolism with the third-generation oral contraceptives could have been detected some years earlier, sufficient for society to take decisions on the benefit-to-harm balance of these drugs before so much needless injury was incurred.

The third-generation oral contraceptives: a judicial assessment

It was extraordinary to find a major epidemiological dispute regarding drug safety being handled by the High Court in England in late July 2002, when the Court handed down its decision regarding thromboembolic events induced by the third-generation oral contraceptives (1401). Essentially, a group of women who claimed to have been injured as a consequence of having using this latest version of "the pill," based on two new progestogens, had sought to reclaim extensive damages from the manufacturers, since in their view the product did not possess the degree of safety which, in the words of European law, the user was legitimately entitled to expect. Since the safety achieved with the widely used products of the second generation was so widely regarded as acceptable, the Court had to decide whether the newer products had significantly failed to meet that standard. Faced with a long procession of expert epidemiological witnesses from both sides, and with some flat contradictions, the judge was obliged to rule on their arguments.

However, that it was an English court in which the issue came to be debated was not surprising, for it was in England that the Committee on Safety of Medicines had written to prescribers in 1995 stating that three unpublished studies on the safety of combined oral contraceptives in relation to venous thromboembolism had indicated about "a two-fold increase in the risk of such conditions" compared with the preceding generation of products. This issue of a "two-fold increase" became crucial to the case. "For reasons of causation," as the Judge put it, the claimants had accepted the burden of proving that the increase in risk was not less than two-fold.

In fact, the English authorities, having rejected a vigorous defence of these products by the manufacturers, were by 1999 speaking more precisely of an increase in risk, as compared with the earlier products "of about 1.7–1.8 after adjustment," which was "not fully explained by bias or confounding"; appropriate label warnings were therefore imposed. These new warnings, summarized, said that an increased risk associated with combined oral contraceptives generally was well established, but was smaller than that associated with pregnancy (60 cases per 100 000 pregnancies). In healthy non-pregnant women who were not taking any combined contraceptive it was about five cases per 100 000 woman-years; in those taking the second-generation products it was about 15; and for third-generation products it was about 25. By September 2001, the European Union's Committee on Proprietary Pharmaceutical Products had formed its own view, and here too it was concluded that the "best estimate of the magnitude of the increased risk is in the range of 1.5–2.0."

In Court to support the claimants, Professor Alexander Walker assessed the relative risk of the third-generation products at 2.2, Dame Margaret Thorogood at 2.1, and Professor Klim McPherson at about 1.9. The experts for the defendants took the view that the relative risk was well below two, and could well be zero. As Mr Justice Mackay noted, having listened to these experts: " ··· the debate between them has been unyielding, at times almost rancorous in tone, and with a few honourable exceptions ··· devoid of willingness to countenance that there may be two sides to the question. So, science has failed to give women clear advice spoken with one voice."

There was also fundamental disagreement on confidence intervals when calculating relative risks in such matters: "The Defendants say that to establish causation in the individual, and therefore a relative risk which is greater than two, there must be seen not just a point estimate but also a lower confidence interval which is greater than two in order for the result to be significantly different from two."

The Court was faced with "a series of studies with different point estimates and largely overlapping confidence intervals. Time after time experts have had their attention drawn to point estimates from studies that appear, to the layman's eye, to be very different. Almost invariably they have dismissed those apparent differences by reference to the overlapping confidence intervals, saying that the figures are statistically compatible and there is no significant difference." Confronted with such material, the Court chose to set aside as inexact and theoretical much of the statistical rhetoric. Having done that, the Judge felt himself in a position to emerge "from that forest into broader more open country where the simpler concept of the balance of probabilities rules." Constructing his judgement in that way, Mr Justice Mackay advanced in the course of 100 pages to the conclusion that the claimants had failed to demonstrate a doubling of the risk. In his view, "the most likely figure to represent the relative risk is around 1.7."

This extraordinary and wise judgement merits most careful reading by anyone anxious to understand the safety issues surrounding oral contraceptives. First, because of the insight that it demonstrates into the manner—not always edifying—in which evidence in this vital matter has been adduced, interpreted, and argued over in the course of more than a decade. Secondly, because it arrives, through a process of tight reasoning, at what is for the moment the most reliable conclusion we have. It seems beyond all possible doubt that the third generation of oral contraceptives is primarily characterized by an increased risk of thromboembolic complications. Whether that risk is great enough to warrant financial compensation is a matter for lawyers to decide. But given the lack of any tangible benefit to the user, the risk is clearly significant in human terms, and it is hard to see that there is any valid reason at all for continuing to use these products.

Prognosis

Despite the vast literature on thromboembolic complications of oral contraceptives, little attention has been paid to factors that may determine the ultimate prognosis and risk of death. Data from the Swedish Adverse Reactions Monitoring Bureau and other sources have now been used to study this question in regard to pulmonary embolism, as well as estimating the incidence (1402). Over 36 years (during which the spectrum and usage of oral contraceptives naturally changed) 248 cases of suspected pulmonary embolism were reported. The presence of thromboembolism was confirmed in all fatal cases and 83% of non-fatal cases. The medical records showed that the presence of nausea or abdominal pain, age above 35 years, concomitant treatment with other drugs that increase the risk of thromboembolism, vein or lymph vessel malformations, and a deep vein thrombosis above the knee were positively associated with a fatal outcome. Chest pain and previous use of a combined oral contraceptive were negatively associated with a fatal outcome. Using pharmacy records to estimate sales, the incidence of verified pulmonary embolism was calculated as 1.72 per 100 000 treatment years; the figure for fatal cases was 0.25 per 100 000 treatment years.

Hormonal contraceptives—progestogen implants

Norplant does not alter the blood pressure (1403).

Hormonal replacement therapy—estrogens + androgens

Androgens appear in some respects to counter the desired effects of estrogens. Doppler flowmetry has been used to study the cardiovascular effects of adding an androgen to an estrogen in an open, randomized study in 40 patients over 8 months, all of whom were using transdermal estradiol (50 micrograms/day) and cyclic medroxyprogesterone acetate (10 mg/day) (1404). Half of the subjects then received additional testosterone undecanoate (40 mg/day). The investigators concluded that while the androgen improved sexual desire and satisfaction and had no effect on endometrial thickness, it did in part counteract the beneficial effects of the estrogens on cerebral vascular activity and lipids. The most notable change was a significant increase in the pulsatility index of the middle cerebral artery. The androgen also resulted in a 10% reduction in HDL cholesterol concentration within 8 months. The authors therefore urged caution in using androgens, at least in the manner used in this study.

There are naturally some groups that have worked together with manufacturers to profile the supposed advantages of particular estrogen + androgen regimens, especially when these are available in the form of fixed combination formulations.

The adverse effects of estrogen + androgen therapy include mild hirsutism and acne (1405). One group of workers, who examined the use of "Estratest" (an esterified combination of estrogen and methyltestosterone), concluded that in their experience under 5% of women developed acne or facial hirsutism, a frequency similar to that experienced when using conjugated estrogens 0.625 mg/day. Women had significantly less nausea with the estrogen + androgen treatment than with conjugated estrogen therapy. Cancers, cardiovascular disease, thromboembolism, and liver disease were stated to be rare among users of the combination. The only adverse events exceeding 4% of total reports were alopecia, acne, weight gain, and hirsutism (1406). However, much higher rates of complications with such combinations have been reported from other centres (1407).

The evident disadvantage of a fixed combination is that it renders it impossible to carry out any fine adjustment of dosages, such as might be called for in the light of the clinical response and adverse effects in a given individual.

All in all, it seems very doubtful whether any of the supposed benefits of androgen therapy justify the risks involved, except possibly as a transitional measure in those recently oophorectomized women who have acute symptoms of sudden androgen withdrawal.

Hormone replacement therapy—estrogens

One of the most serious aspects of the thromboembolic complications now widely acknowledged as being associated with HRT is that their emergence coincides with the development of the conclusion that the role of HRT in reducing the risk of coronary heart disease is at best unproven. A form of treatment that was originally viewed as potentially beneficial to the cardiovascular system is at present on balance perhaps harmful (1408).

A US group examined potential risk factors for venous thromboembolic events in women assigned to HRT in the Postmenopausal Estrogen/Progestin Interventions (PEPI) study, a 3-year double-blind study in 875 postmenopausal women designed to assess the effects of HRT on heart disease risk factors (HDL cholesterol, fibrinogen, blood pressure, and insulin) (1409). Women with a history of estrogen-associated venous thromboembolic events were excluded. Ten women, all assigned to HRT, had a venous thromboembolic event during the study. Only baseline fibrinogen varied significantly between those who had a venous thromboembolic event while assigned to HRT event (mean 2.49 g/l) and those who did not have an event (mean 2.81 g/l). Adjusting for covariates did not affect this finding. As the authors remarked, the lower fibrinogen concentrations among women who subsequently reported venous thromboembolic events may be a marker for a specific, but as yet undefined, coagulopathy that is magnified in the presence of exogenous hormones. However, larger studies are needed to confirm this hypothesis.

Since much of the evidence of thromboembolic complications with HRT relates to the use of conjugated equine estrogens, the degree of risk when natural 17-beta-estradiol was used instead has been examined in Norway in a population-based case-control study involving consecutive women, aged 44–70 years, discharged from a University Hospital between 1990 and 1996 with a diagnosis of deep venous thrombosis or pulmonary embolism (1410). Women with cancer-associated thrombosis were excluded. Random controls were used. The material comprised 176 cases and 352 controls, that is two controls for each case. All the women who received HRT had been given estradiol. The frequency of HRT use was 28% (50/176) in cases and 26% (93/352) in controls. The estimated matched crude odds ratio was 1.13 (CI = 0.71, 1.78), which shows no significant association of overall use of estradiol-based HRT and thromboembolism. However, when the duration of exposure to HRT was taken into account by stratification, there was an increased risk of thromboembolism during the first year of use, with a crude odds ratio of 3.54 (CI = 1.54, 5.2). This effect was reduced by extended use to a crude odds ratio of 0.66 (CI = 0.39, 1.10) after the first year of use. The authors concluded that the use of estradiol for HRT was associated with a three-fold increase in the risk of venous thromboembolism, but that this increased risk was restricted to the first year of use. One is bound to wonder whether this shift in risk was genuine or reflects only the limitations of the study.

Among the less common forms of thromboembolism that have been reported is occlusion of the retinal vein, familiar with the oral contraceptives but unusual with HRT (1411).

Hormone replacement therapy—estrogens + progestogens

Despite biologically plausible mechanisms whereby estrogens might be expected to confer cardioprotection in postmenopausal women, as well as observational data suggesting cardiovascular benefit, the literature continues to provide contradictory outcomes on this.

Electrocardiographic work has suggested that not only the estrogen but also the progestogen component of HRT can have some impact on the electrophysiological properties of the heart (1412), the clinical significance of which, if any, is not understood. The picture is further confused by evidence that a particular regimen may initially increase the risk, yet confer long-term benefit, as in the Heart and Estrogen/progestin Replacement Study (HERS), while in other well-planned work, such as the recent Estrogen Replacement and Atherosclerosis trial (ERA), there was no benefit (1413).

Myocardial ischemia

There is a continuing debate about the relation, if any, between HRT and coronary heart disease (1414,1415).

A paradoxical finding in the Women's Health Initiative (WHI) study was that while HRT resulted in an improvement in blood lipid concentrations there was no reduction in the incidence of coronary heart disease. A re-examination of the findings in 2005 generated the hypothesis that the key to the paradox could lie in effects on specific lipid subgroups rather than on lipids as a whole (1416). This hypothesis was tested by an evaluation of differences in coronary calcification, lipids, and lipoprotein subclasses among menopausal HRT users and non-users in a longitudinal study. Lipoprotein subclasses and coronary artery calcification (the latter measured using electron beam computed tomography) were studied in HRT users (49%) and non-users in a total of 243 women from the Healthy Women Study, who were about 8 years postmenopausal. The distribution of calcification scores was not significantly different between users and non-users and neither were there differences between the groups as regards any LDL subclass. However, regardless of HRT use, women with detectable calcification of the coronary arteries had higher concentrations of VLDL and small LDL particles, higher LDL particle concentration, and smaller mean LDL size compared with women with no detectable calcification. The fact that HRT users had higher concentrations of VLDL particles (triglycerides) and did not have a better LDL subclass distribution could explain the fact that HRT was not associated with a difference in coronary calcification in this study or with a reduction in coronary heart disease risk in randomized clinical trials.

There has been a randomized trial in 270 postmenopausal women to evaluate the effects on cardiovascular risk markers of two continuous combined estrogen + progestogen replacement products (17-beta-estradiol 1 mg with or without norethindrone acetate 0.25 or 0.5 mg) compared with unopposed estrogen or placebo (1417). LDL cholesterol was reduced to a similar extent in all those who took the active treatment (10–14% from baseline). Compared with unopposed 17-beta-estradiol, 17-beta-estradiol plus norethindrone acetate 0.5 mg enhanced the reductions in total cholesterol and apolipoprotein B concentrations. The combination of 17-beta-estradiol plus norethindrone blunted or reversed the increases in concentrations of high-density lipoprotein cholesterol, apolipoprotein A-I, and triglycerides produced by 17-beta-estradiol alone. The effects of 17-beta-estradiol plus norethindrone on hemostatic variables were similar to those of 17-beta-estradiol alone, except for factor VII activity, which was significantly reduced by 17-beta-estradiol plus norethindrone acetate 0.25 and 0.5 mg. The combination of 17-beta-estradiol plus norethindrone blunted reductions in C peptide and insulin concentrations produced by unopposed 17-beta-estradiol, but did not affect them compared with placebo. The authors concluded that 17-beta-estradiol plus norethindrone produced favorable changes in most cardiovascular risk markers and had a profile distinct from that of unopposed estrogen.

The findings of the randomized HERS suggested that in women with clinically recognized heart disease, HRT might be associated with early harm but late benefit in terms of coronary events. The findings of that study seem in the meantime to have been confirmed by some further US work. In one study the histories and subsequent course of 981 postmenopausal women who had survived a first myocardial infarct and had thereafter used estrogen or estrogen + progestogen were examined (1418). Relative to the risk in a parallel group of women not currently using hormones there was a suggestion of increased risk during the first 60 days after starting hormone therapy (RR = 2.16; CI = 0.94, 4.95) but of reduced risk with current hormone use for longer than 1 year (RR = 0.76), although the confidence intervals were wide.

However, in a second study, data on 1857 women from the Coumadin Aspirin Reinfarction Study were used to assess the incidence of cardiac deaths or unstable angina as related to the use of HRT. Of the population studied, 524 (28%) had used HRT at some point and 111 of the latter (21%) had started HRT after suffering a myocardial infarct ("new users"). Women who began HRT after their first myocardial infarct had a significantly higher subsequent incidence of unstable angina than women who had never used hormones (39 versus 20%); however, these new hormone users suffered death or recurrence of myocardial infarct at a much lower rate than never-users (4 versus 15%). These differences are striking. Prior/current users had no excess risk of the composite end-point after adjustment. Users of estrogen plus progestogen had a lower incidence of death, infarct, or unstable angina during follow-up than users of estrogen only (RR = 0.56)

(1419). As Grady and Hulley have commented in an editorial, current data seem to make it clear that "postmenopausal hormone therapy should not be used for the purpose of preventing coronary disease unless future data from well-designed randomized trials document such benefit" (1420).

Cardiac dysrhythmias

Hormone replacement therapy can cause prolongation of the QT interval on the electrocardiogram. In a recent prospective study set of the incidence and extent of this effect 3103 women were followed at intervals over 9 years and data were collected on their use of hormonal replacement and the QT interval, prolongation being defined as an increase of not less than 10% (1421). The QT_c interval was moderately but significantly longer in users of hormone replacement therapy, the risk of QT prolongation being nearly twice that in never-users. The consequences of these changes for the user's health could not be assessed and merit further study.

In 2001 the randomized Heart and Estrogen/progestin Replacement Study (HERS) produced some evidence that in women with clinically manifest heart disease HRT might produce some early harm but confer some later benefit as regards coronary events. The picture is still incomplete, but it has been filled in to some extent by a prospective study of the effects of HRT on cardiovascular function over 1 year of treatment in 46 healthy postmenopausal women, mean age 55 years, who took either estrogen replacement therapy alone (n = 23) or progestogen + estrogen replacement therapy (n = 23) (1422). The doses used were 0.625 mg/day of conjugated equine estrogen with or without medroxyprogesterone acetate 2.5 mg/day. The controls were 25 health premenopausal women, mean age 35 years. Long-term estrogen-only replacement increased the QT interval, QT dispersal, and the index of parasympathetic activity; there was also a small non-significant increase in the incidence of dysrhythmias. Long term use of progestogen + estrogen did not affect the QT interval, QT dispersion, or the frequencies of ventricular dysrhythmias or parasympathetic activity, but it did increase the incidence of supraventricular tachycardias. These findings support the idea that estrogen may directly modulate ventricular repolarization and that progestogens do not.

Thromboembolism

The thrombotic complications of combined HRT in a potentially high-risk group have been assessed in a randomized, multicenter study in the USA in 2763 women, average age 67 years (1423). All had some degree of pre-existing coronary heart disease but no previous venous thromboembolism, and none had undergone hysterectomy. They took either conjugated equine estrogens 0.625 mg + medroxyprogesterone acetate 2.5 mg or a placebo. During an average 4.1 years of follow-up, 34 women in the hormone therapy group and 13 in the placebo group had venous thromboembolism (relative risk = 2.7, excess risk = 3.9 per 1000 woman-years). The mean risk for venous thromboembolism was increased among women who had leg fractures (RR = 18) or cancer (RR = 4) and it was also raised several-fold for 3 months after inpatient surgery or non-surgical hospitalization. The risk was approximately halved by the use of aspirin or statins.

Ibuprofen

Apart from the consequences of salt and water retention, ibuprofen does not affect myocardial or vascular function. Congestive heart failure has rarely been reported (1424).

In a small comparison of ibuprofen and indometacin in preterm infants with patent ductus arteriosus there was no apparent difference in the rate of patent ductus arteriosus closure; ibuprofen did not impair cerebral hemodynamics or oxygenation, while indometacin impaired cerebral oxygen delivery (1425).

Ifosfamide

Atrial fibrillation has been attributed to ifosfamide after a dose of only 1800 mg/m^2, with mesna, in a regimen for metastatic breast cancer (1426).

Iloprost

Myocardial ischemia is unusual during infusion of iloprost. It mainly occurs in patients with pre-existing coronary disease, when it is ascribed to a steal phenomenon detrimental to the subendocardial tissue. As a rule it is transient and exceptionally proceeds to infarction. However, such an event has now been reported in a patient with systemic sclerosis (1427).

- A 57-year-old man with a 1-year history of systemic sclerosis and ischemia of several digits received a first infusion of iloprost using the recommended stepwise increasing dosage scheme; he developed sudden chest pain, with inferior ST segment elevation. Emergency coronary angiography showed an occlusion of the circumflex coronary artery, for which a stent was inserted. At angiography 3 years earlier his coronary arteries had been normal. He died 5 months later from cardiogenic pulmonary edema.

Imatinib

It has been suggested that imatinib may have caused severe heart failure and left ventricular dysfunction in 10 patients with pre-existing conditions such as hypertension, diabetes mellitus, and coronary heart disease (1428). Experimental studies have shown that imatinib induces apoptosis in isolated cardiac myocytes (1428). Several trials and a database of six registration trials have therefore been reviewed.

- The Italian Cooperative Study Group—four consecutive studies of imatinib therapy in 833 patients with Philadelphia chromosome-positive chronic myeloid leukemia, observed for a median of 19–64 months (1429). The overall cardiac mortality rate was 0.3%.

- The MD Anderson experience—clinical trials of imatinib from July 1998 to July 2006, with median follow-up of 5 years (1430). In all the imatinib protocols, standard research monitoring procedures were conducted before treatment and at regular intervals. Electrocardiography, echocardiography, and chest radiography were conducted routinely before treatment and as clinically indicated during follow-up. The eligibility criteria excluded patients with cardiac problems (NYHA classes III and IV). After reviewing all reported adverse events, particularly those that could be considered as having a cardiac origin, 22 patients (1.8%) were identified as having symptoms that could be attributed to congestive heart failure, of whom 12 had previously received interferon and three had received anthracyclines. They included nine patients reported elsewhere (1428). Their median age was 70 (range 49–83) years. The median time from the start of imatinib therapy to a cardiac adverse event was 162 (range 2–2045) days. Eighteen patients had previous medical conditions that predisposed them to cardiac disease: congestive heart failure (n = 6), diabetes mellitus (n = 6), hypertension (n = 10), coronary artery disease (n = 8), dysrhythmias (n = 3), and cardiomyopathy (n = 1). Of the 22 patients, 15 underwent echocardiography or multiple gated acquisition (MUGA) scanning at the time of the event: nine of these 15 patients had low ejection fractions, and six of these nine had significant conditions that predisposed them to cardiac disease (three had coronary artery disease, two congestive heart failure, and one a cardiomyopathy). Of the 22 patients with symptoms of congestive heart failure, 11 continued to take imatinib with dosage adjustments and management of congestive heart failure without further complications. However, with the host of confounding factors involved in these patients, the occurrence of congestive heart failure related to the use of imatinib was reasonably unambiguous in only seven of the 1276 patients reviewed (0.5%).
- Novartis clinical database—six registration trials comprising 2327 patients who took imatinib as monotherapy. These trials represented 5595 patient-years of exposure to imatinib (average exposure 2.4 years). Twelve cases of congestive heart failure (0.5%) were considered to be incident cases (with no previous history of congestive heart failure or left ventricular dysfunction) with a possible or probable relation to imatinib. If these cases are related to the 5595 patients-years of imatinib exposure the incidence of congestive heart failure is 0.2% per year across all trials
- In the largest international, randomized phase III study reported to date, 1106 patients with newly diagnosed chronic myeloid leukemia were randomized to either initial therapy with imatinib or the previous standard treatment of interferon plus cytosine arabinoside (1431). Both regimens were examined for cardiac safety according to an analysis of adverse events as described above. The incident cases of cardiac failure and left ventricular dysfunction, possibly or probably related to exposure to the study medication, was 0.04% per year (1 case in 2309 patient-years) for patients taking imatinib versus 0.75% per year (four cases in 536 patient-years of exposure) in patients taking interferon + cytosine arabinoside.

Imatinib therapy as a cause of congestive heart failure seems to be rare. When it occurs, the symptoms most commonly occur in elderly patients with pre-existing cardiac conditions and may often reflect predisposing cardiac compromise compounded by some element of fluid retention. Patients with a previous cardiac history should be monitored closely and treated aggressively with diuretics if they develop fluid retention.

Immunoglobulins

Intravenous immunoglobulin expands the plasma volume and increases blood viscosity, which can lead to volume overload in patients with cardiac insufficiency (1432). Thromboembolic events (1433,1434,1435,1436,1437,1438), myocardial infarction, and stroke have been reported after high-dose treatment with intravenous immunoglobulin. Risk factors are obesity, advanced age, immobilization, hypertension, diabetes mellitus, and a history of vascular disease or thrombosis (1439).

Stroke has been described in 16 patients, 15 of whom had recognized risk factors (1440). These events appear to be related to hyperviscosity of the blood after intravenous immunoglobulin (1441–1442,1443). This has been confirmed by analysis of blood viscosity after intravenous immunoglobulin infusion (1444). Patients with high plasma concentrations of albumin and fibrinogen are considered to be more susceptible to thrombotic complications after intravenous immunoglobulin infusion (1445).

Thromboembolic complications, such as stroke and acute myocardial infarction, have been observed after intravenous immunoglobulin in patients with pre-existing susceptibility factors (1446,1447). Deep venous thrombosis and pulmonary embolism have also been reported (1448). Increased serum viscosity is not always the underlying mechanism, as has been shown retrospectively in seven patients with susceptibility factors who developed such reactions (1449). Cerebral vasospasm was also excluded as mechanism. The authors suggested that the thromboembolic complications are caused by clotting factors and vasoactive cytokines within specific batches of intravenous immunoglobulin formulations.

Thrombosis in elderly patients with an increased risk of thrombosis, such as those with hypertension or previous episodes of infarction, has been described (1450). A few cases of thrombosis subsequent to intravenous immunoglobulin have been reported, including myocardial infarction in five patients, stroke in four cases, and spinal cord ischemia in one (1451). It has been postulated that these events are induced by platelet activation and increased plasma viscosity (1452).

Several cases of thrombosis have been reported after administration of intravenous immunoglobulin (1453).

- A 75-year-old man with idiopathic thrombocytopenia purpura who was treated with intravenous immunoglobulin developed recurrent myocardial ischemia (1454).
- A 54-year-old woman with idiopathic thrombocytopenic purpura received intravenous immunoglobulin 1 g/kg/day for 2 days and had an ischemic stroke with

hemiparesis; 3 days later she had a deep vein thrombosis (1455).

- A 33-year-old woman with Evans' syndrome received intravenous immunoglobulin 400 mg/kg/day and developed a deep vein thrombosis after 1 week (1455). She was treated with warfarin, and 6 months later received an additional course of intravenous immunoglobulin for recurrent hemolytic anemia; 1 day later she died of pulmonary thromboembolism.
- A 70-year-old woman with polycythemia rubra vera and Guillain–Barré syndrome, but no known risk factors for thrombosis, had a cerebral infarction 10 days after receiving intravenous immunoglobulin; the authors wondered whether there was a relation to the polycythemia vera (1456).

In a randomized, controlled study in 56 patients with untreated autoimmune thrombocytopenic purpura, who were treated with intravenous immunoglobulin 0.7 g/kg/day for 3 days, one had a deep vein thrombosis complicated by pulmonary embolism (1457). One of 10 children with toxic epidermal necrolysis, for which they were given intravenous immunoglobulin 0.5 g/kg/day, developed a deep vein thrombosis requiring heparin (1458). Of the 10 children, this child was the only one who received intravenous immunoglobulin for 7 days instead of the standard 4-day course.

- A 38-year-old woman with multiple sclerosis developed a deep vein thrombosis after a course of intravenous immunoglobulin 2 g over 2 days in combination with methylprednisolone (1459). She was not immobilized by her multiple sclerosis.

The authors proposed that the combination of intravenous immunoglobulin and methylprednisolone might have been associated with the thrombotic event, because this patient had tolerated several courses of intravenous immunoglobulin without methylprednisolone. They discussed 27 previously reported cases of thrombotic events after intravenous immunoglobulin. In eight cases a glucocorticoid had also been used. In nine cases, there were no data on glucocorticoid usage, but the clinical conditions made it likely that a glucocorticoid had indeed been used.

Diffuse venous thromboembolism was reported in a patient with streptococcal toxic syndrome after two courses of high-dose intravenous immunoglobulin (0.4 g/kg/day over 5 days and 8 days later 0.4 g/kg/day over 4 days) (1460). Several causes of immunoglobulin-mediated thrombosis have been postulated, including increased plasma and blood viscosity, platelet activation, cytokine-mediated vasospasm, and contamination with factor IX. The authors suggested that combined therapy, including other inhibitors of the inflammatory cascade such as analogues of activated protein C, should be administered to avoid relapse or complications such as thrombosis.

- Transient hypertension occurred in a patient with dermatomyositis during therapy with intravenous immunoglobulin (1461). In the past, his diastolic blood pressure had been 104–106 mm Hg, but he was normotensive with antihypertensive drug medication.

Several mechanisms for this transient hypertension were postulated, for example stimulation of the vascular endothelium to secrete endothelin to inhibit nitric oxide synthesis.

Hypotension after treatment with intravenous immunoglobulin is rare and is due to the presence of IgG dimers in some immunoglobulin formulations (1462).

- A 54-year-old woman developed an acute severe headache, nausea, and difficulty in speech 1 day after receiving intravenous immunoglobulin (21 g) for isolated IgG1 deficiency (1463). She had a transverse sinus thrombosis, at first considered to be a complication of intravenous immunoglobulin. However, she also had primary thrombocythemia, which is also known to cause headache and is associated with a risk of lateral sinus thrombosis.

It has been recommended that patients with cardiac diseases should be monitored during intravenous immunoglobulin therapy, because hypertension and cardiac failure have occurred, presumably as a result of fluid overload or electrolyte shifts (1464).

Indanediones

As with the coumarins, the main complication of the indanediones is bleeding. The incidence of hemorrhage varies, depending on the age of the patient, the intensity of treatment, and the indication for using an anticoagulant. Hematuria, bruising, and gastrointestinal bleeding are the commonest signs. The most common site of fatal hemorrhage is intracranial (1465).

Indinavir

A hypertensive crisis caused a secondary reversible posterior leukoencephalopathy in a patient taking indinavir-containing antiretroviral therapy (1466).

- A 40-year-old man, who had taken stavudine 30 mg bd, lamivudine 150 mg bd, and indinavir 800 mg qds, developed an occipital headache, nausea, and vomiting. His blood pressure was 220/140 mmHg and he had bilateral papilledema. His blood pressure was controlled and his symptoms disappeared. An MRI scan of the brain showed lesions in the periventricular white matter; the nuclei semiovale and occipital asta were most severely affected. Indinavir was withdrawn and replaced by nelfinavir; his blood pressure returned to normal and the MRI white matter lesions disappeared.

In a retrospective analysis of 198 normotensive patients in a protease inhibitor comparison study, 30% of those who took indinavir developed stage I or worse hypertension, compared with none of the patients who took nelfinavir, ritonavir, or saquinavir (1467).

Indometacin

Clinical experience and reports have provided little evidence that indometacin precipitates angina or myocardial infarction. However, an individual angina-provoking effect has been documented (1468), and there are grounds for believing that it can happen.

Intravenous administration of indometacin increases blood pressure, coronary vascular resistance, and myocardial oxygen demands, decreasing coronary flow. A controlled short-term study showed that indometacin increased blood pressure in patients with mild untreated essential hypertension (SEDA-17, 108). In view of the increasing use of parenteral administration, the acute hemodynamic effects of indometacin may now occur more often, especially in the elderly (1469). The mechanism is poorly understood, but apparently a direct action is exerted on the resistance vessels in various regions. This is probably independent of indometacin's action on prostaglandin formation. The clinical relevance is largely unknown, but other NSAIDs should probably be prescribed for patients with occlusive vascular diseases affecting the cerebral and/or coronary vessels.

Other systemic cardiovascular adverse effects are due to salt and fluid retention and also to a reduction in the vasodilator action of circulating prostaglandins E_2 and I_2 (1470).

Unlike other NSAIDs, indometacin acts as a cerebral vasoconstrictor. It reduces cerebral flow by up to 35% and the response to hypercapnia disappears (SEDA-10, 79). It also reduces blood flow in the splanchnic vascular bed, by increasing local vascular resistance, but does not impair circulation in the forearm and leg muscles.

Infliximab

The preliminary results of a phase II trial in patients with moderate to severe congestive heart failure showed a higher incidence of worsening congestive heart failure and death in patients treated with infliximab compared with placebo (1471). This led to warnings from regulatory agencies and to the limited use of infliximab in patients with congestive heart failure.

The ATTACH (Anti-TNF Therapy Against Congestive Heart Failure) trial showed no benefit of short-term infliximab in 150 patients with moderate-to-severe heart failure, and a risk of worsening heart failure with high doses of infliximab (1472). In addition, an analysis of reports to the FDA Adverse Event Reporting System identified 38 cases of new-onset heart failure and nine cases of worsening heart failure in patients with Crohn's disease, rheumatoid arthritis, or related disorders (1473). Of these, 29 cases were attributed to etanercept and 18 to infliximab. There were identifiable susceptibility factors in 19 of the 38 patients who developed new-onset heart failure. All 10 patients aged under 50 years had new-onset heart failure and only three had an underlying susceptibility factor. Withdrawal of the TNF alfa antagonist and heart failure therapy in these 10 patients resulted in complete recovery in three, improvement in six, and death in one.

Venous thrombosis has been associated with infliximab in two patients.

- A 55-year-old woman with psoriatic arthritis and possible systemic lupus erythematosus developed inspiratory pain, slight dyspnea, and left leg pain 1 week after receiving a second infusion of infliximab 3 mg/kg (1474). A respiratory infection was suspected and she recovered. Similar pulmonary symptoms with right leg pain recurred 6 days after her third infusion of infliximab, and a pulmonary embolism was suggested on spiral CT. She also had raised anti-DNA antibodies and slightly raised cardiolipin antibodies.
- A 45-year-old woman with no history of hypertension, hypercholesterolemia, or diabetes received infliximab for Crohn's disease (1475). She had visual changes after her third dose of infliximab and ophthalmoscopy showed retinal vein thrombosis. There was no underlying coagulation disorder.

Although vascular complications have been associated with Crohn's disease or rheumatoid arthritis, there was a close temporal relation with infliximab treatment in both cases. In addition, the second patient had no evidence of susceptibility factors.

Infliximab can precipitate thrombotic events in patients with various underlying diseases (1476,1477,1478,1479,1480).

It is not yet known whether infliximab has a negative procoagulant effect, but experimental data suggest that TNF-alpha has a strong antithrombotic activity in mice (1481). In contrast, prolonged use of infliximab in seven patients with rheumatoid arthritis improved endothelial function assessed by brachial ultrasonography, at least during the first 7 days after infusion (1482).

Death due to worsening of cardiac insufficiency in patients with congestive heart failure has been reported (SEDA-26, 401).

Influenza vaccine

Influenza infection has been a significant problem in cardiac transplant patients; immunization of such patients could therefore be beneficial. However, its use has been limited by concern that stimulation of the immune system might in principle cause an increased risk of cardiac rejection. In the renal transplant experience, influenza infection itself can trigger an immunological response to cause graft rejection, as well as predisposing to other infections. Another concern is whether an immunosuppressed cardiac transplant recipient could seroconvert sufficiently. In a case-control study in 18 cardiac transplant recipients and 18 control patients 6 months or more beyond transplant surgery, there were no differences in the incidence of cardiac rejection or immune responses (1483).

There have been reports of pericarditis (1484,1485) in temporal relation to influenza vaccine.

Insulins

The cardiovascular effects of hypoglycemia include angina pectoris, dysrhythmias, electrocardiographic changes, and coronary thrombosis. Raised concentrations of catecholamines and reduced concentrations of potassium contribute to cardiac damage during hypoglycemia.

In 215 subjects with type 1 diabetes atherosclerosis was assessed using carotid intima media thickness (1486). There was a positive correlation with cumulative short-acting insulin exposure but no correlation with intermediate-acting insulin. There was no power to distinguish between analogues and regular insulin. A review of insulin therapy has suggested that hyperglycemia is important, based on the results of the DCCT study, in which the intensive control group had less progression of atherosclerosis than the conventional group. Most studies have shown a beneficial or neutral effect of exogenous insulin on cardiovascular disease and atherosclerosis, which is different to the epidemiological data on endogenous insulin, which show an increased cardiovascular risk with increasing insulin concentrations (1487).

Interferon alfa

Hypotension or hypertension, benign sinus or supraventricular tachycardia, and rarely distal cyanosis, have been reported within the first days of treatment in 5–15% of patients receiving high-dose interferon alfa (1488). These adverse effects are usually benign, except in high-risk patients with a previous history of dysrhythmias, coronary disease, or cardiac dysfunction.

Cardiac complications

Severe or life-threatening cardiotoxicity is infrequent and mostly reported in the form of a subacute complication in patients with cancer, and in those with pre-existing heart disease or receiving high-dose interferon alfa. Atrioventricular block, life-threatening ventricular dysrhythmias, pericarditis, dilated cardiomyopathy, cardiogenic shock, asymptomatic or symptomatic myocardial ischemia or even infarction, and sudden death have been observed (SED-13, 1091; SEDA-20, 326; SEDA-22, 369; 1489). The combination of high-dose interleukin-2 (IL-2) with interferon alfa enhanced cardiovascular complications, namely cardiac ischemia and ventricular dysfunction (1490).

Cardiomyopathy has been attributed to interferon alfa in an infant (1491).

- A 3-month-old boy was given interferon alfa (2.5–5.5 MU/m^2) for chronic myelogenous leukemia. After 7.5 months he developed progressive respiratory distress, with anorexia, irritability, and nocturnal sweating. A chest X-ray showed cardiomegaly, an echocardiogram showed a markedly dilated left ventricle, and an electrocardiogram showed left ventricular hypertrophy with abnormal repolarization. Viral cultures and serology for cytomegalovirus, parvovirus B19, and enterovirus were negative. Infectious diseases and metabolic disturbances were excluded. Interferon alfa was withdrawn and digoxin, furosemide, and an angiotensin-converting enzyme inhibitor were given. One year later, he was asymptomatic without further cardiac treatment.

Similar, but anecdotal reports were also described in patients without evidence of previous cardiac disease and receiving low-dose interferon alfa (SEDA-20, 326). In chronic viral hepatitis, only seven of 11 241 patients had severe cardiac adverse effects (1492). The exact risk of such cardiovascular adverse effects is unknown. In patients with chronic viral hepatitis, cardiovascular test results were not modified when patients were re-examined after at least 6 months of treatment, even where there was an earlier cardiac history (1493), but there was a potentially critical reversible reduction in left ventricular ejection of more than 10% in another prospective study (1494).

Myocardial dysfunction can completely reverse after withdrawal of interferon alfa and does not exclude further treatment with lower doses (1495).

- A 47-year-old man with renal cancer and no previous history of cardiovascular disease developed gradually worsening exertional dyspnea after he had received interferon alfa in a total dose of 990 MU over 5 years. Echocardiography and a myocardial CT scan confirmed a dilated cardiomyopathy, with left ventricle dilatation and diffuse heterogeneous perfusion at rest. He improved after interferon alfa withdrawal and treatment with furosemide, quinapril, and digoxin. Myocardial scintigraphy confirmed normal perfusion. He restarted low-dose interferon alfa (6 MU/week) 1 year later and had no recurrence of congestive heart failure after a 1-year follow-up period.

Patients with pre-existing cardiac disease are more likely to develop cardiovascular toxicity while receiving interferon alfa, but these complications are rare. Among 89 patients with chronic hepatitis C, 12-lead electrocardiography monthly during a 12-month treatment period and follow-up for 6-months showed only minimal and non-specific abnormalities in five patients (two had right bundle branch block, one left anterior hemiblock, and two unifocal ventricular extra beats) (1496). None of these disorders required treatment withdrawal, and complete non-invasive cardiovascular assessment was normal. Overall, the role of interferon alfa was uncertain and the 5.6% incidence of electrocardiographic abnormalities was suggested to be similar to that expected in the general population. Nevertheless, severe cardiac dysrhythmias are still possible in isolated cases, as illustrated by the development of third-degree atrioventricular block, reversible on withdrawal, in a 57-year-old man with lower limb arteritis but no other cardiovascular disorder (1497).

Acute myocardial infarction occurred about 12 hours after the administration of pegylated interferon alfa-2b 1 microgram/kg in a healthy 76-year-old woman during a single-dose pharmacokinetic study in 24 patients, of whom 18 were over 65 years (1498). Although the case was poorly documented, the close temporal relation strongly suggested a causal role of the drug.

Pulmonary hypertension has previously been reported in one patient (SEDA-26, 393) and was briefly mentioned as a potential complication of interferon alfa in two other patients who also had multiple ulcers involving the feet or toes (1499). Other vascular events including Raynaud's phenomena, digital ulcerations and gangrene, pulmonary vasculitis, and thrombotic thrombocytopenic purpura are occasionally reported (1499).

Pericarditis has been attributed to interferon alfa (1500)

- A 42-year-old woman with a history of atrioventricular septal defect and atrial fibrillation was given interferon alfa-2b for chronic hepatitis C. Pre-treatment echocardiography showed moderate mitral valve disease and left atrial hypertrophy but no pericardial anomalies. Six hours after her first injection of interferon alfa (3 MU) she had an acute episode of thoracic pain, which disappeared spontaneously within 24 hours. Echocardiography showed a moderate non-constrictive pericardial effusion. Similar symptoms recurred 7 hours after a subsequently reduced dose of interferon alfa (1 MU), and echocardiography again showed a reversible pericardial effusion. Autoimmune diseases, cryoglobulinemia, and myocardial infarction were ruled out. Interferon alfa was withdrawn and no recurrence of pericarditis was noted over 5 years of follow-up.

This seems to have been the first case of interferon alfa-induced pericarditis, because lupus-like syndrome and mixed type 2 cryoglobulinemia with vasculitis were deemed to be the direct cause in two previously reported cases.

Pericarditis without the typical features of a lupus-like syndrome has been reported after 4 months of treatment with low-dose interferon alfa-2b in a 40-year-old woman with chronic hepatitis C (1501). She simultaneously developed a polyneuropathy. Both disorders disappeared after interferon withdrawal.

Peripheral vascular complications

Raynaud's phenomenon

Raynaud's phenomenon can occur, particularly in patients with chronic myelogenous leukemia (SEDA-20, 326) (1502), and severe cases were complicated by digital necrosis (SEDA-20, 326) (SEDA-21, 369).

In 24 cases of Raynaud's syndrome, interferon alfa was the causative agent in 14, interferon beta in 3, and interferon gamma in 5 (1503). There was no consistent delay in onset and the duration of treatment before the occurrence of symptoms ranged from 2 weeks to more than 4 years. The most severe cases were complicated by digital artery occlusion and necrosis requiring amputation. Few patients had other ischemic symptoms, such as myocardial, ophthalmic, central nervous system, or muscular manifestations. Severe Raynaud's phenomenon was also reported in a 5-year-old girl with hepatitis C (1504).

Cryoglobulinemia

Although most patients with mild-to-moderate clinical manifestations of hepatitis C virus-associated mixed cryoglobulinemia improved during treatment with interferon alfa, acute worsening of ischemic lesions has been reported in three patients who had prominent cryoglobulinemia-related ischemic manifestations (1505). All three had acute progression of pre-existing peripheral ischemia or leg ulcers within the first month of treatment, and transmetatarsal or right toe amputations were required in two. The lesions healed after interferon alfa withdrawal. It was therefore suggested that the anti-angiogenic activity of interferon alfa may also impair revascularization and healing of ischemic lesions in patients with initially severe ischemic manifestations.

Venous thrombosis

Whereas clinically insignificant coagulation abnormalities have been documented in patients receiving high-dose continuous interferon alfa (1506), isolated cases of venous thrombosis have been observed (SEDA-20, 329; 1507). Interferon alfa can also induce the production of antiphospholipid antibodies (SEDA-20, 329; SEDA-21, 371). In one study, antiphospholipid antibodies were found in five of 12 patients with melanoma treated with interferon alfa alone or with interferon alfa plus interleukin-2; deep venous thrombosis occurred in four patients with antiphospholipid antibodies (1508). Although the underlying neoplasia undoubtedly played a role in the further development of venous thrombosis, the causative role of interferon alfa was suggested by the absence of antiphospholipid antibodies and venous thrombosis in eight patients treated with interleukin-2 alone.

Other vascular complications

Other anecdotal reports included acrocyanosis and peripheral arterial occlusion (SED-13, 1388; SEDA-22, 400). Although the causal relation is unclear, interferon alfa was considered as a possible cause in the triggering of acute cerebrovascular hemorrhage or ischemic neurological symptoms in few patients (SEDA-21, 370; SEDA-22, 400). The pathogenic mechanisms of these vascular effects are still unclear; vasculitis, hypercoagulability, vasospasm, a paradoxical anginal effect of interferon alfa, or an underlying cardiovascular disease have all been suggested as underlying processes.

Interferon beta

Cardiovascular adverse effects of interferon beta include isolated reports of severe Raynaud's phenomenon (SEDA-22, 374) and acute myocarditis (SEDA-21, 374).

Fatal capillary leak syndrome has been reported (1509).

- A 27-year-old woman had an 8-month history of relapsing–remitting neurological symptoms and a monoclonal gammopathy. She started to take interferon beta-1b for multiple sclerosis, but had marked somnolence 30 hours after a single injection. She rapidly became unresponsiveness, and hemodynamic tests showed low central venous and pulmonary capillary wedge pressures with generalized peripheral edema, ascites, and bilateral pleural effusions. She died within 80 hours after injection from multiple organ failure. At postmortem she was found to have C1 esterase inhibitor deficiency.

In the light of the possible effects of interferon beta on cytokine release and complement activation, a cytokine-mediated reaction was discussed as the cause of the capillary leak syndrome in this case.

Interferon gamma

Heart rate, ventricular or supraventricular extra beats, and asymptomatic cardiac events were not significantly different during treatment compared with baseline in 20 patients receiving interferon gamma (1510). Interferon gamma rarely produced cardiovascular adverse effects. Hypotension, dysrhythmias, and possible coronary spasm were sometimes observed, mostly in patients receiving high doses or with previous cardiovascular disorders (SEDA-20, 333; SEDA-22, 405; 1511,1512).

Exacerbation of Raynaud's syndrome occurred in five of 20 patients with systemic sclerosis treated with interferon gamma (SEDA-20, 333).

Interleukin-1

The most significant adverse effect of both forms of interleukin-1 is dose-limiting hypotension resulting from a capillary leak phenomenon with clinical features of septic shock (1513). Although mild weight gain, dyspnea, and pulmonary infiltrates can occur, severe capillary leak syndrome is usually not observed. Shortness of breath requiring oxygen and benign supraventricular dysrhythmias were sometimes noted. The maximum tolerated dose of interleukin-1 with pressors is therefore 0.3 micrograms/kg/day. The most probable mechanism underlying these complications is an interleukin-1-induced increase in nitric oxide production by vascular smooth muscle.

Interleukin 2

Interleukin-2 therapy has infrequently been associated with myocarditis. Two of 57 patients developed myocarditis during high-dose IL-2 treatment for metastatic melanoma and renal cell carcinoma, including one biopsy-proven case with eosinophilic and lymphocytic infiltrate (1514). Both recovered uneventfully.

Interleukin-3

Hypotension, exacerbation or new onset of atrial fibrillation, and dyspnea were infrequently observed in clinical trials (1513). Although weight gain and peripheral edema can develop, only one fatal case, compatible with a capillary vascular leak syndrome, has been reported (SEDA-20, 335).

Thrombophlebitis was recorded in 45% of patients treated for advanced ovarian cancer who also received intravenous interleukin-3 (1515), and deep venous thromboses were reported in children treated with maintenance interleukin-3 for Diamond–Blackfan anemia (1516). In addition, one smoking breast-cancer patient developed severe hypotension and cerebellar and superior mesenteric thrombosis after subcutaneous interleukin-3 administration (SEDA-19, 340). Collectively, these case reports suggest that interleukin-3 may contribute to the development of thrombosis, but a possible increased risk of thrombosis with interleukin-3 remains to be demonstrated.

Interleukin-4

The vascular leak syndrome was observed at a dose of 15 micrograms/kg by bolus or continuous intravenous administration, but a moderate capillary leak syndrome was also noted at lower subcutaneous doses (1517).

Cardiac toxicity, consistent with myocardial infarction, was observed in three of seven patients with metastatic cancer receiving intravenous bolus interleukin-4 to a daily total of 800 micrograms/m^2 (1518). A unique pattern of myocarditis with predominant polymorphonuclear, eosinophil, and mast cell infiltration was the possible cause of death in one case and suggested an allergic inflammatory myocardial process.

Interleukin 11

Low-dose recombinant interleukin-11 therapy was associated with one case of dysrhythmia and one transient ischemic attack in 32 patients with bone marrow failure due to myelodysplastic syndromes, graft failure, chemotherapy, or aplastic anemia (1519).

Iodinated contrast media

Most severe reactions to contrast media are associated with cardiovascular manifestations, causing hypotensive shock and in some cases ventricular fibrillation and cardiac arrest; these events are reversible in most cases in which prompt treatment is given. In a case of hypotensive collapse reported in 1977, and followed by a small number of others, there was disseminated intravascular coagulation (1520). In milder cases there is only hypotension, which can be transient and symptomless; in some cases there is bradycardia (due apparently to vagal overactivity) rather than tachycardia.

In more than 90 000 cardiac angiographies performed in US hospitals during 1991, the overall rate of complications with low-osmolar contrast media was 1.5%, including idiosyncratic reactions in 0.25%, vascular complications in 0.44%, neurological complications in 0.05%, dysrhythmias in 0.31%, and myocardial infarction in 0.06%; the death rate was 0.11%. In percutaneous coronary angioplasty major complications, generally of the same type, occurred in 5% of cases (1521). In one large series of digital subtraction angiography examinations using iopamidol, the overall incidence of reactions was 2.5%; some occurred with a delay of 1 hour or more (1522).

The safety of iodixanol 320 and iohexol 350 has been investigated in Swedish patients undergoing cardiac angiography for suspected coronary artery disease (1523). Of 1020 patients, 502 aged 25–83 years received iohexol (median dose 105 ml, range 20–440) and 518 aged 18–85 years received iodixanol (median dose 115 ml, range 30–400). There were 134 patients with unstable angina in the iohexol group and 167 in the iodixanol group. Cardiac adverse events (angina pectoris, dysrhythmias, and dyspnea) within 24 hours of the examination were reported by 9% of the patients who received iohexol and 7% of patients who received iodixanol. There were two cases of ventricular fibrillation, both after iohexol. Cardiac adverse events in patients aged 65 years or more occurred in 11% with iohexol and 7% with iodixanol. The proportions of patients with unstable angina and cardiac adverse events were 18% with iohexol and 12% with iodixanol. The authors concluded that iodixanol could be advantageous in old patients and in those with unstable angina.

Iodixanol (a non-ionic dimer, 320 mg of iodine per ml) and ioxaglate 320 (a low-osmolar ionic dimer, 320 mg of iodine per ml) have been compared in a randomized study in 110 consecutive patients referred for coronary angiography and ventriculography (1524). The incidence of adverse reactions was significantly higher with ioxaglate (28 versus 3%) but there was no difference in angiographic quality between the two agents. The increase in left ventricular end-diastolic pressure was significantly less with iodixanol than with ioxaglate. The QT interval was significantly prolonged by both agents, but the changes were less marked after iodixanol. The authors concluded that iodixanol and ioxaglate are of comparable diagnostic efficacy in coronary angiography and ventriculography but that iodixanol is better tolerated and has less marked hemodynamic and electrophysiological effects.

The hemodynamic effect of direct intra-arterial injection of contrast agents on capillary perfusion in man has been investigated (1525). This was achieved through continuous recording of perfusion in the nail-fold capillaries of the right hand before and after a bolus injection of 20 ml of iodixanol 270 (a non-ionic dimer) or iopentol 150 (a non-ionic monomer) into the right axillary artery. The high-viscosity contrast agent iodixanol (5.8 mPa.s) caused a significant reduction in erythrocyte velocity, while iopentol, which has a much lower viscosity (1.7 mPa.s), had no effect. The authors concluded that high-viscosity contrast media can cause reduced organ perfusion. This effect could be significant in patients with atherosclerotic disease, as it might lead to reduced perfusion of the myocardium during coronary angiography.

Effects on blood pressure

Rapid peripheral intravenous injection of concentrated ionic contrast media produces a brief rise in systemic arterial pressure followed by a prolonged fall; the diastolic pressure decreases more than the systolic pressure and the heart slows; the pulse contour changes, and the venous pressure rises; the arterial hypotension is more marked if injection is rapid. The electrocardiogram can show flattening, splitting, or T-wave inversion; tachycardia is probably compensatory, as are the concomitant increases in venous pressure and pulmonary arterial pressure. Hypotension associated with a vasovagal reaction probably explained four deaths from acute coronary insufficiency (two each with iodoalphionic acid and iopanoic acid) in patients with ischemic heart disease.

- A 44-year-old man had a CT scan of the head with intravenous contrast enhancement (35 ml of ioversol 350) (1526). He developed severe back pain 90 minutes later and then became acutely unwell, with nausea, vomiting, chills, tremor, and faintness. He rapidly became shocked (systolic blood pressure 80 mmHg, pulse 140/minute) and had a petechial rash over the trunk and upper limbs. He was given intravenous fluids (polygeline 2 liters and crystalloid 2 liters and adrenaline. Blood cultures were negative, and echocardiography, CT scan of the chest and abdomen, and abdominal ultrasound were normal. He continued to deteriorate, developed acute renal insufficiency with disseminated intravascular coagulation, and was given dopamine, aggressive fluid resuscitation, and antibiotics (gentamicin, ceftriaxone, and erythromycin). His general condition gradually improved and he recovered fully.

The authors attributed these events to a severe delayed reaction to the contrast medium, manifesting as prolonged hypotension.

Injection of contrast media into the right side of the heart or pulmonary artery can be followed by transient pulmonary hypertension but systemic hypotension. The pulmonary hypertension is partially due to an increase in the pulmonary vascular resistance from capillary blockage by the altered erythrocytes, which have a reduced elasticity due to the effect of a hypertonic contrast medium. Reduced cardiac output accompanied by cardiac slowing and diminished force of contraction seem to explain the initial systemic hypotension; persistence of hypotension thereafter is probably due to the vasodilator effect of the contrast medium on the systemic vessels. Pulmonary angiography is particularly dangerous when the right ventricular end-diastolic pressure exceeds 20 mmHg. Iohexol appears to be a safer medium for pulmonary angiography (SEDA-14, 423).

Injection into the left ventricle or the proximal aorta is likely to produce more marked effects. Cardiac rate, stroke volume, and cardiac output increase. There is a rise in right and left atrial pressures and left ventricular end-diastolic pressure. The pulmonary arterial pressure is also increased. The blood volume expands and peripheral blood flow increases and then decreases as systemic resistance falls. The hematocrit falls and venous pressure gradually rises. As the systemic arterial pressure falls, the heart rate increases. These responses are largely due to the injection of strongly hypertonic solutions, which promote a rapid expansion of the plasma volume; water shifts from the extravascular fluid spaces to the blood and moves out of the erythrocytes, which shrink and become crenated. Blood viscosity rises, but plasma viscosity does not increase significantly. The erythrocytes give up potassium to the plasma and this might contribute to the observed reduction in peripheral vascular resistance.

In two cases of severe hypotensive collapse with generalized itching after left ventricular angiography with 76% sodium methylglucamine diatrizoate, the hypotension failed to respond to vasoconstrictors, and measurements of right atrial and right ventricular pressures showed marked reduction in filling pressures. Rapid intravenous infusion of isotonic saline caused prompt improvement in the blood pressure. A similar case of hypotension with a beneficial response to plasma expanders occurred in a case of prolonged shock after intravenous urography. Severe and prolonged hypotensive collapse has also been seen after antegrade pyelography through a nephrostomy tube under general anesthesia, but the patient in question had a previous history of an acute reaction (SEDA-18, 445). In two cases, injection of diatrizoate during arteriography under general anesthesia caused severe hypotensive collapse (SEDA-7, 452).

Acetrizoate has a more marked effect in this respect than an equiosmolar solution of diatrizoate. Methylglucamine salts appear to be relatively less vasoactive in the peripheral vessels. All these changes are considerably less marked with low-osmolar media such as ioxaglate, iopamidol, or iohexol, but not necessarily absent. Abdominal aortography with iohexol has been found to produce both a decrease in the systemic blood pressure and an increase in the plasma concentration of atrial natriuretic peptide; this may be due to increased intravascular volume (SEDA-16, 533).

Intra-arterial injection of conventional ionic contrast media results in vasodilatation. This is due mainly to hypertonicity of the medium, but toxicity is also a factor. The vasodilatation may in addition be partly due to an anti-cholinesterase action, since it is partially blocked by atropine. In clinical practice, aortography and peripheral arteriography are usually associated with a slight fall in blood pressure, tachycardia and discomfort in the limbs, such as heat or pain.

During cerebral angiography with either ionic or non-ionic agents, hypotension, bradycardia, and even transient asystole can occur, though there can also be reflex tachycardia (1527), which can result in hypertension. These changes are more marked during vertebral angiography when the posterior cerebral arteries have been filled, suggesting that they are due to involvement of centers in the hypothalamus or brain stem. Visual disturbances can also occur, due to involvement of the occipital cortex (1528). The reflex cardiovascular changes may be more serious in patients with coronary artery disease and can give rise to left ventricular failure. Both electrocardiographic and electroencephalographic changes are less common when methylglucamine salts are used. Premedication with atropine reduces the incidence of the cardiovascular changes, but not that of focal electroencephalographic effects. Their incidence has also been reduced by use of very small doses of contrast agents and by premedication with hypertonic mannitol in patients with raised intracranial pressure. One patient with metrizamide encephalopathy developed severe hypertension (SEDA-11, 415; SED-12, 1526).

Electrocardiographic effects

In a comparison of the effects of iodixanol and ioxaglate during coronary angiography, 22 patients received ioxaglate for the first injection into the left coronary artery and iodixanol for the next injections, and 20 patients received the media in the reverse order (1529). Those who received ioxaglate first received a mean of 102 ml of contrast medium and the iodixanol group 104 ml. The first three injections into the left coronary artery were subjected to electrocardiographic analysis. Deviation from baseline was greater in those who received ioxaglate first. The most pronounced effects of ioxaglate were on the ST segment and T wave: the T wave change vector magnitude increased 11-fold from baseline after ioxaglate and 5-fold after iodixanol; the increase in ST change vector magnitude was 4-fold with ioxaglate and 3-fold with iodixanol. The authors concluded that iodixanol caused less pronounced electrocardiographic changes than ioxaglate. These findings are in accord with experimental evidence that iodixanol is well tolerated by the myocardium.

Certain contrast media, notably the high-osmolar products based on sodium methylglucamine, Renografin-76 and MD-76, have calcium-binding properties and cause more hemodynamic changes and a higher risk of ventricular fibrillation than the calcium-enriched media Angiovist and Hypaque-76 (1530). In nine patients undergoing coronary arteriography, plasma measurements from the coronary sinus, there was a significant reduction in ionized calcium concentrations immediately after injection of Renografin 76 (sodium methylglucamine diatrizoate) into the coronary arteries. The effect was most marked and lasted longer in patients with vascular disease. The reduction in ionized calcium was attributed to the chelating agents (disodium EDTA and sodium citrate) present in some contrast media. This may have been a factor in causing electromechanical dissociation in cardiac muscle (SEDA-11, 412). The hypocalcemic effect of ionic contrast media can potentiate the effect of a calcium blocker such as verapamil (SEDA-8, 429; SEDA-9, 410).

Non-ionic media cause fewer electrocardiographic changes than ionic media (SEDA-10, 424; SEDA-11, 412). Non-ionic contrast media are almost sodium-free; they have little tendency to cause ventricular fibrillation or to depress cardiac contraction, and this small risk might (if in vitro animal studies are dependable) be further reduced by the addition of a very small amount of sodium (1531). Following studies in dogs in which methylglucamine salts of diatrizoate or iotalamate produced only temporary T wave changes while sodium salts produced more marked changes with a fall in blood pressure and increase in coronary flow, the US manufacturers of a product containing sodium methylglucamine diatrizoate removed virtually all of its sodium content, but without announcing the change (1532). Radiologists using the altered contrast medium noticed an unexpected increase in the incidence of ventricular fibrillation after coronary arteriography. Subsequent animal experiments confirmed that the new medium caused this effect, apparently by prolonging the time of depolarization. The addition of small quantities of sodium to the medium lessened this prolongation of the depolarization phase.

Prinzmetal angina with electrocardiographic changes has been seen 10 minutes after a dose of iodipamide (SEDA-2, 373). In another case an anaphylactoid reaction after left ventriculography was associated with electrocardiographic changes apparently due to coronary artery spasm (1533).

In a comparison of the effects of ioxaglate (a low-osmolar ionic dimer) with iopamidol (a non-ionic monomer), iopamidol caused fewer electrocardiographic changes and a reduction in ventricular excitability compared with ioxaglate (1534).

The electrocardiographic effects of different types of non-ionic low-osmolar contrast media have been investigated in 41 patients undergoing left ventricular angiography (1535). There was transient prolongation of the QT interval in all of the patients. The effect did not cause important cardiac events and was less than 60 ms in most cases. The authors concluded that this effect was too brief to present any significant risk.

Left ventricular failure

Of 65 patients being investigated for intermittent claudication under general anesthesia, five developed pulmonary edema after retrograde aortic injection of sodium iotalamate (Conray 325 or 420). Three of the five had a history of myocardial disease and another had received 200 ml of Conray 420.

Although it is now widely acknowledged that low-osmolar non-ionic contrast media are better tolerated than high-osmolar ionic media, the choice of contrast agent does not affect the early results of percutaneous transluminal coronary angioplasty (1536). However, the authors acknowledged that high-osmolar ionic agents carry higher risks of acute left ventricular failure. They retrospectively reviewed 401 patients who underwent percutaneous transluminal coronary angioplasty, 220 of whom received high-osmolar ionic media and 181 of whom received non-ionic contrast media. Acute left ventricular failure occurred more often in the high-osmolar group (1.4 versus 0%). There were no differences in the incidences of acute myocardial infarction (3.3% in each group) or urgent surgical intervention (0.5% with the high-osmolar agents and 0.6% with the low-osmolar non-ionic media). There were two cases of mild and transient central nervous system complications (loss of orientation and transient hemiparesis) with the high-osmolar contrast media. The authors concluded that in the majority of cases, the type of contrast medium used does not influence the early results of percutaneous transluminal coronary angioplasty in relation to its efficacy, the degree of revascularization, and residual narrowing. However, they acknowledged that the use of high-osmolar ionic media increases the risk of acute left ventricular failure after angioplasty. They attributed the finding that non-ionic contrast media increased the risk of abrupt vessel closure (4.5 versus 1.5%) to intravascular clotting. The suggestion that non-ionic agents are procoagulant is contentious, and there is no conclusive evidence to support this view (SEDA-22, 501).

Myocardial infarction

In spite of the safety of the non-ionic contrast media cardiac arrest can complicate the infusion of these agents.

- A 47-year-old man with chest pain and a myeloproliferative disorder had a CT scan of the abdomen with contrast enhancement (the type of contrast medium was not stated) (1537). He had no significant past medical history or history of allergy. During a later CT scan of the abdomen infusion of 60 ml of the non-ionic monomer iohexol (iodine 300 mg/ml) caused a sudden cardiac arrest. Resuscitation was ineffective and post-mortem examination showed intramural acute and old organizing infarctions in the entire left ventricular wall.

Although the authors suggested that this event was an adverse effect of the contrast medium, it is possible that the cardiac arrest in this patient was secondary to an acute coronary event independent of the contrast agent.

Iodine-containing medicaments

Cardiac dysrhythmias have been seen after accidental ingestion of a large amount of potassium iodide solution (1538). In one case administration of iodide was associated with pulmonary edema and iododerma (SEDA-7, 190).

Ipecacuanha, emetine, and dehydroemetine

Cardiotoxicity is the most serious and dangerous adverse effect of emetine. The clinical signs are tachycardia, dysrhythmias, and hypotension. Deaths have been described. Electrocardiographic abnormalities occur in 60–70% of cases; increased T wave amplitude, prolongation of the PR interval, ST segment depression, and T wave changes are all common. It seems possible that emetine influences the cell permeability of sodium and calcium ions, and this could be the basis of its effect on cardiac automaticity and contractility and on the electrocardiogram (SED-11, 594). The symptoms of emetine toxicity suggest that an effect on intracellular magnesium concentrations could be another possible explanation, but there are no data to support this hypothesis (SED-11, 594).

Iron salts

Intramuscular iron

The major hazard of the intramuscular use of iron sorbitex consists of severe systemic reactions with cardiac involvement, which may be fatal; they occur in up to 1% of cases, they start 10–30 minutes after injection and a patient who has received an injection must be monitored for an hour. Nausea, chest pain, profuse sweating, cardiac dysrhythmias, and loss of consciousness can occur. Cardiac complications include complete atrioventricular block, ventricular tachycardia, and ventricular fibrillation.

Intravenous iron

Non-transferrin-bound iron, which increases after intravenous ferric saccharate, has been suggested to act as a catalytic agent in oxygen radical formation in vitro, and may therefore contribute to endothelial impairment in vivo (1539). The effect of ferric saccharate infusion 10 mg has been investigated in 20 healthy volunteers. Ferric saccharate caused a greater than four-fold increase in non-transferrin-bound iron and transient significant reduction in flow-mediated dilatation 10 minutes after infusion of ferric saccharate. The generation of superoxide in whole blood increased significantly 10 and 240 minutes after infusion of ferric saccharate by 70 and 53% respectively. Thus, infusion of iron leads to increased oxygen radical stress and acute endothelial dysfunction.

Isoflurane

Although atrial dysrhythmias have been reported in 3.9% of patients and ventricular dysrhythmias in 2.5% (1540), the dysrhythmogenicity of isoflurane is less pronounced than that of halothane (1541). Indeed, the incidence of dysrhythmias due to catecholamines in cardiovascular anesthesia and during oral surgery is reduced by using isoflurane rather than the other agents.

Isoflurane can cause marked hypertension during induction of anesthesia. Of 26 patients who were anesthetized with 0.5% isoflurane in oxygen, increased to 4% in 2 minutes, nine had increases in systolic blood pressure by more than 10 mmHg (mean 26) (1542). Tracheal intubation markedly increased the blood pressure in all patients, but there was a negative correlation between the isoflurane-induced increase and that induced by intubation. Tracheal intubation produced a larger increase in blood pressure in the isoflurane-induced hypertensive patients.

The most controversial adverse effect of isoflurane is its potential to cause coronary steal in patients with critical stenosis in the coronary circulation. Most recent work suggests that the risk of myocardial ischemia is not increased, as long as the hemodynamics, especially heart rate, are well controlled (1543). However, there are still isolated reports, suggesting that the issue is not settled. In some cases isoflurane has caused a specific coronary steal even with good hemodynamic control (1544).

Itraconazole

Ventricular fibrillation has been attributed to itraconazole-induced hypokalemia (1545).

- Pleural and subsequent pericardial effusion developed in a woman treated with itraconazole 200 mg bd for a localized pulmonary infection with *Aspergillus fumigatus* (SEDA-18, 282). After more than 9 weeks of treatment she developed a pericardial effusion, which necessitated drainage. Itraconazole was withdrawn. Six weeks later, and 2 weeks after the resumption of itraconazole, she developed signs of pulmonary edema and cardiac enlargement. These signs disappeared rapidly on discontinuation of itraconazole.

Studies in dogs and healthy human volunteers have suggested that itraconazole has a negative inotropic effect; the mechanism is unknown. A systematic analysis of data from the FDA's Adverse Event Reporting System (AERS) identified 58 cases suggestive of congestive heart failure in patients taking itraconazole (1546). A simultaneous search did not identify any cases of congestive heart failure in patients taking fluconazole and ketoconazole, ruling out the possibility of a class effect. In consequence, the labeling of itraconazole has been revised. Itraconazole is now contraindicated for the treatment of onychomycosis in patients with evidence of ventricular dysfunction. For systemic fungal infections, the risks and benefits of itraconazole should be reassessed if signs or symptoms of congestive heart failure develop.

Ivermectin

Supine and postural tachycardia with postural hypotension can occur; in one large study, such effects were found in three of 40 patients (SEDA-14, 262) (SEDA-22, 327). In another there was hypotension in 13 of 69 cases (SEDA-20, 280), but in some series these effects have not been observed at all (SEDA-17, 356). A massive community study in Ghana noted hypotension in only 37 of nearly 15 000 patients treated (1547). Transient electrocardiographic changes are sometimes seen.

Ketamine

Tachycardia and hypertension are common after anesthetic induction with ketamine, although the hypertension can be limited by the addition of diazepam (1548). Nodal dysrhythmias can also occur (1549). Because of possible reduced cardiac and pulmonary performance, ketamine should be avoided in critically ill patients (1550). Pulmonary vasoconstriction and increased ventricular preload secondary to ketamine can be deleterious (1551).

The effects of intramuscular premedication with either clonidine 2 micrograms/kg or midazolam 70 micrograms/kg on perioperative responses to ketamine anesthesia have been assessed in a placebo-controlled study in 30 patients (1552). Clonidine significantly reduced intraoperative oxygen consumption, mean arterial pressure, and heart rate compared with midazolam and placebo. Thus, clonidine was as effective as midazolam, the standard drug used for this purpose, in reducing the undesirable sympathetic stimulation of ketamine.

Oral clonidine, 2.5 or 5.0 micrograms/kg, 90 minutes before ketamine 2 mg/kg has been compared with placebo in 39 patients (1553). In those given clonidine 2.5 micrograms/kg, heart rate responses were reduced compared with placebo (maximum heart rate 97 versus 76 beats/minute). In those given clonidine 5 micrograms/kg, heart rate responses were less (maximum heart rate 97 versus 77 beats/minute) and mean arterial pressure was lower (121 versus 141 mmHg), and there were fewer nightmares and less drooling.

Subanesthetic low-dose ketamine is being used increasingly often for day-case surgery, acute pain, and chronic pain. Angina pectoris has been reported.

- Subcutaneous low-dose ketamine precipitated angina in an elderly man with metastatic bladder cancer and venous gangrene of a leg, in whom antianginal medication had been withdrawn (1554).

Ketoconazole

Ketoconazole has been reported to be prodysrhythmic without concomitant use of drugs that cause prolongation of the QT interval.

- A 63-year-old woman with coronary artery disease developed a markedly prolonged QT interval and torsade de pointes after taking ketoconazole for a fungal infection (1555). Her QT interval returned to normal on withdrawal of ketoconazole. There were no mutations in her genes that encode cardiac IK_r channel proteins.

The authors concluded that because it blocks inward rectifier potassium channels (IK_r) channels, ketoconazole alone can prolong the QT interval and induce torsade de pointes. This calls for attention when ketoconazole is given to patients with risk factors for the long QT

Ketolides

In one study telithromycin did not prolong the QT_c interval in healthy men and women (1556). However, telithromycin can cause prolongation of the QT interval, especially in elderly patients with predisposing conditions (1557).

Torsade de pointes has also been reported (1082).

Lapatinib

The incidence of cardiac toxicity with lapatinib appears to be low; in one study only 37 of 2812 women (1.3%) had a fall in left ventricular ejection fraction (LVEF) of at least 20% from baseline (1558). The onset of reduced LVEF occurred within 9 weeks of treatment in 68% of cases and was rarely symptomatic and generally reversible and non-progressive. The duration of reduction in LVEF averaged 42 days.

Latanoprost

Two patients in their seventies developed hypertension during treatment with topical latanoprost (dosage not stated) for open-angle glaucoma; both were also taking tocopherol (vitamin E) supplements. Neither had a previous history of hypertension (1559). The authors commented that it is likely that systemic absorption of topical latanoprost could cause hypertension. Self-medication with vitamin E has been reported to aggravate or precipitate hypertension.

Leflunomide

The incidence of hypertension in patients with rheumatoid arthritis taking leflunomide 25 mg/day was 11% in a phase II trial (1560). During phase III trials, there was new-onset hypertension in 2.1–3.7% (1561,1562). Increased sympathetic drive has been implicated in its pathogenesis, because leflunomide-induced hypertension is accompanied by an increased heart rate (1563). However, this hypothesis remains to be tested.

Pulmonary hypertension has been described in association with leflunomide (1564).

Levamisole

Hypotension has been attributed to levamisole (1565).

Levobupivacaine

Levobupivacaine is less cardiotoxic than racemic bupivacaine (1566). In seven sheep, racemic bupivacaine caused mild cardiac depression, which was superseded by central nervous system toxicity and then proceeded to severe ventricular dysrhythmias, which were fatal in three sheep, at doses of 125, 150, and 200 mg (1567). Levobupivacaine was consistently less toxic than bupivacaine, and higher doses were needed to produce adverse effects. Convulsions were less severe and of shorter duration, and although levobupivacaine produced QRS prolongation and ventricular dysrhythmias, there were no deaths.

Several animal studies have shown that levobupivacaine on an equivalent dose basis is safer than bupivacaine; 32–57% more levobupivacaine is required to cause death (1568). In sheep, the mean lethal dose of levobupivacaine was 78% higher than that of bupivacaine; the author suggested that there may be a similar trend in humans and concluded that levobupivacaine should be used in preference to bupivacaine, based on safety data alone.

Levodopa and dopa decarboxylase inhibitors

Postural hypotension has been estimated to occur in some 15% of patients who take plain levodopa during the first year of treatment, and in 10% of patients who take co-careldopa; it is doubtful whether the difference is significant. On the other hand, some patients become hypertensive; they may be individuals who absorb or metabolize the drug at an abnormal rate. Levodopa can cause ventricular dysrhythmias in patients with pre-existing cardiac disorders. Transient flushing of the skin is common; palpitation is unusual.

Levofloxacin

Preclinical and clinical trial data and data from phase IV studies have suggested that levofloxacin causes prolongation of the QT interval (1569). There were cardiovascular problems in 1 in 15 million prescriptions compared with 1–3% of patients taking sparfloxacin, who had QT_c prolongation to over 500 ms. Polymorphous ventricular tachycardia with a normal QT interval has been associated with oral levofloxacin in the absence of other causes (1570,1571,1569,1572).

Among 23 patients who took levofloxacin 500 mg/day there was prolongation of the QT_c interval by more than 30 ms in four patients and 60 ms in two patients (1573). There was absolute QT interval prolongation to over 500 ms in four patients, one of whom developed torsade de pointes.

Phlebitis can occur during parenteral administration of levofloxacin. High concentrations of levofloxacin (5 mg/ml) significantly reduced intracellular ATP content in cultured endothelial cells and reduced ADP, GTP, and GDP concentrations (1574). These in vitro data suggest that high doses of levofloxacin are not compatible with maintenance of endothelial cell function and may explain the occurrence of phlebitis. Commercial formulations should be diluted and given into large veins.

Levosalbutamol (levalbuterol)

See also Salbutamol

Levosalbutamol is the R-enantiomer of racemic salbutamol (albuterol). Levosalbutamol 0.63 mg is equipotent to salbutamol 2.5 mg, but with a lower risk of adverse effects, except for the potassium-lowering effect (1575).

Changes in heart rate before and after inhalation of nebulized salbutamol (2.5 mg) or levosalbutamol (0.63 mg) on days 1 and 3 have been retrospectively compared by chart review in 35 patients with acute airflow obstruction (1576). On day 3, heart rate was increased by 2.7 bpm (CI = 0.02, 5.4) after inhalation of salbutamol compared with levosalbutamol. However, levosalbutamol did not provide clinical benefit with respect to drug-associated tachycardia.

In a study of the effect of racemic nebulized salbutamol 2.5 mg and nebulized levosalbutamol 1.25 mg on heart rate in 20 intensive care patients with (n = 10) and without (n = 10) baseline tachycardia, the patients were randomized to receive at least two consecutive doses of salbutamol or levosalbutamol 4 hours apart (1577). Patients with a baseline tachycardia had a mean increase in heart rate of 1.4 bpm with salbutamol and 2 bpm with levosalbutamol over the following 2 hours after drug inhalation. Patients without baseline tachycardia had a mean increase of heart rate of 4.4 bpm with salbutamol and 3.6 bpm with levosalbutamol. Short-term use of nebulized salbutamol and levosalbutamol was therefore associated with similar changes in heart rates in intensive care patients with or without baseline tachycardia.

Lidocaine

Lidocaine can cause dysrhythmias and hypotension. The dysrhythmias that have been reported include sinus bradycardia, supraventricular tachycardia (1578), and rarely torsade de pointes (1589). There have also been rare reports of cardiac arrest (1580) and worsening heart failure (1581). Lidocaine can also cause an increased risk of asystole after repeated attempts at defibrillation (1582). Lidocaine may increase mortality after acute myocardial infarction, and it should be used only in patients with specific so-called warning dysrhythmias (that is frequent or multifocal ventricular extra beats, or salvos) (1583).

Sinus bradycardia has been seen after a bolus injection of 50 mg, atrioventricular block after a dose of 800 mg given over 12 hours, and left bundle branch block after a mere subconjunctival injection of 2% lidocaine.

- High-grade atrioventricular block has been reported in a 14-day-old infant who was given lidocaine 2 mg/kg intravenously (SED-12, 255; 1584).

A death due to ventricular fibrillation after 50 mg and another due to sinus arrest after 100 mg have been reported (SED-12, 255; 1585). Two cases of ventricular fibrillation and cardiopulmonary arrest occurred after local infiltration of lidocaine for cardiac catheterization (SEDA-21, 136).

Lidocaine does not usually cause conduction disturbances, but two cases have been reported in the presence of hyperkalemia (1586).

- A 57-year-old man with a wide-complex tachycardia was given lidocaine 100 mg intravenously and immediately became asystolic. Resuscitation was unsuccessful.
- A 31-year-old woman had a cardiac arrest and was resuscitated to a wide-complex tachycardia, which was treated with intravenous lidocaine 100 mg. She immediately became asystolic but responded to calcium chloride.

In both cases there was severe hyperkalemia, and the authors suggested that hyperkalemia-induced resting membrane depolarization had increased the number of inactivated sodium channels, thus increasing the binding of lidocaine and potentiating its effects.

The degree of hypotension occurring after epidural anesthesia with alkalinized lidocaine (with adrenaline) was greater than with a standard commercial solution (SED-12, 255; 1587).

In 23 patients there was a significant dose-dependent reduction in blood pressure following submucosal infiltration of lidocaine plus adrenaline compared with saline plus adrenaline for orthognathic surgery (1588). The study was randomized but small; larger studies are needed to confirm effects that could easily have been due to multifactorial causes in patients undergoing general anesthesia.

Lincosamides

Rapid intravenous infusion of large doses of lincomycin (600 mg in 5–10 minutes) can cause flushing and a sensation of warmth for about 10 minutes.

- A patient who received 200 mg/kg of lincomycin experienced nausea, vomiting, hypotension, dyspnea, and electrocardiographic changes for 20 minutes (SED-7, 389; 1589).

Rapid intravenous infusion of lincomycin 1–2 g can cause phlebitis.

Clindamycin can prolong the QT interval and cause ventricular fibrillation (1590). Cardiac arrest associated with rapid intravenous administration of clindamycin has been reported (SEDA-8, 258; 1591).

Linezolid

In a placebo-controlled, crossover study in 12 healthy men who took one oral dose of linezolid 600 mg or a placebo tablet followed by an intravenous tyramine pressor test until the systolic blood pressure increased by at least 30 mmHg above baseline, there was a significant difference in the pressor response to intravenous tyramine between linezolid and placebo (1592).

Severe bradycardia with an increased blood pressure has been attributed to linezolid (1596).

- A 49-year-old woman with cancer of the biliary tree developed a fever and jaundice. Dilatation of the right biliary tract was confirmed by ultrasound and nuclear magnetic resonance. She was given ceftazidime 4 g/day and oral linezolid 600 mg/day. Unexpectedly, 2 days later her blood pressure rose to 170/90 mmHg and was associated with severe bradycardia 37–40/minute. Linezolid was withdrawn. The pulse rate became normal after 48 hours and the blood pressure fell.

The mechanism underlying this effect is unknown, but it may have been related to the fact that linezolid is a monoamine oxidase inhibitor.

Lithium

Cardiovascular disease is not a contraindication to lithium, but the risks may be greater, in view of factors such as fluid and electrolyte imbalance and the use of concomitant medications. Close clinical and laboratory monitoring is necessary, and an alternative mood stabilizer may be preferred. While long-term tricyclic antidepressant therapy may be more cardiotoxic than lithium, the newer antidepressants (SSRIs and others) seem to be safe.

In two studies of 277 and 133 patients taking long-term lithium, there was no evidence of increased cardiovascular mortality compared with the general population (1594,1595). While the latter study reported on 16-year mortality, it did not provide information about which patients continued to take lithium after the first 2 years.

Blood pressure

Lithium does not affect blood pressure adversely, nor does it benefit blood pressure, although lithium hippurate was used to treat arterial hypertension in the 1920s.

There is a higher mortality from cardiovascular diseases among patients with bipolar affective disorder than in the general population. In a study of 81 patients taking lithium monotherapy, 40 were studied in detail; one had hypothyroidism and six had hypertension (1596). Of the 81 patients 13 were taking antihypertensive drugs, suggesting a high prevalence of hypertension. One of the points of the study was to assess if lithium was a factor in cardiovascular risk in these patients, but there was no correlation between the duration of lithium treatment or the duration of bipolar disorder and the presence of hypertension.

Two patients who were taking lithium carbonate for mood disorders and who underwent coronary artery bypass grafting developed refractory hypotension during cardiac surgery, which responded to methylthioninium chloride (1597). The authors suspected that chronic lithium therapy had caused cardiac embarrassment and recommended that lithium be withdrawn before cardiac surgery.

Cardiac dysrhythmias

Non-specific, benign ST-T wave electrocardiographic changes are the most common cardiovascular effects of lithium.

- A 13-year-old boy taking lithium developed a "pseudo-myocardial infarct pattern" on the electrocardiogram; this may have been an overinterpretation of nonspecific T-wave changes (1598).

A very uncommon adverse effect involves sinus node dysfunction (extreme bradycardia, sinus arrest, sinoatrial block), which can be associated with syncopal episodes, perhaps due to hypothyroidism (1599,1600). In such cases, lithium must either be withdrawn or continued in the presence of a pacemaker. At therapeutic concentrations, other cardiac conduction disturbances have been reported, sometimes in conjunction with hypercalcemia (1601), but are uncommon.

Two reviews of the cardiac effects of psychotropic drugs briefly mentioned lithium and dysrhythmias, with a focus on sinus node dysfunction (1602,1603), reports of which, as manifested by bradycardia, sinoatrial block, and sinus arrest, continue to accumulate in association with both toxic (1604) and therapeutic (1605,1606) serum lithium concentrations. The rhythm disturbance normalized in some cases when lithium was stopped (1604,1606), persisted despite discontinuation (1605), or was treated with a permanent cardiac pacemaker (1606). Of historical interest is the observation that the first patient treated with lithium by Cade developed manifestations of toxicity in 1950, including bradycardia (1607).

There have been several reports of bradycardia and sinus node dysfunction.

- During an episode of lithium toxicity (serum concentration 3.86 mmol/l), a 42-year-old woman developed sinus bradycardia that required a temporary pacemaker (1604). There was marked prolongation of sinus node recovery time. Lithium was withdrawn and the patient underwent hemodialysis once daily for 3 days; sinus node recovery time normalized. The presence of nontoxic concentrations of carbamazepine may have contributed to the condition.
- A 65-year-old man taking lithium for 2 years, with therapeutic concentrations, developed sinus bradycardia (30 beats/minute), which remitted when the drug was stopped and recurred when it was restarted (1608). Implantation of a permanent pacemaker allowed lithium to be continued.
- Asymptomatic bradycardia occurred in three of 15 patients treated for mania with a 20 mg/kg oral loading dose of slow-release lithium carbonate (1609).
- A 9-year-old boy whose serum lithium concentration was 1.29 mmol/l had a sinus bradycardia with a junctional escape rhythm (40 beats/minute), which normalized at a lower lithium concentration (1610).
- A 58-year-old woman with lithium toxicity developed an irregular bradycardia (as low as 20 beats/minute), which resolved during hemodialysis; persistent sinoatrial conduction delay suggested that she was predisposed to the bradydysrhythmia (1611).
- A 52-year-old man took an overdose of lithium (serum concentration 4.58 mmol/l) and developed asymptomatic sinus bradycardia with sinus node dysfunction and multiple atrial extra beats, which resolved after hemodialysis (1612).
- A 66-year-old woman with pre-existing first-degree AV block, developed sinus bradycardia, a junctional rhythm, a prolonged QT interval, and syncopal episodes (serum lithium concentration 1.4 mmol/l in a 40-hours sample) about 2 weeks after beginning lithium therapy. She was treated successfully with a pacemaker and a lower dose of lithium (1613).
- A 36-year-old man became hypomanic after lithium was withdrawn because of symptomatic first-degree atrioventricular block (although, how first-degree block could have caused symptoms is unclear) (1614).
- A 44-year-old woman developed atropine-resistant but isoprenaline-sensitive bradycardia (36 beats/minute), thought to be due to sinus node dysfunction related to lithium, fentanyl, and propofol (1600).
- A 52-year-old man with a serum lithium concentration of 4.58 mmol/l had sinus node dysfunction with multiple atrial extra beats and an intraventricular conduction

delay, which normalized following hemodialysis (1612). Two patients, a 58-year-old woman and a 74-year-old woman, developed sick sinus syndrome while taking lithium but were able to continue taking it after pacemaker implantation (1615,1616).

- A 59-year-old woman with syncope and sick sinus syndrome, which remitted when lithium was withdrawn, recurred when lithium was restarted, and then persisted despite lithium withdrawal; after a pacemaker was implanted she was treated successfully with lithium for 7 years (1615).

Lithium can also occasionally cause tachycardia.

- A 59-year-old man was noted to have tachycardia, a shortened QT interval, and nonspecific ST-T changes, while hospitalized with lithium-associated hypercalcemia (1617).

An extension of a previously published study (1601) added a third comparator group of 18 hypercalcemic non-lithium treated patients and compared them with 12 hypercalcemic lithium patients, 40 normocalcemic lithium patients, and 20 normocalcemic bipolar patients taking anticonvulsant mood stabilizers (1618). Both hypercalcemic groups had more conduction abnormalities than the other two groups, but did not differ from each other in this regard. While the authors concluded that both lithium and calcium played important roles in the dysrhythmias, their data suggested that hypercalcemia alone was the critical factor.

Cardiac dysrhythmias associated with lithium intoxication in the elderly included sinus node dysfunction and junctional bradycardia (1619). A retrospective chart review of patients on lithium who had mild but persistent hypercalcemia ($n = 12$) showed a greater frequency of cardiographic conduction disturbances compared with normocalcemic patients taking lithium ($n = 40$) and normocalcemic bipolar patients taking anticonvulsant mood stabilizers ($n = 20$), although the overall frequency of cardiographic abnormalities did not differ significantly among the groups (1601). When 21 patients without cardiovascular disease (mean serum lithium 0.66 mmol/l) were compared with healthy controls using standard electrocardiography, vector cardiography, and electrocardiographic body surface potential mapping, the only abnormality was a reduction in the initial phase of depolarization, a finding of questionable clinical significance (1620).

Abnormalities of the QT_c interval have been explored in 495 psychiatric patients (87 taking lithium, but many of them also taking other drugs) and 101 healthy controls (1621). There was no association of lithium with QT_c prolongation but it was associated with nonspecific T-wave abnormalities (odds ratio 1.9) and increased QT dispersion (odds ratio 2.9). Caution was suggested if lithium is used with drugs associated with QT_c prolongation, such as tricyclic antidepressants, droperidol, and thioridazine.

Sudden death has been reported in 14 psychiatric patients and the literature has been reviewed regarding occult cardiac problems, psychotropic drugs, and sudden death (1622).

- A 57-year old man with bipolar disorder taking olanzapine, lithium, and other drugs had underlying mitral valve prolapse, left ventricular hypertrophy, and His bundle anomalies; he died suddenly, probably because of a cardiac dysrhythmia.

The authors suggested that cardiac pathology should be systematically evaluated in patients who take psychotropic drugs.

Cases of lithium toxicity, its cardiac effects, and issues of cardiac dysfunction in children have been reviewed in the light of a cardiac dysrhythmia in a child.

- A 10-year-old boy developed abdominal pain, diarrhea, and vomiting over 2 days (1623). He had a history of bipolar disorder, with psychotic features, a schizoaffective disorder, an intermittent explosive disorder, and attention deficit hyperactivity disorder. He had several other medical problems, including hypothyroidism, asthma, and seizures. He was taking many drugs, including methylphenidate, escitalopram, oxcarbazepine, clonidine, Depakote, thyroid hormone, and lithium. The serum lithium concentration was 3.1 mmol/l. Electrocardiography showed a broad-complex tachydysrhythmia, which persisted despite treatment with intravenous adenosine and lidocaine. The cardiac rhythm was interpreted as a ventricular tachycardia. He was given intravenous procainamide, resulting in temporary slowing of his cardiac rhythm, and a continuous procainamide infusion produced stable sinus rhythm. Over the next 36 hours, he continued to have treatment for his lithium toxicity and procainamide for his ventricular dysrhythmia, and improved. At follow-up a 24-hour Holter monitor showed first-degree atrioventricular block.

The diagnosis in this patient may not have been correct. He obviously had a severe behavioral disturbance, which required treatment; however, it is not clear if his polypharmacy was appropriate for his condition.

Cardiomyopathies

A study of cardiomyopathies found a specific cause in 614 of 1230 patients (the remainder were diagnosed as idiopathic). One was attributed to lithium but no details were provided (1624).

In a study of 1230 patients with initially unexplained cardiomyopathies, lithium was implicated in one case (1595). Using a data-based mining Bayesian statistical approach to the WHO database of adverse reactions to examine antipsychotic drugs and heart muscle disorders, a significant association was found between lithium and cardiomyopathy, but not myocarditis (1625). The authors acknowledged that further study is needed to determine if the association is causal.

- A 78-year-old woman developed a cardiomyopathy while taking lithium, imipramine, amineptine, levomepromazine, and lorazepam; it resolved when the medications were withdrawn (1626). Whether lithium was causally involved is not known.

Electrocardiography

In 30 patients there were only minimal electrocardiographic changes during long-term treatment with lithium using the method of body surface electrocardiographic mapping (1627). In contrast, a tricyclic antidepressant showed dose-related effects.

However, in two studies lithium treatment was associated with prolongation of the corrected QT interval (QT_c). A retrospective analysis of the records of 76 patients taking lithium showed that intervals of over 440 msec were significantly more common in subjects with lithium concentrations over 1.2 mmol/l than in those with concentrations in the usual target range (55% versus 8%); T wave inversion was also more common in subjects with high lithium concentrations (73% versus 17%) (1628).

Similar results have been reported in 39 in-patients with either bipolar illness or schizophrenia; the duration of the QT_c interval correlated significantly with lithium concentrations and those with the longest QT_c intervals had the highest lithium concentrations (1629).

Local anesthetics

See also individual compounds

Cardiovascular complications are not uncommon in the course of local anesthesia; however, most changes are moderate, involving mild peripheral vasodilatation and reduced cardiac output with a change in heart rate.

Local anesthetics reduce myocardial contractility and rate of conduction (1630). They also cause direct vasoconstriction or vasodilatation of vascular smooth muscle (1631) and central stimulation of the autonomic nervous system (1632).

Cardiac arrest and marked myocardial depression, in which hypoxia plays a critical role, have been reported.

Cardiovascular collapse can be severe and refractory to treatment; most fatal cases involve bupivacaine.

The cardiovascular system is more resistant to the toxic effects of local anesthetics than the nervous system. Mild circulatory depression can precede nervous system toxicity, but seizures are more likely to occur before circulatory collapse. The intravenous dose of lidocaine required to produce cardiovascular collapse is seven times that which causes seizures. The safety margin for racemic bupivacaine is much lower. The stereospecific levorotatory isomers levobupivacaine and ropivacaine are less cardiotoxic, and have a higher safety margin than bupivacaine, but not lidocaine; in the case of ropivacaine this may be at the expense of reduced anesthetic potency (1633,1634). Toxicity from anesthetic combinations is additive.

A comparison of the cardiotoxicity of the two stereoisomers of ropivacaine and bupivacaine on the isolated heart showed that both compounds had negative inotropic and negative chronotropic effects irrespective of the stereoisomer used, but bupivacaine had greater effects compared with ropivacaine at equal concentrations (1635). Atrioventricular conduction time showed stereoselectivity for bupivacaine at clinical concentrations; the R(+) isomer had a greater effect in lengthening atrioventricular conduction time, but the less fat-soluble ropivacaine only showed stereoselectivity at concentrations far greater than those used clinically. Similar to the negative inotropic and chronotropic effects, bupivacaine produced greater effects on atrioventricular conduction time than ropivacaine at equal concentrations. This important study has confirmed speculations that not only the stereospecificity of ropivacaine but also its physicochemical properties contribute to its cardiac safety.

Current concepts of resuscitation after local anesthetic cardiotoxicity have been reviewed (1636). Vasopressin may be a logical vasopressor in the setting of hypotension, rather than adrenaline, in view of the dysrhythmogenic potential of the latter. Amiodarone is probably of use in the treatment of dysrhythmias. Calcium channel blockers, phenytoin, and bretyllium should be avoided. In terms of new modes of therapy targeted at the specific action of local anesthetics, lipid infusions, propofol, and insulin/glucose/potassium infusions may all have a role, but further research is necessary.

Prolongation of the QT interval can predispose to dysrhythmias with local anesthetics.

- Intraoperative cardiac arrest occurred in a 9-year-old child with Pfeiffer syndrome (craniosynostosis, mild syndactyly of hands and feet, and dysmorphic facial features) undergoing reversal of a colostomy (1637). All previous anesthetics had been uneventful. The child received an epidural catheter at the L3/4 interspace. A test dose of 2 ml of lidocaine 1% with adrenaline 1: 200 000 was administered and aspiration for spinal fluid was negative. One minute after the first dose of bupivacaine 0.25% 3 ml with adrenaline 1: 200 000 he developed cardiac dysrhythmias and 3 minutes later, and before surgical incision, ventricular fibrillation. After chest compression, 100% oxygen, adrenaline, and sodium bicarbonate, sinus rhythm returned. Blood was aspirated from the epidural catheter. Postoperative investigation showed a long QT syndrome.

Prolongation of the QT interval predisposes to ventricular dysrhythmias and can be triggered by adrenaline. In this case the authors concluded that accidental intravascular injection of bupivacaine and adrenaline may have triggered the dysrhythmia.

Conduction disturbances have been attributed to bupivacaine.

- A 60-year-old woman with pre-existing heart failure awaiting surgery for a fractured humerus was accidentally given a mixture of bupivacaine 75 mg and clonidine 15 micrograms intravenously (1638). She developed a nodal rhythm with extreme bradycardia, severe shock, and convulsions. Seizures were controlled with thiopental, and suxamethonium and adrenaline partially restored the blood pressure to 50/30 mmHg and the heart rate to 60/minute (nodal rhythm). After clonidine 75 micrograms intravenously, her blood pressure rose to 90/70 mmHg and her heart rate to 70/minute. Her cardiac rhythm reverted to sinus rhythm with first degree atrioventricular block.

The authors concluded that clonidine had reversed bupivacaine-induced conduction disturbances.

Successful resuscitation after systemic ropivacaine toxicity during peripheral nerve block has been described.

- A 15-year-old girl was given 18 ml of ropivacaine 0.75% to the sciatic nerve after negative aspiration, and developed convulsions, immediately followed by ventricular fibrillation (1639). Oxygen was delivered by face mask and she received two DC shocks of 200 J. The convulsions stopped and sinus rhythm returned. Postoperatively, there was no evidence of sciatic block. She did not remember the episode and was discharged the next day.

The authors emphasized the importance of electrocardiographic monitoring during nerve block for early identification of complications, the effectiveness of appropriate resuscitation measures in ropivacaine toxicity, and the potential usefulness of low-dose adrenaline in the test dose to detect inadvertent intravascular injection. However, there is evidence that ropivacaine-induced cardiac toxicity is not nearly as troublesome as bupivacaine toxicity (1640).

The longer-acting, more lipophilic agents, such as bupivacaine, can cause cardiovascular toxicity at serum concentrations that are not much greater than those required to cause nervous system toxicity.

- A 65-year-old man had 15 ml of plain bupivacaine 0.5% infiltrated before a planned radiofrequency ablation of a lumbar sympathetic ganglion (1641). He immediately developed respiratory arrest with bradycardia and hypotension (54/40 mmHg). Asystolic cardiac arrest was treated successfully but he subsequently developed pulmonary edema after a hypotensive episode. Angiography showed left anterior descending artery ischemia and his electrocardiographic T waves normalized 7 months later.

The authors report this case as bupivacaine-induced cardiovascular collapse with several novel features. Firstly, it developed after the administration of a relatively low dose of bupivacaine, less than 1.1 mg/kg. Secondly, the presentation was that of mixed cardiogenic and vasomotor shock. Finally, he developed an unexplained delayed cardiographic finding of symmetrically inverted anterior T waves. The authors thought that drug-drug interactions may also have contributed; since he was taking amitriptyline and carbamazepine, each of which is potentially cardiotoxic and may have lowered the threshold for bupivacaine toxicity.

Cardiopulmonary bypass has been used to successfully treat bupivacaine-induced cardiovascular collapse (1642).

- A 39-year-old woman (72 kg, 165 cm) with congenital clubfoot presented for total right ankle arthroplasty. She had no history of syncope, seizures, coronary artery disease, or congenital heart disease. After induction of anesthesia she received a popliteal nerve block with 30 ml of 0.5% bupivacaine using a nerve stimulator device. Communication was maintained with the patient throughout the injection, and she denied any neurological symptoms suggestive of intravascular injection. About 30 seconds after the block, she had a generalized tonic-clonic seizure and soon afterwards developed ventricular fibrillation. Advanced cardiac life support was begun. She was given adrenaline 2 mg and bretylium 1000 mg intravenously, as well as six attempts at electrical defibrillation. Bupivacaine cardiotoxicity because of inadvertent intravascular injection was suspected and cardiopulmonary bypass was begun and continued for 30 minutes. She was extubated on the second postoperative day and discharged home on postoperative day 10 with no neurological sequelae.

The prolonged duration of cardiac support needed after refractory drug-induced cardiotoxicity may make cardiopulmonary bypass, although invasive, the best option for successful resuscitation of such patients. Notwithstanding the practical and technical limitations of staff and equipment availability, the authors argued that cardiopulmonary bypass should become first-line therapy after unsuccessful basic resuscitation in such cases.

Brugada syndrome, right bundle branch block and raised ST segments, can cause sudden cardiac death, potentially hastened by class I antidysrhythmic drugs. Intravenous sodium channel blockers such as local anesthetics can therefore unmask Brugada syndrome.

- A 77-year old man with no previous symptoms of ischemic heart disease underwent elective gastrectomy for carcinoma of the stomach (1643). Preoperative electrocardiography showed partial right bundle branch block. An epidural catheter was inserted at interspace T9/10 before induction. Aspiration of the catheter was negative for blood and cerebrospinal fluid. Bupivacaine 0.25% 10 ml was given in 2 ml increments, and an infusion of 0.125% bupivacaine and fentanyl 2.5 μg/ml was begun at 8 ml/hour. The operation was uneventful. Three epidural bolus doses were given postoperatively over 11 hours, consisting of 0.125% bupivacaine with fentanyl 2.5 μg/ml, 8 ml, 5 ml, and 5 ml. After the last dose, his systolic blood pressure fell to 80 mmHg. An electrocardiogram showed right bundle branch block with new convex-curved ST segment elevation in V1-V3. Acute myocardial infarction was ruled out and a diagnosis of Brugada syndrome was made. Bupivacaine was withdrawn after a total infusion time of 17 hours (total dose of bupivacaine 443 mg). The patient made a complete and uneventful recovery.

As Class Ib drugs such as lidocaine do not induce the characteristic electrocardiographic changes, the authors suggested that bupivacaine causes greater inhibition of the rapid phase of depolarization in Purkinje fibers and ventricular muscle, and remains bound to sodium channels for longer than lidocaine.

Adverse cardiovascular effects of brachial plexus anesthesia have been reported.

- A 34-year-old man undergoing acromioplasty of the right shoulder had a sudden cardiac arrest after an interscalene brachial plexus block with a mixture of ropivacaine 150 mg and lidocaine 360 mg (1644). After successful resuscitation, severe hypotension persisted, necessitating the use of an adrenaline infusion.

The patient developed pulmonary edema and was mechanically ventilated for 22 hours. He eventually made a good recovery.

A similar report with the use of a combination of lidocaine and levobupivacaine has been published (1645). Tachycardia was the only cardiovascular symptom, while seizures were easily treatable. Both reports are in line with the improved cardiovascular safety reported with enantiomer-specific local anesthetics as discussed below.

In three cases cardiac arrest was related to the administration of ropivacaine for lower limb block.

- A 76-year-old woman underwent foot osteotomy under combined femoral and sciatic nerve block (1646). A femoral nerve block using 20 ml of mepivacaine 1.5% with 1: 400 000 adrenaline was followed by sciatic nerve block with 32 ml of ropivacaine 0.5% and 1: 400 000 adrenaline. The injection was stopped as the patient became less responsive, developed twitching of the hand and face, and had a tonic-clonic seizure, which was terminated with intravenous propofol. She was intubated, developed a bradycardia with wide QRS complexes, and subsequently developed ventricular fibrillation. She was given adrenaline and sinus rhythm returned. She made a complete recovery and was discharged on the next day. The total ropivacaine concentration was 3.2 μg/ml, the unbound ropivacaine concentration 0.5 μg/ml, and the mepivacaine concentration 0.22 μg/ml 5 minutes after the injection.
- A 66-year-old woman was admitted for foot surgery and underwent sciatic nerve block with 25 ml of ropivacaine 0.75% (1647). The block was deemed inadequate for surgery and a further 15 ml of ropivacaine 0.75% was used to block the tibial and peroneal nerves at the ankle, resulting in a total ropivacaine dose of 300 mg (6.7 mg/kg). After 1 hour she became agitated and confused and then unresponsive with abnormal oculogyric movements. An electrocardiogram showed wide QRS complexes with worsening bradycardia, despite treatment with atropine and ephedrine. She then had an asystolic cardiac arrest and cardiopulmonary resuscitation was started. More ephedrine was given intravenously. Sinus rhythm was rapidly restored and return of cardiac output was accompanied by return of spontaneous respiration. The ropivacaine concentration was 1.88 μg/ml 70 minutes after the adverse event. She made a full recovery.
- A 66-year-old man scheduled for hip arthroplasty received a lumbar plexus block with 25 ml of ropivacaine 0.75% (total dose 187.5 mg, 1.88 mg/kg) (1648). Two minutes after the injection he had a tonic-clonic seizure, for which diazepam was given. He became asystolic and cardiopulmonary resuscitation was begun. After 5 minutes of cardiopulmonary resuscitation and intravenous adrenaline, cardiac activity was restored. An electrocardiogram showed sinus bradycardia with wide QRS complexes, but this normalized after a further 10 minutes. He was extubated 2 hours later and made a full recovery. The ropivacaine concentration 55 minutes after the episode was 5.61 μg/ml.

In two of these cases the onset of adverse effects was within moments of injection of ropivacaine; the authors concluded that inadvertent intravascular injection was likely to have occurred, despite negative aspiration of blood. In the other case there was a delay of 1 hour between nerve block and cardiac arrest, implying ropivacaine toxicity due to absorption, and the authors acknowledged that the dose of ropivacaine had been excessive.

While cardiac arrest after administration of other local anesthetic agents, such as bupivacaine, is often reported, these are the first cases associated with the use of ropivacaine. In the first case a combination of mepivacaine and ropivacaine was used; however, it is reasonable to conclude that cardiac arrest was due to inadvertent intravascular administration of ropivacaine, as ropivacaine concentrations were high after the episode.

On all three occasions cardiac arrest was immediately preceded by loss of consciousness and a seizure or seizure-like activity. All cases were also associated with bradycardia and wide QRS complexes. On all three occasions cardiac massage was rapidly successful and sinus rhythm was restored without defibrillation or antidysrhythmic drugs. These findings are in stark contrast to cardiac arrest associated with bupivacaine toxicity, which is particularly refractory to treatment and often requires prolonged resuscitation. This suggests that cardiac arrest in the context of ropivacaine toxicity may not only be less likely than with equal doses of bupivacaine, but also more easily treated. Ropivacaine also had significantly less myotoxic potential than bupivacaine in minipigs, when it was injected through femoral nerve catheters (1649).

An 11-month-old child consumed about 2 ml of a benzocaine anesthetic gel 20% accidentally (1650). He developed a tachycardia (200/minute), which resolved over 24 hours. The authors explained that although the cardiotoxicity of benzocaine is milder than that of other local anesthetics, it can cause life-threatening effects and so pediatricians should counsel parents about the potential hazard of anesthetic teething gels; formulations that contain benzocaine should be in a childproof container.

The effect of a lipid emulsion infusion on bupivacaine-induced cardiac toxicity has been studied in dogs (1651). Bupivacaine 10 mg/kg was given intravenously over 10 minutes to fasted dogs under general anesthesia. Resuscitation included 10 minutes of internal cardiac massage followed by either saline or 20% lipid infusion, as a 4 mg/kg bolus followed by a continuous infusion of 0.5 ml/kg/minute for 10 minutes. Electrocardiography, arterial blood pressure, myocardial pH, and myocardial PO_2 were continuously monitored. All six lipid treated dogs survived after 10 minutes of cardiac massage, but there were no survivors among the six dogs who were given saline. Hemodynamics, PO_2, and myocardial pH were also improved in the treatment group. This study supports the need for further investigation of lipid-based resuscitation to treat bupivacaine toxicity in order to determine the optimum dosage regimen.

Intra-operative hypotension is common and potentially dangerous in elderly patients undergoing spinal anesthesia for repair of hip fractures. Combining an intrathecal opioid with a local anesthetic allows a reduction in the dose of local anesthetic and causes less sympathetic block and hypotension, while still maintaining adequate anesthesia. In a double-blind, randomized comparison in 40 patients of glucose-free bupivacaine 9.0 mg with added fentanyl 20 micrograms with glucose-free bupivacaine 11.0 mg alone, the incidence and frequency of hypotension was reduced by the addition of fentanyl (1652). Similarly, falls in systolic, diastolic, and mean blood pressures were all less. However, there were four failed blocks in those given fentanyl compared with one in those given bupivacaine alone.

There have been two reports of lidocaine-induced vasospasm after intra-arterial injection to overcome constriction of a brachial artery after vascular surgical repair and after digital nerve block for hand surgery (1653). While this sounds paradoxical, the authors pointed out that local anesthetics regulate vascular tension in a biphasic manner, and that lower concentrations cause vasoconstriction.

Lopinavir and ritonavir

Two of 16 patients taking lopinavir + ritonavir developed so-called inflammatory edema, which resolved on withdrawal and recurred after rechallenge (1654). In three of eight patients inflammatory edema occurred 1–4 weeks after they started to take regimens that contained lopinavir + ritonavir (1655). The edema affected the feet, ankles, and calves and was associated in one case with fever and in another with a transient rash; in one case the left shoulder and groin were also affected. All three recovered completely within 1–4 weeks despite continued drug treatment, but 7 months later one had a relapse that required withdrawal of lopinavir + ritonavir.

From a case in which there was positive dechallenge and rechallenge it has been concluded that edema of the lower limbs can be an adverse effect of ritonavir in some HIV-positive patients (1656). The authors suspected a relation to the drug's vasodilatory activity. However, it should also be borne in mind that ritonavir has caused reversible renal insufficiency, which should be looked for in any patient who develops edematous changes.

Loratadine

In contrast to astemizole and terfenadine, loratadine is generally believed to be free of adverse cardiac effects. However, there is some evidence that it could be associated with atrial dysrhythmias. In human atrial myocytes loratadine rate-dependently inhibited the transient outward potassium current at therapeutic concentrations, possibly providing a basis for supraventricular dysrhythmias (1657). However, during prolonged exposure (3 months) of healthy adult men to four times the recommended daily dose (40 mg/day) there was no change in the electrocardiogram that would suggest QT_c prolongation or any evidence of cardiac dysrhythmias (1658).

Prolonged QT interval and symptomatic ventricular tachycardia has been described in a patient who took loratadine plus quinidine, but is most likely to have been attributable to the quinidine and the patient's cardiac condition (SEDA-19, 174).

The pharmacokinetics, electrocardiographic effects, and tolerability of loratadine syrup have been studied in 161 children aged 2–5 years (1659). A single-dose open study was performed to characterize the pharmacokinetic profiles of loratadine and its metabolite desloratadine, and a randomized, double-blind, placebo-controlled, parallel-group study was performed to assess the tolerability of loratadine syrup 5 mg after multiple doses. Electrocardiographic parameters were not altered by loratadine compared with placebo. There were no clinically important changes in other tolerability assessments.

In healthy adults loratadine 10 mg/day had no effects on the electrocardiogram when co-administered for 10 days with therapeutic doses of ketoconazole or cimetidine (1660).

Lysergide

Vasoconstriction, affecting both cerebral and peripheral circulations, has been associated with LSD (1661), but it is not usually significant at ordinary doses in people with a normal circulatory system.

Macrolide antibiotics

See also individual drugs

Cardiovascular reactions are rare if macrolide antibiotics are used in the absence of susceptibility factors, which include drug interactions, increasing age, female sex, concomitant diseases, and co-morbidity (1662).

Of the currently available antimicrobial drug classes, the macrolides appear to be associated with the greatest degree of QT interval prolongation and risk of torsade de pointes (1082).

In a prospective study in 47 healthy subjects, azithromycin (3 g total dose given during 5 days) resulted in a small, non-significant prolongation of the QT interval (1663).

Macrolides with at least one published report of torsade de pointes include azithromycin (1664), clarithromycin, erythromycin, roxithromycin, spiramycin, and troleandomycin.

- A 51-year-old woman took azithromycin 500 mg for an upper respiratory tract infection shortly after a dose of over-the-counter pseudoephedrine (1665). Two hours later she had two syncopal events due to polymorphous ventricular tachycardia without QT interval prolongation. Azithromycin was withdrawn and the ventricular tachycardia abated after 10 hours. She was symptom-free 1 year later.

In the Tennessee Medicaid cohort, during 1 249 943 person-years of follow-up there were 1476 cases of sudden death from cardiac causes; the multivariate adjusted rate of sudden death from cardiac causes among patients currently using erythromycin was twice as high (1666). There was no significant increase in the risk of sudden death among former users of erythromycin. The adjusted rate of sudden death from cardiac causes was five times as high among those who concurrently used CYP3A inhibitors and erythromycin as among those who had used neither CYP3A inhibitors nor any of the study antibiotic medications.

- A 6-year-old girl with complex cyanotic heart disease developed torsade de pointes after taking roxithromycin 10 mg/kg/day (1667).
- An 83-year-old woman developed torsade de pointes after taking oral roxithromycin 300 mg/day for 7 days (1668).

Magnesium salts

Circulatory collapse can occur as a symptom of severe hypermagnesemia if magnesium-containing products are taken in large amounts in high-risk patients (1669).

In a retrospective, case-control study in 150 women treated for preterm labor with magnesium sulfate, the susceptibility factors for pulmonary edema included greater magnesium sulfate and intravenous fluid infusion rates, use of a less concentrated solution of magnesium sulfate, infection, multiple gestations, concomitant tocolytics, large positive net fluid balances, and maternal transport (1670). The mean latency period to diagnosis was 1.96 days. Six percent of patients had recurrence if magnesium sulfate tocolysis was continued.

Maprotiline

Maprotiline is among the antidepressants with the lowest reported incidence of tachycardia and postural hypotension.

Mazindol

In another study, an increase in heart rate, probably due to an amphetamine-like stimulatory effect, was ascribed to mazindol (1671).

Medroxyprogesterone

To date medroxyprogesterone does not seem to have been associated with thrombosis, but an authoritative review has pointed out that this remains a possible risk (1672).

Medroxyprogesterone has been reported to have variable effects on blood pressure. In some studies it had no effect on blood pressure at all (1673) or reduced diastolic pressure slightly (1674). In 24 women (21 normotensive and three hypertensive) aged 16–35 years who received injections of medroxyprogesterone acetate 150 mg for contraception mean blood pressure fell from 124/79 to 120/75 mmHg; the change was attributable to effects in the women with hypertension (1675).

Mefloquine

Sinus bradycardia was seen in 18% of patients taking mefloquine (SEDA-12, 693) (1676), occurring some 4–7 days after administration; the bradycardia was asymptomatic and lasted about 3–4 days. Transient sinus arrhythmia was also reported, without a need for treatment (SEDA-12, 808). Asymptomatic dysrhythmias were also recorded in a dosage comparison trial (SEDA-16, 308).

Meglitinides

The meglitinides bind with high affinity to a site, distinct from the sulfonylurea receptor site, on the ATP-sensitive potassium channels in pancreatic beta cells and stimulate insulin secretion. After binding, the ATP-dependent potassium channels are closed, reducing potassium efflux and depolarizing the cell membrane. The meglitinides do not have to be internalized in the membrane, in contrast to the sulfonylureas. This may explain their rapid onset of the action and the end of that action when glucose concentrations are falling. MgADP potentiates the effect in beta cells but not in cardiac cells (1677), which may explain the reduced cardiovascular adverse effects of repaglinide in vivo. Meglitinide-stimulated insulin secretion depends on the glucose concentration; insulin secretion is not stimulated in vitro or in fasted animals. Nateglinide (and perhaps repaglinide) reduces the secretion of glycated insulin, which has poor activity, from islet cells; this may contribute to its hypoglycemic action (1678).

Repaglinide (rINN) is a carbamoylmethyl benzoic acid derivative, which contains the non-sulfonylurea moiety of glibenclamide, and nateglinide (rINN) is a phenylalanine derivative. Nateglinide acts more quickly than repaglinide, and both act more quickly than sulfonylureas, which stimulate insulin secretion independent of blood glucose concentrations (1679). It has been stated (1680) that earlier studies (SEDA-22, 478) (SEDA-23, 462) (1681) showed that the efficacy in lowering HbA_{1c} is almost equivalent for sulfonylureas and repaglinide and is slightly lower for nateglinide. When a lunch-time meal was omitted (SEDA-23, 462) patients taking glibenclamide had the lowest blood glucose concentrations, often within the hypoglycemic range, in contrast to patients taking repaglinide. Both drugs can cause hypoglycemia.

Repaglinide has been reviewed (1682–1684). It stimulates glucose-dependent insulin secretion, amplifying insulin bursts without changing burst frequency, but it does not restore disrupted pulsatile secretion in type 2 diabetes (1685). It is rapidly absorbed. It is metabolized by CYP3A4 in the liver and is 90% excreted in the feces. It has a half-life of 32 minutes. Its pharmacokinetics do not differ in young or older healthy persons (1686). It can be given as monotherapy (1687), and the effective dose is 0.5–8 mg/day, starting with 0.5 mg/day. It has to be given before each meal and can be adapted to irregular food intake or missed meals (1688). When given preprandially it improves glucose control without increasing the risk of adverse effects (1689). There were no differences in action in healthy younger or older volunteers (1690).

Repaglinide is short-acting and seems to be associated with significantly fewer episodes of serious hypoglycemia (1691). In a short review of a number of clinical studies the following contraindications were reported (1692):

- known hypersensitivity to repaglinide or one of the constituents of Novonorm®;
- type 1 diabetes;
- renal or hepatic impairment.

The most frequent adverse effect of meglitinides is hypoglycemia. The overall incidence of hypoglycemia with repaglinide is similar to that reported with sulfonylureas, but the incidence of serious hypoglycemia is lower. Other adverse effects are respiratory tract infections and headache. Cardiovascular events and cardiovascular mortality are not different from those in users of sulfonylureas. In Europe, repaglinide is contraindicated in patients with severe liver dysfunction and it is not recommended in people over 75 years old; in America the advice is to use repaglinide cautiously in patients with impaired liver function and there is no restriction on its use in elderly patients. In renal impairment, the half-life of repaglinide is prolonged. Reasons for withdrawal are hyperglycemia, hypoglycemia, and myocardial infarction (1693).

Repaglinide has a short duration of action and improves postprandial hyperglycemia, a potential risk factor for cardiovascular changes (1694). In a double-blind, multiple-dose, parallel-group study repaglinide stimulated mealtime insulin secretion (1695). Bouts of hypoglycemia were equally frequent with placebo and repaglinide. When repaglinide was added to NPH monotherapy in patients with HbA_{1c} over 7.1% for 3 months, 38% of the patients had an HbA_{1c} below 7.1% (1696). The incidence of hypoglycemia did not change.

When glipizide was compared with repaglinide in 75 patients there were no major hypoglycemic events; minor events were the same in both groups, but after the start of therapy the events occurred much later with repaglinide than glipizide (1697).

The effect of a missed meal during repaglinide and glibenclamide therapy has been compared in 83 randomized patients (1698). During two meals there were six separate hypoglycemic events in those taking glibenclamide. Blood glucose fell from 4.3 mmol/l to 3.4 mmol/l in those taking glibenclamide when lunch was omitted. There were no changes in blood glucose in those taking repaglinide.

Factitious hypoglycemia has been attributed to repaglinide in an 18-year-old man who had bouts of hypoglycemia for 2 months (1699). After glucose administration he recovered promptly and was sent home, but the next night his glucose was 1 mmol/l with high concentrations of insulin (395 pmol/l), C-peptide (2966 pmol/l), and proinsulin (81 pmol/l). The plasma concentrations of repaglinide in three specimens were 4.8–21 ng/ml. Metformin was below the detection limit. He finally admitted to taking repaglinide 4 mg regularly.

Melatonin

There was an increase in blood pressure throughout 24 hours in a double-blind, placebo-controlled, crossover study in 47 hypertensive patients who were also taking nifedipine (1700). This finding differs from other studies in which melatonin had a mild hypotensive effect (1701) and may indicate an interaction between melatonin and nifedipine. Tachycardia, chest pain, and cardiac dysrhythmias have also been reported, although the relation to melatonin was not clearly established (1702).

Mepacrine

Mepacrine inhibits phospholipase A2 and subsequently leukotrienes, which are calcium ionophores. Mepacrine also has an inhibitory effect on phospholipase C and subsequent inositol phosphate formation, which mobilizes cytosolic calcium from intracellular stores. It has therefore been suggested that this might influence myocardial contractile function (SEDA-13, 241). However, mepacrine is not cardiotoxic, although problems could arise in the presence of a "sick" myocardium.

Mepivacaine

See also Local anesthetics

Severe bradypnea and bradycardia requiring external ventricular pacing occurred in a previously asymptomatic 30-year-old woman with a known cardiac conduction defect 85 minutes after a paracervical block with mepivacaine 400 mg (1703). First-degree atrioventricular block has been reported (1704).

Mercury and mercurial salts

Symptoms of chronic mercury poisoning have been reviewed (1705–1707). The symptoms are listed in Table 4. The urinary tract is very sensitive to poisoning by all forms of mercury, a sensitive indicator of early injury being a rise in the urinary excretion of *N*-acetyl-beta-D-glucosaminidase (NAG) (1708).

In 70 patients with psoriasis treated with an ointment containing ammoniated mercury, symptoms and signs of mercury poisoning were detected in 33 (1709): albuminuria, headache, gingivitis, erythroderma, nausea, dizziness, precordial pain, contact dermatitis, conjunctivitis, epistaxis, keratitis, tremor, neuritis, hematological changes, metallic taste in mouth, and purpura.

Acrodynia ("pink disease") is thought to be a particular form of mercury hypersensitivity, which can be caused by organic or inorganic mercury; it formerly occurred in young children exposed to teething formulations containing mercury compounds. Typical signs include pink scaling palms and soles, flushed cheeks, pruritus, photophobia, profuse irritability, and insomnia. A modern case of acrodynia involved a patient with congenital agammaglobulinemia who had received merthiolate-containing gammaglobulin injections for 15 years (SEDA-6, 225).

Mercury poisoning has been reported in a child who took a Chinese herbal medicine (1710).

Table 4 The symptoms of chronic mercury poisoning

System	Symptom
Cardiovascular	Hypertension, hypotension
Nervous system	Emotional disturbances, irritability, hypochondria, psychosis, impaired memory, insomnia, tremor, dysarthria, involuntary movements, vertigo, polyneuropathy, paresthesia, headache
Sensory systems	Corneal opacities and ulcers, conjunctivitis, hypacusis
Endocrine	Hyperthyroidism
Hematologic	Hypochromic anemia, erythrocytosis, lymphocytosis, neutropenia, aplastic anemia
Mouth	Loose teeth, discoloration of the gums and oral mucosa, mouth ulcers, fetor
Gastrointestinal	Anorexia, nausea, vomiting, epigastric pain, diarrhea, constipation
Urinary tract	Nephrotic syndrome
Skin	Tylotic eczema, dry skin, skin ulcers, erythroderma
Musculoskeletal	Acrodynia, arthritis in the legs
Reproductive system	Dysmenorrhea

- A 5-year-old Chinese boy developed oral ulceration, mainly affecting the left lateral aspect of his tongue, and 5 weeks later motor and vocal tics. Herpetic ulceration was diagnosed and confirmed by the isolation of *Herpes simplex* virus type 1 from a tongue swab. The lesion improved with oral aciclovir (200 mg five times a day for 5 days), but relapsed a few days later. A local pharmacist prescribed a Chinese medicinal herb mouth spray called "Watermelon Frost", said to be useful in controlling pain and healing difficult mucosal wounds. Over the following weeks his oral symptoms improved but he became irritable and cleared his throat frequently. He developed a transient skin rash on his trunk and motor tics (eye blinking, head turning, and shoulder shrugging). His blood lead concentrations was 0.31 μmol/l (normal below 1.5) and his blood manganese concentration was 246 nmol/l (normal 70–280); his urine arsenic concentration was 10 nmol/mmol creatinine (normal <68). His blood mercury concentration was 83 nmol/l (normal for adults <50). The mercury content of the spray was 878 ppm (2% methyl mercury and 98% inorganic mercury). There was a significant difference in mercury content between different brands and between batches of the same brand of the Chinese medicinal herb. The spray was withdrawn and his tics completely resolved within 4 weeks.)

Peripheral polyneuropathy as a result of chronic ammoniated mercury poisoning has been studied and followed over 2 years (1711).

- A 36-year-old man developed peripheral polyneuropathy after chronic perianal use of an ammoniated mercury ointment. He had very high blood and urine mercury concentrations. Sural nerve biopsy showed mixed axonal degeneration/demyelination. His symptoms improved progressively over 2 years after withdrawal of the ointment, but neurophysiological recovery was incomplete.
- A 4-month-old boy was hospitalized with a weeping eczema covering more than the half of the body surface and complicated by skin hemorrhage and infection by *Klebsiella pneumoniae* and *Proteus mirabilis*. Apart from general therapy, a compound zinc oxide ointment containing, among other ingredients, yellow mercuric oxide 2 g per 40 g of base was applied daily. After 12 days, he rapidly developed cardiovascular collapse, acute pulmonary edema, and coma Stage II, with right hemiparesis, generalized hypertonia, and muscular tremor. The mercury concentrations in blood, urine, and CSF were respectively 120 ng/ml (normal <10 ng/ml), creatinine 260 micrograms/g (normal <5 micrograms/g), and 4.8 ng/ml (normal <0.1 ng/ml). Despite therapy with dimercaprol and vigorous supportive measures, the child's condition deteriorated, he developed *Klebsiella aerogenes* septicemia, and died 6 weeks later (1712).

Metamizole

In one case, intravenous injection of metamizole caused arterial hypotension (1713).

Methadone

A variety of complications following parenteral self-administration of oral methadone were noted, including regional thrombosis, often associated with shock and multiorgan failure (1714).

The use of methadone/dihydrocodeine has been linked to an acute myocardial infarction (1715).

- A 22-year-old man with a 6-year history of intravenous heroin use was maintained on methadone 60 mg/day and dihydrocodeine 0.5 g/day. He had an extensive anterior myocardial infarction as a result of occlusion of the left anterior descending coronary artery, which was reopened by percutaneous transluminal coronary angioplasty.

This case presents circumstantial evidence only, and the association was probably not a true one.

There has been a report of five cases of episodes of syncope and an electrocardiogram showing ventricular tachydysrhythmias with prolonged QT intervals and episodes of torsade de pointes; all the patients were taking high doses of methadone (270–660 mg/day) with no previous history of cardiac disease (1716). Torsade de pointes also occurred when high doses (3 mg/kg) of the long-acting methadone derivative, levomethadyl acetate HCl (LAAM), were given to a 41-year-old woman with a history of heroin dependence (1707). She was also taking fluoxetine and intravenous cocaine, which can prolong the QT interval, and fluoxetine and marijuana, which inhibit the activity of CYP3A4, which is responsible for the metabolism of LAAM and its active metabolite.

In a retrospective case study in methadone maintenance treatment programs in the USA and a pain management center in Canada, 17 methadone-treated patients developed torsade de pointes during 5 years (1718). The dose of methadone was 65–1000 mg/day. Six patients had had an increase in methadone dose in the months just before the onset of torsade de pointes. One patient had taken nelfinavir, a potent inhibitor of CYP3A4, begun just before the development of torsade de pointes. The above two risk factors (increased drug dosage and drug interactions) are important when eliciting the cause of torsade de pointes in patients taking methadone.

Methohexital

Vasodilatation and depressed myocardial contractility are possible hemodynamic consequences of high-dose methohexital anesthesia (1719).

Methotrexate

Cardiovascular adverse effects of methotrexate are extremely rare.

- There has been one detailed report of ventricular dysrhythmias and myocardial infarction, with recurrence of frequent ventricular extra beats on each readministration of methotrexate in a 36-year-old man (1720).

It has been suggested that methotrexate increases mortality in patients with rheumatoid arthritis with cardiovascular co-morbidity (1721). This assumption was based on a retrospective analysis of 632 patients with rheumatoid arthritis, of whom 73 died. The simultaneous presence of methotrexate and evidence of cardiovascular disease was an independent predictor of mortality. There was no such association with other DMARDs. The authors suggested that this effect may result from a methotrexate-induced increase in serum homocysteine, encouraging atherosclerosis.

Methylenedioxymetamfetamine (MDMA ecstasy)

See also

Cardiotoxicity following ecstasy use has been reported (1722). amphetamines

- A 16-year-old boy took three tablets of ecstasy and amfetamine 0.3 g and several hours later had convulsions and a temperature of 40.9°C. His heart rate was 210/minute and his blood pressure 100/75 mmHg. His creatine kinase activity was raised and he had myoglobinuria, renal impairment, hyperkalemia, and hypocalcemia. An electrocardiogram showed ventricular and supraventricular tachycardias but no myocardial ischemia. A diagnosis of serotonin syndrome due to ecstasy ingestion with associated hyperpyrexia and rhabdomyolysis was made. Following active treatment, his condition stabilized, with restoration of sinus rhythm and normal urine output. However, 12 hours later he developed jaundice, raised liver enzymes, and coagulopathy, suggesting acute liver failure due to ecstasy. With supportive treatment, his liver function improved. However, another 12 hours later, he developed shortness of breath associated withsided chest signs and X-ray changes compatible with aspiration pneumonia, and required emergency intubation 4 days later. He developed pulmonary edema, his pulmonary artery was occluded, and an echocardiogram showed globally impaired left ventricular function with an ejection fraction of 30–35%; there was electrocardiographic T wave inversion. Primary myocardial damage causing cardiac dysfunction was investigated using serial creatine kinase and troponin measurements. He recovered completely with treatment and an echocardiogram showed an ejection fraction of 60%.

The authors reported that this was the first case report of clinical, radiological, biochemical, and echocardiographic evidence of myocardial damage and cardiac dysfunction following ecstasy and amfetamine use.

Transient myocardial ischemia associated with ecstasy has been reported (1723).

- A 25-year-old man, a regular alcohol drinker with a history of asthma, had been out drinking 8 pints of lager and 4 gins. His last drink was spiked with a tablet that was presumably ecstasy. He scooped the tablet out, and even though some of the tablet may have dissolved, he finished the drink. He awoke 3 hours later with restlessness, nausea, and abdominal cramps. In the emergency room, his temperature was 37.2°C, and he was sweating. His heart rate was 120/minute and his blood pressure 130/70 mmHg. He had some abdominal discomfort. A diagnosis of ecstasy ingestion was made, although urine MDMA concentrations were not measured. His electrocardiogram on admission showed sinus tachycardia, with T wave inversion in leads I, aVL, and V4–6 and the voltage criteria for left ventricular hypertrophy. The next day the electrocardiogram had returned to normal. An echocardiogram was within normal limits. He was well on discharge.

The authors suggested that this was the first report of transient myocardial ischemia after ecstasy. They reasoned that the myocardial ischemia did not proceed to necrosis or a dysrhythmia because the amount of drug exposure was low.

Cardiovascular autonomic functioning during MDMA use has been investigated in 12 MDMA users and a matched group of non-users (1724). Resting heart rate variability (an index of parasympathetic tone) and heart rate response to the Valsalva maneuver (Valsalva ratio, an index of overall autonomic responsiveness) were both reduced in the drug users. Thus, seemingly healthy users of MDMA had autonomic dysregulation, comparable to that seen in diabetes mellitus. In several users there was a total absence of post-Valsalva release bradycardia, a sign of parasympathetic dysfunction. Since no cardiac data were available for these patients before their use of ecstasy, and since all were multidrug users, the findings must be interpreted with caution.

Extensive aortic dissection with cardiac tamponade and mesenteric ischemia has been attributed to ecstasy (1725).

- A 29-year-old man who took ecstasy and alcohol at a rave had no immediate adverse effects, slept well later on, and was in good health until he suddenly collapsed to the floor about 2 hours after waking. When seen 36 hours after the last dose of ecstasy he was short of breath and had abdominal pain, diarrhea, and vomiting. He had a loose bloody bowel movement but refused further investigations. He was discharged with a diagnosis of gastroenteritis, only to be readmitted 8 hours later after sudden deterioration and hypertension. Despite extensive efforts, his condition deteriorated and he died 5 hours later. At autopsy, there was a type I aortic dissection, starting at the root and spreading to the bifurcation, which had resulted in cardiac tamponade. The dissection had involved the mesenteric arteries, resulting in bowel ischemia.

Since this condition is rare in young adults, diagnosis can be difficult. The authors believed that this was the first case report of aortic dissection secondary to ecstasy.

Ecstasy has been associated with sudden death and cardiovascular complications. Eight healthy self-reported ecstasy users participated in a four-session, ascending-dose, double-blind, placebo-controlled comparison of the echocardiographic effects of ecstasy and those of dobutamine (1726). Ecstasy 1.5 mg/kg increased the mean heart rate by 28 beats/minute, systolic blood pressure by 25 mmHg, diastolic blood pressure by 7 mmHg, and cardiac output by 2 l/minute. The effects of ecstasy were similar to those produced by dobutamine (40 μg/kg/minute), except that ecstasy had no measurable inotropic effects. Thus, ecstasy increases systolic and diastolic blood pressures in the absence of a significant change in cardiac contractility and end-systolic wall thickness. The resulting increase in the tension of the ventricular wall leads to disproportionately higher myocardial oxygen consumption than would be expected from the observed changes in the heart rate and blood pressure. The authors commented that the behavioral and environmental factors accompanying the use of ecstasy—sustained exercise from dancing, often in crowded nightclubs with high ambient temperature and humidity—could further potentiate toxicity. They recommended a combination of beta-blockers and vasodilators for the emergency treatment of ecstasy-associated vascular instability.

Atrial fibrillation has been reported after the use of ecstasy (1727).

- A 17 year-old previously healthy man had a generalized tonic-clonic seizure. He denied drug abuse, but his urine drug screen was positive for ecstasy. He had an irregular heart rhythm with a normal blood pressure. An electrocardiogram showed atrial fibrillation with a ventricular rate of 102/minute. His routine laboratory investigations, a brain CT scan, and an electroencephalogram were normal. There were no underlying cardiac lesions. He was stabilized with medical treatment and was doing well 6 months later.

Since atrial fibrillation is unusual in young people, the authors speculated that ecstasy may have contributed in this case even though this adverse effect has not been reported before.

Methylergonovine

Fatal cardiac arrest has been reported in a woman with hypertension who was given methylergonovine after termination of pregnancy (1728).

- A 38-year-old Taiwanese woman, with a history of hypertension treated with verapamil and valsartan, was given intravenous methylergonovine 0.2 mg and intramuscular oxytocin 10 IU after termination of pregnancy at 5 weeks of gestation. Five minutes later she complained of chest pain and then had a cardiac arrest. Attempted resuscitation was unsuccessful.

The authors acknowledged that methylergonovine should have been avoided in this case, and in any case is best given intramuscularly. They also noted that six of the seven published cases of ergot alkaloid-induced post-partum myocardial infarction occurred in Asians, suggesting increased susceptibility.

Methylphenidate

Significant increases in blood pressure and/or pulse rate after methylphenidate are more frequent when it is given parenterally (1729) (SEDA-4, 9).

Cardiac dysrhythmias, shock, cardiac muscle pathology, and liver pathology have all been reported (1730).

Ambulatory blood pressure monitoring showed changes in blood pressure and heart rate in boys aged 7–11 years taking stimulant therapy (1731). This preliminary study with chronic methylphenidate or Adderall® (dexamfetamine + levamfetamine) for ADHD showed alterations in awake and asleep blood pressures, with profound nocturnal dipping. Modified-release formulations of methylphenidate and Adderall® now allow more sustained blood concentrations in children. The effects of these newer formulations on cardiovascular indices should be evaluated.

Methysergide

Methysergide can cause tachycardia, postural hypotension, and angina pectoris.

Aortic and mitral valvular fibrosis can lead to congestive cardiac failure; fibrosis rarely affects the endocardium more extensively (extending into the myocardium) or the pericardium (resulting in constrictive pericarditis). Vasospastic effects can occasionally be as severe in susceptible subjects as with ergotamine; especially dangerous are combinations with ergotamine tartrate, as are combinations of ergot alkaloids with beta-blockers (SEDA-9, 128).

Metoclopramide

Complete heart block (SEDA-12, 939; 1732) and supraventricular tachycardia (1733), presumably due to a vagolytic effect, have been described with metoclopramide. If there are predisposing conditions, heart failure can be precipitated (SEDA-16, 418). In one study in which 56 patients were treated with metoclopramide, a single case of second-degree atrioventricular heart block was observed (SEDA-20, 316).

- Complete heart block has been reported in a 65-year-old woman with type 2 diabetes, hypertension, and ischemic heart disease, who was given intravenous metoclopramide 20 mg (1734).

In a double-blind, randomized, placebo-controlled study of metoclopramide as a single intravenous dose of 10 mg in 10 healthy men, metoclopramide caused an increase in QT slope and QT variance 30 minutes after the dose (1735). Caution is recommended with other drugs that affect the QT interval.

A hypertensive crisis has been precipitated by metoclopramide in patients with pheochromocytoma (1736).

Metocurine

Metocurine has significantly less ganglion-blocking activity than D-tubocurarine (1737) and much less of a tendency to provoke histamine release (1738,1739). It does not block cardiac muscarinic receptors at neuromuscular blocking doses (1740). Cardiovascular stability is therefore to be expected (1741). Slight tachycardia and fall in blood pressure have been reported in one-third of patients given larger doses (0.4 mg/kg) rapidly, probably as a result of histamine release, but no bronchospasm was seen (1742).

Metronidazole

Thrombophlebitis can occur after intravenous administration of metronidazole; an incidence of 6% has been cited (SEDA-7, 297; 1743).

Mianserin

Cardiotoxic effects are relatively uncommon with mianserin (1744). In a placebo-controlled study in 50 patients with a variety of cardiac conditions who were taking anticoagulants, mianserin (up to 30 or 60 mg) had no effects on electrocardiography, blood pressure, or pulse rate after 3 weeks. In a second phase, mianserin (up to 60 mg/day) was compared with amitriptyline (up to 150 mg/day) and placebo in 18 healthy volunteers. Measurements included systolic time intervals, electrocardiography at rest and during exercise, echocardiography, and blood pressure. Amitriptyline had a negative inotropic effect; mianserin increased ejection fraction. The results of both these experiments led the authors to conclude that mianserin is an antidepressant with very low cardiac toxicity.

In another study of the electrocardiographic effects of mianserin and a review of previous experiments with doses up to 120 mg/day, there were no consistent cardiovascular effects, although four patients who took 40 mg/day for 13 months had significant increases in pulse rate, with prolonged PR intervals and reduced T-wave amplitude (1745). Fainting and persistent bradycardia were noted in one woman after a single dose of mianserin 60 mg, and her symptoms recurred on rechallenge (SEDA-17, 21).

There have been two instances of possible cardiac effects due to mianserin in elderly patients with pre-existing cardiovascular disorders (1746).

- A 71-year-old man with hypertension who took 30 mg/day developed cardiac failure.
- A 66-year-old woman with mitral regurgitation and atrial fibrillation developed hypokalemia and a variety of rhythm disturbances at dosages of 40 mg or more per day.

Miconazole

Local phlebitis is not uncommon after intravenous miconazole (1747); the type of intravenous solution used is of importance.

Collapse after rapid intravenous injection has been described, as have some cases of tachycardia, ventricular tachycardia, and even, in a few instances, cardiorespiratory arrest, attributable to the histamine-releasing properties of Cremophor (SED-12, 679; 1748).

Midazolam

The incidence of hypotension with the use of midazolam for pre-hospital rapid-sequence intubation of the trachea has been assessed in a retrospective chart review of two aeromedical crews (1749). The rapid-sequence protocols were identical, except for the dose of midazolam. Both crews used 0.1 mg/kg, but one crew had a maximum dose of 5 mg imposed. This meant that patients over 50 kg received lower doses of midazolam; they also had a higher incidence of hypotension. This relation was also present in patients with traumatic brain injury, implying that cerebral perfusion could be compromised at a critical time in those without dosage restriction.

Midazolam depresses both cardiovascular and respiratory function, especially in elderly patients (1750). As little as 0.01 mg/kg can obtund the response to hypoxia and hypercapnia (1751). The simultaneous use of opiates (such as fentanyl) commonly produces hypoxia (1752).

Hypertension and tachycardia during coronary angiography can cause significant problems. In a double-blind, randomized, placebo-controlled study during coronary angiography in 90 patients, midazolam with or without fentanyl under local anesthesia provided better hemodynamic stability than placebo (1753).

Midodrine

Supine hypertension occurs in up to 3% of subjects, although it is not clear how the underlying pathology of the hypotension influences the likelihood of this reaction.

Milrinone

The main limitations of the short-term use of milrinone is the risk of hypotension and ventricular dysrhythmias (9–12) and sudden death has occurred (1754). Acute exacerbation of chronic heart failure has also been reported (1755).

In a retrospective view of 63 patients who received intravenous milrinone for more than 24 hours for advanced cardiac failure, the mean dose was 0.43 microgram/kg/minute and the mean duration of therapy 12 (range 1–70) days (1756). After 24 hours of therapy there was significant improvement in pulmonary artery pressure, pulmonary capillary wedge pressure, and cardiac index. Because of the nature of the study, which was not placebo-controlled, it is impossible to be sure what events could have been attributed to the milrinone. However, the authors reported five cases of asymptomatic, non-sustained ventricular tachycardia, six of symptomatic ventricular tachycardia, and three deaths, one in ventricular tachycardia and two in heart failure. There was no difference in the incidence of these adverse events in patients who received milrinone for more than 7 days compared with the others.

Milrinone can cause tachycardia, partly because of its vasodilatory effects and partly perhaps by a direct effect on the heart.

- A 74-year-old man had a tachycardia of 145/minute during infusion of milrinone after an operation for repair of an abdominal aortic aneurysm (1757). The tachycardia was controlled by esmolol on one occasion and more impressively by metoprolol on a second occasion. However, the hemodynamic effects of milrinone were not altered by beta-blockade.

Presumably the beneficial effects of beta-blockade in this case were non-specific, since milrinone does not affect beta-adrenoceptors.

Of 19 children (12 infants and 7 children aged 1–13 years) who were given either two boluses of 25 micrograms/kg followed by an infusion of 0.5 microgram/kg/minute, or a bolus of 50 micrograms/kg followed by a bolus of 25 micrograms/kg followed by an infusion of 0.75 microgram/kg/minute, two infants developed junctional ectopic tachycardia during infusion of milrinone.

Vasopressin has been used to treat hypotension due to milrinone. In seven patients with congestive heart failure who developed hypotension (systolic arterial pressure below 90 mmHg), vasopressin 0.03–0.07 units/minute increased the systolic arterial pressure to 127 mmHg (1758). This effect was due to peripheral vasoconstriction, since the systemic vascular resistance increased from 1112 to 1460 dyne.s/cm^5 with no change in cardiac index. Urine output also improved significantly. In three patients in whom milrinone caused hypotension, vasopressin (0.03–0.07 U/minute) increased the systolic arterial pressure from 90 to 130 mmHg and reduced the dosages of catecholamines that were being used (1759). The authors hypothesized that vasopressin may have inhibited the milrinone-induced accumulation of cyclic AMP in vascular smooth muscle.

Mivacurium chloride

Benzylisoquinolinium compounds have a tendency to evoke histamine release, the main source of the cardiovascular changes seen with mivacurium. These have been reported as minimal up to and including twice the ED_{95} dose in several studies (1760,1761–1763). At higher dosages (0.2 mg/kg and over) transient hypotension, often associated with facial flushing and lasting only some 2–5 minutes, which correlated significantly with increases in plasma histamine, has been described (1761). Reducing the speed of injection to 30 or 60 seconds reduced the degree of hypotension to insignificant levels. Similar findings in 50% of patients given doses of 0.2 and 0.25 mg/kg have been described elsewhere (1760). In yet another series, mean arterial pressure fell by more than 20% (24–61%) in seven of 15 patients given rapid bolus injections of 0.2–0.25 mg/kg (1762). Pretreatment with oral antihistamines has been suggested as an additional option to reduce histamine-related adverse effects after administration of high-dose mivacurium (1764).

In patients scheduled for coronary artery bypass grafting or valve replacement (1763), significant hypotension was seen in two patients (out of 27 given higher doses), even when mivacurium was injected slowly (over 60 seconds); the mean arterial pressure fell by 24 and 50% after the injection of 0.2 and 0.25 mg/kg respectively. Beta-blockers, calcium channel blockers, and nitrates were not discontinued preoperatively in this study. The authors concluded that "doses larger than 0.15 mg/kg are probably unnecessary and may contribute to hemodynamic instability at least in cardiac patients."

Mizolastine

Erythromycin and ketoconazole both increase mizolastine plasma concentrations and patients with hepatic or renal impairment have altered pharmacokinetics of mizolastine, indicating the need for caution in its use in such individuals (1765).

Moclobemide

There have been two reports of single cases of a rise in blood pressure after moclobemide: in one there was a 40 mmHg increase in systolic pressure after one dose only (1766), while the other report referred only to "episodes of hypertension" without further comment (1767). Interactions were not suggested in either case.

In a prospective drug utilization study in 13 741 patients who received the reversible type A MAO inhibitor moclobemide in a variety of settings, including general practice, psychiatric out-patients, and psychiatric in-patients, there were few episodes of hypertension (0.11%; 95% CI = 0.06, 0.18%) or hypotension (0.04%; 95% CI = 0.02, 0.10%) and episodes were mostly associated with underlying cardiovascular disease (1768). These findings suggest that moclobemide in usual therapeutic doses carries a low risk of producing significant changes in blood pressure relative to conventional MAO inhibitors.

Modafinil

In an open, randomized, crossover study in healthy men, concomitant administration of single oral doses of modafinil (200 mg) and dexamfetamine (10 mg), each given separately, produced a slight increase in blood pressure and the cardiovascular effects were more pronounced after concomitant administration (1769). These changes were considered not to be clinically relevant and would not necessarily preclude short-term administration of the two drugs together. The most frequent adverse events with modafinil or dexamfetamine were headache, dizziness, insomnia, and dry mouth.

Modafinil was effective in narcolepsy in a 9-week, randomized, placebo-controlled, double-blind, 21-center trial in 271 patients (1770). During treatment withdrawal, the patients did not have symptoms associated with amphetamine withdrawal. Nausea and rhinitis were significantly more common in the treatment group; in contrast, in a previous multicenter study in the USA there was a higher incidence of headache (1771). Modafinil was also effective in the treatment of somnolence due to pramipexole in a patient with Parkinson's disease (1772).

Once-daily modafinil for an average of 4.6 weeks has been evaluated in an open trial in 11 children aged 5–15 years with ADHD (1773). This pilot study, with non-blinded ratings, a small number of subjects, and a short duration of treatment, showed significant improvement. Adverse events were responsible for drug withdrawal in one child. The most common adverse event was delayed onset of sleep or sleep disruption, which, in two of three cases, responded to a reduction in dosage. No patient taking modafinil lost weight or had a reduced appetite. A larger-scale, double-blind, placebo-controlled study will be needed to further substantiate the efficacy and safety of modafinil.

In a 9-week, single-blind, placebo-controlled pilot study in 72 patients with multiple sclerosis who took modafinil 200 mg/day for 2 weeks, there was significant improvement in fatigue compared with placebo run-in treatment (1774). The most frequent adverse effects were headache, nausea, and anxiety, and these were rated as either mild or moderate.

Monoamine oxidase (MAO) inhibitors

When MAO is inhibited, the concentrations of noradrenaline, dopamine, and serotonin increase in the central nervous system and heart. The concentrations of the precursors of these amines (dihydroxyphenylalanine and 5-hydroxytryptophan) are greatly increased. The effects are similar to those of the postganglionic blocking amines: postural hypotension occurs and cardiac output is reduced, possibly due to the accumulation of a pseudo-transmitter.

Hypotension

The hypotensive effects of the MAO inhibitors differ from those of the tricyclic antidepressants, inasmuch as the former affect supine as well as orthostatic blood pressure. This was confirmed both in a study (1775) involving tranylcypromine and in an evaluation of blood pressure changes in 14 patients taking phenelzine averaging 65 mg/day for at least 3 weeks (1776). The drug-free mean supine systolic blood pressure was 127 mmHg and it fell significantly (mean drop 5 mmHg) by the end of the first week. By contrast, an increase in the orthostatic drop from the mean predrug baseline of 2 mmHg did not reach significance until the end of the second week of treatment, after which the blood pressure continued to fall. Two patients developed profound orthostatic falls of up to 50 mmHg after more than 2 weeks of treatment. In one patient the hypotension and related symptoms (light-headedness and ataxia) improved with fludrocortisone, but not in the other, who required drug withdrawal. The authors commented that because orthostatic hypotension can develop late, cautious long-term monitoring is required.

Hypertension

Hypertensive crises usually occur when MAO inhibitors are combined with other drugs or foods that cause interactions (see next subsection).

Autopotentiation of hypertensive effects

In a few cases, a hypertensive crisis appears not to have been provoked by known drug or dietary precipitants (1777). From a review of 12 reports of spontaneous hypertension in patients taking tranylcypromine or phenelzine, it has emerged that a family history of hypertension may be a risk factor (SEDA-18, 14). A significant increase in supine blood pressure without similar changes in standing blood pressure after the administration of tranylcypromine has also been described (SEDA-17, 17).

Cardiac arrhythmias

MAO inhibitors can cause bradycardia. A report of two cases of interactions of monoamine oxidase inhibitors with beta-blockers (nadolol and metoprolol) is of interest, and several possible mechanisms were discussed (1778).

Moxifloxacin

See also fluoroquinolone antibiotics

Moxifloxacin blocks the rapid-component delayed-rectifier potassium channel in the heart, and thus prolongs the QT_c interval by 6 minutes after oral administration and 12 minutes after intravenous administration (1779). Moxifloxacin carries a greater risk of QT interval prolongation than ciprofloxacin, levofloxacin, and ofloxacin (1780), and although the risk of moxifloxacin-induced torsade de pointes is expected to be minimal when the drug is given in the recommended dosage (400 mg/day) (1781), moxifloxacin should be used with caution in patients with prodysrhythmic conditions and avoided in patients taking antidysrhythmic drugs, such as quinidine, procainamide, amiodarone, and sotalol (1782).

- A man developed a hypertensive crisis and transient left bundle branch block with QT interval prolongation after taking moxifloxacin (1783).
- Sinus tachycardia (120/minute) associated with moxifloxacin has been reported in a 49-year-old man about 45 minutes after he took a single 400 mg dose of moxifloxacin (1784).

The underlying mechanism may have been vasodilatation, either directly or indirectly, owing to release of histamine, with reflex tachycardia. These effects have been described for other fluoroquinolones.

Muromonab–CD3

Cardiovascular manifestations are usually observed during the cytokine-release syndrome. They mostly included tachycardia and transient hypertension or hypotension (1785). Fluid overload or volume depletion can also play a role. Chest pain or severe dysrhythmias are infrequent (1786).

Muromonab has been thought to exert procoagulant activity and activate fibrinolysis, probably as a result of TNF-alfa release of other cytokines and cellular adhesion to the vascular endothelium (1787,1788). Accordingly, retrospective studies have pointed to an increased risk of intragraft thromboses with thromboses in the renal artery, renal vein, or glomerular capillaries, or thrombotic microangiopathy, leading to a higher incidence of rejection episodes in patients receiving high or conventional doses of muromonab (1787,1789). Among 231 kidney transplant recipients who received muromonab prophylaxis, intragraft thromboses were found in 13 (5.6%), and the use of very high-dose methylprednisolone (30 mg/kg) was suggested to be a major risk factor (1790). In contrast, and despite the evidence of procoagulant activity, some investigators have failed to identify evidence that muromonab increases the risk of thromboembolic complications (1788,1791).

Nalmefene

Pulmonary edema has been described after nalmefene (SEDA-22, 104).

Naloxone

Doses of naloxone over 1 pg/kg should be given with caution, especially to patients with hypertension. Massive release of catecholamines in response to pain after administration of naloxone can trigger left ventricular failure, partly by causing a shift in fluid from the intravascular to the interstitial Thus, alpha-blockers such as phentolamine have been postulated to be beneficial in its management (SEDA-17, 88). A fatal case of pulmonary edema followed the use of naloxone in a young man (1792), although the causal link was disputed (1793).

Severe hypertension and multiple atrial extra beats have been reported after the administration of naloxone, especially in patients with coronary heart disease (1794).

Neuroleptic drugs

See also individual compounds

Neuroleptic drugs can reduce exercise-induced cardiac output as a result of drug-induced increases in plasma catecholamine concentrations and concurrent alpha-adrenoceptor blockade (1795). Cardiomyopathy has been associated with neuroleptic drugs, including clozapine (1796–1797).

Hypotension

Neuroleptic drugs with alpha-blocking activity can cause hypotension. Those of high and intermediate potency, such as haloperidol and loxitane, have minimal alpha-blocking effects and should be less likely to cause hypotension, although in one report orthostatic changes (a fall of 30 mmHg) were reported with these drugs in 27% and 22% of cases respectively (SED-11, 106) (1798). An exception to the relatively safe use of high-potency agents has been noted in the combination of droperidol with the narcotic fentanyl, which can cause marked hypotension (1799).

Postural hypotension is particularly hazardous in susceptible patients, such as the elderly and those with depleted intravascular volume or reduced cardiovascular output. The risk of orthostatic hypotension is markedly increased after parenteral administration. The combination of alpha-adrenoceptor blockade and sedative effects may explain the increased risk of falling when taking neuroleptic drugs (SEDA-12, 52).

Significantly more low blood pressures were documented by 24-hour ambulatory blood pressure monitoring compared with typical blood pressure measurement obtained an average of 3.6 times a day in patients treated with psychotropic drugs ($n = 12$), most of which were neuroleptic drugs (1800). This finding may be of clinical relevance, in view of the potential hemodynamic consequences of hypotension, especially in older patients taking more than one psychotropic drug.

Cardiac dysrhythmias

Neuroleptic drugs are vagolytic and can increase resting and exercise heart rates. Cardiac dysrhythmias have been reported, and include atrial dysrhythmias, ventricular tachycardia, and ventricular fibrillation. Bradycardia is unusual. The risk of cardiac dysrhythmias is dose-related and is increased by pre-existing cardiovascular pathology (SEDA-2, 48), interactions with other cardiovascular or psychotropic drugs (particularly the highly anticholinergic tricyclic antidepressants), increased cardiac sensitivity in the elderly, hypokalemia, and vigorous exercise.

In elderly people it is advisable to avoid low-potency neuroleptic drugs, such as thioridazine, which produce significantly more cardiographic changes than high-potency agents, such as fluphenazine (1801). In any patient with pre-existing heart disease, a pretreatment electrocardiogram with routine follow-up is recommended.

QT interval prolongation due to neuroleptic drugs has been reviewed (1802). It is not a class effect: among currently available agents, thioridazine and ziprasidone are associated with the greatest prolongation. Dysrhythmias are more likely to occur if drug-induced QT prolongation co-exists with other risk factors, such as individual susceptibility, congenital long QT syndromes, heart failure, bradycardia, electrolyte imbalance, overdose of a QT interval-prolonging drug, female sex, restraint, old age, hepatic or renal impairment, and slow metabolizer status; pharmacokinetic or pharmacodynamic interactions can also increase the risk of dysrhythmias. Prolongation of the QT interval occurs more often in patients taking more than 2000 mg of chlorpromazine equivalents daily (1803).

In an open study in 164 patients with schizophrenia of the effect of several neuroleptic drugs on the QT interval, the study drugs were given for 21–29 days and three separate electrocardiograms were obtained after steady state had been achieved and drug concentrations were at their maximum (1804). The mean changes in the QT_c interval were:

(a) thioridazine 36 ms (95% CI = 31, 41; n = 30);
(b) ziprasidone 20 ms (95% CI = 14, 26; n = 31);
(c) quetiapine 15 ms (95% CI = 9.5, 20; n = 27);
(d) risperidone 12 ms (95% CI = 7.4, 16; n = 20);
(e) olanzapine 6.8 ms (95% CI = 0.8, 13; n = 24);
(f) haloperidol 4.7 ms (95% CI = −2, 11; n = 20).

Mechanism

Torsade de pointes, first described in 1966 by Dessertenne (1805), is a potentially fatal ventricular tachydysrhythmia with a characteristic pattern of polymorphous QRS complexes, which appear to twist around the isoelectric line. It is often associated with ventricular extra beats immediately before the dysrhythmia. Torsade de pointes typically occurs in the setting of a prolonged QT interval, which includes depolarization and repolarization times. Conditions or agents that delay ventricular repolarization, causing long QT interval syndromes, can trigger the dysrhythmia. It has been hypothesized that prolongation of the QT interval is caused by early after-depolarization, which in turn may develop in response to abnormal ventricular repolarization (1806–1808). Evidence linking QT interval prolongation, potassium channel function, and torsade de pointes to neuroleptic drugs has been reviewed (1809).

Among the non-cardiac drugs that can cause QT interval prolongation, torsade de pointes, and sudden death (1810) the neuroleptic drugs have particularly been associated with conduction disturbances, torsade de pointes being one of the most worrisome (SED-14, 141) (1811,1812) (SEDA-20, 36) (SEDA-21, 43) (SEDA-22, 45) (SEDA-23, 49) (1813,1814). Although the role of neuroleptic drugs in sudden death is controversial (SEDA-18, 47) (SEDA-20, 36), and although there are other non-cardiac causes of this syndrome, including asphyxia, convulsions, or hyperpyrexia, QT interval prolongation and torsade de pointes provide a plausible mechanism of sudden death.

QT_c prolongation has been proposed as a predictor of sudden death, and psychiatrists are encouraged to perform electrocardiograms in patients taking high-dose neuroleptic drugs to detect conduction abnormalities, especially QT_c prolongation. It has been suggested that QT_c prolongation in itself is not necessarily an indicator of the risk of sudden death (1815). Instead, QT_c dispersion, the difference between the longest and the shortest QT_c interval on the 12-lead electrocardiogram, is an indication of more extreme variability in ventricular repolarization, which could be regarded as a better predictor of the risk of dysrhythmias.

Epidemiology

Electrocardiographic changes are relatively common during treatment with neuroleptic drugs, but there is a lack of unanimity regarding the clinical significance of these findings. The changes that are generally considered benign and non-specific are reversible after withdrawal. Potentially more serious changes include prolongation of the QT interval, depression of the ST segment, flattened T waves, and the appearance of U waves. Non-specific T wave changes are commonly seen during the mid-afternoon, and may be related to the potassium shift and other changes that result after meals, so that a prebreakfast cardiogram may be more desirable.

The prevalence of QT_c prolongation in psychiatric patients has been estimated, to assess whether it is associated with any particular neuroleptic drug (1816). Electrocardiograms were obtained from 101 healthy control individuals and 495 psychiatric patients (aged 18–74 years) in various inpatient and community settings in North-East England. Exclusion criteria were atrial fibrillation, bundle-branch block, and a change in drug therapy within the previous 2 weeks (3 months for depot formulations). The threshold for QT_c prolongation (456 ms) was defined as 2 standard deviations above the mean value in the healthy controls. Values were abnormal in 40 patients. Significant independent predictors of QT_c prolongation in psychiatric patients after adjustment for potential confounding effects were aged over 65 years and the use of tricyclic antidepressants, droperidol (RR = 6.7; CI = 1.8, 25), or thioridazine (RR = 5.3; CI = 2.0, 14). Increasing neuroleptic drug dosage was also associated with an increased risk of QT_c prolongation. Abnormal QT dispersion or T wave abnormalities were not significantly associated with neuroleptic drug treatment. Based on these results, the authors recommended electrocardiographic screening, not only in patients taking high doses of any neuroleptic drug but also in those taking droperidol or thioridazine, even at low doses. In other studies, dose and age were similarly found to be predictive factors (1424), as was co-administration of carbamazepine (1817).

Management

Adverse effects related to QT interval prolongation can be prevented by avoiding higher doses and by avoiding the following patients at risk: those with organic heart disease, particularly congestive heart failure; those with metabolic abnormalities (such as hypokalemia and hypomagnesemia); and those with sinus bradycardia or heart block. Concomitant administration of drugs that inhibit drug metabolism should also be avoided, and the potassium concentration should be controlled. If torsade de pointes is suspected, neuroleptic drugs must be withdrawn.

Magnesium sulfate 2 g (20 ml of a 10% solution intravenously) suppresses torsade de pointes (1818). Most patients report flushing during this injection. A case of torsade de pointes related to high dose haloperidol and treated with magnesium has been reported (1819).

- A 41-year-old woman, with liver lacerations, rib fractures, and pneumothorax after a motor vehicle accident, was given haloperidol for agitation on day 7. During the first 24 hours she received a cumulative intravenous dose of 15 mg, 70 mg on day 2, 190 mg on day 3, 160 mg on days 4 and 5, and 320 mg on day 6. An hour after the first dose of 80 mg on day 7, she had ventricular extra beats followed by 5-beat and 22-beat runs of ventricular tachycardia. The rhythm strips were consistent with polymorphous ventricular tachycardia or torsade de pointes and the QT_c interval was 610 ms (normally under 450 in women). She received intravenous magnesium sulfate 2 g. Concurrent medications included enoxaparin, famotidine, magnesium hydroxide, ampicillin/sulbactam, nystatin suspension, midazolam, and 0.45% saline with 20 mmol/l of potassium chloride. She had no further dysrhythmias after haloperidol was withdrawn. Eight days after the episode of torsade de pointes she had a QT_c interval of 426 ms.

Nicotine replacement therapy

Nicotine can aggravate cardiovascular disease through the hemodynamic consequences of sympathetic neural stimulation, or systemic release of catecholamines, or both (1820). Although several cardiovascular events have been ascribed to nicotine patches, including cerebral arterial narrowing with severe headaches and transient neurological deficits (1821), relatively few adverse effects have been reported (1820). Suicide attempts by means of overdosage using nicotine patches, presumably with the intention of precipitating myocardial infarction, have failed (1822), and it would seem that the risk of nicotine replacement therapy is likely to be much less than that of cigarette smoking, even in patients with coronary heart disease (1820).

Nicotine can provoke angina in patients with coronary artery disease and can also trigger myocardial infarction. It has been hypothesized that activities associated with increased catecholamine concentrations can both trigger plaque rupture and promote an occlusive thrombus formation that can result in myocardial infarction or sudden cardiac death. By evoking the release of catecholamines (stimulating platelet aggregation and increasing platelet/vessel wall interactions), nicotine could therefore be a potent trigger of myocardial infarction. However, a randomized, double-blind, placebo-controlled trial in 584 patients with at least one diagnosis of cardiovascular disease clearly showed that transdermal nicotine does not cause a significant increase in cardiovascular events in high-risk patients with cardiovascular disease. On the other hand, the efficacy of transdermal nicotine as an aid to smoking cessation in such patients has proved to be very limited and may not be sustained over time (1823). In another study (1824) from the Pharmacovigilance Division of the Federal Ministry for Health and Consumer Products of Austria, 41 cases of myocardial infarction were associated with the use of nicotine patches.

Atrial fibrillation was described in association with normal or large doses (1825,1826), suggesting that accumulation can occur when nicotine gum is taken daily.

Nicotinic acid and derivatives

Symptoms observed after high doses of nicotinic acid (such as giddiness, faintness, and vasovagal attacks) result from dilatation of the small arteries and arterioles, resulting in reduced peripheral resistance and blood pressure (1827). Flushing occurs transiently in all patients and persists in 10–15% of cases. When starting with a high dose, hypotension can occur (1828). The flushing of the skin with pruritus is transitory.

Coronary steal with worsening of myocardial ischemia related to vasodilatation has been reported in a patient taking nicotinic acid (1829), which should therefore perhaps be withheld in patients with unstable angina.

Niridazole

Minor electrocardiographic abnormalities, especially T wave changes, are common but are probably of no functional significance of patients (1830). Dysrhythmias with prolongation of the QT interval occur in a small minority of patients (1831).

Nitric oxide

There have been several case reports that high doses of nitric oxide (40–80 ppm) reduce pulmonary vascular resistance and increase pulmonary capillary wedge pressure in some patients with left ventricular dysfunction.

The acute increase in left ventricular filling pressure that ensues can cause or exacerbate pulmonary edema. The author concluded that in patients with severe left ventricular dysfunction (pulmonary capillary wedge pressure over 25 mmHg) it would be prudent to avoid the use of inhaled nitric oxide.

Systemic hypotension has been reported after the introduction of inhaled nitric oxide (1832).

- A 72-year-old woman, who underwent emergency resection of a giant left atrial myxoma, had pulmonary hypertension (pulmonary artery pressure 40 mmHg) and a low cardiac output (2.2 l/minute). Inhaled nitric oxide, 40 ppm, before cardiopulmonary bypass resulted in pulmonary vasodilatation and a fall in pulmonary artery pressure from 39 to 31 mmHg. This was accompanied by a fall in cardiac output from 2.4 to 1.5 l/minute and a fall in mixed venous oxygen saturation. After bypass, inhaled nitric oxide improved pulmonary and systemic hemodynamics and resulted in a rise in cardiac output from 3.0 to 3.5 l/minute.

The authors concluded that inhaled nitric oxide had reduced pulmonary hypertension and pulmonary vascular resistance, but that obstruction by the atrial myxoma had caused the low cardiac output, due to a net reduction in transpulmonary pressure.

Nitrofurantoin

Except for cardiovascular collapse in anaphylactic shock, adverse cardiovascular events seem to be extremely rare with nitrofurantoin (1833,1834). In experimental animals, cardiotoxic effects have been described (1835).

Nitrous oxide

Although myocardial depression has been described in healthy volunteers after the use of 40 or 50% nitrous oxide in oxygen, it is usually mild. It is likely that nitrous oxide can worsen myocardial ischemia in patients with critical coronary stenosis, although this may not be of clinical significance (1836,1837).

Homocysteine concentrations

Nitrous oxide inhibits methionine synthase, thereby preventing the conversion of homocysteine to methionine. A high homocysteine concentration has also been identified as an independent risk factor for coronary artery and cerebrovascular disease.

The effect of nitrous oxide on homocysteine concentrations and perioperative myocardial ischemia/infarction has been extensively reviewed (1838). Nitrous oxide causes acute rises in postoperative homocysteine concentrations temporally associated with postoperative myocardial ischemia. Preoperative oral folate and vitamins B_6 and B_{12} blunt nitrous oxide-induced postoperative increases in plasma homocysteine (1839).

The effect of nitrous oxide on homocysteine concentrations and myocardial ischemia has been studied in a randomized controlled study in 90 patients, ASA grades 1–3, who received a standardized anesthetic consisting of propofol induction, an opioid, and either inhalational isoflurane or isoflurane + 50% nitrous oxide (1840). They underwent carotid endarterectomy (average operation duration 3.3 hours). Electrocardiographic monitoring consisted of a three-channel Holter monitor (leads II, V2, and V5), which was later examined for periods of ischemia by a physician blinded to treatment group. Myocardial enzyme activities were not measured. Baseline homocysteine concentrations (12.7 μmol/l) were significantly increased in the recovery room and at 48 hours to 15.5 and 18.8 μmol/l respectively. The concentrations did not increase in those given nitrous oxide. Periods of preoperative and intraoperative ischemia did not differ. Postoperatively the nitrous oxide group had more patients with ischemia (19 versus 11), longer ischemic events in the first 24 hours (54 minutes versus 17 minutes), and more episodes of ischemia lasting more than 30 minutes (23 versus 14). The authors concluded that nitrous oxide is associated with increased myocardial ischemia. However, they conceded that they had not shown causality. Previous studies have not shown this outcome, but were less sensitive, using only two-channel or once-daily 12-lead electrocardiography, or not monitoring patients postoperatively. The subject warrants a major study before a firm conclusion can be drawn.

Non-steroidal anti-inflammatory drugs (NSAIDs)

See also COY-2 inhibitors

Hypertension

NSAIDs can cause or aggravate hypertension and interact negatively with the effects of antihypertensive drugs, including diuretics, although contrasting data from experimental and clinical studies have been published (1841).

Data from a randomized trial have suggested that ibuprofen significantly increases blood pressure in patients taking ACE inhibitors, but that celecoxib and nabumetone do not (1842).

Compared with placebo, ibuprofen was associated with significantly greater increases in both systolic and diastolic blood pressure, whereas blood pressure increases with nabumetone and celecoxib were not significantly different to placebo. In addition, the proportion of patients with systolic blood pressure increases of clinical concern was significantly greater in those taking ibuprofen (17%) than in those taking nabumetone (5.5%), celecoxib (4.6%), or placebo (1.1%). However, the results of this study must be confirmed in a larger population of hypertensive patients on the basis of relevant clinical outcomes.

The mechanisms by which NSAIDs affect cardiovascular function are complex and controversial. They may include reduced blood flow, a reduction in the filtered load of sodium, an increase in tubular reabsorption of sodium, and a reduction in the synthesis of PGE-1, which may be associated with raised blood viscosity and increased peripheral vascular resistance. This is perhaps the primary mechanism, which is due to increased renal synthesis of endothelin-1.

Meta-analyses

In a meta-analysis of the hypertensive effects of NSAIDs or aspirin (1.5 g/day or greater) in short-term intervention studies; 54 studies and 123 NSAID treatment arms were included (1843). Of the 1324 participating subjects, 92% were hypertensive; they had a mean age of 46 years and none was over 65 years. The major outcome studied was the change in mean arterial pressure. The effects of NSAIDs on blood pressure were found solely in hypertensive subjects; among these, the increase in mean arterial pressure, after adjusting for possible confounders (for example dietary salt intake) was different among different NSAIDs. The increase in mean arterial pressure was 3.59 mmHg for indometacin (57 treatment arms), 3.74 mmHg for naproxen (4 arms), and 0.45 mmHg for piroxicam (4 arms). However, mean arterial pressure fell by 2.59 mmHg with placebo (10 arms), by 0.83 mmHg with ibuprofen (6 arms), 1.76 mmHg with aspirin (4 arms), and 0.16 mmHg with sulindac (23 arms).

Overall, only the effects of indometacin on mean arterial pressure were statistically significantly different from those found with placebo, showing that in this population the effects of NSAIDs on blood pressure were modest and varied considerably among different drugs. However, one must take into account the important limitations of the analysis: the patients were mostly young, the studies included in the meta-analysis were small and short-term, and information on possible confounders was incomplete. The significance of the results of this study is therefore doubtful.

Another meta-analysis provided more complete and useful results (1844). Its primary aim was to produce an estimate of the overall effect of NSAIDs on blood pressure, and its secondary aims were to evaluate the mechanisms by which NSAIDs alter blood pressure and to determine susceptibility factors. Moreover, as NSAIDs have been associated with raised blood pressure in normotensive individuals and in both treated and untreated hypertensive subjects, the authors tried to discover different effects in these subgroups. Finally, they studied whether different NSAIDs alter blood pressure to the same degree.

In all, 50 randomized placebo-controlled trials and 16 randomized comparisons of two or more NSAIDs met the selection criteria. These studies included 771 young volunteers or patients aged 47 years or younger. The studies were small (the mean sample size per trial was 16); many different NSAIDs and antihypertensive drugs were used, but indometacin was used in more than half of all the trials; the duration of therapy with NSAIDs or antihypertensive drugs was 1 week or longer, but in most studies it was less than 3 months. NSAIDs raised supine mean blood pressure by 5.0 mmHg (95% CI = 1.2, 8.7), but had no significant effect on variables measured to assess possible mechanisms (such as body weight, daily urinary sodium output, creatinine clearance, or urinary prostaglandin excretion). Overall, the data suggested that NSAIDs do not appear to increase blood pressure primarily by increasing salt and water retention, because weight and urinary sodium were not altered by NSAIDs and inhibition of blood pressure control was not more marked in patients taking diuretics compared with other antihypertensive drugs. In addition NSAIDs did not significantly alter plasma renin activity or 24-hour urinary excretion of prostaglandin E_2 and 6-ketoprostaglandin $F_{1\alpha}$. Other factors may therefore contribute to the increase in blood pressure caused by NSAIDs. In particular, a potential effect of NSAIDs on peripheral vascular resistance should be considered (1845,1846). There is good evidence to suggest an important role for prostaglandins in the modulation of two major determinants of blood pressure, vasoconstriction of arteriolar smooth muscle and control of extracellular volume.

NSAIDs inhibited the effects of all antihypertensive drug categories. However, in patients taking beta-blockers and vasodilators, NSAIDs produced a greater increase in supine mean blood pressure than in patients taking diuretics, but only the pooled inhibitory effect of NSAIDs on the effects of beta-blockers achieved statistical significance. When the data were analysed by type of NSAID the meta-analysis showed that all NSAIDs increased supine blood pressure, and that piroxicam, indometacin, and ibuprofen produced the most marked increases. However, only piroxicam had a statistically significant effect with respect to placebo. Aspirin, sulindac, and flurbiprofen caused the smallest increases in blood pressure.

In conclusion this meta-analysis has provided more clear evidence that, as a group, NSAIDs significantly increase arterial pressure and can antagonize the blood pressure-lowering effect of some antihypertensive drugs, by mechanisms that are still unclear. Although the hypertensive effect of NSAIDs was more marked in hypertensive subjects taking antihypertensive drugs than in normotensive subjects not taking antihypertensive drugs, the difference was not statistically significant and its clinical relevance is unclear. It is worth noting that the effects of NSAIDs on blood pressure were similar in patients taking antihypertensive drugs for months or only a few days.

This study also had two main limitations: first, most of the trials were small, which precluded definitive conclusions about the effects of individual NSAIDs or individual antihypertensive drug classes; secondly, in most studies, therapy was short term and the patients were relatively young, making generalization of the results difficult, as NSAIDs are most often prescribed long term and for elderly people.

Prospective studies

A case-control study of the effects of NSAID therapy on arterial pressure has been performed in subjects aged 65 years and over, drawn from a large database (the State of New England Medicaid Program) to determine whether NSAIDs affect blood pressure (1847). The investigators calculated the odds ratio (OR) for the initiation of antihypertensive therapy in patients taking NSAIDs relative to non-users after adjusting for possible confounding factors. The 9411 patients had started taking antihypertensive drugs between 1981 and 1990, and a similar number of controls were randomly selected. The date of the first prescription for an antihypertensive drug was defined as the index date. Of those who took antihypertensive drugs, 41% had taken an NSAID during the year before the index date, compared with 26% of the control subjects. This risk increased with the recency of NSAID therapy, and was greatest among recent users (those with a supply of NSAIDs ending more than 60 days before the index date) (OR = 2.10; 95% CI = 1.95, 2.26). For former users (those with a supply of NSAIDs ending more than 60 days before the index date) the adjusted OR compared with non-users was 1.66 (CI = 1.54, 1.80). There was a dose-response relation, with adjusted ORs of 1.55 (CI = 1.38, 1.74), 1.64 (CI = 1.44, 1.87), and 1.82 (CI = 1.62, 2.05) for low, medium, and high daily doses of NSAIDs respectively. The unadjusted ORs for ibuprofen, piroxicam, meclofenamate, and indometacin, were separately calculable, and for each of these drugs the OR increased with increasing dose. The relation between cumulative duration of NSAID use and the initiation of antihypertensive therapy was also examined in recent users. The risk was greatest in those who had used an NSAID for 30–90 days and was less for those who had used an NSAID for less than 30 days or for more than 90 days.

The results of this study suggest that the effects of NSAIDs on blood pressure in older patients taking NSAIDs may be clinically important. Given that 15% of the control group were recent users of NSAIDs, and assuming that the adjusted OR of 1.66 represents a causal association of these drugs with the initiation of antihypertensive therapy, the proportion of cases attributable to the use of these drugs in this sample of elderly population was nearly one in ten.

Despite the high prevalence of the use of minor analgesics (aspirin and paracetamol) there is little information available on the association between the use of these analgesics and the risk of hypertension. A prospective cohort study in 80 020 women aged 31–50 years has provided some useful information (1848). The women had participated in the Nurses' Health Study II and had no previous history of hypertension. The frequency of use of paracetamol, aspirin, and NSAIDs was collected by mailed questionnaires and cases of physician-diagnosed hypertension were identified by self-report. During 164 000 person-years of follow-up, 1650 incident cases of hypertension were identified. Overall 73% of the cohort had used paracetamol at least 1–4 days/month, 51% had used aspirin, and 77% had used an NSAID. Compared with non-users of paracetamol the age-adjusted relative risk (RR) of hypertension was significantly increased even in women who had used paracetamol for only 1–4 days/month (RR = 1.22; CI = 1.07, 1.39). There seemed to be a dose-response relation, as the RR of hypertension compared with non-users was 2.00 (CI = 1.52, 2.62) in women who had taken paracetamol for 29 days/month or more. For women using aspirin or NSAIDs at a frequency of 1–4 days/month the RRs were 1.18 (CI = 1.02, 1.35) and 1.17 (CI = 1.02, 1.36) respectively. However, after adjusting for age and other potential risk factors, only paracetamol and NSAIDs, but not aspirin, remained significantly associated with a risk of hypertension. In summary, the data from this study support the view that paracetamol and NSAIDs are strongly associated with an increased risk of hypertension in women, the risk increasing with increasing frequency of use. Aspirin did not seem to be associated with an increased risk. This conclusion contrasts with the results of some short-term studies that have shown no effect of paracetamol on blood pressure (1849,1850).

However, the results from this study must be interpreted with caution, as there were some limitations: the assessments of analgesic use and hypertension were made using a self-reported questionnaire; relative risk can be influenced by many potentially confounding variables; the results are relevant only for young women and cannot be extrapolated to the general population.

The impact of NSAIDs on blood pressure in elderly people has been evaluated in three epidemiological studies, with similar findings (SEDA-19, 92). The use of NSAIDs was significantly associated with hypertension or the use of antihypertensive drugs. Reliable data are available for hypertension in the elderly. Recent users of NSAIDs have a 1.7-fold increase in the risk of initiating antihypertensive therapy compared with non-users, and the use of NSAIDs significantly predicts the presence of hypertension (OR = 1.4; 95% CI = 1.1, 1.7) (1841).

Conclusions

The overall results of these studies have provided convincing evidence that NSAIDs and paracetamol can raise arterial blood pressure in a dose-related fashion, interfere with the actions of antihypertensive drugs, and prompt the need for new antihypertensive therapy.

Even if the increase in mean blood pressure is probably modest (less than 5.0 mmHg) the clinical relevance of such an increase can be large, especially in elderly people. In fact, an overview of randomized clinical trials of antihypertensive treatment has shown that a 5–6 mmHg increase in diastolic blood pressure over a few years can be associated with a 67% increase in the incidence of strokes and a 15% increase in coronary heart disease (1851). These effects are apparent in both normotensive and hypertensive patients.

Whether these results apply with certainty to patients taking NSAIDs is not known, because these studies included patients not taking NSAIDs, but it is wise to consider this probability. The type and dose of NSAID may be important, but more studies are needed to document this. Hypertensive and elderly patients seem to be particularly at risk. In patients taking long-term NSAIDs, or even paracetamol, periodic monitoring of blood pressure appears to be warranted.

Congestive heart failure

Much less is known about the risk of congestive heart failure with NSAIDs. The rate of hospitalization for congestive heart failure in more than 10 000 patients over 55 years of age during exposure to both diuretics and NSAIDs was compared with the rate in those exposed to diuretics alone (1852). At mean follow up of 4.7 years, there was an increased risk of hospitalization when diuretics and NSAIDs were used concomitantly (RR = 1.8; 95% CI = 1.4, 2.4).

In 600 elderly patients with documented congestive cardiac failure there was a possible or probable link between NSAIDs and heart failure in 27 cases (1853). In some, the mechanism was apparently a reduction in the effect of furosemide. In others the NSAID may have caused an imbalance in circulatory homeostasis. Pre-existing renal impairment was not observed in any of the 27 cases. This study suggests that in elderly people congestive heart failure may be a complication of NSAIDs.

In a matched case-control study the relation between the recent use of NSAIDs and hospitalization with congestive heart failure in elderly patients has been analysed (1854). Cases (n = 365) were patients admitted to hospital with a primary diagnosis of congestive heart failure; controls (n = 658) were patients without congestive heart failure. Structured interviews were used to obtain information on several possible risk factors, such as a history of heart disease and the type and dosage of NSAIDs used. The use of NSAIDs (other than low-dose aspirin) in the previous week was associated with a doubling of the chance of a hospital admission with congestive heart failure (adjusted OR = 2.5, CI = 1.2, 3.3). The risk was even higher in patients with a history of heart disease, in whom the use of NSAIDs was associated with an increased risk of a first admission with congestive heart failure (OR = 11; CI = 0.7, 45), but not in those without a history of heart disease (OR = 1.6; CI = 0.7, 3.7). In contrast to the results of a previous study (1852), the risk of admission to hospital with congestive heart failure was positively related to the dose of NSAID consumed in the previous week and was higher with long half-life NSAIDs than with short half-life NSAIDs. The authors estimated that, assuming that these relations were causal, NSAIDs might be responsible for about 19% of hospital admissions with congestive heart failure. If confirmed, this burden of illness resulting from NSAID-related congestive heart failure may rival that resulting from damage to the gastrointestinal tract and would represent an under-recognized public health problem. In any case, this study reinforces the timely suggestion that NSAIDs should be used with great caution in patients with a history of cardiovascular disease. This recommendation must also include the use of the selective COX-2 inhibitors, until more information is available.

Noradrenaline (norepinephrine)

Like adrenaline, noradrenaline is also sometimes added to local anesthetics (for example in a 1:250 000 concentration) to prolong their effect; it should not be injected into extremities (finger, penis) for this purpose, since dangerous ischemia can result (1855).

When infusing noradrenaline the infusion should always be ended very gradually, since otherwise a catastrophic fall in blood pressure can occur.

Dynamic left ventricular outflow tract obstruction associated with catecholamine therapy in a patient *without* evidence of hypertrophic cardiomyopathy has been reported, possibly for the first time (1856).

- A 31-year-old man developed acute pancreatitis with hypovolemia and shock. Apart from large volumes of fluid and colloid, he was given dopamine 20 micrograms/kg/minute and noradrenaline 0.1 micrograms/kg/minute. After initial improvement he became severely hypotensive, with a blood pressure of 65/40 mmHg, a pulse of 134/minute, and a new loud systolic murmur at the left sternal edge. Echocardiography showed severe left ventricular outflow obstruction with a maximal pressure gradient of 106 mmHg. There was rapid hemodynamic improvement after the withdrawal of catecholamines and he survived, with type 1 diabetes and several pancreatic pseudocysts.

The authors pointed out that this problem can occur in the presence of catecholamine therapy and hypovolemia even in patients with an entirely normal heart, and that early diagnosis and correct management is vital.

Nucleoside analogue reverse transcriptase inhibitors (NRTIs)

Cardiomyopathy is a rare adverse effect that has been observed in patients treated with didanosine, zidovudine, and zalcitabine (1857). In a retrospective, case-control study, cardiomyopathy was 8.4 (95% CI = 1.7, 42) times more likely to develop in children who had previously used zidovudine than in children who had never been exposed to it (1858).

Octreotide

Bradycardia, hypotension, and heart block have been attributed to octreotide (1859).

- A 67-year-old man who was receiving subcutaneous octreotide 100 micrograms bd underwent abdominal laparotomy for metastatic carcinoid. He was given an intravenous bolus of octreotide 100 micrograms 10 minutes after induction and immediately after surgical incision. His heart rate fell to 35/minute and his blood pressure to 85/40 mmHg. He was given ephedrine 20 mg intravenously and recovered. He was given a further bolus of octreotide 100 micrograms 30 minutes later. His heart rate immediately fell to 45/minute and he developed complete heart block. He was given ephedrine 20 mg and glycopyrrolate 0.2 mg intravenously and recovered.

There have been previous reports of bradycardia with octreotide. The authors suggested that intravenous octreotide should be infused slowly when possible.

The records of 21 children who received infusions of octreotide (1-2 micrograms/kg/hour) for 35 gastrointestinal bleeds have been reviewed (1860). There was one case of asymptomatic bradycardia and one case of a sudden unexplained cardiac event in a patient who appeared to be hemodynamically stable just before the event 5 hours after starting octreotide.

Octreotide increases systemic vascular resistance, and bradycardia may be a baroreceptor-induced response. Octreotide also has direct effects on the heart, the main effects being reduced heart rate, reduced myocardial contractility, and slowing of the propagation velocity along the cardiac conduction system.

Olanzapine

Olanzapine can cause QT interval prolongation (SEDA-25, 64).

- A 66-year-old woman taking chlorpromazine and quetiapine had QT_c interval prolongation, which improved when these drugs were withdrawn (1861). However, prolongation later recurred while she was taking high-dosage olanzapine (60 mg/day), which had not occurred with a smaller dosage (40 mg/day).
- QT_c prolongation occurred in a 28-year-old woman while she was taking olanzapine 40 mg/day; after olanzapine withdrawal, the QT_c interval returned to normal (1862).
- A 61-year-old woman with Wolff–Parkinson–White syndrome, who had previously had QT_c interval prolongation with both sulpiride 1200 mg/day and clozapine 50 mg/day, had a QT_c interval of 390 ms, which increased to 466 ms when she took olanzapine 5 mg/day and returned to 395 ms in 2 days when olanzapine was withdrawn (1863). When she was given olanzapine again in the same dose, the QT_c interval increased to 473 ms in 2 weeks and returned to baseline when olanzapine was withdrawn. She was also taking daily valproate 1500 mg/day, lithium 300 mg/day, lorazepam 2 mg/day, and propranolol 40 mg/day.

However, in an analysis of electrocardiograms obtained as part of the safety assessment of olanzapine in four controlled randomized trials (n = 2700) the incidence of maximum QT_c prolongation beyond 450 ms during treatment was approximately equal to the incidence of prolongation of the QT_c beyond 450 ms at baseline (1864). The authors therefore suggested that olanzapine does not contribute to QT_c prolongation. This has been supported by the report of a patient in whom QT_c prolongation while he was taking clozapine reversed when he switched to olanzapine (1865).

An episode of asystole (at which time olanzapine was withdrawn), followed 6 days later by a brain stem stroke, occurred during a double-blind parallel study for 2 weeks in 39 demented patients with agitation (mean age 83 years) (1866); olanzapine (n = 20, mean daily dose 6.65 mg, modal dose 10 mg) or risperidone (n = 19, mean daily dose 1.47 mg, modal dose 2 mg) were given once a day at bedtime.

Two cases of light-headedness or "fainting" in patients taking olanzapine have been reported (1867). Electrocardiograms showed first-degree heart block and AV conduction delay, which normalized after dosage reduction.

- A 44-year-old man took clozapine 10 mg/day and the dose was increased to 50 mg/day, after which he complained of episodes of light-headedness. An electrocardiogram showed prolongation of the PR interval to 227 ms. The dosage was reduced to 30 mg and the PR interval returned to 187 ms within 2 days.
- A 36-year-old man took clozapine 20 mg/day and within 2 weeks began to complain of intermittent, unpredictable "fainting" attacks. An electrocardiogram showed a prolonged PR interval at 230 ms. The dose of olanzapine was reduced to 17.5 mg/day, and a repeat electrocardiogram 1 week later was normal.

Subclinical cases of increased blood pressure related to olanzapine have previously been reported.

- A 29-year-old man developed transient rises in systolic and diastolic blood pressures (160/90; previous blood pressure 130/84 mmHg) with raised transaminases and dependent pitting edema of both feet (1868).

Peripheral edema might be more frequent than expected in patients taking olanzapine. In a recent open, non-randomized study in 49 subjects taking olanzapine, 28 reported edema, which was severe in five (1869). There were no significant differences regarding sex, dose of olanzapine or duration of treatment, concomitant diagnoses, or other psychotropic drugs, but there was a tendency toward greater frequency of thyroid abnormalities and older age in those with edema, in whom there was a positive correlation between age and severity.

Acute massive pulmonary thromboembolism has been attributed to olanzapine in two patients (1870):

- a 64-year-old woman taking bromperidol chronically, who also took risperidone (last dose 6 mg) for 40 days before the event;
- a 48-year-old woman who took risperidone for 6 days (last dose 2 mg).

The first patient died and the second survived. In the same series from a Japanese Emergency Center, seven patients (two men and five women, aged 23–70) who took chlorpromazine, levomepromazine, or propericiazine also had thromboembolic disorders; none of them had any known risk factor for thrombotic disease. Suggested mechanisms are: reduced movement at night as a consequence of the sedative effect of neuroleptic drugs (symptoms developed in all cases in the early morning); an effect of anticardiolipin antibodies, which can occur in some patients taking phenothiazines; or increased $5HT_{2A}$-induced platelet aggregation.

The debate about the need to restrict drug therapy in relation to the risk of cardiovascular events is open, and some authors have already expressed agreement (1871) or disagreement about it (1872), as well as pointing to the need for individual patient meta-analyses and significant changes to clinical trial methods in order to better assess the effectiveness, in contrast to the efficacy, of risperidone and other drugs in dementia (1873).

Opioid analgesics

Orthostatic hypotension can occur and is common after intravenous administration of opioid analgesics. Histamine release sometimes contributes to this.

Levacetylmethadol

The synthetic opioid levacetylmethadol, a metabolite of methadone, can cause torsade de pointes, and its use requires electrocardiographic screening before treatment and during titration (1874). It has been removed from the market throughout the European Union and in the USA production has ceased.

Methadone

Opioids block the cardiac human ether-a-go-go-related gene (HERG) potassium current in susceptible patients without any apparent heart disease and can thus prolong the QT interval (1874). Two cases of QT_c interval prolongation and torsade de pointes have been reported in patients taking methadone (1875).

- A patient was found unconscious with plasma concentrations of bromazepam 277 μg/ml and methadone 3500 μg/ml, both of which were above the toxic threshold. The QT_c interval was 688 ms. After initial improvement, torsade de pointes occurred and the patient was treated with DC shock 200 J, isoprenaline, magnesium, and potassium. The QT_c interval improved to 440 ms after 3 days.
- The second patient was admitted to the hospital in a comatose state. Sinus bradycardia was present with a QT_c interval of 736 ms. The plasma concentration of methadone was 1740 μg/ml. Ventricular bigeminy was followed by torsade de pointes. The patient was treated with DC shock 200 J, lidocaine, and magnesium. By the fifth day after the episode, the QT_c interval had improved to 502 ms.

This potentially life-threatening dysrhythmia has been reported previously in association with methadone and is probably under-recognized in this population. The authors did not provide details about the sex of the patients and reasons why methadone concentrations were high.

Methadone-related torsade de pointes has been reported in a patient with chronic bone and vaso-occlusive pain due to sickle cell disease (1876).

- A 40-year-old man with sickle cell disease, hypertension, congestive heart failure, and a past history of cocaine and marihuana abuse, was given a large dose of oral methadone 560 mg/day, following hydromorphone 170 mg intravenously and by PCA for progressive back and leg pain. On day 2, he developed asymptomatic bradycardia and QT_c prolongation (454–522 msec). On day 3, he developed profuse sweating and non-sustained polymorphous ventricular tachycardia consistent with torsade de pointes. He had hypokalemia and hypocalcaemia. Echocardiography showed normal bilateral ventricular function, mild pulmonary hypertension, and trivial four-valve regurgitation. Methadone was replaced by modified-release morphine and a continuous epidural infusion of hydromorphone + bupivacaine. Daily electrocardiography showed a heart rate of 50–69/minute, a QT_c interval of 375–463 msec, and no further dysrhythmias.

This case highlights the importance of very careful monitoring especially when prescribing such large doses of methadone. The effects of methadone on cardiac function are potentially fatal.

Another report has highlighted the potential risks of combining prodysrhythmic drugs on cardiovascular function (1877).

- A 39-year-old man had recurrent episodes of sinus tachycardia at 115/minute, with no other abnormalities. He was taking methadone 120 mg/day for opioid dependency and doxepin 100 mg/day for anxiety, and was given metoprolol 50 mg/day. During the next few weeks he had episodes of recurrent syncope with sinus bradycardia (47/minute) and prolongation of the QT interval (542 ms). The QT interval and heart rate normalized after withdrawal of all treatment.

In this case it is likely that the myocardial repolarization potential of methadone and doxepin may have been influenced or triggered by bradycardia induced by metoprolol. This shows the importance of cardiac monitoring in patients receiving combination therapy with potential adverse cardiac effects. Patients with co-morbidities are at high risk.

The association between methadone treatment and QT_c interval prolongation, QRS widening, and bradycardia has been explored prospectively in 160 patients with at least a 1-year history of opioid misuse (1878). The QT_c interval increased significantly from baseline at 6 months (n = 149) and 12 months (n = 108). The QRS duration and heart rate did not change. There were no cases of torsade de pointes, cardiac dysrhythmias, syncope, or sudden death. There was a positive correlation between methadone concentration and the QT_c interval.

There have been another 11 cases showing a direct link between QT interval prolongation and oral methadone maintenance treatment at doses of 14–360 micrograms/day (1879,1880). QT interval prolongation can lead to arrhythmias such as torsade de pointes, especially when high doses of methadone are given intravenously and associated with concomitant use of cocaine and/or medications that inhibit the hepatic clearance of methadone (e.g. antidepressants and antihistamines).

Following reports similar to those mentioned above, changes in the QT_c interval were studied in 132 heroin-dependent patients as they were starting treatment with methadone (1881). After baseline electrocardiography methadone 30-150 mg/day was given and a second electrocardiogram was obtained 2 months later. Across all doses of methadone, the QT_c interval increased significantly by a mean of 11 ms over the first 2 months of treatment. No episodes of torsade de pointes were reported. Male sex and methadone doses over 110 mg/day were associated with the greatest prolongation. The average follow-up QT_c interval was 428 ms. Clinical significance is generally attributed to an increase in QT_c interval of 40 ms or greater or a value above 500 ms. None of these patients had an increase that was above this threshold. While these results were statistically significant, the authors were not sure of their clinical significance.

There is a linear relation between QT interval prolongation and the dose of methadone (1882). It is prudent to avoid concomitant use of other drugs that prolong the QT interval. This is of significance in individuals infected with HIV, who might have not only viral cardiomyopathies or autonomic neuropathies but also be taking macrolides, quinolones, clindamycin, trimethoprim, fluconazole, pentamidine, and other drugs that are closely associated with torsade de pointes and other cardiac dysrhythmias (1883). Several case have been described in which methadone in a dose of more than 300 mg/day was associated with an increased risk of cardiac dysrhythmias, especially if prescribed with other drugs that inhibit CYP3A4 (1884).

Morphine

Adverse cardiac effects due to morphine are rare. They comprise inappropriate heart rate responses to hypotension, rather than conduction defects. They are not especially associated with inferior myocardial infarction, as was previously thought (SED-11, 142; 1885).

Opioid antagonists

Nalbuphine

The effects of nalbuphine 5 mg on the cardiovascular and subjective effects of cocaine have been studied in a randomized controlled trial in seven patients (1886). The combination of nalbuphine and cocaine was safe and did not have synergistic effects on heart rate and blood pressure or subjective effects. Nalbuphine was safe and well tolerated and its acute administration moderately attenuated the abuse-related effects of cocaine.

Orphenadrine

Non-sustained ventricular tachycardia has been attributed to orphenadrine (1887).

- A 57-year-old woman had been taking a formulation containing orphenadrine 15 mg and paracetamol 450 mg bd for musculoskeletal pain. She was also taking propafenone 600 mg/day for paroxysmal atrial fibrillation. After 5 days she developed severe palpitation. Holter monitoring showed frequent brief episodes not only of atrial fibrillation but also of non-sustained ventricular tachycardia. After the orphenadrine was withdrawn the palpitation ceased.

The authors pointed out the potential problems of anticholinergic drugs like orphenadrine in patients taking antidysrhythmic drugs.

Oxamniquine

Electrocardiographic and electroencephalographic changes have been reported as rare adverse effects of oxamniquine (SEDA-11, 598; 1888).

Oxybuprocaine

An episode of severe bradycardia, with no perceptible cardiac output, was reported in a previously healthy patient after one drop of 0.4% oxybuprocaine was applied to each eye (1889).

Oxygen-carrying blood substitutes

A modified hemoglobin solution, DCLHb, has been associated with an increase in mean arterial pressure. This has been proposed to be due to binding to nitric oxide, stimulation of the production of endothelin, and sensitization and/or potentiation of beta-1 and beta-2 adrenoceptor responses to catecholamines (1890).

It has been suggested that cell-free hemoglobin, particularly low molecular weight hemoglobin, such as tetrameric hemoglobin, has the ability to come more closely to the endothelial cell lining of blood vessels than erythrocytes and to extravasate into the subendothelial There, hemoglobin might scavenge nitrous oxide and so induce vasoconstriction and hypertension (1891,1892,1893). A randomized, controlled phase II study has suggested that hemoglobin raffimer (Hemolink) is safe and effective in patients undergoing coronary artery bypass surgery. The incidence of hypertension was higher in the treated group, but blood pressure management prevented serious hypertension-related events (1894).

Another hemoglobin substitution product, liposome-encapsulated tetrameric hemoglobin, can exacerbate the manifestations of septic shock (1890).

Oxytocin and analogues

Tachycardia and a fall in blood pressure are common and usually short-lived after oxytocin administration during labor. There has been one reported maternal death after a hypovolemic woman was given a bolus dose of oxytocin 10 units (1895).

In 34 women undergoing cesarean section at full term under spinal anesthesia, heart rate and cardiac output increased significantly within 2 minutes of the rapid administration of either 5 or 10 units of oxytocin, with an associated 10 mmHg fall in mean arterial pressure in those who received 10 units (1896). There were significant ST segment changes in 11 of 26 women undergoing cesarean section, with raised concentrations of troponin I in two; however, the relationship to oxytocin administration was not clear in this report (1897).

- A previously fit 19-year-old woman had severe ST segment depression and increased troponin concentrations after a bolus dose of oxytocin 5 units (1898).

Ventricular tachycardia has been reported in two patients with pre-existing prolongation of the QT interval, immediately after oxytocin was begun (1899).

Paclitaxel

Paclitaxel causes disturbances in cardiac rhythm, but the relevance of these effects has not been fully elucidated. Originally, all patients in trials of paclitaxel were under continuous cardiac monitoring, owing to the risk of hypersensitivity reactions, and cardiac disturbances were therefore more likely to be detected. Many trials limited eligibility to patients without a history of cardiac abnormalities and to those who were not taking medications likely to alter cardiac conduction. The incidence of cardiac dysrhythmias in the population under study not treated with paclitaxel is unknown, and it is therefore not always possible to attribute dysrhythmias to paclitaxel in these patients. The Cremophor EL vehicle does not appear to be implicated in the incidence of dysrhythmias, although hypotension associated with hypersensitivity reactions may occur (1900).

The most common effect of paclitaxel is asymptomatic bradycardia, which occurred in 29% of patients in one phase 2 trial (1901) and in 9% of patients in a further assessment of 402 patients in phase 2 trials (1902). One phase 1 trial showed no significant cardiac dysrhythmias (1903), while another reported cardiac toxicity in 14% of patients, 74% of these being due to asymptomatic bradycardia (1904). Bradycardia is not an indication for discontinuation of treatment, unless it is associated with atrioventricular conduction disturbances or clinically significant effects (for example symptomatic hypotension). More significant bradydysrhythmias and atrioventricular conduction disturbances have been reported during clinical trials, including Mobitz I (Wenckebach syndrome) and Mobitz II atrioventricular block (1901,1905).

One patient died in heart failure 7 days after receiving paclitaxel by infusion; this patient had no prior history of cardiac problems, apart from mild hypertension (1906).

The authors of a review of the cardiac toxicity associated with paclitaxel in a number of studies concluded that the overall incidence of serious cardiac events is low (0.1%) (1907). Heart block and conduction abnormalities occurred infrequently and were often asymptomatic. Sinus bradycardia was the most frequent, occurring in 30% of patients. The causal relation of paclitaxel to atrial and ventricular dysrhythmias and cardiac ischemia was not entirely clear. There did not appear to be any evidence of cumulative toxicity or augmentation of acute cardiac effects of the anthracyclines.

In an attempt to clarify further the cardiotoxicity of paclitaxel, its effect on cardiovascular autonomic regulation has been investigated in 14 women (1908). The authors concluded that autonomic modulation of heart rate is impaired by paclitaxel, but they were unable to say whether it would return to normal on withdrawal. They also investigated the effect of docetaxel on neural cardiovascular regulation in women with breast cancer, previously treated with anthracyclines (1909). They concluded that docetaxel did not impair vagal cardiac control. The changes that they observed in blood pressure suggest that docetaxel changes sympathetic vascular control, although these changes seemed to be related to altered cardiovascular homeostasis rather than peripheral sympathetic neuropathy.

Continuous cardiac monitoring is recommended for patients with serious conduction abnormalities; however, routine cardiac monitoring is considered unnecessary in patients without a history of cardiac conduction abnormalities (1910). Further studies are needed to determine the risk in patients treated with paclitaxel with predisposing cardiac risk factors.

Panax ginseng (Amaranthaceae)

A 64-year-old previously normotensive man presented with amaurosis fugax and hypertension (blood pressure 220/130 mmHg) after taking Ginseng Forte-Dietisa 500 mg/day for 13 days (1911). All other tests were normal. He was advised to stop taking ginseng, and 1 week later his blood pressure was normal (140/90 mmHg).

Pancuronium bromide

Cardiovascular adverse effects are minimal with pancuronium. Ganglion blockade does not occur. Slight dose-dependent rises in heart rate, blood pressure, and cardiac output are common (1912), but are often masked by the actions of other co-administered agents, such as fentanyl or halothane, which cause bradycardia or hypotension. These adverse effects of pancuronium are thus often beneficial and can be deliberately harnessed. Several mechanisms contribute: vagal blockade via selective blockade of cardiac muscarinic receptors (1913), release of noradrenaline from adrenergic nerve endings (1914), increased blood catecholamine concentrations (1915), inhibition of neuronal catecholamine reuptake (1916–1918), and direct effects on myocardial contractility (1919). These have been reviewed (1920–1922).

Occasionally nodal rhythm, atrioventricular dissociation, and tachydysrhythmias (such as ventricular extra beats or even bigeminy) develop, but these usually occur in association with halothane.

Supraventricular tachycardia has been reported after 8 mg pancuronium in a patient taking aminophylline (800 mg/day) (1923).

Nodal rhythm can occur after injection of pancuronium. This dysrhythmia and bradycardia appear to be more common when neostigmine (plus atropine) is given for reversal of pancuronium-induced neuromuscular blockade than for reversal of D-tubocurarine or alcuronium (1924); cholinesterase inhibition by pancuronium may contribute to the bradycardia in these circumstances.

Pantothenic acid derivatives

Eosinophilic pleural effusion with eosinophilic pericardial tamponade has been attributed to concomitant use of pantothenic acid and biotin (1925).

- A 76-year-old woman developed chest pain and difficulty in breathing. She had no history of allergy and had been taking biotin 10 mg/day and pantothenic acid 300 mg/day for 2 months for alopecia. Chest X-rays showed pleural effusions and cardiac enlargement. Blood tests showed an inflammatory syndrome, with an erythrocyte sedimentation rate of 51 mm/hour and an eosinophil count of 1.2–1.5 × 10^9/l. Pericardiotomy showed an eosinophilic infiltrate. There was no evidence of vasculitis. Serological studies were negative for antinuclear antibodies, rheumatoid factor, viruses, bacteria, and Lyme disease. Stool examination and parasitological serologies were negative. A malignant tumor was excluded by mammography, thoracoscopy, and a CT scan. Myelography, a biopsy specimen of the iliac crest bone, and the concentrations of IgE, lysozyme, and vitamin B_{12} were also normal. A week after withdrawal of pantothenic acid and biotin she improved dramatically and her eosinophilia resolved.

Papaverine

A study in neonates has confirmed the hypothesis that continuous infusion of papaverine-containing solutions in peripheral arterial catheters reduces the catheter failure rate and increases the functional duration of the catheter (1926). There was no difference in the incidence of intraventricular hemorrhage and no evidence of hepatotoxicity.

Paracetamol

Despite the high prevalence of the use of minor analgesics (aspirin and paracetamol) there is little information available on the association between the use of these analgesics and the risk of hypertension. A prospective cohort study in 80 020 women aged 31–50 years has provided some useful information (1927). The women had participated in the Nurses' Health Study II and had no previous history of hypertension. The frequency of use of paracetamol, aspirin, and NSAIDs was collected by mailed questionnaires and cases of physician-diagnosed hypertension were identified by self-report. During 164 000 person-years of follow-up, 1650 incident cases of hypertension were identified. Overall, 73% of the cohort had used paracetamol at least 1–4 days/month, 51% had used aspirin, and 77% had used an NSAID. Compared with non-users of paracetamol the age-adjusted relative risk (RR) of hypertension was significantly increased even in women who had used paracetamol for only 1–4 days/month (RR = 1.22; CI = 1.07, 1.39). There seemed to be a dose–response relation, as the RR of hypertension compared with non-users was 2.00 (CI = 1.52, 2.62) in women who had taken paracetamol for 29 days/month or more. For women using aspirin or NSAIDs at a frequency of 1–4 days/month the RRs were 1.18 (CI = 1.02, 1.35) and 1.17 (CI = 1.02, 1.36) respectively. However, after adjusting for age and other potential risk factors, only paracetamol and NSAIDs, but not aspirin, remained significantly associated with a risk of hypertension. In summary, the data from this study support the view that paracetamol and NSAIDs are strongly associated with an increased risk of hypertension in women, the risk increasing with increasing frequency of use. Aspirin did not seem to be associated with an increased risk. This conclusion contrasts with the results of some short-term studies that have shown no effect of paracetamol on blood pressure (1928,1929).

This study suggests that paracetamol can raise arterial blood pressure in a dose-related fashion, interfere with the actions of antihypertensive drugs, and prompt the need for new antihypertensive therapy. However, these results must be interpreted with caution, as there were some limitations: the assessments of analgesic use and hypertension were made using a self-reported questionnaire; relative risk can be influenced by many potentially confounding variables; the results are relevant only for young women and cannot be extrapolated to the general population.

Paraldehyde

Paraldehyde caused microembolization in a neonate immediately after the injection of 0.3 ml/kg (1930). The skin below the waist became red, large purple vesicles formed, and there was loss of skin and sloughing of two toes.

Parathyroid hormone and analogues

Parathyroid hormone lowers the blood pressure by a direct effect on vascular smooth muscle, and there are isolated instances of hypotension or tachycardia (1931). Pre-injection blood pressure was normal in 1093 women randomized to parathyroid hormone (PTH_{1-34}), and dizziness was reported infrequently in 541 women taking 20 micrograms/day but not in 552 taking 40 micrograms/day (1932).

Parenteral nutrition

Infusion phlebitis presents a problem in parenteral nutrition. Various alternative techniques of administration have been compared in order to identify means of countering this problem (1933). Mechanical trauma appears to be a causative factor; it can be reduced by limiting the time of exposure of the vein wall to nutrient infusion and by minimizing the amount of prosthetic material within the vein (1934). This is likely to be even more important in small veins. In one study the addition of heparin (500 U/l) and hydrocortisone (5 micrograms/ml) significantly reduced the risk of thrombophlebitis from 0.43 to 0.11 episodes per patient-day, and a reduction in osmolality of the solution resulted in a further ten-fold fall in the incidence of thrombophlebitis (1935). Other work has concluded that the incidence of infusion phlebitis is minimized during parenteral nutrition by cyclic infusion of nutrient solutions and by rotation of venous access sites (1936).

Paroxetine

Electrocardiographic changes, with a prolonged QT_c interval and bradycardia, have been reported with paroxetine (1937).

Pemoline

There has been a report of pemoline-induced acute movement disorders after presumed overdose (1938).

- Two 3-year-old identical male twins were found playing with an empty bottle of pemoline that had originally contained 59 tablets. They had a history of attention deficit disorder previously unsuccessfully treated with methylphenidate, but no history or family history of movement disorders. Choreoathetoid movements began 45 minutes to 1 hour after ingestion. The children received gastrointestinal decontamination and high doses of intravenous benzodiazepines but continued to have choreoathetosis for about 24 hours and were discharged at 48 hours.

Five children who took excessive amounts of pemoline have been described (1939). They had a relatively benign course and their symptoms appeared to be primarily accentuated pharmacological effects on the central nervous and cardiovascular systems. Sinus tachycardia, hypertension, hyperactivity, choreoathetoid movements, and hallucinations were most commonly observed, consistent with previous reports.

Possible rhabdomyolysis occurred after overdose, accompanied by raised serum creatine kinase activity in three of four patients in whom it was measured; this appears to be common in acute pemoline poisoning. After ingestion, symptoms occurred within 6 hours and lasted up to 48 hours in all cases.

Gastric lavage and activated charcoal are considered to be effective decontamination measures, whereas ipecac-induced emesis should be avoided after massive ingestion, because of the risk of seizures. Aggressive use of benzodiazepine is a reasonable first choice to treat associated involuntary movements, tremor, hyperactivity, and agitation. Chlorpromazine or haloperidol can also be used, especially for serious, life-threatening symptoms, including hypertensive crises and severe hyperthermia, and labetalol or sodium nitroprusside are reasonable choices for rapid stabilization of blood pressure.

Penicillamine

Penicillamine has no direct effect on the cardiovascular system. However, penicillamine-associated polymyositis can involve cardiac muscle and cause dysrhythmias, Adams–Stokes attacks, and death. Necrotizing vasculitis can occur as an immunological reaction to penicillamine (1940). The effect of penicillamine on collagen and elastin fibers, which causes characteristic skin lesions, also includes the vascular wall, but effects of vascular insufficiency have not been reported.

Penicillins

In nearly half of the cases, the course of anaphylactic shock, especially that induced by penicillin and other small molecular substances, is that of a cardiovascular reaction without any other effects suggestive of an allergic mechanism (1941–1943). There is an extensive list of articles on anaphylactic shock to penicillins (1944,1941,1942–1947). General anesthesia does not inhibit the development of anaphylactic shock in penicillin allergy (1948).

Embolic-toxic reactions to penicillin depot formulations were first described in patients with syphilis (1949). The symptoms include fear of death, confusion, acoustic and visual hallucinations, and possibly palpitation, tachycardia, and cyanosis (SEDA-8, 559; 1943,1949–1953). Generalized seizures or twitching of the limbs have been observed in children and adults (1952,1954–1958). As a rule, the symptoms abate and disappear within several minutes to an hour. They rarely persist for up to 24 hours. If a cardiovascular reaction with a fall in blood pressure occurs simultaneously with typical symptoms, a combination with anaphylactic shock must be considered (1959,1960).

Such reactions have been called "pseudo-anaphylactic reactions" or "acute non-allergic reactions" (1942,1950–1952,1961–1965), "panic attack syndrome," and "acute psychotic reactions" (1954,1966). In several countries, the term "Hoigné syndrome" is used.

The frequency of such reactions is about 1–3 reactions per 1000 intramuscular injections of penicillin G procaine, the usual dose being about 0.6–1.2 million units (1946,1955,1967–1969). Eight of 920 patients with venereal diseases had a definite toxic-like reaction with a dose of 4.8 million units of penicillin G procaine, corresponding to about one in 120 patients (1970). In a series of 7700 intramuscular injections with only 400 000 units of penicillin G procaine, there was not one episode (1963).

The mechanism is probably embolic (1949,1961,1964,1967,1971), as has been shown in one case at autopsy, in which emboli of benzathine penicillin crystals were found in the lungs (1961).

Some reports suggest that the procaine component may be especially important (1955,1964). Plasma procaine esterase activity was low in patients with systemic toxic reactions (1955). The same symptoms occurred in three patients after erroneous administration of penicillin G procaine by intravenous infusion (1964), but also in two after procaine-free antihistamine penicillin was injected intramuscularly (1967, 1971). However, observations of similar symptoms with a procaine-free antihistamine penicillin argue against a central role of the procaine component (SED-8, 560; 1943,1959,1962,1963,1967).

Pentagastrin

Pentagastrin causes small transient increases in blood pressure and pulse (1972), autonomic effects that are prevented by inhibiting cholecystokinin receptors (1973). Atrial fibrillation has been observed (1974).

Pentamidine

Severe hypotension can occur after a single intramuscular injection of pentamidine or with rapid intravenous administration, but has been seen with slow infusion as well. Infusing the drug over 60 minutes or more may reduce this risk. Facial flushing, breathlessness, dizziness, and nausea and vomiting can occur at the same time.

Cardiac dysrhythmias, including ventricular tachycardia, have been reported during treatment (SEDA-13, 824; SEDA-16, 331; 1975,1976). Prolongation of the QT interval, which usually precedes the development of ventricular dysrhythmias with pentamidine, occurs in one-third of patients, usually within 2 weeks of starting therapy. Torsade de pointes has been described. Any dysrhythmia can recur many days after the pentamidine has been discontinued, which is not surprising, in view of the long half-life and tissue accumulation. Electrolyte abnormalities, including low serum magnesium concentrations, have been noticed at times of dysrhythmias (SEDA-16, 315; SEDA-17, 331; 1975–1977).

- Torsade de pointes has been reported in a 48-year-old HIV-positive woman treated with intravenous pentamidine (1978).

Local thrombophlebitis can occur after injection of pentamidine, but problems are more often seen at the injection site after intramuscular injection.

Pentamorphone

There was no effect on blood pressure or heart rate with pentamorphone doses of 0.015–0.48 micrograms/kg (1979).

Pergolide

Symptoms suggestive of dose-related angina pectoris were observed in a number of patients, either early or late in treatment, but they were easily controlled by dose reduction without sequelae (1980).

Pergolide can cause severe hypotension in patients already receiving antihypertensive agents (1981).

Pergolide

The profibrotic effects of pergolide on heart valves have been evaluated echocardiographically in 78 patients with Parkinson's disease who were taking pergolide (mean age 71 years; 45 men) and 18 who had never taken an ergot-derived dopamine receptor agonist (mean age 73 years; nine men) (1982). There was detectable restrictive valve disease in 33% of those taking pergolide, and it was considered significant in 19%. The patients were divided into a high-dose group, taking pergolide 5 mg/day or more, and a low-dose group, but the incidence of valvular lesions was not clearly dose-dependent. The mitral valve was most commonly affected (n = 20), followed by the aortic valve (n = 7) and tricuspid (n = 6) valve. In six patients the pergolide was withdrawn because of the valve lesions and in two of these there was regression of the mitral valve lesions. There were no lesions detected in patients who had never been exposed to ergot drugs. The authors suggested that pergolide should be replaced by a non-ergot drug if valve lesions are detected. This was subsequently supported by correspondents from Glasgow, who had switched 88 of their 99 patients from ergot-based to non-ergot-based dopamine receptor agonists without clinical problems and having established dosage equivalence (1983). As a result of this accumulation of evidence pergolide is now not recommended for first-line therapy in Parkinson's disease and its use should now be accompanied by intensive monitoring, which may be a further deterrent to its administration.

Pethidine

- A 70-year-old patient with a metastatic carcinoid tumor of the liver presented with a hypertensive crisis after being given pethidine 10 mg/hour by continuous intravenous infusion (1984). The patient remained hypertensive with a systolic blood pressure of 210 mmHg, even after chemoembolization of the tumor. The blood pressure fell when pethidine was withdrawn and nitroprusside was given. The serum 5-HT concentration was 15 μmol/l (reference range 0.17–0.26) and the urine 5-hydroxyindoleacetic acid concentration was 1311 mg/g of creatinine (reference range less than 10).

The authors postulated that the hypertensive crisis had occurred from the release of 5-HT from the tumor and blockade by pethidine of 5-HT re-uptake.

When used for sedation in children undergoing esophagogastroduodenoscopy, hypoxia with dysrhythmias was more likely to occur with a combination of pethidine and diazepam than with pethidine and midazolam (SEDA-18, 81).

Phenmetrazine and phendimetrazine

A dilated cardiomyopathy has been associated with chronic consumption of phendimetrazine (1985).

Phenols

Phenol is cardiotoxic, and various cardiac dysrhythmias have been noted after application to the skin, or less commonly when it has been used for neurolysis. Ventricular extra beats occurred during topical application of phenol and croton oil in hexachlorophene soap and water for chemical peeling of a giant hairy nevus (1986). Three of sixteen children treated with motor point blocks for cerebral palsy with a phenolic solution under halothane anesthesia developed cardiac dysrhythmias (1987). Severe cardiac dysrhythmias followed by circulatory arrest occurred in an elderly patient with pancreatic cancer, injected with a phenolic solution to produce splanchnic neurolysis (1988). The authors recommended that ethanol should replace phenol for this purpose.

In New Zealand, a patient died with brain damage after a cardiac arrest after being exposed to a chemical face peeling solution containing 64% phenol, Exoderm (1989). The New Zealand Ministry of Health issued a public statement concerning the safety of phenol solutions.

Phenoperidine

Intracranial hypertension occurred within 1 minute in a patient with a severe head injury who received phenoperidine 1 mg intravenously. It was associated with a reduction in arterial blood pressure. A similar reaction occurred when a second 1 mg bolus was given 8 hours later (SED-11, 146; 1990).

Phenoxybenzamine

In patients with myocardial infarction, in whom phenoxybenzamine has been used to improve circulation, it can cause or aggravate pulmonary edema; this could be explained by severe hyponatremia during treatment with phenoxybenzamine (SEDA-13, 113).

Phentermine

In a systematic review of 1279 patients taking fenfluramine, dexfenfluramine, or phentermine, evaluated in seven uncontrolled cohort studies, 236 (18%) and 60 (5%) had aortic and mitral regurgitation respectively (1991). Pooled data from six controlled cohort studies yielded, for aortic regurgitation, a relative risk ratio of 2.32 (95% CI = 1.79, 3.01) and an attributable rate of 4.9% and, for mitral regurgitation, a relative risk ratio of 1.55 (95% CI = 1.06, 2.25) with an attributable rate of 1.0%. Only one case of valvular heart disease was detected in 57 randomized controlled trials, but this was judged unrelated to drug therapy. The authors concluded that the risk of valvular heart disease is significantly increased by the appetite suppressants. Nevertheless, valvulopathy is much less common than suggested by previous less methodologically rigorous studies.

Spontaneous rupture of a retroperitoneal aneurysm occurred in a 70-year-old woman who had been taking phentermine hydrochloride, 30 mg/day, for about 1 month (1992). Other long-term medications included fluoxetine and amitriptyline, and she had no history of coronary artery disease, hypertension, diabetes, or complications of pregnancy. Although it is plausible that phentermine could have contributed to the ruptured aneurysm, other possibilities should be considered, particularly rupture of an anomalous retroperitoneal blood vessel.

Fatal pulmonary hypertension occurred in a 32-year-old man who had been taking phentermine in unknown doses for 4 months (SED-9, 16).

Phenylephrine

The FDA and the National Registry of Drug-induced Ocular Side Effects (Casey Eye Institute, Portland, Oregon) have received 11 reports of adverse systemic reactions to a single dose of topical ocular phenylephrine 10% applied in pledget form (1993). There were eight men and three women, aged 1–76 years. Most of the patients noted systemic effects within minutes of applying phenylephrine, and the adverse systemic reactions included severe hypertension, pulmonary edema, cardiac dysrhythmias, cardiac arrest, and subarachnoid hemorrhage. Ophthalmologists should be warned not to apply phenylephrine in this way, which is believed to be contraindicated in ophthalmic surgery, especially when other medications may be used (SED-14, 2061).

A 10% solution of phenylephrine has sometimes caused extremely severe cardiovascular complications, including myocardial infarction.

In newborn infants the benefit of accurate assessment of gestational age by examination of the anterior vascular capsule of the lens and the value of funduscopic examination in ill premature babies must be weighed against the possible risks of the associated increase in blood pressure produced by the pupillary dilators. Since there is no increase in mydriatic effect with repeated instillation or increasing concentration, and their small body mass places premature neonates at increased risk of phenylephrine overdose, it is prudent to use the lowest possible concentration, as well as the most effective combination of mydriatics for indirect ophthalmoscopy in premature infants when such examination is absolutely necessary. The hypertensive effect is likely to be maximal at some time within the first 20 minutes, and whenever possible (or when risk factors are present) the blood pressure should be monitored.

- A child developed cardiac dysrhythmias, severe hypertension, and pulmonary edema after the intraoperative administration of ocular phenylephrine (1994).
- A 2-month-old child given perioperative phenylephrine drops during cataract extraction developed ventricular extra beats, very severe hypertension, and pulmonary edema requiring intensive therapy (1995). Extubation was possible within 3 hours, and she recovered with no untoward consequences.

The authors commented that changes in arterial blood pressure are well described with phenylephrine eye-drops, especially in infants. Clearly precise dosage is difficult in these very young patients and they suggested that microdrops might be a safer mode of administration.

Cardiographic U waves have been reported to have been caused by phenylephrine (1996).

- A 71-year-old woman suddenly developed global aphasia. She was hypertensive and a heavy smoker and had previously undergone left carotid endarterectomy. There were multiple ischemic areas in the brain on the left side and total occlusion of the common, internal, and external carotid arteries on that side. Perhaps surprisingly, it was thought that hypoperfusion of the language areas of the cortex could be overcome by increasing the arterial pressure. This was done by an intravenous infusion of phenylephrine up to a dose of 180 micrograms/minute, which produced a blood pressure of 220/100 mmHg. Unfortunately, this did not produce any clinical improvement but positive U waves developed in the chest leads of the electrocardiogram; these disappeared after the infusion was discontinued.

The authors noted that while U waves are usually considered to be pathological and may reflect a tendency to the development of ventricular dysrhythmias, particularly torsade de pointes, the mechanisms by which they are formed are still conjectural. In any case, this approach to therapy does not appear very promising.

Phenylpropanolamine (norephedrine)

Much of the concern over the free sale of phenylpropanolamine relates to its cardiovascular effects. Although the commonly recommended 75 mg dose has no significant effect on blood pressure in healthy volunteers, 150 mg causes significant albeit transient hypertension lasting several hours (1997). Hypertensive crises have been reported, for example after a 600 mg overdose (SEDA-18, 158). However, field experience suggests that, because of individual variations in sensitivity, the risks to some patients are greater than these findings suggest. In single doses of 50 mg, phenylpropanolamine increased diastolic blood pressure to over 100 mmHg in 12% of adults.

Phenylpropanolamine-induced myocardial injury has been reported in a young woman using the recommended dose for weight control (1998).

- A 25-year-old woman with a negative medical history developed severe retrosternal chest tightness of sudden onset lasting for 30 minutes after taking phenylpropanolamine 25 mg bd for 3 days. She had never smoked and had no family history of heart disease. The electrocardiogram and cardiac markers were consistent with a non-Q wave myocardial infarct. She was treated with aspirin, intravenous glyceryl trinitrate, oral diltiazem, and intravenous eptifibatide and made a rapid recovery.

The author reviewed other cases of phenylpropanolamine-induced myocardial injury, all in younger women. It is not known why women are especially susceptible to this adverse effect of phenylpropanolamine (1999).

Phenylpropanolamine can also, on occasion, cause cardiac dysrhythmias in mild overdosage.

A transient cardiomyopathy without hypertension occurred in a girl of 14 who had taken only a small overdose (2000) and another in an adult woman (SEDA-17, 163).

A warning was issued by the Swiss Pharmaceutical Association that phenylpropanolamine can cause severe sympathomimetic adverse effects, including hypertensive crises, dysrhythmias, and tachycardia (2001) (SEDA-5, 11) (SEDA-7, 12).

Phenytoin and fosphenytoin

Intravenous phenytoin can cause cardiac dysrhythmias, hypotension, and potentially fatal cardiovascular collapse, especially if the highest recommended infusion rate (50 mg/minute or 1 mg/kg/minute in children) is exceeded. One case of hypersensitivity myocarditis was probably initiated by phenytoin, although carbamazepine may have contributed (SED-13, 142) (2002).

After intravenous use, the most common vascular complication is the so-called purple-glove or purple-limb syndrome, defined as the progressive development of edema, discoloration, and pain in the limb; sequelae include soft-tissue necrosis and limb ischemia. Retrospective analysis of data from 152 patients treated with intravenous phenytoin identified nine (6%) who developed this syndrome: they had received a greater mean initial dose of phenytoin (500 versus 300 mg) and a larger dose over 24 hours (800 versus 500 mg), and they tended to be older (72 versus 49 years) than those without the complication (2003). In one case surgical therapy was required; the others resolved conservatively within 1 month. Extravasation of intravenously injected phenytoin has caused tissue necrosis requiring amputation (SEDA-18, 67). Purple glove syndrome has generally been reported after intravenous phenytoin.

- In a 49-year-old otherwise healthy woman undergoing craniotomy for aneurysm clipping, inadvertent overdose with phenytoin (1500 mg) by rapid infusion caused intraoperative sinus arrest, which was managed successfully with standard resuscitative measures (2004).

This report highlights the cardiovascular risk of intravenous phenytoin, particularly when high infusion rates are used.

However, the purple-glove syndrome can also occasionally occur after oral administration.

- A 10-year-old boy took phenytoin 100 mg/day and his seizures were well controlled (2005). However, a pharmacist gave him about 1000 mg of phenytoin instead of the prescribed dose, and several hours later he became drowsy and his hands and feet turned dark purple with marked swelling. Phenytoin was withdrawn after 4 days and the swelling and discoloration of his hands and feet improved gradually and disappeared 11 days later.

The incidence of purple-glove syndrome associated with intravenous phenytoin has been assessed in a prospective review of 179 consecutive exposures (2006). There were only three mild cases (1.7%).

However, the purple-limb syndrome was recorded in 20 of 67 patients who received intravenous phenytoin over a 5-month period (2007). Affected cases tended to be older (median age 70 years versus 57 years in non-affected cases), and all resolved spontaneously within 3 weeks. These data suggest that the incidence of the syndrome may have been underestimated in the past, possibly owing to its delayed onset, frequent occurrence in patients with impaired communication abilities, and its usually mild self-limiting course.

In 775 patients who received intravenous phenytoin, valproate, or placebo, intravenous site reactions occurred in 25% of patients who received phenytoin (2008). Most of the events (70%) occurred in the first intravenous site, and all occurred in peripheral administration sites. When patients who received the drug by central line were excluded, the estimated incidence was 30%. There were fewer adverse events when phenytoin was given alone than when it was given together with valproate.

Fosphenytoin-induced QT interval prolongation has been reported (2009).

- A 23-year-old man was given intravenous fosphenytoin (equivalent to phenytoin 1500 mg or 20.5 mg/kg) over 85 minutes. He was normocalcemic before the infusion. During the infusion he had prolongation of the QT interval and reductions in the concentrations of total and ionized serum calcium. Plasma phenytoin concentrations were within the target range during the electrocardiographic changes, and the blood pressure was stable.

Fosphenytoin is metabolized by phosphatases to yield phenytoin plus inorganic phosphate. Binding of calcium by phosphate could have lowered the serum concentration of ionized calcium.

Photochemotherapy (PUVA)

A rise in ambient temperature can have cardiovascular effects. In high-risk conditions, cardiovascular monitoring during treatment has been advised (SEDA-6, 148), although patients with cardiovascular disease and hypertension are also reported to tolerate PUVA therapy without evidence of cardiovascular stress (2010).

Edema of the legs has been noted occasionally.

Pilocarpine

Pilocarpine is an antagonist at acetylcholine receptors (2011). While it is mainly used in the eye in the treatment of glaucoma, pilocarpine is still occasionally used for other purposes, for example for treating salivary gland hypofunction and xerostomia. When used in this way, cardiovascular tolerance was good but there was a high incidence of sweating, flushing, increased frequency of micturition, increased nasal secretion, and lacrimation (SEDA-18, 174).

Pimozide

Pimozide can cause QT interval prolongation and torsade de pointes; a total of 40 reports (16 deaths) of serious cardiac reactions, mainly dysrhythmias, were reported to the Committee on Safety of Medicines in the UK from 1971 to 1995 (2012).

Pipecuronium bromide

No histamine release has been reported with pipecuronium, and vagolytic or sympathomimetic effects are not seen in the usual dose range. Rarely, significant hypotension has been reported (2013), but this was transient and occurred during an unstable phase of anesthesia. Bradycardia has also been seen (2013) but is usually mild (2014), and probably due to the vagotonic effects of co-administered drugs, as is seen with vecuronium and atracurium (that is a minor disadvantage of the relaxant's lack of vagolytic or sympathomimetic effects). Usually, no significant changes in heart rate or blood pressure are seen (2015–2017), even with doses up to three times the ED_{95} (2018,2019). Cardiovascular stability has also been reported in cardiac patients (2020), including patients in ASA classes II and III about to undergo coronary artery bypass grafting who received doses up to 0.15 mg/kg (2021) and those who received high-dose fentanyl anesthesia (2022). The absence of tachycardia in these high-risk cardiac patients, in whom any increase in myocardial oxygen demand is unwanted, was considered an advantage of pipecuronium.

Piper methysticum (Piperaceae)

Tachycardia and electrocardiographic abnormalities have been reported in heavy users of kava (2023). Whether these observations represent true adverse effects or are due to other factors is unclear.

Piperazine

Cardiac conduction defects have been described in patients taking piperazine (2024).

Plasma products

Hypotension is a sometimes severe adverse effect of the administration of plasma proteins. It may be associated with the presence of a potent prekallikrein activator, the Hageman factor degradation product, which is thought to initiate the production of kinin from fibrinogen in the recipient's blood. The kinin then produces systemic hypotension.

In five of eight patients treated with fresh frozen plasma to achieve rapid correction of anticoagulation after warfarin-related intracranial hemorrhage, there were complications of fluid overload (2025). In two of these patients congestive heart failure developed, resulting in myocardial infarction and renal insufficiency respectively. Two patients had supraventricular tachydysrhythmias and one developed pulmonary edema.

The administration of Solvent-Detergent (SD)-plasma during plasma exchange in the management of thrombotic thrombocytopenic purpura, was associated with risks of venous thromboembolism (2026).

Platinum-containing cytostatic drugs

Asymptomatic sinus bradycardia (for example 30–40/minute) is observed within 30 minutes to 2 hours after the start of cisplatin infusion. When cisplatin is withdrawn normal rhythm is restored. Because patients who receive platins are not routinely monitored, drug-induced sinus bradycardia may not be detected in practice. However, several case reports have included heavily pretreated patients, which makes a direct relation between cisplatin administration and the onset of cardiotoxic symptoms much more difficult to assess. In conclusion, no dosage adjustment appears to be warranted in patients with cisplatin-induced sinus bradycardia; however, attention should be paid to patients with resting bradycardia or those using medications known to slow the heart rate (2027,2028).

- A 60-year-old woman with a squamous cell lung carcinoma developed a paroxysmal supraventricular tachycardia during administration of cisplatin 20 mg/m^2 and etoposide 75 mg/m^2. The dysrhythmia appeared to be related to cisplatin since normal rhythm was restored after cisplatin was withdrawn (2029).

Orthostatic hypotension was reported in "several" of 126 patients given cisplatin 50 mg/m^2 on days 1, 8, 29, and 36 as part of treatment for lung cancer in combination with etoposide and chest radiotherapy (2030).

There have been 21 reports of life-threatening disease affecting large arteries in patients treated with cisplatin, bleomycin, and vinblastine in combination for germ cell tumors (2031,2032). Five patients died during or after therapy, three from acute myocardial infarction, one from rectal infarction, and one from cerebral infarction. Other patients who developed major vascular disease, including coronary artery and cerebrovascular disease, have been reported. Symptoms occurred acutely in some (within 48 hours of starting therapy), and after months or years had elapsed in others.

Reduced peripheral circulation, Raynaud's phenomenon, and polyneuropathy have been described after the combined use of cisplatin, bleomycin, and vinblastine for testicular tumors. Of eight cases with polyneuropathy that were investigated, it was not possible to confirm a causative association between Raynaud's phenomenon and the chemotherapy (2033).

Platinum compounds have rarely been described to cause phlebitis after intravenous administration (2034).

Poliomyelitis vaccine

Myopericarditis has been attributed to Td-IPV vaccine (2035).

- A 31-year-old man developed arthralgia and chest pain 2 days after Td-IPV immunization and had an acute myopericarditis. He recovered within a few days with high-dose aspirin.

The authors discussed two possible causal mechanisms, natural infection or an immune complex-mediated mechanism. Infection was excluded by negative bacterial and viral serology and a favorable outcome resulted within a few days without antimicrobial drug treatment.

Polymyxins

During sepsis, toxins (for example released from bacteria) can cause shock, disseminated intravascular coagulation, multiorgan dysfunction, and death. Apheresis may be a way of reducing the amounts of toxins and other harmful compounds in the circulation, and polymyxin B may serve as an adsorber. In three patients with septic shock, direct hemoperfusion using a polymyxin B-immobilized fiber column was carried out after antibacterial and antishock therapy. As a result, cardiovascular instabilities improved without increasing the supply of catecholamines (2036). Furthermore, in seven patients with endotoxic shock after laparotomy undergoing hemoperfusion with the polymyxin B-immobilized fiber, there was an early increase in urine volume, attributable to increased glomerular filtration independent of systemic hemodynamic factors (2037).

Polytetrafluoroethylene

Teflon injected in a young woman for urinary incontinence migrated to the pulmonary vascular system (2038).

Particles of Teflon can detach from cardiac valve prostheses, producing embolic complications (2039).

Potassium chloride

When potassium chloride is given by intravenous infusion for the treatment of potassium depletion (for example in diabetic ketoacidosis) there is a risk of cardiac dysrhythmias if the infusion is too rapid. The rate of infusion should be no greater than 20 mmol/hour.

Pramipexole

Orthostatic hypotension is common during oral pramipexole therapy (2040).

Peripheral edema has occasionally been described as an adverse effect of dopamine agonist therapy and has been reported in 17 of 300 patients treated with pramipexole (2041). The mean dose at onset of edema was 1.7 mg/day and the time after initiation of therapy was 2.6 months. In all cases the edema disappeared after the drug was withdrawn but reappeared on rechallenge. Although the condition was dose-dependent in affected individuals its occurrence was idiosyncratic with no obvious predisposing features. Response to diuretics was minimal.

Probucol

Prolongation of the QT interval has been seen and 16 cases of tachydysrhythmias, especially torsade de pointes, have been reported in association with probucol, 15 cases in women (SEDSA-13, 1331) (2042,2043).

Progestogens

Progestogens with a degree of mineralocorticoid activity will tend to cause water and salt retention and to increase blood pressure in susceptible subjects. However, effects on blood pressure can be variable; for example, medroxyprogesterone has been variously reported to cause a fall in blood pressure in some initially hypertensive patients while rapidly increasing diastolic pressure in some other women, or to have no effect on blood pressure at all.

Since thromboembolic complications can occur with progestogens, there may be a danger in using them in patients in whom there are other risk factors for thromboembolism. A Spanish group had to deal with patients with AIDS in whom megestrol acetate seems to be helpful in countering AIDS-related anorexia or cachexia. However, advanced HIV infection is itself a risk factor for thromboembolism, as is tuberculosis, which readily occurs in this population. Of 199 patients with AIDS followed for 2 years, 25 took megestrol 320 mg/day. Deep vein thrombosis occurred in seven patients in the entire series, four of them being in the megestrol group and three having tuberculosis. The duration of hormonal therapy up to the moment of thrombosis averaged 98 days. Tuberculosis was an independent risk factor. Statistical analysis led to the conclusion that in this high-risk population the use of megestrol had increased the risk of thrombosis by a factor of 7.6 (2044).

Thromboembolism and third-generation progestogens

In October 1995 the UK Committee on Safety of Medicines (CSM) issued a warning that oral contraceptives containing the third-generation progestogens gestodene or desogestrel carry a higher risk of venous thromboembolism, and that women using these should consider changing to another brand (2045).

The warning was based on three unpublished studies that had not at the time been formally peer reviewed. All three showed about double the risk of venous thromboembolism with gestodene and desogestrel products compared with oral contraceptives containing other progestogens. The first, a large WHO collaborative case-control study of cardiovascular disease and oral contraceptives was undertaken in 17 countries. It was completed in July 1995, and involved 829 cases of venous thromboembolism and 2641 controls from nine countries in which third-generation oral contraceptives had been used (2046). The second, from the Boston Collaborative Drug Surveillance Program, was an analysis of the occurrence of venous thromboembolism in a UK general practice cohort of 238 130 women who had received a prescription for an oral contraceptive containing levonorgestrel, gestodene, or desogestrel (2047). The third was a transnational case-control study conducted in five European countries, funded by Schering, the leading manufacturer of gestodene (2048).

The WHO study was undertaken because the association between oral contraceptive use and venous thromboembolism had not been examined since the 1970s, when oral contraceptives contained higher doses of estrogen and progestogen than now, and because none of the earlier studies had been done in developing countries. The finding of a higher risk of venous thromboembolism with gestodene and desogestrel came as a surprise, and this, together with publicity in the media, prompted the CSM to commission the UK cohort study. The transnational study was set up at the request of the German regulatory authority to follow up German spontaneous reporting data that in 1990 had strongly suggested higher risks of thromboembolism with gestodene-containing oral contraceptives. This led to much controversy in the press and television. Schering had argued that highly publicized deaths had stimulated selective reporting, and that the claimed higher risk was an artefact.

The sudden announcement by the CSM in October 1995 caught prescribers and users of oral contraceptives completely unprepared. What some General Practitioners described as the "pill panic" in the media (2049) drove thousands of women to consult their doctors (2049–2051), who had not been briefed. A heated debate ensued over whether the CSM's decision was justified, whether its announcement should have been delayed until the data were published, and over the dramatic and confusing way in which it was issued. Before long there was general agreement that the decision was necessary and correct, but that its communication had been handled badly, and the Health Minister said that the Government would review the incident to learn from it. Important suggestions for getting this kind of communication correct have subsequently been made (2052–2054). Official actions in other countries ranged from the imposition of stronger restrictions (in Germany) to decisions in the Netherlands and Canada to wait and consider the published studies, and in the USA not to advise switching to other products (2055). The European Union's Committee on Proprietary Medicinal Products also decided to wait and see.

Although doubts about the relative safety of gestodene emerged in 1990, regulators did not publicly acknowledge them. Nor did they help independent scientists to examine all the relevant data (including prescribing figures, essential to provide denominators for the calculation of risk), perhaps because they were insufficient and might have been wrongly interpreted. In Germany, the regulatory authority privately asked Schering to undertake a case-control study; in the UK the Medicines Control Agency considered the available evidence inadequate and dismissed the doubts. Whether the CSM was consulted at that time or only later remains a minor official secret. The attitude changed early in July 1995, when the CSM saw the early results of the WHO Collaborative Study, and a highly critical UK television programme about gestodene-containing oral contraceptives was broadcast (Granada, "World in Action," 10 July). The CSM asked workers from the transnational case-control study to "expedite their results," and commissioned the cohort study using data routinely collected from General Practitioners. The analysis of venous thromboembolism in the transnational study was completed on 8 October, and was promptly discussed with members of the Medicines Control Agency and the CSM. The CSM quickly made its decision (2048) and announced it on 18 October. The relevant data from the WHO study and the GP cohort study were published on 16 December; those from the transnational study were published on 13 January 1996 (2056). A searching independent analysis of these three studies, and of a fourth (2057) that the CSM had not considered, supported the decision (2058).

In retrospect, what could the CSM and other authorities have done to alert the world to the potential problem during the 5 years of growing doubts about the relative safety of gestodene? To know about the doubts would have helped women and their doctors. They could then have begun to discuss among themselves whether to act on them or not, in order to weigh any real advantages of their gestodene oral contraceptive against the doubt and potential risk. A joint statement by regulators and manufacturers about their plans for further investigations would have made it clear that there was a problem to be resolved, instead of appearing to ignore or deny it. This would have allowed prescribers and independent experts to rethink their prescribing policies, for example deciding in what circumstances gestodene oral contraceptives should no longer be the first choice. That would not have pleased the manufacturers, since it would almost certainly have reduced the sales of their leading product in an uncontrollable way. The position of desogestrel creates additional complications, since its relative safety was not questioned until the data from the WHO study appeared in July 1995. If doubts had been officially expressed about the gestodene-containing products Femodene and Minulet, these oral contraceptives would meanwhile have lost some market share to Marvelon and Mercilon, the desogestrel-containing products. But greater openness from the start could have minimized the alarm and confusion.

Propafenone

Cardiovascular adverse effects have been reported in 13–27% of patients taking propafenone and ventricular dysrhythmias in 8–19% in small studies. However, in large studies the risk has been reported to be about 5%.

Conduction disturbances are common with propafenone and can result in sinus bradycardia, sinoatrial block, sinus arrest, any degree of atrioventricular block, and right or left bundle-branch block (SEDA-10, 151; SEDA-15, 179).

The adverse effects of a single oral dose of propafenone for cardioversion of recent-onset atrial fibrillation have been evaluated in a systematic review (2059). The adverse effects were transient dysrhythmias (atrial flutter, bradycardia, pauses, and junctional rhythm), reversible widening of the QRS complex, transient hypotension, and mild non-cardiac effects (nausea, headache, gastrointestinal disturbances, dizziness, and paresthesia).

Dysrhythmias can occur; these include ventricular tachycardia, ventricular flutter, and atrial fibrillation (2060–2062). Hypotension and worsening of heart failure have occasionally been reported (SEDA-10, 151; 2063).

- Wide-complex tachycardias occurred in two elderly patients (a 74-year-old man and an 80-year-old woman) who had taken propafenone for atrial fibrillation (2064). In the first case the dysrhythmia was due to atrial flutter with 1:1 conduction.

Although drugs of class Ic, such as propafenone, can slow atrial and atrioventricular nodal conduction in patients with atrial fibrillation or atrial flutter, they do not alter the refractoriness of the atrioventricular node, and this allows 1:1 atrioventricular conduction as the atrial rate slows. This happens despite prolongation of the PR interval.

Class Ic drugs can also convert atrial fibrillation to atrial flutter, reportedly in 3.5–5% of patients. Of 187 patients with paroxysmal atrial fibrillation who were treated with flecainide or propafenone, 24 developed atrial flutter, which was typical in 20 cases (2065). These patients underwent radiofrequency ablation, which failed in only one case. All the patients continued to take their pre-existing drugs, and during a mean follow-up period of 11 months, the incidence of atrial fibrillation was higher in patients who were taking combined therapy than in those taking monotherapy. The authors suggested that in patients with atrial fibrillation who developed typical atrial flutter due to class Ic antidysrhythmic drugs, combined catheter ablation and continued drug treatment is highly effective in reducing the occurrence and duration of atrial tachydysrhythmias. They did not report adverse effects.

In controlled trials of oral propafenone (450–600 mg as a single dose) in patients with recent-onset atrial fibrillation without heart failure, atrial flutter with 1:1 atrioventricular conduction occurred in only two of 709 patients (0.3%) who received propafenone (2066).

Propofol

Propofol is a cardiodepressant and resets the baroreflex set-point, with a tendency to bradycardia (which occurs in some 5% of cases), hypotension (16%), or both (1.3%) (2067). The hypotension may be brought about by peripheral vasodilatation, reduced myocardial contractility, and inhibition of sympathetic nervous system outflow (2068). Four deaths due to cardiovascular collapse during induction have been reported in patients aged 78–92 years given propofol 1.1–1.8 mg/kg (2069). The patients were of ASA classes 3 or 4.

Total intravenous anesthesia with propofol resulted in a reduced heart rate and a higher frequency of oculocardiac reflex bradycardia than thiopental/isoflurane anesthesia, with a higher sensitivity of children younger than 6 years in all groups (2070).

Hemodynamic effects

The cardiovascular effects of propofol have been examined in a randomized trial in 40 healthy subjects using transthoracic echocardiography (2071). Propofol was given to the same total dose (2.5 mg/kg) at two different rates, 2 mg/second or 10 mg/second. In both groups, global and segmental ventricular function was unchanged, but propofol caused a markedly reduced end-systolic quotient, presumably related to reduced afterload. With the higher infusion rate, there was a significant reduction in fractional shortening, thought to be related principally to reduced preload.

There has been a prospective, double-blind, controlled comparison of propofol, midazolam, and propofol + midazolam for postoperative sedation in 75 patients who received low-dose opioid-based anesthesia for coronary bypass grafting (2072). Mean induction doses of propofol and midazolam used alone were 2.5 times higher than when both were used together. The single agents caused significant reductions in blood pressure, left atrial filling pressure, and heart rate after induction. These hemodynamic changes returned to normal after 15 minutes with midazolam and after 30 minutes with propofol, except for the bradycardia, which remained for the duration of the sedation. The combination of propofol + midazolam had no significant hemodynamic effects, but was also associated with bradycardia lasting the duration of the sedation. There was a greater than 68% reduction in maintenance doses with the combination. Propofol and propofol + midazolam were associated with comparable times to awakening and extubation, while with midazolam alone recovery was slower. This study clearly showed a reduction in adverse effects from exploiting the sedative synergism between propofol and midazolam.

In a placebo-controlled study of induction of anesthesia with a combination of propofol + fentanyl in 90 patients aged over 60 years, prophylactic intravenous ephedrine 0.1 or 0.2 mg/kg given 1 minute before induction of anesthesia significantly attenuated the fall in blood pressure and heart rate that is usually observed (2073). Prophylactic use of ephedrine may be useful in preventing the occasional instances of cardiovascular collapse recorded after induction of anesthesia using these agents in elderly people.

The hemodynamic effects of combining ephedrine with propofol in an effort to prevent hypotension and bradycardia have been investigated in 40 elderly patients of ASA grades III and IV, who received ephedrine 15, 20, or 25 mg added to propofol 200 mg (2074). The hypotensive response to propofol was effectively prevented, but marked tachycardia in the majority of patients meant that the technique may not be beneficial, given the high incidence of ischemic heart disease in this age group.

The effects of giving calcium chloride 10 mg/kg after induction of anesthesia with propofol, fentanyl, and pancuronium have been investigated in 58 patients undergoing elective coronary artery bypass grafting (2075). Calcium chloride reduced the fall in arterial blood pressure and prevented the reductions in heart rate, stroke volume index, cardiac index, and cardiac output, compared with placebo. Propofol reduces the availability of calcium to the myocardial cells, and calcium chloride effectively minimizes the hemodynamic effects of propofol. However, given that intravenous calcium can be locally toxic when given via peripheral veins, the technique may have limited applicability.

Cardiac dysrhythmias

Propofol causes bradydysrhythmias by reducing sympathetic nervous system activity.

- A four-year-old patient developed a nodal bradycardia while receiving propofol 6 mg/kg/hour + remifentanil 0.25 microgram/kg/minute (2076). The bradycardia responded to atropine 0.3 mg.
- Complete atrioventricular heart block occurred in a 9-year-old boy with Ondine's curse who received a single bolus injection of propofol (2077).

The authors questioned the safe use of propofol in congenital central hypoventilation syndrome, which is a generalized disorder of autonomic function.

- Propofol caused marked prolongation of the QT_c interval in a 71-year-old woman with an acute myocardial infarction who required ventilatory support (2078). Substituting midazolam for propofol was associated with normalization of the QT_c interval. Rechallenge with propofol was associated with further prolongation. There were no malignant ventricular dysrhythmias.

Pain on injection

Propofol can cause severe pain on injection, especially when injected into a small vein (2079); the incidence is 25–74% (2080). Administration of the lipid solvent in which propofol dissolved has confirmed that the solvent is responsible for this adverse effect (2081).

- Severe pain on injection of propofol occurred in a 36-year-old man with severe Raynaud's phenomenon, including a history of skin ulceration when he was given a 2% propofol + lidocaine mixture into a vein on the back of his hand (2082).

The author suggested that selecting a larger antecubital vein might be a wiser choice in these patients.

The effectiveness of lidocaine in preventing pain on injection has been confirmed, and a concentration of 0.1% was optimal (2083). The kallikrein inhibitor nafamostat mesilate was as effective as lidocaine.

A controlled study in 100 women showed that pretreatment with intravenous ketamine 10 mg reduced the incidence of injection pain from 84 to 26% of patients (2084).

Warming propofol to 37°C had no effect on the incidence of pain (2085,2086).

The effects of different doses of ketorolac, with or without venous occlusion, on the incidence and severity of pain after propofol injection have been studied in a randomized, double-blind study in 180 patients (2087).

Pretreatment with intravenous ketorolac 15 mg and 30 mg reduced the pain after propofol injection. A lower dose of ketorolac 10 mg with venous occlusion for 120 seconds achieved the same effect.

Ondansetron, tramadol, and metoclopramide were less effective than lidocaine in preventing pain on injection (2088–2090). Ondansetron and tramadol have been compared in patients being given propofol in a randomized, double-blind study in 100 patients (2091). Tramadol 50 mg intravenously was as effective as ondansetron 4 mg intravenously with 15 seconds of venous occlusion at preventing propofol injection pain. However, there was significantly less nausea and vomiting in those given ondansetron.

Prostaglandins

Both PGE_2 and $PGF_{2\alpha}$ commonly cause a fall in blood pressure and a degree of bradycardia (2092,2093). PGE_2 can cause vasodilatation of small vessels and $PGF_{2\alpha}$ can cause vasoconstriction (2094). These changes are common but often mild. However, angina pectoris and myocardial infarction have been reported with prostaglandins of all types, particularly after inadvertent intramyometrial injection (2095–2098). A single case of pulmonary edema after the infusion of PGE_1 has been reported (2099). In patients with pre-existing cardiovascular disease, the risk of serious aggravation is very real, and both pre-existing hypertension and states of shock can be worsened. A severe rise in maternal blood pressure occurred in a few cases in which fetal death was associated with unresolved pre-eclampsia.

Coronary spasm has been attributed to latanoprost (2100).

- A 58-year-old man with stable angina pectoris started to use latanoprost eye drops and over the next few days his angina worsened and occurred at rest. After 15 days, he had syncope during physical exercise. Angiography showed coronary spasm.

Protamine

Severe acute pulmonary vasoconstriction with cardiovascular collapse was identified in 1983 and subsequent studies showed that the incidence of protamine-induced pulmonary artery vasoconstriction was about 1.5% after cardiac surgery (2101). Protamine-induced pulmonary artery vasoconstriction is accompanied by generation of large quantities of thromboxane A_2, a potent vasoconstrictor (2101).

The precise mechanisms that explain protamine-mediated systemic hypotension are unknown, but there is evidence that it may be mediated by the endothelium and dependent on nitric oxide/cyclic guanosine monophosphate; it has been suggested that methylthioninium chloride (methylene blue) may be of use in treating hemodynamic complications caused by the use of protamine after cardiopulmonary bypass (2102).

Protease inhibitors

Metabolic changes that protease inhibitors can cause after prolonged therapy include raised serum lactate, hypogonadism, hypertension and accelerated cardiovascular disease, reduced bone density, and avascular necrosis of the hip. Two large prospective studies in 1207 patients (2103) and 3191 patients (2104) have clarified the spectrum and incidence of metabolic changes in HAART and have explored the relative importance of protease inhibitors. In addition, data on fat redistribution from a postmarketing review of HIV-infected individuals taking indinavir have been published (2105).

Prothrombin complex concentrate

Thromboembolic events in relation to activated prothrombin complex products include deep venous thrombosis and pulmonary embolism (2106–2108). So far, the exact mechanism of how thromboembolic events are stimulated by prothrombin complex products is not known. It has been postulated that the presence of activated clotting factors VII, IX, and X, as well as procoagulatory phospholipids in prothrombin complex products, are responsible (2109).

In 42 patients who required immediate reversal of oral anticoagulant therapy by prothrombin complex concentrate, no laboratory and clinical evidence for coagulation activation was found (2110). The authors suggested that the concentration of protein C within the prothrombin complex concentrate is an important factor for preventing thrombosis. The activated form of protein C is an important natural anticoagulant.

Myocardial necrosis has been described after the administration of a prothrombin complex products, probably because of excessive kininogen in the product.

Protirelin

A transient rise in blood pressure occurs immediately after the administration of protirelin (2111). In very rare cases, transient amaurosis and bronchospasm have been reported, thought to be due to either vasopressor syncope or cardiac arrhythmias (2112–2114).

Proton pump inhibitors

Reversible peripheral edema has been reported in five women taking the proton pump inhibitors omeprazole, lansoprazole, or pantoprazole for 7–15 days for peptic disorders in recommended standard doses (2115). Edema disappeared within 2–3 days of withdrawal and reappeared in all five patients after re-exposure. High-dose intravenous infusions of omeprazole and pantoprazole (8 mg/hour) caused peripheral edema in three of six young female volunteers and two of six female volunteers respectively. The edema disappeared within 24 hours of stopping the infusion. Similar high doses of omeprazole did not produce edema in male volunteers. Subsequent studies performed on 10 female volunteers to elucidate the cause of the edema did not show any changes in concentrations of serum hormones or C1 esterase inhibitor.

Proxyphylline

In 12 patients with chronic obstructive pulmonary disease, intravenous proxyphylline 16 mg/kg lowered airway resistance by 30% in 9 patients and by 40% in 7 (2116). There were no serious adverse effects and, in particular, no adverse cardiovascular effects or muscle tremor.

In a short-term, double-blind, crossover study in 10 adult asthmatics, a modified-release formulation of proxyphylline (2400 mg/day) was compared with theophylline (800 mg/day) and placebo (2117). There were no significant differences between the treatments with regard to relief of asthma symptoms, the need for additional medication, or the incidence and intensity of adverse effects. The adverse reactions included loss of appetite, palpitation, headache, nausea, stomach ache, muscle tremor, and sleep disturbances.

Pseudoephedrine

In subjects with impaired baroreflex function, oral pseudoephedrine 30 mg or phenylpropanolamine 12.5 mg and 25 mg produced significant increases in blood pressure. When they were taken with water the increase in blood pressure was greater. The maximal increase in systolic blood pressure occurred after 60 minutes and the pressure returned to baseline by 2 hours (2118).

Acute myocardial infarction has been attributed to pseudoephedrine (2119).

- A 19-year-old male cigarette smoker with normal coronary arteriography had an upper respiratory infection, for which he brought Gripex®, each tablet of which contains paracetamol 325 mg, pseudoephedrine HCl 30 mg, and dextromethorphan HBr 10 mg. He took four tablets twice, about 3 hours apart, a total of 240 mg of pseudoephedrine within 3 hours. He had an acute myocardial infarction 12 hours later. Other drugs, such as cocaine and amphetamines, were excluded. Subsequent coronary angiography and echocardiography showed a non-Q-wave myocardial infarction with normal coronary arteries.

About 5% of patients with a myocardial infarction have normal coronary arteries at angiography. In many cases there is coronary artery spasm and/or thrombosis, perhaps with underlying endothelial dysfunction of the epicardial arteries. It is important in such cases to obtain a complete history of the use of drugs, including over-the-counter drugs.

In a meta-analysis of 24 studies involving 1285 subjects the authors concluded that in healthy individuals pseudoephedrine significant increased heart rate (2.83/minute; CI = 2.0, 3.6) and increased systolic blood pressure (0.99 mmHg; 95% CI = 0.08, 1.9), with no effect on diastolic pressure (2120). In patients with controlled hypertension there was an increase of similar magnitude in systolic blood pressure (1.2 mmHg; CI = 0.56, 1.84). Higher doses and immediate-release formulations were associated with greater increases in blood pressure, although it was difficult to be certain what the average dose was in these studies; from the data presented it was usually 60 mg once or twice daily. The authors reported isolated cases of much greater rises in blood pressure (up to 20 mmHg systolic) but again it was difficult to determine the doses used. They concluded that immediate-release formulations had a greater effect than modified-release formulations, and that a dose-response relation could be discerned. Evidently many cases of pseudoephedrine-related cardiovascular effects involved higher doses than have been used in formal studies or are recommended by manufacturers. There appeared to be a smaller effect in women than in men. Shorter duration of use was associated with greater increases in systolic and diastolic blood pressures. There were no clinically significant adverse outcomes. However, a rare adverse event may not be seen with such a small sample size.

Coronary vasospasm has been attributed to pseudoephedrine (2121).

- A previously healthy 32-year-old man from Nigeria developed substernal chest pain at rest associated with nausea and sweating 45 minutes after taking two tablets of an over-the-counter cold remedy containing pseudoephedrine 30 mg and paracetamol 500 mg per tablet. He recalled a similar but less severe episode 1 week earlier after taking the same medication. There was no relevant past medical history or family history of coronary artery disease. An electrocardiogram showed ST elevation in the inferolateral leads and the plasma creatine kinase activity and troponin I concentration were both raised. Coronary angiography showed normal arteries.

The authors concluded that this episode had been caused by coronary vasospasm initiated by pseudoephedrine and warned of the dangers of this type of medication, even in otherwise healthy individuals. The temporal association between ingestion of pseudoephedrine and the myocardial infarction suggested a causal relation. The absence of coronary artery disease at catheterization combined with the cardiac magnetic resonance imaging findings were consistent with an acute myocardial infarction caused by vasospasm due to pseudoephedrine.

Hypertensive strokes have been attributed to pseudoephedrine in 22 patients (2122). The effects of pseudoephedrine may be important when considered on a population basis, given their widespread use as decongestants. Although marked rises in blood pressure were uncommon, there were rises above 140/90 mmHg in nearly 3% of the patients. The benefit to harm balance should therefore be evaluated carefully before pseudoephedrine is used in individual patients most at risk of rises in blood pressure and heart rate.

Pyrazinamide

Acute symptomatic hypertension consistently followed the administration of pyrazinamide to a 65-year-old woman with pulmonary tuberculosis (2123).

Pyridoxine

In letter to the editor (2124) it was mentioned that the suspicion of partiality about the Committee on Toxicity becomes more plausible when one considers the issue of homocysteine. This intermediate metabolite may well turn out to be of greater importance as a risk factor for cardiovascular disease than cholesterol and blood pressure. Raised homocysteine concentrations appear to be accessible to treatment with pyridoxine (100 mg/day) together with vitamin B_{12} and folic acid (2125). Furthermore, the statement that there is no good evidence for the efficacy of pyridoxine in any disease, apart from depression, was criticized, because this ignores important studies in autism, pregnancy outcome, asthma, and sickle-cell anemia (2124).

Quinine

Quinine can cause atrioventricular conduction disturbances. In sensitive patients, such changes can occur with normal dosages given over a prolonged period; however, in most cases cardiac effects are due to overdosage.

Electrocardiographic changes, such as prolongation of the QT interval, widening of the QRS complex, and T wave flattening, can be seen with plasma concentrations above 15 μg/ml (SEDA-11, 590) (SEDA-14, 239).

Quinine, and more profoundly quinidine, its diastereomer, can cause ventricular tachycardia, torsade de pointes, and ventricular fibrillation by prolonging the QT interval (SEDA-20, 261).

- An 8-year-old child given an incorrect dose of quinine had ventricular tachycardia and status epilepticus after 48 hours; the plasma quinine concentration was 20 μg/ml (SEDA-18, 288), compared with the target range of 1.9–4.9 μg/ml.

Radioactive iodine

In a series of thyrocardiac patients, of those dying primarily from thyrotoxicosis more than 21% did so within 3 weeks of ^{131}I treatment (2126), presumably reflecting too sudden a change in metabolic activity for patients with existing cardiac complications.

Radiopharmaceuticals

Dipyridamole-^{201}thallium imaging has been used to assess cases of suspected coronary disease (SEDA-15, 506) (SEDA-16, 537) (SEDA-17, 539). It is difficult to distinguish the adverse effects of dipyridamole from those of the pharmaceutical agent (dipyridamole-^{201}thallium), since dipyridamole can cause bronchospasm. In one series of 400 examinations, there was severe chest pain due to myocardial ischemia in 9% of cases, milder chest pain (probably not associated with cardiac events) in 21%, and severe hypotension in 2.5% (2127). Others have reported instances of cardiovascular collapse (SEDA-15, 506).

Raloxifene

In one published study there was a two- to three-fold rise in the incidence of thromboembolism, similar to that seen with estrogen treatment. There were also some cases of leg cramps and hot flushes (2128).

There have been conflicting reports on the incidence and severity of symptoms such as hot flushes (also known as hot flashes) during long-term treatment with raloxifene for the prevention of osteoporosis. In fact the difference between raloxifene and placebo does not seem to be very great. In a review of three identical randomized trials in which raloxifene 60 mg was given for long periods to healthy postmenopausal women of various ages it was concluded that after 30 months the cumulative incidence of hot flushes was 21% for placebo and 28% for raloxifene, but the difference in frequency was confined to the first 6 months of therapy (2129). There was no difference between placebo and raloxifene in the maximum severity of symptoms or the rate of early discontinuation, while the period during which hot flushes continued was only a little shorter in the raloxifene group. In a US study in more than 1100 postmenopausal women who took raloxifene 30–150 mg/day the only significant adverse effect of therapy was hot flushes (25% with 60 mg/day and 18% in the placebo group) (2130).

The effects of raloxifene on the vascular endothelium have been studied in 19 subjects who underwent endothelial function testing at baseline and after treatment with placebo or raloxifene (60 mg/day for 6 weeks) (2131). The findings in this small short-term study were entirely positive. Brachial artery diameter change (flow-mediated dilatation) increased 5.0% with placebo and 8.6% with raloxifene in response to a hyperemic stimulus. The ratio of AUC response to AUC reference with the use of laser Doppler measures was 1.18 for placebo and 1.28 for raloxifene. Flow-mediated dilatation and AUC ratio correlated significantly. The authors concluded that raloxifene enhanced endothelial-mediated dilatation in brachial arteries and digital vessels in these women, and they discussed the drug's possible cardioprotective effect.

Ranitidine

Atrioventricular block has been attributed to ranitidine (2132).

- A 20-year-old man taking ranitidine 300 mg/day had a brief episode of syncope. The only abnormal finding was first-degree atrioventricular block, which disappeared after withdrawal of ranitidine. Rechallenges on two separate occasions produced recurrence of asymptomatic first-degree atrioventricular block, but cimetidine 400–800 mg/day and famotidine 40–80 mg/day caused no electrocardiographic abnormalities.

The authors hypothesized that this patient may have been abnormally susceptible to the cholinergic or cholinergic-like effect of ranitidine, unrelated to its histamine H_2 blocking action. However, the ability of ranitidine to release histamine may also have contributed.

Bradycardia occurred in a 4-day-old full-term male neonate 2 hours after the intravenous injection of ranitidine and resolved over 24 hours (2133).

Ranunculaceae

There is experimental evidence that berberine can cause arterial hypotension (2134,2135).

Rapacuronium

A major adverse effect of rapacuronium is an increase in heart rate (2136). Plasma histamine concentrations may increase after rapacuronium injection, but this was not correlated with changes in blood pressure or heart rate (2137).

Electrolyte balance

Hyponatremia has been reported with reboxetine (2138,2139).

- A 72-year-old man with diabetes mellitus and cardiovascular disease developed major depression. He was taking aspirin (100 mg/day), enalapril (20 mg/day), and glibenclamide (5 mg/day). His serum sodium was 133 mmol/l (reference range 134–146 mmol/l). He started to take reboxetine (4 mg/day) and after 8 days experienced malaise and nausea, at which time his serum sodium had fallen to 118 mmol/l. The reboxetine was withdrawn, and both his symptoms and the low serum sodium remitted over the next 6 days. Rechallenge with reboxetine produced a recurrence of both the low sodium and the accompanying symptoms.

It appears that, like SSRIs, reboxetine can cause sodium depletion in elderly people. However, in this case the contributions of concomitant general medical illness and its treatment were uncertain.

Remifentanil

Three groups of 20 women due to undergo elective surgery were recruited into a randomized, double-blind study (2140). Group 1 received a bolus dose of remifentanil 1 microgram/kg and an infusion of remifentanil (0.5 micrograms/kg/minute); groups 2 and 3 received remifentanil 0.5 micrograms/kg and an infusion of 0.25 micrograms/kg/minute. Groups 1 and 2 received pretreatment with glycopyrrolate 200 micrograms whilst group 3 did not. Cardiovascular responses to laryngoscopy and orotracheal intubation were measured. There were no significant differences in the three groups, except that there was a significantly lower heart rate in group 3 after induction of anesthesia and after intubation.

The hemodynamic effects of bolus intravenous remifentanil 0.2, 0.33, and 1 microgram/kg/minute have been studied in patients scheduled for coronary artery bypass grafting (2141). The study was terminated after only eight patients had been recruited, because of severe hypotension, bradycardia, and/or evidence of myocardial ischemia. The authors concluded that remifentanil should not be given as a bolus dose of 1 microgram/kg but as an infusion at a low rate. An editorial response to this article suggested that the hemodynamic instability reported may have resulted from other contributing factors, such as hypovolemia, impairment of venous return, or excessive anesthesia due to remifentanil toxicity (2142).

In a prospective study in 12 men undergoing elective coronary artery bypass grafting, remifentanil 0.5 and 2.0 micrograms/kg/minute combined with propofol preserved hemodynamic stability and reduced myocardial blood flow and metabolism to a similar extent (2143).

Asystole has been attributed to remifentanil (2144).

- A 78-year-old man with laryngeal cancer developed asystole 1 minute after an intravenous bolus of remifentanil 0.5 micrograms/kg followed by a continuous infusion of 0.5 micrograms/kg/hour. The asystole was unresponsive to intravenous atropine 1 mg. The remifentanil infusion was stopped and cardiac sinus rhythm resumed after two precordial thumps.

The authors postulated that rapid-sequence induction of anesthesia with sevoflurane had blunted sympathetic tone and allowed uncompensated parasympathetic activation by remifentanil.

Opioids are used routinely to eliminate the stress response in the pre-bypass phase of pediatric cardiac surgery. Remifentanil was used in a double-blind, randomized trial in 49 infants and children under 5 years old, who were given one of four infusion rates (0.25, 1.0, 2.5, or 5.0 micrograms/kg/minute) (2145). Blood glucose, cortisol, and neuropeptide Y concentrations were used as indicators of stress. An infusion rate of 1 microgram/kg/minute was considered a suitable starting rate. Of the 49 patients, nine had significant bradycardia or hypotension requiring intervention. Four of these were neonates with complex cardiac anatomy, and remifentanil should be used with caution in these cases.

In another randomized, double-blind, placebo-controlled study, 25 cardiac surgical patients aged 55–70 years received either remifentanil 0.5 micrograms/kg/minute or placebo during surgery (2146). The patients who were given remifentanil had a lower cardiac output, a lower left ventricular stroke work index (LVSWI), and lower mixed venous oxygen saturation (SvO_2) in the early postoperative phase, suggesting postoperative cardiac depression. This occurred within the first 2 hours after termination of the remifentanil infusion, rather than during remifentanil administration. A possible explanation of this unexpected result is an opioid-related alteration in cardiac responsiveness to sympathetic discharge.

Remifentanil can cause bradycardia and hypotension. In 40 children, cardiac effects were monitored after the administration of remifentanil with or without atropine (2147). Remifentanil reduced blood pressure, heart rate, and cardiac index, even when atropine was added (2148). Glycopyrrolate 6 micrograms/kg prevented the bradycardia caused by remifentanil + sevoflurane anesthesia for cardiac catheterization in children with congenital heart disease (2149). Remifentanil attenuated the rapid rise in systolic blood pressure after electroconvulsive therapy. This sympathetic response can be harmful to patients who already have cardiac problems. Thus the cardiac effects of remifentanil may be beneficial in patients with compromised cardiac function.

Remifentanil-induced bradycardia and asystole may be useful as a protective effect in patients with atrial fibrillation (2150).

- A 90-year-old woman with atrial fibrillation was given digoxin for 3 days before surgery for a pelvic mass. Following anesthesia, induced with thiopental and atracurium, her heart rate rose to 105/minute and her electrocardiogram showed fast atrial fibrillation and ST segment depression. She was given remifentanil 0.25 micrograms/kg/minute and her heart rate fell to 95/minute.

Rifamycins

Shock and a flu-like illness (fever, chills, and myalgia) have been observed most often in patients taking intermittent rifampicin, dosages over 1000 mg/day, or on restarting treatment (2151,2152). Shock and cerebral infarction have been reported in an HIV-positive patient after re-exposure to rifampicin (2153).

Local thrombophlebitis can occur during prolonged intravenous administration (2154).

Risperidone

Cardiac arrest was attributed to risperidone in a patient with no history of cardiac disease (2155).

One of two children aged 29 and 23 months with autistic disorder developed a persistent tachycardia and dose-related QT_c interval prolongation while taking risperidone (2156).

Cardiotoxicity of risperidone has been discussed (2157) in the light of a death in a patient taking a therapeutic dose (SEDA-22, 68).

Risperidone is an alpha-adrenoceptor antagonist and can cause hypotension. This has been reported during anesthesia (2158).

- A 32-year-old woman, who had had a previous uncomplicated vaginal delivery with epidural anesthesia at age 21 and a cesarean delivery with a spinal anesthetic at 28, was at week 39 of her third gestation on treatment with risperidone 2 mg/day and lithium 1200 mg/day, which she had taken throughout pregnancy. In preparation for cesarean section she was given a spinal anesthetic with hyperbaric 0.75% bupivacaine 12 mg, fentanyl 10 micrograms, and preservative-free morphine 0.2 mg. Two minutes later, her blood pressure fell from 120/50 to 70/30 mmHg and did not resolve with 50 mg of ephedrine or 2 liters of Ringer's lactate solution, but did rise with phenylephrine 600 micrograms.

Elderly patients and those predisposed to orthostatic hypotension should be instructed in non-pharmacological interventions to reduce hypotension (for example, rising slowly to the seated position and sitting on the edge of the bed for several minutes before trying to stand up in the morning), and they should avoid circumstances that accentuate hypotension, including sodium depletion and dehydration (2159).

Long-acting depot risperidone can cause cardiac failure (2160).

The possible association of risperidone with QT_c interval prolongation is controversial. There were no significant changes in 73 patients with schizophrenia (mean age 34 years; 59% men) who took risperidone for 42 days (mean dose 3.7, range 4–6 mg/day) (2161).

Ritodrine

Ritodrine can cause bradycardia instead of the expected tachycardia (SEDA-8, 145); an unexpected hypertensive crisis has also been reported (SEDA-8, 145). ST segment depression is a consistent finding in patients during ritodrine infusion, and should therefore not always be interpreted as an indication of myocardial ischemia. The electrocardiographic changes are unrelated to the more generally accepted changes in heart rate, glucose, and potassium concentrations and are probably an intrinsic effect of the drug.

Rituximab

Cardiac dysrhythmias have been reported in 8% of patients treated with rituximab in patients with lymphomas (2162).

Acute coronary syndrome has been reported during rituximab therapy. It was speculated that release of cytokines during rituximab infusion caused vasoconstriction, platelet activation, or rupture of an atherosclerotic plaque (2163).

- A 71-year-old man, with a history of type II diabetes mellitus, hypertension, a myocardial infarction 12 years before, and percutaneous transluminal coronary angioplasty 11 years before, was given chlorambucil for B cell chronic lymphocytic leukemia without a response. Rituximab 375 mg/m^2 was therefore, started at a rate of 50 mg/hour intravenously after premedication with paracetamol, diphenhydramine, and methylprednisolone. Four hours later he developed substernal pain radiating to the neck and left arm with a sinus tachycardia; the pain responded to glyceryl trinitrate; the cardiac enzymes were not raised. Rituximab 30 mg/hour was restarted after an interval of 2 hours. He developed identical symptoms and rituximab was withdrawn.

Rivastigmine

Based on an electrocardiographic analysis of pooled data from four 26-week, phase III, multicenter, double-blind, placebo-controlled trials of rivastigmine ($n = 2791$), there were no adverse effects on cardiac function (2164). Rivastigmine can therefore be safely given to patients with Alzheimer's disease, without the need for cardiac monitoring.

Rocuronium bromide

Rocuronium has virtually no cardiovascular adverse effects (2165–2167). Minor increases in heart rate can occur with higher doses owing to its mild vagolytic properties.

There are several reports of pain during injection of rocuronium (2168,2169). Eight of 10 patients complained of severe pain, one complained of moderate pain, and another reported an unpleasant sensation (2168). This suggests that rocuronium will almost invariably cause pain. The mechanism of this phenomenon is not clear, but there appear to be some similarities to propofol injection pain. Several authors have suggested that rocuronium should not be given to awake patients (2168,2169). On the other hand, small doses of rocuronium have been used, with some success, to prevent fasciculations and myalgia after suxamethonium (2170–2173). With regard to the severity of injection pain, rocuronium pretreatment in awake patients does not seem advisable.

Ropinirole

Supraventricular extra beats have rarely been reported after low doses of ropinirole and have also been reported after pergolide and levodopa (2174). Symptomatic postural hypotension has occurred after even low oral doses of ropinirole (2–5), related to peripheral dopaminergic activity. Hypotensive effects occur within 3 minutes of standing, usually between 2 and 4 hours after an oral dose, associated with non-specific malaise (2175). Dizziness occurred in up to 40% of patients in clinical trials. Related symptoms include faintness, malaise, and yawning (2175). Bradycardia has occasionally accompanied postural hypotension (2176). Syncope has been reported.

Ropivacaine

See also Local anesthetics

The effects on the cardiovascular system of ropivacaine are similar to those of bupivacaine, although direct cardiotoxicity is less severe with ropivacaine than bupivacaine in both man and animals (SEDA-22, 143). Hypotension and bradycardia are prominent adverse effects when ropivacaine is used epidurally, particularly with concentrations of ropivacaine over 0.5% (SEDA-20, 129) (SEDA-22, 143); in one series, hypotension was observed in 30% of patients who received ropivacaine, but in only 13% of those given an equivalent dose of bupivacaine (2177).

Roxithromycin

- Torsade de pointes has been reported in an 83-year-old man who developed severe prolongation of the QT interval after taking roxithromycin 300 mg/day for 4 days (2178).

Salbutamol (albuterol)

See also Levosalbutamol

Peripheral vasodilatation and palpitation have been reported in patients using salbutamol. These effects usually appear with rapid administration of high doses. Inhalation of 0.4 mg did not affect the heart rate, whereas inhalation of 5 mg raised it by 15/minute (2179). Infusion of salbutamol at 0.025 mg/minute had a similar effect. Slight tachycardia can result in a mild rise in blood pressure, whereas at higher doses vasodilatation was more marked and the blood pressure fell. Infusion at 0.125 mg/minute reduced blood pressure by 30 mmHg. Salbutamol 10 puffs, administered from a pressurized aerosol using a spacer, produced significant bronchodilatation without adverse effects in mechanically ventilated patients. Heart rate rose significantly after 28 puffs (SEDA-21, 182).

Patients with severe hypoxemia and low serum potassium have an increased risk of developing extra beats and cardiac dysrhythmias (SEDA-2, 121).

Salbutamol causes an increase in the activity of the MB isoenzyme of creatine kinase, which has been interpreted as meaning that it might be cardiotoxic (2181).

The systemic vascular changes produced by the combination of hypoxia and inhaled salbutamol in eight healthy men with mild asthma in a randomized, double-blind, placebo-controlled, crossover trial have been briefly reported (2182). Forearm blood flow was measured non-invasively to obtain a measure of forearm vascular resistance (equating with systemic vascular resistance) after inhalation of placebo or salbutamol 800 micrograms under normoxic or hypoxic conditions. Both salbutamol alone and hypoxia alone produced small non-significant falls in forearm vascular resistance, but the combination of salbutamol and hypoxia resulted in a dramatic fall (30%). The authors pointed out that this fall in forearm vascular resistance was equivalent to the effects of glyceryl trinitrate 0.6–0.9 mg sublingually or felodipine 10 mg orally. This study emphasizes the marked cardiovascular changes that can occur with even small doses of salbutamol in people with mild asthma when hypoxia is present and confirms the importance of supplementary oxygen in these patients.

Intravenous and intracoronary salbutamol (10–30 and 1–10 micrograms/minute respectively) and intravenous isoprenaline (1–5 micrograms/minute), a mixed $beta_1$/$beta_2$-adrenoceptor agonist, were infused in 85 patients with coronary artery disease and 22 healthy controls during fixed atrial pacing (2183). Both salbutamol and isoprenaline produced large increases in QT dispersion (QT_{onset}, QT_{peak}, and QT_{end}), more pronouncedly in patients with coronary artery disease. Dispersion of the

QT interval is thought to be a surrogate marker for cardiac dysrhythmia (2184). The authors concluded that $beta_2$-adrenoceptors mediate important electrophysiological effects in human ventricular myocardium and can trigger dysrhythmias in susceptible patients.

In a blind, randomized study, 29 children aged under 2 years, with moderate to severe acute exacerbations of hyper-reactive airways disease, were treated with either a standard dose of nebulized salbutamol (0.15 mg/kg) or a low dose of nebulized salbutamol (0.075 mg/kg) plus nebulized ipratropium bromide 250 micrograms (2185). Standard and low-dose nebulized salbutamol was given three times at intervals of 20 minutes and nebulized ipratropium bromide was given once. Clinical improvement, measured as oxygen saturation and respiratory distress, was similar in both groups. QT dispersion was measured at baseline and after treatment and was significantly increased only by the standard dose of nebulized salbutamol.

Salbutamol-mediated activation of cardiac and peripheral $beta_2$-adrenoceptors, inducing positive chronotropic and inotropic effects and redistribution of coronary blood flow, have been discussed as possible causes of myocardial damage (2180).

- An 84-year-old woman with no history of cardiac disease and no risk factors developed an acute transmural anteroseptal myocardial infarction about 12 hours after beginning salbutamol and ipratropium bromide rescue therapy for acute exacerbated COPD. She was given five doses of salbutamol 5 mg plus ipratropium bromide 500 micrograms at 2-hour intervals. Arterial blood gas analysis showed a pH of 7.27, PaO_2 of 6.13 kPa (46 mmHg) and $PaCO_2$ of 8.53 kPa (64 mmHg) when breathing room air. Urgent coronary angiography showed no obstructive coronary artery disease or thrombosis. Salbutamol was suggested to be the probable cause of myocardial infarction in this patient.

As a note of caution, the hyperadrenergic state with acute dyspnea and severe hypoxemia should be considered as alternative causes of myocardial injury in this case.

Salmeterol

At a dose of 0.1 mg of salmeterol twice a day no change in heart rate or rhythm was seen on a 24-hour electrocardiogram (Holter). No change in blood pressure was seen and no electrocardiographic abnormality related to myocardial ischemia was recorded during 24-hour monitoring (2186).

Adverse events data have been pooled from seven randomized, double-blind, parallel-group, multiple-dose studies, to study the cardiovascular safety of salmeterol 50 micrograms bd in chronic obstructive pulmonary disease in 2853 patients for a median of 24 weeks (2187). The incidence of cardiovascular events (8%) was similar in the two groups. There were no episodes of sustained ventricular tachycardia or differences in heart rate, QT interval, or extra beats. The incidence of cardiovascular events increased with age, concurrent cardiovascular disease, and concurrent treatment with antidysrhythmic/bradycardic agents in both groups.

In January 2003, the US Food and Drug Administration (FDA) announced the premature termination of a placebo-controlled evaluation of the risk of serious adverse effects of salmeterol (2188,2189). An interim analysis of 26 000 asthmatic patients showed an excess of serious asthma attacks and asthma deaths in patients using salmeterol. The difference was significant in the subgroup of patients who were not using inhaled glucocorticoids. The high proportion of patients who were not using inhaled glucocorticoids (53% of the study population) is remarkable and not in agreement with current guidelines. This might have biased the results.

Sclerosants

Vascular thrombosis due to sclerosants is due to direct damage to vascular endothelium and red cells, aggregation of platelets, and aggregation of granulocytes at the venous wall endothelium. Where sodium morrhuate is concerned, these effects probably derive from its surfactant properties and its high arachidonate content (2190). Other sclerosant agents in current use (lauromacrogol 400, monoethanolamine oleate, and sodium tetradecylsulfate) cause effects similar to sodium morrhuate.

Cardiac arrest has been reported in a child who received lauromacrogol 400 to sclerose a symptomatic peripheral venous malformation (2191).

- A 5-year-old child with Klippel–Trenaunay syndrome, a cutaneous capillary malformation in the right leg, and a venous malformation of the lateral and posterior aspects of the right thigh and buttock had an injection of 4 ml of 1% lauromacrogol 400 into the malformation in the leg after oral premedication with midazolam 5 mg and atropine 0.5 mg and anesthesia with thiopental 80 mg and vecuronium bromide 2 mg for tracheal intubation. Shortly after the injection the patient developed rapidly progressive sinus bradycardia with eventual asystolic cardiac arrest. Anesthesia was discontinued and cardiopulmonary resuscitation, with 100% oxygen, external cardiac message, and intravenous orciprenaline 0.05 mg, was successful.

Selegiline

Unwanted effects of selegiline on cardiovascular regulation have been investigated as a potential cause for the unexpected mortality in the UK Parkinson's Disease Research Group trial (2192). Head-up tilt caused selective and often severe orthostatic hypotension in nine of 16 patients taking selegiline and levodopa, but had no effect on nine patients taking levodopa alone. Two patients taking selegiline lost consciousness with unrecordable blood pressures and another four had severe symptomatic hypotension. The normal protective rises in heart rate and plasma noradrenaline were impaired. The abnormal response to head up tilt was reversed by withdrawal of selegiline. The authors proposed that these findings might be due to either non-selective inhibition of monoamine oxidase or effects of amfetamine and metamfetamine.

Sertindole

Sertindole was associated with 27 deaths (16 cardiac) in 2194 patients who were enrolled in premarketing studies; further fatal cases have been collected, and the Committee on Safety of Medicines has described reports of 36 deaths (including some sudden cardiac deaths) and 13 serious but non-fatal dysrhythmias also associated with sertindole (2193).

Sertraline

Angina occurred in an elderly woman shortly after she started to take sertraline and on rechallenge (2194).

Sevoflurane

In 28 subjects given either sevoflurane + nitrous oxide or enflurane + nitrous oxide anesthesia, sevoflurane caused fewer cardiodepressant effects than enflurane (2195). Nevertheless, in 10 healthy subjects atrial contraction and left ventricular diastolic function, including active relaxation, passive compliance, and elastic recoil were impaired by sevoflurane (1 MAC) (2196).

Sevoflurane has a similar effect on regional blood flow to other halogenated anesthetics, although it is perhaps slightly less of a coronary artery vasodilator than isoflurane. It reduces myocardial contractility and does not potentiate adrenaline-induced cardiac dysrhythmias (2197). It also reduces baroreflex function, and in that respect is similar to other halogenated anesthetics. Coronary artery disease is not a risk factor for the use of these agents (2198).

In contrast to isoflurane and desflurane, sevoflurane tends not to increase the heart rate, and is usually well tolerated for induction of anesthesia in young children. However, profound bradycardia was reported in four unpremedicated children aged 6 months to 2 years during anesthesia induction with sevoflurane 8% and nitrous oxide 66% (2199). The episodes were not associated with loss of airway or ventilation. In three of the children there was spontaneous recovery of heart rate when the sevoflurane concentration was reduced; the other child received atropine because of evidence of significantly reduced cardiac output. In a previous study of sevoflurane induction of anesthesia in children with atropine premedication there was also a low incidence of this complication (2200), which is probably due to excessive sevoflurane concentrations.

The effects of sevoflurane on cardiac conduction have been studied in 60 healthy unpremedicated infants (2201). They received sevoflurane either as a continuous concentration of 8% from a primed circuit or in incrementally increasing doses. Nodal rhythm occurred in 12 cases. The mean duration of the nodal rhythm was 62 seconds in the incremental group and 90 seconds in the 8% group. All of the dysrhythmias were self-limiting and there were no ventricular or supraventricular dysrhythmias. No adverse events occurred as a result of the dysrhythmias. This study highlights the importance of using electrocardiographic monitoring when inducing anesthesia with volatile agents.

- Complete atrioventricular block occurred in a 10-year-old child with a history of hypertension, severe renal dysfunction, incomplete right bundle branch block, and a ventricular septal defect that had been repaired at birth (2202). After slow induction with sevoflurane and nitrous oxide 66%, complete atrioventricular block occurred when the inspired sevoflurane concentration was 3% and reverted to sinus rhythm after withdrawal of the sevoflurane. The dysrhythmia recurred at the end of the procedure, possibly caused by lidocaine, which had infiltrated into the abdominal wound, and again at 24 hours in association with congestive cardiac failure following absorption of peritoneal dialysis fluid.

Congenital or acquired forms of the long QT syndrome can result in polymorphous ventricular tachycardia (torsade de pointes). Many drugs, including inhalational anesthetics, alter the QT interval, and sevoflurane prolongs the rate-corrected QT interval (QT_c). In a randomized study of whether sevoflurane-associated QT_c prolongation was rapidly reversed when propofol was used instead, 32 patients were randomly allocated to one of two groups (2203). All received sevoflurane induction and maintenance for the first 15 minutes. In one group, sevoflurane was then withdrawn, and anesthesia was maintained with propofol for another 15 minutes; the other group continued to receive sevoflurane for 30 minutes. Sevoflurane-associated QT_c prolongation was fully reversed within 15 minutes when propofol was substituted.

Life-threatening dysrhythmias during anesthesia have been reported in patients with increased QT dispersion (QT_d), the difference between the longest and shortest QT intervals in any of the 12 leads of the electrocardiogram. Sevoflurane prolongs the QT_c and QT_d. In a prospective randomized study of the QT interval, the QT_c, the QT_d, and the QT_{cd} in preoperative, perioperative, and postoperative electrocardiograms in 90 adults undergoing non-cardiac surgery under general anesthesia, sevoflurane, desflurane, and isoflurane all prolonged QT_c, QT_d, and QT_{cd}, but there were no significant intergroup differences (2204).

The effects of single-breath vital capacity rapid inhalation with sevoflurane 5% on QT_c has been assessed in comparison with propofol in 44 adults undergoing laparoscopic surgery in a blind, randomized study (2205). Sevoflurane significantly prolonged the QT_c and seven patients developed ventricular dysrhythmias.

In a randomized trial the QT_c interval was measured in pre-, peri-, and post-operative electrocardiograms in 36 infants aged 1–6 months scheduled for inguinal or umbilical hernia repair (2206). Anesthesia was by either sevoflurane or halothane. There was prolongation of the QT_c interval during sevoflurane anesthesia (mean 473 ms) and 60 minutes after emerging from anesthesia (433 ms) compared with infants who received halothane. The JT_c

interval was analogously affected. The authors suggested that despite sevoflurane's shorter half-life, electrocardiographic monitoring until the QT_c interval has returned to preanesthetic values may increase safety after sevoflurane anesthesia.

Life-threatening dysrhythmias during anesthesia have been reported in patients with increased QT dispersion (the difference between the longest and shortest QT intervals in any of the 12 leads of the electrocardiogram). The effects of sevoflurane on QT dispersion have been compared with those of halothane in 50 children aged 5-15 years in a blind randomized study (2207). Neither sevoflurane nor halothane caused a significant increase in QT dispersion compared with baseline.

The effects of propofol and sevoflurane on the corrected QT (QT_c) and transmural dispersion of repolarization have been investigated in 50 unpremedicated children aged 1–16 years (2208). Sevoflurane significantly prolonged the preoperative QT_c; propofol did not. Neither anesthetic had any significant effect on the preoperative transmural dispersion of repolarization.

A case of torsade de pointes has been attributed to sevoflurane anesthesia (2209).

- A 65-year-old woman, who had had normal preoperative serum electrolytes and a normal QT interval with sinus rhythm, received hydroxyzine and atropine premedication followed by thiopental and vecuronium for anesthetic induction. Endotracheal intubation was difficult and precipitated atrial fibrillation, which was refractory to disopyramide 100 mg. Anesthesia was then maintained with sevoflurane 2% and nitrous oxide 50%. Ten minutes later ventricular tachycardia ensued, refractory to intravenous lidocaine, disopyramide, and magnesium. DC cardioversion resulted in a change to a supraventricular tachycardia, which then deteriorated to torsade de pointes. External cardiac massage and further DC cardioversion were initially unsuccessful, but the cardiac rhythm reverted to atrial fibrillation 10 minutes after the sevoflurane was switched off. Two weeks later she had her operation under combined epidural and general anesthesia, with no changes in cardiac rhythm.

In this case the role of excessive sympathetic drive as a result of the difficult intubation and the lack of opioid use during induction must be considered, even if sevoflurane played a role in precipitating the dysrhythmia.

Severe bradycardia has been described after sevoflurane induction for adeno-tonsillectomy in three children aged 42, 26, and 5 months with trisomy 21 (2210). Two had normal electrocardiography and echocardiography. The third had had a complete AV canal repaired early in life, and had first degree heart block but normal echocardiography. Severe bradycardia (40–44/minute from a baseline of 110–130) and hypotension occurred on induction of anesthesia. Two children responded to atropine and glycopyrrolate, but the third required adrenaline for resuscitation. The authors suggested that children with trisomy 21 should be premedicated with an anticholinergic agent either orally or intramuscularly.

Sibutramine

Cardiovascular risk factors associated with obesity, including dyslipidemia, particularly raised triglyceride concentrations and reduced high-density lipoprotein concentrations, can be improved with weight loss during sibutramine treatment (2211,2212,2213). In general, improvements in serum lipids are proportional to the degree of body weight loss, whether that weight loss occurs with sibutramine or with placebo (2213,2214).

In a study of the efficacy and safety of sibutramine in obese white and African Americans with hypertension, the most common adverse event resulting in withdrawal among those taking sibutramine was hypertension (5.3 versus 1.4% of patients taking placebo) (2215).

The observation that sibutramine, which blocks the reuptake of noradrenaline and serotonin and to a lesser extent dopamine (2216), causes a raised blood pressure has been a cause of concern (2217,2218). Some insight into this problem and its magnitude comes from two recent studies (2219,2220). Most studies have shown a positive relation between blood pressure and weight (2221). The failure of the blood pressure to fall with weight loss in normotensive and hypertensive patients treated with sibutramine differs from the fall seen with orlistat (2222–2224) or weight loss induced by life-style modifications (2225,2226). In the case of sibutramine, the potentially detrimental effect due to the failure of the blood pressure to fall with weight loss may be offset by the reductions in lipids, insulin, and uric acid that occur with weight loss (2227).

Intermittent use of sibutramine has been proposed to reduce potential concerns about its effect on blood pressure. An alternative strategy would be to identify those patients who respond to sibutramine with weight loss but who have minimal changes in blood pressure (2227).

The effects of sibutramine on weight loss, blood pressure, and pulse rate in hypertensive obese patients, whose blood pressure was well controlled with a beta-blocker either alone or with a thiazide diuretic, have been evaluated in a 12-week, double-blind, placebo-controlled, parallel-group, randomized study in 69 patients (2219). Sibutramine was effective and well tolerated and did not exacerbate pre-existing hypertension controlled with beta-blockers. Despite the presence of apparently effective beta-blockade, there were modest increases in pulse rate in those who took sibutramine, suggesting that mechanisms other than increased sympathetic tone may, at least in part, mediate this effect. Based on the potential for increased blood pressure and pulse rate, obese patients with well-controlled hypertension who are taking sibutramine should be monitored periodically (2219).

In a 52-week, multicenter, randomized, double-blind, placebo-controlled, parallel-group study in 220 hypertensive patients with obesity (BMI 27–40 kg/m^2), whose hypertension was well controlled with an angiotensin converting enzyme (ACE) inhibitor with or without a thiazide diuretic, sibutramine 20 mg/day safely and effectively achieved weight loss without compromising blood pressure control (2220). Blood pressure remained in the target range in patients who took sibutramine or placebo, although sibutramine was associated with a small mean increase in blood pressure and a modest increase in pulse rate.

Sibutramine did not affect heart valves or pulmonary artery pressure during 24 weeks in 106 obese patients (51 men and 55 women) with minimal tricuspid regurgitation on echocardiographic examination (2228). There were significant increases in blood pressure and heart rate.

Reversible cardiomyopathy has been reported in a patient taking sibutramine (2229).

- A month after starting to take sibutramine 15 mg/day, a 36-year-old obese man (BMI 38 kg/m^2) developed an upper respiratory tract infection. He developed progressively increasing fatigue and weight gain, and congestive heart failure was diagnosed. He had normal coronary arteries, a dilated left ventricle, and a low ejection fraction. Echocardiography showed some hypokinetic segments. Sibutramine was withdrawn and he was given diuretics, metoprolol, digoxin, ramipril, and low-dose aspirin. His clinical status improved but he continued to have NYHA class 1 symptoms.

A cause-and-effect relation could not be substantiated in this case, because there were other confounders, such as the upper airway infection 3 weeks before the onset of heart failure.

Silver salts and derivatives

A chronic inflammatory reaction in a prosthetic valve has been attributed to silver (2230).

- A prosthetic mitral valve (St Jude Medical Silzone) that had been implanted in a 72-year-old woman became partially detached 4 months later, causing acute cardiac failure. The mitral annulus was ulcerated and there were multiple erosions in the tissues in contact with the valve. Histology showed chronic inflammation, with hemosiderin deposits and giant cells. She was not allergic to silver.

The silver-coated sewing cuff had caused a chronic inflammatory reaction because of a toxic reaction to silver. The Silzone valve was withdrawn from the market in January 2000.

Sirolimus (rapamycin)

In a large multicenter trial, pre-emptive sirolimus therapy was evaluated after liver transplantation. The incidence of hepatic artery thrombosis was six of 110 patients taking sirolimus + tacrolimus and one of 112 taking tacrolimus monotherapy. Whether the increased incidence of hepatic artery thrombosis can be attributed to sirolimus is unclear, but the Food and Drug Administration released a mandated drug warning (2231).

After kidney transplantation or kidney–pancreas transplantation, 13 of 368 recipients (3.5%) developed biopsy-proven thrombotic microangiopathy in the absence of vascular rejection (2232). The incidence was highest in those who used ciclosporin + sirolimus (21%). The relative risk was 16 (95% CI = 4.3, 61) for patients who used ciclosporin + sirolimus compared with those who used tacrolimus + mycophenolate mofetil. The combination of ciclosporin + sirolimus was the only regimen that had both pronecrotic and antiangiogenic effects on arterial endothelial cells, suggesting that this combination produces thrombotic microangiopathy by dual effects on endothelial cell death and repair.

Two intestinal transplant recipients developed thrombotic microangiopathy while taking tacrolimus + sirolimus soon after starting to take sirolimus or increasing the dose; improvement occurred only after withdrawal (2233).

Smallpox vaccine

Acute myocarditis after vaccination against smallpox has been reported (2234). Fatal myocarditis is rare, but electrocardiographic evidence of myocarditis has been found more frequently; this adverse effect is probably not always noticed (2235–2237). Pericarditis after smallpox vaccination has also been described (2238).

All case reports of myocarditis/pericarditis after smallpox vaccination have been carefully evaluated. It was concluded that the data are consistent with a causal relation between myocarditis/pericarditis and smallpox vaccination; however, no causal association between ischemic cardiac events and smallpox has been identified (2239,2240).

- A 57-year-old woman with a history of hypertension, a transient ischemic attack, and carotid endarterectomy died 22 days after smallpox vaccination. Histopathological evaluation showed no evidence of cardiac inflammation.

Pericarditis after smallpox vaccination has been described (2238).

The Advisory Committee on Immunization Practices (ACIP) has recommended that people who have underlying heart disease, with or without symptoms, or who have three or more known major cardiac risk factors (that is hypertension, diabetes, hypercholesterolemia, heart disease at age 50 years in a first-degree relative, and smoking) should be excluded from the pre-event smallpox vaccination program (2241).

During the period of routine smallpox vaccination, only rare reports of cardiac inflammation (pericarditis, fatal myocarditis, and electrocardiographic evidence of myocarditis) were published in the world literature (SED-8, 709). To determine the risk of cardiac death after smallpox vaccination, death certificates were analysed from a period in 1947 when 6 million New York City residents were vaccinated after a smallpox outbreak; the incidence of cardiac deaths did not increase after the vaccination campaign (2242).

Somatostatin and analogues

Both somatostatin and octreotide cause transient increases in mean arterial pressure and mean pulmonary pressure when given intravenously to patients with cirrhosis, more marked with bolus administration than with continuous infusion (2243). This may be either direct or mediated by inhibition of gut vasodilatory peptides (SEDA-24, 505) (2244) and is not usually associated with significant clinical effects.

Severe hypertension with associated headache, nausea, and vomiting was reported within 2 weeks of administration of octreotide LAR 20 mg in a 26-year-old diabetic woman with autonomic neuropathy (2244). Rechallenge with octreotide 75 μg resulted in a transient hypertensive episode lasting 3 hours.

- Exacerbation of pre-existing hypertension was also reported in a 22-month-old boy during octreotide infusion (2245).

There has been one report of acute pulmonary edema during octreotide and intravenous fluid therapy for variceal bleeding (2246).

Sinus bradycardia (less than 50/minute) is reported in up to 25% of acromegalic patients taking octreotide, and conduction abnormalities are also commonly reported in these patients. This adverse effect is reported only rarely in other recipients of somatostatin or octreotide, probably reflecting the high rate of cardiac abnormalities due to acromegaly (2247).

Somatropin

The effect of somatropin on cardiovascular risk is complex, as growth hormone reduces visceral fat and total cholesterol and increases HDL cholesterol concentrations (2248–2250).

Edema, both generalized and peripheral, is common in adults given somatropin, as is hypertension. Symptoms are usually mild and resolve in many patients despite continuing treatment (2251). Increased left ventricular wall thickness has been reported in both adults and children, although in children this is thought to reflect an increase in overall mass and is not thought to be of clinical significance (2252,2253).

In an open study in five patients with severe dilated cardiomyopathy given high-dose somatropin 4 IU/day (1.3 mg/day) for 3 months, ventricular dysrhythmias worsened in all patients during treatment, from Lown class 2 or 3 to 4A or 4B, and returned to baseline when treatment was stopped (2254).

- A 7-year-old boy developed cardiomegaly and edema within a month of starting somatropin, 0.7 IU/kg/week; when the dose was reduced to 0.35 IU/kg/week his heart size returned to normal (2255).

This is a reminder that the adverse effects of somatropin are dose-related within the therapeutic range and that dose escalation should be gradual.

Sorafinib

Grade 3 hypertension is very common in patients taking sorafenib. The median time of onset in patients taking the anti-VEGF antibody bevacizumab was 131 (range 7–316) days. Twelve of 20 patients who took with sorafenib had a rise in systolic blood pressure of at least 20 mmHg compared with baseline, with a median change of 21 mmHg after 3 weeks (2256). There as a significant inverse relation between increased systolic blood pressure and a reduction in catecholamines, suggesting a secondary response to the increase in blood pressure.

Tyrosine kinase inhibitors of the VEGF and PDGF receptor pathways may target the VHL hypoxia-inducible gene pathway, which results in inhibition of hypoxia-inducible factor (HIF)-induced gene products. The latter mediate physiological responses of the myocardium to ischemia, including myocardial remodelling, peri-infarct vascularization, and vascular permeability.

Tyrosine kinase inhibitor-induced inhibition of HIF may be associated with more severe myocardial damage than previously expected. Of 73 patients with advanced renal cell carcinoma and normal creatine kinase MB fraction and cardiac troponin T at baseline, 23% had a significant increase in creatine kinase and troponin T after 2–32 weeks, with symptoms in seven (2257). One patient had an acute coronary artery occlusion and myocardial infarction. Electrocardiographic changes and biochemical markers are important indicators, and both should be measured regularly irrespective of whether sunitinib or sorafenib is used.

Sparfloxacin

See also Fluoroquinolone antibiotic

Some quinolones can prolong the QT interval, with a risk of cardiac dysrhythmias.

Sparfloxacin causes greater prolongation of the QT interval than other quinolones (2258), as has been shown in in vitro comparisons. Compared with grepafloxacin, moxifloxacin, and ciprofloxacin, sparfloxacin caused the greatest prolongation of the action potential duration (2259), and in an in vitro comparison of sparfloxacin, ciprofloxacin, gatifloxacin, gemifloxacin, grepafloxacin, levofloxacin, moxifloxacin, sitafloxacin, tosufloxacin, and trovafloxacin, sparfloxacin caused the largest increase in QT interval (2260). In an in vivo study in conscious dogs with stable idioventricular automaticity and chronic complete atrioventricular block, oral sparfloxacin 60 mg/kg caused torsade de pointes, leading to ventricular fibrillation within 24 hours, while 6 mg/kg did not (2261). In halothane-anesthetized dogs, intravenous sparfloxacin 0.3 mg/kg prolonged the effective refractory period, and an extra 3.0 mg/kg reduced the heart rate and prolonged the effective refractory period and ventricular repolarization phase to a similar extent, suggesting that a backward shift of the relative repolarization period during the cardiac cycle may be the mechanism responsible for the dysrhythmogenic effect of sparfloxacin. Data from in vitro electrophysiological studies of the effect of sparfloxacin, ofloxacin, and levofloxacin on repolarization of rabbit Purkinje fibers indicate that the prolongation of the action potential is observed only with sparfloxacin (2262).

In a double-blind, randomized, placebo-controlled, crossover study of a single oral dose of sparfloxacin in

15 healthy volunteers, prolongation of the QT interval was about 4% greater with sparfloxacin than with placebo (2263). An independent Safety Board concluded from the results of various phase I and phase II studies that the increase in the QT_c interval associated with sparfloxacin is moderate, averaging 3%, and that the few serious adverse cardiovascular events that have been reported during postmarketing surveillance all occurred in patients with underlying heart disease (2264).

In 25 patients taking sparfloxacin for multiresistant tuberculosis, there were six cases of moderate prolongation of the electrocardiographic QT interval (30–40 ms compared to baseline) without clinical symptoms (2265).

- A 37-year-old woman who was taking sparfloxacin as part of modified antituberculosis drug therapy developed torsade de pointes (2266).

Spiramycin

QT interval prolongation (2267) and torsade de pointes (2268) occur rarely in patients taking spiramycin.

Sufentanil

The incidence of hypotension with sufentanil is 7% and that of hypertension 3%. In a double-blind comparison of morphine, pethidine, fentanyl, and sufentanil in balanced anesthesia, patients who received sufentanil had the least hemodynamic disturbances. In high doses, adverse effects such as bradycardia and hypotension can lead to complications in some patients (SEDA-16, 85). Sudden hypotension occurred on induction of anesthesia with sufentanil in four patients, in whom the dose was 8.4–22.7 micrograms/kg (2269). Other workers noted similar findings at doses of 1 and 1.5 micrograms/kg.

Clinically significant bradycardia or asystole occurred on induction when sufentanil was used in conjunction with vecuronium (2270,2271).

The effect of sufentanil was examined in 10 healthy men to find out whether it has the same hemodynamic and sensory effects as when it is used in women in labor. Details of the method of recruitment of volunteers for this double-blind study were not provided, but they received either saline or sufentanil 10 micrograms intrathecally (2272), and blood pressure, heart rates, oxyhemoglobin saturation, cold and pinprick sensation, motor block, and visual analogue scales for sedation pruritus and nausea were all measured. Pruritus and sensory changes to pinprick and cold occurred only in the sufentanil group and there were no significant hemodynamic changes in either group. In view of the frequency and severity of pruritus when sufentanil is used in labor, it is interesting that all five of the male volunteers experienced this symptom, three of them severely. These findings suggested that the hypotension observed with the use of intrathecal sufentanil during labor and the sensory changes may not be mediated by the same pathway. The authors proposed that the hypotension observed in such studies is a direct result of pain relief, which is not an issue in the pain-free men in this investigation.

Sulfonamides

Of 98 patients with drug-induced long QT interval, one taking sulfamethoxazole carried a single-nucleotide polymorphism (SNP; found in about 1.6% of the general population) in KCNE2, which encodes MinK-related peptide 1 (MiRP1), a subunit of the cardiac potassium channel I_{Kr} (2273). Channels with the SNP were normal at baseline but were inhibited by sulfamethoxazole at therapeutic concentrations, which did not affect wild-type channels.

Cardiovascular reactions can be due to sulfonamide myocarditis or systemic vascular collapse, owing to severe adverse events such as widespread skin disease. Sulfonamide myocarditis has been described in relation to earlier sulfonamides and occurs in combination with other hypersensitivity reactions (2274).

Sulfonylureas

Early studies suggested that tolbutamide caused excessive cardiovascular deaths (2275) as reported in the University Group Diabetes Program (UGDP) (SEDA-4, 301). This result generated vigorous debate on the outcome and the methods used (SED-8, 917; SEDA-1, 317), and the results of the UGDP study are now widely regarded as impossible to be unanimously interpreted. The UK Prospective Diabetes Study did not find different mortality rates in patients treated with insulin, glibenclamide, or chlorpropamide (2276).

However, the question of whether sulfonylureas have a negative effect during myocardial infarction and survival during infarction has lingered. New data on the sulfonylurea receptor as part of the ATP-dependent potassium channel (SUR-1 in the beta pancreatic cells, involved in insulin secretion, and SUR-2 in the myocardium, involved in cardiac adaptation during ischemia) has still not yielded a definitive answer. The available experimental and clinical data have been systematically reviewed (2277). The conclusion was that experimentally the effects of sulfonylureas on heart muscle are both deleterious and protective for glibenclamide while tolbutamide, glimepiride, and gliclazide have no effects. There seem to be no adverse cardiac consequences of chronic treatment with sulfonylureas.

This effect of tolbutamide has been attributed to prevention of ischemic preconditioning, a protective manoeuvre that reduces myocardial damage after temporary stoppage of coronary blood flow (2278). Transient myocardial ischemia augments post-ischemic myocardial function and prevents dysrhythmias. K_{ATP} channels play a role in this so-called ischemic preconditioning. In 48 atrial trabeculae, obtained during catheterization, the recovery of the developed muscle force in patients treated with a sulfonylurea was only half of what was found in nondiabetics or diabetics treated with insulin. This suggests that inhibition of K_{ATP} channels by oral sulfonylureas might contribute to increased cardiovascular mortality (2279). The question of whether the findings obtained mainly with glibenclamide can be generalized to all sulfonylureas has been discussed, since sulfonylureas differ greatly in their ability to interfere with vascular or cardiac K_{ATP} channels (2280). Both glimepiride and glibenclamide (2281) improved glycemia in 29 patients in a randomized, double-blind, placebo-controlled, crossover study of a number of susceptibility factors for ischemic heart disease (plasminogen activator inhibitor activity, plasminogen activator inhibitor antigen, LDL cholesterol, C peptide, proinsulin, des-31,32 proinsulin, etc.), but K_{ATP} channels were not investigated.

Glibenclamide prevented the increase in tolerance to myocardial ischemia normally observed during the second of two sequential exercise tests (2282).

In a randomized study of 48 patients with type 2 diabetes, mean age 58 years, those who took glibenclamide for 8 weeks had a mean increase in systolic blood pressure of 3.1 mmHg (2283). In the same study the systolic and diastolic blood pressures fell in those who took rosiglitazone. The authors speculated whether changes in insulin concentrations and sympathetic activity were responsible.

Sulindac

Several cardiac abnormalities have been reported, including congestive heart failure, dysrhythmias, and palpitation (2284).

Sulpiride

Eight patients with schizophrenia who had been free of medication for at least 2 weeks took sulpiride 15 mg/kg for 2 weeks and then haloperidol 0.25 mg/kg for another 2 weeks (1220). QT_c intervals during sulpiride treatment were significantly prolonged by 5.1% and 8.5% compared with haloperidol and no treatment.

Sulprostone

See also Prostaglandins

Several experimental studies have provided support for the hypothesis that coronary spasm plays a major role in the pathophysiology of myocardial infarction during the administration of sulprostone. However, the possibility of myocardial infarction is not mentioned in the product information.

- Two cases of myocardial infarction (one fatal) have been reported in patients receiving sulprostone with mifepristone (2285,2286).
- Myocardial infarction has been reported in a woman aged 35 years with normal coronary arteries and good left ventricular function (2287).
- A 30-year-old woman developed uterine atony and bleeding after induced abortion because of fetal death at 17 weeks of gestation (2288). Sulprostone was given intravenously at a rate of 500 micrograms/hour. When additional sulprostone was injected into the uterine cervix, the patient sustained a myocardial infarction, with ventricular fibrillation and cardiocirculatory arrest, most probably due to coronary artery spasm. She was resuscitated and recovered completely.

Sulprostone should be used with care, particularly in patients with cardiac risk factors, and only in settings equipped to manage complications.

Cardiac dysrhythmias have been reported after the administration of misoprostol.

- A 38-year-old woman developed complete heart block, ventricular fibrillation, and subsequent asystole about 7 minutes after intravenous sulprostone 30 micrograms over 5 minutes, after she had previously been given a total dose of intramyometrial sulprostone 500 micrograms at seven different points for postpartum hemorrhage after cesarean section (2289).

The time-course suggested that the most likely cause of the arrest was the intravenous sulprostone. Contributory causes may have been hemorrhagic shock, electrolyte abnormalities, and hypothermia (from massive blood transfusion).

- Cardiac arrest occurred in a 39-year-old woman 3.5 hours after the administration of sulprostone 250 micrograms directly into the uterine wall for postpartum hemorrhage after manual removal of the placenta (2290). She had specific contraindications to sulprostone, as formulated by the French authorities: age over 35 years, heavy cigarette smoking, and cardiovascular risk factors.

In the Netherlands, sulprostone is registered for intravenous administration only. The authors strongly advised against administration directly into the uterine wall.

Suramin

The use of suramin plus hydrocortisone and androgen deprivation and the use of multiple courses of suramin have been assessed in 59 patients with newly diagnosed metastatic prostate cancer (2291). Suramin (doses aimed at plasma concentrations between a trough of 150 μg/ml and a peak of 250 μg/ml) was given in a 78-day fixed dosage schedule (one cycle) and suramin treatment cycles were repeated every 6 months to a total of four cycles. There was significant broad-spectrum toxicity throughout the study, leading to withdrawal of treatment in 33 patients. Cardiovascular events (dysrhythmias, hypotension, and congestive heart failure), neurotoxic effects, and respiratory effects were more frequent than expected. In consequence, repeated courses of suramin could be given in a minority of cases only. The authors felt that in the light of the relatively non-toxic palliation achieved with standard hormonal therapy, suramin in this dosage schedule has only limited use in patients with newly diagnosed metastatic prostate cancer.

The effects of fixed-dose suramin plus hydrocortisone have been studied in 50 patients with hormone-refractory prostate cancer (2292). Suramin was initially given as a

30-minute test infusion of 200 mg. In the absence of allergic reactions, additional 24-hour intravenous infusions of 500 mg/m^2 were given daily for the next 5 days. Thereafter, 2-hour intravenous infusions (350 mg/m^2) were given weekly on an outpatient basis for 12 weeks or until disease progression. The median duration of response was 16 weeks and the median time to disease progression 13 weeks. Fatigue and lymphopenia were the most commonly reported adverse effects, in 27 patients (54%) and 39 patients (78%) respectively. Skin rash occurred in 12 patients (24%). Suramin was withdrawn in three patients because of acute renal insufficiency ($n = 2$) and Stevens–Johnson syndrome ($n = 1$).

In a randomized study in 390 patients suramin has been given in a fixed low dose (3.192 g/m^2), intermediate dose (5.320 g/m^2), or high dose (7.661 g/m^2) to determine whether its efficacy and toxicity in the treatment of patients with hormone-refractory prostate cancer is dose-dependent (2293). There was no clear dose–response relation for survival or progression-free survival, but toxicity increased especially with the higher dose. There were neurological adverse effects in 40% of the patients and cardiac adverse effects in 15%. This raises questions about the usefulness of suramin, particularly in high doses, in advanced prostate cancer. However, in another phase I study of suramin with once- or twice-monthly dosing in patients with advanced cancer, suramin was relatively safely administered without using plasma concentrations to guide dosing (2294). Dose-limiting toxic effects included fatigue, neuropathy, anorexia, and renal toxicity. Diffuse colitis, erythema multiforme, and hemolytic anemia were reported as unusual effects.

Suxamethonium

Bradycardia and other dysrhythmias are common (80% in some series) and occur after the first and subsequent injections of suxamethonium in infants and children. In adults, these effects are seen more commonly after second or later injections, particularly when the interval between the doses is 2–5 minutes. However, it has been suggested that bradycardia and asystole may now be more frequently seen than previously in adults after a single injection of suxamethonium, as a result of the increased use of fentanyl or the omission of atropine beforehand (2295). Nodal rhythm and wandering pacemaker are frequent. The bradycardia is sometimes extreme (asystolic periods of 15–30 seconds duration have been reported). Usually these minor dysrhythmias revert to normal after a few minutes. Halothane can prolong their presence. The incidence of bradycardic asystole is not known, as atropine (the effective therapy) is usually quickly given.

Over the years cardiac arrest in apparently healthy children has occurred unexpectedly, most cases having been attributed to suxamethonium-induced hyperkalemia in patients with previously undetected myopathies (2296–2306). Several children have died of this complication. A diagnosis of Duchenne dystrophy or another unspecified progressive myopathy was made in 80% of the patients reported to the American Malignant Hyperthermia Registry who were subsequently tested for myopathies (2307). Pointing out that hyperkalemia was detected in 72% of the patients from whom blood samples were taken, the authors suggested that calcium, sodium bicarbonate, hyperventilation, and glucose and insulin should be considered for the treatment of anesthesia-related cardiac arrest in children. This is certainly good advice. Standard resuscitative efforts in such cases are often ineffective, as severe hyperkalemia prevents the restoration of a stable cardiac rhythm. It should be stressed that resuscitative efforts should not be stopped until hyperkalemia has been aggressively treated. Excessive doses of adrenaline, calcium, sodium bicarbonate, and glucose/insulin may be required. Peritoneal dialysis (2308), hemodialysis (2309), and cardiopulmonary bypass (2310) have been used successfully to treat suxamethonium-induced hyperkalemic cardiac arrest.

Regarding the risk of this rare but life-threatening complication in children with undetected myopathy or muscular dystrophy, it has been suggested that the routine use of suxamethonium in pediatric anesthesia be abandoned. It should be reserved for emergency intubation or when immediate securing of the airway is necessary.

Tachycardia and a rise in blood pressure are occasionally seen. Other supraventricular and ventricular dysrhythmias are much less common. Ventricular fibrillation associated with suxamethonium is usually the result of hyperkalemia, but has also been reported in hypercalcemia (2311) and is often seen in the course of malignant hyperthermia. Atropine, especially when given intravenously just before suxamethonium, is the most effective agent for the prevention of dysrhythmias. Hexafluorenium, D-tubocurarine, pancuronium, and other non-depolarizer blockers have also been reported as being effective in prevention. Severe hypotension can occur in patients with anaphylactoid reactions.

On theoretical grounds, suxamethonium, being akin to acetylcholine, should produce effects not only at the neuromuscular junction, but also at autonomic ganglia, at muscarinic receptors, and at postganglionic parasympathetic receptors. However, these other types of cholinoceptors are not so sensitive to its action. Nevertheless, stimulation of sympathetic ganglia has been invoked as being possibly responsible for the tachycardia and rise in blood pressure that sometimes occur transiently after its use. Likewise, stimulation of parasympathetic ganglia or direct stimulation of cardiac muscarinic receptors may be responsible for bradycardia. Differences in resting sympathetic and vagal tone have been said to account for the more frequent occurrence of tachycardia in "vagotonic" adults and bradycardia in "sympathotonic" children. The transient mild rise in blood pressure is possibly the result of the initial fasciculation, inducing an increase in venous return, which may also reflexly result in a slowing of the heart rate. Stimulation of afferent receptors in the carotid sinus has also been claimed to cause reflex bradycardia. Small doses (20–25 mg) are said to convert nodal to sinus rhythm, and larger doses to depress the sinoatrial node and so to cause bradycardia and nodal rhythm. Fasciculation probably produces an increase in afferent discharge from muscle spindles, which may account for the reported arousal pattern on the electroencephalogram; this in turn is postulated as a cause of tachycardia and a rise in blood pressure.

It has been hypothesized that suxamethonium modulates noradrenaline release from postganglionic sympathetic nerve terminals by presynaptic nicotine (+) and muscarinic (−) receptors on these nerve terminals (2312). The refractory period of these presynaptic nicotinic receptors is postulated as being longer than that of the muscarinic receptors, which results in a net muscarinic effect (bradycardia) after a second injection of suxamethonium within 4–5 minutes of the first. To explain the occurrence of bradycardia after an initial injection of suxamethonium in young children, it is postulated that sympathetic nerve terminals mature later, so that muscarinic (bradycardic) effects are unopposed by noradrenaline secretion in younger patients.

Some controversial correspondence has followed the report of four cases of fatal cardiac arrest among 150 patients who were given suxamethonium by paramedics in out-of-hospital emergencies (2313). The authors suggested that this might militate against suxamethonium-facilitated endotracheal intubation in this setting. Others, however, have argued that there was no evidence for a causal role of suxamethonium in those cases (2314). Patients with critical conditions, such as respiratory failure requiring endotracheal intubation, may have a cardiac arrest without being given suxamethonium. Furthermore, undetected esophageal intubation was considered to be an alternative explanation of cardiac arrest. Indeed, when endotracheal intubation is attempted in these often dramatic and stressful circumstances by health-care providers who have no routine experience in this, there may be a high rate of esophageal intubation. In one study 18 of 108 patients who had been intubated by paramedics were found to have the tube in their esophagus (2315). So the role of suxamethonium in the above report is questionable. On the other hand, suxamethonium is part of the protocol for emergency intubation in many centers worldwide and suxamethonium-associated cardiac arrest, apart from anecdotal instances, has not been reported to be a relevant problem (2316). Suxamethonium may increase the success rate of emergency intubations while reducing the incidence of traumatic intubations (2317). Therefore, rapid-sequence intubation with an induction agent such as etomidate and suxamethonium is probably still the technique of choice for airway management in emergencies. Whoever uses this technique must be aware of contraindications to suxamethonium and must have frequent practice in endotracheal intubation.

Tacrolimus

See also Sirolimus

The incidence of hypertension and the prevalence of antihypertensive drug use are lower with tacrolimus than with ciclosporin (2318,2319).

Severe recurrent, but usually reversible hypertrophic cardiomyopathy has been infrequently reported, both in adults and children (SEDA-19, 352; SEDA-20, 346). Based on experimental data and one additional case report, the interaction of tacrolimus with calcium channel blockers in the cardiac muscle has been suggested as a possible mechanism (SEDA-21, 390). However, the role of tacrolimus in the development of cardiomyopathy is still hypothetical. Echocardiographic abnormalities were relatively common before and after liver transplantation in 12 adult patients, and there was no clear evidence that oral tacrolimus specifically alters cardiac function (2320). Other investigators did not show differences in heart weight, ventricular thickness, or valve circumferences between 67 liver transplant recipients treated with tacrolimus and 72 non-transplanted patients who died from end-stage liver disease (2321). In addition, more than 80% of patients in both groups had left ventricular hypertrophy.

Other isolated reports and preliminary studies have suggested a possible risk of life-threatening dysrhythmias, including sinus bradycardia or sinus arrest, asymptomatic but significant mean QT/QT_c interval prolongation in 33 patients (over 500 msec in seven patients), and recurrent episodes of ventricular tachycardia or torsade de pointes in two patients (SEDA-21, 391; SEDA-22, 390; 2322,2323).

Several reports have previously focused on the possible occurrence of cardiomyopathy in tacrolimus-treated transplant patients, particularly children. In two further liver transplant children aged 2.5 and 14 years who died from multiorgan system failure due to sepsis and end-stage liver failure, pathological examination showed prominent concentric left ventricular hypertrophy (2324). Although tacrolimus was regarded as a possible cause of asymptomatic hypertrophic cardiomyopathy in these patients, a direct causal relation was difficult to establish. The cause is probably multifactorial, and potential confounding factors (for example hypertension, glucocorticoids) are numerous in this population. In a retrospective review of 89 pediatric heart transplant patients who had survived for at least 6 months, repeated echocardiography showed signs of cardiac hypertrophy, particularly early after transplantation and in very young infants (2325). However, there was no evidence of progressive hypertrophy on follow-up examinations, and no significant differences in the degree of cardiac hypertrophy between patients aged over 1 year at the time of transplantation who received ciclosporin ($n = 26$) or tacrolimus ($n = 41$).

In a retrospective study, the prevalence of hypertension 2 years after adult liver transplantation was significantly lower in patients treated with tacrolimus (64% of 28 patients) than in patients treated with ciclosporin (82% of 131 patients) (2326). In addition, hypertension occurred later with tacrolimus. A similar benefit of tacrolimus over ciclosporin was found in a randomized, comparative trial in 85 heart transplant patients, and 41% of 39 tacrolimus-treated patients developed new-onset hypertension requiring treatment, compared with 71% of 46 ciclosporin-treated patients (2327).

In 37 patients with liver transplants there was no difference between the pre- and post-transplant QT interval in the 25 taking oral ciclosporin and the 12 taking oral tacrolimus (2328).

Cardiac symptoms manifesting as myocardial ischemia are uncommon, but can occur through tacrolimus toxicity (2329).

- A 20-year-old woman with chest pain, dyspnea, and protracted electrocardiographic ST depression had very high blood tacrolimus concentrations (45 ng/ml). Subsequent coronary angiography ruled out any significant organic lesions, but showed vasospastic coronary arteries. She had no other cardiac symptoms when tacrolimus was restarted with careful surveillance of serum concentrations.

When ciclosporin was replaced by tacrolimus in 22 adult renal recipients with serum total cholesterol concentrations greater than 5.2 mmol/l (200 mg/dl) more than 1 year after kidney transplantation, there was a significant improvement in fibrinogen, total cholesterol, and low-density lipoprotein cholesterol after conversion (2330). Low-dose tacrolimus may be preferable to ciclosporin for chronic maintenance immunosuppression because it improves the overall cardiovascular risk profile without any apparent adverse effects.

Tacrolimus-associated thrombotic microangiopathy resolved after conversion to sirolimus (2331).

- A 29-year-old man received a kidney–pancreas transplant for end-stage diabetic nephropathy. After induction with basiliximab, immunosuppression consisted of prednisone + tacrolimus + mycophenolate mofetil. Thrombotic microangiopathy occurred 24 days after transplantation. The patient was successfully converted from tacrolimus to sirolimus and was treated with plasma infusion. Allograft biopsy showed focal glomerular and arteriolar acute thrombosis without evidence of rejection.

Talipexole

Sinus bradycardia and hypotension have been attributed to talipexole (2332).

- About 4 hours after a 65-year-old man with Parkinson's disease took talipexole hydrochloride 0.8 mg, he acutely developed sleepiness, delusion, akinesia, and faintness associated with hypotension and sinus bradycardia. A similar episode occurred when he took talipexole hydrochloride 1.2 mg/day in combination with co-careldopa (levodopa 200 mg/day plus carbidopa 20 mg/day). These symptoms persisted for 12 hours and abated gradually without any specific treatment.

The authors suggested that talipexole had caused bradycardia and hypotension by stimulating both D_2 dopamine receptors and alpha$_2$-adrenoceptors.

Tamoxifen

Both deep vein thrombosis and pulmonary embolism have been described with tamoxifen.

- Cerebral sinus thrombosis, progressing to hemorrhagic cerebral infarction, occurred in a 52-year-old woman (2333).

Although the authors pointed to the absence of risk factors other than the drug, it must be remembered that cerebral venous thrombosis is a recognized complication of various malignancies. In this case the breast tumor had been treated with various cytostatic drugs and stem cell transplantation, and tamoxifen had been given as an adjuvant, and it was believed that the tumor had been eliminated. Nevertheless, in this complex case one should perhaps be hesitant in attributing the complication solely to the drug.

In the light of three further cases of thrombosis, it has been suggested that there may be a particular predisposition to this complication in patients with high circulating concentrations of homocysteine, and that these should be checked for in advance of treatment (2334).

However, some such incidents may have been attributable to the primary condition being treated, and the risk must not be over-estimated (2335).

The effect of tamoxifen 20 mg/day on the incidence of venous thromboembolism has been assessed in a placebo-controlled breast cancer prevention trial for 5 years in 5408 hysterectomized women (2336). There were 28 incidents of thromboembolism on placebo and 44 on tamoxifen (hazard ratio = 1.63; 95% CI = 1.02, 2.63), 80% of which involved only superficial phlebitis, which accounted for all of the excess due to tamoxifen within 18 months from randomization. Compared with placebo, the risk of venous thromboembolism with tamoxifen was higher in women aged 55 years or older, those with a body mass index of 25 kg/m^2 or more, those with a raised blood pressure or a total cholesterol of 6.50 mmol/l (250 mg/dl) or greater, current smokers, and those with a family history of coronary heart disease, all familiar risk factors for venous complications. Of the 685 women with a coronary heart disease risk score of 5 or greater, one in the placebo arm and 13 in the tamoxifen arm developed venous thromboembolism. In a multivariate regression analysis, age in excess of 60 years, height of 165 cm or more, and a diastolic blood pressure of 90 mmHg or higher all had independent detrimental effects on the risk of venous thromboembolism during tamoxifen therapy, whereas transdermal estrogen therapy concomitant with tamoxifen was not associated with any excess risk (HR = 0.64; 95% CI = 0.23, 1.82). The authors concluded that the increased risk of venous thromboembolism during the use of tamoxifen was largely associated with the well-known risk factors for this condition, and that this information should be part of pre-treatment counselling.

In one case tamoxifen was associated with myocardial infarction (2337).

Taxaceae

Of 11 197 exposures to the berries of *Taxus* species children under 12 years of age were involved in 96% (under 6 years 93%; 6–12 years 3.7%) (2338). When the final outcome of the exposure was documented ($n = 7269$), there were no adverse effects in 93% and minor effects in 7.0%. There were moderate (more pronounced, but not life-threatening) effects in 30 individuals and major (life-threatening) effects in four. There were no deaths. Decontamination therapy had no impact on outcome compared with no therapy. When symptoms occurred after exposure to *Taxus*, the most frequent were gastrointestinal (66%), followed by skin reactions (8.3%), nervous system effects (6.0%), and cardiovascular effects (6.0%).

Taxanes

In 58 patients with breast cancer there were no significant changes in cardiac function after a single dose of docetaxel 100 mg/m^2, but supraventricular extra beats occurred in those who received docetaxel 75 mg/m^2 and epirubicin 75 mg/m^2 (2339). However, the authors were unable to attribute any real clinical significance to these observations.

During a phase 2 study, a weekly regimen containing paclitaxel and trastuzumab had beneficial activity in anthracycline- and taxane-pretreated patients with HER2 overexpression. Since cardiac dysfunction (grade 3) was observed frequently, monitoring of cardiac function is warranted during combined use of these two drugs (2340).

Teicoplanin

Hypotension has been observed as part of the manifestations of a hypersensitivity reaction to teicoplanin (2341).

Terfenadine

Terfenadine can cause prolongation of the QT interval in overdose or during interactions with drugs that inhibit its metabolism, such as erythromycin (2342), ketoconazole (2343), and grapefruit juice (2344). Torsade de pointes can result (2345).

Terodiline

A retrospective analysis of data from 19 patients has shown that the effect of terodiline on the QT interval is dose-related. The authors commented that terodiline has structural similarities to prenylamine, an antianginal drug that was previously shown to have prodysrhythmic properties, so that such problems should really not have been unexpected (SEDA-20, 147).

Tetanus toxoid

Myopericarditis has been attributed to Td-IPV vaccine (2346).

- A 31-year-old man developed arthralgia and chest pain 2 days after Td-IPV immunization and had an acute myopericarditis. He recovered within a few days with high-dose aspirin.

The authors discussed two possible causal mechanisms, natural infection or an immune complex-mediated mechanism. Infection was excluded by negative bacterial and viral serology and a favorable outcome occurred within a few days without antimicrobial treatment.

Tetracaine

See also Local anesthetics

- An 18-month-old child undergoing cardiac surgery developed discoloration of the hand, consistent with severe bruising, after application of 4% tetracaine gel, which was inadvertently left under an occlusive dressing for about 24 hours (2347). There were no long-term sequelae and no treatment was required.

The authors blamed a combination of the vasodilatory properties of tetracaine and the fact that the child was heparinized for surgery, causing capillary leak at the area of application.

Tetrachloroethylene

In cases of poisoning with tetrachloroethylene, hypotension can occur; sympathomimetic drugs should not be used to treat it, since ventricular fibrillation can be precipitated.

Tetracyclines

Cardiovascular reactions to tetracyclines have often been associated with other symptoms of hypersensitivity, such as urticaria, angioedema, bronchial obstruction, and arterial hypotension (2348,2349). Such reactions occurred in patients who had tolerated tetracyclines previously and were therefore considered as anaphylactic.

Thalidomide

Bradycardia, hypotension, orthostatic hypotension, and dizziness have been reported with thalidomide; these effects may have been due to its central sedative effects or to vasovagal activation (2350).

Of 96 patients taking thalidomide 52 had a heart rate below 60/minute at some time during follow-up and 10 developed symptomatic bradycardia; the symptoms abated with reduction of the dose in most cases (2351).

Diabetic microvascular disease has been attributed to thalidomide (2352).

- A 57-year-old man with type I diabetes mellitus started taking thalidomide after failed stem cell transplantation for multiple myeloma and after 10 months developed a sensory neuropathy with loss of pain sensation and ischemic changes in the legs.

The authors suggested that the antiangiogenic properties of thalidomide were implicated in the pathogenesis of diabetic foot disease in this patient.

Thromboembolic disease

> DoTS classification (BMJ 2003;327:1222–5)
> Adverse effect: Deep venous thrombosis with thalidomide
> Time-course: intermediate
> Susceptibility factors: genetic; diseases (multiple myeloma, lupus erythematosus); drugs (chemotherapy, especially doxorubicin; darbepoetin)

Deep venous thrombosis is a common complication of thalidomide, especially when it is used in combination with chemotherapy and in patients with multiple myeloma (2353–2355).

Arterial thrombosis has also been reported. Of 23 patients with myeloma who took thalidomide 150 mg/day for 142 patient-months there were seven cases of thrombosis, five venous and two arterial; in a historical control group of 18 similar patients who did not take thalidomide there was only one case over 289 months (2355,2356).

Incidence

Of 50 patients with multiple myeloma, 14 developed a deep venous thrombosis after taking thalidomide 400 mg/day, compared with two of 50 patients who did not take it (2356) All the episodes occurred during the first 3 cycles of therapy. One patient taking thalidomide had a pulmonary embolus. Most of the patients continued to take thalidomide with the addition of low molecular weight heparin followed by warfarin and there was no progression of deep venous thrombosis.

Of 23 men with advanced androgen-dependent prostatic cancer who received docetaxel alone, none developed a venous thrombosis, compared with nine of 47 men who received docetaxel + thalidomide (2357).

Susceptibility factors

The risk of thrombosis from thalidomide may be increased in certain conditions, such as malignancies, cicatricial pemphigoid (2358), and systemic lupus erythematosus. Four episodes of thrombosis (two arterial, two venous) occurred in three patients with systemic lupus erythematosus and one with severe atopic dermatitis within 10 weeks of starting treatment with thalidomide 50–100 mg/day (2359). All four had at least one risk factor for thrombosis, but none had thrombosis before or after treatment with thalidomide.

The incidence of deep venous thrombosis is increased by the co-administration of doxorubicin, as suggested by a study in 232 patients with multiple myeloma who received a combination of thalidomide and chemotherapy in two protocols that differed only by the inclusion of doxorubicin in one: DT-PACE (dexamethasone + thalidomide + cisplatin + doxorubicin + cyclophosphamide + etoposide) and DCEP-T (dexamethasone + cyclophosphamide + etoposide + cisplatin + thalidomide) (2354). There was an increased risk of deep venous thrombosis in those who received DT-PACE but not in those who received DCEP-T. Multivariate analysis confirmed that those who received thalidomide + doxorubicin had an increased risk of deep venous thrombosis. In two separate trials in patients taking thalidomide for multiple myeloma, deep venous thrombosis occurred in four of 15 patients who received concomitant treatment with doxorubicin + dexamethasone compared with three of 45 who received dexamethasone only (2355).

Of 535 patients who received thalidomide with or without cytostatic chemotherapy, 82 developed a deep venous thrombosis (2360). Multivariate analysis showed that the combination of thalidomide with chemotherapy that contained doxorubicin was associated with the highest odds ratio (OR = 4.3). Newly diagnosed disease (OR = 2.5) and chromosome 11 abnormalities (OR = 1.8) were also independent predictors. After a median period of 2.9 years, survival was worse in those with chromosome 13 abnormalities, aged over 60 years, with a raised lactate dehydrogenase activity, and a raised serum creatinine concentration.

Three episodes of thrombosis occurred in patients who had taken thalidomide (25–100 mg/day) for up to 2 years (2361). However, all had other risk factors (heterozygous protein C resistance in one and surgical intervention or trauma in the others), so a causal role of thalidomide was debatable.

In a trial of thalidomide in Behçet's disease there was superficial thrombophlebitis in 10 of 32 patients taking thalidomide 100 mg/day, two of 31 patients taking thalidomide 300 mg/day, and three of 32 patients taking placebo (2362). The fact that this apparent increase in the incidence of thrombophlebitis in those taking 100 mg/day was not reproduced at the higher dosage suggests that the effect occurred by chance.

Management

Of 256 patients with myeloma randomized to thalidomide or not, 221 received no prophylactic anticoagulation and 35 received low-dose warfarin 1 mg/day (2360). The incidence of deep venous thrombosis was higher in those who took thalidomide (hazard ratio 4.5). Warfarin did not reduce the risk, and prophylactic subcutaneous enoxaparin 40 mg/day was therefore introduced in 68 patients of a subsequent group of 130 patients who received thalidomide. This intervention eliminated the difference in the incidence of deep venous thrombosis between those who took thalidomide and those who did not.

Theaceae

Camellia sinensis (green tea) contains caffeine and antioxidant polyphenols. It has been touted as being useful in a wide variety of conditions, including cancer prevention, mostly on relatively slim epidemiological evidence (2363), cardiovascular disorders, and AIDS.

Theophylline

The relation between toxicity and excessive serum theophylline concentrations has been confirmed, tachycardia being a frequent indication of toxic symptoms (SED-9, 3). It appears that the cardiac and metabolic effects of theophylline are at least partly related to catecholamine release (SEDA-13, 1). The importance of appropriate dosage has been stressed in several papers, all indicating the particular risk of ventricular fibrillation in subjects with respiratory distress (2364). The authors have pointed to the potentially lethal effect of serum theophylline concentrations in excess of 20 μg/ml. In another study, half the patients with serum concentrations greater than 35 μg/ml had life-threatening dysrhythmias (SEDA-7, 7). In one study of the effect of oral aminophylline on cardiac dysrhythmias in 15 patients with chronic obstructive pulmonary disease, aminophylline had both dysrhythmogenic and chronotropic effects but did not change the grade of dysrhythmia (SEDA-7, 7). In another study of 16 patients there was a relation between the use of theophylline and multifocal atrial tachycardia (2365). Whereas raised blood theophylline concentrations are likely in themselves to produce cardiovascular complications, patients with pre-existing cardiac disease are at greater risk.

Thiazolidinediones

Reductions in VLDL cholesterol, LDL cholesterol, and chylomicrons may contribute to a reduction in cardiac complications. Pioglitazone reduced both lipoprotein(a) and the remnant particles (cholesterol-rich particles after the release of triglycerides from the chylomicrons), whereas troglitazone caused increases in lipoprotein(a) (2366).

- A 74-year-old man with long-standing type 2 diabetes and compensated systolic dysfunction taking glibenclamide was given rosiglitazone 4 mg/day, increasing to 8 mg/ml after 1 month (2367). After 2 weeks he had weight gain of 5 kg, increased jugular venous pressure, shortness of breath, bibasal crackles, and a gallop rhythm. Subsequently he had a total weight gain of 17 kg and worse symptoms. When rosiglitazone was withdrawn his weight fell within 12 days to the pretreatment weight and the edema disappeared.

Cardiac dysfunction increases insulin resistance, suggesting that thiazolidinediones, which reduce insulin resistance might be a good choice in patients with diabetes and cardiac dysfunction. However, this case suggests that they can worsen fluid retention, perhaps by vasodilatation (2368).

The American Heart Association and the American Diabetes Association have published a consensus statement in which they stated that in patients with signs or symptoms of NYHA class III or IV cardiac failure thiazolidinediones should not be used and in class I or II they should be used cautiously, starting with a very low dosage (rosiglitazone 2 mg or pioglitazone 15 mg) (2369). Gradual dose escalation is warranted, with careful observation to identify weight gain, edema, or exacerbation of cardiac failure. Even if there are no signs of chronic heart failure, it can develop when thiazolidinediones are begun. When thiazolidinediones are combined with insulin, edema is more common.

In a retrospective study in diabetic patients taking various oral hypoglycemic agents from January 1995 to March 2001 in a large insurance company there were 5441 thiazolidinedione users and 28 103 non-thiazolidinedione users (2370). Those taking thiazolidinediones were younger but more likely to have coronary artery disease or complications of diabetes. The adjusted incidence of heart failure after 40 months was 8.2% in those taking thiazolidinediones and 5.3% in those not taking thiazolidinediones. The validity of the conclusion was discussed (2371) but other commentators supported the need to be vigilant for heart failure when prescribing thiazolidinediones (2372).

In 40 ambulatory hemodialysis patients, 25 of whom were taking pioglitazone and 15 rosiglitazone, there were no increases in intravascular volume, anemia, edema, or chronic heart failure in a retrospective study (2373). It may be that dialysis obviates any increase in intravascular volume. The use of these drugs during dialysis seems to be safe, although there were reductions in systolic and diastolic blood pressures.

Thionamides

Aplasia cutis congenita has been attributed to carbimazole, or its active metabolite thiamazole, given during early pregnancy (SEDA-14, 367) (2374), and a review revealed 16 cases of solitary skin defects associated with intrauterine exposure to thiamazole (2374). The defects can be restricted to a region of the body or can be widespread. Several causes have been documented, including chromosomal abnormalities (for example trisomy 13) and single gene mutations, such as Goltz syndrome. A few cases are believed to result from in utero exposure to teratogens, including thionamides. To date, 16 cases of solitary skin defects associated with intrauterine exposure to thiamazole have been reported. Additional cases of aplasia cutis congenita in thionamide-exposed infants associated with other congenital abnormalities, such as bilateral atresia of the nasal choana, esophageal atresia, imperforate anus, and cardiovascular defects have also been reported (2374).

Additional cases of aplasia cutis congenita in thionamide-exposed infants associated with other congenital abnormalities, such as choanal atresia, esophageal atresia, imperforate anus, and cardiovascular defects, were also reviewed. This pattern of abnormalities has previously led to the term "methimazole [thiamazole] embryopathy" (2375).

- A 3-year-old child, whose mother had been treated for Graves' hyperthyroidism with thiamazole throughout pregnancy, had two scalp lesions and other abnormalities of tissues of ectodermal origin, including dystrophic nails and syndactyly.

The authors suggested that a history of in utero exposure to thiamazole should be sought in all children with aplasia cutis congenita, as well as other ectodermal tissue abnormalities, to allow better definition of the "methimazole embryopathy." However, cautious interpretation of the literature is required, given the small number of thiamazole-associated cases of aplasia cutis congenita compared with the widespread prescription of this drug in pregnant women with hyperthyroidism. On the other hand, the absence of an apparent association with the use of the alternative thionamide, propylthiouracil, argues in favor of using the latter in pregnant patients.

In a prospective study in 241 women referred to a teratology service because of exposure to thiamazole during pregnancy, congenital abnormalities were compared with those found in offspring of 1089 controls referred because of exposure to non-teratogenic drugs or radiography (2375). There were no statistically significant differences between the two groups in terms of major abnormalities, gestational age at delivery, neonatal weight, or head circumference, but among the thiamazole-exposed infants two had a major malformation consistent with "methimazole embryopathy." One had choanal atresia (exposed at 4–7 gestational weeks) and the other had esophageal atresia (exposed at 0–16 gestational weeks) (2376). These are very rare malformations, and the number of cases in the cohort was insufficient to reach statistical significance, so the possibility of a chance association with thiamazole exposure cannot be excluded. These cases do, however, lend support to the view that thiamazole may be teratogenic, although thyrotoxicosis itself may be the associated factor. Until further data are available, treatment of thyrotoxicosis with propylthiouracil may be preferable in women who are planning a pregnancy.

- Scalp atresia has been described in an infant whose mother had taken carbimazole in a high dose (60 mg/day) during the first 12 weeks of pregnancy and propylthiouracil thereafter (2377). The infant had other dysmorphic features (a flat face, low-set ears, upper lip retraction, and a low-set fifth finger) in addition to transient hypothyroidism.
- Choanal atresia has been described in an infant whose mother presented in early pregnancy with Graves' hyperthyroidism and who took carbimazole in doses up to 60 mg/day in the first trimester (2378). She was also clinically and biochemically severely hyperthyroid at this time.

Propylthiouracil crosses the placenta as readily as the thioimidazoles, but the rare and probably real association between thioimidazoles and fetal anomalies makes the thioimidazoles less attractive first-line alternatives (2379–2381). When propylthiouracil is used cautiously in minimal amounts and with frequent dose adjustments, it is probably the safest form of treatment of hyperthyroidism during pregnancy (SEDA-8, 373) (SEDA-11, 357).

Thiopental sodium

Cardiovascular depression is a well-documented complication of thiopental. However, the plasma concentrations necessary to produce loss of corneal reflex and trapezius muscle tone were only minimally depressant to the heart (2382). Problems can in any case be reduced or avoided by proper fluid administration before induction of anesthesia, as well as by cautious choice of dosage and administration in patients with uncompensated cardiac failure.

Surgery for cerebral artery aneurysms sometimes requires cardiopulmonary bypass and deep hypothermic circulatory arrest if they are to be operated on safely. During such bypass procedures patients with such aneurysms often receive large doses of thiopental, in the hope of providing additional cerebral protection. In 42 non-cardiac patients thiopental loading to the point of suppressing electroencephalographic bursts caused only negligible cardiac impairment and did not impede withdrawal of cardiopulmonary bypass; however, there were no data on patients with cardiac disease (2383).

Thioridazine

Cardiac dysrhythmias

Thioridazine has been associated with QT interval prolongation (2384,2385) and several cases of torsade de pointes have been reported (2386,2387).

Two types of T wave changes have also been described after treatment with thioridazine: type I (with rounded, flat, or notched T waves) and type II (with biphasic T waves) (2388).

On July 7, 2000, doctors and pharmacists in the USA were notified about the addition of extensive new safety warnings, including a boxed warning to the professional product label for the neuroleptic drug thioridazine (Melleril, Novartis Pharmaceuticals). The text of the new boxed warning read: "Melleril (thioridazine) has been shown to prolong the QT_c interval in a dose-related manner, and drugs with this potential, including Melleril, have been associated with torsade de pointes-type arrhythmias and sudden death. Due to its potential for significant, possibly life-threatening, prodysrhythmic effects, Melleril should be reserved for use in the treatment of patients with schizophrenia who fail to show an acceptable response to adequate courses of treatment with other neuroleptic drugs, either because of insufficient effectiveness or the inability to achieve an effective dose due to intolerable adverse effects from those drugs."

The new labelling changes were based primarily on the FDA's review of three published studies. The first of these showed increased blood concentrations of thioridazine in patients with a genetic defect, resulting in the slow inactivation of debrisoquine (2389). In this study, 19 healthy subjects (six poor and 13 extensive metabolizers of debrisoquine) took a single oral dose of thioridazine 25 mg. The poor metabolizers reached higher blood concentrations of thioridazine 2.4 times more quickly than the extensive metabolizers. There was a 4.5-fold increase in the systemic availability of the drug in the poor metabolizers, in whom thioridazine remained in the blood twice as long. The second study showed dose-related prolongation of the QT_c interval from 388 (range 370–406) to 411 (range 397–425) ms 4 hours after an oral dose of thioridazine 50 mg (2390). The average maximal increase was 23 ms. This change was statistically greater than that for either placebo or thioridazine 10 mg. In the third study the effect of the selective serotonin re-uptake inhibitor (SSRI) fluvoxamine, 25 mg bd for 1 week, on thioridazine blood concentrations was evaluated in 10 hospitalized men with schizophrenia (2391). The concentrations of thioridazine and its two active breakdown products, mesoridazine and sulforidazine, increased three-fold after the administration of fluvoxamine.

The possibility that some of the cardiac effects of thioridazine may be related to these metabolites as well as the parent compound has been explored (2388,2392) but needs further investigation.

Several regulatory measures (www.mca.gov.uk; www.medsafe.govt.nz; www.imb.ie; www.hc-sc.gc.ca; www.bpfk.org; www.fda.gov; www.who.int/medicines) have been adopted in different countries with regard to thioridazine.

The role of cardiac effects of neuroleptic drugs in sudden death is controversial (SEDA-18, 47; SEDA-20, 26; SEDA-20, 36). There may be multiple non-cardiac causes, including asphyxia, convulsions, or hyperpyrexia. However, some cases of sudden death in apparently young healthy individuals may be directly attributable to cardiac dysrhythmias after treatment with thioridazine or chlorpromazine (2393), and from time to time, cases of sudden death are reported (SEDA-20, 36; SEDA-22, 46), including four cases in which thioridazine in standard doses was implicated as the cause of death or as a contributing factor (2394).

- A 68-year-old man with a 5-year history of Alzheimer's disease was treated with thioridazine 25 mg tds because of violent outbursts (2395). His other drugs, temazepam 10–30 mg at night, carbamazepine 100 mg bd for neuropathic pain, and droperidol 5–10 mg as required, were unaltered. Five days later, he was found dead, having been in his usual condition 2 hours before. Post-mortem examination showed stenosis of the coronary arteries, but no coronary thrombosis, myocardial infarction, or other significant pathology. The certified cause of death was cardiac dysrhythmia due to ischemic heart disease. Thioridazine was considered as a possible contributing factor.

Hypotension

Hypotension is the most commonly observed cardiovascular adverse effect of neuroleptic drugs, particularly after administration of those that are also potent alpha-adrenoceptor antagonists, such as thioridazine. A central mechanism involving the vasomotor regulatory center may also contribute to the lowering of blood pressure.

Severe orthostatic hypotension has been observed in older volunteers ($n = 14$; aged 65–77 years) who participated in a randomized, double-blind, three-period, crossover study, in which they took single oral doses of thioridazine (25 mg), remoxipride (50 mg), or placebo (2396). Compared with placebo, there were falls in supine and erect systolic and diastolic blood pressures after thioridazine, but not remoxipride. Standing systolic blood pressures fell by a maximum of 26 mmHg. There were similar falls in blood pressure in young volunteers.

Hypotension has been reported in two other elderly patients who were taking thioridazine (2397). The patients, men aged 68 and 70 years, had traumatic brain injury and were taking oral thioridazine 25 mg/day for agitation. A few days later they developed mild hypotension (100/50 and 100/60 mmHg respectively).

Thrombolytic agents

Transient hypotension can occur after thrombolysis, and the incidence after infusion of streptokinase in myocardial infarction is high. Hypotension has been related to plasminemia, which leads to bradykinin generation from kallikrein, activation of complement, and potentially endothelial prostacyclin release (2398). During treatment of myocardial infarction, some other mechanisms may explain hypotension: vagal reflexes precipitated by posterior wall reperfusion (2399) and left ventricular dysfunction associated with the infarct process. Hypotension occurs shortly after the start of infusion and can be accompanied by bradycardia, flushing, and anxiety. It is reversible if fluids are given and the infusion temporarily stopped (2398); vasopressors may be needed. The magnitude of this hypotensive reaction is directly related to the rate of infusion of streptokinase. The incidence of such reactions does not seem to be affected by premedication with glucocorticoids. Like streptokinase, anistreplase can also cause transient hypotension (2400). Alteplase seems to be associated with a lower risk of hypotension than streptokinase or anistreplase. In the ISIS-3 trial, the incidences of profound hypotension necessitating drug therapy were: 6.8% with streptokinase, 7.2% with anistreplase, and 4.3% with alteplase (2401).

Embolic detachment of components of venous or mural thrombi can sometimes be involved in the development of thromboembolism or cholesterol embolization. There is a 10% incidence of pulmonary embolism during thrombolysis, lethal in 0–5% (2402), pointing to the risk of detachment of white components of venous thrombi, especially if large veins, such as those in the pelvis, are involved (2403). However, the risk has not been proven to exceed that reported in patients treated with heparin and/or oral anticoagulants.

Cholesterol embolization is thought to occur after removal of mural thrombi covering atherosclerotic plaques leading to direct exposure of the soft lipid-laden core of these plaques to the arterial circulation. The contents of the soft lipid core, including crystallized cholesterol, shower the downstream circulation. Cholesterol crystals are impervious to dissolution or lysis and they lodge in small arterioles, causing obstruction. Cholesterol embolization shortly after thrombolytic treatment is characterized by livedo reticularis or multiple necrotic lesions of the skin of both legs (2404) but can also be associated with lower limb ischemia, gangrene, visceral ischemia, or pseudovasculitis (2405). Cholesterol embolization is difficult to diagnose, and may therefore be more common than suspected. It can cause acute renal insufficiency (2406), and renal biopsy shows acute tubular necrosis, in some instances with arteriolar clefts representing cholesterol crystals (2407).

Reperfusion dysrhythmias can occur when thrombolytic drugs are used to treat myocardial infarction. The most common is transient ventricular tachycardia. However, controlled trials have failed to show an increase in serious ventricular dysrhythmias, the incidence of ventricular fibrillation being actually reduced by thrombolytic therapy in myocardial infarction (2408).

- A coronary artery aneurysm occurred in a 49-year-old man 1 month after successful percutaneous transluminal coronary recanalization with streptokinase (2409).

Phlebitis at the site of infusion of streptokinase has been reported.

Thyroid hormones

Overdosage of thyroid hormones causes tachycardia or palpitation but can also cause several types of dysrhythmia, for example atrial fibrillation. Evidence from the Framingham population that suppression of serum TSH is a susceptibility factor for atrial fibrillation has heightened concern that subclinical hyperthyroidism secondary to levothyroxine can also cause atrial fibrillation (2410). Five women who reported frequent bouts of palpitation were investigated while taking levothyroxine and again after levothyroxine withdrawal (2411). There was a clear increase in mean 24-hour heart rate during levothyroxine treatment, as well as an increase in atrial extra beats and the number of episodes of re-entrant atrioventricular nodal tachycardia. Four of these patients had evidence of abnormal conduction pathways, even when they were not taking levothyroxine, as evidenced by a short PR interval, but exacerbation of atrial dysrhythmias in these predisposed subjects is consistent with the view that thyroid hormones increase atrial excitability and may increase the risk of cardiac morbidity, especially if given in doses sufficient to suppress serum TSH.

Pre-existing cardiac disease, always to be suspected in elderly people or after long-standing hypothyroidism, can be severely aggravated by sudden thyroid substitution, resulting in severe angina pectoris, myocardial infarction, or sudden cardiac death (SEDA-13, 375). In such patients, the initial dosage should be low and the stepwise increase should be spaced out over a prolonged period and with careful clinical and cardiographic monitoring (2412). In some circumstances, for example three-vessel disease, substitution should be postponed until after coronary bypass surgery (SEDA-6, 363). Cardiac decompensation can also result from the increased circulatory demand induced by thyroid hormone substitution or overtreatment. Long-term treatment with levothyroxine in doses that suppress thyrotrophin can have significant effects on cardiac function and structure, especially in patients with hyperthyroid symptoms (2413,2414), but large prospective studies are needed to assess cardiovascular risk in these patients.

In two cases of hypothyroidism myocardial infarction occurred at the start of thyroxine therapy in the absence of evidence of significant coronary artery disease (2415).

- A 58-year-old man with a previous smoking history and a history of hypertension was severely biochemically hypothyroid (serum TSH 221 mU/l) and was given thyroxine, initially in a low dose (25 micrograms/day), increasing to 100 micrograms/day after 2 weeks. A month later he sustained a subendocardial myocardial infarction associated with only minor abnormalities on coronary angiography.
- A 61-year-old woman with severe hypothyroidism (serum TSH 115 mU/l) had an acute myocardial infarction (but no demonstrable abnormality on coronary angiography) 1 month after a thyroxine dosage increase from 50 to 100 micrograms/day.

Cautious introduction of thyroxine, especially in elderly people and those with severe or long-standing hypothyroidism, is prudent.

Myocardial infarction can also occur during long-term use of levothyroxine.

- A 71-year-old woman who had undergone total thyroidectomy with subsequent irradiation because of follicular carcinoma 3 years before (2416). Since then, she had taken oral levothyroxine 0.15 mg and 0.2 mg on alternate days. When latent hypothyroidism became evident despite replacement therapy, the dose of levothyroxine was increased to 0.3 mg/day. Three weeks later, she had formed an acute posterior myocardial infarction, although she had no previous history of coronary artery disease. Subsequent coronary arteriograms revealed no evidence of disease of the major vessels. Myocardial scintigraphy 3 weeks after infarction still showed a persistent perfusion defect.

In the US Coronary Drug Project (2417) a formulation of dextrothyroxine with a so-called low levothyroxine content was used. The study had to be terminated because of an excessive number of cardiac deaths and non-fatal infarcts.

Tiabendazole

Bradycardia, hypotension, and syncope can occur with tiabendazole, even to the point of collapse (2418).

Tiapride

Torsade de pointes has been attributed to tiapride.

- A 76-year-old man developed torsade de pointes while taking tiapride 300 mg/day; the QT_c interval 1 day after starting treatment was 600 ms; the dysrhythmia resolved when tiapride was withdrawn (2419).

Tiotropium bromide

In a randomized, double-blind study, 470 patients with stable COPD (mean FEV_1 38.6% predicted) received tiotropium 18 micrograms or placebo as a once-daily medication via a lactose-based dry-powder inhaler device (6). Spirometry was measured on days 1 and 8 and at regular intervals for 3 months. Tiotropium produced significant improvements in trough FEV_1 and FVC (measured immediately before the next dose), averaging 12% greater than baseline on day 8. The bronchodilatation resulting from tiotropium did not diminish over the 3 months of the study. Upper respiratory tract infections were reported in 15% of the patients in each of the treatment groups and exacerbations of COPD were reported in 22% of patients taking placebo and 16% of patients taking tiotropium. Dry mouth was significantly more common with tiotropium (9.3 versus 1.6%). There was a 6.8% incidence of serious adverse events or events leading to withdrawal (2.5% with tiotropium and 5.8% with placebo). A patient with a long history of cardiovascular disease, who was randomized to receive tiotropium, was found dead and was suspected to have died of a cardiac dysrhythmia. There were no differences in electrocardiograms between the treatment groups or changes in laboratory values.

In a multicenter, double-blind, placebo-controlled trial in patients with stable COPD, tiotropium 18 micrograms/day for 1 year (n = 550) significantly improved lung function (mean trough FEV_1, FVC, PEFR) and reduced dyspnea compared with placebo (n = 371) (7). Moreover, tiotropium recipients had better health status scores, fewer exacerbations of COPD, and fewer hospitalizations. There was no evidence of tachyphylaxis over one year.

Electrocardiographic and 24-hour Holter monitoring in 196 patients with COPD at baseline and after 8 and 12 weeks of treatment with tiotropium 18 micrograms daily was performed in a randomized, double-blind, placebo-controlled study (2420). Tiotropium was not associated with changes in heart rate, heart rhythm, QT interval, or conduction.

Tocainide

Adverse cardiovascular effects have been reported in 6–55% of cases. After a single dose of tocainide the most common effect is hypotension with bradycardia (2421). Angina pectoris has also been reported (2422).

During repeated administration cardiovascular adverse effects are relatively uncommon. Increasing heart failure (2423,2424), worsening dysrhythmias (2424,2425), pericarditis (2424,2426,2427), and sinus arrest with sinoatrial block (2428) have all been reported. Tocainide can worsen ventricular tachycardia (2429).

Tolmetin

Increased blood pressure can occur after long-term treatment with tolmetin (2430).

Tonazocine

Tonazocine is a partial opioid receptor agonist that has not been reported to have adverse effects on the cardiovascular system or to cause clinically significant respiratory depression. When single doses of tonazocine 2, 4, and 8 mg were compared in 150 adults postoperatively, drowsiness was the most frequent adverse effect and visual hallucinations occurred in two patients (2431).

Topiramate

It is unclear whether topiramate played any role in rare cardiovascular events. These included symptoms of Raynaud's phenomenon in three patients, and third-degree atrioventricular block requiring emergency cardiac pacemaker implantation in one patient with pre-existing right bundle branch block (SEDA-21, 76).

Topoisomerase inhibitors

There are reports of myocardial infarction in patients who have received combination chemotherapy containing etoposide. The mechanisms have not been clearly elucidated.

- A 28-year-old man with a non-seminomatous retroperitoneal germ-cell cancer received etoposide (180 mg/day intravenously on days 1–5), bleomycin, and cisplatin (2432). He had no cardiac risk factors and no history of cardiac symptoms. On day 3, during infusion of bleomycin, he developed chest pain and dyspnea. The infusion was discontinued and he was given glyceryl trinitrate and diazepam; his symptoms resolved. On day 4 he was given etoposide as scheduled, but four hours later developed severe angina. The electrocardiogram and raised cardiac enzymes were consistent with an acute posterolateral myocardial infarction. He was given heparin, aspirin, and nitrates, and the chemotherapy was discontinued. Within 20 hours his chest pain completely disappeared and his electrocardiogram became normal.

If hypotension occurs during drug administration, it usually subsides when the infusion ends and intravenous fluids or other supportive agents are given. Elderly patients may be particularly susceptible to etoposide-induced hypotension. During a phase I trial of etoposide by continuous infusion, 17 patients were given 75 mg/m^2/day for 5 days and later courses of 100 mg/m^2/day and 150 mg/m^2/day (2433). Two patients with pre-existing cardiovascular disease developed myocardial infarctions, one at the 100 mg/m^2/day dose and the other at the 150 mg/m^2/day dose. Another patient developed congestive heart failure at the end of the 5-day infusion and died on day 8; however, this patient also received a saline load of 1500 ml/day for 5 days during etoposide administration and had had previous episodes of congestive cardiac failure. The authors concluded that in patients with underlying cardiovascular disease etoposide must be administered cautiously and that extensive saline loading should be avoided in patients with a history of previous congestive heart failure.

Tramadol

Pericarditis has been attributed to tramadol (2434).

- An 88-year-old man took tramadol and 2 days later developed precordial chest pressure radiating to the scapula and increasing with inspiration and movement. His blood pressure was 75/45 mmHg and his heart rate 60/minute. There was a soft pericardial rub, mild cardiomegaly, diffuse ST segment elevation and PR interval shortening, normal left ventricular systolic function, and no wall motion abnormalities or effusion on echocardiography.

The temporal relation between the development of acute pericarditis and the resolution of symptoms on withdrawal of tramadol suggested a causal link.

Trastuzumab

Cardiotoxicity is a major concern with trastuzumab, particularly as it is often used in patients who are receiving or who have previously received anthracycline antibiotics (2435). It occurs in 5% of patients given trastuzumab alone, in 13% of patients given trastuzumab with paclitaxel, and in 27% of patients given trastuzumab in combination with anthracyclines and cyclophosphamide (2436).

The efficacy and safety of trastuzumab have been evaluated in 235 women with metastatic breast cancer receiving standard chemotherapy (2437). The most important adverse event was cardiac failure, which occurred in 27% of those who were given anthracycline, cyclophosphamide, and trastuzumab, 8% of those who were given anthracycline and cyclophosphamide alone, 13% of those who were given paclitaxel and trastuzumab, and 1% of those who were given paclitaxel alone. The incidence of cardiac failure of New York Heart Association class III or class IV was highest among patients who had received an anthracycline, cyclophosphamide, and trastuzumab. The mechanism is unknown. In a retrospective analysis of all patients with cardiac failure by an independent review and evaluation committee, the only significant risk factor was older age. Although the cardiotoxicity was severe, and in some cases life-threatening, the symptoms improved with standard medical management.

Seven phase II and III clinical trials of trastuzumab in patients with metastatic breast cancer have been reviewed (2438). The physiopathology of trastuzumab-associated cardiac disease is poorly understood, and baseline MUGA scanning is recommended to identify cardiac disease. The rates of cardiac disease were highest when trastuzumab was given in combination with anthracyclines or when there had been previous exposure to anthracyclines. Given the 25% improvement in overall survival in these patients with metastatic disease, the use of trastuzumab is justified.

- A 60-year-old woman with coronary heart disease and hypertension developed breast cancer and a core biopsy showed ER+/PR+/HER2–3+ (2436). She enrolled in an institutional study with neoadjuvant trastuzumab with docetaxel. A pretreatment blood pool radionuclide angiography (MUGA) scan revealed a left ventricular ejection fraction of 56%. She had a good clinical response to treatment at 3 months. Before surgery a repeat MUGA showed a dilated left ventricle and a reduced ejection fraction (35%). She underwent surgery and 2 months later her ejection fraction was 44%.
- An overweight 59-year-old woman with hypertension and asthma developed breast cancer (ER+/PR+/HER2–2+) (2436). Her MUGA scan showed a left ventricular ejection fraction of 57%. She was given trastuzumab with docetaxel and had a good response after 4 months. Before surgery she became dyspneic, and a MUGA scan showed an ejection fraction of 24%; even 7 months after surgery her MUGA scan showed no improvement.

The efficacy and tolerability of trastuzumab in clinical trials has been reviewed (2439). Of the first 48 patients treated in Sweden with or without chemotherapy, two had serious cardiac events and both had previously been treated with an anthracycline. As not all patients who received trastuzumab had echocardiography, the number of cardiac events was probably underestimated. It has been postulated that HER2 pathways may be involved in myocyte repair, and that concomitant administration of trastuzumab may interfere with the repair of anthracycline-damaged myocytes (2440). Evidence from 20 patients has shown that cardiotoxicity is related to trastuzumab uptake in the myocardium, suggesting that the extent of HER2 receptor expression, or a related cross-reactive antigen in the myocardium, may be the underlying mechanism.

Trazodone

Based on animal research and restricted experience in overdosage (SEDA-7, 19–21), early attempts to differentiate trazodone from tricyclic antidepressants suggested that it might be relatively free of cardiotoxic effects. However, a preliminary report of a study of the effects of trazodone on the cardiovascular system in 20 subjects mentioned two patients who had ventricular dysrhythmias (2441). Others have reported ventricular tachycardia (2442–2444), atrial fibrillation (2445), and complete heart block (2446).

Additive hypotensive effects of trazodone and phenothiazines have been reported (2447).

Trazodone can cause peripheral edema, as outlined in a report of 10 cases (2448).

Tricyclic antidepressants

The cardiac toxicity of tricyclic antidepressants in overdose has been a source of continued concern. Undesirable cardiovascular effects, besides representing a major therapeutic limitation for this category of drugs, delineate an area in which tricyclic compounds with novel structures, as well as second-generation antidepressants, may have significant advantages. The cardiovascular effects of tricyclic antidepressants and the new generation of antidepressants have been reviewed (SEDA-18, 16; 2449).

Since the inception of SEDA in 1977, each volume has included a review of the evolving literature, focusing on specific aspects, including direct myocardial actions (SEDA-12, 13), hypotension (SEDA-12, 13), and the incidence, severity, and management of overdosage in adults and children (SEDA-10, 19) (SEDA-11, 16; 2450,2451–2455).

Direct myocardial actions

Tricyclic antidepressants are highly concentrated in the myocardium; this may account for the vulnerability of the heart as a target organ as well as for inconsistently and inconclusively reported relations between plasma drug concentrations and specific manifestations of cardiac toxicity. These drugs interfere with the normal rate, rhythm, and contractility of the heart through actions on both nerve and muscle that are mediated by at least four different mechanisms (singly, in combination, or due to imbalance), including an anticholinergic action, interference with re-uptake of catecholamines, direct myocardial depression, and alterations in membrane permeability due to lipophilic and surfactant properties.

Acute experiments in dogs have shown a negative inotropic effect sufficient to cause congestive cardiac failure, but a carefully conducted long-term study in man did not show impaired left ventricular function in depressed patients with concurrent congestive failure (SEDA-9, 18). In another study (2456) nortriptyline (mean dose 76 mg/day, mean plasma concentration 107 ng/ml) was given to 21 depressed patients with either congestive heart failure or enlarged hearts. In this study, nortriptyline was effective and well tolerated, producing only one episode of intolerable hypotension.

The most readily observable change in cardiac function is sinus tachycardia, which occurs to a greater or lesser extent in more patients and which correlates weakly or inconsistently with plasma concentrations (2450–2452). The mechanism may be related to both central and peripheral effects on cholinergic and adrenergic systems, but is not simply a reflex response to hypotension (2450). The presence of tachycardia can serve as an indirect measure of compliance (2450), but it is seldom a cause for concern, except in individuals who anxiously monitor their own physiological functions.

The complex changes that occur in cardiac rhythm have been intensively studied using 24-hour high-speed and high-fidelity cardiographic tracings, His bundle electrocardiography (2457,2458), and cardiac catheterization (2459,2460). Changes in conduction and repolarization cause prolongation of the PR, QRS, and QT intervals and flattening or inversion of T-waves on routine electrocardiograms; conduction delay occurs distal to the atrioventricular node and is apparent as a prolonged HV interval (the time from activation of the bundle of His to contraction of the ventricular muscle). This effect resembles that due to type I cardiac antidysrhythmic drugs, such as quinidine and procainamide. These conduction changes can cause atrioventricular or bundle branch block and can predispose to re-entrant excitation currents with ventricular extra beats, tachycardia, or fibrillation.

The implications and complications of these changes in cardiac function and rhythm are less clear; knowledge of their existence provoked concern about the incidence of sudden death in patients with cardiovascular disease, but the evidence from epidemiological sources is equivocal (2453). General guidelines for the use of these drugs in the elderly have been discussed above, and a review of studies on the cardiovascular effects of therapeutic doses of tricyclic antidepressants has supported their use in elderly patients and those with pre-existing cardiovascular disease, provided precautions are taken. Atrial fibrillation has been reported in predisposed elderly subjects (SEDA-18, 19). In children taking desipramine there have been reports of sudden death (2461,2462) and tachycardia, and cardiographic evidence of an intraventricular conduction defect (SEDA-18, 18; 2463).

The antidysrhythmic effect of imipramine was first reported in 1977 during treatment of two depressed patients whose ventricular extra beats improved during treatment (2464). Tricyclic compounds can trigger serious dysrhythmias at high doses and perhaps also when the myocardium is sensitized.

Care should be taken in patients with a recent myocardial infarction who show evidence of impaired conduction (first-degree heart block, bundle-branch block, or prolongation of the QT_c interval), since tricyclic antidepressants can theoretically add to the already increased risk of ventricular fibrillation in such patients (2465). Reviews in earlier editions of Meyler's *Side Effects of Drugs* discussed these effects and gave practical guidelines on the use of tricyclic antidepressants in patients with heart disease (2466).

There is sometimes a clear-cut correlation between cardiac toxicity and high plasma concentrations (SEDA-18, 18; 2450,2467,2468), but this may not always be so in individuals who are highly sensitive to the drug or in whom prolonged treatment may have led to drug accumulation in the myocardium, despite plasma concentrations in the usual target range. Routine plasma concentration monitoring does not seem indicated, since plasma concentrations account for only a small part of the variance in cardiac effects (2451). If an individual shows significant changes clinically or cardiographically, a spot measurement may show a high plasma concentration, requiring dosage reduction.

To date there have been no prospective studies that clearly show increased mortality in cardiac patients who use tricyclic antidepressants. It has been suggested that overall mortality due to cardiac disease may be higher in depressed patients who remain untreated than in those who receive either an antidepressant or electroconvulsive therapy (2469). Even in patients who have chronic heart disease, the risks of effective treatment with a tricyclic appear to be minimal (2470).

Hypotension

As many as 20% of patients taking adequate doses of a tricyclic antidepressant experience marked postural hypotension. This effect is not consistently correlated with plasma concentrations and tolerance does not develop during treatment (2471–2473). The mechanism for this effect is uncertain; it has been attributed to a peripheral antiadrenergic action, to a myocardial depressant effect, and to an action mediated by alpha-adrenoceptors in the central nervous system (2474). Studies of left ventricular function in man are conflicting. One study of systolic time intervals showed a decrement in left ventricular function with therapeutic doses (2475), while two in which cardiac function was observed directly during cardiac catheterization after overdosage showed no evidence of impaired myocardial efficiency, whereas the hypotension persisted after left ventricular filling pressures and cardiac output had returned to normal (2476,2477).

Postural hypotension can lead to falls. Its occurrence may be predictable, since patients who have raised systolic pressures and who have a pronounced postural drop before treatment are most likely to experience drug-induced hypotension (2473). Such patients should be cautioned to rise slowly from sitting positions and, since elderly people are especially at risk of falling at night, single large bedtime doses of sedative tricyclic drugs should be avoided. These preventive measures are the most helpful, since the hypotensive effect is not directly related to plasma drug concentrations and may not improve with dosage reduction. The wisdom of using sympathomimetic drugs to counter this undesirable effect is questionable (2478).

Use in patients with cardiac disease

The diagnosis of depression and the use of antidepressant medication are both associated with an increased risk of myocardial infarction. The relative contribution of these two factors is uncertain. In a case-control study of 2247 subjects, taking antidepressants was associated with a 2.2-fold (CI = 1.3, 3.7) increase in the risk of myocardial infarction (2479). This increased risk seemed to be accounted for entirely by the use of tricyclic antidepressants, because selective serotonin re-uptake inhibitors were not associated with an increased risk, although the confidence intervals were wide (relative risk 0.8; CI = 0.2, 3.5). These findings support the usual clinical advice that tricyclic antidepressants are best avoided in those with known cardiovascular disease or significant risk factors.

Cardiovascular complications of overdosage

The relation between the dosage of a tricyclic antidepressant and the development of life-threatening cardiovascular complications is unclear and individually variable (2480,2481), although plasma concentrations above 1000 ng/ml give cause for serious concern. Plasma concentrations vary widely, and absorption can be delayed or deceptive, owing to gastric stasis and enterohepatic recycling (2482). Dysrhythmias can occur for the first time up to 36 hours after drug ingestion or admission to hospital (2482). The frequency of serious cardiac conditions in one series of 68 cases (2483) was 46% in patients who took over 2000 mg of imipramine or its equivalent, almost twice that of those who took less (25%). An intensive study of cardiovascular complications among 35 overdose patients showed that 51% had significant hypotension and 80% had abnormal electrocardiograms (2476). The latter consisted of sinus tachycardia (71%) and various abnormalities that reflect impaired conduction, including prolongation of the QT_c interval (86%), QRS complex (29%), and PR interval (11%). The ST segments and T-waves were abnormal in 28% of patients. Despite these manifestations of disturbed conduction and repolarization, there were relatively few dysrhythmias; 13 patients (37%) had ventricular extra beats, which lasted up to 72 hours after admission and subsided within 36 hours in 10 cases. No patients developed sustained repeated ventricular tachydysrhythmias, and the authors speculated that bizarre wide QRS complexes seen in aberrantly conducted supraventricular tachycardia (present in some cases) may sometimes be misinterpreted as ventricular tachycardia.

The basic principles of intensive supportive care should be applied early, and artificial ventilation is often necessary, since respiratory depression is more frequent than is commonly supposed (2484,2485). Patients should be monitored for 24 hours if the initial (or subsequent) electrocardiogram shows a dysrhythmia. Among 75 patients with overdose, none who had a normal electrocardiogram and level of consciousness for 24 hours went on to develop any significant dysrhythmia (2486). However, a case was subsequently reported of a patient who died an acute cardiac death 57 hours after admission and 33 hours after normalization of the electrocardiogram (2487). The authors suggested that prolonged monitoring may be justified in individuals who have taken antidepressants for prolonged periods, compared with those who overdose early in treatment.

More specific treatment to combat cardiotoxic effects is usually necessary in only a minority of instances; in the series reported above (2476), five patients (14%) had marked hypotension. Initial low left ventricular filling pressures were corrected within 3 hours by infusion of isotonic saline. Systemic hypotension persisted and was corrected by infusion of sympathomimetic amines. Routine insertion of a pulmonary artery catheter, with continuous monitoring of blood gases, pulmonary arterial pressure, left atrial wedge pressure, and cardiac output have been recommended (2476). Volume expansion is suggested for low left atrial pressure, with dopamine infusion to improve myocardial contractility if cardiac output remains low.

The management of ventricular extra beats is based on recognition of the quinidine-like basis of the conduction defect. In the series reported above (2476) all 13 patients with ventricular arrhythmias responded to intravenous infusion of lidocaine (mean dose 2.0 mg/minute). This alone might account for the absence of deaths in this series.

Another indirect method of benefiting the patient with cardiotoxic effects has been the alkalinization of plasma to a pH of 7.50–7.55 using sodium bicarbonate infusion (2488). This enhances plasma protein binding, making the drug less available to the tissues. It is claimed that this technique reverses both hypotension and cardiac dysrhythmias without the risk of the undesirable effects of antidysrhythmic drugs (2482). Recommendations for the management of poisoning with tricyclic antidepressants given in a recent review have been summarized (SEDA-16, 8).

Differences among tricyclic compounds

There is evidence that doxepin is significantly less cardiotoxic than other tricyclic compounds (2489). However, a complete review of all the animal and clinical data has suggested that doxepin overdose can still cause lethal dysrhythmias in man, probably by producing more marked respiratory depression.

Three different studies have shown that there is less risk of hypotension in patients treated with nortriptyline (2490–2492) than with other tricyclic compounds. However, a similar claim has been made for doxepin (2493).

Increased pulse rate and blood pressure have been associated with desipramine in the treatment of bulimia nervosa (SEDA-17, 18).

Trihexyphenidyl

Paradoxical bradycardia has been reported; the reaction was specific to trihexyphenidyl and was not observed when the patient took other anticholinergic drugs (SEDA-12, 125).

Trimethoprim and co-trimoxazole

An early review of newer case reports and placebo-controlled trials involving several hundred patients did not show an increase in fetal abnormalities (2494). However, the relative risks of cardiovascular defects and oral clefts in infants whose mothers were exposed to dihydrofolate reductase inhibitors, such as trimethoprim, during the second or third month after the last menstrual period, compared with infants whose mothers had no such exposure, are 3.4 (95% CI = 1.8, 6.4) and 2.6 (1.1, 6.1) respectively (2495). Multivitamin supplements containing folic acid reduced the adverse effects of dihydrofolate reductase inhibitors. There have been two reports of severe spinal malformations in the fetuses of HIV-positive women treated with combination antiretroviral therapy and co-trimoxazole (2496).

Triptans

Particularly when injected, sumatriptan can cause a transient increase in both systolic and diastolic blood pressures; oral doses, which are usually higher, have this effect to a lesser extent. It is unwise to use sumatriptan in patients with uncontrolled severe hypertension.

Some of the minor discomfort often experienced (tingling, flushing, sensations of heat or cold) may also reflect cardiovascular effects.

Coronary vasoconstriction is a potential risk of all triptans, but the risk is minimal in the absence of coronary artery disease or uncontrolled hypertension. Chest tightness and pain are reported in up to 15% of patients taking sumatriptan and are presumed to be due to vasoconstriction of the coronary arteries. Myocardial infarction has been reported and as a consequence sumatriptan should not be used in patients with cardiovascular disease (2497). Intravenous sumatriptan causes some coronary vasoconstriction during diagnostic angiography. Myocardial infarction occurred in a patient who had received sumatriptan 6 mg subcutaneously, but the causal association was not clear.

Cardiac dysrhythmias can occur occasionally.

- A 41-year-old otherwise perfectly healthy woman developed the typical symptoms of a migraine attack after an aerobics class followed by a sauna and took sumatriptan 100 mg (2498). She collapsed half an hour later and a resuscitation team diagnosed ventricular fibrillation, which was converted to ventricular tachycardia and later to sinus rhythm. Neither the electrocardiogram nor enzyme changes suggested myocardial infarction. She remained comatose and died 6 weeks later. At autopsy there was no evidence of atherosclerotic coronary disease or any other underlying cardiac disorder.
- A 34-year-old man with migraine had palpitation after taking sumatriptan by nasal spray for a severe headache (2499). A similar episode had occurred after he had previously taken sumatriptan. He had atrial fibrillation with a rapid ventricular rate. Sinus rhythm returned spontaneously within a few hours. No structural cardiac abnormality was detected.

Myocardial ischemia secondary to coronary spasm was the putative trigger of atrial fibrillation in the latter case.

Vasoconstriction leading to organ ischemia has been repeatedly reported with triptans. The risk is very low in the absence of pre-existing arterial disease. Nevertheless, cases of myocardial infarction, mesenteric ischemia, and ischemic colitis have been described, as has splenic infarction (2500).

- A 48-year-old woman with a low risk of atherosclerotic disease who was not taking oral contraception was admitted with sudden heavy pain in the left hypochondrium. Routine tests were normal, but there was a triangular hypodensity in the spleen on computerized tomography consistent with a splenic infarct. There was a history of postoperative venous thrombosis 15 years before, but tests for thrombophilia were all normal. No source of embolism was identified in the heart or proximal arteries. She had repeatedly taken zolmitriptan over the previous few months for migraine, and the symptoms started 3 hours after the most recent dose.

The temporal relation and the absence of an alternative cause to explain the infarct led the authors to conclude that zolmitriptan had been causative.

Tropicamide

A transient ischemic attack occurred in a 64-year-old patient with cardiovascular risk factors who used topical tropicamide (2501).

Tubocurarine

D-Tubocurarine commonly causes a fall in blood pressure, associated with a slight tachycardia and a reduction in total peripheral resistance; cardiac output is not affected. The frequency of hypotension is reported as being 20–90%. This wide range probably reflects the methods of measurement, the anesthetic agents used, and the general condition of the patients in the various studies, as well as the criteria for diagnosing hypotension. The magnitude of the fall in blood pressure is generally about 20%, and it occurs within 5 minutes of injection. Histamine release is considered to be the principal cause (2502), but blockade of autonomic ganglia may also contribute. It has been suggested that prostacycline, released by histamine acting on H_1 receptors, is the final mediator; intravenous administration of aspirin or an H_1 receptor antagonist beforehand affords some protection (SEDA-15, 126; 2503). Ganglion blockade may also contribute, particularly if high doses are used (2504). Reduction in venous return secondary to muscle relaxation and alterations in intrathoracic and intra-abdominal pressures may also play a role. The fall in blood pressure can be greatly exaggerated in hypovolemic patients, in the elderly, and in others with reduced sympathetic tone. Tubocurarine should be used very cautiously in such patients or another relaxant should be chosen. Concurrent administration of agents known to cause circulatory depression aggravates the problem. The higher the halothane concentration, for example, the greater the fall in blood pressure after D-tubocurarine (2505). Since the degree of hypotension seems to be linked to dose (2502) and the rate of injection (2506), it seems reasonable to use the smallest dose that produces adequate relaxation (under 0.5 mg/kg) and to inject it slowly (over at least 180 seconds) (SEDA-7, 141).

Tumor necrosis factor alfa

Hypotension, perhaps nitrous oxide-mediated, is a dose-limiting adverse effect of tumor necrosis factor alfa (2507).

Severe hypophosphatemia with myocardial dysfunction was noted in patients receiving tumor necrosis factor alfa by continuous hepatic arterial infusion (SEDA-17, 433).

Congestive cardiomyopathy has been attributed to tumor necrosis factor alfa in isolated patients (SED-13, 1110) (2508).

Ultrasound contrast agents

In a post-marketing analysis of more than 157 838 studies of the ultrasound contrast agent Sonovue® there were three (0.002%) fatal and 19 (0.012%) severe, non-fatal adverse reactions; of the latter, nine were cardiac events (four bradycardia, two tachycardia, three myocardial ischemia); 11 of these 19 patients received Sonovue for echocardiography (2509). These results led to the addition of several contraindications to the use of Sonovue. Although a strong relation was established between the non-fatal cases and administration of Sonovue, a causal relation with the fatal cases is debatable. Therefore, the risk associated with the use of Sonovue should be judged carefully, taking into consideration the prevalence of adverse effects of other contrast media and diagnostic procedures used, particularly in cardiology.

The ultrasound contrast agent ECHOVIST-200 has been used for hysterosonography with diagnostic success in 195 cases (2510). The most common adverse effect during the procedure was abdominal pain, which was well tolerated by the patients. There were no other important adverse effects.

Valproate sodium and semisodium

The effect on the blood pressure of intravenous valproate sodium at larger doses and more rapid rates of infusions than recommended in the original labelling has been investigated in a randomized study (2511). The original labelling recommended that valproate sodium injection be administered over 60 minutes, at rates up to 20 mg/min, and, for adjunctive therapy, that doses exceeding 250 mg be given in divided doses. In this study 112 adults and children with epilepsy (age range 13 months to 79 years) were randomized in a 2:1 ratio to receive up to 15 mg/kg of valproate sodium at 3.0 or 1.5 mg/kg/minute. Up to four doses were allowed within 24 hours to achieve target plasma concentrations. There were no significant differences in blood pressures among the two groups of patients.

Vancomycin

Vancomycin can cause phlebitis if infused peripherally (2512,2513).

Vasopressin and analogues

Dose-related adverse effects of vasopressin include skin pallor, hypertension, cardiac dysrhythmias, and myocardial ischemia or infarction; treatment has to be stopped in 20–30% of patients because of these effects (2514).

Intestinal and peripheral vasoconstriction can follow prolonged infusion, resulting in gangrene of intestinal segments or of skin, fingers, or limbs. This has been fatal in several cases, and vasopressin should be withdrawn if skin necrosis occurs (SEDA-13, 1310; 2515).

Desmopressin (N-deamino-8-D-arginine vasopressin, DDAVP)

People with hemophilia can develop atherosclerosis, but they are usually protected from ischemic coronary events.

However, such events can occur when desmopressin is given.

- A 59-year-old man with mild hemophilia A was given a test dose of desmopressin 30 micrograms (0.19 micrograms/kg) in 100 ml of saline by intravenous infusion over 30 minutes (2516). Shortly afterwards, having had a cigarette, he developed chest pain. An electrocardiogram showed ST elevation, and a myocardial infarction was confirmed.

Cardiovascular evaluation may be appropriate before using desmopressin.

Ornipressin

Of 10 patients with hepatorenal syndrome one had to be withdrawn 2 hours after receiving 12 units of ornipressin (6 IU/hour) because of a ventricular tachydysrhythmia, and an infusion of dopamine 2–3 micrograms/kg/minute was started (2517).

Terlipressin

Terlipressin has similar, but less pronounced, systemic hemodynamic effects to vasopressin, including increases in mean arterial pressure and reduced heart rate (2518). Of 105 patients who had continuous terlipressin infusions for variceal bleeding in a multicenter study, lower limb ischemia developed in two and cardiac ischemia in one (2519).

Of 86 cirrhotic patients treated with terlipressin 10 developed a tachycardia, four developed atrial fibrillation, and one developed ventricular tachycardia. Four patients in the same study developed hypertension (2520). In another study, tachycardia occurred in 23% of patients randomized to pitressin for acute variceal bleeding and 8% developed transient hypertension (2521).

Myocardial ischemia has been associated with terlipressin (2522).

- A 61-year-old man with coronary artery stenosis became hypotensive during elective surgery, refractory to ephedrine (cumulative dose of 36 mg over 45 minutes). Immediately after terlipressin 1 mg, he developed hypertension and bradycardia, with evidence of myocardial ischemia.

One of 21 patients with hepatorenal syndrome developed finger ischemia on the fourth day of intermittent intravenous terlipressin and recovered after terlipressin was stopped (2523).

In 32 patients undergoing carotid endarterectomy treated with renin–angiotensin inhibitors hypotension developed under general anesthesia (2524). They were randomized to received terlipressin 1 mg (n = 16) or noradrenaline infusion. Compared with baseline those who received terlipressin had reduced gastric mucosal perfusion for at least 4 hours. There was also reduced oxygen delivery and oxygen consumption index at 30 minutes and 4 hours in those who received terlipressin.

Vecuronium bromide

The expected cardiovascular stability of vecuronium has been confirmed in man (2525–2528). Even doses as large as 0.28 mg/kg in patients undergoing coronary artery bypass grafting produced negligible effects (2529). Bradycardia is the only cardiovascular adverse effect reported, and this is seen in association with opioids such as fentanyl (2530) and sufentanil (SEDA-11, 125) (2531) or other drugs that are themselves capable of producing bradycardia. The lack of vagolytic and sympathomimetic activity of vecuronium means that it does not counteract the bradycardia or the hypotensive effects of other drugs or surgical manipulations. It is an ideal relaxant for patients with pheochromocytoma (2532).

Venlafaxine

A meta-analysis of the effect of venlafaxine on blood pressure in patients studied in randomized placebo-controlled trials of venlafaxine, imipramine, and placebo showed that at the end of the acute phase (6 weeks) the incidence of a sustained rise in supine diastolic blood pressure (over 90 mmHg) was significantly higher in both active treatment groups: venlafaxine 4.8% (135/2817), imipramine 4.7% (15/319), and placebo 2.1% (13/605) (2533). The effect of venlafaxine in causing a rise in diastolic blood pressure appeared to be dose related, with an incidence of 1.7% in patients taking under 100 mg/day and 9.1% in those taking over 300 mg/day.

These data confirm that venlafaxine, particularly in higher dosages, can significantly increase blood pressure. At high doses, venlafaxine inhibits the re-uptake of noradrenaline as well as that of serotonin, which probably accounts for the pressor effect.

In physically healthy subjects venlafaxine has a generally benign cardiovascular profile, although hypotension and dose-related hypertension have been reported (SEDA-23, 20).

Venlafaxine is often used in high doses in patients with treatment-resistant depression. If there is continuing failure to respond, electroconvulsive therapy might be used, often in combination with venlafaxine. A 73-year-old woman taking venlafaxine (112.5 mg/day) had sustained hypertension for several hours after her first treatment (2534). However, electroconvulsive therapy can cause transient hypertension, and the patient had essential hypertension controlled by bendroflumethiazide. It is therefore possible that the reaction might have occurred had she not been taking venlafaxine. Nevertheless, the fact that venlafaxine can cause hypertension when used as sole treatment suggests that blood pressure should be monitored carefully in patients receiving electroconvulsive therapy and venlafaxine together, particularly if there is a history or current evidence of hypertension.

Venlafaxine has not been studied systematically in patients with cardiovascular disease, although there are reports that older patients can have clinically significant disturbances of cardiac rhythm (2535).

- A 69-year-old woman with stable angina and mild single-vessel coronary artery disease developed acute myocardial ischemia within a week of starting venlafaxine (75 mg/day).

Taken together with the information that the authors cited in their review, the current data suggest that venlafaxine should be used with caution in patients with established cardiovascular disease.

Viloxazine

Although hypotension and tachycardia can occur, an extensive review of its animal and clinical pharmacology suggested that viloxazine is relatively free of direct cardiotoxic effects (2536). In a controlled comparison of doxepin and viloxazine (150–450 mg/day), one patient taking viloxazine developed chest pain after 26 days; an electrocardiogram confirmed changes compatible with ischemia but there were no progressive electrocardiographic or enzyme changes, and the patient recovered fully after being dropped from the study (2537).

Vinca alkaloids

Vinca alkaloid-associated myocardial infarction, angina pectoris, and transient electrocardiographic changes related to coronary ischemia are limited to case reports (2538–2540). In addition, patients with these adverse effects received combination cancer chemotherapy, containing drugs such as bleomycin and cisplatin, which both have been suggested to cause cardiovascular adverse effects.

- A 64-year-old Japanese man developed chest pain with concomitant lateral ST segment depression after treatment with intravenous cisplatin 70 mg/m^2, vincristine 1.2 mg/m^2, and bleomycin 12 U/m^2. There was no history of predisposing factors and the symptoms disappeared quickly with glyceryl trinitrate. During the following cycle, containing cisplatin and verapamil as an antianginal agent, chest pain did not reoccur (2539).
- A 46-year-old man developed a Q-wave inferior and a right ventricular myocardial infarct with postinfarction angina after the third cycle of vincristine + doxorubicin for multiple myeloma. The patient had no risk factors for ischemic heart disease, except for a positive smoking history, nor for hyperviscosity (2540).

There is some evidence that both vincristine and doxorubicin are more often associated with ischemic heart disease than other cancer chemotherapeutic agents. Whether drug-induced platelet activation, altered clotting, or endovascular damage are responsible for vascular toxicity is still unclear. The risk of ischemic heart disease must therefore be kept in mind when patients receive a combination of doxorubicin and vincristine, especially when potential risk factors have been identified. Whether the structural similarity between vinca alkaloids and ergot alkaloids, which are vasospastic, is relevant is highly speculative.

After administration of vinorelbine, chest pain occurs in up to 5% of patients. However, subsequent analysis showed that most patients had underlying cardiovascular disease or a tumor in the chest, making interpretation difficult (2541,2542). Three patients developed acute cardiopulmonary toxicity after vinorelbine therapy (2543). The symptoms mimicked acute cardiac ischemia, but with no electrocardiographic changes or raised cardiac enzymes. In two patients, tachypnea, râles, wheezing, and severe dyspnea responded to inhaled salbutamol. One patient developed pulmonary edema and bilateral pleural effusions, which contained no malignant cells when drained.

Venous discomfort and venous chemical phlebitis have been reported with vinorelbine 30 mg/m^2, with an incidence of up to 31% (2544,2545). The risk is increased in obesity (2545). Pain along the vein occurred in five of 43 patients receiving vinorelbine 30 mg/m^2/week; none developed extravasation, but their symptoms were very similar (2546). A similar rate of toxicity, 4.5%, has been reported in a review of the use of vinorelbine in 321 patients with breast cancer (2547).

Viscaceae

In 14 cases of accidental exposure to *Phoradendron flavescens* (*Phoradendron serotinum*, American mistletoe) there were no symptoms, but there was one death from intentional ingestion of an unknown amount of an elixir brewed from the berries (2548). In 92 patients aged 4 months to 42 years (median 2 years), 14 were symptomatic, 11 related to mistletoe exposure. All the symptomatic cases had onset of symptoms within 6 hours. The symptoms included gastrointestinal upsets ($n = 6$), mild drowsiness ($n = 2$), eye irritation ($n = 1$), ataxia ($n = 1$), and seizures ($n = 1$). Treatment included gastrointestinal decontamination in 54 patients, ocular irrigation in one, and an intravenous benzodiazepine in one; decontamination did not affect the outcome. The amount ingested ranged from one berry or leaf to more than 20 berries or five leaves. Eight of ten patients who had taken at least five berries were symptom-free. Of 11 patients who took only leaves (range 1–5 leaves), three had gastrointestinal upsets. There were no cardiovascular effects in any case.

Vitamin A: Carotenoids

It is not therefore surprising that prolonged use of alcohol, drugs, or both not only results in reduced dietary intake of retinoids and carotenoids, but also accelerates the breakdown of retinol, through cross-induction of degradative enzymes. There is also competition between ethanol and retinoic acid precursors. Depletion of retinol and retinoic acid precursors is associated with hepatic and extrahepatic pathology, including carcinogenesis and fetal defects. Unfortunately, correction of vitamin A deficiency by supplementation is complicated by the intrinsic hepatotoxicity of retinol, which is potentiated by concomitant alcohol consumption. Furthermore, the precursor of vitamin A, beta-carotene, interacts with ethanol, interfering with the conversion of beta-carotin to retinol, while the combination of beta-carotene with ethanol results in hepatotoxicity.

In smokers who also consume alcohol, beta-carotene supplementation promotes pulmonary cancer and possibly cardiovascular complications.

It was already known in 1987 that foods that provide large amounts of retinol increase the risk of cancer of the esophagus (2549,2550), and in an epidemiological study the increased risk of cancer associated with the use of cigarettes and alcohol was enhanced by the ingestion of foods containing retinol (2551).

The effect of alcohol abuse, one of the most common aggravating factors in vitamin A toxicity, has been elucidated. Vitamin A toxicity was potentiated in patients who took 10 000 IU/day for sexual dysfunction, and this effect was attributed to excess alcohol consumption (2552). In animals, potentiation of vitamin A toxicity by ethanol resulted in striking hepatic inflammation and necrosis accompanied by a rise in serum glutamate dehydrogenase and aspartate transaminase (2553).

Alcohol may also act indirectly by causing liver disease, which in turn can affect the capacity of the liver to export vitamin A, thereby enhancing its local toxicity. In alcoholics the carrying capacity of retinol binding protein was increased, even in those with low serum retinol concentrations (2554). In such cases, caution in the amount of vitamin A used for therapy is recommended. Similarly, diets that are severely deficient in protein can affect the capacity of the liver to export vitamin A and enhance its hepatotoxicity.

In contrast to retinoids, carotenoids were considered non-toxic, even when taken chronically in large amounts, until recently, when it was found that ethanol interacts with carotenoids, interfering with their conversion to retinol. In baboons, the consumption of ethanol together with beta-carotene resulted in more striking hepatic injury than consumption of either compound alone (2555). This interaction occurred at a total dose of 7.2–10.8 mg of beta-carotene per Joule of diet. This dose is common in people who take supplements and is the same order of magnitude used in the Beta-Carotene and Retinol Efficacy Trial (CARET) (30 mg/day) (2556) and in another study (20 mg/day for 12 weeks) (2557). The amount of alcohol given to the baboons was equivalent to that taken by an average alcoholic. The well-known toxicity of ethanol was potentiated by large amounts of beta-carotene, and the concomitant administration of both beta-carotene and alcohol resulted in striking liver lesions, characterized by increased activity of plasma liver enzymes, an inflammatory response, and striking autophagic vacuoles and alterations in the endoplasmic reticulum and mitochondria (2558).

Besides its hepatotoxic effects, beta-carotene supplementation can also cause cardiovascular complications in smokers and potentiate carcinogenicity. The Alpha-Tocopherol, Beta-Carotene and Cancer Prevention Study (ATBC) (2559) and CARET (2556) showed that supplementation of beta-carotene in smokers increased the incidence of death from coronary artery disease. Recent results suggest that beta-carotene participates as a pro-oxidant in the oxidative degradation of LDL, and that raised LDL concentrations may cancel the protective effect of alpha-tocopherol (2560).

The two trials also showed that beta-carotene supplementation increased the incidence of pulmonary cancer in smokers. Because heavy smokers are commonly heavy drinkers it was supposed that alcohol might have contributed to the increased incidence of lung cancer. Subsequent analysis showed that there was indeed a relation between the incidence of pulmonary cancer and the amount of alcohol consumed.

As detrimental effects result from deficiency as well as an excess of retinoids and carotenoids, and since both have similar adverse effects in terms of fibrosis, carcinogenesis, and possibly embryotoxicity, therapeutic measures must pay attention to the narrow therapeutic window, especially in drinkers, in whom alcohol narrows the therapeutic window even further by promoting the depletion of retinoids and by potentiating their toxicity (2561).

Vitamin A: Retinoids

Although the various retinoids have similar toxicity profiles, they differ in the extent to which they affect various body systems. Cutaneous and mucous membrane symptoms (up to 70%) are by far the most prominent adverse effects.

Acitretin

Capillary leak syndrome with associated edema has been reported following acitretin therapy.

- A 79-year-old man with extensive psoriasis and joint involvement was given acitretin 10 mg/day for 12 days (2562). He developed myalgia and non-pitting edema associated with large hemorrhagic lesions, and weight gain of 13 kg. Laboratory values were normal, except for moderate hypoproteinemia and hypoalbuminemia, and raised creatine kinase, myoglobin, and aspartate transaminase. The clinical and laboratory findings were suggestive of capillary leak syndrome. Withdrawal of acitretin resulted in slow regression of the edema over 2 months.

Isotretinoin

Patients who use isotretinoin have a 50% incidence of conjunctivitis and irritation of the eyes. Musculoskeletal symptoms occur in up to 15% of users. Hypersensitivity reactions are rare and consist of occasional drug rashes. The occurrence of sarcomas in patients treated with isotretinoin may well be a chance finding (SEDA-21, 164); retinoids may prevent or even cure certain malignancies (2563). All the retinoids are strongly teratogenic (2564–2566).

The incidence and time-course of adverse events during a 4-month course of oral isotretinoin (1 mg/kg) for severe acne have been studied prospectively in 189 patients (2567). Most of the adverse events were most often reported during the first 3 months of treatment. However, only a few patients were seen every month as scheduled and only 50 of 189 filled in the questionnaire at 4 months.

Electrocardiographic changes have been attributed to isotretinoin (2568).

- An 18-year-old Caucasian man had been taking isotretinoin 1 mg/kg/day for severe acne for 3 months. He developed a fever, headache, rigidity of the neck and limbs, masseter twitch, and diffuse myalgia mainly localized to the buttocks and the lumbar region. He had a fever (38°C), a sinus tachycardia (140/minute) that did not correlate with the fever, and urinary retention. Renal stones and urinary tract infection were excluded. His liver enzymes were increased four-fold. An electrocardiogram showed sinus tachycardia with right branch bundle block. Electromyography showed a diffuse myopathy, but a muscle biopsy was not specific. A pulmonary CT scan showed interstitial pneumonitis with a small amount of encysted pleural effusion. A whole body bone scan showed dense radioactive areas at the upper wedge of the scapula and at vertebrae T4 to T7. There was hyperostosis in the vertebrosternal joint areas. All other laboratory tests were normal or negative. Isotretinoin was withdrawn and all the symptoms and signs disappeared after 3 weeks. Reintroduction 10 days later of isotretinoin 0.5 mg/kg/day for 1 week caused myalgia and right bundle branch block, but without sinus tachycardia. Withdrawal again produced resolution.

The authors recognized that all the adverse effects of isotretinoin noted in this case have been reported before, except the combination of right bundle branch block with sinus tachycardia and urinary retention. They suggested that electrocardiography should be performed in patients with systemic illnesses who are taking oral isotretinoin.

Tretinoin

In a retrospective study, 22 children taking tretinoin (median age 9.3 years, range 1.8–16.3) for a median of 38 (6–138) days were compared with 22 taking conventional therapy (median age 12.3 years, range: 3.2–16.7) (2569). Overall, 12 of 22 patients had symptoms associated with tretinoin (Table 5). Three developed the retinoic acid syndrome.

Retinoic acid syndrome

The retinoic acid syndrome is a generalized severe capillary leakage syndrome with leukocyte activation, which results in weight gain, pulmonary infiltrates or pleural effusions with acute respiratory distress, and fever without infection. It occurs in up to 25% of patients. Postmortem reports of patients who have died from this syndrome have shown infiltration of maturing myeloid cells and edema in the lungs (2570). The myeloid cells are considered to release various types of cytokines responsible for symptoms such as fever, weight gain, and heart failure. Improvement can be obtained by glucocorticoid treatment, which suppresses the effects of cytokines. Chest CT provides an accurate assessment of the size, number, and distribution of pulmonary opacities associated with the syndrome, as has been demonstrated anecdotally (2571).

Table 5 Features of tretinoin toxicity in 22 patients

Organ system	*N (%)*	*Symptoms*	*N (%)*
Tretinoin-related toxicity	12 (2155)	Retinoic acid syndrome	3 (2151)
Cardiovascular	5 (2140)	Pericardial effusion	1 (2152)
		Weight gain	5 (2140)
		Arterial hypotension	2 (2153)
Respiratory	3 (2151)	Adult respiratory distress syndrome	2 (2153)
		Pleural effusion	1 (2152)
		Pulmonary infiltrates	1 (2152)
Nervous system	6 (2144)	Headache	6 (2144)
		Raised intracranial pressure	1 (2152)
Liver and metabolism	12 (2155)	Raised transaminases	6 (2144)
		Raised bilirubin	1 (2152)
		Raised triglycerides	2 (2153)
Hematologic	4 (2154)	Leukocytosis	4 (2154)
Skin andmusculoskeletal	4 (2154)	Dry skin	1 (2152)
		Itching	1 (2152)
		Joint/bone/muscle pain	3 (2151)
		Osteonecrosis	1 (2152)

Of 69 patients with acute promyelocytic leukemia treated with tretinoin for 5 years, 15 developed retinoic acid syndrome (2572). The following features were found on chest radiographs: an increased cardiothoracic ratio, an increased pedicle width, pulmonary congestion in 13, pleural effusion in 11, ground-glass opacities, septal lines, and peribronchial cuffing in 9, consolidation and nodules in 7, and an air bronchogram in 5. Three patients had pulmonary hemorrhages and bilateral, diffuse, poorly delineated nodules and ground-glass opacities on radiography. Lung infiltrates cleared completely within 8 days after administration of prednisolone.

- A 69-year-old man developed dyspnea, hypoxia, and heart failure 4 days after starting to take tretinoin 70 mg/day. His highest white blood cell count was 72 × 109/l. A plain chest X-ray showed two pulmonary opacities, increased attenuation in the left lower lobe, and bilateral pleural effusions, but a chest CT also showed multiple irregular-shaped opacities localized in the centrilobular and subpleural regions. He improved over 10 days with prednisolone (total dose 5750 mg) and daunorubicin (total dose 360 mg).

The retinoic acid syndrome, its incidence and clinical course, has been investigated in 167 patients taking tretinoin as induction and maintenance therapy for acute promyelocytic leukemia (2573). The syndrome did not occur during maintenance therapy. During induction it occurred in 44 patients (26%) at a median of 11 (range 2–47) days. The major manifestations included respiratory distress (84%), fever (81%), pulmonary edema (54%), pulmonary infiltrates (52%), pleural or pericardial effusions (36%), hypotension (18%), bone pain (14%), headache (14%), congestive heart failure (11%), and acute renal insufficiency (11%). The median white blood cell count was 1.45 × 10^9/l at diagnosis and 31 × 10^9/l (range 6.8–72 × 10^9/l) at the time the syndrome developed. Tretinoin was continued in eight of the 44 patients, with subsequent resolution in 7. It was withdrawn in 36 patients and then reintroduced in 19, after which the syndrome recurred in 3, with one death attributable to reintroduction of the drug. Ten of these 36 patients received chemotherapy without further tretinoin, and 8 achieved complete remission. Of seven patients in whom tretinoin was not reintroduced and who were not given chemotherapy, five achieved complete remission and two died. Two deaths were definitely attributable to the syndrome.

In 63 patients with acute promyelocytic leukemia taking tretinoin (60 mg/day) the rates of leukocytosis, intracranial hypertension, and retinoic acid syndrome were 57%, 9.5%, and 3.2% respectively; the death rate was 11% (2574). The authors suggested that progressive leukocytosis during tretinoin therapy should be an indication for chemotherapy (for example, with homoharringtonine); if the white cell count excceds 10 × 10^9/l before treatment, the patient should be given homoharringtonine only; if it is below 5.0 × 10^9/l homoharringtonine plus tretinoin should be used.

A syndrome similar to that of the retinoic acid syndrome occurred after 10 days of tretinoin therapy in a patient with a relapse of acute myeloblastic leukemia (2575).

- A 75-year-old woman whose acute myeloblastic leukemia relapsed was treated with one dose of intravenous idarubicin (10 mg/m^2), cytarabine 20 mg subcutaneously for 10 days, and oral tretinoin 45 mg/m^2/day. Ten days later she developed a persistent fever. A chest X-ray and a CT scan showed bilateral pleural effusions and interstitial infiltrates, but no pulmonary embolus. Tretinoin was withdrawn and she was given intravenous dexamethasone 10 mg every 12 hours. Her fever disappeared within 24 hours and her respiratory distress gradually improved during the next 24–48 hours. A chest X-ray 7 days later showed total resolution.

The incidence, clinical features, and outcome of retinoic acid syndrome have been analysed in 413 cases of newly diagnosed acute promyelocytic leukemia (2570). Patients under 65 years old with a white blood cell count below 5 x 10^9/l were initially randomized to tretinoin followed by chemotherapy or to tretinoin with chemotherapy started on day 3. In patients with white cell counts over 5 × 10^9/l chemotherapy was rapidly added if the white cell count was greater than 6, 10, and 15 × 10^9/l by days 5, 10, and 15 of tretinoin treatment. The retinoic acid syndrome occurred during induction treatment in 64 of 413 patients (15%). Clinical signs developed after a median of 7 (range 0–35) days. In two cases they were present before the start of treatment; in 11 they occurred on recovery from the phase of aplasia due to the addition of chemotherapy. Respiratory distress (98% of patients), fever (81%), pulmonary infiltrates (81%), weight gain (50%), pleural effusion (47%), renal failure (39%), pericardial effusion (19%), cardiac failure (17%), and hypertension (12%) were the main clinical signs. Mechanical ventilation was required in 13 patients and dialysis in 2. A total of 55 patients (86%) who experienced the retinoic acid syndrome achieved complete remission, compared with 94% of patients who had no retinoic acid syndrome, and nine died of the syndrome. None of the patients with complete remission who received tretinoin for maintenance had recurrence of the syndrome. The syndrome was associated with a lower event-free survival and survival at 2 years.

Retinoic acid syndrome has been reported in a patient who developed diffuse alveolar hemorrhage while being treated with tretinoin for acute promyelocytic leukemia (2576).

- An 18-year-old woman developed promyelocytic leukemia and was given tretinoin and dexamethasone. At 15 days she developed significant hemoptysis and respiratory failure, requiring mechanical ventilation. Her temperature was 39.1°C and she had disseminated intravascular coagulation. A lung biopsy showed diffuse interstitial neutrophilic infiltration, interstitial fibrinoid necrosis, and diffuse alveolar hemorrhage; pulmonary capillaritis was diagnosed. She was given intravenous methylprednisolone 1 g/day for 3 days followed by a tapering dose of oral prednisolone. She subsequently completed a full 45-day course of tretinoin.

Vitamin B_{12} (cobalamins)

Syncope was induced by intramuscular injection of hydroxocobalamin in a paraplegic patient (2577).

- An 81-year-old man had been a prisoner of war 58 years before and had been fed mainly rice. At that time he had developed complete loss of power and sensation in the right leg, weakness in the left leg, and weakness and impaired sensation in the right arm; he had become almost blind and deaf in both ears. When he was given vitamin B tablets his sight became almost normal within 1 month, but recovery of motor power was slower. In 1945 he returned home and was given injections of vitamin B_{12} on and off without adverse effects. In 1997 he had a serum vitamin B_{12} concentration of 91 ng/l (reference range 223–1132), a normal folate concentration, a hemoglobin of 15.7 g/dl, and an MCV of 95 fl. He was given intramuscular vitamin B_{12} 1 mg every 4–6 weeks. In 1998 he fainted about 10 minutes after an injection of vitamin B_{12}. His blood pressure was 60/30 mmHg. Soon after the next injection the patient fainted again and developed red patches and blisters over his arms, chest, and legs. Oral vitamin B_{12} without intrinsic factor resulted in 3.47% excretion in 24 hours. Vitamin B_{12} with intrinsic factor led to a urinary excretion of 1.49%. No reactions to oral vitamin B_{12} were reported.

Vitamin D analogues

Some individuals develop hypertension in response to vitamin D, which in some of them may be directly related to hypercalcemia and which may be reversible when renal function is normalized (2578). Metastatic calcification is observed in various tissues (2579), but arterial calcification is the most usual. In some patients undergoing dialysis, calcification of the blood vessels has been so extensive that cannulation could not be performed (2580).

Hypercalcemia due to vitamin D intoxication with chest pain and electrocardiographic changes mimicking acute myocardial infarction has been reported (2581).

- A 78-year old man who had taken alfacalcidol 1.0 μg/day for osteoporosis and also made an effort to maximize his dietary calcium intake by drinking milk and mineral water developed general fatigue, anorexia and chest pain. An electrocardiogram showed tachycardia, ST elevation in leads V_1-V_3, and diffuse T wave flattening. Emergency coronary angiography showed no significant stenoses. His laboratory results then showed hypercalcemia (4.1 mmol/l). He was given intravenous saline and furosemide and alfacalcidol was withdrawn. His symptoms gradually improved and the electrocardiogram returned to normal as the serum calcium fell to normal. On re-evaluation of the electrocardiogram extreme shortening of the QT_c interval was noted.

Chest pain is not a common symptom of hypercalcemia, but in patients with suspected acute myocardial infarction, when emergency calcium assays are not available, a shortened QT_c interval is a valuable tool to diagnose hypercalcemia.

Most of the adverse effects of alfacalcidol previously reported have not been serious and usually disappear shortly after withdrawal. However, potentially life-threatening congestive heart failure and axonal polyneuropathy has been reported in a patient with systemic lupus erythematosus (2582).

- A 29-year old women with systemic lupus erythematosus, who had taken alfacalcidol 0.5 μg/day for 9 months to prevent glucocorticoid-induced osteoporosis, presented with painful swellings in the legs, followed by numbness and muscle weakness in both arms and legs. She then developed shortness of breath and bilateral leg edema. Blood tests showed no abnormalities to suggest exacerbation of systemic lupus erythematosus. A chest X-ray showed a right pleural effusion and cardiac enlargement. Alfacalcidol was withdrawn. She recovered from the congestive heart failure within 2 months and the symptoms of axonal polyneuropathy had almost completely disappeared after 4 months.

The mechanisms of toxicity leading to congestive heart failure and axonal polyneuropathy are unknown. However, there is a vitamin D receptor in human heart (2583) and murine peripheral nerves (2584), and toxicity could be related to polymorphism of vitamin D receptor genes.

Vitamin E

Vitamin E 300 mg/day can increase serum cholesterol by an average of 1.9 mmol/l (74 mg/100 ml) (2585). It is therefore advisable to use vitamin E with caution in patients with a family history of heart disease. Some studies have shown that vitamin E can prevent atherosclerosis and serious adverse events were not reported in these studies (2586–2588).

A possible link between long-term use of vitamin E and an increased risk of heart failure has been identified from an extensive analysis of experimental and epidemiological data related to the possible prevention of cancer and cardiovascular events by vitamin E supplementation (2589). In fact, clinical trials have generally failed to confirm benefits, possibly because of their relatively short duration. The objective of this study was to evaluate whether long-term supplementation with vitamin E reduces the risks of cancer, death from cancer, and major cardiovascular events. The study, HOPE-TOO (HOPE–The Ongoing Outcomes), was an extension of a randomized, double-blind, placebo-controlled international study (the Heart Outcomes Prevention Evaluation, HOPE, study) in patients aged at least 55 years with vascular disease or diabetes mellitus. Of the initial 267 centers that had enrolled 9541 patients in HOPE, 174 centers participated in HOPE-TOO. Of 7030 patients enrolled at these centers, 916 were already dead at the beginning of the extension study, 1382 refused to participate, 3994 continued to take part, and 738 agreed to passive follow-up (median duration 7.0 years). The intervention comprised a daily dose of natural source vitamin E 400 IU or matching placebo. The primary outcome measures were the incidences of cancer, cancer deaths, and major cardiovascular events (myocardial infarction, stroke, and cardiovascular deaths). Secondary outcomes included heart failure, unstable angina, and revascularization. Among all the HOPE patients, there were no significant differences in the primary analysis between vitamin E and placebo:

- for cancer incidence, 552 (11.6%) versus 586 (12.3%) respectively (RR = 0.94; 95% CI = 0.84, 1.06);
- for cancer deaths 156 (3.3%) versus 178 (3.7%) respectively (RR = 0.88; 95% CI = 0.71, 1.09);
- for major cardiovascular events 1022 (21.5%) versus 985 (20.6%) respectively (RR = 1.04; 95% CI = 0.96, 1.14).

However, one significant finding was that patients who took vitamin E had a higher risk of heart failure (RR = 1.13; 95% CI = 1.01, 1.26) and hospitalization for heart failure (RR = 1.21; 95% CI = 1.00, 1.47). Similarly, among patients enrolled at the centers that participated in HOPE-TOO, there were no differences in cancer incidence, cancer deaths, or major cardiovascular events, but higher rates of heart failure and hospitalization for heart failure. The authors concluded that in patients with vascular disease or diabetes mellitus, long-term vitamin E supplementation does not prevent cancer or major cardiovascular events and may increase the risk of heart failure.

Vitamin K analogues

Results of intravenous administration of 100 sequential doses of vitamin K in 45 patients (34 men, 11 women) have been retrospectively reviewed (2590). Vitamin K_1 was administered as Aquamephyton. The dose averaged 9.7 (range 1–20) mg. The mean age was 44 (range 18–66) years. Many of the patients were seriously ill and had significant underlying surgical and medical problems, as evidenced by the fact that 16 died. However, none of the deaths was attributable to vitamin K. In the 11 instances in which the duration of administration was noted, the mean time was 49 (range 12.5–60) minutes. Only one patient had an episode of transient hypotension.

- A 37-year-old man with a history of ethanol abuse presented with hepatic failure and non-cardiogenic pulmonary edema after an overdose of paracetamol, codeine, ibuprofen, and diazepam. He received two doses of Aquamephyton (phytonadione) over 2 days and 10 minutes after the second dose his blood pressure fell to 79/40 mmHg.

In this study the incidence of adverse effects was 1%, contrasting with the impression that one obtains from reading the medical literature on vitamin K, which suggests an overall incidence of less than 1%.

Voriconazole

QT interval prolongation has been attributed to voriconazole.

- A 15-year-old girl with acute lymphoblastic leukemia, who was taking voriconazole for a Fusarium infection, developed asymptomatic bradycardia, QT interval prolongation, and non-sustained, polymorphic ventricular tachycardia, which recurred on rechallenge (2591). Voriconazole concentrations and metabolism were within expected values.

This patient was also taking several potentially arrhythmogenic drugs and had hypokalemia during the first episode; however, during rechallenge the electrolytes were within normal values and the only co-administered drugs included vancomycin and amikacin. The authors concluded that this observation suggested that closer attention should be paid to cardiac rhythm monitoring during voriconazole treatment.

Xenon

Xenon is a heavy gas (symbol Xe; atomic no 54) that is normally present in the atmosphere. It has been used as an anesthetic and as a diagnostic tool in functional neuroimaging (2592).

Xenon has many characteristics of the ideal anesthetic (2593). It has no effects on the cardiovascular system and has low solubility, enabling faster induction of and emergence from anesthesia. Although its high cost limits its use, the development of closed rebreathing systems has led to further interest.

Yellow fever vaccine

Two episodes have been reported from the Ivory Coast (1974) and Ghana (1982) (2594). In the Ivory Coast, there were 39 cases of severe reactions with eight deaths following a mass campaign, in which 730 000 persons were immunized. The clinical features were uniform: a few hours after immunization the vaccinees developed signs of local inflammation. In severe cases, edema and inflammation were followed by cardiovascular collapse. Bacterial contamination could have been the cause: during the campaign, five-dose vaccine ampoules were pooled to prepare 50 and 100 doses for use in jet injectors. In Ghana, six vaccinees developed fulminant reactions 2–6 hours after immunization, including two deaths. The clinical features resembled those in the Ivory Coast episode. In 2001, during a mass vaccination campaign against yellow fever in Abidjan, the Ivory Coast, more than 2.6 million doses were administered and 87 adverse events were notified, of which 41 were considered to be vaccine-related. There was one case of anaphylaxis and 26 cases of urticaria, five of which were generalized (2595).

People who are known to be suffering from allergy must be tested intradermally before immunization.

Yohimbine

In 25 unmedicated subjects with hypertension yohimbine 22 mg increased mean blood pressure by an average of 5 mm Hg, plasma noradrenaline by 66%, and plasma dihydroxyphenylglycol by 25% at 1 hour after administration (2596). The magnitude of the pressor response was unrelated to baseline pressure but correlated positively with baseline noradrenaline concentration and with the yohimbine-induced increment in plasma noradrenaline.

In 25 healthy volunteers and 29 sex- and age-matched untreated hypertensive patients yohimbine 10 mg caused a significant increase in diastolic pressure only in the hypertensive patients (2597).

In patients taking tricyclic antidepressants, hypertension can occur at a dose of 4 mg tds. The toxicity of yohimbine can be enhanced by other drugs, such as phenothiazines.

Angina pectoris has been attributed to yohimbine (2598).

- A patient with CREST syndrome (calcinosis, Raynaud's phenomenon, esophageal dysfunction, sclerodactyly, and telangiectasia) paradoxically experienced worsening of Raynaud's phenomenon when using yohimbine for erectile dysfunction (2599).

Zinc

- A very old woman accidentally ingested a formulation containing zinc and copper sulphate (2600). At 90 minutes after ingestion, the peak plasma concentrations were 20 mg/ml for zinc and 2 μg/ml for copper. The major complications were gastric and bronchial inflammation, due to the corrosive properties of these compounds. There were also systemic manifestations, with cardiovascular failure and renal insufficiency, but the patient made a complete recovery.
- A schizophrenic patient died after ingesting 461 coins, the first reported case of a death associated with zinc intoxication (2601). The patient presented with clinical manifestations consistent with the local corrosive as well as systemic effects of zinc intoxication and died 40 days after admission with multisystem organ failure. Many British post-1981 pennies, which contain mostly zinc, were severely corroded through prolonged contact with gastric juice. Tissue samples of the kidneys, pancreas, and liver obtained at autopsy contained high concentrations of zinc and showed acute tubular necrosis, mild fibrosis, and acute massive necrosis respectively.

Ziprasidone

Three extensive reviews of ziprasidone have devoted particular attention to the possibility of QT interval prolongation (2602–2604). Ziprasidone up to 160 mg/day prolongs the QT_c interval on average 5.9–9.7 ms (data from 4571 patients); a QT_c interval of over 500 ms was seen in two of 2988 ziprasidone recipients and in one of 440 placebo recipients. In an open study in 31 patients with schizophrenia, ziprasidone given for 21–29 days prolonged the QT_c interval by 20 ms (95% CI = 14, 26).

- A 38-year-old woman with a psychosis who took 4020 mg of ziprasidone had borderline intraventricular conduction delay (QRS duration 111 ms); the QT_c interval was 445 ms (2605). She oscillated between being drowsy and calm, and alert and agitated; her blood pressure fell from 129/81 to 99/34 mmHg 4 hours later. She also had diarrhea and urinary retention.

Reports of cardiovascular adverse events led the FDA to include a "black box" warning in the official labelling of the product. Furthermore, in a warning letter the FDA stated that the manufacturers, Pfizer Inc, had promoted the product in a misleading manner, because they had minimized the greater capacity of ziprasidone to cause QT interval prolongation, as well as its potential to cause torsade de pointes and sudden death (2606). The Indications and Usage section of the manufacturers' approved product labelling stated that ziprasidone has a "greater capacity to prolong the QT/QT_c interval compared with several other antipsychotic drugs" and that this effect "is associated in some other drugs with the ability to cause torsade de pointes-type arrhythmia, a potentially fatal polymorphic ventricular tachycardia, and sudden death. Whether ziprasidone will cause torsade de pointes or increase the rate of sudden death is not yet known." Although torsade de pointes was not observed in pre-marketing studies, experience is too limited to rule out an increased risk. Furthermore, the FDA has received several spontaneous reports of QT interval prolongation greater than 500 ms, all suggestive of a potential risk of this dysrhythmia. There are also reports of sudden death of unknown cause, which could have been due to unrecognized torsade de pointes.

The question of the effect of ziprasidone on the QT_c interval has been analysed in an extensive review (2607). It is generally accepted that 440 ms is the upper limit of normality, and the authors concluded that ziprasidone clearly prolongs the QT_c interval, but that the clinical consequences of this effect are uncertain, and that so far no direct association with torsade de pointes, sudden death, or increased cardiac mortality has been observed. However, they provided recommendations about its use.

1. Before starting treatment, conditions that might predispose to a higher risk of QT_c interval prolongation or

torsade de pointes (either cardiac, metabolic, or others) should be ruled out; a careful medical history is recommended.

2. In people with stress, shock, and extreme or prolonged physical exertion already taking ziprasidone, therapy can be continued, but electrocardiographic monitoring is advised when episodes are severe or prolonged.
3. Although no special precautions are suggested when ziprasidone is co-prescribed with metabolic inhibitors, drugs that do not inhibit hepatic enzymes should be preferred.
4. Ziprasidone should be avoided in the following cases:
   a. when there is evidence of long QT syndrome, a history of myocardial infarction or ischemic heart disease, and persistent or recurrent bradycardia; a cardiologist should be consulted if uncertain.
   b. in conditions often associated with electrolyte disturbance, including anorexia or bulimia; electrolyte disturbance should be always ruled out by means of a blood sample, and low concentrations of calcium, potassium, or magnesium should be corrected before treatment.
   c. in patients taking other drugs that prolong the QT interval, including certain antidysrhythmic drugs, antidepressants, antihistamines, antimicrobial drugs, and others; if co-prescription is unavoidable, electrocardiographic monitoring by a specialist is advised.
   d. in patients taking thioridazine, droperidol, sertindole, or pimozide; these drugs should be always withdrawn before starting ziprasidone.
   e. in patients taking other antipsychotic drugs; this is permissible only when cross-tapering and excessive doses of antipsychotic drugs should not be used in combination.

Zolpidem

The effect of hypnotics during spaceflight on the high incidence of post-flight orthostatic hypotension has been studied in astronauts who took no treatment (n = 20), temazepam 15 or 30 mg (n = 9), or zolpidem 5 or 10 mg (n = 8) (323). Temazepam and zolpidem were only taken the night before landing. On the day of landing, systolic pressure fell significantly and heart rate increased significantly in the temazepam group, but not in the control group or in the zolpidem group. Temazepam may aggravate orthostatic hypotension after spaceflight, when astronauts are hemodynamically compromised. It should not be the initial choice as a sleeping aid for astronauts; zolpidem may be a better choice.

References

1. Hunt TL, Cramer M, Shah A, Stewart W, Benedict CR, Hahne WF. A double-blind, placebo-controlled, dose-ranging safety evaluation of single-dose intravenous dolasetron in healthy male volunteers. J Clin Pharmacol 1995;35(7):705–12.
2. Arsura EL, Brunner NG, Namba T, Grob D. Adverse cardiovascular effects of anticholinesterase medications. Am J Med Sci 1987;293(1):18–23.
3. Bjerke RJ, Mangione MP. Asystole after intravenous neostigmine in a heart transplant recipient. Can J Anaesth 2001;48(3):305–7.
4. Okun MS, Charriez CM, Bhatti MT, Watson RT, Swift TR. Asystole induced by edrophonium following beta blockade. Neurology 2001;57(4):739.
5. Cain JW. Hypertension associated with oral administration of physostigmine in a patient with Alzheimer's disease. Am J Psychiatry 1986;143(7):910–2.
6. Svartling N, Lehtinen AM, Tarkkanen L. The effect of anaesthesia on changes in blood pressure and plasma cortisol levels induced by cementation with methylmethacrylate. Acta Anaesthesiol Scand 1986;30(3):247–52.
7. Esemenli BT, Toker K, Lawrence R. Hypotension associated with methylmethacrylate in partial hip arthroplasties. The role of femoral canal size. Orthop Rev 1991;20(7):619–23.
8. Arac SS, Ercan ZS, Turker RK. Prevention by aprotinin of the hypotension due to acrylic cement implantation into the bone. Curr Ther Res 1980;28:554.
9. Tryba M, Linde I, Voshage G, Zenz M. Histaminfreisetzung und kardiovaskulare Reaktionen nach Implantation von Knochenzement bei totalem Huftgelenkersatz. [Histamine release and cardiovascular reactions to implantation of bone cement during total hip replacement.] Anaesthesist 1991;40(1):25–32.
10. Worthen M, Placik B, Argano B, MacCanon DM, Luisada AA. On the mechanism of epinephrine-induced pulmonary edema. Jpn Heart J 1969;10(2):133–41.
11. Ersoz N, Finestone SC. Adrenaline-induced pulmonary oedema and its treatment. A report of two cases. Br J Anaesth 1971;43(7):709–12.
12. Johnston SL, Unsworth J, Gompels MM. Adrenaline given outside the context of life threatening allergic reactions. BMJ 2003;326:589–90.
13. Douglass JA, O'Hehir RE. Adrenaline and non-life threatening allergic reactions. BMJ 2003;327:226–7.
14. Davis CO, Wax PM. Prehospital epinephrine overdose in a child resulting in ventricular dysrhythmias and myocardial ischemia. Pediatr Emerg Care 1999;15(2):116–8.
15. Jackman WM, Friday KJ, Anderson JL, Aliot EM, Clark M, Lazzara R. The long QT syndromes: a critical review, new clinical observations and a unifying hypothesis. Prog Cardiovasc Dis 1988;31(2):115–72.
16. Cucchiaro G, Rhodes LA. Unusual presentation of long QT syndrome. Br J Anaesth 2003;90:804–7.
17. Kunimatsu T. Two cases of atrioventricular junctional rhythm induced by administration of regional injection of epinephrine. Acta Anesthesiol Scand 2004;48:928.
18. Yang JJ, Wang QP, Wang TY, Sun J, Wang ZY, Zuo D, Xu JG. Marked hypotension induced by adrenaline contained in local anesthetic. Laryngoscope 2005;115(2):348–52.
19. Sia S, Sarro F, Lepri A, Bartoli M. The effect of exogenous epinephrine on the incidence of hypotensive/bradycardic events during shoulder surgery in the sitting position during interscalene block. Anesth Analg 2003;97:583–8.
20. Hahn I-H, Schnadower D, Dakin RJ, Nelson LS. Cellular phone interference as a cause of acute epinephrine poisoning. Ann Emerg Med 2005;46:298–9.
21. Myerson SG, Mitchell ARJ. Mobile phones in hospitals. BMJ 2003;326:460–1.

22. Derbyshire SWG, Burgess A. Use of mobile phones in hospitals. BMJ 2006;333:767–8.
23. Adams MC, McLaughlin KP, Rink RC. Inadvertent concentrated epinephrine injection at newborn circumcision: effect and treatment. J Urol 2000;163(2):592.
24. Galoo E, Godon P, Potier V, Vergeau B. Fibrillation auriculaire compliquant une hémostase endoscopique rectale par injection d' adrénaline. [Atrial fibrillation following a rectal endoscopic injection using epiphedrine solution.] Gastroenterol Clin Biol 2002;26(1):99–100.
25. Mazzocca AD, Meneghini RM, Chhablani R, Badrinath SK, Cole BJ, Bush-Joseph CA. Epinephrine-induced pulmonary edema during arthroscopic knee surgery. J Bone Joint Surg 2003;85A:913–5.
26. Jensen KH, Werther K, Stryger V, Schultz K, Falkenberg B. Arthroscopic shoulder surgery with epinephrine saline irrigation. Arthroscopy 2001;17:578–81.
27. Cartwright MS, Reynolds PS. Intracerebral hemorrhage associated with over-the-counter inhaled epinephrine. Cerebrovasc Dis 2005;19:415–6.
28. Chelliah YR, Manninen PH. Hazards of epinephrine in transsphenoidal pituitary surgery. J Neurosurg Anesthesiol 2002;14(1):43–6.
29. Armbruster C, Robibaro B, Griesmacher A, Vorbach H. Endothelial cell compatibility of trovafloxacin and levofloxacin for intravenous use. J Antimicrob Chemother 2000;45(4):533–5.
30. Drummond GB, Ludlam CA. Is albumin harmful? Br J Haematol 1999;106(2):266–9.
31. Tjoeng MM, Bartelink AK, Thijs LG. Exploding the albumin myth. Pharm World Sci 1999;21(1):17–20.
32. Pandit SK, Dundee JW, Stevenson HM. A clinical comparison of pancuronium with tubocurarine and alcuronium in major cardiothoracic surgery. Anesth Analg 1971;50(6):926–35.
33. Coleman AJ, Downing JW, Leary WP, Moyes DG, Styles M. The immediate cardiovascular effects of pancuronium, alcuronium and tubocurarine in man. Anaesthesia 1972;27(4):415–22.
34. Baraka A. A comparative study between diallylnortoxiferine and tubocurarine. Br J Anaesth 1967;39(8):624–8.
35. Brandli FR. Pancuronium und Alcuronium: Ein klinischer Vergleich. [Pancuronium and alcuronium: a clinical comparison.] Prakt Anaesth 1976;11:239.
36. Tammisto T, Welling I. The effect of alcuronium and tubocurarine on blood pressure and heart rate: a clinical comparison. Br J Anaesth 1969;41(4):317–22.
37. Hughes R, Chapple DJ. Effects on non-depolarizing neuromuscular blocking agents on peripheral autonomic mechanisms in cats. Br J Anaesth 1976;48(2):59–68.
38. Sosman JA, Kohler PC, Hank JA, Moore KH, Bechhofer R, Storer B, Sondel PM. Repetitive weekly cycles of interleukin-2. II. Clinical and immunologic effects of dose, schedule, and addition of indomethacin. J Natl Cancer Inst 1988;80(18):1451–61.
39. Richards JM, Barker E, Latta J, Ramming K, Vogelzang NJ. Phase I study of weekly 24-hour infusions of recombinant human interleukin-2. J Natl Cancer Inst 1988;80(16):1325–8.
40. Nasr S, McKolanis J, Pais R, Findley H, Hnath R, Waldrep K, Ragab AH. A phase I study of interleukin-2 in children with cancer and evaluation of clinical and immunologic status during therapy. A Pediatric Oncology Group Study. Cancer 1989;64(4):783–8.
41. Groeger JS, Bajorin D, Reichman B, Kopec I, Atiq O, Pierri MK. Haemodynamic effects of recombinant interleukin-2 administered by constant infusion. Eur J Cancer 1991;27(12):1613–6.
42. Citterio G, Pellegatta F, Lucca GD, Fragasso G, Scaglietti U, Pini D, Fortis C, Tresoldi M, Rugarli C. Plasma nitrate plus nitrite changes during continuous intravenous infusion interleukin 2. Br J Cancer 1996;74(8):1297–301.
43. Vial T, Descotes J. Clinical toxicity of interleukin-2. Drug Saf 1992;7(6):417–33.
44. White RL Jr, Schwartzentruber DJ, Guleria A, MacFarlane MP, White DE, Tucker E, Rosenberg SA. Cardiopulmonary toxicity of treatment with high dose interleukin-2 in 199 consecutive patients with metastatic melanoma or renal cell carcinoma. Cancer 1994;74(12):3212–22.
45. Citterio G, Fragasso G, Rossetti E, Di Lucca G, Bucci E, Foppoli M, Guerrieri R, Matteucci P, Polastri D, Scaglietti U, Tresoldi M, Chierchia SL, Rugarli C. Isolated left ventricular filling abnormalities may predict interleukin-2-induced cardiovascular toxicity. J Immunother Emphasis Tumor Immunol 1996;19(2):134–41.
46. Fragasso G, Tresoldi M, Benti R, Vidal M, Marcatti M, Borri A, Besana C, Gerundini PP, Rugarli C, Chierchia S. Impaired left ventricular filling rate induced by treatment with recombinant interleukin 2 for advanced cancer. Br Heart J 1994;71(2):166–9.
47. Junghans RP, Manning W, Safar M, Quist W. Biventricular cardiac thrombosis during interleukin-2 infusion. N Engl J Med 2001;344(11):859–60.
48. Lenihan DJ, Alencar AJ, Yang D, Kurzrock R, Keating MJ, Duvic M. Cardiac toxicity of alemtuzumab in patients with mycosis fungoides/Sezary syndrome. Blood 2004;104(3):655–8.
49. Basquiera AL, Berretta AR, Garcia JJ, Palazzo ED. Coronary ischemia related to alemtuzumab therapy. Ann Oncol 2004;15(3):539–40.
50. Ananthanarayan C. Sinus arrest after alfentanil and suxamethonium. Anaesthesia 1989;44(7):614.
51. Nathan HJ. Narcotics and myocardial performance in patients with coronary artery disease. Can J Anaesth 1988;35(3Pt 1):209–13.
52. de Vries JH, van Dorp WT, van Barneveld PW. A randomized trial of alcohol 70% versus alcoholic iodine 2% in skin disinfection before insertion of peripheral infusion catheters. J Hosp Infect 1997;36(4):317–20.
53. Clark JA, Gross TP. Pain and cyanosis associated with $alpha_1$-proteinase inhibitor. Am J Med 1992;92(6):621–6.
54. Johnston PS, Lebovitz HE, Coniff RF, Simonson DC, Raskin P, Munera CL. Advantages of alpha-glucosidase inhibition as monotherapy in elderly type 2 diabetic patients. J Clin Endocrinol Metab 1998;83(5):1515–22.
55. Standl E, Schernthaner G, Rybka J, Hanefeld M, Raptis SA, Naditch L. Improved glycaemic control with miglitol in inadequately-controlled type 2 diabetics. Diabetes Res Clin Pract 2001;51(3):205–13.
56. Scott LJ, Spencer CM. Miglitol: a review of its therapeutic potential in type 2 diabetes mellitus. Drugs 2000;59(3):521–49.
57. Hasegawa T, Yoneda M, Nakamura K, Ohnishi K, Harada H, Kyouda T, Yoshida Y, Makino I. Long-term effect of alpha-glucosidase inhibitor on late dumping syndrome. J Gastroenterol Hepatol 1998;13(12):1201–6.
58. Lindstrom J, Tuomilehto J, Spengler MThe Finnish Acargbos Study Group. Acarbose treatment does not change the habitual diet of patients with type 2 diabetes mellitus. Diabet Med 2000;17(1):20–5.

59. Holman RR, Cull CA, Turner RC. A randomized double-blind trial of acarbose in type 2 diabetes shows improved glycemic control over 3 years (U.K. Prospective Diabetes Study 44) Diabetes Care 1999;22(6):960–4.
60. Nishii Y, Aizawa T, Hashizume K. Ileus: a rare side effect of acarbose. Diabetes Care 1996;19(9):1033.
61. Odawara M, Bannai C, Saitoh T, Kawakami Y, Yamashita K. Potentially lethal ileus associated with acarbose treatment for NIDDM. Diabetes Care 1997;20(7):1210–1.
62. Azami Y. Paralytic ileus accompanied by pneumatosis cystoides intestinalis after acarbose treatment in an elderly diabetic patient with a history of heavy intake of maltitol. Intern Med 2000;39(10):826–9.
63. Oba K, Kudo R, Yano M, Watanabe K, Ajiro Y, Okazaki K, Suzuki T, Nakano H, Metori S. Ileus after administration of cold remedy in an elderly diabetic patient treated with acarbose. J Nippon Med Sch 2001;68(1):61–4.
64. Maeda A, Yokoi S, Kunou T, Murata T. [A case of pneumatosis cystoides intestinalis assumed to be induced by acarbose administration for diabetes mellitus and pemphigus vulgaris.]Nippon Shokakibyo Gakkai Zasshi 2002;99(11):1345–9.
65. Piche T, Raimondi V, Schneider S, Hebuterne X, Rampal P. Acarbose and lymphocytic colitis. Lancet 2000;356(9237):1246.
66. Ranieri P, Franzoni S, Trabucchi M. Alprazolam and hypotension. Int J Geriatr Psychiatry 1999;14(5):401–2.
67. Fujita M, Hatori N, Shimizu M, Yoshizu H, Segawa D, Kimura T, Iizuka Y, Tanaka S. Neutralization of prostaglandin E1 intravenous solution reduces infusion phlebitis. Angiology 2000;51(9):719–23.
68. Lewis GB, Hecker JF. Infusion thrombophlebitis. Br J Anaesth 1985;57(2):220–33.
69. Vale JA, Maclean KS. Amantadine-induced heart-failure. Lancet 1977;1(8010):548.
70. Patterson RN, Herity NA. Acute myocardial infarction following bupropion (Zyban). Q J Med 2002;95(1):58–9.
71. Johansson SA. Acute right heart failure during treatment with epsilon amino caproic acid (E-ACA). Acta Med Scand 1967;182(3):331–4.
72. Martin PD. ECG change associated with streptomycin. Chest 1974;65(4):478.
73. Kristensen KS, Hoegholm A, Bohr L, Friis S. Fatal myocarditis associated with mesalazine. Lancet 1990;335(8689):605.
74. Vayre F, Vayre-Oundjian L, Monsuez JJ. Pericarditis associated with longstanding mesalazine administration in a patient. Int J Cardiol 1999;68(2):243–5.
75. Sentongo TA, Piccoli DA. Recurrent pericarditis due to mesalamine hypersensitivity: a pediatric case report and review of the literature. J Pediatr Gastroenterol Nutr 1998;27(3):344–7.
76. Ishikawa N, Imamura T, Nakajima K, Yamaga J, Yuchi H, Ootsuka M, Inatsu H, Aoki T, Eto T. Acute pericarditis associated with 5-aminosalicylic acid (5-ASA) treatment for severe active ulcerative colitis. Intern Med 2001;40(9):901–4.
77. Oxentenko AS, Loftus EV, Oh JK, Danielson GK, Mangan TF. Constrictive pericarditis in chronic ulcerative colitis. J Clin Gastroenterol 2002;34(3):247–51.
78. Asirvatham S, Sebastian C, Thadani U. Severe symptomatic sinus bradycardia associated with mesalamine use. Am J Gastroenterol 1998;93(3):470–1.
79. Amin HE, Della Siega AJ, Whittaker JS, Munt B. Mesalamine-induced chest pain: a case report. Can J Cardiol 2000;16(5):667–9.
80. Reid J, Holt S, Housley E, Sneddon DJ. Raynaud's phenomenon induced by sulphasalazine. Postgrad Med J 1980;56(652):106–7.
81. Ahmad J, Siddiqui MA, Khan AS, Afzall S. Raynaud's phenomenon induced by sulphasalazine in a case of chronic ulcerative colitis. J Assoc Physicians India 1984;32(4):370.
82. Pedrosa Gil F, Grohmann R, Ruther E. Asymptomatic bradycardia associated with amisulpride. Pharmacopsychiatry 2001;34(6):259–61.
83. Ngouesse B, Basco LK, Ringwald P, Keundjian A, Blackett KN. Cardiac effects of amodiaquine and sulfadoxine-pyrimethamine in malaria-infected African patients. Am J Trop Med Hyg 2001;65(6):711–6.
84. Karch SB, Billingham ME. The pathology and etiology of cocaine-induced heart disease. Arch Pathol Lab Med 1988;112(3):225–30.
85. Ellenhorn DJ, Barceloux DG. Amphetamines. In: Medical Toxicology: Diagnosis and Treatment of Human Poisoning. New York: Elsevier Science Publishers, 1988:625.
86. Call TD, Hartneck J, Dickinson WA, Hartman CW, Bartel AG. Acute cardiomyopathy secondary to intravenous amphetamine abuse. Ann Intern Med 1982;97(4):559–60.
87. Findling RL, Biederman J, Wilens TE, Spencer TJ, McGrough JJ, Lopez FA, Tulloch SJ, on behalf of the SL1381.301 and .302 Study Groups. Short- and long-term cardiovascular effects of mixed amphetamine salts extended release in children. J Pediatr 2005;147:348–54.
88. Findling RL, Short EJ, Manos MJ. Short-term cardiovascular effects of methylphenidate and adderall. J Am Acad Child Adolescent Psychiatry 2001;40:525–9.
89. Gutgesell H, Atkins D, Barst R, Buck M, Franklin W, Humes R, Ringel R, Shaddy R, Taubert KA. AHA scientific statement. Cardiovascular monitoring of children and adolescents receiving psychotropic drugs. J Am Acad Child Adolesc Psychiatry 1999;38:1047–50.
90. Brennan K, Shurmur S, Elhendy A. Coronary artery rupture associated with amphetamine abuse. Cardiol Rev 2004;12:282–3.
91. Hung MJ, Kuo LT, Cherng WJ. Amphetamine-related acute myocardial infarction due to coronary artery spasm. Int J Clin Pract 2003;57:62–4.
92. Angrist B, Sanfilipo M, Wolkin A. Cardiovascular effects of 0.5 milligrams per kilogram oral d-amphetamine and possible attenuation by haloperidol Clin Neuropharmacol 2001;24(3):139–44.
93. Waksman J, Taylor RN Jr, Bodor GS, Daly FF, Jolliff HA, Dart RC. Acute myocardial infarction associated with amphetamine use. Mayo Clin Proc 2001;76(3):323–6.
94. Costa GM, Pizzi C, Bresciani B, Tumscitz C, Gentile M, Bugiardini R. Acute myocardial infarction caused by amphetamines: a case report and review of the literature. Ital Heart J 2001;2(6):478–80.
95. Zaidat OO, Frank J. Vertebral artery dissection with amphetamine abuse. J Stroke Cerebrovasc Dis 2001;10:27–9.
96. Turnipseed SD, Richards JR, Kirk D, Diercks DB, Amsterdam EA. Frequency of acute coronary syndrome in patients presenting to the emergency department with chest pain after methamphetamine use. J Emerg Med 2003;24:369–73.

97. Takasaki T, Nishida N, Esaki R, Ikeda N. Unexpected death due to right-sided infective endocarditis in a methamphetamine abuser. Legal Med 2003;5:65–8.
98. Hung M-J, Kuo L-T, Cherng W-J. Amphetamine-related acute myocardial infarction due to coronary artery spasm. Intl J Clin Pract 2003;57:62–4.
99. Pavese N, Rimoldi O, Gerhard A, Brooks D, Piccini P. Cardiovascular effects of methamphetamine in Parkinson's disease patients. Movement Disord 2004;19 (3):298–30.
100. Le Y, Rana KZ, Dudley MN. Amphotericin B-associated hypertension. Ann Pharmacother 1996;30(7–8):765–7.
101. Groll AH, Piscitelli SC, Walsh TJ. Clinical pharmacology of systemic antifungal agents: a comprehensive review of agents in clinical use, current investigational compounds, and putative targets for antifungal drug development. Adv Pharmacol 1998;44:343–500.
102. Rowles DM, Fraser SL. Amphotericin B lipid complex (ABLC)-associated hypertension: case report and review. Clin Infect Dis 1999;29(6):1564–5.
103. Levy M, Domaratzki J, Koren G. Amphotericin-induced heart-rate decrease in children. Clin Pediatr (Phila) 1995;34(7):358–64.
104. el-Dawlatly AA, Gomaa S, Takrouri MS, Seraj MA. Amphotericin B and cardiac toxicity—a case report. Middle East. J Anesthesiol 1999;15(1):107–12.
105. Johnson MD, Drew RH, Perfect JR. Chest discomfort associated with liposomal amphotericin B: report of three cases and review of the literature. Pharmacotherapy 1998;18(5):1053–61.
106. DeMonaco HJ, McGovern B. Transient asystole associated with amphotericin B infusion. Drug Intell Clin Pharm 1983;17(7–8):547–8.
107. Danaher PJ, Cao MK, Anstead GM, Dolan MJ, DeWitt CC. Reversible dilated cardiomyopathy related to amphotericin B therapy. J Antimicrob Chemother 2004;53:115–7.
108. Zernikow B, Fleischhack G, Hasan C, Bode U. Cyanotic Raynaud's phenomenon with conventional but not with liposomal amphotericin B: three case reports. Mycoses 1997;40(9–10):359–61.
109. Bichel T, Steinbach G, Olry L, Lambert H. Utilisation de l'amrinone intraveineux dans le traitement du choc cardiogenique. [Use of intravenous amrinone in the treatment of cardiogenic shock.] Agressologie 1988;29(3):187–92.
110. Reckelhoff JF, Granger JP. Role of androgens in mediating hypertension and renal injury. Clin Exp Pharmacol Physiol 1999;26(2):127–31.
111. Barry RE, Ford J. Sodium content and neutralising capacity of some commonly used antacids. BMJ 1978;1(6110):413.
112. Wiseman LR, Spencer CM. Dexrazoxane. A review of its use as a cardioprotective agent in patients receiving anthracycline-based chemotherapy. Drugs 1998;56(3):385–403.
113. Okuma K, Ariyoshi Y, Ota K. [Clinical study of acute cardiotoxicity of anti-cancer agents—analysis using Holter ECG monitoring.]Gan To Kagaku Ryoho 1988;15(6):1893–900.
114. Barendswaard EC, Prpic H, Van der Wall EE, Camps JA, Keizer HJ, Pauwels EK. Right ventricle wall motion abnormalities in patients treated with chemotherapy. Clin Nucl Med 1991;16(7):513–6.
115. Viniegra M, Marchetti M, Losso M, Navigante A, Litovska S, Senderowicz A, Borghi L, Lebron J, Pujato D, Marrero H, et al. Cardiovascular autonomic function in anthracycline-treated breast cancer patients. Cancer Chemother Pharmacol 1990;26(3):227–31.
116. Launchbury AP, Habboubi N. Epirubicin and doxorubicin: a comparison of their characteristics, therapeutic activity and toxicity. Cancer Treat Rev 1993;19(3):197–228.
117. Chabner BA, Longo DL. Cancer Chemotherapy and Biotherapy: Principles and Practice. 2nd ed.. Lippincott Williams and Wilkins;. 2001.
118. Souhami RL, Tannock I, Hohenberger P, Horiot JC. Oxford Textbook of Oncology. 2nd ed.. Oxford: Oxford University Press;. 2002.
119. Shan K, Lincoff AM, Young JB. Anthracycline-induced cardiotoxicity. Ann Intern Med 1996;125(1):47–58.
120. Von Hoff DD, Layard MW, Basa P, Davis HL Jr, Von Hoff AL, Rozencweig M, Muggia FM. Risk factors for doxorubicin-induced congestive heart failure. Ann Intern Med 1979;91(5):710–7.
121. Casper ES, Gaynor JJ, Hajdu SI, Magill GB, Tan C, Friedrich C, Brennan MF. A prospective randomized trial of adjuvant chemotherapy with bolus versus continuous infusion of doxorubicin in patients with high-grade extremity soft tissue sarcoma and an analysis of prognostic factors. Cancer 1991;68(6):1221–9.
122. Mortensen SA, Olsen HS, Baandrup U. Chronic anthracycline cardiotoxicity: haemodynamic and histopathological manifestations suggesting a restrictive endomyocardial disease. Br Heart J 1986;55(3):274–82.
123. Swain S, Whaley FS, Ewer MS. Congestive heart failure in patients treated with doxorubicin—a retrospective analysis of three trials. Cancer 2003;97:2869–79.
124. Sorensen K, Levitt GA, Bull C, Dorup I, Sullivan ID. Late anthracycline cardiotoxicity after childhood cancer—a prospective longitudinal study. Cancer 2003;97:1991–8.
125. Levitt GA, Dorup I, Sorensen K, Sullivan I. Does anthracycline administration by infusion in children affect late cardiotoxicity? Br J Haematol 2004;124:463–8.
126. Pein F, Sakiroglu O, Dahan M, Lebidois J, Merlet P, Shamsaldin A, Villain E, DeVathaire F, Sidi D, Hartmann O. Cardiac abnormalities 15 years and more after adriamycin therapy in 229 childhood survivors of a solid tumour at the Institut Gustave Roussy. Br J Cancer 2004;91:37–44.
127. Pihkala J, Saarinen UM, Lundstrom U, Virtanen K, Virkola K, Siimes MA, Pesonen E. Myocardial function in children and adolescents after therapy with anthracyclines and chest irradiation. Eur J Cancer 1996;32A(1):97–103.
128. Hesseling PB, Kalis NN, Wessels G, van der Merwe PL. The effect of anthracyclines on myocardial function in 50 long-term survivors of childhood cancer. Cardiovasc J South Afr 1999;89(Suppl 1):C25–8.
129. Ibrahim NK, Hortobagyi GN, Ewer M, Ali MK, Asmar L, Theriault RL, Fraschini G, Frye DK, Buzdar AU. Doxorubicin-induced congestive heart failure in elderly patients with metastatic breast cancer, with long-term follow-up: the M.D. Anderson experience Cancer Chemother Pharmacol 1999;43(6):471–8.
130. Lipshultz SE, Colan SD, Gelber RD, Perez-Atayde AR, Sallan SE, Sanders SP. Late cardiac effects of doxorubicin therapy for acute lymphoblastic leukemia in childhood. N Engl J Med 1991;324(12):808–15.
131. Steinherz LJ, Steinherz PG, Tan CT, Heller G, Murphy ML. Cardiac toxicity 4 to 20 years after completing anthracycline therapy. JAMA 1991;266(12):1672–7.
132. Drug news. Anthracycline cardiotoxicity uncovered. Drug Ther 1991;57Dec.
133. Fahey J. Cardiovascular function in children with acquired and congenital heart disease. Curr Opin Cardiol 1992;7:111–5.

134. Mather FJ, Simon RM, Clark GM, Von Hoff DD. Cardiotoxicity in patients treated with mitoxantrone: Southwest Oncology Group phase II studies. Cancer Treat Rep 1987;71(6):609–13.
135. McQuillan PJ, Morgan BA, Ramwell J. Adriamycin cardiomyopathy. Fatal outcome of general anaesthesia in a child with adriamycin cardiomyopathy. Anaesthesia 1988;43(4):301–4.
136. Huettemann E, Junker T, Chatzinikolaou KP, Petrat G, Sakka SG, Vogt L, Reinhart K. The influence of anthracycline therapy on cardiac function during anesthesia. Br J Cancer 2004;98:941–7.
137. Schneider HB, Becker H. Is prolongation of the QTc interval during isoflurane anaesthesia more prominent in women pretreated with anthracyclines for breast cancer? Br J Anaesthesia 2004;92:658–61.
138. Coukell AJ, Faulds D. Epirubicin. An updated review of its pharmacodynamic and pharmacokinetic properties and therapeutic efficacy in the management of breast cancer. Drugs 1997;53(3):453–82.
139. Camaggi CM, Comparsi R, Strocchi E, Testoni F, Angelelli B, Pannuti F. Epirubicin and doxorubicin comparative metabolism and pharmacokinetics. A cross-over study. Cancer Chemother Pharmacol 1988;21(3):221–8.
140. Lahtinen R, Kuikka J, Nousiainen T, Uusitupa M, Lansimies E. Cardiotoxicity of epirubicin and doxorubicin: a double-blind randomized study. Eur J Haematol 1991;46(5):301–5.
141. Plosker GL, Faulds D. Epirubicin. A review of its pharmacodynamic and pharmacokinetic properties, and therapeutic use in cancer chemotherapy. Drugs 1993;45(5):788–856.
142. Torti FM, Bristow MM, Lum BL, Carter SK, Howes AE, Aston DA, Brown BW Jr, Hannigan JF Jr, Meyers FJ, Mitchell EP, et al. Cardiotoxicity of epirubicin and doxorubicin: assessment by endomyocardial biopsy. Cancer Res 1986;46(7):3722–7.
143. Booser DJ, Hortobagyi GN. Anthracycline antibiotics in cancer therapy. Focus on drug resistance. Drugs 1994;47(2):223–58.
144. Buckley MM, Lamb HM. Oral idarubicin. A review of its pharmacological properties and clinical efficacy in the treatment of haematological malignancies and advanced breast cancer. Drugs Aging 1997;11(1):61–86.
145. Petti MC, Mandelli F. Idarubicin in acute leukemias: experience of the Italian Cooperative Group GIMEMA. Semin Oncol 1989;16(1 Suppl 2):10–5.
146. Ota K. Clinical review of aclacinomycin A in Japan. Drugs Exp Clin Res 1985;11(1):17–21.
147. Wilke A, Hesse H, Gorg C, Maisch B. Elevation of the pacing threshold: a side effect in a patient with pacemaker undergoing therapy with doxorubicin and vincristine. Oncology 1999;56(2):110–1.
148. Lai KH, Tsai YT, Lee SD, Ng WW, Teng HC, Tam TN, Lo GH, Lin HC, Lin HJ, Wu JC, et al. Phase II study of mitoxantrone in unresectable primary hepatocellular carcinoma following hepatitis B infection. Cancer Chemother Pharmacol 1989:23(1):54–6.
149. Wassmuth R, Lentzsch S, Erdbruegger U, Schulz-Menger J, Doerken B, Dietz R, Friedrich MG. Subclinical cardiotoxic effects of anthracyclines as assessed by magnetic resonance imaging—a pilot study. Am Heart J 2001;141(6):1007–13.
150. Zambetti M, Moliterni A, Materazzo C, Stefanelli M, Cipriani S, Valagussa P, Bonadonna G, Gianni L. Long-term cardiac sequelae in operable breast cancer patients given adjuvant chemotherapy with or without doxorubicin and breast irradiation. J Clin Oncol 2001;19(1):37–43.
151. Dey HM, Kassamali H. Radionuclide evaluation of doxorubicin cardiotoxicity: the need for cautious interpretation. Clin Nucl Med 1988;13(8):565–8.
152. Solymar L, Marky I, Mellander L, Sabel KG. Echocardiographic findings in children treated for malignancy with chemotherapy including adriamycin. Pediatr Hematol Oncol 1988;5(3):209–16.
153. Nakamura K, Miyake T, Kawamura T, Maekawa I. [Prospective monitoring of adriamycin cardiotoxicity with systolic time intervals.]Nippon Gan Chiryo Gakkai Shi 1988;23(8):1633–7.
154. Schwartz CL, Hobbie WL, Truesdell S, Constine LC, Clark EB. Corrected QT interval prolongation in anthracycline-treated survivors of childhood cancer. J Clin Oncol 1993;11(10):1906–10.
155. Vici P, Ferraironi A, Di Lauro L, Carpano S, Conti F, Belli F, Paoletti G, Maini CL, Lopez M. Dexrazoxane cardioprotection in advanced breast cancer patients undergoing high-dose epirubicin treatment. Clin Ter 1998;149(921):15–20.
156. Rowan RA, Masek MA, Billingham ME. Ultrastructural morphometric analysis of endomyocardial biopsies. Idiopathic dilated cardiomyopathy, anthracycline cardiotoxicity, and normal myocardium. Am J Cardiovasc Pathol 1988;2(2):137–44.
157. Weesner KM, Bledsoe M, Chauvenet A, Wofford M. Exercise echocardiography in the detection of anthracycline cardiotoxicity. Cancer 1991;68(2):435–8.
158. Suzuki T, Hayashi D, Yamazaki T, Mizuno T, Kanda Y, Komuro I, Kurabayashi M, Yamaoki K, Mitani K, Hirai H, Nagai R, Yazaki Y. Elevated B-type natriuretic peptide levels after anthracycline administration. Am Heart J 1998;136(2):362–3.
159. Nony P, Guastalla JP, Rebattu P, Landais P, Lievre M, Bontemps L, Itti R, Beaune J, Andre-Fouet X, Janier M. In vivo measurement of myocardial oxidative metabolism and blood flow does not show changes in cancer patients undergoing doxorubicin therapy. Cancer Chemother Pharmacol 2000;45(5):375–80.
160. Cassidy J, Merrick MV, Smyth JF, Leonard RC. Cardiotoxicity of mitozantrone assessed by stress and resting nuclear ventriculography. Eur J Cancer Clin Oncol 1988;24(5):935–8.
161. Brusamolino E, Bertini M, Guidi S, Vitolo U, Inverardi D, Merante S, Colombo A, Resegotti L, Bernasconi C, Ferrini PR, et al. CHOP versus CNOP (N = mitoxantrone) in non-Hodgkin's lymphoma: an interim report comparing efficacy and toxicity Haematologica 1988;73(3):217–22.
162. Saini J, Rich MW, Lyss AP. Reversibility of severe left ventricular dysfunction due to doxorubicin cardiotoxicity. Report of three cases. Ann Intern Med 1987;106(6):814–6.
163. Moreb JS, Oblon DJ. Outcome of clinical congestive heart failure induced by anthracycline chemotherapy. Cancer 1992;70(11):2637–41.
164. Ferrari S, Figus E, Cagnano R, Iantorno D, Bacci G. The role of corrected QT interval in the cardiologic follow-up of young patients treated with Adriamycin. J Chemother 1996;8(3):232–6.
165. Goenen M, Baele P, Lintermans J, Lecomte C, Col J, Ponlot R, Schoevardts JC, Chalant C. Orthotopic heart transplantation eleven years after left pneumonectomy. J Heart Transplant 1988;7(4):309–11.

166. Puccio CA, Feldman EJ, Arlin ZA. Amsacrine is safe in patients with ventricular ectopy. Am J Hematol 1988;28(3):197–8.
167. Weiss RB, Grillo-Lopez AJ, Marsoni S, Posada JG Jr, Hess F, Ross BJ. Amsacrine-associated cardiotoxicity: an analysis of 82 cases. J Clin Oncol 1986;4(6):918–28.
168. Legha SS, Benjamin RS, Mackay B, Ewer M, Wallace S, Valdivieso M, Rasmussen SL, Blumenschein GR, Freireich EJ. Reduction of doxorubicin cardiotoxicity by prolonged continuous intravenous infusion. Ann Intern Med 1982;96(2):133–9.
169. Workman P. Infusional anthracyclines: is slower better? If so, why? Ann Oncol 1992;3(8):591–4.
170. Working PK, Dayan AD. Pharmacological–toxicological expert report. CAELYX. (Stealth liposomal doxorubicin HCl). Hum Exp Toxicol 1996;15(9):751–85.
171. Dezube BJ. Safety assessment: Doxil (doxorubicin HCl liposome injection) in refractory AIDS-related Kaposi's sarcoma. Doxil Clinical Series, Vol. 1, No. 2. Menlo Park, California: SEQUUS Pharmaceuticals Inc, 1996
172. Goebel FD, Goldstein D, Goos M, Jablonowski H, Stewart JSThe International SL-DOX Study Group. Efficacy and safety of Stealth liposomal doxorubicin in AIDS-related Kaposi's sarcoma. Br J Cancer 1996;73(8):989–94.
173. Harrison M, Tomlinson D, Stewart S. Liposomal-entrapped doxorubicin: an active agent in AIDS-related Kaposi's sarcoma. J Clin Oncol 1995;13(4):914–20.
174. Uziely B, Jeffers S, Isacson R, Kutsch K, Wei-Tsao D, Yehoshua Z, Libson E, Muggia FM, Gabizon A. Liposomal doxorubicin: antitumor activity and unique toxicities during two complementary phase I studies. J Clin Oncol 1995;13(7):1777–85.
175. Muggia FM, Hainsworth JD, Jeffers S, Miller P, Groshen S, Tan M, Roman L, Uziely B, Muderspach L, Garcia A, Burnett A, Greco FA, Morrow CP, Paradiso LJ, Liang LJ. Phase II study of liposomal doxorubicin in refractory ovarian cancer: antitumor activity and toxicity modification by liposomal encapsulation. J Clin Oncol 1997;15(3):987–93.
176. Gabizon A, Martin F. Polyethylene glycol-coated (pegylated) liposomal doxorubicin. Rationale for use in solid tumours. Drugs 1997;54(Suppl 4):15–21.
177. Kanter PM, Klaich G, Bullard GA, King JM, Pavelic ZP. Preclinical toxicology study of liposome encapsulated doxorubicin (TLC D-99) given intraperitoneally to dogs. In Vivo 1994;8(6):975–82.
178. Kanter PM, Bullard GA, Pilkiewicz FG, Mayer LD, Cullis PR, Pavelic ZP. Preclinical toxicology study of liposome encapsulated doxorubicin (TLC D-99): comparison with doxorubicin and empty liposomes in mice and dogs. In Vivo 1993;7(1):85–95.
179. Kanter PM, Bullard GA, Ginsberg RA, Pilkiewicz FG, Mayer LD, Cullis PR, Pavelic ZP. Comparison of the cardiotoxic effects of liposomal doxorubicin (TLC D-99) versus free doxorubicin in beagle dogs. In Vivo 1993;7(1):17–26.
180. Harris L, Winer E, Batist G, Rovira D, Navari R, Lee Lthe TLC D-99 Study Group. Phase III study of TLC D-99 (liposome encapsulated doxorubicin) vs. free doxorubicin in patients with metastatic breast cancer (Abstract 26). Proc Am Soc Clin Oncol 1998;17:A474.
181. Shapiro CL, Ervin T, Welles L, Azarnia N, Keating J, Hayes DFTLC D-99 Study Group. Phase II trial of high-dose liposome-encapsulated doxorubicin with granulocyte colony-stimulating factor in metastatic breast cancer. J Clin Oncol 1999;17(5):1435–41.
182. Girard PM, Bouchaud O, Goetschel A, Mukwaya G, Eestermans G, Ross M, Rozenbaum W, Saimot AG. Phase II study of liposomal encapsulated daunorubicin in the treatment of AIDS-associated mucocutaneous Kaposi's sarcoma. AIDS 1996;10(7):753–7.
183. Gill PS, Espina BM, Muggia F, Cabriales S, Tulpule A, Esplin JA, Liebman HA, Forssen E, Ross ME, Levine AM. Phase I/II clinical and pharmacokinetic evaluation of liposomal daunorubicin. J Clin Oncol 1995;13(4):996–1003.
184. Harris L, Batist G, Belt R, Rovira D, Navari R, Azarnia N, Welles L, Winer ETLC D-99 Study Group. Liposome-encapsulated doxorubicin compared with conventional doxorubicin in a randomized multicenter trial as first-line therapy of metastatic breast carcinoma. Cancer 2002;94(1):25–36.
185. Fassas A, Buffels R, Anagnostopoulos A, Gacos E, Vadikolia C, Haloudis P, Kaloyannidis P. Safety and early efficacy assessment of liposomal daunorubicin (DaunoXome) in adults with refractory or relapsed acute myeloblastic leukaemia: a phase I–II study. Br J Haematol 2002;116(2):308–15.
186. Boumpas DT, Furie R, Manzi S, Illei GG, Wallace DJ, Balow JE. A short course of BG9588 (anti-CD40 ligand antibody) improves serologic activity and decreases hematuria in patients with proliferative lupus glomerulonephritis. Am Coll Rheumatol 2003;48:719–27.
187. Wang PS, Levin R, Zhao SZ, Avorn J. Urinary antispasmodic use and the risks of ventricular arrhythmia and sudden death in older patients. J Am Geriatr Soc 2002;50(1):117–24.
188. Olsen KM, Martin SJ. Pharmacokinetics and clinical use of drotrecogin alfa (activated) in patients with severe sepsis. Pharmacotherapy 2002;22(12 Pt 2):S196–205.
189. Rinkenberger RL, Prystowsky EN, Jackman WM, Naccarelli GV, Heger JJ, Zipes DP. Drug conversion of nonsustained ventricular tachycardia to sustained ventricular tachycardia during serial electrophysiologic studies: identification of drugs that exacerbate tachycardia and potential mechanisms. Am Heart J 1982;103(2):177–84.
190. Velebit V, Podrid P, Lown B, Cohen BH, Graboys TB. Aggravation and provocation of ventricular arrhythmias by antiarrhythmic drugs. Circulation 1982;65(5):886–94.
191. Thibault B, Nattel S. Optimal management with Class I and Class III antiarrhythmic drugs should be done in the outpatient setting: protagonist. J Cardiovasc Electrophysiol 1999;10(3):472–81.
192. Wellens HJ, Smeets JL, Vos M, Gorgels AP. Antiarrhythmic drug treatment: need for continuous vigilance. Br Heart J 1992;67(1):25–33.
193. Morganroth J. Early and late proarrhythmia from antiarrhythmic drug therapy. Cardiovasc Drugs Ther 1992;6(1):11–4.
194. Morganroth J. Proarrhythmic effects of antiarrhythmic drugs: evolving concepts. Am Heart J 1992;123(4 Pt 2):1137–9.
195. Hilleman DE, Mohiuddin SM, Gannon JM. Adverse reactions during acute and chronic class I antiarrhythmic therapy. Curr Ther Res 1992;51:730–8.
196. Hilleman DE, Larsen KE. Proarrhythmic effects of antiarrhythmic drugs. PT 1991;520–4June.
197. Libersa C, Caron J, Guedon-Moreau L, Adamantidis M, Nisse C. Adverse cardiovascular effects of anti-arrhythmia drugs. Part I: Proarrhythmic effects. Therapie 1992;47(3):193–8.

198. Podrid PJ, Fogel RI. Aggravation of arrhythmia by antiarrhythmic drugs, and the important role of underlying ischemia. Am J Cardiol 1992;70(1):100–2.
199. Follath F. Clinical pharmacology of antiarrhythmic drugs: variability of metabolism and dose requirements. J Cardiovasc Pharmacol 1991;17(Suppl 6):S74–6.
200. Cowan JC, Coulshed DS, Zaman AG. Antiarrhythmic therapy and survival following myocardial infarction. J Cardiovasc Pharmacol 1991;18(Suppl 2):S92–8.
201. Friedman L, Schron E, Yusuf S. Risk-benefit assessment of antiarrhythmic drugs. An epidemiological perspective. Drug Saf 1991;6(5):323–31.
202. Furberg CD, Yusuf S. Antiarrhythmics and VPD suppression. Circulation 1991;84(2):928–30.
203. Luderitz B. Möglichkeiten und Grenzen der Arrhythmiebehandlung. [Possibilities and limitations of treatment for arrhythmia.] Z Gesamte Inn Med 1991;46(12):425–30.
204. Podrid PJ. Safety and toxicity of antiarrhythmic drug therapy: benefit versus risk. J Cardiovasc Pharmacol 1991;17(Suppl 6):S65–73.
205. Zimmermann M. Antiarrhythmic therapy for ventricular arrhythmias. J Cardiovasc Pharmacol 1991;17(Suppl 6):S59–64.
206. Fauchier JP, Babuty D, Fauchier L, Rouesnel P, Cosnay P. Les effets proarythmiques des antiarythmiques. [Proarrhythmic effects of antiarrhythmic drugs.] Arch Mal Coeur Vaiss 1992;85(6):891–7.
207. Leenhardt A, Coumel P, Slama R. Torsade de pointes. J Cardiovasc Electrophysiol 1992;3:281–92.
208. Fujiki A, Usui M, Nagasawa H, Mizumaki K, Hayashi H, Inoue H. ST segment elevation in the right precordial leads induced with class IC antiarrhythmic drugs: insight into the mechanism of Brugada syndrome. J Cardiovasc Electrophysiol 1999;10(2):214–8.
209. Boriani G. New options for pharmacological conversion of atrial fibrillation. Card Electrophysiol Rev 2001;5:195–200.
210. Donahue TP, Conti JB. Atrial fibrillation: rate control versus maintenance of sinus rhythm. Curr Opin Cardiol 2001;16(1):46–53.
211. Chaudhry GM, Haffajee CI. Antiarrhythmic agents and proarrhythmia. Crit Care Med 2000;28(Suppl 10):N158–64.
212. Witchel HJ, Hancox JC. Familial and acquired long QT syndrome and the cardiac rapid delayed rectifier potassium current. Clin Exp Pharmacol Physiol 2000;27(10):753–66.
213. Brembilla-Perrot B, Houriez P, Beurrier D, Claudon O, Terrier de la Chaise A, Louis P. Predictors of atrial flutter with 1:1 conduction in patients treated with class I antiarrhythmic drugs for atrial tachyarrhythmias. Int J Cardiol 2001;80(1):7–15.
214. Feldman AM, Bristow MR, Parmley WW, Carson PE, Pepine CJ, Gilbert EM, Strobeck JE, Hendrix GH, Powers ER, Bain RP, et alVesnarinone Study Group. Effects of vesnarinone on morbidity and mortality in patients with heart failure. N Engl J Med 1993;329(3):149–55.
215. Fuster V, Rydèn LE, Asinger RW, Cannom DS, Crijns HJ, Frye RL, Halperin JL, Kay GN, Klein WW, Levy S, McNamara RL, Prystowsky EN, Wann LS, Wyse DG, Gibbons RJ, Antman EM, Alpert JS, Faxon DP, Fuster V, Gregoratos G, Hiratzka LF, Jacobs AK, Russell RO, Smith SC Jr, Klein WW, Alonso-Garcia A, Blomstrom-Lundqvist C, de Backer G, Flather M, Hradec J, Oto A, Parkhomenko A, Silber S, Torbicki A. American College of Cardiology/American Heart Association Task Force on Practice Guidelines; European Society of Cardiology Committee for Practice Guidelines and Policy Conferences (Committee to Develop Guidelines for the Management of Patients with Atrial Fibrillation); North American Society of Pacing and Electrophysiology. ACC/AHA/ESC Guidelines for the Management of Patients with Atrial Fibrillation: Executive Summary. A Report of the American College of Cardiology/American Heart Association Task Force on Practice Guidelines and the European Society of Cardiology Committee for Practice Guidelines and Policy Conferences (Committee to Develop Guidelines for the Management of Patients with Atrial Fibrillation) Developed in Collaboration with the North American Society of Pacing and Electrophysiology. Circulation 2001;104(17):2118–50.
216. Fuster V, Ryden LE, Asinger RW, Cannom DS, Crijns HJ, Frye RL, Halperin JL, Kay GN, Klein WW, Levy S, McNamara RL, Prystowsky EN, Wann LS, Wyse DG. American College of Cardiology; American Heart Association; European Society of Cardiology; North American Society of Pacing and Electrophysiology. ACC/AHA/ESC Guidelines for the Management of Patients with Atrial Fibrillation. A report of the American College of Cardiology/American Heart Association Task Force on Practice Guidelines and the European Society of Cardiology Committee for Practice Guidelines and Policy Conferences (Committee to Develop Guidelines for the Management of Patients with Atrial Fibrillation) developed in collaboration with the North American Society of Pacing and Electrophysiology. Eur Heart J 2001;22(20):1852–923.
217. Banai S, Tzivoni D. Drug therapy for torsade de pointes. J Cardiovasc Electrophysiol 1993;4(2):206–10.
218. Brembilla-Perrot B, Houriez P, Claudon O, Yassine M, Suty-Selton C, Vancon AC, Abo el Makarem Y, Makarem E, Courtelour JM. Les effets proarythmiques supraventricularires des antiarythmiques de classe IC sont-ils prévenus par l'association avec des bétabloquants?. [Can the supraventricular proarrhythmic effects of class 1C antiarrhythmic drugs be prevented with the association of beta blockers?.] Ann Cardiol Angeiol (Paris) 2000;49(8):439–42.
219. Scholz H. Antiarrhythmischer und Kardiodepressive Wirkungen antiarrhythmischer Substanzen. [Anti-arrhythmic and cardiodepressive effects of anti-arrhythmia agents.] Z Kardiol 1988;77(Suppl 5):113–9.
220. Luderitz B, Manz M. Hämodynamic bei ventrikularen Rhythmusstörungen und bei ihrer Behandlung. [Hemodynamics in ventricular arrhythmias and in their treatment.] Z Kardiol 1988;77(Suppl 5):143–9.
221. Schlepper M. Cardiodepressive effects of antiarrhythmic drugs. Eur Heart J 1989;10(Suppl E):73–80.
222. Seipel L, Hoffmeister HM. Hemodynamic effects of antiarrhythmic drugs: negative inotropy versus influence on peripheral circulation. Am J Cardiol 1989;64(20):J37–40.
223. Hammermeister KE. Adverse hemodynamic effects of antiarrhythmic drugs in congestive heart failure. Circulation 1990;81(3):1151–3.
224. Blumenthal RS, Baranowski B, Dowsett SA Cardiovascular effects of raloxifene: the arterial and venous systems. Am Heart J 2004;147:783–9.
225. Deitcher SR, Gomes MPV. The risk of venous thromboembolic disease associated with adjuvant hormone therapy for breast carcinoma: a systematic review. Cancer 2004;101:439–49.

226. Wong AR, Rasool AHG. Hydroxyzine-induced supraventricular tachycardia in a nine-year-old child. Singapore Med J 2004;45:90–2.
227. Walsh GM, Annunziato L, Frossard N, Knol K, Levander S, Nicolas JM, Taglialatela M, Tharp MD, Tillement P, Timmerman H. New insights into the second generation antihistamines. Drugs 2001;61:207–36.
228. Honig P, Baraniuk JN. Adverse effects of H1-receptor antagonists in the cardiovascular system. In: Simons FER, editor. Histamine and H_1-receptor Antagonists in Allergic Disease. New York: Marcel Dekker Inc, 1996:383–412.
229. Simons FE, Kesselman MS, Giddins NG, Pelech AN, Simons KJ. Astemizole-induced torsade de pointes. Lancet 1988;2(8611):624.
230. Davies AJ, Harindra V, McEwan A, Ghose RR. Cardiotoxic effect with convulsions in terfenadine overdose. BMJ 1989;298(6669):325.
231. Craft TM. Torsade de pointes after astemizole overdose. BMJ (Clin Res Ed) 1986;292(6521):660.
232. Passalacqua G, Bousquet J, Bachert C, Church MK, Bindsley-Jensen C, Nagy L, Szemere P, Davies RJ, Durham SR, Horak F, Kontou-Fili K, Malling HJ, van Cauwenberge P, Canonica GW. The clinical safety of H1-receptor antagonists. An EAACI position paper. Allergy 1996;51(10):666–75.
233. Barbey JT, Anderson M, Ciprandi G, Frew AJ, Morad M, Priori SG, Ongini E, Affrime MB. Cardiovascular safety of second-generation antihistamines. Am J Rhinol 1999;13(3):235–43.
234. Walsh GM, Annunziato L, Frossard N, Knol K, Levander S, Nicolas JM, Taglialatela M, Tharp MD, Tillement JP, Timmerman H. New insights into the second generation antihistamines. Drugs 2001;61(2):207–36.
235. Woosley RL. Cardiac actions of antihistamines. Annu Rev Pharmacol Toxicol 1996;36:233–52.
236. Monahan BP, Ferguson CL, Killeavy ES, Lloyd BK, Troy J, Cantilena LR Jr. Torsades de pointes occurring in association with terfenadine use. JAMA 1990;264(21):2788–90.
237. Taglialatela M, Castaldo P, Pannaccione A, Giorgio G, Genovese A, Marone G, Annunziato L. Cardiac ion channels and antihistamines: possible mechanisms of cardiotoxicity. Clin Exp Allergy 1999;29(Suppl 3):182–9.
238. DuBuske LM. Second-generation antihistamines: the risk of ventricular arrhythmias. Clin Ther 1999;21(2):281–95.
239. Chaufour S, Caplain H, Lilienthal N, L'heritier C, Deschamps C, Dubruc C, Rosenzweig P. Study of cardiac repolarization in healthy volunteers performed with mizolastine, a new H1-receptor antagonist. Br J Clin Pharmacol 1999;47(5):515–20.
240. Moss AJ, Chaikin P, Garcia JD, Gillen M, Roberts DJ, Morganroth J. A review of the cardiac systemic side-effects of antihistamines: ebastine. Clin Exp Allergy 1999;29(Suppl 3):200–5.
241. de Abajo FJ, Rodriguez LA. Risk of ventricular arrhythmias associated with nonsedating antihistamine drugs. Br J Clin Pharmacol 1999;47(3):307–13.
242. Khalifa M, Drolet B, Daleau P, Lefez C, Gilbert M, Plante S, O'Hara GE, Gleeton O, Hamelin BA, Turgeon J. Block of potassium currents in guinea pig ventricular myocytes and lengthening of cardiac repolarization in man by the histamine H1 receptor antagonist diphenhydramine. J Pharmacol Exp Ther 1999;288(2):858–65.
243. Taglialatela M, Timmerman H, Annunziato L. Cardiotoxic potential and CNS effects of first-generation antihistamines. Trends Pharmacol Sci 2000;21(2):52–6.
244. Huang MY, Argenti D, Wilson J, Garcia J, Heald D. Pharmacokinetics and electrocardiographic effect of ebastine in young versus elderly healthy subjects. Am J Ther 1998;5(3):153–8.
245. Hey JA, del Prado M, Sherwood J, Kreutner W, Egan RW. Comparative analysis of the cardiotoxicity proclivities of second generation antihistamines in an experimental model predictive of adverse clinical ECG effects. Arzneimittelforschung 1996;46(2):153–8.
246. Rijal S, Chappuis F, Singh R, Boelaert M, Loutan L, Koirala S. Sodium stibogluconate cardiotoxicity and safety of generics. Trans R Soc Trop Med Hyg 2003;97:597–8.
247. Costa JML, Garcia AM, Rebelo JMM, Guimaraes KM, Guimaraes RM, Nunes PMS. Fatal case during treatment of American tegumentary leishmaniasis with sodium stibogluconate bp 88® (Shandong Xinhua). Rev Soc Bras Med Trop 2003;36:295–8.
248. Baranwal AK, Mandal RN, Singh R, Singhi SC. Sodium stibogluconate and polymorphic ventricular tachycardia. Indian J Pediatr 2005;72:269–71.
249. Cesur S, Bahar K, Erekul S. Death from cumulative sodium stibogluconate toxicity on kala-azar. Clin Microbiol Infect 2002;8(9):606.
250. Sundar S, Sinha PR, Agrawal NK, Srivastava R, Rainey PM, Berman JD, Murray HW, Singh VP. A cluster of cases of severe cardiotoxicity among kala-azar patients treated with a high-osmolarity lot of sodium antimony gluconate. Am J Trop Med Hyg 1998;59(1):139–43.
251. Thakur CP. Sodium antimony gluconate, amphotericin, and myocardial damage. Lancet 1998;351(9120):1928–9.
252. Seaton RA, Morrison J, Man I, Watson J, Nathwani D. Out-patient parenteral antimicrobial therapy—a viable option for the management of cutaneous leishmaniasis. QJM 1999;92(11):659–67.
253. Wey HE, Richards D, Tirmenstein MA, Mathias PI, Toraason M. The role of intracellular calcium in antimony-induced toxicity in cultured cardiac myocytes. Toxicol Appl Pharmacol 1997;145(1):202–10.
254. Tirmenstein MA, Mathias PI, Snawder JE, Wey HE, Toraason M. Antimony-induced alterations in thiol homeostasis and adenine nucleotide status in cultured cardiac myocytes. Toxicology 1997;119(3):203–11.
255. Straus SM, Sturkenboom MC, Bleumink GS, Dieleman JP, van der Lei J, de Graeff PA, Kingma JH, Stricker BH. Non-cardiac QT_c-prolonging drugs and the risk of sudden cardiac death. Eur Heart J 2005;26:2007–12.
256. Roden DM. Drug-induced prolongation of the QT interval. N Engl J Med 2004;350:1013–22.
257. Justo D, Prokhorov V, Heller K, Zeltser D. Torsade de pointes induced by psychotropic drugs and the prevalence of its risk factors. Acta Psychiatr Scand 2005;111:171–6.
258. Stöllberger C, Huber JO, Finsterer J. Antipsychotic drugs and QT prolongation. Int Clin Psychopharmacol 2005;20:243–51.
259. Copetti R, Proclemer A, Pillinini PP. Brugada-like ECG abnormalities during thioridazine overdose. Br J Clin Pharmacol 2005;59:608.
260. Arizona CERT. Center for Education and Research in Therapeutics. http://www.torsades.org/medical-pros/drug-lists/drug-lists.htm
261. Herrero Hernández R, Cidoncha Gallego M, Herrero de Lucas E, Jiménez Lendínez M. Haloperidol por vía intravenosa y torsade de pointes. Med Intensiva 2004;28:89–90.
262. Akers WS, Flynn JD, Davis GA, Green AE, Winstead PS, Strobel G. Prolonged cardiac repolarization after

tacrolimus and haloperidol administration in the critically ill patient. Pharmacotherapy 2004;24:404–8.
263. Schmid C, Grohmann R, Engel RR, Rüther E, Kropp S. Cardiac adverse effects associated with psychotropic drugs. Pharmacopsychiatry 2004;37 Suppl 1:S65–9.
264. Mathis AS, Rao V. Deep vein thrombosis during rabbit antithymocyte globulin administration. Transplant Proc 2004;36(10):3250–1.
265. Westaby S, Katsumata T. Aprotinin and vein graft occlusion—the controversy continues. J Thorac Cardiovasc Surg 1998;116(5):731–3.
266. Bevan DH. Cardiac bypass haemostasis: putting blood through the mill. Br J Haematol 1999;104(2):208–19.
267. Lemmer JH Jr, Stanford W, Bonney SL, Breen JF, Chomka EV, Eldredge WJ, Holt WW, Karp RB, Laub GW, Lipton MJ, et al. Aprotinin for coronary bypass operations: efficacy, safety, and influence on early saphenous vein graft patency. A multicenter, randomized, double-blind, placebo-controlled study. J Thorac Cardiovasc Surg 1994;107(2):543–51.
268. Havel M, Grabenwoger F, Schneider J, Laufer G, Wollenek G, Owen A, Simon P, Teufelsbauer H, Wolner E. Aprotinin does not decrease early graft patency after coronary artery bypass grafting despite reducing postoperative bleeding and use of donated blood. J Thorac Cardiovasc Surg 1994;107(3):807–10.
269. Kalangos A, Tayyareci G, Pretre R, Di Dio P, Sezerman O. Influence of aprotinin on early graft thrombosis in patients undergoing myocardial revascularization. Eur J Cardiothorac Surg 1994;8(12):651–6.
270. Lass M, Simic O, Ostermeyer J. Re-graft patency and clinical efficacy of aprotinin in elective bypass surgery. Cardiovasc Surg 1997;5(6):604–7.
271. Levy JH, Pifarre R, Schaff HV, Horrow JC, Albus R, Spiess B, Rosengart TK, Murray J, Clark RE, Smith P. A multicenter, double-blind, placebo-controlled trial of aprotinin for reducing blood loss and the requirement for donor-blood transfusion in patients undergoing repeat coronary artery bypass grafting. Circulation 1995;92(8):2236–44.
272. Hayes A, Murphy DB, McCarroll M. The efficacy of single-dose aprotinin 2 million KIU in reducing blood loss and its impact on the incidence of deep venous thrombosis in patients undergoing total hip replacement surgery. J Clin Anesth 1996;8(5):357–60.
273. Anonymous. Ginseng—hypertension. WHO SIGNAL 2002;.
274. Huang CH, Chen WJ, Wu CC, Chen YC, Lee YT. Complete atrioventricular block after arsenic trioxide treatment in an acute promyelocytic leukemic patient. Pacing Clin Electrophysiol 1999;22(6 Pt 1):965–7.
275. Unnikrishnan D, Dutcher JP, Varshneya N, Lucariello R, Api M, Garl S, Wiernik PH, Chiaramida S. Torsades de pointes in 3 patients with leukemia treated with arsenic trioxide. Blood 2001;97(5):1514–6.
276. Lefevre G, Carpenter P, Souppart C, Schmidli H, Martin JM, Lane A, Ward C, Amakye D. Interaction trial between artemether–lumefantrine (Riamet) and quinine in healthy subjects. J Clin Pharmacol 2002;42(10):1147–58.
277. Bindschedler M, Lefevre G, Degen P, Sioufi A. Comparison of the cardiac effects of the antimalarials co-artemether and halofantrine in healthy participants. Am J Trop Med Hyg 2002;66(3):293–8.
278. Kaukinen S, Eerola R, Eerola M, Kaukinen L. A comparison of carticaine and lidocaine in spinal anaesthesia. Ann Clin Res 1978;10(4):191–4.
279. Pal B, Keenan J, Misra HN, Moussa K, Morris J. Raynaud's phenomenon in idiopathic carpal tunnel syndrome. Scand J Rheumatol 1996;25(3):143–5.
280. Heit JA. Risk factors for venous thromboembolism. Clin Chest Med 2003;24:1–12.
281. Ikeda S, Oka H, Matunaga K, Kubo S, Asai S, Miyahara Y, Osaka A, Kohno S. Astemizole-induced torsades de pointes in a patient with vasospastic angina. Jpn Circ J 1998;62(3):225–7.
282. Srivastava S. Angioneurotic oedema following atracurium. Br J Anaesth 1984;56(8):932–3.
283. Siler JN, Mager JG Jr, Wyche MQ Jr. Atracurium: hypotension, tachycardia and bronchospasm. Anesthesiology 1985;62(5):645–6.
284. Lynas AG, Clarke RS, Fee JP, Reid JE. Factors that influence cutaneous reactions following administration of thiopentone and atracurium. Anaesthesia 1988;43(10):825–8.
285. Beemer GH, Dennis WL, Platt PR, Bjorksten AR, Carr AB. Adverse reactions to atracurium and alcuronium. A prospective surveillance study. Br J Anaesth 1988;61(6):680–4.
286. Basta SJ, Ali HH, Savarese JJ, Sunder N, Gionfriddo M, Cloutier G, Lineberry C, Cato AE. Clinical pharmacology of atracurium besylate (BW 33A): a new non-depolarizing muscle relaxant. Anesth Analg 1982;61(9):723–9.
287. Basta SJ, Savarese JJ, Ali HH, Moss J, Gionfriddo M. Histamine-releasing potencies of atracurium, dimethyl tubocurarine and tubocurarine. Br J Anaesth 1983;55(Suppl 1):S105–6.
288. Scott RP, Savarese JJ, Ali HH, et al. Atracurium: clinical strategies for preventing histamine release and attenuating the hemodynamic response. Anesthesiology 1984;61:A287.
289. Shorten GD, Goudsouzian NG, Ali HH. Histamine release following atracurium in the elderly. Anaesthesia 1993;48(7):568–71.
290. Hughes R, Chapple DJ. The pharmacology of atracurium: a new competitive neuromuscular blocking agent. Br J Anaesth 1981;53(1):31–44.
291. Healy TE, Palmer JP. In vitro comparison between the neuromuscular and ganglion blocking potency ratios of atracurium and tubocurarine. Br J Anaesth 1982;54(12):1307–11.
292. Guggiari M, Gallais S, Bianchi A, Guillaume A, Viars P. Effets hémodynamiques de l'atracurium chez l'homme. [Hemodynamic effects of atracurium in man.] Ann Fr Anesth Reanim 1985;4(6):484–8.
293. Carter ML. Bradycardia after the use of atracurium. BMJ (Clin Res Ed) 1983;287(6387):247–8.
294. McHutchon A, Lawler PG. Bradycardia following atracurium. Anaesthesia 1983;38(6):597–8.
295. Woolner DF, Gibbs JM, Smeele PQ. Clinical comparison of atracurium and alcuronium in gynaecological surgery. Anaesth Intensive Care 1985;13(1):33–7.
296. Kinjo M, Nagashima H, Vizi ES. Effect of atracurium and laudanosine on the release of ^{3}H-noradrenaline. Br J Anaesth 1989;62(6):683–90.
297. Sudhaman DA. Atracurium and hypoxaemic episodes. Anaesthesia 1990;45(2):166.
298. Dauchot P, Gravenstein JS. Effects of atropine on the electrocardiogram in different age groups. Clin Pharmacol Ther 1971;12(2):274–80.
299. Gravenstein JS, Ariet M, Thornby JI. Atropine on the electrocardiogram. Clin Pharmacol Ther 1969;10(5):660–6.

300. Allen GD, Everett GB, Kennedy WF Jr. Cardiorespiratory effects of general anesthesia in outpatients: the influence of atropine. J Oral Surg 1972;30(8):576–80.
301. Lunde P. Ventricular fibrillation after intravenous atropine for treatment of sinus bradycardia. Acta Med Scand 1976;199(5):369–71.
302. Brady WJ, Perron AD. Administration of atropine in the setting of acute myocardial infarction: potentiation of the ischemic process? Am J Emerg Med 2001;19(1):81–3.
303. Bouhnik Y, Lemann M, Mary JY, Scemama G, Tai R, Matuchansky C, Modigliani R, Rambaud JC. Long-term follow-up of patients with Crohn's disease treated with azathioprine or 6-mercaptopurine. Lancet 1996;347(8996):215–9.
304. Sandborn WJ. A review of immune modifier therapy for inflammatory bowel disease: azathioprine, 6-mercaptopurine, cyclosporine, and methotrexate. Am J Gastroenterol 1996;91(3):423–33.
305. Hollander AAMJ, Van der Woude FJ. Efficacy and tolerability of conversion from cyclosporin to azathioprine after kidney transplantation. A review of the evidence. BioDrugs 1998;9:197–210.
306. Arellano-Rodrigo E, Garcia A, Mont L, Roque M. Torsade de pointes y parada cardiorrespiratoria inducida pot azitromicina en una paciente con sindrome de QT largo congenito. [Torsade de pointes and cardiorespiratory arrest induced by azithromycin in a patient with congenital long QT syndrome.] Med Clin (Barc) 2001;117(3):118–9.
307. Strle F, Maraspin V. Is azithromycin treatment associated with prolongation of the Q-Tc interval? Wien Klin Wochenschr 2002;114(10–11):396–9.
308. Damm O, Briheim G, Hagstrom T, Jonsson B, Skau T. Ruptured mycotic aneurysm of the abdominal aorta: a serious complication of intravesical instillation Bacillus Calmette–Guerin therapy. J Urol 1998;159(3):984.
309. Eastwood GL. ECG abnormalities associated with the barium enema. JAMA 1972;219(6):719–21.
310. Stremple J, Montgomery C. Nonspecific electrocardiographic abnormalities: the EKG during the barium enema procedure. Marquette Med Rev 1961;27:20–4.
311. Yigitbasi O, Sari S, Kiliccioglu B, Nalbantgil I. Recherche par l'ECG dynamique des modificiations cardiaques pouvant survenir pendant l'administration de lavements opaques. Effets protecteurs des bloqueurs des beta-recepteurs. [Dynamic ECG studies on possible cardiac modifications during the course of barium enemas. Protective effect of beta-receptor blockers.] J Radiol Electrol Med Nucl 1978;59(2):125–8.
312. Fleming GF, Vokes EE, McEvilly JM, Janisch L, Francher D, Smaldone L. Double-blind, randomized crossover study of metoclopramide and batanopride for prevention of cisplatin-induced emesis. Cancer Chemother Pharmacol 1991;28(3):226–7.
313. Ribera JR, Munoz RC, Fernando NA, Sahun NB, Cels AC, Capmany RP. Valvulopatia cardiaca associada al uso de benfluorex. Rev Esp Cardiol 2003;56:215–26.
314. Van Camp G, Flamez A, Cosyn B, Goldstein J, Perdaens C, Schoors D. Heart valvular disease in patients with Parkinson's disease treated with high dose pergolide. Neurology 2003;61:859–61.
315. Pritchett AM, Morrison JF, Edwards JD, Schaff HV, Connolly HM, Espinosa RE. Valvular heart disease in patients taking pergolide. Mayo Clin Prof 2002;77:1280–6.
316. Serratrice J, Disdier P, Habib G, Viallet F, Weiller PJ. Fibrotic valvular heart disease subsequent to bromocriptine treatment. Cardiol Rev 2002;10:334–6.
317. Donaldson D, Gibson G. Systemic complications with intravenous diazepam. Oral Surg Oral Med Oral Pathol 1980;49(2):126–30.
318. Ng E, Klinger G, Shah V, Taddio A. Safety of benzodiazepines in newborns. Ann Pharmacother 2002;36(7–8):1150–5.
319. Glaser JW, Blanton PL, Thrash WJ. Incidence and extent of venous sequelae with intravenous diazepam utilizing a standardized conscious sedation technique. J Periodontol 1982;53(11):700–3.
320. Mikkelsen H, Hoel TM, Bryne H, Krohn CD. Local reactions after i.v. injections of diazepam, flunitrazepam and isotonic saline Br J Anaesth 1980;52(8):817–9.
321. Jensen S, Huttel MS, Schou Olesen A. Venous complications after i.v. administration of Diazemuls (diazepam) and Dormicum (midazolam) Br J Anaesth 1981;53(10):1083–5.
322. Reves JG, Fragen RJ, Vinik HR, Greenblatt DJ. Midazolam: pharmacology and uses. Anesthesiology 1985;62(3):310–24.
323. Shi SJ, Garcia KM, Keck, JV. Temazepam, but not zolpidem, causes orthostatic hypotension in astronauts after spaceflight. J Cardiovasc Pharmacol 2003;41:31–9.
324. Blondheim DS, Levi D, Marmor AT. Mild sedation before transesophageal echo induces significant hemodynamic and respiratory depression. Echocardiography 2004;21(3):241–5.
325. Gershanik J, Boecler B, George W, et al. Gasping syndrome: benzyl alcohol poisoning. Clin Res 1981;29:895a.
326. Gershanik J, Boecler B, Ensley H, McCloskey S, George W. The gasping syndrome and benzyl alcohol poisoning. N Engl J Med 1982;307(22):1384–8.
327. Gershanik J, Boecler B, George W, et al. Neonatal deaths associated with use of benzyl alcohol—United States. Munch Med Wochenschr 1982;31:290.
328. Jardine DS, Rogers K. Relationship of benzyl alcohol to kernicterus, intraventricular hemorrhage, and mortality in preterm infants. Pediatrics 1989;83(2):153–60.
329. Cronin CM, Brown DR, Ahdab-Barmada M. Risk factors associated with kernicterus in the newborn infant: importance of benzyl alcohol exposure. Am J Perinatol 1991;8(2):80–5.
330. Lievre M, Morand S, Besse B, Fiessinger JN, Boissel JP. Oral Beraprost sodium, a prostaglandin I(2) analogue, for intermittent claudication: a double-blind, randomized, multicenter controlled trial. Beraprost et Claudication Intermittente (BERCI) Research Group. Circulation 2000;102(4):426–31.
331. Bristow MR. Beta-adrenergic receptor blockade in chronic heart failure. Circulation 2000;101:558–69.
332. Salpeter SR, Ormiston TM, Salpeter EE. Cardiovascular effects of beta-agonists in patients with asthma and COPD: a meta-analysis. Chest 2004;125:2309–21.
333. Kallergis EM, Manios EG, Kanoupakis EM, Schiza SE, Mavrakis HE, Klapsinos NK, Vardas PE. Acute electrophysiologic effects of inhaled salbutamol in humans. Chest 2005;127:2057–63.
334. Waring WS, Leigh RB. Haemodynamic responses to salbutamol and isometric exercise are altered in young adults with mild asthma. Eur J Clin Pharmacol 2005;61:9–14.
335. Burggraaf J, Westendorp RG, in't Veen JC, Schoemaker RC, Sterk PJ, Cohen AF, Blauw GJ. Cardiovascular side effects of inhaled salbutamol in hypoxic asthmatic patients. Thorax 2001;56(7):567–9.
336. Lowe MD, Rowland E, Brown MJ, Grace AA. Beta(2) adrenergic receptors mediate important

electrophysiological effects in human ventricular myocardium. Heart 2001;86(1):45–51.
337. Pye M, Quinn AC, Cobbe SM. QT interval dispersion: a non-invasive marker of susceptibility to arrhythmia in patients with sustained ventricular arrhythmias? Br Heart J 1994;71(6):511–4.
338. Yuksel H, Coskun S, Polat M, Onag A. Lower arrythmogenic risk of low dose albuterol plus ipratropium. Indian J Pediatr 2001;68(10):945–9.
339. Friedland JS, Warrell DA. The Jarisch–Herxheimer reaction in leptospirosis: possible pathogenesis and review. Rev Infect Dis 1991;13(2):207–10.
340. Maloy AL, Black RD, Segurola RJ Jr. Lyme disease complicated by the Jarisch–Herxheimer reaction. J Emerg Med 1998;16(3):437–8.
341. Young EJ, Weingarten NM, Baughn RE, Duncan WC. Studies on the pathogenesis of the Jarisch–Herxheimer reaction: development of an animal model and evidence against a role for classical endotoxin. J Infect Dis 1982;146(5):606–15.
342. Negussie Y, Remick DG, DeForge LE, Kunkel SL, Eynon A, Griffin GE. Detection of plasma tumor necrosis factor, interleukins 6, and 8 during the Jarisch–Herxheimer reaction of relapsing fever. J Exp Med 1992;175(5):1207–12.
343. Fekade D, Knox K, Hussein K, Melka A, Lalloo DG, Coxon RE, Warrell DA. Prevention of Jarisch–Herxheimer reactions by treatment with antibodies against tumor necrosis factor alpha. N Engl J Med 1996;335(5):311–5.
344. Emmanouilides C, Pegram M, Robinson R, Hecht R, Kabbinavar F, Isacoff W. Anti-VEGF antibody bevacizumab (Avastin) with 5FU/LV as third line treatment for colorectal cancer. Tech Coloproctol 2004;8 Suppl 1:s50-s52.
345. Howes LG, Sundaresan P, Lykos D. Cardiovascular effects of oral hypoglycaemic drugs. Clin Exp Pharmacol Physiol 1996;23(3):201–6.
346. Olsson J, Lindberg G, Gottsater M, Lindwall K, Sjostrand A, Tisell A, Melander A. Increased mortality in Type II diabetic patients using sulphonylurea and metformin in combination: a population-based observational study. Diabetologia 2000;43(5):558–60.
347. Debourdeau PM, Djezzar S, Estival JL, Zammit CM, Richard RC, Castot AC. Life-threatening eosinophilic pleuropericardial effusion related to vitamins B5 and H. Ann Pharmacother 2001;35(4):424–6.
348. Nwokolo CU, Prewett EJ, Sawyer AA, et al. Lack of bismuth absorption from bismuth subnitrate (Roter) tablets. Eur J Gastroenterol Hepatol 1989;5:433.
349. Nwokolo CU, Mistry P, Pounder RE. The absorption of bismuth and salicylate from oral doses of Pepto-Bismol (bismuth salicylate). Aliment Pharmacol Ther 1990;4(2):163–9.
350. Teepker M, Hamer HM, Knake S, Bandmann O, Oertel WH, Rosenow F. Myoclonic encephalopathy caused by chronic bismuth abuse. Epileptic Disord 2002;4(4):229–33.
351. Harris RA, Poole A. Beware of bismuth: post maxillectomy delirium. ANZ J Surg 2002;72(11):846–7.
352. Martin-Bouyer G. Intoxications par les sels de bismuth administrés par voie orale. [Poisoning by orally administered bismuth salts.] Gastroenterol Clin Biol 1978;2(4):349–56.
353. Menge H, Gregor M, Brosius B, Hopert R, Lang A. Pharmacology of bismuth. Eur J Gastroenterol Hepatol 1992;4(Suppl 2):41–7.
354. Jungreis AC, Schaumburg HH. Encephalopathy from abuse of bismuth subsalicylate (Pepto-Bismol). Neurology 1993;43(6):1265.
355. Friedland RP, Lerner AJ, Hedera P, Brass EP. Encephalopathy associated with bismuth subgallate therapy. Clin Neuropharmacol 1993;16(2):173–6.
356. Sharma RR, Cast IP, Redfern RM, O'Brien C. Extradural application of bismuth iodoform paraffin paste causing relapsing bismuth encephalopathy: a case report with CT and MRI studies. J Neurol Neurosurg Psychiatry 1994;57(8):990–3.
357. White DA, Schwartzberg LS, Kris MG, Bosl GJ. Acute chest pain syndrome during bleomycin infusions. Cancer 1987;59(9):1582–5.
358. Baniel J, Foster RS, Rowland RG, Bihrle R, Donohue JP. Complications of post-chemotherapy retroperitoneal lymph node dissection. J Urol 1995;153(3 Pt 2):976–80.
359. Davis JM, Rowley SD, Braine HG, Piantadosi S, Santos GW. Clinical toxicity of cryopreserved bone marrow graft infusion. Blood 1990;75(3):781–6.
360. Blackmore M. Minimising bruising in the antecubital fossa after venepuncture. BMJ (Clin Res Ed) 1987;295:332.
361. Volz HP, Gleiter CH, Moller HJ. Brofaromine versus imipramine in in-patients with major depression—a controlled trial. J Affect Disord 1997;44(2–3):91–9.
362. Hoppe UC, Beuckelmann DJ, Bohm M, Erdmann E. A young mother with severe chest pain. Heart 1998;79(2):205.
363. Matas O, Coppere B, Dupin AC, Richalet F, Ninet J. Accidents ischémiques artériels associés à la bromocriptine. [Ischemic arterial accidents associated with bromocriptine.] JEUR 1999;1:32.
364. Dutt S, Wong F, Spurway JH. Fatal myocardial infarction associated with bromocriptine for postpartum lactation suppression. Aust NZ J Obstet Gynaecol 1998;38(1):116–7.
365. Lindner M, Rosenkranz S, Deutsch HJ, Erdmann E. Ergotamininduzierter postpartaler Myokardinfarct. [Ergotamin-induced post partum myocardial infarction.] Herz Kreisl 2000;32:65–8.
366. Dasgupta D, Garasia M, Gupta KC, Satoskar RS. Randomised double-blind study of centbucridine and lignocaine for subarachnoid block. Indian J Med Res 1983;77:512–6.
367. Samsi AB, Bhalerao RA, Shah SC, Mody BB, Paul T, Satoskar RS. Evaluation of centbucridine as a local anesthetic. Anesth Analg 1983;62(1):109–11.
368. Nolte H. Zur Problematik der Cardiotoxizität von Bupivacain 0.75%. [The problem of the cardiotoxicity of bupivacaine 0.75%.] Reg Anaesth 1986;9(3):57–9.
369. Marx GF. Bupivacaine cardiotoxicity-concentration or dose? Anesthesiology 1986;65(1):116.
370. Hurley R, Feldman H. Toxicity of local anesthetics in obstetrics. I. Bupivacaine. Clin Anaesthesiol 1986;4:93.
371. Heavner JE. Cardiac toxicity of local anesthetics in the intact isolated heart model: a review. Reg Anesth Pain Med 2002;27(6):545–55.
372. Bacsik CJ, Swift JQ, Hargreaves KM. Toxic systemic reactions of bupivacaine and etidocaine. Oral Surg Oral Med Oral Pathol Oral Radiol Endod 1995;79(1):18–23.
373. Tanaka M, Nitta R, Nishikawa T. Increased T-wave amplitude after accidental intravascular injection of

lidocaine plus bupivacaine without epinephrine in sevoflurane-anesthetized child. Anesth Analg 2001;92(4):915–917.

374. Exler U, Nolte H, Milatz W. Die Überwachung der Herzleitung bei Anwendung von Bupivacain 0.75% mit Hilfe der Ventrikulographie (^{99m}Tc). [Monitoring of cardiac output during use of bupivacaine 0.75% by ventriculography (^{99m}Tc).] Reg Anaesth 1986;9(3):68–73.
375. Pape R, Ammer W. Holter-EKG-Überwachung bei Periduralanaesthesie mit Bupivacain 0.75%. [Holter ECG monitoring during peridural anesthesia with bupivacaine 0.75%.] Reg Anaesth 1986;9(3):74–8.
376. Pham-Dang C, Beaumont S, Floch H, Bodin J, Winer A, Pinaud M. Accident aign toxique après bloc du plexus lombaire à la bupivacaine. [Acute toxic accident following lumbar plexus block with bupivacaine.] Ann Fr Anesth Reanim 2000;19(5):356–9.
377. Fujita Y, Endoh S, Yasukawa T, Sari A. Lidocaine increases the ventricular fibrillation threshold during bupivacaine-induced cardiotoxicity in pigs. Br J Anaesth 1998;80(2):218–22.
378. Matsuda T, Kurata Y. Effects of nicardipine and bupivacaine on early after depolarization in rabbit sinoatrial node cells: a possible mechanism of bupivacaine-induced arrhythmias. Gen Pharmacol 1999;33(2):115–25.
379. Pickering AE, Waki H, Headley PM, Paton JF. Investigation of systemic bupivacaine toxicity using the in situ perfused working heart-brainstem preparation of the rat. Anesthesiology 2002;97(6):1550–6.
380. Gouny P, Gaitz JP, Vayssairat M. Acute hand ischemia secondary to intraarterial buprenorphine injection: treatment with iloprost and dextran-40—a case report. Angiology 1999;50(7):605–6.
381. Cracowski JL, Mallaret M, Vanzetto G. Myocardial infarction associated with buprenorphine. Ann Intern Med 1999;130(6):536–7.
382. Weinberger A, Pinkhas J, Sandbank U, Shaklai M, de Vries A. Endocardial fibrosis following busulfan treatment. JAMA 1975;231(5):495.
383. Massin F, Fur A, Reybet-Degat O, Camus P, Jeannin L. La pneumopathie du busulfan. [Busulfan-induced pneumopathy.] Rev Mal Respir 1987;4(1):3–10.
384. Pachter IJ, Evens RP. Butorphanol. Drug Alcohol Depend 1985;14(3–4):325–38.
385. Weinmann S, Siscovick DS, Raghunathan TE, Arbogast P, Smith H, Bovbjerg VE, Cobb LA, Psaty BM. Caffeine intake in relation to the risk of primary cardiac arrest. Epidemiology 1997;8(5):505–8.
386. Mann JI, Thorogood M. Coffee-drinking and myocardial infarction. Lancet 1975;2(7946):1215.
387. International Coffee Organization. United States of America: Coffee Drinking Study. 1989.
388. LeGrady D, Dyer AR, Shekelle RB, Stamler J, Liu K, Paul O, Lepper M, Shryock AM. Coffee consumption and mortality in the Chicago Western Electric Company Study. Am J Epidemiol 1987;126(5):803–12.
389. Hemminki E, Pesonen T. Regional coffee consumption and mortality from ischemic heart disease in Finland. Acta Med Scand 1977;201(1–2):127–30.
390. Yano K, Rhoads GG, Kagan A. Coffee, alcohol and risk of coronary heart disease among Japanese men living in Hawaii. N Engl J Med 1977;297(8):405–9.
391. Cannon ME, Cooke CT, McCarthy JS. Caffeine-induced cardiac arrhythmia: an unrecognised danger of healthfood products. Med J Aust 2001;174(10):520–1.
392. Gennari C, Fischer JA. Cardiovascular action of calcitonin gene-related peptide in humans. Calcif Tissue Int 1985;37(6):581–4.
393. Institute of Medicine. Marijuana and HealthWashington, DC: National Academy Press;. 1982.
394. Avakian EV, Horvath SM, Michael ED, Jacobs S. Effect of marihuana on cardiorespiratory responses to submaximal exercise. Clin Pharmacol Ther 1979;26(6):777–81.
395. Relman AS. Marijuana and health. N Engl J Med 1982;306(10):603–5.
396. Mittleman MA, Lewis RA, Maclure M, Sherwood JB, Muller JE. Triggering myocardial infarction by marijuana. Circulation 2001;103(23):2805–9.
397. Papp E, Czopf L, Habon T, Halmosi R, Horvath B, Marton Z, Tahin T, Komocsi A, Horvath I, Melegh, B, Toth K. Drug-induced myocardial infarction in young patients. Int J Cardiol 2005;98:169–70.
398. Lindsay AC, Foale RA, Warren O, Henry JA. Cannabis as a precipitant of cardiovascular emergencies. Int J Cardiol 2005;104:230–2.
399. Rezkalla SH, Sharma P, Kloner RA. Coronary no-flow and ventricular tachycardia associated with habitual marijuana use. Ann Emerg Med 2003;42:365–9.
400. Fisher BAC, Ghuran A, Vadamalai V, Antonios TF. Cardiovascular complications induced by cannabis smoking: a case report and review of the literature. Emerg Med J 2005;22:679–80.
401. Kosior DA, Filipiak KJ, Stolarz P, Opolski G. Paroxysmal atrial fibrillation following marijuana intoxication: a two-case report of possible association. Int J Cardiol 2001;78(2):183–4.
402. Singh GK. Atrial fibrillation associated with marijuana use. Pediatr Cardiol 2000;21(3):284.
403. Mathew RJ, Wilson WH, Davis R. Postural syncope after marijuana: a transcranial Doppler study of the hemodynamics. Pharmacol Biochem Behav 2003;75:309–18.
404. Schneider HJ, Jha S, Burnand KG. Progressive arteritis associated with cannabis use. Eur J Vasc Endovasc Surg 1999;18(4):366–7.
405. Ducasse E, Chevalier J, Dasnoy D, Speziale F, Fiorani P, Puppinck P. Popliteal artery entrapment associated with cannabis arteritis. Eur J Vasc Endovasc Surg 2004;27(3):327–32.
406. Frickhofen N, Beck FJ, Jung B, Fuhr HG, Andrasch H, Sigmund M. Capecitabine can induce acute coronary syndrome similar to 5-fluorouracil. Ann Oncol 2002;13:797–801.
407. Gary NE, Byra WM, Eisinger RP. Carbamazepine poisoning: treatment by hemoperfusion. Nephron 1981;27(4–5):202–3.
408. Schwartau M, Wahl G, Bucking J. Intramyokardialer Block bei Carbamazepin-Intoxication. [Intramyocardial block in carbamazepine poisoning.] Dtsch Med Wochenschr 1983;108(48):1841–3.
409. Macnab AJ, Robinson JL, Adderly RJ, D'Orsogna L. Heart block secondary to erythromycin-induced carbamazepine toxicity. Pediatrics 1987;80(6):951–3.
410. Boesen F, Andersen EB, Jensen EK, Ladefoged SD. Cardiac conduction disturbances during carbamazepine therapy. Acta Neurol Scand 1983;68(1):49–52.
411. Jette N, Veregin T, Guberman A. Carbamazepine-induced hypertension. Neurology 2002;59(2):275–6.
412. Marini AM, Choi JY, Labutta RJ. Transient neurologic deficits associated with carbamazepine-induced hypertension. Clin Neuropharmacol 2003;26:174–6.

413. Terrence CF, Fromm G. Congestive heart failure during carbamazepine therapy. Ann Neurol 1980;8(2):200–1.
414. Takayanagi K, Hisauchi I, Watanabe J, Maekawa Y, Fujito T, Sakai Y, Hoshi K, Kase M, Nishimura N, Inoue T, Hayashi T, Morooka S. Carbamazepine-induced sinus node dysfunction and atrioventricular block in elderly women. Jpn Heart J 1998;39(4):469–79.
415. Hennessy MJ, Tighe MG, Binnie CD, Nashef L. Sudden withdrawal of carbamazepine increases cardiac sympathetic activity in sleep. Neurology 2001;57(9):1650–4.
416. De Chadarevian JP, Legido A, Miles DK, Katsetos CD. Epilepsy, atherosclerosis, myocardial infarction, and carbamazepine. J Child Neurol 2003;18:150–1.
417. Everitt NJ. Effect of prolonged infusion on vein calibre: a prospective study. Ann R Coll Surg Engl 1999;81(2):109–12.
418. Smirniotis V, Kotsis TE, Antoniou S, Kostopanagiotou G, Labrou A, Kourias E, Papadimitriou J. Incidence of vein thrombosis in peripheral intravenous nutrition: effect of fat emulsions. Clin Nutr 1999;18(2):79–81.
419. Muckart DJ, Neijenhuis PA, Madiba TE. Superior vena caval thrombosis complicating central venous catheterisation and total parenteral nutrition. S Afr J Surg 1998;36(2):48–51.
420. Gluszek S, Kot M, Matykiewicz J. Cardiac tamponade as a complication of catheterization of the subclavian vein—prevention and principles of management. Nutrition 1999;15(7–8):580–2.
421. Hassan NA, Gunaid AA, Abdo-Rabbo AA, Abdel-Kader ZY, al-Mansoob MA, Awad AY, Murray-Lyon IM. The effect of qat chewing on blood pressure and heart rate in healthy volunteers. Trop Doct 2000;30(2):107–8.
422. Al-Hadrani AM. Khat induced hemorrhoidal disease in Yemen. Saudi Med J 2000;21(5):475–7.
423. Burke AP, Saenger J, Mullick F, Virmani R. Hypersensitivity myocarditis. Arch Pathol Lab Med 1991;115(8):764–9.
424. Beghetti M, Wilson GJ, Bohn D, Benson L. Hypersensitivity myocarditis caused by an allergic reaction to cefaclor. J Pediatr 1998;132(1):172–3.
425. Mazarakis A, Koutsojannis CM, Kounis NG, Alexopoulos D. Cefuroxime-induced coronary artery spasm manifesting as Kounis syndrome. Acta Cardiol 2005;60(3):341–5.
426. Estelle F, Simons R. Prospective, long-term safety evaluation of the H_1-receptor antagonist cetirizine in very young children with atopic dermatitis. Allergologie 2000;23:244–55.
427. Simons FE, Silas P, Portnoy JM, Catuogno J, Chapman D, Olufade AO. Safety of cetirizine in infants 6 to 11 months of age: a randomized, double-blind, placebo-controlled study. J Allergy Clin Immunol 2003;111(6):1244–8.
428. Lietman PS. Chloramphenicol and the neonate—1979 view. Clin Perinatol 1979;6(1):151–62.
429. Nahata MC. Lack of predictability of chloramphenicol toxicity in paediatric patients. J Clin Pharm Ther 1989;14(4):297–303.
430. Brown RT. Chloramphenicol toxicity in an adolescent. J Adolesc Health Care 1982;3(1):53–5.
431. Nikaido S, Tanaka M, Yamoto M, Minami T, Akatsuka M, Mori H. [Anaphylactoid shock caused by chlorhexidine gluconate.]Masui 1998;47(3):330–4.
432. Terazawa E, Shimonaka H, Nagase K, Masue T, Dohi S. Severe anaphylactic reaction due to a chlorhexidine-impregnated central venous catheter. Anesthesiology 1998;89(5):1296–8.
433. Stephens R, Mythen M, Kallis P, Davies DW, Egner W, Rickards A. Two episodes of life-threatening anaphylaxis in the same patient to a chlorhexidine-sulphadiazine-coated central venous catheter. Br J Anaesth 2001;87(2):306–8.
434. Pittaway A, Ford S. Allergy to chlorhexidine-coated central venous catheters revisited. Br J Anaesth 2002;88(2):304–5.
435. Nightingale SL. Hypersensitivity to chlorhexidine-impregnated medical devices. JAMA 1998;279:1684.
436. Veenstra DL, Saint S, Sullivan SD. Cost-effectiveness of antiseptic-impregnated central venous catheters for the prevention of catheter-related bloodstream infection. JAMA 1999;282(6):554–60.
437. Raad I, Hanna H. Intravascular catheters impregnated with antimicrobial agents: a milestone in the prevention of bloodstream infections. Support Care Cancer 1999;7(6):386–90.
438. Allen RW, Fee JP, Moore J. A preliminary assessment of epidural chloroprocaine for day procedures. Anaesthesia 1993;48(9):773–5.
439. Veinot JP, Mai KT, Zarychanski R. Chloroquine related cardiac toxicity. J Rheumatol 1998;25(6):1221–5.
440. Guedira N, Hajjaj-Hassouni N, Srairi JE, el Hassani S, Fellat R, Benomar M. Third-degree atrioventricular block in a patient under chloroquine therapy. Rev Rhum Engl Ed 1998;65(1):58–62.
441. Cervera A, Espinosa G, Font J, Ingelmo M. Cardiac toxicity secondary to long term treatment with chloroquine. Ann Rheum Dis 2001;60(3):301.
442. Charlier P, Cochand-Priollet B, Polivka M, Goldgran-Toledano D, Leenhardt A. Cardiomyopathie a la chloroquine revelée par un bloc auriculo-ventriculaire complete. A propos d'une observation. [Chloroquine cardiomyopathy revealed by complete atrio-ventricular block. A case report.] Arch Mal Coeur Vaiss 2002;95(9):833–7.
443. Freihage JH, Patel NC, Jacobs WR, Picken M, Fresco R, Malinowska K, Pisani BA, Mendez JC, Lichtenberg RC, Foy BK, Bakhos M, Mullen GM. Heart transplantation in a patient with chloroquine-induced cardiomyopathy. J Heart Lung Transpl 2004;23(2):252–5.
444. Clemessy JL, Favier C, Borron SW, Hantson PE, Vicaut E, Baud FJ. Hypokalaemia related to acute chloroquine ingestion. Lancet 1995;346(8979):877–80.
445. Axelsson R, Aspenstrom G. Electrocardiographic changes and serum concentrations in thioridazine-treated patients. J Clin Psychiatry 1982;43(8):332–5.
446. Dahl SG. Active metabolites of neuroleptic drugs: possible contribution to therapeutic and toxic effects. Ther Drug Monit 1982;4(1):33–40.
447. Risch SC, Groom GP, Janowsky DS. The effects of psychotropic drugs on the cardiovascular system. J Clin Psychiatry 1982;43(5 Pt 2):16–31.
448. Jardine AG. Assessing the relative risk of cardiovascular disease among renal transplant patients receiving tacrolimus or cyclosporine. Transpl Int 2005;18(4):379–84.
449. Kramer BK, Zulke C, Kammerl MC, Schmidt C, Hengstenberg C, Fischereder M, Marienhagen J; European Tacrolimus vs. Cyclosporine Microemulsion Renal Transplantation Study Group. Cardiovascular risk factors and estimated risk for CAD in a randomized trial comparing calcineurin inhibitors in renal transplantation. Am J Transplant 2003;3:982–7.
450. Kasiske BL, Guijarro C, Massy ZA, Wiederkehr MR, Ma JZ. Cardiovascular disease after renal transplantation. J Am Soc Nephrol 1996;7(1):158–65.

451. Kronenberg F, Lhotta K, Konigsrainer A, Konig P. Renal artery thromboembolism and immunosuppressive therapy. Nephron 1996;72(1):101.

452. Textor SC, Canzanello VJ, Taler SJ, Wilson DJ, Schwartz LL, Augustine JE, Raymer JM, Romero JC, Wiesner RH, Krom RA, et al. Cyclosporine-induced hypertension after transplantation. Mayo Clin Proc 1994;69(12):1182–93.

453. Taler SJ, Textor SC, Canzanello VJ, Schwartz L. Cyclosporin-induced hypertension: incidence, pathogenesis and management. Drug Saf 1999;20(5):437–49.

454. Ventura HO, Mehra MR, Stapleton DD, Smart FW. Cyclosporine-induced hypertension in cardiac transplantation. Med Clin North Am 1997;81(6):1347–57.

455. Sturrock ND, Lang CC, Struthers AD. Cyclosporin-induced hypertension precedes renal dysfunction and sodium retention in man. J Hypertens 1993;11(11):1209–16.

456. Campistol JM, Romero R, Paul J, Gutierrez-Dalmau A. Epidemiology of arterial hypertension in renal transplant patients: changes over the last decade. Nephrol Dial Transplant 2004;19 Suppl 3:iii62-iii66.

457. Thami GP, Bhalla M. Erythromelalgia induced by possible calcium channel blockade by ciclosporin. BMJ 2003;326:910.

458. Davenport A. The effect of renal transplantation and treatment with cyclosporin A on the prevalence of Raynaud's phenomenon. Clin Transplant 1993;7:4–8.

459. Rottenberg Y, Fridlender ZG. Recurrent infusion phlebitis induced by cyclosporine. Ann Pharmacother 2004;38(12):2071–3.

460. Fortin MC, Raymond MA, Madore F, Fugere JA, Paquet M, St Louis G, Hebert MJ. Increased risk of thrombotic microangiopathy in patients receiving a cyclosporin–sirolimus combination. Am J Transplant 2004;4(6):946–52.

461. Bastani B, Galli D, Gellens ME. Cimetidine-induced climacteric symptoms in a young man maintained on chronic hemodialysis. Am J Nephrol 1998;18(6):538–40.

462. Owens RC Jr, Ambrose PG. Torsades de pointes associated with fluoroquinolones. Pharmacotherapy 2002;22(5):663–8discussion 668–72.

463. Singh H, Kishore K, Gupta MS, Khetarpal S, Jain S, Mangla M. Ciprofloxacin-induced QTc prolongation. J Assoc Physicians India 2002;50:430–1.

464. Vandenplas Y, Benatar A, Cools F, Arana A, Hegar B, Hauser B. Efficacy and tolerability of cisapride in children. Paediatr Drugs 2001;3(8):559–73.

465. Benatar A, Feenstra A, Decraene T, Vandenplas Y. Cisapride and proarrhythmia in childhood. Pediatrics 1999;103(4 Pt 1):856–7.

466. Chhina S, Peverini RL, Deming DD, Hopper AO, Hashmi A, Vyhmeister NR. QTc interval in infants receiving cisapride. J Perinatol 2002;22(2):144–8.

467. Tamariz-Martel Moreno A, Rodrigo A B, Sanchez Bayle M, Montero Luis C, Acuna Quiros MD, Cano Fernandez J. Efectos de la cisaprida sobre el intervalo QT en ninos. Rev Esp Cardiol 2004;57(1):89–93.

468. Committee on Safety of Medicines Medicines Contol Agency. Cisapride (Prepulsid) withdrawn. Curr Probl Pharmacovig 2000;26:9–10.

469. Ng KS, Tham LS, Tan HH, Chia BL. Cisapride and torsades de pointes in a pacemaker patient. Pacing Clin Electrophysiol 2000;23(1):130–2.

470. Committee on Safety of Medicines. Cisparide: licences withdrawn. Curr Probl Pharmacvig 2004;30:3.

471. Ramirez-Mayans J, Garrido-Garcia LM, Huerta-Tecanhuey A, Gutierrez-Castrellon P, Cervantes-Bustamante R, Mata-Rivera N, Zarate-Mondragon F. Cisapride and QTc interval in children. Pediatrics 2000;106(5):1028–30.

472. Levy J, Hayes C, Kern J, Harris J, Flores A, Hyams J, Murray R, Tolia V. Does cisapride influence cardiac rhythm? Results of a United States multicenter, double-blind, placebo-controlled pediatric study. J Pediatr Gastroenterol Nutr 2001;32(4):458–63.

473. Cools F, Benatar A, Bougatef A, Vandenplas Y. The effect of cisapride on the corrected QT interval and QT dispersion in premature infants. J Pediatr Gastroenterol Nutr 2001;33(2):178–81.

474. Benatar A, Feenstra A, Decraene T, Vandenplas Y. Cisapride plasma levels and corrected QT interval in infants undergoing routine polysomnography. J Pediatr Gastroenterol Nutr 2001;33(1):41–6.

475. Wysowski DK, Corken A, Gallo-Torres H, Talarico L, Rodriguez EM. Postmarketing reports of QT prolongation and ventricular arrhythmia in association with cisapride and Food and Drug Administration regulatory actions. Am J Gastroenterol 2001;96(6):1698–703.

476. Lien CA, Belmont MR, Abalos A, Eppich L, Quessy S, Abou-Donia MM, Savarese JJ. The cardiovascular effects and histamine-releasing properties of 51W89 in patients receiving nitrous oxide/opioid/barbiturate anesthesia. Anesthesiology 1995;82(5):1131–8.

477. Doenicke A, Soukup J, Hoernecke R, Moss J. The lack of histamine release with cisatracurium: a double-blind comparison with vecuronium. Anesth Analg 1997;84(3):623–8.

478. Schramm WM, Jesenko R, Bartunek A, Gilly H. Effects of cisatracurium on cerebral and cardiovascular hemodynamics in patients with severe brain injury. Acta Anaesthesiol Scand 1997;41(10):1319–23.

479. Schramm WM, Papousek A, Michalek-Sauberer A, Czech T, Illievich U. The cerebral and cardiovascular effects of cisatracurium and atracurium in neurosurgical patients. Anesth Analg 1998;86(1):123–7.

480. Reich DL, Mulier J, Viby-Mogensen J, Konstadt SN, van Aken HK, Jensen FS, DePerio M, Buckley SG. Comparison of the cardiovascular effects of cisatracurium and vecuronium in patients with coronary artery disease. Can J Anaesth 1998;45(8):794–7.

481. Searle NR, Thomson I, Dupont C, Cannon JE, Roy M, Rosenbloom M, Gagnon L, Carrier M. A two-center study evaluating the hemodynamic and pharmacodynamic effects of cisatracurium and vecuronium in patients undergoing coronary artery bypass surgery. J Cardiothorac Vasc Anesth 1999;13(1):20–5.

482. Favre MP, Sztajzel J, Bertschy G. Bradycardia during citalopram treatment: a case report. Pharmacol Res 1999;39(2):149–50.

483. Catalano G, Catalano MC, Epstein MA, Tsambiras PE. QTc interval prolongation associated with citalopram overdose: a case report literature review. Clin Neuropharmacol 2001;24(3):158–62.

484. Bunker JP, Bendixen HH, Murphy AJ. Hemodynamic effects of intravenously administered sodium citrate. Nord Hyg Tidskr 1962;266:372–7.

485. Vallejo Camazon N, Rodriguez Pardo D, Sanchez Hidalgo A, Tornos Mas MP, Ribera E, Soler Soler J. Taquicardia Ventricular y QT largo asociades a la administracion de claritromycina en un patiente afectado de infeccion por el VIH. [Ventricular tachycardia and long

QT associated with clarithromycin administration in a patient with HIV infection.] Rev Esp Cardiol 2002;55(8):878–81.

486. Cousins D, Upton D. Beware bolus clarithromycin. Pharm Pract 1996;4:443–5.
487. de Dios Garcia-Diaz J, Santolaya Perrin R, Paz Martinez Ortega M, Moreno-Vazquez M. Flebitis relacionada con la administracion intravenosa de antibioticos macrolidos. Estudio comparativo de eritromicina y claritromicina. [Phlebitis due to intravenous administration of macrolide antibiotics. A comparative study of erythromycin versus clarithromycin.] Med Clin (Barc) 2001;116(4):133–5.
488. Hoffman RJ, Hoffman RS, Freyberg CL, Poppenga RH, Nelson LS. Clenbuterol ingestion causing prolonged tachycardia, hypokalemia, and hypophosphatemia with confirmation by quantitative levels. J Toxicol Clin Toxicol 2001;39(4):339–44.
489. Baselt RC. Clenbuterol. In: Disposition of Toxic Drugs and Chemicals in Man. 5th ed.. Foster City, CA: Chemical Toxicology Institute, 2000:189–90.
490. Eikmeier G, Kuhlmann R, Gastpar M. Thrombosis of cerebral veins following intravenous application of clomipramine. J Neurol Neurosurg Psychiatry 1988;51(11):1461.
491. Symes MH. Cardiovascular effects of clomipramine (Anafranil). J Int Med Res 1973;1:460.
492. Malan TP Jr, Nolan PE, Lichtenthal PR, Polson JS, Tebich SL, Bose RK, Copeland JG 3rd. Severe, refractory hypotension during anesthesia in a patient on chronic clomipramine therapy. Anesthesiology 2001;95(1):264–6.
493. Killian JG, Kerr K, Lawrence C, Celermajer DS. Myocarditis and cardiomyopathy associated with clozapine. Lancet 1999;354(9193):1841–5.
494. Shiwach RS. Treatment of clozapine induced hypertension and possible mechanisms. Clin Neuropharmacol 1998;21(2):139–40.
495. Donnelly JG, MacLeod AD. Hypotension associated with clozapine after cardiopulmonary bypass. J Cardiothorac Vasc Anesth 1999;13(5):597–9.
496. Low RA Jr, Fuller MA, Popli A. Clozapine induced atrial fibrillation. J Clin Psychopharmacol 1998;18(2):170.
497. Rechlin T, Beck G, Weis M, Kaschka WP. Correlation between plasma clozapine concentration and heart rate variability in schizophrenic patients. Psychopharmacology (Berl) 1998;135(4):338–41.
498. Varma S, Achan K. Dysrhythmia associated with clozapine. Aust NZ J Psychiatry 1999;33(1):118–9.
499. Vaddadi KS, Soosai E, Vaddadi G. Low blood selenium concentrations in schizophrenic patients on clozapine. Br J Clin Pharmacol 2003;55:307–9.
500. Krentz AJ, Mikhail S, Cantrell P, Hill GM. Pseudophaeochromocytoma syndrome associated with clozapine. BMJ 2001;322(7296):1213.
501. Bredbacka PE, Paukkala E, Kinnunen E, Koponen H. Can severe cardiorespiratory dysregulation induced by clozapine monotherapy be predicted? Int Clin Psychopharmacol 1993;8(3):205–6.
502. Kang UG, Kwon JS, Ahn YM, Chung SJ, Ha JH, Koo YJ, Kim YS. Electrocardiographic abnormalities in patients treated with clozapine. J Clin Psychiatry 2000;61(6):441–6.
503. Cohen H, Loewenthal U, Matar MA, Kotler M. Reversal of pathologic cardiac parameters after transition from clozapine to olanzapine treatment: a case report. Clin Neuropharmacol 2001;24(2):106–8.
504. Tie H, Walker BD, Singleton CB, Bursill JA, Wyse KR, Campbell TJ, Valenzuela SM, Breit SN. Clozapine and sudden death. J Clin Psychopharmacol 2001;21(6):630–2.
505. Chatterton R. Eosinophilia after commencement of clozapine treatment. Aust NZ J Psychiatry 1997;31(6):874–6.
506. Leo RJ, Kreeger JL, Kim KY. Cardiomyopathy associated with clozapine. Ann Pharmacother 1996;30(6):603–5.
507. Juul Povlsen U, Noring U, Fog R, Gerlach J. Tolerability and therapeutic effect of clozapine. A retrospective investigation of 216 patients treated with clozapine for up to 12 years. Acta Psychiatr Scand 1985;71(2):176–85.
508. Warner B, Schadelin J. Clinical safety and epidemiology. Leponex/Clozaril and myocarditisBasel, Switzerland: Novartis Pharm AG;. 1999.
509. Walker AM, Lanza LL, Arellano F, Rothman KJ. Mortality in current and former users of clozapine. Epidemiology 1997;8(6):671–7.
510. Merrill DB, Dec GW, Goff DC. Adverse cardiac effects associated with clozapine. J Clin Psychopharmacol 2005;25:32–41.
511. Reid P, McArthur M, Pridmore S. Clozapine rechallenge after myocarditis. Aust NZ J Psychiatry 2001;35(2):249.
512. Catalano G, Catalano MC, Frankel Wetter RL. Clozapine induced polyserositis. Clin Neuropharmacol 1997;20(4):352–6.
513. Murko A, Clarke S, Black DW. Clozapine and pericarditis with pericardial effusion. Am J Psychiatry 2002;159(3):494.
514. Kay SE, Doery J, Sholl D. Clozapine associated pericarditis and elevated troponin I. Aust NZ J Psychiatry 2002;36(1):143–4.
515. Boot E, De Haan L, Guzelcan Y, Scholte WF, Assies H. Pericardial and bilateral pleural effusion associated with clozapine treatment. Eur Psychiatry 2004;19:65–6.
516. Zornberg GL, Jick H. Antipsychotic drug use and risk of first-time idiopathic venous thromboembolism: a case-control study. Lancet 2000;356(9237):1219–23.
517. Hagg S, Spigset O, Soderstrom TG. Association of venous thromboembolism and clozapine. Lancet 2000;355(9210):1155–6.
518. Knudson JF, Kortepeter C, Dubitsky GM, Ahmad SR, Chen M. Antipsychotic drugs and venous thromboembolism. Lancet 2000;356(9225):252–3.
519. Kortepeter C, Chen M, Knudsen JF, Dubitsky GM, Ahmad SR, Beitz J. Clozapine and venous thromboembolism. Am J Psychiatry 2002;159(5):876–7.
520. Ihde-Scholl T, Rolli ML, Jefferson JW. Clozapine and pulmonary embolus. Am J Psychiatry 2001;158(3):499–500.
521. Pan R, John V. Clozapine and pulmonary embolism. Acta Psychiatr Scand 2003;108:76–7.
522. Hem E. Clozapine and pulmonary embolism: invited comment to letter to the editor. Acta Psychiatr Scand 2003;108:77.
523. Selten J-P, Büller H. Clozapine and venous thromboembolism: further evidence. J Clin Psychiatry 2003;64:609.
524. Patel S, Robinson R, Burk M. Hypertensive crisis associated with St. John's wort. Am J Med 2002;112(6):507–8.
525. Bolton-Maggs PH. The management of factor XI deficiency. Haemophilia 1998;4(4):683–8.
526. Giangrande PL. Adverse events in the prophylaxis of haemophilia. Haemophilia 2003;9 Suppl 1:50–6.
527. Collins PW. Management of acquired haemophilia A—more questions than answers. Blood Coagul Fibrinolysis 2003;14 Suppl 1:S23–7.
528. Delgado J, Jimenez-Yuste V, Hernandez-Navarro F, Villar A. Acquired haemophilia: review and meta-analysis

focused on therapy and prognostic factors. Br J Haematol 2003;121:21–35.

529. Dejgaard A. Update on Novo Nordisk's clinical trial programme on NovoSeven. Blood Coagul Fibrinolysis 2003;14 Suppl 1:S39–41.

530. O'Connell NM, Perry DJ, Hodgson AJ, O'Shaughnessy DF, Laffan MA, Smith OP. Recombinant FVIIa in the management of uncontrolled hemorrhage. Transfusion 2003;43:1711–16.

531. Laffan M, O'connell NM, Perry DJ, Hodgson AJ, O'Shaughnessy D, Smith OP. Analysis and results of the recombinant factor VIIa extended-use registry. Blood Coagul Fibrinolysis 2003;14 Suppl 1:S35–8.

532. European Medicines Agency (EMEA). Core Summary of Product Characteristics (SPC) for human plasma-derived and recombinant coagulation factor IX products. CPMP/BPWG/1625/99.

533. Mariani G, Dolce A, Marchetti G, Bernardi F. Clinical picture and management of congenital factor VII deficiency. Haemophilia 2004;10 Suppl 4:180–3.

534. Salomon O, Seligsohn U. New observations on factor XI deficiency. Haemophilia 2004;10 Suppl 4:184–7.

535. Ragni MV. Hemophilia gene transfer: comparison with conventional protein replacement therapy. Semin Thromb Hemost 2004;30(2):239–47.

536. Kerkhoffs JL, Atsma DE, Oemrawsingh PV, Eikenboom J, van der Meer FJ. Acute myocardial infarction during substitution with recombinant factor VIII concentrate in a patient with mild haemophilia A. Thromb Haemost 2004;92(2):425–6.

537. Basso IN, Keeling D. Myocardial infarction following recombinant activated factor VII in a patient with type 2A von Willebrand disease. Blood Coagul Fibrinolysis 2004;15(6):503–4.

538. Najaf SM, Malik A, Quraishi AU, Kazmi K, Kakepoto GN. Myocardial infarction during factor IX infusion in hemophilia B: case report and review of the literature. Ann Hematol 2004;83(9):604–7.

539. Goodnough LT, Hewitt PE, Silliman CC. Transfusion Medicine. Joint ASH and AABB educational session. Hematology Am Soc Hematol Educ Program 2004;(1):457–72.

540. Roberts HR, Monroe DM, White GC. The use of recombinant factor VIIa in the treatment of bleeding disorders. Blood 2004;104(13):3858–64.

541. Aledort LM. Comparative thrombotic event incidence after infusion of recombinant factor VIIa versus factor VIII inhibitor bypass activity. J Thromb Haemost 2004;2(10):1700–8.

542. Abshire T, Kenet G. Recombinant factor VIIa: review of efficacy, dosing regimens and safety in patients with congenital and acquired factor VIII or IX inhibitors. J Thromb Haemost 2004;2(6):899–909.

543. Roberts HR, Monroe DM, III, Hoffman M. Safety profile of recombinant factor VIIa. Semin Hematol 2004;41 (1 Suppl 1):101–8.

544. Luu H, Ewenstein B. FEIBA safety profile in multiple modes of clinical and home-therapy application. Haemophilia 2004;10 Suppl 2:10–16.

545. Von Depka M. Immune tolerance therapy in patients with acquired hemophilia. Hematology 2004;9(4):245–57.

546. Tjonnfjord GE, Brinch L, Gedde-Dahl T, Brosstad FR. Activated prothrombin complex concentrate (FEIBA) treatment during surgery in patients with inhibitors to FVIII/IX. Haemophilia 2004;10(2):174–8.

547. Uhlmann EJ, Eby CS. Recombinant activated factor VII for non-hemophiliac bleeding patients. Curr Opin Hematol 2004;11(3):198–204.

548. Stieltjes N, Altisent C, Auerswald G, Negrier C, Pouzol P, Reynaud J, Roussel-Robert V, Savidge GF, Villar A, Schulman S. Continuous infusion of B-domain deleted recombinant factor VIII (ReFacto) in patients with haemophilia A undergoing surgery: clinical experience. Haemophilia 2004;10(5):452–8.

549. Lethagen S, Carlson M, Hillarp A. A comparative in vitro evaluation of six von Willebrand factor concentrates. Haemophilia 2004;10(3):243–9.

550. Roberts HR. Recombinant factor VIIa: how safe is the stuff? Can J Anaesth 2005;52(1):8–11.

551. MacLaren R, Weber LA, Brake H, Gardner MA, Tanzi M. A multicenter assessment of recombinant factor VIIa off-label usage: clinical experiences and associated outcomes. Transfusion 2005;45(9):1434–42.

552. Mayer SA, Brun NC, Begtrup K, Broderick J, Davis S, Diringer MN, Skolnick BE, Steiner T; Recombinant Activated Factor VII Intracerebral Hemorrhage Trial Investigators. Recombinant activated factor VII for acute intracerebral hemorrhage. N Engl J Med 2005;352(8):777–85.

553. Cregler LL. Cocaine: the newest risk factor for cardiovascular disease. Clin Cardiol 1991;14(6):449–56.

554 Kloner RA, Hale S, Alker K, Rezkalla S. The effects of acute and chronic cocaine use on the heart. Circulation 1992;85(2):407–19.

555. Thadani P. Cardiovascular toxicity of cocaine: underlying mechanisms. NIDA Res Monogr 1991;108:1–238.

556. Boehrer JD, Moliterno DJ, Willard JE, Snyder RW 2nd, Horton RP, Glamann DB, Lange RA, Hillis LD. Hemodynamic effects of intranasal cocaine in humans. J Am Coll Cardiol 1992;20(1):90–3.

557. Stein R, Ellinwood EH Jr. Medical complication of cocaine abuse. Drug Ther 1990;10:40.

558. Cohen S. Reinforcement and rapid delivery systems: understanding adverse consequences of cocaine. NIDA Res Monogr 1985;61:151–7.

559. Cregler LL, Mark H. Medical complications of cocaine abuse. N Engl J Med 1986;315(23):1495–500.

560. Pitts WR, Vongpatanasin W, Cigarroa JE, Hillis LD, Lange RA. Effects of the intracoronary infusion of cocaine on left ventricular systolic and diastolic function in humans. Circulation 1998;97(13):1270–3.

561. Sofuoglu M, Nelson D, Dudish-Poulsen S, Lexau B, Pentel PR, Hatsukami DK. Predictors of cardiovascular response to smoked cocaine in humans. Drug Alcohol Depend 2000;57(3):239–45.

562. Allred RJ, Ewer S. Fatal pulmonary edema following intravenous "freebase" cocaine use. Ann Emerg Med 1981;10(8):441–2.

563. Feldman JA, Fish SS, Beshansky JR, Griffith JL, Woolard RH, Selker HP. Acute cardiac ischemia in patients with cocaine-associated complaints: results of a multicenter trial. Ann Emerg Med 2000;36(5):469–76.

564. Kushman SO, Storrow AB, Liu T, Gibler WB. Cocaine-associated chest pain in a chest pain center. Am J Cardiol 2000;85(3):394–6.

565. Villota JN, Rubio LF, Flores JS, Peris VB, Burguera EP, Gonzalez VB, Banuls MP, Escorihuela AL. Cocaine-induced coronary thrombosis and acute myocardial infarction. Int J Cardiol 2004;96:481–2.

566. Meltser H, Bhakta D, Kalaria VG. Multivessel coronary thrombosis secondary to cocaine use successfully treated

with multivessel primary angioplasty. Int J Cardiovasc Interv 2004;1:39–42.

567. Doshi SN, Marmur JD. Resolution of intracoronary thrombus with direct thrombin inhibition in a cocaine abuser. Heart 2004;90:501.

568. Erwin MB, Hoyle JR, Smith CH, Deliargyris EN. Cocaine and accelerated atherosclerosis: insights from intravasacular ultrasound. Int J Cardiol 2004;93:301–3.

569. Osula S, Stockton P, Abdelaziz MM, Walshaw MJ. Intratracheal cocaine induced myocardial infarction: an unusual complication of fibreoptic bronchoscopy. Thorax 2003;58:733–4.

570. Bizzarri F, Mondillo S, Guerrini F, Barbati R, Frati G, Davoli G. Spontaneous acute coronary dissection after cocaine abuse in a young woman. Can J Cardiol 2003;19:297–9.

571. Mochizuki Y, Zhang M, Golestaneh L, Thananart S, Coco M. Acute aortic thrombosis and renal infarction in acute cocaine intoxication: a case report and review of literature. Clin Nephrol 2003;60:130–3.

572. Stollberger C, Kopsa W, Finsterer J. Resolution of an aortic thrombus under anticoagulant therapy. Eur J Cardiothorac Surg 2001;20:880–2.

573. Inyang VA, Cooper AJ, Hodgkinson DW. Cocaine induced myocardial infarction. J Accid Emerg Med 1999;16(5):374–5.

574. Carley S, Ali B. Towards evidence based emergency medicine: best BETs from the Manchester Royal Infirmary. Acute myocardial infarction in cocaine induced chest pain presenting as an emergency. Emerg Med J 2003;20:174–5.

575. Lange RA, Hillis RD. Cardiovascular complications of cocaine use. New Engl J Med 2001;345:351–8.

576. Mittleman MA, Mintzer D, Maclure M, Tofler GH, Sherwood JB, Muller JE. Triggering of myocardial infarction by cocaine. Circulation 1999;99:2737–41.

577. Qureshi AI, Suri MF, Guterman LR, Hopkins LN. Cocaine use and the likelihood of nonfatal myocardial infarction and stroke: data from the Third National Health and Nutrition Examination Survey. Circulation 2001;103:502–6.

578. Office of Applied Statistics. Year end 2000 emergency department data from the Drug Abuse Warning Network. DAWN Series D-18, Rockville, Md: Substance Abuse and Mental Health Services Administration, 2001 (DHHS Publication No. 01-03532).

579. Brody SC, Slovis CM, Wren KD. Cocaine-related medical problems: consecutive series of 233 patients. Am J Med 1990;88:325–31.

580. Hollander JE. The management of cocaine associated myocardial ischemia. New Engl J Med 1995;333:1267–72.

581. Hockstra JW, Gibler WB, Levy RC, Sayre M, Naber W, Chandra A, Kacich R, Magorien R, Walsh R. Emergency department diagnosis of acute myocardial infarction and ischemia: a cost analysis of two diagnostic protocols. Acad Emerg Med 1994;1:103–10.

582. Weber JE, Shofer FS, Larkin GL, Kalaria AS, Hollander JE. Validation of a brief observation period for patients with cocaine-associated chest pain. New Engl J Med 2003;348:510–7.

583. Hollander JE, Hoffman RS. Cocaine-induced myocardial infarction: an analysis and review of the literature. J Emerg Med 1992;10(2):169–77.

584. Yuen-Green MS, Yen CK, Lim AD, Lull RJ. Tc-99m sestamibi myocardial imaging at rest for evaluation of cocaine-induced myocardial ischemia and infarction. Clin Nucl Med 1992;17(12):923–5.

585. Tanenbaum JH, Miller F. Electrocardiographic evidence of myocardial injury in psychiatrically hospitalized cocaine abusers. Gen Hosp Psychiatry 1992;14(3):201–3.

586. Eisenberg MJ, Mendelson J, Evans GT Jr, Jue J, Jones RT, Schiller NB. Left ventricular function immediately after intravenous cocaine: a quantitative two-dimensional echocardiographic study. J Am Coll Cardiol 1993;22(6):1581–6.

587. Rejali D, Glen P, Odom N. Pneumomediastinum following Ecstasy (methylenedioxymetamphetamine, MDMA) ingestion in two people at the same "rave". J Laryngol Otol 2002;116(1):75–6.

588. Morgan MJ, McFie L, Fleetwood H, Robinson JA. Ecstasy (MDMA): are the psychological problems associated with its use reversed by prolonged abstinence? Psychopharmacology (Berl) 2002;159(3):294–303.

589. Fox HC, McLean A, Turner JJ, Parrott AC, Rogers R, Sahakian BJ. Neuropsychological evidence of a relatively selective profile of temporal dysfunction in drug-free MDMA ("ecstasy") polydrug users. Psychopharmacology (Berl) 2002;162(2):203–14.

590. Kelly RF, Sompalli V, Sattar P, Khankari K. Increased TIMI frame counts in cocaine users: a case for increased microvascular resistance in the absence of epicardial coronary disease or spasm. Clin Cardiol 2003;26:319–22.

591. Minor RL Jr, Scott BD, Brown DD, Winniford MD. Cocaine-induced myocardial infarction in patients with normal coronary arteries. Ann Intern Med 1991;115(10):797–806.

592. Virmani R. Cocaine-associated cardiovascular disease: clinical and pathological aspects. NIDA Res Monogr 1991;108:220–9.

593. Amin M, Gabelman G, Karpel J, Buttrick P. Acute myocardial infarction and chest pain syndromes after cocaine use. Am J Cardiol 1990;66(20):1434–7.

594. Sharkey SW, Glitter MJ, Goldsmith SR. How serious is cocaine-associated acute chest pain syndromes after cocaine use. Cardiol Board Rev 1992;9:58–66.

595. Ross GS, Bell J. Myocardial infarction associated with inappropriate use of topical cocaine as treatment for epistaxis. Am J Emerg Med 1992;10(3):219–22.

596. Kontos MC, Jesse RL, Tatum JL, Ornato J. Coronary angiographic findings in patients with cocaine-associated chest pain. J Emerg Med 2003;24:9–13.

597. Kuczkowski KM. Chest pain and dysrhythmias in a healthy parturient: Wolff–Parkinson–White syndrome vs cocaine-induced myocardial ischemia? Anaesth Intens Care 2004;32:143–4.

598. Weber JE, Shofer FS, Larkin L, Kalaria AS, Hollander JE. Validation of brief observation period for patients with cocaine-associated chest pain. New Engl J Med 2003;348:510–7.

599. Honderick T, Williams D, Seaberg D, Wears R. A prospective, randomized, controlled trial of benzodiazepines and nitroglycerine or nitroglycerine alone in the treatment of cocaine-associated acute coronary syndromes. Am J Emerg Med 2003;21:39–42.

600. Baumann BM, Perrone J, Hornig SE, Shofer FS, Hollander JE. Randomized, double-blind, placebo-controlled trial of diazepam, nitroglycerine, or both for treatment of patients with potential cocaine-associated acute coronary syndromes. Acad Emerg Med 2000;7:878–86.

601. Anonymous. Second American Heart Associations International Evidence Evaluation Conference, Part 6. Advanced cardiovascular life support: Section 1.

Introduction to ACLS from the Guidelines 2000 Conference. Circulation 2000;102 Suppl 1:186–9.

602. Shrma AK, Hamwi SM, Garg N, Castagna MT, Suddath W, Ellahham S, Lindsay J. Percutaneous intervention in patients with cocaine-associated myocardial infarction: a case series and review. Catheter Cardiovasc Intervention 2002;56:346–52.
603. Hollander JE. The management of cocaine associated myocardial ischemia. New Engl J Med 1995;333:1267–72.
604. Cohle SD, Lie JT. Dissection of the aorta and coronary arteries associated with acute cocaine intoxication. Arch Pathol Lab Med 1992;116(11):1239–41.
605. Berry J, van Gorp WG, Herzberg DS, Hinkin C, Boone K, Steinman L, Wilkins JN. Neuropsychological deficits in abstinent cocaine abusers: preliminary findings after two weeks of abstinence. Drug Alcohol Depend 1993;32(3):231–7.
606. Hsue PY, Salinas CL, Bolger AF, Benowitz NL, Waters DD. Acute aortic dissection related to crack cocaine. Circulation 2002;105(13):1592–5.
607. Riaz K, Forker AD, Garg M, McCullough PA. Atypical presentation of cocaine-induced type A aortic dissection: a diagnosis made by transesophageal echocardiography. J Investig Med 2002;50(2):140–2.
608. Neri E, Toscano T, Massetti M, Capannini G, Frati G, Sassi C. Cocaine-induced intramural hematoma of the ascending aorta. Tex Heart Inst J 2001;28(3):218–9.
609. Jaffe BD, Broderick TM, Leier CV. Cocaine-induced coronary-artery dissection. N Engl J Med 1994;330(7):510–1.
610. Eskander KE, Brass NS, Gelfand ET. Cocaine abuse and coronary artery dissection. Ann Thorac Surg 2001;71(1):340–1.
611. Steinhauer JR, Caulfield JB. Spontaneous coronary artery dissection associated with cocaine use: a case report and brief review. Cardiovasc Pathol 2001;10(3):141–5.
612. Kumar PD, Smith HR. Cocaine-related vasculitis causing upper-limb peripheral vascular disease. Ann Intern Med 2000;133(11):923–4.
613. Balbir-Gurman A, Braun-Moscovici Y, Nahir AM. Cocaine-induced Raynaud's phenomenon and ischaemic finger necrosis. Clin Rheumatol 2001;20(5):376–8.
614. Morrow PL, McQuillen JB. Cerebral vasculitis associated with cocaine abuse. J Forensic Sci 1993;38(3):732–8.
615. Martin K, Rogers T, Kavanaugh A. Central nervous system angiopathy associated with cocaine abuse. J Rheumatol 1995;22(4):780–2.
616. Wang RY. pH-dependent cocaine-induced cardiotoxicity. Am J Emerg Med 1999;17(4):364–9.
617. Taylor D Parish D, Thompson L, Cavaliere M. Cocaine induced prolongation of the QT interval. Emerg Med J 2004;21:252–3.
618. Silvain J, Maury E, Qureshi T, Baudel JL, Offenstadt G. A puzzling electrocardiogram. Intensive Care Med 2004;30:340.
619. Grigorov V, Goldberg L, Foccard JP. Cardiovascular complications of acute cocaine poisoning: a clinical case report. Cardiol J S Africa 2004;15:139–42.
620. Kajdasz DK, Moore JW, Donepudi H, Cochrane CE, Malcolm RJ. Cardiac and mood-related changes during short-term abstinence from crack cocaine: the identification of possible withdrawal phenomena. Am J Drug Alcohol Abuse 1999;25(4):629–37.
621. Laffey JG, Neligan P, Ormonde G. Prolonged perioperative myocardial ischemia in a young male: due to topical intranasal cocaine? J Clin Anesth 1999;11(5):419–24.
622. Liao BS, Hilsinger RL Jr, Rasgon BM, Matsuoka K, Adour KK. A preliminary study of cocaine absorption from the nasal mucosa. Laryngoscope 1999;109(1):98–102.
623. Walker SR, MacSweeney ST. Plasma expanders used to treat or prevent hypotension can themselves cause hypotension. Postgrad Med J 1998;74(874):492–3.
624. McKinnon RP, Sinclair CJ. Angiotensinamide in the treatment of probable anaphylaxis to succinylated gelatin (Gelofusine). Anaesthesia 1994;49(4):309–11.
625. Hast R, Hellstrom-Lindberg E, Ohm L, Bjorkholm M, Celsing F, Dahl IM, Dybedal I, Gahrton G, Lindberg G, Lerner R, Linder O, Lofvenberg E, Nilsson-Ehle H, Paul C, Samuelsson J, Tangen JM, Tidefelt U, Turesson I, Wahlin A, Wallvik J, Winquist I, Oberg G, Bernell P. No benefit from adding GM-CSF to induction chemotherapy in transforming myelodysplastic syndromes: better outcome in patients with less proliferative disease. Leukemia 2003;17:1827–33.
626. Darie C, Boutalba S, Fichter P, Huret JF, Jaillot P, Deplus F, Gerenton S, Zenone T, Moreau JL, Grand A. Aortite après injections de G-CSF. Rev Med Interne 2004;25:225–9.
627. Kang HJ, Kim HS, Zhang SY, Park KW, Cho HJ, Koo BK, Kim YJ, Soo Lee D, Sohn DW, Han KS, Oh BH, Lee MM, Park YB. Effects of intracoronary infusion of peripheral blood stem-cells mobilised with granulocyte-colony stimulating factor on left ventricular systolic function and restenosis after coronary stenting in myocardial infarction: the MAGIC cell randomised clinical trial. Lancet 2004;363:751–6.
628. Deeren DH, Zachee P, Malbrain ML. Granulocyte colony-stimulating factor-induced capillary leak syndrome confirmed by extravascular lung water measurements. Ann Hematol 2005;84(2):89–94.
629. Kirchgatterer A, Schwarz CD, Holler E, Punzengruber C, Hartl P, Eber B. Cardiac tamponade following acupuncture. Chest 2000;117(5):1510–1.
630. White AR, Abbot NC, Barnes J, Ernst E. Self-reports of adverse effects of acupuncture included cardiac arrhythmia. Acupunc Med 1996;14:121.
631. Maxwell SR, Moots RJ, Kendall MJ. Corticosteroids: do they damage the cardiovascular system? Postgrad Med J 1994;70(830):863–70.
632. Iuchi T, Akaike M, Mitsui T, Ohshima Y, Shintani Y, Azuma H, Matsumoto T. Glucocorticoid excess induces superoxide production in vascular endothelial cells and elicits vascular endothelial dysfunction. Circ Res 2003;92:81–7.
633. Sato A, Funder JW, Okubo M, Kubota E, Saruta T. Glucocorticoid-induced hypertension in the elderly. Relation to serum calcium and family history of essential hypertension. Am J Hypertens 1995;8(8):823–8.
634. Thedenat B, Leaute-Labreze C, Boralevi F, Roul S, Labbe L, Marliere V, Taieb A. Surveillance tensionnelle des nourrissons traites par corticotherapie generale pour un hemangiome. [Blood pressure monitoring in infants with hemangiomas treated with corticosteroids.] Ann Dermatol Venereol 2002;129(2):183–5.
635. Stewart IM, Marks JSECG. Abnormalities in steroid-treated rheumatoid patients. Lancet 1977;2(8050):1237–8.
636. Baty V, Blain H, Saadi L, Jeandel C, Canton P. Fatal myocardial infarction in an elderly woman with severe ulcerative colitis. what is the role of steroids? Am J Gastroenterol 1998;93(10):2000–1.

637. Machiels JP, Jacques JM, de Meester A. Coronary artery spasm during anaphylaxis. Ann Emerg Med 1996;27(5):674–5.
638. Sato O, Takagi A, Miyata T, Takayama Y. Aortic aneurysms in patients with autoimmune disorders treated with corticosteroids. Eur J Vasc Endovasc Surg 1995;10(3):366–9.
639. Kotha P, McGreevy MJ, Kotha A, Look M, Weisman MH. Early deaths with thrombolytic therapy for acute myocardial infarction in corticosteroid-dependent rheumatoid arthritis. Clin Cardiol 1998;21(11):853–6.
640. Ellis SG, Semenec T, Lander K, Franco I, Raymond R, Whitlow PL. Effects of long-term prednisone (>=5 mg) use on outcomes and complications of percutaneous coronary intervention. Am J Cardiol 2004;93:1389–90.
641. Yunis KA, Bitar FF, Hayek P, Mroueh SM, Mikati M. Transient hypertrophic cardiomyopathy in the newborn following multiple doses of antenatal corticosteroids. Am J Perinatol 1999;16(1):17–21.
642. Gill AW, Warner G, Bull L. Iatrogenic neonatal hypertrophic cardiomyopathy. Pediatr Cardiol 1996;17(5):335–9.
643. Pokorny JJ, Roth F, Balfour I, Rinehart G. An unusual complication of the treatment of a hemangioma. Ann Plast Surg 2002;48(1):83–7.
644. Kothari SN, Kisken WA. Dexamethasone-induced congestive heart failure in a patient with dilated cardiomyopathy caused by occult pheochromocytoma. Surgery 1998;123(1):102–5.
645. Balys R, Manoukian J, Zalai C. Left ventricular hypertrophy with outflow tract obstruction-a complication of dexamethasone treatment for subglottic stenosis. Int J Pediatr Otorhinolaryngol 2005;69(2):271–3.
646. Kucukosmanoglu O, Karabay A, Ozbarlas N, Noyan A, Anarat A. Marked bradycardia due to pulsed and oral methylprednisolone therapy in a patient with rapidly progressive glomerulonephritis. Nephron 1998;80(4):484.
647. Schult M, Lohmann D, Knitsch W, Kuse ER, Nashan B. Recurrent cardiocirculatory arrest after kidney transplantation related to intravenous methylprednisolone bolus therapy. Transplantation 1999;67(11):1497–8.
648. Feuillet L, Guedj E, Laksiri N, Philip E, Habib G, Pelletier J, Cherif AA. Deep vein thrombosis after intravenous immunoglobulins associated with methylprednisolone. Thromb Haemost 2004;92:662–5.
649. Brumund MR, Truemper EJ, Lutin WA, Pearson-Shaver AL. Disseminated varicella and staphylococcal pericarditis after topical steroids. J Pediatr 1997;131(1 Part 1):162–3.
650. Kaiser H. Cortisonderivate in Klink und Praxis. 7th edn.. Stuttgart: G.Thieme;. 1977.
651. Hussain RM, McIntosh SJ, Lawson J, Kenny RA. Fludrocortisone in the treatment of hypotensive disorders in the elderly. Heart 1996;76(6):507–9.
652. Bhattacharyya A, Tymms DJ. Heart failure with fludrocortisone in Addison's disease. J R Soc Med 1998;91(8):433–4.
653. Hiraishi S, Iwanami N, Ogawa N. Images in cardiology. Enlargement of cardiac rhabdomyoma and myocardial ischaemia during corticotropin treatment for infantile spasm. Heart 2000;84(2):170.
654. Lang D, Muhler E, Kupferschmid C, Tacke E, von Bernuth G. Cardiac hypertrophy secondary to ACTH treatment in children. Eur J Pediatr 1984;142(2):121–5.
655. Kusse MC, van Nieuwenhuizen O, van Huffelen AC, van der Mey W, Thijssen JH, van Ree JM. The effect of non-depot ACTH(1–24) on infantile spasms. Dev Med Child Neurol 1993;35(12):1067–73.
656. Hrachovy RA, Frost JD Jr, Glaze DG. High-dose, long-duration versus low-dose, short-duration corticotropin therapy for infantile spasms. J Pediatr 1994;124(5 Pt 1):803–6.
657. Starc TJ, Bierman FZ, Pavlakis SG, Challenger ME, De Vivo DC, Gersony WM. Cardiac size and function during adrenocorticotropic hormone-induced systolic systemic hypertension in infants. Am J Cardiol 1994;73(1):57–64.
658. Akle CA, Joiner CL. Purple toe syndrome. J R Soc Med 1981;74(3):219.
659. Feder W, Auerbach R. "Purple toes": an uncommon sequela of oral coumarin drug therapy. Ann Intern Med 1961;55:911–7.
660. Burton JL, Pennock P. Anticoagulants and "feeling cold". Lancet 1979;1(8116):608.
661. Bruns FJ, Segel DP, Adler S. Control of cholesterol embolization by discontinuation of anticoagulant therapy. Am J Med Sci 1978;275(1):105–8.
662. Aronson JK. The NSAID roller coaster: more about rofecoxib. Br J Clin Pharmacol 2006;62:257–9.
663. Fitzgerald GA. Coxibs and cardiovascular disease. N Engl J Med 2004;351:1709–11.
664. Linton MRF, Fazio S. Cyclooxygenase-2 and inflammation in atherosclerosis. Curr Opin Pharmacol 2004;4:116–23.
665. Clark DW, Layton D, Shakir SA. Do some inhibitors of COX-2 increase the risk of thromboembolic events? Linking pharmacology with pharmacoepidemiology. Drug Saf 2004;27:427–56.
666. Hankey GJ, Eikelboom JW. Cyclooxygenase-2 inhibitors: are they really atherothrombotic, and if not why not? Stroke 2003;34:2736–40.
667. Cipollone F, Fazia M, Iezzi A, Zucchelli M, Pini B, De Cesare D, Ucchino S, Spigon··· F, Bajocchi G, Bei R, Muraro R, Artese L, Piattelli A, Chiarelli F, Cuccurullo F, Mezzetti A. Suppression of the functionally coupled cyclooxygenase-2/prostglandin E synthase as a basis of simvastatin-dependent plaque stabilization in humans. Circulation 2003;107:1479–85.
668. Chenevard R, Hurlimann D, Bechir M, Enseleit F, Spiker L, Hermann M, Riesen Gay S, Gay RE, Neidhart M, Michel B, Luscher TF, Noll G, Ruschitzka F. Selective COX-2 inhibition improves endothelial function in coronary artery disease. Circulation 2003;107:405–9.
669. Armstrong PW. Balancing the cyclooxygenase portfolio. CMAJ 2006;174:1581–2.
670. Mukherjee D, Nissen SE, Topol EJ. Risk of cardiovascular events associated with selective COX-2 inhibitors. JAMA 2001;286(8):954–9.
671. Bombardier C, Laine L, Reicin A, Shapiro D, Burgos-Vargas R, Davis B, Day R, Ferraz MB, Hawkey CJ, Hochberg MC, Kvien TK, Schnitzer TJVIGOR Study Group. Comparison of upper gastrointestinal toxicity of rofecoxib and naproxen in patients with rheumatoid arthritis. N Engl J Med 2000;343(21):1520–8.
672. Sanmuganathan PS, Ghahramani P, Jackson PR, Wallis EJ, Ramsay LE. Aspirin for primary prevention of coronary heart disease: safety and absolute benefit related to coronary risk derived from meta-analysis of randomised trials. Heart 2001;85(3):265–71.
673. Fleming M. Cardiovascular events and COX-2 inhibitors. JAMA 2001;286(22):2808.
674. Burnakis TG. Cardiovascular events and COX-2 inhibitors. JAMA 2001;286(22):2808.
675. Konstam MA, Demopoulos LA. Cardiovascular events and COX-2 inhibitors. JAMA 2001;286(22):2809.

676. Grant KD. Cardiovascular events and COX-2 inhibitors. JAMA 2001;286(22):2809.
677. Haldey EJ, Pappagallo M. Cardiovascular events and COX-2 inhibitors. JAMA 2001;286(22):2809–10.
678. McGeer PL, McGeer EG, Yasojima K. Cardiovascular events and COX-2 inhibitors. JAMA 2001;286(22):2810.
679. White WB, Whelton A. Cardiovascular events and COX-2 inhibitors. JAMA 2001;286(22):2811–2.
680. Konstam MA, Weir MR, Reicin A, Shapiro D, Sperling RS, Barr E, Gertz BJ. Cardiovascular thrombotic events in controlled, clinical trials of rofecoxib. Circulation 2001;104(19):2280–8.
681. FitzGerald GA, Cheng Y, Austin S. COX-2 inhibitors and the cardiovascular system. Clin Exp Rheumatol 2001;19(6 Suppl 25):S31–6.
682. Wooltorton E. What's all the fuss? Safety concerns about COX-2 inhibitors rofecoxib (Vioxx) and celecoxib (Celebrex). CMAJ 2002;166(13):1692–3.
683. Van Hecken A, Schwartz JI, Depre M, De Lepeleire I, Dallob A, Tanaka W, Wynants K, Buntinx A, Arnout J, Wong PH, Ebel DL, Gertz BJ, De Schepper PJ. Comparative inhibitory activity of rofecoxib, meloxicam, diclofenac, ibuprofen, and naproxen on COX-2 versus COX-1 in healthy volunteers. J Clin Pharmacol 2000;40(10):1109–20.
684. Wright JM. The double-edged sword of COX-2 selective NSAIDs. CMAJ 2002;167(10):1131–7.
685. Solomon DH, Glynn RJ, Levin R, Avorn J. Nonsteroidal anti-inflammatory drug use and acute myocardial infarction. Arch Intern Med 2002;162(10):1099–104.
686. Watson DJ, Rhodes T, Cai B, Guess HA. Lower risk of thromboembolic cardiovascular events with naproxen among patients with rheumatoid arthritis. Arch Intern Med 2002;162(10):1105–10.
687. Rahme E, Pilote L, LeLorier J. Association between naproxen use and protection against acute myocardial infarction. Arch Intern Med 2002;162(10):1111–5.
688. Ray WA, Stein CM, Daugherty JR, Hall K, Arbogast PG, Griffin MR. COX-2 selective non-steroidal anti-inflammatory drugs and risk of serious coronary heart disease. Lancet 2002;360:1071–3.
689. Solomon DH, Schneeweiss S, Glynn RJ, Kiyota Y, Levin R, Mogun H, Avorn J. Relationship between selective cyclooxygenase-2 inhibitors and acute myocardial infarction in older adults. Circulation 2004;109:2068–73.
690. Graham DJ, Campen D, Hui R, Spence M, Cheetham C, Levy G, Shoor S, Ray WA. Risk of acute myocardial infarction and sudden cardiac death in patients treated with cyclo-oxygenase 2 selective and non-selective non-steroidal anti-inflammatory drugs: nested case-control study. Lancet 2005;365:475–81.
691. Cleland JG. No reduction in cardiovascular risk with NSAIDs-including aspirin? Lancet 2002;359(9301):92–3.
692. Dalen JE. Selective COX-2 Inhibitors, NSAIDs, aspirin, and myocardial infarction. Arch Intern Med 2002;162(10):1091–2.
693. Ray WA, Stein CM, Hall K, Daugherty JR, Griffin MR. Non-steroidal anti-inflammatory drugs and risk of serious coronary heart disease: an observational cohort study. Lancet 2002;359(9301):118–23.
694. Steering Committee of the Physicians' Health Study Research Group. Final report on the aspirin component of the ongoing Physicians' Health Study. N Engl J Med 1989;321(3):129–35.
695. Topol EJ. Failing the public health—rofecoxib, Merck, and the FDA. N Engl J Med 2004;351:1707–9.
696. Bresalier RS, Sandler RS, Quan H, Bolognese JA, Oxenius B, Horgan K, Lines C, Riddell R, Morton D, Lanas A, Konstam MA, Baron JA, Adenomatous Polyp Prevention on Vioxx (APPROVe) Trial Investigators. Cardiovascular events associated with rofecoxib in a colorectal adenoma chemoprevention trial. N Engl J Med 2005;352:1092–102.
697. Jüni P, Nartey L, Reinchenbach S, Sterchi R, Dieppe PA, Egger M. Risk of cardiovascular events and rofecoxib: cumulative meta-analysis. Lancet 2004;364:2021–29.
698. Nissen SE. Adverse cardiovascular effects of rofecoxib. N Engl J Med 2006;355:203–4.
699. Lévesque LE, Brophy JM, Zhang B. Time variations in the risk of myocardial infarction among elderly users of COX-2 inhibitors. CMAJ 2006;174:1563–9.
700. Aronson JK, Ferner RE. Joining the DoTS. New approach to classifying adverse drug reactions. BMJ 2003;327:1222–5.
701. Konstam MA, Weir MR, Reicin A, Shapiro D, Sperling RS, Barr E, Gertz BJ. Cardiovascular thrombotic events in controlled, clinical trials of rofecoxib. Circulation 2001;104:2280–8.
702. Reicin AS, Shapiro D, Sperling RS, Barr E, Yu Q. Comparison of cardiovascular thrombotic events in patients with osteoarthritis treated with rofecoxib versus nonselective nonsteroidal anti-inflammatory drugs (ibuprofen, diclofenac, and nabumetone). Am J Cardiol 2002;89:971–2.
703. Mamdani M, Rochon P, Juurlink DN, Anderson GM, Kopp A, Naglie G, Austin PC, Laupacies A. Effect of selective cyclooxygenase 2 inhibitors and naproxen on short-term risk of acute myocardial infarction in the elderly. Arch Intern Med 2003;163:481–6.
704. Rahme E, Pilote L, LeLorier J. Association between naproxen use and protection against acute myocardial infarction. Arch Intern Med 2002;162:1111–5.
705. Solomon DH, Glynn RJ, Levin R, Avorn J. Nonsteroidal anti-inflammatory drug use and acute myocardial infarction. Arch Intern Med 2002;162:1099–104.
706. Watson DJ, Rhodes T, Cai B. Guess HA. Lower risk of thromboembolic cardiovascular events with naproxen among patients with rheumatoid arthritis. Arch Intern Med 2002;162:1105–10.
707. Kearney P, Baigent C, Godwin J, Halls H, Emberson JR, Patrono C. Do selective cyclo-oxygenease-2 inhibitors and traditional non-steroidal anti-inflammatory drugs increase the risk of atherothrombosis? Meta-analysis of randomised trials. BMJ 2006;332:1302–8.
708. McGettigan P, Henry D. Cardiovascular risk and inhibition of cyclooxygenase: a systematic review of the observational studies of selective and nonselective inhibitors of cyclooxygenase 2. JAMA 2006;296:1633–44.
709. Solomon SD, McMurray JJV, Pfeffer MA, Wittes J, Fowler R, Finn P, Anderson WF, Zauber A, Hawk W, Bertagnolli M; Adenoma Prevention with Celecoxib (APC) Study Investigators. Cardiovascular risk associated with celecoxib in a clinical trial for colorectal adenoma prevention. N Engl J Med 2005;352:1071–80.
710. Brophy JM. Cardiovascular risk associated with celecoxib. N Engl J Med 2005;352:2648–50.
711. White WB, Faich G, Whelton A, Maurath C, Ridge NJ, Verburg KM, Geis GS, Lefkowith JB. Comparison of thromboembolic events in patients treated with celecoxib, a cyclooxygenase-2 specific inhibitor, versus ibuprofen or diclofenac. Am J Cardiol 2002;89(4):425–30.
712. White WB, Faich G, Borer JS, Makuch RW. Cardiovascular thrombotic events in arthritis trials of the

cyclooxygenase-2 inhibitor celecoxib. Am J Cardiol 2003;92:411–8.

713. Silverstein FE, Faich G, Goldstein JL, Simon LS, Pincus T, Whelton A, Makuch R, Eisen G, Agrawal NM, Stenson WF, Burr AM, Zhao WW, Kent JD, Lefkowith JB, Verburg KM, Geis GS. Gastrointestinal toxicity with celecoxib vs nonsteroidal anti-inflammatory drugs for osteoarthritis and rheumatoid arthritis: the CLASS study: a randomized controlled trial. Celecoxib Long-term Arthritis Safety Study. JAMA 2000;284:1247–55.

714. Ray WA, Stein CM, Hall K, Daugherty JR, Griffin MR. Non-steroidal anti-inflammatory drugs and risk of serious coronary heart disease: an observational cohort study. Lancet 2002;359:118–23.

715. Kimmel SE, Berlin JA, Reilly M, Jaskowiak J, Kishel L, Chittams J, Strom BL. Patients exposed to rofecoxib and celecoxib have different odds of nonfatal myocardial infarction. Ann Intern Med 2005;142:157–64.

716. Shaya FT, Blume SW, Blanchette CM, Weir MR, Mullins CD. Selective cyclooxygenase-2 inhibition and cardiovascular effects: an observational study of a Medicaid population. Arch Intern Med 2005;165:181–6.

717. White WB, Strand V, Roberts R, Whelton A. Effects of the cycloxygenase-2 specific inhibitor valdecoxib versus nonsteroidal antinflammatory agents and placebo on cardiovascular thrombotic events in patients with arthritis. Am J Ther 2004;11:244–50.

718. Ott E, Nussmeier NA, Duke PC, Feneck RO, Alston RP, Snabes MC, Hubbard RC, Hsu PH, Saidman LJ, Mangano DT. Efficacy and safety of the cyclooxygenase 2 inhibitors parecoxib and valdecoxib in patients undergoing coronary artery bypass surgery. J Thorac Cardiovasc Surg 2003;125:1481–92.

719. Nussmeier NA, Whelton AA, Brown MT, Langford RM, Hoeft A, Parlow JL, Boyce SW, Verburg KM. Complications of the COX-2 inhibitors parecoxib and valdecoxib after cardiac surgery. N Engl J Med 2005;352:1081–91.

720. Furberg CD, Psaty BM, FitzGerald GA. Parecoxib, valdecoxib, and cardiovascular risk. Circulation 2005;111:249.

721. Ray WA, Griffin MR, Stein CM. Cardiovascular toxicity of valdecoxib. N Engl J Med 2004;361:2767.

722. Eisen GM, Goldstein JL, Hanna DB, Rublee DA. Meta-analysis: upper gastrointestinal tolerability of valdecoxib, a cyclooxygenase-2-specific inhibitor, compared with nonspecific nonsteroidal anti-inflammatory drugs among patients with osteoarthritis and rheumatoid arthritis. Aliment Pharmacol Ther 2005;21:591–8.

723. Farkouh ME, KirshnerH, Harrington RA, Ruland S, Verheugt FWA, Schnitzer TJ, Burmster GR, Mysler E, Hochberg MC, Doherty M, Ehrsam E, Gitton X, Krammer G, Mellein B, Gimona A, Matchaba P, Hawkey CJ, Chesebro J, on behalf of the TARGET Study Group. Comparison of lumiracoxib with naproxen and ibuprofen in the therapeutic arthritis research and gastrointestinal event trial (TARGET), cardiovascular outcomes: randomised controlled trial. Lancet 2004;364:675–84.

724. Topol EJ, Falk GW. A coxib a day won't keep the doctor away. Lancet 2004;364:639–40.

725. Schnitzer T, Burmester GR, Mysler E, Hochberg MC, Doherty M, Ehrsam E, Gitton X, Krammer G, Mellein B, Matchaba P, Gimona A, Hawkey CJ, on behalf of the TARGET Study Group. Comparison of lumiracoxib with naproxen and ibuprofen in the therapeutic arthritis research and gastrointestinal event trial (TARGET), reduction in ulcer complications: randomised controlled trial. Lancet 2004;364:665–74.

726. Ramey DR, Watson DJ, Yu C, Bolognese JA, Curtis SP, Reicin AS. The incidence of upper gastrointestinal adverse events in clinical trials of etoricoxib vs. non-selective NSAIDs: an updated combined analysis. Curr Med Res Opin 2005;21:715–22.

727. Aldington S, Shirtcliffe P, Weatherall M, Beasley R. Systematic review and meta-analysis of the risk of major cardiovascular events with etoricoxib therapy. N Z Med J 2005;118:U1684.

728. Altman R, Luciardi HL, Muntaner J, Del Rio F, Berman SG, Lopez R, Gonzalez C. Efficacy assessment of meloxicam, a preferential cyclooxygenase-2 inhibitor, in acute coronary syndromes without ST-segment elevation. Non-steroidal Anti-Inflammatory Drugs in Unstable Angina Treatment (NUT-2) pilot study. Circulation 2002;106: 191–5.

729. Mandani M, Juurlink DN, Lee DS, Rochon PA, Kopp A, Naglie G, Austin PC, Laipacis A, Stukel TA. Cyclo-oxygenase-2 inhibitors versus non-selective non-steroidal anti-inflammatory drugs and congestive heart failure outcomes in elderly patients: a population-based cohort study. Lancet 2004;363:1751–6.

730. Hudson M, Richard H, Pilote L. Differences in outcomes of patients with congestive heart failure prescribed celecoxib, rofecoxib, or non-steroidal anti-inflammatory drugs: population based study. BMJ 2005;330:1370.

731. Solomon DH, Scheneeweiss S, Levin R, Avorn J. Relationship between COX-2 specific inhibitors and hypertension. Hypertension 2004;44:140–5.

732. Aw TJ, Haas SJ, Liew D, Krum H. Meta-analysis of cyclooxygenase-2 inhibitors and their effects on blood pressure. Arch Intern Med 2005;165:490–6.

733. Wolfe F, Zhao S, Pettitt D. Blood pressure destabilization and edema among 8538 users of celecoxib, rofecoxib, and nonselective nonsteroidal anti-inflammatory drugs (NSAID) and nonusers of NSAID receiving ordinary clinical care. J Rheumatol 2004;31:1035–7.

734. Mamdani M, Juurlink DN, Kopp A, Naglie G, Austin PC, Laupacis A. Gastrointestinal bleeding after the introduction of COX 2 inhibitors: ecological study. BMJ 2004;328:1415–6.

735. Johnson AG, Nguyen TV, Day RO. Do nonsteroidal anti-inflammatory drugs affect blood pressure? A meta-analysis. Ann Intern Med 1994;121(4):289–300.

736. Whelton A. Renal and related cardiovascular effects of conventional and COX-2-specific NSAIDs and non-NSAID analgesics. Am J Ther 2000;7(2):63–74.

737. Collins R, Peto R, MacMahon S, Hebert P, Fiebach NH, Eberlein KA, Godwin J, Qizilbash N, Taylor JO, Hennekens CH. Blood pressure, stroke, and coronary heart disease. Part 2. Short-term reductions in blood pressure: overview of randomised drug trials in their epidemiological context. Lancet 1990; 335(8693):827–38.

738. Whelton A, Maurath CJ, Verburg KM, Geis GS. Renal safety and tolerability of celecoxib, a novel cyclooxygenase-2 inhibitor. Am J Ther 2000;7(3):159–75.

739. Whelton A, Fort JG, Puma JA, Normandin D, Bello AE, Verburg KMSUCCESS VI Study Group. Cyclooxygenase-2-specific inhibitors and cardiorenal function: a randomized, controlled trial of celecoxib and rofecoxib in older hypertensive osteoarthritis patients. Am J Ther 2001;8(2):85–95.

740. Nietert PJ, Ornstein SM, Dickerson LM, Rothenberg RJ. Comparison of changes in blood pressure measurements and antihypertensive therapy in older, hypertensive,

ambulatory care patients prescribed celecoxib or rofecoxib. Pharmacotherapy 2003;23:1416–23.

741. Pathak A, Boveda S, Defaye P, Mansourati J, Mallaret M, Thebault L, Galinier M, Blanc JJ, Montastruc JL. Celecoxib-associated torsade de pointes. Ann Pharmacother 2002;36(7–8):1290–1.
742. Goldberg MA, Antin JH, Guinan EC, Rappeport JM. Cyclophosphamide cardiotoxicity: an analysis of dosing as a risk factor. Blood 1986;68(5):1114–8.
743. Mugitani A, Yamane T, Park K, Im T, Tatsumi N, Tatsumi Y. Cardiac complications after high-dose chemotherapy with peripheral blood stem cell transplantation. J Jpn Soc Cancer Ther 1996;31:255–62.
744. Nakamae H, Tsumura K, Hino M, Hayashi T, Tatsumi N. QT dispersion as a predictor of acute heart failure after high-dose cyclophosphamide. Lancet 2000;355(9206): 805–6.
745. Hansen DD, Fernandes A, Skovsted P, Berry P. Cyclopropane anaesthesia for renal transplantation. Report of 100 cases. Br J Anaesth 1972;44(6):584–9.
746. Wong KC. Sympathomimetic drugs. In: Smith NT, Miller RD, Corbascio AN, editors. Drug Interactions in Anesthesia. Philadelphia: Lea & Febiger, 1981:66.
747. Seaman H, de Vries C, Farmer R. Venous thromboembolism associated with cyproterone acetate in combination with ethinyloestradiol (Dianette): observational studies using the UK General Practice Research Database. Pharmacoepidemiol Drug Saf 2004;13:427–36.
748. Hermans C, Straetmans N, Michaux JL. Pericarditis induced by high-dose cytosine arabinoside chemotherapy. Ann Hematol 1997;75:55–7.
749. Gähler A, Hitz F, Hess U, Cerny T. Acute pericarditis and pleural effusion complicating cytarabine chemotherapy. Onkologie 2003;26:348–50.
750. Boos C.J, Dawes M, Jones R, Farrell T. Danazol treatment and acute myocardial infarction. J Obstet Gynecol 2003;23:327–8.
751. Saltzman LS, Kates RA, Corke BC, Norfleet EA, Heath KR. Hyperkalemia and cardiovascular collapse after verapamil and dantrolene administration in swine. Anesth Analg 1984;63(5):473–8.
752. Rubin AS, Zablocki AD. Hyperkalemia, verapamil, and dantrolene. Anesthesiology 1987;66(2):246–9.
753. Cowan RE, Wright JT. Dapsone and severe hypoalbuminaemia in dermatitis herpetiformis. Br J Dermatol 1981;104(2):201–4.
754. Berdoukas V, Bentley P, Frost H, Schnebli HP. Toxicity of oral iron chelator L1. Lancet 1993;341(8852):1088.
755. Fosburg MT, Nathan DG. Treatment of Cooley's anemia. Blood 1990;76(3):435–44.
756. McCarthy JT, Milliner DS, Johnson WJ. Clinical experience with desferrioxamine in dialysis patients with aluminium toxicity. Q J Med 1990;74(275):257–76.
757. Jacobs P, Wood L, Bird AR, Ultmann JE. Pseudo deep-vein thrombosis following desferrioxamine infusion: a previously unreported adverse reaction? Lancet 1990;336(8718):815.
758. Stephens AD. Cystinuria and its treatment: 25 years experience at St. Bartholomew's Hospital. J Inherit Metab Dis 1989;12(2):197–209.
759. Hamdy NA, Andrew SM, Shortland JR, Boletis J, Raftery AT, Kanis JA, Brown CB. Fatal cardiac zygomycosis in a renal transplant patient treated with desferrioxamine. Nephrol Dial Transplant 1989;4(10):911–3.
760. Daly AL, Velazquez LA, Bradley SF, Kauffman CA. Mucormycosis: association with deferoxamine therapy. Am J Med 1989;87(4):468–71.
761. Arizono K, Fukui H, Miura H, Hayano K, Otsuka Y, Tajiri M. [A case report of rhinocerebral mucormycosis in hemodialysis patient receiving deferoxamine.]Nippon Jinzo Gakkai Shi 1989;31(1):99–103.
762. Voest EE, Neijt JP, Keunen JE, Dekker AW, van Asbeck BS, Nortier JW, Ros FE, Marx JJ. Phase I study using desferrioxamine and iron sorbitol citrate in an attempt to modulate the iron status of tumor cells to enhance doxorubicin activity. Cancer Chemother Pharmacol 1993;31(5):357–62.
763. Rodig G, Wild K, Behr R, Hobbhahn J. Effects of desflurane and isoflurane on systemic vascular resistance during hypothermic cardiopulmonary bypass. J Cardiothorac Vasc Anesth 1997;11(1):54–7.
764. Warltier DC, Pagel PS. Cardiovascular and respiratory actions of desflurane: is desflurane different from isoflurane? Anesth Analg 1992;75(Suppl 4):S17–31.
765. Bunting HE, Kelly MC, Milligan KR. Effect of nebulized lignocaine on airway irritation and haemodynamic changes during induction of anaesthesia with desflurane. Br J Anaesth 1995;75(5):631–3.
766. Bennett JA, Lingaraju N, Horrow JC, McElrath T, Keykhah MM. Elderly patients recover more rapidly from desflurane than from isoflurane anesthesia. J Clin Anesth 1992;4(5):378–81.
767. Geffroy JC, Gentili ME, Le Pollès R, Triclot P. Massive inhalation of desflurane due to vaporizer dysfunction. Anesthesiology 2005;103(5):1096–8.
768. Smelt WL. Cardiac arrest during desflurane anaesthesia in a patient with Duchenne's muscular dystrophy. Acta Anaesthesiol Scand 2005;49(2):267–9.
769. Meltzer EO, Prenner BM, Nayak AThe Desloratadine Study Group. Efficacy and tolerability of once-daily 5 mg desloratadine, an H_1-receptor antagonist, in patients with seasonal allergic rhinitis: assessment during the spring and fall allergy seasons Clin Drug Invest 2001;21:25–32.
770. Ring J, Hein R, Gauger A, Bronsky E, Miller B, Breneman D, Conneley M, Corren J, Ceuppens J, Fierlbeck G, Friday G, Goldberg P, Graft D, Holst T, Honsinger R, Hornmark A-M, Kaiser H, Kaplan R, Kempers S, Lockey R, Miller SD, Nayak A, Nayak N, Pariser D, Prenner B, Ruzicka T, Stewart GE II, Thompson M, Wein M. Once-daily desloratadine improves the signs and symptoms of chronic idiopathic urticaria: a randomized, double-blind, placebo-controlled study. Int J Dermatol 2001;40(1):72–6.
771. Banfield C, Hunt T, Reyderman L, Statkevich P, Padhi D, Affrime M. Lack of clinically relevant interaction between desloratadine and erythromycin. Clin Pharmacokinet 2002;41(Suppl 1):29–35.
772. Gill JC, Ottum M, Schwartz B. Evaluation of high concentration intranasal and intravenous desmopressin in pediatric patients with mild hemophilia A or mild-to-moderate type 1 von Willebrand disease. J Pediatr 2002;140(5): 595–599.
773. D'Alauro FS, Johns RA. Hypotension related to desmopressin administration following cardiopulmonary bypass. Anesthesiology 1988;69(6):962–3.
774. Israels SJ, Kobrinsky NL. Serious reaction to desmopressin in a child with cyanotic heart disease. N Engl J Med 1989;320(23):1563–4.
775. Ruggeri ZM, Mannucci PM, Lombardi R, Federici AB, Zimmerman TS. Multimeric composition of factor VIII/ von Willebrand factor following administration of DDAVP: implications for pathophysiology and therapy of von Willebrand's disease subtypes. Blood 1982;59(6):1272–8.

776. Bond L, Bevan D. Myocardial infarction in a patient with hemophilia treated with DDAVP. N Engl J Med 1988;318(2):121.
777. van Dantzig JM, Duren DR, Ten Cate JW. Desmopressin and myocardial infarction. Lancet 1989;1(8639):664–5.
778. McLeod BC. Myocardial infarction in a blood donor after administration of desmopressin. Lancet 1990;336(8723):1137–8.
779. Hartmann S, Reinhart W. Fatal complication of desmopressin. Lancet 1995;345(8960):1302–3.
780. Anonymous. Desmopressin and arterial thrombosis. Lancet 1989;1(8644):938–9.
781. Stratton J, Warwicker P, Watkins S, Farrington K. Desmopressin may be hazardous in thrombotic microangiopathy. Nephrol Dial Transplant 2001;16(1):161–2.
782. Levi M, Cromheecke ME, de Jonge E, Prins MH, de Mol BJ, Briet E, Buller HR. Pharmacological strategies to decrease excessive blood loss in cardiac surgery: a meta-analysis of clinically relevant endpoints. Lancet 1999;354(9194):1940–7.
783. Grunwald Z, Sather SD. Intraoperative cerebral infarction after desmopressin administration in infant with end-stage renal disease. Lancet 1995;345(8961):1364–5.
784. Wieting JM, Dykstra DD, Ruggiero MP, Robbins GB, Galusha K. Central nervous system ischemia after *Varicella* infection and desmopressin therapy for enuresis. J Am Osteopath Assoc 1997;97(5):293–5.
785. Mannucci PM, Lusher JM. Desmopressin and thrombosis. Lancet 1989;2(8664):675–6.
786. Mannucci PM, Carlsson S, Harris AS. Desmopressin, surgery and thrombosis. Thromb Haemost 1994;71(1):154–5.
787. Golan A, Siedner M, Bahar M, Ron-El R, Herman A, Caspi E. High-output left ventricular failure after dextran use in an operative hysteroscopy. Fertil Steril 1990;54(5):939–41.
788. Gillard P, Laurent M. Dextropropoxyphene-induced cardiogenic shock: treatment with intra-aortic balloon pump and milrinone. Intensive Care Med 1999;25(3):335.
789. Pickar D, Dubois M, Cohen MR. Behavioral change in a cancer patient following intrathecal beta-endorphin administration. Am J Psychiatry 1984;141(1):103–4.
790. O'Brien JJ, Benfield P. Dezocine. A preliminary review of its pharmacodynamic and pharmacokinectic properties, and therapeutic efficacy. Drugs 1989;38(2):226–48.
791. Joist A, Tibesku CO, Neuber M, Frerichmann U, Joosten U. Fingergangrän nach akziden teller intraarterieller Injektion von Diazepam. [Gangrene of the fingers caused by accidental intra-arterial injection of diazepam.] Dtsch Med Wochenschr 1999;124(24):755–8.
792. Derakshan MR. Amputation due to inadvertent intra-arterial diazepam injection. Iran J Med Sci 2000;25: 84–6.
793. Pavone-Macaluso M, de Voogt HJ, Viggiano G, Barasolo E, Lardennois B, de Pauw M, Sylvester R. Comparison of diethylstilbestrol, cyproterone acetate and medroxyprogesterone acetate in the treatment of advanced prostatic cancer: final analysis of a randomized phase III trial of the European Organization for Research on Treatment of Cancer Urological Group. J Urol 1986;136(3):624–31.
794. Lessard E, Yessine MA, Hamelin BA, Gauvin C, Labbe L, O'Hara G, LeBlanc J, Turgeon J. Diphenhydramine alters the disposition of venlafaxine through inhibition of CYP2D6 activity in humans. J Clin Psychopharmacol 2001;21(2):175–84.
795. Sype JW, Khan IA. Prolonged QT interval with markedly abnormal ventricular repolarization in diphenhydramine overdose. Int J Cardiol 2005;99:333–5.
796. Thakur AC, Aslam AK, Aslam AF, Vasavada BC, Sacchi TJ, Khan IA. QT interval prolongation in diphenhydramine toxicity Int J Cardiol 2005;98:341–3.
797. Lopez-Barbeito B, Lluis M, Delgado V, Jimenez S, Diaz-Infante E, Nogue-Xarau S, Brugada J. Diphenhydramine overdose and Brugada sign. Pacing Clin Electrophysiol 2005;28:730–2.
798. Joshi AK, Sljapic T, Borghei H, Kowey PR. Case of polymorphic ventricular tachycardia in diphenhydramine poisoning. J Cardiovasc Electrophysiol 2004;15(5):591–3.
799. Boccara F, Benhaiem-Sigaux N, Cohen A. Acute myopericarditis after diphtheria, tetanus, and polio vaccination. Chest 2001;120(2):671–2.
800. Chang WT, Lin LC, Yen RF, Huang PJ. Persistent myocardial ischemia after termination of dipyridamole-induced ventricular tachycardia by intravenous aminophylline: scintigraphic demonstration. J Formos Med Assoc 2000;99(3):264–6.
801. Bielen M, Karsera D, Melon P, Kulbertus H. Bradyarythmies sevères au cours d'ure scintigraphie myocardique de perfusion avec injection de dipyridamole (persantine). [Severe bradyarrhythmia during myocardial perfusion scintigraphy with injection of dipyridamole (Persantine).] Rev Med Liege 1999;54(2):105–8.
802. Nedeljkovic MA, Ostojic M, Beleslin B, Nedeljkovic IP, Stankovic G, Stojkovic S, Saponjski J, Babic R, Vukcevic V, Ristic AD, Orlic D. Dipyridamole–atropine-induced myocardial infarction in a patient with patent epicardial coronary arteries. Herz 2001;26(7):485–8.
803. Wartski M, Caussin C, Lancelin B. Spasme coronaire plurifocal declenche par l'injection de dipyridamole. Med Nucl 2001;25:153–9.
804. Whiting JH Jr, Datz FL, Gabor FV, Jones SR, Morton KA. Cerebrovascular accident associated with dipyridamole thallium-201 myocardial imaging: case report. J Nucl Med 1993;34(1):128–30.
805. Lake CR, Major LF, Ziegler MG, Kopin IJ. Increased sympathetic nervous system activity in alcoholic patients treated with disulfiram. Am J Psychiatry 1977;134(12):1411–4.
806. Savas MC, Gullu IH. Disulfiram–ethanol test reaction: significance of supervision. Ann Pharmacother 1997;31(3):374–5.
807. Lattanzi F, Picano E, Adamo E, Varga A. Dobutamine stress echocardiography: safety in diagnosing coronary artery disease. Drug Saf 2000;22(4):251–62.
808. Chenzbraun A, Khoury Z, Gottlieb S, Keren A. Impact of age on the safety and the hemodynamic response pattern during high dose dobutamine echocardiography. Echocardiography 1999;16(2):135–42.
809. Roldan FJ, Vargas-Barron J, Espinola-Zavaleta N, Keirns C, Romero-Cardenas A. Severe dynamic obstruction of the left ventricular outflow tract induced by dobutamine. Echocardiography 2000;17(1):37–40.
810. Previtali M, Repetto A, Scuteri L. Dobutamine induced severe midventricular obstruction and mitral regurgitation in left ventricular apical ballooning syndrome. Heart 2005;91:353–5.
811. Datino T, Garcia-Fernandez MA, Martinez-Selles M, Quiles J, Avanzas P. Cardiac rupture during contrast-enhanced dobutamine stress echocardiography. Int J Cardiol 2005;98:349–50.
812. Lu D, Greenberg MD, Little R, Malik Q, Fernicola DJ, Weissman NJ. Accelerated dobutamine stress testing: safety and feasibility in patients with known or suspected coronary artery disease. Clin Cardiol 2001;24(2):141–5.

813. Previtali M, Scelsi L, Sebastiani R, Lanzarini L, Raisaro A, Klersy C. Feasibility, safety, and prognostic value of dobutamine stress echocardiography in patients > or = 70 years of age early after acute myocardial infarction Am J Cardiol 2002;90(7):792–5.
814. Vecchia L, Ometto R, Finocchi G, Vincenzi M. Torsade de pointes ventricular tachycardia during low dose intermittent dobutamine treatment in a patient with dilated cardiomyopathy and congestive heart failure. Pacing Clin Electrophysiol 1999;22(2):397–9.
815. Varga A, Picano E, Lakatos F. Fatal ventricular fibrillation during a low-dose dobutamine stress test. Am J Med 2000;108(4):352–3.
816. Kahn JH, Starling MR, Supiano MA. Transient dobutamine-mediated pulsus alternans. Can J Cardiol 2001;17(2):203–5.
817. Cox DE, Farmer LD, Hoyle JR, Wells GL. Prognostic significance on nonsustained ventricular tachycardia during dobutamine stress echocardiography. Am J Cardiol 2005;96:1293–8.
818. Burger AJ, Elkayam U, Neibaur MT, Haught H, Ghali J, Horton DP, Aronson D. Comparison of the occurrence of ventricular arrhythmias in patients with acutely decompensated congestive heart failure receiving dobutamine versus nesiritide therapy. Am J Cardiol 2001;88(1):35–9.
819. Poldermans D, Rambaldi R, Bax JJ, Cornel JH, Thomson IR, Valkema R, Boersma E, Fioretti PM, Breburda CS, Roelandt JR. Safety and utility of atropine addition during dobutamine stress echocardiography for the assessment of viable myocardium in patients with severe left ventricular dysfunction. Eur Heart J 1998;19(11):1712–8.
820. Elhendy A, Valkema R, van Domburg RT, Bax JJ, Nierop PR, Cornel JH, Geleijnse ML, Reijs AE, Krenning EP, Roelandt JR. Safety of dobutamine-atropine stress myocardial perfusion scintigraphy. J Nucl Med 1998;39(10):1662–6.
821. Lin SS, Roger VL, Pascoe R, Seward JB, Pellikka PA. Dobutamine stress Doppler hemodynamics in patients with aortic stenosis: feasibility, safety, and surgical correlations. Am Heart J 1998;136(6):1010–6.
822. Burger AJ, Notarianni MP, Aronson D. Safety and efficacy of an accelerated dobutamine stress echocardiography protocol in the evaluation of coronary artery disease. Am J Cardiol 2000;86(8):825–9.
823. Garcimartin I, San Roman JA, Vilacosta I, Munoz JC, de la Torre M, Fernandez-Aviles F. Complicaciones de la ecocardíografia de estrés transesofágica con dobutamina. [Complications of transesophageal echocardiography with dobutamine.] Rev Esp Cardiol 2000;53(8):1136–9.
824. Burger AJ, Horton DP, LeJemtel T, Ghali JK, Torre G, Dennish G, Koren M, Dinerman J, Silver M, Cheng ML, Elkayam U. Prospective Randomized Evaluation of Cardiac Ectopy with Dobutamine or Natrecor Therapy. Effect of nesiritide (B-type natriuretic peptide) and dobutamine on ventricular arrhythmias in the treatment of patients with acutely decompensated congestive heart failure: the PRECEDENT study. Am Heart J 2002;144(6):1102–8.
825. Friart A, Hermans L, De Valeriola Y. Unusual side-effect of a dobutamine stress echocardiography. Am J Noninvasive Cardiol 1993;7:63–4.
826. Kawano H, Fujii H, Motoyama T, Kugiyama K, Ogawa H, Yasue H. Myocardial ischemia due to coronary artery spasm during dobutamine stress echocardiography. Am J Cardiol 2000;85(1):26–30.
827. Pfleger S, Scherhag A, Latsch A, Dempfle CE, Simonis B, Haux P, Voelker W, Gaudron P. Safety of dobutamine echocardiography: no signs of myocardial cell damage or activation of the coagulation system. Dis Manag Clin Outcomes 2001;3:15–9.
828. Beckmann S, Bocksch W, Muller C, Schartl M. Does dobutamine stress echocardiography induce damage during viability diagnosis of patients with chronic regional dysfunction after myocardial infarction? J Am Soc Echocardiogr 1998;11(2):181–7.
829. Hoff JV, Peatty PA, Wade JL. Dermal necrosis from dobutamine. N Engl J Med 1979;300(22):1280.
830. Roussak JB, Carey P, Parry H. Cardiac arrest after treatment with intravenous domperidone. BMJ (Clin Res Ed) 1984;289(6458):1579.
831. Osborne RJ, Slevin ML, Hunter RW, Hamer J. Cardiac arrhythmias during cytotoxic chemotherapy: role of domperidone. Hum Toxicol 1985;4(6):617–26.
832. Anonymous. Sale of drug for cancer patients suspended due to side effects. Asahi Evening News (Tokyo) 1985;.
833. Calvo-Romero JM, Ramos-Salado JL. Bradycardia sinusal sintomatica associada a donepecilo. [Symptomatic sinus bradycardia associated with donepezil.] Rev Neurol 1999;28(11):1070–2.
834. McLaren AT, Allen J, Murray A, Ballard CG, Kenny RA. Cardiovascular effects of donepezil in patients with dementia. Dement Geriatr Cogn Disord 2003;15:183–8.
835. Stahl SM, Markowitz JS, Gutterman EM, Papadopoulos G. Co-use of donepezil and hypnotics among Alzheimer's disease patients living in the community. J Clin Psychiatry 2003;64:466–72.
836. Bordier P, Lanusse S, Garrigue S, Reynard C, Robert F, Gencel L, Lafitte A. Causes of syncope in patients with Alzheimer's disease treated with donepezil. Drugs Aging 2005;22:687–94.
837. Goldstein BJ, Claghorn JL. An overview of seventeen years of experience with dothiepin in the treatment of depression in Europe. J Clin Psychiatry 1980;41(12 Pt 2):64–70.
838. Prentice A, Brown R. Fetal tachyarrhythmia and maternal antidepressant treatment. BMJ 1989;298(6667):190.
839. Basta SJ, Savarese JJ, Ali HH, Embree PB, Schwartz AF, Rudd GD, Wastila WB. Clinical pharmacology of doxacurium chloride. A new long-acting nondepolarizing muscle relaxant. Anesthesiology 1988;69(4):478–86.
840. Reich DL. Transient systemic arterial hypotension and cutaneous flushing in response to doxacurium chloride. Anesthesiology 1989;71(5):783–5.
841. Murray DJ, Mehta MP, Choi WW, Forbes RB, Sokoll MD, Gergis SD, Rudd GD, Abou-Donia MM. The neuromuscular blocking and cardiovascular effects of doxacurium chloride in patients receiving nitrous oxide narcotic anesthesia. Anesthesiology 1988;69(4):472–7.
842. Stoops CM, Curtis CA, Kovach DA, McCammon RL, Stoelting RK, Warren TM, Miller D, Abou-Donia MM. Hemodynamic effects of doxacurium chloride in patients receiving oxygen sufentanil anesthesia for coronary artery bypass grafting or valve replacement. Anesthesiology 1988;69(3):365–70.
843. Reich DL, Konstadt SN, Thys DM, Hillel Z, Raymond R, Kaplan JA. Effects of doxacurium chloride on biventricular cardiac function in patients with cardiac disease. Br J Anaesth 1989;63(6):675–81.
844. Scott RP, Norman J. Doxacurium chloride: a preliminary clinical trial. Br J Anaesth 1989;62(4):373–7.

845. Stephen CR, Talton I. Effects of doxapram on the electrocardiogram during anesthesia. Anesth Analg 1966;45(6):783–9.
846. Wengel SP, Roccaforte WH, Burke WJ, Bayer BL, McNeilly DP, Knop D. Behavioral complications associated with donepezil. Am J Psychiatry 1998;155(11):1632–3.
847. De Villiers GS, Walele A, Van der Merwe PL, Kalis NN. Second-degree atrioventricular heart block after doxapram administration. J Pediatr 1998;133(1):149–50.
848. Weitzenblum E, Parini JP, Roeslin N. The effects on ventilation, gas exchange and haemodynamics of the respiratory stimulation of doxapram in cases of chronic respiratory failure. J Med Strasb 1973;4:1063.
849. Huon C, Rey E, Mussat P, Parat S, Moriette G. Low-dose doxapram for treatment of apnoea following early weaning in very low birthweight infants: a randomized, double-blind study. Acta Paediatr 1998;87(11):1180–4.
850. Barrington KJ, Muttitt SC. Randomized, controlled, blinded trial of doxapram for extubation of the very low birthweight infant. Acta Paediatr 1998;87(2):191–4.
851. D'Adamo DR, Anderson SE, Albritton K, Yamada J, Riedel E, Scheu K, Schwartz GK, Chen H, Maki RG. Phase II study of doxorubicin and bevacizumab for patients with metastatic soft-tissue sarcomas. J Clin Oncol 2005;23(28):7135–42.
852. Warner JP, Barnes TR, Henry JA. Electrocardiographic changes in patients receiving neuroleptic medication. Acta Psychiatr Scand 1996;93(4):311–3.
853. Iwahashi K. Significantly higher plasma haloperidol level during cotreatment with carbamazepine may herald cardiac change. Clin Neuropharmacol 1996;19(3): 267–70.
854. Reilly JG, Ayis SA, Ferrier IN, Jones SJ, Thomas SH. QTc-interval abnormalities and psychotropic drug therapy in psychiatric patients. Lancet 2000;355(9209): 1048–52.
855. Michalets EL, Smith LK, Van Tassel ED. Torsade de pointes resulting from the addition of droperidol to an existing cytochrome P450 drug interaction. Ann Pharmacother 1998;32(7–8):761–5.
856. Kao LW, Kirk MA, Evers SJ, Rosenfeld SH. Droperidol, QT prolongation, and sudden death: what is the evidence? Ann Emerg Med 2003;41:546–58.
857. Habib AS, Gan TJ. Food and drug administration black box warning on the perioperative use of droperidol: a review of the cases. Anesth Analg 2003;96:1377–9.
858. Gan TJ. "Black box" warning on droperidol: a report of the FDA convened expert panel. Anesth Analg 2004;98:1809.
859. van Zwieten K, Mullins ME, Jang T. Droperidol and the black box warning. Ann Emerg Med 2004;43:139–40.
860. Habib AS, Gan TJ. Safety of patients reason for FDA black box warning on droperidol. Anesth Analg 2004;98:551–2.
861. Shafer SL. Safety of patients reason for FDA black box warning on droperidol. Anesth Analg 2004;98:551–2.
862. Thase ME, Tran PV, Curtis W, Pangallo B, Mallinckrodt C, Detke MJ. Cardiovascular profile of duloxetine, a dual re-uptake inhibitor of serotonin and norepinephrine. J Clin Psychopharmacol 2005;25:132–40.
863. Gillen MS, Miller B, Chaikin P, Morganroth J. Effects of supratherapeutic doses of ebastine and terfenadine on the QT interval. Br J Clin Pharmacol 2001;52(2):201–4.
864. Moss AJ, Chaikin P, Garcia JD, Gillen M, Roberts DJ, Morganroth J. A review of the cardiac systemic side-effects of antihistamines: ebastine. Clin Exp Allergy 1999;29(Suppl 3):200–5.
865. Moss AJ, Morganroth J. Cardiac effects of ebastine and other antihistamines in humans. Drug Saf 1999;21(Suppl 1):69–80; discussion 81–7.
866. Shimosato S, Iwatsuki N, Carter JG. Cardio-circulatory effects of enflurane anesthesia in health and disease. Acta Anaesthesiol Scand Suppl 1979;71:69–70.
867. Rifat K. Effets cardiovasculaires de l'enflurane. Med Hyg 1979;37:3602.
868. Reves JG, Samuelson PN, Lell WA, McDaniel HG, Kouchoukos NT, Rogers WJ, Smith LR, Carter MR. Myocardial damage in coronary artery bypass surgical patients anaesthetized with two anaesthetic techniques: a random comparison of halothane and enflurane. Can Anaesth Soc J 1980;27(3):238–45.
869. Saarnivaara L. Comparison of halothane and enflurane anaesthesia for tonsillectomy in adults. Acta Anaesthesiol Scand 1984;28(3):319–24.
870. Willatts DG, Harrison AR, Groom JF, Crowther A. Cardiac arrhythmias during outpatient dental anaesthesia: comparison of halothane with enflurane. Br J Anaesth 1983;55(5):399–403.
871. Chander S. Isorhythmic atrioventricular dissociation during enflurane anesthesia. South Med J 1982;75(8): 945–50.
872. Okun MS, Charriez CM, Bhatti MT, Watson RT, Swift TR. Asystole induced by edrophonium following beta blockade. Neurology 2001;57(4):739.
873. Bjerke RJ, Mangione MP. Asystole after intravenous neostigmine in a heart transplant recipient. Can J Anaesth 2001;48(3):305–7.
874. Aspirin Myocardial Infarction Study Research Group. A randomized, controlled trial of aspirin in persons recovered from myocardial infarction. JAMA 1980;243(7):661–9.
875. Esemenli BT, Toker K, Lawrence R. Hypotension associated with methylmethacrylate in partial hip arthroplasties. The role of femoral canal size. Orthop Rev 1991;20(7):619–23.
876. Laursen LC, Johannesson N, Fagerstrom PO, Weeke B. Intravenous administration of enprofylline to asthmatic patients. Eur J Clin Pharmacol 1983;24(3):323–7.
877. Berlin I, Warot D, Aymard G, Acquaviva E, Legrand M, Labarthe B, Peyron I, Diquet B, Lechat P. Pharmacodynamics and pharmacokinetics of single nasal (5 mg and 10 mg) and oral (50 mg) doses of ephedrine in healthy subjects Eur J Clin Pharmacol 2001;57(6–7): 447–455.
878. Zahn KA, Li RL, Purssell RA. Cardiovascular toxicity after ingestion of "herbal ecstacy". J Emerg Med 1999;17(2):289–91.
879. Jacobs KM, Hirsch KA. Psychiatric complications of Mahuang. Psychosomatics 2000;41(1):58–62.
880. Kluger MT. Ephedrine may predispose to arrhythmias in obstetric anaesthesia. Anaesth Intensive Care 2000;28(3):336.
881. McBride BF, Karapanos AK, Krudysz A, Kluger J, Coleman CI, White CM. Electrocardiographic and hemodynamic effects of a multicomponent dietary supplement containing *Ephedra* and *caffeine*. A randomized controlled trial. JAMA 2004;291:216–21.
882. Gardner SF, Frank AM, Gurley BJ, Haller CA, Singh BK, Mehta JL. Effect of a multicomponent, *Ephedra*-containing dietary supplement (Metabolife 356) on

Holter monitoring and hemostatic parameters in healthy volunteers. Am J Cardiol 2003;91:1510–3.

883. Shekelle PG, Hardy ML, Morton HC, Maglione M, Mojica WA, Suttorp MJ, Rhodes SL, Jungvig L, Gagne J. Efficacy and safety of *Ephedra* and ephedrine for weight loss and athletic performance: a meta-analysis. JAMA 2003;289:1537–45.
884. Traub SJ, Hoyek W, Hoffman RS. Dietary supplements containing ephedra alkaloids. N Engl J Med 2001;344(14):1096.
885. Wahl A, Eberli FR, Thomson DA, Luginbuhl M. Coronary artery spasm and non-Q-wave myocardial infarction following intravenous ephedrine in two healthy women under spinal anaesthesia. Br J Anaesth 2002;89(3):519–23.
886. Mourand I, Ducrocq X, Lacour JC, Taillandier L, Anxionnat R, Weber M. Acute reversible cerebral arteritis associated with parenteral ephedrine use. Cerebrovasc Dis 1999;9(6):355–7.
887. Naik SD, Freudenberger RS. *Ephedra*-associated cardiomyopathy. Ann Pharmacother 2004;38:400–3.
888. Sola S, Helmy T, Kacharava A. Coronary dissection and thrombosis after ingestion of *Ephedra*. Am J Med 2004;116:645–6.
889. Fedotin MS, Hartman C. Ergotamine poisoning producing renal arterial spasm. N Engl J Med 1970;283(10):518–20.
890. Pajewski M, Modai D, Wisgarten J, Freund E, Manor A, Starinski R. Iatrogenic arterial aneurysm associated with ergotamine therapy. Lancet 1981;2(8252):934–5.
891. Lambert DH. Transient neurologic symptoms when phenylephrine is added to tetracaine spinal anesthesia—an alternative. Anesthesiology 1998;89(1):273.
892. Poldermans D, Rambaldi R, Bax JJ, Cornel JH, Thomson IR, Valkema R, Boersma E, Fioretti PM, Breburda CS, Roelandt JR. Safety and utility of atropine addition during dobutamine stress echocardiography for the assessment of viable myocardium in patients with severe left ventricular dysfunction. Eur Heart J 1998;19(11):1712–8.
893. Elhendy A, Valkema R, van Domburg RT, Bax JJ, Nierop PR, Cornel JH, Geleijnse ML, Reijs AE, Krenning EP, Roelandt JR. Safety of dobutamine-atropine stress myocardial perfusion scintigraphy. J Nucl Med 1998;39(10):1662–6.
894. Lin SS, Roger VL, Pascoe R, Seward JB, Pellikka PA. Dobutamine stress Doppler hemodynamics in patients with aortic stenosis: feasibility, safety, and surgical correlations. Am Heart J 1998;136(6):1010–6.
895. Hahne T, Balda BR. Fingerkuppennekrosen nach Dihydroergotaminmedikation bei limitierter systemischer Sklerodermie. [Finger tip necroses after dihydroergotamine medication in limited systemic scleroderma.] Hautarzt 1998;49(9):722–4.
896. Rommel JD, Klee P, Burkard A, Ratthey KP. Normalisierung des Gefäbildes durch Sympathikusblockade bei schwerer arterieller Durchblutungsstörung durch Ergotismus. [Normalization of the vascular picture with sympathetic block in severe arterial ischemia from ergotism.] Anästhesiol Intensivmed Notfallmed Schmerzther 1999;34(9):578–81.
897. Safar HA, Alanezi KH, Cina CS. Successful treatment of threatening limb loss ischemia of the upper limb caused by ergotamine. A case report and review of the literature. J Cardiovasc Surg (Torino) 2002;43(2):245–9.
898. Mousa HA, McKinley CA, Thong J. Acute postpartum myocardial infarction after ergometrine administration in a woman with familial hypercholesterolaemia. BJOG 2000;107(7):939–40.
899. Tsui BC, Stewart B, Fitzmaurice A, Williams R. Cardiac arrest and myocardial infarction induced by postpartum intravenous ergonovine administration. Anesthesiology 2001;94(2):363–4.
900. Yoshitomi Y, Kojima S, Sugi T, Matsumoto Y, Yano M, Kuramochi M. Coronary artery spasm induced by ergonovine in an infarct related coronary artery late after primary angioplasty. J Interv Cardiol 2000;13:31–4.
901. Zavaleta EG, Fernandez BB, Grove MK, Kaye MD. St. Anthony's fire (ergotamine induced leg ischemia)—a case report and review of the literature. Angiology 2001;52(5):349–56.
902. Kernan WN, Viscoli CM, Brass LM, Broderick JP, Brott T, Feldmann E, Morgenstern LB, Wilterdink JL, Horwitz RI. Phenylpropanolamine and the risk of hemorrhagic stroke. N Engl J Med 2000;343(25):1826–32.
903. Rubinstein E. Comparative safety of the different macrolides. Int J Antimicrob Agents 2001;18(Suppl 1):S71–6.
904. Orban Z, MacDonald LL, Peters MA, Guslits B. Erythromycin-induced cardiac toxicity. Am J Cardiol 1995;75(12):859–61.
905. Drici MD, Knollmann BC, Wang WX, Woosley RL. Cardiac actions of erythromycin: influence of female sex. JAMA 1998;280(20):1774–6.
906. Kdesh A, McPherson CA, Yaylali Y, Yasick D, Bradley K, Manthous CA. Effect of erythromycin on myocardial repolarization in patients with community-acquired pneumonia. South Med J 1999;92(12):1178–82.
907. Mishra A, Friedman HS, Sinha AK. The effects of erythromycin on the electrocardiogram. Chest 1999;115(4):983–6.
908. Brixius B, Lindinger A, Baghai A, Limbach HG, Hoffmann W. Ventrikuläre Tachykardie nach Erythromycin-Gabe bei einem Neugeborenen mit angeborenem AV-Block. [Ventricular tachycardia after erythromycin administration in a newborn with congenital AV-block.] Klin Padiatr 1999;211(6):465–8.
909. Washington JA 2nd, Wilson WR. Erythromycin: a microbial and clinical perspective after 30 years of clinical use (1). Mayo Clin Proc 1985;60(3):189–203.
910. Washington JA 2nd, Wilson WR. Erythromycin: a microbial and clinical perspective after 30 years of clinical use (2). Mayo Clin Proc 1985;60(4):271–8.
911. Lanbeck P, Odenholt I, Paulsen O. Antibiotics differ in their tendency to cause infusion phlebitis: a prospective observational study. Scand J Infect Dis 2002;34(7):512–9.
912. Johansson JE, Wersall P, Brandberg Y, Andersson SO, Nordstrom LEPO-Study Group. Efficacy of epoetin beta on hemoglobin, quality of life, and transfusion needs in patients with anemia due to hormone-refractory prostate cancer—a randomized study. Scand J Urol Nephrol 2001;35(4):288–94.
913. El Haggan W, Vallet L, Hurault de Ligny B, Pujo M, Corne B, Lobbedez T, Levaltier B, Ryckelynck JP. Darbepoetin alfa in the treatment of anemia in renal transplant patients: a single-center report. Transplantation 2004;77(12):1914–15.
914. Stasi R, Brunetti M, Terzoli E, Abruzzese E, Amadori S. Once-weekly dosing of recombinant human erythropoietin alpha in patients with myelodysplastic syndromes unresponsive to conventional dosing. Ann Oncol 2004;15(11):1684–90.
915. Gouva C, Nikolopoulos P, Ioannidis JP, Siamopoulos KC. Treating anemia early in renal failure patients slows the

decline of renal function: a randomized controlled trial. Kidney Int 2004;66(2):753–60.

916. Frei U, Nonnast-Daniel B, Koch KM. Erythropoietin und Hypertonie. [Erythropoietin and hypertension.] Klin Wochenschr 1988;66(18):914–9.
917. Ifudu O, Dawood M, Homel P. Erythropoietin-induced elevation in blood pressure is immediate and dose dependent. Nephron 1998;79(4):486–7.
918. Buemi M, Allegra A, Aloisi C, Corica F, Frisina N. Hemodynamic effects of recombinant human erythropoietin. Nephron 1999;81(1):1–4.
919. van den Bent MJ, Bos GM, Sillevis Smitt PA, Cornelissen JJ. Erythropoietin induced visual hallucinations after bone marrow transplantation. J Neurol 1999;246(7):614–6.
920. Kuriyama S, Tomonari H, Hosoya T. Antiplatelet therapy decreases the incidence of erythropoietin-induced hypertension in predialysis patients. Clin Exp Hypertens 1999;21(3):213–22.
921. Kang DH, Yoon KI, Han DS. Acute effects of recombinant human erythropoietin on plasma levels of proendothelin-1 and endothelin-1 in haemodialysis patients. Nephrol Dial Transplant 1998;13(11):2877–83.
922. Brandt JR, Avner ED, Hickman RO, Watkins SL. Safety and efficacy of erythropoietin in children with chronic renal failure. Pediatr Nephrol 1999;13(2):143–7.
923. Locatelli F, Canaud B, Giacardy F, Martin-Malo A, Baker N, Wilson J. Treatment of anaemia in dialysis patients with unit dosing of darbepoetin alfa at a reduced dose frequency relative to recombinant human erythropoietin (rHuEpo). Nephrol Dial Transplant 2003;18:362–9.
924. Raine AE. Seizures and hypertension events. Semin Nephrol 1990;10(2 Suppl 1):40–50.
925. Levin N. Management of blood pressure changes during recombinant human erythropoietin therapy. Semin Nephrol 1989;9(1 Suppl 2):16–20.
926. Eschbach JW, Kelly MR, Haley NR, Abels RI, Adamson JW. Treatment of the anemia of progressive renal failure with recombinant human erythropoietin. N Engl J Med 1989;321(3):158–63.
927. Cases A. Recombinant human erythropoietin treatment in chronic renal failure: effects on hemostasis and vasculature. Drugs Today (Barc) 2000;36(8):541–56.
928. Savonije JH, Spanier BW, van Groeningen CJ, Giaccone G, Pinedo HM. Afname van de transfusiebehoefte bij oncologiepatienten door het gebruik van epoetine tijdens cisplatinebevattende chemotherapie. [Decline in the need for blood transfusions in cancer patients due to the use of epoietin alfa during cisplatin based chemotherapy.] Ned Tijdschr Geneeskd 2001;145(18):878–81.
929. Macdougall IC. Adverse reactions profile 4. Erythropoietin in chronic renal failure. 1992.
930. Tomson CRV, Venning MC, Ward MK. Blood pressure and erythropoietin. Lancet 1988;1(8581):351–2.
931. Taylor J, Pahl M, Rajpoot D. Erythropoietin-induced hypertensive encephalopathy in a child: possible mechanisms. Dial Transplant 2002;31:170–88.
932. Sikole A, Spasovski G, Zafirov D, Polenakovic M. Epoetin omega for treatment of anemia in maintenance hemodialysis patients. Clin Nephrol 2002;57(3):237–45.
933. Faught C, Wells P, Fergusson D, Laupacis A. Adverse effects of methods for minimizing perioperative allogeneic transfusion: a critical review of the literature. Transfus Med Rev 1998;12(3):206–25.
934. Mingoli A, Sapienza P, Puggioni A, Modini C, Cavallaro A. A possible side-effect of human erythropoietin therapy: thrombosis of peripheral arterial reconstruction. Eur J Vasc Endovasc Surg 1999;18(3):273–4.
935. Bohlius J, Langensiepen S, Schwarzer G, Seidenfeld J, Piper M, Bennett C, Engert A. Recombinant human erythropoietin and overall survival in cancer patients: results of a comprehensive meta-analysis. J Natl Cancer Inst 2005;97(7):489–98.
936. Stasi R, Amadori S, Littlewood TJ, Terzoli E, Newland AC, Provan D. Management of cancer-related anemia with erythropoietic agents: doubts, certainties, and concerns. Oncologist 2005;10(7):539–54.
937. Grzeszczak W, Sulowicz W, Rutkowski B, de Vecchi AF, Scanziani R, Durand PY, Bajo A, Vargemezis V; European Collaborative Group. The efficacy and safety of once-weekly and once-fortnightly subcutaneous epoetin beta in peritoneal dialysis patients with chronic renal anaemia. Nephrol Dial Transplant 2005;20(5):936–44.
938. Sasaki N, Ando Y, Kusano E, Asano Y. A case of erythropoietin induced hypertension in a bilaterally nephrectomized patient. ASAIO J 2003;49:131–5.
939. Cogar AA, Hartenberger CH, Ohls RK. Endothelin concentrations in preterm infants treated with human recombinant erythropoietin. Biol Neonate 2000;77(2):105–8.
940. Tsukahara H, Hori C, Tsuchida S, Hiraoka M, Fujisawa K, Mayumi M. Role of endothelin in erythropoietin-induced hypertension in rats. Nephron 1998;79(4):499–500.
941. Kuriyama S, Kobayashi H, Tomonari H, Tokudome G, Hayashi F, Kaguchi Y, Horiguchi M, Ishikawa M, Hosoya T. Circulating adrenomedullin in erythropoietin-induced hypertension. Hypertens Res 2000;23(5):427–32.
942. Ksiazek A, Zaluska WT, Ksiazek P. Effect of recombinant human erythropoietin on adrenergic activity in normotensive hemodialysis patients. Clin Nephrol 2001;56(2):104–10.
943. Kuriyama S, Tomonari H, Tokudome G, Kaguchi Y, Hayashi H, Kobayashi H, Horiguchi M, Ishikawa M, Hara Y, Hosoya T. Association of angiotensinogen gene polymorphism with erythropoietin-induced hypertension: a preliminary report. Hypertens Res 2001;24(5):501–5.
944. Engert A. Recombinant human erythropoietin as an alternative to blood transfusion in cancer-related anaemia. Dis Manage Heath Outcomes 2000;8:259–72.
945. Tong EM, Nissenson AR. Erythropoietin and anemia. Semin Nephrol 2001;21(2):190–203.
946. Sifakis S, Angelakis E, Vardaki E, Koumantaki Y, Matalliotakis I, Koumantakis E. Erythropoietin in the treatment of iron deficiency anemia during pregnancy. Gynecol Obstet Invest 2001;51(3):150–6.
947. Wakeen M, Zimmerman SW. Association between human recombinant EPO and peripheral vascular disease in diabetic patients receiving peritoneal dialysis. Am J Kidney Dis 1998;32(3):488–93.
948. Cases A. Recombinant human erythropoietin treatment in chronic renal failure: effects on hemostasis and vasculature. Drugs Today (Barc) 2000;36(8):541–56.
949. Mingoli A, Sapienza P, Puggioni A, Modini C, Cavallaro A. A possible side-effect of human erythropoietin therapy: thrombosis of peripheral arterial reconstruction. Eur J Vasc Endovasc Surg 1999;18(3): 273–274.
950. MacKinnon GE, Singla D. Epoetin alfa in chronic renal failure. P&T 1998;23:437–46.
951. Cameron JS, Barany P, Barbas J, Carrera F, Chanard J. European best practice guidelines for the management of anaemia in patients with chronic renal failure. Working Party for European Best Practice Guidelines for the

Management of Anaemia in Patients with Chronic Renal Failure. Nephrol Dial Transplant 1999;14(Suppl 5):1–50.
952. Finelli PF, Carley MD. Cerebral venous thrombosis associated with epoetin alfa therapy. Arch Neurol 2000;57(2):260–2.
953. Schwartzberg LS, Yee LK, Senecal FM, Charu V, Tomita D, Wallace J, Rossi G. A randomized comparison of every-2-week darbepoetin alfa and weekly epoetin alfa for the treatment of chemotherapy-induced anemia in patients with breast, lung, or gynecologic cancer. Oncologist 2004;9(6):696–707.
954. Lavey RS, Liu PY, Greer BE, Robinson WR 3rd, Chang PC, Wynn RB, Conrad ME, Jiang C, Markman M, Alberts DS. Recombinant human erythropoietin as an adjunct to radiation therapy and cisplatin for stage IIB-IVA carcinoma of the cervix: a Southwest Oncology Group study. Gynecol Oncol 2004;95(1):145–51.
955. Rosenzweig MQ, Bender CM, Lucke JP, Yasko JM, Brufsky AM. The decision to prematurely terminate a trial of R-HuEPO due to thrombotic events. J Pain Symptom Manage 2004;27(2):185–90.
956. Henry DH, Bowers P, Romano MT, Provenzano R. Epoetin alfa. Clinical evolution of a pleiotropic cytokine. Arch Intern Med 2004;164(3):262–76.
957. Bokemeyer C, Aapro MS, Courdi A, Foubert J, Link H, Osterborg A, Repetto L, Soubeyran P. EORTC guidelines for the use of erythropoietic proteins in anaemic patients with cancer. Eur J Cancer 2004;40(15):2201–16.
958. Perez Gutthann S, Garcia Rodriguez LA, Castellsague J, Duque Oliart A. Hormone replacement therapy and risk of venous thromboembolism: population based case-control study. BMJ 1997;314(7083):796–800.
959. The Coronary Drug Project Research Group. The Coronary Drug Project. Findings leading to discontinuation of the 2.5-mg day estrogen group JAMA 1973;226(6):652–7.
960. Christodoulakos G, Lambrinoudaki I, Panoulis C, Papadias C, Economou E, Creatsas G. Effect of hormone therapy and raloxifene on serum VE-cadherin in postmenopausal women. Fertil Steril 2004;82:634–8.
961. Watson B. Transient hypotension following intravenous ethamsylate (Dicynene). BMJ 1977;1(6077):1664.
962. Kaysen G, Noth RH. The effects of alcohol on blood pressure and electrolytes. Med Clin North Am 1984;68(1): 221–46.
963. Thomas SHL. Drug induced systemic hypertension. Adv Drug React Bull 1993;150:559–62.
964. Schulze VH, Berlin-Buch VH. Plasmaersatzstoffe: Dextran und Hydroxyethylstärke im Vergleich. Krankenhauspharmazie 1991;12:551.
965. Colvin MP, Savege TM, Newland PE, Weaver EJ, Waters AF, Brookes JM, Inniss R. Cardiorespiratory changes following induction of anaesthesia with etomidate in patients with cardiac disease. Br J Anaesth 1979;51(6):551–6.
966. Price ML, Millar B, Grounds M, Cashman J. Changes in cardiac index and estimated systemic vascular resistance during induction of anaesthesia with thiopentone, methohexitone, propofol and etomidate. Br J Anaesth 1992;69(2):172–6.
967. Flach AJ. Systemic toxicity. Associated with topical ophthalmic medications. J Fla Med Assoc 1994;81(4): 256–60.
968. Morgan JF. Complications associated with contact lens solutions. Ophthalmology 1979;86(6):1107–19.
969. Mondino BJ, Salamon SM, Zaidman GW. Allergic and toxic reactions of soft contact lens wearers. Surv Ophthalmol 1982;26(6):337–44.
970. Sigurjonsdottir HA, Franzson L, Manhem K, Ragnarsson J, Sigurdsson G, Wallerstedt S. Liquorice-induced rise in blood pressure: a linear dose-response relationship. J Hum Hypertens 2001;15(8):549–52.
971. Scharrer I. Recombinant factor VIIa for patients with inhibitors to factor VIII or IX or factor VII deficiency. Haemophilia 1999;5(4):253–9.
972. Barthels M. Clinical efficacy of prothrombin complex concentrates and recombinant factor VIIa in the treatment of bleeding episodes in patients with factor VII and IX inhibitors. Thromb Res 1999;95(4 Suppl 1):S31–8.
973. Smith MP, Ludlam CA, Collins PW, Hay CR, Wilde JT, Grigeri A, Melsen T, Savidge GF. Elective surgery on factor VIII inhibitor patients using continuous infusion of recombinant activated factor VII: plasma factor VII activity of 10 IU/ml is associated with an increased incidence of bleeding. Thromb Haemost 2001;86(4):949–53.
974. Santagostino E, Morfini M, Rocino A, Baudo F, Scaraggi FA, Gringeri A. Relationship between factor VII activity and clinical efficacy of recombinant factor VIIa given by continuous infusion to patients with factor VIII inhibitors. Thromb Haemost 2001;86(4):954–8.
975. Van der Planken MG, Schroyens W, Vertessen F, Michiels JJ, Berneman ZN. Distal deep venous thrombosis in a hemophilia A patient with inhibitor and severe infectious disease, 18 days after recombinant activated factor VII transfusion. Blood Coagul Fibrinolysis 2002;13(4):367–70.
976. Roberts HR. Clinical experience with activated factor VII: focus on safety aspects. Blood Coagul Fibrinolysis 1998;9(Suppl 1):S115–8.
977. Hedner U. Use of high dose factor VIIa in hemophilia patients. Adv Exp Med Biol 2001;489:75–88.
978. Mannucci PM, Chediak J, Hanna W, Byrnes J, Ledford M, Ewenstein BM, Retzios AD, Kapelan BA, Schwartz RS, Kessler CAlphanate Study Group. Treatment of von Willebrand disease with a high-purity factor VIII/von Willebrand factor concentrate: a prospective, multicenter study. Blood 2002;99(2):450–6.
979. Makris M, Colvin B, Gupta V, Shields ML, Smith MP. Venous thrombosis following the use of intermediate purity FVIII concentrate to treat patients with von Willebrand's disease. Thromb Haemost 2002;88(3):387–8.
980. Chowdary P, Dasani H, Jones JA, Loran CM, Eldridge A, Hughes S, Collins PW. Recombinant factor IX (BeneFix) by adjusted continuous infusion: a study of stability, sterility and clinical experience. Haemophilia 2001;7(2):140–5.
981. Hughes R, Chapple DJ. Effects on non-depolarizing neuromuscular blocking agents on peripheral autonomic mechanisms in cats. Br J Anaesth 1976;48(2):59–68.
982. Marshall IG. The ganglion blocking and vagolytic actions of three short-acting neuromuscular blocking drugs in the cat. J Pharm Pharmacol 1973;25(7):530–6.
983. Hughes R, Payne JP, Sugai N. Studies on fazadinium bromide (AH 8165): a new non-depolarizing neuromuscular blocking agent. Can Anaesth Soc J 1976;23(1):36–47.
984. Schuh FT. Clinical neuromuscular pharmacology of AH 8165 D, an azobis-arylimidazo-pyridinium-compound. Anaesthesist 1975;24(4):151–6.
985. Lyons SM, Clarke RS, Young HS. A clinical comparison of AH8165 and pancuronium as muscle relaxants in patients undergoing cardiac surgery. Br J Anaesth 1975;47(6):725–9.
986. Lienhart A, Tauvent A, Guggiari M. Effets hemodynamiques des curares. [Hemodynamic effects of curares.]. In: Curares et Curarisation 1. Paris: Librairie Arnette, 1979:384.

987. Pinaud M, Arnould F, Souron R, Nicolas F. Influence of cardiac rhythm on the haemodynamic effects of fazadinium in patients with heart failure. Br J Anaesth 1983;55(6):507–12.
988. Sobel RM. Ruptured retroperitoneal aneurysm in a patient taking phentermine hydrochloride. Am J Emerg Med 1999;17(1):102–3.
989. Fowles RE, Cloward TV, Yowell RL. Endocardial fibrosis associated with fenfluramine–phentermine. N Engl J Med 1998;338(18):1316.
990. Wee CC, Phillips RS, Aurigemma G, Erban S, Kriegel G, Riley M, Douglas PS. Risk for valvular heart disease among users of fenfluramine and dexfenfluramine who underwent echocardiography before use of medication. Ann Intern Med 1998;129(11):870–4.
991. Khan MA, Herzog CA, St Peter JV, Hartley GG, Madlon-Kay R, Dick CD, Asinger RW, Vessey JT. The prevalence of cardiac valvular insufficiency assessed by transthoracic echocardiography in obese patients treated with appetite-suppressant drugs. N Engl J Med 1998;339(11):713–8.
992. Jick H, Vasilakis C, Weinrauch LA, Meier CR, Jick SS, Derby LE. A population-based study of appetite-suppressant drugs and the risk of cardiac-valve regurgitation. N Engl J Med 1998;339(11):719–24.
993. Murthy TH, Weissman NJ. Diet-drug valvulopathy. ACC Curr J Rev 2002;11:17–20.
994. Palmieri V, Arnett DK, Roman MJ, Liu JE, Bella JN, Oberman A, Kitzman DW, Hopkins PN, Morgan D, de Simone G, Devereux RB. Appetite suppressants and valvular heart disease in a population-based sample: the HyperGEN study. Am J Med 2002;112(9):710–5.
995. Tomita T, Zhao Q. Autopsy findings of heart and lungs in a patient with primary pulmonary hypertension associated with use of fenfluramine and phentermine. Chest 2002;121(2):649–52.
996. Abenhaim L, Moride Y, Brenot F, Rich S, Benichou J, Kurz X, Higenbottam T, Oakley C, Wouters E, Aubier M, Simonneau G, Begaud BInternational Primary Pulmonary Hypertension Study Group. Appetite-suppressant drugs and the risk of primary pulmonary hypertension. N Engl J Med 1996;335(9):609–16.
997. Anchors M. Fluoxetine is a safer alternative to fenfluramine in the medical treatment of obesity. Arch Intern Med 1997;157(11):1270.
998. Griffen L, Anchors M. Asymptomatic mitral and aortic valve disease is seen in half of the patients taking "phen-fen". Arch Intern Med 1998;158(1):102.
999. Devereux RB. Appetite suppressants and valvular heart disease. N Engl J Med 1998;339(11):765–6.
1000. Weissman NJ, Tighe JF Jr, Gottdiener JS, Gwynne JTSustained-Release Dexfenfluramine Study Group. An assessment of heart-valve abnormalities in obese patients taking dexfenfluramine, sustained-release dexfenfluramine, or placebo. N Engl J Med 1998;339(11):725–32.
1001. Parisi AF. Diet-drug debacle. Ann Intern Med 1998;129(11):903–5.
1002. Weissman NJ. Appetite suppressant valvulopathy: a review of current data. Cardiovasc Rev Rep 1999;20:146–55.
1003. Adams C, Cohen A. Appetite suppressants and heart valve disorders. Arch Mal Coeur Vaiss 1999;92(9):1213–9.
1004. Ewalenko M, Richard C, Vandenbossche JL. Fenfluramines and cardiac valvular lesions. Rev Med Brux 1999;20(5):419–26.
1005. Cannistra LB, Cannistra AJ. Regression of multivalvular regurgitation after the cessation of fenfluramine and phentermine treatment. N Engl J Med 1998;339(11):771.
1006. Day A. Lessons in women's health: body image and pulmonary disease. CMAJ 1998;159(4):346–9.
1007. Anonymous. Cardiac valvulopathy associated with exposure to fenfluramine or dexfenfluramine: U.S. Department of Health and Human Services interim public health recommendations MMWR Morb Mortal Wkly Rep November 1997;46(45):1061–6.
1008. Anonymous. Statement of the American College of Cardiology on recommendations for patients who have used anorectic drugsBethesda MD: American College of Cardiology;. October 18, 1997.
1009. Connolly HM, Crary JL, McGoon MD, Hensrud DD, Edwards BS, Edwards WD, Schaff HV. Valvular heart disease associated with fenfluramine–phentermine. N Engl J Med 1997;337(9):581–8.
1010. Redmon B, Raatz S, Bantle JP. Valvular heart disease associated with fenfluramine–phentermine. N Engl J Med 1997;337(24):1773–4.
1011. Burger AJ, Sherman HB, Charlamb MJ, Kim J, Asinas LA, Flickner SR, Blackburn GL. Low prevalence of valvular heart disease in 226 phentermine–fenfluramine protocol subjects prospectively followed for up to 30 months. J Am Coll Cardiol 1999;34(4):1153–830.
1012. Graham DJ, Green L. Further cases of valvular heart disease associated with fenfluramine–phentermine. N Engl J Med 1997;337(9):635.
1013. Kurz X, Van Ermen A. Valvular heart disease associated with fenfluramine–phentermine. N Engl J Med 1997;337(24):1772–3.
1014. Rasmussen S, Corya BC, Glassman RD. Valvular heart disease associated with fenfluramine–phentermine. N Engl J Med 1997;337(24):1773.
1015. Teramae CY, Connolly HM, Grogan M, Miller FA Jr. Diet drug-related cardiac valve disease: the Mayo Clinic echocardiographic laboratory experience. Mayo Clin Proc 2000;75(5):456–61.
1016. Gardin JM, Schumacher D, Constantine G, Davis KD, Leung C, Reid CL. Valvular abnormalities and cardiovascular status following exposure to dexfenfluramine or phentermine/fenfluramine. JAMA 2000;283(13):1703–9.
1017. Burger AJ, Charlamb MJ, Singh S, Notarianni M, Blackburn GL, Sherman HB. Low risk of significant echocardiographic valvulopathy in patients treated with anorectic drugs. Int J Cardiol 2001;79(2–3):159–65.
1018. Shively BK, Roldan CA, Gill EA, Najarian T, Loar SB. Prevalence and determinants of valvulopathy in patients treated with dexfenfluramine. Circulation 1999;100(21):2161–7.
1019. Weissman NJ, Tighe JF Jr, Gottdiener JS, Gwynne JT. Prevalence of valvular-regurgitation associated with dexfenfluramine three to five months after discontinuation of treatment. J Am Coll Cardiol 1999;34(7):2088–95.
1020. Mangion JR, Habboub AA, Kamat BR, Tam SK. Transesophageal echocardiography with pathological correlation in severe valvular disease associated with fenfluramine–phentermine. Echocardiography 1999;16(1):27–30.
1021. Tovar EA, Landa DW, Borsari BE. Dose effect of fenfluramine–phentermine in the production of valvular heart disease. Ann Thorac Surg 1999;67(4):1213–4.
1022. Garon CF, Oury JH, Duran CM. Virus-like particles in the mitral and tricuspid valves explanted from a patient

treated with fenfluramine–phentermine. J Heart Valve Dis 1999;8(2):232.
1023. Lepor NE, Gross SB, Daley WL, Samuels BA, Rizzo MJ, Luko SP, Hickey A, Buchbinder NA, Naqvi TZ. Dose and duration of fenfluramine–phentermine therapy impacts the risk of significant valvular heart disease. Am J Cardiol 2000;86(1):107–10.
1024. Jollis JG, Landolfo CK, Kisslo J, Constantine GD, Davis KD, Ryan T. Fenfluramine and phentermine and cardiovascular findings: effect of treatment duration on prevalence of valve abnormalities. Circulation 2000;101(17):2071–7.
1025. Steffee CH, Singh HK, Chitwood WR. Histologic changes in three explanted native cardiac valves following use of fenfluramines. Cardiovasc Pathol 1999;8(5):245–53.
1026. Roldan CA, Gill EA, Shively BK. Prevalence and diagnostic value of precordial murmurs for valvular regurgitation in obese patients treated with dexfenfluramine. Am J Cardiol 2000;86(5):535–9.
1027. Bonow RO, Carabello B, de Leon AC Jr, Edmunds LH Jr, Fedderly BJ, Freed MD, Gaasch WH, McKay CR, Nishimura RA, O'Gara PT, O'Rourke RA, Rahimtoola SH, Ritchie JL, Cheitlin MD, Eagle KA, Gardner TJ, Garson A Jr, Gibbons RJ, Russell RO, Ryan TJ, Smith SC Jr. Guidelines for the management of patients with valvular heart disease: executive summary. A report of the American College of Cardiology/American Heart Association Task Force on Practice Guidelines (Committee on Management of Patients with Valvular Heart Disease). Circulation 1998;98(18):1949–84.
1028. Levine HJ, Gaasch WH. Vasoactive drugs in chronic regurgitant lesions of the mitral and aortic valves. J Am Coll Cardiol 1996;28(5):1083–91.
1029. Kancherla MK, Salti HI, Mulderink TA, Parker M, Bonow RO, Mehlman DJ. Echocardiographic prevalence of mitral and/or aortic regurgitation in patients exposed to either fenfluramine–phentermine combination or to dexfenfluramine. Am J Cardiol 1999;84(11):1335–8.
1030. Mast ST, Jollis JG, Ryan T, Anstrom KJ, Crary JL. The progression of fenfluramine-associated valvular heart disease assessed by echocardiography. Ann Intern Med 2001;134(4):261–6.
1031. Weissman NJ, Panza JA, Tighe JF, Gwynne JT. Natural history of valvular regurgitation 1 year after discontinuation of dexfenfluramine therapy. A randomized, double-blind, placebo-controlled trial. Ann Intern Med 2001;134(4):267–73.
1032. Prasad A, Mehra M, Park M, Scott R, Uber PA, McFadden PM. Cardiac allograft valvulopathy: a case of donor-anorexigen-induced valvular disease. Ann Thorac Surg 1999;68(5):1840–1.
1033. Dahl CF, Allen MR. Regression and progression of valvulopathy associated with fenfluramine and phentermine. Ann Intern Med 2002;136(6):489.
1034. Singh JP, Evans JC, Levy D, Larson MG, Freed LA, Fuller DL, Lehman B, Benjamin EJ. Prevalence and clinical determinants of mitral, tricuspid, and aortic regurgitation (the Framingham Heart Study). Am J Cardiol 1999;83(6):897–902.
1035. Klein AL, Burstow DJ, Tajik AJ, Zachariah PK, Taliercio CP, Taylor CL, Bailey KR, Seward JB. Age-related prevalence of valvular regurgitation in normal subjects: a comprehensive color flow examination of 118 volunteers. J Am Soc Echocardiogr 1990;3(1):54–63.
1036. Bella JN, Devereux RB, Roman MJ, O'Grady MJ, Welty TK, Lee ET, Fabsitz RR, Howard BV. Relations of left ventricular mass to fat-free and adipose body mass: the strong heart study. The Strong Heart Study Investigators. Circulation 1998;98(23):2538–44.
1037. Brenot F, Herve P, Petitpretz P, Parent F, Duroux P, Simonneau G. Primary pulmonary hypertension and fenfluramine use. Br Heart J 1993;70(6):537–41.
1038. Mark EJ, Patalas ED, Chang HT, Evans RJ, Kessler SC. Fatal pulmonary hypertension associated with short-term use of fenfluramine and phentermine. N Engl J Med 1997;337(9):602–6.
1039. Bruce CJ, Connolly HM. Valvular heart disease, pulmonary hypertension and fenfluramine–phentermine use. Cardiol Rev 1998;15:17–9.
1040. Delcroix M, Kurz X, Walckiers D, Demedts M, Naeije R. High incidence of primary pulmonary hypertension associated with appetite suppressants in Belgium. Eur Respir J 1998;12(2):271–6.
1041. Fisher EA, Ruden R. Pulmonary artery pressures and valvular lesions in patients taking diet suppressants. Cardiovasc Rev Rep 1998;19:13–6.
1042. Hellermann J, Salomon F. Appetite depressants and pulmonary hypertension. Ther Umsch 1998;55(9):548–50.
1043. Goldstein SE, Levy Y, Shoenfeld Y. Development of pulmonary hypertension and multi-valvular damage caused by appetite depressants. Harefuah 1998;135(11):489–92568.
1044. Heuer L, Benoit W, Heydrich D, Kummer D, Schick J. Pulmonale Hypertonie durch Appetitzügler (Mirapront). [Pulmonary hypertension caused by appetite suppressants (Mirapront).] Chir Praxis 1978;23:497.
1045. Kramer MS, Lane DA. Aminorex, dexfenfluramine, and primary pulmonary hypertension. J Clin Epidemiol 1998;51(4):361–4.
1046. Rich S, Rubin L, Walker AM, Schneeweiss S, Abenhaim L. Anorexigens and pulmonary hypertension in the United States results from the surveillance of North American pulmonary hypertension. Chest 2000;117(3):870–4.
1047. Rothman RB. The age-adjusted mortality rate from primary pulmonary hypertension, in age range 20–54 years, did not increase during the years of peak "phen/fen" use. Chest 2000;118(5):1516–7.
1048. Simonneau G, Fartoukh M, Sitbon O, Humbert M, Jagot JL, Herve P. Primary pulmonary hypertension associated with the use of fenfluramine derivatives. Chest 1998;114(Suppl. 3):195S–9S.
1049. Nichols WC, Koller DL, Slovis B, Foroud T, Terry VH, Arnold ND, Siemieniak DR, Wheeler L, Phillips JA 3rd, Newman JH, Conneally PM, Ginsburg D, Loyd JE. Localization of the gene for familial primary pulmonary hypertension to chromosome 2q31–32. Nat Genet 1997;15(3):277–80.
1050. Morrell NW, Sarybaev AS, Alikhan A, Mirrakhimov MM, Aldashev AA. ACE genotype and risk of high altitude pulmonary hypertension in Kyrghyz highlanders. Lancet 1999;353(9155):814.
1051. Lullmann H, Lullmann-Rauch R, Wassermann O. Drug-induced phospholipidoses. II. Tissue distribution of the amphiphilic drug chlorphentermine. CRC Crit Rev Toxicol 1975;4(2):185–218.
1052. Hruban Z. Pulmonary and generalized lysosomal storage induced by amphiphilic drugs. Environ Health Perspect 1984;55:53–76.

1053. Halliwell WH. Cationic amphiphilic drug-induced phospholipidosis. Toxicol Pathol 1997;25(1):53–60.

1054. Archer SL, Djaballah K, Humbert M, Weir KE, Fartoukh M, Dall'ava-Santucci J, Mercier JC, Simonneau G, Dinh-Xuan AT. Nitric oxide deficiency in fenfluramine- and dexfenfluramine-induced pulmonary hypertension. Am J Respir Crit Care Med 1998;158(4):1061–7.

1055. Weir EK, Reeve HL, Huang JM, Michelakis E, Nelson DP, Hampl V, Archer SL. Anorexic agents aminorex, fenfluramine, and dexfenfluramine inhibit potassium current in rat pulmonary vascular smooth muscle and cause pulmonary vasoconstriction. Circulation 1996;94(9):2216–20.

1056. Barman SA, Isales CM. Fenfluramine potentiates canine pulmonary vasoreactivity to endothelin-1. Pulm Pharmacol Ther 1998;11(2–3):183–7.

1057. Tuder RM, Radisavljevic Z, Shroyer KR, Polak JM, Voelkel NF. Monoclonal endothelial cells in appetite suppressant-associated pulmonary hypertension. Am J Respir Crit Care Med 1998;158(6):1999–2001.

1058. Mabadeje AF. Fenfluramine-associated hypertension. West Afr J Pharmacol Drug Res 1975;2(2):145–52.

1059. Barancik M. Inadvertent diagnosis of pheochromocytoma after endoscopic premedication. Dig Dis Sci 1989;34(1):136–8.

1060. Eseverri J. Proyeccion de los nuevos antihistaminicos. [Projection of new antihistamines.] Allergol Immunopathol (Madr) 2000;28(3):143–52.

1061. Gupta S, Banfield C, Kantesaria B, Marino M, Clement R, Affrime M, Batra V. Pharmacokinetic and safety profile of desloratadine and fexofenadine when coadministered with azithromycin: a randomized, placebo-controlled, parallel-group study. Clin Ther 2001;23(3):451–66.

1062. Anonymous. Severe cardiac arrhythmia on fexofenadine? Prescrire Int 2000;9(45):212.

1063. Graft DF, Bernstein DI, Goldsobel A, Meltzer EO, Portnoy J, Long J. Safety of fexofenadine in children treated for seasonal allergic rhinitis. Ann Allergy Asthma Immunol 2001;87(1):22–6.

1064. Kram HB, Reuben BI, Fleming AW, Shoemaker WC. Use of fibrin glue in hepatic trauma. J Trauma 1988;28(8):1195–201.

1065. Solov'ev GM, Suprunov MV, Khorobrykh TV. Fibrinovyi klei v serdechno-sosudistoi khirurgii. [Fibrin glue in cardiovascular surgery.] Kardiologiia 2003;43(4):4–5.

1066. Spotnitz WD, Dalton MS, Baker JW, Nolan SP. Reduction of perioperative hemorrhage by anterior mediastinal spray application of fibrin glue during cardiac operations. Ann Thorac Surg 1987;44(5):529–31.

1067. Vaiman M, Eviatar E, Shlamkovich N, Segal S. Effect of modern fibrin glue on bleeding after tonsillectomy and adenoidectomy. Ann Otol Rhinol Laryngol 2003;112(5):410–4.

1068. Brennan M. Fibrin glue. Blood Rev 1991;5(4):240–4.

1069. Jessen C, Sharma P. Use of fibrin glue in thoracic surgery. Ann Thorac Surg 1985;39(6):521–4.

1070. Hammond TM, Grahn MF, Lunniss PJ. Fibrin glue in the management of anal fistulae. Colorectal Dis 2004;6(5):308–19.

1071. Vaiman M, Martinovich U, Eviatar E, Kessler A, Segal S. Fibrin glue in initial treatment of epistaxis in hereditary haemorrhagic telangiectasia (Rendu-Osler-Weber disease). Blood Coagul Fibrinolysis 2004;15(4):359–63.

1072. Moon MC, Molnar K, Yau L, Zahradka P. Perivascular delivery of losartan with surgical fibrin glue prevents neointimal hyperplasia after arterial injury. J Vasc Surg 2004;40(1):130–7.

1073. Berruyer M, Amiral J, Ffrench P, Belleville J, Bastien O, Clerc J, Kassir A, Estanove S, Dechavanne M. Immunization by bovine thrombin used with fibrin glue during cardiovascular operations. Development of thrombin and factor V inhibitors. J Thorac Cardiovasc Surg 1993;105(5):892–7.

1074. Margerger MJ. Long-term effects of finasteride in patients with benign prostatic hyperplasia; a double-blind, placebo-controlled multicenter study. Urology 1998;51:677–86.

1075. Marchioli R, Barzi F, Bomba E, Chieffo C, Di Gregorio D, Di Mascio R, Franzosi MG, Geraci E, Levantesi G, Maggioni AP, Mantini L, Marfisi RM, Mastrogiuseppe G, Mininni N, Nicolosi GL, Santini M, Schweiger C, Tavazzi L, Tognoni G, Tucci C, Valagussa F, on behalf of the GISSI-Prevenzione Investigators. Early protection against sudden death by n-3 polyunsaturated fatty acids after myocardial infarction: time-course analysis of the results of the Gruppo Italiano per lo Studio della Sopravvivenza nell'Infarto Miocardico (GISSI)-Prevenzione. Circulation 2002;105(23):1897–903.

1076. Raitt MH, Connor WE, Morris C, Kron J, Halperin B, Chugh SS, McClelland J, Cook J, MacMurdy K, Swenson R, Connor SL, Gerhard G, Kraemer DF, Oseran D, Marchant C, Calhoun D, Shnider R, McAnulty J. Fish oil supplementation and risk of ventricular tachycardia and ventricular fibrillation in patients with implantable defibrillators: a randomized controlled trial. JAMA 2005;293(23):2884–91.

1077. Tholakanahalli VN, Potti A, Hanley JF, Merliss AD. Fluconazole-induced torsade de pointes. Ann Pharmacother 2001;35(4):432–4.

1078. Wassmann S, Nickenig G, Bohm M. Long QT syndrome and torsade de pointes in a patient receiving fluconazole. Ann Intern Med 1999;131(10):797.

1079. Khazan M, Mathis AS. Probable case of torsades de pointes induced by fluconazole. Pharmacotherapy 2002;22(12):1632–7.

1080. Isetta C, Garaffo R, Bastian G, Jourdan J, Baudouy M, Milano G. Life-threatening 5-fluorouracil-like cardiac toxicity after treatment with 5-fluorocytosine. Clin Pharmacol Ther 2000;67(3):323–5.

1081. Van Besien K, Devine S, Wickrema A, Jessop E, Amin K, Yassine M, Maynard V, Stock W, Peace D, Ravandi F, Chen Y.-H, Hoffman R, Sossman J. Regimen-related toxicity after fludarabine–melphalan conditioning: a prospective study of 31 patients with hematologic malignancies. Bone Marrow Transplant 2003;32:471–6.

1082. Owens RC Jr. QT prolongation with antimicrobial agents: understanding the significance. Drugs 2004;64(10):1091–124.

1083. Patmore L, Fraser S, Mair D, Templeton A. Effects of sparfloxacin, grepafloxacin, moxifloxacin, and ciprofloxacin on cardiac action potential duration. Eur J Pharmacol 2000;406(3):449–52.

1084. Hagiwara T, Satoh S, Kasai Y, Takasuna K. A comparative study of the fluoroquinolone antibacterial agents on the action potential duration in guinea pig ventricular myocardia. Jpn J Pharmacol 2001;87(3):231–4.

1085. Owens RC Jr, Ambrose PG. Torsades de pointes associated with fluoroquinolones. Pharmacotherapy 2002;22(5):663–8discussion 668–72.

1086. Jaillon P, Morganroth J, Brumpt I, Talbot G. Overview of electrocardiographic and cardiovascular safety data for sparfloxacin. Sparfloxacin Safety Group. J Antimicrob Chemother 1996;37(Suppl A):161–7.
1087. Clark DW, Layton D, Wilton LV, Pearce GL, Shakir SA. Profiles of hepatic and dysrhythmic cardiovascular events following use of fluoroquinolone antibacterials: experience from large cohorts from the Drug Safety Research Unit Prescription-Event Monitoring database. Drug Saf 2001;24(15):1143–54.
1088. Frothingham R. Rates of torsades de pointes associated with ciprofloxacin, ofloxacin, levofloxacin, gatifloxacin, and moxifloxacin. Pharmacotherapy 2001;21(12):1468–72.
1089. Yoo BK, Triller DM, Yong CS, Lodise TP. Gemifloxacin: a new fluoroquinolone approved for treatment of respiratory infections. Ann Pharmacother 2004;38(7-8):1226–35.
1090. Amankwa K, Krishnan SC, Tisdale JE. Torsades de pointes associated with fluoroquinolones: importance of concomitant risk factors. Clin Pharmacol Ther 2004;75(3):242–7.
1091. Noel GJ, Natarajan J, Chien S, Hunt TL, Goodman DB, Abels R. Effects of three fluoroquinolones on QT interval in healthy adults after single doses. Clin Pharmacol Ther 2003;73:292–303.
1092. Saravolatz LD, Leggett J. Gatifloxacin, gemifloxacin, and moxifloxacin: the role of 3 newer fluoroquinolones. Clin Infect Dis 2003;37:1210–5.
1093. Sable D, Murakawa GJ. Quinolones in dermatology. Dis Mon 2004;50(7):381–94.
1094. Lacroix P, Crumb WJ, Durando L, Ciottoli GB. Prulifloxacin: in vitro (HERG current) and in vivo (conscious dog) assessment of cardiac risk. Eur J Pharmacol 2003;477:69–72.
1095. Takahara A, Sugiyama A, Satoh Y, Hashimoto K. Effects of mexiletine on the canine model of sparfloxacin-induced long QT syndrome. Eur J Pharmacol 2003;476:115–22.
1096. Fukuda H, Morita Y, Shiotani N, Mizuo M, Komae N. [Effect of pazufloxacin mesilate, a new quinolone antibacterial agent, for intravenous use on QT interval]. Jpn J Antibiot 2004;57(4):404–12.
1097. Keam SJ, Perry CM. Prulifloxacin. Drugs 2004;64(19):2221-34; discussion 2235–6.
1098. Prabhakar M, Krahn AD. Ciprofloxacin-induced acquired long QT syndrome. Heart Rhythm 2004;1(5):624–6.
1099. Daya SK, Gowda RM, Khan IA. Ciprofloxacin- and hypocalcemia-induced torsade de pointes triggered by hemodialysis. Am J Ther 2004;11(1):77–9.
1100. Fteha A, Fteha E, Haq S, Kozer L, Saul B, Kassotis J. Gatifloxacin induced torsades de pointes. Pacing Clin Electrophysiol 2004;27(10):1449–50.
1101. Farooqi IS, Aronson JK. Iatrogenic chest pain: a case of 5-fluorouracil cardiotoxicity. QJM 1996;89(12):953–5.
1102. Clavel M, Simeone P, Grivet B. Toxicité cardiaqué du 5-fluorouracile. Revue de la litterature, cinq nouveaux cas. [Cardiac toxicity of 5-fluorouracil. Review of the literature, 5 new cases.] Presse Méd 1988;17(33):1675–8.
1103. Robben NC, Pippas AW, Moore JO. The syndrome of 5-fluorouracil cardiotoxicity. An elusive cardiopathy. Cancer 1993;71(2):493–509.
1104. Lomeo AM, Avolio C, Iacobellis G, Manzione L. 5-Fluorouracil cardiotoxicity. Eur J Gynaecol Oncol 1990;11(3):237–41.
1105. Rezkalla S, Kloner RA, Ensley J, al-Sarraf M, Revels S, Olivenstein A, Bhasin S, Kerpel-Fronious S, Turi ZG. Continuous ambulatory ECG monitoring during fluorouracil therapy: a prospective study. J Clin Oncol 1989;7(4):509–14.
1106. Collins C, Weiden PL. Cardiotoxicity of 5-fluorouracil. Cancer Treat Rep 1987;71(7–8):733–6.
1107. Freeman NJ, Costanza ME. 5-Fluorouracil-associated cardiotoxicity. Cancer 1988;61(1):36–45.
1108. McKendall GR, Shurman A, Anamur M, Most AS. Toxic cardiogenic shock associated with infusion of 5-fluorouracil. Am Heart J 1989;118(1):184–6.
1109. Patel B, Kloner RA, Ensley J, Al-Sarraf M, Kish J, Wynne J. 5-Fluorouracil cardiotoxicity: left ventricular dysfunction and effect of coronary vasodilators. Am J Med Sci 1987;294(4):238–43.
1110. Misset B, Escudier B, Leclercq B, Rivara D, Rougier P, Nitenberg G. Acute myocardiotoxicity during 5-fluorouracil therapy. Intensive Care Med 1990;16(3):210–1.
1111. Eskilsson J, Albertsson M, Mercke C. Adverse cardiac effects during induction chemotherapy treatment with cisplatin and 5-fluorouracil. Radiother Oncol 1988;13(1):41–6.
1112. Mortimer JE, Higano C. Continuous infusion 5-fluorouracil and folinic acid in disseminated colorectal cancer. Cancer Invest 1988;6(2):129–32.
1113. Coronel B, Madonna O, Mercatello A, Caillette A, Moskovtchenko JF. Myocardiotoxicity of 5 fluorouracil. Intensive Care Med 1988;14(4):429–30.
1114. Chaudary S, Song SY, Jaski BE. Profound, yet reversible, heart failure secondary to 5-fluorouracil. Am J Med 1988;85(3):454–6.
1115. Jakubowski AA, Kemeny N. Hypotension as a manifestation of cardiotoxicity in three patients receiving cisplatin and 5-fluorouracil. Cancer 1988;62(2):266–9.
1116. de Forni M, Malet-Martino MC, Jaillais P, Shubinski RE, Bachaud JM, Lemaire L, Canal P, Chevreau C, Carrie D, Soulie P, Roche H, Boudjema B, Mihura J, Martino R, Bernadet P, Bugat R. Cardiotoxicity of high-dose continuous infusion fluorouracil: a prospective clinical study. J Clin Oncol 1992;10(11):1795–801.
1117. Ensley J, Kish J, Tapazoglou E, et al. 5-FU infusions associated with an ischaemic cardiotoxicity syndrome. Proc Am Soc Clin Oncol 1986;5:142.
1118. Gradishar WJ, Vokes EE. 5-Fluorouracil cardiotoxicity: a critical review. Ann Oncol 1990;1(6):409–14.
1119. Aziz SA, Tramboo NA, Mohi-ud-Din K, Iqbal K, Jalal S, Ahmad M. Supraventricular arrhythmia: a complication of 5-fluorouracil therapy. Clin Oncol (R Coll Radiol) 1998;10(6):377–8.
1120. Wang WS, Hsieh RK, Chiou TJ, Liu JH, Fan FS, Yen CC, Tung SL, Chen PM. Toxic cardiogenic shock in a patient receiving weekly 24-h infusion of high-dose 5-fluorouracil and leucovorin. Jpn J Clin Oncol 1998;28(9):551–4.
1121. Doll DC, Yarbro JW. Vascular toxicity associated with chemotherapy and hormonotherapy. Curr Opin Oncol 1994;6(4):345–50.
1122. Papamichael D, Amft N, Slevin ML, D'Cruz D. 5-Fluorouracil-induced Raynaud's phenomenon. Eur J Cancer 1998;34(12):1983.
1123. Kleiman NS, Lehane DE, Geyer CE Jr, Pratt CM, Young JB. Prinzmetal's angina during 5-fluorouracil chemotherapy. Am J Med 1987;82(3):566–8.
1124. Liss RH, Chadwick M. Correlation of 5-fluorouracil (NSC-19893) distribution in rodents with toxicity and chemotherapy in man. Cancer Chemother Rep 1974;58(6):777–86.
1125. Suzuki T, Nakanishi H, Hayashi A, et al. Cardiac toxicity of 5-FU in rabbits. Jpn J Pharmacol 1972;27(Suppl):137.

1126. Matsubara I, Kamiya J, Imai S. Cardiotoxic effects of 5-fluorouracil in the guinea pig. Jpn J Pharmacol 1980;30(6):871–9.

1127. Orditura M, De Vita F, Sarubbi B, Ducceschi V, Auriemma A, Infusino S, Iacono A, Catalano G. Analysis of recovery time indexes in 5-fluorouracil-treated cancer patients. Oncol Rep 1998;5(3):645–7.

1128. Martin M, Diaz-Rubio E, Furio V, Blazquez J, Almenarez J, Farina J. Lethal cardiac toxicity after cisplatin and 5-fluorouracil chemotherapy. Report of a case with necropsy study. Am J Clin Oncol 1989;12(3):229–34.

1129. Thyss A, Gaspard MH, Marsault R, Milano G, Frelin C, Schneider M. Very high endothelin plasma levels in patients with 5-FU cardiotoxicity. Ann Oncol 1992;3(1):88.

1130. Labianca R, Beretta G, Clerici M, Fraschini P, Luporini G. Cardiac toxicity of 5-fluorouracil: a study on 1083 patients. Tumori 1982;68(6):505–10.

1131. Jeremic B, Jevremovic S, Djuric L, Mijatovic L. Cardiotoxicity during chemotherapy treatment with 5-fluorouracil and cisplatin. J Chemother 1990;2(4):264–7.

1132. Käfer G, Achtnich M, Willer A, Weiss A, Queißer W. Cardiotoxicity in 5-fluorouracil/folinic acid treatment for metastatic colorectal cancer. Onkologie 1998;21:324–7.

1133. Anand AJ. Fluorouracil cardiotoxicity. Ann Pharmacother 1994;28:374–8.

1134. Lemaire L, Malet-Martino MC, de Forni M, Lasserre M. Cardiotoxicity of commercial 5-fluorouracil vials stems from the alkaline hydrolysis of this drug. Br J Cancer 1992;66:119–27.

1135. Lieutaud T, Brain E, Golgran-Toledano D, Vincent F, Cvitkovic E, Leclercq B, Escudier B. 5-Fluorouracil cardiotoxicity: a unique mechanism for ischaemic cardiopathy and cardiac failure. Eur J Cancer 1996;32A(2):368–9.

1136. Südhoff T, Enderle Md, Pahlke M, Petz C, Teschendorf C, Graeven U, Schmiegel W. 5-Fluorouracil induces arterial vasocontractions. Ann Oncol 2004;15:661–4.

1137. Cianci G, Morelli MF, Cannita K, Morese R, Ricevuto E, Rocco ZC Di, Porzio G, Lanfiuti Baldi P, Ficorella C. Prophylactic options in patients with 5-fluorouracil-associated cardiotoxicity. Br J Cancer 2002;88:1507–9.

1138. Dent RG, McColl I. Letter: 5-Fluorouracil and angina. Lancet 1975;1(7902):347–8.

1139. Pottage A, Holt S, Ludgate S, Langlands AO. Fluorouracil cardiotoxicity. BMJ 1978;1(6112):547.

1140. Gamelin E, Gamelin L, Larra F, Turcant A, Alain P, Maillart P, Allain YM, Minier JF, Dubin J. Toxicité cardiaque aiguë du 5-fluorouracile: correlation pharmacocinétique. [Acute cardiac toxicity of 5-fluorouracil: pharmacokinetic correlation.] Bull Cancer 1991;78(12):1147–53.

1141. Keefe DL, Roistacher N, Pierri MK. Clinical cardiotoxicity of 5-fluorouracil. J Clin Pharmacol 1993;33(11):1060–70.

1142. Lynch TJ Jr, Kass F, Kalish LA, Elias AD, Strauss G, Shulman LN, Sugarbaker DJ, Skarin A, Frei E 3rd. Cisplatin, 5-fluorouracil, and etoposide for advanced non-small cell lung cancer. Cancer 1993;71(10):2953–7.

1143. Burger AJ, Mannino S. 5-Fluorouracil-induced coronary vasospasm. Am Heart J 1987;114(2):433–6.

1144. Kohne CH, Thuss-Patience P, Friedrich M, Daniel PT, Kretzschmar A, Benter T, Bauer B, Dietz R, Dorken B. Raltitrexed (Tomudex): an alternative drug for patients with colorectal cancer and 5-fluorouracil associated cardiotoxicity. Br J Cancer 1998;77(6):973–7.

1145. Akpek G, Hartshorn KL. Failure of oral nitrate and calcium channel blocker therapy to prevent 5-fluorouracil-related myocardial ischemia: a case report. Cancer Chemother Pharmacol 1999;43(2):157–61.

1146. Wernicke JF. The side effect profile and safety of fluoxetine. J Clin Psychiatry 1985;46(3 Part 2):59–67.

1147. Roose SP, Glassman AH, Attia E, Woodring S, Giardina EG, Bigger JT Jr. Cardiovascular effects of fluoxetine in depressed patients with heart disease. Am J Psychiatry 1998;155(5):660–5.

1148. Amsterdam JD, Garcia-Espana F, Fawcett J, Quitkin FM, Reimherr FW, Rosenbaum JF, Beasley C. Blood pressure changes during short-term fluoxetine treatment. J Clin Psychopharmacol 1999;19(1):9–14.

1149. Rudnick A, Modai I, Zelikovski A. Fluoxetine-induced Raynaud's phenomenon. Biol Psychiatry 1997;41(12):1218–21.

1150. Varriale P. Fluoxetine (Prozac) as a cause of QT prolongation. Arch Intern Med 2001;161(4):12.

1151. Anderson J, Compton SA. Fluoxetine induced bradycardia in presenile dementia. Ulster Med J 1997;66(2):144–5.

1152. Roos JC. Cardiac effects of antidepressant drugs. A comparison of the tricyclic antidepressants and fluvoxamine. Br J Clin Pharmacol 1983;15(Suppl. 3):439S–45S.

1153. Robinson JF, Doogan DP. A placebo controlled study of the cardiovascular effects of fluvoxamine and clovoxamine in human volunteers. Br J Clin Pharmacol 1982;4(6):805–8.

1154. Cazzola M, Imperatore F, Salzillo A, Di Perna F, Calderaro F, Imperatore A, Matera MG. Cardiac effects of formoterol and salmeterol in patients suffering from COPD with pre-existing cardiac arrhythmias and hypoxemia. Chest 1998;114(2):411–5.

1155. Centanni S, Carlucci P, Santus P, Boveri B, Tarricone D, Fiorentini C, Lombardi F, Cazzola M. Non-pulmonary effects induced by the addition of formoterol to budesonide therapy in patients with mild or moderate persistent asthma. Respiration 2000;67(1):60–4.

1156. Kovarik JM, Schmouder RL, Barilla D, Buche M, Rouilly M, Berthier S, Wang Y, Van Saders C, Mayer T, Gottlieb AB. FTY720 and cyclosporine: evaluation for a pharmacokinetic interaction. Ann Pharmacother 2004;38(7-8):1153–8.

1157. Tedesco-Silva H, Mourad G, Kahan BD, Boira JG, Weimar W, Mulgaonkar S, Nashan B, Madsen S, Charpentier B, Pellet P, Vanrenterghem Y. FTY720, a novel immunomodulator: efficacy and safety results from the first phase 2A study in de novo renal transplantation. Transplantation 2004;77(12):1826-33. Corrected and republished in: Transplantation. 2005;79(11):1553–60.

1158. Kroft LJ, de Roos A. Blood pool contrast agents for cardiovascular MR imaging. J Magn Reson Imaging 1999;10(3):395–403.

1159. Rubin DL, Desser TS, Semelka R, Brown J, Nghiem HV, Stevens WR, Bluemke D, Nelson R, Fultz P, Reimer P, Ho V, Kristy RM, Pierro JA. A multicenter, randomized, double-blind study to evaluate the safety, tolerability, and efficacy of OptiMARK (gadoversetamide injection) compared with Magnevist (gadopentetate dimeglumine) in patients with liver pathology: results of a Phase III clinical trial. J Magn Reson Imaging 1999;9(2):240–50.

1160. Coche EE, Hammer FD, Goffette PP. Demonstration of pulmonary embolism with gadolinium-enhanced spiral CT. Eur Radiol 2001;11(11):2306–9.

1161. Eisele JH, Marta JA, Davis HS. Quantitative aspects of the chronotropic and neuromuscular effects of gallamine in anesthetized man. Anesthesiology 1971;35(6):630–3.

1162. Stoelting RK. Hemodynamic effects of gallamine during halothane–nitrous oxide anesthesia. Anesthesiology 1973;39(6):645–7.
1163. Kennedy BR, Farman JV. Cardiovascular effects of gallamine triethiodide in man. Br J Anaesth 1968;40(10):773–80.
1164. Riker WF Jr, Wescoe WC. The pharmacology of Flaxedil, with observations on certain analogues. Ann NY Acad Sci 1951;54(3):373–94.
1165. Brown BR Jr, Crout JR. The sympathomimetic effect of gallamine on the heart. J Pharmacol Exp Ther 1970;172(2):266–73.
1166. Vercruysse P, Bossuyt P, Hanegreefs G, Verbeuren TJ, Vanhoutte PM. Gallamine and pancuronium inhibit pre- and postjunctional muscarine receptors in canine saphenous veins. J Pharmacol Exp Ther 1979;209(2):225–30.
1167. Reitan JA, Fraser AI, Eisele JH. Lack of cardiac inotropic effects of gallamine in anesthetized man. Anesth Analg 1973;52(6):974–9.
1168. Marshall IG. Pharmacological effects of neuromuscular blocking agents: interaction with cholinoceptors other than nicotinic receptors of the neuromuscular junction. Anest Rianim 1986;27:19.
1169. Bowman WC. Non-relaxant properties of neuromuscular blocking drugs. Br J Anaesth 1982;54(2):147–60.
1170. Bowman WC. Pharmacology of Neuromuscular Function. 2nd ed.. London/Boston/Singapore/Sydney/Toronto/Wellington: Wright;. 1990.
1171. Grasela DM. Clinical pharmacology of gatifloxacin, a new fluoroquinolone. Clin Infect Dis 2000;31(Suppl 2):S51–8.
1172. Lannini PB, Circiumaru I. Gatifloxacin-induced QTc prolongation and ventricular tachycardia. Pharmacotherapy 2001;21(3):361–2.
1173. Owens RC Jr, Ambrose PG. Torsades de pointes associated with fluoroquinolones. Pharmacotherapy 2002;22(5):663–8discussion 668–72.
1174. Bertino JS Jr, Owens RC Jr, Carnes TD, Iannini PB. Gatifloxacin-associated corrected QT interval prolongation, torsades de pointes, and ventricular fibrillation in patients with known risk factors. Clin Infect Dis 2002;34(6):861–3.
1175. Humphreys BD, Sharman JP, Henderson JM, Clark JW, Marks PW, Rennke HG, Zhu AX, Magee CC. Gemcitabine-associated thrombotic microangiopathy. Cancer 2004;100(12):2664–70.
1176. De Pas T, Curigliano G, Franceschelli L, Catania C, Spaggiari L, de Braud F. Gemcitabine-induced systemic capillary leak syndrome. Ann Oncol 2001;12(11):1651–2.
1177. Pulkkanen K, Kataja V, Johansson R. Systemic capillary leak syndrome resulting from gemcitabine treatment in renal cell carcinoma: a case report. J Chemother 2003;15(3):287–9.
1178. Schulte-Sasse U. Life threatening myocardial ischaemia associated with the use of prostaglandin E1 to induce abortion. BJOG 2000;107(5):700–2.
1179. Wadleigh M, Richardson PG, Zahrieh D, Lee SJ, Cutler C, Ho V, Alyea EP, Antin JH, Stone RM, Soiffer RJ, DeAngelo DJ. Prior gemtuzumab ozogamicin exposure significantly increases the risk of veno-occlusive disease in patients who undergo myeloablative allogeneic stem cell transplantation. Blood 2003;102:1578–82.
1180. Zwaan CM, Reinhardt D, Corbacioglu S, et al. Gemtuzumab ozogamicin: first clinical experiences in children with relapsed/refractory acute myeloid leukemia treated on compassionate-use basis. Blood 2003;101:3868–71.
1181. Nabhan C, Rundhaugen LM, Riley MB, Rademaker A, Boehlke L, Jatoi M, Tallman MS. Phase II pilot trial of gemtuzumab ozogamicin (GO) as first line therapy in acute myeloid leukemia patients age 65 or older. Leuk Res 2005;29(1):53–7.
1182. Mckinney MS, Fee JP. Cardiovascular effects of 50% nitrous oxide in older adult patients anaesthetized with isoflurane or halothane. Br J Anaesth 1998;80(2):169–73.
1183. Michaloudis D, Fraidakis O, Petrou A, Gigourtsi C, Parthenakis F. Anaesthesia and the QT interval. Effects of isoflurane and halothane in unpremedicated children. Anaesthesia 1998;53(5):435–9.
1184. Blayney MR, Malins AF, Cooper GM. Cardiac arrhythmias in children during outpatient general anaesthesia for dentistry: a prospective randomised trial. Lancet 1999;354(9193):1864–6.
1185. Viitanen H, Baer G, Koivu H, Annila P. The hemodynamic and Holter-electrocardiogram changes during halothane and sevoflurane anesthesia for adenoidectomy in children aged one to three years. Anesth Analg 1999;89(6):1423–5.
1186. Guler N, Bilge M, Eryonucu B, Kati I, Demirel CB. The effects of halothane and sevoflurane on QT dispersion. Acta Cardiol 1999;54(6):311–5.
1187. Weber S, Genevray B, Pasquier G, Chapsal J, Bonnin A, Degeorges M. Severe coronary spasm during drug-induced immediate hypersensitivity reaction. Lancet 1982;2(8302):821.
1188. Chin DT. Myocardial ischemia induced by glucagon. Ann Pharmacother 1996;30(1):84–5.
1189. Wright JS, Newman DC. Complications with glutaraldehyde-preserved bioprostheses. Med J Aust 1980;1(11):542–3.
1190. Lindley G. Was it something you ate? BMJ 2003;326:87.
1191. Hill C, Pile K, Henderson D, Kirkham B. Neurological side effects in two patients receiving gold injections for rheumatoid arthritis. Br J Rheumatol 1995;34(10):989–90.
1192. Gottlieb NL, Gray RG. Diagnosis and management of adverse reactions from gold compounds. J Anal Toxicol 1978;2:173.
1193. Yim SF, Lau TK, Sahota DS, Chung TK, Chang AM, Haines CJ. Prospective randomized study of the effect of "add-back" hormone replacement on vascular function during treatment with gonadotropin-releasing hormone agonists. Circulation 1998;98(16):1631–5.
1194. Lindemann A, Rumberger B. Vascular complications in patients treated with granulocyte colony-stimulating factor (G-CSF). Eur J Cancer 1993;29A(16):2338–9.
1195. Conti JA, Scher HI. Acute arterial thrombosis after escalated-dose methotrexate, vinblastine, doxorubicin, and cisplatin chemotherapy with recombinant granulocyte colony-stimulating factor. A possible new recombinant granulocyte colony-stimulating factor toxicity. Cancer 1992;70(11):2699–702.
1196. Kuroiwa M, Okamura T, Kanaji T, Okamura S, Harada M, Niho Y. Effects of granulocyte colony-stimulating factor on the hemostatic system in healthy volunteers. Int J Hematol 1996;63(4):311–6.
1197. Vij R, Adkins DR, Brown RA, Khoury H, DiPersio JF, Goodnough T. Unstable angina in a peripheral blood stem and progenitor cell donor given granulocyte-colony-stimulating factor. Transfusion 1999;39(5):542–3.
1198. Oeda E, Shinohara K, Kamei S, Nomiyama J, Inoue H. Capillary leak syndrome likely the result of granulocyte colony-stimulating factor after high-dose chemotherapy. Intern Med 1994;33(2):115–9.

1199. Dereure O, Bessis D, Lavabre-Bertrand T, Exbrayat C, Fegueux N, Biron C, Guilhou JJ. Thrombotic and necrotizing panniculitis associated with recombinant human granulocyte colony-stimulating factor treatment. Br J Dermatol 2000;142(4):834–6.
1200. Vial T, Descotes J. Clinical toxicity of cytokines used as haemopoietic growth factors. Drug Saf 1995;13(6):371–406.
1201. Stephens LC, Haire WD, Schmit-Pokorny K, Kessinger A, Kotulak G. Granulocyte macrophage colony stimulating factor: high incidence of apheresis catheter thrombosis during peripheral stem cell collection. Bone Marrow Transplant 1993;11(1):51–4.
1202. Tolcher AW, Giusti RM, O'Shaughnessy JA, Cowan KH. Arterial thrombosis associated with granulocyte–macrophage colony-stimulating factor (GM-CSF) administration in breast cancer patients treated with dose-intensive chemotherapy: a report of two cases. Cancer Invest 1995;13(2):188–92.
1203. Arning M, Kliche KO, Schneider W. GM-CSF therapy and capillary-leak syndrome. Ann Hematol 1991;62(2–3):83.
1204. Emminger W, Emminger-Schmidmeier W, Peters C, Susani M, Hawliczek R, Hocker P, Gadner H. Capillary leak syndrome during low dose granulocyte–macrophage colony-stimulating factor (rh GM-CSF) treatment of a patient in a continuous febrile state. Blut 1990;61(4):219–21.
1205. Kaczmarski RS, Mufti GJ. Hypoalbuminaemia after prolonged treatment with recombinant granulocyte macrophage colony stimulating factor. BMJ 1990;301(6764):1312–3.
1206. Suzuki T, Kato Y, Sasabe H, Itose M, Miyamoto G, Sugiyama Y. Mechanism for the tissue distribution of grepafloxacin, a fluoroquinolone antibiotic, in rats. Drug Metab Dispos 2002;30(12):1393–9.
1207. Yamamoto H, Koizumi T, Hirota M, Kaneki T, Ogasawara H, Yamazaki Y, Fujimoto K, Kubo K. Lung tissue distribution after intravenous administration of grepafloxacin: comparative study with levofloxacin. Jpn J Pharmacol 2002;88(1):63–8.
1208. Gibaldi M. Grepafloxacin withdrawn from market. Drug Ther Topics Suppl 2000;29:6.
1209. Carbon C. Comparison of side effects of levofloxacin versus other fluoroquinolones. Chemotherapy 2001;47(Suppl 3):9–14discussion 44–8.
1210. Zhanel GG, Ennis K, Vercaigne L, Walkty A, Gin AS, Embil J, Smith H, Hoban DJ. A critical review of the fluoroquinolones: focus on respiratory infections. Drugs 2002;62(1):13–59.
1211. Landin-Wilhelmsen K, Nilsson A, Bosaeus I, Bengtsson BA. Growth hormone increases bone mineral content in postmenopausal osteoporosis: a randomised placebo-controlled trial. J Bone Mineral Res 2003;18:393–405.
1212. Calao A, Di Somma C, Rota F, Di Maio S, Salerno M, Klain A, Spiezia S, Lombardi G. Common carotid intima–media thickness in growth hormone (GH) deficient adolescents: a prospective study after GH withdrawal and restarting GH replacement. J Clin Endocrinol Metab 2005;90:2659–65.
1213. Toivonen L, Viitasalo M, Siikamaki H, Raatikka M, Pohjola-Sintonen S. Provocation of ventricular tachycardia by antimalarial drug halofantrine in congenital long QT syndrome. Clin Cardiol 1994;17(7):403–4.
1214. Sowunmi A, Falade CO, Oduola AM, Ogundahunsi OA, Fehintola FA, Gbotosho GO, Larcier P, Salako LA. Cardiac effects of halofantrine in children suffering from acute uncomplicated falciparum malaria. Trans R Soc Trop Med Hyg 1998;92(4):446–8.
1215. Olivier C, Rizk C, Zhang D, Jacqz-Aigrain E. Allongement de l'espace QT_c compliquant la prescription d'halofantrine chez deux enfants presentant un accès palustre a *plasmodium falciparum*. [Long QT_c interval complicating halofantrine therapy in 2 children with *Plasmodium falciparum* malaria.] Arch Pediatr 1999;6(9):966–70.
1216. van Agtmael M, Bouchaud O, Malvy D, Delmont J, Danis M, Barette S, Gras C, Bernard J, Touze JE, Gathmann I, Mull R. The comparative efficacy and tolerability of CGP 56697 (artemether + lumefantrine) versus halofantrine in the treatment of uncomplicated falciparum malaria in travellers returning from the Tropics to The Netherlands and France Int J Antimicrob Agents 1999;12(2):159–69.
1217. Malvy D, Receveur MC, Ozon P, Djossou F, Le Metayer P, Touze JE, Longy-Boursier M, Le Bras M. Fatal cardiac incident after use of halofantrine. J Travel Med 2000;7(4):215–6.
1218. Herranz U, Rusca A, Assandri A. Emedastine–ketoconazole: pharmacokinetic and pharmacodynamic interactions in healthy volunteers. Int J Clin Pharmacol Ther 2001;39(3):102–9.
1219. Hatta K, Takahashi T, Nakamura H, Yamashiro H, Asukai N, Matsuzaki I, Yonezawa Y. The association between intravenous haloperidol and prolonged QT interval. J Clin Psychopharmacol 2001;21(3):257–61.
1220. Sharma ND, Rosman HS, Padhi ID, Tisdale JE. Torsades de pointes associated with intravenous haloperidol in critically ill patients. Am J Cardiol 1998;81(2):238–40.
1221. Su K-P, Shen WW, Chuang C-L, Chen K-P, Chen CC. A pilot cross-over design study on QT_c interval prolongation associated with sulpiride and haloperidol. Schizophr Res 2002;59:93–4.
1222. Desai M, Tanus-Santos JE, Li L, Gorski JC, Arefayene M, Liu Y, Desta Z, Flockhart DA. Pharmacokinetics and QT interval pharmacodynamics of oral haloperidol in poor and extensive metabolizers of CYP2D6. Pharmacogenomics J 2003;3:105–13.
1223. Jackson T, Ditmanson L, Phibbs B. Torsade de pointes and low-dose oral haloperidol. Arch Intern Med 1997;157(17):2013–5.
1224. LLerena A, Berecz R, de la Rubia A, Dorado P. QT_c interval lengthening and debrisoquine metabolic ratio in psychiatric patients treated with oral haloperidol monotherapy. Eur J Clin Pharmacol 2002;58(3):223–4.
1225. Huyse F, van Schijndel RS. Haloperidol and cardiac arrest. Lancet 1988;2(8610):568–9.
1226. Iwahashi K. Significantly higher plasma haloperidol level during cotreatment with carbamazepine may herald cardiac change. Clin Neuropharmacol 1996;19(3):267–70.
1227. Douglas PH, Block PC. Corrected QT interval prolongation associated with intravenous haloperidol in acute coronary syndromes. Catheter Cardiovasc Interv 2000;50(3):352–5.
1228. Iglesias E, Esteban E, Zabala S, Gascon A. Tiapride-induced torsade de pointes. Am J Med 2000;109(6):509.
1229. Remijnse PL, Eeckhout AM, van Guldener C. Plotseling overlijden na eenmalige orale toediening van haloperidol. [Sudden death following a single oral administration of haloperidol.] Ned Tijdschr Geneeskd 2002;146(16):768–71.

1230. Perrault LP, Denault AY, Carrier M, Cartier R, Belisle S. Torsades de pointes secondary to intravenous haloperidol after coronary bypass grafting surgery. Can J Anaesth 2000;47(3):251–4.
1231. Johri S, Rashid H, Daniel PJ, Soni A. Cardiopulmonary arrest secondary to haloperidol. Am J Emerg Med 2000;18(7):839.
1232. Merin RG. Physiology, pathophysiology and pharmacology of the coronary circulation with particular emphasis on anesthetics. Anaesthesiol Reanim 1992;17(1):5–26.
1233. Maze M, Mason DM. Ąetiology and treatment of halothane-induced arrhythmias. Clin Anaesthesiol 1983;1:301.
1234. Schmidt U, Schwinger RH, Bohm S, Uberfuhr P, Kreuzer E, Reichart B, Meyer L, Erdmann E, Bohm M. Evidence for an interaction of halothane with the L-type Ca^{2+} channel in human myocardium. Anesthesiology 1993;79(2):332–9.
1235. McKinney MS, Fee JP, Clarke RS. Cardiovascular effects of isoflurane and halothane in young and elderly adult patients. Br J Anaesth 1993;71(5):696–701.
1236. Yokoyama K. Arrhythmias due to halothane anesthesia. Jpn J Anesthesiol 1978;27:64.
1237. Lindgren L. E.C.G changes during halothane and enflurane anaesthesia for E.N.T. surgery in children Br J Anaesth 1981;53(6):653–62.
1238. Reinoso-Barbero F, Gutierrez-Marquez M, Diez-Labajo A. Prevention of halothane-induced bradycardia: is intranasal premedication indicated? Paediatr Anaesth 1998;8(3):195–9.
1239. Saghaei M, Mortazavian M. Pulsus alternans during general anesthesia with halothane: effects of permissive hypercapnia. Anesthesiology 2000;93(1):91–4.
1240. Anbari KK, Garino JP, Mackenzie CF. Hemoglobin substitutes. Eur Spine J 2004;13 Suppl 1:S76–82.
1241. Bloomfield EL, Rady MY, Esfandiari S. A prospective trial of diaspirin cross-linked hemoglobin solution in patients after elective repair of abdominal aortic aneurysm. Mil Med 2004;169(7):546–50.
1242. Creteur J, Vincent JL. Hemoglobin solutions. Crit Care Med 2003;31(12 Suppl):S698–707.
1243. Bjoraker DG, Ketcham TR. Hemodynamic and platelet response to the bolus intravenous administration of porcine heparin. Thromb Haemost 1983;49(1):1–4.
1244. Bowler GM, Galloway DW, Meiklejohn BH, Macintyre CC. Sharp fall in blood pressure after injection of heparin containing chlorbutol. Lancet 1986;1(8485):848–9.
1245. Schimpf K, Barth P. Heparinshock mit Verbrauchsreaktion des Blutgerinnungssystems?. [Heparin shock with consumption reaction of the blood coagulation system?.] Klin Wochenschr 1966;44(10):544–7.
1246. Gayer W. Seltene Beobachtungen von Heparin-Unvertreglichkeit. [Unusual case of heparin intolerance.] Gynaecologia 1968;166(1):25.
1247. Winslow LC, Kroll DJ. Herbs as medicines. Arch Intern Med 1998;158(20):2192–9.
1248. Miller LG. Herbal medicinals: selected clinical considerations focusing on known or potential drug–herb interactions. Arch Intern Med 1998;158(20):2200–11.
1249. Mashour NH, Lin GI, Frishman WH. Herbal medicine for the treatment of cardiovascular disease: clinical considerations. Arch Intern Med 1998;158(20):2225–34.
1250. Saller R, Reichling J, Kristof O. Phytotherapie-Behandlung ohne Nebenwirkungen? [Phytotherapy—treatment without side effects?] Dtsch Med Wochenschr 1998;123(3): 58–62.
1251. Bateman J, Chapman RD, Simpson D. Possible toxicity of herbal remedies. Scott Med J 1998;43(1):7–15.
1252. Ernst E. Harmless herbs? A review of the recent literature. Am J Med 1998;104(2):170–8.
1253. Shaw D. Risks or remedies? Safety aspects of herbal remedies in the UK. J R Soc Med 1998;91(6):294–6.
1254. Marrone CM. Safety issues with herbal products. Ann Pharmacother 1999;33(12):1359–62.
1255. Ko RJ. Causes, epidemiology, and clinical evaluation of suspected herbal poisoning. J Toxicol Clin Toxicol 1999;37(6):697–708.
1256. Ernst E. Phytotherapeutika. Wie harmlos sind sie wirklich? Dtsch Arzteblatt 1999;48:3107–8.
1257. Calixto JB. Efficacy, safety, quality control, marketing and regulatory guidelines for herbal medicines (phytotherapeutic agents). Braz J Med Biol Res 2000;33(2):179–89.
1258. De Smet PA. Herbal remedies. N Engl J Med 2002;347(25):2046–56.
1259. Ernst E, Pittler MH. Risks associated with herbal medicinal products. Wien Med Wochenschr 2002;152(7–8):183–9.
1260. Ali MS, Uzair SS. Natural organic toxins. Hamdard Medicus 2002;XLIV:86–93.
1261. Gee BC, Wilson P, Morris AD, Emerson RM. Herbal is not synonymous with safe. Arch Dermatol 2002;138(12):1613.
1262. Matthews HB, Lucier GW, Fisher KD. Medicinal herbs in the United States: research needs. Environ Health Perspect 1999;107(10):773–8.
1263. Hussain SH. Potential risks of health supplements—self-medication practices and the need for public health education. Int J Risk Saf Med 1999;12:167–71.
1264. Fontana RJ. Acute liver failure. Curr Opin Gastroenterol 1999;15:270–7.
1265. Valli G, Giardina EG. Benefits, adverse effects and drug interactions of herbal therapies with cardiovascular effects. J Am Coll Cardiol 2002;39(7):1083–95.
1266. Chitturi S, Farrell GC. Herbal hepatotoxicity: an expanding but poorly defined problem. J Gastroenterol Hepatol 2000;15(10):1093–9.
1267. Haller CA, Dyer JE, Ko R, Olson KR. Making a diagnosis of herbal-related toxic hepatitis. West J Med 2002;176(1):39–44.
1268. Ernst E. Adverse effects of herbal drugs in dermatology. Br J Dermatol 2000;143(5):923–9.
1269. Holsen DS. Flora og efflorescenser—om planter som arsak til hudsykdom. [Plants and plant produce—about plants as cause of diseases.] Tidsskr Nor Laegeforen 2002;122(17):1665–9.
1270. Ernst E. Adverse effects of unconventional therapies in the elderly: a systematic review of the recent literature. J Am Aging Assoc 2002;25:11–20.
1271. Ernst E. Herbal medicinal products during pregnancy: are they safe? BJOG 2002;109(3):227–35.
1272. Hodges PJ, Kam PC. The peri-operative implications of herbal medicines. Anaesthesia 2002;57(9):889–99.
1273. Cheng B, Hung CT, Chiu W. Herbal medicine and anaesthesia. Hong Kong Med J 2002;8(2):123–30.
1274. Bartsch H. Gefahrliche Naturprodukte: sind Karzinogene im Kräutertee? [Hazardous natural products. Are there carcinogens in herbal teas?] MMW Fortschr Med 2002;144(41):14.
1275. Pies R. Adverse neuropsychiatric reactions to herbal and over-the-counter "antidepressants". J Clin Psychiatry 2000;61(11):815–20.

1276. Tomlinson B, Chan TY, Chan JC, Critchley JA, But PP. Toxicity of complementary therapies: an eastern perspective. J Clin Pharmacol 2000;40(5):451–6.
1277. Bensoussan A, Myers SP, Drew AK, Whyte IM, Dawson AH. Development of a Chinese herbal medicine toxicology database. J Toxicol Clin Toxicol 2002;40(2):159–67.
1278. Ernst E. Ayurvedic medicines. Pharmacoepidemiol Drug Saf 2002;11(6):455–6.
1279. Boullata JI, Nace AM. Safety issues with herbal medicine. Pharmacotherapy 2000;20(3):257–69.
1280. Shapiro R. Safety assessment of botanicals. Nutraceuticals World 2000;52–63 July/August.
1281. Saller R, Iten F, Reichling J. Unerwünschte Wirkungen und Wechselwirkungen von Phytotherapeutika. Erfahrungsheilkunde 2000;6:369–76.
1282. Pennachio DL. Drug-herb interactions: how vigilant should you be? Patient Care 2000;19:41–68.
1283. Ernst E. Possible interactions between synthetic and herbal medicinal products. Part 1: a systematic review of the indirect evidence. Perfusion 2000;13:4–68.
1284. Ernst E. Interactions between synthetic and herbal medicinal products. Part 2: a systematic review of the direct evidence. Perfusion 2000;13:60–70.
1285. Blumenthal M. Interactions between herbs and conventional drugs: introductory considerations. Herbal Gram 2000;49:52–63.
1286. Ernst E. Herb–drug interactions: potentially important but woefully under-researched. Eur J Clin Pharmacol 2000;56(8):523–4.
1287. De Smet PAGM, Touw DJ. Sint-janskruid op de balans van werking en interacties. Pharm Weekbl 2000;135:455–62.
1288. Abebe W. Herbal medication: potential for adverse interactions with analgesic drugs. J Clin Pharm Ther 2002;27(6):391–401.
1289. Mason P. Food–drug interactions: nutritional supplements and drugs. Pharm J 2002;269:609–11.
1290. Scott GN, Elmer GW. Update on natural product—drug interactions. Am J Health Syst Pharm 2002;59(4):339–47.
1291. Rahman SZ, Singhal KC. Problems in pharmacovigilance of medicinal products of herbal origin and means to minimize them. Uppsala Rep 2002;17:1–4.
1292. De Smet PA. Health risks of herbal remedies. Drug Saf 1995;13(2):81–93.
1293. De Smet PAGM, D'Arcy PF. Drug interactions with herbal and other non-orthodox drugs. In: D'Arcy PF, McElnay JC, Welling PG, editors. Mechanisms of Drug Interactions. Heidelberg: Springer Verlag, 1996
1294. Stockley I. Drug Interactions. 4th ed. London: The Pharmaceutical Press, 1996.
1295. Newall CA, Anderson LA, Phillipson JD. Herbal medicines. A guide for health-care professionals. London: The Pharmaceutical Press, 1996.
1296. Barnes J, Mills SY, Abbot NC, Willoughby M, Ernst E. Different standards for reporting ADRs to herbal remedies and conventional OTC medicines: face-to-face interviews with 515 users of herbal remedies. Br J Clin Pharmacol 1998;45(5):496–500.
1297. Ernst E. Competence in complementary medicine. Comp Ther Med 1995;3:6–8.
1298. Wong ALN, Chan JTS, Chan TYK. Adverse herbal interactions causing hypotension. Ther Drug Monit 2003;25:297–8.
1299. Dujovne CA. Side effects of statins: hepatitis versus "transaminitis"-myositis versus "CPKitis". Am J Cardiol 2002;89(12):1411–3.
1300. Black DM, Bakker-Arkema RG, Nawrocki JW. An overview of the clinical safety profile of atorvastatin (Lipitor), a new HMG-CoA reductase inhibitor. Arch Intern Med 1998;158(6):577–84.
1301. Jimenez-Alonso J, Osorio JM, Gutierrez-Cabello F, Lopez de la Osa A, Leon L, Mediavilla Garcia JD. Atorvastatin-induced cholestatic hepatitis in a young woman with systemic lupus erythematosus. Grupo Lupus Virgen de las Nieves. Arch Intern Med 1999;159(15):1811–2.
1302. Farmer JA, Torre-Amione G. Comparative tolerability of the HMG-CoA reductase inhibitors. Drug Saf 2000;23(3):197–213.
1303. Illingworth DR, Crouse JR 3rd, Hunninghake DB, Davidson MH, Escobar ID, Stalenhoef AF, Paragh G, Ma PT, Liu M, Melino MR, O'Grady L, Mercuri M, Mitchel YBSimvastatin Atorvastatin HDL Study Group. A comparison of simvastatin and atorvastatin up to maximal recommended doses in a large multicenter randomized clinical trial. Curr Med Res Opin 2001;17(1):43–50.
1304. Spreckelsen U, Kirchhoff R, Haacke H. Cholestatischer Iikterus Wahrend Lovastatin-Einnahme. [Cholestatic jaundice during lovastatin medication.] Dtsch Med Wochenschr 1991;116(19):739–40.
1305. Ballare M, Campanini M, Catania E, Bordin G, Zaccala G, Monteverde A. Acute cholestatic hepatitis during simvastatin administration. Recenti Prog Med 1991;82(4):233–5.
1306. Muggeo M, Travia D, Querena M, Zenti MG, Bagnani M, Branzi P, et al. Long term treatment with pravastatin, simvastatin and gemfibrozil in patients with primary hypercholesterolaemia, a controlled study. Drug Invest 1992;4:376–85.
1307. Assmann G, Huwel D, Schussman KM, Smilde JG, Kosling M, Withagen AJ, Wunderlich J, Stoel I, Van Dormaal JJ, Neuss J, et al. Efficacy and safety of atorvastatin and pravastatin in patients with hypercholesterolemia. Eur J Intern Med 1999;10:33–9.
1308. Nakad A, Bataille L, Hamoir V, Sempoux C, Horsmans Y. Atorvastatin-induced acute hepatitis with absence of cross-toxicity with simvastatin. Lancet 1999;353(9166):1763–4.
1309. A preliminary communication to the Medical Research Council by a Subcommittee. Risk of thromboembolic disease in women taking oral contraceptives. BMJ 1967;2(548):355–9.
1310. Miwa LJ, Edmunds AL, Shaefer MS, Raynor SC. Idiopathic thromboembolism associated with triphasic oral contraceptives. DICP 1989;23(10):773–5.
1311. Lamy AL, Roy PH, Morissette JJ, Cantin R. Intimal hyperplasia and thrombosis of the visceral arteries in a young woman: possible relation with oral contraceptives and smoking. Surgery 1988;103(6):706–10.
1312. Scolding NJ, Gibby OM. Fatal pulmonary embolus in a patient treated with Marvelon. J R Coll Gen Pract 1988;38(317):568.
1313. Porter JB, Hunter JR, Jick H, Stergachis A. Oral contraceptives and nonfatal vascular disease. Obstet Gynecol 1985;66(1):1–4.
1314. Realini JP, Goldzieher JW. Oral contraceptives and cardiovascular disease: a critique of the epidemiologic studies. Am J Obstet Gynecol 1985;152(6 Pt 2):729–98.
1315. Skouby SOCommittee Chairman. Consensus Development Meeting: metabolic aspects of oral

contraceptives of relevance for cardiovascular diseases. Am J Obstet Gynecol 1990;162:1335.
1316. Lidegaard O. Thrombotic diseases in young women and the influence of oral contraceptives. Am J Obstet Gynecol 1998;179(3 Pt 2):S62–7.
1317. Anonymous. Oral contraceptives containing gestodene or desogestrel-up-date: revised product information. WHO Pharm Newslett 1999;7/8:3.
1318. Farmer RD, Lawrenson RA. Oral contraceptives and venous thromboembolic disease: the findings from database studies in the United Kingdom and Germany. Am J Obstet Gynecol 1998;179(3 Pt 2):S78–86.
1319. Burnhill MS. The use of a large-scale surveillance system in Planned Parenthood Federation of America clinics to monitor cardiovascular events in users of combination oral contraceptives. Int J Fertil Womens Med 1999;44(1):19–30.
1320. Jick SS, Myers MW, Jick H. Risk of idiopathic cerebral haemorrhage in women on oral contraceptives with differing progestagen components. Lancet 1999;354(9175):302–3.
1321. Vessey MP, Villard-Mackintosh L, McPherson K, Yeates D. Mortality among oral contraceptive users: 20 year follow up of women in a cohort study. BMJ 1989;299(6714):1487–91.
1322. Schwingl PJ, Ory HW, King TDN. Modeled estimates of cardiovascular mortality risks in the US associated with low dose oral contraceptives. 1995 Unpublished draft.
1323. Thorogood M, Mann J, Murphy M, Vessey M. Is oral contraceptive use still associated with an increased risk of fatal myocardial infarction? Report of a case-control study. Br J Obstet Gynaecol 1991;98(12):1245–53.
1324. Rosenberg L, Palmer JR, Shapiro S. Use of lower dose oral contraceptives and risk of myocardial infarction. Circulation 1991;83:723.
1325. Hannaford PC, Croft PR, Kay CR. Oral contraception and stroke. Evidence from the Royal College of General Practitioners' Oral Contraception Study. Stroke 1994;25(5):935–42.
1326. Lidegaard O. Oral contraception and risk of a cerebral thromboembolic attack: results of a case-control study. BMJ 1993;306(6883):956–63.
1327. Thorogood M, Mann J, Murphy M, Vessey M. Fatal stroke and use of oral contraceptives: findings from a case-control study. Am J Epidemiol 1992;136(1):35–45.
1328. Thorogood M, Mann J, Murphy M, Vessey M. Risk factors for fatal venous thromboembolism in young women: a case-control study. Int J Epidemiol 1992;21(1):48–52.
1329. Longstreth WT, Nelson LM, Koepsell TD, van Belle G. Subarachnoid hemorrhage and hormonal factors in women. A population-based case-control study. Ann Intern Med 1994;121(3):168–73.
1330. Gerstman BB, Piper JM, Tomita DK, Ferguson WJ, Stadel BV, Lundin FE. Oral contraceptive estrogen dose and the risk of deep venous thromboembolic disease. Am J Epidemiol 1991;133(1):32–7.
1331. Rosendaal FR, Helmerhorst FM, Vandenbroucke JP. Oral contraceptives, hormone replacement therapy and thrombosis. Thromb Haemost 2001;86(1):112–23.
1332. Zreik TG, Odunsi K, Cass I, Olive DL, Sarrel P. A case of fatal pulmonary thromboembolism associated with the use of intravenous estrogen therapy. Fertil Steril 1999;71(2):373–5.
1333. Chopard JL, Moulin T, Bourrin JC, et al. Contraception orale et accident vasculaire cérébral ischémique. Semin Hop (Paris) 1988;64:2075.
1334. Lindberg MC. Hepatobiliary complications of oral contraceptives. J Gen Intern Med 1992;7(2):199–209.
1335. Bohler J, Hauenstein KH, Hasler K, Schollmeyer P. Renal vein thrombosis in a dehydrated patient on an oral contraceptive agent. Nephrol Dial Transplant 1989;4(11):993–5.
1336. Hassan HA. Oral contraceptive-induced mesenteric venous thrombosis with resultant intestinal ischemia. J Clin Gastroenterol 1999;29(1):90–5.
1337. Hagglund H, Remberger M, Klaesson S, Lonnqvist B, Ljungman P, Ringden O. Norethisterone treatment, a major risk-factor for veno-occlusive disease in the liver after allogeneic bone marrow transplantation. Blood 1998;92(12):4568–72.
1338. Grimes DA. The safety of oral contraceptives: epidemiologic insights from the first 30 years. Am J Obstet Gynecol 1992;166(6 Pt 2):1950–4.
1339. Hirvonen E, Idanpaan-Heikkila J. Cardiovascular death among women under 40 years of age using low-estrogen oral contraceptives and intrauterine devices in Finland from 1975 to 1984. Am J Obstet Gynecol 1990;163(1 Pt 2):281–4.
1340. Cirkel U, Schweppe KW. Fettstoffwechsel und orale Kontrazeptiva. Arztl Kosmetol 1985;15:253.
1341. Fruzzetti F, Ricci C, Fioretti P. Haemostasis profile in smoking and nonsmoking women taking low-dose oral contraceptives. Contraception 1994;49(6):579–92.
1342. Thorogood M, Villard-Mackintosh L. Combined oral contraceptives: risks and benefits. Br Med Bull 1993;49(1):124–39.
1343. Machin SJ, Mackie IJ, Guillebaud J. Factor V Leiden mutation, venous thromboembolism and combined oral contraceptive usage. Br J Fam Planning 1995;21:13–4.
1344. Rosenberg L, Palmer JR, Sands MI, Grimes D, Bergman U, Daling J, Mills A. Modern oral contraceptives and cardiovascular disease. Am J Obstet Gynecol 1997;177(3):707–15.
1345. Vessey MP, Doll R, Fairbairn AS, Glober G. Postoperative thromboembolism and the use of oral contraceptives. BMJ 1970;3(715):123–6.
1346. Greene GR, Sartwell PE. Oral contraceptive use in patients with thromboembolism following surgery, trauma, or infection. Am J Public Health 1972;62(5):680–5.
1347. Whitehead EM, Whitehead MI. The pill, HRT and postoperative thromboembolism: cause for concern? Anaesthesia 1991;46(7):521–2.
1348. Beller FK. Cardiovascular system: coagulation, thrombosis, and contraceptive steroids is there a link? In: Goldzieher JW, Fotherby K, editors. Pharmacology of the Contraceptive Steroids. New York: Raven Press, 1994:309.
1349. Harlap S, Kost K, Forrest JD. Preventing Pregnancy, Protecting Health: a New Look at Birth Control Choices in the United StatesNew York: The Alan Guttmacher Institute;. 1991.
1350. Mant D, Villard-Mackintosh L, Vessey MP, Yeates D. Myocardial infarction and angina pectoris in young women. J Epidemiol Community Health 1987;41(3):215–9.
1351. Frederiksen H, Ravenholt RT. Thromboembolism, oral contraceptives, and cigarettes. Public Health Rep 1970;85(3):197–205.
1352. Rich-Edwards JW, Manson JE, Hennekens CH, Buring JE. The primary prevention of coronary heart disease in women. N Engl J Med 1995;332(26):1758–66.

1353. Croft P, Hannaford PC. Risk factors for acute myocardial infarction in women: evidence from the Royal College of General Practitioners' oral contraception study. BMJ 1989;298(6667):165–8.
1354. Jain AK. Cigarette smoking, use of oral contraceptives, and myocardial infarction. Am J Obstet Gynecol 1976;126(3):301–7.
1355. Beral V. Mortality among oral-contraceptive users. Royal College of General Practitioners' Oral Contraception Study. Lancet 1977;2(8041):727–31.
1356. Petitti DB, Wingerd J. Use of oral contraceptives, cigarette smoking, and risk of subarachnoid haemorrhage. Lancet 1978;2(8083):234–5.
1357. Bruni V, Rosati D, Bucciantini S, Verni A, Abbate R, Pinto S, Costanzo G, Costanzo M. Platelet and coagulation functions during triphasic oestrogen–progestogen treatment. Contraception 1986;33(1):39–46.
1358. Kjaeldgaard A, Larsson B. Long-term treatment with combined oral contraceptives and cigarette smoking associated with impaired activity of tissue plasminogen activator. Acta Obstet Gynecol Scand 1986;65(3):219–22.
1359. Von Hugo R, Briel RC, Schindler AE. Wirkung oraler Kontrazeptiva auf die Blutgerinnung bei rauchenden und nichtrauchenden Probandinnen. Aktuel Endokrinol Stoffwechsel 1989;10:6.
1360. Adachi T, Sakamoto S. Thromboembolism during hormone therapy in Japanese women. Sem Thromb Hemostas 2005;31:272–80.
1361. Inman WH, Vessey MP. Investigation of deaths from pulmonary, coronary, and cerebral thrombosis and embolism in women of child-bearing age. BMJ 1968;2(599):193–9.
1362. Arthes FG, Masi AT. Myocardial infarction in younger women. Associated clinical features and relationship to use of oral contraceptive drugs. Chest 1976;70(5):574–83.
1363. Koenig W, Gehring J, Mathes P. Orale Kontrazeptiva und Myokardinfarkt bei jungen Frauen. Herz Kreisl 1984;16:508.
1364. Zatti M. Contraccettivi orali: alterazioni delle variabili fisiologiche. C Ital Chim Clin 1983;8:249.
1365. Leone A, Lopez M. Rôle du tabac et de la contraception orale dans l'infarctus du myocarde de la femme: description d'un cas. [Role of tobacco and oral contraception in myocardial infarction in the female. Description of a case.] Pathologica 1984;76(1044):493–8.
1366. Gevers Leuven JA, Kluft C, Bertina RM, Hessel LW. Effects of two low-dose oral contraceptives on circulating components of the coagulation and fibrinolytic systems. J Lab Clin Med 1987;109(6):631–6.
1367. Poller L. Oral contraceptives, blood clotting and thrombosis. Br Med Bull 1978;34(2):151–6.
1368. Plowright C, Adam SA, Thorogood M, Beaumont V, Beaumont JL, Mann JI. Immunogenicity and the vascular risk of oral contraceptives. Br Heart J 1985;53(5):556–61.
1369. Syner FN, Moghissi KS, Agronow SJ. Study on the presence of abnormal proteins in the serum of oral contraceptive users. Fertil Steril 1983;40(2):202–9.
1370. Ernst E. Oral contraceptives, fibrinogen and cardiovascular risk. Atherosclerosis 1992;93(1–2):1–5.
1371. Ylikorkala O, Puolakka J, Viinikka L. Oestrogen containing oral contraceptives decrease prostacyclin production. Lancet 1981;1(8210):42.
1372. Gris JC, Jamin C, Benifla JL, Quere I, Madelenat P, Mares P. APC resistance and third-generation oral contraceptives: acquired resistance to activated protein C, oral contraceptives and the risk of thromboembolic disease. Hum Reprod 2001;16(1):3–8.
1373. Marque V, Alhenc-Gelas M, Plu-Bureau G, Oger E, Scarabin PY. The effects of transdermal and oral estrogen/progesterone regimens on free and total protein S in postmenopausal women. Thromb Haemost 2001;86(2):713–4.
1374. Rosing J, Middeldorp S, Curvers J, Christella M, Thomassen LG, Nicolaes GA, Meijers JC, Bouma BN, Buller HR, Prins MH, Tans G. Low-dose oral contraceptives and acquired resistance to activated protein C: a randomised cross-over study. Lancet 1999;354(9195):2036–40.
1375. Porter JB, Hunter JR, Danielson DA, Jick H, Stergachis A. Oral contraceptives and nonfatal vascular disease—recent experience. Obstet Gynecol 1982;59(3):299–302.
1376. Bailar JC 3rd, Byar DP. Estrogen treatment for cancer of the prostate. Early results with 3 doses of diethylstilbestrol and placebo. Cancer 1970;26(2):257–61.
1377. Badaracco MA, Vessey MP. Recurrence of venous thromboembolic disease and use of oral contraceptives. BMJ 1974;1(901):215–7.
1378. Jick H, Dinan B, Herman R, Rothman KJ. Myocardial infarction and other vascular diseases in young women. Role of estrogens and other factors. JAMA 1978;240(23):2548–52.
1379. Stadel BV. Oral contraceptives and cardiovascular disease (first of two parts). N Engl J Med 1981;305(11):612–8.
1380. Bottiger LE, Boman G, Eklund G, Westerholm B. Oral contraceptives and thromboembolic disease: effects of lowering oestrogen content. Lancet 1980;1(8178):1097–101.
1381. Jung-Hoffmann C, Kuhl H. Interaction with the pharmacokinetics of ethinylestradiol and progestogens contained in oral contraceptives. Contraception 1989;40(3):299–312.
1382. Guengerich FP. Mechanism-based inactivation of human liver microsomal cytochrome P-450 IIIA4 by gestodene. Chem Res Toxicol 1990;3(4):363–71.
1383. Rosing J, Tans G, Nicolaes GA, Thomassen MC, van Oerle R, van der Ploeg PM, Heijnen P, Hamulyak K, Hemker HC. Oral contraceptives and venous thrombosis: different sensitivities to activated protein C in women using second- and third-generation oral contraceptives. Br J Haematol 1997;97(1):233–8.
1384. Mellemkjaer L, Sorensen HT, Dreyer L, Olsen J, Olsen JH. Admission for and mortality from primary venous thromboembolism in women of fertile age in Denmark, 1977–95. BMJ 1999;319(7213):820–1.
1385. Vandenbroucke JP, Bloemenkamp KW, Helmerhorst FM, Rosendaal FR. Mortality from venous thromboembolism and myocardial infarction in young women in the Netherlands. Lancet 1996;348(9024):401–2.
1386. Herings RM, Urquhart J, Leufkens HG. Venous thromboembolism among new users of different oral contraceptives. Lancet 1999;354(9173):127–8.
1387. Hannaford PC, Kay CR. The risk of serious illness among oral contraceptive users: evidence from the RCGP's oral contraceptive study. Br J Gen Pract 1998;48(435):1657–62.
1388. Bruni V, Croxatto H, De La Cruz J, Dhont M, Durlot F, Fernandes MT, Andrade RP, Weisberg E, Rhoa M; Gestodene Study Group. A comparison of cycle control and effect on well-being of monophasic gestodene-, triphasic gestodene- and monophasic

desogestrel-containing oral contraceptives. Gynecol Endocrinol 2000;14(2):90–8.

1389. Marks LV. Sexual chemistry. In: A History of the Contraceptive Pill. New Haven: Yale University Press, 2001:138–57.
1390. Sartwell PE, et al. Oral contraceptives and relative risk of death from venous and pulmonary thromboembolism in the United States; an epidemiologic case-control study. Am J Epidemiol 1969;90:365.
1391. Royal College of General Practitioners. Oral contraception and thrombo-embolic disease. J R Coll Gen Pract 1967;13(3):267–79.
1392. Vessey MP, Doll R. Investigation of relation between use of oral contraceptives and thromboembolic disease. BMJ 1968;2(599):199–205.
1393. Vessey MP, Doll R. Investigation of relation between use of oral contraceptives and thromboembolic disease. A further report. BMJ 1969;2(658):651–7.
1394. Creatsas G, Koliopoulos C, Mastorakos G. Combined oral contraceptive treatment of adolescent girls with polycystic ovary syndrome. Lipid profile. Ann NY Acad Sci 2000;900:245–52.
1395. Jick H, Kaye JA, Vasilakis-Scaramozza C, Jick SS. Risk of venous thromboembolism among users of third generation oral contraceptives compared with users of oral contraceptives with levonorgestrel before and after 1995: cohort and case-control analysis. BMJ 2000;321(7270):1190–5.
1396. Bloemenkamp KW, Rosendaal FR, Helmerhorst FM, Buller HR, Vandenbroucke JP. Enhancement by factor V Leiden mutation of risk of deep-vein thrombosis associated with oral contraceptives containing a third-generation progestagen. Lancet 1995;346(8990):1593–6.
1397. Kemmeren JM, Algra A, Grobbee DE. Third generation oral contraceptives and risk of venous thrombosis: meta-analysis. BMJ 2001;323(7305):131–4.
1398. Bonnar J, Daly L, Carroll E. Blood coagulation with a combination pill containing gestodene and ethinyl estradiol. Int J Fertil 1987;32(Suppl):21–8.
1399. Nicolaes GA, Thomassen MC, Tans G, Rosing J, Hemker HC. Effect of activated protein C on thrombin generation and on the thrombin potential in plasma of normal and APC-resistant individuals. Blood Coagul Fibrinolysis 1997;8(1):28–38.
1400. Meijers JC, Middeldorp S, Tekelenburg W, van den Ende AE, Tans G, Prins MH, Rosing J, Buller HR, Bouma BN. Increased fibrinolytic activity during use of oral contraceptives is counteracted by an enhanced factor XI-independent down regulation of fibrinolysis: a randomized cross-over study of two low-dose oral contraceptives. Thromb Haemost 2000;84(1):9–14.
1401. High Court. XYZ and others (Claimants) versus (1) Schering Health Care Limited, (2) Organon Laboratories Limited and (3) John Wyeth & Brother Limited. Judgement by the Hon. Mr Justice Mackay. London, 29 July 2002. Case No: 0002638. Neutral Citation No: (2002) EWHC 1420 (QB)
1402. Hedenmalm K, Samuelsson E, Spigset O. Pulmonary embolism associated with combined oral contraceptives: reporting incidences and potential risk factors for a fatal outcome. Acta Obstet Gynecol Scand 2004;83:576–85.
1403. Davies GC, Newton JR. Subdermal contraceptive implants—a review: with special reference to Norplant. Br J Fam Plann 1991;17:4.
1404. Penotti M, Sironi L, Cannata L, Vigano P, Casini A, Gabrielli L, Vignali M. Effects of androgen supplementation of hormone replacement therapy on the vascular reactivity of cerebral arteries. Fertil Steril 2001;76(2):235–40.
1405. Cameron DR, Braunstein GD. Androgen replacement therapy in women. Fertil Steril 2004;82(2):273–89.
1406. Barrett-Connor E. Efficacy and safety of estrogen/androgen therapy. Menopausal symptoms, bone, and cardiovascular parameters. J Reprod Med 1998;43(Suppl 8):746–52.
1407. Kaunitz AM. The role of androgens in menopausal hormonal replacement. Endocrinol Metab Clin North Am 1997;26(2):391–7.
1408. Rossouw JE. Hormone replacement therapy and cardiovascular disease. Curr Opin Lipidol 1999;10(5):429–34.
1409. Whiteman MK, Cui Y, Flaws JA, Espeland M, Bush TL. Low fibrinogen level: A predisposing factor for venous thromboembolic events with hormone replacement therapy. Am J Hematol 1999;61(4):271–3.
1410. Hoibraaten E, Abdelnoor M, Sandset PM. Hormone replacement therapy with estradiol and risk of venous thromboembolism—a population-based case-control study. Thromb Haemost 1999;82(4):1218–21.
1411. Cahill M, O'Toole L, Acheson RW. Hormone replacement therapy and retinal vein occlusion. Eye 1999;13(Pt 6):798–800.
1412. Haseroth K, Seyffart K, Wehling M, Christ M. Effects of progestin–estrogen replacement therapy on QT-dispersion in postmenopausal women. Int J Cardiol 2000;75(2–3):161–5.
1413. Wenger NK. Hormonal and nonhormonal therapies for the postmenopausal woman: what is the evidence for cardioprotection? Am J Geriatr Cardiol 2000;9(4):204–9.
1414. Petitti D. Hormone replacement therapy and coronary heart disease: four lessons. Int J Epidemiol 2004;33:461–3.
1415. Lawlor DA, Smith GD, Ebrahim S. The hormone replacement–coronary heart disease conundrum: is this the death of observational epidemiology? Int J Epdemiol 2004;33:464–7.
1416. Mackey R.H, Kuller L.H, Sutton-Tyrrell K, Evans R.W, Holubkov R, Matthews K.A. Hormone therapy, lipoprotein subclasses, and coronary calcification: The Healthy Women Study. Arch Int Med 2005;165:510–5.
1417. Davidson MH, Maki KC, Marx P, Maki AC, Cyrowski MS, Nanavati N, Arce JC. Effects of continuous estrogen and estrogen-progestin replacement regimens on cardiovascular risk markers in postmenopausal women. Arch Intern Med 2000;160(21):3315–25.
1418. Heckbert SR, Kaplan RC, Weiss NS, Psaty BM, Lin D, Furberg CD, Starr JR, Anderson GD, LaCroix AZ. Risk of recurrent coronary events in relation to use and recent initiation of postmenopausal hormone therapy. Arch Intern Med 2001;161(14):1709–13.
1419. Alexander KP, Newby LK, Hellkamp AS, Harrington RA, Peterson ED, Kopecky S, Langer A, O'Gara P, O'Connor CM, Daly RN, Califf RM, Khan S, Fuster V. Initiation of hormone replacement therapy after acute myocardial infarction is associated with more cardiac events during follow-up. J Am Coll Cardiol 2001;38(1):1–7.
1420. Grady D, Hulley SB. Postmenopausal hormones and heart disease. J Am Coll Cardiol 2001;38(1):8–10.
1421. Carnethon MR, Anthony MS, Cascio WE, Folsom AR, Rautaharju PM, Liao D, Evans GW, Heiss G. A prospective evaluation of the risk of QT prolongation with

hormone replacement therapy: the atherosclerosis risk in communities study. Ann Epidemiol 2003;13:530–6.

1422. Gokce M, Karahan B, Yilmaz R, Orem C, Erdol C, Ozdemir S. Long term effects of hormone replacement therapy on heart rate variability, QT interval, QT dispersion and frequencies of arrhythmia. Int J Cardiol 2005;99:373–9.

1423. Grady D, Wenger NK, Herrington D, Khan S, Furberg C, Hunninghake D, Vittinghoff E, Hulley S. Postmenopausal hormone therapy increases risk for venous thromboembolic disease. The Heart and Estrogen/progestin Replacement Study. Ann Intern Med 2000;132(9):689–96.

1424. Schooley RT, Wagley PF, Lietman PS. Edema associated with ibuprofen therapy. JAMA 1977;237(16):1716–7.

1425. Patel J, Marks KA, Roberts I, Azzopardi D, Edwards AD. Ibuprofen treatment of patent ductus arteriosus. Lancet 1995;346(8969):255.

1426. Ingle JN, Krook JE, Mailliard JA, Hartmann LC, Wieand HS. Evaluation of ifosfamide plus mesna as first-line chemotherapy in women with metastatic breast cancer. Am J Clin Oncol 1995;18(6):498–501.

1427. Marroun I, Fialip J, Deleveaux I, Andre M, Lamaison D, Cabane J, Piette JC, Eschalier A, Aumaitre O. Infarctus du myocarde sous iloprost chez un patient atteint de sclérodermie. [Myocardial infarction and iloprost in a patient with scleroderma.] Therapie 2001;56(5):630–2.

1428. Kerkelä R, Grazette L, Yacobi R, Iliescu C, Patten R, Beahm C, Walters B, Shevtsov S, Pesant S, Clubb FJ, Rosenzweig A, Salomon RN, Van Etten RA, Alroy J, Durand JB, Force T. Cardiotoxicity of the cancer therapeutic agent imatinib mesylate. Nat Med 2006;12:908–16.

1429. Rosti G, Martinelli G, Baccarani M. In reply to "Cardiotoxicity of the cancer therapeutic agent imatinib mesylate" Nat Med 2007;13:15–16.

1430. Atallah E, Kantarjian H, Cortes J. In reply to "Cardiotoxicity of the cancer therapeutic agent imatinib mesylate". Nat Med 2007;13:14–16.

1431. Hatfield A, Owen S, Pilot PR. In reply to "Cardiotoxicity of the cancer therapeutic agent imatinib mesylate". Nat Med 2007;13:13–16.

1432. Stangel M, Hartung HP, Marx P, Gold R. Intravenous immunoglobulin treatment of neurological autoimmune diseases. J Neurol Sci 1998;153(2):203–14.

1433. Katz KA, Hivnor CM, Geist DE, Shapiro M, Ming ME, Werth VP. Stroke and deep venous thrombosis complicating intravenous immunoglobulin infusions. Arch Dermatol 2003;139:991–3.

1434. Mohaupt M, Krueger T, Girardi V, Mansouri Taleghani B. Stroke after high-dose intravenous immunoglobulin. Transfus Med Hemother 2003;30:186–8.

1435. Okuda D, Flaster M, Frey J, Sivakumar K. Arterial thrombosis induced by IVIg and its treatment with tPA. Neurology 2003;60:1825–6.

1436. Butler KS, Zeitlin DS. Pulmonary embolism associated with intravenous immunoglobulin therapy. Ann Pharmacother 2003;37:1530.

1437. Brown HC, Ballas ZK. Acute thromboembolic events associated with intravenous immunoglobulin infusion in antibody-deficient patients. J Allergy Clin Immunol 2003;112:797–9.

1438. Dalakas MC. High-dose intravenous immunoglobulin in inflammatory myopathies: experience based on controlled clinical trials. Neurol Sci 2003;24 Suppl 4:S256–9.

1439. Knezevic-Maramica I, Kruskall MS. Intravenous immune globulins: an update for clinicians. Transfusion 2003;43:1460–80.

1440. Caress JB, Cartwright MS, Donofrio PD, Peacock JE, Jr. The clinical features of 16 cases of stroke associated with administration of IVIg. Neurology 2003;60:1822–4.

1441. Reinhart WH, Berchtold PE. Effect of high-dose intravenous immunoglobulin therapy on blood rheology. Lancet 1992;339(8794):662–4.

1442. Machkhas H, Harati Y. Side effects of immunosuppressant therapies used in neurology. Neurol Clin 1998;16(1):171–88.

1443. Nishikawa M, Ichiyama T, Hasegawa M, Kawasaki K, Matsubara T, Furukawa S. Safety from thromboembolism using intravenous immunoglobulin therapy in Kawasaki disease. Study of whole-blood viscosity. Pediatr Int 2003;45:156–8.

1444. Steinberger BA, Ford SM, Coleman TA. Intravenous immunoglobulin therapy results in post-infusional hyperproteinemia, increased serum viscosity, and pseudohyponatremia. Am J Hematol 2003;73:97–100.

1445. Ben-Ami R, Barshtein G, Mardi T, Deutch V, Elkayam O, Yedgar S, Berliner S. A synergistic effect of albumin and fibrinogen on immunoglobulin-induced red blood cell aggregation. Am J Physiol Heart Circ Physiol 2003;285:H2663–9.

1446. Bierling P, Godeau B. Intravenous immunoglobulin and autoimmune thrombocytopenic purpura: 22 years on. Vox Sanguinis 2004;86(1):8–14.

1447. Hefer D, Jaloudi M. Thromboembolic events as an emerging adverse effect during high-dose intravenous immunoglobulin therapy in elderly patients: a case report and discussion of the relevant literature. Ann Hematol 2005;84(6):411–5.

1448. Hommes OR, Sorensen PS, Fazekas F, Enriquez MM, Koelmel HW, Fernandez O, Pozzilli C, O'Connor P. Intravenous immunoglobulin in secondary progressive multiple sclerosis: randomised placebo-controlled trial. Lancet 2004;364(9440):1149–56.

1449. Vucic S, Chong PS, Dawson KT, Cudkowicz M, Cros D. Thromboembolic complications of intravenous immunoglobulin treatment. Eur Neurol 2004;52(3):141–4.

1450. Woodruff RK, Grigg AP, Firkin FC, Smith IL. Fatal thrombotic events during treatment of autoimmune thrombocytopenia with intravenous immunoglobulin in elderly patients. Lancet 1986;2(8500):217–8.

1451. Alliot C, Rapin JP, Besson M, Bedjaoui F, Messouak D. Pulmonary embolism after intravenous immunoglobulin. J R Soc Med 2001;94(4):187–8.

1452. Sherer Y, Levy Y, Langevitz P, Rauova L, Fabrizzi F, Shoenfeld Y. Adverse effects of intravenous immunoglobulin therapy in 56 patients with autoimmune diseases. Pharmacology 2001;62(3):133–7.

1453. Wiles CM, Brown P, Chapel H, Guerrini R, Hughes RA, Martin TD, McCrone P, Newsom-Davis J, Palace J, Rees JH, Rose MR, Scolding N, Webster AD. Intravenous immunoglobulin in neurological disease: a specialist review. J Neurol Neurosurg Psychiatry 2002;72(4):440–8.

1454. Crouch ED, Watson LE. Intravenous immunoglobulin-related acute coronary syndrome and coronary angiography in idiopathic thrombocytopenic purpura–a case report and literature review. Angiology 2002;53(1):113–7.

1455. Emerson GG, Herndon CN, Sreih AG. Thrombotic complications after intravenous immunoglobulin therapy in two patients. Pharmacotherapy 2002;22(12):1638–41.

1456. Byrne NP, Henry JC, Herrmann DN, Abdelhalim AN, Shrier DA, Francis CW, Powers JM. Neuropathologic findings in a Guillain-Barré patient with strokes after IVIg therapy. Neurology 2002;59(3):458–61.

1457. Godeau B, Chevret S, Varet B, Lefrère F, Zini JM, Bassompierre F, Chèze S, Legouffe E, Hulin C, Grange MJ, Fain O, Bierling P; French ATIP Study Group. Intravenous immunoglobulin or high-dose methylprednisolone, with or without oral prednisone, for adults with untreated severe autoimmune thrombocytopenic purpura: a randomised, multicentre trial. Lancet 2002;359(9300):23–9.

1458. Tristani-Firouzi P, Petersen MJ, Saffle JR, Morris SE, Zone JJ. Treatment of toxic epidermal necrolysis with intravenous immunoglobulin in children. J Am Acad Dermatol 2002;47(4):548–52.

1459. Feuillet L, Guedj E, Laksiri N, Philip E, Habib G, Pelletier J, Ali Cherif A. Deep vein thrombosis after intravenous immunoglobulins associated with methylprednisolone. Thromb Haemost 2004;92(3):662–5.

1460. Geller JL, Hackner D. Diffuse venous thromboemboli associated with IVIg therapy in the treatment of streptococcal toxic shock syndrome: case report and review. Ann Hematol 2005;84(9):601–4.

1461. Keohane SG, Kavanagh GM, Gordon PM, Hunter JAA. Transient hypertension during infusion of intravenous gammaglobulin for dermatomyositis. J Dermatol Treat 1999;10(4):289–92.

1462. Kroez M, Kanzy EJ, Gronski P, Dickneite G. Hypotension with intravenous immunoglobulin therapy: importance of pH and dimer formation. Biologicals 2003;31:277–86.

1463. Evangelou N, Littlewood T, Anslow P, Chapel H. Transverse sinus thrombosis and IVIg treatment: a case report and discussion of risk-benefit assessment for immunoglobulin treatment. J Clin Pathol 2003;56:308–9.

1464. Dahl MV, Bridges AG. Intravenous immune globulin: fighting antibodies with antibodies. J Am Acad Dermatol 2001;45(5):775–83.

1465. Lacroix P, Portefaix O, Boucher M, Ramiandrisoa H, Dumas M, Ravon R, Christides C, Laskar M. Condition de survenue des accidents hémorragiques intracraniens des antivitamines K. [The causes of intracranial hemorrhagic complications induced by antivitamins K.] Arch Mal Coeur Vaiss 1994;87(12):1715–9.

1466. Giner V, Fernandez C, Esteban MJ, Galindo MJ, Forner MJ, Guix J, Redon J. Reversible posterior leukoencephalopathy secondary to indinavir-induced hypertensive crisis: a case report. Am J Hypertens 2002;15(5):465–7.

1467. Cattelan AM, Trevenzoli M, Sasset L, Rinaldi L, Balasso V, Cadrobbi P. Indinavir and systemic hypertension. AIDS 2001;15(6):805–7.

1468. Golding D. Angina and indomethacin. BMJ 1970;4(735):622.

1469. Wennmalm A, Carlsson I, Edlund A, Eriksson S, Kaijser L, Nowak J. Central and peripheral haemodynamic effects of non-steroidal anti-inflammatory drugs in man. Arch Toxicol Suppl 1984;7:350–9.

1470. Dzau VJ, Packer M, Lilly LS, Swartz SL, Hollenberg NK, Williams GH. Prostaglandins in severe congestive heart failure. Relation to activation of the renin–angiotensin system and hyponatremia. N Engl J Med 1984;310(6):347–52.

1471. Weisman MH. What are the risks of biologic therapy in rheumatoid arthritis? An update on safety. J Rheumatol Suppl 2002;65:33–8.

1472. Chung ES, Packer M, Lo KH, Fasanmade AA, Willerson JT. Randomized, double-blind, placebo-controlled, pilot trial of infliximab, a chimeric monoclonal antibody to tumor necrosis factor-alfa, in patients with moderate-to-severe heart failure: results of the anti-TNF Therapy Against Congestive Heart Failure (ATTACH) trial. Circulation 2003;107:3133–40.

1473. Kwon HJ, Cote TR, Cuffe MS, Kramer JM, Braun MM. Case reports of heart failure after therapy with a tumor necrosis factor antagonist. Ann Intern Med 2003;138:807–11.

1474. Eklund KK, Peltomaa R, Leirisalo-Repo M. Occurrence of pulmonary thromboembolism during infliximab therapy. Clin Exp Rheumatol 2003;21:679.

1475. Puli SR, Benage DD. Retinal vein thrombosis after infliximab (Remicade) treatment for Crohn's disease. Am J Gastroenterol 2003;98:939–40.

1476. Grange L, Nissen MJ, Garambois K, Dumolard A, Duc C, Gaudin P, Juvin R. Infliximab-induced cerebral thrombophlebitis. Rheumatology 2005;44:260–1.

1477. Ryan BM, Romberg M, Wolters F, Stockbrugger RW. Extensive forearm deep venous thrombosis following a severe infliximab infusion reaction. Eur J Gastroenterol Hepatol 2004;16:941–2.

1478. Settergren M, Tornvall P. Does TNF-alpha blockade cause plaque rupture? Atherosclerosis 2004;173:149.

1479. Shoukeir H, Awaida R, Gupta K, Kodsi R, Oswaldo B, Scileppi T, Tenner S. Myocardial infarction precipitated by infliximab infusion: report of case. Am J Gastroenterol 2004;Suppl:S165.

1480. Sobkeng Goufack E, Mammou S, Scotto B, De Muret A, Maakaroun A, Socie G, Bacq Y. Thrombose des veines hépatiques au cours d'un traitement par infliximab (Remicade®) révélant une hémoglobinurie paroxystique nocturne. Gastroenterol Clin Biol 2004;28:596–9.

1481. Cambien B, Bergmeier W, Saffaripour S, Mitchell HA, Wagner DD. Antithrombotic activity of TNF-alpha. J Clin Invest 2003;112:1589–96.

1482. Gonzalez-Juanatey C, Testa A, Garcia-Castelo A, Garcia-Porrua C, Llorca J, Gonzalez-Gay MA. Active but transient improvement of endothelial function in rheumatoid arthritis patients undergoing long-term treatment with anti-tumor necrosis factor alpha antibody. Arthritis Rheum 2004;51:447–50.

1483. Kobashigawa JA, Warner-Stevenson L, Johnson BL, Moriguchi JD, Kawata N, Drinkwater DC, Laks H. Influenza vaccine does not cause rejection after cardiac transplantation. Transplant Proc 1993;25(4):2738–9.

1484. Medearis DN Jr, Neill CA, Markowitz M. Influenza and cardiopulmonary disease. II. Med Concepts Cardiovasc Dis 1963;32:813–6.

1485. Zanettini MT, Zanettini JO, Zanettini JP. Pericarditis. Series of 84 consecutive cases. Arquivos Brasileiros de Cardiologia 2004;82(4):360–9.

1486. Muis MJ, Bots ML, Bilo HJG, Hoogma RPLM, Hoekstra JBL, Grobbee DE, Stolk RP. High cumulative insulin exposure: a risk factor of atherosclerosis in type 1 diabetes? Atherosclerosis 2005;181:185–92.

1487. Gerstein HC, Rosenstock J. Insulin therapy in people who have dysglycemia and type 2 diabetes mellitus: can it offer both cardiovascular protection and beta-cell preservation? Endocrinol Metab Clin N Am 2005;34:137–54.

1488. Vial T, Descotes J. Clinical toxicity of the interferons. Drug Saf 1994;10(2):115–50.
1489. Sonnenblick M, Rosin A. Cardiotoxicity of interferon. A review of 44 cases. Chest 1991;99(3):557–61.
1490. Kruit WH, Punt KJ, Goey SH, de Mulder PH, van Hoogenhuyze DC, Henzen-Logmans SC, Stoter G. Cardiotoxicity as a dose-limiting factor in a schedule of high dose bolus therapy with interleukin-2 and alpha-interferon. An unexpectedly frequent complication. Cancer 1994;74(10):2850–6.
1491. Angulo MP, Navajas A, Galdeano JM, Astigarraga I, Fernandez-Teijeiro A. Reversible cardiomyopathy secondary to alpha-interferon in an infant. Pediatr Cardiol 1999;20(4):293–4.
1492. Fattovich G, Giustina G, Favarato S, Ruol A. A survey of adverse events in 11,241 patients with chronic viral hepatitis treated with alfa interferon. J Hepatol 1996;24(1):38–47.
1493. Kadayifci A, Aytemir K, Arslan M, Aksoyek S, Sivri B, Kabakci G. Interferon-alpha does not cause significant cardiac dysfunction in patients with chronic active hepatitis. Liver 1997;17(2):99–102.
1494. Sartori M, Andorno S, La Terra G, Pozzoli G, Rudoni M, Sacchetti GM, Inglese E, Aglietta M. Assessment of interferon cardiotoxicity with quantitative radionuclide angiocardiography. Eur J Clin Invest 1995;25(1):68–70.
1495. Kuwata A, Ohashi M, Sugiyama M, Ueda R, Dohi Y. A case of reversible dilated cardiomyopathy after alpha-interferon therapy in a patient with renal cell carcinoma. Am J Med Sci 2002;324(6):331–4.
1496. Colivicchi F, Magnanimi S, Sebastiani F, Silvestri R, Magnanimi R. Incidence of electrocardiographic abnormalities during treatment with human leukocyte interferon-alfa in patients with chronic hepatitis C but without pre-existing cardiovascular disease. Curr Ther Res Clin Exp 1998;59:692–6.
1497. Parrens E, Chevalier JM, Rougier M, Douard H, Labbe L, Quiniou G, Broustet A, Broustet JP. Apparition d'un bloc auriculo-ventriculaire du troisième degré sous interféron alpha: à propos d'un cas. [Third degree atrio-ventricular block induced by interferon alpha. Report of a case.] Arch Mal Coeur Vaiss 1999;92(1):53–6.
1498. Gupta SK, Glue P, Jacobs S, Belle D, Affrime M. Single-dose pharmacokinetics and tolerability of pegylated interferon-alfa2b in young and elderly healthy subjects. Br J Clin Pharmacol 2003;56:131–4.
1499. Al-Zahrani H, Gupta V, Minden MD, Messner HA, Lipton JH. Vascular events associated with alfa interferon therapy. Leuk Lymphoma 2003;44:471–5.
1500. Wisniewski B, Denis J, Fischer D, Labayle D. Péricardite induite par l'interféron alfa au cours d'une hépatite chronique C. Gastroentérol Clin Biol 2004;28:315–16.
1501. Gressens B, Gohy P. Pericarditis due to interferon-alpha therapy during treatment for chronic hepatitis C. Acta Gastroenterol Belg 2004;67:301–2.
1502. Creutzig A, Caspary L, Freund M. The Raynaud phenomenon and interferon therapy. Ann Intern Med 1996;125(5):423.
1503. Schapira D, Nahir AM, Hadad N. Interferon-induced Raynaud's syndrome. Semin Arthritis Rheum 2002;32(3):157–62.
1504. Iorio R, Spagnuolo MI, Sepe A, Zoccali S, Alessio M, Vegnente A. Severe Raynaud's phenomenon with chronic hepatis C disease treated with interferon. Pediatr Infect Dis J 2003;22(2):195–7.
1505. Cid MC, Hernandez-Rodriguez J, Robert J, del Rio A, Casademont J, Coll-Vinent B, Grau JM, Kleinman HK, Urbano-Marquez A, Cardellach F. Interferon-alpha may exacerbate cryoblobulinemia-related ischemic manifestations: an adverse effect potentially related to its anti-angiogenic activity. Arthritis Rheum 1999;42(5):1051–5.
1506. Mirro J Jr, Kalwinsky D, Whisnant J, Weck P, Chesney C, Murphy S. Coagulopathy induced by continuous infusion of high doses of human lymphoblastoid interferon. Cancer Treat Rep 1985;69(3):315–7.
1507. Durand JM, Quiles N, Kaplanski G, Soubeyrand J. Thrombosis and recombinant interferon-alpha. Am J Med 1993;95(1):115–6.
1508. Becker JC, Winkler B, Klingert S, Brocker EB. Antiphospholipid syndrome associated with immunotherapy for patients with melanoma. Cancer 1994;73(6):1621–4.
1509. Schmidt S, Hertfelder HJ, von Spiegel T, Hering R, Harzheim M, Lassmann H, Deckert-Schluter M, Schlegel U. Lethal capillary leak syndrome after a single administration of interferon beta-1b. Neurology 1999;53(1):220–2.
1510. Friess GG, Brown TD, Wrenn RC. Cardiovascular rhythm effects of gamma recombinant DNA interferon. Invest New Drugs 1989;7(2–3):275–80.
1511. Sonnenblick M, Rosin A. Cardiotoxicity of interferon. A review of 44 cases. Chest 1991;99(3):557–61.
1512. Yamamoto N, Nishigaki K, Ban Y, Kawada Y. Coronary vasospasm after interferon administration. Br J Urol 1998;81(6):916–7.
1513. Vial T, Descotes J. Clinical toxicity of cytokines used as haemopoietic growth factors. Drug Saf 1995;13(6):371–406.
1514. Eisner RM, Husain A, Clark JI. Case report and brief review: IL-2-induced myocarditis. Cancer Invest 2004;22:401–4.
1515. Biesma B, Willemse PH, Mulder NH, Sleijfer DT, Gietema JA, Mull R, Limburg PC, Bouma J, Vellenga E, de Vries EG. Effects of interleukin-3 after chemotherapy for advanced ovarian cancer. Blood 1992;80(5):1141–8.
1516. Gillio AP, Faulkner LB, Alter BP, Reilly L, Klafter R, Heller G, Young DC, Lipton JM, Moore MA, O'Reilly RJ. Treatment of Diamond–Blackfan anemia with recombinant human interleukin-3. Blood 1993;82(3):744–51.
1517. Prendiville J, Thatcher N, Lind M, McIntosh R, Ghosh A, Stern P, Crowther D. Recombinant human interleukin-4 (rhu IL-4) administered by the intravenous and subcutaneous routes in patients with advanced cancer—a phase I toxicity study and pharmacokinetic analysis. Eur J Cancer 1993;29A(12):1700–7.
1518. Trehu EG, Isner JM, Mier JW, Karp DD, Atkins MB. Possible myocardial toxicity associated with interleukin-4 therapy. J Immunother 1993;14(4):348–51.
1519. Tsimberidou AM, Giles FJ, Khouri I, Bueso-Ramos C, Pilat S, Thomas DA, Cortes J, Kurzrock R. Low-dose interleukin-11 in patients with bone marrow failure: update of the M D Anderson Cancer Center experience. Ann Oncol 2005;16(1):139–45.
1520. Zeman RK. Disseminated intravascular coagulation following intravenous pyelography. Invest Radiol 1977;12(2):203–4.
1521. Johnson LW, Krone R. Cardiac catheterization 1991: a report of the Registry of the Society for Cardiac

Angiography and Interventions (SCA&I). Cathet Cardiovasc Diagn 1993;28(3):219–20.

1522. Gross-Fengels W, Beyer D, Fischbach R, Lanfermann H. Akute Nebenwirkungen und Komplikationen der zentralvenösen DSA. Ergebnisse bei 2600 Untersuchungen. [Acute adverse effects and complications of central venous digital subtraction angiography (DSA). Results of 2,600 studies.] Med Klin (Munich) 1991;86(11):561–5.

1523. Flinck A, Gottfridsson B. Experiences with iohexol and iodixanol during cardioangiography in an unselected patient population. Int J Cardiol 2001;80(2–3):143–51.

1524. Roriz R, de Gevigney G, Finet G, Nantois-Collet C, Borch KW, Amiel M, Beaune J. Comparaison de l'iodixanol (Visipaque) et de l'ioxaglate (Hexabrix) en coronaro-ventriculographie: une étude randomisée en double aveugle. [Comparison of iodixanol (Visipaque) and ioxaglate (Hexabrix) in coronary angiography and ventriculography: a double-blind randomized study.] J Radiol 1999;80(7):727–32.

1525. Spitzer S, Munster W, Sternitzky R, Bach R, Jung F. Influence of iodixanol-270 and iopentol-150 on the microcirculation in man: influence of viscosity on capillary perfusion. Clin Hemorheol Microcirc 1999;20(1):49–55.

1526. Burton PR, Jarmolowski E, Raineri F, Buist MD, Wriedt CH. A severe, late reaction to radiological contrast media mimicking a sepsis syndrome. Australas Radiol 1999;43(3):360–2.

1527. Mitsumori M, Hayakawa K, Abe M. ECG changes during cerebral angiography; a comparison of low osmolality contrast media. Eur J Radiol 1991;13(1):55–8.

1528. Wishart DL. Complications in vertebral angiography as compared to non-vertebral cerebral angiography in 447 studies. Am J Roentgenol Radium Ther Nucl Med 1971;113(3):527–37.

1529. Flinck A, Selin K. Vectorcardiographic changes during cardioangiography with iodixanol and ioxaglate. Int J Cardiol 2000;76(2–3):173–80.

1530. Matthai WH Jr, Hirshfeld JW Jr. Choice of contrast agents for cardiac angiography: review and recommendations based on clinically important distinctions. Cathet Cardiovasc Diagn 1991;22(4):278–89.

1531. Baath L, Almen T, Oksendal A. Cardiac effects from addition of sodium ions to nonionic contrast media for coronary arteriography. An investigation of the isolated rabbit heart. Invest Radiol 1990;25(Suppl 1):S137–40.

1532. Snyder CF, Formanek A, Frech RS, Amplatz K. The role of sodium in promoting ventricular arrhythmia during selective coronary arteriography. Am J Roentgenol Radium Ther Nucl Med 1971;113(3):567–71.

1533. Druck MN, Johnstone DE, Staniloff H, McLaughlin PR. Coronary artery spasm as a manifestation of anaphylactoid reaction to iodinated contrast material. Can Med Assoc J 1981;125(10):1133–5.

1534. Altun A, Ozbay G. Effects of ionic versus non-ionic contrast agents on dispersion of ventricular repolarization. Turk Kardiyol Dernegi Ars 1998;26:362–7.

1535. Wiggins J, Beckmann R, Weinmann HJ, Lehr R. Electrocardiographic effects of diagnostic imaging agents. Acad Radiol 2002;9(Suppl 2):S444–6.

1536. Lesiak M, Grajek S, Pyda M, Skorupski W, Mitowski P, Cieslinski A. Percutaneous transluminal coronary angioplasty: the influence of non-ionic and high osmolar ionic contrast media on the results and complication of the procedure. Kardiol Pol 1999;50:311–21.

1537. Fukuda N, Shinohara K, Shimohakamada Y, Cochran ST, Bomyea K. Fatal cardiac arrest during infusion of nonionic contrast media in a patient with essential thrombocythemia. Am J Roentgenol 2002;178(3):765–6.

1538. Tresch DD, Sweet DL, Keelan MH Jr, Lange RL. Acute iodide intoxication with cardiac irritability. Arch Intern Med 1974;134(4):760–2.

1539. Rooyakkers TM, Stroes ES, Kooistra MP, van Faassen EE, Hider RC, Rabelink TJ, Marx JJ. Ferric saccharate induces oxygen radical stress and endothelial dysfunction in vivo. Eur J Clin Invest 2002;32(Suppl 1):9–16.

1540. Levy WJ. Clinical anaesthesia with isoflurane. A review of the multicentre study. Br J Anaesth 1984;56(Suppl 1):S101–12.

1541. Rodrigo MR, Moles TM, Lee PK. Comparison of the incidence and nature of cardiac arrhythmias occurring during isoflurane or halothane anaesthesia. Studies during dental surgery. Br J Anaesth 1986;58(4):394–400.

1542. Kobayashi Y. [Pressor responses to inhalation of isoflurane during induction of anesthesia and subsequent tracheal intubation.] Masui 2005;54(8):869–74.

1543. Slogoff S, Keats AS. Randomized trial of primary anesthetic agents on outcome of coronary artery bypass operations. Anesthesiology 1989;70(2):179–88.

1544. Inoue K, Reichelt W, el-Banayosy A, Minami K, Dallmann G, Hartmann N, Windeler J. Does isoflurane lead to a higher incidence of myocardial infarction and perioperative death than enflurane in coronary artery surgery? A clinical study of 1178 patients. Anesth Analg 1990;71(5):469–74.

1545. Nelson MR, Smith D, Erskine D, Gazzard BG. Ventricular fibrillation secondary to itraconazole induced hypokalaemia. J Infect 1993;26(3):348.

1546. Ahmad SR, Singer SJ, Leissa BG. Congestive heart failure associated with itraconazole. Lancet 2001;357(9270):1766–7.

1547. De Sole G, Awadzi K, Remme J, Dadzie KY, Ba O, Giese J, Karam M, Keita FM, Opoku NO. A community trial of ivermectin in the onchocerciasis focus of Asubende, Ghana. II. Adverse reactions. Trop Med Parasitol 1989;40(3):375–82.

1548. Zsigmond EK, Kothary SP, Kumar SM, Kelsch RC. Counteraction of circulatory side effects of ketamine by pretreatment with diazepam. Clin Ther 1980;3(1):28–32.

1549. Cabbabe EB, Behbahani PM. Cardiovascular reactions associated with the use of ketamine and epinephrine in plastic surgery. Ann Plast Surg 1985;15(1):50–6.

1550. Waxman K, Shoemaker WC, Lippmann M. Cardiovascular effects of anesthetic induction with ketamine. Anesth Analg 1980;59(5):355–8.

1551. Tarnow J, Hess W. Pulmonale Hypertonie und Lungenödem nach Ketamin. [Pulmonary hypertension and pulmonary edema caused by intravenous ketamine.] Anaesthesist 1978;27(10):486–7.

1552. Taittonen MT, Kirvela OA, Aantaa R, Kanto JH. The effect of clonidine or midazolam premedication on perioperative responses during ketamine anesthesia. Anesth Analg 1998;87(1):161–7.

1553. Handa F, Tanaka M, Nishikawa T, Toyooka H. Effects of oral clonidine premedication on side effects of intravenous ketamine anesthesia: a randomized, double-blind, placebo-controlled study. J Clin Anesth 2000;12(1):19–24.

1554. Ward J, Standage C. Angina pain precipitated by continuous subcutaneous infusion of ketamine. J Pain Symptom Manage 2003;25:6–7.

1555. Mok NS, Lo YK, Tsui PT, Lam CW. Ketoconazole induced torsades de pointes without concomitant use of QT interval-prolonging drug. J Cardiovasc Electrophysiol 2005;16(12):1375–7.

1556. Demolis JL, Vacheron F, Cardus S, Funck-Brentano C. Effect of single and repeated oral doses of telithromycin on cardiac QT interval in healthy subjects. Clin Pharmacol Ther 2003;73:242–52.

1557. Shain CS, Amsden GW. Telithromycin: the first of the ketolides. Ann Pharmacother 2002;36(3):452–64.

1558. Terkola R. Lapatinib ditosylate (Tykerb). Eur J Oncol Pharm 2007;1:13–17.

1559. Peak AS, Sutton BM. Systemic adverse effects associated with topically applied latanoprost. Ann Pharmacother 1998;32(4):504–5.

1560. Rozman B. Clinical experience with leflunomide in rheumatoid arthritis. Leflunomide Investigators' Group. J Rheumatol Suppl 1998;53:27–32.

1561. Strand V, Cohen S, Schiff M, Weaver A, Fleischmann R, Cannon G, Fox R, Moreland L, Olsen N, Furst D, Caldwell J, Kaine J, Sharp J, Hurley F, Loew-Friedrich I. Treatment of active rheumatoid arthritis with leflunomide compared with placebo and methotrexate. Leflunomide Rheumatoid Arthritis Investigators Group. Arch Intern Med 1999;159(21):2542–50.

1562. Smolen JS, Kalden JR, Scott DL, Rozman B, Kvien TK, Larsen A, Loew-Friedrich I, Oed C, Rosenburg REuropean Leflunomide Study Group. Efficacy and safety of leflunomide compared with placebo and sulphasalazine in active rheumatoid arthritis: a double-blind, randomised, multicentre trial. Lancet 1999;353(9149):259–66.

1563. Rozman B, Praprotnik S, Logar D, Tomsic M, Hojnik M, Kos-Golja M, Accetto R, Dolenc P. Leflunomide and hypertension. Ann Rheum Dis 2002;61(6):567–9.

1564. Martinez-Taboada VM, Rodriguez-Valverde V, Gonzalez-Vilchez F, Armijo JA. Pulmonary hypertension in a patient with rheumatoid arthritis treated with leflunomide. Rheumatology (Oxford) 2004;43(11):1451–3.

1565. Holcombe RF, Li A, Stewart RM. Levamisole and interleukin-2 for advanced malignancy. Biotherapy 1998;11(4):255–8.

1566. Thomas JM, Schug SA. Recent advances in the pharmacokinetics of local anaesthetics. Long-acting amide enantiomers and continuous infusions. Clin Pharmacokinet 1999;36(1):67–83.

1567. Huang YF, Pryor ME, Mather LE, Veering BT. Cardiovascular and central nervous system effects of intravenous levobupivacaine and bupivacaine in sheep. Anesth Analg 1998;86(4):797–804.

1568. Gristwood RW. Cardiac and CNS toxicity of levobupivacaine: strengths of evidence for advantage over bupivacaine. Drug Saf 2002;25(3):153–63.

1569. Owens RC Jr, Ambrose PG. Torsades de pointes associated with fluoroquinolones. Pharmacotherapy 2002; 22(5):663–8; discussion 668–72.

1570. Carbon C. Comparison of side effects of levofloxacin versus other fluoroquinolones. Chemotherapy 2001; 47(Suppl 3):9–14; discussion 44–8.

1571. Kahn JB. Latest industry information on the safety profile of levofloxacin in the US. Chemotherapy 2001; 47(Suppl 3):32–7; discussion 44–8.

1572. Paltoo B, O'Donoghue S, Mousavi MS. Levofloxacin induced polymorphic ventricular tachycardia with normal QT interval. Pacing Clin Electrophysiol 2001;24(5):895–7.

1573. Carbon C. Tolérance de la lévofloxacine, dossier clinique et données de pharmacovigilance. [Levofloxacin adverse effects, data from clinical trials and pharmacovigilance.] Therapie 2001;56(1):35–40.

1574. Armbruster C, Robibaro B, Griesmacher A, Vorbach H. Endothelial cell compatibility of trovafloxacin and levofloxacin for intravenous use. J Antimicrob Chemother 2000;45(4):533–5.

1575. Handley DA, Tinkelman D, Noonan M, Rollins TE, Snider ME, Caron J. Dose-response evaluation of levalbuterol versus racemic albuterol in patients with asthma. J Asthma 2000;37:319–27.

1576. Scott VL, Frazee LA. Retrospective comparison of nebulized levalbuterol and albuterol for adverse events in patients with acute airflow obstruction. Am J Ther 2003;10:341–7.

1577. Lam S, Chen J. Changes in heart rate associated with nebulized racemic albuterol and levalbuterol in intensive care patients. Am J Health-Syst Pharm 2003;60:1971–5.

1578. Ziegelbaum M, Lever H. Acute urinary retention associated with flecainide. Cleve Clin J Med 1990;57(1):86–7.

1579. Krikler DM, Curry PV. Torsade de pointes, an atypical ventricular tachycardia. Br Heart J 1976;38(2):117–20.

1580. Pfeifer HJ, Greenblatt DJ, Koch-Weser J. Clinical use and toxicity of intravenous lidocaine. A report from the Boston Collaborative Drug Surveillance Program. Am Heart J 1976;92(2):168–73.

1581. Gottlieb SS, Packer M. Deleterious hemodynamic effects of lidocaine in severe congestive heart failure. Am Heart J 1989;118(3):611–2.

1582. Weaver WD, Fahrenbruch CE, Johnson DD, Hallstrom AP, Cobb LA, Copass MK. Effect of epinephrine and lidocaine therapy on outcome after cardiac arrest due to ventricular fibrillation. Circulation 1990;82(6):2027–34.

1583. Tisdale JE. Lidocaine prophylaxis in acute myocardial infarction. Henry Ford Hosp Med J 1991;39(3–4):217–25.

1584. Garner L, Stirt JA, Finholt DA. Heart block after intravenous lidocaine in an infant. Can Anaesth Soc J 1985;32(4):425–8.

1585. Hansoti RC, Ashar PN. Atrioventicular block and ventricular fibrillation due to lidocaine therapy. Bombay Hosp J 1975;17:26.

1586. McLean SA, Paul ID, Spector PS. Lidocaine-induced conduction disturbance in patients with systemic hyperkalemia. Ann Emerg Med 2000;36(6):615–8.

1587. Parnass SM, Curran MJ, Becker GL. Incidence of hypotension associated with epidural anesthesia using alkalinized and nonalkalinized lidocaine for cesarean section. Anesth Analg 1987;66(11):1148–50.

1588. Enlund M, Mentell O, Krekmanov L. Unintentional hypotension from lidocaine infiltration during orthognathic surgery and general anaesthesia. Acta Anaesthesiol Scand 2001;45(3):294–7.

1589. Vacek V, Tesarova-Magrova J, Stafova J. Prevention of adverse reactions in therapy with high doses of lincomycin. Arzneimittelforschung 1970;20(1):99–101.

1590. Gabel A, Schymik G, Mehmel HC. Ventricular fibrillation due to long QT syndrome probably caused by clindamycin. Am J Cardiol 1999;83(5):813–5A11.

1591. Aucoin P, Beckner RR, Gantz NM. Clindamycin-induced cardiac arrest. South Med J 1982;75(6):768.

1592. Cantarini MV, Painter CJ, Gilmore EM, Bolger C, Watkins CL, Hughes AM. Effect of oral linezolid on the pressor response to intravenous tyramine. Br J Clin Pharmacol 2004;58(5):470–5.

1593. Tartarone A, Gallucci G, Iodice G, Romano G, Coccaro M, Vigliotti ML, Mele G, Matera R, Di Renzo N. Linezolid-induced bradycardia: a case report. Int J Antimicrob Agents 2004;23(4):412–3.

1594. Kallner G, Lindelius R, Petterson U, Stockman O, Tham A. Mortality in 497 patients with affective disorders attending a lithium clinic or after having left it. Pharmacopsychiatry 2000;33(1):8–13.

1595. Brodersen A, Licht RW, Vestergaard P, Olesen AV, Mortensen PB. Sixteen-year mortality in patients with affective disorder commenced on lithium. Br J Psychiatry 2000;176:429–33.

1596. Klumpers UMH, Boom K, Janssen FMG, Tulen JHM, Loonen AJM. Cardiovascular risk factors in outpatients with bipolar disorder. Pharmacopsychiatry 2004;32:211–6.

1597. Sparicio D, Landoni G, Pappalardo F, Crivellari M, Cerchierini E, Marino G, Zangrillo A. Methyline blue for lithium-induced refractory hypotension in off-pump coronary artery bypass graft: report of two cases. J Thorac Cardiovasc Surg 2004;127:592–3.

1598. Rosebraugh CJ, Flockhart DA, Yasuda SU, Woosley RL. Olanzapine-induced rhabdomyolysis. Ann Pharmacother 2001;35(9):1020–3.

1599. Terao T, Abe H, Abe K. Irreversible sinus node dysfunction induced by resumption of lithium therapy. Acta Psychiatr Scand 1996;93(5):407–8.

1600. Uchiyama Y, Nakao S, Asai T, Shingu K. [A case of atropine-resistant bradycardia in a patient on long-term lithium medication.]Masui 2001;50(11):1229–31.

1601. Wolf ME, Moffat M, Ranade V, Somberg JC, Lehrer E, Mosnaim AD. Lithium, hypercalcemia, and arrhythmia. J Clin Psychopharmacol 1998;18(5):420–3.

1602. Goodnick PJ, Jerry J, Parra F. Psychotropic drugs and the ECG: focus on the QT_c interval. Expert Opin Pharmacother 2002;3(5):479–98.

1603. Chong SA, Mythily, Mahendran R. Cardiac effects of psychotropic drugs. Ann Acad Med Singapore 2001;30(6):625–31.

1604. Lai CL, Chen WJ, Huang CH, Lin FY, Lee YT. Sinus node dysfunction in a patient with lithium intoxication. J Formos Med Assoc 2000;99(1):66–8.

1605. Convery RP, Hendrick DJ, Bourke SJ. Asthma precipitated by cessation of lithium treatment. Postgrad Med J 1999;75(888):637–8.

1606. Numata T, Abe H, Terao T, Nakashima Y. Possible involvement of hypothyroidism as a cause of lithium-induced sinus node dysfunction. Pacing Clin Electrophysiol 1999;22(6 Part 1):954–7.

1607. Davies B. The first patient to receive lithium. Aust NZ J Psychiatry 1999 1983;33:s32–4.

1608. Kahkonen S, Kaartinen M, Juhela P. Permanent pacing-aid to carry out long-term lithium therapy in manic patient with symptomatic bradycardia. Pharmacopsychiatry 2000;33(4):157.

1609. Keck PE Jr, Strakowski SMI, Hawkins JM, Dunayevich E, Tugrul KC, Bennett JA, McElroy SL. A pilot study of rapid lithium administration in the treatment of acute mania. Bipolar Disord 2001;3(2):68–72.

1610. Moltedo JM, Porter GA, State MW, Snyder CS. Sinus node dysfunction associated with lithium therapy in a child. Tex Heart Inst J 2002;29(3):200–2.

1611. Slordal L, Samstad S, Bathen J, Spigset O. A life-threatening interaction between lithium and celecoxib. Br J Clin Pharmacol 2003;55(4):413–4.

1612. Newland KD, Mycyk MB. Hemodialysis reversal of lithium overdose cardiotoxicity. Am J Emerg Med 2002;20(1):67–8.

1613. Delva NJ, Hawken ER. Preventing lithium intoxication. Guide for physicians. Can Fam Physician 2001;47:1595–600.

1614. Montes JM, Ferrando L. Gabapentin-induced anorgasmia as a cause of noncompliance in a bipolar patient. Bipolar Disord 2001;3(1):52.

1615. Terao T. Lithium therapy with pacemaker. Pharmacopsychiatry 2002;35(1):35.

1616. Luby ED, Singareddy RK. Long-term therapy with lithium in a private practice clinic: a naturalistic study. Bipolar Disord 2003;5(1):62–8.

1617. Rifai MA, Moles JK, Harrington DP. Lithium-induced hypercalcemia and parathyroid dysfunction. Psychosomatics 2001;42(4):359–61.

1618. Wolf ME, Ranade V, Molnar J, Somberg J, Mosnaim AD. Hypercalcemia, arrhythmia, and mood stabilizers. J Clin Psychopharmacol 2000;20(2):260–4.

1619. Slavicek J, Paclt I, Hamplova J, Kittnar O, Trefny Z, Horacek BM. Antidepressant drugs and heart electrical field. Physiol Res 1998;47(4):297–300.

1620. Reilly JG, Ayis SA, Ferrier IN, Jones SJ, Thomas SH. QTc-interval abnormalities and psychotropic drug therapy in psychiatric patients. Lancet 2000;355(9209): 1048–52.

1621. Felker GM, Thompson RE, Hare JM, Hruban RH, Clemetson DE, Howard DL, Baughman KL, Kasper EK. Underlying causes and long-term survival in patients with initially unexplained cardiomyopathy. N Engl J Med 2000;342(15):1077–84.

1622. Frassati D, Tabib A, Lachaux B, Giloux N, Daléry J, Vittori F, Charvet D, Barel C, Bui-Xuan B, Mégard R, Jenoudet LP, Descotes J, Vial T, Timour Q. Hidden cardiac lesions and psychotropic drugs as a possible cause of sudden death in psychiatric patients: a report of 14 cases and review of the literature. Can J Psychiatry 2004;491:100–5.

1623. Francis J, Hamzeh RK, Cantin-Hermoso MR. Lithium toxicity-induced wide-complex tachycardia in a pediatric patient. J Pediatr 2004;145:235–40.

1624. Coulter DM, Bate A, Meyboom RH, Lindquist M, Edwards IR. Antipsychotic drugs and heart muscle disorder in international pharmacovigilance: data mining study. BMJ 2001;322(7296):1207–9.

1625. Cruchaudet B, Eicher JC, Sgro C, Wolf JE. [Reversible cardiomyopathy induced by psychotropic drugs: case report and literature overview.] Ann Cardiol Angéiol (Paris) 2002;51(6):386–90.

1626. Jefferson JW. Lithium toxicity in the elderly. In: Nelson JC, editor. Geriatric Psychopharmacology. New York: Marcel Dekker, 1998:273–83.

1627. Paclt I, Slavicek J, Dohnalova A, Kitzlerova E, Pisvejcova K. Electrocardiographic dose-dependent changes in prophylactic doses of dosulepine, lithium and citalopram. Physiol Res 2003;52:311–7.

1628. Hsu CH, Liu PY, Chen JH, Yeh TL, Tsai HY, Lin LJ. Electrocardiographic abnormalities as predictors for over-range lithium levels. Cardiology 2005;103:101–6.

1629. Mamiya K, Sadanaga T, Sekita A, Nebeyama Y, Yao H, Yukawa E. Lithium concentration correlates with QT_c in patients with psychosis. J Electrocardiol 2005;38:148–51.

1630. Lynch C 3rd. Depression of myocardial contractility in vitro by bupivacaine, etidocaine, and lidocaine. Anesth Analg 1986;65(6):551–9.

1631. Ashley EM, Quick DG, El-Behesey B, Bromley LM. A comparison of the vasodilatation produced by two topical anaesthetics. Anaesthesia 1999;54(5):466–9.
1632. Zaugg M, Schulz C, Wacker J, Schaub MC. Sympatho-modulatory therapies in perioperative medicine. Br J Anaesth 2004;93(1):53–62.
1633. Polley LS, Columb MO, Naughton NN, Wagner DS, van de Ven CJ. Relative analgesic potencies of ropivacaine and bupivacaine for epidural analgesia in labor: implications for therapeutic indexes. Anesthesiology 1999;90(4):944–50.
1634. Capogna G, Celleno D, Fusco P, Lyons G, Columb M. Relative potencies of bupivacaine and ropivacaine for analgesia in labour. Br J Anaesth 1999;82(3):371–3.
1635. Graf BM, Abraham I, Eberbach N, Kunst G, Stowe DF, Martin E. Differences in cardiotoxicity of bupivacaine and ropivacaine are the result of physicochemical and stereoselective properties. Anesthesiology 2002;96(6):1427–34.
1636. Weinberg GL. Current concepts in resuscitation of patients with local anesthetic cardiac toxicity. Reg Anesth Pain Med 2002;27(6):568–75.
1637. Cucchiaro G, Rhodes LA. Unusual presentation of long QT syndrome. Br J Anaesth 2003;90:804–7.
1638. Favier JC, Da Conceicao M, Fassassi M, Allanic L, Steiner T, Pitti R. Successful resuscitation of serious bupivacaine intoxication in a patient with pre-existing heart failure. Can J Anaesth 2003;50:62–6.
1639. Gielen M, Slappendel R, Jack N. Successful defibrillation immediately after the intravascular injection of ropivacaine. Can J Anaesth 2005;52(5):490–2.
1640. Finucane BT. Ropivacaine cardiac toxicity—not as troublesome as bupivacaine. Can J Anaesth 2005;52(5):449–53.
1641. Levsky ME, Miller MA. Cardiovascular collapse from low dose bupivacaine. Can J Clin Pharmacol 2005;12(3):e240–5.
1642. Soltesz EG, van Pelt F, Byrne JG. Emergent cardiopulmonary bypass for bupivacaine cardiotoxicity. J Cardiothorac Vasc Anesth 2003;17:357–8.
1643. Phillips N, Priestley M, Denniss AR, Uther JB. Brugada-type electrocardiographic pattern induced by epidural bupivacaine. Anesth Analg 2003;97:264–7.
1644. Reinikainen M, Hedman A, Pelkonen O, Ruokonen E. Cardiac arrest after interscalene brachial plexus block with ropivacaine and lidocaine. Acta Anaesthesiol Scand 2003;47:904–6.
1645. Khan H, Atanassoff PG. Accidental intravascular injection of levobupivacaine and lidocaine during the transarterial approach to the axillary brachial plexus. Can J Anaesth 2003;50:95.
1646. Klein SM, Pierce T, Rubin Y, Nielsen KC, Steele SM. Successful resuscitation after ropivacaine-induced ventricular fibrillation. Anesth Analg 2003;97:901–3.
1647. Chazalon P, Tourtier JP, Villevielle T, Giraud D, Saissy JM, Mion G, Benhamou D. Ropivacaine-induced cardiac arrest after peripheral nerve block: successful resuscitation. Anesthesiology 2003;99:1449–51.
1648. Huet O, Eyrolle LJ, Mazoit JX, Ozier YM. Cardiac arrest after injection of ropivacaine for posterior lumbar plexus blockade. Anesthesiology 2003;99:1451–3.
1649. Zink W, Seif C, Bohl JRE, Hacke N, Braun PM, Sinner B, Martin E, Fink RH, Graf BM. The acute myotoxic effects of bupivacaine and ropivacaine after continuous peripheral nerve blockades. Anesth Analg 2003;97:1173–9.
1650. Calello DP, Muller AA, Henretig FM, Osterhoudt KC. Benzocaine: not dangerous enough? Pediatrics 2005;115(5):1452.
1651. Weinberg G, Ripper R, Feinstein DL, Hoffman W. Lipid emulsion infusion rescues dogs from bupivacaine-induced cardiac toxicity. Reg Anesth Pain Med 2003;28:198–202.
1652. Martyr JW, Stannard KJ, Gillespie G. Spinal-induced hypotension in elderly patients with hip fracture. A comparison of glucose-free bupivacaine with glucose-free bupivacaine and fentanyl. Anaesth Intensive Care 2005;33(1):64–8.
1653. Azma T, Okida M. Does lidocaine provoke clinically significant vasospasm? Acta Anaesthesiol Scand 2003;47:1174–5.
1654. Lascaux AS, Lesprit P, Bertocchi M, Levy Y. Inflammatory oedema of the legs: a new side-effect of lopinavir. AIDS 2001;15(6):819.
1655. Eyer-Silva WA, Neves-Motta R, Pinto JF, Morais-De-Sa CA. Inflammatory oedema associated with lopinavir-including HAART regimens in advanced HIV-1 infection: report of 3 cases. AIDS 2002;16(4):673–4.
1656. Dol L, Geffray L, el Khoury S, Cevallos R, Veyssier P. Oedèmes des members inférieurs chez un patient seropositif pour le VIH: effet secondaire du ritonavir? [Edema of the lower extremities in a HIV seropositive patient: secondary effect of ritonavir?] Presse Méd 1999;28(2):75.
1657. Crumb WJ Jr. Rate-dependent blockade of a potassium current in human atrium by the antihistamine loratadine. Br J Pharmacol 1999;126(3):575–80.
1658. Affrime MB, Brannan MD, Lorber RR, Danzig MR, Cuss F. A 3-month evaluation of electrocardiographic effects of loratadine in healthy individuals. Adv Ther 1999;16:149–57.
1659. Salmun LM, Herron JM, Banfield C, Padhi D, Lorber R, Affrime MB. The pharmacokinetics, electrocardiographic effects, and tolerability of loratadine syrup in children aged 2 to 5 years. Clin Ther 2000;22(5):613–21.
1660. Kosoglou T, Salfi M, Lim JM, Batra VK, Cayen MN, Affrime MB. Evaluation of the pharmacokinetics and electrocardiographic pharmacodynamics of loratadine with concomitant administration of ketoconazole or cimetidine. Br J Clin Pharmacol 2000;50(6):581–9.
1661. Lieberman AN, Bloom W, Kishore PS, Lin JP. Carotid artery occlusion following ingestion of LSD. Stroke 1974;5(2):213–5.
1662. Shaffer D, Singer S, Korvick J, Honig P. Concomitant risk factors in reports of torsades de pointes associated with macrolide use: review of the United States Food and Drug Administration Adverse Event Reporting System. Clin Infect Dis 2002;35(2):197–200.
1663. Ruuskanen O. Safety and tolerability of azithromycin in pediatric infectious diseases: 2003 update. Pediatr Infect Dis J 2004;23(2 Suppl):S135–9.
1664. Matsunaga N, Oki Y, Prigollini A. A case of QT-interval prolongation precipitated by azithromycin. N Z Med J 2003;116:U666.
1665. Kim MH, Berkowitz C, Trohman RG. Polymorphic ventricular tachycardia with a normal QT interval following azithromycin. Pacing Clin Electrophysiol 2005;28(11):1221–2.
1666. Ray WA, Murray KT, Meredith S, Narasimhulu SS, Hall K, Stein CM. Oral erythromycin and the risk of sudden death from cardiac causes. N Engl J Med 2004;351(11):1089–96.

1667. Promphan W, Khongphatthanayothin A, Horchaiprasit K, Benjacholamas V. Roxithromycin induced torsade de pointes in a patient with complex congenital heart disease and complete atrioventricular block. Pacing Clin Electrophysiol 2003;26:1424–6.
1668. Justo D, Mardi T, Zeltser D. Roxithromycin-induced torsades de pointes. Eur J Intern Med 2004;15(5):326–7.
1669. Ali A, Walentik C, Mantych GJ, Sadiq HF, Keenan WJ, Noguchi A. Iatrogenic acute hypermagnesemia after total parenteral nutrition infusion mimicking septic shock syndrome: two case reports. Pediatrics 2003;112(1 Pt 1):e70–2.
1670. Samol JM, Lambers DS. Magnesium sulfate tocolysis and pulmonary edema: the drug or the vehicle? Am J Obstet Gynecol 2005;192(5):1430–2.
1671. Hedges A. AN 448 on critical flicker frequency and heart rate in man. S Afr Med J 1972;46(6):139.
1672. Anonymous. Medroxyprogesterone and palliative care: new indication. No impact on quality of life. Prescrire Int 2001;10(51):3–4.
1673. Lelli G, Angelelli B, Zanichelli L, Strocchi E, Mondini F, Monetti N, Piana E, Pannuti F. The effect of high dose medroxyprogesterone acetate on water and salt metabolism in advanced cancer patients. Chemioterapia 1984;3(5):327–9.
1674. Harvey PJ, Molloy D, Upton J, Wing LM. Dose response effect of cyclical medroxyprogesterone on blood pressure in postmenopausal women. J Hum Hypertens 2001;15(5):313–21.
1675. Black HR, Leppert P, DeCherney A. The effect of medroxyprogesterone acetate on blood pressure. Int J Gynaecol Obstet 1978;17(1):83–7.
1676. Kofi Ekue JM, Ulrich AM, Rwabwogo-Atenyi J, Sheth UK. A double-blind comparative clinical trial of mefloquine and chloroquine in symptomatic falciparum malaria. Bull World Health Organ 1983;61(4):713–8.
1677. Dabrowski M, Wahl P, Holmes WE, Ashcroft FM. Effect of repaglinide on cloned beta cell, cardiac and smooth muscle types of ATP-sensitive potassium channels. Diabetologia 2001;44(6):747–56.
1678. Lindsay JR, McKillop AM, Mooney MH, O'Harte FP, Flatt PR, Bell PM. Effects of nateglinide on the secretion of glycated insulin and glucose tolerance in type 2 diabetes. Diabetes Res Clin Pract 2003;61(3):167–73.
1679. Hu S, Boettcher BR, Dunning BE. The mechanisms underlying the unique pharmacodynamics of nateglinide. Diabetologia 2003;46(Suppl 1):M37–43.
1680. Cohen RM, Ramlo-Halsted BA. How do the new insulin secretagogues compare? Diabetes Care 2002;25(8):1472–3.
1681. Kahn SE, Montgomery B, Howell W, Ligueros-Saylan M, Hsu CH, Devineni D, McLeod JF, Horowitz A, Foley JE. Importance of early phase insulin secretion to intravenous glucose tolerance in subjects with type 2 diabetes mellitus. J Clin Endocrinol Metab 2001;86(12):5824–9.
1682. Parulkar AA, Fonseca VA. Recent advances in pharmacological treatment of type 2 diabetes mellitus. Compr Ther 1999;25(8–10):418–26.
1683. Ratner RE. Repaglinide therapy in the treatment of type 2 diabetes. Today's Ther Trends 1999;17:57–66.
1684. Culy CR, Jarvis B. Repaglinide: a review of its therapeutic use in type 2 diabetes mellitus. Drugs 2001;61(11):1625–60.
1685. Juhl CB, Porksen N, Hollingdal M, Sturis J, Pincus S, Veldhuis JD, Dejgaard A, Schmitz O. Repaglinide acutely amplifies pulsatile insulin secretion by augmentation of burst mass with no effect on burst frequency. Diabetes Care 2000;23(5):675–81.
1686. Hatorp V, Huang WC, Strange P. Repaglinide pharmacokinetics in healthy young adult and elderly subjects. Clin Ther 1999;21(4):702–10.
1687. Gomis R. Repaglinide as monotherapy in Type 2 diabetes. Exp Clin Endocrinol Diabetes 1999;107(Suppl 4):S133–5.
1688. Damsbo P, Marbury TC, Hatorp V, Clauson P, Muller PG. Flexible prandial glucose regulation with repaglinide in patients with type 2 diabetes. Diabetes Res Clin Pract 1999;45(1):31–9.
1689. Moses RG, Gomis R, Frandsen KB, Schlienger JL, Dedov I. Flexible meal-related dosing with repaglinide facilitates glycemic control in therapy-naive type 2 diabetes. Diabetes Care 2001;24(1):11–5.
1690. Hatorp V, Huang WC, Strange P. Pharmacokinetic profiles of repaglinide in elderly subjects with type 2 diabetes. J Clin Endocrinol Metab 1999;84(4):1475–8.
1691. Massi-Benedetti M, Damsbo P. Pharmacology and clinical experience with repaglinide. Expert Opin Investig Drugs 2000;9(4):885–98.
1692. Bouhanick B, Barbosa SS. Repaglinide: Novonorm, une alternative chez le diabétique de type 2. [Rapaglinide: Novonorm, an alternative in type 2 diabetes.] Presse Méd 2000;29(19):1059–61.
1693. Schatz H. Preclinical and clinical studies on safety and tolerability of repaglinide. Exp Clin Endocrinol Diabetes 1999;107(Suppl 4):S144–8.
1694. Schmitz O, Lund S, Andersen PH, Jonler M, Porksen N. Optimizing insulin secretagogue therapy in patients with type 2 diabetes: a randomized double-blind study with repaglinide. Diabetes Care 2002;25(2):342–6.
1695. Van Gaal LF, Van Acker KL, De Leeuw IH. Repaglinide improves blood glucose control in sulphonylurea-naive type 2 diabetes. Diabetes Res Clin Pract 2001;53(3):141–8.
1696. de Luis DA, Aller R, Cuellar L, Terroba C, Ovalle H, Izaola O, Romero E. Effect of repaglinide addition to NPH insulin monotherapy on glycemic control in patients with type 2 diabetes. Diabetes Care 2001;24(10):1844–5.
1697. Madsbad S, Kilhovd B, Lager I, Mustajoki P, Dejgaard A. Scandinavian Repaglinide Group. Comparison between repaglinide and glipizide in Type 2 diabetes mellitus: a 1-year multicentre study. Diabet Med 2001;18(5):395–401.
1698. Damsbo P, Clauson P, Marbury TC, Windfeld K. A double-blind randomized comparison of meal-related glycemic control by repaglinide and glyburide in well-controlled type 2 diabetic patients. Diabetes Care 1999;22(5):789–94.
1699. Hirshberg B, Skarulis MC, Pucino F, Csako G, Brennan R, Gorden P. Repaglinide-induced factitious hypoglycemia. J Clin Endocrinol Metab 2001;86(2):475–7.
1700. Lusardi P, Piazza E, Fogari R. Cardiovascular effects of melatonin in hypertensive patients well controlled by nifedipine: a 24-hour study. Br J Clin Pharmacol 2000;49(5):423–7.
1701. Arangino S, Cagnacci A, Angiolucci M, Vacca AM, Longu G, Volpe A, Melis GB. Effects of melatonin on vascular reactivity, catecholamine levels, and blood pressure in healthy men. Am J Cardiol 1999;83(9):1417–9.
1702. Herxheimer A, Petrie KJ. Melatonin for preventing and treating jet lag. Cochrane Database Syst Rev 2001;(1):CD001520.
1703. Ayestaran C, Matorras R, Gomez S, Arce D, Rodriguez-Escudero F. Severe bradycardia and bradypnea following vaginal oocyte retrieval: a possible toxic effect of paracervical mepivacaine. Eur J Obstet Gynecol Reprod Biol 2000;91(1):71–3.

1704. Griebenow R, Saborouski F, Matthes H, Wald-Oloumier H. EKG-Veränderungen bein Infiltrationsanesthesie mit Mepivacain. [ECG changes after spinal anesthesia using mepivacaine.] Intensivmed 1979;16:163.

1705. LeClercq A, Melennec J, Proteau J. Intoxication mercurielle. Concours Med 1973;95:6055.

1706. Ciaccio EI. Mercury: therapeutic and toxic aspects. Semin Drug Treat 1971;1(2):177–94.

1707. Ward OC, Hingerty D. Pink Disease from cutaneous absorption of mercury. J Ir Med Assoc 1967;60(357):94–5.

1708. Boogaard PJ, Houtsma AT, Journee HL, Van Sittert NJ. Effects of exposure to elemental mercury on the nervous system and the kidneys of workers producing natural gas. Arch Environ Health 1996;51(2):108–15.

1709. Young E. Ammoniated mercury poisoning. Br J Dermatol 1960;72:449–55.

1710. Li AM, Chan MH, Leung TF, Cheung RC, Lam CW, Fok TF. Mercury intoxication presenting with tics. Arch Dis Child 2000;83(2):174–5.

1711. Deleu D, Hanssens Y, al-Salmy HS, Hastie I. Peripheral polyneuropathy due to chronic use of topical ammoniated mercury. J Toxicol Clin Toxicol 1998;36(3):233–7.

1712. De Bont B, Lauwerys R, Govaerts H, Moulin D. Yellow mercuric oxide ointment and mercury intoxication. Eur J Pediatr 1986;145(3):217–8.

1713. Zoppi M, Hoigne R, Keller MF, Streit F, Hess T. Blutdruckabfall unter Dipyron (Novaminsulfon-Natrium). [Reducing blood pressure with Dipyron (novaminsulfone sodium).] Schweiz Med Wochenschr 1983;113(47):1768–70.

1714. Nathan HJ. Narcotics and myocardial performance in patients with coronary artery disease. Can J Anaesth 1988;35(3 Pt 1):209–13.

1715. Backmund M, Meyer K, Zwehl W, Nagengast O, Eichenlaub D. Myocardial Infarction associated with methadone and/or dihydrocodeine. Eur Addict Res 2001;7(1):37–9.

1716. Hays H, Woodroffe MA. High dosing methadone and a possible relationship to serious cardia arrhythmias. Pain Res Manag 2001;6(2):64.

1717. Deamer RL, Wilson DR, Clark DS, Prichard JG. Torsades de pointes associated with high dose levomethadyl acetate (ORLAAM). J Addict Dis 2001;20(4):7–14.

1718. Krantz MJ, Lewkowiez L, Hays H, Woodroffe MA, Robertson AD, Mehler PS. Torsade de pointes associated with very-high-dose methadone. Ann Intern Med 2002;137(6):501–4.

1719. Todd MM, Drummond JC, Sang H. The hemodynamic consequences of high-dose methohexital anesthesia in humans. Anesthesiology 1984;61(5):495–501.

1720. Kettunen R, Huikuri HV, Oikarinen A, Takkunen JT. Methotrexate-linked ventricular arrhythmias. Acta Derm Venereol 1995;75(5):391–2.

1721. Landewe RB, van den Borne BE, Breedveld FC, Dijkmans BA. Methotrexate effects in patients with rheumatoid arthritis with cardiovascular comorbidity. Lancet 2000;355(9215):1616–7.

1722. Barrett PJ, Taylor GT. "Ecstasy" ingestion: a case report of severe complications. J R Soc Med 1993;86(4):233–4.

1723. McCann UD, Ricaurte GA. Lasting neuropsychiatric sequelae of (+-)methylenedioxymethamphetamine ("ecstasy") in recreational users. J Clin Psychopharmacol 1991;11(5): 302–5.

1724. Duxbury AJ. Ecstasy—dental implications. Br Dent J 1993;175(1):38.

1725. Ames D, Wirshing WC. Ecstasy, the serotonin syndrome, and neuroleptic malignant syndrome—a possible link? JAMA 1993;269(7):869–70.

1726. Campkin NJ, Davies UM. Treatment of "ecstasy" overdose with dantrolene. Anaesthesia 1993;48(1):82–3.

1727. Madhok A, Boxer R, Chowdhury D. Atrial fibrillation in an adolescent—the agony of ecstasy. Pediatr Emerg Care 2003;19:348–9.

1728. Lin Y-H, Seow K-M, Hwang J-L, Chen H-H. Myocardial infarction and mortality caused by methylergonovine. Acta Obstet Gynecol Scand 2005;84:1022.

1729. Witton K. On the use of parenteral methylphenidate: a follow-up report. Am J Psychiatry 1964;121:267–8.

1730. Chernoff RW, Wallen MH, Muller OF. Cardiac toxicity of methylphenidate: report of two cases. Nord Hyg Tidskr 1962;266:400–1.

1731. Stowe CD, Gardner SF, Gist CC, Schulz EG, Wells TG, Felin JF. 24-hour ambulatory blood pressure monitoring in male children receiving stimulant therapy. Ann Pharmacother 2002;36(7–8):1142–9.

1732. Vidal Company A, Rodriguez Martin A, Barrio Merino A, Lorente Garcia-Maurino A, Garcia Llop LA. Bloqueo A-V por intoxicación con metoclopramida. [Atrioventricular block caused by metoclopramide poisoning.] An Esp Pediatr 1991;34(4):313–4.

1733. Bevacqua BK. Supraventricular tachycardia associated with postpartum metoclopramide administration. Anesthesiology 1988;68(1):124–5.

1734. Huerta Blanco R, Hernandez Cabrera M, Quinones Morales I, Cardenes Santana MA. Bloqueo cardiaco completo inducido por metoclopramida intravenosa. [Complete heart block induced by intravenous metoclopramide.] An Med Interna 2000;17(4):222–3.

1735. Ellidokuz E, Kaya D. The effect of metoclopramide on QT dynamicity: double-blind, placebo-controlled, crossover study in healthy male volunteers, Aliment Pharmacol Ther 2003;18:151–5.

1736. Sheridan C, Chandra P, Jacinto M, Greenwald ES. Transient hypertension after high doses of metoclopramide. N Engl J Med 1982;307(21):1346.

1737. Savarese JJ, Ali HH, Antonio RP. The clinical pharmacology of metocurine: dimethyltubocurarine revisited. Anesthesiology 1977;47(3):277–84.

1738. Hughes R, Chapple DJ. Effects on non-depolarizing neuromuscular blocking agents on peripheral autonomic mechanisms in cats. Br J Anaesth 1976;48(2):59–68.

1739. McCullough LS, Stone WA, Delaunois AL, Reier CE, Hamelberg W. The effect of dimethyl tubocurarine iodide on cardiovascular parameters, postganglionic sympathetic activity, and histamine release. Anesth Analg 1972;51(4):554–9.

1740. Basta SJ, Savarese JJ, Ali HH, Moss J, Gionfriddo M. Histamine-releasing potencies of atracurium, dimethyl tubocurarine and tubocurarine. Br J Anaesth 1983;55(Suppl 1):S105–6.

1741. Hughes R, Chapple DJ. Cardiovascular and neuromuscular effects of dimethyl tubocurarine in anaesthetized cats and rhesus monkeys. Br J Anaesth 1976;48(9):847–52.

1742. Hughes R, Ingram GS, Payne JP. Studies on dimethyl tubocurarine in anaesthetized man. Br J Anaesth 1976;48(10):969–74.

1743. Stranz MH, Bradley WE. Metronidazole (Flagyl IV, Searle). Drug Intell Clin Pharm 1981;15(11):838–46.

1744. Kopera H, Klein W, Schenk H. Psychotropic drugs and the heart: clinical implications. Prog Neuropsychopharmacol 1980;4(4–5):527–35.

1745. Goldie A, Edwards JG. Electrocardiographic changes during treatment with maprotiline and mianserin. Neuropharmacology 1984;23(2B):273–5.

1746. Whiteford H, Klug P, Evans L. Disturbed cardiac function possibly associated with mianserin therapy. Med J Aust 1984;140(3):166–7.

1747. Stevens DA. Miconazole in the treatment of coccidioidomycosis. Drugs 1983;26(4):347–54.

1748. Drouhet E, Dupont B. Evolution of antifungal agents: past, present, and future. Rev Infect Dis 1987;9(Suppl 1):S4–S14.

1749. Davis DP, Kimbro TA, Vilke GM. The use of midazolam for prehospital rapid-sequence intubation may be associated with a dose-related increase in hypotension. Prehosp Emerg Care 2001;5(2):163–8.

1750. Anonymous. Midazolam—is antagonism justified? Lancet 1988;2(8603):140–2.

1751. Alexander CM, Gross JB. Sedative doses of midazolam depress hypoxic ventilatory responses in humans. Anesth Analg 1988;67(4):377–82.

1752. Bailey PL, Pace NL, Ashburn MA, Moll JW, East KA, Stanley TH. Frequent hypoxemia and apnea after sedation with midazolam and fentanyl. Anesthesiology 1990;73(5):826–30.

1753. Baris S, Karakaya D, Aykent R, Kirdar K, Sagkan O, Tur A. Comparison of midazolam with or without fentanyl for conscious sedation and hemodynamics in coronary angiography. Can J Cardiol 2001;17(3):277–81.

1754. Simonton CA, Chatterjee K, Cody RJ, Kubo SH, Leonard D, Daly P, Rutman H. Milrinone in congestive heart failure: acute and chronic hemodynamic and clinical evaluation. J Am Coll Cardiol 1985;6(2):453–9.

1755. Takei K. Case report: three cases of acute exacerbation of chronic heart failure treated with milrinone. Ther Res 1999;20:245–51.

1756. Milfred-LaForest SK, Shubert J, Mendoza B, Flores I, Eisen HJ, Pina IL. Tolerability of extended duration intravenous milrinone in patients hospitalized for advanced heart failure and the usefulness of uptitration of oral angiotensin-converting enzyme inhibitors. Am J Cardiol 1999;84(8):894–9.

1757. Alhashemi JA, Hooper J. Treatment of milrinone-associated tachycardia with beta-blockers. Can J Anaesth 1998;45(1):67–70.

1758. Gold J, Cullinane S, Chen J, Seo S, Oz MC, Oliver JA, Landry DW. Vasopressin in the treatment of milrinone-induced hypotension in severe heart failure. Am J Cardiol 2000;85(4):506–8.

1759. Gold JA, Cullinane S, Chen J, Oz MC, Oliver JA, Landry DW. Vasopressin as an alternative to norepinephrine in the treatment of milrinone-induced hypotension. Crit Care Med 2000;28(1):249–52.

1760. Choi WW, Mehta MP, Murray DJ, Sokoll MD, Forbes RB, Gergis SD, Abou-Donia M, Kirchner J. Neuromuscular and cardiovascular effects of mivacurium chloride in surgical patients receiving nitrous oxide–narcotic or nitrous oxide–isoflurane anaesthesia. Can J Anaesth 1989;36(6):641–50.

1761. Savarese JJ, Ali HH, Basta SJ, Scott RP, Embree PB, Wastila WB, Abou-Donia MM, Gelb C. The cardiovascular effects of mivacurium chloride (BW B1090U) in patients receiving nitrous oxide-opiate-barbiturate anesthesia. Anesthesiology 1989;70(3):386–94.

1762. Caldwell JE, Heier T, Kitts JB, Lynam DP, Fahey MR, Miller RD. Comparison of the neuromuscular block induced by mivacurium, suxamethonium or atracurium during nitrous oxide–fentanyl anaesthesia. Br J Anaesth 1989;63(4):393–9.

1763. Stoops CM, Curtis CA, Kovach DA, McCammon RL, Stoelting RK, Warren TM, Miller D, Bopp SK, Jugovic DJ, Abou-Donia MM. Hemodynamic effects of mivacurium chloride administered to patients during oxygen–sufentanil anesthesia for coronary artery bypass grafting or valve replacement. Anesth Analg 1989;68(3):333–9.

1764. Doenicke A, Moss J, Lorenz W, Mayer M, Rau J, Jedrzejewski A, Ostwald P. Effect of oral antihistamine premedication on mivacurium-induced histamine release and side effects. Br J Anaesth 1996;77(3):421–3.

1765. Lebrun-Vignes B, Diquet B, Chosidow O. Clinical pharmacokinetics of mizolastine. Clin Pharmacokinet 2001;40(7):501–7.

1766. Baumhackl U, Biziere K, Fischbach R, Geretsegger C, Hebenstreit G, Radmayr E, Stabl M. Efficacy and tolerability of moclobemide compared with imipramine in depressive disorder (DSM-III): an Austrian double-blind, multicentre study. Br J Psychiatry Suppl 1989;6:78–83.

1767. Realini R, Mascetti R, Calanchini C. Efficacité et tolerance du moclobemide (Ro 11-1163 Aurorix) en comparaison avec la maprotiline chez des patients ambulatoires présentant un épisode dépressif majeur. [Effectiveness and tolerance of moclobemide (Ro 11-1163 Aurorix) in comparison with maprotiline in ambulatory patients presenting with a major depressive episode.] Psychol Med 1989;21:1689.

1768. Delini-Stula A, Baier D, Kohnen R, Laux G, Philipp M, Scholz HJ. Undesirable blood pressure changes under naturalistic treatment with moclobemide, a reversible MAO-A inhibitor – results of the drug utilization observation studies. Pharmacopsychiatry 1999;32(2):61–7.

1769. Wong YN, Wang L, Hartman L, Simcoe D, Chen Y, Laughton W, Eldon R, Markland C, Grebow P. Comparison of the single-dose pharmacokinetics and tolerability of modafinil and dextroamphetamine administered alone or in combination in healthy male volunteers. J Clin Pharmacol 1998;38(10):971–8.

1770. Becker PM, Jamieson AO, Jewel CE, Bogan RK, James DS. Randomized trial of modafinil as a treatment for the excessive daytime somnolence of narcolepsy: US Modafinil Narcolepsy Multicenter Study Group. Neurol 2000;54(5):1166–75.

1771. Anonymous. Randomized trial of modafinil for the treatment of pathological somnolence in narcolepsy. US Modafinil Narcolepsy Multicenter Study Group. Ann Neurol 1998;43(1):88–97.

1772. Hauser RA, Wahba MN, Zesiewicz TA, McDowell Anderson W. Modafinil treatment of pramipexole-associated somnolence. Mov Disord 2000;15(6):1269–71.

1773. Rugino TA, Copley TC. Effects of modafinil in children with attention-deficit/hyperactivity disorder: an open-label study. J Am Acad Child Adolesc Psychiatry 2001;40(2):230–5.

1774. Rammohan KW, Rosenberg JH, Lynn DJ, Blumenfeld AM, Pollak CP, Nagaraja HN. Efficacy and safety of modafinil (Provigil) for the treatment of fatigue in multiple sclerosis: a two centre phase 2 study. J Neurol Neurosurg Psychiatry 2002;72(2):179–83.

1775. Razani J, White KL, White J, Simpson G, Sloane RB, Rebal R, Palmer R. The safety and efficacy of combined amitriptyline and tranylcypromine antidepressant treatment. A controlled trial. Arch Gen Psychiatry 1983;40(6):657–61.

1776. Kronig MH, Roose SP, Walsh BT, Woodring S, Glassman AH. Blood pressure effects of phenelzine. J Clin Psychopharmacol 1983;3(5):307–10.

1777. Linet LS. Mysterious MAOI hypertensive episodes. J Clin Psychiatry 1986;47(11):563–5.

1778. Blackwell B. Clinical and pharmacological observations of the interactions of monoamine oxidase inhibitors, amines and foodstuffs. Cambridge University: MD Thesis, 1966

1779. White CM, Grant EM, Quintiliani R. Moxifloxacin does increase the corrected QT interval. Clin Infect Dis 2001;33(8):1441–4.

1780. Moxifloxacin: new preparation. A me-too with more cardiac risks. Prescrire Int 2002;11(62):168–9.

1781. Demolis JL, Kubitza D, Tenneze L, Funck-Brentano C. Effect of a single oral dose of moxifloxacin (400 mg and 800 mg) on ventricular repolarization in healthy subjects Clin Pharmacol Ther 2000;68(6):658–66.

1782. Culley CM, Lacy MK, Klutman N, Edwards B. Moxifloxacin: clinical efficacy and safety. Am J Health Syst Pharm 2001;58(5):379–88.

1783. Salvador Garcia Morillo J, Stiefel Garcia-Junco P, Vallejo Maroto I, Carneado de la Fuente J. Crisis hipertensiva y bloqueo transitorio de rama izquierda con prolongacion del intervalo qt asociados a moxifloxacino. [Hypertensive crisis and transitory left brunch block with QT interval prolongation associated to moxifloxacin.] Med Clin (Barc) 2001;117(5):198–9.

1784. Siepmann M, Kirch W. Drug points: tachycardia associated with moxifloxacin. BMJ 2001;322(7277):23.

1785. Jeyarajah DR, Thistlethwaite JR Jr. General aspects of cytokine-release syndrome: timing and incidence of symptoms. Transplant Proc 1993;25(2 Suppl 1):16–20.

1786. Hall KA, Dole EJ, Hunter GC, Zukoski CF, Putnam CW. Hyperpyrexia-related ventricular tachycardia during OKT3 induction therapy. Transplantation 1992;54(6):1112–3.

1787. Abramowicz D, Pradier O, Marchant A, Florquin S, De Pauw L, Vereerstraeten P, Kinnaert P, Vanherweghem JL, Goldman M. Induction of thromboses within renal grafts by high-dose prophylactic OKT3. Lancet 1992;339(8796):777–8.

1788. Raasveld MH, Hack CE, ten Berge IJ. Activation of coagulation and fibrinolysis following OKT3 administration to renal transplant recipients: association with distinct mediators. Thromb Haemost 1992;68(3):264–7.

1789. Gomez E, Aguado S, Gago E, Escalada P, Alvarez-Grande J. Main graft vessels thromboses due to conventional-dose OKT3 in renal transplantation. Lancet 1992;339(8809):1612–3.

1790. Abramowicz D, Pradier O, De Pauw L, Kinnaert P, Mat O, Surquin M, Doutrelepont JM, Vanherweghem JL, Capel P, Vereerstraeten P, et al. High-dose glucocorticosteroids increase the procoagulant effects of OKT3. Kidney Int 1994;46(6):1596–602.

1791. Hollenbeck M, Westhoff A, Bach D, Grabensee B, Kolvenbach R, Kniemeyer HW. Doppler sonography and renal graft vessel thromboses after OKT3 treatment. Lancet 1992;340(8819):619–20.

1792. Wride SR, Smith RE, Courtney PG. A fatal case of pulmonary oedema in a healthy young male following naloxone administration. Anaesth Intensive Care 1989;17(3):374–7.

1793. Allen T. No adverse reaction. Ann Emerg Med 1989;18(1):116.

1794. Pallasch TJ, Gill CJ. Naloxone-associated morbidity and mortality. Oral Surg Oral Med Oral Pathol 1981;52(6):602–3.

1795. Carlsson C, Dencker SJ, Grimby G, Häggendal J. Noradrenaline in blood-plasma and urine during chlorpromazine treatment. Lancet 1966;1:1208.

1796. Leo RJ, Kreeger JL, Kim KY. Cardiomyopathy associated with clozapine. Ann Pharmacother 1996;30(6):603–5.

1797. Juul Povlsen U, Noring U, Fog R, Gerlach J. Tolerability and therapeutic effect of clozapine. A retrospective investigation of 216 patients treated with clozapine for up to 12 years. Acta Psychiatr Scand 1985;71(2):176–85.

1798. Petrie WM, Ban TA, Berney S, Fujimori M, Guy W, Ragheb M, Wilson WH, Schaffer JD. Loxapine in psychogeriatrics: a placebo- and standard-controlled clinical investigation. J Clin Psychopharmacol 1982;2(2):122–6.

1799. Mandelstam JP. An inquiry into the use of Innovar for pediatric premedication. Anesth Analg 1970;49(5):746–50.

1800. Yanovski A, Kron RE, Townsend RR, Ford V. The clinical utility of ambulatory blood pressure and heart rate monitoring in psychiatric inpatients. Am J Hypertens 1998;11(3 Pt 1):309–15.

1801. Branchey MH, Lee JH, Amin R, Simpson GM. High- and low-potency neuroleptics in elderly psychiatric patients. JAMA 1978;239(18):1860–2.

1802. Haddad PM, Anderson IM. Antipsychotic-related QT_c prolongation, torsade de pointes and sudden death. Drugs 2002;62(11):1649–71.

1803. Warner JP, Barnes TR, Henry JA. Electrocardiographic changes in patients receiving neuroleptic medication. Acta Psychiatr Scand 1996;93(4):311–3.

1804. FDA Psychopharmacological Drugs Advisory Committee. Briefing Document of Zeldox Capsules (Ziprasidone HCI) 19 July 2000; www.fda.gov/ohrms/dockets/ac00/backgrd/3619b1a.pdf 19/07/2000.

1805. Dessertenne F. La tachycardie ventriculaire a deux foyers opposés variables. [Ventricular tachycardia with 2 variable opposing foci.] Arch Mal Coeur Vaiss 1966;59(2):263–72.

1806. Buckley NA, Sanders P. Cardiovascular adverse effects of antipsychotic drugs. Drug Saf 2000;23(3):215–28.

1807. Moss AJ. The QT interval and torsade de pointes. Drug Saf 1999;21(Suppl 1):5–10.

1808. Viskin S. Long QT syndromes torsade de pointes. Lancet 1999;354(9190):1625–33.

1809. Glassman AH, Bigger JT Jr. Antipsychotic drugs: prolonged QT_c interval, torsade de pointes, and sudden death. Am J Psychiatry 2001;158(11):1774–82.

1810. Yap YG, Camm J. Risk of torsades de pointes with non-cardiac drugs. Doctors need to be aware that many drugs can cause qt prolongation. BMJ 2000;320(7243):1158–9.

1811. Kiriike N, Maeda Y, Nishiwaki S, Izumiya Y, Katahara S, Mui K, Kawakita Y, Nishikimi T, Takeuchi K, Takeda T. Iatrogenic torsade de pointes induced by thioridazine. Biol Psychiatry 1987;22(1):99–103.

1812. Connolly MJ, Evemy KL, Snow MH. Torsade de pointes ventricular tachycardia in association with thioridazine therapy: report of two cases. New Trends Arrhythmias 1985;1:157.

1813. Sharma ND, Rosman HS, Padhi ID, Tisdale JE. Torsades de pointes associated with intravenous haloperidol in critically ill patients. Am J Cardiol 1998;81(2):238–40.

1814. Michalets EL, Smith LK, Van Tassel ED. Torsade de pointes resulting from the addition of droperidol to an existing cytochrome P450 drug interaction. Ann Pharmacother 1998;32(7–8):761–5.

1815. Barber JM. Risk of sudden death on high-dose antipsychotic medication: QT_c dispersion. Br J Psychiatry 1998;173:86–7.

1816. Reilly JG, Ayis SA, Ferrier IN, Jones SJ, Thomas SH. QT_c-interval abnormalities and psychotropic drug therapy in psychiatric patients. Lancet 2000;355(9209):1048–52.

1817. Iwahashi K. Significantly higher plasma haloperidol level during cotreatment with carbamazepine may herald cardiac change. Clin Neuropharmacol 1996;19(3):267–70.

1818. Tzivoni D, Banai S, Schuger C, Benhorin J, Keren A, Gottlieb S, Stern S. Treatment of torsade de pointes with magnesium sulfate. Circulation 1988;77(2):392–7.

1819. O'Brien JM, Rockwood RP, Suh KI. Haloperidol-induced torsade de pointes. Ann Pharmacother 1999;33(10):1046–50.

1820. Benowitz NL. Nicotine patches. BMJ 1995;310(6991):1409–10.

1821. Jackson M. Cerebral arterial narrowing with nicotine patch. Lancet 1993;342(8865):236–7.

1822. Engel CJ, Parmentier AH. Suicide attempts and the nicotine patch. JAMA 1993;270(3):323–4.

1823. Joseph AM, Norman SM, Ferry LH, Prochazka AV, Westman EC, Steele BG, Sherman SE, Cleveland M, Antonnucio DO, Hartman N, McGovern PG. The safety of transdermal nicotine as an aid to smoking cessation in patients with cardiac disease. N Engl J Med 1996;335(24):1792–8.

1824. Anonymous. Nicotine patches and chewing gum-cardiac risks with concomitant smoking. WHO Newslett 1996;5,6:6.

1825. Stewart PM, Catterall JR. Chronic nicotine ingestion and atrial fibrillation. Br Heart J 1985;54(2):222–3.

1826. Ottervanger JP, Stricker BH, Klomps HC. Transdermal nicotine: clarifications, side effects, and funding. JAMA 1993;269(15):1940–1.

1827. Britton ML, Bradberry JC, Letassy NA, Mckenney JM, Sirmans SM. ASHP Therapeutic Position Statement on the safe use of niacin in the management of dyslipidemias. American Society of Health-System Pharmacists. Am J Health Syst Pharm 1997;54(24):2815–9.

1828. Carlson LA. The broad spectrum hypolipidaemic drug nicotinic acid. J Drug Dev Suppl 1990;3:223–6.

1829. Pasternak RC, Kolman BS. Unstable myocardial ischemia after the initiation of niacin therapy. Am J Cardiol 1991;67(9):904–6.

1830. Abdallah A, Saif M, Abdel-Meguid M, Badran A, Abdel-Fattah F, Aly IM. Treatment of urinary and intestinal bilharziasis with Ciba 32644-Ba (Ambilhar). A preliminary report. J Egypt Med Assoc 1966;49(2):145–63.

1831. Katz N, Bittencourt D, Oliveira CA, Dias RP, Ferreira H, Grinbaum E, Dias CB, Pellegrino J. Clinical trials with Ciba 32,644-Ba (Ambilhar®) in schistosomiasis mansoni. Folha Med 1966;53(4):561–7.

1832. Rovira I, Fita G, Suarez S, Gomar C, Cartana R. Effects of inhaled nitric oxide in a patient with pulmonary hypertension and left heart failure secondary to a giant left atrial myxoma. J Cardiothorac Vasc Anesth 1999;13(6):726–8.

1833. Lubbers P. Allergische Reaktion gegen Furadantin. Dtsch Med Wochenschr 1962;87:2209.

1834. Holmberg L, Boman G, Bottiger LE, Eriksson B, Spross R, Wessling A. Adverse reactions to nitrofurantoin. Analysis of 921 reports. Am J Med 1980;69(5):733–8.

1835. Biel B, Younes M, Brasch H. Cardiotoxic effects of nitrofurantoin and tertiary butylhydroperoxide in vitro: are oxygen radicals involved? Pharmacol Toxicol 1993;72(1):50–5.

1836. Kozmary SV, Lampe GH, Benefiel D, Cahalan MK, Wauk LZ, Whitendale P, Schiller NB, Eger EI 2nd. No finding of increased myocardial ischemia during or after carotid endarterectomy under anesthesia with nitrous oxide. Anesth Analg 1990;71(6):591–6.

1837. Lampe GH, Donegan JH, Rupp SM, Wauk LZ, Whitendale P, Fouts KE, Rose BM, Litt LL, Rampil IJ, Wilson CB, et al. Nitrous oxide and epinephrine-induced arrhythmias. Anesth Analg 1990;71(6):602–5.

1838. Badner NH, Spence JD. Homocyst(e)ine, nitrous oxide and atherosclerosis. Balliere's Best Pract Res Clin Anesthesiol 2001;15:185–93.

1839. Badner NH, Freeman D, Spence JD. Preoperative oral B vitamins prevent nitrous oxide-induced postoperative plasma homocysteine increases. Anesth Analg 2001;93(6):1507–10.

1840. Badner NH, Beattie WS, Freeman D, Spence JD. Nitrous oxide-induced increased homocysteine concentrations are associated with increased postoperative myocardial ischemia in patients undergoing carotid endarterectomy. Anesth Analg 2000;91(5):1073–9.

1841. Johnson AG. NSAIDs and blood pressure. Clinical importance for older patients. Drugs Aging 1998;12(1):17–27.

1842. Palmer R, Weiss R, Zusman RM, Haig A, Flavin S, MacDonald B. Effects of nabumetone, celecoxib, and ibuprofen on blood pressure control in hypertensive patients on angiotensin converting enzyme inhibitors. Am J Hypertens 2003;16:135–9.

1843. Pope JE, Anderson JJ, Felson DT. A meta-analysis of the effects of nonsteroidal anti-inflammatory drugs on blood pressure. Arch Intern Med 1993;153(4):477–84.

1844. Johnson AG, Nguyen TV, Day RO. Do nonsteroidal anti-inflammatory drugs affect blood pressure? A meta-analysis. Ann Intern Med 1994;121(4):289–300.

1845. Nowak J, Wennmalm A. Influence of indomethacin and of prostaglandin E1 on total and regional blood flow in man. Acta Physiol Scand 1978;102(4):484–91.

1846. Johnson AG, Nguyen TV, Owe-Young R, Williamson DJ, Day RO. Potential mechanisms by which nonsteroidal anti-inflammatory drugs elevate blood pressure: the role of endothelin-1. J Hum Hypertens 1996;10(4):257–61.

1847. Gurwitz JH, Avorn J, Bohn RL, Glynn RJ, Monane M, Mogun H. Initiation of antihypertensive treatment during nonsteroidal anti-inflammatory drug therapy. JAMA 1994;272(10):781–6.

1848. Curhan GC, Willett WC, Rosner B, Stampfer MJ. Frequency of analgesic use and risk of hypertension in younger women. Arch Intern Med 2002;162(19):2204–8.

1849. Radack KL, Deck CC, Bloomfield SS. Ibuprofen interferes with the efficacy of antihypertensive drugs. A randomized, double-blind, placebo-controlled trial of ibuprofen compared with acetaminophen. Ann Intern Med 1987;107(5):628–35.

1850. Chalmers JP, West MJ, Wing LM, Bune AJ, Graham JR. Effects of indomethacin, sulindac, naproxen, aspirin, and paracetamol in treated hypertensive patients. Clin Exp Hypertens A 1984;6(6):1077–93.

1851. Collins R, Peto R, MacMahon S, Hebert P, Fiebach NH, Eberlein KA, Godwin J, Qizilbash N, Taylor JO, Hennekens CH. Blood pressure, stroke, and coronary heart disease. Part 2. Short-term reductions in blood pressure: overview of randomised drug trials in their epidemiological context. Lancet 1990;335(8693):827–38.

1852. Heerdink ER, Leufkens HG, Herings RM, Ottervanger JP, Stricker BH, Bakker A. NSAIDs associated with increased risk of congestive heart failure in elderly patients taking diuretics. Arch Intern Med 1998;158(10):1108–12.

1853. Van den Ouweland FA, Gribnau FW, Meyboom RH. Congestive heart failure due to nonsteroidal anti-inflammatory drugs in the elderly. Age Ageing 1988;17(1):8–16.

1854. Page J, Henry D. Consumption of NSAIDs and the development of congestive heart failure in elderly patients: an underrecognized public health problem. Arch Intern Med 2000;160(6):777–84.

1855. Coffman JD, Cohen RA. Intra-arterial vasodilator agents to reverse human finger vasoconstriction. Clin Pharmacol Ther 1987;41(5):574–9.

1856. Auer J, Berent R, Weber T, Lamm G, Eber B. Catecholamine therapy inducing dynamic left ventricular outflow tract obstruction. Int J Cardiol 2005;101:325–8.

1857. Herskowitz A, Willoughby SB, Baughman KL, Schulman SP, Bartlett JD. Cardiomyopathy associated with antiretroviral therapy in patients with HIV infection: a report of six cases. Ann Intern Med 1992;116(4):311–3.

1858. Domanski MJ, Sloas MM, Follmann DA, Scalise PP 3rd, Tucker EE, Egan D, Pizzo PA. Effect of zidovudine and didanosine treatment on heart function in children infected with human immunodeficiency virus. J Pediatr 1995;127(1):137–46.

1859. Dilger JA, Rho EH, Que FG, Sprung J. Octreotide-induced bradycardia and heart block during surgical resection of a carcinoid tumour. Anesth Analg 2004;98:318–20.

1860. Eroglu Y, Emerick KM, Whitingon PF, Alonso EM. Octreotide therapy for control of acute gastrointestinal bleeding in children. J Ped Gastroenterol Nutr 2004;38:41–7.

1861. Gurovich I, Vempaty A, Lippmann S. QT_c prolongation: chlorpromazine and high-dosage olanzapine. Can J Psychiatry 2003;48:348.

1862. Dineen S, Withrow K, Voronovitch L, Munshi F, Nawbary MW, Lippmann S. QT_c prolongation and high-dose olanzapine. Psychosomatics 2003;44:174–5.

1863. Su K-P, Lane H-Y, Chuang C-L, Chen K-P, Shen WW. Olanzapine-induced QT_c prolongation in a patient with Wolff–Parkinson–White syndrome. Schizophr Res 2004;66:191–2.

1864. Czekalla J, Beasley CM Jr, Dellva MA, Berg PH, Grundy S. Analysis of the QT_c interval during olanzapine treatment of patients with schizophrenia and related psychosis. J Clin Psychiatry 2001;62(3):191–8.

1865. Cohen H, Loewenthal U, Matar MA, Kotler M. Reversal of pathologic cardiac parameters after transition from clozapine to olanzapine treatment: a case report. Clin Neuropharmacol 2001;24(2):106–8.

1866. Fontaine CS, Hynan LS, Koch K, Martin-Cook K, Svetlik D, Weiner MF. A double-blind comparison of olanzapine versus risperidone in the acute treatment of dementia-related behavioral disturbances in extended care facilities. J Clin Psychiatry 2003;64:726–30.

1867. Kosky N. A possible association between high normal and high dose olanzapine and prolongation of the PR interval. J Psychopharmacol 2002;16(2):181–2.

1868. Farooque R. Uncommon side effects associated with olanzapine. Pharmacopsychiatry 2003;36:83.

1869. Ng B, Postlethwaite A, Rollnik J. Peripheral oedema in patients taking olanzapine. Int Clin Psychopharmacol 2003;18:57–9.

1870. Kamijo Y, Soma K, Nagai T, Kurihara K, Ohwada T. Acute massive pulmonary thromboembolism associated with risperidone and conventional phenothiazines. Circ J 2003;67:46–8.

1871. Qureshi N. Atypical antipsychotics and dementia: some reflections! http: //bmj.bmjjournals.com/cgi/eletters/328/7450/1262-b.

1872. Mowat D, Fowlie D, MacEwan T. CSM warning on atypical antipsychotics and stroke may be detrimental for dementia. http: //bmj.bmjjournals.com/cgi/eletters/328/7450/1262-b.

1873. Schneider L, Dagerman K. Meta-analysis of atypical antipsychotics for dementia patients: balancing efficacy and adverse events. http: //ipa.confex.com/ipa/11congress/techprogram/paper_4201.htm.

1874. Krantz MJ, Mehler PS. Synthetic opioids and QT prolongation. Arch Intern Med 2003;163:1615; author reply 1615.

1875. De Bels D, Staroukine M, Devriendt J. Torsades de pointes due to methadone. Ann Intern Med 2003;139:E156.

1876. Porter BO, Coyn PJ, Smith WR. Methadone-related torsade de pointes in a sickle cell patient treated for chronic pain. Am J Hematol 2005;78(4):316–7.

1877. Rademacher S, Dietz R, Haverkamp W. QT prolongation and syncope with methadone, doxepin, and a beta-blocker. Ann Pharmacother 2005;39(10):1762–3.

1878. Martell BA, Arnsten JH, Krantz MJ, Gourevitch MN. Impact of methadone treatment on cardiac repolarisation and conduction in opioid users. Am J Cardiol 2005;95(7):915–8.

1879. Piquet V, Desmeules J, Enret G, Stoller R, Dayer P. QT interval prolongation in patients on methadone with concomitant drugs. J Clin Psychopharmacol 2004;24:446–8.

1880. Decerf JA, Gressens B, Brohet C, Liolios A, Hantson P. Can methadone prolong the QT interval? Intensive Care Med 2004;30:1690–1.

1881. Martell BA, Arnsten JH, Ray B, Gourevitch MN. The impact of methadone induction on cardiac conduction in opiate users. Ann Intern Med 2003;139:154–5.

1882. Kornick CA, Kilborn MJ, Santiago-Palma J, Schulman G, Thaler HT, Keefe DL, Katchman AN, Pezzullo JC, Ebert SN, Woosley RL, Payne R, Manfredi PL. QT_c interval prolongation associated with intravenous methadone. Pain 2003;105:499–506.

1883. Gil M, Sola M, Anguera I, Chapinal O, Cervantes M, Guma JR, Sequra F. QT prolongation and torsades de pointes in patients infected with human immunodeficiency virus and treated with methadone. Am J Cardiol 2003;8:995–7.

1884. Walker PW, Klein D, Kasza L. High dose methadone and ventricular arrhythmia: a report of three cases. Pain 2003;103:321–4.

1885. Semenkovich CF, Jaffe AS. Adverse effects due to morphine sulfate. Challenge to previous clinical doctrine. Am J Med 1985;79(3):325–30.

1886. Dilaveris P, Pantazis A, Vlasseros J, Gialafos J. Non-sustained ventricular tachycardia due to low-dose orphenadrine. Am J Med 2001;111(5):418–9.

1887. Mello NK, Mendelson JH, Sholar MB, Jaszyna-Gasior M, Goletiani N, Siegel AJ. Effects of the mixed mu/kappa opioid nalbuphine on cocaine-induced changes in subjective and cardiovascular responses in men. Neuropsychopharmacology 2005;30, 618–32.

1888. Anonymous. Drugs for parasitic infections. Med Lett Drugs Ther 1986;28(706):9–16.

1889. Christensen C. Bradycardia as a side-effect to oxybuprocaine. Acta Anaesthesiol Scand 1990;34(2):165–6.

1890. Remy B, Deby-Dupont G, Lamy M. Red blood cell substitutes: fluorocarbon emulsions and haemoglobin solutions. Br Med Bull 1999;55(1):277–98.

1891. Stowell CP. Hemoglobin-based oxygen carriers. Curr Opin Hematol 2002;9(6):537–43.

1892. Chang TM. Oxygen carriers. Curr Opin Investig Drugs 2002;3(8):1187–90.

1893. Jahr JS, Nesargi SB, Lewis K, Johnson C. Blood substitutes and oxygen therapeutics: an overview and current status. Am J Ther 2002;9(5):437–43.

1894. Hill SE, Gottschalk LI, Grichnik K. Safety and preliminary efficacy of hemoglobin raffimer for patients undergoing coronary artery bypass surgery. J Cardiothorac Vasc Anesth 2002;16(6):695–702.

1895. Why mothers die 1997–1999. The confidential enquiries into maternal deaths in the United KingdomLondon: RCOG Press;. 2001.

1896. Pinder AJ, Dresner M, Calow C, Shorten GD, O'Riordan J, Johnson R. Haemodynamic changes caused by oxytocin during caesarean section under spinal anaesthesia. Int J Obstet Anesth 2002;11(3):156–9.

1897. Moran C, Ni Bhuinneain M, Geary M, Cunningham S, McKenna P, Gardiner J. Myocardial ischaemia in normal patients undergoing elective Caesarean section: a peripartum assessment. Anaesthesia 2001;56(11):1051–8.

1898. Spence A. Oxytocin during Caesarean section. Anaesthesia 2002;57:710–1.

1899. Liou SC, Chen C, Wong SY, Wong KM. Ventricular tachycardia after oxytocin injection in patients with prolonged Q-T interval syndrome—report of two cases. Acta Anaesthesiol Sin 1998;36(1):49–52.

1900. Rowinsky EK, Eisenhauer EA, Chaudhry V, Arbuck SG, Donehower RC. Clinical toxicities encountered with paclitaxel (Taxol). Semin Oncol 1993;20(4 Suppl 3):1–15.

1901. McGuire WP, Rowinsky EK, Rosenshein NB, Grumbine FC, Ettinger DS, Armstrong DK, Donehower RC. Taxol: a unique antineoplastic agent with significant activity in advanced ovarian epithelial neoplasms. Ann Intern Med 1989;111(4):273–9.

1902. Onetto N, Canetta R, Winograd B, Catane R, Dougan M, Grechko J, Burroughs J, Rozencweig M. Overview of Taxol safety. J Natl Cancer Inst Monogr 1993;(15):131–9.

1903. Schiller JH, Storer B, Tutsch K, Arzoomanian R, Alberti D, Feierabend C, Spriggs D. Phase I trial of 3-hour infusion of paclitaxel with or without granulocyte colony-stimulating factor in patients with advanced cancer. J Clin Oncol 1994;12(2):241–8.

1904. Trimble EL, Adams JD, Vena D, Hawkins MJ, Friedman MA, Fisherman JS, Christian MC, Canetta R, Onetto N, Hayn R, Arbuck S. Paclitaxel for platinum-refractory ovarian cancer: results from the first 1,000 patients registered to National Cancer Institute Treatment Referral Center 9103. J Clin Oncol 1993;11(12):2405–10.

1905. Rowinsky EK, McGuire WP, Guarnieri T, Fisherman JS, Christian MC, Donehower RC. Cardiac disturbances during the administration of taxol. J Clin Oncol 1991;9(9):1704–12.

1906. Alagaratnam TT. Sudden death 7 days after paclitaxel infusion for breast cancer. Lancet 1993;342(8881):1232–3.

1907. Arbuck SG, Strauss H, Rowinsky E, Christian M, Suffness M, Adams J, Oakes M, McGuire W, Reed E, Gibbs H, Greenfield R, Montello M. A reassessment of cardiac toxicity associated with Taxol. J Natl Cancer Inst Monogr 1993;(15):117–30.

1908. Ekholm EM, Salminen EK, Huikuri HV, Jalonen J, Antila KJ, Salmi TA, Rantanen VT. Impairment of heart rate variability during paclitaxel therapy. Cancer 2000;88(9):2149–53.

1909. Ekholm E, Rantanen V, Bergman M, Vesalainen R, Antila K, Salminen E. Docetaxel and autonomic cardiovascular control in anthracycline treated breast cancer patients. Anticancer Res 2000;20(3B):2045–8.

1910. Bristol-Myers Squibb Pharmaceuticals. Taxol (paclitaxel). ABPI Data Sheet Compendium. 1995.

1911. Martínez-Mir I, Rubio E, Morales-Olivas FJ, Palop-Larrea V. Transient ischemic attack secondary to hypertensive crisis related to *Panax ginseng*. Ann Pharmacother 2004;38:1970.

1912. Coleman AJ, Downing JW, Leary WP, Moyes DG, Styles M. The immediate cardiovascular effects of pancuronium, alcuronium and tubocurarine in man. Anaesthesia 1972;27(4):415–22.

1913. Saxena PR, Bonta IL. Mechanism of selective cardiac vagolytic action of pancuronium bromide. Specific blockade of cardiac muscarinic receptors. Eur J Pharmacol 1970;11(3):332–41.

1914. Domenech JS, Garcia RC, Sastain JM, Loyola AQ, Oroz JS. Pancuronium bromide: an indirect sympathomimetic agent. Br J Anaesth 1976;48(12):1143–8.

1915. Cardan E, Nana A, Domokos M. Blood catecholamine changes after pancuronium bromide administration. Xth Congress of the Scandinavian Society of Anesthesiologists. Lund 1971;57:.

1916. Quintana A. Effect of pancuronium bromide on the adrenergic reactivity of the isolated rat vas deferens. Eur J Pharmacol 1977;46(3):275–7.

1917. Docherty JR, McGrath JC. Potentiation of cardiac sympathetic nerve responses in vivo by pancuronium bromide. Br J Pharmacol 1977;61(3):P472–3.

1918. Docherty JR, McGrath JC. Sympathomimetic effects of pancuronium bromide on the cardiovascular system of the pithed rat: a comparison with the effects of drugs blocking the neuronal uptake of noradrenaline. Br J Pharmacol 1978;64(4):589–99.

1919. Seed RF, Chamberlain JH. Myocardial stimulation by pancuronium bromide. Br J Anaesth 1977;49(5):401–7.

1920. Bowman WC. Pharmacology of Neuromuscular Function. 2nd ed. London/Boston/Singapore/Sydney/Toronto/Wellington: Wright, 1990.

1921. Bowman WC. Non-relaxant properties of neuromuscular blocking drugs. Br J Anaesth 1982;54(2):147–60.

1922. Marshall IG. Pharmacological effects of neuromuscular blocking agents: interaction with cholinoceptors other than nicotinic receptors of the neuromuscular junction. Anest Rianim 1986;27:19.

1923. Belani KG, Anderson WW, Buckley JJ. Adverse drug interaction involving pancuronium and aminophylline. Anesth Analg 1982;61(5):473–4.

1924. Heinonen J, Takkunen O. Bradycardia during antagonism of pancuronium-induced neuromuscular block. Br J Anaesth 1977;49(11):1109–15.

1925. Debourdeau PM, Djezzar S, Estival JL, Zammit CM, Richard RC, Castot AC. Life-threatening eosinophilic pleuropericardial effusion related to vitamins B5 and H. Ann Pharmacother 2001;35(4):424–6.

1926. Griffen MP, Siadaty MS. Papaverine prolongs patency of peripheral arterial catheters in neonates. J Pediatr 2005;146(1) 62–5.

1927. Curhan GC, Willett WC, Rosner B, Stampfer MJ. Frequency of analgesic use and risk of hypertension in younger women. Arch Intern Med 2002;162(19):2204–8.

1928. Radack KL, Deck CC, Bloomfield SS. Ibuprofen interferes with the efficacy of antihypertensive drugs. A randomized, double-blind, placebo-controlled trial of ibuprofen compared with acetaminophen. Ann Intern Med 1987;107(5):628–35.

1929. Chalmers JP, West MJ, Wing LM, Bune AJ, Graham JR. Effects of indomethacin, sulindac, naproxen, aspirin, and paracetamol in treated hypertensive patients. Clin Exp Hypertens A 1984;6(6):1077–93.

1930. Wait RB, Greenhalgh D, Gamelli RL. Vascular injury in the neonate associated with intra-arterial injection of paraldehyde. Clin Pediatr (Phila) 1984;23(6):324.

1931. Neer RM, Arnaud CD, Zanchetta JR, Prince R, Gaich GA, Reginster JY, Hodsman AB, Eriksen EF, Ish-Shalom S, Genant HK, Wang O, Mitlak BH. Effect of parathyroid hormone (1–34) on fractures and bone mineral density in postmenopausal women with osteoporosis. N Engl J Med 2001;344(19):1434–41.

1932. Morley P, Whitfield JF, Willick GE. Parathyroid hormone: an anabolic treatment for osteoporosis. Curr Pharm Des 2001;7(8):671–87.

1933. Kane KF, Lowes JR. Peripheral parenteral nutrition and venous thrombophlebitis. Nutrition 1997;13(6):577–8.

1934. May J, Murchan P, MacFie J, Sedman P, Donat R, Palmer D, Mitchell CJ. Prospective study of the aetiology of infusion phlebitis and line failure during peripheral parenteral nutrition. Br J Surg 1996;83(8):1091–4.

1935. Madan M, Alexander DJ, Mellor E, Cooke J, et al. A randomised study of the effects of osmolality and heparin with hydrocortisone on thrombophlebitis in peripheral intravenous nutrition. Clin Nutr 1991;10:309–14.

1936. Kerin MJ, Pickford IR, Jaeger H, Couse NF, et al. A prospective and randomised study comparing the incidence of infusion phlebitis during continuous and cyclic peripheral parenteral nutrition. Clin Nutr 1991;10:315.

1937. Erfurth A, Loew M, Dobmeier P, Wendler G. EKG – Veran derungen nach faroxetineo Drei Fallberichte. [ECG changes after paroxetine. Three case reports.] Nervenarzt 1998;69(7):629–31.

1938. Stork CM, Cantor R. Pemoline induced acute choreoathetosis: case report and review of the literature. J Toxicol Clin Toxicol 1997;35(1):105–8.

1939. Nakamura H, Blumer JL, Reed MD. Pemoline ingestion in children: a report of five cases and review of the literature. J Clin Pharmacol 2002;42(3):275–82.

1940. Pless M, Sandson T. Chronic internuclear ophthalmoplegia. A manifestation of D-penicillamine cerebral vasculitis. J Neuroophthalmol 1997;17(1):44–6.

1941. Blanca M, Perez E, Garcia J, Miranda A, Fernandez J, Vega JM, Terrados S, Avila M, Martin A, Suau R. Anaphylaxis to amoxycillin but good tolerance for benzyl penicillin. In vivo and in vitro studies of specific IgE antibodies. Allergy 1988;43(7):508–10.

1942. Hunziker I, Kunzi UP, Braunschweig S, Zehnder D, Hoigné R. Comprehensive hospital drug monitoring (CHDM): adverse skin reactions, a 20-year survey. Allergy 1997;52(4):388–93.

1943. Hoigne R. Akute Nebenreaktionen auf Penicillinpräparate. [Acute side-reactions to penicillin preparations.] Acta Med Scand 1962;171:201–8.

1944. Idsoe O, Guthe T, Willcox RR, de Weck AL. Art und Ausmass der Penizillinnebenwirkungen unter besonderer Berücksichtigung von 151 Todesfällen nach anaphylaktischem Schock. [Nature and extent of penicillin side effects with special reference to 151 fatal cases after anaphylactic shock.] Schweiz Med Wochenschr 1969;99(33):1190–7contd.

1945. Hoffman DR, Hudson P, Carlyle SJ, Massello W 3rd. Three cases of fatal anaphylaxis to antibiotics in patients with prior histories of allergy to the drug. Ann Allergy 1989;62(2):91–3.

1946. Lin RY. A perspective on penicillin allergy. Arch Intern Med 1992;152(5):930–7.

1947. Spark RP. Fatal anaphylaxis due to oral penicillin. Am J Clin Pathol 1971;56(3):407–11.

1948. Cullen DJ. Severe anaphylactic reaction to penicillin during halothane anaesthesia. A case report. Br J Anaesth 1971;43(4):410–2.

1949. Batchelor RC, Horne GO, Rogerson HL. An unusual reaction to procaine penicillin in aqueous suspension. Lancet 1951;2(5):195–8.

1950. Hoigné R, Schoch K. Anaphylaktischer Schock und akute nichtallergische Reaktionen nach Procain-Penicillin. [Anaphylactic shock and acute nonallergic reactions following procaine-penicillin.] Schweiz Med Wochenschr 1959;89:1350–6.

1951. Dry J, Leynadier F, Damecour C, Pradalier A, Herman D. Réaction pseudo-anaphylactique à la procaine-pénicilline G. Trois cas de syndrome de Hoigné. [Pseudo-anaphylactic reaction to procaine-penicillin G. 3 cases of Hoigné's syndrome.] Nouv Presse Med 1976;5(22):1401–3.

1952. Schmied C, Schmied E, Vogel J, Saurat JH. Syndrome de Hoigné ou réaction pseudo-anaphylactique à la procaine pénicilline G: un classique d'actualité. [Hoigné's syndrome or pseudo-anaphylactic reaction to procaine penicillin G: a still current classic.] Schweiz Med Wochenschr 1990;120(29):1045–9.

1953. Lewis GW. Acute immediate reactions to penicillin. BMJ 1957;(5028):1151–2.

1954. Silber TJ, D'Angelo L. Psychosis and seizures following the injection of penicillin G procaine. Hoigné's syndrome. Am J Dis Child 1985;139(4):335–7.

1955. Downham TF 2nd, Cawley RA, Salley SO, Dal Santo G. Systemic toxic reactions to procaine penicillin G. Sex Transm Dis 1978;5(1):4–9.

1956. Silber TJ, D'Angelo LJ. Panic attack following injection of aqueous procaine penicillin G (Hoigné syndrome). J Pediatr 1985;107(2):314–5.

1957. Berger H, Juchinka H, Tomczyk D, et al. Pseudo-anaphylactic syndrome after procaine penicillin in children. In: Abstracts, 10th Jubilee Congress. Bialystok: Polish Neurology Society, 1977:82.

1958. Menke HE, Pepplinkhuizen L. Acute non-allergic reaction to aqueous procaine penicillin. Lancet 1974;2(7882):723–4.

1959. Hoigné R, Krebs A. Kombinierte anaphylaktische und embolisch-toxische Reaktion durch akzidentelle intravaskuläre Injektion von Procain-Penicillin. [Combined anaphylactic and embolic-toxic reaction caused by the accidental intravascular injection of procaine penicillin.] Schweiz Med Wochenschr 1964;94:610–4.

1960. Kryst L, Wanyura H. Hoigné's syndrome—its course and symptomatology. J Maxillofac Surg 1979;7(4):320–6.

1961. Ernst G, Reuter E. Nicht-allergische tödliche Zwischenfälle nach depot-Penicillin. Beitrag zur Pathogenese und Prophylaxe. [Nonallergic fatal incidents following depot penicillin. Pathogenesis and prevention.] Dtsch Med Wochenschr 1970;95(12):618.

1962. Bornemann K, Schulz E, Heinecker R. Akute, nicht-allergische Reaktionen nach i.m. Gabe von Clemizol-Penicillin G und Streptomycin. [Acute, non-allergic reactions following i.m. administration of clemizole-penicillin G and streptomycin.] Munch Med Wochenschr 1966;108(15):834–7.

1963. Bredt J. Akute nicht-allergische Reaktionen bei Anwendung von Depot-Penicillin. [Acute non-allergic reactions in the use of depot-penicillin.] Dtsch Med Wochenschr 1965;90:1559–63.

1964. Galpin JE, Chow AW, Yoshikawa TT, Guze LB. "Pseudoanaphylactic" reactions from inadvertent infusion of procaine penicillin G. Ann Intern Med 1974;81(3):358–9.

1965. Kraus SJ, Green RL. Pseudoanaphylactic reactions with procaine penicillin. Cutis 1976;17(4):765–7.

1966. Ilechukwu ST. Acute psychotic reactions and stress response syndromes following intramuscular aqueous procaine penicillin. Br J Psychiatry 1990;156:554–9.

1967. Clauberg G. Wiederbelebung bei embolisch-toxischer Komplikation. [Resuscitation in embolic and toxic complication caused by intravascular administration of a depot-penicillin.] Anaesthesist 1966;15(8):284–5.

1968. Utley PM, Lucas JB, Billings TE. Acute psychotic reactions to aqueous procaine penicillin. South Med J 1966;59(11):1271–4.

1969. Randazzo SD, DiPrima G. Psicosi allucinatoria acuta da penicillina-procaina in sospensione acquosa. [Acute hallucinatory psychoses caused by procaine penicillin in aqueous suspension.] Minerva Dermatol 1959;34(6):422–8.

1970. Green RL, Lewis JE, Kraus SJ, Frederickson EL. Elevated plasma procaine concentrations after administration of procaine penicillin G. N Engl J Med 1974;291(5):223–6.

1971. Markowitz M, Kaplan E, Cuttica R, et alInternational Rheumatic Fever Study Group. Allergic reactions to long-term benzathine penicillin prophylaxis for rheumatic fever. Lancet 1991;337(8753):1308–10.

1972. Ewers HR, Brouwers HP, Merguet P, Hengstebeck W. Nebenwirkungen nach Stimulation der Magensekretion mit Pentagastrin. [Side effects after stimulation of gastric secretion with pentagastrin.] Med Klin 1976;71(1):19–23.

1973. Lines C, Challenor J, Traub M. Cholecystokinin and anxiety in normal volunteers: an investigation of the anxiogenic properties of pentagastrin and reversal by the cholecystokinin receptor subtype B antagonist L-365,260. Br J Clin Pharmacol 1995;39(3):235–42.

1974. Drucker D. Atrial fibrillation after administration of calcium and pentagastrin. N Engl J Med 1981;304(23):1427–8.

1975. Masur H. Prevention and treatment of *Pneumocystis* pneumonia. N Engl J Med 1992;327(26):1853–60.

1976. Ryan C, Madalon M, Wortham DW, Graziano FM. Sulfa hypersensitivity in patients with HIV infection: onset, treatment, critical review of the literature. WMJ 1998;97(5):23–7.

1977. Gradon JD, Fricchione L, Sepkowitz D. Severe hypomagnesemia associated with pentamidine therapy. Rev Infect Dis 1991;13(3):511–2.

1978. Kroll CR, Gettes LS. T wave alternans and torsades de Pointes after the use of intravenous pentamidine. J Cardiovasc Electrophysiol 2002;13(9):936–8.

1979. Glass PS, Camporesi EM, Shafron D, Quill T, Reves JG. Evaluation of pentamorphone in humans: a new potent opiate. Anesth Analg 1989;68(3):302–7.

1980. Ahlskog JE, Muenter MD. Pergolide: long-term use in Parkinson's disease. Mayo Clin Proc 1988;63(10):979–87.

1981. Kando JC, Keck PE Jr, Wood PA. Pergolide-induced hypotension. DICP 1990;24(5):543.

1982. Van Camp G, Flamez A, Cosyns B, Weytjens C, Muyldermans L, Van Zandijcke M, De Sutter J, Santens P, Decoodt P, Moerman C, Schoors D. Treatment of Parkinson's disease with pergolide and relation to restrictive valvular disease. Lancet 2004;363:1179–83.

1983. Grosset KA, Grosset DG. Pergolide in Parkinson's disease: time for a change? Lancet 2004;363:1997–8.

1984. Balestrero LM, Beaver CR, Rigas JR. Hypertensive crisis following meperidine administration and chemoembolization of a carcinoid tumor. Arch Intern Med 2000;160(15):2394–5.

1985. Rostagno C, Caciolli S, Felici M, Gori F, Neri Serneri GG. Dilated cardiomyopathy associated with chronic consumption of phendimetrazine. Am Heart J 1996;131(2):407–9.

1986. Warner MA, Harper JV. Cardiac dysrhythmias associated with chemical peeling with phenol. Anesthesiology 1985;62(3):366–7.

1987. Morrison JE Jr, Matthews D, Washington R, Fennessey PV, Harrison LM. Phenol motor point blocks in children: plasma concentrations and cardiac dysrhythmias. Anesthesiology 1991;75(2):359–62.

1988. Gaudy JH, Tricot C, Sezeur A. Troubles du rythme cardiaque graves après phénolisation splanchnique peropératoire. [Serious heart rate disorders following perioperative splanchnic nerve phenol nerve block.] Can J Anaesth 1993;40(4):357–9.

1989. Anonymous. Failure to provide the necessaries of life. NZ Med J 2003;116(1168):1.

1990. Grummitt RM, Goat VA. Intracranial pressure after phenoperidine. Anaesthesia 1984;39(6):565–7.

1991. Loke YK, Derry S, Pritchard-Copley A. Appetite suppressants and valvular heart disease—a systematic review. BMC Clin Pharmacol 2002;2(1):6.

1992. Sobel RM. Ruptured retroperitoneal aneurysm in a patient taking phentermine hydrochloride. Am J Emerg Med 1999;17(1):102–3.

1993. Fraunfelder FW, Fraunfelder FT, Jensvold B. Adverse systemic effects from pledgets of topical ocular phenylephrine 10%. Am J Ophthalmol 2002;134(4):624–5.

1994. Baldwin FJ, Morley AP. Intraoperative pulmonary oedema in a child following systemic absorption of phenylephrine eyedrops. Br J Anaesth 2002;88(3):440–2.

1995. Greher M, Hartmann T, Winkler M, Zimpfer M, Crabnor CM. Hypertension and pulmonary edema associated with subconjunctival phenylephrine in a 2-month-old child during cataract extraction. Anesthesiology 1998;88(5):1394–6.

1996. Hefer D, Bukharovich I, Nasrallah EJ, Plotnikov A. Prominent positive U waves appearing with high-dose intravenous phenylephrine. J Electrocardiol 2005;38:378–82.

1997. Lake CR, Zaloga G, Clymer R, Quirk RM, Chernow B. A double dose of phenylpropanolamine causes transient hypertension. Am J Med 1988;85(3):339–43.

1998. Pilsczek FH, Karcic AA, Freeman I. Case report: Dexatrim (phenylpropanolamine) as a cause of myocardial infarction. Heart Lung 2003;32:100–4.

1999. Oosterbaan R, Burns MJ. Myocardial infarction associated with phenyl-propanolamine. J Emerg Med 2000;18:55–9.

2000. Chin C, Choy M. Cardiomyopathy induced by phenylpropanolamine. J Pediatr 1993;123(5):825–7.

2001. Anonymous. Norephedrin (Phenylpropanolamin) statt Nor-d-pseudoephedrin als Appetitzügler? Schweiz Apoth Ztg 1977;115:430.

2002. Taliercio CP, Olney BA, Lie JT. Myocarditis related to drug hypersensitivity. Mayo Clin Proc 1985;60(7):463–8.

2003. O'Brien TJ, Cascino GD, So EL, Hanna DR. Incidence and clinical consequence of the purple glove syndrome in patients receiving intravenous phenytoin. Neurology 1998;51(4):1034–9.

2004. Berry JM, Kowalski A, Fletcher SA. Sudden asystole during craniotomy: unrecognized phenytoin toxicity. J Neurosurg Anesthesiol 1999;11(1):42–5.

2005. Yoshikawa H, Abe T, Oda Y. Purple glove syndrome caused by oral administration of phenytoin. J Child Neurol 2000;15(11):762.

2006. Burneo JG, Anandan JV, Barkley GL. A prospective study of the incidence of the purple glove syndrome. Epilepsia 2001;42(9):1156–9.

2007. Meara FM, O'Brien TJ, Cook MJ, Vajda FJ. Prospective study of the incidence of local cutaneous reactions (the purple limb syndrome) in patients receiving i.v. phenytoin Epilepsia 1999;40(Suppl 7):145.

2008. Anderson GD, Lin Y, Temkin NR, Fischer JH, Winn HR. Incidence of intravenous site reactions in neurotrauma patients receiving valproate or phenytoin. Ann Pharmacother 2000;34(6):697–702.

2009. Keegan MT, Bondy LR, Blackshear JL, Lanier WL. Hypocalcemia-like electrocardiographic changes after administration of intravenous fosphenytoin. Mayo Clin Proc 2002;77(6):584–6.

2010. Chappe SG, Roenigk HH Jr, Miller AJ, Beeaff DE, Tyrpin L. The effect of photochemotherapy on the cardiovascular system. J Am Acad Dermatol 1981;4(5):561–6.

2011. Wiseman LR, Faulds D. Oral pilocarpine: a review of its pharmocological properties and clinical potential in xerostomia. Drugs 1995;49(1):143–55.

2012. Committee on Safety of Medicines—Medicines Control Agency. Cardiac arrhythmias with pimozide (Orap). Curr Probl Pharmacovigilance 1995;21:1.

2013. Wittek L, Gecsenyi M, Barna B, Hargitay Z, Adorjan K. Report on clinical test of pipecurium bromide. Arzneimittelforschung 1980;30(2a):379–83.

2014. Boros M, Szenohradszky J, Marosi G, Toth I. Comparative clinical study of pipecurium bromide and pancuronium bromide. Arzneimittelforschung 1980;30(2a):389–93.

2015. Alant O, Darvas K, Pulay I, Weltner J, Bihari I. First clinical experience with a new neuromuscular blocker pipecurium bromide. Arzneimittelforschung 1980;30(2a):374–9.

2016. Bunjatjan AA, Miheev VI. Clinical experience with a new steroid muscle relaxant: pipecurium bromide. Arzneimittelforschung 1980;30(2a):383–5.

2017. Newton DE, Richardson FJ, Agoston S. Preliminary studies in man with pipecurium bromide (Arduan), a new steroid neuromuscular blocking agent. Br J Anaesth 1982;54:P789.

2018. Larijani GE, Bartkowski RR, Azad SS, Seltzer JL, Weinberger MJ, Beach CA, Goldberg ME. Clinical pharmacology of pipecuronium bromide. Anesth Analg 1989;68(6):734–9.

2019. Foldes FF, Nagashima H, Nguyen HD, Duncalf D, Goldiner PL. Neuromuscular and cardiovascular effects of pipecuronium. Can J Anaesth 1990;37(5):549–55.

2020. Barankay A. Circulatory effects of pipecurium bromide during anaesthesia of patients with severe valvular and ischaemic heart diseases. Arzneimittelforschung 1980;30(2a):386–9.

2021. Tassonyi E, Neidhart P, Pittet JF, Morel DR, Gemperle M. Cardiovascular effects of pipecuronium and pancuronium in patients undergoing coronary artery bypass grafting. Anesthesiology 1988;69(5):793–6.

2022. Stanley JC, Carson IW, Gibson FM, McMurray TJ, Elliott P, Lyons SM, Mirakhur RK. Comparison of the haemodynamic effects of pipecuronium and pancuronium during fentanyl anaesthesia. Acta Anaesthesiol Scand 1991;35(3):262–6.

2023. Ulbricht C, Basch E, Boon H, Ernst E, Hammerness P, Sollars D, Tsourounis C, Woods J, Bent S. Safety review of kava (*Piper methysticum*) by the Natural Standard Research Collaboration. Expert Opin Drug Saf 2005;4(4):779–94.

2024. Gouffault J, Van den Driessche J, Pony JC, Courgeon P, Thomas R. Les troubles de conduction induits par la pipérazine: étude clinique et expérimentale. [Conduction disorders induced by piperazine; clinical and experimental study.] Arch Mal Coeur Vaiss 1973;66(10):1289–95.

2025. Boulis NM, Bobek MP, Schmaier A, Hoff JT. Use of factor IX complex in warfarin-related intracranial hemorrhage. Neurosurgery 1999;45(5):1113–9.

2026. Yarranton H, Cohen H, Pavord SR, Benjamin S, Hagger D, Machin SJ. Venous thromboembolism associated with the management of acute thrombotic thrombocytopenic purpura. Br J Haematol 2003;121:778–85.

2027. Tassinari D, Sartori S, Drudi G, Panzini I, Gianni L, Pasquini E, Abbasciano V, Ravaioli A, Iorio D. Cardiac arrhythmias after cisplatin infusion: three case reports and a review of the literature. Ann Oncol 1997;8(12):1263–7.

2028. Altundag O, Celik I, Kars A. Recurrent asymptomatic bradycardia episodes after cisplatin infusion. Ann Pharmacother 2001;35(5):641–2.

2029. Fassio T, Canobbio L, Gasparini G, Villani F. Paroxysmal supraventricular tachycardia during treatment with cisplatin and etoposide combination. Oncology 1986;43(4):219–20.

2030. Albain KS, Rusch VW, Crowley JJ, Rice TW, Turrisi AT 3rd, Weick JK, Lonchyna VA, Presant CA, McKenna RJ, Gandara DR, et al. Concurrent cisplatin/etoposide plus chest radiotherapy followed by surgery for stages IIIA (N2) and IIIB non-small-cell lung cancer: mature results of Southwest Oncology Group phase II study 8805. J Clin Oncol 1995;13(8):1880–92.

2031. Samuels BL, Vogelzang NJ, Kennedy BJ. Vascular toxicity following vinblastine, bleomycin, and cisplatin therapy for germ cell tumours. Int J Androl 1987;10(1):363–9.

2032. Samuels BL, Vogelzang NJ, Kennedy BJ. Severe vascular toxicity associated with vinblastine, bleomycin, and cisplatin chemotherapy. Cancer Chemother Pharmacol 1987;19(3):253–6.

2033. Heier MS, Nilsen T, Graver V, Aass N, Fossa SD. Raynaud's phenomenon after combination chemotherapy of testicular cancer, measured by laser Doppler flowmetry. A pilot study. Br J Cancer 1991;63(4):550–2.

2034. Dorr RT. Managing extravasations of vesicant chemotherapy drugs. In: Lipp HP, editor. Anticancer Drug Toxicity; Prevention, Management and Clinical Pharmacokinetics. New York-Basel: Marcel Dekker Inc, 1999:279–318.

2035. Boccara F, Benhaiem-Sigaux N, Cohen A. Acute myopericarditis after diphtheria, tetanus, and polio vaccination. Chest 2001;120(2):671–2.

2036. Yuasa J, Naya Y, Tanaka M, Amakasu M, Yamaguchi K. [Clinical experiences of endotoxin removal columns in septic shock due to urosepsis: report of three cases.]Hinyokika Kiyo 2000;46(11):819–22.

2037. Terawaki H, Kasai K, Kobayashi H, Hirano K, Hamaguchi A, Kase Y, Horiguchi T, Yokoyama K, Yamamoto H, Nakayama M, Kawaguchi Y, Hosoya T. [A study on the mechanism of enhanced diuresis following direct hemoperfusion with polymyxin B-immobilized fiber.]Nippon Jinzo Gakkai Shi 2000;42(5):359–64.

2038. Claes H, Stroobants D, Van Meerbeek J, Verbeken E, Knockaert D. Pulmonary migration following periurethral polytetrafluoroethylene injection for urinary incontinence. J Urol 1991;145:839–40.

2039. Weingarten J, Kauffman SL. Teflon embolization to pulmonary arteries. Ann Thorac Surg 1977;23(4):371–3.

2040. Weiner WJ, Factor SA, Jankovic J, Hauser RA, Tetrud JW, Waters CH, Shulman LM, Glassman PM, Beck B, Paume D, Doyle C. The long-term safety and efficacy of pramipexole in advanced Parkinson's disease. Parkinsonism Relat Disord 2001;7(2):115–20.

2041. Tan EK, Ondo W. Clinical characteristics of pramipexole-induced peripheral edema. Arch Neurol 2000;57(5):729–32.

2042. Gohn DC, Simmons TW. Polymorphic ventricular tachycardia (torsade de pointes) associated with the use of probucol. N Engl J Med 1992;326(21):1435–6.

2043. Matsuhashi H, Onodera S, Kawamura Y, Hasebe N, Kohmura C, Yamashita H, Tobise K. Probucol-induced QT prolongation and torsades de pointes. Jpn J Med 1989;28(5):612–5.

2044. Force L, Barrufet P, Herreras Z, Bolibar I. Deep venous thrombosis and megestrol in patients with HIV infection. AIDS 1999;13(11):1425–6.

2045. Carnall D. Controversy rages over new contraceptive data. BMJ 1995;311:1117–8.

2046. World Health Organization Collaborative Study of Cardiovascular Disease and Steroid Hormone Contraception. Effect of different progestagens in low oestrogen oral contraceptives on venous thromboembolic disease. Lancet 1995;346(8990):1582–8.

2047. Jick H, Jick SS, Gurewich V, Myers MW, Vasilakis C. Risk of idiopathic cardiovascular death and nonfatal venous thromboembolism in women using oral contraceptives with differing progestagen components. Lancet 1995;346(8990):1589–93.

2048. Spitzer WO. Data from transnational study of oral contraceptives have been misused. BMJ 1995;311(7013):1162.

2049. Armstrong JL, Reid M, Bigrigg A. Scare over oral contraceptives. Effect on behaviour of women attending a family planning clinic. BMJ 1995;311(7020):1637.

2050. Seamark CJ. Scare over oral contraceptives. Effect on women in a general practice in Devon. BMJ 1995;311(7020):1637.

2051. Davies AW, York JR, Jones SR. [Scare over oral contraceptives] ··· and south Wales.]BMJ 1995;311(7020):1637–8.

2052. Ketting E. Third generation oral contraceptives. CSM's advice will harm women's health worldwide. BMJ 1996;312(7030):576.

2053. Stewart-Brown S, Pyper C. Third generation oral contraceptives. CSM should rethink its approach for such announcements. BMJ 1996;312(7030):576.

2054. Smith C. Third generation oral contraceptives. How one clinic's practice conforms with CSM's advice. BMJ 1996;312(7030):576–7.

2055. Carnall D, Karcher H, Lie LG, Sheldon T, Spurgeon D, Josefson D, Zinn C. Third generation oral contraceptives—the controversy. BMJ 1995;311(7020):1589–90.

2056. Spitzer WO, Lewis MA, Heinemann LA, Thorogood M, MacRae KD. Third generation oral contraceptives and risk of venous thromboembolic disorders: an international case-control study. Transnational Research Group on Oral Contraceptives and the Health of Young Women. BMJ 1996;312(7023):83–8.

2057. Bloemenkamp KW, Rosendaal FR, Helmerhorst FM, Buller HR, Vandenbroucke JP. Enhancement by factor V Leiden mutation of risk of deep-vein thrombosis associated with oral contraceptives containing a third-generation progestagen. Lancet 1995;346(8990):1593–6.

2058. McPherson K. Third generation oral contraception and venous thromboembolism. BMJ 19967;312(7023):68–9.

2059. Khan IA. Single oral loading dose of propafenone for pharmacological cardioversion of recent-onset atrial fibrillation. J Am Coll Cardiol 2001;37(2):542–7.

2060. Antman EM, Beamer AD, Cantillon C, McGowan N, Friedman PL. Therapy of refractory symptomatic atrial fibrillation and atrial flutter: a staged care approach with new antiarrhythmic drugs. J Am Coll Cardiol 1990;15(3):698–707.

2061. Escande M, Diadema B, Maarek-Charbit M. Étude a long terme de la propafénone dans l'extrasystolie ventriculaire grave du sujet agé. [Long-term study of propafenone in severe ventricular extrasystole in elderly subjects.] Ann Cardiol Angeiol (Paris) 1989;38(9):555–60.

2062. Colas A, Maarek-Charbit M. Propafenone per os dans les troubles du rythme ventriculaire. [Propafenone per os in ventricular arrhythmia.] Cah Anesthesiol 1989;37(4):241–4.

2063. Cobbe SM, Rae AP, Poloniecki JDUK Propafenone PSVT Study Group. A randomized, placebo-controlled trial of propafenone in the prophylaxis of paroxysmal supraventricular tachycardia and paroxysmal atrial fibrillation. Circulation 1995;92(9):2550–7.

2064. Mackstaller LL, Marcus FI. Rapid ventricular response due to treatment of atrial flutter or fibrillation with Class I antiarrhythmic drugs. Ann Noninvasive Electrocardiol 2000;5:101–4.

2065. Schumacher B, Jung W, Lewalter T, Vahlhaus C, Wolpert C, Luderitz B. Radiofrequency ablation of atrial flutter due to administration of class IC antiarrhythmic drugs for atrial fibrillation. Am J Cardiol 1999;83(5):710–3.

2066. Chopra IJ, Baber K. Use of oral cholecystographic agents in the treatment of amiodarone-induced hyperthyroidism. J Clin Endocrinol Metab 2001;86(10):4707–10.

2067. Hug CC Jr, McLeskey CH, Nahrwold ML, Roizen MF, Stanley TH, Thisted RA, Walawander CA, White PF, Apfelbaum JL, Grasela TH, et al. Hemodynamic effects of propofol: data from over 25,000 patients. Anesth Analg 1993;77(Suppl 4):S21–9.

2068. Searle NR, Sahab P. Propofol in patients with cardiac disease. Can J Anaesth 1993;40(8):730–47.

2069. Warden JC, Pickford DR. Fatal cardiovascular collapse following propofol induction in high-risk patients and dilemmas in the selection of a short-acting induction agent. Anaesth Intensive Care 1995;23(4):485–7.

2070. Wilhelm S, Standl T. Bietet Propofol Vorteile gegeniiber Isofluran für die sufentanil-supplementierte Auästhesie bei kindern in des strabismuschirurgie? [Does propofol have advantages over isoflurane for sufentanil supplemented anesthesia in children for strabismus surgery?] Anasthesiol Intensivmed Notfallmed Schmerzther 1996;31(7):414–9.

2071. Bilotta F, Fiorani L, La Rosa I, Spinelli F, Rosa G. Cardiovascular effects of intravenous propofol administered at two infusion rates: a transthoracic echocardiographic study. Anaesthesia 2001;56(3):266–71.

2072. Carrasco G, Cabre L, Sobrepere G, Costa J, Molina R, Cruspinera A, Lacasa C. Synergistic sedation with propofol and midazolam in intensive care patients after coronary artery bypass grafting. Crit Care Med 1998;26(5):844–51.

2073. Michelsen I, Helbo-Hansen HS, Kohler F, Lorenzen AG, Rydlund E, Bentzon MW. Prophylactic ephedrine attenuates the hemodynamic response to propofol in elderly female patients. Anesth Analg 1998;86(3):477–81.

2074. Gamlin F, Freeman J, Winslow L, Berridge J, Vucevic M. The haemodynamic effects of propofol in combination with ephedrine in elderly patients (ASA groups 3 and 4). Anaesth Intensive Care 1999;27(5):477–80.

2075. Tritapepe L, Voci P, Marino P, Cogliati AA, Rossi A, Bottari B, Di Marco P, Menichetti A. Calcium chloride minimizes the hemodynamic effects of propofol in patients undergoing coronary artery bypass grafting. J Cardiothorac Vasc Anesth 1999;13(2):150–3.

2076. Bagshaw O. TIVA with propofol and remifentanil. Anaesthesia 1999;54(5):501–2.

2077. Sochala C, Deenen D, Ville A, Govaerts MJ. Heart block following propofol in a child. Paediatr. Anaesth 1999;9(4):349–51.

2078. Sakabe M, Fujiki A, Inoue H. Propofol induced marked prolongation of QT interval in a patient with acute myocardial infarction. Anesthesiology 2002;97(1):265–6.

2079. Tan CH, Onsiong MK. Pain on injection of propofol. Anaesthesia 1998;53(5):468–76.

2080. Sear JW, Jewkes C, Wanigasekera V. Hemodynamic effects during induction, laryngoscopy, and intubation with eltanolone (5 beta-pregnanolone) or propofol. A study in ASA I and II patients. J Clin Anesth 1995;7(2):126–31.

2081. Nakane M, Iwama H. A potential mechanism of propofol-induced pain on injection based on studies using nafamostat mesilate. Br J Anaesth 1999;83(3):397–404.

2082. Gilston A. Raynaud's phenomenon and propofol. Anaesthesia 1999;54(3):307.

2083. Ho CM, Tsou MY, Sun MS, Chu CC, Lee TY. The optimal effective concentration of lidocaine to reduce pain on injection of propofol. J Clin Anesth 1999;11(4):296–300.

2084. Tan CH, Onsiong MK, Kua SW. The effect of ketamine pretreatment on propofol injection pain in 100 women. Anaesthesia 1998;53(3):302–5.

2085. Uda R, Kadono N, Otsuka M, Shimizu S, Mori H. Strict temperature control has no effect on injection pain with propofol. Anesthesiology 1999;91(2):591–2.

2086. Ozturk E, Izdes S, Babacan A, Kaya K. Temperature of propofol does not reduce the incidence of injection pain. Anesthesiology 1998;89(4):1041.

2087. Huang YW, Buerkle H, Lee TH, Lu CY, Lin CR, Lin SH, Chou AK, Muhammad R, Yang LC. Effect of pretreatment with ketorolac on propofol injection pain. Acta Anaesthesiol Scand 2002;46(8):1021–4.

2088. Ambesh SP, Dubey PK, Sinha PK. Ondansetron pretreatment to alleviate pain on propofol injection: a randomized, controlled, double-blinded study. Anesth Analg 1999;89(1):197–9.

2089. Pang WW, Huang PY, Chang DP, Huang MH. The peripheral analgesic effect of tramadol in reducing propofol injection pain: a comparison with lidocaine. Reg Anesth Pain Med 1999;24(3):246–9.

2090. Mok MS, Pang WW, Hwang MH. The analgesic effect of tramadol, metoclopramide, meperidine and lidocaine in ameliorating propofol injection pain: a comparative study. J Anaesthesiol Clin Pharmacol 1999;15:37–42.

2091. Memis D, Turan A, Karamanlioglu B, Kaya G, Pamukcu Z. The prevention of propofol injection pain by tramadol or ondansetron. Eur J Anaesthesiol 2002;19(1):47–51.

2092. Dusting GJ, Moncada S, Vane JR. Prostaglandins, their intermediates and precursors: cardiovascular actions and regulatory roles in normal and abnormal circulatory systems. Prog Cardiovasc Dis 1979;21(6):405–30.

2093. Lee JB. Cardiovascular-renal effects of prostaglandins: the antihypertensive, natriuretic renal "endocrine" function. Arch Intern Med 1974;133(1):56–76.

2094. Olsson AG, Carlson LA. Clinical, hemodynamic and metabolic effects of intraarterial infusions of prostaglandin E1 in patients with peripheral vascular disease. Adv Prostaglandin Thromboxane Res 1976;1:429–32.

2095. Bugiardini R, Galvani M, Ferrini D, Gridelli C, Tollemeto D, Mari L, Puddu P, Lenzi S. Myocardial ischemia induced by prostacyclin and iloprost. Clin Pharmacol Ther 1985;38(1):101–8.

2096. Fliers E, Duren DR, van Zwieten PA. A prostaglandin analogue as a probable cause of myocardial infarction in a young woman. BMJ 1991;302(6773):416.

2097. Lennox CE, Martin J. Cardiac arrest following intramyometrial prostaglandin E2. J Obstet Gynaecol 1991;11:263–4.

2098. Meyer WJ, Benton SL, Hoon TJ, Gauthier DW, Whiteman VE. Acute myocardial infarction associated with prostaglandin E2. Am J Obstet Gynecol 1991;165(2):359–60.

2099. White JL, Fleming NW, Burke TA, Katz NM, Moront MG, Kim YD. Pulmonary edema after PGE1 infusion. J Cardiothorac Anesth 1990;4(6):744–7.

2100. Marti V, Guindo J, Valles E, Domínguez de Rozas JM. Angina variante asociada con latanoprost. Med Clin (Barc) 2005;125(6):238–9.

2101. Lowenstein E, Zapol WM. Protamine reactions, explosive mediator release, and pulmonary vasoconstriction. Anesthesiology 1990;73(3):373–5.

2102. Viaro F, Dalio MB, Evora PR. Catastrophic cardiovascular adverse reactions to protamine are nitric oxide/cyclic

guanosine monophosphate dependent and endothelium mediated: should methylene blue be the treatment of choice? Chest 2002;122(3):1061–6.
2103. Bonfanti P, Valsecchi L, Parazzini F, Carradori S, Pusterla L, Fortuna P, Timillero L, Alessi F, Ghiselli G, Gabbuti A, Di Cintio E, Martinelli C, Faggion I, Landonio S, Quirino T. Incidence of adverse reactions in HIV patients treated with protease inhibitors: a cohort study. Coordinamento Italiano Studio Allergia e Infezione da HIV (CISAI) Group. J Acquir Immune Defic Syndr 2000;23(3):236–45.
2104. Thiebaut R, Dabis F, Malvy D, Jacqmin-Gadda H, Mercie P, Valentin VD. Serum triglycerides, HIV infection, and highly active antiretroviral therapy, Aquitaine Cohort, France, 1996 to 1998. Groupe d'Epidemiologie Clinique du Sida en Aquitaine (GECSA). J Acquir Immune Defic Syndr 2000;23(3):261–5.
2105. Benson JO, McGhee K, Coplan P, Grunfeld C, Robertson M, Brodovicz KG, Slater E. Fat redistribution in indinavir-treated patients with HIV infection: a review of postmarketing cases. J Acquir Immune Defic Syndr 2000;25(2):130–9.
2106. Penner JA. Management of haemophilia in patients with high-titre inhibitors: focus on the evolution of activated prothrombin complex concentrate AUTOPLEX T. Haemophilia 1999;5(Suppl 3):1–9.
2107. Roberts HR. The use of agents that by-pass factor VIII inhibitors in patients with haemophilia. Vox Sang 1999;77(Suppl 1):38–41.
2108. Shapiro AD. Recombinant factor VIIa: a viewpoint. BioDrugs 1999;12:78.
2109. Barthels M. Clinical efficacy of prothrombin complex concentrates and recombinant factor VIIa in the treatment of bleeding episodes in patients with factor VII and IX inhibitors. Thromb Res 1999;95(4 Suppl 1):S31–8.
2110. Preston FE, Laidlaw ST, Sampson B, Kitchen S. Rapid reversal of oral anticoagulation with warfarin by a prothrombin complex concentrate (Beriplex): efficacy and safety in 42 patients. Br J Haematol 2002;116(3):619–24.
2111. Devlieger R, Vanderlinden S, de Zegher F, Van Assche FA, Spitz B. Effect of antenatal thyrotropin-releasing hormone on uterine contractility, blood pressure, and maternal heart rate. Am J Obstet Gynecol 1997;177(2):431–3.
2112. McFadden RG, McCourtie DR, Rodger NW. TRH and bronchospasm. Lancet 1981;2(8249):758–9.
2113. Drury PL, Belchetz PE, McDonald WI, Thomas DG, Besser GM. Transient amaurosis and headache after thyrotropin releasing hormone. Lancet 1982;1(8265):218–9.
2114. Cimino A, Corsini R, Radaeli E, Bollati A, Giustina G. Transient amaurosis in patient with pituitary macroadenoma after intravenous gonadotropin and thyrotropin releasing hormones. Lancet 1981;2(8237):95.
2115. Brunner G, Athmann C, Boldt JH. Reversible pheripheral edema in female patients taking proton pump inhibitors for peptic acid diseases. Dig Dis Sci 2001;46(5):993–6.
2116. Geisler LS, Thiel H, Forster OB, Rohner HG. Proxyphyllinwirkung bei chronisch-obstruktiven Atemweg-serkrankungen. [Effectiveness of proxyphylline in chronic obstructive pulmonary disease.] Dtsch Med Wochenschr 1980;105(25):894–7.
2117. Mosbech H, Paulsen H, Soborg M. Controlled-release theophylline and proxyphylline in asthmatics: a comparative study. Pharmatherapeutica 1984;3(9):626–30.
2118. Jordan J, Shannon JR, Diedrich A, Black B, Robertson D, Biaggioni B. Water potentiates the pressor effect of Ephedra alkaloids. Circulation 2004;109:1823–5.
2119. Grzesk G, Polak G, Grabczewska Z, Kubica J. Myocardial infarction with normal coronary arteriogram: the role of ephedrine-like alkaloids. Med Sci Monit 2004;10:CS 15–21.
2120. Salerno SM, Jackson JL, Berbano EP. Effect of oral pseudoephedrine on blood pressure and heart rate. A meta-analysis. Arch Intern Med 2005;165:1686–94.
2121. Manini AF, Kabrehl C, Thomsen TW. Acute myocardial infarction after over-the-counter use of pseudoephedrine. Ann Emerg Med 2005;45:213–6.
2122. Cantu C, Arauz A, Murillo-Bonilla LM, Lopez M, Barinagarrementeria F. Stroke associated with sympathomimetics contained in over-the-counter cough and cold drugs. Stroke 2003;34:1667–72.
2123. Goldberg J, Moreno F, Barbara J. Acute hypertension as an adverse effect of pyrazinamide. JAMA 1997;277(17):1356.
2124. Downing D. Debate continues on vitamin B6. Lancet 1998;352(9121):63.
2125. Selhub J, Jacques PF, Wilson PW, Rush D, Rosenberg IH. Vitamin status and intake as primary determinants of homocysteinemia in an elderly population. JAMA 1993;270(22):2693–8.
2126. Shani J, Atkins HL, Wolf W. Adverse reactions to radiopharmaceuticals. Semin Nucl Med 1976;6(3):305–28.
2127. Perper EJ, Segall GM. Safety of dipyridamole–thallium imaging in high risk patients with known or suspected coronary artery disease. J Nucl Med 1991;32(11):2107–14.
2128. Compston JE. Selective oestrogen receptor modulators: potential therapeutic applications. Clin Endocrinol (Oxf) 1998;48(4):389–91.
2129. Cohen FJ, Lu Y. Characterization of hot flashes reported by healthy postmenopausal women receiving raloxifene or placebo during osteoporosis prevention trials. Maturitas 2000;34(1):65–73.
2130. Johnston CC Jr, Bjarnason NH, Cohen FJ, Shah A, Lindsay R, Mitlak BH, Huster W, Draper MW, Harper KD, Heath H 3rd, Gennari C, Christiansen C, Arnaud CD, Delmas PD. Long-term effects of raloxifene on bone mineral density, bone turnover, and serum lipid levels in early postmenopausal women: three-year data from 2 double-blind, randomized, placebo-controlled trials. Arch Intern Med 2000;160(22):3444–50.
2131. Sarrel PM, Nawaz H, Chan W, Fuchs M, Katz DL. Raloxifene and endothelial function in healthy postmenopausal women. Am J Obstet Gynecol 2003;188:304–9.
2132. Allegri G, Pellegrini K, Dobrilla G. First-degree atrioventricular block in a young duodenal ulcer patient treated with a standard oral dose of ranitidine. Agents Actions 1988;24(3-4):237–42.
2133. Nahum E, Reish O, Naor N, Merlob P. Ranitidine-induced bradycardia in a neonate—a first report. Eur J Pediatr 1993;152(11):933–4.
2134. Sabir M, Bhide NK. Study of some pharmacological actions of berberine. Indian J Physiol Pharmacol 1971;15(3):111–32.
2135. Chun YT, Yip TT, Lau KL, Kong YC, Sankawa U. A biochemical study on the hypotensive effect of berberine in rats. Gen Pharmacol 1979;10(3):177–82.
2136. Osmer C, Wulf K, Vogele C, Zickmann B, Hempelmann G. Cardiovascular effects of Org 9487 under isoflurane anaesthesia in man. Eur J Anaesthesiol 1998;15(5):585–9.

2137. Levy JH, Pitts M, Thanopoulos A, Szlam F, Bastian R, Kim J. The effects of rapacuronium on histamine release and hemodynamics in adult patients undergoing general anesthesia. Anesth Analg 1999;89(2):290–5.

2138. Ranieri P, Franzoni S, Trabucchi M. Reboxetine and hyponatremia. N Engl J Med 2000;342(3):215–6.

2139. Schwartz GE, Veith J. Reboxetine and hyponatremia. N Engl J Med 2000;342:216.

2140. Hall AP, Thompson JP, Leslie NA, Fox AJ, Kumar N, Rowbotham DJ. Comparison of different doses of remifentanil on the cardiovascular response to laryngoscopy and tracheal intubation. Br J Anaesth 2000;84(1):100–2.

2141. Elliott P, O'Hare R, Bill KM, Phillips AS, Gibson FM, Mirakhur RK. Severe cardiovascular depression with remifentanil. Anesth Analg 2000;91(1):58–61.

2142. Michelsen LG. Hemodynamic effects of remifentanil in patients undergoing cardiac surgery. Anesth Analg 2000;91(6):1563.

2143. Kazmaier S, Hanekop GG, Buhre W, Weyland A, Busch T, Radke OC, Zoelffel R, Sonntag H. Myocardial consequences of remifentanil in patients with coronary artery disease. Br J Anaesth 2000;84(5):578–83.

2144. Kurdi O, Deleuze A, Marret E, Bonnet F. Asystole during anaesthetic induction with remifentanil and sevoflurane. Br J Anaesth 2001;87(6):943.

2145. Weale NK, Rogers CA, Cooper R, Nolan J, Wolf AR. Effect of remifentanil infusion rate on stress response to the pre-bypass phase of paediatric cardiac surgery. Br J Anaesth 2004;92:187–94.

2146. Pleym H, Stenseth R, Wilseth R, Karevola A, Dale O. Supplemental remifentanil during coronary artery bypass grafting is followed by a transient post-operative cardiac depression. Acta Anaesthesiol Scand 2004;48:1155–62.

2147. Van Zijl DH, Gordon PC, James MF. The comparative effects of remifentanil or magnesium sulphate versus placebo on attenuating the hemodynamic responses after electroconvulsive therapy. Anesth Analg 2005;101(6):1651–5.

2148. Chanavaz C, Tirel O, Wodey E, Bansard JY, Senhadji L, Robert JC, Ecoffey C. Haemodynamic effects of remifentanil in children with and without intravenous atropine. an echocardiographic study. Br J Anaesth 2005;94(1):74–9.

2149. Reyntjens K., Foubert L, De Wolf D, Vanlerberghe G, Mortier E. Glycopyrrolate during sevoflurane–remifentanil-based anaesthesia for cardiac catheterization of children with congenital heart disease. Br J Anaesth 2005;95(5):680–4.

2150. Williams H, Spoelstra C. Use of remifentanil in fast atrial fibrillation. Br J Anaesth 2002;88(4):614.

2151. Ramachandran A, Bhatia VN. Rifampicin induced shock—a case report. Indian J Lepr 1990;62(2):228–9.

2152. Martinez E, Collazos J, Mayo J. Shock and cerebral infarct after rifampin re-exposure in a patient infected with human immunodeficiency virus. Clin Infect Dis 1998;27(5):1329–30.

2153. Kissling M, Xilinas M. Rimactan parenteral formulation in clinical use. J Int Med Res 1981;9(6):459–69.

2154. Mandell GL, Sande MA. Antimicrobial agents: drugs used in the chemotherapy of tuberculosis and leprosy. In: Goodman Gilman A, Rall TW, Nies AS, Taylor P, editors. Goodman and Gilman's The Pharmacological Basis of Therapeutics 8th ed. New York: Pergamon Press, 1990:1146.

2155. Ravin DS, Levenson JW. Fatal cardiac event following initiation of risperidone therapy. Ann Pharmacother 1997;31(7–8):867–70.

2156. Posey DJ, Walsh KH, Wilson GA, McDougle CJ. Risperidone in the treatment of two very young children with autism. J Child Adolesc Psychopharmacol 1999;9(4):273–6.

2157. Henretig FM. Risperidone toxicity acknowledged. J Toxicol Clin Toxicol 1999;37:893–4.

2158. Williams JH, Hepner DL. Risperidone and exaggerated hypotension during a spinal anesthetic. Anesth Analg 2004;98:240–1.

2159. Food and Drug Administration. Risperdal Consta. Long-acting injection. Summary of product characteristics. http://www.fda.gov/medwatch/safety/2005/aug_PI/Risperdal_%20Consta_PI.

2160. Harrison TS, Goa KL. Long-acting risperidone. A review of its use in schizophrenia. CNS Drugs 2004;18:113–32.

2161. Chiu CC, Chang WH, Huang MC, Chiu YW, Lane HY. Regular-dose risperidone on QT_c intervals. J Clin Psychopharmacol 2005;25:391–3.

2162. Foran JM, Rohatiner AZ, Cunningham D, Popescu RA, Solal-Celigny P, Ghielmini M, Coiffier B, Johnson PW, Gisselbrecht C, Reyes F, Radford JA, Bessell EM, Souleau B, Benzohra A, Lister TA. European phase II study of rituximab (chimeric anti-CD20 monoclonal antibody) for patients with newly diagnosed mantle-cell lymphoma and previously treated mantle-cell lymphoma, immunocytoma, and small B cell lymphocytic lymphoma. J Clin Oncol 2000;18(2):317–24.

2163. Garypidou V, Perifanis V, Tziomalos K, Theodoridou S. Cardiac toxicity during rituximab administration. Leuk Lymphoma 2004;45(1):203–4.

2164. Weber JE, Chudnofsky CR, Boczar M, Boyer EW, Wilkerson MD, Hollander JE. Cocaine-associated chest pain: how common is myocardial infarction? Acad Emerg Med 2000;7(8):873–7.

2165. Levy JH, Davis GK, Duggan J, Szlam F. Determination of the hemodynamics and histamine release of rocuronium (Org 9426) when administered in increased doses under N_2O/O_2–sufentanil anesthesia. Anesth Analg 1994;78(2):318–21.

2166. McCoy EP, Maddineni VR, Elliott P, Mirakhur RK, Carson IW, Cooper RA. Haemodynamic effects of rocuronium during fentanyl anaesthesia: comparison with vecuronium. Can J Anaesth 1993;40(8):703–8.

2167. Hudson ME, Rothfield KP, Tullock WC, Firestone LL. Haemodynamic effects of rocuronium bromide in adult cardiac surgical patients. Can J Anaesth 1998;45(2):139–43.

2168. Borgeat A, Kwiatkowski D. Spontaneous movements associated with rocuronium: is pain on injection the cause? Br J Anaesth 1997;79(3):382–3.

2169. Steegers MA, Robertson EN. Pain on injection of rocuronium bromide. Anesth Analg 1996;83(1):203.

2170. Demers-Pelletier J, Drolet P, Girard M, Donati F. Comparison of rocuronium and d-tubocurarine for prevention of succinylcholine-induced fasciculations and myalgia. Can J Anaesth 1997;44(11):1144–7.

2171. Findlay GP, Spittal MJ. Rocuronium pretreatment reduces suxamethonium-induced myalgia: comparison with vecuronium. Br J Anaesth 1996;76(4):526–9.

2172. Motamed C, Choquette R, Donati F. Rocuronium prevents succinylcholine-induced fasciculations. Can J Anaesth 1997;44(12):1262–8.

2173. Tsui BC, Reid S, Gupta S, Kearney R, Mayson T, Finucane B. A rapid precurarization technique using rocuronium. Can J Anaesth 1998;45(5 Pt 1):397–401.

2174. Acton G, Broom C. A dose rising study of the safety and effects on serum prolactin of SK&F 101468, a novel

dopamine D_2-receptor agonist. Br J Clin Pharmacol 1989;28(4):435–41.
2175. de Mey C, Enterling D, Meineke I, Yeulet S. Interactions between domperidone and ropinirole, a novel dopamine D_2-receptor agonist. Br J Clin Pharmacol 1991;32(4):483–8.
2176. Vidailhet MJ, Bonnet AM, Belal S, Dubois B, Marle C, Agid Y. Ropinirole without levodopa in Parkinson's disease. Lancet 1990;336(8710):316–7.
2177. Morrison LM, Emanuelsson BM, McClure JH, Pollok AJ, McKeown DW, Brockway M, Jozwiak H, Wildsmith JA. Efficacy and kinetics of extradural ropivacaine: comparison with bupivacaine. Br J Anaesth 1994;72(2):164–9.
2178. Haffner S, Lapp H, Thurmann PA. Unerwunschi te Arzneimittel wirkungen Der konkrete Fall. [Adverse drug reactions—case report.] Dtsch Med Wochenschr 200210;127(19):1021.
2179. Scherrer M, Bachofen H. Vergleich der Wirkung einer 4.5 Minuten dauernden Aerosolinhalation von Salbutamol und von Trimetoquinol mit derjenigen einer 10–15 Minuten dauernden Tacholoquin-Orciprenalin-Inhalation bei Bronchialasthma. [Comparison of the effect of aerosol inhalation with salbutamol and trimetolquinol of four and a half minutes duration with that of tacholiquin–orciprenalin of fifteen minutes duration in bronchial asthma.] Schweiz Med Wochenschr 1972;102(26):909–14.
2180. Fisher AA, Davis MW, McGill DA. Acute myocardial infarction associated with albuterol. Ann Pharmacother 2004;38:2045–9.
2181. Chazan R, Tadeusiak W, Jaworski A, Droszcz W. Creatine kinase (CK) and creatine kinase isoenzyme (CK-MB) activity in serum before and after intravenous salbutamol administration of patients with bronchial asthma. Int J Clin Pharmacol Ther Toxicol 1992;30(10):371–3.
2182. Burggraaf J, Westendorp RG, in't Veen JC, Schoemaker RC, Sterk PJ, Cohen AF, Blauw GJ. Cardiovascular side effects of inhaled salbutamol in hypoxic asthmatic patients. Thorax 2001;56(7):567–9.
2183. Lowe MD, Rowland E, Brown MJ, Grace AA. Beta$_2$-adrenergic receptors mediate important electrophysiological effects in human ventricular myocardium. Heart 2001;86:45–51.
2184. Pye M, Quinn AC, Cobbe SM. QT interval dispersion: a non-invasive marker of susceptibility to arrhythmia in patients with sugtained ventricular arrhythmias? Br Heart J 1994;71:511–4.
2185. Yuksel H, Coskun S, Polat M, Onag A. Lower arrythmogenic risk of low dose albuterol plus ipratropium. Indian J Pediatr 2001;68(10):945–9.
2186. Tranfa CM, Pelaia G, Grembiale RD, Naty S, Durante S, Borrello G. Short-term cardiovascular effects of salmeterol. Chest 1998;113(5):1272–6.
2187. Ferguson GT, Funck-Brentano C, Fischer T, Darken P, Reisner C. Cardiovascular safety of salmeterol in COPD. Chest 2003;123:1817–24.
2188. FDA Safety Information. http://www.fda.gov/medwatch/SAFETY/2003/serevent.htm.
2189. Excess mortality with salmeterol as single-agent therapy. Prescrire Int 2003;12:142.
2190. Stroncek DF, Hutton SW, Silvis SE, Vercellotti GM, Jacob HS, Hammerschmidt DE. Sodium morrhuate stimulates granulocytes and damages erythrocytes and endothelial cells: probable mechanism of an adverse reaction during sclerotherapy. J Lab Clin Med 1985;106(5):498–504.
2191. Marrocco-Trischitta MM, Guerrini P, Abeni D, Stillo F. Reversible cardiac arrest after polidocanol sclerotherapy of peripheral venous malformation. Dermatol Surg 2002;28(2):153–5.
2192. Churchyard A, Mathias CJ, Boonkongchuen P, Lees AJ. Autonomic effects of selegiline: possible cardiovascular toxicity in Parkinson's disease. J Neurol Neurosurg Psychiatry 1997;63(2):228–34.
2193. Committee on Safety of Medicines-Medicines Control Agency. Cardiac arrhythmias with pomizode (Orap). Curr Probl Pharmacovigilance 1995;21:1.
2194. Sunderji R, Press N, Amin H, Gin K. Unstable angina associated with sertraline. Can J Cardiol 1997;13(9):849–51.
2195. Kikura M, Ikeda K. Comparison of effects of sevoflurane/nitrous oxide and enflurane/nitrous oxide on myocardial contractility in humans. Load-independent and noninvasive assessment with transesophageal echocardiography. Anesthesiology 1993;79(2):235–43.
2196. Kitahata H, Tanaka K, Kimura H, Saito T. [Effects of sevoflurane on left ventricular diastolic function using transesophageal echocardiography.]Masui 1993;42(3):358–64.
2197. Ebert TJ, Harkin CP, Muzi M. Cardiovascular responses to sevoflurane: a review. Anesth Analg 1995;81(Suppl 6):S11–22.
2198. Malan TP Jr, DiNardo JA, Isner RJ, Frink EJ Jr, Goldberg M, Fenster PE, Brown EA, Depa R, Hammond LC, Mata H. Cardiovascular effects of sevoflurane compared with those of isoflurane in volunteers. Anesthesiology 1995;83(5):918–28.
2199. Townsend P, Stokes MA. Bradycardia during rapid inhalation induction with sevoflurane in children. Br J Anaesth 1998;80(3):410.
2200. Sigston PE, Jenkins AM, Jackson EA, Sury MR, Mackersie AM, Hatch DJ. Rapid inhalation induction in children: 8% sevoflurane compared with 5% halothane. Br J Anaesth 1997;78(4):362–5.
2201. Green DH, Townsend P, Bagshaw O, Stokes MA. Nodal rhythm and bradycardia during inhalation induction with sevoflurane in infants: a comparison of incremental and high-concentration techniques. Br J Anaesth 2000;85(3):368–70.
2202. Maruyama K, Agata H, Ono K, Hiroki K, Fujihara T. Slow induction with sevoflurane was associated with complete atrioventricular block in a child with hypertension, renal dysfunction, and impaired cardiac conduction. Paediatr Anaesth 1998;8(1):73–8.
2203. Kleinsasser A, Loeckinger A, Lindner KH, Keller C, Boehler M, Puehringer F. Reversing sevoflurane-associated Q-Tc prolongation by changing to propofol. Anaesthesia 2001;56(3):248–50.
2204. Yildrim H, Adanir T, Atay A, Kataricioglu K, Savaci S. The effects of sevoflurane, isoflurane and desflurane on QT interval of the ECG. Eur J Anaesthesiol 2004;21:566–70.
2205. Sen S, Ozmert G, Boran N, Turan H, Caliskan E. Comparison of the effects of single-breath vital capacity rapid inhalation with sevoflurane 5% and propofol induction on QT interval and haemodynamics for laparoscopic surgery. Eur J Anaesthesiol 2004;21:543–6.
2206. Loeckinger A, Kleinsasser A, Maier S, Furtner B, Keller C, Kuehbacher G, Lindner KH. Sustained prolongation of the QT_c interval after anesthesia with sevoflurane in

infants during the first 6 months of life. Anesthesiology 2003;98:639–42.

2207. Gurkan Y, Canatay H, Agacdiken A, Ural E, Toker K. Effects of halothane and sevoflurane on QT dispersion in paediatric patients. Paediatr Anaesth 2003;13:223–7.

2208. Whyte SD, Booker PD, Buckley DG. The effects of propofol and sevoflurane on the QT interval and transmural dispersion of repolarization in children. Anesth Analg. 2005;100(1):71–7.

2209. Abe K, Takada K, Yoshiya I. Intraoperative torsade de pointes ventricular tachycardia and ventricular fibrillation during sevoflurane anesthesia. Anesth Analg 1998;86(4):701–2.

2210. Roodman S, Bothwell M, Tobias JD. Bradycardia with sevoflurane induction with patients with trisomy 21. Paediatr Anaesth 2003;13:538–40.

2211. Smith IG, Goulder MAOn Behalf of the Members of the Sibutramine Clinical Study 1047 Team. Randomized placebo-controlled trial of long-term treatment with sibutramine in mild to moderate obesity. J Fam Pract 2001;50(6):505–12.

2212. Fanghanel G, Cortinas L, Sanchez-Reyes L, Berber A. Second phase of a double-blind study clinical trial on sibutramine for the treatment of patients suffering essential obesity: 6 months after treatment cross-over. Int J Obes Relat Metab Disord 2001;25(5):741–7.

2213. Luque CA, Rey JA. The discovery and status of sibutramine as an anti-obesity drug. Eur J Pharmacol 2002;440(2–3):119–28.

2214. Dujovne CA, Zavoral JH, Rowe E, Mendel CMSibutramine Study Group. Effects of sibutramine on body weight and serum lipids: a double-blind, randomized, placebo-controlled study in 322 overweight and obese patients with dyslipidemia. Am Heart J 2001;142(3):489–97.

2215. Toubro S, Hansen DL, Hilsted JC, Porsborg PA, Astrup AVSTORM Study Group. Effekt af sibutramin til vaegttabsvedligeholdelse: en randomiseret klinisk kontrolleret undersogelse. [The effect of sibutramine for the maintenance of weight loss: a randomized controlled clinical trial.] Ugeskr Laeger 2001;163(21):2935–40.

2216. McMahon FG, Fujioka K, Singh BN, Mendel CM, Rowe E, Rolston K, Johnson F, Mooradian AD. Efficacy and safety of sibutramine in obese white and African American patients with hypertension: a 1-year, double-blind, placebo-controlled, multicenter trial. Arch Intern Med 2000;160(14):2185–91.

2217. James WP, Astrup A, Finer N, Hilsted J, Kopelman P, Rossner S, Saris WH, Van Gaal LFSTORM Study Group. Effect of sibutramine on weight maintenance after weight loss: a randomised trial. Sibutramine Trial of Obesity Reduction and Maintenance. Lancet 2000;356(9248): 2119–25.

2218. Sramek JJ, Leibowitz MT, Weinstein SP, Rowe ED, Mendel CM, Levy B, McMahon FG, Mullican WS, Toth PD, Cutler NR. Efficacy and safety of sibutramine for weight loss in obese patients with hypertension well controlled by beta-adrenergic blocking agents: a placebo-controlled, double-blind, randomised trial. J Hum Hypertens 2002;16(1):13–9.

2219. McMahon FG, Weinstein SP, Rowe E, Ernst KR, Johnson F, Fujioka KSibutramine in Hypertensives Clinical Study Group. Sibutramine is safe and effective for weight loss in obese patients whose hypertension is well controlled with angiotensin-converting enzyme inhibitors. J Hum Hypertens 2002;16(1):5–11.

2220. Cutler JA. Randomized clinical trials of weight reduction in nonhypertensive persons. Ann Epidemiol 1991;1(4):363–70.

2221. Rossner S, Sjostrom L, Noack R, Meinders AE, Noseda GEuropean Orlistat Obesity Study Group. Weight loss, weight maintenance, and improved cardiovascular risk factors after 2 years treatment with orlistat for obesity. Obes Res 2000;8(1):49–61.

2222. Davidson MH, Hauptman J, DiGirolamo M, Foreyt JP, Halsted CH, Heber D, Heimburger DC, Lucas CP, Robbins DC, Chung J, Heymsfield SB. Weight control and risk factor reduction in obese subjects treated for 2 years with orlistat: a randomized controlled trial. JAMA 1999;281(3):235–42.

2223. Hauptman J, Lucas C, Boldrin MN, Collins H, Segal KR. Orlistat in the long-term treatment of obesity in primary care settings. Arch Fam Med 2000;9(2):160–7.

2224. Stamler R, Stamler J, Grimm R, Gosch FC, Elmer P, Dyer A, Berman R, Fishman J, Van Heel N, Civinelli J, Mc Donald A. Nutritional therapy for high blood pressure. Final report of a four-year randomized controlled trial—the Hypertension Control Program. JAMA 1987;257(11):1484–91.

2225. Anonymous. The Hypertension Prevention Trial: three-year effects of dietary changes on blood pressure. Hypertension Prevention Trial Research Group. Arch Intern Med 1990;150(1):153–62.

2226. Bray GA. Sibutramine and blood pressure: a therapeutic dilemma. J Hum Hypertens 2002;16(1):1–3.

2227. Fanghanel G, Cortinas L, Sanchez-Reyes L, Berber A. A clinical trial of the use of sibutramine for the treatment of patients suffering essential obesity. Int J Obes Relat Metab Disord 2000;24(2):144–50.

2228. Guven A, Koksal N, Cetinkaya A, Sokmen G, Ozdemir R. Effects of the sibutramine therapy on pulmonary artery pressure in obese patients. Diabetes Obesity Metab 2004;6:50–5.

2229. Sayin T, Gindal M. Sibutramine: possible cause of a reversible cardiomyopathy. Int J Cardiol 2005;99:481–2.

2230. Tozzi P, Al-Darweesh A, Vogt P, Stumpe F. Silver-coated prosthetic heart valve: a double-bladed weapon. Eur J Cardiothorac Surg 2001;19(5):729–31.

2231. Neff GW, Montalbano M, Tzakis AG. Ten years of sirolimus therapy in orthotopic liver transplant recipients. Transplant Proc 2003;35 Suppl 3A:209S-216S.

2232. Fortin MC, Raymond MA, Madore F, Fugere JA, Paquet M, St Louis G, Hebert MJ. Increased risk of thrombotic microangiopathy in patients receiving a cyclosporin-sirolimus combination. Am J Transplant 2004;4(6):946–52.

2233. Paramesh AS, Grosskreutz C, Florman SS, Gondolesi GE, Sharma S, Kaufman SS, Fishbein TM. Thrombotic microangiopathy associated with combined sirolimus and tacrolimus immunosuppression after intestinal transplantation. Transplantation 2004;77(1):129–31.

2234. Baldini G, Bani E. Sulle complicanze cardiache in corso di vaccinazione jenneriana. (Contributo clinico ed ec-grafie.). [Cardiac complications in Jennerian vaccination (Clinical and electrocardiographic studies).] Minerva Pediatr 1979;31(1):35–9.

2235. Finlay-Jones LR. Fatal myocarditis after vaccination against smallpox. Report of a case. N Engl J Med 1964;270:41–2.

2236. Mead J. Serum transaminase and electrocardiographic findings after smallpox vaccination: case report. J Am Geriatr Soc 1966;14(7):754–6.

2237. Bessard G, Marchal A, Avezou F, Pont J, Rambaud P. Un nouveau cas de myocardite après vaccination antivariolique. [A new case of myocarditis following smallpox vaccination.] Pediatrie 1974;29(2):179–84.

2238. Price MA, Alpers JH. Acute pericarditis following smallpox vaccination. Papua N Guinea Med J 1968;11:30.

2239. Centers for Disease Control and Prevention (CDC). Supplemental recommendations on adverse events following smallpox vaccine in the pre-event vaccination program: recommendations of the Advisory Committee on Immunization Practices. MMWR Morb Mortal Wkly Rep 2003;52(13):282–4.

2240. Centers for Disease Control and Prevention (CDC). Cardiac deaths after a mass smallpox vaccination campaign—New York City, 1947. MMWR Morb Mortal Wkly Rep 2003;52(39):933–6.

2241. Centers for Disease Control and Prevention (CDC). Update: adverse events following civilian smallpox vaccination—United States, 2003. MMWR Morb Mortal Wkly Rep 2003;52(34):819–20.

2242. Centers for Disease Control and Prevention (CDC). Update: adverse events following smallpox vaccination—United States, 2003. MMWR Morb Mortal Wkly Rep 2003;52(13):278–82.

2243. Hadengue A. Somatostatin or octreotide in acute variceal bleeding. Digestion 1999;60(Suppl 2):31–41.

2244. Pop-Busui R, Chey W, Stevens MJ. Severe hypertension induced by the long-acting somatostatin analogue Sandostatin LAR in a patient with diabetic autonomic neuropathy. J Clin Endocrinol Metab 2000;85(3):943–6.

2245. Beckman RA, Siden R, Yanik GA, Levine JE. Continuous octreotide infusion for the treatment of secretory diarrhea caused by acute intestinal graft-versus-host disease in a child. J Pediatr Hematol Oncol 2000;22(4):344–50.

2246. Jenkins SA, Shields R, Davies M, Elias E, Turnbull AJ, Bassendine MF, James OF, Iredale JP, Vyas SK, Arthur MJ, Kingsnorth AN, Sutton R. A multicentre randomised trial comparing octreotide and injection sclerotherapy in the management and outcome of acute variceal haemorrhage. Gut 1997;41(4):526–33.

2247. Herrington AM, George KW, Moulds CC. Octreotide-induced bradycardia. Pharmacotherapy 1998;18(2):413–6.

2248. Weaver JU, Monson JP, Noonan K, John WG, Edwards A, Evans KA, Cunningham J. The effect of low dose recombinant human growth hormone replacement on regional fat distribution, insulin sensitivity, and cardiovascular risk factors in hypopituitary adults. J Clin Endocrinol Metab 1995;80(1):153–9.

2249. Attanasio AF, Lamberts SW, Matranga AM, Birkett MA, Bates PC, Valk NK, Hilsted J, Bengtsson BA, Strasburger CJAdult Growth Hormone Deficiency Study Group. Adult growth hormone (GH)-deficient patients demonstrate heterogeneity between childhood onset and adult onset before and during human GH treatment. J Clin Endocrinol Metab 1997;82(1):82–8.

2250. Rosenfalck AM, Fisker S, Hilsted J, Dinesen B, Volund A, Jorgensen JO, Christiansen JS, Madsbad S. The effect of the deterioration of insulin sensitivity on beta-cell function in growth-hormone-deficient adults following 4-month growth hormone replacement therapy. Growth Horm IGF Res 1999;9(2):96–105.

2251. Cuneo RC, Judd S, Wallace JD, Perry-Keene D, Burger H, Lim-Tio S, Strauss B, Stockigt J, Topliss D, Alford F, Hew L, Bode H, Conway A, Handelsman D, Dunn S, Boyages S, Cheung NW, Hurley D. The Australian Multicenter Trial of Growth Hormone (GH) Treatment in GH-Deficient Adults. J Clin Endocrinol Metab 1998;83(1):107–16.

2252. Daubeney PE, McCaughey ES, Chase C, Walker JM, Slavik Z, Betts PR, Webber SA. Cardiac effects of growth hormone in short normal children: results after four years of treatment. Arch Dis Child 1995;72(4):337–9.

2253. Crepaz R, Pitscheider W, Radetti G, Paganini C, Gentili L, Morini G, Braito E, Mengarda G. Cardiovascular effects of high-dose growth hormone treatment in growth hormone-deficient children. Pediatr Cardiol 1995;16(5):223–7.

2254. Frustaci A, Gentiloni N, Russo MA. Growth hormone in the treatment of dilated cardiomyopathy. N Engl J Med 1996;335(9):672–3.

2255. Oczkowska U. Przejsciowe powiekszenie sylwetki serca jako nietypowe powiklanie leczenia hormonen wzrostu. [Transient cardiac enlargement: an unusual adverse event associated with growth hormone therapy.] Endokrynol Diabetol Chor Przemiany Materii Wieku Rozw 2001;7(1):53–6.

2256. Veronese ML, Mosenkis A, Flaherty KT, Gallagher M, Stevenson JP, Townsend RR, O'Dwyer PJ. Mechanisms of hypertension associated with BAY 43-9006. J Clin Oncol 2006;24:1363–9.

2257. Schmidinger M, Vogl UM, Schukro C, Bojic A, Bojic M, Schmidinger H, Zielinski CC. Cardiac involvement in patients with sorafenib or sunitinib treatment for metastatic renal cell carcinoma. ASCO Annual Meeting Proceedings. J Clin Oncol 2007;25 (June 20 Suppl):5110.

2258. Stahlmann R. Clinical toxicological aspects of fluoroquinolones. Toxicol Lett 2002;127(1–3):269–77.

2259. Patmore L, Fraser S, Mair D, Templeton A. Effects of sparfloxacin, grepafloxacin, moxifloxacin, and ciprofloxacin on cardiac action potential duration. Eur J Pharmacol 2000;406(3):449–52.

2260. Hagiwara T, Satoh S, Kasai Y, Takasuna K. A comparative study of the fluoroquinolone antibacterial agents on the action potential duration in guinea pig ventricular myocardia. Jpn J Pharmacol 2001;87(3):231–4.

2261. Chiba K, Sugiyama A, Satoh Y, Shiina H, Hashimoto K. Proarrhythmic effects of fluoroquinolone antibacterial agents: in vivo effects as physiologic substrate for torsades. Toxicol Appl Pharmacol 2000;169(1):8–16.

2262. Adamantidis MM, Dumotier BM, Caron JF, Bordet R. Sparfloxacin but not levofloxacin or ofloxacin prolongs cardiac repolarization in rabbit Purkinje fibers. Fundam Clin Pharmacol 1998;12(1):70–6.

2263. Demolis JL, Charransol A, Funck-Brentano C, Jaillon P. Effects of a single oral dose of sparfloxacin on ventricular repolarization in healthy volunteers. Br J Clin Pharmacol 1996;41(6):499–503.

2264. Jaillon P, Morganroth J, Brumpt I, Talbot G. Overview of electrocardiographic and cardiovascular safety data for sparfloxacin. Sparfloxacin Safety Group. J Antimicrob Chemother 1996;37(Suppl A):161–7.

2265. Lubasch A, Erbes R, Mauch H, Lode H. Sparfloxacin in the treatment of drug resistant tuberculosis or intolerance of first line therapy. Eur Respir J 2001;17(4):641–6.

2266. Kakar A, Byotra SP. Torsade de pointes probably induced by sparfloxacin. J Assoc Physicians India 2002;50:1077–8.

2267. Stramba-Badiale M, Guffanti S, Porta N, Frediani M, Beria G, Colnaghi C. QT interval prolongation and

cardiac arrest during antibiotic therapy with spiramycin in a newborn infant. Am Heart J 1993;126(3 Pt 1):740–2.
2268. Verdun F, Mansourati J, Jobic Y, Bouquin V, Munier S, Guillo P, Pages Y, Boschat J, Blanc JJ. Torsades de pointes sous traitement par spiramycine et mequitazine. A propos d'un cas. [Torsades de pointe with spiramycine and metiquazine therapy. Apropos of a case.] Arch Mal Coeur Vaiss 1997;90(1):103–6.
2269. Spiess BD, Sathoff RH, el-Ganzouri AR, Ivankovich AD. High-dose sufentanil: four cases of sudden hypotension on induction. Anesth Analg 1986;65(6):703–5.
2270. Starr NJ, Sethna DH, Estafanous FG. Bradycardia and asystole following the rapid administration of sufentanil with vecuronium. Anesthesiology 1986;64(4):521–3.
2271. Dobson JAR, Davies JM, Hodgson GH. Bradycardia after sufentanil and vecuronium. Can J Anaesth 1988;35:S121.
2272. Riley ET, Hamilton CL, Cohen SE. Intrathecal sufentanil produces sensory changes without hypotension in male volunteers. Anesthesiology 1998;89(1):73–8.
2273. Sesti F, Abbott GW, Wei J, Murray KT, Saksena S, Schwartz PJ, Priori SG, Roden DM, George AL Jr, Goldstein SA. A common polymorphism associated with antibiotic-induced cardiac arrhythmia. Proc Natl Acad Sci USA 2000;97(19):10613–8.
2274. French AJ, Weller CV. Interstitial myocarditis following the clinical and experimental use of sulfonamide drugs. Am J Pathol 1942;18:109.
2275. Leibowitz G, Cerasi E. Sulphonylurea treatment of NIDDM patients with cardiovascular disease: a mixed blessing? Diabetologia 1996;39(5):503–14.
2276. Schwartz TB, Meinert CL. The UGDP controversy: thirty-four years of contentious ambiguity laid to rest. Perspect Biol Med 2004;47(4):564–74.
2277. Riveline JP, Danchin N, Ledru F, Varroud-Vial M, Charpentier G. Sulfonylureas and cardiovascular effect: from experimental data to clinical use. Available data in humans and clinical applications. Diabetes Metab 2003;29:207–22.
2278. Cleveland JC Jr, Meldrum DR, Cain BS, Banerjee A, Harken AH. Oral sulfonylurea hypoglycemic agents prevent ischemic preconditioning in human myocardium. Two paradoxes revisited. Circulation 1997;96(1):29–32.
2279. Wascher TC. Sulfonylureas and cardiovascular mortality in diabetes: a class effect? Circulation 1998;97(14):1427–8.
2280. Britton ME, Denver AE, Mohamed-Ali V, Yudkin JS. Effects of glimepiride vs glibenclamide on ischaemic heart disease risk factors and glycaemic control in patients with type 2 diabetes mellitus. Clin Drug Invest 1998;16:303–17.
2281. Tomai F, Danesi A, Ghini AS, Crea F, Perino M, Gaspardone A, Ruggeri G, Chiariello L, Gioffre PA. Effects of K(ATP) channel blockade by glibenclamide on the warm-up phenomenon. Eur Heart J 1999;20(3):196–202.
2282. Turner RC, Holman RR, Cull CA, Stratton IM, Matthews DR, Frighi V, Manley E, Neil A, McElroy H, Wright D, Kohner E, Fox C, Hadden D. Intensive blood-glucose control with sulphonylureas or insulin compared with conventional treatment and risk of complications in patients with type 2 diabetes (UKPDS 33). UK Prospective Diabetes Study (UKPDS) Group. Lancet 1998;352(9131):837–53.
2283. Yosefy C, Magen E, Kiselevich A, Priluk R, London D, Volchek L, Viskoper JR. Rosiglitazone improves, while glibenclamide worsens blood pressure control in treated hypertensive diabetic and dyslipidemic subjects via modulation of insulin resistance and sympathetic activity. J Cardiovasc Pharmacol 2004;44:215–22.
2284. Anonymous. Clinoril sulindac: cardiac abnormalities. ADR Highlights 1979;July 3.
2285. Anonymous. A death associated with mifeprostone/sulprostone. Lancet 1991;337:969–70.
2286. Ulmann A, Silvestre L, Chemama L, Rezvani Y, Renault M, Aguillaume CJ, Baulieu EE. Medical termination of early pregnancy with mifepristone (RU 486) followed by a prostaglandin analogue. Study in 16,369 women. Acta Obstet Gynecol Scand 1992;71(4):278–83.
2287. Feenstra J, Borst F, Huige MC, Oei SG, Stricker BH. Acuut myocardinfarct na toediening van sulproston. [Acute myocardial infarct following sulprostone administration.] Ned Tijdschr Geneeskd 1998;142(4):192–5.
2288. Kulka PJ, Quent P, Wiebalck A, Jager D, Strumpf M. Myocardial infarction after sulprostone therapy for uterine atony and bleeding: a case report. Geburtshilfe Frauenheilk 1999;59:634–7.
2289. Chen FG, Koh KF, Chong YS. Cardiac arrest associated with sulprostone use during caesarean section. Anaesth Intensive Care 1998;26(3):298–301.
2290. Beerendonk CC, Massuger LF, Lucassen AM, Lerou JG, van den Berg PP. Circulatiestilstand na gebruik van sulproston bij fluxus post partum. [Circulatory arrest following sulprostone administration in postpartum hemorrhage.] Ned Tijdschr Geneeskd 1998;142(4):195–7.
2291. Hussain M, Fisher EI, Petrylak DP, O'Connor J, Wood DP, Small EJ, Eisenberger MA, Crawford ED. Androgen deprivation and four courses of fixed-schedule suramin treatment in patients with newly diagnosed metastatic prostate cancer: a Southwest Oncology Group Study. J Clin Oncol 2000;18(5):1043–9.
2292. Calvo E, Cortes J, Rodriguez J, Sureda M, Beltran C, Rebollo J, Martinez-Monge R, Berian JM, de Irala J, Brugarolas A. Fixed higher dose schedule of suramin plus hydrocortisone in patients with hormone refractory prostate carcinoma a multicenter Phase II study. Cancer 2001;92(9):2435–43.
2293. Small EJ, Halabi S, Ratain MJ, Rosner G, Stadler W, Palchak D, Marshall E, Rago R, Hars V, Wilding G, Petrylak D, Vogelzang NJ. Randomized study of three different doses of suramin administered with a fixed dosing schedule in patients with advanced prostate cancer: results of intergroup 0159, cancer and leukemia group B 9480. J Clin Oncol 2002;20(16):3369–75.
2294. Ryan CW, Vokes EE, Vogelzang NJ, Janisch L, Kobayashi K, Ratain MJ. A phase I study of suramin with once- or twice-monthly dosing in patients with advanced cancer. Cancer Chemother Pharmacol 2002;50(1):1–5.
2295. Sorensen M, Engbaek J, Viby-Mogensen J, Guldager H, Molke Jensen F. Bradycardia and cardiac asystole following a single injection of suxamethonium. Acta Anaesthesiol Scand 1984;28(2):232–5.
2296. Linter SP, Thomas PR, Withington PS, Hall MG. Suxamethonium associated hypertonicity and cardiac arrest in unsuspected pseudohypertrophic muscular dystrophy. Br J Anaesth 1982;54(12):1331–2.
2297. Genever EE. Suxamethonium-induced cardiac arrest in unsuspected pseudohypertrophic muscular dystrophy. Case report. Br J Anaesth 1971;43(10):984–6.
2298. Henderson WA. Succinylcholine-induced cardiac arrest in unsuspected Duchenne muscular dystrophy. Can Anaesth Soc J 1984;31(4):444–6.

2299. Solares G, Herranz JL, Sanz MD. Suxamethonium-induced cardiac arrest as an initial manifestation of Duchenne muscular dystrophy. Br J Anaesth 1986;58(5):576.
2300. Sullivan M, Thompson WK, Hill GD. Succinylcholine-induced cardiac arrest in children with undiagnosed myopathy. Can J Anaesth 1994;41(6):497–501.
2301. Bush GH. Suxamethonium-associated hypertonicity and cardiac arrest in unsuspected pseudohypertrophic muscular dystrophy. Br J Anaesth 1983;55(9):923.
2302. Schaer H, Steinmann B, Jerusalem S, Maier C. Rhabdomyolysis induced by anaesthesia with intraoperative cardiac arrest. Br J Anaesth 1977;49(5):495–9.
2303. Seay AR, Ziter FA, Thompson JA. Cardiac arrest during induction of anesthesia in Duchenne muscular dystrophy. J Pediatr 1978;93(1):88–90.
2304. Parker SF, Bailey A, Drake AF. Infant hyperkalemic arrest after succinylcholine. Anesth Analg 1995;80(1):206–7.
2305. Schulte-Sasse U, Eberlein HJ, Schmucker I, Underwood D, Wolbert R. Sollte die verwendung von succinylcholin in der kinderanästhesie neu uberdacht werden? [Should the use of succinylcholine in pediatric anesthesia be re-evaluated?] Anaesthesiol Reanim 1993;18(1):13–9.
2306. Farrell PT. Anaesthesia-induced rhabdomyolysis causing cardiac arrest: case report and review of anaesthesia and the dystrophinopathies. Anaesth Intensive Care 1994;22(5):597–601.
2307. Larach MG, Rosenberg H, Gronert GA, Allen GC. Hyperkalemic cardiac arrest during anesthesia in infants and children with occult myopathies. Clin Pediatr (Phila) 1997;36(1):9–16.
2308. Jackson MA, Lodwick R, Hutchinson SG. Hyperkalaemic cardiac arrest successfully treated with peritoneal dialysis. BMJ 1996;312(7041):1289–90.
2309. Lin JL, Huang CC. Successful initiation of hemodialysis during cardiopulmonary resuscitation due to lethal hyperkalemia. Crit Care Med 1990;18(3):342–3.
2310. Lee G, Antognini JF, Gronert GA. Complete recovery after prolonged resuscitation and cardiopulmonary bypass for hyperkalemic cardiac arrest. Anesth Analg 1994;79(1):172–4.
2311. Smith RB, Petruscak J. Succinylcholine, digitalis, and hypercalcemia: a case report. Anesth Analg 1972;51(2):202–5.
2312. Nigrovic V. Succinylcholine, cholinoceptors and catecholamines: proposed mechanism of early adverse haemodynamic reactions. Can Anaesth Soc J 1984;31(4):382–94.
2313. Pace SA, Fuller FP. Out-of-hospital succinylcholine-assisted endotracheal intubation by paramedics. Ann Emerg Med 2000;35(6):568–72.
2314. Menegazzi JJ, Wayne MA. Succinylcholine-assisted endotracheal intubation by paramedics. Ann Emerg Med 2001;37(3):360–1.
2315. Katz SH, Falk JL. Misplaced endotracheal tubes by paramedics in an urban emergency medical services system. Ann Emerg Med 2001;37(1):32–7.
2316. Zink BJ, Snyder HS, Raccio-Robak N. Lack of a hyperkalemic response in emergency department patients receiving succinylcholine. Acad Emerg Med 1995;2(11):974–8.
2317. Dronen SC, Merigian KS, Hedges JR, Hoekstra JW, Borron SW. A comparison of blind nasotracheal and succinylcholine-assisted intubation in the poisoned patient. Ann Emerg Med 1987;16(6):650–2.
2318. Hohage H, Bruckner D, Arlt M, Buchholz B, Zidek W, Spieker C. Influence of cyclosporine A and FK506 on 24 h blood pressure monitoring in kidney transplant recipients Clin Nephrol 1996;45(5):342–4.
2319. Dollinger MM, Plevris JN, Chauhan A, MacGilchrist AJ, Finlayson ND, Hayes PC. Tacrolimus and cardiotoxicity in adult liver transplant recipients. Lancet 1995;346(8973):507.
2320. Jain AB, Fung JJ. Cyclosporin and tacrolimus in clinical transplantation. A comparative review. Clin Immunother 1996;5:351–73.
2321. Johnson MC, So S, Marsh JW, Murphy AM. QT prolongation and torsades de pointes after administration of FK506. Transplantation 1992;53(4):929–30.
2322. Sanoski CA, Vasquez EM, Bauman JL. QT interval prolongation associated with the use of tacrolimus in transplant recipients. Pharmacotherapy 1998;18:427.
2323. Chang RK, Alzona M, Alejos J, Jue K, McDiarmid SV. Marked left ventricular hypertrophy in children on tacrolimus (FK506) after orthotopic liver transplantation. Am J Cardiol 1998;81(10):1277–80.
2324. Scott JS, Boyle GJ, Daubeney PE, Miller SA, Law Y, Pigula F, Griffith BP, Webber SA. Tacrolimus: a cause of hypertrophic cardiomyopathy in pediatric heart transplant recipients? Transplant Proc 1999;31(1–2):82–3.
2325. Canzanello VJ, Textor SC, Taler SJ, Schwartz LL, Porayko MK, Wiesner RH, Krom RA. Late hypertension after liver transplantation: a comparison of cyclosporine and tacrolimus (FK 506). Liver Transpl Surg 1998;4(4):328–34.
2326. Taylor DO, Barr ML, Radovancevic B, Renlund DG, Mentzer RM Jr, Smart FW, Tolman DE, Frazier OH, Young JB, VanVeldhuisen P. A randomized, multicenter comparison of tacrolimus and cyclosporine immunosuppressive regimens in cardiac transplantation: decreased hyperlipidemia and hypertension with tacrolimus. J Heart Lung Transplant 1999;18(4):336–45.
2327. Gonzalez MG, Hernandez-Madrid A, Sanroman AL, Monge G, De Vicente E, Barcena R. Comparison of post-liver transplantation electrocardiographic alterations between cyclosporine- and tacrolimus-treated patients. Transplant Proc 1999;31(6):2423–4.
2328. Uchida N, Taniguchi S, Harada N, Shibuya T. Myocardial ischemia following allogeneic bone marrow transplantation: possible implication of tacrolimus overdose. Blood 2000;96(1):370–2.
2329. Pham SM, Kormos RL, Hattler BG, Kawai A, Tsamandas AC, Demetris AJ, Murali S, Fricker FJ, Chang HC, Jain AB, Starzl TE, Hardesty RL, Griffith BP. A prospective trial of tacrolimus (FK 506) in clinical heart transplantation: intermediate-term results. J Thorac Cardiovasc Surg 1996;111(4):764–72.
2330. Baid-Agrawal S, Delmonico FL, Tolkoff-Rubin NE, Farrell M, Williams WW, Shih V, Auchincloss H, Cosimi AB, Pascual M. Cardiovascular risk profile after conversion from cyclosporine A to tacrolimus in stable renal transplant recipients. Transplantation 2004;77(8):1199–202.
2331. Gutierrez dlF, Sola E, Alferez MJ, Navarro A, Cabello M, Burgos D, Gonzalez Molina M. Sindrome emolitico urémico de novo en el postoperatorio de un transplante renopancreatico. [De novo hemolytic uremic syndrome in a kidney-pancreas recipient in the postoperative period.] Nefrologia 2004;24 Suppl 3:3–6.

2332. Sakai T, Ii Y, Kuzuhara S. [Sinus bradycardia induced by talipexole hydrochloride in a patient with Parkinson disease.] Rinsho Shinkeigaku 1998;38(8):771–5.

2333. Finelli PF, Schauer PK. Cerebral sinus thrombosis with tamoxifen. Neurology 2001;56(8):1113–4.

2334. Tisman G. Thromboses after estrogen hormone replacement, progesterone or tamoxifen therapy in patients with elevated blood levels of homocysteine. Am J Hematol 2001;68(2):135.

2335. Goldhaber SZ. Tamoxifen: preventing breast cancer and placing the risk of deep vein thrombosis in perspective. Circulation 2005;111:539-

2336. Decensi A, Maisonneuve P, Rotmensz N, Bettega D, Costa A, Sacchini V, Salvioni A, Travaglini R, Oliviero P, D'Aiuto G, Gulisano M, Gucciardo G, Del Turco MR, Pizzichetta MA, Conforti S, Bonanni B, Boyle P, Veronesi U. Effect of tamoxifen on venous thromboembolic events in a breast cancer prevention trial. Circulation 2005;111:650–6.

2337. Ludwig M, Tolg R, Richardt G, Katus HA, Diedrich K. Myocardial infarction associated with ovarian hyperstimulation syndrome. JAMA 1999;282(7):632–3.

2338. Krenzelok EP, Jacobsen TD, Aronis J. Is the yew really poisonous to you? J Toxicol Clin Toxicol 1998;36(3):219–23.

2339. Syvanen K, Ekholm E, Anttila K, Salminen E. Immediate effects of docetaxel alone or in combination with epirubicin in cardiac function in advanced breast cancer. Anticancer Res 2003;23:1869–74.

2340. Gori S, Colozza M, Mosconi AM, Franceschi E, Basurto C, Cherubini R, Sidoni A, Rulli A, Bisacci C, De Angelis V, Crino L, Crino L, Tonato M. Phase II study of weekly paclitaxel and trastuzumab in anthracycline- and taxane-pretreated patients with HER2-overexpressing metastatic breast cancer. Br J Cancer 2004;90:36–40.

2341. Paul C, Janier M, Carlet J, Tamion F, Carlotti A, Fichelle JM, Daniel F. [Erythroderma induced by teicoplanin.] Ann Dermatol Venereol 1992;119(9):667–9.

2342. Paris DG, Parente TF, Bruschetta HR, Guzman E, Niarchos AP. Torsades de pointes induced by erythromycin and terfenadine. Am J Emerg Med 1994;12(6):636–8.

2343. Monahan BP, Ferguson CL, Killeavy ES, Lloyd BK, Troy J, Cantilena LR Jr. Torsades de pointes occurring in association with terfenadine use. JAMA 1990;264(21):2788–90.

2344. Clifford CP, Adams DA, Murray S, Taylor GW, Wilkins MR, Boobis AR, Davies DS. The cardiac effects of terfenadine after inhibition of its metabolism by grapefruit juice. Eur J Clin Pharmacol 1997;52(4):311–5.

2345. Mathews DR, McNutt B, Okerholm R, Flicker M, McBride G. Torsades de pointes occurring in association with terfenadine use. JAMA 1991;266(17):2375–6.

2346. Boccara F, Benhaiem-Sigaux N, Cohen A. Acute myopericarditis after diphtheria, tetanus, and polio vaccination. Chest 2001;120(2):671–2.

2347. Hewitt T, Eadon H. Check long contact with Ametop. Pharm Pract 1998;8:47–8.

2348. Sastre Dominguez J, Sastre Castillo A, Marin Nunez F. Anafilaxia sistémica a tetraciclinas. [Systemic anaphylaxis caused by tetracyclines.] Rev Clin Esp 1984;174(3–4):135–6.

2349. Pollen RH. Anaphylactoid reaction to orally administered demethylchlortetracycline. N Engl J Med 1964;271:673.

2350. Clark TE, Edom N, Larson J, Lindsey LJ. Thalomid (thalidomide) capsules: a review of the first 18 months of spontaneous postmarketing adverse event surveillance, including off-label prescribing. Drug Saf 2001;24(2):87–117.

2351. Fahdi IE, Gaddam V, Saucedo JF, Kishan CV, Vyas K, Deneke MG, Razek H, Thorn B, Bissett JK, Anaissie EJ, Barlogie B, Mehta JL. Bradycardia during therapy for multiple myeloma with thalidomide. Am J Cardiol 2004;93(8):1052–5.

2352. Pitini V, Arrigo C, Aloi G, Azzarello D, La Gattuta G. Diabetic foot disease in a patient with multiple myeloma receiving thalidomide. Haematologica 2002;87(2):ELT07.

2353. Zangari M, Anaissie E, Barlogie B, Badros A, Desikan R, Gopal AV, Morris C, Toor A, Siegel E, Fink L, Tricot G. Increased risk of deep-vein thrombosis in patients with multiple myeloma receiving thalidomide and chemotherapy. Blood 2001;98(5):1614–5.

2354. Zangari M, Siegel E, Barlogie B, Anaissie E, Saghafifar F, Fassas A, Morris C, Fink L, Tricot G. Thrombogenic activity of doxorubicin in myeloma patients receiving thalidomide: implications for therapy. Blood 2002;100(4):1168–71.

2355. Osman K, Comenzo R, Rajkumar SV. Deep venous thrombosis and thalidomide therapy for multiple myeloma. N Engl J Med 2001;344(25):1951–2.

2356. Bowcock SJ, Rassam SM, Ward SM, Turner JT, Laffan M. Thromboembolism in patients on thalidomide for myeloma. Hematology 2002;7(1):51–3.

2357. Horne MK 3rd, Figg WD, Arlen P, Gulley J, Parker C, Lakhani N, Parnes H, Dahut WL. Increased frequency of venous thromboembolism with the combination of docetaxel and thalidomide in patients with metastatic androgen-independent prostate cancer. Pharmacotherapy 2003;23(3):315–8.

2358. Howell E, Johnson SM. Venous thrombosis occurring after initiation of thalidomide for the treatment of cicatricial pemphigoid. J Drugs Dermatol 2004;3(1):83–5.

2359. Flageul B, Wallach D, Cavelier-Balloy B, Bachelez H, Carsuzaa F, Dubertret L. Thalidomide et thromboses. [Thalidomide and thrombosis.] Ann Dermatol Venereol 2000;127(2):171–4.

2360. Zangari M, Barlogie B, Thertulien R, Jacobson J, Eddleman P, Fink L, Fassas A, Van Rhee F, Talamo G, Lee CK, Tricot G. Thalidomide and deep vein thrombosis in multiple myeloma: risk factors and effect on survival. Clin Lymphoma 2003;4(1):32–5.

2361. Pouaha J, Martin S, Trechot P, Truchetet F, Barbaud A, Schmutz JL. Thalidomide et thromboses: trois observations. [Thalidomide and thrombosis: three observations.] Presse Méd 2001;30(20):1008–9.

2362. Hamuryudan V, Mat C, Saip S, Ozyazgan Y, Siva A, Yurdakul S, Zwingenberger K, Yazici H. Thalidomide in the treatment of the mucocutaneous lesions of the Behçet syndrome. A randomized, double-blind, placebo-controlled trial. Ann Intern Med 1998;128(6):443–50.

2363. Bushman JL. Green tea and cancer in humans: a review of the literature. Nutr Cancer 1998;31(3):151–9.

2364. Chaithiraphan S. Fatal complication associated with intravenous use of aminophylline. J Med Assoc Thai 1976;59(11):507–9.

2365. Levine JH, Michael JR, Guarnieri T. Multifocal atrial tachycardia: a toxic effect of theophylline. Lancet 1985;1(8419):12–4.

2366. Nagai Y, Abe T, Nomura G. Does pioglitazone, like troglitazone, increase serum levels of lipoprotein(a) in diabetic patients? Diabetes Care 2001;24(2):408–9.

2367. Page II RL, Gozansky WS, Ruscin JM. Possible heart failure exacerbation associated with rosiglitazone: case report and literature review. Pharmacotherapy 2003;23:945–54.

2368. Walker AB, Naderali EK, Chattington PD, Buckingham RE, Williams G. Differential vasoactive effects of the insulin sensitizers rosiglitazone (BRL 49653) and troglitazone on human small arteries in vitro. Diabetes 1998;47:810–4.
2369. Nesto RW, Bell D, Bonow RO, Fonseca V, Grundy SM, Horton ES, Winter ML, Porte D, Semenkovich CF, Smith S, Young LH, Kahn R. Thiazolidinedione use, fluid retention and congestive heart failure. Diabetes Care 2004;27:256–63.
2370. Delea TE, Edelsberg JS, Hagiwara MH, Oster G, Phillips LS. Use of thiazolidinediones and risk of heart failure in people with type 2 diabetes; a retrospective cohort study. Diabetes Care 2003;26:2983–9.
2371. Karter AJ, Ahmed AT, Liu J, Moffet HH, Parker MM, Ferrara A, Selby JV. Use of thiazolidinediones and risk of heart failure in people with type 2 diabetes; a retrospective cohort study. Diabetes Care 2004;27:850–1.
2372. Delea TE, Edelsberg JS, Hagiwara MH, Oster G, Phillips LS. Use of thiazolidinediones and risk of heart failure in people with type 2 diabetes; a retrospective cohort study. Diabetes Care 2004;27:852.
2373. Manley HJ, Allcock NM Thiazolidinedione safety and efficacy in ambulatory patients receiving hemodialysis. Pharmacotherapy 2003;23:861–5.
2374. Martin-Denavit T, Edery P, Plauchu H, Attia-Sobol J, Raudrant D, Aurand JM, Thomas L. Ectodermal abnormalities associated with methimazole intrauterine exposure. Am J Med Genet 2000;94(4):338–40.
2375. Clementi M, Di Gianantonio E, Pelo E, Mammi I, Basile RT, Tenconi R. Methimazole embryopathy: delineation of the phenotype. Am J Med Genet 1999;83(1):43–6.
2376. Di Gianantonio E, Schaefer C, Mastroiacovo PP, Cournot MP, Benedicenti F, Reuvers M, Occupati B, Robert E, Bellemin B, Addis A, Arnon J, Clementi M. Adverse effects of prenatal methimazole exposure. Teratology 2001;64(5):262–6.
2377. Bihan H, Vazquez MP, Krivitzky A, Cohen R. Aplasia cutis congenita and dysmorphic syndrome after antithyroid therapy during pregnancy. Endocrinologist 2002;12:87–91.
2378. Barwell J, Fox GF, Round J, Berg J. Choanal atresia: the result of maternal thyrotoxicosis or fetal carbimazole? Am J Med Genet 2002;111(1):55–6.
2379. Mandel SJ, Cooper DS. The use of antithyroid drugs in pregnancy and lactation. J Clin Endocrinol Metab 2001;86(6):2354–9.
2380. Burrow GN. The management of thyrotoxicosis in pregnancy. N Engl J Med 1985;313(9):562–5.
2381. Momotani N, Noh J, Oyanagi H, Ishikawa N, Ito K. Antithyroid drug therapy for Graves' disease during pregnancy. Optimal regimen for fetal thyroid status. N Engl J Med 1986;315(1):24–8.
2382. Becker KE Jr, Tonnesen AS. Cardiovascular effects of plasma levels of thiopental necessary for anesthesia. Anesthesiology 1978;49(3):197–200.
2383. Stone JG, Young WL, Marans ZS, Khambatta HJ, Solomon RA, Smith CR, Ostapkovich N, Jamdar SC, Diaz J. Cardiac performance preserved despite thiopental loading. Anesthesiology 1993;79(1):36–41.
2384. Warner JP, Barnes TR, Henry JA. Electrocardiographic changes in patients receiving neuroleptic medication. Acta Psychiatr Scand 1996;93(4):311–3.
2385. Reilly JG, Ayis SA, Ferrier IN, Jones SJ, Thomas SH. QT_c-interval abnormalities and psychotropic drug therapy in psychiatric patients. Lancet 2000;355(9209): 1048–52.
2386. Kiriike N, Maeda Y, Nishiwaki S, Izumiya Y, Katahara S, Mui K, Kawakita Y, Nishikimi T, Takeuchi K, Takeda T. Iatrogenic torsade de pointes induced by thioridazine. Biol Psychiatry 1987;22(1):99–103.
2387. Connolly MJ, Evemy KL, Snow MH. Torsade de pointes ventricular tachycardia in association with thioridazine therapy: report of two cases. New Trends Arrhythmias 1985;1:157.
2388. Axelsson R, Aspenstrom G. Electrocardiographic changes and serum concentrations in thioridazine-treated patients. J Clin Psychiatry 1982;43(8):332–5.
2389. von Bahr C, Movin G, Nordin C, Liden A, Hammarlund-Udenaes M, Hedberg A, Ring H, Sjoqvist F. Plasma levels of thioridazine and metabolites are influenced by the debrisoquin hydroxylation phenotype. Clin Pharmacol Ther 1991;49(3):234–40.
2390. Hartigan-Go K, Bateman DN, Nyberg G, Martensson E, Thomas SH. Concentration-related pharmacodynamic effects of thioridazine and its metabolites in humans. Clin Pharmacol Ther 1996;60(5):543–53.
2391. Carrillo JA, Ramos SI, Herraiz AG, Llerena A, Agundez JA, Berecz R, Duran M, Benitez J. Pharmacokinetic interaction of fluvoxamine and thioridazine in schizophrenic patients. J Clin Psychopharmacol 1999;19(6):494–9.
2392. Dahl SG. Active metabolites of neuroleptic drugs: possible contribution to therapeutic and toxic effects. Ther Drug Monit 1982;4(1):33–40.
2393. Risch SC, Groom GP, Janowsky DS. The effects of psychotropic drugs on the cardiovascular system. J Clin Psychiatry 1982;43(5 Pt 2):16–31.
2394. Timell AM. Thioridazine: re-evaluating the risk/benefit equation. Ann Clin Psychiatry 2000;12(3):147–51.
2395. Thomas SH, Cooper PN. Sudden death in a patient taking antipsychotic drugs. Postgrad Med J 1998;74(873):445–6.
2396. Swift CG, Lee DR, Maskrey VL, Yisak W, Jackson SH, Tiplady B. Single dose pharmacodynamics of thioridazine and remoxipride in healthy younger and older volunteers. J Psychopharmacol 1999;13(2):159–65.
2397. Rampello L, Raffaele R, Vecchio I, Pistone G, Brunetto MB, Malaguarnera M. Behavioural changes and hypotensive effects of thioridazine in two elderly patients with traumatic brain injury: post-traumatic syndrome and thioridazine. Gaz Med Ital Arch Sci Med 2000;159:121–3.
2398. Lew AS, Laramee P, Cercek B, Shah PK, Ganz W. The hypotensive effect of intravenous streptokinase in patients with acute myocardial infarction. Circulation 1985;72(6):1321–6.
2399. Wei JY, Markis JE, Malagold M, Braunwald E. Cardiovascular reflexes stimulated by reperfusion of ischemic myocardium in acute myocardial infarction. Circulation 1983;67(4):796–801.
2400. Been M, de Bono DP, Muir AL, Boulton FE, Fears R, Standring R, Ferres H. Clinical effects and kinetic properties of intravenous APSAC—anisoylated plasminogen-streptokinase activator complex (BRL 26921) in acute myocardial infarction. Int J Cardiol 1986;11(1):53–61.
2401. ISIS-3 (Third International Study of Infarct Survival) Collaborative Group. ISIS-3: a randomised comparison of streptokinase vs tissue plasminogen activator vs anistreplase and of aspirin plus heparin vs aspirin alone among 41,299 cases of suspected acute myocardial infarction. Lancet 1992;339(8796):753–70.

2402. Meissner AJ, Misiak A, Ziemski JM, Scharf R, Rudowski W, Huszcza S, Kucharski W, Wislawski S. Hazards of thrombolytic therapy in deep vein thrombosis. Br J Surg 1987;74(11):991–3.
2403. Grimm W, Schwieder G, Wagner T. Todliche Lungenembolie bei Bein-Beckenvenenthrombose unter Lysetherapie. [Fatal pulmonary embolism in venous thrombosis of the leg and pelvis during lysis therapy.] Dtsch Med Wochenschr 1990;115(31–32):1183–7.
2404. Queen M, Biem HJ, Moe GW, Sugar L. Development of cholesterol embolization syndrome after intravenous streptokinase for acute myocardial infarction. Am J Cardiol 1990;65(15):1042–3.
2405. Blankenship JC. Cholesterol embolisation after thrombolytic therapy. Drug Saf 1996;14(2):78–84.
2406. Gupta BK, Spinowitz BS, Charytan C, Wahl SJ. Cholesterol crystal embolization-associated renal failure after therapy with recombinant tissue-type plasminogen activator. Am J Kidney Dis 1993;21(6):659–62.
2407. Gurwitz JH, Gore JM, Goldberg RJ, Barron HV, Breen T, Rundle AC, Sloan MA, French W, Rogers WJ. Risk for intracranial hemorrhage after tissue plasminogen activator treatment for acute myocardial infarction. Participants in the National Registry of Myocardial Infarction 2. Ann Intern Med 1998;129(8):597–604.
2408. Cairns JA, Kennedy JW, Fuster V. Coronary thrombolysis. Chest 1998;114(5 Suppl):S634–57.
2409. Chen MF, Liau CS, Lee YT. Coronary arterial aneurysm after percutaneous transluminal coronary recanalization with streptokinase. Int J Cardiol 1990;28(1):117–9.
2410. Sawin CT, Geller A, Wolf PA, Belanger AJ, Baker E, Bacharach P, Wilson PW, Benjamin EJ, D'Agostino RB. Low serum thyrotropin concentrations as a risk factor for atrial fibrillation in older persons. N Engl J Med 1994;331(19):1249–52.
2411. Biondi B, Fazio S, Coltorti F, Palmieri EA, Carella C, Lombardi G, Sacca L. Clinical case seminar. Reentrant atrioventricular nodal tachycardia induced by levothyroxine. J Clin Endocrinol Metab 1998;83(8):2643–5.
2412. Toft AD, Boon NA. Thyroid disease and the heart. Heart 2000;84(4):455–60.
2413. Ching GW, Franklyn JA, Stallard TJ, Daykin J, Sheppard MC, Gammage MD. Cardiac hypertrophy as a result of long-term thyroxine therapy and thyrotoxicosis. Heart 1996;75(4):363–8.
2414. Kohno A, Hara Y. Severe myocardial ischemia following hormone replacement in two cases of hypothyroidism with normal coronary arteriogram. Endocr J 2001;48(5):565–72.
2415. Locker GJ, Kotzmann H, Frey B, Messina FC, Strez FR, Weissel M, Laggner AN. Factitious hyperthyroidism causing acute myocardial infarction. Thyroid 1995;5(6):465–7.
2416. The Coronary Drug Project Research Group. The coronary drug project. Findings leading to further modifications of its protocol with respect to dextrothyroxine. JAMA 1972;220(7):996–1008.
2417. Biondi B, Fazio S, Cuocolo A, Sabatini D, Nicolai E, Lombardi G, Salvatore M, Sacca L. Impaired cardiac reserve and exercise capacity in patients receiving long-term thyrotropin suppressive therapy with levothyroxine. J Clin Endocrinol Metab 1996;81(12):4224–8.
2418. Bagheri H, Simiand E, Montastruc JL, Magnaval JF. Adverse drug reactions to anthelmintics. Ann Pharmacother 2004;38(3):383–8.
2419. Iglesias E, Esteban E, Zabala S, Gascon A. Tiapride-induced torsade de pointes. Am J Med 2000;109(6):509.
2420. Covelli H, Bhattacharya S, Cassino C, Conoscenti C, Kesten S. Absence of electrocardiographic findings and improved function with once-daily tiotropium in patients with chronic obstructive pulmonary disease. Pharmacotherapy 2005;25:1708–18.
2421. Greenspon AJ, Mohiuddin S, Saksena S, Lengerich R, Snapinn S, Holmes G, Irvin J, Sappington E, et al. Comparison of intravenous tocainide with intravenous lidocaine for treating ventricular arrhythmias. Cardiovasc Rev Rep 1989;10:55–9.
2422. Winkle RA, Anderson JL, Peters F, Meffin PJ, Fowles RE, Harrison DC. The hemodynamic effects of intravenous tocainide in patients with heart disease. Circulation 1978;57(4):787–92.
2423. Maloney JD, Nissen RG, McColgan JM. Open clinical studies at a referral center: chronic maintenance tocainide therapy in patients with recurrent sustained ventricular tachycardia refractory to conventional antiarrhythmic agents. Am Heart J 1980;100(6 Pt 2):1023–30.
2424. Cheesman M, Ward DE. Exacerbation of ventricular tachycardia by tocainide. Clin Cardiol 1985;8(1):47–50.
2425. Winkle RA, Meffin PJ, Harrison DC. Long-term tocainide therapy for ventricular arrhythmias. Circulation 1978;57(5):1008–16.
2426. Roden DM, Reele SB, Higgins SB, Carr RK, Smith RF, Oates JA, Woosley RL. Tocainide therapy for refractory ventricular arrhythmias. Am Heart J 1980;100(1):15–22.
2427. Gould LA, Betzu R, Vacek T, Muller R, Pradeep V, Downs L. Sinoatrial block due to tocainide. Am Heart J 1989;118(4):851–3.
2428. Van Natta B, Lazarus M, Li C. Irreversible interstitial pneumonitis associated with tocainide therapy. West J Med 1988;149(1):91–2.
2429. Perlow GM, Jain BP, Pauker SG, Zarren HS, Wistran DC, Epstein RL. Tocainide-associated interstitial pneumonitis. Ann Intern Med 1981;94(4 Pt 1):489–90.
2430. Maibach E. European experiences with tolmetin in the treatment of rheumatic diseases. Curr Ther Res Clin Exp 1976;19(3):350–62.
2431. Lippmann M, Mok MS, Farinacci JV, Lee JC. Tonazocine mesylate in postoperative pain patients: a double-blind placebo controlled analgesic study. J Clin Pharmacol 1989;29(4):373–8.
2432. Schwarzer S, Eber B, Greinix H, Lind P. Non-Q-wave myocardial infarction associated with bleomycin and etoposide chemotherapy. Eur Heart J 1991;12(6):748–50.
2433. Aisner J, Van Echo DA, Whitacre M, Wiernik PH. A phase I trial of continuous infusion VP16–213 (etoposide). Cancer Chemother Pharmacol 1982;7(2–3):157–60.
2434. Krantz MJ, Garcia JA, Mehler PS. Tramadol-associated pericarditis. Int J Cardiol 2005;99(3):497–8.
2435. Seidman AD, Fornier MN, Esteva FJ, Tan L, Kaptain S, Bach A, Panageas KS, Arroyo C, Valero V, Currie V, Gilewski T, Theodoulou M, Moynahan ME, Moasser M, Sklarin N, Dickler M, D'Andrea G, Cristofanilli M, Rivera E, Hortobagyi GN, Norton L, Hudis CA. Weekly trastuzumab and paclitaxel therapy for metastatic breast cancer with analysis of efficacy by HER2 immunophenotype and gene amplification. J Clin Oncol 2001;19(10):2587–95.
2436. Tham YL, Verani MS, Chang J. Reversible and irreversible cardiac dysfunction associated with trastuzumab in breast cancer. Breast Cancer Res Treat 2002;74(2):131–4.

2437. Slamon DJ, Leyland-Jones B, Shak S, Fuchs H, Paton V, Bajamonde A, Fleming T, Eiermann W, Wolter J, Pegram M, Baselga J, Norton L. Use of chemotherapy plus a monoclonal antibody against HER2 for metastatic breast cancer that overexpresses HER2. N Engl J Med 2001;344(11):783–92.
2438. Seidman A, Hudis C, Pierri MK, Shak S, Paton V, Ashby M, Murphy M, Stewart SJ, Keefe D. Cardiac dysfunction in the trastuzumab clinical trials experience. J Clin Oncol 2002;20(5):1215–21.
2439. Andersson J, Linderholm B, Greim G, Lindh B, Lindman H, Tennvall J, Tennvall-Nittby L, Pettersson-Skold D, Sverrisdottir A, Soderberg M, Klaar S, Bergh J. A population-based study on the first forty-eight breast cancer patients receiving trastuzumab (Herceptin) on a named patient basis in Sweden. Acta Oncol 2002;41(3):276–81.
2440. Leonard DS, Hill AD, Kelly L, Dijkstra B, McDermott E, O'Higgins NJ. Anti-human epidermal growth factor receptor 2 monoclonal antibody therapy for breast cancer. Br J Surg 2002;89(3):262–71.
2441. Janowsky D, Curtis G, Zisook S, Kuhn K, Resovsky K, Le Winter M. Trazodone-aggravated ventricular arrhythmias. J Clin Psychopharmacol 1983;3(6):372–6.
2442. Vlay SC, Friedling S. Trazodone exacerbation of VT. Am Heart J 1983;106(3):604.
2443. Vitullo RN, Wharton JM, Allen NB, Pritchett EL. Trazodone-related exercise-induced nonsustained ventricular tachycardia. Chest 1990;98(1):247–8.
2444. Aronson MD, Hafez H. A case of trazodone-induced ventricular tachycardia. J Clin Psychiatry 1986;47(7): 388–9.
2445. White WB, Wong SH. Rapid atrial fibrillation associated with trazodone hydrochloride. Arch Gen Psychiatry 1985;42(4):424.
2446. Rausch JL, Pavlinac DM, Newman PE. Complete heart block following a single dose of trazodone. Am J Psychiatry 1984;141(11):1472–3.
2447. Asayesh K. Combination of trazodone and phenothiazines: a possible additive hypotensive effect. Can J Psychiatry 1986;31(9):857–8.
2448. Barrnett J, Frances A, Kocsis J, Brown R, Mann JJ. Peripheral edema associated with trazodone: a report of ten cases. J Clin Psychopharmacol 1985;5(3):161–4.
2449. Glassman AH, Preud'homme XA. Review of the cardiovascular effects of heterocyclic antidepressants. J Clin Psychiatry 1993;54(Suppl):16–22.
2450. Ziegler VE, Co BT, Biggs JT. Plasma nortriptyline levels and ECG findings. Am J Psychiatry 1977;134(4):441–3.
2451. Veith RC, Friedel RO, Bloom V, Bielski R. Electrocardiogram changes and plasma desipramine levels during treatment of depression. Clin Pharmacol Ther 1980;27(6):796–802.
2452. Spiker DG, Weiss AN, Chang SS, Ruwitch JF Jr, Biggs JT. Tricyclic antidepressant overdose: clinical presentation and plasma levels. Clin Pharmacol Ther 1975;18(5 Part 1):539–46.
2453. Burrows GD, Vohra J, Hunt D, Sloman JG, Scoggins BA, Davies B. Cardiac effects of different tricyclic antidepressant drugs. Br J Psychiatry 1976;129:335–41.
2454. Hallstrom C, Gifford L. Antidepressant blood levels in acute overdose. Postgrad Med J 1976;52(613):687–8.
2455. Petit JM, Spiker DG, Ruwitch JF, Ziegler VE, Weiss AN, Biggs JT. Tricyclic antidepressant plasma levels and adverse effects after overdose. Clin Pharmacol Ther 1977;21(1):47–51.
2456. Roose SP, Glassman AH, Giardina EG, Johnson LL, Walsh BT, Woodring S, Bigger JT Jr. Nortriptyline in depressed patients with left ventricular impairment. JAMA 1986;256(23):3253–7.
2457. Bigger JT, Kantor SJ, Glassman AH, et al. Cardiovascular effects of tricyclic antidepressant drugs. In: Lipton MA, Dimascio A, Killam KF, editors. Psychopharmacology: a Generation of Progress. New York: Raven Press, 1978:1033.
2458. Burrows GD, Vohra J, Dumovic P, et al. Tricyclic antidepressant drugs and cardiac conduction. Prog Neuropsychopharmacol 1977;1:329.
2459. Brorson L, Wennerblom B. Electrophysiological methods in assessing cardiac effects of the tricyclic antidepressant imipramine. Acta Med Scand 1978;203(5):429–32.
2460. Scherlag BJ, Lau SH, Helfant RH, Berkowitz WD, Stein E, Damato AN. Catheter technique for recording His bundle activity in man. Circulation 1969;39(1):13–8.
2461. Anonymous. Sudden death in children treated with a tricyclic antidepressant. Med Lett Drugs Ther 1990;32(819):53.
2462. Riddle MA, Nelson JC, Kleinman CS, Rasmusson A, Leckman JF, King RA, Cohen DJ. Sudden death in children receiving Norpramin: a review of three reported cases and commentary. J Am Acad Child Adolesc Psychiatry 1991;30(1):104–8.
2463. Biederman J, Baldessarini RJ, Wright V, Knee D, Harmatz JS, Goldblatt A. A double-blind placebo controlled study of desipramine in the treatment ADD. II. Serum drug levels and cardiovascular findings. J Am Acad Child Adolesc Psychiatry 1989;28(6):903–11.
2464. Bigger JT, Giardina EG, Perel JM, Kantor SJ, Glassman AH. Cardiac antiarrhythmic effect of imipramine hydrochloride. N Engl J Med 1977;296(4):206–8.
2465. Wasylenki D. Depression in the elderly. Can Med Assoc J 1980;122(5):525–32.
2466. Todd RD, Faber R. Ventricular arrhythmias induced by doxepin and amitriptyline: case report. J Clin Psychiatry 1983;44(11):423–5.
2467. Avery D, Winokur G. Mortality in depressed patients treated with electroconvulsive therapy and antidepressants. Arch Gen Psychiatry 1976;33(9):1029–37.
2468. Veith RC, Raskind MA, Caldwell JH, Barnes RF, Gumbrecht G, Ritchie JL. Cardiovascular effects of tricyclic antidepressants in depressed patients with chronic heart disease. N Engl J Med 1982;306(16):954–9.
2469. Ziegler VE, Taylor JR, Wetzel RD, Biggs JT. Nortriptyline plasma levels and subjective side effects. Br J Psychiatry 1978;132:55.
2470. Reisby N, Gram LF, Bech P, Nagy A, Petersen GO, Ortmann J, Ibsen I, Dencker SJ, Jacobsen O, Krautwald O, Sondergaard I, Christiansen J. Imipramine: clinical effects and pharmacokinetic variability. Psychopharmacology (Berl) 1977;54(3):263–72.
2471. Glassman AH, Bigger JT Jr, Giardina EV, Kantor SJ, Perel JM, Davies M. Clinical characteristics of imipramine-induced orthostatic hypotension. Lancet 1979;1(8114):468–72.
2472. van Zwieten PA. The central action of antihypertensive drugs, mediated via central alpha-receptors. J Pharm Pharmacol 1973;25(2):89–95.
2473. Taylor DJ, Braithwaite RA. Cardiac effects of tricyclic antidepressant medication. A preliminary study of nortriptyline. Br Heart J 1978;40(9):1005–9.

2474. Langou RA, Van Dyke C, Tahan SR, Cohen LS. Cardiovascular manifestations of tricyclic antidepressant overdose. Am Heart J 1980;100(4):458–64.
2475. Thorstrand C. Cardiovascular effects of poisoning with tricyclic antidepressants. Acta Med Scand 1974;195(6): 505–14.
2476. Sternon J, Owieczka J. La prophylaxie de l'hypotension orthostatique induite par les antidepresseurs tricycliques. [The prophylaxis of orthostatic hypotension caused by tricyclic antidepressants.] Ars Med (Brux) 1979;34:641.
2477. Cohen HW, Gibson G, Alderman MH. Excess risk of myocardial infarction in patients treated with antidepressant medications: association with use of tricyclic agents. Am J Med 2000;108(1):2–8.
2478. Siddiqui JH, Vakassi MM, Ghani MF. Cardiac effects of amitriptyline overdose. Curr Ther Res Clin Exp 1977;22:321.
2479. O'Brien JP. A study of low-dose amitriptyline overdoses. Am J Psychiatry 1977;134(1):66–8.
2480. Hoffman JR, McElroy CR. Bicarbonate therapy for dysrhythmia hypotension in tricyclic antidepressant overdose. West J Med 1981;134(1):60–4.
2481. Serafimovski N, Thorball N, Asmussen I, Lunding M. Tricyclic antidepressive poisoning with special reference to cardiac complications. Acta Anaesthesiol Scand Suppl 1975;57:55–63.
2482. Crome P, Newman B. Fatal tricyclic antidepressant poisoning. J R Soc Med 1979;72(9):649–53.
2483. Nogue Xarau S, Nadal Trias P, Bertran Georges A, Mas Ordeig A, Munne Mas P, Milla Santos J. Intoxicacion agoda grave por antidepresivos triciclcos. Estudio retrospective de is casos. [Severe acute poisoning following the ingestion of tricyclic antidepressants.] Med Clin (Barc) 1980;74(7):257–62.
2484. Goldberg RJ, Capone RJ, Hunt JD. Cardiac complications following tricyclic antidepressant overdose. Issues for monitoring policy. JAMA 1985;254(13):1772–5.
2485. McAlpine SB, Calabro JJ, Robinson MD, Burkle FM Jr. Late death in tricyclic antidepressant overdose revisited. Ann Emerg Med 1986;15(11):1349–52.
2486. Brown TC, Barker GA, Dunlop ME, Loughnan PM. The use of sodium bicarbonate in the treatment of tricyclic antidepressant-induced arrhythmias. Anaesth Intensive Care 1973;1(3):203–10.
2487. Pinder RM, Brogden RN, Speight TM, Avery GS. Doxepin up-to-date: a review of its pharmacological properties and therapeutic efficacy with particular reference to depression. Drugs 1977;13(3):161–218.
2488. Reed K, Smith RC, Schoolar JC, Hu R, Leelavathi DE, Mann E, Lippman L. Cardiovascular effects of nortriptyline in geriatric patients. Am J Psychiatry 1980;137(8):986–9.
2489. Roose SP, Glassman AH, Siris SG, Walsh BT, Bruno RL, Wright LB. Comparison of imipramine- and nortriptyline-induced orthostatic hypotension: a meaningful difference. J Clin Psychopharmacol 1981;1(5):316–9.
2490. Thayssen P, Bjerre M, Kragh-Sorensen P, Moller M, Petersen OL, Kristensen CB, Gram LF. Cardiovascular effect of imipramine and nortriptyline in elderly patients. Psychopharmacology (Berl) 1981;74(4):360–4.
2491. Neshkes RE, Gerner R, Jarvik LF, Mintz J, Joseph J, Linde S, Aldrich J, Conolly ME, Rosen R, Hill M. Orthostatic effect of imipramine and doxepin in depressed geriatric outpatients. J Clin Psychopharmacol 1985;5(2):102–6.
2492. Kantor SJ, Glassman AH, Bigger JT Jr, Perel JM, Giardina EV. The cardiac effects of therapeutic plasma concentrations of imipramine. Am J Psychiatry 1978;135(5):534–8.
2493. Kantor SJ, Bigger JT Jr, Glassman AH, Macken DL, Perel JM. Imipramine-induced heart block. A longitudinal case study. JAMA 1975;231(13):1364–6.
2494. Brigg GG, Freedman RK, Jaffe SJ. In: A Reference Guide to Fetal and Neontal Risk: Drugs in Pregnancy and Lactation. 3rd ed.. Baltimore-Hong Kong-London-Sydney: Williams and Wilkins, 1990:621.
2495. Hernandez-Diaz S, Werler MM, Walker AM, Mitchell AA. Folic acid antagonists during pregnancy and the risk of birth defects. N Engl J Med 2000;343(22):1608–14.
2496. Richardson MP, Osrin D, Donaghy S, Brown NA, Hay P, Sharland M. Spinal malformations in the fetuses of HIV infected women receiving combination antiretroviral therapy and co-trimoxazole. Eur J Obstet Gynecol Reprod Biol 2000;93(2):215–7.
2497. Committee on Safety of Medicines. Sumatriptan (Imigran) and chest pain. Curr Probl 1992;34:2.
2498. Hicklin LA, Ryan C, Wong DK, Hinton AC. Nose-bleeds after sildenafil (Viagra). J R Soc Med 2002;95(8):402–3.
2499. Morgan DR, Trimble M, McVeigh GE. Atrial fibrillation associated with sumatriptan. BMJ 2000;321(7256):275.
2500. Bellaiche G, Radu B, Boucard M, Ley G, Slama JL. Infarctus splénique associé à la prise de zolmitriptan. [Splenic infarction associated with zolmitriptan use.] Gastroenterol Clin Biol 2002;26(3):298.
2501. Vicedo CMF, Garcia MB, Bellver MJG, Bustamante AP. Conjunctival tropicamide and transitory ischemic accident (TIA). Farm Clin 1998;15:115–8.
2502. Moss J, Rosow CE, Savarese JJ, Philbin DM, Kniffen KJ. Role of histamine in the hypotensive action of d-tubocurarine in humans. Anesthesiology 1981;55(1):19–25.
2503. Hatano Y, Arai T, Noda J, Komatsu K, Shinkura R, Nakajima Y, Sawada M, Mori K. Contribution of prostacyclin to D-tubocurarine-induced hypotension in humans. Anesthesiology 1990;72(1):28–32.
2504. Marshall IG. Pharmacological effects of neuromuscular blocking agents: interaction with cholinoceptors other than nicotinic receptors of the neuromuscular junction. Anest Rianim 1986;27:19.
2505. Munger WL, Miller RD, Stevens WC. The dependence of a d-tubocurarine-induced hypotension on alveolar concentration of halothane, dose of d-tubocurarine, and nitrous oxide. Anesthesiology 1974;40(5):442–8.
2506. Stoelting RK, McCammon RL, Hilgenberg JC. Changes in blood pressure with varying rates of administration of d-tubocurarine. Anesth Analg 1980;59(9):697–9.
2507. Hanson DS, Leggette CT. Severe hypotension following inadvertent intravenous administration of interferon alfa-2a. Ann Pharmacother 1997;31(3):371–2.
2508. Hegewisch S, Weh HJ, Hossfeld DK. TNF-induced cardiomyopathy. Lancet 1990;335(8684):294–5.
2509. Dijkmans PA, Visser CA, Kamp O. Adverse reactions to ultrasound contrast agents: is the risk worth the benefit? Eur J Echocardiogr 2005;6:363–6.
2510. Tamasi F, Weidner A, Domokos N, Bedros RJ, Bagdany S. ECHOVIST-200 enhanced hystero-sonography: a new technique in the assessment of infertility. Eur J Obstet Gynecol Reprod Biol 2005;121:186–90.
2511. Ramsay RE, Cantrell D, Collins SD, Walch JK, Naritoku DK, Cloyd JC, Sommerville K. Safety and tolerance of

rapidly infused Depacon. A randomized trial in subjects with epilepsy. Epilepsy Res 2003;52:189–201.

2512. Garrelts JC, Smith DF Jr, Ast D, LaRocca J, Peterie JD. Phlebitis associated with vancomycin therapy. Clin Pharm 1988;7(10):720–1.

2513. Hadaway L, Chamallas SN. Vancomycin: new perspectives on an old drug. J Infus Nurs 2003;26(5):278–84.

2514. Burroughs AK, Planas R, Svoboda P. Optimizing emergency care of upper gastrointestinal bleeding in cirrhotic patients. Scand J Gastroenterol Suppl 1998;226:14–24.

2515. Moreno-Sanchez D, Casis B, Martin A, Ortiz P, Castellano G, Munoz MT, Vanaclocha F, Solis-Herruzo JA. Rhabdomyolysis and cutaneous necrosis following intravenous vasopressin infusion. Gastroenterology 1991;101(2):529–32.

2516. Virtanen R, Kauppila M, Itala M. Percutaneous coronary intervention with stenting in a patient with haemophilia A and an acute myocardial infarction following a single dose of desmopressin. Thromb Haemost 2004;92:1154–6.

2517. Gulberg V, Bilzer M, Gerbes AL. Long-term therapy and retreatment of hepatorenal syndrome type 1 with ornipressin and dopamine. Hepatology 1999;30(4):870–5.

2518. Romero G, Kravetz D, Argonz J, Bildozola M, Suarez A, Terg R. Terlipressin is more effective in decreasing variceal pressure than portal pressure in cirrhotic patients. J Hepatol 2000;32(3):419–25.

2519. Escorsell A, Ruiz del Arbol L, Planas R, Albillos A, Banares R, Cales P, Pateron D, Bernard B, Vinel JP, Bosch J. Multicenter randomized controlled trial of terlipressin versus sclerotherapy in the treatment of acute variceal bleeding: the TEST study. Hepatology 2000;32(3):471–6.

2520. Bruha R, Marecek Z, Spicak J, Hulek P, Lata J, Petrtyl J, Urbanek P, Taimr P, Volfova M, Dite P. Double-blind randomized, comparative multicenter study of the effect of terlipressin in the treatment of acute esophageal variceal and/or hypertensive gastropathy bleeding. Hepatogastroenterology 2002;49(46):1161–6.

2521. Zhang HB, Wong BC, Zhou XM, Guo XG, Zhao SJ, Wang JH, Wu KC, Ding J, Lam SK, Fan DM. Effects of somatostatin, octreotide and pitressin plus nitroglycerine on systemic and portal haemodynamics in the control of acute variceal bleeding. Int J Clin Pract 2002;56(6): 447–51.

2522. Medel J, Boccara G, Van de Steen E, Bertrand M, Godet G, Coriat P. Terlipressin for treating intraoperative hypotension: can it unmask myocardial ischemia? Anesth Analg 2001;93(1):53–5.

2523. Ortega R, Gines P, Uriz J, Cardenas A, Calahorra B, De Las Heras D, Guevara M, Bataller R, Jimenez W, Arroyo V, Rodes J. Terlipressin therapy with and without albumin for patients with hepatorenal syndrome: results of a prospective, nonrandomized study. Hepatology 2002;36(4 Pt 1):941–8.

2524. Morelli A, Tritapepe L, Rocco M, Conti G, Orecchioni A, De Gaetano A, Picchini U, Pelaia P, Reale C, Pietropaoli P. Terlipressin versus norepinephrine to counteract anesthesia-induced hypotension in patients treated with renin–angiotensin system inhibitors: effects on systemic and regional hemodynamics. Anesthesiology 2005;102:12–19.

2525. Mirakhur RK, Ferres CJ, Clarke RS, Bali IM, Dundee JW. Clinical evaluation of Org NC 45. Br J Anaesth 1983;55(2):119–24.

2526. Barnes PK, Smith GB, White WD, Tennant R. Comparison of the effects of Org NC 45 and pancuronium bromide on heart rate and arterial pressure in anaesthetized man. Br J Anaesth 1982;54(4):435–9.

2527. Gregoretti SM, Sohn YJ, Sia RL. Heart rate and blood pressure changes after ORG NC45 (vecuronium) and pancuronium during halothane and enflurane anesthesia. Anesthesiology 1982;56(5):392–5.

2528. Lienhart A, Desnault H, Guggiari M, et al. Vecuronium bromide: dose response curve and haemodynamic effects in anaesthetized man. In: Agoston S, editor. Clinical Experiences with Norcuron. Amsterdam: Excerpta Medica, 1982:46.

2529. Morris RB, Cahalan MK, Miller RD, Wilkinson PL, Quasha AL, Robinson SL. The cardiovascular effects of vecuronium (ORG NC45) and pancuronium in patients undergoing coronary artery bypass grafting. Anesthesiology 1983;58(5):438–40.

2530. Salmenpera M, Peltola K, Takkunen O, Heinonen J. Cardiovascular effects of pancuronium and vecuronium during high-dose fentanyl anesthesia. Anesth Analg 1983;62(12):1059–64.

2531. Starr NJ, Sethna DH, Estafanous FG. Bradycardia and asystole following the rapid administration of sufentanil with vecuronium. Anesthesiology 1986;64(4):521–3.

2532. Gencarelli PJ, Roizen MF, Miller RD, Joyce J, Hunt TK, Tyrrell JB. ORG NC45 (Norcuron) and pheochromocytoma: a report of three cases. Anesthesiology 1981;55(6):690–3.

2533. Thase ME. Effects of venlafaxine on blood pressure: a meta-analysis of original data from 3744 depressed patients. J Clin Psychiatry 1998;59(10):502–8.

2534. West S, Hewitt J. Prolonged hypertension: a case report of a potential interaction between electroconvulsive therapy and venlafaxine. Int J Psychiatr Clin Pract 1999;3:55–7.

2535. Reznik I, Rosen Y, Rosen B. An acute ischaemic event associated with the use of venlafaxine: a case report and proposed pathophysiological mechanisms. J Psychopharmacol 1999;13(2):193–5.

2536. Pinder RM, Brogden RN, Speight TM, Avery GS. Viloxazine: a review of its pharmacological properties and therapeutic efficacy in depressive illness. Drugs 1977;13(6):401–21.

2537. Pinder RM, Brogden RN, Speight TM, Avery GS. Doxepin up-to-date: a review of its pharmacological properties and therapeutic efficacy with particular reference to depression. Drugs 1977;13(3):161–218.

2538. Cargill RI, Boyter AC, Lipworth BJ. Reversible myocardial ischaemia following vincristine containing chemotherapy. Respir Med 1994;88(9):709–10.

2539. Dixon A, Nakamura JM, Oishi N, Wachi DH, Fukuyama O. Angina pectoris and therapy with cisplatin, vincristine, and bleomycin. Ann Intern Med 1989;111(4):342–3.

2540. Calvo-Romero JM, Fernandez-Soria-Pantoja R, Arrebola-Garcia JD, Gil-Cubero M. Ischemic heart disease associated with vincristine and doxorubicin chemotherapy. Ann Pharmacother 2001;35(11):1403–5.

2541. Budman DR. New vinca alkaloids and related compounds. Semin Oncol 1992;19(6):639–45.

2542. Furuse K, Kubota K, Kawahara M, Takada M, Kimura I, Fujii M, Ohta M, Hasegawa K, Yoshida K, Nakajima S, Ogura T, Niitani HJapan Lung Cancer Vinorelbine Study Group. Phase II study of vinorelbine in heavily previously treated small cell lung cancer. Oncology 1996;53(2):169–72.

2543. Karminsky N, Merimsky O, Kovner F, Inbar M. Vinorelbine-related acute cardiopulmonary toxicity. Cancer Chemother Pharmacol 1999;43(2):180–2.

2544. Feun LG, Savaraj N, Hurley J, Marini A, Lai S. A clinical trial of intravenous vinorelbine tartrate plus tamoxifen in the treatment of patients with advanced malignant melanoma. Cancer 2000;88(3):584–8.

2545. Yoh K, Niho S, Goto K, Ohmatsu H, Kubota K, Kakinuma R, Nishiwaki Y. High body mass index correlates with increased risk of venous irritation by vinorelbine infusion. Jpn J Clin Oncol 2004;34(4):206–9.

2546. Frasci G, Comella G, Comella P, Salzano F, Cremone L, Della Volpe N, Imbriani A, Persico G. Mitoxantrone plus vinorelbine with granulocyte-colony stimulating factor (G-CSF) support in advanced breast cancer patients. A dose and schedule finding study. Breast Cancer Res Treat 1995;35(2):147–56.

2547. Fumoleau P, Delozier T, Extra JM, Canobbio L, Delgado FM, Hurteloup P. Vinorelbine (Navelbine) in the treatment of breast cancer: the European experience. Semin Oncol 1995;22(2 Suppl 5):22–8.

2548. Spiller HA, Willias DB, Gorman SE, Sanftleban J. Retrospective study of mistletoe ingestion. J Toxicol Clin Toxicol 1996;34(4):405–8.

2549. Tuyns AJ, Riboli E, Doornbos G, Pequignot G. Diet and esophageal cancer in Calvados (France). Nutr Cancer 1987;9(2–3):81–92.

2550. Decarli A, Liati P, Negri E, Franceschi S, La Vecchia C. Vitamin A and other dietary factors in the etiology of esophageal cancer. Nutr Cancer 1987;10(1–2):29–37.

2551. Graham S, Marshall J, Haughey B, Brasure J, Freudenheim J, Zielezny M, Wilkinson G, Nolan J. Nutritional epidemiology of cancer of the esophagus. Am J Epidemiol 1990;131(3):454–67.

2552. Worner TM, Gordon GG, Leo MA, Lieber CS. Vitamin A treatment of sexual dysfunction in male alcoholics. Am J Clin Nutr 1988;48(6):1431–5.

2553. Leo MA, Lieber CS. Hepatic fibrosis after long-term administration of ethanol and moderate vitamin A supplementation in the rat. Hepatology 1983;3(1):1–11.

2554. Chapman KM, Prabhudesai M, Erdman JW Jr. Vitamin A status of alcoholics upon admission and after two weeks of hospitalization. J Am Coll Nutr 1993;12(1): 77–83.

2555. Leo MA, Kim C, Lowe N, Lieber CS. Interaction of ethanol with beta-carotene: delayed blood clearance and enhanced hepatotoxicity. Hepatology 1992;15(5):883–91.

2556. Omenn GS, Goodman GE, Thornquist MD, Balmes J, Cullen MR, Glass A, Keogh JP, Meyskens FL, Valanis B, Williams JH, Barnhart S, Hammar S. Effects of a combination of beta carotene and vitamin A on lung cancer and cardiovascular disease. N Engl J Med 1996;334(18): 1150–5.

2557. Rust P, Eichler I, Renner S, Elmadfa I. Effects of long-term oral beta-carotene supplementation on lipid peroxidation in patients with cystic fibrosis. Int J Vitam Nutr Res 1998;68(2):83–7.

2558. Leo MA, Aleynik SI, Aleynik MK, Lieber CS. beta-Carotene beadlets potentiate hepatotoxicity of alcohol. Am J Clin Nutr 1997;66(6):1461–9.

2559. Baker DL, Krol ES, Jacobsen N, Liebler DC. Reactions of beta-carotene with cigarette smoke oxidants. Identification of carotenoid oxidation products and evaluation of the prooxidant/antioxidant effect. Chem Res Toxicol 1999;12(6):535–43.

2560. Bowen HT, Omaye ST. Oxidative changes associated with beta-carotene and alpha-tocopherol enrichment of human low-density lipoproteins. J Am Coll Nutr 1998;17(2):171–9.

2561. Leo MA, Lieber CS. Alcohol, vitamin A, and beta-carotene: adverse interactions, including hepatotoxicity and carcinogenicity. Am J Clin Nutr 1999;69(6):1071–85.

2562. Estival J, Dupin M, Kanitakis J, Combemale P. Capillary leak syndrome induced by acitretin. Br J Dermatol 2004;150:150–2.

2563. Peck GL. Therapy and prevention of skin cancer. In: Saurat JH, editor. Retinoids. Basel: S. Karger, 1985:345.

2564. Dai WS, LaBraico JM, Stern RS. Epidemiology of isotretinoin exposure during pregnancy. J Am Acad Dermatol 1992;26(4):599–606.

2565. Teelmann K. Retinoids: toxicology and teratogenicity to date. Pharmacol Ther 1989;40(1):29–43.

2566. Chan A, Hanna M, Abbott M, Keane RJ. Oral retinoids and pregnancy. Med J Aust 1996;165(3):164–7.

2567. Hull PR, Demkiw-Bartel C. Isotretinoin use in acne: prospective evaluation of adverse events. J Cutan Med Surg 2000;4(2):66–70.

2568. Charalabopoulos K, Papalimneou V, Charalabopoulos A, Hatzis J. Two new adverse effects of isotretinoin. Br J Dermatol 2003;148:593.

2569. Mann G, Reinhardt D, Ritter J, Hermann J, Schmitt K, Gadner H, Creutzig U. Treatment with all-trans retinoic acid in acute promyelocytic leukemia reduces early deaths in children. Ann Hematol 2001;80(7):417–22.

2570. De Botton S, Dombret H, Sanz M, Miguel JS, Caillot D, Zittoun R, Gardembas M, Stamatoulas A, Conde E, Guerci A, Gardin C, Geiser K, Makhoul DC, Reman O, de la Serna J, Lefrere F, Chomienne C, Chastang C, Degos L, Fenaux P. Incidence, clinical features, and outcome of all trans-retinoic acid syndrome in 413 cases of newly diagnosed acute promyelocytic leukemia. The European APL Group. Blood 1998;92(8):2712–8.

2571. Amano Y, Tajika K, Mizuki T, Amano M, Dan K, Kumazaki T. All-trans retinoic acid syndrome: chest CT assessment. Eur Radiol 2001;11(8):1516–7.

2572. Jung JI, Choi JE, Hahn ST, Min CK, Kim CC, Park SH. Radiologic features of all-trans-retinoic acid syndrome. Am J Roentgenol 2002;178(2):475–80.

2573. Tallman MS, Andersen JW, Schiffer CA, Appelbaum FR, Feusner JH, Ogden A, Shepherd L, Rowe JM, Francois C, Larson RS, Wiernik PH. Clinical description of 44 patients with acute promyelocytic leukemia who developed the retinoic acid syndrome. Blood 2000;95(1):90–5.

2574. Han ZP, Lu HB, Shen ZS. [Severe side effects of the treatment of acute promyelocytic leukemia with all-trans retinoic acid.] Hunan Yi Ke Da Xue Xue Bao 2000;25(3):283–4.

2575. Lehmann S, Paul C. The retinoic acid syndrome in non-M3 acute myeloid leukaemia: a case report. Br J Haematol 2000;108(1):198–9.

2576. Nicolls MR, Terada LS, Tuder RM, Prindiville SA, Schwarz MI. Diffuse alveolar hemorrhage with underlying pulmonary capillaritis in the retinoic acid syndrome. Am J Respir Crit Care Med 1998;158(4):1302–5.

2577. Vaidyanathan S, Soni BM, Oo T, Watt JW, Sett P, Singh G. Syncope following intramuscular injection of hydroxocobalamin in a paraplegic patient: indication for oral administration of cyanocobalamin in spinal cord injury patients. Spinal Cord 1999;37(2):147–9.

2578. Blum M, Kirsten M, Worth MH Jr. Reversible hypertension. Caused by the hypercalcemia of hyperparathyroidism, vitamin D toxicity, and calcium infusion. JAMA 1977;237(3):262–3.

2579. Okada J, Nomura M, Shirataka M, Kondo H. Prevalence of soft tissue calcifications in patients with SLE and effects of alfacarcidol. Lupus 1999;8(6):456–61.

2580. Tsuchihashi K, Takizawa H, Torii T, Ikeda R, Nakahara N, Yuda S, Kobayashi N, Nakata T, Ura N, Shimamoto K. Hypoparathyroidism potentiates cardiovascular complications through disturbed calcium metabolism: possible risk of vitamin D(3) analogue administration in dialysis patients with end-stage renal disease. Nephron 2000;84(1):13–20.

2581. Ashizawa N, Arakawa S, Koide Y, Toda G, Seto S, Yano K. Hypercalcemia due to vitamin D intoxication with clinical features mimicking acute myocardial infarction. Intern Med 2003;42:340–4.

2582. Kikuchi H, Aramaki K, Hirohata S. Congestive heart failure and axonal polyneuropathy induced by alfacacidol in a patient with systemic lupus erythematosus. Mod Rheumatol 2003;13:277–80.

2583. O'Connell TD, Simpson RU. Immunochemical identification of the 1,25-dihydroxyvitamin D3 receptor protein in human heart. Cell Biol Int 1996;20:621–4.

2584. Coret A, Baudet C, Neveu I, Baron-Van Evercooren A, Brachet P, Naveihan P. 1,25-dihydroxyvitamin D3 regulates the expression of VDR and NGF genes in Schwann cells in vitro. J Neurosci Res 1998;15:742–6.

2585. Nako Y, Fukushima N, Tomomasa T, Nagashima K, Kuroume T. Hypervitaminosis D after prolonged feeding with a premature formula. Pediatrics 1993;92(6):862–4.

2586. Stampfer MJ, Hennekens CH, Manson JE, Colditz GA, Rosner B, Willett WC. Vitamin E consumption and the risk of coronary disease in women. N Engl J Med 1993;328(20):1444–9.

2587. Rimm EB, Stampfer MJ, Ascherio A, Giovannucci E, Colditz GA, Willett WC. Vitamin E consumption and the risk of coronary heart disease in men. N Engl J Med 1993;328(20):1450–6.

2588. Stephens NG, Parsons A, Schofield PM, Kelly F, Cheeseman K, Mitchinson MJ. Randomised controlled trial of vitamin E in patients with coronary disease: Cambridge Heart Antioxidant Study (CHAOS). Lancet 1996;347(9004):781–6.

2589. Lonn E. Effects of long-term vitamin E supplementation on cardiovascular events and cancer: a randomized controlled trial. JAMA 2005;293:1338–47.

2590. Bosse GM, Mallory MN, Malone GJ. The safety of intravenously administered vitamin K. Vet Hum Toxicol 2002;44(3):174–6.

2591. Alkan Y, Haefeli WE, Burhenne J, Stein J, Yaniv I, Shalit I. Voriconazole-induced QT interval prolongation and ventricular tachycardia: a non-concentration-dependent adverse effect. Clin Infect Dis 2004;39:e49–52.

2592. Taber KH, Zimmerman JG, Yonas H, Hart W, Hurley RA. Applications of xenon CT in clinical practice: detection of hidden lesions. J Neuropsychiatry Clin Neurosci 1999;11(4):423–5.

2593. Leclerc J, Nieuviarts R, Tavernier B, Vallet B, Scherpereel P. Anesthésie an xénon: du mythe à la réalité. [Xenon anesthesia: from myth to reality.] Ann Fr Anesth Reanim 2001;20(1):70–6.

2594. World Heath Organization. Prevention and Control of Yellow Fever in Africa. Geneva: WHO, 1986.

2595. Fitzner J, Coulibaly D, Kouadio DE, Yavo JC, Loukou YG, Koudou PO, Coulombier D. Safety of the yellow fever vaccine during the September 2001 mass vaccination campaign in Abidjan, Ivory Coast. Vaccine 2004;23(2):156–62.

2596. Grossman E, Rosenthal T, Peleg E, Holmes C, Goldstein DS. Oral yohimbine increases blood pressure and sympathetic nervous outflow in hypertensive patients. J Cardiovasc Pharmacol 1993;22(1):22–6.

2597. Musso NR, Vergassola C, Pende A, Lotti G. Yohimbine effects on blood pressure and plasma catecholamines in human hypertension. Am J Hypertens 1995;8(6):565–71.

2598. Epelde Gonzalo F. Angor inducido por yohimbina. [Yohimbine-induced angina pectoris.] An Med Interna 1998;15(12):676.

2599. Johnson S, Iazzetta J, Dewar C. Severe Raynaud's phenomenon with yohimbine therapy for erectile dysfunction. J Rheumatol 2003;30(11):2503–5.

2600. Hantson P, Lievens M, Mahieu P. Accidental ingestion of a zinc and copper sulfate preparation. J Toxicol Clin Toxicol 1996;34(6):725–30.

2601. Bennett DR, Baird CJ, Chan KM, Crookes PF, Bremner CG, Gottlieb MM, Naritoku WY. Zinc toxicity following massive coin ingestion. Am J Forensic Med Pathol 1997;18(2):148–53.

2602. Stimmel GL, Gutierrez MA, Lee V. Ziprasidone: an atypical antipsychotic drug for the treatment of schizophrenia. Clin Ther 2002;24(1):21–37.

2603. Gunasekara NS, Spencer CM, Keating GM. Ziprasidone: a review of its use in schizophrenia and schizoaffective disorder. Drugs 2002;62(8):1217–51.

2604. Caley CF, Cooper CK. Ziprasidone: the fifth atypical antipsychotic. Ann Pharmacother 2002;36(5):839–51.

2605. House M. Overdose of ziprasidone. Am J Psychiatry 2002;159(6):1061–2.

2606. FDA warning letter. Pfizer Inc. Geodon (ziprasidone HCl). http: //www.pharmcast.com/WarningLetters/Yr2002/.

2607. Taylor D. Ziprasidone in the management of schizophrenia. The QT interval issue in context. CNS Drugs 2003;17:423–30.

Index of drug names

Note: The letter '*t*' with the locater refers to tables.

Printed and bound by CPI Group (UK) Ltd, Croydon, CR0 4YY
03/10/2024
01040329-0020